The Immune Response

The Immune Response

Basic and Clinical Principles

Tak W. Mak and Mary E. Saunders

Contributors

Maya R. Chaddah

Wendy L. Tamminen

AMSTERDAM • BOSTON • HEIDELBERG • LONDON • NEW YORK • OXFORD
PARIS • SAN DIEGO • SAN FRANCISCO • SINGAPORE • SYDNEY • TOKYO
Academic Press is an imprint of Elsevier

Cover Image:
Kazuo Nakamura
Inner Structure, 1956
oil on masonite.
Gift of Charles McFaddin, Toronto, 1985.
© 2004 Art Gallery of Ontario

Acquisitions Editor and Publisher: Tessa Picknett
Development Editor: Victoria Lebedeva
Project Manager: Dawnmarie Simpson
Marketing Manager: Katherine Clevenger
Cover Design: Anthony Coulson
Interior Design: Philip Patenall

Elsevier Academic Press
30 Corporate Drive, Suite 400, Burlington, MA 01803, USA
525 B Street, Suite 1900, San Diego, California 92101–4495, USA
84 Theobald's Road, London WC1X 8RR, UK

This book is printed on acid-free paper. ∞

Library of Congress Cataloging-in-Publication Data
Mak, Tak W., 1945–
    The immune response : basic and clinical principles / Tak W. Mak, Mary E. Saunders.
      p.   cm.
    Includes bibliographical references and index.
    ISBN-13 : 978-0-12-088451-3 (casebound : alk. paper)
    ISBN-10 : 0-12-088451-8 (casebound : alk. paper)
    1. Immunology.   2. Immunity.   I. Saunders, Mary E., Ph.D.
  II. Title.
    [DNLM: 1. Immune System.   2. Immunity.   3. Immune System
  Diseases.   QW 504 M235i 2006]
  QR181.M325 2005
  616.07'9—dc22

                                                                          2005024292

British Library Cataloguing in Publication Data
A catalogue record for this book is available from the British Library

ISBN 13: 978-0-12-088451-3
ISBN 10: 0-12-088451-8

CD ISBN 13: 978-0-12-369532-1
CD ISBN 10: 0-12-369532-5

For all information on all Elsevier Academic Press publications
visit our Web site at www.books.elsevier.com

Printed in China
05   06   07   08   09   10   9 8 7 6 5 4 3 2 1

Pre-publication Review Quotes

". . .Tak and everyone else on the publication team are to be congratulated . . ."
Brian Barber, University Health Network, Toronto, ON, Canada

". . .excellent. . ."
Douglas Fearon, Department of Medicine, University Cambridge School Clinical Medicine, Addenbrookes Hospital, Cambridge, England, UK

". . .illustrations are excellent. . .
. . .delightfully written . . . the text is unique in its style. . ."
Lionel G. Filion, Department of Biochemistry Microbiology and Immunology, Faculty of Medicine, University of Ottawa, Ottawa, ON, Canada

"The illustrations are terrific. . . One of the things I especially like is the use of the 'Introducing. . .' and 'More about . . .' sections. These give students a taste of what is to come without overwhelming them with details. . ."
Douglas R. Green, Department of Immunology, St. Jude Children's Research Hospital, Memphis, TN, USA

". . .I particularly like the clear and conversational style of the prose . . . chapters are a delight to read. . .
I like the format of presentation with Boxes for the specialists and plenty of illustration. This will be an important contribution to the field of Immunology."
Eddy Liew, Division of Immunology, Infection and Inflammation, University of Glasgow, Glasgow, UK

". . .the authors have done a remarkable job of integrating diverse subject matter. . ."
Gary W. Litman, University South Florida, Children's Research Institute, St. Petersburg, FL, USA

". . .A very readable and enjoyable text. . . Historical background is a nice idea that is essentially overlooked in many competing texts . . . contemporary and would be easy to recommend. . ."
Richard M. Locksley, Department of Medicine, University of California in San Francisco, HHMI, San Francisco, CA, USA

". . .very well written; it will be easily accessible to a broad readership, including undergraduate and graduate students. . ."
Ruslan Medzhitov, Section of Immunobiology, HHMI, Yale University School of Medicine, New Haven, CT, USA

"I particularly like the clear and conversational style of the prose . . . chapters are a delight to read. . . I like the format of presentation with boxes for the specialists and plenty of illustration. . . .
This will be an important contribution to the field of immunology."
Noel R. Rose, Director, Center for Autoimmune Disease Research, the Johns Hopkins University, Baltimore, MD, USA

". . .I like the conversational style and the way various topics are integrated. . ."
Warren Strober, NIAID, NIH, Bethesda, USA

"I find the manuscript to be very erudite and quite complete and accurate in its coverage. . ."
Peter A. Ward, University of Michigan Health Systems, Department of Pathology, Ann Arbor, MI, USA

". . .crystal clear definitions of complex immunological concepts. . . Uncomplicated style, effortless for the reader to move from one section to the next, comfortable, unforced learning experience. . ."
Gillian E. Wu, Dean, Faculty of Science and Engineering, York University, Toronto, ON, Canada

". . .this textbook from Mak and Saunders is very timely, reads very well, and should be of great interest to those wanting to get a start in understanding immunology. . ."
Juan-Carlos Zúñiga-Pflücker, Sunnybrook & Women's Research Institute, Toronto, ON, Canada

Preface

As multidisciplinary studies gain value in myriad areas of medical and biological sciences, it has become highly desirable for an individual studying or working in one area to quickly and easily gain knowledge in a complementary field. Immunology is just such a field, since the immune system impinges on virtually every tissue in the body. While immune responses were originally associated only with infectious disease, they were eventually connected with the causation or exacerbation of cancer, transplant rejection, allergies, and autoimmune diseases such as multiple sclerosis, rheumatoid arthritis, and systemic lupus erythematosus. In more recent years, it has become increasingly clear that immunological processes are also linked to heart disease and many neurological disorders. Our goal in writing *The Immune Response: Basic and Clinical Principles* was to produce a book useful to a wide audience, including individuals with undergraduate, graduate, or post-graduate levels of education in diverse areas of medical science. We have taken the approach of establishing a solid grounding in the basic concepts of immunology before progressing to advanced levels of information close to the "cutting edge" of research in this field.

The Immune Response is divided into two major sections: Part I, Basic Immunology, and Part II, Clinical Immunology. To effectively reach a diverse readership, we have started Part I with four chapters (Chs. 1–4) that can serve on their own as a short course in immunology. Newcomers to the field are provided with a gentle introduction to a fascinating science whose specialized terminology and basic paradigms present a unique challenge. Seasoned immunologists may also appreciate the "big picture" presented in these initial chapters, as this overview may be useful in developing a framework in which to present their area of specialization to more general audiences. Chapters 5–21 continue Part I with an in-depth examination of concepts and mechanisms in Basic Immunology. Part II encompasses Chapters 22–30 and covers how the basic elements described in Part I either combine to preserve good health or malfunction to cause disease.

We felt it was critical to write all chapters in a single voice so as to maximize coherence and consistency and make the most of the interconnectedness of many of the topics in this field. While the book begins from the point of view of the intelligent, inquisitive non-immunologist, subsequent chapters gradually build on preceding concepts to bring all readers to the same level of appreciation of the complex workings of the immune system. Immunologists looking for an overview of any topic outside their own area of expertise will find a historical perspective, basic concepts, and more advanced information in any chapter of interest.

We were fortunate to have two contributors who were instrumental in helping us successfully bring *The Immune Response* to fruition. We would like to thank them and acknowledge their importance in the writing of this book.

As an educational consultant, contributor Wendy Tamminen brought undergraduate teaching experience to this project, including an understanding of which approaches and definitions work for novices to the field who are trying to make sense of complicated situations. Dr. Tamminen tirelessly edited every chapter and pored over the illustrations with a view to keeping the authors on the straight and narrow path of accuracy and readability throughout this project. She also contributed endless hours of literature searching, problem solving, and information distillation.

The Immune Response is brought to life by the clean, clear, and distinctive illustrations of contributor Maya Chaddah. With her M.Sc. degree in Immunology and her attention to detail, Ms. Chaddah was able to deftly take the intricate processes described in the text and turn them into beautifully rendered, elegant drawings that capture the plethora of cellular and molecular players participating in immunological events. Shades of two colors were deliberately chosen to reduce the distraction to the reader that full color can sometimes bring. As she developed the illustrations, Ms. Chaddah also reviewed the chapter content, performed numerous literature searches, and sourced the photographs for the book.

The Immune Response has also benefited greatly from the input of many experts on a vast array of immunological topics. These experts, who are listed in the Acknowledgments pages, gave freely of their valuable time and perceptive insights to improve the quality and accuracy of both the text and the illustrations. Any remaining errors are solely the responsibility of the authors. As this is the first edition of *The Immune Response*, there will inevitably be further improvements that readers will want to suggest. We welcome all input that will make this book even more useful for its intended audience.

We now invite our readers to embark on this journey into immunology and to share with us our delight in its elegant mechanisms and our pique that some of its secrets stubbornly resist elucidation. As the reader finishes the last chapter, we are confident that he or she will join us in our conviction that the immune system is among the most vital and intriguing elements of the human body.

Tak W. Mak and Mary E. Saunders

Biographies

Authors

Tak Wah Mak

Tak W. Mak is the Director of the Campbell Family Institute for Breast Cancer Research in the Princess Margaret Hospital, Toronto, and a University Professor of the Departments of Medical Biophysics and Immunology, University of Toronto. He was trained at the University of Wisconsin in Madison, the University of Alberta, and the Ontario Cancer Institute. His research interests center on immune recognition and regulation as well as cell survival and cell death in normal and malignant cells. He gained worldwide prominence in 1983 as the leader of the group that first cloned the genes of the human T cell antigen receptor. His more recent work includes the creation of a series of genetically altered mice that have proved critical to understanding intracellular programs governing the development and function of the immune system, and to dissecting signal transduction cascades in various cell survival and apoptotic pathways. Dr. Mak holds honorary doctoral degrees from universities in North America and Europe, is an Officer of the Order of Canada, and has been elected a Foreign Associate of the National Academy of Sciences (U.S.) as well as a Fellow of the Royal Society of London (U.K.). Dr. Mak has won international recognition as the recipient of the Emil von Behring Prize, the King Faisal Prize for Medicine, the Gairdner Foundation International Award, the Sloan Prize of the General Motors Cancer Foundation, the Novartis Prize in Immunology, the Paul Ehrlich Prize, and the Ludwig Darmstaedter Prize.

Mary Evelyn Saunders

Mary E. Saunders holds the position of Scientific Editor for the Campbell Family Institute for Breast Cancer Research, Toronto. She completed her B.Sc. degree in Genetics at the University of Guelph, Ontario, and received her Ph.D. in Medical Biophysics at the University of Toronto. Dr. Saunders works with Dr. Mak and members of his laboratory on the writing and editing of scientific papers for peer-reviewed journals as well as on various book projects. She takes pride and pleasure in producing concise, clear, highly readable text and making complex scientific processes readily understandable.

Contributors

Wendy Lynn Tamminen

Wendy L. Tamminen completed her B.Sc. degree in Chemistry and Biochemistry at McMaster University, Hamilton, Ontario, and received her Ph.D. in Immunology from the University of Toronto. She taught immunology at the undergraduate level for several years to students in both the biomedical sciences and medicine at the University of Toronto, where her teaching skills were recognized with an Arts and Science Undergraduate Teaching Award. As a writer, editor, and lecturer, Dr. Tamminen's main interest is the communication of scientific concepts to both science specialists and non-specialists. www.wendytamminen.com

Maya Rani Chaddah

Maya R. Chaddah graduated with a double major in Human Biology and Spanish, followed by an M.Sc. in Immunology at the University of Toronto. In 1996, Ms. Chaddah started a business focused on the writing and editing of scientific and medical publications. Her expertise has grown to include scientific and medical illustration, and she continues to produce a variety of communications for diverse audiences in the public and private sectors. www.mayachaddah.com

Acknowledgments

The authors are indebted to the following individuals for the reviewing of one or more draft chapters of this book.

Brian Barber
University Health Network
Toronto, Ontario, Canada

Wiebke Bernhardt
Campbell Family Institute for Breast Cancer Research
Toronto, Ontario, Canada

Paul Cameron
Centre for Clinical Immunology and Biomedical Statistics, Royal Perth Hospital, Perth, Western Australia

James R. Carlyle
Sunnybrook & Women's Research Institute, University of Toronto
Toronto, Ontario, Canada

Vijay K. Chaddah
Grey Bruce Regional Health Centre
Owen Sound, Ontario, Canada

Dominique Charron
Laboratoire d'Immunogénétique Humaine, Hôpital Saint Louis
Paris, France

Irvin Y. Chen
UCLA AIDS Institute, David Geffien School of Medicine at UCLA
Los Angeles, California, USA

Charlotte Cunningham-Rundles
Mount Sinai School of Medicine
New York, New York, USA

Hans-Michael Dosch
The Hospital for Sick Children, University of Toronto
Toronto, Ontario, Canada

Douglas Fearon
Addenbrookes Hospital, University of Cambridge School of Clinical Medicine, Cambridge, England, UK

Lionel G. Filion
University of Ottawa
Ottawa, Ontario, Canada

Amanda Fisher
Hammersmith Hospital, Imperial College School of Medicine
London, England, UK

Dale Godfrey
University of Melbourne
Parkville, Australia

Douglas R. Green
St. Jude Children's Research Hospital
Memphis, Tennessee, USA

David Hafler
Harvard Medical School
Boston, Massachusetts, USA

William Heath
The Walter and Eliza Hall Institute
Victoria, Australia

Peter Henson
National Jewish Medical and Research Center
Denver, Colorado, USA

Jules Hoffmann
CNRS Institute of Molecular and Cellular Biology, University Louis Pasteur, Strasbourg, France

Kristin Ann Hogquist
Center for Immunology, University of Minnesota
Minneapolis, Minnesota, USA

Robert D. Inman
Division of Rheumatology, Toronto Western Hospital, University of Toronto, Toronto, Ontario, Canada

Stefan H.E. Kaufmann
Max-Planck-Institute for Infection Biology
Berlin, Germany

Vijay Kuchroo
Harvard Medical School
Boston, Massachusetts, USA

Robert Lechler
Guy's, King's and St. Thomas' School of Medicine
London, England, UK

Eddy Liew
Division of Immunology, Infection, and Inflammation, University of Glasgow, Glasgow, Scotland, UK

Gary W. Litman
Children's Research Institute, University of South Florida
St. Petersburg, Florida, USA

Richard M. Locksley
Howard Hughes Medical Institute, University of California in San Francisco, San Francisco, California, USA

Bernard Malissen
Centre d'Immunologie de Marseille-Luminy
Marseille, France

Simon Mallal
Centre for Clinical Immunology and Biomedical Statistics, Royal Perth Hospital, Perth, Western Australia

Patty Martinez
Centre for Clinical Immunology and Biomedical Statistics, Royal Perth Hospital, Perth, Western Australia

Ruslan Medzhitov
Howard Hughes Medical Institute, Yale University School of Medicine
New Haven, Connecticut, USA

Mark Minden
Princess Margaret Hospital
Toronto, Ontario, Canada

Thierry Molina
Hôtel Dieu
Paris, France

Pamela Ohashi
Ontario Cancer Institute, University Health Network
Toronto, Ontario, Canada

Christopher Paige
Ontario Cancer Institute, University Health Network
Toronto, Ontario, Canada

Jean-Yves Perrot
Laboratoire INSERM, Universitaire Paris
Paris, France

Michael Reth
Max-Planck-Institute for Immunobiology, University of Freiburg
Freiburg, Germany

Noel R. Rose
Centre for Autoimmune Disease Research, The Johns Hopkins
University, Baltimore, Maryland, USA

Lawrence E. Samelson
Laboratory of Cellular and Molecular Biology, Center for Cancer
Research, NIH, Bethesda, Maryland, USA

Daniel N. Sauder
The Johns Hopkins University
Baltimore, Maryland, USA

Barry W. Sawka
Grey Bruce Health Services
Owen Sound, Ontario, Canada

Hans Schreiber
University of Chicago
Chicago, Illinois, USA

Alan Sher
Laboratory of Parasitic Diseases, National Institute of Allergy and
Infectious Disease, NIH, Bethesda, Maryland, USA

Warren Strober
President of the Society for Mucosal Immunology and Editor of
Current Protocols in Immunology, USA

Megan Sykes
Transplantation Biology Research Center, Massachusetts General
Hospital and Harvard Medical School,
Boston, Massachusetts, USA

Giorgio Trinchieri
Laboratory of Immunological Research,
Schering-Plough Research Institute,
Dardilly, France

Peter A. Ward
University of Michigan Health Systems
Ann Arbor, Michigan, USA

Hans Wigzell
Microbiology and Tumor Biology Centre, Karolinska Institute,
Stockholm, Sweden

Gillian E. Wu
York University
Toronto, Ontario, Canada

Rae Yeung
Division of Rheumatology, University of Toronto
Toronto, Ontario, Canada

Juan-Carlos Zúñniga-Pflücker
Sunnybrook & Women's Research Institute,
University of Toronto
Toronto, Ontario, Canada

The authors are also most grateful to the following individuals
for their helpful suggestions and/or generous gifts of
materials used in the preparation of this book.

Avrum I. Gotlieb
University of Toronto and University Health Network
Toronto, Ontario, Canada

John B. Hay
University of Toronto
Toronto, Ontario, Canada

Steven G.E. Marsh
Anthony Nolan Research Institute, HLA Informatics Group
London, England, UK

Mary Shago
Cytogenetics Laboratory, The Hospital for Sick Children,
University of Toronto, Toronto, Ontario, Canada

Andrea Shugar
Division of Clinical and Metabolic Genetics, The Hospital
for Sick Children, University of Toronto, Toronto, Ontario, Canada

Douglas Tkachuk
Princess Margaret Hospital and University of Toronto
Toronto, Ontario, Canada

Heddy Zola
Child Health Research Institute
North Adelaide, Australia

Contents

Preface   VIII
Biographies   IX
Acknowledgments   XI

PART I: BASIC IMMUNOLOGY   1

Chapter 1: Perspective on Immunity and Immunology   3

A. WHAT IS IMMUNOLOGY?   4
B. WHY HAVE AN IMMUNE SYSTEM AND WHAT DOES IT DO?   7
C. TYPES OF IMMUNE RESPONSES: INNATE AND ADAPTIVE   7
D. WHAT IS "INFECTION"?   10
E. PHASES OF HOST DEFENSE   11
F. HOW ARE ADAPTIVE AND INNATE IMMUNITY RELATED?   12
G. LEUKOCYTES: CELLULAR MEDIATORS OF IMMUNITY   13
H. WHERE DO IMMUNE RESPONSES OCCUR?   14
I. CLINICAL IMMUNOLOGY: WHEN THE IMMUNE SYSTEM DOES NOT WORK PROPERLY   15

Chapter 2: Introduction to the Immune Response   17

A. GENERAL FEATURES OF INNATE IMMUNITY   18
B. GENERAL FEATURES OF ADAPTIVE IMMUNITY   19
   I.   SPECIFICITY   19
   II. IMMUNOLOGIC MEMORY   19
   III. DIVERSITY   20
   IV. TOLERANCE   21
   V.   DIVISION OF LABOR   21
C. ELEMENTS OF IMMUNITY COMMON TO THE INNATE AND ADAPTIVE RESPONSES   22
   I.   INTRODUCING CYTOKINES   22
   II. INTRODUCING INTRACELLULAR SIGNALING   23
D. ELEMENTS OF IMMUNITY EXCLUSIVE TO THE ADAPTIVE RESPONSE   24
   I.   ANTIGENS VERSUS IMMUNOGENS   24
   II.   INTRODUCING SPECIFIC ANTIGEN RECOGNITION: B CELLS   26
   III. INTRODUCING SPECIFIC ANTIGEN RECOGNITION: T CELLS AND THE MAJOR HISTOCOMPATIBILITY COMPLEX   26
   IV.   INTRODUCING ANTIGEN PROCESSING   27
   V.   INTRODUCING CORECEPTORS AND COSTIMULATORY MOLECULES   29
   VI.   INTRODUCING B CELL EFFECTOR FUNCTIONS   29
   VII. INTRODUCING T CELL EFFECTOR FUNCTIONS   30
   VIII. INTRODUCING PRIMARY AND SECONDARY ADAPTIVE IMMUNE RESPONSES   31

Chapter 3: Cells and Tissues of the Immune Response   35

A. CELLS OF THE IMMUNE SYSTEM   36
   I.   TYPES OF HEMATOPOIETIC CELLS   36
   II. CELLS OF THE MYELOID LINEAGE   36
   III. CELLS OF THE LYMPHOID LINEAGE   41
   IV. DENDRITIC CELLS   46
   V.   HEMATOPOIESIS   47
B. LYMPHOID TISSUES   51
   I.   PRIMARY LYMPHOID TISSUES   53
   II. SECONDARY LYMPHOID TISSUES   56

Chapter 4: Innate Immunity   69

A. MECHANISMS OF INNATE IMMUNITY   70
   I.   ANATOMICAL AND PHYSIOLOGICAL BARRIERS TO INFECTION   70
   II. CELLULAR INTERNALIZATION MECHANISMS THAT FIGHT INFECTION   73
   III. INFLAMMATION AS A RESPONSE TO INFECTION OR INJURY   78
B. PATTERN RECOGNITION IN INNATE IMMUNITY   84
   I.   PATTERN RECOGNITION BY PRRs   85
   II. PATTERN RECOGNITION BY RECEPTORS OF NK, NKT, AND γδ T CELLS   88
   III. PATTERN RECOGNITION BY SOLUBLE MOLECULES   89

Chapter 5: B Cell Receptor Structure and Effector Function   93

A. HISTORICAL NOTES   94
B. THE STRUCTURE OF IMMUNOGLOBULINS   97
   I.   GENERAL STRUCTURAL PROPERTIES OF IMMUNOGLOBULIN MOLECULES   97
   II. CHANGES TO Ig STRUCTURE ASSOCIATED WITH FUNCTION   103
   III. THE B CELL ANTIGEN RECEPTOR COMPLEX   108
   IV. Fc RECEPTORS   108
C. EFFECTOR FUNCTIONS OF ANTIBODIES   113
   I.   NEUTRALIZATION   113
   II. OPSONIZATION   113
   III. ANTIBODY-DEPENDENT CELL-MEDIATED CYTOTOXICITY   114
   IV. COMPLEMENT-MEDIATED CLEARANCE OF ANTIGEN   115
D. IMMUNOGLOBULIN ISOTYPES IN BIOLOGICAL CONTEXT   116
   I.   NATURAL DISTRIBUTION OF ANTIBODIES IN THE BODY   116
   II. MORE ABOUT IgM   116
   III. MORE ABOUT IgD   118
   IV. MORE ABOUT IgG   118
   V.   MORE ABOUT IgA   118
   VI. MORE ABOUT IgE   119

Chapter 6: The Nature of Antigen–Antibody Interaction   121

A. THE NATURE OF B CELL IMMUNOGENS   122
   I.   WHAT MOLECULES CAN FUNCTION AS IMMUNOGENS?   122
   II. IMMUNOGENS IN THE HUMORAL RESPONSE   122
   III. PROPERTIES OF Td IMMUNOGENS   125

Contents

B. B CELL–T CELL COOPERATION IN THE HUMORAL
   IMMUNE RESPONSE 130
   I.   THE DISCOVERY OF B–T COOPERATION: RECONSTITUTION
        EXPERIMENTS 130
   II. THE MOLECULAR BASIS OF B–T COOPERATION: THE
        HAPTEN–CARRIER EXPERIMENTS 131
   III. HAPTEN–CARRIER COMPLEXES *in vivo* 134
   IV. THE RATIONALE FOR LINKED RECOGNITION 136

C. THE MECHANICS OF ANTIGEN–ANTIBODY INTERACTION 137
   I.   IDENTIFICATION OF B EPITOPE STRUCTURAL REQUIREMENTS .. 137
   II. WHERE ANTIBODY AND ANTIGEN INTERACT: THE
        COMPLEMENTARITY-DETERMINING REGIONS 137
   III. FORCES AT WORK IN SPECIFIC ANTIGEN–ANTIBODY BINDING . 139
   IV. AFFINITY AND AVIDITY OF ANTIBODY BINDING 140
   V.   ANTIBODY CROSS-REACTIVITY 143

Chapter 7: Exploiting Antigen–Antibody Interaction 147

A. SOURCES OF ANTIBODIES 149
   I.   ANTISERA 149
   II. HYBRIDOMAS AND MONOCLONAL ANTIBODIES 150

B. TECHNIQUES BASED ON IMMUNE COMPLEX FORMATION 155
   I.   CROSS-LINKING AND THE FORMATION OF
        IMMUNE COMPLEXES 155
   II. TECHNIQUES BASED ON THE PRECIPITIN REACTION 156
   III. TECHNIQUES BASED ON AGGLUTINATION 159
   IV. TECHNIQUES BASED ON COMPLEMENT FIXATION 159

C. ASSAYS BASED ON UNITARY ANTIGEN–ANTIBODY
   PAIR FORMATION 162
   I.   GENERAL CONCEPTS 162
   II. DETECTION OF ANTIGEN BY TAG ASSAYS 167
   III. ISOLATION AND CHARACTERIZATION OF ANTIGEN USING
        ANTIBODIES 172

Chapter 8: The Immunoglobulin Genes 179

A. HISTORICAL NOTES 180

B. CHROMOSOMAL ORGANIZATION OF Ig GENES 183
   I.   THE Ig LOCI 183
   II. GENERAL STRUCTURE OF Ig LOCI 184
   III. FINE STRUCTURE OF LIGHT CHAIN GENES 185
   IV. FINE STRUCTURE OF HEAVY CHAIN GENES 186

C. Ig GENE REARRANGEMENT 188
   I.   THE ROLE OF V(D)J RECOMBINATION 188
   II. VDJ JOINING IN THE *Igh* LOCUS 188
   III. VJ JOINING IN THE *Igl* AND *Igk* LOCI 190
   IV. PRODUCTIVITY TESTING 191
   V.   ALLELIC EXCLUSION 192
   VI. KAPPA/LAMBDA EXCLUSION 193
   VII. INTRODUCING KNOCKOUT MICE 194

D. MOLECULAR MECHANISMS OF Ig GENE REARRANGEMENT 196
   I.   HOW DO VDJ SEGMENTS JOIN IN THE RIGHT ORDER? THE RSS . 196
   II. THE RECOMBINASE ENZYMES: RAG-1 AND RAG-2 197

   III. SYNAPSIS, SIGNAL JOINTS, AND CODING JOINTS 198
   IV. MUTATIONS OF V(D)J RECOMBINATION 200

E. ANTIBODY DIVERSITY GENERATED BY GENE
   REARRANGEMENT 201
   I.   MULTIPLICITY OF GERM-LINE GENE SEGMENTS 201
   II. COMBINATORIAL DIVERSITY 201
   III. JUNCTIONAL DIVERSITY 202
   IV. HEAVY–LIGHT Ig CHAIN PAIRING 204

F. CONTROL SEQUENCES IN THE Ig LOCI 204
   I.   ENHANCERS 204
   II. DNA BINDING MOTIFS AND NUCLEAR TRANSCRIPTION
        FACTORS 206

Chapter 9: The Humoral Response: B Cell Development and Activation 209

A. THE MATURATION PHASE OF B CELL DEVELOPMENT 211
   I.   PRO-B CELLS 211
   II. PRE-B CELLS 213

B. THE DIFFERENTIATION PHASE OF B CELL DEVELOPMENT 217
   I.   THE THREE-SIGNAL MODEL OF B CELL ACTIVATION 218
   II. CELLULAR INTERACTIONS DURING B CELL ACTIVATION 229
   III. PROLIFERATION AND SOMATIC HYPERMUTATION 231
   IV. AFFINITY MATURATION 234
   V.   ISOTYPE SWITCHING: GENERATING FUNCTIONAL DIVERSITY . 235
   VI. DIFFERENTIATION OF MEMORY B CELLS AND PLASMA CELLS . 239

Chapter 10: MHC: The Major Histocompatibility Complex 247

A. HISTORICAL NOTES 248
   I.   DISCOVERY OF THE MHC 248
   II. MHC INVOLVEMENT IN T CELL RECOGNITION 249
   III. ELUCIDATION OF THE ANTIGEN-PRESENTING FUNCTION
        OF THE MHC 249

B. GENERAL ASPECTS OF THE MHC IN HUMANS AND MICE 254
   I.   OVERVIEW OF THE MHC PROTEINS 254
   II. OVERVIEW OF THE MHC LOCI 255
   III. INTRODUCING MULTIPLICITY AND POLYMORPHISM IN
        THE MHC LOCI 255
   IV. INTRODUCING HAPLOTYPES 259

C. MHC PROTEINS 260
   I.   MHC CLASS I PROTEINS 260
   II. MHC CLASS II PROTEINS 262

D. MHC GENES 264
   I.   DETAILED ORGANIZATION OF THE H-2 COMPLEX 264
   II. DETAILED ORGANIZATION OF THE HLA COMPLEX 267

E. EXPRESSION OF MHC MOLECULES 269
   I.   THE SXY–CIITA REGULATORY SYSTEM 270
   II. TNF- AND IFN-γ-INDUCED EXPRESSION OF MHC CLASS I ... 271
   III. EXPRESSION OF MHC CLASS Ib GENES 272
   IV. OTHER REGULATORY PATHWAYS GOVERNING MHC
        CLASS II GENE EXPRESSION 272

F. PHYSIOLOGY OF THE MHC — 272

   I. POLYMORPHISM AND THE BIOLOGICAL ROLE OF THE MHC — 272

   II. ALLOREACTIVITY — 273

   III. MHC AND IMMUNE RESPONSIVENESS — 273

   IV. INTRODUCING MHC AND DISEASE PREDISPOSITION — 274

Chapter 11: Antigen Processing and Presentation — 279

A. THE EXOGENOUS OR ENDOCYTIC ANTIGEN PROCESSING PATHWAY — 281

   I. HISTORICAL NOTES — 281

   II. CELLS THAT CAN FUNCTION AS APCs — 282

   III. MECHANISM OF ANTIGEN PROCESSING BY APCs — 289

   IV. FACTORS AFFECTING ANTIGEN PROCESSING AND PRESENTATION BY APCs — 294

B. THE ENDOGENOUS OR CYTOSOLIC ANTIGEN PROCESSING PATHWAY — 295

   I. HISTORICAL NOTES — 295

   II. CELLS THAT CAN FUNCTION AS TARGET CELLS — 295

   III. MECHANISM OF ANTIGEN PROCESSING BY TARGET CELLS — 296

C. OTHER PATHWAYS OF ANTIGEN PRESENTATION — 302

   I. CROSS-PRESENTATION — 302

   II. ANTIGEN PRESENTATION BY NON-CLASSICAL AND MHC-LIKE MOLECULES — 305

Chapter 12: The T Cell Receptor: Structure of Its Proteins and Genes — 311

A. HISTORICAL NOTES — 312

   I. INTRODUCTION — 312

   II. DISCOVERY OF THE GENES AND PROTEINS OF THE TCR — 313

   III. A SECOND T CELL RECEPTOR — 315

B. THE STRUCTURE OF T CELL RECEPTOR PROTEINS — 317

   I. OVERVIEW — 317

C. GENOMIC ORGANIZATION OF THE TCR AND CD3 LOCI — 319

   I. THE TCRα LOCUS — 319

   II. THE TCRβ LOCUS — 320

   III. THE TCRγ LOCUS — 321

   IV. THE TCRδ LOCUS — 321

   V. THE CD3 GENES — 322

D. EXPRESSION OF TCR GENES — 322

   I. MECHANISM OF V(D)J RECOMBINATION IN THE TCR LOCI — 322

   II. TCR GENE TRANSCRIPTION AND PROTEIN ASSEMBLY — 324

E. DEVELOPMENTAL ASPECTS OF V(D)J RECOMBINATION IN THE TCR LOCI — 325

   I. TCRβ LOCUS REARRANGEMENT — 325

   II. TCRα LOCUS REARRANGEMENT — 326

   III. TCRγ AND δ LOCUS REARRANGEMENT — 327

   IV. TCR LOCUS KNOCKOUT MICE — 327

F. GENERATION OF DIVERSITY OF THE T CELL RECEPTOR REPERTOIRE — 327

   I. MULTIPLICITY OF GERMLINE GENE SEGMENTS — 327

   II. COMBINATORIAL DIVERSITY — 328

   III. JUNCTIONAL DIVERSITY — 328

   IV. CHAIN PAIRING — 329

G. REGULATION OF TCR GENE EXPRESSION — 329

H. STRUCTURE AND FUNCTION OF THE CD3 COMPLEX — 330

   I. CD3 PROTEIN STRUCTURE — 330

   II. FUNCTIONS OF THE CD3 COMPLEX — 331

   III. CD3 KNOCKOUT MICE — 331

I. THE CD4 AND CD8 CORECEPTORS — 332

   I. DISCOVERY OF CD4 AND CD8 — 332

   II. WHAT IS A "CORECEPTOR"? — 332

   III. STRUCTURE OF CD4 — 334

   IV. STRUCTURE OF CD8 — 334

   V. FUNCTIONS OF CD4 AND CD8 — 334

J. PHYSICAL ASPECTS OF THE INTERACTION OF THE TCR WITH ANTIGEN — 336

   I. STUDYING TCR–PEPTIDE–MHC INTERACTION — 336

   II. BINDING AFFINITY OF TCR$\alpha\beta$ FOR ITS LIGAND — 336

   III. X-RAY CRYSTAL STRUCTURE OF T CELL RECEPTORS — 337

Chapter 13: T Cell Development — 341

A. HISTORICAL NOTES — 342

B. CONTEXT AND OVERVIEW — 344

   I. COMPARISON OF B AND T CELL ONTOGENY — 344

   II. OVERVIEW OF T CELL DEVELOPMENT — 346

C. T CELL DEVELOPMENT IN THE BONE MARROW — 347

D. T CELL DEVELOPMENT IN THE THYMUS — 347

   I. OVERVIEW — 347

   II. DN (TN) PHASE (TCR$^-$CD4$^-$CD8$^-$) — 351

   III. THE DP PHASE (CD4$^+$CD8$^+$) — 356

   IV. THE SP PHASE: CD4/CD8 LINEAGE COMMITMENT — 364

E. MATURE SP THYMOCYTES IN THE PERIPHERY — 368

F. T CELL DEVELOPMENT IN AN EMBRYOLOGICAL CONTEXT — 368

Chapter 14: T Cell Activation — 373

A. BRINGING T CELLS AND APCs TOGETHER — 375

   I. L-SELECTIN — 376

   II. LFA-1 — 376

B. SIGNAL ONE: BINDING OF PEPTIDE–MHC TO THE TCR — 378

   I. MODELS OF TCR TRIGGERING — 378

   II. FORMATION OF THE IMMUNOLOGICAL SYNAPSE (SMAC) — 380

   III. TCR DOWNREGULATION — 382

   IV. TCR SIGNAL TRANSDUCTION — 383

C. SIGNAL TWO: COSTIMULATION — 392

   I. CD28–B7 — 393

   II. ICOS–ICOSL — 396

   III. PD-1–PDL1/PDL2 — 397

   IV. ???–B7-H3 — 397

   V. THE TNF/TNFR-RELATED COSTIMULATORY MOLECULES — 397

   VI. CD27/CD70 — 398

   VII. OTHER MINOR COSTIMULATORY CONTACTS — 398

Contents

D. SIGNAL THREE: CYTOKINES ... 398

   I. THE IL-2/IL-2R SYSTEM ... 399

   II. CONTROL OF TRANSCRIPTION OF THE IL-2 GENE ... 399

Chapter 15: T Cell Differentiation and Effector Function ... 403

A. HISTORICAL NOTES ... 404

B. Th CELL DIFFERENTIATION AND EFFECTOR FUNCTION ... 405

   I. WHAT ARE "Th1 AND Th2 RESPONSES"? ... 405

   II. PROCESS OF Th CELL DIFFERENTIATION ... 406

   III. ACTIVATION OF EFFECTOR Th1 AND Th2 CELLS ... 413

   IV. EFFECTOR FUNCTIONS OF Th CELLS ... 415

C. Tc CELL DIFFERENTIATION AND CTL EFFECTOR FUNCTION ... 417

   I. GENERATION OF EFFECTOR CTLs ... 418

   II. MECHANISMS OF TARGET CELL DESTRUCTION BY $CD8^+$ CTLs ... 419

D. COMPARISON OF NAIVE AND EFFECTOR T CELLS ... 422

   I. TRAFFICKING AND ADHESION ... 422

   II. ACTIVATION ... 423

   III. FUNCTIONS AND PRODUCTS ... 424

E. ELIMINATION OF EFFECTOR T CELLS ... 424

   I. ACTIVATION-INDUCED CELL DEATH (AICD) ... 424

   II. T CELL EXHAUSTION ... 427

F. MEMORY T CELLS ... 427

   I. GENERATION OF MEMORY T CELLS ... 428

   II. MEMORY T CELL MARKERS ... 428

   III. MEMORY T CELL DISTRIBUTION ... 429

   IV. MEMORY T CELL ACTIVATION ... 429

   V. MEMORY T CELL DIFFERENTIATION AND EFFECTOR FUNCTION ... 430

   VI. MEMORY T CELL LIFE SPAN ... 430

Chapter 16: Immune Tolerance in the Periphery ... 433

A. CONTEXT OF PERIPHERAL TOLERANCE ... 434

   I. HISTORICAL NOTES ... 434

   II. EVIDENCE FOR PERIPHERAL SELF-TOLERANCE MECHANISMS ... 436

B. T CELL PERIPHERAL SELF-TOLERANCE ... 437

   I. THE IMPORTANCE OF DCs AND "DANGER SIGNALS" ... 437

   II. MECHANISMS OF PERIPHERAL T CELL SELF-TOLERANCE ... 439

C. B CELL PERIPHERAL SELF-TOLERANCE ... 447

   I. DEMONSTRATION OF B CELL ANERGY ... 447

   II. MECHANISMS OF PERIPHERAL B CELL TOLERANCE ... 448

D. REGULATORY T CELLS ... 450

   I. $CD4^+CD25^+$ T_{reg} CELLS ... 450

   II. Th3 AND Tr1 CELLS ... 452

   III. $CD8^+$ Ts CELLS ... 452

   IV. INFECTIOUS TOLERANCE AND LINKED SUPPRESSION ... 453

E. EXPERIMENTAL TOLERANCE ... 454

   I. CHARACTERISTICS OF EXPERIMENTAL TOLERANCE ... 454

   II. CHARACTERISTICS OF TOLEROGENS ... 454

   III. DEGREE AND PERSISTENCE OF TOLERANCE ... 457

   IV. "SPLIT" TOLERANCE ... 457

   V. SPECIAL SITUATIONS ... 457

Chapter 17: Cytokines and Cytokine Receptors ... 463

A. HISTORICAL NOTES ... 464

B. OVERVIEW ... 465

   I. GENERAL PROPERTIES OF CYTOKINES ... 465

   II. GENERAL PROPERTIES OF CYTOKINE RECEPTORS ... 467

C. FUNCTION, PRODUCTION, AND RECEPTORS OF CYTOKINES ... 469

   I. THE INTERFERONS ... 469

   II. THE INTERLEUKINS ... 474

   III. TUMOR NECROSIS FACTOR AND RELATED MOLECULES ... 496

   IV. TRANSFORMING GROWTH FACTORS ... 501

   V. HEMATOPOIETIC GROWTH FACTORS ... 502

   VI. CHEMOKINES AND THEIR RECEPTORS ... 505

D. CYTOKINES AND THEIR RECEPTORS IN PHYSIOLOGICAL CONTEXT ... 508

Chapter 18: Bridging Innate and Adaptive Immunity: NK, γδ T, and NKT Cells ... 517

A. NATURAL KILLER (NK) CELLS ... 518

   I. HISTORICAL NOTES ... 518

   II. WHAT AND WHERE ARE NATURAL KILLER CELLS? ... 520

   III. FUNCTIONS OF NK CELLS ... 520

   IV. ACTIVATORY AND INHIBITORY NK RECEPTORS ... 524

   V. NK CELL DEVELOPMENT, INHIBITORY RECEPTOR EXPRESSION, AND TOLERANCE ... 532

B. γδ T CELLS ... 537

   I. WHAT ARE γδ T CELLS? ... 537

   II. ANTIGEN RECOGNITION ... 539

   III. ACTIVATION ... 541

   IV. EFFECTOR FUNCTIONS AND MEMORY ... 542

   V. DEVELOPMENT ... 544

C. NKT CELLS ... 547

   I. WHAT ARE NKT CELLS? ... 547

   II. ANTIGEN RECOGNITION AND ACTIVATION ... 548

   III. EFFECTOR FUNCTIONS ... 549

   IV. DEVELOPMENT ... 549

Chapter 19: Complement ... 553

A. HISTORICAL NOTES ... 554

B. OVERVIEW OF THE COMPLEMENT SYSTEM ... 555

   I. SOURCES OF SOLUBLE COMPLEMENT COMPONENTS ... 555

   II. THREE PATHWAYS OF COMPLEMENT ACTIVATION ... 555

   III. AMPLIFICATION AND SENSITIVITY ... 556

   IV. CONTROL ... 556

   V. COMPLEMENT SYSTEM NOMENCLATURE ... 556

C. THE CLASSICAL PATHWAY OF COMPLEMENT ACTIVATION ... 557

   I. C1 ... 557

   II. C4 ... 558

III. C2 559

IV. C3 559

D. THE ALTERNATIVE PATHWAY OF COMPLEMENT ACTIVATION 561

  I. C3 AND FACTOR B 562

  II. FACTOR D AND THE ALTERNATIVE C3 CONVERTASE 562

  III. PROPERDIN AND THE ALTERNATIVE C5 CONVERTASE 562

E. THE LECTIN PATHWAY OF COMPLEMENT ACTIVATION 562

  I. MBL 562

  II. CRP 563

F. TERMINAL COMPONENTS OF THE COMPLEMENT CASCADE AND FORMATION OF THE MAC 564

  I. C5 564

  II. C6 AND C7 564

  III. C8 AND C9 565

  IV. TARGETS OF THE MAC 566

G. REGULATION OF THE COMPLEMENT SYSTEM 566

  I. REGULATION OF THE CLASSICAL PATHWAY 567

  II. REGULATION OF THE ALTERNATIVE PATHWAY 570

  III. REGULATION OF THE LECTIN PATHWAY 571

  IV. REGULATION OF TERMINAL COMPONENTS 571

H. COMPLEMENT RECEPTORS AND THEIR BIOLOGICAL ROLES 572

  I. CR1 573

  II. CR2 574

  III. CR3 574

  IV. CR4 575

  V. C1q "RECEPTORS" 575

  VI. RECEPTORS FOR C3a, C4a, AND C5a 575

I. COMPLEMENT DEFICIENCIES 575

  I. DEFICIENCY OF C1, C4, OR C2 576

  II. DEFICIENCY OF C3 577

  III. DEFICIENCY OF ALTERNATIVE PATHWAY COMPONENTS 578

  IV. DEFICIENCY OF LECTIN PATHWAY COMPONENTS 578

  V. DEFICIENCY OF TERMINAL COMPONENTS (C5–C9) 578

  VI. DEFICIENCY OF REGULATORY PROTEINS 578

  VII. DEFICIENCY OF COMPLEMENT RECEPTORS 579

J. NEW ROLES FOR COMPLEMENT? 579

Chapter 20: Mucosal and Cutaneous Immunity 583

A. MUCOSAL IMMUNITY 584

  I. GALT 586

  II. BALT/NALT 591

  III. IMMUNE RESPONSES IN THE MAJOR MALT 593

  IV. IMMUNE RESPONSES IN THE MINOR MALT 599

B. CUTANEOUS IMMUNITY 600

  I. COMPONENTS OF SALT 600

  II. IMMUNE RESPONSES IN SALT 605

Chapter 21: Comparative Immunology 611

A. OVERVIEW 612

  I. REPRISE OF THE ANIMAL KINGDOM 612

  II. FORCES SHAPING THE EVOLUTION OF IMMUNE RECOGNITION 614

B. ELEMENTS OF INNATE IMMUNITY 618

  I. ANATOMICAL AND PHYSIOLOGICAL BARRIERS 618

  II. INNATE IMMUNE RESPONSE CELLS AND THEIR RECOGNITION STRUCTURES 618

  III. CYTOKINES 620

  IV. PATHOGEN ELIMINATION 621

C. ELEMENTS OF ADAPTIVE IMMUNITY 626

  I. LYMPHOID TISSUES 626

  II. BCR 628

  III. MHC 632

  IV. TCR 634

PART II: CLINICAL IMMUNOLOGY 629

Chapter 22: Immunity to Pathogens 641

A. OVERVIEW 642

  I. WHAT IS A "PATHOGEN" 642

  II. WHAT IS "DISEASE" 644

  III. INNATE DEFENSE AGAINST PATHOGENS 645

B. IMMUNITY TO EXTRACELLULAR BACTERIA 647

  I. WHAT ARE EXTRACELLULAR BACTERIA? 647

  II. EFFECTOR MECHANISMS 652

  III. EVASION STRATEGIES 654

C. IMMUNITY TO INTRACELLULAR BACTERIA 656

  I. WHAT ARE INTRACELLULAR BACTERIA? 656

  II. EFFECTOR MECHANISMS 658

  III. EVASION STRATEGIES 662

D. IMMUNITY TO VIRUSES 664

  I. WHAT ARE VIRUSES? 664

  II. EFFECTOR MECHANISMS 669

  III. EVASION STRATEGIES 672

E. IMMUNITY TO PARASITES 680

  I. WHAT ARE PARASITES? 680

  II. EFFECTOR MECHANISMS 683

  III. EVASION STRATEGIES 686

F. IMMUNITY TO FUNGI 688

  I. WHAT ARE FUNGI? 688

  II. EFFECTOR MECHANISMS 689

  III. EVASION STRATEGIES 690

G. THE MYSTERIOUS PRIONS 691

Chapter 23: Vaccines and Clinical Immunization 695

A. VACCINATION: PUBLIC HEALTH SUCCESSES AND CHALLENGES 696

B. HISTORICAL NOTES 698

Contents

C. GENERAL PRINCIPLES OF VACCINE DESIGN — 703
- I. BIOLOGICAL PURPOSE OF A VACCINE — 703
- II. VACCINE EFFICACY AND SAFETY — 704

D. TYPES OF VACCINES — 705
- I. LIVE, ATTENUATED VACCINES — 705
- II. KILLED VACCINES — 708
- III. TOXOIDS — 709
- IV. SUBUNIT VACCINES — 709
- V. PEPTIDE VACCINES — 710
- VI. RECOMBINANT VECTOR DNA VACCINES — 711
- VII. "NAKED DNA" VACCINES — 713
- VIII. ANTI-IDIOTYPIC VACCINES — 716

E. FACTORS AFFECTING VACCINATION — 718
- I. SCHEDULING OF VACCINE ADMINISTRATION AND BOOSTING — 718
- II. ROUTES OF VACCINE ADMINISTRATION — 722
- III. ADJUVANTS AND DELIVERY VEHICLES — 722

F. PROPHYLACTIC VACCINES — 727
- I. ANTHRAX — 727
- II. CHOLERA — 729
- III. DIPHTHERIA, TETANUS, AND PERTUSSIS — 729
- IV. *HAEMOPHILUS INFLUENZAE* TYPE b — 730
- V. HEPATITIS A — 731
- VI. HEPATITIS B — 731
- VII. INFLUENZA VIRUS — 731
- VIII. JAPANESE B ENCEPHALITIS — 732
- IX. MEASLES, MUMPS, AND RUBELLA — 732
- X. MENINGOCOCCUS — 734
- XI. PLAGUE — 734
- XII. POLIO — 734
- XIII. RABIES — 735
- XIV. *STREPTOCOCCUS PNEUMONIAE* — 735
- XV. TUBERCULOSIS — 735
- XVI. TYPHOID FEVER — 736
- XVII. VARICELLA ZOSTER (CHICKEN POX) — 738
- XVIII. VARIOLA (SMALLPOX) — 738
- XIX. YELLOW FEVER — 739

G. THE "DARK SIDE" OF VACCINES — 739
- I. ADVERSE EFFECTS OF VACCINES — 739
- II. FAILURE TO VACCINATE AND OPPOSITION TO VACCINATION — 740

H. PROPHYLACTIC VACCINES OF THE FUTURE — 740

I. PASSIVE IMMUNIZATION — 741

J. THERAPEUTIC VACCINES — 744
- I. VACCINES TO COMBAT TUMORS — 744
- II. VACCINES TO CURE CHRONIC VIRAL DISEASES — 745
- III. VACCINES TO MITIGATE INDIRECT EFFECTS OF INFECTIOUS DISEASE — 745
- IV. VACCINES TO MITIGATE AUTOIMMUNITY — 745
- V. VACCINES TO MITIGATE ALLERGY — 746
- VI. VACCINES TO SUPPRESS FERTILITY — 746
- VII. VACCINES TO SUPPRESS ADDICTION — 747

Chapter 24: Primary Immunodeficiencies — 751

A. GENERAL CONCEPTS — 752
- I. DIAGNOSIS OF PIs — 752
- II. TREATMENT OF PIs — 753
- III. FOCUS OF PI RESEARCH — 754

B. PRIMARY IMMUNODEFICIENCIES DUE TO DEFECTS IN ADAPTIVE IMMUNE RESPONSES — 754
- I. COMBINED IMMUNODEFICIENCIES — 754
- II. T CELL-SPECIFIC IMMUNODEFICIENCIES — 763
- III. B CELL-SPECIFIC IMMUNODEFICIENCIES — 765
- IV. ADAPTIVE IMMUNODEFICIENCIES DUE TO DEFECTS IN DNA REPAIR — 768
- V. LYMPHOPROLIFERATIVE IMMUNODEFICIENCY SYNDROMES — 773
- VI. OTHER ADAPTIVE IMMUNODEFICIENCIES — 774

C. PRIMARY IMMUNODEFICIENCIES DUE TO DEFECTS IN THE INNATE IMMUNE RESPONSE — 776
- I. IMMUNODEFICIENCIES AFFECTING PHAGOCYTE RESPONSES — 776
- II. COMPLEMENT DEFICIENCIES — 780

Chapter 25: HIV and Acquired Immunodeficiency Syndrome — 785

A. HISTORICAL NOTES — 786

B. WHAT IS HIV? — 788
- I. OVERVIEW OF THE HIV-1 LIFE CYCLE — 788
- II. OVERVIEW OF HIV STRUCTURE — 789

C. STRUCTURE AND FUNCTION OF HIV PROTEINS — 791
- I. VIRAL PROTEINS DERIVED FROM THE *gag* GENE — 792
- II. VIRAL PROTEINS DERIVED FROM THE *pol* GENE — 792
- III. VIRAL PROTEINS DERIVED FROM THE *env* GENE — 793
- IV. REGULATORY PROTEINS — 794
- V. ACCESSORY PROTEINS — 795

D. HIV INFECTION AND AIDS — 796
- I. CLINICAL VIEW OF HIV INFECTION — 796
- II. MOLECULAR EVENTS UNDERLYING HIV INFECTION — 798

E. THE IMMUNE RESPONSE DURING HIV INFECTION — 802
- I. Th RESPONSES — 802
- II. CTL RESPONSES — 803
- III. ANTIBODY RESPONSES — 803
- IV. CYTOKINES — 804
- V. CEM-15 (APOBEC3G) — 805
- VI. NK CELLS — 805
- VII. COMPLEMENT — 805

F. HOST FACTORS INFLUENCING THE COURSE OF HIV INFECTION — 805
- I. RESISTANCE TO HIV INFECTION — 806
- II. CLINICAL COURSE VARIABILITY — 806

G. EPIDEMIOLOGY AND SOCIOLOGY OF HIV INFECTION — 808
- I. TRANSMISSION OF HIV — 808
- II. EPIDEMIOLOGY — 808
- III. SOCIETAL AND ECONOMIC IMPACT OF AIDS — 812

H.	ANIMAL MODELS OF AIDS	813
	I. PRIMATE MODELS OF AIDS	813
	II. MOUSE MODELS OF AIDS	814
I.	HIV VACCINES	815
	I. BARRIERS TO HIV VACCINE DEVELOPMENT	815
	II. EXPERIMENTAL AIDS VACCINES IN ANIMALS	816
	III. EXPERIMENTAL AIDS VACCINES IN HUMANS	818
	IV. PASSIVE IMMUNIZATION	819
J.	TREATMENT OF HIV INFECTION WITH ANTI-RETROVIRAL DRUGS	819
	I. PROTEASE INHIBITORS	820
	II. NUCLEOSIDE RT INHIBITORS	820
	III. NON-NUCLEOSIDE RT INHIBITORS	820
	IV. FUSION INHIBITORS	820
	V. IMMUNE RESPONSES AND ANTI-RETROVIRAL THERAPY	821
	VI. OTHER DRUG THERAPY ISSUES	821

Chapter 26: Tumor Immunology — 825

A.	HISTORICAL NOTES	826
B.	TUMOR BIOLOGY	829
	I. WHAT IS A TUMOR AND WHAT IS A CANCER?	829
	II. CARCINOGENESIS	832
	III. TUMORIGENIC GENETIC ALTERATIONS	834
	IV. CARCINOGENS	839
C.	IMMUNE RESPONSES TO CANCER	842
	I. THE CONCEPT OF IMMUNOSURVEILLANCE	842
	II. TUMOR ANTIGENS	842
	III. IMMUNE RESPONSES TO TUMOR CELLS	845
	IV. HURDLES TO EFFECTIVE ANTI-TUMOR IMMUNITY	849
D.	CANCER THERAPY	851
	I. RADIATION THERAPY	851
	II. CHEMOTHERAPY	852
	III. TUMOR HYPOXIA	853
	IV. IMMUNOTHERAPY	856

Chapter 27: Transplantation — 873

A.	HISTORICAL NOTES	874
B.	THE MOLECULAR BASIS OF ALLORECOGNITION	880
	I. DIRECT ALLORECOGNITION	881
	II. INDIRECT ALLORECOGNITION	881
C.	MINOR HISTOCOMPATIBILITY ANTIGENS	881
D.	TYPES OF CLINICAL REJECTION AND THEIR MECHANISMS	885
	I. HYPERACUTE GRAFT REJECTION	885
	II. ACUTE GRAFT REJECTION	886
	III. CHRONIC GRAFT REJECTION	888
	IV. THE ROLE OF CYTOKINES AND CHEMOKINES IN GRAFT REJECTION	888
	V. A ROLE FOR TLRs IN GRAFT REJECTION?	888
	VI. GRAFT-VERSUS-HOST DISEASE (GvHD) IN SOLID ORGAN TRANSPLANTS	890

E.	HLA TYPING	890
	I. SEROLOGICAL TECHNIQUES	891
	II. TYPING BY CELLULAR RESPONSE: MLR	892
	III. TYPING AT THE DNA LEVEL	893
F.	IMMUNOSUPPRESSION	896
	I. AZATHIOPRINE	896
	II. CYCLOSPORINE A	896
	III. TACROLIMUS	897
	IV. MYCOPHENOLATE MOFETIL	897
	V. SIROLIMUS	897
	VI. MALONONITRILAMIDES (MNAs)	898
	VII. FTY720	898
	VIII. FLUDARABINE	898
	IX. ANTI-LYMPHOCYTE ANTIBODIES	898
	X. CYTOKINES	899
G.	INDUCTION OF GRAFT TOLERANCE	899
	I. BONE MARROW MANIPULATION	900
	II. THYMIC MANIPULATION	901
	III. COSTIMULATORY BLOCKADE	901
	IV. REGULATORY T CELLS	903
	V. REGULATION BY NKT CELLS	904
	VI. TOLEROGENIC DCs	905
	VII. CAVEATS	905
H.	XENOTRANSPLANTATION	906
	I. CHOICE OF SPECIES FOR XENOTRANSPLANTATION	907
	II. XENOGRAFT REJECTION	907
	III. TRANSMISSION OF ZOONOTIC DISEASES	909
	IV. REGULATORY AND LEGAL OBSTACLES	910
	V. ETHICAL AND MORAL CONSIDERATIONS	910
I.	BLOOD TRANSFUSIONS	910
	I. ALLOREACTIVITY IN BLOOD TRANSFUSIONS	910
	II. ENSURING THE SAFETY OF THE BLOOD SUPPLY	913
J.	HEMATOPOIETIC CELL TRANSPLANTATION (HCT)	913
	I. GRAFT REJECTION IN HCT	914
	II. GRAFT-VERSUS-HOST DISEASE (GvHD) IN HCT	915
	III. GRAFT-VERSUS-LEUKEMIA (GvL) EFFECT	916
	IV. BENEFICIAL EFFECTS OF ALLOGENEIC NK CELLS	916
	V. INFECTION CONTROL	917
K.	GENE THERAPY IN TRANSPLANTATION	917

Chapter 28: Allergy and Hypersensitivity — 923

A.	HISTORICAL NOTES	924
B.	TYPE I HYPERSENSITIVITY: IMMEDIATE OR IgE-MEDIATED	925
	I. WHAT IS TYPE I HS?	925
	II. MECHANISMS UNDERLYING TYPE I HS	927
	III. EXAMPLES OF TYPE I HS	930
	IV. ROLES OF FcεR AND IgE IN TYPE I HS	934
	V. ALLERGEN BIOLOGY	937
	VI. DIAGNOSIS AND THERAPY OF TYPE I HS	940

Contents

C. TYPE II HYPERSENSITIVITY: DIRECT ANTIBODY-MEDIATED
CYTOTOXICITY 949

   I.   WHAT IS TYPE II HS? 949

   II.   MECHANISMS UNDERLYING TYPE II HS 949

   III.   EXAMPLES OF TYPE II HS: CYTOTOXICITY AGAINST
MOBILE CELLS 949

   IV.   EXAMPLES OF TYPE II HS: CYTOTOXICITY AGAINST
FIXED TISSUES 953

D. TYPE III HYPERSENSITIVITY: IMMUNE
COMPLEX-MEDIATED INJURY 954

   I.   WHAT IS TYPE III HS? 954

   II.   MECHANISM UNDERLYING TYPE III HS 954

   III.   EXAMPLES OF TYPE III HS 956

E. TYPE IV HYPERSENSITIVITY: DELAYED-TYPE OR
CELL-MEDIATED HYPERSENSITIVITY 957

   I.   WHAT IS TYPE IV HS? 957

   II.   CHRONIC DTH REACTIONS 957

   III.   CONTACT HYPERSENSITIVITY (CHS) 958

   IV.   HYPERSENSITIVITY PNEUMONITIS (HP) 960

Chapter 29: Autoimmune Disease 963

A. OVERVIEW 964

   I.   HISTORICAL NOTES 964

   II.   WHAT CHARACTERIZES AN AUTOIMMUNE DISEASE? 965

   III.   OUR APPROACH TO DISCUSSING AUTOIMMUNITY 969

B. EXAMPLES OF HUMAN AUTOIMMUNE DISEASES 969

   I.   SYSTEMIC LUPUS ERYTHEMATOSUS (SLE) 970

   II.   RHEUMATOID ARTHRITIS (RA) 971

   III.   RHEUMATIC FEVER (RF) 972

   IV.   TYPE 1 DIABETES MELLITUS (T1DM) 973

   V.   MULTIPLE SCLEROSIS (MS) 974

   VI.   ANKYLOSING SPONDYLITIS (AS) 977

   VII.   SJÖGREN SYNDROME (SS) 977

   VIII.   AUTOIMMUNE THYROIDITIS 978

   IX.   AUTOIMMUNE THROMBOCYTOPENIC PURPURA 980

   X.   SCLERODERMA (SD) 981

   XI.   MYASTHENIA GRAVIS (MG) 982

   XII.   KAWASAKI DISEASE (KD) 982

   XIII.   POLYMYOSITIS (PM) 983

   XIV.   GUILLAIN-BARRÉ SYNDROME (GBS) 984

   XV.   PSORIASIS (PS) 984

   XVI.   ANTI-PHOSPHOLIPID SYNDROME (APS) 985

   XVII.   INFLAMMATORY BOWEL DISEASE (IBD): CROHN'S
DISEASE (CD) AND ULCERATIVE COLITIS (UC) 985

   XVIII.   PEMPHIGUS (PG) 986

   XIX.   GOODPASTURE'S SYNDROME (GS) 986

   XX.   IMMUNODYSREGULATION, POLYENDOCRINOPATHY,
ENTEROPATHY X-LINKED (IPEX) SYNDROME 986

C. ANIMAL MODELS OF AUTOIMMUNE DISEASE 986

   I.   NZB/W F1 MICE 988

   II.   *MRL/lpr* AND *gld* MICE 988

   III.   NOD MICE 989

   IV.   *Scurfy* MICE 990

   V.   THYMECTOMY OF NEONATAL MICE 991

   VI.   IBD MODELS 991

   VII.   EXPERIMENTAL AUTOIMMUNE ENCEPHALITIS (EAE) 991

   VIII.   COLLAGEN-INDUCED ARTHRITIS (CIA) 992

   IX.   EXPERIMENTAL AUTOIMMUNE MYASTHENIA GRAVIS (EAMG) 992

   X.   EXPERIMENTAL AUTOIMMUNE THYROIDITIS (EAT) 993

   XI.   EXPERIMENTAL AUTOIMMUNE UVEORETINITIS (EAU) 993

   XII.   EXPERIMENTAL AUTOIMMUNE MYOCARDITIS (EAM) 994

   XIII.   OTHER POTENTIAL AID MODELS 994

D. DETERMINANTS OF AUTOIMMUNE DISEASE 995

   I.   GENETIC PREDISPOSITION 995

   II.   ENVIRONMENTAL TRIGGERS 998

   III.   HORMONAL INFLUENCES 1000

   IV.   REGIONAL/ETHNIC DIFFERENCES 1000

E. MECHANISMS UNDERLYING AID 1001

   I.   PATHOGEN-RELATED MECHANISMS 1002

   II.   INHERENT DEFECTS IN IMMUNE SYSTEM CELLS 1006

   III.   ALTERATIONS TO CYTOKINE EXPRESSION 1010

   IV.   DEFECTS IN THE COMPLEMENT SYSTEM 1011

   V.   EPITOPE SPREADING 1011

F. THERAPY OF AID 1012

   I.   CONVENTIONAL TREATMENTS 1012

   II.   IMMUNOTHERAPEUTICS 1014

G. RELATIONSHIP BETWEEN AID AND CANCER 1020

Chapter 30: Hematopoietic Cancers 1025

A. HISTORICAL NOTES 1026

B. OVERVIEW OF HEMATOPOIETIC CANCER BIOLOGY 1027

   I.   WHAT ARE HEMATOPOIETIC CANCERS? 1027

   II.   HEMATOPOIETIC CANCER CARCINOGENESIS 1029

C. TERMS USED IN CLINICAL ASSESSMENT AND
TREATMENT OF HC 1030

D. LEUKEMIAS 1035

   I.   ACUTE MYELOID LEUKEMIA (AML) 1035

   II.   CHRONIC MYELOGENOUS LEUKEMIA (CML) 1039

   III.   ACUTE LYMPHOBLASTIC LEUKEMIA (ALL) 1041

   IV.   CHRONIC LYMPHOCYTIC LEUKEMIA (CLL) 1044

   V.   OTHER LEUKEMIAS 1045

E. PLASMA CELL DYSCRASIAS 1046

   I.   MONOCLONAL GAMMOPATHY OF UNDETERMINED
SIGNIFICANCE (MGUS) 1047

   II.   WALDENSTRÖM'S MACROGLOBULINEMIA (WM) 1047

   III.   MYELOMA 1047

F. LYMPHOMAS 1050

   I.   HODGKIN'S LYMPHOMA (HL) 1050

   II.   NON-HODGKIN'S LYMPHOMAS 1054

Appendix: CD Molecules 1065

Glossary 1119

Index 1179

Part I

Basic Immunology

Perspective on Immunity and Immunology

1

Louis Pasteur 1822–1895

Germ theory of disease

© Musée Pasteur Paris

CHAPTER 1

A. WHAT IS IMMUNOLOGY?

B. WHY HAVE AN IMMUNE SYSTEM AND WHAT DOES IT DO?

C. TYPES OF IMMUNE RESPONSES: INNATE AND ADAPTIVE

D. WHAT IS "INFECTION"?

E. PHASES OF HOST DEFENSE

F. HOW ARE ADAPTIVE AND INNATE IMMUNITY RELATED?

G. LEUKOCYTES: CELLULAR MEDIATORS OF IMMUNITY

H. WHERE DO IMMUNE RESPONSES OCCUR?

I. CLINICAL IMMUNOLOGY: WHEN THE IMMUNE SYSTEM DOES NOT WORK PROPERLY

"Science knows no country, because knowledge belongs to humanity, and is the torch which illuminates the world"–Louis Pasteur

A. What Is Immunology?

Immunology is a relatively young science: the word "immunology" did not even appear in the *Index Medicus* until 1910. Immunology has its roots in several other disciplines, including microbiology, biochemistry, genetics, and pathology. Simply put, *immunology is the study of the immune system*. The immune system is *the body's defense system against invasion by non-self entities*, including infectious and inert agents, and tumor cells. The normal functioning of the immune system gives rise to *immunity*, a word derived from the Latin word *immunitas*, meaning "to be exempt from." Originally, this term referred to the fact that certain individuals could be immune to, or exempt from, public service, in the same way as the Church was "immune" from civil control. It had been noticed for centuries that although many people died after exposure to a particular disease, some survived and did not get sick when exposed a second time. In the 1500s, before the causes of disease were understood, these survivors were said to have become "exempt from" the disease, or "immune."

Thucydides of Athens in 430 BC is often cited as being the first to articulate the concept of immunity, in his observations that a first exposure to plague rendered those individuals resistant to a second infection.

> Yet it was with those who had recovered from the disease that the sick and the dying found most compassion. These knew what it was from experience, and had now no fear for themselves; for the same man was never attacked twice—never at least fatally. (*Thucydides, historian—Description of a plague, Athens, 430 BC*)

Even 1000 years later, the same observations were being made, but the causes of diseases and resistance to them remained mysteries.

Still at a later time it came back; then those who dwelt roundabout this land, whom formerly it had afflicted most sorely, it did not touch at all. (*Procopius, historian—Description of a plague, AD 541*)

By the passage of an additional 10 centuries, scientists and physicians of the day were speculating as to whether immunity could be manipulated, although there was still no clear understanding of the mechanisms of immunity.

> Moreover, I have known certain persons who were regularly immune, though surrounded by the plague-stricken, and I shall have something to say about this in its place, and shall inquire whether it is impossible for us to immunize ourselves against pestilential fevers. (*Giralamo Fracastoro—From the book, "On Contagion," 1546*)

Crude practical attempts at inducing immunity by *variolation* (the inhalation of material from smallpox pustules or its insertion into superficial wounds) were recorded by the Chinese and Turks of the 1400s and were put into practice by Lady Mary Wortley Montagu (the wife of the British ambassador to Constantinople) on her own children in 1718. However, it was not until the experiments of the English physician Edward Jenner, in 1796, that immunology as an independent science was born. Jenner had observed that dairymaids and farmers were "fair-skinned" and lacked the pock-marked complexions of their fellow citizens. This observation led Jenner to conclude that the cattle workers had somehow become immune to smallpox, a severe and often fatal disorder that decimated whole villages (Plate 1-1). Jenner wondered whether the cattle workers' resistance to smallpox was related to their close contact with cows. Cows of that era were often afflicted with cowpox disease, a disease similar to smallpox but much less severe. In an experiment that would be prohibited on ethical grounds today, Jenner deliberately exposed an 8-year-old boy to fluid from a cowpox lesion, then 2 months later

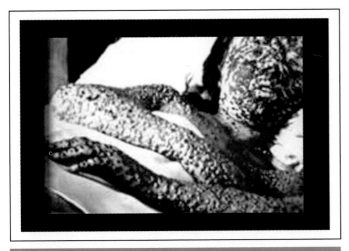

Smallpox
Reproduced courtesy of the Jenner Museum, England
(http://www. jennermuseum.com).

inoculated the same boy with infectious material from a small-pox patient (Plate 1-2). The boy did not develop smallpox, and the first step towards the structured examination of the body's response to foreign material had been taken. Jenner's approach to smallpox prevention was subsequently published and was quickly adopted in countries throughout Europe. The complete story of smallpox immunization, which is one of the most successful public health endeavors in history, is discussed in more detail in Chapter 23.

In Jenner's time, the cause of infectious disease was still a mystery. Theories of the day did not envision the transmission of disease-causing germs; the body was thought to be a passive

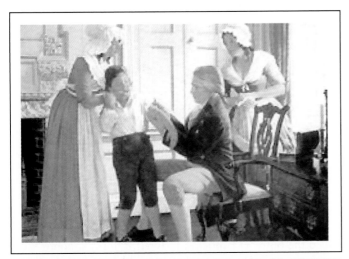

The First Immunization by Edward Jenner
In this painting, Edward Jenner (seated) is depicted vaccinating 8-year-old James Phipps with fluid taken from a cowpox pustule on the hand of local milkmaid Sarah Nelmes (right). The vaccination took place on May 14, 1796. Several weeks later, Jenner inoculated Phipps with live smallpox. The boy did not develop the disease, proving the vaccination was a success. Reproduced with permission from Pfizer.

target, rather than an active combatant. Almost a century passed before Robert Koch proposed the *germ theory of disease* in 1884, which stated that pathogenic microbes invisible to the naked eye (bacteria, viruses, fungi, and parasites) were responsible for specific illnesses. Louis Pasteur had already been thinking very similar thoughts, and had taken Jenner's immunization technique for smallpox and applied it to various animal diseases in the late 1800s. Although Jenner did not know it at the time, he had used the weaker cowpox virus present in cowpox lesions to protect against the development of smallpox caused by the related but much more virulent virus, *variola*. Working in 1870 with diseases of chickens and sheep, and in 1884 with the rabies virus, Pasteur demonstrated that inoculation with a weakened, or *attenuated*, pathogen could protect against a subsequent exposure to the naturally occurring (*virulent*) form of the pathogen. In honor of Jenner's work, Pasteur coined the term *vaccination* (from the Latin *vaccinus*, "derived from cows") for this procedure. Slowly, vaccination became the accepted practice for inducing immunity to infectious disease (see Box 1-1).

Although scientists of Pasteur's time could show that vaccination worked, they could not explain why. Human pathogens were first identified in the late 1800s, and the study of their effects on the body engendered the evolution of immunology from the established fields of microbiology, pathology, and histology. For some time, pathologists observing sites of injury or disease had noted the appearance of cellular infiltrates that stimulated healing and minimized the spread of infection. The importance of these cells in immunity was recognized after the discovery in 1883 by Elie Metchnikoff that certain white blood cells could engulf and digest a variety of pathogenic organisms. These *phagocytes* ("eating cells") appeared to provide non-specific protection to the host and to be more active in immunized animals than in nonimmunized ones, leading Metchnikoff to propose that immunity was *cell-mediated*.

Meanwhile, others were arguing that a different mechanism was responsible for immunity. Emil von Behring and Shibasaburo Kitasato demonstrated, in work that won Behring the first Nobel Prize for medicine in 1901, that the serum fraction of blood from immunized animals could confer immunity on unimmunized animals. Because this form of immunity was derived from body fluids (known as "body humors"), it was called *humoral immunity*. Controversy erupted over whether the true basis of immunity was cellular or humoral (see Box 1-2). By the 1930s, intensive biochemical study by those in the humoral camp had shown that serum contained substances (then called *serum antitoxins*) that could clump and lyse bacteria, and precipitate and neutralize toxins and viruses. Biochemists soon demonstrated that one type of protein, termed an *antibody*, was responsible for all of these activities, and the elucidation of its basic molecular structure was quickly undertaken. The "humoral vs. cell-mediated" debate raged back and forth until the discovery in the 1950s of *lymphocytes*, a class of blood cells also found in lymph and lymph nodes. These cells were shown to mediate activities essential to the immune response, including antibody production. It became clear that both humoral and cell-mediated

Box 1-1. Smallpox vaccine on the move

Edward Jenner was both the inventor of the smallpox vaccine and an outspoken proponent of making it available to people throughout the world as quickly as possible. Two obstacles stood in the way of this noble cause. Firstly, Jenner encountered doubts and skepticism about the safety and effectiveness of the vaccine from people inside and outside the medical establishment, and on many occasions he had to fiercely defend his position. Secondly, he was working during the dawn of the 1800s, so that medical advances of any sort were hindered by communication and travel times that, by today's standards, would be unimaginably slow. With these considerations in mind, the speed at which smallpox vaccination became part of standard medical practice is astounding. To appreciate just how quickly this procedure became routine throughout Europe, consider the accompanying photograph of an original "Certificate of Cowpox Inoculation." This document was found on a Danish island by a descendant of Gomme Andersen, who, as a 2-month-old boy, was vaccinated against smallpox in 1814. This record, dated just 18 years after Jenner's first experiments, attests to the rapid geographical spread of smallpox immunization as a routine medical procedure. The document reads as follows:

Certificate of Cowpox Inoculation
Gomme Andersen, born in Nordby and living in the parish of Nordby, 2 months of age, has by me, the undersigned, been inoculated with cowpox on November 15 of the year 1814. By careful follow-up examination between the 7th and 9th day after inoculation, there were found all the signs showing these to be true cowpox: they were whole, intact and filled with a clear fluid, depressed in the middle and surrounded by a red circle. Gomme Andersen has therefore successfully been exposed to the true cowpox which will protect him against small pox in the future, to which I attest in honour and good faith.

Nordby, February 22, year 1815
H. S. Heegaard

Cases of smallpox decreased steadily throughout the 1800s wherever vaccination was practiced. By 1940, the disease had vanished from England and many parts of Europe. By 1950, smallpox cases were no longer reported in North America, and by 1960, the same was true in China. At that point, the World Health Organization recognized the need to focus on smallpox eradication in developing countries. By 1980, smallpox eradication was complete worldwide. Plate courtesy of Dr. Anders Bennick, University of Toronto.

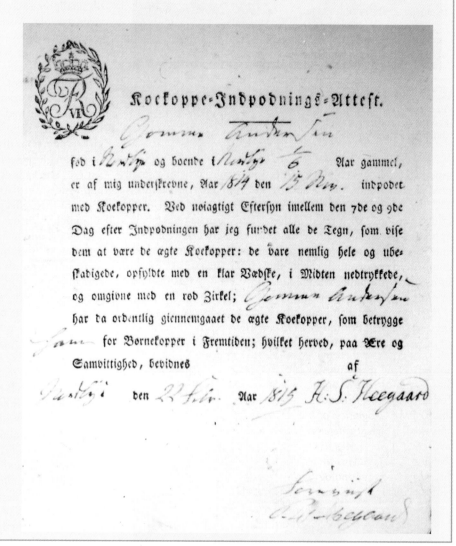

immunity contributed to a complete immune response, and that both types of immunity were dependent on lymphocytes.

The discovery of lymphocytes sustained a surge of interest in cellular immunology into the 1960s. Two types of lymphocytes capable of responding specifically to foreign entities were identified: the *B lymphocytes*, responsible for antibody production, and the *T lymphocytes*, responsible for cell-mediated immunity. The B cells were so labeled because they were originally identified in birds in a unique avian organ called the Bursa of Fabricius; fortunately for taxonomy, B cell development in humans takes place almost entirely in the *bone marrow*. T cell precursors also originate in the bone marrow but migrate to the thymus as immature cells and complete their development there. Today, it is accepted that both the "humoralists" and "cellularists" were correct, that the cellular (T cell) and humoral (B cell) forms of immunity are interrelated and interdependent, and that both are essential to the body's defense.

Box 1-2. Is immunity humoral or cell-mediated?

In the 1800s, it was recognized that immunity to disease could be transferred via the blood of an immune individual. Towards the end of the 19th century, a heated debate was raging over which component of blood—the fluid portion (*plasma*) or the cellular portion (the red and white blood cells)—was responsible for immunity. In 1883, Elie Metchnikoff observed that white blood cells were capable of engulfing fungal spores (a process he called *phagocytosis*, or "cell-eating"). He further noted that these phagocytic cells were more active in immunized animals, as compared to unimmunized animals, and he proposed that immunity to microbes thus had to be *cell-mediated*. However, Emil von Behring and Shibasaburo Kitasato demonstrated in 1890 that inoculation with the blood *serum* (the fluid remaining after plasma is allowed to clot) from animals immunized against diphtheria could protect unimmunized animals. The factor conferring this immunity was called an *antitoxin* (later shown to contain antibodies) and its discovery clearly demonstrated that immunity could be *humoral*, that is, a

property of a body fluid, or "humor." The humoral immunity hypothesis was further bolstered by the discovery of the *complement* system (an antimicrobial enzymatic system in the blood) in 1894 by Jules Bordet, and the demonstration of the *precipitin reaction* (later shown to be due to large antigen–antibody complexes which are insoluble and precipitate out of solution) in 1897 by Rudolf Kraus. Of course, none of this could explain Metchnikoff's observation. The debate was cooled somewhat by Sir Almroth Wright's demonstration in 1903 that antibodies could facilitate phagocytosis by coating microbes (*opsonization*).

Extensive studies of antibodies and the cellular components of blood to determine the agents responsible for both forms of immunity proceeded throughout the first half of the 20th century. In the 1940s, Albert Coons used *immunofluorescence* (the tagging of antibody molecules with fluorescent markers) to show that antigens and antibodies could be identified *in situ* inside cells, and Merrill Chase and Karl Landsteiner demonstrated that at least

one form of hypersensitivity (an immune response to a normally innocuous antigen) was mediated by cells and not by serum. Astrid Fagraeus discovered in 1948 that antibodies were synthesized by *plasma cells*. The controversy was forever put to rest in the 1950s when it was determined that the lymphocyte was ultimately responsible for both humoral and cell-mediated immune responses. That the humoral and cellular elements of the immune response were interdependent was shown in 1966, when it was demonstrated by Henry Claman and his colleagues that there were two types of lymphocytes: the *B cells* (primarily responsible for antibody production) and the *T cells* (primarily responsible for either cytokine release or the cellular destruction of certain pathogens). These researchers determined that B cells and T cells had to cooperate to generate a specific humoral response. Today, the immune response is viewed as a multifaceted event with both humoral and cell-mediated components, and the nature of B–T cell cooperation and interdependence are the focus of much ongoing research.

As immunology overlaps more and more with what were thought to be the strictly delineated fields of pathology, biochemistry, and genetics, we realize that science must be multidisciplinary, just as living organisms depend on the coordination of multiple body systems. It is crucial to understand not only the individual elements of the immune system, but also how the system operates in the organism as a whole, and for that, we must draw on knowledge derived from every field. Today, the core of modern immunology can be defined as *the study of the recognition of non-self entities; the activation, development, and specific defense functions of lymphocytes; the interactions of lymphocytes with other cells; and the investigation of the genes and proteins underlying these interactions.* Landmark discoveries in immunology and immunologists who have won Nobel prizes for their work are featured in Boxes 1-3 and 1-4.

Our environment teems with pathogens of viral, microbial, parasitic, and fungal origin, each of which furnishes many different antigens; yet most infections in healthy normal individuals have few lasting adverse consequences. This happy state is due entirely to the existence of the *immune response*, a coordinated action by numerous cellular and soluble components in a network of tissues and circulating systems. Thus, the immune system is a system of cells, tissues, and their soluble products that recognizes, attacks, and destroys that which is "foreign" to the body. Immunity is the ability to rid the body successfully of a foreign entity. However, it is important to keep in mind at this juncture that, as powerful as it is, the immune system is not able to conquer all infections. The epidemic proportions and persistence of chronic infections such as hepatitis B, hepatitis C, and human immunodeficiency virus (HIV) remind us that certain pathogens are able to circumvent the immune response and cause long-term and debilitating consequences for their hosts.

B. Why Have an Immune System and What Does It Do?

The immune system has evolved to counteract assault on the body by non-self entities, or *antigens*. Such entities include infectious agents, such as viruses, bacteria, and parasites, but also inert injurious materials, such as splinters, and, in some situations, host-generated threats, such as cancers. Since microbes are literally everywhere on Earth, the immune system is primarily occupied with containing attacks from this quarter. Accordingly, we will often focus on this aspect of the immune response to illustrate its properties and functions.

C. Types of Immune Responses: Innate and Adaptive

Any immune response by an organism to an invading foreign entity can be broken down into two stages: (1) the *recognition* of the entity as being non-self and (2) an *effector response*, leading to the elimination of the entity. Immune responses also have the property of being self-limited; that is, the response subsides as the antigen provoking it in the first place is eliminated.

Box 1-3. An abbreviated chronology of landmark discoveries in immunology

1798 Edward Jenner: Cowpox vaccination to prevent smallpox infection.

1880 Louis Pasteur: Attenuated vaccines.

1884 Elie Metchnikoff: Phagocytic theory of immune defense.

1888 Emile Roux and Alexandre Yersin: Isolation of diphtheria toxin.

1889 Hans Buchner: Discovery of complement.

1890 Emil von Behring and Shibasaburo Kitasato: Discovery of antitoxins (antibodies) and serum therapy.

1891 Robert Koch: Koch phenomenon and the tuberculin skin test.

1894 Richard Pfeiffer: Discovery of immune bacteriolysis.

1896 Herbert Durham and Max von Gruber: Bacterial agglutination reactions.

1897 Paul Ehrlich: Side-chain theory of antibody formation.

1897 Rudolph Kraus: Precipitin reaction.

1900 Karl Landsteiner: Discovery of A, B, and O blood groups.

1901 Jules Bordet and Octave Gengou: Complement fixation.

1902 Paul Portier and Charles Richet: Anaphylaxis.

1903 Maurice Arthus: The Arthus reaction.

1903 Almroth Wright and S. R. Douglas: Opsonization reactions.

1904 Julius Donath and Karl Landsteiner: Discovery of autoimmune disease (paroxysmal cold hemoglobinuria).

1905 Clemens von Pirquet and Bela Schick: Serum sickness.

1910 Dale and Barger: Discovery of histamine.

1910 Peyton Rous: Experimental viral cancer immunology.

1920 Karl Landsteiner: Hapten inhibition.

1921 Calmette and Guérin: First use of bacillus Calmette-Guérin (BCG) for tuberculosis vaccination.

1921 Carl Prausnitz and Heinz Küstner: Cutaneous reactions and the passive transfer of allergy.

1923 Gaston Ramon: Formaldehyde treatment of diphtheria toxin to produce toxoid.

1923 Michael Heidelberger and Oswald Avery: Pneumococcal polysaccharides.

1929 Michael Heidelberger and Forrest Kendall: Quantitative precipitin reaction.

1929 Louis Dienes and E. W. Schoenheit: Delayed hypersensitivity to simple proteins.

1930 F. Breinl and Felix Haurowitz: Template ("instructive") theory of antibody formation.

1935 Alexandre Besredka: Local immunity and oral immunization.

1936 Peter Gorer: Identification of H-2 antigen in mice.

1939 Arne Tiselius and Elvin Kabat: Demonstration that antibodies are gamma globulins.

1940 Karl Landsteiner and Alexander Wiener: Rh blood group system.

1942 Albert Coons et al.: Fluorescein labeling of antibodies; immunofluorescence.

1942 Jules Freund and K. McDermott: Freund's adjuvant.

1942 Lloyd Felton: Immunological unresponsiveness.

1942 Karl Landsteiner and Merrill Chase: Passive cellular transfer of delayed type hypersensitivity.

1944 Peter Medawar: Transplantation immunology.

1945 Ray Owen: Chimerism in dizygotic cattle twins.

1946 Jacques Oudin: Precipitin reaction in gels.

1948 Örjan Ouchterlony and S. D. Elek: Immunodiffusion in agarose gels.

1948 Astrid Fagraeus: Antibodies produced by plasma cells.

1952 Ogden Bruton: Description of agammaglobulinemia in humans.

1953 Pierre Grabar and Curtis Williams: Immunoelectrophoresis and the heterogeneity of antibodies.

1953 Peter Medawar, F. MacFarlane Burnet, and Frank Fenner: Theories and demonstration of immunological tolerance.

1955 Niels Jerne: Natural selection theory of antibody formation.

1956 Bruce Glick et al.: Discovery of the bursa of Fabricius.

1956 Elvin Kabat: Size of the antibody combining site.

1957– F. MacFarlane Burnet, David
1959 Talmage, and Joshua Lederberg: Clonal selection theory of antibody formation.

1958 Jean Dausset and Rapaport: Histocompatibility antigens on leukocytes.

1958 Rodney Porter: Enzymatic cleavage of antibody molecules.

1959 Joseph Heremans et al.: Discovery of IgA.

1959 James Gowan: Lymphocyte circulation.

1960 Rosalyn Yalow and S. A. Berson: Radioactive labeling and radioimmunoassays.

1961 Jacques F. A. P. Miller: The role of the thymus in immunity.

1961 James Till and Ernest McCulloch: Existence of hematopoietic stem cells.

1963 Niels Jerne and Albert Nordin: Hemolytic plaque assay.

1963 Henry Kunkel et al.: Discovery of antibody idiotypes.

1966 Barry Bloom and B. Bennet: Discovery of lymphokines.

1966 S. Avrameas et al.: Antigen and antibody labeling with enzymes.

1966 Henry Claman, E. A. Chaperon, and R. F. Triplett: B cell/T cell cooperation.

1966 Kimishige Ishizaka and Teruko Ishizaka: Discovery of IgE.

1969 N. Avrion Mitchison: Helper T cells.

1969 Gerald Edelman: Primary sequence of an immunoglobulin molecule.

1969 Leonard Herzenberg: Invention of cell sorting based on fluorescence.

1970 Elvin Kabat and T. T. Wu: Hypervariable regions of Ig molecules.

1971 Donald Bailey: Recombinant inbred mouse strains.

1972 Baruj Benacerraf and Hugh McDevitt: Histocompatibility-linked immune response genes.

1972 Hans Muller-Eberhard: Characterization of the membrane attack complex of the complement cascade.

1974 Rolf Zinkernagel and Peter Doherty: Major histocompatibility complex (MHC) restriction of immune responses.

1975 Georges Köhler and César Milstein: Development of hybridoma technology to produce monoclonal antibodies.

1976 Nobu Hozumi and Susumu Tonegawa: Discovery of V(D)J gene segments in the generation of antibody diversity.

1977 Jean Borel: Demonstration of cyclosporine A as an immunosuppressive drug.

Continued

Box 1-3. An abbreviated chronology of landmark discoveries in immunology—*cont'd*

1978–
1981 *Growth and maintenance of T lymphocyte clones in long-term culture.

1980 World Health Organization: Worldwide eradication of smallpox.

1980 Ralph Steinman: Characterization of dendritic cells and their role in immunity.

1983 Luc Montagnier: Isolation of the human immunodeficiency virus (HIV) from an AIDS patient.

1983 Melvin Bosma: First report of a natural mouse mutant having severe combined immunodeficiency (SCID).

1984 Tak Mak and Mark Davis: Cloning of the $\alpha\beta$ T cell receptor genes.

1984–
1985 Susumu Tonegawa: Cloning of the γ T cell receptor gene.

1984–
1986 Eugene Butcher and Irving Weissman: Lymphoctye migration and lymphocyte homing receptors.

1985–
1986 Emil Unanue, Alain Townsend, and Howard Grey: Presentation of antigenic peptide in the context of self MHC molecules.

1985–
1986 Donald Metcalf: Characterization and cloning of hematopoietic growth factors.

1985–
1986 *Role of CD4 and CD8 in T cell signal transduction.

1985–
1988 *Recognition of stress proteins by the immune system.

1985–
1989 *Immune recognition of stress proteins.

1986 Timothy Mosmann and Robert Coffman: Discovery of Th1/Th2 cellular subsets.

1987 Pamela Bjorkmann and Donald Wiley: First MHC crystal structure.

1987 Yueh-Hsiu Chien, Mark Davis, *et al.*: Cloning of the T cell δ receptor gene.

1989 Christopher Goodnow: Experimental induction of self-tolerance in peripheral B cells.

1989 John Kappler and Philippa Marrack: Superantigen stimulation of T cells.

1989 Mario Capecchi, Martin Evans, and Oliver Smithies: Generation of first knockout mice.

1990–
1992 Greg Winter: Phage display technology for *in vitro* antibody selection and production.

1991 Harald von Boehmer: First demonstration of thymic selection of T cells.

1991 World Health Organization–International Union of Immunological Societies (WHO–IUIS): Formal adoption of nomenclature guidelines for secreted regulatory proteins of the immune system (interleukins).

1995 Lewis Lanier, Erik Long, and others: Identification and cloning of MHC-binding membrane receptors on NK cells.

1997 Charles Janeway, Jr. and Ruslan Medzhitov: Cloning and characterization of a mammalian Toll-like receptor (TLR).

2001 *Multigroup collaboration: Cloning of the human genome. Reported in *Science* **291**, 5507 (2001) and *Nature* **409**, 6822 (2001).

* A number of researchers contributed concurrently to advances in these areas.

Box 1-4. Nobel Prize winners in immunology

1901 **Emil von Behring**, for his discovery of serum antitoxins (antibodies) and serum therapy, and its application to the treatment of diphtheria.

1905 **Robert Koch**, for his investigations and discoveries in regard to tuberculosis. Koch developed tuberculin reactivity tests which later were important in the development of our current understanding of cellular immunity.

1908 **Elie Metchnikoff**, for his discovery of phagocytosis, and **Paul Ehrlich**, for his work on fundamental immunology.

1913 **Charles Robert Richet**, for the discovery of anaphylaxis.

1919 **Jules Bordet**, for his studies in regard to immunology, particularly complement-mediated lysis.

1930 **Karl Landsteiner**, for his discovery of the human blood groups.

1951 **Max Theiler**, for his development of vaccines against yellow fever.

1957 **Daniel Bovet**, for his development of antihistamines in the treatment of allergy.

1960 **Frank MacFarlane Burnet** and **Peter Brian Medawar**, for the discovery of acquired immunological tolerance.

1972 **Gerald Maurice Edelman** and **Rodney Robert Porter**, for their discoveries concerning the chemical structure of antibodies.

1977 **Rosalyn Yalow** (shared with **Roger Guillemin** and **Andrew Schally**), for the development of radioimmuno-assays for peptide hormones.

1980 **Jean Dausset, George Davis Snell**, and **Baruj Benacerraf**, for their discoveries of the histocompatibility antigens on human and animal cells and their role in tissue and blood transplantation rejection (Dausset and Snell), and for work on the genetic control of immune responses (Benacerraf).

1984 **Georges J. F. Köhler** and **César Milstein**, for their development of cell hybridization as a technique to produce monoclonal antibodies, and **Niels K. Jerne**, for his many fundamental contributions to theoretical immunology.

1987 **Susumu Tonegawa**, for his work on the immunoglobulin genes and the mechanism of generating antibody diversity.

1990 **Joseph E. Murray** and **E. Donnall Thomas**, for their work on organ and bone marrow transplantation.

1996 **Peter C. Doherty** and **Rolf M. Zinkernagel**, for their discovery of the MHC restriction of T cell responses.

Immunity and Defense against Pathogens
Pathogens that manage to penetrate the body's physical barriers (skin, mucous membranes, enzymes, etc.) are first met by cells of the innate immune system. Using broadly specific control mechanisms, these cells attempt to limit the pathogen's spread. If the innate immune system cells are unable to eradicate the threat, immune system cells called lymphocytes (which recognize unique antigens) generate humoral and cell-mediated adaptive responses to pathogens. The interplay between innate and adaptive immunity occurs via chemical messengers and through direct contact between cells of the innate and adaptive responses. By cooperating in this way, innate and adaptive immunity mount an optimum defense against pathogens.

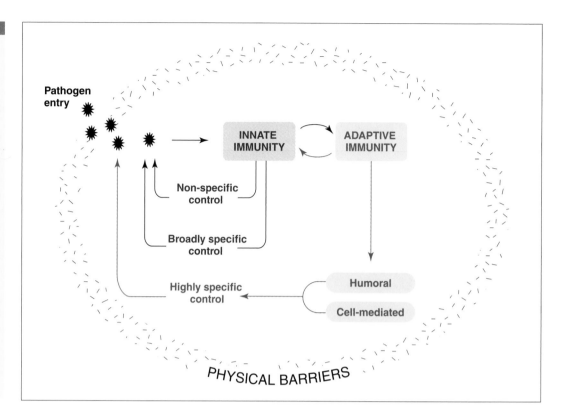

Vertebrates are capable of two types of immune responses, *innate* and *adaptive*. Both the innate and adaptive immune responses distinguish between self and non-self antigens, but the degree of specificity and the mechanisms underlying the recognition of the foreign entity are very different. The innate immune response has two components: (1) a collection of physical, chemical, and molecular barriers that distinguish and exclude antigens in a totally non-specific way and (2) an army of cells possessing receptors that recognize a limited number of molecular patterns common to a wide variety of pathogens. In other words, the recognition and effector stages of the innate response are only broadly specific and are not designed to recognize unique antigens. In addition, should the same pathogen attack a second time, the innate response is exactly the same as during the first encounter. In contrast, the adaptive (also called the *specific* or *acquired*) immune response involves recognition and effector actions that are highly specific to the precise foreign entity that has triggered the response. Unique antigens derived from the foreign entity bind specifically to receptors expressed by B and T lymphocytes, which then take effector action against the entity. Furthermore, a second attack by the same pathogen induces an adaptive response that is much faster and stronger than the first, such that the host does not get sick: the host has "adapted" its immune system such that the pathogen cannot even gain a toehold.

The innate and adaptive immune responses do not operate in isolation, and each is dependent on or enhanced by elements of the other (Fig. 1-1). To achieve optimum defense against pathogens, both the adaptive and innate systems must be functioning normally. The lymphocytes and antibodies of the adaptive response require the involvement of cells and chemical messengers of the innate system to initiate and complete their task of pathogen elimination. Similarly, the barrier functions of the innate response are made more effective by the adaptive immune response cells that reside at or can travel to the barrier surfaces.

D. What Is "Infection"?

Our bodies are under constant assault by less than benign microbes. Over 150 types of viruses are known to invade human cells. It has been estimated that, when breathing indoor air, we inhale an average of eight microbes per minute, and yet, most of the time, these assaults are successfully repelled and disease is prevented. This resistance is due to the sophisticated immune systems developed by vertebrates to combat the pathogenicity of microbes. However, it is a continual horse race as to which will be the more effective mechanism: the vertebrate's immune surveillance or the microbe's strategies to invade the body and establish an infection.

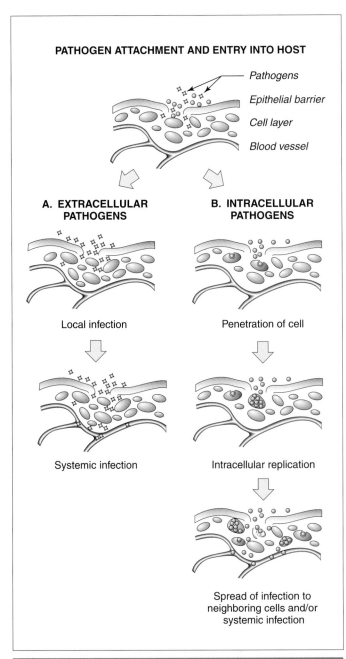

PATHOGEN ATTACHMENT AND ENTRY INTO HOST

Pathogens
Epithelial barrier
Cell layer
Blood vessel

A. EXTRACELLULAR PATHOGENS

Local infection

Systemic infection

B. INTRACELLULAR PATHOGENS

Penetration of cell

Intracellular replication

Spread of infection to neighboring cells and/or systemic infection

Figure 1-2

Establishment of Infection
(A) Extracellular pathogens (e.g., *Staphylococcus aureus, Streptococcus pyogenes*) that are not stopped by epithelial barriers will gain access to the underlying tissues and replicate to cause a local infection. Further spread may result in the pathogen entering the bloodstream to cause a systemic infection. (B) After gaining entrance into the host, intracellular pathogens (e.g., hepatitis C virus, *Mycobacterium tuberculosis*) must penetrate a host cell in order to establish an infection. Replication within the host cell leads to the death of that cell and spread of the pathogen to new host cells and/or to the bloodstream.

Like all species, microorganisms live to reproduce. However, in order to reproduce, many must gain access to the interior of a host cell, and others must at least penetrate the host's body. *Infection* is defined as the attachment and entry of a pathogen into the host's body, which can occur only if the pathogen is successful in avoiding the body's cleansing secretions and penetrating its anatomical defenses (Fig. 1-2). Once inside the body or the cell, the pathogen replicates, spreading its descendants in a localized or systemic fashion. *Extracellular pathogens* are those that do not enter cells, like certain bacteria or their toxins. These first collect in the interstitial fluid in the tissues and then may disseminate in the blood, overwhelming normal body system function and making it "sick." In contrast, *intracellular pathogens*, such as viruses and certain bacteria and parasites, enter a host cell and take over its metabolic machinery. These pathogens subvert the cell and cause it to churn out new virus particles, bacteria or parasites. The host cell is then unable to perform its normal function: it is "sick." The job of the immune response is to "clean up" the infected blood, interstitial fluid, and tissues and to destroy infected host cells so that neighboring host cells do not share their fate. This task requires several strategies, starting with the natural barriers of the innate immune system which help to prevent or contain microbial multiplication. Firstly, certain soluble factors and innate immune cells attempt to destroy foreign macromolecules, microbes, or infected cells. These cells also secrete protein growth factors that attract and stimulate lymphocytes. Secondly, lymphocytes activated by antigen respond with the effector functions of adaptive immunity: the production of antigen-specific antibodies and lytic cells. The invading microbe is successful only if it evades all of these defenses long enough to reproduce, and only if its offspring exit from the body and are transmitted to fresh hosts. Often the microbe's replicative cycle, and occasionally the immune response induced to counter it, may cause or induce tissue damage to the host, which is manifested as particular clinical symptoms of illness. The host can be considered successful if it survives the pathogenic assault with minimal damage.

E. Phases of Host Defense

Following attack by an infectious agent, a vertebrate defends itself in three phases, two of which are innate and the last of which involves adaptive immunity. Although these phases are sequential for an individual pathogen, such as a single influenza virus particle, the actual assault on a host involves billions of identical individual organisms, not all of which will pass through the phases at the same rate. As a result, in the host as a whole, and in the invading pathogen population as a whole, all three phases may be happening concurrently.

Taking immediate effect are the *non-specific, non-inducible* elements of the innate system, involving inherent features of the host's physiology, such as skin pH or proteases in the saliva (Fig. 1-3). The second phase, which occurs about 4–96 hours following a pathogen attack, is marked by innate host defense mechanisms that are *induced* by the infection but which mediate recognition of a broad range of pathogens. Effector elements of this second phase (such as cells that can secrete proteases and growth factors, and phagocytes that can

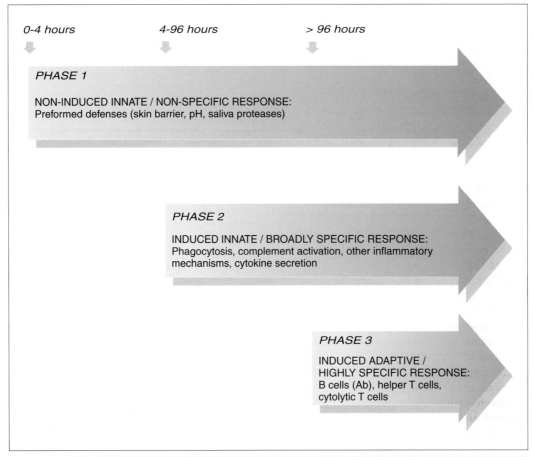

0-4 hours   4-96 hours   > 96 hours

PHASE 1

NON-INDUCED INNATE / NON-SPECIFIC RESPONSE:
Preformed defenses (skin barrier, pH, saliva proteases)

PHASE 2

INDUCED INNATE / BROADLY SPECIFIC RESPONSE:
Phagocytosis, complement activation, other inflammatory
mechanisms, cytokine secretion

PHASE 3

INDUCED ADAPTIVE /
HIGHLY SPECIFIC RESPONSE:
B cells (Ab), helper T cells,
cytolytic T cells

Figure 1-3

The Three Phases of Host Defense
The three phases of host defense are defined on the basis of whether the elements involved in protecting the host are continually ready to act or need to be called into action (induced) by a pathogen, and the degree to which a pathogen is treated as a unique entity relative to other pathogens. During each phase, a new defense tactic takes action against the pathogen and continues to function throughout the infection. Phase 1 represents the pre-existing body barriers that work in a constant but passive way to prevent invaders from gaining a foothold within the body. It is during phase 2 that the innate immune system cells first begin the active fight against the infection by employing mechanisms such as inflammation, phagocytosis and complement activation. These induced and broadly specific defenses kill pathogens and recruit other immune system cells to the site of damage. After 4 days, B and T lymphocytes that uniquely recognize a particular pathogen are called into action for the adaptive response. At this point it is a "primary" response because the host has not encountered the particular invader before. Effector T cells and specific antibodies (Ab) are produced and aggressively work to eliminate the pathogen. Upon future infection with the same pathogen, memory B and memory T lymphocytes remaining in the body after the primary response are induced to respond as early as 24–72 hours after the second infection.

engulf and digest antigens) depend on the recognition of proteins or features common to a wide variety of pathogens. If these T and B cell-independent mechanisms are successful in removing the pathogen, the lymphocytes specific for that pathogen are not fully activated and a complete adaptive immune response is not evoked. Only if the pathogen is processed by cells of the innate response so as to "present" foreign antigens in a particular way to T lymphocytes is the adaptive immune system fully induced in the third phase of host defense. However, this process takes time, and full effector functions, such as the production of cytolytic T cells and specific antibodies, are not observed until 4–5 days after the initial infection. After the infection is resolved, memory T and

B lymphocytes generated during the primary response remain in the host to provide a more rapid response in a repeat exposure to the same pathogen.

F. How Are Adaptive and Innate Immunity Related?

The innate immune response is crucial to the immunity of the host organism because it takes effect immediately and provides an early defense until the activated lymphocytes of the adaptive immune response can play their role. The usefulness of the innate system is not, however, limited to "playing for time" to allow the adaptive immune response to gear up. In many cases, an infection is completely controlled by innate mechanisms before adaptive immunity is even triggered. Furthermore, molecules essential for the induction of the adaptive response, such as messenger molecules called *cytokines* (see Ch.2), are synthesized by cells of the innate system. In fact, one can think of the adaptive response as being a more sophisticated (and more recently evolved) extension of the innate system, in that the complex recognition and memory cascades of the adaptive system trigger many of the same effector cells employed by the innate system to remove pathogens. Cells of the innate system recognize certain conserved antigens on a wide variety of pathogens and work to lyse these invaders, but can do so only in limited numbers and in limited ways. The evolution of the adaptive response has meant that the recognition of a pathogen can trigger the specific proliferation of lymphocytes directed against that particular pathogen. These activated lymphocytes not only undergo differentiation into effector lymphocytes capable of destroying the pathogen, but also secrete products that activate large numbers of innate response cells. To illustrate the overall need for an adaptive immune system, consider that humans lacking normal lymphocyte function are left defenseless against opportunistic infections and are very ill indeed, as in cases of severe combined immunodeficiency disease (SCID; no normal B or T cell functions) and acquired immunodeficiency syndrome (AIDS; the HIV virus destroys T cells) (see Ch.25).

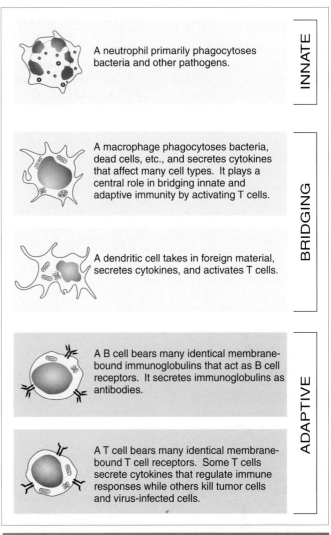

A neutrophil primarily phagocytoses bacteria and other pathogens.

INNATE

A macrophage phagocytoses bacteria, dead cells, etc., and secretes cytokines that affect many cell types. It plays a central role in bridging innate and adaptive immunity by activating T cells.

A dendritic cell takes in foreign material, secretes cytokines, and activates T cells.

BRIDGING

A B cell bears many identical membrane-bound immunoglobulins that act as B cell receptors. It secretes immunoglobulins as antibodies.

A T cell bears many identical membrane-bound T cell receptors. Some T cells secrete cytokines that regulate immune responses while others kill tumor cells and virus-infected cells.

ADAPTIVE

Figure 1-4

Some Important Cells of the Immune System

G. Leukocytes: Cellular Mediators of Immunity

Those cells mediating the innate and adaptive responses are collectively called *leukocytes* ("leuko," white; "cyte," small body, i.e., a cell), or *white blood cells*, as they can be found in the blood, but are distinct from red blood cells. However, "blood cell" is a bit of a misnomer, since a majority of leukocytes reside in tissues and specialized organs, and move around the body through both the blood circulation and a system of vessels called the *lymphatic system* (see later).

A key type of leukocyte important in the innate immune response is the *neutrophil* (Fig. 1-4). Neutrophils use *phagocytosis* to non-specifically engulf and destroy bacteria and other pathogens. Another type of leukocyte is the *macrophage*. These cells are also powerful phagocytes, engulfing not only pathogens but also dead host cells, cellular debris and macromolecules. Macrophages also secrete a vast array of proteases, cytokines, and growth factors, including many molecules crucial for the activation of cells in the adaptive immune response. Indeed, after phagocytosing foreign material, macrophages play a key role in presenting this material so that it can be recognized by cells of the adaptive immune response. In this way, macrophages can be said to be cells "bridging" the innate and adaptive responses. *Dendritic cells* are leukocytes that also engulf foreign material and play a key role in activating the adaptive immune response.

B and T lymphocytes are the leukocytes responsible for adaptive immune responses. The adaptive response was originally named the "specific response" because each B or T cell bears on its cell surface thousands of copies of a single type of specialized receptor that recognizes only one antigen (or occasionally a few very closely related antigens). The antigen receptors on the surface of a B cell are called *B cell receptors (BCR)*, whereas those on the T cell surface are called *T cell receptors (TCR)*. In each case, the antigen receptor is a complex of several proteins. Some of these proteins interact directly and specifically with antigen, and others allow the binding of the antigen to send activation signals into the lymphocyte. The engagement of lymphocyte antigen receptors by molecules of specific antigen is critical for triggering the activation of the lymphocyte and the generation of a particular effector response that eliminates the antigen.

H. Where Do Immune Responses Occur?

Innate immune responses occur all over the body, primarily at its outer surfaces and on its mucous membrane interfaces, the places where microbes attempt to enter the body. Wherever infection or injury to a tissue occurs, an *inflammatory response* is mounted. *An inflammatory response is an influx into the site of attack or injury of innate response leukocytes which fight infections using broadly specific recognition mechanisms* (see Ch.4). Those foreign entities evading these innate measures will be handled by different components of the adaptive response, depending on the chemical nature or route of entry of the incoming threat. The vertebrate immune system includes specialized tissues and organs collectively called the *lymphoid tissues* (Fig. 1-5). The cells of both the innate and adaptive immune responses are generated in the *primary* lymphoid tissues. In mammals, these are the bone marrow and the thymus. The *secondary* lymphoid tissues include the tonsils, lymph nodes, and spleen, and are the sites where leukocytes commonly encounter foreign entities and in which a vast majority of

adaptive immune responses occur. The secondary lymphoid tissues also contain the *accessory cells*, which assist lymphocytes during an immune response. The function of the accessory cells is to collect and trap antigens and to display them for recognition by lymphocytes.

Host defense mediated by the secondary lymphoid tissues starts with the diffuse collections of lymphocytes that are located in common areas of antigen penetration, normally under the skin and adjacent to the mucous membranes lining the respiratory and gastrointestinal tracts. Antigens that are not eliminated here may penetrate further into the tissues and enter the *lymphatic system*, a system of vessels that drains excess fluid from the tissues and in so doing, conveys antigens to the *lymph nodes*. The lymph nodes are organized cellular structures filled with T and B lymphocytes and are the most important sites of adaptive immune response action. The spleen is another organ composed to a large extent of lymphocytes. It functions in the elimination of antigens that have gained access to the blood circulation either by direct introduction (such as by an insect bite) or by "spillover" into the blood (such as in a case of overwhelming systemic infection).

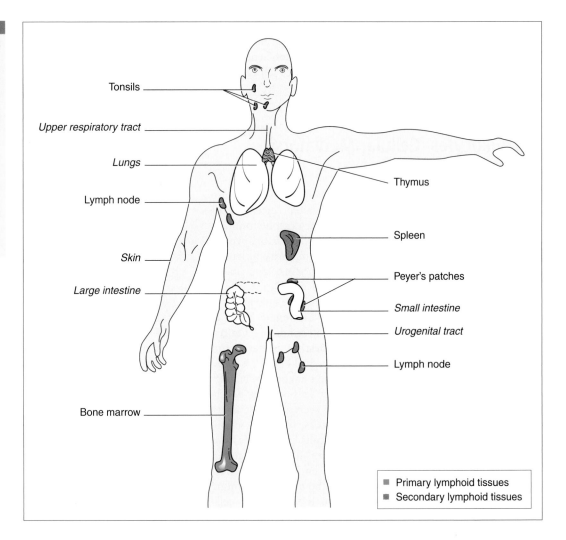

Figure 1-5

Principal Organs and Tissues of the Immune System
Primary lymphoid organs are shown in blue (bone marrow and thymus). Although only a femur is shown here, other bones in the body also contain bone marrow. Secondary lymphoid tissues are shown in gray (spleen, lymph nodes, and tonsils). Not all lymph nodes or tonsils are shown. Locations where pathogens commonly enter the body and where diffuse collections of lymphoid cells are present are noted in italics.

Tonsils

Upper respiratory tract

Lungs

Lymph node

Skin

Large intestine

Bone marrow

Thymus

Spleen

Peyer's patches

Small intestine

Urogenital tract

Lymph node

■ Primary lymphoid tissues
■ Secondary lymphoid tissues

I. Clinical Immunology: When the Immune System Does Not Work Properly

When the immune system is functioning normally, harmful antigens are recognized, and the host is protected from external attack by pathogens and internal attack by cancers. In the process of normal immune defense, some localized tissue damage may result from the inflammatory response that develops as immune system cells work to eliminate the threat, but the effect is limited and controlled. In addition, adaptive immune responses possess the property of *tolerance*: that is, the immune system does not attack normal host tissues and is therefore said to be "tolerant to self."

What happens when the immune system malfunctions, or when its normal functioning has undesirable effects? There are several instances in which actions of the immune system can result in pathologic consequences for the individual. Firstly, the normal attack of a healthy immune system on a transplant of foreign tissue meant to preserve life results in *transplant rejection* and deleterious consequences for the transplant patient. Secondly, when the tolerance of an individual's immune system fails, self tissues are attacked, resulting in *autoimmune disease*. Thirdly, when an individual's immune system responds inappropriately to an antigen that is gener-

ally harmless, the immune response is manifested as an *allergy*. Fourthly, an immune response that is unregulated or too strong can result in *hypersensitivity* to an antigen. In this situation, the inflammation that accompanies the response does not end when it should and causes considerable collateral tissue damage. Lastly, the immune system itself may be incomplete, either because of an inborn error (*primary immunodeficiency*) or because of external factors (*acquired immunodeficiency*, as in AIDS). In either case, individuals lacking a component of the immune system are said to be *immunodeficient*, and their bodies have an increased susceptibility to infection and tumors. These outcomes are illustrated in Figure 1-6.

Thus, the immune system is not entirely benign or completely efficient in its effects. It is a system that is never static, since pathogens are constantly evolving mechanisms to evade or block immune system defenses. Some pathogens are successful and generate chronic conditions that the immune system is unable to mitigate. In other cases, the existence of an efficient immune response is a double-edged sword that may result in either self-defense or self-destruction. For example, removal of pathogens is obviously highly desirable, but rejection of a life-saving transplant or graft is not. As clinicians and scientists, we attempt to capitalize on the positive aspects of the immune response and alleviate its negative aspects, in

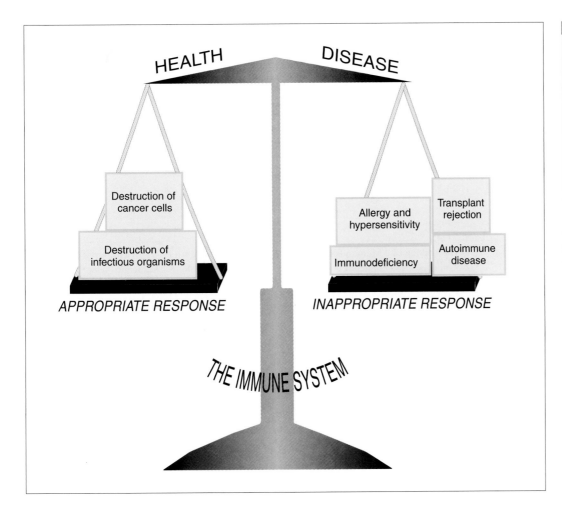

Figure 1-6

The Immune Response Mediates Health and Disease The way in which the immune system mounts a response can maintain a healthy body or lead to disease. In a healthy individual, the immune system mounts an appropriate response that leads to the death of infectious organisms or cancerous cells. States of disease can occur, however, when the immune system mounts an inappropriate response, such as an unwanted attack leading to transplant rejection, an attack on self tissues during autoimmune disease, an overzealous attack resulting in allergy and hypersensitivity, or a weak or absent attack resulting from immunodeficiency.

order to preserve human health. For example, vaccines are developed that prime the immune system, allowing it to prevent disease in the case of exposure to a natural pathogen. To be able to enhance and direct the cancer-fighting properties of the normal immune response would be of great benefit, and much research is dedicated to determining how this might be achieved. However, when the immune response actually causes or exacerbates disease, as in cases of autoimmunity and allergy, researchers attempt to understand what triggers the inappropriate response, and then work to prevent or decrease the intensity of the response. For example, intervention in immune function by the use of immunosuppressive drugs or antihistamines has offered relief to countless individuals victimized by their own immune responses.

Immunology is complex and challenging, but it is also essential and fascinating. We hope we can imbue the reader with some of our enthusiasm for this most exciting science. The remainder of this book will take the reader on a logical tour of the immune response. Part I, Basic Immunology, begins with the introduction of key concepts in Chapter 2 (Introduction to the Immune Response). Chapter 3 presents detailed descriptions of the various cells and tissues involved in mediating immune responses, while Chapter 4 focuses on the body's first line of defense, the innate immune response. In Chapters 5–20, each component of the immune system is described in depth. Chapter 21 closes this part of the book with a discussion of comparative immunology among species. In Part II, Clinical Immunology, Chapters 22–30 address immunological principles not only of practical value to the more medically oriented reader but also of interest to all who seek to understand the pivotal role of the immune system in health and disease. The involvement of the immune system in defense against pathogens and cancers is described, as well as its roles in promoting allergy, transplant rejection, and autoimmunity. The "manufacture" of immunity to bolster the body's defenses is explored in the chapter on vaccination. We also discuss inherited and acquired defects in immunity as well as tumors arising in immune system cells. It is our hope that the reader will come away with a solid understanding of the cellular and molecular mechanisms underlying immunity, how these mechanisms can go awry to cause disease, and how researchers seek to manipulate these mechanisms with the goal of ensuring good health for all.

SUMMARY

Our current understanding of immunity has grown from studies in the related fields of microbiology, pathology, and biochemistry, as well as from clinical observation. The immune response arises from the coordinated action of blood cells called leukocytes, which travel and function throughout the body, recognizing and eliminating extracellular and intracellular pathogens, toxins, and transformed cells. Protection from and clearance of unwanted entities involve both innate and adaptive immunity. Some innate immune mechanisms require no induction and are completely non-specific, whereas others are inducible and involve broadly specific recognition. In contrast, adaptive immunity requires lymphocyte activation and is directed only against the particular inciting entity. The interplay between innate and adaptive immune mechanisms is complex and relies heavily on intercellular messengers called cytokines, which are secreted by leukocytes. When functioning properly, the immune system maintains our health such that we are barely aware of its existence. However, malfunction of the immune system underlies many aspects of human disease, including cancer, autoimmune disorders, allergy, hypersensitivity, transplantation rejection, and immunodeficiency. The immune system is therefore the central player in the maintenance of human health and disease.

Selected Reading List

Abbas A. and Janeway C., Jr. (2000) Immunology: improving on nature in the twenty-first century. *Cell* **100**, 129–138.

Bachmann M. and Kopf M. (2002) Balancing protective immunity and immunopathology. *Current Opinion in Immunology* **14**, 413–419.

Gallagher R., Gilder J., Nossal G. and Salvatore G., eds. (1995) "Immunology: The Making of a Modern Science." Academic Press Limited, London.

Mazumdar P., ed. (1989) "Immunology 1930–1980: Essays on the History of Immunology." Wall and Thompson, Toronto.

Rajeswsky K. (1996) Clonal selection and learning in the antibody system. *Nature* **381**, 751–758.

Silverstein A. (1989) "A History of Immunology." Academic Press, San Diego, CA.

Silverstein A. (2003) Darwinism and immunology: from Metchnikoff to Burnet. *Nature Immunology* **4**, 3–6.

Talmage D. (1986) The acceptance and rejection of immunological concepts. *Annual Review of Immunology* **4**, 1–12.

Introduction to the Immune Response

2

Emil Von Behring 1854–1917

Serum therapy
Nobel Prize 1901
© Nobelstiftelsen

CHAPTER 2

A. GENERAL FEATURES OF INNATE IMMUNITY

B. GENERAL FEATURES OF ADAPTIVE IMMUNITY

C. ELEMENTS OF IMMUNITY COMMON TO
    THE INNATE AND ADAPTIVE RESPONSES

D. ELEMENTS OF IMMUNITY EXCLUSIVE TO
    THE ADAPTIVE RESPONSE

"Satan finds an opening to the universe within
Confusion heard his voice . . .
Till, at his second bidding, Darkness fled, Light shone, and
Order from disorder sprung . . .
Each had his place appointed, each his course;
The rest in circuit walls this Universe"
—From Paradise Lost III by John Milton

The purpose of this chapter is to introduce in a general way important concepts of the immune system, including elements that are common to the innate and adaptive immune responses, and elements that distinguish them. Each of these elements is examined in depth in later chapters, building on the foundation of information presented here. As mentioned in Chapter 1, vertebrates are capable of two types of immune responses, the *innate response*, involving *non-specific* and *broadly specific* recognition and action to eliminate the entity, and the *adaptive response* (or *acquired* or *specific response*), involving *specific* recognition and effector functions. Table 2-1 summarizes and compares the general characteristics of innate and adaptive immune responses. Most foreign antigens are eliminated by the mechanisms of innate immunity. Only those antigens that succeed in penetrating the innate defenses evoke adaptive immune responses.

A. General Features of Innate Immunity

Those immune response mechanisms that do not make fine distinctions among invading entities and do not undergo fundamental permanent change as a result of exposure to a pathogen confer *innate immunity*. A second exposure to an infectious agent provokes exactly the same magnitude and character of innate response as the first exposure; that is, there is no "memory" of a previous exposure to that agent. Elements of an innate immune response can be found in all multicellular organisms, whereas the mechanisms of the later-evolved adaptive immune response (thought to be 400 million years old) are present only in fish, reptiles, amphibians, birds, and mammals.

Non-inducible, non-specific innate defense in vertebrates is mediated by *anatomical barriers* and *physiological barriers*, while inducible, broadly specific innate defense results from *cellular internalization mechanisms* and *inflammation*. An example of an anatomical barrier is intact skin. The low pH of stomach acid and the hydrolytic enzymes in body secretions are physiological barriers. The term "cellular internalization mechanisms" refers to the fact that, while there are various ways that any host cell can ingest certain antigens, specialized phagocytic cells such as macrophages and neutrophils use a more sophisticated method of recognition and engulfment of pathogens and their molecules. Phagocytes express cell surface receptors that can bind to a broad range of molecules common to certain classes of invading entities. In addition, there are *natural killer cells* (NK cells) that use receptors to target and cause the lysis of virally infected cells or tumor cells in a broadly specific way. Inflammation is the influx into a site of injury or infection of leukocytes that phagocytose and digest antigens, produce chemical signals that promote wound

Table 2-1 Comparison of Innate and Adaptive Immune Responses

Innate	Adaptive
Preexisting and induced mechanisms	All mechanisms induced
Broad range of pathogens recognized by a small number of recognition structures of limited diversity	Broad range of pathogens recognized by an almost infinite number of extremely diverse and adaptive receptors
No memory of pathogens	Pathogen remembered by memory cells
Defense is the same with repeated exposure to pathogens	Defense is more vigorous and more finely tuned with repeated exposure

healing, and secrete cytokines summoning lymphocytes of the adaptive response to the site of injury or attack. The innate immune system is discussed in detail in Chapter 4.

B. General Features of Adaptive Immunity

Adaptive immune responses differ from innate immune responses in five key aspects: *specificity, memory, diversity, tolerance*, and the "*division of labor*" that has evolved to share the load among various cells, tissues, and soluble products. The quest to understand the cellular and molecular mechanisms underlying these features drives the bulk of modern immunologic research. We will briefly summarize each of these important concepts in turn, as the ideas involved will recur throughout this book.

I. SPECIFICITY

Specificity in the adaptive response context means that all phases of the response are specific to a unique antigen, from the recognition of the antigen either by antibody (humoral response) or T lymphocyte (cell-mediated response), through lymphocyte activation, to effector function (action by antibody or T lymphocyte to eliminate antigen) and the development of *immunologic memory* (see later). This specificity is mediated by the existence on T and B lymphocytes of antigen-specific receptors that must interact with antigen before a lymphocyte can be activated. Although this is often interpreted as "one lymphocyte recognizes one antigen," specificity is more accurately defined at the molecular level as "one lymphocyte has one antigen receptor gene that is expressed to generate thousands of identical copies of one receptor protein." The distinction made by the molecular-level definition is important, since a lymphocyte can in fact recognize more than one antigen: different molecules closely related in shape or amino acid sequence may all bind to a given antigen receptor with varying degrees of strength. Nevertheless, the range of binding specificities of an individual lymphocyte is very narrow compared to the much broader recognition capacity of an individual cell of the innate system. Recognition structures of the innate system are designed to respond to antigens shared by a wide variety of pathogens, while binding to a lymphocyte antigen receptor is much more restricted and usually hinges on the presence of a unique feature unlikely to appear on more than one pathogen (Fig. 2-1).

II. IMMUNOLOGIC MEMORY

In contrast to the innate immune response, which greets a given pathogen the same way each time it enters the body, the adaptive immune response can "remember" that it has seen

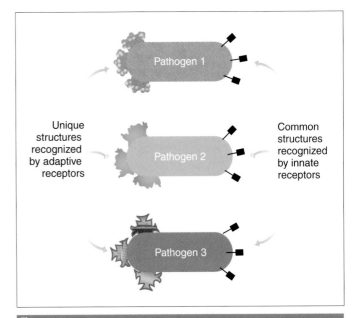

Figure 2-1

Antigens Recognized by the Receptors of the Innate and Adaptive Responses
Unique antigens found on pathogens are bound by the adaptive antigen receptors. In general, these types of antigens are rarely found on more than one type of pathogen. In contrast, common surface antigens may be found on many different pathogens and are bound by recognition molecules of the innate immune system.

a particular antigen before. This "immunologic memory" permits an enhanced adaptive response upon subsequent exposure to the same pathogen, so that signs of clinical illness are mitigated or prevented: the body has effectively "adapted" its defenses and "acquired" the ability to exclude this pathogen.

Immunologic memory arises in the following way. A constant supply of resting lymphocytes is maintained throughout the body, each displaying its unique antigen receptor. When an antigen enters the body, only those clones of lymphocytes bearing receptors specific for that antigen will be triggered to respond: this process is called the *clonal selection* of lymphocytes. Only those lymphocyte clones "selected" by the antigen will leave the resting state and proliferate, undergoing *clonal expansion*. As the selected clone of lymphocytes expands, some daughter cells bearing the identical receptor differentiate into short-lived effector cells equipped to eliminate the antigen, while others differentiate into long-lived "memory" cells that will remain in a resting state until subsequent exposure to antigen at a later time (Fig. 2-2).

The results of clonal expansion and differentiation are manifested as the *primary* and *secondary immune responses*. The primary response occurs upon the first exposure to antigen and results in the specific elimination of that antigen by the formation of antigen-specific effector cells. In addition, large numbers of antigen-specific memory cells are formed and persist in the tissues. The secondary immune response occurs the next time that particular antigen enters the body, because it is met by an existing army of antigen-specific

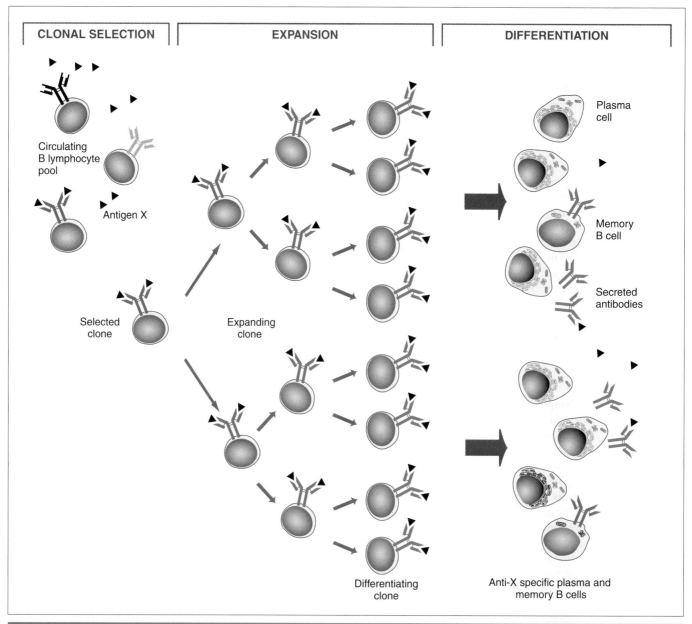

| CLONAL SELECTION | EXPANSION | DIFFERENTIATION |

Circulating B lymphocyte pool

Antigen X

Selected clone

Expanding clone

Plasma cell

Memory B cell

Secreted antibodies

Differentiating clone

Anti-X specific plasma and memory B cells

Figure 2-2

Clonal Selection and Generation of Memory and Effector Cells
From the existing pool of circulating lymphocytes, very few are ever selected by antigen to expand clonally and differentiate into effector memory cells. The scheme shown here has been drawn for B cells, but T cells also undergo this process, giving rise to effector and memory T cells. Note that all plasma cells derived from the selected B cell secrete anti-X specific antibodies. The mechanisms that determine whether a clone becomes a plasma cell or memory cell are not clear.

lymphocytes which undergo rapid differentiation into antigen-specific effector cells. This enables the organism to mount a stronger, faster response, which prevents or mitigates disease, and, in many cases, confers long-term or even lifelong immunity to that pathogen.

III. DIVERSITY

The adaptive immune response also differs from the innate response in its degree of diversity. Whereas the innate immune system exhibits a fixed and limited capacity for antigen recognition, the capacity of the adaptive immune system is nearly limitless. In fact, our bodies are equipped to recognize antigens that do not occur in nature at all: it was shown in the early 20th century that immune responses can be provoked by totally synthetic antigens. It has been estimated from the structure of the antigen receptor genes that more than 10^{12} distinct antigens can theoretically be recognized by the collective cells of the adaptive response. This huge number is derived from the combinatorial effects of several genetic mechanisms, some of which occur before a lymphocyte encounters antigen, and some after.

The first source of diversity is a process called *somatic recombination* or *somatic gene rearrangement*, which occurs during the development of B and T cells. The genes encoding the B cell receptor (BCR) and T cell receptor (TCR) are actually assembled from smaller genetic units selected from a large array of pre-existing *gene segments*. Random combinations of these gene segments generate hundreds of thousands of DNA sequences encoding different antigen receptor proteins. Additional mutational mechanisms also operate that result in further structural diversification of the receptor protein, ensuring that complex recognition molecules are generated. These molecules have the unique antigen-binding site sequences required to facilitate specific antigen recognition, but also share certain common structural features necessary for antigen removal. The sheer numbers of adaptive immune cell clones generated guarantee that there will be a lymphocyte clone with a receptor sequence that can bind any antigen encountered during the host's life span. These clones, with their array of antigen receptor specificities, are collectively called the individual's lymphocyte *repertoire*.

IV. TOLERANCE

The generation of lymphocyte clones that can theoretically recognize any antigen in the universe leads to the question of how we avoid having lymphocytes that attack our own cells and tissues. This brings us to tolerance, the fourth aspect in which the adaptive response exhibits much more refinement than the innate response. The antigenic specificity of each lymphocyte receptor is randomly generated early in development by the somatic gene rearrangement process described in the preceding section. By chance, some of those protein sequences will generate receptors that recognize antigens making up host tissues. Tolerance depends on the identification of lymphocytes carrying receptors that recognize "self antigens," and on the mechanisms that ensure that such lymphocytes are either removed from the system or inactivated. In other words, "tolerance" refers to the development of an active lymphocyte repertoire that does not attack self. This concept was first elaborated in 1953 by Peter Medawar, who showed that the establishment of a lymphocyte repertoire that focuses on "nonself" occurs primarily during early lymphocyte development. For example, in an elaborate screening process that takes place in the thymus, the majority of those immature lymphocytes whose randomly generated antigenic specificity would result in the recognition of self antigens are not permitted to mature and are destroyed in the thymus: this process results in *central tolerance*. Similarly, maturing B cells undergo a central tolerance process in the bone marrow to establish self-tolerance. The B and T lymphocytes that escape the screening of central tolerance are functionally silenced by additional *peripheral tolerance* mechanisms that operate outside the thymus and bone marrow and exert their effects in the body's periphery. Peripheral tolerance is explored in depth in Chapter 16.

V. DIVISION OF LABOR

As we shall see in subsequent chapters, not all pathogens, or even different parts of the same pathogen, are dealt with in the same fashion or by the same component of the immune system. The onerous task of countering every conceivable unique antigen in the universe has been divided up among the effector lymphocytes of the adaptive response. The *intracellular* or *extracellular* location of an antigen determines which cell of the adaptive response will be activated. Extracellular, or *exogenous*, antigens are those that originate outside of cells, such as most bacteria, fungi, and parasites. Intracellular, or *endogenous*, antigens are harmful entities that originate inside a cell or gain access to its interior, such as viruses, cancer cell antigens, certain bacteria, and the surface antigens present on transplanted tissues.

Extracellular threats that succeed in breaching the innate defenses, such as certain bacteria or bacterial toxins, are recognized by the B lymphocytes of the *humoral* response. (T lymphocytes are unable to recognize soluble antigens.) Activated B cells synthesize and secrete antibodies directed against these antigens (Fig. 2-3), antibodies that are in fact a soluble form of the B lymphocyte antigen receptor used to recognize the extracellular entity in the first place. Binding of the antibody protein marks these antigens for elimination by one of several destruction mechanisms. However, antibodies are unable to penetrate cell membranes, so that those antigens that are intracellular in origin must be addressed by *cell-mediated* immunity. T lymphocytes are specialized for this purpose because they recognize certain host membrane proteins that bind pieces of antigen present inside the cell and display them at the cell surface, allowing detection of cells containing foreign entities. The cell-mediated immune response results in the killing of the affected host cell and the concomitant protection of the surrounding host cells. A type of T lymphocyte called a *cytotoxic, cytolytic*, or *killer T cell* specifically recognizes, attacks, and lyses a host cell altered by either malignant transformation or viral infection. A second type of T lymphocyte called a *helper T cell* secretes products that assist most B cells in antibody production and assist cytotoxic T cells in their destruction of targeted host cells.

How a given antigen is dealt with by the adaptive immune system depends on its origin (intracellular or extracellular) and its chemical nature. Complex pathogens may simultaneously present both intra- and extracellular antigens to the immune system during the course of infection. Thus, simultaneous responses by all three types of effector cells—B cells, helper T cells, and cytotoxic T cells—may be evoked by a single foreign entity of sufficient complexity. In order to coordinate the response, B cells, T cells, and non-lymphocytic cells recruited to assist them all secrete small soluble molecules called *cytokines* (see the following section) that influence the behavior of nearby cells, including cells of the innate response. Cytokines induce cells to carry out defense functions particular to that cell type, which may include cellular proliferation, antibody production, the synthesis of antibacterial or antiviral substances, or target cell lysis.

Figure 2-3

B and T Lymphocyte Division of Labor

Lymphocyte subsets have evolved unique effector functions that are particularly suited to ridding the body of different pathogens. Helper T cells act indirectly to rid the body of both extracellular and intracellular threats by secreting cytokines that activate cytotoxic T cells and help B cells to make antibody. By secreting antibodies, B cells are able to mark the surface of extracellular pathogens and soluble toxins for destruction by innate system mechanisms. The spread of infection and disease symptoms caused by toxins are thus blocked. Cytotoxic T lymphocytes clear intracellular pathogens by lysing infected host cells that harbor pathogens.

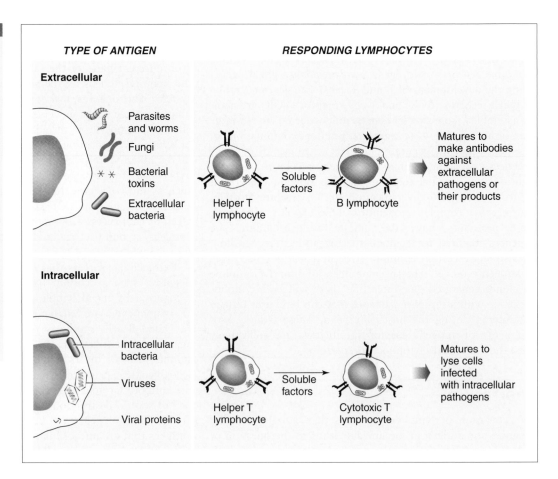

C. Elements of Immunity Common to the Innate and Adaptive Responses

I. INTRODUCING CYTOKINES

In both innate and adaptive immunity, there are many situations when leukocytes must communicate with one another in order to achieve a response. This communication is often carried out by members of a group of low molecular weight proteins called cytokines. The youth of immunology as a science is emphasized when one realizes that although the first cytokines were identified in the 1950s, the definitive isolation, identification, and characterization of the vast majority occurred only in the mid-1980s. Well over 100 cytokines have been identified in the past 10 years.

Cytokines are secreted primarily, but not exclusively, by leukocytes, and regulate not only immune responses, but also the generation of red and white blood cells, wound healing, and other physiological processes (see Table 2-2). During an immune response, cytokines act to intensify or dampen the response by stimulating or inhibiting the secretory activity, activation, proliferation, or differentiation of other cells. This regulation is achieved by *intracellular signaling* (see later) that is triggered when a cytokine binds to its receptor on the membrane of the cell to be influenced. Cytokines are induced in both the adaptive and innate immune responses, and cytokines secreted in one type of response frequently stimulate the secretion of cytokines influential in the other type. For example, the binding of bacterial products to a phagocyte can directly induce the secretion of cytokines that will attract cells of the adaptive immune response for additional defense. Certain cytokines secreted by cells of the adaptive immune response can induce the macrophages, which bridge the innate and adaptive responses, to produce chemical attractants necessary to draw additional immune system cells to the site of pathogen attack.

Cytokines are structurally distinct and genetically unrelated; however, many appear to be functionally redundant. The same biological effect may result from the action of more than one cytokine, ensuring that a critical function is preserved should a particular cytokine be defective or absent. A given cytokine can be produced by multiple cell types, may be *pleiotropic* (act on many different cell types), or have multiple different effects of varying duration on the same target cell. Cytokines can also affect the action of other cytokines, and may behave in a manner that is *synergistic* (two cytokines acting together achieve a result that is greater than additive) or *antagonistic* (one cytokine inhibits the effect of another). Cytokines are discussed in detail in Chapter 17.

Table 2-2 Origin and Action of Well-Studied Cytokines*

Cytokine	Cell Source	Principal Cell Target	Principal Actions
IFNγ	T cells, NK cells	Lymphocytes, monocytes, tissue cells	Immunoregulation, B cell differentiation, some antiviral action
IL-1α, IL-1β	Macrophages, monocytes, dendritic cells, some B cells, fibroblasts, epithelial cells, endothelium, astrocytes	Thymocytes, neutrophils, T and B cells, tissue cells	Immunoregulation, inflammation, fever
IL-2	T cells, NK cells	T cells, B cells, monocytes	Proliferation, activation
IL-3	T cells	Stem cells, progenitors	Colony-stimulating factor
IL-4	T cells	B cells, T cells	Division and differentiation
IL-5	T cells	B cells, eosinophils	Differentiation
IL-6	Macrophages, T cells, fibroblasts, some B cells	T cells, B cells, thymocytes, hepatocytes	Differentiation, acute-phase protein synthesis
IL-8	Macrophages, skin cells	Granulocytes, T cells	Chemotaxis
TNF	Macrophages, lymphocytes	Fibroblasts, endothelium	Inflammation, catabolism, production of other cytokines (IL-1, IL-6, GM-CSF) and adhesion molecules

*Note the multiple sources and pleiotropic actions of cytokines. Only the most important targets and actions are shown.
IFN, interferon; IL, interlukin; TNF, tumor necrosis factor; GM-CSF, granulocyte/macrophage colony-stimulating factor.

II. INTRODUCING INTRACELLULAR SIGNALING

The binding of a ligand such as a cytokine or a growth factor to a cell surface receptor on a leukocyte is the initiating signal indicating that a response is required. A cell surface receptor generally features an extracellular domain, a transmembrane domain, and a cytoplasmic domain. Usually two or three receptor molecules are grouped together in the membrane to form a dimeric or trimeric complex that can respond to the binding of aggregated ligand molecules. The ligands bind to the extracellular domains of the receptor complex chains, inducing a change (most likely conformational) in the cytoplasmic domains of the receptor complex chains. This change in turn induces the phosphorylation (or dephosphorylation) of other proteins in a cytoplasmic kinase/phosphatase cascade, and the release from intracellular stores of mediators such as calcium (Fig. 2-4). When the response required by the cell takes the form of the reorganization of the actin cytoskeleton or the release of pre-formed granules, the end point of the signaling pathway is in the cytoplasm. However, sometimes the proteins needed to carry out an effector action are not routinely present as part of a cell's "housekeeping" metabolism, so that new synthesis of specialized proteins is required. New protein synthesis in turn requires new transcription, so that the signal that a foreign entity has been encountered must be conveyed from the cell surface to its nucleus. The cytoplasmic kinase/phosphatase cascade therefore continues from protein to protein until one or more *nuclear transcription factors* capable of entering the nucleus are stimulated. The activated nuclear transcription factors induce the transcription of the previously silent genes that code for the proteins responsible for effector action.

For example, in many cell surface receptors (both on innate and adaptive immune cells), certain polypeptide chains are associated with cytoplasmic *protein tyrosine kinases* (PTKs). PTKs are enzymes that, when activated, phosphorylate (add phosphorus to) the tyrosine residues of substrate proteins. Within seconds of the binding of molecules of a ligand or growth factor to such a receptor complex, the PTKs are activated. This activation initiates a guanosine triphosphate (GTP)-dependent phosphorylation/dephosphorylation cascade, which involves other kinases and additional substrates that regulate a plethora of cellular activities. These intracellular signals transiently trigger transcription, resulting in the synthesis of new proteins that will be used either in the cell, on the surface of the cell, or outside the cell upon secretion.

The antigen receptors of lymphocytes (the BCR and the TCR) operate in a similar fashion, except that the ligand is, of course, the antigen. New receptors appearing on the cell surface in response to the signaling induced by antigen binding may include those for cytokines that stimulate cellular proliferation and the differentiation of effector cells. In this way, the signal that a foreign entity has been encountered is translated into the production or activation of cells equipped to eliminate the entity. However, intracellular signaling pathways in lymphocytes are among the most complex identified, involving multiple receptor and coreceptor proteins, and the recruitment of multiple PTKs and other molecules. This complexity has no doubt evolved because of the fine degree of control necessary for regulating an immune response and harnessing the power of effector lymphocytes.

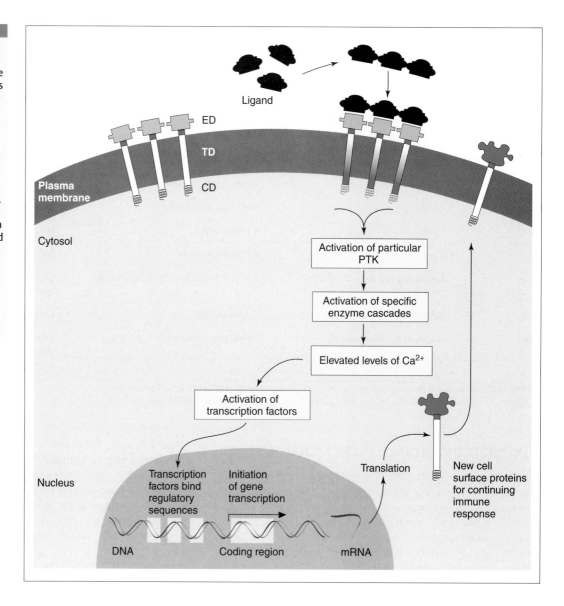

Intracellular Signaling
Ligand binding to cell surface receptors initiates one or more intracellular signaling cascades that ultimately lead to the activation of transcription factors, the initiation of gene transcription, and the expression of new proteins. In this illustration, new cell surface proteins are produced as a result of receptor aggregation at the cell surface. For simplicity, the synthesis of the new membrane protein on the endoplasmic reticulum and its subsequent delivery to the cell surface via transport vesicles has not been shown. ED, Extracellular domain; TD, transmembrane domain; CD, cytoplasmic domain; PTK, protein tyrosine kinase.

D. Elements of Immunity Exclusive to the Adaptive Response

I. ANTIGENS VERSUS IMMUNOGENS

It behooves us to say a word about antigens, without which there would be no need for an immune response. The word "antigen" was coined in the 1800s to describe substances "against which <u>anti</u>bodies were <u>gen</u>erated." Thus, strictly speaking, "antigen" is a word that applies only to the humoral branch of the adaptive immune response. In Chapter 1, we defined an antigen as "any foreign entity eliciting an immune response," accepting that cancer cells (which may no longer resemble normal host cells) are also considered "foreign." This definition applies well to innate immunity and takes into account the fact that both injury and infection can provoke immune responses. However, with respect to

adaptive immunity, a further conundrum must be dealt with. Some foreign entities can bind to the specific antigen receptors on T and B cells but do not induce lymphocyte activation (i.e., cannot lead to the production of antibodies or effector T cells to eliminate the antigen). This has necessitated the coining of the term *immunogen* in order to distinguish between those substances that both bind to lymphocyte antigen receptors and cause lymphocyte activation, and those that merely bind. Thus, an *antigen* in the context of the adaptive response is *any substance that is specifically recognized and bound by receptors of T and B lymphocytes*, a definition that covers both the humoral and cell-mediated branches of adaptive immunity. In contrast, an *immunogen* is defined as *any antigen that elicits an adaptive immune response*. Both binding and activation must occur for the adaptive response to be induced. In other words, all immunogens are antigens, but not all antigens are immunogens. The historical use of the term "antigen" and the irrelevance of lymphocyte activation to the biochemical study of proteins important in immunology have meant that "antigen" is often used when "immunogen" would be more appropriate.

The antigen-binding site of a B or T cell antigen receptor can usually accommodate an entity of a size that might be equivalent to only a small portion of an antigen. Macromolecular antigens are much larger than the binding site and are sometimes, particularly in the case of a microorganism, larger than the entire receptor. The small region of a macromolecule that binds to the antigen receptor is called an *antigenic epitope* or *determinant*. While there may be many different epitopes on a large macromolecule, a particular epitope may also be present more than once (Fig. 2-5). A whole pathogen displays many macromolecules on its exterior, each of which contains multiple antigenic determinants. Thus, a given pathogenic entity, present in thousands of copies, may simultaneously engage the receptors of several lymphocytes with differing specificities. In the case of humoral immunity, each epitope exhibited on a microorganism can bind to the antigen receptor of a different B cell, stimulating the production of antibodies to several different sites on the microbe at once. Similarly, in the case of cell-mediated immunity, T cells may respond to many different fragments of antigenic protein derived from the pathogen. The result is the simultaneous activation of multiple T cell clones of varied specificity. It is also important for the reader to remember that the surfaces of both B cells and T cells are covered with thousands of identical copies of their specific antigen receptors (Fig. 2-6). <u>The interaction of multiple receptors with multiple copies of the same antigenic determinant is required to trigger lymphocyte activation and precipitate an adaptive immune response.</u> The binding of a single molecule of antigen to a single antigen receptor in a lymphocyte membrane is necessary, <u>but not sufficient,</u> for full lymphocyte activation. It is only after the generation of sufficient antigen–receptor multimers on the cell surface that the activation threshold is achieved, and the B or T cell can proceed to proliferate and differentiate into effector and memory cells. Thus, the concentration of an immunogen greatly influences the initiation of an immune response and acts as a control on it: as the immunogen

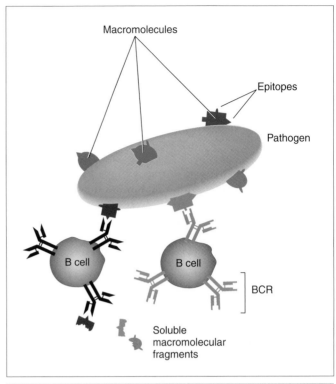

Figure 2-5

Antigenic Epitopes
The surface of a pathogen is studded with many antigenic macromolecules that are composed of small regions of antigenicity called epitopes. Epitopes may be unique to a particular macromolecule or may be shared by other macromolecules on the pathogen's surface. The B cell antigen receptor (BCR) can bind to epitopes that are part of macromolecules on the pathogen surface as well as to epitopes that are part of soluble fragments of macromolecules.

is eliminated and its concentration falls, the number of antigen–receptor entities drops, the lymphocytes are no longer able to reach their activation thresholds, and the response subsides.

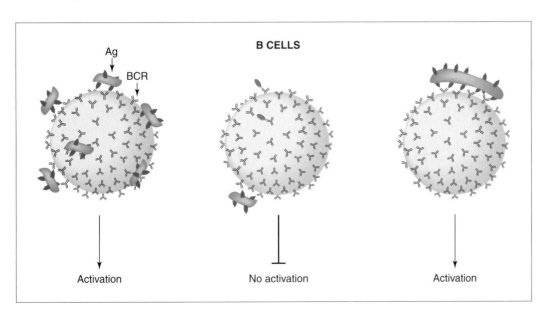

Figure 2-6

B Cell Surface Receptor Density and Activation
B lymphocytes are covered with about 10^5 identical membrane-bound B cell receptors (BCRs). B cell activation is precipitated when a critical threshold of BCR aggregation is achieved. For T cells, which are covered with about 10^4 identical membrane-bound T cell receptors (TCRs), a critical threshold of antigen–TCR interaction is achieved by a different mechanism.

II. INTRODUCING SPECIFIC ANTIGEN RECOGNITION: B CELLS

Although their antigen receptors are similar in structure, B and T lymphocytes "see" antigens differently. B cells provide antigen receptors that can recognize soluble or cell-bound antigens without prior modification. Thus, collectively, the B lymphocyte population is able to recognize an almost infinitely broad range of molecules, including proteins in their native or denatured conformations, simple chemical groups, lipids, carbohydrates, nucleic acids, and other molecules (Fig. 2-7). However, most B cells cannot proceed beyond recognition of the antigen without "help" from T cells. The

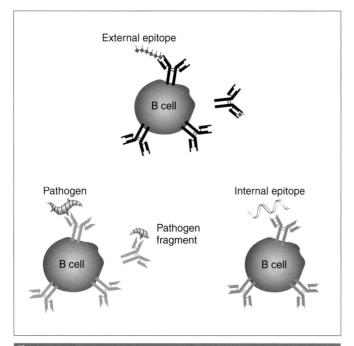

Figure 2-7

B Cell Receptor Antigen Recognition
Both cell-bound and secreted forms of the B cell receptor can recognize and bind to epitopes found on pathogen surfaces as well as internal and external cellular components.

term *T cell help* is used to describe the cooperation between helper T cells and B cells that is necessary for most B cell responses. T cell help takes the form of direct intercellular T–B cell contacts, and the binding to the B cell of specific cytokines produced by a T cell responding to the same antigen. Without T cell help, the majority of B cells (especially those directed against viruses) are unable to achieve complete activation and fail to proliferate and differentiate into memory cells and antibody-secreting *plasma cells*. Antigens that bind to B cell receptors but cannot activate B cells without T cell help are called *T-dependent* (Td) antigens. For some non-protein antigens, interaction between the B cell receptors and the antigen alone is sometimes sufficient to activate the B cell. These antigens, which are often polymers, are called *T-independent* (Ti) antigens, because no T cell help is required for lymphocyte activation.

III. INTRODUCING SPECIFIC ANTIGEN RECOGNITION: T CELLS AND THE MAJOR HISTOCOMPATIBILITY COMPLEX

Antigen receptors on T cells interact with antigen in a different way. The TCR does not recognize non-protein antigens or protein antigens that are soluble or in their native conformation. In order for a T cell to respond to an antigen, a separate host cell (which can be any one of a number of different cell types) must first digest it internally into peptide fragments, and then attach the individual peptides to specialized cell surface molecules called *major histocompatibility complex* (MHC) molecules. It is the overall shape of the peptide–MHC complex, <u>presented as a unit</u> by the host cell to the T cell, that is critical for recognition by the TCR (Fig. 2-8).

The major histocompatibility complex refers to a genetic region encoding a particular set of related cell surface proteins. The function of most of the proteins encoded in the MHC region is to combine with antigenic peptides, both self and nonself, and display them on the surface of host cells for perusal by T cells. There are two structurally similar classes of MHC molecules with this function: MHC class I and MHC class II. <u>MHC class I molecules are expressed on</u>

Figure 2-8

T Cell Receptor Antigen Recognition
T cell receptors (TCRs) recognize the overall shape of a processed peptide combined with an MHC molecule. The TCR on a helper cell (Th) binds to peptide associated with MHC class II on the surface of an antigen-presenting cell (APC). The TCR on a cytolytic T cell (Tc) binds to peptide associated with MHC class I on the surface of a nucleated host cell (HC).

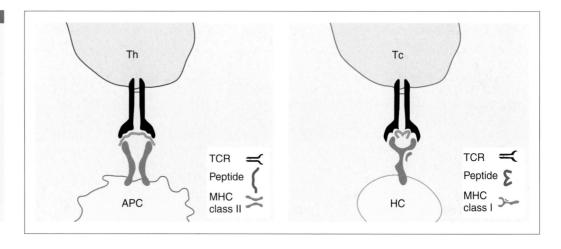

most nucleated cells and present peptides to *cytolytic T cells (Tc)*. MHC class II molecules are expressed only on specialized antigen-presenting cells that present peptides to *helper T cells (Th)*. It is important to note that the host cell displaying the peptides makes no distinction as to the self or nonself nature of the peptides that bind to its MHC molecules: that discrimination is left to the T cell repertoire as determined by central and peripheral tolerance mechanisms.

Although MHC proteins bind antigens, they do so in a much less specific way than the antigen receptors on B cells and T cells. Only one peptide per MHC molecule binds at any one time, but the antigen-binding cleft in the MHC protein is structured such that it is capable of binding an array of different peptides derived from a variety of antigens (Fig. 2-9). When a binding site can accommodate several different ligands with similar affinity, the site is said to exhibit "promiscuous" binding. In addition, while a given lymphocyte expresses its

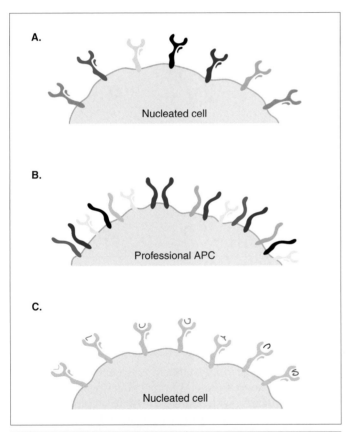

Figure 2-9

MHC Expression and Peptide Binding
MHC molecules are both codominantly expressed and highly polymorphic. On the surface of a single cell, an individual can display proteins encoded by as many as 6–14 MHC alleles. MHC class I molecules are found on the surface of most nucleated cells (A) while MHC class II molecules are expressed only on specialized host cells called professional antigen-presenting cells (APC) (B). Although only one peptide binds to the MHC molecule at any given time, the peptide-binding groove encoded by each MHC class I or II allele is designed to accommodate a wide variety of peptides. This type of binding is called "promiscuous" and is exhibited by the proteins encoded by all MHC alleles (class I and II). An example of promiscuous binding is shown for one MHC class I protein in (C).

antigen receptor protein from a single rearranged functional gene, a given host cell simultaneously expresses MHC proteins derived from multiple MHC gene sequences. Not only are there multiple class I and class II MHC gene loci in an individual's genome, but each MHC locus has an unusually high level of polymorphism, meaning that many different alleles for each gene occur within the total population. The two parental alleles present at each MHC locus in an individual are codominantly expressed at high copy number ($\sim 5 \times 10^5$ per cell) on the appropriate cell type. Thus, the presence of a nonself entity results in many different *peptide–MHC (pMHC)* combinations being displayed for inspection by T cells, increasing the probability of recognition by a T cell and the triggering of an immune response.

IV. INTRODUCING ANTIGEN PROCESSING

Since T cells recognize only peptides (and not native proteins) bound to MHC class I or II molecules, foreign proteins must be degraded into small peptides in a process called *antigen processing*, in order to induce the activation of T cells. Whether a peptide binds to an MHC class I or II molecule often depends on whether it is derived from a protein produced inside a cell (an endogenous antigen) or outside a cell (an exogenous antigen), respectively.

When a somatic cell is infected with a virus, viral proteins are produced endogenously and processed into peptides as part of normal cell metabolism. The peptides are intracellularly complexed to MHC class I molecules and displayed on the cell surface (Fig. 2-10A). Such a "target cell" can be recognized only by T cells expressing a surface protein called CD8, because this molecule facilitates a required interaction with the MHC class I molecule (see later). CD8+ T cells are usually cytolytic, so target cell recognition results in the destruction of the target. Target cells can also be altered self cells, such as cancer cells, transplanted cells, or aging body cells; in short, any cell in which nonself proteins are generated internally and peptides derived from them are displayed on the cell surface in association with MHC class I. As mentioned previously, MHC class I molecules are expressed on almost all nucleated cells, so almost any cell can function as a target cell for CD8+ T cell recognition. This makes evolutionary sense, since all nucleated cells are subject to infection and malignant transformation and may therefore require clearance involving cell lysis.

Exogenous antigens are those that assault host cells from the exterior, such as bacterial toxins or proteins released as debris during the destruction of either pathogens or host cells. In contrast to endogenous antigens, exogenous antigens must be phagocytosed and processed to peptides by the endocytic pathway associated with cellular internalization mechanisms (see Ch.4). These peptides are then complexed intracellularly with MHC class II molecules and displayed on the cell surface (Fig. 2-10B). Although almost all cells have phagocytic/endocytic capability, only a few cell types express MHC class II molecules. By convention, scientists reserve the term *professional antigen-presenting cells* for those few specialized cell

Figure 2-10

Antigen Processing and Presentation

(A) Cells infected with a virus (1) produce endogenous viral peptides (2), which associate with MHC class I in the endoplasmic reticulum (ER) (3). The class I–viral peptide complex is transported through the ER and released from the Golgi apparatus in an exocytic vesicle (4), which is transported to the cell surface (5) and presented as a unit to CD8$^+$ cytolytic T cells (Tc) (6). (B) Exogenous antigens are phagocytosed into the cell (1) and upon entering the endocytic pathway (2) antigen is broken down by acidic granules in endosomes (3a) and lysosomes (3b). Meanwhile, MHC class II molecules are synthesized in the ER and then processed through the Golgi apparatus (4). Peptides associate with MHC class II molecules in an endosome (5) and the two are transported to the cell surface as a unit (6), where they are presented to CD4$^+$ helper T cells (Th) (7).

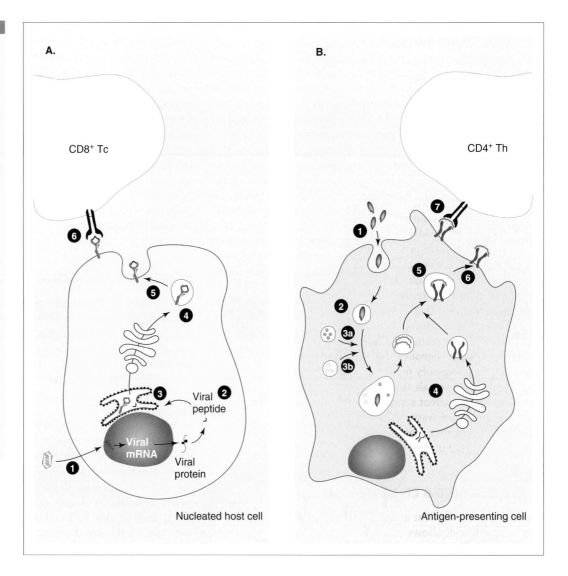

A. CD8$^+$ Tc

Viral mRNA
Viral peptide
Viral protein

Nucleated host cell

B. CD4$^+$ Th

Antigen-presenting cell

types that process exogenous antigen and combine it with MHC class II. Antigen-presenting cells (APCs) are recognized almost exclusively by T cells expressing the CD4 surface protein, since this molecule interacts only with MHC class II molecules. On activation, most CD4$^+$ T cells function as "helper T cells," releasing cytokines to promote the further development and activation of B cells, macrophages, and cytolytic T cells. Limiting the expression of MHC class II molecules to a small number of cell types specialized for presentation of exogenous antigen makes evolutionary sense in two ways. Firstly, since the APCs are usually found in lymphoid tissue, this ensures the necessary co-localization of antigen and CD4$^+$ T cells. Secondly, it ensures that the epitopes from a relatively diffuse extracellular source are concentrated on the surface of only a small number of cells. This has the practical effect of increasing the density of peptide–MHC II complexes on the surface of the APC, so the CD4$^+$ T lymphocyte activation threshold will be reached and "help" will then

be available to fully activate other leukocytes specialized for antigen clearance.

To reiterate: all cells routinely and continuously process both self and nonself proteins into peptides, which then interact intracellularly with expressed MHC molecules. These peptide–MHC complexes of both self- and nonself-peptides are constantly displayed for perusal by passing T cells. If the T cell receptor of a passing T cell recognizes the shape of the pMHC, it can be triggered to activate, proliferate, differentiate, and take effector action that will result in elimination of the antigen or the target cell. These actions are desirable in the case of peptides derived from an organism that has invaded the host's body, but are highly undesirable in the case of proteins derived from the host's own tissues. Fortunately, in a healthy individual, the mechanisms of central and peripheral tolerance ensure that T cells circulating through the periphery do not initiate damaging immune responses against self-peptide–MHC complexes.

V. INTRODUCING CORECEPTORS AND COSTIMULATORY MOLECULES

The activation of any T lymphocyte depends first and foremost on the binding of sufficient pMHC–TCR pairs on the cell surface. However, complete activation of a T cell requires additional intercellular contacts. In other words, target cells and APCs do more than just present foreign antigens. Interactions between additional *accessory molecules* on T cells and their counterparts on the cell surfaces of APCs and target cells are also necessary. Many accessory molecules promote adhesion between the T cell and the APC or target cell, since the binding of the TCR to the pMHC is relatively weak. This adhesion allows the T cell to scan the surface of the APC or target cell and determine if a pMHC that it recognizes is present. The CD4 and CD8 surface proteins mentioned previously are special accessory molecules called *coreceptors*, since they bind to the same MHC molecule (on an APC or target cell) that is complexed with peptide, but at a site that is distant from the antigen-binding cleft (Fig. 2-11). This interaction helps to stabilize the intercellular contact between the T cell and the APC or target cell and also recruits PTKs required for the transduction of the intracellular signaling cascade initiated by pMHC–TCR binding. Other accessory molecules are *costimulatory* in nature, since they deliver signals that promote or regulate intracellular signaling. Much current research is directed at the dissection of these signaling pathways, in order to elucidate their components and the factors that control them. Costimulatory molecules are discussed in detail in Chapter 14.

VI. INTRODUCING B CELL EFFECTOR FUNCTIONS

The binding of sufficient antigen–receptor pairs on the surface of a mature B cell allows it (in the presence of T cell help, if necessary) to become a fully activated lymphocyte. After such clonal selection, this cell and its similarly activated sister cells undergo rapid clonal proliferation and differentiation into both memory B cells and cells that can carry out B cell effector functions. *Effector functions* are defined as those actions taken by effector lymphocytes that result in the elimination of nonself antigens. In the case of B lymphocytes, the effector cells are fully differentiated *plasma cells* that secrete antibody.

As mentioned previously, the antigen receptor of a B cell is a multimolecular complex called the B cell receptor (BCR). The molecule in the BCR that interacts specifically with antigen is the membrane-bound form of a protein called an immunoglobulin (Ig). The antibody secreted by an activated B cell after it differentiates into a plasma cell is a soluble form of the immunoglobulin molecule and has the same antigen specificity. Immunoglobulins are proteins with a Y-shaped structure (Fig. 2-12). The arms of the Y contain "variable regions" with amino acid sequences that differ greatly from Ig to Ig. These sequences are identical on both arms of the Y and encode two identical antigen-binding sites that are unique to that Ig. In contrast, the "stem" of the Y varies very little among Ig molecules and contains "constant region" sequences that dictate which effector function is used to remove the antigen (see Ch.5). Thus, the Ig molecule is "bifunctional," with one end for antigen binding and the other end for antigen elimination. There are five types of constant regions that influence how the Ig interacts with the effector cells and molecules that do the actual work of antigen elimination. Based on these constant regions, Igs are divided into five classes, or *isotypes*, called IgM, IgG, IgA, IgD, and IgE. Each isotype confers slightly different physical and functional characteristics on the Ig molecule. In addition, Ig isotypes also differ in their anatomical distribution, in their ability to form multimers with high antigen-binding potential,

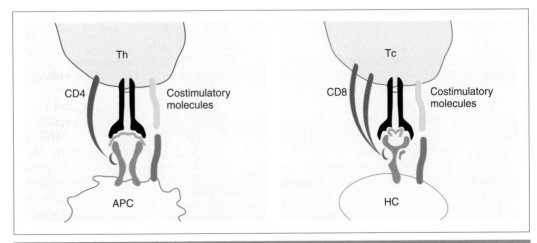

Figure 2-11

CD4 and CD8 Coreceptors and Costimulatory Molecules
Additional intercellular contacts between T cells and their targets are also required for full T cell activation. CD4 and CD8 coreceptors located on T cell surfaces bind to MHC molecules at a site other than the antigen binding cleft, to stabilize the interaction between the TCR and the antigen-presenting cell (APC) or target cell. CD4 is a monomeric molecule that binds to MHC class II, while CD8 is dimeric and binds to MHC class I. Other costimulatory molecules on the surface of T cells make contact with specific ligands on the host cell (HC) or APC to further promote T cell activation.

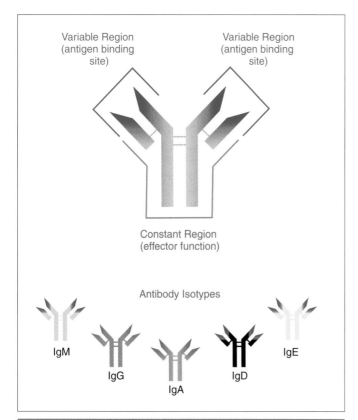

Figure 2-12

The Basic Structure of Immunoglobulins and Soluble Antibodies
Membrane-bound immunoglobulins and soluble antibodies are Y-shaped, bifunctional molecules. At the tips of the Y are identical variable regions that bind to antigen. It is in the variable region that antibody structure differs greatly among B cell clones. The stem of the Y forms the constant region that confers the antibody's effector function. The type of constant region in the Ig molecule determines its class (isotype). At different times in the life of a given B cell, the variable region can be synthesized in combination with any of the five different constant regions that determine effector function, ensuring that there will be an array of different clones recognizing the same antigen but clearing it by a variety of mechanisms.

binding site (the arm of the Y) with any one of the several sequences defining an isotype (the "stem" of the Y). Thus, over its lifetime, a B cell clone may produce Igs of the same antigenic specificity, but of different isotype; however, at any given moment, a given daughter B cell of that clone produces Ig of a single isotype. These matters are examined in detail in Chapter 9.

VII. INTRODUCING T CELL EFFECTOR FUNCTIONS

Unlike B cells, T cells recognize antigen only when it is presented in cell-bound (not soluble) form. Furthermore, T cells do not produce a secreted form of their receptors. Although they mature in the thymus, the effector cells of the T lineage function in the periphery, following their activation in the secondary lymphoid tissues. Effector T cells either work at the site where they are first activated (often a lymph node) or they may travel through the blood and lymphatics to reach distant sites where the triggering antigen is being encountered by the body.

Mature peripheral T cells fall into two categories, depending on their coreceptor expression (CD4$^+$ or CD8$^+$) and mode of cell–cell interaction (Fig. 2-13). Because the

and in the timing of their appearance in primary and secondary immune responses.

All Ig isotypes, when secreted as antibody, can "neutralize" specific antigens and physically prevent them from attaching to and invading host cells. However, the most important function of humoral immunity is the clearance of these antigen–antibody complexes from the body. The antibody acts as a "tag" that marks an antigen for destruction by enzymes or cells that recognize a complex composed of antigen bound to an antibody of a particular isotype. Different Ig isotypes interact with components of different antigen-clearance mechanisms, so that, depending on the isotype of the Ig that has bound the antigen, the antigen is removed by either phagocytosis, destruction by specialized lytic cells, or destruction by the lytic enzymes of the complement system (see Chs.4 and 5).

The variable and constant regions of an Ig molecule are encoded by separate gene segments. An elegant mechanism of *isotype switching* that operates in activated lymphocytes permits the joining of a sequence defining a specific antigen-

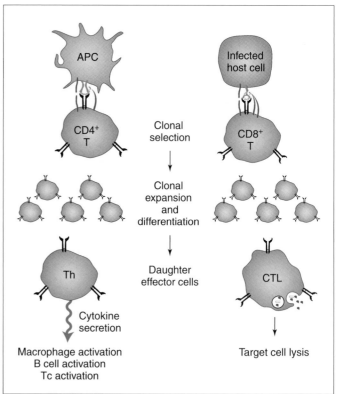

Figure 2-13

T Cell Effector Functions
Circulating CD4$^+$ and CD8$^+$ T cells bind to specific peptide–MHC complexes and undergo clonal selection, expansion, and differentiation into daughter Th cells and cytotoxic T lymphocytes (CTLs), respectively. Effector Th cells secrete cytokines that activate macrophages, B cells, and Tc cells. CTLs lyse infected host cells and tumor cells.

CD4 coreceptor binds exclusively to MHC class II molecules, a CD4$^+$ T cell can scan only the surfaces of passing APCs in search of antigenic peptide–MHC class II complexes that match the shape of its T cell receptor antigen-binding site. Binding of an appropriate peptide–MHC II complex by the T cell activates it to proliferate and differentiate into daughter effector cells that are helper T cells. Their effector action is the *secretion of cytokines* that are crucial for the activation of other cells, including cytolytic T cells, B cells, and other non-lymphocytes. Under the influence of different constellations of cytokines, CD4$^+$ Th cells can further differentiate into *Th1* and *Th2* cells, which in turn secrete cytokines having differential effects on different aspects of the adaptive immune response. For example, among other functions, Th1 cells are involved in defense against intracellular bacteria and viruses, while Th2 cells secrete a "cytokine profile" that aids in defense against parasites and extracellular bacteria.

Because the CD8 coreceptor interacts with MHC class I molecules, which are widely expressed on almost all cells, a CD8$^+$ T cell can survey all passing cells in search of those displaying a peptide–MHC class I complex that matches the shape of the binding site of its T cell receptor. After lymphocyte activation, proliferation, and differentiation, daughter effector CD8$^+$ T cells are generated; these *cytotoxic T lymphocytes* (CTLs) can induce the lysis of other host target cells. A single effector CTL can go from target to target (bearing the same peptide–MHC I complex) delivering "lethal hits" that result in target cell death. The killing of a host cell may not seem at first to be a desirable event, but, in the case of a viral infection, the host cell has become a "virus factory" that is functioning more for the good of "nonself" than "self." The action of the CTL to permanently "shut down" virus production is of overall benefit to the host. Similarly, a malignantly transformed host cell is detrimental to the health of the host, so its destruction by a CTL becomes highly desirable. The mechanisms of CTL-mediated cytolysis are explained in detail in Chapter 15.

It is pertinent to note that CD4$^+$ and CD8$^+$ T cells cannot be absolutely equated to Th and Tc cells, respectively. Some helper T cells have been found to express CD8, and some cytolytic T cells bear CD4. However, it is absolute that CD4 always binds to MHC class II, and CD8 always binds to MHC class I (see Ch.14). In general, cytotoxic T cells respond to peptides of intracellular origin (such as those derived from viruses or tumors), while peptides of extracellular origin (i.e., not produced inside the antigen-presenting cell) are recognized by the receptors of helper T cells.

VIII. INTRODUCING PRIMARY AND SECONDARY ADAPTIVE IMMUNE RESPONSES

In generating the smallpox vaccine, Jenner capitalized (albeit unknowingly) on the ability of an organism's immune system to react faster and with greater intensity following a second exposure to a foreign entity. Both the humoral and cell-mediated branches of the adaptive immune response exhibit these characteristics, the presence of which can be demonstrated by monitoring either serum antibody levels (in humoral immunity) or the speed of rejection of transplanted tissues (in cell-mediated immunity). The kinetics of both types of responses are, in general, identical, but due to the greater technical ease of studying antibodies, more detailed information is available on primary and secondary humoral responses than on primary and secondary cell-mediated responses. Accordingly, we shall use the context of the humoral immune response in the following description.

Let's say an individual is exposed to antigen X for the first time, in what is called a *priming event* (Fig. 2-14). From the individual's lymphocyte repertoire, the small number of B lymphocytes having receptors that can recognize and bind antigen X are stimulated during clonal selection. After a *lag phase* of about 5 days, a short-lived, relatively weak increase in serum antibodies against antigen X can be detected. Next, in the *log phase*, or *exponential phase*, there is a rapid increase in the concentration of anti-X antibody, due to clonal expansion of anti-X-producing lymphocytes, until a *steady-state* level is reached (when antibody is degraded as fast as it is formed). This stage is sometimes called the *stationary phase*. As antigen X is eliminated, the stimulus for production of anti-X is removed, and the antibody concentration gradually drops (the *declining phase*) until it reaches its original pre-exposure level. However, included among the cells generated in the clonal expansion phase of the primary response are X-specific memory cells. These long-lived antigen-specific lymphocytes take up residence in large numbers in the lymphoid tissues after a first exposure, where they can remain dormant for very long periods. This completes the *primary immune response*. This process is repeated for all the antigens our body encounters that are not dealt with by the innate system. Thus, the serum of a healthy adult contains a wide variety of antibodies, each generally present in relatively low concentration.

A second exposure of our hypothetical individual to antigen X sparks a *secondary immune response*, which exhibits the same phases as the primary response, but with some important quantitative differences. The lag phase is reduced to 1–2 days from 5 days, due to the expanded presence of the pre-existing antigen-specific memory cells. These cells swiftly and efficiently differentiate into anti-X-producing plasma cells, as well as more memory cells. As a result, the rate of increase in the serum concentration of anti-X is much faster, the steady-state concentration of anti-X in the serum is much higher, and anti-X can be detected in the serum for much longer periods (months to years) after the second exposure. There are qualitative differences in the antibody produced as well, generated by subtle changes introduced into the secondary-response anti-X antibody protein. Due to the capacity in B cells for rearrangement of the immunoglobulin genes, antibodies of different isotypes may be secreted in the secondary response, a fact that has implications for the effector mechanisms used in antigen clearance. In addition, somatic point mutations in the

Figure 2-14

Features of Primary and Secondary Immune Responses
The graph plots the change in serum antibody concentration after a first and then a second exposure to antigen X. The accompanying chart compares the various properties of primary and secondary immune responses.

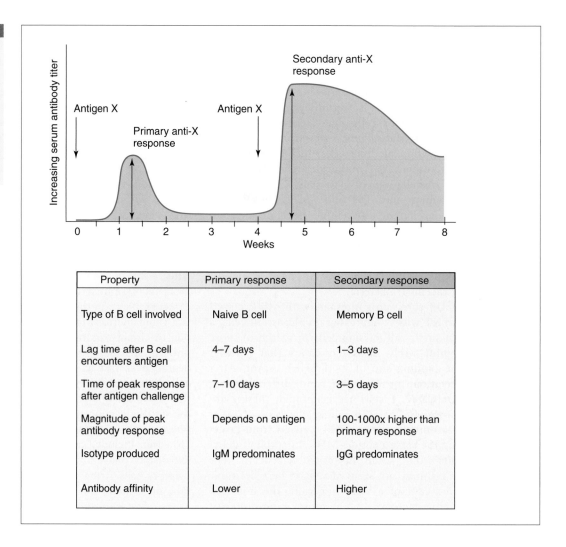

Property	Primary response	Secondary response
Type of B cell involved	Naive B cell	Memory B cell
Lag time after B cell encounters antigen	4–7 days	1–3 days
Time of peak response after antigen challenge	7–10 days	3–5 days
Magnitude of peak antibody response	Depends on antigen	100-1000x higher than primary response
Isotype produced	IgM predominates	IgG predominates
Antibody affinity	Lower	Higher

Ig genes may produce a mutated anti-X antibody that has an increased affinity for antigen X, compared to primary-response anti-X (see Ch.9). These changes often contribute to the increased efficiency of the secondary immune response.

The existence of the secondary response forms the basis for *vaccination*, whereby an individual is immunized against a serious disease by a priming event with an antigen that resembles the disease-causing pathogen. This antigen induces production of the same anti-pathogen antibody and memory cells but does not (generally) cause illness (see Ch.23). Some vaccines consist of a whole pathogen that has been killed, inactivated, or weakened ("attenuated"). Other "subunit" vaccines, composed of only part of a pathogen, are effective because antibodies and effector T cells are able to recognize individual epitopes and do not require the complete molecule or microbe for binding and activation. In other words, the vaccine retains the *immunogenicity*, but not the pathogenicity, of the disease-causing pathogen. A subsequent natural infection stimulates the memory cells produced during the primary response to the vaccine, so that the antigen is eliminated quickly and efficiently in a secondary response and disease is prevented. Some antigens result in the development of memory cells that last a lifetime; others (for reasons still unknown) do not. Thus, protection against certain diseases requires periodic *booster* vaccinations, to replenish the body's army of memory cells.

The reader has now completed an overview of the immune system, and is no doubt curious to learn more about its mechanics. In Chapter 3, we examine the "players" in this game of host defense—the cells and tissues of the immune system.

SUMMARY

Innate immunity involves both pre-existing physical barriers that show little or no pathogen specificity and induced cellular responses of broad specificity. In contrast, adaptive immune responses must be induced and require the activation of B and T lymphocytes. Each lymphocyte clone expresses cell surface antigen receptors of a single specificity, and each clone is activated only upon the interaction of these receptors with complementary antigen. For B cells, this antigen may be a native or denatured protein or carbohydrate of almost any conformation. For T cells, the antigen is a complex of peptide bound to an antigen presentation molecule encoded by genes localized in the major histocompatibility complex (MHC). Clonal selection by antigen makes lymphocyte responses highly specific. The repertoire of lymphocyte specificities available for induction by antigen is almost infinitely diverse because the antigen receptor proteins are encoded by genes that undergo random somatic rearrangement during lymphocyte development. Such diversity inevitably generates some antigen receptors recognizing self tissues. However, lymphocytes with the potential for autoreactivity are either eliminated during development (central tolerance) or regulated in the periphery, so they do not cause harm (peripheral tolerance). Once a lymphocyte is activated by its specific antigen, it undergoes clonal expansion and differentiation into effector and memory cells. Effector lymphocytes are short-lived and mediate the primary response to antigen, while memory cells are long-lived and mediate a stronger, faster secondary response if the same antigen is encountered again. Activated B lymphocytes differentiate into plasma cells, which secrete antibodies critical for defense against extracellular pathogens. Activated Tc lymphocytes differentiate into cytotoxic effectors (CTLs) capable of lysing tumor cells and cells infected with intracellular pathogens. Activated Th lymphocytes differentiate into cytokine-secreting Th effectors that support B cell and Tc cell functions.

Selected Reading List

Alberts B. (2002) The immune system. In "Molecular Biology of the Cell," 4th edn. Garland Science, New York, NY. Ch.23.

Chaplin D. (2003) Overview of the immune response. *Journal of Allergy and Clinical Immunology* 111, S442–S459.

Karp G. (2005) The immune response. In "Cell and Molecular Biology," 4th edn. John Wiley, Hoboken, NJ. Ch.17.

Nossal G. (1987) The basic components of the immune system. *New England Journal of Medicine* 316, 1320–1325.

Parkin J. and Cohen B. (2001) An overview of the immune system. *The Lancet* 357, 1777–1789.

Cells and Tissues of the Immune Response

Paul Ehrlich 1854–1915

Early understanding of immunity
Nobel Prize 1908

© Nobelstiftelsen

CHAPTER 3

A. CELLS OF THE IMMUNE SYSTEM

B. LYMPHOID TISSUES

"We shall not fail or falter; we shall not weaken or tire.
Give us the tools and we will finish the job.
—Sir Winston Churchill (attributed)

I n the first two chapters, we have made reference to T and B lymphocytes, to macrophages and neutrophils, and to lymph nodes, thymus, and bone marrow. It is now time to examine in more detail these and other cells and tissues whose actions and products result in the elimination of foreign entities and the preservation of our good health.

A. Cells of the Immune System

I. TYPES OF HEMATOPOIETIC CELLS

As noted in Chapter 1, mammalian blood is made up of a plasma fluid phase containing oxygen-carrying erythrocytes (red blood cells) and infection-fighting leukocytes (white blood cells). These cells are known as *hematopoietic* cells. All hematopoietic cells are derived in the bone marrow from a common precursor called the *hematopoietic stem cell* (HSC), in a process called *hematopoiesis* ("the generation of blood"), which is described later in this chapter. Hematopoietic cells are categorized as being of either the *myeloid* or *lymphoid* lineage. Myeloid cells include *erythrocytes* and nonlymphocyte leukocytes such as *neutrophils, monocyte/macrophages, eosinophils, basophils,* and *megakaryocytes.* Cells considered to be of the lymphoid lineage include *T and B lymphocytes, natural killer* (NK) *cells,* and *natural killer T* (NKT) *cells. Dendritic cells* (DCs) can arise from either myeloid or lymphoid lineage cells (see Ch.11). The physical characteristics of these cell types are illustrated in Figure 3-1.

Different hematopoietic cell types tend to reside and function in different compartments of the body (Fig. 3-2). Hematopoietic stem cells and early hematopoietic progenitors are generated and remain primarily in the bone marrow. Lymphocytes can be found in both the blood and the peripheral tissues. Other hematopoietic cell types tend

to concentrate in either the blood or the tissues, but may move between these compartments when infection or injury occurs.

We will first explore the features and functions of myeloid cells, followed by those of lymphoid cells. We then move to a discussion of how these cells are generated in the first place, and how and why they die. The second section of this chapter focuses on lymphoid tissues, which are the regions of the body where adaptive immune responses are initiated.

II. CELLS OF THE MYELOID LINEAGE

i) Introducing Myeloid Cells

Among the largest blood cells in the circulation are the *monocytes,* phagocytes that possess a regularly shaped nucleus. The principal functional features of these cells are their cytoplasmic lysosomes, which contain hydrolytic enzymes, and their abundance of organelles, which are required for the synthesis of secreted and membrane-bound proteins. Monocytes circulate in the blood at low density (3–5% of all blood leukocytes) for approximately 1 day before entering the tissues and serous cavities and maturing further to become *macrophages.* Macrophages (from the Greek, "big eaters") are large, powerful phagocytes that function primarily in the tissues, tirelessly engulfing and digesting not only foreign entities, but also spent host cells and cellular debris: they are the "garbage collectors" of our tissues. In addition to their function in innate immunity as key phagocytes during inflammation, macrophages are also important antigen-presenting cells (APCs) for the T cells of the adaptive immune response. The functions of macrophages are discussed later in this chapter in more detail.

Neutrophils, basophils, and eosinophils are all considered to be *granulocytes,* which are myeloid leukocytes that harbor large intracellular granules containing microbe-destroying hydrolytic enzymes. Neutrophils are the most common white

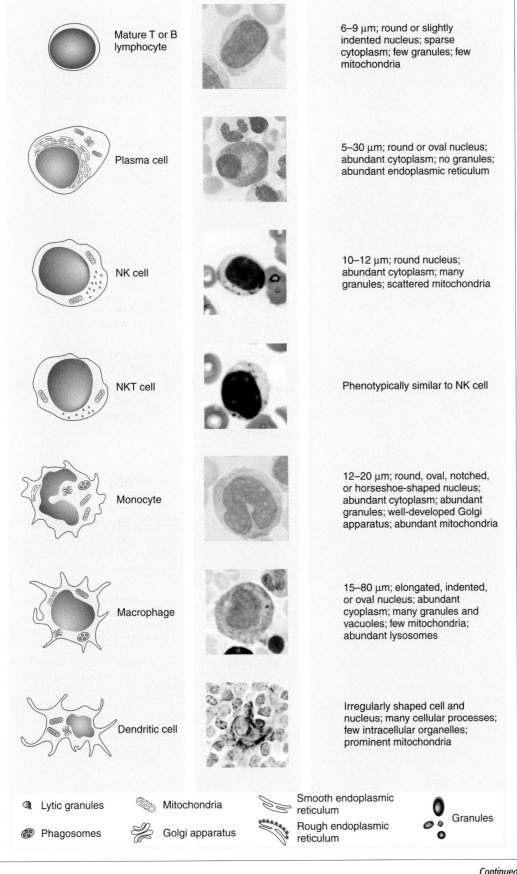

Figure 3-1

Hematopoietic Cells
Micrographs of cells courtesy of Dr. Doug Tkachuk, Princess Margaret Hospital, University Health Network, Toronto; from Tkachuk D. *et al.* (2002) "Atlas of Clinical Hematology." W. B. Saunders Company, Philadelphia.

Mature T or B lymphocyte — 6–9 μm; round or slightly indented nucleus; sparse cytoplasm; few granules; few mitochondria

Plasma cell — 5–30 μm; round or oval nucleus; abundant cytoplasm; no granules; abundant endoplasmic reticulum

NK cell — 10–12 μm; round nucleus; abundant cytoplasm; many granules; scattered mitochondria

NKT cell — Phenotypically similar to NK cell

Monocyte — 12–20 μm; round, oval, notched, or horseshoe-shaped nucleus; abundant cytoplasm; abundant granules; well-developed Golgi apparatus; abundant mitochondria

Macrophage — 15–80 μm; elongated, indented, or oval nucleus; abundant cyoplasm; many granules and vacuoles; few mitochondria; abundant lysosomes

Dendritic cell — Irregularly shaped cell and nucleus; many cellular processes; few intracellular organelles; prominent mitochondria

Lytic granules Mitochondria Smooth endoplasmic reticulum

Phagosomes Golgi apparatus Rough endoplasmic reticulum Granules

Continued

Figure 3-1 *cont'd*

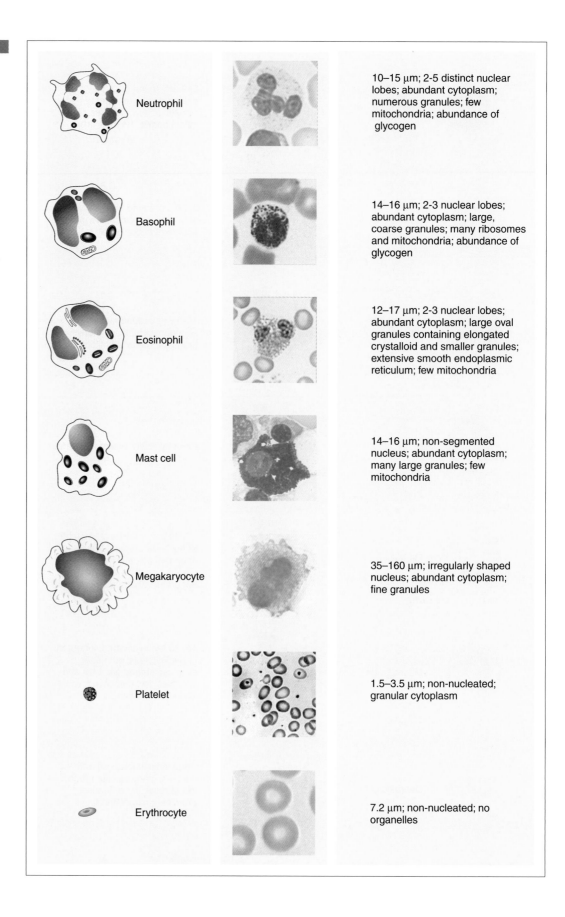

Neutrophil — 10–15 μm; 2-5 distinct nuclear lobes; abundant cytoplasm; numerous granules; few mitochondria; abundance of glycogen

Basophil — 14–16 μm; 2-3 nuclear lobes; abundant cytoplasm; large, coarse granules; many ribosomes and mitochondria; abundance of glycogen

Eosinophil — 12–17 μm; 2-3 nuclear lobes; abundant cytoplasm; large oval granules containing elongated crystalloid and smaller granules; extensive smooth endoplasmic reticulum; few mitochondria

Mast cell — 14–16 μm; non-segmented nucleus; abundant cytoplasm; many large granules; few mitochondria

Megakaryocyte — 35–160 μm; irregularly shaped nucleus; abundant cytoplasm; fine granules

Platelet — 1.5–3.5 μm; non-nucleated; granular cytoplasm

Erythrocyte — 7.2 μm; non-nucleated; no organelles

Bone Marrow	Blood	Tissues
Stem cells Early progenitor cells	Monocytes Erythrocytes Neutrophils Basophils Eosinophils	Macrophages Dendritic cells Mast cells NK cells
	B lymphocytes T lymphocytes	

Figure 3-2

Compartmentalization of Hematopoietic Cells
In the absence of an immune response, hematopoietic cells can generally be found in particular anatomical locations within the body.

cells in the body and respond immediately in great numbers to tissue injury. In addition to being granulocytes, neutrophils are phagocytes capable of engulfing macromolecules and particles. Neutrophils were originally called *polymorphonuclear* (PMN) *leukocytes* because of the appearance under the light microscope of their irregularly shaped, multilobed nuclei. The term "neutrophil" arose when it was observed that the cytoplasm and cytoplasmic granules of these cells stained neutrally with certain dyes. Neutrophil functions are discussed in more detail later in this chapter.

Eosinophils and *basophils* acquired their names from the colors taken on by their cytoplasmic granules (as viewed under the microscope) when these cells were originally stained and examined in blood smears. Eosinophils are leukocytes with bilobed nuclei and large cytoplasmic granules that stain reddish with certain acidic dyes (such as eosin). These granules are filled with highly basic proteins and enzymes that are effective in the killing of large parasites. The eosinophil's primary function is the removal of such pathogens, although it can also carry out phagocytosis. The vast majority (99%) of mature eosinophils reside in the connective tissues and so constitute less than 4% of all leukocytes in the blood. Eosinophils also have a role in allergy, and will be discussed in that context in Chapter 28. Basophils are circulating leukocytes with irregularly shaped nuclei and cytoplasmic granules that react with basic dyes (such as hematoxylin), thus staining a dark blue color. Basophils are important for inflammation, since their granules contain heparin and vasoactive amines, as well as many enzymes. Basophils are present in the body in very low numbers, residing primarily in the blood until they move into the tissues during an inflammatory response (see Ch.4).

Mast cells (from the German word "mast," meaning "food" or "feed") were named by Paul Ehrlich, who found this cell type to be increased in the tissues of fattened animals. This observation is consistent with the role of mast cells in countering attacks by worms and other parasites. The cytoplasmic granules of mast cells stain in a way similar to the staining of basophilic granules, and also contain heparin and

histamines. However, mast cells are derived from a separate cell lineage, mast cell nuclei are not lobed, and mast cell granules tend to be more numerous and of a smaller size compared to those of basophils. Unlike basophils, mast cells are rarely found in the blood, preferring to reside in the connective tissues and, in some cases, the gastrointestinal mucosa. The *degranulation* (release of granule contents) of mast cells is rapidly triggered by tissue injury, causing the release of histamine and dozens of cytokines and growth factors that initiate the inflammatory response and, occasionally, hypersensitivity and allergy (see Ch.28).

Megakaryocytes are multinucleate cells (containing up to 20 nuclei) from which platelets are derived. Platelets are small, colorless, irregularly shaped nonnucleated cells. The primary role of platelets is to mediate blood clotting, but activated platelets also secrete cytokines that influence the functions and migration of other leukocytes during an inflammatory response. These cytokines are produced from preformed messenger RNAs (mRNAs) that are translated upon platelet activation. *Erythrocytes* are myeloid cells that are very important physiologically; however, they contribute minimally to the immune response and we will not discuss them further here.

ii) More about Neutrophils

As previously mentioned, neutrophils constitute the majority of cells activated in the inflammatory response. These cells grow to full maturity in the bone marrow where they become terminally differentiated and incapable of further cell division. Mature neutrophils then enter the blood circulation in great numbers such that the average adult human possesses approximately 50 billion neutrophils in the circulation at any given moment. This large number is necessary because each neutrophil has a life span of only 1–2 days.

Neutrophils are drawn from the circulation into an injured or infected tissue by the presence of specific *chemotactic factors*, which are molecules derived from either host or bacterial metabolism (see Box 4-4 in Ch.4). As early as 30 minutes after an acute injury or the onset of infection, the first neutrophils can be detected at the site and accumulation continues until the production of chemotactic factors falls off. As the neutrophils reach the site of pathogen attack, they immediately begin to phagocytose foreign cells or macromolecules, sequestering the captured entities in intracellular vesicles called *phagosomes* (see Ch.4). Cytoplasmic granules in the neutrophil then fuse with the phagosomes, and within seconds, granule contents are released into these vesicles. The phagosomes are flooded with hydrogen ions, lowering the phagosomal pH and resulting in the direct hydrolysis of the foreign entity, as well as the optimization of reaction conditions for enzymes stored in the granule. Included with the numerous proteases released by a degranulating neutrophil are the *defensins* (small antimicrobial peptides that permeabilize bacterial or fungal cell walls), *lysozyme*, and *lactoferrin* (an iron-chelating protein that inhibits bacterial growth). Besides the hydrolytic action of granule enzymes, the *oxidation* of target molecules is also an important antimicrobial defense. After degranulation, *NADPH-dependent oxidases* attached to the granule membranes stimulate both the generation of inorganic radicals and *myeloperoxidase* action, resulting in the formation of hypochlorous acid. This end-product quickly oxidizes crucial components of the microbe, including nucleic acids, amino acids, and thiols. In order to sustain this antimicrobial mechanism, the neutrophil undergoes a significant increase in oxygen utilization called the *respiratory burst* (see Box 4-2 in Ch.4).

Neutrophil action as described in the preceding paragraph is a component of the innate immune response, since the cells respond with only broad specificity to their target pathogens. However, neutrophils are linked to the adaptive immune response through the role they play in two antibody-mediated antigen clearance mechanisms. Firstly, neutrophils express surface receptors for IgG antibody molecules that allow the neutrophils to bind and phagocytose complexes formed by IgG antibody bound to its specific antigen. Secondly, receptors for certain complement proteins that coat antigens after complement activation (see Ch.5) are also present on the neutrophil surface. Like IgG–antigen complexes, the resulting complement–antigen complexes can then be phagocytosed by the neutrophils. In cases in which complement activation has been initiated by the presence of antigen–antibody complexes, this antigen clearance process represents another way in which neutrophils are linked to the antibody responses of adaptive immunity.

Although neutrophils are critical for the initial stages of infection control, and are highly effective in removing certain pathogens, they are limited in their range of action and can cause collateral damage to host tissues. If a tissue injury provokes a prolonged or intense inflammatory response, enzymes released from neutrophils countering a microbial invader start to liquefy nearby healthy host cells. The accumulation of dead host cells, degraded microbial material, and fluid forms *pus*, a familiar sign of clinical infection and inflammation. "Bystander" damage to host tissues can also arise when the neutrophil releases its granule contents extracellularly in an effort to dissolve a particle too large to engulf, a situation sometimes called "frustrated" phagocytosis. Another limitation to the neutrophil as a host defender is that it can phagocytose only those antigens that bind to the specialized receptors on its surface that initiate phagocytosis. Certain viruses, bacteria, and toxins may not bind to these receptors, thereby escaping clearance by neutrophils. In addition, the neutrophil population does not "learn" from an encounter with a foreign entity and is not capable of strengthening its response upon a second exposure to a pathogen; that is, no immunological memory is developed. Over the course of evolution, these limitations have encouraged vertebrate hosts to develop additional defenses that are more specific and capable of adapting to provide enhanced responses upon repeated exposure to a particular threat.

iii) More about Macrophages

Macrophages are long-lived mononuclear phagocytic cells that reside in all organs and tissues, usually in the locations where they are most likely to encounter foreign entities or unwanted host materials. Regardless of location, macrophages have a multifaceted role in the innate immune response: not only do they have a phagocytic function, but they also secrete cytokines that draw neutrophils and other immune system cells to the site of inflammation, and produce growth factors that stimulate cells involved in wound healing. Their importance, however, reaches even beyond their roles in the innate immune response. Macrophages function as APCs for T cells and secrete cytokines that activate lymphocytes of the adaptive immune response.

Macrophages are generally several times larger than the monocytes from which they are derived and display further enhancements of the cellular protein synthesis and secretion machinery. Subtle differences in the morphology and function of macrophages develop as result of the influence of a particular microenvironment on a particular monocyte, and these differences have given rise to the use of tissue-specific names (Fig. 3-3). For example, macrophages that develop in the liver sinusoids are known as *Kupffer cell*s, whereas those developing in the kidney, brain, connective tissues, or bone are known, respectively, as *mesangial phagocytes*, *microglia*, *histiocytes*, or *osteoclasts*. A tissue macrophage has a life span of 2–4 months, during which it can remain fixed in one spot or can constantly explore the tissue by amoeboid movement, using its many different cell surface receptors to monitor its immediate environment for antigens.

The *activation* of the tissue macrophage by an encounter with a foreign entity is one of the pivotal moments in the immune response. An activated macrophage undergoes metabolic changes that allow it to move faster, phagocytose more rapidly, and kill microbes more effectively. Not only is the phagocytic capability of the cell enhanced, but it also commences secretion of numerous cytokines, hydrolytic enzymes, chemoattractants, complement components, coagulation factors, and reactive oxygen species. A whole host of biological reactions ensues, including the killing of microbes, the recruiting of cells of the adaptive immune response, and the instigation of wound healing. Macrophages are activated by direct contact with certain microbes or macromolecules, or by the binding of a broad range of ligands to *pattern recognition receptors* (PRRs) on the surface of the macrophage (see Ch.4). Molecules that can activate macrophages include host tissue breakdown products, proteins of the complement or coagulation systems, bacterial endotoxins, and carbohydrate-containing structures commonly found on the surfaces of a broad range of pathogens. Macrophage activation is further stimulated by cytokines produced in the adaptive immune response, especially *interferon-gamma* (IFNγ) secreted by helper T cells and NK cells. Conversely, activated macrophages have an enhanced ability to secrete interleukins that activate additional T cells and NK cells. We see here a reciprocity between the innate and adaptive systems, in that the innate response (macrophage activation) is enhanced by products of the adaptive system (cytokines secreted by helper T cells), and that the adaptive response (recruitment and activation of additional T cells) is enhanced by a "bridging" cell of the innate system (cytokines secreted by the activated macrophage). More detail on macrophage activation is given in Box 3-1.

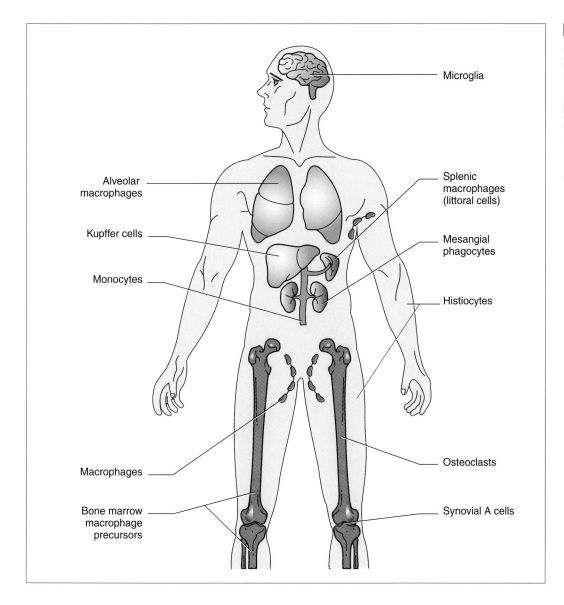

Figure 3-3

Macrophages throughout the Body
Once mature blood monocytes migrate to the tissues, they are called macrophages. Tissue macrophages in different locations in the body have different names, often reflecting where they reside in the body.

Microglia

Splenic macrophages (littoral cells)

Mesangial phagocytes

Histiocytes

Osteoclasts

Synovial A cells

Alveolar macrophages

Kupffer cells

Monocytes

Macrophages

Bone marrow macrophage precursors

III. CELLS OF THE LYMPHOID LINEAGE

As mentioned previously, cells of the lymphoid lineage include T and B lymphocytes, NK cells, and NKT cells. An adult human body contains about 10^{12} lymphocytes, accounting for up to 40% of the body's white blood cells. In human peripheral blood, about 55% of all lymphocytes are CD4$^+$ T cells, 20% are CD8$^+$ T cells, and 10% are B cells. [The remaining 15% is made up of a subset of unconventional T cells called the *T cells* plus NK cells and NKT cells (discussed in detail in Chs.12, 13, and 18)]. We will now describe the physical characteristics and general functions of T and B lymphocytes, NK, and NKT cells.

i) T and B Lymphocytes
ia) T and B lymphocytes and their identification. *Resting* B and T lymphocytes (those in the G_0 phase of the cell cycle) are classed histologically as "small lymphocytes" (refer to Fig. 3-1). Morphologically, small lymphocytes are round cells with a large nucleus surrounded by a narrow rim of cytoplasm. Comparatively few intracellular organelles are contained in the cytoplasmic rim. Resting B and T cells that have not interacted with specific antigen are said to be *virgin, naive,* or *unprimed.* They have a short life span (up to a few weeks) and undergo apoptosis unless they encounter their specific antigen.

Binding of specific antigen *activates* a lymphocyte and stimulates it to progress through cell division. The transcription of numerous genes is triggered, causing the progeny cells to undergo morphologic and functional changes that result in the production of *lymphoblasts.* This conversion of the resting lymphocyte into a lymphoblast occurs within 18–24 hours after antigen receptor activation. Lymphoblasts are larger and display more cytoplasmic complexity than their resting cell counterparts. They undergo rapid cell division and differentiation into long-lived memory and short-lived effector cells; in fact, effector cells can be detected after the first cell division. For B cells, the fully differentiated effector cell is an antibody-secreting plasma cell. T cell effectors of the helper type (Th) become cytokine

Box 3-1. The multiple roles of macrophages

Macrophages occupy a position of central importance in the immune response, since they are key effectors in both the innate and adaptive systems. Four clearly separate roles can be ascribed to activated macrophages. Firstly, they play a role in the inflammatory response during the early stages of host *innate* defense, phagocytosing foreign entities and secreting chemotactic factors and cytokines that draw other inflammatory cells to the site of tissue injury and promote tissue repair. Secondly, macrophages function as *accessory* cells to the lymphocytes of the adaptive immune response, not only secreting pro-inflammatory cytokines such as *tumor necrosis factor* (TNF) and IL-1, which affect lymphocyte proliferation and differentiation, but also processing and presenting antigen to T cells. Thirdly, macrophages act as effectors in the *humoral* immune response, since they possess receptors for immunoglobulin molecules on their cell surfaces and can avidly phagocytose antigens bound by soluble antibody. Fourthly, macrophages act as effectors in *cell-mediated* immunity, in response to cytokines secreted by activated T cells. These "hyperactivated" macrophages gain enhanced antimicrobial and antiparasitic activities and new cytolytic capacities, particularly against tumor cells. Thus, the mutual stimulation of T cells and macrophages is vital to the immune response, since it provides a means of amplifying both the innate effector mechanisms of macrophage action, and the adaptive effector mechanisms of the cell-mediated and humoral responses.

The activation of macrophages is a multifaceted process (see Figure A), with the definition of activation depending to some extent on which macrophage function is under examination. The binding of a particular macrophage activator, which could be either a cytokine, a bacterial product, or an extracellular matrix (ECM) molecule, may stimulate the transcription of those genes responsible for one activity, but may have no effect on the genes responsible for a different function. The activation of macrophages has been described as occurring in four stages. The *resting* or *immature* macrophage receives an initial signal from a site of inflammation that causes it to enter the *responsive* stage, becoming a cell that responds particularly well to chemotactic signals. On receipt of a second signal such as that achieved by low levels of cytokines, in particular IFNγ, the responsive macrophage becomes *activated* or *primed*, and capable of most of the tasks associated with fully

functional macrophages. Morphological, biochemical, and functional changes accompany this evolution, with cells becoming faster, larger, and filled with greater numbers of the intracellular organelles necessary for increased phagocytosis. These cells also excel at *antigen presentation* and *costimulation*. Finally, upon receipt of a third level of signal, such as high levels of IFNγ plus bacterial endotoxin, the activated cell becomes a *hyperactivated* or *triggered cytolytic* macrophage, exhibiting enhanced antimicrobial and antiparasitic activities and the capacity to kill tumor cells. These cells are incapable of proliferation and have a reduced ability to present antigen. All these stages of activation are clearly defined by the presence or absence of specific cell surface markers; for example, MHC class II proteins appear on activated and hyperactivated macrophages, but not on responsive macrophages. Similarly, only

hyperactivated macrophages express the CD11a marker.

The receipt of signals by macrophages at any stage of their development is facilitated by the huge array of specific receptors expressed on the cell membranes. Ligands bound by macrophage receptors include chemoattractant factors, growth factors, and cytokines (especially interferons and interleukins), hormones and related mediators, and other proteins (including fibronectin, lactoferrin, and transferrin). Macrophages also bear receptors for various *opsonins*, which are host proteins that can coat pathogens or macromolecules and make them easier to phagocytose (see Ch.5). Both antibodies and complement components are effective opsonins. Among the cytokines, IFNγ is the best-studied activator of macrophages as well as a key modulator of their functions. Phagocytosis and opsonization are increased

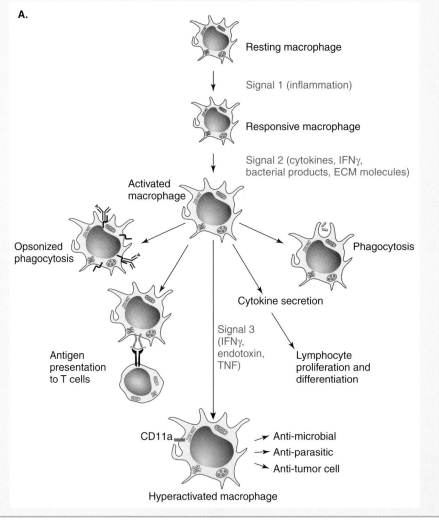

A.

Resting macrophage

Signal 1 (inflammation)

Responsive macrophage

Signal 2 (cytokines, IFNγ, bacterial products, ECM molecules)

Activated macrophage

Opsonized phagocytosis

Phagocytosis

Cytokine secretion

Antigen presentation to T cells

Signal 3 (IFNγ, endotoxin, TNF)

Lymphocyte proliferation and differentiation

CD11a

Anti-microbial
Anti-parasitic
Anti-tumor cell

Hyperactivated macrophage

Continued

Box 3-1. **The multiple roles of macrophages**—*cont'd*

in macrophages treated with IFNγ, and IFNγ also activates the transcription of the MHC class II genes, enhancing the ability of the activated macrophage to present antigen to helper T cells. Although the binding of IFNγ enables a macrophage to kill some types of bacteria, IFN binding alone is not sufficient to hyperactivate the macrophage such that it can kill intracellular bacteria that are resistant to phagosomal digestion. By itself, IFNγ is able to induce the nitric oxide pathway, but an additional signal in the form of TNF supplied by activated T cells is also required to trigger the actual release of the nitric oxide and turn the macrophage into a deadly killer. Similarly, the secretion of cytokines by macrophages is not induced by IFNγ alone, but is also dependent on a secondary signal, such as that derived from encountering a bacterial cell wall component such as endotoxin.

As well as being roving banks of receptor proteins, activated macrophages are secretory cells, flooding the surrounding cells and tissues with over 100 secretory products (see Figure B). Some molecules are produced in response to specific signals, and others are released as part of a more general response. Antimicrobial secretions include lysozyme and various complement components, as well as elastases and collagenases that digest the extracellular matrix to facilitate cell migration. Macrophages also secrete numerous cytokines and chemotactic molecules, influencing the migration, growth, and development of other leukocytes, both myeloid and lymphoid. Which products are secreted depends on the stage of macrophage activation and which signal has triggered the activation. Macrophages responding to acute inflammation produce lipid inflammatory mediators including *platelet-activating factor* (PAF) and *leukotrienes*. Additional binding by cytokines, including IFNγ, enhances the capacity of the macrophage to synthesize and secrete enzymes (such as *thrombin*) that are

necessary for tissue repair and angiogenesis. Secretion of colony-stimulating factors by activated macrophages can promote hematopoiesis. Macrophages also determine the direction of differentiation of helper T cell subsets by secreting either IL-10 (directs development of Th2 cells) or IL-12 (directs development of Th1 cells). Hyperactivated macrophages have significant tumoricidal activity, partly attributable to their secretion of TNF (see Ch.17).

The reciprocal influences of macrophages and T cells on one another are necessary to contain the effect of an innate immune response. The presence of constantly activated

macrophages in the tissues would be highly undesirable. Among their other tasks, activated macrophages take on the destruction of foreign entities that are too large to be phagocytosed, such as parasitic worms. Macrophages accomplish this by releasing destructive compounds into the extracellular milieu in the vicinity of the invader, an action that causes a certain degree of "bystander" damage to host cells. Without the tight controls that T cells exert on the activation of macrophages, the cost to the host in terms of "friendly fire" tissue damage, and the large energy expenditures required to sustain activation, would be prohibitive.

B.

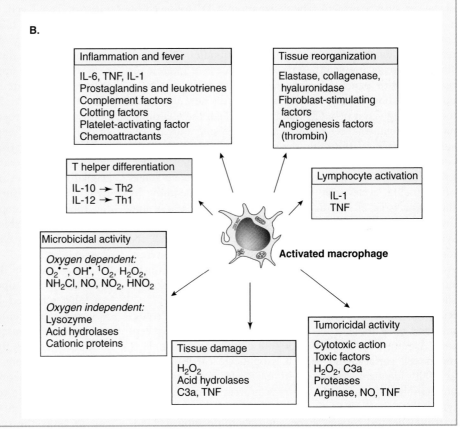

Inflammation and fever
IL-6, TNF, IL-1
Prostaglandins and leukotrienes
Complement factors
Clotting factors
Platelet-activating factor
Chemoattractants

Tissue reorganization
Elastase, collagenase, hyaluronidase
Fibroblast-stimulating factors
Angiogenesis factors (thrombin)

T helper differentiation
IL-10 → Th2
IL-12 → Th1

Lymphocyte activation
IL-1
TNF

Microbicidal activity
Oxygen dependent: $O_2^{\bullet-}$, $OH^{\bullet}$, $^1O_2$, H_2O_2, NH_2Cl, NO, NO_2, HNO_2
Oxygen independent: Lysozyme, Acid hydrolases, Cationic proteins

Activated macrophage

Tissue damage
H_2O_2
Acid hydrolases
C3a, TNF

Tumoricidal activity
Cytotoxic action
Toxic factors
H_2O_2, C3a
Proteases
Arginase, NO, TNF

secretors, while those of the cytotoxic type (Tc) develop a mechanism allowing them to directly lyse target cells.

Morphology can be used to distinguish some forms of lymphocytes. For example, under the microscope, a mature antibody-producing plasma cell, with its dense cytoplasmic concentration of endoplasmic reticulum and Golgi apparatus, is clearly different from a resting B cell. Similarly, assays for functional capability (e.g., determination of antibody or cytokine secretion) can be used to tell active B cells from active T cells. However, consistent with the theory that they develop from the same lymphoid precursor, B and T lympho-

cytes share many features during their early stages of maturation. In fact, prior to activation by antigen, a resting B cell cannot be distinguished from a resting T cell by examination of morphology or assay for effector function alone. Fortunately, the B cell receptors (BCRs) carried by resting B cells can be clearly distinguished from the T cell receptors (TCRs) present on resting T cells. T cells also express either the *CD4* or *CD8 coreceptor* proteins which are not found on B cells. In addition, the sequential appearance of unique cell surface proteins indicates the stage of maturity or activation of a given lymphocyte, and permits the discrimination of early precursors or resting cells from their more mature or activated counterparts.

ib) What is a CD designation? We have made reference previously to distinguishing lymphocyte subsets by their expression of CD4 or CD8. In fact, the CD system is useful for identifying many different types of leukocytes. A cell surface protein whose presence identifies a cell type or a stage of differentiation is known as a *cell surface marker* or a *CD marker*. CD stands for *cluster of differentiation*, a term coined before the elucidation of the functions of many cell surface proteins or protein complexes. Upon immunization of a host animal with the cells of another species, a given protein on the surface of an introduced cell can provoke the development of several different antibodies, each specific for a different *epitope* on the protein molecule. Thus, individual proteins on the surface of an introduced cell were originally identified by a "cluster" of different antibodies that would bind to some part of the cell surface protein. The designation "CD" followed by a unique number indicates that the protein bears a specific collection of *serological epitopes*, that is, epitopes that are recognized by a particular group of antibodies. New cell surface proteins are identified by screening cells of interest for reactivity with panels of new and existing antibodies. If a novel pattern of antibody binding is defined, the new membrane molecule is given a new CD number. Although the CD designation system was originally established using human cell surface molecules, the same designations are applied to the corresponding molecules in other species, particularly mice. Like other leukocytes, T and B lymphocytes at different stages of their life cycles can be recognized experimentally by their expression of specific CD markers. The appearance and disappearance of such markers, indicating cycles of protein synthesis and degradation, enable the scientist to track the identity, maturation, and state of activation of a cell of interest. A more detailed explanation of CD designations is given in Box 3-2 and a complete description of known CD markers is shown in the Appendix.

ic) Lymphocyte specificity and memory. Lymphocytes recognize and bind antigens using their cell surface antigen receptors. A mature T cell bears ~30,000 copies of a single TCR on its cell surface, and each B lymphocyte carries close to 150,000 identical copies of its BCR. In both cases, the antigen-binding site of the receptor protein on an individual lymphocyte is designed to bind with high affinity to a very narrowly defined antigen shape, and only those antigens binding to a sufficient number of receptors (estimated to be less than 20) with adequate affinity will trigger activation of the lymphocyte and initiate an immune response. Molecules that are very closely related in shape to the "ideal antigen" for that receptor will fit to some degree into the antigen-binding cleft of the receptor, and will therefore also be "recognized," or bound, with a strength proportional to the quality of the fit. Such an antigen is said to be *cross-reacting* with the receptor.

Before a lymphocyte ever comes into contact with any antigen, its specificity is determined by the structure of its BCR or TCR. Unlike any other known proteins, the antigen receptors on lymphocytes are encoded by genes generated by the random rearrangement of pre-existing gene segments (see Chs.8 and 12). This *somatic recombination* or *somatic gene rearrangement* occurs early in lymphocyte development, prior to the expression of the antigen receptors on the cell surface and before the cells are released into the peripheral circulation to begin their surveillance. Because of the vast numbers of lymphocytes produced during hematopoiesis, and because of the combinatorial possibilities arising from the process of somatic gene rearrangement, it has been estimated that a pool, or *repertoire*, of more than 10^{12} different antigen specificities is genetically possible in human lymphocytes at any given time. However, many naive lymphocytes comprising this repertoire will never be triggered, for one of the following reasons: (1) The randomly generated antigen receptor sequence of the lymphocyte may be directed against a self antigen, triggering the deletion of that cell in the bone marrow (B cell) or thymus (T cell) by the mechanisms of central tolerance; (2) if the lymphocyte in question is a T cell, its antigen receptor may fail to recognize the specific major histocompatibility complex (MHC) molecules displayed by host cells in the thymus, halting lymphocyte development; (3) the host may complete its life without being exposed to the appropriate antigens; (4) the appropriate mature lymphocyte may die before it can react to the antigen; (5) the lymphocyte may fail to meet up with the appropriate accessory cells necessary for full activation of that lymphocyte; or (6) the lymphocyte may be functionally silenced by mechanisms of peripheral tolerance (see Ch.16). Thus, the huge cellular generation of hematopoiesis is balanced by both regulatory mechanisms and the short life span of these cells. Evolution has ensured that, in spite of these adjustments to the functional receptor repertoire, the number of antigenic specificities available at any one time is still sufficiently large for the body as a whole to react to almost any foreign entity imaginable.

The size of the B cell repertoire is affected by additional mechanisms that operate after a B lymphocyte has been successfully activated by an encounter with specific antigen. Mutational mechanisms operating exclusively in these cells can further increase the diversity of the Ig receptor proteins produced (see Ch.9), so that the total number of antigenic specificities present in the B cell repertoire over the life of the host may theoretically approach 10^{30}. (A caveat: it is almost impossible to know the precise number of antigenic specificities in a given repertoire, since it is obviously difficult to assay the specificity of every lymphocyte clone; the numbers that appear in most texts are necessarily only estimates.)

If only a small clonal population of identical lymphocytes specific for an antigen (let's call it "X") is present in the circulation, how are enough anti-X lymphocytes generated to eliminate the potentially large numbers of X molecules entering the body in a pathogenic invasion? When lymphocytes recognizing antigen X encounter that antigen in the body, they bind X to their cell surface receptors, triggering clonal selection. Should sufficient antigen X–receptor pairs be formed, the lymphocytes are activated and a complex sequence of intracellular signaling events leads to their rapid cell division, resulting in clonal expansion. Only the clone specific for

Box 3-2. CD designations

Since leukocytes were first discovered, immunologists have been searching for ways to tell them apart in the absence of morphological differences. Originally, this was done by raising antibodies, in two different inbred mouse strains, to a cell surface protein. Such antibodies were called *alloantibodies* and were components of an *alloantiserum*. This approach led to the isolation of antibodies capable of identifying mouse T cells in general (e.g., anti-Thy-1), and distinguishing between functionally different subsets of T cells. Because the proteins recognized by these antibodies marked a cell as being of a particular type, or at a specific stage of differentiation, these cell surface molecules were called *differentiation antigens*. However, the use of alloantisera as a method of identification was limited to only those cell surface markers that occurred in allelic forms. Another problem developed from the practice of routinely naming cell surface proteins for the alloantisera that reacted with them. Since different sera from different research laboratories sometimes reacted to the same cell surface protein, the same molecule was often unknowingly called many different names by individual scientists. In addition, although certain physical properties of these proteins were known, the biological functions of few of them had been defined.

In order to resolve this confusion, the First International Workshop on Human Leukocyte Differentiation Antigens was held in 1982 to establish a standard and consistent nomenclature for these cell surface protein markers. It was helpful that in the early 1980s, César Milstein and Georges Köhler had invented the technology that led to the development of *monoclonal antibodies*, which are antibodies produced by a single B cell clone such that the antibody preparation contains a single type of antibody protein with a single specificity (see Ch.6). Workshop participants decided that a surface molecule that had a clearly defined structure, identified a particular stage of cellular differentiation or lineage, and was specifically recognized by a group (cluster) of monoclonal antibodies would be given a unique CD (*cluster of differentiation*) number. Subsequent International Workshops have assigned CD designations

to molecules other than the leukocyte antigens (including some in nonlymphoid tissues), and in species other than humans. New antigens are examined using standard antibodies to determine if they fall within an existing CD designation, and if not, are assigned a provisional "CDw" (for "Workshop") designation until their status is confirmed.

The CD designations are useful for identifying specific cell types, either by lineage, species, stage of maturation, or state of activation. The functions most often ascribed to molecules with CD designations are intercellular adhesion or other intercellular interactions, and the transduction of signals leading to lymphocyte activation. As an example of a marker specific to a cell lineage, consider CD3, expressed only on T cells and representing a complex of five transmembrane proteins that are coordinately expressed with the TCR chains. These proteins are intimately involved with TCR function and do not appear on other cell types. Similarly, almost all B cells carry the CD19 marker (a potential coreceptor for the BCR), and many carry CD20 (a protein that may be involved in controlling intracellular calcium flux) and CD22 (a B cell activation regulator). Some markers are shared with other cell types; all T cells carry the marker CD5 and the intercellular adhesion molecule CD2, but CD2 also appears on NK cells, and CD5 can be found on some B cells. Natural killer cells usually express CD16 and CD56, but lack CD3. Other markers distinguish species. For example, mouse (but not human) T cells express CD90 (the Thy-1 marker), whereas human (but not mouse) T cells bear the CD7 glycoprotein. Similarly, B220 (a high molecular weight isoform of the transmembrane phosphatase CD45) identifies principally (but not exclusively) murine B cells.

Stage of maturation can often be assessed by determining the presence or absence of specific CD markers. These markers are usually transiently expressed during development, and a marker that appears only at one stage in one cell type may be routinely present on another cell type. For example, CD1 occurs on most B cells and dendritic cells, but in the T cell lineage is found only on thymocytes developing in the thymus and not on

mature T cells in the periphery. CD25 (the IL-2 receptor α chain) is absent early in T cell maturation, but CD44 is present; later, both are present, and still later, CD25 appears and CD44 disappears. With respect to the coreceptors, immature T cells start in the thymus as CD4$^-$CD8$^-$ (double negative) cells, which then mature through a CD4$^+$ CD8$^+$ (double positive) stage. In both mice and humans, mature MHC class II-restricted T cells carry a species-specific version of CD4 (but not CD8), whereas cytotoxic T cells express CD8, but not CD4. Mature T cells are also almost unique in expressing the costimulatory molecule CD28. Immature B cells express high levels of CD10 (a peptidase enzyme) and CD20, while mature B cells are distinguished by high levels of CD22 and the costimulatory molecules CD40, CD80, and CD86. Mature B cells also express (but not exclusively) CD35 and CD21 (receptors for complement components C3b and C3d, respectively), and CD32 (FcγRII, the Fc receptor for soluble IgG).

Activation of a cell often induces the transient expression of a CD marker, although this protein may be present on other cell types in their resting states. Increased levels of CD25 (the IL-2 receptor) and the activation marker CD69 can be found at the early stages of T cell activation, whereas the marker CD29 appears only very late in activation. Activated, but not resting, T cells also express the intercellular adhesion molecule CD54, the activation regulator CD27, and the costimulatory molecule CD28. Activated B cells show increased levels of CD124 (the IL-4 receptor) and CD126 (the IL-6 receptor), among others. Molecules that facilitate cell–cell interaction also mark activation, as in the enhanced expression of the CD2 and CD22 on T and B cells, respectively. Still other markers serve to distinguish memory or effector cells from naive cells. For example, the CD29 marker is found on memory T cells but not naive T cells, and CD20 is expressed on resting B cells, but not plasma cells. A comprehensive list of CD designations is contained in Appendix 1. The reader should note that not all molecules cited in this book, even those expressed on cell surfaces, have been assigned a CD number. Examples include GlyCAM-1, LPAM, and MAdCAM.

antigen X is stimulated to multiply, creating a sufficient number of X-specific lymphocytes to mount an effective immune response. Within the expanding clone of X-specific lymphocytes, some will differentiate into effector cells (eliminating antigen X) and others into memory cells. Without the stimulatory presence of X, the effector cells soon die off, but the memory cells remain quiescent until selectively provoked to proliferate again by the binding of fresh X during a subsequent exposure to the same X-bearing pathogen. Because each lymphocyte activated in the initial response gives rise to many memory cells, there will remain an increased proportion of the body's lymphocyte population dedicated to fighting the pathogen that expresses antigen X. If we now imagine a later exposure to X in which the individual is also exposed for the first time to a different antigen Y, we observe a secondary response level of anti-X antibody because anti-X memory cells have been generated, but only a primary response level of anti-Y antibody. These principles have already been illustrated in Figures 2-2 and 2-14.

ii) Natural Killer Cells, Natural Killer T Cells, and γδ T Cells

One might be tempted to think that myeloid ancestry is equated with innate immunity, and lymphoid with adaptive immunity, but as with most processes in biology, it is not that simple. During the evolution of the immune system, two separate classes of lymphoid cells called *natural killer cells* (NK) and *natural killer T cells* (NKT) developed, as well as an unconventional subset of T lymphocytes called *T cells*. These cell types do not have the highly diverse and specific receptors found on B cells and conventional T cells and do not generate memory cells upon activation.

NK cells are large cells found in the blood and lymphoid tissues that morphologically resemble effector T lymphocytes (refer to Fig. 3-1), but which carry out their effector actions without the fine specificity of a T lymphocyte. Two interesting features of NK cells are their cytoplasmic granules, which are present in resting cells as they are in a granulocyte, and their capacity to recognize and kill certain virus-infected and tumor cells. Because this killing is carried out using a broad level of ligand recognition, this type of NK activation is considered part of the innate, rather than the adaptive, immune response. Furthermore, in the presence of sufficient quantities of interleukin-2 (IL-2), NK cells can be induced to differentiate into *lymphokine-activated killer* (LAK) *cells*, which have even broader powers of target cell recognition and cytolysis. Since a prime source of IL-2 is secretion by T cells, the development of NK cells into LAK cells may be another example of the interplay between innate and adaptive immune mechanisms. In addition, NK cells carry surface receptors for immunoglobulin that allow them to interact with and lyse cells that have been coated with antibody as part of an adaptive immune response. More on the specific mechanisms involved in NK killing can be found in Chapter 18.

NKT cells are lymphoid lineage cells that combine several features of both T lymphocytes and NK cells. Like T cells, NKT cells carry a TCR, but unlike T cells, the TCR repertoire is much less diverse such that the NKT cell population "sees" only a small subset of antigens. Once activated, NKT cells quickly secrete cytokines that can help support the activation and differentiation of B and T cells. Like NK cells, NKT cells may also carry out target cell cytolysis. The rapidity of the NKT response and the limited range of antigens to which these cells respond mark them as cells contributing primarily to the innate immune response. However, the influence of NKT-secreted cytokines on B and T lymphocyte functions indicates that NKT cells constitute an important link between innate and adaptive immunity. More on the characteristics and functions of NKT cells can be found in Chapter 18.

As we shall see later in this book, T lymphocytes bear one of two types of antigen receptors: either the *TCR* (expressed on the conventional CD4$^+$ and CD8$^+$ T cells described previously), or the *TCR* (expressed on γδ T cells). Unlike the enormous diversity of αβ TCRs, γδ TCRs bind to only a limited set of broadly expressed antigens. Thus, like NKT cells, γδ T cells can be considered as part of the innate immune system. More on the characteristics and functions of γδ T cells can be found in Chapter 18.

IV. DENDRITIC CELLS

Dendritic cells (DCs) are irregularly shaped cells that exhibit long, fingerlike membrane processes that resemble the dendrites of nerve cells (refer to Fig. 3-1). DCs can arise from myeloid or lymphoid precursors in either the thymus or bone marrow, or from more differentiated cell types in the blood or tissues (see Ch.11). In mice, several distinct subtypes of DCs have been identified which differ with respect to surface markers, cytokine production, cytokines responded to, and influence on T cell responses. Markers of particular interest for murine DC subsets include CD8αα homodimers and a lectin called DEC-205 (CD205). A similar spectrum of DC subsets has been identified in humans but exact parallels between particular mouse and human DC subsets have been difficult to establish.

In general, DCs newly produced in the bone marrow migrate through the blood to the peripheral tissues, where they take up residence as immature DCs and wait for antigen. (DCs generated in the thymus remain there as *thymic DCs*.) As in the case of macrophages, different locations dictate subtle differences in DC morphology and function, leading to a varied nomenclature. Dendritic cells in nonlymphoid organs are called *interstitial DCs*, while those in areas of lymphoid organs (see later) populated primarily by T cells are sometimes called *interdigitating DCs*. *Langerhans cells* are DCs in the epidermis. *Blood DCs* are found in the circulation, but undergo morphological changes as they enter the interstitial fluid in the tissues to become *veiled cells*. Dendritic cells are notable because they are one of the few cell types expressing MHC class II and, as such, are important APCs. Immature DCs in the tissues primarily use receptor-mediated endocytosis to internalize antigen extremely efficiently. Upon antigen uptake, the immature DCs are activated, commence maturation, and migrate to the nearest secondary lymphoid tissue, usually the local lymph node. Here, the mature DCs process the antigen intracellularly and display it in association with MHC class II to resident helper T cells, which then initiate effector action. Indeed, DCs are the only APCs capable of presenting peptide antigens to resting, naive T cells such that they achieve full activation. Nevertheless, for all their power, DCs comprise <1% of the total cell populations of the lymph nodes and spleen. A detailed discussion of DC biology and antigen presentation appears in Chapter 11.

It should be noted that there is another type of cell called a *follicular dendritic cell* (FDC) which is, in spite of its name, unrelated to the antigen-presenting DCs just described. Although FDCs are morphologically similar to the other DCs, their lineage derivation appears to be different. In addition, rather than having a close association with T cells, FDCs are found in areas of lymphoid organs populated mainly by B cells. FDCs also differ in function from antigen-presenting DCs in that they do not internalize antigen; instead, they trap antigen–antibody complexes (for example, a microbe with antibody molecules bound to it) on their cell surfaces, displaying them for extended periods of time. An adaptive B cell response against the microbe may then be initiated when the antigen receptors on the passing B cell bind to appropriate epitopes on the surface of the trapped microbe. The FDCs may also play a role in reactivating memory B cells, although this has not been definitively established because the FDCs are intimately integrated into the structure of the lymphoid organs and are difficult to study in isolation.

V. HEMATOPOIESIS

All of the cell types described above are generated from the hematopoietic stem cell (HSC) during hematopoiesis. Where and how does this tremendous generation of blood cells occur?

i) Sites of Hematopoiesis

There is some controversy over where the mesodermal cells that give rise to the first HSCs are produced. In some studies in both humans and mice, HSCs are initially detected in an actively remodeling embryonic tissue called the *aortagonadmesonephros* (AGM). In other studies, the HSCs appear to arise first in the yolk sac surrounding the embryo. In any case, HSCs are first detected in human embryonic structures at 3–4 weeks of gestation (Fig. 3-4). The HSCs migrate to the fetal liver at 5 weeks of gestation and liver hematopoiesis completely replaces embryonic hematopoiesis by 12 weeks of gestation. At 10–12 weeks of gestation, some stem cells commence migration to the spleen and bone marrow. The fetal spleen transiently produces blood cells between the third and seventh months of gestation. The bone marrow of the long bones assumes an increasingly important role by about the fourth month and takes over as the major site of hematopoiesis during the second half of gestation. By birth, virtually all hematopoiesis occurs in the bone marrow of the long bones. After birth, the activity of the long bones steadily declines and is replaced in the adult by the production of hematopoietic cells in the axial skeleton—pelvis, sternum, ribs, vertebrae, and skull. In situations of injury to the bone marrow, it is possible for the liver and spleen to resume hematopoiesis in the adult. In mice, hematopoiesis in embryonic structures commences at about embryonic day 10.5. Adult hematopoiesis occurs primarily in the bone marrow with some contribution by the spleen.

ii) HSCs and Their Differentiation

Hematologists say that the HSC is *pluripotent* because its progeny are capable of *differentiating* (specializing) into any one of a variety of other cell types. Cells that reach the end of an irreversible maturation pathway are said to be *terminally differentiated*. Terminally differentiated cells do not undergo mitosis and are usually very specialized in their functions. Although the precise steps of hematopoiesis have not been clarified *in vivo*, much evidence for the capacity of HSCs to give rise to multiple intermediates and then to mature hematopoietic cell types has been gained from *in vitro* studies (see Box 3-3).

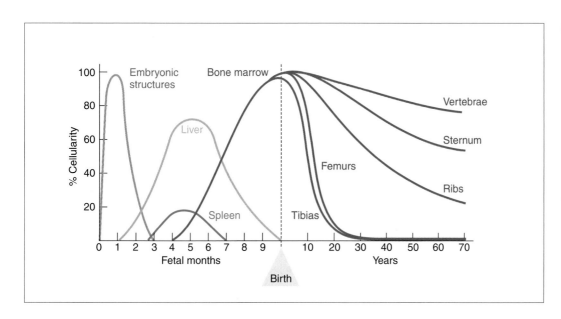

Figure 3-4

Sites of Hematopoiesis in Humans
The site of hematopoiesis (blood cell formation) shifts during fetal and adult life, as shown by changes in the relative number of blood cells (% cellularity) in different tissues at different times. As an individual ages, hematopoiesis becomes less vigorous. Note that the contribution from bone marrow prior to birth is primarily from the long bones (tibias, femurs). Adapted from Klein J. and Hořejší V. (1997). "Immunology," 2nd edn. Blackwell Science, Osney Mead, Oxford.

Box 3-3. The discovery of the hematopoietic stem cell and hematopoietic progenitors

The discovery of the hematopoietic stem cell (HSC) was a serendipitous by-product of irradiation experiments undertaken by James Till and Ernest McCulloch at the University of Toronto and the Ontario Cancer Institute in the early 1960s. Their original objective was to study the sensitivity of proliferating bone marrow cells to irradiation. Fortuitously, these experiments led to the discovery of a cell that could give rise to all of the myeloid cell lineages that populate the blood. In their original experiment, Till and McCulloch first lethally irradiated a mouse in order to destroy all of its bone marrow cells. They then injected this mouse with 10^4–10^5 bone marrow cells from an untreated mouse. Several days later, it was found that the irradiated mouse had developed visible nodules on its spleen (see plate). Till and McCulloch called these nodules *colony-forming units of the spleen* (CFU-S). Further experiments showed that

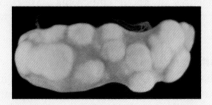

the number of colonies was proportional to the number of bone marrow cells injected (approximately 1 colony per 10^4 injected bone marrow cells), implying that each colony arose from a single progenitor cell. What was startling about these colonies, however, was the fact that each contained cells of multiple morphologies. That is, in a single colony, there were erythrocytes, megakaryocytes, and various granulocytic cells of the myeloid lineage. The interpretation of this result was that the spleens of the recipient animals had been colonized by individual cells that possessed the ability to give rise to all of the various differentiated cell types found in the blood. In other words, each CFU-S was derived from the progeny of what was originally called a "multipotent progenitor cell."

Definitive proof that stem cells existed came out of more complicated experiments in which unique cells could be identified by the presence of particular *karyotypic markers* on their chromosomes. It was eventually demonstrated that all of the cell types found in the blood, including lymphoid cells, can be derived from a single type of pluripotent HSC originating in the bone marrow.

The development of *in vitro* colony formation assays that assess the ability of hematopoietic progenitors to proliferate and respond to growth factors has helped scientists to categorize progenitors and better define the multiple steps of myelopoiesis and lymphopoiesis. Hematopoietic progenitors form colonies *in vitro* when they are immobilized in a semi-solid medium such as agar or methylcellulose, and various cytokines and other factors can be tested for their influence on this process. This methodology has propagated use of the term "colony-forming unit" as an experimental convenience. Within the *in vitro* continuum of cell development, the conventional designations begin with CFU and are followed by a letter indicating the specific cell types appearing in the culture, as defined by their characteristic morphologies (e.g., CFU-E = colony-forming unit, erythrocytes; a colony made up of erythrocytes). Depending on the species and the stage of differentiation, progenitor cells mature *in vitro* at different rates. In mice, both immature and more advanced progenitors take 7–10 and 2 days, respectively, to form colonies, whereas comparable human progenitors require 14 and 7 days. It should be noted that CFU assays measure proliferative and differentiation potentials of cells *in vitro*, and retrospectively assign these characteristics to cells *in vivo*. The precise relationship of the CFU precursors to their analogous cell types *in vivo* may still be unclear.

Scientists have now constructed a working model of *in vivo* hematopoiesis based on CFU assays and other more recently developed cell culture systems. HSCs are thought to give rise to common myeloid progenitors (CMPs) and common lymphoid progenitors (CLPs), which have less self-renewal potential and are committed to producing cells of the myeloid or lymphoid lineages, respectively. CMPs and CLPs in turn give rise to precursors and mature hematopoietic cells that can be identified in *in vitro* CFU assays. For example, in the presence of IL-7 and the growth factor known as *stem cell factor* (SCF), cultures containing CLPs generate B cell progenitors and NK/T progenitors (discussed in detail in Chs.9, 13, and 18). Addition of IL-15 to such a culture favors the development of the NK/T progenitors into NK cells rather than T cells.

Similarly, in the presence of SCF and granulocyte/monocyte colony-stimulating factor (GM-CSF), cultures containing CMPs give rise *in vitro* to a myeloid progenitor called the CFU-GEMM (colony-forming unit, granulocytes, erythrocytes, monocytes, macrophages). CFU-GEMM cells in turn generate two slightly more restricted progenitors: the CFU-BME (CFU, basophils, megakaryocytes, erythrocytes) and the CFU-GM (CFU, granulocytes, monocytes). As the name indicates, the progeny of CFU-BME cells can develop into either basophils, megakaryocytes, or erythrocytes, depending on the action of transcription factors and cytokines in the immediate microenvironment. For example, the presence of *erythropoietin* (Epo) favors the differentiation of CFU-E precursors and eventually mature erythrocytes. Similarly, a CFU-GM cell exposed to *granulocyte colony-stimulating factor* (G-CSF) becomes a committed granulocyte precursor (CFU-G), whereas a CFU-GM cell developing in the absence of G-CSF becomes a committed monocyte precursor (CFU-M). CFU-M cells, in the presence of GM-CSF and monocyte colony-stimulating factor (M-CSF), can give rise to mature monocytes.

It is now clear that the redundancy and pleiotropy of cytokines make them unlikely to be the sole definitive agents of HSC differentiation, with rare exceptions. Thus, research laboratories have turned to the study of other conditions and signals that induce an HSC to differentiate into one cell type rather than another. Many of these signals are now thought to arise from the regulated expression of specific transcription factors that act at various stages during differentiation, while others take the form of intercellular contacts supplied by stromal cells within bone marrow. Transcription factors determining B cell development are discussed in Chapter 9; while those governing T cell development are described in Chapter 13.

Much work remains to be done to identify hematopoietic intermediates *in vivo*, along with the transcription factors, environmental stimuli, and cytokines that influence the decision of a hematopoietic stem cell to self-renew or commit along a particular pathway of differentiation. Photograph of spleen reproduced with permission from McCulloch E. and Till J. (1961). A direct measurement of the radiation sensitivity of normal mouse bone marrow cells. *Radiation Research* **14**(2), 213–222.

Hematopoietic stem cells are difficult to study because they occur at a frequency of just 0.01% of all bone marrow cells, and because they do not grow readily under laboratory conditions. They can be identified by certain characteristic cell surface markers; however, many of these are not exclusive to stem cells. We do know that HSCs are unusual in that they are *self-renewing* (divide into more HSCs) and are capable of tremendous proliferation and differentiation in response to an increased demand for hematopoiesis. This capacity has been demonstrated in experiments that have exploited another unusual property of HSCs, that of extreme radiation sensitivity. Bone marrow cells, and especially HSCs, are particularly susceptible to damage by irradiation. For example, a normal mouse can be subjected to a level of radiation such that all of its bone marrow cells ($\sim 3 \times 10^8$ in number) are destroyed, but its other body cells are unaffected. Such a mouse is said to be "lethally" irradiated because the inability to generate cells of the immune system means it will eventually die of opportunistic infection, unless it receives a bone marrow transplant. However, the proliferative capacity of HSCs is such that a transplant to the mouse's marrow of only 10^4 nonirradiated bone marrow cells (of which the HSCs are only a tiny proportion) can replace all the necessary cell types and rescue the mouse.

When stimulated to divide, a particular HSC in the bone marrow will either self-renew or commit to differentiating into cells of the myeloid lineage (*myelopoiesis*) or cells of the lymphoid lineage (*lymphopoiesis*) (Fig. 3-5). We emphasize that the stages leading from the HSC to mature blood cell types *in vivo* are not yet totally defined, and controversy remains as to the identity of the many intermediates in both lineage paths. Development from an HSC to a mature blood cell is a continuum, and the identifiable intermediates, which are called *progenitor* and *precursor* cells, progressively lose their capacity to self-renew and differentiate into multiple cell types. Progenitor cells are unipotent, bipotent, or multipotent, depending on the number of mature cell types they can generate. Progenitors typically have few distinguishing features under the light microscope and are present in the bone marrow in moderate numbers. Precursors are more often unipotent and tend to be easily recognized when stained and viewed under the light microscope. These cells are present in the bone marrow in large numbers as a result of increased mitosis, but retain little self-renewal capacity compared to HSCs. Mature cells, which are highly differentiated and can no longer generate new precursors, are easily identified under the microscope (refer to Fig. 3-1). In the case of mature T and B lymphocytes, activation induces further proliferation and differentiation that results in the generation of effector cells such as plasma cells, Th cells, and cytotoxic T lymphocytes. Long-lived memory cells, which have the same morphology as mature, naive lymphocytes, are also produced.

Many models of hematopoiesis have been postulated, but none has yet to capture the unanimous approval of the scientific community. According to one widely accepted model (Fig. 3-6), the pluripotent HSC differentiates into two early multipotent progenitors: the *common myeloid progenitor* (CMP) and the *common lymphoid progenitor* (CLP). To date, CMPs and CLPs have been positively identified *in vivo* only in mice. CMPs and CLPs retain some self-renewal capacity and the ability to differ

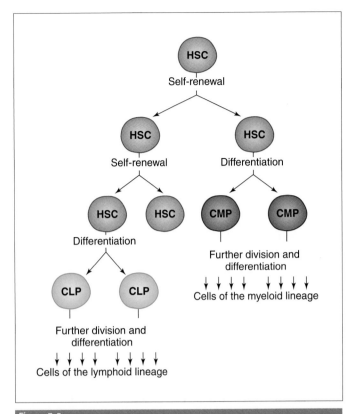

Figure 3-5

Model of Stem Cell Decisions
At the time of cell division, a hematopoietic stem cell (HSC) will choose one of two pathways: self-renewal or differentiation. Self-renewal duplicates the HSC, maintaining it in its primitive state, whereas differentiation drives the HSC toward a more mature state. During differentiation, either two myeloid precursors (common myeloid progenitor; CMP) or two lymphoid precursors (common lymphoid progenitor; CLP) are generated, each committed to producing cells of the myeloid or lymphoid lineage, respectively. The scheme shown here portrays one possible outcome of stem cell decision-making in the adult. The probability that a stem cell will self-renew must be greater than 50% in order to maintain the stem cell population over the life of an individual.

entiate into the full spectrum of either myeloid cells or lymphoid cells, respectively. *In vitro* experiments in adult mice have identified progeny of CMPs and CLPs that are more restricted in their potential. For example, cells giving rise only to neutrophils or monocytes have been isolated, as well as intermediates restricted to producing T cells or NK cells, and still others that are limited to generating granulocytes, erythrocytes, monocytes, or macrophages. It should be noted that, for unknown reasons, the situation appears to be different in fetal mice, in which a precursor capable of giving rise to both B cells and macrophages has been isolated.

The growth and maturation of the progeny of the hematopoietic precursors are dependent on the activation of specific transcription factors required for gene expression, cytokines and growth factors in the immediate microenvironment, and contacts and products furnished by the bone marrow *stroma*. The stroma is a cellular matrix composed of a variety of nonhematopoietic cell types such as fibroblasts, fat cells, and endothelial cells. The stroma provides structural support for developing

Figure 3-6

Model of Hematopoiesis
A hematopoietic stem cell is influenced by its microenvironment and the activity of various transcription factors and cytokines to produce stem cells committed to the myeloid or lymphoid lineages. All red blood cells (erythrocytes) and white blood cells (leukocytes) in the body arise from these two lineages. As differentiation progresses, the capacity of the cells to self-renew decreases, but the absolute cell numbers increase due to increased cell division.

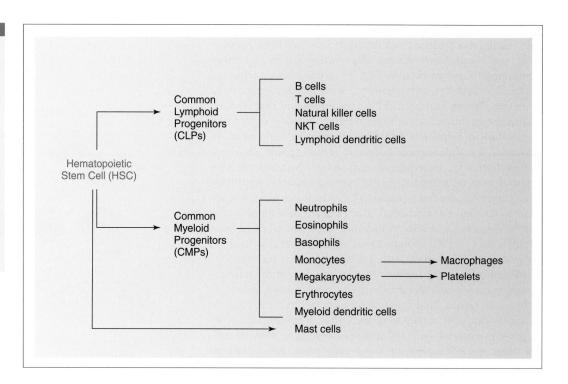

hematopoietic precursors and a collection of diffusible and membrane-bound growth factors that foster maturation. In particular, the *interleukins* (ILs) and *colony-stimulating factors* (CSFs) act to promote the growth and differentiation of hematopoietic cells in the bone marrow. Each of the ILs and CSFs has multiple effects, and a complex pattern of redundancy has been built into the system, so that the development of most cell types is not totally dependent on any one factor. Some ILs and CSFs act to support the survival of HSCs and hematopoietic progenitors and precursors, while others work synergistically (with other cytokines) to promote the proliferation and differentiation of cells at various stages (Table 3-1). For example, IL-7 is essential for the generation of T and B lymphocytes, while NK development requires IL-15 and neutrophil development depends on granulocyte CSF (G-CSF). More detail on the transcription factors and cytokines required for the development of B cells appears in Chapter 9; T cell development is covered in Chapter

13 and the development of NK, NKT, and γδ T cells is discussed in Chapter 18.

The origin of mast cells has remained particularly obscure. Mast cells were originally thought to be related to basophils, or even derived from mature T cells, fibroblasts, or macrophages. However, more recent studies have shown that mast cells originate from less differentiated precursors that are apparently distinct from CMPs and CLPs. Although most descendants of the HSCs remain in the bone marrow until they have fully differentiated, the unidentifiable progenitors of mast cells appear to migrate to the tissues.

iii) The Role of Apoptosis in Homeostasis

Hematopoiesis, the generation of red and white blood cells, is one side of the equation our bodies require to maintain *homeostasis* (a steady state). The other side of the equation is, naturally enough, cell death. Each of the differentiated cell types arising from the HSCs during hematopoiesis has a defined life span, often consistent with its function. Neutrophils die after 1–2 days while eosinophils die after about 5 days. Erythrocytes circulate throughout the body for about 120 days, whereas macrophages roam the tissues for 60–120 days. Most mature lymphocytes in the periphery die after a few weeks if they are not specifically stimulated by antigen. If antigenic stimulation does occur, the memory cells that are formed as part of the response may remain viable for decades or even years.

The sheer numbers of blood cells required for normal function of the body are staggering; for example, the average healthy adult human maintains an estimated 10^{12} lymphocytes, and an estimated 5×10^{10} circulating neutrophils. In order to meet this demand, the bone marrow is a prodigious cell-generating factory, producing stem cells, progenitors, and precursors at a constant and rapid rate. However,

Table 3-1 Examples of Cytokines That Play a Role during Hematopoiesis*

Early-Acting Cytokines	Synergistic Cytokines	Lineage-Specific Cytokines (Acting on Cell Type)
SCF	IL-6	IL-4 (B cells, mast cells)
	IL-11	IL-7 (B and T cell precursors)
		Epo (erythrocytes)
		M-CSF (macrophages)
		G-CSF (neutrophils)
		GM-CSF (neutrophils/macrophages)

*See Ch.17 for further details.
SCF, stem cell factor; IL, interleukin; Epo, erythropoietin; CSF, colony-stimulating factor.

without the limits imposed on the life spans of individual cells, lympho- or myeloproliferative disorders can develop, sometimes leading to a *leukemic* state (a detrimentally high level of leukocytes). Many leukemias result from a situation in which leukocytes continue to pour out of the bone marrow but normal leukocyte death does not occur. More on leukemias and how they arise appears in Chapter 30.

How does a cell know when it is time to die? In other words, how is the tremendous and constant proliferation of cells in the bone marrow balanced with cell death to maintain homeostasis? The answer is *programmed cell death*, also called *apoptosis* (see Box 3-4). Briefly, apoptosis is a process by which a cell dies because it is triggered to activate certain intracellular proteases that cause the orderly breakdown of the cell nucleus and its DNA. Degraded cellular structures are compartmentalized into smaller vesicles which are then phagocytosed and digested by other cells, principally macrophages. Apoptosis occurs when it is advantageous to the body for a cell to die, such as during embryonic development, or in order to remove expended cells. In the case of the hematopoietic system, apoptosis removes aged and spent blood cells as quickly as the bone marrow creates new ones, maintaining a steady-state level of healthy blood cells.

B. Lymphoid Tissues

The immune system is not a discrete organ, like a liver or a kidney. It is an integrated partnership, with contributions by the circulatory system, lymphatic system (see later), various lymphoid organs and tissues, and the specialized hematopoietic cells moving among these. A *lymphoid tissue* is simply a tissue in which lymphocytes are found. Lymphoid tissues range in organization from diffuse arrangements of individual cells to encapsulated organs (Fig. 3-7). *Lymphoid follicles* are organized cylindrical clusters of lymphocytes that, when gathered into groups, are called *lymphoid patches*. *Lymphoid organs* are groups of follicles that are surrounded, or encapsulated, by specialized supporting tissues and membranes.

A photograph of a dissected mouse showing the positions of the major lymphoid organs and tissues as they sit in the

---

Box 3-4. **Apoptosis**

Apoptosis (from the Greek, *apo*, meaning "off," and *ptosis*, meaning "a dropping or falling") is defined as *programmed cell death*. It is distinguished from *necrosis*, which is cell death caused by tissue injury or pathogenic attack, in several important ways. A cell that has been injured lyses in an uncontrolled way and becomes necrotic. The necrotic cell loses the integrity of its nucleus and releases its intracellular contents, which damage nearby cells and provoke an inflammatory response. In contrast, apoptosis is a controlled form of cell death that follows specific biochemical paths and results in specific morphological changes to only those cells marked for "suicide." An apoptotic cell undergoes an orderly demise in which it degrades its own chromosomal DNA, fragments its nucleus, shrinks its cytoplasmic volume, and forms *apoptotic bodies*, which are small vesicles containing intact organelles such as lysosomes and mitochondria. The rapid phagocytosis of the apoptotic bodies by macrophages prevents the leakage of deadly contents (such as hydrolytic enzymes or oxidative molecules) that would otherwise precipitate neighboring cell breakdown and an inflammatory response.

The function of apoptosis is to maintain the balance between cell growth and cell death in a controlled way. Apoptosis is triggered both when selective cell suicide is of advantage to the organism, such as during development and in adult homeostasis, and when certain very specific phenomena (such as the presence of certain chemicals or damage by certain types of radiation) lead to the triggering of the apoptotic pathway. In the immunological context, apoptosis is important during the normal development of bone marrow progenitor cells into just the right number ofzcells of the various hematopoietic lineages, and to remove unwanted effector cells following an immune response. Apoptosis is also important for the *negative selection* process involved in the generation of *central tolerance*. Many immature lymphocytes with random antigenic specificity that could lead them to attack self-antigens are prevented from emerging from the primary lymphoid tissues, because they are induced to undergo apoptosis by an encounter with their specific self-antigens. Apoptosis can also be used as an offensive weapon; some cytotoxic T cells kill their target cells by inducing their apoptosis.

The exact physiological mechanisms of apoptosis are not fully understood, but it is apparent that induction in appropriate cells can occur in response to the binding of specific ligands to the cell surface in a variety of situations. For example, after high affinity binding of the TCR or BCR by self-antigen during lymphocyte development, or following the engagement of these receptors during an immune response, certain new "death-inducing" cell surface molecules appear. The interaction of these surface proteins with their corresponding ligands induces the transcription of several genes, including those for the *caspase* family of proteases. Once activated by the products of other genes transcribed in response to the binding signal, the caspase proteases initiate the nuclear and membrane changes associated with apoptosis. Certain chemicals and types of radiation can also trigger the caspases, resulting in chemical- and radiation-induced apoptosis. Mutations in the "death-inducing" cell surface receptors, or in the ligands that activate this cascade, result in a failure of apoptosis and an accumulation of cells that otherwise would have been deleted. Should these cells happen to be those reacting to self-antigens, the failure in apoptosis may be manifested as autoimmunity (see Ch.29). Similarly, if apoptosis fails to control leukocyte numbers during hematopoiesis, a lymphoproliferative or leukemic state can develop (see Ch.30).

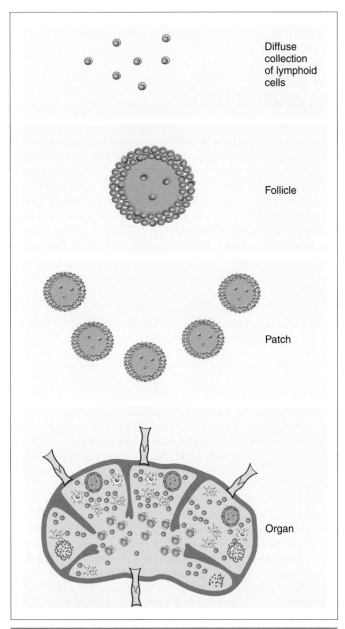

Lymphoid Tissue Organization
Lymphoid tissues are organized into structures of varying complexity depending on the function they serve. In order of increasing structural complexity, there are diffuse lymphoid cells, follicles, patches, and organs.

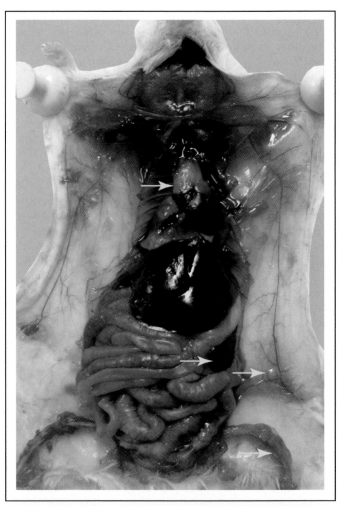

Plate 3-1

Mouse Primary and Secondary Lymphoid Organs and Tissues
Top to bottom: yellow arrows indicate thymus, spleen, superficial lymph node, bone marrow. Courtesy of Dr. Alejandro Ceccarelli, Animal Resource Center, University Health Network, Toronto.

body is shown in Plate 3-1. The *primary* (or *generative*) lymphoid organs are the sites of lymphocyte development and maturation, a process that includes the initial expression of antigen receptors. At these sites, most cells recognizing self antigens are deleted and cells recognizing foreign antigens are permitted to mature. In mammals, the *bone marrow* is the primary lymphoid tissue where all lymphocytes arise and B cells mature (Table 3-2). The *thymus* is the other primary lymphoid tissue, in which T cells mature. Newly produced lymphocytes leave the primary lymphoid tissues and migrate through the blood and lymph (see later) to the secondary lymphoid tissues.

Secondary lymphoid tissues are those in which mature lymphocytes recognize antigen and become activated, undergoing clonal selection and proliferation followed by differentiation into effector and memory cells. Secondary lymphoid tissues in mammals include the *spleen*, the *lymph nodes*, the *mucosa-associated lymphoid tissue* (MALT), and the *skin-associated lymphoid tissue* (SALT). The MALT encompasses the *gut-associated lymphoid tissue* (GALT), the *bronchi-associated lymphoid tissue* (BALT), and the *nasopharynx-associated lymphoid tissue* (NALT) systems. The GALT consists of the *Peyer's patches*, the *appendix*, and the cells in the lining of the intestine. The BALT includes lymphoid patches and the cells in the lining of the *lower respiratory system*, while the NALT includes the *tonsils* and *adenoids*. The SALT comprises the diffuse collections of lymphocytes that reside in the skin. The secondary lymphoid tissues serve to trap antigen moving through the body

Table 3-2 Primary and Secondary Lymphoid Tissues

Primary Lymphoid Tissues	Secondary Lymphoid Tissues
Bone marrow	Lymph nodes
Thymus	Spleen
	MALT (mucosa-associated lymphoid tissue)
	• BALT (bronchi-associated lymphoid tissue)
	• GALT (gut-associated lymphoid tissue)
	• NALT (nasopharynx-associated lymphoid tissue)
	SALT (skin-associated lymphoid tissue)

via the circulation (spleen), through the lymphatic system (lymph nodes), through the mucosal membranes (MALT), or through the skin (SALT).

I. PRIMARY LYMPHOID TISSUES

i) More about the Bone Marrow

The bone marrow is the site of hematopoiesis and HSC differentiation in the adult human, but much remains unknown about how the bone marrow carries out its vital work. Lymphoid cells and their progenitors account for less than 10% of all bone marrow cells; the remainder is composed of 60–70% myeloid lineage cells, 20–30% erythroid lineage cells, and less than 10% other cells, such as reticular cells, adipocytes, and mast cells. Many of these cells are vital for hematopoiesis because they secrete the cytokines and growth factors (CSFs, in particular) that are required for differential blood cell maturation.

The compact outer matrix of a bone surrounds a *central* (also called *medullary*) *cavity*. Within the central cavity is the marrow, which may appear red or yellow in color. The *yellow marrow* is usually hematopoietically inactive but contains significant numbers of adipocytes (fat cells) that act as an important energy reserve. *Red marrow* is hematopoietically active tissue that gains its color from the vast numbers of erythrocytes produced. The bones of an infant contain virtually only red marrow, but with maturation into an adult, the demand for hematopoiesis slackens such that the number of bones with active red marrow declines. In the adult, only a few bones retain red marrow, including the sternum, ribs, pelvis, and skull.

Underneath the outer layer of compact bone, the central cavity is composed of an inner layer of "spongy" bone with a honeycomb structure. The honeycomb is made up of thin strands of bone called *trabeculae*. The red marrow lies within the cavities created by the trabeculae. A hematopoietically active bone is nourished by one or more nutrient arteries that enter the shaft of the bone from the exterior, as well as by arteries that penetrate at the ends of the bone. Branches of nutrient arteries thread through the *Haversian canals* of the compact bone to reach the trabeculae in the central cavity (Fig. 3-8A). The circulatory loop is completed when the branches of the nutrient arteries make contact with venous sinuses that eventually feed into *nutrient veins* that exit the shaft of the bone. Within a trabecular cavity, the network of blood vessels separates groups of developing hematopoietic cells into hematopoietic *cords*. It is in the cords that the HSCs either self-renew or differentiate into red and white blood cells of all lineages (Fig. 3-8B). Tucked into the spaces between blood vessels are interconnecting *adventitial reticular cells* and fibers that form the stromal framework supporting the developing hematopoietic cells. As the hematopoietic cells reach maturity, they leave the cords by squeezing between the endothelial cells lining the venous sinuses. From there, the cells enter a nutrient vein and finally the blood circulation. Erythrocytes remain in the circulation, while leukocytes are distributed between the blood and the tissues.

ii) More about the Thymus

The thymus is the site of T cell maturation, where only those T cells that survive thymic selection go on to mature and become immunocompetent (see later in this chapter and Ch.13). Very immature $CD4^-CD8^-$ T cell precursors known as *thymocytes* arise from hematopoiesis in the fetal liver and bone marrow and pass into the blood vessels. These cells migrate through the circulatory system to the *thymus*, a small bilobed organ located above the heart. Chemoattractant peptides secreted by specialized epithelial cells of the rudimentary thymus in the first 2–3 months of fetal life are thought to draw $CD4^-CD8^-$ thymocytes from the circulation into the thymus, but the precise stimulus and route are unclear.

Each lobe of the thymus is encapsulated and is composed of multiple lobules, which are separated by connective tissue strands called *trabeculae* or *septae* (Fig. 3-9). The thymocytes are organized into the densely packed outer *cortex*, and the sparsely populated inner *medulla*. Arteries enter the capsule surrounding the thymus through the trabeculae, running adjacent to the cortex and then looping back along the corticomedullary junction. Branches of the arteries run across the cortex as numerous capillaries but are fewer in number across the medulla. Some capillaries connect with interlobular veins that run parallel to the arteries, while others travel back across the cortex before becoming postcapillary venules that ultimately drain into interlobular veins. The arterial blood supply to the cortex is much richer than that to the medulla. However, blood cells do not readily pass through the endothelial cells lining the vessels in the cortex, a phenomenon called the *blood–thymus barrier*. This barrier is not absolute; small soluble molecules can pass into the cortex. The vascular barrier in the medulla is not so tight, and larger molecules and even cells can pass into the medulla. It is possible that thymocyte precursors access the organ in this region. There are no *lymphatic vessels* (see later) entering the thymus from the exterior of the organ, but some lymphatics begin in the medulla and connective tissue of the trabeculae, running out of the thymus and draining into the local lymph nodes.

The thymus contains four distinct subsets of maturing T lineage cells, each of which resides in a different region of the thymus: $CD4^-CD8^-$ thymocytes are present in the outer cortex; more mature $CD4^+CD8^+$ thymocytes (which make up 80% of the thymocyte population) occupy the inner cortex; and the mature T cells, either $CD4^+CD8^-$ or $CD4^-CD8^+$,

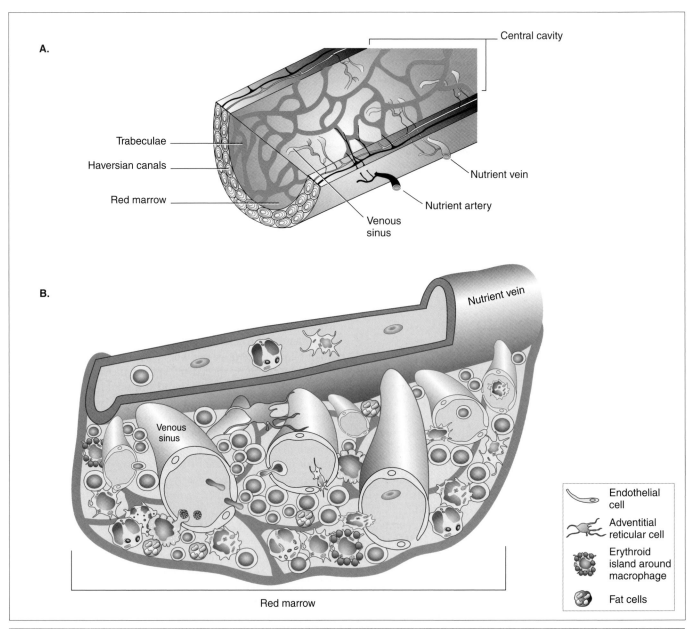

A.

Central cavity

Trabeculae

Haversian canals

Red marrow

Nutrient vein

Nutrient artery

Venous
sinus

B.

Nutrient vein

Venous
sinus

Red marrow

Endothelial
cell

Adventitial
reticular cell

Erythroid
island around
macrophage

Fat cells

Figure 3-8

The Bone Marrow
(A) Cross-section of central cavity of a long bone. (B) Developing hematopoietic cells of the red marrow. A variety of cell types provide the proper microenvironment to nurture developing hematopoietic cells. Hematopoietic cells (icons as in Fig. 3-1) leave the central cavity through pores in the endothelial cell wall of venous sinuses and then drain into the central vein. Megakaryocyte processes puncture the endothelial cell walls of venous sinuses into which they shed platelets. Erythropoiesis seems to take place within clusters called erythroid islands surrounding macrophages. Adapted from Sandorama (1987).

are located in the medulla. As they pass from the cortex through the medulla, the developing thymocytes interact with and are supported by a stromal framework of epithelial cells, dendritic cells, and macrophages. Specialized epithelial cells in the outer cortex, called *nurse cells*, form large multicellular complexes by enveloping up to 50 thymocytes within their long processes. At the junction of the cortex and medulla, interdigitating DCs begin to take on the nurturing role. Epithelial cells that have exhausted their supportive function within the thymus degenerate, forming the *Hassall's corpus-*

cles found in the medulla. Interestingly, it now appears that the maturation of a thymocyte into a mature T cell is more complicated than just moving from the exterior to the interior of the thymus. While most mature T cells appear to exit the thymus via the postcapillary venules near the medulla, some cortical thymocyte subpopulations are thought to mature and exit the thymus without ever reaching the medulla, and other thymocytes may never see the cortex.

Both tremendous T cell proliferation and T cell apoptosis take place in the thymus. The vast majority of T cells

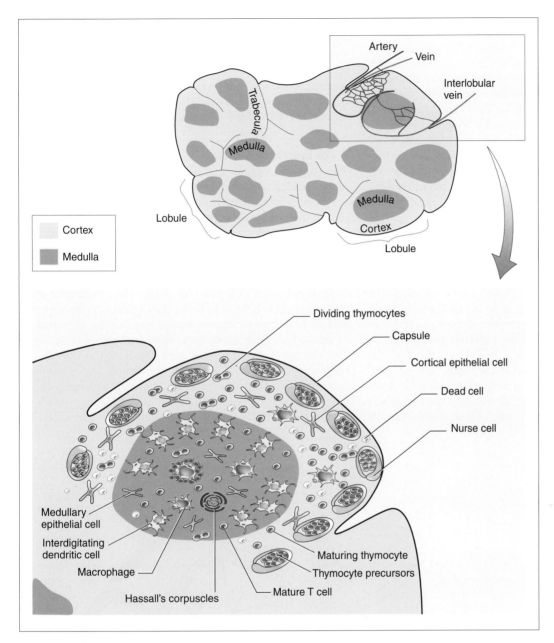

Figure 3-9

The Thymus
The upper panel shows a cross-section of the multiple lobules of the thymus, each with its own medulla and cortex, as well as the blood supply across one lobule. The lower panel depicts the cellular detail within the cortex and medulla of a thymic lobule. Macrophages are more prevalent in the medulla. Cortical epithelial cells, nurse cells, and dendritic cells nurture developing thymocytes. Exhausted epithelial cells form Hassall's corpuscles. With information from Klein J. and Hořejší V. (1997). "Immunology," 2nd edn. Blackwell Science, Osney Mead, Oxford.

generated during early lymphopoiesis will bear TCRs that do not recognize the particular MHC molecules displayed on the thymic stromal cells of the host. Since T cells cannot recognize antigen unless it is presented on "self" MHC molecules, further development of these cells is stymied. Of the remaining T cell clones that are able to bind to the host's MHC molecules, the mechanisms of central tolerance ensure that those cells with TCRs that recognize self-peptides are induced to undergo apoptosis, a process resulting in *clonal deletion*. This elimination of self-reactive cells is called *thymic negative selection* (see Ch.13). Only those thymocytes with the potential to recognize foreign antigenic peptides complexed to self MHC molecules are *positively selected* and go on to complete their maturation. Perhaps a mere 1% of all thymocytes meet these stringent criteria and survive their transit through the thymus, entering the circu-

lation as mature, naive T cells. So efficient is this selection process that the thymus starts to regress after puberty (a process called *thymic involution*), and its lymphoid components are eventually replaced to a large extent with fatty connective tissue. The thymus was previously believed to generate sufficient naive T cells early in life to cope with the entire spectrum of antigens likely to be encountered in an organism's life span. For many years, this was a conundrum, since mature T cells were thought to survive only a few weeks at most if they did not encounter their specific antigen. However, it has been discovered that mature T cells survive longer than first thought. In addition, not all T cells mature in the thymus; certain subsets develop *extrathymically*, such as in particular sites in the intestine. The development of T cells is described in more detail in Chapters 13 and 18.

II. SECONDARY LYMPHOID TISSUES

The generation of cells of the adaptive immune response occurs in the primary lymphoid tissues, but it is in the secondary lymphoid tissues such as the spleen and the numerous lymph nodes that lymphocytes undergo activation, proliferation, and differentiation into effector cells. The effector cells subsequently migrate from the secondary lymphoid tissues to infected organs or tissues where war is waged against invading pathogens. The secondary lymphoid tissues are widely distributed throughout the body in its so-called *periphery*, an evolutionarily favorable arrangement since endogenous antigens can arise within any tissue, and exogenous antigens can enter the body from many access points.

To reach optimal efficiency, cells of the adaptive immune response make use of "gathering places" in which all the necessary partners can come together and collaborate. This co-localization facilitates lymphocyte activation, proliferation, and differentiation into effector cells, which then migrate to the infected tissue to combat the pathogen. For example, in order to make antibody against a T-dependent antigen, several cells must meet in close physical and/or temporal proximity: the pathogen itself (or part thereof), a B cell specific for that pathogen, a T cell specific for that pathogen, and an APC presenting the appropriate antigen to the T cell. If these components were constantly drifting anywhere and everywhere throughout the body, or if there was only a single centrally

located secondary lymphoid organ, the chances of all the required partners being in the same place at the same time would be very low, and the frequency of lymphocyte activation would be correspondingly decreased. Instead, antigen is trapped, collected, and concentrated in dispersed secondary lymphoid tissues, which are also the sites of the greatest concentrations of B cells, T cells, and accessory cells.

i) The Lymphatic System

The lymphoid organs and tissues are connected by both the blood circulation and the *lymphatic system*. All cells in the body are bathed by nutrient-rich *interstitial fluid*, which is blood plasma that, under the pressure of the circulation, leaks from the capillaries into spaces between cells. Ninety percent of this fluid returns to the circulation via the venules, but 10% filters slowly through the tissues and eventually enters a network of tiny blind-ended channels known as the *lymphatic capillaries*, where it becomes known as *lymph* (Fig. 3-10). The overlapping structure of the endothelial cells lining the capillaries creates specialized pores that allow high molecular weight substances such as proteins, fats, and even leukocytes and microbes to enter into the lymphatic capillaries with the lymph. Valves in the lymphatic capillaries ensure that the lymph and its cellular contents move only forward as the lymphatic capillaries collect into progressively larger *lymphatic vessels*. These vessels in turn connect with one of two large *lymphatic trunks* called the

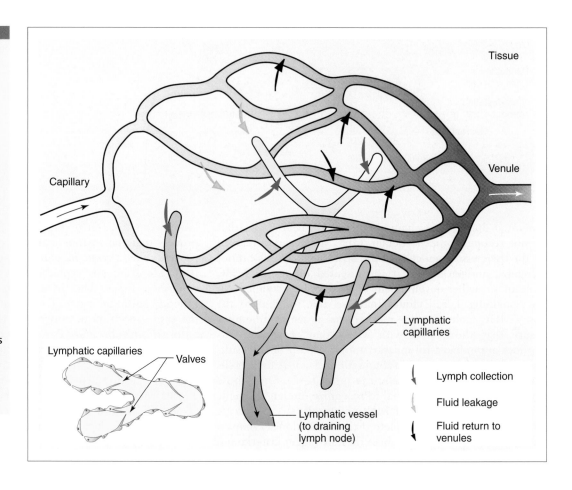

Figure 3-10

The Collection of Lymph in Peripheral Tissues
Most of the fluid that leaks from the capillaries into the tissues is taken up by the venules, but the remainder seeps back into the lymphatic vessels, which are blind-ended, fingerlike protrusions intertwined with the blood capillaries. The overlapping structure of the endothelial cells creates specialized pores that allow high molecular weight substances (proteins, fats, leukocytes, and even microbes) to pass into the lymphatic capillaries. Valves are present in all lymphatic channels to ensure the unidirectional flow of lymph into the collecting lymphatics (insert). The lymphatic vessels lead to local draining lymph nodes and ultimately flow into the thoracic duct, which empties the lymph back into the blood circulation.

right lymphatic duct and the *thoracic* (or *left lymphatic*) *duct* (Fig. 3-11). The right lymphatic duct drains the right upper body, while the entire lower body drains into the *cisterna chyli* at the base of the thoracic duct. Lymph from the left upper body also enters the thoracic duct. The right lymphatic duct empties the lymph into the *right subclavian vein* of the blood circulation, while the thoracic duct connects with the *left subclavian vein*.

As lymph flows through the lymphatic vessels, it passes through the *lymph nodes*, which are bean-shaped, encapsulated structures 2–10 mm in diameter containing the concentrations of lymphocytes, macrophages, and other accessory cells required to deal with foreign antigens. Over the course of a day, up to 72% of a human adult's blood volume will filter through the tissues and lymph system and come under the scrutiny of the lymph nodes. Lymph nodes occur along the entire length of the lymphatic system but are clustered in a few key regions: the *cervical* lymph nodes drain the head and neck; *axillary* nodes in the armpits drain the upper trunk from the

arms; the *thoracic* nodes drain the lungs and respiratory passages; the *intestinal* and *mesenteric* nodes drain the digestive tract; the *inguinal* nodes drain the lower trunk from the legs; the *abdominal* nodes drain the urinary and reproductive systems; and the *popliteal* nodes drain the lower legs.

ii) More about Lymph Nodes

The major site for the activation of lymphocytes during an adaptive immune response is the lymph node. Antigen that finds its way past the body's innate defenses and into the tissues is collected in the lymph, then enters a lymph node and interacts with lymphocytes, DCs, and macrophages within the node. DCs take up antigen, process it, and present it to naive T cells within the node, activating these cells and initiating a primary response.

Lymph enters a lymph node through several *afferent* lymphatic vessels. It then passes through the exterior *cortex* of the node on its way through the interior *medulla*, and exits on

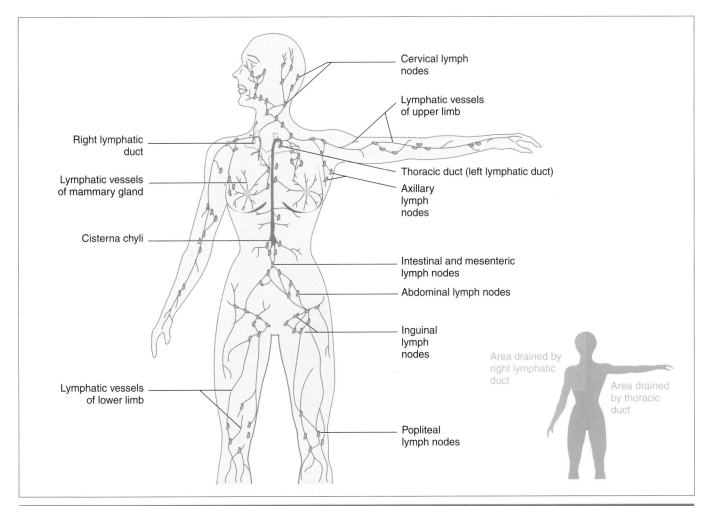

Figure 3-11

Major Vessels and Nodes of the Lymphatic System
Lymphatic vessels and lymph nodes are named according to their location in the body. Lymph collected by smaller lymphatic vessels throughout the body eventually drains into larger lymphatic vessels called the thoracic duct (left lymphatic duct) and right lymphatic duct. The cisterna chyli is the base of the thoracic duct and drains the lumbar and intestinal lymphatic regions. Lymph flows from the thoracic duct and right lymphatic duct into the left and right subclavian veins, respectively, of the heart to re-enter the blood circulation. (Not all lymphatic vessels or lymph nodes are shown.)

the opposite side of the node through a single *efferent* lymphatic vessel (Fig. 3-12). The cortex is considered rich in B cells, FDCs, and macrophages, while the paracortex is home to many T cells and interdigitating DCs, and the medulla is well-stocked with antibody-secreting plasma cells. In a human lymph node, the lymphocyte compartment is composed of approximately 25% B lymphocytes, 60% CD4$^+$ T cells, and 15% CD8$^+$ T cells; NK cells are virtually absent.

Only a small proportion (10%) of lymphocytes arrives in the lymph node via the afferent lymphatics. Newly produced lymphocytes or others that are recirculating throughout the body enter the paracortex of the node from the blood via specialized *postcapillary venules* (see later). Lymphocytes that fail to encounter specific antigen while in the node exit directly via the efferent lymphatic that drains the medulla. However, some lymphocytes may encounter specific antigen that has entered the lymph node via the afferent lymphatics and has been deposited in the cortex or paracortex. The antigen may be held by FDCs for inspection by B cells in the cortex and/or internalized nonspecifically by interdigitating DCs in the paracortex and then processed to a form recognizable by T cells. Engagement of the TCR by antigen induces T cell activation and proliferation. The presence of antigen-activated helper T cells fosters the initial

stages of activation of any antigen-specific B cell clone in the small population of B cells located in the paracortex. Antigen-specific B cells that interact with antigen-stimulated Th cells in small foci at the edges of the paracortex are triggered to proliferate and commence differentiation into plasma cells. About 4–7 days after the initial encounter with antigen, some of the activated B cells and Th cells migrate in response to an unknown stimulus into the cortex of the lymph node.

Prior to the arrival of antigen, the lymph node cortex contains many *primary follicles*, which are spherical aggregates of (mainly) resting mature B lymphocytes, macrophages, and follicular dendritic cells. As mentioned previously, antigen entering a lymph node is trapped very tightly on the membranes of the FDCs for relatively long periods of time. Antigen-specific B cells and Th cells migrating into the cortex from the paracortex move into the primary follicles. Further interactions between the antigen-laden FDCs and the primed antigen-specific B and Th cells lead to the formation of *secondary follicles* containing *germinal centers* (see Ch.9). The germinal centers are aggregations of rapidly proliferating B cells and differentiating memory B and plasma cells. The mature plasma cells migrate back through the node to the medulla, which is home to a mixed population of B and

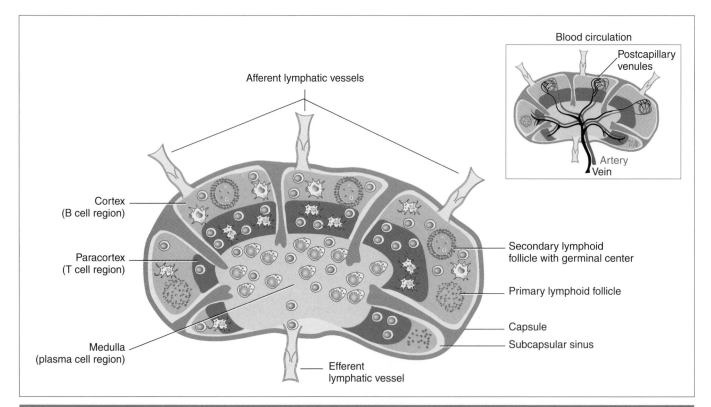

Figure 3-12

The Lymph Node
The lymph node is composed of an outer capsule, subcapsular region, and specific microenvironments called the cortex, paracortex, and medulla. Lymphocytes enter a node through the afferent lymphatic vessels or by squeezing through specialized postcapillary venules. Once lymphocytes are in the node, they can either leave directly via the efferent lymphatic vessels or penetrate deeper into the node and interact with the different cell types that predominate in each region: B cells, macrophages, and follicular dendritic cells in the cortex; T cells and interdigitating dendritic cells in the paracortex; and plasma cells, T cells, and B cells in the medulla. Primary follicles composed of B cells, macrophages, and follicular dendritic cells are found in the cortex in the absence of antigen whereas secondary follicles with germinal centers develop only after lymphocyte activation by antigen.

T cells, macrophages, and plasma cells. The copious quantities of newly synthesized IgM (and later IgG) antibodies secreted by the plasma cells enter the lymph by the efferent lymphatic and are eventually conveyed to the blood, in which they circulate through the body. Effector T cells may also leave the node via the efferent lymphatic and migrate to a site of infection where they are needed, or they may remain in the node to combat a pathogen within the node.

We are all familiar with the examination of our cervical lymph nodes by our doctors when we complain of an incipient infection. A significant infection can result in a swelling of the lymph nodes responsible for draining that area of the body. This swelling, which can be palpated and is often referred to as "swollen glands," helps to confirm the existence of the infection. The swelling of a node is due to several factors. Firstly, the presence of antigen within a node increases the migration of leukocytes into the node. Secondly, lymphocytes recognizing the antigen become activated and are stimulated to proliferate vigorously within the node. Lastly, the proliferation of activated cells triggers a temporary cessation of cellular traffic in and out of the node (called a "shutdown"), resulting in an accumulation of lymphocytes and other cells that causes a clearly discernible swelling of the node.

iii) More about MALT and SALT

The mucosa-associated lymphoid tissues and the skin-associated lymphoid tissues can be viewed as the body's second lines of defense after its passive anatomical and physiological barriers. The innate and adaptive immune system cell populations that make up the MALT and SALT are situated at the most common points of antigen entry, behind the mucosal membranes of the respiratory, gastrointestinal, and urogenital tracts, and just below the skin (Fig. 3-13). Large populations of T and B lymphocytes, macrophages, and plasma cells are positioned in these locations to either dispose of pathogens as they breach a body surface or process the pathogens so as to activate B and T lymphocytes.

The MALT is subclassified as either BALT, GALT, or NALT based on the location of the given mucosal membranes in the body. Thus, cells of BALT protect the bronchial regions of the lower respiratory tract, GALT cells defend the lining of the gut, and NALT cells protect the nasopharynx and upper respiratory tract. Throughout most of the MALT, lymphoid cells are dispersed in diffuse masses under a mucosal epithelial layer. In some cases, slightly more organized collections of cells exist, such as the Peyer's patches, which capture foreign entities that penetrate the intestinal

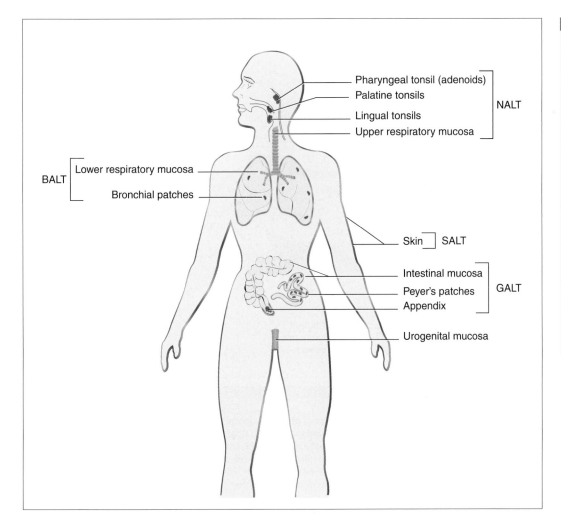

Figure 3-13

Mucosa- and Skin-Associated Lymphoid Tissues
The diffuse system of lymphoid tissues that protect mucosal surfaces lining the digestive, respiratory, and urogenital systems is collectively known as MALT (mucosa-associated lymphoid tissue). Tissue within the MALT that guards a particular region of the body may be further categorized. For example, BALT (bronchi-associated lymphoid tissue) and NALT (nasopharynx-associated lymphoid tissue) together defend the air passages from foreign attack. GALT (gut-associated lymphoid tissue) protects the gastrointestinal tract. SALT is the diffuse collection of immune cells that protect the skin.

Labels in figure:
Pharyngeal tonsil (adenoids)
Palatine tonsils
Lingual tonsils
Upper respiratory mucosa
} NALT

BALT {
Lower respiratory mucosa
Bronchial patches

Skin] SALT

Intestinal mucosa
Peyer's patches
Appendix
} GALT

Urogenital mucosa

epithelium. In other cases, the lymphoid cells are organized into discrete structures such as the tonsils, which are responsible for trapping pathogens passing through the membranes of the nose and throat area.

The SALT is comprised of small populations of immune system cells scattered in the epidermis and dermis of the skin. Dendritic cells known as *Langerhans cells* are scattered throughout the epidermis; these cells both function as APCs and secrete cytokines, drawing lymphocytes to the area. Below the epidermis is a layer of dermis where memory T cells and macrophages await the invasion of foreign antigens. Antigens that escape elimination in these tissues enter the lymph and are channeled to the local subcutaneous lymph node. The net effect of the MALT and SALT systems is to protect the body at all surfaces where an interface with the external environment exists. The MALT and SALT systems are examined in detail in Chapter 20.

iv) More about the Spleen

Most antigens escaping the innate immune response and the diffuse secondary lymphoid tissues make their way into the tissues and are collected in the lymphatic system. However, there are some cases in which an antigen is introduced directly into the blood, such as in the cases of bites (e.g., the malaria parasite via a mosquito bite; snake venom via a snake bite) or injections (e.g., drugs, causing hypersensitivity). In addition, overwhelming local infection at skin and mucosal sites can result in penetration of underlying blood vessels by the pathogen. Finally, systemic infection that cannot be contained by the lymph nodes pours into the lymph and is eventually dumped into the blood. Fortunately, we have the *spleen*, an abdominal organ that traps *blood-borne antigens* (Fig. 3-14). The structural framework of the spleen is created by a network of reticular fibers and tunnels of trabeculae that branch out of the hilus of the spleen and connect with a thin exterior capsule. Antigen carried in the blood circulation enters the spleen by a single artery called the *splenic artery*. The splenic artery branches into a network of smaller arteries that travel through the hollow spaces inside the organ created by the trabeculae. These smaller arteries in turn branch, forming *arterioles* that leave the trabeculae and penetrate the reticular fibers. Encasing each arteriole is a *periarteriolar*

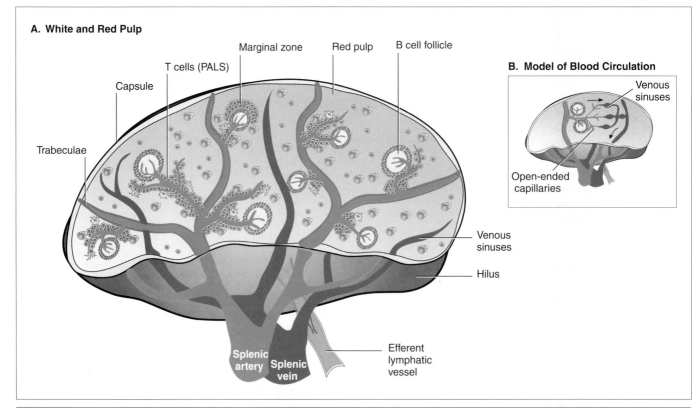

A. White and Red Pulp

Marginal zone · Red pulp · B cell follicle · T cells (PALS) · Capsule · Trabeculae · Venous sinuses · Hilus · Splenic artery · Splenic vein · Efferent lymphatic vessel

B. Model of Blood Circulation

Venous sinuses · Open-ended capillaries

Figure 3-14

The Spleen
The structural framework of the spleen is created by a network of reticular fibers and tunnels of trabeculae that branch out of the hilus of the spleen and connect with the capsule (A) The white pulp is composed of the T cell-rich periarteriolar lymphoid sheath (PALS) and B cell-rich follicles. The PALS, mostly lymphocytes, along with a few plasma cells, macrophages, and interdigitating dendritic cells, surrounds the arteries as they exit the trabeculae. Primary and secondary B cell follicles are located along the PALS, usually at arterial branch points. The follicles have a well-defined marginal zone, composed mostly of memory B cells. These are in close proximity to the red pulp, which forms about 75% of the splenic volume. The red pulp is composed of splenic cords, a meshwork of macrophages, lymphocytes, fibroblasts, and collagen fibers, and the venous sinuses, which are red in appearance due to the abundance of erythrocytes. (B) A model of the blood flow within the spleen.

lymphoid sheath (PALS), a cylindrical arrangement of tissue populated primarily by mature T cells but also containing low numbers of plasma cells, macrophages, and interdigitating DCs. Antigen is taken up by resident DCs, which present it to T cells in the PALS, resulting in T cell activation. Within the PALS, T cells predominate closest to the arterioles, but further away, both T and B cells are present. Surrounding the PALS are the regions containing the primary follicles (similar to those in lymph nodes) filled with B cells and macrophages. When antigen is trapped in proximity to these B cells, they are stimulated to form germinal centers, which then become filled with activated B cells. Surrounding the follicles is the *marginal zone*, which also contains some B cell subsets.

The total lymphocyte compartment in the spleen is composed of ~40% B cells, 50% $CD4^+$ T cells, 10% $CD8^+$ T cells, and less than 5% NK cells. Roughly one-third of all circulating lymphocytes in the body are found in the spleen, and more lymphocytes pass through the spleen than all of the lymph nodes combined. *White pulp* is the name given to the splenic regions containing the splenic arterioles with their periarteriolar lymphoid tissue sheaths, the follicles, and the marginal zone. The white pulp makes up 20% of the spleen and is dominated by lymphocytes. The *red pulp*, named for its abundance of erythrocytes, surrounds the white pulp and functions chiefly in the filtering of particulate material from the blood and in the disposal of senescent or defective erythrocytes and leukocytes. The red pulp consists of the *splenic cords* and the *venous sinuses*. The splenic cords are composed of a collagen-containing lattice surrounding collections of erythrocytes, reticular cells, fibroblasts, macrophages, and lymphocytes. The venous sinuses are small vascular channels, some of which randomly end in the cords as blind-ended fingers.

The exact route of blood flow within the spleen is unclear. The central artery distributes blood to the white pulp and marginal zones, where some capillaries appear to connect directly with the venous sinuses. Other capillaries are thought to open into the splenic cords, allowing blood to percolate through these structures before entering the venous sinuses. The splenic capillaries are unique in that some of them are sheathed at their termini in macrophages. The function of these cells is to survey the passing blood and remove spent erythrocytes, defective cells, and foreign antigens. The filtered blood collects in the venous sinuses and is emptied into small veins that connect with the *splenic vein* exiting the spleen.

Lymphocyte traffic within the spleen is also not completely understood. The spleen has no afferent lymphatic, so that lymphocytes must enter the spleen via the splenic artery. The cells exit the blood via capillary beds in the marginal zone, and then migrate to the follicles and PALS. If not activated by antigen collected by macrophages in the red pulp, the lymphocytes exit the spleen via small efferent lymphatics that start as blind-ended sacs around the germinal centers in the PALS. These lymphatics converge in the trabeculae and drain into the efferent lymphatic located in the splenic hilus, allowing the cells to eventually rejoin the circulation.

Every day, an adult's blood volume courses through the spleen four times, and is thus subject to the watchful perusal of resident lymphocytes and macrophages. The importance of the spleen to the immune system is illustrated by patients who have had to undergo a splenectomy (removal of the spleen), usually because of damage sustained in an accident. Splenectomized patients who no longer enjoy this vigilance survive, but appear to have high numbers of imperfect circulating red blood cells and increased morbidity due to multiple infections. Such consequences are often more severe in the very young or old and in immunosuppressed patients.

v) Leukocyte Extravasation

One of the unique features of the immune system is that its constituent cells are not fixed in a single organ; they move throughout the body to various locations in order to carry out their work efficiently. Although leukocytes circulate in the blood, it is within the tissues that the services of these cells are required. How do leukocytes migrate out of the blood and gain access to the tissues that are injured or under pathogen attack? As they travel through postcapillary venules, leukocytes have the chance to leave the blood via a migration process called *extravasation* (Fig. 3-15). Extravasation is made possible when *cellular adhesion molecules* (CAMs) expressed on the surfaces of leukocytes interact with complementary CAMs on the endothelial cells lining the postcapillary venules. These molecules include members of the *selectin* and the *integrin* adhesion protein families, with different members being expressed on different cell types. For example, L-selectin (CD62L) is expressed on most leukocytes, while P- and E-selectin (CD62P, CD62E) are expressed exclusively on endothelial cells. Expression of selectins and integrins is upregulated on leukocyte and endothelial cell surfaces by inflammatory molecules induced locally by trauma or infection, ensuring that leukocytes will gain access to tissues whenever and wherever they are needed.

The process of extravasation can be divided into two phases: *margination*, in which leukocytes adhere to endothelial cells in the postcapillary venules, and *diapedesis*, in which leukocytes pass between the endothelial cells and through the basement membrane, emigrating into the tissues. The margination phase can be broken down into three steps: selectin-mediated *tethering*, integrin-mediated *activation*, and *activation-dependent arrest and flattening*. In a postcapillary venule, the blood flow is relatively slow, shear forces are decreased, and the electrical charge on the endothelial surface is low compared to that at other vessel sites, allowing intermolecular binding to occur. In the first step, CAMs expressed on an endothelial cell transiently bind to complementary CAMs on a passing leukocyte, "tethering" it to the venule wall. P- and E-selectins on the endothelial cell can then forge contacts with carbohydrate moieties of proteins expressed on the leukocyte cell surface. Similarly, L-selectin molecules on the tethered leukocyte can bind to glycoproteins on the endothelial cell surface. However, these bonds are relatively weak and can be reversed in seconds, so that the leukocyte "rolls" across several endothelial cells, increasingly slowed by the sequential binding of carbohydrate moieties. The interactions sustained during tethering activate the leukocyte and trigger the second, integrin-mediated step of the margination process. The tethered

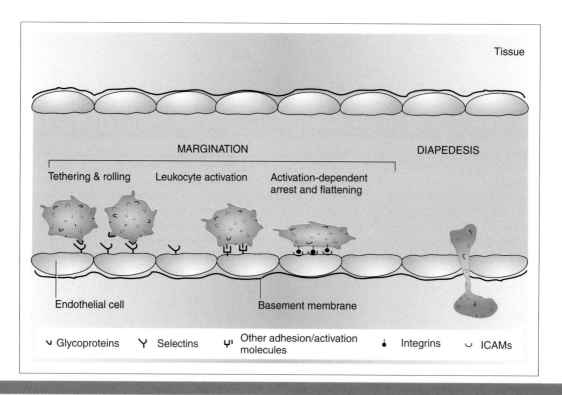

Figure 3-15

Leukocyte Extravasation
Adhesion molecules present on both leukocytes and activated postcapillary venule endothelium allow leukocytes to bind endothelial cells and pass through to the tissues during a process called extravasation. Extravasation is divided into two phases: margination and diapedesis. Margination consists of three parts: tethering and rolling, leukocyte activation, and activation-dependent arrest and flattening. Tethering is initiated when selectins found on the surface of the endothelium latch onto complementary adhesion molecules on the surface of leukocytes. This interaction is a loose one and the leukocyte is able to roll along the endothelium. An unknown trigger rapidly activates leukocytes, and integrin-mediated leukocyte–endothelial contacts result in activation-dependent arrest and flattening of the leukocyte. This prepares the leukocyte for the final step, called diapedesis, when it squeezes between endothelial cells to enter the tissues.

leukocyte rapidly deploys small intracellular vesicles which have membranes containing presynthesized β2-integrin molecules such as LFA-1 (leukocyte function antigen-1) and Mac-1 (*macrophage antigen-1*; CD11b/CD18). The vesicles fuse with the leukocyte plasma membrane and fix the β2-integrins onto the leukocyte cell surface in a position to bind to specific endothelial *intercellular adhesion molecules* (ICAMs) such as ICAM-1 (CD54) and ICAM-2 (CD102). Integrin-mediated binding is strong enough to flatten the leukocyte onto the endothelial surface, resulting in a state of so-called *activation-dependent arrest*.

Once the cell has achieved activation-dependent arrest, it is ready for the second phase of migration, diapedesis. The flattening of the leukocyte permits it to insert pseudopodia between the endothelial cells of the venule and squeeze between them. The leukocyte then secretes enzymes that digest the basement membrane of the endothelium, allowing the leukocyte to complete its migration through the venule wall into the tissue. Once the leukocyte has emigrated into the tissue, it depends on the binding of β1-integrin molecules (called *very late antigens*, or VLAs) for continued migration. VLA-3 (CD49c/CD29), VLA-4 (CD49d/CD29), and VLA-5 (CD49e/CD29) have affinity for the extracellular matrix protein fibronectin, while VLA-2 (CD49b/CD29) and VLA-3 bind to collagen and other components of the extracellular

matrix. In addition, the leukocyte downregulates its expression of L-selectin, since this molecule is of no value in the tissues. The leukocyte has now completed its transformation from a cell moving in the blood to one adapted for moving in the tissues.

vi) Lymphocyte Recirculation: Helping Lymphocytes and Antigen Meet
The ability to extravasate from blood into inflamed tissues is a general characteristic of leukocytes. However, lymphocytes are the only leukocytes that continually migrate from the tissues back into the blood, using the lymphatic system to create a complete blood → tissue → lymph → blood circuit. Because the majority of lymphocytes complete this circuit repeatedly during their life spans, this process has come to be called *lymphocyte recirculation* (Fig. 3-16). This strategy helps to solve the problem of getting the right lymphocyte to the right place at the right time to meet the right antigen. The total cellular mass of lymphocytes in the human body (approximately 10^{12} cells) is the same as that of the liver or the brain. However, because the number of lymphocyte specificities is so large, only a very small number of individual cells (one naive T cell in 10^5) exist to deal with any particular foreign substance. Nevertheless, because this naive cell recirculates from blood to the secondary lymphoid

Figure 3-16

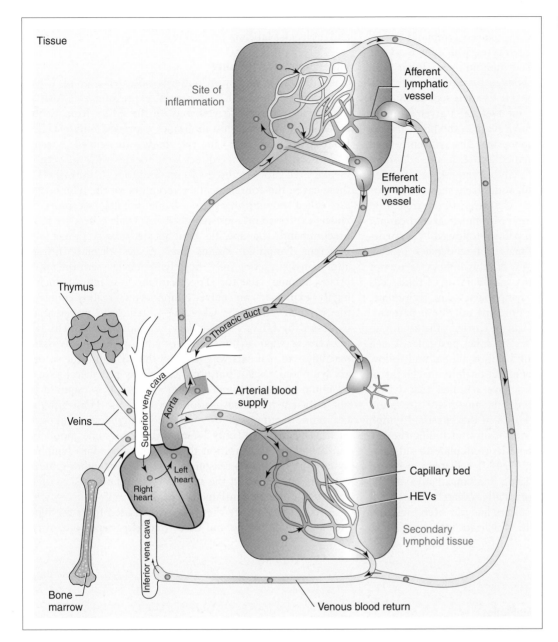

Schematic Representation of Lymphocyte Circulation
Lymphocytes from the bone marrow and thymus enter the circulation via the superior vena cava. The heart pumps lymphocytes through arterial blood into the capillary beds that supply all the tissues to the body. Lymphocytes enter secondary lymphoid tissues and sites of inflammation by squeezing through either HEVs or activated endothelium lining the capillaries, respectively. Excess fluid and lymphocytes are collected by the lymphatic capillaries and filtered through regional lymph nodes draining all tissues prior to being returned to the blood circulation via the left (thoracic duct) and right subclavian veins.

organs to lymph and back to blood on average once or twice each day, the chance that it will meet its antigen is greatly increased.

Lymphocytes do not migrate randomly into all tissues during recirculation. Instead, when mature, naive lymphocytes are released from the thymus or bone marrow into the blood, they begin the cycle of recirculation by extravasating preferentially into the secondary lymphoid tissues, where they can easily sample antigens that have been collected and concentrated at these sites. The migration of lymphocytes into secondary lymphoid sites requires the presence in these tissues of specialized postcapillary venules called *high endothelial venules* (HEVs). Molecules and other factors expressed by the endothelial cells lining these venules facilitate the adherence and transendothelial migration of naive lymphocytes. With the exception of the spleen (where the

site of lymphocyte extravasation appears to be the blood sinusoids in the marginal zone), HEVs are found in all secondary lymphoid tissues, including lymph nodes, Peyer's patches, appendix, tonsils, and the adenoids, as well as the small aggregates of lymphoid tissue in the stomach and large intestines. The endothelial cells in HEVs have a plump, cuboidal appearance rather than the flat morphology of endothelial cells from other vessel sites, hence the term "high" was originally used in describing them. More important than their unique morphology, however, is the expression by HEVs of specialized adhesion molecules that can bind to complementary cell surface molecules on naive lymphocytes, thereby mediating their extravasation. The specialized features of HEVs develop and are maintained under the influence of cells and/or factors present in the afferent lymph coming into lymphoid tissues.

Recirculation works as follows: at any given moment, as blood passes through the HEV in a given lymphoid tissue, about 25% of the passing lymphocytes will adhere to the endothelium and migrate into the lymphoid tissue. It has been estimated that, throughout the body, a total of 5×10^6 lymphocytes exit the blood through HEVs each second. Such a high migration rate means that, on average, a given lymphocyte can travel through every lymph node in the body at least once a day. Thus, naive lymphocytes can efficiently and thoroughly survey sites of antigen collection in the host on a continuous basis. Studies that analyze lymphocyte "traffic" moving through individual lymph nodes have shown that various lymphocyte subsets display characteristic recirculation patterns. Although the results are only statistical and cannot necessarily predict the migration pattern followed by a particular lymphocyte, they do reveal migration tendencies. Indeed, it appears that approximately 40% of all lymphocytes are not capable of recirculating at all, the vast majority of these cells being B lymphocytes. Most T lymphocytes can recirculate, although different subsets have different migration patterns. Molecular labeling experiments can be carried out (usually in sheep) that allow one to isolate and label particular T cell subsets from the lymph nodes of a donor animal and follow their subsequent migration after injection into the blood of a recipient animal. Such studies have revealed that CD8$^+$ T cells isolated from lymph nodes in any tissue are more likely to exit the blood in the HEVs in the intestinal lymph nodes rather than in other secondary sites, including the site of origin. Similarly, γδ T cells tend to recirculate to subcutaneous lymph nodes. In contrast, CD4$^+$ T cells tend to return preferentially to nodes in the same anatomical area as the node from which they were originally isolated. The small percentage of B cells that recirculates has not been analyzed in the same way because B cells, which are inherently more fragile than T cells, are less able to survive current isolation and molecular labeling procedures.

vii) Lymphocyte Homing

The differential migration patterns observed among T lymphocytes are due to the differential expression of CAMs on the surfaces of lymphocytes and on the HEVs from various tissue sites. Cell adhesion molecules expressed only on HEVs from particular sites in the body are called *vascular addressins*, because they direct lymphocytes with the complementary receptors to those specific locations ("addresses"). In these cases, the complementary receptors on the lymphocytes are called *homing receptors*. Some homing receptors are widely expressed on most lymphocytes, while others are found predominantly on specific lymphocyte subsets (Table 3-3). Different lymphoid organs and tissues display different addressins, attracting only those lymphocytes with the appropriate homing receptor. For example, L-selectin, which is highly expressed on naive lymphocytes, readily binds to the vascular addressin GlyCAM-1, which is preferentially expressed by HEVs in secondary lymphoid tissues. The lymphocyte is then in position to employ its extravasation machinery to make its way through the HEV into the secondary lymphoid tissues (but not into healthy peripheral tissues). It should be noted that L-selectin expression is required but not sufficient for extravasation across HEVs. Thus, although L-selectin is expressed on many nonlymphocyte leukocytes such as neutrophils, only lymphocytes and some monocytes can complete extravasation through the secondary lymphoid tissue HEVs. For this reason, neutrophils and other granulocytes are not found in the secondary lymphoid tissues unless these tissues become inflamed (see later).

A naive lymphocyte that has extravasated from the blood into a secondary lymphoid tissue may experience one of two

Table 3-3 Lymphocyte Homing Molecules

Lymphocyte Homing Receptor	Endothelial Vascular Addressins	Role in Extravasation	Cell Type	Primary Homing Pathway
L-Selectin	MAdCAM-1	Tethering	Naive lymphocytes	Peyer's patches
L-Selectin	Peripheral lymph node addressins, GlyCAM-1	Tethering, rolling	Naive lymphocytes	Lymph nodes
INTEGRINS*				
LPAM-1	MAdCAM-1	Tethering, rolling, adhesion	Naive lymphocytes Memory lymphocytes	Peyer's patches and appendix Nonpulmonary mucosal sites
VLA-4	VCAM-1	Tethering, rolling, adhesion	Memory lymphocytes Effector lymphocytes	Extraintestinal inflammatory sites Site of severe inflammation
LFA-1	ICAM-1, ICAM-2	Adhesion	Lymphocyte blasts or memory cells	Intestinal mucosa

LPAM-1, lymphocyte Peyer's patch adhesion molecule-1; MAdCAM-1, mucosal addressin cell adhesion molecule-1; VLA-4, very late antigen-4; LFA-1, leukocyte function-associated molecule-1; ICAM-1, -2, intercellular adhesion molecule-1 and -2.

basic outcomes at this site: no activation, or activation by specific antigen. Let's take the case of a naive T cell that has extravasated into a lymph node through its HEVs. The naive T cell migrates to the lymph node cortex and transiently binds many different APCs in succession in order to survey the different antigenic peptides displayed on the APC surfaces. These transient interactions are mediated by several cell surface proteins, including CD2 and LFA-1 on T cells, and ICAM-1, ICAM-2, and ICAM-3 on APCs (see Ch.14). These antigen-independent contacts weakly hold a T cell to an APC long enough to determine if the peptide fits in the TCR binding cleft. If the cell does not encounter an APC displaying the antigen for which it is specific, the lymphocyte exits the node via an efferent lymphatic vessel, and moves via the lymphatics to other secondary lymphoid tissues. If activation is not invoked at any point along the way, the cell is eventually returned to the blood circulation by the lymphatic system and the recirculation process begins again.

Alternatively, the naive T lymphocyte that has entered the lymph node may be stimulated by appropriately presented antigen. Stimulation of its TCR induces the expression and/or upregulation of adhesion proteins such as VLA-4, VLA-5, VLA-6, and LFA-1. Some of these increase the strength of binding of the T cell to the APC, while others increase adhesion to extracellular matrix molecules. In addition, conformational changes in some cell surface proteins induced by the binding of antigen to the TCR increase their affinity for their counterpart ligands. As a result, the activated T cell is "stuck" in the lymph node and no longer leaves to recirculate. As mentioned previously, the lymph node experiences a temporary shutdown, during which the activated T cell proliferates extensively and cellular migration in and out of the node is sharply reduced. The clone of activated cells in the node then differentiates, generating both effector and memory cells that display an array of adhesion molecules with a lower affinity for lymph node addressins. B cells are thought to undergo a similar process of activation, proliferation, and differentiation, resulting in plasma cell and memory cell generation.

In the case of B lymphocytes, most effector and memory cells generated in the lymph node tend to remain there. The antibodies secreted by the plasma cells diffuse through the medulla of the node and exit it via the efferent lymphatic, eventually entering the blood via the left or right subclavian veins and traveling to distant sites of infection or injury. In contrast, effector and memory T lymphocytes exit the lymph nodes via the lymphatics and rejoin the blood circulation. However, unlike naive lymphocytes, effector and memory T cells express very little L-selectin, and so no longer recirculate through lymph nodes. Instead, these cells feature different cell surface molecules, such as a new isoform of the *signaling phosphatase CD45* (CD45RO) and increased levels of LFA-1 and CD44, as well as new homing receptors (such as VLA-4), that mediate their migration to peripheral (nonlymphoid) tissues. Although effector T cells generally die in the peripheral tissue in which they have been sent to combat an invader, the memory cells establish new recirculation patterns back to those peripheral tissues to which they were first dispatched.

During the primary response, the expression of tissue-specific homing receptors is induced in a new memory cell, which remains fixed in its progeny, causing these cells to return repeatedly to the peripheral tissue in which they first encountered antigen. In other words, there is a constant background level of extravasation of memory lymphocytes through postcapillary venules into specific peripheral tissues, and this migration occurs even in the absence of inflammation. These cells enter the lymph in the peripheral tissues and return to the circulation via the lymphatics as described previously. Since the triggering antigen is likely to enter the body in the same fashion in a subsequent infection, this local homing of memory cells increases their chances of being in the right place at the right time to encounter the specific antigen, contributing to the shorter lag time of the secondary immune response.

Memory lymphocytes of the MALT and SALT have their own recirculation patterns; mucosal and cutaneous addressins draw MALT and SALT lymphocytes, respectively, into these types of tissues all over the body. For example, the expression of the homing receptor LPAM-1 (*lymphocyte Peyer's patch adhesion molecule-1*; also known as integrin α4β7) allows MALT lymphocytes to collect in all elements of the GALT, since endothelial cells in these regions express the vascular addressin MAdCAM-1 (*mucosal addressin cell adhesion molecule-1*), which binds to LPAM-1. More on mucosal and cutaneous recirculation patterns appears in Chapter 20.

In a peripheral tissue where inflammation has been initiated, the postcapillary venules are altered to permit the extravasation of all types of leukocytes. Inflammatory cytokines induce the upregulation or expression of new vascular addressins such as VCAM-1 (CD106) by the now "activated" endothelial cells, allowing them to signal "we're under attack" to passing neutrophils, monocytes, and effector and memory lymphocytes in the blood. In particular, the addressins VCAM-1, ICAM-1, ICAM-2, and hyaluronate expressed by the activated endothelial cells then serve as ligands for the VLA-4, LFA-1, and CD44 homing receptors expressed on effector and memory lymphocytes. These effector and memory cells are thus able to extravasate through the activated endothelium into the inflamed tissue to combat the threat. Because of the influx of effector and memory lymphocytes, such sites of inflammation are sometimes called *tertiary lymphoid tissues*. After completing their tasks, most effector lymphocytes in a site of inflammation undergo apoptosis in this location. However, some nonlymphoid leukocytes that have extravasated into a site of inflammation may end up collected in the lymph and are thus conveyed to the local lymph node. For unknown reasons, these cells are unable to exit the node and cannot continue through the lymphatic system to re-enter the blood as do lymphocytes. Non-lymphoid leukocytes trapped in this way die within the node. If specific antigen is no longer present at a particular site of inflammation, the "unused" lymphocytes can be taken up from the interstitial space by the lymphatic vessels. The cells are conveyed back to the lymph nodes through the afferent lymphatics and exit again through the efferent lymphatics and re-enter the blood. Both effector and memory lymphocytes can then be sent out to survey other sites of inflammation. If, however, the inflammation is not resolved and becomes chronic, the postcapillary venules in the affected site take on the characteristics of HEVs, inducing lymphocytes to traffic through the area on a constant basis.

The control of immune cell migration is not only important for bringing lymphocytes and antigen together efficiently, but also for maintaining homeostasis. An inappropriate or damaging accumulation or depletion of lymphocyte subsets can be detrimental to the host. With new immune system cells constantly being produced by the primary lymphoid organs, a balance must be maintained between incoming new cells and those already in a given tissue niche. Competition for access to specific microenvironments containing factors necessary for the survival of particular lymphocytes may determine the balance between cell proliferation and differentiation, and apoptosis. The contribution of homing receptors to the maintenance of lymphocyte homeostasis is explored in Box 3-5.

Thus, the immune system exercises fine control over lymphocyte deployment by requiring specific microenvironments and molecular interactions between cell surfaces for both cell migration and cell differentiation. The binding of a lymphocyte to another cell or to an extracellular matrix molecule is influenced by the controlled sequence of reversible receptor–ligand binding events, ensuring that the appropriate cells are recruited and that they carry out their responses to antigen where they should. Each step offers the cell a chance to control the process, and whether the next step is taken depends on the local microenvironment and the matching of lymphocyte homing receptors with addressins on tissue cells. The variability in type, expression, and affinity of cell surface molecules, and the requirement for a specified sequence of expression, provide the necessary regulation of lymphocyte homing using a relatively small number of distinct molecules.

Now that we are familiar with the cells and tissues of the immune system, we return in the next chapter to innate immunity, to describe in detail how most of the foreign entities attacking our bodies are defeated.

Box 3-5. The role of homing receptors in homeostasis

How does the body control the balance (achieve *homeostasis*) between the massive generation of immune system cells by the bone marrow, cell circulation through the blood and lymph, and cell death? Because there must be a limited pool size for lymphocytes in each compartment of the body, there must be a mechanism to restrict entry to these pools. New naive T and B cells emerging from the bone marrow must compete with existing naive cells for entry into that pool: the body can tolerate only so many cells in the circulation. Similarly, if memory cells are produced after each exposure to antigen and remain in the tissues, how does the body accommodate the accumulating total of new memory cells as more and more antigens are encountered throughout life? Recent work has suggested that newly produced lymphocytes must compete with pre-existing ones for compartmental space, and that this competition is based on homing to specific microenvironments within the body.

Lymphocytes require certain cytokines and growth and regulatory factors in order to maintain expression of various receptors, or in order to carry out activities such as proliferation, differentiation, and cytolysis. Those cells able to bind the limited supply of these factors will succeed; those that do

not are culled, thus maintaining the balance. A given lymphocyte subset will require different factors at different times in its development, and thus a different microenvironment. Admission to a given microenvironment is limited by the expression of the appropriate addressins and homing receptors. Interaction of an addressin with one receptor on a particular cell type may deliver a "stay here and prosper" signal, whereas its binding to another ligand on a different cell type may induce that cell's apoptosis.

Lymphocytes move continuously from their preferred microenvironments to the general circulation, ensuring that the body's lymphocyte repertoire as a whole must compete regularly for necessary factors. Since the presence of antigen can influence lymphocyte homing and the type of cell subset activated, only those clones best suited to handling the current situation in the body will succeed in re-entering their microenvironments, and undergo the interactions necessary for continued survival. The specific homing of different lymphocyte subsets can also prevent inappropriate competition between subsets of very different function, such as naive versus memory cells.

When a particular niche is depleted of lymphocytes (due to a malfunction in the generation of new lymphocytes), very few cells are competing for the available growth factor resources in the microenvironment, a situation that may permit the survival of lesser-suited cells, such as those few self-reactive clones that were not eliminated in the thymus during the generation of central tolerance. Such a mechanism has been postulated to account for the occurrence of autoimmunity in patients with low numbers of circulating T and B cells, or in cases of "graft versus host" disease observed after bone marrow transplantation (see Ch.27). Similarly, if too many cells are activated at once, and all are competing simultaneously for the limited resources in a given niche, none may get enough resources to survive. This mechanism could account for the rare phenomenon of the complete depletion of responding T cells observed during some viral infections. Thus, not only the fate of an individual cell is determined by its homing characteristics, but also the fate of the lymphocyte recirculating pool as a whole, and by extension, the homeostasis of the organism as a whole.

SUMMARY

Hematopoietic stem cells in the bone marrow give rise to all cells of the myeloid and lymphoid lineages involved in both innate and adaptive immune responses. Hematopoiesis occurs throughout life, and is balanced by the ongoing programmed cell death of spent cells to maintain the health of the host. The primary lymphoid organs are the anatomical sites where T and B lymphocytes develop. Like most leukocytes, B cells complete their maturation in the bone marrow. T cells originate in the bone marrow but complete their maturation in the thymus. While in the thymus, T cells are selected for their ability to recognize self-MHC complexed to peptide (but not self-MHC alone). Once mature, leukocytes of all lineages migrate throughout the body via the blood circulation and lymphatic system. Although different leukocytes tend to concentrate in either the blood or tissues, they can move between these compartments when infection or injury occurs. At such times, inflammation activates endothelial cells to allow an influx of leukocytes, including effector and memory lymphocytes, into the damaged tissue. To facilitate lymphocyte activation, secondary lymphoid tissues are distributed throughout the body to serve as junctions where lymphocytes, antigens, and antigen-presenting cells can co-localize. The major secondary lymphoid structures are the lymph nodes, the spleen, and the mucosa- and skin-associated lymphoid tissues.

Selected Reading List

Akashi K., Reya T., Dalma-Weiszhausz D. and Weissman I. (2000) Lymphoid precursors. *Current Opinion in Immunology* **12**, 144–150.

Ansel K. and Cyster J. (2001) Chemokines in lymphopoiesis and lymphoid organ development. *Current Opinion in Immunology* **13**, 172–179.

Banchereau J., Briere F., Caux C., Davouse J., Lebecque S., Liu Y.-J., Pulendran B. and Palucka K. (2000) Immunobiology of dendritic cells. *Annual Review of Immunology* **18**, 767–811.

Burkitt H., Young B. and Heath J., eds. (1993) "Wheater's Functional Histology: A Text and Colour Atlas," Churchill Livingstone, Edinburgh, Scotland.

Butcher E. and Picker L. (1996) Lymphocyte homing and homeostasis. *Science* **272**, 60–66.

Chadburn A. (2000) The spleen: anatomy and function. *Seminars in Hematology* **37**(Suppl.1), 13–21.

Chaddah M., Wu D. and Phillips R. (1996) Variable self-renewal of reconstituting stem cells in long-term bone marrow cultures. *Experimental Hematology* **24**, 497–508.

Cormack D. H. (2001) "Essential Histology," 2nd edn. Lippincott Williams & Wilkins, Baltimore, MD.

Enver T., Heyworth C. and Dexter T. (1998) Do stem cells play dice? *Blood* **92**, 348–352.

Fu Y.-X. and Chaplin D. D. (1999) Development and maturation of secondary lymphoid tissues. *Annual Review of Immunology* **17**, 399–433.

Fulop G., Wu D. and Phillips R. (1989) The scid mouse as a model to identify and quantify myeloid and lymphoid stem cells. *Current Topics in Microbiology and Immunology* **152**, 173–180.

Girard J.-P. and Springer T. (1995) High endothelial venules (HEVs): specialized endothelium for lymphocyte migration. *Immunology Today* **16**, 449–457.

Handin R., Lux S. and Stossel T., eds. (2003) "Blood: Principles and Practice of Hematology," 2nd edn. Lippincott Williams & Wilkins, Philadelphia, PA.

Kawamoto H., Ohmura K. and Katsura Y. (1997) Direct evidence for the commitment of hematopoietic stem cells to T, B and myeloid lineages in murine fetal liver. *International Immunology* **9**, 1011–1019.

Klinger M. (1997) Platelets and inflammation. *Anatomy and Embryology* **196**, 1–11.

Kondo M., Weissman I. and Akashi K. (1997) Identification of clonogenic common lymphoid progenitors in mouse bone marrow. *Cell* **91**, 661–627.

Lee G., Bithell T., Foerster J., Athens J. and Lukens J. (1993) "Wintrobe's Clinical Hematology," 9th edn., vol. 1. Lea and Febiger, Philadelphia, PA.

Metcalfe D. (1998) Lineage commitment and maturation in hematopoietic cells: the case for extrinsic regulation. *Blood* **92**, 345–348.

Peng C.-A., Koller M. R. and Palsson B. (1996) Unilineage model of hematopoiesis predicts self-renewal of stem and progenitor cells based on ex vivo growth data. *Biotechnology and Bioengineering* **52**, 24–33.

Randolph G. J., Inaba K., Robbiani D. F., Steinman R. M. and Muller W. (1999) Differentiation of phagocytic monocytes into lymph node dendritic cells in vivo. *Immunity* **11**, 753–761.

Reis E., Sousa C., Sher A. and Kaye P. (1999) The role of dendritic cells in the induction and regulation of immunity to microbial infection. *Current Opinion in Immunology* **11**, 392–399.

Stekel D., Parker C. and Nowak M. (1997) A model of lymphocyte recirculation. *Immunology Today* **18**, 216–221.

Till J. E. and McCulloch E. (1961) A direct measurement of the radiation sensitivity of normal mouse bone marrow cells. *Radiation Research* **14**, 213–222.

Weissman I. L. (2000) Stem cells: units of development, units of regeneration, and units in evolution. *Cell* **100**, 157–168.

Whetton A. and Spooncer E. (1998) Role of cytokines and extracellular matrix in the regulation of haemopoietic stem cells. *Current Opinion in Immunology* **10**, 721–726.

Young A. (1999) The physiology of lymphocyte migration through the single lymph node in vivo. *Seminars in Immunology* **11**, 73–83.

Young A. and Hay J. (1999) Lymphocyte migration in development and disease. *Seminars in Immunology* **11**, 71.

Zandstra P., Lauffenburger D. and Eaves C. (2000) A ligand–receptor signaling threshold model of stem cell differentiation control: a biologically conserved mechanism applicable to hematopoiesis. *Blood* **96**, 1215–1222.

Innate Immunity

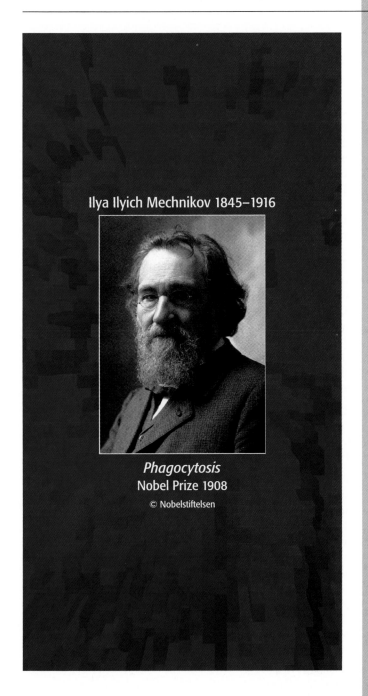

Ilya Ilyich Mechnikov 1845–1916

Phagocytosis
Nobel Prize 1908

© Nobelstiftelsen

CHAPTER 4

A. MECHANISMS OF INNATE IMMUNITY

B. PATTERN RECOGNITION IN INNATE IMMUNITY

"All the flowers of all the tomorrows are in the seeds of today"
—Indian Proverb

The popular press, much of modern research, and, indeed, most of this book, focus on adaptive immunity. However, it is important to keep in mind that host defense is an integrated process, in which adaptive immune recognition, while fascinating and of major importance, is only one of several mechanisms implemented in response to invasion by a pathogen. Furthermore, to achieve an optimal adaptive immune response, the concerted participation of the innate immune response is essential. *Innate immunity* is conferred by the interaction of certain cells, cell surface receptors, and soluble molecules that act either non-specifically or recognize invading entities in a broadly specific way. The host can use innate immunity to rapidly generate an initial response to almost any pathogen, either eliminating it, or at least containing it until the slower adaptive immune response can develop. In addition, the innate response can influence the type of adaptive immune response that occurs, depending on which antigens are present.

The innate immune response is phylogenetically older than the adaptive response. Elements of an innate immune response can be found in all multicellular organisms, whereas the mechanisms of the adaptive immune response (thought to be 400 million years old) are present only in fish, reptiles, amphibians, birds, and mammals (see Ch.21). We will now examine in detail the innate immune response, which accounts for the elimination of the vast majority of foreign antigens we encounter every day.

A. Mechanisms of Innate Immunity

There are three types of fundamental innate defenses in vertebrates: *anatomical and physiological barriers, cellular internalization mechanisms,* and *inflammation.* (Mechanisms such as the cytotoxicity mediated by NK cells are considered to "bridge" the innate and adaptive responses and are dealt with in Chapter 18.) Fundamental innate defenses pre-exist as natural features of the body and do not undergo permanent change as a result of exposure to a pathogen. A second exposure to an infectious agent provokes exactly the same magnitude and character of response as the first exposure; that is, there is no "memory" of a previous exposure to that agent.

Each innate mechanism plays its part in sequence to defend the host against attack by potential pathogens in the environment. The anatomical barriers, some physiological barriers, and the cellular internalization mechanisms fall among the elements of the first phase of the immune response: those that are non-specific and non-inducible, meaning that they are always present or are ongoing activities. Other physiological barriers and the inflammatory response are not pre-existing and are induced only by the presence of the antigen, marking the second phase of the immune response. The recognition of antigen in this case is broadly specific. These two innate phases are followed by the adaptive phase of the immune response that must be induced and is highly antigen-specific.

I. ANATOMICAL AND PHYSIOLOGICAL BARRIERS TO INFECTION

The body of an organism features several anatomical and physiological barriers that substantially decrease the likelihood of infection and reduce its intensity should it occur (Fig. 4-1). By anatomical barriers, we mean structural elements such as the skin and mucous membranes that physically prevent access through the body surfaces and orifices. By physiological barriers, we mean the actions of body structures (such as the waving of nasal cilia) or substances produced by tissues (such as tears and mucous) that work to reinforce the anatomical barriers. The anatomical and physiological barriers cooperate to provide a first line of defense against the penetration of body surfaces, be they external like the skin, or internal like the gut lining.

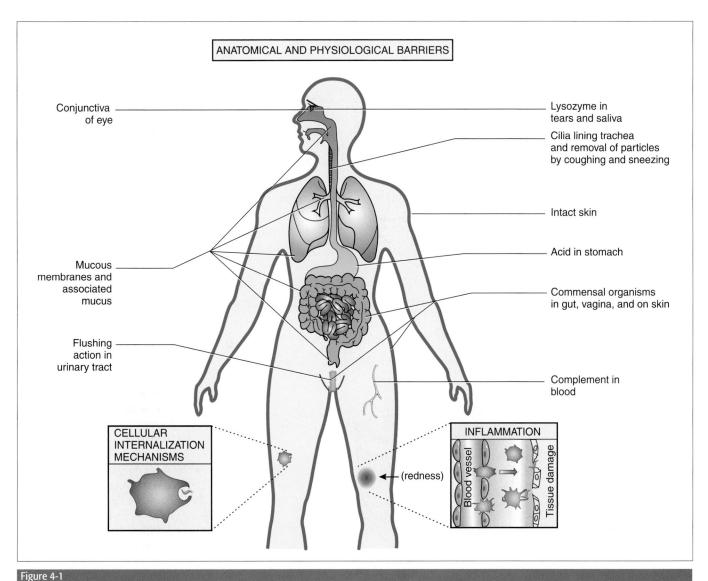

ANATOMICAL AND PHYSIOLOGICAL BARRIERS

Conjunctiva of eye

Mucous membranes and associated mucus

Flushing action in urinary tract

Lysozyme in tears and saliva

Cilia lining trachea and removal of particles by coughing and sneezing

Intact skin

Acid in stomach

Commensal organisms in gut, vagina, and on skin

Complement in blood

CELLULAR INTERNALIZATION MECHANISMS

INFLAMMATION

Blood vessel

Tissue damage

(redness)

Figure 4-1

Innate Defenses in Invertebrates
A variety of anatomical and physiological body barriers prevent the majority of infectious agents from penetrating the body's external and internal surfaces. Those pathogens that do manage to overcome these initial barriers are met by innate cells that provide the next line of defense. These cells circulate throughout the body and immediately respond to invaders by employing mechanisms of cellular internalization and inflammation.

The primary anatomical barrier against infection in vertebrates is the *skin*, a surprisingly large organ that can weigh up to 5 kg in humans. When intact, the skin provides a tough surface that is relatively difficult for microorganisms to penetrate. The skin is composed of three layers, the outer *epidermis*, the underlying *dermis*, and the fatty *hypodermis* (Fig. 4-2A, B). Cells at the exterior surface of the epidermis are, in fact, dead and filled with *keratin*, a protein that confers water resistance. Other features of the epidermis that discourage infection are its lack of blood vessels and its rapid turnover: complete renewal of the outer skin occurs every 15–30 days. Below the epidermis lies the dermis, which contains all of the blood vessels and other tissues necessary to support the epidermis. Among these are the *sebaceous glands*, which produce *sebum*, an oily secretion with a pH of 3–5. Sebum production gives the skin a level of

acidity of about pH 5.5, which inhibits the multiplication of most microbes. The hypodermis provides an additional barrier of fat that pathogens must cross to access the underlying tissues. Failure of the skin as a defense occurs when it is breached, as in the case of insect bites or wounds. More on immune defense of the skin appears in Chapter 20.

Other areas of the body that must come into contact with the outside world have developed different anatomical barriers. Instead of dry skin, which would not permit the passage of air or food, the topologically exterior surfaces of the gastrointestinal, respiratory, and urogenital tracts are covered with *mucosal epithelial layers* that are often called *mucous membranes* (Fig. 4-2C, D). Similarly, the eye possesses delicate membranes called *conjunctiva*, which line the eyelids and protect the exposed surface of the eye while allowing the passage of light. These membranes are further

Figure 4-2

The Skin and Mucosal Epithelium as Innate Barriers

(A) The intact skin, a tough outer layer bathed in acid secretions, is normally an effective barrier to most pathogens. (B) Damaged skin allows pathogens easy access to the blood supply within the dermal layer and the tissues underlying the hypodermis (subcutaneous layer). (C) Invading pathogens are prevented from adhering to epithelial cells in the lung by mucus and cilia covering the surface of the epithelial cells. Cilia sweep the pathogens up to an area where they can be expelled, which in the case of the lung is the upper respiratory tract. (D) Where mucus is decreased, microbes can easily adhere to and infect the epithelium.

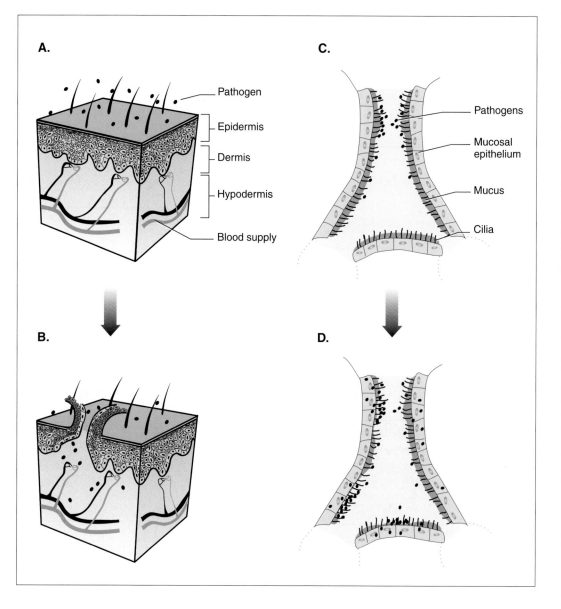

defended from microbial penetration by the flushing action of various secretions, including *mucus, saliva,* and *tears.* Mechanical defense is derived from the sweeping action of tiny oscillating hairs called *cilia* on the surface of epithelial cells in the lower respiratory and gastrointestinal tracts. Coughing and sneezing expel the mucus-coated microbes. Failure of these barriers can occur when a microorganism possesses a cell surface molecule that facilitates penetration of the mucus and allows the microbe to bind to and enter the underlying epithelial cells.

A microorganism that manages to evade the external and internal barriers of the body may succumb to a hostile *physiological environment,* such as a body temperature too high or a stomach pH too low to permit replication. As well as their cleansing action, body fluids may also contain antibacterial or antiviral substances that can kill or inhibit pathogens. For example, *lysozyme,* an enzyme that hydrolyzes a common component of some bacterial cell walls, is found in tears and other mucus secretions. Body surfaces, such as

the skin or intestinal epithelia, are also protected by antimicrobial peptides, including the *defensins* (a product of neutrophil degranulation mentioned in Chapter 3) and the *cathelicidins.* In response to bacterial infection, keratinocytes in the skin or intestinal epithelial cells initiate production of these antimicrobial peptides which then block bacterial replication. Mice lacking genes regulating the production of these peptides are highly susceptible to invasive skin infections with *Staphylococcus,* and mice that express very high levels of an intestinal defensin are resistant to *Salmonella* infection. Infections with viruses induce a different family of soluble antipathogen molecules called the *interferons.* Interferons protect cells surrounding a virally infected cell by causing the bystanders to adopt an *antiviral state,* in which they undergo biochemical changes that make them less susceptible to viral attack (see Ch.17). In addition, interferon-γ (IFNγ) activates macrophages and stimulates NK cells to lyse target cells, particularly those infected with virus (see Ch.18).

Another physiological barrier to infection by harmful pathogens is provided by the billions of *commensal* bacteria and fungi that inhabit the mouth, digestive system, and skin surface. These beneficial microbes are collectively known as the *normal flora*. The normal flora appear to have a protective effect with respect to infection by noncommensal microbes; elimination of subpopulations of the normal flora (say, with antibiotic therapy) renders a patient more susceptible to infection by pathogenic bacteria. The precise mechanisms of this protection by normal flora have yet to be clearly delineated, but may involve competition for available resources, activation of immune system cells, or the secretion of selectively toxic molecules.

The *complement system* represents a physiological barrier of key importance. The complement system is a collection of proteins and proenzymes that circulates in the blood in the inactive state. Activation of the complement system results in a systematic cascade of enzyme activation and the production of biologically active by-products. These products induce a broad range of inflammatory effects which result in containment of the invader (see later). The complement system can be triggered either directly by a microbe (the *nonclassical*, or *alternative*, activation path), by specific antibody that has coated the cell surface of the microbe (the *classical* activation path), or by certain soluble carbohydrate recognition proteins that can bind to a wide range of microbes and can trigger complement activation without the involvement of antibody (the *lectin* activation path). Membrane-encapsulated pathogens (including certain bacteria, parasites, and viruses) can be directly destroyed by complement-induced damage to the pathogen's membrane, or may be controlled by a complement-induced increase in activity of the host's phagocytes (see Box 4-1). Note that the interaction of complement and antibody in the classical complement activation path is an example of the linkage between the second (inducible innate) and third (adaptive) phases of the immune response. The contribution of the complement cascade to adaptive immunity is discussed in Chapter 5 and the complement system as a whole is examined in detail in Chapter 19.

II. CELLULAR INTERNALIZATION MECHANISMS THAT FIGHT INFECTION

Engulfment and destruction via internalization by a cell is the mechanism responsible for the disposal of the vast majority of unwanted materials (either foreign or host generated) encountered in the body. There are three distinct processes for importing extracellular macromolecules and particles into a cell. Two of these, *macropinocytosis* ("cell drinking") and *receptor-mediated endocytosis* ("specific importing of a ligand into a cell"), involve the internalization of soluble macromolecules and are ongoing activities that can be carried out by any cell. Only specialized phagocytic leukocytes, including neutrophils, tissue macrophages, and blood monocytes, are capable of *phagocytosis* ("cell eating"), a related process for the handling of large extracellular particles such as certain viruses and bacteria.

i) Macropinocytosis

Macropinocytosis (Fig. 4-3A) refers to a process by which a cell can "gulp" extracellular fluid by ruffling its plasma membrane and forming a vesicle around a relatively large volume of fluid. These vesicles, called *macropinosomes*, are highly variable in volume (0.2–5 μm in diameter) and contain fluid phase solutes, rather than insoluble particles. Foreign macromolecules enter the vesicle with the extracellular fluid on a gradient of passive diffusion. The number of macromolecules caught in the vesicle depends entirely on their concentration. This process provides an efficient but non-specific means of sampling the extracellular environment.

ii) Receptor-Mediated Endocytosis

In *receptor-mediated endocytosis* (Fig. 4-3B), uptake of the foreign macromolecule depends on its interaction as a soluble ligand with an appropriate receptor on the surface of the endocytic cell. Binding of the macromolecule to a receptor triggers the polymerization of *clathrin*, a protein component of the microtubule network located on the cytoplasmic side of the plasma membrane. Invagination of clathrin-coated "pits" internalizes the receptor and its bound foreign ligand into a "coated" vesicle of a size in the range of 0.15–0.45 μm. Clathrin-coated vesicles are much more uniform in size than those created by macropinocytosis.

The uptake of foreign materials by the mechanisms of macropinocytosis and receptor-mediated endocytosis does more than just sequester antigens: internalization also initiates the processing of these antigens that leads to recognition by cells of the adaptive immune system, another example of the inextricable linkage of the innate and adaptive responses.

iii) Phagocytosis

Phagocytosis, the ingestion of particles more than 0.5 μm in diameter, is initiated when multiple cell surface receptors on a phagocyte bind sequentially in a "zippering" manner to ligands on a foreign particle, or even to ligands on the surface of a whole microbe or spent host cell (Fig. 4-4). These interactions between receptors and ligands induce the polymerization of the cellular skeletal protein *actin* at the site of internalization. The plasma membrane then invaginates, probably via an actin-based mechanism (although other "molecular motors," including myosins I and II, may also be involved), and forms a relatively large vesicle called a *phagosome* around the material to be removed. The morphology of the phagosome can vary depending on which phagocyte receptor has mediated the internalization.

Whether a particle binding to a phagocyte cell surface receptor is disposed of by endocytosis or phagocytosis appears to depend on signals issued by specific domains in the cytoplasmic tails of the transmembrane receptor molecules. Certain domains mediate signals to the actin cytoskeleton, causing it to initiate the internalization associated with phagocytosis; other domains are responsible for signaling the microtubule system to start endocytosis. The guanosine triphosphate (GTP)-binding proteins Rab-1, Rab-5, and Rab-7 are involved in phagocytosis but not endocytosis, while Rab-4 appears to function in endocytosis exclusively.

Box 4-1. Introduction to the complement system

Complement is not a single substance, but a collection of about 30 serum and membrane proteins with a vital role in host defense. In the late 1800s, microbiologist Jules Bordet identified a heat-labile activity in serum that was required for antibacterial antibody to succeed in lysing bacteria. This activity was deemed to assist, or to be "complementary" to, antibody; hence the name "complement." We now know that complement is a complex system of functionally related enzymes that are sequentially activated in a tightly regulated cascade. Complement activation has six principal outcomes: (1) *lysis* of pathogens; (2) *opsonization* (coating) of bacteria and other foreign entities, which enhances phagocytosis; (3) *clearance* of potentially damaging immune (antigen–antibody) complexes; (4) *enhancement of antigen presentation*; (5) *enhancement of B cell activation*; and (6) the *generation of peptide by-products* that are involved in vasodilation, phagocyte adhesion, phagocyte chemotaxis, signaling an inflammatory response, and regulating complement activation.

There are three pathways by which the complement system can be activated (see Figure). In the *classical* pathway, which was discovered first but is probably more recently evolved, the initial enzymatic step follows the binding of the first complement component, C1q, to antibody that is complexed either to soluble antigen (to form an immune complex) or to antigen on the surface of a pathogen. The *nonclassical*, or *alternative*, pathway is triggered by the direct binding (no antibody) of complement component C3b to an appropriate ligand on the surface of a pathogen. Most recently identified is the *lectin* pathway, in which soluble, C1q-like proteins bind to carbohydrates on a pathogen surface and then trigger the complement cascade without the involvement of antibody. The molecules known to trigger the lectin pathway, such as mannose-binding lectin (MBL), are members of a family of soluble molecules called *collectins*, which play a role in the innate recognition of pathogens. Thus, the alternative and lectin paths are elements of the innate immune response, while the classical path represents a means by which effectors of the innate immune response can be activated by the immune complexes of the adaptive response.

Many enzymes of the complement system rest in the circulation in proenzyme or *zymogen* (inactive) form, until the cascade is triggered. The activation of the first component causes it to activate the next, and so on, in a strictly regulated sequence: deficiency of any one component halts the cascade in its tracks. Because of the potential for tissue damage should the complement system be continuously activated, the enzymatic products are short-lived, and enzyme inhibitors generated as the by-products of earlier steps are present to inactivate the cascade once the initiating stimulus is removed.

Although their initial steps are different, the various complement activation pathways eventually result in generation of a common enzymatic component called the *C3 convertase*, which cleaves the complement component *C3* into the smaller peptides C3a and C3b. *C3b*, the larger cleavage fragment of C3, opsonizes bacteria, promoting their uptake by phagocytes expressing complement receptors. C3b also participates in a complex called the C5 convertase, which can cleave C5, the next component in the cascade. The cleavage of C5 results in the generation of the smaller peptides C5a and C5b. C5b can be deposited on the surface of the triggering pathogen or the infected host cell where it serves as a platform for the progression of the cascade, ultimately leading to the generation of a structure called the *membrane attack complex* (MAC). Subsequent enzymatic events take place right on the surface of the target cell, so that the MAC is assembled directly in position to force the opening of a pore through the plasma membrane, causing the pathogen or target cell to lyse.

Cleavage of C3 is pivotal in all three complement activation pathways. This step amplifies the number of C3b molecules available for forming C5 convertase, thus triggering the alternative pathway and opsonizing antigen for disposal by phagocytes that express C3b receptors. Cleavage of C5 is also a key step in the complement cascade, because the C5b product allows MAC formation to proceed. However, C3b and C5b are not the only biologically active molecules produced. *C3a*, the small fragment resulting from C3 cleavage, and *C5a*, the small fragment of C5, are pro-inflammatory molecules that signal an inflammatory response. *C5a* is also a particularly effective chemoattractant for phagocytic neutrophils. A detailed discussion of the biochemistry and effector functions of the complement system appears in Chapter 19.

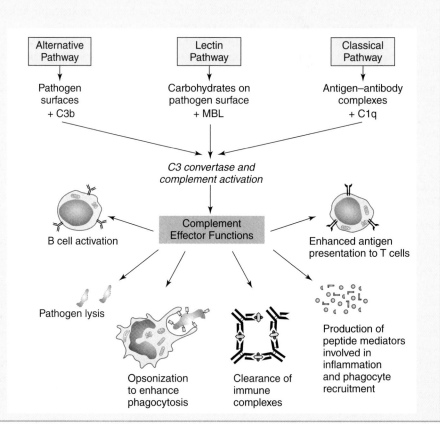

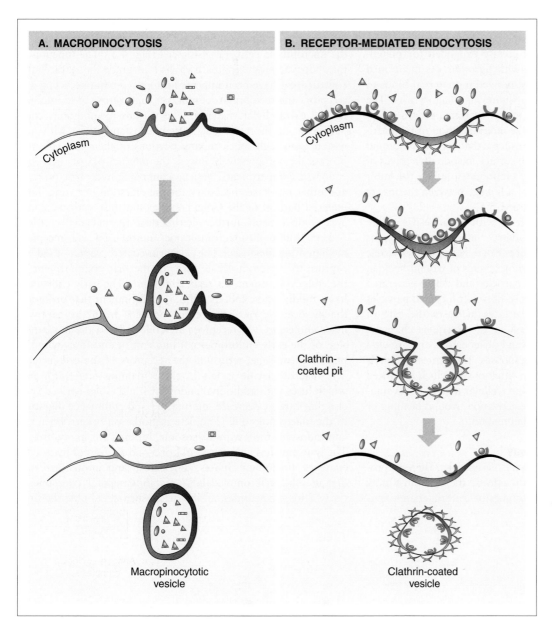

A. MACROPINOCYTOSIS

Cytoplasm

Macropinocytotic
vesicle

B. RECEPTOR-MEDIATED ENDOCYTOSIS

Cytoplasm

Clathrin-
coated pit

Clathrin-coated
vesicle

Figure 4-3

Macropinocytosis and Receptor-Mediated Endocytosis

(A) Macropinocytosis occurs when ruffles of the cell membrane close back against the plasma membrane, indiscriminately engulfing fluid and solutes in a macropinocytic vesicle. A receptor is not required. (B) Extracellular material can also be internalized by receptor-mediated endocytosis. Soluble molecules are bound by specific cell surface receptors and ligand–receptor pairs are targeted to clathrin-coated pits for internalization.

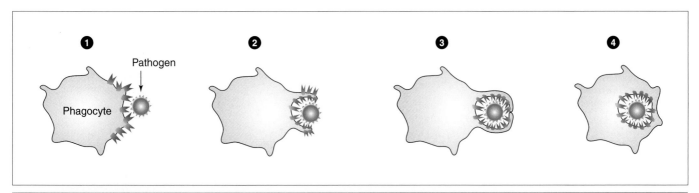

1 Pathogen

Phagocyte

2

3

4

Figure 4-4

Phagocytosis

(1) Pathogen attachment mediated by phagocyte receptors. (2) Extension of pseudopodia around the pathogen and sequential binding of adjacent receptors. (3) "Zippering up" of the pathogen followed by internalization via cytoskeletal-driven mechanisms. (4) Fusion of the phagocyte membrane to form a vesicle, completing phagosome generation.

Microbes and particles that do not interact with cell surface receptors on the phagocyte cannot be phagocytosed directly; however, the range of targets is greatly expanded for phagocytes when microbes are coated with *opsonins*. As mentioned in Chapter 3, an opsonin is a host-derived protein that binds to the exterior of a microbe and facilitates its phagocytosis. The opsonin on the microbe binds to an appropriate opsonin receptor on the phagocyte, triggering the internalization of the particle (Fig. 4-5). Common opsonin proteins that are readily bound by receptors on the phagocyte cell surface include derivatives of the complement proteins as well as immunoglobulins, the latter case being yet another example of a linkage between the innate and adaptive immune responses (see Ch.5).

Integrins are cell surface proteins with an adhesive nature that facilitates cell–cell or cell–matrix interactions. Integrins expressed on phagocytic cell surfaces can enhance phagocytosis. For example, *fibronectin* is important at sites of wound healing, and functions as an opsonin for particles and debris generated as a result of trauma. The binding of fibronectin-coated particles to integrins on phagocyte cell surfaces increases the rate of clearance of the debris. As we shall see in subsequent chapters, integrin functions extend well beyond phagocytosis and are vital to both the adaptive and innate responses. The expression levels of integrins are often increased in situations in which enhanced adhesion to other cells or to the intercellular matrix is advantageous, such as during lymphocyte activation, wound healing, or cell migration associated with inflammation.

iv) Endocytic Processing Pathway

Having engulfed the macromolecule or microbe, thereby protecting nearby cells from harmful effects, the cell must now dispose of its burden. Macropinosomes or clathrin-coated vesi-cles enter the *endocytic processing pathway*, an internal trafficking system for membrane-bound vesicles that is supported by the microtubule network of the cell (Fig. 4-6). The function of the endocytic processing pathway is to degrade the extracellular contents that have been imported into a cell while recycling any receptors and membrane components that were used to internalize the extracellular material. The various membranous compartments of the endocytic system constantly fuse with and fission from each other, mixing portions of their contents in a process that progressively lowers the internal pH of successive endocytic compartments. Vesicles containing hydrolytic proteins and other molecules necessary for the digestion of extracellular material bud off of the Golgi complex and fuse with the endocytic compartments, further altering their internal environments.

Let's follow a macromolecule internalized via receptor-mediated endocytosis. The clathrin-coated vesicle used to capture the macromolecule fuses to the first compartment of the endocytic system, an *early endosome*. The early endosome has a mildly acidic (about pH 6.5) environment that promotes the dissociation of the receptor proteins from the captured macromolecule. The receptors are collected in a tubular extension of the early endosome that buds off of and fuses with the plasma membrane, returning the receptors to the cell surface. The macromolecule is left in the remaining *late endosome*, which fuses with additional endosomes and Golgi-derived vesicles that further decrease internal pH and commence digestion of the macromolecule. Complete degradation occurs when the late endosome fuses with a lysosome, forming an *endolysosome*. The lysosome has an internal pH of 5 and a cargo of hydrolytic enzymes (including lipases, nucleases, and proteases) that degrade the macromolecule to its fundamental components (fatty acids, nucleotides, amino acids, sugars, etc.). At the final

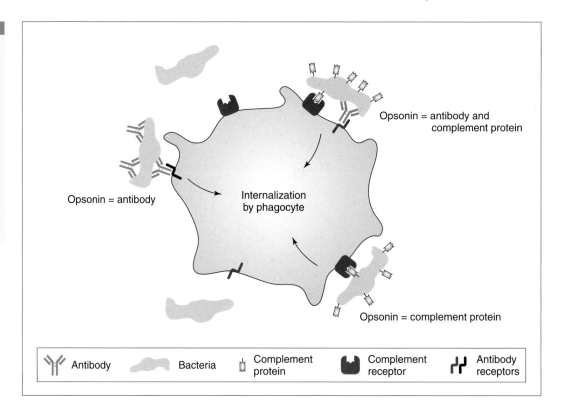

Figure 4-5

Opsonization
Opsonins are host-derived proteins that enhance phagocytosis by coating foreign targets destined for internalization. Antibodies and complement components can act as opsonins, either on their own or in combination, when the appropriate receptors are expressed on the phagocyte. The ability of different opsonins to enhance phagocytosis varies: antibody and complement protein > antibody > complement.

Opsonin = antibody

Opsonin = antibody and complement protein

Internalization by phagocyte

Opsonin = complement protein

Antibody   Bacteria   Complement protein   Complement receptor   Antibody receptors

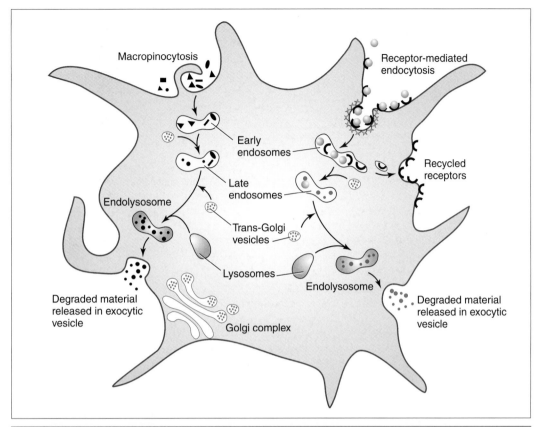

Figure 4-6

Model of the Endocytic Pathway
A variety of pathways exist for bringing material from outside the cell into the endocytic pathway. The endocytic pathway is composed primarily of early endosomes, late endosomes, and lysosomes. These compartments exist in a dynamic state, constantly fusing with macropinocytic or clathrin-coated vesicles, shuttling contents to the next compartment, and recycling internalized membrane receptors back to the cell surface. Many intracellular proteins are involved in the transfer of materials from one compartment to the next, and enzymes contained in vesicles budding from the Golgi complex contribute to the maturation of the compartments. In both macropinocytosis and receptor-mediated endocytosis, internalized vesicles are constantly fusing with early endosomes, in which the mildly acidic environment is sufficient to dissociate ligands from their receptors. These receptors are then recycled back to the cell membrane for further use. Intravesicular contents are concentrated as they make their way from mildly acidic early endosomes (pH 6.3–6.8) to late endosomes, to lysosomes (pH 5), where ultimate degradation occurs. The products of degradation are then released at the cell surface in exocytic vesicles.

stage, the products of digestion are collected in an *exocytic vesicle* which buds off of the endolysosome. In a process called *exocytosis*, the degraded material is expelled from the cell into the extracellular fluid by the fusion of the vesicle membrane with the plasma membrane.

Phagosomes are thought to be handled in a similar way, apparently joining the endocytic processing pathway at the late endosome stage (Fig. 4-7). After internalization of a particle in a phagosome, the actin-based framework is depolymerized via an unknown mechanism. The phagosome uses the microtubule-supported endocytic system to undergo a stepwise "maturation" to eventually form a *phagolysosome*. Maturation involves a poorly understood series of fusion and fission events with various Golgi-derived vesicles, intracellular vacuoles and granules, endosomal compartments, and lysosomes. These events are facilitated by the *annexin* proteins, which bind to membrane phospholipids, and by members of the Rab family

of small GTPases. The mixing of the contents of successively recruited vesicles contributes to changes in the internal protein composition of the maturing phagolysosome, influencing its ability to degrade the foreign entity. For example, the protein that functions as a "proton pump," a component necessary for complete microbial killing, is transferred to the maturing phagolysosome through fusion with a particular intracellular vacuole. As a result, the internal pH of the phagolysosome decreases to pH 5 or below lower. Failure of this transfer can result in decreased acidification of the interior of the phagosome and consequent survival of a phagocytosed bacterium.

In the case in which a phagocyte has engulfed an entire microbe and sequestered it in a phagosome, the microbe must be killed before the phagocyte can digest it. Killing within the phagosome results from the action of various enzymes and the action of inorganic chemicals with antibacterial activity, such as oxygen radicals and nitric oxide (see Box 4-2).

Figure 4-7

Model for the Maturation of Phagolysosomes
Phagosome maturation is a dynamic process of multiple fusion and fission events which are still not completely understood. The phagosome is transformed into a functional phagolysosome by receiving hydrolytic enzymes and other destructive molecules from intracellular vacuoles and the Golgi complex. At some point during this process, the phagosome traffics through the endocytic pathway and merges with the lysosome to become a mature phagolysosome. Degraded contents within the phagolysosome are released in exocytic vesicles.

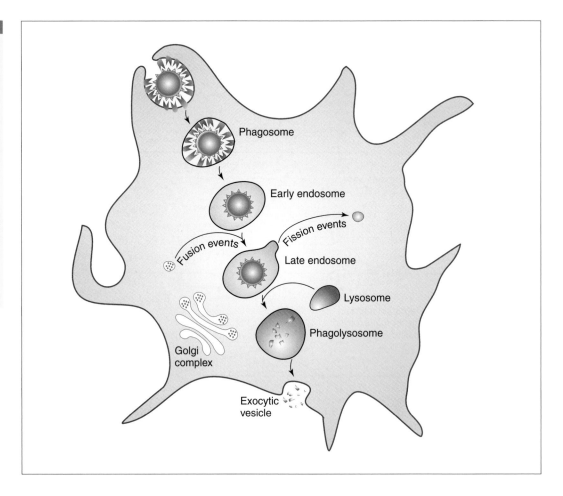

Microbes are continually evolving ways to avoid the defense mechanisms of the host cells, and escape from destruction by phagocytosis is no exception (see Ch.22). Certain pathogens are engulfed by phagocytes and contained in phagosomes, but are not killed by this process. Other pathogens, such as *Listeria monocytogenes*, can actually escape from the phagosome. There has been a surge of interest in research in this area because it has become clear that the recent emergence of drug-resistant *Mycobacterium tuberculosis* strains may stem from their acquired capacity to modify a phagosome from within and retard its maturation. A *Mycobacterium* with this capacity is thought to prevent the phagosome from fusing with vacuoles containing components needed to destroy the microbe; consequently, the microbe builds itself a safe and drug-free hiding place.

III. INFLAMMATION AS A RESPONSE TO INFECTION OR INJURY

Inflammation is a complex series of soluble and cellular events that occurs in response not only to pathogen attack, but also to tissue injury. Since it occurs non-specifically, and must be induced by either infection or cell damage, it constitutes part of the second phase of the immune response. It is important to note, however, that the induction of inflam-

mation and its concomitant antigen-clearing effects can also occur in response to cytokines released by cells of the adaptive immune response, so that temporally, inflammation is an element of both the second and third phases of the immune response. Therefore, inflammation is an integral part of both innate and adaptive immunity.

i) Clinical Signs of Inflammation

We are all familiar with the outward *clinical signs* of inflammation at the site of infection or injury: redness, swelling, heat, and pain. The localized heat and redness result from *vasodilation* (an increase in the diameter of local blood vessels), which allows increased blood flow into the affected areas. An increase in the *adhesion* of a variety of white blood cells to local blood vessel walls and increased *permeability* of the capillaries encourage an *influx of leukocytes* into the tissue site, causing swelling. The pain associated with inflammation results not only from the swelling, but also from the stimulation of pain receptors in the skin by the peptide mediator *bradykinin* (see later), which becomes activated in situations of tissue injury.

The early stages of inflammation are an initial attempt by the body to control a pathogen non-specifically (Fig. 4-8). Damaged tissue cells release molecules that attract leukocytes of the innate response that are circulating in the blood. These cells push through the blood vessel walls and enter the tissues

Box 4-2. Respiratory burst and killing by reactive intermediates

The principal antimicrobial weapon in the innate immune response is phagocytosis, and a key method for the killing of engulfed parasites is the *oxidation* of molecules on and in the pathogen. Oxidation is chemically defined as "the acceptance of electrons," and an oxidizing agent is an entity that "donates an electron." The oxidizing agents present within a phagosome are "radicals," which are chemically unstable atoms or groups of atoms with an unpaired electron. Microbial proteins present within the phagosome stabilize the radicals by accepting the unpaired electrons, but this oxidation event alters the properties of the protein so that it can no longer carry out its normal function. As a result, the microbe dies. As shown in the figure, inside a phagocyte, the formation of a phagosome around an engulfed "agent" results in the (1) activation of *protein kinase C* (PKC), which rapidly (2) phosphorylates *NADPH-dependent oxidases* located in the phagosomal membrane. The NADPH-dependent oxidases catalyze the (3) conversion of molecular oxygen (O_2) to the highly reactive *superoxide anion* (O_2^-) (4), a molecule that is extremely toxic to engulfed microbes. The explosive activation of the NADPH-dependent oxidases is called the *respiratory*, or *metabolic, burst*. The respiratory burst is characterized by an intense, rapid, and short-lived (up to 3 hours) increase in overall oxygen consumption (and concomitant generation of O_2^-) that occurs immediately after phagocytosis of a microbe. The term "respiratory" is used to describe the burst because it is accompanied by metabolic changes usually associated with mitochondrial respiration; however, respiratory mitochondrial enzymes are not involved.

The superoxide anion generated from molecular oxygen by the NADPH-dependent oxidase can be further reduced to reactive oxygen intermediates (ROIs) such as *hydrogen peroxide* (H_2O_2), *singlet oxygen* ($^1O_2$), and *hydroxyl radicals* (OH$^\bullet$), all potent oxidizing agents. *Myeloperoxidase* released from intracellular granules in the phagocyte reacts with Cl^- and H^+ ions in the presence of H_2O_2 to produce *hypochlorite* (HOCl), the key ingredient in bleach. The action of hypochlorite on organic amine residues in microbial proteins *halogenates* them and kills the pathogen. Singlet oxygen is effective in the breaking of double bonds in hydrocarbons, releasing products that can attack other microbial compounds by cross-linking

membrane proteins, inactivating enzymes, and generating more toxic intermediates. Some of these intermediates oxidize the thiol groups that are important in the enzymes of microbial biosynthesis and metabolism. Hydroxyl radicals can break hydrogen bonds, resulting in the degradation of nucleic acids.

Another powerful antimicrobial agent produced within phagocytes is *nitric oxide* (NO). Mouse macrophages activated by exposure to bacterial cell wall products such as lipopolysaccharide and stimulated by the cytokines interferon-γ or tumor necrosis factor express high levels of *inducible nitric oxide synthetase* (iNOS). This enzyme converts L-arginine to citrulline in a reaction producing large amounts of NO. This NO then reacts with other chemical species, such as the superoxide anion, to generate additional toxic molecules, including *reactive nitrogen intermediates* (RNIs). Some of these compounds are thought to interfere with aconitase, an enzyme of the citric acid cycle, and other microbial enzymes containing iron and sulfur atoms. Nitric oxide has also

been shown to inhibit viral replication in mouse cells.

How does the phagocyte protect itself from the reactive species it has loosed upon the ingested microbe, should the phagosome leak? For example, about 80% of the superoxide anion generated by the respiratory burst reacts at the cell membrane or in the phagosome; the remainder diffuses into the cytosol, where, theoretically, it could attack phagocyte proteins. The phagocyte has three enzymes that neutralize escaped reactive intermediates: *superoxide dismutase* (SOD; converts superoxide anion into hydrogen peroxide), *catalase* (degrades hydrogen peroxide to water and oxygen), and *glutathione peroxidase* (reduces hydrogen peroxide to water and oxidized glutathione). In macrophages, SOD is induced by the presence of increased oxygen partial pressure, ensuring that the control path is ready when the respiratory burst is induced. With information from Roitt I. and Peter J. (2001) "Essential Immunology," 10th edn. Blackwell Scientific Publ., Oxford.

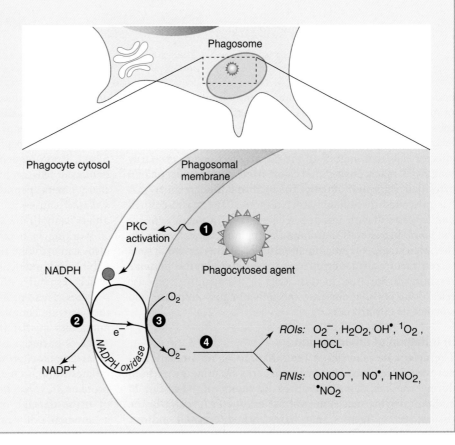

The Inflammatory Response
Tissue damage resulting from bacterial infection or trauma signals mediators of the plasma enzyme systems to increase endothelial cell permeability and adhesion and influx of leukocytes. A chemotactic gradient facilitates the migration of neutrophils, followed by macrophages, to the damaged site. Both types of innate cells non-specifically engulf and degrade foreign particles and secrete soluble mediators that recruit cells of the adaptive immune response to the site of tissue injury. The chemoattractant C5a is derived from complement activation.

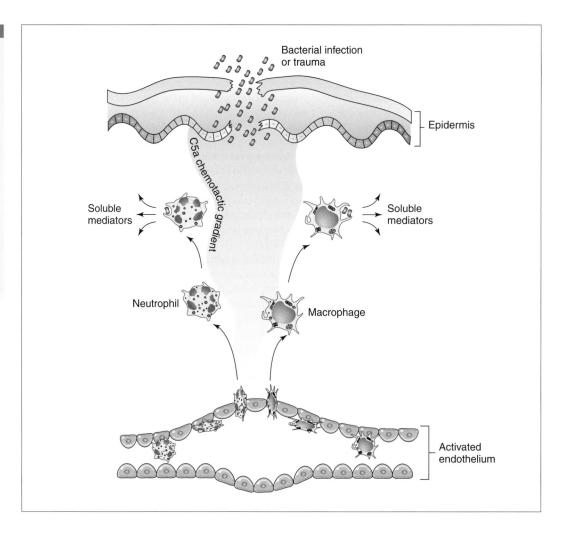

underlying the site of attack. Subsequent migration to the site of injury or infection requires *chemotaxis*, a process whereby cells move along a concentration gradient of a *chemotactic molecule*, or attractant factor (see Box 4-3). The innate effector cells that migrate to the site of infection are primarily long-lived macrophages and short-lived neutrophils, phagocytes that vigorously attempt to engulf the foreign entity that has provoked the inflammatory response. The cream-colored *pus* associated with infected wounds is an accumulation of the white blood cells that have died fighting the infection. In the later stages of inflammation, cells of the innate system secrete factors attracting effector cells of the adaptive immune response to the area. In turn, cells of the adaptive response release factors that increase the inflammatory response and enhance its effectiveness.

ii) Initiation of Inflammation

The actual mechanism of the *initiation* of inflammation is not well understood. It can be triggered by proteins released during the cell damage induced by injury or infection. In addition, roving macrophages that encounter bacteria in the tissues secrete factors that have both direct and indirect effects. Major inflammatory mediators, their sources, and their effects are listed in Table 4-1. Some factors act directly

and locally to induce the initial changes in blood vessel diameter and permeability, signaling that an inflammatory response is under way. Among these factors are the *kinins*, a family of small peptides that circulate in inactive form in the blood. Upon sequential activation by enzymes released from damaged cells or by enzymes of the blood clotting system, these potent peptide mediators (including *bradykinin* and *kallidin*) cause vascular dilation, increased vascular permeability, smooth muscle contraction, and pain.

Working in a more indirect way, activated macrophages also send cytokine signals in the form of tumor necrosis factor (TNF) and interleukin-6 (IL-6) to hepatocytes in the liver, which respond with the systemic production of *acute-phase proteins*. These proteins act somewhat like antibodies, in that they activate complement to destroy bacteria, but differ from antibodies in that they are non-specific and bind to a wide range of bacterial agents (see later). The serum concentration of acute-phase proteins rises rapidly in the early stages of inflammation. One of the most important acute-phase proteins is *C-reactive protein*, which appears within hours of tissue damage or infection. C-Reactive protein binds to a common cell wall component of bacteria and fungi and activates the complement system, resulting in the destruction of the microbe.

Box 4-3. Chemotaxis

The word "chemotaxis" is derived from the Greek *chemos*, meaning "juice" or "liquid," and *taxis*, meaning "movement." In the immunological context, chemotaxis is defined as the *directed* movement of cells in response to "juice," i.e., to chemical signals. (Compare this with *chemokinesis*, the stimulation of *random* leukocyte movement by chemical signals.) The site of an inflammatory response becomes filled with a soup of chemical messengers as injured cells release molecules signaling distress, and the inflammation results in the production of *chemotactic factors*. As these molecular signals diffuse into the tissues surrounding the site of inflammation, a *concentration gradient* is created, with the highest concentration occurring at the site of injury or attack and decreasing with distance from this area. Phagocytic cells (in particular) are equipped to recognize these signals and leave the tissues or blood circulation, traveling up the concentration gradient to the site of injury. By extending a pseudopod into the gradient, a migrating cell can detect a difference of as little as 0.1% in the concentration of chemoattractant at the cell's leading and trailing surfaces. The cell continues to move in the direction of the gradient as long as it detects an increase in factor concentration. Once at the site of injury or invasion (and at the peak of the gradient), these cells of the innate immune response secrete additional molecules that chemotactically attract lymphocytes of the adaptive immune response.

What are these *chemotactic factors*? A wide variety of low molecular weight molecules can act as *chemoattractants*, some of which affect only specific leukocyte populations. For example, the 77-amino acid *C5a* fragment, derived from the enzymatic cleavage of protein C5 of the complement system, is a powerful chemoattractant for certain white blood cells, especially neutrophils, eosinophils, and macrophages. Activation of the complement system by bacterial invasion, or by the release of proteases from injured cells, leads to the production of copious quantities of C5a. In the course of damage to cell membranes, blood coagulation factor XII stimulates the conversion of prekallikrein to *kallikrein*, another early inflammatory chemoattractant. Other molecules generated during blood clotting, such as *platelet-activating factor* (*PAF*), *fibrin peptide B*, and *thrombin*, attract phagocytes. The activation of macrophages,

Chemokine	Source	Chemotactic for
C-C SUBGROUP		
Macrophage activating factor (MAF)	Monocytes, macrophages, fibroblasts	Monocytes, macrophages, T cells
Monocyte chemotactic protein-1 (MCP-1)	Monocytes, macrophages, T cells	Monocytes
Macrophage inflammatory protein-1α (MIP-1α)	Monocytes, macrophages, neutrophils, endothelium	Monocytes, macrophages, T cells, B cells, basophils, eosinophils
Macrophage inflammatory protein-1β (MIP-1β)	Monocytes, macrophages, neutrophils, endothelium	Monocytes, macrophages, naive Tc cells, B cells
RANTES	T cells, platelets	Monocytes, memory T cells, eosinophils, basophils
C-X-C SUBGROUP		
Interleukin-8 (IL-8)	Monocytes, macrophages, endothelium, fibroblasts, neutrophils	Neutrophils, basophils, T cells
Neutrophil-activating protein (NAP-2)	Platelets	Neutrophils, basophils
Platelet factor-4 (PF-4)	Platelets	Neutrophils, fibroblasts

basophils, and mast cells leads to the increased metabolism of arachidonic acid, resulting in the production of *leukotriene B4* (*LTB4*), also chemotactic for neutrophils, eosinophils, and macrophages. Bacteria inherently supply chemotactic attraction, since neutrophils and macrophages bear receptors for the *N-formyl-methionine tripeptide* that is present at the N-termini of prokaryotic, but not eukaryotic, proteins. Release of these proteins during lysis of the bacterial cell wall sends the unmistakable and unique signal that bacteria are present.

The *chemokines* (<u>chemo</u>tactic cyto<u>kines</u>) are a group of over 40 small (~8–10 kDa) structurally homologous cytokines that have chemotactic activity. Specific chemokines are secreted by specific types of leukocytes, resulting in the attraction of a variety of cell types to the inflammatory site. These monomeric proteins can be categorized in several subgroups based on amino acid sequence. Members of the *cys-X-cys* (C-X-C) subgroup (see Table) are produced mainly by activated monocytes and macrophages (but also by endothelial cells, fibroblasts, and megakaryocytes). The CXC chemokines stimulate the migration primarily of neutrophils. Best studied among these is interleukin-8 (*IL-8*), released by activated monocytes, which attracts neutrophils and basophils to sites of inflammation. The *cys-cys* (C-C) chemokines are synthesized principally by activated T cells. Members of the CC subgroup tend to act more selectively on particular leukocyte subsets; for example, the *RANTES* factor acts preferentially on certain memory T cells and monocytes. In

Continued

Box 4-3. **Chemotaxis—cont'd**

general, neutrophils do not respond to CC chemokines. Other examples of selective action by chemokines are the stimulation of cytotoxic (but not helper) T cells by *macrophage inflammatory protein-1β* (MIP-1β), and the chemoattraction of monocytes alone by *monocyte chemotactic protein-1* (MCP-1), a chemokine synthesized by several different types of leukocytes. All chemokines bind to heparan sulfate in the host cell extracellular matrix and in the endothelial cell wall, which has the effect of

displaying these chemoattractant molecules where leukocytes can recognize them. Several different chemokine receptors have been identified on various leukocyte subsets, some of which appear to bind more than one chemokine. More on chemokines appears in Chapter 17.

It remains a mystery as to why there should be so many different chemotactic signals. Particularly in the case of the selectively acting chemokines, the organism may need to exert fine control over both the leukocyte traffic to

the site of injury and the release of chemoattractants acting in the later stages of inflammation. When there is no further need for immune response cells to be drawn to the site of injury or infection, the concentration gradient of chemoattractant is removed, both by inhibition of the chemotactic molecules (by factors released in the complement cascade, among other inhibitors) and by termination of their production by inflammatory cells (since the stimulus furnished by the injury or pathogen has been removed).

Table 4-1 **Major Inflammatory Mediators**

Mediator Type	Origin	Action
PLASMA ENZYME SYSTEMS		
Bradykinin, kallidin	Kinin system	Induces vasodilation, smooth muscle contraction, increased capillary permeability, pain
Fibrin peptides, heparin	Clotting system	Increases vascular permeability, neutrophil and macrophage chemotaxis
C3a	Complement system	Promotes mast cell degranulation, smooth muscle contraction
C5a	Complement system	Promotes mast cell degranulation, neutrophil and macrophage chemotaxis, neutrophil activation, smooth muscle contraction, increased capillary permeability
ACUTE-PHASE PROTEINS		
C-reactive protein Mannose-binding lectin	Hepatocytes	Induce complement fixation, opsonization
CYTOKINES		
IL-1	Mononuclear phagocytes	Activates inflammation and coagulation, stimulates liver to produce acute-phase proteins
IL-6	Mononuclear phagocytes, endothelial cells, monocytes, lymphocytes	Increases acute-phase proteins
IL-8	Mononuclear phagocytes, endothelial cells, monocytes, lymphocytes	Increases acute-phase proteins, neutrophil chemotaxis
TNF	Macrophages, T cells	Activates inflammation and coagulation, increases acute-phase proteins
LIPOXYGENASE PATHWAY		
Leukotrienes	Phospholipids derived from macrophage, monocyte, neutrophil, and mast cell membranes	Promotes neutrophil chemotaxis, increased vascular permeability, smooth muscle contraction
Prostaglandins	Phospholipids derived from macrophage, monocyte, neutrophil, and mast cell membranes	Promotes vasodilation, increased vascular permeability, neutrophil chemotaxis
LEUKOCYTE PRODUCTS		
Histamine	Mast cells, basophils	Increases vascular permeability, smooth muscle contraction, chemotaxis
5-Hydroxytryptamine	Platelets, mast cells (rodents)	Increases vascular permeability, smooth muscle contraction
Platelet-activating factor	Platelets, neutrophils, monocytes, macrophages, mast cells	Induces release of platelet mediators; increases vascular permeability, smooth muscle contraction, neutrophil activation

The *leukotrienes* and *prostaglandins* comprise another family of powerful inflammatory mediators. These molecules are originally derived from the lipoxygenase-mediated degradation of phospholipids in the membranes of macrophages, monocytes, neutrophils, and mast cells. The products of phospholipid degradation are subsequently converted to arachidonic acid, which can be metabolized in two ways. Metabolism of arachidonic acid via the cyclooxygenase pathway produces prostaglandins, whereas processing via the lipoxygenase pathway yields leukotrienes.

iii) Cellular Adhesion and Migration during Inflammation

Inflammation also involves changes in leukocyte migration patterns. During the inflammatory response, the extravasation of leukocytes between the blood and the tissues is significantly increased as large numbers of leukocytes react to chemical messengers emanating from the injured tissues and rush to the body's defense (Fig. 4-9A). In response to tissue damage or to cytokines released during an inflammatory response (including TNF and IL-1), the endothelial cells are activated, resulting in the "upregulation" of the expression of selectins and other cell

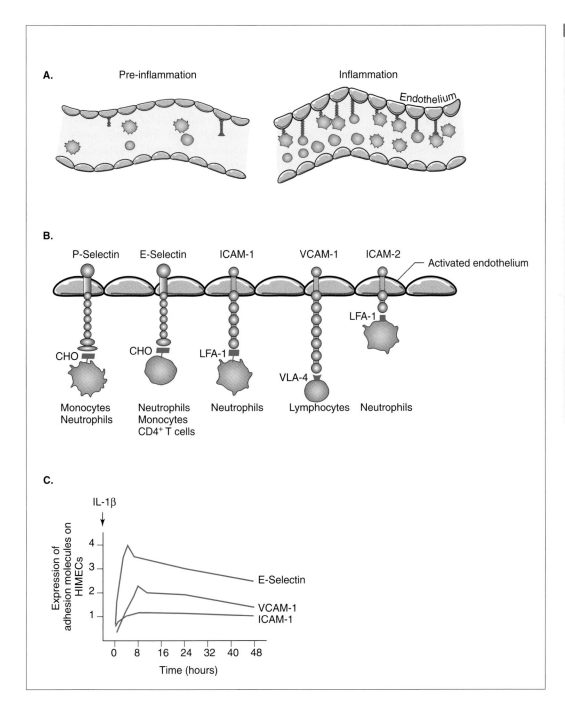

Figure 4-9

Endothelial Adhesion Molecules during Inflammation
(A) The extent of leukocyte adhesion and expression of adhesion molecules on endothelial cells before and after inflammation. (B) Endothelial adhesion molecules and the moieties to which they bind on leukocytes. (C) The changes over time in expression of particular endothelial adhesion molecules following endothelial stimulation. CHO, Carbohydrate; HIMECs, human intestinal microvascular endothelial cells. Part C Haraldsen G. *et al.* (1996). Cytokine-regulated expression of E-selectin, intercellular adhesion molecule-1 (ICAM-1), and vascular cell adhesion molecule-1 (VCAM-1) in human intestinal microvascular endothelial cells. With information from *The Journal of Immunology* **156**, 2558–2565.

adhesion molecules (CAMs) (described in Ch.3) (Fig. 4-9B). The "stickiness" of the endothelial cells is thereby enhanced, increasing the chance that passing leukocytes will stick to a capillary wall (margination) in an area where they are needed. For example, activated endothelial cells increase their expression of P-selectin (CD62P) and E-selectin (CD62E); these membrane proteins bind oligosaccharide moieties that are often present as part of adhesion proteins expressed on the surfaces of leukocytes (Fig. 4-9C). Similarly, exposure to chemotactic factors greatly increases the expression of L-selectin on neutrophils. Tissue damage also increases the production of *platelet-activating factor* (PAF) on the endothelial cell surface. When a neutrophil that is rolling on the surface of an activated endothelial cell encounters PAF, there is a stimulation of the adhesive properties of the LFA-1 (CD11a/CD18) and Mac-1 (CD11b/CD18) molecules positioned in the neutrophil membrane. These β2-integrins then bind tightly to intercellular adhesion molecules ICAM-1 (CD54) and ICAM-2 (CD102) on the endothelial cell surface. Note that because this stimulation of β2-integrin adhesion occurs only in PAF-activated neutrophils, the neutrophils are directed only to locations in the venule close to sites of tissue injury, and are less likely to bind to endothelial cells that have not responded to inflammatory signals. New sets of adhesion molecules that are not constitutively expressed, including the vascular cell adhesion molecule VCAM-1 (CD106), which binds to VLA-4 (CD49d/CD29), are also expressed on the activated endothelium.

As mentioned previously, leukocytes that have extravasated through the endothelium follow a concentration gradient of chemoattractants to the site of injury or infection. Receptors on the surfaces of phagocytic cells commence binding to ligands on the pathogen, and pathogen destruction ensues. Pathogens that are neither directly destroyed by complement, nor bound by acute-phase proteins, nor phagocytosed, must be targeted by the recognition mechanisms of the adaptive immune response, and macrophages begin the production of factors that will attract cells of the adaptive immune response to the area.

iv) Tissue Repair during Inflammation

The increased *permeability* of blood vessels that occurs in an inflammatory response is particularly important for the control of infection and for the repair of damage, since large molecules such as antibodies arising from the adaptive response and enzymes of the blood clotting system are thus better able to enter the tissues during the latest stages of inflammation when they are required. Upon entering a damaged tissue, enzymes of the *plasmin* or *fibrinolytic* system (the blood clotting cascade) can begin depositing *fibrin* to form a blood clot and commence wound healing. The fibrinolytic system shares several activation steps with both the complement and the kinin systems.

v) Auxiliary Cells in Inflammation

Certain *auxiliary cells* are also important in the inflammatory response. These include *platelets* and *mast cells*, which are drawn to the site of tissue injury or invasion by factors secreted by macrophages or damaged cells. Platelets are necessary for the completion of wound healing, while mast cells bear cytoplasmic granules containing heparin and vasoactive amines. During an inflammatory response, mast cells are triggered to degranulate by the presence of kinins. Heparin is a blood clotting inhibitor that helps maintain the required influx of soluble factors and leukocytes into the affected area in the early stages of the response. Vasoactive amines, such as *histamine* and *5-hydroxytryptamine*, induce vasodilation and increase vascular permeability. The vasoactive amines can mediate their effects immediately upon mast cell degranulation because they are "preformed" and do not require new metabolic synthesis by the cell; the fast action of these preformed mediators is crucial for initiating the inflammatory response. On the other hand, *leukotrienes*, which are responsible for attracting and activating other cells, do not pre-exist and must be newly synthesized, and so function later in the inflammatory response.

Mast cells and their vasoactive amines also play a prominent role in a specialized type of immune response involved in both antiparasite immunity and allergic reactions. Cellular and vascular responses similar to those observed in an inflammatory response, and involving many of the same mediators, are induced in these situations. However, the mast cells are triggered in a different process that involves IgE antibody, making mast cells participants in both adaptive and innate immunity. An in-depth discussion of the IgE-mediated responses in antiparasite immunity and allergy can be found in Chapters 22 and 28, respectively.

B. Pattern Recognition in Innate Immunity

It was not until the late 1990s that it became clear how the inducible innate response recognizes a broad range of target molecules and particles as foreign. Like the adaptive response, the induced innate response distinguishes between self and nonself, and often relies on cell-mediated effector mechanisms to eliminate the nonself entities. However, rather than binding to unique antigenic structures or shapes as do the receptors of lymphocytes, the molecules of the induced innate response exhibit a broad *pattern-based recognition* of evolutionarily conserved structures derived from foreign entities (Fig. 4-10). Furthermore, these *pattern recognition molecules* (PRMs) may be expressed by many different cell types, such as NK cells, macrophages, and neutrophils. Some PRMs are present as surface receptors, and so are called *pattern recognition receptors* (PRRs). PRRs recognize patterns of structures derived from components or products that are common to a wide variety of microbes but are not usually present in host cells. These ligands for PRRs are sometimes called *pathogen-associated molecular patterns*, or PAMPs. Other PRMs are soluble (free-floating) proteins that may either activate the complement system or act as opsonins to enhance phagocytosis of the antigen to which they bind. Certain receptors expressed by NK, NKT, and γδ

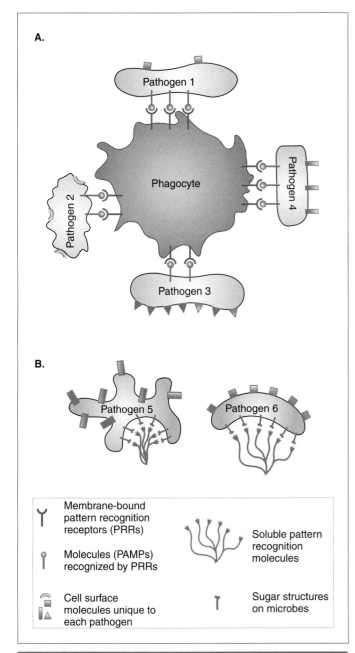

A.

Pathogen 1

Phagocyte

Pathogen 2

Pathogen 4

Pathogen 3

B.

Pathogen 5

Pathogen 6

Y Membrane-bound
pattern recognition
receptors (PRRs)

↑ Molecules (PAMPs)
recognized by PRRs

🖰 Cell surface
molecules unique to
each pathogen

⋔ Soluble pattern
recognition
molecules

⋎ Sugar structures
on microbes

Figure 4-10

General Pattern Recognition
Pattern recognition molecules can be membrane-bound receptors
(A) or soluble proteins (B) that bind to structures found on a variety of
pathogens. In the case of PRRs, the binding of the pathogen initiates
internalization and destruction of the invader. In contrast, soluble pattern
recognition molecules activate the complement system and serve as
opsonins for phagocytosis.

T cells can also be considered PRMs, in that ligand recognition
by these proteins lacks the fine specificity of αβ TCRs and
BCRs. Several examples of PRMs are presented in Table 4-2.
Antigen clearance strategies based on pattern recognition are
the subject of much current research, and several examples are
discussed in the following sections.

I. PATTERN RECOGNITION BY PRRs

The structures recognized by PRRs include pathogen cell wall
components, viral DNA and RNA genomes, bacterial DNA
and proteins, and microbial lipoproteins. In addition to being
found on mammalian phagocytes, some PRRs are also found
on the functionally equivalent hemocytes of invertebrates,
reinforcing the concept that the innate immune system is phylogenetically very old (see Ch.21). Why is recognition by
PRRs of the innate system more broad than that of the antigen receptors of the adaptive immune system? The genes
encoding the B and T cell antigen receptor proteins are not
single entities, but instead are made up of gene segments that
are rearranged and recombined during B and T cell development: this is the process of somatic gene rearrangement. No
such somatic gene rearrangement occurs in the cells expressing PRR genes, so that the amino acid sequences of PRRs are
derived from genes specified inflexibly in the germ line. Thus,
the cells of the innate system exhibit a comparatively limited
repertoire of receptor specificities, usually directed against
conserved features that are likely to be shared by many classes
of pathogens. In contrast, the flexibility of lymphocyte antigen receptor genes enables the adaptive immune system to
develop a huge repertoire of receptors, including those that
will also recognize protein antigens, no matter how rare.

i) CD14 Coreceptor

Before the discovery and characterization in the 1990s of the
Toll-like receptors (TLRs; see next section), there was evidence
that a macrophage could be activated by the binding of bacterial
lipopolysaccharide (LPS) to the CD14 cell surface receptor.
CD14 is a 55-kDa protein expressed primarily on monocytes
and macrophages. It lacks a transmembrane domain and is
instead usually bound to the plasma membrane by a glycophos-
phoinositol (GPI) anchor, although it does occur in soluble
forms. The LPS-mediated stimulation of macrophages results in
dramatically upregulated transcription of the genes encoding
many proinflammatory cytokines, including TNF and IL-1. [See
Box 17-2 (Ch.17) for more on LPS and its inflammatory
effects.] Historically, this interaction of CD14 with LPS was the
first example of a cell surface receptor able to recognize and
respond to a structural pattern common to multiple pathogens.
Until the late 1990s, it was a mystery how CD14, which lacks a
transmembrane domain needed for intracellular signaling, could
deliver the signals required for transcriptional activation. It was
suspected that CD14 must also interact with a transmembrane
domain-containing coreceptor to mediate LPS recognition and
initiate intracellular signaling. The major PRR, known as TLR4,
was then found to be the molecule fulfilling this role.
Interestingly, CD14 does not actually bind bacterial LPS directly.
Instead, a lipid transfer protein called *LPS-binding protein*
(LBP), which is found in the blood plasma, facilitates LPS binding to a site on CD14 prior to interaction of CD14 with TLR4.

ii) Toll-like Receptors

The *Toll* gene was originally discovered in *Drosophila*, where
it was first shown to control dorsal–ventral patterning during
embryonic development of the fly. The protein product of the

Table 4-2 Selected Pattern Recognition Molecules of the Innate System

Molecule	Location	Ligand	Function
CD14	Monocytes, macrophages, PMNs	LPS; numerous microbial cell wall components	LPS sensitivity; clearance of microbes; proinflammatory cytokine induction
Toll-like receptor family	Splenic and peripheral blood leukocytes	TLR2 = bacterial peptidoglycan, techoic acid TLR3 = viral dsRNA TLR4 = bacterial LPS TLR5 = bacterial flagellin TLR7/8 = viral ssDNA TLR9 = bacterial DNA with unmethylated CpG	Induction of IL-1, inflammatory cytokines, and costimulatory molecules
Scavenger receptors (A, B, and C)	Tissue macrophages, hepatic endothelial cells, high endothelial venules	Bacterial and yeast cell walls	Clearance of LPS and microbes; adhesion
NK activatory receptors	NK cells	Self or non-self protein antigens induced on stressed, transformed, or virus-infected host cells	Killing of infected, stressed, or transformed host cells; production of inflammatory cytokines
γδ TCRs	γδ T cells	Native proteins or unprocessed peptides derived from microbial, viral, or damaged host cells	Production of cytokines activating NK cells and macrophages; induction of antimicrobial compounds; production of chemokines
NKT semi-invariant TCR	NKT cells	Glycolipids presented on CD1d	Cytokine secretion and perhaps cytolysis
Collectins (MBl, lung surfactant proteins A and D)	Plasma proteins produced by hepatocytes	Microbial cell wall polysaccharides	Binding of C1q receptor; activation of complement; promotion of phagocytosis; modulation of CD14-induced cytokine production
Acute-phase protein (C-reactive protein)	Plasma protein produced by hepatocytes	Microbial polysaccharides	Activation of complement; enhancement of phagocytosis
Natural antibodies	Secreted by CD5$^+$ B cells	Bacterial, viral, and fungal components; host nucleic acids and other self-components	Enhancement of phagocytosis; clearance of pathogens; protection of fetus and neonate

LPS, lipopolysaccharide; PMNs, polymorphonuclear cells; dsRNA, double-stranded RNA; ssDNA, single-stranded DNA; CpG, cytosine-phosphate-guanine dinucleotide; MBL, mannose-binding lectin.

Toll gene was then identified as a transmembrane receptor important for antifungal immunity in adult *Drosophila*. The ligand for the *Drosophila* Toll receptor was found to be an invertebrate protein encoded by the Spätzle gene. In the mid-1990s, mammalian researchers took an interest in the *Drosophila* Toll receptor and its ligand because the cytoplasmic tail of the *Drosophila* Toll receptor is homologous to the cytoplasmic region of the IL-1 receptor in vertebrates. IL-1 is a cytokine that promotes the inflammatory response, and is a key inducer of the fever component of that response in vertebrates (see later). In addition, the *Drosophila* Toll receptor uses an intracellular signaling pathway that is homologous in many respects to a key signal transduction pathway in vertebrates that results in the activation of the transcription factor NF-κB. Engagement of the IL-1 receptor in vertebrates triggers signal transduction using the NF-κB pathway, which results in inflammation and pleiotropic effects on other immune system cells (see Ch.17). More on the evolution of Toll receptors can be found in Chapter 21, 'Comparative Immunity.'

At least 11 human homologues of Toll (designated TLRs, for "Toll-like receptors") have been identified to date. In both mice and humans, the TLRs are broadly expressed on a range of leukocytes and on some nonleukocyte cell types. In general, monocytes/macrophages and neutrophils show the highest expression of TLRs, with lower levels present on immature DCs and NK cells. Some T and B cell subsets may also express TLRs. Different TLRs recognize different microbial structures. For example, TLR2 recognizes the lipoteichoic acid and peptidoglycan of gram-positive bacterial cell walls, while TLR3 binds to the double-stranded

RNA of viruses. As mentioned previously, TLR4 is activated by the LPS of gram-negative bacterial cell walls, whereas TLR5 responds to bacterial flagellin. TLR7 and TLR8 bind to viral single-stranded RNA, and TLR9 recognizes stretches of bacterial DNA containing unmethylated CpG dinucleotides. Intriguingly, it seems that TLR7, TLR8, and TLR9 may also occur in the membranes of endosomal vesicles within cells, perhaps to assist in the detection of intracellular pathogens. TLR1 and/or TLR6 may cooperate with TLR2, and TLR10 may have a function related to that of TLR9. The ligand of TLR11 is unknown but knockout mice lacking this receptor are highly susceptible to infections with uropathogenic bacteria. In any case, the binding of the correct class of pathogen components to a TLR, alone or in tandem with another TLR, stimulates the receptor and initiates the intracellular signaling that leads to an inflammatory response.

The TLRs that have been studied in depth are transmembrane receptors with *leucine-rich repeat* (LRR) domains in their extracellular portions (like many other PRRs). The intracellular portions of the TLRs contain regions homologous to the IL-1 receptor and interact with many of the same signal transducers. Engagement of the TLRs leads to NF-κB activation and the production of the proinflammatory cytokines IL-1, IL-6, IL-8, IL-12, and TNF. TLR engagement on APCs also leads to the upregulation of costimulatory molecules and the migration of these cells to the lymph nodes for participation in adaptive immune responses. For example, when the TLRs of an immature DC are engaged, the cell becomes activated, commences maturation, and migrates with its load of antigen to the lymph nodes. The mature DC then processes and presents peptide antigen to naive T cells. TLR engagement on macrophages also stimulates inducible nitric oxide synthase (iNOS) induction and the production of reactive nitrogen intermediates involved in microbial killing. In addition, just as the stimulation of Toll in *Drosophila* results in the production of antifungal peptides, activation of TLR2 or TLR4 in human respiratory or intestinal epithelial cells leads to the production of β-defensin-2 (an antimicrobial peptide).

Other types of pathogen ligands appear to be capable of binding to TLR2 and TLR4. TLR2 is the PRR that detects the GPI anchor linking many proteins to the surface of the parasite *Trypanosoma cruzi* (see Ch.22), whereas TLR4 recognizes the fusion protein of the respiratory syncytial virus (see Ch.22). Significantly, TLR4 and TLR2 may also be able to respond to self-molecules induced when the host is under attack. For example, the stress-induced heat shock protein HSP60 is recognized by both TLR4 and TLR2, and a form of fibronectin expressed in situations of tissue injury binds to TLR4. Two other mammalian stress proteins, HSP70 and a molecular chaperone protein known as gp96, also induce intracellular signaling and cytokine secretion by mechanisms involving TLR2, TLR4, and possibly CD14. It is thought that this recognition of endogenous host stress ligands released from dying cells may be the means by which the innate immune system detects traumatic tissue injury (when no pathogen is directly involved).

iii) Scavenger Receptors

"Scavenger receptors" (SRs) are expressed primarily on phagocytes and also recognize foreign entities by pattern recognition. SRs were originally defined by their ability to bind modified low-density lipoproteins (LDLs), such as LDL molecules that have been oxidized or acetylated during cell injury or apoptosis. Scavenger receptors have since been shown to bind a wide variety of lipid-related ligands. In addition to clearance of microbial pathogens, the function of scavenger receptors in vertebrates may be to facilitate wound recognition and phagocytic disposal of host cell debris by macrophages at wound sites. "Nonself" forms of lipids appear not only on microbial cells, but also on damaged host cells, apoptotic cells, and senescent (old) cells. In all of these cases, the host cells have undergone deleterious changes to their membranes such that internal phospholipids have become exposed. In this context, the receptors recognize molecular patterns that are not present on normal healthy host cells.

Several classes of scavenger receptors have been identified, including the A, B, and C classes (Fig. 4-11). Scavenger receptors of the A class (SR-A) were the first macrophage SRs to be identified, and although they are found primarily on monocytes and tissue macrophages, they can also occur on some dendritic cells and specialized endothelial and smooth muscle cells. SR-A molecules are trimeric glycoproteins containing a collagenous extracellular domain (ED) that mediates binding to a broad array of polyanionic ligands, including those that occur on the surfaces of both gram-positive and gram-negative bacteria. In some cases, different class A receptors can be generated by alternative splicing of mRNA from the same gene. Mutant mice lacking SR-A molecules are more susceptible to infections with *Staphylococcus aureus* and other bacteria.

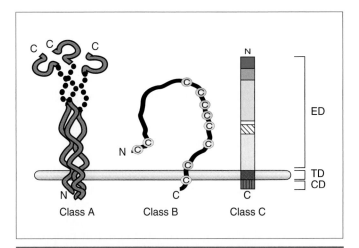

Figure 4-11

Scavenger Receptor Structures
ED, Extracellular domain; TD, transmembrane domain; CD, cytoplasmic domain; C, carboxy terminal; N, amino terminal; ©, cysteine residue.
Adapted from Pearson A. M. (1996). Scavenger receptors in innate immunity. *Current Opinion in Immunology* **8**, 20–28.

The class B scavenger receptors (SR-B) are found on monocytes, macrophages, hemocytes, B lymphocytes, capillary endothelial cells, platelets, and adipocytes. SR-B molecules recognize a different group of primarily lipid ligands, including phosphatidylserine and a *Plasmodium* parasite that infects red blood cells. Recognition by SR-B molecules may depend, in some cases, on the presence of negatively charged chemical groups in the ligand. An interesting feature of the large extracellular domain of SR-B molecules is the presence of many cysteine residues.

To date, class C scavenger receptors (SR-C) have been identified only on *Drosophila* hemocytes. Like the other SRs, the one SR-C class molecule isolated to date binds many polyanionic ligands with high affinity; however, it is unusual in that it can also recognize uncharged microbial glucan molecules. Scientists working in the scavenger receptor field feel that there are many more SRs to be identified, and probably many more ligands and modes of action to be defined.

The intracellular signaling pathways used by scavenger receptors to transduce the message that a ligand has been bound are just beginning to be dissected. Some scientists believe that the signaling initiated by SRs is limited to inducing endocytosis and phagocytosis, but others feel that more is going on, because there is some evidence that SRs can use many of the same elements that function in the more completely elucidated B and T lymphocyte signaling pathways (see Chs.9 and 14, respectively). For example, the binding of a particular LDL molecule to an SR-A molecule results in the rapid phosphorylation of the Lyn kinase (known to be involved in lymphocyte signaling), triggering its activation. Lyn may also be physically associated with the SR-A cytoplasmic tail. The activity of another central intracellular signaling molecule called MAP kinase (mitogen-activated protein kinase) has also been shown to be stimulated by the binding of LDL to SR on macrophage-like cells. Similarly, engagement of an SR-B molecule has been associated with the activation of another signal transducer called Ras. These matters remain under investigation.

II. PATTERN RECOGNITION BY RECEPTORS OF NK, NKT, AND $\gamma\delta$ T CELLS

i) NK Activatory Receptors

It is germane in the context of pattern recognition to remind the reader of NK cells, which are cytolytic lymphoid cells with effector actions that are deemed to be part of the innate response. The NK activatory receptors recognize certain self and nonself protein ligands that are upregulated or induced on virus-infected cells, tumor cells, and host cells undergoing stress responses. In keeping with what might be their intermediate evolutionary position as primitive forebears of cells of the adaptive immune response, the ligands for some of these NK activatory receptors are MHC class I proteins and other very similar proteins. NK cells are discussed in detail in Chapter 18.

ii) $\gamma\delta$ T Cell Receptors

We would like to expand slightly at this point on the nature of the antigen receptors of $\gamma\delta$ T cells. The majority of T cells in the body carry antigen receptors that are composed in part of two polypeptides, known as the "T cell receptor α (TCR)" and "T cell receptor β (TCR)" chains (see Ch.12). These "$\alpha\beta$ T cells" (which may be either CD4$^+$ or CD8$^+$ cells) recognize specific antigenic peptides complexed to MHC molecules, as described briefly in Chapter 2. As alluded to in Chapter 3, a minority of T lymphocytes are called "$\gamma\delta$ T cells" because they bear TCRs in which the antigen-binding site is made up of two different polypeptides, the TCRγ and TCRδ chains; the TCRα and TCRβ chains are never present on these cells. Unlike the wealth of information available on the development and functions of $\alpha\beta$ T cells, much less is known about the development or function of $\gamma\delta$ T cells. Recent evidence suggests that they may have evolved prior to the evolution of the highly specific $\alpha\beta$ T cells, since $\gamma\delta$ T cells appear to recognize certain antigens that are more broadly distributed over a relatively wide range of pathogens. Furthermore, unlike $\alpha\beta$ T cells, $\gamma\delta$ T cells do not require the involvement of MHC molecules or the processing of protein antigens by APCs to recognize ligands. $\gamma\delta$ T cells have been shown to be capable of binding intact proteins, non-MHC-associated peptides, and even some nonpeptide antigens, such as phosphorylated nucleotides. In view of these antigen recognition properties, some researchers consider $\gamma\delta$ T cells to be a class of cells intermediate between B cells, which recognize native protein antigens, and $\alpha\beta$ T cells, which recognize only processed peptides in association with MHC molecules.

The $\gamma\delta$ TCR apparently lacks the fine specificity of the $\alpha\beta$ TCRs and recognizes ligands in a way that is similar to ligand recognition by PRRs of conventional innate immune system cells. Ligands recognized by $\gamma\delta$ T cells include peptides of microbial or viral origin and those generated by damaged host cells, similar to the range of ligands recognized by PRRs. Other evidence has suggested that $\gamma\delta$ T cells may also express some conventional PRRs, such as CD14, and may carry inhibitory receptors similar to those on NK cells. Thus, $\gamma\delta$ T cells may represent another type of "bridging" cell, constituting a link between the innate and adaptive immune responses. What is known about the development and function of $\gamma\delta$ T cells is described in detail in Chapter 18.

iii) NKT Cell TCR

We introduced NKT cells briefly in Chapter 3. The reader will recall that these cells have characteristics of both NK and T cells. However, while NKT cells express an $\alpha\beta$ TCR, it is almost always made up of the same TCRα chain coupled to one of a small collection of TCRβ chains: the enormous diversity of the $\alpha\beta$ TCRs expressed on CD4$^+$ and CD8$^+$ T cells is not seen. The restricted TCRs of the NKT cell population are thought to recognize glycolipid ligands presented not by MHC class I or II molecules, but rather by an MHC-like antigen-presenting molecule called CD1d. The natural ligands activating NKT cells *in vivo* may include glycolipids released from damaged host cells, or from bacterial, viral, or parasitic pathogens. Because of the rapidity of their response and the fact that their TCRs act much like PRRs, NKT cells are also considered sentinels of the innate response. The development and function of NKT cells are described in Chapter 18.

III. PATTERN RECOGNITION BY SOLUBLE MOLECULES

i) Collectins
The collectins are soluble effector proteins that also possess the property of pattern recognition. These molecules are free-floating in the blood and other body fluids. The name "collectins" is derived from the fact that these proteins feature two types of domains fused together: a tail domain characteristic of <u>col</u>lagen molecules and a multimeric globular <u>lectin</u> (carbohydrate-binding) domain (Fig. 4-12). Examples of these soluble defenders include *lung surfactant protein-A*

and -D (SP-A, SP-D) and mannose-binding lectin (MBL). [MBL was formerly known as *serum mannose-binding protein* (sMBP, or just MBP).] The collectins are thought to recognize and interact preferentially with carbohydrate structures that are arranged in distinct patterns on the microbial cell surface. The amino acid sequence of the carbohydrate recognition domain in a given collectin protein defines the repertoire of carbohydrate structures (and hence, the spectrum of pathogens) it "sees" and can bind. This feature probably evolved because carbohydrates in microbial cell walls are clearly distinct from those in eukaryotic cells, leading to the ready identification of nonself molecules. For example, mammalian glycoproteins usually exhibit side chains with galactose and sialic acid as their most distal residues. The chemical structure of these side chains does not fit within the recognition domain of MBL. Hence, cells displaying mammalian glycoproteins (self) are ignored by MBL,

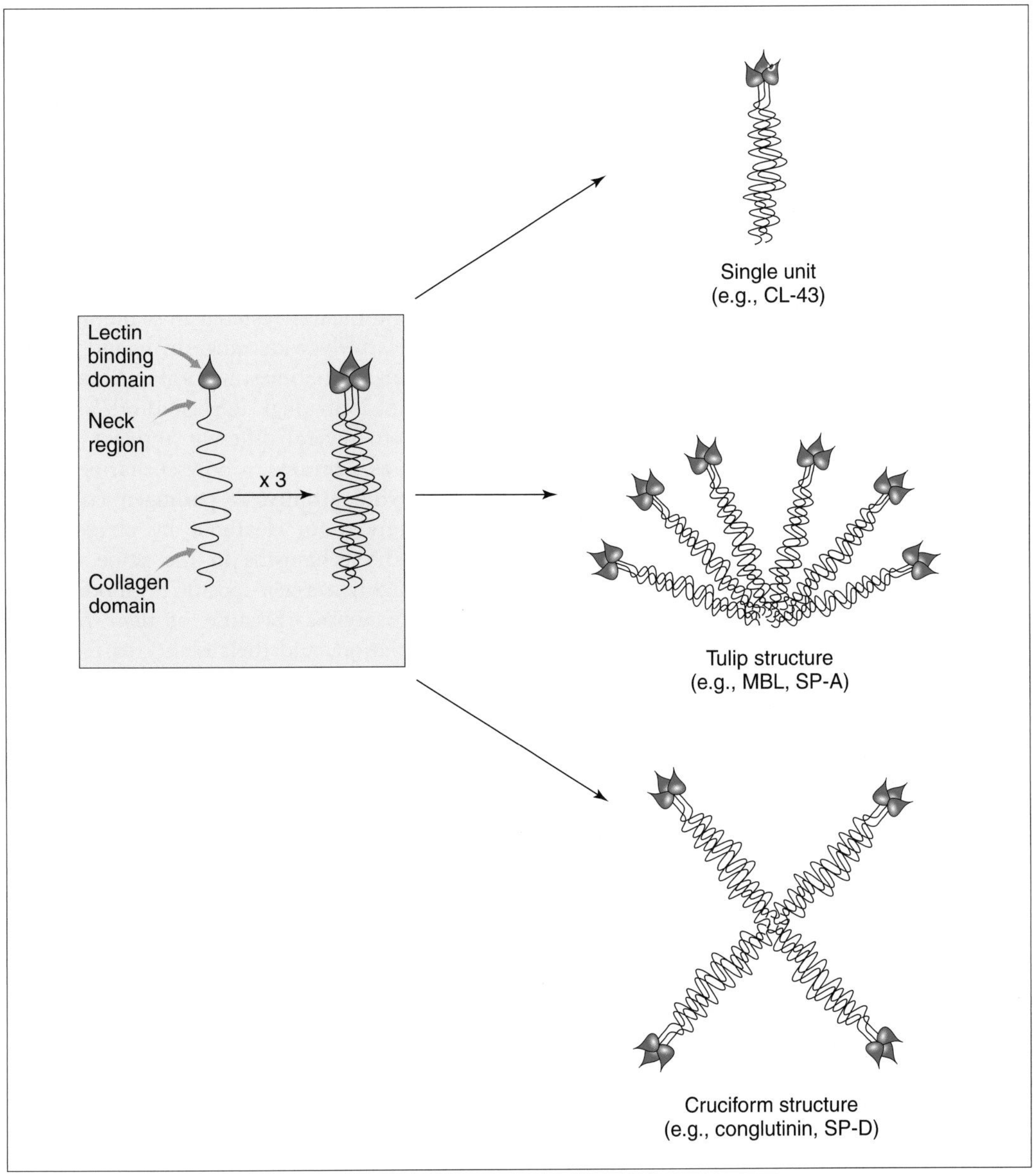

Figure 4-12

Structure of Collectins
Collectins are soluble proteins with a collagen domain, a neck region, and a globular carboxy-terminal lectin binding domain. They exist as multimers of the basic trimer unit of three very similar polypeptides that, when viewed by electron microscopy, are arranged in different structural shapes. The simplest collectin is collectin-43 (CL-43). Mannose-binding lectin (MBL) and surfactant protein-A (SP-A) exist as hexamers of the basic trimer and associate with each other to form "tulip" structures. The basic trimer units are arranged into a "cruciform" structure for surfactant protein-D (SP-D) and conglutinin. Adapted from Jinha L. (1997) Collectins: collectors of microorganisms of the innate immune system. *BioEssays* **19**(**6**), 509–518.

Single unit (e.g., CL-43)

Lectin binding domain / Neck region / Collagen domain / x 3

Tulip structure (e.g., MBL, SP-A)

Cruciform structure (e.g., conglutinin, SP-D)

while effector action is exerted on microbial cells (nonself) with cell surface proteins bearing sugar side chains accommodated by the carbohydrate recognition site of MBL. The determination of self/nonself in this context is also dependent on the fact that microbes (but not mammalian cells) are generally covered in dense arrays of repeating binding sites composed of carbohydrates. The geometry of the multimeric collectins is such that multiple carbohydrate recognition domains are easily lined up to bind to multiple carbohydrates on the microbe, increasing the overall strength of the binding between the collectin and the microbe.

Collectins generally mediate pathogen clearance via complement activation, by opsonization, or by aggregating cells together (*agglutination*; see Ch.6). The multimeric structure of a collectin molecule is reminiscent of the first protein in the classical complement cascade, C1q (see Ch.19). Indeed, the first collectins were identified in 1967 by their ability to trigger the classical complement system, the first non-antibody proteins shown to do so. For example, although their amino acid sequences are unrelated, MBL resembles C1q closely enough in shape to replace it *in vitro*, interacting directly with the early components of the classical complement cascade, C1r and C1s (see Ch.19). *In vivo*, proteases associated with MBL assume the function of C1r and C1s and activate the early components of the classical cascade. Ultimately, this lectin pathway of complement activation leads to production of the key enzyme C3 convertase (see Box 4-1) without the involvement of any specific antibody. The activation of complement can then lead to lysis of the pathogenic cell. Some collectins, including surfactant protein A and MBL, interact directly with bacterial cells to serve as opsonins, enhancing the removal of these cells by phagocytosis. Other collectins, such as surfactant protein D, can agglutinate bacteria or form intermolecular networks that trap fungal cells. The bovine collectin-43 (CL-43) is unusual in two ways: it always appears as a monomer (in contrast to all other collectins described to date) and it completely lacks complement-activating properties.

It is thought that the major impact of the collectins is felt during the lag phase of the adaptive immune response, while lymphocytes are going through the process of activation, proliferation, and differentiation. Similarly, the collectins may play a role in the prevention of recurrent infections in individuals whose adaptive immune response is compromised or not yet fully developed, such as children under 2 years of age.

ii) Acute-Phase Proteins

Early in a local inflammatory response, activated macrophages secrete cytokines that trigger hepatocytes to produce the acute-phase proteins. At least some of these soluble proteins, particularly *C-reactive protein*, bind to a wide variety of bacteria and fungi through recognition of common cell wall components that are absent from host cells. Once deposited on the surface of a microbe, these soluble pattern recognition molecules can activate complement, leading to the surface deposition of complement component C3b. C3b then acts as an opsonin to enhance phagocytosis of the foreign entity

via the cell surface receptors for C3b present on phagocytes. This is another example of how soluble molecules of the innate immune system can focus an antigen clearance mechanism onto "nonself" without requiring the high degree of antigen specificity displayed by antibodies.

iii) Natural Antibodies

The term "natural antibody" refers to a type of immunoglobulin that arises in the body independently of external antigenic stimulation. Antibodies that bind to various bacterial, viral, fungal, and other lower order components, as well as antibodies to self-elements such as nucleic acids, erythrocytes, insulin, and other cellular components, are present in the vertebrate circulation regardless of whether the host has been immunized against them. Natural antibodies also account for the observation that a significant population of B cell clones produce antibacterial antibodies even when the research animals studied have been raised in a germ-free environment. Natural antibodies are mostly of the IgM isotype (but IgG and IgA have also been identified) and are produced primarily by a subpopulation of B cells called CD5$^+$ B cells (also known as B1 cells; see Ch.9). Natural antibodies are encoded by immunoglobulin genes that have undergone the initial somatic rearrangement of their gene segments, but not the subsequent additional mutational mechanisms that are necessary to generate the diversity of antibodies produced in the adaptive response (see Ch.9).

The maintenance of Ig nucleotide sequences that are close to the germline configuration has resulted in a binding cleft that accommodates a relatively wide range of ligands. Thus, in contrast to conventional antibodies, natural antibodies are *polyreactive*, in that an individual IgM natural antibody can recognize and bind to several different antigens of very different structure (such as a protein, a polysaccharide, and a phospholipid) with varying affinity. A pathogen coated in natural antibodies is marked for clearance by phagocytosis or complement-mediated mechanisms in the same way as pathogens marked by highly antigen-specific antibodies during adaptive immune responses. Because of their presence prior to antigenic stimulation, and their ability to recognize more than one antigen, natural antibodies are considered part of the innate response and may be important in the host's initial response to common pathogens. Natural antibodies also account for most of the B cell repertoire in the fetus and neonate (in which the B lymphocyte compartment is dominated by CD5$^+$ B cells), and, although the underlying mechanisms are still unclear, may play a role in shaping the development of the mature B cell repertoire.

The polyreactivity of natural antibodies constitutes a double-edged sword, which may be why antigen-specific antibodies developed as evolution progressed. The common molecular patterns recognized by natural antibodies are not only present on the exogenous antigens provided by bacteria and viruses, but also on host tissues, so that natural antibodies are often self-reactive. It is interesting to note that CD5$^+$ B cells have been associated with several autoimmune conditions.

iv) NOD Proteins

We previously learned how TLRs are, for the most part, cell surface transmembrane receptors that recognize the PAMPs of pathogens. NOD1 and NOD2, members of the *NOD* (nucleotide-binding oligomerization domain) family of proteins, appear to perform an analogous function within a cell's cytoplasm to detect products of intracellular pathogens. The NOD1 and NOD2 proteins are soluble sensor proteins that are structurally related to the TLRs in that they contain leucine-rich repeat (LRR) domains. However, the NOD1 and NOD2 LRRs are thought to mediate binding to unique bacterial peptidoglycan-derived structures that are not recognized by the TLRs. Ligand binding to NOD1 or NOD2 triggers intracellular signaling that leads to NF-κB activation by a pathway slightly different from that utilized by the TLRs. Nevertheless, the end result is the same, in that the expression of genes associated with inflammation (particularly TNF, IL-1, and IL-18) is induced. [It should be noted that researchers studying autoimmunity refer to NOD1 as CARD4 (caspase recruitment domain-4), and NOD2 as CARD15. Mutations in CARD15 have been associated with the autoimmune disorder Crohn's disease, which may involve inappropriate immune responses to normal enteric bacteria (see Ch.29).]

The reader has now completed the introductory section of this book and should be well armed for an in-depth discussion of all aspects of the immune response. We have learned that there is an interdependence and overlapping of innate and adaptive elements in the temporal and functional flow of the immune response. The non-inducible, non-specific anatomical and physiological barriers act in the first phase to repel most foreign entities. Those antigens succeeding in evading these defenses may be engulfed by ever-vigilant phagocytes. The second phase of the immune response features elements that are broadly specific in their recognition function, but that act only in the presence of antigen; examples are the induction of the inflammatory response and the non-classical triggering of complement-mediated destruction. It is in the third phase of the immune response that the adaptive and innate mechanisms become intertwined, with products secreted by inflammatory cells having an effect on lymphocytes, and vice versa. Such interdependence and shared use of effector mechanisms ensure that host defense is integrated, responsive, and effective.

SUMMARY

Innate immunity encompasses anatomical and physiological barriers, cellular internalization mechanisms, and inflammatory responses that are rapidly induced by the presence of antigen. Innate immune mechanisms inhibit pathogen entry, prevent the establishment of infection, and clear both host and microbial debris. Some innate mechanisms are completely antigen non-specific, while others involve broadly specific pattern recognition molecules (PRMs) that play a role in clearing a limited range of pathogens. Some PRMs are pattern recognition receptors (PRRs) expressed on effector cell surfaces, whereas others are soluble molecules that mark pathogens for clearance. Innate immunity either succeeds in eliminating the pathogen, or helps to hold infection in check until the slower, lymphocyte-mediated adaptive immune responses can develop. In addition, cells of the innate immune response release cytokines that are critical in lymphocyte activation and differentiation, influencing both the extent and the type of adaptive immune response to a given pathogen.

Selected Reading List

Aderem A. and Underhill D. M. (1999) Mechanisms of phagocytosis in macrophages. *Annual Review of Immunology* **17**, 593–623.

Allen L.-A. H. and Aderem A. (1996) Mechanisms of phagocytosis. *Current Opinion in Immunology* **8**, 36–40.

Barton G. and Medzhitov R. (2002) Control of adaptive immune responses by Toll-like receptors. *Current Opinion in Immunology* **14**, 380–383.

Bendelac A. and Fearon D. T. (1997) Innate pathways that control acquired immunity. *Current Opinion in Immunology* **9**, 1–3.

Beutler B. and Rietschel E. (2003) Innate sensing and its roots: the story of endotoxin. *Nature Reviews. Immunology* **3**, 169–176.

Boman H. G. (2000) Innate immunity and the normal flora. *Immunological Reviews* **173**, 5–16.

Casali P. and Schettino E. (1996) Structure and function of natural antibodies. *Current Topics in Microbiology and Immunology* **210**, 167–178.

Epstein J., Eichbaum Q., Sheriff S. and Ezekowitz R. A. (1996) The collectins in innate immunity. *Current Opinion in Immunology* **8**, 29–35.

Fearon D. and Locksley R. (1996) The instructive role of innate immunity in the acquired immune response. *Science* **272**, 50–54.

Gadjeva M., Thiel S. and Jensenius J. C. (2001) The mannan-binding-lectin pathway of the innate immune response. *Current Opinion in Immunology* **13**, 74–78.

Galluci S. and Matzinger P. (2001) Danger signals: SOS to the immune system. *Current Opinion in Immunology* **13**, 114–119.

Greenberg S. and Grinstein S. (2002) Phagocytosis and innate immunity. *Current Opinion in Immunology* **14**, 136–145.

Hewlett L. J., Prescott A. R. and Watts C. (1994) The coated pit and macropinocytic pathways serve distinct endosome populations. *The Journal of Cell Biology* **124**, 689–703.

Hoffman J. and Reichhart J. (2002) *Drosophila* innate immunity: an evolutionary perspective. *Nature Immunology* **3**, 121–126.

Holmskov U., Malhotra R., Sim R. and Jensenius J. C. (1994) Collectins: collagenous C-type lectins of the innate immune defense system. *Immunology Today* **15**, 67–73.

Hunziker W. and Geuze H. J. (1995) Intracellular trafficking of lysosomal membrane proteins. *BioEssays* **18**, 379–389.

Iwasaki A. and Medzhitov R. (2004) Toll-like receptor control of the adaptive immune response. *Nature Immunology* **5**, 987–995.

Janeway C., Jr. and Medzhitov R. (2002) Innate immune recognition. *Annual Review of Immunology* **20**, 197–216.

Kawai T., Adachi O., Ogawa T., Takeda K. and Akira S. (1999) Unresponsiveness of MyD88-deficient mice to endotoxin. *Immunity* **11**, 115–122.

Krutzik S. R., Sieling P. A. and Modlin R. L. (2001) The role of Toll-like receptors in host defense against microbial infection. *Current Opinion in Immunology* **13**, 104–108.

Kwiatkowska K. and Sobota A. (1999) Signaling pathways in phagocytosis. *BioEssays* **21**, 422–431.

Lu J. (1997) Collectins: collectors of microorganisms for the innate immune system. *BioEssays* **19**, 509–518.

Mak T. W. and Ferrick D. A. (1998) The gamma-delta bridge: linking innate and acquired immunity. *Nature Medicine* **4**, 764–765.

Medzhitov R. and Janeway C., Jr. (1997) Innate immunity: the virtues of a non-clonal system of recognition. *Cell* **91**, 295–298.

Medzhitov R. and Janeway C. A. J. (1997) Innate immunity: impact on the adaptive immune response. *Current Opinion in Immunology* **9**, 4–9.

Pearson A. M. (1996) Scavenger receptors in innate immunity. *Current Opinion in Immunology* **8**, 20–28.

Peiser L., Mukhopadhyay S. and Gordon S. (2002) Scavenger receptors in innate immunity. *Current Opinion in Immunology* **14**, 123–128.

Rabinovitch M. (1995) Professional and non-professional phagocytes: an introduction. *Trends in Cell Biology* **5**, 85–87.

Rock F. L., Hardiman G., Timans J. C., Kastelein R. A. and Bazan J. J. (1998) A family of human receptors structurally related to *Drosophila* Toll. *Proceedings of the National Academy of Sciences U.S.A.* **95**, 588–593.

Silverstein S. C. (1995) Phagocytosis of microbes: insights and prospects. *Trends in Cell Biology* **5**, 141–142.

Storrie B. and Desjardins M. (1996) The biogenesis of lysosomes: is it kiss and run, continuous fusion and fission process? *BioEssays* **18**, 895–903.

Strzelecka A., Kwaitkowska K. and Sobota A. (1997) Tyrosine phosphorylation and Fc gamma receptor-mediated phagocytosis. *FEBS Letters* **400**, 11–14.

Swanson J. and Baer S. (1995) Phagocytosis by zippers and triggers. *Trends in Cell Biology* **5**, 89–93.

Travis S. M., Singh P. K. and Welsh M. J. (2001) Antimicrobial peptides and proteins in the innate defense of the airway surface. *Current Opinion in Immunology* **13**, 89–95.

Underhill D. and Ozinsky A. (2002) Toll-like receptors: key mediators of microbe detection. *Current Opinion in Immunology* **14**, 103–110.

Wright S. (1999) Toll, a new piece in the puzzle of innate immunity. *Journal of Experimental Medicine* **11**, 115–122.

Yang R.-B., Mark M., Gray A., Huang A., Xie M., Zhang M., Goddard A., Wood W., Gurney A. and Godowski P. (1998) Toll-like receptor-2 mediates lipopolysaccharide-induced cellular signaling. *Nature* **395**, 284–288.

B Cell Receptor Structure and Effector Function

5

Gerald Edelman 1929–

Antibody structure
Nobel Prize 1972

© Nobelstiftelsen

Rodney Porter 1917–1985

Antibody structure
Nobel Prize 1972

© Nobelstiftelsen

CHAPTER 5

A. HISTORICAL NOTES

B. THE STRUCTURE OF IMMUNOGLOBULINS

C. EFFECTOR FUNCTIONS OF ANTIBODIES

D. IMMUNOGLOBULIN ISOTYPES IN BIOLOGICAL CONTEXT

"Regard your soldiers as your children, and they will follow you into the deepest valleys. Look on them as your own beloved sons, and they will stand by you even unto death"—Sun Tzu

We learned in Chapter 2 that the specific recognition of antigen by B and T cells depends on the interaction between epitopes on the antigen and antigen receptors on the lymphocyte cell surface. This chapter focuses on the structure and function of the *immunoglobulins* (Igs), since membrane-bound Igs constitute an integral part of the antigen receptors of B cells. Soluble Igs function as secreted antibody, clearing pathogens during the effector phase of an adaptive immune response.

A. Historical Notes

In the late 1800s, Emil Behring and Shibasaburo Kitasato discovered that the serum component of blood could transfer immunity to microbial toxins from an immunized animal to an unimmunized one. The active agent in the "antiserum" was first called an "antitoxin," and then an "antibody" by the 1930s, after it was shown that the same substance was capable of several activities in addition to toxin neutralization. The discovery of blood group antigens by Landsteiner in 1900 then ushered in the age of <u>serology</u> (the study of <u>ser</u>um), when immunology became the study of "antigens," or those molecules that could <u>gen</u>erate an <u>antibody</u> response. At the same time, debate began concerning the theoretical mechanism by which antigens could stimulate the production of antibodies (see Box 5-1). Studies during this period made routine use of electrophoresis as a technique for characterizing proteins: a mixture of proteins was introduced into a semisolid medium (such as an agar or polyacrylamide gel) and separated on the basis of charge by application of an electric current. Because antibodies were identified in the third peak, or "γ peak," of electrophoretically fractionated serum globulin proteins (Fig. 5-1), they also became known as "gamma globulins."

In the 1940s, immunochemistry allowed scientists to qualitatively and quantitatively assess the structural characteristics

of antibodies and antigens. The question of how a given antigen induced the production of a specific antibody was not resolved until the development of the *clonal selection* theory by F. MacFarlane Burnet in the 1950s (refer to Box 5-1). Burnet postulated that a multitude of antibody-producing cells pre-existed in an organism, each one having on its cell surface an antibody or Ig molecule specific for a different antigen. Encountering its antigen (being "selected") would activate an antibody-producing cell, inducing it to proliferate into a *clone* (a large number of identical cells) secreting antibodies identical to those on the surface of the original cell. This hypothesis has since been demonstrated to be correct, and is now one of the basic tenets of modern immunology.

Early explorations of the structure of antibodies were carried out using limited proteolysis. From the results of these experiments, scientists deduced that all Ig molecules shared a common basic structure responsible for certain functional features. Brief digestion of the immunoglobulin IgG with papain produced two identical *Fab* fragments (each of molecular mass 45 kDa), so-called because they retained the antibody's <u>a</u>ntigen-<u>b</u>inding ability (Fig. 5-2). The remainder of the antibody molecule was called the *Fc* fragment (50 kDa), because it was found to <u>c</u>rystallize at low temperatures. Similarly, partial digestion of IgG with pepsin resulted in a large (100 kDa) fragment with antigen-binding ability, called the $F(ab)_2$ fragment, and a partial Fc fragment designated *Fc*. These structural divisions were reflected in the bifunctional nature of the antibody molecule. While the Fab region of the antibody was found to carry out the recognition function and would specifically bind to a microbe or macromolecule, the Fc region mediated the effector functions of the humoral response that resulted in antigen clearance. Complement was found to be activated by the binding of the Fc region of circulating antibody, and it was observed that receptors for the Fc region present on the surfaces of phagocytic cells allowed antibodies to enhance phagocytosis by acting as opsonins.

Box 5-1. Theories of antibody formation

Antibodies were first identified in the 1890s by scientists studying the body's response to bacterial toxins. Up until about 1960, many competing theories were put forward to explain how antibodies were generated. Concurrently, advances made in understanding the molecular biology of proteins, genes, and DNA influenced the acceptability of various theories. However, the favored theories also shifted as the key concerns of immunologists shifted.

In the earliest years, the immunological community was most concerned with explaining antibody specificity and the size of the antibody repertoire; little attention was paid to explaining the more biological aspects of immunity such as the stronger, higher affinity antibody responses seen after secondary exposure to a foreign entity. This bias led to the rejection of the "side chain" theory proposed in 1897 by Paul Ehrlich. Ehrlich reasoned that since certain cell types had specific receptors facilitating the entrance of particular nutrients into the cell, other specialized cells might have surface receptors that were specific for particular foreign entities (such as toxins); he called these receptors *side chains*. A foreign entity (the term "antigen" had yet to be coined) would bind to a specific side chain, be taken inside the cell, and trigger the release into the blood of structurally identical side chains that were observed as circulating antibodies. This was the first example of an *antibody selection* theory, in which antigen merely selects a complementary preformed antibody from the many available and somehow drives the production of the selected antibody.

An inherent difficulty with any antibody selection theory was explaining how receptors with different antigen specificities could come to preexist in the host prior to antigen exposure. When Ehrlich first introduced the side chain theory, only a limited number of antibodies were known, all of them pathogen specific. It was therefore reasonable to imagine that natural selection would result in the existence of a varied but limited number of antigen-specific receptors conferring disease resistance. However, it was soon shown that antibodies could be made to antigens that would never be encountered in nature, such as laboratory-derived antigens and cells from other species. Furthermore, in the 1920s, Karl Landsteiner demonstrated that antibodies could be raised to inert

substances that pose no natural threat to survival. Despite its elegance, Ehrlich's theory was incompatible with such a large antibody repertoire, and so it fell into disfavor and was ignored for the next 60 years.

During the 1930s and 1940s, most immunologists were convinced that the information to generate an apparently limitless antibody repertoire could not come from within the host but must be controlled in some way by antigen. Various *antigen incorporation* theories proposed that specific antibody was derived directly from the antigen. However, these theories were inconsistent with the observation that the concentration of antigen-specific antibody in serum does not decline after fresh antigen is no longer introduced into the body. Equally incompatible was the finding that a small amount of antigen stimulated production of extremely large amounts of antibody. The failure of antigen incorporation theories led to the rise of various *antibody instruction* theories, which held that antibody was derived from the host but that its final specificity was somehow "taught" to it through exposure of the host to the antigen. In early instruction theories proposed in the 1930s by Friedrich Breinl and Felix Haurowitz, an antibody molecule would obtain its specific shape by using the antigen directly as a template for folding. However, by the end of the 1930s, it was known that proteins (including antibodies) were synthesized by the sequential addition of amino acids. This forced Linus Pauling and others in the 1940s to adapt their "direct template" instruction theories such that the antigen now would direct the actual synthesis of the antibody, and not just its folding.

Direct template instruction theories were compatible with what was known about antibody specificity, the unlimited antibody repertoire, and the generation of copious amounts of a particular antibody in response to a minimal amount of antigen. However, between 1940 and 1960, there was increasing discomfort with the failure of these theories to explain the increased production and quality of antibody observed during a secondary response. F. MacFarlane Burnet extended an earlier hypothesis of Niels Jerne and argued that these features were evidence that antibody production was not only carried out by the original antigen-stimulated cell, but also by its descendants. Thus arose the "indirect template"

instruction theories, in which the influence of the antigen on antibody structure and specificity was somehow passed from the original antibody-producing cell to its daughter cells. Burnet initially argued that enzymes involved in protein synthesis might be modified by antigen, and that these modified enzymes could both affect specificity and be passed on to progeny cells. However, by 1949, this idea lost its appeal as it became clear that protein synthesis was controlled, not by enzymes, but by genetic material.

The discovery in the 1950s that the flow of information in cells went irreversibly from DNA to RNA to protein swept aside the instruction theories. Immunologists now had to accept that antigen could not carry into a cell information that would influence either antibody structure or antibody genes. In 1957, Burnet, David Talmage, and Joshua Lederberg returned at last to Ehrlich's idea that antibodies might act as receptors on the surface of lymphoid cells and that antigen was simply a signal triggering these cells to produce antibodies. They argued that a large number of cells, each displaying only a single specificity, could be available at all times. Antigen would then "select" cells bearing antibody receptors with a complementary fit, triggering them to proliferate and differentiate into antibody producing cells. An increased number of antigen-specific antibody-producing cells would then be present upon subsequent exposure to the antigen, giving rise to higher levels of secondary antibody. Increased antibody affinity could be accounted for by minor somatic mutations in the DNA of the antibody gene. Burnet's *clonal selection theory* of antibody formation quickly became accepted as one of the underlying principles of the immune response.

The road leading from the original idea of Ehrlich to that of Burnet was very long and winding, but the overall distance between them was, in the end, very small. Interestingly, clonal selection theory was quickly and widely accepted despite having the same major flaw that was fatal for Ehrlich's theory. Burnet made no attempt to explain how the information for a limitless repertoire of antibody specificities could be stored in the relatively limited vertebrate genome. Immunologists of the day chose to address this problem separately while retaining the basic tenets of the clonal selection theory. A historical overview of the important debate surrounding the generation of antibody diversity can be found in Chapter 8.

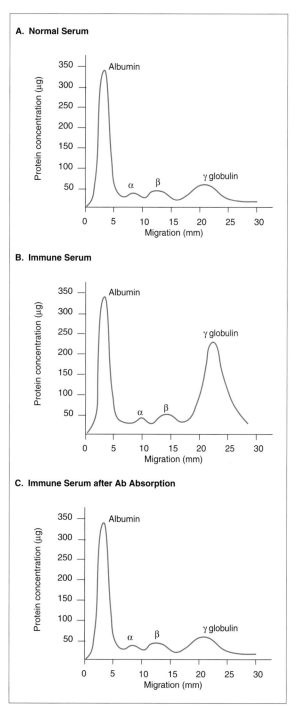

Figure 5-1

Identification of Antibodies as Serum Globulin Proteins
The components of a serum sample can be separated based on charge using electrophoresis. (A) Normal serum separates into a large albumin fraction and three smaller globulin fractions known as the α, β, and γ globulins. (B) When serum from rabbits immunized from antigen X is electrophoresed, a dramatic increase in the γ globulin fraction is observed, suggesting that anti-X antibodies reside in the γ globulin fraction. (C) Pretreatment of the immune serum with antigen X prior to electrophoresis results in a return of the γ globulin peak to normal levels because the anti-X antibodies have been absorbed out of the serum. Adapted from Tisselius A. and Kabat E. A. (1939). An electrophoretic study of immune sera and purified antibody preparations. *Journal of Experimental Medicine* **69**, 119–131.

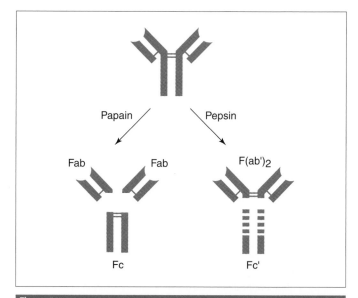

Figure 5-2

Antibody Fragments Produced by Limited Enzymatic Digestion
Limited digestion of IgG with papain results in characteristic fragments that depend on the location of proteolysis sites relative to interchain disulfide bonds. Papain digestion yields two Fab fragments, each containing one antigen-binding site, plus an intact Fc fragment. Pepsin digestion yields one F(ab')$_2$ fragment containing two antigen-binding sites, plus various smaller Fc' fragments derived from digestion of Fc. The prime symbol (') is used to indicate a variation in structure.

The fine structure of the antibody (Ig molecule) was first determined in the laboratories of Gerald Edelman and Rodney Porter in the 1950 and 1960s. Antibody molecules were shown to have a Y-shaped structure composed of four proteins: two light chains and two heavy chains. The N-terminal regions of the branches of the Y contained the two identical antigen-binding sites, and the stem of the Y contained the Fc effector region. Further experimentation revealed that each Fab antibody-binding region of the Ig contained one light chain and part of one heavy chain, and that the effector Fc portion contained parts of two heavy chains and no light chains.

With the rapid progress in protein chemistry in the 1960s and 1970s, it became possible to examine antibody molecules at the amino acid level. However, the elucidation of the actual sequence of an individual Ig molecule initially eluded scientists for several reasons. Because the injection of an antigen stimulates the production of a collection of antibodies with slightly different amino acid sequences (see later discussion of *polyclonal*), an antiserum could not be a source of pure antibody of a single specificity, as was needed for protein sequencing studies. Furthermore, attempts to culture individual B cells long enough for them to produce sufficient quantities of a particular antibody failed because of the very limited life span (only a few days) of plasma cells. The discovery of *myelomas*, or plasma cell cancers, revolutionized the study of antibody structure and specificity. Like all cancers, myeloma cells grow uncontrollably because they have become resistant to the cell death signals that limit the life span of a normal cell; they are said to have become *immortalized*. It was observed that patients in whom a clone of B cells had undergone malignant transformation secreted abnormally large quantities of a structurally homogeneous Ig protein in their blood and urine. Isolation of these myeloma proteins (which are normal Igs in their specificity and sequence) provided the first source of a single antibody in amounts sufficient for protein sequencing. The first complete amino acid sequence of an Ig molecule was reported by Gerald Edelman and colleagues in 1969.

When the sequences of heavy and light chains present in different antibodies were compared, it became obvious that there was a distinct structural pattern to these proteins. In both light and heavy Ig chains, certain regions had amino acid sequences that were very similar among many different antibodies (the *constant*, or *C, regions*), whereas other regions exhibited great variability in amino acid sequence (the *variable*, or *V, regions*). These observations provided the first insights as to how the diversity required to respond to a multitude of antigens could be generated. Recombinant DNA technologies that were developed in the 1970s opened the door to detailed examination of the genes encoding the proteins. How the body manages the huge amount of genetic information required to respond to the universe of antigens was a problem that was not solved until the late 1970s and 1980s. In 1978, work done by Nobumichi Hozumi and various co-workers in the lab of Susumu Tonegawa at the Basel Institute in Switzerland demonstrated that Ig gene segments in individual B cells underwent an intricate DNA rearrangement process, resulting in an array of B cells producing antibody proteins with an almost infinite spectrum of antigen-binding sequences. Research in the late 1980s and 1990s then revealed the complex mechanics underlying gene rearrangement (see Ch.8).

The Polyclonal Nature of Natural Antisera

The reader will recall that the B cell repertoire of antigenic specificities pre-exists so as to equip the host with at least one clone of cells able to react to each molecule in the universe of antigens. An antigen entering the body is an entity exposing a number of both T and B epitopes to immune system cells. Each epitope on the antigen molecule will correspond to the antigen receptor on a particular clone of B cells, so that the activation and differentiation of several different B cell clones, one for each antigenic epitope, can be triggered. Each activated B cell is induced to clonally expand and differentiate into progeny plasma cells secreting antibody recognizing the original epitope. The presence in the host's plasma of all of these different antibodies specific for different regions of the same antigen makes a natural antiserum *polyclonal*—that is, containing antibodies produced by many different B cell clones. Such antigen-specific antibodies in a polyclonal antiserum preparation have a wide variety of amino acid sequences, but they all act against the same antigen. In addition, as the immune response progresses, antibodies to a given epitope are produced, which have the same specificity, but which differ in isotype, providing a further layer of heterogeneity (see *isotype switching* later in this chapter and in Ch.9). It was the polyclonal character of the natural antibody response to antigen, and the fact that a single type of antibody protein is present in only very small amounts, that made the initial sequencing of antibodies such a challenge.

In 1975, Georges Köhler and César Milstein developed the *hybridoma technology*, which allows scientists to create immortalized B cells secreting immense amounts of pure antibody of a single known specificity. Such Igs are known as *monoclonal antibodies* (see Ch.7). This groundbreaking technique, which garnered for its creators the Nobel Prize in Physiology or Medicine in 1984, has been used to great advantage in generating tools for both basic and industrial research and for clinical applications. Studies using monoclonal antibodies, or *mAbs*, have greatly refined our knowledge of the function and structure of Ig molecules.

B. The Structure of Immunoglobulins

Immunoglobulins are specialized glycoproteins synthesized only by B lymphocytes. These proteins are encoded by a genetic locus whose almost unique features confer the unusual structural characteristics of the Ig polypeptide chains. Immunoglobulins are first expressed in a membrane-bound form on the surface of B cells, where they function as part of the B cell's antigen receptor. After lymphocyte activation, proliferation, and differentiation, an Ig protein that has the same antigenic specificity is synthesized in a form that is secreted as circulating antibody by a clone of mature plasma cells. The separation of the recognition and effector functions of the Ig molecule is reflected in its intrinsic structure, since recognition capability and sequence diversity are located in the N-terminal end of the Ig polypeptide chain, while sequences interacting with antigen clearance mechanisms are located in the much less variable C-terminal end.

I. GENERAL STRUCTURAL PROPERTIES OF IMMUNOGLOBULIN MOLECULES

i) What Is the Basic Structure of an Immunoglobulin Molecule?

Immunoglobulins are large glycoproteins of total molecular mass of approximately 150–190 kDa. All Ig molecules have the same basic core structure: two *heavy*, or *H, chains* (55–77 kDa), and two *light*, or *L, chains* (25 kDa) (Fig. 5-3A). This structure is often abbreviated as "H_2L_2." Genetically, an individual B cell can produce only one kind of H chain and only one kind of L chain (see Ch.8), so that the two H chains in a given Ig molecule are identical, and the two L chains are identical. In general, the two H chains are joined to each other by two or more covalent disulfide bonds, and each heavy chain is joined to an L chain by an additional disulfide bond. The number and position of disulfide bonds vary by Ig isotype and subclass. Carbohydrate moieties are attached at particular sites on the H chains. Noncovalent forces in the form of hydrophobic interactions between residues also help to hold the four chains together in a single molecule. Although the H_2L_2 structure defines an Ig unit and represents the structure of Igs functioning in antigen receptors, secreted molecules of some isotypes contain more than

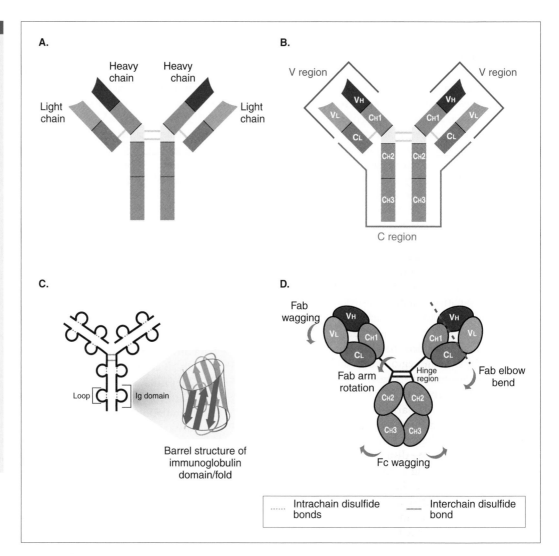

Figure 5-3

The Basic Structure of the Immunoglobulin Molecule
(A) Two heavy (H) chains and two light (L) chains make up the Ig molecule. (B) The Ig variable (V) and constant (C) regions with domain structure are shown schematically. (C) Schematic representation of the characteristic Ig domain loop structure, with inset showing the three-dimensional barrel-like structure of the loop. (D) While the hinge region holds the central area of the Ig molecule rigid, each Fab arm is able to rotate around the V_H–C_H1 axis. Flexibility in the secondary structure outside the hinge region also allows both the Fab and Fc portions of the molecule to move back and forth, to some degree, in a "wagging" motion. The Fab can also bend like an elbow along the interface between its V and C domains. Part D: With information from Brekke O. H. (1996) The structural requirements for complement activation by IgG: does it hinge on the hinge? *Immunology Today* **16**, 85–90.

one of these basic units. For example, a single pentameric IgM molecule contains five basic Ig units, making it an $(H_2L_2)_5$ structure displaying a total of 10 antigen-binding sites. These considerations are discussed in more detail later in this chapter.

ii) Constant and Variable Domains

If the basic structure of all Igs is the same, how do the receptors of different B cells recognize different antigenic structures? Comparison of the complete amino acid sequences of individual Ig molecules reveals a vast variability in the Ig domain at the N-terminus of any Ig polypeptide chain, the *variable*, or *V*, *domain*. The immense sequence diversity required to build a comprehensive B cell repertoire is confined to the V domain, whereas the *constant*, or *C*, domains are relatively conserved (Fig. 5-3B). Each L chain contains one V and one C domain, denoted V_L and C_L, respectively. Similarly, each H chain contains one V_H and three or four C_H domains, depending on its isotype. Thus, each tetrameric Ig molecule contains a total of 12–14 domains: 4 V domains and 8–10 C domains, or 2 V_L, 2 C_L, 2 V_H, and 6–8 C_H domains.

The light and heavy chains are encoded by separate genetic loci, and the sequence of a light chain V or C region is never found in a heavy chain, or vice versa. Studies of pure immunoglobulins have shown that, within an Ig molecule, the H and L polypeptides are aligned such that the V domains and C domains in the L chain are positioned directly opposite their counterparts in the H chain. Thus, each of the two individual antigen-combining sites is formed by the pairing of the V_L domain of an L chain with the V_H domain of the H chain to which it is linked. (The reader should note that the disulfide bond linking an H chain to an L chain is positioned outside the V domains.)

All V and C domains in Ig polypeptide chains are based on a common structural unit known as an *Ig domain* (Fig. 5-3C). An Ig domain is about 70–110 amino acids in length (~12 kDa), with cysteine residues at either end that form an intrachain disulfide bond. The amino acid sequences of Ig domains are not identical, but are similar enough that each forms an identifiable and characteristic cylindrical structure known as the *immunoglobulin barrel* or the *immunoglobulin fold*. Many other proteins involved in the immune system and in intercellular interactions also contain Ig-like domains with a similar barrel structure, leading to coining of the term *Ig superfamily* (see Box 5-2). It is thought that members of the Ig superfamily, including the Ig chains, all evolved from a common ancestral gene containing the sequence of the basic Ig domain. By repeated duplication of this gene and subsequent divergent mutations, a family of proteins related in sequence but involved in diverse molecular recognition functions was generated. Some scientists speculate that an evolutionary advantage was conferred on those proteins exhibiting the Ig-fold, since they appear to have enhanced structural stability in hostile environments. Antibody proteins are remarkably stable under conditions of extreme pH, such as occur in the gut, and are resistant to proteolysis under natural conditions.

The C_L domain in the L chain is paired with and interacts with the first C_H domain of the H chain, and it is in this region that the disulfide bond joining the L chain to its H chain partner is located. In most Ig isotypes, the region of the H chain between the first and second C_H domains (where the Fab region joins the Fc region at the top of the stem of the Y) is somewhat extended and is known as the *hinge* region (Fig. 5-3D). The hinge region contains many proline residues, which impart a rigidity to the top of the stem. However, glycine residues in the hinge region create a flexible secondary structure such that the arms of the Ig molecule can move independently of each other around the proline-stabilized anchor point. The hinge region also contains cysteine residues that interact to form the disulfide bonds holding the H chains together. Many of the Ig carbohydrate moieties as well as the sites on the Ig molecule that bind to complement component C1 are generally located within the C_H2 domain of Igs with three C_H domains, and in the C_H3 domain of Igs with four C_H domains. In contrast to other C_H domains, the C_H2 (or

C_H3) domain on one H chain does not appear to interact with the C_H2 (or C_H3) domain on the other H chain, due to the presence of interfering sugar groups. This lack of close contact may permit molecules associated with the effector response to more easily access this region of the Ig molecule. Continuing down the stem of the Y, the C_H3 (or C_H4) constant domain on one H chain is paired noncovalently with its corresponding partner on the other H chain.

iii) Structural Variation in the V Region

Within the V domains of both the L and H chains are three short *hypervariable regions* that exhibit extreme amino acid variability (Fig. 5-4A). These regions of five to seven amino acids are responsible for the diversity that allows the total repertoire of Igs to recognize almost any molecule in the universe of antigens. The antigen-binding sites are formed by three-dimensional juxtaposition of the three hypervariable regions in the V_L domain with those in the V_H domain. Because these sequences result in a structure complementary to an antigenic epitope, the hypervariable regions are also called *complementarity-determining regions* (CDRs). The hypervariable regions make up about 15–20% of a variable domain, with the remaining 80–85% being made up of four *framework regions* (FRs) of restricted (only 5%) variability (Fig. 5-4B). The framework regions, which are largely composed of the β-strands of the Ig domain, are thought to position and stabilize the hypervariable regions in the correct conformation for antigen binding. X-Ray crystallography studies (see Box 6-1 in Ch.6) have demonstrated that the six CDRs of the bivalent Ig molecule are displayed as loops of varying sizes and shapes, projecting from the relatively flat surfaces of the framework regions, and it is these loops that interact with specific antigen. The CDR loops contribute to unique regions within the antigen-binding site that can be

---

Box 5-2. Ig superfamily

An Ig protein has a modular structure composed of the variable and constant domains of both the H and L chains. Each of the variable (V) and constant (C) domains has a unique amino acid sequence, but each is folded into a barrel-shaped structure common to all Ig domains called the *immunoglobulin fold* (Ig-fold). This fold has a characteristic structure in which the amino acids of the Ig domain are looped back on themselves to form "sandwiches" of two β-pleated sheets. The first such domain was identified in the β2-microglobulin (β2m) molecule that functions as a subunit of the MHC class I molecule. Over the past 25 years, researchers have discovered that there is a plethora of proteins that make use of the basic Ig-fold structure. Although many of the genes that encode these proteins are important for

immune function, some of them encode proteins that function in the central nervous system, or in cell adhesion or in cellular proliferation. Together, these proteins constitute the *immunoglobulin superfamily*.

The working definition of a "superfamily of proteins" is that members must share at least 15% amino acid sequence homology. To be a member of the Ig superfamily, a protein must exhibit at least 15% amino acid sequence homology to Ig proteins <u>and</u> must contain at least one *Ig-like domain*. An Ig-like domain is a region that is homologous to the V or C domains in Ig proteins and assumes a three-dimensional conformation analogous to the Ig-fold. Although the existence of the Ig-fold may not have been confirmed by X-ray crystallography for each Ig superfamily member, examination of the amino acid

sequence is often sufficient to confirm the presence of the correct elemental structure. [It is important to note that, although the Ig-fold structure is a shared feature of Ig superfamily members, only the antigen receptor (TCR and Ig) genes in T and B cells, respectively, undergo the process of somatic gene rearrangement that generates diversity at the protein level (see Ch.8).]

The structure of the Ig-fold is shown in the main text in Figure 5-3. An Ig domain in an Ig molecule varies from 70 to 110 amino acids in length. The domain folds itself into two sheets of antiparallel β-strands joined across the middle by an internal disulfide bond. A sheet is composed of three to five antiparallel β-strands, and 5–10 amino acids make up one β-strand. Variable domains are slightly longer than constant domains, and

Continued

Box 5-2. **Ig superfamily**—*cont'd*

so V domain sheets contain two additional β-strands. One β-strand is joined to the next strand in a sheet by amino acid loops of varying lengths. These loops are made up of very short stretches of highly variable amino acids (the "hypervariable regions") surrounded by the less variable "framework" regions. It is the hypervariable regions in the V_H and V_L domains that are responsible for the diversity of the antigen-binding sites of Ig proteins. The two β-strand sheets of the domain form a "sandwich" such that the hydrophobic amino acid side chains are buried in the core, and the hydrophilic side chains protrude outward, forming the characteristic topology of the Ig-fold. The Ig-fold structure is one of remarkable stability, due both to the internal disulfide bond between the β-strand sheets and to the interactions between the hydrophobic side chains in its core.

Most members of the Ig superfamily are transmembrane molecules involved in some form of molecular recognition in which interaction between protein domains is required to carry out specific functions. By definition, all of these molecules contain Ig-like domains in their extracellular portions (see figure). The Ig-fold structure of Ig-like domains has an extraordinary capacity to promote interactions with other Ig-like domains, either within a multimeric protein or between proteins. As an example of <u>intramolecular</u> interaction, consider the H_2L_2 Ig monomer: C_H domains on one H chain pair with the C_H domains on the other to form the Fc region, while the V_H domains interact with V_L domains to form the antigen binding site. An example of <u>intermolecular</u> interaction is the association of Ig-like domains in MHC molecules with Ig-like domains in the CD4 or CD8 coreceptors, promoting interaction between the T cell and the APC or target cell. Another example is the facilitation of leukocyte extravasation by the interaction of Ig-like domains in the β2-integrins with Ig-like domains in the CAMs. The list of Ig superfamily members is very large and is still growing rapidly, as is the range of their functions. The intramolecular stability of the Ig-fold structure and its ability to facilitate intermolecular associations are the basis for its extraordinarily broad utility and evolutionary conservation.

The similar extracellular regions of the members of the Ig superfamily most likely arose from duplication of a primordial gene

containing a single exon which encoded an early form of the Ig-fold. The product of this gene may have served as a primitive cell surface receptor. Divergent mutation of duplicated copies of this primordial gene and their random distribution and integration across the genome are thought to have produced a multitude of distinct, but related, genes serving widely varied functions. The products of these genes have shared extracellular features, but highly divergent intracellular features that are reflected in each molecule's specialized function. Different intracellular domains trigger different intracellular signaling paths, so that the consequences of ligand recognition at the cell surface vary according to the nature of the signal transduction machinery associated with the intracellular

portion of the protein. By using a modular assembly of the gene sequences for particular protein domains, evolution can link domains for various effector functions to domains for specific ligand recognition, creating entirely new proteins. To illustrate, imagine that evolution duplicates the gene for an extracellular Ig-like domain specifying nerve cell recognition, and couples it via a transmembrane domain to a cytoplasmic domain involved in intracellular calcium release. The result might eventually be a new type of membrane-bound receptor whose engagement by a nerve cell results in intracellular calcium release. The remarkable adaptability of the Ig-fold underlies its selection to serve a wide variety of functions in cellular recognition at the cell surface.

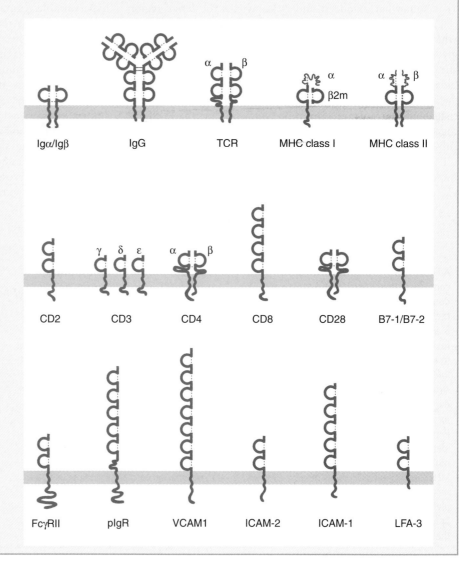

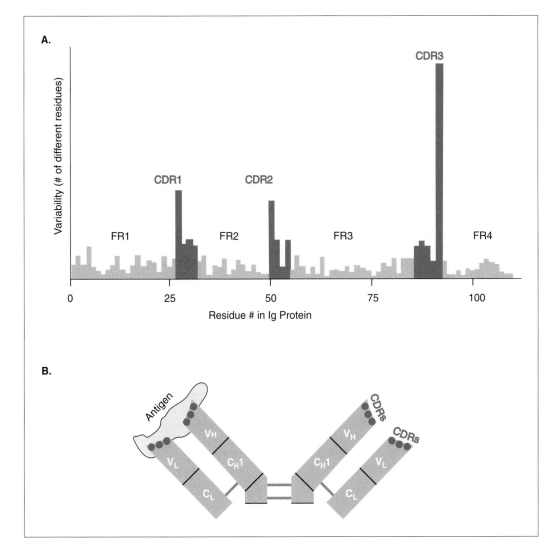

Figure 5-4
Immunoglobulin Hypervariable Regions
(A) The amino acid sequences of many different purified immunoglobulins were compared, allowing production of a variability plot showing the number of different amino acid residues appearing in each given position along the protein. For both heavy and light chains, such a plot reveals three areas of extremely high variability in the V region domains, called complementarity-determining regions (CDRs). Less variable regions between the CDRs are known as framework regions (FRs). (B) When the Ig molecule folds into its native conformation (only the Fab region is shown), the CDRs are grouped together to form an antigen-binding region that is highly variable from antibody to antibody. Part A adapted from Wu T. T. and Kabat E. A. (1970) An analysis of the sequences of the variable regions of Bence Jones proteins and myeloma light chains and their implications for antibody complementarity. *Journal of Experimental Medicine* **132**, 211–250.

recognized by other individual antibodies. These epitopes are called *idiotopes*, and a collection of idiotopes makes up an *idiotype*. These somewhat confusing terms are discussed in more detail in Box 5-3.

iv) Structural Variation in the Constant Region

In contrast to the V region, the amino acid sequence of the Fc region shows very little variation among antibodies, as expected for an entity that triggers the same effector action in response to a wide variety of antigens. It makes evolutionary sense to confine the variability required to generate a complete repertoire of antibodies to one domain (V) at one end of Ig polypeptides, while maintaining a relatively constant protein conformation in those regions (C) that must interact with other proteins involved in antigen clearance mechanisms. Were the variability of the hypervariable regions to extend to the entire Ig protein, not only would antigen clearance mechanisms have to somehow be able to recognize innumerable different proteins, but the stability inherent in the Ig-fold structure might be compromised. Nevertheless, the constant regions of the Igs of a given antigenic specificity can vary in

two major ways: in their isotypes and in their structural isoforms.

v) Isotypes

As well as the major amino acid differences that occur in the variable domains of Ig proteins, there are other differences between Ig molecules in their constant regions, which give rise to different Ig *isotypes*. These amino acid differences can affect the size, charge, solubility, and structural features of a particular Ig, which in turn influence where an Ig goes in the body and how it interacts with surface receptors and other molecules. A mechanism called *isotype switching* (see Ch.9) occurs later in the life span of a B cell clone that allows its individual members to produce Igs of different constant region sequences. All of these differences can influence how a given antibody will clear its antigen.

va) Light chain isotypes. There are two different isotypes, or constant region classes, for L chain Ig polypeptides: the *kappa* (κ) *light chain* and the *lambda* (λ) *light chain* (Fig. 5-5). All antibodies containing the κ light chain have the same C_L sequence (Cκ), which is distinct from the C_L sequence found

Box 5-3. Anti-immunoglobulin antibodies: isotypes, allotypes, and idiotypes

Because Igs are large, globular proteins, they can act as antigens, a fact that immunologists have used to great advantage. By immunizing laboratory animals with Igs, antibodies that recognize different parts of the Ig molecule can be prepared, or "raised." Such "anti-Ig" antibodies have proved to be excellent tools for the isolation and analysis of various Igs. The three major types of anti-Ig molecules that can be produced are called *anti-isotypic, anti-allotypic,* and *anti-idiotypic,* based on which part of the antigenic Ig molecule they recognize. The type of anti-Ig antibody that will be produced depends on the phylogenetic relationship between the animal immunized and the animal from which the antigenic Ig is derived, as this will determine which parts of the Ig appear "foreign" to the responding host. Animals commonly immunized to raise anti-Ig antibodies are those from which it is convenient to obtain antiserum, such as rabbits, sheep, and goats.

Interspecies immunization and anti-isotypic antibodies. Interspecies immunization involves immunizing one species, such as a rabbit, with Ig from another species, such as a mouse. In this case, the overall degree of "foreignness" of the antigenic Ig is high and the host rabbit's antibody response would theoretically involve recognition of epitopes from all constant regions of the mouse Ig. However, the Ig preparation used as the immunogen is usually a polyclonal mixture of antibody specificities and isotypes. Epitopes from specific variable regions of the mouse Ig are present in amounts too low to stimulate a host response, so that the rabbit antibodies will therefore focus mostly on epitopes in the more highly represented constant domains. Among the antibodies raised to the constant regions will be subsets, each of which recognizes epitopes specific to one of the five H chain isotypes or one of the two L chain isotypes. These are called *anti-isotypic antibodies.* If desired, this mixture can be purified so that all antibodies are removed except those recognizing one particular isotype. This yields an anti-isotypic antiserum that is *isotype-specific,* such as rabbit antimouse IgG antiserum or rabbit antimouse kappa antiserum.

Intraspecies immunization and anti-allotypic antibodies. Alleles are multiple forms of the same gene, defined by slightly different nucleotide differences, that are found in different individuals of the same species. Polymorphism is the term used to describe the occurrence of multiple alleles in a

population. Several allelic polymorphisms exist in the various Ig constant region genes of both mice and humans. This genetic variation gives rise to differences in protein structure that make the Ig of one individual immunogenic in some members of the same species. The epitopes that are antigenic in this case are called *allotopes* and the collection of allotopes found in a particular Ig constant region define the *allotype* of the antibody. Thus, a particular isotype of antibody common to all members of a species may exist in different allotypic forms. Anti-allotypic antibodies are usually raised by immunizing one inbred mouse strain with antibodies from a second inbred mouse strain that has allotypic differences in one of the Ig constant regions. Outside the allotypic regions, the Ig constant regions of the two strains will be identical, so the antibody response of the immunized mouse will focus exclusively on the allotypic epitopes, generating a high concentration of *anti-allotypic antibodies.* Interspecies immunizations cannot be used as an effective means to generate anti-allotypic antibodies because allotopes would represent only a very small fraction of the total number of foreign epitopes present, making it impractical to detect the anti-allotypic response separately.

Intrastrain immunization and anti-idiotypic antibodies. In addition to isotypic and allotypic determinants present in the constant region, each Ig molecule has unique determinants (not shared by any other member of the species) called *idiotopes* that are defined by the hypervariable sequences in the variable region. The collection of idiotopes on a given Ig defines its *idiotype.*

In order to generate an *anti-idiotypic antibody* response that is detectable, an immunization protocol must be contrived such that idiotopes are the sole focus of anti-Ig response. Usually this is done by immunizing an inbred mouse with Ig isolated from another member of the <u>same</u> inbred strain, which is, by definition, essentially genetically identical. This will ensure that the response to the relatively minor idiotypic differences is not overwhelmed by the simultaneous production of anti-isotypic or anti-allotypic antibodies. In addition, the response can be focused onto a single idiotype by immunizing with a monoclonal antibody preparation (derived from a single B cell clone), ensuring that the binding sites (and hence the idiotopes) contained in the immunogen are all identical. It has been suggested that these anti-idiotypic antibodies may have a physiological function, such as the regulation of immune responses, but their significance remains unclear.

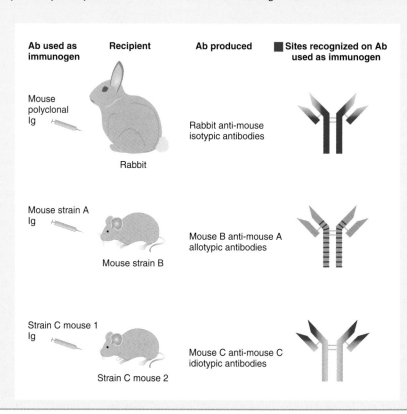

Ab used as immunogen	Recipient	Ab produced	Sites recognized on Ab used as immunogen
Mouse polyclonal Ig	Rabbit	Rabbit anti-mouse isotypic antibodies	
Mouse strain A Ig	Mouse strain B	Mouse B anti-mouse A allotypic antibodies	
Strain C mouse 1 Ig	Strain C mouse 2	Mouse C anti-mouse C idiotypic antibodies	

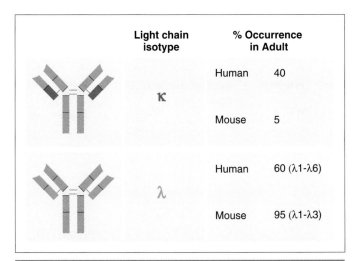

Light chain isotype		% Occurrence in Adult
κ	Human	40
	Mouse	5
λ	Human	60 (λ1-λ6)
	Mouse	95 (λ1-λ3)

Figure 5-5

Light Chain Isotypes
The presence of one of two possible light chain constant regions (κ or λ) determines the light chain isotype of the immunoglobulin molecule. With information from Gorman J. R. and Alt F. W.,(1998) Regulation of Ig Light chain Isotype Expression. *Advances in Immunology* **69**, 157.

in all antibodies containing the λ light chain (Cλ). In fact, human κ chains are more closely related in sequence to mouse κ chains than they are to human λ chains. Single amino acid differences in λ polypeptide sequences have permitted the classification of human λ chains into six subgroups, λ1–λ6, and mouse λ chains into three subgroups, λ1, λ2, and λ3.

Both L chain isotypes can associate with any H chain isotype, but a single B cell expresses antibodies containing only one of the L chain isotypes (see Ch.8). Interestingly, in an individual human's repertoire of antibody-producing B cells, 60% produce antibodies with κ light chains, and 40% produce antibodies with λ chains. In the mouse, due to the nature of its Ig locus (see Ch.8), 95% of antibodies contain κ light chains, and only 5% of mouse B cells make use of the λ chain. There are no known functional differences between κ- and λ-containing antibodies, so the significance of these observations is unclear. However, as is detailed in Chapter 9, the κ L chains are always the first to be expressed during B cell development. Overall, L chains contribute to the diversity rather than to the effector function of antibodies, and so will be discussed in Chapter 8.

vb) Heavy chain isotypes. Minor sequence variation also occurs in the constant regions of H chains. As we learned in Chapter 2, Igs (in most mammals) are categorized into five classes based on differences in their C_H domains. The five major H chain sequences are designated as μ, δ, γ, ε, and α, corresponding to the five major isotypes of immunoglobulins, which are called IgM, IgD, IgG, IgE, and IgA, respectively. The structural differences between H chains are such that the μ and ε chains of human IgM and IgE antibodies contain four C_H domains, whereas IgA, IgD, and IgG antibodies contain only three C_H domains in their shorter α, δ, and γ H chains (Fig. 5-6). The amino acid sequences of C_H1, C_H3, and C_H4 domains in IgM and IgE correspond to the amino acid

sequences of C_H1, C_H2, and C_H3 domains in IgA, IgD, and IgG. IgM and IgE antibodies lack a classical hinge region and rely instead on the pairing of their C_H2 domains (which are unrelated to the C_H2 domains in IgA and IgG) to confer flexibility to the Fab arms. The human IgD and certain IgG immunoglobulins feature an extended hinge region. Further variations among the sequences of the α and γ H chains in humans, and among γ sequences in mice, have given rise to Ig *subclasses*: IgA1 and IgA2, and IgG1, IgG2, IgG3, and IgG4 in humans; and IgG1, IgG2a, IgG2b, and IgG3 in mice. Thus, a total of nine Ig H chain isotypes can be found in humans, and eight in mice.

vi) Carbohydrate Content of Igs
All Igs contain carbohydrate in the form of oligosaccharide side chains, which are generally attached to amino acids in the C_H domains, but not in the V_H, V_L, or C_L domains. Carbohydrate content varies by isotype, ranging from 12–14% for IgD, IgE, and IgM, down to 2–3% for IgA and IgG. Number and location of oligosaccharides can also vary among different molecules secreted by the same B cell clone, depending on host physiological factors at the time of protein synthesis. The precise function of the carbohydrate moieties associated with Igs is not clear.

II. CHANGES TO Ig STRUCTURE ASSOCIATED WITH FUNCTION

i) Introduction to Isotype Switching
When a given B cell clone encounters its T cell-dependent (Td) antigen and is induced to proliferate and differentiate, its first progeny plasma cells produce only IgM antibodies. This is why the majority of antibodies raised in a primary response are found to be of the IgM isotype. However, late in the primary response, proliferating progeny B cells derived from the original antigen-activated clone can differentiate into plasma cells that can produce Igs of isotypes other than IgM. These C-region changes are generated as a result of *isotype switching*, a phenomenon of late B cell clone maturation in which the variable region sequence of the Ig gene is combined at the DNA level with H chain constant region sequences other than those specifying IgM and IgD (see Ch.9). Under the influence of specific collections of cytokines secreted by local antigen-stimulated T helper (Th) cells, a mature activated B cell (in a secondary lymphoid tissue) that encounters fresh antigen toward the end of the primary response can switch to making either IgG, IgE, or IgA (Fig. 5-7). Both memory and plasma cells capable of making these new isotypes will be generated. In subsequent responses to this antigen, a memory B cell of this clone can be triggered to proliferate and differentiate into plasma cells that secrete antibodies of the new isotype. Indeed, progeny cells of the memory B cell (but not fully differentiated plasma cells) may switch isotypes again as the cycle of proliferation and differentiation repeats. In this way, all nine H chain constant region sequences can potentially be employed, resulting in the appearance of a range of isotypes and their subclasses in the collective cells of the clone. In other

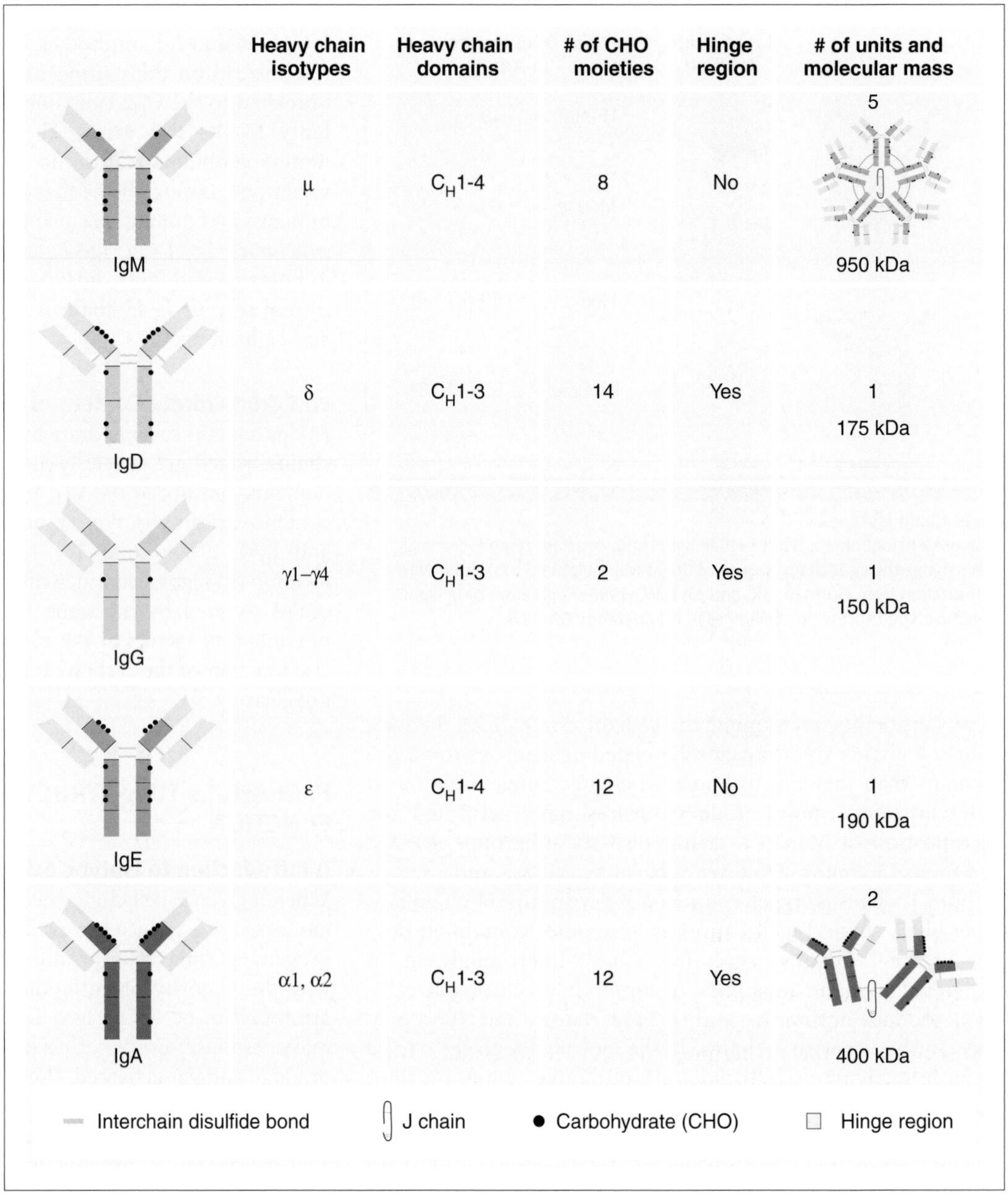

Figure 5-6

Human Heavy Chain Isotypes

The presence of one of five possible heavy chain constant regions (μ, δ, γ, ε, or α) determines the heavy chain isotype of the immunoglobulin molecule. While a single type of μ chain occurs in all IgM antibodies, four slightly different γ chains define the four subtypes of IgG antibodies, and two slightly different α chains define IgA1 and IgA2. The basic monomeric H_2L_2 structures of both IgM and IgA can oligomerize around a small auxiliary protein called the J chain to generate polymeric antibody molecules. IgM is most commonly found as a pentamer whereas IgA usually occurs as a dimer. Murine heavy chain isotypes are similar except that there is only one type of IgA heavy chain and the IgG subgroup chains are called γ1, γ2a, γ2b, and γ3.

	Heavy chain isotypes	Heavy chain domains	# of CHO moieties	Hinge region	# of units and molecular mass
IgM	μ	C_H1-4	8	No	5 950 kDa
IgD	δ	C_H1-3	14	Yes	1 175 kDa
IgG	γ1–γ4	C_H1-3	2	Yes	1 150 kDa
IgE	ε	C_H1-4	12	No	1 190 kDa
IgA	α1, α2	C_H1-3	12	Yes	2 400 kDa

— Interchain disulfide bond   ∫ J chain   ● Carbohydrate (CHO)   ☐ Hinge region

words, all B cells in this clonal population have the same antigenic specificity, but as the primary response progresses, each individual progeny B cell may express this specificity in association with any one of the Ig H chain isotypes, depending on the amount of isotype switching that particular cell and its predecessors have undergone. This capacity of the host to produce antibodies of all isotypes ensures that all effector mechanisms can be brought to bear to eliminate an antigen.

In a healthy adult human, antibodies of all nine Ig subclasses directed against a single antigen can be found simultaneously, although an individual B cell expresses only one isotype of Ig at any one time except for the case of IgM/IgD coexpression in early B cell development (see Ch.9). Each isotype retains the same antigenic specificity in its Fab region but will trigger its preferred effector mechanism by its Fc region,

so that all body compartments are protected from the antigen, no matter how it enters the body. For example, pathogens most often gain access via injured skin, or by penetrating the epithelial barriers lining the respiratory, gastrointestinal, and urogenital tracts. Novel antigens introduced in this way that are not dealt with by the innate barrier defenses provoke a primary adaptive immune response in which predominantly IgM antibodies directed against the pathogen are produced. In a subsequent invasion, memory B cells mediating the secondary response may make specific IgG, IgE, or IgA antibodies against the pathogen, with the possibility that one isotype may predominate. Thus, the pathogen may encounter specific IgA antibodies in the external secretions and specific IgE antibodies on mast cells, as well as specific IgG antibodies in the blood. In a third attack, the same or a different isotype may predominate.

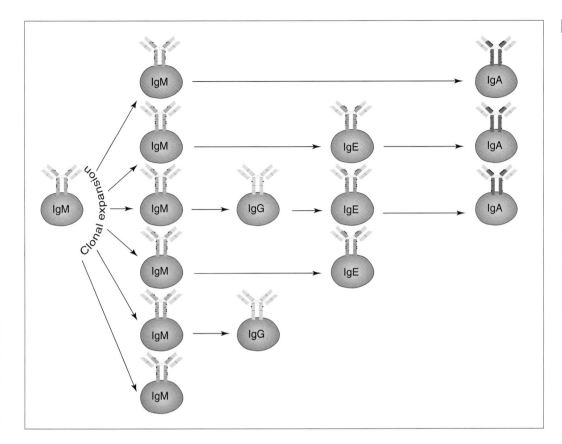

Figure 5-7

Isotype Switching during B Cell Development
During the initial stages of a B cell's primary response to antigen, it produces and secretes IgM. Later in the primary response or during subsequent responses, different heavy chain isotypes may be expressed by the progeny of the original IgM-producing clone. Such "switching" occurs at the DNA level, resulting in the production of an Ig protein with the same V region but a different C region. Thus, over the lifetime of a B cell clone, it may produce antibodies of the same specificity but different isotypes.

ii) Structural Isoforms of Igs

We have noted previously that the basic core structure of any Ig is H_2L_2. Depending on the stage of activation of a given B cell and the cytokine cues it receives from its immediate microenvironment, three different isoforms of the basic Ig can be produced. These structural isoforms, which vary in amino acid sequence at their C-terminal ends, perform distinct functions in the body. *Membrane-bound* Igs serve as part of the B cell antigen receptor. *Secreted* (or *serum*) Igs serve as circulating antibody in the blood. *Secretory* Igs are secreted antibodies that undergo further posttranslational structural modifications that enable them to function in the external secretions of the body, such as in tears and mucus. The details of Ig gene transcription and translation that give rise to the C-terminal differences among the Ig isoforms produced by a given B cell clone will be discussed in Chapter 9. However, it is important to note here that although their antigenic specificities are identical, a membrane-bound Ig protein does not "turn into" a secreted one. The production of a membrane-bound or secreted Ig is regulated by the B cell at the level of transcription. RNA transcripts of the Ig H chain gene that are slightly different at their 3′ (C-terminal) ends are produced, resulting in different structural forms of the Ig protein. This control step occurs prior to protein synthesis.

iia) Membrane-bound Igs.

The BCR on the B cell surface that recognizes antigen is actually a multichain complex, of which the Ig molecule is only one component (the other component chains are discussed later). The membrane-bound Ig protein (often denoted "mIg") in the BCR is very similar in form to the antibody its progeny cells will later secrete. However, compared to the secreted antibody, mIg has an extended C-terminal tail of about 40–60 extra amino acids and lacks a short *tailpiece* (Fig. 5-8A). Membrane-bound Ig is ultimately positioned in the plasma membrane such that its extracellular N-terminal end containing the Fab regions is displayed to the exterior of the cell; the C-terminal tail sequences extending beyond the last C_H Ig domain span the plasma membrane and dangle the C-terminal end of the protein into the cytoplasm. The mIg becomes fixed in the membrane because the Ig H chain gene is transcribed such that the resulting mRNA includes sequences that encode a *transmembrane domain* (TM). The TM characteristically contains 18–22 acidic amino acids followed by a region of 26 residues with hydrophobic side chains that interact with the lipid bilayer. As is detailed in Chapter 9, when the mRNA for the mIg is translated and the Ig protein is synthesized on the rough endoplasmic reticulum (rER), these hydrophobic residues cause the protein to become anchored in the ER membrane. In association with two accessory proteins (Igα and Igβ) needed for the complete BCR (see later), the anchored mIg passes through the Golgi apparatus and into a *secretory vesicle*. The secretory vesicle then fuses with the plasma membrane of the cell, and the Ig protein becomes inserted into the plasma membrane of the B cell. (Note: the use of the term "secretory" here refers to intracellular protein processing and **not** to secretory forms of Igs. Also, the term "Igα" should not be confused with the symbol indicating the IgA H chain constant region, Cα.)

Figure 5-8

Immunoglobulin Structural Isoforms
(A) Membrane-bound Ig is anchored in the B cell membrane by a TM domain. (B) Secreted Ig lacks a TM domain and features a tailpiece. Secreted Ig can polymerize around a J chain protein. (C) Secretory Ig acquires secretory component after passing through a mucosal epithelial cell.

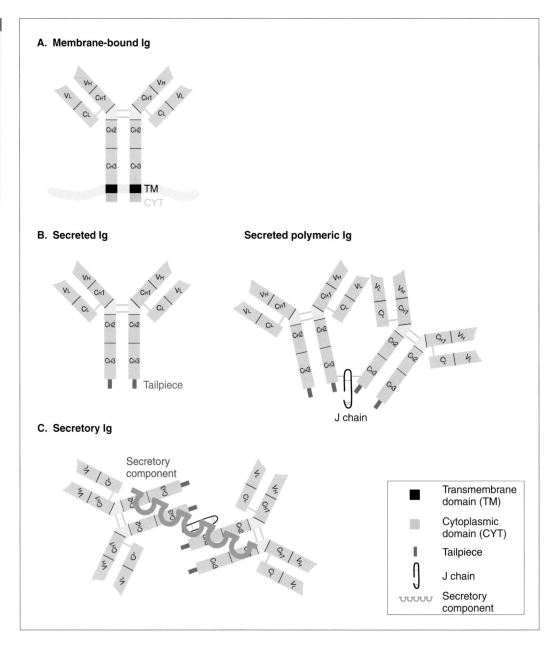

Figure 5-8

Immunoglobulin Structural Isoforms
(A) Membrane-bound Ig is anchored in the B cell membrane by a TM domain. (B) Secreted Ig lacks a TM domain and features a tailpiece. Secreted Ig can polymerize around a J chain protein. (C) Secretory Ig acquires secretory component after passing through a mucosal epithelial cell.

Between the TM and the C-terminus of an Ig H chain is the *cytoplasmic domain*. This region usually contains basic amino acids that interact with the phospholipids on the cytoplasmic side of the membrane. The length of the cytoplasmic region of mIgs varies by isotype: mIgG and mIgE cytoplasmic regions are considerably longer (28 amino acids) than are those of mIgA (14 residues) or mIgM and mIgD (only 3 amino acids).
iib) Secreted Igs: tailpieces and J chains. The mature plasma cell progeny of an activated B cell secrete into the blood the shortest and simplest form of an Ig as antibody, often denoted "sIg" (for "secreted"). An sIg antibody has the same antigenic specificity (same N-terminal sequences) as the receptor mIg on the original B cell that was activated by encountering its antigen, but its C-terminal sequences end shortly after the last C_H Ig domain and it is synthesized without the transmembrane and cytoplasmic sequences. Instead, a short amino acid sequence C-terminal to the last C_H domain, called a *tailpiece* (Fig. 5-8B), is present that facilitates secretion. (It should be noted that "sIg" is sometimes used as an abbreviation for "surface Ig," meaning "membrane Ig." It will not be used in this way in this book, to avoid confusion.)

The basic H_2L_2 unit of an Ig is sometimes referred to as the Ig "monomer." In most cases, mIg is present in the B cell membrane in monomeric form; however, work in the early 2000s has suggested that multimeric structures may exist. On the other hand, it has been known for some time that secreted IgM and IgA frequently assume quaternary structures. These soluble multimers can form because the tailpieces in sIgM and sIgA molecules appear to allow interactions between the L and H chains of several IgM or IgA monomers. Typically, five to six IgM monomers congregate to form a pentamer or hexamer, whereas two to three IgA monomers form dimers or trimers, respectively. All secreted multimeric forms of IgA contain *joining* (J) *chains*, small acidic polypeptides (15 kDa) that bond to the tailpieces of μ and α H chains by disulfide linkages. The J chains are not structurally related to Igs and are encoded by a separate genetic locus. While a single J chain appears to be able to stabilize the component monomers of secreted IgA, it is not crucial for the joining event. Indeed, pentameric sIgM molecules contain the J chain, but hexameric sIgM molecules do not.

iic) Secretory antibodies: the poly-Ig receptor and secretory component. Most antibodies of the IgA isotype occur in the external secretions (such as tears, mucus, breast milk, and saliva) rather than in the blood circulation. The IgA molecules present in these secretions usually take the form of dimers and trimers that contain not only tailpieces and a J chain, but also an additional polypeptide of molecular mass 70 kDa called the *secretory component* (Fig. 5-8C). This heavily glycosylated protein, which is encoded by a gene outside the Ig loci, is actually a piece of a receptor synthesized not by B cells, but rather by mucosal epithelial cells. Mucosal epithelial cells constitute the physiological barrier between tissues such as the gastrointestinal and respiratory tracts and the external environment. (We often call these interfaces the "mucous membranes," or *mucosae*.) Newly synthesized IgA antibodies are released into the tissues underlying the mucosae by IgA-secreting plasma cells that have homed to this location (see Ch.20). Multimeric sIgA molecules gain access to the external secretions by binding to a receptor called the *poly-Ig receptor* ("polymeric immunoglobulin receptor," or pIgR), which is present on that surface of mucosal epithelial cells facing the tissues (rather than the surface facing the lumen of the intestine or respiratory tract,

etc.) (Fig. 5-9). The poly-Ig receptor (whose structure is described in more detail later) recognizes the J chain present in multimeric sIgA molecules. The binding of structural elements in the C-terminal domain of sIgA to pIgR triggers receptor-mediated endocytosis of the Ig multimer into a vesicle, for transport through the epithelial cell to its opposite (lumen-facing) side. This transport process across the cell is called *transcytosis*. However, as the IgA molecule is expelled from the epithelial cell at its luminal surface, a part of the pIgR molecule is enzymatically released from the vesicle membrane and remains attached to the IgA polymer as it is pushed into the external secretions: this piece is the secretory component. Although very few IgM antibodies enter the external secretions, those that do also contain secretory component. It is thought that the J chain in a pentameric IgM molecule is recognized by pIgR in the same way, triggering the same process of receptor-mediated endocytosis and receptor cleavage. Polymeric Igs that do not contain the J chain cannot bind to pIgR and thus do not enter the secretions.

To review the three structural isoforms of Ig proteins: The membrane-bound Ig contains the antigen-binding site and its framework regions in the V region, the C domains, a transmembrane region, and a cytoplasmic region (no tailpiece). It functions in the B cell antigen receptor complex. Secreted antibodies in the blood or tissues contain the same antigen-binding site sequences as are found in the antigen receptor Ig, the same C domains, a tailpiece (no TM region), and a J chain if the antibody is multimeric. Antibodies in the external secretions contain the same antigen-binding site sequences as those in the antigen receptor Ig, the same C domains, a tailpiece, a J chain, and the secretory component. It must be emphasized that the V_L and V_H sequences of all three forms of a given Ig

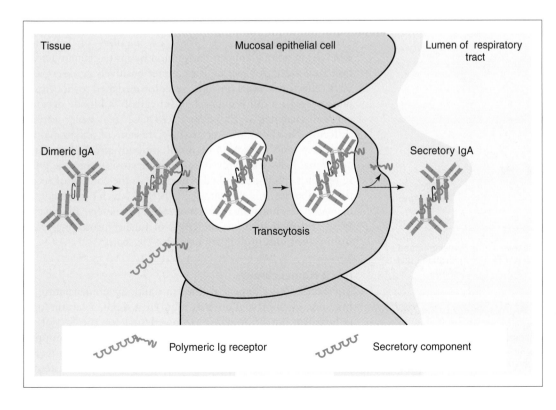

Tissue	Mucosal epithelial cell	Lumen of respiratory tract

Dimeric IgA

Transcytosis

Secretory IgA

Polymeric Ig receptor   Secretory component

Figure 5-9

Generation of Secretory Antibody
In this example of secretory Ab generation, dimeric IgA is secreted into the tissues underlying the mucosal epithelium lining the respiratory tract. The dimeric IgA binds via its J chain to a polymeric Ig receptor (pIgR) expressed on the surface of a mucosal epithelial cell. The IgA is internalized via receptor-mediated endocytosis into a vesicle and transported across the cell (transcytosis). Upon fusion of the vesicle with the side of the epithelial cell facing the lumen of the respiratory tract, the pIgR molecule is enzymatically cleaved. The polymeric IgA and part of the pIgR molecule that remains attached (secretory component) are released into the lumen as secretory IgA.

are identical, so that the antigen-binding sites of the mIg, sIg, and secretory Ig produced by members of one B cell clone are identical. In other words, the form of the Ig protein has no effect on antigenic specificity.

III. THE B CELL ANTIGEN RECEPTOR COMPLEX

Although mIgs provide the specificity for antigen recognition, the C-terminal cytoplasmic sequences of the Ig protein are so short that the mIg molecules need other molecules to help convey the intracellular signals indicating that antigen has been bound. The tyrosine kinases and phosphatases that carry out the actual enzymatic reactions of intracellular signaling simply cannot bind to these short tails. In addition, the Ig cytoplasmic tails contain none of the molecular motifs generally associated with intracellular signaling. Instead, the mIg molecule relies on the Igα and Igβ chains with which it is cosynthesized and co-inserted into the rER (see Ch.9). These chains form the heterodimeric *Igα/Igβ* transmembrane molecule, which is expressed only in B cells (Fig. 5-10). The Igα and Igβ chains are held to each other by a disulfide bond located just N-terminal to the TM region. The extracellular portions of the Igα and Igβ proteins each consist of one Ig-like domain, allowing the Igα/Igβ heterodimer to associate with the membrane-proximal C domains of the H chains in the mIg molecule (of any isotype). In other words, the complete BCR has the stoichiometry of mIg:Igα:Igβ. Recognition of antigen depends on the mIg component, but antigen binding is not perceived by the B cell without the signals conveyed by the Igα/Igβ heterodimer.

How does Igα/Igβ facilitate intracellular signaling? The Igα (42 kDa) and Igβ (37 kDa) polypeptides are glycoproteins

with long tyrosine-containing cytoplasmic tails (61 amino acids in Igα, 4 tyrosine residues; 48 amino acids in Igβ, 2 tyrosines). These tyrosine residues are located in amino acid sequence motifs called "immunoreceptor tyrosine-based activation motifs" (ITAMs). This name was given to these sequences because they were first identified in studies of signal transduction from the BCR and TCR. ITAMs are described fully in Box 9-1 in Chapter 9. Suffice it to say here that the presence of one or more ITAM sequences in the cytoplasmic tail of a molecule confers upon it the ability to recruit intracellular signaling kinases and other molecules needed to trigger a signaling pathway, resulting in the activation of transcription factors in the nucleus. These factors initiate new gene transcription, leading to cellular activation and the triggering of effector functions. The Igα and Igβ chains may have slightly different roles in the BCR signaling process, but this has yet to be completely defined.

IV. Fc RECEPTORS

Antibodies are sophisticated molecular tags which function by binding to a pathogen or other foreign entity in the body and marking it for eventual destruction by various effector functions. The binding of antibody to antigen is therefore only the first step in the elimination of an invading entity. Additional mechanisms (such as phagocytosis) are required for physical removal of the antibody–antigen complex, and these depend on the recognition of the Fc region of the Ig molecule by *Fc receptors* (FcRs). FcRs are expressed on the surfaces of various effector cells (such as macrophages) that actually eliminate the antigen. There are FcRs for all antibody isotypes except IgD. The FcRs binding to IgG, IgA, and IgE are the best characterized.

Molecules in the FcR family can be broadly classified into two functional categories. The vast majority of FcRs are involved in linking antigen recognition by Igs to cellular activation and/or effector mechanisms, either positively or negatively. Such effector actions include receptor-mediated endocytosis, phagocytosis, a cell lysis mechanism called "antibody-dependent cell-mediated cytotoxicity" (ADCC; see next section, 'Effector Functions of Antibodies'), release of inflammatory mediators, and regulation of B cell antibody production. Two additional FcRs are involved in the transcytosis of Igs: the poly-Ig receptor (pIgR), described previously in the context of secretory antibodies, and neonatal FcR (FcRn), which plays a role in neonatal immunity in many mammals (see later). Binding to FcRn and pIgR also has the effect of further protecting an Ig from proteolysis, prolonging its useful life span.

i) FcR Nomenclature
The Fc receptors appear to have a daunting nomenclature at first, but when broken down symbol by symbol, a certain logic is revealed. An Fc receptor is named first for the Ig isotype to which it binds, indicated by a Greek letter. For example, the Fcγ receptors, or FcγRs, are the molecules binding to the Fc region of IgG antibodies. If major subtypes of an Fc receptor protein exist, they are denoted by Roman numerals. For

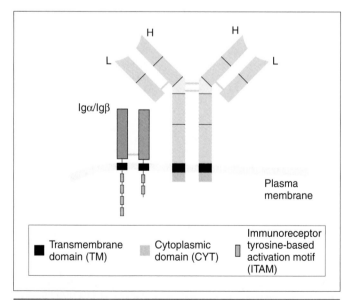

Figure 5-10

The B Cell Antigen Receptor Complex
Noncovalent association between membrane-bound Ig and the transmembrane Igα/Igβ heterodimer forms the B cell antigen receptor complex. The complex allows transduction of signals to the inside of the cell upon antigen–antibody binding.

example, there are three major FcR subtypes that bind to IgG: FcγRI, FcγRII, and FcγRIII. These proteins are derived from independent genes and differ significantly in their amino acid sequences and in their affinities for the various IgG subtypes. They also differ in their cell type distribution, which determines the effector functions they mediate. Within each subtype, receptors with structures that are closely related but which are still derived from distinct genes are indicated with a Roman capital letter. For example, FcγRIII occurs in two slightly different forms: the FcγRIIIA receptor, which is expressed on NK cells, and the FcγRIIIB receptor, expressed exclusively on neutrophils. In a few cases, even more subtle variations in protein sequence exist within an FcR subtype, arising from different alleles of an FcR gene. These minor subtypes are indicated by lowercase letters (e.g., FcγRIIIAa and FcγRIIIAb). Finally, alternative splicing of an FcR allele can give rise to very similar but distinguishable FcR proteins, which are denoted by an Arabic number (e.g., FcγRIIIAa1 and FcγRIIIAa2).

ii) FcR structure

Molecules binding to the Fc regions of Igs vary surprisingly widely in structure (Fig. 5-11). Most FcRs display several Ig-like extracellular domains (making them members of the Ig superfamily), a transmembrane domain, and a cytoplasmic domain that often contains ITAMs. Many membrane-bound FcRs are multisubunit complexes, whereas others are single polypeptide chains. Several FcRs are composed of one subunit that confers Fc region binding specificity (often designated α) plus two other subunits involved in either transport to the cell surface or intracellular signal transduction in response to Ig binding. An "FcR common γ chain," which contains a single ITAM but occurs naturally in FcR complexes as a dimer, conveys intracellular signals for a number of FcRs. The outcome of FcR engagement by antibody depends on interactions between FcRs on the effector cell surface, the signal transduction molecules involved, and the cell type bearing the FcRs. The type of antigen bound to the Ig does not appear to have an influence. Some FcRs occur in soluble forms, derived by alternative splicing of the transmembrane exon or by proteolytic cleavage. Soluble FcRs are able to bind Igs but their function remains unclear. Some observers speculate that soluble FcRs may compete with membrane-bound FcRs for antibody, thereby modulating antigen clearance functions.

Within the class of FcRs that mediates effector functions, the outcome of FcR binding is determined by the presence or absence of the short conserved amino acid sequences that are important for promoting or downregulating intracellular signal transduction. Most effector FcRs display one or more ITAM motifs in their cytoplasmic tails. In contrast, the FcγRIIB family of effector FcRs does not possess an ITAM but instead harbors a six-amino acid motif called an immunoreceptor tyrosine-based inhibitory motif (ITIM), which inhibits signal transduction (particularly calcium mobilization) and acts as a brake on cellular activation. The FcγRIIIB molecules contain neither ITAMs nor ITIMs but can transduce signals by coaggregating with the ITAM-containing common γ chain. The transporter FcRs, which have no need

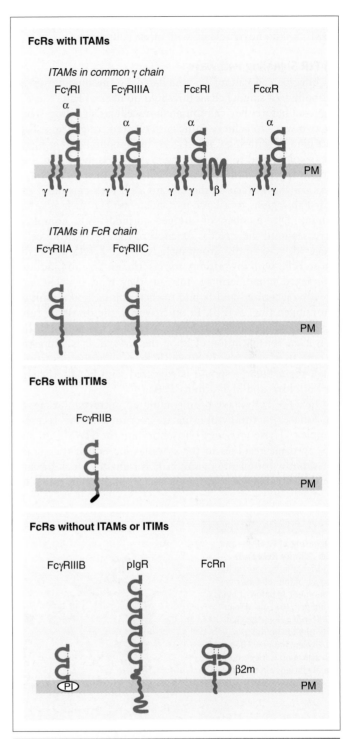

Figure 5-11

Examples of Fc Receptors
Selected examples of Fc receptors, grouped according to the presence or absence of motifs involved in signaling. Most FcRs are anchored in the membrane by a transmembrane domain. FcγRIIIB lacks a transmembrane domain and instead is anchored by a phosphatidylinositol domain. Receptor functions are described in the text. ITAM, Immunoreceptor tyrosine-based activation motif; ITIM, immunoreceptor tyrosine-based inhibitory motif; PM, plasma membrane; β2m, β2-microglobulin; PI, phosphatidylinositol membrane linkage. Adapted from Daeron M. (1997) Fc receptor biology. *Annual Review of Immunology* 15, 203–234.

to transduce intracellular signals, possess neither ITAMs nor ITIMs and do not associate with other FcR chains.

iii) FcR Signaling Pathways

An effector cell cannot be stimulated by the binding of one antibody to a single FcR: a threshold number of FcRs must be triggered before they can initiate intracellular signaling. There are two ways to achieve sufficient FcR triggering. Those FcRs that bind Ig with high affinity can bind monomeric Ig <u>before</u> it has complexed to antigen (Fig. 5-12A). As more and more antigen-specific antibodies become fixed by their Fc regions to FcRs on the surface of the effector cell, they can bind with their Fab regions to a large particulate antigen such as a bacterium. The number of antigen–antibody–FcR complexes increases until the effector function of the FcR-bearing cell is triggered. FcRs in this group include FcγRI, FcεRI, and FcαR. Those FcRs with low affinity for Igs are not able to bind non-complexed monomeric Ig. Instead, they bind Igs that have already recognized and bound to their epitopes on a multivalent antigen (Fig. 5-12B). In other words, the threshold number of stimulated FcR complexes is reached via a pre-existing high local concentration of Ig on the antigen. It is interesting to note that, once the Ig-bound FcRs are successfully stimulated, the cellular response is triggered with equal efficiency for both low- and high-affinity FcRs.

Once the FcRs have been stimulated, intracellular signaling kinases in the cytoplasm are recruited to the receptor complex. These enzymes phosphorylate the ITAMs on specific tyrosine residues in the cytoplasmic tail of the signaling subunit (often the common FcRγ chain). Docking sites are thus created for additional kinases and signal transduc-

tion molecules, which in turn join the receptor complex and are phosphorylated. Various cytoplasmic signaling pathways are triggered, leading to the rapid release of calcium, the reorganization of the cytoskeleton in preparation for phagocytosis, and degranulation. Additional signals that are conveyed to the nucleus of the effector cell activate new gene transcription. It has thus become clear that the function of FcR stimulation is to bring enzymes essential for signal transduction, and their substrates, into comfortable proximity.

iv) Functions of Effector FcRs

Once triggered by Ig binding, all effector FcRs are capable of inducing endocytosis by the effector cell, if the antigen is small enough. Similarly, phagocytes upon which sufficient effector FcRs of almost any type have become stimulated readily phagocytose larger particles without the need for cellular activation. However, only those FcRs that possess ITAMs (or those FcRs lacking ITAMs or ITIMs that coaggregate with an ITAM-containing chain) are capable of delivering the necessary signals for full effector cell activation. Such signals induce neutrophils to degranulate and to mount a respiratory burst, or trigger macrophages to secrete inflammatory cytokines. Mast cells are also induced to degranulate, releasing inflammatory mediators, and NK cells and macrophages begin to lyse target cells using ADCC. However, if an FcR containing an ITAM happens to coaggregate with an FcγRIIB molecule containing an ITIM, the signaling pathways leading to complete activation are disrupted, due to the recruitment of phosphatases that reverse ITAM phosphorylation. The cell is then restricted to endocytosing or phagocytosing the antigen.

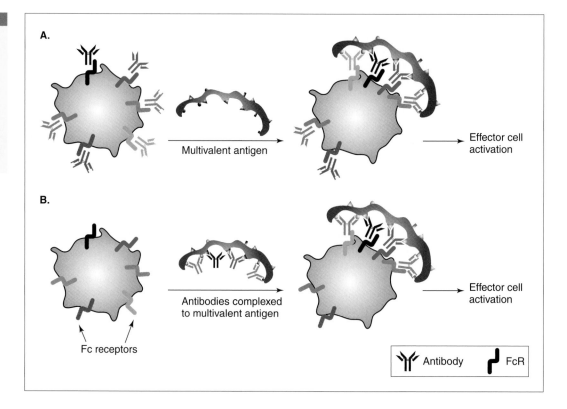

Figure 5-12

Triggering of Fc High- and Low-Affinity Receptors
(A) High-affinity FcRs on the surface of the cell bind monomeric Ig before it binds to antigen. (B) Low-affinity FcRs bind multiple Igs that have already bound to a multivalent antigen. Effector cell activation is triggered with equal efficiency in both cases.

A.

Multivalent antigen

Effector cell activation

B.

Antibodies complexed to multivalent antigen

Effector cell activation

Fc receptors

Antibody   FcR

The isotype of the antibody bound to the antigen determines which effector cells are activated, since different collections of FcRs are expressed on the surfaces of different effector cells. For example, eosinophils generally lack FcγR but express FcεR and FcαR, and so would ignore antigens coated in IgG but respond to those covered in IgA or IgE. However, most effector cells, and especially neutrophils, carry several different types of FcRs, allowing them to bind antigens coated in antibody of almost any isotype. Furthermore, effector cells can often execute more than one type of effector action. The diversity of mechanisms available through the existence of different 18 antibody isotypes, FcRs, and effector actions thus enhances the effectiveness of the antibody response. We will now examine in more detail the specific characteristics and functions of each class of effector FcR (Table 5-1).

iva) FcγR. The FcγRs are arguably the best studied of the Fc receptors, and their diversity in form reflects their diverse functions. The three major subtypes of FcRγ, namely FcγRI, FcγRII, and FcγRIII, differ both in their possession of ITAMs/ITIMs (and therefore in their effects on effector cell activation) and in their affinities for IgG.

FcγRI (CD64) is known as the high-affinity IgG receptor ($K = 10^8 \ M^{-1}$). (K represents a measure of binding affinity. How the values for binding affinities are determined is covered in Chapter 6.) FcγRI expression is restricted to activated granulocytes and is upregulated by the presence of inflammatory cytokines. FcγRI is composed of an IgG-binding α chain and an FcRγ homodimer containing ITAMs (refer to Fig. 5-11). The FcγRI complex binds to the C-terminal region of only one of the H chains making up the Fc region of

the IgG molecule. FcγRI binds to IgG1 or IgG3 antibodies with the greatest affinity, making these Ig isotypes very good opsonins. In other words, pathogens coated with IgG1 or IgG3 are more firmly bound to FcγRI on phagocytes than are those pathogens coated with IgG4 or IgG2, which bind to FcγRI with much lower affinity. FcγRI stimulation by IgG also triggers the secretion of inflammatory cytokines and ADCC-mediated target cell lysis by macrophages and monocytes, and superoxide production by neutrophils. The high affinity of FcγRI ensures that it can trigger effector responses even at the low concentrations of IgG prevalent in the early stages of an immune response.

FcγRII (CD32) is expressed more broadly than FcγRI but the affinity of FcγRII for noncomplexed IgG molecules is only 1/100th that of FcγRI. Instead, FcγRII interacts only with IgG that has already been aggregated by binding to a multivalent antigen. Structurally, the FcγRII molecule consists of a single transmembrane chain. The FcγRIIA isoform (ITAM-containing) is expressed primarily on phagocytes but also on megakaryocytes and platelets. Stimulation of FcγRIIA by IgG binding is associated with the induction of the respiratory burst in neutrophils and the secretion of inflammatory mediators by platelets. The FcγRIIB isoform does not contain an ITAM sequence but rather an ITIM. FcγRIIB appears on lymphoid and myeloid cells. It is thought that the primary function of this receptor on B cells may be to provide feedback signals regarding antibody production levels, damping down B cell activation when sufficient antibody has bound to surface FcγRIIB molecules. In fact, as mentioned previously, if the FcγRIIB molecule on any effector cell type (including mast cells and T cells) is

Table 5-1 Fc Receptor Expression and Main Effector Function

Fc Receptor	Expression	Main Effector Function
FcγRI (CD64)	Mo, Ma, Neu, Gr	Activation, endocytosis, phagocytosis, ADCC
FcγRIIA (CD32)	Ma, Mo, Neu, Meg, Pl, B	Activation, endocytosis, phagocytosis
FcγRIIB (CD32)	Hematopoietic cells (myeloid and lymphoid)	Inhibits cell activation; endocytosis, phagocytosis
FcγIIC (CD32)	Ma, Mo, Neu, Meg, Pl	Activation, endocytosis, phagocytosis
FcγRIIIA (CD16a)	Mo, Ma, NK, M	Activation, endocytosis, phagocytosis, ADCC
FcγRIIIB (CD16b)	Neu	Contributes to signaling by association with other FcRs
FcεRI	Ms, Ba, LC, *Eo, *Mo	Activation, endocytosis, phagocytosis; triggers mast cell/basophil degranulation
FcεRII (CD23)	Mo, Ma, Eo, Neu, T, B	Enhances Ab response; regulation of IgE synthesis
FcαRI (CD89)	Ma, Mo, Neu, Eo	Activation, endocytosis, phagocytosis, ADCC
Poly-Ig receptor	Epithelial cells of gastrointestinal, respiratory, and urogenital tracts	Transport of IgA and IgM
FcRn	Placenta, fetal yolk sac, gut epithelial cells, hepatocytes	IgG transport from mother to newborn

ADCC, antibody-dependent cell-mediated cytotoxicity; B, B cell; Ba, basophil; Eo, eosinophil; Gr, granulocyte; LC, Langerhans cell; M, mast cell; Meg, megakaryocyte; Mo, monocyte; Ma, macrophage; Neu, neutrophil; Pl, platelet; T, T cell.
*In allergic patients.

coaggregated with ITAM-containing chains, the activation of that cell is inhibited. The FcγRIIC isoform is expressed on macrophages, monocytes, neutrophils, megakaryocytes, and platelets. Although it binds IgG with low affinity and contains an ITAM, it does not appear to mediate any specific effector cell action. Its function is so far unknown.

The FcγRIII receptors (CD16) also have low affinity [$K = (1–2) \times 10^5 \ M^{-1}$] for IgG. Similar in structure to the FcγRI, the FcγRIIIA isoform consists of an IgG-binding α chain and a dimer of the ITAM-containing FcRγ chain. FcγRIIIA is expressed principally on phagocytes and NK cells but also on immature B cells, mast cells, Langerhans cells, and γδ T cells. Interestingly, FcγRIIIA is the only FcR expressed on NK cells. FcγRIIIA, which exhibits a typical FcR transmembrane region, mediates ADCC by NK cells and mast cell degranulation. In contrast, the FcγRIIIB isoform, which is expressed exclusively on neutrophils, is a single-chain protein lacking both transmembrane chains and ITAMs. Instead, FcγRIIIB is anchored in the membrane by linkage to phosphatidylinositol residues. In order to carry out intracellular signaling, the FcγRIIIB is thought to coaggregate with an FcγRIIA molecule expressed on the same cell and to take advantage of its FcRγ chain. Neutrophils do not use their FcγRIII molecules to participate in ADCC as do NK cells; instead, they use these receptors to signal the need for vigorous phagocytosis. Expression of FcγRIII is upregulated by IFNγ.

ivb) FcεR. The Fc receptors for IgE have also been reasonably well studied. There are two known subtypes, FcεRI and FcεRII. The high-affinity receptor FcεRI is expressed mostly on basophils in the blood and mast cells in the tissues, but also on Langerhans cells and eosinophils and monocytes of allergic individuals. FcεRI is a tetrameric molecule composed of α and β polypeptides and a dimer of the common (ITAM-containing) FcRγ chain in an $\alpha\beta(\gamma)_2$ structure (refer to Fig. 5-11). The β chain also contains an ITAM in its cytoplasmic domain. Because of its high affinity ($K = 10^9–10^{10} \ M^{-1}$), FcεRI is capable of binding to free monomeric IgE. (Only one of the two CH4 domains of the IgE molecule interacts with the receptor.) Thus, in individuals making low levels of circulating IgE antibodies to specific allergens (such as pollens or animal dander), the FcεRI molecules effectively concentrate the body's IgE on mast cell and basophil surfaces. When multivalent antigen binds to several immobilized IgE molecules on a mast cell, the associated FcεRI molecules aggregate, triggering the mast cell to degranulate and release its vasoactive amines. The result appears as a local inflammatory response and symptoms of allergy (see later, this chapter, and Ch.28).

FcεRII (CD23), which has a much lower affinity for IgE than does FcεRI, consists of a single transmembrane chain. It also has an unusual lectin-binding domain that allows it to bind to certain carbohydrate structures on cells. FcεRII is expressed on many hematopoietic cells, including monocytes, eosinophils, and B and T cells, but does not appear on mast cells or basophils. The FcεRIIA isoform occurs primarily on B cells but its function here is obscure. FcεRIIA may be involved in binding to the important B cell coreceptor molecule CD19, and some observers have suggested that FcεRIIA may play a role in regulating IgE synthesis by B cells. The FcεRIIB

isoform appears on monocytes and eosinophils, allowing these cells to destroy IgE-coated target cells via ADCC. FcεRIIB engagement also appears to stimulate NO production and IL-10 synthesis by macrophages and monocytes. *In vitro*, IL-10 downregulates FcεRII expression on B cells.

ivc) FcαR. Two FcRs exist for monomeric IgA molecules: FcαR (CD89) and FcαRb. FcαR is the high-affinity ($K = 5 \times 10^7 \ M^{-1}$) IgA receptor, capable of binding noncomplexed monomeric IgA. FcαR consists of an IgA-binding α chain and a dimer of the FcRγ chain. FcαR is expressed in five splice variants that are differentially expressed on hematopoietic cells, including eosinophils, monocytes, neutrophils, and alveolar macrophages. Neutrophils are known to use their FcαR to participate in ADCC, whereas macrophages are induced to produce inflammatory mediators such as TNF and IL-6. FcαR expression is upregulated by TNF and downregulated by transforming growth factor β (TGFβ).

FcαRb lacks the transmembrane and cytoplasmic domains of FcαR and instead contains sequences that facilitate secretion. Its precise function is as yet unknown. Expression of mRNA for FcαRb has been detected in neutrophils and eosinophils.

ivd) FcμR. Fc receptors have been identified for IgM but have not been as well characterized as yet.

v) Functions of Transporter FcRs

va) Neonatal immunity mediated by FcRn. The immune system of neonatal mammals is not fully developed at birth and is limited in its ability to eliminate certain microbes. In human infants, although independent IgM synthesis starts at birth, it may take as long as 6–12 months for adequate levels of serum and secretory Igs to be produced. To compensate, evolution has provided protection to fetuses and neonates through two types of *passive immunity*, which is defined as "protection by preformed antibodies transferred to a recipient." The offspring of some species, including humans, other primates, rabbits, and guinea pigs, are protected primarily by *antenatal* passive immunity. In this case, maternal sIgG antibodies of all subclasses (but no other isotypes) cross the placenta and enter the fetal circulation, preparing it for the moment of birth when it enters a pathogen-filled environment. By about 9 months after birth, maternal sIgG is no longer detectable in a human infant's blood. However, human infants and the offspring of certain other species (including ruminants, pigs, and horses) are protected after birth by *neonatal* passive immunity, in which maternal secretory antibodies concentrated in breast milk are taken into the neonatal gut during nursing. Maternal secretory IgA consumed by the neonate remains in the lumen of its gut and protects the digestive tract by neutralizing any ingested pathogens. In addition, in rats (and probably other species), maternal IgG secreted into the breast milk can be taken up by cells lining the neonatal gut in a process that is essentially the reverse of secretory IgA generation. This process is mediated by FcRn (originally called FcRB, for the "Brambell receptor").

FcRn is an IgG Fc receptor with an unusual composition. It consists of an IgG-binding α subunit related in structure to the MHC class I α chain, and the β2-microglobulin chain. However, the counterpart of the MHC class I binding groove in FcRn is too narrow to bind peptides; no function has been ascribed to this structure. In studies in rats, when the breast milk enters the acid environment of the neonatal gut, monomeric FcRn on the surface of the gut epithelial cells facing the gut lumen are able to bind the maternal monomeric sIgG with reasonably high affinity [$K = (2–5) \times 10^7\ M^{-1}$]. The epithelial cells then use receptor-mediated endocytosis and intracellular vesicle transport to transcytose the IgG through the cell, across the membrane, and into the neonatal blood vessels. In the relatively neutral pH of the blood, the affinity of FcRn for IgG drops sharply ($K = 10^5\ M^{-1}$) and the IgG is released into the circulation of the neonate. The large difference in affinity ensures that the transport of the IgG is unidirectional, and the use of differential pH as a release mechanism allows the FcRn molecule to be used for another round of transport. The passively acquired IgG antibodies last just long enough to protect the rat neonate until its own system is functioning at full strength. The astute reader will wonder "How did the sIgG get into the breast milk in the first place, since it is not normally a secretory antibody?" The answer to that question remains unclear, although scientists speculate that, depending on the species, there may be a specialized FcγR on mammary cells in the breast tissue that can take up IgG either directly from the plasma, or from IgG-secreting plasma cells located in the mammary tissue. FcRn has now been implicated in IgG transport in additional tissues, including human placenta, rat fetal yolk sac, and rat hepatocytes.

vb) Poly-Ig receptor. The function of the poly-Ig receptor in generating secretory IgA and IgM was discussed earlier in this chapter. pIgR is expressed on the basal plasma membranes of epithelial cells in the urogenital, respiratory, and gastrointestinal tracts. pIgR is a single-chain molecule containing five extracellular Ig-like domains that can associate with the Fc regions of polymeric IgA and IgM molecules. A disulfide bond between the Fc region on one IgA H chain and the fifth Ig domain of the poly-Ig receptor provides additional stability. pIgR binds dimeric IgA with an affinity of $K = 10^9\ M^{-1}$ and pentameric IgM with an affinity of $K = 10^8\ M^{-1}$, leading to the transcytosis of these J chain-containing polymeric Igs across the epithelium. The reader will note that the direction of transport in this case is opposite to that of the FcRn, moving Igs from the blood to the external secretions. Because proteolysis of the secretory component is used to release the Ig into the external secretions, each pIgR molecule can be used for only one round of transport.

vi) Other Signal Transduction Mediated by FcR Chains

Fc receptors or their components can also participate in what appears to be unrelated intracellular signal transduction, since their cytoplasmic portions have been found to associate with the relevant signaling subunits of non-Fc receptors. For example, FcγRIIIA has been identified on some T cells, and stimulation of FcγRIIIA by antigen-bound IgG is thought to initiate a signal transduction cascade mediated by the association of the cytoplasmic tail of the Fc receptor with part of the T cell receptor complex (specifically, the CD3ζ chain; see Ch.12). Similarly, in some members of a subclass of γδ T cells known as intraepithelial lymphocytes (γδ IELs), the TCR conveys its intracellular signals by coupling to the common FcRγ chain rather than the CD3ζ chain used in conventional T cell signaling (see Ch.12).

C. Effector Functions of Antibodies

The binding of antibody to antigen leads to clearance or destruction of the latter and the protection of the host. As mentioned previously, most of the effector functions of antibody are mediated by the Fc region of the Ig molecule. By interacting with Fc receptors on the cell surfaces of other leukocytes (usually cells of the innate immune response), the antibody signals that the antigen is to be eliminated. Since different isotypes can trigger different effector activities, the means of elimination available for a given antigen will therefore depend on the isotype of the antibodies to which it is bound.

Four types of effector functions can be ascribed to antibodies (1) *neutralization*, (2) *opsonization*, (3) *antibody-dependent cell-mediated cytotoxicity* (ADCC), and (4) *complement activation*. The first of these depends solely on the Fab region of the Ig molecule and so is "isotype independent"; the others involve interactions with the Fc region and are "isotype dependent." Some of these were briefly introduced in earlier chapters. We shall now examine the effector functions of antibodies in detail and how each of these mechanisms contributes to antigen elimination.

I. NEUTRALIZATION

Antigen neutralization is carried out by secreted or secretory antibodies. Certain viruses, bacterial toxins, and the venom of insects or snakes cause disease by binding to proteins on the host cell surface and using them to enter host cells. In this situation, a neutralizing antibody that can recognize and bind to the virus, toxin, or venom can physically prevent it from binding, thereby protecting the cell (Fig. 5-13). In the case of the initial stages of a viral infection, if preformed neutralizing antibodies exist, the spread of the virus can be averted. However, once an infection is entrenched, neutralizing antibodies are no longer sufficient and the action of cytotoxic T cells is vital for successful virus elimination.

II. OPSONIZATION

As was introduced in Chapter 4, *opsonization* is the process by which an antigen is coated with a host protein (the opsonin) in order to enhance recognition by phagocytic cells. Opsonization is important for two reasons: (1) To depend on the random engulfment of antigen by phagocytes would be very inefficient as an immune defense. (2) Unlike lymphocytes, phagocytes do not specifically recognize and bind to unique epitopes on pathogens and antigens, and do not possess the lymphocyte's powers of clonal expansion or mechanisms of diversity generation. Instead, as we learned in Chapter 4, the major phagocytic cell types rely on pattern recognition receptors and Fc receptors expressed on their cell surfaces to mount an effective, nonspecific, defense. The FcγRs bind particularly strongly to the Fc regions of IgG1 and IgG3 antibodies that have combined with the pathogen or antigen, resulting in a stimulation of the phagocytosis of the antigen–antibody complex

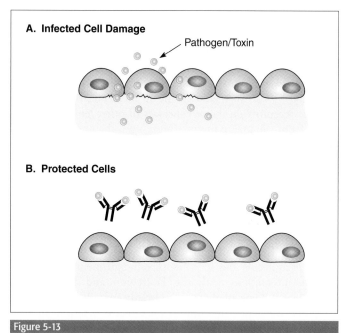

Figure 5-13

Neutralization of Pathogen/Toxin by Antibody
(A) In the absence of specific antibody, a pathogen or pathogen-derived toxin can freely bind to and enter host cells. (B) If specific antibody is present, the pathogen/toxin is blocked from binding and entering the cells. That is, the potential of the pathogen to cause damage is "neutralized."

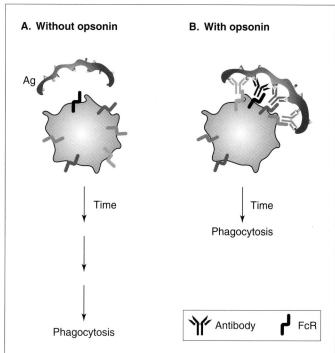

Figure 5-14

Opsonization by Antibody

(Fig. 5-14). As much as a 4000-fold increase in the rate of phagocytosis of an antigen has been observed *in vitro* when specific antibody is added to act as an opsonin.

III. ANTIBODY-DEPENDENT CELL-MEDIATED CYTOTOXICITY

Opsonization is a means of disposing of antibody-coated antigens that are small enough to be engulfed by a phagocyte. This includes most molecular antigens and some very small organisms. When an antigen is antibody coated but too large to be engulfed, as is the case for many cellular pathogens, an alternative mechanism called *antibody-dependent cell-mediated cytotoxicity* (ADCC) becomes important. ADCC is carried out by certain leukocytes that express Fc receptors and have cytolytic capability, such as NK cells, eosinophils, and, to a lesser extent, neutrophils, monocytes, and macrophages. These properties allow these "lytic" cells to lyse Ig-coated targets directly. Such targets include certain bacteria and viruses, certain tumor cells, grafted cells (the effect is one aspect of transplant rejection), and, in cases of autoimmune disease, self cells. Once a target is coated by antibody, the projecting Fc regions can bind to the FcRs of a lytic cell. This binding stimulates the metabolism of the latter, resulting in increased synthesis of the hydrolytic enzymes stored in cytoplasmic granules. Stimulation of the Ig-bound FcR on the surface of the lytic cell triggers degranulation in close proximity to the target. The release of the granule contents, including the membrane-puncturing enzyme *perforin* and the protease *granzyme*, damages the target's membrane such that its internal salt balance is disrupted and it lyses (Fig. 5-15). NK cells and activated monocytes and macrophages whose FcγR are engaged also synthesize and secrete tumor necrosis factor (TNF) and IFNγ, which hasten the demise of the target.

NK cells (discussed in detail in Ch.18) are the most important mediators of ADCC. These cells express FcγRIIIA molecules (but no other FcRs) on their surfaces, which recognize monomeric IgG1 and IgG3 molecules with low affinity. This low affinity is important because it means that monomeric IgG is effectively ignored, and that only IgG complexed to a multivalent antigen such that the Ig molecules are aggregated will bind efficiently to the FcR. Thus, only targets that are well coated with IgG will trigger the release of the NK cell's damaging contents; without precoating of the target by IgG, ADCC cannot occur. Eosinophils carry receptors for IgE and IgA antibodies (FcεRI and FcαR, respectively) that can bind to IgE- or IgA-coated parasitic targets, particularly helminth worms. These pathogens are resistant to the cytotoxic mediators released by neutrophils and NK cells in response to FcγR stimulation, but are susceptible to the contents of eosinophilic granules activated by binding to FcεRI and FcαR.

It is important to note that, while similar, ADCC is distinct both from complement-mediated lysis and from the lytic mechanism used by cytotoxic T cells. In the case of complement-mediated lysis, the lytic signal is delivered by a molecular complex deposited directly on the target cell surface. The complex itself creates a pore in the target cell membrane: no

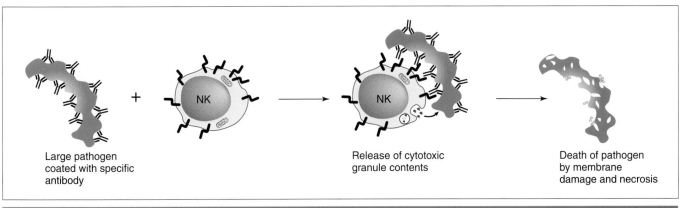

Large pathogen
coated with specific
antibody

Release of cytotoxic
granule contents

Death of pathogen
by membrane
damage and necrosis

Figure 5-15

Antibody-Dependent Cell-Mediated Cytotoxicity (ADCC)
Specific antibodies bound to a cellular antigen (such as a pathogen or tumor cell) bind via their Fc regions to FcRs on a lytic cell. The lytic cell is then stimulated to release granule contents that kill the pathogen at close range via membrane destruction. The lytic cell in this example is an NK cell, but could also be an activated eosinophil, neutrophil, monocyte, or macrophage.

host effector cell is involved. In the case of lysis mediated by cytotoxic T cells, the T cell receptor itself recognizes epitopes on the target cell and does not rely on mediation by antibody for specificity.

IV. COMPLEMENT-MEDIATED CLEARANCE OF ANTIGEN

The complement system was introduced in Chapter 4 as an element of the innate immune response. The reader will recall that "complement" is in fact a system of close to 30 proteins in the blood plasma that sequentially activate each other by proteolytic cleavage or other modifications. In some cases, the complement cascade is triggered directly by the interaction of microbes with particular complement components (especially C3b), a process called the *alternative pathway* of complement activation (refer to figure in Box 4-1). In other situations, antibody that is bound to antigen can interact with the complement component C1 that initiates complement activation via the *classical pathway*. More recently, it has been found that certain lectins can also activate complement via the *lectin pathway*. This route does not involve antibody but does make use of downstream elements of the classical pathway. Since the antigen clearance that results from classical complement activation is considered an antibody effector function, the mechanisms involved are discussed briefly here. A more detailed look at the structure, function, and interaction of complement system components can be found in Chapter 19.

Complement activation can lead to antigen clearance in six ways (Table 5-2). The first way requires the cascade to reach its final end point, but several other important antigen clearance mechanisms are invoked by the various by-products of the many enzymatic reactions that occur as the cascade proceeds.

i) Lysis of Pathogens

A pathogen coated with specific antibody of certain isotypes can initiate the classical pathway of complement activation.

Antibodies that can trigger complement activation in this way (IgM, IgG1, IgG2, and IgG3 in humans) are sometimes called "complement-fixing" antibodies. The binding of antigen to an antibody opens a site in the Fc region that allows it to bind to the classical initiating complement component C1q. However, the C1q molecule must bind simultaneously to two C1q-binding sites (and thus two separate Fc regions) in order for it to activate the cascade. A pathogen must therefore be well coated with Ig molecules supplying Fc regions in close proximity before the cascade can commence. As was briefly described in Box 4-1, the end result of the cascade is the assembly of the *membrane attack complex* (MAC) on the surface of the pathogen. The MAC is actually a group of complement proteins that are deposited on the pathogen surface in a rigorously regulated series of enzymatic reactions. One enzymatic component of the MAC forms a pore in the membrane of the pathogenic cell, causing it to swell and lyse.

ii) Opsonization of Pathogens

Key complement components C3b, iC3b, and C4b (see Ch.19 for nomenclature) are generated as by-products of the classical pathway of complement activation. These low molecular weight proteins can be deposited on the surface of antibody–antigen complexes, where they can act as important opsonins. Macrophages and neutrophils carry receptors for

---

Table 5-2 Complement-Mediated Clearance of Antigen

Lysis of pathogens by MAC deposition

Opsonization of pathogen to promote phagocytosis

Clearance of immune complexes from circulation

Enhancement of antigen presentation

Enhancement of B cell activation

Induction of local inflammation

MAC, membrane attack complex.

these complement proteins, so that the coating of an antigen–antibody complex by one or more of them promotes binding to phagocytes and subsequent phagocytosis of the complex (provided that the complexes are not too large to be engulfed). The greatest increases in the binding of microbes by phagocytes are observed when both antibody and C3b simultaneously opsonize an organism.

iii) Clearance of Immune Complexes from the Circulation

Complement activation helps to clear antigen from the blood circulation in two ways. First, complement components can bind to the Fc regions of antigen-bound antibodies, preventing the antibodies from interacting noncovalently with each other to form extended networks of antigen–antibody complexes. Such large complexes are insoluble and can become deposited in blood vessel walls, leading to inflammation and tissue damage. Complement proteins therefore play an important role in keeping the complexes soluble so that they can be removed. Second, red blood cells also carry receptors for complement component C3b. Thus, red blood cells entering a local area can bind to any C3b-coated antigen–antibody complexes present and transport them through the circulation to the liver and spleen. Phagocytic cells in these locations can then engulf the antigen–antibody complexes during the routine disposal of red blood cells.

iv) Enhancement of Antigen Presentation

By acting as opsonins, complement by-products C3b and iC3b enhance the uptake of antigen by macrophages. Not only does this remove antigen by facilitating phagocytosis, it is also the first step in the processing and presentation of antigen to T cells by macrophages acting as APCs. These same complement components have also been implicated in enhancement of antigen uptake by B cells and follicular dendritic cells, two other cell types important in antigen presentation. The resulting increase in stimulation of antigen-specific lymphocytes is thought to further aid the clearance of the antigen.

v) Enhancement of B Cell Activation

The complement by-products iC3b and C3dg can enhance the activation of B cells by binding to a complement receptor expressed on the B cell surface called complement receptor 2 (CR2). The amount of antigen required to reach the threshold of activation for a B cell is greatly decreased when its Ig receptors are stimulated by antigen, and its CR2 receptors are engaged by C3 products.

vi) Induction of Local Inflammation

Several of the by-products of classical complement activation (C3a, C4a, and C5a) can induce local inflammatory responses (see Ch.19). As a result, increased numbers of leukocytes are drawn into the affected area, increasing the efficiency of the response to the antigen. This is an indirect means by which complement activation helps to clear antigen.

This concludes our discussion of antibody effector mechanisms. We now move to a synthesis of our previous two topics, and relate the structure of Igs and their effector mechanisms to their biological roles.

D. Immunoglobulin Isotypes in Biological Context

Immune responses are divided into two parts: the *recognition* phase (in which antigen is recognized and bound by the antigen receptor) and the *effector* phase (in which cells initiate specific actions to eliminate the antigen). As we learned previously, Igs fixed in the membranes of B cells function as part of its antigen receptor complex in the recognition phase of the humoral response, whereas Igs of the same antigenic specificity in their soluble form are the antibodies secreted by the activated and differentiated progeny B cells in the effector phase. Binding of antigen to the mIg molecules of a B cell stimulates (usually with T cell help) the intracellular signaling cascade that triggers the activation, proliferation, and differentiation of that B cell, causing the generation of plasma cells. One clone of plasma cells produces millions of antibody protein molecules with exactly the same specificity (V region sequences) as that of the original Ig in the antigen receptor complex.

We have also seen that the five major isotypes of Igs are distinguished by the H chain constant region sequences of the Ig molecule. These domains are critical for effector actions since they interact with the initiating protein C1q of the classical complement pathway as well as with the Fc receptors on phagocytes that mediate opsonized phagocytosis, and on lytic cells that carry out ADCC. In addition, each of the five major Ig isotypes has distinct physical properties and biological activities that depend on its total amino acid sequence and the carbohydrate content of its side chains. All of these isotype-dependent properties influence where a given antibody may travel in the body and how it will be involved in host defense.

I. NATURAL DISTRIBUTION OF ANTIBODIES IN THE BODY

The bulk of Ig proteins in the body are in the form of secretory IgA in the external secretions, guarding the mucosal surfaces where pathogens are likely to attempt entry. Next in relative abundance are the secreted antibodies present in the plasma, circulating throughout the body. Size considerations dictate that the pentameric sIgM molecule remains primarily in the blood vessels, but other isotypes are small enough to diffuse freely from the blood into the tissues. Occasionally a low concentration of antibodies can also be found in the interstitial fluid of tissues in which B cell activation, proliferation, and differentiation have occurred. In addition, since they bind to the Fc receptors of NK cells, mast cells, monocytes, and macrophages, antibodies can be found in the tissues where these cell types are distributed. Interestingly, no Ig is normally detected in the brain. A chart summarizing selected biological characteristics for each Ig isotype is shown in Table 5-3.

II. MORE ABOUT IgM

Monomeric mIgM is always the first form and isotype of Ig generated by naive B cells. Following its initial activation by

Table 5-3 Major Functional Properties of Human Ig Isotypes*

	IgG1/2/3/4	IgM	IgA1/2	IgD	IgE
Serum concentration (mg/ml)	3–20	0.1–1.0	1–3	0.001–0.01	0.0001–0.001
Half-life in serum (days)	2–4	1	1	<1	<1
Intravascular distribution (%)	45	80	42	75	50
Tailpiece	No	Yes	Yes	No	No
Secreted form	Monomer	Pentamer	Monomer (IgA1) Monomer, dimer, trimer (IgA2)	Monomer	Monomer
Classical complement activation	++/+/+++/−	+++	−/−	−	−
Placental crossing	+	−	−	−	−
FcR binding	+	+	+	−	+
Triggers mast cells/basophil degranulation in allergy	−	−	−	−	+++

*Adapted from "Monoclonal Antibodies: Principles & Practice," 3rd edn. (1996) Academic Press Ltd., London, with permission.

antigen, the naive B cell proliferates and differentiates, and its progeny produce the pentameric secreted form of sIgM. sIgM is the "default" antibody, since all other antibody isotypes (except IgD) are generated by isotype switching, a process that commences only late in a primary response, and sometimes not until the secondary response. Thus, it is sIgM antibodies that are expressed first in any primary immune response, and those that are synthesized first in the newborn (which has just begun to encounter antigens on its own). The detection of increased sIgM levels in an adult indicates that he or she has recently been exposed to a novel antigen. It is also sIgM antibodies that react to foreign blood group proteins, because these are T-independent antigens that can activate a B cell to secrete sIgM, but which, due to the absence of the cytokines of T cell help, do not induce isotype switching.

Because the bulk of the expanded memory B cell population continues to undergo isotype switching during the secondary response, sIgM antibodies comprise only about 5–10% of normal serum Ig. However, the low absolute numbers of IgM molecules are balanced by the number and properties of their binding sites. Because of its pentameric nature, the IgM antibody displays 10 Fab sites that can theoretically bind to a pathogen; in practice, however, steric hindrance usually prevents the IgM molecule from binding to more than five antigenic epitopes at once. Nevertheless, this number is sufficient for the IgM antibody to bind a large antigen or pathogen displaying multiple copies of the same antigenic determinant and to reduce its infectivity (by neutralization or complement activation) much more efficiently (using fewer molecules) than a monomeric Ig molecule can. In addition, the binding sites are of relatively low affinity, with correspondingly higher levels of cross-reactivity to related epitopes. This allows the host to "cast a broad net" in the primary response, maximizing the number of different B cell clones stimulated by a particular antigen.

As mentioned previously, because the sIgM antibodies are large pentamers of total molecular mass 970 kDa (5 × 190 kDa), they are generally concentrated in the intravascular pool. Since IgM is the most efficient isotype with respect to activating complement, it makes perfect sense that evolution has selected it to be preferentially located where the complement is: in the blood. The secreted pentameric form of IgM <u>when bound to a surface such as a pathogen</u> is ideally suited for complement fixation via the classical pathway, since multiple Fc regions are already juxtaposed in the pentamer and provide the necessary two C1q-binding sites in close proximity. (When in its free, soluble form, the IgM pentamer assumes a conformation in which the Fc regions are not as readily accessible.) The classical cascade can thus be triggered by a single molecule of antigen-bound IgM, in contrast to the many more molecules of IgG that must coat a pathogen before two C1q-binding sites will occur close enough together for classical complement activation to be possible.

Although the bulk of IgM is found in the blood, if vascular permeability has been increased in a local area by the release of vasoactive compounds during an inflammatory response, sIgM antibodies can exit the blood and enter the tissues to reach sites of infection. In addition, because of the presence of the J chain in the pentameric form of sIgM, these antibodies can occasionally acquire the secretory component by passage through epithelial cells and thus enter the external secretions as secretory IgM. Although the concentration of IgM antibodies in the external secretions is very low compared to that of IgA, secretory IgM antibodies do play an important role in mucosal humoral immunity (see Ch.20). IgM is not an isotype prominent in either opsonization or ADCC. While Fc receptors for IgM have been identified on mononuclear cells, very little is known about them.

III. MORE ABOUT IgD

IgD is a mysterious entity at this point. This monomeric, richly glycosylated Ig (180 kDa) is barely detectable in the blood (0.001% of total serum Ig) and is rarely secreted. It is observed only on the surface of immature, peripheral B cells that already express IgM. IgD is therefore the second Ig isotype to be synthesized by a B cell, and first appears on its surface early in B cell development (see Ch.9). However, the expression of IgD is transient; upon activation of the B cell by the binding of antigen, the surface expression of mIgD, but not mIgM, is abrogated.

The function of IgD is unknown. However, mIgD is capable of sending signals to the cell nucleus via its associated Igα/Igβ signaling heterodimer. Since mIgD seems to appear on B cells at specific times during their early development, mIgD may be important in regulating B cell maturation. Controversy currently exists over whether mIgD is expressed on memory B cells. Studies *in vitro* have shown that the secreted form of IgD does not possess complement-fixing activity.

IV. MORE ABOUT IgG

Immunoglobulin G is the "workhorse" of systemic humoral immunity, since it is the isotype most commonly found in the circulation and tissues. In the blood of normal adult humans, 70–75% of serum Ig is sIgG, which occurs in the monomeric H_2L_2 form (146 kDa, on average). The approximate proportions (which vary by individual) of its subclass molecules are: sIgG1, 67%; sIgG2, 22%; sIgG3, 7%; sIgG4, 4%. The small differences in amino acid sequence and disulfide bonding between subclasses confer small differences in physical properties, which in turn are associated with variable longevity in the serum and differences in effector functions. Because they are smaller in size than the pentameric sIgM antibodies, sIgG antibodies are distributed equally between the intra- and extravascular pools. Their diffusibility and high serum concentration make sIgG the prevalent antibodies in the extracellular fluid. This localization is logical for an antibody that is a key opsonin and important in ADCC, since access to FcγR-bearing phagocytes and lytic cells is readily available. The IgG1 and IgG3 subclasses are particularly good opsonins and mediators of ADCC because FcγRs bind sIgG1 and sIgG3 antibodies with high affinity. FcγRs generally bind less well to sIgG4, and hardly at all to sIgG2. These differences among subclasses may be related to length of the hinge region in the C_H2 domain of the Ig structure, since IgG3 has an extended hinge, and IgG4 has a very short hinge.

While not as efficient as sIgM antibodies, sIgG antibodies are also important activators of complement. Free sIgG molecules display readily accessible Fc regions, but because this Ig is monomeric, it supplies only one C1q-binding site per antibody. Two sIgG molecules must be brought together by mutual binding to antigen in order to furnish two C1q-binding sites in close enough proximity to trigger complement activation. Thus, the power of the classical cascade is not unleashed by IgG unless it is warranted due to the presence of sufficient antigen. IgG3 is the most efficient complement-fixing IgG subclass, while IgG1 is somewhat less efficient, and IgG2 is even less so. IgG4 is unable to bind C1q and so cannot fix complement at all in the classical pathway. As well as opsonization, ADCC, and complement fixation, IgG antibodies have an additional unique protective function. Only IgG antibodies can cross the mammalian placenta, and maternal IgG1, IgG3, and IgG4 molecules have been found to be crucial for conferring the passive immunity that protects the developing human fetus and newborn in the first few months of life. IgG2 crosses the placenta with much lower efficiency.

V. MORE ABOUT IgA

More secretory IgA is produced per day in an adult human than all of the other Ig isotypes combined. In humans, 85–90% of IgA antibodies are found in the secretory form in the external secretions of the body, which include tears; saliva; mucous secretions of the gastrointestinal, urogenital, and respiratory tracts; breast milk; and prostatic fluid, among others. Secretory IgA antibodies are of enormous importance because they facilitate antigen removal right at the mucosal surface, the most common site of initial pathogen attack. The predominant effector mechanism carried out by secretory IgA antibodies is neutralization.

The remaining 10–15% of IgA antibodies occurs in blood, produced by B cells in the lymph nodes, bone marrow, and spleen. Two subclasses of human serum IgA exist. IgA1 tends to be monomeric, is present at a fivefold higher concentration than IgA2, and has greater flexibility in its hinge region. sIgA2 tends to be present more often in polymeric forms. No functional differences of note have been observed between sIgA1 and sIgA2 antibodies. In humans, serum IgA antibodies are predominantly (80%) in the monomeric (160 kDa) form; 20% are in the dimeric or trimeric (or more rarely, tetrameric, pentameric, or even hexameric) forms. However, in other mammals, the polymeric forms of sIgA tend to predominate over the monomeric. Memory B cells located in the submucosal diffuse lymphoid tissues, the Peyer's patches, and the tonsils produce sIgA antibodies (usually the J chain-containing dimeric or trimeric forms) almost exclusively, because specific cytokines secreted by Th cell subsets in these microenvironments promote successive switching to the IgA isotype.

Evolution has matched effector action to location for IgA as well. Secretory IgA molecules, which operate primarily at mucosal surfaces where complement and phagocytes do not occur, are poor complement fixers and opsonins. Neither are they good inducers of inflammatory responses, which makes evolutionary sense, since inflammation at a mucosal surface might damage it and compromise an important physical barrier. Secretory IgA is often said to have "antiviral" activity, since the polymeric nature of this antibody allows it to easily bind repeating epitopes on large antigens such as virus particles, impeding their attachment to mucosal cell surfaces. The expulsion of these particles from the body is then carried out by mechanical means, such as the movement of mucus by the undulating cilia of the respiratory tract. Thus, antigens trapped by binding to secretory IgA are removed primarily by mechanisms of the innate immune response, often making it unnecessary to invoke a further secondary adaptive immune response. Since secretory IgA is also passed along in breast milk to the neonatal gut mucosa, it contributes to the passive immunity that protects neonates of several species. sIgA plays a role in the elimination of helminth worms, in that IgA-coated parasites can be dispatched by ADCC carried out by eosinophils bearing FcαR.

VI. MORE ABOUT IgE

Antibodies of the IgE isotype circulate in the blood as monomers of molecular mass 190 kDa. Each ε H chain has five Ig domains (V_ε and C_ε1–4), but no hinge region. Secreted IgE is present in the serum at the lowest concentration of all isotypes, a mere 0.00005 mg/ml, or 0.000003%, due to the infrequency of switching to this isotype. In peripheral blood, IgE molecules have a half-life of 1–5 days. Human serum IgE levels are low at birth but rise to a peak at about age 15 years, after which they decline. IgE antibodies do not cross the placenta, cannot fix complement, and do not function as opsonins. Nevertheless, IgE antibodies have a clinical impact that far outweighs their actual numbers and limited effector options. The serum concentration of IgE rises dramatically in response to worm infections. The processing and presentation of worm antigens by APCs draws to the local area Th2 cells that secrete certain cytokines (principally IL-4 and IL-13), which promote isotype switching in local memory B cells to IgE. Eosinophils, the only lytic cell type competent to combat large parasitic worms, carry FcεRI. The pathogen-induced increase in the local supply of specific IgE antibodies ensures that the pathogen is well coated with IgE and thus is recognized and bound by the cell type best-suited to its elimination.

sIgE antibodies are also responsible for the symptoms experienced in allergic reactions such as hay fever, and more severe conditions such as asthma and anaphylactic shock. These conditions are all manifestations of a type of immune reaction called *immediate hypersensitivity* (see Ch.28). Immediate hypersensitivity is in fact an intense inflammatory response (usually local) that is mediated by mast cells situated close to the allergen's site of entry into the body, often at a mucosal surface or in a local lymph node. In individuals producing IgE recognizing particular allergens, mast cells can exist in a "pre-armed" or "sensitized" state; that is, the IgE molecules can bind to high-affinity FcεRI receptors present on the mast cells <u>prior</u> to binding antigen (Fig. 5-16). The low amounts of circulating IgE antibody are thus effectively concentrated on the surfaces of these cells, which then display the IgE molecules such that they act like antigen receptors. Should a sufficient number of allergen X molecules bind to and cross-link X-specific IgE molecules protruding from the pre-armed mast cell, the mast cell is immediately triggered to deliver its inflammatory response. The mast cell degranulates and releases the histamines and cytokines that induce the allergic symptoms. The supply of pre-existing cytoplasmic granules in a mast cell ensures that no lag is displayed between triggering of the FcR and the mast cell response. The reader will note that the role of FcεRI in allergic responses dif-

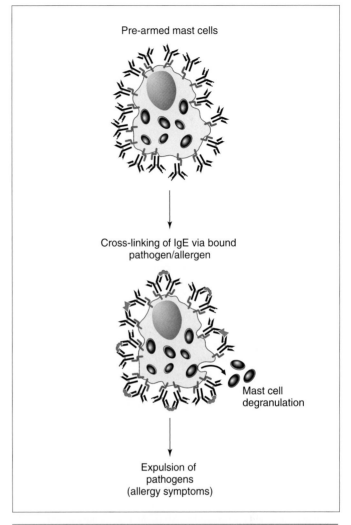

Pre-armed mast cells

Cross-linking of IgE via bound pathogen/allergen

Mast cell degranulation

Expulsion of pathogens (allergy symptoms)

Figure 5-16

IgE-Mediated Mast Cell Degranulation
The FcεRs on the surface of a mast cell bind IgE that is not yet complexed with antigen. On exposure to a specific allergen or pathogen, the antigen–antibody binding that occurs cross-links IgE and thus the FcεRs. These receptors mediate intracellular signals that lead to mast cell degranulation. The contents of the granules attack the pathogen/allergen but also cause allergy symptoms.

fers in two ways from the role of mIg in the BCR. Firstly, unlike an individual B cell that carries a BCR capable of recognizing only one (or a very small number) of antigens, FcεRI recognizes any IgE molecule, regardless of antigenic specificity. As a result, an individual pre-armed mast cell is able to respond to many allergens. Secondly, after an antigen binds to the BCR, a lag is experienced while the B cell undergoes activation, proliferation, and differentiation into plasma cells that can produce antibodies capable of effector actions. No such lag is observed between the stimulation of FcεRI on the surface of a mast cell and the degranulation response.

Because of their role in stimulating the release of inflammatory mediators in response to innocuous (rather than harmful) antigens, IgE antibodies are sometimes called *reaginic* (pathogenic) antibodies. Presumably the original evolutionary purpose of IgE was not to make allergy sufferers miserable but to provide protection. Mast cell degranulation provokes sneezing, coughing, mucus production, and sometimes vomiting or diarrhea, which are reactions that have the underlying function of expelling foreign entities. Mast cell degranulation also releases molecules thought to promote the accumulation of other cells (such as eosinophils) that fight parasitic infec-

tions. These are normal responses to appropriate antigens such as pathogens, but appear as symptoms of allergy in individuals who happen to make IgE recognizing the allergen.

This completes our discussion of Ig structure and biological effector function. In the next two chapters, we examine the mechanics of antigen–antibody interaction and how antibodies serve as important tools in the laboratory.

SUMMARY

Immunoglobulins (Igs) are antigen-binding proteins produced by B cells. All Igs made by a given B cell have the same antigenic specificity. Each Ig molecule has an H_2L_2 structure in which two identical heavy (H) chains and two identical light (L) chains are held together by disulfide bonds to form a Y-shaped structure. At the N-terminal end of each Ig chain is a variable (V) domain that differs widely in amino acid sequence among B cells. The C-terminal end of each light chain contains a single constant (C) domain of relatively invariant amino acid sequence, whereas each H chain contains three or four constant domains. Each H chain pairs with its L chain such that the two V domains combine to form identical antigen binding sites (Fab region), making the Ig molecule bivalent. The constant domains of the H chains form the Fc region. Five isotypes of Igs are distinguished based on the amino acid sequence in the heavy chain constant domains: IgM, IgG, IgA, IgE, and IgD. Immunoglobulins expressed on the B cell surface (mIg) have a transmembrane domain and recognize antigen but are incapable of signaling. The Igα/Igβ molecule associates with mIg in the membrane of naive B cells to form the B cell receptor (BCR) complex that transduces activation signals in response to antigen binding. Once the B cell is activated and differentiates into a plasma cell, it secretes copious amounts of an antibody which is a soluble form (sIg) of the Ig molecule in its BCR. The antibody binds to its specific antigen whether it is present in a soluble or cell-associated state. Antibody-bound antigens are eliminated from the body via interaction of the Fc region with either complement components or Fc receptors (FcRs) on cell types mediating phagocytosis or antibody-dependent cell-mediated cytotoxicity (ADCC). Complement components and FcRs are isotype-specific, so that the antibody isotype determines which antibody effector functions operate to clear bound antigen. The isotype of an antibody also controls where it is found in the body because access to various anatomical locations is Fc-dependent.

Selected Reading List

Bachmann M. F. and Zinkernagel R. M. (1997) Neutralizing antiviral B cell responses. *Annual Review of Immunology* **15**, 235–270.

Bengten E., Wilson M., Miller N., Clem L., Pilstrom L. and Warr G. (2000) Immunoglobulin isotypes: structure, function, and genetics. *Current Topics in Microbiology and Immunology* **248**, 189–219.

Buck C. (1992) Immunoglobulin superfamily: structure, function and relationship to other receptor molecules. *Seminars in Cell Biology* **3**, 179–188.

Burton D. R. and Woof J. M. (1992) Human antibody effector function. *Advances in Immunology* **51**, 1–84.

Daeron M. (1997) Fc receptor biology. *Annual Review of Immunology* **15**, 203–234.

Davies D. and Metzger H. (1983) Structural basis of antibody function. *Annual Review of Immunology* **1**, 87–117.

Frutiger S., Hughes G. J., Packquet N., Luthy R. and Jaton J.-C. (1992) Disulfide bond assignment in human J chain and its covalent pairing with immunoglobulin M. *Biochemistry* **31**, 12643–12647.

Goding J. (1996) "Monoclonal Antibodies: Principles and Practice," 3rd edn. Academic Press, London.

Harris L., Larson S. and McPherson A. (1999) Comparison of intact antibody structures and the implications for effector function. *Advances in Immunology* **72**, 191–208.

Junghans R. (1997) Finally! The Brambell receptor (FcRB). *Immunologic Research* **16**, 29–57.

Koshland M. E. (1985) The coming of age of the immunoglobulin J chain. *Annual Review of Immunology* **3**, 425–453.

Leatherbarrow R., Stedman M. and Wells T. (1991) Structure of immunoglobulin G by scanning tunnelling microscopy. *Journal of Molecular Biology* **221**, 361–365.

Mestcky J. and McGhee J. R. (1987) Immunoglobulin A (IgA): molecular and cellular interactions involved in IgA biosynthesis and immune response. *Advances in Immunology* **40**, 153–245.

Raghavan M. and Bjorkman P. J. (1996) Fc receptors and their interactions with immunoglobulins. *Annual Review of Cell Development and Biology* **12**, 181–220.

Ravetch J. (1997) Fc receptors. *Current Opinion in Immunology* **9**, 121–125.

Roux K. (1999) Immunoglobulin structure and function as revealed by electron microscopy. *International Archives of Allergy and Immunology* **120**, 85–99.

Sutton B. and Gould H. (1993) The human IgE network. *Nature* **366**, 421–428.

Tiselisu A. and Kabat E. A. (1938) An electrophoretic study of immune sera and purified antibody preparations. *Journal of Experimental Medicine* **69**, 119–131.

van de Winkel J. and Hogarth P, eds. (1998) "The Immunoglobulin Receptors and Their Physiological and Pathological Roles in Immunity." Kluwer Academic Publishers, Dordrecht.

Wu T. T. and Kabat E. (1970) An analysis of the sequences of the variable regions of Bence Jones proteins and myeloma light chains and their implication for antibody complementarity. *Journal of Experimental Medicine* **132**, 211–250.

Youinou P., Jamin, C. and Lydard P. (1999) CD5 expression in human B-cell populations. *Immunology Today* **20**, 312–316.

The Nature of Antigen–Antibody Interaction

6

César Milstein 1927–2002

Hybridomas and monoclonal antibodies
Nobel Prize 1984
© Nobelstiftelsen

George Köhler 1946–1995

Hybridomas and monoclonal antibodies
Nobel Prize 1984
© Nobelstiftelsen

CHAPTER 6

A. THE NATURE OF B CELL IMMUNOGENS

B. B CELL–T CELL COOPERATION IN THE HUMORAL IMMUNE RESPONSE

C. THE MECHANICS OF ANTIGEN–ANTIBODY INTERACTION

"Snowflakes are one of nature's most fragile things, but just look at what they can do when they stick together"–Vesta Kelly

In Chapter 5, we discussed the structure of immunoglobulin proteins, both in their membrane-bound form as part of the B cell antigen receptor, and as secreted and secretory antibodies responsible for B cell effector functions. In this chapter, we examine the nature of B cell epitopes, how the antibody recognizes and binds to these epitopes on antigens, and the role of T cell help in B cell activation. In so doing, we gain insight into the engagement of the BCR on the B cell surface by antigen. We also explore the assays used in the laboratory that are based on these interactions. T cell epitopes and their recognition will be discussed in Chapter 12.

A. The Nature of B Cell Immunogens

As was introduced in Chapter 2, an *antigen is any substance that is specifically recognized and bound by the antigen receptors of T or B lymphocytes. An immunogen is any substance that elicits an adaptive immune response.* The reader will recall that *all immunogens are antigens, but not all antigens are immunogens*, since some antigens bind to a receptor but do not induce the activation of the lymphocyte. Despite this very real distinction, we will continue to use these terms interchangeably for convenience (like most of the immunological community). The antigen that is used to immunize the host and induce antibody production is called the *cognate* antigen. An *epitope* or *determinant* is a small region of an immunogenic molecule that binds to an antigen receptor. While B and T cells carry receptors that are similar from the functional point of view, in that they both bind antigen, these receptors recognize epitopes in fundamentally different ways. The T cell receptor recognizes an epitope composed of an antigenic peptide presented in association with an MHC molecule (see Ch.10), while B cell receptors recognize simpler epitopes on intact pathogens (living or not), on macromolecules in solution, or on macromolecules bound to cell surfaces.

I. WHAT MOLECULES CAN FUNCTION AS IMMUNOGENS?

Almost every kind of organic molecule (including proteins, carbohydrates, lipids, and nucleic acids) can be an antigen, but only macromolecules (primarily proteins and polysaccharides) have the size and properties necessary to be physiological immunogens. The reasons for this restriction, which have to do with the ability of the molecule to furnish the epitopes required for lymphocyte activation, are discussed below. Inorganic molecules are by themselves not generally immunogenic, mostly because of their small molecular size but probably for other reasons as well: large inorganic crystals (such as kidney stones) also fail to provoke an immune response. The most powerful immunogens are proteins because they supply both the T and B cell epitopes necessary to generate the T cell help required in the majority of humoral immune responses. In addition, since the activation of T cells depends on the recognition of specific peptide fragments of antigens, proteins are the only immunogens capable of provoking a cell-mediated adaptive immune response. Polysaccharides generally supply B cell epitopes (only) and can provoke a humoral response from B cells without involving T cells. In a natural infection, invading bacteria and viruses appear as collections of proteins, polysaccharides, and other macromolecules to the host immune system. Within this collection, there are many immunogens, each inducing its own adaptive immune response.

II. IMMUNOGENS IN THE HUMORAL RESPONSE

The reader will recall from Chapter 2 that two classes of antigens (or more properly, immunogens) exist: "T-dependent" (Td) antigens and "T-independent" (Ti) antigens. Td antigens can bind to the BCRs of B cells to initiate activation but cannot induce differentiation and Ig secretion unless direct intercellular contacts are established with a primed helper

T cell. To prime the Th cell, its specific T cell epitope must be provided on the Td immunogen. In contrast, Ti antigens do not require T cell help in the form of B cell–T cell intercellular contacts to induce B cell activation and antibody production; that is, a primed Th cell is not required for primary Ti responses. Because no cognate Th cell help is required, B cell responses to a Ti antigen represent a rapid, frontline defense against invaders.

Secondary Ti responses do not appear to exist. Even after repeated exposure of a host to a Ti antigen, the B cells recognizing most Ti antigens react in only early primary immune response mode, with the production of mainly IgM antibodies. Isotype switching is minimal and the germinal centers and memory cells needed for immunological memory are not usually generated. We now know that the generation of memory in the vast majority of cases depends on the specific intercellular costimulatory contacts furnished only by cognate Th cells, making the secondary response a characteristic of antigens that possess T cell epitopes (see Ch.9). Purified Ti antigens generally do not contain T cell epitopes. Thus, substances such as synthetic polymers or antigens composed solely of repeating carbohydrate units are unable to induce germinal center formation and produce only very limited isotype switching in experimental situations. When isotype switching does occur in response to a Ti antigen (say, a natural component of a pathogen), it can usually be traced to the presence of T cell epitopes present in a protein component of the pathogen. In some cases, the cytokines necessary for isotype switching may be produced by non-T cells in the local microenvironment. Evidence exists suggesting that many

Ti antigens have the potential to directly stimulate NK cells and other non-T cells such as macrophages, mast cells, and even B cells to secrete cytokines.

i) Classes of Ti Antigens

Ti antigens generally are large polymeric proteins or polysaccharides (and sometimes lipids or nucleic acids) whose structure is composed of repetitive elements, as occur in many bacterial and viral products and structural elements. Ti antigens fall into two classes: the *Ti type 1* (Ti-1) and the *Ti type 2* (Ti-2) antigens. The properties of these antigens are listed along with those of Td antigens in Table 6-1. B cell responses to Ti-1 antigens are unaffected by and occur in the complete absence of either interactions with cognate Th cells or cytokine help. In contrast, while B cell responses to Ti-2 antigens do not require intercellular contacts with a cognate Th cell, some help in the form of cytokines derived from surrounding cells is required.

ia) Ti-1 antigens. At high concentrations, Ti-1 antigens act as polyclonal B cell activators, stimulating all B cells regardless of their antigenic specificity. Because B cell activation leads to proliferation, Ti-1 antigens are also considered to be *B cell mitogens*; a mitogen is a substance that can stimulate DNA synthesis and cell division. (Mitogens are discussed in detail in Chapter 14.) The fact that Ti-1 antigens act as mitogens implies that they bind to a site on the B cell separate from the antigen-binding site of the BCR. (Thus, strictly speaking, these molecules appear to fall outside our definition of "antigen.") Some Ti-1 antigens bind to an entirely separate receptor, such as occurs in the case of LPS binding to the

Table 6-1 Features of Td, Ti-1, and Ti-2 Antigens

Property	Td Antigens	Ti-1 Antigens	Ti-2 Antigens
Require MHC class II-restricted T cell help	+	−	−
Repetitive epitopes	−	+	+
Engagement of BCR antigen-binding site	+	−	+
Polyclonal B cell activators (mitogens)	−	+	−/+ (some cases)
Stimulate mature B cells	+	+	+
Stimulate immature B cells	−	+	−
Relative response time	Slow	Rapid	Rapid
Source of cytokine help used	Cognate Th	None	Th, NK, B
Secondary response characteristics (memory, isotype switching)	+	−	−/+
Activate complement	−	−	+
General chemical nature	Soluble proteins or peptides	Bacterial cell wall components	Polysaccharides, polymeric proteins
Examples	Influenza nucleoprotein, monomeric bacterial flagellin	LPS, TNP–*Brucella abortus*	Type III pneumoccocal polysaccharide, polymeric bacterial flagellin

CD14/TLR4 complex. Interestingly, low concentrations of Ti-1 antigens induce what appears to be an antigen-specific, rather than a mitogenic, response. That is, only some B cell clones are stimulated and polyclonal activation is no longer observed. Why this phenomenon occurs is not clear, but it may be due to competition between B cell clones for limiting amounts of the Ti-1 antigen. Those B cells with BCRs whose antigen-binding sites happen to bind to the Ti-1 antigen may be preferentially stimulated. This activation at low antigen concentration is still considered to be T-independent, because no cognate Th cell help is required, but it is not yet clear how cellular activation actually occurs. Unlike Td or Ti-2 antigens, Ti-1 antigens are also generally able to stimulate immature B cells, further supporting the idea that the molecular interaction of a Ti-1 antigen with a B cell is unconventional in some way.

ib) Ti-2 antigens. Ti-2 antigens are often linear molecules of large molecular weight with highly organized and repetitive antigenic determinants. Because they tend to be poorly degraded *in vivo*, Ti-2 antigens can persist on the surfaces of macrophages for long periods of time and are displayed as multiple repeats of the same antigenic fragment. The Ti-2 antigen is then acting as a multivalent antigen that binds with high avidity to multiple mIg molecules on a nearby B cell. The BCRs are said to be *cross-linked* because the Fab regions of two separate Ig molecules (of identical specificity) are linked together by virtue of their binding to the same very large immunogen. The requirement that repeating epitopes on the Ti-2 antigen be identical is underscored by a study showing that the amino acid polymer poly(Pro-Gly-Pro) elicits a Ti-2 response, while a randomly ordered polymer with the same overall amino acid composition acts as a Td antigen. While the exact molecular mechanisms underlying Ti-2 responses are not completely understood, it is thought that the extensive cross-linking of BCRs by Ti-2 antigens invokes prolonged and persistent signaling within the B cell that bypasses the need for intercellular contacts with a cognate Th cell. With a few exceptions, Ti-2 antigens do not act as polyclonal B cell activators, evidence that, unlike Ti-1 antigens, the Ti-2 antigen interacts in an antigen-specific way with the binding site of the BCR.

It has become clear in recent years that cytokine help can regulate Ti-2 responses. In the absence of intercellular contacts with a cognate Th cell, cytokine help derived from other sources can increase the magnitude of a Ti-2 B cell response as well as allow some degree of isotype switching and even memory generation in some cases. Non-cognate Th cells, NK cells, macrophages, and B cells may all secrete cytokines supporting Ti-2 responses. It may be that, during Td responses, the contribution from these "ancillary" helper cells is normally eclipsed by the large quantities of cognate Th cell-derived cytokines in the milieu. In addition to cytokine help, B cell responses to polysaccharide Ti-2 antigens may receive a boost from the complement system. These polysaccharides can directly activate the alternative complement pathway, and the C3d that is generated can bind to the CR2 receptor expressed on the mature B cell surface, enhancing the responsiveness of the B cell (see Ch.19).

ic) Other Ti-like antigens. Some viruses contain antigenic epitopes that don't fall readily into either the Ti-1 or Ti-2 category. For example, the surface glycoprotein of vesicular stomatitis virus (VSV) is Ti-1-like in that it can generate an antibody response in the complete absence of intercellular contacts and cytokines. However, like a Ti-2 antigen, VSV is not a polyclonal activator of B cells even at high concentration. Further studies of epitopes from VSV and other viruses have shown that the rigidity of antigenic determinants and their degree of repetition have a significant effect on the efficiency of the B cell response. In other words, depending on its degree of repetition and rigidity of organization in a particular structure, a given B cell epitope may be able to act as part of a Ti-1, Ti-2, or even a Td antigen.

ii) Td Antigens and the Rationale for T Cell Help

Ti antigens are only a small fraction of the immunogens that assault a host's body. A high proportion of the molecules making up a pathogen are proteins of unique amino acid sequences that lack the large repetitive structures needed to cross-link BCRs and trigger B cell activation. Some of these proteins are situated on the surface of the intact pathogen, while others are internal proteins exposed as the pathogen is attacked by the cells and destructive mechanisms of the innate immune response. A protein of unique sequence might present many different epitopes recognized by the BCRs of many different B cell clones. However, because only a single epitope of any one kind is likely to be furnished on a given protein molecule, it is rare that any one B cell will form enough BCR–epitope pairs to trigger complete B cell activation. In other cases, even though the protein may be present in thousands of copies on the surface of an intact pathogen, steric hindrance forestalls extensive BCR stimulation, preventing complete B cell activation. Enter "T cell help," in the form of specific costimulatory intercellular contacts and secreted cytokines supplied by antigen-activated cognate Th cells. The Th cells are activated by peptide derived from the Td antigen that is presented either by an APC or the B cell itself. The costimulatory contacts prepare the B cell for activation, while the cytokines bind to specific cytokine receptors present on the surface of the B cell. Engagement of the cytokine receptors replaces the requirement for direct cross-linking of the BCRs with signals that complete the stimulation of the intracellular signaling pathway leading to complete B cell activation, proliferation, and antibody production (Fig. 6-1). Without these signals, activation events such as protein kinase activation and increases in intracellular Ca^{2+} may be initiated, but neither cellular proliferation nor antibody production can occur. Instead, the occupied BCRs are internalized and the B cell assumes an APC function in that bound antigen is processed and presented on MHC class II molecules.

To summarize: efficient B cell activation by Td antigens requires three steps: (1) the binding of a sufficient number of B cell epitopes of the Td antigen by the BCRs, which makes the B cell "receptive" to T cell help; (2) the formation of costimulatory intercellular contacts between the B cell and the cognate Th cell; and (3) the binding of cytokines secreted by the cognate Th cell.

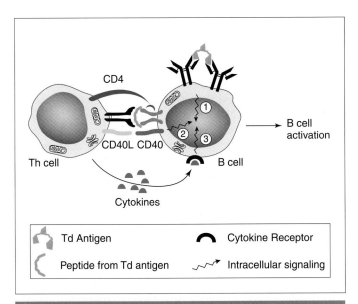

Figure 6-1

B Cell Activation by Td Antigens
Three signals are involved in B cell activation. When the Td antigen binds to its cognate BCR, "Signal 1" is delivered to the B cell, triggering surface expression of CD40 and cytokine receptors that make the B cell receptive to T cell help. When the Td antigen-derived peptide is presented to a T cell by an APC (in this case, the B cell itself), the activated T cell expresses CD40 ligand (CD40L), providing the intercellular contacts needed for "Signal 2." The activated T cell also secretes cytokines that deliver "Signal 3" to the B cell, allowing the completion of B cell activation.

III. PROPERTIES OF Td IMMUNOGENS

Large numbers of repetitive structures on a Ti antigen appear to be sufficient to confer immunogenicity, but what makes a Td antigen immunogenic? Scientists have attempted to dissect and define these properties individually, but the reader should keep in mind that, in a natural infection by a pathogen, immunogenicity will involve a combination of these properties.

Obviously a Td antigen must contain protein, since it must supply at least one peptide that can act as a T cell epitope.

However, the immunogenicity of a Td antigen also depends on other physical properties such as its *foreignness, conformation,* and *molecular complexity.* The nature of the antigenic challenge must also be taken into account: the immunogenicity of a particular molecule may be affected by its *dosage* and its *route of entry or administration.* Finally, *host factors* may influence the immunogenicity of a molecule. Each step in the immune response is controlled by the expression of the host's genes, and subtle allelic differences in the *genetic constitution* of an individual can alter the type, as well as the intensity, of the immune response to a given immunogen (Table 6-2).

i) Foreignness

First and foremost, to be immunogenic, the antigen must be perceived as foreign by the host's immune system. The more an antigen deviates structurally from "self" proteins, the more easily it is recognized as "nonself." A protein that is widely conserved among different species will have a structure similar to that of a molecule that exists in the host and therefore will not be very immunogenic. However, the introduction of a protein unlike anything in the host will usually trigger an intense immune response. Whether an antigen is perceived as nonself by an individual depends on whether there is a lymphocyte in the individual's repertoire whose antigen receptor recognizes the antigen. Lymphocytes whose receptors react with self-antigens are usually eliminated during lymphocyte maturation or are silenced in the periphery, resulting in a state of *tolerance* to self (see Chs.13 and 16). In other words, a lymphocyte clone whose antigen receptors can recognize a very foreign antigen is generally guaranteed to be present and to react, but in the case of proteins that are similar to self-proteins, the lymphocyte clone whose receptors might have recognized that antigen should have already been removed or rendered non-responsive.

For example, bovine serum albumin (BSA) is a protein found in identical form in all cows. Injection of BSA from one cow into another does not elicit an immune response because the immune system of every cow sees BSA as self, regardless

Factor	More Immunogenic	Less Immunogenic
Foreignness	Very different from self	Very similar to self
Molecular complexity:		
Size	Large	Small
Subunit composition	Many	Few
Conformation	Denatured, particulate	Native, soluble
	Intermediate charge	Highly charged
Charge		
Processing potential	High	Low
Dose	Intermediate	High or low
Route of entry	Subcutaneous > intraperitoneal > intravenous or gastric	
Host factors (genetics)		
MHC and other antigen processing molecules	Efficient presentation and peptide binding	Inefficient presentation and peptide binding

Table 6-2 **Factors Affecting Immunogenicity**

of the source. However, injection of BSA into a goat provokes an immune response because BSA is sufficiently different from the goat's own serum albumin (GSA) to be perceived as foreign; that is, some BSA epitopes are recognized by certain goat lymphocytes remaining after tolerance to GSA has developed in the goat. Introduction of BSA into a rabbit results in an even greater response, since rabbits are farther removed phylogenetically from cows than are goats, and rabbit serum albumin is even more different from BSA. Another example of the response to foreignness can be found in attempts to transplant organs from one species to another: the rejection of a pig organ by a human body is much more vigorous than the rejection of a human organ (see Ch.27).

There are some situations in which the presence of a foreign protein does not trigger an immune response. Certain tissues, such as the cornea of the eye, are normally sequestered from the immune system by physiological barriers. Cornea transplants are thus freely possible between individuals despite the presence of intraspecies antigenic differences, because lymphocytes with the potential to react to these corneal antigens are physiologically barred from activation. However, should these local barriers break down, proteins from the cornea can be immunogenic even in the individual from which they came, because self-reactive lymphocytes recognizing corneal components will not have been eliminated.

ii) Molecular Complexity

The molecular complexity of an immunogen encompasses the properties of *molecular size, subunit composition,* *conformation, charge,* and *processing potential*. As we have seen, molecular size and subunit composition are important determinants of immunogenicity for Td antigens, since these properties increase the likelihood that a molecule will contain both the T and B epitopes necessary for lymphocyte activation. It has been found that, with rare exceptions, an antigen must be at least 4 kDa to be immunogenic, and that many of the best immunogens have molecular masses of close to 100 kDa. However, just because a molecule is large does not necessarily mean that it will be a good Td immunogen. Homopolymers (composed of a single type of amino acid) are not effective Td immunogens, but co-polymers of two different amino acids are, because a co-polymer provides more possibilities for a variety of T and B epitopes. Macromolecules with greater numbers of different subunits are generally more effective Td immunogens than are those of a single subunit type, for the same reason.

The *conformation* of a molecule can affect immunogenicity. The evolutionary rationale for having antibodies in the first place is to provide a means of recognizing and counteracting intact pathogens invading the body. In keeping with this strategy, most B cell epitopes are structures that project from the external surfaces of macromolecules or pathogens. The majority of these B cell epitopes are *conformational determinants*, structures in which the contributing amino acids may be located far apart in their linear sequence but which become juxtaposed when the protein is folded in its natural or *native* shape (Fig. 6-2). These epitopes depend on native state folding, so that they disappear when the protein is *denatured*;

Figure 6-2
Effect of Protein Conformation on Antigenic Determinants Conformational B cell determinants (or epitopes) are lost upon denaturation of the antigenic protein. Linear determinants may or may not be accessible when the protein is in its native conformation, but are always accessible upon denaturation of the protein. Adapted from Abbas A. K. *et al.* (1997) "Cellular and Molecular Immunology," 3rd edn. W. B. Saunders Company, Philadelphia.

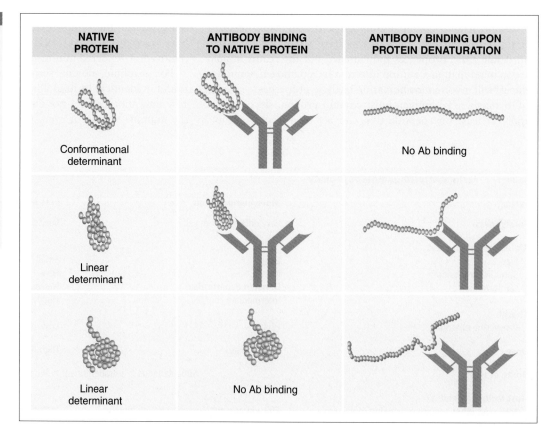

that is, when the intramolecular bonds maintaining the folding are broken. Other epitopes are *linear*; that is, they are defined by a particular stretch of consecutive amino acids. These determinants are present whether the protein is in its native or denatured state, but are often buried deep within the molecule and, if so, are accessible to antibody only when the protein is denatured. Certain epitopes are said to be *immunodominant*; that is, even though there may be numerous epitopes within an antigen, the majority of antibodies are raised to only a few of them. The epitopes that are immunodominant for B cells tend to be on the surface of a macromolecule because they are, by definition, more available than internal epitopes. Structural considerations that affect the ease of binding between the epitope and the antigen

receptor binding site make one surface epitope immunodominant over another.

The most detailed information on B epitope structure has been derived from studies of purified antibody crystallized with and without antigen. The positions of the various atoms in the molecule of interest can be determined quite precisely by analyzing how X-rays interact with the crystal, in a method called *X-ray crystallography* (see Box 6-1). X-Ray crystallographic analyses of the sites on immunogens that combine with antibody have shown that B cell epitopes are generally composed of about 15 residues arranged in irregularly shaped structures of approximately 7 nm^2 in area. Large antigens may display several different B cell epitopes, each recognized by a different specific antibody.

Box 6-1. X-ray crystallography

X-Ray crystallography is a technique that can be used to deduce the three-dimensional structure of a protein. To generate the data for analysis, a crystallized protein is bombarded with monochromatic X-rays from either a rotating anode X-ray generator or a synchrotron source. The latter generates a more powerful X-ray beam that gives better diffraction patterns with shorter exposure times. As the X-rays strike the crystal, most will go right through it, but some will strike the atoms making up the protein and be deflected into a detectable and reproducible pattern. The pattern provides information about the position and orientation of the atoms in the crystal.

The success of structure determinations by X-ray crystallography depends on having the protein of interest available in a well-ordered crystal that can diffract the X-rays strongly enough to give interpretable data. Growing protein crystals can be very difficult, and requires a protein sample that is very pure (>97% purity). Crystals are most often grown using the *hanging drop* method, in which a concentrated droplet of the protein solution is brought to the point of supersaturation by allowing the water to evaporate out of the droplet. Ideally, this high protein concentration results in the crystallization of the protein into a structure in which all of the protein molecules are packed together in a regular, repeating lattice. The successful crystallization of a protein depends on the protein concentration, the temperature and pH of the solution, the nature of the solvent and precipitant, and the ions or ligands that are added to the protein solution. Much trial and error is often required to find appropriate crystallization conditions for a given protein.

The regular arrangement of atoms in the crystal leads to a pattern of positive and negative interference of the diffracted X-rays, which appears as a series of dots on the X-ray detector. This diffracted X-ray pattern is the raw data from which the locations of the atoms in the protein can be calculated, using a mathematical tool called the *Fourier transform*, after the French mathematician Jean Baptiste Joseph Fourier. However, in order to use the X-ray diffraction pattern to deduce the structure of the protein, the primary amino acid sequence of the protein must be known. In addition, a resolution of 3 Å or less (i.e., the ability to distinguish individual atoms that are as little as 3 Å apart) is generally needed to fit the known primary amino acid sequence onto the electron density map. Given these data, a model of the α-carbon backbone of the protein can be constructed.

The first protein structure solved by X-ray crystallography was that of myoglobin by John Kendrew at Cambridge University in 1958. Analysis of the myoglobin structure led to the realization that proteins use a remarkably wide variety of folding patterns. This was a disappointment to scientists in light of the solution 5 years earlier of the structure of DNA by James Watson and Frances Crick. Scientists had expected that the simple elegance of the DNA structure would be repeated in the structure of proteins, but it was discovered that proteins, unlike DNA, required many different structures in order to interact with their many different substrates and ligands. After the first excitement in the late 1950s, the science of X-ray crystallography developed slowly. With more recent technological advances, there has been

an explosion of new protein structure solutions over the past 25 years.

Knowing the structure of a protein facilitates the understanding of its function. In addition, it is possible to co-crystallize a protein with another molecule, such as an enzyme with its inhibitor, or a receptor with its ligand. A number of immunologically important crystal structures have been elucidated in this way, providing scientists with important data on antigen–antibody interaction and T cell receptor–MHC association. For example:

- Solving the crystal structures of Fab fragment structures has led to an understanding of how antibodies fold, and how CDR loops combine to form the antigen-binding site.
- A combination of electron microscopy and X-ray diffraction was used to help explain how antibody binding can neutralize a virus. The hemagglutinin molecule from the influenza virus capsid was bound to antibody and the complex was crystallized. Resolution of this structure showed that the virus did not greatly alter its shape when bound, but that the antibody used the flexibility of its hinge region to bind the virus with both Fab regions, increasing the avidity of the binding and reducing the numbers of hemagglutinin molecules available for binding to host cells. Additional studies showed how viruses in turn mutated the structure of antigenically important regions of their receptors to escape antibody neutralization.
- The resolution of the crystal structure of the human MHC class I molecule HLA-A2 furnished many important insights into the recognition of antigenic peptides by the TCR. The HLA-A2 molecule, like any MHC

Continued

Box 6-1. **X-ray crystallography**–*cont'd*

class I or class II molecule, is unstable without a peptide in the antigen-binding groove (see Ch.10), so the sample co-crystallized with a broad range of peptides in the groove. This analysis revealed how MHC molecules bind to a wide variety of peptides with high affinity, and confirmed that the TCR recognizes both the antigen and the MHC molecule simultaneously. MHC class II structures have since been solved as well.

Modern determinations of protein structure still depend on X-ray crystallography as one of a battery of complementary techniques, in concert with new methods of deciphering the primary amino acid sequence needed to interpret the X-ray data. The primary amino acid sequence can be solved by laborious protein sequencing,

but the more common and easier approach is to sequence the DNA of the gene that encodes the protein. Working from the amino acid sequences predicted from the nucleotide sequence, projected secondary structures are developed that can be compared to the X-ray diffraction data.

The plate shows the crystal structure of hemagglutinin (HA) in light blue and an anti-HA antibody in green. The site in HA that normally binds to host cells is shown in dark blue. A complementarity-determining loop (red) is directly involved in binding HA in a neutralizing fashion.

Plate reproduced with permission from Bizebard T. *et al.* (1995) Structure of influenza virus haemagglutin complexed with a neutralizing antibody. *Nature* **376**, 92–94.

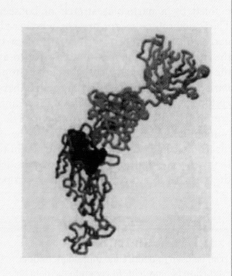

The *charge* or electronegativity of a molecule is also important for immunogenicity. Since B cell epitopes are most often those exposed on the surface of a macromolecule in the aqueous environment of the tissues, those areas of globular proteins acting as B cell epitopes are often hydrophilic in nature. Proteins containing a high ratio of aromatic to non-aromatic amino acid residues often have numerous hydrophilic epitopes and are very effective Td immunogens. However, very highly charged molecules elicit poor antibody production, possibly due to electrostatic interference between the surface of the B lymphocyte and the antigen.

A protein's *potential for intracellular processing* by APCs is also important for immunogenicity of a Td antigen, since the antigenic epitopes recognized by T lymphocytes during activation are in the form of peptides. Macromolecules that are large and insoluble (properties that encourage phagocytosis by APCs), and those that are easily digested by APC enzymes, are more immunogenic than small, soluble molecules.

iii) Dosage

When a pathogen invades the body, it is able to <u>replicate</u> and so delivers a substantial dose of immunogenic molecules to the host. The exact magnitude of a dose in a natural infection is hard to measure and difficult to analyze. However, experimental immunizations with various purified immunogens have provided some insight into the effects of dosage on immunogenicity.

For each immunogen, there appears to be a threshold dose below which no adaptive immune response can occur because insufficient numbers of lymphocytes are activated. This makes evolutionary sense, since most infections our bodies fend off every day involve only low numbers of pathogens that are easily handled by the innate immune response. The adaptive response is reserved for those invasions in which sufficient

organisms are present to constitute a dose of antigen that exceeds the threshold. As the dose of antigen increases above the threshold, the production of antibodies rises until a broad plateau is reached; this is the *optimal range* of doses for immunization with this antigen (Fig. 6-3). Note that maximal

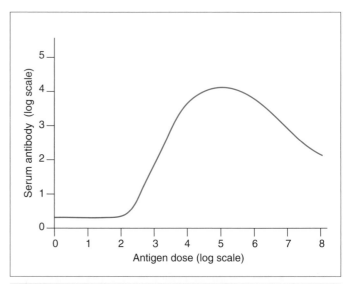

Figure 6-3

Primary Antibody Response to Different Antigen Doses
The antigen dose must exceed a certain threshold before B cells can be activated to produce antibody. Once this threshold is exceeded, antibody production increases exponentially until a plateau is reached where an increase in antigen does not result in a further increase in serum antibody titer. As antigen dose increases still further, antibody production is impeded, perhaps because of tolerance mechanisms. Adapted from Janeway C. A. *et al.* (1997) "Immunobiology," 3rd edn. Current Biology Ltd., London.

antibody production can be achieved with a fairly wide range of antigen concentrations. At very high antigen doses, antibody production declines. The precise reason for this effect is unknown. One cause may be simple steric hindrance, but some researchers also speculate that it may be related to mechanisms responsible for tolerance to self-proteins (which are naturally present at high concentrations). Because of the presence and properties of memory cells, secondary responses tend to be induced by a lower dose of antigen than that which triggered the primary response, and antibody production reaches higher levels. The mechanisms behind these attributes are discussed in Chapters 8 and 9. For reasons that are still unclear, very high or very low doses of antigen can result in decreased antibody production in the secondary response. At these doses, the memory cells appear to adopt a non-responsive state known (respectively) as *high zone* or *low zone tolerance* or *unresponsiveness*. These aspects of tolerance are discussed in detail in Chapter 16.

iv) Route of Entry or Administration

In the natural situation, pathogens most often enter the body by penetrating the skin or the mucosae protecting the respiratory, gastrointestinal, and urogenital tracts. Having gained access to the body's cells and tissues, the pathogens commence replicating. In some instances, the innate immune response is insufficient to hold a pathogen in check. The large amounts of pathogen immunogen generated in such situations trigger a vigorous primary response that generates effector and memory cells. Once the effector cells have vanquished the pathogen, the memory cells remain on guard to counter the pathogen if it enters the body a second time. This "naturally immunized" host is then protected from the effects of the pathogen and shows no clinical signs of disease.

Despite the general competence of our immune systems, a parent rarely wants to take the chance that his/her children might not beat their first exposure to certain pathogens, such as the bacterium that causes diphtheria or the virus that causes polio. Instead, prior to any natural infection with these pathogens, a primary response to a weakened, or *attenuated*, version of the pathogen is induced by vaccination. In a vaccination (or experimental) situation, the immunogen usually does not have the full capacity to replicate, and the immunogenicity of a given molecule can depend surprisingly heavily on its route of administration. For example, the food we eat contains many antigens, but because these antigens are "dead" (cannot replicate) and are introduced via the digestive tract, enzymatic degradation usually destroys them before an immune response is provoked. For this reason, antigens used for vaccinations or experimental immunizations are usually introduced by a route other than the digestive tract, that is, *parenterally* (from the Greek *para*, "beside"; *enteron*, "the intestine") (Table 6-3). One exception to this rule is the Sabin oral polio vaccine, developed because of the terror young children have of hypodermic needles. The poliovirus is an enteric virus, meaning it normally enters the body through the gut. Accordingly, this virus has evolved various mechanisms, such as its tough proteinaceous coat, that help it to survive in the harsh environment of the digestive tract. The

immunogen in the oral polio vaccine is a live poliovirus that has been attenuated by altering those genes that are involved in viral replication (see Ch.23). The attenuated virus is therefore able to penetrate and survive in the gut in the same manner as the wild-type virus, but does not cause significant disease because its replication is impaired; its *virulence* is reduced. Contrast this with the Salk polio vaccine, which contains dead virus totally incapable of both replication and penetration. This vaccine would not survive digestion in large enough quantities to initiate an immune response and so is administered by subcutaneous injection.

At the site of either *subcutaneous* or *intramuscular* immunization, the immunogens are conveyed rapidly to the regional lymph nodes, provoking vigorous immune responses. Immunogens introduced *intravenously* or *intraperitoneally* are channeled to the spleen, inducing immune responses of moderate intensity. Many immunogens introduced via the respiratory tract (*intranasally*) provoke allergic responses (inappropriate immune responses) in susceptible individuals.

The dose of a molecule that provokes a strong immune response when injected by one route of administration may induce only a minimal response when introduced by another. These considerations dictate that, for each immunogen, a dose-response curve must be established, and an assessment of the most effective route of administration be carried out, in order to determine how to achieve maximum immunogenicity. Why should the route of administration matter so much? Different routes of administration bring an immunogen into contact with different local lymphoid cell populations, which may result in differential processing of the antigen and thus varying effects on the immune response observed. Much is still unknown about the localized secondary lymphoid tissues known as the SALT (*skin-associated lymphoid tissue*), MALT (*mucosa-associated lymphoid tissue*), GALT (*gut-associated lymphoid tissue*), and BALT (*bronchial-associated lymphoid tissue*) (see Ch.20). These regionalized cell populations may

Table 6-3 Modes of Immunogen Administration

Mode	Description
Oral	By mouth; channeled to MALT
Parenteral*	
Intravenous (i.v.)	Into a blood vessel; channeled to spleen
Intraperitoneal (i.p.)	Into the peritoneal cavity; channeled to spleen
Intramuscular (i.m.)	Into a muscle; channeled to regional lymph node
Intranasal (i.n.)	Into the nose; channeled to MALT
Subcutaneous (s.q.)	Into the fatty hypodermis layer beneath the skin; channeled to regional lymph node
Intradermal (i.d.)	Into the dermis layer of the skin; channeled to SALT

*Parenteral: by means other than through the digestive tract.

exert significant local effects on the processing and presentation of antigens they encounter.

Since the whole idea of a vaccine is to allow rapid secondary responses when the natural pathogen does attack, it might seem desirable to create a vaccination scenario as close as possible to a natural infection. However, whereas natural pathogens can attack and penetrate mucosal surfaces, dead pathogens or their component macromolecules cannot. Modern medicine has thus had to abandon optimum localization in favor of relatively large doses of non-replicating immunogens administered by more distant routes. Fortunately for us, this seems to work, and the required memory lymphocytes are ready to respond when and where they are needed. Vaccination is discussed in detail in Chapter 23.

v) Adjuvants: Compensating for Non-replication

It should be clear after the preceding discussion that most isolated proteins are not very good immunogens on their own. A live, intact pathogen bearing, for example, a protein X on its surface is a splendid immunogen for three reasons. Because the pathogen multiplies during the course of infection, copious amounts of protein X are generated such that the threshold of lymphocyte activation is readily achieved. Secondly, the pathogen enters the body via its natural route of infection, which is usually an area in which leukocytes either reside or to which they can be recruited quite easily. Finally, the presence of the replicating pathogen stimulates the inflammatory response, which both draws APCs and lymphocytes to the area of attack and promotes the phagocytic uptake of the pathogen and its subsequent transport through the lymph system to the regional lymph nodes. In contrast, a dead pathogen or purified protein X injected into an animal generates only a limited dose of immunogen, which begins to randomly diffuse in the tissues. Far less inflammation is provoked, resulting in a lower level of recruitment and antigen-presenting capacity of immune system cells and decreased drainage to the local lymph nodes. The result is manifested as a poor immune response. In an effort to compensate to some degree for the non-replicating nature of many immunogens, scientists have devised agents called *adjuvants* (from the Latin *adjuvare*, "to help"). Adjuvants are substances that, when mixed with an isolated antigen, increase its immunogenicity. Basically, the presence of the adjuvant can provoke local inflammation and draw large numbers of immune system cells to the site of injection, where they can interact with the antigen. Different adjuvants work in different ways, matters that will be discussed in relation to vaccination in Chapter 23.

B. B Cell–T Cell Cooperation in the Humoral Immune Response

A fundamental tenet in modern immunology is that a T-dependent humoral immune response can occur only if antigen-specific B cells and antigen-specific T cells both respond to the same antigen at the same time. The road

of historical experiments that led to this finding was long and tortuous indeed, and filled with puzzling observations that defied clarification until the 1970s and 1980s. The reader's understanding of the results of the experiments described in the following sections will be enhanced if he/she keeps in mind the following points: (1) Although T cell progenitors are generated in the bone marrow, the bone marrow contains no mature T cells. The majority of T cell progenitors migrate from the bone marrow to the thymus in order to mature into functional effector Th or Tc cells. Therefore, a mouse without a thymus has very few mature T cells. (2) The spleen contains a mixture of mature B and T cells. The mature T cells in this organ arrive there by migrating from the thymus after thymic selection. A spleen lacking mature T cells can be "seeded" with exogenous mature T cells taken from a normal donor. (3) B cells both originate and mature in the bone marrow before migrating to the spleen.

I. THE DISCOVERY OF B–T COOPERATION: RECONSTITUTION EXPERIMENTS

By the early 1960s, researchers had shown that if a mouse had its thymus removed ("thymectomy") at birth, it could mount neither cell-mediated nor humoral immune responses. In studies of newborn chicks in which the bursa (the avian organ in which B cells originate) was removed, cell-mediated immunity was unaffected but antibody production was lost. The erroneous conclusion drawn at that time from these experiments was that the bursa (or its mammalian equivalent, the bone marrow) was necessary for some aspects of antibody production, but that the antibody-forming cells themselves were actually derived from the thymus. However, within a few years this view was challenged by "reconstitution" experiments in which researchers irradiated mice to abolish their immune systems and rebuilt them with various leukocyte populations taken from normal mice. The reconstituted mice were tested for their ability to make specific antibody upon antigenic challenge. When bone marrow cells alone (no mature T cells) were given to the irradiated mice, these cells did not proliferate in response to antigenic challenge and no antigen-specific antibody was produced. Similarly, if the mice were reconstituted with thymic cells alone (no B cells), these cells responded to antigenic challenge by proliferating, but there was still no antibody production. On the other hand, when both bone marrow and thymic cell populations were used to reconstitute the mice, both cell populations proliferated in response to antigen and antibody was produced. It was concluded that both the "thymus-derived" (T cells) and "bone-marrow-derived" cells (B cells) had to cooperate in some unknown way to produce the humoral response.

To identify which cell populations actually produced and secreted the antibody molecules, researchers took advantage of the fact that MHC class I molecules are expressed on all nucleated cells in the body, and that inbred mice with different genetically defined MHC types were available. They first thymectomized neonatal mice expressing an MHC class I allele we'll call type P, leaving the animals severely depleted of

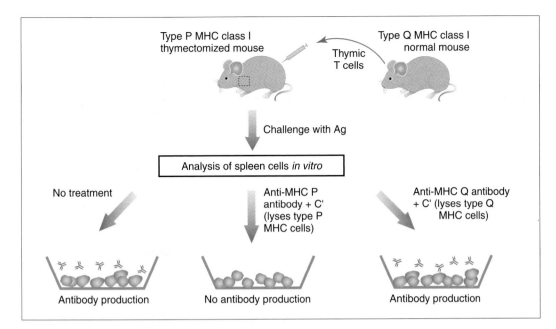

Figure 6-4

Identification of the B Lymphocyte as the Antibody-Producing Cell
The thymectomized mouse has type P MHC B cells (bone marrow-derived) and is reconstituted with thymus-derived type Q MHC T cells. The mouse is challenged *in vivo* with Ag and its spleen cells are isolated 6 days later. Treating the antigen-stimulated spleen cells from this mouse with specific anti-MHC antibodies plus complement (C′) allows selective depletion of either B cells or T cells. Only depletion of B cells prevents antibody production *in vitro*.

T cells (Fig. 6-4). These recipient mice were then reconstituted with thymus-derived cells from donor mice expressing a different MHC class I allele (say, type Q) such that the spleen contained endogenous B cells of type P MHC from the recipient but T cells of type Q MHC seeded from the donor. The reconstituted mice were challenged with specific antigen and their spleens were removed several days later and analyzed *in vitro*. The mixed population of antigen-stimulated spleen cells was shown to be capable of producing antigen-specific antibody in a culture dish. This result set the stage for researchers to selectively remove cell populations from the mixture in the dish to determine exactly which cell type was secreting the antibody. They found that if MHC P-bearing cells (recipient-derived B cells) were lysed by treating the mixture with anti-MHC P antibody plus complement, no antibody was produced. Conversely, if they treated the mixture with anti-MHC Q antibody plus complement, the MHC Q-bearing cells (donor-derived T cells) were removed but antibody production was not affected. This meant that the antibody-forming cells in the mixture were of recipient origin and not derived from the donor thymus.

While it was logical to assume that the recipient cells responsible for antibody production had matured in the bone marrow (as opposed to the thymus), additional experiments were done in which irradiated mice were reconstituted with normal thymic cells and bone marrow cells that carried an identifiable chromosomal aberration. After antigenic challenge, the antibody-producing spleen cells were analyzed and were all found to carry the same chromosomal aberration as the bone marrow cells used for reconstitution. Since the publication of these results, it has been accepted that antibody is produced by lymphoid cells that mature in the bone marrow (B cells). Moreover, these experiments showed that to become activated, B cells required "help" from cells that mature in the thymus (T cells).

II. THE MOLECULAR BASIS OF B–T COOPERATION: THE HAPTEN–CARRIER EXPERIMENTS

i) The Hapten–Carrier System

With the concept of B–T cell cooperation in antibody formation established, attention turned to the as yet unknown mechanism by which the two cell types interacted. Advances were slowly made by exploiting an important type of experimental model known as the "hapten–carrier" system.

In the early 1900s it was thought that antibodies were directed only against pathogenic agents in the environment, and that the repertoire of antibody specificities had been determined by evolutionary pressure favoring animals that had developed antibodies against microorganisms that threatened host survival. This theory was overturned by Karl Landsteiner in a series of landmark experiments carried out in the 1920s and 1930s. Landsteiner demonstrated that antibodies could in fact be raised against small molecules (<10 kDa) that were not even found in nature. However, he observed an antibody response only when these low molecular weight molecules were covalently coupled or "fastened" to larger macromolecules. Landsteiner called the small molecules that were not immunogenic on their own *haptens* (from the Greek *haptien*, "to fasten"). He called the large macromolecules to which the hapten had to be conjugated to stimulate an antibody response *carriers*. By the mid-1960s, many insights into the nature of antibody formation and recognition had been provided by experimentation with small inorganic haptens, but several puzzling observations still remained. The most perplexing was: why did hapten–carrier conjugates provoke antibody production when free haptens did not? Once produced, the anti-hapten antibodies were perfectly able to bind to the hapten alone, so that there did not appear to be an intrinsic problem with the binding site of the anti-hapten antibody. There had to be another element in the puzzle whose role had not yet been clarified.

ii) The Role of the Carrier

It was clear that haptens had to supply B cell determinants, since it had been established that it was the B cells that produced the antibodies that specifically bound to the haptens. However, from Landsteiner's time until the mid-1960s, it was thought that the carrier was simply acting as an inert but necessary support for the hapten. In 1963, Zoltan Ovary and Baruj Benacerraf took a closer look at the behavior of hapten–carrier systems and discovered a more complex role for the carrier that began to shed light on the mechanism of B–T collaboration. These researchers examined primary and secondary antibody responses to various hapten–carrier combinations, as illustrated in Table 6-4. As expected from Landsteiner's work, Ovary and Benacerraf found that the hapten and carrier had to be physically linked together to get an antibody response. However, they also noted that if the same hapten was used for a primary and then a secondary challenge, a secondary antibody response to the hapten was seen only if the hapten was conjugated to the *same* carrier for both immunizations. In other words, the identity of the carrier appeared to be playing a significant role in the development of the memory response to the hapten. This so-called *carrier effect* was very surprising at the time and had no ready explanation.

In order to explore the carrier effect more thoroughly, N. Avrion Mitchison and his colleagues at University College in London, UK reproduced the carrier effect in an artificial situation by a technique called *adoptive transfer*. In this method, cells from one animal (the "donor") are transferred to and function in another animal (the "recipient") whose own lymphocytes have been inactivated by irradiation. Mitchison primed mice with a defined hapten–carrier conjugate (H–C1) and then 6–12 weeks later transferred primed spleen cells to irradiated mice. Each recipient mouse was then challenged with a particular hapten–carrier combination. As expected from Ovary and Benacerraf's observations of the carrier effect, if a recipient mouse received spleen cells primed to a hapten on a particular carrier, a secondary anti-hapten response was observed only if the mouse was challenged later with the hapten conjugated to the same carrier (H–C1). If the secondary challenge used the original hapten conjugated to a different ("heterologous") carrier (H–C2), no secondary antibody response was obtained (Fig. 6-5A).

The development of this adoptive transfer system provided an easy means of manipulating spleen cell populations. Using this methodology, Mitchison made a critical observation that led to the identification of the roles of various cell types participating in the immune response. He found that, if along with the H–C1-primed spleen cells, spleen cells primed to carrier 2 alone (C2) were also transferred into the recipient mouse, then a secondary challenge with H–C2 did stimulate a secondary anti-H antibody response. In other words, the carrier effect could be overcome by "carrier priming" with the heterologous carrier (Fig. 6-5B). The existence of the carrier effect and the fact that it could be overcome by carrier priming suggested that during the secondary response, some form of cooperation was occurring between separate hapten-primed and carrier-primed cell populations. Furthermore, just as the hapten-primed cells distinguished among various haptens, the carrier-primed population recognized the carrier in a specific manner. In other words, adoptive transfer of H–C1-primed and C2-primed spleen cells followed by challenge of the mouse with hapten conjugated to a third carrier (H–C3) would not result in a secondary response.

iii) The Linkage of B and T Cell Determinants

As mentioned previously, Landsteiner's early hapten–carrier experiments indicated that the hapten and carrier had to be physically linked together in order for an immunized animal to make antibody responses to the hapten. This also held true in adoptive transfer hapten–carrier experiments. A mouse reconstituted with spleen cells from an H–C1-primed mouse and a C2-primed mouse would only produce a secondary antibody response upon challenge with H–C2 if H and C2 were linked together, a phenomenon called *linked recognition* (Fig. 6-5C). If challenged with H and C2 as separate entities, no secondary anti-hapten response was observed.

In 1970, Martin Raff, also of University College in London, took the hapten–carrier adoptive transfer experiments one step further to clarify the roles played by B cells and T cells in this system. He treated H–C1-primed spleen cells with anti-Thy1 antibody plus complement to remove T cells, and transferred the remaining cells to an irradiated mouse along with C2-primed spleen cells derived from a separate mouse that had not been immunized with the hapten. When the recipient

Table 6-4 **Carrier Effect**			
Primary Immunization*	**Primary Ab Response to H**	**Secondary Immunization***	**Secondary Response to H**
H + C1	−	H + C1	−
H–C1	+	H + C1	−
H–C1	+	H–C1	+++
H–C1	+	H–C2	−

*H, Hapten; C, carrier; −, covalent linkage.

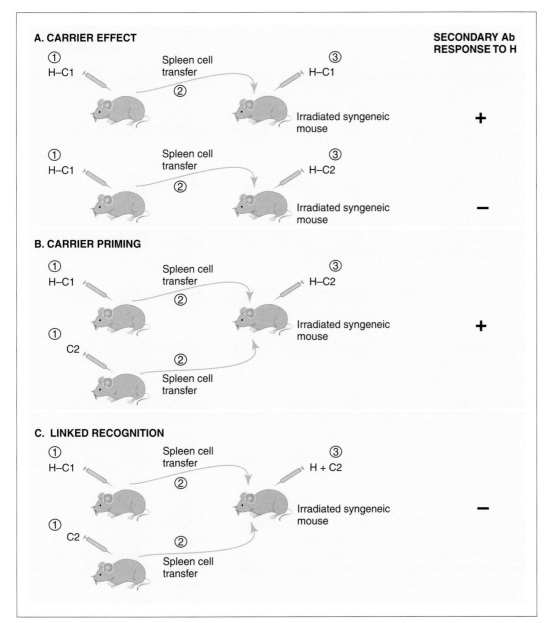

Figure 6-5

Use of Adoptive Transfer to Explore the Hapten–Carrier System
(A) The carrier effect. A secondary anti-hapten response is only seen in the recipient mouse if it is challenged with hapten (H) bound to the same carrier (C1) used to stimulate the donor spleen cells. The anti-hapten response is not observed when the recipient mouse is challenged with a different carrier (C2) linked to H. (B) Carrier priming. The lack of an anti-hapten response due to the carrier effect is overcome if the recipient mouse is also given donor spleen cells primed to the heterologous carrier, C2. (C) Linked recognition. Carrier priming is not successful unless the recipient mouse is challenged with hapten and heterologous carrier (C2) that are covalently linked.

mouse was challenged with H–C2, a secondary response to the hapten was still generated, suggesting that T cells were not involved in hapten recognition (Fig. 6-6A). On the other hand, if the C2-primed spleen cells were treated to remove T cells, the carrier effect could no longer be overcome; the reconstituted mouse could not mount a hapten-specific secondary response after challenge with H–C2. This indicated that the carrier supplied determinants that were specifically recognized by T cells (Fig. 6-6B).

Thus, the use of hapten–carrier systems elegantly demonstrated that clearly defined haptenic and carrier determinants are both required for an immune response, and that these determinants must be physically linked together. Coupling a hapten to a large carrier molecule results in the display of the hapten on the carrier surface in such a way that the hapten

becomes just another epitope on a macromolecular antigen. An antibody response is induced because the hapten supplies a B cell epitope and the carrier protein supplies a T cell epitope. This explains why the free hapten on its own is not immunogenic: it cannot supply the T cell epitope necessary for the generation of T cell help required to complete B cell activation. Furthermore, the hapten cannot act as a Ti antigen because it is small and lacks a repetitive structure capable of cross-linking BCRs.

To summarize, three elements are essential for a secondary response to a hapten: the hapten must be physically linked to the carrier; B cells primed against the hapten must be present; and T cells primed against the carrier used in the secondary immunization must be present. These requirements are illustrated for different hapten–carrier combinations in

Figure 6-6

Roles Played by B and T Cells in the Antibody Response
In hapten–carrier experiments using the adoptive transfer system, selective T cell depletion from donor spleen cell populations is accomplished by treatment with anti-Thy-1 antibody in the presence of complement. (A) When T cells primed to H–C1 are depleted, the recipient mouse can still respond to the hapten (H) via carrier priming, suggesting that B cells are the lymphocytes responding to the hapten. (B) When T cells primed to the heterologous carrier, C2, are depleted, carrier priming no longer works, showing that T cells are the lymphocytes responding to the carrier.

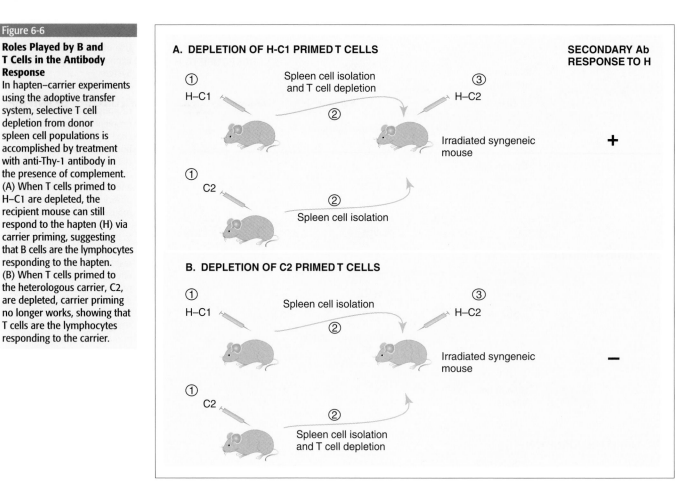

Table 6-5. In the case of a secondary challenge with a hapten conjugated to a <u>different</u> carrier from that used in the primary immunization, the B cell epitope in the hapten is the same, but the T cell epitope has changed. Primed T cells recognizing the first carrier are present, but are not stimulated by the new carrier, and thus do not secrete the cytokines needed to activate the memory B cells. Any T cells present that recognize the second carrier are not primed, and are thus undergoing a primary response to the second carrier and will produce only primary response levels of T cell help. The primary response amount of T cell help is not adequate to induce antibody production of secondary response levels or characteristics. These principles can be extrapolated to all T-dependent B cell immunogens, in that each immunogen can be viewed as a complex molecule containing distinct B cell and T helper cell epitopes.

Scientists have taken advantage of the fact that immunization with a hapten–carrier complex results in a certain proportion of all antibodies in the antiserum being specific for the hapten, even though injection of the unconjugated hapten is non-immunogenic. Antibodies to many small molecules (both natural and synthetic) and even to metallic ions have been raised through conjugation to carriers. Pure nucleic acids and lipids cannot function as Td antigens on their own (depending on their size, some are able to function as Ti antigens), but

will activate lymphocytes when conjugated to protein, enabling the scientist to raise antibodies against nucleoproteins and lipoproteins. It is germane to point out here that protein carrier molecules conjugated to haptens differ from the adjuvants discussed earlier in this chapter. Adjuvants are not covalently attached to the antigen, merely mixed with it. The carrier molecule is important because it supplies a T epitope; the adjuvant is important because it provokes inflammation and/or promotes lymphocyte stimulation.

III. HAPTEN–CARRIER COMPLEXES *in vivo*

Natural examples of hapten–carrier structures can also be found *in vivo*. Inorganic nickel that is absorbed through the skin can form conjugates with skin proteins, causing an immune response that results in a localized skin rash (see Ch.28). In the disease *systemic lupus erythematosus* (see Ch.29), the serum of patients contains anti-DNA antibodies. The immunogen in this disease may be a naturally formed complex of DNA and a host or viral protein, although large DNA molecules (whose structure is inherently repetitive) may also be acting as Ti antigens.

The dilemma of the carrier effect also has its equivalents in clinical application. For example, current vaccines against

Table 6-5 Requirements for the Induction of Secondary Responses by Hapten–Carrier Combinations

	Primary Immunization			Secondary Immunization			
ANTIGEN	B CELLS PRIMED	T CELLS PRIMED	ANTIGEN	PRIMED NIP-SPECIFIC B CELLS PRESENT?	PRIMED CARRIER-SPECIFIC T CELLS PRESENT?	SPECIFIC CARRIER LINKED TO THE HAPTEN?	PRODUCTION OF ANTI-NIP ANTIBODIES
(1) NIP–KLH	B_{NIP} B_{KLH}	T_{KLH}	NIP–KLH	+	+	+	+++
(2) NIP + KLH	– B_{KLH}	T_{KLH}	NIP + KLH	–	+	–	–
(3) NIP–KLH	B_{NIP} B_{KLH}	T_{KLH}	NIP + KLH	+	+	–	–
(4) NIP + KLH	– B_{KLH}	T_{KLH}	NIP–KLH	–	+	+	–
(5) NIP–KLH	B_{NIP} B_{KLH}	T_{KLH}	NIP–BSA	+	–	+	
(6) NIP–KLH + BSA	B_{NIP} B_{KLH} B_{BSA}	T_{KLH} T_{BSA}	NIP–BSA	+	+	+	+++
(7) NIP–KLH + BSA	B_{NIP} B_{KLH} B_{BSA}	T_{KLH} T_{BSA}	NIP + BSA	+	+	–	–
(8) NIP–KLH	B_{NIP} B_{KLH}	T_{KLH}	NIP–BSA + KLH	+	–	–	–

*NIP is the hapten 5-nitrophenyl acetic acid; KLH is the carrier protein keyhole limpet hemocyanin; BSA is the carrier protein bovine serum albumin. NIP–KLH indicates that the hapten is physically linked to the carrier protein. NIP + KLH indicates that the hapten is mixed with, but not physically linked to, the carrier protein.

(1) The hapten is linked to the carrier in the primary immunization, so that B cells are primed against both the hapten and the carrier. T cells are primed against the carrier. In the secondary immunization, the same hapten–carrier combination is used and the hapten is physically linked to the carrier. Both the appropriate primed B cells and T cells are present, so that the secondary response can proceed. (2) Since the hapten is not linked to the carrier in the primary immunization, B cells are primed only to B cell epitopes on the carrier and not to the hapten. T cells are primed to T cell epitopes on the carrier. A secondary response does not occur because no primed anti-hapten B cells are present. (3) The hapten is not linked to the carrier in the secondary immunization. A secondary response does not occur. (4) The hapten is not linked to the carrier in the primary immunization, so that no B cells are primed to the hapten. Even though the same hapten is conjugated to the same carrier in the secondary immunization, the lack of primed B cells prevents a secondary response. (5) In the secondary immunization, the carrier used is not the same as that used in the primary immunization. Primed anti-KLH T cells are present, but are not activated by the BSA. No primed anti-BSA T cells are present, so that a secondary response cannot proceed, only a primary response. This is an example of the *carrier effect*. (6) The hapten is linked to KLH in the primary immunization and mixed with BSA. T cells are primed to both KLH and BSA. The secondary response can occur because the hapten is physically linked to BSA in the secondary immunization, and both primed anti-hapten B cells and primed anti-BSA T cells are present. (7) Though the appropriate primed T and B cells are present, the hapten is not physically linked to the carrier. Therefore a secondary response does not occur. (8) As for (5), the carrier conjugated to the hapten in the secondary immunization is different from that conjugated to it in the primary immunization. T cells primed to KLH are present but those primed to BSA are not. Even though KLH is part of the secondary immunization, it is not conjugated to the hapten. This is a demonstration of *linked recognition*: the B cell epitope and the T cell epitope recognized in the secondary response <u>must</u> be physically linked.

the *Haemophilus influenzae b* bacterium (Hib vaccine) were developed using a polyribosyl phosphate carbohydrate structure of the bacteria as the hapten. However, the carrier used is not part of the bacterium, so that when a vaccinated individual encounters the live bacteria, the hapten appears to the immune system to be attached to a heterologous carrier; the immune reaction is only at primary response levels. A carrier that is naturally present as part of the bacterium itself is under investigation.

Another example is the influenza virus and a phenomenon called "original antigenic sin." A glycoprotein called hemagglutinin (HA) is a protein prominent on the surface of the influenza viral particle. Different strains of influenza virus carry HA molecules with slightly different carbohydrate components. In this situation, certain internal components of the virus (such as nucleoproteins that are relatively invariant from strain to strain) provide the T cell epitopes, while the HA moiety acts as the hapten (attached to the virus particle) and supplies the B cell determinants. In early flu epidemics, people were exposed to certain HA structures that differed from those on viruses attacking in later epidemics. Because the hapten (the HA) was no longer the same, no more than a primary response to these new strains was expected. However, some of the B cell clones that expanded in the early epidemic were able to cross-react with the HA molecules of the later viruses. Since the cross-reacting HA (hapten) was linked to the same carrier (the invariant internal viral components) in the later assault, primed T cell clones were also present, and a secondary response, although weak, was observed to the later flu strains when none was expected. The term "original antigenic sin" refers to the fact that the B cell clones triggered by the original HA antigens were reactivated in response to the new cross-reacting HA molecules due to the presence of the same carrier components in the different strains.

IV. THE RATIONALE FOR LINKED RECOGNITION

Although an in-depth discussion of B cell activation appears in Chapter 9, let's briefly summarize B cell activation events in order to examine the rationale for linked recognition.

To mount a primary humoral response to a Td antigen (or invading pathogen), the BCRs of a B cell must first recognize and bind a B cell epitope, making the B cell receptive to T cell help. Concurrently, an APC in the same vicinity must internalize a molecule of the same antigen (or pathogen), process it, and present a helper T cell epitope (which is different from the B cell epitope) on its surface. The TCR on a helper T cell in the same vicinity must recognize and bind the presented T cell epitope and must receive sufficient costimulatory signals from the APC to become fully activated and commence its proliferation and differentiation into daughter effector cells. An antigen-stimulated daughter Th cell must establish costimulatory contacts with the receptive antigen-specific B cell in a bicellular structure called a *conjugate*. To be fully activated, the B cell must then bind cytokines secreted by the Th cell to specific cell surface receptors whose cytoplasmic tails trigger intracellular signaling. Intracellular signaling stimulates nuclear transcription, facilitating B cell proliferation and differentiation into progeny plasma cells that secrete antibody recognizing the B epitope on the invader.

The formation of the required B–T conjugate is made possible by the structural connection of the B cell and T cell epitopes either <u>on the same macromolecular antigen</u> or <u>"attached" on the same pathogen</u>. Because of this connection, the B cell not only binds the B cell epitope of a Td antigen to its BCR, but can also internalize the protein by receptor-mediated endocytosis, process it in association with MHC molecules, and present the T cell epitopes on its surface for the perusal of nearby antigen-specific helper T cells. In so doing, the B cell facilitates the MHC–peptide:TCR interaction that leads to conjugate formation and establishment of the necessary costimulatory contacts between the B and T cell (Fig. 6-7).

In Chapter 3, we learned of the importance of <u>cellular</u> co-localization in initiating a primary humoral response; that is, an appropriate B cell, T cell, and APC must all come together in the same place at the same time. Now we see that to complete B cell activation in either the primary or secondary response, the B and T cell must actually co-localize

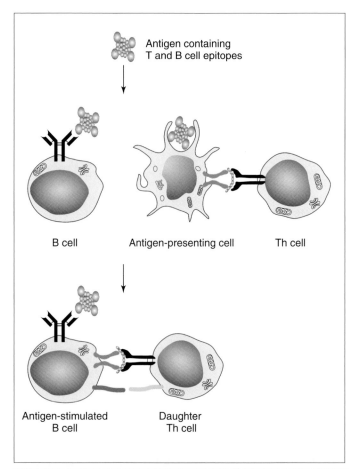

Antigen containing T and B cell epitopes

B cell   Antigen-presenting cell   Th cell

Antigen-stimulated B cell   Daughter Th cell

Figure 6-7

Connectivity of B and T Cell Epitopes
The BCR on the B cell recognizes one epitope on the antigen (gray) while the antigen-presenting cell processes and presents a different epitope (blue) to the T cell. Note that in a secondary response, the B cell itself often acts as the APC. In both primary and secondary responses, the activated Th cell proliferates, giving rise to daughter cells that must then make intercellular contact with the antigen-specific B cell in order to allow its full activation. Due to the connection of the B and T cell epitopes within the same molecule, the B cell will display the peptide–MHC class II epitope that will facilitate formation of a conjugate between the B cell and daughter T cell specific for the same antigen, ensuring activation that is both efficient and specific. Note that not all costimulatory molecules are shown.

to the point of "bridging" at the molecular level so their cell surface molecules are in contact. It is the connectivity of B and T cell epitopes on the same Td antigen that allows this necessary <u>molecular</u> co-localization between B and T cells. Because the T and B cell epitopes must be linked for lymphocyte activation to proceed, smaller molecules containing only one or the other epitope are not likely to be immunogenic, and only proteins of a certain size and complexity will be effective Td antigens. It also follows from molecular co-localizaton that antigen-stimulated B cells do not generally receive T cell help from a Th cell specific for a different antigen. On the rare occasion that a B cell is activated by a Th cell responding to an unrelated antigen, "bystander activation" is said to have occurred.

C. The Mechanics of Antigen–Antibody Interaction

Apart from the nature of their epitopes, the interaction at the molecular level of an antigen with its antigen receptor is fundamentally the same in B and T cells. However, it has been easier to study binding to the BCR because B cells secrete a soluble form of antigen receptor as antibody. In addition, the structure of a B cell epitope is simple compared to the peptide–MHC complex that serves as a T cell epitope. Accordingly, the next section will focus on the binding of antigen to antibody to illustrate the interaction between an antigen and its specific receptor. Of course, one must bear in mind that, *in vivo*, antibody is free in the circulation whereas the antigen receptor is fixed in the B cell membrane.

I. IDENTIFICATION OF B EPITOPE STRUCTURAL REQUIREMENTS

Landsteiner used antibodies specifically directed against haptens to demonstrate that the structure of an epitope is often as important as its chemical nature in determining its recognition by an antibody. He devised a battery of haptens composed of aminobenzene rings substituted with different chemical groups in varying positions and on chains of varying length (Fig. 6-8). Each hapten was coupled to the same carrier and used to immunize mice. In each case, antisera were raised in which a proportion of the antibodies was hapten-specific. He then tested antibodies specific for one hapten for cross-reactivity to the other haptens. The results showed that, while the actual chemical groups on the hapten influenced binding, the overall conformation of the antigen was more important. For example, if the anti-hapten antibody was raised against a hapten with a functional group in a particular molecular orientation, this antibody failed to recognize other haptens, or recognized them only weakly, if the same functional groups were present in a different orientation. In other words, an anti-hapten antibody recognized only the hapten conformation against which it was raised (apart from weak cross-reactions), even though the atomic content of a panel of related haptens might be identical. Landsteiner (and others) therefore concluded that specificity was based on both the atomic content and arrangement of chemical groups. Further investigation revealed that antibodies raised against non-ionic haptens and haptens of identical orientation but of varying chain lengths were more broadly cross-reactive. For example, a hapten displaying a chemical group of substantially different size or shape from that of the hapten used in the original immunization, even if it was in the same orientation, was recognized only weakly by the anti-hapten antibody, if at all. However, this antibody could react with molecules that, at first glance, appeared to be distant relatives of the original hapten, but which exhibited an overall conformation very similar to that of the original hapten. These studies were invaluable in demonstrating the importance of epitope structure for antibody binding and elucidating the forces and contacts required for antigen to bind to antibody (discussed in detail in the following sections).

II. WHERE ANTIBODY AND ANTIGEN INTERACT: THE COMPLEMENTARITY-DETERMINING REGIONS

As we learned in Chapter 5, early studies of antibody structure using limited proteolysis showed that antigens interact with the two Fab sites formed by the V_L and V_H regions at the tips of the Y-shaped Ig structure. When sequenced, these V regions were each found to contain three short stretches of amino acids that were highly variable between Ig molecules, the so-called hypervariable (HV) or complementarity-determining regions (CDRs). The less variable regions in the V domains between the CDRs became known as the framework regions. In laboratory tests, mutations of amino acids in the CDRs, either natural or engineered, altered the ability of an antibody to bind to its antigen. These results led to the deduction that CDR variability was responsible for the precise specificity of antibody–antigen binding. Further indirect evidence was obtained from early X-ray crystallographic studies of crystallized antibody proteins which showed that, in nature, the antibody protein is folded so as to group the CDRs on each arm of the Ig "Y" into one site at the tip of each Fab region, forming loops projecting in exactly the most advantageous position for antigen-binding. Direct evidence of antigen–antibody binding emerged from the analysis of crystallized complexes of antibody bound to small carbohydrates or haptens. These early studies suggested that the N-terminal ends of both the V_L and V_H domains were involved in forming an antigen-binding "pocket" in a naturally folded antibody molecule. The primary function of the framework regions appeared to be to ensure the correct folding of the V regions to form the pocket. Ten to twelve amino acids spread over several CDRs were found to be involved in securing small antigens within the pocket. This discovery led to a long-held misconception that antigen was necessarily nestled in an antibody pocket.

More recent studies of antibody–antigen crystal structures have shown that, while the antibody pocket concept appears to be true for small hapten molecules and peptides, larger macromolecular antigens (such as whole pathogens or globular proteins in their native conformations) not only project into the V_L–V_H pocket but also form specific surface–surface interfaces with the framework regions outside the CDR loops. Although 95% of binding specificity is determined by the CDRs, disruption of the framework contacts can abrogate the binding of antigen, confirming the essential role of the framework regions in keeping antigen and antibody together. Extensive antigen–antibody contacts outside the pocket are particularly well demonstrated by X-ray crystallographic studies of the binding of lysozyme by anti-lysozyme Fab fragment (Plate 6-1).

Close to 100 X-ray crystal structures for Fab fragments or antibodies have now been reported. This larger sample size has provided new insights into many aspects of antibody–antigen interaction, as well as conferring statistical validity on

Reactivity with

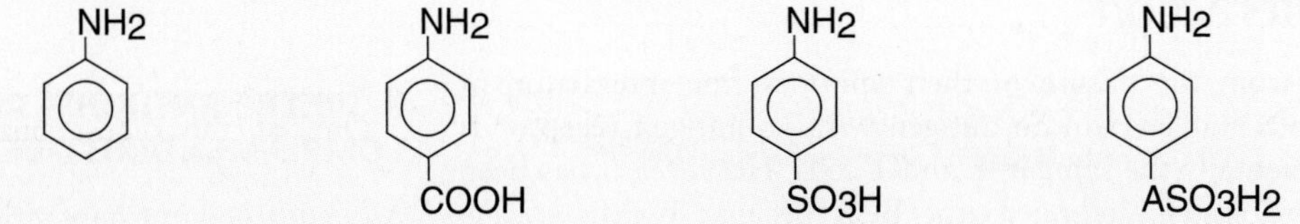

Antiserum Against	Aminobenzene	p-Aminobenzoic Acid	p-Aminobenzene Sulfonic Acid	p-Aminobenzene Arsenic Acid
Aminobenzene	+ + +	0	0	0
p-Aminobenzoic Acid	0	+ + ±	0	0
p-Aminobenzene Sulfonic Acid	0	0	+ + ±	0
p-Aminobenzene Arsenic Acid	0	0	0	+ + ±

Reactivity with

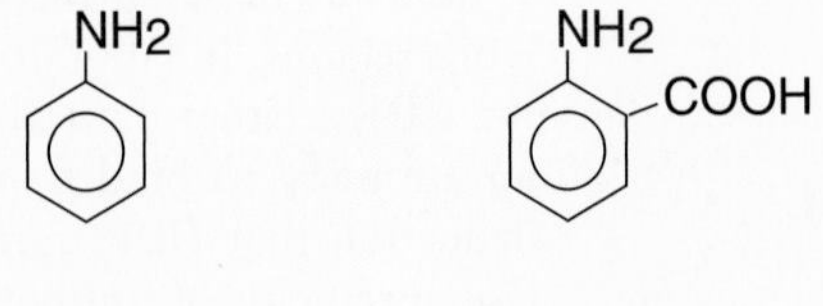

Antiserum Against	Aminobenzene	o-Aminobenzoic Acid	m-Aminobenzoic Acid	o-Aminobenzoic Acid
Aminobenzene	+ + +	0	0	0
o-Aminobenzoic Acid	0	+ + +	0	0
m-Aminobenzoic Acid	0	0	+ + +	0
o-Aminobenzoic Acid	0	0	0	+ + ±

Reactivity with

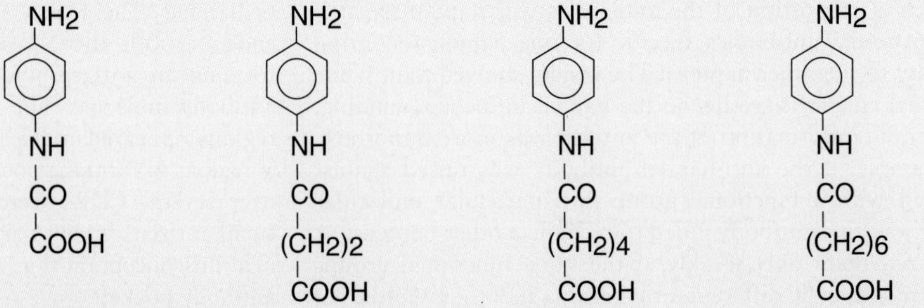

Antiserum Against	p-Aminooxalinic Acid	p-Aminosuccinanilic Acid	p-Aminodipanilic Acid	p-Aminosuberanilic Acid
p-Aminoxanilic Acid	+ +	0	0	0
p-Aminosuccinanilic Acid	0	+ + +	0	0
p-Aminodipanilic Acid	0	±	+ +	+ ±
p-Aminosuberanilic Acid	0	±	+ + ±	+ + +

Figure 6-8

Specificity of Hapten–Antibody Reactions
Effect of the atomic content of hapten (upper panel): Antisera do not cross-react with haptens bearing different chemical groups in the same position. Effect of atomic orientation (middle panel): Antisera do not cross-react with haptens bearing the same chemical group in different positions on the molecule. Effect of chain length (lower panel): Some cross-reactivity is seen with antisera against haptens of similar atomic content and position but varying size. Adapted from K. Landsteiner (1962) "The Specificity of Serologic Reactions." Dover Press, New York, NY.

earlier conclusions. Studies of the lysozyme/anti-lysozyme Fab co-crystal showed that the critical points for antigen–antibody binding in this case were 17 amino acids (spread over all six CDRs) that interacted with 16 amino acids on the lysozyme molecule. In fact, the interaction between most large protein antigens and their antibodies follows the lysozyme model and depends on approximately the same number of amino acid contacts. The antigen contact residues are often discontinuous

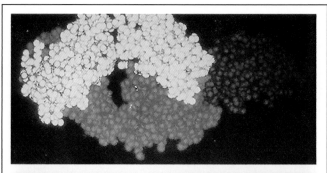

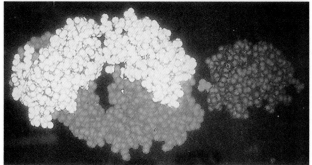

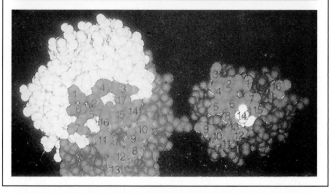

Plate 6-1

X-Ray Crystallography of Fab Lysozyme Complex
Top panel: Anti-lysozyme Fab (light chain in yellow, heavy chain in blue) binds to lysozyme (green). A critical residue of lysozyme, Gln 121, is shown in red. Center panel: Computer-generated image of separated components. Bottom panel: Rotation of separated components to show the contact residues in red (16 in each case) involved in Fab/lysozyme binding. Gln 121 (white) projects slightly from the lysozyme contact area, but is accommodated by a small cleft formed by the CDRs on the Fab binding surface. Reproduced with permission from Amit A. G. *et al.* (1986) Three-dimensional structure of an antigen–antibody complex at 2.8 Å resolution. *Science* **233**, 747–753.

peptide antigen. In general, an antigen will associate with the greatest number of amino acids in CDR3 of the Ig heavy chain. This CDR does not share the canonical structure exhibited by the other CDRs, and displays the greatest degree of sequence variability and conformational adaptability.

One of the oldest puzzles regarding antibody binding was whether an antibody and an antigen fit together in an inflexible lock-and-key mechanism, or whether the antibody and antigen could influence each other's conformation in an "induced fit" mechanism. The data derived from early studies of lysozyme/anti-lysozyme Fab co-crystals favored the former hypothesis and still describes the binding of most antigens to their antibodies: the antibody and antigen contact surfaces show a high degree of complementarity prior to binding. However, more recent evidence has suggested that there may be much more flexibility in the antigen-combining site than had been originally thought. Some antibodies apparently undergo conformational modifications to their CDR3 loops and alter the orientations of certain side chains to form a better bond with the antigen. That is, the structure of the antibody is "induced" to fit by the antigen to some extent. In other antibodies, modifications to the elbow angle allow more than one Fab to bind to the antigen at once, increasing the strength of the bond. The reader will recall from Chapter 5 that the elbow angle is defined by that region in the Ig molecule connecting the V region to the C_H1 domain (not to be confused with the hinge region located between C_H2 and C_H3).

III. FORCES AT WORK IN SPECIFIC ANTIGEN–ANTIBODY BINDING

The antigen–antibody bond is the result of multiple non-covalent intermolecular forces. None of these forces is itself very strong, but because they are all working simultaneously at the various sites defining the antigenic epitope in question, they combine to forge a very tight bond. Nevertheless, the bond is reversible, and the partners can be separated without alteration to either molecule by the application of high salt concentration, detergent, or non-physiological pH. Disruption experiments have demonstrated that four types of weak non-covalent forces can be identified in the antigen–antibody bond (Fig. 6-9).

1. *Hydrogen bonding* is established when the positive charge surrounding a hydrogen atom belonging to a residue in one molecule shares the negative charge of a chemical group in a residue in another molecule. Recent high-resolution analyses of antibody–antigen crystals have revealed the prominent role of bound water molecules in forming hydrogen bond networks between the antigen and the antibody. As many as 50 water molecules were identified in one antigen–antibody pairing.

2. *Van der Waals forces* result when polarities oscillate in the outer electron clouds of two neighboring atoms, creating attractive/repulsive forces between them. The high frequency of aromatic amino acids in the pockets of antibody

in sequence but contiguous in space, involving a surface area of about 600–900 square angstroms ($Å^2$). However, since every antibody has a different sequence in its CDRs (which is, after all, the basis of antigenic specificity), the pattern of contact between one antibody and its antigen is slightly different from that of the next antibody with its antigen. Furthermore, while both the V_L and V_H domains of an antibody are involved in forming the antigen-binding site, not every antibody will use all of its six CDRs to interact with its antigen, particularly if the antibody binds a polysaccharide or small

Figure 6-9

Non-covalent Forces in Antibody–Antigen Binding
The indicated four types of non-covalent bonds contribute to antigen–antibody binding. Adapted from Klein J. and Hořejší V. (1997) "Immunology," 2nd edn. Blackwell Sciences Ltd., Oxford.

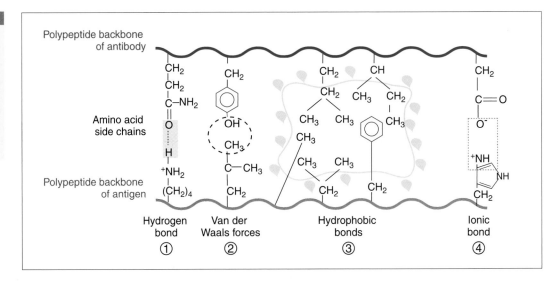

proteins increases the charge in this region, promoting both hydrogen and Van der Waals bonding.

3. *Hydrophobic* ("afraid of water") *bonds* are formed in aqueous solution when polar water molecules force hydrophobic, non-polar chemical groups together in an effort to generate the minimum non-polar surface area possible (and maximize the entropy of the water molecules). The larger the hydrophobic regions involved, the stronger the hydrophobic association between them.

4. *Electrostatic* or *ionic bonds* are the attraction between charged residues with opposite polarities. These forces are thought to play a more minor role in antigen–antibody binding.

The contribution of each type of force to the overall binding depends on the identity and location of the amino acids or other chemical groups in both the antibody and antigen molecules. The more closely the relevant chemical groups can approach one another, the more efficient the binding will be. Similarly, the more complementary the shapes of the antigenic epitope and the antigen-binding site on the antibody, the more contact sites will simultaneously be brought into close proximity, increasing the number of non-covalent bonds of all types and resulting in a stronger overall binding. Mapping of binding sites on antibodies has shown that cavities, grooves, or planes are often present that correspond to complementary structural features on the antigen. If an antigen approaches whose shape is less complementary to the conformation of the binding site on the antibody, fewer bonds are formed and steric hindrance and repulsion by competing electron clouds are more likely to "push" the prospective antigen away. This principle has been shown to be crucial for the antigenic specificity of antibodies: only those antigens shaped such that they make a sufficient number of contacts of the required strength will succeed in binding to the antigen-combining site of the antibody molecule.

IV. AFFINITY AND AVIDITY OF ANTIBODY BINDING

Scientists have devised terms to express how tightly biological molecules bind to each other. Immunologists define the *affinity* of an antibody for its antigen as the strength of the non-covalent association between one antigen-binding site (thus, one Fab arm of an antibody molecule) and one antigenic epitope. In contrast, the *avidity* of an antibody for its antigen is a measure of the total strength of all the Fab–epitope associations possible between the two molecules. Avidity relates to the fact that all antibodies are *polyvalent*; that is, all have more than one antigen-binding site and those that exist in polymeric forms have more than two. In addition, most natural antigens are whole pathogens that are naturally polyvalent (although some of their component individual proteins are not). In the vast majority of cases, more than one contact is established between the antigen and an antibody binding to it. We will now examine these two binding measures in more detail.

i) Antibody Affinity

The affinity of an antibody for an antigen is a measure of how many antibodies in a pool of identical Ig molecules have antigen bound to an antigen-binding site at any given moment (assuming that antigen is not limiting). The greater the affinity, the more complexes will exist at any one time, and the longer these complexes will endure. Affinity is defined mathematically according to the precepts of the Law of Mass Action as applied to a reversible chemical reaction. For the purposes of this discussion, we will describe the wholly unnatural situation in which molecules of a single pure antigen are combining with molecules of a single homogeneous antibody with a single antigen-combining site.

When our hypothetical antibody and antigen solutions are mixed, antibody and antigen begin to find each other and to form complexes at a rate dependent both on their relative concentrations, and on how easily the non-covalent forces holding them together form. The rate of complex formation is known as the k_{on} constant. Because the binding of antigen to antibody is fully reversible, the complexes will also dissociate at a rate that depends on how resistant the non-covalent binding forces are to being disrupted. The rate of complex dissociation is known as the k_{off} constant. After a certain period of time, chemical equilibrium will be established; that is, complexes will be formed at the same rate as they are dissociated, so that the amount of unbound antigen remains constant. The ratio of the concentrations of free and bound antibody at equilibrium is a measure of the affinity of the antibody for the antigen. In other words, when the rate of association between antigen and antibody (k_{on}) is divided by the rate of their dissociation (k_{off}), we obtain a measure of the affinity of an antibody for antigen represented by the *equilibrium constant K* (Fig. 6-10).

In practical terms, an antibody with a very small k_{off} (meaning that the antigen and antibody stick tightly together for long periods) might have K in the order of $10^{11}\ M^{-1}$. In contrast, an antibody with a fast off rate (large value for k_{off}) might have a K of $10^6\ M^{-1}$ and would be considered a low-affinity antibody that binds its antigen weakly and only for short periods. Studies of the relative contribution of rates of association compared to rates of dissociation in determining complex formation have shown that k_{off} is more important than k_{on}. The K value for a small antigen is relatively easy to determine experimentally, as illustrated in Box 6-2.

It is important to note that, in a natural immune response to invasion by a complex antigen, a <u>polyclonal</u> antibody response will be induced in which the antibodies generated

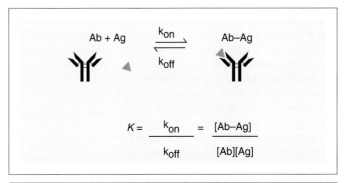

Figure 6-10

Kinetics of Antigen–Antibody Interaction
The binding of antibody and antigen can be viewed as a chemical reaction where antibody (Ab) and antigen (Ag) are the reactants and the antigen–antibody complex (Ab–Ag) is the product. The equilibrium constant, K, is a measure of the affinity of the antibody for the antigen; k_{on}, forward rate constant, k_{off}, reverse rate constant.

will be heterogeneous with respect to both specificity and affinity. Not only will antibodies to different epitopes on the antigen be raised, but a single epitope will induce the activation of various B cells whose receptors display a variable accuracy of fit with the epitope. The result is the production of a collection of antibodies with a range of affinities. Interestingly, for reasons that are still unclear, these affinities do not appear to follow a normal distribution: proportionally more antibodies have low or high affinities for a given antigenic epitope than have an intermediate affinity for it

---

Box 6-2. Determination of antibody affinity

How do we actually arrive at a value for the affinity of a particular antibody for an antigen? The affinity of the antibody binding site for the antigenic epitope is represented by the *equilibrium constant K*, which is equal to the rate of association (k_{on}) of the antigen–antibody pair divided by their rate of dissociation (k_{off}) as follows:

$$K = k_{on}/k_{off}$$

For the reaction Ab + Ag $\leftrightarrow$ Ab · Ag, one can apply the Law of Mass Action to obtain the following relation:

$$k_{on}\,[Ab]\,[Ag] = k_{off}\,[Ab \cdot Ag]$$

So, at equilibrium,

$$K = [Ab \cdot Ag]_{eq}/[Ab]_{eq}\,[Ag]_{eq} \quad (1)$$

where $[Ab \cdot Ag]_{eq}$, $[Ab]_{eq}$, and $[Ag]_{eq}$ are the equilibrium concentrations of antigen–antibody complexes, free antibody,

and free antigen, respectively. Applying laws of thermodynamics that are beyond the scope of this book, Eq. (1) can be expressed as

$$\frac{[AB \cdot Ag]_{eq}/[Ab]_{initial}}{[Ag]_{eq}} = -K([Ab \cdot Ag]_{eq}/[Ab]_{initial}) + K_n$$

or

$$r/c = -K_r + K_n \quad (2)$$

where r is the ratio of complexed antigen concentration to total antibody concentration (i.e., $[Ab \cdot Ag]_{eq}/[Ab]_{initial}$), c is the equilibrium concentration of free antigen (i.e., $[Ag]_{eq}$), and n is the number of binding sites on the antibody (valency).

If one considers Eq. (2) in its more familiar form of $y = mx + b$, it follows that plotting r/c (the "y" value) against r (the "x" value) yields a straight line with

a slope of $-K$ and an x-intercept ("b") of K_n. Such a graph is called a Scatchard plot and can be constructed from a series of experiments in which $[Ab]_{initial}$ is known and constant, $[Ag]_{initial}$ is known and varied, and $[Ab \cdot Ag]_{eq}$ and $[Ag]_{eq}$ are measured. Values for r/c and c can then be calculated and plotted, yielding a value for K.

Experimentally, the equilibrium concentrations of free and complexed antigen ($[Ag]_{eq}$ and $[Ab \cdot Ag]_{eq}$, respectively) can be measured over time using *equilibrium dialysis*. To understand how this system works, consider a dialysis apparatus in which a known concentration of antigen of small molecular size is added to one side ("the outside") of a semi-permeable membrane dividing two chambers. This antigen has been radiolabeled to allow easy assessment of its concentration. As shown in the Figure, in the control experiment (1),

Continued

Box 6-2. **Determination of antibody affinity**–*cont'd*

no antibody is present at first in the "inside" chamber (1a). The labeled antigen diffuses across the membrane until it reaches a state of equilibrium in which both sides of the chamber have the same concentration of antigen (1b); 1c shows a graphical representation of this process. To generate an experimental point (2) for the Scatchard plot, an amount of radiolabeled antigen (that varies for each point of the plot) is added to the "outside" chamber while a known amount of unlabeled antibody is added to the "inside" chamber (2a). The size of the pores in the semi-permeable membrane separating the chambers must be such that antibody (and therefore antigen complexed to antibody) cannot cross the membrane, but free antigen can diffuse from one side of the membrane to the other. Initially, all the labeled antigen is in the "outside" chamber. With time, antigen diffuses to the other side of the membrane in an attempt to equalize its concentration on both sides. However, a certain proportion of antigen arriving on the antibody side of the membrane is bound in an immune complex, effectively taking a fraction of the antigen molecule population out of the equilibrium equation. The greater the affinity of the antibody for the antigen, the more antigen is "removed" as it is drawn from the "outside" to the "inside" (2b). Once equilibrium is established (about 4 hours), the amount of free antigen will be equal in each chamber, while some amount of antigen will be bound to the antibody on the "inside." Thus, the radioactivity on the "outside" of the chamber represents $[Ag]_{eq}$, while the radioactivity on the "inside" represents $([Ab \cdot Ag]_{eq} + [Ag]_{eq})$. The difference in radioactivity between the two chambers ("inside" – "outside") allows calculation of the concentration of complexed antigen, $[Ab \cdot Ag]_{eq}$. Values for r and r/c can then be calculated as described above and the Scatchard plot (3) constructed such that the equilibrium constant K (slope of the line), representing the affinity of the antibody for this antigen, can be determined.

The x-intercept of the Scatchard plot can be used to determine the number of antigen-binding sites on the antibody (which may be monomeric, polymeric, or even a single Fab). When $[Ag]_{eq}$ becomes very large, the amount

of antigen bound by antibody at equilibrium $[Ab \cdot Ag]_{eq}$ becomes increasingly small, so that mathematically, r/c approaches 0. Since

$$r/c = -K_r + K_n \quad (3)$$

then, as r/c approaches 0,

$$0 = -K_r + K_n$$

Therefore

$$n = rK/K$$

and

$$n = r$$

That is, the value of r as the line intersects the x-axis gives a value for the valency of the antibody.

As mentioned previously, this method of K determination only works for small antigens able to pass through a semi-permeable dialysis membrane. For macromolecular antigens, a more difficult determination of the actual concentrations of free antibody, free antigen, and complexes under non-equilibrium conditions must be undertaken. In addition, if the antibodies used were of mixed affinities, a curved line will be generated with a variable slope, reflecting the varying affinities in the antibody preparation.

1. Control

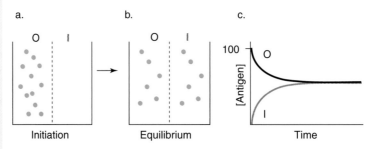

a.   b.   c.

Initiation   Equilibrium   Time

O = outside
I = inside

2. Experimental Point

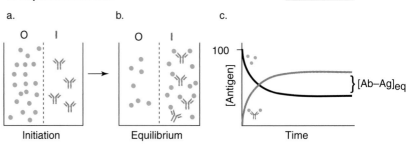

a.   b.   c.

Initiation   Equilibrium   Time

$[Ab–Ag]_{eq}$

3. Scatchard Plot

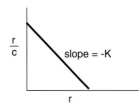

slope = -K

(Fig. 6-11). It is also important to keep in mind that affinity is not an absolute quality for an antibody: a given antibody may display different affinities for a spectrum of related but slightly different antigens (see below). In addition, the affinity of antibodies raised during a secondary immune response to a given epitope is increased compared to antibodies induced in the first encounter, a phenomenon called *affinity maturation*, which is discussed in Chapter 9.

ii) Antibody Avidity

As introduced previously, avidity is defined as the overall strength of the bond between a multivalent antibody and a multivalent antigen. Microorganisms and eukaryotic cells are nothing more than large multivalent antigens when they feature multiple identical protein molecules on their cell surfaces. Other examples of multivalent antigens are macromolecules composed of multiple identical subunits and macromolecules containing a structural epitope that occurs repeatedly along the length of the molecule.

All antibodies are multivalent because they have two or more identical antigen-binding sites. sIgG, sIgD, and sIgE molecules have only two binding sites, but polymeric sIgA molecules can exhibit 4, 6, or even 8 identical antigen-combining sites, and sIgM molecules are naturally found as pentamers with 10 binding sites. Binding at one Fab site on a polymeric antibody holds the antigen in place so that other binding sites on the antibody are more likely to bind as well. With each additional binding site engaged by a polymeric antibody, the probability of the simultaneous release of all bonds drops exponentially, resulting in a lower chance of antigen–antibody dissociation and therefore a stronger overall bond (Fig. 6-12). For example, in the case of pentameric IgM, each site in itself may not have a very strong affinity for antigen, but because all 10 sites may be bound simultaneously, IgM antibodies have considerable <u>avidity</u> for their antigens. At any given

moment, dissociation at any one antibody–antigen combining site may occur, but the complete release of antigen from antibody would require that all 10 binding sites be dissociated from the antigen simultaneously. The practical result of high avidity in an antibody is that the immune complex between antigen and antibody is formed more rapidly and stably, and the antigen is eliminated from the body more efficiently.

V. ANTIBODY CROSS-REACTIVITY

The specificity of an antiserum as a whole is defined by its component antibodies. As we have seen, any one epitope will

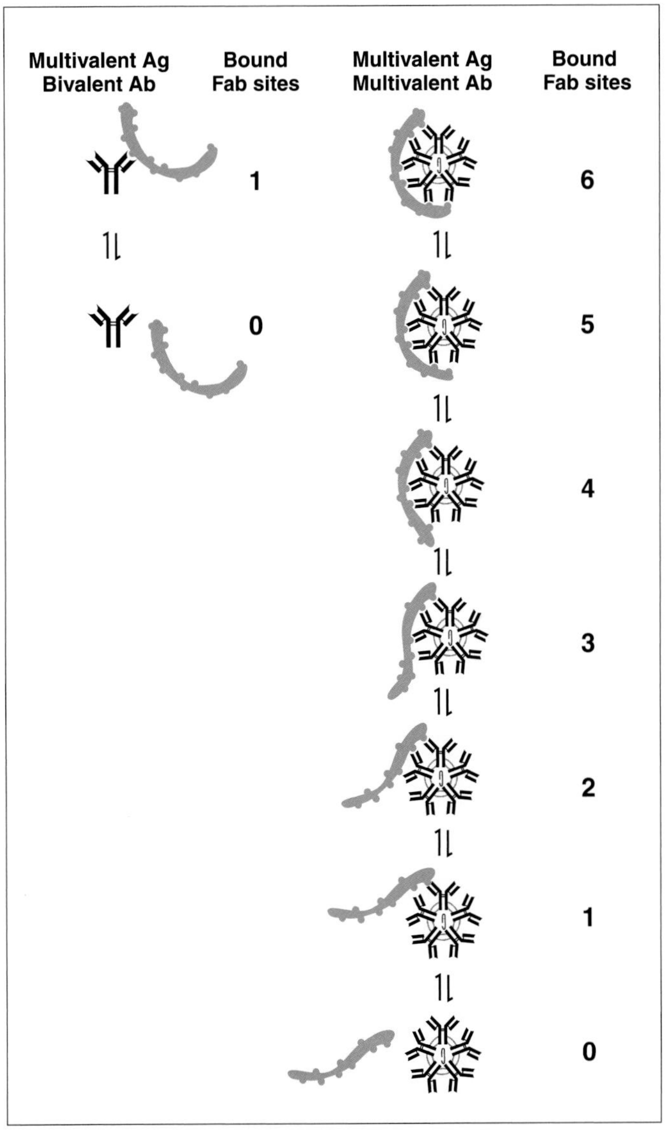

Multivalent Ag Bivalent Ab	Bound Fab sites	Multivalent Ag Multivalent Ab	Bound Fab sites

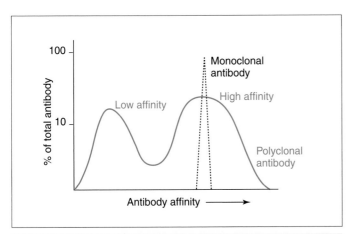

Figure 6-12

Antibody Avidity
A comparison of antigen binding by monomeric vs. pentameric antibody. Overall binding strength (avidity) between an antibody and multivalent antigen increases with the number of binding sites on the antibody that can be engaged. Many more dissociation steps must simultaneously occur before the antigen can be released from the pentameric antibody.

Figure 6-11

Affinity Distribution Curve for a Polyclonal Antibody Response
Polyclonal antibodies are derived from a variety of B cells responding to the same antigen. The variety of antigen binding sites among such antibodies leads to a range of antibody affinities with a distribution that is bi-modal between low and high affinity. In contrast, monoclonal antibodies have the same binding site and therefore display a uniform affinity.

induce the production of a collection of antibodies with a range of affinities for that epitope. As well, an antigen usually exhibits more than one epitope, so that antibodies exhibiting a spectrum of epitope specificities will also be included in the antiserum. A *cross-reaction* is said to have occurred when an antiserum reacts to an antigen other than the cognate antigen. *Cross-reactivity results either when one epitope is shared by two antigens, or when two epitopes on separate antigens are similar.*

For example, let us suppose that we immunize a mouse with antigen X, which contains epitopes A, B, and C; in this context, antigen X is the cognate antigen (Fig. 6-13). A polyclonal antiserum called anti-X will be recovered from this mouse which contains a mixed population of antibodies, including anti-A, anti-B, and anti-C. Let us also suppose that we have a second antigen Y, which contains epitopes A, Q, and R, and a third antigen Z, which contains epitopes S and B* (which is conformationally similar to B). The anti-X antiserum will recognize antigen Y, since a proportion of the antibodies (anti-A) will bind strongly to the shared epitope A on antigen Y. Anti-X antiserum will also recognize antigen Z, because anti-B antibodies will bind to epitope B*, although probably with a different affinity. Anti-C antibodies will make no contribution to cross-reactivity in either case since they fail to recognize any epitope on antigens Y or Z. Anti-X antiserum is said to cross-react with antigen Y because of a shared epitope, and with antigen Z because of the presence of an epitope that is similar. In the same way, anti-Y antiserum and anti-Z antiserum will cross-react with antigen X because of the presence of anti-A and anti-B* antibodies, respectively.

We have seen that a complementarity between the conformation of the epitope and that of the antigen-combining site is necessary to bring sites of critical contact in sufficiently close proximity to permit stable antigen–antibody binding. The differing degrees to which cross-reacting antigens bind

to antibody can be understood in the light of the molecular binding forces described above. An antigen of a shape slightly different from that of the cognate antigen may not be able to form one or more of the multiple types of bonds required for tight binding to antibody; the antibody is thus more easily dissociated from the cross-reacting antigen. To continue the previous example, the more closely B* resembles the shape of B, the greater the possible molecular contacts between B* and the binding site of anti-B, and the more likely cross-reactive binding will occur. However, in some cases, the antibody may undergo an "induced fit" modification to better accommodate a non-cognate antigen. In this case, the cross-reactive epitope may be more different in conformation than one would expect, but the antibody adjusts to it and binds with surprisingly high affinity. In other situations, the antibody actually has a better fit with the non-cognate antigen and will bind it with higher affinity than it will the cognate antigen. Although the antigen–antibody bond is generally a sum of multiple forces at multiple sites on an epitope, in some cases a single amino acid may furnish a critical binding contact, and mutation of this residue alone can be sufficient to prevent antibody binding. We see here evidence of an apparent paradox: some antibodies can distinguish similar antigens with exquisite specificity (one amino acid change can completely abolish antibody binding) while others appear to be broadly cross-reactive to a range of antigens that are related but not identical. These factors result from the individual idiosyncrasies of each antibody–antigen interface, and make cross-reactivity difficult to predict.

The reader may be more familiar with the phenomenon of cross-reaction than he or she realizes. In our previous discussion of haptens, we made reference to the population of antibodies induced by immunization with hapten–carrier complexes. In this population will be antibodies recognizing epitopes on the carrier protein itself, and those recognizing

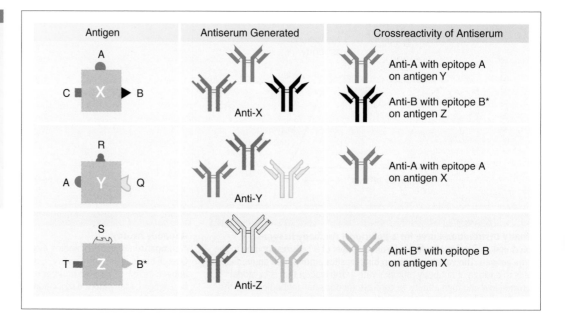

Figure 6-13

Antibody Cross-reactivity
Antigen X, Y, or Z is used to immunize a mouse, generating anti-X, anti-Y, and anti-Z antisera as shown. Each of these polyclonal antisera is then tested for cross-reactive binding to the other two antigens. A, B, C, Q, R, S, and T represent various distinct antigenic epitopes on X, Y, and Z. B* represents an epitope that is conformationally similar to B.

the conjugated hapten as just another epitope. If this anti-serum is mixed with free hapten, that subset of antibodies that was induced in response to the hapten as epitope will bind to free hapten. In other words, there is a cross-reaction to a "shared" epitope: the hapten is "shared" by itself and the hapten carrier conjugate. However, the free hapten often has a shape slightly different from the shape it assumes in the hapten–carrier conjugate, so that the affinity of these anti-bodies for the isolated hapten is often lower than that for the hapten–carrier cognate antigen. The same is true of that subset of antibodies that recognizes epitopes on the carrier portion of the conjugate: their affinity for free carrier is often decreased.

For another familiar example of cross-reaction, the reader need look no further than Jenner's vaccination experiments (described in Ch.1). B cells originally stimulated by the cowpox virus in an immunized individual recognized certain epitopes on the invading smallpox virus and commenced production of antibodies. These antibodies were able to cross-react and bind to the structurally related smallpox virus particles. The immune complexes were subsequently eliminated from the host, saving the individual from a deadly disease.

This type of immunization scheme is unusual in that the vast majority of modern vaccinations do not rely on cross-reaction to a different pathogen, but rather on exposure to the natural pathogen that has been killed or attenuated. Such a vacci-nation process usually generates antibodies that recognize identical, cognate epitopes on the virulent pathogen in a sub-sequent natural infection (see Ch.23). However, should the chemical treatment used to attenuate the natural pathogen slightly alter the epitope in question, cross-reaction could play a role in this type of vaccination as well.

We have now examined the mechanics of antigen–antibody binding at the molecular level, which also gives us a reason-able idea of the interactions that occur on the cell surface between an antigen and an antigen receptor. In the next chapter, we investigate technologies developed in the labora-tory to visualize and quantitate antigen–antibody binding, and explore how scientists have taken advantage of the specificity and reversibility of this remarkable bond. These tools have proved invaluable for the precise manipulation of many bio-logical molecules not only in immunology, but in many other fields as well.

SUMMARY

T-independent (Ti) antigens activate B cells in the absence of MHC class II-restricted T cell help. Ti antigens are generally large poly-meric molecules (such as polysaccharides) that have repetitive subunits capable of cross-linking mIg on the B cell surface. In contrast, T-dependent (Td) antigens are proteins of non-repetitive amino acid sequence that bind to mIg but that cannot fully activate the antigen-specific B cells on their own. The activation of these B cells cannot occur until the Td antigen is processed and presented by APCs that stimulate T helper cells specific for that antigen. The activated T helper cells then secrete the cytokines required by the B cells for full activation. To become receptive to the T cell cytokines, the B cell must also establish intercellular, costimulatory contacts with the antigen-specific T cell. A Td anti-gen thus contains both B cell and T cell epitopes that facilitate the necessary co-localization of the antigen-specific lymphocytes.

When the mIg component of the BCR or the corresponding solu-ble antibody binds to antigen, specificity is ensured by the unique surface created by the combined CDRs of the V_H and V_L domains of the Ig. Intermolecular binding at this interface results from the additive effect of non-covalent binding forces between the antibody and antigen and is therefore influenced by the atomic content of the antigen as well as the position of its chemical groups. The affinity of antibody binding to a given antigen is the strength of the antigen–antibody interaction at one antigen binding site, while antibody avidity measures the total strength of binding in cases in which the multivalent antibody binds multivalent antigen. A given antibody may cross-react with more than one antigen if the recognized epitope is present on different macromolecules, or if similar epitopes are present on separate antigens.

Selected Reading List

Amit A. G., Mariuzza A. and Phillips R. J. P. (1986) Three-dimensional structure of an antigen–antibody complex at 2.8 angstroms resolution. *Science* **233**, 747–753.

Arevalo J. H., Taussig M. J. and Wilson I. A. (1993) Molecular basis of crossreactivity and the limits of antibody–antigen complementarity. *Nature* **365**, 859–863.

Armitage R. J. and Alderson M. R. (1995) B-Cell stimulation. *Current Opinion in Immunology* **7**, 243–247.

Braden B. and Poljak R. (1995) Structural features of the reactions between antibodies and protein antigens. *FASEB Journal* **9**, 9–16.

Braden B. C., Souchon H., Eisele J.-L., Bentley G. A., Bhat T. N., Navaza J. and Poljak R. J. (1994) Three-dimensional structures of the free and the antigen-complexed Fab from monoclonal anti-lysozyme antibody D44.1. *Journal of Molecular Biology* **243**, 767–781.

Brownstone A., Mitchison N. and Pitt-Rivers R. (1966) Chemical and serological studies with an iodine-containing synthetic immunological determinant 4-hydroxy-3-iodo-5-nitrophenyl-acetic acid (NIP) and related compounds. *Immunology* **10**, 465–479.

Claman H. N., Chaperon E. A. and Triplett R. F. (1966) Thymus–marrow cell combinations. Synergism in antibody production. *Proceedings of the Society for Experimental Biology and Medicine* **122**, 1167–1171.

Colman P. (1988) Structure of antibody–antigen complexes: implications for immune recognition. *Advances in Immunology* **43**, 99–132.

Howard J. (1985) Immunological help at last. *Nature* **314**, 494–495.

Landsteiner K. (1963) "The Specificity of Serological Reactions." C. C. Thomas, Springfield, IL.

Macario A. and Conway de Macario E. (1975) Antigen-binding properties of antibody molecules: time-course dynamics and biological significance. *Current Topics in Microbiology and Immunology* **71**, 125–170.

Metzger H. (1974) Effect of antigen binding on the properties of antibody. *Advances in Immunology* **18**, 169–207.

Mian I., Bradwell A. and Olson A. (1991) Structure, function and properties and antibody binding sites. *Journal of Molecular Biology* **217**, 133–151.

Mitchell G. F. and Miller J. F. A. P. (1968) Immunological activity of thymus and thoracic-duct lymphocytes. *Proceedings of the National Academy of Sciences* **59**, 296–303.

Mitchison N. (1968) Recognition of antigen. *Symposia of the International Society for Cell Biology* **7**, 21–42.

Mond J. J., Lees A. and Snapper C. M. (1995) T cell-independent antigens Type 2. *Annual Review of Immunology* **13**, 655–692.

Mongini P. K., Blessinger C. A., Highet P. F. and Inman J. K. (1992) Membrane IgM-mediated signaling of human B cells. *The Journal of Immunology* **148**, 3892–3901.

Mosier D. and Subbarao B. (1982) Thymus-independent antigens: complexity of B-lymphocyte activation revealed. *Immunology Today* **3**, 217–222.

Ovary Z. and Benacerraf B. (1963) Immunological specificity of the secondary response with dinitrophenylated proteins. *Proceedings of the Society for Experimental Biology and Medicine* **114**, 72–76.

Porter R. (1967) The structure of the heavy chain of immunoglobulin and its relevance to the nature of the antibody-combining site. *Biochemical Journal* **105**, 417–426.

Raff M. F. (1970) Role of thymus-derived lymphocytes in the secondary humoral immune response in mice. *Nature* **226**, 1257–1258.

Schumaker V., Phillips M. and Hanson D. (1991) Dynamic aspects of antibody structure. *Molecular Immunology* **28**, 1347–1360.

Sulzer B. and Perelson A. A. (1997) Immunons revisited: binding of multivalent antigens to B cells. *Molecular Immunology* **34**, 63–74.

Sutton B. (1989) Immunoglobulin structure and function: the interaction between antibody and antigen. *Current Opinion in Immunology* **2**, 106–113.

von Bulow G.-U., van Deursen J. M. and Bram R. J. (2001) Regulation of the T-independent humoral response by TACI. *Immunity* **14**, 573–582.

Wilson I. A. and L S. R. (1994) Antibody–antigen interactions: new structure and new conformational changes. *Current Biology* **4**, 857–867.

Exploiting Antigen–Antibody Interaction

7

Rosalyn Yalow 1921–

Radioimmunoassay
Nobel Prize 1977

© Nobelstiftelsen

CHAPTER 7

A.   SOURCES OF ANTIBODIES

B.   TECHNIQUES BASED ON IMMUNE COMPLEX
     FORMATION

C.   ASSAYS BASED ON UNITARY
     ANTIGEN–ANTIBODY PAIR FORMATION

"The important thing in science is not so much to obtain new facts as to discover new ways of thinking about them"–Sir William Lawrence Bragg

Chapters 5 and 6 have described the Ig proteins and their interaction with antigenic epitopes *in vivo*. The extreme specificity and exquisite sensitivity of antibodies have made them valuable tools for use not only in basic research but also for clinical and even industrial applications. In this chapter, we examine some aspects of how the interaction between antigen and antibody can be exploited in research and clinical laboratories. In Section A, we begin our discussion with a look at sources of antibodies. In the remainder of the chapter, we describe two broad categories of techniques that use antibodies to detect or isolate antigen. Section B covers those assays based on detection of large, visible *immune complexes* containing antibody and antigen trapped in a network. Examples of these techniques are the *precipitin reaction, agglutination*, and *complement fixation*. In Section C, we discuss techniques based on the formation of *individual antigen–antibody pairs* in which detection relies on a "tag" chemically introduced onto either the antigen or the antibody molecule. Such tags are usually radioactive, enzymatic, or fluorescent and give rise to easily detectable assay signals. These assays tend to be more sensitive than those based on immune complex formation. Techniques of this type include *radioimmunoassay* (RIA), *enzyme-linked immunosorbent assay* (ELISA), *immunofluorescence, flow cytometry*, and *Western blotting*. Both categories of assays can be used qualitatively to characterize antigens, and quantitatively (to varying degrees) to arrive at a numerical measure of the number of antigen or antibody molecules or immune complexes present. Common antigen–antibody assays and their relative detection sensitivities are shown in Figure 7-1.

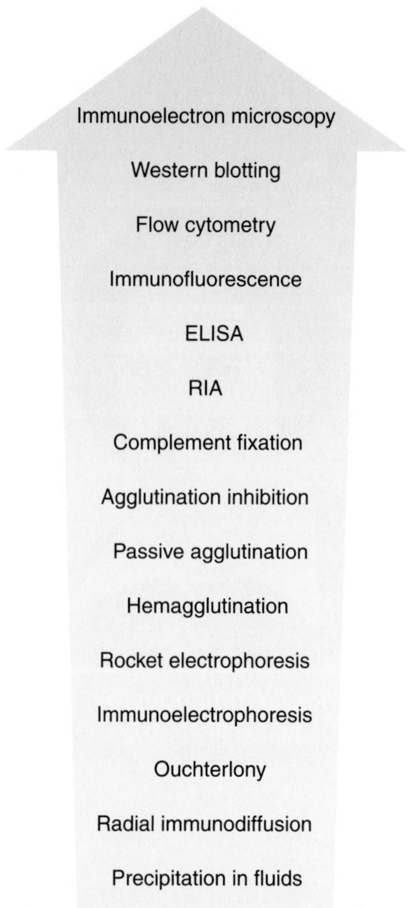

Why Use Antibodies?

The dissection of any biological system requires experimental techniques and assays capable of detecting interactions that will test the scientist's hypothesis of how it works. Because immunology evolved from the sciences of biochemistry, genetics, histology, and pathology (among others), many of the techniques used to characterize features of the immune system are derived

Immunoelectron microscopy

Western blotting

Flow cytometry

Immunofluorescence

ELISA

RIA

Complement fixation

Agglutination inhibition

Passive agglutination

Hemagglutination

Rocket electrophoresis

Immunoelectrophoresis

Ouchterlony

Radial immunodiffusion

Precipitation in fluids

Figure 7-1

Relative Immunoassay Sensitivities
Various types of immunoassays are placed in increasing order of sensitivity, from bottom to top. Note that the hierarchy is not absolute. For example, many researchers consider RIA and ELISA to be of equal sensitivity.

Scientists have come up with a way to use the immune system to create novel antibodies that can substitute for enzymes that have been created by evolution. One common mechanism that enzymes use to catalyze chemical reactions is the stabilization of a high-energy *transition state* between the reactants and the products. This stabilization has the effect of lowering the energetic barrier for a chemical reaction, thereby increasing its rate, often by many orders of magnitude. In 1986, it was shown that it is possible to immunize an animal with a *transition state analogue* of a chemical reaction. The resulting antibodies often cross-reacted with the real transition state species and stabilized them, thus acting as catalysts for the reaction in question. This discovery led to technology employing immune selection to create novel "catalytic antibodies," dubbed *abzymes*.

What advantages do catalytic antibodies have over enzymes that have been created by nature? Antibodies are well-characterized molecules that are easy to purify and produce in relatively large amounts. More importantly, it is possible to custom-design catalytic antibodies for many different types of chemical reactions, even ones for which there is no known natural enzyme. Many early catalytic antibodies had catalytic rates that were substantially lower than the enzyme they were modeled after, but more sophisticated design of the original immunogen has led to the creation of abzymes that rival natural enzymes. In addition, novel uses for the catalytic antibodies have been developed. For instance, catalytic antibodies have been created which hydrolyze cocaine into harmless by-products. It might become

possible, therefore, to treat drug addicts by injecting them with abzymes, which would neutralize a drug soon after it was injected (Ch.23). Alternatively, one could simply immunize a patient with the appropriate transition state analogue, and perhaps they might make their own cocaine-hydrolyzing antibodies. Another attractive application is the development of antibodies that catalyze the breakdown of an environmental pollutant. By expressing this antibody as an Fab fragment in bacteria, it has become possible to create designer bacteria able to clean up environmentally damaging chemical spills or pollutants. The use of abzymes in a wide variety of industrial and non-industrial applications is likely to become ever more widespread as the technique grows in complexity and sophistication.

from the methodologies of these fields. General methodologies appropriate for examining cellular, enzymatic, and genetic components are frequently used. However, immunologists can also take advantage of a unique feature of their science, the antigen–antibody bond, which has the specificity of an enzymatic reaction without its sometimes inconvenient permanent alteration of the bound substrate. This bond has formed the basis for highly sensitive techniques that have made the leap to many other scientific fields. In situations where standard physicochemical techniques cannot distinguish between very closely related molecules, specific antibodies for distinct epitopes on those molecules can do so with ease. Identification and purification of a single component from a complex mixture becomes a ready possibility. Antibodies also make invisible antigens "visible," or at least detectable. Under the appropriate conditions, one antigenic molecule among 10^8 can be detected by specific antibody. For these reasons, techniques employing antibodies can be used to purify antigens, characterize them, quantitate them, and to pinpoint their expression in cells or tissues. In short, antibodies offer custom-designed solutions to many clinical and research needs. Antibodies may even act as highly specific mediators of catalysis, giving them the potential for a broad range of uses (see Box 7-1).

A. Sources of Antibodies

I. ANTISERA

i) What Is an "Antiserum"?

Serology is the branch of science devoted to the study of antibodies present within a given *antiserum*—the clear liquid serum fraction of clotted blood obtained from an immunized individual or an individual exposed to a foreign substance or infectious agent. The antiserum preparation is then tested for its *titer*, or relative concentration of antigen-specific antibodies, by serially diluting samples of the antiserum until binding to specific antigen can no longer be detected. An antiserum that contains a large number of antibodies specific for a given antigen (i.e., can be diluted extensively and still shows binding activity) is said to have a "high titer." The concentration of antibody proteins in such an antiserum may be in the range of 10–20 mg/ml. If further purification of the antibody is required, non-Ig proteins can be removed by ammonium sulfate precipitation followed by a chromatographic method that relies on differences in size, charge, or binding affinity between the Ig and non-Ig proteins in the mixture.

ii) Advantages and Disadvantages of Antisera

As we learned in Chapter 5, antisera are polyclonal, meaning that in a natural immunization, many B cell clones respond to a whole collection of epitopes, producing many different antibody specificities. The resulting complex mixture contains a plethora of antibodies of different amino acid sequences, each present in a relatively small quantity. This mixture is an advantage to an organism *in vivo* because it offers multiple ways to attack a pathogen. Similarly, a researcher who wants to use an antibody preparation to identify an antigen as a whole (as opposed to one particular epitope of an antigen) is well-advised to use an antiserum. Even if the antigen has been mutated at one or more epitopic sites, the mixed population of antibodies will still likely contain at least some antibodies capable of recognizing one or more epitopes on the antigen.

However, the heterogeneity of an antiserum can be problematic when one is trying to identify a specific epitope with as few variables as possible. Removal of undesired antibodies to other epitopes involves adsorption procedures that can be time-consuming, expensive, less than 100% effective (leaving cross-reacting antibodies behind), and resulting in a significant decrease in the concentration of the desired antibody. In addition, antisera vary in composition and titer from one animal to the next even in genetically identical individuals (such as inbred mice), and from immunization to immunization even when the same protocol is followed. Tied to this difficulty is that of limited supply: the amount of antiserum recovered from one individual animal may be insufficient to complete an extensive series of clinical or experimental assays. These considerations, depending on the experiment, can make it difficult to establish a reproducible result. In situations such as these, the use of a homogeneous antibody preparation of a single specificity is more appropriate.

II. HYBRIDOMAS AND MONOCLONAL ANTIBODIES

The only natural bulk sources of Igs of a single specificity are the antibodies secreted by myeloma cells. Myelomas are B cell cancers that arise from the malignant transformation of a single plasma cell *in vivo*. A myeloma clone secretes antibodies of a single specificity like any B cell clone. However, unlike a normal plasma cell clone that dies after a few days, a myeloma clone has an unlimited life span: it is said to be "immortal." This means that unlimited numbers of cells producing massive amounts of a single antibody can be grown in culture and manipulated in the laboratory. However, the antigenic specificities of myeloma antibodies are for the most part unknown, being established during B cell development prior to the transformation of the plasma cell. Thus, myeloma proteins were first used primarily as sources of homogeneous antibody for elucidating the fine structural features of the Ig molecule. However, in 1975, César Milstein and Georges Köhler succeeded in physically uniting a myeloma cell with a B cell whose antigenic specificity was known. Using techniques of *cell fusion* in which the plasma membranes of two cells are joined such that the cytoplasms are combined, they artificially joined a myeloma cell to an activated B cell to create a hybrid cell with the immortality of the myeloma and the known antibody specificity and production capacity of the antigen-specific B cell. These cells are known as *hybridomas*. Why use a myeloma line for fusion if its Ig-producing capacity is irrelevant? Why not use an immortalized cell line derived from a completely different tissue? For some reason, cells of the same lineage fuse more easily and with greater stability than cells from different lineages. More importantly, Ig expression requires specific enhancers and transcriptional regulators that are present primarily in cells of the B lineage. The advent of hybridoma technology allowed the establishment of hybrid cell lines that could be grown indefinitely to produce large amounts of a single Ig protein of desired specificity. Such was the importance of this work that Milstein and Köhler were awarded the Nobel Prize for Physiology or Medicine in 1984.

i) Generation of Somatic Cell Hybrids

How does one go about creating a hybrid cell line? Almost any two cells can be induced to fuse by treatment with either polyethylene glycol (PEG) or inactivated Sendai virus. A mixture of two different parental cell types can be treated with either of these agents such that the plasma membranes of two or more adjacent cells fuse, combining the cytoplasms and incorporating the separate nuclei into a single large cell called a *heterokaryon* (from *hetero*, "different" and *karyon*, "chromosome") (Plate 7-1). As the heterokaryon undergoes cell division, the nuclei are fused to each other, combining the two (or more) sets of chromosomes in one large nucleus and forming a cell called a *somatic cell hybrid*. The fused cells of interest (those hybrids containing genetic material from both parental cell types) occur in small numbers, both because the absolute percentage of cells that undergo fusion in a treated mixed cell culture is quite low, and because they are hidden among hybrids derived from two (or more) cells of the same type. The latter cells are called *homokaryons*, multinucleate fused cells with the "same chromosomes." For example, after a fusion procedure involving parental cell type Q with parental cell type R, five different types of cells can be found in the culture in various proportions: unfused Q cells, unfused R cells, Q–Q homokaryons, R–R homokaryons, and finally, the desired Q–R heterokaryons. How then to isolate the desired hybrids from this mixed population? Let's suppose that only parental cell Q can survive under certain selection conditions, and that only parental cell R can survive under a different set of selection conditions. A mixture of the Q and R cells can be cultured under conditions in which the selection pressures are combined, so that only those Q–R hybrids that have acquired the ability to survive under both sets of conditions will be recovered. This is an example of an event called *complementation*, and its exploitation under these circumstances is called *somatic hybrid selection*. However, the desired Q–R fusion product is not yet a "normal" cell because of the presence of extra sets of chromosomes acquired during fusion. This extra DNA makes the hybrid cell unstable, and it loses chromosomes randomly

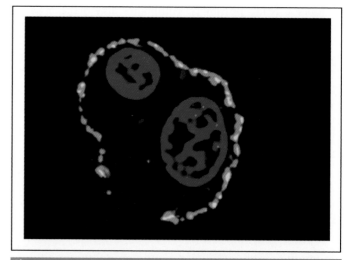

Plate 7-1

Heterokaryon Formation
Fluorescence staining shows a single membrane surrounding the cytoplasm of two fused cells. The separate nuclei are shown in blue. Reproduced with permission from Gottfried E. *et al.* (2002) Characterization of cells prepared by dendritic cell–tumor cell fusion. *Cancer Immunity* **2**, 15.

during subsequent cell divisions until it returns to the normal diploid state of one set of paired chromosomes. The maintenance of the hybrids under the appropriate selection conditions ensures that only the desired ones survive.

The most common selection system used for isolating somatic cell hybrids is called *HAT selection*, a technique that takes advantage of the fact that normal mammalian cells employ two pathways to synthesize the nucleotides making up their DNA. The *de novo pathway* builds new nucleotides from amino acids, while the *salvage pathway* builds new nucleotides from the routine degradation of spent nucleic acids. Thus, when the *de novo* pathway is blocked, as in the presence of the inhibitor *aminopterin*, the cells can survive by using their salvage pathway to synthesize nucleotides. The salvage pathways depend on the enzymes *hypoxanthine–guanine phosphoribosyl transferase* (HGPRT), which initiates the generation of guanosine monophosphate (GMP) and adenosine monophosphate (AMP) from hypoxanthine, and *thymidine kinase* (TK), which utilizes thymidine to generate thymidylate (Fig. 7-2). If cells are cultured in HAT medium (containing <u>a</u>minopterin but supplemented with <u>h</u>ypoxanthine and <u>t</u>hymidine), only those cells able to express both HGPRT and TK will survive. Conversely, cells lacking either of these enzymes will die in the presence of aminopterin, since neither the *de novo* nor salvage pathways of biosynthesis are functioning. Thus, as outlined in the next section, the judicious choice of fusion partners will ensure survival of the desired hybrids cells.

ii) Generation of Murine B Cell Hybridomas

Köhler and Milstein used the general principles of somatic cell hybrid generation to create the first B cell hybridomas. To generate a hybridoma, a mouse is first injected with the antigen to which one wants to raise a specific antibody (Fig. 7-3). The mice are then immunized a second time (boosted) 5–7 days prior to the fusion procedure so that the B cells of interest are fully activated. (For unknown reasons, activated B cells fuse more readily than resting B cells.) The spleen cells of these mice are harvested as a source of activated B cells to serve as one parent of the hybrid. These B cells are all HGPRT$^+$ and TK$^+$ but they are also terminally differentiated cells with a fixed and limited life span, and so are unable to proliferate in culture. The other parental cell type is a cultured mouse myeloma cell line selected for mutations that render both its HGPRT and Ig genes non-functional.

After fusion of the B cell–myeloma mixture, the culture is placed in HAT medium. Unfused B cells and B–B homokaryons die in a few days because they cannot survive in culture under any conditions. The unfused myeloma cells or myeloma homokaryons lack HGPRT and thus cannot use their salvage pathways; these cells die in the presence of aminopterin. However, somatic cell hybrids arising from a fusion of activated splenic B cells (HGPRT$^+$, TK$^+$) and myeloma cells (HGPRT$^-$) survive in HAT medium because they have acquired HGPRT from the B cell parent and the ability to grow indefinitely in culture from the myeloma parent. Over several cycles of cell division, parental chromosomes from the two genomes are <u>randomly</u> deleted until the diploid state is recovered; these hybrids are then called *hybridomas*. Somewhere buried in this collection of hybridomas are those that not only retain the B cell chromosome harboring the Ig locus but also produce the Ig of the desired specificity. To isolate these rare cells, the total population of fused cells is subjected to *limiting dilution*; that is, the cells are distributed in multi-well culture plates at very low density such that single cells, each secreting an Ig of a different specificity, grow into small colonies in isolation. Using assay techniques described later in this chapter, the culture fluids of these individual colonies are screened for the presence of the desired Ig. Those colonies found to be secreting antibody

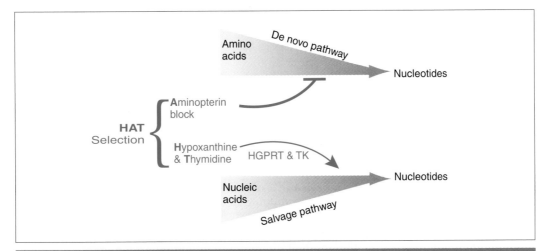

Figure 7-2

Nucleotide Synthesis and HAT Selection
The *de novo* and salvage pathways function simultaneously to produce nucleotides in the cell. The aminopterin present in HAT medium (hypoxanthine, aminopterin, thymidine) blocks the *de novo* pathway by interfering with required enzymes. Cells grown in HAT must therefore generate nucleotides via the salvage pathway. This is only possible if the cells express both HGPRT (hypoxanthine–guanine phosphoribosyl transferase) and TK (thymidine kinase), as these enzymes must be present for the salvage pathway to function.

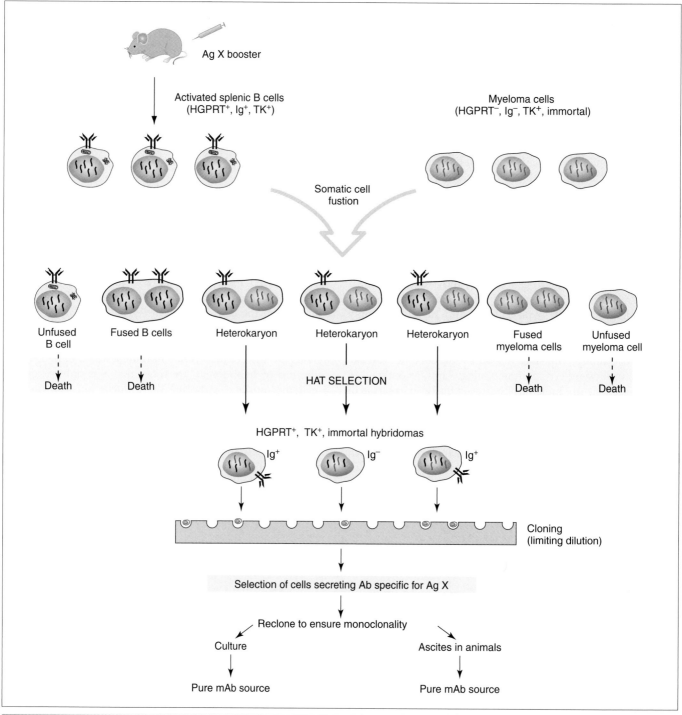

Figure 7-3

Making Hybridomas for Monoclonal Antibody Production
The production of hybridomas secreting monoclonal antibodies (mAb) specific for antigen Ag X is shown schematically. Activated B cells isolated from the spleens of mice immunized with Ag X undergo somatic cell fusion with an immortal myeloma cell line selected for lack of HGPRT expression. A mixed population of cell types results, including heterokaryons formed when a B cell fuses with a myeloma cell. These "hybridomas" are the only segment of the cell population with potential to be both immortal and secrete antibody of the desired specificity. "HAT selection" refers to the post-fusion growth of all cell types in medium containing hypoxanthine, aminopterin, and thymidine, conditions under which only heterokaryons survive. After HAT selection, the hybridomas, some of which may not express Ig due to chromosomal loss, are diluted and seeded into culture wells at a density that statistically predicts that no more than one cell is seeded per well ("cloning by limiting dilution"). Clones are then screened for production of anti-X antibody and positive producers are recloned. Bulk amounts of pure mAb can be obtained by growth of the hybridoma to high density in culture. Alternatively, the hybridoma can be injected into the abdominal cavity of a mouse and the secreted mAb collected at high concentration in the resulting ascites (fluid that accumulates in the abdominal cavity).

of the relevant specificity are then grown up into bulk cultures. Since the antibody secreted by the selected hybridoma was derived originally from a single splenic B cell clone, the antibody is said to be a *monoclonal antibody*, often abbreviated as *mAb*, and is of a single isotype and a single specificity.

iii) What about Human or Non-B Cell mAbs?

The preceding description is of the production of murine monoclonal antibodies in a mouse system. In cases where the goal is to use mAbs in clinical treatment, a problem arises because, not unexpectedly, the introduction of mouse mAbs into humans provokes an immune response to the mouse antibody. It would therefore be desirable to produce human mAbs, if possible. Unfortunately, although human B cell/mouse myeloma hybridomas have been created, for unknown reasons all the human chromosomes are eventually lost from the hybrid, making it impossible so far to produce human Igs in this system. Human B cell–human myeloma cell hybridomas are constructed only with difficulty (not too many individuals would be willing to endure immunization followed by splenectomy; the much less efficient peripheral blood B cells must be used instead) and, unlike their murine counterparts, human myelomas tend not to be immortal in culture. In addition, the few human myeloma cell lines that do grow well in culture produce their own Ig molecules, causing background difficulties during screening. As an alternative to human hybridomas, researchers have created "humanized" mouse antibodies, chimeric Ig molecules in which the mouse V region specific for an antigen of interest in humans is coupled to a human C region. The presence of the human C region greatly decreases the likelihood of an immune response against the antibody when used clinically. The potential therapeutic benefit of monoclonal antibodies for use in humans is tremendous and the subject of much ongoing research (see Box 7-2).

Box 7-2. Producing monoclonal antibodies for clinical use

As long as 100 years ago, Emil von Behring predicted that it should someday be possible to generate antibodies entirely outside an animal's body. Modern immunologists have devoted much energy toward proving him right, in an effort to apply the benefits of antibody specificity to clinical treatment. Although hybridoma technology has liberated scientists from the constraints of polyclonal antibodies, a number of technical and biological drawbacks exist that restrict the clinical application of monoclonal antibodies. First, the kinds of human antibodies that can be produced by hybridoma technology are limited: it is difficult to generate human antibodies against human antigens due to self-tolerance, and it is ethically impossible to immunize humans with harmful antigens such as toxins or cancers. Secondly, because of their non-human C regions, rodent monoclonal antibodies function sub-optimally in humans. Their half-life in serum is short and only some retain a structure capable of triggering the effector functions needed for a complete immune response in humans. In addition, the introduction of rodent antibodies into a human patient may result in the generation of human anti-rodent antibodies, leading to antibody clearance and hypersensitivity reactions. Lastly, from an ethical standpoint, a means of producing a mAb that does not require animals would be highly desirable. These inherent difficulties prodded immunologists into thinking of ways to make rodent antibodies, which are easier to produce, more like human antibodies, which are more clinically useful.

"Humanized" Antibodies

The first "humanized" antibodies were mouse–human chimeras; that is, proteins that were partly mouse Ig and partly human Ig. Using techniques of genetic engineering, the variable domains of mouse mAbs that recognized a human antigen of interest were combined with the constant domains of human antibodies. In brief, the sequences encoding these variable domains were obtained from the DNA of a mouse B cell hybridoma and assembled with the sequences encoding the human constant domains. The chimeric Ig gene was then expressed in cultured mammalian cells to ensure correct glycosylation of the engineered antibody. A humanized antibody has three advantages over the original mouse mAb: the immunogenicity of the mAb is reduced (since much of the immune response occurs against the mouse Ig constant region); the human C region allows for human effector functions to take place; and the serum half-life of the mAb in humans is significantly increased. To make these antibodies even more human, researchers have teased out the mouse CDRs directed against a human antigen of interest and have built them into a human antibody framework. Although the binding affinities of chimeric antibodies are often decreased compared to the original mAb, they can be increased by making framework substitutions and by varying CDR sequences. When different framework regions are combined with the same CDRs, chimeric antibodies specific for the same antigen can elicit different effector functions, extending their therapeutic benefits.

Phage Display

Another approach to deriving mAbs more suitable for use in humans involves artificially constructing soluble Fab fragments. Fab fragments have the advantages of being able to penetrate tissues very efficiently and having less stringent assembly requirements than full-length antibodies (processing through the rER is not essential). This means that signal sequences can be artificially added to direct an Fab to an extracellular or intracellular location. The practical applications arising from these characteristics range from neutralization of drugs, to imaging tumors when the Fab is coupled to radioactive markers. The technology used to create Fab fragments, called "phage display," was invented in the early 1990s and is based on the selection of antigen-specific antibodies from a repertoire library in a manner that mimics that of the immune system *in vivo*.

A phage display library of Fab fragments is constructed by joining each of the variable region gene sequences of a collection of hybridoma clones to a gene encoding a bacteriophage minor coat protein, pIII. (A bacteriophage, or "phage," is a virus that infects bacteria.) Each phage then expresses or "displays" a fusion protein on its coat surface consisting of a mAb variable domain joined to the N-terminus of the pIII coat protein. Selection of the desired Fab fragment is carried out by "panning" the recombinant phage library over the

Continued

immobilized antigen of interest. Low-affinity antibodies that do not bind are simply washed away and successive rounds of selection ensure that the Fabs that remain are of the highest binding affinity. Finally, bacteria are infected with the selected phage, which multiplies to yield an abundance of the desired Fab fragment.

The collection of V region sequences required for phage display can also be obtained by making cDNA copies of IgM RNAs obtained from the blood, bone marrow, or spleen leukocytes of a naive (unimmunized) individual. With this repertoire, only one library is necessary to derive antibodies to all antigens, but these Fabs are of low affinity and only a minute fraction will have specificities of interest. Alternatively, a library of immune repertoire V region sequences can be constructed using the IgG mRNAs from an individual immunized with an antigen of interest. Since affinity maturation (see Ch.9) has probably already occurred, the Fabs produced are likely to be of higher affinity and more easily isolated. However, ethically one might not be able to immunize with the antigen of interest. Another drawback to this latter approach is that for every antigen, a new phage library must be constructed, which is a time-consuming process. A better option is to derive a repertoire of synthetic V region sequences by artificially introducing known

random mutations into the CDR3 loops of either germline or rearranged V sequences. This method allows much greater control over the content and diversity of the library. A more clinically significant problem with the phage display method is that the resulting product is not a full-length antibody. The constant region responsible for effector functions is not present, Ig glycosylation does not occur, and the serum half-life of the Fab fragment falls short of the month-long half-life of full-length antibodies. Nevertheless, phage display allows a specific Fab fragment free of mouse constant region sequences to be produced completely outside a natural host, in some cases bypassing both immunization and immune reaction.

"Plantibodies"

Antibodies can also be produced in transgenic plants (plants carrying foreign genes in their DNA); the resulting mAbs have been dubbed "plantibodies." Plants have an advantage over phage display in that they not only assemble Fab fragments but also full-length antibodies and even multimeric antibodies. Full-length plantibodies preserve the antibody constant region that houses sites specific for glycosylation, complement activation, phagocyte binding, J chain, and secretory component. Because plants are eukaryotic organisms like mammals, the antibodies they

produce have glycosylation patterns similar to those of hybridoma-produced antibodies, and thus tend to be long-lived *in vivo*. Such longevity makes them ideal for therapeutic use. For example, a mAb that can be used to target cancers during chemotherapy has been genetically engineered in soybeans. Other uses for plants include the production of mAbs that at present cannot be produced in sufficient quantities by conventional liquid culture methods.

The potential advantages of using plants as vehicles for monoclonal antibody and vaccine production are enormous (see Ch.23). Most of the genetic transformations engineered in plants result in stable integration of the transgene into the plant DNA, making crosses quite straightforward. As well, transgenic lines can be stored as seeds almost indefinitely and grown more economically than any liquid culture-based technology in practice today. Large-scale production is particularly cost-effective, increasing the economic feasibility of worldwide immunization and control of infectious diseases. Some problems must still be addressed (such as ensuring that the unique glycans in plants do not immunogenically challenge humans), but we may someday see a return to von Behring's time, when most medicinal compounds came from plants.

The reader also may be wondering about the application of hybridoma technology to other cell lineages. T cell hybridomas tend to be more difficult to make than B cell hybridomas because the starting cell population is smaller and antigen binds less efficiently to a TCR than to a BCR. Nevertheless, mouse T cell hybridomas have been produced using normal T cells from immunized mice and T cell tumor cell lines (*T lymphomas*) that are sensitive to HAT selection. These hybrids do not secrete antibodies or T cell receptors, but can be used to examine antigen-specific T cell activation in a well-defined, homogeneous, and permanent T cell population. There are now many stable mouse cytotoxic T and helper T cell hybridomas available. Human T–T hybridomas have also been created.

iv) Advantages and Disadvantages of Monoclonal Antibodies

Because they are of known specificity, monoclonal antibodies are very useful anywhere the presence of an antigen must be detected. They can be used to identify a specific protein marker on a cell surface or in a tissue or serum sample, or to map individual epitopes on macromolecular and cellular antigens. In addition, large quantities of mAbs are useful for purifying proteins of interest to be used in research studies, or in

industrial or clinical applications. Finally, because hybridomas can be clonally expanded and maintained indefinitely, they provide a permanent and uniform source of antibody. In cell cultures, hybridomas secrete antibody at a concentration of about 10–100 μg/ml. However, if the peritoneal cavity of an immunologically compatible mouse is injected with hybridoma cells, antibody will accumulate in the fluid in the cavity (termed the *ascites fluid*) at a concentration of 1–25 mg/ml. Modern techniques of cell culture and biotechnology have now made it possible to buy many mAbs commercially at high concentrations and in essentially unlimited quantities (refer to Box 7-2).

For all the benefits of monoclonal antibodies, it behooves the reader to keep in mind that there are limitations to their use. For example, because mAbs recognize only a single determinant, a virus that manages to mutate that precise epitope can escape detection, whereas a polyclonal antiserum containing antibodies to several determinants on that viral protein would likely still detect the presence of the protein as a whole. In addition, as we shall see in the following discussions, the cross-linking of antibody–antigen complexes is required for many laboratory techniques. Because a mAb preparation recognizes only a single epitope, which may be represented only once on an antigen molecule that occurs naturally as a monomer, immune complex networks between multiple antigen and antibody molecules may not be readily formed. As a result, tests based on agglutination or the formation of a large precipitate may not work if the epitope in question is sparsely distributed and the antibody used is monoclonal. Note, however, that mAbs can be successfully used to cross-link receptors in receptor stimulation studies where the receptors are expressed at sufficient density on the cell surface.

B. Techniques Based on Immune Complex Formation

Assays described in this section make use of the fact that large antigen–antibody complexes precipitate out of solution, or appear as the visible clumping of bacteria or other cells, or activate complement.

I. CROSS-LINKING AND THE FORMATION OF IMMUNE COMPLEXES

The kinetics of antigen–antibody binding are driven by relative concentration, as are all chemical reactions. When complementary antibody and antigen (i.e., a binding pair) are mixed in a fluid in approximately equal amounts, non-covalent bonds are rapidly formed between individual molecules, resulting in a small <u>soluble</u> complex. However, because of bivalency, a single antibody molecule may use one antigen-combining site to bind to its epitope on one molecule of antigen molecule, and the other antigen-combining site to bind the identical epitope on a second antigen molecule. Each of these two antigen molecules

may possess additional epitopes for additional antibody-binding, so that different antibody molecules mutually binding to this antigen are said to be *cross-linked*. Further cross-connections between additional antigen and antibody molecules result in the formation of an *immune complex* or *lattice*. As more and more antigen and antibody molecules become cross-linked, they form lattices large enough to precipitate out of solution and to become visible. These are the properties that were first exploited to examine antigen–antibody interactions, in the form of the *precipitin* and *immunodiffusion* assays. The collection of precipitated immune complexes by centrifugation also provided the first means of isolating and purifying antigens.

As illustrated in the *precipitin curve* shown in Fig. 7-4, when increasing amounts of soluble antigen are added to a fixed amount of soluble complementary antibody, the amount of lattice or immune complex precipitated increases up to a broad peak called the *zone of equivalence* and then slowly declines again. In the *zone of equivalence*, the number of antigenic epitopes and antibody-combining sites is approximately equal and the binding is optimal. Enough antigenic epitopes are present that each individual antibody Fab site binds to a <u>different</u> antigen molecule, but the antigen concentration is still low enough that each antigen molecule is shared between (on average) two antibodies. Under these conditions, cross-linking readily occurs, and neither free antigen nor antibody can be detected. Where the amount of antigen added is very low and an excess of antibody over antigen is present, on average only one Fab site of an antibody is in use, meaning that cross-linking does not occur and the immune complex cannot form. Free antibody remains in the supernatant after the unitary antigen–antibody pairs are removed by centrifugation. This area of the precipitin curve is known as the *zone of antibody excess*. When the antigen concentration increases past the zone of equivalence, a point is reached at which every available Fab site is bound to a separate antigen molecule: no

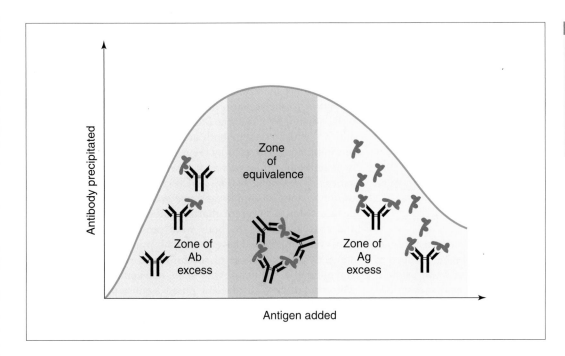

Figure 7-4

Precipitin Curve
To generate the precipitin curve, varying concentrations of antigen are added to a fixed concentration of antibody and the amount of antibody precipitated is measured in each case. Ab, antibody; Ag, antigen.

Antibody precipitated

Zone
of
equivalence

Zone of
Ab
excess

Zone of
Ag
excess

Antigen added

antigen molecule is shared between two antibodies and a large immune complex cannot form. Free antigen remains in the supernatant after the unitary antigen–antibody pairs are removed by centrifugation, giving rise to the area of the precipitin curve known as the *zone of antigen excess*.

The formation of large immune complexes is a property that was used to advantage in early studies of antigen–antibody interaction. However, clinical complications can arise when such complexes occur *in vivo*. The function *in vivo* of immune complexes is to trap antigen in preparation for elimination of the complex by mechanisms of the innate immune response. If the network between antigen and antibody molecules is too extensive, the immune complexes will be too large to remain soluble and will precipitate in the tissues. Quantities of immune complexes deposited in otherwise healthy tissues (instead of being cleared from them) mark them as targets for antibody effector functions, provoking an inflammatory reaction and possible tissue destruction. An example of this kind of reaction is "immune-complex glomerulonephritis," in which circulating antigen–antibody complexes become trapped in the glomeruli of the kidneys during the renal filtration process. The accumulation of immune complexes in the glomeruli and the subsequent inflammatory response results in damage to the glomeruli due to complement activity and cytokines released by neutrophils (see Ch.28).

II. TECHNIQUES BASED ON THE PRECIPITIN REACTION

The precipitin reaction was the first assay developed for the quantitation of antibody. It has since been superseded in research laboratories by more sensitive techniques, but some precipitin reaction-based assays are still used in clinical diagnostics.

i) Precipitin Ring

The *precipitin ring assay*, a quick qualitative test for the presence of antigen–antibody complexes, is based on the principles of the precipitin curve described in Section B.I. A solution containing the antiserum to be investigated is placed in the bottom of a series of clear test tubes (Fig. 7-5A). Solutions containing increasing concentrations of a known antigen are gently layered above the antibody solution. If the test antiserum contains antibody recognizing the antigen at a concentration representing the zone of equivalence, immune complexes will form and be visible as a whitish ring at the interface. Semi-quantitative results can be obtained by comparing the experimental results to those derived from a standard curve constructed using known quantities of antigen and antibody.

ii) Immunodiffusion

The precipitin reaction also works when antigen and antibody diffuse toward each other within a semi-solid medium such as an agar gel, a phenomenon given the name *immunodiffusion*. If immunodiffusion is carried out in a test tube, as first described in the mid-1940s by Jacques Oudin, it is called "one-dimensional immunodiffusion" (Fig. 7-5B). If the assay is set up on a plate, as first devised by Mancini, "radial immunodiffusion" is said to

be occurring (Fig. 7-5C). In each case, comparison to a standard curve generated by using known amounts of antigen and antibody in the assay permits the quantitation of an unknown antigen or antibody.

In the 1950s, Örjan Ouchterlony extended these techniques to *double immunodiffusion*, in which two wells are carved into an agar surface (such as agar allowed to set on a microscope slide) so that they are only a few millimeters apart. Antibody is added to one well and antigen to the other. Each component diffuses away from its well toward the other, forming two concentration gradients. The concentration gradients begin to overlap such that, at the point where the concentrations of antibody and antigen reach the zone of equivalence, a visible precipitin line forms on the slide surface (Fig. 7-5Di). A variation of the double immunodiffusion technique has been particularly useful for qualitatively ascertaining whether two antigens share identical epitopes, and for quickly assessing the purity of either antigen or antibody. This modification requires that the two antigens under investigation be placed in separate wells, with the antiserum placed in a third well so as to form a triangle (Fig. 7-5Dii–v). As the antigens and antibody diffuse out from their respective wells, the distinct patterns of precipitin lines that are generated can be interpreted to assess the relatedness or purity of the antigens. For example, suppose we set up one well containing antigen X, a second well containing antigen Y, and a third well containing anti-X antiserum (Fig. 7-5Dii). A straight precipitin line containing X:anti-X complexes will form between the X and anti-X wells as X and anti-X diffuse toward each other. No precipitin line forms between the diffusing antigen Y and anti-X because anti-X antibody does not recognize Y at all; X and Y share <u>no</u> epitopes.

Suppose now that, instead of antigens X and Y, identical samples of antigen X are placed in the two antigen wells, and anti-X antiserum is placed in the third well (Fig. 7-5Diii). As the antigen samples diffuse from their wells, they each meet the anti-X antibodies and start to form precipitin lines containing X:anti-X immune complexes. The concentration gradients formed as X diffuses from each X well begin to overlap each other, effectively increasing the concentration of X between them. Because of this relative increase in X, the zone of equivalence shifts into an arc shape so that the precipitin lines appear as a single broad curved line in a *pattern of identity*. If, however, we return to the scenario in which antigen X is in one antigen well and antigen Y is in the other, but this time use an antiserum anti-XY which contains both anti-X and anti-Y antibodies, the precipitin lines generated by the reaction of each antigen with its own antibody diffuse past each other freely and cross, forming a *pattern of non-identity* (Fig. 7-5Div). If we knew only that there were two antigen samples and one antiserum sample, we could deduce from this pattern that the antigens shared no epitopes and that antibodies to both were present in the antiserum. However, suppose we have the situation in which two different antigens share only one of several epitopes, as in antigen XY and antigen XZ. Because they share one epitope, these antigens will cross-react with antisera raised against the other. Picture antigen XY in one well, antigen XZ in the other, and anti-XY antiserum in the third well (Fig. 7-5Dv). A pattern of identity will form due to the presence of the shared X epitopes of each

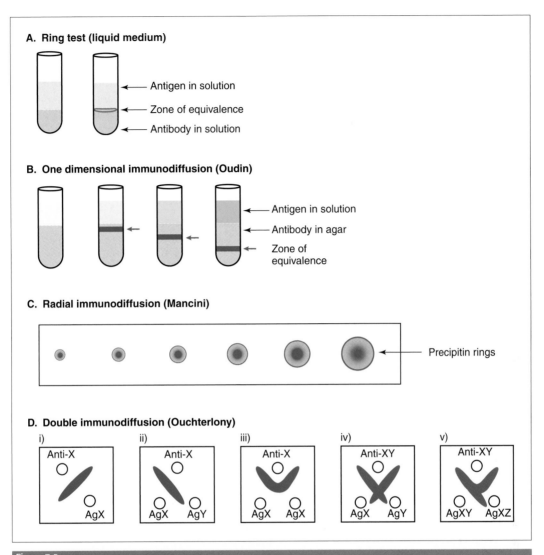

A. Ring test (liquid medium)

Antigen in solution
Zone of equivalence
Antibody in solution

B. One dimensional immunodiffusion (Oudin)

Antigen in solution
Antibody in agar
Zone of equivalence

C. Radial immunodiffusion (Mancini)

Precipitin rings

D. Double immunodiffusion (Ouchterlony)

i)
Anti-X
AgX

ii)
Anti-X
AgX   AgY

iii)
Anti-X
AgX   AgX

iv)
Anti-XY
AgX   AgY

v)
Anti-XY
AgXY   AgXZ

Figure 7-5

Assays Based on the Precipitin Reaction
(A) The ring test. Left-hand tube, no complementary antigen and antibody are present. Right-hand tube, complementary antigen and antibody cause a precipitin line to form at the solution interface (zone of equivalence). (B) One-dimensional immunodiffusion (Oudin) assay. With antibody-impregnated agar as the medium, increasing concentrations of antigen (left to right) form a series of precipitin lines after diffusion into the agar. (C) Radial immunodiffusion (Mancini) assay. Various amounts of antigen are placed in wells in antibody-impregnated agar. The greater the antigen concentration, the greater the diameter of the precipitin ring formed. (D) Double immunodiffusion (Ouchterlony) assay. Antibody and antigen samples diffuse toward each other from wells made in agar. A precipitin line will form wherever a zone of equivalence exists between a complementary antibody and antigen (i). In panels ii–v, a single antiserum is used to give information about the relationship between test antigens. The resulting patterns indicate non-identity (ii, no epitopes shared between antigens), identity (iii), non-identity (iv), and partial identity (v).

antigen and anti-X antibodies in the antiserum. However, super-imposed on this pattern will be the straight precipitin line formed by complexes of Y epitope (on only one antigen) and anti-Y antibodies in the antiserum. The combined precipitin patterns will resemble an incomplete cross or an arc with a *spur*, and constitute a *pattern of partial identity*. Such a result indicates that the antigens cross-react or share one or more epitopes.

iii) Immunoelectrophoresis

The precipitin reaction is useful in revealing some qualitative aspects of an antigen and its interaction with antibody but provides very little detail about its structural or chemical characteristics. In the early 1950s, a coupling of electrophoresis and double immunodiffusion gave rise to *immunoelectrophoresis*, which combines the separation of proteins by electrophoresis with the determination of protein identity by immunodiffusion. During electrophoresis, molecules are separated in an electric field on the basis of their overall charge. In immunoelectrophoresis, a complex mixture of antigens can be analyzed by first separating its component proteins by electrophoresis in a non-denaturing agar gel, so that negatively-charged proteins move toward the positive electrode and positively-charged ones move toward the negative electrode. A channel is then carved in the agar along the axis of separation, and is filled with antibody solution that is allowed to diffuse toward the line of separated antigens (Fig. 7-6A). Where specific antibody meets an antigen in the zone of equivalence, a precipitin arc is formed which can be visualized by appropriate staining. In this way, two or more antigens of differing charge that interact with the antiserum used in the assay can still be viewed individually. This technique is of only limited value because of its relative insensitivity (detection limit approx. 3–20 μg/ml antibody concentration). However, it can be useful in clinical situations where the relative proportions of the various immunoglobulins in a patient's serum are very obviously abnormal.

iv) Rocket Electrophoresis

A modification of immunoelectrophoresis known as *rocket electrophoresis* was a significant improvement, boosting the sensitivity of the assay to approximately 0.2–0.5 μg antibody/ml. The rocket technique involves the immunoelectrophoresis of antigens in an antibody-impregnated gel of a pH at which the antibody remains neutrally-charged and therefore immobile, while the antigens become negatively-charged and migrate

A. Immunoelectrophoresis

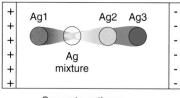

Separate antigens → Add antiserum to trough → Let antiserum diffuse and record antigen/antibody precipitin arc

B. Rocket electrophoresis

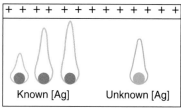

Electrophorese antigens on gel impregnated with antibody

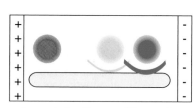

Calculate concentration of unknown antigen

C. Two-dimensional immunoelectrophoresis

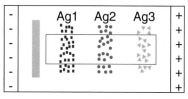

Electrophorese antigen mixture and cut out strip from gel

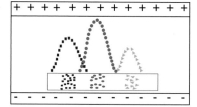

Overlay strip on gel impregnated with Ab

Figure 7-6

Types of Immunoelectrophoresis
(A) Immunoelectrophoresis. Separation on the basis of charge is followed by antigenic characterization using antiserum. The first panel shows some residual antigen mixture in the well of the electrophoresis plate; in reality, all antigen would be gone from the well by the time electrophoresis was complete. (B) Rocket electrophoresis. Electrophoresis of an antigen sample in an antibody-impregnated gel results in a precipitin line whose height is proportional to the antigen concentration (left panel). Using known antigen concentrations, a standard curve can be constructed to determine the sample concentration (right anel). (C) If a sample contains a mixture of antigens, they can first be separated by immunoelectrophoresis and then subjected to analysis by rocket electrophoresis.

toward the positive electrode (Fig. 7-6B). This turns out to be fairly easy to arrange because the amino acid composition of immunoglobulin proteins dictates that they have a relatively high *isoelectric point* (the pH at which the net charge on the molecule is zero), meaning an Ig protein will usually be neutrally-charged at an alkaline pH where most antigens are negatively-charged. As an antigen migrates through the gel and the immobile antibody molecules, the shape of the precipitin line formed by the antigen–antibody complexes resembles a rocket. Because antigen molecules are successively lost in precipitated immune complexes, the concentration of antigen drops at the sides of its leading edge and a progressively narrower precipitin point forms. The height of the rocket is proportional to the concentration of antigen. The concentration of an unknown antigen can be determined from a standard curve based on known antibody and antigen concentrations, but only one antigen at a time can be quantitated using this method.

v) Two-Dimensional Immunoelectrophoresis

Two-dimensional immunoelectrophoresis is a variation of the rocket technique that allows the quantitation of several antigens in a complex mixture in one assay. The antigen mixture is first electrophoresed in one dimension under non-denaturing conditions to separate its component antigens by charge (Fig. 7-6C). The antigen gel is then overlaid at 90° onto a second gel impregnated with antibody, and the gel sandwich is electrophoresed in the other dimension. Individual rockets form for each separated antigen that can react with the antibody. The height of each rocket allows the quantitation of each component antigen by comparison to a standard curve.

III. TECHNIQUES BASED ON AGGLUTINATION

The techniques described in the preceding sections are all based on the ability of antigen–antibody complexes to form a visible lattice. Several other analytical methods to detect interactions between antibody and antigens on the surfaces of cells depend on the clumping of the target cells. Antibody molecules binding directly to antigen molecules on the surfaces of certain bacterial, fungal, or mammalian cells may cross-link them, causing the microbial cells to *agglutinate*, or stick together and form a visible particle. Thus, for a polyclonal antiserum, the presence of antibodies to one of these antigens can be tested by adding the appropriate cells to serial dilutions of the antiserum and looking for agglutination. The *titer* of the antibody is the greatest dilution that still results in the agglutination of the test cells.

Agglutination assays can be classified as *active* or *passive*, and *direct* or *indirect* (Fig. 7-7). In active assays, the epitope of interest occurs naturally on the target cells to be agglutinated. In passive assays, the epitope of interest does not occur naturally on the cells or particles to be agglutinated and must be chemically fixed to them. In direct assays, the antibody to the epitope of interest can agglutinate the test particles on its own. In indirect assays, the antibody is able to bind to the epitope on the cells or particles but fails to agglutinate them. Another *secondary antibody* that recognizes the first or *primary*

antibody is then required. As was described in Box 5-3, secondary antibodies are raised by immunizing a heterologous species with the primary antibody. Among other antibodies generated will be anti-isotypic antibodies that recognize the common constant region of any primary antibody of that isotype. These antibodies can be purified and used to detect primary antibodies of different antigenic specificities. For example, a goat antibody to mouse IgG (termed "goat anti-mouse IgG") recognizes the constant region of the mouse IgG heavy chain and is able to cross-link and agglutinate test cells coated in any mouse IgG antibody. In clinical parlance, antibodies that bind to antigen on cells but fail to agglutinate them are known as "incomplete antibodies." Incomplete antibodies are often of the IgG isotype, since IgG is a monomeric, bivalent molecule. In contrast, the pentameric, multivalent IgM antibody is 700-fold more efficient in cross-linking erythrocytes than the IgG antibody; no secondary antibody is required for agglutination in this case.

When the target cells clumping in an agglutination reaction are erythrocytes (red blood cells, or RBCs), the assay is known as *hemagglutination*. Hemagglutination can be used for the rapid assessment of human ABO blood group types by using human RBCs as the test cells. The ABO antigens are glycoproteins displayed on the surfaces of human RBCs. Samples of blood from the individual to be typed are mixed separately with antisera to each of the ABO antigens. The RBCs of someone with blood type A will cross-link and clump in the presence of antiserum containing anti-A antibodies, but not antiserum containing anti-B antibodies. This is an example of an active, direct assay. Similarly, the hemagglutination of other antigens occurring naturally on RBCs can be used to evaluate serum antibody titers to those antigens. In passive, direct hemagglutination, a soluble antigen of interest is fixed to the surface of a sheep red blood cell (denoted SRBC) by the use of glutaraldehyde, chromic chloride, or tannic acid. Serial dilutions of the antiserum to be assayed are added and the presence of specific antibody to the antigen of interest is detected by the agglutination of the SRBC. Passive agglutination reactions can also be carried out on antigens conjugated to particles other than SRBCs, such as latex or bentonite microbeads. *Reverse agglutination* is the coating of inert particles with purified specific antibody for the detection of an antigen of interest. Solutions containing the antigen of interest show clumping of the test particles. The *inhibition of agglutination* is often used to detect antigens present in blood or other body fluids at concentrations in the range of 0.1–10 µg/ml. The principle of this assay is that agglutination by specific antibody can be inhibited if the antigen of interest is present in the test sample.

IV. TECHNIQUES BASED ON COMPLEMENT FIXATION

When antibody binds to an antigen on the surface of a cell, the antigen–antibody complexes can trigger the classical complement cascade, resulting in the lysis of the cell. As mentioned in Chapter 5, the consumption of the various complement components during this process is called *complement fixation*.

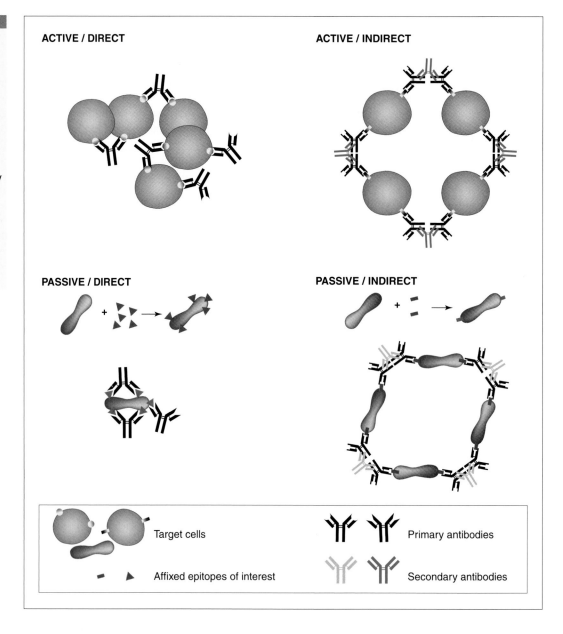

Figure 7-7

Types of Agglutination
Generalized examples of agglutination, in which the binding of specific antibody mediates the visible clumping of target cells (shown) or test particles (not shown). The epitope recognized by the specific antibody may occur naturally on the target cells (active) or may need to be affixed to target cells artificially (passive). The specific antibody itself may be able to agglutinate the target cells (direct). If not, a secondary antibody specific for the primary antibody must be used to mediate agglutination (indirect).

ACTIVE / DIRECT

ACTIVE / INDIRECT

PASSIVE / DIRECT

PASSIVE / INDIRECT

Target cells

Primary antibodies

Affixed epitopes of interest

Secondary antibodies

Since the end result of complement fixation on a cell surface is cell lysis, the complement-mediated lysis of an "indicator" cell can be used to directly assay a sample for the presence of either an antigen or antibody of interest, or to determine if the complement components are intact. One advantage to complement fixation assays is that they can be used to test components of different species origin because the antibodies of one species can interact with the complement proteins of another species to trigger the lysis of a test cell.

Complement fixation assays may be direct or indirect. *Direct complement fixation* assays are one-step assays in which the end point is the lysis of the indicator cell (Fig. 7-8A). Indicator cells may be bacterial (*bacteriolytic* assays), nucleated (*cytolytic* assays), or RBCs (*hemolytic* assays). In bacteriolytic or cytolytic assays, the antigen of interest is present naturally on the cell to be lysed. Lysed cells are identified either by their absorption of vital dyes, which viable, intact cells are able to exclude, or

by the release of an internalized radioisotope. In the latter method, cells are preincubated with a radioisotope such as ^{51}Cr, which is easily absorbed and trapped intracellularly. Viable cells are not able to release the radioisotope; however, when a cell is lysed, ^{51}Cr leaks from the punctured membrane. Detection of ^{51}Cr in the extracellular fluid signals that cytolysis has occurred. In hemolytic complement fixation assays, sheep red blood cells (SRBCs) are used as the indicator cells. Antigens of interest can be readily attached to the erythrocyte surface, and the lysis of the SRBCs releases hemoglobin, which is easily measured using a spectrophotometer.

As an example of a direct complement fixation assay, consider the situation in which we want to ascertain whether a human serum sample contains antibody to an antigen of interest. The antigen of interest is chemically fixed to SRBCs. Serial dilutions of the test antiserum are then incubated with the modified SRBCs. The cells are washed thoroughly to

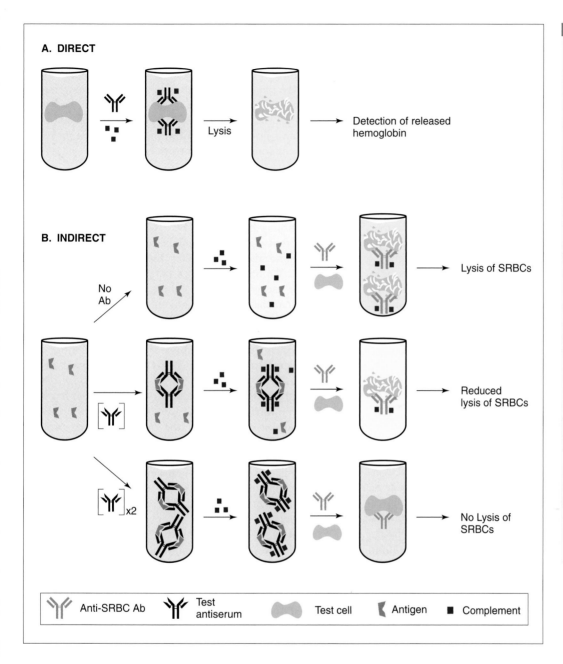

Figure 7-8

Assays Based on Complement Fixation
Complement-mediated lysis of cells can be used to evaluate the presence of specific antibody. In these examples, sheep red blood cells (SRBCs) serve as the test cells so that measurement of released hemoglobin (hemolysis) can be used to evaluate target lysis. Target cells other than SRBCs can also be used, in which case evaluation of lysis can be measured by the uptake of dyes or by the release of an internalized radioisotope. (A) Direct assay; the antigenic epitope of interest occurs on the target cell. (B) Indirect assay; the antigenic epitope of interest does not occur on a target cell. After allowing the antigen and antibody to interact and fix complement, the amount of complement remaining unfixed is measured by adding anti-SRBC antibodies and SRBCs. The amount of hemolysis is inversely proportional to the amount of antibody–antigen binding occurring in the first step of the assay. To simplify the illustration, the antigen–antibody complexes present in the indirect assay are not shown in the final tube, as they play no direct role at this point.

A. DIRECT

Lysis → Detection of released hemoglobin

B. INDIRECT

No Ab → Lysis of SRBCs

Reduced lysis of SRBCs

[Y]x2 → No Lysis of SRBCs

Anti-SRBC Ab   Test antiserum   Test cell   Antigen   ■ Complement

remove unbound antibody and other serum components and an exogenous source of complement is added. If the human serum sample contains antibodies to the antigen of interest, they will have bound to the SRBCs. Complement proteins will in turn bind to the human antibodies and the complement cascade will be triggered, causing the SRBCs to lyse and release hemoglobin. The amount of the hemoglobin released by the lysed SRBCs can be measured spectrophotometrically and is directly proportional over a limited concentration range to the titer of specific antibody in a series of test samples.

Indirect complement fixation assays are based on the fact that the formation of antigen–antibody complexes consumes complement. The amount of intact complement left over in the sample is then measured by the addition of an indicator

cell and an antibody specific for an antigen on the indicator cell (Fig. 7-8B). Such assays can be useful where it is not possible or desirable to fix the antigen of interest to SRBCs. For example, to ascertain whether an antiserum sample contains antibody to antigen A, serial dilutions of antiserum are incubated with a fixed amount of antigen A in test tubes followed by the addition of a known amount of complement in the form of guinea pig serum. If antibody recognizing antigen A is present in the test sample, immune complexes are formed, to which complement components can bind and trigger fixation (even in the absence of indicator cells). The amount of complement fixed depends directly on the amount of specific antibody in the sample. The amount of the <u>unfixed</u> complement left over can then be determined by adding indicator SRBC and anti-SRBC antibody, and then measuring the complement-mediated lysis that results. In other words, if anti-A antibody is present in the original antiserum sample, it forms an immune complex with antigen A, which triggers the complement cascade and causes complement components to be used up in proportion to the amount of anti-A present. The more anti-A in the test sample, the more complement is fixed by antigen A/anti-A complexes, the less complement is left to react with the SRBC/anti-SRBC complexes, and the fewer SRBCs are lysed in the last stage of the assay. It should be noted that considerable utilization of complement in the initial antigen A/anti-A incubation must occur before any significant changes in hemolytic activity are detectable.

While complement fixation assays can be useful for the detection of antibodies, antigens, or complement, considerable care must be taken in their use. These assays by definition depend on the integrity of all nine components of the complement cascade, some of which are very sensitive to poor sample handling. In addition, complement fixation can be inhibited by subtle changes in reaction components or by contaminants in sample preparations. Finally, hemolysis can be triggered by some antibodies or antigens in the absence of complement, so that extensive controls for non-specific cell lysis are essential.

C. Assays Based on Unitary Antigen–Antibody Pair Formation

The preceding section discussed techniques exploiting the fact that antigen–antibody immune complexes are relatively large in size, and as such are easily isolated or readily make themselves "visible" in assays. In contrast, unitary antibody–antigen pairs by definition are not found in extensive complexes. Thus, they cannot be isolated or made visible for quantitation or characterization unless the pair is somehow labeled with an easily detected *tag*. Techniques that make use of detectable tags to track antigens or antibodies of interest have greatly increased the scope of antibody-based assays.

I. GENERAL CONCEPTS

As mentioned at the start of this chapter, the tags used to make unitary antigen–antibody pairs detectable are generally radioisotopes, enzymes, or fluorochromes that are covalently bound to either the antigen or the antibody. In addition, the tag assay principle requires that one partner of the antigen–antibody pair be immobilized in some way so that any tag that is not part of an antigen–antibody pair can be removed from the assay system by washing. The amount of tag detected will then be proportional to the presence of antigen–antibody pairs in the sample, allowing quantitation. Tag assays can either be *direct* or *indirect*.

i) Direct Tag Assays
Direct tag assays refer to single step procedures in which a tagged antigen (or antibody) is used to detect the presence of its untagged antibody (or antigen) binding partner. All direct tag assays follow the same basic principle: antigen or antibody is labeled with a tag and then incubated with the test sample to allow antigen–antibody pairs to form. Unbound tagged molecules are removed by washing and the remaining tagged molecules are quantitated by measuring the amount of tag present. In many cases, the absolute number of antigen–antibody pairs present is not as important as the relative amount of tag (pairs) present compared to controls.

ii) Indirect Tag Assays
Sometimes the antibody or antigen of interest is not available in pure form or is chemically difficult to tag. In these situations, antigen–antibody binding can be detected by tagging a third component that binds to the unlabeled antigen–antibody pair of interest and indirectly makes the presence of this pair measurable. We call these types of experiments *indirect tag assays*. The use of the third component offers the distinct advantage that the scientist can avoid the task of individually purifying and labeling each antibody or antigen required in the research project. Three reagents commonly used as the third component in indirect assays are secondary antibodies, *Staphylococcus aureus* Protein A or G, and the biotin–avidin system (Fig. 7-9).

iia) Secondary antibodies. A secondary antibody that has been tagged can be used to bind to an untagged primary antibody, allowing indirect detection of the original antigen–primary antibody pair (Fig. 7-9A). As previously described, secondary antibodies are often raised in another species and include those recognizing constant domain epitopes on the primary antibody. Secondary antibodies may also be purified to yield a third component that detects only primary antibodies of a particular isotype.

iib) Protein A or G. Proteins other than immunoglobulins can be used to detect the primary antibody (Fig. 7-9B). The Fc region of antibody proteins is recognized by two proteins called "staphylococcal Protein A" and "staphylococcal Protein G," which are found naturally on the surface of *Staphylococcus aureus* bacteria. In the early days of tag assay development, labeled antigen–antibody complexes could be precipitated from solution by the addition of formalin-treated *Staphylococcus aureus* cells followed by centrifugation. Subsequently, labeled purified Protein A and Protein G became commercially available. These preparations are often used as the tagged third component in indirect tag assays. Protein A (42 kDa) has five binding sites and binds best to IgG antibodies but will weakly bind to other isotypes. Protein G (23 kDa) has two binding sites but recognizes only IgG antibodies (Table 7-1). Protein G binds to antibodies in a broader range of species than Protein A.

iic) The biotin–avidin system. Biotin is a small molecule (244 Da) that serves in nature as a covalently-bound cofactor in carboxylase enzymes; it is also known as Vitamin H. The biotin molecule is a useful tool in the laboratory partly because it is easily coupled via amide linkages to practically any protein, but also because of its remarkably strong binding to two proteins called *avidin* and *streptavidin*. Avidin (68 kDa) is a basic glycoprotein derived from egg white

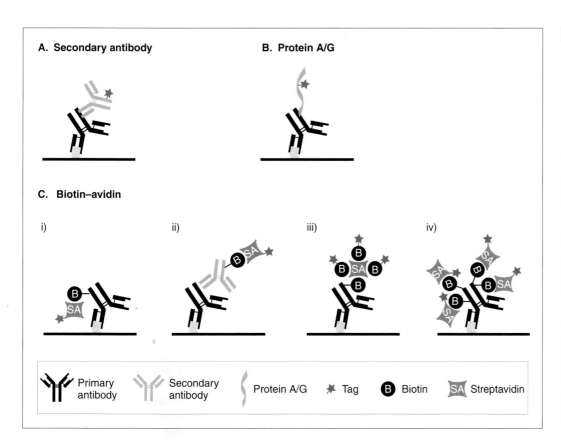

A. Secondary antibody

B. Protein A/G

C. Biotin–avidin

i)　　　　　ii)　　　　　iii)　　　　　iv)

Primary antibody　　Secondary antibody　　Protein A/G　　★ Tag　　**B** Biotin　　SA Streptavidin

Third Components in Antigen–Antibody Assays
Third components can be tagged and used to detect the presence of an antigen–antibody pair. (A) A tagged secondary antibody is specific for the Fc region of the primary antibody. (B) Tagged *Staphylococcus aureus* Protein A or Protein G binds to the Fc region of primary antibodies of certain isotypes (refer to Table 7-1). (C) A primary (i) or secondary (ii) antibody can be conjugated with biotin before use in an assay so that the presence of antigen–antibody pairs can be later detected using tagged streptavidin (SA) or avidin (not shown). The assay sensitivity can be increased by using SA as a multivalent bridge between the biotinylated antibody and tagged biotin (iii), or by ensuring that the biotinylated antibody has undergone biotin conjugation at multiple sites (iv).

while streptavidin (66 kDa) is a bacterial glycoprotein. Both avidin and streptavidin are tetrameric molecules made up of four identical subunits. Each subunit has a binding site with an extraordinarily high affinity for biotin in the order of $K \sim 10^{15}\ M^{-1}$. Such an affinity is considered to approach the strength of a covalent bond and is much stronger than the average antibody–antigen bond.

A variety of indirect assay methods have been developed that take advantage of the high-affinity binding of avidin or streptavidin (A/SA) to biotin (Fig. 7-9C). Let us suppose that we "biotinylate" a primary antibody of interest by coating it with covalently-bound biotin. We then incubate the biotinylated antibody with a mixture potentially containing the antigen of interest so that antigen–antibody pairs form. Tagged A/SA is then added as a third component. The tagged A/SA binds to the biotin on the primary antibody, facilitating detection of the antigen–antibody pair. If it is difficult to conjugate the primary antibody to biotin, a biotinylated secondary antibody directed against the primary antibody can be used, followed by incubation with tagged A/SA as a fourth step. The sensitivity of the system can be further increased by using a variation of these methods in which unlabeled A/SA is added to bind to the biotinylated antibody (primary or secondary), followed by the addition of tagged biotin. Since three tagged biotin molecules can bind to each antibody-bound A/SA molecule, the assay signal is instantly amplified three-fold.

Despite its extra steps, there are distinct advantages to the biotin–A/SA system over other detection methods: (1) Because biotin is so small and chemically innocuous, it can be conju-

gated to proteins without compromising the specificity or affinity of their interactions with other proteins; (2) non-specific binding to A/SA is virtually non-existent, decreasing assay background levels; (3) A/SA will bind to biotin-labeled antibodies of any species or isotype; (4) tag assay signals can be amplified many-fold by successive rounds of biotin–A/SA binding. In addition, a large antibody or antigen molecule can accommodate multiple biotin molecules at once, so that several A/SA molecules can bind to a single biotinylated assay component simultaneously. The strength of the signal associated with formation of a single antibody–antigen pair is thus amplified, greatly increasing the sensitivity of the assay. For example, the multivalency and very high binding affinity of the biotin–A/SA bond have been estimated to increase the sensitivity of the standard RIA by 1000-fold. Because of this increased sensitivity, some types of assays that formerly required the use of radioactive tags for even minimal detection have now been rendered less hazardous by the use of the biotin/avidin system in conjunction with a tag enzyme rather than a radioisotope.

The fact that the bond between biotin and A/SA is practically indestructible has turned out to be a double-edged sword. The system has been used to great advantage in quantitation and detection assays where no further manipulation or analysis of the identified protein of interest is required. However, the harsh conditions needed to break the biotin–A/SA bond may thwart any attempts to recover either the antibody or antigen of interest for further use after the assay. In most cases, even harsh detergents such as sodium dodecyl sulfate (SDS) are

Table 7-1 Antibody Binding by Protein A and Protein G

Ig Isotype	Protein A	Protein G
Human IgG1	+ + + +	+ + + +
Human IgG2	+ + + +	+ + + +
Human IgG3	–	+ + + +
Human IgG4	+ + + +	+ + + +
Human IgA	+ +	–
Human IgM	+ +	–
Human IgE	+ +	–
Mouse IgG1	+	+ +
Mouse IgG2a	+ + + +	+ + + +
Mouse IgG2b	+ + +	+ + +
Mouse IgG3	+ +	+ + +
Mouse IgM	+/–	–
Rat IgG1	–	+
Rat IgG2a	–	+ + + +
Rat IgG2b	–	+ +
Rat IgG2c	+	+ +
Rat IgM	+/–	–
Sheep IgG1	+	+ + +
Sheep IgG2	+ + +	+ + +
Cow IgG1	+	+ + +
Cow IgG2	+	+ + +

Total Ig	Protein A	Protein G
Human	+ + + +	+ + + +
Mouse	+ +	+ +
Rat	+/–	+ +
Rabbit	+ + +	+ + +
Hamster	+	+ +
Guinea Pig	+ + + +	+ +
Cow	+ +	+ + + +
Sheep	+	+ +
Goat	+/–	+ +
Pig	+ + +	+ + +
Chicken	–	+
Horse	+ +	+ + + +
Donkey	+ +	+ + + +
Dog	+ + + +	+ +
Cat	+ + + +	+ +

+ + + +, strongest binding; –, no binding.

ineffective in separating A/SA from a biotinylated protein of interest, precluding a determination of the properties of that protein of interest in isolation.

iii) Molecular Tags

First and foremost, the tags used in tag assays must be readily detectable and easily attached to the antigen or antibody of interest without causing significant alterations to the latter's binding properties. As mentioned previously, the tags most commonly used in immunoassays are *radioisotopes*, certain *enzymes* whose reaction products are colored, and *fluorochromes*. *Electron-dense materials* can also be used where tissues are being examined using electron microscopy. Each tag has its own advantages, disadvantages, and favored applications, all stemming from a tag's individual characteristics (Table 7-2).

iiia) Radioisotopes. The radioactive isotopes of inorganic molecules most commonly used as molecular tags are ^{125}I and ^{131}I (different radioisotopes of iodine), ^{57}Co (cobalt), ^{75}Se (selenium), and ^{32}P (phosphorus). These radioisotopes can be covalently attached to protein (either antigen or antibody) to generate tagged molecules of high specific activity that are easily quantitated by specialized detection instruments. The principal advantage to using radioisotope assays is that they are extremely sensitive, detecting antigen or antibody in the range of 0.001–0.050 μg/ml. The main disadvantage of isotope use is the hazard to the researcher inherent in exposure to radioactivity. Considerable care must be taken to handle assay materials deftly and to work behind the appropriate shielding. Other disadvantages include the short shelf life of many isotopes (e.g., ^{125}I has a half-life of only 60 days), the high cost, and the necessity of isolating each sample and counting its radioactivity for the minimal 1 minute needed to secure a reliable result. In a situation where hundreds of samples must be tested, the time expended in adequately counting all the samples may approach the impractical.

iiib) Enzymes. An enzyme whose action converts a substrate into an easily detected end-product of a particular color can also be used as a molecular tag. The enzyme can be covalently attached to either antigen or antibody. The enzymes most often used in this way are *horseradish peroxidase*, *alkaline phosphatase*, and *β-galactosidase*. Antibody (or, less commonly, antigen) tagged with the enzyme is added to the experimental sample so that the desired antigen–antibody complexes form. After washing away unbound components, the binding of the enyzme-tagged antigen or antibody is detected by the addition of a *chromogen*, a substrate (often colorless) for the conjugated enzyme. The enzyme acts on the chromogenic substrate and causes it to either gain, or change to, a particular color. This color can often be detected by eye for a quick qualitative screen, or its optical density can be precisely quantitated using a spectrophotometer. The intensity of the color produced is proportional to the amount of enzyme present; that is, proportional to the amount of enzyme-linked antibody (or antigen) bound, which is in turn proportional to the amount of untagged antigen (or antibody) present in the experimental sample. Tag enzymes are generally stable and retain high activity at the pH used for antigen– antibody binding reactions. The kinetics of substrate conversion by these enzymes are close to linear with respect to time and concentration of enzyme, so that a quantitative linear response is readily obtained.

Assays using enzymes as tags have distinct advantages over those employing radioisotopes. Tag enzymes are more stable, less hazardous to work with, and cheaper to obtain

Table 7-2 Molecular Tags Used in Antigen–Antibody Assays

Tag	Bound to	Method of Detection	Advantages	Disadvantages
RADIOISOTOPES ^{125}I, ^{131}I, ^{57}Co, ^{75}Se, ^{32}P	Antibody or antigen	Count radioactivity in scintillation counter	Extremely sensitive	Hazardous Costly Short shelf-life Time-consuming
ENZYMES Horseradish peroxidase, alkaline phosphatase, β-galactosidase	Antibody or antigen	Measure intensity of chromogenic reaction product by eye or spectrophotometer	Stable Less hazardous Cheap Quick	End point difficult to pinpoint Background noise Can be difficult to reproduce results Endogenous enzyme interference
FLUOROCHROMES FITC (fluorescein isothiocyanate), PE (phycoerythrin), Rh (rhodamine)	Antibody	Detect fluorescence under fluorescence microscope or by flow cytometry	Can examine multiple antigens simultaneously Less hazardous, Less costly	Less sensitive
ELECTRON DENSE MATERIALS Ferritin, colloidal gold, uranium	Antibody	View under electron microscope	Extremely sensitive	Time-consuming Expensive

than radioisotopes. In addition, because spectrophotometric assessment is often faster than measuring radioactivity, much larger numbers of samples can be evaluated in a shorter period of time, a distinct advantage in a clinical setting. Disadvantages sometimes cited for the tag enzyme assays are that the end point of the reaction can be difficult to pinpoint, and that the assay is more easily affected than isotopic assays by non-specific background "noise" from molecules in a biological sample that affect color development. The temperature and timing of the reaction must also be carefully controlled to achieve a reproducible result. In addition, some enzyme assay reagents are highly sensitive to contamination, poor handling, and light, and some enzyme substrates are mutagenic and/or carcinogenic. Tag enzymes are said by their supporters to be just as sensitive as assays employing radioisotopes, but others claim that this is not quite true in practice. When a tag enzyme assay involves mammalian cells, there is always the possibility that "endogenous activity" (present in the cells themselves) of the enzyme used for the assay could distort the results. For example, peroxidase is found at high levels in macrophages, erythrocytes, and bone marrow, while alkaline phosphatase occurs endogenously in placenta and intestine. Tag assays of these cells and tissues should therefore use β-galactosidase as the indicator enzyme, since there is almost no endogenous β-galactosidase activity in mammalian cells. The drawback to this solution is that β-galactosidase is also much less stable than the other enzymes available.

iiic) Fluorochromes. In some situations, the presence or precise localization of an antigen on an individual cell is desired. The signal visualization methods used with radioisotope and enzyme tags do not provide the necessary degree of resolution. However, certain dye molecules called *fluorochromes* can be visualized exactly in the locations where they bind. Scientists have exploited this property to generate easily detected tags that also have the advantage of not impairing the viability of cells to which they are bound.

Fluorochromes absorb light at a particular wavelength and subsequently emit that light at a longer wavelength (i.e., at a lower energy); these molecules are said to *fluoresce*. Three fluorochromes commonly used to tag antibodies in immunological studies are *fluorescein, rhodamine*, and *phycoerythrin*. Fluorescein and phycoerythrin absorb light at a wavelength of approximately 490 nm. Fluorescein re-emits a bright yellow-green signal at 517 nm, while phycoerythrin re-emits a reddish-orange signal at 575 nm. Rhodamine absorbs light of 550 nm and emits light of an intense red color at 580 nm. The colors emitted by these fluors are visible under a fluorescence microscope, which operates using UV light and interchangeable filters that allow only the appropriate excitation wavelength to reach the fluorochrome-tagged sample. Similarly, an appropriate barrier filter that permits the passage only of emitted light of the optimal wavelength is used to visualize the fluorescent signal. An example of two-color immunofluorescence is shown in Plate 7-2.

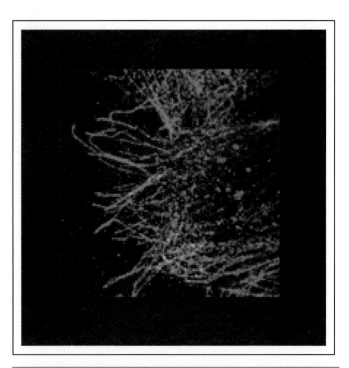

Immunofluorescence of Early Endosomes
Early endosomes (green) are aligned on microtubules (red) within a cell.
Courtesy of Erik Nielson and Marino Zerial, EU Research Training Network.

Fluorochromes can be covalently conjugated to amino groups in the antibody protein that one wishes to label. The tagging process is a bit of an art, so that after the conjugation process, the concentration of the antibody protein itself and the concentration of dye molecules per antibody protein must be determined spectrophotometrically before the tagged antibody can be used for any kind of quantitative application. By labeling antibody of one specificity with one fluor and another antibody of different specificity with a second fluor, the presence of two different antigens can be investigated at once in the same tissue or cell sample.

The intensity of color that one actually sees in a fluorescence assay depends on several factors, so that immunofluorescent tagging cannot be said to be quantitative in the same way as radioactive labeling. The fluorescence observed will depend not only on the concentration of the antigen of interest in the tissue sample (binding the fluorochrome-tagged antibody), but also on the intensity of the irradiating light, and the efficiency of its absorption and conversion into emitted light. In addition, many biological samples contain molecules that can scatter light or act as natural fluorochromes, causing non-specific fluorescent effects. Emission filters must be carefully chosen to minimize these sources of background "noise." Despite these caveats, assays based on fluorochromes have about the same sensitivity (detection limit 10–100 pg/ml) as assays employing either radioisotopes or enzymes. Fluorochrome assays also have the advantage that multiple

antigens can be examined simultaneously by the use of different fluorescent tags, and that the analyzed cells can be manipulated further (see later). Like tag enzyme assays, fluorochrome-based assays are less costly and hazardous than radioisotope assays.

iiid) Electron-dense materials. Antibodies (more often than antigens) can also be conjugated to tags that can be visualized under the electron microscope. Cellular components to which antibodies have been raised can be identified by chemically coupling the antibody to electron-dense materials such as *ferritin* (a protein that contains iron), *colloidal gold*, or *uranium*. The tagged antibody is then used to detect the presence of the cellular component for which it is specific. Electron-dense tags appear as black dots of varying sizes in tissue sections viewed under the electron microscope (see later). Assays employing electron-dense tags cannot be matched for the level of sub-cellular detail they can reveal but are time-consuming to set up. As well, sophisticated and expensive equipment is required to prepare and analyze the samples.

iv) Standard Procedures for Tag Assays
iva) The importance of blocking. In all tag assays, care must be taken to control for non-specific sticking of proteins either to the antibody or antigen of interest, or to any support used to immobilize a component of the assay. For example, suppose an antigen of interest has been bound to the surface of a plastic culture dish. Before adding the specific tagged antibody, the culture dish is flooded with an unrelated protein such as bovine serum albumin (BSA), a process known as "blocking." BSA binds non-specifically to any site in or on the dish that protein will stick to, including to the antigen of interest. The massive amount of BSA used effectively ensures that the specific tagged antibody is prevented from binding to non-specific sites at any significant level. However, the affinity of the antibody for the antigen is so much greater than that of BSA for the antigen that the antibody displaces the BSA, resulting in specific antigen–antibody pair formation. Blocking is particularly important in situations where plastics are involved since these surfaces tend to bind proteins irreversibly.

ivb) The importance of washing. At each step in a tag assay, there must be a way to separate bound from unbound components. Components may be unbound either because they are present in excess, or because they have too low an affinity for the other components of the assay mixture. If these unbound components are not removed prior to the next step in the assay, they may bind inappropriately to subsequently added components and interfere with the remaining assay steps. The result of such an assay may not then properly reflect the concentration of the component that is being evaluated. In particular, if unbound molecules of the tagged assay component are not washed away, a false positive result will be generated.

The requirement for washing to remove unbound components means that the antibody–antigen pair whose binding is being measured must be immobilized in some way so that it will remain behind after the washing is completed. Either antigen or antibody can be immobilized at the start of the assay by attachment to a solid support. A commonly used means of immobilization takes advantage of the fact that proteins will spontaneously bind to the wells of plastic assay dishes. For an assay carried out in a modern multi-well culture plate, assay fluids containing unbound components are easily removed using the ordinary suction of a mini-vacuum. The bound complexes, which are left intact, are then washed in a buffered solution containing detergent. The wash fluid is then removed by suction and the washing procedure is repeated to ensure that only the bound complexes of interest remain in the wells. Antibodies (or antigens) of interest can also be conjugated to insoluble beads, which can then be washed after the tagged partner has had a chance to bind. Such beads are washed either by centrifugation or by being packed into a tube through which fluids, but not particulate matter, may pass. In the case of cell surface antigens on cells in solution, the solid support may be considered the cells themselves, since the cells can be resuspended in buffer and washed repeatedly using a centrifuge. The cells spin down into a pellet, allowing the removal of unbound assay components in the supernatant by ordinary suction. In contrast, internal cellular antigens are effectively immobilized by being part of a tissue or cell sample chemically fixed onto a microscope slide.

These assays, in which one component is bound to a solid support, are called *solid-phase* assays, as distinct from their earlier counterparts in which antibody–antigen complexes were removed by precipitation in solution. There is another topological consideration to be noted with respect to solid-phase assays. Antibodies bind to the surface of a microtiter well in random orientation: some Ig molecules will bind "face down" such that the Fab sites are stuck to the dish and are not available to bind to antigen. Fortunately, a sufficient proportion of molecules bind in the correct orientation so that this factor does not impede the use of these assays.

ivc) Standard curves. Tag assays are not intrinsically quantitative: standard curves generated by the binding of known amounts of antigen and antibody must be established first if quantitation is a goal. The standard curve is constructed graphically from values obtained using known amounts of antigen and antibody in that particular assay system. Generally speaking, there is an approximately linear range in the curve where the signal generated by the presence of the tag is directly proportional to the concentration of antigen–antibody complexes (Fig. 7-10). This relationship continues until a plateau is reached at about 20- to 100-fold background binding levels, where no further increases in the signal associated with the bound tag are achieved. The numerical value of an experimental result is compared to the linear portion of the standard curve and extrapolation of this result produces a quantitative measure of the component of interest. It should be noted that the solution used to dilute the binding partners must be taken into account, as the nature of a given diluent can affect the standard curve.

One last consideration must be borne in mind when designing and using tag assays. The addition of a tag to a protein changes that protein, by definition. A specific antibody

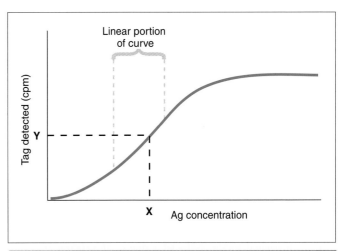

Figure 7-10

Standard Curves for Assay Quantitation
The standard curve is generated by measuring the amount of tag detectable when using a fixed, excess concentration of one component (Ag or Ab) and various known concentrations of the other component. For example, for an RIA in which the concentration of radiolabeled Ab is held constant, the concentration of Ag is varied (X-axis) and the amount of tag detected in counts per minute (cpm) is plotted (Y-axis). In the linear portion of the resulting curve, the amount of radiolabel detected is proportional to the concentration of Ag. A sample of unknown Ag concentration can then be tested in the same assay system and if the amount of radiolabel measured is in the linear portion of the curve (at Y), then the unknown Ag concentration can be read off the graph (at X).

therefore usually shows a slight difference in its binding affinity for an antigen when either it or the antigen is tagged, making it advisable to re-test the binding activity of the antibody–antigen pair after conjugation. It is rare that the binding affinity is reduced enough to affect the overall result of the assay.

II. DETECTION OF ANTIGEN BY TAG ASSAYS

i) Binder–Ligand Tag Assays: Radioimmunoassay and Enzyme-Linked Immunosorbent Assay

A binder–ligand assay is often the method of choice for determining the concentration of one of two binding partners. For example, an antibody can be tagged and used to bind to its specific antigen in such a way that the concentration of unknown specific antigen in an experimental sample can be determined using a standard curve. Alternatively, the antigen can be tagged so as to allow quantitation of an unknown specific antibody in a sample. The first type of immunological binder–ligand assay to be developed was the *radioimmunoassay* (RIA), in which the tag is a radioisotope. The RIA was devised in 1959 by endocrinologists Rosalyn Yalow, Solomon Berson, and their colleagues in their studies of the binding of human anti-insulin antibodies to human insulin. Consistent with its eventual revolutionary effect not only on studies in endocrinology and immunology but also in a wide variety of biological sciences, this work was recognized with the Nobel Prize for Physiology or Medicine in 1977.

Although the first RIAs were done in test tubes, the increasing use of plastics engendered the adaptation of the RIA to multi-well microtiter plates. Components of interest were readily adsorbed to these plates and the required washing steps were much more easily carried out. The soonubiquitous use of microtiter plates for RIA led to their slang designation as "plate assays." However, the safety hazards associated with the handling of radioactivity and the desire for ever-increasing speed in handling large numbers of samples spurred the development of binder–ligand assays using tag enzymes. These assays became known as ELISAs, for *enzyme-linked immunosorbent assay*, in which the sample putatively containing the untagged antigen (usually) is adsorbed to the wells of a plastic microtiter plate or fixed on a microscope slide to immobilize it, and the enzyme-tagged antibody is added to the plate or slide after blocking. After washing to remove unbound antibody, a chromogenic substrate is added. The presence of colored enzyme product indicates the presence of the antigen of interest. Special spectrophotometric "plate readers" have been developed to simultaneously analyze the absorbance of all 96 wells of a standard microtiter plate within a few seconds. While such ELISA assays can be used quantitatively to determine the concentration of an antigen or antibody (using a standard curve), they are very often used to do quick qualitative screenings for the mere presence of the antigen or antibody of interest.

Both radioisotopes and tag enzymes are used essentially the same way in the myriad variations of plate assays: only the tag and the method of its detection differ. Examples of these assays are illustrated for ELISAs in Fig. 7-11. In a *direct* binder–ligand assay (Fig. 7-11A), one partner of the binding pair (say, antibody) is tagged and is incubated with a plated sample which may or may not contain the other partner (the antigen). The presence of immobilized tag after washing indicates that the sample did indeed contain the antigen of interest. In a *competitive* binder–ligand assay (Fig. 7-11B), tagged and untagged forms of the component of interest compete to bind to the specific binding partner that has been immobilized. This type of assay is based on the principle that the presence of an untagged component X in an experimental sample can be detected by observing its displacement of tagged component X of known concentration. *Indirect* binder–ligand assays (Fig. 7-11C) employ secondary antibodies, Protein A or G, or the biotin–avidin system as an extra step and are often useful where one of the binding partners is impure, in short supply, or difficult to tag. The immobilization of the tagged third component indicates that the primary antibody has bound to antigen, and that antigen was indeed in the original sample. The *sandwich* assay is a variation of the indirect binder–ligand assay in which the antigen is "sandwiched" between two antigen-specific antibodies, the second of which is tagged (Fig. 7-11D). If the antigen is multivalent, the two antibodies involved can be identical. If the antigen is monovalent, the two antibodies used must bind to different

non-overlapping sites on the antigen; in this case, the assay is sometimes called a *two-site* assay.

It should be noted that there is considerable latitude within the immunological community in the naming of tag assays. The terms "direct," "indirect," and "competitive" are commonly used, but upon closer examination, one researcher's "direct" assay may be another's "indirect" assay. The reader should perhaps concentrate on the actual mechanics of a given detection process, rather than its name.

ii) Immunohistochemistry

Antibodies are not only sensitive tools for the manipulation and examination of the physical properties of antigens, they have also revolutionized the analysis of antigen expression in intact tissues and on cell surfaces. In short, antibodies have allowed scientists to "see" where antigens play their normal roles *in vivo*. *Histochemical* analyses of cells or tissue samples involve the *staining* of cellular structures of interest with tags that are easily identified using the appropriate type of microscopy. In *immunohistochemical* or *immunostaining assays*, the presence and location of specific molecules are marked by the binding of specific antibody that has been chemically tagged to allow its visualization. Briefly, the cell or tissue sample is incubated in solution or on a microscope slide with tagged antibody recognizing the antigen of interest. After removing non-specific proteins by washing, the appropriate procedure for the detection of the tag is carried out. Immunohistochemical assays can be carried out on either live cells or cells that are non-viable due to chemical fixation. If live cells are used, perhaps because further subculture of particular subpopulations is desired, only surface structures can be characterized, as tagged antibodies will not readily penetrate the viable cell membrane. On the other hand, both internal and external structures can be characterized in fixed cell samples as the cell membrane becomes permeable to antibody under fixation conditions.

iia) Immunohistochemical ELISA and RIA.

Enzymes or radioisotopes can be used as immunohistochemical tags in cases where cell viability after the assay is not an issue and an extremely precise level of resolution is not required. For example, often one wants a more rapid screening of a large number of samples than is possible with fluorochrome-tagged samples. In an immunohistochemical ELISA or RIA, a microscope slide on which a cell culture or tissue sample containing an antigen of interest has been fixed is incubated with antibody carrying an enzyme or radioisotope tag. In the case of the ELISA, after washing away unbound components, the slide bearing the antigen–antibody complexes is incubated with the ELISA chromogenic substrate. The colored product of the ELISA reaction is deposited on the slide in the location of the antigen of interest, allowing the scientist to determine its general localization. If a radioisotope is used as a tag, the presence of the antigen of interest can be determined by autoradiography. Although the autoradiographic image may give some information about the relative cellular position of the detected antigen, the signal emitted by most radioisotopes is broad, making it more difficult to pinpoint the exact position of the component of interest.

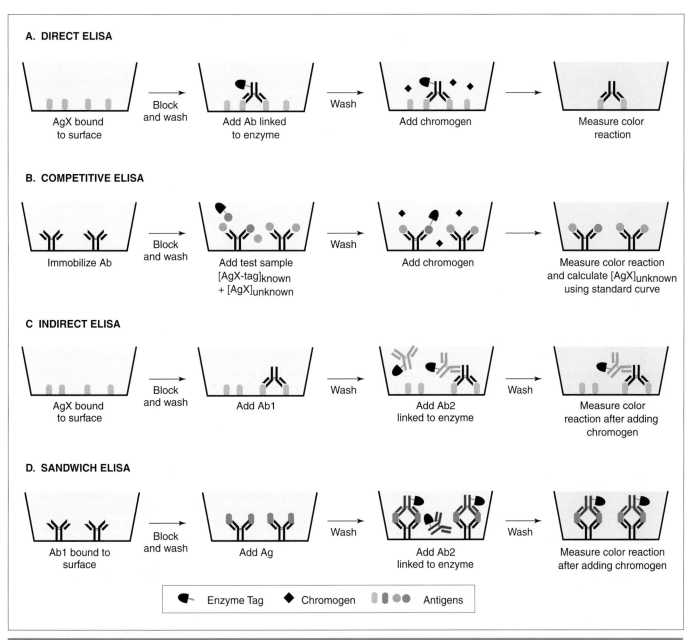

A. DIRECT ELISA

AgX bound to surface → Block and wash → Add Ab linked to enzyme → Wash → Add chromogen → Measure color reaction

B. COMPETITIVE ELISA

Immobilize Ab → Block and wash → Add test sample [AgX-tag]$_{known}$ + [AgX]$_{unknown}$ → Wash → Add chromogen → Measure color reaction and calculate [AgX]$_{unknown}$ using standard curve

C INDIRECT ELISA

AgX bound to surface → Block and wash → Add Ab1 → Wash → Add Ab2 linked to enzyme → Wash → Measure color reaction after adding chromogen

D. SANDWICH ELISA

Ab1 bound to surface → Block and wash → Add Ag → Wash → Add Ab2 linked to enzyme → Wash → Measure color reaction after adding chromogen

Enzyme Tag   ◆ Chromogen   Antigens

Figure 7-11

ELISAs as Examples of Binder–Ligand Assays
In all binder–ligand assays, one component of interest, either an antigen or antibody, is immobilized so that antigen–antibody pairs formed during the course of the assay are not lost during subsequent washing steps. If a radioisotope is used, the assay is an RIA. In an ELISA, the tag is an enzyme that acts on a chromogen to yield a detectable colored product. In the direct ELISA shown in (A), antigen X of unknown concentration is coated onto the surface of a microtiter well and tagged antibody of known concentration is added to detect it. In a competitive ELISA (B), the presence of AgX in the test sample leads to a proportional displacement of labeled AgX from the immobilized antibody, decreasing the intensity of the final color reaction proportionally. For the indirect ELISA shown in (C), a secondary antibody has been used as the third component; however, Protein A, Protein G, or the biotin–avidin system could also be used for this step. The sandwich ELISA (D) is a variation of the indirect ELISA.

iib) Immunoelectron microscopy. Structures at the sub-cellular or even molecular level can be examined using immunohistochemical tags made of electron-dense materials. Specific antibody is tagged with either ferritin, colloidal gold, or uranium and allowed to bind to the structure of interest in an ultra-thin tissue section prepared for electron microscopy (EM). The tissue section is then bombarded with electrons. The electron-dense tag absorbs the electrons and appears in an electron micrograph as a black dot. Different sizes of electron-dense particles can be used to label different antibodies, which appear as black dots of different sizes and allow the simultaneous localization of different antigens in the same tissue section (Plate 7-3). This technique has been of great use in the elucidation of intracellular protein-processing

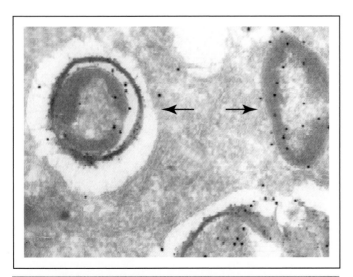

Plate 7-3

Immunoelectron Microscopy
Antibodies specific for two proteins are tagged with two different sizes of gold particles. Using electron microscopy, these antibodies detect co-localization of the proteins within macrophage endocytic compartments (arrows) important for antigen processing. Reproduced with permission from Kleijmeer M. J. *et al.* (1996). Characterization of MHC class II compartments by immunoelectron microscopy. *Methods: A Companion to Methods in Enzymology* **10**, 191–207.

pathways. For example, immunogold particles of different sizes conjugated to antibodies recognizing a test antigen or MHC molecule, respectively, have permitted scientists to follow the internalization of the antigen and its subsequent intracellular processing and association with MHC. Similar methodology has been used to confirm that MHC class I and II molecules operate in distinct intracellular processing paths.

iic) Immunofluorescence. Conjugation of a fluorochrome to the constant domain of an antibody allows the scientist to locate specific antigens on the surface or within immobilized cells using *immunofluorescence* (Fig. 7-12). The nature of

the fluorochrome signal is such that it remains tightly confined to the site where the antigen is located and has bound the antibody, providing a finer degree of visual resolution than antibodies tagged with either an enzyme or a radioisotope. Some useful applications of immunofluorescence techniques are: to detect immune complexes and autoantibodies in cases of autoimmune disease; to identify microbial species in patient tissues, and tumor-specific antigens on tumors; and to investigate transplantation antigens on tissues and organs.

Staining assays in which antigens are detected using fluorochrome-tagged antibodies can be based on either *direct immunofluorescence* or *indirect immunofluorescence*. In direct assays, the tagged antibody is added directly to the cells or tissue section under examination and allowed to bind. Application of UV light excites the fluorochrome conjugated to the antibody, and the position of the antigen is indicated by the fluorescence of the bound antibody as viewed through the fluorescence microscope. In indirect immunofluorescent staining, the primary antibody recognizing antigen is not tagged, but a secondary antibody that recognizes the primary antibody, or *Staphylococcus* Protein A or G, is conjugated to the fluorochrome. This approach obviates the need for purifying and conjugating a different primary antibody for each antigen under investigation.

While fluorescence microscopes can indicate the general cellular location of a fluorochrome tag signal, they do not reveal the particular cellular structures to which the tagged antibodies have bound. More advanced microscopic techniques have therefore been developed in which *phase contrast* microscopy is coupled with fluorescence to better define the cellular features which "light up." Alternatively, *confocal* microscopy, in which a computer generates a very thin optical section of a stained cell or tissue, can be used. While immunofluorescence is primarily thought of as a qualitative technique, *microfluorometers* have also been developed which can accurately measure the amount of light of a given wavelength emitted from a fluorescent sample, permitting the quantitation of bound antibody by reference to a standard curve.

Figure 7-12

Immunofluorescence
A blood cell sample can be tested for the presence of lymphocytes using direct immunofluorescence. The cell sample fixed on the microscope slide is incubated with FITC-conjugated anti-lymphocyte antibody. After washing away excess unbound antibody, the slide is viewed under a fluorescence microscope. Fluorescing sites indicate the presence and location of lymphocytes on the slide. FITC, fluorescein isothiocyanate.

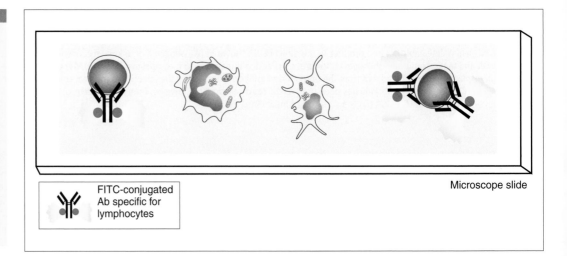

FITC-conjugated Ab specific for lymphocytes

Microscope slide

The sensitivity of immunofluorescence assays can be enhanced by the implementation of the biotin–avidin system. Each molecule of the antibody to be used for the detection of the antigen of interest can be conjugated to many biotin molecules, each of which can bind to fluorochrome-tagged avidin. A single antibody molecule is thus coated with numerous fluorochromes, increasing the total fluorescent signal of that antibody. Antigens present in concentrations of <10 pg/ml have been clearly detected using fluorochrome-tagged avidin.

iii) Flow Cytometry

Cell populations bearing fluorochrome-tagged antibodies can be counted or otherwise evaluated according to their fluorescence using a *flow cytometer*. Flow cytometers are electronic instruments that can analyze various physical and chemical characteristics of individual cells as they are forced through a small opening at high speed. Any cell type that can be prepared as a "single-cell suspension" can be studied using the flow cytometer. The term "single-cell suspension" does not imply that there is a only a single cell in the sample, but that individual cells have been resuspended at relatively low density in culture media *in vitro* and are not clumped together or present as part of an organized tissue. Unlike most other cell types, leukocytes are able to survive quite nicely as single-cell suspensions and are not impeded or made less viable by the surface binding of molecules bearing a fluorochrome tag. These characteristics have made them the mammalian cell type most intensely analyzed by flow cytometry. Flow cytometry can also be used to study non-mammalian cells such as bacteria, yeast, and plant cells, and many marine research vessels have a flow cytometer on board for the study of plankton.

For flow cytometric analysis, the single-cell suspension is incubated with tagged antibody specific for a cell surface molecule of interest and then added to a chamber in the flow cytometer known as the "flow cell." Pressure forces the cells out the tip of a nozzle, which vibrates ultrasonically. The vibration causes each cell to become encased in an individual droplet of buffer. The droplets fall one by one past a laser beam, which excites the fluorochrome tags on the cell surface-bound antibodies. The intensity and wavelength (color) of the fluorescent light emitted by each antibody's tag is analyzed instantaneously by the computer-assisted fluorescence detectors of the flow cytometer. This analysis yields information on the number of cells binding the tagged antibody and the average fluorescent intensity of each, which is a reflection of the level of expression of the molecule of interest.

It should be noted that flow cytometry is a technique that can be very broadly applied, and thus can reveal important information about cells in a sample in addition to that provided by the binding of fluorochrome-conjugated antibodies. Features such as size, DNA and RNA content, and calcium flux can be measured by analyzing how parameters such as viscosity and refractive index are affected when the laser beam hits an individual cell. For example, the flow cytometer determines the size of each cell by examining the "forward light scatter" associated with it, and the granularity of each cell by analyzing its "90° light scatter." The flow cytometer can be calibrated such that cells larger or smaller than those of interest are ignored, meaning that data on only a particular subpopulation of cells (for example, the lymphocytes in an unfractionated leukocyte suspension) can be obtained. Those cells that do not fit the specified parameters are "gated out" and ignored in the flow cytometer's analysis of the test population. Thus, the output of the flow cytometer is a representation of the numbers of cells fitting a particular set of light scatter and fluorescence parameters.

Immunofluorescence in conjunction with flow cytometry can be used to enumerate, characterize, and even isolate particular subsets of living cells in a sample. As a specific example of how antibodies are used in flow cytometric studies of leukocytes, let's imagine we have mouse thymus cells and we want to determine what percentage of that population is expressing the CD8 co-receptor molecule that is expressed on T cells and interacts with MHC class I molecules. A thymus cell suspension is prepared and incubated with fluorescein-tagged antibody recognizing CD8. The cells are then washed free of unbound antibody and passed through the flow cytometer. The flow cytometer analyzes the relative fluorescence intensity of each cell as it passes by the fluorescence detector, providing a relative measure of the amount of fluorescein-conjugated anti-CD8 antibody bound to each thymus cell. The data can be reported as a graph plotting the number of cells versus fluorescence intensity (Plate 7-4A). By also running a control sample with a fluorescein-tagged antibody that is not expected to bind to thymus cells, one can determine a background level of fluorescence above which the cells can be said to be "CD8 positive." From this result, the percentage of $CD8^+$ thymus cells in the sample can be calculated. As an example of the use of two-color fluorescence analysis, let's now suppose that in addition to analyzing the thymus cells expressing CD8, we also wish to simultaneously analyze the thymus cells expressing the CD4 co-receptor that is expressed on T cells and binds to MHC class II. In this case, the thymus cell suspension is stained by incubation with both fluorescein-tagged anti-CD8 antibody and phycoerythrin-tagged anti-CD4 antibody. The flow cytometer assesses each cell as it passes through the fluorescence detectors for the presence of either the fluorescein or the phycoerythrin signal and keeps track of the numbers of each. The data are shown on an attached oscilloscope organized in a graphical representation consisting of four quadrants (Plate 7-4B). The relative percentages of cells bearing only CD4, cells bearing only CD8, cells bearing both CD4 and CD8, and cells bearing neither are plotted in the appropriate quadrants. More sophisticated analyses can involve staining with up to 16 different fluorescent markers and plotting for cell size as well as number and fluorescence. In addition, new dyes that change color upon binding to K^+, Na^+, and Ca^{2+} ions within a living cell allow a researcher to closely follow subtle intracellular reactions such as membrane pore changes and Ca^{2+} flux. Some examples of dyes used in these types of analyses are given in Table 7-3.

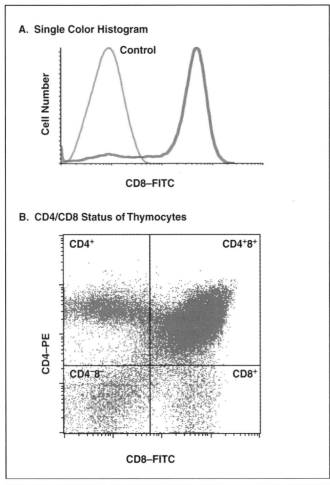

A. Single Color Histogram

Control

Cell Number

CD8–FITC

B. CD4/CD8 Status of Thymocytes

CD4+ | CD4+8+

CD4–PE

CD4 8 | CD8+

CD8–FITC

Analysis of Thymocytes by Flow Cytometry
(A) Thymocytes have been incubated with a non-specific Ab (curve on left) and an anti-CD8–FITC tagged Ab (curve on right). The number of cells at a given level of fluorescence intensity is measured by flow cytometry. (B) Thymocytes are incubated with anti-CD4–PE and anti-CD8–FITC antibodies. Thymocytes expressing different levels of CD4 and CD8 are shown in the various quadrants. Reproduced courtesy of Dr. Juan-Carlos, Zúñiga-Pflücker, Department of Immunology, University of Toronto and Sunnybrook & Women's Research Institute, Toronto.

Table 7-3 **Examples of Dyes Used in Immunofluorescence Analyses**

Name of Dye	Fluorescing Color	Molecules Labeled/Detected
Alexa Fluor 430	Yellow-green	DNA, proteins
Alexa Fluor 405	Blue	DNA, proteins
Alexa Fluor 488	Green	DNA, proteins
Alexa Fluor 546	Orange	DNA, proteins
Alexa Fluor 594	Red	DNA, proteins
Fluoroscein	Green	DNA, proteins
Rhodamine	Red	DNA, proteins
Phycoerythrin	Red-orange	DNA, proteins
Texas Red	Red	DNA, proteins
Hoechst	Blue	Nucleic acids
Acridine Orange	Red-orange	Nucleic acids
DAPI	Blue	Nucleic acids
Fura-2	Red	Calcium
Fluo-4 AM	Green	Calcium
Indo-1	Blue	Calcium
Sodium Green	Green	Sodium
Corona Red	Red-orange	Sodium
PBFI	Blue	Potassium
MQAE	Blue	Chloride

phycoerythrin-tagged anti-CD4 antibody were identified in a thymus cell population. By specifying the collection parameters of the flow cytometer, subpopulations of thymus cells bearing only CD4 or CD8, or both or neither, can be physically separated from each other. In addition, cells that stain very intensely can be isolated from those that stain moderately, or those that stain very weakly. Each of these populations can then be recovered from its collection tube, cultured separately, and manipulated further, providing us with a very fine resolution with which to purify and study developing T cells in the thymus that are very closely related.

III. ISOLATION AND CHARACTERIZATION OF ANTIGEN USING ANTIBODIES

Antigen–antibody interaction is often used to purify antigens and characterize their physical properties. A big advantage to the use of antibodies in this regard is that their specificity often makes extensive preliminary purification of the antigen unnecessary. As in the case of quantitative tag assays, these antigen isolation/characterization techniques lend themselves to both direct and indirect approaches. Molecular tags are used regularly to track the progress of the antigen or antibody through a purification process. While a quantitative determination of the numbers of antigen–antibody pairs formed during the isolation/characterization may not be possible or desirable, the location of the tagged component of interest in the assay system can often provide valuable qualitative information.

As well as identifying and counting particular cell populations, a flow cytometer can also sort these cells for subsequent manipulation. In fact, a great advantage of flow cytometric analysis is that fluorochrome-labeled cells can pass through the instrument without compromising either sterility or cell viability. The fluorescence detectors of the flow cytometer can be connected to computer-operated electromagnetic deflector plates that are programmed to direct a cell with a fluorescent signal of a given color and intensity to a particular collection tube, a process known as "cell-sorting." One might hear a group of immunologists talking about the results of a "FACS" analysis: FACS™ is one biotechnology company's abbreviation for "fluorescence-activated cell sorter."

Let's return to our example, in which cells bearing fluorescein-tagged anti-CD8 antibody and cells bearing

i) Immunoprecipitation

The technique of *immunoprecipitation* by specific antibody can be used to isolate or just detect antigens present at low concentrations in complex mixtures of proteins. Briefly, specific antibody is added to the protein mixture (such as a cell extract), and the antigen–antibody complex is caused to precipitate out of solution by adding an insoluble agent to which the complexes will bind (Fig. 7-13). Techniques currently used to "fish out" the desired primary antibody–antigen complex from solution often employ a secondary binding protein permanently coupled to small agarose or dextran beads. If the primary antibody is of the IgG isotype, purified *Staphylococcus* Protein A coupled to beads can be used to bring down the primary antibody–antigen complex. *Staphylococcus* Protein G coupled to beads can be used where other isotypes need to be retrieved. Less commonly, heterologous secondary anti-Ig antibodies that bind the Fc regions of the primary antibody can be covalently attached to the beads. The beads spin down into a pellet during centrifugation, precipitating the attached complexes from solution. The unwanted supernatant containing the assay fluid and unbound components (including other proteins) is removed and the beads are washed again with neutral buffer. The antigen is then released from the primary antibody–bead structure by disruption of the antigen–antibody bond with chemical compounds such as guanidine-HCl, sodium thiocyanate, or diethylamine that drastically raise or lower the pH (to pH 3.0 or pH 11.5), or increase the ionic strength of the chemical environment. These changes reduce the affinity of binding between antigen and antibody and allow the recovery of the purified antigen from the supernatant by the use of conventional protein chemistry techniques such as dialysis or electrophoresis.

A similar strategy can be used in situations where the cells under study have been labeled metabolically with a radioisotope such that the antigen of interest becomes tagged. An extract is made of the labeled cells and primary antibody conjugated to beads (or primary antibody followed by the addition of a secondary protein conjugated to beads) is incubated with the extract. Binding of the labeled antigen to the primary antibody followed by centrifugation of the beads and washing can be used to "bring down" the tagged antigen. The quantity of tag in the pellet is then determined to yield a measure of the amount of antigen in the extract. Alternatively, the mixture can be separated on a gel by electrophoresis and autoradiographed to detect a band representing the antigen of interest.

ii) Affinity Chromatography

Chromatography is defined as the separation of the components of a mixture by slow passage over or through a material that absorbs the components differently. *Affinity chromatography* is a technique in which the difference in absorption depends on the specific affinity between a substance fixed in the separation material (the *absorbent*) and the desired component in the mixture (the *ligand*). Typically the absorbent is covalently linked to agarose beads, which are then confined in a column over which the protein mixture to be separated is continuously passed. The ligand is absorbed by the absorbent as the mixture passes over the column, while other components fail to bind and exit the column. After thorough washing, the ligand can then be retrieved using

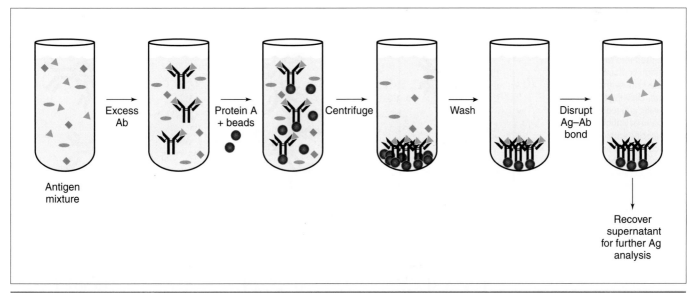

Figure 7-13

Immunoprecipitation

Immunoprecipitation can be used to isolate one specific component from an antigenic mixture in solution. Incubation of the solution with specific antibody is followed by precipitation of the antigen–antibody complexes with Protein A-coated beads. After washing away extraneous antigens, the desired antigen is released from the antibody–bead using altered pH and/or high salt. The antigen can then be analyzed further by methods such as gel purification or thin-layer chromatography (TLC).

chemical elution that disrupts the bond between the absorbent and the ligand. In the immunological context, the absorbent coupled to the agarose beads is called the *immunoabsorbent*, and is usually either a primary antibody (where large quantities of purified antibody are available) or a secondary isotype-specific antibody or Protein A or G (Fig. 7-14). In cases where primary antibodies are used, the antigen mixture is passed slowly over the primary antibody affinity column such that the antigen is retained on the column. The column is washed extensively with a neutral buffer to remove any non-specifically bound components. The purified antigen is then eluted from the column using chemicals that disrupt the antigen–antibody bond as previously described. Where secondary antibody or Protein A or G is adsorbed to the affinity column, the protein mixture to be separated is preincubated with specific primary antibody to allow antigen–antibody pairs to form. The mixture is then passed over an affinity column composed of the secondary binding protein coupled to the agarose beads. The secondary binding protein pulls the primary antibody and its associated antigen out of the mixture. Elution techniques are then used to release the primary antibody from the secondary binding protein, and the antigen from the primary antibody. Subsequent purification by electrophoresis may be required to distinguish between the antigen and the primary antibody.

For the purification of large quantities of antigen, affinity chromatography is the technique of choice. The structure of the column allows the researcher to wash away non-specific proteins from a large number of beads more easily than can be achieved by repeated centrifugation of beads in solution. The relatively cumbersome procedure required to set up the column apparatus is thus well worth the effort for a single large-scale purification. Moreover, the column can be re-used repeatedly for additional purifications of the same antigen. Affinity chromatography is not generally used to separate antigen–antibody complexes from free components in RIA or ELISA situations because it is simply impractical to set up and manipulate so many small-scale columns. Affinity chromatography can also be used to purify specific antibody by coupling purified antigen to the agarose beads. Antiserum recovered from an animal immunized with the antigen of interest is passed over the column. Repeated washes ensure that only antibody specific for the fixed antigen binds to the beads and that all other proteins, including antibodies to other antigens, pass through. The purified antibody can be eluted from the column by applying conditions that reduce its affinity for antigen as described previously.

The biotin–avidin system can also be used for antigen isolation in an affinity chromatography system. For example, a biotinylated antibody can be incubated with a complex protein mixture containing the antigen of interest. The mixture, now containing antigen–antibody–biotin complexes, can be passed over a column of beads to which avidin has been covalently attached. These biotin-containing complexes will

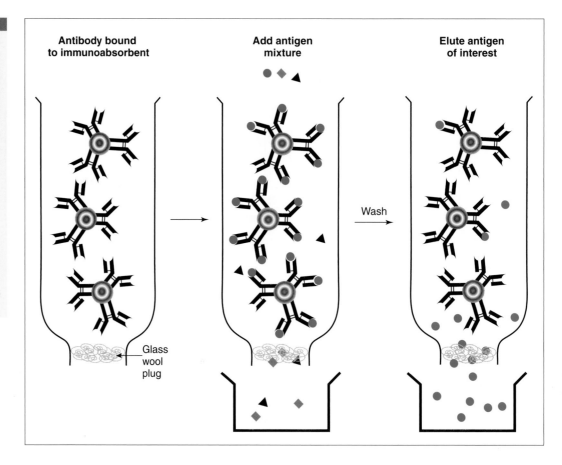

Figure 7-14

Affinity Chromatography
The affinity chromatography column is fitted at the bottom with a plug such as glass wool that is permeable to liquid. Immunoabsorbent beads bearing immobilized antibody are poured into the column. A solution containing antigen is passed slowly through the column, allowing the binding of specific antigen to the immobilized antibody. Non-specific entities are removed by extensive washing. A solution with the appropriate pH and salt concentration to disrupt antigen–antibody binding is then passed through the column to elute (wash off) the antigen of interest.

Antibody bound to immunoabsorbent

Add antigen mixture

Elute antigen of interest

Wash

Glass wool plug

stick to the column. The strength of the avidin–biotin bond means that the antigen will be released from the biotinylated antibody long before the biotinylated antibody is released from the column. In fact, due to the almost irreversible nature of the avidin–biotin interaction, the column can only be used once.

iii) Western Blots

The Western blot uses the specificity of antigen–antibody binding in conjunction with sodium dodecyl sulfate poly-acrylamide gel electrophoresis (SDS-PAGE) to detect very small quantities of a protein of interest in a complex mixture. Scientists often initially characterize a protein mixture by determining the sizes of its constituent proteins. By adding a negatively-charged ionic detergent such as SDS to a protein mixture, all its component proteins become denatured and

uniformly negatively-charged. The mobility of these proteins through a polyacrylamide gel in response to an electric current then depends solely on their size: larger proteins tend to move more slowly toward the positive electrode ("down" the gel) than do smaller proteins. The gel can be stained either with Coomassie blue dye or by using a silver reduction process to visualize the separated proteins. The position of the separated proteins relative to a set of standard size markers electrophoresed on the same gel can be used to determine their size.

A Western blot can be used to determine the actual identity of any one of the proteins separated by SDS-PAGE (Fig. 7-15). Let us assume that we have already separated the total proteins of a tissue preparation on an SDS-PAGE gel. A nitro-cellulose membrane is placed over the gel, allowing the denatured proteins to transfer onto (are "blotted" onto) the

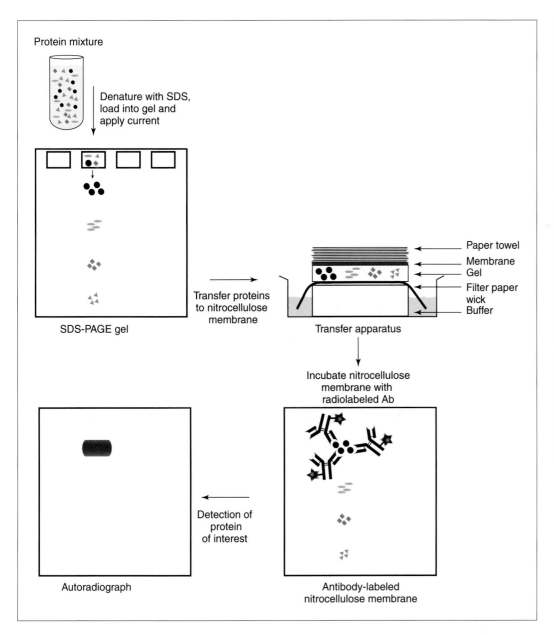

Figure 7-15

Western Blotting
Denatured proteins are separated by charge using polyacrylamide gel electrophoresis (PAGE) in the presence of sodium dodecyl sulfate (SDS). The subsequent transfer of the proteins onto a nitrocellulose membrane is achieved by the "wicking" or capillary action of buffer that is drawn up through filter paper wick toward absorbent paper towel. As the buffer moves through the gel and the nitrocellulose, the proteins are drawn upward but adhere to the nitrocellulose rather than passing through. The antigenic protein of interest is then detected on the membrane by the binding of tagged antibody. If the tag is a radioisotope (as shown in this example), its presence is detected by subjecting the membrane to autoradiography. If the tag is an enzyme, the membrane is exposed to the appropriate chromogenic substrate. A colored band develops where the protein is bound to the membrane.

membrane by capillary action, sometimes accelerated by the application of an electric current. The membrane with its transferred protein bands is incubated with antibody that is specific for a protein of interest and tagged with either a radioisotope or an ELISA enzyme. The membrane is washed to remove non-specific proteins, and treated to reveal the presence and position of the antigen of interest. If the tag is a radioisotope, autoradiography is used to visualize any resulting bands. If an enzyme has been used, the blot is incubated with the enzyme substrate and any bands of interest appear colored. If tagged primary antibody is not available, a tagged secondary protein such as an anti-isotype antibody or Protein A/G can be used to indirectly visualize an antigen–primary antibody pair of interest. Biotinylated proteins separated on a Western blot can be visualized using antibodies labeled with either ^{125}I- or enzyme-tagged avidin.

Western blots can also be used to rapidly detect the presence in a serum sample of antibodies specific for antigens of interest. For example, to confirm a case of HIV infection, the patient's serum can be tested by incubating it with a prepared Western blot of proteins derived from the HIV virus. The various viral antigens are fixed on the blot according to their known molecular weights. If the patient is indeed HIV-positive, antibodies to several viral antigens will be present in the serum sample and will bind to those antigens on the nitrocellulose membrane. The primary antibody can then be visualized by further incubating the washed blot with tagged anti-human Ig. The presence of several bands that "light up" at the molecular weights expected for HIV antigens confirms that the patient is HIV-positive. However, the development of antibodies to HIV can take as long as 3–6 months after acute infection, meaning that a patient might indeed be HIV-positive but negative for anti-HIV antibodies if tested as described here. Fortunately, Western blotting can also be used to test a patient's serum for the presence of the HIV antigens themselves, by electrophoresing and blotting the patient's serum proteins and visualizing the bands with tagged anti-HIV antibodies.

The name "Western" actually has very little to do with geography, and everything to do with parallel technologies. Ed Southern was the scientist who invented nitrocellulose membrane transfer for the analysis of DNA fragments separated on agarose gels (see Ch.8, Box 8-3). Accordingly, these DNA transfers became known as "Southern blots." The same technology applied to the analysis of mRNAs was called "Northern blotting," just for fun. Naturally, when the technique was applied to the separation of proteins, scientists looked toward another point of the compass. (Perhaps a reader of this text will develop an Eastern blot!)

We have now discussed antibodies as effectors and antibodies as laboratory tools. It is time to discuss the Ig genes from which antibodies are expressed and the mechanisms responsible for the almost infinite diversity of the antibody repertoire.

SUMMARY

The specificity and sensitivity of antibodies make them useful biochemical tools. Antibodies can discriminate between two closely related structures, and can "fish out" a desired molecule from a complex mixture. The development of hybridoma technology and monoclonal antibodies in the late 1970s resulted in an explosion of technical applications. Antibodies are used in two main categories of assays: those based on immune complex formation, and those based on unitary antigen–antibody pair formation. In the first category, the assays exploit the fact that large immune complexes are insoluble structures that can either be made visible or can be detected by their activation of complement. Such assays include precipitin reactions, immunodiffusion, immunoelectrophoresis, rocket electrophoresis, two-dimensional immunoelectrophoresis, agglutination of target cells, and complement fixation. Assays based on unitary pair formation rely on the "tagging" of antigens or antibodies to detect binding. Such tags include radioisotopes, enzymes, fluorochromes, and electron-dense materials. Quantitation requires comparison to a standard curve. Direct tag assays involve the antigen and the primary antibody, one of which is tagged. Indirect tag assays involve the antigen, the primary antibody, and a third component that is labeled and used to detect the antigen–antibody pair. Such third components may be Proteins A and G, secondary antibodies, or the biotin–avidin system. Frequently used antibody tag assays are radioimmunoassay (RIA) and enzyme-linked immunosorbent assay (ELISA). Immunohistochemical assays can detect antigens in intact tissues or cells using radioisotopes, enzymes, or electron-dense materials as tags. Immunofluorescence uses fluorescent tags to visualize antigens in an intact tissue or cell under the microscope. Flow cytometry allows the examination and isolation of viable cell populations of interest that are identified using fluorochrome-tagged antibodies. Antibodies can be used to characterize antigens by immunoprecipitation, affinity chromatography, and Western blotting.

Selected Reading List

Barbas III C. F. (1995) Synthetic human antibodies. *Nature Medicine* **1**, 837–838.

Borresbaeck C. (2000) Antibodies in diagnostics—from immunoassays to protein chips. *Immunology Today* **21**, 379–382.

Bruggenmann M. and Neuberger M. S. (1996) Strategies for expressing human antibody. *Immunology Today* **17**, 391–397.

Butler J. (2000) Enzyme-linked immunosorbent assay. *Journal of Immunoassay* **21**, 165–209.

Chord T. (1987) "An Introduction to Radioimmunoassay and Related Techniques," 3rd edn. Elsevier, Amsterdam.

Goding J. (1996) "Monoclonal Antibodies: Principles and Practice," 3rd edn. Academic Press, London.

Golinelli-Pimpaneau B. (2000) Novel reactions catalysed by antibodies. *Current Opinion in Structural Biology* **10**, 697–708.

Hage D. (1999) Immunoassays. *Analytical Chemistry* **71**, 294R–340R.

Hay F. and Westwood O. (2002) "Practical Immunology," 3rd edn. Blackwell Scientific, Oxford.

Hoogenboom H. R. (1997) Designing and optimizing library selection strategies for generating high-affinity antibodies. *Trends in Biotechnology* **15**, 62–70.

Kumar V. (2000) Immunofluorescence and enzyme immunomicroscopy methods. *Journal of Immunoassay* **21**, 235–253.

Kuntz R. and Saltzman W. (1997) Polymeric controlled delivery for immunization. *Trends in Biotechnology* **15**, 364–369.

Laurino J., Shi Q. and Ge J. (1999) Monoclonal antibodies, antigens and molecular diagnostics: a practical overview. *Annals of Clinical and Laboratory Science* **29**, 158–166.

Ma J. K.-C. and Hein M. (1995) Immunotherapeutic potential of antibodies produced in plants. *Trends in Biotechnology* **13**, 522–527.

Maeda M. (2003) New label enzymes for bioluminescent enzyme immunoassay. *Journal of Pharmaceutical and Biomedical Analysis* **30**, 1725–1734.

Mason H. S. and Arntzen C. J. (1995) Transgenic plants as vaccine production systems. *Trends in Biotechnology* **13**, 388–392.

Moffat A. S. (1995) Exploring transgenic plants as a new vaccine source. *Science* **268**, 658–660.

Plested J., Coull P. and Gidney M. (2003) ELISA. *Methods in Molecular Medicine* **71**, 243–261.

Samuelsson G. (1999) What's happening? Protein A columns: current concepts and recent advances. *Transfusion Science* **21**, 215–217.

Scanziani E. (1998) Immunohistochemical staining of fixed tissues. *Methods in Molecular Biology* **104**, 133–140.

Schultz P. and Lerner R. (1995) From molecular diversity to catalysis: lessons from the immune system. *Science* **269**, 1835–1842.

Self C. and Cook D. (1996) Advances in immunoassay technology. *Current Opinion in Biotechnology* **7**, 60–65.

Shaw C. and Zheng J. (1998) Western immunoblot analysis. *Methods in Molecular Biology* **105**, 295–306.

Shiuan D., Wu C., Chang Y. and Chang R. (1997) Competitive enzyme-linked immunosorbent assay for biotin. *Methods in Enzymology* **279**, 321–326.

Vitetta E., Thorpe P. and Uhr J. (1993) Immunotoxins: magic bullets or misguided missiles? *Immunology Today* **14**, 252–259.

Winter G. and Harris W. J. (1993) Humanized antibodies. *Immunology Today* **14**, 243–246.

Winter G., Griffiths A. D., Hawkins R. E. and Hoogenboom H. R. (1994) Making antibodies by phage display technology. *Annual Review of Immunology* **12**, 433–455.

The Immunoglobulin Genes

Susumu Tonegawa 1939–

Immunoglobulin gene rearrangement
Nobel Prize 1987

© Nobelstiftelsen

CHAPTER 8

A.   HISTORICAL NOTES

B.   CHROMOSOMAL ORGANIZATION OF Ig GENES

C.   Ig GENE REARRANGEMENT

D.   MOLECULAR MECHANISMS OF Ig GENE REARRANGEMENT

E.   ANTIBODY DIVERSITY GENERATED BY GENE REARRANGEMENT

F.   CONTROL SEQUENCES IN THE Ig LOCI

"Great acts are made up of small deeds"–Lao Tzu

In Chapter 5, we described the immunoglobulin proteins in detail, and in Chapters 6 and 7, we discussed how these remarkable proteins interact with antigen, both as receptors and as antibodies, and how this bond can be exploited in the laboratory. We have made reference in Chapters 2 and 4 to the almost infinite diversity of antibodies produced to protect us from the universe of potentially hostile antigens. We have hinted at the conundrum of how this huge number of antibody proteins could be derived from the amount of DNA that can physically fit inside the nucleus of a B cell. In this chapter, we examine the genes from which antibody proteins are derived. We will also commence our examination of how antibody diversity is generated, beginning with the fascinating process of *somatic recombination*, a key aspect of the science of *immunogenetics*.

A. Historical Notes

In the first half of the 20th century, before the structures of either the Ig protein or its genes were known and B cells had been identified, scientists had observed some fundamental characteristics of the humoral immune response. First, it was apparent that the body could produce a huge number of different antibodies; secondly, that a humoral response could be mounted to almost any antigen; and thirdly, that only plasma cells were capable of producing secreted immunoglobulins. The total collection of antibody specificities that could be produced in an individual's lifetime came to be called that individual's *antibody repertoire*, and the fact that this repertoire apparently included specificities for essentially all antigens made extreme "diversity" its hallmark. What unique mechanism allowed the plasma cell population to produce a plethora of different antibody proteins, each capable of binding to what appeared to be a single specific antigen?

Before 1965, three types of theories were propounded by scientists to answer this question: the *instructional* theories, the *germline variation* theories, and the *somatic mutation* theories. The *instructional theories* (reviewed in Box 5-1) generally proposed that antibodies were formed by the Ig protein "wrapping around" an antigen, generating a shape complementary to the antigen. However, these theories were disproved when it was shown that specific Ig sequences already existed on the surfaces of Ig-producing cells <u>prior</u> to antigen exposure, and that clonal selection by antigen was responsible for determining which cell was activated to produce and secrete antibody. In other words, antigen played no role in shaping the specificity of these pre-existing antibodies.

The *germline variation* camp pointed out that since these pre-existing antibodies were proteins, the antibody repertoire had to be encoded in the DNA of the Ig-producing cells; that is, had to be present in the germline of the individual. They felt this observation was most easily explained by a model in which the DNA needed to encode every single antibody that would ever be needed pre-existed in every single Ig-producing cell (since every cell of an individual is derived from the original zygote and must carry exactly the same DNA sequences in its genome). This theory was plausible in the days when it was thought that antibodies were directed against microbial antigens only. However, as it was shown over and over that antibodies could be produced against any antigen, not just microbes, the estimates of repertoire size grew to a point where the germline theorists had to account for a possible 10^9–10^{11} different antibody sequences. This number implied that an estimated 3 billion base pairs (or 1 billion codons, even excluding introns) were needed to code for the Ig genes alone, a most unlikely situation since this amount of DNA would have been many times more than that known to be possible in a single cell. Since the nucleus of an Ig-producing cell simply did not have sufficient volume to contain both vast numbers of Ig genes and the amount of DNA required to code for all the other non-Ig genes involved in cell metabolism and function, the germline variation theory began to look less and less tenable.

The *somatic mutation* theorists accounted for the huge size of the antibody repertoire another way. They proposed that a special mechanism of "somatic mutation" existed in Ig-producing cells such that a small number of germline Ig genes could readily undergo mutation to generate antibody diversity, but only in somatic cells; that is, there was no alteration of the Ig genes in the germline of sex cells passed on to the next generation. The somatic mutation theories were gaining ascendance when confounding information about antibody structure emerged. Examination of the Ig light and heavy chains had shown that antibodies were composed of N-terminal variable (V) regions containing highly diverse sequences (which accounted for antigenic specificity) attached to C-terminal constant (C) regions which were of very limited variability indeed. If antigenic specificity was determined by somatic mutation, what cellular mechanism allowed the V region to change, but kept the C region constant?

In 1965, before the exon/intron structure of mammalian genomic genes (described in Box 8-1) was known, William Dreyer and J. Claude Bennett proposed a radical idea for the time—namely, that the Ig protein was in fact derived from two genes, one for the V region, and one for the C region, and that these somehow came together to generate one mRNA. To account for the pre-existence of an almost infinite number of antibody specificities in a limited germline capac-

ity, Dreyer and Bennett hypothesized that multiple V region genes were present in the germline but that only single copies of C region genes were present. Furthermore, they hypothesized that the C genes were physically separated from the V region genes, so that the C genes would escape any mechanism (such as somatic mutation) that increased diversity in the V genes. This model combined the best ideas of both the germline variation and the somatic mutation camps, and adequately explained many of the observations of Ig structure and effector function of the day. However, the Dreyer–Bennett model also went squarely against two accepted dogmas, one that said "one gene, one mRNA, one protein," and another that said "all cells in an organism contain exactly the same DNA sequence." Because initially only indirect experimental proof of this model could be obtained, there was significant resistance to the acceptance of Dreyer and Bennett's proposal by the scientific community. It was not until after the development of techniques of molecular biology in the 1970s that much of what Dreyer and Bennett proposed would be proved to be essentially correct.

In 1976, Nobumichi Hozumi, Susumu Tonegawa, and their colleagues at the Basel Institute in Switzerland finally provided the direct proof that the genome in fact does not contain any intact Ig genes at all. These workers compared Ig sequences in DNA preparations from whole embryos (representing the

---

Box 8-1. Exons, introns, and transcription

A prokaryote's genome contains "simple" genes (see Figure, panel A) whose sequences correspond continuously and linearly to the amino acid sequence of the proteins they produce; all the information in the nucleotide sequence is "coding." Transcription of a prokaryotic gene gives rise to an RNA transcript that can be translated with minimal further processing. As transcription proceeds, a 7-methylguanosine "cap" required for translation by the cytoplasmic polyribosomes is added to the 5′ end of the transcript. In addition, the transcript of any translatable gene contains at least one highly conserved *polyadenylation signal* (AAUAAA). Once transcription has been completed, the transcript is cleaved 15–30 nucleotides downstream of this signal and the polyA polymerase enzyme catabolizes ATP and adds 200–300 adenosine residues to the 3′ end of the transcript. The polyadenylated prokaryotic transcript is now in a form called *messenger RNA* (mRNA), which can be translated directly into protein without further processing.

In contrast, most genes in a eukaryotic genome are not continuous but instead are broken up into *exons* and *introns* (see Figure, panel B). The amino acid coding

information is distributed among the *exons*, which are separated from one another by intervening sequences of non-coding DNA called *introns*. While introns do not contain amino acid coding information, they may contain regulatory sequences necessary for the control of gene expression. Very often a single exon will correspond to a discrete domain in a protein, and some scientists speculate that the presence of exons and introns in eukaryotic genes facilitates the duplication or "swapping" of protein domains, resulting in the creation of new members of protein superfamilies.

A single exon on its own is not transcribable and cannot be translated. Instead, when eukaryotic genes are transcribed, a long RNA molecule called a *primary transcript* is made which includes both the exon and intron sequences. This primary RNA is capped and polyadenylated but before it can leave the nucleus to be translated by the polyribosomes in the cytoplasm, it must undergo a further modification. The introns must be "spliced out" of the RNA structure, joining the exons together in a continuous linear coding sequence that is the mRNA. RNA splicing is carried out by a multi-component complex

called a *spliceosome* and is controlled by *small nuclear RNAs (snRNAs)* in the nucleus, a process which is fascinating but, unfortunately, beyond the scope of this book. Suffice it to say that the splicing machinery in the cell recognizes specific short nucleotide sequences called *splice donor* and *splice acceptor* sites positioned on the 5′ and 3′ ends of introns, respectively. The splicing enzyme brings the transcribed splice donor site that occurs just after the 3′ end of one exon to the splice acceptor site just before the 5′ end of the next exon, looping out and excising the intron between them, and joining the exons together at the *splice junctions* to form a continuous mRNA. The mRNA is now able to move out of the nucleus into the cytoplasm, where it is translated into the nascent protein by the polyribosomes.

In some eukaryotic genes, a splice donor site can pair with more than one splice acceptor site. Instead of choosing the splice acceptor site immediately 3′ of a given intron, the splicing machinery sometimes chooses (in a mechanism still not completely understood) to process the primary transcript such that a splice acceptor site 3′ to a different intron further downstream is used. This has the effect of "skipping" any exons between the

Continued

Box 8-1. Exons, introns, and transcription—*cont'd*

two introns, and deletes them from the mRNA as well as the introns. This process of *alternative splicing* creates an mRNA whose sequence is different at its 3' end, and which thus encodes a protein with a different C-terminal amino acid sequence (see Figure, panel C). The transcription of this gene produces thousands of these primary transcripts. Each of these transcripts may be spliced in either of the two alternatives, so that expression of both proteins simultaneously in one cell is possible.

Proteins with different C-terminal ends can also be encoded by one gene through another mechanism called *differential RNA processing* (described in more detail in Ch.9). Eukaryotic genes often contain multiple polyadenylation sites, positioned after each exon encoding a potential C-terminal end of the protein. Depending on which site is chosen for cleavage and polyadenylation, primary RNA transcripts of different lengths (and thus including different subsets of exons and introns) will be generated (see Figure, panel D). These differentially polyadenylated primary transcripts then undergo standard intronic splicing to generate translatable mRNAs. The resulting population of two mRNAs is translated to produce two subsets of proteins with alternative C-terminal amino acid sequences.

As an example of differential RNA processing, consider the generation of the membrane-bound and secreted forms of the immunoglobulin molecules produced by activated B cells, discussed in detail in Chapter 9. Primary transcripts containing both TM and tailpiece sequences (refer to Ch.5) can be cleaved and polyadenylated after either of these components. Those transcripts cleaved after the TM domain sequence undergo splicing such that the tailpiece is lost; these transcripts specify an mIg. In transcripts that are cleaved and polyadenylated after the tailpiece sequence, the TM sequences are subsequently removed by splicing, leading to an mRNA translated into an sIg.

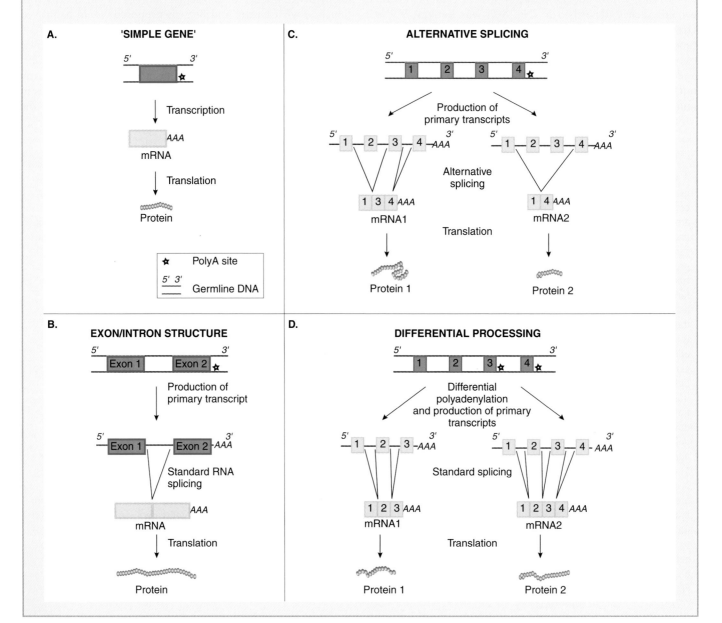

germline sequence maintained in non-B cells) and myeloma cells (representing B lineage cells). As is outlined in more detail in Box 8-2, it was found that the Ig V and C regions were clearly separated in the germline DNA, but had changed position in the myeloma DNA to become closely linked. Subsequent work by these and other researchers showed that the Ig V region is encoded by a V exon that is intricately assembled from small DNA fragments called *gene segments*, and that the gene segments comprising the V exon are physically brought together in the genome of each B cell precursor as it differentiates. Furthermore, the V gene segments are randomly brought together in each B cell precursor by a mechanism called *somatic recombination, somatic gene rearrangement,* or *V(D)J recombination* (see later), so that a huge variety of unique variable exons is generated within the B cell precursor population. The V exon sequences are then spliced to the C exon sequences in the primary transcript and a unique mIg is expressed. We emphasize here that somatic recombination results in a permanent change in the DNA of a given B cell and its clonal progeny. Indeed, without this rearrangement at the DNA level, there would never be any Ig transcripts: the various gene segments and exons making up the gene are too far apart in the germline to permit transcription.

The germline variation and somatic mutation camps started out at the opposite ends of the theoretical spectrum, but, perhaps not surprisingly, neither theory turned out to be entirely right or wrong. The germline theory was partly right, in that multiple V gene segments exist in the germline. The somatic mutation theory was also partly right, in that a mechanism known as "somatic hypermutation" introduces further mutations into the V exon of a mature B cell after it has undergone activation by antigen (see Ch.9), greatly increasing the diversity of the antibody repertoire. What no one had anticipated was the process of somatic recombination. We now know that the enzymatic mechanisms that underlie somatic recombination are only functional in developing lymphocytes and do not affect the DNA of non-lymphocytes. Furthermore, the same enzymatic machinery used in B cell precursors to rearrange the genes coding for the Ig receptor is also used in T cell precursors to rearrange the genes coding for the T cell receptor (TCR) (see Ch.12). The molecular mechanism of somatic recombination is discussed in detail later in this chapter.

B. Chromosomal Organization of Ig Genes

I. THE Ig LOCI

The Ig proteins of most mammalian species (including mice and humans) are encoded by nucleotide sequences residing in three large genetic loci: *Igh* (heavy chain locus), *Igk* (kappa

---

Box 8-2. Discovery of Ig gene rearrangement

In 1965, William Dreyer and J. Claude Bennett first hypothesized that the immunoglobulin protein was encoded in the genome in separate variable and constant segments that joined together in highly variable ways. At the time, this explanation of antibody diversity was met with much skepticism. Experimentally, the tools were not yet available for anyone to put this radical notion to the test. Genes were not yet being studied in isolation because it was not possible to cut DNA at specific points that would allow one to obtain a reproducible sample containing the gene of interest. However, over the next 10 years, molecular biologists discovered a scientific gold mine—the *restriction endonucleases*, sometimes called *restriction enzymes*. These bacterial enzymes cut double-stranded DNA at short (usually 4–8 bp long) specific recognition sequences called *restriction sites*. Each enzyme recognizes a different restriction site, so that digestion of DNA with any one of these enzymes results in the cutting of the DNA into a specific and reproducible pattern of fragments of varying sizes. Using agarose electrophoresis, the collection of fragments generated by digestion with a particular enzyme can be separated by

size to form a ladder of bands. The gel can then either be sliced into individual pieces for analysis (considered a historical method today), or the ladder can be visualized by Southern blotting of the agarose gel (described in Box 8-3).

Technological developments in DNA handling using restriction endonucleases opened the door for Nobumichi Hozumi and other members of Susumu Tonegawa's group to set up an experiment that would test the Dreyer–Bennett hypothesis. The goal was to determine if the gene encoding the Ig light chain in the mouse was in the same configuration in the germline as it was in a cell producing light chains. To do this, light chain mRNA was purified from a myeloma cell line and used to make two different preparations. One was the entire light chain mRNA containing both the variable and constant region coding sequences. The other mRNA preparation, which was generated by cutting the full-length mRNA, consisted mainly of the constant region coding sequences. After radiolabeling these mRNA preparations, they used them as (V + C) and (C-only) *probes* that would hybridize in a

detectable way to corresponding sequences in DNA samples. The DNA samples to be tested were prepared by using a particular restriction endonuclease to digest myeloma and embryonic DNA. In each case, the resulting DNA fragments were separated by agarose gel electrophoresis and gel slices were excised to obtain various fractions containing DNA of different sizes. The DNA fractions were then hybridized to the radiolabeled mRNA probes to determine which fractions contained DNA encoding V regions, and which fractions contained DNA encoding C regions. The crucial finding was that, in the embryonic DNA, the (V + C) probe and the (C-only) probes hybridized to two separate DNA fragments, whereas in the myeloma DNA, both the (V + C) and (C-only) probes hybridized to a single DNA fragment. This work provided convincing evidence that the DNA encoding the variable regions and constant regions of antibodies were separated in the germline, but brought into close proximity in an antibody-producing cell. Such was the importance of this work that Tonegawa was awarded the Nobel Prize for Physiology or Medicine in 1987.

light chain locus), and *Igl* (lambda light chain locus). Each locus is located on a different chromosome (Table 8-1) but exhibits remarkably similar organization. The mRNA for any immunoglobulin polypeptide chain is transcribed from a complex gene made up of a variable exon encoding the variable region of the protein, and a constant exon encoding its constant region. So far, this sounds like fairly standard eukaryotic gene structure. However, as introduced in the preceding section, the Ig genes are unusual in that the variable exon is assembled from smaller nucleotide sequences called *gene segments*. The *Igk* and *Igl* loci contain two types of gene segments called the *V (variable)* and *J (joining)* segments, while the *Igh* locus contains not only V and J segments but also *D (diversity)* gene segments. Each type of gene segment contributes amino acid coding information for a particular set of amino acids in the variable region of an Ig protein. The gene segments of a given locus are randomly recombined by V(D)J recombination in individual developing B cells, to collectively give rise to the large, diverse repertoire of Ig proteins. Although gene segments are separated from each other on the chromosome by non-coding intervening sequences, gene segments are not exons (refer to Box 8-1). They are not bounded by the splice donor and acceptor sites recognized by the cell's RNA splicing machinery and are therefore not joined by splicing of RNA transcribed from unrearranged (germline) DNA. Instead the genomic DNA itself is rearranged by specialized recombinase enzymes unique to lymphocytes (see later), so that the gene segments are physically and permanently juxtaposed at the DNA level to generate a complete variable exon.

In contrast to the variable exon, the constant exons are not formed by the rearrangement of smaller gene segments and are true exons separated by introns. Thus, the constant region sequences are joined to the variable exon at the RNA level by conventional splicing to produce a translatable Ig mRNA. The constant regions of light chain polypeptides are encoded by intact exons that are not subdivided into smaller components. However, the constant region of the longer heavy chain polypeptide is derived from one of the several C_H exons in the *Igh* locus. Each C_H exon is subdivided into several smaller *sub-exons*. The constant region of an Ig heavy chain is encoded by a collection of three to four sub-exons, with additional sub-exons specifying the transmembrane and cytoplasmic regions. The sub-exons are true exons (and not gene segments) because they contain splice donor and acceptor sites and are separated by proper introns that must be removed by conventional RNA splicing. Each C_H sub-exon encodes an Ig domain, and each

encodes about the same number of amino acids as is encoded by a single light chain constant exon (see later).

The nomenclature for the genetic elements of the Ig and TCR genes cited in many sources can be confusing and imprecise. "V gene" can refer either to a specific V gene segment, or to the entire recombined V(D)J exon. The V, D, and J sequences are actually gene segments, but are sometimes referred to as exons. Finally, the constant sequences are often referred to as genes, even though the complete Ig gene also includes the leader and recombined V(D)J exons. The unique complexity of the Ig and TCR loci can be bewildering if the structure and function of these genes and their component elements are not clearly understood.

II. GENERAL STRUCTURE OF Ig LOCI

With the preceding introduction, we can examine the overall structure of the Ig loci in more detail. Each Ig locus or complex is composed of distinct regions, which in turn contain distinct gene segments, or exons and sub-exons. The organization of the Ig complexes, which is remarkably similar for humans and mice, is shown in Figures 8-1 and 8-2, respectively. A light chain Ig complex contains the V gene segments plus their associated leader (L) sequences (see later), the J gene segments, and the C_L exons, while a heavy chain complex contains the V (plus L), D, and J gene segments, and the C_H exons. At the 5′ end of an Ig complex are the V gene segments, each of which is ~300 base pairs (bp) in length and separated from the next V gene segment by a variable stretch (usually about 5000 bp) of non-coding DNA. Immediately adjacent to each V gene segment on its 5′ side is a *leader sequence* of about 120–150 bp (bisected by an intron) that travels with the V gene segment during gene rearrangement. Just 5′ of each L sequence is a promoter sequence that also travels with its associated L and V gene sequences through DNA rearrrangement. From its 5′ end, the L sequence contains a sub-exon of about 45 bp followed by an intron of about 100 bp, and a second sub-exon of just 4 bp immediately 5′ of the start of the V segment. Each L sequence specifies a hydrophobic *leader* or *signal* peptide of about 15–18 amino acids that facilitates the transport of the Ig protein through the endoplasmic reticulum as it is synthesized on ribosomes in the cytoplasm. This leader peptide is clipped off before the Ig chain is assembled with its three counterparts into the basic H_2L_2 monomer (and possibly before translation is even finished), meaning that leader region amino acids do not appear in completed Ig molecules. In the heavy chain complex, the D gene segments (each 9–23 bp in length) are located 3′ to the V region and spread over about 50 kilobase pairs (kb). The J gene segments (each 30–40 bp) are generally located 3′ to V gene segments in *Igk* and *Igl*, and 3′ to the D segments in *Igh*. In the *Igh* and *Igk* loci, the J gene segments are organized in a block between the V gene segments and the C exons located close to the 3′ end of the complex. The *Igl* locus is different in that, rather than separate blocks of J segments and C exons, the Jλ segments tend to be paired with individual Cλ exons. The numbers of different gene segments

Table 8-1 Chromosomal Location of Ig Genes

Genetic Locus	Chromosome	
	Human	Mouse
Igl	22	16
Igk	2	6
Igh	14	12

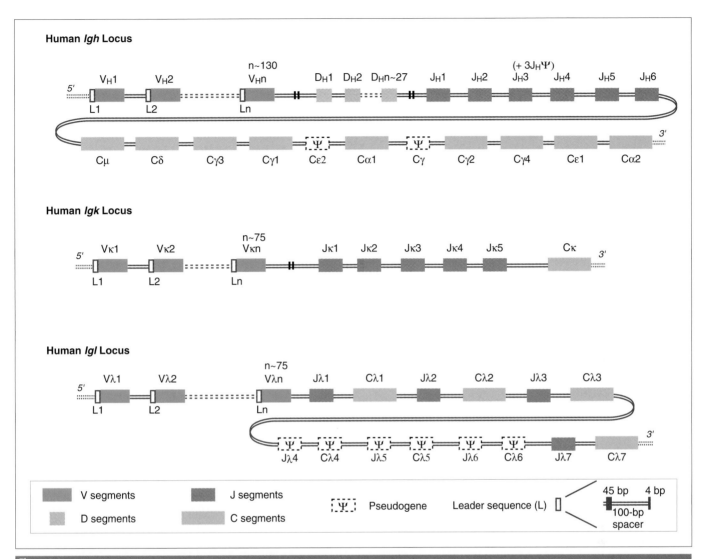

Figure 8-1

Genetic Organization of the Immunoglobulin Loci (Human)
Genomic organization schemes for the *Igh, Igk,* and *Igl* loci are shown for humans. Additional gene segments occur in areas represented by dashed lines, but are not shown. Long stretches of DNA that occur between groups of gene segments are indicated by two vertical strokes. Dotted lines at the 5′ and 3′ ends of the Ig loci indicate the continuation of the chromosome. Note that the sub-exon structure of the constant exons is not shown.

and exons identified in the human and murine Ig complexes are shown in Table 8-2.

Among the V gene segments, families have been identified in which member V gene segments share 70–80% homology of their nucleotide sequences; these families are thought to have arisen by extensive gene duplication of several "founding" V gene segments. In fact, the actual number of V, D, and J gene segments in the *Igh, Igk,* and *Igl* loci can vary slightly among individuals, resulting in genetic polymorphisms for the Ig genes. Some DNA sequences in an Ig complex closely resemble functional gene segments or exons but contain mutations such that they cannot be expressed; these are called *pseudogenes*, often written (for example) as Jψ or Cψ. Since each individual segment or exon can be separated from its neighbors by anywhere from 1.2 to 55 kb, an entire Ig gene complex may cover 1000–2000 kb of genomic DNA.

III. FINE STRUCTURE OF LIGHT CHAIN GENES

The murine *Igk* locus that encodes the Ig κ light chains is shown in Figure 8-2. Since there is only one C region exon, Cκ, the assembly of this polypeptide is the most straightforward. This single exon also explains why all κ light chains have the same amino acid sequence in the constant region of the polypeptide. About 250–300 Vκ gene segments (each with an accompanying L sequence and promoter) and 5 Jκ gene segments (one of which is a pseudogene) have been identified in the murine *Igk* locus. The human *Igk* locus is similar to that in the mouse, being composed of about 75 Vκ and 5 Jκ gene segments (all of which are functional), and one Cκ exon.

The *Igl* locus in both humans and mice has a more complex organization, since multiple Cλ exons are present, giving rise to the known isotype subclasses of λ Ig light chains. As

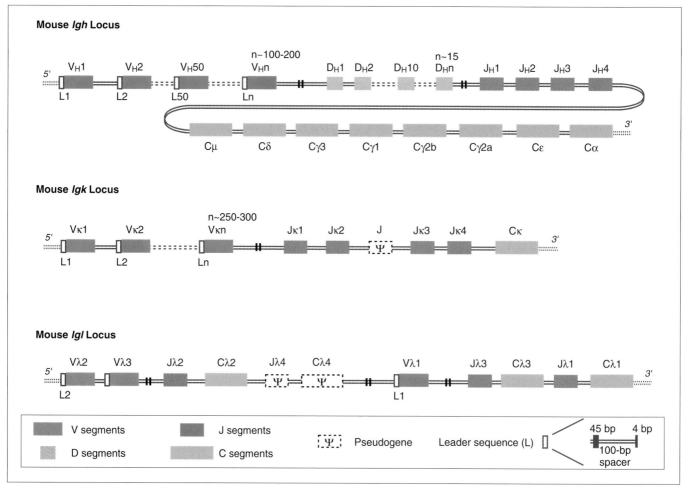

Figure 8-2

Genetic Organization of the Immunoglobulin Loci (Mouse)
Genomic organization schemes for the *Igh*, *Igk*, and *Igl* loci are shown for mice. Additional gene segments occur in areas represented by dashed lines, but are not shown. Long stretches of DNA that occur between groups of gene segments are indicated by two vertical strokes. Dotted lines at the 5′ and 3′ ends of the Ig loci indicate the continuation of the chromosome. Note that the sub-exon structure of the constant exons is not shown.

Table 8-2 Estimates of Numbers of Rearranging Gene Segments in Mouse and Human Ig Loci

Germline Gene Segments	Number of Gene Segments*		
	Heavy Chain	**Light Chain**	
MOUSE		**KAPPA**	**LAMBDA**
V	100–200	250–300	3
D	15	0	0
J	4	4(1)	3(1)
HUMAN			
V	130	75	75
D	27	0	0
J	6(3)	5	7(3)

*Numbers in parentheses are the numbers of these gene segments that are pseudogenes.

noted previously, the organization of this locus is also very different from that found in the *Igh* and *Igk* loci, in that the Jλ gene segments and Cλ exons occur as J–C units. In the case of the mouse, there are 3 Vλ and 4 Jλ gene segments, and 4 Cλ exons. The Jλ4 gene segment and the Cλ4 exon are unexpressed pseudogenes. In humans, there are approximately 75 Vλ gene segments, 7 Jλ gene segments (3 of which are pseudogenes), and 7 Cλ exons (3 of which are pseudogenes).

IV. FINE STRUCTURE OF HEAVY CHAIN GENES

The *Igh* locus in both humans and mice (shown in Figs. 8-1 and 8-2, respectively) is more complex than either the *Igl* or the *Igk* locus. Overall, the organization of the *Igh* resembles that of the *Igk*, in that its V, J, D, and C regions are organized in distinct blocks and individual segments or exons of different types are not intermixed. As previously noted, unlike the light chains, the rearranged variable exon of the heavy chain is composed of three gene segments. A sequential cutting and rejoining process first brings a D_H gene segment together with a J_H segment, and then the $J_H D_H$ entity with a V_H segment to complete the variable exon. The role of the D segment was uncovered somewhat later than that of the J segment. The fact that the amino acids encoded by the D_H segment fall within CDR3 of the Ig heavy chain variable domain gave rise to the designation "D" for "diversity" for this region of the *Igh* complex.

In the human *Igh*, there are about 130 V_H gene segments, 27 D_H segments, 9 J_H segments (3 are pseudogenes), and 11 C_H exons (2 are pseudogenes). Each C_H exon, which corresponds to a C_H isotype, is made up of clustered sub-exons (Fig. 8-3). For example, the $C\mu$ exon includes 4 coding sub-exons called $C\mu1$, $C\mu2$, $C\mu3$, and $C\mu4$, which correspond to the four domains of the IgM heavy chain constant region. In the mouse *Igh*, there are between 100 and 200 V_H gene segments, 15 D_H segments, 4 J_H segments, and 8 C_H exons (composed of sub-exons). In both species, immediately contiguous to the 3' end of the last coding sub-exon in each C_H exon, is a very small sub-exon called "Se" (for "secreted") which encodes the tailpiece, a stretch of hydrophilic amino acids specifying an sIg. A short intron separates the tailpiece exon from two additional sub-exons encoding the transmembrane region and the cytoplasmic region that specify an mIg. The M1 sub-exon encodes the transmembrane region (TM) of the Ig protein, while the M2 sub-exon encodes the cytoplasmic tail (CYT). To emphasize this key point: each C_H exon includes its own Se, M1, and M2 sub-exons, allowing the synthesis of transmembrane and secreted versions of each heavy chain isotype.

Interestingly, the C_H exons are organized consecutively along the chromosome more or less in the order of the appearance of Igs during an immune response. The most 5' of the human C_H exons is the $C\mu$ exon, followed by $C\delta$, $C\gamma3$, $C\gamma1$, $C\epsilon2$, $C\alpha1$, $C\gamma$, $C\gamma2$, $C\gamma4$, $C\epsilon1$, and $C\alpha2$. Similarly, the order of murine C_H exons is $C\mu$, $C\delta$, $C\gamma3$, $C\gamma1$, $C\gamma2b$, $C\gamma2a$, $C\epsilon$, and $C\alpha$. Note that the $C\mu$ and $C\delta$ exons are first in order, corresponding to the production of IgM and IgD first in any immune response, while $C\alpha$ is the most 3' C_H exon, corresponding to the (often) much later production of IgA during subsequent responses. During the isotype switching that occurs in individual B cells late in the primary response, the variable exon ($V_H D_H J_H$) becomes associated with any one of the C_H exons located 3' to $C\delta$ by a mechanism that involves the looping out and deletion of intervening DNA (see Ch.9). As a result of this process, the antibody produced has the same antigenic specificity as the original IgM antibody (same $V_H D_H J_H$ exon) but is of a different isotype (different C_H exon). Each B cell can switch to only one isotype at one particular time, but within an expanded clone of a given specificity, many different isotypes will be produced.

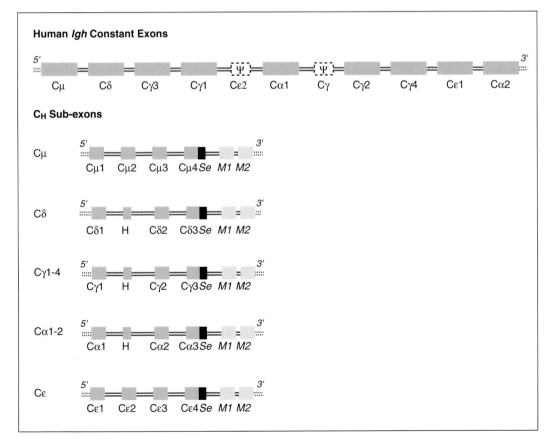

Human *Igh* Constant Exons

$C\mu$ $C\delta$ $C\gamma3$ $C\gamma1$ $C\epsilon2$ $C\alpha1$ $C\gamma$ $C\gamma2$ $C\gamma4$ $C\epsilon1$ $C\alpha2$

C_H Sub-exons

$C\mu$ — $C\mu1$ $C\mu2$ $C\mu3$ $C\mu4$ *Se* *M1* *M2*

$C\delta$ — $C\delta1$ H $C\delta2$ $C\delta3$ *Se* *M1* *M2*

$C\gamma1-4$ — $C\gamma1$ H $C\gamma2$ $C\gamma3$ *Se* *M1* *M2*

$C\alpha1-2$ — $C\alpha1$ H $C\alpha2$ $C\alpha3$ *Se* *M1* *M2*

$C\epsilon$ — $C\epsilon1$ $C\epsilon2$ $C\epsilon3$ $C\epsilon4$ *Se* *M1* *M2*

Figure 8-3

Sub-exons in the Human *Igh* Constant Exons
The human *Igh* constant exons with the structures of the C_H sub-exons shown below. The sub-exon organization of mouse *Igh* constant exons is similar. Each of the numbered C sub-exons and H (hinge) sub-exons encodes a corresponding extracellular domain in the constant region of the final Ig heavy chain protein (see Fig. 5-3 for the domain structure of human isotypes). The Se sub-exon specifies a domain allowing secretion of the immunoglobulin from the B cell, while heavy chain transmembrane and cytoplasmic domains are encoded by the M1 and M2 sub-exons, respectively.

C. Ig Gene Rearrangement

I. THE ROLE OF V(D)J RECOMBINATION

Although every cell in the body contains the DNA of the Ig loci, Ig proteins are produced only by B lineage cells. If one compares the arrangement of the Ig genes in non-lymphoid somatic cells or even pluripotent stem cells with that in mature antibody-producing B cells, one finds marked differences in the content and order of Ig DNA sequences in the V regions. (Historical and contemporary techniques used to identify gene rearrangement are described in Box 8-3.) The fact is that Ig proteins cannot be produced from the string of multiple V, D, and J gene segments that exists in germline Ig DNA. Only precursor B lymphocytes in the bone marrow are capable of rearranging their Ig germline DNA to generate variable exons using V(D)J recombination. Non-lymphocytes never have the capacity for V(D)J recombination, and mature B cells lose it. The gene rearrangement that both generates the variable exon and brings the V and C exons close enough together to permit transcription occurs at the DNA level <u>prior to the transcription</u> of the complete Ig gene that results in an Ig polypeptide. Special *recombinase* enzymes (see later) combine selected V, D, and J gene segments in an enzymatic cutting and rejoining of the DNA strand that excludes the non-selected V, D, and J gene segments and other intervening sequences, and brings the VJ or VDJ exon close to the C exons. This juxtaposition ultimately makes transcription of the complete Ig gene possible. V(D)J recombination is strictly controlled and its DNA excision and rejoining events occur in a precise sequence. This control even reaches to which locus is rearranged first, which in turn influences the maturational steps a B cell precursor follows. Most often, the *Igh* locus is rearranged first, followed by the *Igk* locus, and lastly, the *Igl* locus.

II. VDJ JOINING IN THE *Igh* Locus

Figure 8-4 illustrates the sequence of events that results in the generation of the heavy chain variable exon. A precursor B cell in the bone marrow first rearranges its germline DNA so that a randomly chosen D_H gene segment is joined by the recombinase to a randomly chosen J_H gene segment. The genomic sequences between the selected segments are "looped out" and deleted, but the J_H gene segments 3′ of the selected J_H, and the D_H gene segments 5′ of the selected D_H, are not affected by this process in the majority of cases. Another round of V(D)J recombination joins a randomly selected V_H segment to the $D_H J_H$ joined segment, creating an intact $V_H D_H J_H$ exon and deleting all D_H segments 5′ of the DJ segment, and all V_H plus leader segments 3′ of the selected V_H segment. The V_H, D_H, and J_H segments are covalently and immutably fixed side by side on the rearranged chromosome, forming a single exon. Thus, after two rounds of V(D)J recombination, we are left with a gene composed of the following components (from the 5′ end): a short leader sub-exon, a leader intron, the 3′ leader sub-exon, a rearranged $V_H D_H J_H$ variable exon, an intron containing unrearranged J_H segments, and a series of C_H exons (each composed of sub-exons including the M1 and M2 sub-exons) separated by introns. Unless the most 5′ V_H segment is selected for inclusion, unrearranged V_H segments may still be in place 5′ of the leader of the rearranged VDJ exon. The *Igh* gene is now ready to be transcribed using the promoter of the rearranged V_H gene segment.

At this point in the life of a maturing B cell, *Igh* primary transcripts include at least the $C\mu$ and $C\delta$ sequences (and possibly the other C_H sequences as well; see Ch.9) The mRNA for a μ heavy chain is derived by conventional alternative splicing of the rearranged $V_H D_H J_H$ sequences to the $C\mu$ sequence. Similarly, the mRNA for a δ heavy chain is derived

---

Box 8-3. Detection of Ig rearrangement: from Southern blotting to PCR

During variable exon rearrangement, the gene segments encoding the Ig genes in B cells recombine in a process that permanently alters the genomic locus architecture. The detection of rearrangement depends on comparing the structure of the genomic Ig region before and after V(D)J recombination has occurred. Since the 1970s, techniques to locate or detect a gene of interest have been developed that can be used to visualize these changes to genomic Ig sequences. As we have seen, DNA can be digested by restriction enzymes to generate DNA fragments whose pattern of sizes constitutes a fingerprint for a particular DNA sequence. In the experiments of Hozumi and other members of Tonegawa's lab in the 1970s (refer to Box 8-2), physical slicing of the agarose gel used for the fractionation of restriction enzyme digests was used to determine the size of the DNA

fragments of interest. However, this was a tedious and time-consuming method. Those routinely using restriction enzyme digestions for DNA analysis rejoiced when Ed Southern invented a technique eliminating the need for gel slicing, which came to be called *Southern blotting*.

In Southern blotting, instead of slicing the agarose gel containing the separated DNA fragments of interest, the gel is treated to denature the DNA *in situ* (separate it into its single strands). The treated gel is then covered with a nitrocellulose membrane. By capillary action (sometimes assisted with an electrical current), the denatured DNA fragments transfer to the nitrocellulose membrane in the pattern in which they appeared in the gel ("blotting"). To visualize the DNA fragments transferred to the membrane, *nucleic acid hybridization* is used.

In this technique, a short piece of RNA or DNA that is complementary to the DNA sequence of interest is selected to be the *probe*. The probe is radioactively labeled, denatured by exposure to temperatures just under 100 °C, and incubated with the nitrocellulose membrane containing the denatured DNA fragments of interest. As the temperature of incubation is decreased, double-stranded DNA can start to reform, and the probe uses the rules of DNA complementarity to bind, or *anneal*, to single-stranded DNA fragments on the membrane whose sequences are complementary. Autoradiography of the hybridized membrane shows a characteristic band or bands corresponding to the restriction fragments containing the gene of interest (or parts of it) identified by the probe. The higher the salt concentration in the buffer used for the

Continued

annealing step, the more *stringent* the hybridization conditions, meaning that the probe sequence must match the sequence of a band almost completely in order to stick to it and reveal its presence after autoradiography. If the salt concentration is decreased, the probe will begin to bind to DNA sequences of less than a perfect match. Hybridization under conditions of low stringency is used to detect the presence of genes with DNA sequences similar to the gene of interest, such as homologous genes of the same family in other species.

Variations on the Southern blotting technique have since become common. In a debatable attempt at humor, these techniques have been named Northern and Western blotting. Western blotting (the transfer of denatured proteins to a nitrocellulose membrane) was described in Chapter 7. In Northern blotting, cellular RNA rather than DNA is transferred to the blot. Using the Northern technique, the steady-state level of expression of a gene in a cell, rather than just its presence in the DNA, can be determined.

However, as wonderful as Southern blotting is, immunologists in the early 1980s were faced with the difficulty that the lymphocytes they wanted to analyze were few in number and thus supplied very limited amounts of DNA. This problem was solved in 1983 by the development of a technique called the *polymerase chain reaction*, or *PCR*. Kary Mullis, a scientist at Cetus Corporation in California, USA, devised a technology capable of amplifying a single starting DNA molecule into more than a billion near-identical copies in just a few hours. The beauty of PCR is that absolutely any piece of DNA can be amplified as long as one knows the sequences of short stretches of DNA flanking the gene fragment of interest. Oligonucleotides complementary to these short flanking sequences are first assembled to provide initiation points for the synthesis of the copied DNA strands. A sample of the DNA of interest is heated to denature the double-stranded structure into two single strands, which are then combined with an excess of these oligonucleotides. Once the oligonucleotides bind to their complementary sequences on the template, the template DNA is "primed" and ready to be copied. Incubation of the primed template with the four deoxynucleotide precursors (A, C, G, T) and a heat-resistant DNA polymerase called *Taq* results in the

extension of the primers one nucleotide at a time, all the way toward the end of each DNA template. By repeating the melting-annealing procedure for multiple "cycles," amplification of the "target" DNA, that small stretch of DNA of interest between the primers, is readily achieved. Current preparations of Taq polymerase preparations allow the production of PCR products of 20 kb.

The Table shows the exponential increase in DNA target copies produced during a 32-cycle PCR. Copies containing this region only (as opposed to longer copies incorporating parts of the original template DNA outside the target area) start to be made after the third cycle of PCR. It takes only 32 cycles (a total running time of about 3 hours) for over a billion copies of the target DNA to be made.

By combining PCR with nucleotide sequencing, all types of molecular genetic analyses have been revolutionized. Immunologists, like other scientists, have wasted no time in capitalizing on PCR to study the genes that encode important immune system proteins such as the Ig, TCR, and MHC molecules. Ig rearrangement is now more easily detected, since if the gene segment of interest is not rearranged in a certain cell, the primers designed to amplify the target area of interest will be too far apart and the PCR reaction will fail. However, if rearrangement has occurred, the primers will bind at sites much closer together, permitting PCR to proceed. This technique can have diagnostic applications. For example, in certain follicular lymphomas, part of the immunoglobulin J_H locus on chromosome 14 is translocated to the locus for the Bcl-2 oncogene on chromosome 18. This translocation can be detected with great precision (in one cancer cell in 10^6) using PCR, thus allowing for either primary identification of the cancer or post-treatment monitoring.

Similar applications of PCR have made great strides possible in other areas as well. New MHC alleles have been amplified and sequenced without ever isolating the original MHC gene from genomic DNA. Use of this methodology is therefore streamlining the process of tissue typing prior to transplantation. Scientists have also taken full advantage of the fact that DNA (in the absence of nucleases) is stable for remarkably long periods. A wealth of otherwise inaccessible information has been gathered by performing PCR on cervical

carcinoma biopsies embedded in paraffin for over 40 years, on 4000-year-old Egyptian mummies, and even on 18-million-year-old plant fragments preserved in shale fossil beds. Forensic analyses have been revolutionized by scientists' ability to obtain useful information from miniscule amounts of DNA left, sometimes for extended periods, at a crime scene. Based on the revolutionary contribution of the PCR technique to the analysis of DNA in a broad range of scientific disciplines, Kary Mullis was awarded the Nobel Prize in Physiology or Medicine in 1993.

Power of PCR Amplification	
Cycle Number	**Number of Duplex DNA Targets (copies of the original DNA)**
1	0
2	0
3	2
4	4
5	8
6	16
7	32
8	64
9	128
10	256
11	512
12	1024
13	2048
14	4096
15	8192
16	16,384
17	32,768
18	65,536
19	131,072
20	262,144
21	524,299
22	1,048,476
23	2,097,152
24	4,194,304
25	8,388,608
26	16,777,216
27	33,544,432
28	67,108,864
29	134,217,728
30	268,435,456
31	536,870,912
32	1,073,741,824

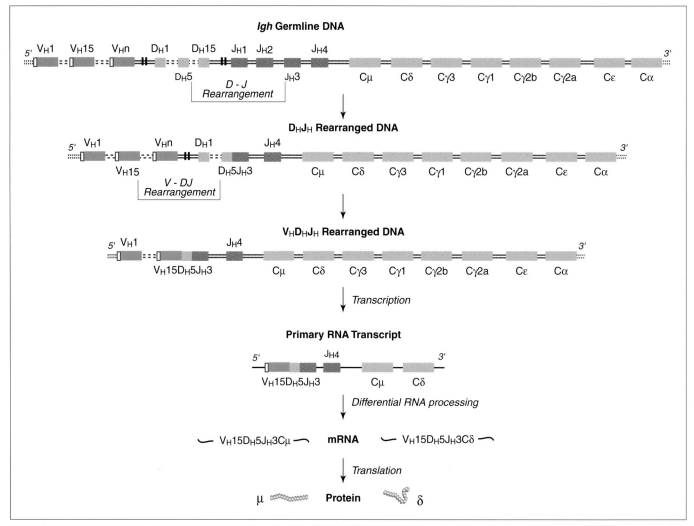

V-D-J Joining in the Mouse *Igh* Locus
During heavy chain V(D)J recombination, D to J joining occurs first, followed by V to DJ joining. In the hypothetical example shown, D_H5 has randomly joined to J_H3, and V_H15 has randomly joined to D_H5J_H3. After V(D)J recombination is completed, transcription of the Ig gene can proceed and primary RNA transcripts are produced. (For reasons made clear in Chapter 9, only the $C\mu$ and $C\delta$ exons are included in the primary transcripts of a maturing naive B cell.) The primary transcripts undergo differential RNA processing and RNA splicing such that the resulting messenger RNAs encode the variable regions linked to either the $C\mu$ or $C\delta$ constant region. Translation of these mRNAs then produces a mixture of both μ and δ heavy chain proteins. Despite their isotypic differences, these heavy chains have identical antigen binding sequences.

by conventional alternative splicing of the rearranged $V_HD_HJ_H$ sequences to the $C\delta$ sequence. Both μ and δ proteins are thus produced in immature B cells. In mature B cells, the other C_H exons can undergo additional rearrangement such that primary transcripts are produced that are translated into heavy chains of the γ, ε, and α isotypes. These issues are fully discussed in Chapter 9.

As we shall see below, the assembly of a functional VDJ exon and the synthesis of a functional heavy chain polypeptide is a key checkpoint in Ig synthesis. The appearance of the newly-translated μ protein correlates with a cessation of further rearrangement of the heavy chain loci and the stimulation of rearrangement of the κ and λ loci.

III. VJ JOINING IN THE *Igl* AND *Igk* LOCI

The $V\kappa J\kappa$ light chain variable exon is assembled by the joining of any $V\kappa$ segment (with its accompanying leader and promoter) to any $J\kappa$ segment (Fig. 8-5). The rearrangement of $V\kappa J\kappa$ brings the variable exon close enough to the single $C\kappa$ constant exon to permit synthesis of a primary RNA transcript. This is followed by splicing out of introns (including those containing unrearranged $J\kappa$ segments), addition of the polyA tail, translation, and clipping of the leader peptide, as for the heavy chain. The synthesis of a functional κ light chain appears to shut down further rearrangement of both the *Igk* and *Igl* loci. However, if a functional κ light chain is not

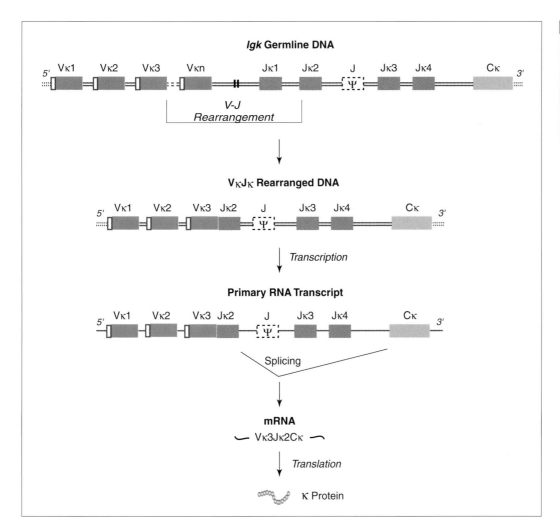

Figure 8-5

V–J Joining in the Mouse *Igk* Locus
Light chain gene rearrangement involves the joining of a V gene segment to a J gene segment. In this hypothetical example of mouse *Igk* locus recombination, Vκ3 joins to Jκ2.

produced from either chromosome, a functional λ chain may be produced by *Igl* rearrangement.

The reader may recall that, in contrast to the *Igk* and *Igh* loci, in which the gene segments are organized in blocks, the J gene segments and C exons of the *Igl* locus occur in J–C units. As a result, less than random selection of J and C sequences occurs. For example, in the mouse, Jλ1 joins only to Cλ1, Jλ2 only to Cλ2, and Jλ3 only to Cλ3. In addition, because the Vλ1 gene segment is positioned 3' of the Jλ2–Cλ2 unit in the genome, it can only join with the Jλ1–Cλ1 or Jλ3–Cλ3 units (Fig. 8-6). In the human *Igl* locus, each Vλ gene segment can pair with any of the Jλ–Cλ units. Transcription, polyA addition, splicing, translation, and clipping of the leader peptide proceed as for the heavy and κ chains.

IV. PRODUCTIVITY TESTING

We have discussed how a rearranged VDJ exon must be functional in order for the cell to proceed any further in Ig synthesis, and how the rearrangement of one Ig locus influences whether another will even start V(D)J recombination. How does the cell carry out such a "productivity test"? V(D)J recom-

bination is a bit of a gamble for the cell, since, although the double-stranded DNA is cut exactly at specified junction points (described later), the joining of the new ends is less precise. Deletions, point and frameshift mutations, and the random insertion of nucleotides often occur. Two-thirds of the time these events generate stop codons that terminate the translation of the Ig polypeptide. During rearrangement, the cell therefore tests whether the joining of a particular combination of V_H, D_H, and J_H segments has produced a functional protein by transcribing the DNA, translating the resulting RNA into protein, and translocating that Ig protein to the B cell surface. Only a rearranged V_H plus leader segment can initiate transcription of a complete primary transcript, so that the RNA polymerase molecule essentially ignores adjacent unrearranged V_H segments and binds preferentially and strongly to the promoter of the rearranged V_H segment. If the primary transcript of the candidate rearranged Ig gene is successfully spliced, polyadenylated, and translated so that a functional μ heavy chain is made which can be expressed on the surface, the entire process has been *productive*. If the cycle of joining has not led to a viable μ chain, it is considered *nonproductive* or *nonfunctional* and the cell tries a different combination of V, D, and J segments on the same chromosome. If, however, no

Figure 8-6

V–J Joining in the Mouse *Igl* Locus

In this hypothetical example of mouse *Igl* locus recombination, Vλ1 joins to Jλ1. In both the *Igk* and *Igl* loci, successful rearrangement is followed by transcription to produce primary RNA transcripts. The primary transcripts are then spliced to remove introns, giving rise to the mRNA transcripts that are then translated into light chain proteins.

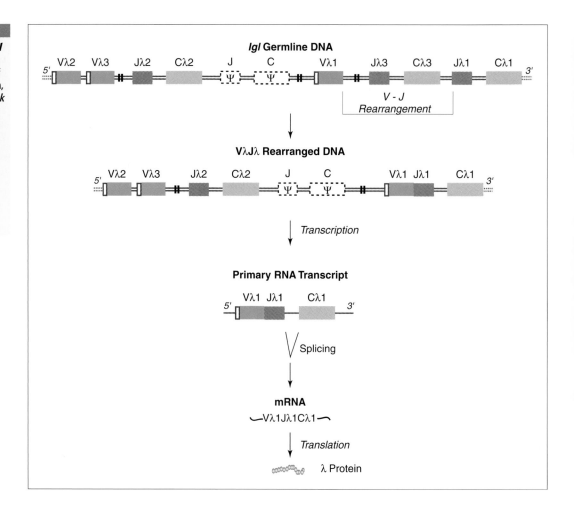

functional Ig heavy chain polypeptide is produced from the first allele, the cell continues the same random joining cycle using gene segments on the other allele on the other chromosome. If neither allele can produce a functional Ig heavy chain polypeptide, the cell dies by apoptosis.

V. ALLELIC EXCLUSION

D–J joining is initiated simultaneously in the *Igh* locus of both chromosomes of a developing B cell. However, once one allele generates a productive V–DJ rearrangement encoding a functional μ chain, further rearrangement of the *Igh* locus of both chromosomes appears to be inhibited, a phenomenon called *allelic exclusion*. Thus, in the vast majority of mature B cells, one finds that one *Igh* allele is functionally rearranged, while the other *Igh* allele is either unproductively rearranged or remains close to germline configuration.

Two types of models have been proposed to account for allelic exclusion: the *stochastic* and the *regulatory*. One *stochastic* model rests on the hypothesis that the probability of both a functional D–J joint and a V–DJ joint occurring on the same chromosome in the same cell is very low, and that the probability that functional DJ and VDJ joints could occur simultaneously on both chromosomes in the same cell is that

much lower. Thus, chance dictates that a cell expresses a complete, in-frame VDJ protein from one chromosome only. Because this model does not invoke any specific regulatory mechanism to shut down rearrangements in the *Igh* locus, one would expect to see almost all mature B cells carrying one productively rearranged *Igh* locus plus an unproductively rearranged *Igh* locus. However, about 60% of mature B cells show one *Igh* locus that is productively rearranged and one near-germline *Igh* locus. This purely stochastic model therefore has fallen out of favor as an explanation for allelic exclusion. Another modified stochastic model proposes that regulatory sequences called *enhancers* (see later) may control the accessibility of the *Igh* locus on a given chromosome, and that random changes to enhancer activities or other factors controlling locus accessibility determine how far rearrangement proceeds on a given chromosome.

Regulatory models propose that a significant degree of control is exercised by the cell, and that rearrangement occurs in a tightly ordered fashion. In some regulatory models, random rearrangement commences on both chromosomes, but if a productive combination is achieved at one *Igh* allele, the presence of the functional μ chain sends some kind of feedback regulatory signal inhibiting any further rearrangement on the other chromosome (Fig. 8-7). Support for this type of model has come from studies of mutant mice engineered to

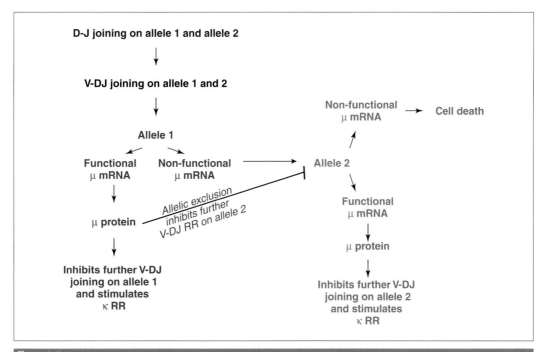

Figure 8-7

Regulatory Model of Productivity Testing and Allelic Exclusion Test
In a given immature B cell, random rearrangements (RRs) of D and J gene segments initiate simultaneously on both *Igh* alleles. As soon as a V–DJ joining occurs on either allele, an attempt is made at transcription, translation, and cell surface expression of the corresponding μ protein. If the μ protein is successfully expressed on the cell surface, the rearrangement is said to be "productive." Any further attempts at V–DJ rearrangement on both alleles are inhibited and V–J rearrangements in the κ locus are stimulated. If no productive μ chain results from rearrangement on the first heavy chain allele, V–DJ joining is stimulated on the second allele, followed by the same productivity testing process. Failure at this stage results in the death of this B cell.

express genes encoding pre-rearranged and functional μ chains. The expression of these modified genes inhibited the rearrangement of the cell's own *Igh* loci. While there is strong support for regulatory models, the actual mechanism by which a productively rearranged heavy chain gene might inhibit further heavy chain rearrangement is still under debate. It is possible that the chromatin of the second allele is altered such that it is no longer accessible to the recombination machinery. Alternatively, there may be direct effects on the recombinases themselves. Other researchers argue that chromosomal accessibility is regulated from the start such that one chromosome is chosen very early in B cell development to eventually undergo V–DJ joining, and that the other chromosome is actively inhibited from doing so. However, there are situations in which individual B cells express transcripts with different VDJ combinations in the cytoplasm, but only one of these transcripts is actually translated and makes it to the cell surface as part of the BCR. This observation suggests that allelic exclusion may occur instead at the level of surface expression, preserving the principle of "one sequence of Ig protein produced per B cell." The precise mechanism underlying allelic exclusion remains under investigation.

The reader may ask "Why have allelic exclusion?" Without it, the expression of both *Igh* alleles would lead to the appearance of Igs of two different specificities, or individual Ig molecules with two different antigen-binding sites, on the surface of

one B cell. In this situation, the number of receptors able to aggregate and respond to the binding of antigen is reduced, which could compromise intracellular signaling leading to an effective response. In addition, activation of the classical complement pathway, which depends on the proximity of antigen-bound antibody Fc regions (see Ch.19), would be diminished. Disruption of normal B cell development and the establishment of a functional, nonself-reactive B cell repertoire (see Ch.9) might also be compromised. From an evolutionary point of view, allelic exclusion conserves the cell's resources: the production of multiple Ig proteins, only one of which recognizes the attacking antigen, would be wasteful.

VI. KAPPA/LAMBDA EXCLUSION

In cells that have undergone productive *Igh* rearrangement, the presence of the μ chain also apparently conveys a regulatory signal stimulating V(D)J recombination of the *Igk* locus. Productivity testing of the reading frame of candidate VκJκ joints is carried out at the *Igk* loci, just as was described for the *Igh* loci. When rearrangement at one *Igk* allele results in the generation of a functional Ig κ chain, it participates in the formation of a mature H_2L_2 Ig molecule. Perhaps using one of the same mechanisms proposed for *Igh* allelic exclusion, the presence of the H_2L_2 monomer inhibits further rearrangement at the other *Igk* allele (*Igk* allelic exclusion). Furthermore, since the cell only needs one type of light chain to make a functional Ig molecule, a regulatory signal is apparently sent, inhibiting rearrangement of the *Igl* locus, a phenomenon called *kappa/lambda exclusion*. If, however, V(D)J recombination at the first *Igk* allele fails to produce a functional light chain, rearrangement at the second *Igk* allele is continued. If neither *Igk* allele is productively rearranged, V(D)J recombination of the *Igl* locus proceeds in an attempt to generate a functional λ light chain. These observations are consistent with the finding that κ-expressing B cells have *Igl* loci that are almost always in germline configuration, whereas both *Igk* alleles in λ-expressing B cells are either non-productively rearranged or deleted, and are rarely found in germline configuration. If neither *Igl* allele can productively rearrange, the cell can produce no light chain at all and undergoes apoptotic death.

The mechanism underlying kappa/lambda exclusion is not yet fully understood. There remains debate as to whether rearrangement on the two light chain loci occurs sequentially or simultaneously. Some researchers believe that the *Igk* rearrangement process completely precedes *Igl* rearrangement. In this case, there would have to be a signal associated with unsuccessful κ chain production that would stimulate subsequent rearrangement of the *Igl* locus. Such a signal might affect DNA demethylation, which governs locus accessibility, or perhaps components of the recombination machinery. According to this model, successful *Igk* rearrangement would result in an absence of this stimulation signal and thus inhibition of the onset of *Igl* rearrangement. Other researchers suggest that *Igk* rearrangement simply appears to precede *Igl* rearrangement because the recombination process occurs much more rapidly and efficiently in the *Igk* locus than in the *Igl* locus. Even if initiation of recombination occurred simultaneously in the *Igk* and *Igl* loci, such a disparity could lead to completion of rearrangement of the *Igk* locus before the *Igl* locus had had a chance to move out of germline configuration. In this case,

successful production of a κ chain would signal a shutdown of further recombination at the *Igl* locus. If *Igk* rearrangement failed on both chromosomes, *Igl* rearrangement would then continue in an attempt to generate a productive light chain.

Exclusionary controls explain why a mature Ig-producing B cell almost never contains more than one functional $V_HD_HJ_H$ sequence, and one functional V_LJ_L sequence. Indeed, only about 1 in 12 pre-B cells in the bone marrow matures to full Ig-producing capacity and exits to the body's periphery. However, it should not be assumed that once rearrangement has occurred, the sequence of the Ig gene is never altered. Under certain circumstances, the $V_HD_HJ_H$ region of the gene may undergo further minor rearrangements in a process called "receptor editing" (see Ch.9).

VII. INTRODUCING KNOCKOUT MICE

Most of the preceding information has come from empirical observations of the DNA of B cells compared to the germline configuration of the DNA of non-lymphocytes. How do we know these events are relevant in the whole animal, and how can we examine each of these events *in vivo*? Advances in the culture and manipulation of mouse embryonic stem (ES) cells, which are capable of differentiating into any cell or tissue in the body (including its germ cells), have made it possible to produce "knockout mouse mutants" deficient for a single specific gene (see Box 8-4). The genome of an ES cell can be engineered *in vitro* to delete, say, the Cκ exon. The mutant ES cell clone is then introduced into a developing mouse embryo and is incorporated within it. After a series of breeding steps, a homozygous mouse mutant missing only the Cκ sequences is generated. The production of Ig proteins by the B cells of these mice is then examined. Studies of Cκ knockout mutants (often denoted as Cκ null or Cκ$^{-/-}$ mice) showed that these animals produced about half the normal number of mature

---

Box 8-4. Gene-targeted "knockout" mice

Many important discoveries in immunology have come about by investigating the molecular basis of diseases of the immune system that arise randomly in humans and mice. Sometimes it has been possible to uncover the role of a particular gene product in immune system development or function by observing what happens when a mutation renders that gene non-functional. However, waiting for a random mutation to arise in every gene of interest is not practical. In the mid- to late 1980s, Mario Capecchi, Martin Evans, and Oliver Smithies independently developed a technique to promote and detect homologous recombination in mammalian cells at non-random, pre-selected sites in the genome. This technology allowed the development of techniques for manipulating a developing mouse embryo such that an endogenous gene

of interest could be disrupted in a targeted manner, allowing the establishment of a mouse strain with the desired genetic deficiency. These *gene knockout mutant mice* have since played a crucial role in the elucidation of myriad different molecular interactions (see Figure). Although knockout mice have been important to many other fields of biology, they are particularly well-suited to the study of immunology, because many immunologically important genes can be disrupted without killing the mouse.

The ability to create gene knockout mice was made possible by the discovery of *pluripotent embryonic stem* (ES) *cells*. These cells are different from hematopoietic pluripotent stem cells in that they have the potential to give rise to not only cells of the blood and immune system, but to <u>any</u> cell type

found in the body, including germ cells. In addition, mouse ES cells can be cultured indefinitely *in vitro* under certain growth conditions, and plasmid DNA can readily be introduced into them. To create ES cells bearing a targeted gene disruption, it is necessary to first build a targeting vector containing the genomic clone of the gene of interest. Two "arms of homology," sequences from the genomic region surrounding the gene of interest, are also included in the construct to facilitate insertion of the vector into the desired part of the ES cell genome. The copy of the gene in the targeting vector is then disrupted by inserting (into a part of the gene essential to its function) a selectable drug resistance marker such as the neomycin resistance gene (*neo*). The targeting vector plasmid is introduced into the ES cells by *electroporation*, a method of

Continued

Box 8-4. **Gene-targeted "knockout" mice**—*cont'd*

transfection that uses electric current to make small pores in the ES cell membrane through which exogenous DNA can enter. Some of the cells will integrate the plasmid DNA into their genomes randomly (outside the sequence of the gene of interest), but a small proportion will incorporate the plasmid using *homologous recombination* between the arms of homology in the plasmid and the gene sequences of interest in the ES cell genome. The end result is a targeted disruption of one allele of the endogenous gene of interest by the insertion of the *neo* gene.

Neomycin-resistant clones are then isolated by *positive drug selection*, in which neomycin treatment kills all cells that have failed to incorporate *neo*. Among the neomycin-resistant clones will be both cells that have integrated *neo* randomly and a small number of cells in which *neo* has been integrated into the gene of interest, disrupting it. These latter cells are correctly *gene-targeted* cells or *homologous recombinants*. Homologous recombinants can then be distinguished from random integrants by restriction mapping and Southern blot screening. However, it is more efficient to distinguish between these two groups of neomycin-resistant cells by including a gene in the targeting plasmid <u>outside</u> the arms of homology that confers sensitivity to a second drug. This drug selection gene is then deleted from the ES cell only if homologous recombination occurs. Thus, if the targeting plasmid is randomly integrated, the second drug selection marker is retained in the ES cell and the cell dies in the presence of the second drug (negative selection). However, homologous recombination of the targeting vector into the gene of interest deletes the second drug selection marker, rendering correctly gene-targeted ES cells resistant to the second drug. The gene-targeted ES cells are then cultured to establish cell lines mutated at one allele of a single gene.

The gene knockout ES cell lines can be used to create *chimeric* ("blended") knockout mice in the following way. The targeted ES cells are injected into a wild-type blastocyst (a 32-cell stage embryo) or cocultured with wild-type morulae (8- to 16-cell stage embryos). The targeted ES cells become incorporated into the embryo as it develops. These "blended" embryos are implanted into pseudopregnant female mice which have been hormonally treated so as to create body conditions resembling pregnancy. The

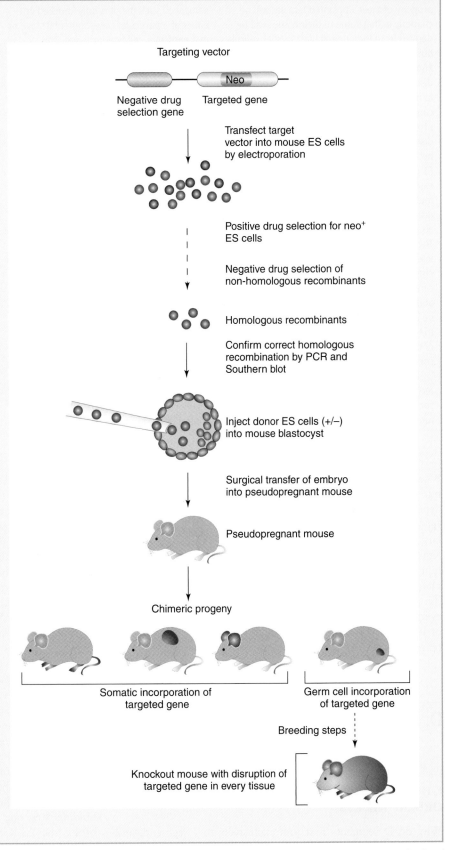

Targeting vector

Neo

Negative drug selection gene   Targeted gene

Transfect target vector into mouse ES cells by electroporation

Positive drug selection for neo⁺ ES cells

Negative drug selection of non-homologous recombinants

Homologous recombinants

Confirm correct homologous recombination by PCR and Southern blot

Inject donor ES cells (+/−) into mouse blastocyst

Surgical transfer of embryo into pseudopregnant mouse

Pseudopregnant mouse

Chimeric progeny

Somatic incorporation of targeted gene

Germ cell incorporation of targeted gene

Breeding steps

Knockout mouse with disruption of targeted gene in every tissue

Continued

progeny of these implanted females will be chimeric in that some of their tissues will be derived from the original blastocyst or morula, but others will be derived from the injected knockout ES cells and will contain the disrupted allele. If a chimeric mouse happens to have germ cells derived from the ES cells, it can be bred to wild-type mice to generate progeny heterozygous for a targeted disruption in the endogenous gene. These heterozygotes are then interbred to produce F1 progeny, one-quarter of which are expected to be homozygous for the targeted mutation. The existence of the genetic disruption is confirmed by Southern blot analysis of the DNA of the knockout mice. These mice will be deficient for the gene of interest in all their cells and tissues, and the effect of the loss of function of a single gene can thus be analyzed in a whole animal. In addition, different knockout mouse strains can be interbred to produce double and even triple knockout mice, allowing the dissection of complex systems in which one gene may compensate for the loss of another.

Recent advances in gene knockout technology have allowed the generation of *conditional* knockout mice; that is, a modification of the gene that is restricted in some way such that it is only knocked out either at a particular developmental stage or in a particular cell type or tissue. This technology depends on the use of a site-specific recombinase called *Cre* ("causes recombination") which was originally derived from bacteriophage P1. The Cre recombinase recognizes the *loxP* ("locus of crossing over") site, a specific sequence of 34 base pairs. Using standard gene manipulation techniques, loxP sites can be introduced into DNA such that they flank a gene of interest. When Cre encounters a segment of DNA flanked on either side by a loxP site, it circularizes the DNA between them and excises it, creating a deletion in the DNA. Cre is extremely efficient, meaning that the chances of obtaining the desired deletion in a given construct are very high. The Cre recombinase can be placed under the control of an unrelated promoter that is inducible or tissue-specific, meaning that the Cre/loxP modification of an appropriate DNA target (a "floxed" target gene—"flanked by loxP sites") will only occur when or where that promoter is activated. By crossing a mouse strain carrying the floxed target gene with another mouse strain carrying the Cre recombinase under the control of the promoter of interest, a conditional mutant mouse can be produced in which the deletion of the target gene occurs only in those tissues in which the promoter is activated. For example, a mouse strain can be created in which Cre is placed under the control of the CD19 promoter. CD19 is a gene expressed exclusively in B cells. In the B cells (only) of the conditional knockout, the gene of interest will be efficiently deleted when the CD19 promoter is activated and Cre is expressed. A similar approach can be used to produce mice in which Cre is placed under the control of an inducible promoter, such as the promoter of a gene whose expression is induced in response to interferon. In this case, the gene of interest is deleted by Cre-recombination only when interferon is applied. This technique allows the study of the absence of the gene of interest at a specific time or developmental stage. Another advantage to Cre-recombination is that it is independent of DNA replication, so that genes can be deleted irrespective of the cell cycle status of the cell. This feature is particularly important in tissues such as the brain, where little cell division occurs. Thus, the use of the Cre/loxP system allows researchers to routinely inactivate a gene *in vivo* in either a cell type-specific or inducible manner.

B cells. VJ rearrangement of the *Igk* locus was much reduced but not absent, showing that the rearrangement process itself depends on more than the Cκ exon. Rearrangement of the *Igl* locus was able to proceed and gave rise to a surprisingly large number of functional λ light chain-expressing B cells. Knockout mice in which the J$_H$ region was deleted had a much more severe phenotype, since these animals were completely devoid of mature B cells in both the bone marrow and the periphery and failed to produce antibodies. Interestingly, even though *Igh* rearrangement was blocked, a low number of κ chains was produced, showing that a base level of *Igk* rearrangement can occur independently of *Igh* rearrangement.

D. Molecular Mechanisms of Ig Gene Rearrangement

I. HOW DO VDJ SEGMENTS JOIN IN THE RIGHT ORDER? THE RSS

Why do D segments first join exclusively to J segments and never to V segments? Why are D–D or V–V fusions almost never found in the B cell genome? How does the cell know where to cut and rejoin its Ig DNA, and what are the mechanisms that carry this out? Much research has been devoted to understanding the enzymatic machinery of V(D)J recombination, and it is now clear that the same recombination mechanism lies behind the repertoire diversity of both B and T cells.

V(D)J recombination, in which gene segments are repositioned within a genetic locus, is an enzymatic process involving the juxtaposition of gene segments, the removal of intervening DNA, and the rejoining of double-stranded DNA. The usual enzymes involved in DNA synthesis and modification, such as the ligases, polymerases, terminal transferases, and exonucleases found in every cell, play their usual roles. However, the juxtaposition and rejoining in the Ig and TCR loci depend on specialized *recombinase* enzymes called *RAG-1* and *RAG-2*. These enzymes are encoded by the *recombination activation genes* (RAG) that are expressed only in lymphocytes. RAG-1 and RAG-2 cooperate to recognize specific *recombination signal sequences* (RSSs) that flank germline V segments on their 3′ sides, J segments on their 5′ sides, and D segments on both sides. The two types of RSSs are called the "12-RSS" and the "23-RSS" (Fig. 8-8). Both sequences are recognized by the recombinase complex that contains two binding sites, one recognizing the 12-RSS and the other recognizing the 23-RSS. The 12-RSS in genomic DNA contains a highly conserved heptamer, a spacer of 12 nucleotides of non-conserved sequence (hence, the name 12-RSS), and a

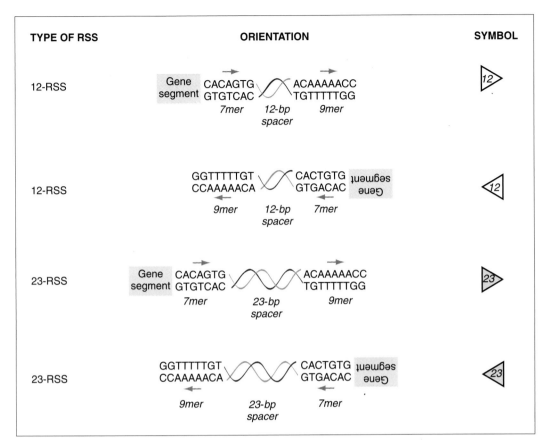

Figure 8-8
Recombination Signal Sequences
All recombination signal sequences (RSSs) have a conserved heptamer and a conserved nonamer base pair (bp) sequence separated by a non-conserved spacer. That spacer is either 12 bp (12-RSS) or 23 bp (23-RSS) in length. The 12-RSS and 23-RSS correspond to one and two turns of the DNA double helix, respectively. The heptamer–spacer–nonamer sequence is always read moving away from the side of the gene segment where it occurs.

conserved nonamer. The 23-RSS contains a conserved nonamer whose sequence is complementary to the 12-RSS nonamer, an unconserved spacer of 23 nucleotides, and a conserved heptamer whose sequence is complementary to the 12-RSS heptamer. Only when one gene segment is flanked on one side by a 12-RSS, and the other gene segment is flanked by a 23-RSS, can the pair be recognized by RAG-1 and RAG-2 and participate in V(D)J joining (Fig. 8-9). This requirement has been called "the 12/23 rule."

In addition to the 12/23 rule, it appears that the recombinase complex recognizes and joins some RSS pairs together more readily than others. This phenomenon arises because there are variations in the heptamer and nonamer sequences of different RSS. These distinctions may help to ensure that DJ joining in the *Igh* locus proceeds before V–DJ joining.

II. THE RECOMBINASE ENZYMES: RAG-1 AND RAG-2

The RAG enzymes carrying out V(D)J recombination are both cell type- and stage-specific: they function only in lymphoid cells (both T and B cells) and only at their very early maturational stages. Once a functional TCR (see Ch.12) or BCR is made, the RAG recombinases are down-regulated. With the assistance of the DNA-binding proteins HMG-1 and HMG-2 (high mobility group), molecules of RAG-1 and

RAG-2 come together in a complex that both recognizes RSS-flanked gene segments according to the 12/23 rule and cleaves double-stranded DNA between an RSS and the coding sequence. The precise structure of the "12/23 recombinase complex" involving the two RSSs is unclear, but it seems that a dimer of RAG-1 and either a dimer or monomer of RAG-2 can bind to a single RSS. In any case, analyses in *in vitro* assay systems have shown that the only proteins needed to produce correctly-targeted double-stranded DNA breaks during V(D)J recombination are RAG-1 and RAG-2. RAG-1 is primarily responsible for DNA nicking and hairpin formation, while RAG-2 appears to promote the binding of the recombinase complex to the DNA and serves as an essential cofactor for RAG-1 activity.

One puzzle that has yet to be solved is "Why do RAG-1 and RAG-2 only act on Ig genes in precursor B cells, and only on TCR genes in precursor T cells?" In other words, why do precursor lymphocytes destined to be B cells rearrange only their Ig genes and not their TCR genes, while precursors destined to be T cells rearrange their TCR genes, but not their Ig genes? Clues to this mystery may lie in factors controlling the accessibility of the Ig and TCR loci to the recombinase enzymes. Local remodeling of chromatin structure is regulated by histone acetylation and deacetylation, and cell type-specific transcription factors or enzymes might govern which gene segments in which locus might be available for V(D)J recombination.

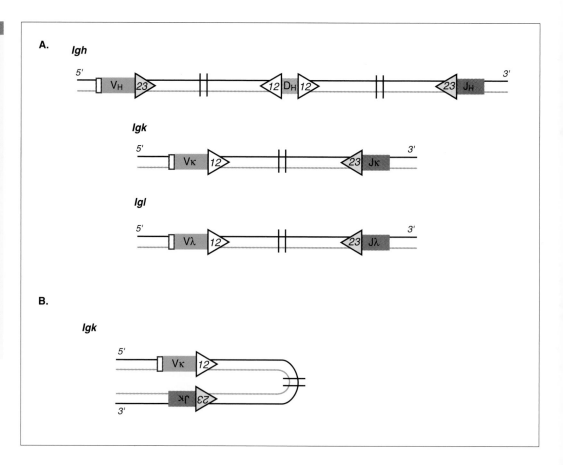

III. SYNAPSIS, SIGNAL JOINTS, AND CODING JOINTS

In the germline DNA of the *Igh* locus, we find each V_H gene segment flanked on its 3′ end by a 23-RSS, each J_H segment flanked on its 5′ end by a 23-RSS, and each D_H gene segment flanked on both sides by a 12-RSS. V(D)J recombination commences when the RAG proteins recognize the 12-RSS on the 3′ side of the D segment and the 23-RSS on the 5′ side of the J segment. (It is not known why the cell starts with the D and J segments and not the V segments.) The RAG proteins then come together to form a recombinase complex that aligns the DNA strands in a structure called a *synapse*. In the synapse, the DNA between the selected D_H and J_H segments is looped out (Fig. 8-10). The double-stranded DNA immediately adjacent to the aligned RSS is then cleaved by the endonuclease activity of the recombinase complex in two steps. First, a nick is made on one strand exactly at the border between the coding ends of the gene segments and the heptamers of the RSS. Secondly, the free 3′ OH ends on either side of the synapse attack the complementary 5′ ends on the opposite strand, forming two hairpin structures (one containing D coding sequences and the other containing J coding sequences), and two blunt 5′-phosphorylated RSS ends. The covalently sealed hairpins consisting of the D and J segments are recognized and bound by the *Ku complex* (composed of the Ku70 and Ku80 proteins). The Ku complex serves as the DNA end-binding component of a multi-subunit DNA repair complex called *DNA-dependent protein kinase* (DNA-PK). The role of the DNA-PK complex and the precise order in which Ku and the catalytic elements of DNA-PK are recruited to the recombinase complex are not entirely clear. Scientists speculate that when Ku binds to the D and J hairpins, DNA-PK is then able to phosphorylate the hairpin structures. It is believed that an endonuclease called *Artemis* (which interacts with DNA-PK and is involved in DNA repair) is a key player in the opening of the hairpins. This enzyme, likely in concert with other components of the recombination machinery, may nick the hairpins at the borders of the D and J coding sequences, creating single-stranded ends on both the D and J gene segments. Non-complementary nucleotides in these ends are trimmed by exonuclease and repaired by the enzymes of a DNA repair pathway present in all cells called the *non-homologous end-joining* (NHEJ) *pathway*. As its name suggests, the NHEJ pathway can repair double-stranded DNA breaks without the need for DNA sequence homology between the ends to be joined. In an Ig gene, the D and J segments are joined together, restoring properly complementary double-stranded DNA in what is called the *coding joint*. The blunt RSS ends are also joined together in a *signal joint*. The formation of the signal joint creates a small circular product containing the DNA excised between the selected D and J segments; this product is subsequently lost during cell division. The final sealing of both the coding and signal joints is carried out by *DNA ligase IV* complexed to a protein called *XRCC4*. (XRCC stands for "X-ray cross-complementation." The XRCC gene is mutated in a particular subgroup of X-ray sensitive rodents.)

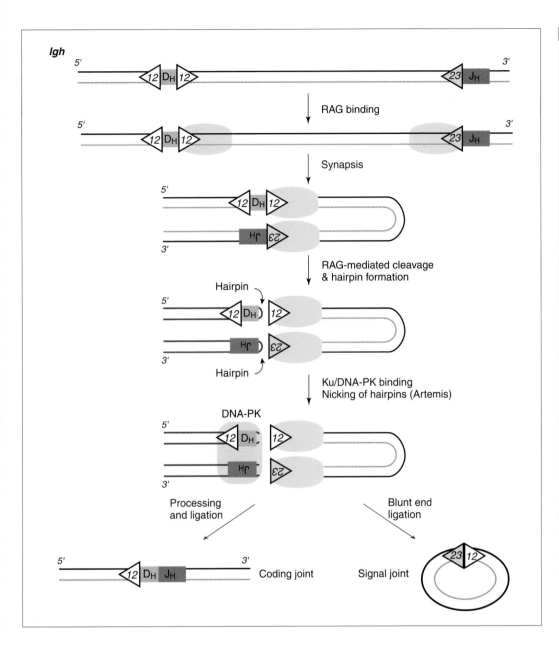

Figure 8-10

An Example of V(D)J Recombination: D–J Joining
When elements of the RAG recombinase complex (indicated by the large blue oval) bind to the appropriate gene segments, they form a synapse structure in which the RSSs align themselves and the DNA between the gene segments is looped out. The RAG complex then mediates DNA cleavage between each gene segment and its heptamer such that the ends of the DNA of each gene segment form a hairpin structure. The hairpins are then bound by the Ku protein of the DNA repair complex enzyme DNA-dependent protein kinase (DNA-PK) and are opened by nicking, most likely by Artemis. Processing and ligation at the opened hairpins result in the formation of a double-stranded DNA join between the D and J segments. In the course of processing prior to ligation, single-stranded DNA ends may be trimmed and the resulting gaps filled in with new nucleotides, resulting in the generation of P and N junctional diversity (see Fig. 8-12) in the coding joint. The blunt RSS ends are ligated to each other to form a circular signal joint containing DNA that is lost from the genome.

At the conclusion of this round of V(D)J recombination, the fused DJ segment has a 12-RSS on its 5′ end, and no RSS on its 3′ end, which prevents any further rearrangement 3′ to the J, but invites fusion with the 23-RSS 5′ of a V gene segment. Thus, the RSS 12/23 pairing rule ensures that V does not fuse to V (nor D to D, nor J to J) and that the gene segments are assembled in the correct order. RAG-1 and RAG-2 then cooperate in the same way to make VJ coding joints for the Ig light chains. In the *Igk* locus, the Vκ gene segments are flanked on the 3′ side by 12-RSS, while the Jκ segments have 23-RSS on their 5′ sides. In contrast, in the *Igl* locus, the Vλ segments have a 23-RSS on their 3′ sides, while 12-RSS are located 5′ to the Jλ segments.

We have described the formation of the coding joints for the situation where the hairpin loops arising from RSS alignment are nicked exactly at the ends of the coding sequences,

blunt ends are formed, and the coding joint is exact. If, however, the loops are nicked at another position such that 5′ or 3′ overhangs are generated, the rejoining may be imprecise. Coding nucleotides may be deleted, or additional nucleotides called *P* and *N nucleotides* may appear in the coding sequence, possibly giving rise to a new amino acid sequence in the Ig protein. These methods of increasing diversity in the antibody repertoire are addressed in Section D, 'Antibody Diversity Generated by Gene Rearrangement.'

In addition, the reader should note that there are actually two methods of joining gene segments together, depending on the orientation of the gene segments in the germline. The process described above is called *deletional joining* and applies when the two gene segments to be brought into apposition are in the same transcriptional orientation. However, in some instances, the two segments to be fused are in opposite transcriptional orientations in the germline. In these cases, joining occurs by a mechanism called *inversion*. As can be seen in Fig. 8-11, the coding joint is formed in the usual way but the signal joint and intervening DNA in the loop do not form a circular product, are not excised, and instead are retained in the rearranged Ig locus.

IV. MUTATIONS OF V(D)J RECOMBINATION

The existence of mutants for genes involved in the very specialized processes of V(D)J recombination has provided scientists with experimental models with which to analyze the diverse roles of B and T cells, and to explore human immunodeficiencies (see Ch.24). For example, a natural mouse mutant strain called *scid*, for "severe combined immuno deficiency," lacks both B cells and T cells. Lymphoid cells in *scid* mice have intact RAG-1 and RAG-2 and so can carry out the nicking and hairpin formation steps that start V(D)J recombination. However, these mice have a defect in the gene encoding the catalytic subunit of the DNA-PK complex, so that hairpins are not opened and a proper coding joint cannot be formed. Blunt signal ends and hairpins accumulate, blocking the development of both mature B and T cells. Similarly, knockout mice lacking Artemis show impaired coding joint formation. In contrast to *scid* and Artemis-deficient mice,

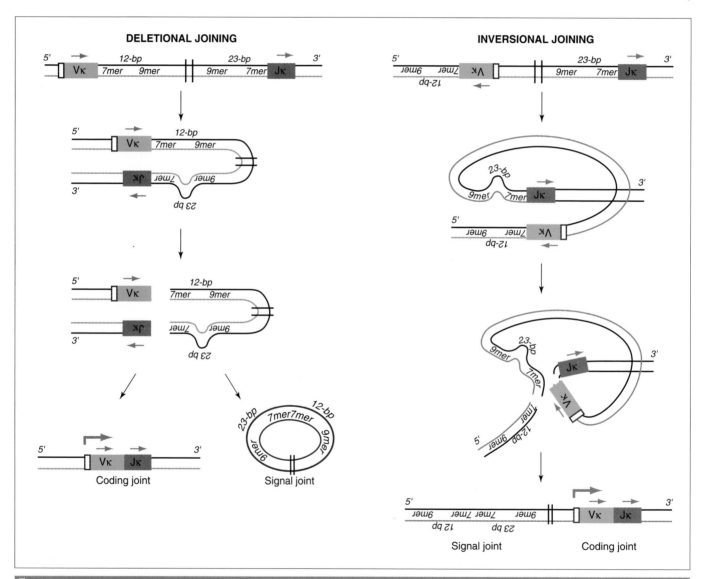

Figure 8-11

Deletional and Inversional Joining of Gene Segments
In this figure, the 12-RSS and 23-RSS symbols are shown in "long form" to illustrate how alignment occurs during deletional and inversional joining. Deletional joining occurs when the gene segments to be joined are in the same transcriptional orientation in the germline. In this case, the relevant RSSs are in opposite order to each other. Inversional joining occurs when the gene segments are in the opposite transcriptional orientation in the germline with their corresponding RSSs in the same order. A coding joint and signal joint result in both cases, but the signal joint formed by inversional joining is retained in the germline DNA rather than being excised as in deletional joining. In each case, synapse formation and processing to form the signal and coding joints occur as described in the text and as shown in Figure 8-10.

knockout mice deficient in either RAG-1 or RAG-2 cannot even start V(D)J recombination of their Ig or TCR genes, and are completely devoid of mature B or T cells. (This phenotype further demonstrates the importance of gene rearrangement to the development of lymphoid cells.) While they do exhibit severe combined immunodeficiencies, *scid*, RAG-1$^{-/-}$, and RAG-2$^{-/-}$ mice can survive quite nicely under conditions in which pathogens are strictly controlled.

E. Antibody Diversity Generated by Gene Rearrangement

The diversity in the antibody repertoire is derived from (1) diversity in antigen recognition (which depends on variation in antigen-binding site sequence, a function of the V region of the antibody protein) and (2) diversity in antibody effector function (which depends on the antibody's isotype, derived from its C region). The majority of antibody diversity involved in antigen recognition arises as a result of the rearrangement of V region elements of the genome of the immature B cell prior to antigenic stimulation. The four sources of pre-antigen variability are: *multiple germline gene segments, combinatorial joining, junctional diversity*, and *heavy–light chain pairing* (Table 8-3). These sources of diversity, which create a unique BCR sequence for each B cell <u>prior</u> to its release into the periphery, are discussed in this chapter. Another process called *somatic hypermutation* contributes to diversity in antigenic recognition <u>after</u>

the naive B cell is released to the periphery and encounters antigen. Point mutations in the BCRs of progeny of the original B cell are generated that subtly increase the variation in antigen-binding site sequences. Isotype switching, which is induced by antigen in the periphery, contributes to C region diversity and multiplicity of effector function rather than to diversity in antigen recognition. Somatic hypermutation and isotype switching are fully discussed in Chapter 9.

We will now describe the four sources of antibody diversity that exist in developing B cells in the bone marrow prior to antigenic stimulation.

I. MULTIPLICITY OF GERMLINE GENE SEGMENTS

The first source of diversity in antigen recognition is that derived from the existence of multiple gene segments in the germline; that is, the hundreds of different V, D, and J gene segments. The fact that there are so many different germline gene segments increases the variation that arises from the second source of diversity in antigen recognition, combinatorial diversity.

II. COMBINATORIAL DIVERSITY

Combinatorial diversity stems from the fact that single V, D, and J segments are joined to form a variable exon. Because there are hundreds of V, D, and J gene segments available to choose from, there are thousands of potential combinations that can be used to create the variable exons in a given B cell.

Table 8-3 Generation of Antibody Diversity in the Mouse

Mechanism of Diversity*	Heavy Chain	Kappa Light Chain	Lambda Light Chain
		Variable Region	
Multiple germline gene segments			
V	100–200	250–300	3
D	15	0	0
J	4	4	3
Combinatorial V–D–J or V–J joining	$100 \times 15 \times 4 = 6.0 \times 10^3$	$250 \times 4 = 1 \times 10^3$	$(3 \times 3) - 1 = 8$[†]
P nucleotide addition	✓	✓	✓
N nucleotide addition	✓	Rare	Rare
Imprecise joining	✓	✓	✓
Somatic hypermutation	✓	✓	✓
Receptor editing	Rare	✓	✓
Heavy and light chain association	$>6.0 \times 10^3 \times (1000 + 8) = >6.0 \times 10^6$		
Total estimated diversity (including junctional diversity and somatic mutation)	$\sim 10^{11}$		

*Numbers in combinatorial joining row are approximate.
[†]Due to its position in the genome, Vλ1 cannot join with Jλ2–Cλ2.

We can now start to see how the amazing diversity of antibody proteins arises. To illustrate the power of combinatorial joining, consider the generation of the variable domain of an Ig κ chain in a mouse: each B cell precursor would have at its disposal any one of at least 250 Vκ gene segments × 4 (functional) Jκ gene segments = 1000 variable exon sequences. The B cell precursor chooses one combination of Vκ and Jκ segments to fuse covalently and irreversibly into its variable exon, which remains fixed thereafter in the cell's genomic DNA and determines its antigenic specificity and that of its progeny cells for life. The selected and fused variable exon then participates with the single Cκ exon in transcripts of complete Ig κ chains of this one specificity.

Similar combinatorial joining can occur in the *Igh* locus, but with an added multiplier effect attributable to the D gene segments. Although they encode only a small number of amino acids, the presence of the D gene segments in the *Igh* locus greatly increases the diversity of heavy chains that can be achieved. Consider the following for a mouse Ig heavy chain: at least 100 V_H segments × 15 D_H segments × 4 J_H segments = at least 6000 different variable exons, as compared to the 400 possible without the D_H segment.

III. JUNCTIONAL DIVERSITY

Junctional diversity arises when the fusion of V, D, and J segments is not exact. Such "imprecise joining" of gene segments can result from the deletion of nucleotides in the joint region, and/or the addition of P and/or N nucleotides during V(D)J recombination (Fig. 8-12).

i) Deletion
The reader will recall that, in the last stages of V(D)J recombination, the strands of the nicked hairpins are paired and then trimmed by exonuclease prior to repair by DNA ligases. Very often, the exonuclease trims into the coding sequence itself, eliminating "templated" nucleotides. After DNA repair is complete, the sequence of the rearranged Ig gene in this region differs from that predicted from the germline sequence. In fact, 80% of adult human Ig genes show junctional diversity due to these sorts of deletions.

ii) P Nucleotide Addition
The reader will also recall that, if the hairpin loops arising from RSS alignment are nicked exactly at the ends of the coding sequences, blunt ends are formed and the coding joint is exact. However, if the loops are nicked in the intervening DNA between gene segments, one strand on each gene segment will have a recessed end while the other strand will have an overhang. The gaps on both strands are filled by standard exonuclease trimming, DNA repair, and ligation to form an intact coding joint. As a result of this process, there are new nucleotides in the coding joint that did not appear in the original sequence. The insertion of these so-called *P nucleotides* into the coding joint can change the amino acid sequence of the Ig protein, making this process a minor source of antibody repertoire diversity. According to studies of the sequences of adult human antibodies, about 5% of joints in Ig molecules contain P nucleotides.

iii) N Nucleotide Addition
Another source of junctional diversity is *N nucleotide* addition, which occurs almost exclusively in the heavy chain genes. If one examines the nucleotides surrounding the coding joints of VDJ variable exons, one can find up to 15 extra residues that do not appear in the germline sequence and are not accounted for by P nucleotides (refer to Fig. 8-12). These N nucleotides ("non-template nucleotides") are added randomly by the terminal dideoxy transferase (TdT) enzyme on the ends of the nicked hairpins before final repair and ligation. TdT is likely recruited to the site of recombination by Ku, since Ku-deficient mice lack N nucleotides in their *Igh* genes. N nucleotide addition occurs predominantly in heavy chain but not in light chain gene rearrangement because the TdT is essentially "turned off" at that point in B cell development when light chain gene rearrangement occurs.

It is worth reiterating here that because N nucleotide addition occurs at the VDJ coding joints, it creates additional diversity in the CDR3 hypervariable region of the Ig protein, the region involved in antigen binding. To illustrate the power of this mechanism to generate diversity, consider that, on average, 6 random nucleotides are added at each joint in the heavy chain variable exon. Two new amino acids may thus appear between the V_H and D_H sequences, and two between the D_H and J_H sequences in this region of the polypeptide. Since the nucleotides are chosen at random, any one of the 20 amino acids can be encoded at each site of N nucleotide addition. Thus, on average, $20 \times 20 \times 20 \times 20 = 160,000$ possibilities for additional amino acids can arise in a given VDJ joint. Evidence for N nucleotide addition can be found in about 50% of adult human antibodies.

To further illustrate the concept of junctional diversity, let's suppose we have individual developing B cells X and Y that have coincidentally used the same V, D, and J segments during successful V(D)J recombination. Imagine that B cell X expresses an Ig heavy chain containing the amino acid usually encoded by the last nucleotide triplet of the coding sequence of a V gene segment, followed in sequence by the amino acid usually encoded by the first triplet of the adjacent J gene segment. Because of imprecise joining during V(D)J recombination, the developing B cell Y may express an Ig protein in which the first amino acid may be derived instead from a different in-frame fusion of the nucleotides originally present at the ends to be joined, which may or may not represent a deletion and/or include P or N nucleotides. The second amino acid may also therefore be encoded by a different nucleotide triplet. As long as the new nucleotide sequence can be translated into a functional Ig protein, developing B cell Y will proceed to produce this new Ig polypeptide, which may have subtle differences in its amino acid sequence affecting its antigenic specificity. Such variation due to junctional diversity would also be expected to occur in rearranged light chain genes having the same V–J combinations. For example, the extreme variability of the amino acid at position 96 in the murine κ light chain has been shown to be due to imprecise joining.

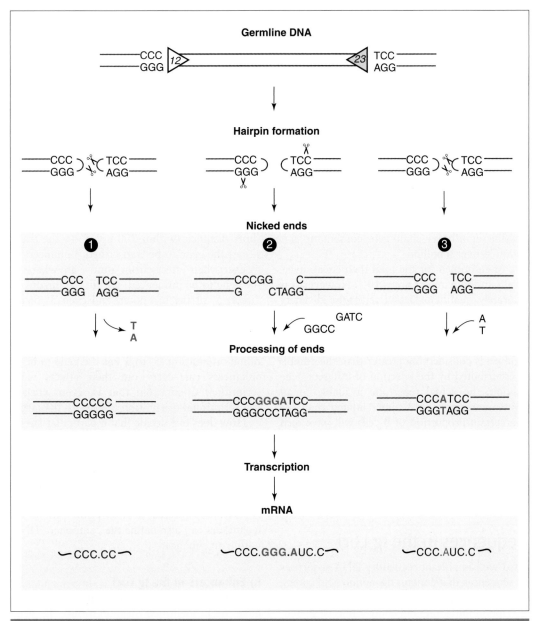

Figure 8-12

Junctional Diversity

A schematic representation of three ways in which imprecise joining of gene segments leads to binding site diversity in antibodies: (1) deletion, (2) P nucleotide addition, and (3) N nucleotide addition. These mechanisms are not mutually exclusive and may occur simultaneously during a single rearrangement event. Note that in this figure the gene segments have been shown in a linear configuration (rather than in synapsis) to highlight alterations to the germline base pair sequence. Sites of cleavage are indicated by the scissors symbol and nucleotides lost or gained relative to the original sequence are shown in blue. (1) Deletions from the germline sequence are caused by exonuclease trimming during processing of the coding joint. (2) When the DNA of a gene segment in synapse is nicked at a site located several base pairs away from the end of the hairpin structure, unpaired nucleotides are left in the uneven ends. These nucleotides serve as templates for the addition of complementary nucleotides inserted during the final processing of the coding joint ends, resulting in the addition of P nucleotides to the original sequence. (3) Non-templated N nucleotides are added when the enzyme terminal dideoxy transferase (TdT) adds random nucleotides to the ends of nicked hairpins prior to their final ligation. N nucleotide addition occurs almost exclusively during heavy chain gene rearrangement.

IV. HEAVY–LIGHT Ig CHAIN PAIRING

A final source of diversity in antigen recognition prior to antigenic stimulation can be found in the random pairing of heavy and light chains within a given B cell. A mature Ig protein molecule's antigen binding site is composed of the variable domains of both the light and heavy chains. Our previous calculations of VJ and VDJ combinatorial joining resulted in a total of at least 1000 variable exons possible for a light chain gene, and at least 6000 variable exons possible for a heavy chain gene (not including diversity contributed by deletion and N and P nucleotide addition). Any one B cell synthesizes only one sequence of heavy chain, and one sequence of light chain, but if one assumes that any of the 6000 heavy chain genes can occur in the same B cell as any of the 1000 light chain genes, the number of possible light/heavy chain gene combinations is at least $1000 \times 6000 =$ over 6 million.

The figure most often quoted for the total theoretical diversity of an individual's antibody repertoire is 10^{11}. However, it is impossible to accurately quantitate actual repertoire diversity because of the variations introduced by junctional diversity and (later on) somatic hypermutation. All that one can safely say is that the number of different antibody molecules that an individual can produce is considerably greater than the combinatorial diversity contributed by the selection of different gene segments. In addition, the actual repertoire available to an individual to counter antigens will be more limited than the theoretical, since a certain proportion of B cells will have died before ever encountering specific antigen and undergoing activation. This process is called *negative selection* and is described in detail in Chapter 9.

F. Control Sequences in the Ig Loci

The Ig loci are filled with important regulatory DNA sequences in addition to the sequences that contain the amino acid coding information. These control sequences regulate V(D)J recombination, govern transcription initiation, and control isotype switching (see Ch.9). *Enhancers* constitute one such class of control elements, and specific enhancers buried within the Ig loci have proved critical for the optimal expression of the Ig genes. Enhancers are short regions of DNA containing collections of even shorter DNA sequences called *DNA binding motifs*. DNA binding motifs are important because they are the binding sites for DNA-binding proteins called *nuclear transcription factors*. By binding to motifs within the enhancer, the nuclear transcription factors can influence the transcription of the gene, either positively or negatively. Consequently, nuclear transcription factors exert a high degree of control over gene expression and cellular development. Although the DNA specifying an enhancer is, by definition, present in every cell of an organism, it may be active in only a single or limited number of cell types or at a certain time during development, and therefore contributes to the tissue- or temporally-specific expression of the gene that it influences. The Ig loci enhancers are examples of just such tissue-specific sequences, since they operate exclusively in B cells.

I. ENHANCERS

i) General Characteristics

Enhancers are short nucleotide sequences in genomic DNA that have been found to influence the rate of transcription of particular target genes. Like promoters, enhancers are *cis-acting* in that they influence only genes on the same DNA molecule: they cannot increase the transcription of genes on a different chromosome. However, enhancers are unlike promoters in three ways. First, the sole function of a promoter is to initiate transcription, while enhancers may either increase or decrease transcription of their target genes. An enhancer that primarily suppresses transcription of a gene is often called a *silencer*. Secondly, enhancers have greater freedom of position than do promoters. Promoters must be located at a short distance (within 200 bp) 5′ to the transcriptional start site of the gene to be transcribed. Enhancers have been found to exert their effects from almost anywhere in the vicinity of the gene to be influenced: within the introns of the genes they affect, 5′ of the promoter, and even 3′ of all coding exons. Thus, some enhancers exert their effects on promoters that are physically located comparatively great (2–25 kb) distances away. Thirdly, while promoters must be in the same transcriptional orientation (5′ to 3′) as the gene to be controlled, many enhancers can carry out their effects regardless of transcriptional orientation: that is, even when in the opposite transcriptional orientation relative to the target gene.

How does one decide that a particular stretch of nucleotides constitutes an enhancer? Several techniques routinely used to identify enhancer sequences are described in Box 8-5. By methodically mutating the nucleotide sequence of the putative enhancer, and assaying *in vitro* whether the transcription of an easily detected test target gene is affected, the precise sequence necessary to increase transcription can be delineated. These techniques can also define the component DNA binding motifs within an enhancer.

ii) Enhancers in the Ig Loci

To date, nine enhancers contributing to Ig gene expression have been identified in the mouse Ig genes: two in the *Igk* locus, three in the *Igl* locus, and five in the *Igh* locus (Fig. 8-13). Each Ig locus has an *intron enhancer* located in an intron upstream of all its C exons; these are denoted $E_{i\kappa}$, $E_{i\lambda}$, and E_{iH}. E_{iH} was originally called $E\mu$ ("enhancer-μ chain") and was the first Ig enhancer to be discovered. The terms $E\mu$ and E_{iH} are still sometimes used interchangeably. Each Ig locus also has at least one enhancer 3′ of the last C exon. In the light chain loci, these enhancers are denoted $E_{3'\kappa}$, $E_{3'\lambda}$, and $E_{3'\lambda}3$-1. The *Igh* locus has four enhancers 3′ of the last C_H exon, called 3′Cα, 3′E, HS3, and HS4. Such positioning ensures that not all the benefits of the increase in transcription provided by enhancers are lost during the DNA rearrangements that occur in heavy chain isotype switching (see Ch.9). In the human Ig loci, there are two enhancers in the *Igk* locus (one on either side of the Cκ exon), one enhancer in the *Igl* locus (3′ of the J–C units), and three in the *Igh* locus (one 5′ of the Cμ exon, one 3′ of the Cα1 exon, and one 3′ of the Cα2 locus).

Box 8-5. Identification of enhancer sequences and nuclear transcription factor binding sites

Enhancers are *cis*-acting DNA sequences that influence gene transcription (either positively or negatively). An enhancer is actually nothing more than a collection of specific binding sites for nuclear transcription factors. Assays used to analyze the binding sites or *motifs* in an enhancer take advantage of the fact that the binding of a nuclear transcription factor to its motif protects that stretch of DNA from cleavage or modification.

DNase footprinting relies on the use of the enzyme DNase I, which non-specifically cuts a single strand of a double-stranded DNA helix that is not protected by chromatin. In this technique, a DNA fragment containing the putative DNA binding site is radioactively end-labeled with ^{32}P, and half of this probe preparation is mixed with a nuclear extract containing DNA binding proteins; the control half is not mixed with the extract. Both preparations are then exposed briefly to DNase I, such that each DNA fragment is nicked once on average. DNA fragments in the control preparation will be randomly nicked at each nucleotide in the total pool of DNA molecules, so that when the DNA is electrophoretically separated on a gel, a uniform "ladder" of DNA fragments is observed. DNA fragments that were exposed to the nuclear proteins may have been bound by a specific transcription factor, and the presence of this protein bound to the DNA results in the protection of that particular portion of the DNA fragment from DNase I cleavage. Thus, this portion of the DNA ladder will be missing when the reaction is electrophoresed and autoradiographed, resulting in a "footprint" in the region of the transcription factor binding site. Mutagenesis of that region can

then be used for detailed analysis of the recognition sequence. In a similar assay known as *methylation interference*, the chemical methylation of the DNA fragments renders them sensitive to cleavage by a specific enzyme. If a nuclear factor is bound to the DNA fragment at a specific site, it protects that region of DNA from methylation, resulting in a footprint similar to that just described.

Another assay used to identify sequences that serve as binding sites for transcription factors is the *gel mobility shift assay*. In this assay, a radiolabeled DNA fragment is incubated *in vitro* with a nuclear or cellular extract, allowing the interaction of DNA binding proteins with a putative binding site. The DNA fragment is then electrophoresed on a gel and compared to the profile of the same DNA fragment not incubated with the extract. If a protein binds to the DNA, the resulting complex will be larger than the DNA alone, and will migrate through the gel more slowly. Thus, compared to the control, the DNA fragment will have been shifted upward on the gel, indicating the presence of a bound protein.

The *reporter gene assay* is another widely used technique for the detection of enhancer sequences or other *cis*-elements influencing transcription rates. In this assay, a plasmid is constructed in which a copy of the putative enhancer element is placed in front of a gene whose expression can easily be detected (the reporter gene). Ideally, the gene chosen as a reporter is not found in cells of the species under study. A well-known reporter gene of the 1970s was the bacterial chloramphenicol acetyl transferase (CAT) gene, which catalyzes the addition of the acetyl group of acetyl CoA to

chloramphenicol. β-Galactosidase, which can be present at low levels in mammalian cells, can also be used as a reporter. This enzyme converts a colorless *o*-nitrophenol galactopyranoside substrate into a yellow *o*-nitrophenol product. These days, most reporter gene assays take advantage of the properties of firefly luciferase (the enzyme responsible for the oxidation of luciferin that results in the eerie yellow-green glow of these insects on a summer night). Reporter assays based on luciferase are highly sensitive.

To carry out any reporter gene assay, the construct is transfected into recipient cells, the cells are incubated for a period to allow the expression of the putative enhancer–reporter gene construct, and an extract is prepared from the cells. Substrates (which may be naturally chromogenic or tagged with radioisotope or fluorochrome) for the enzyme encoded by the reporter gene are then added to the extract and the level of reporter enzyme activity assessed. If the putative enhancer element is truly a transcriptional activation site, it will drive expression of the reporter gene such that reporter mRNA is translated into active reporter enzyme protein. This enzyme converts the substrate into a product that can be easily detected and quantitated. The amount of product correlates with the level of reporter enzyme activity, which in turn is related to the rate of transcription of the reporter gene. If the genetic element placed in proximity to the reporter gene is a positively-acting enhancer, the level of reporter enzyme expression will correspondingly increase relative to a control plasmid containing the reporter gene alone.

DNA sequences that suppress Ig gene transcription (silencers) are also present in the Ig loci. Two have been identified in the murine *Igh* locus, one on either side of the E_{iH} intronic enhancer. Two others have been located in the *Igk* locus: one just 5′ of the κ intronic enhancer, and the other just 3′ of the 3′κ enhancer. No silencers have been identified yet in the *Igl* locus. It is thought that, in non-B cells, a nuclear transcription factor called NF-μNR can bind to the silencer sequences, inhibiting the function of the neighboring enhancers and thereby shutting down Ig transcription. However, since not every enhancer sequence has a silencer nearby, other means of controlling Ig gene transcription (which have yet to be elucidated) must exist in non-B cells.

iii) Ig Enhancer Functions

Ig locus enhancers are important for the transcription of the Ig genes because RNA polymerase II binds only weakly to Ig promoters in the variable region. Prior to V(D)J recombination, the distance between the Ig variable promoters and the Ig enhancers is in the order of 200–300 kb. This distance is reduced to 2–3 kb after V(D)J recombination, bringing the promoters within the range of enhancer influence. The effect of bringing the enhancer within range of the promoter is to increase the efficiency of transcription initiation, but the precise mechanism is still unknown. It is thought to be related to the change in chromatin structure that occurs when a nuclear DNA binding factor binds to its specific motif in the enhancer region.

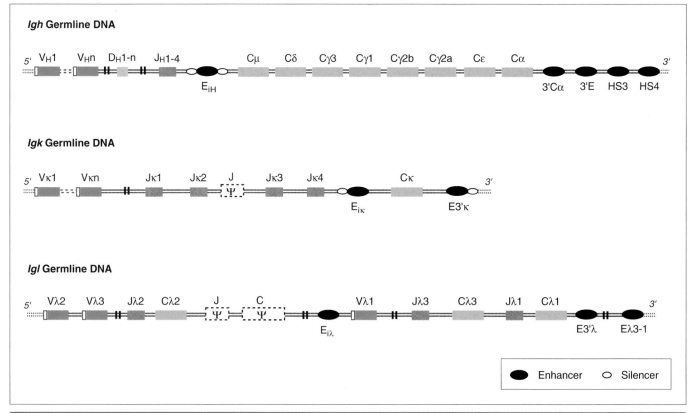

Figure 8-13

Igh Locus Enhancers
The positions of sequences in the murine *Igh, Igk,* and *Igl* loci that can act as enhancers or silencers of Ig transcription are shown. Similar sites of transcriptional control exist in the human Ig loci. The 3′Cα enhancer is also called HS3a and the 3′E enhancer is also called HS1,2.

This structural change may increase the stability of the transcription initiation complex that assembles on the promoter.

The E_{iH} enhancer was identified originally as being required for early Ig expression but is also now thought to be involved in V(D)J recombination. The binding of nuclear factors to E_{iH} may affect the state of the chromatin in the area and increase the accessibility of the locus to recombinase activity. However, studies of $E\mu^{-/-}$ knockout mice which lack E_{iH} have shown that V(D)J recombination is impaired but not abolished in the absence of E_{iH}. It is possible that one or more of the enhancers located 3′ to the coding exons may be able to compensate for a deficiency of E_{iH}.

The importance to Ig gene expression of enhancers located in the introns is highlighted by the fact that knockout mutant $iE\kappa^{-/-}$ mice, which lack the intronic enhancer of the *Igk* locus ($E_{i\kappa}$) but retain the Cκ exon, fail to carry out VκJκ rearrangement in their B cells. About 50% of the normal number of mature B cells is produced in these mice, and there is a dramatic increase in the proportion of B cells expressing λ chains. These results demonstrate that the function of $E_{i\kappa}$ is critical for *Igk* rearrangement, but that *Igk* rearrangement *per se* is not critical for *Igl* rearrangement and λ light chain expression. Contrast this blockage of VJ rearrangement with the phenotype of the $C\kappa^{-/-}$ knockout mice discussed earlier in this chapter. These animals lack the constant exon of the kappa chain gene

and cannot produce a functional κ light chain, but still are able to carry out VκJκ rearrangement, albeit at a reduced level.

II. DNA BINDING MOTIFS AND NUCLEAR TRANSCRIPTION FACTORS

As stated previously, enhancers are collections of DNA motifs that are binding sites for nuclear transcription factors. DNA binding motifs can be found in either gene promoters or enhancer regions, and some are found in both. Nuclear transcription factors are proteins or complexes of protein subunits that are made in the cytoplasm but that translocate to the nucleus to bind to DNA motifs. Nuclear transcription factors are *trans-acting*, meaning that they are free molecules capable of binding to their specific motifs on any chromosome. These molecules have been identified by a battery of assays that depend on the association of the nuclear transcription factor with the correct DNA binding motif (refer to Box 8-5).

In response to an extracellular stimulus, such as the engagement of an antigen receptor by antigen, an intracellular signaling pathway is activated that ultimately results in the binding of one or more transcription factors to specific motifs located in a promoter and/or enhancer of one or more genes. By binding to its motif, the transcription factor protein alters the conformation of the DNA at that site in some way such that transcription from the promoter of the target gene is influenced, either positively or negatively. Since more than one factor likely binds to and influences a given gene at one time, total enhancer activity and the net level of transcription will depend on the competing influences of these factors. Similarly, the deletion of a single binding site rarely has a great influence on the function of an enhancer, because it is just one of a number of such sites. Some nuclear transcription factors are expressed in a wide variety of cells and induce the transcription of multiple genes; others expressed only in a particular cell type at a particular stage in cellular maturation. Two nuclear transcription factors that are important in Ig gene expression are discussed in the following sections; six others that play roles in B cell development and activation are described in Chapter 9.

i) NF-κB

NF-κB is a heterodimeric protein that was first identified as a nuclear factor stimulating the transcription of the κ chain gene in B cells. In developing B cells which do not yet express mature Ig chains, NF-κB is a dimer of two DNA-binding subunits: p50 ("protein of 50 kDa") and p65 ("protein of 65 kDa"). It should be noted that NF-κB is expressed in the cytoplasm of a wide variety of cells, and can be present as a heterodimer of p50/p65, or as a homodimer of p65 subunits. In order to control its stimulatory activity, NF-κB is regulated by the binding of several inhibitory proteins, known as IκB ("inhibitor of κB") proteins. Prior to external stimulation of the cell, the inactive NF-κB heterodimer is complexed in the cytoplasm with an IκB molecule (Fig. 8-14). Certain stimuli, such as the binding of a cytokine to its receptor or an antigen to a BCR, trigger an intracellular signaling pathway resulting in the activation of a cytoplasmic complex with kinase activity known as the IKK (IκB kinase) complex. The IKK complex contains two kinase enzymes, IKK1 (IKKα) and IKK2 (IKKβ), and a modulatory subunit called NEMO (NF-κB essential modulator; also known as IKKγ). The kinase activity of the complex phosphorylates IκB, causing it to be released from NF-κB. Thus freed, the NF-κB molecule becomes active and can enter the nucleus through the nuclear membrane pores. Once in the nucleus, NF-κB seeks out its binding motif of 10 nucleotides known as the κB sequence on the chromosomal DNA, binds to it, and in so doing, dramatically increases the efficiency of transcription of the target gene. In antigen-stimulated mature murine B cells, NF-κB is constitutively activated and always present in the nucleus, so that the transcription of the κ chain gene (which has undergone V(D)J recombination such that the enhancer regions are positioned within range of the variable region promoters) is constant and vigorous, as one would want in these cells. The κB motif also appears within the promoters and/or enhancers associated with many other genes important for lymphocyte development, cell cycle, cell survival, cell adhesion, and inflammatory and immune responses, including the genes encoding MHC class I and several cytokines.

ii) OCTA-Binding Transcription Factors

One key DNA binding motif that is active almost exclusively in B cells is an octamer known as OCTA (nucleotide sequence

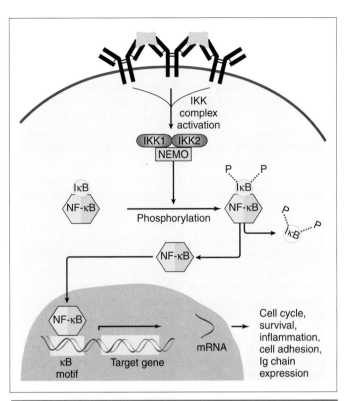

Figure 8-14

NF-κB (p65/p50) Regulation of Gene Transcription
This figure illustrates the sequence of cellular events leading to stimulation of gene transcription by the nuclear transcription factor NF-κB. In a resting cell, NF-κB remains inactive in the cytoplasm due to binding of a group of inhibitory proteins called IκB. An external stimulus, such as receptor cross-linking, causes activation of the IKK complex made up of the modulator protein NEMO and the kinases IKK1 and IKK2. The activated kinases phosphorylate IκB, triggering its degradation and the release of NF-κB. Free NF-κB moves into the nucleus and binds to κB motifs located in the promoters of its many target genes, inducing their transcription. Examples of cell surface receptors that trigger specific gene transcription via NF-κB include TNFR, the TLRs, and IL-1R.

ATGCAAAT). OCTA motifs are utilized in both the mouse Ig heavy chain and κ light chain promoters, and in the internal enhancers of these genes (as well as in a number of non-Ig genes). The presence of an intact OCTA motif in the promoter has been shown to be crucial for Ig gene transcription. Two DNA-binding proteins recognize and bind the OCTA motif: Oct-1, a ubiquitously-expressed transcriptional activator, and Oct-2, which is expressed only in lymphoid cells but at all stages of B cell differentiation. However, knockout mutant Oct-2$^{-/-}$ mice, which lack any expression of the Oct-2 transcription factor, are still able to express their Ig genes. Perhaps surprisingly, these mice do show a block in the maturation of Ig-bearing B cells into Ig-secreting B cells. This result indicates that Oct-2 may not be strictly required for the expression of the Ig genes themselves but is essential for another maturational step.

This brings us to the end of our discussion of Ig gene structure. In the next chapter, we examine the ontogeny of B cells, their activation by antigen, the generation of antibody diversity following activation, and the biological expression of these marvelously complex immunoglobulin genes.

SUMMARY

Antibodies are made up of heavy and light immunoglobulin chains, and these chains are derived from the Ig genes located in the *Igh*, *Igl*, and *Igk* loci. The Ig genes contain variable and constant exons but are unusual in that the variable (V) exons are assembled at the DNA level from essentially random combinations of V (variable), D (diversity), and J (joining) gene segments. There are several *Igh* constant (C) exons, each of which is made up of several smaller sub-exons. A single intact *Igk* constant exon exists, as well as multiple intact *Igl* constant exons. Using recombination signal sequences (RSSs) flanking each V, D, and J gene segment, the RAG recombinase enzymes carry out V(D)J recombination to join an essentially randomly chosen D segment in the *Igh* locus to a randomly chosen J segment. The DJ entity is then joined to a random V segment to assemble a complete *Igh* V exon. V and J segments are similarly joined in the *Igk* and *Igl* loci to form light chain V exons. In a developing B cell, the *Igh* locus is rearranged first. To produce a functional Ig heavy chain, a primary transcript that includes the assembled V, $C\mu$, and $C\delta$ exons is processed by con-ventional mRNA splicing. Productivity testing relies on the expression of a completed *Igh* chain on the cell surface. Successful rearrangement on one chromosome invokes allelic exclusion which shuts down V(D)J rearrangement on the other chromosome. Igk locus rearrangement follows that of the *Igh* locus and appears to be stimulated by the successful production of a functional μ chain. If a functional κ light chain is not produced from either chromosome, recombination of the *Igl* locus may then result in generation of a functional λ chain. Productive rearrangement of at least one heavy and one light chain must occur in order for the B cell to survive. Much of the diversity in the antibody repertoire is derived from the existence and random recombination of the multiple germline V, D, and J gene segments as well as from random heavy–light chain pairing. Additional junctional diversity arises from the imprecise joining of V, D, and J gene segments and can result from nucleotide deletion and the addition of N and P nucleotides. Control sequences in the Ig loci include enhancers containing multiple DNA binding motifs for nuclear transcription factors.

Selected Reading List

Bain G., Maandag E. C., Izon D. J., Amsen D., Kruisbeek A. M. *et al.* (1994) E2A proteins are required for proper B cell development and initiation of Ig gene rearrangements. *Cell* **79**, 885–892.

Blackwell T. K. and Alt F. W. (1989) Mechanism and developmental program of immunoglobulin gene rearrangement in mammals. *Annual Review of Genetics* **23**, 605–636.

Bogue M. and Roth D. R. (1996) Mechanism of V(D)J recombination. *Current Opinion in Immunology* **8**, 175–180.

Brandt V. and Roth D. (2002) A recombinase diversified: new functions of the RAG proteins. *Current Opinion in Immunology* **14**, 224–229.

Butler J. E. (1997) Immunoglobulin gene organization and the mechanism of repertoire development. *Scandinavian Journal of Immunology* **45**, 455–462.

Cook G. P. and Tomlinson I. M. (1995) The human immunoglobulin VH repertoire. *Immunology Today* **16**, 237–242.

Dang W., Nicolajczyk B. S. and Sen R. (1998) Exploring functional redundancy in the immunoglobulin u heavy-chain gene enhancer. *Molecular and Cellular Biology* **18**, 6870–6878.

Grawunder U. and Harfst E. (2001) How to make ends meet in V(D)J recombination. *Current Opinion in Immunology* **13**, 186–194.

Grawunder U., West R. B. and Lieber M. R. (1998) Antigen receptor gene rearrangement. *Current Opinion in Immunology* **10**, 172–180.

Hozumi N. and Tonegawa S. (1976) Evidence for somatic rearrangement of immunoglobulin genes coding for variable and constant regions. *Proceedings of the National Academy of Sciences U.S.A.* **73**, 3628–3632.

Kuhn R. and Schwenk F. (1997) Advances in gene targeting methods. *Current Opinion in Immunology* **9**, 183–188.

Lieber M. R., Chang C.-P., Gallo M., Gauss G., Gerstein R. and Islas A. (1994) The mechanism of V(D)J recombination: site-specificity, reaction fidelity and immunologic diversity. *Seminars in Immunology* **6**, 143–153.

Livak F. and Petrie H. (2001) Somatic generation of antigen-receptor diversity: a reprise. *Trends in Immunology* **22**, 608–612.

Mak T. and Simard J. (1998) "Handbook of Immune Response Genes." Plenum Press, New York, NY.

Matsuda F. and Honjo T. (1996) Organization of the human immunoglobulin heavy-chain locus. *Advances in Immunology* **62**, 1–29.

May M. J. and Sankar G. (1998) Signal transduction through NF-kappa B. *Immunology Today* **19**, 80–88.

Mostoslavsky R., Alt F. and Rajewsky K. (2004) The lingering enigma of the allelic exclusion mechanism. *Cell* **118**, 539–544.

Mullis K. B. (1990) The unusual origin of the polymerase chain reaction. *Scientific American* **April**, 56–65.

Oettinger M. A. (1999) V(D)J recombination: on the cutting edge. *Current Opinion in Cell Biology* **11**, 325–329.

Radic M. and Zouali M. (1996) Receptor editing, immune diversification, and self-tolerance. *Immunity* **5**, 505–511.

Ramsden D. A., van Gent D. C. and Gellert M. (1997) Specificity in V(D)J recombination: new lessons from biochemistry and genetics. *Current Opinion in Immunology* **9**, 114–120.

Rashtchian A. (1995) Novel methods for cloning and engineering genes using the polymerase chain reaction. *Current Opinion in Biotechnology* **6**, 30–36.

Saiki R. K., Gelfand D. H., Stoffel S., Scharf S. J., Higughi R. *et al.* (1988) Primer-directed enzymatic amplification of DNA with a thermostable DNA polymerase. *Science* **239**, 487–491.

Sambrook J., Fritsch E. and Maniatis T. (1989) In "Molecular Cloning: A Laboratory Manual," 2nd edn. Cold Spring Harbor Laboratory Press, Cold Spring Harbor, NY. 16.57–16.62.

Schatz D. (2004) V(D)J recombination. *Immunological Reviews* **200**, 5–11.

Taylor G. R. and Logan W. P. (1995) The polymerase chain reaction: new variations on an old theme. *Current Opinion in Biotechnology* **6**, 24–29.

Tonegawa S. (1983) Somatic generation of antibody diversity. *Nature* **302**, 575–581.

Watson J., Tooze J. and Kurtz D. "Recombinant DNA: A Short Course." W. H. Freeman, New York, NY.

Williams A. F. and Barcla A. N. (1988) The immunoglobulin superfamily—domains for cell surface recognition. *Annual Review of Immunology* **6**, 381–405.

The Humoral Response: B Cell Development and Activation

9

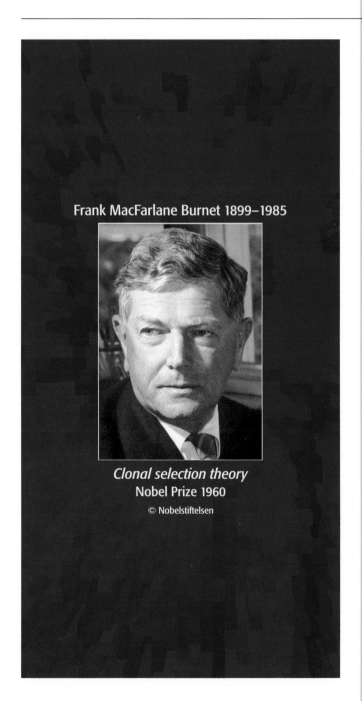

Frank MacFarlane Burnet 1899–1985

Clonal selection theory
Nobel Prize 1960

© Nobelstiftelsen

CHAPTER 9

A. THE MATURATION PHASE OF B CELL DEVELOPMENT

B. THE DIFFERENTIATION PHASE OF B CELL DEVELOPMENT

"The military don't start wars. Politicians start wars"
—William Westmoreland

In Chapter 8, we examined in detail the immunoglobulin genes and the mechanisms of gene segment shuffling and combinatorial pairing that contribute to the diversity of the antibody repertoire evident in naive B cells. In this chapter, we discuss B cell development from the stem cell to mature memory cells and Ig-producing plasma cells, and how the developmental pathway of the B cell dictates the humoral immune response. The expression patterns of the genes involved in B cell development determine the fate of each newly-produced B cell clone, and whether it survives the selection mechanisms of central B cell tolerance that distinguish self-reactive from nonself-reactive cells. The mechanisms of peripheral tolerance (described in Ch.16) then silence any self-reactive lymphocytes that escaped elimination during the establishment of central tolerance. Thus, it is largely B cells directed against nonself antigens that are left to mediate the antibody response. We will also continue our examination of antibody diversity, focusing on those mechanisms that come into play after a naive B cell is activated by an encounter with antigen, namely somatic hypermutation and isotype switching.

Overview of B Cell Ontogeny

As we learned in Chapter 3, all B lymphocytes are derived from hematopoietic stem cells (HSCs). During gestation, B lineage cells are generated first by the embryo, and then by the fetal liver, spleen, and bone marrow. After birth, B cell lymphopoiesis occurs almost exclusively in the bone marrow. Regardless of species, the life span of the resulting mature naive B cell is short, the average half-life being 1–4 days. In contrast, the life span of an antigen-stimulated B cell in the periphery may be in the order of weeks.

Direct contact with stromal bone marrow cells as well as various growth factors and cytokines secreted by stromal and other hematopoietic cells influence a stem cell to direct its descendants down the path of B cell development. How can one tell whether a developing cell is destined to become a mature B cell? In the mouse, the B220 cell surface protein

(also known as CD45R) has long been regarded as a pan-B cell marker; that is, present on all B cells regardless of developmental stage. However, some NK cell precursors have also been found to express B220. A better choice as an indicator of B lineage in mice is the CD19 marker, which appears on all B cell populations (except the earliest precursors) but not on NK cells. The human version of the CD19 protein also marks a human cell as being of B lineage, although it has yet to be definitively confirmed whether the earliest committed human B precursors also express this protein.

B cell development takes place in a series of well-defined stages that can be grouped into two phases: the *maturation phase* (hematopoietic stem cell to mature naive B cell) and the *differentiation phase* (antigen-activated mature B cell to antibody-secreting plasma cells and memory B cells). Under the influence of the appropriate cytokines and growth factors but independently of foreign antigen, an HSC divides and evolves into mature naive B cells through at least 10 developmental stages (Fig. 9-1). These stages are as follows (where "pro" indicates "progenitor," and "pre" indicates "precursor"): the HSC, the *common lymphoid progenitor* (CLP; discussed in Ch.3), the *pro-B cell*, the *pre-B-I cell*, the *pre-B-II cell* (three different subtypes), the *immature* or *early B cell*, the *mature naive B cell* in the *bone marrow*, and the *mature naive B cell* in the *periphery*. Because the morphology of these different types of B cells is almost identical, these stages are identified *in vitro* by the presence or absence of specific cell surface marker proteins (including the Ig chains). The maturation phase ends when mature naive B cells exit the bone marrow and fan out into the secondary lymphoid tissues in the body's periphery, including the lymph nodes, tonsils, Peyer's patches, and spleen. The interaction of the mature naive B cell with specific antigen in these peripheral locations initiates the differentiation phase, and the complete activation and differentiation of a mature naive B cell depends on the continued presence of specific antigen. Two stages of the differentiation phase of B cell development can be identified: (1) the activation of the

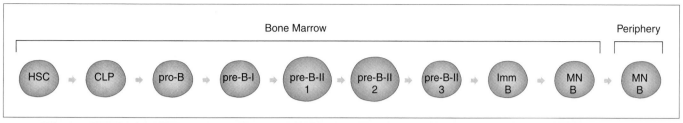

Figure 9-1

Developmental Stages of B Cell Maturation
A hematopoietic stem cell (HSC) gives rise to a common lymphoid progenitor (CLP). The CLP proceeds through the progenitor ("pro-B") and multiple precursor ("pre-B") steps to become an immature B cell (Imm B) and then a mature, naïve B cell (MN B) in the bone marrow. The MN B cells then leave the bone marrow via the circulation and spread throughout the secondary lymphoid tissues in the periphery, where they await interaction with specific antigen.

mature B cell, and (2) the generation of effector *plasma cells* and *memory B cells*. Plasma cells gain morphological features that make them readily recognizable under the light microscope, but activated mature and memory B cells must again be distinguished by the presence or absence of specific cell surface marker proteins.

A. The Maturation Phase of B Cell Development

We will now discuss in detail the maturation of mouse B cells in the bone marrow from the stem cell to mature naïve B cell stage. The key proteins that exhibit differential expression during this phase of B cell development in the mouse are c-kit, IL-7Rα, CD25, CD40, CD43, surrogate light chain (see later), TdT, RAG-1 and RAG-2, and transcripts from the Igk locus. The role of each of these is described in the following sections, and the stage of development at which each appears is illustrated in Figure 9-2. In humans, the list of proteins known to show differential expression during B cell development includes CD10, CD34, surrogate light chain, CD40, RAG-1 and RAG-2, and TdT. As well as genes expressed in the B cells themselves, particular proteins expressed by non-B cells in the near vicinity have profound and essential effects on B cell development.

The discussion that follows focuses primarily on B cell development in the mouse, since this is a species in which experimental work can readily be carried out and about which most is known. Where possible, references to human B cell development have been included.

I. PRO-B CELLS

i) Pro-B Cell Markers

Pro-B cells are the very earliest clearly defined stage of B cell development. The earliest pro-B cells can be recognized by their expression of B220 and by the fact that all their Ig genes are still in germline configuration; that is, the stage just prior to *Igh* locus rearrangement. Since no germline rearrangement has commenced in these cells, they appear in normal

numbers in mice lacking elements of the V(D)J recombination machinery, including *scid* mice and RAG-1/2 knockout mice. Pro-B cells do not yet express CD19, CD25, CD40, or transcripts from the *Igk* locus but do express c-kit, CD43 (leukosialin, an adhesion molecule), RAG-1/2, and TdT. As we saw in Chapter 8, these enzyme activities are required to commence rearrangement of the *Igh* locus. In the latest pro-B cells, the first attempts at $D_H J_H$ joining in the *Igh* locus are initiated, such that the $D_H J_H$ sequences on at least one chromosome are clearly distinct from the germline *Igh* sequence. Pro-B cells also express a specialized heterodimer called the *surrogate light chain* (SLC) and the IL-7R. The SLC pairs with newly synthesized μ heavy chains later in B cell development. Since there are no μ heavy chains yet in pro-B cells, the SLC appears on the cell surface in association with a non-Ig complex of glycoproteins known as gp130/gp35–65.

Cells equivalent to pro-B cells in mice have been identified in humans but have yet to be extensively characterized.

ii) The Role of Bone Marrow Stromal Cells

The stromal cells of the bone marrow are absolutely required for the development of pro-B cells in the mouse. It is thought that the stromal cells are vital for two reasons: (1) they secrete various chemokines and cytokines that are critical for hematopoietic development and (2) they make necessary direct cell–cell contacts with both pro-B and pre-B cells. Early B cells adhere to stromal cells through the interactions between the adhesion molecule *VLA-4* on the B cells with the ligand *fibronectin* on stromal cells (Fig. 9-3). The earliest contribution by stromal cells is their secretion of *SDF-1* (stromal cell derived factor). SDF-1 is a chemokine (also designated CXCL-12) that binds to its receptor CXCR4 on pro-B cells. SDF-1 not only induces the proliferation of pro-B cells but also causes them (and their successors) to remain in the fetal liver or bone marrow until development is completed. In addition, a pro-B cell developing in the correct bone marrow microenvironment expresses the adhesion molecule *CD44*, which can bind to *hyaluronic acid* (one of its many ligands) present on the surface of the bone marrow stromal cell. Stimulation of CD44 in this way promotes the interaction of the receptor *c-kit* on the pro-B cell with its ligand *stem cell factor* (SCF) produced by stromal cells. The c-kit receptor has intrinsic tyrosine kinase activity and so is able to convey an intracellular signal to the nucleus of

	pro-B	pre-B-I	pre-B-II 1	pre-B-II 2	pre-B-II 3	Imm B	MN B	MN B
				Bone Marrow				Periphery
c-kit	+	+						
CD43	+	+	+/−	+/−				
B220	+	+	+	+	+	+	+	+
CD19		+	+	+	+	+	+	+
CD25			+	+	+			
CD40					+	+	+	+
IL-7R	+	+	+	+	+			
RAG-1/2	+	+			+	+	+	
TdT	+	+						
SLC	+	+	+					
μ chain			+	+	+	+	+	+
κ chain					+	+	+	+
Igα/β		+	+	+	+	+	+	+
Rearranged D$_H$J$_H$	+ (late)	+	+	+	+	+	+	+
Rearranged V$_H$D$_H$J$_H$		+	+	+	+	+	+	+
Rearranged Vκ					+	+	+	+
κ transcription				+*	+	+	+	+
pre-BCR complex			+	+/−				
IgM						+	+	+
IgD						+/−	+	+
MHC class II							+	+
LFA-1, ICAM-1, CD22							+	+
L-selectin							+	+
CD23							+	+

Figure 9-2

Major Changes in Markers of Mouse B Cell Maturation
The expression of various proteins as well as the rearrangement status of the immunoglobulin genes can be used to identify and distinguish cells at various stages of B cell maturation. The asterisk (*) indicates the production of sterile (non-translatable) transcripts. SLC, surrogate light chain. With information from Hardy R. and Hayakawa K. (2001) B cell developmental pathways. *Annual Reviews in Immunology* **19**, 595–621.

the pro-B cell, telling it both to undergo mitosis and to commence expression of the IL-7Rα chain. Related to c-kit is *Flk2/Flt3*, another receptor tyrosine kinase on the B cell surface whose stimulation promotes early B lymphopoiesis. Gene knockout mice lacking Flk2/Flt3 have a partial deficiency of B cell progenitors, and mice deficient for both c-kit and Flk2/Flt3 lack almost all hematopoietic cells.

An interesting aspect of murine B cell development is highlighted by comparing the phenotypes of IL-7$^{-/-}$ and IL-7Rα$^{-/-}$ knockout mice. IL-7 is an essential growth factor for murine B cell precursors (but is much less important for human B cell development). In IL-7Rα$^{-/-}$ mice, a block in B cell development

occurs in the stem cell to pro-B cell transition; in IL-7$^{-/-}$ animals, it occurs slightly later, at the pro-B to pre-B transition. This effect arises because, in addition to the IL-7 receptor, the IL-7Rα chain participates in the receptor for a widely-expressed growth factor called *thymic stromal lymphopoietin* (TSLP). TSLP has activities that are similar to those of IL-7, but signaling through its receptor does not apparently involve either the common γc chain or any of the Jak kinases (see Ch.17). The loss of TSLP signaling seems to impede B cell development prior to the point at which IL-7 acts.

It should be noted that bone marrow cells other than the stromal cells may also produce cytokines influencing mouse

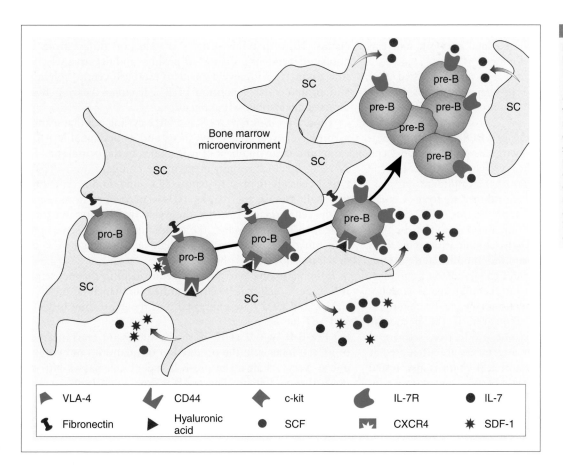

Figure 9-3

Role of Stromal Cells in B Cell Development
Bone marrow stromal cells (SCs) support early B cell development by expressing the cell surface adhesion molecules and secreting the chemokines and cytokines shown in the key. IL-7 produced by the SCs may eventually downregulate the expression of adhesion molecules on pre-B cells, allowing their release from the SC network. Abbreviations are defined in the text.

▶ VLA-4	CD44	c-kit	IL-7R	● IL-7
Fibronectin	Hyaluronic acid	● SCF	CXCR4	✳ SDF-1

B cell development. For example, bone marrow macrophages secrete both IFNα and IFNβ, which inhibit IL-7-induced B cell growth, and IL-1, which can act either positively or negatively on B cell growth. IL-3, a cytokine with a pleiotropic role in hematopoiesis, binds to its receptor on both pro- and pre-B cells, inducing cellular proliferation in the latter case.

II. PRE-B CELLS

i) The Role of the Surrogate Light Chain
In order to make sense of the pre-B cell progression, we need to digress for a moment to discuss some unusual proteins critical for production of the BCR. The surrogate light chain (SLC) plays a key role in the production of the first μ heavy chains in pre-B cells. The intracellular transport machinery requires that an Ig heavy chain in the cytoplasm be combined with some form of Ig light chain to get the complex to the cell membrane where it is tested for functionality, the "productivity testing" of a particular $V_H D_H J_H$ recombination. Since the *Igl* and *Igk* loci undergo V(D)J recombination only after rearrangement has occurred in the *Igh* locus, no functional mature Ig light chains are present during early B cell development to do the job of getting the first candidate μ heavy chains from the cytoplasm to the membrane. Instead, proteins homologous to the mature Ig light chains come together to form the SLC, and the SLC pairs with a newly synthesized μ heavy chain and conveys it

to the membrane. The proteins composing the SLC are transcribed from genes related to the Ig light chains that are located in the *Igl* locus. In both mice and humans, a gene called V_{preB} is expressed in all pro-B and pre-B cells. V_{preB} encodes a polypeptide whose invariant sequence is homologous to the V regions of λ and κ chains. Similarly, a gene called λ5 supplies a constant region-like polypeptide. The polypeptides encoded by V_{preB} and λ5 associate non-covalently to form a light chain-like structure capable of associating with a few of the cytoplasmic μ chains and facilitating their transient insertion into the membrane of a cell at the so-called pre-B-I stage (see later).

ii) Pre-B Cell Stages
iia) Pre-B-I cells. The pre-B-I cell is the earliest cell in which complete V(D)J rearrangement of *Igh* genes can be detected; that is, a V_H gene segment is joined to the previously joined $D_H J_H$ segment on the same chromosome. However, no μ Ig chains are yet detectable in the cytoplasm. Both the *Igl* and *Igk* loci remain in germline configuration and there is no expression yet of light chain transcripts.

From the point of view of identifying markers, a mouse pre-B-I cell expresses B220, CD19, c-kit, CD43, and SLC in association with gp130/gp35–65, but lacks CD40 and CD25. The presence of the complete IL-7 receptor (CD127) on the pre-B-I cell surface allows the binding of IL-7 secreted by the surrounding stromal cells and signaling that supports continued pre-B cell development and proliferation. Eventually, the expression of adhesion molecules is downregulated so that the pre-B cells are released from the stromal network and can proceed to the next stage of development.

A putative human pre-B-I cell has been isolated that does not contain cytoplasmic μ Ig chains but does express SLC, RAG-1/2, and TdT, and the markers CD10, CD34, and CD19.

iib) Pre-B-II type 1 cells and the pre-BCR. Following complete rearrangement of the *Igh* locus, the pre-B-II type 1 cell carries out test production of heavy chains of the μ isotype (only). SLCs combine with copies of the candidate cytoplasmic μ chain and the Igα/Igβ heterodimer to form a small number of *pre-B cell receptor* complexes (*pre-BCR*) that are inserted transiently in the membrane in order to test the productivity of this particular $V_H D_H J_H$ combination (Fig. 9-4). A pre-BCR is not a true immunoglobulin and cannot respond to antigen, but it may bind to ligands on bone marrow stromal cells such that an intracellular signal is delivered, telling the cell that a functional heavy Ig chain has been successfully synthesized. However, if the particular rearranged sequence selected is not productive, sequential $D_H \rightarrow J_H$ and $V_H \rightarrow D_H J_H$ rearrangements are attempted at the second *Igh* allele and the resulting μ chain is also tested in pre-BCR form. Failure on both chromosomes occurs in about half of pre-B-II type 1 cells, pre-empting further maturation and leading to the apoptotic death of most of these cells in the bone marrow.

With the production of a functional pre-BCR, further *Igh* locus rearrangements are terminated (allelic exclusion), and preparations for light chain locus rearrangement commence. The activities of c-kit, RAG-1, RAG-2, and TdT are lost, CD43 is downregulated, and the production of SLC decreases. The synthesis of the successful Ig μ heavy chain proceeds in earnest and results in the accumulation of large numbers of these molecules in the cytoplasm. The expression of the pre-BCR is also correlated with the proliferative expansion of those pre-B-II type 1 cells in which *Igh* locus rearrangement was successful, in effect providing a means

of *positive selection* of those cells with functional heavy chains. Note that this process of selection differs from the antigen-dependent processes of positive and negative selection that occur later in development to establish central tolerance and promote the expansion of B cell clones bearing BCRs directed against nonself antigens.

Thus, a pre-B-II type 1 cell is a large cycling cell that features a completely rearranged *Igh* locus and a functional $V_H D_H J_H$ exon, and large numbers of μ Ig chains in the cytoplasm. The *Igl* and *Igk* loci remain in germline configuration. Pre-B-II type 1 cells are the last to express SLC, and do not yet express true light chains, the marker CD40, or mature Ig molecules on the cell surface. However, these cells are the first to express CD25, the gene encoding the α subunit of the IL-2 receptor. The function of this receptor chain at this stage of B cell development remains obscure.

In human bone marrow, a cell equivalent to the mouse pre-B-II type 1 cell has been identified in which a pre-BCR, CD10, and CD19 are expressed, but the expression of CD34 and TdT has been lost.

iic) Pre-B-II type 2 cells. With the successful rearrangement of its *Igh* locus and the production of quantities of cytoplasmic μ heavy chains, the pre-B-II type 1 cell passes into the pre-B-II type 2 stage. The pre-B-II type 2 cell is also a large cycling cell but it ceases heavy chain rearrangements and the production of SLC. Signaling from the pre-BCR is no longer required and this transient complex vanishes from the cell surface. The cell prepares to commence rearrangement of the *Igk* locus (but does not actually do so) by making sterile (non-functional) κ transcripts of its unrearranged *Igk* locus. Transcription (but not translation) of the RAG-1 and RAG-2 genes resumes.

iid) Pre-B-II type 3 cells. After the signal to induce light chain locus rearrangement is received in the large pre-B-II type 2 cell, it goes through a 2-day transition during which it becomes a smaller resting cell. RAG-1 and RAG-2 proteins and their activities are again detectable. However, since TdT expression remains silenced, the light chains do not undergo

Figure 9-4
Pre-B Cell Receptor Expression
Synthesis and expression of the pre-BCR and BCR are shown at three different stages of B cell development. Not shown are the membrane-bound vesicles that normally surround proteins in transit to the cell surface. SLC (surrogate light chain) = $V_{pre-B}/\lambda 5$.

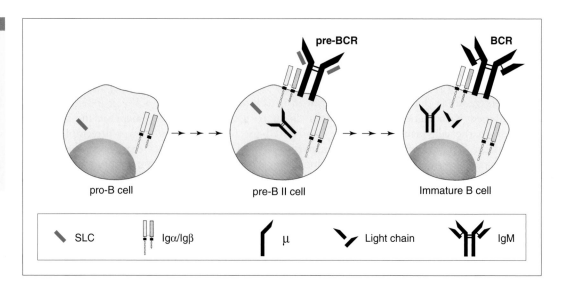

N-nucleotide addition. CD40 is also expressed for the first time, although its function at this stage is unknown.

As was described in Chapter 8, light chain production starts with the initiation of V(D)J recombination at both alleles of the *Igk* locus. Successful rearrangement on one chromosome invokes both allelic and κ–λ exclusion and the production of functional κ light chains that will combine with the cytoplasmic μ heavy chains to form functional monomeric mIgM molecules. However, if productive rearrangement fails at both κ alleles, the cell attempts to productively rearrange one or the other of the *Igl* loci. Complete failure of light chain rearrangement such that neither a functional κ nor a functional λ light chain is produced results in the apoptotic death of the cell.

Humans have cells similar to the small pre-B-II type 3 cells of mice. These human counterparts do not express CD34, SLC, or TdT, but do express CD10 and CD19 and have cytoplasmic μ heavy chains. The *Igl* and *Igk* loci may also be rearranged.

iii) Immature mIg⁺ B Cells in the Bone Marrow

The appearance of the surface IgM molecule in the membrane marks a successful cell as an immature B cell, often denoted as either an "mIg⁺" ("membrane Ig") or sometimes "sIg⁺" ("surface Ig") B cell. A signal is delivered that terminates all Ig gene rearrangements. Immature mIg⁺ B cells are also identified by their minimal IgD expression and suspended mitosis. CD25 is no longer expressed but RAG-1 and RAG-2 expression continue in spite of the cessation of gene rearrangement. These cells reside in the bone marrow and have short half-lives of only 2–3 days. In spite of their functional BCRs, these cells do not proliferate and differentiate in response to antigen binding (cannot take part in immune responses).

iiia) Central B cell tolerance.
Bone marrow cells express "housekeeping" molecules, including surface self-antigens present on all body cells. The expression of the surface IgM on an immature mIg⁺ B cell allows it to determine before it enters the periphery whether it is *autoreactive*; that is, whether the molecular structure recognized by its antigen receptor occurs in the housekeeping molecules present on the surrounding bone marrow cells. (Such structures may be proteins but are frequently carbohydrates or glycolipids.) If released to the body's periphery, cells recognizing housekeeping structures could attack self-tissues if they received T cell help from a Th cell directed against the same self-antigen. Immature B cells in the bone marrow whose BCRs bind with high affinity to these self-antigens (and thus are potentially autoreactive) receive an intracellular signal to halt development. The cell then uses its residual RAG-1 and RAG-2 activities to carry out a process called *receptor editing* (see later) in an attempt to produce a non-autoreactive BCR. If this attempt fails and the B cell is still autoreactive, it receives a further signal that induces it to undergo apoptosis before it ever becomes a mature naive B cell. An apoptotic cell is characterized by reduced cell volume, membrane "blebbing," chromatin condensation, and DNA cleavage. Such a cell sends strong signals that it should be removed by phagocytosis. By using this orderly program of cellular demise to destroy

autoreactive lymphocytes, the body avoids autoimmunity and prevents the inflammation that could result if these cells simply died in an uncontrolled manner. Thus, the establishment of central B cell tolerance relies on *negative selection* by apoptosis to eliminate self-reactive lymphocytes before they are released from the bone marrow.

Two principal pathways of apoptosis are used to effect negative B cell selection. When the mIg on an immature B cell in the bone marrow binds to self-antigen, intracellular signaling is triggered that leads directly to the apoptosis of these cells. In addition, engagement of the BCR of an immature B cell in the bone marrow results in the upregulation of a cell surface protein called *Fas* (also known as APO-1 and CD95). When Fas molecules on the surface of an immature B cell bind to molecules of *Fas ligand* (FasL) protein present on the surface of a nearby Th cell, apoptosis of the B cell is triggered. Because most antigens encountered in the bone marrow will necessarily be "self" in nature, the BCRs of immature B cells directed against nonself antigens will not be engaged, and these cells will be spared. Fas-mediated apoptosis is thought to be a key mechanism for maintaining central B cell tolerance, and also to be responsible for certain aspects of peripheral B cell tolerance (see Ch.16).

Some scientists believe that immature mIg⁺ B cells that are not negatively selected must receive some kind of *positive selection* signal in order to complete their maturation. It is thought that the receipt of this signal may depend on the BCR not recognizing self-antigen (i.e., no engagement of the BCR). In other words, the engagement of the BCR by self-antigen may preclude the receipt by a cell of the positive selection signal. The nature and indeed existence of a B cell positive selection signal remain controversial.

iiib) Receptor editing: avoiding negative selection.
An autoreactive immature mIg⁺ B cell is given a brief window of opportunity (16–18 hours after self-antigen binds to the BCR) to try to rearrange its light chain again and stave off apoptosis by altering its antigenic specificity. The cell does this by a process of secondary gene rearrangement called *receptor editing* (Table 9-1). Receptor editing occurs primarily in the light chain because, after V$_H$D$_H$J$_H$ recombination of the heavy chain is completed, there are no longer any D$_H$ gene segments available to permit further rearrangements that satisfy the structural rules of V(D)J recombination. The light chain locus, on the other hand, can still have available many upstream V$_L$ gene segments and a few downstream J$_L$ segments if an early attempt at rearrangement has been successful. In addition, light chain rearrangement can occur at the *Igl* locus if autoreactivity has resulted from rearrangement at both alleles of the *Igk* locus. It appears that the expression of the RAG genes remains "turned on" after successful L chain rearrangement in order to facilitate receptor editing. (It is also possible that RAG expression is turned off after successful L chain rearrangement, but is somehow turned back on in cells that have specificity for an autoantigen.) If receptor editing fails to achieve success in time, the B cell is negatively selected and undergoes apoptosis. However, if receptor editing is successful, the BCR no longer recognizes self-antigen, can perhaps receive a positive selection signal, and becomes a mature naive B cell in the bone marrow.

Table 9-1 **Receptor Editing**

Immature IgM⁺ B Cell			
Recognizes Self-peptide as a Result of V(D)J Joining	Carries Out Receptor Editing	Recognizes Self-peptide after Receptor Editing	Developmental Status
No	No	No	Positive selection
Yes	Yes	No	Positive selection
Yes	Yes	Yes	Negative selection

iiic) IgM and IgD co-expression. On a daily basis, about 2–5% of the pool of immature mIg⁺ B cells survives the selection processes establishing central B cell tolerance (Fig. 9-5). These cells remain in the bone marrow for 1–3 days, during which new cell surface markers are expressed prior to emigra-tion to the body's periphery. As well as the presence of surface IgM (which is mandatory for a bone marrow B cell to make the transition to a peripheral B cell), higher levels of IgD also start to appear. Both Igs are derived from the same rearranged heavy and light chain genes containing the same $V_H D_H J_H$ and

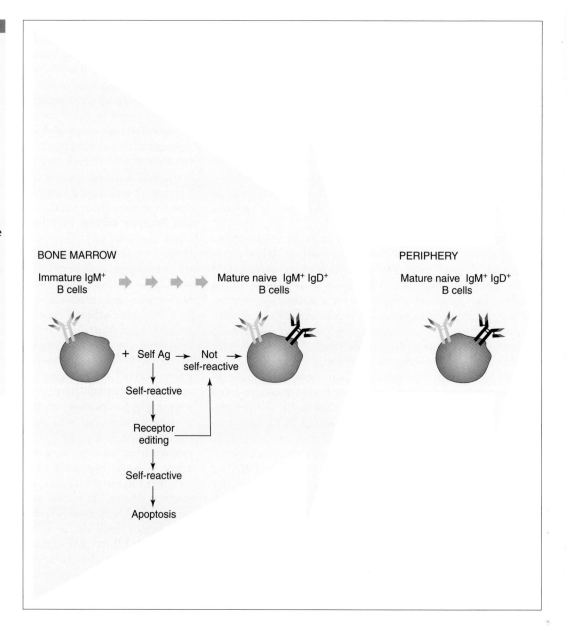

Figure 9-5

Transition from Immature to Mature B Cells
All immature IgM⁺ B cells are exposed to self-antigens present on cells in the bone marrow. Cells that are not self-reactive are positively selected and proceed directly to the mature B cell stage, during which IgD is co-expressed on the cell surface with IgM. Self-reactive immature B cells undergo receptor editing of their light chain genes in an attempt to avoid negative selection. Those that are no longer self-reactive join the mature, self-tolerant B cell pool (<5%) that enters the peripheral lymphoid tissues. The remaining 95% of immature B cells dies via apoptosis. This selection process eliminates most self-reactive B cells before they reach the periphery, establishing central B cell tolerance.

BONE MARROW

Immature IgM⁺ B cells → → → → Mature naive IgM⁺ IgD⁺ B cells

+ Self Ag → Not self-reactive →

Self-reactive
↓
Receptor editing
↓
Self-reactive
↓
Apoptosis

PERIPHERY

Mature naive IgM⁺ IgD⁺ B cells

V_LJ_L exons, respectively, and therefore have the same V domains in their antigen binding sites and the same antigenic specificity. It is only the constant region of the heavy chain protein that differs between the IgM and the IgD molecules.

Co-expression of IgM and IgD occurs because the long (about 15 kb) primary RNA transcripts that are synthesized from the rearranged heavy chain gene contain the sequences encoded by the $V_HD_HJ_H$ exon and <u>both</u> the $C\mu$ and $C\delta$ exons (refer to Fig. 8-3 in Ch. 8). The $C\mu$ and $C\delta$ exons are only 5 kb apart in the genomic DNA and there is no isotype *switch region* (see later) between them, so that this transcription is readily accomplished. This transcript also contains four polyA addition sites, two after the sub-exons in the $C\mu$ exon and two after the sub-exons of the $C\delta$ exon. If the $C\mu$- and $C\delta$-containing primary transcript is cleaved at either of the polyA sites following the $C\mu$ exon, the $C\delta$ sequences are lost and the mRNA is translated into an IgM heavy chain. Alternatively, if either of the polyA sites following the $C\delta$ exon are chosen for cleavage, the splice donor site 3′ to the $V_HD_HJ_H$ exon is joined to the splice acceptor site just 5′ to the $C\delta$ exon, removing the $C\mu$ exon and all introns in between. The resulting mRNA is translated into an IgD heavy chain. Both splicing pathways appear to proceed simultaneously, giving rise to a single B cell that expresses two different isotypes of membrane-bound Ig molecules. The signals dictating which splice junctions and polyadenylation sites are chosen when, and those that trigger the initiation of alternative splicing in the first place, are still unknown.

Other proteins whose presence identifies later stage immature bone marrow B cells are complement receptors; new adhesion molecules, including LFA-1 and ICAM-1; CD23, an Fcε receptor with affinity for oligosaccharides; and lymphocyte homing receptors, including L-selectin. Proteins important for interaction with helper T cells are also now expressed, including MHC class II and CD40. Expression of CXCR4, the receptor for SDF-1 that acts to retain B cells in the bone marrow, is downregulated. Thus, fully equipped to participate in an immune response, these immature naive B cells are released from the bone marrow into the blood.

iv) Naive B Cells in the Periphery: The Transition to Maturity

At this point, immature naive B cells traveling in the blood are known as *transitional type 1 immature B cells*, or T1 cells. T1 cells extravasate first into the red pulp of the spleen and then into the PALS, all the while undergoing further maturation marked by a gradual increase in membrane IgD expression. After about 24 hours in the PALS, T1 cells become *transitional type 2 immature B cells*, or T2 cells. Some of these T2 cells start to colonize the B cell-rich areas of the spleen and acquire the ability to emigrate to other secondary lymphoid tissues. These cells, which will recirculate throughout the lymphoid system, are known as *follicular B cells*, and are considered fully mature naive peripheral B cells. Other T2 cells move from the red pulp of the spleen to its marginal zone, becoming mature *marginal zone (MZ) B cells*. While primarily splenic, some MZ B cells can also be found in lymphoid follicles in the Peyer's patches and mesenteric lymph nodes. These mostly non-recirculating B cells tend to have a lower threshold of activation than do follicular B cells and have longer life spans. MZ B cells respond primarily to Ti antigens (rather than to Td antigens, like follicular B cells do) and do not give rise to memory B cells. How a T2 cell decides to become a follicular or MZ B cell may depend on signals conveyed by the strength of the interaction between the BCR and surrounding self-antigens. Weak avidity interactions appear to favor MZ development, while stronger interactions promote the follicular fate.

The transition of T1 cells to T2 cells depends on survival signals delivered by a molecule called BAFF (*B lymphocyte activating factor belonging to the TNF family*). [BAFF is also known in the literature as BLyS (B lymphocyte stimulator) as well as by several other names.] BAFF is produced principally by macrophages and DCs and occurs in a transmembrane form as well as a biologically active soluble homotrimer created by cleavage of the transmembrane form. The expression of membrane-bound BAFF on macrophages and DCs is upregulated by treatment with IFNγ, IFNα/β, or LPS, and downregulated by IL-4. BAFF binds to its cognate receptor BAFF-R, expressed on T1 and T2 B cells. This engagement triggers survival signals that are thought to lead to NF-κB activation and the upregulation of the anti-apoptotic genes Bcl-2 and Bcl-xL, and the downregulation of various pro-apoptotic genes. Knockout mice lacking BAFF show a complete lack of T2 and mature peripheral B cells, indicating that BAFF is indispensable for continued B cell development in the secondary lymphoid tissues.

While mIgM is essential for all responses of mature, naive B cells to antigen, mIgD does not appear to be required for B cell function. There is some evidence that IgD may prolong the life span of mature naive B cells in the periphery, which may in turn later influence the development of B cell memory in response to Td antigens. RAG-1 and RAG-2 activities are permanently lost in mature B cells, so that no further V(D)J gene segment shuffling can occur in either the mature B cell or its memory and plasma cell progeny. These B cells are now poised to encounter antigen.

B. The Differentiation Phase of B Cell Development

Once mature naive follicular B cells enter the periphery, antigenic stimulation starts to play a critical role in B cell development. It is thought mature naive B cells are induced to undergo apoptosis within a few days unless they are stimulated by specific antigen. In fact, the vast majority of newly-produced B cells die within a few days of entering the periphery: only those few that reach the threshold of receptor stimulation necessary for activation survive. The scale of this "waste" is staggering. The adult human bone marrow continuously churns out new pro-B cells such that about 10^9 mature naive B cells are released into the periphery every day. However, over the course of their brief life span, the chance of these B cells encountering their specific epitope as well as the T lymphocytes and accessory cells necessary to respond is extremely limited: only 1 in 10^5 mature naive B cells in the periphery is activated at any one time. Similarly, an adult mouse bone marrow produces about 5×10^7 mature B cells daily, contributing to a pool of about 5×10^8 mature peripheral B cells. However, at any one time, only $(1–2) \times 10^6$ will encounter antigen and receive a survival signal. It should be noted that the probability of a given B cell becoming stimulated is affected by neither the immunological state nor immunological history of the individual, nor by prior VDJ combinations (which occur randomly). Antigen-stimulated mature B cells have long life spans, often exhibiting half-lives of more than 6 weeks. The small numbers of B cells that actually participate in an immune response are thus compensated for by their increased longevity and their immense proliferation upon activation.

Mature naive follicular B cells in the periphery face different fates depending on the nature of the antigen they encounter (Table 9-2). As mentioned previously, if the BCR fails entirely to bind to antigen, the B cell dies by apoptosis. If the BCR binds to a Td antigen and the B cell is successfully activated by interacting with a Th cell stimulated by its epitope on the same antigen, the B cell is induced to proliferate and differentiate and becomes fully capable of somatic hypermutation and isotype switching. Both memory cells and plasma cells are produced and an effective humoral response is mounted. If the BCR binds to a Ti antigen, the B cell will undergo only limited isotype switching. Plasma cells are generated, the majority of which secrete IgM antibodies, but differentiation into memory B cells does not occur. If the BCR binds to a self-antigen (that was not encountered during negative selection in the bone marrow), the B cell will either be deleted by apoptosis or induced to become *anergic* (non-responsive) by the mechanisms that maintain peripheral tolerance (see Ch.16). Interestingly, overexpression of BAFF in transgenic mice leads to autoimmunity, most likely because the survival signals triggered by BAFF overpower the negative selection signals necessary to remove autoreactive B cells (see Ch.29). BAFF also binds to two other long-windedly named

receptors called BCMA (*B cell maturation antigen*) and TACI (*transmembrane activator and CAML interactor*; CAML, *calcium modulator and cyclophilin ligand*). The binding of BAFF to these receptors does not mediate B cell survival, but may deliver negative regulatory signals. Indeed, knockout mice lacking TACI show greatly elevated B cell numbers and die rapidly with symptoms of severe autoimmunity.

A B cell subpopulation that expresses upregulated levels of the marker CD5 appears to be an exception to the "be stimulated or die" rule. CD5 is a cell surface protein of unknown function that can associate with both the BCR and TCR complexes, and which is phosphorylated following antigen receptor stimulation. CD5 is present on more than half of human fetal B cells, but on only 3% of adult human lymph node B cells; it has also been found on thymocytes and T cells. In mice, $CD5^+$ B cells are known in B cell research circles as "B1 cells," while $CD5^-$ B cells, which comprise the vast majority of B lymphocytes in an individual, have been designated "B2 cells." $CD5^+$ B cells tend to be distributed only in the murine pleural and peritoneal cavities. Interestingly, BAFF deficiency does not affect this subset, suggesting that they do not require the positive signal delivered by BAFF to $CD5^-$ B cells. In addition, $CD5^+$ B cells are resistant to Fas- and mIg-mediated apoptosis, accounting for their extraordinarily long lives even in the absence of antigenic stimulation. Perhaps because of their restricted localization, $CD5^+$ B cells are more likely to produce the "natural antibodies" discussed in Chapter 4, making them a cell type that bridges adaptive and innate immunity. $CD5^+$ B cells are also more likely to bear BCRs recognizing autoantigens (see Ch.29), and almost all cases of chronic lymphocytic leukemia feature transformed cells that are $CD5^+$ (see Ch.30).

I. THE THREE SIGNAL MODEL OF B CELL ACTIVATION

Let's continue to follow the development of a mature naive follicular B cell that has made it into the periphery and whose BCR happens to be directed against a Td antigen. The activation of this resting mature naive B cell clone in response to binding by the antigen occurs in three steps: (1) the binding of antigen molecules to the BCRs, (2) the receipt of costimulatory signals, and (3) the receipt of T cell help in the form of cytokines (Fig. 9-6). We will briefly examine each of these events from a mechanistic point of view before integrating them back into our discussion of the continuum of B cell maturation.

i) Signal One: The Binding of Antigen Molecules to the BCRs

ia) Signal transduction from the BCR. The transduction of signals from the BCR is a large field of study to which a whole chapter, or even a whole book, could be devoted. The following description of this process is necessarily an abbreviated summary.

The complete BCR on a mature naive B cell in the periphery consists of mIgM non-covalently complexed with the

Table 9-2 **Effects of Antigen Type on Mature B Cell Fates**	
Type of Antigen Bound by BCR	**Fate of Mature IgM⁺/IgD⁺ B Cell**
None	Apoptosis
Td	Differentiation • plasma cells • memory cells • somatic hypermutation • isotype switching
Ti	Differentiation • plasma cells • limited isotype switching
Self	Apoptosis or anergy

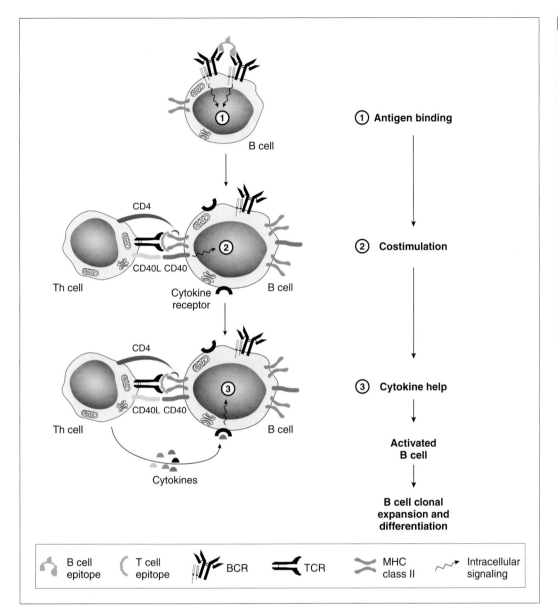

Figure 9-6

The Three Signal Model of B Cell Activation
A mature B cell in the periphery must receive three distinct signals to achieve full activation. Signal 1 is triggered when multiple BCRs are aggregated by specific antigen. Signal 2 is delivered when adhesive and costimulatory molecules on an activated Th cell make direct contact with their complementary ligands on the B cell surface. Cytokine receptors are then upregulated on the B cell, which allows it to receive Signal 3 in the form of cytokines secreted by the Th cell. Note that the Th cell is the daughter effector of a mature peripheral T cell activated by an encounter with the same antigen.

(1) Antigen binding

(2) Costimulation

(3) Cytokine help

Activated
B cell

B cell clonal
expansion and
differentiation

CD4 · CD40L CD40 · Th cell · Cytokine receptor · B cell · Cytokines

| | B cell epitope | | T cell epitope | | BCR | | TCR | | MHC class II | | Intracellular signaling |

accessory signaling proteins Igα and Igβ, also known as the CD79 complex. (While mIgD expressed on this cell is also associated with Igα and Igβ and can be involved in signaling, its role in the actual response to antigen is unclear.) This B cell is said to be in a *cognitive* state, in that it is capable of both recognizing and reacting to a specific antigen that binds to its BCRs. In studies of B cell activation by one Ti antigen, it was determined that about 10–12 immunogen–BCR pairs had to be formed on the B cell surface before activation could occur. Such a restriction seems sensible because it holds the adaptive response in check until significant numbers of immunogen molecules threaten the body. It is unknown whether a similar number of BCRs must be engaged for activation of a B cell by a Td antigen, but it is clear that multiple BCR-immunogen pairs must be engaged to induce intracellular signaling leading to the production of new proteins required for complete cellular activation.

Let us suppose that sufficient molecules of a Td antigen bind specifically to the antigen-combining sites of the B cell such that sufficient BCRs are stimulated and the response threshold is reached. Within a minute of this event, a signal is conveyed down the length of the transmembrane Igα and Igβ proteins from the surface toward the cytoplasm and the nucleus of the B cell. The signal is transduced in the following way (Fig. 9-7). Several protein tyrosine kinases (PTKs) belonging to the *Src family* are habitually clustered around the cytoplasmic base of the BCR. The Src family members involved early in signal transduction in a B cell are the tyrosine kinase enzymes *Fyn, Lyn,* and *Blk* (also known as $p59^{fyn}$, $p56^{lyn}$, and $p55^{blk}$, respectively). In resting cells, these enzymes are held in inactive form by the presence of a negative regulatory phosphate group on their C-terminal tyrosine residues. When antigen-induced stimulation of the BCRs occurs, Fyn, Lyn, and Blk are activated by the enzyme *CD45*

Signal Transduction from the BCR

Upon antigen–BCR binding, CD45 phosphatase removes negative regulatory phosphate groups from Src family tyrosine kinases Fyn, Lyn, and Blk. Activated Fyn, Lyn, and Blk then phosphorylate Igα and Igβ in the ITAMs (see text), an event that recruits other protein tyrosine kinases such as Syk and Btk to the signaling complex. Once phosphorylated by Fyn, Lyn, and Blk, the activated Syk/Btk/BLNK complex triggers signaling pathways initiated by Ras and phospholipase C (PLC-γ). As described in the text, these pathways lead to the production of second-messenger molecules and the release of intracellular calcium that ultimately activate transcription factors. These transcription factors are required for the expression of genes necessary to sustain B cell activation. PIP_2, phosphatidylinositol 4,5-bisphosphate.

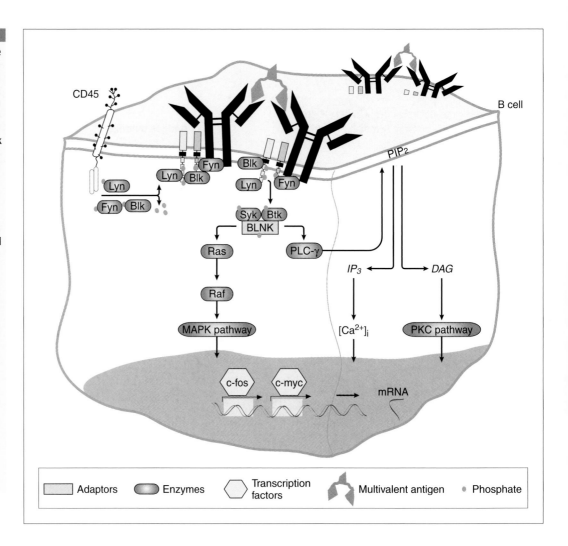

(B220). CD45 is a phosphatase that is expressed in all leukocytes and associates with the BCR in the B cells. In response to engagement of the BCR, CD45 removes the negative regulatory phosphate group from the Fyn, Lyn, and Blk tails, allowing them to phosphorylate specific tyrosine molecules in the cytoplasmic tails of the Igα and Igβ molecules. The targeted tyrosine residues are located in the ITAMs, the *immunoreceptor tyrosine-based activation motifs* introduced in Chapter 5. The ITAMs and their inhibitory counterparts, the ITIMs (*immunoreceptor tyrosine-based inhibitory motifs*), are discussed in the context of antigen receptor signaling in Box 9-1.

Tyrosine phosphorylation in the ITAMs can be detected as early as 30 seconds after BCR stimulation and peaks within 3–5 minutes. In B cells, ITAM phosphorylation results in the recruitment of new PTKs to the growing complex, including members of the *Syk* and *Tec* kinase families such as *Syk* and *Btk* (*Bruton's tyrosine kinase*), respectively (refer to Fig. 9-7). Once recruited, these kinases are also phosphorylated by Fyn, Lyn, and Blk and become activated. The activation of Syk results in the phosphorylation of the adaptor protein BLNK, which in turn serves as the initiator of a complex containing other adaptor proteins, Btk, and the important signaling enzyme *phospholipase C-γ* (PLC-γ; see later).

Signal transduction from the antigen receptor is very similar in B and T lymphocytes, and many of the enzymes are shared or have closely related counterparts. For example, Lyn and Fyn (but not the B cell-specific Blk) also carry out ITAM phosphorylation in T cells. Similarly, the function subserved by Syk and Btk in B cells is carried out by the related kinase *ZAP-70* in T cells, in which this enzyme is exclusively expressed. (ZAP-70 stands for "zeta-associated kinase," indicating that this enzyme associates with the CD3 zeta chain of the TCR complex.) In both B and T lymphocytes, the two main consequences of Syk, Btk, or ZAP-70 phosphorylation appear to be the activation of *Ras* and *PLC-γ*. Ras is an enzyme that occupies a central role in intracellular signaling. Activated Ras activates a serine/threonine kinase called *Raf*, which in turn activates another kinase called *MAPK* (*mitogen-activated protein kinase*). The Ras–Raf–MAPK pathway of signal transduction proceeds through several additional enzymes and ultimately culminates in the activation of various nuclear transcription factors that enter the nucleus and influence the pattern of gene expression in the cell. While all this is transpiring, PLC-γ activated by Syk, Btk, or ZAP-70 cleaves *phosphatidylinositol 4,5 bisphosphate* (PIP2) into the intracellular second-messenger molecules *diacylglycerol* (DAG) and *inositol triphosphate* (IP3). DAG in turn activates *protein kinase* C (PKC), a serine/threonine kinase similar in function to Raf. IP_3 causes an increase in the release of calcium from intracellular stores (an event known as a *calcium flux*), which then activates a calcium-dependent enzyme cascade. Phosphorylated proteins activated by both of these pathways may act as transcriptional inducers, resulting in the synthesis of *c-myc* and *c-fos*. These proteins function as key subunits of important nuclear transcription factors. Binding to DNA motifs by these nuclear transcription factors (see Box 9-2) results in the activation and regulation of the transcription of new genes, the protein products of which control the proliferation and differentiation of the B cell.

The end result of the signal transduction cascade induced by antigen binding is the preparation of the B cell to receive both costimulatory signals and T cell help. Messenger RNAs encoding nuclear transcription factors have been detected in B cells within 60 minutes of antigen binding, and visible changes in the morphology of the B cell can occur within 12 hours of stimulation. The cell expands in size, increases its RNA content, and appears to move from the resting phase

of the cell cycle (G_0) to the G_1 phase. The G_1 phase occurs just prior to the S ("synthesis") phase in which the DNA is replicated prior to cell division. MHC class II molecules are upregulated on the cell surface as are costimulatory molecules that are required for the B cell to establish the intercellular contacts conveying "signal two" from the Th cell. The expression of cytokine receptors necessary for the receipt of "signal three" also commences.

ib) Modulation of the signaling cascade. The complex signal transduction cascade just described is made more complicated by the actions of several proteins that can modulate BCR signaling. As mentioned previously, phosphatases (such as CD45) dephosphorylate the C-terminal regulatory tyrosine residue of Src-PTKs, thereby activating them. Conversely, phosphorylation of the same tyrosine residues by certain kinases (such as Csk kinase) inhibits Src-PTK activity. The balance of these molecules results in a threshold for antigen responsiveness.

CD19 is considered to be a coreceptor of the BCR. This protein is expressed exclusively in B cells from the early pre-B cell stage to the point of differentiation into a plasma cell.

Box 9-2. Nuclear transcription factors important in B cell development and activation

In Chapter 8, we discussed the nuclear transcription factor NF-κB and the OCTA-binding proteins that are important for Ig synthesis. Here, we describe a collection of nuclear transcription factors that are particularly important for the development of B cells (c-myb, PU.1, Ikaros, E2A, EBF, BSAP/Pax5, LEF-1, and OCA-B), and several more that are more directly involved in B cell activation (Spi-B, Aiolos, IRF-4, and ABF-1). The temporal sequence and consequences of the actions of these transcription factors are described in the following sections and summarized in the Figure. [The Figure also contains information on Bcl-6, Blimp-1, and XBP-1, which are discussed later in the main text in the context of plasma cell (PC) and memory cell differentiation].

• Transcription Factors Associated with B Cell Development

i) c-myb

The nuclear transcription factor gene c-myb is thought to encode one of the first regulatory proteins in hematopoiesis, since knockout mice lacking this gene exhibit a very early blockage in stem cell development. Erythroid, myeloid, and lymphoid lineages are all absent, but megakaryocytes are present in normal numbers. C-myb is active in B cell development at both the mature, naive (MN) and activated (Act) B cell stages.

ii) PU.1

The nuclear DNA binding protein PU.1 is a member of the Ets family of transcription factors. PU.1 is specifically expressed in the hematopoietic system, most highly in cells of the monocytic, granulocytic, and B cell lineages. PU.1 is essential for the development and maturation both of the common lymphoid progenitor (CLP) and the common myeloid progenitor (CMP). Embryos from knockout mice deficient for PU.1 lack all cells of both the myeloid and lymphoid lineages, but have normal numbers of cells of megakaryocyte and erythrocyte lineages. More specifically, PU.1 knockout mice have no mature B cells (or macrophages or T cells) and fail to show any evidence of Ig gene rearrangement. At a molecular level, PU.1 appears to exert its effects by regulating the transcription of cytokine receptors needed to receive cytokine-mediated developmental signals. For example, the IL-7Rα gene is a direct target of PU.1. IL-7Rα is required for the assembly of the IL-7R, which in turn is required to receive the IL-7-mediated signals necessary for B (and T) cell differentiation. Low levels of PU.1 activate IL-7Rα expression and promote lymphoid development, while high PU.1 levels block IL-7Rα expression and promote myeloid development. PU.1 also upregulates receptors for M-CSF, G-CSF, and GM-CSF, allowing myeloid progenitors to bind these growth factors and develop into macrophages or granulocytes. In addition, PU.1 is required for the expression of the transcription factors EBF and BSAP/Pax5, which operate later during B cell development.

iii) Ikaros

The expression of the Ikaros gene results in the production of five different transcription factors, Ik-1 to Ik-5, which are generated by differential splicing. Ikaros binding sites can be found in the enhancers of a number of genes involved in both B and T cell development. Knockout mice lacking Ikaros generally die within 1–3 weeks after birth (for unknown reasons) and are characterized by an absence of B and T lymphocytes as well as their progenitors. Since only lymphocytes and some DCs are missing in these mice, Ikaros is thought to act later, or to be "downstream," of PU.1 and to affect later progenitors in addition to the CLPs. As well as by activating transcription of vital genes such as c-kit and Flk2/Flt3 in HSCs, Ikaros has been postulated to control lineage choice in later progenitors by repressing expression of "inappropriate" genes. For example, early in hematopoietic development, Ikaros may repress expression of myeloid-specific genes in the CLP. Later on, Ikaros may repress the expression of T cell-specific genes (such as CD4 and CD8) in common T/B lymphoid progenitors destined to become B cells, and B cell-specific genes (such as CD19 and λ5) in those that become T cells. Mechanistically, Ikaros is thought to repress transcription by recruiting histone deacetylases, which reduce the chromatin accessibility of a locus.

iv) E2A

E2A is the prototypical member of a family of dimeric DNA-binding proteins known as the "basic helix–loop–helix" (bHLH) proteins. Hetero- and homodimeric forms of the E2A proteins bind to sequences called E-box elements, which have the consensus sequence GCAXTG. As well as being found in the promoters of many tissue-specific genes, E-box elements known as the μE1–μE4, μB, and μA sequences are found in the E_{iH} ($E\mu$) enhancer of the Igh locus. Similarly, E-box elements known as κE1 and κE2 are found in

	HSC	CLP	pro-B	pre-B	Imm B	MN B	Act B	PC
c-myb	+		+			+	+	
PU.1	+	+	+			+		
Ikaros		+	+	+				
E2A			+	+			+	
EBF			+/−	+	+	+	+	
BSAP			+/−	+		+		
LEF-1			+	+/−				
Aiolos						+	+	
Oct-2							+	
OCA-B						+	+	
ABF-1							+	
Bcl-6							+	
NF-κB/Rel						+		
Spi-B						+		
IRF-4							+	
Blimp-1								+
XBP-1								+

Continued

the κ light chain gene enhancer of the *Igk* locus. E2A homodimeric gene products have been found to increasingly bind to E-box elements as a B cell matures, suggesting that E2A is a critical regulator of Ig gene expression. In agreement with this observation, DJ rearrangement does not occur in the few B cell progenitors found in knockout E2A$^{-/-}$ mice (which die within 1 week of birth of unknown causes). In addition, the transcription of other germline genes important for Ig expression is drastically reduced, there is a profound deficiency of pro-B cells, and no mature B cells are observed. E2A is thus critical for normal B cell differentiation and Ig production. Interestingly, T cells in neonatal E2A$^{-/-}$ mice are normal, suggesting that E2A may play a role in the differential control of RAG gene expression in B cells and T cells.

The phenotype of the E2A$^{-/-}$ mouse and studies of E2A function indicate that E2A acts downstream of PU.1 and Ikaros, but upstream of EBF. There is also now evidence that, in later stages of development, E2A cooperates with EBF to induce the expression of B cell-specific genes such as V_{preB} and λ5 (which encode the SLC) and the signal transducers Igα and Igβ. Furthermore, studies of E2A function suggest that being a B cell is the default fate of a T/B progenitor. Notch1 is a transmembrane cell fate determination molecule that is critical for early T cell development (see Ch.13). Part of the function of Notch1 is to interfere with E2A activity, preventing development down the B cell path. Similarly, a transcription factor called Id2 interferes with E2A in cells destined to become NK cells. The homeostasis of B cell development is also controlled at the E2A stage: TGFβ signaling induces the expression of the inhibitor Id3, which binds to E2A, arresting progenitor development.

v) EBF

Early B cell factor (EBF) is expressed at all stages of B cell development except in terminally differentiated plasma cells. Most knockout EBF$^{-/-}$ mutant mice are viable, although about 30% of them die by 4 weeks of age for unknown reasons. The mice that survive exhibit a marked deficiency of B cells and certain early B cell progenitors. However, B220$^+$CD43$^+$ pro-B cells are observed in these mice, indicating that EBF acts downstream of PU.1, Ikaros, and E2A. As mentioned previously, expression of EBF

appears to be controlled by E2A and to cooperate with it for the induction of the RAG, V_{preB}, λ5, Igα, and Igβ genes. The B220$^+$ cells of EBF$^{-/-}$ mice do not express RAG (nor the SLC) and thus do not initiate VDJ recombination.

vi) BSAP/Pax5

B cell-specific activator protein (BSAP) is the product of the *Pax5* gene. BSAP/Pax5 exhibits a tissue-specific expression pattern similar to that of EBF, except that BSAP/Pax5 is also expressed in the developing central nervous system. Knockout mutant mice lacking BSAP/Pax5 die within 3 weeks of birth and exhibit defects in B cell development similar to those observed in EBF$^{-/-}$ mice. Pro-B cells are generated but cannot differentiate into pre-B cells. CD19 expression is not induced. BSAP/Pax5 is thought to act downstream of EBF because expression of RAG and SLC is normal in EBF$^{-/-}$ pro-B cells. Expression of the *Pax5* gene in pro-B cells may be regulated by IL-7 signaling, which in turn is controlled by PU.1 control of IL-7Rα chain expression. Interestingly, Pax5$^{-/-}$ B cells resemble HSCs in that they retain the capacity to self-renew; that is, they are unable to make the commitment to become B cells. Thus, although the activities of E2A and EBF are required for B cell commitment, they are insufficient to guarantee further maturation in the absence of BSAP/Pax5. Mechanistically, BSAP/Pax5 appears to simultaneously activate B lineage-specific genes and repress lineage-inappropriate genes. For example, BSAP/Pax5 induces expression of CD19 (part of the BCR coreceptor complex) and BLNK (an adaptor protein crucial for BCR signaling), while repressing transcription of receptors (such as M-CSFR) capable of receiving signals from cytokines promoting myeloid development. BSAP/Pax5 also represses transcription of Notch1 and thus suppresses T cell development.

vii) LEF-1 and Sox-4

LEF-1 (*lymphoid enhancer-binding factor* 1) is required for normal B cell development but functions as a DNA structure-modifying protein rather than as a gene expression activator. Mice lacking LEF-1 have reduced numbers of pro-B and pre-B cells but normal numbers of earlier progenitors that are responsive to IL-7. Expression of Fas and c-myc, both of which can induce apoptosis, is upregulated in LEF-1$^{-/-}$ B cells. In a wild-type cell, LEF-1 acts to bend the DNA helix

and also receives signals from the Wnt intracellular signaling pathway. It is thought that the Wnt pathway may be important for the survival of B cell progenitors. Sox-4 is a transcription factor that is related to LEF-1 and expressed in early B cell lymphopoiesis. However, Sox-4$^{-/-}$ mice have reduced numbers of pro-B cells that cannot respond to IL-7.

viii) Others?

Transcription factors further downstream that control pre-B cell development and beyond remain to be identified for the most part. We do know that OCA-B, a B cell-specific transcriptional co-activator that cooperates with Oct-1 and Oct-2, appears to be required for the emergence of mature, naive B cells from the bone marrow into the periphery.

• Transcription Factors Associated with B Cell Activation

The activation, proliferation, and differentiation of B cells bring new combinations of transcription factors into play. Some of these transcription factors have one function during B cell development and maturation, and another during B cell activation and effector generation. For example, PU.1, in addition to its role in early lymphoid progenitor development, is thought to affect transcription of key signal transduction components needed for BCR signaling during B cell activation. In addition, *Igh* transcription is upregulated when PU.1 binds to a site in the *Igh* 3′ enhancer in mature B cells. Similarly, OCA-B, required for exit of naive B cells to the periphery, is necessary for the formation of structures called *germinal centers* (see later in the main text) during an immune response. OCA-B also functions as a co-activator for the Oct-1 and Oct-2 transcription factors that govern Ig gene expression. BSAP/Pax5, necessary for B lineage commitment and pre-B cell generation, conversely has a negative effect on *Igh* locus expression in mature B cells. BSAP/Pax5 binds to a site in the *Igh* 3′ enhancer that is very close to the PU.1 binding site, presumably interfering with the positive effects of PU.1 on the *Igh* enhancer.

We will now briefly touch on several other transcription factors that have no known function in B cell development but figure prominently in B cell activation and post-activation events.

Continued

i) Spi-B

Spi-B is a member of the Ets transcription factor family and closely related in structure to PU.1. Studies of Spi-B$^{-/-}$ mice suggest that, like the later function of PU.1, Spi-B is required for the expression of molecules participating in the first steps of signal transduction from the BCR. B cells from Spi-B$^{-/-}$ mice fail to proliferate in response to BCR stimulation. Spi-B$^{-/-}$ mice have abnormal germinal centers and cannot mount normal immune responses to Td antigens.

ii) Aiolos

Aiolos is related structurally to Ikaros but is highly expressed in mature B cells rather than in progenitors. Aiolos appears to have a negative regulatory role because in the absence of Aiolos, mature B cells exhibit a lower threshold of activation and proliferate uncontrollably in response to BCR stimulation. Even without immunization, germinal centers can form in Aiolos$^{-/-}$ mice and serum Ig titers are elevated. Aiolos also

influences a T2 cell to become either an MZ or a follicular B cell. Aiolos$^{-/-}$ mice show greatly reduced numbers of MZ B cells because, in the absence of Aiolos, BCR signaling exceeds an important threshold in MZ B cells and induces their negative selection.

iii) IRF-4

IRF-4 is a member of the *interferon regulatory factor* (IRF) family of transcription factors, which were first identified as being involved in the expression of interferon-induced genes. Unlike other IRF family members, IRF-4 expression is restricted to lymphoid cells. In B cells, IRF-4 exerts its effects by interacting with PU.1 and binding to specific motifs in the enhancers of the light chain genes. B cells are present in normal numbers in unimmunized IRF-4$^{-/-}$ mice and low levels of complete Ig molecules containing light chains can be detected. However, IRF-4$^{-/-}$ mice cannot mount antibody responses to either Td or Ti antigens and exhibit severely decreased

serum Ig levels compared to immunized wild-type mice. These results suggest that IRF-4 is not required for the expression of light chain genes during B cell development but is necessary for the upregulated level of *Igl* and *Igk* expression induced by activation of a B cell. Interestingly, IRF-4 may cooperate with the myeloid lineage commitment factor IRF-8 (ICSBP; *interferon consensus sequence binding protein*) to promote the maturation of pre-B cells. These precursors accumulate in IRF-4$^{-/-}$IRF-8$^{-/-}$ double knockout mice, fail to repress SLC expression, and do not initiate rearrangement of the Ig light chain genes.

iv) ABF-1

ABF-1 is a bHLH transcription factor that is not expressed until mature B cells are activated. To be functional, ABF-1 forms either homodimers or heterodimers with E2A proteins. ABF-1 can bind to E boxes in Ig gene enhancers and may participate in the regulation of isotype switching in activated B cells.

The CD19 protein is found on the B cell surface in a complex that contains CD21 and CD81. CD21 is also known as complement receptor 2 (CR2) and its ligand is complement component C3d (see Ch.19). CD81 is a membrane anchor protein whose physiological ligand is a mystery. Tyrosine residues in the cytoplasmic tail of CD19 are phosphorylated in response to the engagement of mIg by antigen, resulting in the association of CD19 with Lyn and Fyn, among other kinases. CD19 can also be stimulated when an antigen to which complement has become attached binds to CD21 in the CD19–CD21–CD81 complex (Fig. 9-8A). If the C3d complement component attached to the antigen binds to the CD21 molecule at the same time as the antigenic epitope binds to the mIg, the antigen receptor and the CD19 coreceptor are effectively cross-linked, or *co-ligated*. Co-ligation in this way appears to lead to a stronger intracellular activation signal. Thus, the CD19 coreceptor complex can enhance B cell responses to opsonized antigens. CD19 may also be required to receive T cell help, since knockout mutant mice lacking CD19 have a phenotype suggesting that their B cells can no longer associate normally with T cells.

Co-ligation events can also result in the suppression of B cell activation (Fig. 9-8B). As introduced in Chapter 5, B cells express an isoform of the Fcγ receptor called FcγRIIB that recognizes the Fc region of antigen-bound IgG antibody molecules. This Fcγ receptor is situated close enough to the BCR on the cell surface that when mIg binds to its epitope on the antigen, the FcγRIIB can bind to the Fc

region of the IgG molecule bound to the antigen molecule, actually incorporating the Fc receptor into the BCR complex. The co-ligation of the mIg and the Fcγ receptor in this way appears to send an intracellular signal indicating that enough antibody against that antigen is present, and that the B cell should not be activated to produce more. The actual mechanism of suppression is based on the interaction between an ITIM sequence in the cytoplasmic tail of the FcγRIIB molecule and SHP-1 tyrosine phosphatase (refer to Box 9-1).

ii) Signal Two: Costimulation of the B Cell

Costimulatory molecules are proteins on the surfaces of lymphocytes whose engagement by specific ligand appears to be necessary for a complete activation response following antigen receptor binding by antigen. Signaling through costimulatory molecules can affect antigen receptor signaling in very important ways. In the case of B cells, if the BCR is engaged by antigen in the absence of costimulation (that is, the cell receives signal 1 alone), the lymphocyte eventually undergoes apoptosis or becomes anergic (unresponsive to its specific antigen). If, however, the lymphocyte receives both signal 1 (the antigen binds to the receptor) and signal 2 (costimulatory molecules on the B cell make intercellular contact with specific ligands on the Th cell), the lymphocyte can accept signal 3 (Th-secreted cytokines) and complete its activation (refer to Fig. 9-6). Only a fully activated cell proceeds beyond proliferation to differentiation and the execution of

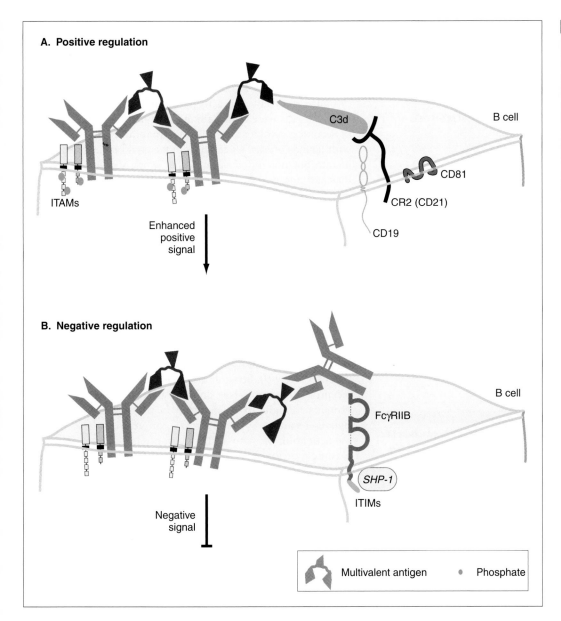

A. Positive regulation

C3d

B cell

CD81

CR2 (CD21)

ITAMs

CD19

Enhanced
positive
signal

B. Negative regulation

FcγRIIB

B cell

SHP-1

ITIMs

Negative
signal

Multivalent antigen • Phosphate

Figure 9-8

Co-ligation of BCR
Co-ligation of the BCR to other cell surface molecules can result in either positive (A) or negative (B) regulation of the signal transduced upon antigen binding. (A) Co-ligation of CD19 and the BCR via antigen complexed to complement component C3d. (B) Co-ligation of FcγRIIB and the BCR via antigen complexed to soluble IgG antibody.

effector functions. Many of the details of how these intricate signal transduction outcomes are controlled are not yet understood.

iia) Adhesive Th–B cell contacts. Before an antigen-stimulated B cell can receive a costimulatory signal, it must make stable adhesive contacts with the antigen-stimulated Th cell delivering that signal. Some of the extensive intercellular adhesive contacts that are established between ligand–receptor pairs on the surfaces of B and Th cells are shown in Figure 9-9. These interactions ensure that the two cells are held together long enough to achieve complete and mutual activation.

When the TCRs of the antigen-specific Th cell are stimulated by binding to specific peptide–MHC complexes (presented either by an APC or by the antigen-specific B cell itself acting as an APC), intracellular signaling is triggered that affects the affinity of LFA-1 on the Th cell for its ligand ICAM-1 (CD54) on the B cell. The reader will recall from Chapter 4 that LFA-1 (leukocyte function antigen-1, a heterodimer also known as CD11a/CD18) is a member of the family of integrin receptors that bind to the intercellular adhesion molecule (ICAM) family of ligands. LFA-1 also binds to ICAM-3 on the B cell in an interaction that has an important costimulatory effect on the Th cell (see Ch.14) as well as an adhesive one. Conversely, ICAM-1 on the Th cell may bind to the heterodimeric β2-integrin molecule CD11c/CD18 expressed on the B cell, further increasing intercellular adhesion. Two additional adhesion molecules on the B cell, CD58 (LFA-3; in humans) and CD48 (CD2 ligand; in mice), interact with CD2 (LFA-2) on the Th cell, resulting in not only adhesion but also activation signals to the Th cell and the induction of IL-5, IL-6, and IL-10 expression by the Th cell.

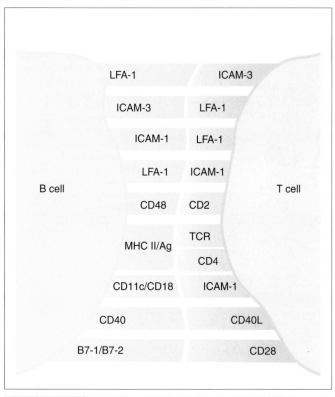

Figure 9-9

Examples of Adhesive and Costimulatory Th Cell–B Cell Contacts in Mice
Molecules on B and T cell surfaces interact to maintain the adhesion and costimulatory contacts necessary for complete activation. See text for details of signals triggered. Note that in the case of LFA-1/ICAM-1 binding and LFA-1/ICAM-3 binding, both members of the binding pair are expressed on B and T cells.

iib) Costimulatory Th–B cell contacts. The importance of costimulatory contacts was established in experiments in which antigen-stimulated B cells were treated with the culture supernatants of Th cells activated *in vitro*. Although these supernatants contained significant concentrations of a variety of cytokines, they failed to activate B cells. In addition, if the activated Th cells were chemically fixed so that they were incapable of secreting cytokines, B cells that were able to make contact with ligands on the Th cell membrane could be induced to proliferate but not differentiate. Other work showed that B cells that had not even bound antigen could sometimes be triggered to proliferate if particular intercellular contacts between it and an activated Th cell were established (the "bystander activation" effect). These observations demonstrated that cytokines alone were insufficient to achieve full B cell activation, and that direct intercellular contact between the antigen-stimulated B cell and the antigen-stimulated Th cell was crucial.

The most important costimulatory contact for B cell activation is the interaction between *CD40* on B cells and *CD40L* (CD40 ligand) on T cells. CD40 is a 50 kDa glycoprotein that is highly expressed on mature naive B cells but not on plasma cells, while CD40L (originally called *gp39* in humans) is a 39 kDa protein transiently expressed on activated Th cells (refer to Fig. 9-9). Structurally, CD40 is a member of the TNF receptor superfamily (see Box 9-3 and Ch.15), a collection of proteins thought to be involved in transducing intracellular signals resulting in cell survival or cell death as well as the activation of many inflammation-related genes. The cytoplasmic portion of CD40 associates with the intracellular signaling proteins TRAF2 and TRAF3 (*TNFR-associated factor 2 and 3*). TRAFs in turn interact with the NF-κB

Box 9-3. TNF receptor superfamily

The *tumor necrosis factor receptor* (TNFR) superfamily (also known as the nerve growth factor receptor family) is a very important class of signal transduction molecules in the immune system. Included in this family are the TNF receptors 1 and 2 (TNFR1 and 2), the low-affinity nerve growth factor receptor (NGFR), CD40, CD30, Fas, and others indicated in the Table. Members of the TNF receptor family are generally transmembrane proteins, but many of them can also be secreted as soluble molecules, derived either by proteolytic cleavage from the membrane or by differential mRNA processing. The TNFR family is characterized by the presence of cysteine-rich motifs of 40 amino acids in the extracellular domain that are involved in ligand binding. The ligands for these receptors are type II transmembrane proteins with a "jelly-roll" β-sandwich structure, many of which can also be secreted.

Some TNF receptor family members are associated with additional signal transduction molecules called *TNFR-associated factors*

(TRAFs). These proteins contain zinc binding domains that are thought to mediate the binding of the protein to DNA, leading to transcriptional activation. Signal transduction is initiated by stimulation of oligomerized receptor complexes upon ligand binding. The ligands, such as tumor necrosis factor (TNF), are often found in multimeric form, and this multivalency enhances the induction of signaling. Other members of the TNFR family (most notably, the apoptosis-inducing molecule Fas) contain *death domains* (DDs) in their cytoplasmic tails. For example, the DD sequence allows the interaction of the membrane-bound Fas molecule with the signal-transducing protein FADD (*Fas-associated death domain*), which in turn delivers signals causing the cell to initiate apoptosis. More recently it has been discovered that several proteins containing the DD sequence are not involved in apoptosis at all but actually promote cell survival. In fact, depending on downstream events, engagement of TNFR1 (which contains a DD sequence) can lead to either cell death or cell survival. Signals

governing these cell fate decisions are explored in more depth in Chapter 15.

The functional consequences of ligand engagement by members of the TNF receptor superfamily are very diverse. Fas binding induces apoptosis, which is important in the maintenance of immune self-tolerance. CD27, CD30, 4–1BB, and CD40 signaling enhance the survival, proliferation, and activation of B or T lymphocytes, often playing critical roles as costimulators or signal modulators. Signaling by other TNFR family members results in the activation of NF-κB in macrophages and the induction of an inflammatory response. The TNF receptor superfamily is thus central to many aspects of both innate and acquired immunity and plays fundamental roles in both cell death and survival. A greater understanding of this receptor family, as well as its corresponding family of ligands, holds great potential for the creation of novel drugs and therapeutics that could be used to control many aspects of autoimmunity and other immunopathologies.

Continued

TNFR Superfamily Members

Name (alternate name)	Ligand	Downstream Signal Transducing Molecules	Involved in
TNFR1 (TNFRp55, CD120a)	TNF, LTα	TRADD, FADD, TRAF1, 2	Inflammation, acute-phase response, cell survival and apoptosis, bacterial toxic shock
TNFR2 (TNFRp75, CD120b)	TNF, LTα	TRAF1, 2, 3	Inflammation, thymocyte proliferation
Fas (CD95, APO-1)	FasL	FADD	Apoptosis
NGFR (p75^{NGFR})	NGF, BDNF, NT-3, NT-4/5	TRAF6	Neuronal growth stimulation; NF-κB activation; apoptosis; sensory neuron development and function
CD40	CD40L (CD154, gp39)	TRAF1, 2, 3, 5, 6	B cell proliferation, development, and costimulation; isotype switching; apoptosis, cytotoxic T cell responses
CD30	CD30L (CD153)	TRAF1, 2, 3, 5	Cell death?
DR3	TWEAK (VEGI)	TRADD, FADD	Apoptosis
DR4 (TRAILR-1, CD261)	TRAIL (CD253)	FADD	Apoptosis
DR5 (TRAILR-2, CD262)	TRAIL (CD253)	FADD	Apoptosis
OPGL-R (RANK, ODAR)	OPGL* (RANKL, TRANCE)	TRAF1, 2, 3, 5, 6	Osteoclast differentiation and/or activation; dendritic cell survival and function
4–1BB (CD137)	4–1BBL (CD137L)	TRAF1, 2	T cell costimulation
CD27	CD70 (CD27L)	TRAF2, 5	T cell development and activation
OX40 (CD134)	OX40L (CD252)	TRAF2, 3	B cell differentiation and secondary Ig responses; adhesion of activated T cells to endothelium

*OPGL is also bound by the antagonist OPG.

LTα, lymphotoxin-α; TRADD, TNF receptor-associated death domain; FasL, Fas ligand; FADD, Fas-associated death domain; NGFR, nerve growth factor receptor; NGF, nerve growth factor; BDNF, brain-derived neurotropic factor; NT, neurotropin; DR, death receptor; TWEAK, TNF-related weak inducer of apoptosis; VEGI, vascular endothelial growth inhibitor; TRAIL, TNF-related apoptosis-inducing ligand; OPGL-R, osteoprotegerin ligand receptor; RANK, receptor activator of NF-κB; ODAR, osteoclast differentiation and activation receptor; TRANCE, TNF-related activation-induced cytokine; OPG, osteoprotegerin.

transcription factor. The expression of CD40L on the Th cell is induced only after it is activated by antigenic stimulation, and only once the intercellular adhesion contacts are firmly established. Studies of CD40 and CD40L knockout mice have shown that this pair of molecules is essential for both isotype switching and the crucial stage of B cell development called *germinal center* formation (see later). That the interaction of CD40 with CD40L is essential in humans for isotype switching is further demonstrated by one form of a congenital defect called hyper-IgM syndrome (see Ch.24). A deficiency of CD40L in these patients leads to a lack of T cell help and consequently an absence of isotype switching, resulting in an over-representation of IgM antibodies. The CD40L knockout mouse has a similar phenotype.

Other intercellular contacts between Th and B cells also have costimulatory effects. Perhaps the most important contact with respect to the costimulation of T cells is the binding of CD28 on the Th cell to either of two ligands, B7-1 (CD80) or B7-2 (CD86), on the B cell. Interaction of these molecules with an activated Th cell stabilizes the mRNA for IL-2, among other cytokines, thus prolonging the proliferative signal. The effects on Th cells of CD28 stimulation are discussed in detail in Chapter 14.

iii) Signal Three: T Cell Help in the Form of Cytokines

The binding of antigen to sufficient BCRs on a B cell coupled with CD40-mediated costimulation triggers the induction of expression of new cytokine receptors (particularly the receptors for IL-1 and IL-4) on the B cell surface. However, without the binding of the relevant cytokines to these receptors, the B cell usually undergoes only limited proliferation. While some of the pertinent cytokines required for B cell activation can be secreted by a nearby macrophage or dendritic cell, most are derived from the antigen-stimulated Th cell that has made contact with the B cell. Shortly after adhesive and costimulatory contacts are established between the B and Th cell, the Golgi apparatus is reorganized within the Th cell such that it is closer to the site where contact has been established with the B cell. The release of the cytokines then becomes directed more precisely toward the B cell, increasing the efficiency of the delivery of "signal three." Such rearrangement of the Golgi is not observed if a Th cell and a B cell come into contact in the absence of specific antigenic stimulation.

The binding of specific antigen by a B cell (either a mature naive B cell or a memory B cell) such that the threshold number of mIg receptors is stimulated prepares the cell to progress from the resting G_0 stage of the cell cycle to the G_1 blast stage (Fig. 9-10). In many (but not all) cases, the binding of IL-4 secreted by the antigen-activated Th cell in contact with the B cell is the primary event that triggers the B cell to enter the G_1 phase of the cell cycle. Continued exposure to IL-4 pushes the B cell further into S phase and induces the replication of DNA in preparation

for cell division. At this point, IL-2, IL-4, IL-5, IL-10, and TGFβ can all act to promote activated B cell proliferation.

While no one cytokine is sufficient on its own for B cell activation, neither is any particular one essential. In other words, blocking the action of any one cytokine has essentially no effect on activated B cell proliferation in response to protein antigens. Knockout mouse mutants lacking any one of the genes for IL-2, IL-4, or IL-10 have essentially normal numbers of all B cell subsets, although the IL-4-deficient mouse does have a defect in IgE production. The IL-5 knockout mouse shows a defect only in the generation of CD5$^+$ B cells.

As the progeny B cells continue to divide, the action of additional cytokines becomes important for inducing differentiation. IL-2, IL-4, IL-5 (in the mouse only), IL-6, IL-10, IFNγ, and TGFβ have all been implicated in stimulating the differentiation of the progeny of the original activated B cell

---

Figure 9-10

Completion of B Cell Activation by Signal 3
A costimulated B cell upregulates cytokine receptor expression and binds Th-derived cytokines that deliver Signal 3. Under the influence of additional cytokines in the milieu, the B cell commences proliferation as described in the text and differentiates into plasma cells (PC) and memory B cells (Mem B). Not all cytokines shown in proliferation and differentiation steps are required simultaneously.

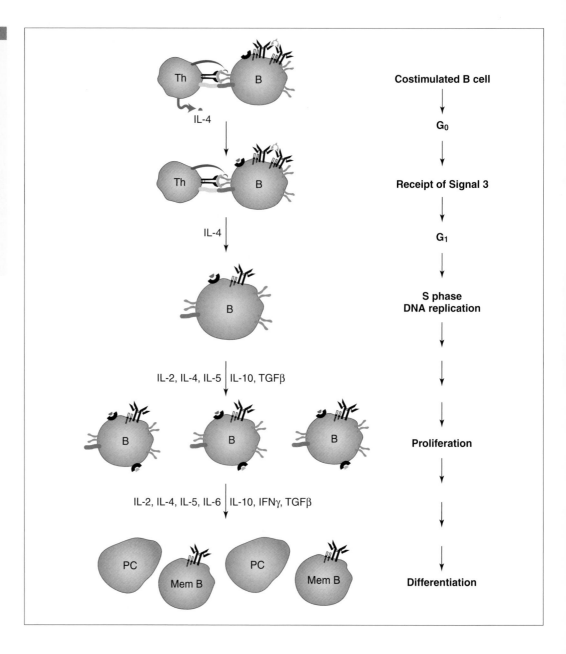

into memory cells or *plasmablasts*, the immediate precursors of plasma cells (see later). Activated murine B cells exposed *in vitro* to IL-4 and IL-5 are strongly induced to begin antibody secretion, while IL-2 and IL-6 appear to be more important in human cells for this function.

It should be noted that B cells not only receive cytokines but also secrete them to a limited extent. IL-6 and TNF secreted by a B cell can increase the efficiency of Th cell activation, which in turn increases its production of cytokines important for sustaining B cell proliferation and differentiation.

II. CELLULAR INTERACTIONS DURING B CELL ACTIVATION

Let us now return to our mature naive follicular B cell to see where in the body the three activation signals are delivered. The anatomy of the secondary lymphoid tissues is crucial for B cell activation and the subsequent maturation of its progeny into memory and plasma cells. As we learned in Chapter 3, primary immune responses take place principally in the secondary lymphoid tissues in the body's periphery, since these are the locations where antigens accumulate and the necessary accessory cells are concentrated. The events that transpire after the recognition of antigen are thought to be similar in the spleen and the lymph nodes, the best-studied secondary tissues: it is the delivery of the antigen to the tissue and certain structural details that differ. Antigens entering the body via the blood are carried to the spleen, while those penetrating the mucosae or skin are conveyed by the afferent lymphatic vessels to the regional lymph nodes (Table 9-3). Mature lymphocytes maintaining immune surveillance in the periphery are stimulated by an encounter with antigen to home to the secondary lymphoid organs in order to complete the immune response.

i) B and T Lymphocytes Encounter the Antigen and Each Other

One of the most common places for antigen and lymphocyte to meet is in the lymph node, since this organ collects a wide variety of antigens. Within 1–2 days of penetrating the skin or mucosae, antigen X is collected by the lymph system and brought to the cortex and paracortex of a regional lymph node. Meanwhile, a mature naive follicular B cell clone with a BCR specific for an epitope on antigen X may be policing the body's periphery by recirculating through the blood and the lymphatic system with its associated tissues and organs. Members of this B cell clone that do not encounter antigen X undergo apoptosis. However, if in passing through the paracortex of the lymph node, molecules of antigen X bind in sufficient numbers to the

BCRs of a member of this clone, activation signal 1 is delivered. The B cell is induced to express both the costimulatory molecule CD40 in anticipation of signal 2 and the apoptotic molecule Fas on its surface. Meanwhile, the proteins of antigen X are digested and processed by the lymph node's various professional APCs, including paracortical interdigitating dendritic cells and the ubiquitous macrophages. (In a secondary response, the memory B cell itself might function as the APC; see Ch.11). The peptides derived from antigen X are presented in association with MHC class II to naive T cells located in the paracortical region of the node.

ii) The Importance of Follicular Dendritic Cells

The follicles of the secondary lymphoid organs are the only sites where follicular dendritic cells (FDCs) are found. Whether these stromal cells originate from hematopoietic progenitors or from other stromal elements is still controversial. FDCs are characterized by extended cellular processes upon which antigen is trapped and retained for extended periods, sometimes for months or years. The mechanism that permits this kind of display without degradation or internalization of the antigen has yet to be elucidated.

Resting B cells recirculating through the body migrate into a lymph node, travel through the network of FDC cells in the cortical region, and pass out of the node again. There is substantial evidence derived from cell culture experiments *in vitro* that FDCs contribute both to the structural organization of the primary follicles and to the survival and activation of peripheral B cells. FDCs express all three types of complement receptors, allowing them to trap antigen coated in complement components. CD21 (CR2) expressed on the FDC has turned out to be the major receptor for trapping of complement-coated antigen and immune complexes. Knockout mouse mutants lacking CD21 have defects in B cell development and in immune responses to Td antigens. Besides mediating antigen retention, CD21 on the FDC may play an adhesive role by binding to CD23 (FcεRII) on the B cell surface. In the presence of T cell help supplied by the antigen-specific Th cell, B cells encountering specific antigen trapped on the FDC are induced to express the adhesive ligands LFA-1 and VLA-4. Only the antigen-stimulated B cells are thus retained among the FDCs because the latter constitutively express ICAM-1 and VCAM-1, ligands that bind with high affinity to LFA-1 and VLA-4, respectively.

FDCs also express Fc receptors for all antibody isotypes, allowing them to facilitate secondary responses in which circulating antibody from the primary response plays an important role in clearing antigen. The Fc receptors bind antigen complexed to antibody and hold it on the FDC surface. An individual FDC can then "bud off" a small piece of its membrane upon which multiple immune complexes are trapped, forming an immune complex-covered particle of about 0.35 μm called an *iccosome* (*immune complex coating*). These iccosomes are released from the FDCs such that a passing B cell whose BCR recognizes the antigen binds the iccosome to its mIg. The B cell can then act on its own as an APC, internalizing the iccosome, digesting the antigen and presenting it in association with MHC class II on its surface for interaction with an antigen-specific T cell.

Table 9-3 **Antigen Localization to Secondary Lymphoid Tissues**	
Location of Entry	**Conveyed to**
Blood	Spleen
Skin	Lymph nodes
Mucosae	Lymph nodes

iii) Influence of Th Subtype

Once the antigen-specific B cell has received signal 1 from its interaction with antigen in the lymph node, what happens next depends on which subtype of antigen-activated Th cell the B cell encounters. Initially, antigen X-derived peptides are displayed in association with MHC class II on APCs in the paracortical region of the node and are recognized by an antigen X-specific CD4$^+$ T cell (either a naive or memory cell). The T cell is then activated and induced to proliferate, resulting in the expansion of the clone and its differentiation into either Th1 or Th2 effectors capable of secreting cytokines (see Ch.15). The expression of the costimulatory ligand CD40L is induced on the surface of Th2 cells, while FasL appears on the Th1 cell surface. If the stimulated B cell happens to meet the activated Th1 cell bearing FasL before it meets the appropriate Th2 cell, FasL binds to Fas expressed on the B cell and apoptosis is induced; the humoral response is downregulated. If, however, the B cell meets an activated antigen-specific Th2 cell first, CD40/CD40L interaction promotes B cell activation. Th2 cells do not express FasL, so they cannot kill B cells during the upregulation of Fas that occurs during B cell activation and CD40 stimulation. In addition, Th2 cells (but not Th1 cells) secrete IL-4, a cytokine that can protect FasL-expressing B cells from apoptosis. Overall, therefore, Th2 activation tends to induce humoral immunity while Th1 cells favor cell-mediated immunity. The immune response can thus be skewed as needed to deal with a given antigen. The balance of the signals delivered by Th1 and Th2 cells may also function to prevent the uncontrolled overexpansion of B cell clones.

In the paracortical region of the lymph node, the contact between CD40L on the activated antigen-specific Th2 cell and CD40 on the stimulated antigen-specific B cell (among other costimulatory and adhesive contacts) delivers signal 2: the B cell is induced to synthesize new cytokine receptors and becomes receptive to T cell help. The Th2 cell secretes copious quantities of cytokines (particularly IL-4) in a polarized manner that deliver signal 3 complete the activation of the B cell, and promote its proliferation.

iv) B Cell Proliferation: Formation of Secondary Follicles and Germinal Centers

The sequence of events that leads to B cell proliferation and differentiation follows a predictable time course (Table 9-4). Within 1–4 days after contact with antigen X, both activated Th and B cells commence expression of CXCR5, a receptor that confers the ability to respond to the chemokine BLC (*B lymphocyte chemoattractant*). The activated B and T cells then migrate into the *primary follicles* located in the B cell-rich cortex of the lymph node. This region of the node contains high levels of BLC and is filled with naive B cells. About 4–6 days post-priming, the activated follicular B cell undergoes one of two fates. The majority of these cells immediately proliferate and terminally differentiate into a population of *short-lived plasma cells* (sometimes called *antibody-forming cells*, AFCs) without carrying out isotype switching or somatic hypermutation. The antibodies produced by these plasma cells are of relatively low affinity and limited diversity, so that although they are of short-term value, they do not contribute much to the complete and efficient long term clearance of a pathogen. A later but more powerful component of the humoral response is derived from the minority of activated follicular B cells that delay their differentiation into *long-lived plasma cells* long enough to allow isotype switching and somatic hypermutation. The antibodies secreted by long-lived plasma cells are of the immense diversity and high affinity necessary to ensure the eventual elimination of most pathogens. Plasma cell differentiation is discussed in detail later in this chapter.

An activated follicular B cell destined to generate long-lived plasma cells undergoes rapid clonal expansion within the primary follicle to convert it into a *secondary follicle* (Fig. 9-11). This process requires the continued presence of antigen X, stimulatory Th2-secreted cytokines, and contact between CD40 and CD40L. By 6–9 days after priming, the uninvolved naive B cells that filled the primary follicle are displaced by the proliferating activated B cells and are compressed at the edges of the follicle to form the *follicular mantle*. By 9–12 days after priming, the secondary follicle polarizes into two distinct areas, a *dark zone* and a *light zone*, and becomes a *germinal*

Table 9-4 **Time Course of Germinal Center Formation**

Day 0 (priming)	Antigen-loaded DCs migrate to the T cell zone of a secondary lymphoid organ and activate naive T cells. Antigen that has been conveyed here in the circulation activates resident naive B cells.
Days 1–4	Activated Th and B cells migrate to the borders of primary follicles containing naive B cells and enter the primary follicles.
Days 4–6	Activated B cells expand to form the secondary follicle.
Days 6–9	Naive B cells in the secondary follicle are pushed to its edges to form the mantle.
Days 9–12	The secondary follicle polarizes into dark and light zones and becomes a germinal center. Somatic hypermutation takes place in proliferating centroblasts in the dark zone. Selection for survival and affinity maturation followed by isotype switching subsequently occur in non-proliferating centrocytes in the light zone.
Days 12–21	Surviving B cells differentiate into memory and plasma cells. The majority leave the germinal center to enter the circulation, but some are retained.
Day 21	The germinal center slowly regresses and returns to the primary follicle stage.

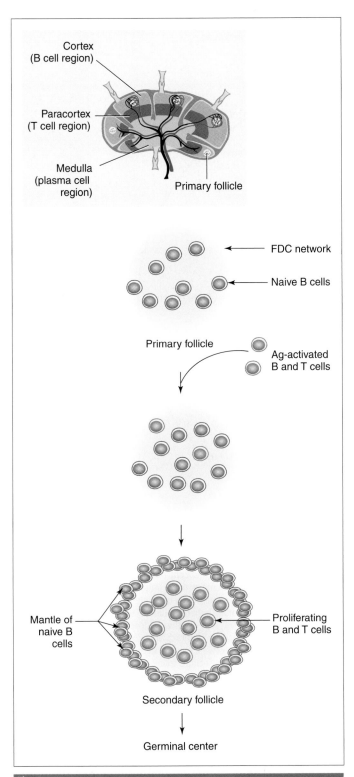

Figure 9-11

Primary and Secondary Follicle Structure in the Lymph Node
After receiving costimulation and Th-derived cytokine help in the paracortical region of the lymph node (top left), the antigen-activated B cell migrates with the antigen-activated T cell into the B cell-rich cortex. The cortex contains many primary follicles made up of an FDC network and a cluster of naive B cells. The activated B and T cells enter a primary follicle where they proliferate rapidly, filling the FDC network to form a secondary follicle. The proliferating cells push the naive B cells out to the perimeter of the follicle to form the follicular mantle. Shortly after, the secondary follicle polarizes to become a germinal center.

center (GC). The dark zone of the GC is where the antibody repertoire undergoes its final diversification by somatic hypermutation of the Ig V_H and V_L exons. The light zone of the GC is where these mutations are either negatively selected to establish peripheral B cell tolerance, or selected for increased affinity for antigen (*affinity maturation*). The light zone is also where the Ig C_H exons undergo isotype switching to increase functional diversity. (Each of these processes is discussed in detail in the following sections.) Because GC formation is tied intimately to all these cellular encounters and events, some researchers and clinicians refer to the appearance of GCs as the *germinal center reaction* ("GC reaction").

The germinal center provides the unique microenvironment for proliferating antigen-activated follicular B cells that is required for their subsequent differentiation into either long-lived plasma cells or memory B cells. The precise mechanisms underlying GC formation and the signals that trigger the subsequent maturation of its B cells have yet to be completely defined, but interactions between activated B cells, activated Th cells, and FDCs are known to be crucial. Indeed, without the FDC, a GC may start to form but then regresses. One function of the FDC network may be to act as an effective antigen depot, maintaining the presence of antigen required to sustain the GC. FDCs may also supply soluble factors and/or intercellular contacts that attract activated B cells and promote the GC reaction. While no individual cytokine has been found to be either essential or sufficient to induce GC formation, CD40–CD40L signaling is critical. Individuals who lack T cells do not have GCs, and knockout mice lacking either CD40 or CD40L fail to develop them as well. CD40 signaling in an activated follicular B cell destined to undergo a GC reaction induces expression of the transcriptional repressor Bcl-6, which suppresses the immediate differentiation of the activated B cell into antibody-producing plasma cells. This suppression allows time for proliferation, selection, affinity maturation, and isotype switching to occur in the GC. The tremendous proliferative activity in GCs throughout the lymph node persists for up to 21 days after priming, after which the number and size of the GCs decrease unless there is a fresh assault by antigen. The development of GCs is accelerated in a secondary response because the memory B cells home more rapidly (due to the expression of additional adhesion molecules) to the secondary lymphoid tissues than do naive B cells.

We will now discuss each of the maturational steps that occurs during the GC reaction.

III. PROLIFERATION AND SOMATIC HYPERMUTATION

i) Centroblasts

The dark zone of a GC is filled with rapidly proliferating, antigen-activated follicular B cells that are called *centroblasts* at this stage (Fig. 9-12). Novel molecules expressed on the FDCs, such as one called 8D6, have been implicated in delivering proliferative signals to these cells. Centroblasts can be identified by the presence of the CD77 cell surface marker and increased expression of c-myc, which, depending on its regulation, can influence cells to either grow or die. The cells

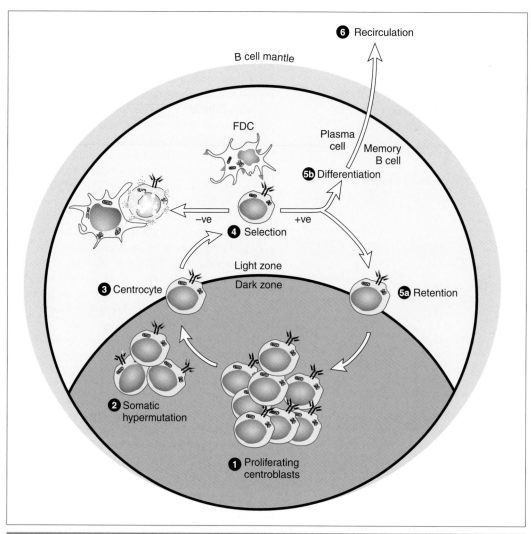

Figure 9-12

Germinal Center Formation and Function
A secondary follicle in the lymph node becomes a germinal center when it polarizes into dark and light zones. Antigen-activated B cells progress from the dark to the light zone, moving through several steps as they complete their differentiation. (1) Rapidly dividing centroblasts undergo apoptosis unless rescued by a survival signal dependent on a minimal level of affinity for the specific antigen. (2) Surviving centroblasts undergo somatic hypermutation of their V_H and V_L exons. (3) Centroblasts diversified by somatic hypermutation give rise to smaller, non-dividing centrocytes along the border of the dark and light zones. 4) The centrocytes migrate into the light zone and interact with FDC displaying the specific antigen. Only clones having relatively high affinity for the antigen receive a survival signal (+ve selection). Those with lower affinity (the majority) undergo death by apoptosis (−ve selection) followed by phagocytosis by tingible body macrophages. Positively selected centrocytes either return to the beginning of the germinal center cycle for further expansion, diversification, and selection (5a) or continue to differentiate into plasma or memory cells (5b) that enter the circulation (6). Adapted from McHeyzer-Williams M. G. and Ahmed R. (1999) B cell memory and the long-lived plasma cell. *Current Opinion in Immunology* **11**, 172–179.

divide at an accelerated rate, doubling every 6–12 hours and producing up to 5000 smaller progeny lymphocytes within 4–5 days. However, centroblasts are very prone to apoptosis unless rescued by a survival signal. This propensity for cell death correlates with their expression of c-myc (which in this case primes the cells to die) and the expression of the apoptosis-inducing genes p53, Bax, and Fas. In addition, centroblasts fail to express the anti-apoptotic molecule Bcl-2. An unknown survival signal appears to be delivered in the dark zone <u>prior</u> to somatic hypermutation that allows only those

clones with BCRs having sufficiently high affinity for antigen X to mature further. Those centroblasts with low affinity for antigen X do not receive the survival signal and die by apoptosis.

ii) Somatic Hypermutation: Generating More V Region Diversity

iia) What is somatic hypermutation? If one compares the sequences of IgM antibodies produced by an activated B cell early in a primary response to a Td antigen with the IgG antibodies produced by its progeny cells in a secondary response, one sees that the nucleotide sequences in the V regions of these Ig proteins are <u>not</u> identical. At about day 9–10 of a primary response, the proliferating centroblasts in the GC dark zone undergo a cycle of DNA replication in which their $V_H D_H J_H$ and $V_L J_L$ exons are particularly subject to the occurrence of random mutations, a process called *somatic hypermutation*. The majority of changes introduced are point mutations rather than insertions or deletions, and only one DNA strand is targeted. As the centroblasts continue to divide (in the presence of specific antigen), mutations continue to accumulate in a step-wise fashion until at least day 18 of the response. As more and more point mutations in the V segments are collected, significant sequence alterations are seen in the CDR1 and CDR2 regions of the Ig protein. These mutations do not usually destroy the ability of the antibody to bind to the antigen but may affect its binding affinity (see later). Interestingly, the efficiency of somatic hypermutation declines with the age of an organism, one reason why older individuals have weaker immune systems.

The rate of point mutation in the V region is estimated to be about 1 change per 1000 base pairs per cell division, about 1000 times the rate of somatic mutation for non-Ig genes (which is why the phenomenon is called "somatic <u>hypermutation</u>"). The V_L and V_H exons that encode the V_L and V_H regions of the mature light and heavy Ig chains expressed by a B cell contain a combined total of about 700 nucleotides, meaning that close to one point mutation occurs at every cell division. This accumulation of mutations means that the nucleotide sequence of an antibody produced by a B cell clone in the secondary response to an antigen may exhibit as much as a 5% divergence from the original germline sequence. Of course, not all nucleotide substitutions result in changes to the amino acid sequence of the Ig protein. On rare occasions, somatic hypermutation can increase the diversity of antigen recognition if a member of a B cell clone is mutated such that its BCR no longer recognizes the original antigen, but gains the ability to respond to an entirely new antigen.

iib) Mechanism of somatic hypermutation: AID. The mechanism of somatic hypermutation has not yet been fully elucidated. Although somatic hypermutation can be observed in the V exons themselves and to a lesser extent in the approximately 2 kb surrounding the rearranged V region, it is not clear why it occurs predominantly in those gene segments that encode the CDR1 and CDR2 regions of the Ig protein (Fig. 9-13). Mutational "hotspots" of the sequence A/G G C/T A/T have been identified that are often, but not always, mutated in mature activated B cells. The CDR3 region of the V exon contains many fewer of these hotspots, and the C_H and C_L exons are entirely unaffected.

How can one explain why somatic hypermutation selectively alters the V regions of the Ig genes but not the V regions of the TCR genes? There is some evidence that active transcription of the Ig genes is necessary to allow hypermutation, so that accessibility of a stretch of DNA may control hypermutability. Interestingly, *in vitro* experimentation has shown that the V gene sequences themselves are not needed to induce hypermutation. If the V region of the Ig gene is replaced with

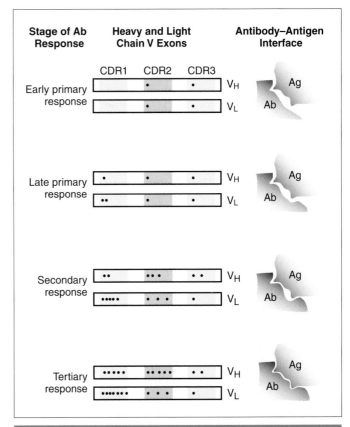

Figure 9-13

Somatic Hypermutation
As a given antigen-specific B cell clone progresses from the early primary response through to the tertiary response (and beyond), successive rounds of somatic hypermutation lead to the accumulation of point mutations primarily in the CDR1 and CDR2 regions of V_H and V_L. Some of these mutations, represented by the black dots, result in altered amino acid sequences in the corresponding heavy and light chain proteins. The affinity-based selection that occurs in the germinal center after each round of somatic hypermutation favors the survival of B cell clones with increased affinity for antigen, so that the antibody–antigen interface becomes increasingly complementary with successive antibody responses.

an unrelated (strong) promoter and an unrelated target gene, but retains the major intronic Ig enhancer and the 3′ Ig enhancer, the target gene will still undergo somatic hypermutation. Other studies have shown that somatic hypermutation in the B cell requires helper T cell interaction, including binding between CD40 and B7-2 on the B cell, to the CD40L and CD28, respectively, on the Th cell.

These observations have been tied together by the work of Tasuku Honjo and colleagues at Kyoto University in Japan. In their pioneering studies of the mechanisms underlying somatic hypermutation and isotype switching, these researchers have shown, using artificial substrates in lymphoid and non-lymphoid cells, that an enzyme called *activation-induced cytidine deaminase* (AID) is essential for both processes (see later for discussion of AID in isotype switching). Expressed exclusively in GC B cells in the lymph nodes and spleen, AID is a small protein of 24 kDa that converts cytidine to uridine.

Honjo and colleagues created fibroblast cell lines bearing a plasmid containing a nucleotide sequence identified as a somatic hypermutation hotspot in the Ig genes. When AID was introduced by retroviral infection into the plasmid-bearing fibroblasts, the plasmid substrate underwent somatic hypermutation to a degree that depended on the dose of AID–retrovirus. The frequency of mutation was similar to that observed after physiological somatic hypermutation in B cells. As occurs in the Ig loci in B cells, somatic hypermutation in the modified fibroblasts proceeded only when the substrate was actively transcribed, and the level of mutation correlated with the level of substrate transcription. In other words, a locus that is inaccessible (not being transcribed) cannot be affected by AID. Furthermore, the introduction of AID alone was sufficient to induce somatic hypermutation in the modified fibroblasts, meaning that all other cofactors or additional enzymes required for somatic hypermutation were constitutively expressed in these non-lymphoid cells. CD40 signaling appears to be required to induce expression of AID in B cells, correlating with the observation that somatic hypermutation is inhibited in the absence of CD40 engagement.

It is not yet clear what the precise targets of AID are or how this enzyme participates in somatic hypermutation and isotype switching. It is possible that the cytidine deamination activity of AID is required to modify a given mRNA so that it encodes an endonuclease in one case and a DNA repair enzyme in another case. Both enzymes may then be required for somatic hypermutation and/or isotype switching, as there is some evidence that these processes share functional elements. There are precedents in plants and protozoa for the use of RNA editing to create multiple different proteins from a limited number of genes in the genome. Examples also exist in mammals in which two proteins of very different functions and expression patterns are derived from differential post-transcriptional processing of the same mRNA. However, it remains to be demonstrated that AID actually acts on RNA. Others have thus proposed that, rather than editing RNA, AID deaminates cytidine to uridine directly in the DNA itself, or is a structural component of a larger protein complex that acts on DNA to carry out recombination and/or mutation.

IV. AFFINITY MATURATION

Let us return to our anti-antigen X B cell centroblasts in the GC that, having undergone somatic hypermutation, are ready to proceed to the next stage of their development. The centroblasts continue to proliferate and eventually give rise to smaller, non-dividing cells called *centrocytes* (refer to Fig. 9-12). Centrocytes have lost the CD77 marker and express high levels of Fas and low levels of Bcl-2. These cells migrate from the dark zone into the light zone of the GC, a region of few T cells but large numbers of FDCs whose extended cellular processes continue to display antigen X. It is at this stage that the fate of the centrocytes, to survive or die, is decided by their interaction with antigen X.

Somatic hypermutation is a random process so that, theoretically, a set of mutations in the antigen binding site of an antibody could increase, decrease, or have no effect on its binding affinity. In reality, the supply of antigen is limited, so that competition for antigen occurs between centrocytes with different somatic mutations. Many mutations introduced by somatic hypermutation will result in centrocyte clones expressing a BCR of unacceptably low affinity for antigen X. These clones do not succeed in binding to antigen X and therefore do not receive a necessary survival signal ("positive selection") that rescues them from apoptosis; the vast majority of centrocytes therefore die. Because of this tremendous amount of B cell death, the light zone is also home to a special type of macrophage called the *tingible-body macrophage*. This cell has the job of phagocytosing all the apoptosed centrocytes that failed to survive this stage.

Those centrocytes whose somatic mutations confer a BCR with higher affinity for antigen X trapped on the FDC are better at presenting antigen to helper T cells, which means they will preferentially receive growth stimulatory signals from those T cells. These clones replicate faster than those clones with somatic mutations that did not increase BCR affinity, and the antibodies produced by these successful clones (those of higher affinity) predominate later in the response. In this way, somatic hypermutation increases the diversity of the antibody repertoire (with respect to antigen recognition) <u>after</u> antigenic stimulation. The positive selection signal conveyed by high-affinity interaction with antigen X in the GC light zone is thought to be related to the interaction of CD23 on the FDC with CD21 on the centrocytes, which leads to the upregulation of Bcl-2 and the downregulation of Fas. Only those positively-selected centrocytes expressing Bcl-2 undergo further maturation, and most positively-selected centrocytes become either memory cells or plasma cells. However, a proportion of these centrocytes is retained within the GC and is "recycled" back into the dark zone to undergo further proliferation and somatic hypermutation. In a secondary response, the positive selection of centrocytes in the light zone is accelerated because pre-existing circulating antibody forms immune complexes with antigen X that are easily accumulated by the FDCs.

Interestingly, those centrocytes showing a significant increase in affinity for antigen X after hypermutation in the primary response are preferentially selected to become long-lived memory cells. These cells retain their mutated Ig sequences, so that when activated in a secondary or subsequent response, their progeny plasma cells produce antibodies which are still specific for the antigen, but of higher affinity. Hence, the phenomenon is called affinity <u>maturation</u>. Those centrocytes with more modest affinities for antigen X during the primary response tend to become plasma cells and <u>not</u> memory cells. Once a centrocyte has differentiated into either a plasma cell or memory cell, it experiences no further somatic hypermutation. However, additional somatic hypermutation can occur in progeny cells the next time antigen enters the body and activates memory B cells. In fact, there is an upper limit to the overall increase in affinity that can be achieved (estimated to be about 100-fold) because continued somatic hypermutation will eventually replace beneficial mutations (those that increase affinity) with deleterious ones. Affinity maturation is a feature only of responses to Td antigens because, as we have seen, signals from Th cells are important for the induction of somatic hypermutation.

V. ISOTYPE SWITCHING: GENERATING FUNCTIONAL DIVERSITY

Those centrocytes displaying a high affinity for antigen are also those that undergo isotype switching. While isotype switching is independent of somatic hypermutation and can occur without it, researchers have shown that most B cells that express Igs of new isotypes have already undergone somatic hypermutation. This order of process not only ensures that the population of memory B cells left to react in the next exposure to antigen can make use of all possible antibody effector functions but may also prevent the expression of new isotypes by low-affinity or autoreactive B cells.

i) What Is Isotype Switching?

V(D)J recombination of V and J gene segments in the *Igk* and *Igl* loci, and of V, D, and J gene segments in the *Igh* locus, generate the variable region exons that are subsequently linked to constant region exon sequences by conventional splicing of primary RNA transcripts. The V_H, V_L, and C_L genomic regions in a given B cell experience no further changes, aside from somatic hypermutation in the V regions, once a B cell matures beyond the pre-B cell stage. However, after antigenic activation, the C_H

region of the B cell genome is capable of undergoing a separate series of cutting/rejoining events <u>at the DNA level</u> that can bring any of the downstream C_H exons next to the $V_H D_H J_H$ exon previously established by V(D)J recombination, with the apparent loss of the intervening C_H exons and other sequences. The effect is one of isotype "switching" in the Ig produced by that B cell clone: the antigenic specificity is the same (because the V exons are unchanged) but this specificity is now attached to a constant region that may confer a different effector function. Such alterations due to isotype switching can govern the outcome of a given antibody response. For example, a switch to an IgE antibody can elicit an allergic reaction to an antigen, whereas a switch to IgA results in a protective response at mucosal surfaces and in bodily secretions.

ii) Molecular Mechanism of Isotype Switching
iia) Switch regions. The actual mechanism of isotype switching, called *switch recombination*, is not yet fully understood. The prevailing hypothesis is that, once the signal to switch is received, the DNA is looped out such that the selected C_H exon is juxtaposed next to the rearranged VDJ exon, and the intervening sequences (containing any C_H exons 5' to the selected C_H exon) are deleted. Switch recombination depends on the presence of highly conserved guanine-rich *switch regions* (S_H), which are tandem repeats of 52 base pairs (total length 1–10 kb) located 2–3 kb upstream in the introns 5' of each C_H exon (except Cδ) (Fig. 9-14). The switch region preceding the

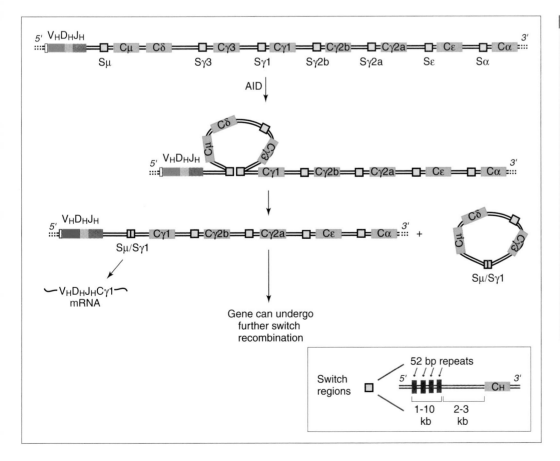

Figure 9-14

Mechanism of Switch Recombination
In this example of switch recombination from IgM to IgG1, the heavy chain DNA loops out such that the Sμ and Sγ1 switch regions are juxtaposed. The DNA is cut and rejoined at a random site within Sμ/Sγ1, excising all C_H exons 5' of Cγ1. The resulting Ig gene has the Cγ1 exon in the correct position for transcription of mRNA specifying IgG1 instead of IgM. This gene may or may not undergo further switch recombination in daughter cells arising from this clone. The DNA/RNA editing enzyme AID is essential for the process of isotype switching. The inset shows details of general switch region structure.

Cμ exon is denoted Sμ, that before the Cγ1 as Sγ1, etc. A double-stranded break is introduced (by an unknown DNA cleavage mechanism) somewhere in the DNA of the Sμ and another in that of the S_H region immediately 5′ to the selected C_H exon. According to standard switch recombination models, the S_H regions are then paired so that circular products formed by the looping out of the intervening DNA are excised, resulting in the deletion of all C_H exons 5′ to the selected C_H exon. The actual joining within the S_H regions as the DNA strands are paired does not have to be precise and appears to happen at some random point in the switch sequence.

During Ig synthesis, the C_H exon closest to the $V_H D_H J_H$ exon is preferentially transcribed, spliced, and translated. Thus, in the case where switch recombination has deleted all C_H exons except Cγ2b, Cγ2a, Cε, and Cα, only the Cγ2b exon is transcribed and spliced to the $V_H D_H J_H$ exon (Fig. 9-15). The final result is the production of IgG2b anti-

bodies and not IgE or IgA antibodies. Each daughter of this IgG2b-producing B cell, however, may subsequently undergo its own randomly-determined switch in response to antigen (remembering that the Cμ, Cδ, and Cγ1 genes were deleted in the switch in the parent cell). For example, in one daughter, Sμ/Sγ2b may be joined to Sε, deleting the Cγ2b and Cγ2a genes and setting up a transcriptional environment that results in the production of IgE antibodies by this daughter cell. Simultaneously, in another daughter cell, Sμ/Sγ2b may join to Sα, so that the Cγ2b, Cγ2a, and Cε genes are all deleted and the cell produces IgA antibodies. This exon deletion mechanism offers an attractive explanation for the unidirectional flavor of isotype switching. That is, cells that switch to IgA production cannot "reverse" and start to make IgG or IgE antibodies again because the Cγ and Cε exons have been lost in the rearrangement that connected the V_H exon to the Cα exon.

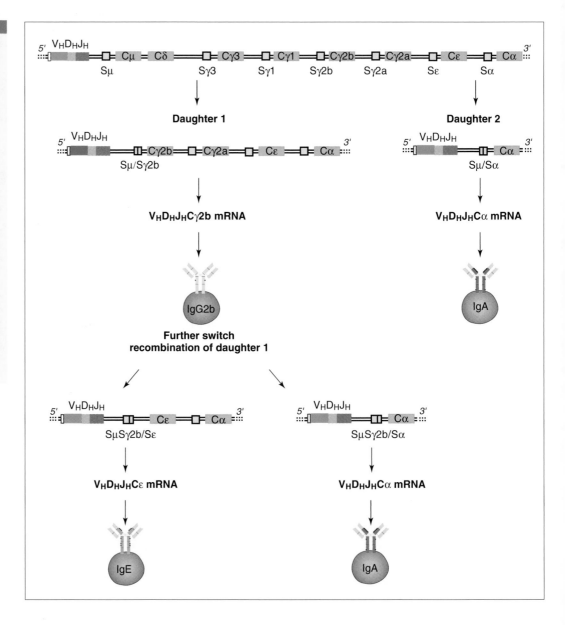

Figure 9-15

An Example of Isotype Switching
As an antigen-activated B cell proliferates and differentiates, various daughter cells may produce different antibody isotypes, depending on the switch recombination experienced in various members of the expanding clone. In this example (described in detail in the text), Daughter 1 undergoes a switch from IgM to IgG2b production while Daughter 2 switches directly to IgA production. Progeny of Daughter 1 can subsequently switch to the production of any isotype whose C_H exon remained in the Ig gene. However, the progeny of Daughter 2 can no longer switch to any other isotype because all other 5′ C_H exons have already been excised from the DNA.

The preceding scheme seems like a neat and tidy model, but there is a puzzling flaw: very occasionally one finds a single antigen-activated B cell that produces antibodies of two different isotypes, including an isotype whose gene has theoretically been deleted! These observations cannot be reconciled with the switch recombination model as it stands, so that some scientists have speculated about the existence of an alternative means of DNA looping that does not involve permanent deletion of C_H exons. The production of the first IgM and IgD antibodies in an immature B cell involves the transcription of a very long primary RNA thought to encompass both the $C\mu$ and $C\delta$ exons. This transcript undergoes alternative splicing to generate Igs of both isotypes in one cell. It is possible that a similar transcriptional mechanism may occasionally operate in a mature activated B cell such that a huge primary transcript of more than 180 kb covering all the C_H exons is made. This transcript might be alternatively spliced to bring any C_H exon in apposition to the $V_HD_HJ_H$ exon. In this model, antibodies of different isotypes could theoretically be produced in any order in different daughter cells because no C_H exons would be physically deleted. However, if this model represented the general process of isotype switching, one would have to invoke a mechanism that accounts for the progressively $3'$ character of the isotypes produced as successive switches occur.

iib) Role of cytokines and germline C_H transcription.

Numerous studies have demonstrated that most isotype switching is promoted by cytokines (particularly TGFβ, IFNγ, and IL-4) in the immediate microenvironment of the activated B cell, and that different cytokines favor switching to different isotypes. *In vitro* experimentation has also shown that isotype switching cannot occur in activated B cells unless there is prior production of short transcripts (which are not translated) of the selected C_H exon in its germline configuration. We now know that cytokines promote isotype switching because they influence this transcription of the *Igh* locus and particularly the selected C_H exon. In fact, the mere process of cytokine-induced C_H exon transcription is not enough to induce isotype switching: the presence of the germline isotype-specific transcript itself is somehow necessary, possibly as a component of the switch recombination machinery. Without the germline C_H transcript, the germline DNA in the activated B cell is not altered and antibodies of the selected isotype cannot be produced. As might be expected for a molecular "machine part" that can be varied, the germline transcripts of all the C_H exons are similar in molecular structure, allowing the switch machinery to recognize any one of them.

The transcription of germline C_H transcripts induced in activated B cells by TGFβ, IFNγ, or IL-4 begins in leader-like sequences in the C_H region called the *I exons*. Each C_H exon has an associated I exon located just $5'$ to its specific S_H site in the germline configuration, leading to the designation Iγ1, Iα, etc. (Fig. 9-16). Each I exon contains a transcriptional start site and an RNA splice donor site, which permit the initiation of C_H exon transcripts in the germline configuration. For example, assume that germline transcription of the $C\gamma$1 exon has been triggered by the presence of a cytokine. The RNA polymerase binds to the transcription start site in the Iγ1 exon $5'$ of the Sγ1 switch site and proceeds to

transcribe through Sγ1, through the $C\gamma$1 exon coding sequence, and ends at the normal polyA sites for the membrane-bound or secreted Ig heavy chain mRNAs of each isotype. Sequences between Iγ1 and the $C\gamma$1 exon, including the Sγ1 region, are spliced out in the normal way. The result is an IgG1 heavy chain constant region transcript that, unlike mature IgG1 chain mRNA, contains no variable exon and does not appear to be translated *in vivo*. In the presence of the germline Iγ1–Cγ1 transcript, the switch recombination machinery aligns the Sμ and Sγ1 switch sites in the germline DNA and loops out the DNA between them, consequently removing the Iγ1 transcriptional start site. As a result, Ig heavy chain transcription in this cell begins at the transcriptional start site upstream of the $V_HD_HJ_H$ variable exon, and proceeds through the $C\gamma$1 coding sequences to produce mature IgG1 heavy chains. The $3'$ transcriptional enhancer located $3'$ of the C_H exons contributes to the efficiency of isotype switching by affecting the levels of germline C_H transcripts. Knockout mice deficient for this enhancer show defects in isotype switching.

Specific cytokines induce the germline transcription of particular C_H exons, and so each promotes switching to a particular isotype (Table 9-5). For example:

(a) IL-4 induces germline $C\gamma$1 and $C\varepsilon$ transcripts in mouse B cells and therefore switching to IgG1 and IgE production. Knockout mouse mutants lacking the IL-4 gene therefore secrete greatly reduced levels of IgE antibodies in response to a Td antigen. This dominant role for IL-4 in inducing switching to IgE has been documented for the B cells of all species examined to date. However, the closely related cytokine IL-13, whose biological activities overlap those of IL-4, can replace IL-4 to a certain extent for switching to IgE and IgG1 (in the presence of CD40 signaling; see later). IL-4 appears to inhibit switching to IgG2a.

(b) IFNγ can induce $C\gamma$2a or $C\gamma$3 germline transcripts in activated murine B cells, which lead to IgGγ2a or IgG3 production, respectively. Knockout mice lacking the IFNγ receptor show decreased production of both IgG2a and IgG3 antigen-specific antibodies. IFNγ also inhibits switching to IgG1, IgE, or IgA in certain *in vitro* systems.

(c) TGFβ promotes germline $C\alpha$ transcription and switching to IgA production in both mouse and human B cells. In mice, TGFβ also increases germline transcription of $C\gamma$2b and subsequent switching to IgG2b. Germline transcription of $C\varepsilon$ and consequent IgE production induced by IL-4 is inhibited in the presence of TGFβ.

(d) IL-5 also appears to increase isotype switching but in an as yet undefined way that does not involve germline transcription. Some scientists speculate that it may be involved in the mechanisms of switch recombination itself. In mucosal lymphoid tissues, a combination of IL-5 and TGFβ promotes switching to IgA production.

(e) Several other cytokines have been implicated in enhancing isotype switching triggered by the major cytokines listed above. For example, IL-2 and IL-6 ensure optimum IgG1

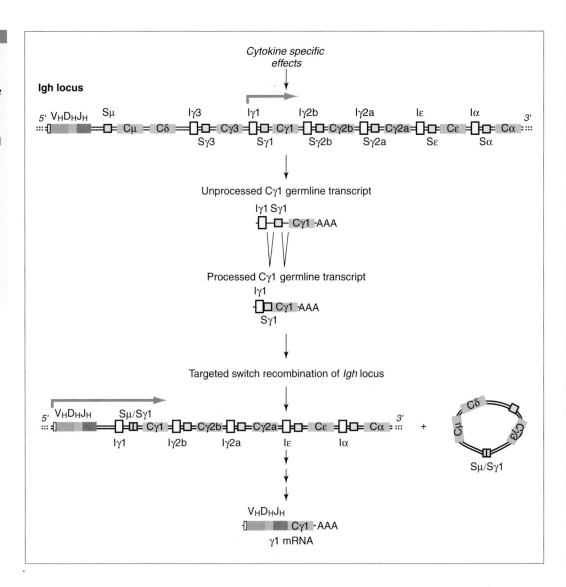

Figure 9-16

Germline Transcripts and Isotype Switching
In the example shown, cytokine-specific effects on the Iγ1 site trigger the transcription of the Cγ1 exon in its germline configuration. The presence of the processed Cγ1 germline transcript then promotes the juxtaposition of the Sμ and Sγ1 switch sites, resulting in the excision of a DNA loop containing the intervening Cμ, Cδ, and Cγ3 exons. Transcription of the Ig heavy chain in this cell will now produce γ1 mRNAs that will be translated into γ1 heavy chains, converting the cell into an IgG1 producer.

Table 9-5 Cytokine Effects on Isotype Switching in Mice

Cytokine	Inhibits Isotype Switch to C_H Exon	Promotes Isotype Switch to C_H Exon	Ig Produced
IL-4 (IL-13)	Cγ2a	Cγ1 Cε	IgG1 IgE
IFNγ	Cγ1, Cε, Cα	Cγ2 Cγ3	IgG2a IgG3
TGFβ	Cε	Cα Cγ2b	IgA IgG2b
IL-5 + IL-1 or TGFβ		Cα	IgA
IL-4 + IL-2 or IL-6		Cγ1	IgG1
IL-10 + TGFβ		Cα	IgA

responses, while IL-1 increases IgA production induced by IL-5. IL-10 increases switching to IgA induced by TGFβ.

The antagonistic effects of cytokines in isotype switching are reflected in the sharply delineated functions of the Th cell subtypes. Th1 cells preferentially secrete a subset of cytokines that includes IFNγ and excludes IL-4. Conversely, Th2 cells secrete a subset that includes IL-4 but excludes IFNγ. Thus, the type of Th cell that makes contact with the B cell will determine which cytokines are secreted in its immediate microenvironment and consequently which antibody isotype will be produced.

iic) Differential requirements for isotype switching. Although the cytokines play important roles in isotype switching, they may be neither strictly necessary nor sufficient, depending on the antigen. Studies *in vitro* of the responses of cultured activated B cells to Td antigens have shown that cytokines alone cannot induce switching. Switching also requires that the mIg be engaged (as would be the case for an activated B cell) and that CD40 signaling occur. Naive B cells from knockout mice lacking either CD40 or CD40L are impaired in switching in response to Td antigens.

Isotype switching to some isotypes can also be observed late in primary responses to Ti antigens. However, the requirements for switching in these cases are less stringent than for Td antigens. For example, CD40 engagement is not essential for switching in response to Ti antigens. Naive B cells from CD40 or CD40L knockout mice cannot switch in response to a Td antigen, but are capable of limited switching to all isotypes except IgE in response to a Ti antigen. The one fundamental requirement for switching in response to Ti antigens is that an activation signal be received, but after this, the necessary factors depend on whether the antigen is of the Ti-1 or Ti-2 type. The Ti-1 antigen LPS induces IgM and IgG3 production in normal B cells *in vitro* even in the absence of cytokines. The addition of IL-4 results in IgG1 and IgE secretion, while the addition of IFNγ promotes switching to IgG2b and IgA production. In contrast, switching in response to Ti-2 antigens does not appear to occur in the absence of cytokines, and depends on their secretion by non-T cells such as NK cells (IL-4, IFNγ), macrophages (TGFβ), B cells, mast cells, and others.

iid) Role of AID in isotype switching. As mentioned previously, AID (activation-induced cytidine deaminase) is essential not only for somatic hypermutation but also for isotype switching. Just as they investigated mechanisms of somatic hypermutation in non-lymphoid cells, Tasuku Honjo and colleagues have shown that AID is necessary and sufficient to induce switch recombination of an artificial switch region construct expressed in fibroblasts. The switch region construct resembled endogenous Ig switch regions in sequence and secondary structure. Again, transcription of the targeted switch region was required for switch recombination of the substrate in the modified fibroblasts, and the level of switching was comparable to that observed for the endogenous *Igh* locus in activated B cells. However, the endogenous Ig genes in the modified fibroblast genome were not affected by transfection of AID, confirming that the locus must be accessible and under active (germline) transcription for switch recombination to be possible. AID expression is restricted to GC B cells, but these experiments showed that all other components necessary for the completion of switching are constitutively available in non-lymphoid as well as lymphoid cells. Filling in another piece of the puzzle, signaling induced by engagement of cytokine receptors and CD40 is likely required for isotype switching because these events induce germline transcription and the expression of AID. It is still unknown whether AID is directly involved in cleaving the DNA in the switch regions, or whether it is a component of a separate recombinase enzyme or complex, or whether it regulates such a complex, or whether it uses its putative RNA-editing function to generate a transcript for a new DNA cleavage enzyme.

VI. DIFFERENTIATION OF MEMORY B CELLS AND PLASMA CELLS

i) Decisions, Decisions—To Be a Plasma Cell or a Memory Cell?

A centrocyte that has enjoyed positive selection, avoided negative selection, and undergone isotype switching makes the decision to become either a memory cell or a plasma cell. The development of memory B cells is entirely dependent on the FDC, the germinal centers, and T cell help in the form of CD40L signaling. Evidence for the importance of T cell help in the development of memory comes from the fact that Ti antigens can stimulate the differentiation of plasma cells but do not induce the generation of memory B cells.

In vitro experimentation has shown that, in the presence of IL-2 and IL-10, engagement of CD40 on the centrocyte by CD40L on the Th cell initiates signaling that blocks the plasma cell differentiation pathway such that the centrocyte becomes a memory B cell instead (Fig. 9-17). Without CD40 signaling (but in the presence of IL-2 and IL-10), the centrocyte becomes a large plasmablast and eventually an antibody-secreting plasma cell. As mentioned previously, CD40 signaling induces the expression of the transcriptional repressor Bcl-6. Bcl-6 in turn inhibits expression of Blimp-1 (*B lymphocyte-induced maturation protein 1*), a transcription factor that is one of two master regulators of plasma cell differentiation. The other factor is called XBP-1 (*X box binding protein-1*). Blimp-1 represses the transcription of proteins necessary for the expression of MHC class II molecules and for proliferation, and activates genes involved in Ig secretion. XBP-1 controls the final steps of plasma cell differentiation.

Some scientists feel that there may be a twist to the above scenario, and that two subtly different subsets of memory B cells are generated that represent separate progenitors of future memory cells and plasma cells. Those primary response memory cells that later give rise to plasma cells express high to intermediate levels of a cell surface marker of unknown function called HSA (CD24), while those destined to become a future generation of memory cells express lower levels of HSA and mIgM, and higher levels of mIgD. These subpopulations can be separated from one another even prior to antigenic stimulation. More markers and *in vitro* studies are needed to definitively test the memory subset theory.

How long can re-stimulated memory cells continue to give rise to new memory cell progeny? *In vitro*, memory cells treated with IL-2 and IL-10 give rise to more plasma cells, and consequently to higher levels of secreted Ig, than do naive B cells treated in the same way. This observation has suggested the so-called "decreasing potential" hypothesis, which holds that, after each round of antigenic stimulation, the progeny of memory cells are more likely to become terminally differentiated plasma cells than new memory cells. *In vivo*, this may translate into an increased number of effector cells in the secondary response and a control on the possible over-expansion of one particular memory B cell clone. On the other hand, it also means that the host may one day no longer have memory cells of this clone to call upon when the relevant pathogen strikes.

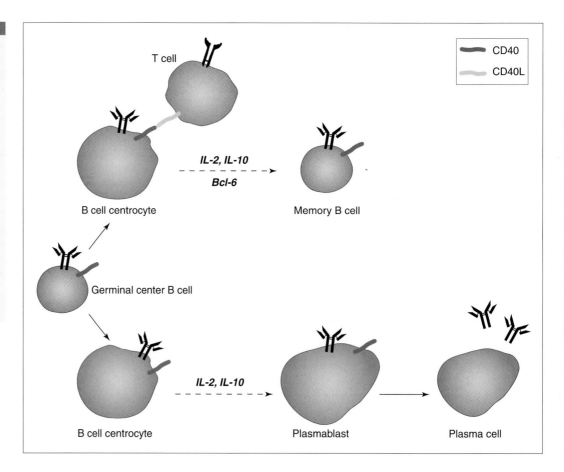

Figure 9-17

Memory Cell vs. Plasma Cell Differentiation Pathway
After somatic hypermutation, selection, and isotype switching, B cells in the germinal center make a decision to become either an antibody-secreting plasma cell or a memory B cell. A germinal center B cell differentiating in the presence of IL-2 and IL-10 becomes a memory B cell if CD40 signaling has induced expression of the transcriptional repressor Bcl-6. Bcl-6 expression blocks the plasma cell differentiation pathway. If CD40 signaling is absent, the cell becomes a plasmablast and eventually a plasma cell.

ii) Plasma Cells

iia) General characteristics. As mentioned earlier in this chapter, activated follicular B cells that do not undergo the GC reaction do not experience sustained CD40 signaling and thus do not upregulate Bcl-6. These cells thus immediately enter the plasma cell terminal differentiation pathway without undergoing isotype switching or somatic hypermutation. Short-lived plasma cells with a limited life span of only 2–3 days are produced that secrete low-affinity IgM antibody. Short-lived plasma cells of follicular origin are concentrated in the secondary lymphoid tissues and peripheral sites.

Short-lived plasma cells can also arise in the spleen from MZ B cells in the marginal zone. An encounter with antigen (usually a Ti antigen) in MZ B cells fails to upregulate Bcl-6 expression. As a result, these B cells immediately proliferate and differentiate into IgM-secreting plasma cells without undergoing isotype switching or somatic hypermutation. It is thought that MZ B cells may play an important role in the very early stages of the splenic adaptive response against blood-borne antigens.

An activated follicular B cell that has undergone a GC reaction (and has completed somatic hypermutation and isotype switching) but is later deprived of CD40 signaling is directed to the plasma cell terminal differentiation path (rather than the memory cell path). Long-lived plasma cells are produced that are capable of secreting antibody for several months in the absence of any cell division or re-exposure to antigen. These cells are concentrated primarily in the bone marrow but also occur in the spleen. The prolonged survival of long-lived plasma cells has been attributed to intercellular contacts with stromal cells that result in their production of IL-6. Long-lived plasma cells secrete high-affinity IgG, IgA, and/or IgE antibodies directed against the original Td antigen.

Regardless of its origin, an activated B cell destined to become a plasma cell first becomes a *plasmablast*, in which internal changes to the rER occur that prepare the cell for its primary task of antibody synthesis. Plasmablasts migrate primarily to the bone marrow but can take up residence in the medulla of the lymph node or, in the case of an immune response occurring in the spleen, in the red pulp. In these sites, the plasmablasts continue to mature and further acquire the features that mark them as plasma cells. Fully differentiated plasma cells express little or no CD19, CD20, CD22, CD40, MHC class II, or mIg on their cell surfaces, can no longer receive T cell help, and are incapable of cell division. Instead, these cells are characterized by enlarged rER and Golgi compartments and increased numbers of ribosomes. Up to 40% of the mature plasma cell's total cellular protein may be immunoglobulin, most of which is released into the blood or lymph as circulating antibody.

iib) Membrane vs. secreted Igs. The reader will recall from Chapters 5 and 8 that a given B cell can make either the membrane-bound form of its Ig protein to serve in its BCR, or it can make the secreted form of its Ig protein to serve as circulating antibody. An individual B cell clone produces varying proportions of these forms as it matures from an activated B cell toward a plasma cell. As we have learned, the amino acid sequences of the antigen-binding sites (that is, of the V regions) of both the secreted and membrane-bound forms of an Ig are identical: their different fates stem from changes in the C region amino acids of the heavy chains. How is the transition from mIg to sIg accomplished? After transcription of the rearranged *Igh* locus, differential RNA processing and alternative splicing (described in Box 8-1) modify the primary Ig RNA transcript at its 3' end. The mRNA produced specifies an amino acid sequence at the C-terminal end of the heavy chain protein that directs whether the Ig is to be secreted or not. The mechanism that controls how this choice is made has yet to be defined but clearly depends on the maturation and activation status of the B cell.

To examine this process in more detail, let's step back to the predecessor of the plasmablast, the resting naive mature B cell. In this cell, the complete rearranged Ig heavy chain gene is composed of the V_H exon and the C_H exons, each of which is followed by its own Se (tailpiece), M1 (transmembrane, TM), and M2 (cytoplasmic, CYT) exons. Let's follow one of the thousands of primary transcripts produced from this gene by the RNA polymerase <u>prior</u> to an encounter with antigen. Under the influence of as yet unknown signals that

indicate that mIgM (as opposed to sIgM) is to be made, the primary RNA transcript containing μ tailpiece, TM, and CYT sequences is processed such that the polyadenylation site located 3' of the M2 exon is used and both the μ TM and CYT sequences remain in the transcript (Fig. 9-18). <u>When these exons are present,</u> the spliceosome is able to make use of a splice donor site located in $C_\mu 4$ just before the Se sequence, and so the tailpiece is lost with the intron in the splicing that joins $C_\mu 4$ to M1 and M2. This mRNA is then translated into a protein with the hydrophobic TM domain that is capable of integrating into the membrane as mIgM. The vast majority of the transcripts produced from the heavy chain Ig gene in this naive cell will undergo the same fate and will be translated into mIgM molecules.

Suppose now that the B cell meets an antigen that activates it. The cell receives signals during activation and maturation that start to alter splicing such that, in some primary transcripts, a polyadenylation site located immediately after the tailpiece sequence is used rather than the site located after M2. By polyadenylating transcripts in this new way, the μ TM and CYT sequences are lost, so that the splice donor site just before the tailpiece is not used, and the tailpiece is retained in the transcript. When translated, this mRNA produces an sIg protein that is incapable of anchoring into the membrane (no TM domain) and is secreted out of the cell in a secretory vesicle that

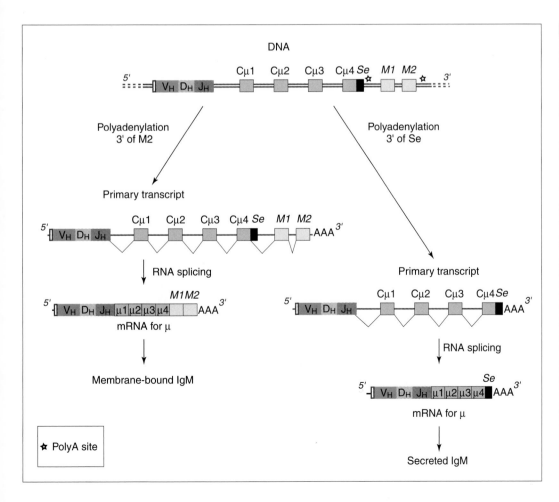

Figure 9-18

Membrane vs. Secreted Immunoglobulin
Each C_H exon is followed by its own Se, M1, and M2 exons that encode the immunoglobulin tailpiece (Se), a transmembrane domain (TM), and a cytoplasmic domain (CYT), respectively. When the polyadenylation site 3' of M2 is used during transcription, the primary transcripts retain the Se, M1, and M2 sequences. RNA splicing removes the Se sequence, leading to heavy chains with transmembrane domains that are incorporated into the membrane as part of the mIg molecule. When the polyadenylation site 3' of Se is used, only a tailpiece is retained at the 3' end of the primary transcript and the Ig is secreted. In the example shown, the B cell has not yet undergone isotype switching to a C_H exon other than C_μ. Either mIgM or secreted IgM can be produced.

fuses with the B cell plasma membrane (see below). This same splicing decision process happens over and over for the thousands of transcripts that are produced by the cell from its heavy chain Ig gene, giving rise to discrete populations of secreted and membrane-bound proteins. However, as the progeny cells differentiate into plasmablasts, more and more primary transcripts are processed so as to produce sIg, and the frequency of transcripts containing the TM domain drops precipitously: secreted antibodies of the IgM isotype are produced in large quantities and the cell surface level of mIgM decreases. Sequences encoding TM domains, CYT domains, and tailpieces are associated with each C_H exon, so that sIgs of all isotypes can be produced after isotype switching is triggered late in the primary response. However, the tailpiece sequences that occur in the $C\delta$ exon appear to be seldom used, since IgD is rarely secreted.

iic) Antibody protein synthesis. Transcripts encoding the sIg heavy chains are translated in the cytoplasm concurrently with, but independently of, transcripts encoding the light Ig chains and the Igα and Igβ chains. The usual co-translational mechanism ensures that all these chains are inserted into the ER membrane as they are synthesized. As the Ig chains pass from the rER into the Golgi apparatus, two heavy chain polypeptides are joined by disulfide bridges to two light chains to form the H_2L_2 core Ig monomer (Fig. 9-19). Surprisingly, how the chains are assembled into the monomer depends on the isotype of the heavy chain. The assembly of the IgG monomer occurs in the sequence H, H_2, H_2L, H_2L_2. In contrast, the IgM H and L chains are linked first to each other to form H–L pairs, and then the two H–L pairs are joined to make the H_2L_2 monomer. The H_2L_2 monomers

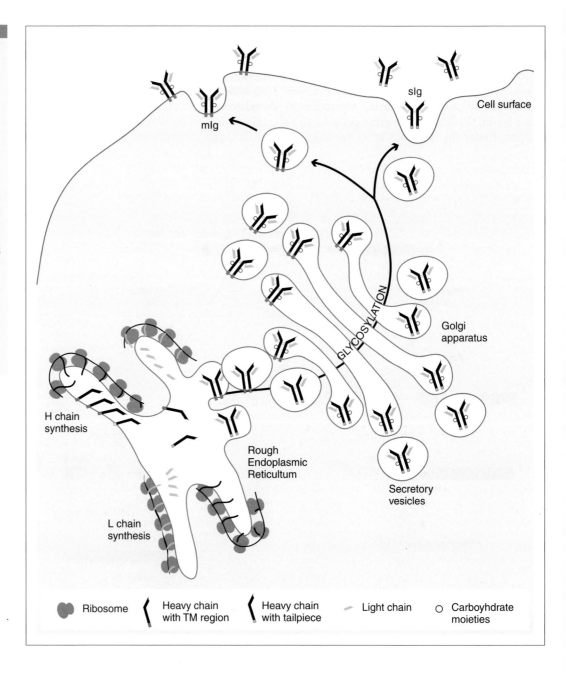

Figure 9-19

Immunoglobulin Synthesis and Glycosylation Pathway
Immunoglobulin molecules that are newly assembled in the rER are targeted differently, depending on the amino acid sequences found at the C-terminus of the heavy chain. Those with a transmembrane (TM) domain are integrated into the membranes of the Golgi apparatus and are transported to the cell surface to form mIg. Those Ig chains with only a tailpiece are sequestered as free molecules inside secretory vesicles and are secreted. Both Ig forms undergo glycosylation as they pass through the Golgi apparatus. Note that for simplicity, Igα/Igβ normally associated with membrane-bound Ig has been omitted.

must also associate with the Igα and Igβ chains in order for the entire Ig complex to pass out of the rER into the Golgi. The glycosylation of all these chains occurs as they pass into the Golgi and the secretory apparatus. Ig heavy chains that have been synthesized using the hydrophobic sequences encoded by the TM exon become anchored in the membrane of intracellular transport vesicles. In a differentiated plasma cell, H_2L_2 monomers synthesized without the TM hydrophobic sequences cannot anchor in membranes and are instead sequestered inside intracellular vesicles as free-floating sIg. When these vesicles undergo reverse fusion with the plasma membrane, the membrane-anchored Ig complexes are expressed on the B cell as surface mIg, while the sIg is released into the extracellular milieu. Certain plasma cells (particularly sIgA-synthesizing cells) further modify their sIg prior to secretion by incorporating the J chain and polymerizing the Ig monomers. Upon release from the cell, interaction of the polymeric Ig with the poly-Ig receptor of a mucosal cell results in the production of secretory Igs, as was described in Chapter 5.

iii) Memory B Cells
iiia) General characteristics. As stated previously, a centrocyte (derived from an activated follicular B cell that underwent the GC reaction) that receives CD40 signaling is diverted from the plasma cell differentiation path and becomes a memory B cell. Although they resemble naive B cells in their small size and general morphology, memory B cells have a longer life span than do naive B cells and do not divide until activated by a subsequent exposure to antigen. Thus, anti-X memory B cells proceed to adopt a resting state as antigen X is cleared, terminating the primary response. Some memory B cells take up residence in the follicular mantle and are ready to react rapidly when a fresh dose of antigen X is conveyed to the lymph node. Other memory B cells leave the GC and enter the circulation, where they recirculate in the periphery and maintain surveillance for a subsequent attack by antigen X. In the spleen, memory B cells congregate in the splenic marginal zones, precisely where blood-borne antigens first collect. Many memory cells appear to localize preferentially to the epithelial tissues, where antigen penetrating the skin might be expected to be encountered. For reasons that are not yet understood, memory cells specific for different antigens have different life spans. We see the tangible results of these limitations when repeated booster shots are required for vaccination against some pathogens but not others (see Ch.23).

Let us follow the fate of an anti-X memory B cell in a secondary response. When antigen X again collects in the secondary lymphoid tissues, anti-X memory B cells, which express increased numbers and types of adhesion molecules, home to the primary follicles more rapidly than did the naive B cells in the primary response. Memory B cells are activated in much the same way as described for naive B cells, but have several properties that make secondary (and subsequent) responses more efficient. First, due to affinity maturation, the BCR of the memory B cell has an increased affinity for antigen so that the cell is stimulated more easily. Secondly, memory B cells are present in expanded numbers and so play a greater role in the presentation of antigen to Th cells. Rather than depending on macrophages or dendritic cells to carry out the APC role, memory B cells readily internalize, process, and present peptide-MHC directly to Th cells (see Ch.11). The increased efficiency in antigen presentation also means that the threshold of antigen required to activate a memory cell is lower than that required for a naive B cell. Thirdly, antigen presentation by a memory B cell is associated with more rapid upregulation of the costimulatory molecules B7-1 and B7-2. These three factors decrease the lag time of the response. In addition, the population of anti-antigen X memory B cells generated in the primary response will have already undergone somatic hypermutation and isotype switching. This means that the progeny cells of the expanding anti-X memory B cell clone that differentiate into second-generation plasma cells may produce anti-X antibodies of altered affinity and different isotype. No further somatic hypermutation is thought to occur in these cells. However, positive selection in favor of higher affinity anti-antigen X clones continues in the germinal centers (assuming antigen is limiting), resulting in further affinity maturation and the production of antibodies that clear antigen X with the increased efficiency characteristic of the secondary response. In contrast, the progeny cells of the activated anti-X memory B cell that go on to be second-generation memory cells may undergo another round of somatic hypermutation and isotype switching. Still more point mutations accumulate in the Ig V regions, which may contribute to increased affinity in a tertiary response while further changes in isotype may influence effector function. With the clearance of antigen X, second-generation plasma cells die off, leaving second-generation anti-X memory B cells to adopt a resting state and resume peripheral surveillance.

iiib) Memory markers. Morphologically, memory cells resemble naive B cells but can be distinguished from them by the expression of mIg isotypes other than IgM and by a lower level of mIgD. Memory B cells in both mice and humans display increased levels of various adhesion molecules, and lower levels of CD24, compared to naive B cells. Memory cells can be readily distinguished from plasma cells by morphology under the light microscope because the former are smaller in size and lack the enhanced protein synthetic machinery of plasma cells. In addition, both human and mouse memory B cells lack expression of CD38 while plasmablasts express this marker. (CD38 is a cell surface enzyme thought to be involved in adhesion or Ca^{2+} signal transduction.) Human memory B cells, but not plasmablasts or mouse B cells, carry the markers CD20 and CD27. (CD20 is a membrane-spanning protein involved in the regulation of Ca^{2+} flux across the membrane, while CD27 regulates B cell proliferation.) More on markers distinguishing various B cell subsets in mice and humans can be found in Tables 9-6 and 9-7 (see pg. 244).

This concludes our discussion of B cell development and activation. The next two chapters deal with the MHC and antigen processing pathways. An in-depth understanding of these indispensable aspects of the T cell immune response is required to fully understand T cell activation, the lynch pin of both humoral and cell-mediated immunity.

Table 9-6 Examples of Mouse B Cell Differentiation Markers

Marker	Naive B Cell	Centroblast	Centrocyte	Memory B Cell	Plasma Cell
sIgD	+				
CD24	++			+/−	
CD38		+	+		+
CD44	+			+	+
Bcl-2	+		−/+	+	+
CD40	−/+	+	+	+	

+/−, downregulation; −/+, upregulation.

Table 9-7 Examples of Human B Cell Differentiation Markers

Marker	Naive B Cell	Centroblast	Centrocyte	Memory B Cell	Plasma Cell
sIgD	+	+	+		
CD10		+	+		
CD38		+	+		++
CD20				+	+
CD24	++			+	
CD27				+	
CD22	++			+	
CD40	−/+	+	+	+	

−/+, upregulation.

SUMMARY

In the bone marrow, pluripotential hematopoietic stem cells (HSCs) give rise to common lymphoid progenitors (CLPs) that eventually produce B lineage cells. The maturation phase of B cell development includes the multipartite pro-B, pre-B, immature B, and mature B cell stages that occur in the bone marrow and are independent of foreign antigen. V(D)J recombination of the Ig genes initiates in the earliest pre-B cells. Immature B cells express a complete surface IgM molecule and undergo negative selection in the bone marrow to remove potentially autoreactive cells. Receptor editing allows a B cell a second chance to produce a non-autoreactive BCR. Mature naive B cells co-expressing IgM and IgD are released to the periphery. The differentiation phase of B cell development takes place in secondary lymphoid tissues and involves the activation of mature B cells by foreign antigen and the generation of antibody-producing plasma cells and memory B cells. Complete activation of mature B cells by foreign antigen requires three signals: the binding of antigen to the BCR, the establishment of costimulatory contacts with activated antigen-specific T helper cells, and the receipt of Th-derived cytokines. Signaling is transduced by Src kinase-mediated phosphorylation of immunoreceptor tyrosine-based activation motifs (ITAMs) in the cytoplasmic tails of the Igα/Igβ molecules in the BCR complex. Activation of ZAP-70 and the PLC-γ1 and Ras/MAPK pathways leads to activation of transcription factors such as c-myc and c-fos that induce the new gene transcription necessary for further differentiation. In the germinal centers of the secondary lymphoid follicles, an activated B cell proliferates and differentiates into centroblasts that undergo somatic hypermutation of the V exons of their Ig genes. The centroblasts mature into centrocytes within the GC and undergo affinity maturation and isotype switching. These cells mature further into antibody-secreting plasma cells and memory B cells that home to the site of infection.

Selected Reading List

Ansel K. and Cyster J. (2001) Chemokines in lymphopoiesis and lymphoid organ development. *Current Opinion in Immunology* **13**, 172–179.

Bartholdy B. and Matthias P. (2004) Transcriptional control of B cell development and function. *Gene* **327**, 1–23.

Benschop R. J. and Cambier J. C. (1999) B cell development: signal transduction by antigen receptors and their surrogates. *Current Opinion in Immunology* **11**, 143–151.

Bishop G. and Hostager B. (2001) B lymphocyte activation by contact-mediated interactions with T lymphocytes. *Current Opinion in Immunology* **13**, 278–285.

Blackwood E. M. and Kadonaga J. T. (1998) Going the distance: a current view of enhancer action. *Science* **287**, 60–63.

Brandt V. and Roth D. (2002) A recombinase diversified: new functions of the RAG proteins. *Current Opinion in Immunology* **14**, 224–229.

Cariappa A. and Pillai S. (2002) Antigen-dependent B-cell development. *Current Opinion in Immunology* **14**, 241–249.

Diaz M. and Casali P. (2002) Somatic immunoglobulin hypermutation. *Current Opinion in Immunology* **14**, 235–240.

Do R. and Chen-Kiang S. (2002) Mechanism of BLys action in B cell immunity. *Cytokine and Growth Factor Reviews* **13**, 19–25.

Good R. and Cain W. (1970) Relationship between thymus-dependent cells and humoral immunity. *Nature* **226**, 1256–1257.

Hardy R. and Hayakawa K. (2001) B cell developmental pathways. *Annual Reviews in Immunology* **19**, 595–621.

Hollowood K. and Goodlad J. (1998) Germinal centre cell kinetics. *Journal of Pathology* **185**, 229–233.

Howard J. (1985) Immunological help at last. *Nature* **314**, 494–495.

Lanier L. (2001) Face-off—the interplay between activating and inhibitory immune receptors. *Current Opinion in Immunology* **13**, 326–331.

Leo A. and Schraven B. (2001) Adapters in lymphocyte signalling. *Current Opinion in Immunology* **13**, 307–316.

Li Z., Woo C., Iglesias-Ussel M. and Scharff M. (2004) The generation of antibody diversity through somatic recombination and class switch recombination. *Genes and Development* **18**, 1–11.

Lieber M. (2000) Antibody diversity: a link between switching and hypermutation. *Current Biology* **10**, R798–R800.

Livak F. and Petrie H. (2001) Somatic generation of antigen-receptor diversity: a reprise. *Trends in Immunology* **22**, 608–612.

MacKay F. and Browning J. (2002) BAFF: a fundamental survival factor for B cells. *Nature Reviews Immunology* **2**, 465–475.

Manis J., Tian M. and Alt F. (2002) Mechanism and control of class-switch recombination. *Trends in Immunology* **23**, 31–39.

McHeyzer-Williams M. G. and Ahmed R. (1999) B cell memory and the long-lived plasma cell. *Current Opinion in Immunology* **11**, 172–179.

McHeyzer-Williams L., Driver D. and McHeyzer-Williams M. (2001) Germinal center reaction. *Current Opinion in Hematology* **8**, 52–59.

Okamura H. and Rao A. (2001) Transcriptional regulation in lymphocytes. *Current Opinion in Cell Biology* **13**, 239–243.

Osmond D. G., Rolink A. and Melchers F. (1998) Murine B lymphopoiesis; towards a unified model. *Immunology Today* **19**, 65–68.

Raff M. (1970) Role of thymus-derived lymphocytes in the secondary humoral immune response in mice. *Nature* **226**, 1257–1258.

Rolink A. and Melchers F. (2002) BAFFled B cells survive and thrive: roles of BAFF in B-cell development. *Current Opinion in Immunology* **14**, 266–275.

Rolink A., Schaniel C., Andersson J. and Melchers F. (2001) Selection events operating at various stages in B cell development. *Current Opinion in Immunology* **13**, 202–207.

Sprent J. (1997) Immunological memory. *Current Opinion in Immunology* **9**, 371–379.

Wang L. and Clark M. (2003) B cell antigen receptor signaling in lymphocyte development. *Immunology* **110**, 411–420.

Wu T. T. and Kabat E. (1970) An analysis of the sequences of the variable regions of Bence Jones proteins and myeloma light chains and their implication for antibody complementarity. *Journal of Experimental Medicine* **132**, 211–250.

Zhang K. (2000) Immunoglobulin class switch recombination machinery: progress and challenges. *Clinical Immunology* **95**, 1–8.

MHC: The Major Histocompatibility Complex

George Snell 1903–1996

MHC complex
Nobel Prize 1980

© Nobelstiftelsen

CHAPTER 10

A. HISTORICAL NOTES

B. GENERAL ASPECTS OF THE MHC IN HUMANS AND MICE

C. MHC PROTEINS

D. MHC GENES

E. EXPRESSION OF MHC MOLECULES

F. PHYSIOLOGY OF THE MHC

"Near or far, hiddenly,
To each other linked are,
That thou canst not stir a flower
Without troubling a star." —Francis Thompson

A s was introduced in Chapters 2 and 3, recognition of antigen by T cells is generally more complex than antigen recognition by B cells. While the B cell receptor binds directly to a unitary epitope on a molecule that is foreign to the host's body, the T cell receptor binds to a two-part antigen. In the normal course of T cell metabolism, peptides derived from both foreign and host proteins are individually complexed to and displayed by "self" cell surface molecules on host cells. When a foreign peptide is bound by a "self" display molecule, the combined shape is a "nonself" entity that is recognized by the T cell receptor, leading to an immune response. These peptide display molecules were originally discovered during studies of tissue rejection, and were therefore named for their role in *histocompatibility*, as described in the following discussion.

What is the Major Histocompatibility Complex?

"Histocompatibility" is a word that says pretty much what it means: "histo" means "tissue" and "compatibility" means "getting along." The *major histocompatibility complex* (MHC) genes were first identified in experiments investigating why the tissues and organs from one individual of a species were destroyed when introduced into another member of the same species. For example, a patch of skin transplanted from a donor mouse to a recipient is sometimes accepted (histocompatible) but at other times rejected (histo-incompatible). If the recipient mouse and the donor are of the same genetic background (same inbred strain), the transplanted skin graft is accepted and melds permanently with the recipient's own skin. However, if the recipient and donor are of genetically different backgrounds (different inbred mouse strains), there are antigens in the graft perceived as foreign to the recipient, resulting in an attack by the recipient's immune system and the destruction of the transplanted tissue. The antigens responsible for this type of reactivity are the peptide display molecules of the donor, its MHC proteins.

The genes controlling the histocompatibility of tissue transplantation were localized to a large genetic region containing multiple loci; hence, the term "complex." The molecules encoded by these genes were found to have striking effects on histocompatibility, and to distinguish them from other molecules (encoded elsewhere in the genome) that had relatively minor effects on histocompatibility, these molecules were called the "major" histocompatibility molecules. Thus, the genes encoding these molecules were dubbed the "major histocompatibility complex." Because of the multiple loci present in the MHC, any one individual was found to express a variety of different MHC molecules on his/her cells. An array of MHC alleles collectively came to be called an MHC *haplotype* (see later).

The involvement of the MHC in tissue acceptance and rejection was appreciated early on, as were its clinical applications with respect to transplantation. However, tissue transplantation is not a natural event, which left scientists wondering what the physiological role of these molecules was within a host's body. Answering this question turned out to be considerably more difficult than first imagined. Only after decades of conjecture and several landmark experiments was the physiological role of the MHC in T cell antigen recognition finally elucidated.

A. Historical Notes

I. DISCOVERY OF THE MHC

In the 1930s, Peter Gorer, a professor of pathology at Guy's Hospital, London, UK, was attempting to understand what factors allowed some mice to resist the growth of transplanted tumors. In 1937, he reported that resistance to transplanted tumors correlated with the presence in the mice of antibodies that would bind to an antigenic molecule on leukocytes from the tumor donors. To be consistent with the blood cell antigen nomenclature of the day, Gorer called this molecule "antigen II." Further individual and collaborative work by Gorer and George Snell revealed that "antigen II" was encoded on mouse chromosome 17 within a genetic locus that Snell had previously called "H" for "histocompatibility." Together, they renamed this locus "H$_2$," which later became "H-2."

In 1944, Peter Medawar at the University of London, UK found that the survival of skin grafts was also controlled genetically by the H-2 locus. More importantly, Medawar also observed that, when a second graft was attempted on a recipient, tissue rejection was faster and more intense if and only if the same host and donor were involved. Thus, the graft rejection process showed both the specificity and memory characteristic of an adaptive immune response. From this result it was concluded that the H-2 locus and its encoded molecules were linked to immunological function in some as yet unknown way. Further studies revealed that the H-2 "locus" was actually a grouping of at least seven independent genetic loci that encoded different cell surface molecules. Several of these MHC loci in mice were defined on the basis of tissue rejection. For example, if the donor and recipient were different in what were first called the *Ir genes* (later known as the I, K, or D regions of the MHC), a rapid immune response resulted in rejection of the transplant. At this point, "H-2" became known as the "major histocompatibility complex," denoting the multi-locus nature of the genes involved and their distinction from unrelated minor histocompatibility loci that caused slower and less dramatic transplant destruction. Similar studies were carried out for the human histocompatibility antigens. In 1965, Jean Dausset and his colleagues defined a system of at least 10 human antigens that coded for human histocompatibility. Numerous sub-loci were identified within each major locus, and each sub-locus was found to specify a number of antigenic alleles. The importance of this body of work was later recognized by the awarding in 1980 of a Nobel Prize jointly to Jean Dausset and George Davis Snell (for their discoveries of histocompatibility antigens on human and animal cells and their role in tissue and blood transplantation) and to Baruj Benacerraf (for his work on the genetic control of immune responses).

II. MHC INVOLVEMENT IN T CELL RECOGNITION

By the 1950s, immunologists understood that lymphocytes were the mediators of immune responses to foreign tissues and proteins as well as pathogens. Over the next two decades, intense research efforts by many laboratories revealed that, in all such responses, an individual's MHC genotype influenced the degree to which his/her immune system would respond to the antigenic challenge. However, the mechanism by which antigenic recognition occurred, and how the MHC was involved, remained murky.

Separate lineages of antigen-specific lymphocytes were identified by the early 1960s, namely B cells and T cells. B cells were the lymphocytes that produced and secreted antibody upon stimulation, and recognized antigen using a membrane-bound form of that antibody molecule. The antigen receptor on the B cell was shown to bind directly to its epitope on a foreign molecule, which could be of any size or chemical composition, and which could exist either in isolation or as part of a larger complex. In comparison, T lymphocytes were poorly understood. It was known that two subpopulations of T cells existed: the "helper" T cells that were necessary to fully stimulate many B cells, and the "cytotoxic" T cells that could directly kill virally-infected or tumor cells. Assays were developed that aided scientists in detecting and monitoring the activation of T lymphocytes (see Box 10-1).

At this point, research into understanding T cell antigen recognition intensified on two different but related fronts. Several studies in the early 1970s had suggested that T cells had Ig-like receptors on their surfaces. However, it had become clear these receptors were not able to bind free antigen and the details of their structure remained obscure. The isolation of the putative T cell receptor became the focus of many research labs (see Ch.12). Other groups chose to study the nature of the structure that was being recognized by T cells. This latter approach, usually involving studies in mice, led to the finding that Th cells could be stimulated to proliferate only if the antigen was associated with "antigen-presenting cells." Furthermore, the T cells and antigen-presenting cells had to be isolated from mice that were histocompatible at the I region (now called the A and E regions) of the MHC. Similarly, for Th cells to fully activate antigen-specific B cells, the Th cell and the B cell had to be derived from mice histocompatible at the I region. In addition, researchers studying the immune responses of mice to various viruses noted that virus-specific CTLs would only kill infected target cells if both the CTLs and target cells were taken from mice that were histocompatible at the K and D regions of the MHC. The influence of histocompatibility in defining the cells with which T lymphocytes could productively interact came to be called *MHC restriction*.

III. ELUCIDATION OF THE ANTIGEN-PRESENTING FUNCTION OF THE MHC

Initially, it was unclear how the molecules encoded by the MHC genes might play a role in the interaction between T cells and their antigens. Most scientists supported the "intimacy hypothesis," which held that antigen recognition occurred when both the T cells and the antigen-bearing host cells carried the same MHC molecules on their surfaces (Fig. 10-1A). In other words, interaction between the T cell and the host cell depended on a direct "like–like" association between MHC molecules on the T cell and MHC molecules on the host cell, perhaps by physically linking them together in some fashion. Recognition of the antigen would then be left to the T cell receptor. However, in an elegant series of experiments (see Box 10-2), Rolf Zinkernagel and Peter Doherty showed that the intimacy hypothesis could not be valid. Rather, their work demonstrated that the MHC molecules expressed on the T cells themselves were not involved in antigen recognition. Zinkernagel and Doherty then surmised

Box 10-1. Assays of T cell activation

How does one check if a T cell is specifically activated in response to an antigen? For a CTL clone, specificity is defined by the killing of target cells of known MHC genotype that are infected with a particular intracellular pathogen. For a Th effector clone, the rapid proliferation of cells followed by cytokine secretion indicates that the clone has responded to a particular antigen. In this Box, we discuss various assays commonly used to measure T cell proliferation and CTL-mediated target cell lysis. In such assays, experimental conditions affecting the T cells, antigens, APCs, or target cells can be manipulated and the resulting effects, if any, can then shed light on the various mechanisms involved in activating the T cell.

Assay of CTL Function

The function of a Tc cell is to generate CTLs, which lyse host cell targets, either tumor cells or cells infected with a virus or intracellular bacterium. Thus, the effector function of a CTL, which depends on the presentation of antigenic peptide in the context of MHC class I, can be assessed by manipulating and measuring target cell lysis. Target cell lysis by CTLs is often measured *in vitro* using the ^{51}Cr (radioactive chromium-51) release test (see Figure, panel A). For this test, one must generate two cell populations: primed responder lymphocytes and labeled target cells. For ethical reasons, one does not inject humans and isolate their spleens, which leaves only the peripheral blood cells as a (rather poor) source of CTLs for human CTL assays. We will therefore describe this test for the mouse situation. Typically, a mouse is infected with the virus or intracellularly replicating bacterium of interest. About 2–3 weeks later (a period that allows a primary response resulting in clonal expansion and the generation of resting memory Tc cells), the spleen is removed from the immunized mouse and a spleen cell suspension is cultured in a plate. This total population of spleen cells is infected *in vitro* with the same bacteria or virus to activate the memory Tc cells in a secondary response, stimulating them to expand and differentiate into effector CTLs over a period of 7–10 days. The culture thus becomes a source of responder cells for the assay. Established cell lines derived from the original Tc cells can also be used as a uniform source of responders to ensure reproducibility between assays.

The target cells of interest, which could include both infected and uninfected controls

from various sources and of various haplotypes, are incubated with ^{51}Cr. The cells take up the radioisotope and become labeled intracellularly because ^{51}Cr binds to soluble macromolecules that cannot cross the membrane. As long as the labeled cells are viable and intact, the ^{51}Cr–macromolecule complexes remain contained by the cell membrane. The labeled cells are washed and plated in a culture dish. The primed responder cells are then added to the culture dish containing the infected, labeled target cells and incubated for approximately 4 hours. If CTLs in the responder population recognize any bacterial or viral peptide–MHC combinations displayed on the target cells, they will attack and lyse them. Lysis of the target cells releases ^{51}Cr into the culture supernatant, where it can be measured by scintillation counting. The level of ^{51}Cr released is proportional to the level of CTL activity.

The ^{51}Cr release assay can be used in a similar way to assess MHC haplotype relatedness. However, for reasons that are explained in the main text of this chapter, this type of recognition does not require any preliminary *in vivo* priming. It is enough to mix together *in vitro* a source of responder Tc cells, such as splenocytes, with "stimulator" cells isolated from another mouse or a cultured cell line. If the responder Tc cells recognize specific MHC class I alleles on the stimulator cells, they will expand and differentiate into specific CTLs that can then be tested for their ability to lyse ^{51}Cr-labeled target cells of various haplotypes, including the stimulator cell type itself.

Assays of Th Function

Since the Th cells that depend on MHC class II for their function do not generally lyse cells, an effector property other than ^{51}Cr release must be used to assess Th function. Activated Th cells undergo rapid proliferation, an activity that can be measured by ^{3}H-thymidine incorporation into the newly synthesized DNA of rapidly dividing cells (see Figure, panel B). For example, let us suppose a mouse is immunized with a potent immunogen that stimulates the expansion of Th clones and the generation of memory cells recognizing this immunogen. Again, 2–3 weeks after immunization of the mouse, the spleen is harvested and the lymphocytes are cultured with isolated APCs (of the appropriate MHC haplotype) that were incubated with immunogen. After several days of culture, ^{3}H-thymidine is added to the culture medium. If the memory Th cells respond to presentation

of peptides by the APCs, they commence cell division and begin to incorporate the ^{3}H-thymidine into new daughter DNA molecules. The dividing cells are harvested after 18–24 hours of incubation with the radioactive label, lysed, and the amount of incorporated ^{3}H-thymidine, which is proportional to the level of proliferation, is measured.

This assay can also be used to detect differences in MHC between individuals in the absence of immunogen. When the spleen cells of one mouse strain are incubated with spleen cells of an MHC-mismatched strain, the Th cells present in both populations react to the presence of a nonself MHC allele and commence proliferation in order to generate effector Th cells. At this point, their ability to incorporate ^{3}H-thymidine can be measured as described above for antigen-specific responses. This response is known as the *mixed leukocyte reaction* (MLR; see Figure, panel B, inset). Because the Th cells of both strains are able to give a proliferative response, this reaction is called a *two-way MLR*. Although the occurrence of proliferation tells you that the two strains differ at their MHC, it is impossible in this situation to quantify the response of either strain individually; the assay measures the total response of both strains.

If, however, the proliferative ability of one strain is abolished by treating the cells derived from it with DNA-damaging X-rays or a mitosis inhibitor such as mitomycin C, only the untreated Th cells will be able to proliferate in response to cell mixing. This reaction, called a *one-way MLR*, can be used to measure the response of one of two strains. In this context, the untreated cells that can respond are called the "responders" and the treated cells that can no longer respond are called the "stimulators." The level of responder cell proliferation in a one-way MLR is often used to judge the degree of genetic relatedness between the responder and stimulator individuals: the greater the number of differences in MHC class II alleles, the greater the proliferation of the responding cells.

When the antigenic specificity of the T cell and the restricting MHC allele are known, Tc and Th activation assays can also be used to measure whether a particular MHC molecule is present and/or functional. Activation of the T cell indicates that the peptide-MHC complex has been successfully presented as nonself and that the MHC molecule in question is indeed functional.

Continued

Box 10-1. **Assays of T cell activation**—*cont'd*

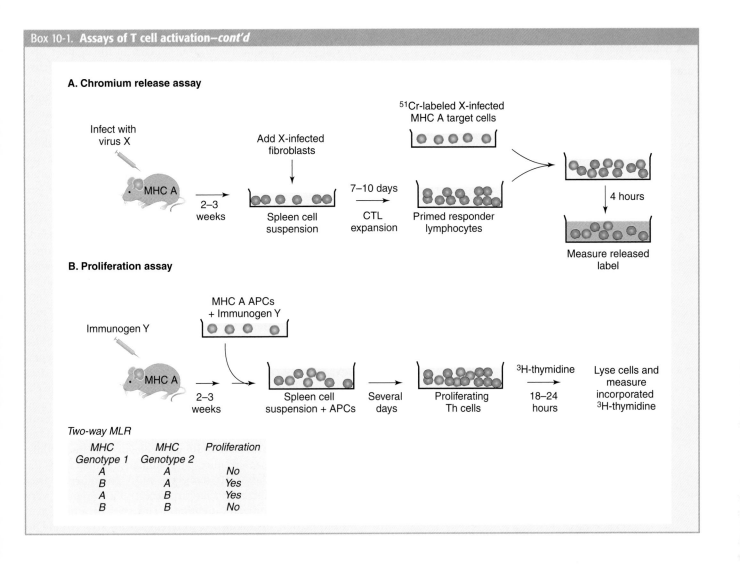

A. Chromium release assay

B. Proliferation assay

Two-way MLR

MHC Genotype 1	MHC Genotype 2	Proliferation
A	A	No
B	A	Yes
A	B	Yes
B	B	No

that each T cell clone most likely expressed a receptor (or receptors) able to recognize both MHC and antigen simultaneously. MHC restriction was observed because, due to the specificity of its receptor, a given T cell clone could only be restimulated by the same antigen when it was associated with the same MHC that led to the original stimulation of the T cell.

What was the mechanism that allowed the T cell to recognize both antigen and MHC at the same time? Did antigen recognition by T cells depend on two separate molecules, one to recognize the MHC molecule and one to recognize the antigen? Or was the TCR a single molecule, and if so, did it possess two binding sites, one for MHC and one for antigen, or a single binding site recognizing some sort of complex between MHC and antigen? After Zinkernagel and Doherty's experiments showed that the intimacy hypothesis could not be true, two theories of the nature of the structure recognized by T cells came into vogue, called the "dual recognition" and "altered-self" hypotheses (Fig. 10-1B, C). Some immunologists believed that antigen X and the MHC molecule remained separate and were recognized by the T cell as

distinct entities, a concept labeled "dual recognition." The "altered-self" theory posited that the MHC (the "self") was somehow altered by the binding of antigen and this alteration was what was recognized by T cells. In other words, this camp felt it more likely that T cells recognized an associative structure in which antigen X and MHC combined physically to form a single (MHC + X) determinant. This idea was actually suggested as early as 1959 by H. Sherwood Lawrence, who had speculated that intracellular pathogens might trigger an immune response by associating with transplantation antigens expressed by the host cell and forming a "self + X" complex.

Dual recognition theories were originally favored because early studies had suggested that the T cell receptor was similar to the Ig molecule and therefore was likely to have two binding sites (perhaps one for MHC and one for X). However, the dual recognition hypothesis could not be reconciled with a growing list of experimental results. For example, several laboratories showed that virally-infected human cells of the incorrect MHC, or uninfected cells of the correct MHC, could not block the CTL-mediated lysis of virally-infected target cells of

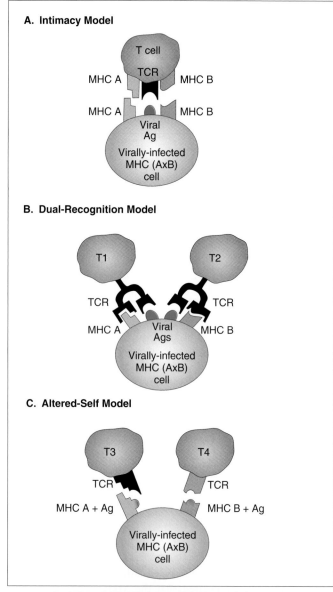

A. Intimacy Model

B. Dual-Recognition Model

C. Altered-Self Model

Figure 10-1

Historical Models of MHC–TCR Interaction
Three hypotheses attempting to account for the restricting role of MHC molecules in the interaction between T cells and their antigens. (A) Intimacy model. A direct "like–like" binding occurs between identical MHC molecules expressed on both the infected cell and the responding T cell. The TCR "sees" the antigen alone. (B) Dual recognition model. Each T cell (T1 and T2) expresses either a single receptor with separate MHC- and antigen-binding sites (as illustrated); or two separate receptors, one that recognizes a specific MHC molecule and one that binds to antigen (not shown). (C) Altered-self model. Each T cell (T3 and T4) expresses one receptor that sees a complex of specific MHC altered by the binding of antigen. Each T cell clone recognizes only one combination of MHC and antigen. Adapted from Zinkernagel R. M. and Doherty P. C. (1974) Immunological surveillance against altered self components by sensitized T lymphocytes in lymphocytic choriomeningitis. *Nature* **251**, 547–548.

evidence contrary to dual recognition was obtained. First, T cell clones specific for a given antigen in the context of a particular self-MHC molecule could often show cross-reactive recognition of a nonself MHC allele. In addition, a T cell clone specific for (MHC "A" + X) could, in some cases, recognize (MHC "B" + Y), a result inconsistent with the notion of separate receptors specific for MHC "A" and antigen X. As findings inconsistent with dual recognition accumulated, those studying the TCR itself found convincing evidence that the TCR was a single molecule unlikely to have the two separate binding sites that the dual recognition theory demanded (see Box 12-1 in Ch.12).

By the mid-1980s, the long-elusive genes for the TCR were cloned (see Ch.12) and were indeed found to encode an entity with a single binding site. This finding confirmed the validity of the altered-self recognition model by showing that the MHC and antigen had to combine into a single structure to be recognized. Further intensive research into the presentation of antigen by both APCs and target cells revealed that the MHC molecules were associated, not with whole antigenic molecules, but with small peptides derived from protein immunogens. The key experimental advances concerning antigen presentation are discussed in Chapter 11. As for the structure of the MHC molecule itself, two decades of work analyzing first the amino acid sequences of MHC proteins and then the DNA sequences of the corresponding genes had already revealed much about the molecular nature of MHC molecules by the mid-1980s. Population studies in mice and humans had shown the genetic loci of the MHC to be the most polymorphic ever observed, allowing a species to successfully present a very wide range of antigenic peptides. The sites of amino acid variation that distinguished different alleles were found to be clustered in the two domains where the antigenic peptides were thought to bind. This model of the structure and function of the MHC molecule was confirmed in 1987 by Don Wiley and colleagues with the resolution of the first X-ray crystal structure of an MHC allele (see later).

Thus, from the early observations of tissue rejection by Gorer and Medawar to the X-ray crystal structure of an isolated MHC molecule, the structure/function portrait of the MHC has been completed. These molecules continually bind a broad range of peptide fragments from the intracellular and extracellular environments, creating complexes of host MHC and peptide that are then displayed on the host cell surface. A given peripheral T cell is specific for a particular self-MHC/peptide X complex. The same T cell does not respond to peptide X in association with any other MHC molecule, providing the molecular basis for the observation of "MHC restriction."

No one knows precisely why T cells require an MHC context for antigen recognition, especially since B lymphocytes are perfectly capable of recognizing foreign antigens without such a context. Some scientists have speculated that there was an evolutionary need for a method of monitoring the health of the intracellular environment; i.e., a way to scan the cell's internal proteins. The spectrum of peptides derived from the processing of all proteins within

the correct MHC, as would be expected if the dual recognition model were valid. Similarly, murine CTLs could not be blocked with purified murine MHC molecules. With the advent of T cell hybridization and cloning techniques, more

Box 10-2. **The discovery of MHC restriction**

Note to readers: For those who have not yet encountered the full definition and description of MHC haplotypes in the text, the generalized notation of "MHC A," "MHC B," etc. is used here to represent inbred mouse strains that differ genetically at their H-2 loci. For those familiar with haplotype notation, the following correlations can be made with respect to the experiments described in the following discussion: MHC A = $H-2^k$, MHC B = $H-2^b$, MHC C = $H-2^d$, MHC (A $\times$ B) = $H-2^{k/b}$, MHC (A $\times$ C) = $H-2^{k/d}$ and MHC (B $\times$ C) = $H-2^{b/d}$.

From 1954 to 1974, many scientists hot on the trail of H-2 transplantation antigens cultivated the seeds of the idea of MHC restriction, but none planted the idea so irrevocably within the immunological community as Peter Doherty and Rolf Zinkernagel. It was through twists and turns of fate that these two young scientists, originally trained in medicine and veterinary science, were thrown together in an Australian laboratory to study the role of cytotoxic T lymphocytes in lethal choriomeningitis in mice, a disease associated with the lymphocytic choriomeningitis virus (LCMV).

At the beginning of their collaboration, Doherty and Zinkernagel developed an assay that showed that the CTL harvested from the cerebrospinal fluid of LCMV-infected mice had antiviral properties that might be responsible for the destruction of brain tissue during LCMV infection. As other investigators had already linked many diseases to H-2 genotype, Doherty and Zinkernagel tested whether the severity of choriomeningitis could be correlated to H-2 genotype. The results puzzled them. While some mouse strains generated virus-specific CTLs that were detectable in their assay, others did not, although all the mice eventually died from LCMV. Further experiments were illuminating. They injected a variety of mouse strains intracerebrally with LCMV virus and tested the spleen cells 7 days later for their ability to kill ^{51}Cr-labeled LCMV-infected MHC A fibroblasts (see Figure, panel A). They found that target cell killing occurred only when the targets and CTLs shared MHC identity. That is, only CTLs from strains that were homozygous or heterozygous for MHC A were able to kill the infected MHC A fibroblasts; CTLs from

strains of other H-2 haplotypes could not. They successfully repeated this experiment using LCMV-infected macrophages to rule out the possibility that the effect they saw was limited to the fibroblasts they initially chose.

These observations were not totally new. Other scientists, such as Berenice Kindred and Donald C. Shreffler, Jean-Claude Leclerc and colleagues, and David H. Lavrin and colleagues, had published experiments using different assay systems that showed the necessity for matching the MHC haplotypes of CTLs and target cells. As described in the text, the paradigm of the day used to explain MHC restriction was the "intimacy" model. After extensive discussions with their colleagues at the John Curtin School of Medical Research in Canberra, Australia, Doherty and Zinkernagel realized they could use their LCMV-based CTL recognition system to design experiments that would put the intimacy model to the test.

They reasoned that if this "like–like" interaction model were true, then all LCMV-specific clones in an F1 MHC (A $\times$ B) hybrid mouse exposed to LCMV should be restimulated by LCMV-infected cells of either parental haplotype, namely MHC A or MHC B. To test this hypothesis, they bred MHC (A $\times$ B) hybrid mice, infected these mice with LCMV, and removed their spleens and lymph node cells 7 days later (see Figure, panel B). These donor LCMV-primed MHC (A $\times$ B) lymphocytes were then injected into recipient mice of different H-2 haplotypes that had been previously irradiated. (Irradiation destroys resident lymphocyte populations that might otherwise compete with the growth of donor lymphocyte populations.) The recipient mice were then infected with LCMV. The success of this protocol hinged on the fact that it was known that primed T cells could not be restimulated by free virus but had to be exposed to virally-infected cells in the recipient. After allowing time for the transferred donor lymphocytes to be restimulated by infected cells in the recipients, the lymphocytes (from either lymph node or spleen) were removed and tested for their ability to kill LCMV-infected target cells of the MHC A haplotype using a standard *in vitro* CTL killing assay (refer to

Box 10-1). If the intimacy model were correct, the LCMV-primed F1 MHC (A $\times$ B) lymphocytes should have been restimulated *in vivo* in any infected recipient expressing either MHC A or MHC B. This would have included recipients that were MHC A, MHC B, MHC (A $\times$ B), MHC (B $\times$ C), or MHC (A $\times$ C). No restimulation or subsequent killing was expected from lymphocytes restimulated in control mice of the MHC C haplotype.

In fact, Zinkernagel and Doherty's results did not fit the intimacy paradigm. They found that, while F1 LCMV-primed T cells injected into and recovered from MHC A or MHC (A $\times$ B) irradiated recipients were able to lyse the MHC A targets, those that passed through the MHC B, MHC C, or MHC (B $\times$ C) recipients could not. It seemed that, in order to get expansion of CTLs that would eventually kill LCMV-infected MHC A target cells *in vitro*, the original primed MHC (A $\times$ B) Tc cell population had to encounter viral antigen presented in the context of MHC A. Doherty and Zinkernagel suggested that, rather than requiring interactions between matching H-2 molecules on the T cell and target cell, the donor cells from the F1 hybrid contained individual LCMV-specific T cell clones, each of which recognized a viral antigen in the context of only one of the parental MHC haplotypes. In other words, the LCMV=specific T cell population in the F1 hybrid was a mixture of clones, each recognizing either (MHC A + LCMV) or (MHC B + LCMV), but not both. Thus, it became clear that a given T cell was simultaneously recognizing both the antigen and one particular MHC molecule on the target cell, and that this recognition requirement was the basis of the MHC restriction observed in T cell responses to a given antigen.

By clearly demonstrating MHC restriction and convincingly disproving the intimacy model of T cell recognition, Zinkernagel and Doherty opened the doors for themselves and others to elucidate the true mechanism of T cell antigen recognition. Such was the importance of Zinkernagel and Doherty's work in establishing many of the fundamental concepts in this field that they were awarded the 1996 Nobel Prize in Physiology or Medicine.

Continued

Box 10-2. **The discovery of MHC restriction**–*cont'd*

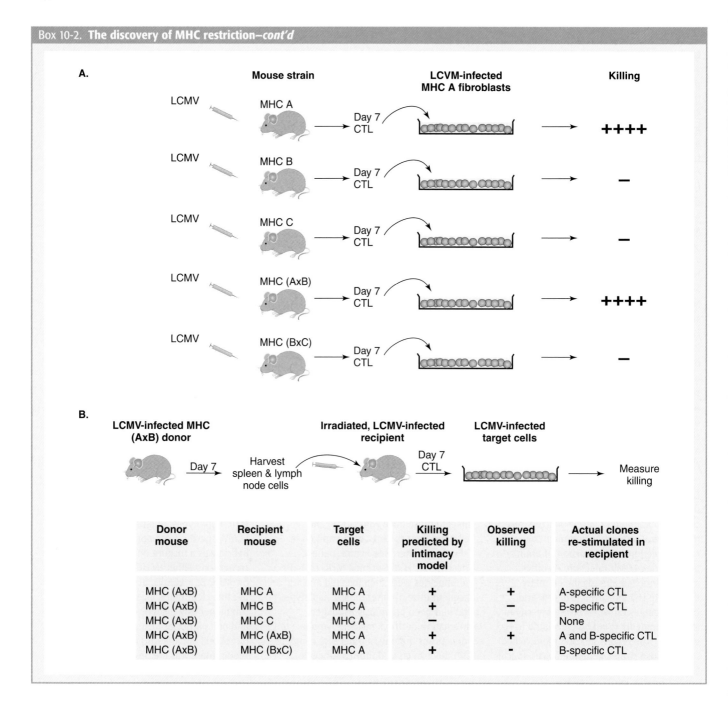

Donor mouse	Recipient mouse	Target cells	Killing predicted by intimacy model	Observed killing	Actual clones re-stimulated in recipient
MHC (AxB)	MHC A	MHC A	+	+	A-specific CTL
MHC (AxB)	MHC B	MHC A	+	–	B-specific CTL
MHC (AxB)	MHC C	MHC A	–	–	None
MHC (AxB)	MHC (AxB)	MHC A	+	+	A and B-specific CTL
MHC (AxB)	MHC (BxC)	MHC A	+	-	B-specific CTL

a cell would be an indicator of whether that cell was truly normal. The display of peptides derived from the processing of proteins associated with intracellular bacteria, viruses, or tumors would be a red flag to the organism that some cells were not normal and that an immune response should be mounted. Similarly, the display of extracellularly-derived peptides by macrophages and other APCs might be a non-invasive means of determining what sort of "garbage" the host's collection system was encountering, and whether the contents of that garbage demanded immune response action. The evolution of the MHC protein as a peptide display molecule may have been the logical result of such pressures.

B. General Aspects of the MHC in Humans and Mice

I. OVERVIEW OF THE MHC PROTEINS

MHC proteins and their antigen-presenting capabilities appear to be a feature of adaptive immunity exclusive to vertebrates, since all vertebrates (but no invertebrates) investigated to date possess some form of MHC. The MHC class I and class II molecules functioning on the surfaces of human and murine cells are very similar and take the form of heterodimeric

Table 10-1 Chromosomal Location of MHC Class I and II Genes

	Chromosome	
	Mouse	**Human**
MHC class Iα	17	6
β2m	2	15
MHC class IIα	17	6
MHC class IIβ	17	6

transmembrane proteins. In both species, the MHC class I molecule is composed of a large MHC-encoded transmembrane α chain and a small non-MHC-encoded chain called *β2-microglobulin (β2m)* that associates non-covalently with the membrane distal region of the α chain. The MHC class II molecule is composed of an α chain and a slightly smaller β chain, both of which are transmembrane proteins encoded by genes in the MHC. The chromosomal locations of the genes encoding the MHC class I and class II proteins in mice and humans are shown in Table 10-1.

II. OVERVIEW OF THE MHC LOCI

The MHC in mice, known as the H-2 complex as noted previously, is spread over 3000 kb on chromosome 17. This area is occupied by 9 loci called the K, P, A, E, S, D, Q, T, and M regions (Fig. 10-2A). Each region contains dozens of genes, most of which have yet to be characterized and which may or may not be involved in antigen recognition. The K and D regions contain single functional genes that encode mouse MHC class I α chains, while the A and E regions contain single functional genes that encode the α and β chains of the mouse MHC class II proteins. The S region of the H-2 complex contains genes encoding the MHC class III proteins, including certain complement proteins, heat shock proteins,

and the cytokines tumor necrosis factor (TNF) and lymphotoxin (LT). The P, Q, T, and M regions contain pseudogenes and genes encoding "non-classical" (also known as *class Ib* or *class IIb*) MHC proteins that in fact make up most of the H-2 complex. All these regions will be discussed in detail below.

In humans, the MHC region is called the HLA complex (for "human leukocyte antigens"). The HLA complex covers about 3500 kb on chromosome 6 and also contains many groups of genes. The most immunologically prominent of these are shown in Figure 10-2B. The single functional genes encoding the human MHC class I α chains are called HLA-A, HLA-B, and HLA-C. Multiple functional genes encoding both MHC class II α and β chains are found in the DP, DQ, and DR regions. Multiple genes in the DM and DO regions and single genes known as the HLA-E, -F, and -G genes encode "non-classical," or class Ib and class IIb MHC proteins. The MHC class III region in the HLA complex is the equivalent of the S region in the murine MHC and contains genes for complement components, heat shock proteins, TNF, and LT.

III. INTRODUCING MULTIPLICITY AND POLYMORPHISM IN THE MHC LOCI

Most proteins in our bodies are unique; that is, there is only one functional gene in the genome that encodes a protein carrying out that particular function. The MHC genes are unusual in that, due to gene duplication during evolution, two to three separate, functional genes encoding exactly the same type of MHC class I or II polypeptide exist, a phenomenon we call *multiplicity*. The genes are named for their region of location and the chain they specify. For example, there is one functional gene called A_a (or A_α) within the A region of the mouse H-2 that encodes a MHC class II α chain. Another MHC class II α chain is encoded by a gene in the E region called E_a (or E_α). A similar situation holds for the MHC class II β chain. These proteins are encoded by both the A_b (or A_β) gene in the A region and the E_b (or E_β) gene in the E region. Figure 10-3A shows examples of how MHC class I and II

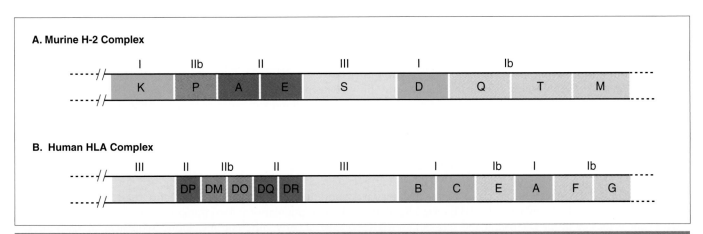

Figure 10-2

General Organization of the Major Histocompatibility Complex in Mice and Humans
(A) Murine H-2 complex on chromosome 17 and (B) human HLA complex on chromosome 6 showing color-coded regions in which MHC class I, Ib, II, IIb, and III genes are found. Regions in the HLA complex containing MHC class III genes do not have letter names.

Figure 10-3

Multiplicity in the MHC Loci
Multiple genes encoding MHC class I and II proteins in the (A) murine H-2 complex and (B) human HLA complex. Examples of how chains derived from each locus can combine to form complete MHC heterodimers are shown. For instance, either Kα (an MHC class Iα chain derived from the K locus) or Dα (an MHC class Iα chain derived from the D locus) can combine with β2m (encoded outside the MHC) to form a murine MHC class I protein. Similarly, in the HLA complex, a DPα chain can combine with a DPβ chain (but not with a β chain derived from either the DR or DQ locus) to form HLA-DP, one of four possible human MHC class II proteins. For clarity, MHC class Ib, IIb, and III loci are not shown. PM, plasma membrane.

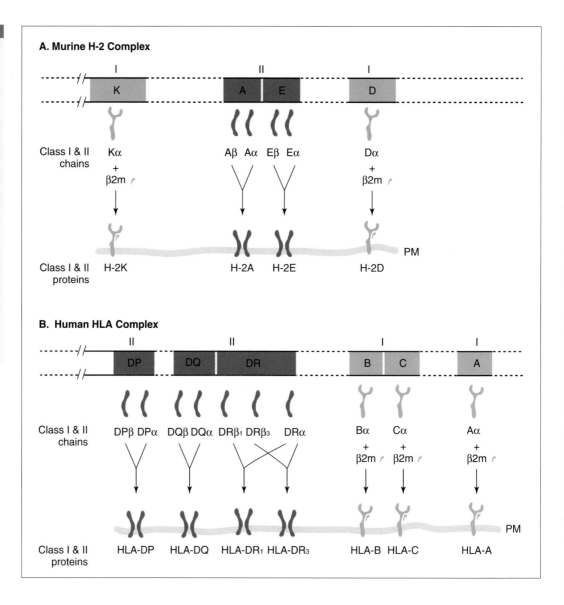

genes in the mouse give rise to complete MHC class I and II molecules (remembering that the β2m chain is encoded outside the MHC region). Note that, for unknown reasons, an α chain derived from the A region gene almost always combines with a β chain derived from the A region and very rarely with a β chain derived from the E region. The same is true for E region-derived α and β chains. In other words, an APC typically expresses an MHC class II molecule composed of H-2A$_a$A$_b$ or H-2E$_a$E$_b$ but not H-2A$_a$E$_b$.

Multiple genes giving rise to MHC class II α and β chains can be found in the HLA-DP, -DQ, and -DR regions. In HLA nomenclature, the letter indicating the chain type is capitalized. Thus, genes in the human DP region encoding MHC class II α chains are called DPA genes, while those in the DP region encoding MHC class I β chains are called DPB genes. Similarly, a DRB gene is located in the DR region and specifies an MHC class II β chain, while a DQA gene is located in the DQ region and specifies an MHC class II α chain. Just as is true in the mouse, however, the αβ pairs are almost always derived from the same region; that is, it is extremely rare to find a mixed

MHC class II molecule composed of, say, HLA-DRA/DQB. Figure 10-3B illustrates how the products of these genes can come together to form complete human MHC molecules.

In addition to the existence of multiple genes, the MHC loci are the most *polymorphic* of all the genes in the vertebrate genome. The vast majority (>90%) of all other vertebrate proteins examined to date are *monomorphic*; that is, almost all individuals in the species share the same nucleotide sequence at that locus. Polymorphism is the existence <u>in a population</u> of several different alleles at one locus. Alleles are slightly different nucleotide sequences of a gene; the protein products of alleles have the same function. For example, more than 290 alleles have been identified for the HLA-A gene, more than 553 for HLA-B and more than 140 for HLA-C in human populations. Two humans, therefore, are very likely to have slightly different nucleotide sequences at each of the HLA loci, meaning that the MHC proteins expressed on their cells will be of slightly different amino acid sequence but of identical function. In addition, because of the abundance of MHC alleles, most members of an outbred population are *heterozygous* at every MHC locus; that is, they have two different alleles for a given MHC locus on the maternal and paternal chromosomes. In inbred populations (such as laboratory mouse strains), only one allele is found in the population at each locus and all members are *homozygous* (have the same allele on both the maternal and paternal chromosomes).

On top of multiplicity and polymorphism, for each MHC locus, the gene on both chromosomes in an individual is expressed independently, or *co-dominantly*. In other words, a human expresses the genes on both the maternal and paternal chromosomes that encode HLA-A class I α, HLA-B class I α, HLA-C class I α, HLA-DP class II α and β, HLA-DQ class II α and β, and HLA-DR class II α and β chains on the appropriate cells. Similarly in mice, the appropriate cells will co-dominantly express two alleles of H-2K class I α chain, two of H-2D class I α chain, two of H-2A class II α and β chains, and two of H-2E class II α and β chains. So, in humans for example, complete MHC class II αβ heterodimers are derived from the products of the D family genes on both parental chromosomes. Therefore, for a given class II molecule such as HLA-DP, there are two choices of α chain and two choices of β chain, resulting in expression of four different HLA-DP heterodimers. For a region like HLA-DR, which has more than one DRB gene in some individuals, the number of possible combinations is even higher.

At this point, we come to some definitions that will be useful both in this and future chapters. As is also the case for identical human twins, members of a particular inbred mouse strain, which are homozygous at all loci and carry the same allele for every gene (including the MHC loci), are said to be *syngeneic*. Conversely, one mouse is said to be *allogeneic* (*allo*, meaning "other") to another if the animals have different alleles at one or more loci in the genome. When immunologists use the term

"allogeneic," they mean that the mice in question are different at a locus or loci in the MHC or at one of the minor histocompatibility loci. (In fact, the mice may also differ at other non-histocompatibility background loci.) The MHC alleles of one individual of an outbred population may be seen as foreign by the immune system of another member of that population, leading to an *alloreactive* immune response (see later in this chapter). Since syngeneic individuals have exactly the same alleles at all major and minor histocompatibility loci, no alloreactivity occurs between them and they do not reject each other's skin grafts. Such inbred mouse strains can be produced by repeated brother–sister mating, as described in Box 10-3.

Scientists carrying out early investigations of the MHC by studying skin graft rejection also used another genetic tool to help decipher the nature of histocompatibility: *congenic* mouse strains (refer to Box 10-3). An inbred mouse strain X is said to be congenic to another inbred mouse strain Y if it carries identical alleles to X *at all but one locus*. (Technically, since the short segment of DNA that is physically being selected for contains millions of base pairs of DNA in addition to the gene of interest, the strains may also differ in other non-selected loci within this short segment.) Thus, MHC congenic mice are identical at all loci other than one MHC locus. MHC congenic mice therefore reject each other's skin grafts, but are essentially identical at all other loci and can be used to study the effects of differences in the MHC alone.

The terms "syngeneic," "allogeneic," and "congenic" refer to comparisons between different strains within a single species. The relationship between members of two different species is said to be *xenogeneic* (*xeno*, meaning "foreign"). More on these terms will be presented in the discussion of transplantation found in Chapter 27.

Box 10-3. Inbred and congenic mouse strains

Experimental animals have played a fundamental role in almost all of the major discoveries in immunology, and foremost among the animals contributing to immunological discoveries is the mouse. Mice were chosen as laboratory subjects for several reasons, but chief among them was the availability of inbred strains. For an experiment to be interpretable, as many sources of variability as possible must be controlled, and one important source of variation in an experiment involving animals is genetic variability. In outbred populations, even though most loci are *monomorphic* (only one allele exists in the population), about 10% of all genes are *polymorphic*; that is, the protein in question may exhibit a slightly different amino acid sequence (and thus a slightly different function or expression pattern) in different individuals. Because of this variability, one cannot be sure that the results of

experiments carried out in outbred populations are due to the experimental variable, or to the influence of differing genetic factors. It is therefore desirable to conduct experiments using *inbred* strains in which all members are genetically identical at virtually all loci.

The development of the inbred mouse as the premier laboratory animal for immunological experiments has an interesting history. For centuries, Chinese and Japanese hobbyists had been keeping and breeding mice that had attractive coat colors, or other unusual characteristics such as crooked tails, obesity, or dwarfism. In the 19th century, "mouse fancying," as it came to be called, spread to England and became quite popular. From England, it spread to America, and some American pet shop owners found that much of their business came from the sale of such mice. In order to maintain the desired characteristic, the pet shop owners bred

siblings together, often for many consecutive generations. In this way, the mice became highly inbred (approaching genetic identity). One such pet store owner, Miss Abbie Lathrop of Grandby, Massachusetts, was the source of the founder mice of many of the important inbred mouse strains used in laboratories around the world to this day (for example, the DBA and C57BL strains).

How are inbred mouse strains developed? The fastest way is to mate brothers with sisters (*sibling mating*) for many successive generations; however, matings between other relatives will also eventually result in genetic homogeneity. As the mice are inbred, heterozygosity is slowly lost, so that after about 20 generations of sibling mating, the progeny are essentially genetically identical; that is, there is homozygosity at all loci. Thus, genetic variability can be eliminated as a factor influencing experimental results. It is important to realize, however, that

Continued

Box 10-3. Inbred and congenic mouse strains–*cont'd*

inbreeding is complicated by the fact that the health and reproductive capacity of the mice often decreases as genetic variability decreases, an effect known as *inbreeding depression*. Inbreeding depression results from the fact that all individuals harbor recessive, deleterious genetic defects that are masked by the normal, dominant allele. As inbreeding progresses, some of the disadvantageous alleles become homozygous, leading to reduced fitness, infertility, or death. Thus, many inbred lines die out before complete inbreeding is achieved.

Another type of mouse strain important in immunological research is the *congenic* mouse. Two congenic strains are identical at every genetic locus except one. Congenic mice are created as illustrated in the Figure. Let us suppose we have two

inbred mouse strains A and B, and B has a trait that we can select for and would like to have expressed in strain A mice; that is, we want a B trait on an A genetic background. [For example, if A has red eyes and an agouti coat, and B has brown eyes and a black coat, we might want to generate a congenic mouse with the brown eyes of B but the agouti coat (and all other characteristics) of A.] We first cross A with B mice to produce AB progeny that are heterozygous at all loci: 50% of their genes are from A, and 50% are from B. The AB progeny with the desired B trait (*b*) are then bred back, or *backcrossed*, to strain A mice and the progeny are again examined for the *b* characteristic. Those progeny with brown eyes are identified as AB (now 75% A and 25% B genes) and are backcrossed again to

strain A mice, and the AB progeny (87.5% A, 12.5% B genes) are again selected based on the presence of the desired *b* trait. Backcrossing is repeated for many generations, until the resulting mice are identical to strain A in every way except for the selected *b* trait. These congenic mice can then be used to study the effect of the B-derived gene, *b*, in the A genetic background.

With the use of such congenic and inbred mouse strains, scientists have been able to dissect important aspects of the major histocompatibility complex, tumor rejection, organ transplantation, and autoimmunity, among other immunological phenomena. Such mice have also been used for similar investigations in other fields when genetic variability would be a confounding factor.

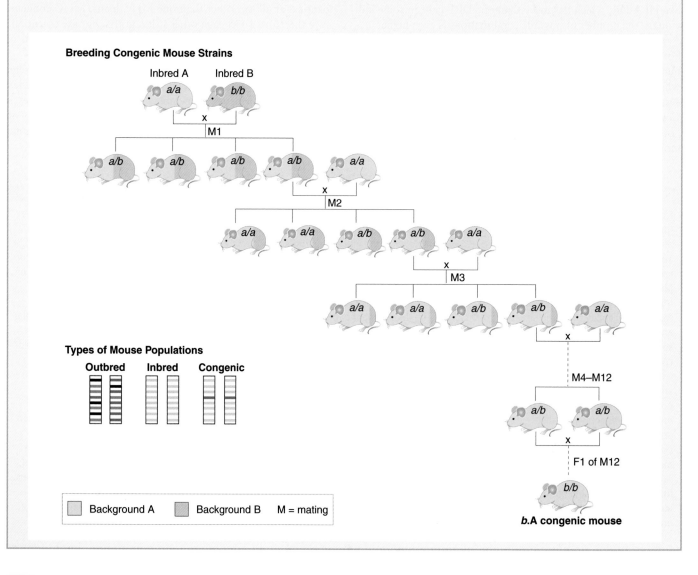

Breeding Congenic Mouse Strains

Types of Mouse Populations

Outbred   Inbred   Congenic

Background A   Background B   M = mating

b. **A congenic mouse**

IV. INTRODUCING HAPLOTYPES

As well as being polymorphic, the MHC loci are closely linked, meaning that the specific set of alleles for all MHC loci is usually passed on to the next generation as an intact block of DNA. This set of alleles on a single chromosome is called a *haplotype*, and any individual inherits two haplotypes from its parents: the MHC block on the paternal chromosome and the MHC block on the maternal chromosome. In an outbred population, the high degree of MHC polymorphism leads to great variation in haplotypes among individuals. In an inbred population (e.g., laboratory mice), both parents have the same allele at each MHC locus, the maternal and paternal haplotypes are the same, and all offspring inherit the same single haplotype on both chromosomes.

It should be noted that, within the human population, researchers have found certain haplotypes that are not only shared within a single family, but among a large number of families. These conserved haplotypes are thought to have been passed from generation to generation, starting from ancestral times when relatively few families were responsible for establishing populations that eventually spread around the world. Over 30 of these *ancestral haplotypes* (also referred to as "common extended haplotypes" or "HLA supratypes") have been described. While ancestral haplotypes were originally defined by serological studies, the complete sequencing of the HLA complex (completed in 1999) has allowed the direct identification of the large, conserved blocks of DNA involved. Because ancestral haplotypes originated in restricted populations that may have ultimately settled in diverse geographic regions, a given ancestral haplotype is often associated with individuals of a particular ethnic background. For example, one such haplotype is found predominantly amongst Basques and Sardinians, while another is specific to Eastern European Jews, and a third is exclusive to South East Asians. In contrast, there are some ancestral haplotypes that are shared amongst geographically diverse populations, perhaps because they shared a common history at some point. For example, an indigenous Taiwanese population shares an ancestral haplotype with the Maori of New Zealand, the Orochon of China, and several Amerindian populations.

With respect to inbred mice, a single term acting as a short form of identification is often used to indicate the haplotype of a particular strain. For example, the short form for the haplotype of the C57BL/6 mouse strain is "H-2^b" where the "b" in H-2^b means that allele number 12 is present at the K locus, allele number 74 is at the Aβ locus, allele number 3 is at the Aα locus, allele number 18 is at the Eβ locus, and so on. Another inbred mouse strain 129 has the same H-2^b haplotype, meaning that the genome of this mouse also contains alleles K-12, Aβ-74, Aα-3, etc., but differs from the C57BL/6 mouse at the so-called "background" loci outside the MHC. Another mouse strain (such as CBA) may have a different haplotype, say H-2^k. The H-2^k haplotype is defined by a different set of alleles: allele number 3 at the K locus, Aβ-22, Aα-54, etc. A fourth mouse strain (C3H) of haplotype H-2^k will have the same collection of MHC alleles (K-3, Aβ-22, Aα-54, etc.) but will differ from CBA mice at the background loci. If an H-2^b mouse is bred to an H-2^k mouse, the F1 offspring will be heterozygous at each locus (H-2$^{b/k}$) and will express both proteins for each locus: K-12, K-3; Aβ-22, Aβ-74; Aα-3, Aα-54; etc. If an H-2$^{b/k}$ mouse is crossed with another heterozygous mouse of haplotype H-2$^{a/d}$, since each progeny mouse inherits one haplotype from the mother and one from the father, the offspring may be either H-2$^{b/a}$, H-2$^{b/d}$, H-2$^{k/a}$, or H-2$^{k/d}$. The immune system of mice inheriting the H-2$^{b/d}$ haplotype will "see" the proteins specified by the H-2^a or H-2^k alleles as foreign.

More detailed short forms are often used to indicate specific alleles in a haplotype. For example, for the community of scientists studying mouse MHC genetics, the term H-2D^b means that the allele at the D locus in this mouse strain is that which occurs in the "b" haplotype. Workers in the field discussing this strain would identify this allele as D^b, and verbalize it as "D of b". Similarly, H-2D^k would come out in conversation as "D of k" and mean that the allele at the D locus is that found in the "k" haplotype. Examples of murine haplotypes are shown in Table 10-2.

Table 10-2 Murine Haplotypes

Sample Strains	Haplotype	K	Ab	Aa	Eb	Ea	D	MHC Protein Expressed Class I	Class II
CBA, C3H/HeJ	H-2^k	k	k	k	k	k	k	K^k, D^k	A^k, E^k
BALB/cJ, DBA/2J	H-2^d	d	d	d	d	d	d	K^d, D^d	A^d, E^d
A, B10.A	H-2^a	k	k	k	k	k	d	K^k, D^d	A^k, E^k
C57BL/6, C57BL/10	H-2^b	b	b	b	b	b	b	K^b, D^b	A^b, E^b
A.SW/Sn, B10.S	H-2^s	s	s	s	s	s	s	K^s, D^s	A^s, E^s
RIIIS/J	H-2^r	r	r	r	r	r	r	K^r, D^r	A^r, E^r
P/J	H-2^p	p	p	p	p	p	p	K^p, D^p	A^p, E^p

With information from Klein J., Figueroa F., and David C. S. (1983) *Immunogenetics* **17**, 553–596.

We will now examine the structure of the MHC proteins in detail before discussing the genes encoding them.

C. MHC Proteins

MHC class I and class II molecules are similar in their overall tertiary structure, in that both are heterodimeric molecules comprising an extracellular N-terminal peptide-binding region, an Ig-like extracellular region, a hydrophobic transmembrane region, and a short C-terminal cytoplasmic region. The structure of the MHC binding site in both class I and class II molecules is such that the affinity of an MHC molecule for peptide is much lower than that of antibody for its cognate antigen. This relaxed binding is a necessity if a single type of MHC molecule is to carry out its task of presenting a wide range of peptides for T cell perusal. However, the component chains of the class I and class II MHC heterodimers are quite different, as is their interaction with antigenic peptide. For the discussion on MHC protein structure and peptide binding that follows, the reader may wish to refer several times to the images of these molecules in Plate 10-1.

I. MHC CLASS I PROTEINS

The reader will recall from Chapter 2 that MHC class I proteins are expressed on almost all nucleated cells, including leukocytes. (One prominent exception can be found in the central nervous system, where neurons have only very low levels of MHC class I.) Most nucleated cells sport a mixed population of MHC proteins derived from the multiple class I genes on both the paternal and maternal chromosomes. Up to 250,000 copies of the products of each MHC class I gene may appear on the cell surface. The peptides bound by MHC class I are generally of endogenous origin; that is, they are derived from degradation of proteins synthesized within the cell (see Ch.11). The vast majority of these peptides will be "self" in nature, because most proteins routinely produced within the cell at any one time are of host origin (as opposed to proteins of nonself origin, such as those generated during a viral infection). The MHC molecule does not discriminate among "self" and "nonself" peptides; that job is left to the TCRs of CD8-bearing T cells (Tc cells), which recognize antigenic peptides complexed to MHC class I molecules.

i) MHC Class I Component Polypeptides

In both mice and humans, an MHC class I molecule contains an α chain associated with the β2-microglobulin chain. MHC

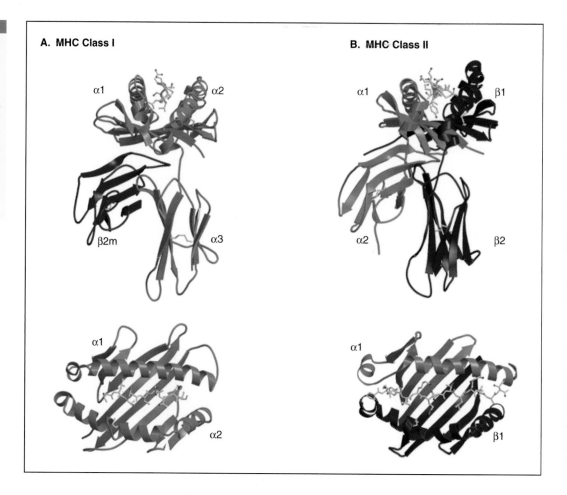

A. MHC Class I

B. MHC Class II

class I α chains are glycoproteins of about 44 kDa (350 amino acids) composed of three extracellular globular domains, each of about 90 amino acids in length (Fig. 10-4). Domains α1 and α2 at the N-terminal end of the chain pair with each other to form the peptide-binding site, while the α3 domain pairs with the non-covalently associated β2m polypeptide to form the Ig-like region. The α chain also supplies the transmembrane domain (approximately 40 amino acids) and the cytoplasmic tail (approximately 30 amino acids). The precise functions associated with the transmembrane and cytoplasmic domains are unknown but are probably related to the regulation of the interaction of the MHC molecule with other membrane proteins and/or the direction of the MHC molecules to their proper membrane locations within the cell.

The α1 domain maintains its shape without disulfide linkage, but the α2 and α3 domains each have an internal disulfide bond in approximately the same relative position. Most of the polymorphism of MHC molecules is localized in the α1 and α2 domains (those comprising the peptide-binding site) while the α3 domain is less polymorphic and more Ig-like. The site

of contact with the coreceptor CD8 is located in an invariant region of the α3 domain. The transmembrane region of MHC class I α chains is highly conserved. While more variation has been found in the cytoplasmic tail, some features are conserved, including certain phosphorylation sites and other sites that may influence the interaction of MHC class I with intracellular elements such as the cytoskeleton. An N-linked oligosaccharide occurs at the junction of the α1 and α2 domains in mice and humans; murine MHC class I α chains have an additional N-linked sugar near the carboxy-terminus.

The other partner in the MHC class I molecule, the β2m protein, is a soluble polypeptide of about 12 kDa (99 amino acids) that was first identified in urine as a serum globulin protein. In contrast to the α chain, β2m exhibits no (in humans) or very little (in mice) polymorphism within a species or even between species. Structurally β2m resembles a single Ig-like domain, and indeed is the prototypical member of the Ig superfamily of proteins. While the MHC class I α chain constitutes part of all four regions of the complete protein, the β2m chain participates primarily in the extracellular Ig-like region by associating non-covalently with the α3 domain. β2m does not have peptide-binding, transmembrane, or cytoplasmic regions of its own and does not make contact with the cell surface.

The considerable homology shared by β2m and the α3 domain allows them to pair up and maintain the overall conformation of the MHC class I molecule. The α3 domain and the β2m chain are structured as anti-parallel β strands of amino acids that interact to form the two β-pleated sheets of the immunoglobulin fold structure described in Chapter 5. β2m also makes extensive contacts with certain regions of the α1 and α2 domains that are important for MHC molecule stability. In addition, β2m has an important function in the intracellular transport of MHC class I molecules. The β2m chain associates with the MHC class I α chain in the endoplasmic reticulum soon after protein synthesis, an event essential for the transportation of the complete heterodimer to the cell surface. Studies of tumor cell lines and knockout mice mutated in the β2m gene have shown that without β2m, the α chain gene is transcribed and the mRNA is translated, but only a very small number of MHC class I molecules appear on the cell surface.

ii) MHC Class I Peptide-Binding Site

Much information about MHC class I binding has come from studies of crystallized MHC molecules. The first structure studied, a crystal composed of millions of molecules of HLA-A2 to which were bound a heterogeneous pool of peptides, was solved in 1987 by Don Wiley and colleagues. These researchers confirmed that the binding site of the MHC molecule was relatively small so that, unlike BCRs, MHC molecules cannot recognize large native antigens. Rather, antigens must be processed to small peptides before they can fit into the MHC groove and be presented to T cells. However, it was also observed that each MHC molecule could bind a spectrum of small peptides with moderately high affinity (keeping in mind that an MHC molecule contains only one peptide-binding site, and can bind only one peptide at a time).

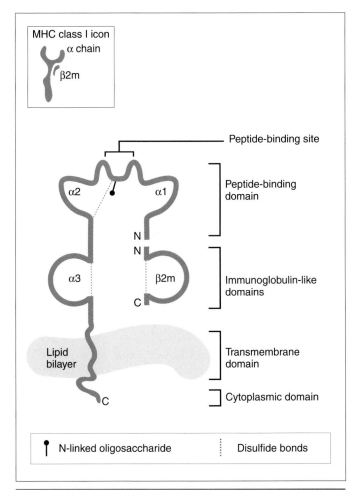

Figure 10-4

Structure of the MHC Class I Protein
Schematic representation showing component chains and domains and the position of the MHC class I protein in the cell membrane. N, amino-terminus; C, carboxy-terminus. With information from Klein J. and Hořejší V. (1997) "Immunology," 2nd edn. Blackwell Science, Oxford.

Thus, a single type of MHC molecule can present a number of quite different peptides to T cells over time, in contrast to the single antigen (or very small number of closely related antigens) that can fit in the binding site of an Ig molecule. In fact, it has been estimated that to scan all the copies of all the intracellular proteins within an average cell, each MHC class I molecule would have to be compatible with up to 500 different peptides, a classic example of promiscuous binding.

The peptide-binding site of the MHC class I molecule is a groove in which a peptide of about 8–10 amino acids is accommodated by anchoring the ends and allowing the middle portion to bulge "upward" (Fig. 10-5A). The groove is formed by the juxtaposition and interaction of the α1 and α2 domains of the α chain. The floor of the groove is a relatively flat region of eight anti-parallel β strands derived from both α1 and α2, while the sides of the groove are made up of two parallel alpha helices, one of which is derived from α1 and the other from α2. The α3 domain can be proteolytically cleaved without affecting the peptide-binding groove or peptide binding, indicating that it plays no role in peptide binding itself. Although it interacts primarily with the α3 domain, β2m also interacts with the amino acids in α1 and α2 that form the floor of the peptide-binding groove. In fact, these interactions are tightened when the groove is occupied by peptide, stabilizing the entire MHC class I structure. Thus, the most energetically favorable form of functional MHC class I is what has come to be called the "MHC class I trimer," made up of the MHC class I α chain, β2m, and peptide. The trimer is more stable than either MHC class I α alone, MHC class I α plus peptide, or MHC class I plus β2m.

The peptide is held in the MHC class I groove by interactions between residues of the MHC class I molecule and the N- and C-termini of the peptide (Fig. 10-5B). Small binding "pockets" bind to certain "anchor residues" at either end of the peptide to hold it in the groove. The peptide anchor residues, usually the first two and last two amino acids, point "down" into the groove between the α1 and α2 helices of the MHC class I α chain. Those peptide residues located between the anchor residues project "up," away from the MHC class I molecule and toward the TCR. A sufficient degree of conformational flexibility exists such that peptides of widely varying amino acid sequence in the region between the anchor residues can fit in the groove, conferring the broad specificity of MHC peptide binding. The ends of the MHC class I groove are closed, which means peptides larger than 8–10 amino acids can fit in only if their central residues can protrude up out of the groove. It is these protruding central residues that determine the degree of affinity with which the TCR binds to the MHC–peptide complex. Different MHC class I genes encode proteins with different pockets able to interact with different anchor residues. In addition, different alleles of a given MHC class I gene can also show subtle differences in anchor residue specificity. In both cases, broad variability in the central region of the peptide is permitted.

In 1992, studies of co-crystals of a particular MHC class I allele bound to a single type of peptide (rather than a collection of different peptides) were independently published by the laboratories of Don Wiley and Ian Wilson. These and subsequent investigations have shown that water molecules play an important role in MHC–peptide binding. The fit of the peptide in the groove is "tightened" when water molecules fill any empty pockets in the complex. In addition, side chains of certain residues in the MHC class I molecule can change their conformation slightly to allow a better fit of a given peptide, similar to the "induced fit" model of antigen–antibody interaction.

II. MHC CLASS II PROTEINS

The reader will recall from Chapter 2 that the expression of MHC class II proteins is much more restricted than that of MHC class I proteins. MHC class II molecules are found primarily on professional antigen-presenting cells such as dendritic cells, macrophages, and mature B cells. The TCRs of CD4-bearing T cells recognize antigenic peptides complexed to MHC class II molecules. The peptides bound by MHC class II are of exogenous origin; that is, derived from the degradation of proteins that have entered the cell from the exterior, usually via phagocytosis (see Ch.11). Again, because the professional APCs also digest spent host proteins of extracellular origin, the vast majority of peptides presented on MHC class II molecules on the APC surface will be "self" in nature and will not trigger a T cell response. However, nonself peptides presented on MHC class II will be identified by

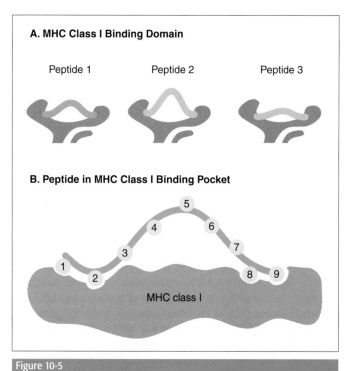

Figure 10-5

MHC Class I Peptide-Binding Site
(A) Schematic representation of how three different peptides might fit within the peptide-binding groove of an MHC class I protein. Note that the N- and C-termini of each peptide are anchored to the ends of the enclosed groove. (B) Detail of peptide binding in an MHC class I protein groove. Numbers represent individual amino acids of the peptide. Peptides are usually 8–9 amino acids in length.

passing antigen-specific Th cells as foreign and, in the presence of the correct costimulatory molecules (such as CD28), will trigger T cell activation.

i) MHC Class II Component Polypeptides

An MHC class II molecule is composed of an α chain non-covalently associated with a β chain (Fig. 10-6). Murine and human MHC class II α and β chains are similar in structure, being glycoproteins of 24–32 and 29–31 kDa, respectively, with an N-terminal domain, an extracellular Ig-like domain, a hydrophobic transmembrane region of about 25 amino acids, and a short cytoplasmic tail. In the MHC class II α chain, the polymorphic N-terminal α1 domain consists of about 90 residues that make up half the peptide-binding site. The α2 domain contains an internal disulfide bond that forms a globular domain of about 90 amino acids that is homologous to the Ig-fold. The α2 domain of MHC class II does not contribute to the peptide-binding site. The MHC class II α chain is glycosylated in both the α1 domain and in the globular

loop of the α2 domain. The MHC class II β chain is also divided into domains. The β1 domain (about 90 amino acids) contributes the other half of the peptide-binding site. The β1 domain also contains an internal disulfide bond, unlike the α1 domain, and bears the single β chain glycosylation site. The Ig-like β2 domain houses the site of contact with the CD4 coreceptor. While the MHC class I molecule has one transmembrane chain plus the extracellular β2m chain, both chains of the MHC class II molecule have transmembrane sequences and are thus anchored in the cell membrane. However, in terms of overall structure, the α2/β2 area of the MHC class II heterodimer closely resembles the α3/β2m area of the class I molecule.

ii) MHC Class II Peptide-Binding Site

The MHC class II peptide-binding groove is very similar in overall structure to that of MHC class I molecules. However, while the groove is formed solely by the α chain in MHC class I, domains of both the α and β chains are required to form the groove in MHC class II molecules. Another important difference from the MHC class I binding groove is the structure of the ends of the cleft: these are closed in MHC class I grooves but open in MHC class II (Fig. 10-7A). These open ends permit the binding of much longer peptides, some as long as 30 amino acids, although the majority of peptides isolated from purified MHC class II molecules are between 13 and 18 amino acids long. The open ends also mean that binding in MHC class II

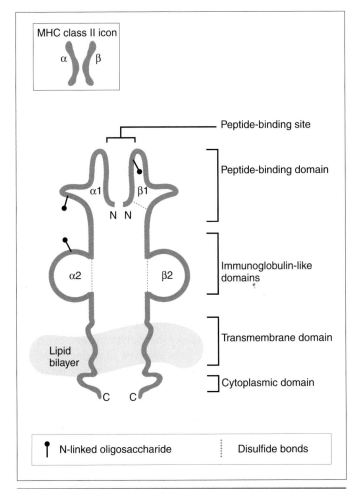

Figure 10-6

Structure of MHC Class II Protein
(A) Schematic representation showing component chains and domains and the position of the MHC class II protein in the cell membrane. N, amino terminus; C, carboxy-terminus. With information from Klein J. and Hořejší V. (1997) "Immunology," 2nd edn. Blackwell Science, Oxford.

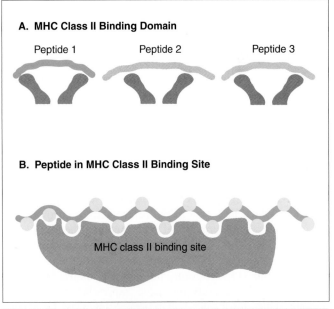

Figure 10-7

MHC Class II Peptide-Binding Site
(A) Schematic representation of how three different peptides might bind to the peptide-binding site of an MHC class II protein. Note that the N- and C-termini of each peptide spill over the ends of the relatively open groove. (B) Detail of peptide binding in an MHC class II protein groove. Blue circles represent individual amino acids of the peptide that must be arranged in a polyproline-like structure to stay bound in the groove. No specific anchor residues are involved. Peptides are usually 13–18 amino acids in length.

grooves does not generally depend on conserved anchor residues in the amino- and carboxy-termini of the peptide. Rather, affinity of binding in the groove appears to depend on hydrogen bonding between the peptide backbone and side chains of certain MHC residues (Fig. 10-7B). Antigenic peptides that are successfully bound to the floor of the groove possess a particular conserved secondary structure (resembling a polyproline chain) in the portion of the peptide that aligns with the critical MHC residues. As a result of this conformational requirement, MHC class II alleles are less promiscuous than MHC class I alleles with respect to the range of peptides they can bind. The conserved MHC residues are acidic and are located in the middle of the MHC class II groove. On the other hand, the peptide ends are permitted greater flexibility compared with peptides binding to MHC class I.

While the overall polyproline shape is conserved in MHC class II-binding peptides, different MHC class II genes and alleles can specify different binding pockets in the floor and walls of the groove. Accordingly, the polymorphism evident in MHC class II molecules is concentrated in the $\alpha1$ and $\beta1$ domains in areas forming the floor and sides of the groove. Crystallization studies have shown that four β-pleated sheet "floor" strands and one α-helical "side panel" are supplied by the MHC class II $\alpha1$ domain, while another four β-pleated sheet strands and another α-helical side panel are contributed by the MHC class II $\beta1$ domain. While the backbone of the peptide is involved in binding to MHC, antigenic specificity is conferred by the interaction of side chains of the peptide protruding up out of the groove toward the TCR.

D. MHC Genes

In this section, we take an in-depth look at the genomic structure of the MHC in mice and humans.

I. DETAILED ORGANIZATION OF THE H-2 COMPLEX

i) Murine MHC Class I Genes

As introduced previously, the class I genes are also known as the "classical" MHC class I genes, or the *class Ia* genes. The K region, which lies close to the centromere of chromosome 17, contains a single functional gene encoding the H-2K class I α chain (Fig. 10-8A). The D region contains one functional class I α gene (the H-2D gene) and several pseudogenes (D2, D3, and D4). In some, but not all, mouse strains, a region called L contains one gene expressing a functional α chain. Thus, most mice will express MHC class I α chains from both the K and D loci but some will also express an MHC class I α chain from the L locus. Each of these chains can combine with $\beta2m$ to form a functional MHC class I heterodimer on the cell surface, so that as many as six different MHC class I molecules (three from the maternal chromosome and three from the paternal chromosome) may appear. In reality, differences in regulatory elements governing each locus are thought to result in a much lower number of L

α chains being expressed compared to K α and D α chains, and lower K α than D α expression.

All MHC class I genes, in both mice and humans, have a similar exon–intron structure (Fig. 10-9A). Class I genes are composed of a 5′ leader exon containing a 5′ untranslated region and encoding a short signal peptide (cleaved off to form the mature MHC class I α chain); three exons encoding domains $\alpha1$, $\alpha2$, and $\alpha3$; an exon encoding a connecting peptide and the transmembrane region; and one or two exons making up the cytoplasmic tail plus a 3′ untranslated region. Some researchers speculate that the multiple cytoplasmic tail exons may reflect as yet unidentified distinct functional roles carried out by different regions of the MHC tail. Nucleotide sequence variability predominates in the exons encoding the $\alpha1$ and $\alpha2$ domains at positions involved in forming the peptide-binding groove.

ii) Murine β2-Microglobulin Gene

The non-polymorphic β2-microglobulin chain is the other part of the MHC class I heterodimer. In the mouse, the gene encoding β2m lies outside the MHC loci on murine chromosome 2 and consists of 3 exons. The first encodes a 5′ untranslated region and the leader while the second and third exons contain the coding sequence. The third exon also contains a 3′ untranslated region.

iii) Murine MHC Class Ib Genes

The products of class Ib genes are "non-classical" MHC molecules. Although these proteins generally resemble the classical MHC class I molecules and many bind to β2m, they exhibit far less polymorphism than do class I molecules, are tissue-restricted in their expression, and are generally not involved in classical antigen presentation. In fact, these genes constitute the majority of genes in the MHC and are found in the Q, T, and M regions spread over about 2500 kb of DNA (refer to Fig. 10-8A). The functions of the products of these genes are under intense investigation (see Box 10-4).

iv) Murine MHC Class II Genes

As introduced previously, murine MHC class II α and β chains are encoded by related genes within the A and E regions of H-2 complex (refer to Fig. 10-8A). The class II α chains are encoded by the A_a and E_a genes while the class II β chains are encoded by the A_b and E_b genes. Considerable polymorphism exists for each of these genes at those sites involved in peptide binding, with the exception of the E_a chain, which is relatively invariant. While there is an E_{b2} gene, it appears to be non-functional.

A vast majority of MHC class II α chain genes have the following structure: a 5′ leader exon, two exons encoding the $\alpha1$ and $\alpha2$ domains, an exon encoding the transmembrane domain and part of the cytoplasmic tail, and a last exon completing the cytoplasmic tail and containing the 3′ untranslated region (Fig. 10-9B). MHC class II β chain genes have a similar structure: a 5′ leader exon, two exons encoding the $\beta1$ and $\beta2$ domains, an exon encoding the transmembrane domain and part of the cytoplasmic tail, and two more 3′ exons completing the cytoplasmic tail and containing the 3′ untranslated region.

Occasionally one will see the murine class II loci referred to as the I-A and I-E loci, a historical leftover from the early

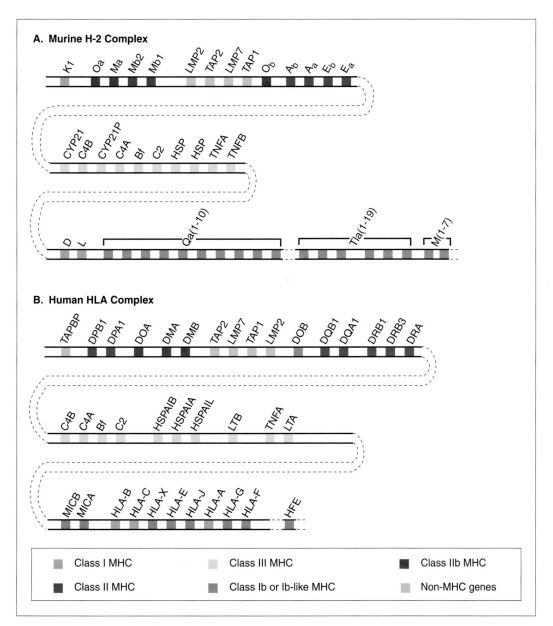

Figure 10-8

Detailed Organization of the Major Histocompatibility Complex
Genes located in the (A) murine H-2 complex and (B) human HLA complex. Class I, Ib, II, IIb, and III genes are color-coded and pseudogenes are not shown. In (A), all genes between Oa and TAP1 are considered to be located in the P region. In (B), all genes between DOA and DOB are considered to be in the DM and DO regions. The individual genotype shown has DRB1 and DRB3 although other DRB gene configurations are possible. See text for gene symbol definitions.

investigations of the influence of MHC haplotype on the level of immune responsiveness to a given antigen. The "I" referred to the "immune response region" of the mouse MHC.

v) Murine MHC Class IIb Genes

The P region located between the K and A/E regions contains several genes that resemble the functional A and E genes but which are much less polymorphic. These genes, known as Oa, Ob, Ma, Mb1, and Mb2, are currently of unknown function (if any) and will not be discussed further.

vi) Murine Non-MHC Genes in the MHC

Telomeric to the presumed pseudogenes in the P region lies a collection of genes that are important for the assembly of MHC class I molecules. Endogenous proteins in the cytosol of

a cell must be degraded in a complex structure called a *proteasome* before the resulting peptides can be combined with MHC class I and presented to CD8 T cells (see Ch.11). Two non-MHC class genes found in this location, LMP2 and LMP7, encode subunits of the proteasome complex. (LMP is pronounced "lamp" and stands for "<u>l</u>arge <u>m</u>ultifunctional <u>p</u>rotease.") Two other genes, TAP1 ("<u>t</u>ransporter in <u>a</u>ntigen <u>p</u>rocessing 1") and TAP2, encode proteins that form an intracellular pump that transports newly-produced peptides from the cytosol into the rER. The transported peptides then bind to newly-synthesized MHC class I molecules and make their way to the cell surface.

vii) Murine MHC Class III Genes

A group of genes known as the MHC class III genes is also located within the MHC, telomeric to the E region (refer to

Figure 10-9

Exon/Intron Structure of MHC Genes

Exon/intron structure of genes encoding (A) MHC class I chains and (B) MHC class II chains. The structures of the MHC genes are very similar in mice and humans. Schematics of the component protein chains are shown as reminders. L, leader sequence; UTR, untranslated region; CP, connecting peptide; TM, transmembrane region; C, cytoplasmic region; CS, coding sequence.

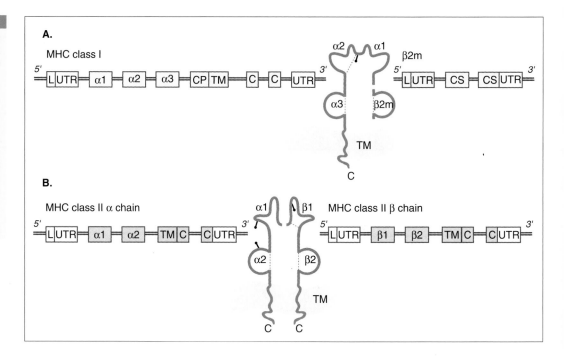

Box 10-4. "Non-classical MHC" and "MHC-like" genes and proteins

When we hear or see the term "MHC," we immediately think of molecules encoded in the MHC genetic region that present peptide antigens to T cells. These are the "classical" MHC proteins performing what is thought of as the normal MHC function. However, other molecules that resemble the classical MHC proteins in structure but which do not appear to present peptides to T cells, or which are encoded outside the MHC, also exist. Those encoded within the MHC loci are called the "non-classical" MHC proteins, and those that are encoded outside the MHC are called "MHC-like" molecules. As more is learned of these molecules, it appears that some may not be as far removed from the classical function as originally thought.

Non-Classical MHC Genes: Class Ib and IIb

The classical MHC class I and class II genes are sometimes called MHC class Ia and class IIa genes, and their non-classical counterparts are called MHC class Ib and class IIb genes, respectively. The MHC class Ib genes in the mouse are located in the H-2Q, -T, and -M regions. Similarly, the human class Ib genes include the HLA-E, -F, -G, -J, and -X genes and the HFE gene (formerly known as HLA-H). Some non-classical MHC I gene products are secreted (unlike the products of classical class I

genes), others are membrane-bound, and still others are unexpressed pseudogenes. Expression level varies by locus and by tissue, although the extracellular domains of all class Ib molecules share considerable sequence identity within a species. This variability in expression appears to be due to the loss by class Ib gene promoters of regulatory sequence motifs conferring ubiquitous expression on class Ia genes. However, mice lacking the Q genes have no apparent immunological defect, and most of the genes in the T region have turned out to be pseudogenes, so that some scientists feel these genes are evolutionary "leftovers" from processes giving rise to the class Ia genes. On the other hand, some T region products have been implicated in antigen presentation to $\gamma\delta$ T cells, and at least one M region gene (H-2M3) encodes a product that appears to be involved in a specialized form of peptide presentation to $\alpha\beta$ T cells. The product of the H-2M3 gene was found to bind to peptides that bear a formylated methionine residue at the N-terminus, a structure that occurs in prokaryotic (but not eukaryotic) proteins. Some scientists have speculated that the H-2M3 product may therefore play a role in presenting peptides derived from bacteria that invade individual cells and replicate intracellularly, such as certain *Mycobacterium* and *Listeria* species.

In humans, the expression patterns and degree of sequence conservation of both HLA-E and HLA-F have suggested that these genes may have important functions related to peptide presentation to $\gamma\delta$ T cells (see Ch.18). HLA-G is expressed in placental cells at specific stages of fetal development. It is thought that its protein product may help protect a fetus, which is allogeneic to its mother, from stimulating a maternal immune response (see Ch.16). The functions of HLA-X and HLA-J remain unknown.

The case of the HFE gene ("high in iron") is interesting. This gene (which is located in the MHC) encodes a protein that, when defective, results in an autosomal recessive disorder called "hereditary hemochromatosis" (HH). HH is characterized by excessive deposition of iron in various organs. The HFE gene product is a cell surface glycoprotein that shares 37% identity with the prototypical MHC class I molecule and associates with β2-microglobulin to form a heterodimer. In fact, β2m is required for HFE surface expression, just as it is required for surface MHC class I expression. However, examination of the crystal structure of HFE has shown that, while the protein folds as expected for an MHC class I molecule, a peptide-binding groove such as occurs in a classical MHC class I molecule does not form. Thus, this non-classical MHC class I gene appears to have a role in iron absorption rather than peptide

Continued

presentation. A putative murine homologue of the human HFE gene has been localized on chromosome 13.

Tucked between the class III and class I genes of the human MHC (specifically between the TNF and HLA-B loci) are the MICA and MICB genes. These genes show an even greater divergence from the classical MHC loci than the class Ib loci. MICA and MICB are stress-induced molecules, like heat shock proteins, and share stress-sensitive regulatory motifs in their promoters. Unlike classical MHC class I molecules, cell surface expression of MICA and MICB requires neither peptide nor β2m. In addition, MICA and MICB expression appears to be restricted to normal intestinal epithelial cells and epithelial tumors. Rather than functioning in antigen presentation, MICA (at least) binds to a particular receptor on NK and γδ T cells; this engagement results in T and NK cell activation and effector functions (see Ch.18). MICA has also been associated with the positive modulation of certain CD8$^+$ αβ T cell responses. Some observers have suggested that the MIC genes be categorized in a new class called the "MHC class Ic" genes, but this designation awaits general acceptance.

In humans, genes of very limited polymorphism in the DO and DM regions encode non-classical MHC class IIb proteins. The role of the HLA-DM protein is discussed in detail in Chapter 11. Suffice it to say here that this molecule, made up of class IIb α and β chains, is localized not on the cell surface but within the endosomal compartment where MHC class II molecules are loaded with peptides. The HLA-DM heterodimer appears to facilitate and regulate peptide loading onto classical MHC class II molecules. The HLA-DO MHC class IIb molecule has recently been found to associate with HLA-DM in the endosomal compartment of B cells and to negatively regulate HLA-DM activity (see Ch.11).

MHC-like Molecules

Several MHC-like molecules that are encoded outside the MHC loci share certain structural similarities (and therefore perhaps functions) with classical MHC molecules. The N-terminal end of all classical MHC class I and II molecules is arranged into a fold containing the peptide-binding groove. The MHC-like molecules also possess the MHC fold, and bind their ligands (which are generally not peptide in nature) either by forming a binding groove similar to the classical MHC peptide binding site, or by using another part of the MHC-like fold as a recognition site. The CD1 gene product (derived from a gene outside the MHC) is found primarily on hematopoietically-derived cells in most mammalian species. The protein associates with β2m but has a binding groove that is deeper and narrower than those found in MHC class I and II molecules. Because of the amino acid composition of the side chains in the groove, the binding site is extremely hydrophobic, a perfect spot for the binding of lipid and glycolipid antigenic fragments. Scientists were bemused when it became clear that cells bearing CD1 could display on their cell surfaces lipid antigens derived from *Mycobacteria*, and that there was some limited recognition of these non-peptide antigens by certain T cell subsets. More on antigen presentation by CD1 molecules appears in Box 11-1 in Chapter 11.

The Fcn receptor (FcRn, described in Chapter 5) was originally named for its involvement in the binding of ingested maternal IgG to intestinal epithelium and the transfer of the IgG to the bloodstream of neonatal rodents. In securing the IgG for transport, FcRn appears to protect the IgG from degradation in the acidic environment of the gut. FcRn structurally resembles an MHC class I molecule and associates with β2m. It possesses an MHC fold, but there is no binding groove. Studies of the crystal structure of FcRn bound to IgG showed that the Fc portion of the IgG molecule interacts with a site on the side of FcRn molecule distinct from the sites of classical MHC contact with TCRs or coreceptors.

Two other MHC class I-like molecules have been identified in humans. MR1 resembles an MHC class Ia molecule but is encoded on chromosome 1 (like human CD1); its function is unknown. Zinc α2-glycoprotein is an MHC class I-like molecule found in plasma and other body fluids; again, its function is unknown.

Other MHC class I-like molecules have evolved to thwart our immune systems. For example, when a cytomegalovirus (CMV) infects a host cell, it induces the downregulation of MHC class I molecules on that cell. This cell cannot then be recognized by CTLs, which would normally bind to the viral peptide complexed to MHC class I, destroy the infected host cell, and protect the host. However, a CMV-infected host cell remains open to attack by NK cells, which seek out and lyse only those cells lacking MHC class I expression (see Ch.18). Cytomegaloviruses have attempted to thwart NK cell attack by synthesizing within the host cell MHC-like molecules that share 25% identity with the prototypical MHC class I protein. These homologues are called UL18 in human CMV and ml44 in murine CMV. Both UL18 and ml44 associate with β2m, but only UL18 has a binding groove similar to an MHC class I molecule. Scientists speculate that the expression of UL18 or ml44 on an infected host cell may allow it to engage the inhibitory receptors of the NK cell that normally recognize MHC class I, thus fooling the NK cell and staving off its attack.

Fig. 10-8A). The class III genes do not appear to encode molecules directly involved in antigen display. The class III region was originally called the "S" region in mice because the complement component C4, isolated from serum, was localized to this area. In fact, the genes encoding the C2, C4a, C4b, and Factor B complement components are all located here. Genes encoding the functional isoform of 21-hydroxylase (21-OH-A, derived from the CYP21 gene) and its non-functional isoform 21-OH-B (derived from the CYP21P pseudogene) are also present. Telomeric to these genes are the heat shock protein genes, including that encoding the HSP70 ("heat shock protein 70") molecule, which may function during the protein degradation required for antigen presentation (see Ch.11). Between the HSP genes and the D region genes lie the TNFA and TNFB genes, which encode the cytokines TNF and lymphotoxin, respectively (see Ch.17).

II. DETAILED ORGANIZATION OF THE HLA COMPLEX

i) Human MHC Class I Genes

As introduced in preceding sections, HLA-A, -B, and -C are single genes encoding the human classical MHC class I α chains (refer to Fig. 10-8B). These genes are equivalent to the functional mouse K, D, and L genes, respectively. Each of the HLA-A, -B, and -C genes on both parental chromosomes encodes an α chain that can combine with the invariant β2m molecule to form complete MHC class I α-β2m heterodimers, meaning that six different MHC class I molecules can be co-dominantly expressed on any given cell. Regulatory differences in the level of expression of different MHC genes occur as they do in the mouse. For example, even though they possess the gene, 20–50% (depending on race) of humans apparently do not express a functional HLA-C protein from one of the two alleles present at the HLA-C locus.

ii) Human β2-Microglobulin Gene
The human β2m gene has the same exon–intron structure as its murine counterpart (refer to Fig. 10-9A) and is located outside the MHC region on chromosome 15.

iii) Human MHC Class Ib Genes
The HLA-X, -E, -J, -G, and -F and HFE (formerly known as HLA-H) genes make up the class Ib genes that encode "non-classical" MHC products analogous to the products of genes in the mouse Q, T, and M regions (refer to Fig. 10-8B). The precise functions of the HLA-X, -E, -F, -G, and -J genes are unknown, while HFE ["high in iron (Fe)"] appears to be involved in iron absorption. These genes have a similar structure to the classical class I genes but lack their polymorphism. Some class Ib gene products combine with β2m, are expressed on cell surfaces, and may be involved in aspects of antigen presentation (refer to Box 10-4 and Ch.11). Many other sequences in the HLA class Ib region appear to be non-functional pseudogenes.

iv) Human MHC Class II Genes
The classical MHC class II α and β chain genes are encoded by separate genes within at least three of the five D families (DP, DQ, DR, DO, and DM; refer to Fig. 10-8B). The HLA-DR genes resemble the murine H-2E genes, while those of HLA-DQ resemble the murine H-2A genes. The HLA-DP region contains two genes, DPA1 and DPA2, which could encode α chains, and two genes, DPB1 and DPB2, which could encode β chains, but only DPA1 and DPB1 appear to be functional. While fewer than 25 DPA1 alleles have been identified, there are more than 100 alleles for DPB1. Similarly, the DQ family contains the DQA1 and -A2 genes, which could encode α chains, and the DQB1, -B2, and -B3 genes, which could encode β chains, but only DQA1 (25 alleles) and DQB1 (more than 50 alleles) are functional.

The DR family is unique in that not every individual carries the same number of DRB loci on his/her chromosomes. The DR α chain is encoded by a single, almost invariant gene designated DRA; this gene is functional and present in every individual. In contrast, within the human population, nine different loci encoding β chains designated DRB1 to DRB9 have been identified. The DRB1 (more than 354 alleles) is much more polymorphic than the other DRB genes. While every individual has the DRB1 and DRB9 loci, different individuals may also have one or more DRB loci selected from

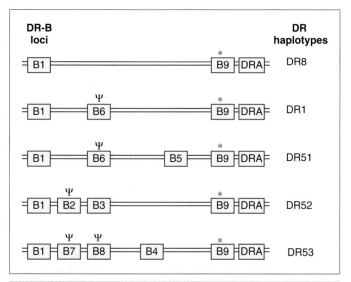

Figure 10-10

Human DR Haplotypes
Multiple DRB loci have been identified in the human HLA complex. All individuals possess the DRB1 and DRB9 loci and many have additional DR loci selected from among seven others. Each collection of DRB loci has been given a haplotype number. Functional MHC class II chains are always expressed from DRB1. Individuals possessing DRB3, -B4, or -B5 may also express MHC class II chains from these genes. *, gene fragment; ψ, pseudogene. With information from Klein J. and Hořejší V. (1997) "Immunology," 2nd edn. Blackwell Science, Oxford.

among the DRB2 to DRB8 genes (Fig. 10-10). However, of the DRB genes, only DRB1, DRB3, DRB4, and DRB5 are functional, DR9 is a gene fragment, and DRB2, B6, B7, and B8 are pseudogenes. To date, of the DRB3, DRB4, or DRB5 genes, only one has been observed in any given genotype examined. Because of this structural and functional variation, some individuals express β chains from two β genes in the DR regions of each chromosome: DRB1 and, if they possess the gene, an additional β chain derived from either DRB3, -B4, or -B5. Thus, a cell may simultaneously express two different forms of MHC class II from this region on one chromosome, for example, HLA-DRA/DRB1 and HLA-DRA/DRB4. As mentioned previously, chains derived from the paternal and maternal chromosomes can be mixed and matched to create heterodimers (for example, an HLA-DRA$_{paternal}$/HLA-DRB1$_{maternal}$ class II molecule). Recall, however, that an α chain from a DR gene almost always combines with a β chain from a DR gene, and only very rarely with a β chain from a DP or DQ gene.

v) Human MHC Class IIb Genes
Genes in the DO and DM regions of the HLA complex (refer to Fig. 10-8B) encode polypeptides that closely resemble classical MHC class II α and β chains. However, while these chains do form heterodimers, these molecules show very little polymorphism, are generally not expressed on the cell surface, and are not directly involved in antigen display. As is discussed in detail in Box 10-4 and Chapter 11, both HLA-DM

and HLA-DO appear to be important for the regulation of antigen processing inside the endocytic pathway.

vi) Human Non-MHC Genes in the MHC

As was true in the mouse, genes important for the assembly of MHC class I molecules are located near the class II genes (refer to Fig. 10-8B). The LMP2, LMP7, TAP1, and TAP2 genes are located between the DP and DQ regions and encode proteins with the same functions as their murine counterparts (see previous discussion).

vii) Human Class III Genes

Between the class II and class I genes of the human MHC lies the class III region, which contains additional genes of immunological importance (refer to Fig. 10-8B). The human class III region is analogous to the S region of the mouse MHC and contains the corresponding (mostly complement-related and proinflammatory) genes. In humans, the lymphotoxins are derived from the LTA and LTB genes.

Table 10-3 summarizes some general characteristics of the products of each gene group discussed in this section.

E. Expression of MHC Molecules

The function of an MHC molecule is to bind a peptide and present it to passing T cells for inspection. In the case of self-peptides, those T cells bearing a TCR that can bind to that peptide–MHC complex with high affinity have most likely been deleted by the mechanisms of central tolerance or rendered non-responsive by peripheral tolerance (see Chs.13 and 16). However, in the case of a nonself peptide, a T cell clone bearing a TCR recognizing that peptide coupled to MHC will probably exist. An immune response occurs when an APC or target cell supplies enough peptide–MHC complexes to occupy sufficient TCRs on the antigen-specific T cell to trigger the activation process. Thus, the level and timing of expression of MHC molecules is a crucial factor in determining whether an immune response will be mounted.

When and where the MHC genes are expressed is controlled to a large extent by cytokines and other stimuli released in the host cell's vicinity. Depending on the type of host cell and the tissue in which it resides, these stimuli may be either constitutively produced or induced during an

Table 10-3 Comparison of Products of MHC Loci*

	Class I	Class II	Class Ib	Class IIb	Class III	MHC-like
Gene encoded in MHC region	Yes	Yes	Yes	Yes	Yes	No
Polypeptides	Class I α plus β2m	Class II α plus class II β	Class I α-like plus β2m	Class II α-like plus class II β-like	Neither class I nor class II chains	Non-MHC chain associates with β2m
Tissue expression	Almost ubiquitous	APCs	Restricted	APCs	Almost ubiquitous	Restricted
Soluble form?	No	No	Some	Some	Yes	Some
Polymorphism	Extreme	Extreme	No	No	No	No
Function	Ag presentation to $CD8^+$ cells	Ag presentation to $CD4^+$ cells	Specialized Ag presentation to some αβ and γδ T cell subsets	Peptide loading of MHC class II	Complement components, chaperone proteins, inflammatory molecules	Ag presentation to γδ cells
Domains contributing to binding site	α1/α2 of α chain	α1/β1 of α and β chains				Presents non-peptides
Contents bound in groove	Peptide (8–9 amino acids long)	Peptide (13–18 amino acids long)	For H2-M, bacterial peptides with formylated methionine			Lipids
Ends of groove	Closed	Open	May not have groove			Groove may be reduced in size or absent
Peptide origin	Intracellular	Extracellular				Microbial or self lipid antigens
Examples	H-2K HLA-B	H-2A HLA-DP	H-2M HLA-E	H-2P HLA-DO	C4, TNF, HSP70	CD1, FcγRn

*See Box 10-4 for a detailed discussion of functions of non-classical MHC gene products.

inflammatory or immune response to injury, pathogens, or tumors. For example, molecules in the walls of invading bacteria stimulate macrophages to produce TNF and lymphotoxins, and viral infection induces the infected cells to synthesize interferons (IFNs). The interaction of these cytokines with specific receptors on a host cell triggers intracellular signaling cascades that terminate in the activation of transcription factors. The activated transcription factors enter the host cell nucleus and bind to 5′ regulatory motifs in the DNA upstream of the MHC genes, either promoting or repressing their expression.

Fundamental differences exist in the expression of MHC class I and class II genes. MHC class I proteins are expressed on almost all nucleated cells, while MHC class II expression is restricted to a few cell types capable of presenting antigen to CD4$^+$ T cells, including B cells, macrophages, Langerhans cells, follicular dendritic cells, and interdigitating cells. MHC class II also appears on epithelial cells in the thymus, functioning in the processes of thymic selection that lead to tolerance (see Ch.13). Despite these differences, it is now known that the MHC class I and II genes share at least one major regulatory pathway dependent on the *SXY regulatory module*.

I. THE SXY–CIITA REGULATORY SYSTEM

The 5′ upstream regions of the promoters of the MHC class Iα, β2m, and MHC class II α and β genes (and other accessory genes, such as that encoding the invariant chain) contain regulatory elements that allow them to be coordinately transcribed. The proteins necessary to form complete MHC class I and II molecules and other proteins functioning in antigen presentation are thus produced together, even though some of the genes may be located outside the MHC. One of the most important of these regulatory motifs is the *SXY module*, which contains (starting from the 5′ end of the regulatory region and proceeding 3′ toward the promoter) the S box (sometimes known as the W/S box); the X box (containing tandem X1 and X2 sequences), and the Y box (Fig. 10-11A). Historically, these regulatory motifs were identified by comparing the sequences of the promoter regions of the murine and human MHC class II genes, and by mutational analyses examining their effects on MHC class II gene expression.

The regulatory DNA sequences of the S, X, and Y boxes are conserved among MHC genes and interact with at least four or five different multi-subunit factors (and therefore at least 8–10 different polypeptides) to form a complex that ultimately controls MHC gene expression. Proper spacing (i.e., correct number of helical turns) between the X and Y boxes has been shown to be important for the binding of these regulatory proteins. Indeed, the X and Y boxes are separated by a 19–20 bp stretch of DNA that can vary in sequence but is conserved in length across all species examined to date. Each of the three elements of the SXY module is bound by at least one member of a giant cooperative binding complex. The heterotrimeric DNA binding factor RFX binds weakly to the S box and strongly to the X1 element of the X box. RFX is composed of the RFX5 protein, the *RXF-associated protein* (RFXAP), and the *RFX ankyrin repeat protein* (RFXANK). The X2 element of the X box is bound by *X2 binding protein* (X2BP), itself a

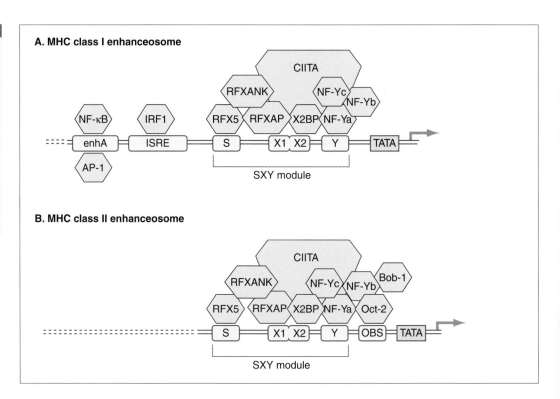

Regulatory Elements of MHC Class I and II Genes
Constitutive and inducible MHC class I and II transcription both depend on the binding of the class I and II enhanceosomes to sites in the class I and II SXY modules, respectively. An enhanceosome is a multi-protein complex that binds to regulatory sites in an active promoter. Details of specific enhanceosome subunits and the binding motifs they recognize are in the text.

A. MHC class I enhanceosome

B. MHC class II enhanceosome

complex of CREB (*cAMP-responsive element binding*) transcription factors. X2BP appears to be required for RFX binding to DNA *in vivo*. The Y box, which scientists speculate may play a role in stabilizing the RFX–X2BP–DNA complex, is bound by the heterotrimeric DNA binding factor NF-Y composed of the NF-Ya, NF-Yb, and NF-Yc subunits. The Y box is the most highly conserved element of the SXY molecule. The NF-Y complex is thought to interact with local chromatin structures, making it easier for transcription to occur. We know the Y box is important because individuals with mutations in the Y box show greatly decreased levels of constitutive MHC class I expression.

The RFX, X2BP, and NF-Y binding factor proteins are all constitutively expressed in most cells and routinely occupy the X and Y boxes of the SXY module. This assembly of the RFX/X2BP/NF-Y complex on the SXY module then serves as a platform for the *class II transactivator* (CIITA) protein. CIITA was originally thought to be expressed only in MHC class II-positive cells (hence its name) but has recently been found to influence MHC class I transcription as well. Low levels of CIITA are required for optimal constitutive and inducible MHC class I expression. CIITA does not bind directly to the promoter DNA but rather binds to the RFX/X2BP/NF-Y complex as a whole (refer to Fig. 10-11A). This type of complicated, multi-protein transactivator structure situated on an active promoter has been dubbed an *enhanceosome*. CIITA is even more important for MHC class II transcription (Fig. 10-11B). Without high levels of CIITA, both constitutive and inducible MHC class II expression are virtually abolished. It has been suggested that CIITA is therefore a crucial "molecular switch" or "master regulator" for class II expression, and that the recruitment of CIITA to the RFX/X2BP/NF-Y complex is absolutely required for MHC class II gene transcription in APCs. Consistent with CIITA's role as the critical factor limiting MHC class II expression, mature plasma cells, which do not express MHC class II, do not express CIITA either. Interestingly, while CIITA$^{-/-}$ mice lack all expression of MHC class II on their cells, humans lacking CIITA do show a low level of MHC class II expression, implying that different compensatory mechanisms may operate in different species. In B cells, CIITA may also interact with another non-DNA binding regulatory protein called Bob-1, also known as OBF-1 (*octamer binding factor-1*) or OCA-B. Bob-1, which is specifically expressed in B cells, interacts with the transcription factor Oct-2 (described in Ch.8) to activate the transcription of genes important for B cell development and activation. Oct-2 binds to the octamer-binding site (OBS) motif located between the Y box and the TATA box. Some researchers think that CIITA and Bob-1 interact synergistically to promote the high levels of MHC class II gene transcription that are observed in most B cells.

CIITA deficiency is one cause of an autosomal recessive human immunodeficiency called "bare lymphocyte syndrome." This disorder got its name when histologists attempting to tissue type these patients could not do so; that is, they could not detect MHC class II antigens on the patient specimens isolated to do the analysis, in this case, a mixed population of B and T lymphocytes. This disorder perhaps should be called "bare APC syndrome," but was discovered before it was known that B cells can act as APCs. More recently, it has been shown that genetic deficiencies for each of the subunits of the RFX factor can also cause bare lymphocyte syndrome.

II. TNF- AND IFNγ-INDUCED EXPRESSION OF MHC CLASS I

The low levels of constitutive MHC class I on almost all cell types are thought to be controlled by RFX expression. However, a further increase in MHC class I expression can be induced by TNF or by IFNα, -β, or -γ. Invasion of the host tissues by gram-negative bacteria induces the production of TNF and lymphotoxins by activated macrophages during both innate and adaptive immune responses. Well upstream of the SXY module in the promoters of MHC class I genes lies a sequence called Enhancer A (refer to Fig. 10-11A), which contains numerous motifs for nuclear transcription factors [particularly NF-κB and *activating protein-1* (AP-1)]. In response to TNF, NF-κB and AP-1 are activated and bind to their appropriate motifs in both Enhancer A and the Y box of the SXY module, stimulating MHC class I transcription.

Another regulatory element called the *interferon-stimulated response element* (ISRE) plays a prominent role during viral infections. In response to viral attack, IFNα is produced by cells of the innate immune system while fibroblasts respond with IFNβ production. IFNγ is secreted by NK cells, CD8$^+$ T cells, and CD4$^+$ Th1 cells activated in the course of the immune response. The IFNs act to mobilize members of a family of regulatory molecules called the IRF (*interferon regulatory factor*) family. Some IRF members are interferon-inducible, while others are constitutively expressed. In any case, mobilized IRF molecules bind to the ISRE regulatory motif located between Enhancer A and the SXY module in the promoters of all MHC class I genes (refer to Fig. 10-11A). Binding of IRFs to the ISRE stimulates the activation of nuclear transcription factors that promote MHC class I transcription, increasing the presentation of viral peptides on the infected cells. Killing of these targets by CD8$^+$ T cells is thus increased and the infection is contained more rapidly. While many IRFs that bind to the ISRE induce the activation of nuclear transcription factors that act to increase MHC class I transcription, negative regulation of MHC class I transcription appears to be mediated by the ISRE-binding protein ICSBP (*interferon consensus sequence binding protein*).

It is the ISRE pathway that may be responsible for the fact that MHC class I expression is only reduced and not abolished in the absence of CIITA. MHC class II genes lack ISRE sequences, meaning that IFNγ cannot directly induce their expression by these means. Instead, in normal APCs, IFNγ induces expression of CIITA mRNA, and this co-activator then interacts with the RFX/X2BP/NF-Y complex to upregulate MHC class II transcription on these cells.

III. EXPRESSION OF MHC CLASS Ib GENES

Unlike MHC class I genes, many class Ib genes are not ubiquitously expressed. Nucleotide sequence differences in the promoter regions of class Ib genes have provided a molecular explanation for the altered expression patterns. For example, the ISRE site in the promoter of the HLA-G gene is partly deleted, meaning that the transcription of this gene cannot be induced by IFNγ. In addition, the κB binding motifs in Enhancer A and the X2 and Y boxes in the SXY module are altered in the HLA-G promoter, meaning that this gene cannot be induced by NF-κB or CIITA either. Indeed, HLA-G molecules are expressed almost exclusively in the very specialized trophoblast cells of the pregnant uterus and likely serve as mediators of maternal–fetal tolerance (see Ch.16). The search is on for regulatory elements that govern expression under these circumstances. Similar alterations to sequences in the HLA-E promoter region block transcription of HLA-E in response to NF-κB. In addition, IFNγ induces HLA-E production, not via the ISRE, but rather via a STAT1 (signal transducer and activator of transcription 1) binding site upstream of the ISRE region. HLA-E can also be induced by CIITA via the SXY module. In the case of HLA-F, NF-κB induction appears to rely partly on interaction with some sequences in or near the ISRE. HLA-F is fully inducible by IFNγ via the ISRE and by CIITA via the SXY module.

IV. OTHER REGULATORY PATHWAYS GOVERNING MHC CLASS II GENE EXPRESSION

MHC class II expression is subject to the influence of several other cytokines in addition to IFNγ, including TGFβ, IL-4, IL-13, IL-10, and IFNβ (see Ch.16). TGFβ inhibits the effects of IFNγ on MHC class II expression by downregulating the expression of CIITA mRNA. IL-4 and IL-13 induce increased levels of MHC class II on many types of APCs, while both IL-10 and IFNβ inhibit IFNγ-mediated stimulation of MHC class II expression in a cell type-specific fashion.

F. Physiology of the MHC

I. POLYMORPHISM AND THE BIOLOGICAL ROLE OF THE MHC

As was stated earlier, while the vast majority of non-MHC genes in a vertebrate are monomorphic (one allele in the entire species), and a few are oligomorphic (two to four alleles at a given locus in a species), the MHC loci in most vertebrates are highly polymorphic. Not only does each individual have several genetic loci encoding different versions of each class of MHC polypeptide (multiplicity), but also, within a species, many different alleles exist for each of these loci. The genes at each MHC locus are co-dominantly expressed

and each individual in an outbred population is therefore heterozygous for most of them. The degree of sequence variation amongst alleles can be astonishing: while oligomorphic alleles outside the MHC differ from each other in usually one and at most a very few residues, MHC alleles can differ from each other by as many as 56 amino acids, with the differences being concentrated in the region comprising the peptide-binding groove. In fact, it has been estimated that as many as 100 alleles may exist for each of the MHC genes in mice, meaning that the theoretical number of different MHC haplotypes that exist for the mouse is [class I (100 H-2K) × (100 H-2D)] × [class II (100 H-2A$_\alpha$) × (100 H-2A$_\beta$) × (100 H-2E$_\alpha$) × (100 H-2E$_\beta$)] = 10^{12} in total.

To understand why the MHC loci are so polymorphic, it is necessary to consider the function of the proteins they encode. As we have seen, both class I and class II MHC molecules serve as antigen presentation structures. Their role is to hold up endogenous and exogenous protein fragments so that T cells can detect and respond to harmful aberrations within the host cell or in its environment. In an ancient, antigenically simple world, one could imagine a peptide receptor of very limited (or non-existent) variability that could bind a small number of foreign peptides with sufficient affinity to form a presentation complex. However, over evolutionary time, as most species faced an increasingly complicated antigenic world with thousands of molecularly distinct antigenic challenges, the presentation molecule may have faced two competing selection pressures. On one hand, an antigen presentation molecule would have had to develop a very broad binding capability to deal with these myriad antigens. A broader range of specificity would have required a concomitant decrease in the affinity between the presentation molecule and each antigen. On the other hand, as a "receptor," the same presentation molecule would have needed to maintain a threshold affinity such that each antigen could still be bound to the MHC molecule and held to it to form a viable complex.

The evolutionary solution to the pressure of requiring broader binding capability while maintaining adequate affinity for each antigen appears to have been two-fold. First, it seems that evolution has allowed multiple gene duplications of a primeval MHC gene, so that more than one gene can be dedicated to antigen presentation in the individual. Secondly, different alleles of each gene have evolved, with selection acting to maintain those alleles showing diversity in the peptide-binding groove of the presentation molecule. The resulting polymorphism at multiple MHC loci ensures that the individual will likely be heterozygous at each locus, with an increased chance of successfully presenting a given antigen. For the species as a whole, polymorphism means a larger catalogue of MHC alleles is spread over the entire population, making it more likely that a significant fraction of the population will be able to respond to a given pathogen and that the species as a whole (but not all individuals) will survive. In other words, any mutation a pathogen can come up with to outwit a particular individual's immune system will likely be recognized by an MHC allele somewhere in the species.

On the basis of this reasoning, one might suppose that in today's world, species with limited MHC diversity might not exist, or at least be very vulnerable to infection. However, in a restricted environment, this may not be true. For example, the MHC loci in the Syrian hamster and cottontop tamarin monkey are virtually monomorphic, but these species survive quite happily in their natural, very isolated, habitats. Presumably these animals have been challenged with only a very limited range of pathogens over evolutionary time. Once removed from their isolated habitat and exposed to new pathogens, these animals are indeed very susceptible to infection, most likely because they lack MHC alleles able to adequately bind peptides derived from the new threats.

II. ALLOREACTIVITY

One of the last mysteries to be solved regarding MHC physiology was that of *alloreactivity*. Alloreactivity (*allo*, meaning "other") describes the situation in which lymphocytes from one member of a species are activated by cells from another member of the species that is allogeneic at the MHC (has different MHC alleles). Most proteins that exist in a species can be transferred from one individual to another without provoking an immune response because most proteins in a species are monomorphic and therefore identical in all members of a species. In other words, such a protein is seen as "self" by all members of that species. However, as we have seen, the MHC is the most polymorphic locus in the genome, and a wide variety of MHC alleles can be expected to be present in the incoming tissue from another individual.

However, prior to the discovery of the role of the MHC in T cell antigen recognition, scientists were puzzled by the sheer magnitude of alloreactive tissue rejection responses. Alloreactive responses are generally very vigorous – much stronger than one sees against a pathogen. Once the molecular basis of antigen recognition by T cells was understood and the concept of MHC molecules as peptide-presenting molecules was accepted, the mystery of the "allo" response began to unravel. Essentially, alloreactivity is a case of molecular cross-reactivity. Allogeneic MHC molecules on allogeneic cells will be complexed for the most part to peptides that are invariant within the species and therefore viewed as "self" in all individuals. However, when analyzed from the point of view of a host receiving allogeneic cells or tissues, many of the allo-MHC plus self-peptide complexes will have the same molecular shape as self-MHC plus foreign peptide; that is, the allogeneic cells mimic self-APCs or target cells whose MHC molecules are presenting foreign peptide. Because of the sheer numbers of different self-peptides presented on the surfaces of the allogeneic cells, a significant number of T cell clones in the host will respond to the perceived threat, viewing the many incoming allo-MHC plus self-peptide structures as various self-MHC plus foreign peptide combinations for which their TCRs are specific. TCR stimulation is highly favored, and the simultaneous activation of this relatively large number of host T cell clones causes an intense alloreactive response.

Table 10-4 MHC Haplotype Correlated with Immune Responsiveness in Mice

Strain	H-2	Response to (T, G-A-L)*
C3H	k	Low
C3H.SW	b	High
A	a	Low
A.BY	b	High
B10	b	High
B10BR	k	Low

*T, G-A-L, branched multichain synthetic polypeptide (poly(L-lysine) backbone, side chains of poly(DL-alanine) and short random sequences of tyrosine and glutamic acid at the ends of the side chains).

III. MHC AND IMMUNE RESPONSIVENESS

From the beginnings of immunology as an organized discipline, it had long been observed that some foreign proteins that provoked strong immune responses in some individuals failed to do so in others. Those individuals failing to mount a response were called *non-responders*, while those that did react were called *responders*. Among responders, subtle differences in the level of the response could be observed, leading to the description of individuals as either *low* or *high* responders. Even before the direct involvement of the MHC in T cell recognition was appreciated, Baruj Benacerraf and Hugh McDevitt had mapped the "immune response" or "Ir" genes to the MHC. The work of each of these researchers showed that mice of different MHC haplotypes responded differently to a given peptide (Table 10-4). Later, when it was shown that MHC molecules interact with antigen to make it "visible" to T cells, the connection between immune responsiveness and the MHC made physical sense. In addition, in light of the extensive polymorphism of the MHC, variations in response levels could be interpreted as variations in the ability of different MHC alleles to be recognized by T cells. For example, pigeon cytochrome *c* (PC*c*) is a very strong immunogen for most inbred mouse strains, but strains lacking a certain MHC class II allele fail to respond to PC*c*. Artificial introduction into the same mice of the gene encoding the responsive MHC class II allele conferred both the expression of the MHC protein and the ability to mount an immune response to PC*c*. Non-responsiveness is a phenomenon rarely observed in outbred populations, since individuals in such a population will be heterozygous at (usually) all MHC loci and will likely have amongst their alleles at least one that can successfully present antigen. However, in an inbred population, individuals are homozygous at all MHC loci and thus more limited in the number of different MHC alleles they have at their disposal. In this case, there is a greater possibility that an individual will lack an MHC allele that leads to specific T cell activation during challenge by a particular antigen.

Two hypotheses, which may not be mutually exclusive, have been proposed to account for non-responsiveness. These theories are the "determinant selection" model and the "hole in the T cell repertoire" model.

i) Determinant Selection Model

Because an immune response to a foreign protein by T cells requires that the TCR recognize a complex of a self-MHC molecule and a peptide derived from the protein, the presence in the host of an MHC allele capable of fitting such a peptide in its binding groove is essential for a response. Should the host have an MHC haplotype that does not include such an allele, no peptides from the foreign peptide are bound, the foreign peptide is never presented to T cells, and an immune response is never triggered. The determinant selection model assumes that responsiveness depends on the strength of binding between a given MHC allele and a given determinant (peptide), which in turn depends on structural compatibility. In other words, the MHC proteins in an individual "select" which determinants will be immunogenic in that individual. Since a foreign protein is usually processed into three to four strongly immunogenic peptides, the outbred individual is very likely to possess an MHC allele capable of binding to at least one of these peptides and provoking an immune response against the immunogen.

The determinant selection model has been supported by several avenues of *in vitro* experimentation. Epitope mapping studies of large protein immunogens have been done to determine which of the component peptides of these proteins can act as T cell epitopes. Component peptides, artificially synthesized such that their sequences overlap slightly, are introduced one type at a time into culture with either APCs or target cells. In the case of MHC class I, this experimental system does not parallel the usual physiological pathway. However, by adding massive amounts of a given peptide directly to the culture dish, MHC class I alleles on the cell surface form peptide–MHC complexes (but only with a peptide they would normally present). Alternatively, peptides of interest can be introduced endogenously by transfecting the corresponding DNA sequences individually into the appropriate, syngeneic target cell lines. The peptide DNA is transcribed, translated, processed, and displayed on MHC class I in the usual way. One then assesses the ability of peptide-loaded APCs or target cells from mice of different MHC haplotypes to either stimulate T cell proliferation *in vitro* (in the case of MHC class II)

or to be killed by CTLs (in the case of MHC class I). Such experiments show that a given peptide may be immunogenic only when coupled to a particular MHC allele. For example, in inbred mice bearing the H-2K^k allele, a peptide containing amino acids 50–63 of the influenza protein NP is the immunodominant epitope—i.e., it provokes an immune response (Table 10-5). However, in mice of the H-2D^b haplotype, peptide 50–63 fails to stimulate T cells and the immunodominant epitope is a peptide containing amino acids 365–380.

The determinant selection model is also consistent with experiments in which the binding affinity of isolated peptides to particular MHC proteins has been measured and correlated with the MHC haplotype of responder mice. Equilibrium dialysis binding studies showed that a given radiolabeled peptide bound with highest affinity to only one MHC allele, and that different alleles favored the binding of different peptides. This result generally correlated with the MHC haplotypes of the responders and non-responders to a given peptide: i.e., if peptide X, which is normally recognized by H-2A-restricted T cells, bound with high affinity to MHC allele H-2A^d, but did not bind to allele H-2A^k, it also provoked a greater immune response in inbred mice of MHC haplotype H-2^d ("responder") than in inbred mice of MHC haplotype H-2^k ("non-responder" for this peptide).

ii) "Hole in the T Cell Repertoire" Model

Immune non-responsiveness may also result from tolerance mechanisms. It may be that, in non-responders, a particular foreign peptide/self-MHC combination very closely resembles the structure of a self-peptide/self-MHC combination, such that the T cell clone capable of recognizing the foreign peptide/self-MHC combination was deleted as autoreactive during thymic selection. In a non-responder, this would result in a missing T cell specificity or a "hole" in the T cell repertoire relative to the repertoire of a "responder." Even if B cell clones with the specificity in question existed, their activation would be unlikely due to the lack of specific T cell help. Thus, a "hole in the T cell repertoire" becomes a "hole" in overall responsiveness. More on the establishment of T cell tolerance appears in Chapters 13 and 16.

IV. INTRODUCING MHC AND DISEASE PREDISPOSITION

As already noted, one's MHC haplotype determines one's responsiveness to immunogens, and whether or not one can mount an immune response to a given immunogen may affect many aspects of one's well-being. The reader will recall from Chapter 1 that immune responses can have six different outcomes, some of which lead to the maintenance of good health (defense against foreign entities and the elimination of cancer cells), but others of which have deleterious consequences (autoimmunity, allergy, hypersensitivity, and transplant rejection). In other words, disease can result when an appropriate immune response to an infection or

Table 10-5 **Determinant Selection by Different MHC Alleles in Mice**		
Allele	**Influenza Peptide Presented**	**Immune Response**
H-2K^k	aa 50–63 NP	Yes
H-2D^b	aa 50–63 NP	No
H-2D^b	aa 365–380 NP	Yes

aa, amino acid; NP, nucleoprotein.

Table 10-6 Examples of HLA-Associated Disorders in Humans

Disease	Examples of Associated HLA Alleles
Acute anterior uveitis	B27
Ankylosing spondylitis	B27
Celiac disease	DR2
Goodpasture's syndrome	DR2
Graves' disease	DR3
Hashimoto's thyroiditis	DR3, DR5
Hodgkin's disease	A1
Type 1 diabetes mellitus	DR3, DR4, DQ2, DQ8
Multiple sclerosis	DR2
Reiter's syndrome	B27
Rheumatoid arthritis	DR4
Sjögren syndrome	DR3
Systemic lupus erythematosus	DR2, DR3

cancer does not occur, or when an inappropriate immune response does occur. Disease may also occur when the appropriate immune response causes collateral tissue damage (*immunopathology*).

Particular MHC haplotypes may predispose individuals to certain disorders, including several infectious diseases and immune system-related disorders (Table 10-6). A link between MHC haplotype and disease susceptibility does not tend to be observed when a relatively common infectious agent causes acute, potentially lethal disease. If the immune protection mounted cannot do the job, the host will likely die prior to reproducing, effectively removing the associated MHC alleles from the population. In contrast, the disorders that are usually observed to have an MHC association tend to be chronic in nature; the host is affected, but not killed or prevented from reproducing. In these cases, individuals possessing particular MHC alleles may fend off an infectious agent less efficiently, and/or mount an immune response that may drag on long enough or be vigorous enough to cause significant immunopathological damage. This is not to say that simple possession of the allele is inextricably linked to disease: other environmental factors and perhaps other genes also influence disease incidence and outcome.

We will cite here two human examples of MHC-related disease predisposition that are particularly striking. The MHC allele HLA-DQ8 is eight times more prevalent in groups of humans suffering from type 1 (insulin-dependent) diabetes than it is in healthy populations. Similarly, 90% of Caucasian patients suffering from a degenerative disease of the spine called "ankylosing spondylitis" carry the HLA-B27 allele, whereas only 9% of healthy Caucasians do. It should be noted that, in many cases, an MHC disease association shows a bias toward a particular ethnic group. This bias arises because, as described earlier in this chapter, a conserved ancestral haplotype is often found predominantly or exclusively in a particular ethnic group. Thus, if a particular MHC allele or combination of alleles within this haplotype is

associated with a given disease, the disease is more likely to occur in individuals of this ethnic background. More on MHC disease association in humans appears in Chapter 29.

An example of MHC-associated disease in mice is susceptibility to lymphocytic choriomeningitis virus (LCMV) infection, in which lethal immunopathology can be caused by the destruction of the brain tissue by CTLs responding to the virus. Different mouse strains of different MHC haplotypes at the H-2D locus exhibit either a weak, strong, or intermediate immune response to LCMV introduced intracerebrally. A strong responder rapidly clears the virus, minimal CTL-mediated damage to the brain tissues ensues, and the mouse remains healthy. A weak responder mounts only a limited CTL response, the brain tissue is largely spared, the infection becomes established chronically but sub-lethally, and the mouse survives. However, in a strain that is an intermediate responder, CTLs responding to the virus are not able to clear it rapidly and summon more CTLs to the scene, causing considerable damage to the brain tissue, which kills the mouse. In this case, an MHC haplotype supporting an intermediate response is more dangerous to have than a haplotype supporting a weak response. Sendai virus infection offers another model of MHC disease association in mice and shows remarkable specificity at the molecular level. The *bm1* mouse strain is susceptible to the lethal pneumonia caused by Sendai virus infection because the animal is a weak responder, while the B6 mouse strain is relatively resistant to this virus due to a strong H-2K^b-restricted CTL response. However, the *bm1* mouse (a mutant strain originally derived from B6) differs from B6 by only three amino acids in its H-2K molecule (H-2K^{bm1}). The *bm1* mouse is apparently impaired only in its ability to mount a CTL response against Sendai virus. T cell proliferation, antibody production, and NK activity all are normal in this strain during Sendai virus infection, and *bm1* mice are not particularly susceptible to any other murine virus commonly used in immunological studies.

Perhaps of more importance to Western societies, many of the disorders associated with specific MHC alleles manifest as autoimmunity; that is, in the "responder," antibodies and T cells reactive to self-components are present. (A complete discussion of MHC and autoimmunity is presented in Ch.29.) For example, type 1 diabetes is thought to arise from an autoimmune attack on antigens of the β cells of the pancreatic islets, resulting in insulin deficiency. The β cell self-antigens are routinely processed and component peptides are displayed on MHC molecules to passing T cells. One of the β cell peptides binds particularly strongly to the HLA-DQ8 allele. If this peptide–MHC complex is recognized by a T cell clone that has somehow escaped self-tolerance mechanisms, an individual expressing this allele may show an increased likelihood of mounting an immune response against the β cells, destroying the islets.

The next chapter, 'Antigen Processing and Presentation,' deals with how the MHC molecule associates with the antigenic peptide and presents it to T cells, and how peptides are derived from immunogens in the first place.

SUMMARY

Multiple genetic loci in the major histocompatibility complex (MHC) encode cell surface proteins that present peptides (both self and nonself) for inspection and potential recognition by T cells. Proteins encoded by MHC class I loci are expressed on almost all cells in the body, while those encoded by MHC class II loci are found only on cells specialized for antigen presentation. Both the class I and class II regions of the MHC contain multiple genes that are extremely polymorphic throughout a species. In addition, the alleles of these genes are co-dominantly expressed, ensuring that every individual expresses a wide array of different MHC alleles, and that few members of a population share the same MHC genotype. MHC differences are responsible for several immune phenomena. In a transplant situation, cross-reactivity between foreign MHC/peptide complexes on allogeneic donor tissues and self-MHC/peptide complexes on tissues of the recipient trigger T cell-mediated transplant rejection in the recipient; this response gives rise to the observed "major histocompatibility" barrier between most individuals. Within a given species, the range of immune responsiveness to a given pathogen results from variation in the ability of different MHC alleles to form pMHC complexes that are recognized by peripheral T cells. The high level of polymorphism in the MHC dramatically decreases the chance that a single pathogen will decimate an entire population. MHC class I and class II proteins are heterodimeric molecules with highly variant N-terminal domains that form a peptide-binding site. An invariant membrane-proximal domain on the class I molecule interacts with the CD8 coreceptor found on Tc cells, while a similar domain on the class II molecule facilitates interaction with the CD4 coreceptor on Th cells. Non-classical MHC class I and class II genes as well as MHC class III genes also constitute part of the MHC. The functions of many of these genes contribute to the overall effectiveness of immune responses.

Selected Reading List

Accolla R. S., Adorini L., Sartoris S., Sinigaglia F. and Guardiola J. (1995) MHC: orchestrating the immune response. *Immunology Today* **16**, 8–11.

Alfonso C. and Karlsson L. (2000) Nonclassical MHC class II molecules. *Annual Review of Immunology* **18**, 113–142.

Benacerraf B. and Katz D. H. The histocompatibility-linked immune response genes. *Advances in Cancer Research* **21**, 121–173.

Bjorkman P., Saper M., Samraoui B., Bennett W., Strominger J. and Wiley D. (1987) The foreign antigen binding site and T cell recognition regions of class I histocompatibility antigens. *Nature* **329**, 512–518.

Bjorkman P., Saper M., Samraoui B., Bennett W., Strominger J. and Wiley D. (1987) Structure of the human class I histocompatibility antigen, HLA-A2. *Nature* **329**, 506–512.

Bodmer J. G., Parham P., Albert E. D. and Marsh S. G. (1997) Putting a hold on "HLA-H." *Nature Genetics* **15**, 234–235.

Boss J. M. (1997) Regulation of transcription of MHC class II genes. *Current Opinion in Immunology* **9**, 107–113.

Campbell R. D. and Trowsdale J. (1993) Map of the human MHC. *Immunology Today* **14**, 349–352.

Doherty P. C. and Zinkernagel R. M. (1975) Capacity of sensitized thymus-derived lymphocytes to induce fatal lymphocytic choriomeningitis is restricted by the H-2 gene complex. *The Journal of Immunology* **114**, 30–33.

Fathman G. and Frelinger J. (1983) T lymphocyte clones. *Annual Review of Immunology* **1**, 633–640.

Fremont D. H., Matsumura M., Stura E. A., Peterson P. A. and Wilson I. A. (1992) Crystal structures of two viral peptides in complex with murine MHC class I H-2Kb. *Science* **257**, 919–927.

Gobin S. and van den Elsen P. (2000) Transcriptional regulation of the MHC class Ib genes HLA-E, HLA-F and HLA-G. *Human Immunology* **61**, 1102–1107.

Gobin S., van Zutphen M., Westerheide S., Boss J. and van den Elsen P. (2001) The MHC-specific enhanceosome and its role in MHC class I and β2-microglobulin gene transactivation. *Journal of Immunology* **167**, 5175–5184.

Groh V. (2001) Costimulation of CD8 alpha beta T cells by NKG2D via engagement by MIC induced on virus-infected cells. *Nature Immunology* **2**, 255–260.

Gunther E. and Walter L. (2000) Comparative genomic aspects of rat, mouse and human MHC class I gene regions. *Cytogenetics and Cell Genetics* **91**, 107–112.

Hennecke J. and Wiley D. (2001) T cell receptor–MHC interactions up close. *Cell* **104**, 1–4.

Hughes A. L., Yeager M., Elshof A. E. and Chorney M. J. (1999) A new taxonomy of mammalian MHC class I molecules. *Immunology Today* **20**, 22–26.

Jones E. Y. (1997) MHC class I and class II structures. *Current Opinion in Immunology* **9**, 75–79.

Kast W. M., Bronkhosrt A. M., De Waal L. P. and Melief C. J. M. (1986) Cooperation between cytotoxic and helper T lymphocytes in protection against lethal sendai virus infection. *Journal of Experimental Medicine* **164**, 722–739.

Kern I., Steimle V., Siegrist C.-A. and Mach B. (1995) The two novel MHC class II transactivators RFX5 and CIITA both control expression of HLA-DM genes. *International Immunology* **7**, 1295–1299.

Klein J. (1986) "Natural History of the Major Histocompatibility Complex." Wiley, New York, NY.

Liebson P. J. (1998) Cytotoxic lymphocyte recognition of HLA-E: utilizing a nonclassical window to peer into classical MHC. *Immunity* **9**, 289–294.

Maenaka K. and Jones E. (1999) MHC superfamily structure and the immune system. *Current Opinion in Immunology* **9**, 745–753.

Masternak K., Barras E., Zufferey M., Conrad B., Corthals G. *et al.* (1998) A gene encoding a novel RFX-associated transactivator is mutated in the majority of MHC class II deficiency patients. *Nature Genetics* **20**, 273–277.

Matzinger P. and Zamoyska R. (1982) A beginner's guide to major histocompatibility complex function. *Nature* **297**, 628.

McDevitt H. O. (2000) Discovering the role of the major histocompatibility complex in the immune response. *Annual Review of Immunology* **18**, 1–17.

Mercier B., Mura C. and Ferec C. (1997) Putting a hold on 'HLA-H.' *Nature Genetics* **15**, 234–235.

Meyer C. G., May J. and Schnittger L. (1997) HLA-DP—part of the concert. *Immunology Today* **19**, 58–61.

Mosyak L., Zaller D. M. and Wiley D. C. (1998) The structure of HLA-DM, the peptide exchange catalyst that loads antigen onto class II MHC molecules during antigen presentation. *Immunity* **9**, 377–383.

Parham P. and Ohta T. (1996) Poplulation biology of antigen presentation by MHC class I molecules. *Science* **272**, 67–73.

Parkkila S., Abdul W., Britton R. S., Bacon B. R., Zhou X. Y. *et al.*(1997) Association of the transferrin receptor in human placenta with HFE, the protein defective in hereditary hemochromatosis. *Proceedings of the National Academy of Sciences U.S.A.* **91**, 13198–13202.

Rached L., McDermott M. and Pontarotti P. (1999) The MHC big bang. *Immunological Reviews* **167**, 33–44.

Sercatz E. E. (1998) Immune focusing vs diversification and their connection to immune regulation. *Immunological Reviews* **164**, 5–10.

Silverstein A. (1989) "A History of Immunology." Academic Press, Inc., San Diego, CA.

Takiguchi M., Nishimura I., Hayashi H., Karaki S., Kariyone A. and Kano K. (1989) The structure and expression of genes encoding serologically undetected HLA-C locus antigens. *The Journal of Immunology* **143**, 1372–1378.

Thorsby E. (1999) MHC structure and function. *Transplantation Proceedings* **31**, 713–716.

Ting J. and Trowsdale J. (2002) Genetic control of MHC class II expression. *Cell* **109**, S21–S33.

Trowsdale J. (1993) Genomic structure and function in the MHC. *Trends in Genetics* **9**, 117–122.

Trowsdale J. (2001) Genetic and functional relationships between MHC and NK receptor genes. *Immunity* **15**, 363–374.

van den Elsen P. J., Peijneneburg A., van Eggermond C. J. A. and Gobin S. J. P. (1998) Shared regulatory elements in the promoters of MHC class I and class II genes. *Immunology Today* **19**, 308–312.

van Ham S. M., Tjin E. P. M., Lillemeier B. F., Gruneberg U., van Jeijgaarden K. E. *et al.* (1997) HLA-DO is a negative modulator of HLA–DM-mediated MHC class II peptide loading. *Current Biology* **7**, 950–957.

Waldburger J.-M., Masternak K., Muhlethaler-Mottet A., Villard J., Peretti M. *et al.* (2000) Lessons from the bare lymphocyte syndrome: molecular mechanisms regulating MHC class II expression. *Immunological Reviews* **178**, 148–165.

Wilson I. and Bjorkman P. (1998) Unusual MHC-like molecules: CD1, Fc receptor, the hemochromatosis gene product, and viral homologs. *Current Opinion in Immunology* **10**, 67–73.

Yu C., Yang Z., Blanchong C. and Miller W. (2000) The human and mouse MHC class III region: a parade of 21 genes at the centromeric segment. *Immunology Today* **21**, 320–328.

Zinkernagel R. and Doherty P. (1974) Immunological surveillance against altered self components by sensitised T lymphocytes in lymphocytic choriomeningitis. *Nature* **251**, 547–548.

Zinkernagel R. and Doherty P. (1979) MHC-restricted cytotoxic T cells. *Advances in Immunology* **27**, 51–177.

Zinkernagel R. M. and Doherty P. C. (1997) The discovery of MHC restriction. *Immunology Today* **18**, 14–17.

Antigen Processing and Presentation

Baruj Benacerraf 1920–

Genetic control of immune response
Nobel Prize 1980

© Nobelstiftelsen

CHAPTER 11

A. THE EXOGENOUS OR ENDOCYTIC ANTIGEN PROCESSING PATHWAY

B. THE ENDOGENOUS OR CYTOSOLIC ANTIGEN PROCESSING PATHWAY

C. OTHER PATHWAYS OF ANTIGEN PRESENTATION

"A flowerless room is a soulless room, to my way of thinking;
but even one solitary little vase of a living flower may redeem it"
—Vita Sackville-West

We learned in previous chapters that T cells, unlike B cells, cannot recognize native antigen and only "see" peptide antigens presented as a complex with an MHC molecule. The presented peptides must therefore be displayed on the surfaces of APCs or target cells for recognition by Th cells or Tc cells, respectively. The purpose of this chapter is to discuss how an antigenic peptide is produced, how the MHC molecule combines with it, and how the complex is displayed for T cell perusal. The actual recognition of the peptide–MHC complex by the T cell is discussed in Chapter 12.

Two Types of Antigen Processing

The production of antigenic peptides from protein antigens is called, not surprisingly, *antigen processing*. Antigen processing exists to provide the host with a means of scanning samples of the proteins constantly being produced and turned over in the body. If a nonself protein is detected in the course of this sampling, the immune response kicks into action to counter the threat. Each APC or target cell displays several hundred thousand peptide–MHC complexes on its surface. Experiments in which the displayed peptides have been eluted from the cell surface and characterized have shown that this population is made up of hundreds of distinct peptides, the vast majority of which are "self" in origin. Hidden within the forest of self-peptide–MHC complexes are the relatively rare antigenic peptide–MHC complexes, present at levels as low as 10–100 copies per cell. Nevertheless, so vigilant and sensitive are T cells that this number is usually sufficient to trigger the activation of the antigen-specific T cell. The important point here is that MHC molecules bind self- and nonself-peptides in exactly the same way: the job of discrimination is left to the exquisitely specific T cell compartment.

As was outlined in Chapter 2, the processing of an antigen, and consequently what type of T cell responds to it, depends on the origin of the antigen. Consider those antigens that are synthesized outside the cell, like those derived from extracel-

lular bacteria, toxins, or parasites (Fig. 11-1, left). These microbes or their component macromolecules are engulfed by APCs with cellular internalization capability and enter the *exogenous* or *endocytic* pathway of processing. Within the endosomes, the antigens are degraded into peptides and eventually brought into contact with MHC class II molecules headed to the cell surface from the ER. This contact facilitates the formation and eventual cell surface display of complexes composed of exogenous peptides and MHC class II molecules. Due to the presence of MHC class II in these structures, a T cell must be equipped with the CD4 coreceptor in order to fully respond to the cell presenting the complex. As CD4$^+$ T cells are generally Th cells, the end result is the activation of that subset of T cells able to provide T cell help to antigen-specific B cells. The activated B cells produce and secrete the antibodies necessary to combat the original extracellular threat. Because constitutive MHC class II expression and the specialized mechanisms of antigen internalization are restricted to only a few cell types, the presentation of exogenous antigen to Th cells is efficiently focused onto a relatively small number of cells, increasing the chance that the density of any given epitope will reach the threshold level required for Th activation.

Now let's consider a foreign antigen that has originated in the cytosol of a host cell, such as the product of intracellular replication by a virus or intracellular bacterium. These pathogens and their products cannot be reached by antibody, so the entire corrupted host cell must be destroyed by CTLs to protect the host. In this situation, the antigenic peptides are produced from the intracellular antigens via the *endogenous* or *cytosolic* pathway (Fig. 11-1, right). Antigens of internal origin are degraded in the cytosol and the resulting peptides are sequestered into the ER such that they come into contact with MHC class I molecules. Because MHC class I forms part of the antigenic complex, the T cell recognizing the affected host cell must be equipped with the CD8 coreceptor in order to

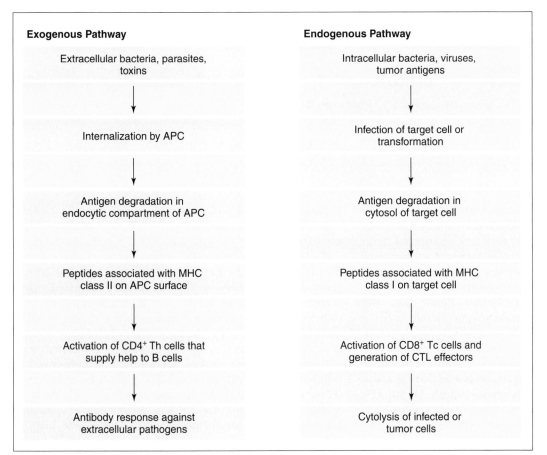

Figure 11-1

Overview of the Exogenous and Endogenous Antigen Processing Pathways Extracellular antigens are captured by APCs and processed via the exogenous pathway. Antigens that arise within a host cell are processed via the endogenous pathway. See text for details.

bind to it and become fully activated. CD8-bearing T cells are generally of the Tc subtype, so that target cells presenting endogenous antigenic peptides on MHC class I usually provoke the generation (often with Th help) of armed CTLs capable of lysing host cells displaying the same peptide–MHC complex. Because any nucleated cell can express MHC class I, the CTLs can monitor all host cells, any one of which may become infected during its life span. In addition, CTLs can eliminate body cells of any type that begin to produce abnormal (perceived as "nonself") proteins. In addition to infected cells, such cells include those that are naturally aging and tumor cells expressing abnormal antigens in the cytosol as a result of host cell oncogenic transformation.

The separate pathways of MHC class I and MHC class II intracellular trafficking allow peptides of two different origins (intracellular or extracellular) to activate two different classes of T cell in a way that correlates with the type of T cell effector function required. Interestingly, the cytosolic and endocytic proteolytic antigen processing pathways appear to be more evolutionarily ancient than MHC function, prompting some observers to suggest that the expression patterns of MHC class I and class II molecules as well as their assembly and peptide loading may have evolved to fit their respective processing paths, rather than the other way around.

We will now discuss each of these pathways in detail.

A. The Exogenous or Endocytic Antigen Processing Pathway

I. HISTORICAL NOTES

At about the time that T and B lymphocytes were identified as distinct entities mediating different types of immune responses, scientists noted that the T cells involved in aiding antibody production seemed to recognize antigen in a different way than the B cells they were helping. Experiments in the late 1950s and early 1960s by Philip G. H. Gell and Baruj Benacerraf were critical in demonstrating this difference. These scientists immunized mice with a protein in its native conformation, eliciting both primary B and T cell responses as expected. They then injected the same mice with either the native protein again or the same protein in a denatured state in which its native conformation was lost. They found that a secondary B cell response occurred only if the native form of antigen was used for both the primary and secondary immunizations, whereas the T cells underwent a proliferative response whether the secondary antigen was native or denatured (Table 11-1). Concurrently, other researchers found that, while antibodies against a native antigen could block the activation of B cells specific for that antigen, such antibodies

Table 11-1 Demonstration of Differences in Antigen Recognition by B Cells and T Cells

Primary Immunization	Primary Response		Secondary Immunization	Secondary Response	
	B cell antibody production	T cell proliferation		B cell antibody production	T cell proliferation
NP	+	+	NP	+	+
			DP	−	+

NP, native protein; DP, denatured protein.

could not block the Th cell response to the antigen. These data strongly suggested that, unlike B cells, which recognized conformational epitopes, T cells somehow recognized linear epitopes present whether the protein antigen was denatured or not. At the time, there was no molecular mechanism envisioned that could explain this observation.

The next piece in the T cell recognition puzzle came with the finding that foreign antigen could activate Th cells only if it was "associated" with host cells bearing MHC class II on their surfaces. In contrast, B cells could bind to and be stimulated by soluble antigen in the absence of such "antigen-presenting cells" (APCs). Subsequent investigations showed that the actual uptake and internalization of exogenous antigen by the APC was essential for Th cell activation. Furthermore, studies using paraformaldehyde to fix or "paralyze" APCs at various times revealed that presentation of the internalized antigen required the continued activity of the APC for a period of about 1 hour after exposure to antigen. More specifically, when APCs were treated with drugs known to block the intracellular degradation of proteins, antigen presentation was inhibited.

The apparent requirement for antigen breakdown or "antigen processing" accounted for the earlier observation that native and denatured antigens were equally stimulatory for Th cells. In addition, if APCs were exposed directly to peptides obtained by digestion of native protein or by direct synthesis *in vitro* from component amino acids, antigen presentation could be achieved even if the APCs were treated with inhibiting drugs or fixatives. Later, it was shown that purified MHC class II molecules that had been inserted into artificial biological membranes or solubilized in detergent solutions could bind isolated peptides. Finally, only MHC class II alleles known to confer on animals "responder" status for a given peptide were found to bind to the peptide in question. The resulting peptide–MHC class II complexes were also shown to stimulate appropriate Th hybridomas in a peptide-specific, MHC-restricted manner. These observations were crucial in establishing that the role of APCs in Th cell activation is to internalize and degrade protein antigen into fragments. The resulting peptides are then combined with MHC class II molecules and displayed at the cell surface for recognition by Th cells.

II. CELLS THAT CAN FUNCTION AS APCs

Only a limited number of cell types have both the cellular internalization properties and the constitutive or inducible expression of MHC class II and costimulatory molecules that allow them to function as professional APCs with the capacity to efficiently activate CD4$^+$ T cells. Professional APCs include dendritic cells, monocytes/macrophages, and B cells (Table 11-2). Exposure of dendritic cells (DCs) and macrophages to IFNγ or TNF increases and sustains their surface expression of MHC class II, while exposure of B cells to IL-4 does the same. These inducible modes of MHC class II expression allow the host to greatly amplify its antigen-presenting capacity when cytokines associated with pathogen attack are present. Several other cell types can be induced to transiently express MHC class II by exposure to IFNγ produced during the course of an inflammatory response, including skin fibroblasts and some epithelial, endothelial, and mesenchymal cells. These cells may be able to act as "non-professional" APCs for a short period of time, but since they do not supply costimulatory molecules to any great extent, their role in T cell activation is limited.

We will now discuss several of the professional APCs before delving into the details of the endocytic processing pathway.

i) Dendritic Cells as APCs

An introduction to the morphology, function, and origin of DCs was given in Chapter 3. Dendritic cells make very efficient use of phagocytosis, macropinocytosis, and receptor-mediated endocytosis to capture antigens and can thus present peptides derived from both particulate and soluble

Table 11-2 Types of Antigen-Presenting Cells

Professional Antigen-Presenting Cells	Non-professional Antigen-Presenting Cells
Interdigitating tissue DCs	Skin fibroblasts
Langerhans cells (epidermal DCs)	Brain glial cells
Thymic DCs	Pancreatic β-islet cells
Monocytes	Thyroid epithelial cells
Macrophages	Vascular endothelial cells (mouse)
B cells	Epithelial cell subsets
Endothelial cell subsets	Mesenchymal cell subsets
Thymic epithelial cells	

Box 11-1. The multi-talented dendritic cells

The more we study dendritic cells, the more we respect their many talents. In this chapter, we focus on the DC as an agent of antigen processing, but it becomes clearer with each passing year that DCs function at the very heart of both adaptive and innate immune responses. In this box, we introduce the many roles of DCs. Detail on these various functions appears in later chapters as indicated.

- Antigen processing and presentation. Immature DCs are exceptionally efficient at capturing exogenous antigens or even whole infected cells derived from the extracellular environment. Only DCs use all three major methods of antigen internalization to a significant degree. In addition to presenting such antigens on MHC class II molecules, DCs can also present some exogenously derived antigens on class I molecules. This latter function is called "cross-presentation" and is discussed later in this chapter. Thus, DCs present antigen on both MHC class I and II to Tc and Th cells, respectively, initiating both humoral and cell-mediated adaptive immune responses.

- Innate immune responses. DCs bear TLRs and complement receptors that trigger their activation in the presence of bacterial and viral products and complement components, meaning that DCs serve an innate immune system function in scanning the body's environment for these molecules. Activated DCs secrete cytokines (such as IFNγ, IL-12, IL-15, and IL-18) that promote the development and activation of both NK cells and the related NKT cells (see Ch.18).

- Lymphocyte activation. DCs are the only APCs with the costimulatory capacity necessary to activate naive T cells. In other words, DCs are the obligatory APCs for primary adaptive immune responses. DCs can also facilitate B cell activation by capturing native antigen and transferring it to the BCRs of B cells. In addition, DCs deliver costimulatory signals and secrete cytokines that support B cell activation and promote differentiation into plasma cells.

- T cell differentiation. Some scientists believe that different subsets of DCs (termed "DC1" and "DC2" in humans) drive T cell differentiation down different paths. In this scenario (explained in detail in Ch.15), engagement of TLRs or CD40 on the surface of DC1 cells induces the production of IL-12 and the IFNs. These cytokines promote differentiation of Th1 cells that function primarily in cell-mediated responses. In contrast, activated DC2 cells produce IL-13, which fosters Th2 differentiation and thus humoral responses.

- T cell tolerance. Presentation of self-peptides in the thymus by thymic DCs is crucial for T cell selection during the establishment of central tolerance. In the periphery, self antigen presented by immature DCs (which lack high levels of costimulatory molecules) can either cause the deletion of antigen-specific peripheral T cell clones or render them nonresponsive. Complete descriptions of central tolerance and peripheral tolerance can be found in Chapters 13 and 16, respectively.

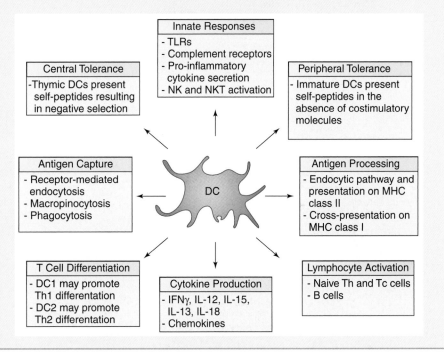

antigens. However, as outlined in Box 11-1, DCs are adept at much more than just antigen presentation. These cells, which were first characterized immunologically by Ralph Steinman in the early 1980s, are now understood to be key players that facilitate T cell activation, differentiation, and tolerance, and also influence innate immune responses.

ia) DC development and subtypes. Multiple subtypes of DCs arising from hematopoietic stem cells have been identified in both mice and humans, defined by marker expression, function, and their location in the body. However, strict parallels between different DC subsets within these species have been difficult to draw. We will therefore concentrate on conveying the current view of mouse DC biology, about which there is more information.

- **Thymic DCs:** *Thymic* DCs are thought to develop from an (unidentified) intrathymic hematopoietic precursor and to remain in this organ throughout their short life span. These DCs most likely participate in the establishment of central tolerance, presenting peptides from self antigens to T cells developing within the thymus. Immature T cells strongly recognizing these pMHCs (self-reactive T cells) are then eliminated.

- **Conventional DCs:** *Conventional* DCs (sometimes called *interdigitating* DCs) are those routinely found in the murine spleen and lymph nodes, and are the APCs that do most of the work in initiating immune responses. Two major subtypes of conventional DCs can be distinguished based on how they gain access to the secondary lymphoid tissues, with "blood-derived" DCs arriving via the blood circulation and "tissue-derived" DCs arriving via lymphatic drainage. Because the spleen has no afferent lymphatic vessels, only blood-derived DCs are found in this location, while both subtypes may be found in lymph nodes, other secondary lymphoid tissues, and the thymus. Both these subtypes can arise from $CD14^+$ monocyte-like cells in the blood or other poorly defined $CD34^+$ bone marrow-derived precursors (Fig. 11-2 and Table 11-3).

There are three subtypes of blood-derived conventional DCs defined by their expression of CD4 and CD8 (although these markers do not appear to have a function on DCs). $CD8^+$ conventional DCs are distinguished by their expression of the $CD8\alpha\alpha$ homodimer (rather than the $CD8\alpha\beta$ coreceptor expressed on T cells) and the lectin CD205. Unlike other DC subsets, $CD8^+$ conventional DCs have only very low expression of the integrin CD11b. Once stimulated, this $CD8^+$ DC subset produces large quantities of IL-12 to drive Th1 differentiation and promote $CD8^+$ T cell proliferation. $CD8^+$ conventional DCs preferentially reside in the T cell-rich areas of lymphoid organs, particularly in splenic PALS. In contrast, $CD4^+$ conventional DCs lack CD8 and CD205 expression but have high surface levels of CD11b. $CD4^+$ conventional DCs are found primarily in the marginal zone of the spleen. The third blood-derived DC subtype expresses high levels of CD11b but not CD4, CD8, or CD205. These $CD4^-8^-$ conventional DCs are found primarily in the lymph nodes. Studies of knockout mice have shown that the transcription factors NF-κB and Ikaros (discussed in Ch.9) are particularly important for the differentiation of various blood-derived conventional DC subsets.

The tissue-derived conventional DCs encompass two major subtypes: *Langerhans cells* (LCs), which are epidermal (skin) DCs, and *interstitial* DCs, which reside in almost all

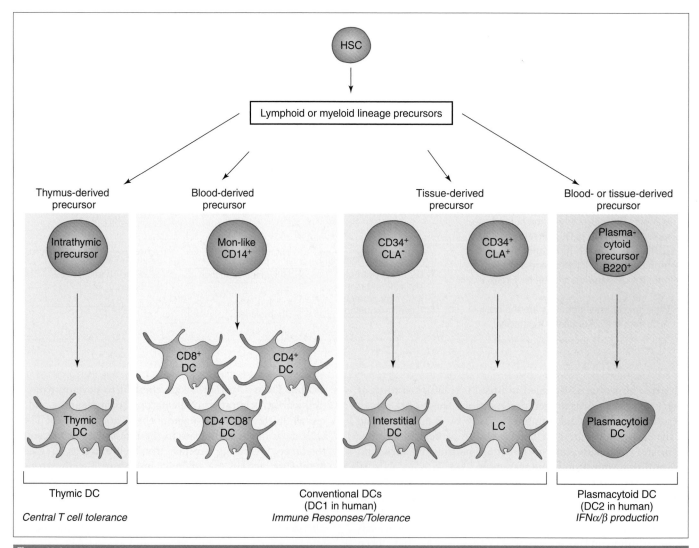

Figure 11-2

Development and Markers of Dendritic Cells in Mice
HSCs in the bone marrow develop into lymphoid or myeloid lineage precursors that eventually give rise to different types of DCs. Mon, monocytes.

Table 11-3 **Murine DC Subtype Characteristics**

DC Subtype	CD4	CD8	CD11b	CD11c	CD205	CD45RA	B220	Birbeck Granules	IFNα/β Production	Tissue Location
CD8⁺	–	+	–	+	+	–	–	–	–	S, LN
CD4⁺	+	–	+	+	–	–	–	–	–	S, LN
CD4⁻CD8⁻	–	–	+	+	–	–	–	–	–	S, LN
Interstitial	–	–	+	+	+	–	–	–	–	NL
LC	–	–/+	+	+	++	–	–	+	–	Skin
Plasmacytoid	+/–	+/–	–	–/+	–	+	+	–	+	B, S, LN

B, blood; LN, lymph nodes; NL, non-lymphoid tissues; S, spleen.

non-lymphoid peripheral tissues. Studies *in vitro* have suggested that there are specific routes of tissue-derived DC production. For example, exposure of hematopoietic bone marrow progenitors *in vitro* to GM-CSF and TNF stimulates the development of a precursor that expresses the marker CD1a, the cutaneous lymphoid antigen (CLA) that directs homing to the skin, and the intercellular adhesion molecule E-cadherin. These precursors subsequently generate LCs as identified under the microscope by the presence of characteristic cytosolic "Birbeck granules." Structurally, these granules are cytoplasmic tubules bound to the plasma membrane; they look striated under the microscope. Birbeck granules contain lectins that may function in endocytosis in the skin.

In humans, other precursors that arise from the treatment of human bone marrow with GM-CSF and TNF develop into interstitial DCs that express CD1a but lack CLA, Birbeck granules, and E-cadherin expression. These cells commence the expression of markers characteristic of DCs resident in tissues other than the skin, and become immature DCs present in peripheral tissues that capture antigen and migrate to the draining lymph node to interact with naive T cells. There is also evidence that at least some interstitial DCs may be derived from the monocyte-like precursor that generates blood-derived DCs.

- **Plasmacytoid DCs:** *Plasmacytoid* DCs are named for the morphology of their precursor. This precursor, which is likely a lymphoid lineage cell, morphologically resembles the B cell-derived plasma cells that produce antibodies. At least *in vitro*, stimulation of this precursor with bacterial products and IL-3 causes it to adopt the usual DC morphology (without proliferating). Plasmacytoid DCs express the B cell marker B220 and contain increased rER (like plasma cells) but do not produce immunoglobulin. Surface markers expressed by these cells include CD4 and/or CD8, IL-3R, and low/absent levels of CD11c and CD11b. Myeloid lineage markers are not expressed. Plasmacytoid DCs are preferentially located in the blood and lymph nodes, but it is not clear that they play a prominent role in immune responses. Instead, upon stimulation by pathogen products, these cells become vigorous producers of the type I interferons IFNα and IFNβ, distinguishing them from other types of DCs.

It should be noted that in humans, conventional blood-derived and tissue-derived interstitial DCs are collectively considered to be DC1 cells, while plasmacytoid DCs are DC2 cells.

ib) Antigen capture by immature DCs. Precursor DCs of various subsets travel through the blood and seed lymphoid and non-lymphoid tissues, developing into immature DCs in these locations. Immature DCs do synthesize MHC class II and some is transported to the cell membrane, but these molecules are quickly internalized and returned to a large intracellular pool. The majority of MHC class II synthesized in immature DCs is directed to the endolysosomes and held there within the endocytic system rather than continuing to the cell surface. As a result, immature DCs have relatively low levels of MHC class II on their cell surfaces, limiting their antigen display and thus their capacity to stimulate Th cells. However, immature DCs are expert at capturing antigen due to their large collections of cell surface receptors that carry out receptor-mediated endocytosis (Fig. 11-3). As well as the TLRs, which generally bind non-protein antigens not presented on MHC, immature DCs express other receptors that facilitate the processing of protein antigens. Two such receptors are FcγRII (a low-affinity IgG receptor) and a ubiquitously expressed mannose receptor (MR). The FcγRII molecule allows the efficient binding of any antigen complexed to IgG (through opsonization), while the mannose receptor, which features multiple extracellular carbohydrate domains, binds to a wide variety of antigens exhibiting exposed mannose residues. There is an interesting difference in the internalization mechanisms of these receptors. An FcγRII molecule that has captured an antigen–IgG complex is internalized into the cell and the receptor protein is degraded in the endosome along with the antigen: only newly synthesized receptors are present on the cell surface and no "recycling" of receptors is possible. In contrast, the mannose receptor is internalized into the cell's early endosomal compartment with its captured antigen but releases the antigen under the influence of the low pH in this organelle. The receptor protein itself returns intact to the plasma membrane to bind fresh antigen. It is probably because of this recycling mechanism that antigens with mannose groups are presented by DCs with an estimated 100-fold greater efficiency than non-glycosylated antigens. DEC-205 (CD205), which is exclusively expressed in DCs and

Antigen Capture by Dendritic Cells
Immature DCs express many different types of cell surface receptors which bind different kinds of antigens, including pathogen macromolecules and apoptotic host cells. See text for details.

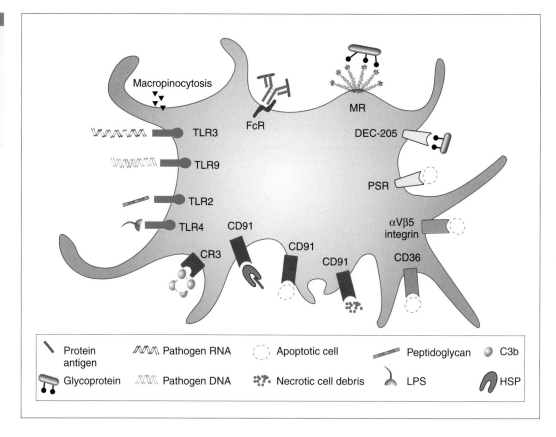

| Protein antigen | Pathogen RNA | Apoptotic cell | Peptidoglycan | C3b |
| Glycoprotein | Pathogen DNA | Necrotic cell debris | LPS | HSP |

thus serves as a marker for these cells, also recognizes glycosylated antigens and undergoes recycling. Additional minor receptors, including FcεRI, FcεRII (CD23), and two novel receptors related to FcαR, may contribute to antibody-complexed antigen internalized by DCs. Immature DCs can also capture antigen by internalizing whole apoptotic cells. The receptors mediating this capture are thought to include αVβ5 integrin protein, the scavenger receptors CD36 and CD91, complement receptor 3 (if the dying cell is coated in complement components), and phosphatidyl serine receptor (PSR). CD91 also mediates the uptake of antigenic proteins released by necrotic cells (see later).

ic) DC migration and maturation. A key property of DCs contributing to their effectiveness as APCs is their ability to migrate, thought to be controlled by constellations of site-specific chemokine receptors and adhesion molecules. For example, many immature DCs express high levels of molecules such as E-cadherin that favor interactions with cells in the non-lymphoid tissues of the periphery; the immature DCs therefore become widely dispersed. From these sites, they can respond rapidly to chemokine gradients and migrate to sites of inflammation or infection, where they collect antigen. The cytokines and bacterial or viral products produced during an inflammatory response (including TNF, IL-1, LPS, viral RNA, and bacterial CpG dinucleotides) influence the immature, antigen-collecting DCs that have gathered in the immediate area to commence maturation.

The precise signal transduction pathways mediating DC maturation remain to be characterized, but engagement of CD40 and the TLRs appears to be involved. For example,

engagement of either TLR3 (which binds double-stranded viral RNA) or TLR9 (which binds to bacterial CpG sequences) promotes DC maturation *in vitro*. Different DC subtypes may express different collections of TLRs: blood-derived conventional DCs tend to express TLR1, 2, 4, 5, and 8, while plasmacytoid DCs express TLR7 and 9. In addition to TLR signaling, intracellular signaling mediated by the MHC class II molecules themselves may make a contribution (Box 11-2). In response to these maturation signals, the cytoskeleton of an immature DC undergoes reorganization. The expression of several of the receptors involved in antigen internalization (such as CD91, αVβ5, and CD36) is downregulated. In contrast, the expression of other molecules (such as certain splice variants of CD44 and chemokine receptors) facilitating the interaction of DCs with cells in the secondary lymphoid tissues is increased. In particular, the expression of the chemokine receptor CCR7 is upregulated while CCR1, CCR5, and CCR6 levels are downregulated. The antigen-loaded DCs then migrate out of the non-lymphoid tissues into the blood or lymph and travel to the secondary lymphoid organs such as the spleen and the local draining lymph nodes. Subpopulations of immature DCs strategically located below the epithelial layer of the intestine capture antigens at particular sites within this zone. The maturing DCs then convey the antigen to the Peyer's patches or mesenteric lymph nodes. Similarly, immature DCs recruited to areas in the hepatic sinusoids can capture antigens moving in the blood through this area and subsequently transport them into the hepatic lymph.

Box 11-2. Intracellular signaling functions of MHC class II molecules

For a long time, MHC class II molecules were thought to function only as antigen-presenting molecules and to be devoid of an independent signaling capacity. This belief was reinforced by the short cytoplasmic tails of the MHC class II α and β chains. Work in the 1990s revealed that the ligation of MHC class II molecules by ligands such as specific mAbs or bacterial superantigens could trigger intracellular signaling in APCs (including monocyte/macrophages, DCs, and B cells) that could affect the maturation, proliferation, cytokine production, or death of these cells, depending on the cell type and the circumstances. Such signaling could then control the direction and duration of the immune response. A natural ligand that seems to bind to MHC class II molecules and produce the same effect as mAb or superantigen binding is a protein called LAG-3 (*lymphocyte activation gene* 3; CD223). LAG-3 is expressed on activated T effector cells and NK cells. *In vitro*, immature DCs stimulated by LAG-3 binding to MHC class II are induced to mature and to express chemokines and chemokine receptors known to draw lymphocytes to the lymph nodes. Naive T cells do not express LAG-3, so that the DC that first interacts with a naive T cell in a lymph node must be induced to mature by other means. However, the LAG-3 expressed on an effector T cell that happens to interact with an immature DC later on in the immune response may promote the maturation of this DC and its migration to the lymph nodes. Such interaction could help amplify the ongoing immune response. In contrast, engagement of MHC class II by LAG-3 can induce activated APCs to undergo cell death, at least *in vitro*. The occurrence of this phenomenon *in vivo* would help to terminate an immune response. The mechanism underlying the cell death induced by MHC class II engagement does not depend on Fas or the caspases, making it an alternative pathway to those instrumental in removing activated T and B cells from the field of battle. MHC class II-mediated cell death has been linked to CD40

signaling since interference with CD40/CD40L interaction increases the susceptibility of an activated APC to this type of death. One could imagine that, as an immune response concludes and less antigen is around to sustain T cell activation and CD40L expression (see Ch.14), CD40/CD40L-mediated signaling within the APC decreases, encouraging the now-redundant APC to succumb to MHC class II/LAG-3-mediated death.

What molecules are activated downstream of MHC class II engagement? The evidence suggests that the MHC class II molecules in activated APCs are located in large multi-protein complexes in the plasma membrane, and that other members of this complex can transduce signals initiated by MHC class II ligation. These signals can lead to death, proliferation, cytokine secretion, or costimulatory molecule expression depending on the specific mediators involved. In activated B cells, the Igα/Igβ heterodimer that facilitates signaling initiated by the binding of antigen to the BCR can also associate with MHC class II molecules. However, instead of supporting activation, this binding promotes B cell death in which the B cell coreceptor components CD19 and CD81 and the B cell marker CD20 have also been implicated. Similarly, in activated monocyte/macrophages, the β2-integrin CD18 associates with certain human MHC class II alleles and supports signaling leading to cell death. In other cases of MHC class II engagement, there is activation of various signaling molecules that we have seen before in reference to BCR signaling (refer to Ch.9) and some that we will encounter in Chapter 14 in our discussions of TCR signaling. Briefly, MHC class II engagement can trigger the tyrosine phosphorylation of various Src kinases (including Lyn) and/or Syk kinase. Activation of these enzymes in turn triggers the eventual activation of several pathways that result in APC cytokine production, maturation, and/or proliferation. Although incompletely defined as yet, these pathways

are thought to involve PLCγ plus two MAPKs prominent in TCR signaling called *ERK* and *p38* (see Ch.14). In contrast, the MHC class II-mediated death of activated APCs results from an activation of the PKC pathway that does not depend on Src activation.

Where in the APC does MHC class II signaling occur? As we shall see in Chapter 14, the interface between the APC presenting pMHC and a responding T cell is a very complex structure that involves numerous adhesion proteins, signaling proteins, and signaling adaptors of the types discussed in Chapter 9. Areas where these molecules are concentrated in the plasma membrane are called *rafts*. Scientists hypothesize that, depending on whether an MHC class II molecule finds itself within or just outside of a raft when it is engaged, it delivers PLCγ- and ERK-mediated signals supporting APC proliferation/maturation, or PKC-mediated signals promoting APC death, respectively.

The multiplicity of MHC class II molecules may also figure in the immune response regulation exerted by APCs. For example, APCs expressing HLA-DR molecules that are bound by ligand show phosphorylation of Src kinases, but those on which HLA-DP is engaged do not. Different HLA molecules also appear to dictate which downstream signaling mediators will be used, and these differences can translate into effects on APC functions. For example, when APCs expressing HLA-DR are stimulated by interaction with a T cell bearing a TCR recognizing the HLA-DR-containing pMHC, both p38 and ERK are activated and the APCs produce IL-1β. In contrast, when APCs expressing HLA-DQ encounter a T cell bearing a TCR that recognizes the HLA-DQ-containing pMHC, only p38 appears to be activated and the APCs secrete IL-10. Although much remains to be determined about the mechanisms involved, it is clear that the effects of MHC class II-mediated signaling on APCs have profound implications for immune response control.

Significantly, contact with an apoptotic cell (whose disintegrating contents remain enclosed within the dying cell) cannot induce DC maturation, but contact with a necrotic cell can. Thus, the type of cell death associated with normal development does not provoke an immune response, but unprogrammed, "messy" death implies injury or pathogen attack and justifies an immune response. In the case of tumors, the transformed

cells contain elevated levels of *heat shock proteins* (see later) whose expression is induced in cells under stress. Once released, these molecules appear to spur the maturation of DCs, enhancing their capacity to initiate immune responses against tumors.

Once settled in their lymphoid locations, DCs continue to mature, losing the capacity to endocytose antigen but gaining the capacity to stimulate Th cells (Table 11-4). Receptor-mediated

Table 11-4 **Comparison of Immature and Mature Dendritic Cells**

	Immature Dendritic Cells	Mature Dendritic Cells
Location	Peripheral tissues	Secondary lymphoid tissues
Surface MHC class II	Low	High
Intracellular MHC class II	High	Low
Endocytosis	Active	Low
Costimulatory molecules	Low	High
CD40, CD25, IL-12	Low	High
Antigen presentation to T cells	Inefficient	Very efficient
Chemokine receptors	High CCR1, CCR5, CCR6; Low CCR7	Low CCR1, CCR5, CCR6; High CCR7
Arrays of actin filaments	Present	Absent

endocytosis is downregulated and MHC class II synthesis is temporarily upregulated. In addition, a decrease in MHC class II internalization and an alteration in the intracellular trafficking of newly synthesized MHC class II molecules result in a 5- to 20-fold net increase in the levels of MHC class II expressed on mature DCs. This increase then allows the rapid presentation of many different antigenic peptides to CD4$^+$ T cells. Note that because of the loss of endocytotic activity induced by maturation, the antigenic peptides presented to a Th cell by a mature DC in the lymph node must have been derived from antigen captured earlier in the periphery by the DC when it was still immature. In other words, the actual surveillance of the non-lymphoid periphery is undertaken by immature DCs acting as outriding scouts, while mature DCs specialize in presenting the collected antigen to T cells in the lymphoid tissues.

Mature DCs are the only APCs able to activate naive T cells; all other APCs must interact with memory or effector T cells. Several features of mature DCs contribute to this talent. First and foremost, mature DCs express high levels of peptide–MHC class II complexes on their cell surfaces. The requisite number of antigenic peptide plus MHC epitopes is easily attained. Secondly, DCs are the only APCs to express a membrane-bound molecule called DC-SIGN (<u>DC</u>-specific, <u>I</u>CAM-3 grabbing <u>n</u>on-integrin), which recognizes mannose residues and is thought to promote initial adhesion to naive T cells. Thirdly, mature DCs have low levels of sialic acid on their surfaces, which may decrease the natural repulsive forces between cells. Mature DCs also express high levels of adhesion molecules such as CD11c that mediate tight binding between the DC and the T cell. Lastly, mature DCs express very high levels (much higher than other APCs) of costimulatory molecules such as B7-2 and CD40. The level of intracellular signaling achieved by the T cell in the presence of elevated concentrations of these molecules is sufficient to lower the activation threshold of naive T cells (see Ch.14). The T cell may reciprocate, stimulating the additional expression of costimulatory molecules on the DC. Once T cell activation is achieved, the DCs have one more role to play. Several lines of evidence suggest that the type of DC the T cell interacts with may influence Th1/Th2

differentiation. For example, in humans, interaction of a CD4$^+$ Th cell with a stimulated, mature DC1 cell preferentially induces the differentiation of Th1 effectors producing IFNγ and IL-2. Conversely, interaction of the Th cell with a stimulated, mature human DC2 cell preferentially induces Th2 differentiation and production of IL-4 and IL-10. Full discussions of T cell activation and Th1/Th2 differentiation appear in Chapters 14 and 15, respectively.

ii) Monocytes/Macrophages as APCs

The cell biology of monocytes and macrophages was discussed in detail in Chapter 3. Macrophages capture mainly particulate and cell-associated antigens using phagocytosis and receptor-mediated endocytosis. Their considerable capacity for phagocytosis enables them to ingest whole bacteria or parasites or other large native antigens and quickly digest them, producing a spectrum of antigenic peptides that are combined with MHC class II and presented to Th cells. However, unlike DCs, which are able to prime naive T cells, macrophages are mainly restricted to acting as APCs for memory or effector T cells. Macrophages are also key players in the amplification of antigen presentation in a localized area. For example, consider the situation in which a macrophage has internalized an antigen, digested it, and presented a peptide bound in the groove of an MHC class II molecule to a specific Th cell such that that Th cell is activated. The activated Th cell produces IFNγ, which not only activates the macrophage so that it becomes a more aggressive phagocyte but also upregulates the expression of MHC class II on additional local macrophages and DCs, and induces MHC class II expression on non-professional APCs in the vicinity. The result is an army of APCs that stimulates many more members of the specific Th cell clone and thus amplifies the adaptive response.

iii) B Cells as APCs

Although best known for their role as antibody secretors, B cells are also considered professional APCs, since this is one of the few cell types that constitutively expresses MHC class II. B cells use their aggregated BCRs to internalize protein antigens by receptor-mediated endocytosis. Within the internalized BCR, the Igα/Igβ moiety becomes phosphorylated on its ITAMs and recruits Syk kinase. Unknown signals delivered by Syk and Igα/Igβ direct the internalized antigen into the endocytic pathway. We know these signals exist because an absence or block of Syk or Igα/Igβ signaling results in profound inhibition of antigen presentation. Antigen confined in the endosomes is processed to peptides, combined with MHC class II molecules, and transported to the cell surface. Peptide–MHC class II complexes appear on the B cell surface 0.5–6 hours after antigen binding to the BCR. The BCRs cannot be recycled, but within 8–24 hours, the B cell replaces the internalized mIg molecules with newly synthesized BCRs ready to bind fresh antigen.

A B cell acting as an APC is one of the most efficient antigen presenters in the body, since the BCR binds specific antigen with high affinity and can thus capture antigens present at much lower concentrations (100–10,000× lower) than are required to engage the lower affinity receptors of other APCs such as macrophages. In this situation, the B cell neatly takes care of all its requirements for activation. A Th cell responding to the peptide–MHC complex presented on the B cell surface is, by definition, close by, and so can rapidly deliver both the costimulatory signals and cytokines necessary for B cell activation.

Several additional aspects of antigen presentation by B cells should be noted. First, although they are effective concentrators of antigen, B cells for any particular epitope are very rare in the unimmunized individual. Thus, in a primary response to antigen X, the chance of a B cell directed against X acting as an APC in close proximity to an equally rare anti-X Th cell is exceedingly small. In addition, resting mature naive B cells express only low levels of the costimulatory molecules required for full Th activation. Thus, B cells do not generally serve as APCs in the primary humoral response. However, in the secondary response, memory B cells against antigen X are present in significantly increased numbers. Memory B cells also express greater numbers of costimulatory molecules than do naive B cells and so are able to promote efficient Th cell activation. Memory B cells therefore quickly predominate as APCs in the secondary response, becoming increasingly prominent upon each subsequent encounter with antigen X. Secondly, unlike macrophages and DCs with their powers of phagocytosis, B cells are unable to internalize ligands that do not bind to their antigen receptors, and so are very limited in the range of antigens that can be captured. However, when the B cell internalizes a whole complex protein antigen that has bound to its BCR, reduces the protein to peptides, and displays them complexed to MHC class II, most, if not all, of the peptides will represent epitopes other than those recognized by the original BCR. This will expand the number of epitopes presented to T cells and increase the number of Th clones activated in the response. Thirdly, whether or not a B cell functions as an APC depends on the antigen captured. Although non-protein, T-independent antigens bind to antigen receptors of B cells and can be internalized, their chemical nature precludes them from undergoing the intracellular processing necessary for association with MHC class II. Consequently, no peptide–MHC class II complex is displayed on the surface of these B cells and no Th cells are stimulated.

III. MECHANISM OF ANTIGEN PROCESSING BY APCs

Having established which cells and receptors are involved in the internalization of an extracellular antigen, we will now follow the fate of such an antigen in the endocytic compartment of an APC. The degradation of the antigen and the combination of its component peptides with MHC class II molecules constitute a fascinating and complex journey.

i) The Invariant Chain

The reader will recall that MHC class II molecules are heterodimers composed of MHC-encoded α and β chains. In professional APCs, these polypeptides are constitutively synthesized on membrane-bound ribosomes and, because of internal signal sequences, are co-translationally inserted into the ER membrane. A third protein called the *invariant chain* (Ii) is coordinately expressed with MHC class II and is also co-translationally inserted into the ER such that its N-terminus remains in the cytosol. It is thought that the function of the Ii molecule is two-fold. First, Ii binds to newly assembled MHC class II molecules and prevents endogenous peptides in the ER from occupying the binding groove, thus "preserving" the MHC class II molecules for the exogenous peptides generated in the endocytic compartment. Secondly, particular amino acid sequences in the cytosolic N-terminus of the Ii chain deflect the transport of the MHC class II–Ii complexes away from the cell's secretory pathway and into the endocytic pathway. Eventually, the acidic environment of the endosome forces the release of Ii from the MHC class II molecule, freeing the MHC class II binding groove for loading with exogenous peptide.

The invariant chain (CD74) is a non-polymorphic 30 kDa protein belonging to the Ig superfamily. In humans, Ii occurs in four different isoforms derived from a single gene by alternative splicing and alternative translation start sites (Fig. 11-4). In mice, two different functional isoforms of Ii have been identified. Ii isoforms appear to have similar functions but are distributed in different ratios in different cell types. All Ii isoforms contain a short N-terminal, di-leucine sequence in the cytoplasmic region, which targets the Ii–MHC class II complex to the endocytic compartment. Two isoforms also contain an N-terminal ER retention signal. The endocytic targeting domain is followed by a short transmembrane (TM) domain. Next comes the CLIP (*class II associated invariant chain peptide*) region, the part of the intact Ii protein that interacts directly with the MHC class II binding groove much like an antigenic peptide (Fig. 11-5A). The C-terminal end of the Ii protein contains a trimerization domain that allows the formation of a nonameric complex consisting of three MHC class II heterodimers (three αβ pairs) and one Ii trimer (three Ii polypeptides) (Fig. 11-5B). Both Ii homotrimers and Ii heterotrimers may be formed using the various Ii isoform chains. Two of the human Ii isoforms contain an extra exon just after the trimerization domain, which encodes a cysteine protease inhibitor peptide (refer to Fig. 11-4). It is speculated that this protease inhibitor function may control the later steps of Ii degradation.

Let's follow the progress of an individual MHC class II molecule from the ribosome to the cell surface, starting at the transcription stage. MHC class II gene expression in the professional

Figure 11-4

Structural Isoforms of Human Ii
Human Ii occurs in four isoforms that are co-expressed in APCs. The functional domains of each isoform are indicated. The Ii trimers that interact with MHC class II αβ heterodimers in the endocytic pathway can be composed of any combination of the Ii isoforms. ER, endoplasmic reticulum; TM, transmembrane domain; CLIP, class II associated invariant peptide. With information from Pieters J. (1997) MHC class II restricted presentation. *Current Opinion in Immunology* **9**, 89–96.

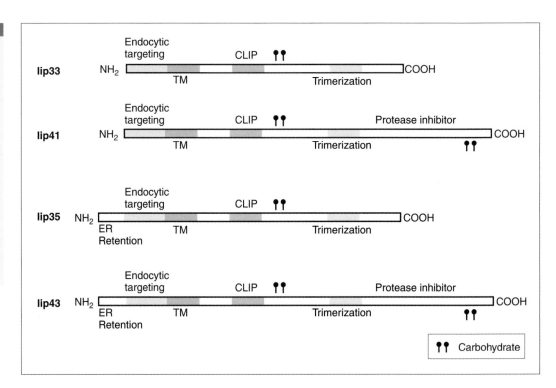

APCs is regulated by the CIITA protein (*class II transactivator protein*) described in Chapter 10. CIITA also controls and coordinates the expression of the Ii gene and that of the HLA-DM/H-2M genes involved in peptide loading (see later). These transcripts are all translated by ER-bound ribosomes. While still within the ER, the MHC class II α and β chains assemble into properly folded heterodimers that associate with an Ii trimer to form the nonameric MHC class II–Ii complex (Fig. 11-6). Within this complex, the CLIP region of each Ii chain is nestled in the MHC binding groove of each MHC class II αβ heterodimer. (In fact, a small number of MHC class II alleles do not bind CLIP, suggesting that other regions of Ii may subserve the "filler" function in some cases.) Once assembled, the MHC class II–Ii nonamer complex exits the ER, enters the secretory pathway, and passes through to the *trans-Golgi network* (TGN). However, the N-terminal targeting signals of the Ii polypeptides divert the complexes from the TGN into the endocytic pathway, where they are sequestered in the endolysosomes. Although the precise molecular signals underlying this targeting have yet to be determined, trimerization of Ii has been found to be critical for correct endolysosomal localization of MHC class II. Like lysosomes, the endolysosomes are home to many proteases and other degradative molecules but, unlike lysosomes, degradation is restricted such that MHC class II–Ii complexes and many antigens can remain intact. The MHC class II–Ii nonamers are then cleaved by members of the cathepsin protease family to yield individual MHC class II molecules bearing CLIP in the peptide-binding groove. The MHC–CLIP complexes then pass from the endolysosomes into specialized late endosomal compartments known as MIICs (*MHC class II compartments; also called CIIVs, class II vesicles*). MIICs are distinguished from most other endolysosomal compartments by the presence of the LAMP-1 protein (*lysosome-associated membrane protein 1*). It is in the MIICs that most peptide loading onto MHC class II occurs, although some peptides appear to be loaded onto MHC class II in the endolysosomes.

ii) Generation of Exogenous Peptides

Let's now consider the other half of the equation, the exogenous antigenic peptide. To process proteins that are captured from the exterior by cellular internalization, proteins that have been recycled from the cell surface, and proteins that have been diverted for some reason from the secretory path into the endosomes, the cell uses hydrolytic proteases and peptidases contained in the membrane-bound compartments of the endocytic pathway. To illustrate, let us suppose that an APC has internalized a whole microbe or macromolecule by phagocytosis, receptor-mediated endocytosis, or macropinocytosis. The antigen is transported across the cell in a membrane-bound vesicle and enters the endocytic pathway (Fig. 11-7). As described in Chapter 4, the transport vesicle is successively fused to a series of endocytic vesicles of increasingly acidic pH and containing a collection of acidophilic proteolytic enzymes. (That acidic pH is important for antigen processing is demonstrated by the fact that chemicals that raise the pH of endolysosomes inhibit the appearance of peptide-loaded MHC class II on the APC surface.) The inter- and intramolecular disulfide bonds that hold many native antigens together must then be broken in order for the protein to "unfold" and become accessible to endolysosomal hydrolases. The reduction of these bonds is thought to be carried out by GILT (*IFNγ-inducible lysosomal thiol reductase*), an endolysosomal enzyme that retains its activity at very low pH. Once the disulfide bonds are reduced by GILT, the antigen undergoes an initial single cleavage by an endopeptidase to completely "unlock" its conformation.

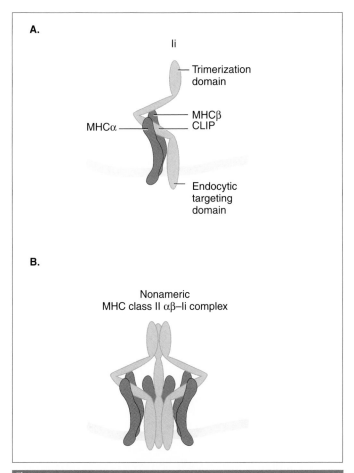

A.

Ii

Trimerization
domain

MHCβ
CLIP

MHCα

Endocytic
targeting
domain

B.

Nonameric
MHC class II αβ–Ii complex

Figure 11-5

Interaction of MHC Class II and Invariant Chain
(A) Model of an MHC class II αβ heterodimer interacting with an Ii
monomer such that the CLIP domain of the Ii chain is positioned in the
peptide-binding groove of the MHC class II molecule. (B) Model of the
nonameric complex formed by three MHC class II molecules (three α and
three β chains) and three Ii chains in the ER. The Ii chains are held in place
by interactions between their trimerization domains. Each MHC class II
binding groove contains an Ii CLIP region. Adapted from Pieters J. (1997)
MHC class II restricted presentation. *Current Opinion in Immunology* **9**,
89–96.

Numerous endolysosomal proteases then attack the unlocked
protein and progressively degrade it into peptides of 10–30
amino acids that can be detected about 1–3 hours after antigen
internalization.

The exact intracellular site where a given antigenic protein
is broken down into peptides partially depends on the nature
of the antigen and its mode of internalization. Some antigens
are processed most efficiently in the less acidic early endosomes
whereas others are better degraded in the later, more acidic
endolysosomes. Cytokines in the immediate microenvironment can regulate the pH of an endolysosomal compartment
and the activity of its hydrolases. The precise method of antigen
capture (mannose receptor, Fc receptor, macropinocytosis, B cell
receptor, phagocytosis, etc.) may also influence the location
and progress of antigenic peptide production. Why the peptides
are not completely degraded into individual amino acids is
not known. It may be purely a matter of timing, in that

peptides of the correct length that happen to be loaded onto
an MHC class II molecule may escape further degradation,
and thus may be those that end up displayed on the cell surface. Alternatively, certain proteins may be intrinsically more
resistant than others to total endosomal degradation, such
that the presentation of peptides from these particular proteins is
favored. This hypothesis is supported by studies of the immunodominant antigens of some common allergens. For example,
antigens in some foods to which many people are allergic (like
peanuts and shrimp) are highly resistant to degradation by
proteases, chemicals, or heat. Similarly, the principal antigens
responsible for allergy to the common house dust mite are the
organism's digestive enzymes, proteins that one would expect to
be resistant to endosomal degradation because of their normal
function in the harsh environment of the gut.

How do the antigenic peptides meet up with MHC class II
molecules in the MIICs within the APC? This question is as
yet largely unresolved. The majority of peptides appear to be
transported by vesicle fusion between the endolysosome and
the MIIC. A few antigens are thought to be degraded directly
in the MIICs. In still other cases, it may be that whole protein
antigens are conveyed to the MIICs where they later bind in
intact form to monomeric MHC class II molecules (see later).
Exopeptidases then trim off overhanging polypeptide pieces
not buried in the peptide-binding groove. Details aside, sequestration within a membrane-bound compartment prior to loading onto an MHC class II molecule is a fundamental element
of exogenous antigen processing.

iii) CLIP Exchange and Peptide Loading

iiia) Ii digestion. Let's return to a nonamer MHC class II–Ii
complex in the endolysosome of an APC that has just been
activated (refer to Fig. 11-6). In order to bind antigenic
peptide, the MHC molecule must be freed from the Ii. The acidophilic proteases *cathepsin S* and *cathepsin L*, which reside in
endolysosomes, progressively degrade the Ii chains into smaller
and smaller fragments. Cathepsin S, which is highly expressed
in DCs and B cells and is inducible by IFNγ, is particularly
crucial for Ii digestion in B cells, some macrophages, and DCs.
Cathepsin L performs the same function in other macrophage
populations and in the thymic epithelial cells that serve as
APCs in the thymus. Several other cathepsin enzymes have
been identified in APCs but their function in antigen processing remains controversial. Cathepsins can be induced and activated by IFNγ and operate in both early and later endosomes
over a wide range of internal pHs. However, these enzymes
stop short of completely digesting the Ii chains: a peptide
encompassing the CLIP region is left in the binding groove of
each of the three original MHC class II heterodimers to prevent premature peptide loading.

Several models have been advanced to explain the dynamics
of MHC class II loading and pMHC expression on the DC
surface. In one mouse model, it has been proposed that cathepsin S activity is controlled in DCs by an intracellular inhibitor
called *cystatin C*. Cystatin C levels are higher in immature
DCs than in mature DCs. It is hypothesized that the increased
cystatin C in immature DCs inhibits cathepsin S so that it does
not degrade Ii bound to MHC class II. The Ii targeting signals

MHC Class II Antigen Presentation Pathway
As shown in the lower left corner of the figure, mRNAs encoding MHC class II α and β chains and the Ii chains are translated on ER-bound ribosomes. Nonameric complexes are formed in which the peptide-binding groove of each MHC class II heterodimer is blocked by the CLIP region of an Ii chain. Due to signaling motifs in the Ii chain, the complexes exit the ER, pass through the trans-Golgi network (TGN), and enter the endolysosomes. When cells or macromolecules are internalized by the APC (upper left of the figure), they enter the endocytic pathway where proteins are eventually degraded to peptides by endolysosomal proteases (see Figure 11-7 for details). These peptides are transferred by vesicle fusion into a MIIC (MHC class II compartment). Meanwhile, back in the endolysosome, the Ii in the nonamers is partially degraded by cathepsin proteases, leaving the CLIP peptide in the MHC class II binding groove. The MHC–CLIP complexes are transferred to the MIIC, where the exchange of CLIP for peptide can be made. The peptide-loaded MHC class II molecule then makes its way to the APC surface.

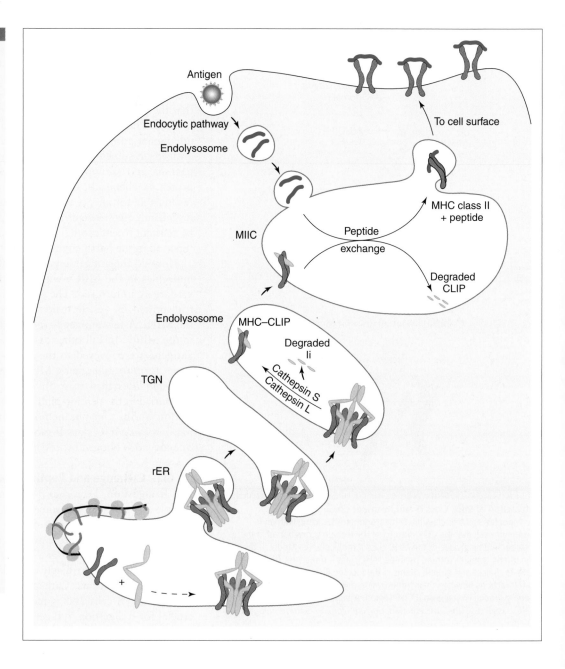

remain intact and the MHC class II–Ii complex is retained in the endolysosome. In this model, as DCs mature, levels of cystatin C drop, cathepsin S cleaves Ii more efficiently, the endolysosomal targeting sequences are lost, and MHC class II is freed to move to the MIICs, accept peptide, and proceed to the cell surface as a pMHC complex. Interestingly, a certain parasitic worm produces a molecule homologous to cystatin C that blocks MHC class II migration to an APC surface. The immune response against this pathogen is thereby crippled.

A different model developed in human DCs suggests that cytokines, rather than a specific inhibitor such as cystatin C, play a major role in controlling cathepsin S activity and thus pMHC expression. Inflammatory cytokines (such IL-1 and TNF and others that would be encountered by an immature DC in the tissues in the course of infection or injury) rapidly upregulate cathepsin S activity in isolated human DCs via a mecha-

nism that is likely independent of transcription. Conversely, the anti-inflammatory cytokine IL-10 blocks this upregulation. The cytokine milieu in which a DC finds itself may thus determine its cathepsin S activity and therefore its level of pMHC expression. (It remains possible that the DC loading mechanism may be slightly different in mice and humans.)

An alternative model concerning the regulation of MHC class II peptide loading and pMHC expression at the DC surface was developed in mice and argues that these processes are independent of cathepsin S-mediated degradation of Ii. Instead, this theory holds that the level of pMHC on the DC surface depends on the ability of the DC to recycle MHC class II from the cell surface, which in turn depends on the state of maturity of the DC. That is, immature DCs actively recycle MHC class II molecules, so that the majority of pMHC complexes assembled appear only transiently on the DC surface and are rapidly

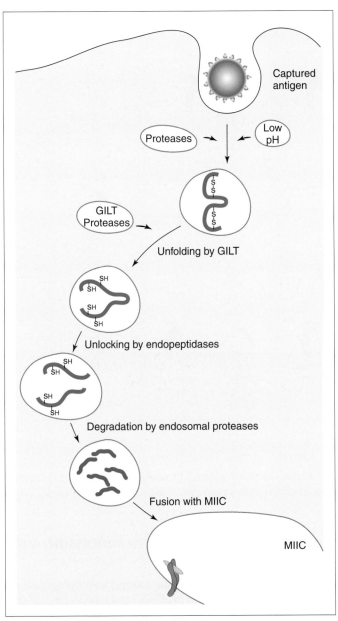

Figure 11-7

Generation of Exogenous Peptides
A pathogen is internalized by an APC and enters the endocytic pathway. Digestion in successively more acidic endosomal compartments produces a collection of pathogen macromolecules. Macromolecular proteins are "unfolded" within endolysosomes by GILT-mediated reduction of intramolecular disulfide bonds and "unlocked" by endopeptidase cleavage. Degradation to peptides by endolysosomal proteases can then proceed. Peptide exchange with CLIP occurs in MIICs (see Fig. 11-6). GILT, IFN-γ-inducible lysosomal thiol reductase.

returned to its interior. In contrast, mature DCs have a dramatically lower recycling rate such that the pMHCs accumulate in large numbers on the DC surface, ready for antigen presentation to CD4$^+$ T cells. Aspects of all three models may contribute to the physiological truth.

iiib) HLA-DM/H-2M. The next step in endocytic antigen processing is the exchange of the CLIP peptide in the MHC binding groove for the exogenous peptide. Originally it was thought that proper MHC class II maturation and Ii degradation were sufficient for antigen presentation, but then researchers discovered the importance of the non-classical HLA-DM/H-2M genes introduced in Chapter 10. Human cell lines deficient for HLA-DM exhibited normal MHC class II maturation and Ii degradation, but defective presentation of whole protein antigens supplied exogenously. The MHC class II molecules on the surfaces of these mutant cells contained only CLIP peptides and no antigenic peptides, suggesting a defect in CLIP exchange. These results were later confirmed in H-2M-deficient knockout mice.

The mechanism by which the HLA-DM/H-2M proteins catalyze CLIP exchange is still unclear. The reader will recall that the HLA-DM and H-2M molecules are MHC class IIb heterodimers closely resembling conventional MHC class II molecules. However, HLA-DM/H-2M proteins do not bind peptides. Rather, scientists speculate that HLA-DM/H-2M associates with the MHC class II–CLIP complex and causes a conformational change that breaks hydrogen bonds and promotes the release of CLIP (Fig. 11-8). HLA-DM/H-2M also has a stabilization function since it remains with the MHC class II heterodimer after CLIP release until peptide is loaded. In the low pH of the endosomes, the temporarily empty MHC class II molecule would denature in the absence of HLA-DM/H-2M. The actual mechanism of peptide loading onto the MHC class II molecule has yet to be delineated. We do know that when a peptide finally does bind stably in the MHC class II groove, the conformation of the MHC class II molecule alters again to force dissociation of the HLA-DM/H-2M molecule. The MHC class II–peptide complex is then transported out of the MIICs by reverse vesicle fusion, inserted into the APC membrane, and displayed to T cells.

Let's spend a moment more on the functions of HLA-DM/H-2M. This molecule has an important role as a "presentation editor" because, as well as stimulating the release of CLIP, it promotes the dissociation of peptides that fail to bind stably to MHC class II and replaces them with peptides that bind with higher stability. HLA-DM/H-2M thus helps to shape the T cell repertoire during T cell development in the thymus (see Ch.13), biasing it in favor of peptides that bind strongly to MHC class II and succeed in being displayed on the cell surface. There is also evidence that HLA-DM/H-2M may be able to carry out this editing function on MHC class II molecules both in recycling compartments and on the surfaces of B cells and DCs. About 10% of the total HLA-DM/H-2M of these cell types is located on the cell surface rather than in the endocytic system. The presence of HLA-DM/H-2M in this location allows lower stability peptides already present in the MHC class II groove to be replaced *in situ* with extracellularly generated, higher stability peptides that have approached the APC surface. Interestingly, HLA-DM has been found to block the surface loading onto MHC class II of a peptide derived from an autoantigen. In this way, HLA-DM may inhibit the activation of T cells recognizing this peptide and thus help to prevent autoimmunity by maintaining peripheral tolerance (see Ch.16).

iiic) HLA-DO/H-2O. In human and murine B cells and thymic epithelial cells, peptide loading is further modulated by a negative regulator of the HLA-DM/H-2M protein called

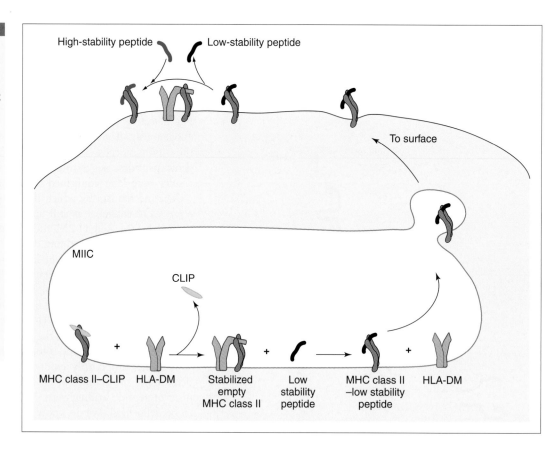

Figure 11-8

Role of HLA-DM/H-2M in CLIP Exchange
Within the MIIC, HLA-DM (H-2M in the mouse) binds to the MHC class II–CLIP complex (lower left) and forces expulsion of the CLIP peptide. HLA-DM then stabilizes the empty MHC class II molecule until it is loaded with exogenous peptide. Conformational changes in the peptide-loaded MHC class II molecule result in the release of HLA-DM and the transport of peptide–MHC class II to the APC surface. While still in the ER, MHC class II loaded with a low-stability peptide can undergo HLA-DM-mediated exchange for another peptide of higher stability. HLA-DM expressed on the APC surface can facilitate the same type of exchange in this location.

HLA-DO/H-2O. The HLA-DO/H-2O molecule is another MHC class IIb protein that does not bind peptides but rather associates with HLA-DM/H-2M in the ER. Indeed, it has been shown that HLA-DO cannot exit from the ER in the absence of HLA-DM. These two proteins are thought to form a complex in which HLA-DO inhibits the HLA-DM-catalyzed release of CLIP from the MHC class II binding groove. Overexpression of transfected HLA-DO increases the number of MHC class II molecules on the cell surface that are still loaded with CLIP rather than peptide, and antigen presentation to T cells is significantly impaired. However, other experiments have shown that, depending on the pH of the endosomal compartment in which peptide loading occurs, HLA-DO/H-2O may actually enhance the loading of some peptides. These contradictory data remain under investigation.

iiid) Minor pathways of MHC class II peptide loading. Antigen processing and the appearance of peptide–MHC class II on the cell surface is very rapid (within hours) in cells that express cathepsin S (or L). In isolated non-professional APCs, which express little or no cathepsin S (or L), there appears to be an alternative path of Ii degradation that allows delayed display of peptide-loaded MHC class II on the cell surface. In this situation, the MHC class II–Ii complexes remain in the endolysosomes, where Ii is very slowly digested, CLIP is released, and the peptides generated in the endolysosome are loaded onto MHC class II under the influence of HLA-DM. The peptide–MHC complex can then slowly migrate through the endocytic pathway to the cell surface without passing through MIICs. The physiological impact of this pathway *in vivo* remains to be determined.

IV. FACTORS AFFECTING ANTIGEN PROCESSING AND PRESENTATION BY APCs

It should be emphasized that many factors can influence antigen processing and presentation by APCs (and target cells, for that matter), and so it is impossible to predict the level of presentation for any single peptide determinant. Several factors have been identified for APCs in *in vitro* experiments in which T cells are stimulated by various APC/peptide/MHC combinations. Different APCs differ in their efficiency or route of antigen uptake (such as macropinocytosis versus phagocytosis); their levels of expression of a particular MHC class II allele, Ii, or HLA-DM; the trafficking of Ii, HLA-DM, and MHC molecules; the localization or nature of the proteases of the intracellular degradative machinery; the ability of these enzymes to cleave intraprotein bonds; and the characteristics of peptide loading, including access to MIICs. Such variation can result in dramatic differences in the presentation of a given antigen by different APCs to the same T cells. For example, the presentation of ovalbumin to T cell hybridomas was found to vary widely among cultured cell lines, freshly isolated splenocytes, purified splenic B cells, and peritoneal macrophages used as APCs. APCs of different lineages may also differ in their responses to environmental stimuli such as cytokines, viral or bacterial infections, or other cellular stressors that induce heat shock protein production. APCs of the same lineage may differ in localization, developmental stage, and state of activation. Lastly, aside from factors affecting the efficiency of generating specific peptide–MHC complexes, some APCs may be more efficient costimulators of T cells than others. For example, the B7-1 and B7-2 molecules, ligands for the costimulatory molecule CD28 present on T cells, are differentially expressed on different types of APCs and with different kinetics. *In vivo*, all of these factors can affect the sets of peptides generated and presented by different APCs or even APCs of the same lineage. This diversity may help to increase the range of peptides produced by the collective APCs of a host, maximizing the number of T cell clones activated and amplifying the intensity of the ensuing immune response.

B. The Endogenous or Cytosolic Antigen Processing Pathway

I. HISTORICAL NOTES

Once it was established that Tc cells recognized virus-infected host cells in an MHC class I-restricted manner (see Ch.10), interest in this field of study focused on understanding the interaction of viral antigens with MHC class I molecules. Initially, the form of antigen recognized by an MHC class I-restricted Tc cell appeared obvious if one considered the molecular nature of the antigens displayed on the corresponding target cells. It was already known that the antigens recognized by alloreactive T cells were allogeneic MHC class I molecules, which are cell surface transmembrane proteins. It was also known that most virus-infected cells expressed virus-derived transmembrane proteins on the host cell surface. A correlation was therefore made between expression of foreign membrane proteins and the ability to be recognized by antigen-specific Tc cells. Researchers then assumed that a transmembrane protein of nonself origin (antigen X) interacted with an MHC class I molecule on the target cell surface to form an associative (MHC class I + native antigen X) structure recognized by MHC class I-restricted Tc cells. At this point, the intermolecular interactions required by this model were far from being clearly understood.

As researchers carried out experiments to confirm and expand this "membrane antigen recognition" model, inconsistencies began to mount. For example, antibodies specific for the membrane proteins of various viruses were not able to block the activity of virus-specific CTLs, as would have been expected if the CTL were recognizing an MHC molecule co-localized on the cell surface with a viral membrane protein in its native form. This observation was difficult to reconcile with the CTL blocking that was readily achieved when various MHC class I-specific antibodies were used. Further futile attempts were made, using antiviral protein and anti-MHC antibodies tagged with fluors or radioisotopes, to show that transmembrane proteins of viral origin could be found in close proximity to MHC molecules on the cell surface. Another inconsistency arose when it was discovered that a large proportion of CTLs specific for particular viruses actually seemed to recognize internal proteins of these viruses; that is, viral proteins that were never expressed on the cell surface, such as the nucleoprotein (NP) of influenza virus or the large T antigen of SV40 virus. Soon after, CTL clones that clearly recognized internal viral proteins were successfully isolated using T cell cloning techniques. However, the question remained: how were these internal viral proteins arriving at the cell surface in a form recognizable by MHC class I-restricted CTLs?

In retrospect, it seems obvious to conclude that, like MHC class II molecules, MHC class I molecules were presenting processed pieces of proteins. However, at the time, the immunological community was not quite ready for the idea that Tc recognition was very similar to Th recognition. It took a landmark series of experiments by Alain Townsend and his colleagues to eventually win acceptance of this basic concept. Townsend's group showed that CTLs specific for the internal influenza protein NP could efficiently recognize uninfected cells of the appropriate haplotype if these cells were transfected with the NP gene. Furthermore, the expression of the entire NP molecule was not even necessary for recognition: the transfection of a small section of the NP gene corresponding to amino acids 365–380 was just as effective as the whole protein at making the target cell recognizable to the NP-specific CTL. Similar results were reported in the SV40 virus system in that cells expressing fragments of the internal large T antigen protein were efficiently lysed by SV40-specific CTLs. It was then discovered that even viral proteins normally expressed on the cell surface, such as influenza hemagglutinin (HA), did not actually need to be on the cell surface to generate target cells recognized by virus-specific CTL. Cells transfected with a version of the HA gene from which the leader sequence had been omitted constituted targets that could still be recognized by HA-specific CTLs. At this point, the immunological community embraced the indisputable evidence that Tc cells recognize peptide fragments that are derived from endogenously synthesized protein antigens and then presented by MHC class I molecules. The field of T cell antigen recognition thus became more streamlined, as the basic molecular mechanisms involved in both MHC class I- and MHC class II-restricted recognition were finally understood to have antigen processing and MHC association at their core.

II. CELLS THAT CAN FUNCTION AS TARGET CELLS

Unlike the limited number of cell types that express MHC class II and so can function as professional APCs, almost any nucleated body cell expresses MHC class I and so can present intracellularly derived peptide antigens to CD8[+] T cells. Thus, almost any body cell that has become aberrant due to old age, cancer, or infection by intracellularly replicating pathogens can alert the immune system that it is in trouble. The affected cells are lost to CTL cytolysis but the host organism is saved.

III. MECHANISM OF ANTIGEN PROCESSING BY TARGET CELLS

The processing and presentation of antigen by target cells is a multifaceted process about which much is still unclear. We have attempted to dissect the system into its known components, indicating where gaps exist in our knowledge.

i) MHC Class I α Chain and Chaperones

The α chain of an MHC class I molecule is synthesized on a membrane-bound ribosome and is co-translationally inserted into the membrane of the ER (Fig. 11-9). As it enters the ER membrane, the MHC class I α chain associates with an 88 kDa "chaperone" protein called *calnexin*. Calnexin is a trans-ER membrane protein that binds to glucosyl groups on the MHC class I α chain and facilitates polypeptide folding and intra-chain disulfide bond formation. Calnexin also mediates the association of the coordinately expressed β2m chain with the MHC class I α chain in the ER. In humans, the calnexin is released from the MHC heterodimer at this point and is replaced by a second chaperone protein, *calreticulin*, a soluble homologue of calnexin. In mice, calnexin remains associated with the heterodimer after association with β2m, perhaps until peptide binding. In addition to these chaperones, the MHC class I molecule is bound to a thiol-dependent oxidoreductase enzyme called *ERp57* (*endoplasmic reticulum protein 57*). ERp57 catalyzes the making and breaking of disulfide bonds. In both mice and humans, the chaperone proteins retain the MHC heterodimer in the ER until its groove is loaded with endogenous peptide. To put it another way, the MHC class I heterodimer cannot be transported to the cell surface unless it is loaded with peptide. Studies of mutant cells with defects in peptide loading have shown that the peptide is crucial for stabilizing the MHC class I heterodimer at this stage: without the bound peptide, the MHC class I heterodimer eventually falls apart, the MHC class I α chain and the β2m chain accumulate as separate entities in the cytoplasm, and no MHC class I can be detected on the target cell membrane.

ii) Generation of Endogenous Peptides: The Proteasomes

iia) The core proteasome. How are endogenous antigenic peptides generated? Viruses or intracellular bacteria that have taken over the host cell force it to use its own protein synthesis machinery to make viral or bacterial proteins. Antigenic proteins are thus derived from the translation of viral or bacterial mRNA on the cytosolic ribosomes in the host cell. Similarly, host-derived tumor antigens synthesized in the cytosol can give rise to proteins that may stimulate an immune response. In order for a Tc response to these antigens to be possible, they must be degraded in the cytosol to peptides of a size that can fit into the MHC class I peptide-binding grooves, and must be transported from the cytosol into the lumen of the ER where the newly assembled MHC class I molecule is waiting. The process used to generate endogenous antigenic peptides is basically the same as that used to deal with misfolded, spent, or damaged host proteins. In the cytosol of every host cell are abundant (up to 1% of total cellular protein), non-lysosomal organelles called *proteasomes*. A proteasome is a huge multi-subunit protease complex of about 700 kDa that possesses multiple catalytic activities, meaning that it can digest a protein at three or four different cleavage sites defined by short conserved amino acid sequences known as *consensus peptide motifs*. The essential role that proteasomes play in MHC class I-mediated antigen presentation has been demonstrated in specific inhibitor experiments. When agents abolishing proteasomal proteolytic activity are introduced into the cytoplasm of target cells, both surface MHC class I expression and the capacity to present antigen are reduced.

There are two types of proteasome: the *standard proteasome* and the *immunoproteasome* (Fig. 11-10). Both types of proteasome have at their core a structure called the *20S core proteasome*, a hollow cylinder of four stacked polypeptide rings surrounding a central channel of 10–20 Å in diameter. The top and bottom rings are composed of seven α-type subunits each while the inner two rings are composed of seven β-type subunits each. The α and β subunits are structurally similar within their groups, but all are distinct gene

Human MHC Class I Chaperones
MHC class I α chains are synthesized on ER-bound ribosomes and are bound in succession by the chaperone proteins calnexin, calreticulin, and ERp57. A complex composed of calreticulin, a complete MHC class I heterodimer, and ERp57 participates in peptide loading.

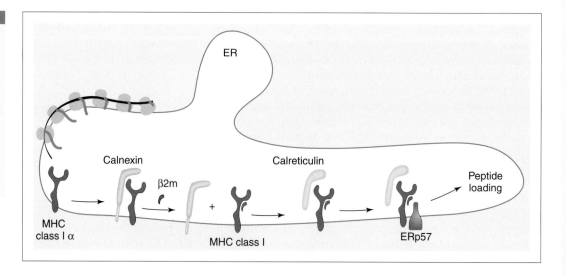

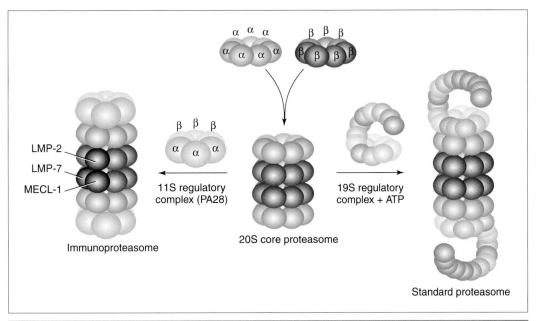

Figure 11-10

Proteasome Structures
The 20S core proteasome contains stacked rings of non-identical α and β subunits. The standard (26S) proteasome is composed of the core proteasome plus two copies of the 19S regulatory complex. The immunoproteasome is composed of the core proteasome plus two copies of the 11S (PA28) regulatory complex. The PA28 regulator contains PA28 α and PA28 β subunits, which are distinct from the core α and β subunits. LMP-2, LMP-7, and MEC-1 are catalytic subunits specific to the immunoproteasome that replace three catalytic β subunits in the 20S core. Adapted from Rivett A. J. (1998) Intracellular distribution of proteasomes. *Current Opinion in Immunology* **10**, 110–114.

products with different amino acid sequences. The α subunits maintain the conformation of the proteasome core, while three of the β subunits (β1, β2, and β5) are its catalytically active components. β1 hydrolyzes peptidylglutamylpeptide bonds, β2 has trypsin-like activity, and β5 is chymotrypsin-like. The β subunits are structured such that the enzymic active sites are positioned on the inner surface of the cylinder. This arrangement ensures that proteins must actually enter the proteasome to be degraded and decreases the chance of accidental degradation of a nearby useful cellular protein.

iib) The standard proteasome. In general, spent and unwanted self proteins are degraded by the standard proteasome (sometimes called the *26S proteasome*) present in all cells. The standard proteasome is composed of the 20S core proteasome plus two copies of a structure called the *19S regulatory complex* (refer to Fig. 11-10). The 19S regulatory complex contains about 20 different proteins, including six ATPase molecules, proteins involved in ubiquitin recognition (see later), and enzymes involved in protein unfolding and translocation. The peptides produced by the standard proteasome range in length from 3–22 residues, although the average is 13–18 residues. About 20% are the size (8–10 amino acids) that fits neatly into an MHC class I binding groove. The degradation of endogenous proteins in this way and subsequent loading of these peptides onto MHC class I allows routine scanning of self-peptides and the monitoring of the cell's internal health.

How does a cell know when proteins in its cytosol should be degraded to peptides by proteasomes? Cells are equipped with a developmental blueprint specifying the half-life of any given self protein. Proteins that are needed for ongoing housekeeping functions (like metabolic enzymes) will have longer half-lives than proteins induced in response to an acute signal (like transcription factors, oncogenes, and cell cycle regulators). The latter types of proteins might go on to harm the cell or organism if not reined in shortly after the stimulus has abated. Several pathways for identifying proteins that have outlived their usefulness exist in all cells, and these pathways also process viral, bacterial, and tumor proteins.

One of the most commonly used paths for identifying and removing unwanted proteins is an ATP-dependent mechanism called the *ubiquitination* pathway (Fig. 11-11). In the cytoplasm of all cells is a small protein of 76 amino acids called *ubiquitin*. Ubiquitin has poorly characterized roles in cell cycle control, heat shock responses, receptor signaling, DNA repair, and transcriptional activation, but its major function appears to be to act as a "garbage tag." Ubiquitinating enzymes present in every cell's cytosol covalently bind polymers of ubiquitin (polyUb) to the NH_2 group of a lysine side chain of the protein to be degraded, creating ubiquitinated targets that are cleaved by the standard proteasome. In a process that is still not well understood, a specific subunit of the standard proteasome appears to be able to recognize the polyubiquitin, take up the tagged protein, and degrade it to peptides.

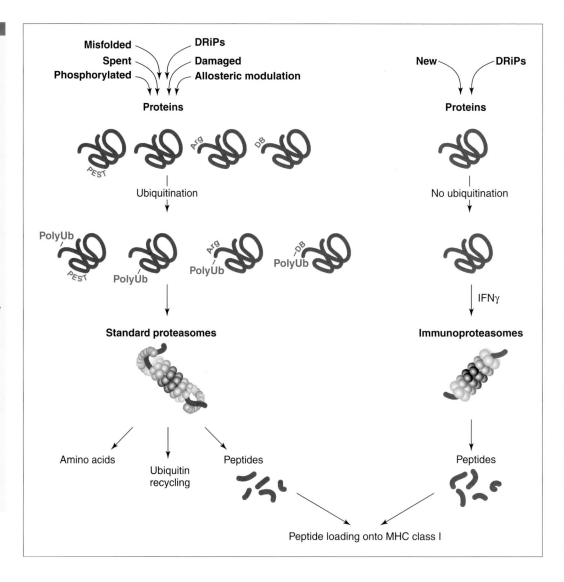

Figure 11-11

Disposal Tags Directing Proteins to the MHC Class I Pathway
In all cells, the disposal of many proteins that are spent, damaged, or misfolded is controlled by covalently adding the small molecule ubiquitin to them (left side of figure). The presence of the ubiquitin directs the protein to standard proteasomes for degradation and the production of peptides. In addition, proteins with limited half-lives contain disposal tag sequences such as PEST or DB that direct them to the ubiquitination pathway. Ubiquitination is also the fate of many newly synthesized "defective ribosomal products" (DRiPs) that never attain the correct form or function. Cells that are exposed to IFNγ due to the presence of a pathogen are induced to assemble immunoproteasomes (right side of figure). Newly synthesized proteins and DRiPs can be directed to the immunoproteasomes in the absence of ubiquitination. If the pathogenic threat is intracellular, a high proportion of the peptides generated will be pathogen-derived.

Ubiquitinating enzymes are thought to identify unwanted proteins by recognizing some type of post-translational modification, such as phosphorylation, dephosphorylation, or glycosylation. For example, many metabolic enzymes, transcription factors, kinases, and phosphatases contain a *PEST sequence*, a variable stretch of amino acids rich in proline, glutamine, serine, and threonine. Phosphorylation of a protein's PEST sequence leads to its rapid destruction via ubiquitination. Similarly, if a protein contains a *cyclin destruction box sequence* (RXALGXIXN) about 40–50 amino acids away from its N-terminus, the degradation of that protein is tied to the cell cycle. It is thought that the destruction box (DB) acts as a docking platform for one of the ubiquitinating enzymes whose expression is also regulated by the cell cycle. In addition, ubiquitinating enzymes recognize substrates based on a hierarchy of N-terminal amino acid sequences. The identity of the N-terminal amino acid of a protein thus determines its half-life and the presentation of its antigenic epitopes to CD8+ T cells. In experiments using viral or bacterial proteins engineered to have different N-terminal amino acids, those with N-terminal residues (such as Arg) that destabilized the

protein and accelerated degradation enhanced antigen presentation to CTL, while those with N-terminal residues (such as methionine) that stabilized the protein and slowed degradation inhibited antigen presentation to CTL.

The *DRiP pathway* (defective ribosomal products) also contributes to the generation of antigenic peptides presented on MHC class I by standard proteasomes. DRiPs are misfolded polypeptides that never assume the correct structure of the native protein. The misfolding may result from a translation error during synthesis by the ribosome or may be due to a problem in post-translational modification processes. It has been estimated that up to 30% of the nascent products of ribosomal synthesis are DRiPs. Since these misfolded proteins are not functional, the cell seeks to dispose of them. Many are rapidly ubiquitinated and digested by standard proteasomes such that DRiP peptides are displayed on MHC class I and contribute to the monitoring of the cell's internal health.

iic) The immunoproteasome. Immunoproteasomes are induced in cells exposed to pro-inflammatory cytokines (such as IFNγ and TNF) that are present at high concentrations during a pathogen attack. Accordingly, immunoproteasomes are responsible for most of the antigen processing associated with immune responses; that is, the degradation of peptides from foreign, rather than self, proteins. Interestingly, immunoproteasomes are found constitutively in cells within lymphoid organs such as the thymus, spleen, and lymph nodes. Indeed, half the proteasomes in immature DCs are immunoproteasomes, while the mature DC contains immunoproteasomes exclusively. The reader may wonder why tissues primarily associated with antigen presentation on MHC class II should have high numbers of immunoproteasomes necessary for antigen presentation on MHC class I. The answer likely has to do with the process of *cross-presentation* (explained later in this chapter), in which peptides from exogenous antigens are presented on MHC class I.

The immunoproteasome contains a modified 20S core: the three catalytically active β subunits of the standard proteasome are replaced by the *LMP-2* and *LMP-7* proteins (encoded by genes located in the MHC; refer to Ch.10), and the *MECL-1* protein (encoded outside the MHC). In addition, the modified 20S core associates not with the 19S regulatory complex but with a different regulatory complex called the *11S regulatory* or *PA28* complex (refer to Fig.11-10). PA28 is composed of three PA28α and three PA28β subunits and is essential for the incorporation of LMP-2, LMP-7, and MECL-1 into the 20S core. These modifications have an effect on the specificity (but not the efficiency) of proteasomal cleavage: overlapping sets of peptides are produced when a given antigenic protein is incubated with immunoproteasomes as opposed to standard proteasomes. In particular, more of the peptides produced by the immunoproteasome are suitable in amino acid length for binding in the groove of MHC class I.

The expression of both PA28 subunits and LMP-2/LMP-7 in a cell are upregulated by TNF and IFNγ, accounting for the fact that immunoproteasome assembly is favored over that of the standard proteasome during an immune response against a pathogen. In the case of a pathogen infection, because of the takeover of the host cell ribosomal machinery by the pathogen mRNA, most newly synthesized proteins will be of pathogen origin. The immunoproteasome processes these newly synthesized proteins to antigenic peptides, many before translation is even completed. Ubiquitination is not required for immunoproteasome function, consistent with the observation that animals with mutations of ubiquitinating enzymes can still present antigenic peptides on MHC class I. Interestingly, while many pathogen proteins are rapidly degraded into antigenic peptides by immunoproteasomes *in vitro*, many self proteins are not. In addition, because the immunoproteasome cleaves some sequences more readily than others, it helps to select which antigenic peptides are loaded onto each MHC class I allele. Due to their constitutive expression in the thymus, these extraordinary organelles exert a regulatory role and influence development of the T cell repertoire of specificities during T cell maturation in the thymus (see Ch.13).

iid) Sampling of organelle proteins. How do those proteins (self and nonself) confined within intracellular organelles such as the mitochondria, nucleus, or ER access the cytosolic processing pathway for sampling? Studies have shown that peptides derived from at least some of the proteins in these organelles are presented in association with MHC class I. Since the degradation of endogenous proteins prior to complexing with MHC class I is known to occur in the cytosol, how do these proteins get from the organelle into the cytosol for processing? Some scientists speculate that a retrieval mechanism for such proteins may exist such that a few intact molecules are plucked out of the ER into the cytosol, thus becoming available for proteasomal degradation. Evidence to support this hypothesis lies in the discovery of MHC-associated peptides that have been post-translationally modified, making them slightly different in some way from the amino acid sequence encoded in the DNA of the gene. For example, an MHC class I-presented peptide derived from a tyrosinase protein in melanoma cells was found to contain an Asp residue instead of the Asn dictated by the gene. It was surmised that a precursor protein was made in the cytosol, glycosylated in the ER, and directed to an organelle, before being exported <u>back</u> to the cytosol (for sampling) where the modification of Asn to Asp occurred. Cysteinylated and phosphorylated derivatives of other peptides have also been identified, consistent with modification in the cytosol. It may be that these modifications serve as tags that tell the cell to process these proteins by proteasomal degradation.

iii) Transport of Peptides into the ER: TAP

How do the peptides generated in the cytosol by proteasomes access the ER and encounter the MHC class I heterodimer? In the membrane of the ER are positioned transporter structures known as TAP (*transporter associated with antigen processing*) (Fig. 11-12). The genes encoding the two subunits of this heterodimeric molecule, TAP-1 and TAP-2, are located in the MHC and were introduced in Chapter 10. TAP gene expression is influenced by the same constellation of cytokines regulating the expression of the MHC class I genes. The TAP-1 and TAP-2 proteins are homologous to a family of *ATP-binding cassette* (ABC) transporter proteins known to convey small molecules across intracellular membranes. Both TAP-1 and TAP-2 are hydrophobic proteins with multiple membrane-spanning segments and a cytosolic nucleotide (ATP) binding domain. These proteins associate non-covalently to assemble the TAP peptide transporter with the cytosolic domains combining to form a single peptide-binding site. Both TAP-1 and TAP-2 are strictly required for the eventual formation of the peptide–MHC class I complex such that mice in which either of the TAP genes is disrupted show greatly reduced surface expression of MHC class I. The importance of TAP to the cell-mediated immune response is illustrated by the fact that some viruses, including herpes simplex viruses 1 and 2 and cytomegalovirus, inhibit CTL killing of infected cells by blocking TAP. A lack of TAP function interferes with the supply of viral peptides that would normally be loaded onto waiting MHC class I heterodimers, resulting in a failure to activate antiviral CD8$^+$ T cells.

How are the peptides produced by proteasomes conveyed to TAP for their rendezvous with MHC class I in the ER? Peptides are not found freely floating in the cytosol after degradation but rather are chaperoned by the heat shock proteins HSP70, HSP90, and HSP110. Heat shock proteins (HSPs) are highly conserved in all organisms studied to date and are a subset of a

Figure 11-12

MHC Class I Antigen Presentation Pathway
At the center left of the figure, MHC class I α and β2m chains are synthesized on ER-bound ribosomes. After MHC class I heterodimer assembly and binding to chaperones, the complex interacts with the TAP peptide transporter (composed of TAP-1 and TAP-2) via binding to tapasin. Meanwhile, at the bottom center of the figure, endogenous proteins have been processed in the cytosol by the proteasomes to generate peptides. These peptides are bound to heat shock proteins (HSPs) that convey them to TAP. A peptide passes into the ER via TAP and is bound by ER chaperone proteins such as BiP. The peptide is transferred from BiP to the MHC class I binding groove, the chaperones are released, and the peptide–MHC class I complex makes its way through the TGN to the cell surface.

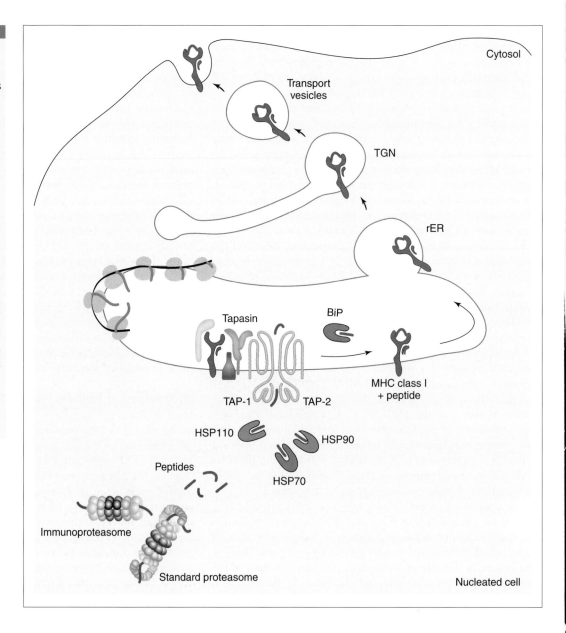

larger group of proteins of unknown function called "stress proteins." HSP expression is sharply increased in cells in response to environmental stresses such as a sudden increase in ambient temperature, cancerous transformation, or an inflammatory response (see Box 11-3). In the context of antigen presentation, HSP70, HSP90, and HSP110 convey peptides from the proteasome to the cytosolic side of TAP, while the ER-resident chaperone molecules gp96, grp170, and BiP take delivery of peptides coming through TAP and bind to them on the ER side of TAP. In *in vitro* experiments, peptides not bound to HSP were loaded onto MHC class I with far less efficiency than peptides bound to HSP. Peptide presentation by MHC class I is also abrogated by specific HSP inhibitors.

The affinity of the TAP heterodimer for peptides in isolation is relatively low (K_d = 1–10 μM), but *in vivo*, transport is facilitated by high local concentrations of peptides created by the action of the HSP. Energy from ATP hydrolysis is not needed for the transfer of the peptide from the HSP to TAP but is required for the import of the peptide into the ER. It has been estimated that TAP can translocate more than 20,000 peptides/min/cell to supply nascent MHC class I molecules generated at a rate of 10–100/min, more than enough to ensure a steady supply for loading. TAP preferentially imports peptides of 8–12 residues in length, although longer peptides can be transported with lower efficiency. Studies in which peptides have been mutated at different positions have shown that the residues crucial for TAP binding seem to be at the ends of the peptide, rather than the middle. Such a structure would allow TAP to transport large peptides by "looping out" their central portions. In general, TAP is far more promiscuous in its peptide-binding profile than is an MHC molecule, which makes sense because TAP has to supply peptides to all the different MHC class I alleles. That being said, the TAP heterodimer does display preferences for certain types of amino acids at certain positions. Some of these preferences are species-specific. For example, human TAP molecules bind equally well to peptides regardless

Box 11-3. Heat shock/stress proteins as peptide chaperones for antigen processing

Cells of all organisms respond to extracellular assaults by inducing new transcription, presumably of genes encoding proteins designed to save the host from destruction. It was first discovered in *Drosophila* that sudden increases in temperature reproducibly led to massive increases in expression of a defined set of proteins originally labeled "heat shock proteins," or HSPs. It was later found that highly conserved HSP homologues were present in all organisms, including pathogens and the earliest multi-cellular hosts, and that HSPs occurred in relatively high abundance even in the absence of stress. Moreover, HSP expression was found to be upregulated not only by heat, but also by a variety of other extracellular stresses, such as changes to the nutritional milieu or chemical microenvironment. Additional families of proteins whose expression was increased in cells in response to a variety of stresses were identified and called "stress proteins." Examples of HSP/stress proteins include the HSP70, HSP90, and HSP110 proteins localized in the cytosol and the gp96, grp170, PDI (protein disulfide isomerase), and BiP (immunoglobulin heavy chain binding protein) chaperone proteins found in the ER.

The story of how HSP/stress proteins were finally connected to antigen processing is a long, convoluted tale. It starts with observations in the 1940s that inbred mice immunized with irradiated, syngeneic cancer cells mounted immune responses such that they became resistant to subsequent challenges with live cells from the cancer originally used for immunization. This resistance was found to be specific to the individual cancer: it did not protect the mice from assault with cells from a different type of cancer, or even the same type of cancer from a different individual. In an effort to identify the source of the resistance, cancer tissues were homogenized and separated into fractions that were assayed for their ability to confer resistance. Each fraction was used to immunize a separate group of mice, and all groups were then challenged with live cells from the original cancer. When characterized, the proteins conferring protection against the cancer turned out to be HSP70, HSP90, and gp96. Furthermore, although these same three proteins appeared in the protective fraction for any

cancer studied, the protective fraction derived from one cancer would not protect a mouse from challenge with a different cancer. The mystery remained: how then did the HSP confer specific immunological resistance to a given cancer?

It was not until the 1990s that the supposedly homogeneous preparations of HSPs isolated from cancers were analyzed in more depth and were found to contain large numbers of peptides of varying sequences. When these peptides were removed from the protective fraction used to immunize the mice, the resistance to subsequent cancer challenge was abrogated. While it proved impossible to find cancer-specific peptides among the vast collection present in the HSP homogenate, it was at least clear that peptides associated with the HSP/stress proteins were responsible for inducing the protective immune response. Eventually it was discovered that the mechanism of this resistance could be attributed to the induction of both cell-mediated and humoral anti-peptide responses. A mound of *in vitro* evidence led to the conclusion that HSP/stress proteins possessed peptide-binding grooves, a finding later confirmed by extensive crystallographic studies of HSP90. Furthermore, functional assays showed that peptides that were non-immunogenic themselves could become so if they were first complexed *in vitro* to HSPs. This effect was unique to the HSP/stress proteins: peptide complexed to a non-stress peptide-binding protein was not immunogenic.

What then is the mechanism of HSP–peptide complex immunogenicity? In other words, what is the function of HSPs outside cells? In the course of their "messy" deaths, necrotic cells release peptides and proteins that have become bound to HSPs and other stress proteins. It has been shown that activated macrophages and DCs (but not B cells or non-professional APCs) express the HSP receptors CD36 and CD91. TLR4 and CD14 (also expressed on macrophages and DCs) have also been shown to bind HSPs. *In vitro*, these receptors facilitate the uptake of exogenously added HSP–peptide complexes and display of these peptides on MHC class I. *In vivo*, it may be that HSPs contribute to the process of *cross-presentation* of exogenous antigens (such as those released by a dying, infected cell) important for antiviral CTL

responses (cross-presentation is explained later in this chapter).

What function do the HSP/stress proteins serve <u>within</u> the cells of a multi-cellular organism, and why should peptides bind to them? Some of these proteins (notably gp96, grp170, and BiP) act as quality control monitors and have been found to bind to misfolded proteins to prevent them from leaving the ER. Others have a more general "chaperoning" function, facilitating polypeptide folding and protecting newly synthesized proteins from intracellular degradation. With respect to proteasomal degradation, some HSPs appear to act as disposal tags, guiding a protein to the proteasome for destruction. At the other end of the process, peptides derived by proteasomal degradation are complexed with HSP70, HSP90, or HSP110 in the cytosol for transport to TAP. On the other side of TAP in the ER, the peptides are complexed to gp96 or PDI. Some scientists have termed this collection of chaperone molecules and structures involving both sides of the ER membrane a *presentosome*. In this model, the HSP/stress proteins form an integral part of the presentosome, passing the peptide from the cytosol through TAP to the ER.

Can we take any clinical advantage of HSP functions? The original work identifying HSPs in immune responses to cancer suggested that there might be therapeutic uses for HSP/stress proteins. Because the HSP/stress proteins are monomorphic, their structure is the same in all individuals of a species and they are not intrinsically immunogenic. In addition, HSP/stress proteins isolated from cells bind to not just one but a wide variety of peptides, meaning that all possible epitopes of a cancer antigen can be presented. Indeed, HSP/stress protein–peptide complexes isolated from cancer cells or virus-infected cells have been shown to provoke immune responses to all identifiable epitopes from the original cancer or virus. Thus, there is no need to identify specific cancer epitopes in a given cancer tissue: immunization of an individual with HSP/stress protein–peptide complexes isolated from the cancer is sufficient to elicit both CTL and antibody responses. The possibilities of using HSP/stress proteins for cancer treatment are under investigation (see Ch.26).

of the hydrophobicity of their C-termini, while mouse TAP preferentially binds peptides with aromatic or aliphatic residues at the C-terminus. The TAP genes are also polymorphic, so that some differences in peptide preference may be a function of different TAP alleles in the same species. Variation at five positions in the membrane-spanning domain of rat TAP was shown to result in functional differences between different alleles, manifested as alterations to the repertoire of peptides transported. Structural polymorphism also exists in mouse and human TAP proteins but to a much more limited extent; no functional differences have yet been established.

iv) Peptide Loading onto MHC Class I

How does the peptide bound to TAP become associated with MHC class I? The details of this transfer are still unclear, but it seems that MHC class I can also transiently associate with TAP, presumably to facilitate peptide loading. A 48 kDa molecule called *tapasin* forms a complex with MHC and TAP in humans (refer to Fig. 11-12). The human tapasin gene is located just outside MHC but within 500 kb of the TAP genes; homologous genes in similar positions have been identified in mice and rats. The tapasin protein is a proline-rich transmembrane glycoprotein and a member of the Ig superfamily due to the presence of an Ig constant region-like extracellular domain. The hydrophobic transmembrane domain contains a lysine residue that interacts with TAP. The short cytoplasmic tail contains a configuration of lysine residues that serves as an ER retention signal. Like other antigen processing genes, the expression of the tapasin gene can be induced by IFNγ.

Since the tapasin molecule has independent binding sites for TAP and the MHC class I α chain, researchers believe that tapasin functions as a physical bridge between MHC and TAP, helping to stabilize empty MHC class I heterodimers in a conformation suitable for peptide loading. In this way, the function of tapasin is parallel to that of the HLA-DM/H-2M molecule that facilitates the loading and stabilization of MHC class II molecules. Analysis of isolated complexes containing TAP, tapasin, MHC class I, and calreticulin, or just TAP and tapasin have shown that one TAP heterodimer is linked to four tapasin molecules and four MHC class I heterodimers (still bound to calreticulin and ERp57). This arrangement quadruples the chance that whatever peptide is bound to TAP at the time will be able to fit in the binding groove of at least one of the four MHC class I molecules bound to TAP. *In vitro*, peptide translocation by TAP is slowed in the absence of tapasin. Tapasin may also be responsible for retaining improperly or suboptimally loaded MHC class I–peptide complexes in the ER, since cells of knockout mice lacking tapasin have abnormalities in the few MHC class I molecules that make it to the cell surface. Moreover, because of reduced MHC class I presentation, CD8-driven T cell selection in the thymus is decreased so that these mutants exhibit a deficiency of mature CD8$^+$ T cells. Exactly how the peptide is transferred from TAP to the chaperone molecules such as BiP and then to the MHC class I binding groove has yet to be determined. The fact that ERp57 is present in the loading complex implies that disulfide bonds have to be broken and reformed. In addition, it is suspected that gp96, which has aminopepti-

dase activity, may trim off any amino acids overhanging the end of the MHC class I groove. In any case, only after the MHC class I heterodimer is loaded with peptide is it stable enough to be transported via the trans-Golgi network to the plasma membrane. With the insertion of the peptide–MHC class I complex into the membrane of the target cell, it is ready for inspection by passing CD8$^+$ Tc cells and CTLs.

C. Other Pathways of Antigen Presentation

As well as the standard exogenous and endogenous pathways of antigen processing and presentation just discussed, there are several other ways by which antigens can end up displayed on a cell surface.

I. CROSS-PRESENTATION

i) Mechanism of Cross-presentation

Until now, we have maintained that peptides from endogenous antigens are produced in the cytosol and are combined with MHC class I molecules for presentation to CD8$^+$ T cells, and that peptides from exogenous antigens are produced within the endosomes and are combined with MHC class II molecules for presentation to CD4$^+$ T cells. Naturally enough, as for most rules in biology, there are exceptions. The first irregularities came to light when researchers noticed that CD8$^+$ CTL responses could be mounted to certain exogenous antigens, and then that peptides from some viral antigens could be presented on MHC class I even when the endogenous processing pathway was blocked. The name eventually given to this process was *cross-presentation*, because the viral protein physically "crossed over" from one cell source to an APC that would then present it on MHC class I as if it had originated in the interior of the APC.

How and where were these peptides being produced and loaded on MHC class I such that they could activate Tc cells? It was discovered that uninfected APCs can acquire viral antigens by internalizing debris from cells that have undergone either necrotic or apoptotic death. There is even evidence that cellular material to be cross-presented can be acquired from portions of the membrane of live cells ("membrane nibbling"). In all cases, because the antigen has entered the APC from the extracellular environment, it is initially directed to the endocytic system in the usual way. Thus, peptides derived from it appear on the surface associated with MHC class II, and a Th response to the antigen can be initiated. However, after the antigen has been processed in the early endosome to a size of 3–20 kDa, a fraction of it is somehow actively transported from the endosomes into the cytosol, where the fragments are taken up by proteasomes and degraded to peptides. These peptides are then loaded onto MHC class I just as if the protein had originated within the APC itself (Fig. 11-13). Interestingly, a tyrosine residue in the tail of MHC class I

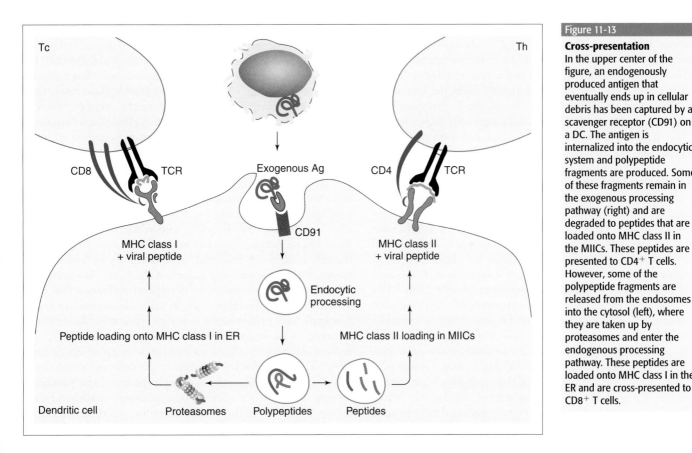

Figure 11-13

Cross-presentation
In the upper center of the figure, an endogenously produced antigen that eventually ends up in cellular debris has been captured by a scavenger receptor (CD91) on a DC. The antigen is internalized into the endocytic system and polypeptide fragments are produced. Some of these fragments remain in the exogenous processing pathway (right) and are degraded to peptides that are loaded onto MHC class II in the MIICs. These peptides are presented to CD4$^+$ T cells. However, some of the polypeptide fragments are released from the endosomes into the cytosol (left), where they are taken up by proteasomes and enter the endogenous processing pathway. These peptides are loaded onto MHC class I in the ER and are cross-presented to CD8$^+$ T cells.

molecules is responsible for the success of cross-presentation: mutation of this residue allows the mutated MHC class I molecule to participate in the endogenous pathway but not the cross-presentation pathway. It should also be noted that cross-presentation is not restricted to viral infections, as peptides from intracellular bacteria and parasites and even purified exogenous proteins introduced into animals in the course of experimentation can also be presented on MHC class I.

There is some evidence that at least some instances of cross-presentation are mediated by an incompletely defined mechanism in which the phagosome containing internalized antigens fuses with a vesicle derived from a portion of the rER membrane to form a new hybrid compartment. This structure contains all the players (including TAP, calreticulin, and tapasin) needed for the loading of peptides onto the newly synthesized MHC class I present in the rER vesicle. To obtain those peptides, the hybrid structure recruits proteasomes to its outer (cytosolic) surface. The antigens, which may be partially digested to a size of 3–20 kDa, are actively transported (in an unknown way) out of the hybrid structure and into the cytosol, where the fragments are taken up by the associated proteasomes and degraded to peptides. Using the TAP of the hybrid structure, the peptides re-enter the hybrid structure, are loaded onto MHC class I, and are transported to the cell membrane <u>without</u> passing through the Golgi. This intriguing hybrid structure and transport mechanism appear to be a feature predominantly of macrophages and DCs, particularly conventional CD8$^+$ CD205$^+$ DCs found in the T cell-rich areas of the lymph nodes and spleen.

Another term often used in the cross-presentation context is *cross-priming*, meaning that an APC has externally acquired an antigen that was originally synthesized in one cell (such as a nearby infected cell) and has used it to prime a CD8$^+$ CTL response against that antigen presented on MHC class I. Researchers studying cross-priming by DCs found that the process was more efficient when whole cells, cell-associated antigens, particulate antigens, or antibody-associated antigens were taken up by the DC than when soluble antigens were internalized. *In vitro* studies indicated that HSP-mediated chaperoning of proteins and peptides from the donor cell to the DC might be important for cross-priming since the presence of an HSP greatly reduces the amount of antigen needed to provoke a CTL response. In addition, the binding of an HSP to an immature DC via CD36 or CD91 can induce DC maturation *in vitro* and the secretion of cytokines and chemokines. However, *in vivo*, chaperones may not be essential, since whole proteins that are stable enough to be efficiently taken up by DCs can successfully prime a CTL response in the absence of chaperone proteins.

Optimal cross-priming of CTL responses by DCs has been found to depend on CD40-CD40L signaling in most instances. Thus, the sequence of cellular events during cross-priming *in vivo* is thought to go something like this: an immature DC is induced to mature by its capture of exogenous antigen either shed by an infected cell (which may be associated with HSP or another chaperone), or in the form of an internalized apoptotic or necrotic infected cell, or via "membrane nibbling" of a live infected cell. The mature DC displays peptides on both MHC class II and MHC class I and thus is able to interact with CD4$^+$ Th cells and CD8$^+$ Tc cells recognizing these pMHCs, respectively. The Th cells are activated and deliver CD40L-mediated signals to the mature CD40$^+$ DC. In the presence of CD40-mediated signals from the stimulated DC plus cytokine help from the Th cell, the Tc generates CTL effectors that attack virally infected target cells displaying the MHC class I–peptide complexes. More detail on the activation of naive Th and Tc cells by APCs appears in Chapter 14.

ii) Rationale for Cross-presentation

Why should cross-presentation exist? One reason may be that this system of antigen acquisition allows a DC to display antigens from pathogens that might not infect DCs at all. Since DCs are the only APCs that can activate naive Tc cells, and the cytolytic T cell response is crucial for eliminating most viruses and intracellular bacteria, it is important to the host that DCs be able to present antigenic peptides on MHC class I. Being able to acquire and present antigens from infected non-immune system cells gives the DC the ability to stimulate the type of immune response that will be most effective. Another possible function may be to counter the protective mechanisms of infecting viruses. Certain viruses express proteins that interfere with the classical endogenous MHC class I pathway (see Ch.22). Cross-presentation enables the cell to get around these blocks and still end up with pathogen peptide displayed on MHC class I. Cross-presentation also circumvents the situation in which a DC is rapidly killed by viral or intracellular bacterial infection or is inhibited from maturing. Uninfected bystander DCs can acquire the pathogen antigen from the infected DC, undergo maturation, and use the cross-presentation pathway to initiate a CTL response. For example, the measles virus is deadly to DCs, but immunity after natural measles infection is very robust. It is likely that uninfected DCs are acquiring measles proteins from the debris of infected cells, and cross-presenting peptides from these antigens on MHC class I to activate anti-measles Tc cells.

Another function of cross-presentation may be to expand the body's development of self-tolerance. As will be described in detail in Chapter 16, peripheral tolerance depends on the inactivation of autoreactive T cell clones that have escaped deletion during the establishment of central tolerance. To inactivate such a clone, the autoreactive T cells must bind to an APC presenting peptides from the specific self antigen in a situation where costimulatory signals are not delivered to the T cell. Consider the case of a self antigen that is expressed only in the periphery (and not endogenously in APCs), such as certain pancreas and kidney proteins. The body must remove all Tc clones recognizing peptides from these self antigens or CTLs will be generated that could attack the pancreas or kidney and cause autoimmune disease. How does the pancreatic or kidney self antigen make it to the secondary lymphoid tissues where the naive Tc cells that need to be inactivated reside? It is thought that small amounts of these antigens may be shed by the pancreas or kidney into the extracellular spaces around the organ. These shed proteins are internalized by tissue-resident immature DCs which then travel to the local lymph node and use the usual endocytic

pathway to present the self antigen on MHC class II. Because self antigens do not induce the production of pro-inflammatory cytokines (and no bacterial or viral products are present), the DC does not receive a signal to mature and only limited numbers of costimulatory molecules are expressed. Thus, rather than activating a naive Th cell recognizing this peptide, the DC causes the Th cell to become inactivated. However, to inactivate naive Tc cells, the DC must use the cross-presentation pathway to shift the externally acquired self antigen from the endosomes into the cytosol and the endogenous antigen processing pathway. An autoreactive Tc cell may recognize a peptide from the self antigen bound to MHC class I but, because no costimulation occurs and no activated Th cell is present to offer T cell help, the Tc cell is also rendered nonresponsive. As a result, neither humoral nor cell-mediated responses to the self antigen are mounted, as is necessary to preserve the health of the pancreas and kidney of the host. This presumed role of cross-presentation in peripheral tolerance has been demonstrated in the laboratory. In mice engineered to express hemagglutinin protein only in the pancreatic β-islet cells (an example of a tissue-restricted self antigen), bone marrow-derived APCs were found to present hemagglutinin to both CD4$^+$ Th cells and CD8$^+$ Tc cells in the draining lymph node. Naive Tc and Th clones recognizing peptides of this processed antigen were either rendered nonresponsive or deleted, establishing tolerance of the T cell repertoire for these otherwise less accessible antigens.

Cross-presentation may lend itself to exploitation for clinical use. There has been a recent surge of interest in DCs as therapeutic agents because they can combine synthetic peptides from pathogen or tumor antigens with MHC class I and present them to Tc cells with high efficiency. Researchers have artificially loaded ("pulsed") DCs with synthetic peptides, peptides from tumor cell lines, or even intact insoluble tumor proteins and induced these DCs to mature *ex vivo*. The mature, loaded DCs were then used to immunize mice, resulting in the mounting of potent anti-tumor CTL responses *in vivo*. Evidence for anti-tumor immunity has also been obtained in situations where DCs were transfected with the gene encoding a tumor antigen and used to immunize mice. Since DCs are relatively abundant and easy to isolate, DCs pulsed with a tumor antigen peptide or transfected with a tumor antigen gene offer the potential of a highly effective means of eliciting anti-tumor immunity. More on tumor immunology and treatment regimens appears in Chapter 26.

iii) Minor Pathways of Cross-presentation on MHC Class I
iiia) Peptide "regurgitation." Another possible mechanism resulting in cross-presentation is the putative "regurgitation" pathway of antigen presentation (Fig. 11-14). Some *in vivo* anti-LCMV and anti-influenza CTL responses have been found to depend on DC function but <u>not</u> on TAP or proteasomes. It is speculated that extracellular viral proteins may be internalized and processed to peptides as usual within the DC endosomes but that the peptides are then released ("regurgitated") back into the extracellular environment. Peptides generated in this manner, or those produced by extracellular enzymatic action or released from lysed or apoptotic cells, may then be able to force

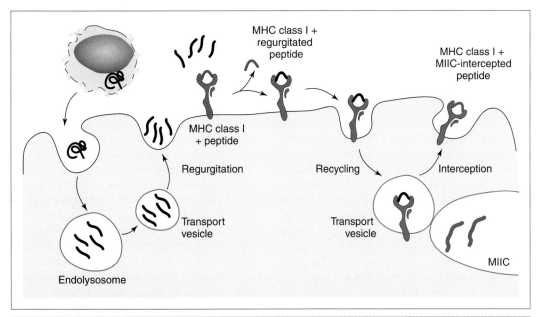

Figure 11-14

Minor Pathways of Cross-presentation on MHC Class I
Regurgitation pathway. At the upper left of the figure, an antigen has been internalized and enters the endocytic system. However, instead of proceeding from the endolysosome to the MIICs, the peptides enter a transport vesicle and are "regurgitated" back out of the cell. The local concentration of peptides on the surface can then force an exchange with a lower stability peptide in the binding groove of an MHC class I molecule already on the surface. Interception pathway. The interception pathway involves peptide MHC class I complexes recycling from the cell surface. By chance, the transport vesicle containing the recycling peptide MHC class I complex is "intercepted" and fuses with a MIIC containing exogenous peptides. The low pH of the MIIC forces the release of the endogenous peptide, and an exogenous peptide is loaded onto the MHC class I molecule in a TAP-independent manner. The new peptide–MHC class I complex returns to the surface.

an exchange with peptides already present on surface MHC class I molecules. Exogenous β2m in the culture medium has been found to increase the level of binding of extracellular peptides, possibly because increased β2m stabilizes the MHC class I chain long enough for surface peptide exchange to occur. It has been estimated that TAP-independent loading of MHC class I molecules is only 1/100 as efficient as the conventional endogenous processing pathway. Interestingly, scientists have developed an experimental equivalent to the regurgitation pathway that can be used in *in vitro* experimentation. By adding huge excesses of exogenous peptide, APCs and target cells in culture can be forced to replace peptides already associated with MHC molecules at the cell surface with an exogenous peptide of interest, allowing the researcher to work with large numbers of cells all presenting vast amounts of the same peptide.

iiib) Peptide "interception." Another means of loading exogenous peptide onto MHC class I appears to involve the MIICs. Some scientists believe that a MIIC containing peptides can sometimes fuse with a vesicle containing MHC class I (complete with an endogenous peptide in its binding groove) recycling from the cell surface. The very low pH (pH 5) in the MIICs forces the endogenous peptide out of the groove of the recycling MHC class I molecule and allows the TAP-independent loading of an exogenous peptide resident in the MIIC. The exogenous peptide then appears on MHC class I instead of MHC class II

and is displayed on the DC surface to Tc cells rather than to Th cells (refer to Fig. 11-14).

II. ANTIGEN PRESENTATION BY NON-CLASSICAL AND MHC-LIKE MOLECULES

We learned in Box 10-3 that two groups of molecules exist that share structural features with MHC proteins: the non-classical (class Ib and IIb) MHC molecules and the MHC-like molecules. Most of these molecules do not function as conventional peptide display molecules, but some do contribute to antigen presentation in specialized ways by interacting with elements of both the cytosolic and endocytic antigen processing paths.

i) Antigen Presentation by Non-classical MHC Proteins
The reader will recall that the class Ib MHC proteins are encoded by the H-2Q, -T, and -M regions in the mouse and the HLA-E, -F, -G, -J, and -X genes and the HFE gene in humans. The glycoproteins encoded by class Ib genes are closely related structurally to the classical MHC class I molecules in both species, but are less polymorphic, are expressed at lower levels, and have a more limited pattern of tissue distribution. The class Ib proteins often have intriguing variations at their C-termini that may confer specialized functions. Several class

Ib molecules occur in both secreted and membrane-anchored forms but do not bind antigenic peptides. Other class Ib proteins can bind to certain subsets of foreign peptides and present them to αβ and γδ T cells in a TAP-dependent manner. This observation suggests that at least some class Ib molecules do indeed present peptides using the standard cytosolic pathway. However, in those class Ib molecules that have been studied, the structure of the peptide-binding groove has turned out to be less accommodating than that of a conventional MHC class I molecule. The binding pocket is partially occluded such that a narrower range of peptides is presented. For example, in mice, the Qa-1 molecule is involved in T cell responses restricted to some microbial peptides, particularly those derived from the intracellular pathogen *Listeria monocytogenes*. Similarly, the H-2M3 molecule has been shown so far to present only *N*-formylated peptides derived from prokaryotic organisms. These peptides are 5–7 amino acids long, just the right size to fit in the smaller groove of the H-2M3 molecule. In humans, examples of non-classical antigen presentation are more rare. However, the HLA-G class Ib molecule involved in maternal-fetal immune interactions (see Box 10-4) has been shown to depend on TAP for its surface expression and the binding of peptides. Similarly, HLA-E has been found to present peptides derived from signal sequences of membrane proteins.

The low level of polymorphism of the non-classical MHC class molecules may offer an attractive clinical opportunity. It may be possible to devise a vaccine that activates those T cells recognizing pathogen peptides presented preferentially on non-classical MHC molecules. Because these MHC molecules are the same in virtually all members of a population, there would be no question of an individual failing to present the pathogen peptide to his/her T cells, as might occur if the individual happened to lack the classical MHC allele favored by a vaccine peptide. Immune protection of the population of a whole would thus be increased. Vaccination is discussed in depth in Chapter 23.

ii) Non-peptide Antigen Presentation by CD1 Molecules

The reader will recall from Chapter 10 that the MHC-like molecules are encoded outside the MHC but feature an MHC-like fold in their structures. The MHC-like CD1 molecules are of particular interest with respect to antigen presentation, since these non-polymorphic proteins were the first molecules shown to be capable of presenting non-peptide lipid and glycolipid antigens to T cells. Structurally, CD1 molecules show very limited sequence homology to MHC class I molecules at the amino acid level but do contain the β2-microglobulin chain. Significantly, the antigen-binding groove in CD1 molecules is much more hydrophobic than that of classical MHC molecules. Structural analyses have indicated that the fatty acid tails of a glycolipid antigen fit neatly into two narrow, deep pockets in the CD1 antigen-binding groove while the polar head group of the glycolipid projects out of the groove and makes contact with the TCR. It is thought that the T cell "sees" a combined epitope composed of amino acids of the CD1 molecule plus a small portion of the carbohydrate head group of the lipid. T cells can be very discriminating in their recognition of CD1-presented antigens, failing to respond if the orientation of even

a single hydroxyl group is changed. For its part, the binding affinity of a CD1 molecule for its ligand approximates that of a classical MHC class I molecule for its peptide.

CD1 molecules are classified into two groups, which differ significantly in their amino acid sequences and tissue distribution. Group 1 includes the CD1a, b, and c molecules, which are expressed on many professional APCs in humans (no mouse homologues to these molecules have yet been identified). The Group 1 genes also include a CD1e gene, but it is not yet clear that a protein is expressed from it. Group 2 includes human CD1d1 and CD1d2 and their murine homologues, which are expressed primarily on conventional DCs in the intestine and to a lesser extent in other non-lymphoid tissues. Initially, researchers were surprised by the finding that CD1a and b molecules could be recognized by T cells, as this was the first indication that these molecules might be involved in antigen presentation or T cell activation. Even more surprising was the subsequent discovery that CD1b could present a branched chain fatty acid group derived from *Mycobacterium* in a restricted fashion to T cells (Fig. 11-15A). Other mycobacterial and even self glycolipid and glycosphingolipid antigens presented to αβ T cells in association with CD1a, CD1b, and CD1c have since been identified, and an unknown ligand bound to CD1c is recognized by γδ T cells (see Ch.18). In contrast to CD1a, b, and c, CD1d does not associate with mycobacterial antigens. Instead, both murine and human CD1d molecules present a restricted set of antigens containing α-galactosylceramide to certain subsets of T cells and NKT cells (see Ch.18). The consequences of CD1-mediated antigen presentation are virtually identical to those of peptide presentation. T and NKT cells recognizing CD1-associated antigens respond by secreting IFNγ and other Th1 cytokines and employing perforin-mediated cytolysis or Fas-mediated apoptosis to kill target cells.

Antigen processing and presentation by CD1 molecules is thought to differ from that mediated by either MHC class I or MHC class II molecules and to take advantage of elements of both the cytosolic and endocytic paths. For extracellular glycolipid antigens, mannose receptors on the APC surface mediate internalization followed by trimming or digestion of the antigen into smaller antigenic fragments in the endosomes. For mycobacteria already captured in a phagosome, lipid degradation products are sorted and directed to different endosomal compartments depending on their molecular structure (Fig. 11-15B). These various types of lipids are taken up by the different CD1 molecules resident in different endosomal compartments. CD1a lacks a tyrosine-based localization signal in its cytoplasmic tail that directs molecules into the later compartments of the endocytic system, and so is found in recycling vesicles and in early endosomes. No specific ligand for CD1a has been isolated to date, but this CD1 molecule is thought to preferentially bind to lipids featuring short tails and unsaturated fatty acids. CD1c contains the tyrosine motif, resides in intermediate endosomes, and binds to a broader range of lipid antigens, including some with only one fatty acid tail. CD1b also contains the tyrosine motif but is found only in late endosomes and endolysosomes. CD1b binds to lipids with branched or long

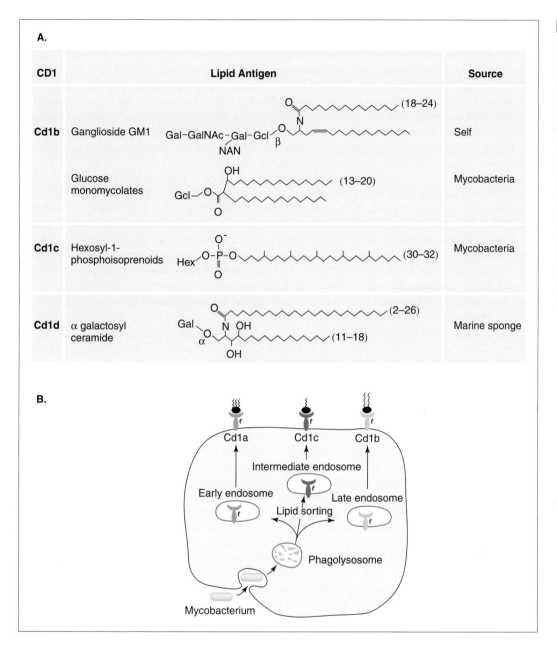

Figure 11-15

Lipid Antigen Presentation by CD1 Molecules
(A) Structures of lipid antigens presented by CD1 molecules. The precise structure of the CD1a antigen is unknown. (B) Sorting of mycobacterial lipids. At the bottom left of the figure, a *Mycobacterium* has been phagocytosed and degraded. The mycobacterial lipids are sorted by structure into different endosomal compartments where they are loaded onto different CD1 molecules for presentation to T cells. Panel A adapted from Matsuda J. L. and Kronenberg M. (2001) *Current Opinion in Immunology* **13**, 19–25.

tails containing unsaturated fatty acids. The highly acidic environment of the late endosome is thought to be crucial for opening the groove of the CD1b molecule, allowing it to combine with its lipid antigen. Agents that block either cellular internalization mechanisms or the acidification of the endosomal compartment have been shown to inhibit CD1b presentation. Like MHC class I, CD1 molecules must be associated with β2m to be transported to the cell surface, but, unlike MHC class I, TAP is not required for antigen loading. Like MHC class II, CD1 molecules are targeted to the endocytic system but no association with either invariant chain or HLA-DM is required. Indeed, no chaperone proteins have been identified that are required for CD1-mediated presentation of mycobacterial antigens. The definition of the steps between antigen loading and presentation on the cell surface is a mystery yet to be solved.

Whether CD1-mediated presentation of pathogen lipid antigens is actually relevant in a physiological immune response against infectious disease remains to be definitively proven. Upregulation of CD1 expression has been observed in patients with *Mycobacterium leprai* infections, and T cells recognizing a lipid antigen from *M. leprai* have been isolated from disease lesions. Many microbial lipids are different from mammalian lipids, making it logical to assume that the cell might monitor these molecules for the appearance of "nonself." However, the vast range of pathogen lipids that researchers expected to come to light as being presented by CD1 molecules has not materialized. In addition, many experiments designed to show direct recognition of bacterial lipids by T and NKT cells *in vitro* or in animal models have shown equivocal results. Some scientists are therefore hypothesizing that the CD1 antigen presentation system is instead a means of monitoring the health of host cell membranes by sampling endogenous lipids and glycolipids. Stress and aging can cause modifications to carbohydrates of membrane components that might help to alert the host as to the necessity of removing this spent cell. It may be that mycobacteria have merely subverted the CD1 system for their own use. Perhaps cells induced to die by the action of T cells or NKT cells activated via CD1 provide a favorable environment for the long-term survival of mycobacteria. Obviously, investigation of additional lipid ligands recognized by CD1-dependent T cells *in vivo* is needed to clarify this matter.

With the identification of the CD1 molecules, we can see that antigen presentation during the adaptive immune response may actually occur via at least four paths: the cytosolic pathway used by MHC class I and non-classical MHC molecules to handle (for the most part) antigens derived from intracellular pathogens and waste host proteins; the endocytic pathway used by MHC class II molecules to handle antigens derived from extracellular pathogens; the cross-presentation pathway used to display extracellularly acquired antigens on MHC class I; and the CD1 pathway used to present non-peptide antigens from extracellular pathogens. That being said, it bears mentioning here that some T cell responses can occur in the complete absence of antigen processing. Unlike αβ T cells, some γδ T cells can directly recognize native (unprocessed) protein antigens. Other γδ T cell clones bind to MHC independent of the specificity of the bound peptide, while still others have been shown to directly recognize small phosphate-containing compounds. Antigen recognition by γδ T cells is discussed in detail in Chapter 18.

This concludes our discussion of antigen processing and presentation. We move now to a set of five chapters (Chs.12–16) describing in detail the biology of T cells, from the structure and genetics of their antigen receptors to their development and activation. In these chapters we discuss how the antigen processed and presented by the APC or target cell interacts with the TCR and sparks the cycle of activation, proliferation, and differentiation that results in the generation of T effector cells crucial for both the cell-mediated and humoral immune responses.

SUMMARY

T cells recognize peptide antigens presented by MHC molecules on the surface of host cells. These peptides are derived from the processing of proteins by either the exogenous (endocytic) or endogenous (cytosolic) pathways. Proteins from entities produced exogenously, such as extracellular bacteria, toxins, and fungi, are processed via the endocytic pathway. These antigens are internalized by specialized antigen-presenting cells (APCs) and degraded within endocytic compartments. Within the endosomes, the resulting peptides bind to MHC class II molecules to form pMHC complexes that are transported to the APC surface for recognition by antigen-specific CD4$^+$ Th cells. The activated Th cells supply help to antigen-specific B cells, facilitating the production of antibodies that combat the extracellular threat. Dendritic cells, macrophages, and B cells are all extremely effective APCs because they take up antigen efficiently and express MHC class II constitutively. Proteins produced endogenously by viruses or intracellular parasites or as a result of host cell transformation are processed via the cytosolic processing pathway. These antigens are degraded by proteasomes in the cytosol of the affected cell and the resulting peptides are actively transported into the endoplasmic reticulum where they bind to newly synthesized MHC class I molecules. When these pMHC complexes are transported to the cell surface, they are recognized by CD8$^+$ CTLs that mediate cytolysis of the affected cell. Since almost all cells express MHC class I molecules, the Tc population can effectively monitor the entire host for infection or transformation. Cross-presentation occurs when some molecules of an exogenously acquired protein are processed in the cytosol instead of in the endosome, and the resulting peptides are consequently presented on MHC class I to Tc cells. Antigens can also be presented to γδ T cells and NKT cells by non-classical MHC molecules and MHC-like molecules such as CD1.

Selected Reading List

Al-Daccak R., Mooney N. and Charron D. (2004) MHC class II signaling in antigen-presenting cells. *Current Opinion in Immunology* 16, 108–113.

Babbitt B. P., Allen P. M., Matsueda G., Habert E. and Unanue E. R. (1985) Binding of immunogenic peptides to Ia histocompatibility molecules. *Nature* 317, 359–361.

Banchereau J., Briere F., Caux C., Davouse J., Lebecque S. *et al.* (2000) Immunobiology of dendritic cells. *Annual Review of Immunology* 18, 767–811.

Boss J. and Jensen P. (2003) Transcriptional regulation of the MHC class II antigen presentation pathway. *Current Opinion in Immunology* 15, 105–111.

Boss J. M. (1997) Regulation of transcription of MHC class II genes. *Current Opinion in Immunology* 9, 107–113.

Brocke P., Garbi B., Momburg F. and Hammerling G. (2002) HLA-DM, HLA-DO and tapasin: functional similarities and differences. *Current Opinion in Immunology* 14, 22–29.

Busch R., Doebele R., Patil N., Pashine A. and Mellins E. (2000) Accessory molecules for MHC class II peptide loading. *Current Opinion in Immunology* 12, 99–106.

Buus S., Sette A., Colon S., Jenis D. and Grey H. (1986) Isolation and characterization of antigen-Ia complexes involved in T cell recognition. *Cell* 47, 1071–1077.

Carbone F., Kurts C., Bennett S., Miller J. and Heath W. (1998) Cross-presentation: a general mechanism for CTL immunity and tolerance. *Immunology Today* 19, 368–373.

Cresswell P. and Lanzavecchia A. (2001) Antigen processing and recognition. *Current Opinion in Immunology* 13, 11–12.

den Haan J. and Bevan M. (2001) Antigen presentation to CD8$^+$ T cells: cross-priming in infectious diseases. *Current Opinion in Immunology* 13, 437–441.

Fonteneau J.-F., Larsson M. and Bhardwaj N. (2002) Interactions between dead cells and dendritic cells in the induction of antiviral CTL responses. *Current Opinion in Immunology* 14, 471–477.

Germain R. N. (1993) The biochemistry and cell biology of antigen processing and presentation. *Annual Review of Immunology* 11, 403–450.

Geuze H. J. (1998) The role of endosomes and lysosomes in MHC class II functioning. *Immunology Today* 19, 282–287.

Goldman, A. (2003) Protein degradation and protection against misfolded or damaged proteins. *Nature* 426, 895–899.

Grandea III A. and Van Kaer L. (2001) Tapasin: an ER chaperone that controls MHC class I assembly with peptide. *Trends in Immunology* 22, 194–199.

Heath W., Belz G., Behrens G., Smith C., Forehan S. *et al.* (1999) Cross-presentation, dendritic cell subsets, and the generation of immunity to cellular antigens. *Immunological Reviews* 199, 9–26.

Hiltbold E. and Roche P. (2002) Trafficking of MHC class II molecules in the late secretory pathway. *Current Opinion in Immunology* 14, 30–35.

Hwang L.-Y., Lieu P., Peterson P. and Yang Y. (2001) Functional regulation of immunoproteasomes and transporter associated with antigen processing. *Immunological Research* 24, 245–272.

Jayawardena-Wolf J. and Bendelac A. (2001) CD1 and lipid antigens: intracellular pathways for antigen presentation. *Current Opinion in Immunology* 13, 109–113.

Lennon-Dumenil A.-M., Bakker A., Wolf-Bryant P., Ploegh H. and Lagaudriere-Gesbert C. (2002) A closer look at proteolysis and MHC class II-restricted antigen presentation. *Current Opinion in Immunology* 14, 15–21.

Li Z., Menoret A. and Srivastava P. (2002) Roles of heat-shock proteins in antigen presentation and cross-presentation. *Current Opinion in Immunology* 14, 45–51.

Maffei A., Papadopoulos K. and Harris P. E. (1997) MHC class I antigen processing pathways. *Human Immunology* 54, 91–103.

Marusina K. and Monaco J. J. (1996) Peptide transport in antigen presentation. *Current Opinion in Hematology* 3, 319–326.

Matsuda J. L. and Kronenberg M. (2001) Presentation of self and microbial lipids by CD1 molecules. *Current Opinion in Immunology* 13, 19–25.

Mellman I. and Steinman R. (2001) Dendritic cells: specialized and regulated antigen processing machines. *Cell* 106, 255–258.

Pamer E. and Cresswell P. (1998) Mechanisms of MHC class I-restricted antigen processing. *Annual Review of Immunology* 16, 323–358.

Pieters J. (1997) MHC class II restricted antigen presentation. *Current Opinion in Immunology* 9, 89–96.

Reimann J. and Kaufmann S. H. (1997) Alternative antigen processing pathways in anti-infective immunity. *Current Opinion in Immunology* 9, 462–469.

Reits E., Vos J., Gromme M. and Neefjes J. (2000) The major substrates for TAP in vivo are derived from newly synthesized proteins. *Nature* 404, 774–778.

Rivett A. J. (1998) Intracellular distribution of proteasomes. *Current Opinion in Immunology* 10, 110–114.

Schneider S. C. and Sercarz E. E. (1997) Antigen processing differences among APC. *Human Immunology* 54, 148–158.

Schubert U., Anton L. C., Gibbs J., Norbury C. C., Yewdell J. W. and Bennink J. R. (2000) Rapid degradation of a large fraction of newly synthesized proteins by proteasomes. *Nature* 404, 770–774.

Shortman K. and Liu Y. (2002) Mouse and human dendritic cell subtypes. *Nature Reviews Immunology* 2, 151–161.

Sikorski R. and Peters R. (1997) Cell shocked. *Science* 278, 2143–2144.

Smith J. D., Lie W.-R., Gorka J., Myers N. B. and Hansen T. H. (1992) Extensive peptide ligand exchange by surface class I major histocompatibility complex molecules independent of exogenous β2-microglobulin. *Proceedings of the National Academy of Sciences USA* 89, 7767–7771.

Srivastava P. K., Menoret A., Basu S., Binder R. J. and McQuade K. L. (1998) Heat shock proteins come of age: primitive functions

acquire new roles in an adaptive world. *Immunity* 8, 657–665.

Steinman R. (2000) DC-SIGN: a guide to some mysteries of dendritic cells. *Cell* 100, 491–494.

Steinman R. M., Inaba K., Turley S., Pierre P. and Mellman I. (1999) Antigen capture, processing, and presentation by dendritic cells: recent cell biological studies. *Human Immunology* 60, 562–567.

Thery C. and Amigorena S. (2001) The cell biology of antigen presentation in dendritic cells. *Current Opinion in Immunology* 13, 25–51.

Uebel S. and Tampe R. (1999) Specificity of the proteasome and the TAP transporter. *Current Opinion in Immunology* 11, 203–208.

Van den Eynde B. and Morel S. (2001) Differential processing of class-I-restricted epitopes by the standard proteasome and the immunoproteasome. *Current Opinion in Immunology* 13, 147–153.

van Ham S. M., Tjin E. P. M., Lillemeier B. F., Gruneberg U., van Jeijgaarden K. E. *et al.* (1997) HLA-DO is a negative modulator of HLA-DM-mediated MHC class II peptide loading. *Current Biology* 7, 950–957.

Villadangos J. and Ploegh H. (2000) Proteolysis in MHC class II antigen presentation: who's in charge? *Immunity* 12, 233–239.

Villadangos J., Cardoso M., Steptoe R., van Berkel D., Pooley J. *et al.* (2001) MHC class II expression is regulated in dendritic cells independently of invariant chain degradation. *Immunity* 14, 739–749.

Watts T. and McConnell H. (1987) Biophysical aspects of antigen recognition by T cells. *Annual Review of Immunology* 5, 461–475.

Wells A. D. and Malkovsky M. (2000) Heat shock proteins, tumor immunogenicity and antigen presentation: an integrated view. *Immunology Today* 21, 129–132.

Yewdell J. W. and Bennick J. R. (2001) Cut and trim: generating MHC class I peptide ligands. *Current Opinion in Immunology* 13, 13–18.

The T Cell Receptor: Structure of Its Proteins and Genes

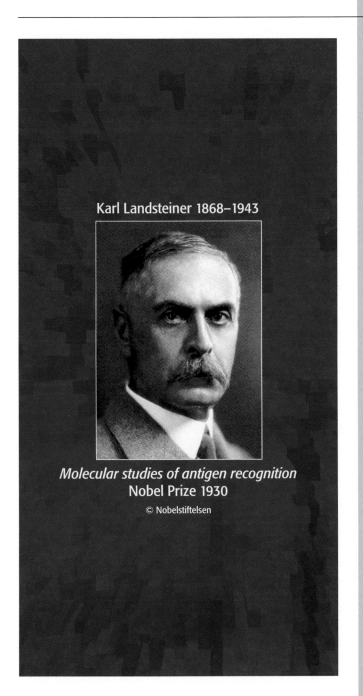

Karl Landsteiner 1868–1943

Molecular studies of antigen recognition
Nobel Prize 1930

© Nobelstiftelsen

CHAPTER 12

A. HISTORICAL NOTES

B. THE STRUCTURE OF T CELL RECEPTOR PROTEINS

C. GENOMIC ORGANIZATION OF THE TCR AND CD3 LOCI

D. EXPRESSION OF TCR GENES

E. DEVELOPMENTAL ASPECTS OF V(D)J RECOMBINATION IN THE TCR LOCI

F. GENERATION OF DIVERSITY OF THE T CELL RECEPTOR REPERTOIRE

G. REGULATION OF TCR GENE EXPRESSION

H. STRUCTURE AND FUNCTION OF THE CD3 COMPLEX

I. THE CD4 AND CD8 CORECEPTORS

J. PHYSICAL ASPECTS OF THE INTERACTION OF THE TCR WITH ANTIGEN

"The best way to have a good idea is to have lots of ideas"
—Linus Pauling

We learned in Chapter 5 that B cells recognize antigen using the BCR, an antigen receptor containing the immunoglobulin molecule that mediates binding specificity. Immunoglobulins are capable of binding to a variety of structures, including soluble or membrane-bound proteins, carbohydrates, and lipids, as well as other organic and even some non-organic compounds. Moreover, immunoglobulins are secreted as antibodies that can destroy the pathogens they bind to by using one of several different effector functions: neutralization, opsonization, ADCC, or complement activation. In this chapter, we discuss the structure and nature of the antigen recognition receptor on T cells, the T cell receptor (TCR). T cell receptors are expressed by all T cells except the earliest precursor cells, so that most *thymocytes* (immature T cells developing in the thymus) and all mature Th and Tc cells in the periphery bear TCRs. Like their B cell counterparts, the antigenic specificities of T cells are clonal in nature, meaning that members of an individual T cell clone carry many identical copies of a receptor protein with a unique binding site. The binding site is complementary to the overall shape created when one of a particular set of antigenic peptides of similar primary structure lodges in the peptide-binding groove of a specific MHC allele. Like the Ig genes, the TCR genes undergo V(D)J recombination of gene segments to produce a repertoire of receptors with immense sequence diversity. Moreover, just as the BCR contains not only the Ig H_2L_2 monomer but also the accessory Igα and Igβ proteins, the TCR $\alpha\beta$ (or $\gamma\delta$) polypeptides are also part of a larger multiprotein complex in which the TCR molecule supplies the specificity of antigen binding but additional proteins called the *CD3 proteins* are required for signal transduction.

Although they have a similar function, major differences exist between TCRs and BCRs. First, while the repertoire of B cell antigen receptors can recognize and bind to virtually any structure, the T cell repertoire is much more restricted. T cells recognize only antigenic peptides complexed to self-MHC molecules that are displayed on the surfaces of antigen-presenting cells (APCs) or target cells. Secondly, unlike B cells, T cells do not secrete their antigen-specific receptors in a form analogous to antibodies. Instead, when the TCR-mediated recognition of a peptide–MHC complex (pMHC) has occurred, the activated T cells proliferate and their progeny differentiate effector functions that do not involve the TCR.

Much interest has been focused on the TCR because of its central role in the adaptive immune response. Obviously, the TCR is fundamental to the recognition of antigen by Th and Tc cells, and a functional Th cell population is required for most humoral and cytotoxic cell-mediated responses. As will be discussed in detail in Chapter 13, the recognition of antigen by the TCR is also central to the processes of positive and negative selection that occur in the thymus and establish immune tolerance. Thus, to fully understand the immune response, a solid comprehension of the structure of the TCR complex and the genes encoding its protein chains is essential.

A. Historical Notes

I. INTRODUCTION

As was described in Chapter 6, the importance of thymus-derived lymphocytes in an immune response was discovered in the early 1960s. Although much of our early understanding of the adaptive immune response was gained from the study of B lymphocytes and how they produce antibodies, it was soon appreciated that the thymus-derived lymphocytes also recognized antigens in a specific manner. This property dictated the presence of an antigen receptor on T cells, which was aptly termed the T cell receptor.

As described in Chapter 10 (refer to Box 10-2), it was in the mid-1970s that Zinkernagel and Doherty discovered that a T cell recognizes antigen and MHC on the surface of a target cell or APC <u>simultaneously</u> as a unit. This revelation sparked a decade of intense and widespread speculation about the molecular nature of the T cell antigen receptor. To be complete, any model of T cell antigen recognition had to somehow account for the interaction of the T cell with both antigen and MHC. A commonly held hypothesis was that two separate receptor molecules existed on the T cell surface: one that interacted with an antigenic epitope and another that bound to the MHC molecule on the stimulatory cell. In 1981, an elegant test of this model was devised in the lab of John Kappler and Phillipa Marrack of the National Jewish Hospital in Denver. First, Kappler and Marrack created T cell hybridomas by fusing antigen-specific, MHC class II-restricted T cells with immortal T cell tumors. Each T cell hybrid therefore recognized one particular known antigen in the context of a particular known MHC molecule. Throughout their experimental work, Kappler and Marrack also ensured that all cells growing in a given culture were derived from the same parental cell and were therefore clonal. In a further step, they developed a technique to produce a "double hybrid," in which cells of one of the cloned T cell hybrids were fused to normal T cells of a different pMHC specificity. In other words, they fused a T cell hybrid recognizing (MHC A + X) with a normal T cell recognizing (MHC B + Y). If it were true that two separate, non-interacting receptors on a T cell accounted for the simultaneous recognition of peptide and MHC, then each double hybrid should bear independent receptors for MHC A, MHC B, peptide X, and peptide Y, and thus be able to recognize any combination of these peptides and MHC molecules. However, none of the hybrids screened displayed the mixed specificity that would have allowed them to recognize cells bearing (MHC A + Y) or (MHC B + X). This result made it clear that T cell recognition did not involve separate receptors specific for MHC and peptide, forcing a critical re-evaluation of the models attempting to describe the molecular nature of the TCR.

While Kappler and Marrack's elegant experimentation at the cellular level ruled out a favored but incorrect model, it did not solve the problem of understanding the TCR structure. Several competing models could still be envisioned. Was the TCR a single receptor molecule with a single binding site? Or was it a single receptor molecule with distinct MHC- and peptide-binding sites? Perhaps it was a multimeric molecule that formed a single binding site with a shape uniquely complementary to a particular peptide/MHC combination? At this point, what was needed was a more direct biochemical or genetic approach to the problem that would allow absolute characterization of the TCR at the molecular level. The obvious solution was to purify the TCR proteins and clone the genes encoding them, tasks that turned out to be extremely difficult despite the intense efforts of many laboratories. Because of its monumental importance, and the long drought without progress, the pursuit of the understanding of the nature of the T cell receptor became known as the "Holy Grail of Immunology."

Several factors contributed to the lack of early success in characterizing the TCR proteins and genes. First, unlike B cells, which secrete an abundance of their antigen receptors, T cells contain relatively small numbers of antigen receptors fixed in the plasma membrane. Secondly, unlike the relatively easy assays in which the specificity of a B cell is determined by the binding of its soluble antibody, T cell specificity has to be determined using more complex functional assays (refer to Box 10-1). For example, to determine whether a given Th cell clone is specific for a test antigen, one measures the results of Th cell activation by assaying cellular proliferation, the secretion of certain cytokines, or the activation of B or Tc cells. Similarly, the specificity of a Tc cell is assayed by monitoring the lysis of target cells or measuring the secretion of IFNγ in response to a given antigen. These somewhat cumbersome assays are often further complicated by the relatively low affinity (compared to BCR affinities) that TCRs have for their antigens.

A third reason for the difficulty in characterizing the TCR proteins and genes was the mistaken belief at that time that the T cell receptor was closely related in structure to the Ig molecule. Several laboratories produced data suggesting that anti-idiotypic antibodies (directed against the variable regions of the Ig protein) could bind to a small percentage of T cells. These claims diverted a considerable amount of effort toward searching for Ig-related or Ig-associated molecules on T cells. It was not until 1983 that scientists were able to demonstrate that no functional rearranged Ig genes were present in T cells and that a separate set of genes therefore had to encode the TCR. Finally, by 1984, the combined efforts of several laboratories with expertise in molecular biology, protein chemistry, and immunology finally provided solutions to the long-standing mystery of the molecular nature of the T cell receptor.

II. DISCOVERY OF THE GENES AND PROTEINS OF THE TCR

Prior to the mid-1980s, the characterization of most proteins occurred via the following steps: identify the protein, sequence it, use this sequence to probe for the gene encoding the protein, and finally, sequence the gene. The discovery of the TCR was unusual for the time because both the gene and the protein were identified and sequenced almost simultaneously. While several researchers were attempting to purify an antigenic recognition structure from the surfaces of T cell hybridomas, other groups were attempting to fish out of the T cell genome a gene sequence that had the properties predicted for a TCR. Because these approaches were concurrent, a chronological description of the discovery of the TCR is necessarily winding and jumps back and forth from proteins to genes.

The development of hybridoma technology in the early 1980s allowed researchers to grow large quantities of T cells that all expressed (but did not secrete) the same TCR. By immunizing mice with these monoclonal T cells, one could raise antibodies recognizing unique TCR structures and use these antibodies to isolate and study the receptor proteins. In 1982–1983, several laboratories (including those of Kappler and Marrack, and James Allison at the University of Texas) generated a series of monoclonal antibodies, each of which bound to structures on the surfaces of cells belonging to one T cell clone only and to no other clones. It was soon realized that each of these "clonotypic" antibodies was binding to a region in the structure that was unique to each T cell, a feature predicted for the TCR. The clonotypic antibodies were used to isolate the clonally specific murine T cell molecules, which were subsequently shown to be composed of two disulfide-bonded glycoprotein subunits, an acidic chain of about 45 kDa and a basic chain of about 40 kDa. These proteins came to be called the *TCRα* and *TCRβ* chains, respectively. Subsequent peptide fragment analyses of these chains isolated from different T cell clones revealed that TCRs had regions of shared polypeptide sequences ("constant" regions) as well as sequences not shared by other T cell receptors ("variable" regions).

At the same time as these unique protein structures were being identified on T cell surfaces, two laboratories were taking a genetic, rather than a protein-based, approach to chasing the Grail. In 1984, Tak Mak and colleagues at the Ontario Cancer Institute in Toronto, and Mark Davis and colleagues at Stanford University, independently used the

techniques of modern molecular biology to directly clone the TCR genes, finally solving the T cell receptor mystery. The successful approaches of Mak and Davis rested on four assumptions: (1) the vast majority of non-receptor genes in B and T lymphocytes should be highly related, if not identical; (2) the TCR genes should be expressed in T cells only; (3) the TCR genes should rearrange at the DNA level to generate repertoire diversity like the Ig genes; and (4) although the Ig and TCR proteins were probably evolutionarily related and therefore might show a low degree of protein sequence similarity, the nucleotide sequences of their genes should be quite distinct and should not share a high enough degree of sequence similarity to hybridize in solution.

Based on the above assumptions, two novel methodologies called *cDNA subtraction* and *differential screening* were devised by Davis and Mak, respectively, to clone the TCR genes (Fig. 12-1). Both methods depended on the synthesis of artificial

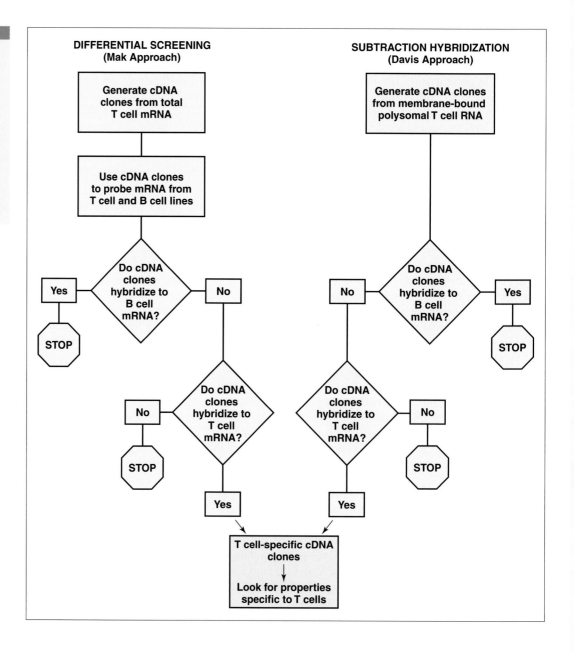

Figure 12-1

Experimental Techniques Used to Clone the TCRβ Gene
The TCRβ gene was cloned independently by the Mak laboratory and the Davis laboratory. The two approaches differ slightly in their initial steps but both ultimately end with the isolation of cDNA clones derived from T cell mRNAs that fail to hybridize to B cell mRNAs.

DNA copies, called *cDNAs*, of all the mRNAs present in a T cell clone. In the cDNA subtraction experiments of Davis and colleagues, the mRNAs of B cells were mixed with the cDNAs of T cells under conditions such that T cell cDNAs of similar sequence to B cell mRNAs would hybridize and anneal, leaving free only those cDNAs that were found in T cells but not in B cells (about 2% of expressed genes). In the case of the differential screening experiments of Mak and colleagues, cDNAs from a T cell line were used to screen the mRNAs present in a B cell line compared with the T cell line. Only those clones that hybridized strongly to T cell mRNA but minimally to B cell mRNA on a Northern blot were selected for further analysis. In both the Davis and the Mak experiments, several cDNA clones were isolated that corresponded to T cell-specific proteins of the appropriate size for the putative receptor chains. The researchers then examined these genes for the properties postulated to exist for T cell receptors. That is, they determined whether the T cell-derived cDNAs contained nucleotide sequences that (1) were analogous to the V, D, and J regions of the Ig genes, and (2) were not in the germline configuration found in non-T cells (suggesting that DNA rearrangement had occurred). Such cDNAs were indeed identified and the amino acid sequences of the putative TCRβ chains were predicted for both humans and mice.

Very shortly after the publication of the deduced protein sequences of the putative TCR genes by Mak and Davis, Ellis Reinherz reported the amino acid sequence of a partial TCRβ polypeptide isolated using the T cell clonotypic antibody approach. The amino acid sequence of this polypeptide was identical to the deduced amino acid sequence of the human TCRβ gene. Thus, the protein studies confirmed that the Mak laboratory was the first to clone the human TCRβ gene while the Davis laboratory had independently identified the mouse TCRβ gene. Several laboratories then went on to clone the TCRα gene. Both components of the αβ TCR were therefore identified at the protein and gene levels by the end of the momentous year 1984 (Fig. 12-2), allowing immunologists to say with confidence that the Grail had indeed been seized.

The availability of the genes encoding the TCRα and TCRβ polypeptides allowed the direct testing of whether the TCRαβ heterodimer itself bound to both antigenic peptide and MHC. The genes encoding the TCRα and β chains of a T cell clone of known specificity could be transferred into another clone of a different antigenic specificity to see if the recipient cell could gain the antigenic specificity of the donor T cell. This was indeed found to be the case, which meant that a single TCRαβ heterodimer contained sufficient structural information to simultaneously recognize <u>both</u> the peptide and the MHC protein. The remaining structural puzzle was solved soon after when it was shown that the antigenic peptide and MHC form a single combined epitope that interacts as a unit with both the TCRα and β chains simultaneously. By the early 1990s, the crystal structure of a TCRαβ heterodimer bound to its specific pMHC antigen revealed in three dimensions the molecular interactions involved in the recognition of ligand by T cells (see later in this chapter). With these experiments, the protracted quest to define the structure of the T cell receptor and to elucidate its recognition of antigens was essentially brought to a close. In more recent years, this knowledge of TCR structure has been applied to the development of tools for the analysis of the T cell specificity, including the quantitation of antigen-specific T cells (Box 12-1).

III. A SECOND T CELL RECEPTOR

During the search for the αβ TCR, two laboratories uncovered evidence for the existence of a second type of antigen receptor present on a distinct subpopulation of T cells. Using a subtraction hybridization approach similar to that used to clone the TCRβ cDNA, the laboratory of Susumu Tonegawa at the Massachusetts Institute of Technology isolated a T cell-specific cDNA which was distinct from that encoding the TCRβ chain, but which also possessed the properties of an antigen receptor. Analysis of this novel cDNA indicated that it appeared to undergo V(D)J rearrangement and exhibited low but recognizable

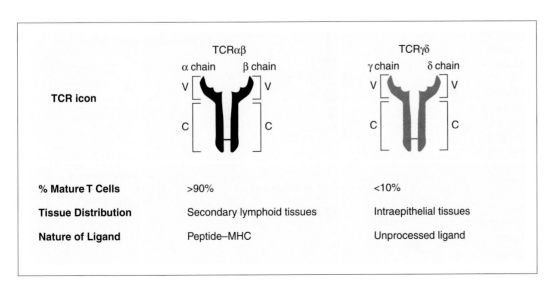

TCR icon

	TCRαβ	TCRγδ
% Mature T Cells	>90%	<10%
Tissue Distribution	Secondary lymphoid tissues	Intraepithelial tissues
Nature of Ligand	Peptide–MHC	Unprocessed ligand

Figure 12-2

Basic Characteristics of TCRαβ and TCRγδ
Two TCRs exist as defined by their component chains. Expression of these TCRs delineates two distinct T cell subsets: αβ T cells and γδ T cells. The proportions of these subsets in the total mature T lymphocyte population in mice are indicated, as are their general tissue distribution and type of ligand recognized.

Box 12-1. Quantitative analysis of antigen-specific T cells

Prior to 1996, limiting dilution analysis (LDA) was the usual method of quantitating the number of T cells generated during an immune response that were specific for an antigen of interest. In LDA, cells harvested from an immunized animal are diluted to a known concentration so low that when the cell suspension is dispensed into multi-well culture plates, individual wells do not contain more than a single T cell each. Individual cells (a proportion of which were primed in the animal) are then cultured and tested for their ability to proliferate in response to the antigen of interest *in vitro*, and the frequency of cells able to respond is calculated. However, the accuracy of LDA is limited by the fact that some antigen-specific T cells might not have been activated during priming in the animal, and would thus fail to proliferate in culture, leading to an underestimate of their frequency. (For technical reasons, T cells cannot be primed *in vitro*; only the proliferation associated with a secondary response can be detected.) Staining of T cells with their cognate ligands followed by flow cytometry, while a logical thought, does not work well in practice for quantitation because of the relatively rapid dissociation of the pMHC from the TCR. Mark Davis and his colleagues overcame this difficulty by constructing tetrameric pMHC complexes able to bind four TCRs simultaneously on a specific T cell. The probability of dissociation of all four TCRs at once from the pMHC multimer is remote, such that immunostaining with cognate antigen became a feasible means of quantitating antigen-specific T cells.

To construct the pMHC tetramers, Davis and colleagues attached a biotin molecule to the C-terminus of a genetically engineered, soluble form of the HLA-A2 α chain. The biotinylated HLA chain was mixed *in vitro* with β2-microglobulin and the peptide of interest to generate complete and loaded single pMHC structures bearing biotin. These structures were then mixed at a ratio of 4:1 with phycoerythrin-labeled avidin molecules, each of which, as the reader will recall, has four biotin-binding sites. By assembling tetramers with specific peptides derived from HIV proteins, a panel of customized staining reagents was created that could be used to identify peptide-specific T cell populations using flow cytometry (see Figure). For example, an HLA-A2-restricted CTL clone specific for a peptide derived from the HIV Pol protein was stained using a labeled A2/Pol tetramer. In contrast, no binding was seen using an HLA-A2/Gag tetramer produced

using a peptide from the HIV Gag protein, confirming the specificity of the staining. Furthermore, the cell populations that stained positively with the tetrameric reagent also exhibited functional killing activity *in vitro*.

Davis and colleagues then tried this staining method on freshly isolated blood samples from HIV-infected patients. Indeed, the tetramer method was able to distinguish and quantitate small populations of T cells directed specifically against either the Gag or Pol epitopes. The highest frequency of T cells found to be antigen-specific obtained in this experiment was 0.77% for CD8+ T cells binding the HLA-A2/Pol tetramer. These cells showed cytolytic activity specifically directed against Pol-loaded target cells *in vitro*. In a different patient, the percentage of T cells binding the HLA-A2/Gag tetramer was in the range of 0.2%. These CTLs lysed Gag-loaded,

but not Pol-loaded, target cells *in vitro*. Analysis of surface markers of the HLA-A2/Pol-specific cells showed that they were a mixture of predominantly memory and some effector T cells. The frequencies obtained using the tetramer methodology are consistent with the previous estimates of frequencies of HIV antigen-specific T cells arrived at by more arduous means. Thus, provided that the relevant peptide epitope has been identified and can be isolated or synthesized, the tetramer methodology represents a very useful approach for monitoring T cells specific for epitopes derived from tumors, pathogens, and autoantigens. Figure panels B and C reproduced with permission from Altman J. D. *et al*. (1996) Phenotypic analysis of antigen-specific T lymphocytes. *Science* **274**, 94–96.

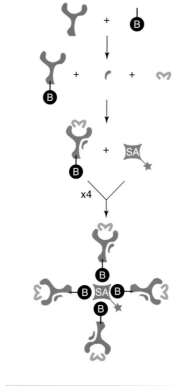

A. Assembly of Tetramer for TCR Staining

⊼ HLA-A2 α chain	↶ β2-microglobulin
∿ Pol peptide	SA Streptavidin
B Biotin	✦ Tag

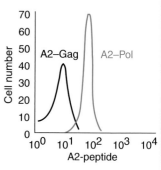

B. Binding of Tetramer to A2–Pol-Specific T Cells

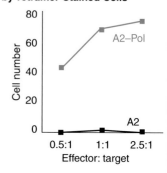

C. Lysis of A2–Pol-Bearing Targets by Tetramer-Stained Cells

sequence similarity to the V, J, and C regions of both the Ig and TCRβ chains. The gene encoding this new receptor chain was initially thought to be TCRα but was subsequently designated TCRγ after the true TCRα chain was cloned.

Meanwhile, the laboratory of Michael Brenner at Harvard University generated a monoclonal antibody that recognized a TCR-like heterodimeric protein present on only some human T cell clones. One of the heterodimer partners turned out to be the TCRγ chain, while its novel partner was designated TCRδ. These experiments, and similar work carried out in the laboratory of Mark Davis on mouse T cells, provided evidence for the existence of a second T cell receptor, designated *TCRγδ* (refer to Fig. 12-2). The detailed molecular structure of TCRγδ has now been established by X-ray crystal structure analysis. TCRγδ receptors are present on γδ *T cells*, a minor subpopulation (usually <10%) of mature T cells that does not express TCRαβ. Although γδ T cells constitute only a minority of T cells in the peripheral lymphoid tissues, they predominate in the intraepithelial tissues of the lung, gut, and reproductive tract where αβ T cells are rare. As was discussed briefly in Chapter 4, γδ T lymphocytes are strikingly different from αβ T cells in that, for the most part, the antigens recognized by their TCRs are not composed of peptides complexed with MHC. Rather, a γδ TCR is designed to bind to a broad range of cell surface molecules encountered in their natural, unprocessed forms. γδ T cells develop independently of αβ T cells and are thought to play a role in bridging innate and adaptive immunity. γδ T cells are discussed in detail in Chapter 18.

B. The Structure of T Cell Receptor Proteins

I. OVERVIEW

An individual T cell generally carries between 10,000 and 30,000 identical copies of a single TCRαβ or TCRγδ antigen receptor protein on its cell surface (never both). Unlike the Ig antigen receptor, which is made up of two light chains and two heavy chains and has two identical antigen binding sites, the T cell antigen receptor is a heterodimeric glycoprotein with a single antigen binding site. Each TCR on an individual T cell is composed of either a TCRα chain (49 kDa) linked via a disulfide bond to a TCRβ chain (43 kDa), or a TCRγ chain (40–55 kDa) linked via a disulfide bond to a TCRδ chain (45 kDa); TCRαδ or γβ structures do not exist in nature. Each TCR chain is composed of an Ig-like variable domain, an Ig-like constant domain, a cysteine-containing connecting sequence, a charged transmembrane portion, and a short cytoplasmic tail (Fig. 12-3). Both the constant and variable domains are arranged in an Ig-fold structure stabilized by an internal disulfide bond, making the TCR a member of the Ig superfamily. The variable region of the TCR, which is composed of the N-terminal ends of both TCR polypeptides, contains the single antigen-binding site that establishes contacts with both the antigenic peptide and the MHC molecule. Unlike the possibility of several different immunoglobulin isotypes in B cells, TCRs feature the same

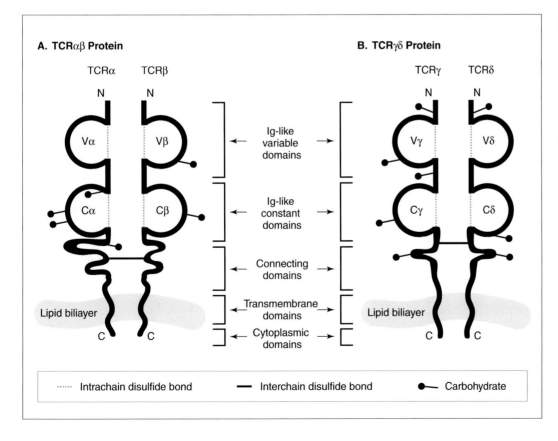

A. TCRαβ Protein

TCRα TCRβ

N   N

Vα   Vβ ← Ig-like variable domains →

Cα   Cβ ← Ig-like constant domains →

← Connecting domains →

Lipid biliayer

← Transmembrane domains →

C   C   ← Cytoplasmic domains →

B. TCRγδ Protein

TCRγ TCRδ

N   N

Vγ   Vδ

Cγ   Cδ

Lipid biliayer

C   C

······ Intrachain disulfide bond   — Interchain disulfide bond   ●— Carbohydrate

Figure 12-3

Schematic Representations of TCRαβ and TCRγδ Proteins
The Ig-like domains present in each of the TCRα, TCRβ, TCRγ, and TCRδ chains are indicated, as are glycosylation sites and disulfide bonds. Note that the two types of TCRs have identical domain structures. N, N-terminus; C, C-terminus. With information from Klein and Hořejší (1997) "Immunology," 2nd ed., Blackwell Science, Oxford.

constant domain throughout the life of the T cell clone, and the constant regions are not involved in "dictating" effector functions in the same way as the C regions of antibodies. A short connecting domain analogous to the Ig hinge appears between the TCR constant domain and the transmembrane domain. As in the mIg proteins, a transmembrane domain of about 21 amino acids anchors the TCR chain in the T cell membrane. TCR chains end with a short cytoplasmic domain (only about 5–12 amino acids) that is considered too short to be capable of signal transduction. The reader will recall that in the BCR, the Ig molecule provided the specificity of antigen recognition, while the Igα-Igβ chains used the ITAM motifs in their cytoplasmic tails to transduce the signal intracellularly. Similarly, while the extracellular domains of the TCRαβ or TCRγδ recognize the antigen, their relatively short cytoplasmic tails make it necessary for them to rely on the *CD3 complex* for signal transduction following antigen binding. The CD3 complex contains variable heterodimeric combinations of five protein chains designated *CD3γ*, CD3δ, CD3ε, CD3ζ (ζ, zeta), and *CD3η* (η, eta) (Fig. 12-4). Like Igα/Igβ, the CD3 chains contain ITAMs that facilitate intracellular signaling. The structure and function of the CD3 complex are discussed in detail later in this chapter.

i) TCRαβ

The V regions of the TCRα and TCRβ chains are about 110 amino acids long, while the C regions of both chains average 159 amino acids. The intrachain disulfide bonds

forming the V and C domain loops span about 60–75 amino acids. The interchain disulfide bond that links the TCRα and TCRβ polypeptides is positioned at the cysteine residue in the 33 amino acid connecting domain between the constant domain and the transmembrane domain (refer to Fig. 12-3A). The constant and transmembrane domains contain amino acids whose side chains establish contacts with the CD3 chains. In addition, the extracellular regions of both the TCRα and TCRβ chains contain multiple N-glycosylation sites.

As is the case for the Ig heavy chain, the variable domain of each TCR polypeptide contains hypervariable (HV) regions similar to the complementarity-determining region (CDRs) present in Ig molecules (Plate 12-1). In the TCR, there are four such regions of increased amino acid variability called (in order from the N-terminus): CDR1, CDR2, HV4, and CDR3. HV4 was identified somewhat later than the CDR regions, which accounts for its different name. The CDR/HV regions are not as tightly defined as are the equivalent CDRs in Ig proteins and tend to involve a slightly greater number of amino acid positions. This property may be responsible for the observation that, in comparison to BCRs, TCRs display a higher degree of cross-reactivity in antigen recognition. All CDRs and HV4 regions of both the TCRα and TCRβ chains (a total of six CDRs and two HV4 regions) are almost always involved in antigen recognition. Functional and structural studies suggest that each CDR and HV4 region can contact a different site on the pMHC complex. In some cases, the CDR1, CDR2, and HV4 regions primarily recognize polymorphic epitopes on MHC class I or class II proteins, while the highly diverse CDR3 regions preferentially make contact with the antigenic peptide nestled within the binding pocket of the MHC molecule. In

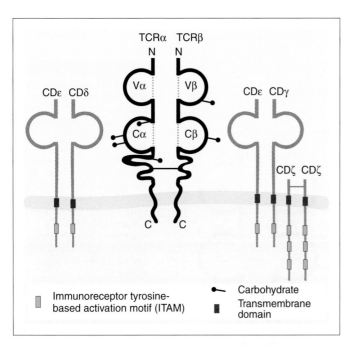

Figure 12-4

The TCRαβ–CD3 Complex
The TCRαβ heterodimer cannot be expressed on the T cell surface in the absence of the CD3 complex. This complex is usually composed of a CDε-CDδ heterodimer, a CDε-CDγ heterodimer, and a CDζ homodimer. Signaling initiated by the binding of pMHC to the TCR antigen binding site is conveyed to the interior of the cell via the ITAMs of the CD3 molecules. Note that although a CDζ homodimer is shown, it can be replaced with a CDζ-CDη heterodimer. The CDη chain contains two rather than three ITAMs.

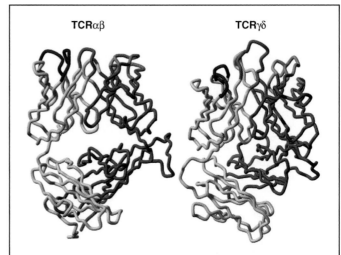

Plate 12-1

Structures of TCRαβ and TCRγδ
Crystal structures showing the carbon backbone of TCRαβ and TCRγδ. TCR α and δ chains are light grey. TCR β and γ chains are dark grey. For the α and δ chains, CDR1 is dark blue, CDR2 is magenta, and CDR3 is green. For the β and γ chains, CDR1 is cyan, CDR2 is pink, CDR3 is yellow, and HV4 is orange. Reproduced with permission from Rudolph M. G. and Wilson I. A. (2002) The specificity of TCR/pMHC interaction. *Current Opinion in Immunology* 14, 52–65.

other cases, the CDR1 and/or CDR2 regions may bind to part of the peptide, while the CDR3 region may make contact with the MHC molecule. Structural studies examining the binding of the TCR to pMHC are discussed later in this chapter.

ii) TCRγδ

In humans, the TCRγ and δ chains may be covalently linked by a disulfide bond or non-covalently linked, depending on which of two exons is used to encode the constant region of the TCRγ chain. In mice, the TCRγ and δ chains are always covalently linked (refer to Fig. 12-3B). The TCRγ and δ chains also feature CDR1, CDR2, HV4, and CDR3 hypervariable regions in their variable domains, but, because TCRγδ ligands are often non-peptides, their precise roles in antigen binding are thought to be slightly different (refer to Plate 12-1). Antigen recognition by γδ TCRs is discussed in Chapter 18.

Table 12-1 **Chromosomal Localizations**

Genetic Locus	Human Chromosome	Mouse Chromosome
TCRA	14	14
TCRB	7	6
TCRG	7	13
TCRD	14	14
CD3γ,δ,ε	11	9
CD3η,ζ	1	1
CD4	12	6
CD8	2	6

domains. The CD3 genes are conventional in structure and do not contain gene segments or undergo V(D)J recombination. The chromosomal locations of the various TCR and CD3 gene loci are listed in Table 12-1 along with those of the CD4 and CD8 loci, which are discussed later in this chapter.

C. Genomic Organization of the TCR and CD3 Loci

Like the genes encoding the Ig proteins, the genes encoding the TCR proteins are composed of a constant exon and a variable exon, with the variable exon being made up of smaller gene segments. The TCR variable exons are assembled at the DNA level by V(D)J recombination of the variable (V), joining (J), and, where present, diversity (D) gene segments, just as occurs in the Ig loci. The TCR constant exons are made up of three to four sub-exons encoding C region, transmembrane, and cytoplasmic

I. THE TCRα LOCUS

i) The Human TCRα Gene

The human TCRα locus is located on the long arm of chromosome 14. It contains about 42 functional Vα gene segments, each of which is composed of a leader sequence encoding a leader peptide of 20 amino acids, a first Vα sub-exon encoding about 15 amino acids, an intron of 100 bp, and a second Vα sub-exon encoding about 98 amino acids (Fig. 12-5A). The Vα gene segments are followed by a stretch

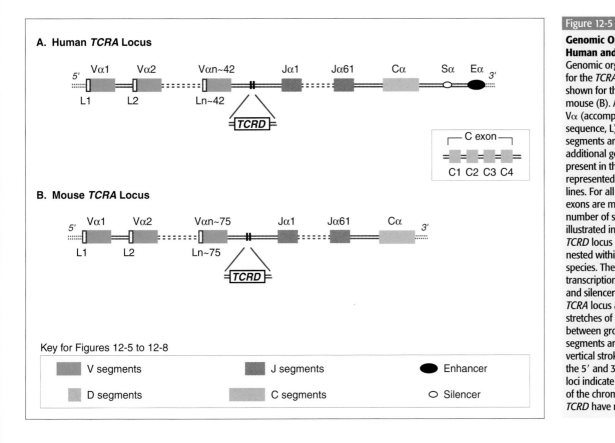

A. Human *TCRA* Locus

Vα1   Vα2   Vαn~42   Jα1   Jα61   Cα   Sα   Eα

L1   L2   Ln~42

TCRD

C exon
C1 C2 C3 C4

B. Mouse *TCRA* Locus

Vα1   Vα2   Vαn~75   Jα1   Jα61   Cα

L1   L2   Ln~75

TCRD

Key for Figures 12-5 to 12-8

| ▨ | V segments | ▨ | J segments | ● | Enhancer |
| ▨ | D segments | ▨ | C segments | ○ | Silencer |

Figure 12-5

Genomic Organization of the Human and Mouse *TCRA* Loci Genomic organization schemes for the *TCRA* (TCRα) locus are shown for the human (A) and mouse (B). A few functional Vα (accompanied by a leader sequence, L) and Jα gene segments are shown, with additional gene segments present in the areas represented by the dashed lines. For all TCR loci, the C exons are made up of a variable number of smaller sub-exons as illustrated in the inset. The *TCRD* locus (see Fig. 12-9) is nested within *TCRA* in both species. The positions of the transcriptional enhancer (Eα) and silencer (Sα) in the human *TCRA* locus are indicated. Long stretches of DNA that occur between groups of gene segments are indicated by two vertical strokes. Dotted lines at the 5′ and 3′ ends of the TCR loci indicate the continuation of the chromosome. *TCRA* and *TCRD* have no D segments.

of 61 Jα gene segments, each of which encodes about 17–21 amino acids. Downstream of the Jα gene segments is a single Cα exon, composed of four sub-exons and three introns. Following the Cα exon are two regions that play a role in regulating TCRα gene transcription: an α silencer (which represses TCRα gene transcription) and an α enhancer (which promotes TCRα gene transcription). There are no Dα gene segments, making the TCRα locus analogous to the Ig light chain loci. If 75% nucleotide sequence similarity is taken to constitute a gene subfamily, the 42 Vα gene segments can be classified into 32 subfamilies, with the majority existing as single member subfamilies. Thus, most Vα gene segments are quite dissimilar in structure, contributing to the substantial variability and potential diversity in antigen recognition in the T cell repertoire. The TCRα locus has one other highly unusual feature: the TCRδ locus is nested within it, positioned between the Vα and Jα gene segments. The TCRδ locus is discussed in detail later in this chapter.

ii) The Mouse TCRα Gene

The TCRα locus in the mouse is found on chromosome 14 and has a genomic organization very similar to that of the human TCRα locus (Fig. 12-5B). The murine TCRα locus has about 75 functional Vα gene segments that can be classified into 25 subfamilies, and 61 functional Jα gene segments. Again, a single constant region exon Cα exists but there are no Dα gene segments. A murine TCRα enhancer and α silencer have yet to be reported. The murine TCRδ locus is also nested within the TCRα locus between the Vα and Jα gene segments.

II. THE TCRβ LOCUS

i) The Human TCRβ Gene

The TCRβ locus in humans is located on the long arm of chromosome 7. It spans about 685 kb and contains a total of 48 Vβ gene segments in 35 subfamilies (Fig.12-6A). Downstream of the Vβ gene segments are two separate clusters of Dβ and Jβ gene segments, each of which is followed by a Cβ constant exon. These two DJ clusters are termed Dβ1/Jβ1 and Dβ2/Jβ2, respectively. The Dβ1/Jβ1 cluster contains one functional Dβ1 and six functional Jβ1 segments, while the Dβ2/Jβ2 cluster contains one functional Dβ2 and seven functional Jβ2 segments. The Jβ gene segments range in length from 32 to 48 bp, while the Dβ gene segments are shorter. Functional V segments can rearrange with DJ units originating from either DJβ cluster. The Cβ1 and Cβ2 exons both contain four sub-exons and three introns. The powerful β core enhancer (Eβ), which lies 3′ of Cβ2, contains binding motifs for several transcription factors important in T cell development (see later).

Cβ1 and Cβ2 are very similar and encode amino acid sequences that differ at only six positions, suggesting that they are most likely derived from relatively recent gene duplication during evolution. After a Vβ–Dβ1/Jβ1 rearrangement, both Cβ1 and Cβ2 exons remain downstream so that either may be transcribed. Which Cβ exon is chosen appears to occur at random. There is no apparent functional reason why a T cell might need to choose one Cβ over the other, and there are no known T cell subsets that preferentially express one Cβ over the other. On the other hand, a Vβ–Dβ2/Jβ2 rearrangement results in the loss of the Cβ1 exon so that only Cβ2 can be transcribed. Correspondingly, in TCRβ proteins, the Cβ1 region is only found with Dβ1/Jβ1 sequences, while the Cβ2 region is found with either Dβ1/Jβ1 or Dβ2/Jβ2 sequences. It should also be noted that no mechanism analogous to isotype switching occurs in T cells, probably because the TCR is not secreted like an Ig that must interface with several different effector mechanisms. The sole function of the TCR constant domains appears to be association with the CD3 complex, and Cβ1 and Cβ2 have equivalent roles in this respect.

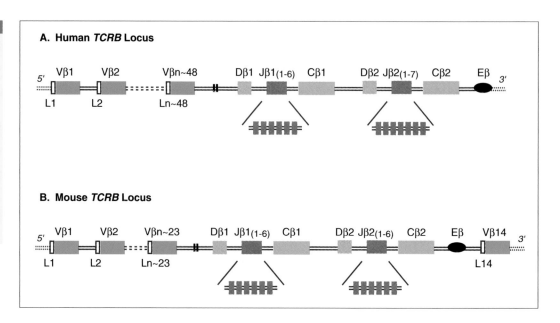

Figure 12-6

Genomic Organization of the Human and Mouse *TCRB* Loci

Genomic organization schemes for the *TCRB* (TCRβ) locus are shown for the human (A) and mouse (B). The Jβ gene segments occur in two clusters in which some gene segments are functional (shown) but others are pseudogenes (not shown). The Jβ clusters are not drawn to scale. The position of the core Eβ enhancer is shown for both species.

A. Human *TCRB* Locus

B. Mouse *TCRB* Locus

ii) The Mouse TCRβ Gene

In the mouse, the TCRβ locus is found on chromosome 6 and is organized like the human TCRβ locus (Fig. 12-6B). A total of 23 functional Vβ gene segments classified into 20 subfamilies is found primarily upstream of two DJβ clusters. The Vβ14 gene segment and its associated leader, however, are found 3′ of Cβ2, and the murine Eβ enhancer is found between Cβ2 and Vβ14. The murine Dβ1/Jβ1 cluster contains one Dβ1 and six Jβ1 gene segments followed by the Cβ1 exon, while the murine Dβ2/Jβ2 cluster contains one Dβ2 and six Jβ2 gene segments followed by the Cβ2 exon. Any functional Vβ segment can rearrange with a DJ unit drawn from either DJβ cluster. As is true in the human locus, the Cβ1 exon is only used in conjunction with Dβ1 and Jβ1 gene segments, while the Cβ2 exon can be transcribed with Dβ and Jβ gene segments from both clusters. Again, the Cβ1 and Cβ2 exons are highly similar in sequence and appear to be functionally equivalent.

III. THE TCRγ LOCUS

i) The Human TCRγ Gene

The human TCRγ locus is located on the short arm of chromosome 7 and occupies about 150 kb. Although there are at least 14 Vγ gene segments belonging to six subfamilies, only six Vγ are apparently functional (Vγ2–Vγ5, Vγ9, and Vγ11) (Fig. 12-7A). Downstream of the Vγ gene segments are two Jγ clusters: the Jγ1 cluster contains three Jγ segments and is followed by the Cγ1 exon, while the Jγ2 cluster contains two Jγ gene segments and is followed by the Cγ2 exon. Each Cγ exon is composed of three sub-exons and two introns, but only Cγ1 encodes the cysteine residue enabling disulfide linkage to the TCRδ chain. TCRγ chains containing Cγ2 use noncovalent bonds to dimerize with TCRδ. A γ enhancer has been identified 3′ of the Cγ2 exon, and at least two γ silencers are located between this enhancer and Cγ2. Like the TCRα

chain, the TCRγ chain resembles an Ig light chain in its lack of D segments.

ii) The Mouse TCRγ Gene

The TCRγ locus in the mouse is located on chromosome 13. Four discrete clusters of gene segments exist, each with its own constant region exon, numbered Cγ1–Cγ4 (Fig. 12-7B). These four clusters contain a total of seven Vγ and four Jγ gene segments. The first cluster contains (in order from the 5′ end of the locus) Vγ5, Vγ2, Vγ4, Vγ3, Jγ1, and Cγ1; the second contains Vγ1.3, Jγ3 (a pseudogene), and Cγ3; the third contains Cγ2 and Jγ2; and the fourth contains Vγ1.2, Vγ1.1, Jγ4, and Cγ4. The gene segments Vγ1.2, Vγ1.1, and Vγ1.3 are highly homologous members of the same subfamily. The Vγ gene segments undergo rearrangement only with Jγ and Cγ sequences that are proximal and in the same orientation. For example, Vγ1.3 joins only with Jγ2 and Cγ2, while Vγ1.1 joins only with Jγ4 and Cγ4, but any of Vγ2–Vγ5 can join with Jγ1 and Cγ1. Because of a defect affecting gene splicing, Cγ3 is a pseudogene and is not expressed, preventing the production of functional chains containing Vγ1.3. Thus, the total numbers of functional TCRγ gene segments and exons in the mouse are six Vγ, three Jγ, and three Cγ. Three γ enhancers (but no silencers) have been identified in the mouse TCRγ locus: one 3′ of Cγ1, one 3′ of Cγ3, and one 5′ of Cγ2.

IV. THE TCRδ LOCUS

In both mice and humans, the TCRδ locus is located on chromosome 14 but is positioned in a highly unusual way. The TCRδ gene segments and exons are nested within the TCRα locus between the Vα and Jα gene segments in both species (Fig. 12-8). The TCRδ locus contains its own Vδ, Dδ, and Jδ gene segments and Cδ exon but also shares the use of some Vα gene segments. However, Vδ gene segments recombine only with Jδ and Cδ sequences and never with Jα or Cα

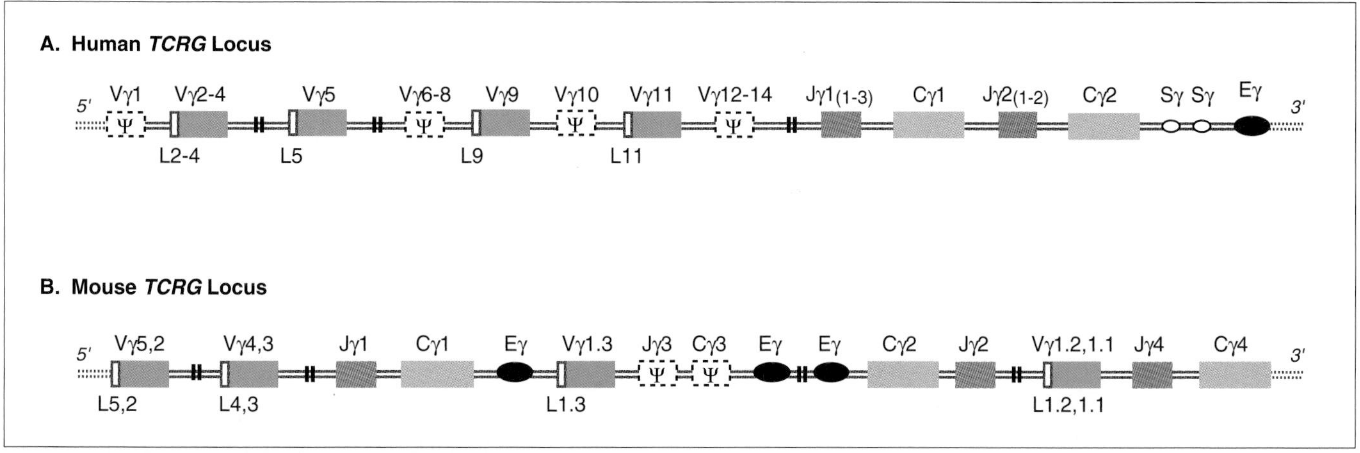

A. Human *TCRG* Locus

B. Mouse *TCRG* Locus

Figure 12-7

Genomic Organization of the Human and Mouse *TCRG* Loci
Genomic organization schemes for the *TCRG* (TCRγ) locus are shown for the human (A) and mouse (B). Gene segments known to be pseudogenes are indicated by Ψ. The positions of the TCRγ enhancer (Eγ) and silencers (Sγ; in the human locus only) are shown. Note that the transcriptional orientation of mouse Cγ2 is inverted.

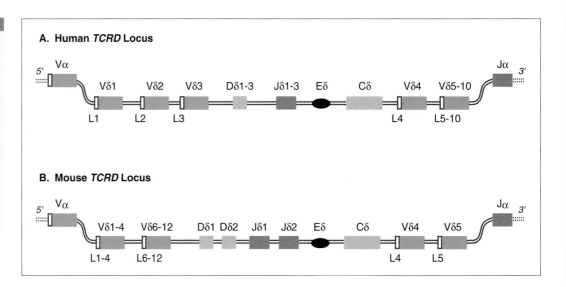

Figure 12-8

Genomic Organization of the Human and Mouse *TCRD* Loci
Genomic organization schemes for the *TCRD* (TCRδ) locus are shown for the human (A) and mouse (B). The *TCRD* locus is positioned between the Vα and Jα gene segments in both species. Only some of the Vδ gene segments shown are considered functional. The position of the TCRδ enhancer (Eδ) is indicated for both species.

A. Human *TCRD* Locus

B. Mouse *TCRD* Locus

sequences. This rather unconventional location of the TCRδ gene prevents the expression of both TCRδ and TCRα on the same T cell, since the recombination of the Vα gene segments with Jα gene segments deletes the entire TCRδ locus.

i) The Human TCRδ Gene

In the human TCRδ locus, there are at least ten Vδ, three Dδ, and three Jδ gene segments, and one Cδ exon, but only five Vδ are considered functional (refer to Fig.12-8A). Sequences encoded by five Vα gene segments have also been found in a small proportion of TCRδ chains. The three Dδ and three Jδ gene segments and the single Cδ exon are clustered together between the Vδ3 and Vδ4 gene segments. Interestingly, rather than the usual VDJ pattern of assembly, the three Dδ gene segments can be used in tandem, occasionally creating a VDDDJ joint. This arrangement increases the variability contributed by the germline repertoire of Dδ segments and, because of the additional joining reactions, dramatically enhances the opportunities for junctional diversity (see Chapter 8 and later). Both of these factors contribute to the overall diversity of the TCRδ chain repertoire. A δ enhancer has been identified between Jδ3 and Cδ.

ii) The Mouse TCRδ Gene

The organization of the TCRδ locus in the mouse is similar to that in humans and contains a total of twelve Vδ, two Dδ, and two Jδ gene segments and one Cδ exon (refer to Fig. 12-8B). The Vδ gene segments recombine only with the Dδ, Jδ, and Cδ gene segments, but sequences encoded by four Vα gene segments can be found in a small number of mouse TCRδ chains. Again, the joining process is unusual in that two Dδ gene segments can appear in tandem in the rearranged gene (VDDJ). A δ enhancer is located between Jδ2 and Cδ.

V. THE CD3 GENES

The genes encoding the CD3γ, δ, and ε chains share a high degree of sequence similarity and are thought to have evolved

by gene duplication. These genes are clustered together on chromosome 11 in humans and on chromosome 9 in mice (refer to Table 12-1). The CD3ζ and CD3η proteins are both derived from a single 10 exon CD3ζ-η gene located on chromosome 1 in both humans and mice. The amino acid sequences of CD3ζ and CD3η are identical for the first 122 residues but then diverge into distinct C-termini. The primary transcript of the CD3ζ-η gene is alternatively spliced to include exon 8 (but not exon 9) in the CD3ζ chain and exon 9 (but not exon 8) in the CD3η chain.

D. Expression of TCR Genes

I. MECHANISM OF V(D)J RECOMBINATION IN THE TCR LOCI

Just as the expressed Ig gene is assembled from V, D, and J Ig gene segments rearranged at the DNA level by V(D)J recombination, the expressed TCR gene is assembled from V, D, and J TCR gene segments brought together by V(D)J joining. The same machinery is thought to be used to assemble both Ig and TCR gene segments into their respective functional genes. Supporting evidence comes from the study of *scid* mice, which have an early defect in the V(D)J recombination process and lack rearrangements of both their Ig and TCR genes. Also, if one transfects germline TCR genes into immature B cell lines *in vitro*, rearrangement of both the exogenous TCR genes and the endogenous Ig genes can occur.

Because the same recombination machinery is used for Ig and TCR V(D)J joining, the TCR gene segments are flanked by the same recombination signal sequences (12-RSS and 23-RSS) that we encountered in Chapter 8 (refer to Figs. 8-8, 8-9, 8-10, and 8-11). That is, among the TCR gene segments, the same conserved heptamer and nonamer sequences containing 12 bp or 23 bp spacers appear on any 5′ or 3′ side that undergoes joining to other gene segments. In addition, the same RAG recombinase enzymes follow the same "12-23 rule" to juxtapose only those RSSs that are not of the same type. Like Ig V gene segments, TCR V segments are flanked on the 3′ side by a 23-RSS. However, while the Ig J_H segments are flanked on their 5′ sides by a 23-RSS (ensuring they cannot join to Ig V_H segments and must join to an Ig D_H segment), TCR J segments are flanked on the 5′ side by a 12-RSS (Fig. 12-9). This organization permits TCR J segments to directly fuse to TCR V segments. More importantly, unlike Ig D_H segments, which are flanked on both sides by a 12-RSS (eliminating the possibility of D–D joining), D segments in the TCR loci are flanked on the 5′ side by a 12-RSS and on the 3′ side by a 23-RSS. This means that a TCR V segment can join directly to a TCR J segment or to a TCR D segment, and that two (in mouse) and three (in humans) TCR D segments can be included in the variable exon before the TCR J segment is added. These modifications have profound effects on the generation of diversity in the T cell repertoire (see later), permitting it to exceed the level of diversity achieved in the Ig repertoire.

One puzzling observation remains: despite the apparent duplication of the V(D)J recombination apparatus in B cells and T cells, the Ig genes are not rearranged in mature T cells, and the TCR genes are not rearranged in mature B cells. Some as yet undefined regulatory mechanism must therefore distinguish between lymphocyte precursors that are destined to become B cells and those that become T cells. A concept called *locus accessibility* has been proposed by some observers to account for this discrimination. Just prior to rearrangement, the structure of the chromatin of a specific antigen receptor locus may change to allow it alone to serve as a substrate for the recombinase complex. Such alterations to local DNA structure have been correlated with transcriptional activity in the germline antigen receptor genes, and particularly with the presence of specific enhancers and promoters (see later). For example, mutant mice lacking the Eβ enhancer

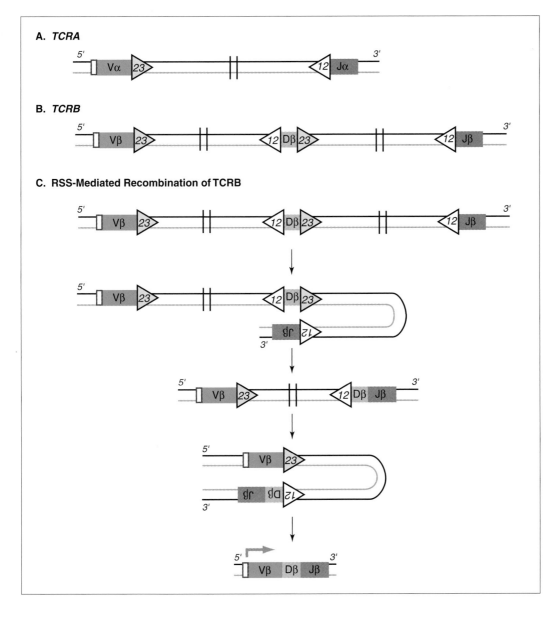

A. TCRA

B. TCRB

C. RSS-Mediated Recombination of TCRB

Figure 12-9

Recombination Signal Sequence Alignment in the TCR Loci
The 23- and 12-recombination signal sequences (RSSs) flanking V, D, and J gene segments in the murine *TCRA* (A) and *TCRB* (B) loci are shown. (C) Somatic recombination of V, D, and J gene segments is shown in the mouse *TCRB* locus. The RSSs flanking each gene segment are aligned following the "12/23" rule. Gene products excised by the action of the recombinase are not shown. The same RSSs occur in the *TCRG* and *TCRD* loci.

are impaired both in germline transcription of the TCRβ locus and in TCRβ gene rearrangement. Elements controlling locus accessibility, germline transcription in the TCR and Ig loci, and the timing and cell type-specific expression of RAG remain under investigation.

II. TCR GENE TRANSCRIPTION AND PROTEIN ASSEMBLY

Following V(D)J recombination to generate the V exon, the rearranged gene undergoes conventional transcription from the promoter associated with the participating V gene segment. A single contiguous primary transcript in which the V exon is linked to the C exon is generated (Fig. 12-10). The introns in the primary transcript are then spliced out to generate mature mRNAs that are translated into the TCR polypeptides. Note that since the TCR is not secreted, there is no need for alternative mRNA transcripts specifying either a membrane-bound or secreted protein.

How do the TCR gene segments correspond to the functional areas of the TCR polypeptides? The CDR1, CDR2, and HV4 hypervariability sites are all located within the V gene segment of a variable exon and contain primarily germline-encoded variation (derived from the random choice of a given V gene segment) (Fig. 12-11). In contrast, the CDR3 region encompasses the junctions of either the VJ (in TCRα and γ) or VDJ (in TCRβ and δ) gene segments so that diversity in CDR3 is contributed by junctional diversity mechanisms as well as by germline-encoded variation of the D and/or J gene segments. This hypervariability site is thus a major contributor to TCR binding site diversity.

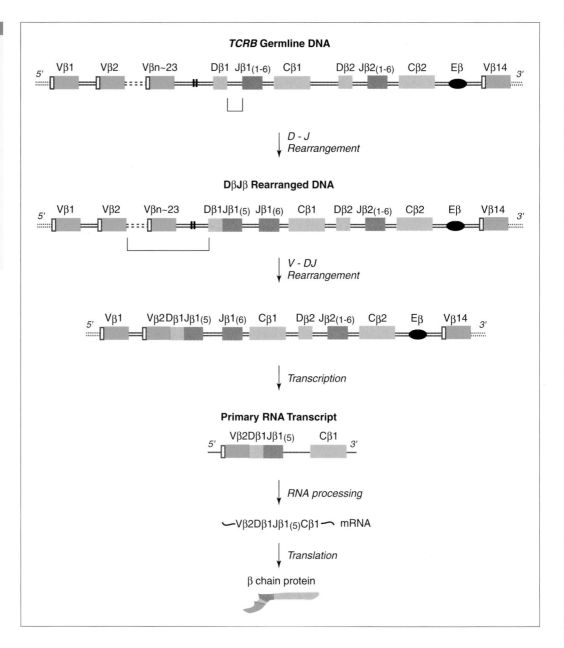

Figure 12-10

V(D)J Rearrangement in the Mouse *TCRB* Locus
An example of V(D)J recombination to generate a rearranged mouse TCRβ gene is shown. Gene segment Dβ1 has been arbitrarily selected to join with the fifth gene segment of the Jβ1 cluster [Jβ1(5)] and then to Vβ2. Transcription results in a primary transcript that contains the V exon [Vβ2:Dβ1:Jβ1(5)] and the Cβ1 exon. After conventional RNA splicing, an mRNA containing the indicated sequences is produced that is translated into a TCRβ protein.

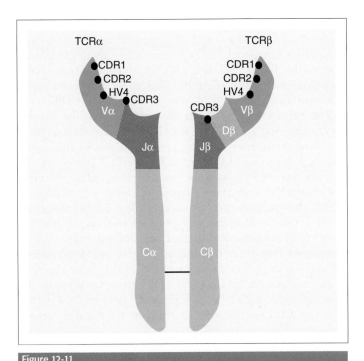

Figure 12-11

Correspondence of TCR Hypervariable Regions to TCR Gene Segments
In this schematic example, the areas of the TCRα and TCRβ proteins derived from the indicated gene segments are shown in different colors. The CDR1, CDR2, and HV4 hypervariable regions are clustered in the variable domains of each chain, while the CDR3 region encompasses the VJ joint in the TCRα chain and the DJ joint in the TCRβ chain. The joints are areas of imprecise joining and non-templated N and P nucleotide addition.

In considering the binding of TCR proteins to peptide–MHC, it is important to note that both MHC class I and II molecules are recognized by products of the same TCR genes. There is only one gene locus encoding the TCRα chain and one encoding the TCRβ chain, so that discrimination between MHC class I and II structures cannot be determined by different classes of TCRs encoded by different sets of genes. The discrimination in binding is determined primarily by the CD4 and CD8 coreceptors, which specifically bind to MHC class II and class I, respectively (see later and Ch.13).

E. Developmental Aspects of V(D)J Recombination in the TCR Loci

V(D)J recombination of the TCR loci is intimately tied to the process of T cell development. Although we discuss T cell development in detail in Chapter 13, we felt it would be helpful for the understanding of the sequence of TCR locus rearrangement to introduce certain elements of thymocyte maturation in this chapter.

When a T lymphocyte progenitor leaves the bone marrow and enters the thymus to become an immature thymocyte, its TCR genes are in the germline configuration. After the decision is made to rearrange TCR genes and not Ig genes, the very immature thymocyte or *pro-T cell* then has to decide whether to mature into a TCRαβ or a TCRγδ *pre-T cell*. There is convincing evidence that αβ pre-T cells and γδ pre-T cells are separate lineages derived from the common pro-T cell precursor that expresses no TCR at all. What influences a hematopoietic progenitor to become a pro-T cell, and a pro-T cell to become an αβ pre-T cell or a γδ pre-T cell, has yet to be precisely determined. It is clear that mice lacking elements of the V(D)J recombination machinery, such as *scid* mice or RAG1/2-deficient mutants, have neither αβ nor γδ mature T cells. Evidence is also accumulating that cell fate determination factors that can alter transcription patterns (like Notch1; refer to Ch.13) play major roles in T lineage commitment decisions. What remains a puzzle is whether rearrangement takes place in cells that are already committed, or whether the rearrangement process itself has a defining role in αβ or γδ commitment.

I. TCRβ LOCUS REARRANGEMENT

Irrevocable commitment of an αβ pre-T cell to the TCRαβ lineage and the continued maturation of the clone depend on V(D)J recombination resulting in a functional TCRβ gene, a process called *β-selection*. Like the *Igh* locus in B cells, in a pre-T cell destined to become an αβ T cell, the TCRβ locus is the first to undergo rearrangement that leads to protein synthesis. Some rearrangement of the TCRγ and δ loci may also occur in αβ pre-T cells at this time but functional chains are not produced. The γ silencer is thought to play a role in preventing the synthesis of TCRγ chains in αβ pre-T cells. In the rearranging TCRβ locus, a Dβ gene segment is joined to a Jβ segment, followed by the joining to the DJ unit of a Vβ segment. In order to test the productivity of the rearranged TCRβ gene, the pre-T cell must make use of a temporary TCR structure called the *pre-TCR*, analogous to the pre-BCR structure in pre-B cells.

The counterpart of the surrogate light chain used in B cells to convey newly rearranged heavy chains to the B cell membrane for productivity testing is the *pre-T alpha chain* (pTα). The pTα chain is expressed only in immature CD4⁻CD8⁻ thymocytes and not in CD4⁺CD8⁺ thymocytes or mature CD4⁺ or CD8⁺ T cells. Murine pTα is an invariant transmembrane receptor protein of 33 kDa with one extracellular Ig-like domain, a transmembrane region, and a 33 amino acid cytoplasmic tail containing two potential phosphorylation sites (Fig. 12-12). Murine (but not human) pTα also has an intracellular signaling domain called the "Src homology 3" domain, or SH3 domain. Expression of pTα is independent of V(D)J recombination. pTα functions as a "surrogate TCRα chain" and covalently associates with a newly rearranged and synthesized TCRβ chain (and the CD3 signaling components) to form a pre-TCR in the T cell membrane.

The pre-TCR acts as a productivity sensor, so that if a particular pTα/TCRβ chain combination succeeds in delivering the required signal (which is unknown and may or may not involve the binding of a ligand of some sort), the rearranged

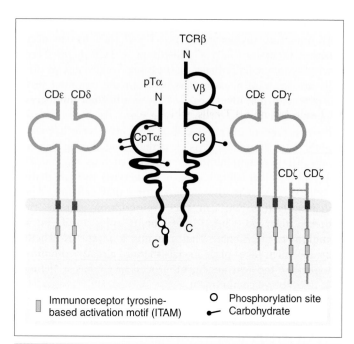

Figure 12-12

Structure of the Pre-TCR
Productivity testing of newly synthesized TCRβ chains requires the participation of the pTα chain and the CD3 complex to form the pre-TCR. The pTα chain differs from the TCRα chain in its lack of a V-like domain. Signaling initiated in an unknown way by the pTα-TCRβ heterodimer is conveyed by the ITAMs of the CD3 chains.

TCRβ gene is considered acceptable and the αβ pre-T cell is β-selected and becomes a committed αβ thymocyte. This commitment means that the mature cell will not later synthesize TCRγ chains even if the TCRγ genes have rearranged. Since the pTα cytoplasmic signaling domain is much larger than its counterpart in any of the TCRα, β, γ, or δ chains, it is tempting to speculate that the pTα tail may mediate a distinct intracellular signal that is not delivered in γδ pre-T cells (which do not express pTα). Mice in which the pTα gene has been knocked out show a 90% reduction in total thymocytes, since 90% of all thymocytes are normally αβ T cells. The development of the remaining 10% of thymocytes, which are γδ T cells, is totally unaffected by the absence of pTα. Curiously, a few pTα$^{-/-}$ thymocytes still manage to express a functional αβ TCR on the cell surface, implying that a parallel "low efficiency" path of β-selection and productivity testing exists.

The successful rearrangement of the TCRβ gene on one chromosome signals to the cell to suppress both V(D)J rearrangement of the TCRβ locus on the other chromosome and rearrangement of both TCRγ loci. This is another example of *allelic exclusion*, which was described in Chapter 9 for the *Igh* locus in B cells. In a pre-T cell in which TCRβ rearrangement on both chromosomes has been unsuccessful, the commitment to αβ lineage is not made, and the cell neither attempts to rearrange its TCRα genes nor becomes a γδ

T cell; instead, it dies by apoptosis. Thus, the synthesis of a functional TCRβ chain and its successful combination with pTα into a functional pre-TCR complex leads to the expression in the T cell of a single TCRβ chain sequence, allelic exclusion of the unrearranged TCRβ allele, and commitment to the αβ T cell lineage.

The existence of pTα and the pre-TCR may be nature's way of ensuring an adequate pool of αβ T cells while weeding out those thymocytes carrying non-productive TCRβ chains. During β-selection, in an event called *pre-TCR activation*, signals are delivered through the pre-TCR that induce the rapid proliferation of only those thymocyte clones with functional TCRβ chains. Since there is no equivalent to the pTα chain in γδ T cells, and apparently no equivalent expansion, the body's total population of thymocytes is biased toward the αβ lineage. As they enter the next stage of T cell development, αβ pre-T cells producing a functional TCRβ chain also start to synthesize both the CD4 and CD8 coreceptor proteins and thus become so-called "double positive" (DP) or CD4$^+$CD8$^+$ cells (see Ch.13). The stage is thus set for the next crucial steps in thymocyte maturation.

II. TCRα LOCUS REARRANGEMENT

Shortly after β-selection and completion of the proliferative burst of DP cells expressing the pre-TCR, RAG gene expression can be detected and V(D)J recombination commences at the TCRα locus on both chromosomes. The synthesis of pTα is downregulated. The juxtaposition of a Vα gene segment and a Jα gene segment excises the entire TCRδ locus (rearranged or not), preventing any possible synthesis of TCRδ chains. The γ silencer is thought to continue to repress the transcription of the TCRγ locus, whether or not it has been functionally rearranged. The rearranged TCRα gene is transcribed and, if the sequence is in-frame, a TCRα chain is synthesized and combined with the TCRβ chain to form a complete TCRαβ heterodimer. However, the productivity of the newly rearranged TCRα chain gene is not tested in the same way as TCRβ or Ig chains; that is, there appears to be no mechanism at this early stage for the cell to determine whether TCRα gene rearrangement was successful and whether the TCR heterodimer is functional. Furthermore, unlike TCRβ, the synthesis of a TCRα protein in a cell does not appear to suppress further TCRα rearrangements on either chromosome. Indeed, RAG1/2 expression and the rearrangement of the TCRα genes on both chromosomes continue for several days. It remains controversial whether receptor editing, which gives unproductively rearranged Ig light chains a "second chance," occurs in the TCR loci. In any case, TCRα rearrangement is maintained during the processes of positive and negative thymic selection. As is explained in detail in Chapter 13, negative selection eliminates self-reactive thymocytes (those binding strongly to self-peptide presented on self-MHC), while positive selection ensures that only thymocytes expressing functional TCRs that recognize self-MHC but not self-peptide are allowed to mature and reach the periphery.

If the TCRα genes on both chromosomes happen to rearrange productively, each may produce a TCRα protein that can combine with the TCRβ chain. In fact, a substantial number of DP thymocytes (10–20%) may exhibit surface expression of two different TCRαβ heterodimers (that share the same TCRβ chain). It is unclear how these cells fare during positive selection. Widely varying estimates of the frequency of dual Vα expression on mature T cells have been reported, but most researchers agree that dual expression happens less often on mature T cells than on thymocytes. Some peripheral T cells appear to exhibit cytoplasmic (but not surface) expression of a second TCRα chain with apparently lower affinity for the TCRβ chain. It is thought that either the positive selection mechanism or a competition for binding to the single TCRβ chain ensures that mature T cells generally express only one type of TCR heterodimer on the surface. Thus, the TCRα locus is said to exhibit "phenotypic" allelic exclusion rather than the "genotypic" allelic exclusion that occurs at the DNA level in the TCRβ locus. If TCRα locus rearrangement is unsuccessful on both chromosomes, the cell cannot be positively selected and dies by apoptosis.

After thymic selection, RAG1/2 expression is extinguished, leading to the cessation of all TCR locus rearrangements. The binding site sequence of the TCR expressed by a given thymocyte is now fixed and, since somatic hypermutation does not occur in T cells, will be the same in mature T cells derived from this thymocyte even after antigenic activation.

III. TCRγ AND δ LOCUS REARRANGEMENT

In pre-T cells destined to become γδ T cells, rearrangement is commenced in the TCRγ, TCRδ, and TCRβ loci at about the same time but completely independently. The TCRα locus does not undergo rearrangement, perhaps due to the regulatory effects of the α silencer (at least in humans). TCRγδ thymocytes and mature T cells that contain in-frame rearrangements of TCRβ have been isolated, but the significance of these rearrangements remains obscure. Thus, mature γδ T cells typically contain rearranged TCRγ and δ loci (and sometimes a rearranged TCRβ locus) but a germline TCRα locus, while mature αβ T cells have rearranged TCRβ, α, and γ loci (since TCRδ was deleted by TCRα rearrangement). VJ joining of TCRγ gene segments occurs in the usual way, but the TCRδ D gene segments in the mouse can be combined with each other to form a tandem DδDδ unit, which is in turn joined to Vδ and then finally to Jδ to form VδDδDδJδ. Another Dδ gene segment can be included in the tandem string in humans. There is apparently no functional equivalent to the pTα chain for the γδ TCR, so that any productivity testing of the TCRγ or δ chains before they are covalently associated and inserted in the pre-T cell membrane as a putative TCRγδ must rely on as yet undefined mechanisms. Similarly, it is not known how newly produced γδ T cell clones bearing non-functional TCRs are weeded out, since thymic positive and negative selection processes are thought to apply only to αβ thymocytes. More on the biology and development of γδ T cells appears in Chapter18.

IV. TCR LOCUS KNOCKOUT MICE

Studies of knockout mice in which the various TCR genes have been disrupted have confirmed the sequence of many of the above events. Mice in which the TCRα gene is mutated (without affecting the TCRδ locus) have normal numbers of γδ T cells but no αβ T cells, which emphasizes that the development of γδ and αβ T cells is completely independent. Moreover, the rearrangement and transcription of the TCRβ genes in TCRα$^{-/-}$ thymocytes is normal. However, thymocyte development is arrested at the DP stage such that thymocytes expressing both CD4 and CD8 as well as TCRβ (but not TCRα) are observed. Mice in which the TCRβ locus is disrupted also have reduced numbers of cells in the thymus, but thymocyte maturation is blocked at the earlier pre-T cell stage. Again, γδ T cell development is unaffected. TCRβ$^{-/-}$ mice continue to show significant amounts of TCRα gene rearrangement, a result that was originally surprising due to the prevailing assumption that TCRα rearrangement depended on TCRβ rearrangement. Similarly, mice in which the TCRδ locus is disrupted without affecting the expression of TCRα show a complete loss of γδ T cells but normal numbers of αβ T cells. These data confirm that the initiation of V(D)J recombination at any one TCR locus is independent of its initiation at any other TCR locus. However, certain functions of the αβ T cells in TCRδ$^{-/-}$ mice were abnormal, suggesting that γδ T cells may be necessary for some aspects of the regulation of αβ T cells during an immune response (see Ch.18).

F. Generation of Diversity of the T Cell Receptor Repertoire

As we learned in Chapter 8, diversity in the B cell receptor repertoire arises from four sources prior to encounter with specific antigen and from two sources after the B cell is activated by antigen. Since the mechanisms of isotype switching and somatic hypermutation do not operate in T cells, the diversity in the T cell repertoire is established entirely by those mechanisms that operate prior to antigenic stimulation. In other words, the sources of diversity that are relevant to T cells are multiplicity of germline segments, combinatorial diversity, junctional diversity, and αβ (or γδ) chain pairing. The theoretical contributions of these various mechanisms to the overall diversity of the TCR repertoire in the mouse are summarized in Table 12-2.

I. MULTIPLICITY OF GERMLINE GENE SEGMENTS

In TCRαβ, the number of different TCR V gene segments is much lower than the number of corresponding Ig V segments in the Ig loci, but the number of TCR J segments is greater than the number of Ig J segments. The overall contribution to the maximum theoretical repertoire from this source is 10- to 30-fold lower than for the Ig repertoire.

Table 12-2 Generation of TCR Diversity in the Mouse

Mechanism of Diversity*	α Chain	β Chain	γ Chain	δ Chain		
Multiple germline gene segments						
V	75	23	6	12		
D	None	2	None	2		
J	61	12	3	2		
C	1	2	3	1		
Combinatorial V-J or V-D-J joining	$75 \times 61 \times 1 = 4575$	$23 \times 2 \times 12 \times 2 = 1104$	$6 \times 3 \times 3 = 54$	V-J $12 \times 2 \times 1 = 24$	V-D-J $12 \times 2 \times 2 \times 1 = 48$	V-D-D-J $12 \times 2 \times 2 \times 2 \times 1 = \underline{96}$ 168
Chain pairing	$>(4.6 \times 10^3) \times (1.1 \times 10^3)$		$>(54) \times (1.1 \times 10^2)$			
N- and P-nucleotide addition	✓	✓	✓	✓		
Imprecise joining	✓	✓	✓	✓		
Total estimated diversity	TCRαβ = $\sim 10^{20}$		TCRγδ = $\sim 10^{18}$			

* Numbers in combinational joining row are approximate.

II. COMBINATORIAL DIVERSITY

The theoretical combinatorial diversity derived from the random juxtaposition of TCR V, D, and J segments during V(D)J recombination can be calculated just as for the Ig genes. For example, for the mouse TCRα chain, the number of possible combinations is 75 Vα × 61 Jα × 1 Cα = 4575, while that for mouse TCRβ is 23 Vβ × 2 Dβ × 12 Jβ × 2 Cβ = 1104. Using this methodology, one might also conclude that there are (considering functional segments only) 6 Vγ × 3 Jγ × 3 Cγ = 54 possible combinations for the mouse TCRγ chain, and 12 Vδ × 2 Dδ × 2 Jδ × 1 Cδ = 48 theoretical combinations for the mouse TCRδ chain. However, this calculation does not take into account either the fact that Cβ1 is found only in conjunction with Dβ1 and Jβ1, or the preference of Vγ segments to rearrange only with segments in the proximal DJγ cluster. In addition, the gene segments that make up γδ TCRs do not appear to be chosen entirely at random. Different γδ TCRs appear to contain specific Vγ and Vδ gene sequences depending on the cellular subset or anatomic location in which they are found (Table 12-3). For example, over 70% of murine γδ peripheral T cells have been found to express TCRδ chains containing Vδ2, with another 30% using Vδ1. While γδ TCRs containing Vδ1 may contain any one of the six Vγ gene

segments, those γδ TCRs containing Vδ2 almost always contain Vγ9, rather than other Vγ sequences. These strictures mean that diversity derived from combinatorial sources is more limited than the theoretical. Fortunately, what is lost in combinatorial diversity is more than compensated for by variable D segment inclusion and concomitant junctional diversity.

The variable inclusion of one or more TCR D segments represents a combinatorial mechanism available only to γδ T cells and not to αβ T cells or B cells. Although the Ig loci contain higher numbers of D gene segments, these segments may join in only one way to an Ig J segment. Diversity in the γδ TCR repertoire is increased because a TCR Dδ segment may, or may not, be included in a given TCRδ chain. For example, in mice, a Vδ segment may join directly with a Jδ segment to form a variable exon of VJ, or to a Dδ segment to form a VDJ variable exon, or even to another Dδ segment to form a VDDJ variable exon; that is, there are three possibilities for a TCRδ chain as opposed to one for an Ig chain. In humans, even more diversity arises from this source because three D segments can appear in tandem in a VDDDJ variable exon.

III. JUNCTIONAL DIVERSITY

The reader will recall from Chapter 8 that junctional diversity arises from the introduction of non-templated nucleotides during imprecise joining of gene segments. Both P nucleotides (inserted as a result of imprecise joining during V(D)J recombination) and N nucleotides (a string of random nucleotides inserted at each VDJ joint by the enzyme TdT) are responsible for the appearance in TCR chains of amino acids that are not germline-encoded. Assuming that an average of six non-germline nucleotides is inserted at each joint, TCRα and TCRγ chains may contain two randomly chosen amino acids at the VJ joint, while TCRβ chains may contain four new

Table 12-3 Use of TCRγ and δ Gene Segments in Murine Tissues

Tissue	Preferential γ Usage	Preferential δ Usage
Skin	Vγ3Jγ1Cγ1, Vγ5Jγ1Cγ1	Vδ1Dδ2 Jδ2Cδ
Uterine	Vγ6Jγ1Cγ1	Vδ1Dδ1Jδ2Cδ
Tongue	Vγ6Jγ1Cγ1	Vδ1Dδ1Jδ2Cδ
Intestine	Vγ7JγnCγn	Vδ4DδnJδnCδ

amino acids over the VDJ joint. Because more than one Dδ segment may be included in tandem in a TCRδ chain, additional opportunities for P and N nucleotide addition occur. TCRδ chains may contain up to four, six, or eight non-germline amino acids depending on whether the joint is VDJ, VDDJ, or VDDDJ, respectively (noting that the mouse TCRδ locus has only two D segments). Since each of these extra residues could be any one of the 20 amino acids, junctional diversity provides at least $20 \times 20 \times 20 \times 20$ possible combinations for each joint plus a further 20×20 possibilities for each additional D segment included. In other words, the contribution of junctional diversity to the TCRα chain is on average 20^4 (1.6×10^5) possible combinations, but its contribution to the TCRδ chain can be as high as 20^8 (2.56×10^{10}) possible combinations.

IV. CHAIN PAIRING

As is true for the Ig genes, a final source of diversity in antigen recognition can be found in the random pairing of TCRα and β chains (or TCRγ and δ chains) within a given αβ (or γδ) T cell. In the case of an αβ T cell, a mature TCR molecule's antigen binding site is composed of the variable domains of both the α and β chains. Any one T cell synthesizes only one sequence of TCRα chain (that appears on the cell surface), and one sequence of TCRβ chain. However, if one assumes that any one of the vast number of possible sequences for a TCRα chain gene can occur in the same T cell as any one of the even more numerous possibilities for a TCRβ gene, the number of possible α/β chain combinations is thought to approach about 10^{20}. Using the same rationale, the repertoire of functional TCRγδ heterodimers is estimated to be about 10^{18}. These numbers compare very favorably to the estimated total diversity of the Ig repertoire, speculated to be in the order of 10^{11} specificities.

The immense potential diversity generated by the random rearrangement and assembly of TCR genes in immature thymocytes is tempered in the repertoire of TCRs of mature T cells in the periphery. Just as maturing B cell clones undergo positive and negative selection in the bone marrow, thymocytes undergo positive and negative selection in the thymus, which limit the TCR sequences available to encounter antigen invading the body. Moreover, because there is no somatic hypermutation of the TCR genes, there are no known opportunities for the affinity maturation process that contributes to the diversity of the B cell repertoire after antigenic stimulation. As a result, the diversity of TCR specificities in the blood of an adult human is thought to be reduced to about 25×10^6 different TCRαβ combinations.

G. Regulation of TCR Gene Expression

The rearrangement and expression of each of the TCR genes are distinct and strictly regulated events. Temporal control can be observed in the mouse embryo, where the rearrangement

of the TCRδ gene in some pre-T cells can be detected as early as embryonic day 14 (E14). Rearrangement of the TCRβ and TCRγ genes follows on E15–16, but rearrangement of the TCRα gene does not commence until E17. In addition to temporal regulation, lineage-specific controls of TCR gene transcription must also exist. As mentioned earlier in this chapter, T cells express either TCRαβ or TCRγδ (but not both) on the cell surface, even though the TCRβ gene may be rearranged in γδ T cells and TCRγ may be rearranged in αβ T cells. Because of these temporal and lineage-specific controls, the earliest thymocytes detected in a mammalian fetus bear TCRγδ, followed a few days later by thymocytes bearing TCRαβ.

Like other events in mammalian development, the expression of a functional TCR requires the coordinated expression of multiple genes. Considerable study has been made of the promoters and enhancers situated in the TCR loci. Coordinated gene expression is often controlled through the response of the promoters of the relevant genes to shared tissue-specific transcription factors. Scientists had speculated that, since the TCR genes are expressed only in T cells, T cell-specific promoters or nuclear transcription factors might be identified. In fact, although each V gene segment has an associated promoter, these promoters do not appear to be sequences unique to T cells. Furthermore, the V promoters are not very active (in either T cells or non-T cells) until V(D)J recombination brings them near to the enhancer located (generally) 3' of the C exon.

The power and specificity of TCR enhancers has been demonstrated in both *in vitro* transcription systems and *in vivo*. Experiments with artificial recombination substrates have shown that the recombination machinery's access to the VDJ gene segments is controlled in a *cis* fashion by the TCR enhancers. For example, the transcription of reporter genes under the control of non-TCR promoters became highly active in a T cell-specific manner when the promoters were spliced to the TCRα enhancers. More recently, the significance of enhancer control has been demonstrated *in vivo* in knockout mouse mutants missing either the TCRα or TCRβ enhancers. Mice heterozygous for a targeted gene disruption of the TCRβ enhancer were unable to rearrange the TCRβ locus on that chromosome, and homozygosity for this mutation led to a complete block in αβ T cell development at the stage of TCRβ locus rearrangement. Similarly, thymocytes lacking the TCRα enhancer were unable to rearrange their TCRα loci and failed to mature.

The human TCRα enhancer contains six consensus binding motifs for known nuclear transcription factors (from the 5' end): ATF/CREB (ATF, *activating transcription factor*; CREB, *cAMP-responsive element binding protein*); the closely related molecules Tcf-1 (*T cell factor*) and Lef-1 (*lymphoid enhancer factor*); CBF (*core binding factor*; also known as PEBP2); Ets-1 (introduced in Box 9-2); GATA-3 (a member of a family of transcription factors named for the sequence of their DNA binding sites); and bHLH (named for their *basic helix-loop-helix* binding target structures) (Fig. 12-13). Studies in genetically engineered mice have indicated that Tcf-1 and Lef-1 are particularly important for TCRα enhancer activity, as are

Figure 12-13

TCR Locus Enhancers
Binding motifs for the indicated transcription factors for various human and mouse TCR loci are shown. Details of the action of selected transcription factors are given in the text.

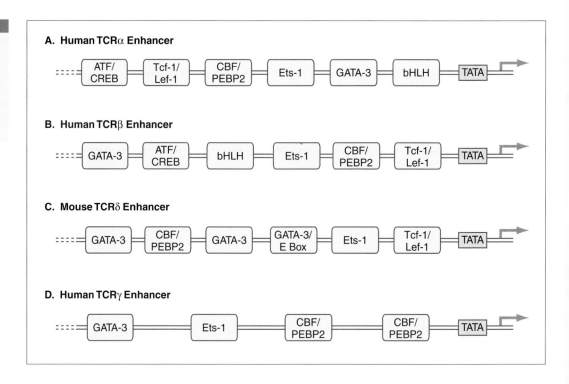

Figure 12-13

TCR Locus Enhancers
Binding motifs for the indicated transcription factors for various human and mouse TCR loci are shown. Details of the action of selected transcription factors are given in the text.

certain Ets family members. The TCRβ, γ, and δ enhancers contain similar collections of binding sites for the same and additional transcription factors. Even though the TCRδ gene is embedded within the TCRα gene, its transcription is controlled by the internal δ enhancer and not the 3′ α enhancer. The recombinational activity of the TCRδ enhancer is thought to be mediated by the transcription factors c-myb (described in Box 9-2) and CBF. Nuclear transcription factors relevant to aspects of T cell development other than TCR gene rearrangement are discussed in detail in Chapter 13.

While the TCR enhancers do appear to be responsible for increased transcription from V promoters in T cells, the nuclear transcription factors that bind to the identified motifs in the enhancers are not T cell-specific and none appears to be essential for TCR gene rearrangement. Furthermore, the enhancers appear to function equally well in αβ and γδ T cells in *in vitro* experiments. So why do only T cells rearrange and transcribe the TCR genes *in vivo*, and why do T cells express either TCRαβ or TCRγδ but not both? A hint may lie in the activity of the locus-specific silencers present in the TCR loci. The TCR silencers are DNA motifs thought to restrict the activity of enhancers to cells of the appropriate lineage. In many αβ T cells, the TCRγ gene may be productively rearranged and yet no functional TCRγ mRNA can be detected. The γ silencer 3′ of the Cγ1 gene segment has been shown to specifically suppress TCRγ gene transcription in αβ T cells and non-T cells in a manner that is independent of position or orientation. The γ silencer is thought to control transcription of the TCRγ gene by dampening the effectiveness of the γ enhancer, without which transcription from the TCRγ V promoters is minimal. Similarly, the α silencer identified in the human TCRα gene 5′ of the α enhancer inhibits the transcription of the TCRα locus in both γδ T cells and

non-T cells. Precisely how silencers exert their effects on enhancers is unknown. Silencers do contain binding motifs for nuclear binding proteins, but mutagenesis experiments have so far failed to show that any of these sites are required for silencer function.

H. Structure and Function of the CD3 Complex

To establish fully functional antigen receptors on the T cell surface capable of conveying the information that antigen has bound, a functional TCRαβ or γδ heterodimer must be assembled with a multi-protein CD3 complex on the T cell surface.

I. CD3 PROTEIN STRUCTURE

In a complete TCR complex, one TCR heterodimer associates non-covalently with at least five and usually six invariant transmembrane proteins. As mentioned earlier in this chapter, the five most commonly included molecules are *CD3γ* (21–28 kDa), *CD3δ* (20–28 kDa), *CD3ε* (20–25 kDa), *CD3ζ*, (16 kDa), and *CD3η* (22 kDa). The γ chain of the FcεRI receptor can occasionally participate in the CD3 complex because it shares certain structural similarities with the CD3ζ molecule. In general, the CD3 complex that clusters around the TCRαβ heterodimer is composed of a CD3γε heterodimer, a CD3δε heterodimer, and either a CD3ζζ homodimer, a CD3ζη heterodimer, or a CD3ζ-FcεRIγ heterodimer (refer to Fig. 12-4). In TCRγδ, the CD3δε heterodimer is replaced by another CD3γε heterodimer. Close to 90% of T cells bear the CD3ζζ homodimer, with just under 10% carrying the CD3ζη heterodimer and a very small fraction expressing CD3ζ-FCεRIγ. Each of the CD3ζζ, CD3ζη, and CD3ζ-FCεRIγ dimers has a disulfide bond linking its component chains, while CD3γε and CD3δε rely on non-covalent bonding to achieve their heterodimeric structure. Thus, the complete TCR–CD3 complex contains at least four dimeric proteins: TCRαβ (or TCRγδ), CD3γε, CD3δε, and CD3ζζ (or, occasionally, CD3ζη or CD3ζ-FCεRIγ).

The CD3γ, CD3δ, and CD3ε chains feature an Ig-like extracellular domain that may interact with the Ig-like extracellular domain in the TCR chains to help to keep the TCR and CD3 complex in close association. The extracellular domain of a CD3 chain is invariant and so does not contribute to the specificity of antigen recognition. The transmembrane regions of CD3γ, CD3δ, and CD3ε all contain an aspartic acid residue that is negatively charged. The side chain of this amino acid is thought to interact with the positively charged side chains of residues in the transmembrane regions of the TCRα and β chains, an interaction that may contribute to either complex integrity or signaling function. That the CD3 complex is physically associated with the TCR has been demonstrated in experiments in which antibodies to either the TCR or the CD3 complex have immunoprecipitated both TCR and CD3 chains from solubilized membrane preparations. Moreover, studies *in vitro* have shown that an isolated TCRα chain preferentially associates with CD3δ or ε, while the TCRβ chain binds to the CD3δ, ε, or γ chains.

II. FUNCTIONS OF THE CD3 COMPLEX

The CD3 chains serve two functions: facilitation of TCR insertion in the plasma membrane, and signal transduction following TCR binding site engagement.

i) TCR Surface Expression

Mutations that affect the expression of CD3 abolish surface expression of the TCR and vice versa, as demonstrated by the lack of surface expression of TCRs in both mouse and human T cell lines deficient for CD3 proteins. The TCR heterodimer is physically associated with a CD3 core structure in the rER before moving to the Golgi for glycosylation and then insertion as a complete TCR–CD3 complex in the membrane. Except for the CD3ζ chain, all component chains (including TCRα and β) of the TCR–CD3 complex are synthesized in excess of the amounts needed to form the number of TCR–CD3 structures observed on the T cell surface. The synthesis and incorporation of the CD3ζ chain into the complex appear to be rate-limiting so that the production of this protein controls the assembly and transport of the entire TCR–CD3 complex.

ii) TCR Signal Transduction

All CD3 chains and the FcεRγ chain have at least one ITAM in their relatively long cytoplasmic tails (about 40 amino acids) (refer to Fig. 12-4). (In the T cell literature, the ITAM is also known as the *ARAM*, or *a*ntigen *r*eceptor *a*ctivation

motif.) Unlike the other CD3 chains, the CD3ζ and CD3η chains lack significant Ig-like extracellular domains (only nine amino acids) and are composed essentially of a transmembrane portion followed by a very long cytoplasmic tail. The cytoplasmic tail in the CD3ζ chain consists of 113 amino acids containing three ITAMs, while the CD3η tail consists of 155 amino acids containing two ITAMs. As was described in Box 9-1, each ITAM features six conserved amino acids appearing in the sequence $D/E-X_7-D/E-X_2-Y-X_2-L/I-X_7-Y-X_2-L/I$, where D/E is either aspartic or glutamic acid, X is any amino acid, Y is tyrosine, and L/I is either leucine or isoleucine. The more ITAMs in a CD3 chain's cytoplasmic tail, the greater its affinity for the intracellular signaling kinases Lck and Fyn. Thus, after the peptide–MHC complex binds to and activates the TCR, the tyrosine residues of these ITAMs are phosphorylated by these kinases, allowing the CD3 chains to bind to the SH2 domains of other signal transduction proteins (such as ZAP-70 kinase) and recruit them to the signaling cascade. The extent of phosphorylation of the CD3 ITAMs may influence the subsequent protein activation and signaling cascades, which in turn may determine whether a T cell is activated or rendered non-responsive (see Chs.14 and 16).

As mentioned previously, the CD3γ, δ, ε, and ζ chains are also found in the pre-TCR in developing thymocytes, associating with the pTα chain and the newly synthesized TCRβ chain. CD3 chains are produced before rearrangement of the TCRβ chain takes place, so that they are available to participate with the TCRβ chain in its productivity test. However, it is not clear whether all these CD3 proteins are in fact required for signaling through the pre-TCR, even though the Lck and Fyn kinases are known to be required for pre-TCR function.

Often one will find references in the immunological literature to T cell activation by anti-CD3 cross-linking *in vitro*. In this experimental situation, the cell has been treated with specific antibody recognizing some component of the CD3 complex. The response of the T cell to anti-CD3 resembles T cell activation induced by binding of specific antigen to the TCR, but the activation is due to a non-specific triggering of the signal transduction function of the CD3 chains. Since any T cell can be stimulated in this way, anti-CD3 antibodies are said to be *polyclonal activators* of T cells.

III. CD3 KNOCKOUT MICE

Several knockout mutants have been created that are deficient for part of the CD3 complex. Mice lacking the CD3ε chain have reduced numbers of thymocytes in the thymus and lack mature αβ T cells in the periphery, emphasizing the role the CD3 complex plays in early T cell development. These mice were able to rearrange their TCRβ, γ, and δ genes but did not rearrange their TCRα genes, most likely due to a lack of pre-TCR signaling. Mice in which the entire CD3ζ-η gene was disrupted so that neither the CD3ζ nor CD3η chain was produced also had decreased thymic cellularity and were blocked in thymocyte development. Thymocytes of these mutants exhibited a drastic decrease in surface levels of TCR complexes, consistent with the essential role of the CD3ζ chain in transporting the TCR to the cell surface. The phenotype of this mouse also demonstrated that the function of the CD3ζ-η gene could not be replaced by the FcεRIγ chain in thymocytes, although a homodimer of FcεRIγ can associate with TCRs in gut intraepithelial lymphocytes that develop independently of the thymus. In addition, CD3ζ function cannot be replaced by CD3η, since mice in which the exon specific for CD3ζ was disrupted without affecting the transcription of CD3η had a phenotype similar to that of the CD3ζ-η$^{-/-}$ mice. However, when CD3η was disrupted without affecting the transcription of the CD3ζ chains, T cell development and immune responses were normal, showing that there is redundancy for CD3η function but not for that of CD3ζ. The specific roles of individual CD chains are under investigation.

I. The CD4 and CD8 Coreceptors

I. DISCOVERY OF CD4 AND CD8

Shortly after the development of monoclonal antibody technology by Köhler and Milstein in 1975, immunologists began to generate mAbs directed against specific markers on different thymocyte and lymphocyte subsets. Two mAbs were identified that bound to a molecule predominantly expressed on T cells with helper function. This molecule was originally called OKT4 in humans and L3T4 in mice; it is now known as CD4 in both species. Another mAb identified a molecule originally called OKT8 in humans and Lyt2 in mice that became known as CD8 in both species. Treatment with antibody that bound to CD4 blocked the stimulation of MHC class II-restricted CD4$^+$ T cells, while treatment of CD8$^+$ T cells with anti-CD8 antibody blocked MHC class I-restricted cytotoxic killing of target cells. These experiments demonstrated that the engagement of these molecules was necessary for T cell effector functions.

As introduced in Chapter 2, we now know that two types of mature T cells patrol the periphery in mice and humans: those bearing the coreceptor CD4 and those bearing CD8. Late-stage αβ thymocytes bearing CD4 engage peptide–MHC class II complexes and generally mature into Th cells, while late-stage αβ thymocytes bearing CD8 interact with peptide–MHC class I complexes and generally mature into Tc cells. About two-thirds of the total population of mature αβ T cells

in the periphery of both humans and mice are CD4$^+$ cells, while one-third are CD8$^+$ cells. Interestingly, CD4 and CD8 do not occur on most mature γδ T cells (some γδ T cells in the gut express CD8), which is consistent with the finding that the particular class of MHC encountered may not be as important for γδ T cell recognition as it is for αβ T cell recognition (see Ch.18).

II. WHAT IS A "CORECEPTOR"?

Why are CD4 and CD8 called "coreceptors"? Studies of the engineered expression of CD4 and CD8 in mature αβ T cells have confirmed that T cell activation is most efficient when the pMHC antigen both binds to the TCR and interacts with CD4 or CD8. Furthermore, unlike costimulatory molecules such as CD28 that bind to ligands somewhat removed from the TCR complex, to be effective CD4 or CD8 must bind to the non-polymorphic region of <u>exactly the same</u> MHC molecule as that engaged by the TCR. It is because both the TCR and CD4 or CD8 on the T cell must bind to the same MHC molecule on the APC/target cell that CD4 and CD8 are termed "<u>coreceptors</u>." Because CD4 and CD8 recognize sites on their respective MHC molecules that are in invariant regions outside the peptide-binding groove, coreceptor binding does not depend on the presence of the antigenic peptide. It is still not known why the presence of CD4 and the recognition of peptide complexed to MHC class II results in the development of T cell clones with helper effector function almost exclusively, while T cells bearing CD8 and recognizing peptide–MHC class I differentiate into Tc clones.

The elucidation of the roles of CD4 and CD8, which are expressed at comparatively low levels on normal T cells, would have been difficult without the development of *transgene* technology. A transgenic animal is created when an isolated gene (the "transgene") is introduced into the genome of a whole embryo and is expressed along with the recipient's own genes. The expression of the transgene can be controlled either by its natural transcriptional promoters and enhancers, or by exogenous regulatory elements engineered into it. "Overexpression" of a molecule of interest can be achieved, making it possible to discern its function and interaction with other molecules. Studies of transgenic mice overexpressing CD4 or CD8 were instrumental in defining the functions of these molecules. A description of transgene technology appears in Box 12-2.

Box 12-2. Construction of transgenic mouse strains

The reader will recall from Box 8–4 in Chapter 8 that gene-targeted knockout mouse mutants are animals manipulated at the embryonic stem cell stage such that a particular gene is disrupted in every cell of the animal, leading to a loss of gene function. Transgenic mice are animals in which the gain of a gene is engineered. To make a transgenic mouse

strain, an excess of cloned foreign DNA (usually ≤50 copies of the sequence of interest) is microinjected into one pronucleus of a fertilized mouse egg (zygote). The injected zygotes are implanted into pseudopregnant females for embryogenesis. After about 20 days of gestation, the pups are born, and their genotypes are analyzed by

Southern blotting or PCR for the presence of the injected transgene.

Transgenic manipulation works because up to a maximum of 50 copies of the microinjected gene of interest can integrate in tandem into a random site in the genomic DNA of the zygote pronucleus and are replicated along with it as the zygote divides

Continued

Box 12-2. **Construction of transgenic mouse strains**—*cont'd*

and grows. The integration of the transgene will occur before the first cell division in 10–25% of the embryos, so that all the cells of these animals (including the germ cells) will carry the transgene. In a smaller percentage of animals, integration occurs after the zygote has already divided, so that the transgene does not integrate into the DNA of some cells. These animals of blended genetic background are said to be *chimeric* rather than transgenic. In the end, because the microinjection procedure is not terribly efficient, only about one to two transgenic mice will be generated for every 100 zygotes microinjected.

Because transgenes integrate at random sites in the genome, in some cases their introduction will disrupt the function of an essential gene; such embryos do not survive. By definition, therefore, in transgenic strains, the positioning of the transgene does not disrupt the normal functioning of the recipient cells. In addition, because the integration often occurs in a single site on one chromosome (even if multiple copies are involved), founder transgenic mice are usually heterozygous. The transgene is passed on in simple Mendelian fashion, so that heterozygotes can be interbred to generate homozygous lines of transgenic mice.

Transgenic mice are useful for examining the expression of a gene of interest in various tissues *in vivo*, because although every cell of the transgenic animal carries the transgenic DNA, cell- and tissue-specific regulatory elements will determine its expression in various cell types. Alternatively, transgenes can be placed under the control of tissue-specific or inducible promoters and the expression of the gene examined in a tissue or at a time of interest. For example, strains have been created using constructs in which the transgene is placed under the control of the T cell-specific Lck kinase promoter, which means that the gene of interest will be expressed only in T lymphocytes. Transgenes can also be cloned next to a promoter conferring IFN inducibility. When the mice are kept under pathogen-free conditions (so as not to provoke a natural immune response), the transgene will not be expressed until exogenous IFN is administered; that is, until the promoter is activated by the binding of IFN. Similarly, inducible control of transgenes can be exerted by placing them under the control of promoters engineered to be regulated by the addition of the antibiotic tetracycline or the synthetic estrogen tamoxifen.

Transgenes are also helpful for probing the relationship between structure and function of a molecule. The transfection of a mutated version of a protein that has been engineered to lack a particular domain may reveal whether or not that domain is essential for function. Transgenes can also be used to rescue phenotypes and thus assess the nature of the original defect. For example, if a mutant in which a developmental process has stalled can be rescued by the overexpression of a molecule known to promote cell survival, it is likely that the original mutation affected a protein with a role in survival. This protein may be quite distinct from the product of the transgene.

A *dominant negative* transgene can be used to interfere with the expression of an endogenous protein. Often the protein derived from a dominant negative transgene is a catalytically inactive version of the molecule of interest, the theory being that large quantities of such a protein (5- to 10-fold excess) will compete with the endogenous protein for essential substrates or co-factors. The promoter used to drive transgene expression in this case must be capable of achieving the necessary high level of expression. It is also essential to determine how much interference is required to completely inactivate the endogenous protein, and how specific an inhibitor the dominant negative molecule truly is. It is not unusual for such an inhibitor to sequester molecular intermediates needed by more than one endogenous enzyme or pathway. This latter characteristic can be an advantage when one wants to overcome redundancy of function and simultaneously disable all isoforms of a protein of interest. Transgenes encoding *bacterial toxins* are valuable tools because of their specificity and high potency at low concentration. For example, transgenes encoding the bacterial cholera, botulism, and pertussis toxins have been used as inhibitors to determine the roles of various intermediates in intracellular signaling pathways. *Constitutively active* transgenes can be constructed to overcome natural control mechanisms within a cell. Mutations of residues in negative control sequences or enzymatic catalytic sites can result in constitutive activation of a protein.

With respect to TCR biology, transgenic mice have been engineered so that their lymphocytes express TCRs specific for known pMHC complexes. In these experiments, the functionally rearranged TCRα and β genes from a T cell clone of known specificity are introduced into a recipient strain. Because the integrated exogenous TCR genes are already productively rearranged, genotypic and phenotypic allelic exclusion are invoked and the expression on the surface of TCRs containing endogenous chains is suppressed in all T cell progenitors. Thus, the TCRs of all T cells of transgenic mice will be directed against one antigenic peptide–MHC allele combination, creating a homogeneous population of TCRs. Such mice are very useful for the examination of thymic selection and T cell activation processes. For example, one could not hope to detect the deletion of a single TCR specificity in a wild-type mouse. However, in a TCR transgenic mouse, essentially all of the T cells express the same TCR. If negative selection of this TCR specificity occurs, almost all the thymocytes in this mouse will be deleted, amounting to an (easily detected) deficit of mature peripheral T cells.

An important caveat pertains to the use of transgenic mice. During "standard" transgenesis, the transgene undergoes integration into the host genome in a random location. Although one may be able to express the gene of interest in amounts, at times, or in tissues where it is not normally expressed, the results obtained may be distinctly unphysiological. Caution should therefore be exercised in drawing conclusions about the natural function of the gene under these circumstances. *Knock-in* technology rectifies some of these defects because the transgene is introduced into a precise location in the genome by homologous recombination; that is, the transgene is designed such that sequences flanking the gene of interest are homologous in sequence to the endogenous locus and thus ensure that the transgene integrates in its natural position. The natural transcription controls of the gene of interest are thus preserved and the danger of overexpression artifacts is reduced. Knock-in mutations can be used to introduce reporter constructs to facilitate detection of a hard-to-monitor gene, or regulatory constructs can be added to alter expression patterns. For example, to investigate control of IL-4 expression *in vivo*, a reporter gene was "knocked into" the IL-4 locus. This study led to the generation of much useful information on the regulation of effector T cell differentiation.

III. STRUCTURE OF CD4

The CD4 gene is located on chromosome 12 in humans and on chromosome 6 in mice (refer to Table 12-1). CD4 is a transmembrane glycoprotein of 55 kDa that is expressed as a single polypeptide chain on the surface of MHC class II-restricted T cells. CD4 is also expressed to a minor extent on macrophages, some monocytes, some DCs and Langerhans cells, some B cells, and brain microglial cells. The CD4 protein contains four extracellular Ig-like domains (making it a member of the Ig superfamily), a hydrophobic transmembrane domain, and a highly basic cytoplasmic tail of 38 amino acids containing three serine residues that can be phosphorylated (Fig. 12-14A). The N-terminal Ig-like domains of CD4 interact with the non-polymorphic α2 and β2 domains of MHC class II molecules, mediating both recognition and adhesion functions (see later). The cytoplasmic tail of CD4 has sites that facilitate physical association with Lck kinase.

In recent times, the CD4 molecule has become identified with more insidious events. The HIV virus has evolved to take advantage of CD4 and exploit it to invade CD4+ T cells. The N-terminal extracellular domains of CD4 provide a perfect attachment site for the gp120 envelope glycoprotein of HIV, such that CD4 acts as part of a high affinity receptor complex that facilitates viral entry into CD4+ cells. The destruction of the CD4+ T cell population that follows HIV infection causes the overwhelming immune dysfunction suffered by AIDS patients (see Ch.25).

IV. STRUCTURE OF CD8

Despite their ostensibly equivalent functions, the CD4 and CD8 proteins show little similarity in either structure or amino acid sequence. In mice, the majority of CD8 molecules expressed on MHC class I-restricted T cells are found as CD8αβ heterodimers composed of a 38 kDa α chain joined by a disulfide link to a 30 kDa β chain (Fig. 12-14B). However, some intestinal intraepithelial T lymphocytes (iIEL cells) express a CD8αα homodimer on their cell surfaces. (CD8ββ homodimers apparently do not exist.) On human T cells, CD8 appears almost exclusively as a CDαα homodimer. In both species, each CD8 chain contains one Ig-like extracellular domain, a hydrophobic transmembrane domain, and a cytoplasmic tail of about 26 residues. CD8 functions as a recognition and adhesion molecule by binding to the α3 domain of the same MHC class I molecule engaged by the TCR. Lck kinase can also associate with the CD8 cytoplasmic tail but does so much more weakly than with the CD4 tail. The human CD8 gene is located on chromosome 2, while the mouse CD8 gene is on chromosome 6. In both cases, the CD8 gene is positioned near the *Igk* locus.

V. FUNCTIONS OF CD4 AND CD8

Why does a T cell need a coreceptor? CD4 and CD8 serve three functions: discrimination between MHC class I and II, intercellular adhesion between T cells and APC/target cells,

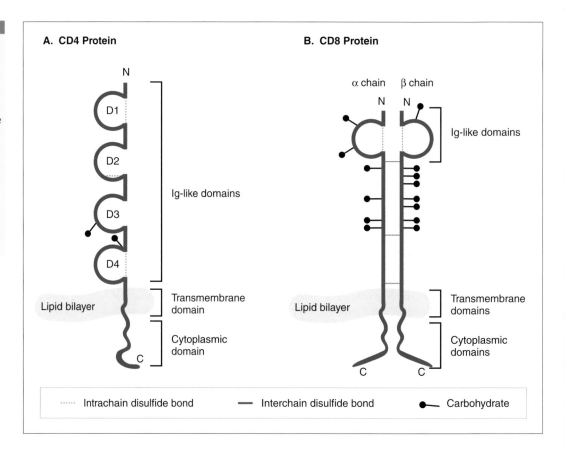

Figure 12-14

CD4 and CD8 Structures
Schematic representations of the structures of the (A) CD4 and (B) CD8 coreceptors are shown, including Ig-like domains and glycosylation sites. CD8αβ heterodimers are shown but CD8αα homodimers also exist. The number of carbohydrate sites in the CD8β chain varies during the developmental stages of a given thymocyte. With information from Klein and Hořejší (1997) "Immunology," 2nd ed., Blackwell Science, Oxford.

A. CD4 Protein

N
D1
D2
D3
D4
Ig-like domains
Lipid bilayer
Transmembrane domain
Cytoplasmic domain
C

B. CD8 Protein

α chain   β chain
N   N
Ig-like domains
Lipid bilayer
Transmembrane domains
Cytoplasmic domains
C   C

········ Intrachain disulfide bond   ——— Interchain disulfide bond   ●— Carbohydrate

and recruitment of Lck to the TCR signaling complex. Discrimination between MHC class I and II is invoked because CD4 binds only to MHC class II, and CD8 only to MHC class I. Why this mechanism should be in place is a mystery probed in some detail in Chapter 13. With respect to the adhesive role, while neither CD4 nor CD8 is absolutely required for the initial engagement of TCRαβ by pMHC, the adhesive and stimulatory contacts these molecules establish with the MHC molecule greatly facilitate the efficiency of TCRαβ receptor–ligand interactions. In other words, coreceptor binding may reduce the "off" rate of the interaction between the T cell and the MHC-expressing APC/target cell, enhancing the affinity of the TCR for its pMHC ligand by as much as 100-fold. In addition, coreceptor binding, particularly that of CD8, may induce a conformational change in the MHC molecule that favors TCR/pMHC interaction. With respect to the recruiting function, because of the positioning of the coreceptors near the TCR in the membrane, Lck associated with the cytoplasmic tails of the coreceptors is brought into close proximity with the signaling tails of the CD3 complex (Fig. 12-15). Lck juxtaposed to the TCR complex in this way may contribute to tyrosine phosphorylation of ITAMs in the CD3 tails and/or the recruitment of additional signaling molecules. Indeed, when the TCR and CD4 molecule are stably associated with one another, TCR signal transduction is optimized. Subsequent phosphorylation of the serine residues in the CD4 tail downregulates the expression of CD4 and dissociates it from Lck, helping to damp down T cell activation after it is no longer needed (see Ch.14).

Initial evidence for a direct role for coreceptors in promoting T cell activation came from studies of *agonist* and *antagonist* peptides. *Agonist peptides* are those that can bind to the TCR and elicit some degree of response from the T cell. Strong agonists result in complete T cell activation (in the presence of the appropriate adhesive and costimulatory interactions), while weak agonists are thought to dissociate from the TCR–CD3 complex too early and achieve only partial activation. For example, T cells interacting with a weak agonist might secrete cytokines but fail to proliferate, or vice versa. *Antagonist peptides* are those that can bind to the TCR but that interfere in some way with the transduction of the activation signal such that T cell activation is completely blocked. These peptides may have a faster "off" rate than agonist peptides (such that the activation signal is not completed), or may somehow inactivate or fail to activate substrates of the signal transduction cascade. In the case of CD4+ T cells, if one engineers a decrease in the level of CD4 expression on the Th cell, a peptide that originally behaved as a weak antagonist (partial inhibition of T cell activation) sometimes acts more like a strong antagonist. Similarly, for CD8+ T cells, whether a peptide is seen as an agonist or an antagonist may depend on the level of expression of CD8β. Some scientists believe that, for weak agonists, the coreceptors play a major role in T cell activation by increasing the avidity of TCR–MHC interaction; however, this role may be dispensable for strong agonists.

Other scientists contend that it is the recruitment of Lck to the TCR signaling complex by the coreceptors that is

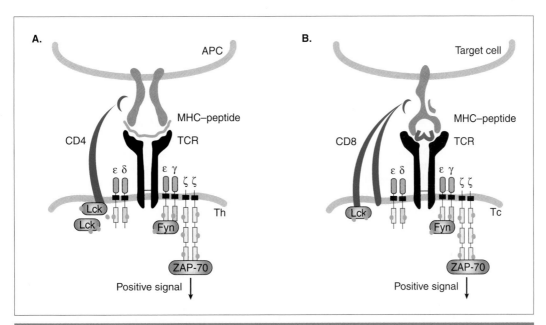

Figure 12-15

Interaction of Coreceptors with Peptide–MHC
Coreceptors promote T cell binding to the appropriate pMHC complex displayed on an APC or target cell. Lck kinase associated with the cytoplasmic tail of the coreceptor triggers a signaling cascade propagated by the kinases Fyn and ZAP-70 associated with the ITAMs of the CD3 chains surrounding TCRαβ. This signaling cascade delivers a stimulatory message to the T cell nucleus. (A) CD4 interacts with a non-polymorphic region of MHC class II. Substantial amounts of Lck are associated with the CD4 cytoplasmic tail. (B) CD8 interacts with a non-polymorphic region of MHC class I. Moderate amounts of Lck are associated with the CD8 cytoplasmic tail.

most important. Indeed, in the antagonist experiments outlined previously, the tyrosine phosphorylation of cellular proteins can be changed from a pattern associated with an agonist peptide to one associated with an antagonist peptide simply by blocking the interaction of CD4 with MHC class II. In addition, there are several examples of antibodies that bind to the TCR signaling complex without triggering it. T cells treated with such antibodies adopt an anergic (non-responsive) state in the presence of agonist peptide. It has been shown that these "non-stimulatory" antibodies do not allow the coreceptor to be brought into the TCR signaling complex, possibly decreasing the activation of Lck and stalling the activation cascade at that point. Other factors that may be involved in distinguishing agonist- from antagonist-initiated TCR signaling are discussed in Chapter 14.

In addition to T cell activation, CD4 and CD8 play crucial roles in the thymic positive and negative selection processes that shape the mature T cell repertoire. How the coreceptors are involved in thymocyte maturation, and other information derived from studies of CD4$^{-/-}$ and CD8$^{-/-}$ knockout mice, are discussed in Chapter 13.

J. Physical Aspects of the Interaction of the TCR with Antigen

The binding of the TCR to its pMHC antigen is a critical event underlying the most fundamental aspects of cell-mediated immunology. Engagement of the TCR results in a series of protein modifications and signal transduction events that deliver signals to the nucleus of the T cell that elicit specific alterations in gene expression. These changes have to be interpreted by the cell so that an appropriate course of action will be taken. For example, the strength of binding between a given thymocyte TCR and a pMHC complex encountered in the thymus determines whether it will be negatively or positively selected; that is, whether the cell initiates apoptosis and dies within the thymus, or proliferates such that its progeny emerge from the thymus to become mature T cells. Similarly, the strength of binding between the TCR on a mature T cell and a foreign peptide-MHC complex on an APC in the periphery determines whether the T cell will be activated to proliferate and differentiate into effector cells, or induced to adopt a state of anergy (see Chs.14 and 16). Thus, the physical act of a TCR binding to its pMHC antigen can result in several different outcomes, and scientists still do not fully understand the molecular interactions underlying how the cell interprets each situation and regulates these cell fate decisions. The intricate involvement of the invariant components of the larger TCR–CD3 complex must also be considered, as well as the influence of costimulatory proteins found on the surface of the T lymphocyte and the APC or target cell. The structural aspects of TCR binding to pMHC are presented here, while those issues more directly associated with the activation process itself are discussed in Chapter 14.

I. STUDYING TCR–PEPTIDE–MHC INTERACTION

As described in Chapter 5, the binding of the Ig molecule to its antigen involves two molecules, and the molecular interaction between the two can be studied relatively easily in solution because the Ig is secreted as antibody. The TCR, however, sees the peptide and the MHC molecule <u>together</u> as its antigen, making the binding tri-molecular in nature. In addition, unlike antibodies, neither the TCRs nor the pMHC complexes are ever secreted, making the study of binding interactions between the TCR and its antigen very difficult. Subcellular experimentation is not a perfect remedy either, since the assembly *in vitro* of isolated TCRα and TCRβ chains rarely gives rise to a functional heterodimer. Researchers have devised various ways of solving these problems. One method of producing soluble TCRs involves engineering TCRα and TCRβ genes such that their cytoplasmic exons are removed and replaced with a sequence encoding the C-terminal end of a protein that can be enzymatically cleaved. Expression of the hybrid protein on a cell and application of the enzyme releases the extracellular portion (the TCR component) of the hybrid protein. Another method for producing soluble TCRs relies on disulfide linkage of isolated Vα and Vβ domains; however, this method does not always result in correct folding of the molecule. The expression of single VαVβCβ chains in which these domains have been connected by peptide linkers can also be engineered; a single chain molecule of this type folds correctly and is often functional *in vitro*. The Cβ domain has been found to be essential for the correct folding of the chain (probably because there are extensive interactions between the Cβ and Vβ domains in the intact molecule) while the Cα domain is dispensable.

II. BINDING AFFINITY OF TCRαβ FOR ITS LIGAND

Studies of solubilized TCRαβ molecules capable of binding to solubilized peptide–MHC complexes have shown that the binding affinity of the TCR for its ligand ($K = \sim 5 \times 10^5$ M^{-1}) is significantly lower than that of an antibody for its antigen ($K = 10^7$–10^{11} M^{-1}). Because this affinity is relatively modest, it is thought that intercellular adhesion in the form of complementary pairs of cellular adhesion molecules (CAMs) plays a large role in establishing the first contacts between T cells and APCs or target cells. Specific TCR–ligand contacts are made only after the cells are held in close enough proximity by the CAMs to permit the T cell to scan the pMHC complexes in the APC/target cell membrane. These intercellular interactions cause the clustering of the CAMs in the membrane region between the two cells, increasing the strength of the bond between them. At this point, contacts between the CD4 or CD8 coreceptors and the MHC molecule also become important for TCR–ligand binding. T cells expressing CD4 solidify their binding to APCs expressing MHC class II, while T cells expressing CD8 become more firmly associated with APCs expressing MHC class I.

III. X-RAY CRYSTAL STRUCTURE OF T CELL RECEPTORS

The resolution of X-ray crystal structures detailing the binding between TCRαβ and pMHC capped a 30 year struggle by immunologists to understand the crucial receptor–ligand interaction that triggers a T cell response. In 1995, the first crystal structure of a TCR Vβ region was analyzed, followed shortly thereafter by the structure of TCR Vα. Major breakthroughs came in 1996, when Ian Wilson and colleagues at the Scripps Research Institute and Donald Wiley and colleagues at Harvard University independently published the crystal structure of a complete TCRαβ heterodimer binding to its antigenic peptide complexed to an MHC class I molecule (Plate 12-2). These studies revealed both the orientation of the TCRα and β chains in the complex and the contacts established between the MHC, the TCR, and the peptide.

The X-ray crystal structure of a Vδ domain of a human TCRγδ was resolved in 1998, while the crystal structure of a complete human TCRγδ known to bind small phosphorylated antigens was resolved in 2001. Analyses of these structures confirmed that TCRγδ is indeed structurally different from TCRαβ. In comparison to the TCRαβ V region, the structure of the TCR Vδ looks more like the Ig V_H than the TCR Vα or Vβ. Indeed, the CDRs of the TCRδ chain have a conformation that incorporates structural elements of both Vα and V_H. In addition, the orientation of the V and C regions is unique in TCRγδ, with the elbow angle between the V and C regions being smaller than in either TCRαβ or Ig Fab. The fact that the V domain of TCRγδ has features also found in BCRs suggests that antigen recognition by TCRγδ may more closely resemble antigen recognition by antibody than by TCRαβ. These results are consistent with the findings of functional studies that suggest that γδ T cells recognize structures other than pMHC complexes (see Ch.18).

i) Contacts between TCRαβ, MHC, and Peptide

Analyses of TCRαβ–peptide–MHC X-ray crystal structures have delineated the extensive interactions between the loops formed by the variable domains of the TCRα and TCRβ chains and the peptide and MHC molecules. The variable domain loops of the TCRαβ heterodimer superficially resemble the variable domain loops of the Ig molecule but distinct differences arise in the interdomain pairing of the constant regions. In addition, in contrast to the situation in Ig molecules, TCR Vβ and TCR Cβ are closely associated within the crystals, which may help to impart a certain degree of inflexibility to the binding region of the TCR that is analogous to the Ig Fab region. A short exposed loop in Cβ is thought to interact with one of the CD3 proteins.

How does the TCR use its variable domain loops to bind to pMHC? The amino acid sequences of TCRβ CDR1 and CDR2 are well-conserved across murine Vβs, allowing their interaction with relatively common anchor residues in most MHC

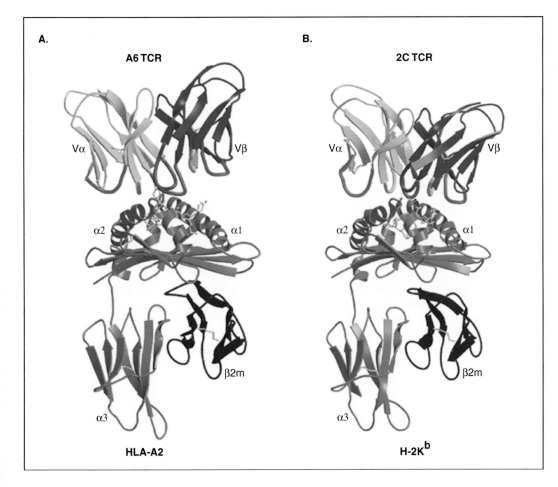

A. A6 TCR
Vα Vβ
α2 α1
β2m
α3
HLA-A2

B. 2C TCR
Vα Vβ
α2 α1
β2m
α3
H-2K^b

Plate 12-2

Structures of Human and Mouse TCR/pMHC Complexes
Crystal structures showing carbon backbones of (A) human A6 TCR binding to peptide (yellow) presented by HLA-A2 (Garboczi *et al.*, 1996) and (B) mouse 2C TCR binding to peptide (yellow) presented by H-2K^b (Garcia *et al.*, 1996). Both the α and β chains and β2-microglobulin of the MHC class I molecules are illustrated, but only the V domains of the TCRs are shown. For both TCRs, CDR1 is yellow; CDR2, green; and CDR3, red. Reproduced with permission from Bjorkman P. J. (1997) MHC restriction in three dimensions: a view of T cell receptor/ligand interactions. *Cell* **89**, 167–170.

α-helices. Except for a deep hydrophilic cavity between the TCRα CDR3 and TCRβ CDR3, the TCR binding site as a whole is relatively flat. When bound, the TCR is oriented in a diagonal position over the pMHC complex, which allows this flat region to interact with the peptide. It is important to note that both the TCRα and β chains are usually involved in binding to both the MHC molecule and the peptide: it is not as if one chain binds only to the MHC, while the other binds only to peptide. Neither is there a temporal separation: the TCR does not bind first to the MHC moiety and then only later to the peptide. In general, the highly variable CDR3 regions of the TCRα and β chains bind to the middle of a peptide lodged between the α- and β-helices that form the MHC-binding groove, as well as to points on the MHC α-helix itself. The less variable CDR1 and CDR2 regions tend to bind the ends of the peptide and to conserved sites on the MHC α-helices.

ii) Stringency of Ligand Recognition by TCRαβ

The area of contact between the TCR and the pMHC is relatively small such that only a few of the residues in the peptide generally make contact with a TCR chain. The diagonal orientation of the TCR central pocket over the middle of the peptide in the groove between the two α-helical regions of the MHC molecule (either class I or class II) is thought to be a general binding mode applicable to most pMHC complexes. The limited opportunity for intermolecular bonding means that the specificity of TCR binding is not as great as that of the Ig binding to antigen, and that TCRs are not as specific for their cognate antigens as originally thought. Such flexibility in binding specificity facilitates thymic selection because one peptide can positively select several thymocyte clones, amplifying the T cell repertoire (see Ch.13). Studies of X-ray crystals of TCR–pMHC complexes have confirmed that promiscuous TCR binding does occur. Indeed, arrays of agonist peptides for a given TCR have been created by introducing mutations into areas of a cognate peptide known to be contacted directly by the TCR. T cell clones ostensibly specific for that cognate pMHC ligand are still able to respond to virtually all members of the agonist panel. At the molecular level, this is manifested as considerable variation in the participation of the CDRs in pMHC binding. Three different TCRs that "see" the same MHC molecule (bearing three different peptides) can show very different binding characteristics at the atomic level with respect to specific peptide residues bound, specific MHC residues bound, positions of side chains, and overall binding affinity.

A primary source of the degeneracy of TCR–pMHC binding is the CDR3 region. The DNA sequence encoding the CDR3 region spans the VJ joint in the rearranged TCRα gene and the VDJ joint in the rearranged TCRβ gene, accounting for its extreme variability in amino acid sequence, length, and conformation. In general, a CDR3 region spans an average of 10 amino acid residues, similar to its size in Ig light chains but much smaller than the equivalent region in Ig heavy chains. Comparisons of the conformation of TCRs that have not bound to pMHC versus TCRs bound to pMHC have demonstrated that the CDR3 region is capable of undergoing an enormous conformational shift in order to achieve the diagonal orientation favored for binding to pMHC. The adoption of this "induced fit" affects only the CDR3 region and does not alter the conformation of either the rest of the TCR molecule or the MHC molecule. Indeed, it has been shown that, even as the CDR3 conformation shifts dramatically, the CDR1 or CDR2 regions undergo only minor positional changes or rotations of the side chains on their component amino acids. Again, there is no specific order in which a CDR binds to its pMHC target: sometimes CDR3 initiates interaction with peptide first, and sometimes CDR1 or CDR2 binding to pMHC is established first. In any case, once a TCR finalizes its contacts with a given pMHC, the entire complex is stabilized and the flexibility of both the TCR and pMHC binding surfaces is lost.

We have come to the end of our discussion of the T cell receptor complex proteins and genes. We move now to the fascinating subject of T cell development, and the crucial role TCRs play in the positive and negative selection processes that shape the T cell repertoire.

SUMMARY

Like B lymphocytes, T lineage cells express antigen receptors that are derived from the V(D)J recombination of gene segments. Unlike B lymphocytes, however, there are two types of T cell receptors (TCRs), TCRαβ and TCRγδ, which are expressed by αβ and γδ T cells, respectively. The TCRαβ and TCRγδ proteins are heterodimers made up of the TCRα and TCRβ chains, and the TCRγ and the TCRδ chains, respectively; αδ or βγ structures do not exist. The TCR chains themselves are incapable of signal transduction. This function is carried out by ITAM-containing CD3 chains that are present in the TCRαβ or TCRγδ complex on the T cell surface. Each TCR chain has an Ig-like variable domain at its N-terminal end and a membrane proximal Ig-like constant domain. The pairing of the variable domains in the TCR heterodimer forms an antigen-binding site for peptide– MHC complexes. Each of the variable domains in the TCRαβ molecule has four hypervariable regions comparable to the three CDRs in the variable domains of the Ig heavy and light chains. However, the TCR is more promiscuous in recognizing antigens than an Ig and can often bind to a small collection of highly similar peptide–MHC complexes. The TCR chains are derived from separate TCR genes that, like the Ig loci, contain multiple V, D, and J gene segments and one or two constant exons containing sub-exons. An unusual feature of the TCR genes is that the TCRδ locus is nested within the TCRα locus. Although isotype switching and somatic hypermutation do not occur in T cells, the diversity of the T cell repertoire is greater than that of the Ig repertoire because of expanded possibilities of junctional diversification. As well as their TCRs, αβ T cells express the CD4 and/or the CD8 coreceptor molecules that bind to non-polymorphic regions of MHC class II or I, respectively. The coreceptors allow the T cell to discriminate between MHC class I and II, increase adhesion between the T cell and its APC or target cell, and recruit the signaling kinase Lck.

Selected Reading List

Acuto O., Fabbi M., Smart J., Poole C. B., Protentis J. *et al.* (1984) Purification and NH2-terminal amino acid sequencing of the β subunit of a human T-cell antigen receptor. *Proceedings of the National Academy of Sciences USA* **81**, 3851–3855.

Allison J. P. (1987) Structure, function, and serology of the T-cell antigen receptor complex. *Annual Review of Immunology* **5**, 503–540.

Allison T., Winter C., Fournie J.-J., Bonneville M. and Garboczi D. (2001) Structure of a human gamma-delta T-cell antigen receptor. *Nature* **411**, 820–824.

Altman J. D., Moss P. A. H., Goulder P. J. R., Barouch D. H., McHeyzer-Williams M. G. *et al* (1996) Phenotypic analysis of antigen-specific T lymphocytes. *Science* **274**, 94–96.

Arden B. (1998) Conserved motifs in T-cell receptor CDR1 and CDR2: implications for ligand and CD8 co-receptor binding. *Current Opinion in Immunology* **10**, 74–81.

Borst J., Jacobs H. and Brouns G. (1996) Composition and function of the T cell receptor and B cell receptor complexes on precursor lymphocytes. *Current Opinion in Immunology* **8**, 181–190.

Brandt V. and Roth D. (2002) A recombinase diversified: new functions of the RAG proteins. *Current Opinion in Immunology* **14**, 224–229.

Carlyle J. R. and Zuniga-Pflucker J. C. (1998) Requirement for the thymus in αβ T lymphocyte lineage commitment. *Immunity* **9**, 187–197.

Chess A. (1998) Expansion of the allelic exclusion principle. *Science* **279**, 2067–2068.

Chien Y.-H., Becker D. M., Lindsten T., Okamura M., Cohen D. I. *et al.* (1984) A third type of murine T-cell receptor gene. *Nature* **312**, 31–35.

Ellemeir W., Sawada S. and Littman D. R. (2000) The regulation of CD4 and CD8 coreceptor gene expression during T cell development. *Annual Review of Immunology* **18**, 523–524.

Fitzsimmons D. and Hagman J. (1996) Regulation of gene expression at early stages of B-cell and T-cell differentiation. *Current Opinion in Immunology* **8**, 166–174.

Fowlkes B. and Pardoll D. (1989) Molecular and cellular events of T cell development. *Advances in Immunology* **44**, 207–264.

Fremont D. H., Matsumura M., Stura E. A., Peterson P. A. and Wilson I. A. (1992) Crystal structures of two viral peptides in complex with murine MHC class I H-2K[b]. *Science* **257**, 919–927.

Fremont D. H., Rees W. A. and Kozono H. (1996) Biophysical studies of T-cell receptors and their ligands. *Current Opinion in Immunology* **8**, 93–100.

Garboczi D. N., Ghosh P., Utz U., Fan Q. R., Biddison W. E. *et al.* (1996) Structure of the complex between human T-cell receptor, viral peptide and HLA-A2. *Nature* **384**, 134–141.

Garcia K., Degano M., Pease L., Huang M., Peterson P. *et al.* (1998) Structural basis of plasticity in T cell receptor recognition of a self peptide-MHC antigen. *Science* **279**, 1166–1172.

Garcia K. C., Degano M., Stanfield R. L., Brunmark A., Jackson M. R. *et al.* (1996) An alpha/beta T cell receptor structure at 2.5 A and its orientation in the TCR-MHC complex. *Science* **274**, 209–219.

Haskins K., Kubo R., White J., Pigeon M., Kappler J. *et al.* (1983) The major histocompatibility complex-restricted antigen receptor on T cells: I. Isolation with a monoclonal antibody. *Journal of Experimental Medicine* **157**, 1149–1169.

Hayday A. C. (2000) Gamma/delta cells: a right time and a right place for a conserved third way of protection. *Annual Review of Immunology* **18**, 975–1026.

Hedrick S., Nielsen E., Kavaler J., Cohen D. and Davis M. (1984) Sequence relationships between putative T-cell receptor polypeptides and immunoglobulins. *Nature* **308**, 153–158.

Hedrick S. J., Cohen D. I., Nielsen E. A. and Davis M. M. (1984) Isolation of cDNA clones encoding T cell-specific membrane-associated proteins. *Nature* **38**, 149–153.

Housset D. and Malissen B. (2003) What do TCR-pMHC crystal structures teach us about MHC restriction and alloreactivity? *Trends in Immunology* **24**, 429–437.

Kappler J. W., Skidmore B., White J. and Marrack P. (1981) Antigen inducible, H-2 restricted, interleukin-2-producing T cell hybridomas. *Journal of Experimental Medicine* **147**, 1198–1214.

Koyasu S., Clayton L. K., Lerner A., Heiken H., Parkes A. *et al.* (1997) Pre-TCR signaling components trigger transcriptional activation of rearranged TCRalpha gene locus and silencing of the pre-TCRalpha locus: implications for intrathymic differentiation. *International Immunology* **9**, 1475–1480.

Mak T., Penninger J. and Ohashi P. (2001) Knockout mice: a paradigm shift in modern immunology. *Nature Reviews Immunology* **1**, 11–20.

Malissen M., Trucy J., Jouvin-Marche E., Cazenave P.-A., Scollay R. *et al.* (1992) Regulation of TCR α and β gene allelic exclusion during T cell development. *Immunology Today* **13**, 315–322.

Margulies D. H. (1997) Interactions of TCRs with MHC-peptide complexes: a quantitative basis for mechanistic models. *Current Opinion in Immunology* **9**, 390–395.

Marrack P., Hannum C., Harris M., Haskins K., Kubo R. *et al.* (1983) Antigen-specific, major histocompatibility complex-restricted T cell receptors. *Immunological Reviews* **76**, 131–145.

Matzinger P. (1981) A one-receptor view of T-cell behaviour. *Nature* **292**, 497–501.

Metzger H. (1992) Transmembrane signaling: the joy of aggregation. *The Journal of Immunology* **149**, 1477–1487.

Meuer S. C., Fitzgerald K. A., Hussey R. E., Hodgdon J. C., Schlossman S. F. *et al.* (1983) Clonotypic structures involved in antigen-specific human T cell function. Relationship to the T3 molecular complex. *Journal of Experimental Medicine* **157**, 705–719.

Reich Z., Boniface J. J., Lyons D. S., Borochov N., Wachtel E. J. *et al.* (1998) Ligand-specific oligomerization of T-cell receptor molecules. *Nature* **387**, 617–620.

Reinherz E., Tan K., Tang L., Kern P., Liu J. *et al.* (1999) The crystal structure of a T cell receptor in complex with peptide and MHC class II. *Science* **286**, 1913–1921.

Rudolph M. and Wilson I. (2002) The specificity of TCR/pMHC interaction. *Current Opinion in Immunology* **14**, 52–65.

Saito H., Kranz D. M., Takagaki Y., Hayday A. C., Eisen H. N. *et al.* (1984) A third rearranged and expressed gene in a clone of cytotoxic T lymphocytes. *Nature* **312**, 36–40.

Sim G. K., Yague J., Nelson J., Philippa M., Palmer E. *et al.* (1984) Primary structure of human T-cell receptor alpha-chain. *Nature* **312**, 771–775.

Sleckman B., Carabana J., Zhong X. and Krangel M. (2001) Assessing a role for enhancer-blocking activity in gene regulation within the murine T-cell receptor alpha/delta locus. *Immunology* **104**, 11–18.

Spada F. M., Grant E. P., Peters P. J., Sugita M., Melian A. *et al.* (2000) Self-recognition of CD1 by gamma/delta T cells: implications for innate immunity. *Journal of Experimental Medicine* **191**, 937–948.

Toyonaga B. and Mak T. W. (1987) Genes of the T-cell antigen receptor in normal and malignant T cells. *Annual Review of Immunology* **5**, 585–620.

Vernooij B., Lenstra J., Wang K. and Hood L. (1993) Organization of the murine T-cell receptor gamma locus. *Genomics* **17**, 566–574.

Ward E. S. and Zadri A. (1997) Biophysical and structural studies of TCRs and ligands; implications for T cell signaling. *Current Opinion in Immunology* **9**, 97–106.

Xiong N., Kang C. and Raulet D. (2002) Redundant and unique roles of two enhancer elements in the TCR gamma locus in gene regulation and gamma-delta T cell development. *Immunity* **16**, 453–463.

Yanagi Y., Yoshikai Y., Leggett K., Clark S. P., Aleksander I. *et al.* (1984) A human T cell-specific cDNA clone encodes a protein having extensive homology to immunoglobulin chains. *Nature* **308**, 145–148.

Yanagi Y., Chan A., Chin B., Minden M. and Mak T. W. (1985) Analysis of cDNA clones specific for human T cells and the alpha and beta chains of the T-cell receptor heterodimer from a human T-cell line. *Proceedings of the National Academy of Sciences USA* **82**, 3430–3434.

T Cell Development

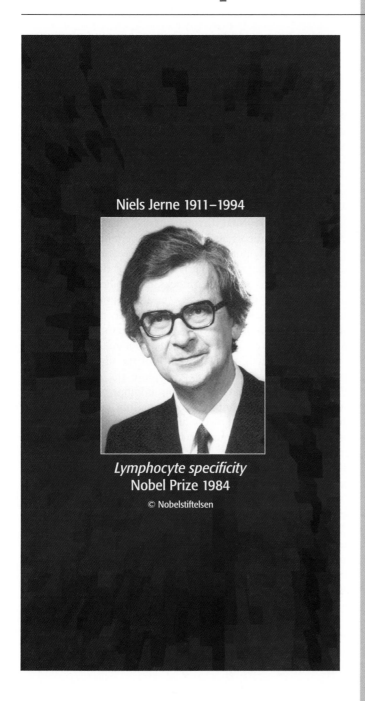

Niels Jerne 1911–1994

Lymphocyte specificity
Nobel Prize 1984

© Nobelstiftelsen

CHAPTER 13

A. HISTORICAL NOTES

B. CONTEXT AND OVERVIEW

C. T CELL DEVELOPMENT IN THE BONE MARROW

D. T CELL DEVELOPMENT IN THE THYMUS

E. MATURE SP THYMOCYTES IN THE PERIPHERY

F. T CELL DEVELOPMENT IN AN EMBRYOLOGICAL CONTEXT

"Tis death that makes life live"—Robert Browning

In the last chapter, we discussed the genes and proteins of the TCR and described the generation of their diversity. It is time to put TCR gene rearrangement events in their developmental context, and also to weave into the story the positive and negative selection of thymocytes that constitutes central T cell tolerance and shapes the mature peripheral T cell repertoire. At least for αβ T cells, the TCR and the thymus play crucial roles in these selection processes, as do the coreceptors CD4 and CD8. The development of γδ T cells is less well defined. There is evidence that the maturation of at least some γδ T cell subsets occurs in tissues other than the thymus, so-called "extrathymic development." The biology of γδ T cells and extrathymic T cell development will be discussed in Chapter 18. The activation by antigen of mature T cells in the periphery, the proliferation and differentiation of progeny cells into effector T cells, and the mechanisms of T effector functions will be addressed in Chapters 14 and 15.

A. Historical Notes

Prior to 1960, the thymus was an organ of unknown function; its role in immunity was not recognized and its connection to lymphocytes completely unknown. In 1960, Jacques F. A. P. Miller was working at Guy's Hospital in London, England. He and his colleagues, not unlike many investigators at the time, were intrigued by the revelation that certain viruses, including what came to be called the Moloney murine leukemia virus (MoMuLV), could induce leukemias and lymphomas when injected into newborn mice. Curious as to the origin of the target cell for transformation by these viruses, and suspecting that it might be a cell resident in the thymus (a *thymocyte*), they thymectomized newborn mice before administering MoMuLV. To their complete surprise, most of the thymectomized animals, including the control mice that had <u>not</u> been injected with virus, succumbed to widespread infections and

died. Obviously, although the thymus itself was not essential for life, its removal had compromised the immune responses of even the control mice. Other investigators reported similar results in chickens, where thymectomy resulted in complete abrogation of cellular immunity. These findings were further bolstered by the observations of Robert Good and Byron Waksman, who found that some primary immunodeficiency patients had abnormal thymi. Physical features similar to those of these patients and increased susceptibility to infection were subsequently noted in the *nude* mouse strain. *Nude* mice get their name from their hairless state, but the same autosomal recessive mutation that blocks hair follicle development also prevents or greatly reduces the development of the thymus. These animals were found to be deficient for a subgroup of lymphocytes (now known as T cells) whose absence correlated with severely impaired cell-mediated immune responses.

After establishing the importance of the thymus in cellular immunity, researchers began to explore the mechanisms by which this organ influenced the immune response. They noticed that a large number of immature cells entered the thymus and underwent a very high degree of proliferation, but only a relatively small number of mature lymphocytes actually emerged from this organ. To account for this huge loss of cells during maturation, Niels Jerne argued that a selection process had to be occurring. This hypothesis meshed nicely with the earlier suggestion by Sir MacFarlane Burnett in the 1950s that self-reactive lymphocytes would have to somehow undergo "clonal deletion" in order to establish the self-tolerance necessary for the health of the host. The idea that clonal deletion of self-reactive T cells was happening in the thymus seemed theoretically sound, but at the time was difficult to demonstrate experimentally in any convincing way.

By the mid-1970s, immunologists knew that the mature T cells that succeeded in exiting the thymus were functionally restricted with respect to the MHC haplotype they could recognize. Rolf Zinkernagel and colleagues, as well as Alfred Singer and colleagues, surmised that the selection occurring in the thymus was somehow related to the MHC restriction of mature T cells. To test this hypothesis, these researchers manipulated mice to create "chimeric" animals, that is, mice in which the thymus was of one MHC haplotype but all other cells, including all bone marrow-derived cells, were of a different haplotype. To construct a chimeric mouse, an adult recipient mouse was thymectomized and irradiated to remove all adult bone marrow cell populations. (Some scientists irradiated the athymic *nude* mice described above and used them as recipients.) A recipient mouse, now devoid of all thymus- and bone marrow-derived cells, was reconstituted with bone marrow from a donor mouse of haplotype A and a grafted irradiated thymus from a donor mouse of haplotype B. The MHC restriction of the T cells maturing in the chimeric animal was then tested; that is, did the mature T cells recognize MHC A or MHC B? It was found that the mature T cells exiting the thymus recognized the haplotype of the thymus (MHC B) rather than that of the bone marrow cells (MHC A), showing that the haplotype of the immature T cell precursors entering the thymus from the bone marrow did not influence the haplotype recognized by mature naive T cells exiting the thymus. It was concluded that the actual transit through and exposure to the thymus imposed MHC restriction on immature T cells, selecting for survival only those cells with appropriate specificity for the MHC molecules expressed in the thymus.

The results of Zinkernagel's and Singer's experiments further implied that only those immature T cells specific for nonself antigen/self-MHC were "positively" selected for survival, resulting in a mature T cell repertoire that appeared to be "educated" to see those entities that would constitute a threat to the host. (In fact, the terms "educated" and "thymic education," although still used today by some immunologists, give a false impression of an individual T cell somehow altering its TCR to achieve the needed specificity. T cells do not change their TCRs; whether a given clone survives to appear in the periphery is simply a matter of being selected or not selected. The overall result of selection of millions of individual T cell clones is that the entire mature T cell repertoire in the periphery appears to have been "educated" to see only nonself and not self antigen.) However, for all the elegance of their design, different chimeric mouse experiments sometimes produced conflicting findings. Furthermore, it was difficult to conceptualize or test how this "positive selection" process could select T cells specific for foreign antigen when no foreign antigen was apparently present in the thymus at the time the selection was supposedly occurring. Something more than just MHC restriction studies was needed to carry the selection hypothesis further, but no techniques to do so were available at the time.

In the 1980s, further steps in understanding thymic selection became possible with the characterization of T cell-specific markers such as CD4, CD8, and CD3. Researchers could then analyze the differences in the phenotype of thymocytes derived from various locations within the thymus. In addition, single cell suspensions of thymic cells were stained with fluorescently tagged antibodies recognizing T cell markers, and flow cytometry was then used to sort the thymocytes into populations based on marker expression. These isolated subpopulations could then be injected into irradiated recipient mice and re-analyzed after varying lengths of residency in the recipient thymus. From such studies, developmental pathways were constructed that took cells from immature thymocytes to mature, peripheral T cells based on changes in surface marker expression as the cells migrated through the thymus.

A technique called *fetal thymic organ culture* (FTOC) then became more widely available. It had long been established that the thymus could be removed from a fetal mouse and cultured in the laboratory for an extended period. Such a cultured thymus was found to be fully capable of all its natural functions with respect to T lymphocyte maturation, and FTOC experiments were used to confirm the selection hypothesis for both MHC class I and MHC class II molecules. Monoclonal antibodies directed against either MHC class I or MHC class II were added to cultures of fetal thymi to block the binding of thymocyte TCRs to the MHC molecules expressed on thymic stromal APCs. Subsequent examination of T cells in these cultures showed that no mature CD4$^+$ T cells could be found in cultures treated with anti-MHC class II mAbs, and no mature CD8$^+$ T cells were found in cultures treated with anti-MHC class I mAbs. Thus, to generate full complements of mature CD4$^+$ and CD8$^+$ T cells, intercellular contacts had to be established between the appropriate MHC molecules on thymic cells and TCRs on T cells. Furthermore, it was clear that the coreceptors had a role in facilitating the contact between MHC and the TCR.

Also during the 1980s, identification of the TCR genes, as well as key technological advances in the generation of transgenic mice (refer to Box 12-1 in Ch.12) and gene-targeted knockout mice (refer to Box 8-4 in Ch.8), allowed controlled investigations of the cellular and molecular events that occur during the passage of a thymocyte through the thymus. The transgenic (Tg) mice that have been of particular use in these studies are those that carry a TCRαβ transgene. Such a TCR transgenic mouse strain is created by introducing into its germline DNA the rearranged genes encoding a particular combination of TCRα and β chains that makes up a TCR recognizing a specific peptide–MHC complex (pMHC). Instead of a varied repertoire of TCR specificities, a large percentage of the T cells in each TCR Tg mouse expresses exactly the same transgenic TCR protein and so exhibits the same antigenic specificity. Such TCR Tg animals have been useful for the examination of thymic selection pathways and the function *in vivo* of T cells recognizing a specific antigenic pMHC. In the case of gene-targeted mouse strains, the disruption of a specific gene creates knockout mutant animals that cannot express a particular protein (or some part of it). The homozygous deletion of an immune system gene allows researchers to specifically study the requirement for that gene's product in the development and function of T cells. However, if the gene of interest plays a role in embryonic development as well as in the immune system, the animal will die during embryogenesis before its immune system is established. In these cases, a technique called *RAG blastocyst complementation* can be used to study immune system development and function in the absence of the gene of interest. An example of this method is given in Box 13-1, and examples of the information gained from early and current studies of transgenic and knockout mice appear throughout this chapter.

Box 13-1. GATA-3 and RAG blastocyst complementation

Targeted disruption of a gene can result in early embryonic lethality, meaning that a viable mouse is not available for immunological study. The GATA-3 knockout mutation is just such a case: the mutant mice die early during embryonic development with multiple developmental anomalies. To study the effect of the GATA-3 knockout mutation on the lymphoid system alone, researchers took advantage of a technique called *RAG$^{-/-}$ blastocyst complementation*.

Mice lacking the RAG gene functions are unable to generate lymphocytes: the mice are viable but totally lack T and B cells. To determine the function of GATA-3 in lymphocytes, one can combine in one embryo the cells of the knockout GATA-3$^{-/-}$ strain, which carries an identifiable mutation in the *Gata3* gene but <u>intact RAG genes</u>, with cells of a RAG$^{-/-}$ strain, which has an intact *Gata3* gene but has lost all RAG function. If one introduces embryonic stem (ES) cells from the GATA-3$^{-/-}$ strain into the blastocyst (8-cell stage) of a developing RAG$^{-/-}$ embryo, the embryo absorbs the new cells and continues normal development after implantation into a pseudopregnant

female (see Figure). However, because its founding cells are now a mixture of RAG$^{-/-}$ and GATA-3$^{-/-}$ genetic backgrounds, and different tissues are derived from different parts of the embryo, some tissues of the newborn mice will be of the GATA-3$^{-/-}$ background while others will not: these mice are genetically *chimeric*. Which tissues are derived from the GATA-3$^{-/-}$ cells will be different among the individual pups. If the lymphoid system of a given embryo happens to develop from GATA-3$^{-/-}$ cells (which have intact RAG genes) rather than RAG$^{-/-}$ cells, the chimeric mouse will be able to produce lymphocytes that can only be of the GATA-3$^{-/-}$ genetic background. That is, any lymphocytes found in that mouse will have been generated from the introduced GATA-3$^{-/-}$ ES cells. The effect of the GATA-3 mutation on lymphocyte development can then be analyzed in the usual way. Figure adapted from Yeung R. S., Ohashi P. and Mak, T. W. (in press) T cell development. In Ochs D., Smith, C. I. and Puck J. M. (eds.) "Primary Immunodeficiency Diseases. A Molecular and Genetic Approach." Oxford University Press, Oxford.

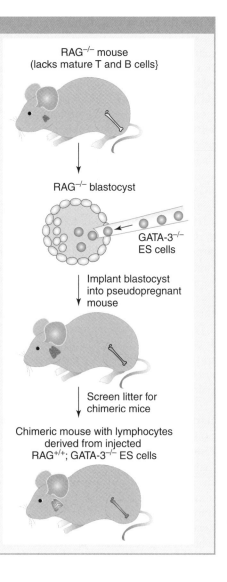

RAG$^{-/-}$ mouse
(lacks mature T and B cells}

RAG$^{-/-}$ blastocyst

GATA-3$^{-/-}$
ES cells

Implant blastocyst
into pseudopregnant
mouse

Screen litter for
chimeric mice

Chimeric mouse with lymphocytes
derived from injected
RAG$^{+/+}$; GATA-3$^{-/-}$ ES cells

Engineered transgenic and gene knockout studies in humans are obviously not ethically feasible, but the available evidence suggests that the majority of events that occur during T cell development are the same in mice and humans. In this chapter, we describe T cell development as it has been determined largely in the mouse.

B. Context and Overview

I. COMPARISON OF B AND T CELL ONTOGENY

While B and T cells are both lymphocytes and are derived from the same hematopoietic progenitor cells, they differ greatly in their differentiation paths and the effector cells and products that result. With respect to ontogeny, the most important difference distinguishing B and T cells is the requirement of the thymus for T cell development. The primary role of the thymus is to foster the development of the majority of

mature peripheral T cells that bear the diverse repertoire of TCRs needed to stock the peripheral lymphoid organs. The unique microenvironment of the thymus enables it to direct T cell development (at least for αβ T cells) in a series of well-defined steps. The TCR is not required for the early steps of αβ T cell development (as is true for the BCR in B cell development) but is essential for the middle and later steps. However, the coreceptors CD4 and CD8 also play increasingly important roles as T cell development progresses, but these molecules have no direct equivalents in B cell development.

The duration of T and B lymphocyte generation is also different. Once the involution of the thymus (see Box 13-2) commences around puberty, new naive T cell production is sharply reduced. Thus, for the rest of its life, the maturing adult is largely dependent on the repertoire of T cells generated in the successive waves of its early existence. This stands in contrast to the situation of B cells, which are freshly produced at essentially the same rate throughout the life of the animal. Thus, while the B cell repertoire continues to evolve and expand throughout adulthood, the T cell repertoire is dominated by those cells created in the early stages of an animal's life.

Up until puberty, the mammalian thymus steadily increases in size and produces wave after wave of the naive mature T cells needed for protective immune responses. After puberty, the thymus starts to lose both its tissue mass and its functionality. Most of the loss is sustained just prior to middle age, followed by a slower rate of involution extending into old age. In humans (and to a lesser extent in mice), the thymic tissue is replaced by fat deposits. The loss of cellularity in the thymic tissue arises because DN thymocytes are prevented from undergoing the proliferation and differentiation necessary to generate the major subpopulations of thymocytes that constitute the bulk of the thymic cellular mass. Thus, the thymus effectively shrinks and contributes less and less to the pool of naive T cells. One effect of this reduced number of naive T cells is that the diversity of the TCR repertoire in an individual progressively becomes more limited. A second effect is related to the homeostatic mechanisms that ensure that the total peripheral T cell pool is maintained at a steady level. Because there are fewer naive T cells entering the periphery of an older individual, peripheral T cells responding to antigen will proliferate more than those in a younger individual in an effort to maintain a steady overall T cell number. As a result, many T cells in the memory T cell pool are close to their replicative limit. Thus, upon reactivation, a T cell in an older individual cannot undergo the same degree of clonal expansion as a T cell in a younger person. Fewer effector cells are generated and the observed immune response is weaker. In addition, peripheral T cells isolated from older individuals respond less efficiently to stimuli *in vitro* than do T cells from younger individuals. Differences in cell surface molecules, intracellular signaling, cell division, and cytokine production have all been implicated. It is believed that this combination of decreased naive T cell production coupled with decreased responsiveness of existing peripheral T cells results in the reduced resistance to infection and tumorigenesis seen in aged individuals.

Many studies have been done to determine the cause of thymic involution, and several hypotheses have been explored and ruled out. Involution is not due to lower numbers of T cell progenitors in the bone marrow of aged mice, because bone marrow samples of aged and young mice contain equal numbers of these cells. Moreover, T cell progenitors from young and aged mice are equally able to repopulate the spleens and thymi of depleted recipients of any age, ruling out an intrinsic T cell progenitor defect. However, examination of transgenic mice expressing a TCR that is already complete and does not require V(D)J recombination has revealed that these mutant animals do not exhibit thymic involution. Thus, one piece of the puzzle may be that DN cells in aged mice are unable to rearrange their TCRβ genes properly, possibly due to inadequate IL-7 expression (see later). Another piece of the puzzle may be related to the role of thymic stromal cells. Reconstitution experiments have shown that the introduction of the bone marrow of a young mouse into an aged mouse cannot reverse involution in the latter, as would be expected if the problem were solely in the T cells. While the total number of stromal cells in the thymus does not appear to change with age, it may be that these cells are also reaching their replicative limits in aged individuals. In addition, altered gene expression patterns may occur in aged thymic stromal cells that could affect thymocyte maturation. Indeed, the expression of MHC class II decreases with age in certain *thymic epithelial cell* (TEC) subpopulations.

Age-dependent alterations to the expression of cytokines and growth factors may also influence the thymic microenvironment such that thymocyte maturation is inhibited. For example, expression levels of IL-2, IL-3, and IL-7 are significantly decreased in the thymi of older mice compared to younger ones. Strikingly, if young mice are treated with anti-IL-7 antibody, they immediately show signs of thymic involution that reverses when the antibody is withdrawn. It is speculated that IL-7 is needed to drive the expression of Bcl-2 required to sustain the survival of DN thymocytes trying to initiate TCRβ chain rearrangement. Without this rearrangement, pre-TCR signaling cannot occur and DN thymocytes cannot proliferate, resulting in decreased thymic cellularity and functionality. Other cytokines, such as IL-4, IL-5, and IL-6, show an increase in thymic expression with age. Indeed, administration of high levels of IL-6 to rodents can induce thymic atrophy. Similar results were obtained with administration of *leukemia inhibitory factor* (LIF), another cytokine whose expression in the thymus increases with age. The precise mechanisms by which these cytokines might promote involution are unknown.

The growth factors *nerve growth factor* (NGF) and *insulin-like growth factor-1* (IGF-1) have been shown to play key roles in supporting thymic functions. Medullary thymic epithelial cells (mTECs) produce NGF, and receptors able to bind NGF (including a molecule called TrkA) are expressed on the mTECs themselves as well as on thymocytes, cortical thymic epithelial cells (cTECs), and thymic DCs. *In vivo*, administration of high levels of NGF to rats causes overexpansion of both thymocytes and TECs, while knockout mice lacking TrkA undergo accelerated involution. *In vitro*, NGF upregulates IL-2R expression on thymocytes. IGF-1 is produced by thymic macrophages, while its receptor IGF-1R is expressed mainly on thymocytes. *In vivo*, administration of IGF-1 to rodents stimulates DN thymocyte production such that thymic mass is significantly increased. In normal mammals, the levels of both NGF and IGF-1 decline with age. It is possible that decreased NGF and IGF-1 contribute to defects in thymocyte and/or TEC production or proliferation, resulting in loss of thymic cellularity.

The involution of the thymus is thought to have clinical consequences in our modern age, since the average life span of a person in a developed country has extended considerably over the last century. With the aging of the population in mind, researchers have been looking at ways to "reactivate" the thymus in older individuals, hoping to reverse the decline in immune responsiveness and the associated susceptibility to potentially lethal infections. Because the thymus is influenced by the neuroendocrine system, researchers have tried to reverse involution by treatments with various hormones, but none has been found to be truly effective. Aged rodents and primates treated with IGF-1 show increased thymic cellularity, but thymocyte production is never fully restored to youthful levels and numbers of mature naive T cells remain low. While SCF is critically important for the earliest stages of thymopoiesis, treatment with SCF does nothing to reverse thymocyte death associated with involution either *in vitro* or *in vivo*. In aged mice, subcutaneous injection of recombinant IL-7 can stop thymic involution and restore some production of naive T cells. However, it remains a challenge to ensure that enough

Continued

Another major difference between B and T cell development paths is the involvement of the MHC. The TCR on a developing T cell must not only be functional (be derived from productive rearrangements of the TCR genes) but also must have an affinity for the host's self-MHC molecules. Thus, although pre-B cells also undergo selection processes that minimize the presence of autoreactive clones in the periphery (see Ch.9), the requirement that there be an affinity for host MHC imposes an additional layer of selection on T cells that developing B cells do not encounter. Positive selection of developing thymocytes promotes the survival of T cell clones whose TCRs weakly recognize self-MHC molecules, while those that fail to recognize self-MHC die of apoptosis. Negative selection of thymocytes then involves the induced deletion of T cell clones whose TCRs strongly recognize self-MHC complexed to self-peptide. Thus, thymic selection leaves as survivors those clones whose TCRs are more likely to recognize self-MHC complexed to non-self-peptide. The result is a mature T cell repertoire that is diverse in its recognition of foreign antigens but tolerant to self-antigens. In other words, *central T cell tolerance* is established by selection in the thymus, just as central B cell tolerance is established by selection in the bone marrow.

While the generation of diversity of the antigen receptor repertoires is largely similar in the B and T cell compartments, there are two major differences. First, diversity is established for T cells entirely in the thymus through the processes of V(D)J recombination, junctional diversification, and heterodimer assembly. Unlike B cells, T cells do not express the RNA-editing enzyme AID essential for somatic hypermutation of the Ig genes. While some workers have reported low levels of receptor editing of TCR chains, others have found the opposite and the issue remains controversial. Thus, T cells generally appear to lack mechanisms that can adjust the antigen receptor's specificity <u>after</u> encounter with antigen. Secondly, there is no collection of constant exons in the TCR genes that can be mobilized for isotype switching, so that there is far less diversity of effector function within a T cell clone than occurs for B cell clones. Even if alternative constant exons were available, T cells do not express the AID activity necessary for isotype switching of Ig constant exons. Finally, while all B cells appear to follow a single developmental program as they mature, functional T cells can result from several different paths. First, there is the divergence of γδ T cells and αβ T cells. Secondly, two major types of mature naive T cells, the Th and Tc subpopulations, differentiate late in αβ T cell development, as opposed to a single major type of mature naive B cell. Thirdly, antigen-activated Th clones further differentiate into either Th1 and Th2 cell clones, which differ in effector functions. Lastly, while B cell maturation appears to occur almost exclusively in the bone marrow, functional subpopulations of T cells have been identified in individuals lacking a thymus, indicating that other as-yet-unidentified tissues can contribute to the T cell compartment.

II. OVERVIEW OF T CELL DEVELOPMENT

It is important to note that the thymus is empty of lymphocytes and their progenitors early in fetal development. The thymus must be "colonized" or seeded with pluripotent hematopoietic stem cells (usually the common lymphoid progenitors or CLPs described in Ch.3) that subsequently proliferate and mature in the thymus into functional naive T cells. The colonization of the thymus by CLPs occurs in a limited number of distinct waves. Before birth, at least two waves of CLPs migrate from the fetal liver to the thymus. For example, in the mouse (gestation period of 20–21 days), the first lymphoid precursor cells can be detected in the fetal thymus between days 10 and 13. These cells have been shown to be the source of the mature T cells emigrating from the thymus within the first week after birth. A second wave of CLPs colonizes the fetal thymus after day 13 of gestation, giving rise to a second generation of mature T cells that predominates after the first week of life. In fact, the first waves of mature T cells exported from the murine thymus to seed the peripheral tissues are γδ T cells rather than αβ T cells. It is not until after birth that αβ T cells gain ascendance and the γδ T cells become a minor population (Table 13-1). Subsequent colonization of the thymus in the very young adult is carried out by waves of CLPs generated in the bone marrow. In general, the developmental path of CLPs to mature T cells exported from the thymus is the same in the adult and the fetus but displays slightly faster kinetics in the fetus. Surface marker expression may also differ slightly. In addition, the TdT enzyme, responsible for much of the junctional diversity generated during TCR gene rearrangement, is not expressed until shortly after birth, meaning that the repertoire of T cell specificities available in the neonate will be significantly less diverse than in the adult.

For all the prodigious production of hematopoietic T cell progenitors in the bone marrow and the vast expansion of T cell precursors within the thymus, it is estimated that more than 98% of all thymocytes generated undergo apoptosis. Some clones fail to produce functional TCR chains during V(D)J recombination, while others bear TCRs that fail to recognize

Table 13-1 Comparison of Murine Thymocyte Development at Different Life Stages

	Fetus	Neonate	Adult
Origin of CLPs	Fetal liver	Fetal liver	Bone marrow
TCRs	Only γδ No αβ	Majority γδ Minority αβ	Minority γδ Majority αβ
Kinetics of progression from CLP to mature T cell	Fast	Fast	Slow
TdT expression	None	Initiated	Fully active
T cell repertoire diversity	Limited	Limited	Fully diversified
Generation of thymocytes	Continuous	Continuous	Minimal after involution of thymus

self-MHC or bind strongly to self-peptide/self-MHC complexes and thus are autoreactive. It is only the remaining 1–2% of all thymocyte clones that secure survival signals during thymic selection. It is then up to these cells to carry out the vital functions of immune responses. We will now follow the development of one particular T cell clone from its generation in the bone marrow through its emigration to the thymus and its subsequent maturation into a mature naive αβ T cell clone.

C. T Cell Development in the Bone Marrow

A T cell starts life in the fetal liver (during embryogenesis) or in the bone marrow (post-natally) much as a B cell does, derived from the same pluripotent HSC. As described in Chapter 3, the HSC proliferates to produce a collection of CLPs that eventually gives rise to B, T, and NK cells depending on the influence of specific transcription factors and cytokines. CLPs receiving both IL-7 signaling and IL-3 signaling become B cell progenitors, while those receiving predominantly IL-7 signaling become NK/T precursors capable of generating T cells, NK cells, and DCs. Still under the influence of IL-7 and other unknown signals, some of the NK/T precursors (and perhaps even some CLPs) leave the fetal liver or bone marrow, travel through the blood, and enter the thymus at the corticomedullary junction. It is likely that such thymus-bound precursors express a distinct receptor(s) on their cell surfaces that enables them to follow a chemotactic gradient of chemokines secreted by thymic stromal cells. One of the most important of these chemokines may be *thymus-expressed chemokine* (TECK), which can bind to chemokine receptor CCR9. The same or different receptors may promote the binding of thymic progenitors to thymic vascular endothelial cells, ensuring that they exit the circulation at the correct site for the next developmental steps. It is also speculated that

the interaction of the adhesion protein CD44 present on thymic progenitors with elements of the *extracellular matrix* (ECM) may assist in the migration of these cells to the thymus. Entry into the thymic cortex itself appears to be governed by signaling initiated by engagement of the chemokine receptor CXCR4, the receptor for SDF-1. In any case, very few T cell progenitors are required to colonize a thymus; it has been estimated that even one lonely cell can do the job.

There are as yet no unique surface markers that identify CLPs, and while NK/T precursors express IL-7Rα and the γ$_c$ chain, these markers are not exclusively expressed by these cells. Interestingly, expression of mRNA for the pre-Tα (pTα) chain has been detected in mouse bone marrow. Since pre-B cells, B cells, and other hematopoietic cells do not express pTα, its expression has allowed researchers to identify those CLPs (or their descendants) that will eventually become T cells.

D. T Cell Development in the Thymus

I. OVERVIEW

i) Thymocytes

Upon their arrival in the thymus, the T cell progenitors evolve into thymocytes. Aside from some size variation, thymocytes at different developmental stages are morphologically very similar and so are usually distinguished by one or both of two experimental methods. First, genomic DNA rearrangements of the TCR genes can be analyzed by restriction mapping or PCR to detect stages of the rearrangement of the TCR genes. Secondly, specific monoclonal antibodies can be used to determine sequential expression patterns of certain surface proteins that act as markers. Markers followed during murine T cell development include the TCR, CD3, CD4, and CD8 at all stages, and c-kit, CD44, and CD25 during the earliest stages. c-kit (CD117) is the tyrosine kinase receptor for stem cell factor (SCF), which, as mentioned in Chapter 9, is also crucial for early B cell development. As stated above and discussed in more detail in Chapter 14, CD44 is a glycoprotein that mediates T cell adhesion to the matrix component hyaluronic acid (among other ligands). CD25 is the α chain of the IL-2 receptor.

In humans, T cell development has not been as clearly characterized due to the more limited options for thymic reconstitution assays. In the 1990s, two techniques became available for the study of human T cell differentiation: human FTOC and the human *scid* mouse model. In the latter, a fragment of a human fetal thymus is transplanted under the kidney capsule of a *scid* mouse (which cannot make its own T or B cells) and seeded with human HSCs. A human thymic microenvironment and T cell developmental pathway is thus reconstituted in the mouse kidney. Analyses of developing T cells in these systems have shown that CD25 is not a useful maturation marker for human T cells. Rather, the loss of expression of CD34[+] (a surface marker associated with bone marrow HSCs) and the gain of expression of first CD7, followed by CD1a, and finally by CD3, CD4, and CD8 serve to define human T cell development.

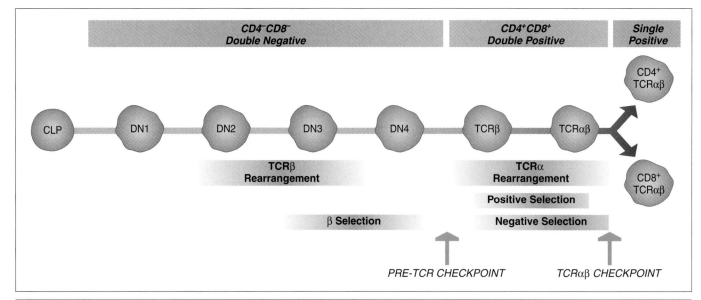

Figure 13-1

Overview of T Cell Development in the Murine Thymus
Under the influence of particular cytokines and intercellular contacts, CLPs pass through the four stages (DN1–4) of the double negative (CD4⁻CD8⁻) phase. The pre-TCR checkpoint marks the end of this phase and entry into the double positive (CD4⁺CD8⁺) phase. After passing through the TCRαβ checkpoint, the developing thymocytes enter the single positive phase and become mature CD4⁺CD8⁻ or CD4⁻CD8⁺ T cells. Adapted from Yeung R. S., Ohashi P. and Mak T.W. (in press) T cell development. In Ochs H. D., Smith, C. I. and Puck, J. M., eds. "Primary Immunodeficiency Diseases. A Molecular and Genetic Approach." Oxford University Press, Oxford.

T cell development in the murine thymus takes place in three broad phases controlled by two developmental checkpoints (Fig. 13-1). The different phases are distinguished on the basis of the CD4/CD8 expression status of the cell. The earliest thymocytes are said to be in the *double negative* or *DN phase*, as they express neither CD4 nor CD8. However, they may sometimes be called *triple negative* or *TN* thymocytes, since they are also negative for TCR expression. Within the DN phase are four subsets of thymocytes labeled DN1–4. As the thymocyte matures and its pre-TCR becomes functional, it gains expression of both the CD4 and CD8 coreceptors, so this phase is called the *double positive* or *DP phase*. DP T cells run the gauntlet of thymic selection, making a final commitment to either the CD4 or CD8 lineage, and emerging in the periphery as mature *single positive* CD4⁺ Th or CD8⁺ Tc cells in the *SP phase*. In other words, thymocytes are generally said to develop from DN (or TN) through DP to SP cells.

The first of the two developmental checkpoints occurs as the DN (TN) phase ends. As we saw in Chapter 12, a process called β-*selection* occurs whereby each candidate TCRβ chain expressed is tested for functionality using the pre-TCR. Thus, the cell is said to have reached the *pre-TCR* or β-*selection checkpoint*. Dramatic expansion of β-selected clones expressing functional pre-TCRs ensues as the expression of both CD4 and CD8 is upregulated. At this point, the DN (TN) phase ends and the cells become DP thymocytes. TCRα gene rearrangement takes place in early DP cells, followed by TCRαβ expression at the cell surface. The cells have then reached the second developmental checkpoint, which is often called the *TCRαβ checkpoint*. At this time, positive selection of DP cells expressing a fully functional TCRαβ promotes the survival of thymocyte clones recognizing self-MHC alleles, while negative selection deletes autoreactive thymocyte clones. Late DP thymocytes that express useful αβ TCRs subsequently lose expression of either CD4 (on CD8⁺ cells) or CD8 (on CD4⁺ cells) to become SP thymocytes. These cells exit to the periphery and become the mature naive CD4⁺ Th and CD8⁺ Tc subsets that patrol the body for antigens.

ii) The Thymic Environment

It is difficult to take the earliest thymocytes out of the thymus and induce them to develop into DP cells in culture in the absence of stromal cell architecture. Indeed, mutations altering thymic structure or stromal development, including defects in various embryologically important transcription factors, have indirect but important effects on T cell development. This observation indicates that T cell maturation is highly dependent on passage through a succession of thymic microenvironments characterized by a different mix of stromal cell types (Fig. 13-2). Among the most important of these cell types are the thymic mesenchymal fibroblasts, cortical thymic epithelial cells (cTECs), and medullary thymic epithelial cells (mTECs). Interestingly, *nude* mice, which are essentially athymic, have a mutation in the gene that encodes the transcription factor Foxn1, which is required for the development of both cTECs and mTECs.

In the 1990s, a method of thymic "reaggregation" *in vitro* was devised that allowed researchers to culture purified murine thymocytes with one of several purified murine tissues (including thymic epithelium, salivary epithelium, gut epithelium, fetal mesenchymal fibroblasts, and thymic DCs) in an attempt to determine the precise function of each stromal population. It

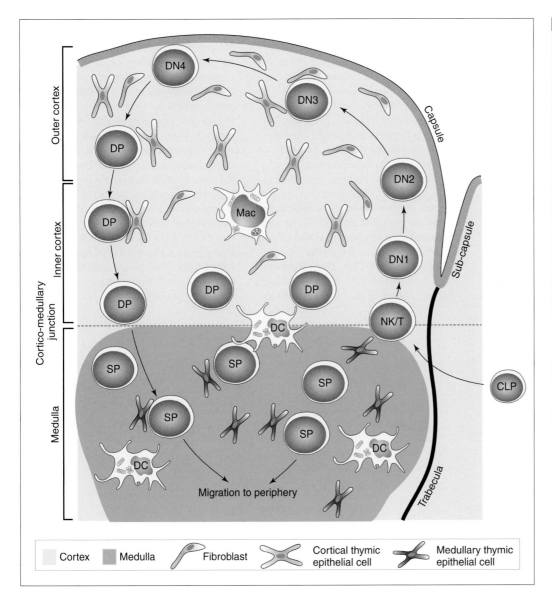

Figure 13-2

Thymic Microenvironment and Location of Developing Thymocytes
T cell progenitors (most likely CLPs or NK/T cells or their descendants) migrate from the bone marrow and enter the thymus at the cortico-medullary junction. Thymocyte development then proceeds in a series of stages that are confined to specific regions of the thymus and influenced by the type of thymic epithelial cell present in that region. DN1 thymocytes are found in the inner cortex but soon migrate toward the outer cortex and become DN2 cells. DN3 cells predominate in the outer cortex and make the transition to the DN4 stage in the subcapsular region. These cells then become DP thymocytes and migrate back down through the cortex to the cortico-medullary junction. In the medulla, only SP thymocytes are found. Adapted from Blackburn C. C. and Manley N. R. (2004) *Nature Reviews Immunology* **4**, 278–289.

Cortex   Medulla   Fibroblast   Cortical thymic epithelial cell   Medullary thymic epithelial cell

was discovered that the DN to DP transition required both thymic epithelial cells and thymic mesenchymal fibroblasts. The critical function of the latter cell type turned out to be production of components of the ECM, such as fibronectin, collagen, and laminin. Furthermore, the fibroblasts had to be organized in a particular three-dimensional structure, since disruption of the fibroblasts and/or their production of ECM components blocked DN maturation in the reaggregated cultures. These observations led to the hypothesis that the fibroblast-secreted ECM serves as a scaffold upon which growth factors and cytokines (such as IL-7) required for thymocyte development can be concentrated. In addition, while some ECM proteins are involved in the adhesion of thymocytes to stromal cells, others are involved in their release, so the production of such components at different times and in different locations may play a role in the directed migration of thymocytes through the thymus.

Thymic stromal cells also supply the ligand(s) for Notch1, a protein crucial for T lineage commitment. The Notch proteins are a family of highly conserved transmembrane receptors that control binary cell fate decisions in diverse organisms (see Box 13-3). The reader will recall from Box 9-2 that Notch1 promotes T cell development at the expense of B cell development. Notch1 regulates transcription by associating directly with nuclear transcription factors. Among the genes activated as a result of this association are the HES genes, transcriptional repressors that act as downstream effectors in the Notch1 signaling pathway. Thus, the interaction of Notch1 expressed on DN cells with its ligand Deltex-1 expressed on thymic stromal cells results in the repression of genes needed for B cell differentiation (like E2A) and the promotion of V to DJ recombination in the TCRβ locus. Accordingly, loss-of-function mutations of Notch1 result in a severe early block in T cell development and the appearance of ectopic B cells in the thymus. (*Ectopic* is a term that refers to an abnormal expression or localization of a protein or structure.) Conversely, expression of a constitutively active form of

Box 13-3. **Cell fate determination via Notch signaling**

Notch is a transmembrane protein that was first identified in *Drosophila* embryos as being involved in determining cell fate. It was noticed that, in a population of seemingly identical *Drosophila* cells grouped together, some developed into epidermal cells while others developed into neural cells. Further work determined that cells expressing slightly more Notch protein received a signal that caused them to develop as epidermal cells, whereas cells that did not express as much Notch followed a default pathway and developed as neurons. A fascinating intercellular interaction was revealed in which a cell apparently assesses its expression level of Notch and Notch ligand relative to those on neighboring cells. If its levels are lower, the cell does not vary from the default developmental path. This type of mechanism is called *lateral inhibition* and occurs in mammalian cells as well, where a member of the Notch family called Notch1 is involved.

Let us suppose we have a collection of developmentally identical progenitor cells in which each cell initially expresses both Notch and one or more of its many ligands. Interaction between Notch and the ligand results in proteolytic cleavage of Notch and the release of its intracellular domain (IC). The Notch-IC translocates to the nucleus and interacts with the transcription factor CBF1 (see Figure, panel A), converting it from a repressor to an activator of genes involved in cell fate decisions. Some of these genes act to increase expression of Notch while others suppress expression of Notch ligands. Small differences in the level of Notch expression within our collection of cells mean that some cells will receive less Notch signaling. As a result, relative to their immediate neighbors, these cells subsequently express less Notch and more Notch ligand (see Figure, panel B). With less Notch expressed on the surface, these cells

then receive even less Notch signaling but become capable of stimulating neighboring cells that have initially received more Notch signal and thus have upregulated expression of Notch. This feedback loop continues to amplify small differences in Notch expression between different members of the progenitor collection. Eventually, two groups of cells are formed: those that receive Notch signaling and those that do not. As a result, neighboring cells tend to adopt opposite fates, and mixed populations of differentiated cell types can be found in the same location. As noted in the text, there is growing speculation that Notch signals may be involved in the developmental divergence of TCRαβ vs. TCRγδ thymocytes (see Ch.18), and CD4 vs. CD8 TCRαβ cells. Figure panel A adapted from Germaine R. N. (2002) T-cell development and the CD4-CD8 lineage decision. *Nature Reviews Immunology* **2**, 309–322.

A. Notch Signaling

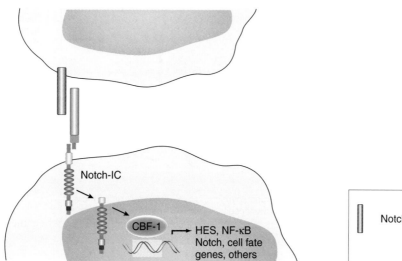

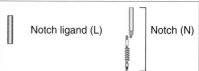

B. Lateral Inhibition

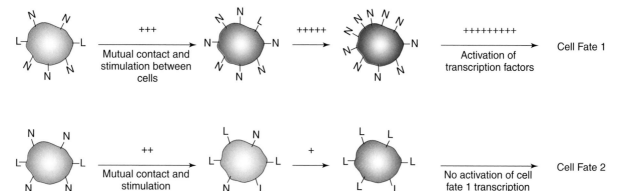

Notch1 in bone marrow stem cells results in the abnormal development of precursor T cells outside the thymus.

The definition of additional intercellular contacts between molecules on thymic stromal cells and immature thymocytes that are necessary to promote T cell development is still in progress. Interaction of SCF (either on the surface of or secreted by thymic stromal cells) with its ligand c-kit on the earliest DN thymocytes may be important for efficient maturation to proceed, although studies of mutant animals suggest that other molecules may also be able to supply the necessary functions. Thymic stromal cells also produce many other growth factors and cytokines, but studies of knockout mice have shown that T cell development can proceed to some extent in most cytokine-deficient mutants. For example, while cytokines such as IL-1, IL-2, and IL-10 have been shown to influence T cell differentiation *in vitro*, studies of IL-1$^{-/-}$, IL-2$^{-/-}$, and IL-10$^{-/-}$ mice proved that they are not essential for thymic development *in vivo*.

The exception is IL-7. Mutants lacking IL-7, components of the IL-7 receptor, or its signal transducers exhibit T cell development that is severely compromised. The IL-7 receptor is made up of a unique IL-7α chain and a common γ chain (the γ_c chain) which is also shared by the receptors for IL-4, IL-9, and IL-15 (see Ch.17). Intracellular signaling from these receptors is transduced by the association of γ_c with the cytoplasmic tyrosine kinase *Jak3*. Mouse mutants lacking Jak3, IL-7, IL-7α, or the γ_c chain have similar phenotypes characterized by a very early block in T cell development prior to TCR gene rearrangement. Furthermore, double mutants lacking both c-kit and the γ_c chain display a complete abrogation of T cell development. The cellular source of IL-7 in the thymus has yet to be precisely identified but is likely to be stromal cells. It is thought that IL-7 signaling may provide DN cells with an essential survival or proliferation stimulus prior to TCRβ gene rearrangement. If a transgene encoding the anti-apoptotic protein Bcl-2 is introduced into IL-7- or IL-7R-deficient mice, T cell development is rescued, implying that an important function of IL-7 signaling is to maintain thymocyte survival. *In vitro*, treatment of progenitor thymocytes with IL-7 results in the upregulation of Bcl-2 and protection of these cells from apoptotic stimuli. The putative signaling pathway between the IL-7 receptor and the Bcl-2 gene remains to be elucidated.

Several other molecules have been identified as being important for thymic stromal cell development and/or function. GHLF, Whn, Pax1, and Hox3a are all transcription/differentiation factors important for cTEC/mTEC development and the establishment of a normal thymic microenvironment. "Sonic hedgehog" is a maturation factor produced by thymic stromal cells that acts on DN thymocytes to hold them at this stage of development. The transition to DP cells cannot occur until an unknown mechanism blocks sonic hedgehog signaling by thymic stromal cells.

We will now examine in detail each phase and checkpoint of T cell development as defined in the mouse, with reference to the situation in humans where possible. Figure 13-3 depicts thymocyte development in detail, including selected influential molecules defined by gene-targeting studies in mice. The reader need not be intimidated by this figure: we include this level of detail merely to illustrate the complexity of the process and the many genes involved. Examples of transcription factors whose activities have effects in each phase are discussed in Box 13-4.

II. DN (TN) PHASE (TCR$^-$CD4$^-$CD8$^-$)

Within cells traversing the DN (TN) phase (TCR$^-$CD4$^-$CD8$^-$) in the mouse, one can define the four discrete subsets (DN1–4) by their differing expression of the markers c-kit, CD44, and CD25. Although early DN thymocytes express the marker Thy-1 (in mice), they do not yet express the TCR, CD3, or coreceptor chains, and cannot bind antigen or carry out effector functions. The components of the IL-7 receptor are expressed throughout all DN stages, as is pTα mRNA. The pTα protein is first expressed on DN3 cells.

i) DN1: c-kit$^+$ CD44$^+$ CD25$^-$ Thymic Progenitors

The first identifiable subpopulation of murine DN thymocytes can be characterized as c-kit$^+$ CD44$^+$ CD25$^-$. (Strictly speaking, these cells are not really "double negative" with respect to CD4/CD8 because they do carry very low levels of CD4 on their cell surfaces. However, the function of CD4 here appears irrelevant to maturation, since very early thymocyte development is normal in CD4$^{-/-}$ mice.) DN1 cells reside near the corticomedullary junction side of the cortex. These cortical cells are small but so closely packed together among the supporting thymic epithelial cells that they are pushed into polyhedral shapes rather than the spherical morphology more typical of lymphocytes. SCF synthesized by thymic stromal cells delivers survival signals to these cells through c-kit, without which maturation ceases and the cells die. The TCR genes remain in germline configuration. Interestingly, DN1 thymocytes retain some capacity for returning to a pluripotential state (Fig. 13-4). For example, when these cells are isolated and transferred intravenously to a recipient mouse, they can give rise not only to T cells but also to B cells, NK cells, and thymic DCs. *In vivo*, however, Notch1 signaling initiated by interaction with Deltex-1 expressed by thymic stromal cells establishes the commitment to the T lineage.

In humans, the earliest thymocytes, which are also multipotent, can be characterized as CD34$^+$, CD7$^+$, CD1a$^-$, CD4$^-$, and CD8$^-$.

ii) DN2: c-kit$^+$ CD44$^+$ CD25$^+$ Pro-T Cells

By the next identifiable thymic DN stage, the developing murine thymocytes have started their migration toward the subcapsule of the thymus and are present primarily in the outer cortex. DN2 thymocytes have lost the low level of CD4 expression but continue to express c-kit and CD44. In addition, these DN2 thymocytes commence expression of CD25. The function of CD25 in early thymocyte development is unknown, as IL-2 is apparently not required. The MHC-like molecule CD1 can also be found on these cells but its function here is unknown. The TCR genes remain in germline configuration. Although they do not yet express TCRα or β chains,

Figure 13-3

**Selected Molecules
Influencing Murine
Thymocyte Development**
The molecules shown in the
blue boxes have been
determined to act at the
indicated stage of thymocyte
development through studies of
cell numbers and phenotypes
in knockout mice lacking these
molecules. For example, in the
absence of RAG1/2, one sees a
decrease in numbers of DN3
cells and an accumulation of
DN2 cells that are "stuck" at this
stage of development. Note that
the effects of a given molecule
may not be restricted to one
stage, and that residual cells
that make it past a given
deficiency may be affected later
on by the same defect. For
example, loss of CD45 affects
both the DN2–DN3 and the
DP–SP transitions. The original
names for the DN stages and
markers distinguishing these
thymocyte subsets are shown
on the left (see text for details).
Adapted from Yeung R. S.,
Ohashi P. and Mak T.W. (in
press) T cell development. In
Ochs H. D., Smith C. I. and
Puck J. M., eds. "Primary
Immunodeficiency Diseases.
A Molecular and Genetic
Approach." Oxford University
Press, Oxford.

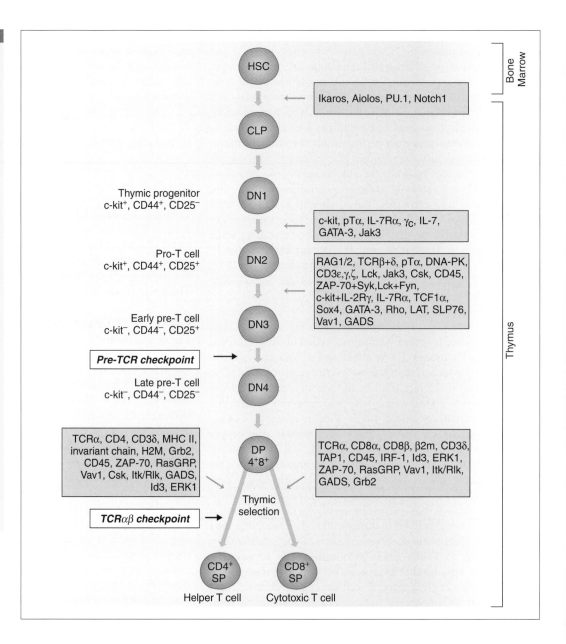

$c\text{-kit}^+$ $CD44^+$ $CD25^+$ DN2 thymocytes do express CD3 chains, which can appear on the DN cell surface in association with the molecular chaperone molecule *calnexin*, although the function of this usually ER-associated protein here is unknown. (Calnexin is not found associated with the TCRs of mature T cells.) Whether these CD3-containing complexes have a signaling function *in vivo* is unclear, but experiments *in vitro* using anti-CD3 antibodies to cross-link these complexes have shown that they are capable of transducing signals that promote the differentiation of DN thymocytes to the DP stage. In any case, DN2 thymocytes start to proliferate rapidly, resulting in a 5-fold increase in cell numbers thought to be sparked by SCF in combination with IL-7. DN2 cells retain less pluripotential capacity than DN1 cells since isolated DN2 thymocytes transferred to a recipient mouse can no longer give rise to B cells. *In vitro* studies have suggested that these

thymocytes can still generate NK cells, but with greatly reduced efficiency compared to cells of the preceding stage.

Human pro-T cells at the equivalent stage can be identified by their expression pattern of $CD34^+$, $CD7^-$, $CD1a^+$, $CD4^-$, and $CD8^-$.

iii) DN3: $c\text{-kit}^-$ $CD44^-$ $CD25^+$ Early Pre-T Cells

In the next developmental step, the maturing murine DN cells are present in the outer cortex or subcapsule. These DN3 cells cease to cycle and downregulate their expression of c-kit and CD44 but continue to express CD25. (The human early pre-T cell has lost expression of CD34 and CD7 but shows high expression of CD1a. These cells may even start to express low levels of CD4 and $CD8\alpha$, but not $CD8\beta$, prior to pre-TCR signaling.) The DN3 stage is critical in T cell development because five key events occur at this time: (1) the thymocytes

Box 13-4. **Transcription factors important in T cell development**

We introduced aspects of transcriptional control in the TCR structural genes in Chapter 12 in the context of regulating TCR gene rearrangement. Here, we expand on the nuclear transcription factors important for T cell development (see figure). Studies of transcriptional enhancers and promoters of both the TCR genes and other genes known to be induced during T cell development have identified specific DNA sequences that constitute binding sites for numerous transcription factors. The fact that many of these genes exhibit binding sites for many of the same factors suggests that their expression is deliberately coordinated during T cell development. Elements have been found that appear to regulate T cell lineage commitment, $\gamma\delta/\alpha\beta$ commitment, thymocyte development prior to CD4/CD8 expression, and thymocyte development after CD4/CD8 expression.

It is important to note that a specific cell fate is the sum of the interactions of an extremely complex network of transcription factor activities. Two different levels of activity of a given transcription factor can have vastly different cellular outcomes. For example, expression of PU.1 in CLPs is necessary to drive thymocyte development, but abnormal extension of PU.1 expression late into the DN phase inhibits T cell development. Similarly, GATA-3 is absolutely required for T cell development from the CLP stage, but overexpression of GATA-3 blocks the development of not only T cells but also B cells and myeloid cells. In addition, a given transcription factor may post-translationally alter the DNA binding ability or transactivation capability of another transcription factor. For example, PU.1 and GATA-3 can mutually interfere with each other's functions. It is therefore thought that important transcription factors are present in limiting concentrations, and that the sum of their interactions with enhancer motifs, and their positive and negative effects on each other's expression and activity, determine an individual cell's developmental fate.

• Transcription Factors Important for the DN Phase
i) PU.1 and c-myb
As was discussed in Box 9-2 of Chapter 9, PU.1 and c-myb are both required for the development of CLPs from HSCs. Since homozygous mutation of the c-myb gene is embryonic lethal in mice, researchers carried out RAG blastocyst complementation using c-myb$^{-/-}$ ES cells and showed that the chimeric animals contained no

detectable immune system cells. Similarly, mature T cells (and macrophages, neutrophils, and B cells) were missing in newborn PU.1$^{-/-}$ mice, leading to their deaths from severe septicemia soon after birth. However, if these PU.1$^{-/-}$ mice were rescued with antibiotics, adult mice were later able to generate both neutrophils and T cells. These findings imply that PU.1 is essential for T cell development during embryogenesis but not in the adult. (However, PU.1 is essential for murine fetal and adult B lymphopoiesis.) As was mentioned in Chapter 9, PU.1 is thought to exert many of its effects by controlling the expression of various cytokine receptor genes. PU.1$^{-/-}$ progenitors in the bone marrow fail to express IL-7Rα, preventing the receipt of the important IL-7 developmental signal. Interestingly, PU.1 expression is almost undetectable in the thymus, implying that the influence of this transcription factor on T cells ends before emigration of thymocyte precursors to the thymus. The expression of IL-7Rα by thymocytes must also therefore be controlled by transcription factors other than PU.1, unlike the situation for B cell progenitors.

ii) Ikaros, Aiolos, and Helios
Ikaros and Aiolos were also discussed in Box 9-2 of Chapter 9. To review briefly, Ikaros is a zinc-finger DNA-binding protein expressed primarily in the hematopoietic system. It is thought to control the entry of progenitor cells into both the B and T lymphoid lineages by repressing the expression of lineage-inappropriate genes. Ikaros$^{-/-}$ mice show severe disruptions of lymphoid organ development and profoundly reduced numbers of T, B, and NK cells. (The few mature T cells that are found in the periphery of these animals are thought to be of extrathymic origin.) Expression of Aiolos is also restricted to lymphoid cells. Aiolos is related to Ikaros in structure and mode of action and may interact with Ikaros to promote CLP survival and establish T lineage commitment. Another related transcription factor called Helios may also contribute to these processes.

iii) GATA-3
GATA-3 is a member of the GATA family of conserved zinc-finger transcription factors, several of which are involved in hematopoiesis. GATA-3 is highly expressed in T cells and a wide variety of other tissues, including the CNS and fetal liver. Studies of GATA-3$^{-/-}$; RAG$^{-/-}$ chimeric mice (refer to Box 13-1) showed that, although these

mice could generate B lymphocytes, not even the earliest DN thymocytes could be detected. GATA-3 is therefore crucial for commitment to the T lineage at the point when hematopoietic CLPs destined to become B cells diverge from those destined to become T cells. As detailed later in this chapter, GATA-3 is also crucial for commitment of DP thymocytes to the CD4 lineage. Finally, as is discussed in Chapter 15, GATA-3 plays an important role still later in the differentiation of Th2 effector cells following T cell activation, another example of a transcription factor functioning at multiple developmental stages.

iv) Id proteins
Like their structural homologues the E proteins (derived from the E2A gene discussed in Ch.9), the Id proteins are members of the bHLH (basic helix–loop–helix) family. In general, the Id proteins are negative regulators that cannot bind to DNA themselves but interfere with the DNA binding of E proteins. Id2 blocks T cell commitment, instead encouraging NK/T precursors to develop into NK cells rather than T cells (see Ch.18). At the DN3 stage, Id3 plays a prominent role in promoting $\gamma\delta$ T cell development at the expense of $\alpha\beta$ T cell development.

• Transcription Factors Important for the DN to DP Transition
In pre-T cells that become $\alpha\beta$ T cells, the expression of a functional TCRβ chain and the process of β-selection constitutes a key developmental checkpoint. In addition to their role in mediating the recombinational activity of the TCRα enhancer, the transcription factors **Tcf-1** and **Lef-1** have virtually redundant roles in pre-TCR activation just prior to TCRα gene rearrangement and the DP transition. Double Lef-1$^{-/-}$; Tcf-1$^{-/-}$ mutants have no DP thymocytes. The transcription factor **Sox4** is related in structure to Tcf-1 and Lef-1 and is also essential for the DN to DP transition. **Ikaros** makes a repeat entrance since this molecule sets the signaling threshold that must be exceeded by pre-TCR signaling for a DN cell to proceed to the DP stage. Products of the **E2A** gene (primarily the E proteins E12 and E47) have also been implicated in the DN to DP transition because E box motifs are present in the enhancers of the TCRα and TCRβ genes. Many other regulatory molecules are likely to be required but have yet to be identified.

Continued

- **Transcription Factors Important for the DP to SP Transition**

Our knowledge of the transcription factors involved in thymic selection and CD4/CD8 commitment is still incomplete. The related transcription factors **Nur-77** and **Nor-1** are required for negative selection, while **MEF-2** regulates the expression of Nur-77.

Curiously, despite its demonstrated role in promoting γδ T cell development, **Id3** has also been implicated in the positive selection of αβ thymocytes. Studies of knockout mice have indicated that **IRF-1** (interferon regulatory factor 1) and members of the **NF-κB** family are important for the development of CD8+ SP cells. Conversely, **CREB** participates in the

development of CD4+ SP cells. **LKLF** is highly expressed by both types of SP thymocytes in the thymic medulla and is essential for the survival of these cells. **Ets-1** is also required for mature SP cell survival.

Transcription factors important in T cell activation and post-activation events are discussed in Chapters 14 and 15.

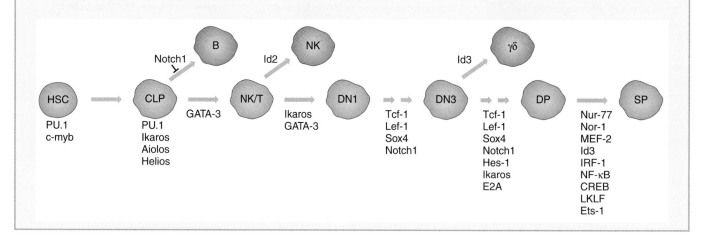

become fully restricted to the T lineage and lose all capacity for pluripotential differentiation; (2) the commitment to either the γδ or αβ T cell lineage is made; (3) the TCRγ, δ, and β genes commence rearrangement; (4) β-selection of thymocytes with a functional pre-TCR occurs; and (5) allelic exclusion occurs at the TCRβ locus, inducing the cessation of further γδ rearrangements. In short, it is during this stage that the pool of αβ pre-T cells constituting a diverse repertoire of TCRβ chains is established.

The levels of expression of RAG1/2, Ku80, DNA-PK, and TdT are upregulated in DN3 pre-T cells to allow for TCRβ, γ, and δ locus rearrangement and N nucleotide addition. The modification of the TCRβ chain gene by N nucleotide addition turns out to be dispensable for T cell maturation because TdT$^{-/-}$ mice exhibit normal thymocyte development. The limited repertoire of T cell specificities in adults of this strain resembled that found in wild-type neonates and fetal mice that had not yet commenced expression of TdT. In spite of this decreased diversity, TdT$^{-/-}$ mice mounted apparently normal immune responses to a wide variety of infectious organisms, and responses to specific peptides were maintained.

The γδ T cell developmental path branches off at this point (see Ch.18), although the pre-γδ T cell precursors cannot be easily distinguished from pre-αβ T cell precursors at this stage. However, as mentioned in Chapter 12, the independence of γδ and αβ pre-T cells is inferred from the fact that TCRβ knockout mice have normal γδ T cell development and TCRδ knockout animals have normal numbers of αβ T cells. The TCRγ silencer has an essential function early in the development of αβ pre-T cells, since deletion of the silencer prevents the appearance of mature αβ T cells in mutant mice.

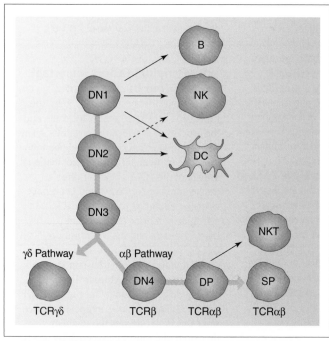

Figure 13-4

Murine Thymocyte Lineage Restriction

Early DN thymocytes have the capacity to develop into other cell types should these cells encounter the right cytokines and intercellular contacts. For example, DN1 thymocytes can give rise to DCs, B cells, and NK cells. DN2 cells are slightly more restricted, giving rise to DCs and possibly NK cells but not B cells. DN3 thymocytes are fully committed to the T lineage but can give rise to thymocytes bearing TCRγδ or DN4 cells bearing the pre-TCR. DN4 cells can give rise to αβ T cells but not to γδ T cells. However, as is described in Chapter 18, DP thymocytes can also give rise to NKT cells.

Unproductive rearrangements may occur in the TCRγ locus in αβ pre-T cells, but these cannot be expressed when the silencer is functional and do not appear to affect the αβ T cell developmental pathway.

TCRβ locus rearrangement soon commences in earnest in DN3 αβ pre-T cells, and the newly synthesized β chains are combined with the pTα chain and expressed on the cell surface in association with CD3 to form the pre-TCR described in Chapter 12. Calnexin continues to appear in the pre-TCR although its function here is unknown. The further maturation of αβ pre-T cells beyond this point requires signaling through the pre-TCR (Fig. 13-5). Accordingly, the expression of certain of the CD3 chains and Src family cytoplasmic protein tyrosine kinases (particularly Lck) becomes important. Mice lacking the CD3γ or ε subunits cannot form a functional pre-TCR and thymocyte maturation is blocked at the DN3 stage (refer to Fig. 13-3). Interestingly, mice lacking CD3δ have normal numbers of DP cells. This observation suggests that CD3δ, although part of the pre-TCR, does not play a critical role at this point. Mice lacking the CD3ζ or η subunits are blocked at slightly later stages of T cell development (see later).

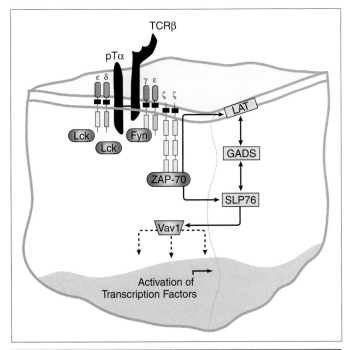

Figure 13-5

Pre-TCR Signaling in DN3 Cells
The pre-TCR (pTα-TCRβ-CD3) associates with Lck and Fyn kinases to phosphorylate the CD3ζ chains of the TCR complex and the downstream kinase ZAP-70. It is unknown what ligand, if any, binds to the pre-TCR to induce this activation. Phosphorylated ZAP-70 then recruits the adaptor proteins LAT, GADS, and SLP76 (discussed in detail in Ch.14), which facilitate activation of the GTP exchange factor Vav1. Activation of Vav1 leads to the activation of transcription factors that induce new gene transcription in the nucleus. This new gene expression drives the proliferation and further maturation of the developing thymocytes.

Signal transduction by kinases associated with the cytoplasmic tails of the CD3γ and ε chains is also crucial for continued maturation. Experiments in mutant mice transgenic for Lck suggest that this kinase may be particularly important for signaling downstream of the pre-TCR. A high level of expression of the Lck transgene in DP cells drives them into proliferation and differentiation even in the absence of a TCRβ chain. However, the function served by Lck may also be covered at least partially by Fyn kinase. Lck$^{-/-}$ mice were only partially blocked at the DN to DP transition, while Fyn$^{-/-}$ mice showed normal T cell development. In contrast to the findings with single mutants, double Fyn/Lck knockout mutant thymocytes were completely blocked at the DN2 pro-T cell stage, implying that at least one function served by these kinases is required even before pre-TCR signaling. Functional overlap between the related Syk and ZAP-70 kinases also exists (at least in mice), since double Syk/ZAP-70 mutant thymocytes, unlike Syk$^{-/-}$ or ZAP-70$^{-/-}$ single mutants, are blocked at the DN3 stage. Downstream of ZAP-70, the molecular adaptors SLP76, LAT, and GADS (discussed in detail in Ch.14 in Box 14-2) and the GTP exchange factor Vav1 (see later) are all essential for pre-TCR signaling and DN3 thymocyte maturation. Knockout mice lacking any one of these molecules show a block in T cell development at this stage.

Studies of V(D)J recombination in αβ pre-T cells have shown that the 10% of DN3 cells that successfully rearrange their TCRβ gene (are β-selected) enter the cell cycle. The remainder exhibit out-of-frame TCRβ gene rearrangements, fail to commence cell division, and eventually die, clearly demonstrating that β-selection is directly linked to the proliferation of thymocytes that can proceed further in maturation. In other words, β-selection is a vital developmental checkpoint: only those cells that have generated a functional pre-TCR receive proliferative signals. No cellular resources are wasted on expanding those thymocyte clones that have non-functional pre-TCRs. After β-selection, RAG-2 appears to be temporarily downregulated to prevent further TCRβ locus rearrangements, ensuring that only one functional TCRβ chain protein is expressed on a given T cell.

The fact that β-selection is an essential checkpoint in T cell development was first experimentally demonstrated in Lck knockout mice. Lck kinase is vital for pre-TCR signaling, and animals missing the gene encoding this enzyme show impaired thymocyte development at the DN stage. In other experiments, a productively rearranged TCRβ transgene was introduced into the immature thymocytes of mutant mice exhibiting a block at the DN to DP transition (such as *scid*, RAG$^{-/-}$, or TCRβ$^{-/-}$ mice). The presence of the functional TCRβ chain allowed a partial rescue of T cell maturation, such that T cells in the transgenic mice were able to reach the DP stage. Finally, pTα$^{-/-}$ mice are almost totally devoid of DP TCRαβ thymocytes, a finding attributed to their inability to carry out the pre-TCR signaling required for β-selection. That being said, the fact that pTα$^{-/-}$ mice do have a few early DP thymocytes indicates the existence of a secondary pathway (mentioned in Ch.12) that some c-kit^{low} CD44^{low} CD25$^+$ thymocytes can take that is independent of pTα signaling and β-selection. Cells following this path do not undergo the rapid cell division of the β-selected DN3 thymocytes and retain partial expression of CD25 in the next stage of development. pTα$^{-/-}$ mice show normal, and sometimes increased, γδ T cell development.

The thymic environment may again play an active role at this point in DN development. Thymic stromal cells appear to be required to supply intercellular contacts necessary for thymocytes to pass through the β-selection checkpoint. It is speculated that the pre-TCR may have to interact with a ligand (presumably displayed on the surface of a thymic stromal cell) to test the productivity of TCRβ gene rearrangement. However, there is as yet no concrete evidence for such a ligand.

iv) DN4: ckit⁻ CD44⁻ CD25⁻ Late Pre-T Cells

In the last DN thymocyte stage, the expression of CD25 and IL-7Rα is downregulated and low levels of CD4 and CD8 start to appear on αβ pre-T cells. Cells in this phase, which are large in size and concentrated in the subcapsule, can also be identified by their functionally rearranged TCRβ genes. Human T cells at this stage are CD1a⁺, CD4⁺, CD8α⁺, and CD8β⁺.

III. THE DP PHASE (CD4⁺CD8⁺)

The DP phase of αβ T cell development is dominated by the thymic selection process necessary to shape the mature αβ T cell repertoire. As αβ thymocytes pass through the DP phase, the expression of the CD4 and CD8 coreceptors is steadily upregulated. These molecules now begin to play increasingly important roles in directing thymocyte development. DP thymocytes also traverse the thymic architecture as they mature, moving from the subcapsule through the outer cortex and back through the inner cortex toward the medulla.

i) TCRα Locus Rearrangements

Early DP cells receive proliferative signals through the pre-TCR such that they rapidly expand, providing the large pool of DP thymocytes necessary to accommodate the generation of a large repertoire of αβ TCRs. It is not known what ligand, if any, is necessary for this signaling, known as *pre-TCR activation*. In fact, thymocytes bearing a truncated version of the pre-TCR which lacks the extracellular domain of the TCRβ chain are still able to continue their maturation. Thus, the only regions of the TCRβ chain that are essential for pre-TCR function are those that associate with the CD3 chains and the pTα chain. Although the pTα protein has a relatively long cytoplasmic tail, it is thought that the key signaling chains in the pre-TCR are the CD3 subunits, just as in the mature TCRαβ complex. Mice lacking the CD3ζ and η chains fail to make the transition from DN4 to DP αβ T cells, implying that although these proteins are not essential for β-selection, they are required for signaling associated with pre-TCR activation.

At this point, RAG expression resumes and rearrangement of the TCRα locus commences, resulting in the deletion of the TCRδ locus. Expression of the pTα gene is gradually downregulated so that the pTα chain is displaced by the TCRα chain. As described in Chapter 12, at this point in development, TCRα chains do not undergo productivity testing like TCRβ chains. TCRα locus rearrangements continue on both chromosomes until positive selection delivers a survival signal to those thymocytes with functional TCRs that weakly recognize self-MHC (see later).

ii) Expansion of the TCRαβ Pool

TCRαβ⁺ DP cells are generally small and are located within the inner cortex of the thymus. This large pool of vigorously dividing thymocytes with different combinations of TCRα and TCRβ chains will establish the entire TCRαβ repertoire. (Note, however, that neither the products of TCRα gene rearrangement nor the newly assembled TCRαβ heterodimers have been tested yet for functionality.) The descendants of these cells comprise a peripheral T cell compartment that recognizes more antigens than the B cell compartment. However, mature T cells in the periphery do not respond to any and all antigenic structures but are limited to epitopes consisting primarily of non-self-peptide/self-MHC combinations. This limitation is established by the multi-faceted process of thymic selection that establishes T cell central tolerance. The onset of selection also signals the end of transcription of pTα chains, a downregulation of RAG1/2 expression, and the cessation of TCRα locus rearrangements.

iii) Thymic Selection

iiia) What is thymic selection? Thymic selection is a process that removes from the T cell repertoire DP cells with non-functional TCRs (*death by neglect* or *non-selection*) and DP cells whose TCRs strongly recognize self-peptides bound to self-MHC (*negative selection*), while preserving those DP cells whose TCRs recognize self-MHC only weakly (*positive selection*). Selection outcome is thought to depend on the strength of TCR signaling, a property determined both by the level of aggregation of TCR and coreceptor molecules and by the affinity of the bond between an individual TCR and a pMHC epitope. In other words, selection is determined primarily by the overall avidity of the interaction between the T cell and the APC and the level of intracellular signaling triggered (Fig. 13-6). This signaling is influenced by myriad elements, including the expression and engagement of the coreceptors CD4 and CD8, the type of APC the thymocyte meets, and the co-stimulatory molecules the APC expresses. Scientists have attempted to dissect the intricacies of thymic selection through the use of several animal and cellular models. Experimental systems that allow the study of thymic selection processes are outlined in Box 13-5.

iiib) Rationale for thymic selection. Why have positive selection? In a given host, the APCs will express only that host's MHC alleles. Therefore, since mature T cells must recognize both MHC and peptide simultaneously on the surface of an APC/target cell in the periphery in order to mount an immune response, TCRs must bind to the host's MHC molecules with at least moderate affinity. Which particular MHC alleles are expressed on any given host's cells obviously depends on the genetic makeup of that host, but the TCRs on all T cells of

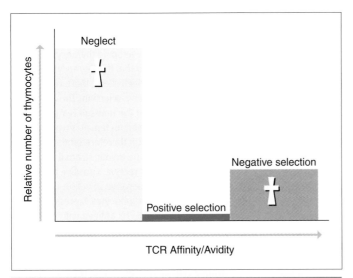

Figure 13-6

Affinity/Avidity Model of Thymic Selection
Developing thymocytes undergo a selection process in the thymus that results in "neglect" (non-selection), positive selection, or negative selection. Selection is based on the overall affinity/avidity of the interaction between thymocyte TCRs and self-peptide–MHC complexes presented by thymic stromal cells. The level of intracellular signaling subsequently induced within the thymocyte determines its fate. The majority of developing thymocytes carry TCRs that fail to bind to self-pMHC or do so very weakly: these cells are non-selected and are induced to undergo apoptosis. About 20% of thymocytes bear TCRs that bind strongly to self-pMHC (are self-reactive); these cells are negatively selected and also die of apoptosis. The remaining 1–2% of thymocytes bear TCRs that interact with self-pMHC with moderate affinity; these cells receive a survival signal and are positively selected. Adapted from Yeung R. S., Ohashi P. and Mak T.W. (in press) T cell development. In Ochs H. D., Smith C. I. and Puck J. M., eds. "Primary Immunodeficiency Diseases. A Molecular and Genetic Approach." Oxford University Press, Oxford.

that host will have to recognize at least one of those host MHC alleles. T cells that fail to produce functional TCRs, or which produce TCRs that fail to recognize host MHC molecules, cannot be activated in the periphery and thus are useless with respect to defending the host; they represent a waste of cellular resources. These cells, which can number as high as 80% of developing thymocytes, are therefore "non-selected" and induced to undergo apoptosis.

Why have negative selection? Peripheral T cells whose TCRs recognize and are activated by a combination of self-peptide and self-MHC could spark an autoimmune response in the host, damaging host tissues and perhaps even causing death. Evolution has provided that those DP cells whose TCRs bind with high affinity/avidity to self-peptide/self-MHC (estimated by some to be about 20% of developing thymocytes) are actively driven to apoptosis and are therefore deleted in the thymus before they ever reach the periphery. Thus, the vast majority of the 1–2% of all thymocytes that survives thymic selection and is released to the periphery bears TCRs that are most likely to recognize nonself-peptide on self-MHC, exactly what an APC or target cell will present when an immune response is required. The few

remaining potentially autoreactive clones that escape thymic deletion, and those that recognize self-antigens that emerge only later in life (after positive selection has been completed), are neutralized by the mechanisms of peripheral tolerance (see Ch.16).

It should be noted that researchers are still debating exactly where in the thymus each type of selection occurs. It is agreed that newly differentiated DP cells first appear in the outer cortex and then migrate toward the medulla. However, some scientists maintain that positive selection takes place primarily in the cortex, with negative selection happening later when the DP cells approach the medulla. Others believe that both positive and negative selection can occur in either the cortex or the medulla. Since negative selection can occur prior to positive selection (on a cell population basis), these processes are believed to be temporally independent as well.

iv) Processes of Thymic Selection

iva) Thymocyte death by neglect (non-selection). As stated previously, at this point in T cell development, the TCRα genes in the earliest DP cells are still undergoing rearrangement on both chromosomes. Over a period of 3–4 days (known as the *DP pause* in some immunological circles), a DP thymocyte produces a TCRα chain from various VαJα combinations, assembles it with the functional TCRβ chain it already expresses, and tests this TCR for binding to the self-peptide–MHC epitopes presented by surrounding thymic DCs and TECs (Fig. 13-7). Interestingly, the physical design and rearrangement mechanism of the TCRα locus allows the cell to examine the various possible combinations for the TCRα chain processively, in that the first candidates tested are a combination of the most 3′ V gene segments with the most 5′ J segments. As time goes on, progressively more upstream Vα gene segments and more downstream Jα gene segments are tested. However, despite this orderly approach, the majority of DP thymocytes cannot interact at all with the epitopes presented by thymic cells. Such cells have TCRs that result from out-of-frame rearrangements of Vα and Jα such that no protein can be produced, or from successful rearrangements that produce TCRs that have a conformation that simply cannot bind to self-MHC with any level of affinity. As a result, TCR signaling is not triggered and these "non-selected" or "neglected" cells proceed down a default path of death by apoptosis and die in the cortex by their 4th day. The apoptotic DP cells are rapidly removed from the thymus by macrophage scavenging.

ivb) Positive and negative selection. Within the sea of nonfunctional thymocytes, a limited number (<22%) bear TCRαβ receptors that are capable of binding, to varying degrees, to self-MHCs whose binding sites are occupied with self-peptides. These self-peptide–MHC complexes are presented by APCs and epithelial cells present mostly in the cortex but also in the cortico-medullary junction and the medulla. As stated previously, evidence from *in vitro* experiments suggests that whether such a DP thymocyte is positively or negatively selected is determined by the overall avidity of the bond. A large number of low affinity contacts between a T cell and

Box 13-5. **Experimental systems for thymic selection**

In normal mice, any given T cell specificity represents only a tiny fraction of the T cell repertoire. Studying the positive or negative selection of such a small subpopulation is difficult because its presence or absence is hard to detect. Several experimental systems have thus been devised to dissect thymic selection under controlled conditions in which experimental manipulations have a detectable effect. These include TCR transgenic mice, fetal thymic organ culture (FTOC), the H-Y system, and activation by superantigens.

TCR Transgenic Mice

Positive and negative selection have been manipulated most easily in TCR transgenic (Tg) mouse strains. The vast majority of T lymphocytes in these strains express only one combination of TCRα and β chains that recognizes a specific (known) combination of antigenic peptide and MHC allele. Thus, the effect on developing thymocytes of changing factors involved in thymic selection can be easily observed because the entire population will be affected in the same way. For example, suppose a series of mouse strains of differing haplotypes has been created, all of which are transgenic for a TCR that recognizes foreign peptide X complexed to H-2D^k. Mature X-specific CD8$^+$ SP lymphocytes will appear only in those animals of haplotype H-2D^k, since the developing TCR Tg thymocytes must interact with thymic stromal cells expressing H-2D^k in order to be positively selected, proliferate, and mature. (Virtually no X-specific CD4$^+$ SP cells will be selected because the transgenic TCR does not bind to MHC class II.) In a strain of a different haplotype, for example, H-2D^b, T cell development proceeds normally until the DP stage (as determined by flow cytometric analysis for surface markers). However, because the transgenic TCR cannot recognize peptide X complexed to H-2D^b on the thymic stromal cells, virtually no X-specific thymocytes can be positively selected, and mature X-specific CD8$^+$ SP cells do not appear in the periphery of these mice.

Fetal Thymic Organ Culture (FTOC)

Both positive and negative selection can be artificially replicated using FTOC derived from mice deficient in β2m or TAP-1. These proteins are required for peptide loading onto MHC class I molecules, so that in such FTOC, the stromal cells express only low levels of MHC class I α chains and thymic selection is blocked. However, with the addition to the FTOC of exogenous β2m and large amounts of purified peptide of interest, the stromal cells are able to express properly assembled MHC class I molecules loaded with the selecting peptide. Normal thymocyte development resumes and the production of easily defined SP T cell populations can be monitored. To study selection for a single TCR specificity, one can use FTOC derived from β2m-deficient, TCR Tg mice in which the TCR specificity is known. Upon addition of exogenous β2m plus the peptide recognized by the transgenic TCR, thymocyte development can proceed. For example, to determine whether a particular peptide is an agent of positive selection for a given transgenic TCR, molecules of the peptide plus exogenous β2m are introduced directly into the β2m-deficient, TCR Tg FTOC. The peptide is taken up by stromal APCs in the FTOC and is displayed on both MHC class I and II for thymocyte perusal. If the avidity of the binding between peptide X–MHC and the transgenic TCR is weak, the positive selection of TCR Tg thymocytes will be triggered and mature TCR Tg CD8$^+$ T cells directed against peptide X–MHC will appear in the culture. Similarly, a peptide Y that combines with MHC in the FTOC to form an epitope to which the transgenic TCR binds with high avidity may trigger negative selection. A failure to find TCR Tg CD8$^+$ SP T cells in the FTOC indicates that clonal deletion did indeed occur and that peptide Y is a negative selector for that transgenic TCR.

The H-Y System

Scientists have also developed elegant models to study negative thymic selection in the whole animal, solving the problem of how to devise a positive control for an assay whose readout is, by definition, an absence of cells. One such system relies on the use of the H-Y male antigen, a protein encoded on the Y chromosome that is highly expressed on many cell types in male mice but not in female mice. In this situation, the H-Y protein will be a "self" antigen for lymphocytes from a male mouse, but a "non-self" antigen for lymphocytes from a female mouse. Let us suppose we have a strain of TCR Tg mice of the H-2^k haplotype in which the TCR is directed against a peptide of the H-Y antigen complexed to the MHC allele H-2D^k. What will happen to the lymphocytes in the males and females of this strain? In female mice, the MHC haplotype expressed on the thymic stromal cells is the same as that recognized by the TCR of the developing thymocytes (both are H-2D^k), so that the thymocytes are not likely to die of neglect and there is no impediment to positive selection. The TCR is also directed against the non-self H-Y peptide, which is not expressed on female thymic stromal cells. The TCR therefore binds with only low avidity to the thymic stromal cells, and the thymocytes receive a positive selection signal and avoid a negative selection signal. Mature CD8$^+$ SP lymphocytes expressing the TCR directed against H-Y antigen will populate the periphery of the female mice and constitute a large proportion of the total CD8$^+$ population. In male mice, H-Y is expressed on the thymic stromal cells, and furthermore, the transgenic TCR binds to it with high avidity. A negative selection signal is delivered to the vast majority of lymphocyte clones because they all recognize H-Y peptide complexed to H-2D^k and are thus deleted in the male thymus. Virtually no mature CD8$^+$ T cells will appear in the periphery of the male Tg mice, a result that is easy to confirm relative to controls.

The H-Y system has been used to determine the effect on negative selection of mutations in individual genes by breeding the H-Y TCR Tg mice to knockout mice lacking the gene of interest. The lymphocyte populations in the periphery of male and female offspring are then examined. For example, a possible role for a cell surface antigen called CD30 in negative selection was uncovered by crossing the H-Y TCR Tg mice described previously to a line of inbred mice that also had the H-2D^k haplotype but were disrupted in the CD30 gene. By carrying out flow cytometric analyses of the lymphocyte populations in the H-Y TCR transgenic CD30$^{-/-}$ males and comparing them with H-Y TCR Tg CD30$^{+/+}$ males, the researchers were able to ask whether the absence of CD30 had an effect on negative selection. That is, if CD30 was involved in negative selection, its absence in the Tg knockout mouse should cause a failure of negative selection and lymphocytes bearing the H-Y TCR should therefore be detected in the periphery of these male mice, as indeed was the case. In contrast, when wild-type CD30 was present, negative selection proceeded normally and no lymphocytes bearing the H-Y TCR appeared in the periphery. In females of both CD30$^{+/+}$ and CD30$^{-/-}$ transgenic strains, the H-Y protein was not present (was not a self-antigen). Lymphocytes bearing the H-Y TCR therefore

Continued

avoided negative selection and appeared in the periphery of both CD30$^{+/+}$ and CD30$^{-/-}$ female transgenic mice. Additional studies have since revealed that CD30 may be required for negative selection of thymocytes in some (but not all) transgenic strains. The fact that the requirement varies with the genetic background of the mice underscores the complexity of the signaling required in thymic selection.

Superantigen Activation

Another system that has been used to demonstrate negative selection relies on *superantigens*. Many bacterial toxins and retroviral products act as superantigens, as is discussed in detail in Box 14-2 in Chapter 14. Briefly, superantigens are proteins that bind simultaneously to invariant sites on the MHC protein outside the peptide binding groove, and to the CDR2 of certain TCRβ chains. CDR2 is a region of the TCRβ chain that does not make contact with MHC in a conventional MHC–peptide binding situation. In other words, these antigens make a connection between the TCR on the T cell and MHC on the APC or target cell that completely ignores whatever peptide is in the MHC peptide-

binding groove. Nevertheless, this connection is sufficient to trigger strong intracellular signaling that results in negative selection in the thymus or T cell activation in the periphery. So far, these interactions have been observed only between MHC class II molecules and TCRβ chains containing particular subfamilies of Vβ gene segments. Superantigens are called "super" because a large number of T cell clones bearing TCRs with the appropriate Vβ segments are non-specifically activated at the same time in response to the presence of superantigen in the host. The immune response that results is thus of unusually high magnitude.

With respect to thymic selection, if a particular superantigen is injected into a mouse such that it binds to MHC class II on thymic stromal cells, any thymocyte clone whose TCR includes the Vβ segment recognized by that superantigen will bind tightly to the stromal cells and a negative selection signal will be delivered. The mouse TCRβ locus includes 23 Vβ gene segments, meaning that at least one in 23 (assuming random usage) thymocyte clones will be deleted in the thymus in response to superantigen. In fact, many superantigens

recognize more than one Vβ sequence, meaning that an even greater proportion of clones will be negatively selected. Such a huge deletion of whole families of T cells creates a major "hole in the T cell repertoire." For example, suppose one injected neonatal mice with sublethal doses of a "toxic shock" antigen called TSST1. TSST1 is a bacterial toxin that, when present in sufficient quantity, leads to systemic shock in the host by inducing an overwhelming release of cytokines throughout the host's body. Injected TSST1 molecules can bind to thymic stromal MHC class II and interact with any thymocyte bearing a TCR containing a murine TCRβ chain with either the Vβ15 or Vβ16 sequence. The binding is such that a negative selection signal is received. Consequently, if one stains for the expression of TCRs on thymocytes and mature SP T cells in these mice using mAbs specific for Vβ15 and Vβ16, one finds that essentially only early DP cells in the thymus are able to bind the mAbs. Almost no peripheral SP T cells stain positively because those that had TCRs containing Vβ15 or Vβ16 sequences were clonally deleted in the thymus by superantigen.

---

an APC may produce the same amount of TCR signaling and thus promote negative selection as readily as a small number of high affinity contacts.

Those TCRαβ thymocytes that bind to self-peptide presented on self-MHC with low avidity (faster dissociation rate) can be said to "not recognize self-antigen." That is, if released to the periphery, the TCRs of these clones will not bind to self-peptide–MHC complexes with sufficient avidity to trigger autoreactivity. The TCR likely disengages from the APC before a complete TCR signaling complex can be assembled, so that only low levels of intracellular signaling are achieved. However, this level of signaling during thymic selection appears to be sufficient to induce the transcription of genes that rescue these clones from apoptosis and promote their subsequent expansion and continued maturation. It is these clones that have been positively selected. Certain members of the anti-apoptotic Bcl-2 family are thought to be candidate genes whose transcription may mediate positive selection. When expression of these genes was measured in different thymocyte populations, Bcl-xL was found to be upregulated in newly selected DP thymocytes, but not in less mature thymocytes in the outer cortex. Similarly, the most mature SP thymocytes in the medulla express elevated levels of Bcl-2.

Those clones that bind with high avidity to self-peptide presented on self-MHC can be said to "recognize self-antigen." That is, if released to the periphery, the TCRs of these clones

might bind to self antigen–self-MHC with sufficient avidity to exceed the activation threshold and therefore trigger autoreactivity. Because of the slow dissociation rate of TCR–pMHC binding in these high affinity/avidity clones, the TCR remains bound to the thymic APC long enough for TCR–pMHC complexes to readily cluster together and assemble a complete TCR signaling complex. Vigorous intracellular signal transduction is triggered that delivers a powerful apoptotic signal, inducing these clones to commit suicide. Presumably the transcription of pro-apoptotic genes is induced whose effects swamp those of the Bcl-2 family members, pushing the cell into apoptosis. It is these clones that are negatively selected. In fact, a thymocyte clone that has been positively selected can still be lost to negative selection if the cells subsequently bind with high avidity to a different self-peptide–MHC combination.

v) Signaling Molecules Involved in Positive and Negative Selection

Much effort has been expended on uncovering the paths of signal transduction during thymic selection. It is now generally accepted that the quantity and perhaps the quality of signaling information transduced through the TCR to its downstream intermediaries largely determine positive and negative thymocyte selection. In this section, we present a simplified version of TCR signaling that highlights molecules known to be

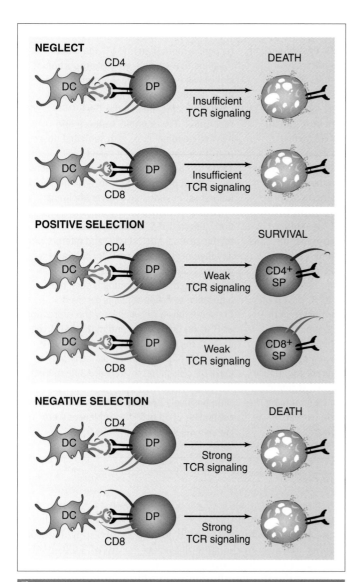

Figure 13-7

Cellular Depiction of Affinity/Avidity Model of Thymic Selection
DP thymocytes encounter thymic DCs (or cortical thymic epithelial cells; not shown) presenting self-peptides on MHC class I or II. The intracellular signaling induced by this interaction determines the thymocyte's fate. It is unclear precisely where in the thymus, either in the cortex or the medulla, each stage of selection takes place.

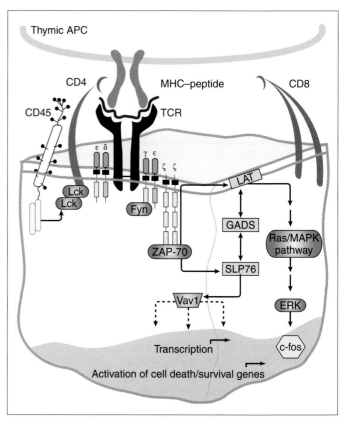

Figure 13-8

Simplified TCR Signaling During Thymic Selection in DP Thymocytes
Upon the interaction of the TCR and pMHC, the phosphatase CD45 activates Lck kinase molecules associated with the cytoplasmic tail of the coreceptor (CD4 and MHC class II are shown in this example). The CD3ζ chains of the TCR complex are phosphorylated, leading to activation of ZAP-70 and recruitment of the adaptors LAT, GADS, and SLP76. These adaptors facilitate activation of both Vav1 and the Ras/MAPK signal transduction pathway (see Ch.14 for detailed description). Ras activation leads in turn to ERK kinase activation and activation of the transcription factor c-fos. New gene transcription initiated by c-fos and by other transcription factors activated via Vav1 leads to cell death or survival. The level of ERK activation is important in determining whether the thymocyte will be positively or negatively selected or neglected. Additional costimulatory signaling may also contribute to selection outcome.

involved in thymic selection based on knockout mouse studies (Fig. 13-8). A complete description of the very complex network of molecules involved in transducing TCR signaling in activated mature T cells is presented in Chapter 14.

va) TCR chains. We described in Chapter 12 the phenotypes of mice deficient for expression of one of the TCR loci. In the context of the scheme of thymocyte development presented previously, we can revisit these animals and state that the small thymus present in TCR$\beta^{-/-}$ mice is due to the block at the DN3 stage, where β-selection is required for the expansion of thymocyte clones. pT$\alpha^{-/-}$ mice have a similar phenotype because of the requirement of pre-TCR signaling for the progression of DN3 cells. In contrast, TCR$\alpha^{-/-}$ mice have

a block at the DP stage, where the expression of a complete TCR$\alpha\beta$ heterodimer is required for thymic selection. As mentioned previously, mice lacking either TCRγ or TCRδ show normal $\alpha\beta$ T cell development.

vb) CD3 chains. Because signaling through the TCR is required for thymic selection, it should come as no surprise that the CD3 signaling chains are also important. There is, however, some evidence that the CD3 molecules and TCR heterodimer may not be as tightly coupled in the receptor complex in thymocytes as in mature T cells. These cell populations differ in the Ca^{2+} flux they show in response to TCR engagement, a factor with the potential to affect intracellular signaling and thus cellular outcome. In addition, although the FcϵRIγ chain can successfully substitute for the CD3ζ chain in mature $\alpha\beta$ T cells, the positive selection of thymocytes is impaired by this alteration, again indicating that subtle differences in signaling requirements exist between thymocytes and mature αβ T cells. It is interesting to note that the ITAM motifs of the CD3 chains in the TCR complex have been shown to play a role in the efficiency of thymic selection. Experiments have been conducted in which the selection for specific TCRs was examined in thymocytes in which one or more individual CD3 ITAM sequences were deleted. While no one ITAM proved to be essential, the greater the number of CD3 ITAMs in the TCR complex, the more efficient was both positive and negative selection.

vc) Lck, Fyn, CD45, and Vav1. The tyrosine kinases Lck and Fyn are required for thymic selection. Lck has a crucial role (that cannot be assumed by Fyn) since engineered deletion of this kinase blocks thymocyte development at the DN3 stage. Similarly, overexpression of a catalytically inactive form of Lck prevents both positive and negative selection. ZAP-70, which depends on Lck for its activation, is also important for both selection and the DP to SP transition. Knockout mice lacking ZAP-70 have normal numbers of DP thymocytes but exhibit impaired positive and negative selection and lack both sets of SP cells. The highly homologous Syk kinase is not able to compensate for an absence of ZAP-70 for these functions. As we learned in Chapter 9, the transmembrane phosphatase CD45 dephosphorylates negative regulatory tyrosine residues of Fyn and Lck, activating these enzymes and promoting signal transduction from the TCR. Studies of CD45$^{-/-}$ mice have shown that CD45 is particularly important for the negative selection of T cells, and for the later DP to SP transition.

Vav1 is a GTP exchange factor involved in intracellular signaling and is widely expressed in hematopoietic cells. Vav1$^{-/-}$ knockout mice have a block in thymocyte differentiation very similar to that observed in CD45$^{-/-}$ mice, in that thymocyte development is inhibited at the DP stage and only a very small number of mature T cells are observed in the periphery. The thymocytes of these mutant mice also exhibit a defect in Ca^{2+} mobilization in response to TCR stimulation. Vav1 regulates peptide-specific apoptosis in thymocytes and is required for both positive and negative selection.

vd) Ras/MAPK pathway and ERK1 activation. There is now convincing evidence that whether a thymocyte is positively or negatively selected may correlate directly with the degree of stimulation of the Ras/MAPK (*mitogen-activated protein kinase*) signaling pathway, which is linked to TCR signaling at several junctures (see Ch.14 for details). TCR stimulation leads to the activation of the membrane-associated signal transducer Ras, which in turn eventually causes the activation of ERK1 (*extracellular signal regulated kinase* 1). ERK1 activation leads to activation of the nuclear transcription factor *c-fos*, which helps to upregulate the transcription of a variety of genes important for cell survival and cell death. Dominant negative transgenes have been used to inhibit various molecules within the Ras/MAPK pathway. The mutant animals show defects in positive thymocyte selection and a significant reduction in the number of mature SP thymocytes. A similar phenotype is observed in ERK$^{-/-}$ mice. Elegant experiments

by Pamela Ohashi and others using *in vitro* selection models and FTOC have confirmed that the level of ERK1 activity influences thymocyte fate to a large degree. Clones that were negatively selected in the model system (driven to apoptosis) exhibited vigorous but short-lived ERK1 activity, while those that were positively selected (proliferated) had only weak ERK1 signaling that was sustained. Furthermore, reduction of ERK1 activity with a specific inhibitor was able to convert a negatively selected clone into a positively selected one, and a positively selected clone into one that was neither positively nor negatively selected in the model system (equivalent to death by neglect *in vivo*).

It is hypothesized that the level of ERK1 signaling in a thymocyte (and thus its selection fate) is controlled by the affinity/avidity of its TCRs for the selecting pMHCs (Table 13-2). TCRs that are engaged by pMHC undergo internalization at a rate determined by the affinity/avidity of the interaction (see Ch.14). That is, ligands with strong affinity/avidity for a given TCR induce rapid internalization of those TCRs, removing them from the pool that can be activated. In contrast, ligands with weak affinity/avidity for a given TCR do not trigger receptor internalization. It is surmised that positively selecting ligands (weak affinity/avidity) sustain the expression of TCRs on the thymocyte surface and continue to trigger TCR signaling.; weak but prolonged ERK activation results. In contrast, strong affinity/avidity ligands (negatively selecting) induce vigorous ERK activation that is rapidly extinguished by TCR internalization. In the absence of ERK signaling, the thymocyte undergoes apoptosis. Despite the enticing simplicity of this scenario, it must be stressed that there are likely to be many other as-yet-undefined pathways involved in controlling selection, and that these pathways probably influence each other in complex ways *in vivo*.

Table 13-2 Molecular Features Leading to Different Selection Outcomes

	Selection		
	Negative	**Positive**	**Neglect**
TCR-pMHC avidity	Strong	Weak	None
TCR internalization	Rapid	Slow	Negligible
ERK1 signaling strength	Strong	Weak	None
ERK1 signaling duration	Short	Sustained	None
Stromal cells	mTECs	cTECs	?
Other potentially influential molecules	CD28, CD40L Itk, Rlk *CD4, CD8	Itk, Rlk TAP, β2m (CD8$^+$ cells) β2m HLA-DO (CD4$^+$ T cells)	?

* Separate from their role in increasing T cell-APC binding.

ve) Costimulatory signaling required for thymic selection.
Examination of knockout mice has indicated that signaling via the co-stimulatory molecules CD28 and CD40L (see Ch.14 for details) may contribute to negative selection. Some (but not all) strains of CD28$^{-/-}$ or CD40L$^{-/-}$ mice show impaired negative selection of DP thymocytes and an accumulation of autoreactive T cells in the periphery. Interestingly, the mechanisms involved in CD28- and CD40-mediated promotion of negative selection appear to differ. Experiments *in vitro* have shown that apoptosis induced by strong, sustained TCR signaling could not be achieved unless CD28 was also engaged. However, in chimeric mice in which CD40L^{+} and CD40L^{-} thymocytes co-existed, both types of thymocytes were negatively selected with equal efficiency. This observation suggests that the CD40L signaling effect necessary for negative selection of a given thymocyte may be delivered by either an intercellular or an intracellular route. The mechanism remains unclear, but it does not appear that CD40L signaling is required for the upregulation of CD28.

vf) Other molecules implicated in signaling required for thymic selection. Several lines of evidence have suggested that thymic selection requires signals in addition to those transduced through TCR engagement and co-stimulation. First, the Tec kinase family members Itk and Rlk have been implicated in the "fine-tuning" of the signaling thresholds necessary for both positive and negative selection. Secondly, the TNF receptor was cited as a co-factor for negative selection in some *in vitro* experiments. However, TNFR$^{-/-}$ mice show normal positive and negative selection, indicating that this molecule does not play a significant role in selection *in vivo*. Lastly, the Fas death receptor was originally suggested by some to be instrumental in negative selection because of its role in the induction of apoptosis. However, Fas is expressed on all thymocytes (including those that are positively selected) and thus is unlikely to be the agent of cell death for only some of them. In addition, thymocytes from natural mouse mutants deficient for either Fas (*lpr* mice) or Fas ligand (*gld* mice) are still negatively selected.

vi) Role of the Coreceptors in Thymic Selection

The binding of the coreceptors CD4 and CD8 to MHC affects thymic selection in two ways. First, the overall avidity of the interaction between the T cell and the APC is increased by coreceptor binding. Secondly, the coreceptors promote TCR signaling because they bring the Lck molecules associated with their cytoplasmic tails into contact with the TCR–CD3 complex. The importance of the coreceptors for T cell development and thymic selection has been demonstrated in a variety of *in vivo* and *in vitro* experiments. Studies of T cell populations in FTOC model systems where anti-CD4 or anti-CD8 mAbs were used to block coreceptor interaction with MHC class II or class I showed a blockage in the development of SP cells. Consistent with these results, CD4$^{-/-}$ mice have no classical helper T cells, and 90% of peripheral αβ T cells in these animals are CD8^{+}. Conversely, mice lacking the CD8α chain (which is required for the surface expression of the CD8β chain) have no CTLs, and the vast majority of mature T cells in the thymus and the periphery are CD4^{+} T cells.

These results also confirm that CD4 is not required for the positive selection of CD8^{+} T cells and vice versa.

Recent evidence from studies of models of negative selection using CD4$^{-/-}$ or CD8$^{-/-}$ thymocytes bearing TCRs specific for various superantigens (refer to Box 13-5) has suggested that coreceptor expression may be a key regulatory checkpoint differentiating positive from negative selection. When cultures of TCR Tg CD4$^{-/-}$ thymocytes were treated with a peptide derived from a superantigen to which the transgenic TCR bound with high affinity, negative selection was able to proceed in the absence of CD4. However, in cases where the superantigen peptide was bound with somewhat lower affinity by the transgenic TCR, negative selection was impaired in the absence of CD4. In other words, interaction with the coreceptor modulated the selection outcome when there was "room for improvement" in the affinity of the TCR for the pMHC. In another set of experiments, thymocytes from mice that were engineered to express decreased levels of CD8 were more often positively selected than were thymocytes expressing higher levels of CD8. In fact, positive selection could be converted to negative selection by the engineered overexpression of CD8 in these cells. Similarly, increased expression of CD4 could convert a peptide that only weakly activated a given T cell clone (low affinity interaction of TCR and pMHC and therefore a "positively selecting" peptide) to one that strongly activated the cells (high affinity interaction and therefore a "negatively selecting" peptide).

vii) Role of Antigen Processing Molecules in Thymic Selection

We have seen that the presence of peptide in the MHC-binding groove is necessary to stabilize the pMHC complex such that a TCR has time to scan it for binding fit; without peptide, the MHC molecule is intrinsically unstable and cannot be "seen" properly by the TCR. Thus, the molecules involved in antigen processing and presentation are also necessary for thymic selection. For example, mice defective in the synthesis or loading of MHC class I molecules (such as those lacking β2m or TAP) showed no positive selection of CD8^{+} T cells but did produce CD4^{+} T cells, implying that the MHC molecule must be loaded with peptide in order to achieve selection. With respect to MHC class II, the HLA-DM and HLA-DO molecules encountered in Chapter 10 are important. The reader will recall that HLA-DM is involved in peptide exchange in the MHC class II groove, displacing CLIP and facilitating the binding of diverse peptides. In the course of driving peptide loading, HLA-DM is thought to promote the dissociation of peptides that do not bind stably in the MHC groove. Immunologists say that HLA-DM is "helping to shape the T cell repertoire" because DP cells undergoing selection in the thymus encounter only those self-peptides that bind with reasonable stability to MHC. HLA-DO acts to inhibit HLA-DM activity and thereby blocks CLIP exchange. The balance between HLA-DM and HLA-DO activities may limit the array of self-peptides displayed, influencing the outcome of thymic selection and thus the peripheral CD4^{+} T cell repertoire. An equivalent "shaping" mechanism has yet to be identified for the CD8[+] repertoire.

viii) Role of pMHC in Thymic Selection

The concentration of a given peptide governs the kinetics of MHC binding to that peptide and consequently TCR–pMHC binding. Thus, the same peptide can induce either positive or negative selection of thymocytes bearing a defined TCR depending on the level of TCR signaling triggered. However, peptide plays an additional role beyond MHC stabilization and TCR triggering. In FTOC experiments in which the cultured thymus was artificially loaded with a single type of peptide, the positive selection of a surprisingly large number of CD4[+] and CD8[+] T cells with diverse TCR specificities was observed. The large range of T cell specificities selected in the presence of a single peptide suggested that the TCR–pMHC interaction that occurs during thymic selection is relatively promiscuous. On the other hand, if peptide made only a structural contribution to the proper display of MHC for TCR perusal, then the T cell repertoire selected in this experiment using a single selecting peptide should have contained a complete rather than partial range of specificities. This was not found to be the case, implying that the peptide itself influences the nature of the resulting T cell repertoire.

Further exploration of the relationship between selecting peptides in the thymus and subsequent mature T cell–peptide interactions in the periphery has been achieved using TCR Tg FTOC systems. Such experiments employ thymic lobes from mice that are transgenic for a defined TCR, making it easy to quantify and track the fate of thymocytes as they mature. In this system, more than one peptide was found to positively select thymocytes of a single TCR specificity, again confirming the relative promiscuity of the TCR–pMHC interaction in the thymus. More surprisingly, a given peptide that was able to activate mature T cells that had developed in such thymic cultures often exhibited little or no amino acid homology to the peptide used for selection during thymocyte development. Furthermore, mature T cells that developed during the course of such an FTOC experiment could not later be activated by the same peptide that mediated their thymic selection.

The above evidence strongly indicates that positively selecting peptides in the thymus can be very different from the antigenic peptides that trigger mature T cell activation in the periphery. What mechanism would allow a peptide that positively selects T cells in the thymus to be later ignored in the periphery? In 1996, Pamela Ohashi and colleagues used an elegant FTOC-based approach to investigate this question. They established FTOC from mice that were transgenic for a TCR that recognized an LCMV-derived peptide p33 presented by the H-2D[b] MHC class I molecule. In this FTOC environment, p33–H-2D[b] constituted a "self" epitope, and p33 added to the FTOC triggered the negative selection (elimination) of the transgenic thymocytes. The system also involved a related peptide A4Y that acted as an agonist in this system, in that A4Y was also able to activate mature transgenic T cells recognizing p33–H-2D[b]. Because

of its efficacy in stimulating the mature T cells, A4Y was expected to negatively select developing thymocytes with specificity for p33–H-2D[b] in the FTOC environment. However, A4Y acted instead as a positively selecting peptide, and even very high concentrations of A4Y failed to trigger clonal deletion of the maturing thymocytes. An important clue to the mechanism responsible for this observation was the finding that the mature T cells arising from the thymocyte clones that were positively selected by A4Y were unable to respond to A4Y itself, but could still respond to the cognate peptide p33. Ohashi interpreted these results to mean that a positively selecting peptide somehow "modifies" the thymocytes it interacts with, establishing in them a "resting threshold" for mature T cell activation that is above the threshold in thymocytes. As a result, the T cells are unresponsive to any further encounters with the selecting peptide, although they may respond to structurally similar peptides that bind in a manner that pushes the T cell above the resting threshold. This model implies that the positively selecting peptide itself acts on the thymocyte to "fine-tune" its resting state by either modifying the avidity of its TCR or modulating an intrinsic biochemical pathway. It should be kept in mind that, in a physiological setting, the thymocytes are positively selected most often by self-peptides, so that the setting of the resting threshold is a mechanism that prevents peripheral autoimmunity but allows responses to foreign peptides of an appropriate specificity that exceed the resting threshold.

ix) Role of Thymic Stromal Cells in Positive Thymic Selection

While many researchers were and still are occupied with the study of thymocytes themselves to elucidate the underlying mechanisms of thymic selection, others have probed the contributions to this process of various types of thymic stromal cells. In the mid-1990s, thymic reaggregation experiments showed that positive selection critically depends on the presence of thymic epithelial cells. When purified murine DP thymocytes were cultured with thymic epithelium, salivary epithelium, gut epithelium, fetal mesenchymal fibroblasts, or thymic DCs, only thymic epithelial cells were capable of supporting the differentiation of the DP cells into mature SP thymocytes. This result indicated that the stromal cells were supplying a unique signal necessary for positive selection and continued maturation of the DP cells. Subsequent analyses demonstrated that it is the cTECs, which present thymic self-antigens almost exclusively (Table 13-3), that are predominantly responsible for positive selection. How they accomplish this task remains a mystery: cTECs do not present a specialized set of pMHC ligands, and do not induce any of the known cell survival signaling pathways. There is some evidence that cTECs may be able to induce a type of TCR receptor editing (similar to that described in Ch.9 for the BCR) that allows a thymocyte to produce a non-self-reactive TCR and thus avoid death. Alternatively, it may be that cTECs express a particular (unknown) combination of accessory and/or co-stimulatory molecules that stimulates signaling, resulting in positive selection.

Table 13-3 Examples of Selected Self-antigens Expressed by Thymic Epithelial Cells (TECs)

Gene Expressed	cTECs	mTECs
Albumin	–	+
α1 anti-trypsin	–	+
C reactive protein	+/–	+
Erythropoietin	–	+
α-Fetoprotein	–	+
Glutamic acid decarboxylase	–	+
Haptoglobin	–	+
Insulin	+/–	+
Intestinal fatty acid binding protein	+/–	+
Lactalbumin	+/–	+
Myelin oligodendrocyte glycoprotein	–	+
Nicotinic acetylcholine receptor	–	+
Serum amyloid P component	–	+
Somatostatin	+/–	+
Trypsin	+/–	+
Tyrosinase	–	+

Adapted from Kyewski B. *et al.* (2002) *Trends in Immunology* **23**(7), 364–371.

x) Role of Thymic Stromal Cells in Negative Thymic Selection

A long-standing puzzle in the field of thymic selection was the following: how can the vast array of self-peptides present throughout the host's body influence a selection process that is localized in a relatively isolated organ such as the thymus? Unlike other lymphoid organs, the thymus has no afferent lymphatic vessels, so that non-thymic antigens are not easily conveyed to this location. The cortex in particular is a sequestered region. It was thus originally believed that mature T cells recognizing tissue-specific self proteins were not inactivated during the establishment of central tolerance and were controlled instead by peripheral tolerance mechanisms. However, work in the 1990s showed that the thymic medulla was more open to the body's circulation than the cortex, and that the medulla was packed with bone marrow-derived APCs. A few scattered reports then claimed that tissue-specific proteins shed into the circulation by the tissue in question could be taken up and displayed by recirculating DCs and macrophages able to access the thymus. In addition, there were sporadic reports that expression of tissue-specific proteins such as myelin and cytokeratin could be detected in the thymus itself. Experiments with transgenic mice expressing genes under the control of tissue-specific promoters often showed expression in the thymus as well as in the tissue of interest. However, most scientists remained skeptical about these overexpression systems and pointed out that the results could be due to promoters with leaky expression due to their site of integration.

These doubts were assuaged somewhat by studies in the early 2000s that identified a population of mTECs in the thymus that can transiently express a wide range of non-thymic, tissue-specific molecules. Such expression would allow for the deletion of reactive T cell clones before their release to the periphery. mTECs have been shown to express RNAs encoding hormones, secreted proteins, membrane proteins, and transcription factors, among other molecules (refer to Table 13-3). Experiments in transgenic and non-transgenic mice have confirmed that ectopic expression of a protein in mTECs can lead to later T cell tolerance of that protein. It has yet to be determined whether all mTEC clones can express all tissue-specific proteins, or whether individual mTEC clones express different subsets of tissue-specific proteins. The latter situation would enable an individual mTEC cell to express a greater percentage of its cell surface proteins as the tissue-specific protein in question, increasing epitope density and the likelihood of a sufficiently strong TCR signal being delivered to the T cell to induce negative selection.

It is not yet clear whether expression of tissue-specific genes in mTECs is the result of random derepression of transcription or a more directed activation process. Similarly, the influence of external factors on mTEC gene expression remains a mystery. However, new evidence has implicated a transcription factor called AIRE (*autoimmune regulator*) in mTEC-induced central tolerance. Both humans and mice deficient for AIRE suffer from autoimmune symptoms that affect multiple organs. Analysis of transcription by cells of AIRE-deficient mice showed that mTEC expression of several tissue-specific proteins associated with various autoimmune symptoms was reduced in the absence of AIRE. Studies of chimeric animals in which AIRE expression was missing either from hematopoietic cells or from stromal cells showed that autoimmunity was present only when AIRE was missing from the stromal cells. AIRE is known to interact with the transcriptional co-activator CREB-binding protein, and it has been suggested that AIRE may participate in a multi-subunit complex in mTECs that activates the transcription of multiple genes encoding tissue-specific proteins. In the absence of AIRE, negative selection of T cells recognizing these proteins does not appear to occur, and tissue-reactive T cell clones escape to the periphery.

IV. THE SP PHASE: CD4/CD8 LINEAGE COMMITMENT

It is at this point in T cell development that crucial events leading to the appearance of mature naive SP T cells in the thymic medulla occur. Either during or just after positive selection, the class of MHC recognized by a given TCR on a late, surviving DP cell is fixed by the loss of expression of either CD4 or CD8 from that cell, generating early SP CD4$^+$ or CD8$^+$ thymocytes. From this point on, the TCRs of the descendants of an SP CD4$^+$ cell can bind only to MHC class II, and those of an SP CD8$^+$ cell only to MHC class I. It is worth reiterating that any TCR on a CD4$^+$ CD8$^+$ thymocyte can potentially interact with either MHC class I or class II: it is the coreceptor molecule retained during development that determines which class of MHC is recognized by the mature naive T cell, which in turn defines the eventual effector function (T cell help or cytotoxicity) acquired by that clone. Once commitment to the CD4 or CD8 lineage is established, an SP thymocyte upregulates its expression of CCR9 and CCR4, which enable it to follow gradients of TECK and *macrophage-derived chemokine* (MDC) emanating from stromal cells in the medulla.

Why bother to lose expression of a coreceptor? If both coreceptors were retained, the mature T cell would have a double specificity, able to recognize not only peptide A–MHC class I but also a conformationally similar combination of peptide B–MHC class II. Since the recognition of MHC class I or class II determines whether the mature naive T cell will be a Th or a Tc clone, and thus affects the effector functions brought to bear on an invader, a developing T cell with double specificity might not develop into either because of a conflict in developmental programming. The immune response of the host might therefore be compromised.

i) Theories of CD4/CD8 Lineage Commitment

The precise sequence of events causing one DP cell to downregulate CD4 expression and become a CD8$^+$ SP cell and another DP cell to downregulate CD8 expression and become a CD4$^+$ SP cell is unknown, but two types of theories involving multiple signaling pathways have been advanced to explain this phenomenon (Table 13-4). Those models based on *stochastic* principles propose that chance is in charge, and that loss of either CD8 or CD4 occurs randomly during selection. According to these models, those cells whose TCRs interact with peptide–MHC class II and randomly retain CD4 receive some kind of survival or stimulatory signal, while cells that interact with MHC class II but randomly retain CD8 are induced to undergo apoptosis. Conversely, cells whose TCRs recognize peptide–MHC class I survive only if they retain expression of CD8 and not CD4. In contrast, *instructive* models hold that some kind of downstream intracellular signaling induced by TCR–pMHC or coreceptor–pMHC interaction determines which class of MHC a thymocyte will recognize and therefore its SP status. For example, if a thymocyte's TCR engages an epitope that includes an MHC class II allele, CD4 (but not CD8) on the DP surface will also bind to this MHC class II molecule. Once this binding is established, the interaction between the MHC class II molecule and the CD4 coreceptor may send a signal to the T cell to downregulate the synthesis of CD8, so that the progeny of this clone become SP CD4$^+$ T cells. A similar scenario can be envisioned to account for the development of CD8$^+$ SP T cells.

One theory, the *coreceptor reversal* model, is a meld of the stochastic and the instructive philosophies. This model posits that CD8 expression is first downregulated in all DP thymocytes regardless of their MHC restriction, so that all thymocytes initially become CD4$^+$ cells. Those thymocytes that subsequently receive both an IL-7 signal and a weak signal delivered by binding to MHC class I resume CD8 expression and downregulate CD4 expression, effectively "reversing" their prior CD4$^+$ phenotype to a CD8$^+$ phenotype. However, those thymocytes that receive a signal through MHC class II maintain their CD4 expression and do not resume CD8 expression.

Table 13-4 **Models and Mechanisms of CD4/CD8 Commitment**

Model	CD4 SP Development	CD8 SP Development
Stochastic	Randomly lose CD8	Randomly lose CD4
Instructive	Signal directs loss of CD8	Signal directs loss of D4
Coreceptor reversal	All thymocytes become CD4$^+$; signal delivered through MHC class II maintains CD4 expression	All thymocytes become CD4$^+$; weak signals delivered through MHC class I plus IL-7 downregulate CD4 and upregulate CD8
Asymmetric	Only if high concentration of peptide	Default*
Signaling		
Lck	Strong	Weak
TCR	Strong	Weak
ERK	Strong	Weak*
Notch1	Weak	Strong
DNA interactions	GATA-3 activation	Runx-3 binding to CD4 silencer
TCR Vα sequence	AV2	AV8

* Evidence for the reverse situation also exists.

Another theory is the *asymmetric* model of CD4/CD8 commitment. In this scenario, the default developmental path favors the expression of one coreceptor over the other unless a strong signal is received by the DP cell. One mouse model used to investigate this possibility used TCR Tg mice specific for ovalbumin (OVA) peptide. The TCR expressed by these mice is specific for a combination of OVA–MHC class II, so if this pMHC combination was encountered by thymocytes in the DP stage in these mice, CD4 would be involved and its expression would therefore be retained. However, researchers found that CD4$^+$ T cells did not appear unless relatively high concentrations of OVA peptide were administered to the mice. When lower doses of OVA were used, only CD8$^+$ T cells were produced, suggesting that CD8$^+$ lineage commitment occurred by default. In another *in vitro* system in which DP thymocytes were transiently stimulated with mitogens to induce positive selection in suspension culture, CD8$^+$ SP cells emerged after stimulation of a moderate duration and intensity, while CD4$^+$ SP cells required stimulation of much longer duration and higher intensity. Work in still other systems has suggested that CD4$^+$ lineage commitment is the default path and CD8$^+$ SP cells emerge only when CD8 is strongly engaged on the DP cell during selection. These conflicting findings remain to be reconciled.

ii) Signaling Leading to CD4/CD8 Commitment

What is the nature of the intracellular signals delivered to the thymocyte nucleus that cause it to commit to either the SP CD4 or CD8 fate? Several different signaling mechanisms have been implicated, making it likely that the cell assesses multiple factors before making its decision.

iia) Lck signaling. Because Lck associates differentially with the cytoplasmic tails of both coreceptors, Lck signaling has been proposed as a mechanism of CD4/CD8 commitment. The model holds that strong Lck signaling favors the differentiation of DP cells into CD4$^+$ SP cells, while little or no signaling results in CD8$^+$ SP cells. The amount of Lck signaling is determined by the MHC restriction of the TCR: MHC class I-restricted TCRs bind to CD8, which is associated with low levels of Lck activity, while MHC class II-restricted TCRs bind to CD4, which is associated with about 20-fold higher levels of Lck activity. In one set of experiments, mutated CD8 molecules were engineered that were capable of binding different amounts of Lck. It was found that the greater the amount of Lck bound to CD8, the more CD4$^+$ T cells emerged from the DP population. In another system, SP development was monitored following the binding of a TCR to either wild-type MHC class II (fully capable of binding Lck) or mutated MHC class II lacking the CD4 binding site (decreased Lck binding). In the absence of significant Lck binding (and signaling), thymocyte differentiation was skewed to the CD8$^+$ subset. Additional evidence supporting a commitment model based on quantitative differences in Lck signaling has come out of transgenic studies in which constitutively active or dominant negative mutants of Lck were used. The former induced MHC class I-restricted thymocytes to develop into CD4$^+$ T cells, whereas the latter

drove MHC class II-restricted thymocytes down the CD8$^+$ path. Confirmation of these results has been obtained by introducing a transgene for inducible Lck expression into Lck$^{-/-}$ cells. When high levels of Lck expression were induced in the mutant thymocytes, only CD4$^+$ T cells emerged in the culture.

iib) TCR signaling. Despite the results of the Lck signaling experiments, work using cells from TCR Tg mice lacking the coreceptors has shown that CD4 and CD8 do not in fact convey signals that are essential for SP cell maturation. Rather, the level of TCR signaling itself, which the coreceptors enhance by stabilizing T cell–APC interactions, was found to correlate with lineage commitment. Strong TCR signaling was associated with the production of CD4$^+$ SP cells while weaker TCR signaling resulted in the appearance of CD8$^+$ SP cells. Some researchers speculate that individual CD3 subunits of the TCR complex may have specialized signaling functions that result in a bias toward CD4 or CD8 SP development, an attractive theory to explain how one receptor structure can mediate multiple outcomes. Studies of knockout mice have shown that mutants in which the CD3γ gene is ablated show a greater decrease in the numbers of CD8$^+$ SP T cells (relative to the decrease in CD4$^+$ SP cells), while those in which the CD3ζ gene is disrupted show a greater relative decrease in CD4$^+$ SP T cells. Mutations of the CD3ζ/η gene or CD3δ gene appeared to affect both subsets equally. A difficulty with these studies, however, is that all CD3 mutations drastically reduce the expression of TCRs (and hence, all CD3 components) on the cell surface.

iic) ERK signaling. The role of ERK signaling in CD4/CD8 lineage commitment is under debate. In one study, high levels of ERK signaling appeared to favor CD4$^+$ over CD8$^+$ SP development. Transfection of mice with a transgene encoding a constitutively active form of ERK resulted in increased numbers of CD4$^+$ thymocytes but decreased numbers of CD8$^+$ cells. Consistent with this result, CD4 maturation was blocked by chemical inhibition of ERK signaling while CD8 development was enhanced. However, other studies have shown that CD8 differentiation can be blocked by deletion, inhibition, or interference with ERK activity.

iid) Notch1 signaling. As for ERK signaling, the involvement of Notch1 signaling in CD4/CD8 lineage commitment is unclear. Early studies demonstrated that expression of a constitutively active form of Notch1 in developing T cells resulted in an increase in CD8$^+$ lineage cells and a decrease in CD4$^+$ lineage cells, even if MHC class I was absent. Conversely, when Notch1 signaling was blocked with antibodies, CD4$^+$ SP development was favored. Complementary studies revealed that a silencer (see later) in the CD4 gene that prevents CD4 expression in DN and CD8 SP thymocytes has an HES-1 binding site, and HES genes are known to be regulated by Notch1. However, CD4/CD8 differentiation is normal in HES-1-deficient mice. Other data have suggested that, rather than promoting only the development of CD8$^+$ SP cells, Notch1 signaling is involved in the maturation of both CD4 and CD8 SP thymocytes. Still others believe that Notch1 may exert its effect indirectly, by reducing the strength of signal delivered through the TCR. Cells receiving high Notch1 signals would have reduced signaling through the TCR and would consequently adopt the CD8 cell fate, whereas cells receiving lower Notch1 signals would have stronger TCR signals and become CD4 cells. The situation is complicated by the fact that there are three other Notch genes operating in mammalian cells. The products of these genes can apparently compensate for Notch1 deficiency since Notch1-deficient mice show normal CD4/CD8 differentiation.

iii) Interactions at the DNA Level Associated with CD4/CD8 Commitment

The ultimate goal of any or all of the signaling paths outlined previously may be to affect transcription of the CD4 and CD8 genes. Evidence accumulated in the early 2000s has implicated a transcriptional regulator called Runx-3 in CD8 commitment, and the transcription factor GATA-3 in CD4 commitment.

iiia) Runx-3 and CD8 commitment. As mentioned in Chapter 12, there are two CD8 proteins, CD8α and CD8β, that form CD8αβ heterodimers (and sometimes CD8αα homodimers) constituting the CD8 coreceptor. The CD8α and CD8β chains are derived from two closely linked genes, CD8A and CD8B, in the CD8 locus. The promoters of each of these genes are linked to several enhancers that are active only in CD8-expressing cells such as DP thymocytes and SP CD8$^+$ T cells. Different enhancers govern CD8 expression at different stages of thymocyte maturation. To date, however, the elements that bind to these enhancers and promote CD8 expression have yet to be identified. Indeed, the CD8 T cell fate appears to depend more on what happens in the CD4 locus than in the CD8 locus.

The CD4 locus contains three enhancers that are positioned 5′ and 3′ of the CD4 coding exons. These enhancers are differentially required for the expression of CD4 at different stages of T cell development. Significantly, the CD4 enhancers are also active in cells destined to become SP CD8$^+$ T cells. To block CD4 expression in DN thymocytes and SP CD8$^+$ T cells, the CD4 gene contains a silencer in the intron between its first two coding exons. When the DNA binding protein Runx-3 binds to the CD4 silencer in a DP thymocyte, it recruits *chromatin remodeling complexes* (CRCs). CRCs are large, variable multi-component enzymatic complexes capable of reorganizing and bending DNA such that its transcription is either enhanced or repressed. In the case of Runx-3 binding to the CD4 silencer, the CRC causes the chromatin of the CD4 locus to become "closed" and inaccessible to the transcriptional machinery. CD4 expression is shut down but CD8 expression is intact. Mutation of the CD4 silencer sequence results in abnormal CD4 expression by CD8$^+$ T cells, and conditional mutation of the Runx-3 gene leads to production of CD4 by mature CD8$^+$ T cells. (The lineage of the CD8 cells in these cases was established by examining the expression of other CD8-specific cytolytic genes such as perforin; see Ch.15.) It is possible that TCR engagement by MHC class I (but not

MHC class II) triggers signaling in DP thymocytes that upregulates expression of Runx-3 and thus promotes CD8 lineage commitment.

iiib) GATA-3 and CD4 commitment. As mentioned in Box 13-1, GATA-3 is a transcription factor that plays vital roles at several steps in T cell development and differentiation. With respect to CD4/CD8 lineage commitment, GATA-3 is implicated because its expression is increased in CD4 SP thymocytes compared to DP and CD8 SP thymocytes. In addition, in experiments designed to manipulate GATA-3 expression in reaggregated thymic cultures, a bias toward the production of CD4 SP T cells was associated with increased GATA-3 expression, while CD8 SP T cell production dominated when GATA-3 expression was inhibited. It may be that TCR engagement of MHC class II (but not MHC class I) on a DP thymocyte triggers strong or sustained signaling that both upregulates GATA-3 and downregulates CD8, driving CD4 SP differentiation. So far, there is no evidence for a silencer in the CD8 locus, ruling out a mechanism parallel to that described for the effects of Runx-3 on the CD4 locus. However, there is some evidence that Runx-3 upregulation in DP thymocytes may block GATA-3 expression, inhibiting CD4 SP differentiation.

iv) Influence of TCR V Segment Usage on CD4/CD8 Commitment

Although the TCRs present on Tc and Th cells of an individual are derived from exactly the same TCR genes, there is evidence that some portions of the TCR sequence can influence CD4/CD8 lineage commitment. Analysis of populations of mature CD4$^+$ and CD8$^+$ peripheral T cells has shown that there is a bias in the usage of TCR Vα gene segments in the TCRs of SP cells. Certain families of Vα sequences preferentially occur in the TCRs of CD4$^+$ cells, while other Vα sequences are found more frequently in the TCRs of CD8$^+$ cells. For example, the AV2 family of TCR Vα sequences is often expressed by CD4$^+$ cells, while the AV8 family turns up on CD8$^+$ cells. Since different Vα sequences specify different binding site sequences on the TCR, some scientists have speculated that Vα bias results due to the propensity of a given Vα sequence type to bind to an MHC molecule of a particular class. For example, if the Vα region in a TCR on a given DP thymocyte interacts better with an MHC class I molecule than a class II molecule, cells with that Vα will be preferentially selected on MHC class I. Presumably a signal (perhaps delivered by Lck, ERK, or Notch1?) is then delivered that results in the downregulation of CD4 synthesis and the production of more CD8$^+$ cells than CD4$^+$ cells bearing this Vα sequence. A mechanism such as this could account for the AV8/CD8 bias that is observed. Similarly, if a thymocyte expresses a TCR containing an AV2 gene segment, interaction of the TCR with MHC class II may be favored, resulting in the subsequent promotion of CD4$^+$ cell development. In experiments using mouse TCR transgenes in which a Vα sequence favoring the development of CD4$^+$ cells was altered by engineered mutation to a sequence preferentially expressed in CD8$^+$ cells, the developmental bias was correspondingly affected. A relative increase in the number of

CD8$^+$ T cells expressing the transgenic TCR was detected in the periphery of the mouse.

E. Mature SP Thymocytes in the Periphery

The generation of SP thymocytes marks the last stage of thymic T cell development. The SP cells loiter in the medulla for a time (about 2–3 days in the mouse) before they receive a final (unknown) proliferative signal and expand once more in the thymic medulla. These cells also become fully functional, and exit into the bloodstream to take up residence in the periphery as naive mature SP T cells. Once in the periphery, the naive SP T cells adopt a resting state (G$_0$ of the cell cycle) and recirculate through the blood and peripheral lymphoid organs every 12–24 hours. The expression of IL-7Rα resumes in these cells, but it is unclear what effect IL-7 has on mature peripheral T cells *in vivo*. Interestingly, the exit of both thymocytes from the thymus and mature T cells from the peripheral lymphoid organs during recirculation is controlled by expression of a single receptor called *S1P receptor 1* (S1P$_1$). The ligand for this receptor is a phospholipid called S1P, which is present at high concentrations in the blood. It may be that the presence of high levels of S1P$_1$ on a T cell's surface allows it to follow a chemotactic gradient of S1P from the lymphoid organs into the blood, or that signaling via S1P$_1$ overcomes a lymphoid organ retention signal. Knockout mice deficient for expression of S1P$_1$ show an accumulation of naive SP cells in the thymus but a complete lack of these cells in the periphery.

Until recently, it was thought that naive mature SP cells were doomed to apoptotic death in the periphery within a few days if they were not activated by binding to antigenic pMHC. However, it is now thought that mature resting T cells are prevented from undergoing apoptotic death in the absence of specific antigen by *TCR tickling*. The TCRs of resting naive T cells are continually but very subtly stimulated in the lymph nodes by the presentation of self-pMHC complexes that may be the same self-pMHC complexes that were encountered during thymic selection. The stimulation is not enough to activate the T cell or cause it to enter the cell cycle, but is enough to stave off apoptosis and permit survival for at least 5–7 weeks. It is thought that TCR tickling may promote the expression of the transcription factor *lung Krüppel-like factor* (LKLF), which, from studies of LKLF$^{-/-}$ mice, appears to be essential for the maintenance of mature naive SP T cells.

The generation of new T cell precursors in the bone marrow is prodigious, and when a vertebrate is young, the emigration of naive T cells from the thymus (even taking into account the impact of thymic selection) is substantial. It has been estimated that about 1% of total thymocytes leaves the thymus daily to seed the secondary lymphoid organs and tissues. TCR tickling sends a survival signal to large numbers of these cells, preventing them from undergoing apoptosis in the absence of antigen, so that the natural mechanism that disposes of B cells unless specific antigen is in the immediate vicinity does not operate for T cells in the same way. The maintenance of steady-state numbers of naive T cells, activated T cells, and memory T cells requires a complex balance of lymphocyte survival and death, and the precise mechanisms by which the overall homeostasis of T cells is preserved are not known. However, researchers are beginning to elucidate elements of some of the signaling pathways thought to control peripheral T cell death. These pathways are described in detail in Chapter 14.

F. T Cell Development in an Embryological Context

Now that the reader is familiar with the stages of T cell development for an individual clone in an individual wave, it is appropriate to follow a wave in an embryological context. In reality, waves of T cell development overlap and it is difficult to pinpoint where one ends and another begins. Scientists have maneuvered around this problem in two ways. First, development can be monitored in early FTOC such that the development of the very first T cell wave can be followed. Alternatively, the DNA of dividing cells can be metabolically labeled with the molecule bromo-deoxyuridine (BrdU). Injected or ingested BrdU is incorporated into the DNA of the earliest T cell precursors of a wave. The maturation of these cells even in adult mice can then be tracked using anti-BrdU antibodies and flow cytometry.

In mice, an organized thymus is in place by embryonic day 11 (E11), when the first thymocyte precursors arriving from the fetal liver can be identified within it. V(D)J recombination of the TCRγ and δ loci first commences in DN3 (CD44$^-$CD25$^+$) pre-T cells at about E12–13 (Fig. 13-9). In the small proportion of pre-T cells destined to become γδ T cells, functional mRNAs for both the TCRγ and δ chains as well as CD3 molecules can be detected at about E13–14. A functional γδ TCR associated with CD3 appears on the mouse thymocyte surface by about E14–15. Thus, the γδ TCR is the first to be detected during murine embryogenesis. At about E15–16, productive TCRβ rearrangements can be detected in the DNA of αβ pre-T cells, and full-length TCRβ transcripts are produced. The expression of CD4 and CD8 is spontaneously and greatly upregulated on DN thymocyte clones with functional pre-TCRs at this stage (the DN to DP transition). Complete TCRαβ molecules associated with CD3 start to be expressed at about E17–18. The vigorous cycling of DP cells ensures the expansion of TCRαβ clones such that the TCRαβ thymocyte population quickly outstrips the TCRγδ population. Thymic selection of DP cells then takes place over about 3 days at this stage such that by E19, mature naive CD4$^+$ and CD8$^+$ SP T cells

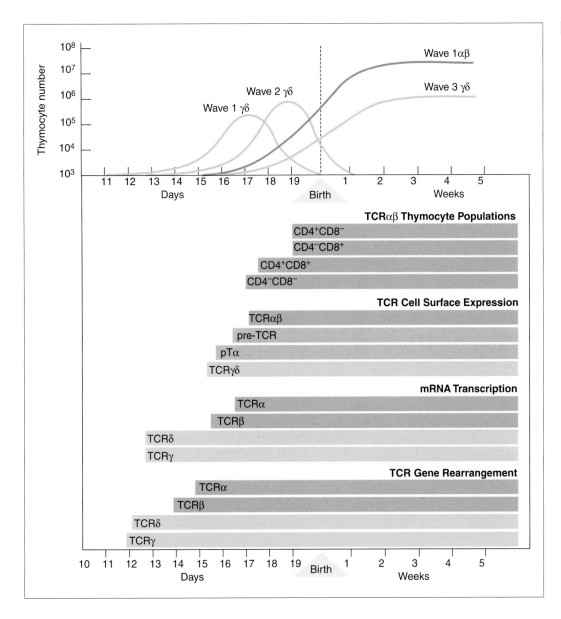

Figure 13-9

TCR and Coreceptor Gene Expression during Murine Embryogenesis
An organized thymus is not present in a mouse until day 11 of embryogenesis. The top graph shows a time course of individual waves of newly generated thymocytes. Note that the first waves are composed of γδ T cell populations. The bottom graph shows the order and timing of TCR gene rearrangement, transcription, and cell surface expression, and when DN, DP, and SP αβ thymocytes first appear. TCRγ is the first TCR gene to be expressed, followed by TCRδ and TCRβ. TCRα is not expressed on the cell surface until after β-selection mediated by the pre-TCR. Top panel adapted from Klein J. and Hořejší V. (1997) "Immunology," 2nd edn. Blackwell Science, Oxford. Bottom panel adapted from Fowlkes B. J. and Pardoll D. M. (1989) Molecular and cellular events of T cell development. *Advances in Immunology* **44**, 207–264.

can be detected (DP to SP transition). The mature SP cells linger in the thymus for 2–3 more days before exiting to the periphery. The mouse is born at about E21 with a full complement of functional T lymphocytes, which are overwhelmingly (>90%) αβ T cells. (Note that there are substantial species differences in the proportion of γδ T cells generated: while 0.5–5% of T cells are also of the γδ lineage in humans, the percentage is significantly higher in both chickens and sheep.) Overall, in a given wave of T cell development in the fetal mouse, the DN stage lasts for about 4 days, the DP stage for about 3 days, and the SP stage for about 2–3 days. The kinetics of an adult wave are slower. For example, the proliferation phase just prior to TCRβ gene rearrangement expands from about 3–4 days to about 1 week, so that it might take from 2 to 3 weeks for CLP precursors to completely mature into SP αβ T cells in the adult mouse.

In humans, a primitive thymic structure can be detected at about 7 weeks gestation. At this time, the first thymocyte precursors come from the fetal liver and reach the thymic primordium. V(D)J recombination of the γ and δ loci in DN3 cells commences so that the first mature γδ T cells appear at about 8 weeks of fetal life. Complete αβ TCRs associated with CD3 start to be expressed at 10 weeks. The thymus becomes a well-defined organ at 14–15 weeks. At 15–16 weeks, the structure of the thymus appears as it will in the neonate. The first Hassall's corpuscles can be detected in the medulla, indicating the presence of degenerating lymphocytes. After week 16, fetal liver precursors also colonize the bone marrow, and by week 22, most of the thymic progenitors are derived from the bone marrow. This situation stands in contrast to that of the mouse, in which all thymic progenitors seeding the thymus during the entire course of embryogenesis are derived from the fetal liver.

In both adult humans and mice, successive waves of T cell precursors continue to enter the thymus even as it gradually involutes, so that thymocytes at all stages of development can be identified in the adult thymus (Fig. 13-10). For example, at any one time in the mouse, the DN subpopulations of thymocytes together constitute about 5% of all thymocytes, while 80% are DP cells. About half of DP cells bear TCRαβ, and half have yet to express a TCR. Rounding out the thymocyte compartment are 12% SP CD4$^+$ cells and 3% SP CD8$^+$ cells. Close to 99% of the SP CD4$^+$ cells present in the adult murine thymus are TCRαβ$^+$, while half of the SP CD8$^+$ cells are mature cells bearing an αβ TCR. The other half of the SP CD8$^+$ population consists of cells in transition from the DN to DP stage; CD8 is upregulated first in most mouse strains, and these cells will shortly become fully DP in phenotype, complete with TCR. The 0.5% of total thymocytes that bear TCRγδ are almost all CD4$^-$CD8$^-$. There is also a late-developing population of about 5% of CD4$^-$CD8$^-$ thymocytes that are TCRαβ$^+$. It is thought that these cells may arise by loss of CD8 or CD4 from thymocytes that were originally SP cells.

We have come to the end of our discussion of naive T cell development. At this point, mature SP T cells are quietly patrolling the periphery of the body, subjected now and then to TCR tickling. Each cell is awaiting the encounter with foreign antigenic peptide that will activate it enough to trigger proliferation and differentiation into memory and effector helper and cytotoxic T cells. The activation process, how effector T cells are generated, and how they function are described in Chapters 14 and 15.

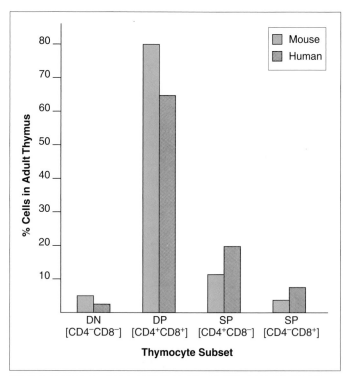

Figure 13-10

Relative Numbers of Different Thymocyte Populations in Adult Murine and Human Thymi
DP cells constitute the vast majority of thymocytes in both species, followed by CD4$^+$ SP cells. Lower numbers of CD8$^+$ SP cells and DN cells are present in both species.

SUMMARY

T cell development starts with pluripotential hematopoietic stem cells (HSCs) in the bone marrow. Under the influence of specific stromal cell contacts and cytokines, HSCs give rise to common lymphoid progenitors (CLPs), which in turn give rise to the more restricted NK/T precursors. The NK/T precursors migrate to the thymus, where thymocyte development proceeds in three multipartite phases: the DN phase (neither CD4 nor CD8 appears on the thymocyte surface), the DP phase (both CD4 and CD8 are present), and the SP phase (either CD4 or CD8 is present on the surface of a mature T cell). Thymocyte development into αβ T cells is controlled by two checkpoints: the β-selection checkpoint and the TCRαβ checkpoint. β-selection occurs in thymocytes during the DN stage and controls irreversible commitment to either the αβ or γδ T lineages. Rearrangement of the TCRβ, γ, and δ genes begins, and thymocytes with successful TCRβ gene rearrangement and production of a functional TCRβ chain are committed to the αβ lineage. The developing αβ thymocytes then pass into the DP phase. Rearrangement of the TCRα locus begins, newly formed TCRαβ heterodimers are expressed, and the pool of αβ thymocytes undergoes thymic selection at the TCRαβ checkpoint. Thymocytes whose TCRs do not recognize self-pMHC presented by surrounding thymic stromal cells do not receive survival signals and die of neglect. Thymocytes whose TCRs bind strongly to self-pMHC would likely be autoreactive in the periphery. These cells receive death signals and are negatively selected. Central tolerance is thus established. Thymocytes whose TCRs bind weakly to self-pMHC are likely to bind strongly to foreign pMHC. These thymocytes receive survival signals and are positively selected. Positively selected thymocytes represent only 1–2% of the original pool of TCRαβ thymocytes. The surviving thymocytes then pass into the SP phase, commence coreceptor expression, and exit to the periphery as mature T cells.

Selected Reading List

Adkins B. (1999) T-cell function in newborn mice and humans. *Immunology Today* **20**, 330–335.

Alam S. M. and Gascoigne N. R. J. (1998) Posttranslational regulation of TCR Vα allelic exclusion during T cell differentiation. *The Journal of Immunology* **160**, 3883–3890.

Anderson G. and Jenkinson E. (2001) Lymphostromal interactions in thymic development and function. *Nature Reviews Immunology* **1**, 31–40.

Anderson M., Venanzi E., Klein L., Chen Z., Berzins S. *et al.* (2002) Projection of an immunological self shadow within the thymus by the Aire protein. *Science* **298**, 1395–1401.

Ashwell J. D., Lu F. W. M. and Vacchio M. S. (2000) Glucocorticoids in T cell development and function. *Annual Review of Immunolgy* **18**, 304–345.

Aspinall R. and Andrew D. (2000) Thymic involution in aging. J*ournal of Clinical Immunology* **20**, 250–256.

Benoist C. and Mathis D. (1997) Positive selection of T cells: fastidious or promiscuous? *Current Opinion in Immunology* **9**, 245–249.

Berg L. and Kang J. (2001) Molecular determinants of TCR expression and selection. *Current Opinion in Immunology* **13**, 232–241.

Blom B., Res P. C. M. and Spits H. (1998) T cell precursors in man and mice. *Critical Reviews in Immunology* **18**, 371–388.

Bommhardt U., Beyer M., Hunig T. and Reichardt H. (2004) Molecular and cellular mechanisms of T cell development. *Cellular and Molecular Life Sciences* **61**, 263–280.

Bosselut R. (2004) CD4/CD8-lineage differentiation in the thymus: from nuclear effectors to membrane signals. *Nature Reviews Immunology* **4**, 529–540.

Brandt V. and Roth D. (2002) A recombinase diversified: new functions of the RAG proteins. *Current Opinion in Immunology* **14**, 224–229.

Brugnera E., Bhandoola A., Cibotti R., Yu Q., Guinter T. *et al.* (2000) Coreceptor reversal in the thymus: signaled CD4+ CD8+ thymocytes initially terminate CD8 transcription even when differentiating into CD8+ T cells. *Immunity* **13**, 59–71.

Cantrell D. (2002) Transgenic analysis of thymocyte signal transduction. *Nature Reviews Immunology* **2**, 20–27.

Chen F., Rowen L., Hood L. and Rothenberg E. (2001) Differential transcriptional regulation of individual TCR Vbeta segments before gene rearrangement. *Journal of Immunology* **166**, 1771–1780.

Chidgey A. and Boyd R. (2001) Thymic stromal cells and positive selection. *Acta Pathologica, Microbiologica, et Immunologica Scandinavica* **109**, 481–492.

Claman H. N. Chaperon E. A. and Triplett R. F. (1966) Thymus-marrow cell combinations. Synergism in antibody production. *Proceedings of the Society for Experimental Biology and Medicine* **122**, 1167–1171.

Clevers H. and Ferrier P. (1998) Transcriptional control during T-cell development. *Current Opinion in Immunology* **10**, 166–171.

Cyster JG. (2003) Lymphoid organ development and cell migration. *Immunological Reviews* **195**, 5–14.

Di Santo J. P. and Rodewald H.-R. (1998) In vivo roles of receptor tyrosine kinases and cytokine receptors in early thymocyte development. *Current Opinion in Immunology* **10**, 196–207.

Elliot J. I. (1998) Selection of dual Valpha T cells. *European Journal of Immunology* **28**, 2115–2123.

Ellmeier W., Sawada S. and Littman D. R. (1999) The regulation of CD4 and CD8 coreceptor gene expression during T cell development. *Annual Review of Immunology* **17**, 523–554.

Fehling H. J. and von Boehmer H. (1997) Early αβT cell development in the thymus of normal and genetically altered mice. *Current Opinion in Immunology* **9**, 263–275.

Glimcher L. and Singh H. (2001) Transcription factors in lymphocyte development—T and B cells get together. *Cell* **96**, 13–23.

Gordon J., Wilson V., Blair N., Sheridan J., Farley A. *et al.* (2004) Functional evidence for a single endodermal origin for the thymic epithelium. *Nature Immunology* **5**, 546–553.

Haynes B. F., Markert M. L., Sempowski G. D., Patel K. D. and Hale L. P. (2000) The role of the thymus in immune reconstitution in aging, bone marrow transplantation, and HIV-1 infection. *Annual Review of Immunology* **18**, 529–560.

Hogquist K. (2001) Signal strength in thymic selection and lineage commitment. *Current Opinion in Immunology* **13**, 225–231.

Hozumi K., Tanaka T., Sato T., Wilson A. *et al.* (1998) Evidence of stage-specific element for germ-line transcritpion of the TCR alpha gene lcoated upstream of Jalpha49 locus. *European Journal of Immunology* **28**, 1368–1378.

Jacobs H. (1997) Pre-TCR/CD3 and TCR/CD3 complexes: decamers with differential signalling properties? *Immunology Today* **18**, 565–569.

Jameson S. C. and Bevan M. J. (1998) T-cell selection. *Current Opinion in Immunology* **10**, 214–219.

Killeen N., Irving B. A., Pippig S. and Zingler K. (1998) Signaling checkpoints during the development of T lymphocytes. *Current Opinion in Immunology* **10**, 360–367.

Klein L. and Kyewski B. (2000) Self-antigen presentation by thymic stromal cells: a subtle division of labor. *Current Opinion in Immunology* **12**, 179–186.

Leiden J. M. (1993) Transcriptional regulation of T cell receptor genes. *Annual Review of Immunology* **11**, 539–570.

Malek T., Porter B. and He Y.-W. (1999) Multiple gamma c-dependent cytokines regulate T-cell development. *Immunology Today* **20**, 71–76.

Miller J. F. A. P., Leuchars E., Cross A. M. and Dukor P. (1964) Immunologic role of the thymus in radiation chimeras. *New York Academy of Sciences* **120**, 205–216.

Miller J. F. A. P. and Osoba D. (1967) Current concepts of the immunological function of the thymus. *Physiological Reviews* **47**, 437–520.

Mitchell G. F. and Miller J. F. A. P. (1968) Immunological activity of thymus and thoracic-duct lymphocytes. *Proceedings of the National Academy of Sciences USA* **59**, 296–303.

Moroy T. and Karsunky H. (2000) Regulation of pre-T-cell development. *Cellular and Molecular Life Sciences* **57**, 957–975.

Osborne B. (2000) Transcriptional control of T cell development. *Current Opinion in Immunology* **12**, 301–306.

Palmer E. (2003) Negative selection—clearing out the bad apples from the T-cell repertoire. *Nature Reviews Immunology* **3**, 383–391.

Playfair J. H. L. and Papermaster B. W. (1965) Focal antibody production by transferred spleen cells in irradiated mice. *Science* **149**, 998–1000.

Quong M., Romanow W. and Murre C. (2002) E protein function in lymphocyte development. *Annual Review of Immunology* **20**, 301–322.

Raff M. F. (1970) Role of thymus-derived lymphocytes in the secondary humoral immune response in mice. *Nature* **226**, 1257–1258.

Ramiro A. R., Trigueros C., Marguez C., San Millan J. L. and Toribio M. L. (1996) Regulation of pre-T cell receptor (pTalpha-TCRbeta) gene expression during human thymic development. *Journal of Experimental Medicine* **184**, 519–530.

Robey E. and Fowlkes B. (1998) The alpha/beta versus gamma/delta T-cell lineage choice. *Current Opinion in Immunology* **10**, 181–187.

Rothenberg E. and Anderson M. (2002) Elements of transcription factor network design for T-lineage specification. *Developmental Biology* **246**, 29–44.

Saito T. and Watanabe N. (1998) Positive and negative thymocyte selection. *Critical Reviews in Immunology* **18**, 359–370.

Sant'Angelo D. B., Lucas B., Waterbury P. G., Cohen B., Brabb T. *et al.* (1998) A molecular map of T cell development. *Immunity* **9**, 179–186.

Savino W., Mendes-Da-Cruz D., Smaniotto S., Silva-Monteiro E. and Villa-Verde D. (2004) Molecular mechanisms governing thymocyte migration: combined role of chemokines and extracellular matrix. *Journal of Leukocyte Biology* **75**, 951–961.

Schlesinger M. (1970) Anti-theta antibodies for detecting thymus-dependent lymphocytes in the immune response of mice to SRBC. *Nature* **226**, 1254–1259.

Sebzda E., Kundig T. M., Thomson C. T., Aoki K., Mak S.-Y. *et al.* (1996) Mature T cell reactivity altered by peptide agonist that induces positive selection. *Journal of Experimental Medicine* **183**, 1093–1104.

Shores E. W. and Love P. E. (1997) TCR zeta chain in T cell development and selection. *Current Opinion in Immunology* **9**, 380–389.

Shortman K. and Wu L. (1996) Early T lymphocyte progenitors. *Annual Review of Immunology* **14**, 29–47.

Sim B.-C., Lo D. and Gascoigne N. R. J. (1998) Preferential expression of TCR Valpha regions in CD4/CD8 subsets: class discrimination or coreceptor recognition. *Immunology Today* **19**, 276–282.

Staal F., Weerkamp F., Langerak A., Hendriks R. and Clevers H. (2001) Transcriptional control of T lymphocyte differentiation. *Stem Cells* **19**, 165–179.

Starr T., Jameson S. and Hogquist K. (2003) Positive and negative selection of T cells. *Annual Review of Immunology* **21**, 139–176.

Williams O., Tanaka Y., Tarazona R. and Kioussis D. (1997) The agonist-antagonist balance in positive selection. *Immunology Today* **18**, 121–126.

T Cell Activation

Rolf Zinkernagel 1944–

MHC restriction of T cell responses
Nobel Prize 1996

© Nobelstiftelsen

CHAPTER 14

A. BRINGING T CELLS AND APCs TOGETHER

B. SIGNAL ONE: BINDING OF PEPTIDE–MHC
   TO THE TCR

C. SIGNAL TWO: COSTIMULATION

D. SIGNAL THREE: CYTOKINES

"Sometimes I think it should be a rule of war that you have to see somebody up close and get to know him before you can shoot him"
—Colonel Potter of *M*A*S*H*

n the last chapter, we followed the development of a T lymphocyte from its stem cell progenitor through V(D)J recombination and thymic selection to its exit from the thymus into the periphery as a mature naive peripheral T cell. Be it a CD4$^+$ or a CD8$^+$ T cell, it circulates among the secondary lymphoid tissues via the blood vessels and lymphatics, patrolling for non-self-peptides displayed on the appropriate MHC molecules on APCs or target cells. The T cell is prevented from undergoing immediate apoptosis by occasional "TCR tickling," the stimulation of its TCR by weakly cross-reacting self-peptide coupled to MHC. In this chapter, we examine what happens when that T cell encounters a foreign peptide bound to MHC that is recognized by its TCRs: the phenomenon of T cell activation.

Overview

The activation of the mature naive peripheral Th cell is the defining moment in an adaptive immune response, for it is the activated Th cell that supplies the T cell help usually required by both antigen-specific B cells and Tc cells. The reader will recall that B cell activation requires three signals. Signal 1 results from the intracellular signaling transduced by antigen binding to the BCR. Signal two depends on costimulatory contacts, particularly CD40–CD40L, established between the B cell and an activated antigen-specific effector Th cell. Signal three is T cell help in the form of cytokines secreted by the activated Th cell. The activation of all T cells is similar, in that more than one signal is required to stimulate proliferation and differentiation and to prevent clonal deletion or *anergy* (see Ch.16), a state of non-responsiveness in which the T cell survives but cannot be induced to proliferate or generate effectors in response to its specific antigen. However, the precise sequence of activation events differs slightly between Th cells and Tc cells, as one might expect since the activation of most Tc cells requires Th help.

The activation of naive Th cells involves just two cells: the Th cell and the mature DC that will act as an APC. Sustained

interaction between pMHC and sufficient numbers of TCRs (signal one) plus costimulatory and other intercellular contacts (signal two) must be established between the mature DC and the Th cell. As a result of these contacts, both the Th cell itself and the DC produce cytokines (signal three) that contribute to full Th cell activation. The end result of these multiple interactions is the induction of new gene transcription, particularly of genes involved in cellular survival, proliferation, and differentiation. The activation of most naive Tc cells is more complex, and involves a total of four cells: the Tc cell, the target cell, an activated APC, and an activated, antigen-specific Th effector cell. As introduced in previous chapters, in most cases, the naive Tc cell must have its TCRs engaged by pMHC presented by a target cell and must receive cytokines from Th cells (or sometimes from APCs) as well as costimulatory signals. However, since target cells are not professional APCs, they do not express high levels of costimulatory molecules. How then are naive CD8$^+$ cells co-stimulated? Many scientists now believe that, prior to receiving cytokines and perhaps even before binding to the target cell, most naive Tc cells are stimulated by contact with an activated, mature APC to obtain the necessary costimulation. Such an APC has been activated by presenting the appropriate pMHC to an antigen-specific Th and is now "licensed" to costimulate a naive Tc (see Box 14-1). A co-stimulated naive Tc cell whose TCR then binds with sufficient avidity to antigenic peptide presented on MHC class I of a target cell in the correct cytokine milieu can proceed with proliferation and differentiation into effector cytotoxic T lymphocytes (CTLs). It is only the differentiated CTLs that can interact with and lyse, without any further need for costimulation, any cell in the body that is displaying non-self-peptides coupled to MHC class I.

Whether the costimulatory mechanisms and molecules delivering signals two and three to Tc cells are precisely the same as those for Th cells remains under investigation. In any case, the initial stages of signal transduction in Tc and Th cells are essentially parallel, and we will merely refer to "T cells" and "APCs" in most of the following discussion. We first explore the

The vast majority of naive Tc responses require both costimulation and T cell help. Until relatively recently, these features of Tc activation presented immunologists with two logistical puzzles. First, with respect to costimulation, Tc cells are like Th cells in that they require the interaction of their CD28 molecules with B7 molecules on a costimulatory cell. However, while B7 is expressed on the professional APCs that interact with Th cells, most target cells that are the subject of Tc recognition are non-APCs and so do not express the required B7 molecules. Secondly, with respect to Th involvement in Tc responses, it was originally not clear how the Th cytokines required by the Tc would be delivered, since Tc cells and Th cells do not express surface molecules permitting direct intercellular contact (unlike the interaction between CD40 and CD40L that allows the formation of a B–Th conjugate). In an attempt to circumvent these theoretical problems, models of Tc activation were envisioned in which Tc and Th cells interacted simultaneously with the same APC, with the Th cell recognizing one peptide from antigen X presented on MHC class II and the Tc cell recognizing a different peptide from antigen X cross-presented on MHC class I (refer to Ch.11). Such a scenario would provide the Tc cell with the required activated, cytokine-secreting Th cell and a B7-expressing costimulatory cell. However, the coming together in the exact same place at the exact same time of Tc and Th cells specific for the same antigen, as well as an APC that happened to be displaying both types of antigenic peptides simultaneously, seemed somewhat improbable. More recently, models have been devised in which naive Tc activation requirements are elegantly met by a phenomenon called "APC licensing."

According to the APC licensing model, the role of the Th cell is to stimulate an APC (most likely an immature DC) so that it becomes activated and expresses surface molecules that "license" it to later costimulate the Tc cell. To put this in context, let us suppose that an animal has a viral infection. The virus replicates within target cells which release virus particles and/or debris of viral origin. A passing macrophage may phagocytize some of this debris, process the viral proteins, and present viral peptides on MHC class II which may be recognized by an antigen-specific memory Th cell. The Th cell is activated, starts synthesis of IL-2, and expresses CD40L. (Very few Tc cells express CD40L.) CD40L on the Th surface can in turn interact with CD40 on a nearby immature DC. Engagement of CD40 causes the DC to upregulate its expression of MHC and B7 molecules (and perhaps others). The activated DC can be released from the Th cell but still maintains its elevated expression of B7 molecules. The freed and activated DC is said to be "licensed" because it expresses the costimulatory ligands necessary to engage CD28 on naive Tc cells. We can picture the licensed DC subsequently encountering an antigen-specific naive Tc cell in the area and engaging it via a B7–CD28 connection or perhaps some other costimulatory combination. It is important to note that the Tc cell may or may not have already had its TCRs engaged by peptide presented on MHC class I on a target cell. In some cases, the Tc cell binds to the licensed APC before it encounters the target cell, meaning that signal two may be delivered before signal one. In other cases, the Tc cell may have already initiated interaction with a target cell, and then may be joined by a licensed

APC to form a tri-cellular complex. Either way, in the presence of cytokines or intercellular contacts delivered by an antigen-specific Th cell and/or an activated APC, a Tc cell that has received both signals one and two becomes activated and undergoes proliferation and differentiation into effector CTLs.

Experiments *in vivo* have lent credence to the licensing hypothesis. When Th-depleted mice were injected with an agonist anti-CD40 antibody to cross-link CD40L on the APC surface, "help" was delivered such that a CTL response could be observed. Furthermore, if wild-type mice were injected with antagonist anti-CD40L antibody to block stimulation of the APCs, the mounting of a helper-dependent CTL response was blocked. The specific cytotoxic responses could be restored in these mice simply by adding agonist anti-CD40 antibody. A final piece of supporting evidence has come from studies of knockout mice deficient for either CD40 or CD40L. These mutants are unable to mount helper-dependent CTL responses.

The licensing theory is attractive because it removes the necessity for the APC, Tc, and Th cells to be in exactly the same place at the same time. This model can also account for those cases in which CTL responses appear to be APC dependent but totally Th independent. In addition to being stimulated by Th cells, immature DCs can be activated directly by inflammatory cytokines, bacterial products (via PRRs such as the TLRs), or viral infection such that DC surface expression of MHC and costimulatory molecules is upregulated. These licensed, mature DCs may secrete the cytokines necessary to induce CTL proliferation and differentiation, perhaps obviating the necessity for Th help in these cases.

interaction of naive T cells and APCs and the signaling pathways sparked by pMHC–TCR interaction. It should be noted that T cell priming is only carried out effectively when the APCs involved are DCs, although "APC" and "DC" are often used interchangeably when discussing naive T cell activation.

A. Bringing T Cells and APCs Together

The activation of naive T cells takes place in a well-ordered series of steps, starting with the entry of antigen at the site of infection, injury, or inflammation. The reader will recall

from Chapter 4 that the inflammatory response involves the emanation from the site of injury or attack of chemokines that summon leukocytes, including APCs and T cells, to the area. Antigens, either on their own or within phagocytes, are conveyed to the local lymph node, processed, and displayed on APCs in the node, primarily DCs and macrophages. Cytokines secreted by cells of the innate response induce or enhance the expression of accessory and adhesive molecules on the naive T cell surface that assist it in efficiently getting to the local lymph node where activation would be useful, and in interacting with the antigen-displaying APCs once it gets there. For example, as well as drawing T cells through the HEVs and into the T cell areas of lymph nodes, chemokines such as SLC (*secondary lymphoid tissue chemokine*) induce rapid myosin activation and actin polymerization at the T cell surface. These events cause a polarization in the T cell structure such that "actin protrusions" are formed. These protrusions form the leading edge of a T cell inspecting the pMHC complexes presented by an APC. TCRs in these protrusions engage pMHC faster than do TCRs on other regions of the T cell surface, reflecting a heightened sensitivity to interface formation. Adhesion receptors in the actin protrusions are also more inclined to establish binding pairs with their ligands. We will now discuss several of the major molecules governing the circulatory and adhesive properties of naive T cells (Fig. 14-1).

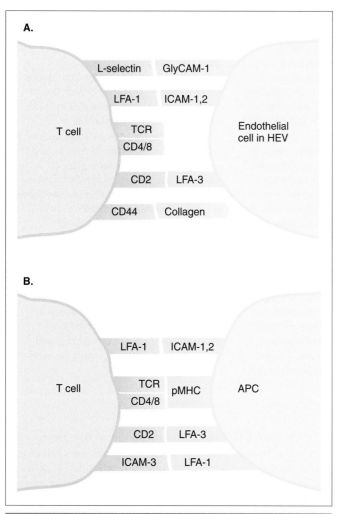

Figure 14-1

T Cell Adhesion Contacts in Humans
(A) Molecules important for the interaction of T cells with endothelial cells in the HEVs are shown. CD44 binds to collagen in the extracellular matrix. Note that either CD4 or CD8 is expressed on the surface of a mature Th or Tc cell, respectively. (B) Molecules important for the adhesion of Th cells to APCs are shown. These binding pairs are also used by Tc cells to carry out their initial inspections of target cells.

I. L-SELECTIN

As mentioned in Chapter 3, L-selectin (also known as CD62L and MEL-14) is a member of the selectin family of adhesion molecules constitutively expressed on all types of leukocytes. Selectins bear multiple sites for N-linked glycosylation and contain an N-terminal "C-type" lectin domain that binds to the carbohydrate moieties of counter-receptors expressed on endothelial cells. In naive lymphocytes, L-selectin functions as a homing receptor, directing the cells to the HEVs of the peripheral lymph nodes. Here, L-selectin interacts with the vascular addressin GlyCAM-1 (which is expressed only on the surfaces of the HEVs) and induces the lymphocytes to extravasate into the lymph nodes. Studies of gene-targeted knockout mice deficient for L-selectin have confirmed that L-selectin has a critical role both in the normal recirculation of lymphocytes and in mediating leukocyte emigration into the tissues at sites of inflammation (where blood vessels take on HEV-like characteristics). Primary T cell-mediated immune responses, T cell proliferation, and cytokine production are all impaired in the absence of L-selectin.

II. LFA-1

LFA-1 is an *integrin*, which means it functions primarily as an adhesion molecule promoting contact either between two cells or between a cell and the extracellular matrix. All integrins are non-covalently linked heterodimers composed of an α polypeptide of 120–200 kDa and a β chain of 90–100 kDa. There are three integrin subfamilies: the β1-, β2-, and β3-integrins. Members of a given subfamily have a unique α chain associated with a shared β chain. The extracellular domains of integrins bind divalent cations that are essential for integrin binding function, while the cytoplasmic tails interact with elements of the cellular cytoskeleton (such as actin). LFA-1 is a β2-integrin expressed on the vast majority of thymocytes and bone marrow cells, and on all mature T cells, B cells, monocytes, and granulocytes. LFA-1 is made up of a unique α chain designated CD11a and a shared β chain called CD18.

LFA-1 binds to three ligands that we have encountered before in our discussions of adhesion: ICAM-1 (CD54), ICAM-2 (CD102), and ICAM-3 (CD50) (Fig. 14-2). ICAM-1, a widely expressed membrane glycoprotein of 80–114 kDa, contains five extracellular Ig-like domains, making it another member of the Ig superfamily. As we have noted before, the level of expression of ICAM-1 on the cell surface is increased in the presence of certain cytokines, particularly during an inflammatory response. The increase in ICAM-1 ensures that passing T cells expressing LFA-1 will be slowed in an area in which they may be needed to scan APC/target cells for foreign pMHC complexes. LFA-1 can also bind to ICAM-2, which has two extracellular Ig-like domains and is also widely expressed but which is not upregulated in response to cytokines. The binding of LFA-1 and ICAM-3 also plays a role in T cell adhesion. In this case ICAM-3 is expressed primarily on lymphoid cells, so it is the LFA-1 expressed on the APC that is involved in the intercellular contact. LFA-1/ICAM

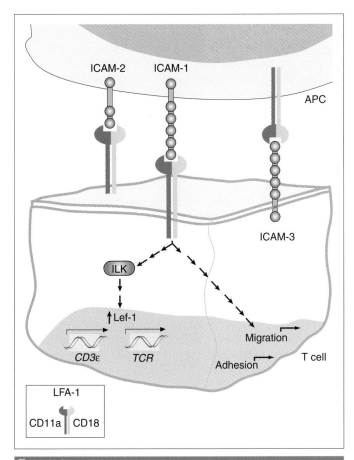

Figure 14-2

Integrin Functions
The binding pairs ICAM-1/LFA-1, ICAM-2/LFA-1, and LFA-1/ICAM-3 mediate adhesion between APCs and T cells. Engagement of LFA-1 also stimulates migration of T cells and induces activation of the kinase ILK associated with the LFA-1 cytoplasmic tail. Intracellular signaling initiated via ILK leads to activation of the transcription factor Lef-1, which may stimulate the transcription of the TCR and CD3ε genes.

binding is increased following stimulation of the TCR, further stabilizing T cell contacts with APC/target cells, thereby facilitating the completion of intracellular signaling and full T cell activation.

Several types of experiments have demonstrated that LFA-1 is important for normal T cell effector function as well as lymphocyte trafficking during inflammatory responses *in vivo*. Blocking the binding of LFA-1 to its ligands with anti-LFA-1 antibodies resulted in decreased T cell proliferation and activation *in vitro*, reduced adherence of lymphocytes to the HEV in the lymph nodes, and the inhibition of lymphocyte migration to the secondary lymphoid tissues. Antibodies against LFA-1 have also been shown to block CTL or NK killing of target cells *in vitro*. However, knockout mice in which the CD11a component of LFA-1 is mutated show normal Tc responses to at least some viruses, and normal activity of isolated Tc cells *in vitro*. Thymic development is normal in CD11a$^{-/-}$ mice, showing that LFA-1 does not have a unique function in this regard. There is an apparent defect in T cell proliferative responses in these mice, which supports the involvement of

LFA-1 in T cell activation in some capacity. One interesting defect noted in CD11a$^{-/-}$ mice was an impairment in the response to and clearing of tumors. A natural human mutation in which the CD18 component of LFA-1 is mutated results in *lymphocyte adhesion deficiency* (LAD), a primary immunodeficiency disease characterized by a pronounced susceptibility to viral infections (see Ch.24).

In addition to their functions in cell adhesion and migration, integrins such as LFA-1 have also been implicated in the regulation of the cell cycle. The cytoplasmic domains of several integrins can associate with a unique serine-threonine kinase called *integrin-linked kinase* (ILK). The activity of this 59 kDa enzyme is modulated by interactions between the cell and the extracellular matrix. These interactions have been shown to influence not only cell shape and motility but also growth, survival, differentiation, and gene expression. ILK may be relevant to T cell activation in that overexpression of ILK results in the enhanced formation of a nuclear transcription factor complex containing Lef-1 (lymphoid enhancer factor-1). As mentioned in Chapter 12, Lef-1 is closely related to, and may be redundant in function with, the transcription factor Tcf-1 (T cell factor-1), which binds to a motif in the enhancers of several genes, including the CD3ε and TCRα, β, and δ genes.

III. CD44

CD44 is an acidic cell surface adhesion protein that is broadly expressed, including in the membranes of B cells, granulocytes, monocytes, and erythrocytes as well as on many thymocytes and mature T cells. CD44 occurs naturally in at least 20 different transmembrane isoforms that are derived from the alternative splicing of 12 exons of the 20-exon CD44 gene. Preferential expression of various isoforms in certain tissues has been observed, with the appearance of some CD44 isoforms being associated with the inhibition of tumor metastasis. In general, CD44 molecules have seven sites available for N-linked and O-linked glycosylation. CD44 has been shown to bind to hyaluronic acid, heparin sulfate, collagen, and fibronectin (all components of the extracellular matrix) and several cytoskeletal proteins. In T cells, CD44 primarily mediates adhesion to the endothelium, promoting lymphocyte rolling during extravasation through the HEV at the site of an inflammatory response. Consistent with this role, CD44 expression is induced by the pro-inflammatory cytokine TNF and is transiently upregulated on T cells during an immune response. This increase in CD44 expression may also help to retain activated T cells within a lymph node by facilitating binding to extracellular proteins acting as structural components of the node.

Knockout mice deficient for CD44 do not show a significant reduction in the total numbers of leukocytes in the peripheral blood or of hematopoietic progenitors in the liver. However, the distribution of hematopoietic progenitors is abnormal, and granuloma formation (see Chs.15 and 22), which requires the migration of monocytes and macrophages, is defective. No evidence has been obtained from CD44$^{-/-}$ mice, which suggests an absolute requirement for CD44 in either T cell costimulation or thymocyte maturation.

IV. CD2

Human CD2 (also known as LFA-2, for *leukocyte function-associated antigen-2*) is a monomeric transmembrane glycoprotein of 45–50 kDa and a member of the Ig superfamily. It has two extracellular Ig-like domains (one V-like and one C-like) and a long, proline-rich cytoplasmic tail. CD2 is expressed early during human thymocyte development and is found on about half of thymocytes and almost all mature peripheral T cells. In mice, CD2 is also expressed on some B cells and NK cells. CD2 functions as an intercellular adhesion molecule, binding with high affinity in humans to its ligand LFA-3 (CD58). Human LFA-3 is a surface glycoprotein of 55–70 kDa that also contains two extracellular Ig-like domains and that is expressed on APCs (including B cells), among many other cell types. Murine CD2 binds to a protein closely related in structure to LFA-3 called CD48.

The binding between CD2 and LFA-3 is thought to stabilize the joining of the T cell to the APC or target cell so that the two cells form a conjugate. The formation of the conjugate gives the TCR a longer interval to scan the various pMHC combinations presented by the APC, determine if a match has been made, and, if so, complete the intracellular signaling necessary for T cell activation. T cells treated *in vitro* with antibodies directed against CD2 or LFA-3 are prevented from forming conjugates and consequently show decreased T helper or CTL activity. In addition, the binding of CD2 on the T cell by LFA-3 on a B cell induces the upregulation of CD40 expression on the B cell, implying that CD2 may play a role in facilitating the costimulation and/or delivery of T help to B cells. Studies *in vitro* have shown that stimulation of CD2 on a T cell by certain anti-CD2 antibodies or soluble LFA-3 can stimulate tyrosine phosphorylation of CD3 chains and, in some cases, can induce T cells to proliferate and secrete cytokines. However, the CD2 knockout mouse has no detectable defects in either T cell development or immune responses. The production of both antibodies and CTLs was found to be completely normal following immunization of these mutants. Thymocyte selection experiments using transgenic variants of these animals showed that both positive and negative selection processes were also normal. Such a phenotype indicates that CD2 function is not essential for the development and function of T cells and that other molecules can compensate for its absence *in vivo*.

B. Signal One: Binding of Peptide–MHC to the TCR

Signal one for the activation of T cells starts with the binding to a TCR of an appropriate pMHC complex of a shape complementary to the binding site of the TCR. It was originally thought that the binding of a pMHC complex to a TCR might cause a conformational change in the TCR that directly relayed the intracellular activation signal. However, studies of both soluble and crystallized TCR–pMHC complexes have found no physical evidence for a significant confor-

mational change in the TCR upon engagement. Instead, the TCR–pMHC binding event results in intracellular events responsible for the activation of the cell. However, as we have learned, the triggering of a single TCR is insufficient to activate the T cell. The importance of the action of adhesion molecules in holding the T cell and the APC together (and the influence of the co-receptors) becomes clear once one realizes that <u>sustained</u> signaling of up to 30 hours, not just a quick hit or a transient aggregation of TCRs, is required to achieve complete naive T cell activation. In fact, transient TCR engagement leads to the triggering of anergy in the T cell rather than activation (see Ch.16). By some estimates, sustained signaling resulting in activation requires the engagement of at least 8000 of the approximately 30,000 TCRs on the T cell by about 50–100 pMHC complexes on the APC. The sustained signal is also associated with a prolonged increase in the concentration of intracellular Ca^{2+} that lasts for at least 60 minutes after initial T cell/APC binding. The elevated Ca^{2+} concentration is necessary for the maintenance of adequate concentrations of transcription factors in the nucleus, particularly NFAT (*nuclear factor of activated T cells*; see later). These transcription factors are required for transcription of the IL-2 gene. Indeed, a maximal IL-2 response occurs only after 10–24 hours of contact between the T cell and the APC.

A T cell that has encountered an APC presenting foreign peptide starts by "crawling" slowly over the surface of the APC. The actin cytoskeleton within the T cell is reorganized so that the surface receptors and secretory machinery become concentrated at the interface between the two cells. If one allows T cells and APCs to begin interacting, but then introduces experimental agents such as anti-MHC antibodies or chemicals blocking Ca^{2+} flux or actin cytoskeleton reorganization, one finds that intracellular signaling ceases after only 2–4 minutes and activation is not achieved. These results demonstrate that the TCR signal produced by any given TCR–pMHC interaction is very short-lived and not self-sustaining. In studies of intermolecular complexes in which one TCR was bound to one pMHC, the "off" rate for TCR–pMHC interaction was estimated to be very high, such that the half-life of such complexes was 5 seconds to a few minutes at most. How then is sustained signaling achieved?

I. MODELS OF TCR TRIGGERING

i) Oligomerization Model

Work in the mid-1990s showed that an intracellular calcium response could be achieved in a naive T cell only if at least three TCRs were brought together. This (and similar observations) led to the *oligomerization* model, which holds that the binding of several pMHC complexes to several TCRs in a small region at the interface between the T cell and the APC induces a structural change in each TCR that promotes oligomerization with neighboring TCR–pMHC–CD3 complexes through transient Vα–Vα interactions. These interactions are thought to build a stable structure of multiple TCR–pMHC–CD3 complexes that restricts the movement of individual TCR–pMHC–CD3 complexes in the T cell membrane. This aggregation of TCR–CD3 cytoplasmic tails in one spot may raise the local concentration of intracellular signaling molecules and their substrates, increasing the efficiency of intracellular signal transduction and promoting cellular activation. In this model, antagonist peptides presumably disrupt T cell activation by somehow interfering with the oligomerization of TCRs necessary to build an activation complex.

Evidence to support the oligomerization model originally came from X-ray crystallographic analyses carried out on the murine Vα domain of a TCRαβ molecule recognizing a particular bacterial peptide in association with MHC class II. The structure of the Vα region in this complex was shown to be very similar to that of an Ig light chain variable region. However, the folding of the β-pleated sheets was altered, forming a space that allowed two Vα domains to pack as a dimer. These findings suggested that, *in vivo*, TCRαβ undergoes oligomerization in response to pMHC binding, an event mediated primarily by interactions between the Vα domains of separate TCRαβ molecules. Other indirect evidence supports the oligomerization hypothesis. First, the MHC molecule HLA-DR was found to crystallize as a dimer of MHCαβ heterodimers, implying that in nature, the MHC structure on the surface of the APC may occur as a dimer that can interact with two TCRs. Secondly, a soluble MHC class II dimer created by chemical cross-linking of two H-2E MHC molecules was able to activate T cells *in vitro* where a monomer could not. Similarly, bivalent experimental antibodies can activate T cells *in vitro*, but monovalent antibody chains cannot. These results imply that at least two TCRs must be present in very close proximity to initiate activation in response to ligand binding.

Kinetic studies by Mark Davis and his colleagues have suggested that more than two TCRs may oligomerize after the initial pMHC binding and just prior to the first detectable intracellular signaling. The kinetics of binding in solution of an isolated TCR of known specificity to its isolated pMHC target were examined by a method that allowed the researchers to determine the size of individual complexes. When increasing concentrations of TCR and pMHC molecules were mixed together, oligomeric complexes formed that correlated in size to entities containing 2–6 TCRs. These complexes formed only if the correct peptide was bound to the correct MHC allele, and not when unrelated peptides, unrelated MHC alleles, or an "empty" molecule of the correct MHC allele was used. However, later studies provided evidence that the formation of TCR oligomers alone is insufficient to generate a sustained signal because the low affinity of the binding between the pMHC and the TCR does not permit the duration of interaction that T cell activation requires. These findings have led to a greater acceptance of a model that takes advantage of this low affinity: *serial TCR triggering*.

ii) Serial Triggering Model

The serial triggering model posits that relatively numerous TCRs can be successively engaged by relatively few pMHC complexes to generate the sustained signaling necessary for complete activation of a naive T cell. By measuring the kinetics

of TCR–pMHC interaction at the single cell level, Antonio Lanzavecchia and colleagues have estimated that about 18,000–20,000 TCRs on a T cell are usually triggered by about 100 pMHC complexes in the course of full cellular activation. Thus, it follows that each pMHC complex may be able to interact sequentially with as many as 180–200 different TCR molecules on the same T cell's surface (Fig. 14-3). Each TCR in succession receives a "hit," delivers intracellular signals such that the CD3 ITAMs are phosphorylated, then disengages, leaving the pMHC free to trigger a fresh TCR recruited to the contact area. The first TCR, meanwhile, becomes associated with small microdomains of the T cell plasma membrane called *rafts*. Membrane rafts are small areas of the fluid bilayer of the plasma membrane that are rich in glycosphingolipids and cholesterol. The rafts can thus be pictured as "floating" in a sea of more fluid membrane phospholipids. In a resting T cell, membrane proteins may transiently "hop" in and out of the rafts and a stable structure is not formed. However, in a triggered T cell, proteins important for the downstream transduction of TCR signaling (see later) preferentially associate with the rafts and a large, relatively stable structure is formed that serves as a platform for the recruitment and activation of additional signaling molecules. The rafts plus their cargo of signaling molecules then move within the membrane to associate with the cytoplasmic portion of the triggered TCR. The antigen-stimulated TCRs are thus brought into close proximity with all the signaling molecules needed to transduce the activation signal (Table 14-1). Indeed, in the absence of rafts, T cell activation is markedly less efficient. The entire membrane complex of antigen-stimulated TCR, CD3 chains, costimulatory and adhesion molecules, co-receptors, and other signaling molecules associated with the rafts has been called an *immunosome* by some researchers.

The necessity for each pMHC to engage multiple TCRs implies that the pMHC must disengage fairly quickly from the TCR once the initial signal to phosphorylate downstream elements is delivered so as to bind a fresh TCR. Evidence consistent with the requirement for dissociation of the pMHC from the TCR for T cell activation comes from *in vitro* studies using anti-CD3 antibodies. In binding to their epitopes in the TCR complex, the antibodies act like very high affinity ligands with a very slow "off" rate. Although in a laboratory setting anti-CD3 can activate any T cell (because it binds to the invariant CD3 components of any TCR), it does so inefficiently because it cannot dissociate easily to trigger fresh TCRs, meaning that a sustained signal is not easily generated. Some antagonist peptides may act in this way as well, refusing to release a TCR (even in the presence of agonist peptide) and interfering with serial triggering. TCRs that have bound a pMHC that fails to activate them are said to be *spoiled* or *paralyzed* because an agonist pMHC can no longer make use of them for activation.

The importance of the balance of the "on-off" rates for pMHC binding to the TCR is highlighted by studies of weak agonist peptides. Weak agonists and some antagonists have been shown to dissociate more quickly from the TCR than do agonist peptides, consistent with a model in which a minimum amount of time (probably a few seconds) of engagement is required for

Figure 14-3

Serial Triggering Model of T Cell Activation
Engagement of the TCR by pMHC leads to phosphorylation of the CD3 chains and recruitment of membrane rafts. The membrane rafts contain many of the signaling molecules required for transduction of the TCR signal (refer to Table 14-1). The pMHC disengages from the first, now-triggered, TCR and engages a fresh resting TCR. The process of CD3 phosphorylation, raft recruitment, and signal initiation is repeated as the pMHC moves from TCR to TCR. Eventually, sufficient TCRs are triggered such that the cumulative intracellular signal is strong enough to induce T cell activation.

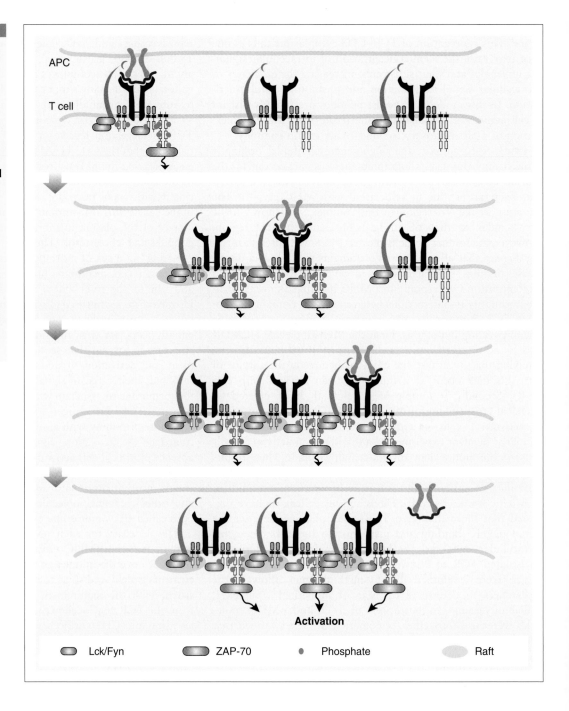

APC

T cell

Activation

Lck/Fyn   ZAP-70   • Phosphate   Raft

the initiation of TCR triggering. Indeed, if a weak agonist exhibits as little as a 5-fold increase in the "off" rate of binding to the TCR, T cell activation cannot be achieved. Abnormal ITAM phosphorylation patterns and a failure of ZAP-70 activation have been demonstrated in T cells treated with a weak agonist having too fast an "off" rate. A "kinetic model" has been proposed in which interaction between a pMHC and the TCR must endure for at least 5 seconds to be productive, that is, initiate TCR triggering. Weak agonists that bind for 3–4 seconds might induce partial activation or anergy, while interaction for less than 3 seconds would not trigger the TCR at all. Thus, if the "off" rate is too slow, serial triggering is frozen, and if the "off" rate is too fast, serial triggering cannot even start.

Another scenario that extends and modifies these observations to account for differences in agonist/antagonist TCR signaling is discussed later in this chapter. Regardless of the model, it is agreed that the degree of TCR triggering can have a profound effect on the cellular response observed (Fig. 14-4).

II. FORMATION OF THE IMMUNOLOGICAL SYNAPSE (SMAC)

The interface or contact zone between the T cell and the APC where stabilized rafts and triggered TCRs accumulate is called the *immunological synapse* or *supramolecular activation*

Table 14-1 Well-Studied Signaling Molecules Associated with Membrane Rafts

Molecule	Presence in Rafts	Function
CD2	Constitutive	Adhesion; minor co-stimulation
CD3ε	Inducible	Substrate for Src PTK phosphorylation
CD3ζ	Inducible	Substrate for Src PTK phosphorylation
CD4, CD8	Inducible	Co-receptors conveying Lck to TCR and increasing adhesion
Fyn	Constitutive	Src PTK that phosphorylates CD3 ITAMs and ZAP-70
IKK	Inducible	Removes inhibitor of NF-κB
LAT	Constitutive	Adaptor linking TCR to Ras/MAPK and PKCγ1 signaling pathways
Lck	Constitutive	Src PTK that phosphorylates CD3 ITAMs and ZAP-70
MEKK2	Inducible	Transduction kinase in SAPK/JNK pathway
PKCθ	Inducible	Links DAG to NF-κB activation
Rac	Constitutive	Substrate of Vav1 that links TCR to cytoskeletal reorganization
Vav1	Inducible	Links TCR and CD28 signals to cytoskeletal movement
ZAP-70	Inducible	Links TCR to Vav1, PLC-γ1, PKCθ, Ras/MAPK, and SAPK/JNK signaling pathways

Note: Pathways are discussed in detail later in this chapter.

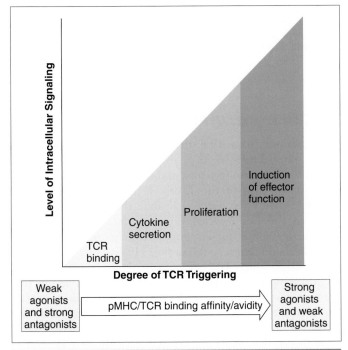

Figure 14-4

Cellular Outcomes of TCR Triggering
The level of intracellular signaling induced by TCR engagement determines the subsequent actions of the T cell. Weak agonists bind only tenuously to the TCR such that little intracellular signaling is generated. The T cell does not respond in any detectable way to the antigen. A similar result occurs when a strong antagonist with a fast off rate interacts with the TCR. As pMHC–TCR binding affinity/avidity increases, TCR triggering increases and more cellular functions are induced, including cytokine secretion and proliferation. When the TCR is engaged by a strong agonist (or the cognate antigen), high levels of TCR triggering lead to complete activation of the T cell and the generation of fully differentiated effector cells. A weak antagonist may permit a level of signaling that can induce some of these outcomes. Adapted from Backman M. F. and Ohashi P. S. (1999) The role of T cell dimerization in T-cell activation. *Immunology Today* **20**(12), 568–576.

complex (SMAC). The changes to the actin cytoskeleton that were initiated by the establishment of adhesive contacts (and extended by costimulatory signaling; see later) become more dramatic. Stable contact is forged between the T cell and APC that is dependent on the actions of Vav1 and molecules that bind to it. An increase in intracellular Ca^{2+} induced by these proteins triggers a pronounced reorganization of the cytoskeleton and the redistribution of receptors and counter-receptors in the membranes of the interfacing cells (Fig. 14-5A and Plate 14-1). TCRs from all over the surface of the T cell congregate at the interface and draw pMHC on the APC into the same area. Seen in cross-section (Fig. 14-5B), the completed SMAC has a concentric triple ring structure in which the innermost ring (sometimes referred to as the cSMAC, for "central" SMAC) contains the engaged or triggered TCRs, the pMHC complexes, the coreceptors (and their associated Lck and Fyn kinases), and the lipid rafts, which include additional enzymes and molecules crucial for signal transduction and costimulation. Thus, the cSMAC may be regarded as a zone specialized for TCR signal transduction, bringing clustered TCRs into close proximity with both pMHCs and signaling molecules so that serial triggering is optimized. The cSMAC is surrounded by the inner pSMAC (for "peripheral" SMAC), which contains many molecules of CD2 bound to its partner LFA-3 (CD58 in humans; CD48 in mice). An adaptor protein called CD2AP (*CD2 associated protein*) that binds to the cytoplasmic tail of CD2 helps to link the signaling complex to the T cell cytoskeleton. The inner pSMAC in turn is encircled by the outer pSMAC in which LFA-1 is bound to ICAM-1 and linked to the actin cytoskeleton-associated structural proteins *talin* and *ezrin*. Other molecules (such as *myosin II*) important for connecting the cytoskeleton to membrane-bound proteins may also appear in the outer pSMAC. The pSMAC layers ensure that the T cell and APC stay fixed together in the correct position long enough to allow signal transduction to be completed within the cSMAC.

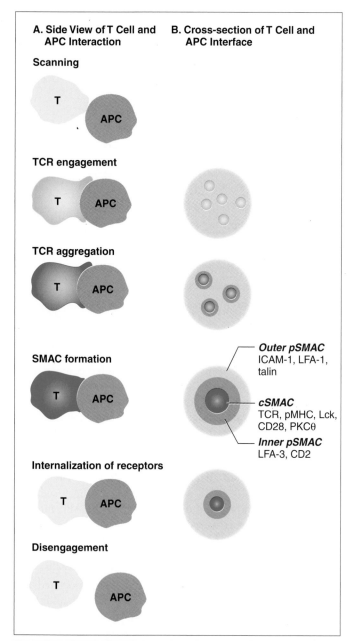

A. Side View of T Cell and APC Interaction

B. Cross-section of T Cell and APC Interface

Scanning

TCR engagement

TCR aggregation

SMAC formation

Outer pSMAC
ICAM-1, LFA-1, talin

cSMAC
TCR, pMHC, Lck, CD28, PKCθ

Inner pSMAC
LFA-3, CD2

Internalization of receptors

Disengagement

Figure 14-5

SMAC Formation

(A) Side view and (B) cross-sectional view of the interface between the T cell and APC during T cell activation. The T cell first scans the surface of the APC to determine if its TCR recognizes the pMHC presented on the APC surface. If sufficient numbers of TCRs are strongly engaged, the cytoskeleton of the T cell reorganizes to generate clusters of signaling molecules in the site of TCR–pMHC interaction. These clusters eventually coalesce into the cSMAC, which contains triggered TCRs and pMHC complexes, and the membrane rafts with their signaling molecules. The cSMAC is surrounded by the inner pSMAC, which is dominated by binding pairs of LFA-3 (or CD48) and CD2. The adhesion molecules that first facilitated the T cell–APC interaction become concentrated in the outer pSMAC, which surrounds the inner pSMAC. Triggered TCRs that have completed their participation in signaling are then internalized, and the SMAC structure slowly dissipates before the T cell disengages from the APC to commence proliferation and effector differentiation. Adapted from Krummel M. F. and Davis M. D. (2002) Dynamics of the immunological synapse: finding, establishing and solidifying a connection. *Current Opinion in Immunology* **14**, 66–74.

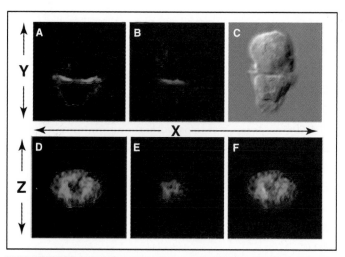

Plate 14-1

SMAC Formation

Three-dimensional reconstruction of a T cell–APC contact site shows the localization of talin (green) and PCKθ (red). (A and D) The T cell–APC conjugate is labeled with talin-specific antibodies. (B and E) The conjugate is labeled with PCKθ-specific antibodies. (A and B) Optical section along the X-Y axis in the center of the cell conjugate. (C) Phase microscopy image of the T cell–APC conjugate showing the interface. (D–F) Three-dimensional view of the entire cell contact site along the X-Z axis. Talin and PCKθ are both present at the cell contact site but in two distinct domains. Reproduced with permission from Monks *et al.* (1998) *Nature* **395**, 82–86.

III. TCR DOWNREGULATION

What happens to a TCR that has done its part to initiate signal transduction? A TCR that is fully phosphorylated is *internalized* or *downregulated* and degraded within the lysosomal compartment of the T cell. In other words, after one encounter with its cognate ligand, the triggered TCR is removed and not recycled to the surface like many other surface receptors. (Unphosphorylated or partially phosphorylated TCRs do return to the surface.) In one study, it was estimated that 70% of TCRs was internalized within 15 minutes of cSMAC formation. This loss of triggered TCRs implies that a continuous flow of new TCRs must make it into the contact site between the T cell and the APC for signaling to be sustained. Studies in which T cells are stimulated in tissue culture have shown that TCRs are internalized only after ligation to specific antigen, so that downregulation is an accurate measure of the number of TCRs that have been triggered in a response to specific ligand. Moreover, T cells treated with weak agonist peptides exhibit less TCR downregulation and a decreased Ca^{2+} flux that correlate with their ability to induce only partial T cell activation. Antagonist peptides fail to disengage, thus blocking the internalization of the TCR and the triggering of a fresh TCR by the pMHC.

The precise mechanism of TCR downregulation has yet to be fully elucidated, but the evidence to date suggests that phosphorylation of the serine 126 residue in the CD3γ chain by *protein kinase* C (PKC; see later) correlates with the TCR internalization observed during activation. The internalization of TCRs appears to depend on the presence of the cSMAC, since mutant mice in which this structure cannot form show abnormally high levels of TCRs on the surfaces of antigen-stimulated T cells. The physiological function of TCR internalization and degradation may be to prevent T cell death due to excessive signaling and overstimulation. The cSMAC would thus function as the master controller of T cell activation, promoting intense TCR signaling for the appropriate time, which is balanced by cSMAC-mediated promotion of TCR degradation.

IV. TCR SIGNAL TRANSDUCTION

The end result of sufficient TCR engagement and SMAC formation is the recruitment of signal transducing proteins and the activation of specific signaling domains within these molecules. It is becoming clear that lymphocyte signal transduction is anything but linear, involving not only the multiple pathways fired up by antigen receptor engagement, but also interlinked pathways induced by coreceptor and costimulatory receptor engagement (Fig. 14-6). These pathways involve shared

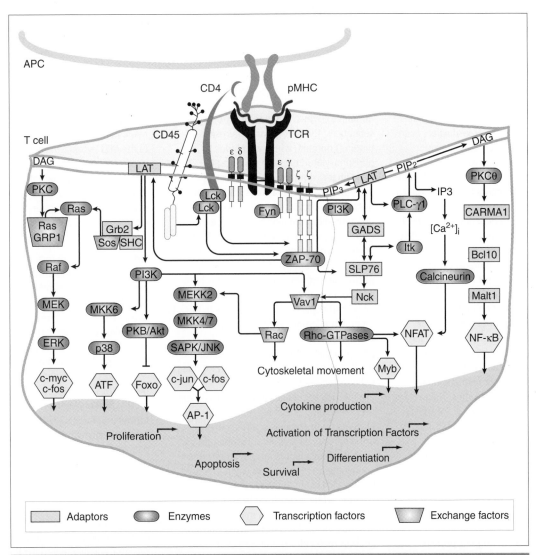

Figure 14-6

TCR Signal Transduction
The various signaling cascades initiated by TCR triggering are shown. The major pathways include the Ras/MAPK pathway (far left) leading to c-fos and c-myc activation; the p38 pathway leading to expression of the transcription factor ATF; the SAPK/JNK pathway leading to activation of the transcription factor c-jun; the activation of ZAP-70 and Vav1, which leads to cytoskeletal movement; the PLC-γ1 pathway, which leads to activation of the transcription factor NFAT; and PKC activation leading to the activation of the transcription factor NF-κB. PI3K has numerous roles in facilitating signal transduction by these major pathways. The transcription factors activated by these pathways lead to new gene transcription that stimulates survival, proliferation, cytokine production, and effector cell differentiation, among other outcomes. Details of the players in each pathway and the interactions between them are given in the text. Adapted from Yeung R. S., Ohashi P. and Mak T. W. (in press) T cell development. In Ochs H. D., Smith C. I. and Puck J. M., eds. "Primary Immunodeficiency Diseases. A Molecular and Genetic Approach." Oxford University Press, Oxford.

molecules that influence each other's concentrations and activities, such that "cross-talk" and "feedback" mechanisms are established. Another layer of complexity to antigen receptor signaling was revealed in the late 1990s with the discovery of *adaptor* molecules. These non-enzymatic proteins allow various signal transduction molecules to physically interact with each other (see Box 14-2). Adaptors are examples of what are often referred to as "scaffold proteins" because they serve as platforms for the binding and interaction of two or more unrelated proteins. Many adaptors are concentrated in the membrane rafts within the cSMAC where triggered TCRs cluster.

Much of the information that follows regarding signal transduction from the TCR has been derived from experiments *in vitro* in which scientists use *mitogens* and *superantigens* to non-specifically activate T cells (see Box 14-3). It is thought that the intracellular signaling that follows such activation mimics that in an antigen-stimulated T cell, and so can be used to investigate T cell activation without the complexities and limitations imposed by the use of specific antigen. In general, the transduction of the signal from the activated TCR complex to the nucleus of the cell is similar in many aspects to transduction of the antigen-binding signal from the BCR.

i) Recruitment of Src Protein Tyrosine Kinases (PTKs) and Activation by CD45

Just as the Igα/β molecules are phosphorylated in response to antigen binding to the BCR (described in Ch.9), the earliest intracellular signaling event that can be detected within a T cell that has bound pMHC is the phosphorylation of the ITAMs in the cytoplasmic tails of the CD3 molecules in the TCR complex.

Box 14-2. Adaptor proteins in antigen receptor-mediated signaling

We have made reference in our descriptions of antigen receptor signal transduction, from both the TCR and the BCR, to signaling complexes that have to be assembled and downstream elements that must be linked to those complexes. Scientists have succeeded in identifying many of the enzymes involved in the cascade of phosphorylations that activates signaling components, but until the late 1990s, it was unknown what mechanism allowed the enzymes and their substrates to come together efficiently enough to sustain activation. The glue in the signaling complexes turns out to be a collection of *adaptor proteins*, so-called because they serve as structural links between different signaling enzymes and their substrates or binding partners. Adaptor proteins possess no enzymatic activity of their own and do not trigger the transcription of genes. Their role is to allow multiple components of the signal transduction machinery to assemble in large complexes, thereby promoting efficient processing of substrates by the component enzymes. Adaptors may be either cytosolic or transmembrane proteins, and those that are transmembrane may or may not be associated with the membrane rafts involved in TCR signaling.

The structure of an adaptor protein resembles a series of plugs or docking bays. Each docking site is a specific domain that can bind to a specific protein motif in a signaling element. Some domains, many of which we have encountered before, facilitate the binding of signaling substrates, for example, the *SH2* domain (Src homology domain 2), which binds to phosphorylated tyrosine residues in signaling proteins such as Lck and Fyn. Other important domains include the *SH3* domain (Src homology

domain 3), which binds to proline-rich sequences; the *PTB* (phosphotyrosine-binding) domain, which also binds to phosphorylated tyrosine residues; the *ZF* (zinc finger) domain, which mediates protein–protein interactions; the *RF* (ring finger) domain, which targets proteins for ubiquitination and degradation; the *LZ* (leucine zipper) domain, which mediates protein–protein interactions; and the *PH* (plekstrin homology) domain, which binds phosphoinositides. Most adaptor proteins also contain one or more protein motifs that allow particular signaling enzymes to dock to them. For example, an adaptor protein containing tyrosines that can be phosphorylated by PTKs is a target for the binding of enzymes containing SH2 domains. Other adaptors contain proline-rich sequences that serve to recruit SH3-containing enzymes. Adaptor proteins often serve as scaffolds that bring an enzyme and its substrate into close proximity. For example, an enzyme containing an SH3 domain could bind to a proline-rich motif at the N-terminus of an adaptor, while a substrate bearing phosphorylated tyrosines might bind to an SH2 domain at the C-terminus of the same adaptor molecule. The efficiency of the conversion of the substrate by the enzyme into product is thus increased.

Signal transduction following antigen receptor engagement is still somewhat of a black box in many areas. While many enzymes and adaptor proteins have been identified, their precise roles and relationships to each other have yet to be completely identified. We offer here a summary of what is known about several important adaptor proteins.

SHC: As we have seen, ligation of the TCR results in the phosphorylation of the ITAMs in the CD3 cytoplasmic tails by Src PTKs, particularly Lck. The phosphorylated ITAMs constitute a binding target for a ubiquitously expressed adaptor protein called *SHC*. SHC is subsequently phosphorylated and becomes capable of binding to the SH2 domain of an adaptor called *Grb2* (see next section).

Grb2: Grb2 (*growth factor receptor binding protein 2*) is a ubiquitously expressed adaptor protein containing two SH3 domains and an SH2 domain. After SHC binds to Grb2's SH2 domain, Grb2 in turn recruits the guanine nucleotide exchange factor *Sos* (*son of sevenless*). Because Sos is linked to the Ras/MAPK signal transduction pathway, Grb2 contributes to the linking of TCR signaling to the Ras/MAPK pathway in T cells. Similarly in B cells, BCR engagement results in the formation of a SHC–Grb2–Sos complex that promotes signaling through the Ras/MAPK pathway. As well as interacting with SHC bound to an antigen receptor, Grb2 can use its SH2 domain to bind to SHC or other adaptor proteins associated with the cytoplasmic tails of growth factor receptors that have been engaged by ligand. Grb2 then uses its SH3 domain to bind to Sos, linking growth factor receptor engagement to the Ras/MAPK pathway.

LAT: The SH2 domain of Grb2 can also associate with an adaptor protein called LAT (*linker for activation of T cells*). LAT is an integral membrane protein concentrated in the membrane rafts of activated T cells. It has a long cytoplasmic tail containing nine tyrosine residues that are phosphorylated by ZAP-70 after TCR engagement. LAT is thus considered to be one of the earliest and

Continued

Box 14-2. **Adaptor proteins in antigen receptor-mediated signaling**–*cont'd*

most important adaptor proteins in the TCR signaling cascade. As well as binding to Grb2, and thus linking the TCR to Ras/MAPK signaling, phosphorylated LAT binds to the SH2 domain of PLC-γ1 and recruits this phospholipase to the membrane to initiate the inositol signaling cascade (see main text). Cells engineered to lack phosphorylated LAT show defective intracellular Ca^{2+} flux and failed T cell activation.

SLP76: The phosphorylation of LAT by ZAP-70 requires the function of the adaptor protein SLP76 (*SH2 domain-containing leukocyte protein of 76 kDa*). Tyrosine phosphorylation of SLP76 by ZAP-70 is necessary for complete signal transduction from the TCR, and in particular for the activation of transcriptional elements needed for IL-2 production. SLP76 binds to LAT via association with additional adaptor proteins such as GADS. Tyrosine-phosphorylated SLP76 also interacts with the SH2 domain of Vav1 and the adaptor molecule Nck (see next section). It is speculated that SLP76 may act as a scaffold on which to bring together Vav1 and effector proteins associated with Nck, resulting in regulation of TCR-mediated actin polymerization and subsequent cytoskeletal changes. In addition, SLP76 may coordinate the interaction of PLC-γ1 and the kinase Itk required to phosphorylate and activate PLC-γ1 (refer to Fig. 14-6). Mice lacking SLP76 show defects in both the PLC-γ1 and Ras/MAPK signaling pathways that lead to profoundly impaired T cell development.

Nck: Nck is a ubiquitously expressed 47 kDa cytosolic adaptor protein containing three SH3 domains and one SH2 domain. Nck is recruited to the SLP76 scaffold and has the very important function of linking the TCR to Vav1 and effector proteins involved in alterations of the actin cytoskeleton that are associated with T cell activation. Nck may serve a parallel linking function for several other receptor complexes, including those assembled in response to BCR engagement and the receptors for PDGF (*platelet derived growth factor*) and EGF (*epidermal growth factor*).

Others: Numerous other adaptor proteins containing SH2, SH3, and other domains have recently been identified, but their functions are unclear. Their names are mentioned here merely to illustrate the complexity of the TCR

signaling web. *SLP* binds to the SH3 domain of Grb2 and appears to be a substrate of PTKs activated in response to TCR engagement. SLP also associates with the SH2 domain of Vav1. The SH2 domain of SLP76 binds to *SLAP*, which has been reported to have both positive and negative effects on T cell activation. SLAP also binds to *SKAP*, another

tyrosine-phosphorylated adaptor protein that contains SH3 and PH domains and has been found in association with the Fyn SH2 domain.

Figure adapted from Peterson E. J. *et al.* (1998) Adaptor proteins in lymphocyte antigen-receptor signaling. *Current Opinion in Immunology* **10**, 337–344.

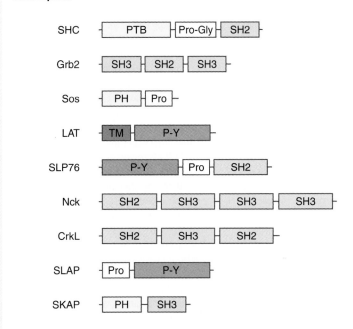

A. Adaptors

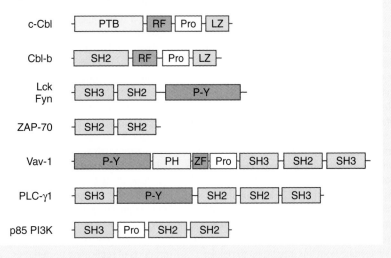

B. Enzymes and Exchange Factors

Box 14-3. Mimicking antigen-induced lymphocyte activation

The complexities and quantitative limitations of lymphocyte activation have made its study problematic both *in vivo* and *in vitro*. Fortunately for immunologists, molecules exist that are relatively easy to use *in vitro* and that mimic antigen-induced lymphocyte activation to a large extent. Molecules that promote cellular mitosis in a non-specific way are known as *mitogens*. In the context of T cells, mitogens activate lymphocytes but bypass antigen receptor specificity. *Superantigens*, which were introduced in Chapter 13, are a sub-class of mitogens. The use of mitogens in experiments permits far greater control over the initiation of T cell activation and consequently more reliable monitoring of downstream responses. The most visible and measurable result of mitogenic activation of a lymphocyte clone is its proliferation. Because a mitogen simultaneously activates a large number of T cell clones, the response is magnified and thus easier to measure.

Mitogens

Many mitogens are *lectins*, carbohydrate-binding proteins usually derived from bacteria or plants. Lectins recognize particular carbohydrate moieties on membrane glycoproteins on the surfaces of host cells. By binding to these moieties, lectins often agglutinate several cells together. In the case of T lymphocytes, the binding of certain lectins to carbohydrates of the TCR or CD3 chains is sufficient to trigger intracellular signaling, resulting in T cell activation and proliferation. Since the carbohydrate elements are usually the same no matter what the antigenic specificity of the TCR, a lectin activates many T cell clones of various specificities and thus functions as a polyclonal activator. Certain anti-CD3 antibodies can also be considered mitogens because they trigger the intracellular signaling path that leads to T cell activation regardless of antigenic specificity. Different mitogens stimulate different cell types (Table A). *Phytohemagglutinin* (PHA), *wheat germ agglutinin* (WGA), and *Concanavalin A* (Con A) promote human and murine T cell proliferation, respectively, whereas pokeweed mitogen (PWM) stimulates both human and murine T and B cells. As was discussed in Chapter 4, LPS is a non-lectin mitogen for murine and human B cells. LPS binds not to the BCR but to CD14 and TLR4 present in the B cell membrane. The signaling path leading from this interaction to cellular

activation and proliferation is described in detail in Chapter 17.

Superantigens

Superantigens are MHC class II-dependent mitogens (Table B). As introduced in Box 13-5, some native (i.e., unprocessed) viral and bacterial proteins can form an "abnormal" linkage between certain MHC class II molecules and TCRs bearing particular subfamilies of Vβ gene segments that leads to non-specific T cell activation. T cell clones whose TCRs do not contain that Vβ segment will not be activated by the superantigen. The superantigen does not undergo antigen processing within the APC; instead, it binds to an MHC class II molecule on an APC at an invariant site outside the peptide-binding groove. The MHC-superantigen structure is then recognized by TCRs bearing a specific Vβ sequence in the CDR2 of the TCRβ chain, which does not usually make contact with MHC. Structural and biological studies have shown that superantigens can bind to either the TCR Vβ region or the MHC class II protein alone but that the binding affinity increases substantially when all three entities form a complex. X-ray crystallographic studies of the bacterial toxin SEB (*Staphylococcus aureus enterotoxin B*) have shown that it binds preferentially to one side of the human class II MHC molecule HLA-DR1. In contrast, TSST-1 (*toxic shock syndrome toxin-1*) covers half of the flanking MHC α-helices of HLA-DR1 and even makes contact with the peptide-binding site. Nevertheless, it is not known whether the peptide-binding groove must be occupied (for conformational reasons) in order for superantigen activation to proceed. In any case, the superantigen can bind to any T cell bearing a TCR containing the appropriate Vβ segment, meaning that a large number (up to 5–20%) of different T cell clones (all sharing the same Vβ segment) can be simultaneously activated. This number of clones greatly exceeds the 0.001% of T cell clones activated during normal antigen presentation, hence the term superantigen. Concomitant with this activation is the release of very large amounts of pro-inflammatory cytokines such as IL-1 and TNF that can lead to acute illness, including *endotoxic shock* (see Ch.17). Finally, very high concentrations of superantigen that induce intense signaling lead to T cell apoptosis rather than activation.

Superantigens can be either exogenous (products of replicating bacteria) or

endogenous (products of host or integrated viral genes). Exogenous superantigens include toxins produced by gram-positive bacteria, such as the enterotoxins produced by *Staphylococcus* species that cause food poisoning, the pyogenic exotoxins of *Streptococcus* species associated with rheumatic fever (see Ch.29), and toxic shock syndrome exotoxins (produced by both staphylococcal and streptococcal species), which cause a serious, potentially fatal condition associated with the incorrect use of internal feminine hygiene products. Crohn's disease (a type of inflammatory bowel disease; see Ch.29) is associated with a relatively new superantigen called SAG (super-Ag) I2. SAG I2, which resembles a transcriptional activator of bacterial genes, is produced by normal bacterial flora in the intestine of Crohn's disease patients. When expressed in the human or murine gut, SAG I2 activates the subset of T cell clones that bears Vβ5 TCRs. It is speculated that these stimulated T cells may then contribute to the pathogenesis of the disease.

Endogenous superantigens are the protein products of genes in the host's genome, such as integrated retroviral genes. For example, certain genes of the mouse mammary tumor virus (MMTV) that can integrate into the mouse genome are translated into protein products that can act as superantigens. These genes were first identified when the lymphocytes from two mice of the same MHC haplotype exhibited an unexpected MLR reaction, leading to the designation of the corresponding proteins as the *minor lymphocyte stimulating* (Mls) antigens. In fact, the MMTV genes had integrated into the genome of one mouse strain but not the other, so that the Mls antigen became an allogeneic antigen. Several different Mls superantigens have been identified to date, each derived from a different MMTV strain and integrated into a different chromosome. Each confers the stable inheritance of a retroviral gene whose product binds to a specific Vβ region. In the host itself, T cells bearing receptors containing this V_β gene segment will have been deleted during the establishment of central tolerance (the retroviral protein becomes perceived as self) and so a response is not detected in that individual. However, if T cells of this host (which express the viral product) are mixed with T cells from an uninfected mouse of the same strain (whose TCRs contain the Vβ

Continued

region in question), a strong mitogenically driven response will be observed.

Although the intense immune response provoked by superantigens provides scientists with a wealth of material with which to study processes downstream of T cell activation, a caveat applies. It is not known how closely the sequence of events triggered by the binding of superantigens to TCR–pMHC complexes *in vitro* parallels events taking place during a physiological immune response. Some scientists feel that superantigen activation obscures the complex mechanisms that modulate T cell activation by antigenic pMHC, making the use of these stimulators problematic in regulatory studies. On the other hand, other scientists feel that responses to superantigens may underlie many autoimmune disorders. For example, in the synovial fluid of patients with rheumatoid arthritis, T cells specific for particular Vβ gene segments are selectively activated, perhaps explaining the infiltration of these lymphocytes into the joint (see Ch.29).

Table A **Selected Mitogens and Target Cells**

Mitogen	Source	Ligand	Primary Target Cells
Phytohemagglutinin (PHA)	Kidney beans	*N*-acetyl-galactosamine	T cells
Wheat germ agglutinin (WGA)	Wheat germ	*N*-acetyl-glucosamine; sialic acid	T cells
Concanavalin A (ConA)	Jack beans	α-D-glucose, α-D-mannose	T cells
Pokeweed mitogen (PWM)	Pokeweed	Di-*N*-acetylchitobiose	T and B cells
Lipopolysaccharide (LPS)	Gram negative bacteria	CD14/TLR4	B cells

Table B **Superantigen Specificity for Certain TCR Vβ Sequences**

Superantigen	Human Vβ Specificity	Murine Vβ Specificity	Disease Caused in Humans
SEA (bacterial)	ND	1, 3, 10, 11, 12, 17	Food poisoning
SEB (bacterial)	3, 12, 14, 15, 17, 20	7, 8.1, 8.3	Food poisoning
SEC$_2$ (bacterial)	12, 13, 14, 15, 17, 20	8.2, 10	Food poisoning
SED (bacterial)	5, 12	3, 7, 8.3, 11, 17	Food poisoning
TSST-1	2	15, 16	Toxic shock
SPE-A (bacterial)	2, 12, 14, 15	ND	Rheumatic fever, shock
SPE-B (bacterial)	8	ND	Rheumatic fever, shock
SPE-C (bacterial)	2	ND	Rheumatic fever, shock
SAG I2 (bacterial)	5	5	Crohn's disease
MMTV-C3H (viral)	NA	14, 15	NA
MMTV-SW (viral)	NA	6, 7, 8.1, 9	NA

SAG I2, superantigen I2; SEA, staphylococcal enterotoxin A; SEB, staphylococcal enterotoxin B; SEC, staphylococcal enterotoxin C; SED, staphylococcal enterotoxin D; SPE, streptococcal pyrogenic exotoxin; MMTV, mouse mammary tumor virus; TSST, toxic shock syndrome toxin; ND, not determined; NA, not applicable because viruses are species-specific.

How does this occur? We learned in Chapter 12 that the CD3 chains do not have inherent catalytic ability but rather serve as substrates for phosphorylation by Lck and Fyn, members of the Src family of protein tyrosine kinases (PTKs). In a resting T cell, the CD3 chains are either incompletely phosphorylated or not phosphorylated at all, and the Src PTKs are maintained in an inactive state by phosphorylation of a particular C-terminal tyrosine residue by members of the Csk family of PTKs. However, within minutes of TCR engagement, the CD4 or CD8 co-receptor and its associated phosphorylated (inactive) Lck molecules are induced to co-aggregate with the TCR and membrane rafts. Similarly, phosphorylated Fyn kinase is recruited to the CD3ε cytoplasmic tail. Lck and Fyn are then activated by dephosphorylation of the C-terminal regulatory tyrosine residue by *CD45*, a very large phosphatase highly expressed on the surfaces of all hematopoietic cells. (There is ongoing debate over whether at least some CD45 molecules are included in the membrane rafts and/or in the cSMAC.) Once dephosphorylated, Lck and Fyn can phosphorylate specific tyrosine residues in the ITAMs in the cytoplasmic tails of the CD3 proteins. Inactivation of CD45 blocks both thymic development and mature T cell activation, demonstrating the importance of this step.

The reader was first introduced in Chapter 9 to CD45 as a regulator of Src PTKs in B cells. In both human and murine T cells, CD45 exists in several different structural isoforms ranging in size from 180–220 kDa, depending on the state of activation of the cell. These isoforms result from alternative splicing of one or more of exons 4, 5, 6, or 7 of the CD45 gene. In human T cells, activation induces a switch in expression of CD45 from its highest molecular weight form CD45R (containing exons 4, 5, 6, and 7) to a low molecular weight form, CD45RO (missing exons 4, 5, and 6 and exhibiting a much lower level of glycosylation). CD45RO is typically observed in the latest stages of activation, several days after the original antigenic stimulation. CD45 isoform expression has also been used to distinguish different subsets of naive T lymphocytes, containing CD45RA (missing exon 4), CD45RB (missing exon 5), or CD45RC (missing exon 6), from memory T cells (CD45RO). Whether these different isoforms actually have subtly different functions is not clear. There is some evidence that CD45RO associates more strongly with CD4 and CD8 than does CD45RA, which may affect the efficiency of dephosphorylation of Lck and Fyn. Alternatively, some scientists speculate that the shift in isoform to one of lower glycosylation decreases the density of the "forest" of CD45 molecules on the cell surface, thereby facilitating the interaction of TCR and CD28 molecules with their respective binding partners.

Despite intense research efforts, the physiological ligand that binds to CD45 and stimulates its phosphatase activity has eluded identification. Some researchers now think that perhaps no intercellular ligands are involved at all, and that CD45 may be activated by binding to a receptor on the very same T cell. Possible candidates include CD4 or CD8, since the co-receptors are associated with the activation of Lck in the TCR complex. Other work has suggested that one ligand for CD45 may be CD45 itself. Studies of the regulation of CD45 activity have indicated that dimerization of CD45 inhibits its function because the active site of one molecule of the pair experiences structural interference exerted by a particular region of the other molecule of the pair. Mice with a mutation of CD45 that blocked dimerization showed autoimmune lymphoproliferation and autoantibody production, confirming the importance of dimerization as a means of turning off CD45 activity *in vivo*. Interestingly, the smaller CD45RO isoform that is observed after lymphocyte activation dimerizes more easily than the CD45A isoform present before activation, consistent with a scenario in which dimerization of CD45 acts to limit stimulatory signaling.

ii) Activation of ZAP-70

After the dephosphorylated Src PTKs activate the ITAM motifs in the CD3ζ chain, the tyrosine kinase *ZAP-70* can be recruited to the signaling complex via its SH2 domains.

ZAP-70 is related to the B cell kinase Syk but is expressed exclusively in T cells. Once ZAP-70 has joined the signaling complex, Lck and Fyn can then phosphorylate it, activating its catalytic activity. Activated ZAP-70 can then phosphorylate itself, providing more docking sites for additional SH2-containing effector molecules. From this point in the cascade, both Lck and ZAP-70 phosphorylate and thereby activate several crucial signal transduction molecules, each of which leads to specific intracellular effects necessary to fully activate the T cell. Among the substrates phosphorylated by ZAP-70 are the adaptor proteins SLP76 and LAT (refer to Box 14-2), which provide vital links to signaling intermediates that transduce the TCR signal and reorganize the actin cytoskeleton.

A powerful kinase such as ZAP-70 must obviously be held under strict control, a function mediated by the SH2-containing tyrosine phosphatase *SHP-1*. When the SH2 domains of SHP-1 bind to ZAP-70, SHP-1 phosphatase activity is stimulated, ZAP-70 is dephosphorylated, and its kinase activity is decreased. The regulatory role of SHP-1 has been demonstrated in experiments in which the overexpression of wild-type SHP-1 was engineered in antigen-stimulated T cells. T cell activation was damped down, and a decrease, rather than an increase, in IL-2 production was observed after antigenic stimulation.

iii) Vav1 and the Cytoskeleton

One of the most important SH2-containing proteins that interacts with ZAP-70 is *Vav1*, which was introduced in Chapter 13. The Vav1 protein contains an intriguing collection of structural motifs in addition to SH2, including a domain known to be involved in binding to the actin cytoskeleton and a region resembling the catalytic domain in certain small GTPase enzymes. Activated Vav1, ZAP-70, and the phosphorylated CD3ζ chain all become associated with the cytoskeleton after TCR triggering, promoting the coalescence of the membrane rafts and the assembly of the SMAC. Accordingly, T cells from Vav1$^{-/-}$ mice show defects in actin polymerization and cytoskeletal reorganization, their TCRs fail to aggregate and form a SMAC, and TCR-mediated Ca^{2+} flux, proliferation, and IL-2 production are all impaired.

How is actin cytoskeletal reorganization induced? Among the proteins recruited to Vav1 and the TCR signaling complex by the phosphorylation of the CD3ε ITAMS are the SH2 domain-containing adaptor proteins SLP76 and Nck (refer to Box 14-2). Phosphorylated Nck interacts via its second SH3 domain with PAK (*p21-activated kinase*) and WIP (*Wiscott-Aldrich syndrome protein-interacting protein*) (Fig. 14-7). WIP in turn binds to WASP (*Wiscott-Aldrich syndrome protein*). WIP and WASP have both been shown to be important for the induction of actin polymerization and cytoskeletal reorganization in lymphocytes. WASP and PAK are target substrates for the Rho-GTPases activated by Vav1, as is another molecule called Rac that is associated with actin polymerization. Thus, the binding of Nck to SLP76 serves to link activated Vav1 with target effector proteins that mediate actin polymerization associated with T cell activation. The Vav1-WASP pathway is subject to negative regulation by the ubiquitination enzyme *Cbl-b*. Cbl-b targets proteins for degradation, including (presumably) proteins necessary for Vav1-mediated functions during T cell

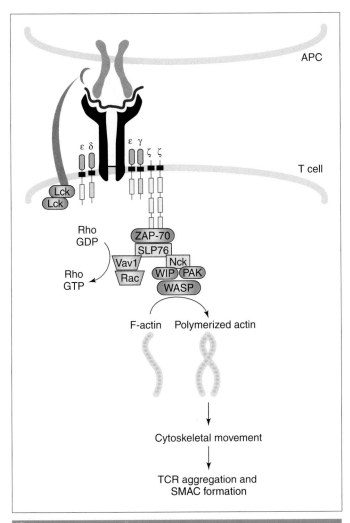

Induction of Cytoskeletal Reorganization by TCR Engagement
Engagement of the TCR triggers signaling transduced by Lck and ZAP-70 that leads to recruitment of adaptor proteins and GTPases (see text for details) that stimulate the polymerization of filamentous actin (F-actin). Once actin is polymerized, cytoskeletal movement takes place that facilitates TCR clustering and SMAC formation.

activation. (Cbl-b does not appear to degrade Vav1 itself.) Cbl-b$^{-/-}$ mice show increased Vav1 activation leading to deregulated T cell activation that can occur in the absence of CD28 costimulation. As a result, these mutant animals spontaneously develop autoimmunity (see Ch.29).

We note here that another member of the Cbl family called *c-Cbl* also negatively regulates antigen receptor signal transduction. c-Cbl is inducibly phosphorylated following BCR or TCR engagement. c-Cbl contains several of the domain elements cited in Box 14-2 that permit direct association with SH2-containing proteins such as Fyn, PI3K, Vav1, and Syk family PTKs. However, instead of promoting signal transduction, c-Cbl is thought to inhibit transduction by impairing the kinase activity of Syk and/or by recruiting another adaptor protein called *CrkL* that ultimately inhibits Ras/MAPK signaling (see later). Interestingly, c-Cbl$^{-/-}$ mice show a marked

elevation of TCRs on the surfaces of their thymocytes, suggesting that c-Cbl may also be involved in the degradation of internalized, triggered TCRs.

iv) PLC-γ1 Activation

Another important enzyme activated by ZAP-70 in T cells is *phospholipase C-γ1* (PLC-γ1) (refer to Fig. 14-6). The recruitment of SLP76 to ZAP-70 allows subsequent association of the adaptors GADS and LAT. PLC-γ1 binds to LAT, while the Tec family kinase Itk binds to SLP76. Itk then activates PLC-γ1 via tyrosine phosphorylation. Activated PLC-γ1 proceeds to hydrolyze the membrane lipid phosphatidyl inositol 4,5-bisphosphate (PIP$_2$) into the intracellular second messengers IP$_3$ (inositol 1,4,5-triphosphate) and DAG (1,2-diacylglycerol). As described for B cells in Chapter 9, intracellular IP$_3$ binds to a calcium channel, causing the release of Ca^{2+} from intracellular stores within the ER. The mobilized Ca^{2+} complexes to and activates the ubiquitous calcium-binding protein *calmodulin*. In the presence of high Ca^{2+} concentrations, the Ca^{2+}–calmodulin complex in turn activates the serine/threonine phosphatase *calcineurin*. Once activated, calcineurin dephosphorylates the cytoplasmic, inactive forms of the members of the NFAT nuclear transcription factor family, particularly *NFATc* ("c," cytoplasmic). Dephosphorylated NFATc is then free to cross the nuclear membrane and to pair with the nuclear transcription factor AP-1 to form an entity sometimes called *NFATn* ("n," nuclear). Like the transcription factors introduced in Chapter 9, NFATn interacts with specific binding motifs in the 5' flanking regions of several early genes needed for T cell activation and stimulates the transcription of these genes. NFATn is subsequently rephosphorylated in the nucleus by the kinase GSK3, which results in the dissociation of NFATc from AP-1 and the nuclear export of phosphorylated NFATc back to the cytoplasm. The immunosuppression drug cyclosporin A, which is used to prevent organ transplant rejection (see Ch.27), acts by inhibiting the dephosphorylation of NFATc, thereby blocking T cell activation.

Meanwhile, intracellular DAG derived from the hydrolysis of PIP$_2$ by PLC-γ1, together with the mobilized Ca^{2+}, activates the serine/threonine kinase PKC (*protein kinase C*). A specific isoform of this enzyme called *PKCθ* is recruited to the rafts in the contact zone between the T cell and the APC. PKCθ in turn starts a phosphorylation cascade involving the adaptor protein CARMA1 and the signal transducers Bcl-10 and MALT1 that culminates in the activation of the transcription factor NF-κB (described in Ch.8). NF-κB activation is required for maximal induction of IL-2 transcription and also results in the expression of the anti-apoptotic Bcl-xL gene, which promotes cell survival. Intracellular DAG also activates the Ras-MAPK pathway, as described later.

v) PI3K Function

When the adaptor protein LAT is phosphorylated by ZAP-70, it becomes capable of recruiting another important enzyme called *phosphatidyl inositol 3-kinase* (PI3K) to the T cell membrane. PI3K is a heterodimer composed of a p110 catalytic subunit and a p85 regulatory subunit. It is the p85 subunit that is thought to bind to LAT and perhaps other adaptor proteins. By acting primarily as a lipid kinase, PI3K serves as a multi-functional regulator of TCR signaling. First, PI3K phosphorylates PIP$_2$ and converts it to phosphatidyl inositol 3,4,5-triphosphate (PIP$_3$), decreasing the production of the second messengers DAG and IP$_3$. Thus, PI3K negatively regulates PLC-γ1 activity. However, PIP$_3$ produced by PI3K can bind to the PH (pleckstrin homology) domain of Vav1 and recruit it to the membrane rafts, promoting TCR signal transduction. PI3K also stimulates the activity of *protein kinase B* (PKB; also called Akt), an enzyme crucial for T cell survival. Activation of PKB/Akt leads to the inhibition of *Foxo* transcription proteins, whose normal function is to drive the expression of cell death genes.

PI3K activates two other pathways that contribute signals favoring T cell activation: the SAPK/JNK pathway and the p38 signaling pathway (refer to Fig. 14-6). (SAPK stands for *stress-activated protein kinase* while JNK stands for *c-jun N-terminal kinase*.) The SAPK/JNK signaling pathway is activated in cells subjected to extracellular stresses such as osmotic shock or treatment with chemical agents, situations that mimic physiological stresses. In such situations, PI3K (and perhaps other enzymes) phosphorylate and activate a cascade of kinases (including MEKK2 and MKK4/7) that act on each other. The activation of the final kinase in the cascade, SAPK/JNK, leads to the activation of the c-jun transcription factor. Similarly, PI3K can stimulate the phosphorylation of the kinase MKK6, which in turn phosphorylates *p38 kinase*. Activation of this enzyme is responsible for activating members of the ATF transcription factor family. c-jun and ATF can combine to drive the transcription of many pro-inflammatory genes, including TNF and E-selectin.

It should be noted that, despite all the details on the effects of PI3K cited above, how this enzyme is actually activated is still not clear. That is, the identity of the receptor that must be engaged to activate PI3K has yet to be definitively proven. Many scientists believe that, at least in T cells, it is CD28 that first recruits and activates PI3K, and that the interaction of PI3K and LAT described previously is indirect.

vi) The Ras/MAPK Pathway

The activation of the Ras/MAPK pathway introduced in Chapter 13 is critical for T cell development and function because it leads to activation of the important signaling kinase ERK (introduced in Ch.13). Ras is a small, monomeric signal transducing protein constitutively associated with the cytoplasmic side of the cell membrane (refer to Fig. 14-6). It is an example of a signal transducer that binds to guanine nucleotides such that its activity is in turn controlled by the nucleotide's phosphorylation status. Proteins known as *nucleotide exchange factors* (GEFs) are responsible for the phosphorylation of such nucleotides. Thus, Ras-GTP is active whereas Ras-GDP is not, and a GEF is required for the conversion of Ras-GDP to Ras-GTP. Conversely, *GTPase activating proteins* (GAPs) stimulate the intrinsic GTPase activity of Ras, which then converts GTP back to GDP as it activates other components of the signaling cascade. In some cell types, Sos (a GEF described in Box 14-2) is responsible for Ras-GDP activation. Early studies provided evidence that Sos mediated Ras activation in T cells as well, because activated Lck phosphorylates not only ZAP-70 but

also the adaptor protein *SHC*. ZAP-70-mediated phosphorylation of LAT results in the recruitment of Grb2 and thus Sos, and phosphorylated SHC can associate with Grb2-Sos. However, work in the early 2000s by James Stone and colleagues showed that the principal route by which Ras is activated in T lineage cells occurs via a GEF called Ras-GRP1 (<u>Ras</u> <u>g</u>uanyl nucleotide <u>r</u>eleasing <u>p</u>rotein-1) (Fig. 14-8). Ras-GRP1 contains domains that bind to DAG and calcium, products of PLC-γ1 activation. Under the influence of PKCθ and DAG, Ras-GRP1 is recruited from the cytoplasm to the membrane, where it interacts with Ras-GDP and triggers its conversion to Ras-GTP. Proof that Ras-GRP is an essential activator of Ras signaling in T lineage cells stems from studies of knockout mice lacking Ras-GRP1. These mutant animals show defects in thymocyte proliferation and differentiation due to decreased ERK activation (which is required for positive selection). Similarly, mature T cells in which Ras-GRP1 is inactivated cannot produce IL-2 or proliferate following stimulation with agonistic anti-CD3 and anti-CD28 antibodies.

Regardless of its route of activation, in almost all cell types, activated Ras-GTP triggers a cascade of serine/threonine MAPK kinases, including Raf, MEK (*MAPK ERK kinase*), and ERK. This cascade results in the phosphorylation and activation of several nuclear transcription factors, including c-myc and c-fos (refer to Fig. 14-6). As is described in more detail in the following section, activation of these transcription factors (among

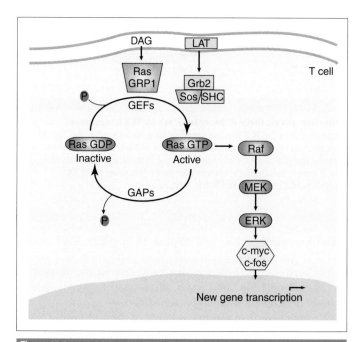

Figure 14-8

The Ras/MAPK Pathway
Following engagement of the TCR by pMHC, PLC-γ1 is activated and hydrolyzes PIP$_2$ into DAG. DAG recruits Ras-GRP1, a guanine nucleotide exchange factor (GEF), which facilitates the conversion of inactive Ras-GDP to active Ras-GTP. This conversion can also be mediated by the GEF Sos. Ras-GTP in turn initiates a cascade of kinase activation that leads to activation of c-myc and c-fos and new gene transcription. The Ras/MAPK pathway is downregulated by the dephosphorylation of Ras-GTP mediated by GAPs (GTPase-activating proteins).

others) leads to the initiation of new gene transcription. To place limits on this transcriptional activation, the Ras/MAPK pathway is subject to regulation by the dephosphorylating activity of the phosphatase *Pac-1*, an enzyme whose expression is rapidly upregulated during T cell activation.

A positive feedback loop involving the Ras/MAPK pathway, Lck, and SHP-1 has been proposed by Ronald Germain and colleagues as the driving force behind the sustained signaling required to complete T cell activation. In this model, pMHC–TCR binding induces the phosphorylation and activation of Lck, which recruits and phosphorylates not only CD3ζ and ZAP-70 but also SHP-1. Phosphorylated SHP-1 then acts to dephosphorylate Lck (and possibly other proteins in the TCR complex), exerting negative feedback inhibition on the TCR signaling cascade. However, if the pMHC that binds to the TCR does so with sufficient avidity, the Ras/MAPK pathway is triggered, leading to the activation of ERK. ERK phosphorylates a particular serine residue in Lck that alters the accessibility of its SH2 domain and thus bars the recruitment of phosphorylated SHP-1 to the TCR complex. As a result, Lck is not dephosphorylated and the TCR signaling cascade continues long enough to complete T cell activation. This model has been proposed to account for the differences in signaling outcomes observed when a T cell is treated with an agonist peptide (activation) as opposed to an antagonist peptide (anergy). Indeed, T cells that encountered APCs bearing an agonist peptide showed a dramatic increase in ERK activity within 60 seconds of contact, whereas ERK activation was only weak and transient in T cells interacting with APCs bearing antagonist peptides. In addition, antagonist peptides induced faster recruitment of SHP-1 to Lck than did agonist peptides, accelerating the shutdown of TCR signaling. This model also accounts for the partial dephosphorylation of CD3ζ and ZAP-70 observed when ERK activity is blocked by an exogenous inhibitor (phosphorylated SHP-1 successfully binds to Lck and is then in a position to dephosphorylate other proteins in the signalsome).

vii) Role of Coreceptors

There is ongoing debate as to the precise roles of the coreceptors CD4 and CD8 in TCR signaling. It is clear that the binding of TCRs to pMHC complexes promotes the recruitment of the coreceptors into the growing SMAC. While the coreceptors definitely contribute an adhesion function, is their primary role to bring their associated cargo of Lck to the TCR signaling complex? Or does the binding of a coreceptor alter some structural aspect of the complex that favors signal transduction? What effect does coreceptor binding to pMHC have on cell fate? The answers to these questions appear to depend on the context. At the biochemical level, Ian Wilson and colleagues have shown that CD8 participates actively during TCR recognition of pMHC, reducing the "off" rate for TCR/pMHC binding. They speculated that CD8 either might be helping the TCR to bind efficiently to pMHC, or was perhaps inducing conformational changes to the MHC molecule that facilitated TCR/pMHC binding. In keeping with an important role for CD8 in T cell activation, the T cells of CD8α$^{-/-}$ mice are completely unable to recognize

MHC class I and the animals mount no CTL responses. However, studies of CD4$^{-/-}$ mice have shown that CD4 is not necessary for all cases of MHC class II recognition, and that many Th-dependent immune responses can proceed in the absence of this coreceptor. In addition, as noted in Chapter 13, not all TCR signaling depends on the function of coreceptors. Indeed, the whole multi-partite DN stage of thymocyte development takes place in the absence of coreceptor expression. In CD4$^-$CD8$^-$ cells, it is thought that the Lck molecules necessary for proximal TCR signaling may associate directly with ZAP-70 via SH2 domain interactions.

Studies of T cell lines have also identified cases of TCR signaling that exhibit coreceptor independence. It has been shown *in vitro* that some T cell clones are still able to respond to antigen even in the presence of anti-CD4 or anti-CD8 antibodies. This property was found to be intrinsic to the individual TCR. Coreceptor-independent TCRs expressed as transgenes in T cell clones lacking the appropriate coreceptor could be still activated by antigen, whereas TCRs isolated from coreceptor-dependent clones could not. It was originally believed that coreceptor independence was a function of the affinity of a TCR for its pMHC ligand: a TCR with very low binding affinity would rely on coreceptors to support the interaction, while a TCR with higher binding affinity would not need to. However, it was discovered that repeated stimulation of a coreceptor-dependent T cell clone *in vitro* could slowly convert it to coreceptor independence, even though the TCR itself (and its affinity for pMHC) had not changed. It is now thought that subtle differences in signal transduction may exist in clones that carry TCRs that are similar in affinity, but differ in their requirement for coreceptor involvement. Comparison of intracellular tyrosine phosphorylation patterns in coreceptor-independent and coreceptor-dependent clones has suggested that a stronger and longer signal is delivered by coreceptor-independent TCRs in response to antigen or anti-CD3 binding. These results imply that TCRs of similar affinity may exhibit differences in sensitivity to antigenic stimulation that dictate whether coreceptor involvement is required. Studies to determine whether the structure of a given TCR protein might underlie the need for coreceptor binding are in progress.

viii) Rationale for T Cell Signal Transduction Complexity

Why have such a complicated signaling system—interactions between ITAMs and PTKs, coreceptors and PTKs, PTKs and SH2-containing effectors, and so on? With a weapon as powerful as the adaptive immune response, the organism has to be sure of employing effector T cells only where they are needed and for only as long as necessary. An overall threshold of T cell activation exists, meaning that all the positive and negative regulatory signals received by the cell are balanced against each other to determine if activation will occur. Some scientists have proposed that the function of the multi-layered activation cascade is to create a lag between the binding of antigen to the receptor and the delivery of the signal. It is likely that evolution has devised a protocol that allows the cell the time to discriminate between ligands of varying affinities. Those ligands of lesser affinity may not stay bound to the TCR long enough for the entire signaling complex to be assembled, meaning that only a partial or perhaps abortive signal is delivered. Such an explanation could account for the striking differences in ZAP-70 and CD3ζ phosphorylation patterns observed when T cells are stimulated with weak versus strong agonists. Indeed, those peptides that induce T cells to become anergic fail to induce the activation of the Ras/MAPK pathway and/or pathways downstream of CD28 signaling (see Ch.16). The fine balancing act between too much and not enough binding to the TCR is amplified by the elements of the signal transduction cascade, making it easier to understand how a variety of dif-

ferent cellular outcomes can be dictated by the initial act of antigen engaging a single receptor.

C. Signal Two: Costimulation

Just as efficient activation of a B cell requires the engagement of the costimulatory molecule CD40 by CD40L as well as the binding of antigen to the mIg, the complete activation of T cells requires costimulatory contacts in addition to the

Table 14-2 **Molecules Affecting T Cell Costimulation**

Receptor	Receptor Present on	Receptor Expression	Ligand	Ligand Present on	Functions
CD28	T cells	Upregulated	B7-1, B7-2	B cells, DCs, monocytes, macrophages	IL-2 and IL-2R transcription, cell survival, proliferation, blocks anergy, prevents T cell apoptosis
CTLA-4	T cells	Segregated*	B7-1, B7-2	B cells, DCs, monocytes, macrophages	Negative regulator of T cell activation due to dephosphorylation of signaling molecules, downregulates IL-2 transcription, blocks T cell proliferation
ICOS	T cells	Inducible	ICOSL	B cells, DCs, monocytes, macrophages	Germinal center formation, isotype switching, IL-4 and IL-13 transcription, Th2 responses, prevents T cell apoptosis
PD-1	Activated T and B cells, myeloid cells	Inducible	PDL1 PDL2	Non-lymphoid cells; inducible on DCs and macrophages	Negative regulator of T cell activation due to dephosphorylation of signaling molecules, downregulates IL-2 transcription, blocks T cell proliferation
???	Activated T cells?	Inducible?	B7-H3	Induced on APCs	Promotes T cell proliferation, IFNγ production, regulates Th1 responses
CD40L	T cells NK cells	Upregulated	CD40	B cells, DCs, monocytes	Promotes APC activation, primary T cell responses, germinal center formation, isotype switching
4-1BB (CD137)	T cells	Upregulated	4-1BBL	B cells, macrophages	Promotes CD8+ T cell proliferation and IFNγ production
OX40	T cells B cells	Inducible	OX40L	Activated T and B cells, DCs, endothelial cells	Promotes adhesion of activated T cells to endothelium, T cell proliferation, cytokine production, B cell antibody secretion
CD27	T cells B cells NK cells	Transiently upregulated	CD70	Activated T and B cells	Enhances T cell proliferation (but not effector cell generation), required for IgG synthesis and generation of T cell memory
LFA-1	T cells	Constitutive	ICAM-1, 2, 3	APCs, activated T and B cells	Mild costimulatory effect on T cell proliferation, does not activate NF-κB, induce IL-2 transcription or prevent anergy
CD2	T cells	Constitutive	LFA-3 (Hu) CD48 (Mu)	Most hematopoietic cells, APCs	Mild costimulatory effect on T cell proliferation, does not activate NF-κB, induce IL-2 transcription or prevent anergy
CD47 (IAP)	T cells, epithelial and endothelial cells, fibroblasts	Upregulated	?	APCs?	Mild costimulatory effect on T cell proliferation

* "Segregated" in this context means that pre-formed CTLA-4 is held in vesicles inside the T cell until activation is under way.

binding of antigen to the TCR. Some of these costimulatory molecules are expressed constitutively while others are upregulated or inducible (Table 14-2). We will also discuss two negative regulators, CLTA-4 and PD-1, that counter costimulatory effects. These molecules ensure that, once the threshold of T cell activation is exceeded, the signaling cascade does not get out of hand.

I. CD28–B7

The most important costimulatory pairing identified to date is that of CD28 on the Th cell with B7 ligands on the APC. Signaling through CD28 sustains the survival of T cells whose TCRs are engaged by pMHC and contributes to proliferation and production of IL-2. In the absence of CD28 costimulation, binding of pMHC to the TCR often induces apoptosis or anergy (see Ch.16).

Controversy exists over whether CD28 plays as large a role in delivering signal two to Tc cells as it does for Th cells. It has been shown that neither CD28 signaling nor a Th cell is required for either primary or secondary CTL responses to certain aggressive viruses, such as LCMV. Infection with LCMV results in the strong, persistent presentation of antigenic peptides on MHC class I molecules of an infected target cell, leading to highly sustained signaling through the TCRs of an engaged naive Tc cell. In this situation of very high TCR occupancy, the Tc cell responding to LCMV may not even require costimulation. However, it is also possible that the IL-2 (or other cytokines) required can be induced by means other than CD28 signaling. For example, stimulation of Vav1 can promote autocrine IL-2 synthesis by a CD28-independent pathway. It is now clear that both Vav1 and CD28 involvement are required for primary Tc responses to weaker viruses. However, while CD28-deficient mice are still able to mount primary Tc responses against LCMV,

Vav1-deficient mice are not. This result suggests that, at least in the case of LCMV infection, signal two leading to the promotion of IL-2 synthesis can be delivered by other means (e.g., via Vav1) rather than by CD28.

We will now discuss the CD28–B7 costimulatory pair and several aspects of the intracellular signaling induced by their interaction.

i) CD28

CD28 is a glycoprotein of 90 kDa and a member of the Ig superfamily. It occurs naturally as a disulfide-linked homodimer of a polypeptide containing a single Ig-like extracellular domain. CD28 is constitutively expressed on both resting and activated T cells, on almost all CD4$^+$ and on at least half of CD8$^+$ T cells in humans, and on 100% of CD8$^+$ cells in mice. Upon the binding of pMHC to the TCR, CD28 expression is sharply upregulated on the T cell, further increasing its interaction with its ligands. When signaling through CD28 is present (in conjunction with TCR signaling), cells are pushed from a resting state to enter the cell cycle, and chromatin remodeling and alterations to DNA methylation are triggered. The expression of the high affinity IL-2 receptor is readily induced, as are the transcription and secretion of IL-2. CD28 costimulation of Th cells also enhances the production of IFNγ, IL-1, TNF, IL-4, IL-5, various chemokines, and receptors for these molecules. Finally, CD28 induces expression or upregulation of several other costimulatory and regulatory molecules, including ICOS, 4-1BB, and CTLA-4 (all defined below), and the CD40L molecules necessary for T–B cell interaction (Table 14-3).

ii) B7-1 and B7-2

CD28 binds with moderate affinity to two cell surface ligands called B7-1 (CD80) and B7-2 (CD86). While both B7-1 and B7-2 bind to CD28 and costimulate T cell proliferation and IL-2 production, they are dissimilar in expression, affinity for

---

Table 14-3 Costimulatory Effects of CD28 Signaling

Activates	Increases	Decreases
• IL-2 and IL-2R transcription and T cell proliferation • IL-4 production and Th2 differentiation • IL-5, IL-10, IFN-γ, TNF, IL-1, IL-6, MIP-1α production • SAPK/JNK signaling leading to c-jun production	• Chromatin remodeling and demethylation • Ras/MAPK signaling and c-fos production • Vav1 hyperphosphorylation and cytoskeletal movement • Coalescence of rafts with TCR and formation of SMAC • Stability of IL-2 mRNA • Bcl-xL expression and cell survival • Expression of CD40L, ICOS, OX40, 4–1BB, IL-12R, CXCR5 • Phosphorylation of CD3 ITAMs and thus duration of TCR signal transduction • NF-κB activation via PKCθ • AP-1 formation • PI3K signaling	• Anergization • Threshold of TCR triggering required for T cell activation • IκB expression • GSK3 activity (NFAT retained in the nucleus)

CD28, and kinetics of action. B7-1 is a 50–60 kDa glycoprotein that contains extracellular Ig-like V and C domains, a transmembrane region, and a short cytoplasmic tail. While B7-2 is similar in overall structure to B7-1, it is only 25% homologous to it at the amino acid level. B7-2 also binds to different determinants than B7-1 on both CD28 and the negative regulator CTLA-4. X-ray crystallography studies have determined that, in their natural state, two B7 molecules come together to form a homodimer, half of which binds to half of a homodimer of CD28 or CTLA-4.

Both B7 ligands are dramatically upregulated on APCs following interaction with T cells but with different kinetics. In response to TCR engagement, B7-2 expression is rapidly upregulated on APCs and its expression can be sustained for up to 96 hours. In contrast, B7-1 does not appear on APCs until 48 hours after activation and declines rapidly after 72 hours. The expression of both B7-1 and B7-2 on B cells is enhanced by the interaction of CD40 on the B cell with CD40L on the Th cell (thereby increasing costimulation of the Th cell). Similarly, B7-1 expression on DCs is upregulated in response to either CD40 engagement or the presence of IFNγ or GM-CSF.

Studies using transfected cells transiently expressing B7-1 or B7-2 and monoclonal antibodies directed against them have shown that B7-1 and B7-2 have at least partially redundant functions. Both molecules are equivalent in costimulating T cell proliferation and the production of IL-2 and IFNγ, and in upregulating the expression of IL-2Rα during the activation of both naive and memory T cells. However, engagement of B7-1 and B7-2 may differentially regulate the production of other cytokines important for the differentiation of the Th1 and Th2 subsets. For example, the induction of IL-4, which drives the development of Th2 effector cells (see Ch.15), is increased more in both naive and memory T cells by engagement of B7-2 as compared to B7-1. Production of lymphotoxin α (LTα), which has effects similar to TNF (see Ch.17), is also increased by B7-2 engagement, while B7-1 engagement favors GM-CSF secretion. Since no differences have been detected in the CD28 signaling pathway that results from engagement by either B7-1 or B7-2, the mechanisms underlying these dissimilarities remain to be identified. It is possible that differences in the strength or timing of the stimulation of CD28 or CTLA-4 may subtly affect the results of downstream signaling.

Studies of knockout mice have shown that B7-mediated signaling is crucial for germinal center formation and isotype switching. B7-2$^{-/-}$ mice showed normal expression of B7-1 and normal antigen-specific IgM production but impaired isotype switching and germinal center formation. B7-1$^{-/-}$ mice have normal germinal centers but impaired humoral responses to antigens. In double knockout B7-1$^{-/-}$; B7-2$^{-/-}$ mice, T cell help was critically decreased such that the entire humoral response was compromised.

iii) CD28 Signaling

The engagement of CD28 has been said to lower the T cell activation threshold, decreasing the number of TCRs that have to be aggregated in the SMAC and internalized (i.e., the number of TCRs encountering specific pMHC) before a proliferative signal can result. For example, when APCs expressing costimulatory molecules were used to stimulate CD4$^+$ T cells in vitro, only 1000–1500 TCRs had to be engaged to achieve activation. In contrast, when the APCs were engineered to lack all costimulatory molecules, it was found that a minimum of 8000 TCRs on a CD4$^+$ T cell had to be engaged by pMHC before the cell could proliferate and differentiate into effectors secreting IFNγ and IL-2. Furthermore, essentially the same result was achieved whether the antigen used was a bacterial superantigen or a monovalent anti-CD3 antibody, indicating that the chemical nature of the activating ligand was irrelevant for TCR triggering. Another way to look at it is to consider the time required for T cell commitment to activation. In the presence of costimulation, a T cell becomes committed to activation after 6 hours of TCR triggering; in the absence of costimulation, 30 hours are required.

How does CD28 signaling reduce the activation threshold? Engagement of CD28 promotes the initial phosphorylation events of signal transduction from the TCR, including the phosphorylation of CD3ζ. At least in vitro, the phosphorylation of downstream signaling molecules is only transient in the absence of CD28. Some researchers believe that CD28 plays a critical role in the coalescence of the TCR and membrane rafts at the T cell–APC interface. Because the TCR signaling complex and its associated signal transducers become concentrated in the rafts, the phosphorylated substrates within the rafts (products of the initial phosphorylation events) are protected from attack by cellular phosphatases. The initial stages of signal transduction may therefore be stabilized. In support of this hypothesis, Mark Davis, Antonio Lanzavecchia, and their respective colleagues have shown that, when rafts of T cells stimulated with plate-bound anti-CD3 antibody were artificially cross-linked in vitro using antibodies to non-signaling components within the raft, the proliferative response achieved was equivalent to that resulting from CD28 engagement. This result suggests that CD28 costimulation is at least partially mediated via the clustering of the rafts and their signaling molecules at the TCR contact site.

At the biochemical level, it has been shown that B7 engagement triggers tyrosine phosphorylation of the 41 amino acid cytoplasmic domain of CD28, possibly by Lck (and/or Fyn). This event results in the recruitment of PI3K to the cytoplasmic tail of CD28 and the activation of this kinase (Fig. 14-9). As we have seen, activated PI3K generates PIP$_3$ and other phospholipid mediators. CD28 engagement also promotes (in an unknown way) the hyperphosphorylation of Vav1; that is, more phosphate residues are added to Vav1 molecules than is observed if the TCR alone is engaged. Rac-mediated reorganization of the actin cytoskeleton results such that membrane rafts readily coalesce around the TCR signaling complex. The rafts contain, among other signaling proteins, elements of the Ras/MAPK and SAPK/JNK pathways (refer to Fig. 14-6). The Ras/MAPK and SAPK/JNK pathways lead to the activation (by phosphorylation) of

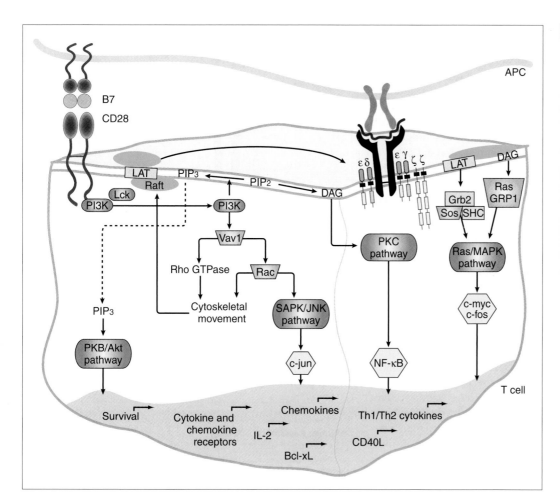

Figure 14-9

CD28 Signal Transduction
Engagement of CD28 by B7 promotes the recruitment of PI3K to the T cell membrane and phosphorylation of Vav1 over and above that induced by TCR engagement alone. Signaling downstream of Vav1 is increased, promoting the cytoskeletal movement and clustering of membrane rafts and their associated signaling molecules around the engaged TCR. In the presence of CD28, signals are transduced via the PKB/Akt, SAPK/JNK, PKC, and Ras/MAPK pathways that lead to complete T cell activation.

nuclear transcription factors, particularly c-fos and c-jun, respectively. These proteins combine to form the AP-1 transcription factor. In particular, AP-1 binds to appropriate motifs in the IL-2 promoter (see later), helping to induce the expression of the IL-2 gene and the stabilization of IL-2 mRNA. AP-1 formation also increases the expression of the anti-apoptotic gene Bcl-xL. In addition, CD28 engagement is thought to contribute to reduced IκB expression and thus the removal of the key negative regulator of NF-κB activation. Signaling via the NF-κB pathway contributes to the expression of numerous cytokines, including IL-2 and IL-6. The expression levels of other cytokines, such as IL-4, IL-6, and IFNγ; chemokines such as MIP-1α; and cytokine/chemokine receptors such as IL-2R, IL-12R, and CXCR5 have also been shown to increase in T cells following CD28 signaling. Finally, CD28 engagement promotes the expression of additional costimulatory molecules, including CD40L, ICOS, OX40, and 4-1BB (all described later). The binding of the appropriate ligands to these molecules makes a complementary contribution to signaling initiated by TCR and CD28 engagement, constituting a "second wave" of signaling that sustains the activation cascade long enough for normal proliferation, effector differentiation, and memory cell generation.

Analyses of CD28$^{-/-}$ mice have revealed that CD28 signaling may not only lower the threshold of TCR triggering but may also act to sustain signaling emanating from the response. Initial studies of CD28$^{-/-}$ mutant mice showed that many T cell responses were impaired in the absence of CD28, but, somewhat surprisingly, not all. While CD28$^{-/-}$ mice could mount a protective antiviral response against certain virulent viruses, they were susceptible to less virulent strains. This observation led to the speculation that the attack by the virulent virus resulted in abnormally persistent antigenic stimulation, which might have stimulated the T cell enough such that it no longer needed CD28 costimulation to become fully activated. In the case of a less virulent virus (less persistent antigenic stimulation), CD28 might have been required to supply signaling to sustain the response. Indeed, if the CD28$^{-/-}$ mice were repeatedly injected with peptides from the less virulent virus (such that antigenic stimulation became persistent), the mice were able to resist the attack. Consistent with this hypothesis, CD28$^{-/-}$ T cells were able to initiate, but not sustain, proliferation *in vitro*. Taken together, these findings indicate that a primary function of CD28 signaling is to sustain intracellular signaling long enough for events leading to cellular proliferation to occur.

The hypothesis that CD28 signaling primarily amplifies the signaling triggered by TCR engagement rather than by inducing unique gene expression is supported by analyses of genomic gene expression. Using a tiny glass chip upon which the genomic DNA of hundreds of known genes was fixed in a "microarray," mRNAs were isolated and quantitated from T cells that had been stimulated either via TCR engagement alone or via TCR engagement plus CD28 engagement. It was found that almost all the same genes were activated in both cases, but that the level of expression of genes induced in response to TCR engagement was increased if CD28 was also engaged. CD28 signaling also inhibited the GSK3 enzyme that promotes NFAT export out of the nucleus, sustaining NFAT-mediated signaling. Looking at the DNA itself, CD28 signaling supports the chromatin remodeling and demethylation that underlies changes to gene expression required for proliferation and differentiation.

iv) Control of TCR and CD28 Signaling by CTLA-4

While the binding of the TCR to pMHC and CD28 to the B7 ligands promotes T cell activation, the binding to the B7 ligands of a negative regulator called *CTLA-4* downregulates T cell activation. CTLA-4 stands for *cytotoxic T lymphocyte-associated molecule 4*, although CTLA-4 has been identified on both Th and Tc cells. CTLA-4 (CD152) is a disulfide-linked homodimeric glycoprotein evolutionarily related to CD28 (sharing about 75% sequence homology) and therefore closely resembling it in structure. However, CTLA-4 binds to the B7 ligands with much higher affinity (10- to 20-fold greater) than does CD28. Unlike CD28, which is constitutively expressed on most T cells, CTLA-4 is expressed in significant amounts on the T cell surface only after the TCR is triggered and the activation cascade is underway. It is thought that the increased affinity of CTLA-4 for the B7 ligands causes it to displace CD28, disrupting costimulatory signaling and, if the disruption is severe enough, blocking cell division at the G1 to S transition. In addition, CTLA-4 can recruit the tyrosine phosphatase *SHP-2* to the T cell membrane. SHP-2 (also known as SYP and PTP-1D) binds to an ITIM motif (refer to Ch.9) in the cytoplasmic tail of CTLA-4 and is thought to reverse T cell activation events by dephosphorylating CD3ζ and possibly SHC, Lck, Fyn, ZAP-70, and/or some of their effector substrates, including Vav1 (Fig. 14-10). An alternative hypothesis is that CTLA-4 is essential for the maintenance or function of certain subsets of regulatory T cells that control immune responses (see Ch.16). In any case, CTLA-4 is essential for maintaining the homeostasis of the T lymphocyte compartment, helping to "deactivate" effector cells and ultimately control their numbers.

The expression of CTLA-4 is tightly controlled, as one might expect. To prevent CTLA-4 from acting too early (such that CD28 would not be able to bind to the B7 molecules and deliver signal two), the majority of CTLA-4 molecules in a resting T cell are sequestered in a subcellular Golgi compartment. Within 2 hours of TCR triggering, CTLA-4 molecules become reoriented toward the T cell–APC interface and continue to accumulate within the vesicle until at least 6 hours after TCR triggering. Only 48–72 hours after TCR triggering do CTLA-4 molecules move out of this compartment and appear on the T cell surface, ready to bind B7 ligands and damp down T cell activation.

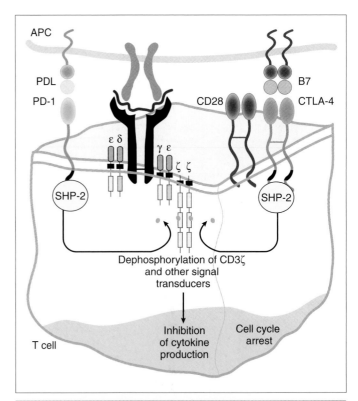

Figure 14-10

Negative Regulatory Signaling
PD-1 (discussed later in the text) and CTLA-4 are negative regulators of TCR signaling. When PD-1 is engaged by its ligand PDL, or CTLA-4 binds to B7 and displaces CD28, the phosphatase SHP-2 is recruited to the ITIM motifs in the cytoplasmic tails of PD-1 and CTLA-4. SHP-2 dephosphorylates the CD3ζ chains and other signaling molecules, halting TCR signaling and subsequent events associated with T cell activation.

The importance of CTLA-4 function is dramatically illustrated by the phenotype of CTLA-4$^{-/-}$ knockout mice. These animals die at 3–4 weeks of age due to a disruption of peripheral lymphocyte homeostasis that leads to massive infiltration of activated T cells into the vital organs. The TCR signaling pathway appears to be constitutively activated in CTLA-4$^{-/-}$ T cells, since CD3ζ, Fyn, Lck, and ZAP-70 are all hyperphosphorylated.

II. ICOS–ICOSL

The phenotype of the CD28$^{-/-}$ mouse was unexpectedly mild for such an important costimulatory molecule. The mutants had no defects in thymic maturation or positive and negative selection of αβ T cells, or in γδ T cell development. These results suggested that other proteins might be able to compensate for the absence of CD28 for some functions. A third member of the CD28 family called *ICOS*, for *inducible costimulator*, was identified in the late 1990s. This homodimeric protein of 55–60 kDa was isolated through its interactions with a panel of monoclonal antibodies that had been screened for exclusive binding to activated T cells (both Th and Tc). ICOS is structurally similar to CD28 and shares some important functional similarities with it. Like CD28,

Table 14-4 Comparison of CD28 and ICOS Costimulation

	CD28	ICOS
Present on	Resting and activated T cells	Only activated T cells
Expression	Upregulated	Inducible
Ligands	B7-1, B7-2	ICOSL
Stabilizes mRNA for	IL-2	CD40L
Critical for production of	IL-2	IL-10, IL-4
Signaling required for	Activation of naive T cells, IL-2 production, T cell proliferation	Germinal center formation, isotype switching, Th2 effector cell differentiation

stimulation of ICOS augments the upregulation of adhesion molecules and CD40L, and promotes T cell proliferation and cytokine secretion (i.e., the provision of T cell help to B cells).

However, as summarized in Table 14-4, ICOS and CD28 exhibit striking differences: (1) Unlike CD28, which is constitutively expressed on T cell surfaces, ICOS is synthesized on the cell surface only during T cell activation induced by foreign antigen and CD28/B7 costimulation. (2) Rather than the promotion of IL-2 production (a prime outcome of CD28 costimulation), ICOS instead induces the production of high levels of IL-10, a cytokine that promotes B cell differentiation. Moreover, ICOS stabilizes the mRNA for CD40L (parallel to the stabilization of IL-2 mRNA by CD28), promoting expression of an important costimulatory ligand for B cells. As we have seen, CD40 signaling is also essential for isotype switching. ICOS is highly expressed *in vivo* on T cells closely associated with B cells in the apical light zone of the germinal centers, where terminal B cell differentiation into plasma cells takes place. ICOS$^{-/-}$ mice show defects in germinal center formation, isotype switching, and T cell-dependent antibody responses. ICOS$^{-/-}$ effector T cells can produce IFNγ but not IL-4. It is thought that the ICOS costimulatory pathway is necessary to promote the production of IL-4 necessary for complete Th2 differentiation. (3) The sequence motif present in both CD28 and CTLA-4 that facilitates binding to the B7 ligands is not present in ICOS, suggesting that, in spite of its structural similarities, ICOS binds to a different ligand. The late 1990s saw the identification of *ICOS ligand* (ICOSL), a molecule that is constitutively expressed on murine APCs and B cells but whose expression is upregulated by inflammatory signals. ICOSL is very similar in structure and sequence to B7-1 and B7-2 but binds only to ICOS and not to CD28 or CTLA-4. Mice deficient for ICOSL show a phenotype identical to that of ICOS$^{-/-}$ mice.

III. PD-1–PDL1/PDL2

PD-1 (*programmed death-1*) is a negative regulatory molecule that is expressed on activated (but not resting) T cells, B cells, and myeloid cells. The name of this protein is derived from the

screening circumstances under which it was identified, and is a bit of a misnomer. Rather than being associated with cell death, PD-1 is associated with the reining-in of cellular activation. The cytoplasmic tail of the monomeric PD-1 protein contains two tyrosine residues that can be phosphorylated, one of which resides in an ITIM (refer to Fig. 14-10). PD-1 binds to two ligands with homology to the B7 molecules, PDL1 (*PD ligand 1*) and PDL2. In contrast to the B7 molecules, however, PDL1 and PDL2 are expressed in both lymphoid and non-lymphoid tissues and bind only to PD-1 and not to members of the CD28 family. Expression of PDL1 and PDL2 is upregulated on macrophages and DCs in response to IFNγ treatment. When both the TCR and PD-1 are engaged, the activation and phosphorylation of SHP-2 recruited to the PD-1 tail results. As described previously for CTLA-4, multiple signaling molecules are then inactivated by SHP-2-mediated dephosphorylation, cytokine mRNA synthesis is downregulated, and T cell proliferation is blocked. Mice deficient for PD-1 show splenomegaly and increased numbers of B cells and myeloid cells, but reduced numbers of mature T cells. Some strains of PD-1$^{-/-}$ mice develop autoimmunity. Experiments in transgenic mouse models have indicated that autoreactive cells that escaped negative selection in the thymus may be kept under control in the periphery partly through the action of PD-1.

IV. ???–B7-H3

B7-H3 is another member of the B7 family. This molecule is not expressed on resting APCs but can be induced on macrophages and DCs by exposure to GM-CSF or IFNγ. Exposure to IL-4 suppresses B7-H3 expression on DCs. Sequence analysis indicates that the most common form of B7-H3 in humans consists of two IgV-like domains alternating with two IgC-like domains. In mice, the protein product has been predicted to have one IgV-like domain and one IgC-like domain. The ligand of B7-H3 is currently unknown but appears to be expressed on activated (but not resting) T cells. B7-H3 does not bind to CD28, CTLA-4, ICOS, or PD-1. The function of B7-H3 was not properly revealed until B7-H3$^{-/-}$ mice were created. The loss of B7-H3 in these mutant animals enhanced their Th1 responses to the appropriate stimuli but had no effect on their Th2 responses. Thus, B7-H3 is a negative regulator that preferentially controls Th1 responses.

V. TNF/TNFR-RELATED COSTIMULATORY MOLECULES

Several molecules that are members of the TNF or TNF receptor (TNFR) superfamilies have costimulatory effects on T cells (refer to Table 14-2). Engagement of TNFR-related costimulatory receptors by membrane-bound ligands with homology to TNF initiates a signaling cascade through the TRAF (TNFR-associated factor) family of molecules that leads to NF-κB activation and cell survival. We have previously discussed the TNF superfamily member CD40 as a costimulator of B cells. From the perspective of T cell activation, the engagement of the TNFR family member CD40L on the T cell by CD40 on an APC not only promotes activation of the APC but also prepares the naive T cell for activation. Two other TNFR-related co-stimulators are *4-1BB* (also known as CD137) and *OX40*. These molecules are induced or upregulated on activated lymphocytes. Their TNF-related ligands are expressed either constitutively or inducibly on APCs. Engagement of 4-1BB or OX40 by their respective ligands 4-1BBL and OX40L enhances T cell proliferation and cytokine production.

VI. CD27/CD70

CD27, a transmembrane homodimeric phosphoglycoprotein of 120 kDa, also appears to have a costimulatory role. CD27 is expressed on the majority of both CD4$^+$ and CD8$^+$ resting T cells in the peripheral blood. Upon activation by antigen, expression of CD27 increases. The ligand for CD27, CD70, is a transmembrane glycoprotein expressed on T and B cells in response to antigen stimulation; it is thus considered a marker of the early stages of activation. *In vitro*, the interaction of CD27 on a T cell and CD70 on a B cell enhances T cell activation in terms of proliferation but only relatively low amounts of IL-2 are secreted. Studies of knockout mice have shown that CD27 plays a minor part in naive T cell activation but is crucial for the generation of T cell memory, a topic addressed in Chapter 15.

VII. OTHER MINOR COSTIMULATORY CONTACTS

In vitro experiments using knockout and transgenic mice have suggested that other molecular pairs may act as minor co-stimulators contributing to T cell activation *in vivo*. In addition to their role in adhesion, LFA-1 and CD2 are considered to have minor costimulatory roles in cell-mediated responses because treatment with agonistic antibodies recognizing these molecules (in conjunction with TCR stimulation) enhances T cell proliferation and/or cytokine production. In addition, *CD47*, also called IAP for *integrin associated protein*, has been reported to stimulate T cell proliferation following TCR engagement *in vitro*. CD47 is expressed on most hematopoietic cells as well as endothelial and epithelial cells and fibroblasts. There is some experimental evidence that CD47 mediates Ca^{2+} flux in response to the binding of adhesion molecules to their ligands. It remains to be determined how relevant these contacts are *in vivo*. It should be noted that, although many of the minor costimulatory molecules support the initial events of T cell activation that are mediated through the TCR, they do not induce IL-2 production, promote T cell survival, or prevent the apoptosis of T cells (unlike CD28 and ICOS).

D. Signal Three: Cytokines

A T cell that has received both signal one and signal two delivered by engagement of the TCR and CD28 is in a position to proceed to the next step of activation: the transcription of the hundreds of genes needed to support T cell proliferation and differentiation. With respect to proliferation, these genes have been classified on the basis of how soon their products appear after TCR triggering (Fig.14-11). The products of the *immediate* genes can be detected in less than 30 minutes of TCR triggering. In general, these genes encode components of nuclear transcription factors necessary to stimulate the transcription of early and late genes involved in proliferation. Products of the *early* genes are detected within 0.5–24 hours after TCR triggering, including IL-2, various other cytokines (such as IL-3, IL-4, IL-5, and IFNγ), and cytokine, chemokine, and growth factor receptors. The *late* genes are expressed more than 48 hours after TCR triggering.

Time after TCR Engagement and Costimulation

	Minutes	
Activation of transcription factors	15 →	c-fos, c-myc
	20 →	NFAT, NF-κB
	30 →	IFNγ
Transcription of cytokines	45 →	IL-2, CD40L, FasL
	60 →	IL-3, LTα, insulin receptor
	Hours	
	2 →	TGFβ, IL-2R
Cell cycle protein synthesis	4 →	Cyclin
	6 →	IL-4, IL-5, IL-6
	16 →	c-myb
	20 →	GM-CSF
	Days	
Proliferation of activated T cell	3 →	Upregulated MHC class II
	4 →	Perforin, granzymes, chemokines, VLA-4
Effector cell generation	7 →	VLA-1, 2, 3, 5
Effector cell death	14	

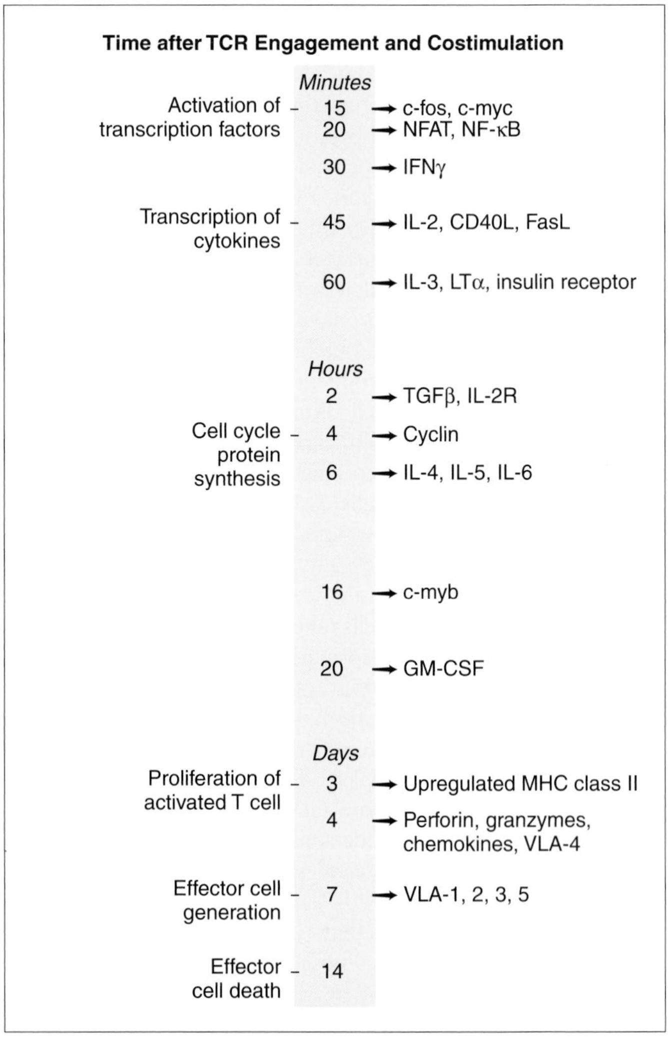

Figure 14-11

Time Course of Gene Expression Following TCR Triggering
The genes for the indicated molecules start to be expressed or are upregulated at the indicated times after TCR engagement and CD28 costimulation. "Immediate" genes are expressed less than 30 minutes after TCR engagement, while the products of "early" genes appear in 0.5–24 hours. "Late" gene expression is detected 2–3 days after TCR engagement. Genes shown are representative examples only; many other genes are also expressed.

New types of adhesion molecules allowing wide-ranging migration in the tissues appear on the cell surface, and the synthesis of cytotoxic and other molecules necessary for T cell effector functions commences.

Activated T cells must first proliferate to be able to generate daughter cells that can differentiate into memory or effector T cells, so that the induction of cytokine and cytokine receptor expression supporting cell survival and division is crucial to the immune response. Prior to the characterization of the IL-2 knockout mouse, most scientists believed that IL-2 was the cytokine critical for T cell activation. Indeed, IL-2 is essential for T cell responses *in vitro*. However, mice lacking IL-2 do not show defects in T cell activation *in vivo*. (In fact, these animals show T cell hyperactivation due to a defect in the regulation of T cell responses; see Ch.16.) The current thinking is that several other cytokines may be capable of contributing to the clonal expansion of an activated T cell in the whole animal. One attractive candidate is IL-15, because the IL-15 receptor shares two of the chains that make up the IL-2 receptor (see Ch.17). A shared signaling pathway may thus account for some of the overlapping biological activities of IL-15 and IL-2.

Because the IL-2 system has been studied in detail, we offer a short description of the role of this molecule and its receptor during T cell activation as an example of cytokine function in this context. We start with an introduction to the IL-2 molecule and its receptor components. This powerful cytokine, its receptor chains, and its signaling path are discussed in detail in Chapter 17.

I. THE IL-2/IL-2R SYSTEM

IL-2 is a 15.5 kDa protein secreted primarily by activated Th cells. Small amounts of IL-2 are also released by activated Tc cells, eosinophils, and B cells. IL-2 delivers proliferative signals by binding to its receptor, IL-2R. The IL-2R is a complex of three proteins: IL-2Rα, IL-2Rβ, and IL-2Rγ. A low affinity heterodimeric receptor containing only IL-2Rβ and IL-2Rγ is constitutively expressed on most mature T cells. However, after the engagement of both the TCR and CD28, the expression of IL-2Rα is upregulated and the three components come together to form a heterotrimeric receptor that binds IL-2 with high affinity. IL-2 can act in either an autocrine or a paracrine fashion to induce Th cell division. Withdrawal of IL-2 after Th cell activation *in vitro* leads to the apoptosis of that cell, suggesting that, *in vivo*, the continued presence of IL-2 (or a compensatory cytokine) may be needed to support T cell proliferation after activation. In Th cells and some Tc cells directed against peptides of aggressive pathogens, an autocrine loop is established in which IL-2 produced by the activated T cell stimulates that same cell to move through the cell cycle. However, although most Tc cells can be induced to express both IL-2 and its receptor, autocrine production of IL-2 is usually <u>not</u> sufficient to push a Tc cell into proliferation. Thus, most Tc cells require "help" in the form of IL-2 secreted by an activated antigen-specific Th cell in order to be able to proliferate. As is described in detail in Chapter 15, cytokines in addition to IL-2 are required for the final differentiation steps that generate effector CTLs with full cytolytic competence.

II. CONTROL OF TRANSCRIPTION OF THE IL-2 GENE

Several layers of regulation are in place to control IL-2 transcription. The enhancer region positioned 5′ of the promoter of the IL-2 gene contains several DNA-binding motifs, and transcription of the gene depends on which of these are bound by the appropriate nuclear transcription factors. Multiple sites for NFAT, AP-1, NF-κB, and Oct-1 have been identified in the IL-2 enhancer region of both the mouse and human genes (Fig. 14-12). To make matters more complicated,

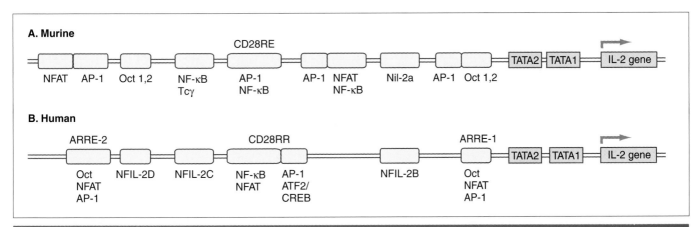

Figure 14-12

Transcriptional Control of Murine and Human IL-2 Genes
(A) Murine IL-2 promoter showing the CD28 response element (CD28RE) and motifs for the indicated transcription factors. (B) Human IL-2 promoter has two antigen-responsive elements (ARRE-1 and ARRE-2) and a CD28 response region (CD28RR). Increased levels of the indicated transcription factors bind to the CD28RE and CD28RR motifs upon CD28 costimulation. The ARRE regions in the human IL-2 gene show increased transcription factor binding in response to TCR stimulation.

many of these transcription factors are composed of more than one subunit, and the activation of these subunits or even transcription of the corresponding genes is under regulatory control as well. For example, AP-1 is composed of two subunits, the c-fos and c-jun factors mentioned earlier, whose activation is controlled independently.

Among the stimulatory motifs in the murine enhancer is a sequence called the *CD28 response element* (CD28RE), which is critical for the induction of IL-2 transcription. The CD28RE is a composite binding site for a complex of NFAT and AP-1. (The human enhancer also contains a CD28RE but its structure is slightly different; refer to Fig. 14-12.) Part of the contribution of costimulation can thus be explained in terms of providing components of this transcription factor complex. TCR engagement alone activates NFAT, but little AP-1. CD28 signaling is required for proper assembly of AP-1, without which the CD28RE cannot be activated, IL-2 transcription is not induced, and anergy ensues. CD28 signaling in the presence of TCR engagement activates the SAPK/JNK pathway and stimulates the Ras/MAPK pathway (refer to Fig. 14-9), resulting in increased phosphorylation of c-jun and c-fos, respectively. These transcription factors combine to generate increased amounts of stable, activated AP-1.

Both NFAT and AP-1 are then available to bind to the CD28RE, the transcription of the IL-2 gene can proceed, and cellular proliferation is induced. As mentioned previously, CD28 signaling is also required for maximal induction of NF-κB initiated by PKCθ activation. Thus, optimal IL-2 transcription leading to T cell proliferation and differentiation *in vitro* requires the synergistic activation of multiple signaling cascades.

At this point in our examination of T cell biology, we have reached the point after encounter with antigen when naive Tc and Th cells have been successfully activated and are proliferating. In Chapter 15, we continue to follow these cells through the differentiation process that produces effector Th cells and CTLs that act to eliminate the antigen. In Chapter 16, we examine the mechanisms of peripheral tolerance and explain how naive self-reactive Tc and Th cells that escaped deletion during the establishment of central tolerance are prevented from attacking self-tissues in the periphery. The subject matter of Chapter 16 has important parallels to the material presented in Chapters 9 and 14 on B and T cell activation: successful lymphocyte activation in the presence of antigen produces an immune response, but absence of activation in the presence of antigen results in anergy and peripheral tolerance.

SUMMARY

T cell activation is a defining moment in the adaptive immune response. Inflammatory cytokines in a site of infection activate immature DCs and increase their uptake of antigen. The DCs migrate to the local lymph node, maturing as they travel so that their powers of adhesion and antigen presentation are enhanced by the time they meet with naive CD4$^+$ T cells in the node. A CD4$^+$ T cell establishes contact with an APC via the interaction of multiple adhesion molecules, followed by interaction between the TCR and peptide presented on MHC class II. If the binding is of sufficient avidity to trigger T cell activation, a supramolecular activation complex (SMAC) is formed between the T cell and the APC that holds the cells together, reorganizes the cytoskeleton, and coalesces signaling molecules in the T cell membrane. The SMAC then allows serial triggering and oligomerization of TCRs that delivers a sustained "signal one." However, signal one alone is insufficient to completely activate a T cell. "Signal two," in the form of costimula-tory contacts such as CD28-B7-1/B7-2 and ICOS-ICOSL, are needed to contribute additional activatory signaling. "Signal three" is delivered by the binding of cytokines such as IL-2 (that may be secreted by the same T cell) to receptors upregulated on the activated T cell. As a result of signals one, two, and three, a complex signaling cascade involving the phosphorylation of ITAMs in the CD3 chains and trans-duction via ZAP-70 and Vav1 leads to the activation of the PLC-γ1, PI3K, PKCθ, SAPK/JNK, and Ras/MAPK signaling pathways. These pathways activate multiple transcription factors, including c-myc, c-fos, c-jun, NFAT, and NF-κB, that initiate the new gene transcription necessary for proliferation and effector cell differentiation. Naive CD8$^+$ T cells are activated after interaction with peptide–MHC class I complexes presented by "licensed" DCs. Licensed DCs express upregulated levels of costimulatory molecules due to prior interaction with an activated CD4$^+$ cell and the establishment of CD40–CD40L contacts.

Selected Reading List

Acuto O. and Cantrell D. (2000) T cell activation and the cytoskeleton. *Annual Review of Immunology* 18, 165–184.

Acuto O. and Michel F. (2003) CD28-mediated co-stimulation: a quantitative support for TCR signalling. *Nature Reviews Immunology* 3, 939–951.

Chambers C. A. and Allison J. P. (1997) Co-stimulation in T cell responses. *Current Opinion in Immunology* 9, 396–404.

Chan A. and Shaw A. (1996) Regulation of antigen receptor signal transduction by protein tyrosine kinases. *Current Opinion in Immunology* 8, 394–401.

Davis M. (2002) A new trigger for T cells. *Cell* 110, 285–287.

Dustin M. and Chan A. (2000) Signaling takes shape in the immune system. *Cell* 103, 283–294.

Frauwirth K. and Thompson C. (2002) Activation and inhibition of lymphocytes by co-stimulation. *Journal of Clinical Investigation* 109, 295–299.

Garcia K. C., Scott C., Brunmark A., Carbone F., Peterson P. *et al.* (1998) CD8 enhances formation of stable T-cell receptor/MHC class I molecule complexes. *Nature* 384, 577–581.

Harder T. (2003) Formation of functional cell membrane domains: the interplay of lipid- and protein-mediated interactions. *Philosophical Transactions of the Royal Society of London Series B: Biological Sciences* **358**, 863–868.

Hořejší V., Zhang W. and Schraven B. (2004) Transmembrane adaptor proteins: organizers of immunoreceptor signaling. *Nature Reviews Immunology* **4**, 603–616.

Hutloff A., Kittrich A., Beier K., Elijzschewitsch B., Kraft R. *et al.* (1999) ICOS is an inducible T-cell co-stimulator structurally and functionally related to CD28. *Nature* **397**, 263–266.

Iezzi G., Karjalainen K. and Lanzavecchia A. (1998) The duration of antigenic stimulation determines the fate of naive and effector T cells. *Immunity* **8**, 89–95.

Konig R. (2002) Interactions between MHC molecules and co-receptors of the TCR. *Current Opinion in Immunology* **14**, 75–83.

Koretzky G. and Myung P. (2001) Positive and negative regulation of T-cell activation by adaptor proteins. *Nature Reviews Immunology* **1**, 95–107.

Kupfer A. and Kupfer H. (2003) Imaging immune cell interactions and functions: SMACs and the immunological synapse. *Seminars in Immunology* **15**, 295–300.

Lanzavecchia A. (1998) Licence to kill. *Nature* **393**, 413–414.

Lanzavecchia A., Iezzi G. and Viola A. (1999) From TCR engagement to T cell activation: a kinetic view of T cell behavior. *Cell* **96**, 1–4.

Leitenberg K., Balamuth F. and Bottomly K. (2001) Changes in the T cell receptor macromolecular signaling complex and membrane microdomains during T cell development and activation. *Immunology/Seminars in Immunology* **13**, 129–138.

Nel A. (2002) T-cell activation through the antigen receptor. Part 1: signaling components, signaling pathways, and signal integration at the T-cell antigen receptor synapse. *Journal of Allergy and Clinical Immunology* **109**, 758–770.

Peterson E., Clements J., Fang N. and Koretzky G. (1998) Adaptor proteins in lymphocyte antigen-receptor signaling. *Current Opinion in Immunology* **10**, 337–344.

Rudd C. (1996) Upstream-downstream: CD28 cosignaling pathways and T cell function. *Immunity* **4**, 527–534.

Sharpe A. (1995) Analysis of lymphocyte co-stimulation in vivo using transgenic and knockout mice. *Current Opinion in Immunology* **7**, 389–395.

Sharpe A. and Freeman G. (2002) The B7-CD28 superfamily. *Nature Reviews Immunology* **2**, 116–126.

Stefanova I., Hemmer B., Vergelli M., Martin R., Biddison W. *et al.* (2003) TCR ligand discrimination is enforced by competing ERK positive and SHP-1 negative feedback pathways. *Nature Immunology* **4**, 248–254.

Sundberg E., Li Y. and Mariuzza R. (2002) So many ways of getting in the way: diversity in the molecular architecture of superantigen-dependent T-cell signaling complexes. *Current Opinion in Immunology* **14**, 36–44.

Valitutti S. and Lanzavecchia A. (1997) Serial triggering of TCRs: a basis for the sensitivity and specificity of antigen recognition. *Immunology Today* **18**, 299–304.

Viola A., Schroeder S., Sakakibara Y. and Lanzavecchia A. (1999) T lymphocyte co-stimulation mediated by reorganization of membrane microdomains. *Science* **283**, 680–682.

Wange R. and Samelson L. (1996) Complex complexes: signaling at the TCR. *Immunity* **5**, 197–205.

Wardenburg J., Pappu R., Bu J.-Y., Mayer B., Chernoff J. *et al.* (1998) Regulation of PAK activation and the T cell cytoskeleton by the linker protein SLP76. *Immunity* **9**, 607–616.

Werlen G. and Palmer E. (2002) The TCR signalsome: a dynamic structure with expanding complexity. *Current Opinion in Immunology* **14**, 299–305.

Veillette A., Latour S. and Davidson D. (2002) Negative regulation of immunoreceptor signaling. *Annual Review of Immunology* **20**, 669–707.

Zamoyska R. (1998) CD4 and CD8: modulators of T-cell receptor recognition of antigen and of immune responses? *Current Opinion in Immunology* **10**, 82–87.

T Cell Differentiation and Effector Function

15

Peter Doherty 1940–

MHC restriction of T cell responses
Nobel Prize 1996

© Nobelstiftelsen

CHAPTER 15

A. HISTORICAL NOTES

B. Th CELL DIFFERENTIATION AND EFFECTOR FUNCTION

C. Tc CELL DIFFERENTIATION AND CTL EFFECTOR FUNCTION

D. COMPARISON OF NAIVE AND EFFECTOR T CELLS

E. ELIMINATION OF EFFECTOR T CELLS

F. MEMORY T CELLS

"Don't fall before you're pushed"—English Proverb

In Chapter 14, we described the activation of resting naive Th and Tc cells by antigen and introduced the resulting proliferation of these cells. In this chapter, we discuss in detail how daughter cells differentiate into Th1, Th2, or CTL effector cells, and the actions that they take to protect the host during an immune response. We then describe how the primary response concludes with the elimination of effector T cells and the generation of memory T cells. We end the chapter with a discussion of the activation and effector functions of memory T cells that underlie the secondary response.

A. Historical Notes

A major breakthrough in understanding the immune system occurred in the mid-1960s when it was discovered that lymphocytes could be subdivided into two types: B cells and T cells. By the 1970s, antisera raised against T lymphocytes were used to define particular surface markers of these cells. In 1975, Harvey Cantor and Edward Boyse reported the existence of two different T cell surface marker profiles that correlated with their differential functions. T cells that were involved in B cell activation and facilitated antibody production were called "helper" T cells (Th) and had the marker phenotype Ly $1^+2^-3^-$, L3T4$^+$, while the marker phenotype of "cytotoxic" T cells (Tc) was Ly $1^-2^+3^+$, L3T4$^-$. At this time, no further distinctions among T cells were made based on surface marker profiles. Over the next few years, controversy bubbled over the possible existence of further functional heterogeneity within the Th subset based on criteria such as nylon wool adherence, MHC restriction, MHC class II expression, and whether the help provided to B cells was antigen-specific or non-specific. However, such avenues of investigation did not prove to be particularly fruitful and this area of immunological study stagnated for a while due to the lack of new approaches.

The next major advance in understanding Th cells came as new research directions were pursued and new technologies implemented. In the 1980s, the focus shifted onto defining the secreted cytokines of Th cells and their effects. For the first time, the availability of purified proteins and recombinant DNA technology allowed researchers to correlate known cytokine activities with well-characterized proteins. In addition, techniques were developed to clone and grow large numbers of functionally active, antigen-specific Th cells, allowing them to be studied *in vitro*. However, a major complication blocked the clear understanding of Th cytokine secretion and its implications: because cytokines themselves have multiple effects on cells (pleiotropy), and most cell types respond to more than one cytokine, it was difficult to develop assays that would measure only the cellular secretion of a particular cytokine. Fortunately, another major technological advance that developed concurrently with this new interest in cytokines was the production of monoclonal antibodies of desired specificity. Monoclonal antibodies specific for the various cytokines were used to develop "monospecific" cytokine assays to measure the levels of one cytokine at a time. This was often done by using specific mAbs to neutralize the biological activity of all cytokines other than that of interest. For example, the presence of IL-2 secreted in a culture was detected using a "T cell growth assay" that measured the amount a T cell clone proliferated in response to a culture supernatant. However, this assay was complicated by the fact that IL-4 in the culture supernatant also induced T cell proliferation. To solve this problem, an excess amount of anti-IL-4 mAb was added to completely deplete the supernatant of IL-4 prior to the assay.

By the mid-1980s, definite patterns in T cell cytokine production had been detected. Tim Mosmann and Robert Coffman studied a large number of Th clones and found that one subtype, which they called "Th1," secreted IL-2, IFNγ, and lymphotoxin. A second subtype, "Th2," did not produce these cytokines but instead secreted IL-4 and IL-5. Not surprisingly, the different cytokine profiles of the

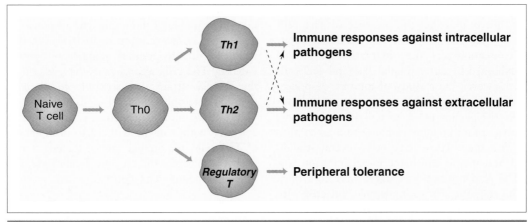

Figure 15-1

Helper T Cell Subsets
Naive T cells that encounter cognate antigen are activated to become Th0 cells. Th0 cells proliferate and, under the influence of particular cytokines, differentiate into one of three subtypes of helper effector T cells. While regulatory T cells secrete cytokines that impose peripheral tolerance, Th2 and Th1 cells supply "help" to B cells and Tc cells in the form of cytokines and intercellular contacts. This "help" aids in the battle against extracellular and intracellular pathogens. Dotted arrows indicate that Th1 and Th2 cells may sometimes supply "help" for responses against intracellular and extracellular pathogens, respectively.

Th subsets correlated with their different functions. Th1 cells became known as the subtype primarily responsible for both cell-mediated immunity and antibody responses against intracellular pathogens, while Th2 cells were found to predominantly facilitate the humoral immune responses needed to eliminate extracellular pathogens (Fig. 15-1). These findings were subsequently confirmed *in vivo* with the isolation of distinct Th1 and Th2 subsets from mice and humans recovering from various infections. In the late 1990s, a third class of T helper cells called "regulatory T cells" was identified. Regulatory T cells are present in comparatively low numbers, and various subsets deliver intercellular contacts that inhibit effector T cells, and/or secrete large amounts of the immunosuppressive cytokines IL-10 and TGFβ. Regulatory T cells are involved primarily in establishing peripheral tolerance and are therefore discussed in Chapter 16.

B. Th Cell Differentiation and Effector Function

The development of a T helper response is best thought of as occurring in three phases: (1) the activation of a naive CD4$^+$ Th cell, (2) its proliferation and differentiation into Th1 and Th2 effector cells, and (3) the activation of the Th1 and Th2 cells that leads to the delivery of "T cell help" to antigen-specific B cells and antigen-specific Tc cells. The differentiation of Th1 and/or Th2 cells derived from an antigen-activated CD4$^+$ T cell depends on the immediate cytokine milieu. The presence of IFNγ and IL-12 favors Th1 development, while the presence of IL-4 promotes Th2 development. This differentiation is fixed, so that Th1 cells cannot later become Th2 cells or vice versa. Once differentiated, the Th1 and Th2 effectors

provide costimulatory contacts (such as CD28 and CD40L) and secrete distinct panels of cytokines that can act either in an autocrine manner (on the cell that produced them) or in a paracrine fashion (on nearby cells). As mentioned previously, the "help" required for Tc cell activation and the production of antibody isotypes effective against intracellular threats is most often provided by Th1 cells, while Th2 cells secrete cytokines that are primarily involved in the production of antibody isotypes needed for defense against extracellular bacteria and large parasites. However, because one type of help is not aimed exclusively at either intracellular or extracellular pathogens, some scientists think of Th1 help as promoting antibody isotypes that engage in opsonization, complement activation, or ADCC (mainly IgG), while Th2 help favors secretory antibodies (IgA) and the IgE that arms mast cells and basophils. In any case, cytokines produced by both types of Th cells also influence the behavior and activation of macrophages, vascular endothelial cells, and other host cells important in both the adaptive and innate immune responses.

I. WHAT ARE "Th1 AND Th2 RESPONSES"?

Among immunologists, it is said that an immune response has either a Th1 or Th2 character or *phenotype*, depending on the predominant Th subset and cytokines observed in the host during that response. An attack on a host by intracellular pathogens stimulates the production by DCs and macrophages of cytokines favoring Th1 development, while invasion of the host by extracellular pathogens most often promotes the development of a Th2 response. Immunological disease states also tend to have either a Th1 or Th2 phenotype. For example, allergies are associated with a prevalence of Th2 cells, while Th1 cells dominate in autoimmune disorders and allograft rejection.

It was originally thought that Th1 and Th2 type responses were mutually exclusive, and that the presence of Th1 cells might preclude the development of Th2 cells, and vice versa. Supporting this contention was the observation of mutual cross-regulation exhibited by differentiated Th1 and Th2 cells (described later). However, close study of immune responses using model systems of infections by various organisms have shown that, unlike the phenotype of an individual Th clone, the overall phenotype of an immune response to a given antigen can change with time. There have been several studies, both in mice and humans, in which the initial infection was countered by a Th1 response but later stages were controlled by a persistent Th2 response. For example, mice infected with the parasite *Plasmodium* first develop a Th1 response in which Th1 cells secrete IFNγ and macrophages are activated to secrete cytotoxic cytokines and produce large quantities of NO. In addition, B cells are induced to switch to the production of specific IgG2a antibodies. However, within 10 days of the initial infection, a Th2 response is observed, characterized by high serum levels of IL-4, IL-10, and specific IgG1 antibodies. Similar patterns have been identified in human HIV and measles infections. It is postulated that a Th1 response may be invoked first in an attempt to deal with the infection at its intracellular stage, whereas after a virus replicates and releases new virus particles extracellularly, or a bacterial toxin is produced, a Th2 response is better suited for defending the host.

Work *in vitro* has demonstrated that within a single immunogen, different immunodominant peptides can elicit opposing responses. While the T cell clones responding to one particular peptide of that immunogen might be Th1 cells, the clones responding to another peptide may be Th2 cells. *In vivo*, the response to the whole immunogen becomes dominated by one subset or the other. Since Th1 cells cannot become Th2 cells or vice versa, a shift in response phenotype must depend on population dynamics such that one subset is downregulated and the other is upregulated. It is thought that the shift in response phenotype is most likely influenced by the specific cellular components of the local microenvironment and their costimulatory capacity, the nature of the immunogen and its concentration, and the cytokines that are present.

Prevalence of APCs of a certain type can influence whether a shift from the Th1 to the Th2 phenotype is observed during the course of an immune response. For example, early in an infection, macrophages are quite likely to be the dominant APC, since they are elements of the innate response and extensive proliferation of antigen-activated B cells has yet to occur. Macrophages present antigen most efficiently to Th1 cells and produce cytokines stimulating their proliferation, promoting the dominance of this subset. However, as the infection is contained toward the end of the primary response, antigen-specific B cells increase in number and start to play a more important role in antigen presentation. Since B cells present antigen to Th2 cells more efficiently than to Th1 cells (see later), a Th2 response is eventually favored.

The genetic background of the host may also bias its Th responses one way or the other. For example, effective clearance of the intracellular pathogen *Leishmania major* requires a Th1 response to activate the macrophages and Tc cells

necessary to efficiently dispose of this invader. Most strains of mice are capable of mounting Th1 responses against *L. major* and therefore are resistant to it. However, the BALB/c mouse strain consistently mounts adaptive responses that are dominated by Th2 cells regardless of the triggering pathogen. Thus, the BALB/c strain is susceptible to *L. major* because the required macrophages and Tc cells are not activated. However, if a BALB/c mouse is treated early in the infection (within 2 weeks) with exogenous IL-12 to promote the differentiation of Th1 cells, the animal can be rescued. Later intervention is unsuccessful, since the progeny T cells have already irreversibly differentiated into Th2 clones.

II. PROCESS OF Th CELL DIFFERENTIATION

i) Th0 Cells

The reader will recall that, for an antigen-specific naive Th cell to respond to antigen, it must engage its TCRs and receive costimulatory signals such that the appropriate cytokine receptors (particularly IL-2R) are expressed; resting Th cells, which lack these receptors, ignore surrounding cytokines. When IL-2 (or another cytokine) induces the proliferation of activated naive Th cells, the resulting progeny are large blast cells called *Th0* cells, precursors to the differentiated effector Th1 and Th2 cells. Detectable numbers of Th0 cells appear in a proliferating Th cell clone within 48 hours of antigenic stimulation. Th0 cells secrete copious quantities of IL-2 as well as low levels of IFNγ and IL-4. However, the cytokines secreted by Th0 cells are insufficient to influence Th1/Th2 differentiation on their own. Larger amounts of additional cytokines, particularly IL-2, IFNγ, IL-27, IL-12 for Th1 differentiation, and IL-4 for Th2 differentiation, must be supplied by other cell types in the immediate area that have been activated by the presence of the pathogen.

In vitro evidence suggests that Th0 cells may have more than just a precursor role. It has been demonstrated that a population of effector Th cells with a Th0 cytokine profile (that is, secreting IL-4 and IFNγ simultaneously) can be generated from naive antigen-specific CD4+ T cells repeatedly exposed to moderate doses of antigen in the continuing presence of moderate concentrations of both IL-4 and IFNγ. In this experiment, the development of the Th0 effector population required time, in that the number of cells secreting multiple cytokines reached its peak only after 9 days in culture. The balance of exogenous IL-4, IFNγ, and IL-12 in the culture was found to be critical for the emergence of these cells, as was an absence of highly costimulatory APCs. In other words, while strongly polarizing influences favor the development of Th1 or Th2 cells, moderate conditions of antigen dose and costimulation, coupled with a particular balance of cytokines, can produce Th0 effector cells. The existence and relevance of this population *in vivo* remain to be determined.

ii) Th1 and Th2 Cells

About 48–72 hours after antigenic stimulation leading to activation and proliferation, most Th0 cells begin to differentiate into functional effectors, either Th1 or Th2 cells (Table 15-1

Table 15-1 Th1 vs. Th2 Differentiation

	Th1 Development	Th2 Development
Inducing pathogen	Intracellular	Extracellular
Costimulation promoting differentiation	CD28/B7-1 CD40/CD40L	CD28/B7-2 ICOS/ICOSL
Major cytokines promoting differentiation	IL-12, IFNγ, IL-27	IL-4, IL-6
Status of receptor chains during differentiation	↑ IL-12Rβ2 ↓ IFNγRβ	IL-4Rα⁺ IL-12Rβ2⁻ IFNγRβ⁺
Major signal transducer	STAT4	STAT6
Major transcription factors	T-bet, IRF-1, ERM	GATA-3, c-maf

and Fig. 15-2). This differentiation requires three elements: (1) antigen, which must be present to keep the progeny cells stimulated as they are differentiating, (2) cytokines, and (3) cytokine receptors. Four cytokines are especially influential: IL-12, IL-27 and IFNγ, which favor Th1 development, and IL-4, which promotes Th2 development.

As mentioned previously, Th0 cells themselves produce only small amounts of IFNγ and IL-4, and no IL-12 or IL-27. Thus, the bulk of the cytokines required for effector differentiation must be obtained from activated non-lymphocytes (principally macrophages and DCs) in the vicinity of the antigen-stimulated Th0 cell. During differentiation, each Th0 cell gains or loses the expression of genes encoding certain cytokines and cell surface receptors that determine which cytokines the cell will respond to and eventually produce. In Th0 cells that later become Th1 cells, the transcription of IL-4

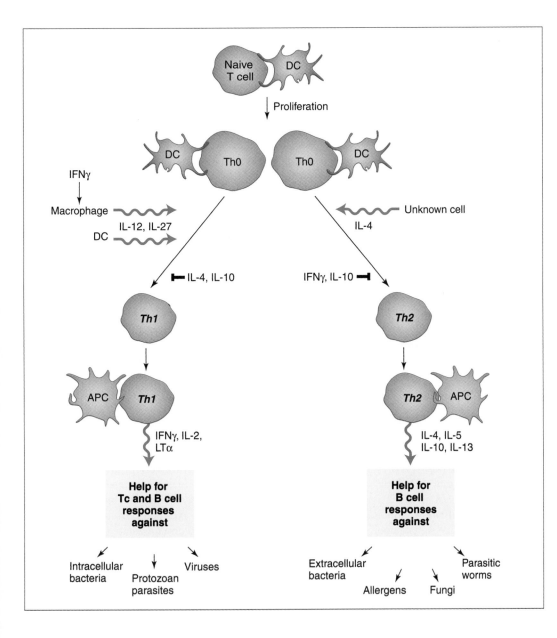

Figure 15-2

Th1/Th2 Differentiation and Effector Functions
A naive T cell that encounters cognate antigen presented by a DC is activated and produces daughter Th0 cells. In the continued presence of antigen and under the influence of IL-12 and IL-27, a given Th0 cell proliferates and differentiates into Th1 effector T cells. This process is stimulated when IFNγ activates macrophages and DCs to produce more IL-12, and is inhibited by a cytokine milieu dominated by IL-4 and IL-10. On the other hand, if a Th0 cell encounters IL-4, it differentiates into Th2 effectors in a process that is inhibited by IFNγ and IL-10. Only the major cytokines influencing Th1/Th2 differentiation are shown (for more details, see Ch 17). Mature Th1 effectors secrete cytokines that aid primarily in the fight against intracellular pathogens, while mature Th2 effectors produce cytokines that mainly bolster defense against extracellular pathogens.

ceases at about the 48 hour point but IL-2 and IFNγ production continues. In the presence of exogenous IL-27 and IFNγ, these cells commence expression of the inducible IL-12Rβ2 subunit that allows the assembly of the complete IL-12R. (The other subunit, IL-12Rβ1, is constitutively expressed.) In the presence of IL-12, these cells then become irreversibly committed to the Th1 phenotype 5–7 days after antigenic stimulation. Conversely, in a cytokine milieu dominated by IL-4, Th0 cells that later become Th2 clones abrogate their production of IFNγ about 48 hours after antigenic stimulation. These cells do not upregulate the IL-12Rβ2 chain and thus do not express a complete IL-12R. In the absence of IL-12 signaling and the presence of IL-4 signaling, these cells become full-fledged Th2 cells 3–5 days later.

iii) Effect of Antigen on Th1/Th2 Differentiation

The nature of the antigen to which a Th0 cell responds largely determines whether it will differentiate into a collection of Th1 or Th2 cells. Pathogens that induce the production of IL-12 and IL-27 by macrophages or DCs preferentially induce the differentiation of Th1 cells, at least initially. Such pathogens tend to be intracellular bacteria and viruses, which are most efficiently eliminated by the cell-mediated response delivered by CTLs and certain antibody isotypes. Pathogens that fail to induce IL-12 and IL-27 production by macrophages or DCs tend to be dealt with by Th2 cells. Such pathogens tend to be extracellular bacteria and large parasites, which are best handled solely by a humoral response.

Studies using transgenic mice and weak agonist peptides have shown that antigen concentration can influence Th1/Th2 differentiation, but the underlying mechanism is not clear. *In vitro*, very high and very low doses of antigen tend to promote Th2 differentiation, while intermediate concentrations of antigen favor Th1 development. However, in a whole animal immunized with bacterial products, low antigen concentrations produce Th1 responses, while intermediate or high concentrations favor Th2 responses. The nature of the antigen rather than its concentration is currently thought to have more influence on which type of Th effector cell differentiates.

iv) Effect of Costimulation on Th1/Th2 Differentiation

The level and type of costimulation a naive Th0 cell receives also appear to bias its differentiation to Th1 or Th2. The engagement of CD28 on a Th0 cell by B7-1 expressed on an APC favors Th1 differentiation, while the interaction of B7-2 with CD28 promotes Th2 cell production (refer to Table 15-1). Experiments *in vitro* have shown that engagement of B7-2 results in a significantly higher level of IL-4 production by naive CD4+ T cells than does engagement of B7-1. The presence of this increased IL-4 may then help to promote Th2 differentiation. Indeed, when the binding of CD28 to B7-2 is abrogated, either *in vivo* or *in vitro*, Th2 differentiation is blocked. In addition, the inducible costimulatory signals conveyed by ICOS/ICOSL interaction are vital for Th2 responses (see later). In general, the evidence suggests that the generation of Th2 cells (but not their activation; see later) is more dependent on costimulation than is the generation of Th1 cells. Conversely, the upregulation of CD40L expression on

activated T cells is important for Th1 differentiation. Interaction of CD40L on T cells with CD40 expressed on DCs induces the latter to produce large amounts of IL-12 that drive a Th0 cell down the Th1 path.

v) Effect of APC Type on Th1/Th2 Differentiation

As we learned in Chapters 3 and 14, DCs are the only cell type capable of activating naive T cells because only these cells deliver sufficient costimulatory signaling to push a T cell over the TCR signaling threshold. As mentioned in previous chapters, the expression of costimulatory ligands such as B7-1, B7-2, and ICOSL is upregulated on DCs in response to engagement of PRRs on the DC surface by pathogens. Thus, the innate immune response is directly linked to Th effector differentiation in the adaptive immune response via DC activation. The engagement of certain TLRs on the DC induces the latter to produce IL-12, thus promoting Th1 (but not Th2) differentiation. The importance of the TLRs in Th1 differentiation has been illustrated by gene-targeting experiments. Mice lacking components of the TLR signaling pathway cannot mount Th1 responses to pathogens and their T cells are unable to produce IFNγ *in vitro*. However, the mutant mice were able to produce antibody isotypes associated with Th2 responses. Thus, it may be that Th2 differentiation is the default development path of a Th0 cell unless IL-12 is produced. Alternatively, PRRs other than TLRs may be engaged that induce the production of an unknown cytokine that initiates Th2 differentiation. A third possibility is that another type of (currently unknown) host cell supplies the initial amounts of IL-4 needed. There is some evidence that NKT cells may be responsible (see Ch.18).

Some researchers believe that different DC subsets influence Th1/Th2 differentiation. For example, there have been reports in mice that CD8+ conventional DCs are primarily associated with Th1 differentiation, while CD8− conventional DCs promote Th2 differentiation. However, in humans, the conventional DC1 cells have been linked to Th1 differentiation, while the plasmacytoid DC2 cells have been associated with the Th2 phenotype. Other scientists believe that there may be two different DC subsets governing Th1/Th2 differentiation but that they are not defined by marker expression or how they evolved from their precursors. In this scenario, one DC subset recognizes pathogens that induce Th1 responses, while the other subset is stimulated by pathogens later eliminated by Th2 responses. Still other researchers believe that there is only one type of DC that is capable of modifying its behavior (to secrete IL-12 or not) depending on the type of pathogen it encounters and the tissue microenvironment. Tissue-specific molecular signals that are constitutively produced by component cells may instruct DCs to bias a response toward Th1 or Th2. For example, a molecule called prostaglandin-E2 (PGE2) is constitutively produced by stromal cells in the intestine. PGE2 causes stimulated DCs to downregulate their production of IL-12, conferring the tendency on the DC to induce activated Th0 cells in the direction of Th2 differentiation.

It should be noted that, regardless of its phenotype or origin, a single DC is insufficient to support an immune response. To complete Th1/2 differentiation, a naive T cell must be activated and proliferate, and its daughter cells must maintain contact with specific antigen. All of this requires sustained TCR signaling, which in turn requires a steady supply of DCs bearing specific pMHC. The first mature DC in the lymph node that interacts with the naive T cell produces chemokines that draw new waves of immature DCs bearing antigen into the lymph node. These DCs then mature and interact with developing effector cells, maintaining TCR signaling and helping to supply the necessary cytokines.

vi) Cytokines Affecting Th1 Differentiation

via) IL-12. Th1 differentiation is driven principally by IL-12. IL-12 is a powerful heterodimeric cytokine secreted primarily by activated macrophages in response to infection by a limited spectrum of agents, including some viruses, protozoa, and intracellular bacteria such as *Listeria* and *Mycobacterium* species. Studies of gene-targeted knockout mice lacking either the genes for the IL-12 subunits or those for the IL-12 receptor chains have confirmed that the differentiation of antigen-specific Th1 cells from naive T cells is largely dependent on IL-12. The binding of IL-12 to its receptor on a Th0 cell triggers intracellular signaling that results in the activation of the kinases Jak2 and Tyk2. These enzymes in turn phosphorylate tyrosine residues of the transcription factor STAT4 (*signal transducer and activator of transcription 4*) (see Box 15-1). Activated STAT4 then binds to specific motifs in the promoters of genes regulated by IL-12, including IFNγ and others required for development along the Th1 pathway. In STAT4-deficient mice, IFNγ induction in response to IL-12 was impaired, and the mice mounted poor Th1 responses to most intracellular pathogens. The development of Th1 cells *in vitro* in response to IL-12 was also defective. However, in these animals, as well as in IL-12$^{-/-}$ and IL-12R$^{-/-}$ mice, immune responses dependent on Th2 cells were unaffected.

It should be noted that in humans (but apparently not in mice) there is an alternative means of inducing Th1 differentiation that is independent of IL-12. A signaling pathway exists in human cells by which IFNα and IFNβ, produced *in vivo* by leukocytes and fibroblasts in response to infection, can activate STAT4 directly and induce transcription of those genes required for Th1 differentiation. In addition, STAT1, which is induced by both IFNα and IL-12, may be able to substitute for STAT4 in human T cells and murine CD8$^+$ T cells.

vib) IL-27. IL-27 is crucial for the initiation of Th1 differentiation because this cytokine is thought to be the primary inducer of T-bet upregulation leading to IL-12Rβ2 expression on the activated Th0 cell surface. Only after the complete IL-12R is assembled on the Th0 cell can it receive the IL-12 signals that drive Th1 differentiation. T-bet also induces IFNγ production and suppresses IL-4 production in these cells. After differentiation is complete, the mature Th1 effector loses expression of IL-27R and ceases to respond to this cytokine. More on IL-27 appears in Chapter 17.

vic) IFNγ. IFNγ contributes both directly and indirectly to Th1 differentiation. As detailed in Box 15-1, IFNγ signaling can contribute to the upregulation of T-bet and thus IL-12R assembly. However, studies of knockout mice deficient for either IFNγ or the IFNγ receptor have shown that IFNγ is also important for Th1 responses because of its effects on APCs. IFNγ is a powerful activator of macrophages, inducing them to produce high levels of IL-12. A positive feedback loop is thus created in which differentiating Th1 cells are stimulated by IL-12 to produce IFNγ, which in turn stimulates APCs to produce more IL-12 (Fig. 15-3). The first molecules of IFNγ that induce macrophages to start production of IL-12 may be secreted by the Th0 cell itself, but additional IFNγ is required to push the production of IL-12 to the level necessary for the completion of Th1 effector cell generation. In an example of the linkage of the innate and adaptive immune responses, NK cells stimulated by bacterial or viral antigens are able to produce IFNγ early in the innate response, even in the absence of other cytokines. In the presence of this IFNγ, macrophages are stimulated to produce the amount of IL-12 needed to direct antigen-activated Th cells down the Th1 differentiation path.

vid) IL-10. As is described in more detail in Chapter 17, IL-10 acts as a brake on inflammatory responses, inhibiting the synthesis of many cytokines by many different cell types. IL-10 interferes with T cell activation by inhibiting the expression of the B7 co-stimulatory ligands by macrophages. A striking downregulation of MHC class II molecules is also induced on macrophages, preventing them from presenting antigen to Th cells. These effects work to dampen the development of both Th1 and Th2 cells. However, because of the reliance of murine Th1 differentiation on IL-12 produced by activated macrophages, the downregulation of these cells by IL-10 makes this cytokine a more powerful inhibitor of Th1 cell development than Th2 development.

vie) IL-4. IL-4 inhibits several activities of macrophages associated with the inflammatory response, including phagocytosis, respiratory burst, and the release of certain pro-inflammatory cytokines. However, IL-4 has also been shown to upregulate IL-12 expression by DCs and macrophages, possibly to counteract IL-4's powerful induction of the Th2 response.

vii) Cytokines Affecting Th2 Differentiation

viia) IL-4. IL-4 is the primary driver of Th2 cell differentiation. The binding of IL-4 to its receptor on a Th0 cell triggers the kinases Jak1 and Jak3 associated with the cytoplasmic region of the receptor complex to phosphorylate STAT6 (refer to Box 15-1). Activated STAT6 in turn binds to the *IL-4 response element* (IL-4RE) in the promoter of the Th2-specific GATA-3 transcription factor and induces its expression. GATA-3 then drives expression of the cluster of closely linked (within 200 kb) genes encoding IL-4, IL-5, and IL-13. STAT6 also activates c-maf, which contributes to the induction of IL-4 transcription. Knockout mice deficient for either STAT6 or IL-4 produce almost no Th2 cells, but responses dependent on Th1 cells remain intact.

What is the source of the IL-4 required to sustain Th2 development? Besides the small amount secreted by activated Th0 cells, IL-4 is produced by only a very few cell types in the

The intracellular signaling pathways leading to Th1/Th2 differentiation have yet to be completely elucidated. Many different transcription factors are likely to be important for Th1/Th2 development as well as for later cytokine production by the fully differentiated effector cells. We are also only beginning to understand how engagement of a cytokine receptor translates into *chromatin remodeling* that allows a locus to be easily transcribed.

DNase Hypersensitivity Sites

Why do some Th0 cells become Th1 cells after encountering IL-12, while others become Th2 cells after encountering IL-4? It seems that the signaling pathways triggered by these cytokines lead both to the activation of specific transcription factors and to changes in the chromatin structure of previously silent cytokine genes that allow them to be transcribed. For example, Anjana Rao and her colleagues found that the IFNγ gene in Th1 cells contains certain sites that are hypersensitive to DNase I digestion, suggesting that the gene has an "open" structure compatible with the binding of transcription factors to the promoter. (In an "open" gene, histones associated with the DNA are acetylated and CG pairs show decreased methylation. A "closed" gene shows low levels of DNA acetylation and high levels of CG methylation. It takes several cell divisions for a gene to convert from the closed to the open state.) DNase hypersensitive sites are not seen in the IFNγ genes of Th2 cells. Conversely, the IL-4, IL-5, and IL-13 genes in Th2 cells (but not Th1 cells) contain specific DNase hypersensitivity sites. Demethylation of the IFNγ gene in differentiating Th1 cells, and the IL-4 and IL-5 genes in differentiating Th2 cells, has been reported. These patterns are stably inherited by daughter cells.

Transcription Factors Associated with Th1 Differentiation

As might be expected from the importance of IL-12 in Th1 effector cell generation, murine Th1 cell differentiation cannot occur in the absence of **STAT4**, the transcription factor through which IL-12 signaling is transduced (see Ch.17 for more on cytokine signal transduction). Mice lacking **IRF-1** also lack Th1 cells because this transcription factor controls expression of the IL-12 p40 gene. However, perhaps the most important transcription factor for Th1 differentiation is **T-bet** ("T-box expressed in T cells"; the *T-box* is a highly conserved DNA-binding domain).

T-bet expression occurs only in differentiating Th1 cells and Th1 effectors and is induced by IL-27 and/or IFNγ via a signaling pathway that usually involves STAT1 (see figure, left side). A major function of T-bet is to induce expression of the IL-12Rβ2 chain on Th0 cells destined to become Th1 effectors, allowing IL-12 to bind and trigger STAT4 activation (see Ch.17). STAT4 then acts on the promoter of the IFNγ gene to induce production of this cytokine. IFNγ expression activates inducible genes important for the Th1 response and operates in a feedback mechanism to increase T-bet expression. In addition to activating IL-12R expression, T-bet may suppress the expression in Th1 cells of either IL-4R or **GATA-3**. GATA-3, a transcription factor that we first encountered during our discussion of the early stages of naive T cell development, is also important for Th2 differentiation (see later). Conversely, T-bet expression is suppressed in developing Th2 cells by GATA-3. In genetically engineered mice, overexpression of T-bet enhances IFNγ expression but suppresses IL-4 and IL-5 production. IFNγ production is almost completely abrogated in T-bet$^{-/-}$ mice and the animals develop symptoms reminiscent of human asthma (a Th2 type disease; see Ch.28). **ERM** is a member of the Ets transcription factor family that is induced by STAT4 and TCR stimulation and is expressed exclusively in Th1 cells. However, unlike overexpression of T-bet, overexpression of ERM in developing Th2 cells cannot induce IFNγ expression.

Transcription Factors Associated with Th2 Differentiation

Th2 differentiation is dependent first and foremost on **STAT6** because this molecule is required for transducing IL-4 signaling (see figure, right side). STAT6 is thought to upregulate the expression of GATA-3, which suppresses T-bet activity and IFNγ expression in differentiating Th2 cells. GATA-3 activity also inhibits differentiating Th0 cells from expressing the IL-12Rβ2 chain necessary to receive IL-12 signaling. GATA-3 expression in Th1 cells is repressed by IL-12/STAT4 signaling as well as by T-bet. In Th2 cells, CD28 signaling has been shown to induce GATA-3 expression independently of IL-4 or STAT6 activation. It is still not clear exactly how GATA-3 works. Curiously, there is no GATA-3 binding site in the proximal IL-4 promoter. Researchers speculate that GATA-3 must therefore bind to some type of more distant enhancer sequence. Functional

GATA-3 binding sites have been identified in the IL-5 gene promoter and in control regions of the GATA-3 gene itself. Forced overexpression of GATA-3 can induce DNase hypersensitivity in loci associated with Th2 responses, suggesting that the principal function of GATA-3 is to promote locus accessibility. In fully differentiated Th2 effectors, GATA-3 is important not only for the expression of IL-4 but also for that of additional Th2 cytokines such as IL-5, IL-10, and IL-13.

The transcription factor **c-maf** is exclusively expressed during the maturation of Th2 cells and in fully differentiated Th2 effectors. Indeed, the transgenic expression of c-maf in non-Th2 cells (and even in non-lymphocytes) activates the IL-4 promoter and induces IL-4 synthesis. Acting in concert with the AP-1 transcription factor component **c-jun**, c-maf binds to its response element located in the IL-4 promoter and induces the synthesis of IL-4 while downregulating IFNγ production. Th2 cells from the c-maf$^{-/-}$ mouse completely lack IL-4 expression, but other cell types that secrete this cytokine (such as mast cells) are unaffected. Unlike GATA-3, c-maf does not affect the transcription of other Th2-associated cytokines.

The transcription factor **C/EBPβ (NF-IL6)** plays a role both in Th2 differentiation and IL-4 promoter activation in fully differentiated Th2 cells. C/EBPβ expression increases dramatically in several different cell types in response to inflammatory stimuli such as LPS. C/EBPβ can activate transcription of the IL-6 gene in macrophages, which in turn may be important for promoting IL-4 production by Th2 cells. Furthermore, C/EBPβ has been shown to specifically increase transcription of the IL-4 gene *in vitro*. However, C/EBPβ$^{-/-}$ knockout mice also show a decrease in IL-12 production, suggesting that this pleiotropic transcription factor may differentially affect both Th2 and Th1 cytokine expression.

The transcriptional repressor **Bcl-6**, first encountered in Chapter 9 in our discussion of plasma cell differentiation, may also influence Th2 differentiation because Bcl-6$^{-/-}$ mice show supranormal Th2 responses. Bcl-6 competes with STAT6 for binding sites in gene promoters, but recruits transcriptional repressor rather than activator complexes. Th2 gene expression is therefore dampened. Another transcriptional repressor, **Socs-1**, downregulates Th2 gene expression in

Continued

Box 15-1. Transcription factors important in Th effector cell differentiation and activation—*cont'd*

response to IFNγ. IFNγ induces the expression of Socs-1, which then binds to the Jak kinases and inhibits the activation of STAT6.

A balance of the various **NFAT** transcription factors also contributes to Th1/Th2 differentiation. NFATp and NFATc are both expressed in Th1 and Th2 cells and can activate the promoters of Th1- and Th2-associated cytokine genes *in vitro*.

However, in some cases, these factors appear to regulate the same cytokine promoter in opposing ways. For example, in studies of knockout mice by Laurie Glimcher and colleagues, it was found that NFATc$^{-/-}$ T lymphocytes were able to produce only very low levels of IL-4, while cells from NFATp$^{-/-}$ mice showed excessive IL-4 production. Cells from mice lacking both

NFATp and another family member, NFAT4, secreted extremely high levels of IL-4, spurred by constitutive localization of NFATc in the nucleus. These data suggest that, at least under certain circumstances, NFATc can function as an activator of IL-4 transcription, while NFATp and NFAT4 control IL-4 expression by acting as repressors.

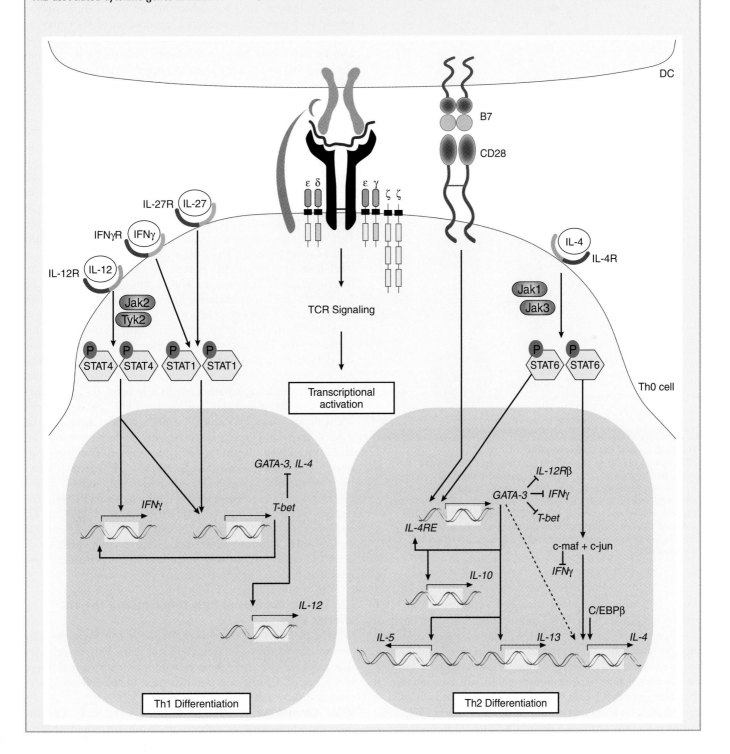

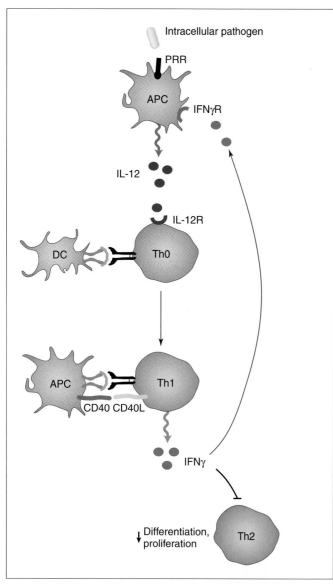

Figure 15-3

Feedback Loop of Th1 Differentiation
APCs that encounter an intracellular pathogen are activated and stimulated to produce IL-12. This IL-12 can bind to the IL-12R expressed by Th0 cells that have been activated by cognate antigen. The Th0 cell differentiates into Th1 effector cells that interact with APCs presenting cognate antigen and supplying the CD40 costimulatory contact. The activated Th1 cell produces copious quantities of IFNγ, which both inhibits Th2 differentiation and promotes additional secretion of IL-12 by activated APCs.

An interesting possibility is that a member(s) of the Notch family of cell fate determination molecules (discussed in Chs.9 and 13) may be involved in generating this IL-4. Notch1 receptors bind to two families of ligands called Delta (or Deltex) and Jagged. It has been proposed that a Th0 cell expressing Notch1 on its surface may interact with an APC expressing Jagged on its surface, and the Th0 cell is thereby induced to differentiate into a Th2 cell. Conversely, Notch1 engagement by Delta on an APC appears to induce Th1 differentiation. Upon engagement by Jagged, the Notch signaling pathway induces GATA-3, which in turn drives transcription of the IL-4 gene. This autocrine production of IL-4 may be sufficient to drive the Th0 cell down the Th2 path. It is not known how Delta engagement might induce Th1 differentiation. Significantly, expression of Jagged in APCs is upregulated under Th2-promoting conditions, while expression of Delta in APCs is enhanced by a microenvironment favoring a Th1 response.

The reader might be wondering about IL-13, the Th2 cytokine whose functions overlap those of IL-4 but are lower in magnitude. The IL-13 receptor contains the IL-4Rα chain (which does not bind IL-13 directly; see Ch.17), and, like IL-4, IL-13 signals through STAT6. Furthermore, IL-13 can substitute for IL-4 to some degree in supporting isotype switching to IgE and IgG1 in B cells (refer to Ch.9). However, IL-13R is not expressed on the surfaces of T cells and thus IL-13 cannot promote Th2 differentiation.

viib) IL-6. It is suspected that IL-6, which is produced by macrophages in response to almost any pathogen, may be important for Th2 differentiation. Exogenous IL-6 can induce CD4$^+$ T cells to produce IL-4, which could operate in an autocrine fashion to promote Th2 differentiation. In this situation, the Th2 pathway of development may operate as the default path, and Th1 development may occur only when macrophages encounter a member of the particular subgroup of intracellular pathogens that induces IL-12 production. Certain products (such as LPS) derived from intracellular bacteria and viruses could activate the macrophages and induce them to secrete large amounts of IL-12, swamping the amount of IL-6 secreted. The differentiation of Th1 cells would then be favored. However, in infections with other organisms that do not induce IL-12 production, IL-6 secretion would dominate and Th2 differentiation might be favored. In IL-6-deficient mice, CTL responses to LCMV infection (which are Th1-dependent) were normal, but antibody responses to vaccinia virus, vesicular stomatitis virus, and *Listeria* (Th2-dependent) were impaired.

viii) Surface Markers of Fully Differentiated Th1/Th2 Effector Cells

When either Th1 or Th2 differentiation is complete, the cells are considered fully functional effector cells. Various cell surface markers and chemokine receptors that distinguish these two types of Th effectors are discussed in the following sections and shown in Figure 15-4.

viiia) Chemokine receptors. Th1 and Th2 cells express different chemokine receptors, implying that they acquire

body: Th2 cells themselves, activated mast cells, and NKT cells (see Ch.18). Since NKT cells produce a huge burst of IL-4 very shortly after anti-CD3 stimulation, it was originally thought that the IL-4 produced by these cells might be required for Th2 differentiation. However, later studies demonstrated that Th2 responses could still be mounted in the absence of NKT cells. Thus, other cellular sources of IL-4 contributing to the support of Th2 differentiation likely exist but their identity remains a mystery to date.

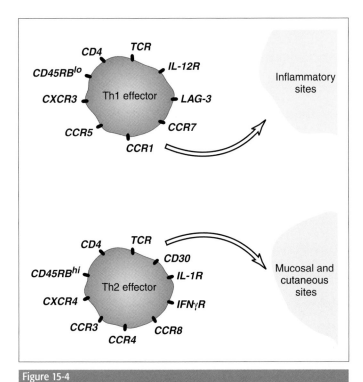

Surface Phenotypes of Th1 and Th2 Effector Cells
Important molecules expressed on the surfaces of Th1 and Th2 effector cells are shown. See text for details.

Interestingly, CCR3 is found not only on Th2 cells but also on eosinophils and basophils. This expression of a common chemokine receptor allows these different cell types to congregate in the same vicinity in response to the appropriate chemokine ligands. For example, alveolar epithelial cells in the airway can secrete eotaxin, a chemokine named for its ability to attract eosinophils. Eotaxin is a ligand for CCR3, such that eosinophils, basophils, and Th2 cells are chemoattracted to the lungs and become co-localized there. This situation favors the initiation of allergic inflammation because the IL-4 and IL-5 secreted by the Th2 cells promote the growth and activation of both eosinophils and basophils.

viiib) IFNγR. IFNγ is a crucial activator of macrophages, but lymphocytes bearing IFNγR experience inhibition of their proliferation in the presence of IFNγ. Th1 cells lose expression of the β chain of the IFNγR early during their differentiation, sparing them from the anti-proliferative effects of the IFNγ that they themselves (and other cell types) secrete. However, the proliferation of Th2 cells, which continue to express a functional IFNγR, is inhibited.

viiic) LAG-3. Th1 cells carry LAG-3 (*lymphocyte activation gene 3*; CD223), a transmembrane protein of the Ig superfamily that structurally resembles CD4 but which is not involved in antigen recognition. LAG-3 expression is induced on activated Th1 cells by IFNγ and enhanced by IL-12. LAG-3 is not found on Th2 cells, probably because LAG-3 expression is inhibited by IL-4. As mentioned in Box 11-2, LAG-3 has been shown to function as a ligand for MHC class II molecules, triggering signaling within a mature DC that promotes the expression of particular chemokine receptors. Engagement of MHC class II by purified LAG-3 *in vitro* can induce the maturation of immature DCs.

viiid) CD30. In Th2 cells (but not Th1 cells), IL-4 induces high surface levels of CD30. We first encountered CD30 in our discussion of negative selection in the thymus; the function of CD30 in the periphery is unknown. It is thought that the IFNγ secreted by Th1 cells blocks their expression of CD30.

III. ACTIVATION OF EFFECTOR TH1 AND TH2 CELLS

Fully differentiated effector Th1 and Th2 cells are ready to be activated by encounter with specific antigen. Each subset has a number of characteristic phenotypic and functional features that are described in the following sections and summarized in Table 15-2.

i) Nature of APCs

With the completion of Th1/Th2 cell differentiation, the site of injury or inflammation in the peripheral tissue under attack becomes filled with competent Th effector cells. Depending on the nature of the insult to the host and the cytokine milieu, cells of the appropriate subset dominate the response. Effector Th cells encounter antigen presented by APCs in the site of attack and are activated in much the same way as naive Th cells but with some important differences. First of all, effector Th cells express higher levels of adhesion molecules

differential trafficking patterns as well as differential cytokine production. While both Th1 and Th2 cells initially express CCR7 (CC chemokine receptor 7), which allows them to home to lymphoid follicles and supply help to B cells, only Th1 cells express CCR1, CCR5, and CXCR3 (CXC chemokine receptor 3), which direct cells to inflammatory sites. In contrast, CCR4, CCR3, and CCR8 are expressed by Th2 cells but not by Th1 cells. These chemokine receptors direct T cells to sites affected by Th2 responses (such as respiratory mucosae irritated by an allergen). Forced expression of a chemokine receptor of the opposite Th type causes inappropriate location of the Th cell and impairment of its function. Chemokine receptor expression also affects APC localization, which in turn can influence the Th1/Th2 balance. For example, Langerhans cells (skin DCs) need CCR2 to migrate properly and contribute to protective Th1 responses against *L. major*. CCR2$^{-/-}$ mice are unable to mount such responses and instead show increased Th2 responses that leave the animal highly susceptible to the disease.

The expression of particular chemokine receptors is directly influenced by the presence of cytokines promoting either Th1 or Th2 development. For example, it has been shown for human T cells that IFNα enhances CXCR3 expression but inhibits that of CCR3. Similarly, TGFβ and other cytokines favoring Th2 development have been shown to enhance the expression of CCR4 and inhibit the expression of chemokine receptors preferentially (but not exclusively) appearing on human Th1 cells.

Table 15-2 Comparison of Th1 vs. Th2 Type Effectors

	Th1 Response	Th2 Response
Pathogens combated	Intracellular	Extracellular
Preferred APCs	Macrophages	B cells
Surface markers of Th effector cells	CD45RB^{lo} CXCR3, CCR5, CCR1, CCR7 LAG-3, IL-12R	CD45RB^{hi} CCR3, CCR4, CCR8 CD30, IFN-γR, IL-1R
Costimulation required for effector activation	CD28/B7^{lo} OX40/OX40L IL-18	CD28/B7$^{very\ lo}$ ICOS/ICOSL IL-1
Cytokines secreted	IFNγ, IL-2, LTα, GM-CSF^{hi}	IL-4, IL-5, IL-13, IL-10, IL-3, IL-6, GM-CSF^{lo}, IL-1
Characteristic Ab promoted	IgG2a, IgG2b (mouse) IgG1, IgG3 (human)	IgA, IgE, IgG1 (mouse) IgG4 (human)
Ab effector functions promoted	ADCC, complement activation, opsonization	Neutralization
Cell functions influenced	Macrophage, DC, Tc cell, Th2 cell	Macrophage, DC, B cell, eosinophil, mast cell, Th1 cell

than do naive T cells. This increased adhesion between the APC and effector Th cells stabilizes their interaction, facilitating TCR triggering. As noted in Chapter 14, an estimated 20–30 hours of sustained TCR signaling is required for a naive T cell to become fully committed to proliferation, and significant quantities of antigen must be presented by DCs. In contrast, only 1 hour of signaling is required to commit an effector Th cell to proliferation. Thus, effector Th cells can respond efficiently to antigen presented by DCs, macrophages, or B cells and may even be stimulated by antigen presented on cells lacking costimulatory capacity, such as keratinocytes. In addition, the concentration of antigen required for activation of Th effectors is lower. As mentioned previously, as the primary response proceeds, fully differentiated Th1 and Th2 effectors display a bias for the type of APC with which they interact. In general, Th2 cells respond preferentially to antigen presented by B cells, while Th1 cells favor antigen presented by macrophages.

ii) Costimulation

iia) CD28/B7. During the activation of both naive and effector Th cells, CD28 signaling increases the efficiency of signal transduction from the triggered TCRs. However, in effector Th cells, the activation threshold is lower and the time to reach it is decreased, so that effector Th cells are less dependent on CD28 costimulation for activation than naive cells. Interestingly, Th2 cells are less susceptible to the induction of anergy (non-responsiveness; see Ch.16) than Th1 cells, perhaps reflecting a relatively lower need for CD28-mediated costimulation during activation (but not differentiation) compared to Th1 cells.

However, the most important effect of CD28 costimulation on effector cells may be protection against *activation-induced*

cell death (AICD). AICD is defined as the induction of apoptosis in response to prolonged antigenic stimulation. (AICD is a means of eliminating effector cells after they are no longer needed and thus is discussed in detail in Section E.) Because of their decreased activation threshold, effector cells can be over-stimulated relatively easily, leading to their elimination by AICD. CD28 engagement reduces the elapsed time required to achieve the activation threshold, thus avoiding prolonged stimulation and preventing the initiation of AICD. In *in vitro* studies, brief exposure to a high concentration of antigen, which can also shorten the time required for activation, had the same protective effect on effector T cells.

CD28 signaling has effects on Th1 and Th2 effectors in addition to costimulation. As was true in naive T cells, engagement of CD28 induces the expression of the anti-apoptotic protein Bcl-xL in effector cells. Bcl-xL also assists in preventing AICD. In addition, CD28 signaling downregulates the expression of chemokine receptors, preventing the effector cell from migrating away from the site where antigen has been encountered.

iib) Supplementary costimulation. Different supplementary costimulatory molecules are influential in Th1 versus Th2 responses (refer to Fig. 15-4 and Table 15-2). OX40 signaling is particularly important for Th1 responses, since OX40$^{-/-}$ mice show a decreased number of IFNγ-producing Th1 cells, and OX40L$^{-/-}$ mice have impaired DTH responses. In contrast, Th2 responses are more dependent on ICOS signaling. ICOS is highly expressed by Th2 effector cells but rarely by Th1 effectors. Experiments using blocking antibodies and in ICOS$^{-/-}$ and ICOSL$^{-/-}$ mice have shown that an absence of ICOS function prevents the production of IL-4 and IL-5 by Th2 cells. Isotype switching to antibody isotypes associated with Th2 responses is inhibited and germinal center formation is impaired. Th2-mediated responses in mouse model systems of asthma involving airway hyper-responsiveness and the infiltration of eosinophils into the lung are also abrogated.

iic) IL-1R/IL-1. While IL-1 has no appreciable influence on naive T cells, both the proliferation and the ability of antigen-stimulated Th2 effector cells to respond to IL-4 are significantly enhanced in the presence of IL-1. IL-1 is a pro-inflammatory cytokine produced by many cell types in response to injury or infection but particularly by mononuclear phagocytes (see Ch.17). In addition, while B cells (the preferred APCs for Th2 cells) do not make IL-1, Th2 cells soon acquire the capacity to make IL-1 after interacting with B cells to engage CD28. Unlike Th2 cells, Th1 cells do not express IL-1R, meaning that they cannot receive the proliferative signals delivered by IL-1. Furthermore, Th1 cells produce IFNγ, which triggers intracellular events interfering with IL-1 signaling.

iid) IL-18. IL-18 is an important stimulator of antigen-activated Th1 cells. IL-18 is synthesized by activated macrophages and has effects similar to those of IL-12 (see Ch.17). While IL-18 does not contribute to Th1 differentiation per se, this cytokine acts synergistically with IL-12 to maximize the production of IFNγ by mature Th1 effectors. IL-18 also promotes the proliferation of Th1 (but not Th2) cells. IL-18 uses a different signaling pathway from that triggered by IL-12 engagement, resulting in the activation of the transcription factor NF-κB rather than STAT4. IL-18 also induces the production of IL-2 by Th1 cells and upregulates the expression of the IL-2Rα chain on Th1 cells (but not on Th2 cells).

iii) Downstream Signal Transduction Pathways

Th1 and Th2 effector cells exhibit striking differences in signaling paths downstream of TCR engagement. Work in the early 1990s showed that Th2 cells have a much lower threshold of activation than Th1 cells. Cross-linking of anti-CD3 antibodies is required to activate isolated Th1 cells *in vitro*, whereas isolated Th2 cells are able to respond to incubation with anti-CD3 antibodies without cross-linking. Secondly, as they differentiate, Th2 cells gradually lose the ability to alter Ca^{2+} flux in response to TCR stimulation, unlike Th1 cells. Thirdly, chemical agents (such as cholera toxin) that are potent inhibitors of the activation of Th1 cells have little or no effect on Th2 activation. Taken together, these results imply that some signaling pathways may be unique to either Th1 or Th2 activation, while other pathways may be common to both but have different levels of importance (Fig. 15-5).

Biochemical studies have confirmed that TCR engagement may emphasize the activation of different signaling paths in Th1 and Th2 cells. Two of the T cell signaling pathways introduced in Chapter 14 are particularly important in Th1 activation: the p38 pathway leading to the activation of various ATF transcription factors, and the SAPK/JNK pathway leading to c-jun activation. The IFNγ promoter contains binding sites for both c-jun and the ATFs. Studies in knockout mice have demonstrated that defects in any one of the signaling intermediaries (MKK3, MKK6; refer to Fig. 14-6) in the p38 pathway block IFNγ production by Th1 cells. The small GTP-binding

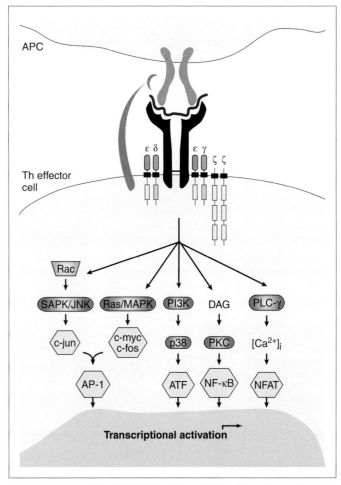

Figure 15-5

TCR Signaling Pathways in Th Cells
Engagement of the TCR by cognate antigen activates several major intracellular signaling pathways: the SAPK/JNK pathway, the Ras/MAPK pathway, the p38 pathway, the PKC pathway, and the PLC-γ1 pathway. As a result, transcription factors such as AP-1, ATF, NF-κB, and NFAT are mobilized to induce new gene transcription. Different pathways appear to have different levels of importance in Th1 and Th2 cells. For example, defects in SAPK/JNK and p38 pathways block Th1 effector functions.

protein Rac plays a starring role, since mice lacking this molecule have reduced IFNγ production. Conversely, transgenic overexpression of Rac2 in T cells enhances IFNγ expression.

IV. EFFECTOR FUNCTIONS OF Th CELLS

As stated previously, the primary effector function of Th cells is to provide "T cell help" in the form of secreted cytokines and costimulatory contacts. These cytokines and contacts then contribute to three cellular outcomes: the activation of B cells, the activation of Tc cells, and the hyperactivation of macrophages. The roles of Th cells in activating macrophages and B cells (particularly the provision of the CD40 contact) have been described in Chapters 3 and 9, respectively. How Th cells supply help for CTL generation is described in Section C of this chapter. Here we briefly review how the cytokines secreted by Th1 and Th2 cells promote these outcomes, as well as how Th1 and Th2 cells cross-regulate each other. A more detailed discussion of the production and broader actions of these cytokines can be found in Chapter 17.

i) Cytokine Secretion by Th1 and Th2 Cells

Th1 cells support cell-mediated immune responses by secreting IFNγ, IL-2, lymphotoxin α (LTα), and GM-CSF (high levels) but not IL-4, IL-5, IL-10, or IL-13 (Fig. 15-6 and refer to Table 15-2). IFNγ and LTα activate macrophages and spur them to secrete additional cytokines, undertake vigorous phagocytosis, and upregulate NO production. The expression of high affinity FcγR molecules is also increased on macrophages by IFNγ, permitting these cells to participate in increased ADCC. Th1 cells are the major producers of IL-2, which promotes the proliferation of both T and B cells and enhances the production of reactive oxygen intermediates by macrophages. Th1 cells also supply the CD40L contact and

IFNγ required for B cell production of Ig isotypes best suited for opsonization, phagocytosis, and complement activation: IgG2a and IgG2b in mice and IgG1 and IgG3 in humans. These antibodies also bind to FcR on macrophages and other phagocytes with high affinity, further promoting ADCC. In addition, Th1 cytokines increase the antigen-presenting potential of macrophages by upregulating MHC class II and TAP. In contrast, B cells encountering Th1 cytokines decrease their expression of molecules required for antigen presentation, and thus become less efficient APCs. It is also primarily Th1 cells that support the activation of Tc cells via production of IL-2 and IFNγ, and provision of the CD40/CD40L contacts that "license" APCs to present antigen to naive Tc cells (refer to Box 14-1).

Th2 cells secrete IL-3, IL-4, IL-5, IL-6, IL-10, IL-13, and GM-CSF (low levels), but not IL-2, IFNγ, or LTα. A major function of Th2 cells is to establish CD40-CD40L contacts with B cells and to secrete IL-4 and IL-5, cytokines that

---

Figure 15-6

Major Th1/Th2 Effector Functions in Mouse
Cytokines secreted by Th1 effector cells have profound effects on B cells, macrophages, and Tc cells. B cells undergo isotype switching to antibody isotypes best suited for opsonization and ADCC, the microbicidal activities of macrophages are stimulated, and Tc cell differentiation into CTLs is promoted. Cytokines secreted by Th2 cells act on B cells, mast cells, eosinophils, and macrophages. B cells undergo isotype switching to isotypes best suited for mucosal defense and neutralization. Mast cells and eosinophils are activated but macrophage activities are suppressed.

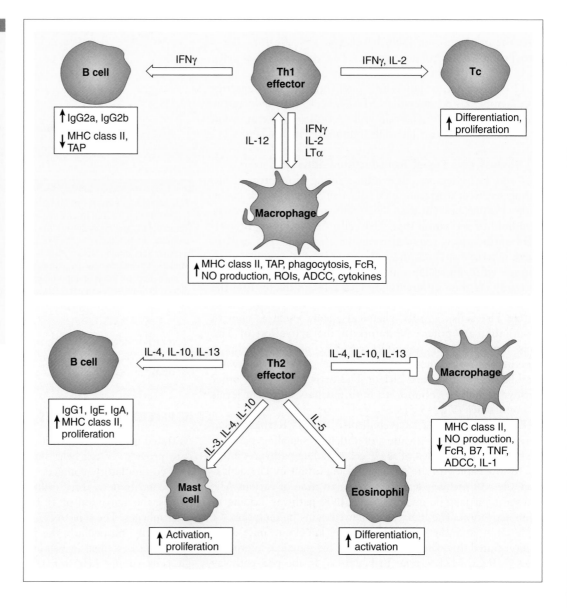

induce isotype switching to IgA, IgE, and IgG1 (in mice) or IgG4 (in humans). These isotypes are less proficient at complement activation and ADCC, which is an advantage in combating pathogens in mucosal sites where neutralization in the absence of inflammation-inducing effector functions is preferred. IL-5 produced by Th2 cells is also important because it promotes the growth, differentiation, and activation of eosinophils crucial for the elimination of large parasites such as helminth worms. Similarly, Th2-derived IL-3, IL-4, and IL-10 work together to stimulate mast cell activation and proliferation. The stimulation of both eosinophils and mast cells and the presence of IgE antibodies can set the stage for various types of allergic inflammation. Another major function of Th2 cells is the regulation of macrophage activation via secretion of IL-4, IL-10, and IL-13. IL-10 inhibits effector functions of macrophages, abrogates their production of IL-12, and downregulates MHC class II expression on these cells. IL-10 also downregulates B7 ligand expression on macrophages and DCs but increases MHC class II expression on B cells. IL-4 and IL-13 inhibit pro-inflammatory cytokine production, downregulate NO production, and decrease FcγR expression on macrophages, blocking ADCC. However, IL-4 upregulates MHC class II expression on APCs such as macrophages, DCs, and B cells, and stimulates B cell proliferation.

ii) Macrophage Hyperactivation and the Delayed Type Hypersensitivity Reaction

Macrophage hyperactivation, described in detail in Box 3-1, is another Th effector function. Hyperactivation of macrophages by Th1 cells enhances the ability of the macrophages to dispose of invaders, but in some cases leads to detrimental consequences for the host in the form of *delayed type hypersensitivity* (DTH) *reactions*. "Hypersensitivity" in the immunological context is a reaction to an antigen that causes clinically recognizable damage to a host previously exposed to that antigen. "Delayed" hypersensitivity is so-named because, in contrast to "immediate" hypersensitivity (which is manifested in minutes), a localized inflammatory reaction is not observed until 2–3 days after the antigen is encountered by a primed individual. Thus, a DTH reaction does not result from Th1 cell effector function per se, but rather from the effector functions of hyperactivated macrophages. However, because hyperactivation of the macrophages depends on the cytokines secreted by antigen-stimulated Th1 cells, some commentators consider DTH to be a Th1 effector function. In any case, as with all hypersensitivities, a DTH reaction can be considered the unfortunate corollary of an overenthusiastic host defense mechanism. More on DTH and other forms of immune hypersensitivity can be found in Chapter 28.

iii) Cross-regulation of Th1 and Th2 Cells

The reader may have already noticed that a collateral function of the Th subsets is the cross-regulation of aspects of each other's responses, either positively or negatively. For convenience, we collect these observations here in summary form. We emphasize that cross-regulation cannot alter the nature of a fully differentiated Th1 cell to Th2, or vice versa.

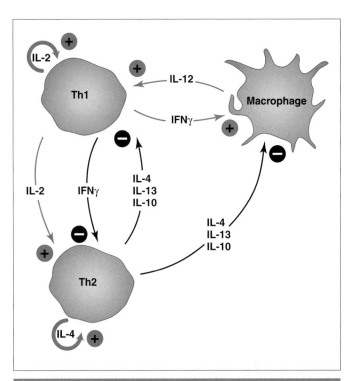

Figure 15-7

Th1/Th2 Cross-regulation
IL-2 produced by Th1 cells stimulates the proliferation of Th2 effectors as well as that of the Th1 effectors themselves. IFNγ produced by Th1 cells inhibits Th2 differentiation but stimulates macrophages to produce IL-12. IL-4 produced by Th2 effectors stimulates their proliferation, but IL-4, IL-13, and IL-10 inhibit both Th1 differentiation and macrophage activation.

While IFNγ is directly associated with the promotion of Th1 differentiation because of its induction of T-bet and its upregulatory effects on macrophages, it also has inhibitory effects on both Th2 development and effector actions (Fig. 15-7). IFNγ has a direct anti-proliferative effect on Th2 cells and interferes with the costimulatory effects of IL-1 on Th2 cell activation. Conversely, Th2 cells secrete IL-4, IL-13, and IL-10, which can downregulate macrophage production of IL-12 and suppress IFNγ and IL-2 secretion by Th1 cells.

We move now to a discussion of the differentiation and effector functions of Tc cells, the cells that monitor threats to the host that arise from intracellular alterations.

C. Tc Cell Differentiation and CTL Effector Function

As was true for Th cells, the development of a cytotoxic T cell response can be pictured as occurring in three phases: (1) the activation of a naive Tc cell in the lymphoid tissue, (2) its proliferation and differentiation into effector cells called CTLs, and (3) the activation of a CTL in the periphery by encounter with specific antigen on a target cell, leading to the lysis of that cell and others displaying an identical pMHC structure. CD8$^+$ Tc cells recognize nonself antigen presented on MHC class I on target cells, which include tumor cells, tissue grafts, and cells infected with virus or intracellularly replicating bacteria. Thus, CTL-mediated immune responses are important in the containment of tumors, the rejection of allografts, and the elimination of intracellular pathogens.

The importance of CD8$^+$ Tc cells to immunity was first demonstrated using mice depleted of CD8$^+$ cells by the injection of anti-CD8 antibodies prior to infection with an intracellular pathogen. The course of the disease induced by the pathogen was considerably more severe in the depleted mice than in the controls. Similarly, the transfer of CD8$^+$ T cells alone from mice immune to a particular intracellular parasite was found to render a recipient (of the correct MHC haplotype) resistant to infection with that parasite. These and other experiments clearly showed that CD8$^+$ T cell-mediated cytotoxic responses are essential for effective clearance of many endogenous pathogens. However, some intracellular parasites and viruses are most effectively cleared by antibodies and/or phagocyte-mediated mechanisms, even though these pathogens also stimulate a CD8$^+$ CTL-mediated response. For example, while CD8$^+$ CTLs are crucial contributors to host defense against *Listeria monocytogenes*, *Toxoplasma gondii*, *Mycobacterium tuberculosis*, and LCMV, they are not as important for protection against *L. major*, vaccinia virus, or VSV (see Ch.22).

We will now examine the process by which effector CD8$^+$ CTLs develop. The reader should note that, although the vast majority of CTLs in the body carry the CD8 coreceptor, a few have been found to express CD4 instead. CD4$^+$ CTLs lyse cells bearing antigenic peptide presented on MHC class II, apparently using the same mechanisms as CD8$^+$ CTLs.

I. GENERATION OF EFFECTOR CTLs

Most of the material described in the following sections has come from studies of CTL clones or T cells derived from TCR Tg mice. These approaches allow the responses of large numbers of CTLs to be dissected with a minimum of confounding background effects.

i) Differentiation of Tc Cells into Armed CTLs

As was discussed in Chapter 14, when naive, resting CD8$^+$ Tc cells resting in a lymph node encounter specific peptide presented by MHC class I on a target cell, and then (usually) interact with a licensed APC capable of delivering costimulatory signals, they become activated and commence expression of the high affinity IL-2 receptor. In the presence of IL-2 either secreted by itself or by an activated, antigen-specific Th1 cell, the activated Tc cell proliferates into daughter cells. (Note that IL-2 cannot influence unrelated Tc cells in the vicinity because they have not been induced to express IL-2Rα. This stricture prevents the indiscriminant activation of Tc cells and the uncontrolled generation of CTLs.) Finally, in the presence of other cytokines secreted by the Th1 cell or an activated APC, these Tc daughter cells differentiate

into effector CTLs, which fan out to function throughout the periphery.

We emphasize that the original Tc cell, even when activated by antigen, has no lytic powers at all; it is only the mature CTLs that develop cytotoxicity. For this reason, Tc cells that have been activated by antigen but have not yet differentiated into CTLs are sometimes called *precursor CTLs* (pCTLs) or CTL-P cells, for *CTL precursors* (Fig. 15-8). The daughter cells of the activated antigen-specific Tc cell undergo a process of differentiation into CTL effectors in which IL-12 plays a particularly important role. IFNγ and IL-6 may also be involved. These pro-inflammatory cytokines are for the most part derived from APCs activated by injury or pathogen invasion. Indeed, culture of activated Tc cells with DCs from IL-12-deficient mice induced only very low levels of CTL generation. This restriction of the development of cytotoxicity to an inflammatory context is another means by which the power of CTLs is reserved for situations in which a pathogenic threat is present.

In contrast to the wealth of information on transcription factors involved in Th1/Th2 differentiation (refer to Box 15-1), comparatively little is known as yet about transcription factors driving pCTL differentiation. T-bet, the transcription factor crucial for CD4$^+$ Th1 differentiation, appears to be involved in CD8$^+$ Tc differentiation but is not sufficient. Differentiating pCTLs also feature upregulated expression of a transcription factor called *eomesodermin* that was first identified as playing an important role in early murine embryological development. Like T-bet, eomesodermin contains the highly conserved DNA-binding T-box domain. It is thought that CTL function depends on the binding of the T-box domain (of perhaps either transcription factor) to sites in genes regulating the expression of the IFNγ and the cytolytic granule protein genes. Indeed, while T-bet$^{-/-}$ mice lack Th1 cells, eomesodermin is able to compensate for the absence of T-bet such that these mutant animals retain relatively normal CD8$^+$ T cell responses. In addition, it has been found that CD8$^+$ T cells from STAT4$^{-/-}$ mice are still able to produce IFNγ. Together, these results suggest that eomesodermin may operate through a different pathway than T-bet to ultimately induce IFNγ expression and Th1 differentiation, and that this pathway may not rely on IL-12R induction and STAT4. These matters remain under investigation.

Within 24–48 hours of TCR stimulation of the original CD8$^+$ Tc cell, cytotoxic granules containing lytic proteins can be found in the cytoplasm of newly generated CTLs. Experiments *in vitro* have shown that the cytotoxic response of a mature CTL does not require further *de novo* protein synthesis, indicating that the proteins that mediate target cell lysis are already in place when the response is induced by encounter with specific antigen. A mature CTL (that has yet to encounter antigen) is said to be "armed" when it has synthesized the chemical mediators that will be used to carry out target cell destruction.

ii) Activation of Armed CTLs and Conjugate Formation

During the course of its differentiation, the armed CTL upregulates homing receptors (including LPAM-1 and LFA-1) that

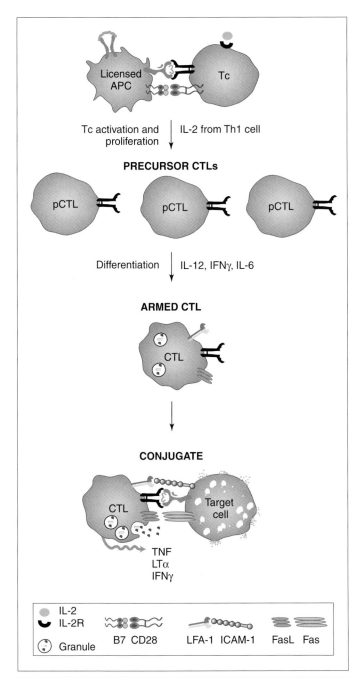

Figure 15-8

CTL Generation and Cytotoxicity
A naive Tc cell encountering a licensed APC presenting cognate antigen in the presence of IL-2 is activated, proliferates, and generates daughter cells called "precursor CTLs." In the correct cytokine milieu, these precursors differentiate into armed CTLs containing cytotoxic granules and expressing FasL. When the TCR of the armed CTL is engaged by cognate antigen, the apoptotic death of the target cell is induced by Fas killing, cytotoxic granule release, or the secretion of cytotoxic cytokines such as TNF. (All three mechanisms are shown in the figure for convenience.) The ultimate death of the target cell occurs after the CTL has detached.

just enough time to scan the pMHC class I complexes on display. The reader will recall that, because of negative selection in the thymus, the CTL most likely bears a TCR that does not recognize self-peptide bound to self-MHC, and so the CTL usually detaches and moves onto the next host cell without incident. However, should the armed CTL encounter a nonself-pMHC for which its TCR is specific (has sufficient affinity), the host cell becomes a target. Stimulation of the TCR of an armed CTL induces a rapid change in the conformation of the adhesion molecule LFA-1 that increases the affinity of its binding to ICAM-1 on the target cell, resulting in a strong bond between the two cells. The bicellular structure formed by the adherence of the CTL and the target cell is called a *conjugate*. The formation of the conjugate not only allows the efficient delivery of the "lethal hit" of chemical mediators causing target cell lysis but also ensures that the activated CTL will kill <u>only</u> those target cells that display the correct nonself-pMHC complex; bystander cells are affected only rarely.

Experiments *in vitro* have shown that extremely low concentrations of agonist peptides can activate armed CTLs, in comparison to the much greater concentrations needed to activate the corresponding naive Tc cells. Further investigation has confirmed that CTL activation requires the engagement of far fewer TCRs than naive Tc cell activation. In fact, only a single TCR on an armed CTL need be engaged by a single specific pMHC ligand to induce a killing response. Consequently, no detectable TCR internalization occurs. Furthermore, the interaction of the TCR with the appropriate pMHC (signal one) is sufficient for armed CTL activation; no costimulation (i.e., no signal two) nor interaction with a third cell is necessary for a CTL killing response. Compare this increased sensitivity with the estimated threshold of 8000 TCRs that must be engaged for most naive CD8+ Tc cells to receive signal one in the absence of costimulation, and the 1500 TCRs that have to be triggered even with costimulation (refer to Ch.14).

II. MECHANISMS OF TARGET CELL DESTRUCTION BY CD8+ CTLs

Target cell destruction mediated by CTLs occurs via three methods: the *granule exocytosis* pathway, the Fas pathway, and the release of cytolytic cytokines such as TNF, LTα, and IFNγ (Fig. 15-9). Which cytotoxic pathway takes precedence in the killing of a particular intracellular pathogen varies with the organism. For example, the granule exocytosis pathway is more important in the control of *L. monocytogenes* than of *L. major*, which is controlled primarily by Fas killing. Overall, granule exocytosis accounts for the majority of target cell killing by CD8+ CTLs.

i) Granule Exocytosis

The granule exocytosis pathway depends in large part on *perforin* and the *granzymes*, cytotoxic proteins contained in the preformed cytoplasmic granules of CTLs and NK cells. Within a few minutes of conjugate formation and

permit easy access to both diffuse and tertiary lymphoid tissues so that the CTL can survey much of the body for antigen. The armed CTL binds weakly to host cells in these tissues, using its adhesion molecules to stick to a host cell for

Figure 15-9

Mechanisms of CTL Cytotoxicity
The figure shows details of the three possible ways a CTL can kill a target cell. In the granule exocytosis pathway, a CTL releases granules containing granzymes and perforin in close proximity to the target cell membrane. Granzymes and perforin enter the cell by an undefined mechanism and are captured in endolysosomal vesicles. Perforin facilitates the release of the granzymes from the vesicles into the target cell cytoplasm. The granzymes then activate apoptosis by either the mitochondrial pathway or an unknown caspase-independent mechanism. Granzymes can also access the nucleus to induce DNA fragmentation directly. In the Fas pathway, FasL expressed on the surface of the armed CTL binds to Fas expressed on the target cell, inducing caspase-8 activation and apoptosis mediated by either the mitochondrial pathway or direct caspase-3 cleavage (see later in text for details). Similarly, the cytotoxic cytokine pathway is invoked when, for example, TNF secreted by the CTL binds to TNFR1 on the target cell and induces caspase-8 activation.

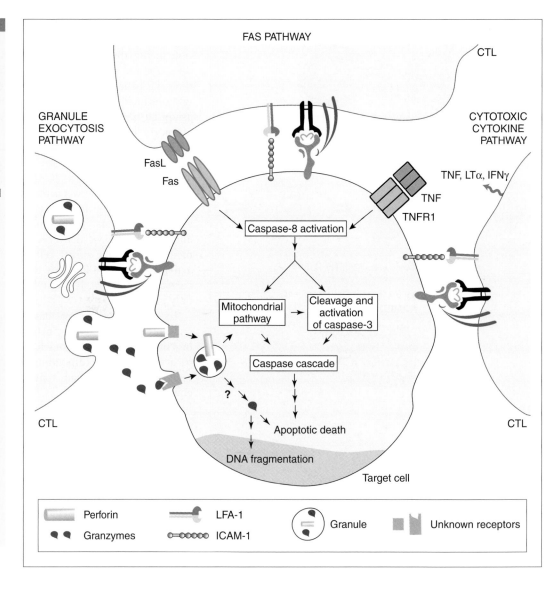

TCR triggering, a change in intracellular energy consumption and Ca^{2+} flux can be detected within an armed CTL. The Golgi reorients and the granules move along the cytoskeleton toward the site of CTL–target cell contact. The granules fuse with the CTL membrane and the cytotoxic contents of the granules are directionally exocytosed and released toward the target cell membrane. Within 2 minutes of formation of a CTL–target cell conjugate, degraded proteins can be detected within the target cell. Such rapid speed is needed to ensure the death of an infected cell before the pathogen can escape to a new host.

ia) Role of perforin. Perforin is a pore-forming protein similar in structure to complement component C9. Perforin is synthesized as monomers of 65 kDa and sequestered within the cytotoxic granules of armed CTLs and NK cells. Early *in vitro* work demonstrated that when perforin is present in the extracellular milieu, contact with the membrane bilayer of a cell causes the conformation of perforin monomers to alter

such that a hydrophobic domain is exposed. This domain permits the insertion of the monomers into the bilayer. In the presence of Ca^{2+} ions, 3–20 inserted monomers polymerize to form a cylindrical perforin pore of 5–20 nm through the membrane, a structure that resembles the membrane attack complex (MAC) of complement killing. Thus, it was originally thought that, once a CTL or NK cell had degranulated, the granzymes accessed the cytoplasm of the target cell through a perforin pore established in the cell's plasma membrane. Indeed, granzyme-mediated target cell death was shown to be impaired in the absence of perforin, and a dependence on Ca^{2+} was originally used to distinguish granule exocytosis-mediated CTL killing from that mediated by the Ca^{2+}-independent Fas pathway. However, subsequent work has demonstrated that granzymes are confined to endolysosomal vesicles after entering target cells (see later). The role of perforin appears to be to facilitate the release of the granzymes from the endolysosomal vesicles. Interestingly, it is now thought that perforin pores are much too small to allow the passage of the relatively large granzyme proteins, making it unlikely that perforin creates a tunnel for their escape. Instead, the perforin must act in some other manner that regulates intracellular trafficking of the endolysosomal vesicles and/or induces granzyme release from them.

Prior to the generation of perforin knockout mice, it was not entirely clear how important perforin actually was in cytotoxic immune responses. Perforin$^{-/-}$ mice show a clear deficit in the killing of Fas-negative target cells, although Fas-positive target cells can still be killed via the Fas pathway. Perforin-mediated cytotoxicity was subsequently found to be crucial for the control of the intracellular pathogen *L. monocytogenes* and the LCMV virus. Another interesting finding was that, in a mouse model of autoimmune diabetes, mice lacking perforin failed to develop the disease. This result implied that, in susceptible mice developing autoimmune diabetes, the attack by CTL on the pancreatic β-islet cells was perforin-mediated. However, CTLs functioning in allogeneic transplant rejection do not depend primarily on the perforin pathway, since allogeneic heart grafts were rejected equally in wild-type and perforin$^{-/-}$ mice.

ib) Role of granzymes. Members of the granzyme family of proteases (also called *fragmentins*) are exocytosed (along with perforin) toward the target cell membrane during CTL and NK cell degranulation. The granzymes are neutral serine proteases of size 27–65 kDa that are found almost exclusively in cytolytic granules. Granzymes are first expressed as inactive precursor proteins with leader sequences that allow them to enter the secretory pathway and ultimately the granules. The granzymes are not active until they are cleaved by proteases (such as cathepsin C) confined to the granules. However, the pH of the granules is more acidic than the optimal operating pH of the granzymes, inhibiting their proteolytic activity within CTLs and NK cells. As well, hematopoietic (but not non-hematopoietic) cells express cytoplasmic protease inhibitors that specifically inhibit granzymes.

Upon degranulation of a CTL or NK cell, the granzymes (and perforin) are taken up by the target cell in a mechanism that remains unclear. Fluid phase pinocytosis and receptor-mediated endocytosis via a cation-independent mannose-6-phosphate receptor have both been cited as potential mechanisms, but this issue is controversial. In any case, upon target cell entry, perforin and the granzymes are immediately confined to the endolysosomal system of the target cell where these molecules can do no damage. As mentioned previously, perforin then facilitates the release of granzymes from the endolysosomal vesicles into the cytoplasm of the target cell. Once freed into the cytoplasm and able to access the nucleus, at least some granzymes participate in the induction of apoptosis and degrade important intracellular substrates, including DNA.

In mice, nine separate enzymes have been identified, called granzymes A, B, C, D, E, F, G, K, and M. The genes encoding these enzymes are located on two separate chromosomes: the granzyme A and M genes are located on mouse chromosome 10 while the granzyme B, C, D, E, F, and G genes are clustered on chromosome 14. In humans, five granzymes have been isolated: A, B, H, K, and M. The genes for human granzymes A and K are located on chromosome 5, while those for granzymes B and H are on chromosome 14. The granzyme M gene is on chromosome 19. Granzymes A and B are equivalent in mice and humans, but human granzyme K is apparently not a counterpart of any of mouse granzymes C–G. Human granzyme H, which is expressed primarily in NK cells, has some similarities to murine granzymes C, D, and F. Only granzymes A and B have been characterized in any detail.

The most important granzyme studied to date is granzyme B, a monomer of 27–29 kDa. Analysis of the crystal structure of granzyme B has confirmed that it is highly similar to the protease chymotrypsin. However, granzyme B cleaves peptide bonds after aspartic acid residues. This specificity is rare and resembles the restricted cleavage patterns of the caspases (which, as the reader will recall from Box 3-4 in Ch.3, are cysteine proteases promoting apoptosis). Indeed, granzyme B appears to have a function closely related to that of the caspases. Purified granzyme B can activate caspases-3, -6, -7, -8, -9, and -10 *in vitro*, and induces the apoptosis of target cells *in vitro*. However, there is also evidence that granzyme B can bypass the caspases and induce apoptosis by an unknown mechanism. Finally, once it accesses the nucleus, granzyme B can directly induce DNA fragmentation and cleave intracellular substrates associated with DNA repair such as DNA-PK and *poly-ADP-ribose polymerase* (PARP).

Granzyme B$^{-/-}$ mice show dramatic defects in the induction of target cell apoptosis even in the presence of normal perforin, indicating that the early induction of apoptosis in target cells is absolutely dependent on granzyme B. The DNA fragmentation and morphological changes characteristic of apoptosis are not evident in allogeneic target cells subjected to attack by granzyme B$^{-/-}$ CTLs. Furthermore, granzyme B$^{-/-}$ CTLs are unable to cleave procaspase-3, indicating an inability to initiate the caspase cascade. The mutation of the granzyme B gene also dramatically disrupts the expression of granzymes C, D, E, F, and G, whose genes lie downstream of the granzyme B gene in the cluster. Whether granzymes C–G play any role(s) in cytotoxicity has yet to be determined; there is some evidence that at least some of their functions may be redundant with those of granzyme B.

Granzyme A is a homodimer of total size about 60 kDa that cleaves peptide bonds after arginine residues. Contrary to original hypotheses, granzyme A does not appear to be particularly important for cytotoxicity. Granzyme A can induce target cell apoptosis *in vitro*, but does so much more slowly than granzyme B. In contrast to granzyme B$^{-/-}$ mice, granzyme A$^{-/-}$ knockout mice exhibit normal CTL- and NK-induced cytolysis of target cells, and are perfectly capable of combating LCMV and *Listeria* infections. Thus, at least some functions of granzyme A may overlap those of granzyme B. However, the substrates and modes of action of granzymes A and B are very different. Granzyme A does not activate caspases or cleave PARP but does degrade the nuclear lamins responsible for nuclear integrity. In addition, rather than the DNA laddering associated with granzyme B action, granzyme A appears to activate a nuclease that causes single-stranded DNA breaks.

ii) Fas Pathway

The role of Fas-mediated apoptosis was discussed in Chapter 13 in the context of negative selection in the thymus. In fact, Fas is a transmembrane "death receptor" that is widely expressed on mammalian cells. Engagement of trimeric Fas on a cell by a trimer of its membrane-bound ligand FasL expressed either on itself or on an adjacent cell triggers the Fas-bearing cell to initiate apoptosis (see later in this chapter). Naive Tc cells do not express FasL, but after activation by encounter with antigen, FasL starts to be synthesized and stored in specialized transport vesicles in differentiating CTL. Upon conjugate formation, the FasL-containing transport vesicles move to the cell surface where fusion with the cell membrane leaves FasL anchored within it. The high levels of FasL expressed on an armed CTL allow it to kill Fas-expressing target cells even in the absence of cytotoxic granules. Some scientists have speculated that there are faster- and slower-acting forms of the cytotoxicity delivered by a given $CD8^+$ CTL, and that the faster form employs the perforin/ granzyme system, while the slower form depends on Fas killing. This conclusion was derived from experiments *in vitro* in which CTLs from perforin$^{-/-}$ mice were unable to induce apoptosis of target cells with normal rapidity, but could eventually induce the lysis of certain types of target cells (presumably those bearing Fas) if the incubation was prolonged. Mice doubly deficient for both perforin and FasL were completely unable to induce target cell lysis, even after lengthy incubations. Of interest, FasL is also expressed on some Th1 and Th2 cells, perhaps accounting for the ability of certain rare $CD4^+$ T cells (which are <u>not</u> CTLs) to kill target cells in the absence of perforin or granzymes.

iii) Cytolytic Cytokines

Although they are not as accomplished secretors of cytokines as Th effector cells, CTLs do produce cytokines that influence host cell killing. When TNF produced by CTLs binds to TNFR1 on the target cell surface, apoptosis of the target is usually induced (see Box 9-3 and refer to Fig. 15-9). In addition, CTLs secrete LTα, which is also cytotoxic, and IFNγ, which has multiple effects that promote cytolysis. For example, IFNγ stimulates B cells to produce antibodies that can lead to death by ADCC or complement activation. As well, IFNγ upregulates MHC class I on nearby host cells, enhancing antigen display and making the target more visible to scanning CTLs. The production and functions of TNF, LTα, and IFNγ are discussed in detail in Chapter 17.

III. DISSOCIATION

About 5–10 minutes after conjugate formation and delivery of the lethal hit to the target cell, the LFA-1 molecules on the CTL resume their low affinity conformation, allowing the CTL to dissociate from the damaged target cell. The target cell succumbs to apoptosis within 15 minutes to 3 hours of dissociation, while the CTL commences synthesis of new cytotoxic granules and moves off to scan other host cells in the area. A single armed (and re-armed) CTL can attach to many host cells in succession, delivering lethal hits without sustaining any damage itself. How the CTL avoids self-destruction by the contents of its granules is not clear. Some scientists speculate that the CTL must actually fuse its granules to the membrane of the host cell, delivering the damaging contents in such a way that none escapes into the surrounding extracellular environment. Alternatively, activated CTLs may possess a mechanism or express surface proteins that counter the re-entry of granzymes from the extracellular milieu.

D. Comparison of Naive and Effector T Cells

Now that the reader is familiar with the process of generation of effector $CD4^+$ and $CD8^+$ T cells and their actions, it might be helpful to step back and review the differences in characteristics between naive and effector T cells. Although they carry the same DNA and respond to the same antigen, naive and effector T cells differ with respect to their homing and adhesion molecules, activation signals, and functions and products. They also differ in their susceptibility to AICD, which is discussed later in Section E. Similarities and differences between naive and effector T cells are summarized in the first two columns of Table 15-3.

I. TRAFFICKING AND ADHESION

As described in Chapter 3, the trafficking of lymphocytes around the body is controlled in large part by the pairing of homing receptors expressed on the surfaces of lymphocytes with their ligands expressed by tissue cells. For example, L-selectin expressed on a T cell's surface allows it to bind to GlyCAM-1 in the HEVs of secondary lymphoid tissues such as the lymph nodes. Naive T cells express high levels of L-selectin but low levels of CD2, LFA-1, and CD44. In contrast, effector T cells increase their expression of these latter molecules (which increases their adhesion to APCs) but downregulate L-selectin, allowing effector cells to move out of the secondary lymphoid tissues.

Effector cells also upregulate their expression of new types of adhesion molecules that permit them to circulate through a wide range of body tissues. For example, to join in the diffuse collections of lymphocytes that exist in the mucosae and just under the skin, and to access inflammatory sites, an effector T cell must carry not only LFA-1 but also LPAM-1. In addition, effector cells take advantage of certain members of the β1-integrin subfamily of six transmembrane proteins called the VLA (for *very late activation*) integrins, which mediate adhesion. Each VLA integrin is composed of one of six unique α chains of 120–210 kDa (designated CD49a–f) non-covalently linked to a common β chain of 130 kDa called CD29. Thus, VLA-1 is made up of CD49a/CD29, VLA-2 of CD49b/CD29, and so on. The VLA integrins are widely expressed on leukocytes and interact with a broad range of ligands, many of which, such as fibronectin, collagen, and laminin, are part of the extracellular matrix. In the case of

Table 15-3 Naive vs. Effector vs. Memory Murine T Cell Characteristics

Property	Naive T Cell	Effector T Cell	Memory T Cell
Preferred tissue location	Lymph nodes, spleen, and other secondary lymphoid tissues	Peripheral tissues such as skin, gut, lung, mucosae; inflammatory sites; diffuse lymphoid tissues	Peripheral tissues such as intestine, lung, skin, mucosae; inflammatory sites; diffuse lymphoid tissues
Chemokine receptors	CCR7	CCR4, CCR5, CCR9, CXCR4	CXCR5 (CD4$^+$ memory cells) CCR7$^+$ ("central" memory) CCR7$^-$ ("effector" memory cells)
Surface markers	CD45RA^{hi}, CD45RB^{hi}	CD45RO^{hi}, CD45RB^{lo} CD25^{hi}, CD69^{hi}, CD40L^{hi}	CD45RB^{lo}, CD45RO^{lo} CD27, Ly6C, IL-2Rβ (CD8$^+$ memory cells) CLA (skin), α4β7 integrin (gut)
Preferred APCs	DCs	Macrophages, B cells, DCs	B cells, macrophages, DCs
Adhesion molecules	L-selectin^{hi} CD2^{lo}, LFA-1^{lo} CD44^{lo}, ICAM-1^{lo}	L-selectin^{lo} CD2^{hi}, LFA-1^{hi} CD44^{hi}, ICAM-1^{hi} LPAM-1, VLA-4, 5, 6	L-selectin^{lo} CD2^{hi}, LFA-1^{hi} CD44^{hi}, ICAM-1^{hi} LPAM-1, VLA-4, 5, 6
Costimulation required	CD28/B7 (high)	CD28/B7 (low)	Minimal to none
Duration of TCR signaling	20–30 hours	<1 hour	<1 hour
Appearance of IL-2Rα chain	6 hours	<1 hour	<1 hour
Cell division	Slow	Rapid	Moderate
Function	Produces daughter cells for effector differentiation	Produces cytokines and costimulatory contacts for Th help to B and Tc cells and macrophages Produces granzymes/perforin for CTL-mediated cytotoxicity	Gives rise to second generation of effector cells within 12 hours
Lifespan	5–7 weeks	2–3 days	Up to 50 years
Sensitivity to AICD	Low	High	High

T cells, VLA-4, -5, and -6 are expressed on daughter cells (but not naive T cells) several days after the initial triggering of the TCR and just at the start of the differentiation process into effector T cells. These molecules undergo increases in expression and affinity for their ligands in response to TCR engagement. In particular, VLA-4 can interact with the ligand VCAM-1 (*vascular cell adhesion molecule-1*), which is upregulated on "activated" endothelium (endothelium at the site of an inflammatory response). This interaction is thought to assist in controlling the trafficking of effector lymphocytes out of the blood vessels into the tissues at sites where they are needed.

Differences in chemokine receptors also contribute to distinct trafficking patterns that result in the wide dispersal of effector T cells throughout the body. Tissues under attack by a pathogen increase their production of various chemokines, drawing effector T cells bearing the appropriate chemokine receptors to the site. For example, effector T cells eventually lose expression of CCR7, which directs T cells to the lymph

nodes, and upregulate expression of CCR9 and CCR4, which facilitate homing to the gut and skin epithelium, respectively.

II. ACTIVATION

Naive and effector cells differ in their requirements for activation, with a higher threshold demanded of the naive cells (refer to Table 15-3). This is physiologically rational because the naive T cell should not be activated unless there is a significant need for it, but once the need has been demonstrated and antigen is present in significant quantities, there should not be a delay in mobilizing the effector cells capable of clearing that antigen. As described in Chapter 14, most naive T cells require multiple signals before complete activation can be achieved and proliferation launched, including sustained TCR triggering, CD28/B7 costimulation, and cytokine binding. Antigen stimulation and costimulation must be in place for a prolonged period (up to 20 hours) for complete activation to occur. In contrast, the activation of effector T cells can be achieved within 1 hour because these cells have a reduced requirement for CD28/B7 costimulation and show more efficient triggering of the crucial signaling kinases Lck and Fyn. Indeed, fully armed CTLs do not require either serial TCR triggering or costimulation for activation, allowing them to lyse the many types of target cells that do not express B7. These differences correlate with differences in crucial signaling molecules expressed by naive and effector T cells. For example, naive murine T cells express the CD45RA and CD45RB isoforms of the phosphatase CD45 at very high density on their surfaces. These isoforms associate with the CD4 and CD8 coreceptors less efficiently than the CD45RO isoform that dominates on the surfaces of murine effector cells.

What makes the activation requirements of naive and effector T cells so different? Antonio Lanzavecchia and colleagues have theorized that perhaps naive cells need a longer time to accumulate the second messengers necessary for intracellular signaling and commitment to proliferation. For example, resting naive Th cells have very few membrane signaling rafts on their surfaces, necessitating a lag while CD28 signaling is deployed to recruit them. Effector cells, in contrast, already have sufficient numbers of rafts at the surface, and so can be activated more quickly with a decreased need for CD28 signaling for this purpose. Alternatively, activation may be preceded by a specific metabolic or nuclear event that takes longer in naive than in effector cells. For example, solid evidence has emerged that the more rapid commitment to activation in effector Th cells correlates with their faster upregulation of expression of the IL-2Rα chain. Newly synthesized chains can be detected in effector cells within 1 hour of antigenic stimulation, permitting a prompt response to IL-2 in the vicinity. In contrast, IL-2Rα appears on naive Th cells only after 6 hours of antigenic stimulation. It is conceivable that other components of the activation machinery make similarly tardy appearances in naive cells.

III. FUNCTIONS AND PRODUCTS

The function of a naive T cell is to proliferate and generate daughter cells that can differentiate into effector cells. The function of an effector cell is to either lyse target cells (CTLs), or secrete cytokines, supply costimulatory contacts, and activate macrophages (Th cells). Effector cells therefore have components that naive cells do not, including soluble cytokines such as IFNγ and IL-4, membrane-bound receptors such as FasL and CD40L, and granules containing perforin and granzymes. These properties were discussed in detail earlier in this chapter.

E. Elimination of Effector T Cells

After the antigenic stimulus sparking a primary immune response has been removed by the actions of effector T cells, there is no further need for their presence. Two mechanisms serve to dispose of effector T cells (both CD4$^+$ and CD8$^+$) and contribute to the maintenance of T cell homeostasis as the primary response concludes: *AICD* and *T cell exhaustion*.

I. ACTIVATION-INDUCED CELL DEATH (AICD)

i) What is AICD?

About 2–3 days after differentiation and execution of their effector functions, most effector T cells are induced to undergo *activation-induced cell death* (AICD) in a tightly controlled manner. AICD is a subset of the normal apoptosis needed to maintain an organism's homeostasis. Many different stimuli can induce cell death, and the antigen that engages the TCR to spark T cell activation is just one of them. AICD of an effector T cell results when signaling downstream of TCR engagement by antigen becomes prolonged.

Apoptosis (including AICD) occurs in two phases: the *induction* phase and the *execution* phase. In the induction phase of AICD in both CD4$^+$ and CD8$^+$ effectors, prolonged TCR stimulation triggers intracellular signaling that induces the transcription of pro-apoptotic genes (e.g., Fas, FasL, Bad, Bax) and decreases the expression of anti-apoptotic molecules (e.g., Bcl-2, Bcl-xL). An intracellular environment is thus created that is permissive for apoptosis. In the execution phase, the caspase cascade is triggered, which inexorably leads to the cleavage of multiple cellular proteins, the condensation of the chromatin, and the other features of apoptotic death described in Box 3-4. The regulatory genes of the induction phase are not transcribed until the TCR is engaged, and gene products of the execution phase, although continually present, depend on the gene products of the induction phase for their activation.

ii) Pathways Leading to AICD

TCR engagement on an activated T cell influences the transcription of members of the TNF and TNFR superfamilies that are key links between the induction and execution phases of apoptosis. For example, TCR stimulation triggers intracellular signaling through ZAP-70 that induces the activation of transcription factors governing the expression of TNFR, FasL, and Fas. The increased expression of TNFR or Fas on a cell's surface increases the chance of that cell binding TNF or FasL, respectively, and receiving intracellular signals that trigger the activation of the caspase cascade. When the cells binding TNF or FasL are activated T cells, elimination of effectors by AICD results. Studies *in vitro* have identified two principal apoptotic pathways called the *extrinsic* or *death receptor* signaling pathway, and the *intrinsic* or *mitochondrial* pathway (Fig. 15-10).

iia) Extrinsic pathway of apoptosis. The engagement of certain cell surface receptors, including Fas and TNFR1, leads to apoptosis via the extrinsic pathway. These receptors have therefore been dubbed "death receptors."

- **Fas-mediated AICD:** Fas and FasL were identified as mediators of AICD through studies of *lpr* mice (which bear a natural mutation in the Fas gene) and *gld* mice (which

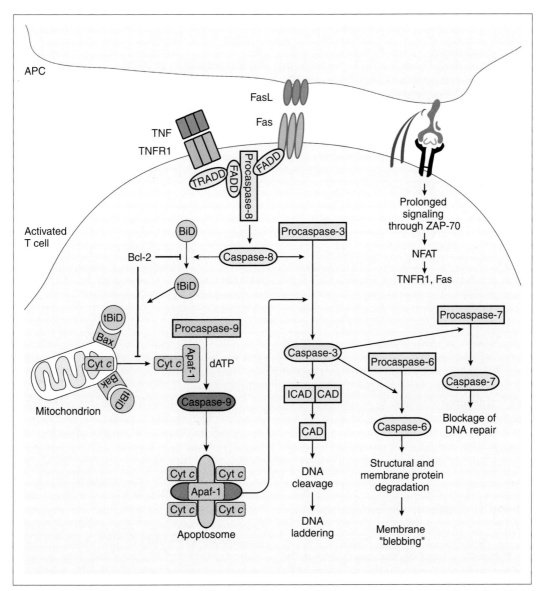

Figure 15-10

Signaling Pathways of AICD
Elements of the extrinsic and intrinsic apoptotic pathways are shown in blue and grey, respectively. **Extrinsic pathway**: prolonged TCR engagement delivers signals via ZAP-70 and NFAT that increase TNFR1 and Fas expression. Upon the binding of several FasL molecules expressed on an APC to Fas expressed on the activated T cell (only one pair is shown for clarity), the adaptor protein FADD is recruited to the cytoplasmic tail of the Fas molecule. FADD in turn recruits procaspase-8, which cleaves itself to generate active caspase-8. Caspase-8 cleaves procaspase-3 to activate this key executioner enzyme, which in turn cleaves procaspase-6 and procaspase-7. Activated caspase-6 and -7 are responsible for many of the morphological hallmarks of apoptotic cells. Caspase-3 also degrades ICAD, releasing CAD, which promotes DNA degradation. The extrinsic pathway is also triggered by the binding of TNF to TNFR1. The adaptor TRADD recruited to the TNFR1 complex can recruit FADD, which proceeds to activate caspase-8 as described previously for Fas killing. **Intrinsic pathway**: external apoptotic stimuli (including granzymes) induce the release of cytochrome c from the activated T cell's mitochondria. In the presence of dATP, cytochrome c associates with Apaf-1 to promote cleavage of procaspase-9. Activated caspase-9 joins Apaf-1 and cytochrome c to form the apoptosome. The apoptosome can then cleave caspase-3, triggering the downstream events of the extrinsic pathway. The intrinsic pathway is promoted when the anti-apoptotic molecule BiD is cleaved by caspase-8. Truncated Bid (tBiD) binds to Bax and Bak in the mitochondrial membrane and induces cytochrome c release.

have a natural mutation in the FasL gene). Both these strains exhibit a lymphoproliferative disorder (uncontrolled proliferation of lymphocytes) that was subsequently shown to be due to a defect in the apoptosis of antigen-stimulated effector T cells. The cytoplasmic tail of Fas contains a *death domain* (DD) sequence of 70–80 amino acids. FasL expressed on the T cell itself, a neighboring T cell, or an APC is able to bind to Fas on the activated T cell surface.

Upon engagement of several Fas receptors by FasL (triggering the induction phase), the Fas molecules aggregate into a structure called the *death-inducing signaling complex* (DISC). The presence of the DDs in the DISC allows it to interact with the DD-containing signal transducer FADD. Once associated with the DISC, FADD can recruit procaspase-8, a member of the caspase protease family and a key initiator of the caspase cascade. Autocatalytic cleavage of procaspase-8 gives rise to activated caspase-8, which in turn triggers the execution phase by cleaving procaspase-3. Activated caspase-3 in turn cleaves procaspase-6 and procaspase-7 and activates them. These proteases degrade proteins essential for cytoskeletal function and DNA repair. Caspase-3 also activates a DNase called *caspase-activated DNase* (CAD). In the absence of an apoptotic stimulus, CAD is held inactive in the nucleus by binding to a protein called *inhibitor of CAD* (ICAD). Activated caspase-3 cleaves ICAD and releases and activates CAD, which then degrades the nuclear DNA. The irreparable damage to the DNA and degradation of cytoskeletal proteins inexorably lead to the orderly death of the cell.

- **TNFR-mediated AICD:** Toward the end of a primary immune response, TNF produced by cells of the innate response engages TNFR1 on activated effector T cells and induces AICD. Like Fas, the cytoplasmic tail of TNFR1 contains the DD sequence, allowing it to interact with another DD-containing protein called <u>TNFR</u>-*associated* <u>death</u> <u>domain</u> (TRADD). TRADD in turn can recruit FADD, which then triggers the induction of apoptosis through procaspase-8 activation as described previously (refer to Fig. 15-10).

iib) Intrinsic pathway of apoptosis. The intrinsic apoptotic pathway also makes a major contribution to the AICD of effector T cells. The intrinsic pathway is triggered by the action of various apoptotic stimuli on the membrane of the mitochondrion. Alterations to this membrane change its permeability characteristics, allowing the escape of cytochrome *c* into the cytoplasm of the cell. In the presence of dATP, cytochrome *c* combines with procaspase-9 and <u>*apoptosis-activating factor-1*</u> (Apaf-1) to cleave and activate procaspase-9 (refer to Fig. 15-10). A complex of cytochrome *c*, activated caspase-9, and Apaf-1 is formed which is called the *apoptosome*. The apoptosome acts on procaspase-3 to activate it, triggering the execution phase as described previously.

In activated T cells, the intrinsic pathway is linked to the extrinsic pathway via activated caspase-8. As well as cleaving procaspase-3, activated caspase-8 cleaves the anti-apoptotic protein BiD (refer to Fig. 15-10). The resulting fragment of BiD (called tBiD, for "truncated BiD") subsequently translocates from the cytoplasm of the doomed cell to its mitochondria. tBiD binds to its receptors Bax and Bak in the mitochondrial membrane, initiating detrimental changes. Cytochrome *c* is released into the cytoplasm and the apoptosome is formed as described previously. The anti-apoptotic protein Bcl-2 can protect cells from apoptosis both by antagonizing BiD cleavage and by directly interfering with the release of cytochrome *c* from the mitochondria.

The induction of the caspase cascade via the mitochondrial route is thus blocked.

iii) Susceptibility to AICD

Naive T cells are relatively resistant to AICD induction, and prolonged TCR engagement leads to (indeed, is required for) activation rather than cell death. In contrast, effector Th cells, which require much less costimulation for activation, and CTLs, which require none, readily succumb to AICD in the face of prolonged antigenic stimulation. Differences in susceptibility to AICD also exist among Th effectors. Although they express similar levels of Fas, Th1 effectors succumb much more readily to Fas-mediated AICD than do Th2 cells. This death may be self-inflicted, induced by the higher upregulation of FasL expression on the Th1 cell surface compared to the Th2 cell surface. Unlike Th1 cells, Th2 cells do not produce the IL-2 required for FasL upregulation. It is not clear whether the relative resistance of Th2 cells to AICD rests in their lower expression of FasL or their higher expression of *Fas-associated phosphatase* (FAP-1), a protein thought to block Fas signaling associated with AICD. Since host tissue damage is more likely to result from the collateral effects of Th1 cytokines than Th2 cytokines, it may be that evolution has provided extra layers of control for the powerful Th1 subset.

The reader may well ask at this point "Do B cells undergo AICD?" The answer appears to be "Not like T cells." As described in Chapter 9, strong signaling through the BCR can trigger the apoptosis of autoreactive B cells during B cell development, but there appears to be no equivalent to the death of activated T cells mediated by Fas or TNFR. In one study, peripheral B cells were stimulated by CD40L, surface IgM cross-linking, or LPS, and the expression of FasL was examined. It was found that peripheral B cells do express low constitutive levels of FasL mRNA but do not upregulate them upon activation. No detectable FasL protein was observed on the surfaces of activated B cells, precluding Fas-mediated AICD. Signals to remove B cells at the conclusion of an immune response are instead thought to be mediated through the BCR itself.

iv) The Pro-/Anti-apoptotic Balance

While engagement of TNFR1 by TNF can lead to apoptosis as described previously, it may startle the reader to learn that it can also promote cell survival. TNF binding to TNFR1 induces the transcription of anti-apoptotic proteins such as the *cellular inhibitor of apoptosis proteins* (cIAPs), which inhibit caspase-3 activation, and signal transducing elements of the NF-κB and SAPK/JNK pathways, which promote cell survival. Which cell fate results, survival or death, appears to depend on which signal transducing molecules are recruited to the cytoplasmic tail of the receptor. We have already described how the cytoplasmic tail of TNFR1 interacts with TRADD. However, TRADD can interact with either FADD, resulting in cell death, or with TRAF2 and *receptor-interacting protein* (RIP), resulting in cell survival (Fig. 15-11). The TNFR1/TRADD/TRAF2/RIP complex activates both the SAPK/JNK and NF-κB pathways, leading to

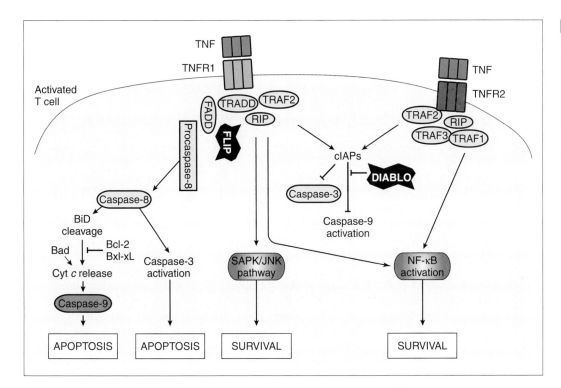

Figure 15-11

Pro/Anti-apoptotic Balance
When TNF engages TNFR1, either survival or death signals are delivered depending on whether FADD, or TRAF2 and RIP are recruited to TRADD. Recruitment of FADD leads to apoptosis via caspase-8 activation, but recruitment of TRAF2 and RIP leads to activation of both the NF-κB and SAPK/JNK pathways and thus cell survival. TRAF2 also activates the cIAPs that block caspase function. When TNF binds to TNFR2, several transducers, including RIP, TRAF2, TRAF1, and TRAF3 but not TRADD, are recruited to the complex. Cell survival via cIAPs and NF-κB activation is promoted. The pro/anti-apoptotic balance is also influenced by FLIP, which inhibits caspase-8 activation, and DIABLO, which blocks the cIAPs.

the transcription of cell survival genes. In addition, the TRAF2 moiety promotes activation of the cIAPs. The push toward survival may be assisted by the binding of TNF to a second receptor called TNFR2. Unlike TNFR1, TNFR2 lacks the DD and so cannot recruit TRADD (and thus not FADD) to its cytoplasmic tail. Rather, the tail of TNFR2 associates with RIP, TRAF2, TRAF1, and TRAF3 and initiates signaling leading to NF-κB activation. Again, TRAF2 recruited to TNFR2 mediates cIAP activation. Thus, on the same cell, the interaction of all these receptor–ligand pairs (and others beyond the scope of this book) sets up a push-pull dynamic between survival and apoptosis. Should the interactions favoring cell death outweigh those favoring cell survival, the cell undergoes AICD.

Two other proteins that play a role in maintaining T lymphocyte homeostasis are FLIP (*FLICE inhibitor protein*; caspase-8 was previously known as FLICE) and DIABLO (*direct IAP binding protein with low isoelectric point*). FLIP is a regulator of both the extrinsic and intrinsic apoptotic pathways. FLIP contains two DDs, is homologous to procaspase-8, and competes with it for binding to FADD. However, FLIP lacks a catalytic domain so that neither BiD nor procaspase-3 can be cleaved and activated, resulting in the inhibition of caspase-8 activity. The expression of FLIP is downregulated as T cell activation progresses, allowing conditions favoring AICD to develop. Significantly, FLIP expression is also downregulated by IL-2. DIABLO is a regulator of the intrinsic apoptotic pathway and is released from the mitochondria in response to apoptotic stimuli. DIABLO acts to curb the effects of the cIAP molecules that inhibit caspase-9 activation, and thus also promotes AICD.

II. T CELL EXHAUSTION

While most T cell effectors succumb to AICD, some activated T cell clones appear to be eliminated by "exhaustion," also called *replicative senescence*. (Senescence means "death due to old age," and age in cells is determined by the number of cell divisions [replications] they have undergone.) In this case, the activated T cells literally become exhausted from continuous exposure to antigen and may simply divide so relentlessly into effectors that they burn out metabolically without generating memory cells. In other words, both the effector cells generated to deal with the initial assault by antigen, and memory cells that normally would have dealt with a subsequent assault, completely disappear from the host. Such an absence of a T cell clone can render the host unresponsive to that antigen due to a "hole in the T cell repertoire," a phenomenon introduced in Chapter 13 and discussed in more detail in Chapter 16.

F. Memory T Cells

Only a small fraction of the antigen-specific T cells generated in a primary response survives the AICD that eliminates the vast majority of effector T cells. It is these cells that are, or give rise to, the memory T cells that survive beyond the primary response and take up residence in the tissues, ready to rapidly combat a given pathogen should it challenge the host again. In general, the memory or secondary response is faster and stronger than the primary response, a difference attributable to the localization, increased numbers, and enhanced capacities of antigen-specific memory cells. Memory T cells recognize the same pMHC structure as naive and effector cells but have properties intermediate between these two cell subsets (refer to the third column of Table 15-3). The vast majority of memory cells adopt a resting state, like naive T cells, but recirculate more widely throughout the periphery of the body, like effector cells. Memory cell activation more closely resembles that of an effector cell than a naive cell, needing only minimal costimulation, but memory cell function mirrors that of naive T cells, i.e., the generation of effector cells that, upon activation, secrete cytokines (Th cells) or lyse target cells (CTLs). The adhesion molecules expressed on memory cells are generally the same as those expressed on effector cells, but the life span of a memory cell far exceeds that of an effector cell. The fact that memory T cells do not fully revert to a naive phenotype has been interpreted to mean that the memory cells must receive some type of stimulus that naive cells do not, perhaps mediated by contact with trace deposits of antigen left behind after conclusion of the primary response (see later).

I. GENERATION OF MEMORY T CELLS

The reader will recall that, in the case of B cells, the generation of memory cells is restricted to a microenvironment favoring anti-apoptotic gene expression, namely the germinal center. However, there is no clearly analogous structure in memory T cell development, and the actual generation of memory T cells is still poorly understood. One school of thought holds that a differentiating progeny T cell becomes either an effector cell or a memory cell depending on the immediate microenvironment in which it finds itself. A location in which antigen is present in low to moderate amounts (so as to avoid exhaustion), where CD28 (and/or ICOS) on the T cell surface is readily engaged and IL-2 (and perhaps other cytokines) is continuously present, may favor the expression of anti-apoptotic molecules such as the cIAPs, Bcl-2, and Bcl-xL over that of pro-apoptotic molecules such as Fas, Bad, and Bax. Such conditions would favor the formation of memory cells. In contrast, differentiating progeny T cells that found themselves in a microenvironment containing higher concentrations of antigen and in which the engagement of IL-2R, CD28, and/or ICOS was weaker might express relatively low levels of the anti-apoptotic molecules. Such cells might become the short-lived effector cells, soon doomed to disappear by exhaustion or AICD.

Another school of thought proposes that memory T cells are derived from among the last proliferative waves of the original expanding clone. Such clones would arrive at the site of the immune response as the antigen was being mopped up. Antigen would therefore be presented by APCs at the relatively low concentrations needed to avoid exhaustion. Presumably where sufficient CD28 and IL-2 signaling still occurred, the expression of Bcl-xL and Bcl-2 would be upregulated, and the cell would avoid AICD and live on as a memory cell. While these hypotheses are attractive, neither

has been formally proven. It therefore remains possible that memory cells evolve in an unknown way from fully differentiated effector T cells.

Two molecules have been pinpointed as being important for memory T cell generation: IL-7 and CD27. IL-7Rα expression is downregulated during naive T cell activation but increases again on at least some subsets of memory CD8$^+$ T cells. IL-7 also appears to be moderately important for certain memory CD4$^+$ T cell populations. It is speculated that IL-7's contribution here may be its ability to upregulate expression of both Bcl-2 and LKLF, a transcription factor that promotes T cell survival (refer to Ch.13). CD27, a member of the TNF superfamily, is a disulfide-linked homodimer that binds to CD70, a member of the TNFR superfamily. Detectable expression of CD27 and CD70 is induced on naive and effector T cells after activation, and CD27/CD70 signaling enhances the expansion of activated effector T cells. However, studies of CD27$^{-/-}$ mice have shown that CD27/CD70 signaling may be most important for memory CTL responses to pathogens. In the absence of CD27, CTL responses to influenza virus infection are delayed and reach only primary response levels. Whether CD27/CD70 signaling is also necessary for the Th memory response is under investigation.

II. MEMORY T CELL MARKERS

Memory T cells can be distinguished from naive and effector T cells by surface marker expression. As was mentioned earlier in this chapter, naive human CD4$^+$ T cells can be identified by their high expression of L-selectin and the CD45RA and RB isoforms, whereas levels of CD44, ICAM-1, LFA-1, and the VLA integrins are relatively low (refer to Table 15-3). On activated effector cells, L-selectin and the CD45 isoforms are downregulated while CD44, ICAM-1, LFA-1, and the VLA integrins are upregulated. Fully activated effector T cells also acquire the expression of CD25, CD69, and CD40L. In mice, resting memory CD4$^+$ T cells have a surface marker phenotype somewhat intermediate between resting naive and fully activated effector CD4$^+$ T cells. Memory CD4$^+$ T cells exhibit high levels of CD44 and low levels of expression of L-selectin and CD45RB but lack expression of CD25, CD69, and CD40L. Murine memory CD8$^+$ T cells also show upregulated expression of the IL-2Rβ chain and the cell surface marker Ly6C, a protein of unknown function that is expressed at much lower levels on CD8$^+$ effectors.

In humans, memory T cells are CD45RA$^-$ but express low levels of CD45RO. The extent of downregulation of the CD45 isoforms is more complete in CD4$^+$ memory T cells than in CD8$^+$ memory T cells. Interestingly, two classes of resting human CD4$^+$ memory cells have been identified, only one of which expresses L-selectin. Reminiscent of the Th1 and Th2 effector subsets, the L-selectin$^-$ subset produces IFNγ, while the L-selectin$^+$ subset can be induced to secrete IL-4 and IL-5 *in vitro*. It is not yet known whether these subsets constitute distinct Th1 and Th2 memory T cells.

III. MEMORY T CELL DISTRIBUTION

Unlike naive cells, the majority of murine memory T cells express low levels of L-selectin and so cannot access lymphoid tissues via the HEV. Rather, memory cells are thought to recirculate by first migrating out of the blood into a non-lymphoid tissue, draining with the lymph into the afferent lymphatics, then passing through the lymph nodes in the lymph and returning to the blood via the efferent lymphatics that join with the thoracic duct. In the absence of antigen, the number of memory cells found in lymph nodes is actually quite small compared to the number of naive T cells; the bulk of the memory cell population (which in total far exceeds the number of naive cells) tends to be dispersed more widely in other secondary and diffuse lymphoid tissues. This modified trafficking pattern is facilitated by the expression of chemokine receptors and adhesion molecules that are more similar to those on effector T cells than those on naive T cells. For example, memory (but not naive) CD4$^+$ T cells express high levels of the chemokine receptor CXCR5, which allows the cell to follow a chemotactic gradient of the chemokine BLC (*B lymphocyte chemoattractant*). BLC, which was originally identified as a chemoattractant for B cells, is produced by the antigen-storing FDC in the lymphoid follicles. In addition, like effector cells, memory T cells express LPAM-1 and the VLAs that allow access to the diffuse lymphoid tissues and inflammatory sites. Regardless of their initial distribution, as they age, memory T cells gradually accumulate in the blood and spleen (as opposed to the lymph nodes or non-lymphoid tissues), meaning that memory responses can often be demonstrated in these compartments long after the original exposure to antigen.

Interestingly, the route of administration of an antigen can dictate the homing pattern of resting memory T cells directed against that antigen. The cytokine milieu in which the original T cell was primed, or perhaps residual antigen trapped close to the site of administration, can influence which adhesion molecules (particularly integrins) are expressed or lost on a given memory T cell. In turn, the presence or absence of particular adhesion molecules determines the access of a memory T cell to a tissue. For example, the expression of the integrin α4β7 is upregulated on memory CD4$^+$ T cells generated during a primary response to antigen in the GALT. This integrin binds to the vascular addressin MAdCAM expressed in the gut. Thus, memory cells bearing α4β7 recirculate preferentially back to the gut, the site of their original formation and precisely where the next assault by that antigen might be anticipated. Similarly, when an animal is immunized intraperitoneally, adhesion molecule expression is altered such that memory Tc cells accumulate preferentially in the spleen rather than the lymph nodes. If the priming of a Tc cell occurs via intranasal administration (a mucosal route), memory CD8$^+$ Tc cells gather preferentially in other mucosal sites, such as in the linings of the digestive, respiratory, and reproductive tracts (see Ch.20).

In a secondary response, tissue-specific recruitment of Th1 and Th2 effectors may affect subsequent memory T cell distribution. During an inflammatory reaction in the skin, the chemokine RANTES produced by stimulated non-lymphoid cells in the site of attack specifically draws antigen-specific Th1 effector cells and memory T cells bearing the appropriate chemokine receptors to the area. The expression of Th1 cytokines by the effector cells induces the expression of CLA (*cutaneous lymphocyte antigen*) on the memory T cells, facilitating their entry into the stricken skin tissue. Similarly, exposure to TGFβ present because of a Th2 response in the gut induces the expression on memory T cells of β integrins promoting retention in the gut mucosae.

Some researchers claim that memory cell distribution patterns and responses reflect the existence of two different memory cell populations: one called *central memory* and the other called *effector memory*. According to this hypothesis, central memory cells express CCR7 (like naive T cells) and thus reside in a resting state in the secondary lymphoid tissues. These cells quickly proliferate and differentiate into a second generation of effector T cells upon a subsequent encounter with specific antigen. In contrast, memory cells deemed to be part of the effector memory population lack CCR7 but express other chemokine receptors that allow them to home to sites of inflammation. These cells do not have to differentiate into effectors: they already possess effector function and can act immediately at the site of attack to combat the pathogen. The existence of these two types of memory cell populations *in vivo* is under debate.

IV. MEMORY T CELL ACTIVATION

The activation of a memory Th cell requires the same basic elements of intracellular signaling as the priming of naive CD4$^+$ T cells; however, there appear to be important differences in how the activation signals are delivered. Because memory Th cells are dispersed in anatomical sites quite different from those occupied by naive Th cells, memory cell activation depends on different types of APCs. Thus, whereas naive Th cells are activated exclusively by pMHC presented by DCs in the lymph nodes, memory Th cells respond to pMHC presented by DCs, B cells, and macrophages almost anywhere in the periphery. Compared to naive Th cells (and much like effector Th cells), the activation of memory Th cells requires a lower concentration of antigen (as much as 10- to 50-fold lower), a shorter period of sustained signaling, and much less costimulation. Memory Th cells have high levels of membrane rafts already in place and apparently do not require costimulation to induce the coalescence of these structures. The increased levels of adhesion molecules and different CD45 isoforms expressed by memory Th cells may also contribute to their decreased threshold of activation. Once activated, memory Th cells proliferate more readily and for longer periods than their naive counterparts, and die more readily due to AICD when subjected to prolonged antigenic stimulation. It remains unresolved whether memory Th cells are more or less sensitive to the induction of anergy than naive Th cells.

Although naive CD8$^+$ T cells often require T cell help from antigen-specific CD4$^+$ T cells to become activated, memory

CD8$^+$ T cells do not appear to have the same requirement. Studies of HY transgenic mice have shown that the primary CTL response to HY is absolutely dependent on the presence of CD4$^+$ T cell help. However, once CD8$^+$ memory cells are generated, their survival and function can be independent of such help. In one experiment, normal female mice were injected with normal male spleen cells (of the same strain and bearing the HY antigen) to prime them and allow the generation of HY-specific CD8$^+$ memory cells. After a 1 month rest period, some of the female mice were depleted of CD4$^+$ T cells by repeated administration of anti-CD4 antibodies over several more months. When later challenged with HY, both anti-CD4-treated and control mice exhibited the same levels and functions of HY-specific CD8$^+$ memory cells for up to 9 months.

V. MEMORY T CELL DIFFERENTIATION AND EFFECTOR FUNCTION

Antigen-stimulated memory CD4$^+$ and CD8$^+$ T cells appear to undergo the same differentiation pathways as naive CD4$^+$ and CD8$^+$ T cells. Unlike the memory B cell situation, there is neither isotype switching of the TCR genes nor somatic hypermutation. However, whereas the generation of effector cells from either CD4$^+$ or CD8$^+$ naive T cells takes about 4–5 days, the differentiation of effectors from antigen-stimulated memory T cells is significantly faster. In fact, memory CD8$^+$ T cells can expand and differentiate into new effector CTLs within 12 hours of encountering antigen. Such readiness to respond accounts for allograft rejection studies in which memory CD8$^+$ T cells destroyed allogeneic tumor cells much more quickly than naive T cells. Some scientists maintain that some memory CD8$^+$ T cells, especially those directed against virulent viruses, are not really resting but instead are maintained in a type of "pre-activation" state (which could perhaps correlate with their intermediate marker phenotype). It is proposed that this pre-activation state makes it easier for the cells to immediately differentiate into new effectors capable of quickly combating the aggressive pathogen, rather than having to progress through a complete activation cycle.

Antigen-stimulated memory CD4$^+$ T cells differentiate into either Th1 or Th2 effector subsets with the same properties as those generated by the activation of naive CD4$^+$ T cells. However, the upregulation of cytokine mRNA transcription is much faster in memory CD4$^+$ T cells. The induction of Th1 differentiation from memory CD4$^+$ T cells is also slightly different. Rather than IL-27, IL-23 (which does not participate in the primary response to a pathogen) induces activated memory CD4$^+$ T cells to commence the differentiation of Th1 effectors (see Ch.17).

The actual functions of Th1, Th2, and CTL effectors derived from memory CD4$^+$ and CD8$^+$ T cells are seemingly identical to those of effectors generated from naive CD4$^+$ and CD8$^+$ T cells. It remains unclear whether memory T cells can give rise to a new generation of memory cells as well as effector cells.

VI. MEMORY T CELL LIFE SPAN

As mentioned in Chapter 9, the life span of memory lymphocytes is clone specific and depends on the nature of the antigen that provoked the primary response. Most memory cells appear to have life spans of at least several months and often years. For example, CD4$^+$ and CD8$^+$ memory T cells specific for vaccinia virus have been found in humans 50 years after the original vaccination. The scientific community is still divided over whether the persistence of memory lymphocytes depends on some sort of periodic low level of stimulation by residual antigen. Such stimulation would be analogous to the "TCR tickling" required to sustain mature, naive T cells in the periphery without activating them (see Ch.13). Presumably a "tickled" memory T cell would express low levels of anti-apoptotic molecules that would permit it to survive. Indeed, in mice, CD8$^+$ memory T cells express higher levels of Bcl-2 than do naive Tc cells, while CD4$^+$ memory T cells upregulate Bcl-2 and Bcl-xL. However, there have also been reports that memory cells can survive in the complete absence of specific antigen, and even in mice completely deficient for MHC. One molecule that has been found to make a crucial contribution to the long-term persistence of both CD8$^+$ and CD4$^+$ memory T cells is IL-15. Numbers of memory CD8$^+$ T cells are increased in IL-15 transgenic mice but decreased in IL-15$^{-/-}$ mice. In addition, IL-15 upregulates the expression of the anti-apoptotic molecule Bcl-xL in memory T cells *in vitro*. These results fit well with the *in vivo* upregulated expression of IL-2Rβ, a component of the IL-15R, on at least some subsets of memory CD8$^+$ T cells. IL-7 may also make a contribution here, since it can promote the survival of both CD4$^+$ and CD8$^+$ T cells in culture in the absence of cell division. IL-7 would be especially helpful for those memory T cell subsets that express low levels of IL-2Rβ.

There has also been debate over whether memory cells, tickled or not, remain in a non-dividing state until the reappearance of antigen, or whether cell division in fact does go on at a low rate. Studies using the DNA synthesis marker BrdU to examine the turnover of memory T cells have revealed that at least some of these cells do divide at a slow, steady pace, a level of cycling that may be sufficient to maintain the memory T cell population for long periods. IL-15 appears to be responsible for driving this low level of proliferation for CD8$^+$ memory T cells, even in the absence of antigen. Indeed, IL-15 can induce the proliferation of CD8$^+$ memory T cells (and CD4$^+$ memory T cells, to a lesser extent) *in vitro* and *in vivo*, and inhibition or loss of IL-15 impairs the proliferation of at least some CD8$^+$ memory T cells. However, IL-15 is not required for the actual functions of memory CD8$^+$ T cell effectors. This matter of IL-15-driven proliferation brings up an interesting question: if some proportion of memory cells is actively dividing, there must be compensation in the form of memory cell death in order to maintain homeostasis in the T lymphocyte compartment. Does IL-15 also have indirect effects that might curtail survival? The regulatory mechanism that may control the balance of memory T cell survival and death has yet to be elucidated.

Can memory T cells protect a host forever? Evidence from studies of the aging of the immune system indicate that memory T cells can be stimulated only so many times before they fail to proliferate in response to antigen due to the natural process of cellular senescence. Theoretically, it is possible that every time an antigen enters the body, new naive T cells specific for that antigen are stimulated as well as the memory T cells generated in the last exposure. Primary responses would therefore occur simultaneously with memory cell-mediated secondary responses, and new batches of memory T cells would be generated. However, as we have seen in Chapter 13, the production of new naive T cells by the thymus declines precipitously after puberty, progressively limiting the number of new naive T cells produced as a host ages. Thus, between a declining supply of naive T cells to generate fresh primary responses and the inexorable replicative senescence of memory T cell clones, effector T cell generation is ultimately limited, and the host becomes increasingly susceptible to pathogens toward the end of its life.

We have now described all the cellular components of an adaptive immune response and how naive cells are activated when sufficient numbers of TCRs or BCRs interact with sufficient avidity with an immunogenic epitope. Such responses constitute the recognition and removal of entities that are nonself, and that are critical for the elimination of pathogens. In the next chapter, we examine *peripheral tolerance*: a collection of mechanisms that controls those mature naive lymphocytes in the periphery whose antigen receptors are directed against self-antigens. For various reasons, these clones escaped culling by the mechanisms of central tolerance that largely rid the developing lymphocyte pool of self-reactive cells. In the healthy host, peripheral tolerance mechanisms prevent these escaped anti-self clones from causing autoimmune disease.

SUMMARY

Once a T cell is activated, it proliferates and differentiates into effector cells that carry out actions designed to eliminate antigens. Memory T cells that mediate secondary immune responses are also produced. Both effector and memory T cells are activated at lower levels of TCR engagement and costimulatory signaling than are naive cells, and express different subsets of adhesion molecules. The activation of naive CD4$^+$ T cells leads to the generation of Th1 and Th2 effector cells that secrete different panels of cytokines. These cytokines can either take direct action against pathogens or support the activation of B cells and CD8$^+$ T cells. If a naive CD4$^+$ T cell is activated in a cytokine milieu dominated by IL-12 and IFNγ, the cell becomes a Th1 effector that secretes IL-2, LTα, GM-CSF, and IFNγ. These cytokines promote inflammation and Tc activation, and favor isotype switching of antibodies to isotypes that readily trigger complement activation. In contrast, if a naive CD4$^+$ T cell is activated in a cytokine milieu dominated by IL-4, the cell becomes a Th2 effector that secretes IL-4, IL-5, IL-10, and IL-13. These cytokines have an anti-inflammatory influence and promote class switching to isotypes that are associated with minimal tissue damage. Specific subtypes of DCs may also bias Th1/Th2 differentiation in a particular direction. The cytokines secreted by a Th1 cell downregulate Th2 differentiation and vice versa, leading to cross-regulation. Th1 and Th2 cells express different subsets of chemokine receptors, resulting in different homing patterns. The activation of naive CD8$^+$ T cells leads to the generation of CTL effectors that kill tumor cells and infected target cells by perforin- and granzyme-mediated cytotoxicity, by Fas ligation, and by secretion of the cytotoxic cytokines IFNγ, LTα, and TNF. The duration of both Th1/2 and CTL responses is controlled by T cell exhaustion and activation-induced cell death (AICD) mediated by both extrinsic and intrinsic apoptotic pathways. Memory T cells persist in the host that can rapidly differentiate into new effectors upon a second exposure to an antigen.

Selected Reading List

Ansel K. M., Lee D. U. and Rao A. (2003) An epigenetic view of helper T cell differentiation. *Nature Immunology* 4, 616–623.

Asnagli H. and Murphy K. (2001) Stability and commitment in T helper cell development. *Current Opinion in Immunology* 13, 242–247.

Barry M. and Bleackley R. (2002) Cytotoxic T lymphocytes: all roads lead to death. *Nature Reviews. Immunology* 2, 401–409.

Barton G. and Medzhitov R. (2002) Control of adaptive immune responses by Toll-like receptors. *Current Opinion in Immunology* 14, 380–383.

Blink E. J., Trapani J. A. and Jans D. A. (1999) Perforin-dependent nuclear targeting of granzymes: a central role in the nuclear events of granule-exocytosis-mediated apoptosis. *Immunology and Cell Biology* 77, 206–215.

Bradley L. M. (2003) Migration and T-lymphocyte effector function. *Current Opinion in Immunology* 15, 343–348.

Budd R. (2001) Activation-induced cell death. *Current Opinion in Immunology* 13, 356–362.

Dutton R. W., Bradley L. M. and Swain S. L. (1998) T cell memory. *Annual Review of Immunology* 16, 201–223.

Effros R. B. (2003) Replicative senescence: the final stage of memory T cell differentiation? *Current HIV Research* 1, 153–165.

Ge Q., Hu H., Eisen H. N. and Chen J. (2002) Naive to memory T-cell differentiation during homeostasis-driven proliferation. *Microbes & Infection* 4, 555–558.

Glimcher L. and Singh H. (2001) Transcription factors in lymphocyte development—T and B cells get together. *Cell* 96, 13–23.

Hendriks J., Gravestein L., Tesselaar K., van Lier R., Schumacher T. *et al.* (2000) CD27 is required for generation and long-term maintenance of T cell immunity. *Nature* 1, 433–440.

Lanzavecchia A. and Sallusto F. (2000) From synapses to immunological memory: the role of sustained T cell stimulation. *Current Opinion in Immunology* **12**, 92–98.

Lezzi G., Karjlalainen K. and Lanzavecchia A. (1998) The duration of antigenic stimulation determines the fate of naive and effector T cells. *Immunity* **8**, 89–95.

Liang L. and Sha W. (2002) The right place at the right time: novel B7 family members regulate effector T cell responses. *Current Opinion in Immunology* **14**, 384–390.

Liu Y. J., Kanzler H. and Soumelis V. (2001) Dendritic cell lineage, plasticity and cross-regulation. *Nature Immunology* **2**, 585–589.

Morel P. A. and Oriss T. B. (1998) Crossregulation between Th1 and Th2 cells. *Critical Reviews in Immunology* **18**, 275–303.

Moser M. and Murphy K. (2000) Dendritic cell regulation of Th1-Th2 development. *Nature Immunology* **1**, 199–205.

Rao A. and Avni O. (2000) Molecular aspects of T-cell differentiation. *British Medical Bulletin* **56**, 969–984.

Reiner S. (2001) Helper T cell differentiation, inside and out. *Current Opinion in Immunology* **13**, 351–355.

Russel J. and Ley T. (2002) Lymphocyte-mediated cytotoxicity. *Annual Review of Immunology* **20**, 323–370.

Sallusto F., Geginat J. and Lanzavecchia A. (2004) Central memory and effector memory T cell subsets: function, generation, and maintenance. *Annual Review of Immunology* **22**, 745–763.

Santana M. A. and Rosenstein Y. (2003) What it takes to become an effector T cell: the process, the cells involved, and the mechanisms. *Journal of Cellular Physiology* **195**, 392–401.

Schluns K. and Lefrancois L. (2003) Cytokine control of memory T-cell development and survival. *Nature Reviews Immunology* **3**, 269–279.

Seder R. A. and Ahmed R. (2003) Similarities and differences in CD4+ and CD8+ effector and memory T cell generation. *Nature Immunology* **4**, 835–842.

Smyth M., Kelly J., Sutton V., Davis J., Browne K. *et al.* (2001) Unlocking the secrets of the cytotoxic granule proteins. *Journal of Leukocyte Biology* **70**, 18–29.

Sprent J. and Surh C. (2002) T cell memory. *Annual Review of Immunology* **20**, 551–579.

Staal F. J., Weerkamp F., Langerak A. W., Hendriks R. W. and Clevers H. C. (2001)

Transcriptional control of t lymphocyte differentiation. *Stem Cells* **19**, 165–179.

Stenger S. and Modlin R. L. (1998) Cytotoxic T cell responses to intracellular pathogens. *Current Opinion in Immunology* **10**, 471–477.

Trapani J. A., Sutton V. R. and Smyth M. J. (1999) CTL granules: evolution of vesicles essential for combating virus infections. *Immunology Today* **20**, 351–356.

Wallach D., Varfolomeev E., Malinin N., Goltsev Y. V., Kovalenko A. *et al.* (1999) Tumor necrosis factor receptor and Fas signaling mechanisms. *Annual Review of Immunology* **17**, 331–367.

Ward S. G., Bacon K. and Westwick J. (1998) Chemokines and T lymphocytes: more than an attraction. *Immunity* **9**, 1–11.

Waterhouse N. and Trapani J. (2002) CTL: caspases terminate life, but that's not the whole story. *Tissue Antigens* **59**, 175–183.

Wong B. and Choi Y. (1997) Pathways leading to cell death in T cells. *Current Opinion in Immunology* **9**, 358–364.

Zamoyska R., Basson A., Filby A., Legname G., Lovatt M. *et al.* (2003) The influence of the src-family kinases, Lck and Fyn, on T cell differentiation, survival and activation. *Immunological Reviews* **191**, 107–118.

Immune Tolerance in the Periphery

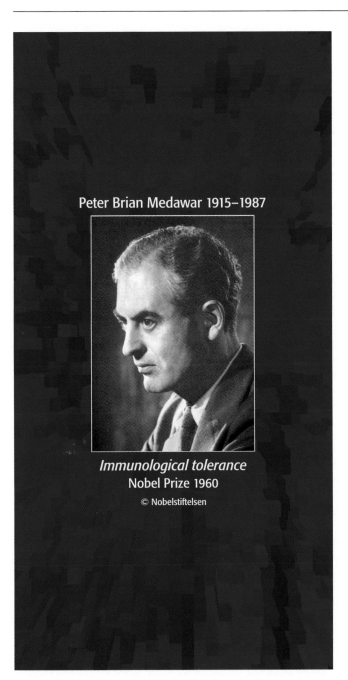

Peter Brian Medawar 1915–1987

Immunological tolerance
Nobel Prize 1960

© Nobelstiftelsen

CHAPTER 16

A.   CONTEXT OF PERIPHERAL TOLERANCE

B.   T CELL PERIPHERAL SELF-TOLERANCE

C.   B CELL PERIPHERAL SELF-TOLERANCE

D.   REGULATORY T CELLS

E.   EXPERIMENTAL TOLERANCE

"War is only a cowardly escape from the problems of peace"
—Thomas Mann

In Chapters 14 and 15, we discussed T cell activation and differentiation into effector cells, a process that occurs when a non-self antigen is recognized with sufficient avidity by antigen receptors and co-stimulatory signals are successfully received. The activation and differentiation of B cells were similarly described in Chapter 9. Because the antigen receptors of B and T cells are randomly generated, a certain proportion of developing lymphocytes will inevitably bear receptors directed against antigens comprising the tissues of a healthy body. In Chapter 9 for B cells and in Chapter 13 for T cells, we described the mechanisms of central tolerance that result in the clonal deletion of self-reactive lymphocytes during their maturation in the bone marrow and thymus, respectively. However, some self-reactive lymphocyte clones are not eliminated during the establishment of central tolerance, most often because the relevant self antigens are not expressed at sufficient levels in the bone marrow or thymus to induce deletion. Instead, control mechanisms in the periphery ensure that when such self-reactive cells mature and leave the bone marrow or thymus they do not attack self tissues. In this chapter, we explore what some have called "the final frontier of immunology," the phenomenon of *peripheral tolerance*.

Peripheral tolerance is manifested when the interaction between a mature peripheral lymphocyte and its cognate antigen does not result in an immune response against the antigen. In this book, we will address two aspects of peripheral tolerance: *self-tolerance* and *experimental tolerance*. Peripheral self-tolerance as it exists in a normal, healthy animal refers to the lack of an immune response to <u>self</u> antigens in the animal's tissues. A failure in peripheral self-tolerance can lead to autoimmunity. Experimental tolerance refers to a lack of an immune response to a <u>foreign</u> antigen, a state that scientists can induce by artificial means. In the sections that follow, we first describe the ways by which T and B cell peripheral self-tolerance is established. We then discuss the methods used to induce experimental tolerance and observations derived from these studies that can be used to gain insight into the mechanisms of peripheral self-tolerance. We conclude the chapter with a description of immune non-responsiveness in special situations: fetal–maternal tolerance and neonatal tolerance.

A. Context of Peripheral Tolerance

I. HISTORICAL NOTES

One of the earliest observations of "non-responsiveness to self" (a failure to mount an immune response) was made by Paul Ehrlich in the early 1900s. He found that almost anything that could be injected into an individual would provoke an immune response, including cells from an unrelated member of the same species. However, the same individual would not respond to cells from its own tissues. Intuitively Ehrlich recognized that an immune attack on self tissues would be incompatible with normal health, and that the immune system would therefore try at all costs to avoid such assaults. Ehrlich called this concept *horror autotoxicus*, meaning that the immune system would have a "horror of being toxic" to its host. However, at the time, the rudimentary understanding of the immune system precluded the development of useful theories or models to formally explain how the immune system could distinguish between self and non-self and only attack the latter. (Note that there was also no distinction between what we now call "central" and "peripheral" tolerance.) In addition, the focus of immunological studies soon shifted to the biochemistry and actions of antibody molecules, contributing to a lapse in progress with respect to understanding the type of self-tolerance observed by Ehrlich. However, in the 1940s, Ray Owen made a seminal contribution to this field. He noticed that when the blood types of dizygotic (fraternal) cattle twins were analyzed, the test results often appeared to indicate that the twins were monozygotic (identical). That is, the immune systems of twins that had different blood types did not recognize each other's blood group antigens as foreign, as was expected. Earlier work in cattle had shown that dizygotic twin fetuses are sometimes so close together physically that the fetuses actually exchange blood *in utero*. Owen therefore concluded that early contact with the blood group antigen of Twin B was what made Twin A non-responsive to it, and vice versa.

Building on Owen's studies, immunologists F. MacFarlane Burnet and Frank Fenner put forth the theory that the self/non-self discrimination displayed by the immune system was established during embryonic development. In support of their model, they drew attention to earlier work in the late 1930s by Erich Traub, who showed that LCMV infection that was passed *in utero* from mother to fetus resulted in mice that could not respond to LCMV antigens after birth. Traub explained his results as the "development of a tolerance to the foreign micro-organism during embryonic life," and his report

is thought to represent the first use of the term "tolerance" in an immunological sense.

Peter Medawar and his colleagues soon extended Owen's observations by showing that dizygotic twin cattle that had shared a single placental circulation could also accept skin grafts from each other. Furthermore, working with mice, they found that if lymphoid cells from strain B were injected into either strain A embryos or strain A neonates, these animals (when adults) would be tolerant of skin grafts from strain B mice but not from strain C mice (Fig. 16-1A). Again, exposure to an alloantigen early in development appeared to induce specific immunological tolerance. This was the first demonstration that self/non-self discrimination could be manipulated experimentally, and the implication of this result was that "self" in an adult individual was defined by the antigens that were present during development. In an interesting test of this hypothesis, Edward L. Triplett reasoned that removal of a self

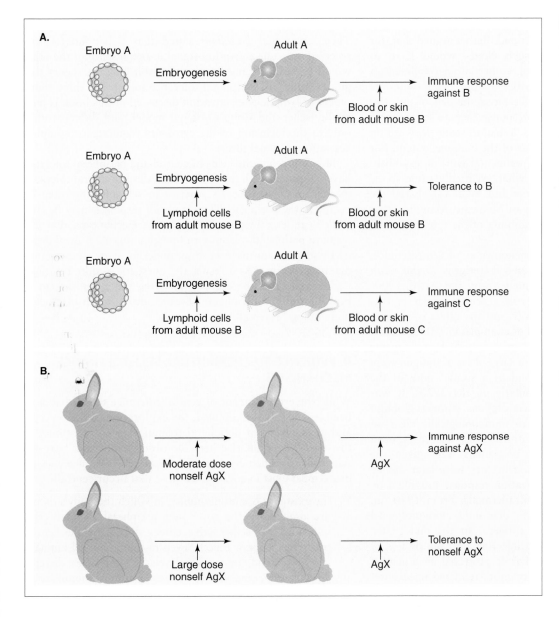

A.

Embryo A → Embryogenesis → Adult A → Blood or skin from adult mouse B → Immune response against B

Embryo A → Embryogenesis (Lymphoid cells from adult mouse B) → Adult A → Blood or skin from adult mouse B → Tolerance to B

Embryo A → Embyrogenesis (Lymphoid cells from adult mouse B) → Adult A → Blood or skin from adult mouse C → Immune response against C

B.

Moderate dose nonself AgX → AgX → Immune response against AgX

Large dose nonself AgX → AgX → Tolerance to nonself AgX

Figure 16-1

Discovery of Immune Tolerance

(A) An early demonstration of self-tolerance. An adult mouse of strain "A" mounts an immune response against transplanted blood or tissue from a mouse of strain "B." However, if lymphoid cells from mouse B are introduced into the developing embryo of mouse A, adult mouse A is tolerant to transplanted blood or tissue from mouse B but still mounts an immune response to blood or tissue from mouse C. (B) An early demonstration of experimental tolerance. A moderate dose of a nonself antigen X (AgX) injected into an adult rabbit induces an immune response against AgX, as expected. However, a large dose of AgX renders the rabbit incapable of responding to a subsequent dose of AgX: tolerance to AgX that does not involve antigen exposure during embryogenesis has been induced.

component at an early age should result in a lack of tolerance to that same component. Triplett removed the pituitary glands of embryonic frogs, growing the glands separately as implants in the dermis of tadpoles, and then grafted the glands back into their original hosts once the frogs had grown to adulthood. The re-introduced pituitaries were rejected as foreign, verifying that an antigen cannot be seen as "self" by an individual if it is not present when self-tolerance is being established during embryonic development. In an important control for this experiment, it was found that if only part of the pituitary was removed, it would be accepted back into the host in adulthood, a result that eliminated the possibility that the transplant procedure itself had somehow made the pituitary tissue appear "foreign."

It was Burnet who first proposed a mechanism to account for these observations of immunological tolerance. As part of his theory of antibody formation published in 1959, Burnet hypothesized that the antibodies made by each individual clone of antibody-producing cells were of a single specificity that was randomly determined prior to antigen exposure. Since this process was random, some anti-self clones would necessarily be generated. Burnet argued that, for tolerance to be established, such clones would have to undergo "clonal deletion" during embryonic development, a model that would later be verified in the 1980s with the development of various transgenic mouse models of negative selection. The concept that non-responsiveness to self is established during development was a major stride forward in understanding the basic workings of the immune system. For their seminal theoretical and experimental work in establishing the general mechanism of what eventually came to be called "central" tolerance, Burnet and Medawar shared the 1960 Nobel Prize for Physiology or Medicine. Many feel that their achievements mark the beginning of the modern era of immunology.

As was the case with the phenomenon of self-tolerance, observations of tolerance to non-self antigens in the adult individual were also documented from the early 1900s onward. Much of the early work generating such reports was focused on studies of allergy and the effect of antigen dose on the antibody response. For example, in 1911, Herbert G. Wells and Thomas B. Osborne described experiments showing that adult animals fed a large dose of antigen were subsequently less able to respond to a second dose of the same antigen (Fig. 16-1B). Similarly, in the 1920s, it was noted that a large dose of heterologous gamma globulin reduced the ability of rabbits to immunologically clear an antigen after a later challenge with the same antigen. Marion Sulzberger then reported that if animals were given the drug neoarsphenamine intravenously, they were later unable to mount the usual skin sensitization response provoked by this drug when it was given intradermally. From 1940 on, there were many similar papers describing immunological tolerance induced to non-self antigens. In the mid-1950s, Lloyd D. Felton noted that the injection of 1 mg or less of pneumococcal capsule polysaccharide protected mice against a later challenge, while 50 mg or more rendered mice unresponsive over a long period of time. At that time, much

debate ensued as to the mechanism underlying this latter observation. Some immunologists felt that the chemically resistant polysaccharide antigen was released from phagocytosed antigen–antibody complexes to repeatedly clear newly-produced specific antibody from the system, while others insisted that specific inhibition of antibody synthesis was involved. By the early 1960s, the form of the antigen used for a challenge had been noted to have an effect on the immune response, and N. Avrion Mitchison published descriptions of the phenomena of low zone and high zone tolerance (see Ch.6). The accumulating body of evidence pointed to the conclusion that tolerance to foreign antigens in the periphery could be acquired, and that immune system cells could adapt to functionally "ignore" certain antigens that existed outside the bone marrow and thymus.

By the early 1970s, despite conceptual advances in understanding the mechanism of self-tolerance, there was as yet no theoretical framework to explain tolerance to foreign antigens in the periphery. Peter Bretscher and Melvin Cohn then put forth the groundbreaking proposal that lymphocytes must receive two separate signals to become activated (see Box 16-1). Their *two-signal paradigm* stated that a lymphocyte that received only one activation signal in the absence of the other would be rendered unresponsive, establishing the theory that guides studies of peripheral tolerance today. The three signal model of lymphocyte activation discussed in this book is built on Bretscher and Cohn's original model, and differs from it only in the inclusion of the cytokines required for complete activation as signal three.

In recent years, the fields of central and peripheral tolerance have converged with the finding that, despite central tolerance mechanisms, lymphocytes reactive to self antigens often do escape to the periphery. Thus, anti-self responses may be controlled, at least in part, by the same mechanisms that can result in peripheral tolerance to foreign antigens. A breakdown in these mechanisms may thus underlie many autoimmune disease processes. As a result, the quest for a more thorough understanding of the events underlying the establishment of both central and peripheral tolerance drives much of modern immunological research.

II. EVIDENCE FOR PERIPHERAL SELF-TOLERANCE MECHANISMS

While the establishment of central tolerance to self has now been reasonably well defined, the exploration of the precepts and mechanisms of peripheral tolerance to self continues to this day. We summarize here the observations that have led scientists to conclude that peripheral self-tolerance mechanisms must exist in addition to those that occur centrally.

1. The existence of autoimmunity, in which the body's tissues are attacked by the host's own peripheral T and B cells, indicates that self-reactive clones do escape the central tolerance process. Indeed, even in unimmunized animals, very low (and innocuous) levels of antibodies directed against self components, such as serum albumin, actin, various complement components, collagen, and even

Box 16-1. The "two-signal" paradigm

The "two-signal paradigm" proposed by Peter A. Bretscher and Melvin Cohn in 1970 held that lymphocyte activation required two signals, one delivered through the antigen receptor and one that we now call co-stimulatory. These researchers were studying B cell activation, attempting to explain Ovary and Benacerraf's observation that an antibody response requires the recognition of two distinct but linked determinants on the same molecule of an antigen. They were also trying to account for Mitchison's finding that different doses of the same antigen could result in either tolerance or immune reactivity. Bretscher and Cohn suggested that if a B cell received only one activation signal, the cell would be tolerized, but if two signals were received, the cell would be activated. This concept made (and still makes) logical sense, defining a mechanism that minimizes unwanted self-reactivity and thus decreases the chance of autoimmune disease. In their original scheme, which was formulated prior to the discovery of T–B cell cooperation, Bretscher and Cohn pictured the construction of an antigen bridge between two B cells; that is, the cells would be linked by the binding of their antigen receptors to the same antigen molecule. Without the second B cell to send the second activation signal, the first B cell would be "tolerized" rather than activated by antigen binding to its BCR. Soon after this model was proposed, it was found that a T cell rather than a second B cell was the source of "help" for B cell activation, with CD40–CD40L interaction being subsequently defined as the crucial second signal.

The two-signal model for T cell activation was slower to emerge. In Bretscher and Cohn's original experiments on T lymphocyte activation *in vitro*, complete Freund's adjuvant (CFA) was found to be necessary to provide the second signal. (CFA is a non-specific stimulator of the innate immune system that causes local inflammation.) In 1975, Kevin J. Lafferty and Alastair J. Cunningham used a grafting model to argue that the second signal was supplied by accessory leukocytes. Today we know that the major molecular interaction delivering the second signal for T cell activation occurs between CD28 on the T cell and either B7-1 or B7-2 on professional APCs. Looking back on Bretscher and Cohn's observations, we can deduce that the CFA upregulated the expression of B7 ligands on the APCs used in their experiment. *In vivo*, the role of CFA is assumed by components of natural pathogens processed by the innate immune system.

DNA, can be found. However, the fact that autoimmune diseases are comparatively rare implies that mature autoreactive clones are normally barred from attacking self tissues by tolerance mechanisms operating beyond the thymus and bone marrow.

2. Negative selection appears to occur only in the thymus and bone marrow. We learned in Chapter 13 that many peripheral tissue antigens are expressed in the thymus by mTECs, or are conveyed to the thymus and bone marrow by cross-presenting APCs migrating from the periphery. Nevertheless, it is highly unlikely that every self antigen in the host's body is presented in the thymus. Furthermore, some developing lymphocytes may escape negative selection even when the relevant self antigen is present because these cells express decreased levels of either the antigen receptor or the coreceptors, resulting in a "mistakenly" low avidity of binding. Thus, some self-reactive clones will inevitably slip through the central tolerance process and into the periphery.

3. Many self antigens are acquired long after birth. For example, mammary tissues develop only in the young adult animal, but these tissues are not normally subjected to immunological attack.

4. Tolerance to foreign antigens can be induced experimentally; that is, one can introduce a foreign antigen into adult animals (in which T cell development is largely complete) under conditions that will result in non-responsiveness rather than immunity.

5. Studies of transgenic mouse models have shown that when the expression of an antigen is engineered to occur specifically <u>outside</u> the thymus and bone marrow, immune tolerance to this antigen can still be clearly demonstrated.

B. T Cell Peripheral Self-Tolerance

I. THE IMPORTANCE OF DCs AND "DANGER SIGNALS"

When a naive, autoreactive T cell in the periphery meets its cognate antigen *in vivo*, what determines whether it will be activated and attack self tissues, or tolerized and rendered harmless? The short answer is: the strength of the TCR signal, the presence or absence of co-stimulation, and the cytokine milieu in which the T–APC interaction occurs. In particular, the DC that interacts with a naive autoreactive T cell plays a pivotal role in translating the context of the antigen encounter into T cell action or inaction. As recently as the mid-1990s, the role of a DC in immunity was deemed to be one solely of naive T cell activation. The job of an immature DC appeared to be to collect antigen and receive activation signals that would induce it to mature and upregulate its expression of MHC class II and B7-1/B7-2, thereby becoming a competent co-stimulator and presenter of antigen. However, the turn of the century saw the generation of a wealth of data implicating immature DCs as the prime inducers of peripheral tolerance as well. Insights into APC behavior that began in the late 1990s revealed the importance of "danger signals" to the induction of immune responses (Box 16-2), and conversely, led to the realization that the interaction of an immature DC and a naive T cell in the absence of "danger signals" could result in tolerization of the T cell.

"Danger signals" are supplied when a host tissue faces a threat such as infection or injury. Examples of danger signals are products from invading bacteria (e.g., certain mannose-containing carbohydrates, LPS, superantigen, CpG DNA) or

Box 16-2. **The "danger signal" model**

In 1994, the work of Charles Janeway and Polly Matzinger (among others) led to the notion that, although a professional APC constitutively expresses low levels of MHC class II, an external "danger signal" of some sort is required to stimulate the APC and induce the elevated expression of MHC class II and the B7 molecules required for lymphocyte activation. In other words, it is not just recognition of antigen through lymphocyte receptors that is needed to activate the immune system, but also the detection of "danger to the host." This danger can take the form of injury, infection, or even transplant surgery because these processes all generate molecular signals (see Table) that either activate an APC directly or stimulate cells of the innate immune system to secrete cytokines activating the APC.

According to the danger signal model, functional tolerance is induced in response to any antigenic peptide presented in the absence of a danger signal. Without engagement of their PRRs, innate response cells do not produce activating cytokines, APCs (including DCs) fail to upregulate co-stimulatory molecules, signal two is not delivered to Th cells that have bound to peptide/MHC class II presented by the DCs, and the Th cells are not activated. Since B cells and most Tc cells depend on Th help, their activation is also blocked by an absence of danger signals. Healthy tissues, free of injury or infection and thus danger signals, may constantly induce tolerance to their component self antigens in this way. The growing appreciation of the role of danger signals has led some scientists to argue that the immune system is designed to identify and destroy that which is <u>dangerous</u> to self, rather than that which is simply non-self. This view is a shift from the traditional belief held since Burnet's time that the role of the immune system is to identify and destroy non-self. Naturally, that which is dangerous to self will usually be non-self, but according to the danger signal model, this is not a prerequisite for triggering an immune response. The traditional self/non-self definition of immune function holds that the presence of both self-reactive clones and corresponding self antigen in the periphery automatically constitutes a pathological situation. However, according to the danger signal model, the key to controlling immune pathology and even immunotherapy is to control the presence/absence of danger signals, rather than the presence/absence of self-reactive lymphocytes. More on the manipulation of DCs and their activation in

clinical settings can be found in various sections of Chapters 26–30.

How does the danger signal model account for autoimmunity, tumor immunology, and transplant rejection? With respect to autoimmunity, the danger signal model holds that self-reactive clones freely meet cognate antigen all the time in the periphery, but are tolerized rather than activated because these encounters occur in the absence of danger signals. Autoimmune disease arises when the very first encounter of an escaped self-reactive lymphocyte with its cognate antigen occurs in the context of danger signals; that is, before tolerization can occur. Indeed, many autoimmune disorders appear to follow an infection with a pathogen of some sort. A pathogen attack normally results in the release of danger signals, activating local DCs, which in turn activate naive T cells recognizing pathogen antigens. Unfortunately, some of these activated DCs may also present cognate self-peptide to a nearby naive self-reactive T cell that has not yet been tolerized. The self-reactive T cell may be activated by the pathogen-activated DC and go on to attack self tissues. More on autoimmunity and its mechanisms appears in Chapter 29.

How does the danger signal model relate to tumor immunology? As is outlined in detail in Chapter 26, there is much ongoing debate about the expression by tumor cells of unique antigens that would represent non-self to a host's immune system. However, even if such unique antigens are expressed by the cells of a tumor, the cancer is generally very small at the start and

does not damage surrounding healthy host tissues. No danger signals are triggered, so that although APCs may be processing and presenting antigens from the tumor, no Th cells recognizing peptides derived from that tumor antigen are activated. Moreover, those circulating reactive Tc cells that recognize tumor-derived antigens and that might be able to destroy the tumor cell are then tolerized rather than activated, because of the lack of Th-mediated co-stimulation. In this way, a tumor might induce tolerance to itself in the same way as a normal self tissue does. When the tumor finally starts to cause collateral damage to healthy tissues and danger signals are released, the cancer is able to grow with impunity because the majority of T cells that might recognize its antigens are tolerized.

In the case of tissue transplantation, the danger signal model argues that the act of surgically preparing the recipient to receive the new tissue damages cells in the transplant site, and these cells release molecules proclaiming "danger." Local APCs (which can be of donor or recipient origin) acquire antigens shed from the incoming graft. In the presence of the danger signals released by the surgically damaged recipient cells, the APCs become capable of activating T cells recognizing graft antigens as foreign. The transplanted tissue is then destroyed. Indeed, as is described in more detail in Chapter 27, transplantation success is increased if the tissue to be transplanted is naturally low in APCs (such as the thyroid gland), or if the donated tissue is first depleted of APCs.

Danger Signals	Examples
Products of invading bacteria	LPS, CpG
Products of infecting viruses	DS and SS RNA
Pro-inflammatory cytokines	IFNα, TNF, IL-6, IL-1
Pro-inflammatory chemokines	RANTES, IL-18
Reactive oxygen intermediates	O_2^-, H_2O_2, OH·
Products of complement activation	C3b, C4b, iC3b, C3d
Heat shock proteins	HSP70, HSP60, gp96

Receptors Receiving Danger Signals	Examples
TLRs	TLR2, TLR4, TLR7, TLR9
Other PRRs	Scavenger receptors, CD14
Complement receptors	CR1, CR2
$\gamma\delta$ TCRs	$\gamma\delta$ TCRs recognizing stress ligands[†]

DS, double-stranded; SS, single-stranded;
[†] See Chapter 18.

viruses (e.g., single- or double-stranded viral RNA); inflammatory cytokines (e.g., IFNα, TNF, IL-6, or IL-1) or chemokines (e.g., RANTES or IL-18) secreted by innate system cells; stress molecules or internal reactive oxygen intermediates generated by physically or metabolically stressed cells; and products of complement activation. Alarm signals may also emanate from sites of tissue injury or necrotic cell death. (Necrotic death is by nature "messy," in that the cell membrane is breached and harmful internal cell contents may spill out into the surrounding tissues.) These danger signals are most likely mediated by host stress molecules such as the heat shock proteins (HSPs). For example, HSP70 released from its sequestration in the cytosol and nucleus during cell necrosis has been shown to activate DCs and macrophages *in vitro*. Similarly, the extracellular release of HSP60, normally concentrated in the mitochondria, or of gp96, normally present in the ER, serves notice to the body that its cells are under attack.

Danger signal molecules operate by engaging the PRRs (such as TLRs, scavenger receptors, complement receptors, and γδ TCRs) of innate system cells, which then produce the cytokines (such as TNF, IL-1β, IFNα, IFNγ, and GM-CSF) necessary to induce NF-κB-mediated gene expression leading to DC maturation and upregulated expression of B7-1/B7-2 and other co-stimulatory molecules (Fig. 16-2). Direct engagement of TLRs and other PRRs on an immature DC by danger signal molecules can also activate it. The activated, mature DC delivers signal one by presenting pMHC derived from the pathogen or host stress antigen to the TCR of a naïve Th cell, and signal two by B7-mediated co-stimulation. In the presence of cytokines delivering signal three, the Th cell proceeds with activation, cytokine secretion, effector generation, and an immune response that will eliminate the threat.

T cell tolerization results when the TCR is engaged by pMHC presented by an immature DC that does <u>not</u> express costimulatory molecules, a situation that arises when the immature DC acquires antigen in the <u>absence</u> of danger signals. Immature DCs are broadly distributed in the peripheral tissues, including in the skin, between vital organs and tucked in various epithelial layers. In these positions, immature DCs can take up self antigens shed by healthy tissues, or engulf apoptotic cells undergoing normal turnover. (During apoptotic cell death, a cell's internal contents are neatly packaged and degraded intracellularly; stress molecules are not released.) Since neither of these situations constitutes injury or attack, "danger signals" are not present. The production of inflammatory cytokines by innate immune system cells is not triggered, and the DC is not induced to mature and upregulate its co-stimulatory molecules or other factors. The immature DC can convey its burden of normal self antigens to the lymph nodes and present them to naïve T cells in the absence of signal two. If the TCR of a naïve autoreactive T cell happens to bind to a pMHC presented by this immature DC, the T cell is either eliminated or inactivated by mechanisms described in the next section. With the functional removal of this autoreactive T cell, peripheral tolerance to this self antigen is maintained.

Some scientists feel that, just as there may be separate subsets of DCs that influence Th1/Th2 differentiation, there may

be DC subsets that are intrinsically more tolerogenic than others. For example, in the mouse spleen, the conventional CD8α⁺ DC subset has been associated with the induction of apoptosis of self-reactive T cells. However, it remains controversial whether distinct subsets of tolerogenic DCs exist, or whether there are myriad states of activation of a single subset. It should also be noted that the induction of tolerance by DCs is not always beneficial. There is some evidence that many tumors are replete with immature DCs, while mature DCs are essentially confined to the periphery of the malignancy. It may be that the immature DCs act to suppress immune responses to tumor antigens, blocking the differentiation of CTLs that could eliminate the tumor cells (see Chs.26 and 30).

II. MECHANISMS OF PERIPHERAL T CELL SELF-TOLERANCE

Peripheral T cell self-tolerance can be induced by several mechanisms that interfere directly with one or more of the molecular signals 1, 2, or 3 needed for complete lymphocyte activation. These mechanisms include *peripheral clonal deletion, anergization, ignorance,* and *immune privilege* (Table 16-1). In addition, tolerance can result from immunoregulation via *immunosuppressive cytokines* and *immune deviation*. *Regulatory T cells,* which are increasingly thought to be important mediators of peripheral self-tolerance, are discussed in Section D later in this chapter.

i) Peripheral Clonal Deletion of T Cells

The most important mechanism by which peripheral tolerance is maintained is by the clonal deletion of self-reactive peripheral T cells in the lymph nodes. Host tissues and organs constantly shed low levels of component proteins or apoptotic cells that provide a source of self antigens. These self antigens are routinely taken up and processed by local immature DCs that then bear them to the draining lymph node. Within the lymph node, pMHCs derived from these self antigens are presented and cross-presented on MHC class II and I to resting naïve Th and Tc cells, respectively. Because there has been no injury or pathogen attack, there are no "danger signals" and the immature DC maintains relatively low levels of MHC class II and costimulatory molecules on its surface. If a T cell whose TCR is specific for one of these self-pMHCs interacts with such an immature DC, the T cell is most often induced to undergo apoptosis.

The precise mechanism by which clonal deletion associated with peripheral tolerance is induced has yet to be fully elucidated, but the cell death does not appear to be AICD and is independent of both Fas and TNFR ligation. *In vitro*, the passive apoptotic death of lymphocytes can be induced by withdrawing IL-2 or decreasing intracellular levels of necessary survival factors such as Bcl-xL. Interestingly, both of these molecules are influenced by signal two. In addition to its role in Ras/MAPK signaling during T cell activation, CD28/B7 co-stimulation both stabilizes IL-2Rα mRNA stability, leading to increased surface expression of this receptor,

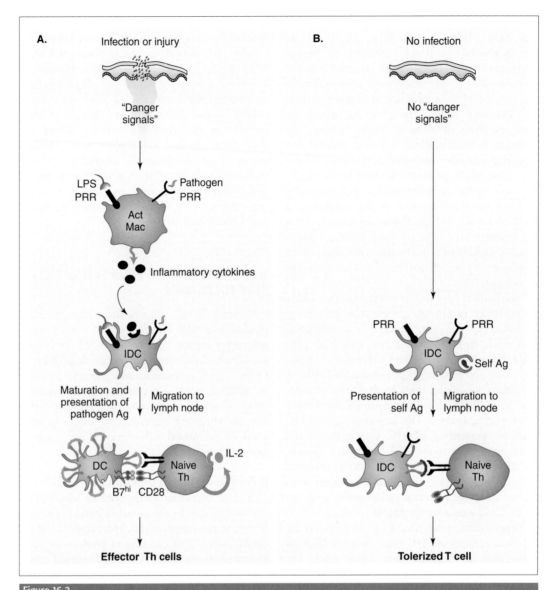

Figure 16-2

Model of T Cell Tolerization in Absence of Danger Signals
(A) Injury or infection of the host generates "danger signals" (in the form of pathogen products) that activate cells of the innate immune system to secrete inflammatory cytokines. These cytokines activate immature DCs (IDCs), which acquire antigen and mature while migrating to the lymph node. In this location, the mature DC presents high concentrations of cognate antigen to a naive T cell and upregulates B7 expression to facilitate costimulation. Upon receipt of signal three in the form of cytokines such as IL-2, effector T cells are generated and an immune response to the pathogen is observed. (B) In the absence of injury or infection, no "danger signals" are generated. Immature DCs capture self antigens and convey them to the lymph node but the DCs do not mature because of the absence of inflammatory cytokines. Cognate antigen is presented at low concentration and B7 is not upregulated on the DC surface. A naive Th cell that interacts with such a DC receives signal one in the absence of signal two, inducing tolerization of that Th cell. No immune response is observed.

and upregulates Bcl-xL transcription. It is thus possible that TCR stimulation in the absence of CD28 engagement depresses both IL-2Rα surface expression and Bcl-xL upregulation such that a T cell is pushed toward apoptosis, but this scenario has yet to be confirmed *in vivo*. Studies *in vitro* have shown that the administration of agonist anti-CD28 mAb increases the expression of both IL-2Rα and Bcl-xL and prevents the T cell death associated with tolerance. However, the expression of other stimulatory and regulatory contacts that depend on CD28/B7 costimulation is also altered, meaning that other molecules probably also play roles in the peripheral deletion pathway *in vivo*.

Table 16-1 Major Mechanisms of Peripheral Tolerance

T Cells	B Cells
Peripheral clonal deletion	Peripheral clonal deletion following anergy induced by:
Induction of anergy	• absence or anergization of a required Th cell
Immunological ignorance	• absence of danger signals required for B7-2 upregulation
Immune privilege	• receptor blockade
Immunosuppressive cytokines	Apoptosis of bystander B cells induced by signal two in the absence of signal one
Immune deviation	Tolerogenic PKCδ signaling
Interaction with regulatory T cells	

ii) Anergization of T Cells

Not all T cells that recognize pMHC presented by immature DCs undergo deletion: some survive but become *anergized*. Anergy is the name used to describe the unresponsiveness of a lymphocyte that is induced by an encounter with its cognate antigen under less than optimal conditions. The anergic lymphocyte survives but fails to proliferate and differentiate into effector cells capable of attacking the source of the antigen. Some limited effector functions may be observed in the primary response, but the anergic lymphocyte cannot respond when it encounters that antigen a second time, even when signals one, two, and three are provided under ideal conditions. The implications of this latter point are important experimentally: scientists cannot determine whether a cell is anergic (received signal one but not signal two) or merely resting (never received signal one) until they re-challenge it under conditions known to supply signals one and two and show that it fails to mount even a primary response. An anergic T cell can maintain its unresponsive state for a considerable period, even several months.

iia) Demonstrations of T cell anergy. The first, now "classical," demonstrations of lymphocyte anergy were carried out in the 1980s using Th cells *in vitro* (Fig. 16-3). Marc K. Jenkins and Ronald H. Schwartz chemically treated APCs such that they were still able to present peptide–MHC complexes to a cognate T cell but could no longer co-stimulate it. When these treated APCs were incubated with antigen-specific CD4$^+$ Th cells, the Th cells became unresponsive. That is, when later incubated with fully competent APCs presenting the same antigen, these Th cells, instead of becoming activated, failed to proliferate and produced minimal IL-2. An identical result was obtained if a normal APC was manipulated to present an antagonistic peptide. The "altered" signal one delivered by the antagonist in the presence of adequate signal two rendered the T cell anergic. Similar observations were made by Tania Watts and her colleagues when they engineered the presentation of antigens to T cells by purified MHC molecules isolated in a lipid bilayer (where no costimulation is possible). The lymphocytes adopted an anergic state that could be maintained for long periods (measured in weeks), but that could also be reversed in isolated cells by the addition of IL-2 or stimulation with a 10-fold higher dose of antigen. These results showed that, although they failed to mount a response in the absence of signal two, the antigen-specific lymphocytes were not clon-

ally deleted but remained alive and capable of responding under certain circumstances.

In vivo demonstration of the requirement for signal two to avoid anergy was then achieved using CD28-deficient mice. In wild-type mice, the injection of certain superantigens leads to lethality from toxic shock (see Box 14-3) because multiple T cell clones are simultaneously activated and release toxic quantities of cytokines. However, in CD28$^{-/-}$ mice, although the injected superantigen can still bind to the TCRs, CD28 co-stimulation is absent. Only low levels of cytokines are produced, and the animals survive. When CD4$^+$ T cells of these mice are later challenged *in vitro* with B7-expressing APCs presenting the superantigen, the T cells fail to respond, indicating that they were anergized *in vivo* by their prior encounter with superantigen in the absence of CD28.

That the anergization of T cells is clonal was convincingly demonstrated by Linda C. Burkly and Richard Flavell in 1990. These researchers used a transgenic (Tg) mouse system in which the transgene encoded a TCR containing a particular Vβ domain recognized by an endogenous superantigen. This superantigen also bound to the MHC class II E molecule outside the peptide-binding groove. It had been previously shown that, in the presence of this superantigen, thymocytes bearing TCRs containing this particular Vβ sequence bound to thymic stromal cells expressing MHC class II E, and that the thymocytes were then clonally deleted to establish central tolerance to the superantigen. Burkly decided to use a variation of this system to explore peripheral tolerance. She placed the expression of an MHC class II E transgene under the control of the insulin promoter and bred mice expressing this construct to MHC class II E-deficient mice. Under these circumstances, the MHC class II E molecule was present on the surfaces of only those cells that express insulin, the pancreatic β-islet cells. Pancreatic β-islet cells are a decidedly non-professional type of APC and do not express co-stimulatory molecules. When the mature T cells in these mice were examined, it was found that T cells expressing TCRs containing the Vβ sequence of interest were present but non-responsive. Presumably the presence of the endogenous superantigen in the pancreas allowed the interaction of the Vβ-containing TCR with the MHC class II E molecule on the β-islet cells in the Tg mouse, delivering signal one. However, because

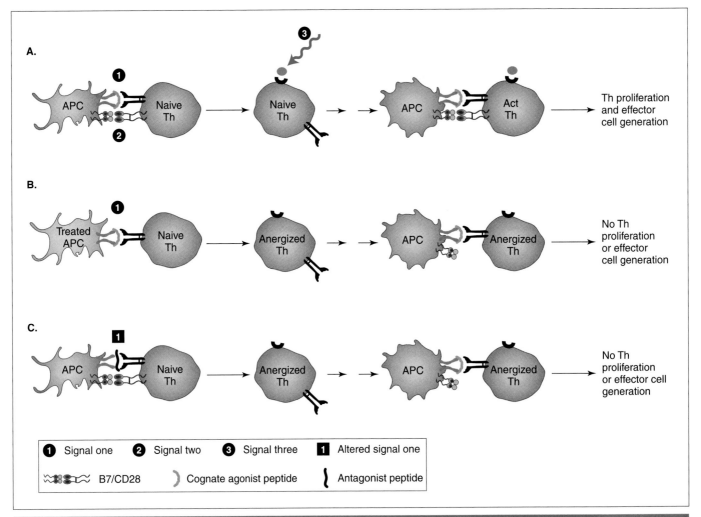

Figure 16-3

Comparison of Th Cell Activation and Anergy
(A) A naive CD4$^+$ T cell that encounters cognate antigen (signal one), costimulation (signal two), and cytokines (signal three) undergoes complete activation. The Th cell proliferates and differentiates into Th effector cells. (B) If the APCs are chemically inactivated, signal two is not delivered and the Th cell is anergized. When this Th cell later encounters a fully functional APC presenting cognate antigen, the anergized Th cell cannot respond and no effector cells are generated. (C) If the APCs present an antagonist peptide to the naive Th cell, signal one is not delivered and the Th cell is anergized. When this Th cell encounters a functional APC presenting cognate peptide, the anergized Th cell cannot respond and no effector cells are generated.

the β-islet cells were incapable of delivering signal two, the T cells were rendered anergic and peripheral tolerance to the superantigen was established.

iib) Induction of T cell anergy. What signaling pathways contribute to T cell anergy? The answer to this question is not completely clear as yet, but several studies of different aspects of intracellular signaling following TCR engagement (plus or minus co-stimulation) may offer some clues. At the very minimum, to be eligible for anergization, the T cell must at least initiate antigen receptor signaling. However, signaling may not proceed much past the earliest stages, as suggested by the findings that anergic T cells show reductions in the activities of the Src kinases Lck (up to five-fold) and Fyn (up to four-fold), and changes in the phosphorylation patterns of

substrates (e.g., CD3ζ) of these enzymes (Table 16-2). In particular, in the absence of CD28/B7 signaling, the phosphatase SHP-1 appears to be recruited to the TCR–Lck complex, correlating with incomplete phosphorylation of the CD3ζ ITAMs and ZAP-70 (Fig. 16-4). Perhaps most importantly, without costimulatory CD28 signaling, the conversion of Ras-GDP to Ras-GTP does not occur in response to TCR binding as it normally would, resulting in incomplete activation of the Ras/MAPK signaling pathway. Without Ras/MAPK activation, the production of c-fos is inhibited. Similarly, incomplete ZAP-70 activation impairs the production of c-jun via the SAPK/JNK pathway. As noted in Chapter 14, c-fos and c-jun are required to form the transcription factor AP-1. The deficit in ZAP-70 activation also impairs the PLC-γ1 pathway such

Table 16-2 Signaling and Outcomes during T Cell Activation vs. Anergy

Activation	Anergy
TCR plus CD28 signaling	TCR signaling only
Immunosome assembly	Tolerosome assembly
Lck and Fyn activated	↓ Lck activity, ↓ Fyn activity
CD3 ITAMs phosphorylated	↓ CD3 ITAM phosphorylation; SHP-1 recruitment; CTLA-4 recruits SHP-2
ZAP-70 activation	↓ ZAP-70 activation
Vav1 phosphorylation and activation	↓ Vav1 activation
Ras/MAPK leads to ↑ c-fos	Expression of a Ras/MAPK pathway inhibitor? ↓ RasGTP leads to ↓ c-fos
SAPK/JNK leads to ↑ c-jun, ↑ AP-1 p38 leads to ↑ ATF	↓ SAPK/JNK leads to ↓ c-jun, ↓ AP-1 ↓ p38 leads to ↓ ATF
PLC-γ leads to ↑ Ca^{2+}, NFAT	↓ Ca^{2+} but normal NFAT leads to ↑ anergy genes?
PKC leads to ↑ CARMA1, ↑ Bcl-10, ↑ NF-κB	↓ CARMA1, ↑ Cbl-b, ↓ NF-κB
↑ IL-2 mRNA stability	↓ IL-2 mRNA stability
Production of IL-2 and other cytokines	↓ Production of IL-2 and other cytokines
↑ Bcl-xL synthesis	↓ Bcl-xL synthesis
Proliferation and effector generation	No proliferation or effector generation

that the cleavage of PIP_2 to generate DAG is decreased. The PKC_1 pathway is thus not properly activated and NF-κB signaling is reduced. As a result of all these alterations to transcription factors, the production of cytokines necessary to support proliferation and differentiation is impaired. For example, IL-2 synthesis may be reduced up to 50-fold, IL-3 production by 10-fold, GM-CSF production by 10-fold, and IFNγ production by 2-fold. Interestingly, IL-4 production appears to be unaffected. Altered Ca^{2+} flux has also been noted in anergized cells, although NFAT activation appears to be normal. Studies of T cells activated *in vitro* with a mitogen that activates the NFAT pathway only (mimicking the absence of CD28 costimulation) have suggested that NFAT activation in the absence of normal AP-1 and NF-κB activation may induce the expression of a specific subset of genes associated with anergy. Such genes include additional transcription factors, signaling kinases and phosphatases, surface receptors, and proteases. Indeed, T cells from NFAT$^{-/-}$ mice are less susceptible to the induction of anergy than are wild-type cells.

Many scientists believe that anergy involves not only differences in intracellular signaling transducers but also differences in the physical structures that are assembled within the lipid rafts. That is, TCR engagement can lead to either T cell activation or anergization based on the collection of various adaptors and enzymes gathered at the cytoplasmic base of TCR complex. The hypothesis is that assembly of the immunosome described in Chapter 14 leads to an immune response, while assembly of an analogous structure called a *tolerosome* results

in anergy. In theory, TCR signaling in the absence of CD28 signaling would favor the assembly of the tolerosome over the immunosome. Both the immunosome and the tolerosome appear to share early components such as the Src kinases, PLC-γ1, ZAP-70, LAT, and Vav1, leading to the designation of this structure as the *early signalosome*. However, an important difference between immunosomes and tolerosomes is the participation of the CARMA1 scaffolding protein (refer to Ch.14). CARMA1 binds to Bcl-10 and is essential for the downstream activation of NF-κB and IL-2 production that occurs in an activated T cell but not in an anergized T cell. In an anergic T cell, CARMA1 is not recruited to the early signalosome. Instead, the ubiquitination enzyme Cbl-b (refer to Ch.14) joins the complex. *In vitro*, Cbl-b can also inhibit the interaction of PLC-γ with Vav1, and CD28 with PI3K. *In vivo*, animals lacking Cbl-b function are predisposed to developing autoimmunity. It is speculated that the action of Cbl-b and molecules that interact with it may promote the formation of the tolerosome and block the formation of the immunosome.

It has also been suggested that the CD28 homologue CTLA-4 may play a role in anergy induction because this molecule binds to B7 ligands but acts as a negative regulator of Th cell activation. As discussed in Chapter 14, CTLA-4 displaces CD28 bound to B7 ligands and indirectly reverses TCR-mediated phosphorylation of substrates via recruitment of the tyrosine phosphatase SHP-2. *In vitro*, CTLA-4 has been shown to inhibit the transcription of IL-2 and the production of several cell cycle components necessary for proliferation. Antigens that are only weakly stimulatory, in that they do not

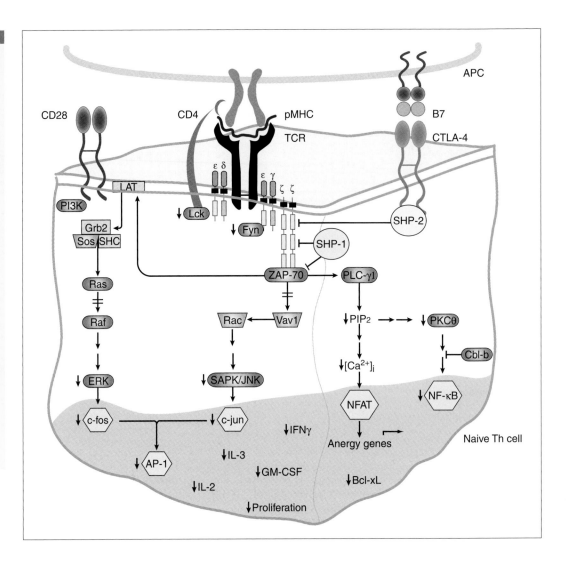

Figure 16-4

Signaling Pathways Associated with T Cell Anergy

If CD28 is not bound by B7, signal transduction via the Ras/MAPK pathway is disrupted and ERK is not activated, leading to reduced c-fos. The phosphatase SHP-1 is recruited to the TCR complex and prevents complete phosphorylation of the CD3 chains and ZAP-70. SHP-2 recruited by CTLA-4 bound to B7 may also contribute to these effects. In the absence of phosphorylated ZAP-70, Vav1 and the SAPK/JNK and PLC-γ1 pathways are not properly activated, leading to reduced c-jun and Ca²⁺ flux. The PKCθ pathway is also affected and may be further inhibited by Cbl-b, leading to decreased NF-κB activation. The transcription of genes required for immune responsiveness is impaired. In addition, the activation of NFAT in the absence of normal AP-1 and NF-κB may lead to the induction of a subset of anergy-specific genes.

trigger sufficient signaling to overcome the CTLA-4-controlled threshold, may therefore trigger anergization rather than activation of the T cell. Other theories about the involvement of CTLA-4 in maintaining peripheral tolerance are discussed in this chapter in Section D, 'Regulatory T Cells.'

Some investigators feel that more than just an absence of co-stimulation is involved in inducing anergy, and that TCR engagement in isolation (signal one alone) delivers a message that leads to the synthesis or activation of an inhibitor that actively impairs cytokine production. Such an inhibitor might block the conversion of Ras-GDP to Ras-GTP, or might act as a transcriptional repressor further downstream. Evidence for such inhibitors has been demonstrated *in vitro* in several ways. In one study, anergic murine T cells were fused *in vitro* to activated human T cells from the Jurkat leukemia cell line to form heterokaryons (Fig. 16-5). The murine T cells were incapable of producing murine IL-2 due to their anergy, but the human T cells were actively producing human IL-2. The cell mixture containing the heterokaryons was then stimulated with a combination of anti-human CD3 and anti-human CD28 antibodies to mimic signals 1 and 2. It was found that the transcription of the active human IL-2 gene was disrupted

in the heterokaryon, implying the presence of a dominant repressor in the anergic murine T cells that was able to interfere with transcription of the human IL-2 gene. When the anergic murine T cells were fused to normal murine T cells and the heterokaryons were subjected to activating stimuli, the production of mRNA for the unrelated lymphotactin gene was normal, although the synthesis of IL-2 remained minimal. This result showed that at least some signaling pathways remained intact in anergic T cells, and that the repressor did not act globally on all genes. The precise nature and mechanisms of action of the putative repressor have yet to be defined. It is speculated that such a repressor might affect signal transduction from the heterokaryon's TCRs to the Ras/MAPK pathway, or might be a nuclear repressor that enters the human nucleus of the heterokaryon to block initiation of IL-2 transcription. In other studies using assays designed to detect DNA-binding proteins, a nuclear repressor protein called Nil-2a (*negative regulator of IL-2a*) was identified. This protein was found to bind to regions upstream of the AP-1 site in the IL-2 promoter and to have the capacity to inhibit transactivation of the IL-2 gene *in vitro*. Whether Nil-2a is the same molecule as the repressor identified in the heterokaryon experiments is unclear.

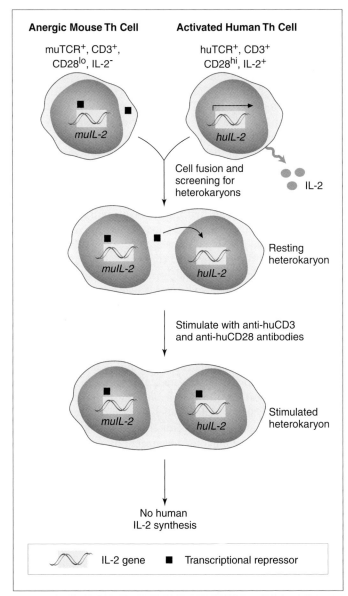

Figure 16-5

Demonstration that an Active Inhibitor May Mediate T Cell Anergy
An anergic mouse Th cell that is incapable of expressing murine IL-2 (muIL-2) is fused to an activated human Th cell secreting human IL-2 (huIL-2). Heterokaryons are isolated and stimulated with anti-human CD3 and anti-human CD28 antibodies to mimic antigen-mediated activation and co-stimulation. However, the heterokaryons fail to express human (or murine) IL-2, indicating that they have become anergic. These results suggest that the mouse cells are incapable of secreting murine IL-2 because of the presence of a diffusible repressor, and that this repressor is able to cross from the mouse nucleus of the heterokaryon to the human nucleus and block human IL-2 synthesis.

In addition to its occurrence in the absence of co-stimulation, anergy has been demonstrated in two other situations for which the molecular bases remain incompletely defined. In the first case, human T cells subjected to very high doses of antigenic pMHC administered in the presence of competent co-stimulation can be anergized. In this case, the block in downstream signaling appears to be at the level of Ca^{2+} and

calcineurin signaling and the production of all cytokines is impeded. In the second instance, as mentioned previously, treatment of T cells *in vitro* with antagonist peptides and APCs fully competent to deliver signal two can result in a later failure to respond to cognate peptide. Despite the presence of normal CD28 signaling, Th cells whose TCRs engage antagonists appear to receive an altered signal one. A reduction in TCR-associated tyrosine kinase activities can easily be demonstrated, suggesting that the block is at the earliest step. Only partial phosphorylation of the CD3 ITAMs is induced, T cell activation does not proceed, and IL-2 secretion is again minimal. Originally, there was some suggestion that the binding of pMHC to the TCR might have some direct influence on TCR conformation that could control whether the binding of antigen resulted in activation or anergy of the T cell. However, the outcome of TCR–pMHC interaction does not correlate with ligand-induced conformational changes to the TCR. This was elegantly shown in experiments by Don Wiley and colleagues in which the interactions of a wild-type viral peptide with a given TCR and MHC molecule were compared with those of three related peptides bearing single amino acid substitutions. The mutations caused the substituted peptides to act as either a weak agonist that resulted in only partial T cell activation, or an antagonist that induced anergy in the T cell. When the three-dimensional structures of the three substituted peptide–TCR–MHC complexes were compared to the wild-type TCR–MHC structure by X-ray crystallography, no major differences in conformation were found. These results clearly show that different TCR-mediated signals resulting in distinct cellular outcomes are not generated by different ligand-induced conformational changes in the TCR. The signaling pathways associated with these instances of anergy are under investigation.

iii) Immunological Ignorance

Sometimes a peripheral T cell appears to be rendered tolerant because it never receives signal one. Many self components are present in such low amounts that immature DCs rarely acquire them and presentation of component peptides to T cells is thus minimal. Because these antigens are present but do not engage the TCR and provoke a response, the immune system is said to be "ignoring" them such that tolerance to these antigens is observed.

Evidence consistent with tolerance due to immunological ignorance was provided by Pamela Ohashi and her colleagues in 1991 using Tg mice and the pancreatic β-islet cell system. The CD8$^+$ T cells in these Tg mice expressed a TCR recognizing a viral antigen engineered to be expressed only on the surface of the β-islet cells (i.e., the viral protein essentially became a "self"-antigen of the β-islet cells). If ignorance did not prevail, CTLs derived from these Tg CD8$^+$ T cells would attack the β-islet cells and destroy them, causing the mice to develop diabetes. However, the β-islet cells remained untouched and the mice stayed healthy. The Tg CD8$^+$ T cells in these animals were shown to be present (i.e., not clonally deleted) and capable of being activated *in vitro*, but were inactive in the whole animal. It was not until the mice were systemically infected with the original virus that diabetes was triggered. The researchers speculated that the β-islet cells were not destroyed until the Tg T cells were activated by the virus in secondary lymphoid tissues such that they generated effector CTLs capable of infiltrating peripheral tissues, including the pancreas. In the non-infected mice, immunological ignorance due to insufficient cross-presentation of the antigen to the naive Tg CD8$^+$ T cells maintained self-tolerance.

These results have been confirmed in mouse models in which the level of antigen expression and means of presentation to antigen-specific T cells can be manipulated. For example, when pancreatic β-islet cells were engineered to express relatively low levels of the experimental antigen ovalbumin, no detectable cross-presentation of this antigen occurred and ovalbumin-specific CD8$^+$ T cells remained ignorant and inactive. However, when the expression of ovalbumin was increased on the β-islet cells (in the absence of inflammation), cross-presentation by DCs became detectable in the draining lymph node and ovalbumin-specific CD8$^+$ T islet cells were clonally deleted. When the mice were manipulated so that their DCs were incapable of presenting ovalbumin, ignorance once more prevailed. These observations emphasize the pivotal role of DCs in determining tolerance, and illustrate that ignorance can indeed be bliss when it comes to lymphocyte recognition of self antigens.

iv) Immune Privilege

Certain anatomical sites in the body are naturally less subject to immune responses than most other sites, a phenomenon known as "immune privilege." Even foreign antigens accessing these sites, which include the central nervous system and brain, anterior chambers of the eyes, and the testes, do not generally trigger immune responses. Thus, grafts are accepted more readily in immune-privileged sites than in other tissues, which is why transplants of eye components can easily be made between unrelated individuals. For a long time, it was thought that immune privilege stemmed from physical barriers that prevented the migration of lymphocytes into these sites. For example, a "blood–testis" barrier is formed by cellular and membranous structures that surround the seminiferous tubules and prevent the infiltration of cells from the blood into the testis. Similarly, a "blood–brain barrier" exists because the walls of the brain capillaries are constructed in such a way that most cells and large molecules cannot pass through from the blood into the brain. Furthermore, there is no lymphatic drainage from the brain to a lymph node, meaning that brain antigens rarely reach a secondary lymphoid tissue where a naive T cell might be activated. Originally, therefore, the tolerance to self antigens in these "privileged" tissues was seen to be due to a failure in the delivery of signal one: naive lymphocytes simply never encountered APCs presenting the relevant antigen. However, studies in the 1990s showed that antigens deliberately introduced into immune-privileged sites could sometimes enter the circulation and activate naive T cells following uptake and processing by APCs. The effector T cells that were subsequently generated started to express adhesion molecules that allowed them to penetrate structures such as the blood–brain barrier. Nevertheless, there was still no immune reactivity in the privileged tissue. Scientists then sought to define the additional mechanisms that had to exist in privileged sites to prevent immune responses.

In the brain, a passive resistance to the initiation of immune responses holds. Neurons express only very low levels of MHC class I molecules even in an inflammatory situation, meaning that antigen presentation to any CD8$^+$ T cells that may have invaded the brain is minimal. In addition, very few DCs are present in the brain, meaning that the presentation of antigen on MHC class II to CD4$^+$ T cells is also unlikely.

In the testis and the anterior chamber of the eye, more direct approaches are taken to terminate both innate and adaptive immune responses. In the eye, cells of the iris and ciliary bodies can secrete the immunosuppressive cytokine TGFβ, which interferes with antigen presentation by APCs, dampens T cell activation, and inhibits Th1 and NK responses. In addition to TGFβ, other factors are present that inhibit neutrophil activation and repress nitric oxide production by macrophages. Non-lymphoid cells in the retina, iris, and cornea of the eye and certain regions of the testis are also equipped to eliminate activated T lymphocytes via Fas killing. Fas is widely expressed on non-lymphoid cells and is induced on T cells after activation, whereas FasL is expressed on only a very few cell types. Significantly, these cell types are often found in immune-privileged sites. For example, the Sertoli cells in the testis constitutively express high levels of FasL. When a T cell whose TCR is specific for a self antigen expressed in the testis manages to enter this tissue, the T cell encounters its antigen, is activated, and starts to express Fas. Interaction of Fas on the T cell with FasL on a Sertoli cell induces the activated T cell to undergo apoptosis before it has a chance to differentiate into effector cells that could attack testicular tissue. Sustained inflammatory reactions are also prevented by this mechanism since T cells are killed before they can secrete the necessary pro-inflammatory cytokines.

As a further illustration of the role of Fas killing in immune privilege, consider the case of testicular graft acceptance. If testicular grafts taken from normal FasL$^{+/+}$ mice are introduced under the kidney capsules of normal, allogeneic recipients, an immune response is not mounted and the graft is not rejected. However, grafts from FasL$^{-/-}$ mice are rejected rapidly. These results have been interpreted to mean that when T cells in the recipient kidney capsule are activated by alloantigens of the normal graft, the activated T cells express Fas, interact with FasL on the normal testicular tissue, and are induced to undergo apoptosis before a graft rejection response can occur. However, when FasL is absent from the testicular graft, T cells in the recipient kidney capsule escape apoptosis and proceed to mount a normal immune response resulting in graft rejection. Interestingly, some tumors have also been found to express FasL, suggesting that they may employ the same Fas killing mechanism to fend off attacks by immune system cells.

v) Immunosuppressive Cytokines

Certain cytokines, in particular IL-10 and TGFβ (see Ch.17), have immunosuppressive effects that can act as a brake on innate and adaptive immune responses, including those initiated against self antigens. Although most often produced by innate system cells activated in the course of a pathogen invasion, these cytokines may also be synthesized in significant amounts by anergic T cells and by certain subsets of regulatory T cells (see later, Section D). IL-10 downregulates Ras/MAPK signal transduction, inhibits macrophage activation and cytokine secretion, blocks APC function, prevents the proliferation of Th1 cells, and destabilizes the mRNAs of many cytokines, including IL-2. The ensuing lack of IL-2 can thus contribute to the T cell anergization. Similarly, TGFβ inhibits macrophage activation, blocks the proliferation and IL-2 production of activated T cells, downregulates Ig synthesis by B cells, and interferes with the stimulatory effects of IL-2 on T and B cells. As noted previously, immunosuppressive cytokine functions are often particularly important for promoting self-tolerance in immune-privileged sites.

vi) Immune Deviation

In the 1960s, scientists occasionally observed a lack of immune reactivity that was popularly referred to as "immune suppression" or "immune deviation." These terms were used to describe a phenomenon in which an adaptive immune response that was harmful, such as an autoimmune reaction, was converted to a less harmful response. This mitigation of damage was viewed as a form of tolerance. Today, we understand these observations to be a function of immunoregulation rather than the establishment of tolerance per se, brought about by the manner in which microenvironments can influence Th1/Th2 differentiation, and how Th cells can cross-regulate each other's effector actions (as discussed in Ch.15). A bias toward one subset of T cells, either Th1 or Th2, will avoid potential damage that might have been caused by an autoimmune attack by the other subset. For example, consider the n̲o̲n̲-o̲b̲ese d̲i̲abetic (NOD) mouse. These mutant mice are genetically more susceptible than the wild-type to developing diabetes caused by an autoimmune attack on the β-islet cells of the pancreas (see Ch.29). This attack is mediated in large part by Th1 cells. However, when immune deviation is induced in NOD mice, Th2 cells prevail in the islets, the number of damaging Th1 cells is reduced, and diabetes is avoided. Phenotypically, the mice appear to have become tolerant to the self antigens responsible for clinical diabetes. In reality, the autoimmune response may continue but it no longer produces visible pathology. These types of effects have been seen *in vitro* as well. Surprisingly, Th0 clones treated in a way expected to render them anergic do not become totally unresponsive but rather adopt a Th2 phenotype at the expense of the development of Th1 cells.

Immune deviation often turns up as an important means of avoiding harmful immune responses in immune-privileged sites. For example, the sensitive tissues of the eye would be badly damaged if Th1-mediated DTH responses predominated rather than Th2-mediated humoral responses.

vii) Peripheral Clonal Exhaustion?

Scientists speculate that the complete loss of fully activated antigen-specific lymphocytes from the periphery may also result in tolerance to some self antigens. As we saw in Chapter 15 in our discussions of the termination of the primary immune response, T cell exhaustion can lead to the removal of all mature antigen-specific T cells that have undergone full activation. Although this primarily homeostatic mechanism appears on the face of it to apply to foreign antigens (and is thus more relevant to experimental tolerance than to self-tolerance), some

scientists think that tolerance to many self antigens which are widely distributed and present in high abundance in the body may be generated during early development in just this way. Prolonged exposure of lymphocytes to specific antigen (such as would occur for a constitutively expressed self antigen) could lead to rapid cell division, the early onset of replicative senescence, and the death of the clone without the generation of memory cells. Although an intriguing and logical possibility on paper, this theory has yet to be substantiated *in vivo*.

C. B Cell Peripheral Self-Tolerance

Self-reactive B cells are also controlled by peripheral tolerance, although the mechanisms differ somewhat from those we have described for T cells. We begin with a description of an experiment considered to be a classic in the field of peripheral tolerance.

I. DEMONSTRATION OF B CELL ANERGY

In 1992, Christopher C. Goodnow and his colleagues carried out a study that was among the first to convincingly demonstrate the existence of peripheral tolerance to self antigens. A Tg mouse strain was engineered to express secreted hen egg lysozyme (HEL) under the control of a zinc-responsive promoter. This promoter had the effect of restricting HEL expression to the periphery, so that central tolerance was not a factor. Another Tg mouse strain carried Ig genes manipulated to encode anti-HEL antibody. In this latter strain, at least 60% of the B cells present expressed anti-HEL BCRs. Goodnow crossed these strains to produce double transgenic (DTg) offspring carrying both transgenes. As is typical in transgenic litters, some of the DTg mice expressed high levels of circulating HEL, while others expressed 10-fold less. Goodnow then examined the specificity and reactivity of peripheral B cells in both groups of DTg offspring. Large numbers of anti-HEL peripheral B cells were found in the spleens and lymph nodes of all DTg mice. However, in DTg mice expressing high levels of HEL, the B cells were anergic. In DTg mice expressing low levels of HEL, very few anti-HEL B cells were detected but these appeared to be functionally responsive. A striking difference in membrane Ig expression was also evident upon a comparison of the B cells of the two groups of DTg mice. While reactive anti-HEL B cells in the periphery of "low HEL" mice expressed normal levels of IgM and IgD, the tolerized anti-HEL B cells of "high HEL" mice lost about 90% of their surface IgM (but retained normal IgD). Furthermore, if the "low HEL" mice were then given drinking water containing high levels of zinc, causing their expression of circulating HEL to increase 70-fold, IgM disappeared from the surface of the anti-HEL B cells in these mice and the cells were tolerized. These data demonstrated convincingly that high levels of a circulating antigen could indeed induce B cell peripheral tolerance, and that the signal was likely conveyed at the level of membrane IgM signaling.

II. MECHANISMS OF PERIPHERAL B CELL TOLERANCE

For B cells, signal one is derived from the stimulation of BCRs by a high-affinity antigen. Neither activation nor anergization can be induced in the absence of signal one. During activation, stimulation of BCRs by antigen triggers the activation of the receptor-associated tyrosine kinases Lyn and Syk. (A review of signaling pathways associated with B cell activation is shown in Fig. 16-6.) These kinases mediate phosphorylation of the ITAMs in the Igα/β components of the BCR that lead to changes in Ca^{2+} flux, activation of the PKCβ isozyme of the PKC family of signaling kinases, and eventually the activation of ERK and NF-κB. In response to these signals (and promoted by cytokines of the innate response induced by "danger signals"), B7-2 is upregulated on the B cell surface, which in turn binds to CD28 molecules on a Th cell whose TCRs have been engaged by antigenic pMHC displayed by the B cell. The delivery of signal one to the Th cell stimulates the upregulation of CD40L on its surface, allowing the delivery of signal two to the B cell in the form of CD40 engagement. In the presence of (usually) IL-4 supplied by the Th cell, the B cell is stimulated to proliferate and differentiate into antibody-secreting plasma cells.

There are several routes (described in the following sections) by which B cells can be anergized. However, unlike anergic T cells, anergic B cells have a very short life span (3–4 days) and clonal deletion occurs rapidly. Even if treated *in vitro* with antigen and supplied with T cell help, anergic B cells nevertheless progress to apoptosis (unlike anergic T cells). It is thought that a lack of both normal BCR signaling and CD40 signaling during the establishment of anergy may prevent the receipt of signals necessary for survival, such as the induction of members of the Bcl-2 family. *In vivo*, anergic B cells apparently cannot compete successfully with activated B cells to enter the secondary follicles. Unable to reach the follicles and receive survival signals normally delivered in this location, the anergic B cells quickly undergo apoptosis. Interestingly, the death of anergic B cells requires not only the continued presence of antigen but also the engagement of CD40 and Fas, suggesting that perhaps an activated T cell (that presumably did not arrive in time to prevent the original anergization) cooperates to supply the necessary CD40L and FasL ligands for killing.

We now discuss the various means by which B cell anergy is induced.

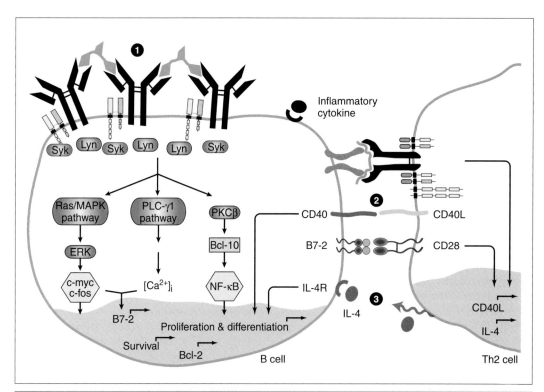

Figure 16-6

Review of Signaling Pathways Associated with B Cell Activation
When antigen cross-links BCRs and aggregates them, signal one is initiated by the activation of the Src kinases Syk and Lyn associated with the Igα/Igβ signaling chains. The Ras/MAPK, PLC-γ1, and PKCβ pathways are activated, leading to transcription factor activation and expression of B7-2. Inflammatory cytokines induced by "danger signals" promote B7-2 expression. The B cell also presents processed antigen on MHC class II to the TCR of a Th cell. The interaction of B7-2 with CD28 on the Th cell upregulates expression of CD40L on the Th surface and induces Th expression of IL-4. CD40/CD40L interaction constitutes signal two for the B cell while IL-4 interaction with IL-4R on the B cell delivers signal three. The B cell proliferates and differentiates into antibody-secreting plasma cells.

i) Lack of an Antigen-Specific Th Cell

Like autoreactive T cells, autoreactive peripheral B cells are anergized by the receipt of signal one in the absence of signal two. If a Th cell of the appropriate specificity is not on hand, there can be no delivery of Th help to a B cell in the periphery whose BCRs have been engaged and B cell anergy will ensue (Fig. 16-7). Because the B cell depends on the Th cell in this way, the host can benefit from the somatic hypermutation that occurs in the Ig genes without undue risk of increased self-reactivity. The reader will recall that somatic hypermutation increases the diversity of the BCR repertoire, but that among these new mutations may be some that cause a B cell to become self-reactive. However, such a B cell will most likely not encounter a Th cell able to offer signal two because self-reactive T cells have been eliminated for the most part during the establishment of central tolerance. Moreover, since the TCR genes do not undergo somatic hypermutation, a non-self-reactive T cell cannot suddenly become self-reactive.

ii) Lack of "Danger Signals"

Even if a self-reactive Th cell is present, without the "danger signals" conveyed by the innate immune response, an antigen-stimulated B cell will not upregulate its expression of B7-2. The Th cell will not be co-stimulated, CD40L will not be upregulated, and the B cell will not receive signal two (refer to Fig. 16-7). Thus, the B cell will be anergized rather than activated, leading to the eventual apoptosis of the cell.

iii) Receptor Blockade

Very large amounts of an antigen that persistently occupy the BCRs without cross-linking them form a "receptor blockade" that can also anergize B cells (refer to Fig. 16-7). It has been

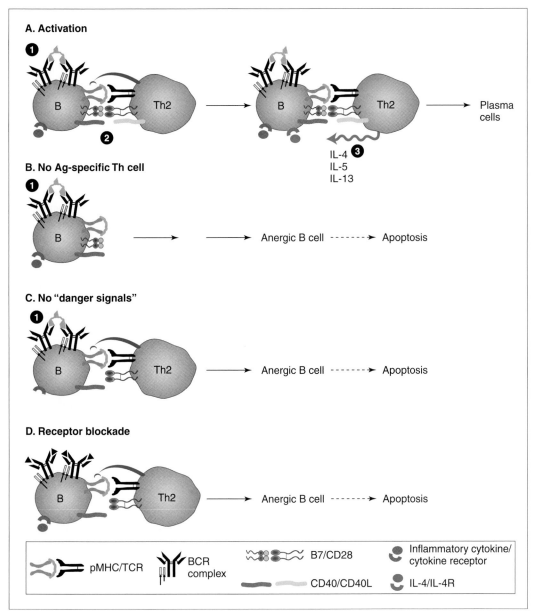

Figure 16-7

Comparison of B Cell Activation and Anergy
(A) Signals one, two, and three for the B cell are delivered as described in Figure 16-6. Plasma cells are produced and an immune response is observed. (B) In the absence of an antigen-specific Th cell, signal one may be received by the B cell but signals two and three are not. The B cell is anergized and rapidly undergoes apoptotic death. (C) In the absence of "danger signals," B7-2 is not upregulated on the B cell surface. The Th2 cell is thus not co-stimulated, CD40L is not upregulated on the Th2 cell surface, and cytokine secretion is blocked. The B cell receives neither signal two nor three, becomes anergized, and undergoes apoptosis. (D) When very large amounts of antigen occupy the BCRs without cross-linking them, signal one is not initiated and B7-2 cannot be upregulated even in the presence of inflammatory cytokines. Anergization and apoptosis of the B cell result.

shown that a receptor blockade affects normal BCR signal one delivery at the level of receptor-associated tyrosine kinase activation. Both Syk kinase and the ITAMs on the Igα/β chains in the BCR complex are incompletely tyrosine-phosphorylated such that an inhibitory signal, perhaps analogous to the Ras/MAPK inhibitor of anergic T cells, is induced. Downstream events such as NF-κB and SAPK/JNK activation are blocked and Ca²⁺ flux is reduced. Consequently, upregulation of B7-2 on the B cell surface does not occur. Without CD28-B7 interaction, any Th cell that manages to form a conjugate with the B cell cannot receive its own signal two and does not upregulate CD40L, so that no signal two is delivered to the B cell. Extensive downregulation of surface IgM expression is also found on these anergic B cells, resulting from the inhibition of IgM transport from the ER to the Golgi. Thus, further signal one delivery is also precluded.

iv) Bystander Apoptosis

Apoptosis of B cells (including self-reactive B cells) can result in cases where signal two is received in the absence of signal one. For example, a bystander B cell whose BCRs are <u>not</u> stimulated by antigen may present an antigenic pMHC to a Th cell whose TCR happens to recognize it. Signal two might still be delivered to the B cell through low levels of CD40–CD40L signaling. However, the stimulation of the TCR also upregulates the expression of FasL on the T cell, which can bind to Fas on the B cell and, in the absence of strong survival signals conveyed by BCR stimulation, condemns the B cell to apoptotic death.

v) Tolerogenic PKCδ Signaling

A relatively recently discovered mechanism that contributes to the maintenance of normal B cell peripheral tolerance appears to be mediated by an isozyme of the PKC signaling family called PKCδ. PKCδ expression is highest in developing B cells and lowest in mature T cells. Knockout mice lacking PKCδ show evidence of autoimmunity and uncontrolled B cell proliferation due to increased IL-6 secretion by the mutant B cells. In wild-type cells, PKCδ may phosphorylate and inactivate a transcription factor called NF-IL6 that drives IL-6 expression (see Ch.17). In the absence of this negative regulation, excess IL-6 production non-specifically promotes the proliferation of B cells, including autoreactive B cells. Other alterations in intracellular signaling, particularly with respect to NF-κB activation, have also been detected in PKCδ-deficient B cells. PKCδ may thus be an example of an enzyme that is specifically involved in tolerogenic, but not immunogenic, B cell signaling responses to an antigen.

D. Regulatory T Cells

In the 1970s, studies of "suppressor" T cells dominated much of cellular immunological research. These regulatory cells were proposed to control immune responses and establish self-tolerance. Many models were generated but none withstood examination at the molecular level. In fact, it was later shown that most of the results of these particular experiments could be explained by immune deviation. Concrete evidence for the existence of regulatory T cells came from studies in the 1990s of *lymphopenic* mice, which are deficient in lymphocytes. Researchers found that a lymphopenic environment could provoke the activation of self-reactive T cells and the initiation of an autoimmune response. These results were interpreted to mean that a subset of T cells needed to suppress self-reactivity was missing. For example, if normal adult mice are thymectomized and treated with the drug cyclophosphamide such that their lymphocyte numbers are drastically reduced, the mice spontaneously develop autoimmune diseases very similar to the human autoimmune disorders type 1 diabetes and Hashimoto's thyroiditis (see Ch.29). Another common manifestation is *autoimmune gastritis*, in which the parietal cells in the lining of the stomach are attacked by T cells. Autoantibodies to antigens expressed by parietal cells appear in the circulation, and the chronic inflammatory reaction that is established destroys the stomach lining. Neonatal mice that are thymectomized between 2 and 4 days after birth also spontaneously develop autoimmune gastritis. In both cyclophosphamide-treated adult mice and thymectomized neonatal mice, the onset of autoimmune disease can be mitigated by the transfer of unfractionated normal thymocytes or T cells. These observations have been interpreted to mean that the transferred T cells or thymocytes contain a sub-population of regulatory T cells or thymocytes required to control autoreactive peripheral T cells and actively establish tolerance.

In fairness, it should be noted that some scientists have suggested that lymphopenic tolerance is not due to regulatory T cells but rather to altered patterns of antigen presentation. In the absence of the thymus or under other conditions inducing lymphopenia, an animal could become subject to spontaneous or infection-induced inflammation in a target organ; such inflammation could constitute "danger signals." The immature DCs acquiring self antigens shed from the target organ would then be induced to mature and become capable of activating previously ignorant autoreactive T cells in the local lymph node. The activated self-reactive T cell would generate the effectors that could migrate to the target organ and mount an autoimmune attack.

While some controversy remains, most researchers are now convinced that subsets of regulatory thymocytes and T cells do exist, and that, under certain circumstances, these subsets can act in various ways to suppress the activation of self-reactive naive lymphocytes. In these cases, the naive cell is usually anergized as a result of its interaction with the regulatory cells.

I. CD4⁺CD25⁺ T_reg CELLS

The best-documented regulatory T cell population has a surface phenotype of CD4⁺CD25⁺ (Table 16-3). That is, unlike resting, conventional mature CD4⁺ T cells, these CD4⁺ T cells express the IL-2Rα chain (CD25) <u>prior</u> to activation. (IL-2Rβ is not expressed.) The T_reg population comprises about 10% of normal mouse or human peripheral CD4⁺ T cells and does not appear to include conventional activated and memory T cells. T_reg cells do not proliferate significantly and show negligible

Table 16-3 **Types of Regulatory T Cells**

	T_{reg}	Th3	Tr1	Ts
Characteristic markers	$CD4^+CD25^{hi}$ $GITR^{hi}$ $CTLA-4^{hi}$	$CD4^+CD25^{lo}$ $GITR^{med}$ $CTLA-4^{med}$	$CD4^+CD25^{lo}$ $GITR^-$ $CTLA-4^{lo}$	$CD8^+$ $CD57^+$ $CD28^-$
Dominant suppressive mechanism	Intercellular contact	Secretion of TGFβ	Secretion of IL-10, TGFβ	Induces upregulation of ILT3, ILT4 on DCs Secretion of IL-2, IL-4, IFN-γ, TGFβ
Regulatory effects	Suppresses activated T cells Induces Th0 cells to produce Tr1 and Th3 cells	Suppresses activated T cells	Suppresses activated T cells	Modulates DCs so that they induce the production of Tr1 and Th3 cells from Th0 cells Suppresses B cell production of certain Ab isotypes
Molecules important for generation	IL-2, FoxP3, CTLA-4, CD28	TGFβ, IL-10, IL-4	TGFβ, IL-10, IFNα	?
Major site of generation	Independent T cell lineage in the thymus	Induced by Th0 contact with modulated DCs or T_{reg} cells in the periphery	Induced by contact with modulated DCs or T_{reg} cells in the periphery	?

cytokine secretion after activation. However, T_{reg} cells whose TCRs are engaged can strongly suppress the proliferation of effector T cells in a non-specific way. The identity of the antigens activating T_{reg} cells remains a mystery.

T_{reg} cells were first identified by their ability to prevent the induction of organ-specific autoimmune disease in mice. That is, if the $CD4^+CD25^+$ population was removed from a $CD4^+$ T cell inoculum injected into susceptible mice, the animals were more likely to develop destruction of one or more organs due to attack by autoreactive $CD4^+CD25^-$ T cells present in the inoculum. If the $CD4^+CD25^+$ population was restored to the inoculum, the attack by the $CD4^+CD25^-$ T cells was prevented. *In vitro* examination of the $CD4^+CD25^+$ cell population from normal mice has shown that development of these regulators from thymic precursors requires contact in the periphery with cognate self antigen (presented by immature DCs) soon after the release of the cells from the thymus. Once mature, subsequent interaction with cognate antigen in the periphery apparently renders the mature T_{reg} cells anergic. However, $CD4^+CD25^+$ cells gain the ability to block the proliferation of $CD4^+CD25^-$ cells of <u>any</u> antigenic specificity and inhibit the latter's IL-2 production. The suppressive effect is not cytokine-mediated but rather requires direct intercellular contact with the target T cell via surface molecules that have yet to be characterized.

As well as by their constitutive expression of CD25, T_{reg} cells can be distinguished by their constitutive expression of L-selectin, OX40, CTLA-4, membrane-bound TGFβ, and a member of the TNFR superfamily of receptors called *GITR* (*glucocorticoid-induced TNFR-related receptor*). However,

whether any of these molecules is actually involved in the suppressive function of T_{reg} cells has been controversial. Blocking mAbs against L-selectin, OX40, or TGFβ do not reduce T_{reg} mediated suppression of conventional T cells. Blocking mAbs against CTLA-4 block suppression in some systems but not in others. In mice, the addition of anti-GITR mAb can abrogate the suppressive activity of T_{reg} cells *in vitro*, and treatment with the same mAb induces autoimmunity *in vivo*. However, activated conventional $CD4^+$ T cells that do not have suppressive activity also express GITR. Moreover, human conventional T cells do not express a ligand for GITR. Thus, some scientists feel that GITR engagement may be necessary for triggering the induction of the suppressive mechanism, but that the suppressive contact itself is not delivered by GITR.

A molecule that appears to be critically important for the generation and function of T_{reg} cells is a transcription factor called FoxP3. FoxP3 is expressed exclusively by T_{reg} cells in the thymus and the periphery and not by conventional $CD4^+$ T cells. Indeed, forced expression of FoxP3 in conventional T cells confers suppressive powers upon them. Some scientists speculate that, just as T-bet and GATA-3 expression defines Th1 and Th2 differentiation, respectively, FoxP3 may govern T_{reg} differentiation. As is discussed in detail in Chapter 29, mice and humans deficient for FoxP3 lack T_{reg} cells and develop autoimmune diseases.

The identification of T_{reg} cells finally provided a rationale for the perplexing phenotype of $IL-2^{-/-}$, $IL-2R\alpha^{-/-}$, and $IL-2R\beta^{-/-}$ mice. Because IL-2 had been shown to be required for *in vitro* T cell responses, it had been assumed that IL-2 was essential for T cell-mediated immune responses *in vivo*.

However, knockout mice lacking either IL-2 or components of the IL-2R showed the opposite phenotype: increased, rather than decreased, numbers of peripheral T cells, and hyperactivation, rather than anergy, of these T cells. These animals all died of multi-organ autoimmunity within a few months of birth. A striking finding was that immune responses to pathogens were almost normal in these mutants. It turns out that while adequate activation of conventional T cells occurs in the absence of IL-2, this cytokine is crucial for the generation of the T_{reg} cell population, and that without an adequate T_{reg} population, autoimmunity cannot be prevented. Indeed, IL-2$^{-/-}$, IL-2R$\alpha^{-/-}$, and IL-2R$\beta^{-/-}$ mice all have greatly reduced numbers of T_{reg} cells, and the development of autoimmunity in these mutant animals can be prevented by timely provision of exogenous T_{reg} cells.

Some scientists speculate that CTLA-4 may contribute to peripheral T cell tolerance not because this molecule blocks CD28-mediated costimulation but because it has an important (still undefined) role in T_{reg} generation. Such a role could explain the autoimmune, lymphoproliferative phenotype of CTLA-4$^{-/-}$ mice. Finally, since the induction of the expression of both IL-2 and CTLA-4 depends on CD28 signaling, CD28 is also important for T_{reg} generation.

II. Th3 AND Tr1 CELLS

Two other CD4$^+$ regulatory T cell subsets that differ in surface phenotype from T_{reg} cells have been identified: Th3 cells and Tr1 cells. Th3 cells express low levels of CD25 and moderate levels of both GITR and CTLA-4. Tr1 cells do not express detectable levels of GITR and only low levels of CTLA-4 and CD25. Both Th3 and Tr1 cells exert suppression of conventional T cells in a non-specific manner by secreting immunosuppressive cytokines. Th3 cells preferentially secrete TGFβ, while Tr1 cells preferentially secrete IL-10 plus low amounts of TGFβ. While Tr1 cells may secrete IFNγ, this cytokine does not appear to play a large role in Tr1-mediated suppression. Neither are intercellular contacts important for Th3- or Tr1-mediated suppression.

Unlike T_{reg} cells, which arise as a separate sub-lineage from T cell precursors in the thymus, Th3 and Tr1 cells are thought to be derived from conventional peripheral Th0 cells that interact with *modulated* DCs. A modulated DC is an immature DC that acquires a self antigen or a benign foreign antigen (such as food) in the absence of "danger signals" but in the presence of certain immunosuppressive molecules, particularly IL-10. Complete maturation does not occur, the expression of co-stimulatory and adhesion molecules and MHC class II is not significantly upregulated, and proinflammatory cytokines are not produced. However, when the modulated DC migrates to the regional lymph node and interacts with a naive conventional Th0 cell, the Th0 cell is not anergized as it would be if it interacted with an immature DC, nor activated to produce Th1 or Th2 effectors as it would be if it interacted with a mature DC. Rather, the Th0 cell is induced to proliferate and differentiate into either Th3 or Tr1 cells (at least *in vitro*) (Fig. 16-8). The mechanism underlying this induction is

poorly defined but appears to involve ICOS signaling and the presence of slightly different panels of cytokines. *In vitro*, Th0 cells can be induced to differentiate into Tr1 cells in the presence of TGFβ, IL-10, and IFNα, while TGFβ, IL-10, and IL-4 favor Th3 differentiation. Once generated, the Th3 and Tr1 cells home back to the tissue where the immature DCs first picked up the self antigen or benign foreign antigen. The Tr1 and Th3 cells can then be activated by the antigen to regulate any immune responses to it in that site. Some scientists think that, as well as by contact with modulated DCs, the generation of Th3 and Tr1 cells from Th0 cells may be induced by intercellular contacts with T_{reg} cells. Others believe that a feedback loop may be in operation between Tr1/Th3 cells and modulated DCs, because the interaction of DC progenitors with these regulatory T cells *in vitro* results in the generation of modulated DCs that can anergize naive Th0 cells.

A word on the nomenclature of regulatory T cells in the literature: because T_{reg} cells arise as an independent lineage in the thymus, some commentators refer to them as "natural," "naturally-occurring," or "intrinsic" regulatory T cells. Because Th3 and Tr1 cells differentiate from mature Th0 cells, they are sometimes called "induced" or "adaptive" regulatory T cells.

III. CD8$^+$ Ts CELLS

Ts (T suppressor) cells are a regulatory subset of CD8$^+$ T cells distinguished by a marker profile of CD8$^+$CD57$^+$CD28$^-$. (CTLs are CD8$^+$CD57$^-$CD28$^+$; CD57 is a surface marker of unknown function found on NK cells and small numbers of peripheral T cells.) Ts cells were originally found to be increased in number after allogeneic transplantation or in patients infected with HIV or cytomegalovirus (where numbers of conventional CD4$^+$ and CD8$^+$ T cells are decreased), and decreased after radiotherapy or in patients suffering from autoimmune disease. Ts cells secrete a Th0-like profile of cytokines (including IL-2, IL-4, IFNγ, and TGFβ, but not IL-10) and can suppress the production of several antibody isotypes. However, rather than acting on effector lymphocytes like the CD4$^+$ regulatory T cell subsets do, the most important effect of the Ts subset appears to be on APCs. Antigen-specific Ts cells that interact with an APC presenting peptides from the relevant antigen alter NF-κB-induced gene expression in the APC, causing it to suppress its expression of B7 molecules but upregulate expression of two surface receptors called ILT3 (immunoglobulin-like transcript 3; see Ch.18) and ILT4. These surface receptors are expressed only by monocytes, macrophages, and DCs, and their appearance is associated with the ability of the APC to anergize, rather than activate, antigen-specific CD4$^+$ Th0 cells, with which it subsequently interacts. This effect is strikingly similar to the modulation of DCs described previously, and indeed, treatment of APCs with IL-10 and IFNα can induce the expression of ILT3 and ILT4 by these APCs. Importantly, a CD4$^+$ Th0 cell that interacts with an ILT3/ ILT4-expressing APC is induced to generate antigen-specific Th3 or Tr1 cells, propagating the tolerance. Studies *in vitro* by the transplantation research community (see Ch.27) have confirmed that the interaction of an immature DC with an activated Ts cell directed

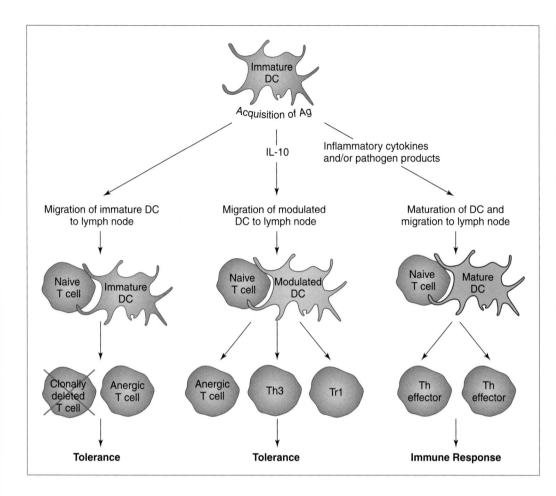

Figure 16-8

Roles of DCs in T Cell Activation, Anergy, and Regulatory T Cell Generation
If an immature DC acquires antigen in the absence of "danger signals" (far left), it does not initiate maturation or upregulate co-stimulatory molecules. Naïve T cells interacting with this DC are either clonally deleted or anergized. If IL-10 is present when the immature DC acquires the antigen, maturation of the DC is "modulated" and the T cells are either anergized or generate Tr1 or Th3 cells. If inflammatory cytokines and pathogen products are present during antigen acquisition, the DC matures fully and the T cells generate Th effector cells.

against an alloantigen induces the upregulation of ILT3 and ILT4 on the DC, rendering it tolerogenic. Conventional alloreactive CD4$^+$ T cells are subsequently anergized when they interact with these DCs presenting peptides from the alloantigen. It is unknown how Ts cells originate or what drives their development, but prolonged culture of peripheral blood cells (containing allogeneic APCs) of a prospective transplant with tissue from a prospective donor results in the generation of Ts cells *in vitro*. *In vivo*, heart transplant recipients that do not reject their new hearts have higher levels of circulating Ts cells than do those that reject their new organ.

IV. INFECTIOUS TOLERANCE AND LINKED SUPPRESSION

Regulatory T cells have been postulated to play a role in two previously mysterious phenomena called *infectious tolerance* and *linked suppression*. Infectious tolerance is said to have occurred when induced tolerance is apparently "passed along" from one population of anergic T cells to a normal T cell population, rendering the normal cells also tolerant. Depletion of CD4$^+$ cells in some situations, and of CD8$^+$ cells in others, abrogates the tolerance and its transmissibility. To illustrate the concept of infectious tolerance, consider the case of the increased acceptance of skin grafts induced by

monoclonal antibodies. Mice treated with blocking mAbs to either CD3, CD4, CD8, LFA-1, CD40L, or IL-2Rα at the same time as skin grafts are applied are able to accept those grafts indefinitely, indicating the induction of tolerance to alloantigens in the graft. (The mAbs used in this case are non-depleting; that is, the binding of the mAb to a cell does not induce cell death.) The mice are still fully capable of mounting immune responses to viral infections, indicating that the tolerance is limited to the T cells and APCs of the lymphoid tissues draining the region containing the grafted tissue. However, normal T cells (which have never seen the antibody) that are subsequently injected into the graft site of the tolerant mice become tolerant. Once recovered from the animal, these cells can then in turn pass on the tolerance to a fresh set of normal T cells. In short, it looks like the original tolerant cells are "infecting" the normal injected cells with their tolerance. In this case, the establishment of tolerance depends both on the continuous presence of antigen as supplied by the graft and a population of CD4$^+$ (but not CD8$^+$) T cells. In fact, if CD4$^+$ T cells are depleted from these mice after tolerance is established, the remaining CD8$^+$ T cells eventually reject the graft. In some experiments, TGFβ, IL-10, and IL-4 appear to be involved in the maintenance of infectious tolerance to skin grafts, suggesting the possible induction and function of Th3 or Tr1 cells. In other examples of infectious tolerance, T$_{reg}$ cells have been implicated as the inducing agents.

While infectious tolerance refers to tolerance for a single specific antigen X that can be transferred from the anti-X lymphocytes of one animal to the anti-X lymphocytes of another animal, linked suppression describes the spread of tolerance, not to other animals, but to other antigens within the same animal. In this case, a variety of T cell clones (anti-X, anti-Y, anti-Z) recognizing a limited set of antigens expressed in close physical proximity are all tolerized together upon administration of antigen X alone. It appears as though the tolerance of anti-X lymphocytes for antigen X has spread to anti-Y and anti-Z clones, abrogating responses to antigens Y and Z even though these antigens were not administered to the animal. T_{reg} cells have been predominantly associated with linked suppression but the precise mechanism remains unclear.

Most workers now believe that the original blocking mAb experiments induced infectious tolerance because they promoted the development of a subset of regulatory T cells that could actively tolerize neighboring T cells. However, there are other possibilities. For example, the involvement of TGFβ and the Th2 cytokines IL-4 and IL-10 in the maintenance of infectious tolerance to skin grafts has been taken by some as evidence of tolerance due to immune deviation. Others have theorized that the blocking mAbs may inhibit the receipt of co-stimulatory signals by Th cells recognizing skin graft antigens, rendering these Th cells anergic. Presumably the anergic Th cells cannot supply T help to nearby $CD8^+$ cells with the potential to attack the graft, resulting in acceptance. These anergic T cells might also compete with competent T cells for access to antigen and APCs, reducing the graft rejection response. Alternatively, mAb treatment might interfere with the ability of Th cells to license APCs for Tc activation, preventing the activation of perhaps multiple Tc clones recognizing unrelated antigens in the same vicinity. Such an effect could neatly account for linked suppression. Furthermore, unlicensed APCs might be incapable of delivering co-stimulation, forcing the anergization of any new T cells attempting to interact with them.

Although still poorly understood, infectious tolerance generated by blocking antibodies could represent an attractive means of therapy to combat transplant rejection, autoimmunity, and allergies, and is under active investigation both to determine its underlying mechanism and assess its therapeutic potential.

E. Experimental Tolerance

Much of what we have learned about self-tolerance has been derived from experimental models in which a lack of responsiveness to a foreign antigen is induced. That is, various experimental contrivances are used to convince the host animal's body that a foreign antigen is "self" such that an immune response is not mounted against it. With respect to definitions, just as an "immunogen" is an antigen that elicits an adaptive immune response, a *tolerogen* is an experimental foreign antigen that is recognized by a T or B lymphocyte but that

induces lymphocytes to become refractory to activation. The understanding of experimental tolerance grew from empirical observations on what molecular characteristics and system parameters had to be in place to induce tolerance to a foreign antigen. Scientists now believe that the mechanisms establishing experimental tolerance are essentially the same as those underlying peripheral self-tolerance, as evidenced by the results of studies of transgenic systems in which self-tolerance can be directly studied.

I. CHARACTERISTICS OF EXPERIMENTAL TOLERANCE

Experimental tolerance is most easily induced in an animal with a lymphoid system that is not at mature full strength, such as in a newborn animal or an animal with an immune system that has been damaged by treatment with drugs or irradiation. Mature animals with a compromised immune system can often be tolerized with either a non-immunogenic (i.e., tolerogenic) form of the antigen of interest using the usual route of immunization, or the immunogenic form of the antigen using a non-immunogenic route of administration. The immunocompromising treatment is then discontinued and the immune system of the animal is allowed to recover completely. Subsequent exposure to the normally immunogenic form of the antigen elicits no response. However, the immune system is not globally suppressed because responses to other, unrelated immunogens remain intact, meaning that, like an immune response, experimental tolerance is antigen-specific. As is discussed in more detail in the following sections, experimental tolerance is not usually permanent and its maintenance depends on continued exposure to the tolerogen in either a persistent or an intermittent fashion. Loss of the tolerogen for a prolonged period slowly restores normal responsiveness to the foreign antigen. This reversal may represent a restoration of the immune reactivity of the previously tolerized cells, and/or may be due to the presence of newly generated T cells with normal reactivity.

II. CHARACTERISTICS OF TOLEROGENS

During early attempts to induce experimental tolerance, several observations were made about the physical and behavioral characteristics of tolerogens (Table 16-4).

i) Nature of Molecules

For reasons that are still unclear, some molecules are naturally more tolerogenic than are others. For example, while L-amino acid polymers are immunogenic at almost any dose, D-amino acid polymers are tolerogenic at the same doses. For other molecules, a slight chemical modification is enough to turn an immunogen into a tolerogen for the same lymphocyte clone. Also important is the density of the antigenic epitope on the molecule. Having a moderate number of epitopes per molecule promotes immunogenicity, but a large number of epitopes per molecule favors tolerance induction, probably for the same reasons as cited in "high zone tolerance" (see later).

Table 16-4 Immunogens vs. Tolerogens

	Immunogen	Tolerogen
Number of antigenic epitopes per molecule Ag	Moderate	High
Size	Large, aggregated, polyvalent	* Small, disaggregated
Solubility	Insoluble	Soluble
Dose	Moderate	Very high or very low
Schedule of Ag administration	One large dose	Many small doses
Route of Ag administration	Subcutaneous	Intravenous, oral

* But larger than a hapten.

ii) Molecular Size

As we learned in Chapters 2 and 6, the best immunogens are large molecules. Small molecules do not often make good immunogens, but they can make good tolerogens. However, a tolerogen is distinguished from a hapten in that a tolerogen is a larger, more complex molecule and often has all the other physical characteristics of an immunogen. The importance of molecule size to immunogenicity and tolerogenicity can be illustrated using a protein that can be prepared in two forms: a large aggregated form, and a smaller, soluble deaggregated form (which is still larger than a hapten). In a naive animal, an immune response to the protein will be induced by a sufficient dose of the aggregate form, but tolerance results when the deaggregated form is administered. However, an immune response to the smaller form can be evoked if the protein is administered with adjuvant. It is speculated that soluble molecules (in the absence of adjuvant) may induce tolerance because their presence does not trigger the "danger signals" necessary to activate the APCs processing them, so that costimulatory molecules are not upregulated. Signal 2 is therefore not delivered to the T cell with a TCR that binds to the presented peptide derived from the small form; only a tolerogenic signal one is received. (Note, however, that if the animal has been previously exposed to the immunogenic form of the molecule, a secondary immune response will result upon presentation of peptide derived from either the larger or smaller forms of the protein, because memory T cells can be activated with much lower levels of co-stimulation.)

For B cells, a tolerogen (like an immunogen) must be capable of stimulating the BCRs such that signal one is delivered. However, the stimulation must not be so complete that the B cell is activated even in the absence of signal two (CD40L) delivered by an activated Th cell, as happens with Ti antigens. For this reason, very few Ti antigens are good tolerogens. Very small soluble molecules, such as monovalent haptens, cannot cross-link BCRs at all and therefore do not tolerize (or activate) B cells.

iii) Dose

It was in the 1960s that immunologists first recognized that exposure of an animal to a very high or very low dose of a foreign antigen (in the absence of an adjuvant) rendered the animal highly susceptible to a later infection with the pathogen from which the antigen was derived. In other words, instead of mounting a secondary immune response, the animal was unresponsive; it had become tolerant to the foreign antigen. Further investigation of the relationship between antigen dose and protective response showed that, in general, whereas intermediate doses of an antigen induce an immune response, very high and very low doses of the same antigen induce tolerance. These types of tolerance, which were briefly introduced in Chapter 6, are called "high zone" and "low zone" tolerance, respectively.

iiia) High zone tolerance. High zone tolerance was a surprise. In the early 1960s, researchers were investigating the possibility of generating a vaccine against bacterial pneumonia. Antigen prepared from pneumococcal bacteria was administered to mice at a high dose or a moderate dose, in an effort to define an effective concentration range. When the mice were subsequently challenged with live pneumococcal bacteria, those that had received the moderate dose mounted an immune response, but those that had received the high dose did not and died. It was subsequently discovered that the animals treated with the high dose of antigen became susceptible to infection because the antibody response had been silenced. Later studies demonstrated that B cells that can respond to a moderate dose of an antigen generally become non-responsive when exposed to 10-fold higher levels. At least for B cells, it appears that high doses of tolerogens can result in a receptor blockade, as described earlier in this chapter. Consequently, in some cases, limited degradation of the antigen with a protease, to lower the effective concentration of the antigen, removes the blockade and restores normal BCR stimulation. The tolerogenic nature of the antigen appears to have been reversed.

The failure to mount an antibody response to an antigen can also be due to T cell tolerization, since B cells require T cell help to respond to Td antigens. Tolerization with high-dose antigen differs between T and B cells. T cells can be tolerized at doses 100- to 1000-fold lower than the doses required to tolerize B cells. In addition, while B cell tolerization is not observed until about 2 days after antigen administration, high zone tolerance for T cells is evident within just hours of exposure. Experiments using superantigens or transgenic mice carrying TCRs recognizing known peptides have indicated that T cell high zone tolerance is often a manifestation of clonal deletion due to AICD. The large numbers of antigen-specific

T cells initially detected appear to be continuously activated such that they soon die from clonal exhaustion, removing the T cells necessary to respond to the antigen.

iiib) Low zone tolerance. In 1964, N. A. Mitchison demonstrated that tolerance to an antigen in an animal can also result if a very low dose (too low to provoke an immune response) is administered over a long period of time. Mitchison repeatedly injected a series of mice with a wide range of doses of BSA over 16 weeks. At 18 weeks, he challenged each of the mice with a dose of BSA known to be immunogenic. He then measured the level of anti-BSA antibodies produced in the serum of each mouse. He deduced that intermediate priming doses of BSA must have induced an immune response, since secondary response levels of anti-BSA antibodies were produced after the final challenge. However, very low priming doses appeared to suppress antibody production, in that the amounts of antibody produced were far less than the primary response levels observed in control mice that had been subjected only to the final immunogenic dose of BSA. Mitchison concluded that, in addition to high zone tolerance, another zone of tolerance could be induced by very low doses of antigen. In general, low zone tolerance can be demonstrated for a wide range of antigens in neonatal animals and immunocompromised adults, but for only a few antigens in healthy adults. The mechanism underlying low zone tolerance is unknown but has been linked in some cases to immune deviation and in others to regulatory T cells.

iv) Route of Administration

As we learned in Chapter 6, the route by which an antigen accesses the body determines how it is processed and affects whether it is immunogenic or not. Some routes of administration of experimental antigens trigger immune responses, while other routes evoke tolerance to the same antigen. The exact reasons for these differences are not known but are probably related to the frequency and types of APCs that take up the antigen. The most immunogenic mode of delivery is the subcutaneous route. Langerhans cells resident in the skin are very effective APCs, meaning that antigen delivered in this way often triggers a robust immune reaction. In contrast, antigen delivered via intravenous or oral routes often results in tolerance to the antigen. Antigen delivered intravenously is conveyed to the spleen, where it is presented primarily by naive splenic B cells (which express only low levels of co-stimulatory molecules in the absence of inflammatory cytokines) and other non-professional APCs, which may not deliver a strong signal two (if any). Responding T cells are therefore generally anergized and the immune response to the antigen is negligible.

iva) Oral tolerance. Oral administration is a very effective way of inducing peripheral tolerance to immunogens. Tolerance in this context involves the GALT (*gut-associated lymphoid tissue*). The GALT is explored in more detail in Chapter 20, but suffice it to say here that the intestinal villi, with their interspersed Peyer's patches, intraepithelial lymphocytes, and γδ T cells, provide an immunological barrier that combats both infection by ingested pathogens and injury by ingested proteins. Most dietary antigens are degraded in the stomach before ever reaching the small intestine, but some

partially digested or even intact molecules do get absorbed into the systemic circulation. In animal models, tolerance to such a systemic antigen can be observed within 5–7 days after the antigen is consumed. T cell tolerance (as measured by a DTH response) after a single feeding can last for a prolonged period (up to 18 months), whereas B cell tolerance (as measured by suppression of a specific antibody response) lasts for 3–6 months. Indeed, oral tolerance has been used successfully to treat autoimmune diseases in animal models and is currently being investigated for the treatment of certain human autoimmune disorders (see Ch.29).

Despite intensive study, it is still not clear how oral tolerance is achieved. Several mechanisms, all of which may make a contribution, have been postulated (Table 16-5). To begin with, oral tolerance appears to be mostly a function of CD4$^+$ T cell tolerization rather than either CD8$^+$ T cell or B cell tolerization, since oral tolerance can be efficiently established in mice lacking either B cells or CD8$^+$ T cells. Large concentrations of an orally administered antigen may allow its uptake by the enterocytes lining the intestine. Enterocytes express only low levels of MHC class II and no co-stimulatory molecules. Thus, mucosal CD4$^+$ T cells responding to antigenic peptides presented by enterocytes may be rendered anergic. In addition, although professional APCs, particularly DCs, are present in the intestinal wall and its draining lymphoid tissues, the consumption of food occurs without "danger signals." Thus, the DCs in this location may remain in an immature state and express only low levels of co-stimulatory molecules. Again, anergization or deletion of responding T cells may result.

When small amounts of an antigen are eaten, it is thought that oral tolerance is established principally by the action of immunosuppressive cytokines rather than by the anergization. The GALT favors immune deviation toward Th2, and the IL-10 produced by Th2 cells has an immunosuppressive effect on nearby T cells. Indeed, a large dose of IFNγ or IL-12 fed to animals abrogates oral tolerance, whereas IL-4 or anti-IL-12 enhances it. However, oral tolerance can be successfully induced in both IL-4$^{-/-}$ and IL-10$^{-/-}$ mice, confirming that more than just deviation to Th2 responses is involved. Indeed, the gut microenvironment also promotes the differentiation of

Table 16-5 **Factors Promoting Oral Tolerance**
Induction of CD4$^+$ T cell anergy by enterocytes with low costimulatory capacity
Lack of DC activation because Ag entering the gut does not trigger "danger signals"
TGFβ secretion by gut Th3 cells
Immune deviation to Th2 response induced by gut environment cytokines
Induction of T$_{reg}$ differentiation
Influence of gut γδ T cells (mechanism unknown)

intraepithelial Th3 cells. Small numbers of these cells may be activated by the ingested antigen and induced to secrete elevated levels of TGFβ. T_{reg} cells may also participate, since oral tolerance induced by low doses of antigen has been associated with the generation of these cells. Finally, the gut mucosae are home to significant concentrations of γδ T cells which appear to have an important, if ill-defined, role in oral tolerance. Treatment of an animal with anti-TCRγδ antibodies prior to antigen feeding results in a failure to establish oral tolerance, as does the feeding of antigen to TCRγδ-deficient mice.

v) Genetic Constitution

The genetic background of an animal may have an impact on whether it can be tolerized to a foreign antigen. Most mouse strains can achieve oral tolerance to a large number of antigens, and differences in MHC haplotype appear to make no difference. However, some inbred mouse strains consistently mount immune responses to a given foreign protein, whereas other strains are much more easily tolerized to the same protein. Genetic variations that affect the ability of the animal to take up antigen, clear it from the circulation, or process and present it to lymphocytes could influence susceptibility to tolerization.

III. DEGREE AND PERSISTENCE OF TOLERANCE

Tolerance is mediated by antigen receptor binding and the subsequent induction of intracellular signaling pathways. Thus, as happens in an immune response, tolerance is influenced by the affinity of antigen binding. Different lymphocyte clones recognizing the tolerogen or its components will bind to it with differing affinities/avidities. Some of these clones may be induced to initiate signal one delivery more readily or strongly than others, resulting in a tolerant state characterized by varying degrees and duration of unresponsiveness among the relevant clones. Thus, the overall ease of tolerance induction to an antigen and the duration of the tolerant state will reflect an average among the clonal population. In general, a constant presence of the antigen is required to maintain experimental tolerance, because newly developed lymphocytes released by the primary lymphoid tissues must continually be rendered anergic or clonally deleted to achieve ongoing non-responsiveness to the antigen. In the absence of persistent antigen, T cell tolerance can still last several months because T cells turn over relatively slowly and the production of new T cells is limited (especially in adults). For example, work in thymectomized adult animals has shown that anergic T cells can persist in a host for an extended period, as long as the host cannot make new T cells that compete with the anergized T cells for space in the secondary lymphoid follicles. In contrast, because anergized B cells have short life spans and the bone marrow continuously generates new B cells throughout life, B cell tolerance generally lasts for only a few weeks in the absence of persistent antigen. It is not long before non-tolerized clones able to respond to the antigen appear in the animal's B cell population, and an immune response is mounted.

IV. "SPLIT" TOLERANCE

"Split" tolerance is a term that unfortunately has three different meanings. In the first context, "split tolerance" refers to experimental situations in which different measures of immune reactivity give different results. That is, while antibodies of a particular isotype directed against an antigen may be present in an animal (evidence for immune reactivity), antibodies of other isotypes, and/or a DTH response representing the cell-mediated arm, may be missing (evidence of tolerance). In this situation, the split tolerance is simply a manifestation of immune deviation, where Th2 cells producing cytokines favoring a humoral response outnumber Th1 cells responsible for DTH reactions. In the second context, "split tolerance" is used to describe situations in which tolerance can be induced to an epitope A on a molecule, but not to another epitope B on the same molecule, or to only some antigens present on a given cell. In this case, tolerance to epitope A and reactivity to epitope B would be observed. In the third context, "split tolerance" refers to a situation where T cell tolerance to a tolerogen is invoked but B cell reactivity is maintained.

V. SPECIAL SITUATIONS

i) Fetal–Maternal Tolerance

A mammalian mother does not normally reject her fetuses, despite the fact that half of the histocompatibility molecules they express are derived from the father and are therefore potentially "foreign" to the mother. This phenomenon represents a specialized form of immune tolerance in that the fetus that implants in the uterus resembles a tissue graft that is not rejected. Tolerance to paternal histocompatibility molecules is maintained during fetal development and antibodies to paternal antigens expressed in the fetus are not generally made until after delivery. At this point, the maternal immune system recovers and immune responses can be mounted against fetal cells shed into maternal tissues during the trauma of delivery.

Maternal–fetal tolerance is not primarily due to physical barriers, since the fetal and maternal tissues coming together in the placentae of (at least) rodents and primates are not separated by basement membranes. Furthermore, although the maternal and fetal circulatory systems are separated by the structures of the highly organized trophoblast, the barrier is not absolute and fetal antigens and cells do enter the maternal circulation and reach the secondary lymphoid organs. Much of maternal tolerance to the fetus has similarities to immune privilege, since the maternal–fetal unit is essentially a localized microenvironment that is somehow protected from damaging lymphocyte activation.

What mechanisms contribute to fetal–maternal tolerance? Reproduction and all of its accompanying hormonal changes may prepare the uterus to accept the "fetal graft" in a way that would not naturally be initiated by a surgically implanted allograft. In other words, the innate immune system's response to the presence of a growing fetus may be qualitatively different from the response to the perceived "injury" of an allogeneic tissue graft (even if the human purpose is to help, not harm). Even prior to gestation, steroid hormones produced in the female have powerful modulatory influences on APCs and lymphocytes. Once pregnancy is established, the unique hormonal and cytokine microenvironment of the placenta favors immunosuppression and immune deviation. For example, certain placental cells produce high levels of progesterone, which regulates potassium channels and dampens Ca^{2+} signaling such that effector T cell responses leading to graft rejection are suppressed (Table 16-6). Other placental cells, including placental γδ T cells, secrete a collection of cytokines that includes TGFβ, IL-10, IL-4, and IL-5. Immunosuppression is thus favored, as is immune deviation toward the differentiation of Th2 lymphocytes rather than toward the Th1 cells largely responsible for graft rejection. The importance of immune deviation to Th2 has been documented in pregnancies of both mice and humans. If circumstances arise during pregnancy that result in an increase in IFNγ production and a decrease in IL-4 or IL-10 secretion, the chances of a spontaneous abortion increase, consistent with a failure in tolerance. Experimental injection of pregnant mice with IL-2, which favors Th1 responses at the expense of Th2 responses, also results in increased fetal loss.

The APCs located in the tissues forming the maternal–fetal interface are generally "non-professional" cells that lack expression of costimulatory molecules. Moreover, pregnancy does not usually represent a "danger" that would induce the production of inflammatory cytokines and the activation of professional APCs. Thus, a maternal T cell encountering fetal antigen presented by placental APC should be tolerized rather than activated. In addition, placental macrophages appear to be a unique cellular subset that has anti-inflammatory rather than pro-inflammatory properties. These cells secrete IL-10 and the anti-inflammatory protein IL-1Ra (*IL-1 receptor antagonist*; see Ch.17). However, these effects must be quite localized in nature because, in experimental animals, pregnancy does not abrogate systemic responses to injected paternal MHC alloantigens, and pregnant animals are quite capable of mounting responses against infectious organisms. Interestingly, maternal–fetal tolerance has been cited as another example of infectious tolerance, due to evidence that a tolerized maternal T cell at the fetal–maternal interface can "pass on" its anergic state to nearby T cells recognizing the same or similar alloantigens,

extending fetal protection. Such observations suggest the possible involvement of regulatory T cells. Indeed, Tr1 and Th3 cells have been implicated in the prevention of miscarriages.

Tolerance is favored by other aspects of placental biology. The placental tissues of many mammals as well as developing fetal tissues are devoid of DCs and do not express MHC class I and II molecules. In fact, the only MHC gene that an early human fetus is known to express is the non-polymorphic HLA-G gene (described in Ch.10) whose product acts as a ligand for an inhibitory receptor on maternal NK cells (see Ch.18). As a result of this interaction, the maternal NK cells are inactivated and prevented from killing the fetal "intruders." Fetal cells also express two proteins that inhibit the activation of maternal complement. These two proteins, called DAF (*decay accelerating factor*) and MCP (*membrane cofactor protein*), are discussed in detail in Chapter 19.

Fas killing also contributes to maternal–fetal tolerance. Like cells in the eye and testis, some placental cells can express FasL. Maternal T lymphocytes activated by an encounter with antigens derived from circulating fetal cells start to express Fas. If these lymphocytes succeed at infiltrating the placental tissues, the interaction of Fas on the T cell with FasL on a placental cell will trigger the apoptosis of the activated T cell, maintaining tolerance to the fetus. Consistent with this scenario, knockout mice lacking expression of FasL exhibit increased fetal resorption and consequently decreased litter sizes.

A mechanism that seems to be particularly important for mouse fetus survival depends on metabolic deprivation. The degradation of the amino acid tryptophan is controlled by the rate-limiting enzyme *indoleamine 2,3-dioxygenase* (IDO). Activated macrophages and DCs resident in the trophoblast upregulate the expression of this enzyme, increasing the local degradation of tryptophan at the maternal–fetal interface. Without this amino acid, activated T cells in the immediate area cannot proliferate or differentiate into effector cells that might attack the fetal tissue. In addition, there is evidence that antigen-stimulated Th0 cells that interact with IDO-expressing DCs develop into T_{reg} cells capable of suppressing the activity of nearby effectors. The converse also appears to be true, in that T_{reg} cells that interact with trophoblastic macrophages and DCs can induce these APCs to upregulate their expression

Table 16-6 Factors Promoting Fetal–Maternal Tolerance

- High levels of progesterone produced by placental cells suppress effector cells
- IL-10, TGFβ secretion by placental cells suppress effector cells
- IL-4, IL-5 secretion by placental cells induces immune deviation to Th2
- Presentation of fetal antigens by non-professional APCs anergizes maternal lymphocytes
- IL-10, IL-1Ra secreted by placental macrophages suppress effector cells and inflammation, respectively
- Infectious tolerance (mediated by Tr1, Th3 cells?) successively anergizes maternal T cells
- Low level of MHC class I and class II on placental cells and developing fetal tissues decreases antigen presentation
- Virtual absence of DCs in placenta decreases antigen presentation
- HLA-G expressed by fetal cells inactivates maternal NK cells
- DAF and MCP produced by fetal cells inhibit maternal complement activation
- FasL expressed by placental cells kills Fas-expressing activated maternal T cells
- Tryptophan deprivation via IDO production leads to decreased maternal T cell activation and increased T_{reg} induction

of IDO. The mechanism in this case is linked in some way to CTLA-4 binding to B7 molecules.

ii) Neonatal Tolerance

In the 1950s, Peter Medawar and his colleagues infused the spleen and bone marrow cells of a mouse strain A into neonatal mice of another strain B, and showed that the procedure allowed strain B mice to later accept skin grafts from strain A mice. Furthermore, the tolerance induced to the allogeneic antigens was lifelong. This result was interesting because the same experiment would not work in adult mice: under the same circumstances, skin grafts of strain A were uniformly rejected by adult mice of strain B. Medawar's results were taken as support for Burnet's theory that self-tolerance was maintained by the clonal deletion of self-reactive lymphocytes during embryogenesis. Other work studying viral infection showed that neonatal animals exposed to a broad spectrum of viruses tended to become tolerant to those viruses rather than mounting immune responses against them. For many years, knowledge gained about the workings of the neonatal immune system suggested that there was "something special" about it, and that there was a period shortly after birth when tolerance to antigens could be "learned," before the animal became fully capable of mounting its own protective immune responses. This concept became known as "neonatal tolerance."

Over the years, several theories were advanced to explain the observations of Medawar and others. The earliest hypothesis held that the T cells in a neonate were too immature to destroy all of the allogeneic donor cells. The surviving allogeneic cells would then migrate to the thymus and cause deletion of reactive T cells during the ongoing establishment of central tolerance, rendering the adult tolerant to an allogeneic skin graft imposed later in life. Alternatively, the injection of allogeneic cells into a neonate might establish a *de facto* state of chimerism in the animal. That is, the foreign cells would essentially become a part of the animal and thus persist for long periods of time. Reactive T lymphocytes might then be clonally deleted, establishing tolerance that was maintained into adulthood. When the same experiment was done with a purified protein antigen, which was cleared from the host much more quickly, tolerance was induced in the neonate for a shorter period. These results were felt to support the requirement of antigen persistence for neonatal tolerance. More recently, several researchers proposed that the results of Medawar and others could be explained if immune deviation to Th2 occurred in neonates. This theory held that neonatal T cells were fully competent to respond but were often spontaneously influenced to differentiate into Th2 cells at the expense of Th1 cells. Since it is Th1 cells that are primarily involved in cell-mediated rejection responses, the destruction of allogeneic cells would be reduced in neonates.

By the mid-1990s, however, more and more conflicting evidence was accumulating to challenge previous theories regarding neonatal tolerance as a special case of tolerance. Various experiments demonstrated that neonatal Th1 responses could indeed be induced under a variety of circumstances, that some (but not all) viruses could provoke neonatal immune responses rather than tolerance, and that the dose rather than the nature of the immunogen was critical to inducing neonatal tolerance or immunity. Some immunologists began to muse that the tolerance mechanisms in neonates might be fundamentally the same as those in adults but were triggered more easily because of maturational differences between the cells of the neonatal and adult immune systems (Table 16-7). For example, FDCs in very young animals are not very efficient at trapping antigen, and newly produced APCs have a reduced

Table 16-7 Comparison of Neonatal vs. Adult Immune Responses

	Neonate	Adult
Original observations	Tolerant to allogeneic skin grafts	Rejected allogeneic skin grafts
FDC trapping capacity	Low	High
APC function capacity	Low	High
Anergizing APCs	Many	Few
B cell differentiation into plasma cells	Delayed	Rapid
B cell responses to Ti Ag	Usually strong	Strong
B cell responses to Td Ag	Weak and delayed	Strong
Relative number of mature naive T cells	1	10,000
Localization of mature naive T cells	Secondary lymphoid tissues and peripheral tissues such as lung and skin	Secondary lymphoid tissues only
T cell cytokine secretion	IL-10, TGFβ	IL-2, IFN-γ (Th1) IL-4, IL-5 (Th2)
Ag dose inducing high zone tolerance	Relatively low	High

antigen presentation capacity, both of which could contribute to a more sluggish immune response. B cells of early neonates, unlike those of adults, are difficult to provoke into mounting antibody responses *in vivo* or *in vitro* and remain highly susceptible to tolerance induction for several days. When antigen exposure does result in a response, these cells simply do not differentiate into plasma cells as quickly as do B cells from adults. Neonatal antibody responses to Ti antigens are generally strong, but responses to Td antigens are weak and often delayed, a state that persists for 2–8 weeks after birth in mice. Numbers of mature naive T cells are 10,000-fold lower in neonatal mice than in adults, although their expression of cytokine receptors and costimulatory molecules is similar to that of naive T cells in adults. The spleen in the newborn contains only a few mature T and B cells, and while the lymph nodes contain adult proportions of mature T and B cells, their absolute numbers are much reduced. Lymphocyte trafficking patterns are also different. Naive T cells can access non-lymphoid peripheral tissues much more easily in neonates than in adults and tend to accumulate in lung and skin. In contrast, as we have learned, naive T cells in adults have more restricted circulation patterns, migrating in large numbers to the spleen and lymph nodes. Finally, T cells from neonates secrete a different cytokine profile compared to T cells from adult mice, such that higher levels of immunosuppressive cytokines are present in neonates than in adults. Some or all of these factors could lead to sub-threshold levels of signal one or two being delivered, resulting in either lymphocyte deletion, anergization, or ignorance.

Ultimately, in 1996, several researchers independently demonstrated that neonatal T cells are fully immunocompetent when antigen is presented in an appropriate way. First, Polly Matzinger and her colleagues set out to replicate Medawar's experiments, but rather than using spleen and bone marrow cells as the inoculum and skin graft rejection as the read-out as Medawar did, Matzinger's group injected neonatal female mice with cells from syngeneic male mice, and looked for CTL responses against the H-Y male antigen as the indicator. When the female neonates were injected with a mixture of spleen and bone marrow cells, CTL responses were not mounted against the H-Y antigen, indicating that tolerance had been invoked just as Medawar had observed. However, when enriched preparations of DCs from male mice were injected into the female neonates, vigorous anti-H-Y CTL responses could be demonstrated, indicating that the female neonates were perfectly capable of mounting immune responses when the foreign antigen was presented in a particular way. Why did the nature of the inoculum make such a difference? Matzinger speculated that the inoculum of spleen and bone marrow cells used by Medawar contained an overwhelming number of non-APC cells that could anergize the few mature CD8$^+$ Tc cells in the neonatal mouse. In other words, the Tc cells could not find a licensed professional APC and its co-stimulatory signals in time to avoid being anergized. When the proportion of professional APCs in the inoculum was increased by purifying the DCs, a Tc cell was more likely to encounter a DC able to cross-present the H-Y peptides. Strong signals

one and two would be received and activation leading to a CTL response would be achieved. To confirm the hypothesis that the proportion of APCs present was critical, Matzinger injected adult mice with a huge inoculum of non-APCs and observed both dramatically decreased CTL responses and an inability to subsequently reject skin grafts, exactly what Medawar had originally observed in his neonates.

About the same time as Matzinger's experiments, another group demonstrated that, as long as antigen was introduced into neonatal mice with an appropriate dose of adjuvant, immune responsiveness could be demonstrated that was equal to that observed in adult mice. A third group examined the often-observed situation in which a neonatal mouse is tolerized by the same dose of virus that is immunogenic in the adult. These researchers kept reducing the amount of leukemia virus introduced into neonatal mice and eventually reached a dose able to prime, rather than tolerize, the newborns. It was postulated that because the number of T cells is so much lower in the neonate than in the adult, a situation of high zone tolerance is easily created. Relatively speaking, the T cells in the neonate meet a much higher concentration of antigen (on a per cell basis), and so are tolerized at a dose that is immunogenic in the adult. As a result of these experiments, most immunologists now believe that the tolerance induction mechanisms in neonates are not substantially different from those in adults, but because of differences in T cell numbers and their access to professional APCs, neonates appear to be tolerized more readily when protocols used to immunize adults are employed. In other words, appropriate adjustments to antigen dose or APC concentration can allow the induction of an immune response to an antigen in a neonate.

One area in which knowledge of the mechanisms of neonatal tolerance is critical is in studies of vaccination. Ideally, we would like to protect babies and children from infection long before adulthood, but if a vaccine induces tolerance instead of protective immunity, the child fails to mount an immune response against the pathogen and might succumb to it. For example, consider the case in which scientists devised an experimental vaccine based on the DNA encoding a protein of *Plasmodium falciparum*, the organism that causes malaria. (See Ch.23 for a discussion of DNA vaccines.) Adult mice immunized with this particular DNA vaccine mount protective anti-malaria Th1 and antibody responses, but neither T nor B cell responses were observed in vaccinated newborn mice. Subsequent exposure to *P. falciparum* showed that the newborn mice had been rendered tolerant to the pathogen and were therefore susceptible to it. The tolerance in this case turned out to be associated with the presence of a subset of CD8$^+$ regulatory T cells, as demonstrated by adoptive transfer experiments in which spleen cells from neonatally tolerized mice were able to transfer the tolerant state to naive syngeneic recipients. This tolerance was abrogated if the spleen cell isolate was depleted of CD8$^+$, but not CD4$^+$, T cells prior to transfer. Furthermore, if CD8$^+$ T cells from the tolerant neonatal mice were mixed *in vitro* with reactive T cells from the immunized adult mice, the tolerant neonatal cells were able to suppress the immune responsiveness of the reactive adult cells. It is not clear whether these CD8$^+$ regulatory T cells were identical to the Ts cells described previously.

As we reach the end of this chapter, we have completed our study of two important cellular components of the adaptive immune response: B cells and αβ T cells. In the next chapter, we examine a vital non-cellular component of both the adaptive and innate immune responses: the network of cytokines that sustains intercellular communication. Without cytokines, many of the cells of our immune system would be ignorant of danger and unable to respond and defend us.

SUMMARY

Multiple peripheral self-tolerance mechanisms ensure that any autoreactive cells that escaped negative selection in the thymus are unable to cause a damaging response to their cognate antigens in the body's periphery. Clonal deletion due to the presentation and cross-presentation of antigen by immature DCs in the absence of "danger signals" is the primary means by which naive autoreactive T cells in local lymph nodes are removed. These cells undergo apoptosis, so their death does not trigger inflammation. Other T cells are tolerized by becoming anergic when they encounter their cognate antigen in the absence of costimulatory signals, or when a modulated DC is involved in the presentation process. The T cell survives, and may or may not proliferate, but it does not generate effector or memory cells. Moreover, the T cell is unable to respond to cognate antigen the next time it meets the antigen under optimal conditions. The complex signaling pathways leading to anergy are only beginning to be unraveled. Immunological ignorance prevails when only very low levels of an antigen are available for presentation to T cells. A site that enjoys immune privilege contains cells that secrete suppressive cytokines or upregulate death receptors to kill invading T cells. Immunoregulation by the immunosuppressive cytokines IL-10 and TGFβ can also block immune responses. Immune deviation may convert a tissue-damaging Th1 response into a less harmful Th2 response. B cell anergy may occur if the BCR is engaged by a Td antigen but the appropriate Th cell is not available, or if the Th cell fails to upregulate CD40L. Receptor blockade can also lead to B cell anergy. Negative regulatory signaling via PKCδ is important for the maintenance of B cell self-tolerance. Regulatory subsets of CD4+ and CD8+ T cells may anergize lymphocytes via intercellular contacts or secretion of IL-10 and/or TGFβ.

Selected Reading List

Acuto O. and Michel F. (2003) CD28-mediated co-stimulation: a quantitative support for TCR signalling. *Nature Reviews Immunology* **3**, 939–951.

Alferink J., Tafuri A., Vestweber D., Hallmann R., Hammerling G. *et al.* (1998) Control of neonatal tolerance to tissue antigens by peripheral T cell trafficking. *Science* **282**, 1338–1341.

Astori M., Finke D., Karapetian O. and Acha-Orbea H. (1999) Development of T–B cell collaboration in neonatal mice. *International Immunology* **11**, 445–451.

Bachmann M. F., Speiser D. E., Mak T. W. and Ohashi P. (1999) Absence of co-stimulation and not the intensity of TCR signaling is critical for the induction of T cell unresponsiveness in vivo. *European Journal of Immunology* **29**, 2156–2166.

Bretscher P. and Cohn M. (1970) A theory of self–nonself discrimination. *Science* **169**, 1042–1049.

Burkly L., Lo D., Kanagawa O., Brinster R. and Flavell R. (1989) T-cell tolerance by clonal anergy in transgenic mice with nonlymphoid expression of MHC class II I–E. *Nature* **342**, 564–566.

Burnet F. M. (1959) "The Clonal Selection Theory of Acquired Immunity." Cambridge University Press, New York, NY.

Curtsinger J., Lins D and Mescher M. (2003) Signal 3 determines tolerance versus full activation of naive CD8 T cells: Dissociating proliferation and development of effector function. *Journal of Experimental Medicine* **197**, 1141–1151.

Ding Y., Baker B., Garboczi D., Biddison W. and Wiley D. (1999) Four A6-TCR/peptide/HLA-A2 structures that generate very different T cell signals are nearly identical. *Immunity* **11**, 45–56.

Erlebacher A. (2001) Why isn't the fetus rejected? *Current Opinion in Immunology* **13**, 590–593.

Forsthuber T., Yip H. C. and Lehmann P. V. (1996) Induction of Th1 and Th2 immunity in neonatal mice. *Science* **271**, 1728–1730.

Galluci S. and Matzinger P. (2001) Danger signals: SOS to the immune system. *Current Opinion in Immunology* **13**, 114–119.

Goodnow C., Crosbie J, Jorgensen H., Brink R. and Basten A. (1989) Induction of self-tolerance in mature peripheral B lymphocytes. *Nature* **342**, 385–391.

Guerder S., Meyerhoff J. and Flavell R. (1994) The role of the T cell costimulator B7-1 in autoimmunity and the induction and maintenance of tolerance to peripheral antigen. *Immunity* **1**, 155–166.

Ichino M., Mor G., Conover J., Weiss W. R., Takeno M. *et al.* (1999) Factors associated with the development of neonatal tolerance after the administration of a plasmid DNA vaccine. *The Journal of Immunology* **162**, 3814–3818.

Janeway C. A., Jr., Goodnow C. C. and Medzhitov R. (1996) Immunological tolerance: danger—pathogen on the premises! *Current Biology* **6**, 519–522.

Jenkins M. and Schwartz R. (1987) Antigen presentation by chemically modified splenocytes induces antigen-specific T cell unresponsiveness in vitro and in vivo. *Journal of Experimental Medicine* **165**, 302–319.

Jonuleit H. and Schmitt E. (2003) The regulatory T cell family: distinct subsets and their interrelations. *Journal of Immunology* **171**, 6323–6327.

Katz J, Benoist C. and Mathis D. (1995) T helper cell subsets in insulin-dependent diabetes. *Science* **268**, 1185–1188.

Kurts C., Kosaka H., Carbone F., Allison J., Miller J. *et al.* (1997) Class I-restricted cross-presentation of exogenous self antigens leads to deletion of autoreactive CD8+ T cells. *Journal of Experimental Medicine* **186**, 239–245.

Kyewski B. and Derbinski J. (2004) Self-representation in the thymus: an extended view. *Nature Reviews. Immunology* **4**, 688–698.

Mahnke K., Schmitt E., Bonifaz L., Enk A. and Jonuleit H. (2002) Immature, but not inactive: the tolerogenic function of immature dendritic cells. *Immunology and Cell Biology* **80**, 477–483.

Matzinger P. (1998) An innate sense of danger. *Seminars in Immunology* **10**, 399–415.

Matzinger P. (2002) The danger model: a renewed sense of self. *Science* **296**, 301–305.

Medzhitov R. and Janeway C., Jr. (2002) Decoding the patterns of self and nonself by the innate immune system. *Science* **296**, 298–300.

Mellor A. and Dunn D. (2000) Immunology at the maternal–fetal interface. *Annual Review of Immunology* **18**, 367–391.

Mitchison N. A. (1964) Induction of immunological paralysis in two zones of dosage. *Proceedings of the Royal Society of London, Series B—Biological Sciences* **161**, 275–292.

Niederkorn J. Y. (1999) The immune privilege of corneal allografts. *Transplantation* **67**, 1503–1508.

Ohashi P., Oehen S., Burki K., Pircher H., Ohashi C. *et al.* (1991) Ablation of "tolerance" and induction of diabetes by virus infection in viral antigen transgenic mice. *Cell* **65**, 305–317.

Owen R. (1945) Immunogenetic consequences of vascular anastomoses between bovine twins. *Science* **102**, 400–401.

Pennisi E. (1996) Teetering on the brink of danger. *Science* **271**, 1665–1667.

Ridge J. P., Fuchs E. and Matzinger P. (1996) Neonatal tolerance revisited; turning on newborn T cell with dendritic cells. *Science* **271**, 1723–1726.

Sakaguchi S. (2004) Naturally arising CD4$^+$ regulatory T cells for immunologic self-tolerance and negative control of immune responses. *Annual Review of Immunology* **22**, 531–562.

Sarzotti M., Robbins D. S. and Hoffman P. M. (1996) Induction of protective CTL responses in newborn mice by a murine retrovirus. *Science* **271**, 1726–1728.

Schonrich G., Kalinke U., Momburg F., Malissen M., Schmitt-Verhulst A.-M. *et al.* (1991) Down-regulation of T cell receptors on self-reactive T cells as a novel mechanism for extrathymic tolerance induction. *Cell* **65**, 293–304.

Schonrich G., Momburg F., Malissen M., Schmitt-Verhulst A.-M., Malissen B. *et al.* (1992) Distinct mechanisms of extrathymic T cell tolerance due to differential expression of self antigen. *International Immunology* **4**, 581–590.

Schwartz R. (1997) T cell clonal anergy. *Current Opinion in Immunology* **9**, 351–357.

Schwartz R. (2003) T cell anergy. *Annual Review of Immunology* **21**, 305–334.

Seddon B. and Mason D. (2000) The third function of the thymus. *Immunology Today* **21**, 95–99.

Shevach E. (2002) CD4$^+$CD25$^+$ suppressor T cells: more questions than answers. *Nature Reviews. Immunology* **2**, 389–400.

Steinman R. and Nussenzweig M. (2002) Avoiding horror autotoxicus: The importance of dendritic cells in peripheral T cell tolerance. *Proceedings of the National Academy of Sciences U.S.A.* **99**, 351–358.

Stockinger B. (1999) T lymphocyte tolerance: from thymic deletion to peripheral control mechanisms. *Advances in Immunology* **71**, 229–265.

Streilein J. and Stein-Streilein J. (2000) Does innate immune privilege exist? *Journal of Leukocyte Biology* **67**, 479–487.

Strobel S. and Mowat A. McI. (1998) Immune responses to dietary antigens: oral tolerance. *Immunology Today* **19**, 173–181.

Thellin O., Coumans B., Zorzi W., Igout A. and Heinen E. (2000) Tolerance to the foeto-placental 'graft': ten ways to support a child for nine months. *Current Opinion in Immunology* **12**, 731–737.

Thellin O. and Heinen E. (2002) Pregnancy and the immune system: between tolerance and rejection. *Toxicology* **185**, 179–184.

Todryk S., Melcher A., Kalgleish A. and Vile R. (2000) Heat shock proteins refine the danger theory. *Immunology* **99**, 334–337.

Traub E. (1936) Persistence of lymphocytic choriomeningitis virus in immune animals and its relation to immunity. *Journal of Experimental Medicine* **63**, 847–861.

Vance R. (2000) A copernican revolution? Doubts about the danger theory. *Journal of Immunology* **165**, 1725–1728.

von Bubnoff D., de la Salle H., Wessendorf J., Koch S., Hanau D. and Bieber T. (2002) Antigen-presenting cells and tolerance induction. *Allergy* **57**, 2–8.

Waldmann H. and Cobbold S. (1998) How do monoclonal antibodies induce tolerance? A role for infectious tolerance? *Annual Review of Immunology* **16**, 619–644.

Webb S., Morris C. and Sprent J. (1990) Extrathymic tolerance of mature T cell: clonal elimination as a consequence of immunity. *Cell* **63**, 1249–1256.

Weiner H. (2001) Oral tolerance: immune mechanisms and the generation of Th3-type TGF-beta-secreting regulatory cells. *Microbes and Infection* **3**, 947–954.

Cytokines and Cytokine Receptors

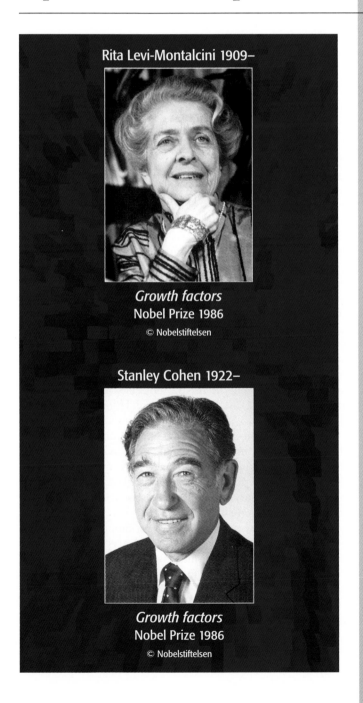

Rita Levi-Montalcini 1909–

Growth factors
Nobel Prize 1986
© Nobelstiftelsen

Stanley Cohen 1922–

Growth factors
Nobel Prize 1986
© Nobelstiftelsen

CHAPTER 17

A. HISTORICAL NOTES

B. OVERVIEW

C. FUNCTION, PRODUCTION, AND RECEPTORS OF CYTOKINES

D. CYTOKINES AND THEIR RECEPTORS IN PHYSIOLOGICAL CONTEXT

"I have yet to see any problem, however complicated, which, when you looked at it in the right way, did not become still more complicated"
—Poul Anderson

Throughout this book the reader has run into countless references to cytokines and their roles in both the adaptive and innate immune responses. These molecules were introduced in Chapter 2 and have turned up in every chapter since. We will now examine a number of these cytokines and their receptors in isolation, focusing on their pleiotropic effects on the many different aspects of the immune system that are subject to their influence. Because more than 100 cytokines have been identified and characterized to date, we shall concentrate on those most directly associated with the immune response. We will also investigate the nature of cytokine receptors, without which a cytokine would have no effect. We conclude the chapter with a series of tables summarizing how certain constellations of cytokines cooperate in important physiological events, including innate and adaptive immune responses.

A. Historical Notes

The effects of cytokines on cells were first recognized in the 1950s by researchers studying cultures of allogeneic lymphocytes. They found that the supernatant of such a culture contained biologically active, soluble material capable of inducing the proliferation and differentiation of other types of cultured cells. Subsequently, scientists studying infectious diseases and immune responses showed that these active substances were produced in response to infection or stimulation of lymphocytes by antigen.

Before the pleiotropy of cytokines was recognized and their purification a real possibility, each separate effect observed in culture was attributed to a different cytokine. Moreover, the same molecule was often unknowingly discovered and independently named by two or more groups, giving rise to great confusion in nomenclature in the 1970s. For example, IFNγ was identified by immunologists as a macrophage activator but by virologists as an antiviral protein. Similarly, IL-5 was variously known at one time as T cell replacement factor, B cell growth factor II (BCGFII), and eosinophil differentiation factor (EDF). Attempts to clarify these issues using biochemical approaches were stymied by the very low quantities of cytokines secreted into culture supernatants and the imprecise specificity of anti-cytokine antibody preparations of the day. Some progress in distinguishing the various molecules responsible for the observed effects was made when tumor cell lines that secrete abnormally large quantities of a given cytokine were isolated. In addition, easy assay systems became available with the isolation of cell lines that depended on a particular cytokine for growth. Ultimately, the molecular characterization of cytokines was made possible by gene cloning and monoclonal antibody technology in the 1980s. Finally, rationalization of the long list of various factors and their effects could be completed, and multiple activities assigned to a much smaller group of cytokines. For example, these types of analyses revealed that the IFNβ2 molecule, B cell differentiation factor (BCDF), and hepatocyte stimulating factor were all in fact IL-6.

Current work on cytokines has centered on the delineation of their precise physiological functions *in vivo* and their potential clinical applications. The roles of cytokines *in vivo* have been addressed using both transgenic and knockout mice, and some surprising differences between *in vivo* and *in vitro* effects have been observed. For example, most cytokines judged by *in vitro* assay to be important in lymphocyte development have proved not to be absolutely necessary in the whole animal. Concomitantly, unexpected fundamental roles for certain cytokines in embryonic development have been uncovered. We should not feel too confident that we have identified either all the cytokines or their effects in different organ systems just yet.

With respect to therapeutic applications, cytokines (and their close cousins, the growth factors) have been used in the clinic for some time, often on the basis of functions defined by *in vitro* studies. For example, both G-CSF and GM-CSF were defined by colony formation studies to be important for the development of the myeloid lineage. Accordingly, patients with damagingly low levels of neutrophils (*neutropenic* patients) have been treated with pharmacological doses of human recombinant GM-CSF or G-CSF with clinically beneficial effects. However, while gene-targeted mice lacking G-CSF show a profound impairment of granulopoiesis (as expected), knockout animals lacking either GM-CSF or a component of its receptor fail to show any dramatic effect on myeloid cell development. Such results reinforce the concept of the redundancy of the cytokine network but also illustrate how non-physiologic doses of effector molecules can often have effects that extend beyond their normal functions *in vivo*.

B. Overview

I. GENERAL PROPERTIES OF CYTOKINES

Cytokines share several properties with hormones and growth factors, the other major "communication" molecules of the body, but also differ from them in important ways (Table 17-1). Like hormones and growth factors, cytokines are soluble proteins present in very low amounts that exert their effects by binding to specific receptors on the cell to be influenced. However, cytokines differ from hormones and growth factors in three ways: site of production, mode of operation, and range of influence. Hormones are induced by specific stimuli and are synthesized in specialized glands. Hormones tend to operate in an endocrine fashion (over substantial distances or systemically) and influence only a very limited spectrum of target cells. Growth factors, on the other hand, are usually produced constitutively and by individual cells (both leukocytes and non-leukocytes) rather than by glands. Many but not all growth factors can be detected at significant levels in the circulation, and a broad range of cell types is influenced. Taking the middle road, cytokines are synthesized under tight regulatory controls by a sizable number of types of leukocytes and non-hematopoietic cells, and generally exert their effects in a paracrine or autocrine fashion.

i) Structure

As introduced in Chapter 2, cytokines are low molecular weight peptides or glycoproteins of diverse structure and function whose primary role is intercellular communication. Several structural families can be discerned among cytokines, although molecules of similar function can have very different structures, and those of similar structure can be very different in function. The largest structural family is the "short chain 4 α-helix" family, which is a shorthand description for a protein structure that includes four α-helical regions of short length connected to each other by simple linear stretches of amino acids (Table 17-2). Very few or no regions of β-pleated sheets are present in these molecules. IL-2 and IFNγ are just two examples of the many functionally diverse members of this structural family. Similarly, several other cytokines, including IFNα/β and IL-6, belong to the "long chain 4 α-helix" family and share a structure in which four longer α-helical stretches are present. In contrast, IL-1 and IL-18 each form a tertiary structure made entirely of 12 β-pleated sheets; this structure is known as the "β-trefoil" structure. Members of the TNF family of molecules exist in nature as homotrimers with a bell-shaped structure referred to as the "β jelly roll." Members of the TGF family of molecules occur as homodimers linked by disulfide bridges in a form known as the "cysteine knot." The reader will recall that chemokines fall into several classes based on the arrangement of their N-terminal cysteine residues. The structure of CC and CXC chemokines was introduced in Box 4-3 in Chapter 4. In CC chemokines, the two cysteine residues are positioned side by side, whereas a single unconserved amino acid separates the terminal cysteines in a CXC chemokine. The lone CX3C chemokine identified to date has three unconserved amino acids between the cysteines. The C chemokines (sometimes called the XC group) contain only one N-terminal cysteine residue.

ii) Production and Half-Life

Most of the cytokines discussed in this chapter are synthesized primarily by activated Th cells in response to antigen,

Table 17-1 **Cytokine Properties in Comparison to Growth Factors and Hormones**

	Cytokine	Growth Factor	Hormone
Solubility	Soluble	Soluble	Soluble
Receptor required?	Yes	Yes	Yes
Site of production	Many cell types with wide tissue distribution	Several cell types with moderate tissue distribution	Specialized cell types in glands
Expression	Induced or upregulated	Constitutive	Induced
Range of effect	Autocrine or paracrine	Paracrine or endocrine*	Endocrine
Cell types influenced	Several	Broad range	Very limited

* In this context, endocrine means "systemic" effects.

Table 17-2 Cytokine Structures

Structural Motif	Examples of Cytokine Family Members
Short chain 4 α-helix	IL-2, IL-15, IL-21* IL-3 IL-4, IL-13 IL-5 IL-7 IL-9 IFNγ
Long chain 4 α-helix	IL-6, IL-11 IL-10, IL-19, IL-20, IL-22, IL-24, IL-26 IL-12, IL-23, IL-27 IFNα/β, IL-28A, IL-28B, IL-29
β-trefoil	IL-1, IL-18
β jelly roll	TNF, LTα, LTβ
Cysteine knot	TGFβ
C (or XC) chemokine	Lymphotactin
CC chemokine	RANTES, MIP-1β, MCP-3, eotaxin
CXC chemokine	IL-8, IP-10, SDF-1, BLC
CX3C chemokine	Fractalkine
Other	IL-14 IL-16 IL-17 IL-25

* Cytokines in same row are more closely related in structure.

or by activated macrophages in response to the presence of microbial or viral products. More modest contributions are made by other leukocytes and some non-leukocyte cell types. Several means are used to regulate cytokine production and effects. The half-lives of cytokines and their mRNAs are generally very short, meaning that new transcription and translation are required when an inducing stimulus is received. The translated cytokine may also be post-translationally processed before it is very quickly secreted out of the producing cell. The effect is one of a transient flurry of cytokine production followed by resumption of a resting state in the absence of fresh stimulus. The influence of cytokines is also curtailed by controls on receptor expression. Cells lacking expression of the appropriate receptor may be sitting in a pool of cytokine but are unable to respond to it. Lastly, since most cytokines act only over a short distance, only those cells that are in the immediate vicinity of a producing cell (and that express the required receptors) will be influenced.

iii) Function

In general, there are three broad categories of cytokine function: regulation of the innate response, regulation of the adaptive response, and regulation of the growth and differentiation of hematopoietic cells. Many cytokines mediate

aspects of all three, resulting in a constellation of effects on the proliferation, differentiation, migration, adhesion, and/or function of a range of cell types. As we have learned in previous chapters, some events that can be attributed at least in part to cytokine action include pro-inflammatory and anti-inflammatory responses, antiviral responses, growth and differentiation responses, cell-mediated immune responses, humoral immune responses, Ig isotype switching, and chemotaxis.

A cytokine exerts its effects by binding to specific receptor complexes expressed in the membranes of particular cell types, triggering intracellular signaling that results in new specific gene transcription and changes to cellular activities. Stimuli in the immediate environment of a cell can induce it to express a particular cytokine receptor, making that cell now able to receive cytokine-mediated signals. The affinity of binding between cytokines and their receptors is high ($K = 10^{10}$–10^{12} M^{-1}), so that only picomolar concentrations of cytokine are often sufficient to produce a physiological effect. A frequent result of cytokine action is the induction of expression of another cytokine or its receptors, creating a cascade of receptivity and a coordinated response. It is therefore not unusual for one cytokine to induce the transcription of a nuclear transcription factor needed to bind to the promoter of another cytokine gene and spark production of this second cytokine. For example, TNF and IL-1 stimulate transcription of NF-κB, which binds to a specific motif in the promoter of the IL-6 gene and initiates IL-6 production. Consequently, a coordinated response of TNF, IL-1, and IL-6 secretion is observed. Alternatively, as we saw for Th1/Th2 cross-regulation in Chapter 15, one cytokine may repress the expression of another cytokine or of its receptors, creating an antagonistic situation.

Because the immune system cannot operate without the signals cytokines deliver, their activities are often redundant. The sharing of protein chains by different cytokine receptors forms the basis for at least some of this observed functional redundancy. In addition, more than one effect of an individual cytokine may be felt by a given "target" cell over a period of time, with some aspects being felt immediately or within minutes, and others delayed by hours or even days. *In vivo*, a target cell may be exposed to a complex mixture of cytokines for an extended period, making the cellular outcome difficult to predict.

iv) Rationale for Cytokine Network Complexity

Why should evolution have created this complex web of cytokine interactions? Cytokines are a remarkably flexible means of controlling the immune response, which can be harmful to the host if too vigorous or too long in duration. Positive and negative feedback loops, agonistic and antagonistic relationships, and redundant functions exist among the cytokines to provide a fine level of control over the powerful cells and effector mechanisms that can be unleashed to contain a pathogen attack. Without such controls, extreme collateral damage to host tissues or even autoimmunity could result. Such multi-level structuring also ensures that, in most cases, no matter what strategy the pathogen uses to invade the host and avoid immune surveillance, a cytokine can be induced that will mobilize an effective response, and that if this cytokine fails, there are others with overlapping functions that can fill the gap.

II. GENERAL PROPERTIES OF CYTOKINE RECEPTORS

i) Structural Families

Like the cytokines themselves, the cytokine receptors come in a variety of shapes and complexities. Nonetheless, each has an extracellular region, a transmembrane region, and a cytoplasmic region. While some receptors are monomeric, most are at least dimeric. It is not unusual for a particular signal transducing chain to be shared by several different receptors (see later). In general, one polypeptide chain of a multimeric receptor (usually designated "α") confers high-affinity ligand binding and specificity, while another β or even γ chain supplies the signal transduction function. Such signal transducing chains usually cannot directly phosphorylate tyrosine residues themselves but associate with intracellular kinases such as Lck, Fyn, or members of the Jak family that do the job. Signal transduction paths then result in the activation of transcription factors such as NF-κB, AP-1, and members of the STAT family. (For a discussion of Jak kinases and STAT transcription factors, see Box 17-1). New, and often coordinated, activation (or suppression) of gene transcription then ensues. While some cell types constitutively express some cytokine receptors, the expression of many must be induced, often by another cytokine.

Cytokine receptor chains are classified into six distinct families on the basis of particular amino acid motifs or structural elements in the extracellular region, as follows and as illustrated in Figure 17-1.

ia) Type I cytokine receptor family.
(Also known as the hematopoietin receptor family.) Chains belonging to this group include IL-2Rβ, IL-2Rγ, IL-3Rα, IL-4Rα, IL-5Rα, IL-6Rα, IL-7Rα, IL-9Rα, IL-11Rα, IL-12Rβ1, IL-12Rβ2, IL-13R1α, IL-15Rα, IL-21Rα, βc, and gp130. The receptors for the growth factors Epo, G-CSF, and GM-CSF (among others) also belong to this group. In the extracellular region close to the transmembrane domain of each of these polypeptides, there is a defining tryptophan-serine-X-tryptophan-serine sequence known as the WSXWS motif. This sequence is critical for correct protein folding, ligand binding, and signal transduction. Four cysteine residues are also conserved in the extracellular domains of several members of this family. Almost all members of this family also contain a sequence called the Box1/Box2 motif, which is found in the cytoplasmic region and is important for recruiting Jak kinases. Several of these receptor chains feature fibronectin domains proximal to the membrane on the extracellular side.

ib) Type II cytokine receptor family.
(Also known as the interferon receptor-like family.) This group contains various chains participating in the IFNγR, the IFNα/βR, IL-10R, IL-20R, and IL-22R. Chains of the latter three receptors are grouped here because the nucleotide sequences of their extra-cellular domains are similar to the sequence found in both subunits of IFNγR. These extracellular regions usually contain tandem fibronectin domains with a characteristic pattern of cysteine and proline residues.

ic) TNF receptor superfamily.
(Also known as the Type III cytokine receptor family.) The TNFR superfamily was reviewed in Box 9-3 in Chapter 9. TNFR1 has six "cysteine repeat" domains, while TNFR2 has five such domains. Only the C-terminal end of TNFR1 features the TNFR "death domain" required to induce apoptosis.

id) TGFβ receptors.
Three types of TGFβ receptors exist, all of which contain cysteine-rich domains in their extracellular regions. Unlike most cytokine receptors, TGFβR types I and II (but not type III) also feature a serine-threonine kinase domain in their cytoplasmic regions.

ie) Ig superfamily cytokine receptors.
The IL-1RI and IL-1RII molecules exhibit three Ig-like domains in their extracellular regions, as does IL-18R. The receptors for several growth factors, including fibroblast growth factor (FGF), platelet-derived growth factor (PDGF), M-CSF, and stem cell factor (SCF), also contain Ig-like domains in their extracellular regions.

if) Chemokine receptor family.
The two receptors that bind IL-8 would be unusual among interleukin receptors in that, instead of a single transmembrane domain, they feature seven transmembrane domains that repeatedly cross the membrane. However, it has turned out that IL-8 is a chemokine rather than an interleukin, and that the structure of its G protein-coupled receptors is standard among chemokine receptors. The receptors formerly known as the IL-8 receptors have thus now been renamed as the chemokine receptors CXCR1 and CXCR2 (see later in this chapter).

ii) Shared Receptor Subunits

Redundancy of effect is a key property of the cytokine network, established by evolution to ensure that vital signaling is not lost if one cytokine ceases to function. The molecular basis for these overlapping functions is the sharing of signal transducing subunits among the cytokine receptors, leading to the activation of the same downstream intracellular signaling elements. Sharing of signaling chains can also account for competitive antagonism between cytokines. The shared chains are called "public" subunits, while those unique chains that dictate binding specificity are called "private" subunits. Several examples of receptor chain sharing are shown among the receptor complexes illustrated in Figure 17-2.

iia) IL-2R chains.
As discussed in Chapter 15, the high-affinity IL-2R is composed of three polypeptides: the unique IL-2Rα chain, the IL-2Rβ chain, and the IL-2Rγ chain. The IL-2Rβ chain is shared by the receptors for IL-2 and IL-15. The IL-2Rγ chain also appears in the receptors for IL-4, IL-7, IL-9, IL-15, and IL-21 and so is often called the γc *chain*, for "common gamma" chain. Both IL-2Rβ and IL-2Rγ belong to the type I receptor family and contain the WSXWS motif.

iib) Gp130 chain.
Gp130 is a component of the IL-6R, IL-11R, and IL-27R, and also appears in the receptors for several

Box 17-1. **Jak/STAT signal transduction pathways**

In Chapters 9 and 12, we described signal transduction pathways from the BCR and TCR that involved the linking of receptor cytoplasmic regions to intracellular kinases such as Lck and Fyn, and thence to downstream elements such as the PKC and Ras/MAPK pathways. While the binding of some cytokines to their receptors can also trigger these signaling pathways, another group of pathways involving members of the Jak/STAT families plays a more prominent role in cytokine signaling. The Jak/STAT signaling pathways were first identified in investigations of IFN signaling. Depending on one's point of view, "Jak" stands either for Janus, a mythical Roman god with two faces, or for "just another kinase." The Jak enzymes are non-receptor tyrosine kinases that contain a kinase domain (JH1), a kinase-like domain (JH2), and several other domains (JH3–7) conserved among Jak family members. It is the tandem juxtaposition of the kinase and kinase-like domains that gave rise to the "two-faced" reference. Jak proteins do not contain SH2 or SH3 domains. The STAT (*signal transducer and activator of transcription*) proteins

are transcription factors frequently involved in downstream cytokine signal transduction. At least four mammalian Jak kinases have been identified to date, Jak1, Jak2, Jak3, and Tyk2 (*tyrosine kinase 2*), and six STAT proteins, STATs 1–6 (see Table).

The cytoplasmic tails of cytokine receptors signaling through Jak/STAT pathways have binding sites for various Jaks and Tyks that associate constitutively but loosely with the receptor tail (see Figure). Upon the binding of a cytokine molecule to its receptor, the clustering of receptor proteins is induced, allowing the associated Jak enzymes to cross-activate each other and phosphorylate several different tyrosine residues in the receptor tails. This phosphorylation both stabilizes the association between the kinase and the receptor protein and triggers the recruitment of specific STAT proteins. For example, STAT6 is essential for IL-4-induced isotype switching to IgE, and STAT4 is crucial for IL-12-mediated induction of Th1 differentiation. Similarly, STAT5 activation is necessary for signaling through both IL-2R and the erythropoietin (Epo) receptor.

Once associated with the receptor protein via SH2 domain interactions, the STAT protein itself is phosphorylated by the bound Jak kinases. Within the cluster of receptor tails, the SH2 domain of one phosphorylated STAT can interact with phosphotyrosine residues of another STAT, and vice versa. In fact, the affinity of the STATs for phosphotyrosines on each other is higher than the affinity of each for phosphotyrosines in the receptor tails, so that two phosphorylated STATS homo- or heterodimerize through their SH2 domains and detach from the receptor. The dimerized STAT molecule is now an activated transcription factor that can make its way to the nucleus to facilitate the initiation of new gene transcription. Meanwhile, the now-vacant STAT binding sites in the receptor tails maintain their phosphorylation (assuming the ligand is still stimulating the receptor) and are free to bind additional STAT molecules, perpetuating the signal transduction cascade.

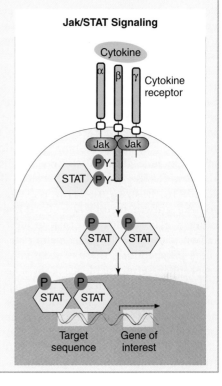

Jak/STAT Signaling

Jaks	Activating Cytokines
Jak1	IFNα/β, IFNγ, IL-2, 4, 5, 6, 7, 9, 10, 13, 15, 27, GM-CSF, G-CSF, (IL-19, 20, 22, 24?)
Jak2	IFNγ, IL-3, 4, 5, 6, 12, IL-15 (IL-15Rx), IL-23, GM-CSF, G-CSF, SCF
Jak3	IL-2, 4, 7, 9, 15, Epo
Tyk2	IFNα/β, IL-6, 10, 12, 13, 23

STAT	Activating Cytokines
STAT1	IFNα/β, IFNγ, IL-6, 9, 10, 11, 26, 27, SCF
STAT2	IFNα/β, IL-17
STAT3	IL-2, 6, 9, 10, 11, 12, 15, 17, 21, 22, 23, 26, 27, G-CSF
STAT4	IL-12, 17, 23
STAT5	IL-2, 3, 5, 6, 7, 9, 10, 15, 21, 22, 23, GM-CSF
STAT6	IL-4, 5, 13, 24

closely related minor cytokines that share the helical structure and signaling pattern of IL-6. These include LIF (*leukemia inhibitory factor*), which inhibits the differentiation of ES cells and promotes the survival and growth of hematopoietic progenitors; CNTF (*ciliary neurotrophic factor*), which promotes

survival of neuronal cell types; and OSM (oncostatin), which inhibits growth of melanoma cells and induces IL-6 secretion by AIDS-related Kaposi's sarcoma cells.
iic) βc chain. The *βc* ("common β") *chain* appears in the receptors for IL-3, IL-5, and GM-CSF. This chain does not

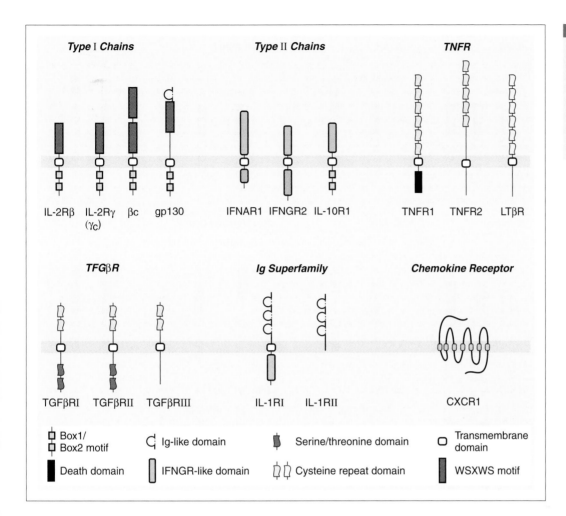

Figure 17-1

Families of Cytokine Receptor Chains
Examples of single cytokine chains belonging to the type I, type II, TNFR, TGFβR, Ig superfamily, and chemokine receptor families are shown. The key identifies domains found in these chains.

Type I Chains

IL-2Rβ   IL-2Rγ   βc   gp130
          (γc)

Type II Chains

IFNAR1   IFNGR2   IL-10R1

TNFR

TNFR1   TNFR2   LTβR

TFGβR

TGFβRI   TGFβRII   TGFβRIII

Ig Superfamily

IL-1RI   IL-1RII

Chemokine Receptor

CXCR1

| ▣ Box1/ ▢ Box2 motif | ⊏ Ig-like domain | ▮ Serine/threonine domain | ▢ Transmembrane domain |
| ▮ Death domain | ▯ IFNGR-like domain | ▯▯ Cysteine repeat domain | ▮ WSXWS motif |

bind to cytokine molecules in the absence of the private sub-unit of each receptor. Furthermore, dimerization of the βc subunit is required for intracellular signaling by these receptors. It is thought that the cross-inhibition between IL-3, IL-5, and GM-CSF (in cells responsive to all three cytokines) stems from competition for limited numbers of βc molecules.

iid) Other shared chains. Several other cytokine receptors chains are shared with one or two other cytokines. The IL-4Rα chain is shared by the receptors for IL-4 and IL-13, a fact that most likely accounts for their overlapping biological effects. Similarly, the IL-10R2 chain is shared by the receptors for IL-10, IL-22, and IL-26. The IL-12Rβ1 chain appears in both the IL-12R and the IL-23R complexes. Both IL-19 and IL-20 bind to the IL-20R composed of IL-20R1 and IL-20R2. IL-20R1 also participates with IL-10R2 in the IL-26R. The IL-20R2 chain can pair with either IL-20R1 or IL-22R1 and bind to either IL-20 or IL-24.

C. Function, Production, and Receptors of Cytokines

The remainder of this chapter is devoted to describing individual cytokines, their function, production, and receptors.

The reader may note that various immunological reference books have information on cytokines organized in different ways. We have chosen not to rely solely on summary charts as we felt this limits the amount of context that can be presented. We have also avoided grouping the cytokines by structural or functional features, because this makes the information about a specific molecule difficult to locate. Furthermore, such groupings may be misleading due to the pleiotropy exhibited by many cytokines. We have instead listed the cytokines individually in an alphanumeric fashion as much as possible, hoping that the reader will find it easy to locate the information on each. Table 17-3 contains a key identifying the various cell type abbreviations used in the figures in this chapter.

I. THE INTERFERONS

The IFNs were among the first cytokines to be discovered and characterized. These amazingly pleiotropic molecules were first identified over 40 years ago by researchers who observed that infection of a cell with a first virus prevented superinfection of that cell by a second virus. This "interference" phenomenon turned out to be due to the fact that infection of the cell by the first virus induced the production

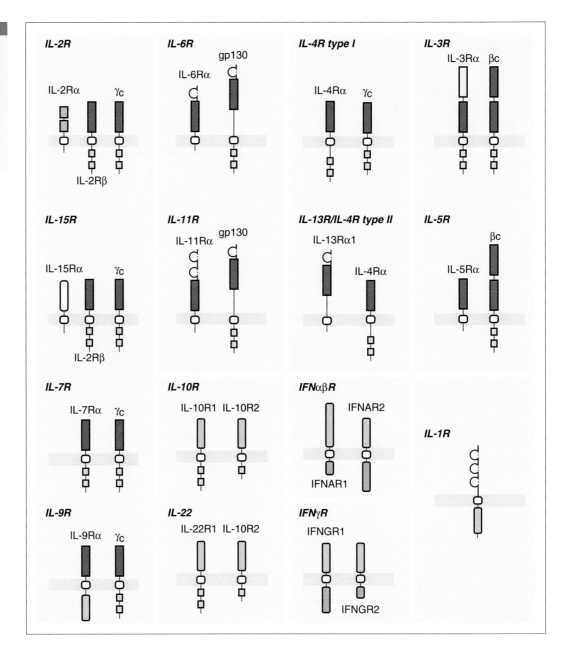

Figure 17-2

Examples of Cytokine Receptor Complexes
Receptors grouped together share at least one component chain. The interferon receptor chains are not shared. The IL-1R is composed of a single unshared chain. Domains are identified as in the key for Figure 17-1.

of IFN molecules that caused the cell to resist infection by the second virus. Hence, the molecules responsible were named "interferons."

Without IFNs, which act long before an adaptive immune response can be mounted, the vertebrate host would rapidly succumb to viral infections. There are two types of IFNs based on similarity in function: type I and type II. Type I IFNs include the long-known IFNα and IFNβ molecules, and the more recently identified IFNδ (in pigs), IFNω (in cattle and humans), and IFNτ (in cattle and sheep). IFNα and IFNβ are evolutionarily related but antigenically distinct types of interferons with identical (as far as can be determined) functions. The other type I IFNs generally appear to have similar functions to IFNα/β. IFNγ is the sole type II or

"immune" interferon. IFNγ is structurally distinct from the type I IFNs and has additional unique immunomodulatory functions.

Functions

i) Functions common to all IFNs. All IFNs have two major functions that are crucial elements of the innate immune response: antiviral activity and antiproliferative activity (Fig. 17-3A). With respect to direct antiviral activity, IFN can interfere with almost any stage of the viral attack, and different stages are targeted in different viruses. For example, IFN blocks the entry of SV40 into the host cell, inhibits the genomic transcription of influenza virus, interferes with the protein translation of adenovirus, and prevents the release of

Table 17-3 **Cell Type Abbreviations**

Abbreviation	Name
αβ T	αβ T cell
Act	Activated
AG	Adrenal gland
APC	Antigen-presenting cell
B	B cell
Br	Brain
Bas	Basophil
BV	Blood vessel
CFU-BME	Colony-forming unit: basophil, megakaryocyte, erythrocyte
CFU-GEMM	Colony-forming unit: granulocyte, erythrocyte, monocyte, macrophage
CFU-GM	Colony-forming unit: granulocyte, monocyte
CMP	Common myeloid progenitor
CLP	Common lymphoid progenitor
CTL	Cytotoxic T lymphocyte
DC	Dendritic cell
Emb	Embryonic cell
Endo	Endothelial cell
Eo	Eosinophil
Epi	Epithelial cell
Fibro	Fibroblast
γδ T	γδ T cell
Hep	Hepatocyte
HSC	Hematopoietic stem cell
IEL	Intraepithelial lymphocyte
LAK	Lymphokine-activated killer cell
Lym	Lymphocyte
Mac	Macrophage
Mast	Mast cell
Meg	Megakaryocyte
Melan	Melanocyte
Mem	Memory
MGP	Monocyte/granulocyte precursor
Mon	Monocyte
Myel prec	Myeloid precursors
Ne	Neuron
Neu	Neutrophil
NHP	Non-hematopoietic cell
NK	NK cell
OB	Osteoblast
OC	Osteoclast
Pre-B	Pre-B cell
Pre-T	Pre-T cell
RBC	Red blood cells
RBC pre	Red blood cell precursors
SM	Smooth muscle
Str	Stromal cell
T	T cell
Th0	Th0 cell
Th1	Th1 cell
Th2	Th2 cell

mature VSV particles from the host cell. IFN also induces host cells near the cell under viral attack to adopt an "antiviral state," which allows these cells to resist viral infection. The antiviral state is a metabolic condition resulting from the action of several factors (Fig. 17-4). First, IFN induces the activation of *PKR*, a dsRNA-dependent serine-threonine kinase. PKR is inactive until it encounters a dsRNA molecule, such as might be derived from a viral genome. Upon activation by dsRNA, PKR phosphorylates itself and other substrates, including a subunit of the translation initiation factor eIF2. eIF2 is inactivated by phosphorylation so that translation of viral (and host) proteins is halted. Secondly, IFN induces the synthesis of a set of *2-5A synthetase* enzymes. These enzymes, which are activated by the presence of dsRNA, use ATP to generate a series of short 2′,5′ oligoadenonucleotides (2-5A) that are required for the dimerization and activation of *RNase L*. RNase L in turn degrades single-stranded RNA, as might be produced by a replicating virus. Thirdly, IFN induces the synthesis of the *Mx proteins*, GTPases that interfere with transcription of viral RNA. Should a virus manage to take hold despite the above, the type I IFNs act to enhance NK-mediated cytolysis of virally infected cells.

As well as antiviral activity, all IFNs have antiproliferative activity, in that they inhibit the growth of many cell types and thus control the expansion of infected or cancerous cells. While the precise genes that IFN acts upon to prevent proliferation have not been identified, both IFNα and IFNγ have been shown to reduce expression of c-myc and to interact with several cell cycle components *in vitro*. However, these effects may not hold for all cell types.

In addition to their important effects on the innate response, IFNs directly and indirectly influence the adaptive immune response. For example, all IFNs share the ability to upregulate MHC class I expression via activation of the transcription factor IRF-1 (*interferon regulatory factor-1*), thereby indirectly promoting the development of CD8$^+$ T cell responses against virally infected cells. It is also now thought that both IFNα/β and IFNγ have direct effects on B cells, governing their maturation, proliferation, antibody secretion, and even isotype switching. In situations where IFNγ is not available, IFNα/β is apparently able to compensate and induce normal switching.

ii) Functions specific to IFNγ. IFNγ was identified in 1965 as an entity in the culture supernatants of activated T cells that not only had the antiviral and antiproliferative activities of the type I IFNs but also had profound effects on several types of immune system cells and on the adaptive immune response (Fig. 17-3B). For example, unlike IFNα/β, IFNγ induces the activation of macrophages, resulting in the production of inflammatory cytokines such as TNF, IL-1β, IL-12, and IL-18. IFNγ is also essential for the expression of iNOS, the inducible nitric oxide synthase enzyme responsible for generating anti-microbial intracellular nitric oxide. IFNγ stimulates the production of reactive oxygen intermediates (ROIs), and induces the expression of Fc receptors suitable for complemented-mediated pathogen destruction and ADCC. In short, IFNγ is a cytokine closely identified with promotion of

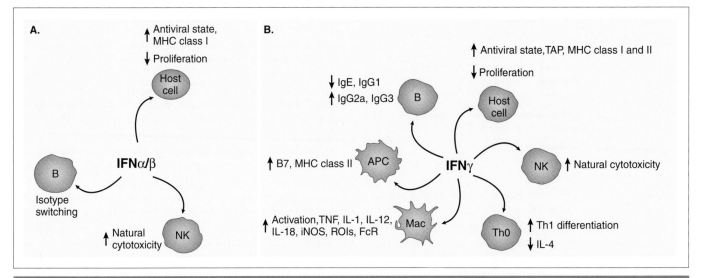

Figure 17-3

Major Functions of Interferons
(A) Functions of IFNα/β. These Th1 cytokines are crucial for innate defense against viruses and other intracellular pathogens. (B) Functions of IFNγ. IFNγ has all the functions of IFNα/β but also has immunoregulatory activities that influence cells of both the innate and specific immune responses. IFNγ is particularly important for stimulating macrophages to secrete the IL-12 necessary for Th1 differentiation.

the inflammatory response and the elimination of intracellularly-replicating pathogens.

With respect to the adaptive response, IFNγ upregulates not only MHC class I expression but also MHC class II on a wide variety of cell types. IFNγ also modulates levels of B7 expression on APCs and thus promotes T cell costimulation. IFNγ regulates the expression of various subunits of proteasomes and the TAP proteins, thereby influencing endogenous antigen processing. Because IFNγ promotes macrophage expression of IL-12, IL-18, and IL-12Rβ2 and blocks IL-4 production, this cytokine favors Th1 over Th2 differentiation. An immune response is thus mounted which is best-suited for assisting in the fight against intracellularly-replicating pathogens. Indeed, many of the powerful actions of IFNγ are inhibited by IL-4, creating a mutually antagonistic form of regulation that helps to keep the immune response under control. In mouse B cells, IFNγ inhibits isotype switching to IgE and IgG1 but supports switching to IgG2a and IgG3, isotypes promoting complement-mediated and phagocytosis-mediated effector mechanisms. In addition to all these effects, IFNγ can either promote or inhibit apoptosis, depending on the cell type, stage of differentiation, and the presence of other apoptotic inducers such as dsRNA and LPS. IFNγ is less important in defense against helminth worms.

Production. In humans, there are about 20 different IFNα proteins (each encoded by its own gene) made primarily by activated macrophages and monocytes with some contribution by activated T cells. Despite their structural variation, the functions and effects of IFNα molecules appear to be identical. The IFNβ class contains a single member in humans and is made primarily by fibroblasts. Fibroblasts in other species make multiple types of IFNβ molecules, again all with essentially the same function. Many human cell types make both IFNα and IFNβ in response to viral infections. IFNγ is produced primarily by activated Th1 cells and to a lesser extent by CTLs and NK cells. As we have learned in previous chapters, IL-12 and IL-18 are important inducers of IFNγ production.

The expression of all IFNs is controlled in part by the IRF family of regulatory factors introduced previously. The IRF proteins bind to specific motifs called *interferon-stimulated regulatory elements* (ISRE) that are found in the promoters of the IFN genes and the promoters of genes induced by the IFNs. Two members of the IRF family, IRF-1 and IRF-2, are thought to cooperate in binding to the same ISREs. Mice deficient for either IRF-1 or IRF-2 are unable to produce IL-12 or IFNγ, and therefore cannot mount Th1 responses even in the presence of exogenous IL-12. The differentiation of Th1 cells (and NK cells) is severely compromised in these mice.

IFNβ production can also be induced by the LPS of gram-negative bacteria via a TLR4-mediated signaling pathway. This pathway is best understood after digesting the information on IL-1 signaling presented in the next section, and will be discussed at that point.

Receptors

i) IFNα/β receptor. IFNα and IFNβ are not really very similar in primary structure. However, perhaps surprisingly, they bind to the same receptor and trigger the same signal transduction pathway, thus accounting for their identical biological effects. IFNα/βR, also known as the type I IFN

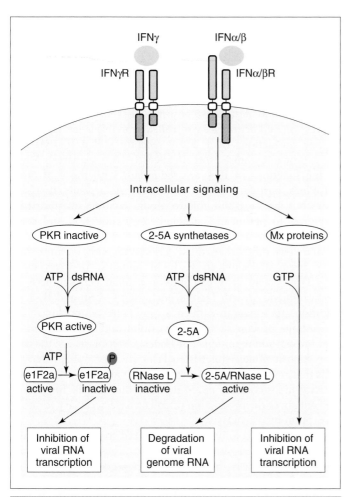

Figure 17-4

Antiviral Mechanisms of IFNs
Upon engagement of an IFN receptor by IFN, intracellular signaling is initiated that leads to the activation of PKR, 2-5A synthetases, and Mx proteins. Each of these molecules plays a role in blocking the viral life cycle, as detailed in the text. IFN thus induces an uninfected host cell to adopt an "antiviral state" that allows it to shut down a virus as soon as it tries to replicate. Adapted from Stark G. R. *et al.* (1998) How cells respond to interferons. *Annual Review of Biochemistry* **67**, 227–264.

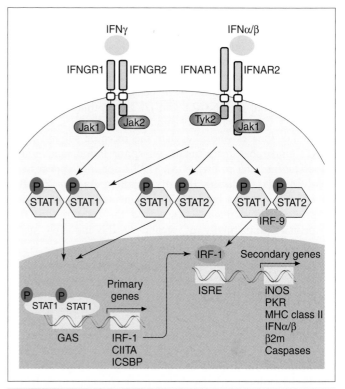

Figure 17-5

IFN Signaling Pathways
When IFNα/β engages IFNαβR, or IFNγ binds to IFNγR, Jak kinases are activated that phosphorylate and activate various STAT transcription factors. Homo- or heterodimers of phosphorylated STATs translocate to the nucleus and bind to specific motifs (GAS, ISRE) in the promoters of "primary" and "secondary" IFN-inducible genes, respectively (see text). The primary IFN-inducible genes include the transcription factors IRF-1, CIITA, and ICSBP (interferon consensus sequence binding protein). The secondary IFN-inducible genes encode molecules necessary for antiviral and microbicidal activity and inflammation. A STAT1/STAT2 heterodimer combined with IRF-9 forms the ISGF-3 transcription factor, which can also bind to ISREs.

receptor, is a heterodimeric protein composed of the IFNAR1 and IFNAR2 subunits (refer to Fig. 17-2). Both subunits are required for high-affinity binding to IFNα and IFNβ. The tyrosine kinase Tyk2 is constitutively associated with the IFNAR1 subunit, while the tyrosine kinase Jak1 is associated with IFNAR2 (Fig. 17-5). Upon IFN binding, Tyk2 is phosphorylated first, followed by cross-activation of Jak1 and reciprocal enhanced activation of Tyk2. STAT dimerization can then occur by one of three possible routes. (1) If STAT1 homodimerizes with another STAT1 molecule, the resulting transcription factor translocates to the nucleus and binds to a DNA motif (5′TTNCNNNAA3′) called GAS (*gamma activating sequence*). GAS, which was originally identified in studies of IFNγ-induced transcription, is present in the promoter regions of many so-called "primary"

IFN-inducible genes, those genes whose expression is required for the transcription of a second subset of genes called the "secondary" IFN-inducible genes. IRF-1 is a primary IFN-inducible gene, and interaction of the STAT1 homodimer with GAS drives IRF-1 transcription. (2) If STAT1 heterodimerizes with STAT2 and makes its way to the nucleus unencumbered, the resulting transcription factor also binds to GAS and the outcome is the same. (3) If, however, a small protein called IRF-9 (or p48) binds to the STAT1/STAT2 heterodimer before it accesses the nucleus, a new transcription factor called ISGF-3 (*interferon stimulatory gene factor-3*) is formed. Rather than the GAS motif, ISGF-3 recognizes the ISRE motif present in the promoters of many "secondary" IFN-inducible genes. IRF-1 also recognizes the ISRE motif and thus helps to drive expression of the secondary IFN-inducible genes. These genes comprise a large subset of immune response genes responsible for a variety of cellular functions, including cytokine production, cell replication, NO production, and antiviral responses.

ii) IFNγ receptor. IFNγR, or the type II IFN receptor, is expressed on all cell types except erythrocytes. IFNγR is a heterodimeric protein containing the IFNGR1 and IFNGR2 subunits, both of which are required for signal transduction. Prior to IFNγ binding, the IFNGR1 and IFNGR2 chains are not associated. Jak1 kinase is associated specifically with the cytoplasmic tail of IFNGR1 and Jak2 with that of IFNGR2. After IFNγ binding, IFNGR1 and IFNGR2 dimerize, the associated Jak kinases phosphorylate the receptor subunits, and STAT1 undergoes recruitment, phosphorylation, and homodimerization (refer to Fig. 17-5). STAT2 does not appear to be involved in transducing IFNγR signaling. As for IFNαβR, STAT1 homodimers bind to GAS elements in the promoters of primary IFN-inducible genes, and the transcription first of primary and finally of secondary IFN-inducible genes proceeds.

II. THE INTERLEUKINS

i) IL-1

Function. IL-1 is a highly inflammatory cytokine prominent in innate defense. Two functionally equivalent isoforms exist: the acidic IL-1α protein and the neutral IL-1β protein, but the IL-1β isoform is present in greater concentration. In the peripheral tissues, low concentrations of IL-1 mediate local inflammation by promoting the expression of adhesion molecules (such as ICAM-1, VCAM-1, and the selectins) on endothelial cells, and inducing macrophages and endothelial cells to produce leukocyte-activating cytokines, particularly IL-6, IL-8, and TNF (Fig. 17-6). However, IL-1 cannot directly activate neutrophils or other inflammatory cells. At moderate concentrations, IL-1 acts on a wide variety of cell types to induce the onset of fever, the acute phase response, and sometimes *cachexia*, a wasting of the body due to uncontrolled cellular catabolism. At high concentrations induced by a massive gram-negative bacterial infection, IL-1 floods out

of the tissues and into the circulation, becoming a systemic inflammatory activator. In so doing, IL-1 is a principal mediator of the sometimes fatal condition known as *endotoxic* or *septic shock*, a collapse of circulatory and metabolic systems induced by overwhelming amounts of cytokines (particularly IL-1 and TNF) released into the circulation (see Box 17-2). This potentially destructive power of IL-1 has dictated that its expression and action are controlled at multiple levels, including by the actions of a decoy receptor called IL-1RII and a receptor antagonist protein called IL-1Ra (see later). Many of the functions of IL-1 are redundant with those of other cytokines, particularly TNF (see later). As a result, IL-1β$^{-/-}$ knockout mice are still able to mount credible inflammatory responses to pathogens, although fever induction and acute phase responses are abnormal.

Activities other than those associated with inflammation have been attributed to IL-1. In humans, IL-1 produced by CD34$^+$ stem cells can induce stromal cell synthesis of hematopoietic growth factors, including G-CSF and IL-3. In addition, IL-1 has an influence on "bone remodeling," the continuous process of bone deposition and bone resorption that sculpts and maintains our bones. Osteoblasts are the cells that deposit bone, while osteoclasts are macrophage-derived cells that resorb bone. IL-1 depresses the alkaline phosphatase activity in osteoblasts necessary for bone matrix formation, and enhances the production of collagenase necessary for bone resorption by osteoclasts.

Production. IL-1α and IL-1β are derived from separate genes of surprisingly dissimilar sequence. These cytokines are expressed primarily by macrophages and neutrophils but also by keratinocytes, epithelial cells, and endothelial cells. Expression of IL-1 is triggered by the presence of TNF, bacterial products such as LPS, IL-1 itself, and, in the case of macrophages, by contact with activated Th cells. Both IL-1α and IL-1β are synthesized as pro-proteins without the usual secretory signal sequence, making their entry into the tissues and circulation an unsolved mystery. The pro-protein form

Figure 17-6

Major Functions of IL-1
This cytokine plays a key role in the inflammatory response and in endotoxic shock.

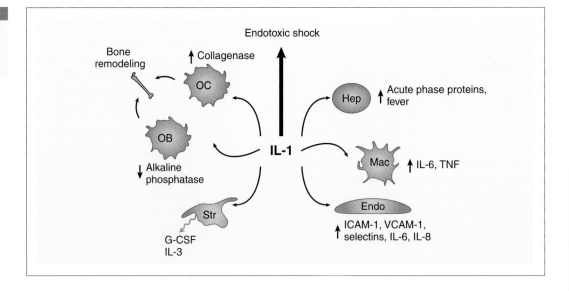

Box 17-2. **Bacterial endotoxic shock**

A little TNF is a very good thing and essential to combat infections, especially those of gram-negative bacteria. However, when an overwhelming amount of the bacterial cell wall component LPS is present, it induces activated macrophages to produce massive amounts of TNF, with catastrophic effects for the host. Very high concentrations of TNF stimulate the overproduction by activated macrophages of the pro-inflammatory cytokine IL-1, which in turn amplifies LPS-induced production of IL-6 and IL-8 by vascular endothelial cells and macrophages. This systemic tidal wave of pro-inflammatory cytokines, termed *endotoxic* or *septic shock*, has devastating effects on several key organ systems.

First, animals suffering from endotoxic shock show evidence of *disseminated intravascular coagulation* (DIC), a massive blockage of capillaries by aggregations of neutrophils and excessive blood clot formation on the walls of vascular endothelial cells. Clotting factors are subsequently depleted, leading to hemorrhage. Secondly, cardiomyocytes in the heart begin to fail due to an accumulation of nitric oxide produced by the cytokine-activated enzyme iNOS. Cytokine-induced slowing of blood circulation also contributes to a drop in blood pressure followed by circulatory collapse. Thirdly, cytokines induced by LPS stimulate muscles to consume glucose at an increased rate but also interfere with glucose synthesis by the liver, leading to metabolic failure. Fever and diarrhea are also features of patients suffering from endotoxic shock. The combination of circulatory failure and metabolic collapse results in irreversible, and often fatal, damage to host organ systems. Death due to endotoxic shock can happen within hours of infection by certain gram-negative bacteria, including those causing enteric disorders, meningitis, and pneumonia. Close to 600,000 people a year experience endotoxic shock in North America, and more than 10% of cases have a fatal outcome. Experimental injection of anti-TNF antibody has been shown to mitigate endotoxic shock in LPS-injected animals, confirming the participation of TNF in initiating this often deadly disease. Unfortunately, injection of anti-TNF or IL-1Ra (to block the binding of IL-1) into humans has

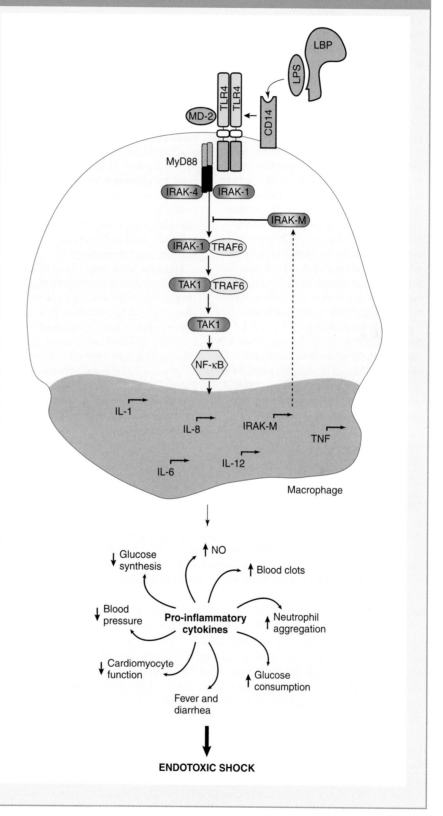

Continued

Box 17-2. **Bacterial endotoxic shock**—*cont'd*

been less effective. Overall, the data indicate that TNF and IL-1 work synergistically to induce bacterial endotoxic shock. Because of their roles in activating macrophages and priming them for a lethal response, IFNγ and IL-12 are also thought to contribute to endotoxic shock. An effective therapeutic strategy must take the redundancy and pleiotropy of all these cytokines into account.

Our growing knowledge of the signaling pathway leading from an encounter with LPS to TNF and IL-1 production may also open up avenues for therapy. As was introduced in Chapter 4, LPS is thought to form a complex with LBP, CD14, an adaptor protein called MD2, and a TLR4 homodimer in order to initiate intracellular signaling leading to the activation of macrophages and cytokine production (see Figure). Work *in vitro* with both mouse and human TLR family members has shown that IRAK-1, the same kinase recruited to the IL-1R in response to IL-1 binding, can associate via the adaptor protein MyD88 with the intracellular portions of TLR4. Under the influence of phosphorylated IRAK-4, signal transduction downstream of IRAK-1 proceeds through TRAF6 recruitment, IκB degradation, and NF-κB (but not MAPK)

activation. Activated NF-κB then drives the expression of many pro-inflammatory cytokines that contribute to endotoxic shock. However, this cascade is subject to control via a feedback loop. Among the genes induced by NF-κB is IRAK-M, an inhibitor that blocks the activation of IRAK-1. We can see the roles of these molecules demonstrated quite nicely in knockout mouse mutants. Mice with deletions of CD14, TLR4, TRAF6, or MyD88 exhibit hyporesponsiveness to LPS treatment and fail to secrete elevated levels of IL-1, IL-6, or TNF; that is, do not exhibit signs of endotoxic shock. Conversely, mice lacking IRAK-M are exquisitely sensitive to LPS, succumbing to endotoxic shock and dying after stimulation with very small amounts of LPS.

Endotoxic shock induced by LPS exposure can be replicated experimentally in a *localized Shwartzman reaction*. Gregory Shwartzman observed that if two sublethal doses of LPS were injected 24 hours apart into a solid tissue in a rabbit, localized coagulation resulted followed by hemorrhaging and necrosis at the tissue site. It was subsequently shown that TNF was a key mediator of this process, and that repeated injections of IL-1 could also induce localized Shwartzman reactions. A

systemic Shwartzman reaction mimics the symptoms of natural bacterial endotoxic shock and was found to result when two sublethal doses of LPS were given intravenously (rather than in solid tissue) 24 hours apart. Administration of anti-IFNγ or anti-IL-12 antibodies prior to treatment with endotoxin was able to prevent mortality associated with a systemic Shwartzman reaction, further suggesting the involvement of IFNγ and IL-12 in endotoxic shock. However, no successful trials of anti-IFNγ or anti-IL-12 as treatments for endotoxic shock have been reported to date. In the early 2000s, pre-clinical trials of inhibitors of TLR signaling were investigated as a future potential therapy for endotoxic shock. In particular, a small chemical compound that selectively blocks the binding between the TIR domains (see text) of IL-1R to MyD88 has shown promise.

Lastly, the reader should note that endotoxic shock, induced by LPS and mediated by TNF and IL-1 secretion, is distinct from bacterial *toxic shock*, the massive cytokine induction mediated by the polyclonal activation of T cells by bacterial superantigen TSST1 (discussed in Box 14-3 in Ch.14).

of IL-1β must be cleaved by caspase-1 (also known as ICE, *IL-1β converting enzyme*) to activate the cytokine. Regulation of IL-1β expression is thus achieved in part by the regulation of caspase-1 expression. Interestingly, caspase-1 does not cleave the IL-1α precursor, a task executed by other membrane proteases. Synthesis of both forms of IL-1 is inhibited by IL-4 and IL-10.

Receptor. Most cell types express two monomeric IL-1 receptors: the type I receptor (IL-1RI), through which most biological activity is mediated, and the type II receptor, which acts as a decoy. Both IL-1α and IL-1β signal through IL-1RI but, in so doing, bind to distinct sites on this molecule. No signal is transduced by the binding of either IL-1α or IL-1β to IL-1RII because this receptor lacks the appropriate cytoplasmic domain. It is thought that the *in vivo* function of IL-1RII is to bind and sequester IL-1, preventing it from binding to IL-1RI. Both IL-1RI and IL-1RII can also be released from the cell membrane, creating a supply of soluble receptors that compete for free IL-1.

The IL-1RI chain is insufficient on its own for IL-1 signal transduction. When IL-1 first binds to IL-1RI, a co-receptor protein called IL-1RAcP (*IL-1R accessory protein*) is brought into close association with the IL-1/IL-1RI complex and greatly increases the affinity of binding. IL-1RAcP does not have any other known role in signal transduction. Rather,

signaling depends on the presence in the IL-1RI chain of a highly conserved cytoplasmic domain called the TIR (*Toll/IL-1R*) domain (also found in the cytoplasmic tails of TLRs). The TIR does not have intrinsic tyrosine kinase activity. However, the TIR facilitates the recruitment of a cytoplasmic adaptor protein called MyD88 (*myeloid differentiation protein 88*), which also contains a TIR and dimerizes as it associates with the IL-1/IL-1RI/IL-RAcP complex (Fig. 17-7). MyD88 in turn recruits both IRAK-1 (*IL-1R associated kinase 1*) and IRAK-4 to the IL-1R complex. When IRAK-4 is brought into close proximity to IRAK-1, it induces the phosphorylation of IRAK-1 and activates it. (IRAK-4 does not bind directly to IRAK-1.) Activated IRAK-1 then hyperphosphorylates itself, which allows the binding of the adaptor protein TRAF6 (*TNFR-associated factor 6*). The IRAK-1/TRAF6 complex then dissociates from the IL-1RI tail and interacts with a multi-protein complex containing TAK1 (*transforming growth factor β activated kinase-1*). As a result of these interactions, IRAK-1 is degraded to yield a complex containing TRAF6 and TAK1 (among other components). TRAF6 is then degraded by ubiquitination, activating the TAK1 complex. TAK1 carries out phosphorylation that activates two pathways. First, TAK1 can phosphorylate components of the IKK (IκB kinase) complex. As introduced in Chapter 8, the IKK complex is composed of three subunits: the kinases IKKα (*IκB kinase α*)

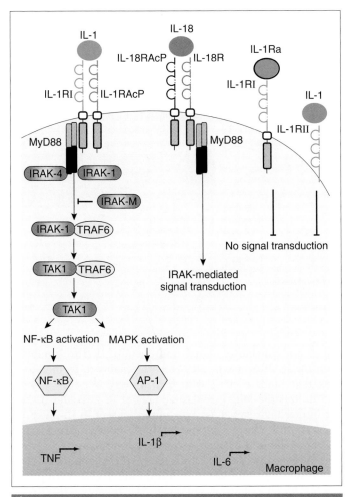

Figure 17-7

IL-1/IRAK Signaling Pathway
The engagement of IL-1RI by IL-1 recruits IL-1RAcP into the complex, followed by recruitment and dimerization of MyD88. MyD88 then recruits IRAK-1 and IRAK-4 to the complex, where IRAK-4 induces the phosphorylation of IRAK-1 in an interaction negatively regulated by IRAK-M. In a series of events described in the text, activated IRAK-1 interacts with TRAF6, which subsequently activates TAK1. TAK1 induces the activation of both the NF-κB and MAPK signaling pathways, leading to the activation of transcription factors including NF-κB and AP-1 and the transcription of the TNF, IL-1β, and IL-6 genes. As is described later in the text, the IL-1-related cytokine IL-18 signals through a similar pathway mediated by IL-18R, IL-18RAcP, and MyD88 that leads to IRAK-1 activation. Signaling through IL-1RI can be blocked by the binding of the antagonist IL-1Ra, which inhibits the recruitment of MyD88 and downregulates IL-1 signaling. The IL-1RII receptor lacks a signaling function and serves to sequester IL-1. Not all signaling proteins are shown.

and IKKβ, and a non-catalytic subunit called IKKγ or NEMO (*NF-κB essential modulator*). Phosphorylation of the IKK complex leads to its activation and the phosphorylation of the inhibitor IκB bound to the transcription factor NF-κB. Phosphorylation of IκB signals for its ubiquitination and degradation, releasing NF-κB and allowing it to translocate into the nucleus and activate transcription. In addition to IKK phosphorylation, TAK1 phosphorylates MEKK3, 4, and 6, which are MAP kinases that facilitate transduction in the p38 and SAPK/JNK pathways (refer to Ch.14). These pathways

lead to the activation of the transcription factors c-jun, AP-1, and ATF. Together, these transcription factors induce the expression of a battery of pro-inflammatory genes, including TNF, IL-6, and IL-1β.

We return here to the TLR4-mediated signaling pathway that leads to IFNβ production. We note in Box 17-2 that LPS binding to TLR4 can lead to recruitment of MyD88 and the IRAK signaling pathway. However, there is also evidence that LPS binding to TLR4 can trigger a signaling pathway that is independent of MyD88. This pathway involves the replacement of MyD88 with an adaptor called *TRIF* and the phosphorylation of the transcription factor *IRF3* by a kinase called *TNK1*. Activated IRF3 dimerizes, translocates to the nucleus, and combines with two coactivators to initiate transcription of type I IFNs, particularly IFNβ. The interaction of TRAF6 with TRIF and the binding of NF-κB to the promoter are also required for complete activation of the IFNβ gene. The IFNβ molecules produced as a result of this activation then drive expression of various IFN-inducible genes. Thus, rather than pro-inflammatory cytokine expression, the MyD88-independent path leads to the expression of IFNβ and IFN-inducible proteins. To date, the MyD88-independent pathway has not been associated with IL-1RI signaling.

IL-1 receptor antagonist. IL-1 is one of the very few cytokines that has a natural competitive inhibitor. IL-1Ra, for "IL-1 receptor antagonist," is produced primarily by mononuclear phagocytes (but also by other cell types) in response to bacterial or viral products and other stimulants. IL-1Ra is therefore considered one of the acute phase proteins induced early during the inflammatory response. *In vitro*, the best inducers of IL-1Ra synthesis are LPS, IL-4, and GM-CSF. IL-1Ra is structurally similar to IL-1 but contains a secretory peptide and so is readily exported out of cells. IL-1Ra binds to both IL-1RI and IL-1RII but does not initiate signal transduction, possibly because it fails to recruit MyD88. Since the amount of IL-1Ra present *in vivo* is 100-fold less than the amount required to achieve total abolition of IL-1 activity *in vitro*, the role of IL-1Ra appears to be the modulation of IL-1 signaling. *In vivo*, studies of IL-1Ra$^{-/-}$ mice showed that IL-1Ra is crucial for the prevention of endotoxic shock. In humans, endogenous IL-1Ra exerts an important anti-inflammatory effect in cases of colitis and arthritis, among other inflammatory disorders. Commercial synthesis and clinical use of IL-1Ra for these disorders are now under way.

ii) IL-2

Function. IL-2 and its receptor were introduced in Chapter 14 in the context of T cell activation. IL-2 stimulates the proliferation of activated CD4$^+$, CD8$^+$, and γδ T cells, promotes the differentiation of CTLs, and functions as a T cell chemoattractant. In addition to its effects on T cells, IL-2 enhances monocyte responses and B cell activation, differentiation, and proliferation (Fig. 17-8). IL-2 also induces the proliferation of NK cells, synergizes with IL-12 to induce IFNγ and TNF production by NK cells, and induces NK cell differentiation *in vitro* into the aggressive, tumor-killing LAK cells.

As elaborated in Chapter 16, studies of IL-2 and IL-2R knockout mice have shown that, unlike other cytokines, IL-2

Figure 17-8

Major Functions of IL-2
This Th1 cytokine has vital roles in promoting the proliferation of lymphocytes, sustaining the generation of T_{reg} cells and inducing the AICD required for peripheral tolerance. Dotted arrow indicates an *in vitro* observation.

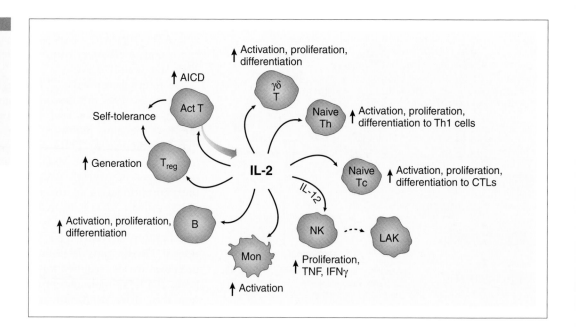

is crucial for the induction of self-tolerance and the maintenance of homeostasis. IL-2$^{-/-}$ and IL-2R$^{-/-}$ animals all have striking autoimmune disorders characterized by spontaneous activation of T cells leading to hemolytic anemia and chronic inflammatory bowel disease. The reader will recall that, in normal mice, activated T cells are controlled toward the end of an immune response both by the actions of regulatory T cells and by AICD resulting from Fas- and/or TNF-mediated apoptosis. In IL-2$^{-/-}$ animals, the generation of T_{reg} cells is compromised, as is Fas-mediated AICD of activated T cells. IL-2 is required for the downregulation of expression of the anti-apoptotic regulator FLIP mentioned in Chapter 15.

Production. The vast majority of IL-2 in the body is produced by activated T cells after binding of the TCR to antigen and the delivery of costimulatory signals via CD28/B7 interaction. IL-2 expression is controlled at the transcriptional level and depends largely on the stability of the IL-2 mRNA. As we learned in Chapter 14, one of the primary functions of CD28 costimulation is to stabilize IL-2 mRNA.

Receptor. The IL-2 high-affinity receptor is composed of the private IL-2Rα chain and the public IL-2Rβ and IL-2Rγ (γc) chains. In humans, a low-affinity receptor consisting only of IL-2Rβ and IL-2Rγ exists that can transduce low levels of signaling in response to IL-2 binding. However, in mice, the IL-2Rα chain is absolutely required for IL-2 signaling.

IL-2 receptor α chain. IL-2Rα (CD25), also known as TAC (T cell activation antigen), is a 55 kDa receptor subunit highly expressed on activated T, B, and NK cells, and to a lesser extent on monocytes and thymocytes. Structurally, IL-2Rα has two extracellular domains (but no WSXWS motif), a transmembrane domain, and a short cytoplasmic tail (refer to Fig. 17-2). Unlike the IL-2Rβ and IL-2Rγ chains, the IL-2Rα chain is not constitutively produced. It is synthesized by T, B, or NK cells only after induction by cellular activation. By itself, IL-2Rα has only a very low affinity ($K = 10^8\ M^{-1}$) for IL-2 and cannot trigger cellular responses because its short

tail is not capable of signal transduction. IL-2Rα combines with the IL-2Rβ and IL-2Rγ receptor chains to form a complex that binds to IL-2 with high affinity ($K = 10^{11}\ M^{-1}$), and utilizes the long cytoplasmic tails of IL-2Rβ and IL-2Rγ for intracellular signaling. Like mice deficient in IL-2 expression, IL-2R$\alpha^{-/-}$ mice exhibit relatively normal T and B cell development but display an increased incidence of certain autoimmune diseases. IL-2Rα is also expressed on thymocytes and is used as a marker (CD25$^+$) during T cell development but its function here is unknown; the thymi of IL-2R$\alpha^{-/-}$ mice are apparently normal.

IL-2 receptor β chain. IL-2Rβ (CD122) is a 75 kDa surface protein containing a WSXWS motif, categorizing it as a member of the type I cytokine receptor family (refer to Fig. 17-1). In humans, IL-2Rβ is expressed constitutively on most mature T lymphocytes and combines with IL-2Rγ to form a signaling receptor of moderate affinity ($K = 10^9\ M^{-1}$). The role of IL-2Rβ in the IL-2 signaling pathway is not well understood. It has been shown that an "acid-rich" domain of the IL-2Rβ cytoplasmic tail binds the adaptor protein SHC (refer to Box 14-2), and that Jak and Lck kinases can associate with this chain. IL-2R$\beta^{-/-}$ knockout mice exhibit an autoimmune reaction characterized by generalized T cell activation and proliferation, and production of autoantibodies. Lymphocyte homeostasis is disrupted and the animals die prematurely. Interestingly, unlike IL-2$^{-/-}$ and IL-2R$\alpha^{-/-}$ mice, Fas-mediated apoptosis appears to be normal in IL-2R$\beta^{-/-}$ mice, implying that the abnormal lymphocyte homeostasis in these animals is due primarily to the deficit in T_{reg} cells rather than a lack of AICD. Whether this difference has anything to do with the fact that the IL-2Rβ chain is also part of the IL-15 receptor (see later) remains to be determined. Engagement of the IL-2 and IL-15 receptors has been shown to result in a number of overlapping biological effects.

IL-2 receptor γ chain. IL-2Rγ (CD132) is a 40 kDa receptor protein expressed on T, B, and NK cells. As mentioned previously, IL-2Rγ (or γc) is also a subunit of IL-4R, IL-7R, IL-9R, and IL-15R (refer to Fig. 17-2). Like IL-2Rβ, IL-2Rγ possesses a WSXWS sequence, and Jak kinases associate with its cytoplasmic tail. The binding of IL-2 to the high-affinity heterotrimeric IL-2R causes its chains to aggregate, bringing the Jak kinases associated with both IL-2Rβ and IL-2Rγ into close proximity. Subsequent phosphorylation of Jak1 and Jak3 results in the activation, dimerization, and nuclear translocation of STAT3 and STAT5. The IL-2Rγ protein also contains a Src homology (SH) domain in its cytoplasmic tail but its significance in IL-2R signaling is not known.

Natural mutations of the IL-2Rγ gene in humans result in a form of severe combined immunodeficiency (called XSCID; see Ch.24) because the affected lymphocytes can no longer receive many essential cytokine signals. Knockout mice in which the IL-2Rγ gene is disrupted exhibit immune defects similar to those found in these human patients. In IL-2Rγ$^{-/-}$ mice, the thymus is hypoplastic and lymphocyte development is severely perturbed. No NK or γδ T cells can be detected in these animals, and the numbers of αβ T cells and B cells are profoundly decreased. However, these αβ T cells are abnormally activated and provoke autoimmune reactions. The development of secondary lymphoid tissues is also impaired in that peripheral lymph nodes and the gut-associated lymphoid tissues, including the Peyer's patches, fail to develop.

iii) IL-3

Function. IL-3 was originally thought to be crucial for stimulating the growth of multiple hematopoietic progenitors. A large number of *in vitro* experiments demonstrated that IL-3 promoted the growth of the earliest hematopoietic progenitor cells and their expansion into all mature cell types. However, the IL-3$^{-/-}$ knockout mouse has surprisingly few abnormalities, indicating that other cytokines can compensate for an absence of IL-3 during early hematopoiesis. *In vivo*, IL-3 does appear to

be an essential growth factor for mast cells and important for basophil and mast cell responses during parasite infections (Fig. 17-9). IL-3 regulates the 5-lipoxygenase/ leukotriene pathway in mast cells and monocytes. IL-3 also has a role in the late inflammatory stages of allergic reactions since it appears to activate and promote recruitment of mature basophils. IL-3 has been shown to upregulate Bcl-2 expression, promoting cell survival, and can induce expression of cell cycle regulators.

Production. IL-3 is produced primarily by activated CD4$^+$ T cells. Mast cells whose Fcε receptors are bound by IgE can also secrete IL-3.

Receptor. The IL-3 receptor is expressed on early hematopoietic cells, most myeloid lineages, and some B cells (but not T cells). IL-3R is composed of a private IL-3Rα chain that binds IL-3 and the non-binding βc chain shared with the receptors for IL-5 and GM-CSF (refer to Fig. 17-2). The expression of βc is upregulated by the inflammatory cytokines IL-1, TNF, and IFNγ. Binding of IL-3 to the IL-3Rα chain of the receptor induces the phosphorylation of the βc chain, recruitment of SH2-containing adaptor molecules such as Vav1, and downstream signal transduction via Jak2/STAT5 and the Ras/ MAPK pathway. In mice (but not humans), there is a minor alternative signaling chain called β$_{IL-3}$ that can bind IL-3.

iv) IL-4

Function. IL-4 is one of the most powerful and pleiotropic cytokines in the body. Its actions are generally antagonistic to those of IFNγ. Because IL-4R is widely expressed, IL-4 influences almost all cell types. In T cells, IL-4 is crucial for the differentiation and growth of the Th2 subset (Fig. 17-10). As such, IL-4 promotes the establishment of the humoral response necessary to combat pathogens that live and reproduce extracellularly. In B cells, IL-4 stimulates growth and differentiation and induces upregulation of MHC class II and FcεRII (CD23). IL-4 also promotes isotype switching in murine B cells to IgG1

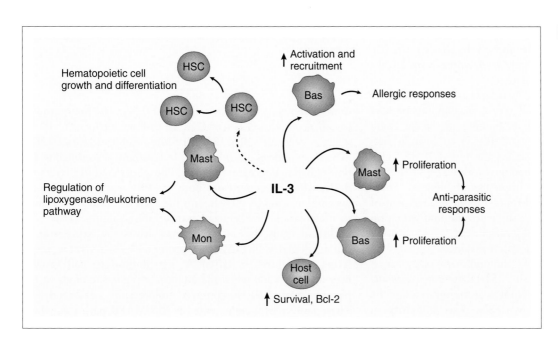

Figure 17-9

Major Functions of IL-3
IL-3's most important role is as a growth factor for mast cells and basophils.

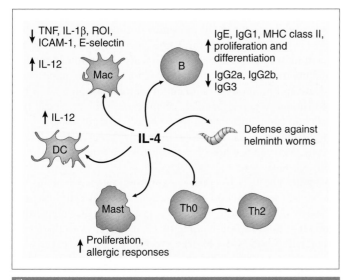

Figure 17-10

Major Functions of IL-4
IL-4 is crucial for Th2 differentiation and Th2 responses. IL-4 also blocks many macrophage effector functions.

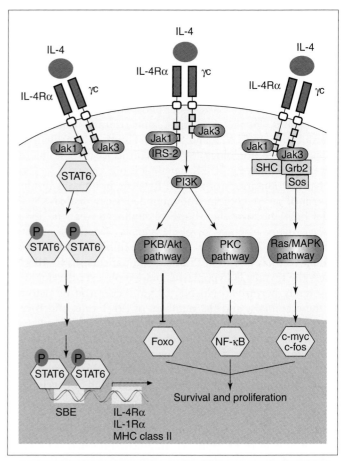

Figure 17-11

IL-4 Receptor Signaling
Engagement of the type I IL-4R complex by IL-4 triggers three separate intracellular signaling pathways, all of which are mediated via Jak1 and Jak3 activation. Immunoregulation occurs when activated STAT6 binds to SBE motifs in the promoters of IL-4-inducible genes. Survival and proliferation are promoted by the two other pathways. Recruitment of IRS-2 to Jak1 results in activation of the PKB/Akt and PKC pathways, which lead to inhibition of Foxo and activation of NF-κB, respectively. The third pathway is triggered when Jak3 recruits the adaptor protein complex Shc–Grb2–Sos and activates the Ras/MAPK pathway. This pathway promotes activation of c-myc and c-fos.

and IgE but inhibits switching to IgG2a, IgG2b, and IgG3. IL-4 is a growth factor for mast cells and plays a major regulatory role in allergic responses since these involve IgE-mediated mast cell degranulation. IL-4 is also important for defense against helminth worms because the IgE production promoted by IL-4 allows eosinophils bearing FcεRIIB to carry out efficient ADCC. In macrophages, IL-4 inhibits the secretion of pro-inflammatory chemokines and cytokines such as TNF and IL-1β, impairs the ability of these cells to produce reactive oxygen and nitrogen intermediates, and blocks IFNγ-induced expression of cellular adhesion molecules such as ICAM and E-selectin. However, IL-4 can also induce DCs and macrophages to upregulate their synthesis of IL-12, supplying a negative feedback mechanism to regulate the Th2 response.

Production. IL-4 is secreted primarily by Th2 CD4+ T cells but also by basophils, mast cells, and NKT cells (see Ch.18). Cross-linking of CD40L upregulates IL-4 synthesis by T cells, while IFNγ and IL-12 downregulate it.

Receptor. The high-affinity IL-4 receptor is expressed in hematopoietic cells. In these cell types, the binding of IL-4 to the IL-4Rα chain results in its binding to the γc chain (shared with IL-2, IL-7, IL-9, and IL-15) to form a heterodimeric receptor sometimes called the type I IL-4R. This receptor has no intrinsic kinase activity although the IL-4Rα chain becomes phosphorylated upon ligand binding. Three signaling pathways are triggered following the engagement of the type I IL-4R by IL-4, one leading to immunoregulation and two leading to cellular proliferation (Fig. 17-11). In the immunoregulatory pathway, IL-4R transduces signals via Jak1, Jak3, and STAT6. STAT6 homodimers recognize a GAS-like sequence called the SBE (*STAT binding element*) that differs by one nucleotide from that recognized by the STAT1 homodimers activated by IFNγ. This similarity in recognized DNA motifs may contribute to the observed

mutual antagonism between IL-4 and IFNγ. The first pathway to cell proliferation is triggered by IL-4 engagement of IL-4R when a molecule called IRS-2 (*insulin receptor substrate-2*) or 4PS (*IL-4-induced phosphotyrosine substrate*) is recruited to the phosphorylated receptor complex. Phosphorylated IRS-2 recruits PI3K, which then activates both the PKC pathway culminating in activation of NF-κB, and the PKB/Akt pathway leading to inactivation of Forkhead transcription factors. (Foxo is a Forkhead transcription factor that drives the expression of cell death genes. PKB/Akt thus promotes cell survival by interfering with Foxo activity.) A contribution to cell growth may also be made when the adaptor protein Shc binds to the phosphorylated IL-4R complex and recruits Grb2 and Sos. These molecules then activate the Ras/MAPK pathway, which may in turn eventually activate c-myc and c-fos transcription.

A low-affinity IL-4 receptor (also called the type II IL-4R) is expressed by non-hematopoietic cells. Upon the binding of IL-4 to the IL-4Rα chain in these cell types, the IL-13Rα chain (rather than the γc chain) is recruited to form the heterodimer. Curiously, this receptor is structurally identical to the IL-13R (refer to Fig. 17-2) whose assembly is induced by the binding of IL-13 to a private chain called IL-13Rα (see later). Evidence from patients lacking γc suggests that IL-4 may be able to exert some of its effects through either the type II IL-4R or IL-13R.

v) IL-5

Function. IL-5 is a homodimeric cytokine that functions principally in the eosinophil arm of the Th2 response, promoting the survival, differentiation, and chemotaxis of these cells (Fig. 17-12). In the presence of IL-5, mature eosinophils are activated such that they become competent to kill helminth worms by degranulation. IL-5 is not involved in IgE production but can act on mast cells to promote the histamine release seen in allergies. IL-5 can also upregulate IL-2Rα on B cells, enhance IgA production, and stimulate the growth and differentiation of B cells and CTLs. However, the importance of IL-5 for these functions *in vivo* is less clear since IL-5$^{-/-}$ mice show no impairment of B or T cell responses. IL-5 is becoming a target of pharmacological interest because of the involvement of eosinophils in Th2-mediated allergic and asthmatic responses (see Ch.28). In one mouse model of asthma, the lung damage normally mediated by IL-5-induced degranulation of eosinophils was significantly decreased in IL-5$^{-/-}$ mice. Since IL-5 appears to be a less pleiotropic cytokine than most, it is conceivable that a drug targeting IL-5 might successfully mitigate asthma attacks caused by eosinophilia without unduly compromising other aspects of host health.

Production. IL-5 is produced primarily by activated Th2 cells, with some contribution by activated mast cells, NK cells, B cells, and eosinophils.

Receptor. The IL-5R is composed of a unique IL-5Rα chain which binds IL-5 and the non-binding βc chain also shared by the receptors for GM-CSF and IL-3. Signal transduction is carried out via the Jak2/STAT5 pathway. IL-5R is expressed predominantly on mast cells, basophils, and eosinophils.

vi) IL-6

Function. IL-6 is the quintessential multi-functional cytokine. It plays roles in the adaptive immune response, inflammation, hematopoiesis, the nervous system, and the endocrine system. However, studies of IL-6$^{-/-}$ mice have shown that, while IL-6 contributes to numerous cellular events, it is crucial for the acute phase response, mucosal production of IgA, and the fever response during inflammation (Fig. 17-13). In particular, IL-6 induces the synthesis of acute phase proteins (particularly fibrinogen) by hepatocytes and stimulates the pathogen clearance functions of neutrophils. IL-6 also appears to have a major influence on the end stages of B cell differentiation. Overproduction of IL-6 in transgenic mice results in the development of plasmacytomas (cancerous plasma cells), while IL-6-deficient mice fail to develop plasmacytomas in response to a chemical inducer. *In vitro*, IL-6 supports the differentiation and/or maturation of B cells, T cells, macrophages, megakaryocytes, certain neurons, and osteoclasts. It can also act as a minor growth factor for numerous other cell types, both normal and transformed. Excessive IL-6 has been associated with estrogen-induced bone loss.

Production. IL-6 is produced primarily by mononuclear phagocytes, fibroblasts, and vascular endothelial cells in response to IL-1 or TNF. Some activated T cells can also secrete IL-6.

Receptor. The IL-6R consists of a unique IL-6Rα chain coupled to gp130 as the signal transducing subunit. Both IL-6Rα and gp130 contain the WSXWS motif typical of a type I

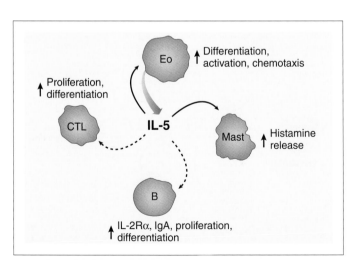

Figure 17-12

Major Functions of IL-5
This Th2 cytokine is particularly important for eosinophil differentiation, activation, and chemotaxis.

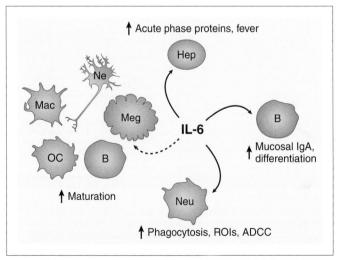

Figure 17-13

Major Functions of IL-6
IL-6 is a supremely multi-functional pro-inflammatory cytokine that also has regulatory roles in the immune system and the neuroendocrine system. In particular, IL-6 is crucial for the acute phase response.

cytokine receptor and an Ig-like extracellular domain, but only the IL-6Rα chain is capable of binding IL-6. The gp130 chain, while not itself intrinsically catalytic, mediates signal transduction because each of Jak1, Jak2, and Tyk2 can associate constitutively with the intracellular Box1/Box2 motif of gp130 (Fig. 17-14). These kinases in turn recruit and activate various STATs (including STAT1, STAT3, STAT5), which translocate to the nucleus of the cell and initiate transcription of the genes encoding additional, more specific transcription factors such as NF-IL6 (*nuclear factor binding to the IL-6 gene*; also known as C/EBPβ). NF-IL6 binds to type I IL-6 response elements (IL-6RE) in the promoters of several genes encoding acute phase response proteins, including the IL-6 gene itself, the haptoglobin gene, and the C-reactive protein gene. A type II IL-6RE serves as a binding site for STAT3 in

the promoters of several other IL-6-inducible genes, including fibrinogen and α₂-macroglobulin. The phosphorylation of different tyrosine residues in the gp130 intracellular tail (and thus the recruitment of different STAT molecules) is associated with the various effects attributed to IL-6, including induction of the acute phase response, growth arrest, anti-apoptosis, and cell proliferation. Engagement of IL-6R can also recruit Shc–Grb2–Sos and thus stimulate the Ras/MAPK pathway. IL-6R is most highly expressed on activated B cells but also on peripheral blood monocytes, mature T cells, neutrophils, and hepatocytes.

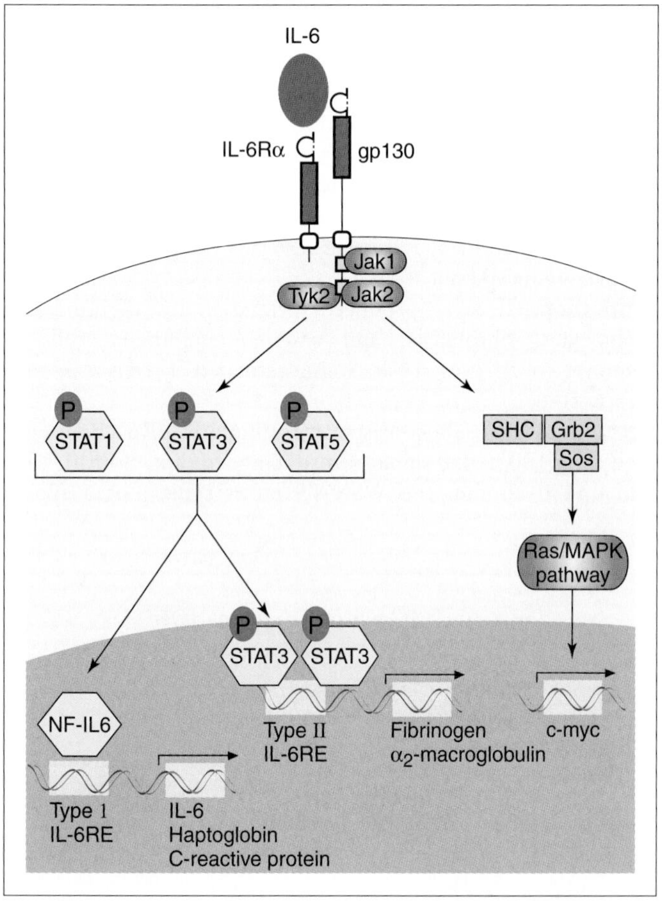

vii) IL-7

Function. As has been discussed in earlier chapters, IL-7 is the only cytokine essential for lymphopoiesis in mice (Fig. 17-15). It acts on the earliest hematopoietic progenitor cells in the bone marrow and on T cell precursors in the thymus. In IL-7-deficient mice, lymphocyte development is blocked at the pro-T cell and pro-B cell stages, and c-kit expression is decreased. IL-7 may also induce the expression of other important cytokine receptors in αβ pre-T cells. IL-7 is particularly critical for the development of both γδ pre-T cells in the thymus and intestinal intraepithelial γδ T cells (iIELs; see Ch.18). Both these cellular subsets are missing in IL-7⁻/⁻ mice, indicating that IL-7 is essential for both the thymic and extrathymic development of γδ T cells. Other cytokines apparently can substitute for IL-7 for the development of αβ iIELs. *In vitro*, IL-7 acts like IL-2 in that it supports the growth and differentiation of both pre-B and pre-T cells. More specifically, IL-7 may enhance the survival of lymphocyte precursors by activating the PI3K survival pathway and by inducing expression of anti-apoptotic proteins such as

Figure 17-14

IL-6 Receptor Signaling

In the major signaling pathway, IL-6 engagement of IL-6R activates the Jak kinases associated with the cytoplasmic tail of gp130. Multiple STATs are activated, leading to the formation of various homo- and heterodimeric combinations. These STATs induce the activation of other transcription factors such as NF-IL6. NF-IL6 binds to type I IL-6RE motifs in the promoters of many acute phase proteins. STAT3 homodimers bind to type II IL-6RE motifs present in the promoters of other IL-6-inducible genes. Engagement of IL-6R can also induce recruitment of the Shc–Grb2–Sos adaptor protein complex and activation of the Ras/MAPK pathway and c-myc transcription.

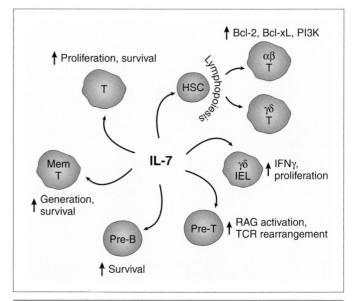

Figure 17-15

Major Functions of IL-7

IL-7 is critical for γδ T cell development and important for αβ T cell and B cell development. IL-7 also maintains memory T cells.

Bcl-2 and Bcl-xL. IL-7 promotes TCR gene rearrangement *in vitro* by sustaining the expression of the RAG genes, and may regulate the accessibility of these loci to transcription factors. IL-7 also stimulates both the expansion of lymphocyte clones (especially γδ iIELs) that infiltrate tumors and the expression of IFNγ by these cells. Lastly, IL-7 contributes to the generation and maintenance of memory T cells.

Production. IL-7 is produced principally by stromal cells in the bone marrow and thymus, as well as by B cells, monocytes, macrophages, FDCs, keratinocytes, and iIELs.

Receptor. The IL-7 receptor consists of a unique IL-7 binding, non-signaling IL-7Rα chain and the signal transducing γc chain. The IL-7Rα chain is a member of the type I cytokine receptor family. Signaling is transduced by Jak1, Jak3, and STAT5. IL-7R is expressed on pro-B and pre-B cells, and on c-kit$^+$ CD44$^+$CD25$^+$ TN pro-T cells. Data from studies of IL-7Rα$^{-/-}$ mice suggested that, in certain cell types, other cytokines may be able to make use of the IL-7R, or that other receptors can substitute for IL-7R. For example, IL-7Rα$^{-/-}$ mice show a complete absence of γδ T cells and a block in B cell development even earlier than that observed in IL-7$^{-/-}$ mice, suggesting that the IL-7Rα chain must be involved in critical interactions that are independent of IL-7. Indeed, it turns out that the IL-7Rα chain also participates in the receptor for a hematopoietic growth factor called *thymic stromal lymphopoietin*. αβ T cells and NK cells are present in greatly reduced numbers in IL-7Rα$^{-/-}$ mice but these few are able to develop normally. Consistent with these observations, IL-7R expression on thymocytes was found to be required for the burst of proliferation that occurs prior to TCR gene rearrangement.

viii) IL-8

Function. As mentioned previously, IL-8 is a CXC chemokine rather than an interleukin. This molecule is a powerful chemoattractant for neutrophils and promotes inflammatory responses dominated by them (Fig. 17-16). IL-8 also upregulates adhesion molecules on neutrophils and induces their degranulation and the respiratory burst. *In vitro* (but possibly not *in vivo*), IL-8 also stimulates the activation of endothelial cells, and the activation and mobility of T cells, eosinophils, basophils, and monocytes. Chemokines are discussed in detail later in this chapter.

Production. IL-8 is not expressed in resting cells but is induced in essentially all cells which encounter TNF, IL-1, or bacterial endotoxin. Such a broad distribution ensures that no matter where tissue injury or infection occurs, cells in the area can produce IL-8 and summon neutrophils. Activated mononuclear phagocytes, fibroblasts, and endothelial cells are particularly good sources of IL-8. IL-8 made by megakaryocytes is stored in platelets.

Receptor. IL-8 binds to two distinct receptors expressed most highly on the surfaces of neutrophils and T cells, and to a lesser extent on basophils and myeloid lineage cells. As mentioned previously, the IL-8 receptors were recently renamed to reflect their true function as chemokine receptors. Thus, IL-8R1 (known formerly as IL-8Rα or IL-8RA) is now called CXCR1, while IL-8R2 (IL-8Rβ or IL-8RB) is called CXCR2. CXCR2 is

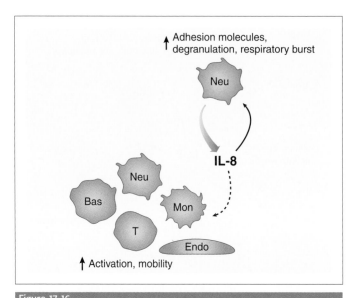

↑ Adhesion molecules, degranulation, respiratory burst

IL-8

↑ Activation, mobility

Figure 17-16

Major Functions of IL-8
This CXC chemokine has powerful effects on neutrophil chemotaxis and function.

able to bind to at least four other chemokines in addition to IL-8, whereas CXCR1 appears to bind only to IL-8 and one other chemokine called GCP-2. CXCR1 and CXCR2 are both G protein-coupled receptors whose α-helices loop back and forth across the membrane seven times. Although CXCR1 and CXCR2 can interact with many different types of G proteins, the binding of IL-8 to either of these receptors induces coupling with the Gi2α G protein only. The receptors phosphorylate the GDP on an inactive Gi2α protein and convert it to GTP, activating the Gi2α molecule. Downstream signaling from this point depends on which receptor IL-8 has bound to: stimulation of either CXCR1 or CXCR2 induces changes in calcium flux and degranulation, but only binding to CXCR1 leads to the activation of the intracellular messenger phospholipase D and the respiratory burst. The signaling intermediaries in these divergent paths have not been identified. As expected, mice deficient for either CXCR1 or CXCR2 show major defects in neutrophil migration.

ix) IL-9

Function. IL-9 has pleiotropic effects on the immune response, hematopoiesis, and tumorigenesis (Fig. 17-17). Its major biological activity appears to be the stimulation of growth and differentiation, particularly of hematopoietic cells. IL-9 promotes the differentiation of erythroid, myeloid, and neuronal precursors, and the proliferation and differentiation of mast cells. Cultured mast cells treated with IL-9 upregulate their expression of IL-6, FcεRα, and granzyme B and other proteases. Although IL-9 on its own does not stimulate Ig production by B cells, B cells treated with a combination of IL-9 and IL-4 show synergistically enhanced production of IgE and IgG1. Mice that are able to resist infection with helminth parasites show elevated levels of IL-9 in mesenteric

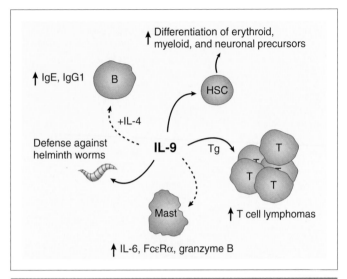

Major Functions of IL-9
IL-9 serves chiefly as a growth factor for hematopoietic cells. In transgenic (Tg) mice, overexpression of IL-9 has been associated with T cell lymphoma development.

lymph nodes. IL-9 has also been shown to stave off apoptosis induced by dexamethasone.

IL-9 transgenic mice spontaneously develop lymphomas and are very susceptible to carcinogens such as γ-irradiation and chemical mutagens. IL-9 is structurally related to IL-2, IL-4, IL-7, IL-13, and IL-15, all of which have profound effects on T cell growth. However, freshly isolated normal T cells do not respond to IL-9, making its influence on normal T cells *in vivo* uncertain.

Production. IL-9 is produced almost exclusively in the later stages of activation of Th2 cells that have encountered IL-2, IL-4, and IL-10. IL-2 is crucial for the induction of IL-4, which in turn works with IL-10 to induce IL-9. Accordingly,

IL-9 secretion is drastically reduced in IL-2$^{-/-}$ mice. Mitogen treatment of T cell populations stimulates production of IL-9 by CD4$^+$ T cells, while CD3/CD28 cross-linking induces IL-9 secretion specifically in memory CD4$^+$ T cells.

Receptor. The IL-9 receptor is composed of a private IL-9-binding IL-9Rα chain (which is a member of the type I cytokine receptor family) and the signal transducing γc chain. The presence of the γc chain means that IL-9R shares downstream transduction elements with IL-2R, IL-4R, IL-7R, and IL-15R. Stimulation of IL-9R by IL-9 catalyzes the phosphorylation and activation of Jak1 (recruited to IL-9Rα) and Jak3 (associated with γc). These kinases phosphorylate IL-9Rα chains at a single tyrosine residue that serves as a docking site for STAT1, STAT3, or STAT5.

x) IL-10

Function. IL-10 has roles in both innate and adaptive immunity that are manifested as either immunosuppressive or immunostimulatory effects on various cell types (Fig. 17-18). Although IL-10 structurally resembles IFNγ, these two cytokines have opposing biological activities. The primary role of IL-10 is to act as a brake during inflammatory responses, targeting principally macrophages, neutrophils, eosinophils, and mast cells. Accordingly, IL-10$^{-/-}$ mice die of chronic inflammatory bowel disease characterized by damaging accumulations of hyperactive CD4$^+$ T cells and macrophages. IL-10 carries out its immunosuppressive role largely by inhibiting NF-κB-activated transcription of genes encoding the pro-inflammatory cytokines, particularly TNF, IL-1, IL-6, IL-8, and IL-12. There is also evidence that IL-10 destabilizes cytokine mRNAs, including the IL-10 mRNA itself. In addition, IL-10 can inhibit the activation of the Ras/MAPK signaling pathway and the tyrosine phosphorylation of Vav1, again downregulating transcriptional activity. Through its regulation of cytokine secretion, IL-10 is important for protection against the massive cytokine release associated with endotoxic shock. IL-10 also inhibits the

Major Functions of IL-10
This Th2 cytokine has powerful immunosuppressive effects on many hematopoietic cell types and acts as a brake on inflammatory responses.

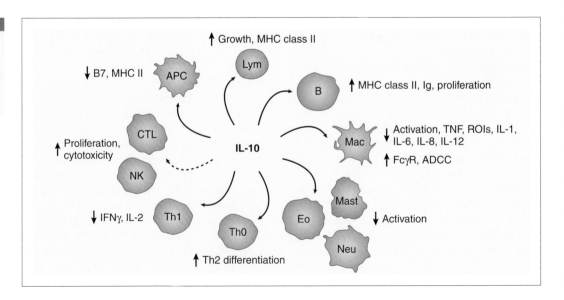

respiratory burst and nitric oxide-dependent killing of microbes by macrophages.

As we saw in Chapter 15, IL-10 promotes a Th2 response by inhibiting the secretion of IFNγ and IL-2 by Th1 cells. In addition, IL-10 downregulates the expression of MHC class II, B7-1, and B7-2 on APCs, thereby further damping down the CD4$^+$ T cell response. There is also evidence that IL-10 may directly inhibit T cells by interfering with TCR signaling. However, *in vitro*, IL-10 has been shown to support the proliferation and cytotoxic activity of CD8$^+$ T cells and to stimulate NK cells. IL-10 also promotes the growth of thymocytes and mature T and B cells in culture, and upregulates the expression of MHC class II and Igs in the latter. Fcγ receptor expression is upregulated on macrophages encountering IL-10, resulting in enhanced ADCC.

Production. IL-10 production is induced not only by invasion of pathogens such as intracellular parasites and fungi but also by the presence of inflammatory cytokines, such as TNF, IL-6, IL-12, and type I IFNs. IL-10 is secreted primarily by activated monocytes, macrophages, and Th2 cells but also by B cells, eosinophils, mast cells, keratinocytes, hepatocytes, and several other cell types. Cellular injury or stress in the form of hypoxia or UV irradiation can also induce IL-10 production. As mentioned earlier, IL-10 self-regulates its synthesis by destabilizing its own mRNA.

Receptor. The heterodimeric IL-10 receptor is expressed mainly on hematopoietic cells. The IL-10R1 chain has an extracellular domain that resembles that in the IFN receptors and a long cytoplasmic domain that associates primarily with Jak1. The IL-10R2 chain has a shorter cytoplasmic domain that associates primarily with Tyk2. Engagement of IL-10R induces phosphorylation of STAT1, STAT3, and STAT5. The cytoplasmic region of an IL-10R1 chain contains four phosphotyrosine residues, each of which appears to be associated with a different signaling function. The most membrane-proximal phosphotyrosine appears to mediate negative regulation, while the membrane-distal phosphotyrosines are associated with the promotion of differentiation and proliferation. Various Jak/STAT pathway proteins are differentially

phosphorylated in response to IL-10R engagement in different cell types, partially accounting for the wide-ranging biological effects of IL-10.

xi) IL-11

Function. IL-11 is a cytokine related to IL-6 that acts primarily to stimulate the growth of hematopoietic cells, with a major effect on the production of megakaryocytes (Fig. 17-19). IL-11 has milder effects on erythropoiesis and myelopoiesis but may contribute to the differentiation of stem cells into committed precursors. IL-11 stimulates the growth of T and B cells and enhances the recovery of neutrophil production after bone marrow transplantation or chemotherapy. IL-11 also exerts anti-inflammatory effects by inhibiting TNF, IL-1, IL-6, IL-12, and NO production by macrophages, and the production of Th1 cytokines, particularly IFNγ, by Th1 cells. IL-11 has been reported to influence non-hematopoietic cells in a variety of ways. This cytokine is an important stimulatory factor for fibroblasts in the connective tissues and can promote collagen deposition. IL-11 may therefore be involved in the airway "remodeling" that occurs in patients with severe asthma (see Ch.28). IL-11 also induces acute phase proteins in hepatocytes *in vitro*, but it is not clear that these effects are relevant in the whole animal. Studies of mice deficient for the ligand binding subunit of the IL-11R have demonstrated that IL-11 is completely redundant (most likely with IL-6) for at least its hematopoietic functions. IL-11 has also been implicated in bone resorption.

Production. IL-11, although apparently widely expressed, is rarely detected in the blood. IL-11 is produced primarily by bone marrow stromal cells but also by osteoblasts and cells in the brain, joints, and testes. IL-11 mRNA has also been detected in kidney, intestine, heart, and lung as well as in thymus and spleen.

Receptor. The IL-11 receptor is composed of a unique IL-11-binding IL-11Rα chain (containing the WSXWS motif) and the gp130 chain shared with the IL-6R. STAT1 and STAT3 are used for downstream signal transduction. A soluble form of the IL-11Rα chain exists but its function is unclear.

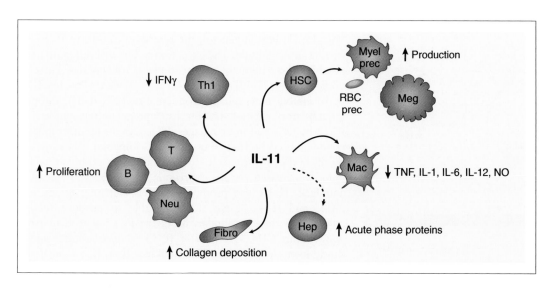

Figure 17-19

Major Functions of IL-11
This IL-6-related cytokine promotes the growth of many hematopoietic cell types but is most important for megakaryocyte production.

xii) IL-12

Function. IL-12 is a cytokine with a central role in the immune response because it links the activation of macrophages by bacterial invaders to activation of both NK (innate response) and Th1 (adaptive response) effector functions (Fig. 17-20). In particular, IL-12 is crucial for the generation of IFNγ required for host defense against a broad range of intracellular pathogens, including many species of intracellularly-replicating bacteria. As we learned in Chapter 15, IL-12 (1) promotes the differentiation of naive Th0 cells (and, in humans but not mice, memory CD4$^+$ T cells) into IFNγ-secreting Th1 cells; (2) directly promotes or synergizes with IL-18 to induce IFNγ secretion by activated Th1 cells and NK cells; (3) enhances the cytolytic capacities of activated (but not resting) CD8$^+$ CTLs and NK cells; and (4) enhances the production of antibody isotypes associated with Th1 responses, such as IgG2a, IgG2b, and IgG3, while suppressing that of Th2-associated IgG1 and IgE. In addition, IL-12 induces macrophage and DC secretion of low concentrations of TNF, GM-CSF, IL-8, and IL-10 but does not apparently regulate IL-2 production. IL-12 may contribute to the pathology of Th1-mediated disorders such as Crohn's disease, type 1 diabetes mellitus, and multiple sclerosis (see Ch.29), since patients with these diseases exhibit elevated serum IL-12. Clinicians are also investigating the treatment of Th2-mediated diseases by administration of IL-12, which has the effect of suppressing the development of the Th2 response.

Production. IL-12 has an unusual structure for a cytokine in that it is a heterodimer rather than a monomer or homodimer. IL-12 is composed of disulfide-linked p35 and p40 subunits that are unrelated in structure. p35 is homologous to regions of IL-6 and G-CSF, while p40 has the structure of a hematopoietin receptor molecule (although no membrane-bound form of IL-12 or p40 has yet been identified). IL-12 is produced primarily by activated macrophages and with some contribution by neutrophils, DCs, monocytes, and B cells. While the p35 subunit is constitutively expressed in most cell types, expression of both the p40 and p35 subunits is upregulated in APCs such as B cells, monocytes, macrophages, and DCs in response to bacterial infection.

CD40 plays a major role in the induction of IL-12 synthesis, at least *in vitro*. Engagement of CD40 on macrophages and DCs by CD40L on activated T cells results in the production of high levels of IL-12 by the APCs. This production is controlled positively by IFNγ, which upregulates CD40 expression on APCs, and negatively by IL-10, which downregulates CD40. Interestingly, while one might expect the Th2 cytokines IL-4 and IL-13 to inhibit the production of IL-12, this is not always the case. There is some evidence suggesting that, in the absence of an ongoing Th1 response or a strong Th1 inducer such as LPS, IL-4 and IL-13 can actually enhance IL-12 production triggered by CD40 engagement.

Receptor. The high-affinity IL-12R is expressed primarily by activated T and NK cells but also by DCs and B cells. IL-12R is a dimeric molecule composed of two very similar subunits, IL-12Rβ1 and IL-12Rβ2. Both these chains resemble gp130 and two copies of each come together to form a complete receptor. Although they are similar in amino acid sequence, it is IL-12Rβ2 that transduces the signal after IL-12 engagement, while IL-12Rβ1 supplies the IL-12 binding function. The expression of IL-12Rβ2 on activated T cells is much more restricted than that of IL-12Rβ1. IL-12Rβ2 is expressed only on Th1 cells, whereas IL-12Rβ1 is present on both Th1 and Th2 cells. The absence of IL-12Rβ2 on Th2 cells correlates with the lack of response of these cells to IL-12. An important role for IFNγ in the Th1 response is to upregulate expression of IL-12Rβ2, thereby promoting IL-12 responsiveness. The signaling path downstream of IL-12R culminates in the activation of STAT4. This signaling path will be discussed later in this chapter when we come to the IL-12-related cytokine IL-23. The IL-23 receptor shares the p40-binding IL-12Rβ1 chain but transduces signals through a unique IL-23R chain.

xiii) IL-13

Function. IL-13 is closely related to IL-4 in biological function (Fig. 17-21) but, in general, the responses of cells to IL-13 are smaller in magnitude. The exception is Th2 cell differentiation, in which IL-13 does not appear to play a role because Th0 cells do not express IL-13R on their surfaces. Nevertheless, like IL-4, IL-13 inhibits macrophage production of TNF, IL-1β, and pro-inflammatory chemokines, but can upregulate the synthesis of IL-12 by DCs and macrophages. Phagocytosis by macrophages is not blocked by IL-13 but ADCC is impaired. Activation of B cells in the presence of IL-13 stimulates their proliferation and induces increased surface expression of CD23, MHC class II, and IgM. IL-13 also promotes isotype switching to IgE and IgG1, and IL-4$^{-/-}$ mice are still able to produce modest amounts of these antibodies using the IL-13-dependent pathway. Mature CD4$^+$ T cells from IL-13$^{-/-}$ mice produce significantly lower levels of the Th2 cytokines IL-4, IL-5, and IL-10. These animals also have lower basal levels of serum IgE.

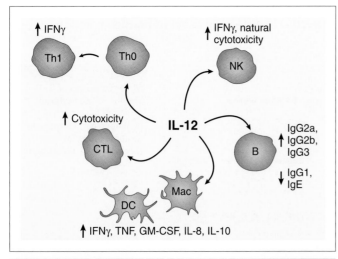

Figure 17-20

Major Functions of IL-12
IL-12 is critical for Th1 differentiation and promotes the production of IFNγ by activated Th1 effectors, macrophages, and NK cells.

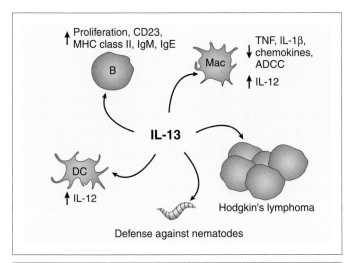

Figure 17-21

Major Functions of IL-13
This Th2 cytokine has many of the same functions as IL-4 but is absolutely critical for immune defense against nematodes. Autocrine secretion of IL-13 by Hodgkin's lymphoma tumor cells may promote expansion of the tumor.

Studies of nematode infections of STAT6$^{-/-}$, IL-4R$\alpha^{-/-}$, and IL-13$^{-/-}$ mice have shown that IL-13 signaling is absolutely required for defense against these parasites. Lastly, studies of antigen-stimulated T cells treated with anti-IL-13 antibodies have suggested that the inhibition of IL-13 function may impair IL-4 expression. The functional relationship between IL-13 and IL-4 remains to be clarified.

Production. The gene for IL-13 in both mice and humans is located in a cluster that includes IL-3, IL-4, and IL-5. IL-13 is normally secreted by activated T cells (particularly Th2 cells), activated mast cells, and basophils stimulated by IL-3. Interestingly, IL-13 is also produced by the B cell-like Reed-Sternberg tumor cells of Hodgkin's lymphoma (see Ch.30). Moreover, IL-13 specifically promotes the proliferation of these tumor cells, suggesting that autocrine secretion of IL-13 may be critical in the etiology of Hodgkin's lymphoma.

Receptor. In most cells, the IL-13R consists of the private IL-13Rα1 chain paired with the shared IL-4Rα chain (refer to Fig. 17-2). This heterodimerization is initiated by the binding of IL-13 to the IL-13Rα1 chain and results in a receptor structurally identical to the type II IL-4R. (A second chain called IL-13Rα2 has turned out to be a decoy receptor.) All these chains belong to the type I receptor family. In some cell types, a structure containing IL-13Rα1 and IL-4Rα and possibly the γc chain has been identified, although this result remains controversial. The IL-13Rα1 chain does not bind IL-4, and IL-13 cannot bind to the IL-4Rα chain directly. However, mutation of the IL-4Rα chain, or treatment with monoclonal anti-IL-4Rα antibodies, blocks IL-13 as well as IL-4 activity, indicating that IL-4Rα is intimately involved in IL-13R function. Interestingly, Th2 development in IL-4R$\alpha^{-/-}$ mice is more severely affected than in IL-4$^{-/-}$ mice, implying some indirect function for IL-13.

Binding of IL-13 to its receptor induces tyrosine phosphorylation of IL-4Rα and at least the IL-13Rα1 chain. The activation of a signaling pathway similar to the primary pathway induced by the binding of IL-4 to its high-affinity receptor ensues, except that IL-13 activates Jak2 (associated with IL-4Rα) and Tyk2 (associated with IL-13Rα1) but not Jak3. STAT6 is the principal downstream signal transducer but STATs 3 and 5 are also phosphorylated in response to IL-13. The secondary signaling pathway involving IRS-2 (refer to Fig. 17-11) can also be activated in response to IL-13 binding to the IL-13R. IL-13R appears principally on the surfaces of vascular endothelial cells, monocyte/macrophages, and B cells. There is some evidence of intracellular (but not surface) expression of IL-13Rα protein in T cells.

xiv) IL-14

Function. Little is currently known about IL-14. *In vitro*, it appears to function primarily as a growth factor for B cells and promotes their proliferation in culture. However, such IL-14-treated B cells become incapable of differentiating into plasma cells and producing antibody. The *in vivo* function of IL-14 has yet to be determined.

Production. IL-14 is secreted by activated T cells and has been found in the culture supernatants of T and B cell lymphomas.

Receptor. Only activated B cells appear to bind IL-14, but the receptor has yet to be characterized in any detail.

xv) IL-15

Function. IL-15's most important functions are exerted on NK cells (Fig. 17-22). NK cells are absolutely dependent on IL-15 for their development (see Ch.18) and continued exposure to IL-15 can promote the differentiation of NK cells into LAKs *in vitro*. IL-15 upregulates IFNγ production by activated NK cells and increases NK cytotoxicity. In addition, IL-15 can synergize with IL-12 to promote even greater production of IFNγ and TNF by NK cells. IL-15 also promotes the development and proliferation of another innate immune system cell, the $\gamma\delta$ T cell. These functions cannot be replaced by IL-2. However, like IL-2, IL-15 stimulates the activation and proliferation of T and B cells, promotes the differentiation of CTLs, and supports memory T cell generation and survival. IL-15 also upregulates adhesion molecules on T cells and serves as a T cell chemoattractant.

Several striking differences between IL-2 and IL-15 have been revealed through study of gene-targeted mice. As expected, mice deficient for the IL-15Rα chain showed sharply decreased numbers of NK cells and $\gamma\delta$ T cells. However, in contrast to the accumulation of autoreactive and spontaneously activated $\alpha\beta$ T cells in IL-2$^{-/-}$ and IL-2R$^{-/-}$ mice, IL-15R$\alpha^{-/-}$ animals exhibit severe reductions in lymphocyte numbers. In particular, CD8$^+$ T cells were profoundly decreased and CD8$^+$ memory T cells were lacking. Moreover, the few lymphocytes present in these mutants did not home properly to the peripheral lymph nodes. These data suggest that IL-15 is heavily involved in lymphocyte recirculation, particularly that of cells destined for the peripheral tissues. Also unlike IL-2, IL-15 has stimulatory effects on skeletal muscle cells and promotes the proliferation of mast cells.

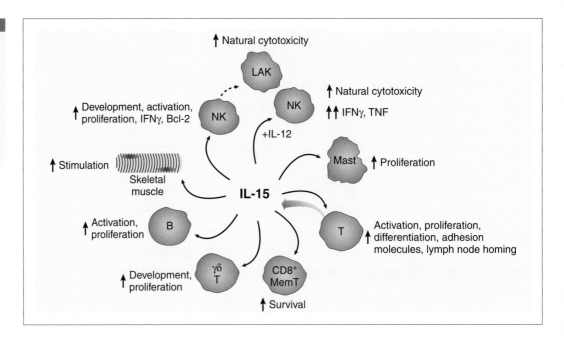

Figure 17-22

Major Functions of IL-15
IL-15 is critical for the development and differentiation of NK cells and important for γδ T cell development. IL-15 greatly increases TNF and IFNγ secretion by activated NK cells and the proliferation of mast cells.

Abnormally elevated IL-15 expression has been noted in cases of rheumatoid arthritis where an inflammatory infiltrate consisting mostly of activated T cells invades the joint tissues (see Ch.29). It is speculated that the chemoattractive properties of the high levels of IL-15 found in such tissues may recruit the activated T cells to the site.

Production. While IL-15 is closely related to IL-2 in overall structure, its amino acid sequence is quite different. IL-15 is secreted primarily by monocytes and macrophages activated by products of viral or bacterial infection. Dendritic cells can be induced to secrete IL-15 *in vitro* by engagement of CD40. Unlike IL-2, IL-15 is not produced by activated T cells. The IL-15 essential for NK cell development is produced by bone marrow stromal cells. Constitutive expression of IL-15 mRNA (but not protein) has been reported in a broad range of tissues, including lung, heart, epithelium, kidney, and skeletal muscle. While synthesis of IL-2 is controlled primarily at the levels of transcription and mRNA stabilization, IL-15 synthesis is controlled at the levels of translation and intracellular transport as well as transcription. The 5′ untranslated region of IL-15 mRNA is unusually long (465 nucleotides in mice) and contains five extra AUGs that can interfere with the efficiency of translation initiation, keeping IL-15 production in check despite the surrounding abundance of IL-15 mRNA. In addition, the secretion signal peptide in the IL-15 protein is a lengthy 48 amino acids, another factor slowing IL-15 production. The rationale for having widespread but untranslated IL-15 mRNA, and what mechanism(s) exist to overcome its inherent translational blocks when IL-15 is needed, are under investigation.

Receptor. IL-15 signals are received through two different receptors that are connected to two different signaling pathways operating in different cell types (Fig. 17-23). In activated T, B, and NK cells, the IL-15 receptor is composed of a unique IL-15α chain and the IL-2Rβ and γc chains. The non-signaling IL-15Rα chain, which binds IL-15 but not IL-2, is not a member of the type I cytokine receptor family. However, because it contains the IL-2Rβ and γc chains, IL-15 signaling transduced through this receptor uses Jak1 and Jak3 coupled to STAT5 and STAT3, just like high-affinity IL-2 signaling. In mast cells, the IL-15 receptor (sometimes referred to as IL-15RX) has a different structure. IL-15RX is composed of the γc chain coupled to one of several unique isoforms of the IL-15Rα chain. These isoforms are smaller than the IL-15Rα chain found in the T cell IL-15 receptor and are derived by alternative splicing of the IL-15Rα gene. IL-15RX contains neither the IL-2Rβ chain nor the full-length IL-15Rα chain. Signals are transduced via Jak2 and STATs 3, 5, and 6.

xvi) IL-16

Function. IL-16 is known primarily for its immunomodulatory effects on CD4+ T cells. IL-16 is also a powerful chemoattractant for both resting and activated cells bearing the CD4 co-receptor, including CD4+ T cells and certain CD4+ eosinophil and CD4+ monocyte subsets (Fig. 17-24). (The function of CD4 on these latter cell types is unclear.) Surface association of IL-16 with CD4 is mandatory for the transduction of IL-16-mediated chemotactic and activation signals. IL-16 initiates cell cycle progression (but not IL-2 synthesis) in a percentage of CD4+ T cells and upregulates IL-2R expression. IL-16 may also protect CD4+ T cells from AICD. IL-16 appears to play a role in asthma by recruiting and activating CD4+ T cells in an antigen-independent way and drawing them to inflammatory sites in the airway. IL-16 also increases eosinophil adhesion to matrix proteins, and MHC class II expression on monocytes.

Production. IL-16 shows no sequence homology to any other known cytokine or chemokine but is highly conserved across

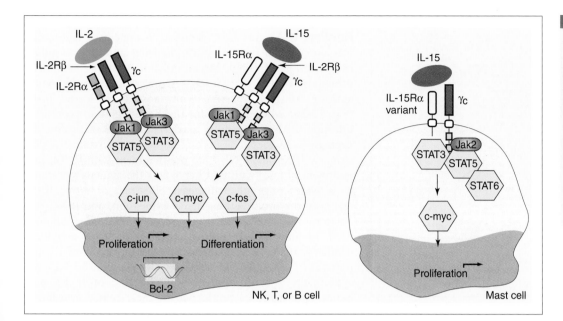

Figure 17-23

IL-15 Receptor Signaling
In T, B, and NK cells, the IL-15R closely resembles the IL-2R in that IL-15Rα combines with IL-2Rβ and γc to form the receptor. Jak/STAT signal transduction leads to the activation of various transcription factors that promote survival, proliferation, and differentiation. In mast cells, the IL-15R (sometimes called IL-15RX) is composed of splice variant isoforms of the IL-15Rα chain linked to γc.

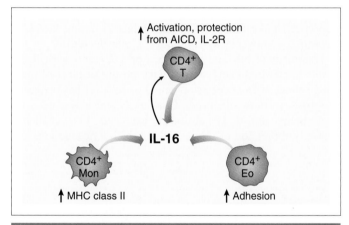

Figure 17-24

Major Functions of IL-16
IL-16 is a potent chemoattractant for CD4-bearing cells and uses CD4 as its receptor.

species. IL-16 mRNA is constitutively expressed in both $CD4^+$ and $CD8^+$ T cells and translated into a 68 kDa precursor protein called pro-IL-16. Resting $CD8^+$ cells contain active caspase-3, which cleaves the pro-IL-16 protein into molecules of a small peptide responsible for all the biological activities of IL-16. However, caspase-3 is inactive in resting $CD4^+$ T cells. Only in response to antigen, T cell mitogens, or vasoactive amines is caspase-3 activated and pro-IL-16 processed into the active IL-16 peptide in $CD4^+$ T cells. This process is accelerated by CD28 costimulation. *In vitro*, IL-16 peptides can autoaggregate into a tetramer able to cross-link the CD4 molecules that serve as the IL-16 receptor. *In vivo*, it is unclear whether autoaggregation into the functional IL-16 tetramer precedes or follows secretion. IL-16 is secreted at low levels

by a variety of other immune system cells, such as DCs, eosinophils, and mast cells, and by airway epithelial cells.

Receptor. IL-16 is unusual among the cytokines because it engages the CD4 molecule as its receptor instead of a unique IL-16R. The sites on the CD4 molecule bound by IL-16 are distinct from those bound by MHC class II (or HIV; see Ch.25), and the TCR is not involved in IL-16 biological activity. Interestingly, the T cell migratory response mediated by IL-16 requires the activity of the Lck kinase associated with the CD4 cytoplasmic tail. However, the cell cycle response induced by IL-16 merely requires the protein recruitment sites of Lck. In other cell types, the situation is slightly different. Certain subsets of eosinophils and monocytes express CD4 but not Lck, implying that IL-16 binding induces a different signal transduction pathway in these cells.

xvii) IL-17

Function. IL-17 functions only during memory $CD4^+$ T cell responses since it is produced only by these cells. IL-17 acts on stromal cells such as keratinocytes, fibroblasts, epithelial cells, and endothelial cells and induces them to secrete pro-inflammatory cytokines such as IL-6, IL-8, and G-CSF (Fig. 17-25). IL-17-mediated release of IL-8 has been shown to increase human neutrophil migration *in vitro*. *In vivo*, injection of IL-17 into mice induces IL-6-dependent neutrophilia (an acute increase in neutrophil numbers in the blood) but cells in the bone marrow are not mobilized. Consistent with the observation of neutrophilia, mice pretreated with IL-17 are resistant to virulent bacterial infections. IL-17 also directly induces the upregulation of ICAM-1 expression on fibroblasts. Under certain conditions, IL-17 can spur the differentiation of $CD34^+$ human hematopoietic stem cells into neutrophils, most likely an indirect effect resulting from the induction of secretion of IL-6 by stromal cells. IL-17 appears to have no detectable direct effect on other hematopoietic cells. High levels of IL-17 are

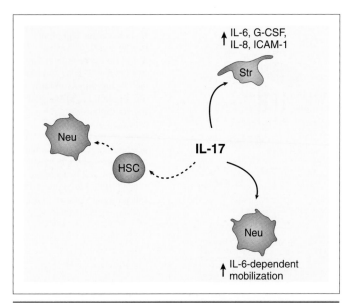

Figure 17-25

Major Functions of IL-17
IL-17 functions during memory CD4$^+$ T cell responses only and induces stromal cells to secrete pro-inflammatory cytokines. In this context, stromal cells are endothelial and epithelial cells and keratinocytes.

found in the synovial tissues of rheumatoid arthritis patients (see Ch.29). The IL-17 in the joint is thought to stimulate the production of IL-6 by non-lymphoid cells in this location, contributing to the inflammatory deterioration of the joint.

Production. IL-17 has one of the most restricted synthesis profiles among the cytokines. In humans, the functional homodimer is produced only by activated memory CD4$^+$ T cells; neither memory CD8$^+$ T cells nor naive CD4$^+$ T cells express IL-17. However, IL-17 constitutes a significant proportion of the cytokine molecules made by memory CD4$^+$ T cells. Extensive testing of a wide variety of fetal and adult tissues has failed to detect IL-17 synthesis by other cell types. Studies of cultured T cell lines have indicated that Th1 but not Th2 effector cells express IL-17. Consistent with this bias in expression, synthesis of IL-17 by T cell-rich synovial explants from rheumatoid arthritis patients was found to be downregulated by the Th2 cytokines IL-4 and IL-13.

Receptor. In contrast to the very limited expression of IL-17, mRNA for the IL-17R is expressed ubiquitously, with the highest levels present in spleen and kidney. The IL-17R is a single-span transmembrane protein of 130 kDa. It has no WSXWS motif and does not resemble members of either the Ig or TNFR superfamilies. The putative receptor protein does contain acidic and serine-rich regions similar to those in the IL-2Rβ, IL-4R, and G-CSFR proteins. *In vitro*, IL-17 induces the tyrosine phosphorylation of Jak1, Jak2, Jak3, and Tyk2 and the activation of STAT1, STAT2, STAT3, and STAT4, suggesting that at least some of these molecules are involved in the transduction of IL-17 signaling. Several other proteins related in structure to IL-17R have been identified that may function as decoy receptors.

xviii) IL-18

Function. IL-18 was first identified in mice as a factor that was upregulated following LPS challenge and able to induce the secretion of IFNγ. Structurally, IL-18 is a monomeric protein of 18 kDa that closely resembles IL-1. The biological activities of IL-18 overlap with those of IL-1 and IL-12 (Fig. 17-26). Mice deficient for IL-18 are highly susceptible to infection with intracellular bacteria and show profoundly decreased synthesis of pro-inflammatory cytokines, particularly IFNγ and TNF. Like pre-treatment with IL-12, pre-treatment of mice with IL-18 enhances the animals' resistance to a wide range of pathogens. However, *in vitro*, IL-18 increases the proliferation, IL-2R expression, and IFNγ production of Th1 cells over that induced by saturating amounts of IL-12 alone. These results indicate that IL-18 acts independently of IL-12 and synergizes with this cytokine to support Th1 responses. In addition, IL-18 induces NK cell cytotoxicity and, in combination with IL-2 or IL-12, promotes vigorous IFNγ production by these cells. Accordingly, mice doubly deficient for both IL-18 and IL-12 show impairments of NK activity and Th1 responses significantly more severe than in either single mutant. Temporally, IL-18 appears to exert its effects later than IL-12: whereas IL-12 is required to initiate Th1 differentiation, IL-18 acts to sustain the Th1 response. Unlike IL-12, IL-18 can induce IL-2 and GM-CSF secretion by antigenically-stimulated T cells in culture.

Treatment of NK and Th1 (but not Th2) cells with IL-18 results in elevated expression of FasL on the surfaces of these cells. Killing of Fas-expressing neighbor cells via Fas-mediated apoptosis is thus enhanced. Conversely, mice treated with neutralizing anti-IL-18 antibody prior to LPS challenge showed suppression of IFNγ, FasL, and TNF production, and

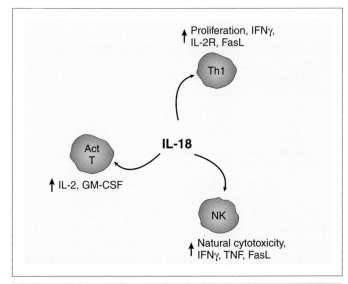

Figure 17-26

Major Functions of IL-18
IL-18 acts independently of IL-12, as shown, but can synergize with it late during the primary Th1 response to sustain IFNγ production by Th1 and NK cells, and to increase NK cytotoxicity.

mitigation of liver damage caused by TNF- and Fas-mediated apoptosis of hepatocytes. Interestingly, IL-18 may also have an indirect role in bone remodeling. Primary osteoblastic stromal cells produce IL-18, which in turn induces T cell production of GM-CSF, a cytokine that inhibits the generation of osteoclasts (thus, an effect opposite to that of IL-1 and TNF). This effect may be important in limiting the occurrence of osteoporosis (brittle bones).

Production. Unlike IL-1, IL-18 is produced by both leukocytes and non-leukocytes. Indeed, IL-18 mRNA has been detected in almost every tissue examined. IL-18 protein is synthesized primarily by activated Kupffer cells in the liver in response to stresses such as LPS or very low temperatures. Like IL-1, IL-18 is synthesized as a precursor protein that requires cleavage by caspase-1 before it becomes biologically active. *In vitro*, IL-18 production can be induced by treating activated macrophages with IFNγ. As mentioned previously, osteoblastic stromal cells also secrete IL-18.

Receptor. The IL-18 receptor is similar in structure to IL-1R, but its extracellular domain is distinct in that it binds only IL-18 and not IL-1α, IL-1β, or IL-1Ra. Like the IL-1RI chain requires IL-1RAcP for high-affinity binding, the IL-18R chain requires an IL-18RAcP co-receptor for optimal binding affinity and signal transduction. The IL-18R complex is expressed widely (but not universally). A soluble form of IL-18R exists that can neutralize IL-18. Although similar in biological effect to IL-12, IL-18 does not activate STAT4 in Th1 cells. Instead, the IL-18R operates through MyD88 and the IRAK pathway previously described for IL-1RI signaling to induce the activation of transcription factors such as NF-κB and AP-1 (refer to Fig. 17-7). IRAK-deficient mice show severe defects in IL-18-induced Th1 responses and NK activation accompanied by impaired c-jun and NF-κB activation. Studies of IL-18R$^{-/-}$ mice have confirmed the importance of the IL-18R chain for IL-18 signaling. IL-18 was unable to bind to IL-18R$^{-/-}$ Th1 cells, and c-jun and NF-κB activation were not detected in these cells. Production of IFNγ by NK cells and NK cytolytic activity induced by IL-18 were impaired. Th1 cell development was also inhibited in these animals.

xix) IL-19

Function. IL-19 is one of a group of IL-10-related cytokines discovered in the early 2000s. The genes encoding IL-10, IL-19, IL-20, and IL-24 (see later) are clustered together in a 200 kb region on human chromosome 1. The physiological function of IL-19 is unknown but it is suspected that it may participate in the regulation of pro-inflammatory cytokine expression. Treatment of mouse monocytes *in vitro* with IL-19 induced the production of IL-6 and TNF, and then led to the apoptosis of these cells.

Production. IL-19 is secreted by monocytes stimulated with LPS or GM-CSF. Due to alternative gene splicing and variable glycosylation, this cytokine exists in multiple isoforms in the range of 30–40 kDa.

Receptor. IL-19 shares the IL-20 receptor composed of IL-20R1 plus IL-20R2 (see later). Curiously, however, the binding of IL-19 to this receptor does not appear to induce a skin phenotype, unlike the binding of IL-20.

xx) IL-20

Function. IL-20 is also an IL-10-related cytokine. IL-20 appears to have a role in skin development, since transgenic mice over-expressing IL-20 die soon after birth with abnormal epidermis due to keratinocyte hyperproliferation. It is suspected that IL-20 may regulate inflammation in the skin.

Production. The precise cell type producing IL-20 is currently unknown.

Receptor. IL-20 binds to both the IL-20R composed of IL-20R1 plus IL-20R2, and an alternative receptor containing IL-20R1 plus IL-22R1.

xxi) IL-21

Function. IL-21 has functional characteristics of both IL-2 and IL-15 and is considered to be a Th1 cytokine. Like IL-2, IL-21 stimulates the proliferation of activated T cells (Fig. 17-27). Like IL-15, IL-21 can promote the maturation and natural cytotoxicity of NK cells. Treatment of mature NK cells with IL-21 *in vitro* upregulates expression of the NK inhibitory receptor CD94/NKG2A. *In vivo* treatment of mice with IL-21 results in NK cell activation and enhanced NK cell-mediated anti-tumor immunity. There is also some evidence that IL-21 may promote the long term survival of tumor-specific CD8$^+$ T cells. In addition, IL-21 enhances the *in vitro* proliferation of both αβ and γδ peripheral T cells induced by IL-2, IL-15, or IL-7, and the production of IFNγ by T cells. Interestingly, IL-21 inhibits the proliferation of B cells induced by IgM engagement or LPS treatment, suggesting that IL-21 may control B cell expansion in response to Ti antigens. However, IL-21 increases the proliferation of B cells responding to Td antigens, and supports isotype switching to IgG1 and IgG3. IL-21 may also inhibit IgE production, since IL-21$^{-/-}$ and IL-21R$^{-/-}$ mice exhibit elevated levels of this antibody.

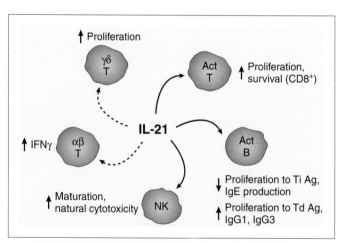

Figure 17-27

Major Functions of IL-21
This Th1 cytokine has functions overlapping those of both IL-2 and IL-15 in that it stimulates the proliferation of activated T cells and increases NK cytotoxicity.

Production. IL-21 is a small protein of 15 kDa which is similar to IL-2 and IL-15 in primary sequence and structure. It is secreted only by activated CD4⁺ T cells. The gene encoding IL-21 lies on human chromosome 4 within 180 kb of the IL-2 gene.

Receptor. Like IL-2 and IL-15, the heterodimeric IL-21R includes the γc chain. Ligand specificity is provided by the unique IL-21Rα chain, which is similar in sequence to the IL-2Rβ chain but closer to the IL-4Rα chain in domain organization. IL-21R is expressed only on T, B, and NK cells. The gene encoding IL-21Rα is located on chromosome 16, near that encoding IL-4Rα.

xxii) IL-22

Function. IL-22 is another IL-10-related cytokine discovered in the early 2000s. Unlike IL-19, IL-20, and IL-24, however, IL-22 is encoded by a gene on human chromosome 12, within 100 kb of the IFNγ gene. The production of acute phase proteins is increased in hepatocytes by administration of IL-22 (Fig. 17-28). Basophil and platelet numbers are increased but erythrocyte numbers are decreased. Unlike IL-10, IL-22 does not inhibit the production of TNF, IL-1, and IL-6 by monocytes treated with LPS, and does not decrease IFNγ production by Th1 cells.

Production. IL-22 is secreted by murine Th1 cells and mast cells that have encountered IL-9. LPS injection *in vivo* can also stimulate IL-22 production by cells in various organs.

Receptor. The IL-22 receptor consists of the IL-22R1 chain plus IL-10R2. Signal transduction involves STAT3 and STAT5. IL-22R is most highly expressed in placenta and spleen but is also present in a wide range of other tissues. A soluble decoy receptor called "IL-22 binding protein" (IL-22BP) binds to IL-22 and specifically inhibits it from binding to IL-22R and inducing STAT activation. IL-22BP does not bind to IL-10, IL-19, IL-20, or IL-24.

xxiii) IL-23

Function. IL-23 is a heterodimeric cytokine composed of the p40 subunit of IL-12 plus a p19 subunit unique to IL-23. The p19 subunit has homology to both IL-6 and the p35 subunit of IL-12. The functions of IL-23 thus partially overlap with those of IL-12. However, unlike IL-12, IL-23 has no direct effect on naive CD4⁺ T cells. Instead, IL-23 acts on memory CD4⁺ T cells to promote their proliferation and support their differentiation into Th1 effectors during a secondary response (Fig. 17-29). IL-23 also stimulates DCs (and possibly NK cells) to produce IFNγ. Some researchers interpret these data to mean that IL-23's primary role is to promote sustained cell-mediated responses designed to clear persistent intracellular infections. Transgenic mice overexpressing p19 exhibit inflammation in multiple organs as well as growth and fertility defects.

Production. While p19 is expressed by macrophages, DCs, T cells, and endothelial cells, only activated APCs also produce the IL-12 p40 subunit required to complete the IL-23 molecule. The majority of the body's IL-23 is produced by activated DCs.

Receptor. The IL-23 receptor is composed of the IL-12Rβ1 chain plus a unique signaling chain called IL-23R that resembles IL-12Rβ2 and gp130 (refer to Fig. 17-2). The gene encoding the IL-23R chain is located very close to that encoding IL-12Rβ2 on human chromosome 1. Memory (but not naive) CD4⁺ T cells, DCs, and NK cells express IL-23R. The IL-23 receptor complex uses many of the same signaling elements as IL-12R (Fig. 17-30). With the binding of IL-12 to the IL-12R complex or IL-23 to the IL-23R complex, the associated kinases Jak2 and Tyk2 are activated. These kinases phosphorylate key sites on IL-12Rβ2 and the IL-23R chain and form docking sites for a variety of STATs, including STATs 1, 3, 4, and 5. The STATs are then phosphorylated and activated by Jak2 and move to the nucleus to activate the transcription of target genes. While IL-12R engagement results in the activation of large amounts of homodimeric STAT4, the binding of IL-23 to the IL-23R complex produces both homodimeric STAT4 and heterodimers of STAT4 and STAT3. Researchers speculate that these transcription activators may bind to the promoters of slightly different collections of target genes.

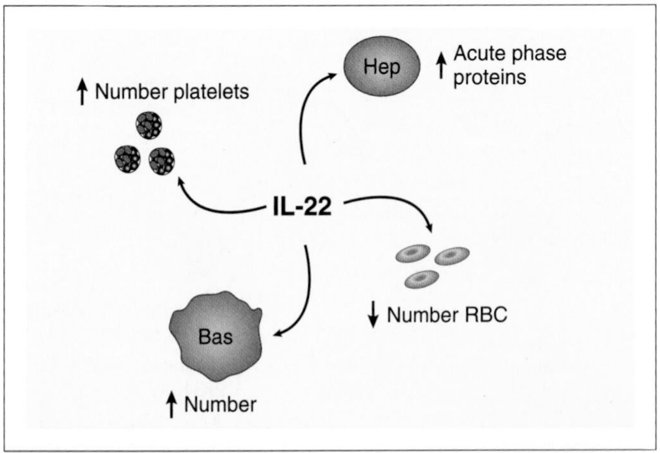

Figure 17-28

Major Functions of IL-22
This IL-10-related cytokine is not immunosuppressive like IL-10 but rather promotes the acute phase response.

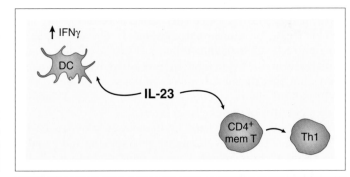

Figure 17-29

Major Functions of IL-23
This IL-12-related cytokine acts on memory CD4+ T cells (but not naive T cells) to induce Th1 differentiation.

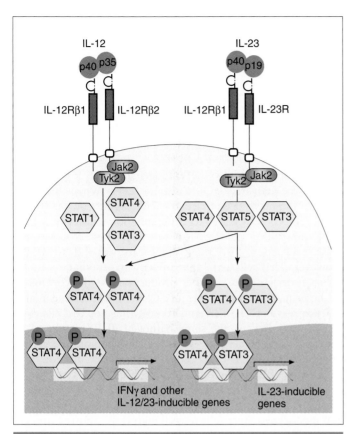

Figure 17-30

IL-12R and IL-23R Signaling
IL-12 and IL-23 share the p40 subunit, while the IL-12R and IL-23R complexes share the IL-12Rβ1 chain. Signaling through IL-12R results in the formation of STAT4 homodimers that can bind to the promoters of the IFNγ gene and other IL-12-inducible genes. Signaling through IL-23R generates STAT4 homodimers and STAT4/STAT3 heterodimers, both of which induce transcription of a unique subset of IL-23-inducible genes.

Receptor. Like IL-20, IL-24 can bind to two receptor complexes: one composed of IL-20R1 plus IL-20R2, and the other consisting of IL-22R1 plus IL-20R2. Signaling induced by binding of IL-24 is thought to be transduced though STAT6.

xxv) IL-25

Function. IL-25 has structural similarities to IL-17 but very different activities (Fig. 17-31). When infused into mice, IL-25 induces the expression of IL-4, IL-5, and IL-13 by unknown non-T cells. The mice develop Th2 responses characterized by the production of IgE, IgG1, and IgA antibodies. IL-25 may also enhance allergic responses in that eotaxin produced in response to this cytokine contributes to airway eosinophilia, increased mucus secretion, and airway hyper-responsiveness (see Ch.28).

Production. The cellular source of IL-25 is still somewhat of a mystery. There is some evidence that activated Th2 cells may produce this highly potent molecule. Other data suggest that IL-25 may be produced by bone marrow stromal cells. Mast cells activated by FcεRI ligation also produce IL-25.

Receptor. In mice, the receptor for IL-25 has been identified as a protein called "thymic shared antigen-1" (TSA-1). Monocyte-derived APCs express both membrane-bound and soluble forms of the IL-25R upon stimulation with Th2 cytokines.

xxvi) IL-26

Function. IL-26 is yet another IL-10-related cytokine. Like IL-22, IL-26 is encoded by a gene on human chromosome 12 that is near the IFNγ gene. IL-26 is thought to occur naturally as a homodimer of a 19 kDa protein that is not highly glycosylated. Its physiological function remains a mystery but, based on where it is produced, IL-26 may act as an autocrine growth factor for transformed and/or virus-infected T cells. Recombinant IL-26 can induce the secretion of IL-10 and IL-8 by some types of epithelial cells, suggesting a function in mucosal or cutaneous immunity. IL-26 mRNA is also expressed in activated normal NK cells and Th1 effectors.

xxiv) IL-24

Function. IL-24 is another IL-10-related cytokine. IL-24 appears to have different functions in different cell types. High levels of IL-24 have been found in some types of tumors. For example, IL-24 was originally identified as "melanoma differentiation-associated gene-7" (mda-7), which is induced during terminal differentiation in human melanoma cells. However, IL-24 can promote the induction of apoptosis in other cancer cells. In rodents, IL-24 has been linked to wound healing, perhaps functioning to increase fibroblast proliferation or to support the inflammation associated with wound healing.

Production. IL-24 is secreted by murine Th2 cells in which both the TCR and the IL-4R have been engaged. IL-24 has also been found in relatively high concentrations in human melanoma and other tumor cells, and in rat fibroblasts at wound healing sites. The IL-24 gene, which is located on chromosome 1 in a cluster with the genes encoding IL-10, IL-19, and IL-20, undergoes alternative splicing.

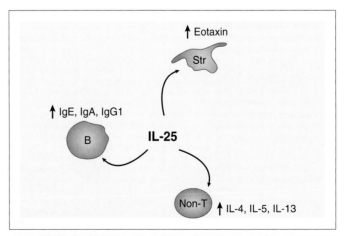

Figure 17-31

Major Functions of IL-25
This putative Th2 cytokine can induce non-T cells to produce Th2 cytokines.

Production. Among normal hematopoietic cells, IL-26 is produced only by activated memory T cells. However, virally transformed T cell lines express high levels of IL-26 and at least some virally transformed B cell lines express low levels of this protein. The gene encoding IL-26 is located on chromosome 12, and the protein is thought to be a homodimer.

Receptor. IL-26 signals through a receptor composed of IL-20R1 plus IL-10R2 that does not bind IL-10, IL-19, IL-20, IL-22, or IL-24. STAT1 and STAT3 are activated.

xxvii) IL-27

Function. IL-27 is a heterodimeric cytokine with many different activities (Fig. 17-32). Structurally, IL-27 is most closely related to IL-12 and IL-23 because it is composed of one subunit that resembles the p40 subunit of IL-12 (called EBI3; "Epstein–Barr induced gene 3") and a smaller subunit called p28 that resembles the p35 subunit of IL-12. The secretion of the complete heterodimer depends on the presence of the EBI3 subunit. The primary function of IL-27 is to induce the expression of the IL-12Rβ2 chain on naive CD4$^+$ T cells. These cells express IL-27R but do not carry the IL-12Rβ2 receptor subunit in the resting state, and so cannot respond to IL-12. Upon activation by pMHC and IL-27R engagement, naive CD4$^+$ T cells commence expression of the transcription factor T-bet, which activates IL-12Rβ2 expression. The naive T cell can then express a complete IL-12R, allowing it to receive the IL-12 signaling necessary to push the cell down the Th1 differentiation path. Importantly, IL-27 can induce this activation of T-bet in the absence of IFNγ, consistent with a role for IL-27 in the earliest stages of an immune response. In addition, like IL-12 and IFNγ, IL-27 suppresses the expression of GATA-3 and thus the expression of IL-4 required for Th2 differentiation. Evidence from the study of knockout mice lacking IL-27R suggests that, unlike IL-12, IL-27 is required only for the initiation of Th1 responses and not for their maintenance. Moreover, unlike IL-12, IL-27 does not promote the differentiation of Th1 effectors from memory CD4$^+$ T cells. However, IL-27 can enhance the production of IFNγ by differentiated Th1 effectors and activated NK cells. Activated B cells that encounter IL-27 are induced to express T-bet and undergo isotype switching to IgG2a. Switching to IgG1 induced by IL-4 is inhibited by IL-27.

How do IL-12, IL-18, IL-23, and IL-27 fit together in inducing Th1 responses? Upon activation by a pathogen, DCs secrete IL-27, which induces the expression of the complete IL-12R on naive CD4$^+$ T cells (Fig. 17-33). At the same time, NK cells stimulated by the pathogen produce IFNγ, which induces the activated DCs to secrete IL-12. When this IL-12 engages the newly assembled IL-12R complex on the T cell, the T cell proceeds with Th1 differentiation and IFNγ synthesis. Expression of the IL-27R is lost on Th1 effectors, so that these cells soon lose their ability to respond to IL-27. However, Th1 cells gain the capacity to respond to IL-18, which further increases IFNγ production. At the conclusion of the primary response, memory CD4$^+$ T cells are generated that express both IL-12R and IL-23R (but not IL-27R). In a secondary attack by the same pathogen, IL-23 produced by activated APCs synergizes with IL-12 to promote the activation of memory CD4$^+$ T cells and the differentiation of a fresh batch of IFNγ-secreting Th1 effectors.

In the early 2000s, it was discovered that IL-27 can induce the expression of pro-inflammatory cytokines by mast cells and monocytes. Mast cells treated with IL-27 showed upregulation of mRNA encoding IL-1, IL-18, and TNF (among several other molecules), while monocytes increased their expression of IL-1 and TNF mRNA. However, there is also evidence that IL-27 can have anti-inflammatory effects late in adaptive responses to acute parasite infections. When knockout mice lacking the IL-27Rα chain (see later) were infected with intracellular parasites, they showed increased immunopathic liver necrosis compared to their infected wild-type counterparts. Indeed, the vigorous Th2 responses required to eliminate helminth worms were mounted much faster in these

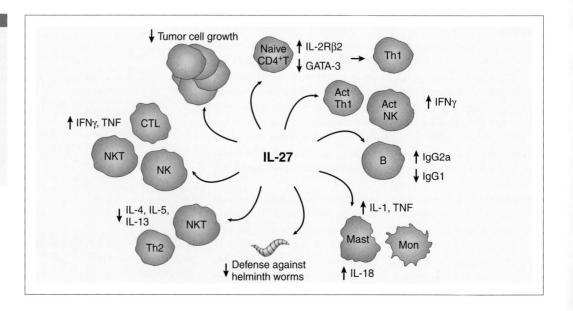

Figure 17-32

Major Functions of IL-27
IL-27 plays a major role in initiating the Th1 response by inducing expression of the IL-12Rβ2 subunit. IL-27 also increases pro-inflammatory cytokine secretion by several cell types and dampens Th2 responses directed against helminth worms. In addition, IL-27 has potent anti-tumor effects.

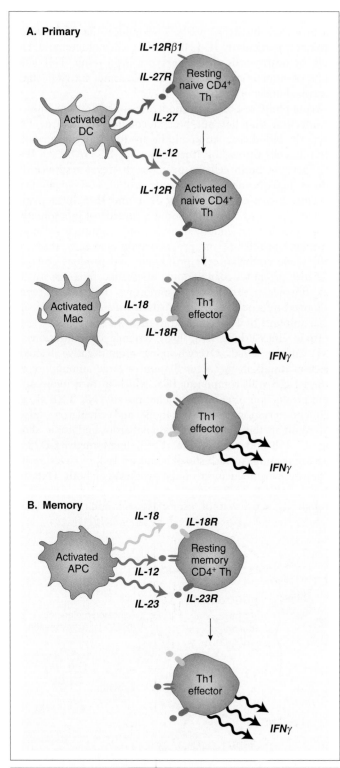

A. Primary

IL-12Rβ1

IL-27R

Activated DC

IL-27

IL-12

Resting naive CD4⁺ Th

IL-12R

Activated naive CD4⁺ Th

Activated Mac

IL-18

IL-18R

Th1 effector

IFNγ

Th1 effector

IFNγ

B. Memory

IL-18

IL-18R

Activated APC

IL-12

Resting memory CD4⁺ Th

IL-23

IL-23R

Th1 effector

IFNγ

Figure 17-33

Relationship of IL-12, -18, -23, and -27 in Th1 Responses
(A) Primary Th1 response. Resting naive T cells express only IL-12Rβ1 and not the complete IL-12R, and so initially cannot respond to IL-12. However, these cells do express IL-27R and can bind IL-27 secreted by activated DCs. IL-27 signaling induces expression of IL-12Rβ2 and assembly of the complete IL-12R. Upon binding of IL-12, Th1 differentiation is initiated and is then sustained by IL-18 signaling. (B) Memory Th1 response. Memory CD4⁺ T cells express IL-12R, IL-18R, and IL-23R (but not IL-27R). IL-18 acts to sustain the Th1 response initiated by IL-12 and IL-23.

mutant animals than in wild-type mice. These results suggest that IL-27 may normally play a role in damping down the anti-parasite adaptive response and preventing it from causing collateral damage in its late stages. Compared to cells from wild-type mice, CTLs, NK, and NKT cells isolated from these infected mutant mice produce more IFNγ and TNF, and Th and NKT cells produce more IL-4, IL-5, and IL-13.

Finally, in 2004 it was reported that IL-27 has potent anti-tumor activity, at least in animals. In one study, mouse colon carcinoma cells (C26) that cause significant and sometimes fatal tumors in mice were transduced with the IL-27 gene. When these IL-27-secreting C26 cells were inoculated into normal mice, minimal tumor growth occurred that later resolved completely. Analysis of these mice showed that IFNγ production was increased and CTLs directed against C26 cells had been generated. Furthermore, mice that had been inoculated with the IL-27-secreting C26 cells and had recovered completely showed tumor-specific immunity to a later challenge with the parental C26 tumor cells. Analyses of IFN⁻/⁻, T-bet⁻/⁻, and CD8⁺ T cell-depleted mice subjected to this protocol demonstrated that the anti-tumor activity of IL-27 depends on both IFNγ and CD8⁺ T cells. Strikingly similar results were obtained in a different study in which mouse neuroblastoma cells (TBJ) were engineered to overexpress IL-27. Mice inoculated with these cells showed diminished tumor growth that almost always resolved completely. Again, these animals exhibited elevated IFNγ and CD8⁺ T cell-mediated tumor-specific immunity. The application of these results to human cancer treatment is under investigation.

Production. IL-27 synthesis is restricted to activated monocytes, DCs, and macrophages serving as APCs during the early stages of the adaptive response.

Receptor. The IL-27R is made up of gp130 plus a unique IL-27Rα chain (also called TCCR or WSX-1) that binds to IL-27. Signal transduction is propagated primarily by phosphorylation and activation of Jak1, STAT1, and STAT3. STAT1 phosphorylation leads to activation of T-bet, which drives IL-12Rβ2 expression. A STAT1-independent pathway to T-bet activation has also been reported, as has IL-27-mediated phosphorylation (at modest levels) of STAT4 and STAT5. IL-27R is expressed most highly on naive CD4⁺ T cells and to a lesser extent on NK cells. As mentioned above, IL-27Rα expression is downregulated once T effector cells are fully differentiated. Memory CD4⁺ T cells do not express IL-27R. With respect to other cell types, mRNA for both chains of IL-27R has been found in B cells, mast cells, DCs, monocytes, and endothelial cells.

xxviii) IL-28 and IL-29

Function. In 2003, three IFNα/β-like cytokines designated IL-28A (also called IFNλ1), IL-28B (IFNλ2), and IL-29 (IFNλ3) were identified that induce an antiviral state in infected cells independently of IFNα/β.

Production. IL-28A, IL-28B, and IL-29 secretion can be induced in various cell lines by viral infection. In humans, the genes encoding these three cytokines are localized in close proximity on chromosome 19.

Receptor. All three cytokines bind to a distinct receptor complex composed of two subunits. One subunit is a unique chain designated IL-28Rα (also called CRF2-12, for "cytokine receptor family 2: protein 12") and the other is the IL-10R2 chain (also known in this context as CRF2-4). Both chains are constitutively expressed on many different cell types and transduce signaling via the Jak/STAT pathway.

III. TUMOR NECROSIS FACTOR AND RELATED MOLECULES

In this section, we will describe in detail TNF and several molecules related to it in structure or function, including the lymphotoxins LTα and LTαβ, and a molecule called OPGL (osteoprotegerin ligand), which has crucial roles in both bone remodeling and the immune system.

i) TNF

Function. Originally known as TNFα, TNF was first identified as a substance capable of inducing the hemorrhagic necrosis of certain tumors in mice. It is now known that TNF is the major inflammatory mediator induced by the presence of gram-negative bacteria and their components. Depending on its concentration, TNF also has potent immunoregulatory, cytotoxic, antiviral, and pro-coagulatory activities, and effects on hematopoiesis.

ia) Inflammatory effects.

Local stimulatory effects: At <u>low</u> concentrations, TNF acts locally to upregulate the expression of adhesion molecules on vascular endothelial cells, neutrophils, macrophages, and lymphocytes, leading to enhanced leukocyte extravasation into the tissue under attack (Fig. 17-34). TNF also directly stimulates

neutrophils, eosinophils, and monocyte/macrophages to carry out their microbicidal activities. TNF induces the production of cytokines, particularly IL-1, IL-6, IFNs, chemokines, and TNF itself, by nearby cells of various types. As a result, TNF has a range of immunomodulatory, cytotoxic, and antiviral effects (described in detail later).

Acute phase response: At <u>moderate</u> concentrations induced by infection that has not been adequately contained, TNF enters the blood circulation and begins to act like a hormone, affecting cells throughout the body (Fig. 17-35A). Like IL-1, TNF acts on brain cells to promote the fever response and induces hepatocytes to commence synthesis of acute phase proteins. Increased synthesis of IL-1 and IL-6 by activated macrophages also occurs, building a cascade of inflammatory cytokines. Vascular endothelial cells, smooth muscle cells, and fibroblasts proliferate and produce additional IL-6, IL-8, and TNF, while epithelial cells proliferate and produce collagen. TNF also promotes coagulation and vascular repair processes, and fibroblast proliferation important for tissue repair. Prolonged infection resulting in sustained production of sublethal amounts of TNF can lead to cachexia.

Toxic effects: Like IL-1, <u>high</u> systemic concentrations of TNF can be lethal. Overwhelming gram-negative bacterial infection results in the accumulation of large amounts of the bacterial cell wall component LPS, which in turn induces the rapid production of enormous quantities of TNF. TNF at very high concentration exerts metabolic and circulatory effects that result in the death of the animal from endotoxic shock (Fig. 17-35B and refer to Box 17-2). Interestingly, CD28$^{-/-}$ mice are resistant to such shock because a lack of CD28 results in drastic downregulation of the synthesis of both TNF and IFNγ. Like IL-1, TNF is implicated in bone matrix destruction through its activation of osteoclasts. In addition, excessive

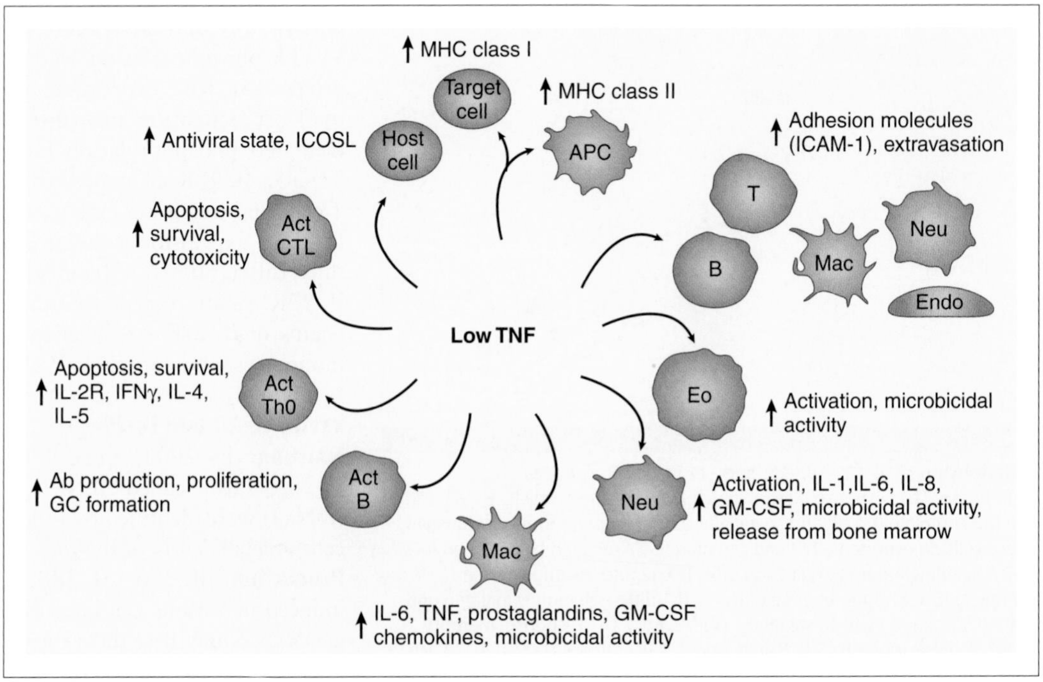

Figure 17-34

Effects of Low Concentrations of TNF
TNF is the major inducer of inflammatory responses to gram-negative bacteria and their components. It also has immunoregulatory, cytotoxic, and antiviral effects.

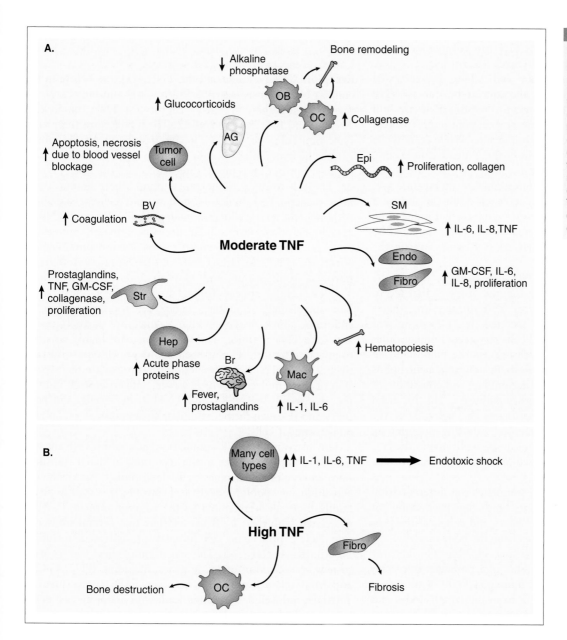

Figure 17-35

Effects of Moderate and High Concentrations of TNF
(A) As TNF concentration increases to a moderate level, this cytokine accesses the circulation and affects a broad range of non-hematopoietic cells. (B) High concentrations of TNF stimulate many cell types (particularly macrophages) to produce vast quantities of IL-1, IL-6, and TNF. Endotoxic shock, among other deleterious consequences, can result.

stimulation of fibroblast proliferation by TNF can result in the fibrosis often associated with chronic inflammation.

ib) Immunoregulatory effects. TNF resembles IL-1 in many of its immunoregulatory actions. TNF is expressed in the thymus and paradoxically can promote both apoptosis and cell survival of T lineage cells, depending on which receptor is engaged (TNFR1 or TNFR2) and the balance of signal transduction molecules recruited to the receptor complex (see Box 9-2 in Ch.9). Although TNF is expressed by thymic stromal cells and has important effects on thymocytes, TNF$^{-/-}$ knockout mice have normal numbers and ratios of the various T cell subsets, indicating that TNF is dispensable for T cell development. TNF promotes increases in adhesion molecules and MHC class I and II on the appropriate cell types, facilitating

B and T cell proliferative responses to antigen. Mature resting T and B cells do not express TNFR and so are not affected by TNF, but antigen-stimulated T and B cells readily upregulate TNFR expression during activation. Receipt of the TNF signal induces upregulated expression of IL-2R and IFNγ by T cells and stimulates antibody production by B cells. Embryonic and cultured fibroblasts subjected to TNF treatment *in vitro* rapidly upregulate expression of the costimulatory molecule ICOSL (refer to Ch.14). ICOSL is also induced in various non-lymphoid tissues in response to LPS treatment (a powerful inducer of TNF) *in vivo*. However, TNF cannot induce ICOSL expression on splenocytes *in vivo* or *in vitro*. As the immune response progresses, activated mature T effectors become more susceptible to TNF-mediated AICD.

ic) Cytotoxic effects. In addition to its ability to induce tumor cell death by necrosis (see later), TNF can kill many types of tumor cells by inducing their apoptosis (refer to Fig. 17-35A). Indeed, a large part of tumor cell killing by activated macrophages and NK cells is thought to be due to TNF secreted by these cells. However, for reasons that are still unclear, normal cells maintaining normal continuous protein synthesis escape TNF-induced cell death. While it may be tempting to think that TNF might be the ultimate cancer cure, the pleiotropy of TNF means that it has severe side-effects on other organ systems, making it unsuitable for therapeutic use.

id) Antiviral effects. TNF at low concentration can induce a cell to adopt an antiviral state very similar to that induced by IFN. CTL killing of virally infected cells is also stimulated by TNF-mediated induction of MHC class I expression on the infected cell.

ie) Pro-coagulatory effects. TNF also controls tumors by inducing their hemorrhagic necrosis, the property that led to the identification and naming of TNF in the first place. Moderate amounts of TNF secreted by LPS-stimulated macrophages induce endothelial cells to secrete (among other molecules) tissue factor III, which triggers the blood clotting cascade and promotes coagulation. In addition, neutrophils drawn to the inflammatory site accumulate and contribute to the stifling of local microcirculatory pathways. This localized blockage of the blood supply initiates the necrosis of the tumor without, for unknown reasons, affecting surrounding normal tissues. This phenomenon can be replicated in the laboratory in the form of a localized *Shwartzman reaction*, discussed in Box 17-2.

if) Hematopoietic effects. TNF stimulates the differentiation of granulo-monocytic precursors and the production of GM-CSF by various cell types *in vitro*, and enhances the production and release of neutrophils from the bone marrow.

For all the wide-ranging activities of TNF, the TNF$^{-/-}$ knockout mouse is relatively robust in a pathogen-free environment. The major phenotypic finding in TNF$^{-/-}$ animals is that germinal centers are missing from the lymph nodes and the FDC networks are disrupted, compromising the humoral response. Interestingly, there is no increased incidence of tumor formation. As expected, TNF$^{-/-}$ animals succumb quickly to infection with intracellular bacteria such as *Listeria monocytogenes* but are resistant to endotoxic shock induced by high doses of LPS.

Production. TNF is synthesized in a pro-protein form with an unusually long secretion signal sequence. This precursor is cleaved to generate a mature protein of 17 kDa that immediately trimerizes, a structure that facilitates binding to TNF receptor complexes and thus the triggering of intracellular signaling. TNF is produced primarily by activated monocytes/macrophages and mast cells but also by T, B, and NK cells and some non-hematopoietic cells (including fibroblasts, hepatocytes, splenocytes, and ovarian, epidermal, and thymic stromal cells). Large amounts of TNF are also stored in the granules of mast cells. The rapid degranulation of activated mast cells, which results in the release of this preformed TNF, represents an important early source of this cytokine during infections. Both IL-1 and GM-CSF can stimulate TNF production by monocytes,

while IFNγ or TNF itself promotes the release of additional TNF by activated macrophages. Signaling through the TCR plus CD28 appears to promote TNF synthesis by T cells.

Receptor. We have discussed the TNF receptors before in the context of apoptosis (see Box 9-3 in Ch.9 and Fig. 15-9). To review briefly, there are two receptors for TNF: the 55 kDa TNFR1 and the 75 kDa TNFR2. The TNFRs are the prototypic molecules for the TNFR superfamily, molecules whose extracellular domains contain multiple repeated cysteine-rich regions. For both receptors, three molecules of either TNFR1 or TNFR2 form a homotrimer that, when bound by a homotrimeric TNF molecule, initiates intracellular signaling. Although the extracellular domains of TNFR1 and TNFR2 are similar, their intracellular domains are not. Studies of knockout mice have shown that most of the physiological effects of TNF, including NF-κB activation, cytotoxicity, fibroblast proliferation, and apoptosis, are mediated through TNFR1. TNFR2 appears to mediate primarily T cell survival. Not surprisingly, TNFR1$^{-/-}$ mice are totally resistant to endotoxic shock, while TNFR2$^{-/-}$ mice are partially resistant. Like TNF$^{-/-}$ mice, TNFR1$^{-/-}$ mice are highly sensitive to infection by intracellular bacteria. Most cell types constitutively express TNFR but induction can upregulate the expression of these molecules. For example, activation itself induces the expression of TNFR1 and TNFR2 on T cells, and both IFNγ and IL-2 can further stimulate the expression of TNFR2 on activated T cells.

Downstream of TNFR stimulation induced by TNF engagement, a complex balance in the recruiting of signal transduction molecules determines cell fate: either survival and activation, or apoptotic death (refer to Fig. 15-9). If a TRAF family molecule is recruited to the adaptor protein TRADD associated with the TNFR cytoplasmic tail, several transcription factors promoting cell survival are activated, including NF-κB, IRF-1, and AP-1. The activation of these factors leads to new transcription of many genes, including IFNβ, G-CSF, and MHC class I and II. If, however, FADD is recruited to TRADD, intracellular signaling leading to apoptotic cell death is induced. Interestingly, TNF receptor molecules can be cleaved from the membranes of activated cells and shed into the extracellular environment. These free TNFRs can still bind TNF and thus may act as competitive inhibitors for TNF binding to TNFR still fixed in the membrane. The free receptors may thus play a role in the regulation of TNF signaling.

ii) Lymphotoxins (LTs)

In humans and mice there are two different lymphotoxins, LTα and LT$\alpha\beta$, that make use of the same subunits. LTα is a secreted homotrimer of the LTα chain and binds to both TNFR1 and TNFR2 to mediate many of the same effects as TNF (Fig. 17-36). LT$\alpha\beta$ is a membrane-bound heterotrimer composed of an LTα subunit plus two copies of a distinct LTβ subunit. LT$\alpha\beta$ binds to a different receptor called LTβR. LT$\alpha\beta$'s primary role is to support peripheral lymphoid tissue development.

LTα function. LTα has often been called TNFβ in the past because it binds to both TNFRs and thus has effects very similar to those of TNF. However, in *in vitro* experiments using

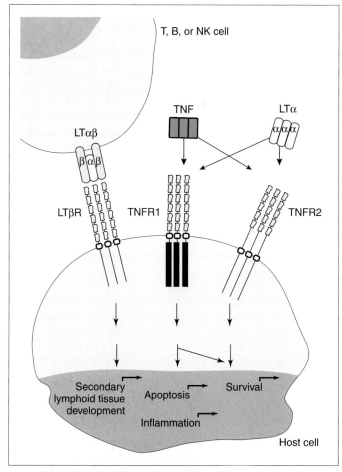

Lymphotoxin Signaling
The LTα chain functions as a component of both the transmembrane protein LTαβ and the soluble cytokine LTa. LTαβ binds to a unique receptor called LTβR and transduces signals important for the development of secondary lymphoid tissues. LTα binds to both TNFR1 and TNFR2 and initiates signaling leading to either apoptosis or survival, depending on the balance of mediators present.

human cell lines, LTα had only 1/10–1/1000 the potency of TNF. Like TNF, LTα promotes inflammation, has antiviral activity, and kills tumor (but not normal) cells by apoptosis. LTα also has been shown to activate the same transcription factors as TNF and to stimulate cellular proliferation and hematopoietic differentiation. LTα is a particularly powerful stimulator of neutrophils, an effect enhanced in the presence of IFNγ. However, because LTα is naturally present at much lower levels than TNF, LTα does not mediate endotoxic shock or other tissue injuries attributed to high levels of TNF. Evidence from LTα$^{-/-}$ knockout mice revealed an unexpected role for the LTα chain in secondary lymphoid organ development that arises from its participation in the LTαβ molecule (see later).

LTα production. As mentioned previously, functional LTα is secreted and occurs as a trimer of LTα chains, reminiscent of the structure of functional TNF. LTα is expressed primarily by

activated Th1, B, and NK cells but, in contrast to TNF, is not produced by activated macrophages. LTα is often produced coordinately with IFNγ.

LTα receptor. Secreted LTα uses TNFR1 and TNFR2 as its receptors, even though TNF and LTα are only 28% homologous at the amino acid level.

LTαβ function. LTαβ lacks the pro-inflammatory, antiviral, proliferative, and cytotoxic properties of TNF and LTα, explained in part by the fact that LTαβ does not bind to TNFR1 or TNFR2. Rather, LTαβ appears to be essential for developmental functions not served by TNF. The role of LTαβ was brought to light by study of the phenotype of LTα$^{-/-}$ and LTβ$^{-/-}$ knockout mice. Lymph nodes, Peyer's patches, and splenic white pulp are all missing in LTα$^{-/-}$ mice and only a few germinal centers are present in the spleen. LTβ$^{-/-}$ mice resemble LTα$^{-/-}$ animals but are less severely affected, lacking Peyer's patches and most lymph nodes but retaining the mesenteric and cervical lymph nodes. The splenic white pulp is present in LTβ$^{-/-}$ mice but scrambled with the red pulp. These defects in lymph node and spleen development present in mice deficient for either one of the LT subunits are not observed in TNF$^{-/-}$ or TNFR$^{-/-}$ mice, indicating that LTαβ has a specialized role in secondary lymphoid organ development.

LTαβ production. Like the LTα chain, the LTβ chain is expressed primarily by activated Th1 cells, B cells, and NK cells. However, LTβ is not secreted and exists as an integral membrane protein. In the LTαβ heterotrimer, two transmembrane LTβ chains serve to anchor the (transmembrane-less) LTα chain to form the membrane-bound complex.

LTαβ receptor. LTαβ binds to a receptor called LTβR, which is similar in structure to TNFR1 and II but distinct from them. The LTβR is found primarily on non-lymphoid cells and transduces signals through TRAF5 that activate NF-κB.

iii) OPGL

Function. OPGL (also known as TRANCE, for <u>T</u>NF-<u>r</u>elated <u>a</u>ctivation-<u>in</u>duced <u>c</u>ytokine; ODF, for <u>o</u>steoclast <u>d</u>ifferentiation <u>f</u>actor; and RANKL, for <u>r</u>eceptor <u>a</u>ctivator of <u>N</u>F-κB <u>l</u>igand) is a member of the TNF superfamily. Both secreted and transmembrane forms of OPGL exist and are functional. OPGL has key regulatory roles in bone homeostasis and immune responses (Fig. 17-37). *In vitro*, OPGL mediates osteoclastogenesis in the presence of CSF-1 (*colony-stimulating factor 1*), and can activate mature bone-resorbing osteoclasts in culture. *In vivo*, OPGL$^{-/-}$ mice suffer from severe osteopetrosis (increased bone density), leading to death from poor nutrition secondary to a failure of the teeth to erupt through the abnormally dense jaw bone. OPGL$^{-/-}$ mice completely lack both immature and mature osteoclasts, indicating that OPGL signaling is essential for osteoclast differentiation. Specifically, it was shown that osteoclast precursors are normal in OPGL$^{-/-}$ mice but that a failure of surrounding osteoblast/stromal cells to produce OPGL blocks any further differentiation. Studies of OPGL$^{-/-}$ mice have also shown that OPGL may play a partial role in T and B cell maturation at the pre-TCR and pre-BCR stages. *In vitro*, OPGL prolongs the survival of mature DCs in culture (correlating with an

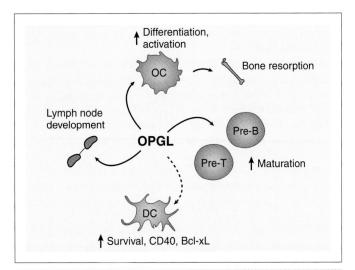

Figure 17-37

Major Functions of OPGL
This member of the TNF superfamily is the master regulator of osteoclastogenesis.

upregulation of the expression of both CD40 and the anti-apoptotic gene Bcl-xL) but does not alter the expression of costimulatory molecules or MHC class II on these APCs. *In vivo*, OPGL is not essential for DC maturation. Finally, OPGL regulates lymph node organogenesis, since, although OPGL$^{-/-}$ mice have normal spleens and Peyer's patches, they lack all lymph nodes.

Production. OPGL is highly expressed by osteoblasts and bone marrow stromal cells. OPGL expression on these cells is upregulated by factors known to promote bone resorption, such as Vitamin D3, parathyroid hormone-related peptide (PTHrP), and glucocorticoids. OPGL is also expressed by activated T cells (particularly Th1 cells) as an early gene and is secreted by these cells. In addition, OPGL expression can be induced by the pro-inflammatory cytokines TNF, IL-1, and IL-11.

Receptor. RANK (*receptor activator of NF-κB*; also known as ODAR, *osteoclast differentiation and activation receptor*) is the cognate receptor for OPGL. Interaction between OPGL expressed on stromal cells and RANK on osteoclast precursors is essential for osteoclastogenesis. RANK is also highly expressed on mature DCs, and is found at lower levels on chondrocytes, mature osteoclasts, endothelial cells of large arteries, hematopoietic precursors, and mature T cells. As a member of the TNFR superfamily, RANK is most closely related in structure to CD40. Several different TRAF molecules (including TRAF2, 3, 5, and 6) can interact with the cytoplasmic tail of RANK, suggesting that this receptor may be able to transduce signals leading to different cellular outcomes. For example, TRAF6$^{-/-}$ mice have normal numbers of osteoclasts (unlike OPGL$^{-/-}$ mice) but the activation and function of these cells are impaired. This result suggests that other TRAFs recruited to RANK can participate in osteoclastogenesis itself, but that signals transduced through TRAF6

are essential for normal functionality. RANK signaling downstream of TRAF6 leads to the activation of NF-κB and c-jun. RANK stimulation has also been shown to result in the activation of Src family kinases and the important cell survival kinase PKB/Akt.

Bone remodeling and bone loss are regulated by a balance between OPGL–RANK interaction and the binding of OPGL to a soluble physiological decoy receptor called OPG (*osteoprotegerin*) (Fig. 17-38). OPG is a member of the TNFR superfamily and is structurally related to RANK. OPG is encoded by a unique gene and, in contrast to other TNFR family members that are generally membrane-bound and functional only as trimers, OPG is produced solely in soluble form and is functional as a dimer. OPG protects bone because it blocks OPGL binding to RANK, thereby inhibiting the maturation of osteoclasts. Interestingly, the hormone estrogen increases OPG expression, providing a molecular explanation for the success of hormone replacement therapy in combating postmenopausal osteoporosis and loss of bone mass. The negative regulatory effect of TGFβ (see later) on osteoclastogenesis is mediated by an upregulation of OPG and a downregulation of OPGL production by osteoblasts. In a T cell-dependent model of rat arthritis, the blocking of OPGL via OPG at the onset of the disease prevented bone and cartilage destruction (but not joint inflammation). Such results

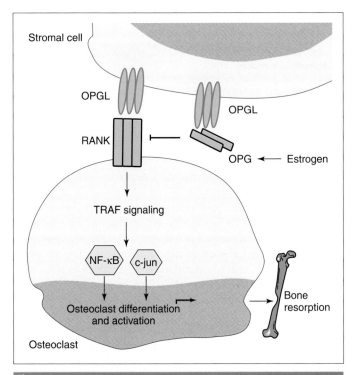

Figure 17-38

OPGL–RANK–OPG Interaction
Binding of OPGL to the RANK receptor initiates TRAF signaling that leads to transcription factor activation and osteoclast differentiation and activation. OPGL activity is regulated by the natural soluble decoy receptor OPG, which binds to OPGL and blocks its interaction with RANK.

hold considerable promise for the therapeutic use of OPG to treat osteoporosis, arthritis, and bone loss in certain types of bone cancers.

iv) BAFF

Function. BAFF (_B lymphocyte activating factor belonging to the TNF family_) was discussed in Chapter 9 in the context of its role in B cell development. To review briefly, BAFF is a survival factor required for the maturation of transitional T1 B cells in the spleen into transitional T2 B cells. In the absence of BAFF, no mature B cells are found in the periphery. It is thought that BAFF signaling activates NF-κB and induces Bcl-2 expression. Overexpression of BAFF leads to autoimmunity in mice, and elevated serum levels of BAFF have been found in human patients with various autoimmune diseases (see Ch.29). Moreover, a direct correlation has been found between serum BAFF levels and the severity of the autoimmune disease symptoms. BAFF has several other names in the literature, including BLyS (_B lymphocyte stimulating factor_), TALL-1 (_TNF- and ApoL-related leukocyte-expressed ligand_), THANK (_TNF homologue that activates apoptosis, NF-κB, and c-jun kinase_), and TNFSF13B (_TNF superfamily member 13B_).

Production. BAFF is produced primarily by monocytes, macrophages, and DCs but not by B cells. Pro-inflammatory stimuli such as LPS, IFNγ, and IFNα/β stimulate BAFF expression by macrophages. Interestingly, the anti-inflammatory cytokine IL-10 has also been reported to stimulate strong BAFF expression by macrophages and weak expression by DCs. Structurally, BAFF has a crystal structure unusual for the TNF superfamily in that a particular polypeptide loop thought to be involved in receptor binding is longer than in related TNF-like ligands. BAFF occurs both as a homotrimeric transmembrane molecule and in a soluble homotrimeric form created by cleavage of the transmembrane form. It is thought that the soluble form is the source of the majority of BAFF biological activity.

Receptor. As noted in Chapter 9, BAFF binds to three receptors, BAFF-R, TACI, and BCMA. BAFF-R is expressed almost exclusively by B cells in the secondary lymphoid tissues. TACI is expressed by B cells and certain subsets of activated T cells. BCMA is preferentially expressed by mature B cells and plasma cells but is also present at low levels in several other cell types, including T cells. However, the majority of BCMA in these latter cell types appears to be localized intracellularly. Only the engagement of BAFF-R by BAFF delivers the B cell survival signals necessary to promote the T1 to T2 transition. Engagement of TACI on B cells by BAFF may deliver negative regulatory signals governing this process. Engagement of BCMA by BAFF appears to deliver a signal supporting B cell proliferation that is redundant with signals triggered by BAFF/BAFF-R interaction. Members of the TRAF family of signal transducers have been identified as acting downstream of TACI and BMCA engagement by BAFF.

In addition to BAFF, a soluble TNF-related molecule called APRIL (_a proliferation-inducing ligand_), which is structurally very similar to BAFF, can bind to BCMA and TACI (but not BAFF-R). APRIL was first identified as a molecule secreted at high levels by some types of cancers, and treatment of tumor cells _in vitro_ with purified APRIL can stimulate their proliferation. In a healthy individual, low levels of APRIL are produced by macrophages and DCs. The interaction of APRIL with TACI expressed on normal B cells is thought to be required for optimal humoral responses to Ti antigens, as responses to these antigens are impaired in TACI-deficient mice. APRIL-deficient mice show upregulated IgG responses to Td antigens but impaired isotype switching to IgA.

IV. TRANSFORMING GROWTH FACTORS

Like many cytokines, the TGF molecules were named for the first (but perhaps not the most important) of their many activities identified by researchers of the day. TGF was the name originally given to a growth factor secreted by certain tumor cells in culture. Normal connective tissue cells exposed to this factor became transiently able to grow in soft agar in an "anchorage-independent" fashion, an ability usually associated with transformed (cancerous) cells. Accordingly, the factor was named "transforming growth factor." The original TGF turned out to be a mixture of two molecules: TGFα, a growth factor for non-hematopoietic cells (and of little immunological interest), and TGFβ. A family of TGFβ proteins is now known to exist, comprising highly pleiotropic molecules with crucial negative regulatory roles in cell growth and the adaptive immune response.

i) TGFβ

Function. Three homodimeric isoforms of TGFβ exist in mammals, derived from the related but distinct TGFβ1, TGFβ2, and TGFβ3 genes. All three of these cytokines signal through the same receptors and act in virtually identical ways on the same range of cellular targets, although each isoform is expressed independently. The TGFβ molecules are very pleiotropic and are expressed in a variety of tissues during embryonic development. TGFβ proteins play roles in adhesion, proliferation, differentiation, transformation, chemotaxis, and immunoregulation. TGFβ also stimulates angiogenesis and promotes the production of extracellular matrix proteins _in vitro_. The effect of TGFβ on a given cell can be strongly influenced by minor adjustments to culture conditions, with proliferation being induced or blocked depending on the circumstances.

TGFβ1 is the isoform most closely associated with the immune system, acting primarily as a brake on lymphocyte function and cytokine synthesis. TGFβ1 inhibits macrophage activation, the proliferation of activated T cells, homing of activated T cells to endothelial sites, Ig synthesis by B cells, MHC class II expression, the growth of bone marrow progenitors, the differentiation of CTLs, and the cytotoxic activities of CTLs and NK cells (Fig. 17-39). TGFβ1 also blocks the effects of pro-inflammatory cytokines on neutrophils and endothelial cells, and downregulates the production of IL-1, IL-2, IL-6, IFNγ, and TNF by various cell types. TGFβ1 counters the stimulatory effects of IL-2 on

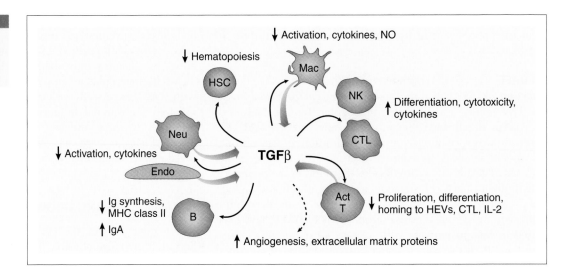

Figure 17-39

Major Functions of TGFβ
TGFβ is a regulatory cytokine with immunosuppressive and anti-inflammatory properties.

T and B cells, and of IL-1 on thymocytes. TGFβ1 induces the synthesis of IL-1Ra, which may account for some of its anti-inflammatory activity. In addition, NO synthesis by macrophages is controlled in part by TGFβ1. TGFβ1 does have some positive effects, inducing isotype switching to IgA in some B cells, and functioning as a potent chemoattractant for T cells, monocytes, and neutrophils. In accordance with the fundamental roles of TGFβ1 in embryonic development and immunoregulation, TGFβ1$^{-/-}$ knockout mice exhibit two phenotypes: embryonic lethality in some mutant mice, and massive multifocal inflammation leading to death within 5 weeks of birth in others. TGFβ3 has very similar effects to TGFβ1 but may also play a unique role during embryonic development. TGFβ3$^{-/-}$ mice die within 20 hours of birth of malformations of the palate and lung that prevent suckling, alterations that are not observed in TGFβ1$^{-/-}$ mice. The TGFβ2$^{-/-}$ mouse has numerous developmental abnormalities not found in TGFβ1$^{-/-}$ or TGFβ3$^{-/-}$ mice, including defects in the cranium, urogenital tract, limbs, spine, eye, and inner ear.

Production. TGFβ molecules are synthesized as pro-proteins whose signal sequences are cleaved to yield small monomers. TGFβ pro-proteins are produced primarily by hematopoietic cells, including activated T, B, and NK cells, mature DCs, macrophages, platelets, and thymic stromal cells, but also by certain cells in non-hematopoietic tissues such as the kidney, CNS, and eye. TGFβ1 constitutes the majority of TGFβ produced by these cells. Non-hematopoietic cells in other tissues have been found to produce significant quantities of TGFβ3. TGFβ in the circulation takes the form of a 25 kDa homodimer non-covalently complexed to the 75 kDa *latency-associated protein (LAP)*. LAP holds TGFβ inactive until a signal delivered in a poorly defined way releases the TGFβ molecule. Some researchers believe the extracellular matrix protein thrombospondin-1 stimulates an unknown extracellular protease to cleave LAP and activate TGFβ. Others have evidence that the integrin protein αvβ6 expressed exclusively on epithelial cells binds to LAP and alters its conformation such that TGFβ is activated.

Receptor. TGFβ1, TGFβ2, and TGFβ3 can each bind to three glycoprotein chains: TGFβRI, TGFβRII, and TGFβRIII. TGFβRI and TGFβRII combine to form the high-affinity TGFβ receptor. The TGFβRIII chain, which binds TGFβ with lower affinity, may interact with the TGFβRI and TGFβRII chains in a regulatory way, serving as a non-signaling "holding facility" for TGFβ molecules. The TGFβR chains are widely expressed on most cell types, can be upregulated upon T cell activation, and are downregulated on monocytes by IFNγ. Unlike most cytokine receptor chains, TGFβRI and TGFβRII both have intrinsic serine-threonine kinase activity, and TGFβRII is capable of autophosphorylation.

To initiate signal transduction, TGFβ binds with high affinity to the TGFβRII chain, which then recruits and phosphorylates the TGFβRI chain. Subsequent signal transduction occurs principally through interactions of the TGFβRI chain with members of a family of signal transducing molecules called the Smads (Fig. 17-40). The Smad family consists of at least nine members with diverse functions. Some Smads (particularly Smad2 and Smad3) interact directly with the TGFβR complex such that these transducers become phosphorylated. Smad4, the most important downstream mediator, is then recruited to form a heterodimeric transcription factor with Smad2. This transcription factor translocates to the nucleus and regulates transcription of TGFβ-induced genes. The Smad-mediated pathway of TGFβ signal transduction is controlled by the activities of the inhibitory Smads (including Smad7) that block recruitment of Smads to the TGFβR complex. In addition to the Smad pathway, TGFβR engagement can activate the p38 and SAPK/JNK signaling pathways via the kinase Tak1. This serine-threonine kinase phosphorylates and activates upstream MAPK enzymes in these pathways.

V. HEMATOPOIETIC GROWTH FACTORS

Because of their importance to the development of cells mediating immune responses, and their similarities to cytokines, a short discussion on selected growth factors has been included

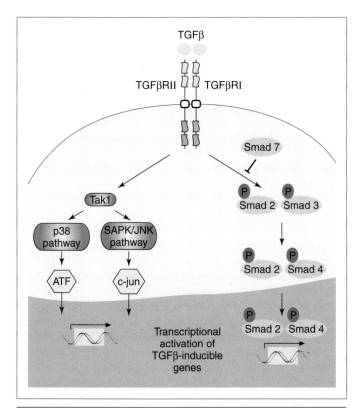

Figure 17-40

TGFβ Receptor Signaling
Engagement of TGFβR triggers signal transduction via Smad recruitment or activation of Tak1 and the p38 and SAPK/JNK pathways. The transcription of various TGFβ-inducible genes is activated. Smad 7 inhibits signaling mediated by other Smads.

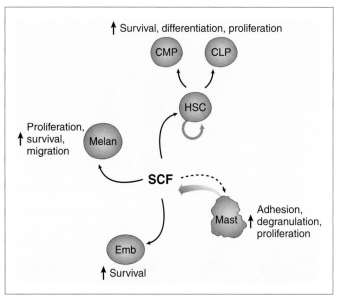

Figure 17-41

Major Functions of SCF
This cytokine is critical for the survival, self-renewal, and differentiation of HSC, and the survival, proliferation, and migration of melanocytes.

in this chapter. The functions of G-CSF, GM-CSF, and M-CSF were introduced in Chapter 3, while the importance of SCF and its receptor c-kit to B and T cell development was described in Chapters 9 and 13, respectively.

i) Stem Cell Factor (SCF)
Function. SCF (also known as c-kit ligand, Steel factor, and mast cell growth factor) supports the survival and growth of the earliest hematopoietic precursors *in vivo* (Fig. 17-41). *In vitro*, SCF is sufficient to support the survival and self-renewal of hematopoietic stem cells. Later during hematopoiesis, and in combination with other growth factors, SCF enhances the proliferation of cells committed to the lymphoid or myeloid lineages. SCF is also crucial for the survival during embryogenesis of primordial germ cells and melanocyte precursors. SCF promotes the growth of mast cells *in vitro*, and stimulates mast cell chemotaxis, adhesion, and secretory and degranulation functions. Mice lacking SCF protein, such as the natural mutant *Sl*, die either before birth or perinatally from severe anemia. It is thought that the major role of SCF is to make stem cells receptive to the influence of other cytokines.
Production. SCF is produced by stromal cells in the fetal liver, bone marrow, and thymus (particularly endothelial

cells, adipocytes, and fibroblasts), in the central nervous system, and in the gut mucosa. Two forms of SCF exist, one membrane-bound and the other secreted. The secreted form is a non-covalently associated homodimer. The membrane-bound form appears to be the more important for hematopoietic cell development. SCF production by marrow stromal cells can be modestly increased by inflammatory cytokines such as IL-1 and TNF, and modestly decreased by TGFβ (at least *in vitro*).
Receptor. The receptor for SCF is c-kit, a protein that was first identified as the product of an oncogene. C-kit is most highly expressed on pluripotent stem cells in the bone marrow but is also found on cells in the central nervous system and gut. Expression of c-kit is downregulated upon exposure of hematopoietic progenitors to IL-4, TNF, or TGFβ.

Unlike many cytokine receptors, c-kit has intrinsic tyrosine kinase activity and can transduce signals via several different pathways (Fig. 17-42). Binding of SCF triggers c-kit homodimerization in the membrane and the autophosphorylation of residues in the receptor protein that supply docking sites for SH2-containing signal transduction proteins, including Vav1 and Jak2. C-kit then mediates tyrosine phosphorylation leading to the activation of the Ras/MAPK, SAPK/JNK, PLC-γ, PI3K, PKB/Akt, Vav1, and p38 pathways, and Jak2/STAT1. Negative regulation of SCF-induced Jak2 phosphorylation is mediated by the phosphatase SHP-1. C-kit activity is also regulated by the binding of adaptor proteins. For example, the *Socs1* protein, which binds to c-kit via SH2 domain interactions, suppresses SCF-dependent mitogenic responses by cells in culture. However, the kinase activity of c-kit is not itself

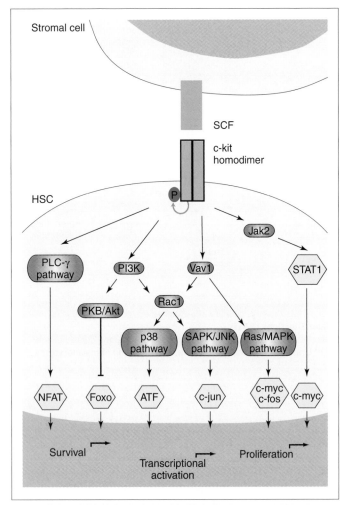

Figure 17-42

SCF Receptor Signaling
C-kit homodimers are formed upon engagement by SCF followed by autophosphorylation. The activated c-kit complex then triggers several different pathways, each of which promotes cellular survival and proliferation (see text).

inhibited and cell survival signals are maintained in the presence of Socs1, indicating that Socs1 may function as a switch determining which response is to occur in response to SCF binding. Socs1 mRNA is rapidly upregulated by SCF, establishing a feedback loop for control of SCF-induced proliferation.

A natural mouse mutation of c-kit exists called the W or "white" mutation. Mild versions of the W mutation (W^v) are characterized by white spotting of the hair coat in these animals, In contrast, mice carrying severe W mutations that result in a complete lack of c-kit protein die before birth or in the perinatal period of severe anemia, like SCF-deficient mice. A human equivalent of the mild W^v mutation exists and is called "piebaldism." People with this disorder are not anemic but display white shocks of hair and pale areas of skin that lack melanocytes and their pigments. The phenotypes of the W^v mice and piebald humans suggest that SCF is required for

melanocyte survival, proliferation, and/or migration during fetal development.

ii) GM-CSF

Function. GM-CSF can act on bone marrow to increase the generation of hematopoietic precursor cells, particularly inflammatory leukocytes, from the most immature precursors (Fig. 17-43). As we saw in Chapter 3, CMPs cultured *in vitro* with GM-CSF give rise to CFU-GEMM cells, which are the progenitors of granulocytes, erythrocytes, monocytes, and macrophages. In the continued presence of GM-CSF, CFU-GEMMs generate CFU-BME (a precursor giving rise to basophils, megakaryocytes, and erythrocytes) and CFU-GM (a precursor giving rise to granulocytes and monocytes). From the CFU-GM cells eventually come the neutrophils that are so important during the earliest stages of any infection. However, GM-CSF has more clinical side effects than G-CSF (see later) and so is only used sparingly to promote neutrophil recovery in patients that have lost bone marrow function. GM-CSF also activates macrophages *in vitro* (although to a lesser extent than IFNγ) and has been reported to be involved in DC differentiation. Interestingly, the main phenotypic defect in GM-CSF$^{-/-}$ mice is one of pathologic lung abnormalities rather than disruption of mononuclear phagocyte homeostasis. These data suggest that, *in vivo*, GM-CSF is essential only for the generation of a certain type of lung macrophage, and possibly not for the whole spectrum of activities observed *in vitro*.

Production. GM-CSF is produced primarily by activated T cells, stromal cells, endothelial cells and macrophages. Unlike G-CSF, GM-CSF action is restricted to cells in the producing cell's immediate vicinity, and therefore is not found in the blood circulation.

Receptor. The GM-CSF receptor is composed of a unique ligand-binding α chain and the non-binding βc common chain shared with the IL-3R and IL-5R. GM-CSF signaling activates the Ras/MAPK and Jak/STAT pathways. Stimulation of the receptor by GM-CSF induces rapid phosphorylation of primarily Jak2 but also Jak1, followed by activation of STAT5.

iii) G-CSF

Function. G-CSF influences CFU-GM cells to differentiate into granulocytes *in vitro* (refer to Fig. 17-43). G-CSF also supports the survival of these cells *in vitro* and *in vivo*. Studies of G-CSF$^{-/-}$ mice confirmed that G-CSF is required for the steady-state production of neutrophils and implicate G-CSF in the "emergency production" of neutrophils that occurs in response to acute injury or infection. Clinicians take advantage of this property of G-CSF to rebuild bone marrow-derived cell populations in patients with neutropenia and in cancer patients who have lost bone marrow function due to chemotherapy or irradiation.

Production. G-CSF is produced both in the bone marrow and by activated T cells, endothelial cells, fibroblasts, and mononuclear phagocytes. G-CSF is a growth factor that is capable of acting at a distance from the cell that secreted it, and thus can be detected in the blood circulation.

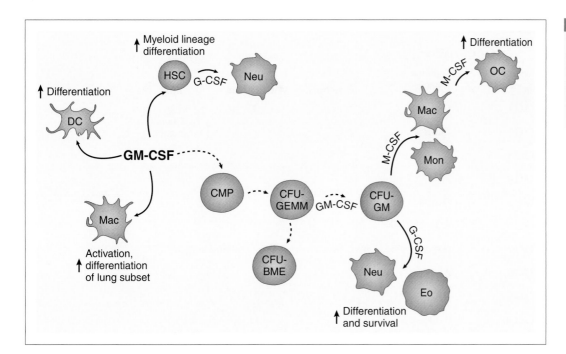

Figure 17-43

Major Functions of GM-CSF, G-CSF, and M-CSF
These growth factors promote the differentiation and activation of many myeloid cell types both *in vivo* and *in vitro*.

Receptor. The G-CSF receptor chain contains the WSXWS motif and is expressed on committed monocyte/granulocyte precursors. Dimerization of two G-CSFR chains is required to form the high-affinity binding site for G-CSF. Intracellular signaling is transduced primarily through Jak2 and STAT3, although Jak1 and STAT1 are also rapidly activated in response to G-CSFR engagement.

iv) M-CSF

Function. M-CSF influences CFU-GM cells to differentiate into monocytes and macrophages *in vitro*. *In vivo*, the role of M-CSF may be more specialized than anticipated from *in vitro* results. A natural form of M-CSF$^{-/-}$ mouse exists in the *op/op* mutants, whose main phenotype is osteoporosis. Studies of these animals have revealed a crucial role for M-CSF in generating osteoclasts (refer to Fig. 17-43).

Production. M-CSF is produced in the bone marrow and by endothelial cells, fibroblasts, and mononuclear phagocytes. Its effects are felt locally; that is, M-CSF is not found in the blood circulation.

Receptor. The M-CSF receptor is related to c-kit in structure, has intrinsic tyrosine kinase activity, and is expressed on committed monocyte/granulocyte precursors.

VI. CHEMOKINES AND THEIR RECEPTORS

As introduced in Box 4-3 in Chapter 4, chemokines constitute a large superfamily of structurally related small molecules (70–100 amino acids) that are characterized by the presence of internal disulfide bridges involving conserved N-terminal cysteine residues. The more than 50 chemokines that have been identified have been classified by structure.

In humans, a standardized system of nomenclature was adopted in 2000 that names the chemokines numerically within their structural group (Table 17-4). For example, the systematic name for the CXC chemokine IL-8 is CXCL8 ("L" for "ligand"), while that for the CC chemokine RANTES (*regulated on activation, normal T cell expressed and secreted*) is CCL5. The best known C chemokine is lymphotactin, systematically known as XCL1, while the CX3C chemokine fractalkine is designated CX3CL1. There are mouse homologues for most (but not all) human chemokines and vice versa.

Function

Chemokines are cytokines that are best known for their chemoattractant effects on particular types of leukocytes (Fig. 17-44), but it should be noted that these molecules can also influence the behavior of endothelial cells and various cell types in the central nervous system. With respect to the immune system, chemokines function primarily to mediate selective trafficking of leukocyte subsets between the blood and various tissues, and the recirculation of lymphocytes between the tissues and the lymphatics. The C (XC) chemokines lymphotactin-α and lymphotactin-β are strongly and preferentially chemotactic for T lymphocytes, while the CC chemokines tend to attract mostly monocytes and T cells, with lesser effects on basophils and eosinophils. In general, neutrophils do not respond to members of the CC chemokine group. In contrast, most CXC chemokines act predominantly on neutrophils and have lesser effects on monocytes and lymphocytes. The CX3C chemokine fractalkine (CX3CL1) contributes to the recruitment of Th effectors to peripheral tissues and lymphoid organs, and to adhesion between monocytes and endothelial cells.

Table 17-4 **Chemokine Nomenclature**

CC Chemokines		CXC Chemokines	
Systematic Name	Common Name	Systematic Name	Common Name
CCL1	I-309	CXCL1	GROα
CCL2	MCP-1	CXCL2	GROβ
CCL3	MIP-1α	CXCL3	GROγ
CCL4	MIP-1β	CXCL4	PF4
CCL5	RANTES	CXCL5	ENA-78
CCL6*	None	CXCL6	GCP-2
CCL7	MCP-3	CXCL7	NAP-2
CCL8	MCP-2	CXCL8	IL-8
CCL9*	None	CXCL9	MIG
CCL10*	None	CXCL10	IP-10
CCL11	Eotaxin	CXCL11	I-TAC
CCL13†	MCP-4	CXCL12	SDF-1
CCL14†	HCC-1	CXCL13	BLC
CCL15†	HCC-2	CXCL14*	BRAK
CCL16	HCC-4	CXCL15	
CCL17	TARC	CXCL16	
CCL18†	PARC		
CCL19	MIP-3β		
CCL20	MIP-3α	**C (XC) Chemokines**	
CCL21	SLC	XCL1	Lymphotactin-α
CCL22	MDC	XCL2	Lymphotactin-β
CCL23†	MPIF-1		
CCL24†	MPIF-2	**CX3C Chemokine**	
CCL25	TECK	CX3CL1	Fractalkine
CCL26†	Eotaxin-3		
CCL27	CTACK		
CCL28			

* Currently found in mice but not humans.
† Currently found in humans but not mice.

In addition to their roles in directing cellular migration, many chemokines appear to have a broader function. For example, as well as inducing neutrophil migration to sites of inflammation, IL-8 (CXCL8) promotes the degranulation and respiratory burst of these cells. IL-8 also promotes intracellular signaling events downstream of the TCR in T cells such as PKC activation and calcium flux. RANTES (CCL5) has also

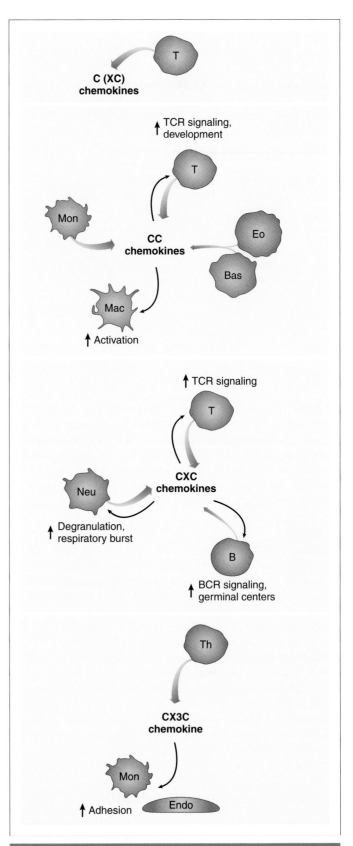

Figure 17-44

Major Functions of Chemokines
Four structural classes of chemokines can be distinguished. Cell types that are only weakly attracted by the indicated chemokine class are indicated by thin gradient arrows. See text for details.

been shown to directly promote T lymphocyte activation and differentiation by enhancing tyrosine phosphorylation of various protein tyrosine kinases involved in intracellular signaling. Similarly, the CC chemokine MCP-1 (*monocyte chemotactic protein-1*; CCL2) stimulates downstream signaling events in macrophages, including alterations to calcium flux and the generation of inositol triphosphate compounds. TARC (*thymus and activation-regulated chemokine*; CCL17) and SLC (*secondary lymphoid tissue chemokine*; CCL21) are among a group of CC chemokines that are constitutively expressed at high levels in the thymus, lymph nodes, and other lymphoid tissues. It is thought that by drawing thymocytes and T cells bearing the appropriate chemokine receptors into these tissues, these chemokines play an important role in the regulation of T cell development.

In contrast to the wealth of information on T cell, macrophage, and neutrophil chemokines, very little is known about B cell chemokines. One study of 10 CXC and 19 CC human chemokines found that only a CXC chemokine called BCA-1 (*B cell attracting chemokine 1*; CXCL13) had a chemotactic effect on B cells. BCA-1 is expressed constitutively in the secondary lymphoid organs and acts specifically on B cells to induce chemotaxis and Ca^{2+} mobilization. BCA-1-mediated B cell migration is necessary for the normal organization of the germinal centers and primary follicles in the secondary lymphoid tissues. BCA-1 acts through the CXCR5 receptor, which is expressed exclusively on B cells. BCA-1 consequently has no effect on T cells, neutrophils, or macrophage/monocytes. Later studies uncovered the existence of the CXCR6 receptor on B cells, implying that the CXCL15 and CXCL16 chemokines that bind to this receptor are also chemotactic for B cells.

Production

The C chemokines are produced by activated T lymphocytes. In humans, these molecules are encoded by a gene on chromosome 1. The human CC chemokines are made primarily by activated T cells and DCs from genes contained in a large cluster on chromosome 17 and two smaller clusters on chromosomes 9 and 16. The CXC chemokines are generally the products of activated macrophages, neutrophils, fibroblasts, and endothelial cells expressed from genes located on chromosome 4 in humans. Several CXC chemokines that act on T cells are clustered in a separate group located on human chromosome 4. The CXC3 chemokine fractalkine, known as neurotactin in mice, is synthesized by a wide variety of cell types from a gene encoded on human chromosome 16. Activated endothelial cells, intestinal epithelial cells, DCs, and neurons in the CNS all express fractalkine. The production of a few chemokines is tissue-specific. For example, CCL27 is exclusively produced by keratinocytes in the skin, serving as a means of drawing a subset of T cells expressing CLA[+] (cutaneous lymphocyte antigen) to areas of injury in the skin.

Receptors

At least 20 chemokine receptors have been identified to date whose genes are generally clustered on human chromosomes 2 and 3. Promiscuity is a feature of chemokine/receptor binding

in that one chemokine may bind to several different chemokine receptors, and a single chemokine receptor may transduce signals from several different chemokines (Fig. 17-45). However, this promiscuity is generally class-restricted: CC receptors (CCR) bind only CC chemokines, while CXC receptors (CXCR) bind only CXC chemokines. The so-called "Duffy antigen" of erythrocytes, which diverges considerably in structure from the classical CC and CXC receptors, is the only chemokine receptor that can bind both CC and CXC chemokines. So far, however, there is no evidence that this receptor in fact initiates any intracellular signaling. The XCR1 and CX3CR1 receptors bind the C and CX3C chemokines, respectively.

As explained in our discussion of IL-8, most chemokine receptors are seven-transmembrane domain G protein-coupled molecules, meaning that the signal initiated by chemokine binding is transduced via activation of G proteins. Intracellular signaling downstream of the G proteins is transduced by protein tyrosine kinase cascades leading to the activation of PLC-γ, RhoGTPases, PI3K, and the PKB/Akt pathway (Fig. 17-46). STAT1 and STAT3 become phosphorylated.

Different chemokines affect different leukocytes, depending on their expression of particular chemokine receptors. For example, the chemokines IP-10 (*IFNγ-inducible protein 10*; CXCL10) and MIG (*monokine induced by IFNγ*; CXCL9), which are chemotactic for T cells, are ligands for CXCR3, which is expressed almost exclusively on activated T cells. Different cell types can express different constellations of these receptors depending on the state of activation of the cells. For example, as we learned in Chapter 14, naive T cells express primarily CXCR4, the receptor for SDF-1 (*stromal cell-derived factor*; CXCL12) which is thought to govern the basal trafficking of these cells. The activation of T cells and B7-mediated stimulation of CD28 causes them to upregulate the expression of CCR1, CCR2, CCR4, and CCR5, and thus become capable of binding RANTES and TARC. In particular, memory T cells express large numbers of CCR5 molecules, the receptor for RANTES which influences the migration of activated and memory T cells. As we have seen in Chapter 15, Th1/Th2 maturation also results in differential chemokine receptor expression. CXCR3 and CCR5 are preferentially found on Th1 cells, while CCR3, CCR4, and CCR8 are prevalent on Th2 cells, allowing the required T cell subset to be drawn into the area by the presence of the appropriate chemokine.

Chemokine receptor expression can also influence the behavior of other immune system cell types. Immature DCs in the tissues are able to respond to macrophage-produced MIP-3α (CCL20) because of the expression of CCR6 on the immature DC surface. However, once these cells acquire antigen and start to mature, expression of CCR6 is replaced by expression of CCR7, the receptor mediating migration to the lymph nodes in response to MIP-3β (CCL19) and SLC. Similarly, the expression of CCR3 on eosinophils draws them to sites of allergic reactions, where production of eotaxin, a key ligand for CCR3, is upregulated. As mentioned previously, CXCR5 and CXCR6 appear to be expressed only on

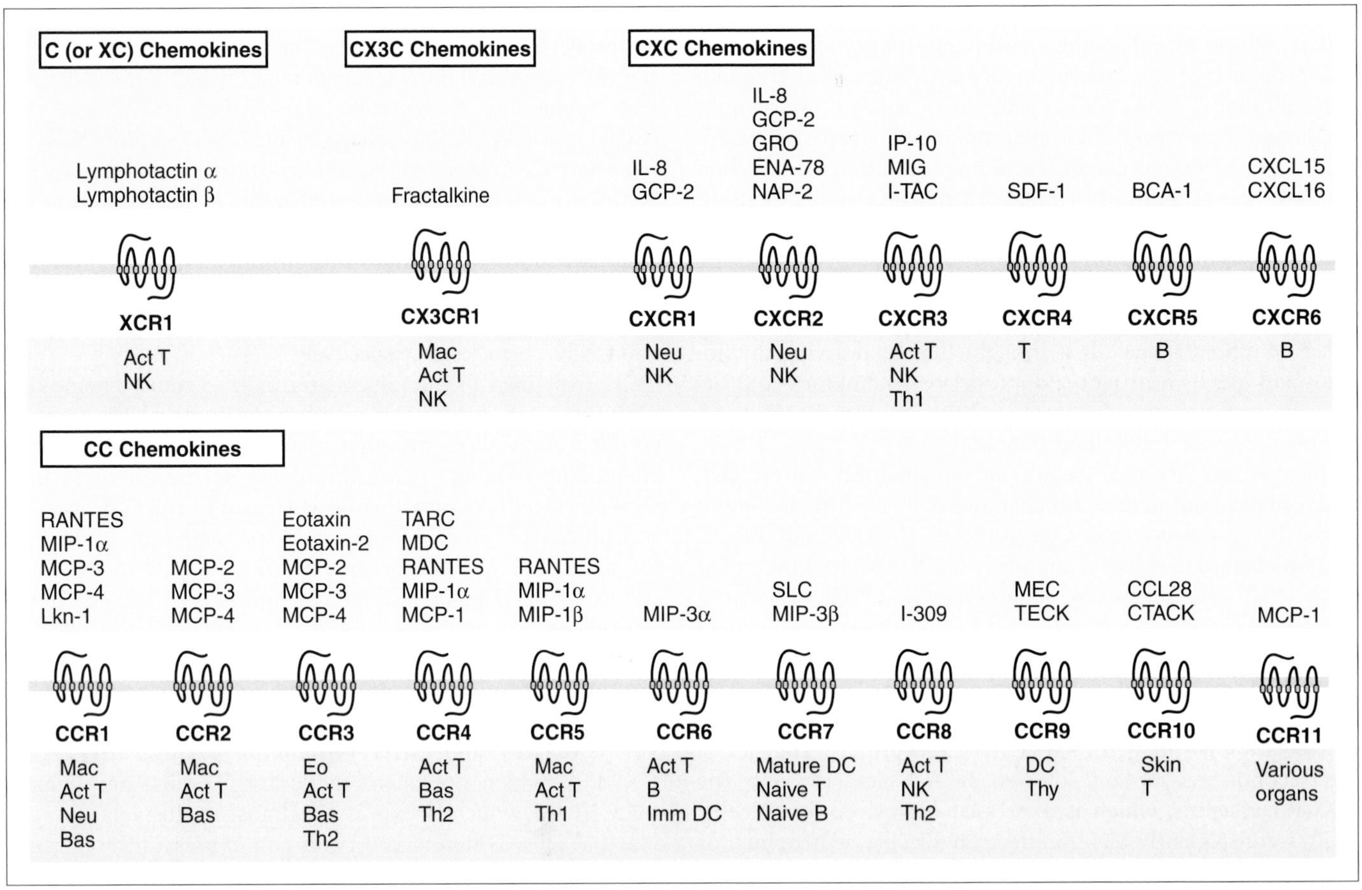

C (or XC) Chemokines

Lymphotactin α
Lymphotactin β

XCR1
Act T
NK

CX3C Chemokines

Fractalkine

CX3CR1
Mac
Act T
NK

CXC Chemokines

IL-8 GCP-2	IL-8 GCP-2 GRO ENA-78 NAP-2	IP-10 MIG I-TAC	SDF-1	BCA-1	CXCL15 CXCL16
CXCR1	**CXCR2**	**CXCR3**	**CXCR4**	**CXCR5**	**CXCR6**
Neu NK	Neu NK	Act T NK Th1	T	B	B

CC Chemokines

RANTES MIP-1α MCP-3 MCP-4 Lkn-1	MCP-2 MCP-3 MCP-4	MCP-2 MCP-3 MCP-4	Eotaxin Eotaxin-2 MCP-2 MIP-1α MCP-1	TARC MDC RANTES MIP-1α MIP-1β	MIP-3α	SLC MIP-3β	I-309	MEC TECK	CCL28 CTACK	MCP-1
CCR1	**CCR2**	**CCR3**	**CCR4**	**CCR5**	**CCR6**	**CCR7**	**CCR8**	**CCR9**	**CCR10**	**CCR11**
Mac Act T Neu Bas	Mac Act T Bas	Eo Act T Bas Th2	Act T Bas Th2	Mac Act T Th1	Act T B Imm DC	Mature DC Naive T Naive B	Act T NK Th2	DC Thy	Skin T	Various organs

Figure 17-45

Chemokine Receptors
A relatively small number of chemokine receptors bind to multiple members of a much larger collection of chemokines (blue shading). Cells expressing the chemokine receptor in question are indicated below the name of the receptor (gray shading). Adapted from Hancock W. W. *et al.* (2002) Chemokines and their receptors in allograft rejection. *Current Opinion in Immunology* **12**, 511–516.

B cells and promote the migration of these cells into the germinal centers.

The importance of selective chemokine receptor expression as a means of directing leukocyte traffic, including the circulation of T cells between different lymphoid compartments, is now widely accepted. In addition, it is becoming clear that some pathogens have evolved to exploit certain chemokine receptors as a means of gaining entry into cells. For example, the malarial parasite *Plasmodium vivax* binds to the Duffy antigen to gain entry to human erythrocytes (see Ch.22). Similarly, HIV binds to CCR5 and CXCR4 to invade T cells and macrophages bearing CD4 (see Ch.25). Humans who carry a mutation of the CCR5 gene such that the receptor does not appear on the T cell surface are resistant to the virus. Other viruses have gone a step further and appropriated genes encoding mammalian cytokines and cytokine receptors. For example, the Epstein–Barr virus appears to have picked up a copy of the human IL-10 gene during evolution and subsequently modified the gene product to suit its nefarious purposes. The protein secreted by EBV has activity similar to that of mammalian IL-10 such that the host's immune responses are inhibited, promoting viral survival. Other examples of the exploitation of various cytokines and their receptors by viruses are given in Box 17-3.

D. Cytokines and Their Receptors in Physiological Context

In this section, we attempt to draw much of the preceding information together in a series of tables capturing the physiological context of cytokine action. The major cytokines involved in the inflammatory response, in responses to intracellular pathogens, in responses to extracellular pathogens, and in hematopoiesis and bone remodeling are collected in Tables 17-5–17-10 detailing the major contribution of each cytokine to the response. It should be borne in mind that chemokines, although not specifically detailed in these tables, are an integral part of any immune response principally because they draw innate and adaptive immune system cells to the sites where they are needed.

Figure 17-46

Chemokine Receptor Signaling
Engagement of a chemokine receptor by a chemokine induces activation of G proteins as well as various non-Src and Src PTKs. Non-Src PTKs act on calcium channels to increase Ca^{2+} flux. This increase, coupled with activation of the PLC-γ1 pathway, leads to vigorous NFAT activation. G protein activation also promotes activation of PI3K and the PKB/Akt pathway, and stimulates Rho-GTPases, which contribute in an unknown way to cellular activation. Src PTKs activated by the chemokine receptor complex transduce signals via ZAP-70 and STATs, which contribute to increased cytokine production.

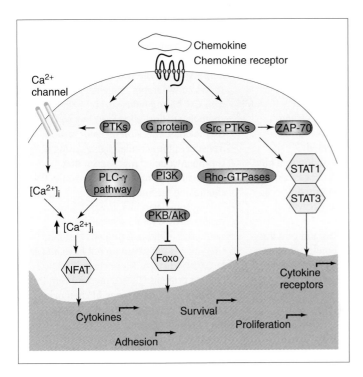

Box 17-3. Viral homologues of cytokines and cytokine receptors

Viruses succeed in infecting us because they escape or subvert the mechanisms of both the innate and adaptive immune responses. Our cells are subsequently forced to churn out virus-specific genomes and proteins. An effective mechanism of subversion employed by certain viruses (most notably members of the herpesvirus and poxvirus groups) involves virus-encoded homologues of host cytokines and cytokine receptors. These viral homologues disrupt host cytokine signaling and thus cripple the host's ability to respond to the viral threat. In some cases, the virus-encoded molecule may be homologous in sequence to a host molecule; in other cases, the resemblance is based on structure or function rather than primary amino acid sequence. Intracellular, transmembrane, or secreted versions of a host protein may be synthesized. We will now describe several examples of these tools of infection, starting with viral cytokines.

- Several types of herpesviruses and poxviruses make viral cytokines resembling IL-10. For example, the BCRF1 gene of Epstein–Barr virus (EBV) encodes a functional protein that is highly homologous to its human and murine counterparts and has most of their functions, including blocking cytokine synthesis (particularly that of IFNγ) by macrophages, NK cells, and T cells. *In vitro*, the BCRF1 product stimulates B cell proliferation and differentiation, like IL-10. Unlike IL-10, however, BCRF1 does not stimulate mast cells or T cells, or induce MHC class II expression on B cells.

- The vaccinia, myxoma, and variola poxviruses have appropriated mammalian growth factor genes for vascular endothelial growth factor (VEGF), transforming growth factor α (TGFα), and epidermal growth factor (EGF). Viral proteins are expressed with at least some of the activities of these molecules. It is thought that these viral growth factors stimulate cell growth inappropriately, promoting viral replication.

- A number of herpesviruses express chemokine-like genes. Some of the products of these genes, like that of the MC148R gene (of the Molluscum contagiosum virus) which resembles the CC chemokine MIP-1β, are competitive inhibitors in that they bind to host cell chemokine receptors but fail to trigger signaling. Others, such as the K4 and K6 ORFs (open reading frames) in KSHV which resemble MIP-1α, appear to be capable of inducing signal transduction. Such signals are thought to induce unwarranted chemotaxis and even cellular activation that benefits the virus.

  The herpesvirus and poxvirus families also make several viral cytokine receptors. Indeed, the virulence of some of these viruses depends on the expression of these molecules.

- The M-T2 gene of the myxoma virus encodes one of the best-studied viral cytokine receptors. Parts of the structure of the protein encoded by this gene closely resemble critical domains in mammalian TNF receptors. The M-T2 viral receptor lacks a transmembrane domain but retains a signal sequence, implying that it functions by being secreted and then binding and neutralizing TNF-related ligands. *In vitro* analyses have confirmed that the M-T2 can compete with cell-bound host receptors for TNF.

- Similar TNF-related viral receptors have been identified in cowpox, variola, and vaccinia viruses. The product of the

Continued

Box 17-3. Viral homologues of cytokines and cytokine receptors—*cont'd*

crmC gene of cowpox virus is a viral receptor homologous to TNFR2. CrmC protein binds specifically to TNF but not to other TNF-related ligands. *In vitro*, crmC can block cytolysis triggered by TNF.

- In addition to the M-T2 protein, the myxoma virus synthesizes the M-T7 protein, which resembles the IFNγ receptor. *In vitro*, M-T7 has been shown to bind to IFNγ and prevent it from engaging its receptor. Establishment of the antiviral state in host cells can thus be inhibited. In addition, M-T7 binds promiscuously to members of the CC, CXC, and C chemokine families. Other IFNγR-like viral receptors of varying functional similarity to M-T7 have been identified in various pox viruses. The tanapox virus produces a viral receptor that is astonishingly broad in its binding activity, sequestering not only IFNγ but also IL-2 and IL-5. Vaccinia virus synthesizes a viral receptor called B18R that is expressed in both secreted and transmembrane forms and is able to bind and inhibit IFNα. Despite this biological activity, however, the B18R protein only vaguely resembles the IFNα/β receptor in

structure. B18R function is essential for viral pathogenesis.

- Members of another class of secreted viral receptors expressed by vaccinia and cowpox viruses mimic the IL-1 receptor. For example, the B15R gene of vaccinia encodes a protein with the structural features of a secreted receptor specific for IL-1β (it does not bind IL-1α or IL-1Ra). *In vitro*, B15R can block the binding of IL-1β to IL-1R on host cells and abrogates the biological activity of IL-1β.
- Members of the herpesvirus and poxvirus families also express viral molecules that mimic certain mammalian chemokine receptors. For example, human cytomegalovirus (CMV) produces the US28 protein, which resembles CCR1 and binds several CC (but no CXC) chemokines with high affinity. Unlike the secreted viral receptors described thus far, US28 is a transmembrane protein capable of transducing intracellular signals and is thought to perhaps facilitate the chemotaxis of infected cells around the body. Like CCR5, US28 can function as a co-receptor for HIV entry into human CD4$^+$ T cells.

Proteins resembling CXC receptors have been identified in certain gamma herpesviruses. The HSV74 product shows considerable sequence homology with CXCR2 and binds several CXC chemokines (including IL-8) but no CC chemokines. This viral receptor also transduces intracellular signals using a G protein-coupled mechanism that appears to be similar to that used by chemokine receptors.

- The myxoma, vaccinia, cowpox, and variola viruses (among others) also synthesize additional proteins that bind chemokines but which do not have a CCR or CXCR structure. These chemokine binding proteins (CBPs) interfere with the activities of a broad range of chemokines, usually of either the CC or CXC class but not both. For example, the M-T1 product of myxoma virus has been shown to block the binding of CC chemokines to host cell receptors. The M-T7 product of myxoma (already described as a viral receptor resembling IFNγR) can also act as a CBP but in a different way, disrupting the interactions of chemokines with the extracellular matrix.

Viral Cytokine	Virus Source	Homologous to	Effect
BCRF1	EBV	IL-10	↓ IFNγ, ↑ B cell proliferation
VGF	Vaccinia	EGF, TGF	↑ Viral replication
MGF	Myxoma	EGF, TGF	↑ Viral replication
MC148R	Molluscum contagiosum	MIP-1β	Competitive inhibitor of MIP-1β
K4, K6 ORFs	KSHV	MIP-1α	↑ Chemotaxis and cellular activation
Viral Cytokine Receptor	**Virus Source**	**Homologous to**	**Effect**
M-T1	Myxoma	CC chemokine receptors	Binds CC chemokines
M-T2	Myxoma	TNFR1	Binds TNF
M-T7	Myxoma	IFNγR	Binds IFNγ and CC, CXC, C chemokines
crmC	Cowpox	TNFR2	Binds TNF
B18R	Vaccinia	IFNαR	Binds IFNα
B15R	Vaccinia	IL-1R	Binds IL-1β
US28	CMV	CCR1; CCR5	Binds CC chemokines; HIV coreceptor
HSV74	Herpesvirus	CXCR2	Binds CXC chemokines (IL-8)

This concludes our discussion of cytokine and cytokine receptor characteristics. However, the reader should not be surprised to see them turn up again and again in future chapters as they have appeared in past chapters. Without the vital intercellular communication network mediated by cytokines, our immune systems could not be mobilized to ward off infection and heal injury. We move now to a detailed discussion of three cell types thought to bridge adaptive and innate immunity and which, until relatively recently, were fairly mysterious in their workings: NK cells, γδ T cells, and NKT cells.

Table 17-5 Summary Table of Cytokines Grouped by Functional Category

Process	Promoted by	Inhibited by
Inflammation	IFNγ, IL-1, 6, 8, 12, 17, 18, 19?, 20?, 22, 24, 27, TNF, LTα	IL-4, 10, 11, 13, TGFβ
Extracellular pathogen responses	IL-2, 3, 4, 5, 6, 9, 10, 11, 13, 15, 16, 25, 27, TNF	IFNγ, IL-10, 12, 27, TGFβ
Hematopoiesis	IL-1, 3, 7, 9, 11, 12, 14, 15, 17, 25, OPGL, SCF, GM-CSF, G-CSF, M-CSF, BAFF	
Intracellular pathogen responses	IFNα/β, IFNγ, IL-2, 4, 12, 13, 15, 18, 21, 23, 27, IL-28A, IL-28B, IL-29, TNF, LTα	IL-10, IL-11, TGFβ
Bone remodeling	OPGL, M-CSF, IL-1, IL-6, IL-11	OPG, IL-18, GM-CSF
Anti-tumor	TNF, LTα, IL-27	
Secondary lymphoid organ development	LTαβ, OPGL	
Leukocyte chemotaxis	IL-8, IL-15, IL-16, TGFβ, chemokines	

Table 17-6 Cytokines Influencing Inflammation

Promoted by	Response	Promoted by	Response
IFNγ	↑ Macrophage activation, FcR, iNOS ↑ Macrophage production of TNF, IL-1, IL-12, IL-18	TNF	↑ Acute phase response by hepatocytes ↑ Macrophage production of IL-1, IL-6, TNF, prostaglandins ↑ Neutrophil production of IL-1, IL-6, IL-8 ↑ Neutrophil and macrophage microbicidal functions ↑ Activation, adhesion, and extravasation of vascular endothelial cells, neutrophils, macrophages, and lymphocytes
IL-1	↑ Acute phase response by hepatocytes ↑ Adhesion molecules on endothelial cells and leukocytes ↑ IL-6, TNF production		
IL-6	↑ Acute phase response by hepatocytes ↑ Neutrophil microbicidal functions	LTα	↑ Acute phase response by hepatocytes ↑ Macrophage production of IL-1, IL-6, TNF, prostaglandins ↑ Neutrophil production of IL-1, IL-6, IL-8 ↑ Neutrophil and macrophage microbicidal functions ↑ Activation, adhesion, and extravasation of vascular endothelial cells, neutrophils, macrophages, and lymphocytes
IL-8	↑ Neutrophil chemotaxis ↑ Neutrophil degranulation and microbicidal functions		
IL-12	↑ DC and macrophage production of TNF, IL-8, GM-CSF, IL-10		

Promoted by	Response	Inhibited by	Response
IL-17 (in memory T cell responses)	↑ IL-6, 8 production ↑ Neutrophil mobilization ↑ Fibroblast adhesion	IL-4	↓ IFNγ functions
IL-18	↑ NK cell function and mobilization ↑ IFNγ, TNF production	IL-10	↓ Macrophage activation and production of TNF, IL-1, IL-6, IL-8, IL-12 ↓ Activation of neutrophils, mast cells, eosinophils
IL-19? 20?	Presumed pro-inflammatory or regulatory roles?	IL-11	↓ Macrophage production of TNF, IL-6, IL-1, IL-12, NO ↓ IFNγ production by Th1 cells
IL-22	↑ Acute phase response by hepatocytes ↑ Basophils and platelets but ↓ erythrocytes	IL-13	↓ Macrophage production of TNF, IL-6, IL-1; ↓ ADCC
IL-24	↑ Wound healing	TGFβ	↓ Production and effects of IL-1, IL-2, IL-6, IFNγ TNF ↓ Macrophage production of NO ↑ Synthesis of IL-1Ra ↑ Angiogenesis and extracellular matrix protein production
IL-27	↑ IFNγ production by Th1 effectors and activated NK cells ↑ IL-1, IL-18, TNF production by mast cells ↑ IL-1, TNF production by monocytes		

Table 17-7 **Cytokines Influencing Responses to Extracellular Pathogens**

Promoted by	Response	Promoted by	Response
IL-2	↑ T cell activation, proliferation, differentiation ↑ B cell activation, proliferation, differentiation	IL-15	↑ Mast cell proliferation
IL-3	↑ Proliferation of mast cells and basophils ↑ T cell production of IL-10, IL-13 ↑ IgE ↑ Anti-parasite response	IL-16	↑ T cell activation Eosinophil chemotaxis CD4$^+$ T cell and monocyte chemotaxis ↑ MHC class II on CD4$^+$ monocytes
IL-4	↑ Th2 differentiation ↓ Macrophage production of TNF, IL-1, ROI ↓ Adhesion molecules ↑ IgE, ↑ IgG1, ↓ IgG2a,b, ↓ IgG3 ↑ MHC class II on B cells ↑ B cell proliferation and differentiation ↑ Mast cell proliferation	IL-25	↑ Production of IL-4, IL-5, and IL-13 by unknown non-T cells ↑ IgE, IgG1, and IgA ↑ Stromal cell production of eotaxin ↑ Allergic responses
		IL-27	↑ IgG2a, ↓ IgG1
IL-5	↑ Eosinophil chemotaxis and activation ↑ Mast cell histamine release ↑ IgA	TNF	↑ ICOSL expression on non-lymphoid cells ↑ MHC class II on APCs ↑ B cell proliferation, GC formation, Ab production ↑ T cell production of IL-4, 5
IL-6	↑ B cell terminal differentiation	OPGL	↑ DC survival (*in vitro*) ↑ DC expression of CD40 (*in vitro*)
IL-9 (in presence of IL-4)	↑ IgE, IgG1 ↑ Anti-helminth worm defense	**Inhibited by**	**Response**
IL-10	↓ IL-2, IFNγ production by T cells (↓ Th1)	IL-12, IFNγ	↓ Th2 differentiation and cytokine production
IL-11	↓ IL-2, IFNγ production by T cells (↓ Th1)	IL-10	↓ Activation of APCs, lymphocytes, eosinophils
IL-13	↑ Th2 cell production of IL-4, IL-5, IL-10 ↓ Macrophage production of IL-1, IL-6, TNF, chemokines ↓ Macrophage ADCC ↑ B cell proliferation, MHC class II, IgM, CD23 ↑ IgE ↑ Anti-nematode defense	IL-27	↓ GATA3 and IL-4 expression ↓ Anti-parasite defense in the late stages
		TGFβ	↓ Activation of T, B, macrophages ↓ Ab production, MHC class II ↓ T cell proliferation and homing

Table 17-8 **Cytokines Influencing Hematopoiesis**

Promoted by	Response	Promoted by	Response
IL-1	↑ BM stromal cell production of G-CSF, IL-3	IL-17	↑ Neutrophil development from HSC (*in vitro*) ↑ Stromal cell production of G-CSF
IL-3	↑ Hematopoietic cell growth and differentiation (*in vitro*) ↑ Bcl-2 (↑ cell survival)	IL-25	↑ Lymphoid (but not myeloid) precursor proliferation
IL-7	↑ Development of αβ T cells, γδ T cells, B cells ↑ Generation and maintenance of memory T cells ↑ Bcl-2, ↑ Bcl-xL, ↑ PI3K	OPGL	Required for pre-T and pre-B cell maturation
		SCF	↑ CMP and CLP generation from HSC ↑ Proliferation of lymphoid and myeloid precursors ↑ Mast cell proliferation, adhesion, and degranulation
IL-9	↑ Erythroid, myeloid, and neuronal precursor differentiation ↑ Erythroid, myeloid, and neuronal precursor proliferation ↑ Mast cell proliferation and differentiation ↑ T cell survival	GM-CSF	↑ Generation of monocyte/granulocyte precursors ↑ Lung macrophage subset differentiation ↑ DC differentiation
IL-11	↑ Erythroid, myeloid, and megakaryocyte precursors ↑ T and B cell and neutrophil proliferation	G-CSF	Acts on monocyte/granulocyte precursors to generate granulocytes ↑ Steady-state production of neutrophils
IL-12	↑ Macrophage production of GM-CSF		
IL-14	↑ B cell proliferation in culture	M-CSF	Acts on monocyte/granulocyte precursors to generate monocytes and macrophages ↑ Osteoclast generation
IL-15	↑ NK development ↑ γδ T cell development ↑ Mast cell proliferation	BAFF	Drives T1 to T2 transition during B cell development

Table 17-9 Cytokines Influencing Responses to Intracellular Pathogens

Promoted by	Response
IFNα/β	Antiviral state induced ↓ Cell proliferation ↑ NK cell function ↑ MHC class I on CTLs ↑ IgG2a, ↑ IgG3, ↓ IgE, ↓ IgG1
IFNγ	Antiviral state induced ↓ Cell proliferation ↑ NK cell function ↑ MHC class I on CTLs and host cells ↑ IgG2a,b, ↑ IgG3, ↓ IgE, ↓ IgG1 ↓ IL-4 production and Th2 differentiation ↑ APC expression of TAP, MHC class II, B7
IL-2	↑ T cell activation, proliferation, differentiation ↑ B cell activation, proliferation, differentiation ↑ NK cell proliferation and production of TNF, IFNγ ↑ LAK differentiation (*in vitro*)
IL-4	↑ DC, macrophage production of IL-12
IL-12	↑ Th1 differentiation ↑ Macrophage production of IFNγ ↑ Activated Th1 production of IFNγ ↑ NK cell production of IFNγ, cytotoxicity ↑ IgG2a,b, ↑ IgG3, ↓ IgE, ↓ IgG1 ↑ CTL cytotoxicity ↑ Memory T cell differentiation into Th1 cells
IL-13	↑ DC, macrophage production of IL-12
IL-15	↑ NK cell proliferation and production of TNF, IFNγ ↑ T cell activation, proliferation, differentiation ↑ T cell homing and adhesion ↑ Memory CD8⁺ T cell survival ↑ B cell activation, proliferation ↑ LAK differentiation, cytotoxicity (*in vitro*) ↑ Bcl-2
IL-18 (with IL-12)	↑ Later stages of Th1 response ↑ Th1 cell production of IFNγ, IL-2R ↑ Th1 cell proliferation and expression of Fas ↑ NK cytotoxicity and production of IFNγ, TNF

Promoted by	Response
IL-21	↑ T cell proliferation ↑ Th1 cell activation and production of IFNγ ↑ Survival of anti-tumor CTLs ↑ B cell proliferation (Td Ag) ↓ B cell proliferation (Ti Ag) ↓ IgE ↑ NK cytotoxicity
IL-23	↑ Memory CD4⁺ T cell proliferation and differentiation into Th1 ↑ IFNγ production by DCs and memory Th1 effectors
IL-27	↑ Early stages of Th1 response (IL-12Rβ2 induction) ↑ Th1 cell activation and production of IFNγ ↑ NK production of IFNγ ↑ Activity of anti-tumor CTLs and NK cells
IL-28A, 28B, 29	Antiviral state induced
TNF	Antiviral state induced ↑ MHC class I on target cells ↑ B cell proliferation, GC formation ↑ Ab production ↑ T cell production of IFNγ, IL-2R ↑ Tumor cell apoptosis
LTα	Antiviral state induced ↑ MHC class I on target cells ↑ B cell proliferation, GC formation, Ab production ↑ T cell production of IFNγ, IL-2R ↑ Tumor cell apoptosis
OPGL	↑ DC survival (*in vitro*) ↑ DC expression of CD40 (*in vitro*)

Inhibited by	Response
IL-10	↓ Th1 cytokine production ↓ APC activation
IL-11	↓ Th1 cytokine production
TGFβ	↓ Macrophage production of IFNγ ↓ CTL differentiation and cytotoxicity

Table 17-10 Cytokines Influencing Bone Remodeling

Bone Resorption		Bone Generation	
CYTOKINE	**INFLUENCE**	**CYTOKINE**	**INFLUENCE**
OPGL (with CSF-1)	↑ Generation of osteoclasts ↑ Osteoclast activation	GM-CSF	↓ Osteoclast generation
IL-1	↓ Osteoblast alkaline phosphatase ↑ Osteoclast collagenase	IL-18	↑ GM-CSF
IL-6	↑ Osteoclast maturation	OPG	↓ Osteoclast maturation
IL-11	↑ Osteoclast generation		
M-CSF	↑ Osteoclast generation		

SUMMARY

Cytokines are structurally diverse, soluble proteins that are synthesized under tight regulatory controls mainly by leukocytes. Cytokines act in an autocrine or paracrine fashion as intercellular messengers, exerting their effects primarily on other leukocytes. Chemokines are cytokines that function as chemoattractants for leukocytes. By binding to their receptors on cell surfaces, cytokines trigger intracellular signaling pathways that induce new gene transcription resulting in proliferation, survival, apoptosis, migration, differentiation, or other cellular outcomes. Cytokine receptors belong to several structural families but can share component chains, accounting for the overlapping functions of some cytokines. Cytokines are necessary for both the initiation and control of normal innate and adaptive immune responses. Inflammation is promoted by cytokines such as IL-1, IFNγ, and TNF, but suppressed by IL-10 and TGFβ. The IFN family has important antiviral effects and can induce an "antiviral state" that protects uninfected cells. IL-12 is the primary cytokine driving Th1 differenti-ation and adaptive immune responses designed to eliminate intra-cellular pathogens. Conversely, IL-4 is crucial for Th2 differentiation and adaptive immune responses designed to eliminate extracellular pathogens. Th1 and Th2 cytokines cross-regulate each other's effects and synthesis, controlling the Th1/Th2 balance. Many other cytokines, such as IL-3, IL-7, IL-9, and GM-CSF, are involved in supporting hematopoiesis and influencing lineage differentiation, while LT$\alpha\beta$ plays a major role in secondary lymphoid organ development. The combined activities of cytokines such as IL-11, IL-18, OPG, and OPGL control bone remodeling, while TNF and IL-27 combat tumors. Cytokines can act as double-edged swords: for example, TNF and IL-1 are critical immunoregulators when present in small amounts but excess TNF and IL-1 cause endotoxic shock. The importance of cytokines and cytokine receptors in immunity is underscored by the fact that various viruses have evolved homo-logues of these molecules to serve as means of evading the immune response.

Selected Reading List

Aarvak T., Chabaud M., Miossec P. and Natwig J. (1999) IL-17 is produced by some proinflammatory Th1/Th0 cells but not by Th2 cells. *Journal of Immunology* **162**, 1246–1251.

Akira S. and Takeda K. (2004) Toll-like receptor signalling. *Nature Reviews Immunology* **4**, 499–511.

Alcami A. (2003) Viral mimicry of cytokines, chemokines and their receptors. *Nature Immunology* **3**, 36–50.

Annes J. (2002) The integrin αvβ6 binds and activates latent TGFβ3. *FEBS Letters* **511**, 65–68.

Ansel K. and Cyster J. (2001) Chemokines in lymphopoiesis and lymphoid organ development. *Current Opinion in Immunology* **13**, 172–179.

Arend W. P., Malyak M., Gunthridge C. J. and Cem G. (1998) Interleukin-1 receptor antagonist: role in biology. *Annual Review of Immunology* **16**, 27–55.

Aronica S. M. and Broxmeyer H. (1996) Advances in understanding the postreceptor mechanisms of action of GM-CSF, G-CSF, and Steel factor. *Current Opinion in Hematology* **3**, 185–190.

Barry M. and McFadden G. (1997) Virus encoded cytokines and cytokine receptors. *Parasitology* **115**, S89–S100.

Bogdan C. (2000) The function of type I interferons in antimicrobial immunity. *Current Opinion in Immunology* **12**, 419–424.

Borish L. and Steinke J. (2003) Cytokines and chemokines. *Journal of Allergy and Clinical Immunology* **111**, S460–S475.

Broudy V. (1997) Stem cell factor and hematopoiesis. *Blood* **90**, 1345–1364.

Bulanova E., Budagian V., Orinska Z., Krause H., Paus R. *et al.* (2003) Mast cells express novel functional IL-15 receptor alpha isoforms. *Journal of Immunology* **170**, 5045–5055.

Burdach S., Nishinakamura R., Kirksen U. and Murray R. (1998) The physiologic role of interleukin-3, interleukin-5, granuloctye–macrophage colony-stimulating factor, and the betac receptor system. *Current Opinion in Hematology* **5**, 177–180.

Callard R. E., Matthews D. J. and Hibbert L. J. (1997) Interleukin 4 and interleukin 13: same response, different receptors. *Biochemical Society Transactions* **25**, 451–455.

Chabaud M., Durand N., Buchs F., Fossiez F., Page G. *et al.* (1999) Human interleukin-17: a T cell-derived proinflammatory cytokine produced by the rheumatoid synovium. *Arthritis and Rheumatism* **42**, 963–970.

Cruickshank W. W., Kornfeld H. and Center D. M. (1998) Signaling and functional properties of interleukin-16. *International Review of Immunology* **16**, 523–540.

Demoulin J.-B. and Renaluld J.-C. (1998) Interleukin 9 and its receptor: an overview of structure and function. *International Review of Immunology* **16**, 145–164.

De Sepulveda P., Okkenhaug K., La Rose J., Hawley R., Dubreuil P. *et al.* (1999) Socs1 binds to multiple signalling proteins and suppresses Steel factor-dependent proliferation. *EMBO Journal* **18**, 904–915.

Dinarello C. A. (1998) Interleukin-1, interleukin-1 receptors and interleukin-1 receptor antagonist. *International Review of Immunology* **16**, 457–499.

Elloso M., Wallace M., Manning D. and Weidanz W. (1998) The effects of interleukin-15 on human gammadelta T cell responses to Plasmodium falciparum in vitro. *Immunology Letters* **64**, 125–132.

Fossiez F., Banchereau J., Murray R., Van Kooten C., Garrone P. *et al.* (1998) Interleukin-17. *International Review of Immunology* **16**, 541–551.

Gately M. K., Renzetti L. M., Magram J., Stern A. S., Adorini L. *et al.* (1998) The interleukin-12/interleukin-12 receptor system: role in normal and pathologic immune responses. *Annual Review of Immunology* **16**, 495–521.

Gillespie M. T. and Horwood N. J. (1998) Interleukin-18: perspectives on the newest interleukin. *Cytokine and Growth Factor Reviews* **9**, 109–116.

Haddad J. (2002) Cytokines and related receptor-mediated signaling pathways. *Biochemical and Biophysical Research Communications* **297**, 700–713.

Hancock W., Gao W., Faia K. and Csizmadia V. (2000) Chemokines and their receptors in allograft rejection. *Current Opinion in Immunology* 12, 511–516.

Hirano T. (1998) Interleukin 6 and its receptor: ten years later. *International Review of Immunology* 16, 249–284.

Holst P. and Rosenkilde M. (2003) Microbiological exploitation of the chemokine system. *Microbes and Infection* 5, 179–187.

Hoshino K., Tsutsui H., Kawai T., Takeda Y. and Akira S. (1999) Generation of IL-18 receptor-deficient mice: evidence for IL-1 receptor-related protein as an essential IL-18 binding receptor. *The Journal of Immunology* 162, 5041–5044.

Kanakaraj P. (1999) Defective interleukin (IL)-18-mediated natural killer and T helper cell type 1 responses in IL-1 receptor-associated kinase (IRAK)-deficient mice. *Journal of Experimental Medicine* 189, 1129–1138.

Karlen S., De Boer M. L., Lipscombe R. J., Lutz W., Mordvinov V. A. *et al.* (1998) Biological and molecular characteristics of interleukin-5 and its receptor. *International Review of Immunology* 16, 227–247.

Kawai T., Adachi O., Ogawa T., Takeda K. and Akira S. (1999) Unresponsiveness of MyD88-deficient mice to endotoxin. *Immunity* 11, 115–122.

Kim C. and Broxmeyer H. (1999) Chemokines: signal lamps for trafficking of T and B cells for development and effector function. *Leukocyte Biology* 65, 6–15.

Kong Y.-Y., Yoshida H., Sarosi I., Tan H.-L., Timms E. *et al.* (1999) OPGL is a key regulator of osteoclastogenesis, lymphocyte development and lymph-node organogenesis. *Nature* 397, 315–323.

Kotenko S. (2002) The family of IL-10-related cytokines and their receptors: related, but to what extent? *Cytokine and Growth Factor Reviews* 13, 223–240.

Kuniyoshi J., Kuniyoshi C., Lim A., Wang F., Bade E. *et al.* (1999) Dendritic cell secretion of IL-15 is induced by recombinant huCD40LT and augments stimulation of antigen-specific cytolytic T cells. *Cellular Immunology* 193, 48–58.

Laan M., Cui Z., Hoshino H., Lotvall, J., Sjostrand M. *et al.* (1999) Neutrophil recruitment by human IL-17 via CXC chemokine release in the airways. *Journal of Immunology* 162, 2347–2352.

Lankford C. and Frucht D. (2003) A unique role for IL-23 in promoting cellular immunity. *Journal of Leukocyte Biology* 73, 49–56.

Legler D., Loetscher M., Stuber Roos R., Clark-Lewis I., Baggiolini M. *et al.* (1998) B cell-attracting chemokine 1, a human CXC chemokine expressed in lymphoid tissues,

selectively attracts B lymphocytes via BLR/CXCR5. *Journal of Experimental Medicine* 187, 655–660.

Letterio J. J. and Roberts A. B. (1998) Regulation of immune responses by TGF-beta. *Annual Review of Immunology* 16, 137–161.

Linnekin D., Sheery M., Deberry C. S., Weiler S., Keller J. R. *et al.* (1997) Stem cell factor, the JAK–STAT pathway and signal transduction. *Leukemia and Lymphoma* 27, 439–444.

Lodolce J. P., Boone D. L., Chai S., Swain R. E., Dassopoulos T. *et al.* (1998) IL-15 receptor maintains lymphoid homeostasis by supporting lymphocyte homing and proliferation. *Immunity* 9, 669–676.

Lomaga M. A., Yen W.-C., Sarosi I., Duncan G. S., Furlonger C. *et al.* (1999) TRAF6 deficiency results in osteopetrosis and defective interleukin-1, CD40, and LPS signaling. *Genes and Development* 13, 1015–1024.

Loughnan M. S. and Nossal G. J. V. (1989) Interleukins 4 and 5 control expression of IL-2 receptor on murine B cells through independent induction of its two chains. *Nature* 340, 76–79.

Luster A. (2002) The role of chemokines in linking innate and adaptive immunity. *Current Opinion in Immunology* 14, 129–135.

Maeurer M. J. and Lotze M. T. (1998) Interleukin-7 (IL-7) knockout mice. Implications for lymphopoiesis and organ-specific immunity. *International Review of Immunology* 16, 309–322.

Mak T. and Simard J. (1998) "Handbook of Immune Response Genes." Plenum Press, New York, NY.

McFadden G., Everett L. H., Nash P. and Xu X. (1998) Virus-encoded receptors for cytokines and chemokines. *Seminars in Cell & Developmental Biology* 19, 359–368.

McKenzie G. J., Emson C. L., Bell S. E., Anderson S., Fallon P. *et al.* (1998) Impaired development of Th2 cells in IL-13 deficient mice. *Immunity* 9, 423–432.

Moseley T., Haudenschild D., Rose L. and Reddi A.. (2003) Interleukin-17 family and IL-17 receptors. *Growth Factor Reviews* 14, 155–174.

Murphy P. (1997) Neutrophil receptors for interleukin-8 and related CXC chemokines. *Seminars in Hematology* 34, 311–318.

Nakanishi K., Yoshimoto T., Tsutsui H. and Okamura H. (2001) Interleukin-18 regulates both Th1 and Th2 responses. *Annual Review of Immunology* 19, 423–474.

Nandurkar H., Robb L., Tarlinton D., Barnett L., Kontgen F. *et al.* (1997) Adult mice with targeted mutation of the interleukin-11 receptor (IL-11Ra) display normal hematopoiesis. *Blood* 90, 2148–2159.

Paludan S. R. (1998) Interleukin-4 and interferon-gamma: the quintessence of a mutual antagonistic relationship. *Scandinavian Journal of Immunology* 48, 459–468.

Parrish-Novak J., Foster D., Holly R. and Clegg C. (2002) Interleukin-21 and the IL-21 receptor: novel effectors of NK and T cell responses. *Journal of Leukocyte Biology* 72, 856–863.

Pflanz S., Timans J., Cheung J., Rosales R., Kanzler H. *et al.* (2002) IL-27, a heterodimeric cytokine composed of EBI3 and p28 protein, induces proliferation of naive CD4[+] T cells. *Immunity* 16, 779–790.

Robinson D. and O'Garra A. (2002) Further checkpoints in Th1 development. *Immunity* 16, 755–758.

Rooke H. and Crosier K. (2001) The SMAD proteins and TGFbeta signalling: uncovering a pathway critical in cancer. *Pathology* 33, 73–84.

Roy B., Bhattacharjee A., Xu B., Ford D., Maizel A. L. *et al.* (2002) IL-13 signal transduction in human monocytes: phosphorylation of receptor components, association with Jaks, and phosphorylation/activation of Stats. *Journal of Leukocyte Biology* 72, 580–589.

Sallusto F., Lenig D., Mackay C. R. and Lanzavecchia A. (1998) Flexible programs of chemokine receptor expression on human polarized T helper 1 and 2 lymphocytes. *Journal of Experimental Medicine* 187, 875–883.

Sandford L., Ormsby I., Gittenberger-de Groot A., Sariola H., Friedman R. *et al.* (1997) TGFbeta2 knockout mice have multiple developmental defects that are non-overlapping with other TGFbeta knockout phenotypes. *Development* 124, 2659–2670.

Schwertschlag U., Trepicchio W., Kykstra K., Keith J., Turner K. *et al.* (1999) Hematopoietic, immunomodulatory and epithelial effects of interleukin-11. *Leukemia* 13, 1307–1315.

Shin I. (1999) Expression of IL-17 in human memory CD45RO[+] T lymphocytes and its regulation by protein kinase A pathway. *Cytokine* 11, 257–266.

Stordeur P. and Goldman M. (1998) Interleukin-10 as a regulatory cytokine induced by cellular stress: molecular aspects.

International Review of Immunology **16**, 501–522.

Subramaniam S., Cooper R. and Adunyah, S. (1999) Evidence for the involvement of JAK/STAT pathway in the signaling mechanism of interleukin-17. *Biochemical and Biophysical Research Communications* **262**, 14–19.

Sugawara I., Yamada H., Kaneko H., Mizuno S., Takeda K. *et al.* (1999) Role of interleukin-18 (IL-18) in mycobacterial infection in IL-18-gene-disrupted mice. *Infection and Immunity* **67**, 2585–2589.

Tagaya Y., Bamford, Richard N, DeFilippis A. P. and Waldmann T. A. (1996) IL-15: a pleiotrophic cytokine with diverse receptor/signaling pathways whose expression is controlled at multiple levels. *Immunity* **4**, 329–336.

Takeda K., Tsutsui H., Yoshimoto T., Adachi O., Yoshida N. *et al.* (1998) Defective NK cell activity and Th1 response in IL-18-deficient mice. *Immunity* **8**, 383–390.

Trinchieri G. (2004) Cytokines and cytokine receptors. *Immunological Reviews* **202**, 5–7.

Trinchieri G., Plfanz S. and Kaselstein R. (2003) The IL-12 family of heterodimeric cytokines: new players in the regulation of T cell responses. *Immunity* **19**, 641–644.

Underhill D. and Ozinsky A. (2002) Toll-like receptors: key mediators of microbe detection. *Current Opinion in Immunology* **14**, 103–110.

Van de Vosse E., Lichtenauer-Katigis E., van Dissel J. and Ottenhoff T. (2003) Genetic variations in the interleukin-12/interleukin-23 receptor (beta1) chain, and implications for IL-12 and IL-23 receptor structure and function. *Immunogenetics* **54**, 817–829.

Waldmann T., Tagaya Y. and Bamford R. (1998) Interleukin-2, interleukin-15 and their receptors. *International Review of Immunology* **16**, 205–226.

Ward S. G., Bacon K. and Westwick J. (1998) Chemokines and T lymphocytes: more than an attraction. *Immunity* **9**, 1–11.

Ware C. F., VanArsdale S. and VanArsdale T. L. (1996) Apoptosis mediated by the TNF-related cytokine and receptor families. *Journal of Cellular Biochemistry* **60**, 47–55.

Wei X., Leung B., Niedbala W., Piedrafita D., Feng G. *et al.* (1999) Altered immune responses and susceptibility to *Leishmania major* and *Staphylococcus aureus* infection in IL-18-deficient mice. *Journal of Immunology* **163**, 2821–2828.

Wong B. R., Josien R. and Choi Y. (1999) TRANCE is a TNF family member that regulates dendritic cell and osteoclast function. *Journal of Leukocyte Biology* **65**, 715–724.

Wright S. (1999) Toll, a new piece in the puzzle of innate immunity. *Journal of Experimental Medicine* **189**, 115–122.

Wu D., Zhang Y., Parada N., Kornfeld H., Nicoll J. *et al.* (1999) Processing and release of IL-16 from CD4+ but not CD8+ T cells is activation dependent. *Journal of Immunology* **162**, 1287–1293.

Yang R., Mark M., Gurney A. and Godowski P. (1999) Signaling events induced by lipopolysaccharide-activated toll-like receptor 2. *Journal of Immunology* **163**, 639–643.

Zlotnik A. and Yoshie O. (2000) Chemokines: A new classification system and their role in immunity. *Immunity* **12**, 121–127.

Bridging Innate and Adaptive Immunity: NK, $\gamma\delta$ T, and NKT Cells

18

Charles Janeway, Jr. 1943–2003

Pattern recognition

© Yale School of Medicine

CHAPTER 18

A. NATURAL KILLER (NK) CELLS

B. $\gamma\delta$ T CELLS

C. NKT CELLS

"If it's natural to kill, why do men have to go into training to learn how?"

—Joan Baez

In Chapter 1, we described macrophages as "bridging" innate and adaptive immunity, because these myeloid cells not only act rapidly to phagocytose invaders but also present pathogen antigens to T and B cells and secrete cytokines that influence lymphocyte function. In this chapter, we explore three other cell types, natural killer (NK) cells, $\gamma\delta$ T cells, and natural killer T (NKT) cells, that are also considered to bridge adaptive and innate immunity. In this case, both the forms and functions of these cell types mark them as belonging to the innate and adaptive responses. NK, $\gamma\delta$ T, and NKT cells are undeniably of the T cell lineage, and $\gamma\delta$ T and NKT cells express TCRs derived from V(D)J recombination. However, the response of these three cell types to infection or injury is rapid and involves broadly specific recognition of antigen independent of classical MHC molecules, making it characteristic of the induced innate immune response. Scientists now believe that NK, $\gamma\delta$ T, and NKT cells are not just primitive ancestors of the $\alpha\beta$ T cells but are distinct and contemporaneous lineages in their own rights, and that their comparative lack of antigen-binding specificity is essential to their primary physiological function as early defenders in the innate response. In addition, NK, NKT, and $\gamma\delta$ T cells have important roles in directly influencing the differentiation of $\alpha\beta$ T cells and B cells and the adaptive responses mediated by these lymphocytes. We discuss the occurrence, function, and development first of NK cells, then of $\gamma\delta$ T cells, and finally of NKT cells.

A. Natural Killer (NK) Cells

I. HISTORICAL NOTES

The reader will recall from Chapter 3 that NK cells have cytolytic activity very similar to that of CTLs of the adaptive immune response. However, unlike CTLs, NK cells lack the TCR and CD3 complex, and, like cells of the innate response, respond rapidly to targets recognized in a much less specific way. NK cells were discovered in the early 1970s as an outgrowth of work on cell-mediated cytotoxicity. At that time, it had only recently been realized that there were two classes of lymphocytes mediating immune responses: T cells and B cells. In addition, the chromium release assay, which provided an easy means of detecting cytolytic activity, had just been optimized by Hans Wigzell of the Karolinska Institute in Sweden. In 1971, the prevailing wisdom was that cell-mediated cytotoxicity was totally T cell-mediated and required prior immunization of an animal with the relevant antigen. However, niggling discrepancies in the literature persisted, such as a report that the lymphocyte population from an <u>unimmunized</u> animal could kill target cells if the latter were coated in antibody. Since killing did not occur in the absence of antibody, this phenomenon became known as "antibody-dependent cell-mediated cytotoxicity" (ADCC; discussed in Ch.5). In late 1971/early 1972, two independent reports asserted that the lymphocytes responsible for ADCC had to be B cells rather than T cells because the cells mediating ADCC lacked the mouse T cell diagnostic surface marker Thy-1. However, when Arnold Greenberg of Ivan Roitt's laboratory in England tested this hypothesis by purifying murine B cells and determining whether they could carry out ADCC on antibody-coated target cells, he found that they could not. Greenberg went on to isolate the cells responsible for the observed ADCC and noted that they constituted 18% of the total lymphocyte population, lacked the properties of macrophages and granulocytes, and did not express surface markers characteristic of either B or T cells. Greenberg therefore called his new class of killer lymphocytes "null cells."

In 1974, a collaboration between Greenberg and John Playfair led to an unexpected result. When syngeneic tumor cells were used as target cells in cytotoxicity assays, splenocytes from unimmunized animals could kill them whether or not they were coated with antibody. Since no immunization was involved, the cells responsible for this "spontaneous" killing could not be CTLs. Rather, they turned out to be large, granular lymphocyte-like cells present at about 10% of the recirculating lymphocyte population. Assessment of surface marker expression suggested that these cells were identical to Greenberg's "null cells." At about the same time, Rolf Kiessling and Eva Klein in Hans Wigzell's group demonstrated that splenocytes of unimmunized mice could spontaneously lyse cells of a Moloney leukemia cell line. Kiessling referred to the non-B, non-T lymphocyte-like cells responsible for this cytolysis as "natural killers." We now know that these cells are identical to the "null cells," but only the term "natural killer" is in common use today.

In 1975, Ron Herberman at the National Institutes of Health in the United States reported that mouse lymphoid cells could carry out spontaneous killing of both syngeneic and allogeneic tumor cells, indicating that the newly discovered phenomenon of MHC restriction did not apply to NK cells in the same way as it did to antigen-specific CTLs. Greenberg and Mark Greene at the University of Manitoba in Canada further showed in 1976 that tumor surveillance by NK cells was T cell-independent, in that mice completely lacking a thymus still showed some evidence of tumor resistance. Kiessling's group subsequently found that a natural mouse mutant called *beige*, which is abnormally sensitive to transplanted tumors and exhibits runaway metastasis, is very deficient in NK activity. This observation was taken as indirect evidence for a role of NK cells in controlling tumor formation. The range of NK cell functions was further expanded in 1978 when it was noted by Magnus Gidlund of Hans Wigzell's group that IFNα/β induced by viral infection enhanced the cytotoxic activity of NK cells, suggesting that these cells also played a role in defense against viruses. Indeed, complementary work by Herberman's group showed that IFN and IFN inducers could augment the residual NK cell activity in *beige* mice.

At this point, it was clear that NK cells could attack tumor and virus-infected cells, but how did they recognize them? In 1986, Klas Kärre and co-workers at the Karolinska Institute made the seminal observation that murine tumors lacking MHC class I expression were susceptible to killing *in vivo* and *in vitro*. Since CD8$^+$ CTLs (which rely on MHC class I recognition) could not be responsible for this phenomenon, they proposed that cell lysis in this instance was being carried out by NK cells. Many investigators subsequently showed both *in vivo* and *in vitro* that NK cells preferentially killed cells that lacked or had abnormally low expression of MHC class I, a characteristic of many tumor cells as well as virus-infected cells. Conversely, self-MHC class I expression engineered on target cells was found to confer resistance to NK killing. In addition, NK cells proved able to kill non-transformed allogeneic cells that expressed MHC molecules other than those found in the host; i.e., "nonself MHC." With these observations came a hint of how the killing power of NK cells could be directed away from uninfected healthy host cells and toward non-host or infected host cells. Kärre and colleagues devised the "missing self" hypothesis to account for these results, postulating that the function of NK cells was to kill cells "missing" the normal expression of self-MHC class I that occurs on almost all healthy host cells. Thus, tumor cells and virus-infected cells deficient in MHC class I expression, and normal cells expressing nonself MHC (i.e., allogeneic cells), would be eliminated by NK cells.

The "missing self" hypothesis was bolstered by the observation of a phenomenon in mice called *F1 hybrid resistance*, in which a bone marrow graft from a H-2$^{k/k}$ donor was rejected by the H-2$^{k/d}$ F1 offspring of an H-2$^{k/k}$ × H-2$^{d/d}$ mating. This rejection was not consistent with the classical solid organ rejection mediated by T and B cells, which depends on the recognition of nonself pMHC structures. Indeed, solid organ grafts from either parental strain were accepted as "self" by the H-2$^{k/d}$ F1 offspring. Since the bone marrow graft did not express any nonself MHC relative to the recipient, it was theorized that the rejection was due to a failure of the graft to express one of the parental MHC alleles; i.e., the engrafted bone marrow cells were "missing some self" and were therefore attacked. (Bone marrow grafts are inherently more sensitive to rejection than solid organ grafts, so that the solid organ grafts in this experiment were able to resist the attack even though they also lacked self-MHC.)

Despite these results, several years of controversy ensued during which conflicting data concerning the role of MHC class I in NK killing were obtained. These discrepancies made it difficult to formulate a molecular model of target recognition by NK cells. It is now clear that some of the early problems in reconciling contradictory reports stemmed from the variety of experimental systems used, MHC class I polymorphisms, and the existence of several different types of receptors that can collectively recognize multiple MHC class I molecules. Finally, in 1990, the molecular basis for NK function was confirmed experimentally. Mark Bix and David Raulet of the University of California at Berkeley showed that NK cells in mice transgenic for a specific MHC class I allele rejected bone marrow grafts from donors lacking either the transgenic MHC allele or β2m (whose absence would also reduce MHC class I expression). Several investigators then proposed that NK cytotoxicity was controlled by "inhibitory receptors" that recognized self-MHC. A lack of self-MHC expression on a target cell would lead to a failure to engage the inhibitory receptors on the NK cell, resulting in activation of NK cytotoxicity and destruction of the target. The question now arose: exactly what molecules or pathways were the inhibitory receptors inhibiting?

It was presumed that the inhibitory receptors were necessary to prevent the lysis of normal self cells, suggesting that there had to be an activatory pathway of some sort that was triggered by ligands that were not just present on nonself, stressed, or infected host cells but also on normal self cells. In the late 1980s/early 1990s, several groups of investigators cross-linked various NK cell surface molecules with specific mAbs and looked for activation of NK cytotoxicity *in vitro*. The results suggested that there was not just one essential activatory pathway but rather several weak pathways that cooperated to induce NK activation. In the early 2000s, significant progress was made in identifying some of these activatory receptors and their ligands (see below).

The current hypothesis for the triggering of NK cytotoxicity holds that activation leading to killing depends on competing signals mediated by two sets of cell surface receptors of broad binding specificity: the *activatory* receptors and the *inhibitory* receptors. In most situations, the NK inhibitory receptors are dominant, in that their binding to the self-MHC class I molecules expressed on most normal host cells conveys a signal that blocks positive signaling initiated by the binding of ligands to the activatory receptors. However, there are examples of cases in which engagement of a single activatory receptor can override inhibitory signaling and activate an NK cell. Fortunately, the ligands that provoke this type of activation appear to be expressed only by cells that should be eliminated (such as virus-infected cells or tumor cells). Ongoing research in the NK field focuses on activatory receptors and ligands, regulation of natural cytotoxicity, and NK self-tolerance.

II. WHAT AND WHERE ARE NATURAL KILLER CELLS?

NK cells are morphologically similar to lymphocytes but are generally larger and more granular in appearance under the microscope. Some histologists consider them to be a class of the *large granular lymphocytes* (LGLs), which comprise about 10–15% of peripheral blood lymphocytes. LGLs contain azurophilic cytoplasmic granules but are not phagocytic (unlike neutrophils) and are non-adherent (unlike macrophages). At the molecular level, NK cells are distinguished from NKT, B, and T cells by their lack of expression of TCRs or BCRs and the germline configuration of their TCR and BCR genes. Although NK cells are found at their highest frequency in the peripheral blood, spleen, and liver, they also tend to congregate in the mucosal epithelium and are important components of the GALT and BALT/NALT (see Ch.20). Moderate numbers of NK cells are found in the bone marrow, lymph nodes, and peritoneum. Very low levels of NK cells occur in the thymus and tonsils such that functional tests do not register their presence. The life span of a peripheral mature NK cell (in the absence of activation) is about 7–10 days.

i) Surface Markers Identifying NK Cells

For a long time, no one cell surface marker was considered diagnostic for NK cells and a species-appropriate panel of molecules had to be considered. Refinements in marker analysis in the late 1990s–2000s have made it easier to distinguish NK cells from both T cells and NKT cells (Table 18-1). In mice, staining with an antibody called DX-5 (which binds to the integrin subunit CD49b of VLA-2) identifies NK cells almost exclusively, while expression of the NK costimulatory molecule NK1.1 (see later) is found on NK cells and NKT cells but not on T cells. In mice and humans, certain members of the NCR (*natural cytotoxicity receptor*; see later) family of NK activatory receptors are exclusively associated with NK cells. In humans, the marker molecule CD56 is expressed much more often on human NK cells than on T or B cells. CD56 is an alternatively spliced form of a *neural cell adhesion molecule* (NCAM) that contains both Ig and fibronectin domains. It may be involved in the adhesion of an NK cell to its target.

In both mice and humans, NK cells share the expression of several other markers with resting αβ T cells, such as IL-2Rβ, CD2, and CD28. IL-2Rα is not expressed on NK cells, allowing them to bind IL-2 with only moderate affinity. Although they do not express TCRs, NK cells do express CD3ζ on their surfaces, and several other CD3 components are present in the cytoplasm. CD3ζ is often found in close association with the FcγRIIIA (CD16) molecule, the Fc receptor for IgG molecules involved in NK-mediated ADCC. FcγRIIIA is not found on T or B cells, but is expressed on NKT cells, neutrophils, and macrophages. NK cells constitutively express CD69, a molecule constitutively expressed by NKT cells and found on T cells following activation. NK cells do not express the MHC class II and CD19 markers found on B cells, nor the T and B cell marker CD5. CD8 can be found on some NK cells. Other markers include members of the NK activatory or inhibitory receptor families such as the CD94 and NKG2 (*natural killer gene 2*) molecules (on human and mouse NK cells), Ly-49 (on mouse NK cells only), and the KIRs (*killer inhibitory receptors*) (on human NK cells only). However, several inhibitory receptors or their subunits have also been found on CD8⁺ T, γδ T, and NKT cell subsets.

III. FUNCTIONS OF NK CELLS

Extensive experimental work has indicated that the primary functions of NK cells are to induce the cytolysis of tumor or virus-infected cells and to secrete cytokines that regulate T and B cell function and differentiation. Importantly, although lymphoid in origin, activated NK cells do not have to take the time to differentiate further into effector cells in order to do their job. This means that, unlike the T and B cell responses, the peak NK response can be detected within the first several hours after a primary infection as part of the induced innate response. Studies of murine cytomegalovirus (CMV) infection in mice have shown that the NK response actually occurs in two phases. Immediately after infection, a broad spectrum of NK cell clones proliferates due to the presence of IL-15 and other cytokines secreted by macrophages activated by viral products. However, the second wave of proliferation that occurs at about 4–6 days after infection involves only those NK clones that express an activatory receptor whose natural ligand resembles a motif found on a CMV protein. This more specific expansion of NK cells that "see" CMV-infected host cells directly helps to ensure their efficient elimination.

Defining the functions of NK cells has posed several challenges to researchers over the years. No known mutant animals lack NK cell functions alone, which is understandable when one considers that the development and immunoregulation of NK cells and T cells are intricately linked. For this reason, a deficiency in one lineage can affect the development and/or function of the other. For example, while *nude*, *scid*, and RAG1/2 mice lack T cell functions, they exhibit not just normal but enhanced NK cell activities, making it difficult to directly evaluate the intrinsic role of NK cells. Immune responses in the *beige* mouse, which shows a decrease in NK killing, were originally studied in an attempt to understand

Table 18-1 Comparison of Major Surface Markers on NK, NKT, αβ, and γδ T Cells*

Marker	NK Cells	αβ T Cells	γδ T Cells	NKT Cells
αβ TCR	−	Very diverse	−	Limited diversity
γδ TCR	−	−	Limited diversity	−
CD2	+	+	+	+
CD3ζ	+	+	+	+
CD4	−	+	−	±
CD5	−	+	+	+
CD8	±	+ (αβ isoform)	+ (αα isoform)	− (mu) ± (hu)
CD11b	+	−	−	−
CD16 (FcRγRIIIA)	+	−	−	+
CD28	+	+	±	+
CD40L	−	+	±	+
CD44		+	±	+
CD45	+	+	±	+
CD49b (DX-5) (mu)	+	−	−	±
CD56 (hu)	+	±	±	±
CD62L (L-selectin)	±	+	−	
CD69	+	+ (activated)	±	+
CD94	+	+	+	±
FcγRIγ	+	+	−	
IL-2Rα	−	+ (activated)	−	+ (neonates)
IL-2Rβ	+	+	±	+
Ly-49 (mu)	+	±	−	±
KIRs (hu)	+	±	±	±
NCRs	+	−	−	−
NK1.1 (mu)	+	−	−	±
NKG2	+	−	+	±

* ±, The marker is sometimes expressed on restricted subpopulations; hu, human; mu, murine.

the physiological role of NK cells in normal animals. Now, however, it is known that the *beige* mutation affects granule function in general. Deficiency of NK function in mice and humans does lead to increased susceptibility to infection with many viruses, including influenza in mice and herpesviruses such as *Epstein–Barr virus* (EBV) in humans. These observations may constitute *in vivo* evidence that, while the down-regulation of MHC class I expression on infected host cells helps viruses to avoid detection by CTLs, it simultaneously invites attack by NK cells. With respect to malignancies,

although there are ample experimental data in animals to support an anti-tumor role for NK cells, the physiological contribution of these cells to the fighting of cancer in humans has yet to be definitively established.

i) Cytolysis
Destruction mediated by NK cells takes the form of either ADCC as described in Chapter 5 or *natural cytotoxicity*. Natural cytotoxicity is a form of target cell cytolysis very similar to that carried out by effector CTLs but independent

of MHC-restriction; indeed, it depends on the <u>absence</u> of self-MHC. The advantage of NK-mediated over CTL-mediated cytolysis is that the NK cell is immediately competent and supplies early protection until activated Tc cells can proliferate and differentiate into CTLs at the 7-day mark.

ia) Natural cytotoxicity. Although no effector differentiation is required, resting NK cells must still be primed through exposure to IFNγ, IL-2, and/or IL-15 to initiate the acquisition of lytic competence (Fig. 18-1A). These cytokines are often present in the local milieu due to the actions of activated innate and adaptive immune system cells in a tissue under attack. The priming cytokines induce upregulation of the expression of multiple activatory receptors on NK cells. These receptors bind to ligands that may be expressed or upregulated on target cells in response to viral infection or malignant transformation but which may also occur on

stressed normal cells. Maximal efficiency of the NK cytotoxic response is achieved upon exposure to IL-12, IL-18, and/or IFNα/β, cytokines arising in the course of virus infections. Mice deficient for IL-12 or IL-18 show marked reductions in IFNγ production and natural cytotoxicity, suggesting that the highest levels of NK functionality require these cytokines. As was mentioned in Chapter 3, activated NK cells exposed in culture to continuous high concentrations of IL-2 are spurred to differentiate into even more powerful killers called *lymphokine-activated killer* (LAK) cells, but whether LAKs represent a distinct subset *in vivo* is controversial.

The initial adhesion between NK cells and target cells relies to a large extent on the interaction between the β2-integrin LFA-1 on the NK cell and ICAM-1 on the potential target (Fig. 18-1B). This interaction allows the NK cell to scan the potential target for normal levels of self-MHC class I

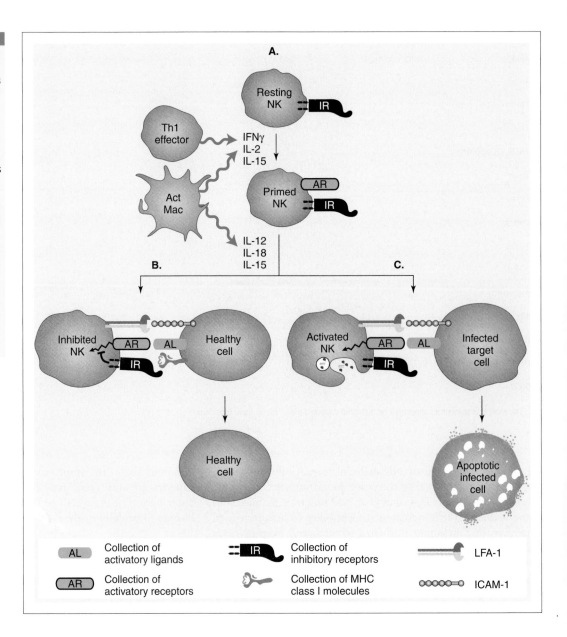

Figure 18-1

Priming and Activation of NK Cells
(A) Th1 cells and macrophages supply cytokines that prime resting NK cells and induce upregulation of NK activatory receptors (AR). (B) If the primed NK cell encounters a healthy host cell, engagement of multiple inhibitory receptors (IR) by MHC class I molecules sends a dominant signal that blocks activatory signaling and prevents NK cell activation. (C) If the primed NK cell encounters an infected (or transformed) cell that has lost expression of MHC class I, activatory signaling is not blocked, the NK cell is activated, and the infected cell is killed by granule-mediated natural cytotoxicity.

molecules. Engagement of the NK inhibitory receptors by self-MHC triggers a dominant signal that blocks NK activation. However, host cells that no longer express self MHC class I (but do express activatory receptor ligands) are killed by natural cytotoxicity (Fig. 18-1C). In other words, a primed NK cell whose activatory receptors bind an appropriate ligand on a host cell, and whose inhibitory receptors <u>fail</u> to bind self-MHC class I, becomes fully activated and delivers a lethal "hit" to the target host cell. The NK "hit" takes the form of perforin/granzyme-mediated apoptosis, the same mechanism used by CTLs for target cell elimination. However, the granules containing perforin and granzymes are preformed in an NK cell and do not have to be synthesized in response to activation, unlike the case with a Tc cell. Upon engagement of the activatory receptors, and in the absence of an inhibitory signal, the cytolytic granules in the NK cytoplasm are mobilized and directed toward the membrane of the target cell at the site of intercellular contact (Plate 18-1). As was described in Chapter 15, perforin and granzymes released from the granules are taken up into the target cell (via pinocytosis or possibly mannose-6-phosphate receptor-mediated endocytosis) and captured in endolysosomal vesicles. Perforin then releases the granzymes from these vesicles, allowing them access to the cytoplasm and nucleus of the target cell. The granzymes then induce target cell apoptosis primarily via the mitochondrial pathway. Knockout mice deficient for either perforin or granzyme B show large deficits in target cell killing by both CTLs and NK cells.

Interestingly, despite their functional similarities, CTL-mediated and natural cytotoxicity result from different signaling pathways. In Tc cells, engagement of the TCR initiates multiple signaling pathways (illustrated in Fig. 14-6), including that transduced via ZAP-70 through Ras to ERK activation.

In NK cells, the initial signaling is transduced by Syk kinase (rather than ZAP-70) and cytotoxicity is independent of Ras. Instead, PI3K is activated, which in turn activates Vav1 and Rac (Fig. 18-2). Activated Rac binds to PAK1 (*p21-activated kinase-1*), which phosphorylates MEK, and phosphorylated MEK activates ERK. Inhibition of PI3K blocks ERK activation in NK cells, and interferes with the positioning of the granules near the target cell attachment site. Such positioning involves actin cytoskeleton reorganization, a process known to require PI3K in many cell types.

ib) ADCC. The majority of NK cells in mice and humans express FcγRIIIA (CD16), which means that these cells can also kill targets by ADCC. The Fc regions of IgG antibodies bound to virus-infected or tumor cells can engage FcγRIIIA, a 70 kDa membrane-anchored glycoprotein that is closely associated on the NK surface with either the CD3ζ or FcεRIγ chain (Fig. 18-3). Both CD3ζ and FcεRIγ contain the ITAM sequences cited in Chapters 9 and 14 as being important for BCR and TCR signaling leading to activation. In cell cultures, ligand-induced stimulation of FcγRIIIA leads to the association of Lck and/or

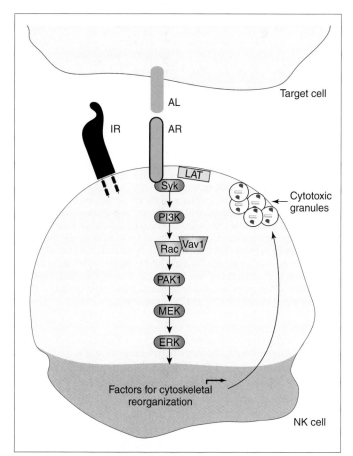

Figure 18-2

Signaling Pathway of ADCC in NK Cells
In the absence of a blocking signal initiated by engagement of the IR, engagement of multiple ARs by multiple activatory ligands (AL) leads to activation of Syk and PI3K (only one AL/AR pair is shown here for clarity). Downstream signal transduction via various mediators results in the cytoskeletal reorganization required to position the NK granules for apoptotic killing of the target cell.

Plate 18-1

NK Cell Killing of a Tumor Cell
(A) Black and white contrast photo of an NK cell killing a tumor cell.
(B) Multi-agent staining of the NK cell and tumor cell in A. Nuclei are stained pale blue in both cells and NK granules are stained red. Note the congregation of the granules at the interface between the NK cell and the tumor cell and the membrane blebbing commencing on the tumor cell. Courtesy of Charles Sentman and Mikael Erikkson, Department of Microbiology and Immunology, Dartmouth Medical School, Lebanon, USA.

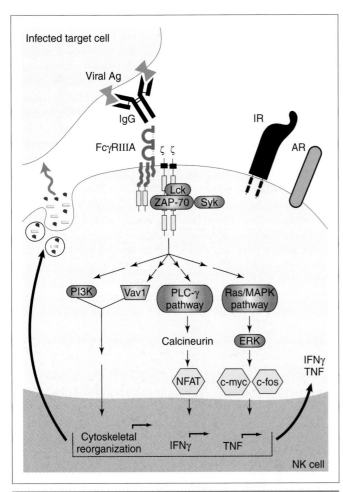

Figure 18-3

Signaling Pathway of ADCC in NK Cells
ADCC occurs when FcγRIIIA molecules on the NK surface are stimulated by the Fc regions of IgG antibodies that have bound to pathogen antigens on the surface of an infected cell. (A single antibody bound to viral antigens is shown here for simplicity.) The ARs do not have to be engaged, and the IRs must not be engaged. Lck recruited to the cytoplasmic tail of FcγRIIIA transduces signals that activate Syk and ZAP-70, which in turn lead to the activation of PI3K, Vav1, PLC-γ, and Ras. Various transcription factors are then activated that promote cytoskeletal reorganization and cytotoxic granule release as well as IFNγ and TNF production.

other Src kinases with the FcγRIIIA/CD3ζ complex, facilitating phosphorylation of the ITAMs and subsequent signal transduction via ZAP-70 and/or Syk. Changes to Ca²⁺ influx occur accompanied by activation of PI3K, NFAT, and ERK (via Ras), leading to new gene transcription and the induction of NK cell activation. The activated NK cell is triggered to degranulate, releasing perforin and granzyme molecules that eliminate the target cell as described previously. Stimulation via FcγRIIIA can also promote NK proliferation and cytokine secretion *in vitro*. Interestingly, like the natural cytotoxicity pathway, the ADCC pathway to NK degranulation can be blocked by engagement of the NK inhibitory receptors.

ic) Fas-killing? Activated NK cells express FasL, which suggests that they may eliminate some target cells by Fas-killing. *In vitro* work has indicated that NK cells can undergo AICD due to

FcR-initiated induction of FasL expression. However, the relevance of these observations to the situation *in vivo* is not clear.

ii) Immunoregulation via Cytokine Secretion
NK cells influence cells of both the innate and adaptive immune responses by their secretion of IFNγ and TNF, proinflammatory cytokines whose properties were described in detail in Chapter 17. In response to infection, phagocytes produce IL-12 and TNF, which induce primed (but not activated) NK cells to synthesize copious quantities of IFNγ (Fig. 18-4). Subsequent activation of primed NK cells by engagement of either the activatory receptors (in the absence of an inhibitory signal) or stimulation of FcγRIIIA/CD3ζ complexes leads to the production of a new battery of cytokines, chemokines, and growth factors, including not only TNF but also IFNα, IL-1β, IL-3, IL-6, MIP-1α, MIP-1β, lymphotactin, RANTES, and GM-CSF. The properties of these molecules were also detailed in Chapter 17.

IV. ACTIVATORY AND INHIBITORY NK RECEPTORS
We'll now take a closer look at the receptors controlling natural cytotoxicity. In general, activatory receptors on NK cells recognize pathogen-derived proteins or carbohydrates on

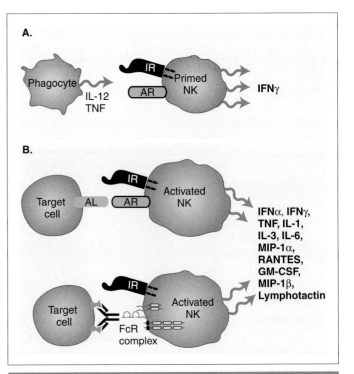

Figure 18-4

NK Cell Cytokine Secretion
(A) In the presence of cytokines secreted by activated phagocytes, primed NK cells contribute to the inflammatory and Th1 responses by producing copious amounts of IFNγ. (B) Should a primed NK cell be activated by AR engagement (top) or by FcγRIIIA stimulation (bottom) in the absence of IR engagement, the NK cell secretes many more inflammatory and immunoregulatory cytokines and chemokines.

the target cell surface, or non-classical MHC molecules. In contrast, inhibitory receptors recognize classical MHC class I molecules independent of the identity of the peptide in the binding groove. Interestingly, many genes related to NK function are clustered together on human chromosome 12 and mouse chromosome 6. These genetic regions constitute the *natural killer gene complex* (NKC) of each species.

i) Activatory Receptors

It has been observed that different NK clones in one individual have different capabilities of lysing a given MHC class I-deficient target, and any one NK cell clone has variable cytotoxic activity against a panel of MHC class I-deficient cells. Thus, it is suspected that, in most situations, a collection or series of receptors has to be engaged to reach a threshold sufficient to trigger NK degranulation and target

cell lysis. One can also picture the activatory receptors as serving a docking function, holding the NK cell and the potential target together long enough for the NK inhibitory receptors to scan the surface of the target and determine whether it expresses self-MHC or not. In the absence of inhibitory receptor engagement, and with contributions by costimulatory molecules, positive signaling initiated by the engagement of multiple activatory receptors triggers NK cell activation.

We will now discuss what is known about the major NK activatory and costimulatory receptors (summarized in Table 18-2) before moving on to the major NK inhibitory receptors (summarized in Table 18-3). Structures of selected activatory and inhibitory receptors are included in Box 18-1, which discusses the occurrence of both activatory and inhibitory receptors in the same structural family.

Table 18-2 Major NK Activatory and Costimulatory Receptors

Receptor	Present on	Associated Signaling Chain	NK Functions Promoted by Engagement	Activatory Ligand
ACTIVATORY NCR CLASS				
NKp46	Activated and resting NK	FcεRIγ or CD3ζ	Natural cytotoxicity against virally infected and tumor cells. Cytokine secretion	Sendai and influenza HAs
NKp44	LAK	DAP12	Cooperates with NKp46 for natural cytotoxicity against virally infected and tumor cells	Sendai and influenza HAs
NKp30	Activated and resting NK	CD3ζ	Natural cytotoxicity against NKp46-resistant targets	?
NKG2 CLASS				
CD94/NKG2C	NK, γδ T, some NKT	DAP12	Promote natural cytotoxicity	HLA-E (hu), Qa-1b (mu)
NKG2D	NK, some T, some NKT	DAP10	Can override inhibitory signaling and trigger natural cytotoxicity and cytokine secretion	MICA (hu), MICB (hu), Rae-1 (mu), H-60 (mu)
KAR CLASS				
KAR p50	NK, some T	DAP12	Promotes natural cytotoxicity	HLA-C
COSTIMULATORY				
2B4	NK, CD8+ T, γδ T, basophils, monocytes	None	Enhances NK natural cytotoxicity	CD48
LAG3	NK	None	Enhances NK natural cytotoxicity?	?
CD2	NK, T, NKT	None	Enhances NK natural cytotoxicity	CD48
CD69	NK, activated T, NKT	None	Promotes NK proliferation, TNF secretion, natural cytotoxicity	Unknown cell surface carbohydrates
NK1.1 (mu)	NK (some strains of mice)	None	Promotes IFNγ production, natural cytotoxicity	Unknown cell surface carbohydrates

HA, Hemaglutinin proteins; hu, human; mu, murine.

Table 18-3 **Major NK Inhibitory Receptors**

Inhibitory Receptor	Present on	NK Functions Inhibited by Engagement	Inhibitory Ligand
C-TYPE LECTIN CLASS			
Ly-49 (mu)	NK, NKT, some CD8$^+$ memory T	Inhibit NK natural cytotoxicity, cytokine secretion, ADCC	Many MHC class I alleles
CD94/NKG2A (mu & hu) CD94/NKG2B (mu & hu)	NK, some T	Inhibit NK natural cytotoxicity, cytokine secretion, ADCC	HLA-E (hu) Qa-1b (mu)
KIR CLASS (hu)			
KIR-2DL	NK, some T	Inhibit NK natural cytotoxicity, cytokine secretion, ADCC	HLA-C (KIR p58) HLA-C (HP3E4/EB6)
KIR-2DS	NK	Inhibit NK natural cytotoxicity, cytokine secretion, ADCC	HLA-C
KIR-3DL	NK, some T	Inhibit NK natural cytotoxicity, cytokine secretion, ADCC	HLA-B (NKB1) HLA-A (p140)
KIR-2DL4	NK	Decreases activation of maternal NK cells against fetal tissue	HLA-G
ILT CLASS (hu)			
ILT2	NK, T, B, myeloid	Inhibits NK natural cytotoxicity, cytokine secretion, ADCC	HLA-A HLA-B HLA-E HLA-G

mu, murine; hu, human.

Box 18-1. Activatory members of inhibitory receptor families, and vice versa

A recurring theme that has been revealed by the study of various receptor families in NK cells is that of structurally related family members that "see" the same ligand but appear to initiate opposing actions. In general, the inhibitory members of a family are transmembrane proteins with long cytoplasmic tails containing one or more ITIM motifs used to recruit SHP tyrosine phosphatases. The action of these enzymes to dephosphorylate key tyrosine residues shuts down activatory signaling. Inhibitory receptors tend to bind with high affinity and fast kinetics to their ligands. The activatory receptors in a family are transmembrane proteins whose extracellular domains closely resemble those of their inhibitory counterparts. However, activatory receptors have much shorter cytoplasmic tails that lack ITIMs, precluding an inhibitory function. Because it does not contain its own ITAMs, to have an activatory function the activatory receptor must associate with an independent ITAM-containing signaling chain such as DAP10, DAP12, FcεRIγ, or CD3ζ. The ITAMs presumably recruit intracellular signaling kinases such as the Src family kinases, ZAP-70, or Syk, which deliver an activatory or costimulatory signal via tyrosine phosphorylation. Activatory receptors tend to bind to the shared ligand with lower affinity and slower kinetics than do the inhibitory receptors.

Examples of activatory receptors can be found in all inhibitory receptor families. For example, the activatory counterparts to the human KIRs (killer inhibitory receptors) are the KARs (*killer activatory receptors*). KARs (usually about 50 kDa) have two or three extracellular Ig-like domains like the KIRs but, unlike them, exhibit a charged amino acid in the transmembrane domain. Most importantly, the KARs have short cytoplasmic tails (about 39 amino acids) that lack the ITIMs present in the long cytoplasmic tails of the KIRs. For example, a KAR called p50 associates with the signaling chain DAP12 to convey signals leading to activation of natural cytotoxicity *in vitro*.

Similarly, the NKG2C and NKG2E molecules in humans and mice have truncated cytoplasmic tails lacking the ITIM motif but associate with CD94 and either DAP10 or DAP12 to form the activatory receptors CD94/NKG2C and CD94/NKG2E. Both of these receptors bind to HLA-E in humans (Qa-1b in mice) like the inhibitory receptor CD94/NKG2A but do so with lower affinity and slower kinetics. The related activatory receptor NKG2D does not associate with CD94 and instead forms homodimers that signal via DAP10. DAP10 contains an SH2-binding site capable of recruiting the p85 subunit of PI3K. In humans, NKG2D binds to the stress-associated MHC class Ib molecules MICA and MICB (see text for details). In mice, the NKG2D homologue binds to two proteins, Rae-1 and H-60, which are only distantly related to MHC class I and are not homologous to MICA and MICB.

While most members of the murine Ly-49 receptor family are inhibitory, Ly-49D, Ly-49H, Ly-49P, and Ly-49W are short proteins that activate rather than inhibit NK cells. Ligation of these receptors with specific mAbs can lead to NK cytotoxicity *in vitro*. While the activatory Ly-49 receptors can bind to various MHC class I ligands *in vitro*, the *in vivo* function of most in terms of defense against pathogens or tumors remains unclear. One exception is the Ly-49H receptor, which has been directly implicated in defense against murine CMV. Mice expressing this receptor are resistant to the virus, whereas

Continued

Box 18-1. Activatory members of inhibitory receptor families, and vice versa—*cont'd*

animals lacking Ly-49H are susceptible. Murine CMV expresses an MHC class I-like protein called m157, which activates natural cytotoxicity if bound to Ly-49H. It is speculated that m157 may closely resemble a self protein that is the natural ligand for Ly-49H.

Activatory analogues of the human ILT (*immunoglobulin-like transcript*) inhibitory receptors have also been identified. Most members of the ILT family are expressed on non-NK cells (see Box 18-2), with only ILT2 being an NK inhibitory receptor. The activatory ILTs are found predominantly on human myeloid cells rather than lymphoid cells. The activatory ILTs contain a charged residue in the transmembrane domain and have short cytoplasmic tails lacking ITIMs. Activatory ILTs are thought to associate with the ITAM-containing FcεRIγ chain rather than with members of the DAP family. For example, ILT1 can participate in a multi-protein complex with FcεRIγ but not with DAP12. Ligation of ILT1 with specific mAb can trigger degranulation and Ca²⁺ release in an ILT1-transfected tumor cell line.

In a reversal of the above, some NK receptors that promote activation have inhibitory counterparts. Structurally, NKR-P1B is a member of the murine NKR-P1 family of costimulatory receptors that includes NK1.1. However, NKR-P1B contains an ITIM in its cytoplasmic tail, suggesting that it may act to inhibit NK cell activation.

The evolutionary rationale for the generation of these paired receptor systems remains unclear. Paradoxically, both activatory and inhibitory members of a family can be expressed simultaneously on a given cell. It is not difficult to imagine that inhibitory receptors might be needed to prevent unwanted attacks by NK cells on healthy host cells, but the physiological role of the activatory counterparts is puzzling. Why should an NK cell express a receptor that recognizes self-MHC but could theoretically activate cytotoxicity, thereby destroying self? It may be that, rather than activation leading to an effector response, these activatory analogues have another function for which their close pairing with the inhibitory families is important. Lewis Lanier has proposed three scenarios exploring the possible purpose of activatory members of the Ly-49 inhibitory receptor family. In the first scenario, the MHC class I molecule recognized by an activatory Ly-49 receptor of a given NK cell is identical to that recognized by one of the Ly-49 inhibitory receptors of that NK cell. In

this case, a regulatory purpose of some kind might be served, such as the recruitment of the unknown kinases required to phosphorylate the ITIM of the inhibitory receptor. Alternatively, the activatory receptor might deliver a growth and/or survival signal during NK development that favors the maturation of NK cells whose inhibitory receptors recognize that self-MHC. Such an action could help establish the "NK repertoire" of specificities (see later in this chapter). In the second scenario, the activatory Ly-49 receptor recognizes a different self-MHC (let's say H-2Dᵏ) than the inhibitory receptor (let's say H-2Dᵈ). In

this case, if infection or transformation causes the loss of H-2Dᵈ on a cell, the remaining H-2Dᵏ molecules could stimulate the activatory Ly-49 receptor, resulting in the death of the altered-self cell by natural cytotoxicity. Lastly, Lanier has speculated that activatory and inhibitory Ly-49 receptors may recognize different types of ligands altogether. It is possible that, *in vivo*, activatory Ly-49 receptors do not respond to classical MHC molecules at all but instead recognize pathogen components that resemble MHC class I. The activation of these receptors could over-ride inhibitory receptor signaling, promoting the death of infected target cells.

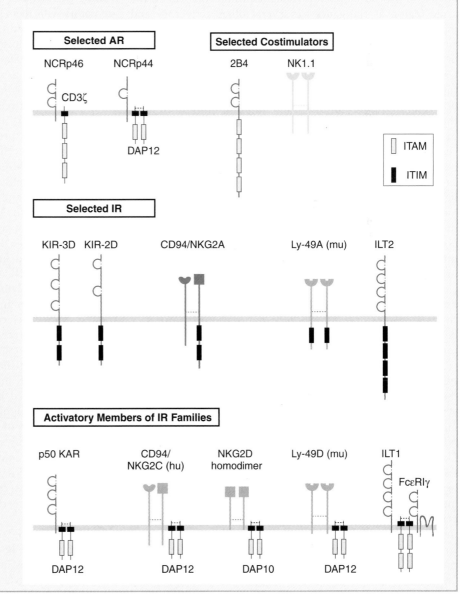

ia) Natural cytotoxicity receptors (NCRs).

In the late 1990s, Lorenzo Moretta and colleagues at the University of Genoa, Italy carried out exhaustive mAb analyses of proteins that were both associated with the induction of natural cytotoxicity and expressed exclusively on NK cells. These efforts resulted in the isolation of three putative activatory receptors called NKp46, NKp44, and NKp30, which were deemed the inaugural members of a growing family called the *natural cytotoxicity receptors* (NCRs). The NCRs, which are expressed in both mice and humans, are the major receptors responsible for the cytolytic destruction of tumor cells.

NKp46 (CD335) is expressed on activated and resting NK cells but on no other cell types examined thus far. NKp46 is a 46 kDa type I transmembrane glycoprotein containing two Ig domains in its extracellular region. It is associated in the NK membrane with ITAM-containing FcεRIγ or CD3ζ chains which become phosphorylated following mAb cross-linking of NKp46. The gene encoding NKp46 has been mapped to human chromosome 19 in the same region as the genes for several other activatory and inhibitory receptors. (This region is not the NKC.) When engaged by specific agonist mAb (and in the absence of action by the inhibitory receptors), NKp46 triggers Ca²⁺ mobilization, cytokine secretion, and NK killing of tumor cells of self or allogeneic origin *in vitro*. It should be noted, however, that not all tumors (even if derived from the same cell type) are equally susceptible to lysis mediated by NKp46. Experiments using blocking mAbs further suggest that at least one ligand for NKp46 is expressed on both normal and abnormal cells, and that ligand expression can vary among tumor cell lines. In 2001, two natural ligands for NKp46 were identified: the hemagglutinin (HA) molecules of influenza and Sendai viruses. Antibodies directed against these pathogen proteins can inhibit NK-mediated cytolysis of infected cells, suggesting that the direct binding of the viral antigen to the NCR activates natural cytotoxicity. It should be noted that not every NK cell clone in an individual expresses the same level of NKp46 or exhibits the same degree of NKp46-mediated cytotoxicity, supporting the current hypothesis that natural cytotoxicity depends on the actions of a collection of activatory receptors rather than a single dominant protein.

NKp44 (CD336) is an NCR expressed only on activated NK cells. NKp44 is a 44 kDa glycoprotein and a member of the Ig superfamily but is not homologous to NKp46. There is evidence that NKp44 may cooperate with NKp46 in mediating natural cytotoxicity against tumor cells, and, like NKp46, NKp44 binds to viral HA molecules. NKp44 has a single V-like extracellular domain, lacks ITAMs in its cytoplasmic domain, and contains a basic amino acid in its transmembrane domain that allows association with the DAP12 signaling chain. DAP12 is a disulfide-linked homodimer that contains an ITAM in its cytoplasmic tail. Phosphorylation of the ITAM by a (unknown) kinase presumably allows downstream interactions with ZAP-70 or Syk that promote activation of the NK cell.

Like NKp46, NKp30 (CD337) is an NCR expressed on both resting and activated NK cells. However, the structure of NKp30 appears to be quite different from that of either NKp46 or NKp44. NKp30 associates with the CD3ζ chain for intracellular signaling. Importantly, although NKp30 may cooperate with NKp46 and NKp44 in killing some human target cells, it also appears to mediate natural cytotoxicity against other targets that can resist NKp46-mediated cytotoxicity. In mice, there appears to be only a pseudogene corresponding to NKp30.

ib) Activatory members of the NKG2 class.

As introduced in Box 18-1, activatory NKG2 molecules resemble their inhibitory counterparts structurally and bind to non-classical MHC class I-like molecules but have no inhibitory properties. For example, the NK receptor composed of CD94 plus NKG2C binds to the non-classical MHC class Ib molecule HLA-E (like the inhibitory receptor CD94/NKG2A) but promotes, rather than inhibits, natural cytotoxicity. The activatory NKG2 proteins contain no ITIMs and associate in the NK cell membrane with ITAM-containing signaling chains such as FcεRIγ, CD3ζ, DAP10, and DAP12. *In vitro*, ligation of many of these receptors with their ligands (if known) or specific mAbs leads to activation of natural cytotoxicity.

Perhaps the most important NKG2 activatory receptor is NKG2D, a 42 kDa transmembrane molecule that forms a homodimer and associates with DAP10. Evidence acquired in the early 2000s has shown that human NKG2D binds to two MHC class I-related molecules called MICA (*MHC class I chain-related A*) and MICB. MICA and MICB are transmembrane stress proteins that are induced on intestinal or thymic epithelial cells in response to heat shock or other cellular stressors. MICA and MICB are also upregulated on virally-infected cells and cells of several types of carcinomas. Human MICA and MICB are tri-domain proteins structurally similar to a MHC class I α chain but with an extremely narrow peptide-binding groove. Neither MICA or MICB binds to the β2m chain. The human MICA and MICB genes are located near HLA-B in the MHC complex. The genes for both MICA and MICB have multiple alleles but the effect of such polymorphism on NK cell recognition has not been explored. In any case, when NKG2D/DAP10 expressed on an NK cell is engaged by the MICA protein, PI3K is recruited to the receptor complex and both NK cytotoxicity and cytokine secretion are triggered. Indeed, the signal delivered is so strong that NKG2D engagement alone can override collective NK inhibitory signaling. Fortunately, as MICA is rarely expressed by normal healthy cells, few targets are inappropriately eliminated.

ic) Costimulators of NK activation: 2B4, LAG-3, CD2, CD69, and NK1.1.

In this section, we consider the effects of ligation of five molecules that appear to contribute in a costimulatory way to NK cell activation. That is, the engagement of these proteins cannot induce cytotoxicity on its own but does enhance the effects of the major NK activatory receptors.

2B4 and LAG-3 are type 1 integral membrane proteins belonging to the Ig superfamily. 2B4 is homologous to the adhesion and costimulatory molecule LFA-3 (CD58) while LAG-3 (*lymphocyte activation gene 3*; CD223), of unknown function, is structurally related to CD4. The ligand for 2B4 is CD48, a molecule that can also bind to CD2, while the ligand for LAG-3 is unknown. CD48 contains two Ig-like domains and is anchored in the target cell membrane via a GPI linkage. Antibody ligation of 2B4 has been shown to enhance NK killing of some targets *in vitro*, while mice deficient for LAG-3 show impaired NK-mediated killing of targets after activation. 2B4 contains four ITAMs in its cytoplasmic tail and is expressed not only on NK cells but also on many CD8$^+$ T cells and γδ T cells, and on all basophils and monocytes. However, when engaged by specific mAb, 2B4 appears to promote only the activation of NK cells.

CD2 is a membrane glycoprotein of 50 kDa that may also act as a costimulatory molecule for NK cell activation. CD2 is expressed on murine and human T and NK cells as well as NKT cells. When engaged by its ligand CD48, CD2 promotes the activation of Src family kinases and PI3K and the induction of natural cytotoxicity. CD2 is not essential for NK activation, however, because knockout mice lacking CD2 have normal NK cell development and function.

CD69, best known as a marker of activated T cells, is also expressed on most mature human NK cells and NKT cells. CD69 is a homodimeric glycoprotein containing an extracellular C-type (calcium-dependent) lectin domain. The gene encoding CD69 in humans is found in the NKC. Whereas the function of CD69 on activated T cells is unknown, the engagement of CD69 on primed NK cells *in vitro* leads to NK proliferation, promotion of cytotoxicity, and the release of TNF. These activities can be blocked by addition of a mAb engaging the inhibitory receptor component CD94 (see later). CD69 appears to recognize as yet undefined cell surface carbohydrates on target cell surfaces.

NK1.1 is a mouse protein with a structure similar to that of CD69, being a disulfide-linked homodimer of 60 kDa subunits that contain integrin and C-type lectin motifs. It is also encoded in the NKC. In strains in which it is expressed, NK1.1 is found on almost all NK cells. Both natural cytotoxicity against target cells and the production of IFNγ can be induced *in vitro* by the cross-linking of NK1.1 on NK cells. The physiological ligand for NK1.1 is believed to be a cell surface carbohydrate but its identity remains obscure. NK1.1 is a member of the NKR-P1 (*natural killer receptor-protein*) family. All mouse NKR-P1 cytoplasmic domains contain a motif shared by the CD4 and CD8 coreceptors that allows interaction with phosphorylated Lck kinase. Engagement of NK1.1 on mouse cells by anti-NK1.1 mAbs induces PI3K activation, generation of arachidonic acid metabolites, and an increase in intracellular Ca^{2+}. The human equivalent of NK1.1 is NKR-P1c (CD161c).

ii) Inhibitory Receptors

The inhibitory receptors of NK cells are much better understood than the activatory receptors. As stated previously, the "missing self" theory of NK recognition holds that host MHC class I molecules represent "self," so that allogeneic cells or cells that express significantly reduced levels of self-MHC class I are subject to NK killing. To put it simplistically, while Tc cells search for peptide–MHC complexes that can be recognized by a specific TCR, NK cells hunt for target cells that lack these same structures. In general, NK inhibitory receptors bind to conserved sites in the extracellular domains of MHC class I molecules that do not involve the peptide-binding groove. Studies of MHC class I recognition by NK inhibitory receptors have shown that the binding of the receptor protein to an MHC class I molecule on a healthy cell requires that there be peptide in the MHC binding groove, but that it does not matter whether this peptide is of self or foreign origin. It is thought that the peptide is needed only to stabilize the conformation of the MHC class I molecule. The inhibitory receptors engaging MHC class I send negative intracellular signals that overrule the positive triggering events of the activatory receptors. In the absence of a signal by the inhibitory receptor (as occurs where self-MHC is missing or only allogeneic MHC is expressed), the activatory signal is unopposed and leads to activation of the NK cell and target lysis. It should be noted that, although the inhibitory signal is normally dominant over the activatory signal, the inhibitory signal can be swamped if enough activatory signals are received at once. In a model system in which multiple mAbs were used to simultaneously activate four activatory receptors while other mAbs were used to stimulate one class of inhibitory receptor, target cell lysis could not be prevented.

Three distinct classes of inhibitory receptors have been identified based on structure. The first class of inhibitory receptors consists of C-type lectin-like proteins. The second class (found only in humans to date) comprises the KIRs (killer inhibitory receptors). The third class contains the ILTs (*immunoglobulin-like transcripts*); only ILT2 is expressed on NK cells. In humans, receptors of all three classes can be simultaneously expressed on one human NK cell, and co-expression of more than one type of C-type lectin or KIR on a given NK cell is common. Indeed, isolated human NK cell clones have been found to express an average of four to five different types of inhibitory receptors spanning all three classes.

Despite their considerable structural differences, once engaged by MHC class I, all three classes of receptors appear to use the same signaling mechanism for delivering the inhibitory signal and blocking both natural cytotoxicity and cytokine secretion. Inhibitory receptors typically are transmembrane proteins with long cytoplasmic tails that contain one or more ITIMs. As we saw in past chapters, ITIMs act to shut down activatory signaling in T and B cells by recruiting phosphatases to the signaling complex. When NK inhibitory receptors are stimulated by the binding of MHC class I, the tyrosine residue in the ITIM is phosphorylated (by kinases yet to be identified), allowing the recruitment of the tyrosine phosphatases SHP-1 or SHP-2 to the inhibitory receptor complex. It is hypothesized that SHP-1 (at least) dephosphorylates signaling intermediates (such as Src kinases and PLC-γ) of the NK activatory receptor complexes, opposing their action and reversing the signal for cellular activation and natural cytotoxicity. In particular, Ca^{2+} flux and the production of IP$_3$ by PI3K are sharply decreased in response to inhibitory receptor engagement, indicating that the inhibition occurs early in the signaling cascade. Experiments with NK cells from *motheaten* mice, natural mutants that lack SHP-1, support the contention that SHP-1 is crucial for delivery of the inhibitory signal.

We shall now discuss each of the inhibitory receptor families in detail. We remind the reader that the key features of inhibitory receptors are summarized in Table 18-3 and that

the structures of selected examples are shown in Box 18-1. In addition, certain members of these inhibitory receptor classes are expressed on non-NK hematopoietic cells, where they are thought to serve a regulatory function. This function was introduced previously and is elaborated upon in Box 18-2.

Box 18-2. Inhibitory receptor families expressed on non-NK cells

This chapter is focused in large part on NK cells and the mechanisms of their inhibitory receptors. Although these molecules were first described for NK cells, it has become apparent that members of inhibitory receptor families (including activatory molecules as described in Box 18-1) can also be found on several other cell types. The expression of inhibitory receptors on γδ T cells is covered separately in the second half of this chapter. In this box, we look at inhibitory receptors expressed on other hematopoietic cell types.

• Ly-49 family

Ly-49 inhibitory receptors have been found on a subset of memory CD8$^+$ T cells in mice. These cells are distinct from NKT cells (see later in this chapter), which are CD4$^+$CD8$^-$ or CD4$^-$CD8$^-$. Ly-49$^+$CD8$^+$ T cells are not present in the thymus but are prominent in the spleen. Interestingly, the presence of Ly-49 on these cells does not appear to significantly inhibit activation of cytotoxicity induced by anti-CD3 ligation *in vitro*. Some scientists speculate that Ly-49 may be present on these cells to modulate cytotoxic activity, rather than block it outright. It should be noted that activatory Ly-49 receptors have not been found on this cell subset.

• CD94/NKG2

The expression of the lectin family inhibitory receptor CD94/NKG2A can be induced on T cells by exposure to IL-15 and TGFβ. An increase in the proportion of cells expressing this receptor has been associated with HIV infection, implying that the inhibition of T cell effector function in this case may have pathological implications.

• KIRs

Parallel to Ly-49 expression on murine CD8$^+$ memory T cells, KIR expression has been detected on subsets of human memory CD8$^+$ T cells. Indeed, KIR$^+$ or KIR$^-$ αβ T cells bearing identical TCRs have been isolated, indicating that KIR induction occurs late

during T cell maturation, certainly well after the completion of TCR gene rearrangement. It is thought that KIR expression may be induced during T cell activation. So far it has not been possible to induce KIR expression on mature lymphoid cells *in vitro*, even using the combination of IL-15 and TGFβ sufficient to induce *in vitro* expression of CD94/NKG2. Nevertheless, engagement of KIRs on CTLs by specific antibody can inhibit cytolysis *in vitro*, suggesting that KIRs on T cells may act as a brake on activation. Such a brake might play a role in the establishment of peripheral tolerance or in the control of AICD such that these memory cells survive. Interestingly, KIR transcripts can also be detected in T cells that do not express the corresponding protein on the cell surface, indicating that translational controls are also in place.

• PIRs

A family of Ig domain-containing proteins called the PIRs (*paired Ig-like receptors*) was identified on murine B and myeloid cells in the late 1990s. The PIRs may be analogous to the human ILTs. There are 12 activatory "A" type PIRs and one inhibitory "B" type PIR. "A" PIRs have short cytoplasmic tails and a basic amino acid in the transmembrane domain that mediates association with the ITAM-containing FcεRIγ. In contrast, the "B" PIR has a long cytoplasmic region containing four ITIMs that recruit the phosphatase SHP-1. It is suspected that the ligands for these receptors are MHC class I-like molecules.

• SIRPs

Another Ig-like pair of activatory and inhibitory receptors that occurs in humans is called SIRP (*signal regulatory proteins*). SIRPα is an ITIM-containing receptor expressed both in myeloid cells, where it recruits SHP-1, and in neurons, where it recruits SHP-2. A mouse homologue of SIRPα has been identified. One ligand for SIRPα is the widely expressed cell surface glycoprotein CD47. It is speculated that a host cell expressing CD47 is protected from host

macrophage phagocytosis because engagement of SIRPα by CD47 shuts down signaling within the macrophage that could lead to engulfment. SIRPβ is an activatory receptor that has a short cytoplasmic domain and associates with DAP12. SIRPβ expression is limited to hematopoietic cells. SIRPβ does not appear to bind to CD47 but no ligand has been identified. No mouse homologue of SIRPβ has been isolated to date.

• ILTs

Various members of the ILT family of Ig-like inhibitory receptors are expressed on a broad range of human leukocytes, including T and B cells, NK cells, monocytes, macrophages, DCs, and granulocytes. As mentioned previously, at least some of the ILTs may be the human counterparts of the murine PIR molecules. The reader will recall that ILT2 inhibits NK cytotoxicity upon engagement by a range of HLA-A, -B, -E, and -G molecules. Similarly, engagement of ILT2 on monocytes by MHC class I inhibits the production of several cytokines by these cells, while ILT2 cross-linking blocks CTL-mediated cytotoxicity in T cells, and Ca^{2+} mobilization in B cells and macrophages. MHC class I-mediated engagement of ILT4, which is expressed exclusively on monocytes/macrophages, blocks tyrosine phosphorylation and Ca^{2+} mobilization in these cells. Moreover, cytokine and chemokine secretion induced by CD40L-stimulation of monocytes can be inhibited by anti-ILT2 or anti-ILT4 mAbs, leading some scientists to speculate that most ILTs function as a brake on cytokine/chemokine secretion by myeloid cells during inflammatory responses. An exception may be the putative activatory receptor ILT1, which is expressed on myeloid cells in association with the ITAM-containing FcεRIγ chain. However, whether this complex actually delivers activatory signals affecting cytokine production, costimulatory molecules, or respiratory burst in myeloid effectors is unclear.

---

iia) C-type lectin-like class
• Ly-49

The best-characterized NK inhibitory receptors are those of the Ly-49 family expressed in mice. The Ly-49 receptors are type II disulfide-linked homodimeric integral membrane glycoproteins with structural similarities to the C-type lectins, meaning that they contain carbohydrate recognition domains in their extracellular regions. Members of the Ly-49 family are expressed on murine NK and NKT cells and on a minor subset of CD8$^+$ memory T cells. Humans and higher primates have a Ly-49-like gene called Ly-49L but it is inactivated by point mutations in these species. Lower primates and cows have a single functional Ly-49L gene. In mice, about a dozen highly homologous Ly-49 genes encode the multiple members of the Ly-49 family (Ly-49A, -B, -C, -D, -E, -F, -G, -H, -I, -P, and -W). These genes are clustered in the mouse NKC.

Different members of the murine Ly-49 family bind to different H-2 proteins, an interaction that can be influenced by variations in the carbohydrate residues on H-2 molecules. The cytoplasmic tails of Ly-49A, -B, -C, -E, -F, -G, and -I all contain the ITIM motif and act as conventional inhibitory receptors. However, Ly-49D, -H, -P, and -W do not possess an ITIM and are putative activatory analogues of the Ly-49 family (refer to Box 18-1). Alternative splicing and allelic polymorphism of the Ly-49 genes increase the diversity in receptors that can be expressed by the NK cell population. Individual NK cells can produce more than one type of Ly-49 receptor, but only one allele of that receptor is expressed per cell—a type of allelic exclusion whose underlying molecular mechanism is unclear.

• CD94/NKG2

Molecules structurally similar (but not homologous) to Ly-49 have been identified on both murine and human NK cells. The CD94 chain is a type II glycoprotein of 30 kDa and is a member of the C-type lectin superfamily. The non-polymorphic CD94 molecule is expressed primarily on NK cells and some T cells. CD94 lacks a cytoplasmic tail, suggesting that it cannot promote inhibition by itself. However, CD94 associates in the membrane with either the NKG2A or NKG2B chain. Both of these proteins, which are about 40 kDa in size, have long cytoplasmic tails containing two ITIM motifs capable of binding to SHP-1. In fact, the appearance of the NKG2A protein in the cell membrane appears to depend on coördinated expression and association with CD94, and the CD94 and NKG2 genes are clustered together in the NKC. In humans, the resulting heterodimeric inhibitory receptor binds specifically to HLA-E molecules. HLA-E is a non-classical MHC class Ib molecule whose peptide-binding groove is occupied by peptides derived from the leader sequences of other MHC class I molecules. One can imagine that the presence of these peptides in the HLA-E groove could facilitate scanning for normal expression of self-MHC class I. Murine CD94/NKG2 heterodimers bind to the non-classical MHC class Ib molecule Qa-1b, a functional homologue of HLA-E. Peptides in the binding groove of Qa-1 molecules are derived from

other H-2 molecules, suggesting evolutionary conservation of this scanning mechanism. CD94 can also form disulfide-linked homodimers but the function *in vivo* of this receptor is unclear.

iib) KIR class. The second class of NK inhibitory receptors was first identified in humans by the ability of these proteins to bind to HLA molecules and block natural cytotoxicity. Despite intensive searching, homologues to the KIRs have not been found in rodents to date. The KIR molecules are not structurally related to either the Ly-49 or CD94/NKG2 families, and are instead type I integral membrane glycoproteins of the Ig superfamily. Different KIR family members are present as either monomers or disulfide-linked homodimers. Like the Ly-49 molecules, the KIRs are expressed on NK cells and certain T subsets, and an individual NK cell can express more than one type of KIR. However, in contrast to Ly-49, carbohydrate does not appear to be involved in KIR recognition of HLA. KIRs bind rapidly and with high affinity to the MHC class I α1 helix at a site distinct from that recognized by the TCR of T cells. The nature of the peptide bound in the groove of HLA molecules recognized by the KIRs may affect the affinity of interaction between the two proteins, but no direct interaction between a KIR and the peptide itself has been observed.

There are at least 14 different KIR genes clustered in a 150 kb region on human chromosome 19. Allelic polymorphism and alternative splicing of these genes make major contributions to KIR diversity. The KIR genes encode at least two distinct types of KIR proteins, distinguished by the number of Ig domains present: two (KIR-2D) or three (KIR-3D). Within the KIR-2D and KIR-3D families, two subgroups are denoted as long (L) or short (S), depending on the length of their cytoplasmic tails and the presence of ITIMs. Those with long (76–84 amino acids) cytoplasmic tails possess two ITIMs able to recruit SHP-1 (and the related SHP-2), and thus act as classical NK inhibitory receptors. For example, p58 is a KIR-2DL molecule of 58 kDa which binds specifically to HLA-C molecules. Similarly, NKB1 (<u>N</u>K receptor for HLA-<u>B</u>) is a KIR-3DL molecule of 70 kDa that interacts with HLA-B. NKB1 can also homodimerize to form a KIR called p140 (140 kDa), which binds specifically to HLA-A. Another KIR called KIR-2DL4 binds to HLA-G, the non-classical HLA molecule expressed on placental trophoblast cells. It is thought that engagement of KIR-2DL4 on maternal NK cells by HLA-G expressed on the trophoblast helps to protect the allogeneic fetus during pregnancy. Other KIR-2D and KIR-3D molecules have short cytoplasmic tails that lack ITIMs and are putative activatory analogues (the KARs mentioned in Box 18-1).

iic) ILT2. Another inhibitory receptor important for human NK cytotoxicity is ILT2. The ILT2 glycoprotein contains four extracellular Ig-like domains and a cytoplasmic tail with multiple ITIMs that can recruit SHP-1. The N-terminus of ILT2 interacts with the α3 domain of the MHC class I protein. Analysis of the crystal structure of ILT2 suggests that the ILT ligand binding site is quite different from that of the KIRs. Engagement of ILT2 by either specific mAb or

MHC class I inhibits the induction of NK cytotoxicity. ILT2 interacts with HLA-A, -B, -E, and -G but apparently not with HLA-C. Intriguingly, ILT2 also binds to the MHC class I-like UL18 protein expressed by human CMV. It is likely that the virus synthesizes UL18 in order to shut down the NK response directed against it (see Ch.22).

As described in Box 18-2, ILT2 belongs to the ILT family of human inhibitory receptors, most of which are expressed on non-NK cells. ILT proteins contain two or four extracellular Ig-like domains homologous to the KIR extracellular domains and are encoded by a cluster of genes near the KIR genes on human chromosome 19. Extensive polymorphism exists for some ILT family members, most notably ILT5. Several ILT family members have been shown to interact with various HLA molecules, but others have no known ligand. Several activatory ILTs have been identified (refer to Box 18-1).

V. NK CELL DEVELOPMENT, INHIBITORY RECEPTOR EXPRESSION, AND TOLERANCE

i) NK Cell Development

NK cells are now considered to be the third major lymphoid lineage after T and B cells, with a particularly close phenotypic relationship to T cells. However, the path of NK development has not been as clearly defined as that of T and B cells and is still under investigation. We do know that NK cells originate from the same bone marrow-derived HSCs that give rise to T and B cell lineages. As we saw in Chapters 9 and 13, the activities of the transcription factors PU.1 and Ikaros and the cell fate determination molecule Notch1 are crucial for the maturation of these precursors. The HSCs give rise in the bone marrow to CLPs, which are capable of generating the T, B, and NK cell lineages (but not myeloid lineages). The CLPs (or another proximal descendant) are thought to then generate a common *NK/T precursor* that can develop into T or NK cells but not B cells. [The term "NK/T precursor" should not be confused with "NKT cells" or "NKT precursor" (see later).] Some of the NK/T precursors arising from the CLPs migrate from the bone marrow to the thymus to give rise to pro-T cells, which go on to generate αβ T cells, NKT cells, and intrathymic γδ T cells. Other NK/T precursors remain in the bone marrow and give rise to NK cells. Some researchers believe that still other NK/T precursors may migrate to the fetal liver, mesenteric lymph nodes, or intestinal cryptopatches to complete their maturation and generate extrathymic γδ T cells (see later in this chapter). (The intestinal cryptopatches are small regions of lymphoid tissue in the lamina propria of the small intestine; see Ch.20.) Other scientists believe that this extrathymic lymphopoiesis occurs only when normal thymic lymphopoiesis cannot proceed, and that the gut epithelium is normally colonized by mature, thymus-derived lymphocytes.

We shall now follow NK development by describing the fate of various markers in mice, keeping in mind that not all of the commonly studied inbred strains express all the following molecules. (Human NK development is thought to be similar to that in the mouse at many stages, but studies lag behind those in mice.) As we saw in Chapter 13, murine HSCs express the c-kit receptor required to receive stem cell factor (SCF) signaling (Fig. 18-5), but not the NK marker NK1.1 or the chains participating in the IL-2 and IL-15 receptors. CLPs generated from these HSCs then commence expression of the IL-7Rα and γc chains required to receive IL-7 signaling. Under the influence of GATA-3 and Notch1 signaling, some of these CLPs become NK/T precursors, while others receive strong IL-7 and IL-3 signaling and go on to be B cells. NK/T precursors commence expression of a variety of cytokine receptors, and cells that receive strong IL-7 signaling and migrate to the thymus go on to become T cells. Other NK/T precursors that remain in the bone marrow become subject to *Flt3 ligand* signaling (see later) followed by IL-15 signaling and eventually develop into NK cells. The inhibitory transcription factor Id2 may also play a role here, because Id2$^{-/-}$ mice lack all NK cells as well as some DC subsets. Id2 is thought to interfere with the function of the E2A transcription factor (important for B cell development) in NK/T precursors, allowing them to become NK cells. We note here that while IL-7 is vital for T and B cell development, it is not as crucial for NK development. Mice with null mutations of the IL-7 or IL-7Rα genes show profoundly decreased numbers of T and B cells but retain near-normal numbers of NK cells. On the other hand, those lacking the γc chain, which is present in the IL-7R and IL-15R (among other receptors), lack all three cell types.

Flt3 ligand (Flt3-L) is a hematopoietic growth factor particularly important for NK development. [Flt3-L is related to MCSF (*macrophage colony-stimulating factor*), and MCSF binds to a receptor known as "fms." Hence, Flt3 is a receptor called "<u>f</u>ms-<u>l</u>ike <u>t</u>yrosine kinase <u>3</u>," and Flt3-L is its ligand.] Flt3-L shows strong homology to SCF and, like SCF, is produced by bone marrow stromal cells. The functions of Flt3-L and SCF may overlap since both work with other growth factors to support early hematopoiesis. However, while NK development is not impaired in SCF-deficient mice, mice deficient for Flt3-L have a profound deficit in NK cell generation. Mice treated with Flt3-L show a significant increase in the number of cells expressing the NK1.1 marker in bone marrow and spleen, an effect replicated *in vitro*. Some scientists speculate that Flt3-L may be involved in inducing or sustaining the expression of IL-2Rβ and/or IL-15Rα required for the receipt of IL-15 signaling by NK/T precursors. Because it ultimately induces an expansion of NK cells that can combat tumor cells, Flt3-L has been touted as an anti-cancer agent and, indeed, has an anti-tumor effect on transplanted tumors in mice.

NK/T precursors were first identified as a subset of early fetal thymocytes that expressed not only the T cell marker Thy-1 but also the NK marker CD16 (FcRγIII) used to distinguish mature NK cells. If these thymocytes were transferred to the thymus of an animal and allowed to develop, they gave rise to αβ T cells; but if they underwent intravenous transfer, they gave rise to NK cells. B cells and

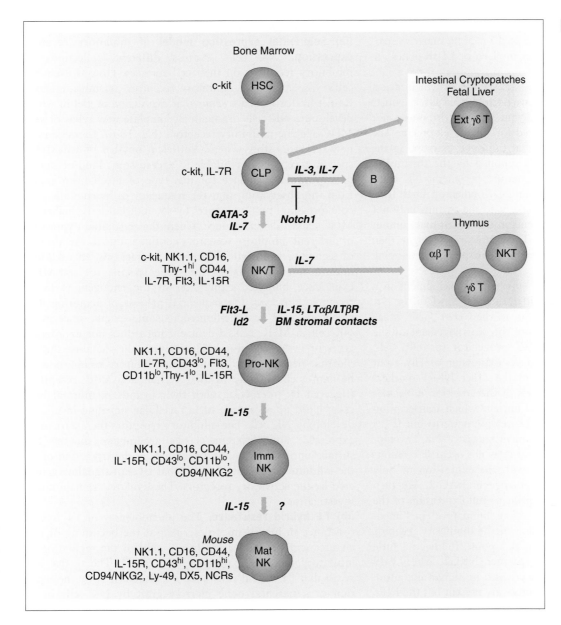

Figure 18-5

Murine NK Cell Development
NK cells are derived from the same HSCs and CLPs that give rise to T and B cells. CLPs that migrate to the intestinal cryptopatches or fetal liver produce extrathymic (Ext) γδ T cells. CLPs that remain in the bone marrow and receive strong Notch1 inhibition as well as IL-7, and IL-3 signaling develop into B cells, while CLPs that receive GATA-3 and IL-7 signaling give rise to NK/T precursors. NK/T precursors that receive strong IL-7 signaling and migrate to the thymus produce αβ T cells and intrathymic γδ T cells and NKT cells. NK/T precursors that remain in the bone marrow and receive strong Id2, Flt3-L, and IL-15 signaling give rise to pro-NK cells. In the continued presence of IL-15, immature NK cells develop and initiate expression of CD94/NKG2 inhibitory receptors. Unknown factors drive the expression of additional IRs such as the Ly-49 receptors. Markers expressed at various developmental stages are indicated.

myeloid cells were never observed. With respect to other surface markers, the NK/T precursor can be characterized as c-kit$^+$, γc$^+$, IL-2Rβ$^+$, IL-15Rα$^+$, NK1.1$^+$, Thy-1$^+$, CD44$^+$ but DX5$^-$, CD3$^-$, CD25$^-$, and Ly-49$^-$. The TCRβ genes in these cells remain in the germline configuration. At this point, IL-15 becomes the key cytokine governing NK development. IL-15 (but not IL-2) is secreted in significant quantities by bone marrow stromal cells during hematopoiesis. When the bone marrow is damaged, the spleen can "take over" the differentiation of several hematopoietic cell types, but not that of NK cells. Some scientists have speculated that NK differentiation may be crippled in secondary lymphoid organs because these tissues cannot match the high local concentrations of IL-15 produced by bone marrow stromal

cells. The addition of exogenous IL-15 can rescue the development of NK cells isolated from mice in which the bone marrow has been ablated. Furthermore, although IL-2R and IL-15R share the IL-2Rβ and γc chains, IL-2 cannot replace IL-15 for the development of lytically-competent NK cells. Knockout mice lacking either IL-2 or IL-2Rα show normal NK development, but mice lacking components of IL-15R or IL-15 itself do not. Jak3 signaling (required to transduce IL-15 signals) must also be intact in NK/T precursors, since knockout mice lacking Jak3 have impaired NK development. Studies of knockout mice lacking the transcription factor IRF-1 have indicated a role for this molecule in the bone marrow secretion of IL-15 and thus in the development of NK cells. As well as their production of IL-15, bone marrow stromal cells may supply intercellular contacts essential for NK development. LTβR is expressed on bone marrow stromal cells, and mice lacking either the LTα or LTβR genes are devoid of NK cells.

Because *nude, scid,* and RAG1/2 mice (all of which lack a thymus and/or mature T cells) have normal NK cells, scientists now believe that neither the thymic microenvironment nor the presence of T cells is required for NK development. The receipt of IL-15 signaling while in the bone marrow, in the absence of later contacts/signals supplied by the thymus, is sufficient for the generation of mature peripheral NK cells. However, particularly in early fetal development, a small fraction of NK/T precursors that make it to the thymus may not be subjected to the inductive events in this organ that commit a precursor to the T lineage. These precursors might then become intrathymic NK progenitors capable of developing into mature NK cells in this site.

It is at about day 12–14 of embryonic development in a mouse when progenitors committed to either the T or the NK lineage can be identified, just prior to the CD44$^+$C25$^+$ stage of T cell development and before the rearrangement of the TCRβ genes. The committed NK progenitor (pro-NK cell) loses expression of c-kit and reduces expression of Thy-1 but retains the expression pattern of the other NK/T markers cited previously. Expression levels of the adhesion molecules CD43 and CD11b are low. It is unclear which transcription factors govern these changes in expression patterns but IL-15 remains important for all subsequent stages of NK development. Somatic gene rearrangement does not occur in committed NK progenitors, consistent with the expression of only low levels of RAG-1 and the absence of RAG-2. We then come to a stage of maturation prior to full expression of the NK inhibitory receptors known as the "immature" or "uneducated" NK cell stage. Expression of the inhibitory receptor component CD94 commences at this point, followed a little later by that of its signaling partner NKG2. The Ly-49 inhibitory receptors are still not present. In human immature NK cells, the CD94/NKG2 receptors are present but the KIRs are not yet expressed.

In the last stage of murine NK cell maturation, we start to see the expression of CD49b (DX-5) and the Ly-49 receptors, in addition to ongoing expression of NK1.1, the IL-15R components, and CD16. The markers CD43 and CD11b are also upregulated. In human NK cells at this stage, expression of the marker CD56 and the KIRs can be detected. Interestingly, an unknown signal in addition to IL-15 appears to be required for the full expression of the KIRs and Ly-49 receptors. Tests of several known cytokines and factors have failed to induce expression of these receptors on isolated precursor populations. A clue may lie in the fact that the expression of Ly-49 in a murine *in vitro* differentiation system was strictly dependent on the presence of stromal cells. What specific contact or cytokine these cells supply at this juncture is not clear. IL-15 continues to be important after the newly produced NK cells leave the bone marrow and home to peripheral tissues. It is thought that IL-15 induces the expression of anti-apoptotic genes such as Bcl-2 that promote the survival of mature NK cells in the periphery.

ii) NK Inhibitory Receptor Expression
iia) Sequential expression model of inhibitory receptor expression.
NK cells express different collections of inhibitory receptors on their cell surfaces. How is each NK cell's "repertoire" of inhibitory receptors assembled? David Raulet has proposed a *sequential expression* model in which developing NK cells gradually accumulate new types of self-MHC-specific inhibitory receptors (Fig. 18-6). Expression of new receptors stops when a sufficient number of self-MHC-specific receptors is available to exceed some kind of signaling threshold. Evidence for this type of mechanism comes from the observation that the frequency of murine NK cells co-expressing two or more Ly-49 molecules is higher in MHC class I-deficient mice, where the sequential expression of different inhibitory receptors continues due to the absence of signaling from MHC–receptor interactions. In addition, mice expressing a Ly-49 transgene specific for self-MHC experience inhibited expression of the endogenous Ly-49 receptor. However, the sequential process of acquiring new receptors must be time-limited to some extent, since every NK cell in MHC class I-deficient mice does not express all types of Ly-49 receptors. What defines the signaling threshold that determines when no new inhibitory receptors need be expressed? It may be that certain activatory receptors, triggered by interaction with their ligands on normal host cells, build up a cascade of intracellular signaling inside the developing NK cell. The inhibitory receptors then start to be expressed, and different classes accumulate on the cell surface until the signaling generated by the triggering of all the inhibitory receptors exceeds that level of signaling generated by the activatory receptors. These matters remain under investigation.

iib) F1 hybrid resistance.
The phenomenon of F1 hybrid resistance in inbred mice described at the beginning of this chapter can be understood in the light of the regulation of inhibitory receptor expression (Fig. 18-7). The reader will recall that F1 hybrid resistance is characterized by the rejection of a parental bone marrow graft by NK cells in an F1 hybrid offspring. While the MHC haplotype of either parent would be "self" from the point of view of a B or T cell in the F1 offspring, let's consider the situation from the NK point of view. Suppose a parent of haplotype MHC X was crossed to a parent of haplotype MHC Y to give rise to F1 offspring of MHC XY. Within the NK population in the offspring, the variable expression of inhibitory receptors would mean that not every NK cell would express the same collection of receptors. In other words, the offspring would have some NK cells bearing a mixed inhibitory receptor population recognizing MHC X and MHC Y, while other NK cells would bear only receptors recognizing MHC X, and still other NK cells would bear only receptors recognizing MHC Y. Self-tissues expressing both X and Y would inactivate all three populations of NK cells and there would be no attack on self. However, upon introduction of parental X bone marrow, those NK cells in the offspring that sported <u>only</u> inhibitory receptors recognizing MHC Y would fail to engage those receptors, thereby allowing activatory receptors to complete signaling for cytotoxicity and

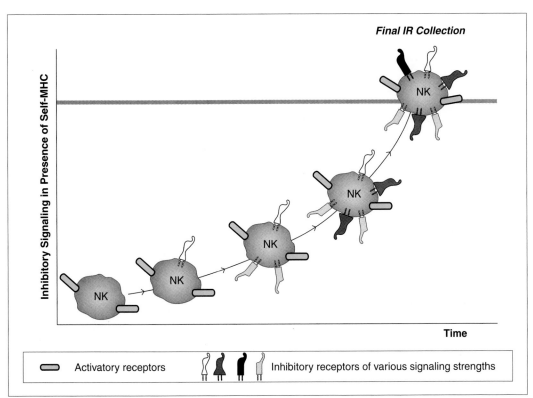

Final IR Collection

Inhibitory Signaling in Presence of Self-MHC

Time

Activatory receptors — Inhibitory receptors of various signaling strengths

Figure 18-6

Sequential Expression Model of IR Regulation
As the NK cell develops, it starts by expressing a single type of IR. Engagement of this IR by self-MHC delivers a level of signaling that allows the cell to continue development and to gradually express more and more different types of IRs. Once the cumulative signal generated by engagement of all new types of IRs exceeds a putative regulatory threshold (gray line), the NK cell terminates expression of new IR types.

kill the grafted cells. (Note that the equivalent phenomenon would not be observed in an outbred population because rejection would occur due to the numerous differences at the MHC loci.)

iii) NK Tolerance

NK cells are sometimes described as "self-tolerant" because they do not attack normal self cells. An NK clone with the potential to attack self would be a clone whose collection of expressed inhibitory receptors failed to recognize self-MHC expressed by the host. That is, the inhibitory receptors would not be engaged when ligands on normal self cells stimulated the NK activatory receptors, resulting in activation of the NK cytolytic program and destruction of the self cells. Thus, during development, an NK cell that does not express appropriate inhibitory receptors must undergo some kind of tolerance process to either delete it or render it harmless. Evidence for such "NK tolerization" was first demonstrated in 1991 by Petter Höglund and colleagues in Kärre's laboratory. These researchers showed that, unlike normal NK cells, NK cells from β2m$^{-/-}$ mice did not destroy lymphoblast targets from β2m$^{-/-}$ mice, despite the lack of engagement of the NK inhibitory receptors. Furthermore, experiments with chimeric mice in which β2m was expressed in the peripheral tissues but not in the bone marrow showed that peripheral expression of MHC class I during development was not enough to restore NK killing of β2m$^{-/-}$ targets. These results indicated that the NK cells had "learned" to tolerate MHC

class I-deficient targets, and that the "education" process responsible was taking place centrally in the bone marrow. Unlike self-reactive B and T cells, however, self-reactive NK clones do not appear to be removed by clonal deletion during development.

We will now briefly discuss three models that have been put forward to account for NK tolerance. These models are not necessarily mutually exclusive.

iiia) "At least one" model. As mentioned previously, most human NK cells express more than one class of inhibitory receptor, and more than one type in each class. Thus, the average NK clone expresses four to five different inhibitory receptors, each of which may bind to a different subset of MHC alleles. The "at least one" model of NK tolerance proposes that each developing NK cell becomes "tolerant to self" by expressing at least one inhibitory receptor specific for self-MHC. Evidence for this model comes from the analysis of a large number of human NK clones from two individuals, in which it was found that each clone expressed at least one receptor that was self-MHC-specific. The model presumes that a survival signal of some sort is delivered during NK maturation so that clones expressing such receptors are selected for further differentiation. Lanier has speculated that the function of those activatory members of inhibitory receptor families that recognize self-MHC may be to promote just this sort of positive selection during NK development. However, it should be noted that some NK cells in the individuals studied may have

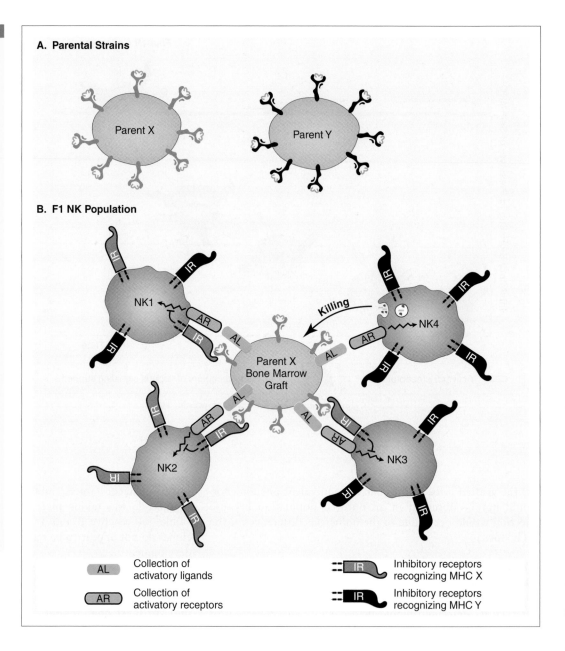

Figure 18-7

NK F1 Hybrid Resistance
(A) Parental mouse strain X expresses MHC class I haplotype X, whereas parental mouse strain Y expresses MHC class I haplotype Y. (B) A bone marrow graft from parental strain X is introduced into an F1 mouse derived from a cross of parental strains X and Y. Different NK cells in the F1 recipient express different collections of IR. For example, F1 NK cell #1 (NK1) expresses one type of IR that recognizes MHC X and another IR that recognizes MHC Y. These F1 NK cells are inhibited from attacking the graft by signaling through the IR that recognizes MHC X. F1 NK2 expresses only IRs that recognize MHC X, so it is also inactivated by the graft. Although F1 NK3 expresses only one type of IR recognizing MHC X (among many recognizing MHC Y), the resulting signaling is sufficient to prevent this cell from attacking the graft. It is F1 NK4 that causes the "resistance of the F1 progeny mouse to transplantation" because this cell expresses only IRs that recognize MHC Y. Since none of the IRs on this cell is engaged by MHC X, an inhibitory signal is not delivered, and F1 NK4 attacks the graft.

A. Parental Strains

Parent X

Parent Y

B. F1 NK Population

NK1

Killing

NK4

Parent X
Bone Marrow
Graft

NK2

NK3

AL	Collection of activatory ligands

AR	Collection of activatory receptors

IR	Inhibitory receptors recognizing MHC X

IR	Inhibitory receptors recognizing MHC Y

avoided self-reactivity through anergy (see later) rather than through expression of one or more self-MHC-specific inhibitory receptors. Such anergized NK cells would be difficult to clone and might therefore have been missed in this particular study.

iiib) "Receptor calibration" model. The "receptor calibration" model proposes that cell surface levels of self-MHC-specific inhibitory receptors are calibrated to ensure that the cumulative signal triggered by the binding of self-MHC to all inhibitory receptors is sufficient to exceed some kind of signaling threshold. If this threshold is not exceeded (meaning that, in the periphery, the inhibitory signals would not be enough to overcome activatory signaling), an unknown mechanism prevents the NK cell from completing normal development. The strength of the signal mediated by any inhibitory receptor is determined by the affinity of interac-

tion between the receptor and a self-MHC molecule. If there is only a weak interaction, a greater surface level of expression of that receptor may occur to ensure an adequate signal (Fig. 18-8). Lower levels of expression may suffice for receptors with higher affinities for self-MHC. *In vitro* experiments have confirmed that surface levels of Ly-49 receptors are higher when the MHC ligands binding to them do so weakly.

iiic) "NK anergy" model. What happens to NK cells that fail to express inhibitory receptors, or express receptors that do not deliver a cumulative signal that will be sufficient to overcome activatory signaling in the periphery? Some observers have suggested that these cells are anergized in an unknown way. Anergization could explain the presence of the hyporesponsive NK cells in mice that are MHC class I-deficient or completely devoid of MHC class I. By

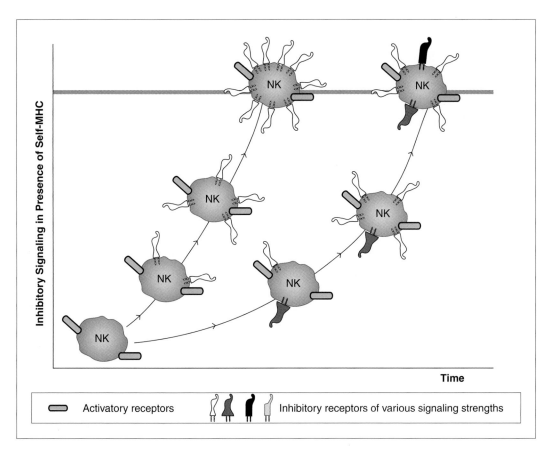

Figure 18-8

Receptor Calibration Model of NK Tolerance
The strength of the signal delivered when each inhibitory receptor binds to self-MHC is combined in an effort to exceed a certain signaling threshold. A large number of IRs with weak affinity for self-MHC, a small number of IRs with strong affinity for self-MHC, or various combinations of both IR types may supply a cumulative signal that overcomes a signaling threshold (gray line). These cells go on to become mature NK cells in the periphery that are activated in the absence of self-MHC.

Inhibitory Signaling in Presence of Self-MHC

Time

⬭ Activatory receptors   Inhibitory receptors of various signaling strengths

extension, it follows that NK cells in the periphery of normal mice may also be rendered anergic under certain conditions if their inhibitory receptors are insufficiently engaged. Indeed, such a tolerance mechanism would be needed to keep NK cells from killing normal host cells that are naturally low in MHC class I expression, such as cells in the brain or trophoblast. The current thinking is that NK anergy may be achieved by the dampening of normal activatory signaling pathways such that lysis of non-transformed, uninfected targets is prevented. The underlying mechanisms remain a mystery.

B. γδ T Cells

We have described in earlier chapters how two distinct types of T lymphocytes can be distinguished based on the polypeptide chains making up their TCRs. The majority of T cells in the body carry TCRs that contain the TCRα and TCRβ chains, making them αβ T cells. Mature αβ T cells carry either the CD4 or CD8 coreceptors. Those bearing CD4 bind exclusively to MHC class II and are usually helper T cells, which can be further subdivided into Th1 or Th2 cells depending on their profile of secreted cytokines. Those αβ T cells carrying the CD8 coreceptor bind to MHC class I and are usually cytotoxic T cells. In this section, we discuss

the biology of γδ T lymphocytes that bear TCRs composed of the TCRγ and TCRδ chains and are generally CD8+ or CD4−CD8−. These differences have profound implications for the development and function of these lymphocytes. A comparison of the properties of αβ and γδ T cells is presented in Table 18-4.

I. WHAT ARE γδ T CELLS?

The history of the discovery of T cells bearing γδ TCRs was outlined in Chapter 12. γδ T cells do not initiate rearrangement of the TCRα locus and never express the TCRα and β chains. In addition, the anatomic localization of γδ T cells is very different from that of αβ T cells, and their developmental paths appear to be surprisingly distinct. The frequency of γδ T cells also differs from that of αβ T cells, although this parameter varies widely among species. For example, in adult humans and mice, γδ T cells comprise only 0.5–5% of total T lymphocytes, while the frequency is much higher in cattle, sheep, and chickens.

The evidence suggests that γδ T cells may have evolved prior to the evolution of the highly specific αβ T cells, in that the γδ TCR recognizes ligands in a way that is similar to ligand recognition by the PRRs of conventional innate immune system cells. That is, γδ T cells can recognize antigens that are broadly distributed over a wide range of pathogens. Curiously,

Table 18-4 **Comparison of $\alpha\beta$ and $\gamma\delta$ T Cells**

	$\gamma\delta$ T Cells	$\alpha\beta$ T Cells
Immune system branch	Innate	Adaptive
Frequency among T cells	0.5–5%[†]	95–99%
Anatomical distribution	Epithelial layer of skin and mucosae of respiratory, digestive, and urogenital tracts (IELs)	Primary and secondary lymphoid organs and tissues
TCR	TCR$\gamma\delta$	TCR$\alpha\beta$
Repertoire diversity	Limited	Extremely diverse
Receptor specificity	Promiscuous	Highly specific
Distribution of epitope recognized	Broadly expressed on wide range of pathogens	Expressed on one pathogen
Ligand recognized	Intact proteins, MICA, MICB, phospho- and glycolipids, phosphonucleotides, pyrophosphate, HSPs	Peptide–MHC complexes
APCs required for antigen presentation	No	Yes
Mature T cell co-receptor expression	CD4$^+$CD8$^-$ CD4$^-$CD8$\alpha\beta^+$	CD4$^-$CD8$\alpha\alpha^+$ CD4$^-$CD8$^-$
Co-stimulation required for activation	No[‡]	Yes
Activation threshold	Very high	High
TCR localizes in rafts	No	Yes
Preferred signaling kinase	Syk	ZAP-70
PLC-γ1 and LAT involved	No	Yes
Pace of effector cell generation	Rapid (1–2 days)	Slow (7 days)
Types of effectors generated	Th1, Th2, CTL, immunoregulatory	Th1, Th2, CTL, immunoregulatory
Capable of memory response	No	Yes

IELs, intraepithelial lymphocytes; HSP, heat shock proteins.
[†] In humans and mice.
[‡] That is, CD28- and CD40L-mediated costimulation are not required.

despite the fact that the array of gene segments in the γ and δ TCR loci has the potential to generate a very large number of unique TCR sequences by V(D)J recombination, the functional repertoire of $\gamma\delta$ TCR antigenic specificities is much more limited than that of $\alpha\beta$ TCRs. In addition, unlike conventional $\alpha\beta$ T cells, $\gamma\delta$ T cells do not require the involvement of conventional MHC nor the processing and presentation of antigens by APCs to recognize their ligands. Intact proteins or peptides from pathogens or stressed host cells, and non-protein antigens such as phosphorylated nucleotides, can serve as $\gamma\delta$ T cell ligands. These ligands may be either soluble or appear on cell surfaces. This pattern of ligand recognition is also reminiscent of B cells, which can recognize native protein and non-protein antigens. Interestingly, analysis of the amino acid sequence of $\gamma\delta$ TCRs (particularly the V region of the TCRδ chain) suggests that they are very closely related to Ig molecules, perhaps more so than to $\alpha\beta$ TCRs. X-ray crystallographic analysis of a human Vγ9Vδ2 TCR has supported this hypothesis, since this molecule was found to share struc-

tural similarities with Ig heavy and light chains, and to have a BCR-like binding site capable of binding small phosphate compounds.

Once activated, $\gamma\delta$ T cells respond by proliferating and differentiating into $\gamma\delta$ Th and CTL effectors much like $\alpha\beta$ T cells, but do so much more rapidly, in smaller numbers, and in the apparent absence of conventional costimulation. Thus, like macrophages, NK, and NKT cells, $\gamma\delta$ T cells can be counted among the "sentinels" of induced innate immunity.

i) Surface Markers
Naturally, the most distinguishing surface feature of $\gamma\delta$ T cells is the $\gamma\delta$ TCR, the expression of which was detailed in Chapters 12 and 13. No other marker exists to date that unequivocally identifies $\gamma\delta$ T cells. It is not uncommon for $\gamma\delta$ T cell clones to express conventional PRRs, and many human clones also express the CD94/NKG2 inhibitory receptors found on NK cells (refer to Table 18-1), while murine $\gamma\delta$

T cells often express NK1.1. In several species, CD2, CD5, 2B4, and CD6 are also present on most γδ T cells (CD6 is involved in thymocyte–epithelial cell adhesion). Less often, IL-2Rβ, CD69, and CD44 can be found on γδ T cell clones. When γδ T cells express CD8, it is present most often as the CD8αα isoform rather than the CD8αβ isoform found on αβ T cells. CD28 and CD40L are expressed on only some γδ T cells and do not appear to act as costimulatory molecules for these cells.

ii) Anatomical Distribution

One of the most intriguing aspects of γδ T cell biology is the anatomical distribution of these cells, which is strikingly different from that of αβ T cells. Only very low numbers of γδ T cells in mice and humans are found in the secondary lymphoid organs or the thymus. Instead, these cells are concentrated in the tissues themselves, interspersed among the epithelial cells comprising the skin and the mucosal linings of various body tracts. In fact, there is a 10-fold higher ratio of γδ:αβ T cells in the human gut mucosa compared to that in a lymph node. However, there are considerable differences among species in γδ T cell anatomical localization. For example, while γδ T cells are very common in chickens, they are not present at detectable levels in the skin of these animals.

In humans and mice, the γδ T cells dispersed among epithelial cells constitute one type of *intraepithelial lymphocyte* (IEL). When IELs are found in the gut, they are called iIELs, for "intestinal IELs." In mice, there are particularly high concentrations of γδ iIELs tucked among the enterocytes forming the intestinal epithelium. In humans, γδ T cells make up a much smaller proportion of the iIEL compartment. When IELs are resident in the top layers of the skin, they are sometimes called *dendritic epidermal T cells* (DETCs). The vast majority of DETCs in adult mice are γδ T cells, but again, γδ T cells occur far less frequently in human skin. High concentrations of γδ T cells are also present in the mucosae of the urogenital and respiratory tracts of mice, while only low numbers are present in these tissues in humans. The localization of γδ T cells at mucosal surfaces allows them to be among the first to confront invading pathogens or injurious substances. The prevalence of γδ T cells in tissues where other lymphoid and myeloid defenders are scarce reflects their importance for defense against certain pathogens. For example, certain intracellular bacteria (e.g., *Nocardia*) that invade the airway epithelium are combated by an influx of neutrophils, which is critical for the removal of necrotic epithelial cells and containment of the pathogen. However, TCRγδ-deficient mice cannot mount this influx and thus succumb to the infection, evidence that cooperation between γδ T cells and neutrophils is essential for protection of the lung.

Unlike αβ T cells, which utilize highly random combinations of V gene segments for their TCRs regardless of their anatomical location, the γδ T cells resident in a particular tissue express a dominant or "canonical" TCR. For example, as was introduced in Table 12-3, γδ T cells in mouse skin predominantly express TCRs containing Vγ5Vδ1 or Vγ3Vδ1, while the genital epithelium and the tongue feature an abundance of Vγ6Vδ1-expressing cells. Vγ7 appears more often than not in the TCRs of iIELs, whereas Vγ2Vδ2 cells dominate in the tonsils, spleen, and peripheral blood. In humans, Vγ1Vδ2 TCRs are prevalent on iIELs. The mechanisms responsible for compartmentalizing the γδ T cell repertoire in this way are unknown, but it is suspected that both genetic and microenvironmental factors are involved. The precise time during embryonic development when a given γδ T cell precursor is generated may also play a large role in defining its TCR and thus its destiny. The shaping of the repertoire continues in later life, as neonates have many more different γδ T cell clones in their blood than do adults.

II. ANTIGEN RECOGNITION

i) General Principles

Conventional MHC molecules are not part of the structure recognized by the γδ TCR, as demonstrated by the fact that MHC class I- and class II-deficient mice have normal γδ T cell development and function. As already mentioned, γδ T cells can recognize low molecular weight non-peptide antigens directly, without the need for presentation by another molecule or cell. However, some subsets of γδ T cells appear to bind to the non-polymorphic CD1 family of MHC-like molecules that present non-peptide antigens to αβ T cells, as discussed in Chapters 10 and 11. Whether small antigens have to be present in the binding site of CD1 molecules for presentation to γδ T cells remains unresolved.

Because the γδ TCRs lack the fine antigen specificity of αβ TCRs, they are often broadly cross-reactive. Work in mice lacking specific Vγ gene segments has shown that it is the overall shape of the γδ TCR that is important for specificity, rather than the particular Vγ segment used. For example, Vγ3-deficient mice compensate for the loss of Vγ3 with higher usage of the Vγ1 segment in their TCRs, and a mAb raised against the Vγ3-bearing γδ T cell population in wild-type mice can recognize the Vγ1-bearing γδ T cell population in the mutant mice. These findings suggest that the overall structures of these TCRs are similar, and imply that the expression of a given TCR on γδ T cells in a given tissue is driven by the need to recognize particular molecules in that tissue. In other words, if one gene segment is missing, another that creates a TCR of similar structure can ultimately take its place. This type of substitution most likely explains why knockout mice in which a particular Vγ or Vδ gene segment has been deleted often show no immunological deficits. These observations also suggest that regions of the γδ TCR outside of the antigen-binding site may make a significant contribution to antigen binding.

ii) Sources of Antigens

It should be noted that most of the information that follows comes from experimental systems, and may only give us a partial indication of the powers of recognition of γδ T cells. Crystal structure determinations and *in vivo* studies are ongoing to more thoroughly define the physiological ligands of γδ TCRs.

iia) Pathogen-derived antigens. γδ T cells respond to a wide variety of bacterial, protozoan, and viral molecules, including those derived from *Listeria*, *Mycobacteria*, *Plasmodium*, *Leishmania*, *Salmonella*, and EBV. Antigenic molecules from these organisms include modified phospholipids, lipoproteins, and phosphorylated oligonucleotides that often feature pyrophosphate-like epitopes. Certain alkyl amines, which are common components of pathogenic bacteria and parasites, have also been shown to stimulate particular γδ T cell subsets. In addition, peptides of viral origin and even whole viral proteins can serve as γδ T cell antigens. For example, in one study, an intact surface glycoprotein of HSV was specifically recognized by a murine γδ T cell clone.

Different γδ T cell subsets are specific for different types of determinants and thus may counter a specific group of pathogens (Table 18-5). For example, human γδ T cells bearing Vγ2Vδ2 recognize small pyrophosphate-like molecules. Phosphorylated metabolites of this type are prevalent during mycobacterial infections and may serve as a signal to the immune system that live bacteria are present. Consequently, the Vγ2Vδ2 population increases in the peripheral blood when a patient suffers from mycobacterial lesions. Vγ2Vδ2 T cells also expand in cases of malaria, leprosy, and EBV infection. In contrast, in cases of HIV infection or Lyme disease (caused by the spirochete *Borrelia burgdorferi*), T cells bearing TCRs containing Vδ1 increase in frequency. These various pathogens tend to enter the body at characteristic locations, providing a possible rationale for the differential localization of various γδ subsets as described previously.

iib) Host stress antigens. γδ T cells also recognize various host "distress" molecules that are expressed by cells suffering injury, infection, or transformation. These stress antigens are not generally found on the surfaces of healthy host cells, so that their presence on a cell is a flag to the immune system indicating that the cell is unhealthy and should be destroyed. The expression of a single stress antigen in response to myriad different infections or injuries would allow a γδ T cell population with a limited

antigen receptor repertoire to monitor a wide range of assaults to the host epithelium.

While some stress molecules able to activate γδ T cells take the form of non-peptidic pyrophosphate-like molecules, others are peptides or whole proteins, such as the evolutionarily conserved heat shock proteins introduced in Chapter 11 (Fig. 18-9A). For example, engineered expression of heat shock protein 58 (HSP58) on a lymphocyte membrane provokes a vigorous response from human Vγ9Vδ2 T cells, and a homologue of mammalian HSP60 that is expressed by *Mycobacteria* (mycobacterial HSP60) can activate murine Vγ1Vd6 cells. Similarly, the Vγ3-expressing γδ T cells that predominate in murine skin are specific for a stress antigen released by injured or infected keratinocytes. Mice experiencing stress, inflammation, or infection upregulate expression by their epithelial cells of the so-called "T10/22 proteins," which are non-classical MHC class Ib molecules encoded by genes in the T region. T10/22 proteins have been shown to bind to cloned epithelial γδ T cells and activate them. In many cases, the presence of a stress antigen is inferred, as in the case of γδ T cell/neutrophil cooperation required to minimize lung damage in mice. It is speculated that airway epithelial cells under attack express a stress antigen on the surface that activates Vγ6 T cells, resulting in the secretion of cytokines and chemokines that draw neutrophils to the area.

• MICA and MICB

The stress antigens MICA and MICB, previously discussed as ligands for the NKG2D activatory receptor of NK cells, have also been identified as important γδ T cell antigens in humans. Cells on which MICA and MICB are expressed become targets for γδ iIELs bearing Vγ1Vδ1 TCRs. These iIELs can then kill the MIC-expressing epithelial cells, at least *in vitro*. Indeed, in one experiment, overexpression of human MICA on B cells stimulated the vigorous activation of Vγ1Vδ1 iIELs. Interestingly, both intestinal and epidermal γδ IELs also express NKG2D. It is unclear if or how the γδ TCR and NKG2D cooperate or oppose each other in intracellular signaling, but some scientists have mused that MIC molecules might be bound by both the γδ TCR and NKG2D on Vγ1Vδ1 T cells (Fig. 18-9B), just as MHC class I is bound by both the αβ TCR and CD8 coreceptor on αβ T cells (Fig. 18-9C). This "co-engagement" might be pivotal for γδ T cell activation, and NKG2D may be serving in this context as a coreceptor. Alternatively, NKG2D may be acting as a costimulatory molecule that supports activation through the γδ TCR. In studies of αβ IELs, which also express NKG2D, it was shown that these αβ CD8+ T cells could be fully activated in the absence of CD28 engagement if NKG2D was ligated instead. Consistent with this hypothesis, PI3K signaling is activated by the engagement of either CD28 or NKG2D.

• CD1c

In humans, some γδ T cell clones can recognize the nonconventional, non-polymorphic antigen presentation molecule CD1c. As discussed in Chapter 11, CD1 molecules on professional APCs can present lipid and glycolipid foreign antigens to αβ T cells. Expression of CD1c is induced on human monocytes,

---

Table 18-5 γδ T Cell Antigens

γδ Usage	Antigen Recognized
Vγ2Vδ2 (hu)	Small pyrophosphates; epitopes derived from EBV, *Mycobacteria*, *Plasmodium*, *Leishmania*, *Salmonella*
Vγ9Vδ2 (hu)	HSP58
Vγ3 (mu)	Stress antigen of keratinocytes
Vδ1 (hu)	Epitopes of HIV, *Borrelia*, CMV; CD1c
Vγ6 (mu)	Stress antigens of lung epithelial cells
Vγ1Vδ1 (hu)	MICA, MICB
Vγ?Vδ? (mu)	T10/22 epithelial cell stress proteins

mu, murine; hu, human.

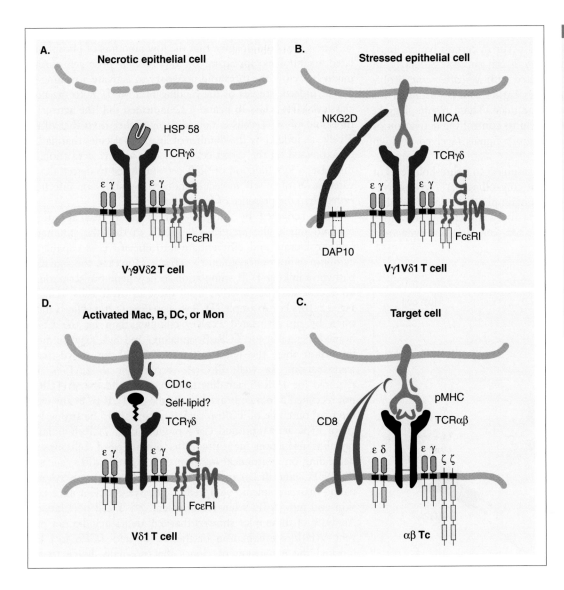

Figure 18-9

Binding of γδ TCRs to Various Antigens
(A) Necrotic epithelial cells leak stress proteins such as the heat shock protein HSP58, which is recognized by the γδ TCR of Vγ9Vδ2 T cells. (B) Recognition of the MHC class I-like molecule MICA on a stressed epithelial cell by the γδ TCR of Vγ1Vδ1 T cells. NKG2D may serve as a coreceptor, much like CD8 functions in the αβ Tc cell in panel D. (C) Recognition of pMHC on a target cell by the αβ TCR of an αβ Tc cell. (D) Recognition of CD1c on activated APCs by the γδ TCR of Vδ1 T cells. The antigen-binding site of the CD1c molecule may contain self-lipid.

macrophages, B cells, and DCs in response to cytokines secreted during stress situations such as inflammation and bacterial invasion. Activated human Vδ1 T cells can respond to CD1c molecules on macrophages, B cells, and DCs *in vitro* (Fig. 18-9D), resulting in proliferation and perforin- or Fas-mediated lysis to destroy CD1c-bearing cells. No foreign antigen need be in the binding site of the CD1c molecule, but it is unclear whether the γδ TCR recognizes CD1c alone, or CD1c bound to a self-lipid molecule. The biological function of such potentially self-reactive clones is also unknown.

III. ACTIVATION

The proliferation of γδ T cells in response to injury or infection is generally very rapid, as befits cells participating in the induced innate response. Indeed, the release of cytokines by γδ T cells can precede the activation of αβ T cells by several days. However, the details of γδ T cell activation remain a mystery.

Small phosphorylated metabolites are able to activate Vγ9Vδ2 clones directly, at least *in vitro*, emphasizing that neither conventional antigen presentation nor costimulation (at least by CD28 or CD40) is required for γδ T cell activation. Nevertheless, γδ T cells proliferate more strongly *in vitro* in the presence of macrophages, monocytes, or DCs than in their absence. Myeloid cells may therefore "present" some antigens to γδ T cells in an MHC-independent manner and/or provide them with cytokines such as IL-1, IL-12, and IL-15. Mature peripheral γδ T cells can also be activated by IL-7, which is interesting in light of the fact that epithelial cells are an important source of this cytokine. Activated CD4+ αβ T cells may also supply stimulatory cytokines, and may even be able to stimulate γδ T cells by direct (unknown) intercellular contacts in those circumstances and locations where the two cell subsets come together (such as in sites of inflammation).

γδ T cell activation may also be influenced by inhibitory receptors. As was true for NK cells, the transgenic expression of MHC class I alleles on target cells inhibits their ability to be lysed by γδ CTL clones. Conversely, aberrant B cells that fail to express MHC class I (such as occur in certain lymphomas) have been shown to activate γδ T cell subsets. Similarly, in MHC class I-deficient mice infected with *Mycobacteria*, proliferation of γδ T cells commences earlier and is sustained for longer than in infected wild-type mice. These results suggest that inhibitory receptors may help to control the activation of γδ T cells, balancing the stimulatory signals received through the γδ TCR (Fig. 18-10).

γδ T cell activation may also be subject to a threshold level of signaling, perhaps similar to that controlling αβ T cell activation. Normal healthy host cells contain phosphorylated metabolites, but these are present at relatively minor concentrations compared to those in an infected cell and are generally sequestered in the cytosol rather than being exposed at the cell surface. One could imagine that the low amounts of phosphorylated metabolites appearing on the normal host cell surface might be below the threshold necessary to activate a passing γδ T cell. Indeed, studies of the binding of a γδ TCR to the nonclassical MHC class Ib ligand T22 indicated that the activation threshold of γδ T cells is considerably higher than that of αβ T cells, as judged by the number of ligand molecules that had to be expressed on the target cell surface to achieve activation. In this case, a key function of γδ T cells might be to detect an elevated level of a self molecule induced by stress or infection, rather than the presence of foreign molecules.

Once activated by antigen, mature γδ T cells appear to respond much like mature αβ T cells in that the stimulated cells proliferate and differentiate into effector cells capable of cytolytic or immunoregulatory activity. However, the signaling pathways linking TCR stimulation to new gene transcription in γδ T cells may be different from those in αβ T cells (refer to Fig. 18-10). For example, TCRγδ signaling is transduced more often by an associated FcεRIγ complex than by the CD3ζ homodimer used for TCRαβ signaling, and Syk may be more important than the related kinase ZAP-70 for downstream transduction. As well, PLC-γ1 recruitment to LAT is not required for TCRγδ signaling transduction, and the γδ TCR is not recruited into membrane rafts as the αβ TCR is. In any case, the final outcome of TCRγδ signaling appears to be strong levels of ERK activation and Ca^{2+} mobilization. TCRγδ-initiated signal transduction may also differ among γδ T cell subsets, depending on anatomical location. For example, IELs (such as the DETCs and iIELs) appear to be more dependent on Syk signaling than are splenic γδ T cells, since Syk-deficient mice lack skin and gut γδ IELs but retain splenic γδ T cell populations. Analyses of these mice showed that Syk deficiency did not prevent V(D)J recombination of the TCRγ and TCRδ loci but blocked the expansion of clones that normally home to the epithelial tissues.

With respect to the termination of activation, γδ T cells appear to be subject to at least some of the same mechanisms used to remove activated αβ T cells. In Fas-deficient mice, clones of γδ T cells in the gut activated by prolonged exposure to antigen proliferate uncontrollably, indicating that Fas-mediated AICD may normally remove excess activated γδ T cells.

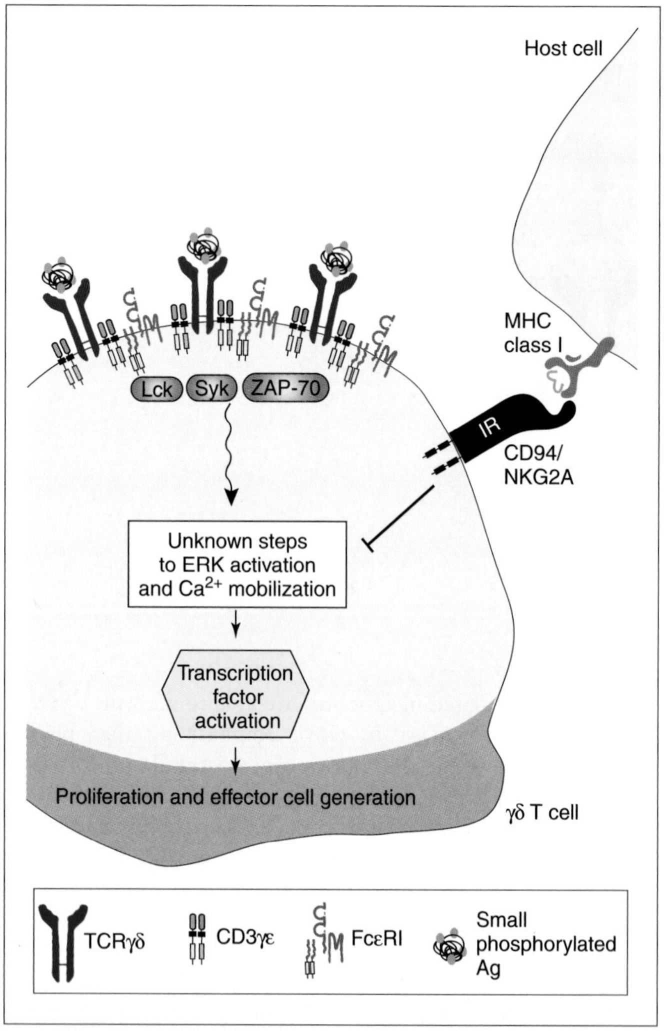

IV. EFFECTOR FUNCTIONS AND MEMORY

i) Effector Functions

In general, γδ T cells have effector functions that are not dissimilar to those of αβ T cells (except that they are not MHC-restricted). Some γδ T cells are cytolytic like αβ Tc cells, developing into CTLs that use perforin-, granzyme-, or Fas-mediated cytolysis to eliminate both infected and transformed cells (Fig. 18-11). Cytolytic Vδ1 iIELs also secrete the antimicrobial protein granulysin. B lymphomas, sarcomas, and carcinomas have all been shown to be subject to γδ T cell-mediated cytotoxicity. Other γδ T cells are like αβ Th cells and secrete cytokines and growth factors that affect other leukocytes. Indeed, Th1 and Th2 subtypes of γδ T cells can be identified

Figure 18-10

Control of Signaling Leading to γδ T Cell Activation
Binding of antigen (in this case, a small phosphorylated antigen) to the γδ TCR complex triggers signaling via Lck, Syk, and ZAP-70, which leads by unknown steps to the ERK activation, Ca^{2+} mobilization, and transcription factor activation required for proliferation and effector cell generation. This signaling path may be inhibited by signals initiated by the engagement of IRs by MHC class I.

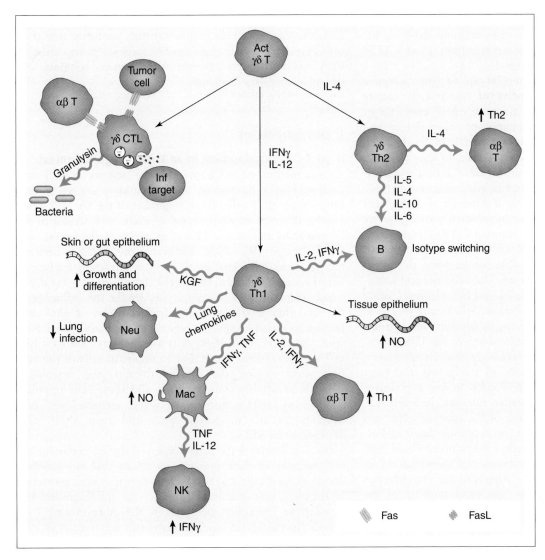

Figure 18-11

Effector and Immunoregulatory Functions of γδ T Cells
Activated γδ T cells generate both γδ CTL and γδ Th1 and Th2 effector cells. γδ CTLs kill infected (Inf) target cells and tumor cells by granule-mediated cytotoxicity and/or Fas killing, and eliminate bacteria by granulysin secretion. γδ CTLs and Th cells exert immunoregulatory control over αβ T cells, the former killing them via Fas ligation or the latter inducing their differentiation into αβ Th1 or Th2 cells via cytokine secretion. Cytokines and chemokines produced by γδ Th cells also influence isotype switching in B cells and have various other effects on epithelial cells and phagocytes (see text). KGF, keratinocyte growth factor.

based on their secretion of IL-2 and IFNγ, or IL-4, IL-5, IL-6, and IL-10, respectively, although the Th1 type predominates. Interestingly, the path to Th1/Th2 differentiation appears to be slightly different for γδ T cells than for αβ T cells. While Th1 γδ T cells are produced in the presence of IL-12 (and thus IFNγ), and Th2 γδ T cells are produced in the presence of IL-4, different transcription factors and cytokine receptors appear to be involved. This pathway has yet to be completely defined.

With respect to immunoregulation of the innate response, IFNγ and TNF secreted by Th1 γδ T cells are very important for the activation of NK cells and macrophages. For example, consider immune responses to the intracellular parasite *Listeria*: the large amounts of IFNγ secreted by NK cells are vital for protection. The NK cells are activated by IL-12 and TNF is produced by macrophages activated in the course of infection. However, studies of TCRδ-deficient mice have suggested that it is the γδ T cells that produce the early IFNγ that first induces the macrophages to produce TNF. This is another example of the collaboration between γδ T cells and other cell types in the provision of innate defense. IFNγ production and cytolytic

activity attributable to γδ T cells have also been cited in the early phase of responses against various viruses.

Other molecules secreted by γδ T cells contribute to the innate response. Chemokines produced by activated γδ IELs influence the trafficking of leukocytes to sites of injury or attack, especially damaged epithelium. We have already noted that γδ T cells play a key role in recruiting neutrophils to sites of invasion of airway epithelial cells. Activated γδ IELs can also induce the production of the anti-microbial compound nitric oxide in neighboring epithelial cells. Activated γδ T cells that have matured in the skin to become DETCs, or in the intestine to become iIELs, can secrete *keratinocyte growth factor* (KGF), a molecule that promotes the growth and differentiation of epithelial cells. αβ T cells and γδ T cells that develop elsewhere in the body do not have this capacity, giving γδ IELs a unique role in tissue repair. Indeed, the intestinal epithelial cells of mice lacking γδ T cells show abnormal differentiation and reduced proliferation. γδ T cells may also provide innate defense independent of the TCR, perhaps via other receptors on their surfaces that are equivalent to those on PRR-bearing cells. For example, LPS is able to activate some mouse skin IEL clones *in vitro* without involving the TCR, implying the presence of a receptor (that is not a TLR) able to transduce LPS-initiated signals.

With respect to immunoregulation of the adaptive response, different subpopulations of γδ T cells secrete cytokines that influence the activity of αβ T cells (up or down) and promote the differentiation of either Th1 or Th2 αβ T cells. In addition, γδ T cells may be able to deliver T cell help to B cells in the absence of αβ T cells. In mice deficient for TCRβ (and therefore deficient for αβ T cells), low levels of germinal center formation and isotype switching to IgG, IgA, and IgE were still able to occur. It is thought that γδ T cells present in these mutant animals compensated for the absence of αβ Th cells and supplied the required CD40L intercellular contact to the activated B cells. Other data have suggested that γδ T cells may be responsible for secreting the initial burst of IL-4 responsible for the αβ Th2 response and subsequent production by B cells of IgG1 and IgE. However, other work with inhaled antigens has indicated that some γδ T cells can inhibit IgE production, perhaps by blocking the maturation of Th2 cells. Interestingly, mice deficient for TCRδ show defective production of IgA, the antibody most closely associated with defense at body surfaces. γδ T cells may also exert direct control over activated αβ T cells. In one experiment studying infection with the bacterium *Borrelia*, γδ T cell subsets proliferating during the course of the infection expressed high levels of FasL and induced the apoptotic death of Fas-bearing CD4+ αβ Th2 cells. Such a mechanism would help regulate an adaptive immune response and contribute to its termination when appropriate. Alternatively, the selective death of Th2 cells could cause unwanted skewing of the immune response to the Th1 phenotype, possibly exacerbating inflammation.

We see a two-pronged approach to T cell-mediated immune defense here: γδ T cells, with their broad specificities and speedy activation, send signals to the rest of the body that a pathogen has been detected, and then act to destroy that pathogen while it is still in the peripheral tissues in small numbers. The αβ T cell response kicks in only if the γδ T cells (and their fellow innate immune system defenders) have been unsuccessful; that is, when significant amounts of inflammatory cytokines permeate the tissues and APCs have been activated to process and present pathogen peptides on MHC. Additional subsets of γδ T cells may then contribute to the termination of the αβ T cell response. The sentinel role of γδ T cells has spurred interest in defining the *in vivo* functions of γδ T cells in disease situations. Some examples of the work demonstrating the importance of this cell type in human diseases and in animal models of human diseases are given in Box 18-3.

ii) Memory

Immunological memory does not appear to be a feature of γδ T cell responses. In some infections, two phases of γδ T cell proliferation are observed: a rapid response immediately after infection with the pathogen, and a later proliferative burst 2–3 days after the primary αβ T cell response is concluding. However, even though these γδ T cells may express memory markers such as CD45R0, they do not appear to be fully functional memory cells. Some scientists speculate that these cells may have a restricted capacity to develop into effectors such that defense remains only at the primary response level during a secondary challenge.

V. DEVELOPMENT

i) γδ T Cell Development in an Embryological Context

As was mentioned in Chapter 12, the first waves of T cells produced in a human or murine embryo are γδ T cells rather than αβ T cells. Rearrangement of the γδ TCR genes can be detected in thymocytes as early as 8 weeks in the human fetus and or day 12.5 of gestation in the mouse. The TdT enzyme that adds additional nucleotides between the V, D, and J segments is not yet activated, leading to V(D)J recombination in these cells that is not very far diversified from the germline sequence. Under the influence of the fetal thymic stromal cells, distinct waves of γδ T cells fan out to populate specific regions of the body. As we have seen, the γδ TCRs of these cells are characterized by a bias in Vγ gene segment usage. For example, in cells of the very first wave exiting the murine thymus, the rearranged γδ TCR invariably contains Vγ3, and these cells populate the skin as DETCs. Subsequent waves include Vγ6+ cells heading for the urogenital tract and then Vγ7+ cells destined to be iIELs.

γδ T cells may have been conserved during evolution not only because of their presence in tissues that are generally devoid of αβ T cells, but also because they may offer protection in the very young animal prior to the full establishment of adaptive immunity mediated by the more powerful αβ T cells. In particular, αβ Th1 responses are weaker in very young animals than in adults. Perhaps not coincidentally, γδ T cells are present at a higher frequency in the neonatal period than in later life, and many are biased toward the Th1 phenotype. Indeed, while neonatal TCRαβ-deficient mice are able to resist infection with the protozoan parasite *Eimeria vermiformis*, neonatal TCRδ-deficient mice are not. Furthermore, this parasite causes severe disease in adults, suggesting that the lower frequency of γδ T cells present later in life makes the host vulnerable. Other evidence supporting an early life function for γδ T cells comes from studies of the fates of particular populations of these cells. For example, the thymus in an adult mouse does not appear to be able to support the development of several γδ T cell subsets, particularly the Vγ3+ subset constituting the DETCs.

ii) γδ T Cell Development and TCR Gene Rearrangement

As introduced in Chapter 13, γδ T cells are thought to arise from the same NK/T precursor that generates αβ T cells in the thymus and NK cells in the bone marrow (Fig. 18-12). Interestingly, IL-7 is more important for the development of γδ T cells than for that of either αβ T cells or NK cells. While

Box 18-3. Roles of γδ T cells in disease

γδ T cells have proved to be important players in both human diseases and animal models of human diseases. In this box, we will briefly consider the role of γδ T cells in infectious diseases, tumor surveillance, and autoimmune disorders. With respect to pathogen infections, the cytokines and chemokines secreted by γδ T cells tend to promote the inflammatory response that attempts to contain the pathogen. Animals lacking γδ T cells have been found to be more susceptible than wild-type animals to malaria, HSV-1-mediated lethal encephalitis, listeriosis, and infections by *Mycobacteria* or *Nocardia*. In humans, populations of (primarily) Vγ9Vδ2 T cells increase following infection with the bacteria causing tuberculosis, salmonellosis, and brucellosis, or with the parasites causing malaria, leishmaniasis, trypanosomiasis, and toxoplasmosis (see Ch.22). HIV-infected individuals show increased numbers of Vδ2$^+$ T cells in their lungs and increased Vδ1$^+$ T cells in their peripheral blood. Clinical researchers speculate that these expanded cell populations may be present to combat the many opportunistic infections suffered by AIDS patients (see Ch.25). Similarly, kidney transplant patients who contract CMV infection after the operation show increased numbers of Vδ1$^+$ and Vδ3$^+$ T cells capable of responding to CMV antigens.

With respect to tumor surveillance, γδ T cells may have an advantage over αβ T cells in that they appear to recognize certain tumor antigens rapidly and directly, without MHC restriction or the need for antigen processing. Different γδ T cell clones can kill different types of tumor cells *in vitro*. Although none of the specific antigens triggering γδ T cell activation has been identified, clones have been isolated that destroy thymic lymphoma cells, erythroleukemia cells, EBV-transformed cells, myeloma cells, and Burkitt's lymphoma cells. Expanded oligoclonal populations of γδ T cells have been found surrounding CNS tumors and renal and lung carcinomas, and γδ T cells have been shown to infiltrate skin melanomas. The γδ T cell antigen MICA appears on many epithelial tumors, inviting recognition by Vδ1$^+$ T cells. Mice transgenic for the TCRγ chain become resistant to the induction of acute T cell leukemia.

In the case of autoimmune diseases, γδ T cells may function as a double-edged sword. In some situations, these cells appear to drive the autoreactive inflammatory response, but in other situations, γδ T cells may act to limit the inflammation promoted by the actions of autoreactive αβ T cells. For example, although the antigens have yet to be identified, numbers of γδ T cells are increased in the affected joints of rheumatoid arthritis patients, in the blood and intestine of inflammatory bowel disease patients, and in the brain and cerebral spinal fluid of multiple sclerosis patients suffering relapses. Monoclonal expansion of γδ T cells has been noted in cases of autoimmunity associated with neutropenia (abnormally low numbers of neutrophils). This observation suggests that the neutrophils in these patients may express a surface antigen recognized by a particular γδ T cell clone. However, in TCRδ-deficient mice, the inflammation associated with autoimmunity is more widespread and aggressive than that seen in wild-type mice. Similarly, in murine models of autoimmune arthritis, lupus nephritis, and allergic encephalomyelitis (see Ch.29), disease is exacerbated in the absence of γδ T cells. In NOD mice, which are susceptible to the development of diabetes, a population of regulatory CD8$^+$ γδ T cells was identified that was necessary for the prevention of diabetes induction in the mutant animals. A similar paradigm has been noted in humans with celiac disease, an autoimmune disorder resulting from an attack by wheat-antigen-specific αβ T cells against self antigens of similar conformation in the gut. Scientists have observed a correlation between disease status and number of γδ T cells in the gut, in that the quiescent phases of the disease are associated with the presence of large numbers of intestinal γδ T cells. On the other hand, disease flare-ups and severe inflammatory damage to the tissues are associated with a striking decrease in numbers of gut γδ T cells. The importance of γδ T cells in controlling autoreactive αβ T cells has also been illustrated using *lpr* mice, which are Fas-deficient and prone to developing autoimmunity. In *lpr* mice lacking TCRδ, there was an elevated autoantibody titer, enhanced Ig deposition in the kidney, and increased numbers of infiltrating CD4$^+$ αβ T cells compared to TCRδ-competent *lpr* mice. TCRδ-deficient *lpr* mice also die prematurely, reinforcing the concept that γδ T cells play a discrete but crucial role in maintaining health.

IL-7 deficiency has only a mild effect on NK maturation and αβ T cell development is blocked at the pro-T cell stage, the generation of γδ T cells is completely abolished. Similarly, IL-7Rα-deficient mice are missing thymic and extrathymic γδ T cell subsets, while NK and αβ T cell populations are present but reduced in number. Some researchers have used the expression of IL-7Rα to distinguish among early thymocyte precursors. In these experiments, IL-7Rα^{hi} thymocytes later developed into γδ T cells while IL-7Rα^{lo} thymocytes went on to become αβ T cells. It was first hypothesized that IL-7 signaling might deliver a survival signal to lymphoid cells, such as upregulating expression of the anti-apoptotic molecule Bcl-2. Indeed, transgenic expression of Bcl-2 can restore αβ T cell development in mice lacking IL-7R components. However, γδ T cell development was still compromised in IL-7-deficient Bcl-2 transgenic animals. It was discovered that, in the absence of IL-7, V–J joining is disrupted in the TCRγ loci during T cell development, a defect that cannot be rescued by Bcl-2. IL-7's most important role in γδ T cells may

thus be to sustain activation of the RAG genes necessary for antigen receptor gene rearrangement.

As we have seen, an NK/T precursor that migrates to the thymus and receives the appropriate stimulation there becomes committed to the T lineage and eventually gives rise to DN3 thymocytes. At the DN3 stage, the TCRγ locus rearranges first followed by the TCRδ locus, as described in detail in Chapter 12. Some rearrangement of the TCRβ locus may also occur. Unlike the TCRβ gene but like the TCRα gene, the TCRγ and TCRδ genes fail to exhibit strong allelic exclusion. Both alleles of the TCRγ and TCRδ genes can show in-frame rearrangements, leading to the possibility of a γδ T cell expressing two types of TCR. In one study of a panel of 27 clonal hybridomas derived from mouse γδ T cells, it was found that 17 clones had V(D)J rearrangements on both TCRδ alleles, and that 6 (35%) of these had two in-frame rearrangements. However, since γδ T cells expressing two TCRs are observed very rarely, it is likely that at least one of the alleles encoded a TCRδ

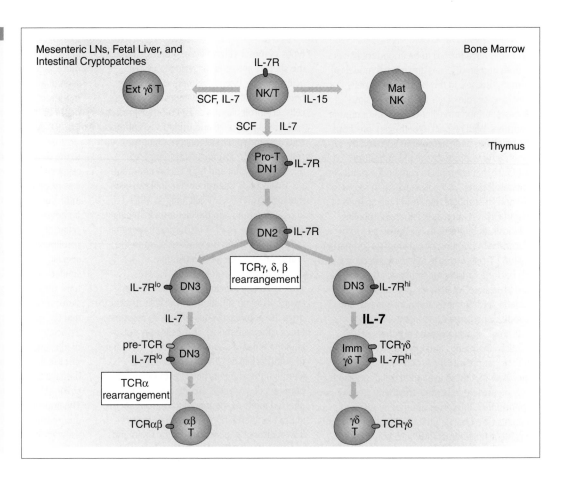

Figure 18-12

Development of γδ T Cells
γδ T cells arise from the same NK/T precursors in the bone marrow that produce αβ T cells and NK cells. NK/T precursors that migrate to the mesenteric lymph nodes, fetal liver, and intestinal cryptopatches give rise to extrathymic (Ext) γδ T cells. NK/T precursors that migrate to the thymus and receive IL-7 signaling eventually give rise to DN3 thymocytes in which the TCR-γ, δ, and β genes have rearranged. Some DN3 cells have relatively low levels of IL-7Rα and subsequently express the pre-TCR. These cells receive moderate IL-7 signaling and undergo the process of β-selection that results in marked expansion of the αβ T cell pool. Other DN3 cells express high levels of IL-7Rα, receive strong IL-7 signaling, and subsequently express the γδ TCR. Only limited expansion of the γδ T cell pool occurs. Mat NK, mature NK cell; Imm γδ T, immature γδ T cell.

chain that failed to pair successfully with a TCRγ chain to form a functional γδ TCR.

At some unknown point prior to TCRα rearrangement, the developing DN3 cell makes a decision to become either an immature γδ T cell or a committed DN3 αβ T cell. We know that αβ/γδ commitment does not occur prior to TCR gene rearrangement because evidence of TCRβ rearrangement can be found in mature γδ T cells, and rearranged TCRγ and δ genes have been detected in mature αβ T cells. It appears that the expression of a complete γδ TCR and signaling through it precludes further TCRβ rearrangement and blocks access to the αβ T cell developmental path. In contrast, in DN3 cells destined for the αβ T cell path, the TCRβ chain is paired with the pTα chain to form the pre-TCR. Engagement of the pre-TCR delivers the β-selection signal crucial for promoting survival and expansion of the αβ T cell pool, an event that does not occur for γδ T cells. Indeed, mice lacking pTα exhibit increased numbers of γδ T cells in the thymus compared to wild-type mice. It is thought that stimulation of the cytoplasmic tail of the pTα molecule activates the transcription factor NF-κB, which may in turn activate genes required to sustain αβ T cell differentiation. Since they lack the pre-TCR, immature γδ T cells do not exhibit this NF-κB activation, and thus most likely cannot express the αβ-determining genes.

Intracellular signaling during γδ T cell development remains largely undefined. Syk kinase signaling, which is present at a higher level than ZAP-70 signaling early in ontogeny, is thought to be important. Signals transduced by Syk from the γδ TCR appear to rescue γδ T cells from apoptosis and induce further differentiation events. It is unknown which ligands in the thymus engage the γδ TCR to initiate such signaling. Another factor that may be relevant to developmental TCRγδ signaling is the lack of the CD3δε heterodimer in the γδ TCR. In DP thymocytes bearing αβ TCRs, an important link is forged between CD3δε and CD4 or CD8 and their associated Lck molecules. TCRγδ thymocytes lack CD3δε and may thus have to recruit different signaling molecules into the receptor complex.

Evidence is accumulating suggesting that many γδ T cells mature extrathymically. In mutant animals with abnormal thymus structure, apparently normal numbers of functional γδ, but not αβ, T cells are able to develop. In particular, the mesenteric lymph nodes, fetal liver, and possibly the intestinal cryptopatches may be important extrathymic sites of γδ TCR rearrangement and cell maturation (although this hypothesis is still controversial). In one study, stem cells and NK/T precursors from the bone marrow were shown to home directly to the cryptopatches, bypassing the thymus, and to develop into precursors giving rise to γδ T cells as long as SCF and IL-7 were present. Interestingly, IL-7 is produced by stromal cells, keratinocytes, and gut epithelial cells, cell types that are known to co-localize and interact with γδ T cells.

iii) Selection and Tolerance

Positive and negative thymic selection appears to occur for γδ T cells maturing in the thymus in a manner similar to selection shaping the αβ T cell repertoire. Recognition of self-MHC obviously does not come into play here, since γδ TCRs do not recognize antigen in conjunction with MHC like αβ T cells. However, the positive selection of at least some γδ T cell clones appears to be driven by the expression of cellular stress antigens and MHC class Ib molecules on cTECs. Interestingly, the signaling downstream of TCR engagement during selection must be slightly different, since the ERK activation required for the positive selection of αβ T cells is not necessary for the positive selection of thymic γδ T cells. Autoreactive γδ T cells are negatively selected or anergized in the thymus in a poorly defined process that may require CD30. Outside of the thymus, positive selection leading to the expansion of certain VγVδ subsets may be induced by an encounter with particular peripheral host ligands. For example, the C57BL6 strain of mice shows a dramatic expansion of Vγ4Vδ7 T cells about a month after birth, without the apparent involvement of either the thymus or pathogens. Similarly, healthy humans show a high frequency of Vγ9Vδ2 cells in peripheral blood. In both cases, the selection of a particular γδ T cell subset appears to be driven by (unknown) extrathymic antigens.

How is peripheral tolerance achieved in γδ T cells? One mechanism may be the presence of the inhibitory receptors on γδ T cell surfaces. Perhaps only very strong activation signals delivered through the γδ TCR can overcome the signal delivered through the inhibitory receptors. The possible co-engagement of the γδ TCR and the activatory molecule NKG2D by an MHC class Ib molecule (such as MICA) as described previously may come into play here (refer to Fig. 18-9B). In the absence of stress, the MHC class Ib molecule might not be expressed, so that co-ligation of the TCR and the putative coreceptor would not occur, and the γδ T cell would not be activated even though it might have bound its self-ligand. Another mechanism resulting in putative γδ T cell tolerance may be exerted by macrophages. These cells are able to establish intercellular contacts with γδ T cells that inhibit the activation of the latter until a strong antigenic stimulus is received. A similar tolerance mechanism may prevent the overactivation of the sentinel Vγ9Vδ2 T cell subset that recognizes a wide selection of host and pathogen non-peptide ligands. Erythrocytes express on their surfaces large amounts of a natural diphosphoglyceric acid that can bind as a weak agonist to the TCRγδ of any Vγ9Vδ2 T cell. A Vγ9Vδ2 T cell encountering this ligand is anergized for several days and shows impaired TCR signaling unless it encounters an antigen providing a strong activation signal. Unrestrained mass activation of one of the most powerful γδ T cell subsets is thus reserved for situations in which it is truly required.

Autoreactive mucosal and cutaneous IELs may be controlled differently. These γδ T cells do not recirculate like systemic T cells and thus encounter a very restricted set of epithelial self antigens. The need to remove these cells may thus not be as great as for cells patrolling the periphery, and clonal deletion may not occur. Cells bearing γδ TCRs specific for self antigens have been identified among IELs, but these cells appear to be anergic.

C. NKT Cells

Investigations in the early 2000s yielded an explosion of information about the once mysterious NKT cells. Although much remains to be confirmed about their physiological roles, it is becoming clear that NKT cells are important players in the first lines of defense against pathogens and in tumor surveillance.

I. WHAT ARE NKT CELLS?

NKT cells are lymphoid lineage cells that share both physical and functional characteristics of NK cells and T cells. Morphologically, NKT cells closely resemble conventional T cells. With respect to anatomical distribution, low numbers of NKT cells are found virtually everywhere T and NK cells are found, in the peripheral blood, spleen, liver, thymus, bone marrow, and lymph nodes. Interestingly, NKT cells make up a larger proportion of the lymphoid cell population found in the livers of unimmunized mice. Following activation, NKT cells migrate to sites of infection or inflammation.

NKT cells carry what is sometimes referred to as a "semi-invariant" αβ TCR. In mice, NKT cells overwhelmingly bear αβ TCRs containing a TCRα chain of the composition Vα14/Jα281 (now known as Vα14/Jα18) and a TCRβ chain containing either Vβ2, Vβ7, or Vβ8. In humans, the TCRs of NKT cells almost always contain Vα24/Jα18 paired with a TCRβ chain containing Vβ11. Many (but not all) subsets of mouse NKT cells express the NK costimulatory molecule NK1.1, while many human NKT cells express the human homologue of NK1.1, NKR-P1c (CD161c). NKT cells of both species also express NK inhibitory receptors. For example, human NKT cells express both the KIRs and inhibitory CD94/NKG2 receptors, while mouse NKT cells express members of the Ly-49 family. Human NKT cells also express the activatory NK receptor NKG2D. With respect to T cell coreceptors, the human NKT cell population includes CD4+, CD8+, and CD4−CD8− subsets, while mouse NKT cells are either CD4+ or CD4−CD8− (refer to Table 18-1). Whether these subsets have true functional differences *in vivo* (such as slightly different cytokine secretion profiles) is under investigation. CD69, the IL-2Rβ chain, and the memory marker CD45R0 are constitutively expressed on naive mouse and human peripheral NKT cells. CD25 is expressed on activated NKT cells of human neonates, but not on activated NKT cells of adults. CD62L (L-selectin), a marker of naive T cells, is not expressed on human or mouse NKT cells.

The literature contains several descriptions of NKT-like cells that show features slightly different than those just described. Some of these unconventional subsets have moderately diverse TCRs instead of the semi-invariant TCR expressed by classical NKT cells. While some of these NKT-like cells seem to differ in their cytokine secretion profiles *in vitro*, little is yet known about their development or physiological relevance.

II. ANTIGEN RECOGNITION AND ACTIVATION

Rather than interacting with the polymorphic peptide-binding MHC molecules recognized by classical T cells, the TCRs of NKT cells generally appear to recognize glycolipid or lipid structures presented on the non-polymorphic CD1d molecule (Fig. 18-13). Although CD1d has greater structural similarity to MHC class I, it behaves more like MHC class II in that the presentation of antigens by CD1d is believed to result primarily via the exogenous pathway. Indeed, endosomal localization and trafficking are required for CD1d-mediated antigen presentation in mice. In addition, like MHC class II, CD1d is preferentially expressed by professional APCs. (Antigen presentation by CD1 was discussed in Chs.10 and 11.)

The first ligand identified as activating murine NKT cells *in vitro* was a glycosphingolipid molecule called α-galactosylceramide (α-GalCer) isolated from a marine sponge by a drug company screening for molecules with anti-tumor activity. Administration of α-GalCer to a mouse potently acti-

vates its NKT cells, protects against infection with various pathogens, and can even promote tumor rejection. However, very few natural ligands for NKT TCRs have been identified. Some scientists have suggested that two natural ligands for murine NKT cells may be the parasite version of the cellular glycolipid *glycosylphosphatidylinositol* (GPI) and mycobacterial *phosphatidylinositol mannoside* (PIM). It is tempting to speculate that NKT cells with slightly different TCRs might recognize these structurally different ligands. Indeed, NKT cells that respond to PIM *in vitro* express a TCR containing Vβ3 rather than Vβ8. However, studies attempting to confirm the physiological relevance of these observations have yielded conflicting results. NKT cell-deficient animals that are infected with various pathogens often have perfectly normal phenotypes. Some researchers speculate that NKT cells are such a powerful weapon that there has been evolutionary pressure on pathogens to develop mechanisms to disarm NKT cells. Indeed, pathogen infection can induce the downregulation of CD1d on APCs, blocking the presentation of pathogen molecules to NKT cells. Other laboratories have found evidence *in vitro* for the existence of NKT TCRs that recognize host stress glycolipid ligands released from cells damaged during an infection. If true *in vivo*, NKT cells could play a defensive recognition role very similar to that of γδ T cells.

In addition to TCR ligation, NKT cells can sometimes be activated via engagement of the NK activatory receptor NKG2D expressed on their cell surfaces. In this situation, activation appears to be independent of any glycolipid antigen, but whether ligands such as MICA are involved is unclear. Finally, it seems that the mere presence of pro-inflammatory cytokines such as IL-12 and IL-18 can

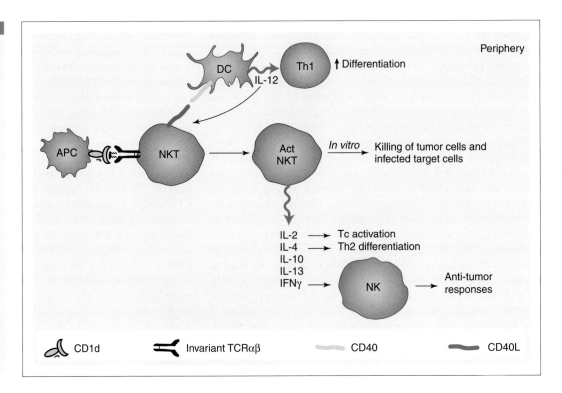

Figure 18-13

Activation and Effector Functions of NKT Cells
NKT cells are activated by the interaction of their invariant αβ TCRs with unknown ligands presented on CD1d by APCs. The activated (Act) NKT cell secretes cytokines that contribute to Th2 differentiation and Tc activation. Activation also upregulates the expression of CD40L on the NKT cell, which allows it to stimulate DCs by engaging CD40. IL-12 produced by the stimulated DCs promotes Th1 differentiation and induces the activated NKT cell to produce IFN-γ. The IFN-γ induces anti-tumor responses by the NK cell. *In vitro*, activated NKT cells kill tumor cells or infected targets via an NK-like mechanism involving activatory receptors.

sometimes activate NKT cells, in the absence of any other (known) stimulus.

The characteristics of induced NKT activation are consistent with a cell type prominent in the innate immune response. Following systemic administration of α-GalCer to a mouse, activated mature NKT cells in the periphery are stimulated to undergo rapid clonal expansion in the spleen, liver, and bone marrow. Expression of NK1.1 is transiently downregulated. Peak numbers of NKT cells, which may reach 10× their levels in unstimulated animals, occur within 3 days of α-GalCer treatment. Massive amounts of cytokines are produced in a sustained fashion, after which NKT numbers return to their resting levels by 9–12 days. Thus, in a natural infection, NKT cells would "cover" that period during which conventional T cells take the time to proliferate and differentiate into the effectors of the more finely tailored adaptive response.

III. EFFECTOR FUNCTIONS

Once activated, NKT cells can immediately (without the need for differentiation) perform functions characteristic of either cytokine-secreting "helper" Th0- or cytolytic NK-like effectors. *In vitro*, activated NKT cells can use FasL- or perforin-dependent mechanisms to lyse the same types of targets as NK cells (i.e., infected cells and tumor cells), and appear to be subject to the same type of control by inhibitory receptors. However, it remains unclear how relevant this cytolytic capacity is *in vivo*. The most important *in vivo* effector function of activated NKT cells appears to be cytokine secretion. NKT cells carry preformed mRNAs for a collection of cytokines, including IL-2, IL-4, IL-13, and IL-10 and IFNγ, similar to that of Th0 αβ T cells. Upon triggering, the NKT cells produce these cytokines within 1–2 hours without the need to differentiate into a separate Th-like effector. It has been speculated that the early burst of IL-4 produced by activated NKT cells contributes to the initiation of the Th2 response, supplying the necessary concentrations of this cytokine required to induce Th2 differentiation of nearby activated T cells. Similarly, the IL-2 produced by activated NKT cells may supply help for Tc activation under circumstances where Th help might be sub-optimal. NKT cells may also contribute cytokine help for humoral responses, at least initially. In one study, CD1d-deficient mice (which lack NKT cells; see below) were infected with parasites and antibody responses to parasite glycolipid antigens were monitored. In the absence of NKT cells, lower levels of such anti-parasite antibodies were observed during the early stages of the infection.

Activated NKT cells also have important effects on DCs and NK cells. Upon CD1d-dependent activation, the NKT cell upregulates its expression of CD40L. Interaction of this CD40L with CD40 on a DC activates the latter, spurring it to produce more IL-12. As we have learned, IL-12 is the key cytokine governing Th1 differentiation. In addition, following exposure to IL-12, activated NKT cells selectively increase their secretion of IFNγ. This IFNγ appears to be required for the anti-tumor responses of NK cells whose effector actions help to prevent tumor metastasis (see Ch.26). Indeed, in *in vivo* experiments, the striking anti-metastatic effects of IL-12 administration have been shown to depend on the presence of NKT cells. However, NKT cells do not act independently of NK cells in fighting tumor cells, since the protective effect of NKT cells is lost in animals depleted of NK cells. A similar mechanism may account for the protection against some viral, bacterial, and parasitic infections offered by NKT cells. Mice genetically engineered to lack NKT cells show higher pathogen loads under some circumstances. Scientists theorize that cytokines produced by NKT cells may stimulate NK cells and promote the conventional T cell responses best-suited to fighting the infection.

As well as responding to pathogens, NKT cells may play a role in preventing autoimmunity. Mice that are prone to the development of type 1 diabetes show abnormally low levels of NKT cells, and these cells show functional defects. Similarly, human diabetic patients show decreased numbers of NKT cells that produce reduced levels of IL-4. If diabetes-prone animals are treated in advance with NKT cells or α-GalCer, the onset of the disease can be prevented. Treatment with α-GalCer can also be of benefit in animal models of colitis and multiple sclerosis, two other autoimmune disorders. The underlying mechanism of protection in these cases remains under investigation. More on the role of NKT cells in autoimmunity appears in Chapter 29.

IV. DEVELOPMENT

Early studies showed that NKT cells had to be derived from committed T lineage cells, since committed NK progenitors could not give rise to NKT cells. Subsequent analyses using CD1d tetramer staining have indicated that the first committed NKT precursors in the thymus are a subset of DP thymocytes that happen to express a TCR containing Vα14/Jα18 (Fig. 18-14). This DP subset expressing the semi-invariant NKT TCR is positively selected by signals received from neighboring CD1d-expressing DP thymocytes (see later) and gives rise to immature CD4$^+$ NK1.1$^-$ NKT cells. Once positive selection has been carried out, a non-hematopoietic cell type (perhaps an mTEC?) appears to play an important role in continued NKT development. Studies of mutant mice have shown that an absence of NF-κB-mediated signaling in these non-hematopoietic cells (which do not have to express CD1d) drastically reduces mature NKT cell numbers. How these mystery cells interact with CD4$^+$ NK1.1$^-$ NKT cells and prepare them for exit from the thymus remains undefined, but impaired NF-κB signaling in thymic stromal cells can reduce their production of IL-15 mRNA. In any case, CD4$^+$ NK1.1$^-$ NKT cells subsequently migrate from the thymus into the periphery, and particularly into the spleen and liver. In these locations, differentiation continues until mature CD4$^+$ NK1.1$^+$ NKT cells are generated. Following activation in the periphery, a mature NKT cell undergoes clonal expansion and

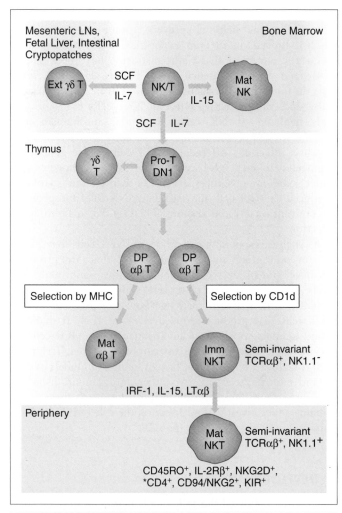

Figure 18-14

Development of NKT Cells
NKT precursors that enter the thymus progress through the DN thymocyte stages. At the DP stage, only thymocytes expressing αβ TCRs that bind to the CD1d expressed on other thymocytes become immature NKT cells. Under the influence of IRF-1, IL-15, and LTαβ, the NKT cells mature, gaining expression of the indicated markers and receptors. *, some NKT subsets.

transiently downregulates its expression of NK1.1. Thus, it should not be assumed that, just because an NKT cell is NK1.1⁻, it is immature.

What cytokines drive NKT development? IRF-1⁻/⁻, IL-15⁻/⁻, IL-15α⁻/⁻, and γc chain knockout mice all lack mature NKT cells, indicating that, like NK cells, the development of NKT cells specifically requires IL-15. Some immature CD4⁺ NK1.1⁻ cells were found in the γc chain knockout mice, confirming the importance of signaling via this chain for the complete maturation of CD4⁺ NK1.1⁺ NKT cells. Interestingly, like the development of NK cells, the development of NKT cells is also absolutely dependent on membrane-bound LTαβ; mice lacking either LTα or LTβ are devoid of NKT cells. Mice lacking IL-7 have a normal proportion of mature NKT cells but reduced numbers.

Cytokine production by NKT cells is also impaired in IL-7⁻/⁻ mice. Mutations in several other molecules that have no effect on conventional T cell development abrogate NKT development, including Ras, Fyn, and Ets1. The mechanisms underlying the involvement of these molecules in NKT development are under investigation.

Like T cells, developing NKT cells undergo thymic selection. Positive selection of developing NK1.1⁻ NKT cells depends on the expression of CD1d by surrounding DP thymocytes (rather than the cortical thymic epithelial cells crucial for conventional T cell selection). It is a mystery what self-ligand, if any, is presented to the developing NKT cells to positively select them. It is also unknown what degree of avidity of TCR-CD1d binding is necessary for positive NKT selection. However, an intact exogenous antigen processing pathway (presumably capable of normal CD1d trafficking) is required for NKT development, since mutant mice lacking either the invariant chain or cathepsin S (refer to Ch.11) lack NKT cells. Negative selection of autoreactive NKT cells can be induced if developing NKT cells are stimulated with CD1d bound to an agonist ligand, or even by high levels of CD1d alone. In this case, the cell type expressing the CD1d is most likely to be a thymic DC. Interestingly, engineered overexpression of CD8 in a mouse blocks NKT cell development. Some scientists have interpreted this observation to mean that CD8 expression on an NKT cell may also be associated with negative selection. The theory is that the combined binding of CD8 and the NKT TCR on the developing NKT cell to CD1d molecules on a DP thymocyte sends an enhanced signal to the NKT cell to undergo deletion. However, the avidity of binding between CD8 and CD1d is quite low, leading to doubts that it could generate a sufficiently strong signal to be relevant. Alternatively, abnormally early expression of CD8 on a developing DN thymocyte interferes with the transition to the DP stage, which would also block NKT development.

Embryologically speaking, the timing of NKT development is controversial. Some investigators have reported that NKT cells (or their precursors) are absent in neonatal mice, and that the full complement of these cells is not present in the thymus until 6–8 weeks of age. Others have found that at least some NKT cells are present in the yolk sac and liver (but not the thymus) of early mouse embryos. NKT cells are present in newborn humans but the cells are not yet fully competent to secrete cytokines. This situation stands in contrast to that of NK cells, which are fully mature and functional in fetal life.

We have now come to the end of our study of the basic cell types of the immune system, the modes of antigen recognition of these cells, and their effector functions. These various cell types, some of which are elements of innate immunity, some of which strictly mediate adaptive immunity, and others that bridge the innate and the adaptive responses, combine to distinguish self from nonself, and danger from benign circumstance. In Chapter 19, we explore the complement system and how it serves a crucial role in host defense by interacting with cells of both the innate and adaptive immune responses.

SUMMARY

The forms and functions of NK, γδ T, and NKT cells mark them as being integral parts of both innate and adaptive immune responses. All three cell types belong to the T cell lineage, but they recognize antigens in a much less specific way than conventional αβ T cells. In addition, as well as having direct effects on pathogens, NK, γδ T, and NKT cells influence the differentiation of B cells and αβ T cells by the cytokines they secrete. NK cells express activatory receptors that bind mostly unknown ligands expressed by most host cells. The activatory pathway is normally held in check by a dominant inhibitory signal initiated by the binding of NK inhibitory receptors to classical MHC class I molecules. Infected cells and tumor cells often have downregulated expression of MHC class I, such that the inhibitory receptors are not engaged. In this case, the activatory pathway causes the NK cell to release the contents of its cytotoxic granules and kill the target via "natural cytotoxicity." Due to their expression of numerous FcRs, NK cells can also kill by ADCC. γδ T cells are prominent in the mucosal and cutaneous epithelial layers of the body and so have a key sentinel function. γδ T cells use their γδ TCRs to recognize non-peptide antigens derived from pathogen proteins and host stress molecules that are presented either directly or on unconventional MHC molecules such as CD1c. Once activated, γδ T cells generate CTLs that kill infected target cells by perforin- and granzyme-mediated cytotoxicity, and Th effectors that secrete cytokines influencing αβ T cells and other leukocytes. NKT cells bear a semi-invariant αβ TCR that recognizes unknown glycolipid and lipid ligands presented on CD1d. Activated NKT cells are capable of target cell cytolysis *in vitro*, but their primary effector function is the rapid secretion of a panel of Th1 and Th2 cytokines that exert immunoregulatory effects. Secretion of IFNγ by NKT cells stimulates NK effector functions and is associated with an important anti-metastatic effect *in vivo*.

Selected Reading List

Akashi K., Reya T., Dalma-Weiszhausz D. and Weissman I. (2000) Lymphoid precursors. *Current Opinion in Immunology* **12**, 144–150.

Allison T. and Garboczi D. (2002) Structure of gammadelta T cell receptors and their recognition of non-peptide antigens. *Molecular Immunology* **38**, 1051–1061.

Biron C. A., Nguyen K. B., Pien G. C., Cousens L. P. and Salazar-Mather T. P. (1999) Natural killer cells in antiviral defense: function and regulation by innate cytokines. *Annual Reviews of Biomedical Sciences* **17**, 189–220.

Blom B., Res P. C. M. and Spits H. (1998) T cell precursors in man and mice. *Critical Reviews in Immunology* **18**, 371–388.

Brumbaugh K., Binstadt B. and Leibson P. (1998) Signal transduction during NK cell activation: Balancing opposing forces. *Current Topics in Microbiology and Immunology* **230**, 103–122.

Brutkiewicz R. and Sriram V. (2002) Natural killer T (NKT) cell and their role in antitumor immunity. *Critical Reviews in Oncology/Hematology* **41**, 287–298.

Carding S. and Egan P. (2002) Gammadelta T cells: functional plasticity and heterogeneity. *Nature Reviews. Immunology* **2**, 336–345.

Chen Z. (2002) Comparative biology of gamma delta T cells. *Science Progress* **85**, 347–358.

De Libero G. (2000) Tissue distribution, antigen specificity and effector functions of gamma delta T cells in human diseases. *Springer Seminars in Immunopathology* **22**, 219–238.

French A. and Yokoyama W. (2003) Natural killer cells and viral infections. *Current Opinion in Immunology* **15**, 45–51.

Godfrey D. and Kronenberg M. (2004) Going both ways: immune regulation via CD1d-dependent NKT cells. *Journal of Clinical Investigation* **114**, 1379–1388.

Greenberg A. H. (1994) The origins of the NK cell, or a Canadian in King Ivan's court. *Clinical and Investigative Medicine* **17**, 626–631.

Hammond K. and Godfrey D. (2002) NKT cells: potential targets for autoimmune disease therapy? *Tissue Antigens* **59**, 353–363.

Hayday A. C. (2000) Gamma/delta cells: a right time and a right place for a conserved third way of protection. *Annual Review of Immunology* **18**, 975–1026.

Hayday A. and Tigelaar R. (2003) Immunoregulation in the tissues by gammadelta T cells. *Reviews in Immunology* **3**, 233–242.

Hayes S., Shores E. and Love P. (2003) An architectural perspective on signaling by the pre-, alpha beta and gamma delta T cell receptors. *Immunological Reviews* **191**, 28–37.

Herberman R. B., Nunn M. E. and Lavrin D. H. (1975) Natural cytotoxic reactivity of mouse lymphoid cells against syngeneic and allogeneic tumors. I. Distribution of reactivity and specificity. *International Journal of Cancer* **16**, 216–229.

Ikawa T., Kawamoto H., Fujimoto S. and Katsura Y. (1999) Commitment of common T/Natural killer (NK) progenitors to unipotent T and NK progenitors in the murine fetal thymus revealed by a single progenitor assay. *Journal of Experimental Medicine* **190**, 1617–1625.

Kärre K., Ljunggren H., Piontek G. and Kiessling R. (1986) Selective rejection of H-2-deficient lymphoma variants suggests alternative immune defence strategy. *Nature* **319**, 675–678.

Kiessling R., Klein E. and Wigzell H. (1975) "Natural" killer cells in the mouse I. Cytotoxic cells with specificity for mouse Moloney leukemia cells. Specificity and distribution according to genotype. *European Journal of Immunology* **5**, 112–117.

Killeen N., Irving B. A., Pippig S. and Zingler K. (1998) Signaling checkpoints during the development of T lymphocytes. *Current Opinion in Immunology* **10**, 360–367.

Lanier L. (2000) Turning on natural killer cells. *Journal of Experimental Medicine* **191**, 1259–1262.

Lanier L. (2003) Natural killer cell receptor signaling. *Current Opinion in Immunology* **15**, 308–314.

Lanier L. L. (1998) NK cell receptors. *Annual Review of Immunology* **16**, 359–393.

Liu C.-C., Perussia B. and Young J. D.-E. (2000) The emerging role of IL-15 in NK-cell development. *Immunology Today* **21**, 113–116.

Ljunggren H.-G. and Kärre K. (1990) In search of the 'missing self': MHC molecules and NK cell recognition. *Immunology Today* **11**, 237–244.

Lopez-Botet M., Llano M., Navarro F. and Bellon T. (2000) NK cell recognition of non-classical HLA class I molecules. *Seminars in Immunology* **12**, 109–119.

Maeurer M. J. and Lotze M. T. (1998) Interleukin-7 (IL-7) knockout mice.

Implications for lymphopoiesis and organspecific immunity. *International Review of Immunology* **16**, 309–322.

Mak T. W. and Ferrick D. A. (1998) The gamma-delta bridge: linking innate and acquired immunity. *Nature Medicine* **4**, 764–765.

Ortaldo J. R. and Herberman R. B. (1984) Heterogeneity of natural killer cells. *Annual Review of Immunology* **2**, 359–394.

Pardoll D. (2001) Immunology. Stress, NK receptors, and immune surveillance. *Science* **294**, 534–536.

Perussia B. (2000) Signaling for cytotoxicity. *Nature Immunology* **1**, 372–374.

Radaev S. and Sun P. (2003) Structure and function of natural killer surface receptors. *Annual Review of Biophysics and Biomolecular Structure* **32**, 93–114.

Raulet D. (2003) Roles of the NKG2D immunoreceptor and its ligands. *Nature Reviews Immunology* **3**, 781–790.

Raulet D. (2004) Interplay of natural killer cells and their receptors with the adaptive immune response. *Nature Immunology* **5**, 996–1002.

Raulet D., Vance R. and McMahon C. (2001) Regulation of the natural killer cell receptor repertoire. *Annual Review of Immunology* **19**, 291–330.

Robey E. and Fowlkes B. (1998) The alpha/beta versus gamma/delta T-cell lineage choice. *Current Opinion in Immunology* **10**, 181–187.

Shortman K. and Wu L. (1996) Early T lymphocyte progenitors. *Annual Review of Immunology* **14**, 29–47.

Snyder M., Weyand C. and Goronzy J. (2004) The double life of NK receptors: stimulation or costimulation? *Trends in Immunology* **25**, 25–32.

Swann J., Crowe N., Hayakawa Y., Godfrey D. and Smyth M. (2004) Regulation of

antitumour immunity by CD1d-restricted NKT cells. *Immunology and Cell Biology* **82**, 323–331.

Tanaka Y., Sano S., Nieves E., De Libero G., Rosa D. *et al.* (1994) Nonpeptide ligands for human gamma/delta T cells. *Proceedings of the National Academy of Sciences U.S.A.* **91**, 8175–8179.

Ugolini S. and Vivier E. (2000) Regulation of T cell function by NK cell receptors for classical MHC class I molecules. *Current Opinion in Immunology* **12**, 295–3000.

Vivier E., Tomasello E. and Paul P. (2002) Lymphocyte activation via NKG2D: towards a new paradigm in immune recognition? *Current Opinion in Immunology* **14**, 306–311.

Williams N. (1998) T cells on the mucosal frontline. *Science* **280**, 198–200.

Complement

Jules Bordet 1870–1961

Complement
Nobel Prize 1919

© Nobelstiftelsen

CHAPTER 19

A. HISTORICAL NOTES

B. OVERVIEW OF THE COMPLEMENT SYSTEM

C. THE CLASSICAL PATHWAY OF COMPLEMENT ACTIVATION

D. THE ALTERNATIVE PATHWAY OF COMPLEMENT ACTIVATION

E. THE LECTIN PATHWAY OF COMPLEMENT ACTIVATION

F. TERMINAL COMPONENTS OF THE COMPLEMENT CASCADE AND FORMATION OF THE MAC

G. REGULATION OF THE COMPLEMENT SYSTEM

H. COMPLEMENT RECEPTORS AND THEIR BIOLOGICAL ROLES

I. COMPLEMENT DEFICIENCIES

J. NEW ROLES FOR COMPLEMENT?

"It is a mistake to try to look too far ahead. The chain of destiny can only be grasped one link at a time"–Sir Winston Churchill

In the preceding chapters, the basic elements of the immune system were described in detail, and the reader should now have a sound understanding of how innate and adaptive immune responses are triggered and function in host protection. This chapter concerns a very important element of host defense—the complement system—that was introduced in Chapters 4 and 5. We felt the reader would benefit from leaving a thorough investigation of the actions of complement until after all the basic elements of the immune system were mastered, because complement interacts with or has an effect on every one of them.

As was described in Chapters 4 and 5, "complement" is a complex system of at least 31 different glycoproteins. Of these, 20 are soluble blood plasma proteins that sequentially activate each other by proteolytic cascade or other modifications. The remaining 11 known components of the complement system are regulatory proteins or complement receptors. The reader will recall that there are three catalytic pathways of complement activation: the *classical* pathway, the *lectin* pathway, and the *alternative* pathway. Regardless of the route of proteolysis, the same six complement effector functions can be achieved: (1) *lysis of pathogenic cells*, (2) *opsonization of pathogens*, (3) *clearance of immune complexes*, (4) *enhancement of antigen presentation*, (5) *enhancement of B cell activation and memory*, and (6) *generation of chemoattractant and regulatory peptides*. As a result, the complement system helps the host to control infection by many species of bacteria and viruses whose component molecules provide triggers for one or more of these pathways.

A. Historical Notes

In the late 1800s, Richard Pfeiffer observed that if cholera bacilli were introduced into the abdominal cavity of a guinea pig that had previously been exposed to cholera bacilli, the bacteria underwent immediate dissolution, or *bacteriolysis*. This phenomenon did not occur in an unimmunized animal,

indicating that the bacteriolysis in this case was immune system-related. If cholera bacilli were then transferred to another unimmunized guinea pig along with the serum from the cholera-primed animal, the bacteria were again destroyed. Since the effect could be transferred with "immune serum" (which contains no cells), the results were attributed to the presence of preformed antibodies induced during the priming exposure to the cholera bacilli. Elie Metchnikoff attempted to dissect out all the components necessary for bacteriolysis by re-creating the phenomenon *in vitro*. He showed that bacteriolysis could still take place if the bacteria were combined in a test tube with immune guinea pig serum and a small amount of fluid from the abdomen of an unimmunized guinea pig. (This latter component was meant to replicate the environment of the live guinea pig's peritoneum.) In 1894, Jules Bordet demonstrated that the abdominal fluid was not required, and the only elements necessary for *in vitro* bacteriolysis were immune serum and bacteria. However, *in vitro* bacteriolysis worked only if the immune serum was perfectly fresh. In other words, if the serum was allowed to stand at room temperature for any length of time, the reaction would not take place. Since it was known by this time that antibodies were heat-stable (survive incubation up to 56°C), it was apparent that something else in fresh serum worked with or "complemented" the specific antibody in the lysis of the bacteria. This newly discovered serum component therefore came to be called "complement." Unlike the antigen-specific antibody, complement was not heat-stable and was present in the serum of unimmunized as well as immunized animals. For this work, among other important observations, Bordet was awarded the Nobel Prize for Medicine or Physiology in 1919.

Throughout the first half of the 20th century, complement was primarily regarded as a tool to be used in the various antibody and antigen detection assays that were under development during this period. While its physiological role and mode of action were certainly of interest to immunologists, an understanding of its biology was difficult to approach experimentally due to the limited biochemical techniques available at that time. However, by the 1950s, advances in protein biochemistry led to the isolation of individual proteins that together were understood to comprise the complement "system." By the 1960s, it became clear that activation of the complement system involved the sequential fragmentation of various protein components that gave rise to inflammatory molecules. Paul Ehrlich had originally speculated decades earlier that this process was enzymatic, and he was proved right by several groups who showed that a cascade of enzymatic cleavages generated products that were themselves proteases. The fact that the cascade was involved in inflammatory processes focused interest on the concept of complement as an important mediator of both immune and allergic reactions. Immunologists began trying to tease apart and understand the individual reactions in which various complement components were involved.

Around this time, Louis Pillemer and colleagues isolated a protein called *properdin* and demonstrated that it was involved in what they assumed was another pathway of complement activation. This group proposed that properdin-based complement activation was a form of innate or "natural" immunity, since no antigen–antibody reactions were involved. However, vigorous debate ensued as others argued that this system actually relied on natural antibody for triggering, making it a variant of the original complement activation pathway. In the mid-1970s, several researchers showed that antibodies against complement component C2 did not inhibit the properdin system. Furthermore, guinea pigs deficient in C4 remained able to trigger the complement cascade through the properdin pathway. Activation involving properdin was finally accepted as a distinct mechanism and came to be called the *alternative complement activation pathway*. Activation involving antigen–antibody interaction was then referred to as the *classical complement activation pathway*. It was not until the late 1990s that a third pathway of complement activation called the *lectin pathway* was defined, revealing a defense mechanism thought to be the most evolutionarily ancient (see Ch.21). The lectin pathway was shown to be triggered when bacterial monosaccharides bind to a host lectin associated with a complement-activating protease.

Complement has turned out to have a role in B cell activation that did not become clear until long after the first clue to this function was discovered. In 1972, Mark Pepys observed that mice that had been depleted of C3 with cobra venom factor were unable to mount normal antibody responses to several different Td antigens, including ovalbumin, human IgG, and various hapten–protein conjugates. Antibody responses to Ti antigens were unaffected. This report attracted little notice at the time because most immunological research of the day was riveted on T and B lymphocytes, their receptors, and their interactions with accessory cells. Investigations of the complement system remained primarily biochemical in nature and restricted to effects within the innate immune system. However, in the late 1980s, the importance of coreceptors and costimulation for T lymphocyte activation and the role of APCs in these functions were recognized. The search then intensified for equivalent co-receptors and co-stimulators for B lymphocytes. In the late 1990s, complement receptor 2 (CR2) was identified as part of the co-receptor for B cell

activation. The earlier result of Pepys suggesting the involvement of C3 in antibody responses was finally put into context with the demonstration that the binding of complement component C3d to CR2 delivers a signal that lowers the threshold of B cell activation and promotes B cell survival. In fact, complement components are now considered to be extremely important for both innate responses and the humoral responses of adaptive immunity.

B. Overview of the Complement System

I. SOURCES OF SOLUBLE COMPLEMENT COMPONENTS

We have noted that the soluble complement components circulate in the blood plasma in inactive *zymogen* or *proenzyme* form. The molecular masses of these soluble components range from 25 to 410 kDa, with serum concentrations from approximately 1 to 1200 µg/ml. In total, complement proteins represent 15% of the plasma globulin fraction and are present in normal plasma at a total concentration of 3 g/liter. Complement proteins are synthesized principally by hepatocytes in the liver but are also secreted by tissue macrophages, blood monocytes, and epithelial cells of the urogenital and gastrointestinal tracts. Macrophages may be particularly significant sources of the complement proenzymes since these cells congregate at the sites of inflammatory responses. Cytokines such as TNF, IL-1, IL-6, and IFNγ can also induce the biosynthesis and secretion of certain complement proteins by specific cell types. As such, the complement components are counted among the "acute phase proteins" rapidly induced in response to tissue injury or invasion.

II. THREE PATHWAYS OF COMPLEMENT ACTIVATION

As shown in Figure 19-1, the complement cascade can be initiated in three different ways which ultimately lead to the production of the same key component, C3b. The classical pathway is triggered by antibody binding to a specific antigen on a pathogen or by antibody participating in soluble immune complexes. The lectin pathway is initiated by the binding of certain carbohydrate-binding proteins (such as *mannose-binding lectin*; MBL) to a range of monosaccharides present on a pathogen surface. After their respective initiation steps, the classical and lectin pathways produce C3b by the same enzymatic cascade. The alternative pathway arrives at C3b by a different route and involves the direct binding of C3b to pathogens without the intervention of either antibody or lectin. Downstream of C3b, all three paths lead to assembly of the *terminal complement components* and the formation of the *membrane attack complex* (MAC). As we learned in Chapter 7, the utilization of complement components during the process of MAC formation is called *complement fixation*.

Why have multiple pathways of complement activation evolved? The classical pathway has the advantage of specificity

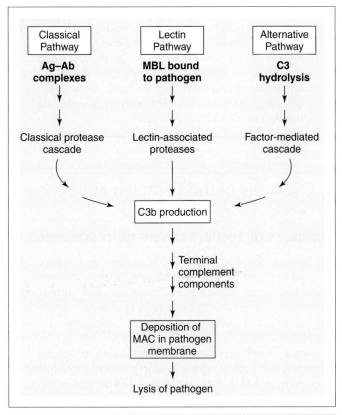

Figure 19-1

Three Pathways for Initiating Complement Activation
In the **classical** pathway, soluble immune complexes or antibodies bound to antigen on a pathogen surface initiate the classical complement protease cascade that eventually generates the complement product C3b. In the **lectin** pathway, MBL binds to carbohydrates on a pathogen, and proteases associated with MBL cleave the second and third components of the classical complement cascade to generate the same C3b product. In the **alternative** pathway, spontaneous hydrolysis of serum C3 leads to the generation of C3b that binds to a pathogen surface and serves as a platform for an enzymatic cascade that generates still more C3b. For all three pathways, the presence of C3b on a membrane surface allows the recruitment of the terminal complement components, assembly of the MAC, and pathogen lysis.

but the disadvantage of tardiness, as an adaptive B cell response is required first. In contrast, what the lectin and alternative pathways lack in pathogen specificity, they gain in rapidity of response. No antibody is needed to trigger either of these pathways, meaning that complement can still be activated to defend the host during the lag phase required for the activation of B cells and synthesis of the specific antibodies used by the classical pathway. Thus, all three pathways work together synergistically to optimize host defense.

III. AMPLIFICATION AND SENSITIVITY

Amplification is a key feature of complement activation. The cascade is made up of a large number of sequential activation steps, some of which generate multiple products. Since many of these products are themselves enzymes that trigger subsequent steps, an exponential increase in activated molecules is generated. A huge amplification of an initial signal can be achieved, making it possible to generate a massive response from a single triggering event within a short period of time. However, the system is also highly sensitive, since a deficiency of any one enzyme in the sequence can halt the progression of the cascade completely and prevent the individual from mounting optimal immune responses.

IV. CONTROL

Because of the tremendous number of players involved and their non-specific destructive capacity, the complement cascade is rigidly organized and tightly regulated to minimize damage to self-tissues. The first control lies in the fact that most of the component enzymes are present in the plasma as zymogens; that is, the molecule is inactive until a specific fragment is cleaved off by the enzyme preceding it in the cascade. In addition, the activated enzymes function for only a short time before being inactivated again. There are also numerous regulatory and inhibitory proteins and cell surface receptors that ensure that the system becomes activated where and when it should (see later in this chapter).

V. COMPLEMENT SYSTEM NOMENCLATURE

The nomenclature of the proteins involved in the complement cascade appears to us now to be profoundly illogical and infuriatingly confusing, but it reflects the historical sequence of identification and biochemical purification of the complement components.

To help keep straight the sequence of enzyme activations and the generation of products, biochemists adopted conventions in naming the complement components. Because they were identified first, the enzymes of the classical path and the terminal components are designated by the letter C and the numbers 1–9 (which reflects their order of discovery, <u>not</u> their order of action). Once a protein is cleaved, the resulting peptide fragments are designated as "a" and "b," where "a" indicates the smaller product and "b" indicates the larger one. Generally speaking, the "b" fragments contribute to the next enzymatic activity in the cascade (e.g., C3b) while the "a" fragments are chemotactic for inflammatory leukocytes (e.g., C5a). The exception is C2, where C2a (not C2b) is the larger fragment and contains the enzymatic domain. An active component is sometimes indicated by drawing a line over its name, which distinguishes it from its inactive precursor; e.g.. C3 is inactive, but $\overline{C3b}$ is active. Often several complement components must associate to form an active complex, e.g., $\overline{C4b2a3b}$. The components of the alternative pathway to C3b formation were discovered later historically and are called "Factors." The Factors are assigned a single letter (e.g., Factor B, Factor D, etc.) and are often abbreviated simply as B, F, D, H, and I. As far as is known, there are no complement components unique to the lectin pathway.

In the sections that follow, we present a description of the path-specific biochemical steps initiating each of the three pathways, followed by a delineation of the biochemistry of the terminal steps common to all three pathways. We then examine the regulation of the initial steps of each pathway and that of MAC formation, focusing on activation in the context of a pathogen attack. Complement can also initiate assaults on host cells, but these are held in check by regulatory mechanisms discussed in Section G. Complement receptors and their physiological functions are explored in Section H. Sections I and J conclude the chapter with brief descriptions of the clinical effects of complement deficiencies and potential new roles for complement components.

C. The Classical Pathway of Complement Activation

The classical pathway of complement activation is totally dependent on the humoral response, since it is triggered by the interaction of the Fc region of an antigen-bound antibody with the first component of the classical pathway, C1. It is important to note that only antibodies that have bound to antigen (either as a soluble immune complex or on the surface of a pathogenic cell) can bind C1. This makes evolutionary sense because one does not want the power of complement-mediated destruction unleashed by free antibodies in the vicinity, but only by those that have recognized and bound a foreign antigen. Antigen binding causes a conformational change in the antibody that exposes a C1-binding site in the Fc region. However, only certain Igs are "complement-fixing antibodies," i.e., can efficiently trigger the complement cascade when bound to antigen. In both humans and mice, complement is most efficiently fixed by IgM and IgG antibodies that use C1-binding sites occurring specifically in the C_H2 domain of IgG and the C_H3 domain of IgM. However, even among these antibodies there are variations in the degree of efficiency of complement fixation. In humans, IgM, IgG1,

and IgG3 activate C1 most efficiently, IgG2 does so less efficiently, and IgG4 cannot activate C1 at all. The presence of a serine residue at position 331 in IgG4, as opposed to Pro at this position in IgG1, prevents IgG4 from fixing complement. In mice, C1 is readily activated by antigen-bound IgM, IgG2a, and IgG2b, but not by IgG1. There are further restrictions. A C1 molecule can be activated only if it binds to two C1-binding sites simultaneously; that is, at least two different Fc regions must be bound at once by a C1 molecule for activation to ensue. A single pentameric IgM molecule bound to antigen furnishes a total of five Fc regions in close proximity, such that C1 can readily bind to two of them at once, facilitating activation. In contrast, two monomeric IgG molecules are required to be positioned side-by-side on the surface of the microorganism or immune complex in order to provide two Fc regions in proximity. Since Ig molecules bind randomly to surfaces, as many as 1000 IgG molecules may have to be involved before two find themselves close enough to bind C1. It is worth noting that the two antibodies supplying the Fc sites for C1q activation do <u>not</u> have to be of the same antigenic specificity: two antibodies, each recognizing a different antigen on the surface of a whole pathogen, can do the job.

In vitro, the classical pathway can be triggered by molecules other than antibodies. Some retroviruses, mycoplasmas, and other microbes bind directly to a subunit of C1 called C1q, causing the activation of the classical cascade in the absence of antibody. The relevance of this pathway *in vivo* remains to be determined. We will now describe in detail the early steps of the classical pathway.

I. C1

The complement component C1 circulates in the plasma as a huge (790 kDa) inactive complex containing three different proteins held together by non-covalent bonds involving Ca^{2+} ions. One *C1q*, two *C1r*, and two *C1s* proteins make up a C1 complex (Fig. 19-2). The C1q protein binds to the

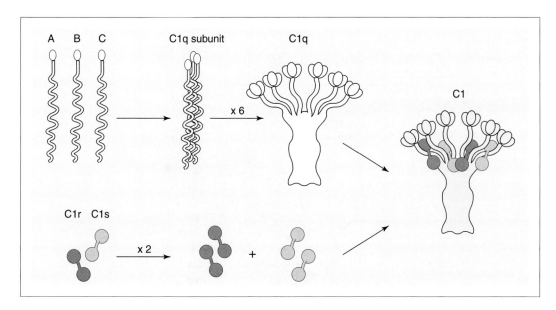

A B C   C1q subunit   C1q

x 6

C1

C1r C1s

x 2   +

Figure 19-2

C1 Structure
Schematic representation of the structure and stoichiometry of the C1 complex and its components C1q, C1r, and C1s. A, B, C are three different polypeptide chains that come together to form the C1q subunit.

Fc region of Igs, while the cleavage activity of C1 as a whole rests in the serine esterase activities of the C1r and C1s components. The C1r and C1s proteins are both single-chain polypeptides of about 85 kDa. Each contains two domains, one that associates with the C1q protein in the complex and another that has protease activity after part of it is cleaved off. C1q itself is a very large protein (462 kDa) made up of six identical subunits in which each subunit is composed of three different polypeptide chains. The six subunits wind around each other to form a stalk that resembles collagen. Projecting radially from the stalk are six globular domains, or "heads"; these heads contain the sequences that bind to the Fc regions of Igs bound to antigen. At least two of these C1q globular domains must be simultaneously bound in order to initiate a conformational change in the C1 complex that leads to the cleavage and activation of one of the C1r subunits by autocatalysis. This one activated $\overline{C1r}$ fragment (28 kDa) can then cleave and activate the other C1r protein, and both can act proteolytically on the C1s proteins to activate them, generating the $\overline{C1s}$ serine esterases that act on the next two components of the classical complement cascade, C4 and C2 (Fig. 19-3).

II. C4

The next substrate in the classical complement pathway is the plasma glycoprotein C4. In its inactive pre-protease form, C4 is a heterotrimer (210 kDa) of three polypeptides labeled α, β, and γ. C4 has structural homology to the C3 protein, the key downstream component of all complement activation pathways, and shares with it the interesting feature of an internal thioester bond. When $\overline{C1s}$ cleaves the C4 molecule, a large C4b fragment containing most of the C4α chain (including the thioester bond) and all of the β and γ chains is generated, as well as a small C4a fragment (8.6 kDa) consisting of the N-terminus of the C4α chain. C4a diffuses away to become a weak *anaphylatoxin*, capable of inducing inflammatory effects (see Box 19-1). Removal of C4a exposes the thioester bond in the C4b fragment to attack by nucleophilic groups, making the C4b protein unstable (written as C4b*). A form of control of the cascade is exerted here, since most of the C4b* molecules are rapidly hydrolyzed by surrounding water molecules

---

Figure 19-3

Early Steps of the Classical Pathway

The binding of C1q to the Fc region of antibodies involved in antigen–antibody complexes (either soluble or fixed on a pathogen surface) activates C1r. C1r activates C1s, which in turn cleaves serum C4 into C4a (an anaphylatoxin) and C4b. The binding of C4b to proteins on the surface of a pathogen is the first step in building a platform upon which the MAC is eventually assembled. Serum C2 binds to C4b and is cleaved by C1s into C2a (protease) and C2b (lost). The C4bC2a structure is known as the classical C3 convertase. The C3 convertase cleaves serum C3, releasing C3a (anaphylatoxin) and C3b* (the common product of all three pathways). The binding of C3b to C4bC2a forms the classical C5 convertase, responsible for cleaving serum C5. C3b can also bind to surface proteins on nearby pathogens or to antibody proteins participating in immune complexes.

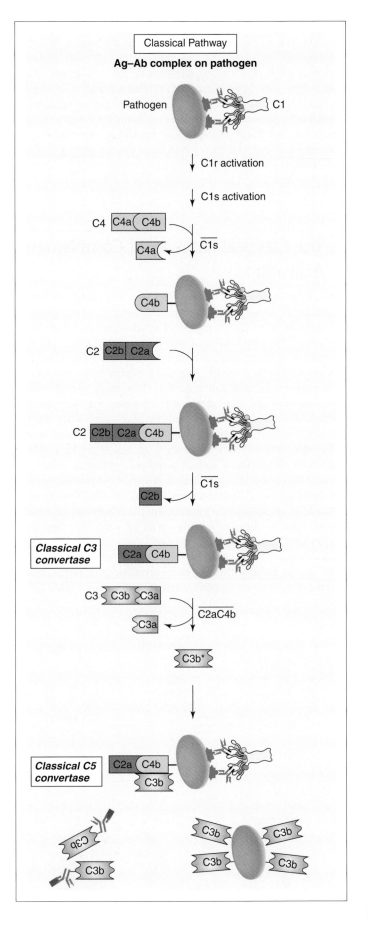

Box 19-1. C3a, C4a, and C5a: the anaphylatoxins

The small fragments that are clipped off the complement proenzymes in the course of the cascade are not just inert bits of detritus; they, too, have biological roles, principally as chemoattractants or as inducers of inflammation. The anaphylatoxins C3a, C4a, and C5a were originally so named because severe bacterial infections that led to systemic complement activation resulted in dramatic cardiovascular and bronchial effects that resembled anaphylaxis (a severe systemic allergic reaction; see Ch.28). All three anaphylatoxins are single-chain polypeptides of about MW 9000, clipped from the N-terminal ends of their respective parent molecules. It is important to note that the anaphylatoxins can also be generated when the complement components are cleaved by enzymes other than those involved in the complement cascade. Kallikrein, plasmin, lysosomal enzymes, and even certain bacterial proteases can directly cleave free C3, C4,

and C5 circulating in the blood, to produce C3a, C4a, and C5a.

C3a, C4a, and particularly C5a trigger the degranulation of mast cells and basophils, which release the vasoactive amines that cause the increased vascular permeability and smooth muscle contraction characteristic of inflammation. The movement of antibodies and accessory cells into the site of entry of the antigen that activated the complement cascade in the first place is thus facilitated. C5a is 2500 times as potent as C4a, and 20 times as potent as C3a, in inducing these effects. C5a is also a powerful chemoattractant, especially for neutrophils, so that complement activation generates compelling signals that draw these inflammatory cells to the site of attack. C5a and C3a upregulate expression of CR3 on neutrophils and ICAM-1 on endothelial cells, increasing adhesion and promoting extravasation into the tissues at sites where an antigen has triggered complement

activation. C5a also stimulates neutrophil degranulation and the respiratory burst. Macrophages and monocytes increase their cell surface expression of adhesion molecules and complement receptors in response to C5a, and secrete increased amounts of IL-1 and IL-6. These cytokines can stimulate the proliferation of activated T cells so that C5a has an indirect effect on adaptive immunity as well.

Like the rest of the complement system, specific regulatory mechanisms exist to curb the power of the anaphylatoxins. *Carboxypeptidase N*, an enzyme circulating in the serum, removes the C-terminal arginine of each anaphylatoxin and abrogates its biological activity. The action of C5a is further controlled by binding to C5a receptors present on the surfaces of circulating leukocytes. These cells immediately endocytose and digest any C5a engaged by the C5aR, removing it from the circulation.

---

to form an inactive intermediate called iC4b. However, some of the thioester bonds will form covalent ester or amide bonds with the nucleophilic side chains of proteins on the pathogen surface. Two isoforms of C4 have been identified: the C4A isoform, whose C4b thioesters preferentially form amide bonds with surface proteins (often the antibody that originally bound C1), and the C4B isoform, whose C4b thioesters form ester links with surface carbohydrates. In any case, the formation of the ester or amide bond prevents hydrolysis by water and stabilizes the C4b on the pathogen surface. By this move, the remainder of the complement cascade is essentially confined to the surface of the invader that has been tagged with antibody, exactly where it is needed.

III. C2

At this point in the classical pathway, C4b is covalently bonded to a surface molecule of a pathogen, carrying out its role as a platform for the next step in the cascade (refer to Fig. 19-3). C2, a single-chain pre-protease of 102 kDa, can bind to surface-immobilized C4b in the presence of Mg^{2+}. Binding to C4b causes C2 to become susceptible to cleavage by the nearby $\overline{C1s}$ molecule, generating the C2b (35 kDa) and C2a (75 kDa) fragments. The smaller C2b fragment diffuses away, while the larger C2a fragment (containing the enzymatic domain) remains associated with C4b to form a new entity: the $\overline{C4b2a}$ complex. Because this new entity gains the capacity to bind, cleave, and activate C3, the next element in the classical complement cascade, it is known as the *classical C3 convertase*. The C4b part binds C3, while the C2a part cleaves it. However, another measure of control is exerted at this point

because the C4b2a complex is quite unstable. The C2a may detach from the C4b and diffuse away as an enzymatically inactive fragment, halting the cascade until a new C2 molecule binds to the surface-bound C4b and a new C4bC2a complex is created.

IV. C3

C3 is arguably the most important complement component. Not only is it present in the serum at a concentration higher than that of any other complement protein, it interacts with a plethora of host and microbial proteins involved in complement activation and effector function. C3 is a 195 kDa glycoprotein composed of two polypeptide chains, the C3α chain (115 kDa) and the C3β chain (75 kDa), which are linked by a single disulfide bond. C3 is synthesized primarily by hepatocytes but also by macrophages, keratinocytes, and endothelial cells. C3 synthesis is upregulated by inflammatory cytokines such as IL-1, IL-6, and IFNγ.

As in the C4α chain, an internal thioester bond between two residues in the C3α chain is hidden within the C3 molecule until it is cleaved by the C3 convertase ($\overline{C4b2a}$), an event that releases an enzymatically inactive C3a fragment (9 kDa) from the C3α chain and exposes the reactive thioester bond in the remaining unstable C3b* fragment (refer to Fig. 19-3). C3a is another anaphylatoxin, while C3b has multiple roles (see later). As in the case of the C4b* fragment, most of the C3b* molecules react with water and are inactivated, halting the cascade. However, in a small proportion of C3b* fragments, the thioester will irreversibly form covalent amide and ester bonds with suitable proteins on a pathogen surface in close proximity, including bound Ig molecules. If C3b* binds to C4b2a, a stable, tripartite C4b2a3b complex known as the *classical C5 convertase* is formed, which can continue the complement cascade by acting on the first of the terminal complement path components, the C5 protein (see later).

The cleavage of C3 links the classical and alternative complement activation pathways and represents an important amplification step. One molecule of classical C3 convertase can produce up to 1000 molecules of C3b*, each of which might succeed in binding to another site on the target surface, to surface proteins on another nearby target, or even to proteins on an uninfected bystander host cell. However, the classical path cannot take advantage of this excess C3b* to form multitudes of classical C5 convertases because the required C4 and C2 molecules have not been similarly expanded in number. Rather, the excess C3b* created by the classical path feeds directly into the alternative path to generate the *alternative C5 convertase* (see later). In other words, the amplification is of the alternative path, not the classical path. With the initiation of the alternative path, increased numbers of alternative C5 convertase complexes can be generated and more MACs assembled to penetrate the target. It is worth reiterating here that these elements of the complement cascade are not reactive unless anchored to a surface protein, confining MAC-mediated destruction to the target alone.

C3b participates in host defense in other ways that do not involve the MAC. The reader will recall that C3b is an important opsonin. Even if it does not participate in a C5 convertase, C3b bound to the surface of an immune complex or pathogen can interact with phagocytes carrying CR1 (*complement receptor 1*; also known as CD35). Phagocytic defense is enhanced as the phagocyte uses its CR1 molecules to ingest the C3b-coated antigen (Fig. 19-4A). As well, C3b deposited on antigen-bound Ig (as a result of the classical path) can mitigate the formation of immune complexes by physically preventing extensive cross-linking of the antibodies, thus limiting the size of the complexes in the plasma (Fig. 19-4B). In addition, C3b can assist in defense against viral infection. Some non-enveloped viruses, while not lysed by the MAC, can be neutralized by the stabilization of C3b* on surface proteins. The coating of C3b sterically hinders the binding of the virus to the host cell viral receptors (Fig. 19-4C). Lastly, membrane-bound C3b can give rise to the important complement product C3d. C3d is produced from C3b when an 88 kDa serine esterase called *Factor I*

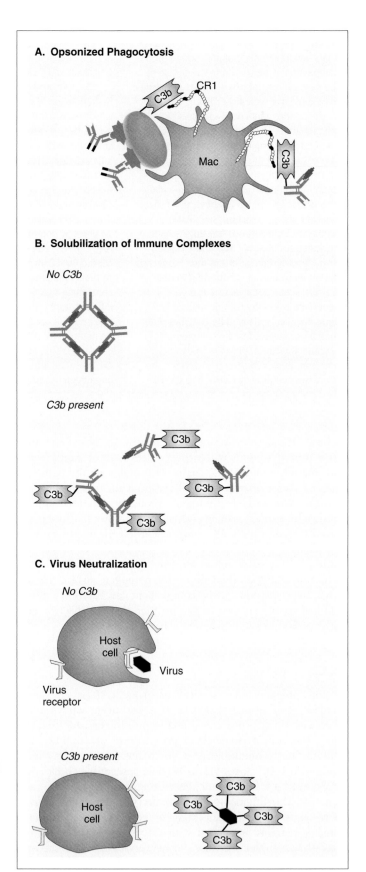

Figure 19-4

Other Functions of C3b
(A) Opsonization. C3b bound to a pathogen or immune complex opsonizes it for uptake via the CR1 molecules of macrophages. (B) Solubilization of immune complexes. Binding of C3b to the Fc region of antibody bound to antigen prevents the networking necessary to form large insoluble immune complexes. (C) Viral neutralization. A coating of C3b on a virus prevents it from binding to receptors on a host cell.

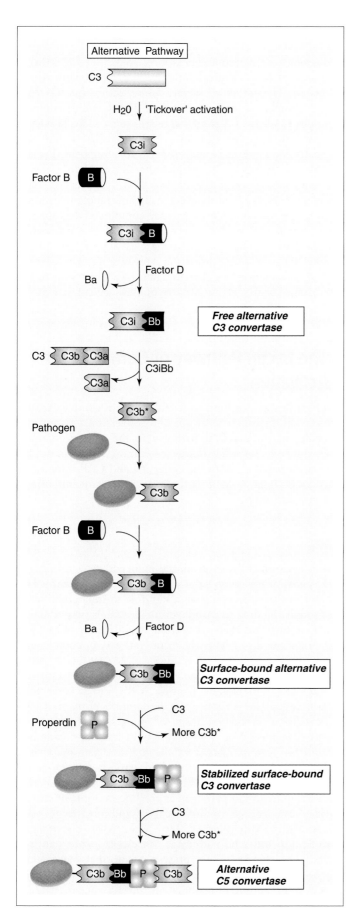

(see later) cleaves a small fragment off membrane-bound C3b, leaving an attached structure called C3d. C3d no longer supports MAC formation (because the formation of the C5 convertase is disrupted) but can still act as an important opsonin and ligand for the B cell coreceptor CR2 (see later).

Returning to the classical path, the generation of the C5 convertase allows the cascade to proceed to the terminal stages of complement activation. However, because the terminal components are common to all three pathways, we will digress to first describe how the alternative and lectin pathways generate C3b and C5 convertases.

D. The Alternative Pathway of Complement Activation

This pathway was outlined in Chapter 4 because it is an integral part of innate immunity. Whereas the classical path is triggered by the interaction of C1 with antibody bound to antigen, the alternative path is triggered by the binding of C3b* to specific surface proteins of a pathogen <u>without the intervention of antibody or the participation of C1, C2, or C4</u>. This C3b* is generated either by previous action of the classical path, or by spontaneous hydrolysis of circulating C3 followed by the action of the factors of the alternative path. Most gram-negative and gram-positive bacteria trigger the alternative pathway because the molecules of lipopolysaccharide and teichoic acid, respectively, in their cell walls readily bind C3b* to the microbial surface in such a way that C3b is stabilized on the pathogen. This C3b then either gives the MAC a place to form (in the case of most gram-negative bacteria and some enveloped viruses), or acts as an opsonin promoting disposal by phagocytosis (in the case of most gram-positive bacteria and some non-enveloped viruses). Similarly, zymosan present in the cell walls of fungal and yeast cells can provide an effective anchor for C3b*. C3b* deposition on the surface of virus-infected host cells, or even attachment to proteins on some non-pathogenic cells (such as heterologous RBCs), can also activate the alternative pathway. The early steps of the alternative pathway are summarized in Figure 19-5.

Figure 19-5

Early Steps of the Alternative Complement Activation Pathway
Serum C3 is spontaneously hydrolyzed to form C3i, which can bind to soluble Factor B. Cleavage of this complex by Factor D yields the free alternative C3 convertase (C3iBb) and the Ba fragment (lost). The free alternative C3 convertase then acts on serum C3 to generate C3b*, which binds to surface proteins on pathogens. Another molecule of Factor B binds to surface-bound C3b and is cleaved by Factor D to generate the surface-bound alternative C3 convertase. This enzyme is capable of generating still more C3b*, amplifying the pathway. Properdin then binds to Bb to form the stabilized surface-bound alternative C3 convertase. The addition of another molecule of C3b* to this complex forms the alternative C5 convertase.

I. C3 AND FACTOR B

The alternative complement activation pathway starts with the freely circulating inactive serum protein C3. A spontaneous low level of hydrolysis (called *C3 tickover activation*) of the internal thioester bond of C3 by water is thought to continuously release tiny amounts of a reactive form of C3 called C3i or $C3(H_2O)$ into the blood. In the presence of Mn^{2+}, C3i binds to the first protein exclusive to the alternative path, Factor B. Factor B is a single-chain protein of 93 kDa that shows homology to C2 of the classical path. There is some evidence that the C2 and Factor B genes may have been generated by a gene duplication event, since these proteins are derived in humans from two genes of similar sequence tandemly arranged on chromosome 6. In a reaction analogous to the binding of C2 to C4b, Factor B associated with C3i (C3iB) becomes susceptible to cleavage by the second alternative path protease, Factor D.

II. FACTOR D AND THE ALTERNATIVE C3 CONVERTASE

Factor D is a 25 kDa serine protease (reminiscent of the C1 serine esterase) that circulates constitutively in the serum at low concentration, apparently in its active form. Two fragments are generated by Factor D-mediated cleavage of C3iB: Ba (33 kDa; unknown function) and $\overline{Bb}$ (63 kDa). Ba diffuses away, leaving the free-floating complex $\overline{C3iBb}$, which is called the *free alternative C3 convertase*. This complex is enzymatically active because the $\overline{Bb}$ component retains its serine protease activity and is capable of cleaving circulating C3 to generate C3b* and C3a. Free C3b* (generated either in this way or by the classical pathway) may randomly bind to side chains of surface proteins through its reactive thioester bond, creating stabilized surface-bound C3b. The binding of another molecule of Factor B to surface-bound C3b creates a surface-bound C3bB complex that can also be cleaved by Factor D, generating a *surface-bound alternative C3 convertase* composed of $\overline{C3bBb}$. This second type of alternative C3 convertase can act on additional C3 molecules to generate still more C3a and C3b*.

Thus, C3b is an initiator of the alternative pathway when Factor B binds to its surface-bound form. However, it is also a product of the same pathway since the alternative C3 convertase formed ultimately generates more C3b*, some of which will become surface-stabilized C3b. In this way, a positive feedback loop is created that amplifies the response. It has been estimated that as many as 2×10^6 molecules of C3b* can be produced and bound to a microbial surface within 5 minutes, quickly rendering the pathogen susceptible to capture by phagocytosis, or lysis by MAC generation.

III. PROPERDIN AND THE ALTERNATIVE C5 CONVERTASE

While both alternative convertases can produce the all-important C3b, only the surface-bound alternative convertase can go on to support MAC formation. Another complement cascade control point comes into play here, however, because

the C3bBb complex is relatively unstable and dissociates within 5 minutes unless a molecule of the serum protein properdin binds to it. Properdin (abbreviated "P"; 220 kDa) is composed of four identical subunits that are non-covalently associated. The *stabilized surface-bound alternative C3 convertase* complex is written $\overline{C3bBbP}$. The $\overline{C3bBbP}$ complex continues to cleave C3 to generate still more C3b. However, the $\overline{C3bBbP}$ complex has an additional role: it can also physically incorporate an additional C3b* fragment, becoming an active $\overline{C3bBbPC3b}$ complex that is capable of cleaving C5; that is, it becomes the *alternative C5 convertase*. This enzymatic complex is both structurally and functionally analogous to the classical C5 convertase complex $\overline{C4b2a3b}$, and brings us to the point of convergence of the alternative and classical pathways, namely, the cleavage of C5 and the terminal stages of MAC formation. However, prior to describing MAC formation, we digress again to a discussion of the initial stages of the lectin pathway of complement activation.

E. The Lectin Pathway of Complement Activation

Two host lectins can bind to residues within complex carbohydrates on microbial cell walls and trigger the activation of molecules of the classical pathway in the absence of antibody: the collectin MBL (*mannose-binding lectin*) and CRP (*C-reactive protein*). Both the collectins and CRP were introduced in Chapter 4. MBL is the more important initiator of the lectin pathway.

I. MBL

MBL is a serum collectin produced by hepatocytes during the acute phase response to attack or injury. As we learned in Chapter 4, the collectins have a key role in innate immunity as opsonins, binding to carbohydrate structures on pathogens such that they are then recognized by PRRs on phagocytic cells. Some can also trigger the lectin pathway of complement activation, the best-studied of these molecules being MBL. MBL is a *C-type* lectin, meaning that calcium must be present in order for it to bind to ligands. MBL uses a C-terminal domain called the *carbohydrate recognition domain* (CRD) to bind to a broad spectrum of monosaccharide hexose structures expressed on a wide variety of microbes, including *Pneumocystis carinii*, *Listeria monocytogenes*, *Neisseria meningitidis*, and the influenza virus and various herpesviruses. Although the affinity of MBL for a single monosaccharide residue is quite low, a pathogen expressing multiple monosaccharides can engage multiple CRDs and greatly increase the avidity of MBL binding. How does MBL discriminate between host cells and pathogens, since both express surface sugars? The surface sugars present on normal mammalian cells are not arranged in the repetitive patterns exhibited by microbes. In addition, mammalian surface sugars tend to terminate in sialic acid, a molecule not recognized by MBL.

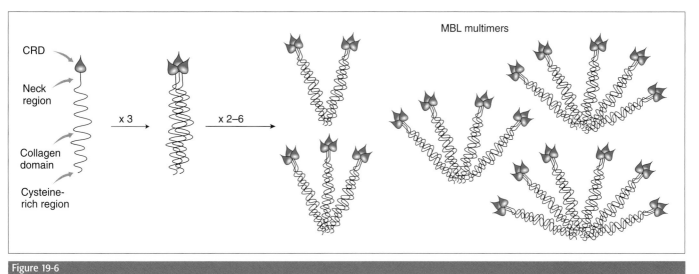

Figure 19-6

MBL Structure
Schematic representation of the structure and stoichiometry of MBL. CRD, Carbohydrate recognition domain. Adapted from Jinhua L. (1997) Collectins: collectors of microorganisms of the innate immune system. *BioEssays* **19** (6), 509–518.

The active form of MBL is a homomultimer of total molecular mass 200–600 kDa. The homomultimer is made up of 2–6 structural units, and each structural unit is a trimer of subunit chains. Each subunit chain (31 kDa) contains an N-terminal cysteine-containing region, a collagen-like region, a neck region, and a globular C-terminal region housing the CRD (Fig. 19-6). The three subunits comprising a structural unit are held together via disulfide bonds between the cysteines in their N-termini. The collagen-like domains can then form a triple-helical stalk from which the carbohydrate binding sites project. In other words, MBL resembles C1q in structure. However, in C1q, binding sites in the C-terminal region recognize the Fc regions of IgM or IgG molecules; in MBL, a range of monosaccharides can be bound to the carbohydrate binding sites, principally mannose but also fucose, glucose, galactose, *N*-acetylmannosamine, and *N*-acetylglucosamine. Any invading microbe expressing abundant amounts of surface molecules containing any of these sugar ligands can expect to find itself quickly coated in MBL and thus subject to opsonization or complement-mediated lysis.

MAC-mediated lysis initiated by MBL occurs because ligand-bound MBL feeds directly into the classical pathway of complement activation at the level of the classical component C4 (Fig. 19-7). MBL is specifically associated with two unique serine proteases, MASP-1 (*MBL-associated serine protease* 1) and MASP-2, and a small non-protease of unknown function called MAp19 or sMAP (*small MBL-associated protein*). The association between MBL, the MASPs, and MAp19 is non-covalent but calcium-dependent. As is true for MBL, the liver is the primary source of the MASPs and MAp19.

The domain structure of the MASPs is essentially identical to that of C1r and C1s, and MASP-2 has enzymatic activity closely resembling that of C1s. It is not entirely clear how the inactive MASP zymogens attain their active forms, but it is possible that one of the MASPs can autoactivate by self-cleavage, and then can cleave the other MASP. In any case, activated MASP-2 then initiates the cleavage of C4 and the deposition of C4b. C2 is then cleaved by MASP-2 as it would be by C1s, and the classical C3 convertase is formed. This enzyme cleaves C3 to generate C3b*, and the addition of C3b to the classical C3 convertase forms the classical C5 convertase, just as is true in the classical pathway. In addition to C3 convertase-mediated cleavage of C3, the MBL/MASP-1/MAp19 complex may be able to activate C3 directly. C3 cleavage products generated by MASP-mediated cleavage may subsequently feed into the alternative pathway, leading to the formation of the alternative C3 convertase and thus the generation of additional C3b*.

II. CRP

CRP, which is also produced by hepatocytes during the acute phase response, has two roles: to battle infections as part of the innate immune response and to aid phagocytes in the removal of membrane and nuclear debris from necrotic cells. CRP is a member of the *pentraxin* family, molecules in which five identical subunits of 20–30 kDa associate non-covalently in a cyclic pentameric structure. Many pentraxins have lectin activity since they bind weakly to bacterial and fungal carbohydrates, and some also recognize a spectrum of small ribonuclear particles, chromatin, histones, and phospholipids. In the case of CRP, each of the five subunits supplies a binding site that weakly recognizes phosphatidylcholine (PC) in the C-polysaccharide found on the surfaces of various pneumococcal species, and similar molecules derived from the membranes of damaged host cells or their nuclear components. Again, a high-avidity interaction develops between CRP and its multivalent ligands. Multiple (but not single) CRP pentamers bound to PC on a pathogen are able to serve

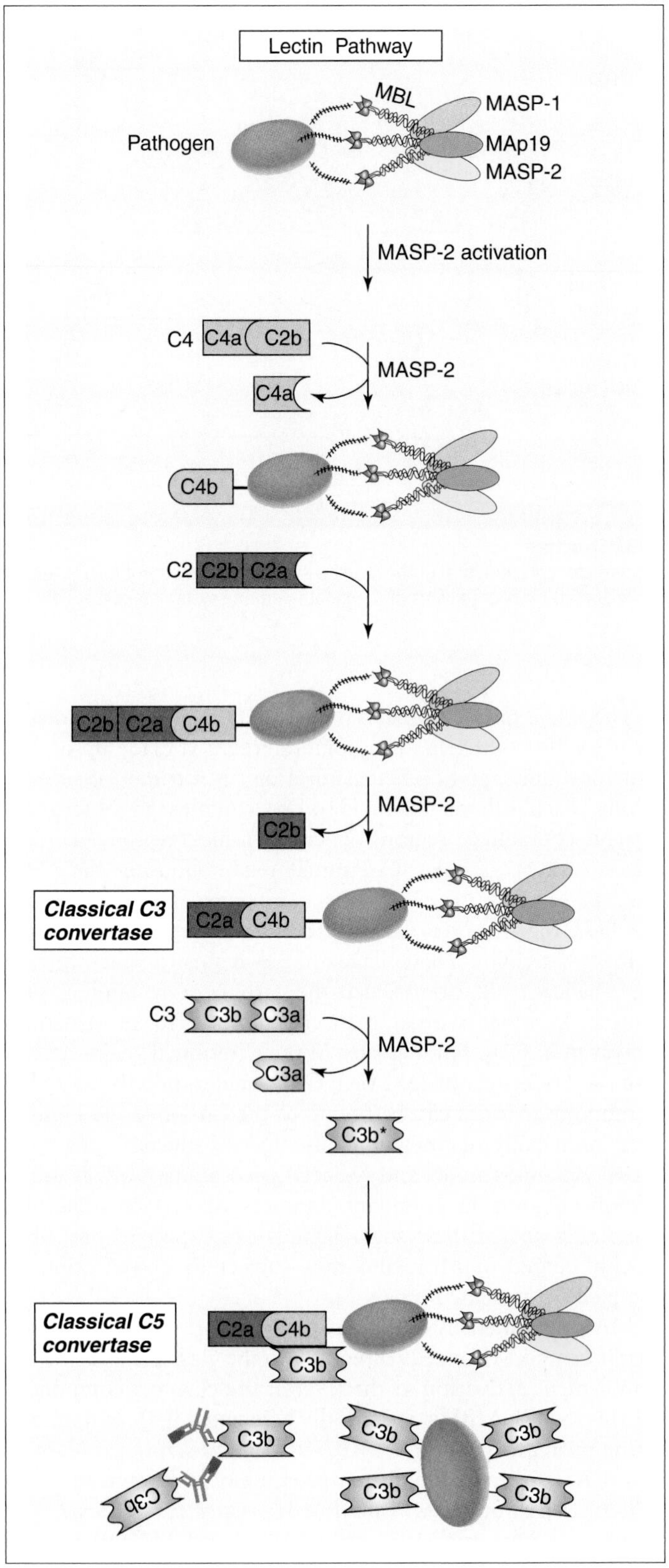

Figure 19-7

Early Steps of the Lectin Pathway
The CRDs of MBL bind to carbohydrate moieties in pathogen proteins. The proteases MASP-1 and MASP-2 and the accessory protein MAp19 are associated with MBL in a complex. MASP-2 cleaves serum C4 to generate C4a (anaphylatoxin) and C4b (binds to proteins on the pathogen surface). MASP-2 then cleaves C2 such that the classical C3 convertase is formed. This convertase cleaves serum C3 to generate C3b* which can both participate in the classical C5 convertase and bind to nearby pathogen surfaces or immune complexes.

as a binding site for C1q, leading to initiation of complement activation by the classical pathway. However, the cascade stops with the building of the C3 convertase. CRP interferes with the assembly of a C5 convertase and so CRP-mediated complement activation is associated with opsonization but not MAC formation.

F. Terminal Components of the Complement Cascade and Formation of the MAC

The first halves of the classical, lectin, and alternative complement pathways are dedicated to the production of vast numbers of C3b* molecules and the assembly of the various C5 convertase complexes. From this point on, the cascade takes a single course leading to MAC formation and cell lysis. These steps are summarized in Figure 19-8.

I. C5

C5 is a 190 kDa serum protein containing two non-identical subunits, α and β, which are linked by a disulfide bond. Although C5 is homologous in some sequences to C3 and C4, it lacks the internal thioester bond. Both the classical $\overline{C4b2a3b}$ and the alternative $\overline{C3bBbPC3b}$ C5 convertases have the function of binding and cleaving C5 to produce two fragments, C5a (11 kDa), a strong anaphylatoxin, and C5b (180 kDa). Upon binding by the C3b element of either C5 convertase, the C5 molecule becomes susceptible to cleavage by either the C2a or the Bb protease element in the convertase. C5a is released from the N-terminus of the α chain, while the C5b fragment (partial α and complete β chain) remains bound to the pathogen surface, anchored by the C5 convertase. This is the last enzymatic step in the complement cascade: the remaining steps in assembling the MAC are non-enzymatic and rely on the conformational arrangement of complement proteins C6–C9 on the C5b platform.

II. C6 AND C7

The C5b fragment associated with either form of the C5 convertase on the cell surface is intrinsically labile and becomes inactive within 2 minutes unless stabilized by binding to C6, a single-chain protein of 128 kDa. (The C6 binding site is not accessible in uncleaved C5.) The C5b6 complex does not have enzymatic activity and serves only to bind C7, another

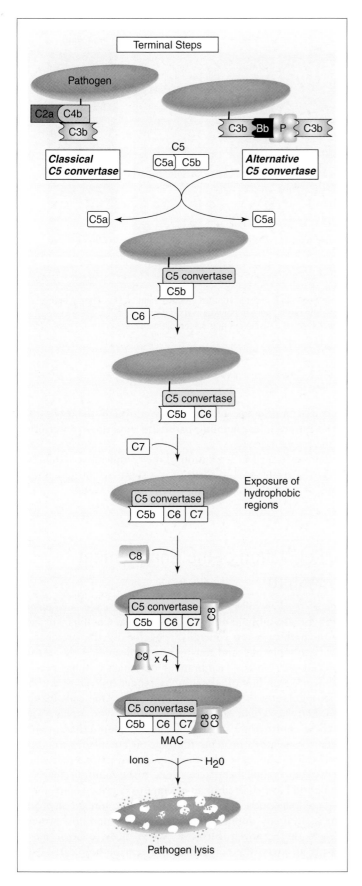

Figure 19-8

Terminal Steps of Complement Activation
Once C5 convertases (regardless of derivation) are fixed on the pathogen surface, they cleave serum C5 to yield C5a (anaphylatoxin) and C5b (remains bound to the convertase). After stabilization of the complex by C6, the addition of C7 exposes hydrophobic regions that facilitate penetration into the pathogen membrane. The addition of C8 stabilizes the complex in the membrane and pore formation is initiated. With the addition of at least four molecules of C9, the pore becomes a tunnel, allowing ion and water molecules to pour through the pathogen membrane into the interior. The pathogen explodes due to osmotic imbalance.

single-chain protein of 121 kDa. The binding of C7 triggers a conformational change in the complex, exposing several hydrophobic regions that, by their chemical nature, recoil from the aqueous environment surrounding the cell. The C5b67 complex is therefore rendered lipophilic and is drawn into the phospholipid bilayer of the cell membrane or viral envelope. If, however, the complex has formed on the surface of an immune complex, there is no bilayer to move into, and the C5b67 complex is released into the fluid phase. The freed complex is very unstable, with a half-life of about 0.1 second. However, should it happen to incorporate itself into the membrane of a pathogen, the complex can proceed to complete the cascade and cause "bystander" lysis of that pathogen.

III. C8 AND C9

Upon insertion into a membrane lipid bilayer, the C5b67 complex acts as a receptor with high affinity for the C8 component. C8 is a 155 kDa heterotrimer composed of an α chain (64 kDa), a β chain (64 kDa), and a γ chain (22 kDa). The C8γ chain is linked to the C8α chain by a disulfide bond. Binding of one molecule of C8 to a C5b67 complex causes another conformational change that permits the insertion of the C8γ chain into the lipid bilayer. This insertion stabilizes the C5b678 complex in the membrane such that it forms a pore of 10 Å in diameter. This small pore causes the cell to become slightly leaky but does not usually lead to overt lysis. Full MAC activity requires the addition of C9 to the C5b678 complex. C9 is a perforin-like single-chain protein of 79 kDa that is capable of polymerizing upon contact with the C5b678 complex. Multiple molecules of C9 may attach to the C5b678 complex, forming a $C5b678(9)_n$ complex we know as the MAC. Only 4 C9 molecules need be present in a MAC for it to have full lytic capabilities, and up to 18 C9 molecules have been identified in a single MAC. In contrast to the small hole formed by the C5b678 complex, the C9 molecules of the $C5b678(9)_n$ complex form a larger cylindrical tunnel of internal diameter approximately 100 Å through the lipid bilayer (Plate 19-1). One such tunnel allows sufficient ion (particularly Ca^{2+}) and water molecule traffic to enter the cell such that the internal osmotic balance is lost and the cell literally explodes from increased internal water pressure.

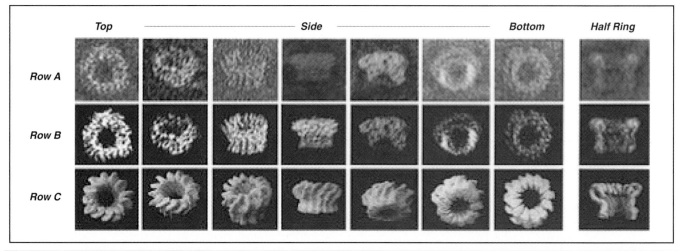

Plate 19-1

Electron Micrographs of Polymerized C9
Row A: Polymerized C9 in several orientations. Row B: The same images are shown after computer-assisted contrast enhancement. Row C: Photographs of a ceramic model of polymerized C9, positioned in different orientations to match Row A. The central pore of polymerized C9 is approximately 100 Å in diameter. Reproduced with permission from DiScipio R. G. and Berlin C. (1999) The architectural transition of human complement component C9 to poly(C9). *Molecular Immunology* **36**, 575–585.

IV. TARGETS OF THE MAC

Any entity that has a membrane in which the MAC can assemble can be destroyed by complement-mediated lysis. Because they have a membrane, most gram-negative bacteria are susceptible to MAC-mediated lysis, whereas gram-positive species, which lack a membrane, are not. Some enveloped virus particles and virally-infected host cells that express viral antigens in their membrane and thus become coated in anti-viral antibodies are also subject to MAC-mediated lysis. Even heterologous erythrocytes introduced into a host by transfusion may be destroyed by MAC formation if the cells bear blood antigens recognized as foreign by host antibodies. However, most uninfected nucleated cells resist MAC-mediated lysis because of the action of host regulatory proteins (discussed later in this chapter). Furthermore, should C3b* happen to randomly bind to a surface protein on a healthy host cell, the metabolic activity of that cell usually results in either endocytosis or exocytosis of any assembled MAC before lysis can occur. Indeed, most vigorously growing cells, including cancer cells, can avoid destruction by endocytosing the MAC before it can instigate lysis. Unfortunately, this means that treating cancer cells with tumor-specific antibodies in the hopes of triggering classical complement-mediated lysis often does not work (see Ch.26).

Another role for the MAC in addition to cell lysis has been proposed. As mentioned previously, most nucleated cells resist MAC-mediated lysis, but it has been shown that MAC formation can lead to the activation of numerous inflammatory cells in the absence of lysis. The activated cells then degranulate, proliferate, and secrete cytokines to assist in the clearance of the pathogen. It remains unclear how the MAC structure could trigger such events, but it is thought that increased trafficking of Ca^{2+} ions through the pore may affect intracellular signaling pathways initiating transcription of the relevant genes.

Before we turn to regulation of the complement system, we direct the reader to Table 19-1 for a summary of the components of the three complement activation pathways.

G. Regulation of the Complement System

One of the key roles of the complement cascade is to ensure step-wise assembly of the MAC. While the complement activation pathways are generally triggered by the presence of pathogen molecules and not by host molecules, MAC formation can occur in any membrane upon which C3b has become stabilized, even that of a healthy bystander cell. Thus, without strict regulation of the complement system, the cells making up our tissues could come under constant lytic attack. Some of these controls are passive in nature, while others involve specific regulatory proteins that act on complement components. Host cells have also developed adaptations that shield them from complement activation. These multiple levels of regulation have evolved in an attempt to confine complement-mediated destruction to entities signaling foreign invasion; that is, antibody–antigen complexes and the non-host proteins present on pathogen surfaces. When these control mechanisms go awry, the complement system blithely carries out its normal destructive function in an indiscriminate manner, destroying host tissues and causing or exacerbating disease (see Box 19-2).

Table 19-1 **Summary of Complement Activation Pathways**

	Classical Pathway	Lectin Pathway	Alternative Pathway
Start point	Ab–Ag complex interacts with C1	Microbial carbohydrates interact with host MBL/CRP	C3 hydrolysis
Initiating components	C1 complex	MBL/CRP and MASP-1, MASP-2, MAp19	C3i
Early components	C4 C2	C4 C2	Factor B Factor D
C3 convertase	C2aC4b	C2aC4b	C3iBb (free) C3bBb (surface-bound) C3bBbP (stabilized surface-bound)
Component that attaches to pathogen surface	C4b	C4b	C3b
C5 convertase	C2aC4bC3b	C2aC4bC3b	C3bBbPC3b
Terminal components	C5–C9	C5–C9	C5–C9

We have already mentioned that most complement serum proteins circulate in inactive zymogen form, providing an initial level of control. In addition, many cleavage products (particularly C3b*) are rapidly degraded, and various complexes are quickly dissociated unless stabilized by binding to a subsequent complement component or to a cell surface protein or carbohydrate. Such limited diffusibility protects nearby self-cells from accumulating C3b and precipitating MAC-induced lysis. Some investigators have suggested that one role of circulating natural antibody is to bind to activated complement products and thereby limit the damage they inflict. In addition, each section of the complement cascade—the classical and lectin paths, the alternative path, and the formation of the MAC—contains its own regulatory elements, as described in the following sections.

I. REGULATION OF THE CLASSICAL PATHWAY

Most of the C1 circulating in the blood is bound to a 105 kDa single-chain glycoprotein called *C1 inhibitor* (C1inh). C1inh belongs to the superfamily of serine proteinase inhibitors known as *serpins*. When C1inh is bound to free C1, the latter cannot undergo spontaneous activation because C1inh acts as a decoy substrate for the activated C1r and C1s serine proteases within the C1 complex (Fig. 19-9). Once C1r and C1s cleave C1inh, covalent links are formed with the resulting fragments that cause the dissociation of C1r and C1s from C1q, and block their ability to cleave C4 or C2. However, if C1 succeeds in binding to an antigen–antibody complex prior to C1inh cleavage, the C1inh molecule is released intact and the cascade can proceed. C1inh also inhibits enzymes of the wound healing cascade.

Further on in the classical cascade, the formation and activity of the classical C3 convertase complex (C4b2a) are regulated by a family of structurally homologous soluble and membrane proteins that bind to C4 and C3 products. They are known as the *regulators of complement activation* (RCA) or *complement control proteins* (CCPs). The RCA proteins are encoded by closely linked genes and are characterized by the presence of multiple (~15–20) copies of 60–70 amino acid sequences known variously as *short consensus repeats* (SCRs), *CCP repeats*, or *Sushi domains*. The SCRs figure prominently in the C4/C3 product binding sites of each RCA. The primary function of the RCA proteins is to prevent the assembly of the C3 convertases on host cell surfaces (thereby pre-empting the formation of the MAC), or to hasten its dissociation. Since different host cell types express different amounts of RCA proteins, the C3 convertases are inactivated more readily in some cells than in others, and thus in some anatomical sites more than in others.

Key RCA proteins in the classical path are the membrane-bound receptors CR1 and DAF (*decay accelerating factor*; CD55), and the soluble serum protein C4bp (*C4 binding protein*). CR1 is described in more detail later in the section on complement receptors. Suffice it to say here that CR1 binds to both C4b (in the classical pathway) and C3b (in the alternative pathway). DAF is a 70 kDa protein anchored by a glycosylphosphatidylinositol (GPI) link in the membranes of vascular endothelial cells, many types of epithelial cells, and peripheral blood cells. DAF also binds to C4b and C3b. C4bp is a large, multimeric, soluble protein composed of seven α chains (75 kDa) and one shorter β chain (45 kDa). The multi-armed structure is held together by non-covalent forces and disulfide bonds. C4bp is synthesized in the liver and its levels increase during infection and inflammation. As its name indicates, C4bp binds to C4b.

How do the RCA proteins control complement activation so as to spare host cells? DAF, C4bp, and CR1 bind to C4b that has managed to attach to a host cell protein. The binding of C2 to C4b is thus competitively inhibited and formation of the classical C3 convertase is blocked (Fig. 19-10). DAF and CR1 also promote the dissociation ("accelerate the decay") of already assembled C3 convertases, releasing C2a from the cell surface.

Box 19-2. Pathological effects of complement activation

Like most powerful weapons, the complement system is a double-edged sword capable of inflicting collateral damage. In the course of its regular protective duties, and despite control by regulatory proteins, complement can have deleterious effects on nearby normal tissues. For example, in the course of attempting to clear immune complexes, complement activation results in the release of the anaphylatoxins, which summon phagocytes into the tissue to remove the coated complexes. Free radicals, proteases, and/or histamines derived from these cells can destroy nearby vascular endothelial cells. A small amount of such destruction can be tolerated in healthy individuals. However, if the collateral damage is too extensive or lasts for too long, pathology can result. A disorder called *immune complex glomerulonephritis* illustrates complement-mediated pathology very clearly. The glomerulus of the kidney is highly susceptible to the accumulation of immune complexes because of its structure and function as a filter of body fluids. The immune complexes both provoke inflammation and invite complement activation via the classical pathway, initiating the destruction of cells of the glomerular capillary walls.

Pathological complement activation is also observed when an autoantibody that has recognized a host tissue triggers MAC assembly in host cells, resulting in manifestations of autoimmunity. For example, complement-mediated destruction of cells in the neuromuscular end plate has been associated with the autoimmune disorder *myasthenia gravis*, in which patients exhibit muscle weakness due to impaired neuromuscular transmission (see Ch.29). Other disorders in which complement activation has been implicated as a cause of tissue injury include hemolytic anemia, type II collagen-induced vasculitis, and multiple sclerosis.

Tissue injury can also occur as a result of inappropriate complement activation secondary to a primary clinical event. In this context, complement activation is often observed in association with cases of stroke, burn injury, or heart attack. The pro-inflammatory effect of the complement cascade triggered by injury to these tissues further exacerbates the damage to the patient. For example, a heart attack (more properly known as a *myocardial infarct*) in humans provokes an acute phase response and the production of CRP. CRP is deposited on myocardial cells in the infarct and is able to initiate complement activation via the

lectin pathway. There is also evidence that oxidative stress may alter cell surface membrane glycosylation such that MBL can bind to myocardial cells, again triggering the lectin pathway. In addition, it is thought that the loss of oxygenation during the infarct causes the myocardial cells to lose expression of the complement regulator MIRL (CD59; see main text), leaving them vulnerable to MAC-mediated damage. In the case of burns, it is unknown what initiates complement activation, but elevated levels of the pro-inflammatory mediators C3a and C5a can be detected both in the plasma and in burn blister fluid very soon after the trauma.

Medical interventions intended to help patients can also result in unwanted complement activation. The mechanical parts of dialysis or heart bypass machinery may activate complement components in patient blood passing through the machinery, causing some patients to develop circulatory and breathing difficulties known as *post bypass syndrome*. The interaction between the polymers making up the bypass machine parts and the patient's complement components are known as *bioincompatibility reactions*. Allograft transplantation can also trigger complement activation, which can then contribute to graft rejection (see Ch.27). The endothelial cells in the grafted tissue may express epitopes recognized by natural antibodies in the recipient, triggering complement activation via the classical pathway.

How then to eliminate complement-induced collateral tissue damage? Most pharmacological agents that inhibit complement activation *in vitro* have toxic side effects or are of uselessly low activity *in vivo*. Clinicians are now starting to exploit the regulatory molecules of the complement system as a means of mitigating complement-induced collateral damage. C1inh (see main text) has been under investigation for many years as a clinical agent able to inhibit not only complement activation and its generation of pro-inflammatory peptides, but also the production of mediators by other enzymatic systems such as the intrinsic coagulation pathway. However, a problem with this type of therapy is that C1inh in inflamed tissues is often rapidly inactivated by proteolytic enzymes released by activated neutrophils in the vicinity. Alternatively, one can target the complement receptors. A soluble recombinant form of CR1 that lacks both transmembrane and cytoplasmic regions and acts as a competitive inhibitor of the natural molecule has shown some promise in

animal models. However, inhibition of the complement pathway at this early stage leaves the animal subject to infections of greater severity and frequency than if only the terminal components of the pathway are blocked. Accordingly, treatments targeting the generation of the MAC have been proposed. No pharmacological drug has yet been discovered that is specific for this part of the pathway, but some researchers are carrying out trials employing natural inhibitors of MAC formation such as soluble, modified forms of regulatory proteins. In another approach, monoclonal antibodies directed against C5, C6, and C8 have proved to be effective inhibitors of MAC formation in model systems *in vivo*. With respect to allografts, experimental manipulation has shown that graft survival can be prolonged if the grafted endothelial cells express high levels of complement regulatory proteins such as MIRL.

Encouraging results have been obtained with a small cyclic 13-residue peptide called *compstatin*, which was discovered in a lengthy search of a random peptide library. Compstatin works by binding to C3 in such a way that the cleavage of C3 into C3a and C3b is blocked reversibly. Despite the structural similarities of C4 and C5 to C3, compstatin inhibits only C3 cleavage. Compstatin is active *in vitro*, in that it is inhibitory in model systems of bioincompatibility, and has also shown promise in examinations of organs cultured outside the body (*ex vivo*). In addition, compstatin has been investigated as a means of controlling complement activation during organ transplantation between species (*xenotransplantation*; see Ch.27). In one experiment, xenotransplantation was mimicked by culturing pig kidneys *ex vivo* and perfusing them with fresh human blood. Compstatin (or a control agent) was added to the human blood, and the survival of the pig kidneys, the accumulation of C3 activation products, and MAC formation were monitored. In the presence of compstatin, C3 activation was blocked, as was MAC formation, and the mean survival time of the kidneys was increased about four-fold over the control. Follow-up studies in primates have shown that compstatin is safe and effective in inhibiting the complement system *in vivo*. Significantly, compstatin binds and inhibits only primate C3 molecules and not those of mice, rats, or rabbits. Compstatin thus represents a class of molecules that may provide significant therapeutic benefits in the myriad situations in which inappropriate complement activation is a threat.

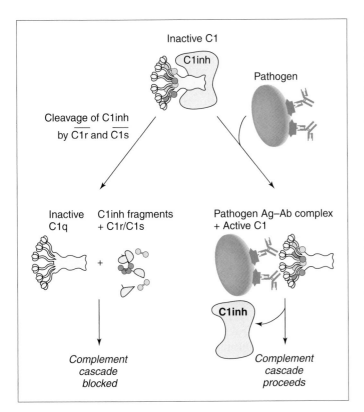

Figure 19-9

Regulation of Classical Complement Activation by the C1 Inhibitor (C1inh)
The C1 complex is held inactive in the blood by the binding of C1inh. In the absence of a pathogen (left), C1inh acts as a decoy substrate and is cleaved by activated C1r and C1s. The resulting fragments bind to C1r and C1s and inactivate them while releasing free (inactive) C1q: the cascade is blocked. However, if the C1–C1inh complex encounters a pathogen sporting a coat of antibody (right), C1q binds to the Fc region of the antibody and releases intact C1inh. C1r and C1s remain activated and the cascade proceeds.

While DAF acts on convertases on the same cell where it is located, CR1 can act on convertases on neighboring cells and immune complexes. Classical C3 convertase formation is also hindered by the proteolytic degradation of C4b carried out by *Factor I*. Factor I is a regulatory protein that was first identified in studies of the alternative pathway. Factor I can cleave membrane-bound C4b into smaller fragments but does so efficiently only when CR1 or C4bp is bound to C4b.

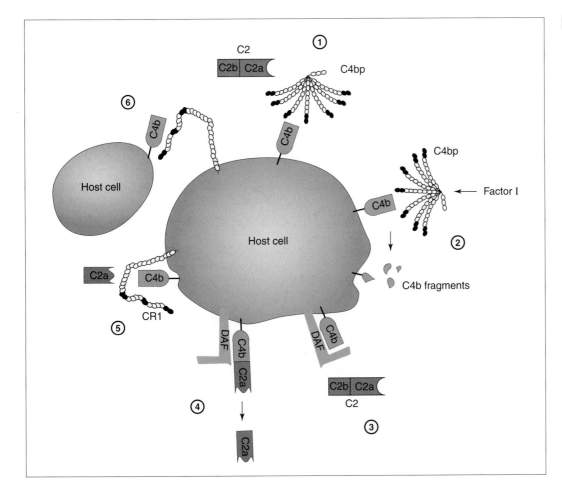

Figure 19-10

RCA Control of Classical Complement Activation
On a healthy host cell, C4bp binding to C4b prevents the subsequent binding of C2 (1) and promotes the degradation of surface-bound C4b by soluble Factor I (2). These two measures block the formation of the classical C3 convertase. DAF binds to C4b to prevent the subsequent binding of C2 (3) and also promotes the release ("decay") of C2a from the C4bC2a complex (4). CR1 blocks access to C4b on both the healthy host cell (5) and neighboring cells (6), promotes the decay of the C4bC2a complex (as for 4), and facilitates degradation of surface-bound C4b by Factor I (as for 2).

II. REGULATION OF THE ALTERNATIVE PATHWAY

Alternative C3 convertase ($\overline{C3bBbP}$) complexes that assemble on microbial cells can continue the fixation process and achieve cell lysis, but those complexes that attach to host cells are prevented from completing the cascade. Several contributing mechanisms have been identified. First, mammalian cell membranes contain high levels of sialic acid that enhance the dissociation of C3b from the host cell surface. Secondly, DAF and CR1 inhibit the formation of the alternative C3 convertase just as they block formation of the classical C3 convertase (Fig. 19-11). DAF and CR1 can bind to C3b (or C3i) and thus block the binding of either Factor B or its fragment Bb to the surface C3b. In addition, DAF can act to accelerate the decay of the alternative C3 convertase, promoting the release of Bb that has bound to C3b.

A soluble regulator unique to the alternative pathway is *Factor H*. Factor H competes with Factor B for binding to C3b, and can displace Bb from the surface-bound alternative C3b convertase. More importantly, the binding of C3b to Factor H makes the C3b molecule susceptible to cleavage by Factor I, which cleaves the cell-bound C3b into inactive fragments. The fragment remaining bound to the cell surface is called *iC3b* (*inactive C3b*). The iC3b fragment cannot support the formation of the alternative C3 convertase so that the cascade is blocked. However, not all biological activity is lost, since iC3b and additional products of its degradation function as opsonins. These molecules can also infiltrate and solubilize immune complexes accumulated in the tissues.

The ability of a molecule to promote the binding of Factor H instead of Factor B to the C3b attached to its surface determines whether that molecule is an "activator" or "non-activator" of the alternative pathway. Microbial cell wall components are generally "activators" because they lack certain surface constituents that allow Factor H to bind to C3b. Factor B binding is thus favored over Factor H, Factor I cannot inactivate C3b, and the complement cascade proceeds. "Non-activators" are generally entities such as host cell surfaces and other structures containing sialic acids or glycosaminoglycans. These molecules provide a microenvironment surrounding the surface-bound C3b molecule that promotes high-affinity interactions with Factor H. Interaction with Factor I can then proceed and Factor B binding is excluded, shutting down the pathway. This reining-in of the alternative pathway is very important, for without control by Factor H (and other regulatory proteins), the alternative C3 convertases would rapidly and continually consume circulating C3, leaving the host with an acquired C3 deficiency and increased vulnerability to infection.

As well as the fluid phase regulator Factor H, mammals possess a membrane-bound regulator specific to the alternative pathway called MCP (*membrane co-factor protein*; CD46). MCP is a 50–70 kDa glycoprotein that structurally resembles DAF. MCP is present in the membranes of virtually all epithelial and endothelial cells, and of all circulating cell types except erythrocytes. Like Factor H, MCP binds to C3b and C3i and promotes Factor I-mediated inactivation of these molecules.

Interestingly, the pathogen world has exploited the RCA proteins on several fronts. The measles virus uses MCP as a coreceptor in gaining access to host cells, binding at a site distinct from the complement binding site. The human herpesvirus 6 (HHV-6) can also exploit MCP for host cell entry. In addition, MCP has been implicated in the adhesion of Group A *Streptococcus* to human cells and in the attachment of *Helicobacter pylori* (ulcer-causing organism) to gastric cells. *Neisseria gonorrhoeae* and *N. meningitidis* express cell surface structures that can bind to MCP-expressing cells. DAF is a receptor for many human picornaviruses and acts as a co-receptor for invasion by many strains of *Escherichia coli*.

Figure 19-11

RCA Control of Alternative Complement Activation
DAF and CR1 can bind to C3b to block formation of the alternative C3 convertase (1) and also accelerate the decay of the C3bBb complex (2). Soluble Factor H blocks the binding of Factor B or its fragment Bb to surface-bound C3b (3), promotes dissociation of Bb from C3b (4), and facilitates the degradation of C3b by Factor I (5). MCP binds to C3b to block Bb binding (6) and promotes the degradation of C3b by Factor I (as for 5).

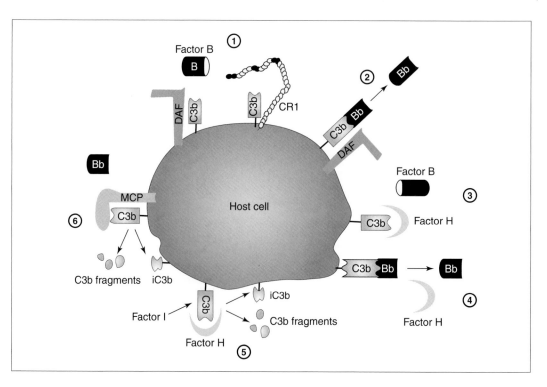

III. REGULATION OF THE LECTIN PATHWAY

The MASP enzymes operating in the lectin pathway have structural and functional homology to C1r and C1s. Accordingly, C1inh can bind to MBL–MASP complexes and inhibit their enzymatic activity. In addition, the MASPs appear to be regulated, at least in part, by the acute phase protein α_2-macroglobulin. This protein has no inhibitory effect on C1r or C1s but does form a complex with the MASPs that blocks their activity. Naturally, since the lectin pathway generates C4b to form the C3 convertase, the regulatory proteins (C4bp, Factor I, etc.) cited previously as acting at this point in the classical pathway also affect the lectin pathway.

IV. REGULATION OF TERMINAL COMPONENTS

Regulation of complement fixation is also exerted at the level of the terminal complement components (Fig. 19-12).

i) Vitronectin and Clusterin

The membrane-penetrating $\overline{C5b67}$ complex can be released into the fluid phase, and if allowed to freely diffuse, could insert itself into the bilayers of neighboring self-cells and bring on their bystander lysis by MAC formation. Fortunately, *vitronectin* (also known as *S protein*) is also present in the fluid phase. Vitronectin, a highly acidic, soluble glycoprotein of 65–75 kDa, is expressed principally by hepatocytes but also by platelets and macrophages. Vitronectin functions primarily as an adhesion molecule, facilitating cell–matrix interactions, but has a secondary function in complement regulation. Vitronectin binds to the free $\overline{C5b67}$ complex and induces it to become less hydrophobic. C8 and C9 can still join the complex, but

C9 polymerization is inhibited and the MAC is no longer able to tunnel through the membranes of nearby host cells. Knockout mice deficient for vitronectin are perfectly normal, indicating that this molecule is redundant in function. Indeed, it has been shown that a multifunctional plasma protein called *clusterin* is also able to bind to soluble complexes of complement terminal components and prevent their insertion into a membrane. Clusterin, a heterodimeric serum glycoprotein of 35–40 kDa, is synthesized primarily in the liver but is also highly expressed by Sertoli cells in the testes. The actions of vitronectin and clusterin are designed to achieve one goal: to ensure that complete MAC formation is essentially confined to the original site of complement activation—the surface of the pathogen.

ii) MIRL and HRF

If $\overline{C5b67}$ released from an assembling MAC is not blocked by vitronectin or clusterin and manages to penetrate the membrane of a bystander cell, the complex is no longer subject to control by mediators in the plasma. Regulation then devolves to two host cell membrane-bound proteins (which are not expressed by pathogens) that specifically block the completion of the MAC in host cells. The *membrane inhibitor of reactive lysis* protein (MIRL, or CD59; also known as *protectin*) binds to C8 and C9 bound on a host cell, preventing membrane penetration and polymerization of C9. MIRL, a heavily glycosylated protein of 18–25 kDa anchored in the membrane by GPI, is highly expressed on sperm, all circulating blood cells, endothelial cells, and most epithelial cells. The *homologous restriction factor* (HRF, also known as *C8 binding protein*; 65 kDa) also interferes with MAC completion, probably at the level of C9 addition. MIRL and HRF are said to feature *homologous restriction*; that is, in *in vitro* experiments, they bind only to a

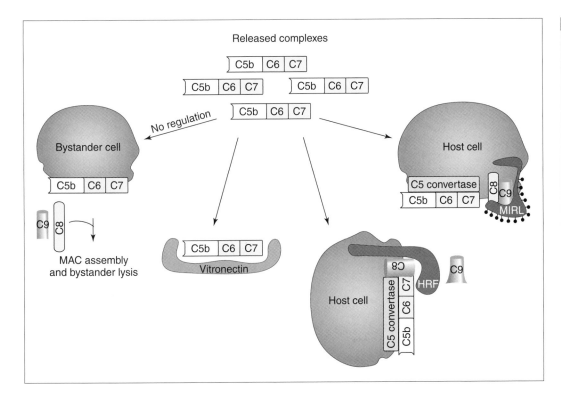

Released complexes

Figure 19-12

Regulation of the Terminal Complement Components
Free C5bC6C7 complexes released from pathogen membranes can penetrate into the membranes of neighboring cells and initiate MAC assembly that, if left unregulated, leads to unwanted bystander lysis. Vitronectin and clusterin (not shown) are soluble RCA proteins in the plasma that bind to and inactivate free C5bC6C7 complexes. Should a C5bC6C7 complex escape vitronectin and penetrate into a host cell membrane, the cellular HRF protein binds to the C8 moiety and blocks the addition of C9. Should regulation by HRF fail, the cellular protein MIRL binds to both C8 and C9 and blocks polymerization of C9 and pore formation.

terminal complement protein that is derived from the <u>same</u> species. The discovery of MIRL and HRF explains a long-standing phenomenon observed in experiments in which hapten-conjugated erythrocytes of an animal are used in a complement fixation assay to test for the presence of anti-hapten antibodies in a serum sample. The hapten-conjugated cells are lysed most effectively by antibody when heterologous complement is used. For example, hapten-bound guinea pig cells succumb to rabbit complement much more readily than to guinea pig complement. We now understand that the guinea pig cells bear guinea pig MIRL and HRF proteins that do not recognize rabbit C8 and C9, and so do not inhibit rabbit complement-mediated lysis. Obviously, guinea pig complement contains guinea pig C8 and C9, so that MIRL and HRF on the guinea pig cells recognize the homologous C8 and C9 proteins, bind them, and inhibit subsequent MAC formation and lysis.

A summary of all three complement activation pathways and the points at which they are regulated by the RCAs is presented in Figure 19-13.

H. Complement Receptors and Their Biological Roles

As we have mentioned in our tour of the complement cascade, there are several points at which products of the complement system initiate host defense actions other than lysis of pathogens. Certain components that remain bound to a pathogen's surface (e.g., C3b and iC3b) act as opsonins, facilitating the uptake of the pathogen by phagocytes. Other fragments generated during the formation of the C3 and C5 convertases are released and serve as anaphylatoxins, provoking an inflammatory response. However, in order for these events to occur, the phagocytes and inflammatory cells involved must express the appropriate complement receptor that binds to the complement product mediating that response. Without the complement receptors, the complement cascade would be much more limited in its range of effects and less efficient in its regulation. Best-characterized of the complement receptors are

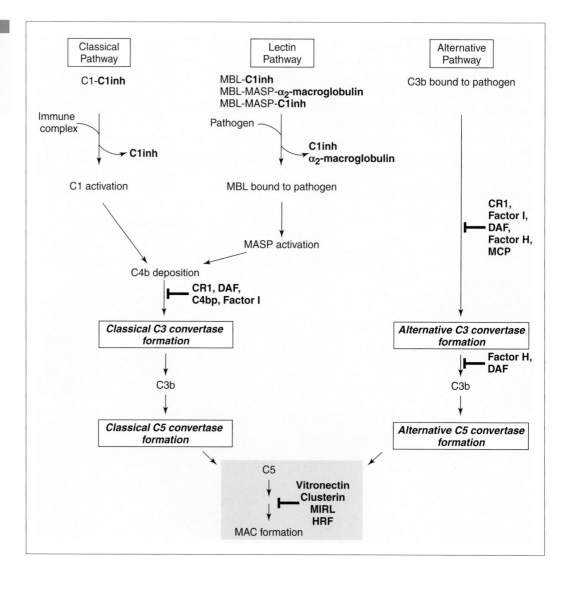

Figure 19-13

Regulation of Complement Activation Pathways
Summary of the regulatory mechanisms controlling the three complement activation pathways.

those that recognize surface-bound complement products. As a group, these products include C3b, iC3b, C3dg (a cleavage product of iC3b), C3d, and C4b. Each is recognized by at least one of a group of receptors that includes CR1, CR2, CR3, and CR4. Each of these receptors is expressed on distinct subsets of immune system cells, and each recognizes a limited number of complement component fragments (Table 19-2).

I. CR1

CR1 (also known as *C3b receptor* and CD35) is a single-chain glycoprotein of 165–280 kDa, depending on the isoform. In mice, CR1 and CR2 are derived from the same gene such that CR1 contains all of the CR2 sequences plus six additional N-terminal SCRs that encode the C3b binding site. In an interesting example of the degree of integration of the complement system, the synthesis of CR1 on phagocytes is stimulated by the proteolytic fragment C5a derived from complement activation. The role of CR1 as an RCA protein regulating the classical complement cascade was discussed previously. CR1 has three additional important functions stemming from its high affinity for both C3b and C4b, and lesser affinity for iC3b.

i) Opsonized Phagocytosis

CR1 expression on phagocytes such as macrophages and neutrophils enables these cells to bind pathogenic cells or viral particles displaying iC3b, C3b, or C4b, promoting the clearance of

these threats by opsonized phagocytosis. In particular, defense against fungal and bacterial (as opposed to viral) infections relies on the phagocytosis of microbes opsonized with C3b and iC3b. Large antigens or immune complexes may also sport a coat of C3b.

ii) Clearance of Immune Complexes

The expression of CR1 on all circulating erythrocytes is particularly important for the clearance of immune complexes. Our bodies contain significant quantities of circulating antibodies in the blood, and a small proportion of these continually encounter antigen and form immune complexes. These complexes are innocuous in small numbers, but the extensive complexes formed during an intense humoral response to a massive infection would have deleterious consequences were it not for CR1 and the complement system. Immune complexes that grow so large that they lose their solubility can be deposited in the tissues, where they activate complement and provoke inflammatory responses that result in significant tissue damage. As mentioned previously, C3b can help dissolve complexes to a certain extent, but should this prove insufficient, the presence of C3b or C4b on a complex allows it to be bound by CR1 on erythrocytes, which then convey the complex to the liver or spleen for disposal by resident phagocytes. Although the density of CR1 on erythrocytes is about 100-fold less than that on a granulocyte, there are 1000-fold more individual erythrocyte cells, so that the vast majority of

Table 19-2 **Major Complement Receptors**

Receptor	Other Names	Ligands	Expression	Major Functions
CR1	CD35, C3b receptor	C3b, C4b, iC3b	Macrophages, neutrophils, erythrocytes, FDCs, B cells, DCs, some T cells, some eosinophils	Inhibition of classical and alternative convertase formation Opsonized phagocytosis Clearance of immune complexes Enhanced Ag processing
CR2	CD21, C3d receptor	C3d, C3dg, iC3b	B cells, FDC, some epithelial cells	Opsonized phagocytosis Clearance of immune complexes Enhanced Ag processing Co-receptor for B cell activation CD5$^+$ B cell development and natural Ab repertoire Memory B cell development
CR3	Mac-1, Cd11b/Cd18 (β2-integrin)	ICAM-1, iC3b, microbial carbohydrates	Neutrophils, monocytes, macrophages, mast cells, NK cells	Extravasation of leukocytes into inflammatory sites Opsonized phagocytosis Clearance of immune complexes Enhanced Ag processing Lectin for microbial carbohydrates
CR4	P150,95, CD11c/CD18 (β2-integrin)	ICAM-1, iC3b, microbial carbohydrates	Neutrophils, monocytes, macrophages, mast cells, NK cells	Extravasation of leukocytes into inflammatory sites Opsonized phagocytosis Clearance of immune complexes Enhanced Ag processing Lectin for microbial carbohytrates

CR1 in the body is actually localized in the erythrocyte compartment.

iii) Enhancement of Antigen Processing

The CR1 expressed on FDCs, B cells, and DCs may contribute to antigen processing and presentation because pathogens or immune complexes coated with C3b, iC3b, or C4b are readily phagocytosed. The pathogen proteins can then be digested to antigenic peptides, which are presented on MHC to host T lymphocytes. For FDCs in particular, CR1 may help to prolong the display of complement-coated pathogens or complexes to cells of the adaptive immune response.

iv) Other

CR1 is also expressed on some T cells and eosinophils, but not on platelets. The precise function of CR1 on these cells is not clear. There is now evidence that CR1 may be involved as a coreceptor in triggering HIV replication in HIV-infected CD4$^+$ T cells.

II. CR2

CR2, also known as *C3d receptor* and CD21, is a single-chain glycoprotein receptor of about 140 kDa that binds iC3b, C3d, and C3dg. The primary functions of CR2 are opsonization, immune complex clearance, and antigen presentation, as described for CR1. However, the cellular distribution of CR2 is much more limited than that of CR1, being confined primarily to B cells, FDCs, and some epithelial cells. Intriguingly, CR2 has two other functions that make it a defining link between the complement system and humoral responses.

i) B Cell Activation

The reader will recall that CR2 expressed on B cells associates with CD19 on these cells to form the B cell coreceptor (discussed in Ch.9). When the Ig receptors of a B cell are stimulated by antigen, and its CR2 receptors are engaged by C3 products, the B cell is more easily activated than if the BCRs alone are engaged (refer to Fig. 9-8). The binding of iC3b or C3dg to CR2 on B cells is thought to lower the threshold of antibody stimulation required to trigger B cell activation. In fact, the amount of antigen required to activate B cells *in vitro* can be reduced by 10- to 100-fold if CR2 is co-ligated with the BCR.

Complement also plays a unique role in the activation of B cells producing natural antibodies. The reader will recall from Chapter 4 that natural antibodies are germline-encoded IgM molecules produced primarily by the unconventional CD5$^+$ subset of B cells. The repertoire of natural antibodies is limited and fixed, and directed against structures common to pathogens such as LPS and phosphatidylcholine. Natural antibodies are present at birth, circulate throughout the body, and are thought to provide immediate antibody protection against a broad range of pathogens without the lag required for the production of specific antibodies by activated B cells. Interestingly, unlike conventional B cells, the clonal selection and expansion of CD5$^+$ B cells (and thus natural antibody production) depend on contact of the clone with antigen coated in C3d. Mice deficient in CR1 or CR2 show decreased numbers of CD5$^+$ B cells and a reduced repertoire of natural antibody specificities. A feedback mechanism exists, because antigen-bound natural antibody activates complement, resulting in the production of more C3d required for the expansion of additional CD5$^+$ B cells.

ii) B Cell Memory Development

CR2 has two roles in promoting B cell memory development. First, the presence of CR2 on the FDC allows the concentration of C3d-coated antigens on these cells and thus the display of those antigens for long periods in the germinal centers. In one experiment, multiple molecules of C3d were coupled to an antigen, and the amount of antigen needed to induce a B cell memory response was assessed. When large amounts of C3d were used, memory responses were induced by significantly lower concentrations of the antigen. Secondly, CR2 expression on B cells is essential for firm contact with the antigens retained on FDCs, an event that delivers a crucial survival signal allowing a stimulated B cell to enter the follicles and initiate a germinal center reaction. Without this CR2-mediated survival signal, the B cell is triggered to undergo apoptosis and memory cell development is compromised. This observation has been confirmed in several different types of experiments. When challenged with a Td immunogen, knockout mice deficient for C3 or CR2 mounted a normal secondary T cell response but a profoundly impaired secondary B cell response characterized by a drastic reduction in the number and size of germinal centers. In addition, in chimeric mice engineered so that their FDCs expressed CR2 but their B cells did not, the defect in the secondary response was still evident, indicating that CR2 on B cells is of critical importance. Lastly, splenic germinal center development was found to be severely impaired in wild-type mice in which C3d binding to CR2 on B cells was blocked by the injection of a soluble form of CR2.

iii) Other

A less benign role for CR2 lies in its function in humans as a convenient route of host cell access for the often tumorigenic Epstein–Barr virus (EBV). Infection with EBV has been strongly linked to certain B cell lymphomas (see Ch.30). EBV displays amino acid homology to the C3dg fragment, causing the virus to be readily taken up by CR2. CR2 has also been shown to bind IFNα, although the physiological significance of this finding is unknown.

III. CR3

CR3 (also known as Mac-1 or CD11b/CD18) was originally characterized as a member of the β2-integrin family of intercellular adhesion molecules introduced in Chapter 3. As such, CR3 on the surfaces of neutrophils and monocytes binds to ICAM-1 on endothelial cells and enhances the extravasation of these leukocytes into a site of inflammation. In addition, like CR1 and CR2, CR3 is important for opsonized phagocytosis and immune complex clearance, and may contribute to antigen processing. However, unlike CR1 and CR2, CR3 binds only to the iC3b product derived from the Factor I-mediated cleavage of C3b. The ability to bind iCb3 is Ca^{2+}-dependent and appears to rest in the α chain. CR3 also differs from both CR1 and CR2 in its cellular distribution. While CR1 is synthesized by a broad range of cell types, and CR2 is found on B cells, FDCs, and some epithelial cells, CR3 is expressed on mast cells, NK cells, and phagocytes. There is evidence that simultaneous engagement of CR3 and Fcγ receptors on a phagocyte by a particle bearing both iC3b and IgG antibody results in a significant stimulation of phagocytosis over the rate observed when either receptor alone is engaged. Finally, CR3 can serve as a surface-bound lectin, specifically binding carbohydrate moieties of some microorganisms and promoting their disposal by phagocytic cells.

IV. CR4

CR4 (also known as p150,95 or CD11c/CD18) is another β2-integrin molecule with cellular distribution and functions in endothelial adhesion and phagocytosis very similar to CR3.

V. C1q "RECEPTORS"

Five molecules have been reported to bind to C1q, but C1q is a notoriously sticky molecule so that these identifications remain tentative. One of these molecules, designated C1qRp (_C1q receptor of phagocytosis_; CD93), is a 126 kDa cell surface receptor expressed on monocytes. C1qRp may be involved in phagocytosis by these cells. Another putative C1qR is a 70 kDa receptor that binds to the collagen-like tail of the C1q component of C1 after the C1 has been dissociated from C1inh. This C1qR occurs on phagocytic cells, many B cells, and some other lymphocytic subsets and may function in the uptake of immune complexes that have bound C1q. CR1 has also been shown to bind to C1q, presumably facilitating opsonized phagoctyosis. Intriguingly, two other proteins that bind to the globular heads of C1q are soluble intracellular proteins: the ER chaperone protein calreticulin (60 kDa) and the mitochondrial matrix protein p33 (33 kDa). Although it is unknown how these internal proteins get to the cell surface to bind C1q, it is speculated that they may have a role in assisting C1q in the clearing of potential autoantigens that leak from damaged cells (see later).

VI. RECEPTORS FOR C3a, C4a, AND C5a

The reader will recall that the C3a, C4a, and C5a fragments are derived from the sequential cleavage of enzymatic components in the classical complement cascade. These fragments were originally called anaphylatoxins because the biological effects they exert on cells, including smooth muscle contraction and the rapid degranulation of basophils and mast cells, resemble extreme allergic reactions. (These biological effects were discussed in detail in Box 19-1.) The functional

responses caused by the anaphylatoxins are mediated by their binding to specific and separate cell surface receptors that are members of the G protein-coupled receptor family. Receptors for C3a, C4a, and C5a occur on smooth muscle cells, lymphocytes, mast cells, basophils, and other granulocytes. The C5a receptor (C5aR; CD88) present on myeloid cells has been examined in some depth and has been found to be homologous to several receptors that transduce chemotactic signals. Studies of knockout mice deficient for C5aR showed that this receptor is critical for the clearance of microbes from lung mucosal surfaces but not from the peritoneum. The phenotype of these mice resembles that of humans with cystic fibrosis, in that neither can clear _Pseudomonas aeruginosa_ infections despite an intense neutrophil-mediated inflammatory response. Both mice and humans in this situation eventually die of pneumonia.

Recent work has also identified receptors for C3a and C5a on various types of CNS cells, including neurons. Mice whose brains are injected with either C3a or C5a alter their drinking and eating behaviors, implying that these molecules can directly or indirectly affect neuronal functions. C5a was also able to induce the release of a neurotransmitter in _in vitro_ studies of crude CNS tissue preparations. Based on a possible analogy with the outcome of engagement of the myeloid C5aR, it is speculated that stimulation of neuronal C5aR might induce chemotaxis of neuroblasts and/or production of adhesion molecules. Alternatively, neuronal complement receptors may act to clear anaphylatoxins that have accessed the CNS.

I. Complement Deficiencies

In this section, we examine what happens when the complement system fails. We have chosen to outline the clinical aspects of complement deficiencies here rather than in the clinical section of this book so as to have a close association between the description of the normally functioning cascade and the consequences of removing a single component. Genetic deficiencies have been defined for almost all complement components (Table 19-3). Most of these conditions are quite rare and are usually inherited in an autosomal recessive pattern. Deficiencies of complement components can also be secondary to an unrelated abnormality, such as hepatic disease. Because all complement components except Factor D are made primarily in the liver, severe liver disease can result in impaired synthesis of normal complement components and thus an acquired deficiency.

Two phenotypes are generally associated with complement deficiencies: recurrent infections and a predisposition to autoimmune disease. Interestingly, and for unknown reasons, autoimmunity is the dominant phenotype in patients lacking early classical pathway components, whereas the dominant phenotype in patients lacking components from C3 on is infection. It should be noted that because there are three pathways of activating complement, while patients with a deficiency in one particular pathway may suffer more than

Table 19-3 Complement Deficiencies

	Component	Numbers of Reported Patients	Phenotype (in Order of Dominant Symptom)
Classical pathway	C1q	<40	SLE; pyogenic infections
	C1r/s	<15	SLE; glomerulonephritis, pyogenic infections
	C4	<30	SLE; glomerulonephritis; other AID; pyogenic infections
	C2	<150	SLE; glomerulonephritis; non-pyogenic infections
All pathways	C3	<25	Severe, recurrent pyogenic infections; glomerulonephritis; SLE; vasculitis; arthritis
Alternative pathway	Factor B	<5	Infections with *Staphylococcus, Meningococcus, Pneumococcus, Neisseria*
	Factor D	<5	Infections with *Staphylococcus, Meningococcus, Pneumococcus, Neisseria*
	Properdin	<80	Infections with *Staphylococcus, Meningococcus, Pneumococcus, Neisseria*
Lectin pathway	MBL	>500	Recurrent infections; SLE; rheumatoid arthritis
Terminal components	C5	<35	Infection with *Neisseria*
	C6	<100	Infection with *Neisseria*
	C7	<100	Infection with *Neisseria*
	C8	<100	Infection with *Neisseria*
	C9	<30	Healthy; infection with *Neisseria*
RCA proteins	C1inh	>500	Hereditary angioedema
	Factor H	<20	Infection with *Neisseria*; glomerulonephritis; SLE
	Factor I	<20	Infection with *Neisseria*; SLE
	MCP	<5	Hemolysis, thrombosis
	DAF	<20	Autoimmune bowel disease; PNH
	HRF	<20	PNH
	MIRL	<20	PNH
Receptors	CR3	<30	Skin infections

SLE, Systemic lupus erythrematosus; AID, autoimmune disease; PNH, paroxysmal nocturnal hemoglobinuria.

normal individuals from serious or frequent infections, very few patients actually succumb to them.

I. DEFICIENCY OF C1, C4, OR C2

Patients genetically deficient for C1q, C1r, C1s, C4, or C2 are predisposed to the development of the autoimmune disorder *systemic lupus erythematosus* (SLE; see Ch.29). SLE is characterized by the accumulation of immune complexes in the tissues and the presence of antibodies directed against nuclear components such as double-stranded DNA. Over 90% of C1-deficient patients, 75% of C4-deficient patients, and 20% of C2-deficient patients exhibit signs of SLE. In humans, the presence of anti-C1q or anti-C1s autoantibodies that induce a deficiency of complement components has also been associated with an increased risk of SLE. These autoantibodies apparently act to produce an acceleration in the cleavage of C4, leading to decreased serum levels of both C4 and C3. These results point to a role for the early classical complement components in preventing autoimmunity, for which several theories have been proposed (see Box 19-3). Genetic deficiency for C1q, C1r, C1s, or C4 also leaves an individual with increased vulnerability to infections, particularly those

Box 19-3. A role for complement in peripheral tolerance?

The high incidence of autoimmunity in patients with genetic deficiencies of the early classical complement components C1, C4, or C2 suggests that these molecules may play roles in maintaining peripheral tolerance. Three theories have been advanced to explain this phenomenon.

The first theory relates to the role of the classical complement pathway in processing immune complexes. In the absence of C1, C4, or C2, not enough C3b would be produced to coat immune complexes. In the absence of C3b, CR1-bearing macrophages and erythrocytes would then be unable to convey the complexes to the liver and spleen where the complexes are normally removed and degraded. As a result, large lattices would accumulate in the peripheral tissues, particularly in the glomeruli of the kidney. This abnormal deposition in the tissues might trigger an inflammatory response, which in turn could result in the release of autoantigens and the subsequent synthesis of autoantibodies. Evidence to support this theory has come from studies of knockout mice deficient for C1q, which suffer from SLE-like symptoms. C1q$^{-/-}$ mice have anti-nuclear antibodies and tend to develop glomerulonephritis characterized by the accumulation of immune complexes.

The second theory for the involvement of the complement system in peripheral tolerance postulates a role for C3d/CR2 and natural antibody in the anergization of self-reactive B cells. The reader will recall that natural antibodies are made by the unconventional CD5$^+$ B cells, are abundantly present at birth, and are broadly specific, meaning that a relatively high proportion may cross-react with self-antigens. A self-antigen recognized by a natural antibody provides a chance for the initiation of the classical complement pathway, resulting in the deposition of C3d on the antigen. The presence of C3d on an antigen enhances its capacity to initiate BCR signaling through its interaction with the BCR coreceptor CR2. If a naive conventional B cell specific for the same self-antigen then met the C3d-coated self-antigen/natural antibody complex, it would most likely be anergized by the encounter, since a strong activation signal (thanks to C3d engagement of the co-receptor) would be delivered to the B cell in the absence of T cell help. This self-reactive B cell would thus be anergized before it could find an appropriate self-reactive T cell to deliver costimulatory signals, preventing the production of autoantibodies by this cell. (In this case, the natural antibody merely facilitates the coating of the antigen with C3d; the natural antibody has no other function in the anergization process.) If, however, an individual was not able to coat antigen in C3d, or was missing the CR2 receptor, the co-receptor could not contribute to the signal received by the naive self-reactive B cell when it bound to self-antigen. As a result, insufficient signal 1 would be received, and the self-reactive cell would be neither anergized nor activated. Should later circumstances arise where T cell help became available or sufficient stimulation was delivered in another manner, the self-reactive B cell, which otherwise would have been rendered harmless by previous anergization, could become activated and attack self-tissues.

More evidence indicating a role for CR2 engagement in maintaining peripheral tolerance of anti-self B cells has come from studies of *lpr* (Fas-deficient) mice, which exhibit autoimmunity. It is speculated that, because *lpr* mice lack Fas expression, the removal of activated self-reactive T cells by AICD is impaired, leading to enhanced antibody production by T-dependent autoreactive B cells. If an *lpr* mouse is crossed to a CR2$^{-/-}$ or C4$^{-/-}$ mouse (C4 being required for the processing of C3 and thus the generation of C3d), the symptoms of autoimmunity are exacerbated, suggesting that self-reactive B cells in these mutant animals are neither anergized nor deleted as they normally would be.

The third theory (sometimes called the "waste disposal hypothesis") for the role of at least one complement component in maintaining peripheral tolerance arises from the observation that apoptotic cells accumulate in the affected glomeruli of C1q$^{-/-}$ mice. It has been shown *in vitro* that, as well as binding to Ig molecules, C1q can bind to apoptotic (but not normal) keratinocytes in the absence of antibody, suggesting that C1q may participate in the clearance of apoptotic self-cells ("waste material") by facilitating receptor-mediated uptake by phagocytes. In one study of SLE mechanisms, SLE autoantigens were shown to accumulate on the surfaces of apoptotic keratinocytes. A failure to clear such apoptotic cells such that they accumulated in the tissues could result in the eventual release of autoantigens (such as double-stranded DNA) and an autoimmune reaction. Experiments investigating skin rashes associated with C1q deficiency in humans have indicated that C1q may indeed act to opsonize autoantigens present on or released by apoptotic or damaged cells. In this context, the binding of C1q by calreticulin and p33 may be relevant for the clearance of C1q–autoantigen complexes. More on the mechanisms of autoimmunity can be found in Chapter 29.

---

that are pyogenic in nature (induce high fever). Individuals lacking C2 also experience more frequent infections but, for unknown reasons, these tend not to be pyogenic.

II. DEFICIENCY OF C3

As might be expected from the central position of C3 in the complement system, a deficiency of C3 is a serious matter. Although C3-deficient patients experience a lower incidence of SLE and related autoimmune disorders than patients lacking C1, C4, or C2, they are subject to severe, recurrent, and potentially life-threatening pyogenic infections caused by a wide variety of encapsulated bacteria. C3 plays a crucial role in host defense against these organisms, facilitating their opsonization, phagocytosis, and lysis. Interestingly, unlike the human situation, autoimmunity is not observed in C3-deficient (or C4-deficient) mice. Instead, these animals show defects in humoral Td responses, probably due to insufficient production of the BCR coreceptor ligand C3d. Why should there be manifestations of autoimmunity in some C3-deficient humans but not in C3-deficient mice? The answer may come down to a matter of degree. It may be that, in mice, the overall decrease in antibody production (due to a lack of CR2 engagement by C3d) reduces the size and volume of immune complexes accumulating, to the extent that deposition in the tissues is limited and does not provoke an inflammatory response of enough magnitude to facilitate the onset of autoimmunity. Of course, one would then have to assume that the deficit in C3d production does not have a significant effect on the solubilization of those complexes that do form. This question awaits definitive resolution.

Low serum levels of C3 can also result from excessive C3 convertase activity induced by an autoantibody. In a human syndrome characterized by glomerulonephritis and partial *lipodystrophy* (a disturbance of fat metabolism), researchers identified what they first called "nephritic factors" that appeared to induce damage to the kidneys. In fact, these nephritic factors turned out to be autoantibodies to the alternative C3 convertase. The autoantibody bound to C3bBb and prevented its dissociation, making the convertase constitutively active. The convertase then cleaved C3 in an unregulated manner, presumably resulting in a level of serum C3 inadequate to sustain the cascade and prevent deposition of the immune complexes in the kidney.

III. DEFICIENCY OF ALTERNATIVE PATHWAY COMPONENTS

In contrast to patients lacking early elements of the classical path, SLE has rarely been found in patients lacking components exclusive to the alternative path of complement activation. Rather, individuals lacking properdin, Factor B, or Factor D of the alternative pathway suffer from frequent infections with aggressive pathogens such as species of *Staphylococcus*, *Meningococcus*, *Pneumococcus*, and *Neisseria*.

IV. DEFICIENCY OF LECTIN PATHWAY COMPONENTS

MBL, the key initiator of the lectin pathway, is a factor in disease susceptibility in situations involving less than fully competent immune systems. Certain variant MBL alleles confer low levels of expression of this collectin in the serum, and partial deficiencies are not uncommon. In immunocompromised individuals or newborns (whose immune systems are still immature), a decreased level of MBL may be enough to tip the balance toward recurrent infections. As well, low serum MBL has been implicated as a risk factor for the development and progression of SLE, HIV infection, and *rheumatoid arthritis* (RA). In the latter case, it is speculated that low levels of MBL in patients susceptible to RA allow the establishment of a viral trigger for RA onset (see Ch.29). A complete absence of serum MBL has been identified in patients who are homo- or heterozygous for point mutations in the first exon of the MBL subunit chain gene. The mutation blocks the formation of the MBL trimers and thus the assembly of the large functional multimer. MBL-mediated complement activation is therefore absent, leading to an increased risk of recurrent infections.

Why would variant alleles of MBL persist in a population if they lead to increased susceptibility to infection? One theory is that, just as heterozygosity for the sickle cell anemia gene

protects against malaria (see Ch.22), heterozygosity for variant alleles of MBL may confer resistance to mycobacterial infections. In one study, at least one variant MBL allele was present at a higher frequency in healthy individuals than in tuberculosis patients.

V. DEFICIENCY OF TERMINAL COMPONENTS (C5–C9)

In contrast to the range of bacteria infecting C3-deficient patients, those patients lacking C5, C6, C7, or C8 appear to suffer predominantly from infection with only one kind of bacterium, the intracellular pathogen *Neisseria*. Such a phenotype emphasizes the importance of MAC formation as a means of destruction for this particular microbe. However, it also reminds us of the redundancy of innate defense, in that non-MAC-dependent methods of microbe disposal, such as opsonization and immune complex formation, and indeed non-complement-mediated mechanisms, remain to protect the host. While some C9-deficient patients have the same phenotypes as C8-deficient individuals, many patients lacking only C9 appear to be perfectly healthy. This observation indicates that MACs can sometimes be assembled sufficiently well in the absence of C9 to allow lysis to proceed and control even neisserial infections.

With respect to autoimmunity, patients lacking the terminal complement components run the same risk as the general population, since the incidence of autoimmune disorders in both groups is generally <5%.

VI. DEFICIENCY OF REGULATORY PROTEINS

A deficiency of C1 inhibitor (C1inh) results in a clinical syndrome called *hereditary angioedema*, in which the face, extremities, and sometimes the mucosae of the respiratory and/or gastrointestinal tracts exhibit swelling. The attacks are episodic and usually harmless, but can cause severe pain in the case of abdominal wall involvement, or (rarely) death if swelling in the upper respiratory tract is severe. Unlike most complement deficiencies, C1inh deficiency is autosomal dominant, meaning that only one allele needs to be disrupted before the mutated phenotype becomes apparent. Without normal levels of C1inh, the cleavage of C1 is uncontrolled, which in turn leads to continuous cleavage of C4 and thus C2. It is the abnormal release of a peptide called *C2 kinin* from C2 that is thought to induce edema. C1inh also inhibits the serine esterases of the wound healing system, such that loss of C1inh results in increased production of bradykinin, another inducer of edema.

Deficiency of either Factor I or H results in uncontrolled activity of the C3 convertases and thus total consumption of C3. The clinical picture thus sometimes resembles that of C3 deficiency. Without the ongoing production of C3b and C3d, inadequate coating of pathogens and immune complexes occurs and the frequency of infections and SLE in these patients is increased. About 50% of patients suffer from neisserial infections, while 15–30% experience autoimmune reactions.

A disorder called *paroxysmal nocturnal hemoglobinuria* (PNH) results whenever DAF, HRF, and MIRL fail to function properly. PNH is characterized by abnormally frequent lysis of erythrocytes. The reader will recall that DAF, HRF, and MIRL are regulatory proteins that depend on a GPI anchoring mechanism, rather than on a membrane-spanning domain, to attach them to the host cell membrane. When the GPI linkage is disrupted, these proteins cannot attach to the host cell surface and control complement activation. In PNH patients, there is a mutation in the gene encoding an early enzyme required for GPI synthesis in hematopoietic progenitors. The GPI anchor is thus absent in the bone marrow precursors that give rise to erythrocytes and leukocytes. Erythrocytes are particularly vulnerable to lysis following MAC assembly, accounting for the observed increase in hemolysis. Studies discriminating between the roles of the various GPI-anchored regulatory proteins have shown that a deficiency of DAF alone is not sufficient to induce PNH. Rather, it is the loss of control of MAC assembly due to a deficiency of MIRL that is the most important factor driving PNH pathogenesis.

VII. DEFICIENCY OF COMPLEMENT RECEPTORS

Deficiencies in complement receptors are extremely rare, and complete deficiency of CR1 or CR2 in humans has yet to be reported. A group of patients lacking CR3 has been described in which infections, particularly of the skin, are increased in frequency. Cells of these individuals have a decreased capacity to bind C3 products, and so exhibit aspects of C3 deficiency such as compromised opsonized phagocytosis. Because CR3 is an integrin, extravasation of phagocytes into the tissues is also impaired, contributing to the increased susceptibility of these individuals to infection by a broad range of organisms.

J. New Roles for Complement?

Work in the late 1990s and early 2000s has revealed possible new roles for complement. As well as being critical for anti-pathogen defense and B cell activation, complement appears to play a part in several developmental processes, including skeletal development, early hematopoiesis, vasculogenesis, limb and organ regeneration, and reproduction. For example, bone marrow stromal cells and osteoblasts have been found to secrete C3, and C3 synergizes with M-CSF in promoting osteoclast differentiation. C3, C5, C9, Factor B, and properdin have all been found within developing cartilage. A molecule strikingly similar to C1q acts as a growth factor promoting the *in vitro* growth and differentiation of mouse chondrocytes (which synthesize cartilage).

With respect to hematopoiesis and vasculogenesis, high levels of expression of the murine homologue of C1qRp have been found in vascular endothelial cells and HSCs. How C1q (or molecules like it) participates in these processes remains undefined. Both C3aR and C5aR are expressed by human HSCs, and stimulation of C3aR promotes the chemotaxis of HSCs. Intriguingly, C3 and C5 are secreted by bone marrow stromal cells, such that complement components may act to promote HSC homing to specific bone marrow niches. *In vitro*, C3a can stimulate the growth of megakaryocyte and erythroid lineages.

In lower vertebrates such as amphibians, limb regeneration is not an uncommon occurrence. During this complex process, epithelial and mesodermal cells undergo de-differentiation to form a cell layer called the *blastema*. Cells of the blastema receive signals from the surrounding microenvironment that cause them to re-differentiate into the cells required to reconstitute the tissues of the missing limb. The link with complement arises from the observation that cells of the blastema secrete C3. Interestingly, the presence of C3 is specifically linked to limb regeneration, since this molecule is not present when embryonic limbs develop "from scratch." Indeed, treatment with cobra venom factor (which depletes C3) delays limb regeneration in experimental amphibians. Exactly how C3 promotes limb regeneration is not clear, but roles in intercellular adhesion or cellular migration have been proposed.

In mammals, the liver is one of the few adult organs capable of significant regeneration. TNF and IL-6 are thought to be the key cytokines priming hepatocytes to proliferate during regeneration. It seems that C5 may also play a role, since mice lacking expression of this complement component are unable to regenerate their livers after injury. In a still murky process, C5 gives rise to C5a, which binds to C5aR and triggers downstream intracellular signaling in macrophages present in the liver (Kupffer cells). These macrophages may then secrete cytokines and growth factors that stimulate hepatocyte proliferation. In addition, C5a stimulates Kupffer cells to release prostaglandins. The prostaglandins promote glucose synthesis by hepatocytes, pushing these cells into the cell cycle.

We have already mentioned the role of RCAs in fetal tolerance (see Ch.16), in that expression of DAF and MCP by fetal cells protects them from maternal complement activation. However, the complement proteins themselves appear to play a much earlier role in reproduction, facilitating the fusion of oocyte and sperm. MCP molecules expressed on the sperm head not only protect the sperm from female complement attack but also may bind to C3b and iC3b present on the plasma membrane of the oocyte. The two gametes are thus held in the close proximity required for fertilization.

This brings us to the end of our discussion of the complement system. How microbes use strategies to defeat complement-mediated immune defense is discussed in Chapter 22, 'Immunity to Pathogens,' while the difficulties that the complement system poses for transplantation are described in Chapter 27, 'Transplantation.' We move now to Chapter 20 and an exploration of mucosal and cutaneous immunity, the front line defenses of the most common ports of pathogen entry.

SUMMARY

Complement is a system of over 30 glycoproteins that are activated in an enzymatic cascade. Various products of the cascade have anti-pathogen and pro-inflammatory effects, and complement activation contributes to both the innate and adaptive immune responses. The complement system can be activated in three ways: the lectin pathway, the alternative pathway, and the classical pathway. The lectin pathway is activated when a carbohydrate present on a pathogen surface binds to the soluble host lectin MBL. The alternative pathway is activated when C3b binds to a pathogen surface protein. The classical pathway is activated when C1q binds to antibody that has bound to a pathogen surface. A common point of convergence for all three pathways is the formation of C3 convertase and subsequent assembly of the membrane attack complex (MAC) in the membrane of the pathogen or infected cell. MAC-induced lysis of these targets then follows. In addition to MAC-mediated lysis, complement activation promotes the clearance of pathogens and their products by means of the following effector functions: opsonization, immune complex clearance, enhancement of antigen presentation, enhancement of B cell activation and memory, and generation of chemoattractant and immunoregulatory peptides. A key feature of the complement system is its potential for tremendous amplification of the anti-pathogen response due to the formation of C3b that drives the alternative pathway in a feedback loop. Complement activation is kept under control both by the expression of the component enzymes in zymogen form and by the production of numerous RCA proteins that inactivate various complement components. Additional new roles for complement components continue to be identified, including in developmental processes, tissue regeneration, and reproduction. Complement may also contribute to the establishment of peripheral tolerance, since patients deficient for C1, C4, or C2 show signs of autoimmunity.

Selected Reading List

Arlaud G., Gaboriaud C., Thielens N., Budayova-Spano M., Rossi V. *et al.* (2002) Structural biology of the C1 complex of complement unveils the mechanism of its activation and proteolytic activity. *Molecular Immunology* **39**, 383–394.

Bellamy R. and Hill A. V. (1998) Genetic susceptibility to mycobacteria and other infectious pathogens in humans. *Current Opinion in Immunology* **10**, 483–487.

Bohana-Kashtan O., Ziporen L., Donin N., Kraus S. and Fishelson Z. (2004) Cell signals transduced by complement. *Molecular Immunology* **41**, 583–597.

Botto M. (1998) C1q knock-out mice for the study of complement deficiency in autoimmune disease. *Experimental Clinical Immunogenetics* **15**, 231–234.

Caliezi C. (2000) C1-esterase inhibitor: an anti-inflammatory agent and its potential use in the treatment of diseases other than hereditary angioedema. *Pharmacological Reviews* **52**, 91–112.

Carroll M. (1998) The role of complement and complement receptors in induction and regulation of immunity. *Annual Reviews of Immunology* **16**, 545–568.

Carroll M. (2004) The complement system in the regulation of adaptive immunity. *Nature Immunology* **5**, 981–986.

Carroll M. C. and Prodeus A. P. (1998) Linkages of innate and adaptive immunity. *Current Opinion in Immunology* **10**, 36–40.

Collard C., Lekowski R., Jordon J., Agah A. and Stahl G. (1999) Complement activation following oxidative stress. *Molecular Immunology* **36**, 941–948.

Davis A. E. 3rd, Cai S. and Liu D. (2004) The biological role of the C1 inhibitor in the regulation of vascular permeability and modulation of inflammation. *Advances in Immunology* **82**, 331–363.

Eisen D. and Minchinton R. (2003) Impact of mannose-binding lectin on susceptibility to infectious diseases. *Clinical Infectious Diseases* **37**, 1496–1505.

Fearon D. T. (1998) The complement system and adaptive immunity. *Seminars in Immunology* **10**, 355–361.

Fiane A., Mollnes T., Videm V., Hovig T. Fontecilla-Camps J. *et al.* (1999) Compstatin, a peptide inhibitor of C3, prolongs survival of ex vivo perfused pig xenografts. *Xenotransplantation* **6**, 52–65.

Gaboriaud C., Thielens N., Gregory L., Rossi V., Fontecilla-Camps J. *et al.* (2004) Structure and activation of the C1 complex of complement: unraveling the puzzle. *Trends in Immunology* **25**, 368–373.

Gadjeva M., Thiel S. and Jensenius J. C. (2001) The mannan-binding-lectin pathway of the innate immune response. *Current Opinion in Immunology* **13**, 74–78.

Griselli M., Herbert J., Hutchinson W., Taylor K., Sohail M. *et al.* (1999) C-reactive protein and complement are important mediators of tissue damage in acute myocardial infarction. *Journal of Experimental Medicine* **190**, 1733–1739.

Kawasaki T. (1999) Structure and biology of mannan-binding protein, MBP, an important component of innate immunity. *Biochimica et Biophysica Acta* **1473**, 186–195.

Lindahl G., Sjobring U. and Johnsson E. (2000) Human complement regulators: a major target for pathogenic microorganisms. *Current Opinion in Immunology* **12**, 44–51.

Mastellos K. and Lambris J. (2002) Complement: more than a "guard" against invading pathogens? *Trends in Immunology* **23**, 485–491.

Matsushita M., Endo Y., Nonaka M. and Fujita T. (1998) Complement-related serine proteases in tunicates and vertebrates. *Current Opinion in Immunology* **10**, 29–35.

Meri S. and Jarva H. (1998) Complement regulation. *Vox Sanguinis* **74**, 291–302.

Miletic V. and Frank M. (1995) Complement–immunoglobulin interactions. *Current Opinion in Immunology* **7**, 41–47.

Mold C., Gewurz H. and Clos T. W. D. (1999) Regulation of complement activation by C-reactive protein. *Immunopharmacology* **42**, 23–30.

Morgan B. P. (1999) Regulation of the complement membrane attack pathway. *Critical Reviews in Immunology* **19**, 173–198.

Morgan B. P. and Harris C. L. (2003) Complement therapeutics: history and current progress. *Molecular Immunology* **40**, 159–170.

Nataf S., Stahel P. F., Davoust N. and Barnum S. R. (1999) Complement anaphylatoxin receptors on neuron: new tricks for old receptors? *Trends in Neuroscience* **22**, 397–402.

Nicholson-Weller A. and Klickstein L. (1999) C1q-binding proteins and C1q receptors. *Current Opinion in Immunology* **11**, 42–46.

Nilsson B., Larsson R., Hong J., Elgue G., Nilsson E. *et al.* (1998) Compstatin inhibits complement and cellular activation in whole blood in two models of extracorporeal circulation. *Blood* **92**, 1661–1667.

Pepys M. (1976) Role of complement in the induction of immunological responses. *Transplantation Reviews* **32**, 93–120.

Reid K. B. M., Colomb M. G. and Loos M. (1998) Complement component C1 and the collectins: parallels between routes of acquired and innate immunity. *Immunology Today* **19**, 56–59.

Rodriguez de Cordoba S., Esparza-Gordillo J., Goicoechea de Jorge E., Lopez-Trascasa M., Sanchez-Corral P. *et al.* (2004) The human factor H: functional roles, genetic variations and disease associations. *Molecular Immunology* **41**, 355–367.

Sahu A., Morikis D. and Lambris D. (2003) Compstatin, a peptide inhibitor of complement, exhibits species-specific binding to complement component C3. *Molecular Immunology* **39**, 557–566.

Sahu A., Sunyer J. O., Moore W. T., Sarrias M. R., Soulika A. M. *et al.* (1998) Structure, functions and evolution of the third complement component and viral molecular mimicry. *Immunologic Research* **17**, 109–121.

Smith L. C., Kaoru A. and Nonaka M. (1999) Complement systems in invertebrates. The ancient alternative and lectin pathways. *Immunopharmacology* **42**, 107–120.

Sullivan K. E. (1998) Complement deficiency and autoimmunity. *Current Opinion in Pediatrics* **10**, 600–606.

Tedesco F., Fischetti F., Pausa M., Dobrina A., Sim R. *et al.* (1999) Complement-endothelial cell interactions: pathophysiological implication. *Molecular Immunology* **36**, 261–268.

Thellin O. and Heinen E. (2003) Pregnancy and the immune system: between tolerance and rejection. *Toxicology* **185**, 179–184.

Thompson D., Pepys M. and Wood S. (1999) The physiological structure of human C-reactive protein and its complex with phosphocholine. *Structure Folding Design* **7**, 169–177.

Turner M. (1998) Mannose-binding lectin in health and disease. *Immunobiology* **199**, 327–339.

Vasta G. R., Quesenberry M., Ahrmed H. and O'Leary N. (1999) C-type lectins and galectins mediate innate and adaptive immune functions: their roles in the complement activation pathway. *Developmental and Comparative Immunology* **23**, 401–420.

Volanakis J. (2001) Human C-reactive protein: expression, structure and function. *Molecular Immunology* **38**, 189–197.

Walport M. (2001) Complement. *New England Journal of Medicine* **344**, 1058–1066.

Ward P. A. (2004) The dark side of C5a in sepsis. *Nature Reviews Immunology* **4**, 133–142.

Mucosal and Cutaneous Immunity

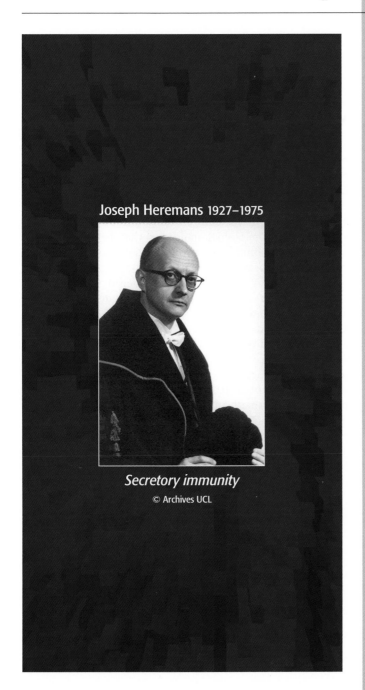

Joseph Heremans 1927–1975

Secretory immunity

© Archives UCL

CHAPTER 20

A. MUCOSAL IMMUNITY

B. CUTANEOUS IMMUNITY

"The problem in defense is how far you can go without destroying from within what you are trying to defend from without"
—Dwight D. Eisenhower

Most human pathogens gain access to the body through attacks on its mucosal surfaces or by penetrating the outer layers of the skin. The integrity of our mucosae and skin is thus the essential "thin red line" separating the vulnerable inner workings of our bodies from the majority of external assaults. In Chapter 3 we introduced the concepts of MALT (mucosa-associated lymphoid tissue) and SALT (skin-associated lymphoid tissue). These arms of the immune system are responsible for what researchers working in this area call "mucosal" and "cutaneous" adaptive immune responses, respectively. The MALT and SALT are made up of diffuse collections of APCs and lymphocytes that can respond to injurious entities that make contact with the mucosae or skin. These responses can be mounted without necessarily involving the draining lymph node. However, a systemic immune response with the potential to attack antigen throughout the body can be induced when antigen-bearing APCs of the MALT or SALT migrate to the local lymph node. Naive T and B cells in the node are activated and generate effectors that fan out into the blood and lymphatics and eliminate antigen that has penetrated beyond the MALT or SALT. We will now discuss in depth how leukocytes are integrated with other elements of the mucosae and skin to provide immune defense at the body portals.

A. Mucosal Immunity

"The mucosae" is the term given to the external surfaces of the epithelial cells that line body passages such as the gut, the respiratory tract, and the urogenital tract. The mucosae get their name from their capacity to produce mucus, a very viscous solution of polysaccharides in water that covers the mucosal surface. The mucus contains various secretory antibodies and anti-microbial molecules that help protect the mucosae from pathogen invasion. In adult humans, the area of all mucosal surfaces exceeds 400 m², a huge expanse that must constantly be defended against pathogen assaults.

MALT is the collective term used to describe all the mucosal lymphoid tissues in the body. Making up MALT are the subsystems of diffuse lymphoid elements associated with each of the body tracts. These subsystems, which can differ in structure and cellular composition, are given their own names by location (Table 20-1). For example, *the gut-associated lymphoid tissue* (GALT) defends the mucosal surfaces of the small and large intestines with a different range and organization of cell types than is used by the *bronchi-associated lymphoid tissue* (BALT) to defend the lungs. Lymphoid tissue associated with the tonsils in the nasopharynx (NALT; *nasopharynx-associated lymphoid*

Table 20-1 The "ALTs"

System Name	Subsystem	Definition	Tissue Defended
SALT		Skin-associated lymphoid tissue	Skin covering body surface
MALT		Mucosa-associated lymphoid tissue	Mucosae of body tracts
	BALT	Bronchi-associated lymphoid tissue	Mucosae of bronchi in lungs
	NALT	Nasopharynx-associated lymphoid tissue	Mucosa of nasopharynx
	GALT	Gut-associated lymphoid tissue	Mucosae of small and large intestines

tissue) is important for upper airway defense, particularly in mice and humans.

Scientists working in the field of mucosal immunity speak of *inductive sites* and *effector sites* (Fig. 20-1). An *inductive* site is a local area where antigen is encountered and a primary adaptive response initiates, assuming that the innate mucosal defenses have been unsuccessful in eliminating the invader. Inductive sites in the GALT include the Peyer's patches (PPs), appendix, and the diffuse collections of lymphocytes and APCs scattered just under the gut epithelium. Similarly, the main inductive site of the NALT is the collection of tonsils in the nasopharynx. Lymphocytes activated in a mucosal inductive site differentiate into effector cells that exert their effector actions (such as antibody production) at mucosal locations called *effector sites*. Such effector sites include the diffuse lymphoid tissues underlying often distant mucosal surfaces of the body, and exocrine glands, such as the salivary and lacrimal glands, which produce protective external secretions (into which antibody can be introduced). Defense at multiple and widely separated effector sites may occur in response to lymphocyte activation in one inductive site. However, the immune responses induced are not of uniform strength, being strongest at the mucosal effector sites closest to the inductive site, or in tissues sharing lymph drainage. In other words, the entire mucosal system may be alerted by an attack at one location, but the resources needed to combat the threat are most concentrated near the site of attack.

How are the inductive and effector sites linked? Specialized cells in inductive sites take up antigen and deliver it to APCs and lymphocytes resident in the underlying sub-epithelial layer. Antigen-activated B and T cells then emigrate from the inductive site into the lymphatic system (without necessarily being further stimulated in the draining lymph node) and home

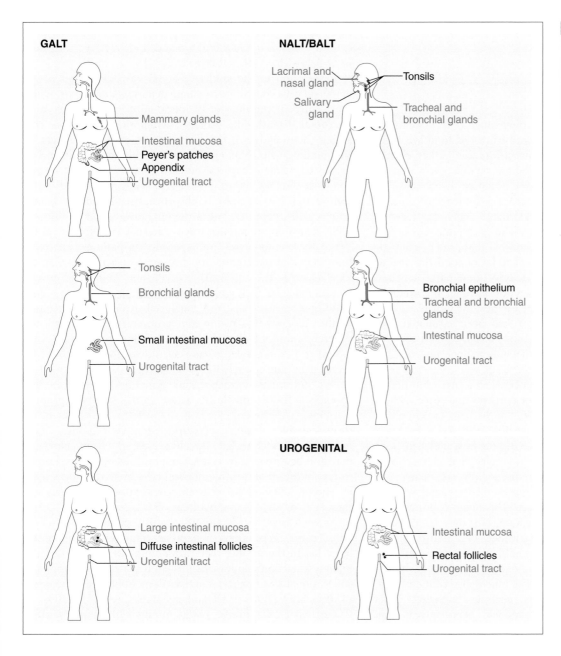

GALT

Mammary glands
Intestinal mucosa
Peyer's patches
Appendix
Urogenital tract

Tonsils
Bronchial glands
Small intestinal mucosa
Urogenital tract

Large intestinal mucosa
Diffuse intestinal follicles
Urogenital tract

NALT/BALT

Lacrimal and nasal gland
Salivary gland
Tonsils
Tracheal and bronchial glands

Bronchial epithelium
Tracheal and bronchial glands
Intestinal mucosa
Urogenital tract

UROGENITAL

Intestinal mucosa
Rectal follicles
Urogenital tract

Figure 20-1

Mucosal Inductive and Effector Sites
An antigen accessing an inductive site (black) provokes detectable immune responses at remote effector sites (blue).

through the blood circulation to multiple mucosal effector sites. It is in the effector sites that the activated lymphocytes complete their differentiation into antigen-specific B, CTL, and Th effector cells. These cells then carry out effector actions to eliminate the threat. To a large degree, defense at the mucosae depends on the copious amounts of secretory IgM and IgA antibodies produced by the high numbers of plasma cells that congregate in tissues underlying the epithelium, but other effector mechanisms also make a contribution. These mechanisms are discussed in depth later in this chapter.

We will now examine the components and organization of the MALT in humans and mice. We focus first on the elements of the major inductive subsystems, the GALT and the BALT/NALT. We then discuss mucosal immune responses in both GALT and BALT/NALT and the concept of a "common mucosal immune system." Immune responses in other mucosal tissues such as the urogenital tract and the middle ear cavity are not as well-studied and are discussed briefly in a separate section.

I. GALT

i) Components of GALT

ia) Gut epithelium. Our gastrointestinal (GI) systems are designed to both absorb nutrients from the food we ingest and repel pathogens and noxious substances. A key element for both of these functions is the gut epithelium, a mucosal surface composed of a single layer of gut epithelial cells. The gut epithelium is not flat, but rather is folded into repeated villus ("hill") and crypt ("cave") structures such that its entire surface area approaches 300 m^2. The gut epithelium is fully exposed to the countless pathogens that enter from the outside world in the estimated 2500 kg of food the average person ingests over a lifetime. Many of these pathogens have developed sophisticated strategies for piercing the epithelial layer, and so the gut mucosae are heavily populated with immune system cells. Indeed, it has been estimated that, overall, the intestinal lining contains more lymphoid cells and produces more antibody than the spleen, bone marrow, and lymph nodes combined.

What does a pathogen encounter when it arrives in the gut lumen? First, it has to compete for nutrients and space with the ~10^{13} *commensal* (from the Latin: "at the same table") organisms that live in the intestinal tract (Fig. 20-2 and Table 20-2). Although these microbes generally do the host no harm, some of them secrete anti-microbial molecules that can seriously impede a pathogen. Pathogens successfully competing with commensal organisms for space must then contend with the mucus coating the mucosal surface. The mucus is produced by *goblet cells* within the gut epithelium. Not only is the mucus sticky enough to physically trap pathogens but it also contains high concentrations of secretory antibodies. These antibodies prevent a wide variety of pathogens from establishing a foothold on the epithelial layer. The mucus also contains molecules with anti-bacterial effects, such as lysozyme, which breaks down the cell wall component peptidoglycan, and lactoferrin, which sequesters iron needed for bacterial growth.

Should a pathogen succeed in penetrating the mucus and approaching the epithelial layer, it still has to negotiate the innate defenses presented by the gut epithelial cells before it can attach, cross the epithelium into the underlying gut tissues, and establish an infection. The apical (lumen-facing) surfaces of gut epithelial cells are covered with dense microvilli collectively known as the *brush border*. The brush border is coated with the *glycocalyx*, a thick (400–500 nm) glue-like layer of mucin-related molecules anchored in the apical membrane. The negative charge of the glycocalyx repels many pathogens. The glycocalyx also has embedded within it several types of hydrolytic enzymes that degrade microbes or macromolecules attempting to make contact with the epithelial cells. Furthermore, most gut epithelial cells are closely joined at their apical poles by structures called *tight junctions*. These junctions virtually seal the epithelium such that the passive diffusion of pathogens, macromolecules, and even peptides between cells is inhibited. It is now clear, however, that the tight junctions are not static and can open in response to cytokine signals (including IFNγ and IL-4) received from cells in or below the epithelial layer.

In addition to the goblet cells, several other types of gut epithelial cells contribute to innate defense. *Paneth cells*, which are located at the bottom of intestinal crypts, produce anti-microbial defensin-like proteins, while *enteroendocrine cells* secrete neuroendocrine molecules with paracrine effects on surrounding cells. Some of these molecules are constitutively expressed, while others are induced by infection. *Enterocytes* comprise the majority of gut epithelial cells, the primary function of which is to actively absorb nutrients such as sugars, amino acids, peptides, and proteins. However, in addition to their transport function, some enterocytes express TLRs. When these enterocytes are exposed to a pathogen or toxin, they secrete pro-inflammatory chemokines (such as IL-8), cytokines (such as IL-1, IL-6, IL-7, IL-11, and TNF), and growth factors (such as SCF and GM-CSF). These molecules recruit additional neutrophils and mast cells from the circulation to the sub-epithelial region and promote the activation and differentiation of leukocytes present within or near the epithelial layer. For example, it has been shown that IL-7 and SCF secreted by gut epithelial cells cooperate to activate a subset of γδ intestinal intraepithelial lymphocytes (iIELs) that co-express IL-7R and IL-2R. These γδ iIELs in turn produce cytokines and chemokines that support the activation of αβ iIELs capable of more powerful adaptive responses. Low numbers of intraepithelial NK cells present in the gut epithelium may also be activated by enterocyte-produced cytokines, leading to the killing of unhealthy mucosal cells by natural cytotoxicity and/or ADCC. NKT cells are found at only extremely low frequency in the iIEL population.

As we learned in Chapter 18, the iIEL population also contains Tc and Th cells directly interspersed among the gut epithelial cells. In the mouse, there are about 5–10 × 10^7 iIELs in the intestinal epithelium, which is equivalent to the number of conventional T lymphocytes in the spleen. In humans, a majority of iIELs are CD8αβ or CD8αα αβ T cells, while 10–20% are CD8αα γδ T cells. In mice, equal numbers of αβ and γδ iIELs are found in the small intestine, and many iIELs in the large

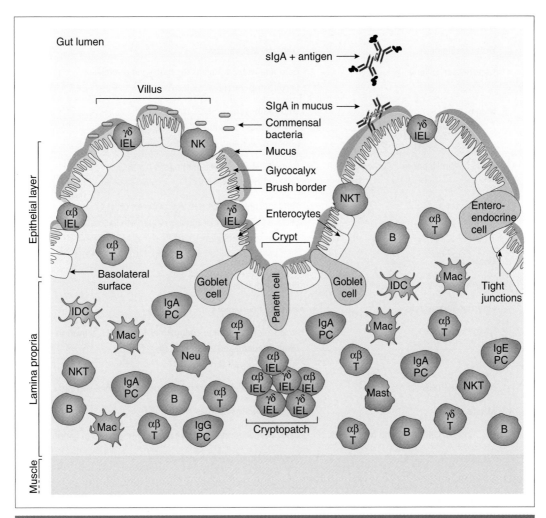

Figure 20-2

GALT Structure

The gut lumen is lined with an epithelial layer overlying the lamina propria and a muscle layer. The epithelium is folded into villi and crypts. An antigen (which could be a whole pathogen or a toxin) approaching the gut lining faces several barriers, including secreted IgA (sIgA), secretory IgA (SIgA) present in the mucus, numerous commensal bacteria, the glycocalyx, and the brush border. Interspersed among the epithelial enterocytes are various types of secretory cells, including enteroendocrine cells, Paneth cells, and mucus-producing goblet cells. Intraepithelial αβ and γδ T lymphocytes (IELs) and NK and NKT cells can also be found tucked between the enterocytes. αβ and γδ IELs may also be present in the cryptopatches. The lamina propria contains memory αβ T cells, mature B cells, immature DCs (IDC), macrophages (Mac), mast cells (Mast), and neutrophils (Neu). Plasma cells (PC) producing IgA antibody typically associate with goblet cells. Adapted from Fagarasan S. and Honjo T. (2002) Intestinal IgA synthesis: regulation of frontline defences. *Nature Reviews Immunology* **3**, 63–72.

intestine express CD4 rather than CD8. These cells are originally generated in the thymus but at least some are thought to mature extrathymically in the mesenteric lymph nodes or fetal liver, or in the intestinal cryptopatches that lie at the far ends of the crypts.

Some scientists have proposed that an important function of γδ iIELs is to monitor specific groups of about 20–100 epithelial cells for the presence of endogenous stress molecules or pathogen antigens, and to take the initial steps to defend them. For example, the repair and growth of epithelial tissue are stimulated by factors secreted by γδ Th iIELs, and γδ Tc iIELs eliminate terminally damaged epithelial cells. In addi-

tion, IFNγ secreted by activated γδ iIELs upregulates the expression of ICAM-1 on the epithelial cell apical surface and unknown adhesion molecules on the basolateral (non-luminal) surface. Neutrophils in the underlying tissue that express the appropriate counter-receptors are drawn between the epithelial cells and held to their apical surfaces in an ideal position to phagocytose invaders. Studies of γδ TCR knockout mice have shown that γδ iIELs are also important in some way for the development of IgA-secreting mucosal B cells, but the mechanism remains unclear. One possibility is that the γδ iIELs may contribute Th2 cytokines that support the humoral response.

Table 20-2 Elements of Mucosal Immunity in the Gut

Region of Gut	Defense Element	Defense Mechanism
Lumen	Commensal bacteria	Compete with pathogens for space and nutrients Secrete anti-microbial molecules
	Mucus	Traps pathogens Blocks access to epithelial cells Contains secretory Abs and anti-microbial molecules
	Epithelial brush border, glycocalyx	Negative charge repels pathogens Contains hydrolases
Gut epithelial layer	Enterocytes	Tight junctions inhibit passage between cells Surface TLRs trigger production of growth factors and pro-inflammatory cytokines Capture some antigens
	Goblet cells	Produce mucus
	Paneth cells	Produce defensins
	Enteroendocrine cells	Produce neuroendocrine mediators
	$\gamma\delta$ iIELs	Produce cytokines and chemokines for $\alpha\beta$ iIEL activation Eliminate damaged epithelial cells Support mucosal B cells
	NK and NKT cells	Destroy infected cells by cytotoxicity or cytokine secretion
	M cells	Capture and transcytose antigen
Lamina propria	$\alpha\beta$ T cells, B cells, DCs, and other APCs	Mount conventional adaptive immune responses in lymphoid follicles
	T_{reg}, Th3	Regulate lymphocyte responses

ib) Lamina propria. Between the basolateral surface of the gut epithelium and the underlying muscle layer is a loose connective tissue called the *lamina propria* (refer to Fig. 20-2). The lamina propria is home to numerous macrophages and neutrophils and lower numbers of NKT cells, mast cells, and immature DCs. Only small numbers of conventional $\gamma\delta$ T cells are present here, and almost no NK cells. However, mature $\alpha\beta$ T cells (both CD4$^+$ and CD8$^+$) and B cells are present in relative abundance. Close to 95% of these lymphocytes have a memory phenotype, and about 10–15% of lamina propria B cells have differentiated into Ig-producing plasma cells. Indeed, about 80% of all human plasma cells are located in the intestinal lamina propria. In mice, over 95% of plasma cells produce IgA, while 85–90% of human plasma cells produce IgA and 10–15% produce IgM or IgG.

Although lymphocytes may be scattered loosely in the lamina propria, they are most often organized into *intestinal follicles* (Fig. 20-3). Intestinal follicles are made up of a germinal center containing B cells and FDCs topped by a "dome" containing a collection of immature DCs, macrophages, CD4$^+$ T cells, and mature B cells. The T cells surrounding the intestinal follicle are mature $\alpha\beta$ cells, of which one-third are CD8$^+$ Tc cells and two-thirds are CD4$^+$ Th cells. As discussed in Chapter 16, small numbers of regulatory T_{reg} and Th3 cells are also present. Intestinal follicles can occur singly, as can be found

scattered along the entire length of the intestine, or in groups, as occur in the PPs and appendix. About 30,000 single follicles monitor the gut lining, while a typical PP contains 30–40 follicles. The average adult human possesses about 200 PPs. Where follicles are grouped, they are separated from one another by *interfollicular regions* that contain high concentrations of mature T cells surrounding the high endothelial venules (HEVs) discussed in Chapter 3. The reader will recall that the HEVs are the sites of extravasation of lymphocytes through which naive B and T cells pass from the blood into the secondary lymphoid tissues.

As mentioned previously, unlike B cells activated in the spleen or lymph nodes, mucosally activated B cells (such as those in the PPs) do not usually differentiate in the inductive site. Rather, the activated B cells exit the inductive site via a local lymphatic channel, pass through the draining lymph node without being further stimulated, and eventually enter the blood. The B cells then use specific homing receptors to extravasate through the endothelium at effector sites in both local and distant mucosal tissues and various exocrine glands (see later). It is not until they reach the effector sites that the activated B cells differentiate into antibody-producing plasma cells. In the case of the gut, the effector sites are generally areas of the lamina propria that do <u>not</u> contain PPs or single follicles.

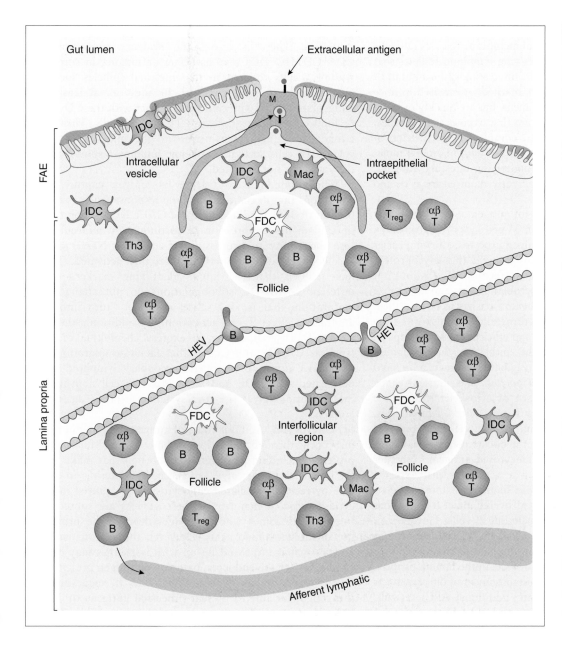

Figure 20-3

Structure of FAE and Intestinal Follicles
The follicle-associated epithelium (FAE) is the region of gut epithelium covering an intestinal follicle. M cells, which have neither a brush border nor a glycocalyx, capture antigen via binding to surface glycoproteins. The antigen is then internalized, confined in an intracellular vesicle, and transcytosed across the M cell cytoplasm. Fusion of the vesicle with the basolateral membrane of the M cell releases the antigen into the intraepithelial pocket. Immature DCs (IDC) of particular subsets and other APCs associated with the M cell take up the expelled antigen and convey it to mature T and B cells resident in the "dome" topping each follicle, in the follicle itself, and elsewhere in the lamina propria. Low numbers of regulatory T cells (Th3, T_{reg}) are present to control lymphocyte activation. Naive T and B cells access the follicles by extravasating from the local high endothelial venule (HEV) located in the interfollicular region between two follicles. Mature T cells and IDC subsets are abundant in the interfollicular region. B cells activated in the intestinal follicles enter the afferent lymphatic and then the blood circulation, differentiating into plasma cells only when they reach an effector site.

ii) Antigen Sampling in the GALT

iia) M cells and the FAE.

The innate defenses of the mucosae, both structural and cellular, are designed to limit pathogen penetration of the epithelial layer. However, nature has apparently realized that despite the best efforts of the innate response, some pathogens or their products are inevitably going to penetrate the mucus and the brush border, slip through the tight junctions linking epithelial cells, and establish infections in the underlying tissues that cannot be eliminated by phagocytes, γδ iIELs, or NK cells. Such pathogens will require the mounting of adaptive immune responses for elimination. To facilitate the collecting of pathogen antigens and the conveying of these antigens to αβ T cells and B cells in the lamina propria, small specialized sections of the gut epithelium have been modified to make antigen capture easier. These sections, which are called the *follicle-associated epithelium* (FAE), lie directly over single or aggregated follicles in the GALT (refer to Fig. 20-3).

The FAE represents only 0.01% of the intestinal mucosal surface but is specialized for a form of macromolecule transport called *transcytosis*. Transcytosis refers to a process in which a macromolecule located extracellularly near one pole of a cell is captured, either by receptor-mediated endocytosis or macropinocytosis, and is internalized and confined to an intracellular vesicle. The vesicle crosses the cell cytoplasm to the opposite pole, fuses with the plasma membrane in this location, and releases its contents into the extracellular space under the basolateral surface. The vast majority of cells in the FAE are enterocytes, but 10–20% are large odd-shaped cells called M (*membranous*) epithelial cells. It is the M cells that are experts at the transcytosis of antigens, particles, and pathogens from the lumen of the gut across the epithelial layer to its basolateral surface. Fusion of the transcytosing intracellular vesicle with the M cell basolateral membrane releases antigens directly into regions of the underlying lymphoid follicles that contain APCs, T cells, and B cells. Interestingly, the amount of FAE and numbers of M cells are markedly decreased in mice reared in a pathogen-free environment, but are quickly restored to normal levels when the mice are first exposed to microbes. This observation implies that the complete differentiation of the FAE and M cells requires interaction with microbial molecules. Signals from cells in the underlying intestinal follicles, particularly B cells, are also apparently required for FAE and M cell formation.

How is the M cell specialized for antigen delivery via transcytosis? The apical surface of an M cell lacks the glycocalyx and thick brush border present on enterocytes, allowing easier sampling of intact antigens. It is thought that glycoprotein receptors expressed on the apical surfaces of M cells (but not enterocytes) can interact with microbial ligands and internalize them, but a specific receptor of this sort has yet to be identified. In any case, once taken up, the transcellular transport of the antigen is facilitated by the large and efficient endocytic apparatus present in M cells. Within the epithelium, the basolateral surface of the M cell is deeply invaginated to create a *de facto* pocket called the *intraepithelial pocket*. This unusual shape is maintained by a thick network of cytoskeletal filaments. Subpopulations of CD4$^+$ memory T cells, naive B lymphocytes, and macrophages can migrate into the M cell intraepithelial pocket and apparently adhere to the pocket membrane, allowing the cells ready access to endocytosed antigens. Whether the antigens are released within a vesicle or just expelled "naked" from the M cell is unclear. In any case, most antigens taken up by M cells are probably delivered intact (i.e., without processing) to waiting APCs, including B cells. However, in addition to the intraepithelial pocket, the M cell has cellular processes that extend down from the base of the M cell into the dome covering the intestinal follicle in the lamina propria. This arrangement may allow the establishment of direct contacts between the M cell and the many additional APCs, T cells, and B cells residing in the dome, and most likely promotes antigen uptake by some of these cells. In this still hypothetical scenario, the M cell acts as a conduit to sweep pathogens into the most immunocompetent regions of the gut.

iib) GALT DCs. Three distinct populations of immature conventional DCs have been identified in intestinal lymphoid follicles. Immature CD4$^+$CD11b$^+$ DCs reside in the dome immediately under the FAE, and so are well positioned to acquire luminal antigens released from M cells. Immature CD8$^+$CD11b$^-$ DCs dominate in the T cell-rich interfollicular regions. A third subset of immature DCs known as DN (double negative) DCs (because they express neither CD8α nor CD11b) is uniquely associated with the GALT and is present not only in the dome and the interfollicular regions but also in the FAE. Under normal, non-threatening conditions, the default action of these mucosal DCs is to tolerize naive T cells, partially accounting for the robust nature of oral tolerance to innocuous antigens such as food. Indeed, the conventional CD8$^+$CD11b$^-$ DCs resident in the mucosae have been shown to constitutively secrete immunosuppressive cytokines

(such as TGFβ and IL-10) that dampen incipient immune responses. There is also some evidence suggesting that CD4$^+$CD11b$^-$ DCs may associate or interact in some way with T$_{reg}$ cells resident in the intestinal follicles such that mucosal tolerance is promoted. In addition, at least some antigen-stimulated Th0 cells that interact with these DCs may be induced to differentiate into Th3 or Tr1 cells. These regulatory T cells may then migrate to the spleen and tolerize splenic Th cells so that a systemic immune response against the antigen is not mounted.

When stimulated to mature by antigen capture under conditions that demand an immune response (such as inflammation or TLR binding), all three GALT DC subpopulations upregulate their expression of costimulatory molecules and lose the ability to process intact antigens. Naive mucosal T cells interacting with these DCs are then activated. (Studies have shown that the initial identifying markers of the different GALT DC subpopulations do not change upon maturation; that is, one subset does not "turn into" into another to effect T cell activation.) In addition, stimulated CD4$^+$CD11b$^+$ DCs start to express chemokine receptors (such as CXCR4 and CCR7) that facilitate migration from the GALT dome to peripheral secondary lymphoid tissues permeated with SDF and SLC. Naive T cell activation in these locations gives rise to a systemic response against the antigen.

iic) Enterocytes. A minor route of antigen sampling in the gut may occur through the actions of enterocytes. Most ordinary proteins (including pathogen components) that make it past the mucus and glycocalyx and succeed in entering an enterocyte by receptor-mediated endocytosis are captured in endocytic vesicles. A majority of these vesicles are subsequently fused to lysosomes whose enzymes degrade the protein to its constituent amino acids. However, those proteins whose vesicles manage to avoid fusion with lysosomes may become candidates for the endocytic pathway of antigen presentation. Some enterocytes do express MHC class II, suggesting that they might be able to present processed antigens to T cells. Indeed, some researchers have reported this activity *in vitro* for certain enterocyte cell lines. However, it remains unclear whether this route exists *in vivo*, and, if so, whether the enterocytes in question present antigen to iIELs or to conventional T cells in the underlying lamina propria. In addition, since most enterocytes appear to lack expression of the B7 costimulatory molecules, it is unclear how such cells could induce immune responses. Interestingly, enterocytes do express CD1d, the MHC class Ib molecule shown to be a ligand for the TCRs of NKT cells, and gp180, a 180 kDa glycoprotein that can bind to CD8 *in vitro* and induce activation of its associated Lck kinase. Some scientists have speculated that perhaps enterocytes use CD1d to present antigens to NKT cells while gp180 binds to CD8 to provide a coreceptor-like function.

iid) Lamina propria macrophages. Another minor means of GALT antigen sampling may rely on macrophages resident in the lamina propria. At least some of these cells appear to express enzymes that allow them to penetrate and reseal the tight junctions between epithelial cells. There is some evidence that these APCs can capture antigens by extending cytoplasmic projections between the epithelial cells directly into the gut lumen.

Antigen sampling leads to immune responses, but before we discuss immune responses in the MALT (the details of which are largely shared in GALT and BALT/NALT), we will briefly describe the components of BALT/NALT and antigen sampling in these tissues.

II. BALT/NALT

i) Components of BALT/NALT

The components of the mucosal system of the mammalian respiratory tract include the airway epithelium lining the nasopharynx; the trachea and lungs; the tonsils and adenoids; and the follicles and diffuse collections of lymphocytes underlying the airway epithelium (Fig. 20-4). Early studies established BALT as being vital for defense of the airway in rabbits, but more recent work has demonstrated that NALT is more important in mice and humans. Significant BALT is present in the lungs of children and adolescents, but in adults, BALT is usually detectable only after induction by pathogen attack or inhalation of a toxin. Similar findings have been recorded in rats and pigs.

How mucosal immunity works in the respiratory system is not as well understood as how it operates in the gut. Like the gut, the mucosae of the respiratory tract represent a large surface area, that of the lung alone being about 100 m². However, depending on location, respiratory tract epithelial layers differ in structure. Some regions, such as parts of the bronchioles, are covered by a single layer of epithelial cells as occurs in the gut. Other regions, such as certain types of tonsils, are covered in several epithelial layers in an arrangement known as "stratified epithelium." Cells of the airway epithelium in the nasopharynx are the first to encounter dubious substances in the 10,000 liters of air inhaled every day, including bacteria and viruses as well as damaging agents such as air pollutants and tobacco smoke. The first lines of defense are physical: the barrier function of the nose hairs, the trapping function of the mucus coating the epithelial cells, the sweeping movement of cilia on respiratory epithelial cells, and coughing in the throat. All of these mechanisms work to clear mucus and eject foreign particles (Table 20-3). The second

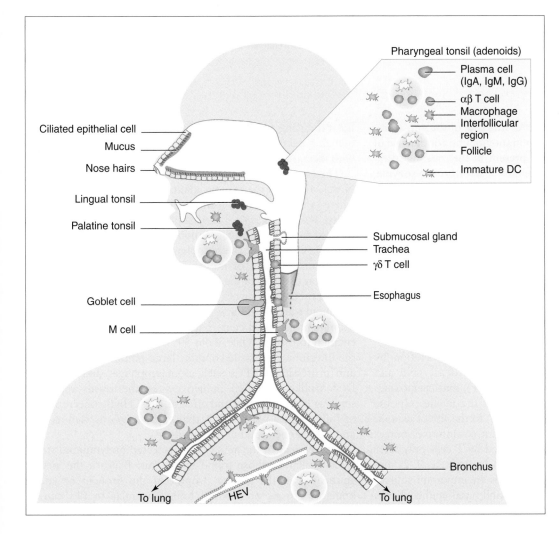

Figure 20-4

BALT/NALT Structure
The airway is defended first by the hairs in the nostrils and the IgA-containing mucus coating the respiratory epithelium. The sweeping motions by the cilia of certain respiratory epithelial cells facilitate the movement to the exterior of mucus that has trapped antigen. The tonsils, which are important inductive sites, are collections of mucosal lymphoid follicles that contain the cell types shown in the inset. Regions of FAE containing M cells and positioned over additional lymphoid follicles are scattered throughout the epithelial layer of the airway as it descends toward the bronchi and ultimately the lungs.

Labels in figure: Pharyngeal tonsil (adenoids); Plasma cell (IgA, IgM, IgG); αβ T cell; Macrophage; Interfollicular region; Follicle; Immature DC; Ciliated epithelial cell; Mucus; Nose hairs; Lingual tonsil; Palatine tonsil; Submucosal gland; Trachea; γδ T cell; Esophagus; Goblet cell; M cell; Bronchus; To lung; HEV; To lung

Table 20-3 **Elements of Mucosal Immunity in the Airway**

Region of the Airway	Defense Element	Defense Mechanism
Nasopharynx	Mucus	Traps pathogens Blocks access to epithelial cells Contains secretory Abs and anti-microbial molecules
	Epithelial brush border, glycocalyx	Negative charge repels pathogens Contains hydrolases
	Nose hairs	Ciliary movement repels pathogens
Respiratory epithelium	Single layer of epithelial cells	Secretes chemokines and pro-inflammatory cytokines Contains interspersed iIELs
	M cells and motile DCs	Capture antigen
Tonsils and adenoids	Stratified epithelial cells	Lymphoid follicles IgG, IgM, and IgA plasma cells
Bronchi	Mucus	Traps pathogens Blocks access to epithelial cells Contains secretory Abs and anti-microbial molecules Contains surfactants A and D
Respiratory epithelium	Epithelial brush border, glycocalyx	Negative charge repels pathogens Contains hydrolases
	Single layer of epithelial cells	Secretes chemokines and pro-inflammatory cytokines Contains interspersed iIELs Carries out limited antigen capture
Lamina propria	$\alpha\beta$ T cells, B cells, DCs, and other APCs	Mount conventional adaptive immune responses in lymphoid follicles

lines of defense are biochemical. As was true in the gut, secretory IgA is the most important immunoglobulin for defense of respiratory tract surfaces and is present in the mucous secretions covering the airway epithelium. Lactoferrin, lysozyme, defensins, anionic peptides, and secreted phospholipases also exert anti-microbial effects. Surfactant proteins A and D produced by the alveoli bind to bacteria and fungi in this location, enhancing removal by phagocytes. The activity of the anti-microbial proteins present in mucus can be affected by the ionic microenvironment. For example, *in vitro*, high salt concentrations can reduce the potency of many of these proteins. This effect may parallel the situation in cystic fibrosis patients, in whom altered ion trafficking leads to higher NaCl concentrations in the mucus concomitant with an increased incidence of airway infections.

Further defense of the airway relies on the secretion by epithelial cells of pro-inflammatory cytokines (IL-1, TNF, and IL-6), growth factors (G-CSF and GM-CSF), and chemokines (IL-8 and eotaxin) that summon cells of the innate response (particularly neutrophils and eosinophils) to the site of assault. Airway epithelial cells are also important constitutive sources of cytokines that are otherwise produced by T cells only after activation. Most airway epithelial cells can secrete IL-2, and bronchial and bronchiolar epithelial cells are important sources of TGFβ, IL-6, and IL-10. Bronchial and nasal epithelial cells also produce IL-5, required for the complete differentiation of

IgA-producing plasma cells. Consistent with this observation, human nasal mucosae treated with IL-5 produce nasal secretions with increased concentrations of secretory IgA.

The tonsils are the most important inductive sites in human NALT, although the salivary glands also make a contribution. There are five tonsils forming a nasopharyngeal structure called *Waldeyer's ring*. The nasopharyngeal tonsil (commonly referred to as the *adenoids*, although this is a single structure) is found in the roof of the nasopharynx, while two *lingual* tonsils lie at the back of the tongue. Two *palatine* tonsils are located at the sides of the back of the mouth. In general, a tonsil is composed of a network of reticular cells that supports typical lymphoid follicles and interfollicular tissue. The follicles contain prominent germinal centers featuring FDCs for antigen presentation to B cells. While about 50% of tonsillar cells are B cells, the interfollicular tissue contains large numbers of αβ T cells and professional APCs such as macrophages and immature DCs. Unlike the GALT inductive sites, the tonsils appear to serve as both inductive and effector sites. High concentrations of fully mature plasma cells that express IgG as well as IgM and IgA can be found in the tonsils.

Other areas of the airway are protected by lymphoid structures similar to those found in the gut. Beneath the airway epithelium in the lamina propria of the bronchi of young humans, groups of mucosal follicles similar to PPs can be found, complete with B cell-containing germinal centers flanked by HEVs. A dome containing CD4$^+$ and CD8$^+$ αβ T cells is present between the epithelium and each follicle. There are no afferent lymphatic vessels in the lungs, so that extravasation through the HEVs in the interfollicular region is the only route by which lymphocytes can access these structures. In the BALT, afferent lymphatics occur near the follicles so that activated lymphocytes can migrate to the draining bronchial lymph node. IgA-secreting plasma cells return to the airway epithelium in effector sites and can even infiltrate within it.

ii) Antigen Sampling

Some scientists have speculated that antigen uptake in the airway may generally be easier than that in the GALT, because the respiratory tract lacks the harsh degradative enzymes and low pH of the gut. Not surprisingly, antigen sampling in BALT/NALT is also influenced by the structure of the epithelial layer. In single epithelial layers of the respiratory tract, antigen sampling can be carried out by M cells, as was described previously for the gut. Indeed, the primary route of the influenza virus to the NALT is thought to be via M cells (see Ch.22). In regions of the airway covered by stratified epithelium, antigens are conveyed to inductive sites by motile DCs. In response to an inhaled antigen, DCs are rapidly recruited in high concentrations to the airway epithelium. The DCs are then able to move through the epithelium to the luminal surface to acquire the inhaled antigens and begin maturation. The activated, mature DCs bearing antigen return to the underlying diffuse lymphoid tissues in the lamina propria or migrate to the more distant draining lymph nodes. These DCs then function as the principal cells supplying antigen to be recognized by T and B cells resident in these locations. It is thought that because antigen-sampling DCs appear at the luminal surface for only brief intervals (unlike the membrane-fixed M cells), they do not generally offer a means of pathogen access.

The adenoids and lingual tonsils are also covered in stratified epithelium and rely on motile DCs for antigen sampling. The DCs move out of the tonsil through the local epithelium to acquire antigen, and then back to the interfollicular region to present the antigen to the resident T cells. In contrast, the palatine tonsils are covered in a single layer of epithelium. The surface area of the palatine tonsils is increased by the presence of up to 20 crypts, some of which appear to contain M cells.

Lastly, some airway epithelial cells express MHC class II and B7 molecules, implying that they may carry out limited antigen presentation. It has been shown that human bronchial and nasal epithelial cells constitutively expressing MHC class II can stimulate T cell proliferation *in vitro*.

III. IMMUNE RESPONSES IN THE MAJOR MALT

The information presented in this section is generalized from studies mainly of GALT but also of BALT/NALT. Less is known about immune responses in other MALT tissues, although they are presumed to share most features. Specific information about immune responses in the urogenital tract, eye, and middle ear cavity is presented later in Section A.IV.

i) Rationale for Th2 Responses in the Mucosae

The mucosae are inherently fragile structures that can easily be damaged by the products and actions of cells activated during inflammation. For example, the mounting of a vigorous Th1 response in the gut, such as a DTH reaction, with its accompanying production of TNF and IFNγ, damages the intestinal villi and thus threatens to interfere with the absorption of nutrients and passive protection mechanisms. Diseases in the human gut characterized by chronic inflammation and tissue damage have been repeatedly associated with aberrant increases in Th1 responses. For this reason, mucosal immunity depends to a large extent on secretory IgA, an antibody that coats mucosal surfaces throughout the body and can protect against pathogens without inducing an inflammatory response. Obviously, production of secretory IgA depends on isotype switching to IgA in B cells, an event that requires the Th2 cytokines IL-4 and IL-10 as well as TGFβ. Thus, a Th2 response at a mucosal surface promotes host defense compatible with mucosal structures because it induces IgA production while downplaying inflammation.

It should be noted, however, that overwhelming and acute infections do induce prominent Th1 responses in the MALT. For example, infections with the aggressive gut pathogen *Salmonella typhimurium* result in the differentiation of Th1 cells in the PPs and the production of IFNγ. *Salmonella typhimurium* is very invasive and hides within macrophages and DCs, necessitating a CTL-mediated response for elimination. These effector T cells rapidly undergo apoptosis after the infection is contained, minimizing the damage to the gut. This situation is reminiscent of oral tolerance (see below and Ch.16), in which a single high oral dose of a protein antigen provokes a Th1 response, while repeated low oral doses of the same antigen induce innocuous Th2 responses and apparent tolerance to the antigen.

ii) Induction of Th2 Responses in the Mucosae

How is the immune response biased toward the Th2 type in the MALT? Scientists investigating mucosal DC populations in the PPs of the GALT have observed intrinsic functional differences between total PP DCs (all subsets) and splenic DCs. Naive T cells primed *in vitro* in the presence of PP DCs were induced to differentiate into Th2 cells producing IL-4 and IL-10. In contrast, naive T cells cultured with splenic DCs became Th1 cells secreting IFNγ. Closer examination of the two populations of DCs suggested that PP DCs may routinely secrete a combination of cytokines (including high levels of IL-10 plus IFNγ and IL-4) that, on balance, promotes Th2 and regulatory T cell differentiation while suppressing Th1 differentiation. Even if a Th1 response is initiated in the GALT, it is usually contained by the actions of various regulatory T cell subsets present in the dome and by immunosuppressive cytokines. A parallel bias toward Th2 responses has also been observed in airway DCs, which preferentially induce Th2 differentiation unless fully activated by highly stimulatory microbial products. Immature mucosal DCs may thus make a decision as to which cytokines to secrete based on the context in which they capture antigen (Fig. 20-5). Most mucosal antigens are encountered under circumstances that do not represent

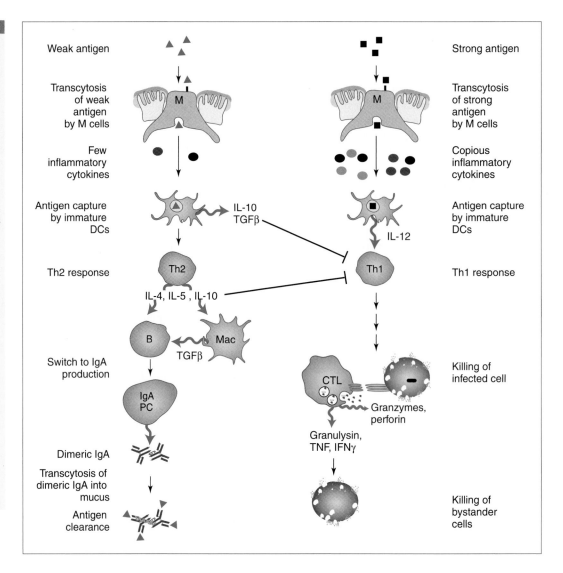

Figure 20-5

Model for the Influence of DCs on Mucosal Immune Responses

On the left side, the antigen breaching the mucosal surface is relatively weak and does not induce the production of high levels of inflammatory cytokines. When such an antigen is transcytosed by an M cell and transferred to an immature mucosal DC in this type of milieu, the DC secretes IL-10 and TGFβ that suppress Th1 differentiation and perhaps other cytokines that promote Th2 differentiation. The Th2 effectors in turn secrete cytokines that maintain the block on Th1 differentiation, activate other hematopoietic cells to secrete TGFβ, and support B cell differentiation into IgA-secreting plasma cells. Secretory IgA enters the mucus and traps antigen without damaging surrounding tissues. On the right side, the antigen attacking the mucosal surface induces vigorous inflammation. The DC that captures the antigen in this type of milieu secretes IL-12 and induces Th1 differentiation. The CTL effectors that are generated to counter the antigen also damage bystander cells.

acute "danger" or provoke significant inflammation (at least at first). The mucosal DC takes up the antigen but may become modulated (see Ch.11) rather than completely activated. Such a DC secretes Th2-inducing cytokines that promote an effective antibody response capable of defending the tissue without collateral damage. It is only if the immature mucosal DC encounters large quantities of pathogen-derived molecules that promote significant inflammation and signal an acute threat that complete DC activation and maturation are triggered. These DCs produce IL-12, which promotes Th1 differentiation and the generation of effectors that are inherently more destructive in nature.

Another hypothesis is that different DC subsets may bias the differentiation of the Th0 cell that they encounter. For example, in one *in vitro* study of PP DCs, treatment with microbial products or IFNγ stimulated conventional CD4$^+$CD11b$^+$ DCs (those usually present in the dome under the FAE) to secrete high levels of IL-10. In contrast, stimulated conventional CD8$^+$CD11b$^-$ DCs (usually present in the interfollicular areas of the PPs) produced IL-12. In keeping with these profiles, the treated CD4$^+$CD11b$^+$ DCs primed antigen-stimulated Th0 cells to secrete IL-4 and IL-10 (favoring a Th2 response) while the treated CD8$^+$CD11b$^-$ subset primed Th0 cells for IFNγ production (promoting a Th1 response.) It may be that relatively small amounts of antigen are mopped up by the CD4$^+$CD11b$^+$ DCs (among the first cells to encounter antigens entering the dome) such that the Th2 response dominates, while large amounts of antigen cannot be captured quickly enough and penetrate further into the interfollicular area, where they are taken up by the CD8$^+$CD11b$^-$ subset. A Th1 response might therefore be initiated.

When mucosal Th2 effectors are generated (the case for most mucosal antigens), they produce IL-4, IL-5, and IL-10. These cytokines in turn facilitate the secretion of the immunosuppressive cytokine TGFβ by activated hematopoietic cells in the area. The targets of all these cytokines are the antigen-activated B cells in the follicular germinal centers that undergo activation culminating in isotype switching to IgA production and affinity maturation. IL-4 and IL-10 suppress Th1 responses and IFNγ production. TGFβ is crucial for transcriptional activation of the IgA genes and isotype switching to Cα exons in activated B cells. TGFβ also upregulates the expression of IL-5R on B cells. In the presence of IL-5 derived from either epithelial or Th2 cells, mucosal B cells that have made the switch to IgA production evolve into fully competent polymeric IgA-secreting plasma cells.

iii) Mucosal Function and Production of Secretory Antibodies

iiia) Secretory antibody function. Secretory IgA is the principal weapon protecting us from pathogens and toxins that might otherwise penetrate mucosal surfaces. These antibodies, whose expression was discussed in Chapter 5, are key components of the mucosal mucus and other body secretions such as saliva and tears. It is important to appreciate how prominent secretory IgA is in immune defense. The majority of the body's entire pool of activated B cells is located near the mucosae and the exocrine glands. Both in the lamina propria of the gut and the exocrine glands, about 80% of all B cells and plasma cells present produce polymeric IgA. This polymeric IgA is converted to secretory IgA as it is introduced into the mucus and body secretions via the poly-Ig receptor (pIgR) (see Ch.5 and later in this chapter). The reader will recall that, in humans, there are two subclasses of secreted IgA: IgA1 and IgA2. While IgA1 predominates in the nasal and bronchial secretions of the human upper respiratory tract and in the upper GI tract, more than 60% of plasma cells in the lamina propria of the large intestine secrete IgA2. The lacrimal glands are associated with 80% IgA1-producing cells versus 20% IgA2-producing cells. In the tonsils, over 90% of plasma cells produce IgA1. The body's production of secretory IgA is prodigious, with about 2–3 grams of this antibody synthesized in the average adult human gut every day. This production rate far exceeds that for any other isotype.

Secretory IgA has several features and functions that make it ideal for mucosal defense. Indeed, secretory IgA antibody alone is sufficient for defense against certain virulent pathogens introduced via the stomach, intestine, or nasal passages. First, this antibody protects the mucosae by a form of neutralization that mucosal immunologists call *immune exclusion*; that is, the antibody defends the host by binding to the pathogen or toxin and "excluding" it from making contact with the epithelial cells. Secondly, independent of antigenic specificity, the carbohydrate moieties of secretory IgA molecules can bind to lectin-like *adhesin* molecules expressed by many pathogens, trapping the invaders on the luminal surface. Thirdly, at least in the gut, about half of all secretory IgA antibodies are unusually cross-reactive, a feature that allows them to cope with "antigenic drift." (Antigenic drift is the term used to describe the subtle changes in antigens that arise over generations of replicating microbes; see Ch.22.) These cross-reactive IgA molecules are derived from the CD5⁺ B1 B cells in the peritoneum (see Ch.9), which, although they do not undergo somatic hypermutation or affinity maturation, generate antibodies with broad powers of pathogen recognition. (Some researchers therefore believe that

the B1 B cell population represents a means of mucosal defense that developed relatively early during evolution.) Fourthly, secretory IgA is not an efficient activator of complement. Thus, even though monocytes, macrophages, and mucosal epithelial cells all synthesize complement pro-enzymes, the bias toward secretory IgA production in the mucosae decreases the chance of activating the complement cascade and initiating damaging inflammation. Fifthly, secretory IgA is highly resistant to a wide variety of host and microbial proteases, making it the isotype of choice for defense where the body has to interface with the exterior world and particularly in the harsh, degradative environment of the gut.

We note here that, in addition to its direct role in anti-pathogen defense, secretory IgA appears to have an indirect developmental function. Mutant mice that cannot produce secretory IgA show increased numbers of non-pathogenic bacteria in the gut and excessive development of single B cell follicles in the lamina propria. Moreover, normal mice reared in a pathogen-free environment have abnormally low numbers of single follicles in the gut lamina propria. Researchers have hypothesized that the presence of commensal bacteria in the gut is necessary for the development of single follicles, and that secretory IgA maintains the homeostasis of these bacterial populations at the correct density for producing normal numbers of single follicles. These various contributions of secretory IgA to mucosal defense and GALT development are summarized in Table 20-4.

Despite the dominance of secretory IgA in mucosal defense, it should be noted that the exocrine secretions may also contain low amounts of secretory IgM or IgG. Like IgA, IgM is made in polymeric form and can bind to the pIgR. However, secretory IgM is less resistant to gut proteases than is secretory IgA. In contrast, IgG is not produced in polymeric form so that it cannot be transcytosed across the epithelial barrier by pIgR. Nevertheless, low levels of IgG antibodies are consistently found in the secretions of the urogenital and respiratory tracts. Less IgG is present in the highly degradative environment of the gut due to this isotype's comparative lack of protease resistance. How IgG accesses gland secretions is not clear: some scientists believe that IgG molecules produced by plasma cells positioned near an endocrine gland can somehow passively diffuse between its epithelial cells and enter the gland secretion. In any case, the relatively low level of IgG antibodies at mucosal

---

Table 20-4 Virtues of Secretory IgA for Mucosal Defense

Excludes pathogens from making contact with epithelial cells

Traps pathogens on the luminal surface via lectin binding

Cross-reacts with a broad range of pathogens

Does not activate complement efficiently

Is resistant to host and microbial proteases

Maintains homeostasis of commensal bacteria required for follicle development

surfaces is generally a good thing, because most IgG isotypes are efficient complement-activating antibodies. The presence of IgG at the mucosae in any significant amount could lead to inflammation and damage to the epithelium, should complement components also be present.

iiib) Secretory antibody production. We will now review the details of how secretory antibodies are produced and how they get into the mucosal and exocrine secretions. As illustrated in Figure 20-6, foreign antigen taken up across the epithelial barrier in an inductive site (such as the appendix or PPs) is processed so as to activate both antigen-specific B cells and helper T cells in the local MALT. The surrounding cytokine milieu induces proliferation and isotype switching to IgA, after which the activated B cells circulate via the lymph and blood systems and exit in effector sites according to the pairing of homing receptors and adhesion molecules upregu-

lated during activation. For example, venules in the lamina propria of effector sites in the human gut express high levels of the mucosal addressin MAdCAM-1, which interacts with the integrin $\alpha 4\beta 7$ (LPAM-1) present on activated mucosal B cells. [One puzzle that was only recently solved concerned why IgA-expressing B cells, but not IgG-expressing B cells, home to gut effector sites. It turns out (at least in mice) that epithelial cells of the small intestine secrete a chemokine called TECK (CCL25), and IgG-producing B cells do not express the CCR9 receptor necessary for TECK recognition.] Upon reaching the effector sites, a majority of activated B cells finally differentiate into mature IgA-producing plasma cells under the influence of IL-5, IL-6, and IL-10. Once fully mature, these plasma cells release polymeric IgA in the vicinity of the basolateral surface of the epithelial cells lining the tract surface or the exocrine gland. Smaller amounts of

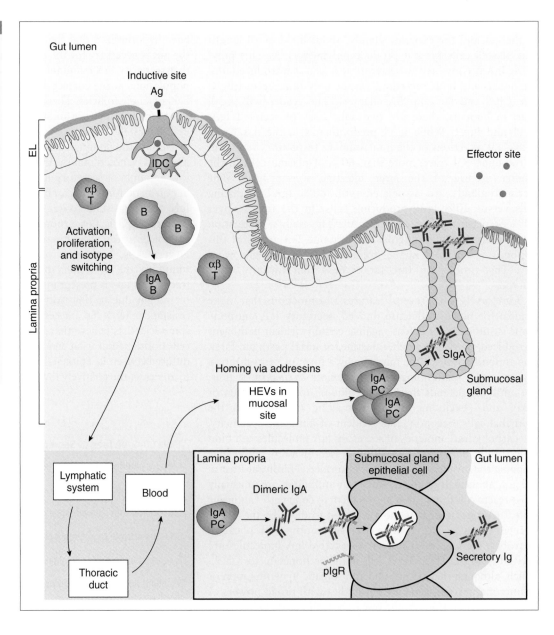

Figure 20-6

From Antigen Uptake to Secretory IgA Production
In this example, antigen taken in at a gut inductive site activates B cells in the follicle below. Intercellular contacts and cytokines in the milieu induce these B cells to generate progeny that undergo isotype switching to IgA. These B cells then travel through the lymphatic system and the blood and, guided by mucosal addressins, home to an effector site elsewhere in the gut. The B cells exit the blood via the HEVs in the effector site and finish their maturation into plasma cells. As shown in the inset, dimeric IgA produced by such a plasma cell binds to pIgR on an epithelial cell of a submucosal gland. Transcytosis of the IgA complexed to pIgR results in the release of secretory IgA into the submucosal gland secretion.

polymeric IgM may be produced in the same area by other plasma cells that do not undergo isotype switching.

The reader will recall from Chapter 5 that mucosal epithelial cells express on their basolateral surfaces a transmembrane receptor called the poly-Ig receptor (pIgR). pIgR, a 100 kDa glycoprotein member of the Ig superfamily, is also expressed on cells in the acinar regions of the exocrine glands. (Acinar tissue is composed of small sack-like dilations that deliver secretions into an exocrine duct.) pIgR binds with high affinity to the polymeric IgA or IgM released by plasma cells that have homed to the mucosae. These antibodies are then imported via pIgR into the epithelial cell using receptor-mediated endocytosis (refer to Fig. 5-9 and inset of Fig. 20-6). During transcytosis of the pIgA- or pIgM-containing intracellular vesicle across the epithelial cell to the apical surface, disulfide bonds form between the pIgR and the antibody molecule. Upon exocytosis, the pIgR is cleaved such that the 80 kDa extracellular domain of the receptor protein known as "secretory component" remains attached to the antibody, resulting in release of the secretory forms of IgA and IgM into the mucus or body secretion. It is the addition of the secretory component that stabilizes the antibody and masks proteolytic sites within it, conferring superior resistance to proteases.

We mentioned previously that secretory antibodies are found in abundance in MALT effector sites but only rarely in inductive sites. The FAE of inductive sites is specialized for antigen uptake and does not express the pIgR, precluding the formation of secretory antibodies. Conversely, in effector sites, the expression of pIgR is upregulated on epithelial cells in response to IFNγ, IL-4, and TNF produced by adaptive and innate immune system cells activated in the course of pathogen attack. These sites thus facilitate the production of secretory antibodies and their distribution on the luminal surfaces of body tracts. Unusual aspects of the transcytotic ability of pIgR that contribute to mucosal defense are described in Box 20-1.

iv) Cell-Mediated Responses: CD8$^+$ αβ CTLs

Production of secretory antibodies is the most important but not the sole means of mucosal defense. Successful penetration of the mucosal epithelium by a virus results in the activation of αβ Tc cells that are resident in mucosal inductive sites. These Tc cells proliferate and mature into CTLs thought to be crucial for the elimination of virus-infected epithelial cells, containing the infection in this relatively superficial layer before the virus can expand into deeper tissues. However, it is not entirely clear how such cells are activated, where they mature, or how they get to effector sites to kill infected epithelial cells. In PPs, resting Tc cells are generally positioned in the interfollicular areas surrounding the intestinal follicles. The M cells of the FAE overlying the follicles express receptors exploited by enteric viruses, particularly reovirus, to gain entry to this cell. Virus particles released by the M cell then infect nearby epithelial cells via receptors expressed on the basolateral surface. Viral antigens acquired from necrotic or apoptotic epithelial cells are taken up by nearby APCs, including macrophages and DCs (particularly the DN DC subset in the dome), and are used to activate antigen-specific Tc cells. It is possible that the activated Tc cells enter the lymphatic system and travel to the nearest lymph node to proliferate and mature into CTLs. The CTLs might then exit the lymph node and enter the systemic circulation, using homing receptors to access infected epithelial tissue in an effector site. In at least some of these sites, virus-specific CTLs appear to infiltrate the epithelial barrier to become part of the iIEL population. We note here that, while NK and NKT cells have also been associated with protection of the mucosae, their contributions have not been well characterized as yet.

v) A Common Mucosal Immune System?

B cells activated in a mucosal inductive site migrate to a large number of effector sites in various mucosal tissues. Thus, pathogen invasion at one location in the intestine can lead to secretory IgA protection not only of mucosae along the entire length of the gut, but also in the upper respiratory tract, salivary glands, lacrimal glands, ocular tissue, middle ear mucosa, and even the lactating mammary glands. Similarly, antigen introduction intranasally can result in detectable secretory IgA responses in the saliva, tonsils, trachea, lung, and gut. This broad dissemination of responding B cells ensures that almost all mucosal surfaces in the body are protected by a wide range of secretory antibodies, and has given rise to the concept of a *common mucosal immune system* (CMIS). However, as alluded to above, the CMIS must be compartmentalized to a degree to account for the fact that the strongest responses occur at effector sites located nearest the inductive site or linked to it by shared lymph drainage. For example, if antigen is encountered in the tonsils of NALT, the most vigorous antigen-specific immune responses are found in the nasal, lacrimal, and salivary glands (NALT) and bronchial glands (BALT), with a lesser response in the mammary glands. Similarly, if the PPs (GALT) are the inductive site, a strong response will be detected in the small intestinal mucosae (GALT) and a weak response observed in the tonsils (NALT).

The basis for this dissemination of immune responsiveness is the differential expression of tissue-specific addressins such as MAdCAM-1 (Fig. 20-7). As we have learned, MAdCAM-1 interacts with the integrin α4β7 expressed preferentially on B cells activated in mucosal inductive sites. One can picture that an antigen or pathogen introduced intragastrically could activate α4β7$^+$ B cells located in the GALT, and that these cells might migrate to local draining lymph nodes and commence proliferation. They might then enter the lymphatics and eventually the blood via the thoracic duct, and from there home to effector tissues expressing MAdCAM-1. MAdCAM-1 is highly expressed on the endothelial cells of HEVs and exocrine glands in the GALT, but is present at much lower levels on the endothelia of effector sites in the NALT/BALT or urogenital tract. Thus, more mucosal B cells will exit the circulation in gut effector sites than in NALT/BALT effector sites, leading to responses of different strengths. Different (unknown) pairs of effector site addressins and B cell adhesion molecules are thought to direct mucosal B cell migration to effector sites in NALT/BALT and the urogenital tract. Systemic peripheral B cells commonly express integrin α4β1, and therefore bypass the mucosae and travel to endothelia where the adhesion molecule VCAM-1 is expressed, typically sites of tissue inflammation.

Box 20-1. Pathogen neutralization via pIgR

In Chapter 5, we discussed how the poly-Ig receptor (pIgR) is essential for the production of secretory antibodies, transcytosing polymeric IgA and IgM (pIgA and pIgM) across the epithelial barrier and into the glands and lumens of body tracts. Nature has taken advantage of the positioning of the pIgR and its ability to transcytose large macromolecules to create other avenues of mucosal defense (see Figure). For example, infection with a virus induces the production of anti-virus pIgA destined to pass through the epithelial barrier via pIgR on its way to the mucosal surface. If the epithelial cell through which the antibody is being transported contains virus particles that are being assembled, and if the anti-virus pIgA molecule recognizes one of the assembly components, there is a chance for <u>intracellular</u> neutralization of the virus by the transiting anti-virus pIgA/pIgR complex. In one experimental system, monolayers of epithelial cells expressing pIgR on their basolateral surfaces were cultured in double-chambered devices such that viruses had access only to the apical surfaces of the epithelial cells, while anti-virus antibodies had access only to the basolateral surfaces. The epithelial cells were then infected on their apical surfaces with either Sendai, parainfluenza, or influenza virus and the production of new virus particles was measured. In all three cases, new virus production was inhibited if specific anti-virus pIgA was added to the culture chamber that allowed access to the basolateral surface and pIgR. These results implied that the pIgR imported the exogenous pIgA, which then blocked virus production intracellularly. Studies using immunofluorescence and immunoelectron microscopy have confirmed that transcytosing pIgA can indeed access subcellular vesicles containing viral proteins. It is speculated that the fusion of vesicles containing the anti-virus pIgA/pIgR complexes with post-Golgi vesicles containing newly synthesized viral envelope proteins allows the antibody to bind to the viral proteins, blocking virus assembly and preventing the spread of infection. This example illustrates that, while we are more accustomed to thinking of antibodies as mediators solely of <u>extracellular</u> defense, we must now broaden our view of how and where at least some antibodies function.

It should not be assumed that intracellular neutralization of viruses by polymeric antibody is a common means of mucosal protection. Several factors likely affect the chance of this type of interaction occurring. First, the virus must rely on a glycosylated component that requires processing through the Golgi (as opposed to a protein synthesized on free ribosomes in the cytoplasm). Secondly, the characteristics and kinetics of the replication cycle of the virus may affect the availability of the antigenic epitope for interaction. Considerable good luck in timing and specificity has to be in place for effective intracellular inhibition of virus production to occur *in vivo*.

In addition to intracellular neutralization of viruses, pIgR may help to dispose of intact noxious antigens (such as toxins and fragments of pathogens) that have passively slipped through the epithelial tight junctions and gained access to the lamina propria. In this location, pIgA of the correct specificity can bind to the escaped antigens and form immune complexes that can be recognized and transcytosed back across the epithelial barrier by pIgR, using the same mechanism that allows transport of free pIgA. The immune complexes are not degraded in transit and are expelled intact into the tract lumen in a type of "immune excretion." This mechanism may not be limited to immune complexes: there is some evidence that particles as large as whole viruses can be eliminated by this route.

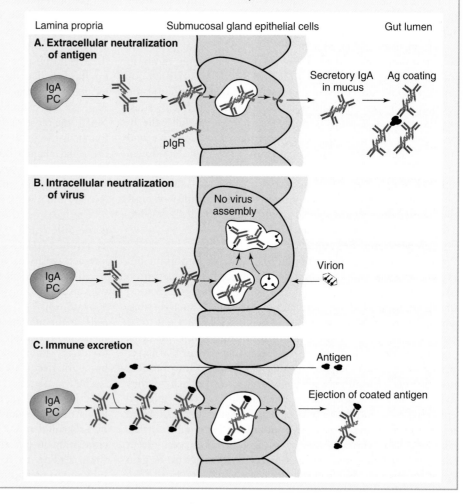

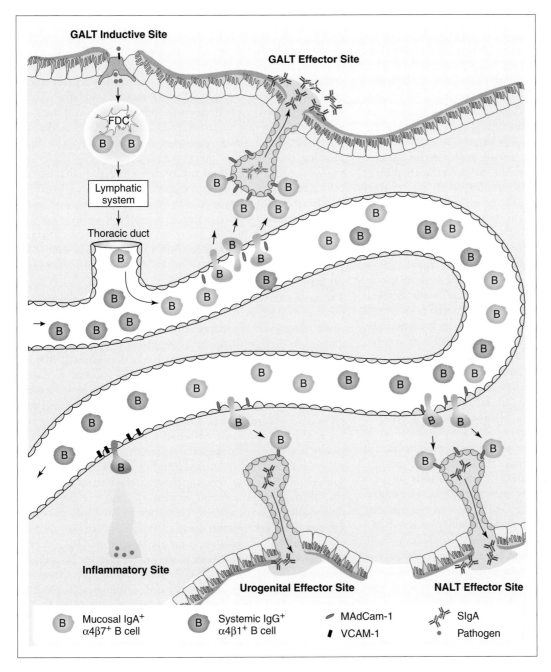

Figure 20-7

A "Common Mucosal Immune System"?
Antigen taken in at a GALT inductive site activates B cells in the follicle below. Progeny mucosal B cells (blue) undergo isotype switching and enter the lymphatic system and blood, where they circulate along with systemic B cells (gray). Mucosal B cells express the integrin $\alpha4\beta7$, which binds specifically to the mucosal addressin MAdCAM-1. The majority of mucosal B cells in this example exit the circulation at GALT effector sites because cells in this area express high levels of MAdCAM-1. The remaining mucosal B cells continue their circulation and exit at more remote mucosal effector sites, such as the NALT and urogenital tract, which express lower levels of MAdCAM-1. A coordinated response in multiple mucosal sites is thus established. Systemic B cells, which express the integrin $\alpha4\beta1$, exit the circulation only in inflammatory sites, where the endothelium has become activated and expresses VCAM-1.

GALT Inductive Site

GALT Effector Site

FDC

B   B

Lymphatic system

Thoracic duct

B

Inflammatory Site

Urogenital Effector Site

NALT Effector Site

	Mucosal IgA+ $\alpha4\beta7^+$ B cell		Systemic IgG+ $\alpha4\beta1^+$ B cell		MAdCam-1		SIgA
					VCAM-1		Pathogen

Much current research in the field of mucosal immunity is devoted to exploitation of the CMIS for oral and intranasal vaccinations (which are easier to administer and receive than injections). The hope is that B cells activated in the gut or nasopharynx in response to the vaccination might initiate antibody production elsewhere, to protect far distant portals of the body. Furthermore, there is evidence that introduction of a vaccine by the oral or intranasal route can induce systemic (as judged by levels of serum antigen-specific antibodies) as well as mucosal immune responses. Vaccination by the intranasal route is more effective than the intragastric route in inducing strong systemic and generalized mucosal responses, while vaccination by the intravaginal route is least effective. More on the issue of vaccination appears in Chapter 23.

IV. IMMUNE RESPONSES IN THE MINOR MALT

i) MALT in the Urogenital Tract

Immune responses in the urogenital tract are less well studied than are those in other MALT. The vagina differs from GALT and BALT/NALT in that it lacks the organized lymphoid structures present in MALT inductive sites. Intraepithelial DCs and macrophages are present in regions of the cervical and vaginal epithelium, but no M cells or lymphoid follicles have been detected in murine genital tracts. In general, introduction of an antigen into the vagina promotes neither mucosal nor systemic immune responses, meaning that the vagina is not a classical inductive site. This lack of responsiveness is evolutionarily desirable because an immune response to incoming sperm could block reproduction. Mucosal IgA responses in the murine vagina can be induced by immunization via the intranasal, intragastric, or even intramuscular routes, allowing protection from pathogen attack by the principle of CMIS. Predominantly secretory IgA (but also some IgG) can be found in vaginal secretions, confirming that the vagina is an effector site. The situation is complicated by the fact that antigen presentation and secretory IgA production in the vagina are regulated by estrous cycle hormones. The expression of MHC class II on various cell types in the female reproductive tract is increased by estradiol, and intraepithelial γδ T cells are present in their greatest numbers when estradiol levels are at their highest. Estradiol also increases the production of IFNγ and IL-6 by uterine non-lymphoid cells. Intrauterine levels of IL-6 affect both antigen presentation and secretory IgA concentration in uterine secretions. In contrast, progesterone suppresses antigen presentation, decreasing immune responses at the time in the estrous cycle when the vagina is most receptive to sperm.

In contrast to the dearth of immunological components in the vagina, the penile urethra contains substantial populations of leukocytes and may therefore function as an inductive site. The epithelial cells lining the urethra produce secretory component, and the lamina propria underlying the urethral mucosae contains many IgM- and IgA-secreting plasma cells. As a result, high concentrations of secretory IgA and secretory IgM can be found in the secretions coating the surface of the urethral mucosae. IgG-producing plasma cells may also be present in the lamina propria. Intraepithelial DCs reside in the mucosae of the distal tip of the urethra, while macrophages predominate in the epithelium and lamina propria of the length of the urethra. Large populations of CD4$^+$ and CD8$^+$ memory T cells are present in the lamina propria and epithelium throughout the urethra. These T cells are characterized by expression of the memory marker CD45RO and the α4β7 integrin homing receptor specific for mucosal tissues. Very few γδ T cells or NK cells reside in the urethral mucosae. In human males, 10-fold increases in specific IgA and IgG antibodies in urine can be induced by intranasal or oral immunization protocols, indicating that CMIS is also at work protecting the male urogenital tract.

ii) MALT in the Ear

The middle ear cavity is lined with a thin covering of mucus that overlies a mucosal epithelium made up of several types of secretory, ciliated, and non-ciliated epithelial cells. The mucus is constantly conveyed toward the Eustachian tube and nasopharynx by the beating of the cilia on the ciliated epithelial cells. This action helps to keep the middle ear cavity sterile because the tide of bacteria trying to access the middle ear from the nasopharynx is continually swept backward. The anti-bacterial molecules present in the mucus also take their toll. Comparatively few pathogens access the middle ear from the exterior through the auditory canal, and those that attempt it are usually thwarted by the tough keratinized layer of squamous epithelium covering the "exterior" side of the tympanic membrane.

There are very few organized lymphoid structures or cells in the normal, healthy middle ear cavity, meaning that it is not an inductive site. However, when infection occurs, the cavity takes on the appearance of a mucosal effector site, complete with the synthesis of secretory component and local production of secretory IgA. Antigen-specific secretory IgA, antibodies of other Ig isotypes, and pro-inflammatory cytokines such as IL-1, IL-6, and TNF can be detected in the middle ear fluid of infected patients, and macrophages, neutrophils, and T cell subsets become associated with the mucosal epithelium. Consistent with the function of the ear cavity as an effector site, immunization with live virus in the respiratory or intestinal tracts leads via CMIS to the synthesis of virus-specific antibodies in the middle ear fluid. Homing receptors that guide lymphocytes specifically to the middle ear have yet to be identified. It is more likely that effector cells generated in the inductive sites of the GALT or BALT are homing to CAMs expressed as a result of inflammation in the infected middle ear. For example, it has been shown that lymphocytes from the PPs can adhere to the mucosa of an inflamed middle ear, but not to the mucosa of a healthy middle ear.

iii) MALT in the Eye

The conjunctiva and anterior ocular surface of the eye are mucosal tissues, and as such are physically delicate structures easily damaged by inappropriate immune responses. Any damage to these structures may be enough to distort the visual axis and cause blindness. Evolution has thus ensured that the eye is an immune-privileged site in which immune responses are stringently controlled. Cells and macromolecules cannot readily pass through the walls of the micro-blood vessels supplying the eye, and the eye is not connected to a draining lymph node. Rather, the purity of blood flowing through the eye is ensured by passage through the spleen. Antigens that do access the eye are captured by intraocular APCs (including specialized subsets of DCs) that migrate from interior structures of the eye, through its trabecular meshwork into the blood, and thence to the spleen, where presentation to lymphocytes occurs. Effector T cells then home back to the eye, where they mediate non-inflammatory adaptive immune responses. To minimize damage to the sensitive eye tissue, these effector T cells are usually of the Th2 type. TGFβ2, produced in high concentration in the aqueous humor of the eye, is thought to confer on the intraocular APCs distinctive properties that promote Th2 development when the APC makes contact with a Th cell.

B. Cutaneous Immunity

As was introduced in Chapter 4, when we say "cutaneous immunity," we are referring to skin-mediated mechanisms that combat infection and injury. The immune system elements that underlie this defense are collectively referred to as SALT (*skin-associated lymphoid tissue*) (Table 20-5).

I. COMPONENTS OF SALT

The skin of an average adult human body has a surface area of about 1.5–1.7 m^2. The dead outer layers of keratinized skin are essential as a first means of physically blocking pathogen entry to the body. However, in the late 1970s, it

Table 20-5 **Elements of Cutaneous Immunity (SALT)**

Region of Skin	Defense Element	Defense Mechanism
Outermost ↓	Commensal bacteria	Compete with pathogens for space
		Produce anti-microbial molecules
		Produce lipases that decrease skin pH
	Keratin layer	Physical barrier when intact
		Takes bacteria with it during sloughing off
↓	Lower epidermis	Keratinocytes produce keratin, growth factors, IL-7, pro-inflammatory cytokines, chemokines, and complement components
		Interspersed γδ and αβ iIELs, Langerhans cells
↓	Dermis	Fibroblasts produce collagen, elastin, growth factors, cytokines, and chemokines
		Macrophages, DCs, mast cells, and αβ T cells mount adaptive responses; B cells are not usually present
↓		Mast cells respond to nerve fiber signals by degranulating and secreting TNF
		Sebaceous glands produce sebum, which decreases skin pH
↓		Nerve termini secrete peptides, stimulating immune responses
Innermost	Hypodermis	Fatty layer provides barrier defense

was recognized that the skin also is equipped with its own lymphoid elements. These elements include the αβ and γδ T lymphocytes and DCs resident among the living epithelial cells of the epidermis, and the mature αβ T cells, fibroblasts, DCs, macrophages, and lymphatic vessels present in the underlying dermis (Fig. 20-8). In general, resting B cells do not reside in the skin. The epidermis, which contains no blood vessels, is separated from the vascularized dermis by the basement membrane. Beneath the dermis is the hypodermis, a fatty layer that functions chiefly as an energy source but which also provides passive barrier defense and support for lymphatics and blood vessels.

i) Keratin Layer

The tough outer layer of the skin that resists penetration by inert stimuli as well as microbes is made up of filaments of a resilient, fibrous protein called *keratin*. Keratin is produced by specialized squamous epithelial cells called *keratinocytes*, which comprise over 90% of the cells in the epidermis. The epidermis is divided into several stratified layers, the outermost representing the oldest keratinocytes. New, less highly differentiated keratinocytes are constantly being produced from beneath in the lower layers of the epidermis such that the skin eternally renews itself from the inside out. Keratinocytes are generated in organized waves, with the cells in each wave being physically connected by specialized intracellular junctions known as *desmosomes*. The desmosomes ensure the formation of regimented horizontal layers of keratinocytes that divide and migrate upward as a unit. As they age and are pushed up to the skin surface by younger cells beneath them, the older keratinocytes increase their production of keratin fibrils. Eventually, the cytoplasm of the outermost keratinocytes is almost completely filled with keratin. As the keratinocytes approach the skin surface, their nuclei disintegrate and their lysosomes burst, releasing contents that both kill the cell and polymerize the keratin. The result is an inanimate layer of long,

tough fibers that can be 10 μm thick. The shells of the dead keratinocytes remain linked by the desmosomes during this process, ensuring that the keratin fibers retain their rigid horizontal alignment. The most exterior layers of keratinized shells finally lose the desmosomes and the shells slough off, which we see as the flaking of dead skin. This constant turnover of the keratinocytes is an effective defense mechanism because microbes generally do not have a chance to get seriously entrenched before they are sloughed off with the dead skin.

In addition to the physical barrier thrown up by the keratin layer, billions of bacteria, present at an average concentration of about $10^6/cm^2$, accumulate on the skin surface. The vast majority of these bacteria are harmless commensal species that actively contribute to defense of the skin against more virulent pathogens, by competing with them for space and nutrients and by secreting anti-microbial substances. Gram-positive commensals secrete lipases that break down fats in the skin into free fatty acids. These acids reduce the pH of the skin surface, discouraging microbial replication such that the commensal population is kept at a manageable level and the chance of a serious pathogen infection is reduced. As mentioned in Chapter 4, the acidity of the skin is also maintained by sebum produced by the sebaceous glands originating in the dermis.

ii) Lower Epidermis

Just below the keratin layer lie the differentiating strata of keratinocytes. These cells do much more than just produce keratin. Keratinocytes constitutively secrete GM-CSF and M-CSF, required for the survival and activation of DCs, and IL-7, which supports T cell maturation. Keratinocytes also constitutively express CD14 and several TLRs, and upregulate expression of these molecules upon exposure to LPS. Within minutes of injury or pathogen exposure, keratinocytes release pro-inflammatory cytokines such as IL-1α, IL-3, IL-6, and TNF, and also synthesize numerous chemokines and complement components. Keratinocyte secretion

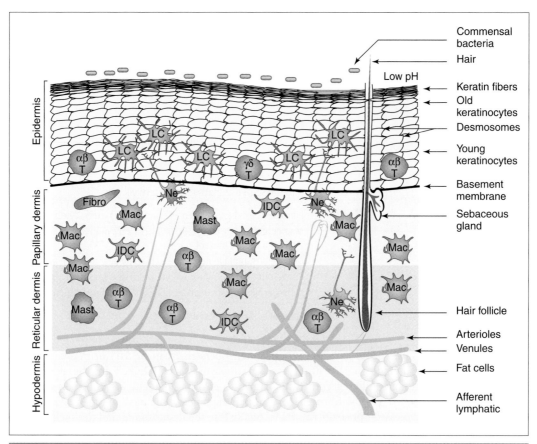

Figure 20-8

SALT Structure
The skin is composed of the epidermis, two dermal layers, and the hypodermis. The surface of the epidermis is protected by commensal bacteria, hair, low pH, and keratin fibers derived from compressed keratinocytes. Below the keratin fibers are multiple layers of differentiating keratinocytes held in horizontal strata by desmosome junctions. Interspersed among the keratinocytes are epidermal DCs (Langerhans cells) and epidermal αβ and γδ T cells. The epidermis is separated from the dermis by the basement membrane. The dermis, which contains the blood supply, is composed of the papillary layer and the more collagenous reticular layer. No keratinocytes are present in the dermis, which is populated by dermal DCs, macrophages, mast cells, and dermal αβ T cells. Nerve endings originating in the dermis extend into the epidermis, and afferent lymphatics take activated T cells to the draining lymph node and beyond. IDC, immature dendritic cell; LC, Langerhans cell; Ne, neuron.

of IL-1 and TNF is then further augmented in response to IFNγ produced by responding innate immune system cells. IFNγ also induces keratinocytes to synthesize MHC class II but not the B7 molecules, allowing these cells to act as anergizing APCs. Such a function may be important to control peripheral T cells that enter the skin toward the late stages of an adaptive immune response, when inflammatory signals are fading (see later). Keratinocytes thus participate in the regulation of both the innate and adaptive immune responses.

Although keratinocytes comprise the vast majority of cells in the epidermis, low numbers of T lymphocytes and motile skin DCs can be found in the most interior stratum of the epidermis, just above the basement membrane. As we learned in Chapter 18, in mice, the skin is home to the DETCs ("dendritic epidermal T cells"). The majority of these cells are γδ T cells exclusively expressing TCRs bearing Vγ3. In contrast, 10 times as many αβ as γδ T cells are present in human epider-

mis. Both human T cell subpopulations are predominantly CD8⁺ and TCR gene segment usage is biased in both cases. Interestingly, epidermal T cells have a memory phenotype, at least in humans.

In both mice and humans, the highly motile DCs present in the skin are bone marrow-derived and of the myeloid lineage. These DCs infiltrate their long, slender processes between keratinocytes in what appear under the microscope to be "veils" (hence, as mentioned in Ch.3, these DCs are sometimes called "veiled cells"). Epidermal DCs are more commonly known as Langerhans cells (LCs) and are the equivalent of the airway DCs described above in the NALT/BALT system. Antigen sampling in the epidermis is carried out primarily by LCs whose numerous dendritic processes allow them to survey a large section of an epidermal stratum. Langerhans cells express CD1 as well as high levels of MHC class I and II, making them ideal presenters of both peptide and glycolipid antigens to αβ or γδ T cells. Human LCs and epidermal T cells both express a particular integrin called αEβ7, which allows them to adhere to keratinocytes expressing E-cadherin. Successive binding interactions between these molecules permit the epidermal T cells and LCs to home upward into the keratinocyte layers. This sharing of a particular adhesion molecule may also explain why T cells are usually found in close proximity to the LCs in the skin, an ideal setup for antigen presentation required for an adaptive immune response.

The lower epidermis is also the site of termination of numerous nerve fibers extending from neurons present in the dermis. Scientists are now convinced that the nervous and immune systems are intricately linked, and that the proximity of these dermal nerve endings to the epidermal T and LC populations is meant to allow the monitoring of immune responses in this area by the nervous system. There is evidence that the nerve endings can secrete bioactive peptides that influence the immune response, as is elaborated in more detail in Box 20-2.

Box 20-2. Neurological components of mucosal and cutaneous immunity

There is growing evidence of concrete connections between the nervous system and the immune system. Stimulation of neural circuits can result in the production of neuropeptides that can have pro-inflammatory effects on systemic immune responses. Conversely, cytokines are able to cross the blood–brain barrier under some circumstances and bind to receptors present on CNS cells, causing them to synthesize molecules that signal the brain. For example, cytokines can stimulate the hypothalamic–pituitary–adrenal (HPA) axis, the neural circuit responsible for the "fight or flight" response to acute stress. HPA stimulation results in the synthesis of glucocorticoids that have potent anti-inflammatory effects. Thus, periods of high stress in an animal are linked to excess glucocorticoid production, suppressed inflammatory responses, and increased susceptibility to infections. Indeed, measurable changes to NK cytolytic activity and antibody titers have been observed in individuals asked to perform something as simple as mental arithmetic.

The status of the host at its interfaces with the external environment is monitored by nerve fibers extending near or into the epithelial layers of the SALT and MALT (see Table). Associations between these nerve fibers and immune system cells then help determine whether there is a local threat that requires an immune response. Disruptions in this monitoring can lead to human disorders that have both neurological and immunological components. For example, dysfunctions in the circuitry linking the HPA axis to the control of immune responses are thought to contribute to the pathology of autoimmune diseases such as diabetes and SLE (see Ch.29). Gut disorders such as inflammatory bowel disease, and skin diseases such as psoriasis and atopic dermatitis (see Ch.28), also fall into this category, as do certain other skin diseases.

With respect to SALT, the lower epidermis is the site of termination of myriad unmyelinated nerve fibers extending from neurons located near the dermal blood vessels. The dendrites spiral upward between the lower ranks of keratinocytes, and indeed make physical contact with them. Human LCs are also often found in very close proximity to the distal ends of these dendrites. These nerve endings emanate from neurons that have two functions: the conduction of electrical signals from the skin to the brain, and the synthesis of various neuropeptides and neurohormones that can participate in intercellular signaling. In the presence of cytokines released by epidermal cells during an inflammatory response, the dermal neurons secrete several factors, including nitric oxide, the vasodilator bradykinin, and the neuropeptides *Substance P* and *calcitonin gene-related peptide* (CGRP). Substance P is an 11-amino-acid peptide belonging to the neurokinin family, and CGRP is a 37-amino-acid peptide found primarily in sensory nerves. Both of these molecules are neurotransmitters that are widely distributed in the central and peripheral nervous systems. Receptors for these molecules are expressed on leukocytes such as T cells and mast cells in SALT and MALT (but not on leukocytes in the peripheral circulation).

The release of neuropeptides by dermal neurons has diverse effects. Vasodilation and permeability of the dermal venules is increased, facilitating extravasation of incoming leukocytes. Immune system cell proliferation, cytokine production, and antigen presentation are modulated. For example, receptors for Substance P are expressed on many cutaneous leukocytes, and their engagement by Substance P induces the production of pro-inflammatory cytokines such as IL-1, IL-6, and TNF. Mast cell degranulation is also induced by Substance P (and other neuropeptides). As mentioned in previous chapters, mast

cells release histamine and other molecules that increase vasodilation and permeability, as well as TNF and other cytokines that promote inflammation. The pro-inflammatory thrust induced by Substance P is counterbalanced by CGRP. Receptors for CGRP are present on cutaneous T and B cells as well as macrophages. *In vitro*, CGRP inhibits T cell proliferation and NK killing and blocks the release of TNF, IL-6, and IL-12 by macrophages. CGRP can also promote IL-10 release by macrophages stimulated with LPS and GM-CSF. Other studies have suggested that LCs can establish direct connections with nerve termini that are actively secreting CGRP, and that CGRP and Substance P may regulate the homing of LCs to the skin.

Another neuropeptide with an immunological connection is α-*melanocyte-stimulating hormone* (α-MSH). α-MSH was originally described as a neuropeptide produced by the pituitary in response to exogenous and endogenous stresses. More recently, however, α-MSH has been found to be secreted by activated epidermal cells such as keratinocytes and LCs, among others, to act as a brake on the inflammatory response. A receptor called MC-1 binds α-MSH and is upregulated on monocytes in response to pro-inflammatory cytokines, mitogens, and endotoxins. Engagement of this receptor by α-MSH inhibits the production of the pro-inflammatory cytokines IL-1, IL-6, IL-12, TNF, and IFNγ, upregulates secretion of the anti-inflammatory cytokine IL-10, and downregulates the expression of the costimulatory B7 molecules. Nitric oxide production by macrophages is blocked by α-MSH and expression of MHC class I is decreased.

One of the seminal observations establishing a link between the neuroendocrine system and cutaneous immunity is that UV-B irradiation (as occurs during prolonged

Continued

Box 20-2. **Neurological components of mucosal and cutaneous immunity—*cont'd***

exposure to the sun) leads to failed *contact hypersensitivity* (CHS) reactions in the skin. Like DTH, CHS is basically a strong secondary immune response characterized by a predominance of Th1 effectors (see Ch.28). These effectors produce cytokines that hyperactivate macrophages to such an extent that inflammatory tissue damage results. In the case of CHS, a chemical or element penetrates the skin and forms complexes with local host proteins, transforming them into immunogens. Because CHS results in recognizable clinical damage on the skin surface after 2–3 days, the CHS reaction is a convenient laboratory "read-out" for the presence of normal cutaneous immunity. Thus, a reduction in CHS observed upon UV-B exposure clearly indicates a loss of cutaneous immunity due to the irradiation. The fact that UV-B damages LCs, which are known to be involved in CHS, might initially suggest a direct immunological reason for the irradiation-induced shutdown of these hypersensitivity reactions. However, exposure to UV-B also causes epidermal nerve endings to release CGRP. CGRP in turn stimulates mast cells to secrete IL-10, which suppresses the inflammatory response that normally leads to CHS. In fact, if a Substance P agonist is later given to the UV-B-treated animals, they recover the ability to mount CHS reactions. In addition, although normal, resting epidermal cells do not express significant quantities of α-MSH, upregulation occurs following exposure to IL-1 or UV irradiation. Intriguingly, like UV-B, α-MSH can block the appearance of CHS in animals, promoting apparent experimental

tolerance to the inducing antigen. These results strongly suggest that signals received by the cutaneous nervous system can be directly translated into effects on SALT. Whether or not an antigen assaulting the skin leads to an immune response may thus be at least partly determined by the actions of nerve endings in the epidermis.

The gut is the best-studied component of MALT with respect to neuro-immune connections. Sensory neurons abound in the gut, linked by neural projections to neurons in other organs (such as the pancreas) that contribute to digestion. The Peyer's patches of the GALT are innervated by enteric neurons, and the lamina propria of the intestinal lining contains numerous afferent nerve endings that extend to the epithelial layer of the intestinal villi. In the lamina propria, membrane-to-membrane contacts have been observed between neuronal axons and immune system cells, particularly mast cells. Like skin damage, damage to the gut mucosae stimulates leukocytes in the GALT to produce mediators such as IL-1, IL-6, bradykinin, and histamine. These mediators act on the nerves, which in turn deliver signals causing alterations to fluid secretion and smooth muscle motility in the gut. The stimulated neurons also release Substance P, CGRP, and *vasoactive intestinal peptide* (VIP), which act on gut mucosal cells (including epithelial cells, macrophages, mast cells, and lymphocytes) bearing receptors for these neurotransmitters. Production of cytokines and other mediators, antigen presentation, MHC class II expression,

antibody synthesis, and NK cytotoxicity have all been shown to be influenced by neuropeptides.

One last note: the interplay between the nervous system and SALT/MALT is not just unidirectional. Keratinocytes and dermal fibroblasts secrete neurotrophins and nerve growth factor (NGF) that are essential for neuronal survival and regeneration as well as for regulation of neuronal responses to external stimuli. Gut fibroblasts and mast cells in an inflammatory milieu also secrete NGF. Activated neutrophils and eosinophils produce molecules such as cAMP and cationic proteins that induce the sensory nerve fibers in the gut and airway epithelial layers to secrete additional neurotransmitters. Furthermore, immune and non-immune system cells resident in the skin can produce their own Substance P, which influences nearby neurons to send signals back to the brain that an inflammatory response is occurring. Conversely, the nervous system can produce its own cytokines. Almost all known cytokines and their receptors have been identified in the resting mammalian brain. In response to infection or LPS administration, several types of neurons in the brain greatly upregulate their production of IL-1, thought to be the primary agent inducing fever in response to endotoxin and the alteration of gastrointestinal functions. The emerging picture is one of intricate interconnectivity between the sensory, structural, and protective elements of the skin and mucosae.

Neuro–Immuno Interactions

Molecule	Secreted by	Receptors Expressed by	Action
Substance P	Dermal neurons	SALT/MALT T cells, mast cells, and other leukocytes	↑ IL-1, IL-6, TNF ↑ Mast cell degranulation ↑ LC homing to skin ↑ Vasodilation and venule permeability
CGRP	Dermal neurons	SALT/MALT macrophages, T, B, and NK cells	↓ Macrophage TNF, IL-6, IL-12, but ↑ IL-10 ↓ NK killing ↓ T cell proliferation ↓ LC homing to skin
α-MSH	Activated keratinocytes, LCs	Monocytes	↓ IL-1, IL-6, IL-12, TNF, IFNγ, but ↑ IL-10 ↓ B7, MHC class I ↓ NO production ↓ CHS reactions
NGF	Keratinocytes, gut and dermal fibroblasts, activated mast cells	Neurons	↑ Survival ↑ Regeneration ↑ Responsiveness

iii) Basement Membrane

The youngest keratinocyte layer of the epidermis is separated from the underlying dermis by the *basement membrane*. The basement membrane is a protective layer composed of collagen, laminin, heparan sulfate, and glycosaminoglycan produced by epidermal keratinocytes in combination with fibronectin produced by dermal fibroblasts. As mentioned previously, there are no blood vessels in the epidermis. Thus, nutrients required to sustain the keratinocytes must exit the circulation in the dermal blood vessels and diffuse across the basement membrane. Similarly, leukocytes that access the epidermis, including the T cell and DC populations, must migrate from the dermal vessels and force their way through the basement membrane before entering the epidermis. Activated leukocytes secrete enzymes that dissolve components of the basement membrane, allowing rapid passage of the migrating cells into the epidermis.

iv) Dermis

Compared to the tightly packed cells of the epidermis, the dermis is a much airier mixture of structural fibers, blood vessels, lymphatic vessels, and low numbers of immune system cells. Nerve fibers also criss-cross the dermis, stretching up through the basement membrane. The dermis is composed of two relatively static layers, the *papillary* dermis (just below the basement membrane) and the *reticular* dermis (below the papillary dermis). Both layers are formed from networks of collagen and elastin fibers embedded in a glue-like hyaluronic acid matrix. The collagen fibers are thicker and more numerous in the reticular dermis than in the papillary dermis. Collagen provides structural support for the skin, elastin gives skin its resilience, and the highly negatively-charged hyaluronic acid traps water molecules that furnish both turgor support and moisture to the skin.

Both dermal layers contain neurons, fibroblasts, and leukocytes such as macrophages, mast cells, DCs, and αβ T cells. (As mentioned previously, B cells are not usually present.) Leukocytes access the dermis by extravasating through the endothelial cell layer lining the dermal post-capillary venules. These endothelial cells assist leukocyte migration by secreting chemokines and providing other contacts necessary to sustain an innate immune response. Macrophages are the most prevalent leukocytes resident in the dermis. Dermal T cells are found clustered around the arterioles and venules penetrating the dermis. Despite the much less cellular nature of the dermis compared to the epidermis, 98% of skin T cells are in the dermis and only 2% are in the epidermis. Most dermal T cells are memory cells expressing the memory marker CD45RO and high levels of CD25 (IL-2Rα). CD4$^+$ and CD8$^+$ cells are equally represented in the dermal population but the TCR repertoire is restricted. The dermis is also home to DCs expressing high levels of CD1 and MHC class I and class II. These dermal DCs appear to be morphologically distinct from epidermal DCs (LCs). Dermal fibroblasts synthesize not only the matrix components collagen, fibronectin, and elastin, but also growth factors and cytokines that promote the survival and differentiation of leukocytes and keratinocytes. Mast cells congregate around the dermal arterioles and venules and frequently make contact with nerve fibers. These contacts allow

the mast cells to receive stimulatory signals that induce the release of their vasoactive contents, resulting in altered blood pressure and vessel permeability. In addition, mast cells are an important source of TNF in the skin. Finally, mast cells are the mediators of IgE hypersensitivity (as introduced in Ch.5), manifested in the skin as hives (urticaria) and/or a rash. More on this latter topic appears in Chapter 28.

II. IMMUNE RESPONSES IN SALT

How do all of the elements just described come together to provide immune defense for the skin? Pathogens or injurious materials breaching the outer keratinized layer and damaging a living keratinocyte's membrane induce that cell to release pre-existing cytoplasmic stores of the inflammatory cytokine IL-1α and TNF (Fig. 20-9). These cytokines induce other keratinocytes to synthesize and release CXC chemokines (IL-8), growth factors (GM-CSF, M-CSF, IL-7, and IL-15), and additional inflammatory cytokines (IL-1β, TNF, and IL-6) within the epidermis. These molecules not only activate resident epidermal cells but also diffuse down through the basement membrane into the dermis. Dermal fibroblasts respond with the synthesis of additional inflammatory cytokines and chemokines. Some of these proteins reach the endothelial cells of the dermal blood vessels, promoting local vasodilation and selectin expression and thus facilitating leukocyte extravasation. Chemokines reaching the basolateral surface of an endothelial cell can also be internalized and conveyed across the cytoplasm by transcytosis. Transcytosed chemokines are then bound on the luminal surface of the endothelial cell through interactions of their C-termini with heparan sulfate, a molecule abundantly expressed on the endothelial cell surface. Chemokine molecules displayed in this way in the lumen of the blood vessel constitute a direct signpost for migrating phagocytes in the blood. Thus, a concentration gradient is created from the site of the assault that first summons neutrophils from the circulation through the endothelium, into the dermis, and upward through the epidermis.

As the neutrophils extravasate through the endothelial walls and pursue the chemokine gradient, they begin expression of adhesion molecules necessary for migration in the tissues. For example, expression of LFA-1 on the neutrophil surface facilitates migration among ICAM-1-expressing dermal fibroblasts, while other adhesion molecules interact with ligands such as fibronectin and collagen within the dermal fibrous matrix. As the neutrophils make contact with the basement membrane, they are induced to produce enzymes capable of degrading its components, allowing the cells to penetrate into the epidermis. The neutrophils continue their migration along the chemokine gradient, squeezing between keratinocytes, which close ranks again behind them. At the site of assault, the presence of pro-inflammatory cytokines induces full activation of the neutrophils and deployment of the respiratory burst.

Activation of neutrophils also upregulates their expression of complement receptors, opening the door to pathogen destruction by complement-mediated opsonization and phagocytosis. There are two sources of complement components in

Figure 20-9

Immune Responses in SALT
An antigen penetrating the skin injures keratinocytes, which produce IL-1α and TNF. These cytokines act on neighboring keratinocytes and induce them to produce a range of inflammatory mediators. A chemokine gradient is formed, which summons neutrophils from the circulation through gaps in the basement membrane created by hydrolases secreted by activated dermal macrophages. Meanwhile, antigen present in the epidermis can directly activate epidermal γδ T cells and macrophages. Antigen captured by LCs may promote activation of epidermal memory αβ T cells, which generate effectors that can act within the epidermis. Alternatively, the antigen may be conveyed to the draining lymph node to induce a systemic response. Systemic effectors bearing the skin-specific addressin CLA travel via the lymph system and blood to the dermis, where they exit the dermal blood vessels and commence battle against the antigen in this tissue. IDC, immature dendritic cell; LC, Langerhans cell (immature dendritic cell); MDC, mature dendritic cell. See text for details.

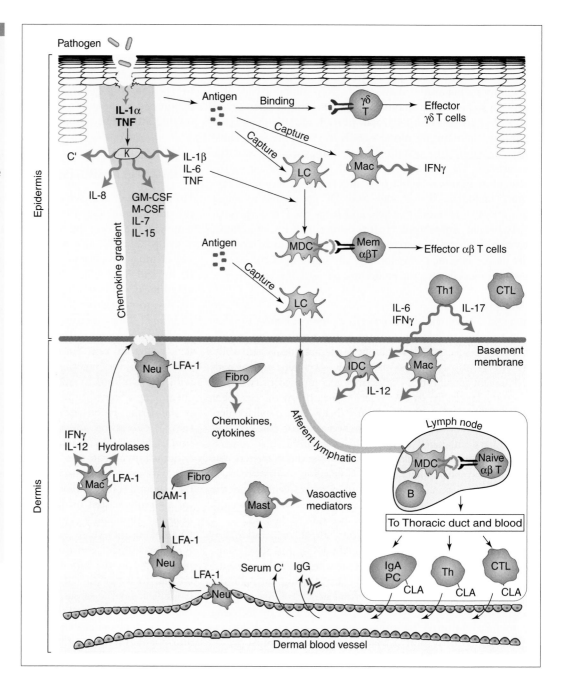

the skin. Keratinocytes can synthesize various complement components, and the vasodilation of inflamed dermal blood vessels allows the leakage of serum complement proteins into the dermis. Should an appropriate microbial component be present, complement can be activated via the alternative pathway. In addition, should there be pre-formed antibodies present in the serum that can bind to a penetrating foreign antigen, classical complement activation may ensue. Such pre-formed antibodies can also participate in pathogen destruction via antibody-mediated opsonization, since Fc receptors are upregulated during neutrophil activation. Mast cells are also stimulated by complement components to degranulate, releasing substances that increase local blood vessel dilation and sustain inflammation.

At the same time as the inflammatory response to injury is being mounted by keratinocytes, dermal fibroblasts, neutrophils, and macrophages, the epidermal T cells and DCs swing into action. Damage to keratinocytes supplies antigens either from the keratinocytes themselves and/or from the microbes leaking through the breached keratinized layer. The reader will recall that epidermal T cells are either γδ T cells or αβ memory T cells. The epidermal γδ T cells immediately recognize the common skin stress or bacterial antigens without the intervention of DCs and are activated upon engagement of their TCRs. Antigens are also taken up by the LCs. If the cytokine milieu is rich enough, the LCs can start to mature within the epidermis without having to go to the local lymph node. Because of the co-localization of LCs and T cells in the epidermis, these maturing DCs can then directly present the stress or bacterial antigens and activate the epidermal αβ T cells. Because these T cells are primarily memory cells, the response is almost as rapid as that of the γδ T cells. Within 24 hours, the memory T cells commence differentiation into effectors. Because of the high local concentrations of IFNγ in the site of assault, these differentiating effectors are biased toward Th1 development and produce still more IFNγ to maintain the inflammatory response. Note here that because of the inherent toughness of the skin as a tissue, Th1 responses are well tolerated (unlike the less sturdy mucosae, which benefit from a Th2 response).

With the activation of the epidermal T cells, more and more pro-inflammatory cytokines are secreted, particularly IL-6 and IFNγ. Epidermal effector T cells also produce IL-17, which acts on keratinocytes to increase their production of IL-8 and IL-6 and expression of ICAM-1. IFNγ and bacterial products diffuse from the epidermis into the dermis and activate dermal macrophages and DCs. IFNγ stimulates these cells to produce IL-12, which promotes further IFNγ production and enhances the inflammatory response. In addition to IFNγ, activated dermal macrophages produce enzymes that clip components of the basement membrane, making it easier for granulocytes to access the epidermis. The dermal macrophages themselves may cross into the epidermis under the influence of chemokines secreted by LCs. Using adhesion molecules such as a surface receptor that serves as an alternate ligand for E-cadherin, the dermal macrophages migrate up through the keratinocyte layers toward the site of the assault.

Traffic across the basement membrane can run the opposite way as well. About 24–36 hours after the initiation of the inflammatory response, some LCs lose their expression of E-cadherin and upregulate integrins that favor binding to ligands on dermal cells. The LCs can then enter the dermis and access the local lymphatics that lead to the draining cutaneous lymph node. In this way, antigen can be transported to reservoirs of naive peripheral T cells to initiate a systemic adaptive immune response. Within the local lymph node, the LCs complete their maturation and further upregulate MHC class II and the B7 molecules in preparation for naive T cell activation. Naive T cells expressing TCRs that recognize the antigens presented by the activated LCs are activated and commence the differentiation process which takes 5–7 days to produce effector cells. Importantly, these effectors are induced to express *cutaneous lymphocyte antigen* (CLA), a sialylated carbohydrate that specifically binds to E-selectin expressed on the post-capillary venules of inflamed dermis. CLA expression thus allows these T cells to migrate through the circulation and home to sites of inflammation in the dermis.

If the invasion has been contained within about 36 hours, IFNγ production subsides, and IL-10 (secreted by phagocytes and keratinocytes) and TGFβ (produced by macrophages and lymphocytes) start to make their anti-inflammatory effects felt. TGFβ depresses the secretion of chemokines by endothelial cells and inhibits secretion of the enzymes used by macrophages and neutrophils to degrade the basement membrane. Differentiating T effectors arriving in the site now find a cytokine milieu low in IFNγ and high in IL-10, promoting

Th2 differentiation. However, very little antigen remains at this point, so that the humoral response supported by Th2 effectors is not really required. Rather, the job of the Th2 cells becomes one of terminating inflammation. The Th2 cells secrete more IL-10 and IL-4, inducing even more secretion of TGFβ by surrounding cells and further suppression of IFNγ expression. As inflammation abates, keratinocytes expressing MHC class II but not costimulatory molecules compete with dwindling numbers of professional APCs for presentation of antigens to local T cells, anergizing the latter. With the resolution of inflammation, TGFβ stimulates dermal fibroblasts and other matrix cells to produce new collagen to repair the damage to the skin.

However, should the assault not be resolved within 36 hours, high levels of IFNγ continue to permeate the site of attack. Endothelial cells of the dermal venules alter their expression of adhesion molecules so as to decrease the neutrophil influx and increase traffic in monocytes and lymphocytes. Epidermal fibroblasts and dermal endothelial cells switch their secretion of chemokines from the CXC class to the CC class, drawing increasing numbers of macrophages and lymphocytes (and decreasing numbers of neutrophils) to the site. In particular, activated Th1 cells are attracted by the CC chemokines RANTES (CCL5) and CTACK (CCL27). In the presence of IFNγ, TNF, and CD40L/CD40 contacts supplied by the Th1 cells, macrophages become hyperactivated and gain increased microbicidal powers, marking the onset of a DTH reaction. Defense is further bolstered by the actions of activated CTLs that employ perforin- and granzyme-mediated destruction to destroy the invader.

Should the CTL response and granulocyte-mediated phagocytosis not prove sufficient to contain the pathogen, a switch is made to Th2 conditions that promote a humoral response. Hyperactivated macrophages that fail to dispose of an invader ("frustrated phagocytosis") start to produce more IL-10 (favors Th2) than IL-12. Dermal mast cells contribute large amounts of IL-4 to the mix, and Th2 differentiation dominates. In the continuing presence of antigen, these Th2 effectors are in a position to deliver help to B cells. However, as mentioned previously, there are very few B cells in the skin, which lacks the follicle structures of the mucosae. It is believed that the Th2 cells recirculate back to the draining lymph node and interact with antigen-stimulated B cells in this location. Antibody-synthesizing plasma cells then migrate from the active germinal centers in the node to sites where the antibodies can enter the blood. Many scientists believe that a substantial proportion of the plasma cells in fact migrate back to the bone marrow, releasing sIgs into the blood circulation from which they disseminate throughout the body. The majority of these Igs are IgG antibodies that facilitate both complement-mediated pathogen destruction and ADCC carried out by NK cells and granulocytes. Although not of major importance in cutaneous immunity, secretory IgA has been found in skin secretions such as perspiration. Epithelial cells in the acinar region of the sweat glands express pIgR and can transcytose polymeric IgA to the skin surface, where secretory IgA neutralizes pathogens as described for mucosal immunity.

If the pathogen is eliminated by the Th2 response, antigen subsides and IL-10 and TGFβ again become the dominant cytokines, suppressing further responses by both lymphocytes and granulocytes. A resting state is again achieved, with only low levels of IL-10 secreted by keratinocytes in the area. If, however, the pathogen resists all efforts to remove it, a chronic DTH reaction sets in to wall the invader off from the rest of the body. The prolonged presence of high levels of IL-10 and TGFβ induces hyperactivated macrophages surrounding the pathogen to fuse together into giant cells.

A granuloma is formed as connective tissue strands produced by local fibroblasts and stromal cells isolate the site of now chronic inflammation. More on DTH reactions and granulomas appears in Chapters 22 and 28.

We have now covered all the basic elements of the immune system and have described their roles in innate and adaptive immune responses. We move now to Chapter 21 and a discussion of comparative immunology, tracing the occurrence and evolution of the elements of the innate and adaptive immune responses from the most primitive organisms to humans.

SUMMARY

Mucosal and cutaneous immune responses are the first lines of defense protecting the body tracts and skin from pathogen attack. The major mucosa-associated lymphoid tissue (MALT) consists of the lymphoid follicles and diffuse collections of lymphocytes in the gut (GALT) and respiratory tract (BALT/NALT). Passive anatomical barriers, IgA-containing mucus, and intraepithelial leukocytes play key roles in protecting the epithelium. However, antigen that bypasses these initial defense mechanisms is taken up by the MALT in inductive sites. M cells in the follicle-associated epithelium (FAE) capture pathogens and convey them by transcytosis to mucosal APCs and T cells resident in the intraepithelial pocket and the "dome" covering the B cell-containing lymphoid follicles. Effector cells then migrate to multiple mucosal effector sites, including the exocrine glands, to establish a common mucosal immune response.

Immune responses in the MALT are generally biased to Th2 to avoid inflammatory damage to the relatively fragile mucosae. In contrast, responses in the skin-associated lymphoid tissue (SALT) tend to be biased to Th1. Keratinocytes both form a physical barrier to pathogen penetration and secrete pro-inflammatory cytokines and chemokines that mobilize and activate phagocytes and DCs in the lower epidermis. Pathogen antigen captured by Langerhans cells either activates epidermal memory T cells or is conveyed through the basement membrane to naive T cells in the local lymph node. Effector T cells are generated that home back to the site of pathogen attack in the skin. Inflammation and cellular activation triggered by a persistent pathogen may initiate a chronic DTH reaction mediated by hyperactivated macrophages. B cells are not prominent in the SALT.

Selected Reading List

Beagley K. W. and Husband A. J. (1998) Intraepithelial lymphocytes: origins, distribution, and function. *Critical Reviews in Immunology* **18**, 237–254.

Bilsborough J. and Viney J. (2002) Getting to the guts of immune regulation. *Immunology* **106**,139–143.

Brandtzaeg P., Farstad I. and Jahnsen F. (1998) Cellular and molecular mechanisms for induction of mucosal immunity. *Developmental Biological Standards* **92**, 93–108.

Brandtzaeg P., Baekkevold E. S., Farstad I. N., Jahnsen F. L., Johansen F.-E. *et al.* (1999) Regional specialization in the mucosal immune system: what happens in the microcompartments? *Immunology Today* **20**, 141–151.

Debenedictis C., Joubeh S., Zhang G., Barria M., Ghohestani R. F. *et al.* (2001) Immune functions of the skin. *Clinics in Dermatology* **19**, 573–585.

Downing J. and Miyan J. (2000) Neural immunoregulation: emerging roles for nerves in immune homeostasis and disease. *Immunology Today* **21**, 281–289.

Fagarasan S. and Honjo T. (2003) Intestinal IgA synthesis: regulation of front-line body defences. *Nature Reviews Immunology* **3**, 63–72.

Fleeton M., Contractor M., Leon F., Wetzel J.D., Dermody T. S. *et al.* (2004) Peyer's patch dendritic cells process viral antigen from apoptotic epithelial cells in the intestine of reovirus-infected mice. *Journal of Experimental Medicine* **200**, 235–245.

Furness J., Kunze W. and Clerc N. (1999) Nutrient tasting and signaling mechanisms in the gut II. The intestine as a sensory organ: neural, endocrine and immune responses. *American Journal of Physiology* **277**, G922–G928.

Guy-Grand D., Azogui O., Celli S., Darche S., Nussenzweig M. C. *et al.* (2003) Extrathymic T cell lymphopoiesis: ontogeny and contribution to gut intraepithelial lymphocytes in athymic and euthymic mice. *Journal of Experimental Medicine* **197**, 333–341

Heijen C. and Kavelaars A. (1999) The importance of being receptive. *Journal of Neuroimmunology* **100**, 197–202.

Iwasaki A. and Kelsall B. L. (1998) Mucosal immunity and inflammation I. Mucosal dendritic cells: their specialized role in initiating T cell responses. *American Physiological Society* **276**, G1074–G1078.

Iwasaki A. and Kelsall B. L. (2001) Unique functions of CD11b+, CD8α+, and double negative Peyer's patch dendritic cells. *Journal of Immunology* **166**, 4884–4890.

Jafarian-Tehrani M. and Sternberg E. (1999) Animal models of neuroimmune interactions in inflammatory diseases. *Journal of Neuroimmunology* **100**, 13–20.

Johansson E.-L., Rask C., Fredriksson M., Eriksoon K., Czerkinsky C. *et al.* (1998) Antibodies and antibody-secreting cells in the female genital tract after vaginal or intranasal immunization with cholera toxin B subunit or conjugates. *Infection and Immunity* **66**, 514–520.

Johnston B. and Butcher E. (2002) Chemokines in rapid leukocyte adhesion triggering and migration. *Seminars in Immunology* **14**, 83–92.

Kagnoff M. F. (1998) Current concepts in mucosal immunity III. Ontogeny and function of γδ T cells in the intestine. *American Journal of Physiology: Gastrointestinal and Liver Physiology* **274**, G455–G458.

Lamm M. E. (1998) Current concepts in mucosal immunity IV. How epithelial transport of IgA antibodies relates to host defense. *American Journal of Physiology: Gastrointestinal and Liver Physiology* **274**(4), G614–G617.

Liu J., Chen M. and Wang X. (2000) Calcitonin gene-related peptide inhibits lipopolysaccharide-induced interleukin-12 release from mouse peritoneal macrophages, mediated by the cAMP pathway. *Immunology* **101**, 61–67.

Luger T., Scholzen T., Brzoska T., Becher E., Slominski A. *et al.* (1998) Cutaneous immunomodulation and coordination of skin stress responses by alpha-melanocyte-stimulating hormone. *Annals of the New York Academy of Sciences* **840**, 381–394.

Matsunaga T. and Rahman A. (1998) What brought the adaptive immune system to vertebrates?—The jaw hypothesis and the seahorse. *Immunological Reviews* **166**, 177–186.

Mayer L. (1998) Current concepts in mucosal immunity I. Antigen presentation in the intestine: new rules and regulations. *American Journal of Physiology: Gastrointestinal and Liver Physiology* **274**(1), G7–G9.

Middleton J., Neil S., Wintle J., Lam C., Auer M. *et al.* (1997) Trancytosis and surface presentation of IL-8 by venular endothelial cells. *Cell* **91**, 385–395.

Misery L. (1997) Skin, immunity and the nervous system. *British Journal of Dermatology* **137**, 843–850.

Mowat A.M. (2003) Anatomical basis of tolerance and immunity to intestinal antigens. *Nature Reviews Immunology* **3**, 331–341.

Neutra M. R. (1998) Current concepts in mucosal immunity V. Role of M cells in transepithelial transport of antigens and pathogens to the mucosal immune system. *American Journal of Physiology: Gastrointestinal and Liver Physiology* **274**(5), G785–G791.

Neutra M. R., Pringault E. and Kraehenbuhl J.-P. (1996) Antigen sampling across epithelial barriers and induction of mucosal immune responses. *Annual Review of Immunology* **14**, 275–300.

Ogra P. L. (2000) Mucosal immune response in the ear, nose and throat. *Pediatric Infectious Disease Journal* **19**, S4–S8.

O'Keefe J., Doherty D., Kenna T, Sheahan K., O'Donoghue D. P. *et al.* (2004) Diverse populations of T cells with NK cell receptors accumulate in the human intestine in health and colorectal cancer. *European Journal of Immunology* **34**, 2110–2118.

Perdue M. H. (1999) Mucosal immunity and inflammation III. The mucosal antigen barrier: cross talk with mucosal cytokines. *American Journal of Physiology: Gastrointestinal and Liver Physiology* **277**(1), G1–G5.

Pivarsci A., Bodai L., Rethi B., Kenderessy-Szabo A., Koreck A. *et al.* (2003) Expression and function of Toll-like receptors 2 and 4 in human keratinocytes. *International Immunology* **15**, 721–730.

Prabhala R. H and Wira C. R. (1995) Sex hormone and IL-6 regulation of antigen presentation in the female reproductive tract mucosal tissues. *The Journal of Immunology* **155**, 5566–5573.

Rameshwar P. (1997) A regulatory neuropeptide for hematopoiesis and immune functions. *Clinical Immunology and Immunopathology* **85**, 129–133.

Rudin A., Ruse G. C. and Holmgren J. (1999) Antibody responses in the lower respiratory tract and male urogenital tract in humans after nasal and oral vaccination with cholera B subunit. *Infection and Immunity* **67**, 2884–2890.

Salvi S. and Holgate S. (1999) Could the airway epithelium play an important role in mucosal immunoglobulin A production? *Clinical and Experimental Allergy* **29**, 1597–1605.

Scholzen T., Armstrong C., Bunnett N., Luger T., Olerud J. *et al.* (1998) Neuropeptides in the skin: interactions between the neuroendocrine and the skin immune systems. *Experimental Dermatology* **7**, 81–96.

Snijdelaar D. (2000) Substance P. *European Journal of Pain* **4**, 121–135.

Spellberg B. (2000) The cutaneous citadel: A holistic view of skin and immunity. *Life Sciences* **67**, 477–502.

Stenfors L.-E. (1999) Non-specific and specific immunity to bacterial invasion of the middle ear cavity. *International Journal of Pediatric Otorhinolaryngology* **49**, S223–S226.

Sternberg E. (2000) Interactions between the immune and neuroendocrine systems. *Progress in Brain Research* **122**, 35–42.

Streilein J. W. (1999) Immunoregulatory mechanisms of the eye. *Progress in Retinal and Eye Research* **18**, 357–370.

Streilein J., Alard P. and Niizeki H. (1999) A new concept of skin-associated lymphoid tissue (SALT): UVB light-impaired cutaneous immunity reveals a prominent role for cutaneous nerves. *The Keio Journal of Medicine* **48**, 22–27.

Strober W. (1998) Interactions between epithelial cells and immune cells in the intestine. *Annals of the New York Academy of Sciences* **859**, 37–45.

Strober W. (2004) Epithelial cells pay a Toll for protection. *Nature Medicine* **10**, 898–900.

Strober W., Fuss I., Nakamura K. and Kitani A. (2003) Recent advances in the understanding of the induction and regulation of mucosal inflammation. *Journal of Gastroenterology* **38**(Suppl. 15), 55–58.

Takizawa H. (1998) Airway epithelial cells as regulators of airway inflammation. *International Journal of Molecular Medicine* **1**, 367–389.

Theodorou V., Fioramonti J. and Bueno L. (1996) Integrative neuroimmunology of the digestive tract. *Veterinary Research* **27**, 427–442.

Toms C. and Powrie F. (2001) Control of intestinal inflammation by regulatory T cells. *Microbes and Infection* **3**, 929–935.

Travis S. M., Singh P. K. and Welsh M. J. (2001) Antimicrobial peptides and proteins in the innate defense of the airway surface. *Current Opinion in Immunology* **13**, 89–95.

Tschernig T. and Pabst R. (2000) Bronchus-associated lymphoid tissue (BALT) is not present in the normal adult lung but in different diseases. *Pathobiology* **68**, 1–8.

Wu H.-Y. and Russell M. W. (1997) Nasal lymphoid tissue, intranasal immunization, and compartmentalization of the common mucosal immune system. *Immunologic Research* **16**, 187–201.

Yuan Q. and Walker W. A. (2004) Innate immunity of the gut: mucosal defense in health and disease. *Journal of Pediatric Gastroenterology and Nutrition* **38**, 463–473.

Comparative Immunology

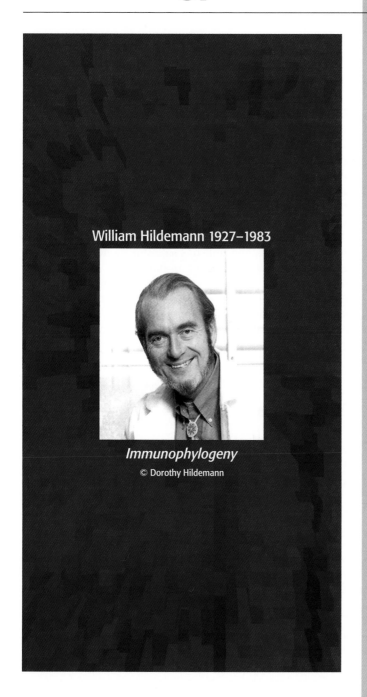

William Hildemann 1927–1983

Immunophylogeny

© Dorothy Hildemann

CHAPTER 21

A.   OVERVIEW

B.   ELEMENTS OF INNATE IMMUNITY

C.   ELEMENTS OF ADAPTIVE IMMUNITY

"I like pigs. Dogs look up to us. Cats look down on us. Pigs treat us as equals"
—Sir Winston Churchill

The preceding chapters in this book have concerned themselves with the adaptive and innate components of the mammalian immune system, focusing on humans and mice. In this chapter, we expand our horizons, examining host defense in organisms ranging from the lowly sponge up through the ranks of the animal kingdom to birds and reptiles. We also take a brief look at methods plants use to defend themselves from pathogens. Our approach will be to compare host defense in these organisms with that in mammals, looking at elements of both innate and adaptive responses. We will also attempt to apply a rationale for each step along the evolutionary trail, speculating as to why certain modes of immunity might have developed in certain organisms as they were subjected to the constant pressure of selection to protect the species against pathogens.

Why bother to explore such topics in a world dominated by a focus on the human condition? One answer is that examination of innate response elements in lower organisms may reveal enormous amounts of information on the very fundamentals of mammalian innate immunity. The innate response plays a hugely important role in human health, both directly on the front lines of host defense, and by facilitating the adaptive immune response. From an evolutionary point of view, the innate response appears to be extremely ancient, much more so than the adaptive response. For example, anti-microbial peptides have been identified in humans, mice, frogs, fish, and even in primitive vertebrates such as tunicates. Studying the variations in the structure and function of these molecules in simpler animals may help us to understand how they work in humans, which in turn could yield clues to new clinical therapies. This type of research is of direct relevance to cystic fibrosis patients, who suffer from repeated airway infections when their anti-microbial peptides malfunction. Another example of a system whose examination in other organisms could be of benefit to humans is the complement cascade. Complement-mediated damage in organ transplantation and autoimmune disease continues to defy effective clinical mitigation, so that study of this long-established system of host defense may lead to the identification and/or design of practical inhibitors for humans.

Another factor to consider in justifying the exploration of immunity in other species is that animals and insects are often important carriers of organisms that can cause disease in humans. Understanding the immune system of the carrier may permit the manipulation of its immune response and perhaps block the transmission of the organism to humans. On the less medical side, knowledge of the immune response in fish is directly applicable to successful aquaculture or "fish farming," just one example of an industry in which knowing how the farmed organism is able to resist pathogens could assist in developing a more resilient and therefore more economically viable product. Similar considerations apply to crop breeding and maintenance, and to preserving threatened species.

A. Overview

I. REPRISE OF THE ANIMAL KINGDOM

A brief review of the hierarchy of the phyla of the animal kingdom is in order to orient the reader with respect to the evolutionary tree (Fig. 21-1). The first animals were protozoans, single-celled organisms that gave rise about 900 million years ago to multicellular ancestral metazoans. The Porifera (sponges) are considered the simplest metazoans. The ancestral metazoans include two other lines, the acoelomates (simple multicellular animals with no body cavity), and the coelomates (simple multicellular animals with a body cavity). The coelenterates (corals and jellyfish), the platyhelminths (flatworms), the nematodes (roundworms), and the nemertines (ribbon worms) were all originally derived from a common acoelomate ancestor. These phyla are considered the lower

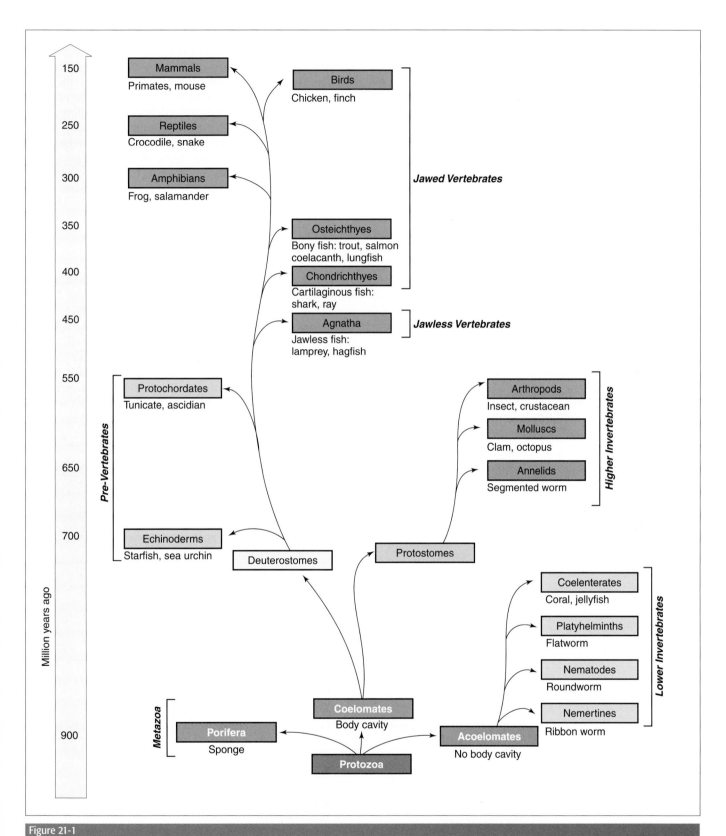

Figure 21-1

Evolutionary Tree of Kingdom Animalia
The major phyla and the approximate times when animals classified in these phyla first appeared are shown, starting with the protozoa at 900 million years ago. Phyla are grouped into the metazoa, lower invertebrates, higher invertebrates, pre-vertebrates, and vertebrates as indicated. Time scale is approximate. Adapted from Sunyer J. O. and Lambris J. D. (1998). Evolution and diversity of the complement system in poikilothermic vertebrates. *Immunological Reviews* 166, 39–57.

invertebrates. Two major lines of more complex animals were derived from the coelomates: the protostomes, which became the higher invertebrates, and the deuterostomes, which eventually gave rise to the vertebrates. About 650 million years ago, the annelids (segmented worms) evolved from a protostome ancestor, followed about 50 million years later by the molluscs (clams, snails, mussels, squid, octopus). This evolutionary line culminates in the arthropods (insects, crustaceans), which emerged about 500 million years ago. The annelids, molluscs, and arthropods constitute the higher invertebrate phyla. Meanwhile, about 700 million years ago, the echinoderms (starfish, sea urchins) evolved from a deuterostome ancestor, followed about 150 million years later by the protochordates (ascidians and tunicates, such as sea squirts). Because the echinoderms and protochordates lack the spinal columns of the animals following them in phylogeny, we have called these animals "pre-vertebrates" (others have called them "deuterostome invertebrates"). About 450 million years ago, the Agnatha (jawless fish) appear, the first of the true vertebrates. Another approximately 50 million years sees the arrival of the first jawed fish, the Chondrichthyes (cartilaginous fish). These animals, which include the skates, sharks, and rays, have a backbone of cartilage rather than of bone. Fish with a backbone of true bone and a swim bladder, which constitute the Osteichthyes (bony fish), emerged about 350 million years ago. There are three distinct classes of bony fish. The most primitive species are the lobe-finned or fleshy-finned fish (coelacanths, lungfish), followed by the intermediate ray-finned fish (sturgeons, garfish). The teleosts (salmon, herring, trout) are the most advanced bony fish. Following the bony fish, the amphibians (frogs, toads, salamanders) evolved about 300 million years ago, and the reptiles (crocodiles, snakes) emerged about 50 million years later. The birds and mammals diverged from the reptiles around the same time, about 100–150 million years ago. To put things in perspective, our species *Homo sapiens* has been around for only about the last 100,000 years.

II. FORCES SHAPING THE EVOLUTION OF IMMUNE RECOGNITION

All forms of life have a mechanism to distinguish self from non-self, and the means to preserve that self in the face of undue competition for nutrients and pathogen attack. These mechanisms and means were shaped by the various evolutionary and developmental pressures experienced by each organism (Table 21-1). Even life forms as primitive as the microorganisms secrete anti-microbial substances that halt or eliminate invaders competing for limited resources. Similarly, restriction enzymes did not evolve to facilitate DNA analysis for modern scientists; rather, these enzymes developed in

Table 21-1 **Evolutionary Pressure and Immune Defense Development**

Organism	Evolutionary Drivers	Adaptation of Immune Defense
Porifera and protochordates	Non-self colony overgrowth	Lectins and phagocytes Primitive PRMs and complement-like components Primitive self-recognition molecules
Lower invertebrates	Circulatory system Multiple body layers Increased mobility and lifespan	Phagocytes expressing PRMs Anti-microbial proteins and peptides Pathogen-trapping mucus
Higher invertebrates	Increased anatomical complexity Increased mobility Open circulatory system Relatively long lifespan High numbers of offspring	Exoskeleton as passive defense Anti-microbial proteins and peptides Lectin-mediated opsonization Ig-like molecule for opsonization Coagulation cascade ProPO system of cytotoxicity (arthropods) Limited cytokines and chemokines
Pre-vertebrates	Increased anatomical complexity Increased mobility	Lectin pathway of complement activation Limited cytokines and chemokines Mucus
Vertebrates	Low numbers of offspring Very long lifespans Great mobility Development of jaws[*] Varied diet and habitat Sun exposure Vascular system Warm-bloodedness Eggs and live young	Skin as passive defense Enzymes that wall off pathogens Anti-microbial proteins and peptides Structured lymphoid tissues Full complement of cytokines and chemokines Adaptive immunity, including distinct T and B cells, Igs, MHC, V(D)J recombination Classical, lectin, and alternative pathways of complement activation Immune tolerance mechanisms

[*] The development of jaws separates the jawless Agnathan vertebrates from the rest of the jawed vertebrates and was a major milestone in vertebrate evolution.

bacteria to cleave the DNA of invading viruses while sparing the host DNA. The most basic of multicellular organisms such as the sponges, coelenterates, and tunicates are also able to distinguish and protect self. These animals grow in colonies of relatively undifferentiated cells, and the need to preserve one's own colony from encroachment by an adjacent colony spurred the evolution of mechanisms that can mediate what appears to be a short-term graft rejection-like response. The overgrowing cells of a non-self colony are killed and the integrity of the self colony is maintained. Since, in this situation, the attacker's DNA is protected within a membrane-enclosed cell, restriction enzymes are of no use; a surface mechanism to identify self is required. The existence of molecules serving what is seen as a histocompatibility function in higher organisms has been postulated from studies of colony overgrowth killing. Lectins and phagocytic cells capable of disposing of non-self can also be detected in these animals, and the existence of at least some complement components (or molecules related to them) has been suggested (Fig. 21-2).

With the evolution in the lowest invertebrates of circulatory systems and multiple body layers to more efficiently manage the digestion and distribution of food and the removal of waste products, the functions of host nutrition and host defense can be carried out by different cell types, some of which travel in the circulation. Increased anatomical complexity is often accompanied by increased mobility, such that the animal is usually exposed to a wider range of pathogens. These animals also have longer life spans that correlate with an increased time to reach reproductive maturity. Thus, to survive as a species, longer-lived animals have to be able to withstand a greater number of viral and bacterial infections. Cells in the circulation become specialized for detecting infected cells via the expression of soluble or membrane-bound pattern recognition molecules (PRMs) of relatively broad specificity (refer to Ch.4). Phagocytic cells patrol the body and engulf invaders, and anti-microbial peptides and proteins are produced. In many early protostomes and deuterostomes, molecules resembling complement components of the lectin pathway can be found, but these proteins function only as opsonins. No true lymphocytes, antibodies, or MHC molecules are present from the annelids through to the arthropods, or in pre-vertebrates. In higher invertebrates, a unique pathway called the ProPO system (*prophenoloxidase-activating system*; discussed later in this chapter) mediates cytotoxicity.

While the range of PRMs expands as one proceeds higher in the pre-vertebrate branch, it seems that adaptive immunity is not needed for the success of species with a limited habitat range, relatively short life spans (short reproductive cycle), and large reproductive capacities. The immune repertoire is limited, as must be true when the genes encoding defense molecules do not undergo somatic recombination and are "hard-wired" in the germline. However, this repertoire is sufficient for survival because a limited habitat means that the range of pathogen types encountered is generally narrow. A short life span means the total number of pathogens encountered is relatively low, and a large reproductive capacity means that the huge numbers of offspring produced offset the loss of

substantial numbers of them to pathogens. However, in vertebrates, we find animals that take a longer time to reach reproductive age than invertebrates and wander over greater distances, meaning that they may encounter a much larger number and broader range of pathogens. These animals also produce many fewer offspring than either invertebrates or pre-vertebrates, such that severe losses to pathogens could threaten the species as a whole. These evolutionary pressures are thought to have promoted the development of adaptive immunity in vertebrates.

While no conventional antibody can be detected in the jawless fish (Agnatha), primitive GALT tissue is present. With the development of hinged jaws, cartilaginous fish like the sharks become better predators and can take advantage of new nutritional opportunities. However, with such a diet comes an increased chance of internal injury and/or infection. Even with innate immunity in place, a strictly germline-encoded repertoire of non-self recognition molecules is no longer sufficiently diverse to counter all the pathogens such vertebrates meet. Strong selection pressure is exerted that results in the emergence of mechanism that can somatically diversify immune system genes and expand the immune repertoire. A distinct thymus and spleen and true lymphocytes expressing forms of Ig and TCR molecules are present. IgM and non-mammalian antibody isotypes (see later) are produced, as are the terminal complement components. MAC lysis is first seen in the cartilaginous fish. In the bony fish, an IgD-like antibody joins IgM as an antibody isotype, and complement components unique to the alternative pathway appear. The anterior kidney serves as a bone marrow equivalent.

With the progression of vertebrates from the sea to the land in the form of amphibians, additional host defense mechanisms are required to cope with the new environment. The limbs necessary to move on land evolve, and with them sophisticated vascular systems containing multiple types of circulating cells. Skins designed to shield the exposed animal from the sun's harmful rays also develop, providing a physical barrier against pathogens. In the amphibians, bone marrow-like tissue serves as a source of distinct T and B cells, and lymphoid tissues of increased complexity and wider distribution are present. By the time we reach the reptiles, we find more advanced lymphoid tissues plus eggs in which the young develop in a self-contained aqueous system enclosed by multiple membranes. This innovation frees these (still) cold-blooded animals from having to return to the water to reproduce and increases barrier protection for the offspring. Because one of the protective membranes surrounding the embryo is called the *amnion*, the reptiles are considered to be "amniotes," a group that also includes the birds and mammals.

The first true forests appear on earth at about the same time that the reptiles diverge from the amphibians, and some scientists hypothesize that it was the new array of food and habitat possibilities that led to this evolutionary event. With land living comes an expansion in the array of immune system cells available for host defense, conferring increased efficiency and division of labor in immune responses. However, the cold-bloodedness of the amphibians and reptiles affects

	Mammals	Birds	Reptiles	Amphibians	Bony fish	Cartilaginous fish	Jawless fish	Protochordates	Echinoderms	Higher invertebrates	Lower invertebrates	Protozoa
PRM Recognition												
PRMs	+	+	+	+	+	+	+	+	+	+	+	+
Anti-microbials	+	+	+	+	+	+	+	+	+	+	+	+
Cytokines	+	+	+	+	+	+	+	+	+	+	+*	+*
Chemokines	+	+	+	+	+	+	+	+	?	+	–	–
ProPO System												
ProPO	–	–	–	–	–	–	–	–	–	+	–	–
Haemolin	–	–	–	–	–	–	–	–	–	+	–	–
Complement System												
CR	+	+	+	+*	+*	+*	+*	+*	+*	–	–	–
C3	+	+	+	+	+	+	+*	+*	+*	–	–	–
MBL/MASP	+	+	+?	+	+*	+*	+*	+*	+*	–	–	–
Terminal C'	+	+	+	+	+	+	–	–	–	–	–	–
Factor I	+	+	+	+	+	+	–	–	–	–	–	–
Factor H	+	+	+	+	+	–	–	–	–	–	–	–
Lectin path	+	+	+	+	+	+	+	+	–	–	–	–
Classical path	+	+	+?	+	+	+	–	–?	–	–	–	–
Alternative path	+	+	+	+	+	+	–	–	–	–	–	–
Lymphoid Tissue												
Lymphocytes	T, B	T, B	T, B	T, B	T, B	T*, B*	T/B	+*	+*	+*	+*	–
GALT	+	+	+	+	+	+	+	–	–	–	–	–
Spleen and thymus	+	+	+	+	+	+	–	–	–	–	–	–
Bone marrow	+	+	+	+*	–	–	–	–	–	–	–	–
Lymph nodes (+ GC)	+	+	–	–	–	–	–	–	–	–	–	–
MALT/SALT	+	+	–	–	–	–	–	–	–	–	–	–
Leydig organ	–	+	–	–	–	+	–	–	–	–	–	–
Epigonal organ	–	+	–	–	–	+	–	–	–	–	–	–
Bursa	–	+	–	–	–	–	–	–	–	–	–	–

Figure 21-2

Elements of Innate and Specific Immunity through Evolution

All animals, no matter how primitive, have some type of pattern recognition molecule (PRM). Many also have at least some components of the complement system. Organized lymphoid tissues are a feature of the vertebrates, but cells with lymphocyte-like functions can be found in the most primitive pre-vertebrates as well as higher and lower invertebrates. Similarly, while recognizable MHC-like molecules are not found in invertebrates or pre-vertebrates, the immune systems of these animals are capable of recognizing and killing non-self cells. Ig responses and TCR genes are found starting with the cartilaginous fish. *, like; ?, not definitively proven; $+/-$, limited; C', complement components; GC, germinal center; F, fast; S, slow; VS, very slow; NAR, novel antigen receptor; NARC, new antigen receptor from cartilaginous fish; BF, class I genes in chicken; BL, class II genes in chicken; BG, non-classical MHC genes in chicken; •, IgW was originally termed IgX and is synonyomous with NARC; ♦, only in lobe-finned class of bony fish; •, non-MHC histocompatibility molecules exist.

Continued

their immune responses, causing measurable seasonal variations in the proliferative capacity of lymphocytes and the production of cytokines. Warm-bloodedness evolves in the birds and mammals, possibly as an adaptation allowing these animals to hunt at night, when the cold-blooded reptiles are less active. However, with a permanently warm body comes an

	Mammals	Birds	Reptiles	Amphibians	Bony fish	Cartilaginous fish	Jawless fish	Protochordates	Echinoderms	Higher invertebrates	Lower invertebrates	Protozoa
Ig Responses												
RAG	+	+	+?	+	+	+	–	–	–	–	–	–
Tdt	+	+	+	+	+	+	–	–	–	–	–	–
Isotype switching	+	+	+	+	–	–	–	–	–	–	–	–
Somatic hypermutation	+	+	+	+	+	+	–	–	–	–	–	–
Affinity maturation	+	+	+?	+/–	+/–	–	–	–	–	–	–	–
Memory	+	+	+/–	+/–	+/–	–	–	–	–	–	–	–
Ig Isotypes												
IgM	+	+	+	+	+	+	–	–	–	–	–	–
IgD	+	+	–	–	+	–	–	–	–	–	–	–
IgX	–	–	–	+	–	–	–	–	–	–	–	–
IgW◆	–	–	–	–	+◆	+	–	–	–	–	–	–
NAR	–	–	–	–	+◆	+	–	–	–	–	–	–
IgY	–	+	+	+	–	–	–	–	–	–	–	–
IgA	+	+	–	–	–	–	–	–	–	–	–	–
IgE	+	–	–	–	–	–	–	–	–	–	–	–
IgG	+	–	–	–	–	–	–	–	–	–	–	–
MHC												
MHC class I	+	BF	+	+	+	+	–•	–•	–	–	–	–
MHC class II	+	BL	+	+	+	+	–	–	–	–	–	–
MHC class III	+	+*	+?	+	–	–	–	–	–	–	–	–
MHC class Ib	+	BG	–	+	+	+	–	–	–	–	–	–
β2-microglobulin	+	+	+	+	+	+	–	–	–	+*	+*	–
Allograft rejection	F	F	F	F	F	S	VS	+*	+*	+*	+*	+*
TCR												
TCR genes	+	+	+?	+	+	+	–	–	–	–	–	–
Td responses	+	+	+	+	+	–	–	–	–	–	–	–
Ti responses	+	+	?	+	+	+	–	–	–	–	–	–

Figure 21-2 *cont'd*

environment favoring pathogen growth, necessitating the development of still more complex immune defenses. Lymphoid tissues differentiate and become more structured, resulting in the presence of distinct germinal centers and lymph nodes. Cell-mediated and humoral responses are better coordinated and controlled to deliver the most efficient response possible with the least amount of collateral damage to the host. In mammals, an additional layer of control develops to prevent rejection of the live young that develop inside the mother rather than in a separate egg.

The preceding completes our sketch of the animal kingdom and the evolutionary pressures that may have forced the development of first innate and then adaptive immune defense systems. We will now examine the elements of innate and adaptive immunity in an evolutionary context, describing in more detail what systems are present in lower organisms and what they can tell us about human immune responses. The reader is urged to continually refer to Figure 21-2 as necessary as he or she reads the rest of this chapter.

B. Elements of Innate Immunity

Evolutionary immunologists have defined innate immunity as "the immediate ability of a host to prevent or limit the consequences of injury or an infectious assault." With this definition in mind, elements similar to those of the mammalian innate immune system can be detected in all multicellular organisms. In fact, much more is known about innate defense in higher invertebrates and pre-vertebrates than has been formally shown for non-mammalian vertebrates, and that slant will be reflected in the following discussion.

I. ANATOMICAL AND PHYSIOLOGICAL BARRIERS

As is true for mammals, the first line of innate defense lies in the anatomical and physiological barriers that deter pathogen invasion. These defenses obviously vary with the complexity of the organism. Many lower invertebrates and pre-vertebrates, such as worms and tunicates, produce mucus that can "wrap up" a pathogen and isolate it. Many higher invertebrates, such as insects and molluscs, possess shells or other forms of hard exoskeletons that resist penetration. In addition, the enzymes involved in exoskeleton formation in insects are thought to play an active role in biochemically helping to wall off invaders. Non-mammalian vertebrates make use of many of the same barrier defenses at mucosal surfaces as those outlined for mammals in Chapters 4 and 20.

II. INNATE IMMUNE RESPONSE CELLS AND THEIR RECOGNITION STRUCTURES

i) Phagocytic and Other "Blood" Cells

The next line of defense involves cell-mediated elimination of pathogens, carried out at its most basic level by some form of phagocytic cell (Table 21-2). Mobile phagocytic cells called *amoebocytes* have been detected in the most primitive organisms. Amoebocytes are multi-functional prowlers in the body, carrying out not only host defense via phagocytosis but also

Table 21-2 "Blood" Cell Evolution

Organism	Form of "Blood" Cells
Coelomates	Amoebocytes
Echinoderms and lower invertebrates	Various specialized amoebocytes
Protochordates	Hemocytes
Higher invertebrates	Phagocytes, granular cells, nutritive cells, lymphocyte-like progenitor cells
Non-mammalian vertebrates	Erythrocytes, macrophages, neutrophils, NK-like cells, T/B lymphocyte-like cells
Mammalian vertebrates	Erythrocytes, macrophages, neutrophils, NK cells, NKT cells, T/B lymphocytes

digestion and excretion. In more advanced protostomes and deuterostomes, whose bodies are bigger and more complex, these functions are carried out by specialized types of amoebocytes. Further up the evolutionary tree, a circulatory system that contains *hemolymph* ferries nutrition and waste products around the body. With the migration of some amoebocytes from the tissues into the circulatory system, the first leukocyte-like cells called *hemocytes* evolve. Indeed, white "blood" cells of various types and functions are present to some degree in the fluids filling the body cavities or the circulation of all but the most primitive of the lower invertebrates and pre-vertebrates. Note, however, that erythrocytes are not generated, and organized lymphoid-like tissues are not present.

As the evolution of invertebrates progresses, the hemocytes evolve into several distinct hematopoietic cell types, some of which are thought to play a role in host defense. *Progenitor cells* in invertebrates resemble vertebrate lymphocytes in appearance, in that they have large nuclei surrounded by a thin rim of cytoplasm. However, these cells do not function like lymphocytes and serve primarily as stem cells that develop into other cell types. *Phagocytic cells* remain ubiquitous, engulfing invaders in much the same way as mammalian macrophages and neutrophils. *Granular* cells (also called *hemostatic* cells) are key players in higher invertebrate defense, carrying out the effector actions of wound healing and the associated coagulation that serves to trap pathogens (see later). *Nutritive* cells, which appear in only a few species of invertebrates, may also contribute to host defense via encapsulation of pathogens.

In higher non-mammalian vertebrates, distinct neutrophils, macrophages, NK-like cells, and cells closely resembling mammalian T and B lymphocytes can be identified. Where they have been examined, the behavior and functions of macrophages and NK cells in non-mammalian vertebrates appear to be very similar to those in mammals. For example, NK-like cells identified in the clawed frog *Xenopus laevis* have been shown to mount immediate cytotoxic responses against tumors or virally infected cells. In the zebrafish, a family of *novel immune-type receptor* (NITR) genes has been identified that putatively encode transmembrane receptors reminiscent of the KIRs expressed by mammalian NK cells. The predicted proteins encoded by these genes contain ITIMs in their cytoplasmic tails. *In vitro*, engineered expression of NITR proteins on human NK cells followed by antibody-mediated cross-linking leads to negative regulation of NK activatory signaling.

The evolution of B and T cells, and the lymphoid tissues with which they are associated, are discussed in more detail later in Section C, "Elements of Adaptive Immunity."

ii) Recognition of Non-self

As mentioned previously, PRMs of one form or another that allow organisms to distinguish between self and non-self can be found in virtually all organisms (refer to Fig. 21-2). For example, homologues of the Toll receptor, first identified in *Drosophila melanogaster* as important for differentiation, have been discovered in organisms as diverse as plants and mammals (see Box 21-1). While some PRMs are constitutively

Box 21-1. Conservation of the Toll receptor pathway

The reader will recall from Chapter 4 that the Toll protein is a PRR that was first identified in *Drosophila*. In that chapter, we described the existence of mammalian homologues to Toll and the roles of the TLRs in signaling in response to the perception of danger in the form of bacterial LPS. It turns out that the signaling pathways mediated by Toll and TLRs are evolutionarily very ancient, homologous elements having been identified in vertebrates, invertebrates, and even in plants (see Box 21-2 later in this chapter). Toll-related molecules generally possess several conserved domains, including the *leucine-rich repeat* (LRR) domain, the serine-threonine kinase domain, and the *Toll/IL-1R homology* (TIR) domain, which resembles a region important for signaling in mammalian IL-1R.

In *Drosophila*, at least eight Toll proteins have been described. The original Toll was identified by its role in early *Drosophila* embryonic development, where it is required to specify dorsal-ventral patterning (see Table). A maternally secreted ventral factor called Spätzle serves in its activated form as the ligand for the transmembrane Toll receptor. Engagement of Toll by activated Spätzle results in dimerization of the receptor protein and the recruitment of an adaptor protein called Tube (see figure). Tube subsequently recruits a serine-threonine kinase called Pelle, which is a member of a conserved group of molecules called the <u>serine/threonine innate immunity kinases</u> (SIIK). Interaction of Pelle with Tube activates the former, which leads to the phosphorylation of Cactus, a molecule homologous to mammalian IκB. Just like IκB, unphosphorylated Cactus is an inhibitor that binds to the nuclear transcription factor Dorsal and prevents its translocation into the nucleus. Phosphorylation of Cactus precipitates its ubiquitin-mediated degradation and the release of Dorsal. Dorsal translocates into the nucleus, where it binds to elements in the upstream regulatory regions of genes that must be transcribed for normal larval development.

Where things become more immune system-related is in the adult fly. The expression of Toll mRNA is complex and implies additional functions for this receptor after embryonic development is complete. Immunity in invertebrates relies greatly on the production of anti-microbial peptides, the synthesis of which must be induced. In *Drosophila*, Toll signaling is required for the synthesis of several (but not all) anti-microbial peptides. It is speculated that pathogen molecules (still to be precisely identified) somehow activate Spätzle and initiate Toll dimerization and signaling in hemocytes. However, in this case, the Dif and Relish transcription factors (which are related to Dorsal) are activated. Dif/Relish move into the hemocyte nuclei and initiate new gene transcription, resulting in the production of anti-microbial peptides and their secretion into the hemolymph.

Why would a fly need eight different Toll receptors? In some cases, the engagement of a particular Toll receptor appears to induce the synthesis of a specific anti-microbial peptide. This finding implies that the recognition of a particular pathogen type induces synthesis of the molecule needed to combat that pathogen. For example, *Drosophila* lacking the original Toll fail to induce synthesis of the anti-fungal

Conservation of the Toll Receptor Pathway

Organism	Receptor	Ligand	TF* Activated	Outcome
Embryonic *Drosophila*	Toll	Spätzle	Dorsal	Dorsal-ventral patterning during development
Adult *Drosophila*	Toll	Pathogen product Spätzle	Dif/Relish	Drosomycin (anti-fungal) production
Mouse	TLR2	Lipotechoic acid and peptidoglycan from gram-negative bacteria	NF-κB	Pro-inflammatory cytokine and chemokine production
	TLR3	Viral double-stranded RNA	NF-κB	Pro-inflammatory cytokine and chemokine production
	TLR4	LPS from gram-negative bacteria	NF-κB	Pro-inflammatory cytokine and chemokine production
	TLR5	Bacterial flagellin	NF-κB	Pro-inflammatory cytokine and chemokine production
	TLR6	May cooperate with TLR2	NF-κB	Pro-inflammatory cytokine and chemokine production
	TLR7	Viral single-stranded RNA	NF-κB	Pro-inflammatory cytokine and chemokine production
	TLR8	Viral single-stranded RNA	NF-κB	Pro-inflammatory cytokine and chemokine production
	TLR9	Bacterial unmethylated CpG	NF-κB	Pro-inflammatory cytokine and chemokine production
	TLR10	May cooperate with or have same function as TLR9	NF-κB	Pro-inflammatory cytokine and chemokine production
Human	TLR11	?	NF-κB	Prevention of uropathogenic bacterial infections

*TF, transcription factor.

Continued

Box 21-1. **Conservation of the Toll receptor pathway**—*cont'd*

protein drosomycin and thus succumb to fungal infections.

In humans, at least 11 TLRs have been identified to date. While the human TLRs were originally discovered as molecules with sequence homology to IL-1R, they are actually closer in sequence to *Drosophila* Toll. As mentioned in Chapter 4, the functions of several mouse TLRs have been clarified (refer to Table). Moreover, the component chains of the TLRs can heterodimerize, likely extending the range of pathogens that can be recognized by this family of receptors. Analyses of the various TLR signaling paths have confirmed that they all use the MyD88/IRAK/IκB/NF-κB pathway to induce new gene transcription. Among these new genes are pro-inflammatory cytokines and chemokines such as IL-1, IL-6, and IL-8. There is also evidence that the important costimulatory molecules B7-1 and B7-2 are upregulated on APCs in response to TLR signaling, linking TLRs directly to adaptive immunity as well.

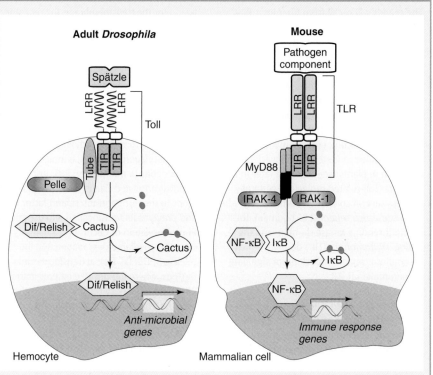

produced, many are induced within a short time of pathogen invasion. This rapid induction is particularly true for insects.

In mammals, PRMs are generally found either as soluble molecules circulating in the plasma (like the collectins) or as membrane-bound receptors expressed on the surface of phagocytes (PRRs). A parallel scenario presumably exists for non-mammalian vertebrates, although this has yet to be formally confirmed. Similarly, some invertebrate PRMs are soluble and present in the hemolymph, while others are PRRs expressed on the surfaces of phagocytic and granular cells. Bacterial or fungal cell wall products such as LPS, peptidoglycan, β1,3 glucan, and various mannose-containing carbohydrates (*mannans*) can be recognized and bound directly by PRMs in most species. Many soluble PRMs in invertebrates are enzymes capable of degrading bound substrates, while others act as opsonins, enhancing phagocytosis of the invader. Still other PRMs activate components of the important host defense cascades, such as complement activation in vertebrates or the ProPO system in higher invertebrates (see later). Engagement of PRRs triggers conserved host defense signaling pathways that lead to new gene transcription essential for an optimal immune response. Like vertebrate macrophages, invertebrate granular cells stimulated by engagement of their PRRs commence aggregation at the site of injury or attack and are triggered to degranulate, releasing factors that initiate or enhance various host defense mechanisms.

In addition to PRMs, non-self is identified by recognition systems involving forms of histocompatibility molecules. Even in some tunicates (protochordates), molecules are expressed that have functions analogous in a general way to mammalian MHC, in that they appear to mediate allorecognition. Early MHC-like proteins and their successors are discussed later in Section C.III, 'MHC.'

III. CYTOKINES

We have seen in mammals that intercellular communication via cytokine secretion is essential for both the innate and adaptive responses, and for coordination between them. There is now evidence that cytokine-like molecules are present even in protozoans, making them one of the earliest elements of immune reactivity to be established. Molecules resembling mammalian cytokines, growth factors, and chemokines have generally been identified in lower organisms by their activity in vertebrate assay systems, so that their precise natural functions have yet to be completely defined. Nevertheless, molecules that functionally and/or structurally equate to IL-1, IL-2, IL-6, IFNγ, and TNF, among others, have been reported in organisms as diverse as arthropods, echinoderms, protochordates and annelids. For example, the invertebrate equivalent of IL-1 stimulates insect hemocytes to proliferate and aggregate, and enhances phagocytosis and encapsulation. Tunicates, fish, and amphibians also produce IL-1-like molecules that can stimulate mouse thymocytes *in vitro*. However, an equivalent to IL-2, which plays an important role in mammalian lymphocyte proliferation, has yet to be identified in an invertebrate species. Amphibians, reptiles, and birds produce forms of IL-2 that are phylum-specific; that is, reptile IL-2 stimulates the proliferation of only reptile T cells and will not stimulate mouse T cells. In echinoderms, hemocytes produce an IL-6-like protein that can stimulate hemocyte aggregation and can also induce the proliferation of mammalian lymphocytes *in vitro*. The equivalent of IFNγ in bony fish (Osteichthyes) can stimulate the respiratory burst of fish macrophages and thus the production of ROI. Interestingly, while birds have been reported to possess IFN-mediated antiviral activity, no such activity has yet been identified in amphibians or reptiles. Molecules related to the leukotrienes (inflammatory molecules in mammals) have also been identified in amphibians and higher fish, where they influence phagocyte migration and lymphocyte proliferation.

Chemokine-like molecules have also been identified in higher invertebrates and protochordates. Damaged or invaded tissue in a higher invertebrate sends out alarm signals thought to be mediated by chemotactic factors that summon both phagocytic and granular cells to the site. Similarly, the migration of bony fish phagocytes can be stimulated by chemokine-like molecules.

IV. PATHOGEN ELIMINATION

i) Phagocytosis and Encapsulation

As mentioned previously, phagocytic cells are ubiquitous throughout the animal kingdom. Chemotaxis, pathogen engulfment, and digestion take place much as we have described for mammals, although the receptors mediating recognition and engulfment differ. From at least the amphibians on up, macrophages and neutrophils mediate phagocytic defense. However, macrophage distribution in the tissues can differ dramatically among the vertebrates. For example, chickens have very few macrophages resident in their respiratory tracts and abdominal cavities compared to mammals, and yet disease resistance is not compromised because the presence of replicating bacteria induces a massive and very rapid influx of macrophages into the affected site. Furthermore, those macrophages that do flood into a site of abdominal infection have particularly efficient bactericidal powers.

In higher invertebrates, various LPS-binding or mannose-binding proteins bind to the surfaces of pathogens, enhancing phagocytosis through opsonization. Pathogens too large to be engulfed are *encapsulated*; that is, the pathogen is surrounded by many phagocytic cells that cooperate to seal the invader off from the invertebrate's open circulatory system, much as hyper-activated macrophages band together to form granulomas in vertebrates. Encapsulated pathogens are then killed by ROI and lysosomal enzymes released by the phagocytic cells. Bactericidal proteins and opsonins released by granular cells enhance both phagocytosis and encapsulation by the phagocytic cells, indicating a degree of cooperation between different invertebrate cell types in host defense.

ii) Coagulation

The wound healing cascade and coagulation offer another means of host defense, particularly in invertebrates. Wound healing involves the coagulation of body fluids in a cascade similar to the blood clotting cascade in vertebrates. In the horseshoe crab, the presence of LPS triggers the activation of a zymogen cascade involving three proteases that results in the cross-linking of the *coagulogen* protein present in the hemolymph. A clot is formed that prevents further leakage of body fluids and walls off any pathogen that has entered the wound. Interestingly, the LPS-binding protease of the coagulation cascade contains domains similar to the SCR repeats found in mammalian complement proteins. The coagulation system is important for another reason. Enzymes activated in the course of coagulation also trigger the activation of another key cascade in invertebrate host defense, namely the ProPO system.

iii) The Prophenoloxidase (ProPO) System

The ProPO system is an important encapsulating mechanism used by many higher invertebrates but, for unknown reasons, not by pre-vertebrates or vertebrates. *Prophenoloxidase* (ProPO) is the zymogen form of the *phenoloxidase* (PO) enzyme and has a molecular mass of 70–80 kDa. In more primitive animals, ProPO is found either in free form in the hemolymph or deposited in the exoskeleton. In more advanced species, ProPO is often sequestered in intracellular vesicles within the granular cells. When microbial LPS, glucans, or peptidoglycans are present, the coagulation cascade is triggered and/or changes to Ca^{2+} concentration or pH occur that activate a series of serine proteases present in the hemolymph (Fig. 21-3). These activated serine proteases cleave ProPO to generate activated PO (molecular mass of 60–70 kDa). Other molecules have been shown to trigger ProPO cleavage, perhaps via associated proteins with MASP-like activities. These triggering molecules include a cockroach agglutinin, certain phospholipids, and soluble molecules in the hemolymph with similarities to vertebrate mannose-binding lectins. PO is an oxidoreductase that oxidizes phenols to produce quinones, which then polymerize non-enzymatically to form the pigment *melanin*. Both melanin and the intermediates in its synthesis are toxic to microbes. A pathogen in the hemolymph caught in this cascade thus ends up as a paralyzed, blackened blob, as the melanin is deposited on it in a process called the *melanization reaction*.

The ProPO system is highly sensitive, such that only a very few picograms per liter of LPS are required to activate the cascade. To prevent premature activation of the ProPO system, the pathway is controlled by the interactions of various proteases and protease inhibitors, including *pacifastin* in crayfish. The activation of the ProPO cascade also generates intermediates that can function as opsonins to enhance phagocytosis and stimulate other proteins participating in host defense. Because of these characteristics, the ProPO system and melanization reaction have been likened to the lectin or alternative complement activation pathways described for mammals in Chapter 19. The activation of the ProPO system also appears to be necessary in some way for the biological activity of *peroxinectin*, a cell adhesion molecule in crayfish hemocytes that can act both as an opsonin and as a peroxidase generating anti-microbial molecules. Peroxinectin and peroxinectin-like molecules in other arthropods are synthesized in granules in the hemocytes and released during degranulation. These proteins have striking functional similarities to mammalian myeloperoxidase, which

The ProPO System of Higher Invertebrates
Microbial insect pathogens shed components such as LPS that are recognized by PRMs on the granular hemocytes of insects. These cells release the contents of their granules, initiating the coagulation cascade in the hemolymph. Among its other activities, this cascade activates serine proteases that convert inactive proPO to active PO. In the presence of O_2, active PO converts phenols to quinone monomers that spontaneously polymerize to form melanin. Melanin deposition is toxic to microbes. Active PO also promotes anti-microbial peroxinectin functions.

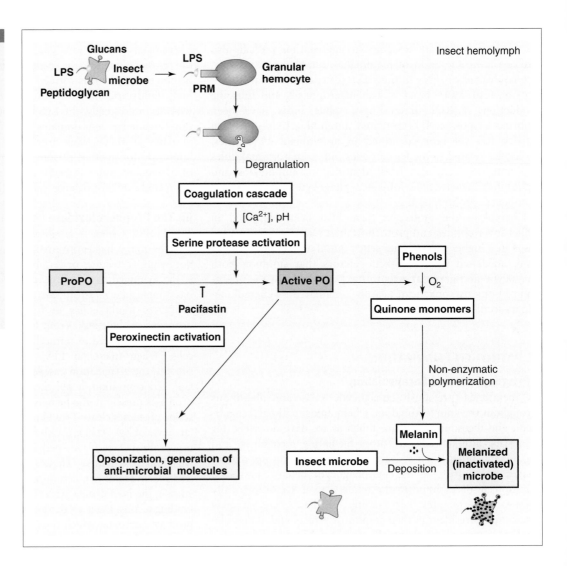

was described in Chapter 4 as contributing to the mammalian innate immune response. Finally, the ProPO molecule has been found associated with an IL-1-like molecule in the tobacco hornworm moth *Manduca sexta*, suggesting that the ProPO system may be linked to invertebrate cytokine activity.

iv) Complement Activation

Elements of a lectin complement activation pathway appear in organisms as primitive as the protochordates and some echinoderms, suggesting that complement activation as a defense mechanism is at least 600–700 million years old (refer to Fig. 21-2). The primordial complement activation pathway appears to have been much reduced in complexity compared to the mammalian cascade covered in Chapter 19. It has been proposed that, in the early deuterostomes (but not the protostomes), the equivalents of MBL, MASP, Factor B, and C3 may have come together in a primitive pathway to facilitate the opsonized phagocytosis of microbial antigens. The evidence suggests that the recognition of foreignness originally fell to an MBL-like molecule able to recognize mannose- or

N-acetylglucosamine-bearing carbohydrate structures on the surfaces of microorganisms. MBL-like molecules first appear among the protochordates, while true MBL homologues have been identified in bony fish (Osteichthyes) and chickens. MBL homologues are presumed to exist in other non-mammalian vertebrates but have not been studied extensively. Homologues of C-reactive protein-like pentraxins have been identified in several pre-vertebrates. While some insects (higher invertebrates) express MBL-like receptors on phagocytic cells present in the hemolymph, genes encoding true homologues of C3 and Factor B have not been found. Thioester proteins vaguely resembling C3 have been found in some arthropods but their function is unclear. Instead, lectins present in insect hemolymph may act as opsonins, coating pathogens resistant to the melanization reaction and promoting their receptor-mediated phagocytosis.

The next step in the primordial complement pathway is thought to have been the interaction of pathogen-bound MBL with a MASP-like serine protease. In theory, the MASP-like enzyme would then act on a Factor B-like protease in an abbreviated cascade. Activated Factor B in turn would cleave a C3-like molecule such that its reactive thioester bond was exposed, allowing the opsonization of foreign entities. As was true for MBL-like molecules, MASP-like molecules are first evident in protochordates. For example, two types of pre-vertebrate MASP proteins with domain structures similar to mammalian MASP-1 and MASP-2 (and thus C1r and C1s) have been found in tunicates. MASP-like proteins have also been found in lampreys (Agnatha), sharks (Chondrichthyes), carp (Osteichthyes), and *Xenopus* (amphibians). Among their several domains, these MASP enzymes contain two SCR domains and a serine protease domain resembling that of mammalian MASP enzymes. The reader will recall that, in mammals, the SCR domains promote interactions between Factor B and C3, and between C2 and C4. It has been proposed that both pre-vertebrate C3 and Factor B could be substrates for pre-vertebrate MASP enzymes. Pre-vertebrate Factor B contains the serine esterase domain necessary for C3 cleavage as well as other domains corresponding to those found in the mammalian enzyme. Interestingly, whereas the mammalian enzyme has three SCRs, the pre-vertebrate molecule has five SCRs. Lastly, in trout (Osteichthyes), a molecule crucial for the activation of both the alternative and classical complement activation pathways has been found that may be the common ancestor of the functionally analogous Factor B and C2 proteins in mammals.

Various forms of pre-vertebrate C3 can be found in the coelomic fluid of sea urchins (echinoderms) and tunicates (protochordates). These molecules generally resemble mammalian C3 in structure and contain reactive thioester sites. However, the C3a sequence is not conserved, suggesting that C3a anaphylatoxin activity evolved in vertebrates. In addition, no separate C4 or C5 genes are present in pre-vertebrates. Functionally, sea urchin C3 promotes phagocytosis via opsonization. Furthermore, injection of a sea urchin with LPS induces the coordinated transcription of homologues of C3 and Factor B in circulating cells called *coelomocytes*. Tunicate hemolymph depleted of C3 by the use of specific antibody can no longer opsonize yeast cells.

In most organisms, C3 is the product of a single gene, but up to five isoforms of vertebrate C3 derived from separate genes have been isolated in the bony fish (Osteichthyes). The proteins encoded by these genes have varying efficiencies of binding to different pathogen surfaces capable of activating the cascade. It is speculated that these animals may have increased their host defense arsenal by duplicating and diverging their C3 genes, expressing multiple forms such that a broader range of pathogens can be destroyed. Such a strategy could compensate for the relatively weak antibody-mediated immunity in bony fish. Only IgM- and IgD-like antibodies are expressed (see later), and these are of weak affinity and low diversity. The lag time for antibody production is also longer than in mammals, making the multiple forms of complement component C3 expressed in these animals that much more important for host defense.

Components specific to the alternative and classical complement activation pathways are thought to have developed some 200–300 million years after the primordial lectin pathway via

sequential gene duplication and divergence. For example, in mammals, C2 and Factor B are closely related, and C1r and C1s are similar in structure and function to MASP-1 and MASP-2. The terminal complement components first appear in the cartilaginous fish (Chondrichthyes), that is, after the evolution of jaws in vertebrates about 500 million years ago. The terminal complement proteins are closely related in structure, suggesting that they also developed via gene duplication and divergence. The cartilaginous fish are thus the first animals in which forms of the alternative and lectin pathways can mediate host defense by direct MAC-mediated lysis of pathogens as well as by opsonization. The presence of the terminal complement components also allows the full utilization of the classical pathway of complement activation.

Effective use of complement activation as a defense mechanism requires the presence of complement receptors on effector cells. Complement receptors have been found on phagocytes of several lower organisms. For example, a CR1-like receptor able to bind to yeast cells coated with mammalian C3b was identified on coelomocytes in echinoderms and on hemocytes in hagfish (Agnatha). In addition, various types of CR3-like complement receptors have been found in echinoderms and right on up through the animal kingdom. The finding of complement receptors in such a wide range of organisms confirms that these molecules evolved soon after phagocytosis as a means of enhancing this method of pathogen disposal. With respect to the RCA proteins, the large number of regulatory factors acting on the mammalian complement system suggests that similar, perhaps simpler, controls might exist for the pre-vertebrate system; however, little evidence of such proteins currently exists.

We include a last note on the synthesis of complement components. The reader will recall that, in vertebrates, the soluble complement components are made primarily in the liver. However, echinoderms lack a liver, and their complement components are thought to be produced almost exclusively by coelomocytes. In tunicates (protochordates), which are more advanced but still lack a liver, the complement components are made by a primordial structure called the *hepatopancreas*.

v) Anti-microbial Proteins and Peptides

In Chapters 3 and 4 we mentioned various anti-microbial proteins and peptides (such as lysozyme and the defensins) that are important for innate immunity in mammals. Lower organisms also express a variety of anti-microbial proteins and peptides that make a major contribution to immune defense. Even plants have developed anti-microbial proteins and peptides to eliminate pathogens, as described in Box 21-2. Some anti-microbial proteins are constitutively present in the hemolymph of lower invertebrates and pre-vertebrates. Examples include lysozyme, which kills gram-positive bacteria, and destructive enzymes such as *hemolysin*. A defensin-like molecule has also been found in a primitive fish. Other anti-microbial molecules are inducible, such as some of the PRMs previously described. Most insects (arthropods) also express small (usually) cationic peptides with short-lived, toxic activity against a broad range of gram-positive and gram-negative bacteria, fungi, protozoa, and parasitic worms. It is thought that many of these peptides work by damaging

Box 21-2. **Immunity in plants**

This chapter and indeed this book are concerned primarily with immunity in the animal kingdom, but it would be remiss of us not to touch at least briefly on defense mechanisms operating in plants. A plant is sedentary and has neither a circulatory system capable of transporting cells nor antigen receptors formed by somatic gene rearrangement. Nevertheless, these organisms have their own complex immune systems capable of recognizing and eliminating non-self. Rather than adaptive immunity, a plant defends itself with a series of innate measures, some of which are constitutive and some of which are inducible.

How does a plant recognize non-self and thus know when to initiate an immune response? Most often, a microbe attacking a plant gives its presence away by secreting the product of an *avirulence (avr)* gene directly into the plant cell. The plant has disease resistance genes known as *R* genes whose products can interact with the products of the microbial *avr* genes to induce an immune response that neutralizes the pathogen. However, different microbes secrete different *avr* products, and different plants have different *R* genes. Only when the *R* product of the plant and the *avr* product of the attacking pathogen "match" can the plant respond by walling off the pathogen and then degrading it.

The successful interaction of an *R* product with an *avr* product induces a series of events in the infected plant cell that allow it to destroy the pathogen (see Figure). Changes to the flux of Ca^{2+} and other ions across the plant cell membrane occur, followed by the activation of various MAP kinases and the production of reactive oxygen intermediates (ROI) by an enzyme that is similar to that mediating the respiratory burst in mammalian neutrophils. These ROI, along with nitric oxide (NO) also released by the cell under attack, then contribute to the induction of the *hypersensitive response* (HR). In an HR, the plant cells and tissues in immediate contact with the pathogen undergo rapid necrosis, releasing the contents of the damaged plant cells. These contents include hydrolytic enzymes (HL) capable of damaging pathogen structures. In addition, dramatic metabolic changes occur in plant cells surrounding the site of the HR that lead to a state of *local acquired resistance* (LAR).

Glycoproteins, polysaccharides, lignin, and other molecules are deposited and cross-linked to fortify the walls of plant cells in the area of LAR and thus trap the pathogen. The synthesis of an army of small *anti-microbial peptides* (AMP) is then induced in LAR cells, including the plant defensins, thionins, lipid transfer proteins, and phytoalexins. These molecules are thought to act directly (in different ways) on the plasma membrane of the pathogen. Various regulatory peptides, proteins, lipids, and other compounds are also induced during the LAR response, including *salicylic acid* (SA), *ethylene* (ET), and lipid-derived molecules called *jasmonates* (JA). Ethylene and jasmonates are plant hormones that are considered analogous to inflammatory mediators in mammals. SA, ethylene, and jasmonates induce the expression of plant *pathogenesis-related* (PR) proteins (see later). Among the PR gene products are chitinases and β1,3 glucanases capable of degrading the pathogen cell wall. Small degradation products are then released that can induce additional plant responses, amplifying the defense. The pathogen also secretes hydrolases that can break down the plant cell wall, resulting in small oligo-saccharides with biological activity that can contribute to a type of "danger" signaling.

The induction of the HR followed by a state of LAR can lead to the eventual establishment of *systemic acquired resistance* (SAR), a state in which the entire plant is immune from attack by a broad range of pathogens. The presence of elevated concentrations of SA and PR proteins characterize SAR. We know that SA is particularly important for the establishment of SAR because transgenic plants expressing the *nahG* gene (which encodes a hydroxylase that inactivates SA) remain susceptible to pathogen attack and fail to induce expression of certain PR proteins. However, it is now believed that NO (rather than SA) is the traveling molecule that conveys the signal to establish SAR throughout the entire plant. A protein called NPR1 that is structurally (but not functionally) related to IκB is involved in promoting transcription of SA-induced PR genes. Other non-overlapping sets of genes encoding anti-microbial proteins are induced by ethylene and jasmonates. The time required to establish SAR varies from plant

to plant, and the level of protection depends on the degree of necrosis in the HR and the level of signaling mediators in the area of LAR. The SAR state can endure for several weeks in some plant species.

How do the *R* genes trigger the HR response? It is thought that most *R* genes encode protein products that trigger intracellular signaling when engaged by pathogen molecules such as the *avr* products. Over 20 *R* genes have been identified whose products fall into five structurally distinct classes of proteins. Some *R* proteins are transmembrane molecules and interact with the pathogen outside the plant cell. Others are cytoplasmic such that the pathogen must actually penetrate the plant cell wall and secrete *avr* peptides in order to be recognized and trigger the HR. The various forms of *R* proteins do share some structural features, notably a region rich in leucine repeats and a nucleotide-binding site. This leucine-rich region is homologous to the LRR domain of Toll-related receptors and appears to be responsible for pathogen recognition. Some *R* proteins also have a serine-threonine kinase domain marking them as belonging to the SIIK group. In addition, the cytoplasmic domains of certain *R* proteins, such as the N protein in tobacco, the Cf-9 protein in tomato, and the Xa-21 protein in rice, contain regions similar to the TIR domain present in the cytoplasmic regions of *Drosophila* Toll and mammalian TLRs. Molecules resembling IκB and various mammalian and/or *Drosophila* enzymes and adaptor proteins have also been identified in plants. Notably, the Pto serine-threonine kinase of tomato is homologous to *Drosophila* Pelle and mammalian IRAK-1. Avr product binding to Pto may induce autophosphorylation of Pto and downstream signal transduction leading to an HR. Finally, proteins interacting with another *R* protein in tomato have been shown to bind to the promoters of genes encoding certain PR proteins. Precise delineation of the signaling pathways downstream of *R* proteins awaits further experimentation.

The PR proteins are also under intense investigation to determine their structures and modes of action. At least 11 major families of PR proteins across multiple plant species have been identified. The PR proteins are grouped according to

Continued

Box 21-2. Immunity in plants—*cont'd*

sequence similarity and biological activity, and include β1,3 glucanases, endochitinases, proteinases, proteinase inhibitors, peroxidases, RNAses, inhibitors of pathogen hydrolases, and others. Most of these enzymatic activities are directed against pathogen structures such as elements of the cell wall or the plasma membrane. Some PR proteins dubbed *permatins* or *osmotins* are thought to create pores through pathogen membranes, altering the internal osmotic balance. This means of defense is particularly effective against fungi.

With the current commercial importance of so many plant species, much research is focused on practical ways of boosting plant immunity. Transgenic manipulation resulting in the constitutive expression of PR genes encoding chitinase or β1,3 glucanase has been shown to protect crops from fungal attack. Furthermore, the expression of more than one PR gene has a synergistic effect on plant disease resistance. These observations have spurred investigators to examine the regulatory elements upstream of PR gene promoters with the hope of identifying signals that might be used to generate a response of increased strength against an expanded range of plant pathogens.

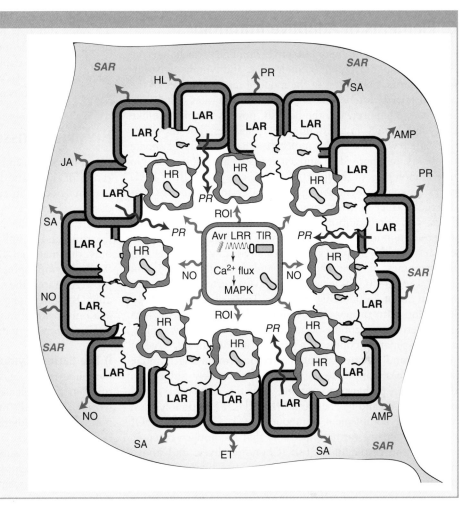

the membranes of microbial but not eukaryotic cells. The cationic peptides are attracted to the anionic phospholipid head groups that are present in microbial membranes, and either bore a hole through the membrane or otherwise interfere with membrane structure. Investigation of the production of these peptides in insects has shown that the expression of many is inducible and regulated. Transcription factors binding to sequences similar to the κB motif recognized by the mammalian transcription factor NF-κB are often involved. Some of these transcription factors are homologous to the mammalian transcription factors regulating the expression of the acute phase response genes in mammals.

What are the major anti-microbial peptides and proteins in invertebrates? The *cecropins* (first discovered in the giant silkmoth *Hyalophora cecropia*) exhibit relatively short-lived anti-microbial activity against a broad range of bacteria. The insect cecropin *haemolin*, which contains regions that structurally resemble Ig domains, is particularly important for the binding of LPS and thus for promoting opsonization and phagocytosis. The *agglutinins* found in insects act as sugar-binding opsonins, while the *attacins* are induced proteins with anti-microbial activity against a limited range of gram-negative bacteria. In *Drosophila*, *drosocin* and *diptericin* combat gram-negative

bacteria, while *drosomycin* and *metchnikowin* are anti-fungal peptides. Silkworms possess a protein called *gram-negative binding protein* (GNBP), which, not surprisingly, binds to gram-negative bacteria. GNBP resembles a glucanase but does not appear to have any actual β1,3 glucanase activity. It may therefore function as a tag on foreign entities, signaling for either opsonization or the recruitment of other bactericidal molecules. *Factor G* in horseshoe crabs has a β1,3 glucan binding subunit that binds to glucans and a serine protease subunit that directly activates the coagulation cascade, stimulating the activation of the ProPO system leading to pathogen melanization. *Lipophorin* found in insect hemolymph binds to LPS and facilitates its removal.

Moving higher up the evolutionary tree, glands in the skin and gut of the amphibian *Xenopus* produce a family of small (20–46 amino acid) anti-microbial peptides called *magainins*, which kill a broad range of fungi, protozoa, and gram-negative and gram-positive bacteria. (The name "magainin" is derived from a Hebrew word meaning "shield.") Peptides similar to the magainins and cecropins have also been identified in a tunicate, and molecules related structurally to the insect cecropins have been found in pig intestine. These observations emphasize the conservation of this ancient mechanism of host defense.

C. Elements of Adaptive Immunity

Elements associated with adaptive immunity, such as true B and T lymphocytes, MHC molecules, and lymphocyte antigen receptors, occur first in the modern representatives of the ancient jawed vertebrates. The jawless vertebrates (Agnatha), such as the lamprey and the hagfish, possess lymphocyte-like cells and mount induced immune responses, but these are associated with lectin-like molecules characteristic of innate responses, as opposed to antigen receptor-like molecules. No true homologues of the Ig, TCR, MHC class I and II, or RAG genes have been identified in Agnatha or in any species that diverged earlier in invertebrate and protochordate phylogeny. Thus, a major evolutionary shift giving rise to conventional adaptive immunity occurred between the emergence of jawless and jawed vertebrates. Another interesting finding is that the genomes of protochordates and invertebrates contain about 15,000–20,000 genes, while jawed vertebrate genomes contain about 25,000–30,000 genes. These observations have led scientists to propose that multiple duplications of regions of the vertebrate genome occurred as the jaw was developing, allowing separate copies of a gene to mutate and diverge in sequence and function. Genomic regions containing immune system genes appear to have been particularly subject to duplication, perhaps accounting for the similarities seen between TCRs and Igs, between MHC class I and II molecules, and among complement components C3, C4, and C5. Of course, it remains formally possible that the antigen receptor genes in the cartilaginous fish (Chondrichthyes) and above have diverged so much from antigen receptor-like genes in the jawless vertebrates that the latter cannot readily be identified using either cross-hybridization techniques or genome scanning. In any case, with the development of cartilaginous fish, recognizable Ig, MHC, and TCR genes can be detected, and somatic rearrangement of variable and constant gene segments can be demonstrated for the BCR and TCR loci in most vertebrates. However, isotype switching and affinity maturation do not occur in pre-vertebrates or fish and are seen only in the amphibians and above.

Many transcription factors and other molecules important in lymphocyte development and activation are also first seen in jawed vertebrates. For example, multiple members of the Src family of signaling kinases, which includes Fyn, Lck, and Lyn, have been identified in various forms in animals as primitive as the cartilaginous fish. Similarly, homologues of the transcription factor Ikaros, required for the lineage differentiation of mammalian NK, T, and B cells, are expressed during the early development of lymphoid cells in animals as diverse as trout, *Xenopus*, and chicken. Various forms of Ikaros and/or the related transcription factor Aiolos have also been found in the skate and lamprey. Skates also possess molecules resembling GATA-3, GATA-1, Pax-5, and EBF-1. Molecules related to the mammalian transcription factor PU.1 occur both in cartilaginous fish and in the lamprey (Agnatha). A homologue of BTK (Bruton's tyrosine kinase), which is involved in lineage discrimination between B and T cells, has also been identified in cartilaginous fish. The reader should note, however, that these homologues may have functions in lower animals that are quite different from the roles they assume in mammals, despite their nucleotide sequence similarities.

We start our exploration of the phylogeny of elements of adaptive immunity with an examination of lymphoid tissues and cells in various organisms.

I. LYMPHOID TISSUES

As we have seen in earlier chapters, the generation and maturation of the T and B cells that mediate adaptive immunity in mammals occur in complex primary and secondary lymphoid tissues. The full complement of lymphoid tissues includes the MALT (GALT plus NALT/BALT), SALT, thymus, spleen, bone marrow, and lymph nodes with germinal centers. However, only the mammals and birds contain all of these features. While T lymphopoiesis almost always occurs in the thymus in pre-vertebrates, the site of B lymphopoiesis can vary considerably. By examining the immune response-related cells and anatomical structures present in lower organisms, we can trace the evolution of modern mammalian lymphoid tissues.

i) The Earliest Lymphoid Elements

Lymphocyte-like cells derived from mesoderm have been identified in the coelenterates (jellyfish), annelids, and nemertine worms. These cells may be involved in distinguishing self from non-self and mediating the primitive forms of graft rejection that occur in these animals. If these lymphocyte-like cells are stimulated with substances known to activate mammalian lymphocytes, such as ConA or LPS, the cells show functional responses reminiscent of both B and T cells. Lymphocyte-like cells in the echinoderms and protochordates carry out recognizable graft rejection and natural killing responses, and in protochordates at least, these cells can be stimulated by mammalian mitogens.

ii) Lymphoid Tissues in Jawless Fish (Agnatha)

We see the beginnings of organized lymphoid tissue in the first true vertebrates, namely jawless fish like the hagfish and the lamprey (refer to Fig. 21-2). In these animals, lymphocyte-like cells that develop from the mesoderm invade tissues derived from the endoderm layer. Such tissues include those associated with the gut so that the first GALT is formed. Note that agnathans still lack both a proper spleen and a thymus, organs that are present in all jawed vertebrates. Agnathans also lack the lymphoid follicles found in the intestinal lamina propria in all other vertebrates. As we saw in Chapter 18, these are crucial sites for mucosal immunity and the thymus-independent development of IELs.

The primitive lymphocytes found in both the GALT and parts of the kidney of the lamprey appear to lack specialization and exhibit some of the functional characteristics of both B and T cells. While these cells do not possess TCR or Ig genes, they do express surface receptors called *variable lymphocyte receptors* (VLRs) that show somatic variation. The VLRs are GPI-anchored cell surface proteins containing a variable number of leucine-rich repeats (LRRs; refer to Box 21-1). The VLR locus contains a large panel of different LRR sequences that can be inserted as a cassette into a basic germline gene framework to give the complete coding sequence of a given receptor. The cassette chosen is different in different lymphocyte-like cells, giving this population recognition powers that faintly echo the adaptive response in jawed vertebrates.

iii) Lymphoid Tissues in Cartilaginous Fish (Chondrichthyes)

As we move up the evolutionary tree to the cartilaginous fish, we find that highly organized GALT is present from here on up. No bone marrow, lymph nodes, or Peyer's patches are yet present, but both spleen and thymus can be found. The spleen possesses both red and white pulp and is important for the generation of both lymphocyte-like cells and erythrocyte-like cells. Lymphomyeloid tissue is also found in the kidney and liver of sharks and rays. The lymphocyte-like cells in sharks show some differentiation of B and T cells, in that recognizable plasma cells can be identified following immunization. However, T cell-dependent immune responses have been difficult to demonstrate. A molecule resembling CD45, required in vertebrates for the regulation of B cell signaling, has been identified in the horned shark.

Many cartilaginous fish also possess two unique lymphoid tissues called the *Leydig organ* and the *epigonal organ*. The Leydig organ is composed of two lobes of tissue positioned on top of and below the esophagus. The epigonal organ is a single tissue mass positioned behind the gonads. Both these organs contain cells resembling mammalian lymphocytes and myeloid cells. Coordinated expression of Rag-1, TdT, IgM, and IgX (see later) as well as Ig light chains can be detected in both the Leydig and epigonal organs in embryos. The TCR genes are not expressed in these organs.

As mentioned earlier in this chapter, scientists speculate that the development of a jaw and a highly predatory lifestyle exposed the first jawed fish to new threats of injury and infection. Organized GALT mediating local immunity may have thus developed in the gut, where a new collection of pathogens was suddenly being encountered. As we learned in Chapter 18, the majority of IELs (intraepithelial lymphocytes) maturing in the GALT (and thus independently of the thymus) are CD8αα IEL T cells bearing γδ TCRs. The reader will recall that γδ TCRs function more like PRRs than conventional antigen receptors. That is, γδ T cells recognize heat shock proteins, and bacterial non-peptide antigens such as phosphorylated nucleotides and certain pyrophosphates, without involving MHC. These receptors and the cells that express them are thus considered to be an evolutionary bridge between local innate immunity and the systemic adaptive immunity mediated in vertebrates by thymus-dependent αβ T cells and bone marrow-dependent B cells.

iv) Lymphoid Tissues in Bony Fish (Osteichthyes)

In the bony fish, lymphocytes are generated primarily in the spleen and kidney. The GALT of bony fish is well developed, with lymphocytes present in the gut epithelium and in follicles in the intestinal lamina propria. Concentrations of macrophages and granulocytes occur in the gut epithelium and lamina propria. The thymus contains mostly T cells but also appreciable numbers of B cells. However, T and B lymphocyte functions have clearly diverged in the bony fish, since a population of thymic lymphocytes can be isolated that responds exclusively to mammalian T cell mitogens. Similarly, among kidney lymphocytes, a different lymphocyte subset can be found that responds only to B cell mitogens. Thymectomy in bony fish inhibits both T and B cell functions, implying the existence of T-B cooperation. Indeed, both Td and Ti responses can be demonstrated in intact bony fish, with Td responses requiring APCs and T cells as is true in mammalian responses. However, only very weak B cell memory responses have been observed in these animals, likely due to the lack of germinal centers in their lymphoid tissues.

Some very interesting work in the seahorse (which is actually an advanced bony fish) has shed more light on a possible relationship between the evolution of the jaw and organized GALT. Seahorses lack jaws and instead use a siphon mechanism to eat. Perhaps not coincidentally, GALT is not well developed in the seahorse. Unlike other bony fish, there are no lymphocytes in the gut epithelium or intestinal lamina propria, and the spleen is missing. Scientists speculate that the loss of the jaw in the seahorse may have reduced the incidence of injury or infection in gut, meaning that GALT may have become dispensable to the animal in a kind of reverse evolution.

v) Lymphoid Tissues in Amphibians

The lymphoid tissues of several amphibians have been relatively well-studied, revealing that the presence and degree of organization of lymphoid components varies among member species. Diffuse and nodular GALT, a spleen, and a well-developed thymus are present in almost all species. Some higher amphibians have evidence of bone marrow-like tissue although lymphopoiesis does not occur here; amphibian lymphopoiesis occurs primarily in regions of the kidney, spleen, and liver. Bone marrow-like tissue in *Xenopus* contains mainly granulocytes, although some antibody-producing B cells can be found. The majority of B cells in *Xenopus* mature in the liver and spleen. Some higher amphibians also have rudimentary lymph node-like structures that can function in a limited way to trap antigen. However, these structures lack true germinal centers. Without functional germinal centers, the higher affinity mutations generated by somatic hypermutation cannot be successfully selected, meaning that antibody responses are often not much stronger in the secondary response than in the primary response. Primary responses in *Xenopus* peak at 4 weeks post-immunization, considerably later than the 5–7 days characteristic of mammalian responses. Upon a second exposure to an immunogen, specific antibody levels are increased only about 10-fold in *Xenopus* (compared to 100- to 1000-fold in mammals) in a response that peaks at 2 weeks post-immunization. In many lower amphibians (such as the axolotl, a Mexican salamander), the levels of antibody produced in the secondary response are no greater than in the primary response, although the peak is reached more quickly.

The amphibian spleen closely resembles the mammalian spleen in that distinct regions of white and red pulp are present, but differs from the mammalian organ in that germinal centers are missing. The splenic white pulp is populated mainly by maturing B cells, while T cells are present in the marginal zone. In the thymus, the epithelium contains cells resembling mammalian T lymphocytes that mature in this organ. Both positive and negative thymic selection have been demonstrated in *Xenopus*. Macrophages, DCs, granular stromal cells, and antibody-producing B cells are also found in the thymus. Thymectomy abolishes all T cell-mediated responses, including allograft rejection and Td antibody production, but not responses to Ti antigens.

vi) Lymphoid Tissues in Reptiles

In contrast to the amphibians, the immune systems of reptiles have not been particularly well studied. Sites of hematopoiesis and the numbers and proportions of leukocytes vary with the species of reptile but the full complement of cell types is present. Both B and T lymphopoiesis occur in the fetal spleen and bone marrow in many reptile species, with the bone marrow becoming the more important organ in the adult. B cells mature in the bone marrow. Reptiles have a well-developed thymus where T cells mature and a fully functional spleen that exhibits definite white and red pulp regions. Functionally distinct T and B cells are present in the spleen but it is unclear whether there are specific T cell-rich and B cell-rich zones in this organ. Rudimentary lymph nodes can be found that still lack germinal centers. GALT is present in most reptiles but develops after the spleen and does not appear to be as important as GALT in pre-vertebrates. In fact, GALT is undetectable in insectivorous (as opposed to carnivorous) lizards. All antibody responses analyzed in reptiles to date appear to be T-dependent.

vii) Lymphoid Tissues in Birds

Birds are the animals closest to mammals in terms of lymphoid tissue development. In addition to GALT, thymus, spleen, and bone marrow, birds have lymph nodes with germinal centers which presumably permit the development of the strong memory responses observed in these animals. T cell precursors arising in the bone marrow mature in the thymus. The spleen, which also contains germinal centers and whose architecture is quite similar to that of a mammalian spleen, has distinct areas in which $\alpha\beta$ T cells, $\gamma\delta$ T cells, and B cells are each concentrated. Rudimentary Peyer's patches have also been identified in some avian species. Birds are unique in having an organ located in the cloaca called the *bursa of Fabricius*. Very early during embryogenesis, stem cells produced in the bone marrow migrate to the bursa, where they mature into B cells. These B cells are continuously released to the periphery until the bursa involutes shortly after hatching.

Having described the lymphoid tissues in which lymphocytes or lymphocyte-like cells develop, we now move to a discussion of antigen receptors, MHC molecules, and the genes encoding them. We examine first the BCR, then the MHC, and finally the TCR.

II. BCR

Immunoglobulins do not occur in invertebrates, although these animals do possess a variety of molecules exhibiting Ig domain structures. For example, in insects, the haemolin molecule is a member of the immunoglobulin superfamily. Haemolin occurs both as a soluble molecule in the hemolymph and as a membrane-bound protein on the surfaces of phagocytic hemocytes. Haemolin expression is induced or upregulated upon infection. Haemolin has been shown to bind to both whole bacteria and LPS, suggesting that it plays a major role in the recognition of non-self. Haemolin also has the intriguing property of self-interaction, or *homophilia*. It is speculated that free haemolin in the hemolymph binds to invading bacteria, and that the Ig domains in this bacteria-bound haemolin can then interact with the Ig domains of haemolin fixed in the membrane of phagocytic cells. Phagocytosis of the haemolin-bound invader is then stimulated. Other surface receptors may also recognize haemolin and further facilitate opsonized phagocytosis. In addition, the binding of free haemolin to membrane-bound haemolin on hemocyte surfaces can prevent the hemocytes from sticking together, inhibiting cellular aggregation that might impede an immune response.

Moving into the vertebrate world, we find that conventional antibodies are not produced in jawless fishes (Agnatha), meaning that neither a conventional adaptive humoral response nor the classical pathway of complement activation is available for host defense in these animals (refer to Fig. 21-2). However, in response to immunization with human RBCs or gram-negative bacteria, jawless fish produce a labile, 310 kDa non-Ig serum protein molecule that appears to recognize immunogens. It is not clear how this protein relates to the VLR proteins mentioned previously. True IgM-like antibodies, in which μ-like heavy chains containing a V region and four constant domains are linked by disulfide bonds to light chains, finally appear in the cartilaginous fish (Chondrichthyes). However, classical rearrangement of the Ig genes does not occur until the bony fish (Osteichthyes), even though the somatic recombinases RAG-1 and RAG-2 are present in all animals from the cartilaginous fish on up. The RSS (recombination signal sequences) and the gene encoding the TdT enzyme (required for junctional diversification) are also remarkably conserved, starting with the cartilaginous fishes. The memory response in cold-blooded vertebrates (amphibians, reptiles) is weak and the secondary antibodies produced are of relatively low affinity. It is not until we get to the warm-blooded vertebrates (birds, mammals) that we observe powerful memory responses characterized by an increased rate and level of production of higher affinity antibodies.

Where do the Ig genes originally come from? The available evidence suggests that the six loci that undergo V(D)J recombination in mammals, namely *Igh*, *Igl*, *TCRA*, *TCRB*, *TCRG*, and *TCRD*, may all have evolved some 300–400 million years ago from a common ancestral gene that originally was not capable of rearranging. Indeed, the genomes of most species of bony fish contain a group of antigen receptor-like genes that cannot rearrange but that contain regions very similar to the V and J regions of the Ig and TCR genes.

i) Immunoglobulin Heavy Chains

ia) Ig heavy chains in cartilaginous fish (Chondrichthyes).

Cartilaginous fish are the first organisms to possess true immunoglobulins. The presence of an IgM-like antibody allows for the first time not only specific B cell-mediated effector functions but also the complete array of complement activation pathways. No isotype switching occurs in these animals so that Ig isotypes corresponding to the IgG, IgA, and IgE of mammals do not appear. However, the cartilaginous fish are equipped with two additional Ig heavy chain genes called IgW (also known in the past as IgX and NARC [*new antigen receptor from cartilaginous fish*]) and NAR (*novel antigen receptor*). NAR is an interesting molecule, in that its V region differs considerably in amino acid sequence and structure from conventional V regions in mammalian Ig and TCR molecules. Unlike IgW, NAR does not appear to form dimers with Ig light chains (see later) but instead functions as an independent entity. Like mammalian Igs, both transmembrane and secreted forms of shark Igs are generated. Both clonal selection of B cells producing specific antibodies and allelic exclusion of Ig heavy chains are thought to exist.

In the shark genome, which is thought to be representative of the genomes of cartilaginous fish, the IgM-like heavy chain genes are organized into about 200 discrete functional clusters of V, D, J, and C segments that are often joined in the germline (Fig. 21-4). A transmembrane (TM) exon is also present. Although RAG and TdT activities are present in cartilaginous fish and all gene segments are surrounded by RSS,

their functions are less important in a species with germline-joined Ig genes. Unlike the situation in mammalian Ig genes, V(D)J recombination at the DNA level between segments of different clusters does not occur and transcription proceeds directly from the germline DNA. However, in the IgM loci of some Chondrichthyes species, there are more D segments per cluster than in mammals, which greatly increases junctional diversity. The other *Igh*-type loci are also of the cluster type but are unlinked and spread over multiple chromosomes. IgW clusters contain V_w, D_w, J_w, and C_w segments homologous to the corresponding V, D, J, and C segments of the IgM locus, plus four additional constant exons. There is also a TM exon.

Much of the diversity in the shark and skate B cell primary repertoires is derived from direct somatic hypermutation of the cluster gene sequences in the germline. The resulting somatic variation is greatest for the shark NAR locus, less so for the IgM locus. Because hypermutation occurs during B cell development (as opposed to during the secondary response, as occurs for the human and murine Ig genes), Chondrichthyes species exhibit neither true memory B cell responses nor antibody affinity maturation. The reader will recall that for affinity maturation to take place in humans and mice, the B lymphocytes involved must be located in the germinal centers of the lymph nodes, structures that are missing in cartilaginous fish. Thus, even if higher affinity antibodies are generated by hypermutation, the lack of germinal centers means they cannot be selected, so that the collection of antibodies produced is effectively no different from that produced in the primary response.

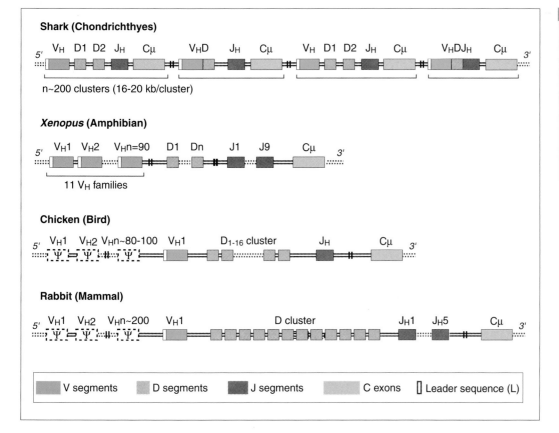

Shark (Chondrichthyes)

$5'$ V_H D1 D2 J_H $C\mu$ V_HD J_H $C\mu$ V_H D1 D2 J_H $C\mu$ V_HDJ_H $C\mu$ $3'$

n~200 clusters (16-20 kb/cluster)

Xenopus (Amphibian)

$5'$ V_H1 V_H2 V_Hn=90 D1 Dn J1 J9 $C\mu$ $3'$

11 V_H families

Chicken (Bird)

$5'$ V_H1 V_H2 V_Hn~80-100 V_H1 D_{1-16} cluster J_H $C\mu$ $3'$

Rabbit (Mammal)

$5'$ V_H1 V_H2 V_Hn~200 V_H1 D cluster J_H1 J_H5 $C\mu$ $3'$

■ V segments   ■ D segments   ■ J segments   ▯ C exons   ▯ Leader sequence (L)

Figure 21-4

Comparison of Loci Producing IgM Chains
The arrangement of V, D, and J segments and C5 exons in loci encoding IgM heavy chains is shown for shark, *Xenopus*, chicken, and rabbit. Not all heavy chain exons are shown. The reader is referred to Figures 8-1 and 8-2 for mammalian *Igh* locus structure.

ib) Ig heavy chains in bony fish (Osteichthyes). As we ascend the evolutionary tree to the higher fish, the organization of the Ig heavy chain genes becomes much more dispersed like that in mammals, rather than the cluster paradigm typical of cartilaginous fish. In concert, RAG-mediated rearrangement becomes more important. In the ray-finned fish (sturgeon, garfish), which evolved between cartilaginous fish and the most advanced bony fish, a single IgM locus is found containing closely linked V_H, D_H, and J_H segments situated within 100 kb of four C_H exons and two TM exons. Antibody diversity derived from combinatorial joining is considerable in these animals because of their relatively large numbers of V_H, D_H, and J_H segments. In some species, another Ig heavy chain locus exists that encodes an IgM-like C_H exon plus six to seven other novel C_H exons. A C_H exon thought to resemble mammalian C_δ is located 3' of the IgM-like C_H exon, and the corresponding heavy chain is co-expressed with the IgM-like chain by some B cells in catfish. However, the hinge region found in mammalian IgD does not appear in catfish immunoglobulins. Further differences between fish and mammalian Ig proteins have been noted in amino acid distribution, degree of glycosylation, hydrophobicity of the constant regions, and splicing patterns of the TM exons. It is in the ray-finned fish that we also find the first evidence for proteins associated with the membrane-bound forms of these Ig molecules that may have the signal transduction functions of $Ig\alpha/\beta$ in mammals. An Igh enhancer-like sequence has also been identified downstream of the last C_H exon in catfish.

Bony fish (Osteichthyes) demonstrate a weak B cell memory response characterized by an increase in antibody levels and a rate of production greater than that exhibited in the primary response but still far less than that observed in mammals. Somatic hypermutation occurs in bony fish but is associated with only a very modest affinity maturation of antibodies. Isotype switching apparently does not occur, since the secondary response antibody is of the IgM isotype. High priming doses of antigen favor effector cell generation over memory development and result in almost negligible memory responses. Only low priming doses, which do not fully activate the effector program and favor the formation and/or development of memory cells, result in memory responses in these fish.

An evolutionary paradox can be found when one examines the Ig heavy chains of the lobe-finned fish, the primitive class of bony fish thought to have emerged from the sea to give rise to the first land vertebrates. Examination of the Ig genes and proteins in the lobe-finned African lungfish has revealed the presence of an IgW gene very similar to that present in the cartilaginous fish. However, neither IgW nor NAR is found in either the intermediate ray-finned fish or the advanced teleost bony fish. It remains a mystery why mammalian Ig heavy genes are more similar to those of the teleosts rather than to those of the lobe-finned fish from which land-based mammals are ultimately derived.

ic) Ig heavy chains in amphibians. The *Igh* locus in amphibians is organized like that in mammals and not in the clusters. In *Xenopus*, a relatively advanced amphibian, there are about 90 V_H segments that can be classified into 11 distinct V_H families. These V_H segments are interspersed among multiple D_H and J_H segments (refer to Fig. 21-4). As in mammals, the order of RAG-mediated segment joining is D_H to J_H, followed by V_H to $D_H J_H$. Regulatory octomer sequences resembling those in mammalian *Igh* loci are present 5' of the *Xenopus* V_H segments. In the axolotl, a lower amphibian, there are fewer gene segments than in *Xenopus*, V(D)J recombination is not as extensive, and antibody affinities are lower than those in higher fish. Somatic hypermutation occurs in different species to different extents.

The isolation of B cells expressing BCRs of a single antigenic specificity has confirmed that allelic exclusion operates in amphibians. More importantly, isotype switching first emerges in the higher amphibians. The *Igh* locus of *Xenopus* contains three distinct constant exons, giving rise to IgM, IgX, and IgY chains. It is thought that the ancestral IgM exon duplicated and diverged to produce the IgX and IgY exons. Both CDRs and framework regions can be identified in each heavy chain. IgX, found principally in the gut, appears to have an IgA-like function, but IgY is not easy to equate to mammalian Ig isotypes. Functionally, IgY appears to be the ancestor of the uniquely mammalian isotypes IgG and IgE, and assumes the workhorse role that IgG performs in amphibians, reptiles, and birds. No IgD-like exon has been found in *Xenopus*. Sequences containing multiple palindromic repeats that resemble the S switch regions in mammals appear to mediate isotype switching in *Xenopus*. Particularly in the secondary response, the *Xenopus* $S\mu$ and Sx regions preceding the IgM and IgX exons, respectively, come together to delete the IgM exon, causing the B cell to express IgX. A secretory form of IgY, but not IgX, has been found in axolotl. In some species of lower amphibians, isotype switching apparently cannot occur at all.

id) Ig heavy chains in reptiles. The analysis of Ig loci in reptiles lags far behind that in amphibians or mammals. Evidence for very large numbers of closely linked V_H genes has been found by Southern blotting in caimans and turtles. There may also be large numbers of J_H gene segments. Northern blotting of mRNA from a turtle species has indicated the presence of at least two μ-like transcripts and two non-μ transcripts. Primary responses are characterized by production of IgM antibodies while secondary responses feature IgY. During reproduction, large concentrations of maternally derived IgY can be found in the egg yolk.

ie) Ig heavy chains in birds. The generation of antibody diversity in birds differs significantly from that in the species described previously and in mammals. The best-studied bird with respect to Ig gene structure is the chicken (refer to Fig. 21-4). The *Igh* locus in chickens contains only a single functional V_H segment, a pool of about 90 V_H pseudogenes positioned 5' of this segment, 16 D_H segments 3' of the functional V segment, a single J_H segment, and 3 C_H exons specifying IgM, IgY, and IgA. It should be noted that this is the first appearance of true IgA in phylogeny. Despite the relatively large number of D segments, very little diversification is due to D segment recombination, and V(D)J recombination in general plays only a minor role. Rather, a mutational mechanism called *somatic gene conversion* introduces diversity into the antigen-binding site (see Box 21-3). During this process, a

Box 21-3. Distinguishing mutational mechanisms

Unlike T cell repertoires, which are generated solely by V(D)J recombination, B cell repertoires are generated by different mechanisms in different species. These mechanisms include V(D)J recombination, somatic hypermutation (both discussed in Ch.9), and somatic gene conversion. Somatic gene conversion does not appear to play a large role in generating diversity in human or murine antigen receptors but is very important in birds, cattle, pigs, sheep, and rabbits. To help the reader distinguish these mechanisms, we offer the following definitions, derived from the work of Jean-Claude Weill and Claude-Agnès Reynaud:

V(D)J recombination: site-specific recombination of gene segments mediated by recognition, cutting, and rejoining of the heptamer-nonamer recombination signal sequences (RSS) flanking those gene segments.

Somatic hypermutation: single or double nucleotide substitution that is not derived from a pre-existing sequence (non-templated alteration).

Somatic gene conversion: homologous recombination in which an acceptor gene is altered by the copying of multiple templated nucleotides from a homologous donor gene (templated alteration), but the donor gene

sequence is not altered (non-reciprocal). (In the case of the antigen receptor genes, the recombination takes place between related gene segments.)

Both somatic hypermutation and somatic gene conversion are thought to be initiated by single-strand (and perhaps double-strand) breaks in the DNA of a gene segment. Hypermutation occurs when there is a lack of effective DNA repair; that is, errors are made during routine repair and ligation of the break that are not fixed. Single nucleotide substitutions are readily introduced into the gene and its sequence is altered. In contrast, gene conversion between an acceptor gene segment (that with the initiating break in the DNA) and a donor gene segment can occur only if a number of conditions are met: (1) The stretch of donor DNA must be physically close to the acceptor DNA and it must be quite homologous to the acceptor gene segment sequence. (2) The donor DNA must also be fairly accessible; that is, not methylated or under the control of a silencer. (3) The break in the acceptor DNA must be expanded (by endonuclease action) such that a free strand end is generated. This end must then be able to invade the donor DNA at the site of homology, initiate templated

copying of the donor DNA, and recombine the copy with the "broken" sequence. In this way, the acceptor gene segment appears to have "converted" to the donor gene segment's sequence while the donor gene segment remains unaltered (see Figure).

Thus, evolution seems to have tried a variety of methods to achieve the same goal: the generation of a collection of BCRs with diverse antigen-binding site sequences. In humans and mice, the diversity of the primary B cell repertoire is generated in large part by V(D)J recombination. Multiple gene segments are present that can be recombined in an extremely large number of combinations. Somatic hypermutation makes a further contribution in the secondary response that results in almost limitless diversity. However, in chickens, pigs, and certain other mammals, the germline contains the elements for only a single functional Ig molecule. In other words, the entire B cell precursor pool expresses the same founding gene after V(D)J recombination. The required diverse population of BCRs is then generated by gene conversion in each B cell precursor using different sequences from a pool of non-functional V gene segments.

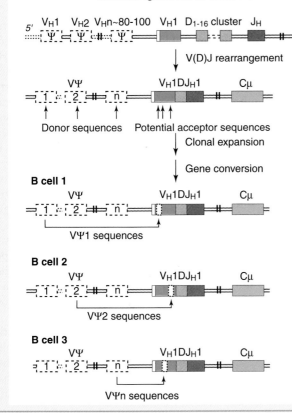

DNA "overwriting" event occurs involving the functional V_H segment and one of the V_H pseudogenes that slightly modifies the sequence of the functional V_H segment but leaves the V_H pseudogene unaltered. Since a different V_H pseudogene is likely to participate in different B cells, different modifications of the functional V_H sequence occur, giving rise to a sufficiently diverse B cell repertoire. Isotype switching of the C_H exons results in IgM giving way to IgA or IgY in the secondary response concomitant with somatic hypermutation leading to an increase in antibody affinity. Studies of Ig gene structure in ducks have confirmed that IgY is probably the common ancestor of mammalian IgG and IgE. In birds, IgY is a skin-sensitizing antibody that can mediate anaphylactic reactions, highlighting its functional similarity to mammalian IgE (see Ch.28).

ii) Immunoglobulin Light Chains

iia) Ig light chains in cartilaginous fish (Chondrichthyes). Ig light chains first appear in the cartilaginous fish. Three types of light chain genes have been identified, called types I, II, and III. The V (but not C) region of the type III light chain resembles the mammalian κ light chain. Types I and II cannot be readily classified as either κ- or λ-like. A typical type I light chain gene contains single V_L, J_L, and C_L segments that are closely linked in a cluster. As was true for the Ig heavy chain genes, multiple clusters of type I light chain genes exist that function independently. In some species but not others, the gene segments of one or more of the three types of light chains may be joined in the germline. The light chain gene products associate at the protein level with the IgM-like heavy chains just described.

iib) Ig light chains in bony fish (Osteichthyes). Different patterns of Ig light chain genes occur in different classes of bony fish, highlighting another evolutionary paradox. In at least some teleosts (advanced bony fish), Ig light chain organization is more similar to that of the cartilaginous fish than to that of the intermediate ray-finned fish. For example, two types of Ig light chains are found in trout. The locus encoding the first type, IgLC1, has its V_L and J_L segments organized in closely linked clusters that are not joined in the germline, like the type I light chain genes in the shark. The second locus, IgLC2, is also organized in clusters but shows very little similarity to IgLC1. Similarly, catfish express two isotypes of light chains called G and F that are also organized in clusters without germline joining. The variable region in F is similar to mammalian $V_κ$, while the variable region in G appears to be closely related to the V region of IgLC1. The constant regions in the Ig light chains of these animals appear to be completely unrelated to mammalian κ or λ sequences. Zebrafish express three types of light chains that are similar to trout IgLC1, trout IgLC2, and catfish F. In contrast, the more primitive sturgeon (a ray-finned fish) has an Ig light chain locus that is not organized in clusters and looks like a mammalian Igκ locus. A single C_L exon exists that recombines with multiple V_L gene segments that are positioned in a block upstream of a block of multiple J_L segments.

iic) Ig light chains in amphibians. Amphibian light chains have been studied most extensively in *Xenopus*. Three types of light chains have been identified that are encoded by separate loci on different chromosomes: the λ chain, the ρ chain, and the σ chain. The λ chain is homologous to the mammalian λ chain, while the ρ chain is κ-like. The σ chain has no apparent mammalian counterpart. Multiple V and J gene segments are present in each locus, and multiple C exons can be found in the λ and σ loci. A single $C_ρ$ exon has been identified. While the diversity of the λ chains is extensive, that of the ρ and σ chains is limited.

iid) Ig light chains in reptiles. Very little is known about light chain genes in reptiles. Two distinct light chain isotypes have been identified in alligators.

iie) Ig light chains in birds. In the bird species studied to date, only a single Ig light chain locus has been identified that contains one (or at most two) functional V_L gene segment(s). A single J_L segment is closely linked to the functional V_L segment, followed by a single C_L exon which is λ-like. No other light chain isotypes have been identified. Because of the profound dearth of V_L segments, V(D)J recombination can make little or no contribution to Ig light chain diversity. However, a pool of about 25 V_L pseudogenes lies 5' of the functional V_L segment(s), such that a repertoire can be generated by the same somatic gene conversion process described previously for the heavy chain. Diversity in the light chains is further increased by somatic point mutations.

III. MHC

As early as the tunicates (protochordates), we start to find evidence of histocompatibility, although the genes controlling this response are not related to the vertebrate MHC. In protochordates, both colony fusion and foreign colony repulsion are controlled by a single gene with multiple alleles that mediates alloreactivity between genetically mismatched colonies. Thus, two protochordate colonies growing toward one another can fuse if they share at least one allele at this locus, but will reject each other via the actions of phagocytic and lymphocyte-like cells if they do not. In this way, the proteins encoded by this ancient form of histocompatibility locus function to maintain the integrity of the "self" by acting as a "self" flag. However, in higher animals such as invertebrates and vertebrates, there is no threat of the self being absorbed into a "non-self" colony. In these cases, histocompatibility is relevant only with respect to transplantation, which is an entirely non-physiological occurrence that could not have driven further evolution of histocompatibility genes. So why did the MHC genes evolve in higher animals?

One rationale for the development of the MHC is that the evolutionary success of a species at this level of complexity becomes increasingly dependent on surviving challenges by intracellular pathogens and cancers that cannot be handled by humoral immune responses. Continued evolution therefore required a mechanism that would detect aberrations within self cells. It is therefore believed that vertebrate MHC class I molecules developed to act as "antigen presentation flags," displaying peptide fragments from all internally produced proteins. The T cell repertoire developed alongside the MHC system to supply the CTL effector cells responsible for specifically destroying cells displaying harmful, "non-self" peptides, that is, those that were aberrant or derived from infection. MHC polymorphism may then have evolved to increase the chances of a species surviving. In a species in which different individuals express different MHC molecules able to present different peptides derived from a pathogen, there are likely to be at least some individuals that can survive a severe pathogen onslaught. In a species in which only a few types of MHC molecules are able to present a limited number of peptides, almost all individuals mount the same type and degree of defense, perhaps leaving the species vulnerable to extinction if a catastrophic pathogen attack occurs. The evolution of the MHC class II molecules may have been driven by the subsequent specialization of T cell functions that resulted in the generation of T helper cells. These lymphocytes required interaction with a slightly different type of MHC molecule able to present exogenous antigen on the surface of antigen-capturing cells.

While the ultimate role of the MHC is clear in animals higher up the phylogenetic tree, there is still some question as to the precise pathway of evolution that led to the development of these antigen presentation molecules. As mentioned previously, the genes controlling cell fusion in protochordates are not related to those in the MHC found in higher vertebrates. In addition, while β2m-like molecules have been found in invertebrates such as the annelids and arthropods, it is not clear that these primeval β2m-like molecules actually play a role in any cell-mediated responses. It thus seems unlikely that an ancestral β2m gene underwent repeated duplication and divergence to give rise to vertebrate MHC genes encoding MHC class I and class II proteins. Moreover, no identifiable β2m or MHC genes have been found in the genome of any jawless fish (Agnatha). Intriguingly, the available evidence suggests that a gene duplication event occurred on a single chromosome of an evolving cartilaginous fish that abruptly established the primordial MHC about 400 million years ago. Indeed, in sharks, a primordial genetic region contains homologues of a proteasome LMP subunit, the TAP peptide transporter, a complement component, and the heat shock protein HSP70. Scientists theorize that this region may have been duplicated and refined to eventually give rise to modern MHC class II α and β genes. Another striking feature of the MHC in cartilaginous fish is that the MHC class I, class II, and class I-like genetic regions are linked together as they are in mammals. Some researchers speculate that this linked arrangement of the MHC may have persisted through evolution because it made immune responses easier to coordinate.

The ancestral vertebrate HSP70-like gene may have been a major contributor to MHC gene evolution. The reader will recall that heat shock proteins are involved in the cellular mechanisms that routinely dispose of incorrectly folded or spent host proteins that have been broken down into peptides. It has been proposed that the MHC class I α chains may have been derived from a hybrid structure containing the peptide-binding domain of the HSP70-related protein combined with the domains of a membrane-bound Ig-like molecule. Further comparative studies of MHC and MHC-like genes in a variety of animals will no doubt shed greater light on the evolutionary path of the MHC. In the meantime, we offer a brief summary of the MHC from the jawless fish to the birds.

i) MHC in Jawless Fish (Agnatha)

As mentioned previously, while there is evidence for the occurrence of very slow allograft rejection in jawless fish such as the lamprey, no genes specifically controlling histocompatibility have been identified. Lymphocyte-like cells in these animals can carry out an MLR response *in vitro*.

ii) MHC in Cartilaginous Fish (Chondrichthyes)

We reiterate that it is in the cartilaginous fish that the first recognizable MHC appears. In sharks, mixed leukocyte reactivity, slow allograft rejection (1–3 months), T-B cell cooperation, and T cell selection have all been shown to be under the control of a single polymorphic MHC-like genetic region. The MHC class I, class II, and class Ib-like regions in these animals show extensive polymorphism and are linked together as they are in mammals.

iii) MHC in Bony Fish (Osteichthyes)

Bony fish are the first group to exhibit rapid allograft rejection (within 2 weeks). Recognizable MHC class I, class II, and class I-like regions have been identified in several species of bony fish. Curiously, however, although the bony fish are considered to be evolutionarily more advanced than the cartilaginous fish, the MHC class I and II regions are not closely linked in teleosts (such as trout) and can even occur on different chromosomes. In addition, although complement component genes are present in teleosts, the genes encoding them are not located near the MHC class I and class II regions; that is, these fish have no readily identifiable equivalent to the MHC class III region. These observations suggest the occurrence of a violent genomic event in this class of fish in which the originally linked MHC class I, II, and III regions may have been torn apart. Whether the MHC class I, II, and III regions are linked in the most primitive bony fish species (such as the lungfish) is not known.

iv) MHC in Amphibians

In the primitive amphibian axolotl, heterodimeric MHC class II molecules are present but are of low polymorphism and wide tissue distribution. MHC class I molecules have yet to be definitively identified. In the more advanced amphibian *Xenopus*, a single functional MHC class I gene is linked to several MHC class II genes in one region of the locus. A large number of non-functional class I-like genes are present outside this region, one of which bears a striking resemblance to HSP70 in its protein-binding domain. *Xenopus* MHC class I-like molecules, which are expressed on all cells, are composed of a three-domain α chain and a β2m-like chain of about 13 kDa. The MHC class II locus in *Xenopus* contains two regions, one for α chain genes and one for β chain genes. MHC class II polymorphism is much more extensive than in axolotl. A form of invariant chain associates with the MHC class II molecule during synthesis. In adult *Xenopus*, MHC class II is expressed constitutively on T and B cells and on skin APCs resembling DCs and Langerhans cells in humans; in *Xenopus* larvae, MHC class II is not expressed on T cells. With respect to MHC function, *Xenopus* represents the first group of non-mammalian organisms in which it appears that T cells have been selected to recognize self-pMHC on an APC. Lastly, complement component genes are linked to the MHC in *Xenopus*, constituting a type of MHC class III collection. Overall, the features of *Xenopus* MHC structure and function make their MHC remarkably similar to that in mammals.

v) MHC in Reptiles

Little is known about the MHC in reptiles. MLR responses, graft rejection, and the generation of cytotoxic T cells have been shown to occur in snakes, caimans, and turtles. Graft rejection is temperature-dependent, the response being less vigorous at lower temperatures. MHC class I and II proteins similar to those in mammals have been identified in some species, and a β2m-like molecule of 19 kDa has been isolated from a snake.

vi) MHC in Birds

In contrast to the dearth of information on reptilian MHC, the first report of a putative MHC locus in chickens (the "B" complex) appeared in 1950. (It should be noted that the chicken may not be representative of the vast majority of avian species; it is, however, by far the best-studied.) The chicken MHC contains (at least) the BF region (class I genes), the BL region (class II genes), and the BG region (genes similar to the non-classical Q genes in mammalian MHC). At least three different BL β chain genes have been identified that are similar to the human HLA-DQ β chain sequence. One interesting feature of the chicken MHC is that its component regions are very compactly clustered such that they occupy only about 400 kb of DNA. (Compare this to the 3000 kb H-2 region in mice and the 3500 kb HLA region in humans.) In particular, the BF and BL genes are separated by very short distances and each gene is comparatively very short with very small introns.

With respect to MHC proteins, a β2m-like chain of 12 kDa combines with an α chain of about 42 kDa to form a BF molecule with clear homology to mammalian MHC class I molecules. The tissue distribution of BF expression is quite broad and includes the liver, spleen, thymus, erythrocytes, and leukocytes. The distribution of the BL (MHC class II) proteins is considerably narrower, with BL expressed only on monocytes and B cells. The BG molecules (molecular mass of about 30–40 kDa) are highly polymorphic and are expressed on erythrocytes, lymphocytes, and thrombocytes. However, in keeping with their similarity to the mammalian non-classical MHC molecules, the BG proteins are not required for graft rejection or MLR, and their functions remain a mystery. Lastly, an MHC class III-like region exists in the chicken genome that contains genes encoding complement components.

IV. TCR

Functional evidence suggests that, like B cells and BCRs, T cells and their TCRs mediate immune responses in all organisms from the cartilaginous fishes (Chondrichthyes) on up. However, unlike the BCR genes, the TCR genes have been much more difficult to characterize. The primary sequences of those TCR genes that have been isolated diverge considerably more than the sequences of comparable Igs, meaning that the DNA cross-hybridization techniques used to identify Ig genes across species do not work well for TCR gene isolation in many cases. This observation tells us that the TCR genes are diverging in sequence at a faster rate than the Ig genes. However, despite their variability in DNA sequence, TCR loci appear to be more strongly conserved in overall organization than Ig loci. This may reflect in part the fact that αβTCRs are much more restricted than BCRs in their binding requirements, "seeing" only MHC molecules bearing small peptides. Similarly, γδ T cells tend to express relatively invariant TCRs specific for a narrow range of antigens, including small pyrophosphate moieties and stress antigens. T cells and their genes and functions are the subject of much study in many commercially and medically important species at present, and much remains to be learned. We next present a sketch of what is currently known about TCR genes in non-mammalian vertebrates.

i) TCR Genes in Cartilaginous Fish (Chondrichthyes)

The genome of the horned shark contains genes that are recognizably TCRβ-like. The Vβ, Jβ, and Cβ gene segments and exons are present in multiple copies, whereas there appears to be a single Dβ gene segment. At least seven Vβ families and 18 Jβ sequences have been identified. The reader will recall that the Ig genes in the shark are organized in germline clusters of V-D-J-C sequences. In contrast, in the TCR locus, the multiple Vβ gene segments are grouped together, as are the multiple Jβ and Cβ gene segments, so that TCRβ organization in the shark resembles that in mammals. Recombination signal sequences very similar in structure to mammalian RSSs flank (at least) the Vβ and Jβ gene segments, and combinatorial joining is important for the creation of a diverse repertoire of TCRβ chains. Two TCRδ sequences have also been identified in the horned shark.

All four TCR loci (α, β, γ, δ) have been identified in the clearnose skate, suggesting that the divergence of TCRα from TCRδ occurred very early in the evolution of jawed vertebrates.

The skate TCRα locus boasts at least four Vα and six Jα families, while the skate TCRβ locus contains six Vβ families, four Jβ families, two Dβ sequences, and a single Cβ exon. The skate TCRγ locus contains five Vγ and two Jγ families, while the TCRδ locus has five Vδ and two Jδ families and three Dδ segments. Interestingly, the TCRδ locus is not nested within the TCRα locus and these genes are relatively far apart. The distances between the V and J gene segments are often very short, and direct VJ joinings have been observed in both TCRα and γ chains.

ii) TCR Genes in Bony Fish (Osteichthyes)

All four TCR loci have also been identified in several of the higher fish. It is in these animals that we first see the nesting of the TCRδ locus within the TCRα locus. However, the genomic structure of these loci differs from that in mammals. In the puffer fish, the TCRα/δ locus contains (in order): 13 Vα/δ gene segments, a Cα exon, 12 Jα gene segments, a Cδ exon, two Jδ gene segments, and a few Dδ gene segments. The V, D, and J segments are flanked by canonical RSSs. Curiously, however, the Vα/δ gene segments are present in inverse orientation to the rest of the locus. This observation implies that, unlike the case in mammals, the complete TCRα and δ genes must be assembled by a gene inversion mechanism. It is unclear whether all Vα/δ gene segments can be used in both the TCRα and TCRδ proteins, or whether certain Vα/δ gene segments are preferentially used to generate the TCRα protein while others are used for the TCRδ protein. Another interesting feature of the puffer fish TCRα/δ locus is its very compact organization. Introns are comparatively short (65–200 nucleotides), the numbers of different V and J segments are relatively low, and no pseudogenes have been identified.

In rainbow trout, six Vα families, 32 Jα gene segments, and a single Cα exon have been identified. Similarly, in the TCRβ locus, a single Cβ exon is accompanied by three Vβ families, 10 Jβ gene segments, and a single short (8 nucleotide) Dβ sequence. The Dβ segment is flanked by RSS very similar to those in mammals. There is a relatively low frequency of junctional diversification in Dβ-Jβ joinings in this animal. In Atlantic salmon, the organization of the TCR loci appears to be very similar to that in trout. The single Cβ exon identified is closely related to that in trout, and the single Dβ segment is identical. Multiple Vβ and Jβ gene segments have also been identified. In catfish, the TCRα locus contains three Vα families, seven Jα segments, and a single Cα exon, while the TCRβ locus contains five Vβ families, seven Jβ segments, two putative Dβ segments, and two Cβ sequences.

iii) TCR Genes in Amphibians

Amphibians possess all four TCR genes but the organization and expression of TCRγ and TCRδ have not been completely characterized as yet. The TCRα and TCRβ loci have been studied in both lower amphibians such as axolotl and higher amphibians such as *Xenopus*. In axolotl, the TCRα locus contains five Vα families and 14 Jα gene segments. The TCRβ locus contains nine Vβ families, nine Jβ gene segments, two Dβ sequences, and two Cβ exons. Preferential joining occurs in this locus, in that different Jβ segments are found associated more often with one Cβ exon than the other. In addition, Dβ1, Jβ1, and Cβ1 join together preferentially, as do Dβ2, Jβ2, and Cβ2. Direct Vβ-Jβ joining has been observed in some transcripts and junctional diversification by N-nucleotide addition is common.

Little has been reported on the TCRα locus in *Xenopus*, and it is not clear whether the TCRδ locus is nested within it. Considerable work has been done on the TCRβ locus. A single Cβ exon associates with elements from 10 very diverse Vβ families, 10 Jβ segments, and two Dβ segments. One of these Dβ sequences is identical to a Dβ sequence found in axolotl, trout, chicken, and human. A third putative Dβ sequence is closely related to a Dβ segment in the horned shark. Typical RSSs surround Vβ, Jβ, and Dβ segments in *Xenopus*.

iv) TCR Genes in Reptiles

Again, the reptiles are the poor cousins in evolutionary studies of the TCR. No information has been reported to date on TCR loci in reptiles, although these genes are presumed to be present.

v) TCR Genes in Birds

Chickens are the best-studied non-mammals with respect to TCR genes and organization. All four TCR loci have been identified. In the chicken TCRα locus, more than 25 Vα gene segments grouped into two Vα families have been identified, as well as multiple Jα segments and a single Cα exon. The Vα and Jα segments are flanked by typical RSSs. In the TCRβ locus, six Vβ1 and four Vβ2 segments are linked to four Jβ segments, a single Dβ segment, and a single Cβ exon. Upregulation of Tdt occurs prior to junctional diversification, just as in mammals. In the TCRγ locus, three Vγ families comprising a total of about 30 segments, three Jγ segments, and a single Cγ exon have been described. Lastly, a putative TCRδ locus has been identified that contains at least one Cδ exon and two Vδ families. As is true in mammals, some Vα gene segments have been found associated with the Cδ exon. Neither the Cγ nor the Cδ sequence is expressed in avian αβ T cell lines. Unlike humans and mice (but like sheep, pigs, and cattle), chickens have high numbers of γδ T cells and express a more complex repertoire of γ and δ rearrangements. As yet, there is no evidence in birds for the early waves of γδ T cells bearing invariant TCRs that precede αβ T cell development in mammals.

We have now completed the 'Basic Immunology' section of this book. In the second section, which is more clinically oriented, we turn our attention back to the mammalian immune system and how our knowledge of it can be applied to human medicine. We commence with 'Immunity to Pathogens' in Chapter 22, which describes how all the elements described in preceding chapters come together to defeat microbial invaders.

SUMMARY

All organisms within the animal kingdom display immune defense mechanisms. As animal anatomy, physiology, and lifestyle became more advanced through evolution, the molecular and cellular systems available to participate in immune responses increased in complexity. At the same time, as mobility and life spans increased, animals encountered a broader range of pathogens that had an expanding window of opportunity to compromise the reproductive success of particular species, creating evolutionary pressure for more sophisticated immune responses. Innate immune mechanisms are extremely ancient compared to those of adaptive immunity. Various defense systems based on pattern recognition molecules such as Toll-like receptors can be found in the protozoa on up through the mammals. In addition, primitive complement-like systems can be found as early as the pre-vertebrates. On the other hand, elements of adaptive immunity, such as lymphoid tissues, lymphocytes, BCRs, and TCRs, do not appear until the jawed vertebrates. These elements become more complex as one moves up the evolutionary tree. Early allograft rejection mechanisms based on self/non-self recognition are found in the protozoans, invertebrates and pre-vertebrates, although structures defined as true MHC molecules are not seen until the jawed vertebrates. Studying the immune responses of organisms on the lower branches of the evolutionary tree is important both academically and clinically because the information it reveals about the fundamentals of mammalian immunity may ultimately apply to human health and disease.

Selected Reading List

Agrawal A., Eastman Q. M. and Schatz D. G. (1998) Implications of transposition mediated by V(D)J-recombination proteins RAG1 and RAG2 for origins of antigen-specific immunity. *Nature* **394**, 744–751.

Baker B., Zambryski P., Staskawicz B. and Dinesh-Kuman S. (1997) Signaling in plant-microbe interactions. *Science* **276**, 726–733.

Cannon J., Haire R., Rast J. and Litman G. (2001) The phylogenetic origins of the antigen-binding receptors and somatic diversification mechanisms. *Immunological Reviews* **200**, 12–22.

Carey C., Cohen N. and Rollins-Smith L. (1999) Amphibian declines; an immunological perspective. *Developmental and Comparative Immunology* **23**, 459–472.

Cerenius L. and Soderhall K. (2004) The prophenoloxidase-activating system in invertebrates. *Immunological Reviews* **198**, 116–126.

Cohn J., Sessa G. and Martin G. B. (2001) Innate immunity in plants. *Current Opinion in Immunology* **13**, 55–62.

Cohn M. (1998) At the feet of the master: the search for universalities. Divining the evolutionary selection pressures that resulted in an immune system. *Cytogenetic Cell Genetics* **80**, 54–60.

Du Pasquier L., Zucchetti I., and De Santis R. (2004) Immunoglobulin superfamily receptors in protochordates: before RAG time. *Immunological Reviews* **198**, 233–248.

Erf G. (2004) Cell-mediated immunity in poultry. *Poultry Science* **83**, 580–590.

Ezekowitz R. A. and Hoffman J. (1998) The blossoming of innate immunity. *Current Opinion in Immunology* **10**, 9–11.

Fritig B., Heitz T. and Legrand M. (1998) Antimicrobial proteins in induced plant defense. *Current Opinion in Immunology* **10**, 16–22.

Fujita T., Matsushita M. and Endo Y. (2004) The lectin-complement pathway—its role in innate immunity and evolution. *Immunological Reviews* **198**, 185–202.

Ganz T. and Lehrer R. I. (1998) Antimicrobial peptides of vertebrates. *Current Opinion in Immunology* **10**, 41–44.

Glinske Z. and Buczek J. (1999) Aspects of reptile immunity. *Medycyna Weterynaryjna* **55**, 574–578.

Gomez-Gomez L. (2004) Plant perception systems for pathogen recognition and defence. *Molecular Immunology* **41**, 1055–1062.

Hoffman J. and Reichhart J. (2002) *Drosophila* innate immunity: an evolutionary perspective. *Nature Immunology* **3**, 121–126.

Hughes A. L. and Yeager M. (1997) Molecular evolution of the vertebrate immune system. *BioEssays* **19**, 777–786.

Johansson M. W. (1999) Cell adhesion molecules in invertebrate immunity. *Developmental & Comparative Immunology* **23**, 303–315.

Kasahara M. (1998) What do the paralogous regions in the genome tell us about the origin of the adaptive immune system? *Immunological Reviews* **166**, 159–175.

Khalturin K., Panzer Z., Cooper M. and Bosch T. (2004) Recognition strategies in the innate immune system of ancestral chordates. *Molecular Immunology* **41**, 1077–1087.

Laing K. and Secombes C. (2004) Chemokines. *Developmental and Comparative Immunology* **28**, 443–460.

Litman G. W., Anderson M. K. and Rast J. P. (1999) Evolution of antigen binding receptors. *Annual Review of Immunology* **17**, 109–147.

Magor K. and Vasta G. (1998) Ancestral immunity comes of age. *Immunology Today* **19**, 54–56.

Matsunaga T. and Rahman A. (1998) What brought the adaptive immune system to vertebrates? The jaw hypothesis and the seahorse. *Immunological Reviews* **166**, 177–186.

Matsushita M., Endo Y., Nonaka M. and Fujita T. (1998) Complement-related serine proteases in tunicates and vertebrates. *Current Opinion in Immunology* **10**, 29–35.

Medzhitov R. and Janeway C. (1998) An ancient system of host defense. *Current Opinion in Immunology* **10**, 12–15.

Medzhitov R. and Janeway C. J. (2000) Innate immune recognition: mechanisms and pathways. *Immunological Reviews* **173**, 89–97.

Nonaka M. (2001) Evolution of the complement system. *Current Opinion in Immunology* **13**, 69–73.

Nurnberger T., Brunner F., Kemmerling B. and Piater L. (2004) Innate immunity in plants and animals: striking similarities and obvious differences. *Immunological Reviews* **198**, 249–266.

Pancer Z., Amemiya C., Ehrhardt G., Ceitlin J. *et al.* (2004) Somatic diversification of variable lymphocyte receptors in the agnathan sea lamprey. *Nature* **430**, 174–180.

Rovet J. (2004) Infectious non-self recognition in invertebrates: lessons from *Drosophila* and other insect models. *Molecular Immunology* **41**, 1063–1075.

Schultz U., Kaspers B. and Staeheli P. (2004) The interferon system of non-mammalian vertebrates. *Developmental and Comparative Immunology* **28**, 499–508.

Smith L. C., Kaoru A. and Nonaka M. (1999) Complement systems in invertebrates. The ancient alternative and lectin pathways. *Immunopharmacology* **42**, 107–120.

Soderhall K. and Cerenius L. (1998) Role of the prophenoloxidase-activating system in invertebrate immunity. *Current Opinion in Immunology* **10**, 23–28.

Sticher L., Mauch-Mani B. and Metraux J. (1997) Systemic acquired resistance. *Annual Review of Phytopathology* **35**, 235–270.

Thomma B. P., Penninckx I. A., Brodkaert W. F. and Cammue B. P. (2001) The complexity of disease signaling in Arabidopsis. *Current Opinion in Immunology* **13**, 63–68.

Trowsdale J. (1995) "Both man & bird & beast": comparative organization of MHC genes. *Immunogenetics* **41**, 1–17.

Warr G., Magor K. and Higgins D. (1995) IgY: clues to the origins of modern antibodies. *Immunology Today* **8**, 392–398.

Weill J.-C. and Reynaud C.-A. (1996) Rearrangement, hypermutation, gene conversion: when, where and why? *Immunology Today* **17**, 92–97.

Yoder J., Mueller M. G., Wei S., Corliss B. *et al.* (2001) Immune-type receptor genes in zebrafish share genetic and functional properties with genes encoded by the mammalian leukocyte receptor cluster. *Proceedings of the National Academy of Sciences USA* **98**, 6771–6776.

Part II

Clinical Immunology

Immunity to Pathogens

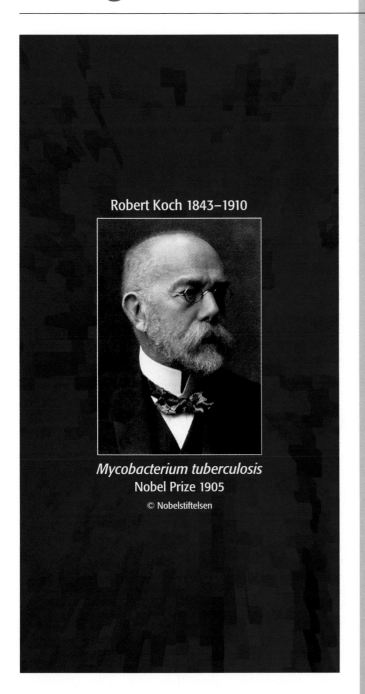

Robert Koch 1843–1910

Mycobacterium tuberculosis
Nobel Prize 1905
© Nobelstiftelsen

CHAPTER 22

A.   OVERVIEW

B.   IMMUNITY TO EXTRACELLULAR BACTERIA

C.   IMMUNITY TO INTRACELLULAR BACTERIA

D.   IMMUNITY TO VIRUSES

E.   IMMUNITY TO PARASITES

F.   IMMUNITY TO FUNGI

G.   THE MYSTERIOUS PRIONS

"Never will the doctrine of spontaneous generation recover from the mortal blow of this simple experiment. No, there is now no circumstance known in which it can be affirmed that microscopic beings came into the world without germs, without parents similar to themselves"—Louis Pasteur

In the first part of this book, we examined each element of innate and adaptive immune responses essentially in isolation, detailing how each developed and functioned. In the second part, we examine the immune response holistically, describing its role in the defense of health against assault from without and within. We discuss what happens when an immune response is unwanted, and the clinical consequences of abnormal or absent immune responses. In this chapter, we focus on defense against pathogens.

Infectious diseases were the leading cause of natural deaths until the early part of the 20th century. In 1900, the three most common causes of death in the United States were pneumonia, tuberculosis, and diarrhea/enteritis. These infectious diseases, along with diphtheria, were responsible for one-third of all deaths. However, due to the late 19th century discovery that specific microorganisms were the cause of such lethal diseases, public health initiatives soon led to dramatic improvements in infectious disease control. Better sanitation and hygiene practices were introduced, including proper sewage disposal, water treatment, and public education about the virtues of hand-washing. The discovery of the first antibiotic—penicillin—in 1928 led to its development as a medical treatment beginning in the 1940s. Since that time, many other antibiotics have come into common use, providing cures in countless cases of otherwise fatal bacterial infections. In addition, in North America, Europe, and much of the Pacific Rim, the development of vaccines and effective public vaccination strategies have all but eliminated potentially lethal diseases such as diphtheria, measles, mumps, and rubella that were once endemic in childhood. The success of these public health measures can hardly be overstated: since 1900, human average life expectancy has increased by 30 years. There has been a corresponding drop in infant and child mortality rates in developed countries such that the proportion of annual deaths in this age group due to infectious disease has fallen to less than 2% from over 30%. Despite these gains, of the approximately 57 million human deaths that occur annually around the world today, about 25% are still due to infectious diseases, considerably more deaths than can be attributed to cancer, cardiovascular disease, or trauma. While new types of antibiotics and vaccines have provided great boons in eradicating the threat posed by some deadly pathogens, others continue to defy all our attempts to control them.

In the following sections, we cover how all the elements detailed in Chapters 1–20 combine to eliminate five major classes of human pathogens. The common physical barriers and innate defense mechanisms encountered by all pathogens are discussed first. We then describe, for each class of pathogen in turn, the immune responses commonly mounted against that class of pathogen, followed by a discussion of the evasion strategies employed by that class to stave off destruction. A short discussion of *prions*, a new category of pathogen that involves infectious proteins, appears at the end of this chapter. Vaccination, which primes a host's immune response to better combat infection, is addressed in Chapter 23.

A. Overview

I. WHAT IS A "PATHOGEN"?

The life objective of any organism is to reproduce, and a pathogen is an organism that can cause disease in its host while attempting to achieve this goal. There are five major types of pathogens: *extracellular bacteria, intracellular bacteria, viruses, parasites,* and *fungi.* Bacteria are microscopic, single-celled organisms that are considered prokaryotic because they do not have the "true" nucleus found in eukaryotes. The prokaryotic nucleus lacks a nuclear membrane and the genetic material of these organisms is usually contained in a single linear chromosome. Extracellular bacteria do not have to enter host cells to reproduce, while intracellular bacteria do. At this point, the reader should take a moment to familiarize him- or herself

Box 22-1. **Classification of bacteria**

Bacteria have been traditionally classified on the basis of their shapes and groupings, their use of oxygen, and their gram stain status. Despite the advent of DNA technology, these terms of reference remain entrenched in the microbiological research community. For our discussions of bacteria in the context of immunology, we have divided the bacteria into the "extracellular" and "intracellular" categories, depending on where they live in the body in relation to host cells. Both these categories contain members of each of the groups described here.

Shapes and Groupings

Bacterial cells come in various shapes and groupings that are associated with specific nomenclature. The most common bacterial shapes are rod-like, which are the *bacilli* ("little staffs"; singular is "bacillus"); spherical or ovoid, which are the *cocci* ("berries"; singular is "coccus"); and *comma forms* and *spirals*. Bacteria may also be star-shaped or square, although these shapes are less common. Cocci are commonly grouped in pairs (*diplo*), chains (*strepto*) or clusters (*staphylo*), giving rise to diplococci, streptococci, and staphylococci. Bacilli are found in pairs and chains, giving rise to diplobacilli and streptobacilli. Spirals are usually found singly, but may occur as *spirilla* (rigid helix) or *spirochetes* ("flexible helix"). *Vibrios* are comma-shaped bacteria.

Use of Oxygen

Microbes in general and bacteria in particular are often described with respect to their use of oxygen. Those that can use oxygen are called *aerobes*, while those that do not are called *anaerobes*. A microbe that uses oxygen when available but can live anaerobically in the absence of oxygen is said to be a *facultative aerobe*. Those microbes that must have oxygen to survive are called *obligate aerobes*, while those that can grow only in the complete absence of molecular oxygen are called *obligate anaerobes*.

Gram Stain Status

Bacteria are also classified into two groups according to how they appear when stained in the laboratory using the multi-step gram stain procedure developed in 1884 by Hans Christian Gram. One group of bacteria has cell walls with many layers of peptidoglycan interwoven with both lipotechoic and techoic acid. These relatively thick cell walls retain the dark purple color imposed by the initial step

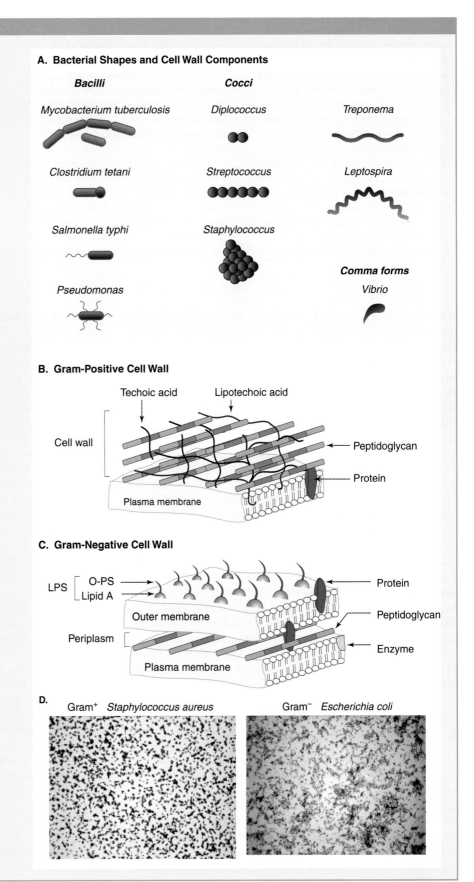

A. Bacterial Shapes and Cell Wall Components

Bacilli

Mycobacterium tuberculosis

Clostridium tetani

Salmonella typhi

Pseudomonas

Cocci

Diplococcus

Streptococcus

Staphylococcus

Treponema

Leptospira

Comma forms

Vibrio

B. Gram-Positive Cell Wall

Techoic acid   Lipotechoic acid

Cell wall

Peptidoglycan

Protein

Plasma membrane

C. Gram-Negative Cell Wall

LPS { O-PS / Lipid A }

Protein

Outer membrane

Peptidoglycan

Periplasm

Enzyme

Plasma membrane

D. Gram⁺ *Staphylococcus aureus*   Gram⁻ *Escherichia coli*

Continued

of the staining procedure despite undergoing a later decolorizing step and counterstaining with the red dye safranin. Bacteria staining purple in this fashion are called "gram-positive" bacteria. The other group, called "gram-negative" bacteria, have a cell wall with a thin peptidoglycan layer that loses its integrity during the staining procedure, allowing decolorization to occur. The purple color vanishes, allowing the cells to stain red with safranin. As well as peptidoglycan, the gram-negative cell wall has an outer plasma membrane containing lipopolysaccharide macromolecules composed of Lipid A (conserved among all gram-negative species)

and an O-linked polysaccharide (O-PS; variable among gram-negative species). Functionally, the O-linked polysaccharide confers virulence (the ability to infect host cells) on the bacterium. It is the Lipid A moiety that functions as an endotoxin capable of inducing the endotoxic shock described in Box 17-2.

Gram-negative and gram-positive bacteria can also be distinguished by their differential susceptibility to physical assaults as well as various enzymes and drugs. For example, gram-positive bacteria resist dry conditions better than gram-negative bacteria and cannot be crushed as readily. Gram-positive bacteria

are more resistant to lysozyme digestion and killing by streptomycin and tetracycline, but are highly susceptible to penicillin, anionic detergents, and basic dyes. In contrast, gram-negative bacteria resist lysozyme digestion and killing by penicillin but dry out easily, are inhibited by basic dyes, and are susceptible to killing by tetracycline.

Panel B adapted from "Microbiology, An Introduction," 6th and 7th edns. (2002, 2004) Gerard J. Tortora, Berdell R. Funke, Christine L. Case, eds. Published by Benjamin Cummings, San Francisco. Panel D courtesy of Dr. Tony Mazzulli, Mount Sinai Hospital, Toronto.

with the common terms and concepts used in classifying or discussing bacteria (Box 22-1). Viruses are not considered to be either prokaryotic or eukaryotic. Viruses are acellular particles consisting of a protein coat (called a *capsid*) encasing a genome of either DNA or RNA. Due to their small size, viruses can only be observed with an electron microscope. To propagate, a virus must enter a host cell possessing protein synthesis machinery that the virus can exploit. Parasites and fungi are eukaryotic organisms that possess several chromosomes contained in a membrane-bounded nucleus. Parasites all share the characteristic of taking advantage of and being dependent on a host organism for both habitat and nutrition at some point in their life cycles, usually damaging the host but not killing it. The major classes of parasites are the protozoa, which are single-celled organisms, and the helminth worms, which are multicellular. While some parasites may be only a few micrometers in size, many are significantly larger, with some worms reaching several meters in length. Fungi are eukaryotic organisms that can exist comfortably outside a host but that will invade and colonize a host if given the opportunity. Fungi may be single-celled, as in the case of yeast, or multicellular, as in bread molds. Prions are curious entities that are infectious but appear to be made entirely of protein and lack any type of nucleic acid. Prions cause disease by altering normal proteins in the brain of the infected host.

II. WHAT IS "DISEASE"?

Disease is not the same as infection. Infection is said to have occurred when an organism successfully avoids innate defense mechanisms and stably colonizes a niche in the body. To establish an infection, the invader must first penetrate the anatomic and physiological barriers that guard the skin and mucosal surfaces of the host. Secondly, the organism must be able to survive in the host cellular milieu long enough to reproduce. This replication may or may not cause visible, clinical damage to the host tissues, symptoms that we call "disease." It should be noted that sometimes the products

released by a pathogen, such as bacterial toxins, are sufficient to cause disease even in the absence of widespread pathogen colonization. An "emerging infectious disease" is a disorder that has come to light only within the past couple of decades, or is changing in character such that it has an increased impact on the global population. Table 22-1 contains short descriptions of several emerging infectious diseases.

In addition to direct destruction by pathogen actions, host tissues are often injured by collateral damage inflicted by the immune response itself as it strives to destroy the pathogen. This damage results in *immunopathic* disease symptoms. CTLs may directly induce the lysis of bystander host cells via cytotoxic cytokine secretion, or phagocytes may release antimicrobial factors (like H_2O_2 or free radicals) that are also toxic to host cells. A very dramatic example of immunopathic damage occurs in mice infected in the brain with *lymphocytic choriomeningitis virus* (LCMV). Elsewhere in the body, this virus is quite harmless to the animals. In the brain, however, the CTL response to the virus results in neuron cytolysis that kills the mice. The endotoxic shock triggered by the LPS of gram-negative bacteria (refer to Box 17-2) is another example of immunopathic damage. Several other examples are given in Table 22-2.

Virulence is the ability to invade host tissues and cause disease in the host; the more virulent a pathogen is, the better able it is to resist the immune system. Many factors affect virulence: how quickly the pathogen can access a host cell and/or replicate, whether the pathogen can express molecules that damage the host or dampen its immune response, and whether the host MHC and PRR molecules can recognize the pathogen's processed peptides and PAMPs, respectively. If the immune system is able to eliminate the organism before it can successfully propagate, the infection is said to have been *abortive*. If, however, the pathogen multiplies despite the actions of the immune system, the infection is *productive*. Whether an infection is abortive or productive depends on the initial dose of the pathogen accessing the host, the virulence of the organism, and the strength and rapidity of the immune responses mounted to

Table 22-1 **Emerging Infectious Diseases**

Disease	Pathogen	Recent Outbreak	Description
Cryptosporidiosis	*Cryptosporidium* (protozoan)	1993	Diarrheal illness transmitted by contaminated water
Pulmonary syndrome	Hantavirus	1993	Fever, cough, rapid respiratory failure; carried by deer mice
Hemorrhagic fever	Ebola virus	1995	Bleeding from all orifices of body, followed by clotting in blood vessels
Necrotizing fasciitis (flesh-eating disease)	Group A *Streptococcus*	1995	Rapid and progressive loss of tissue layers; can require amputation of infected limb
West Nile encephalitis	West Nile virus	1999	Ranges from flu-like illness to potentially fatal brain infection; mosquito-borne
Hamburger disease	*E. coli* 0157:H7	2001	Bloody diarrhea, severe dehydration due to consumption of uncooked meat or contaminated water
Severe acute respiratory syndrome (SARS)	Coronavirus	2003	Respiratory disease resembling atypical pneumonia; can be fatal

combat the invasion. The latter will be determined in part by whether the host has been previously exposed to the pathogen in question. In the last phase of a productive infection, the progeny of the pathogen escape the original host and travel to new ones.

An ongoing struggle or "horse race" is thus established between a pathogen and the immune system: the pathogen tries to replicate and expand its niche, and the immune system tries to eliminate the pathogen, or at least confine it.

Table 22-2 **Examples of Immunopathic Disease Symptoms**

Disease Symptom	Immunopathic Cause
Endotoxic shock	Flood of pro-inflammatory cytokines released in response to LPS of gram-negative bacteria
Flu-associated malaise and fever	Pro-inflammatory cytokines released to fight influenza virus
Hepatitis virus-associated liver damage	Fas-mediated apoptosis of hepatocytes
HTLV-associated spinal cord damage	Chronic inflammatory response to residual viral antigen
LCMV-associated brain pathology	Pro-inflammatory cytokines released to fight LCMV kill neurons in the brain
Measles red rash	Cell-mediated immune response against infected cells in skin
Tuberculosis	Granuloma formation in the lungs caused by hyperactivated macrophage response to bacteria
Ulcers	Cytokines produced by Th1 cells responding to *Helicobacter pylori* infection in the gut

The interval between the time of infection and the onset of disease is called the *incubation time*, which can vary considerably in length depending on the lifestyle of the pathogen. An infection is said to be *acute* when it causes disease symptoms that appear rapidly but remain for only a short time. A *persistent* infection is one in which the pathogen remains in the host's body for prolonged periods or possibly throughout life. Persistent infections may result in *chronic disease* if symptoms are experienced on an ongoing or recurring basis. In other cases, a persistent pathogen may lurk for an extended time without producing any symptoms, in which case the infection is *latent*. A pathogen may or may not be infectious during latency. A person with a latent infection that can be transmitted unknowingly to others is a *carrier*.

The threat of pathogen attack has prompted hosts to evolve immune responses designed to repel and neutralize the invaders and stop their appropriation of host resources. Naturally, the pathogens, whose evolution occurs much more rapidly than that of their hosts, have responded with the development of various "evasion strategies" designed to outwit or compromise the host's immune response. For example, a host population may be resistant to a certain pathogen because of an effective neutralizing antibody response that prevents the pathogen from gaining a foothold on host cells. However, if the pathogen undergoes a mutation that allows it to make novel use of a host surface protein to enter the cell, the pathogen can escape the humoral response. The host population is once again susceptible to disease.

III. INNATE DEFENSE AGAINST PATHOGENS

As described in Chapter 20, the first lines of defense encountered by any pathogen are the intact skin and mucosae. The vast majority of microorganisms cannot penetrate the tough keratin layer protecting the epidermis, and any pathogen arriving on the skin surface has to compete with the commensal flora for space and nutrients. Furthermore, the routine shedding of skin layers often deposes a pathogen before it has a chance to establish a firm foothold. Similarly in the mucosae, pathogens ingested into the gut or inhaled into the respiratory tract are for the most part trapped by mucus and secretory IgA coating the mucosal surfaces. Microbicidal molecules in the body secretions and the low pH and hydrolases in the gut also take their toll. However, should there be a wound in the skin or the mucosae that allows a pathogen access to tissues underlying the epithelium, the consequences can be severe. Some organisms are *opportunistic* pathogens, in that they do not cause disease unless offered an unexpected opportunity by a failure in host defense. *Staphylococcus aureus* is an example of an extracellular bacterium that is not particularly invasive and that normally colonizes the skin surface without harming the host. If inhaled, *S. aureus* is successfully repelled by an intact respiratory epithelial layer. However, should the skin experience a tear or the respiratory epithelium be damaged by trauma or another more invasive pathogen, *S. aureus* can take advantage of the breach and establish a serious infection. When the opportunity arises to invade the respiratory tract, staphylococcal pneumonia can result. Opportunistic invasions are a major problem in hosts immunocompromised either by disease or by design. For example, patients on immunosuppressive therapy following a tissue transplant, or those whose immune systems have been destroyed by HIV, are vulnerable to infection by the opportunistic pathogens *Pneumocystis carinii* and *Candida albicans*.

A small minority of microorganisms are *invasive* pathogens, capable of entering the body even when surface defenses are intact. As we learned in Chapter 20, some viruses (and bacteria, as we shall see later in this chapter) can target the antigen-collecting M cells of the mucosal follicle-associated epithelium (FAE) to effect entry, while other pathogens have the capacity to adhere to host molecules expressed on cutaneous or mucosal epithelial cells. Still other pathogens exploit host surface molecules as receptors, inducing receptor-mediated internalization. Many highly invasive pathogens can be found in the group A *Streptococcus*, bacteria that cause 15,000 severe infections per year in the United States.

Most times, a pathogen that succeeds in penetrating the skin or mucosae either through a wound or via exploitation of host cell receptors meets effective innate defense in the form of subepithelial macrophages and NK cells and intraepithelial γδ T cells (Fig. 22-1). (Table 22-3 contains a list of the abbreviations used in the figures in this chapter.) Molecules on the pathogen surface may be recognized directly by the PRRs of resident macrophages, by the activatory receptors of NK cells, and by the restricted TCRs of γδ T cells and NKT cells. Inflammatory responses are mounted that first draw neutrophils into the site of incursion. Different types of pathogens elicit the production of different classes of chemokines that facilitate the inflammatory response. For example, acute infection by an extracellular gram-positive bacterium such as *Streptococcus pneumoniae* results in the production of predominantly CXC chemokines, particularly IL-8, that summon the neutrophils required to eliminate the bacteria. The lipopolysaccharides of the cell walls present in gram-negative bacteria such as *Escherichia coli* are also powerful inducers of CXC chemokine secretion. In contrast, infection with a bacterial spirochete such as *Borrelia burgdorferi* leads to a more chronic condition eventually requiring action by macrophages and lymphocytes. Infection with this extracellular organism provokes the secretion by activated macrophages of CC (e.g., MCP-1) or C (e.g., lymphotactin) chemokines that attract the required cells.

Inflammatory cells responding to an invading pathogen produce toxic substances such as nitric oxide (NO) and acute phase proteins, and cytotoxic and activatory cytokines such as TNF and IFNγ. Complement components that either have leaked into the tissue from the blood or were produced by cells in the site of attack may be activated by either the alternative or lectin pathways. The pathogen will then be coated in C3b or C3d, enhancing phagocytosis via complement receptors on phagocytes. MAC-mediated destruction of the pathogen may also ensue if a membranous structure is present. Other complement components serve as chemoattractants, drawing more phagocytic cells to the site of attack. After phagocytes have engulfed an invader, molecules such as lactoferrin are introduced into the phagosome to sequester iron atoms, further inhibiting bacterial growth. Despite all these measures, a pathogen may avoid phagocytosis, opsonized or otherwise, and access the blood circulation. *Bacteremia*, *viremia*, and *parasitemia* are terms referring to the presence in the blood of bacteria, viruses, or parasites, respectively. Innate defense in the blood falls to blood monocytes and neutrophils. Should the pathogen reach the liver or the spleen, the macrophages present in these organs attempt to confine the invader. A pathogen tough enough to escape these measures may successfully colonize other organs and cause serious abscesses.

As the innate response proceeds, the pathogen or some of its components interact with the PRRs expressed on macrophages and DCs and are ingested, resulting in activation of these cell types such that they become capable of acting as APCs for the adaptive response. The products of many species of bacteria and protozoa (particularly lipopolysaccharides) are potent activators of DCs, promoting their maturation into fully competent APCs. As well, the DNA of some parasites and bacteria, particularly if it is unmethylated, can activate DCs via binding to TLR9. The DNA of these organisms, but not that of vertebrates, contains stretches of unmethylated cytosine-phosphate-guanosine residues called *CpG motifs*. TLR9 appears to be specific for these motifs. Lipopolysaccharides also induce the upregulation of the B7 molecules, CD40, and MHC class I and II on the DC surface, and promote DC migration to the secondary lymphoid tissues. Cytokine production by DCs, including TNF, IL-12, IL-10, IL-6, and IFNα/β, is induced. The arrival of activated, antigen-bearing macrophages and DCs in the local lymph node

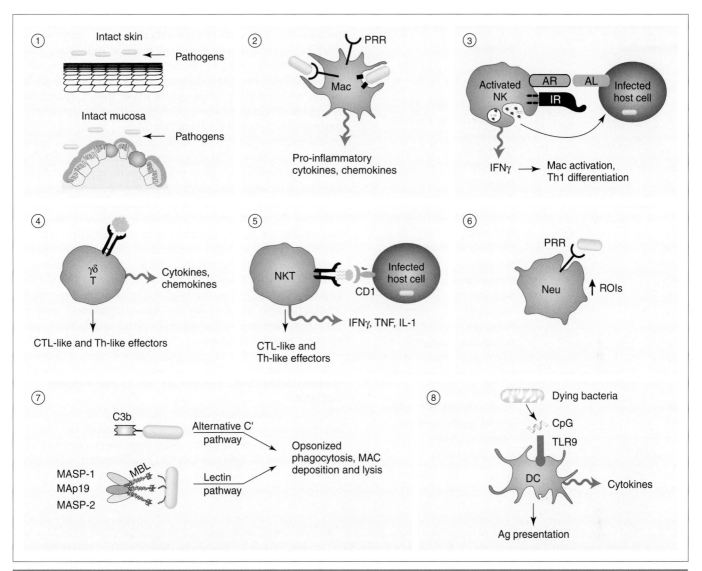

Figure 22-1

Major Mechanisms of Innate Defense against Pathogens
(1) Intact skin and mucosa prevent pathogen penetration. (2) Macrophages activated by PRR-mediated recognition of pathogen PAMPs carry out phagocytosis and secrete cytokines. (3) Infected cells that have downregulated MHC class I activate NK cells and are destroyed by natural cytotoxicity. IFNγ secreted by the activated NK cell activates macrophages and supports Th1 differentiation. (4) γδ T cells activated by soluble non-peptide pathogen antigens differentiate into effector cells and secrete cytokines. (5) Infected cells that present pathogen antigens on CD1 activate NKT cells that differentiate into effector cells and secrete cytokines. (6) Neutrophils activated by PRR-mediated recognition of pathogen PAMPs kill phagocytosed pathogens via respiratory burst and increased ROI production. (7) Pathogens that have bound C3b or MBL are destroyed via the alternative or lectin complement activation pathways, respectively. (8) DCs activated by TLR-mediated recognition of pathogen PAMPs take up pathogen debris, increase their presentation of pathogen peptides to T cells, and secrete cytokines. Abbreviations are defined in Table 22-3 on page 648.

stimulates the activation of naive T and B cells. The adaptive immune response then proceeds in its bid to eliminate the pathogen. As is described later, the adaptive response effector mechanisms most effective in countering a pathogen are determined by the invader's lifestyle and mode of replication.

We move now to a discussion of each class of pathogen, the details of the immune response effector mechanisms employed by the host to combat them, and the evasion tactics used by each to escape the immune system.

B. Immunity to Extracellular Bacteria

I. WHAT ARE EXTRACELLULAR BACTERIA?

As mentioned previously, extracellular bacteria do not have to enter host cells to replicate. Rather, they occupy spaces topologically outside cells: interstitial regions in connective tissue, the blood circulation, and the lumens of the respiratory, urogenital,

Table 22-3 Abbreviations Used in Chapter 22 Figures

Abbreviation	Definition
Act Mac	Activated macrophage
Ag	Antigen
AL	Activatory ligand (NK cells)
AR	Activatory receptor (NK cells)
C'	Complement
Cyt c	Cytochrome c
Eo	Eosinophil
FcR	Fc receptor
Hyper Mac	Hyperactivated macrophage
IR	Inhibitory receptor (NK cells)
Mac	Macrophage
MAC	Membrane attack complex
Mast	Mast cell
MBL	Mannose binding lectin
Neu	Neutrophil
NO	Nitric oxide
PC	Plasma cell
PrP^c	Prion protein, cellular
PrP^{sc}	Prion protein, scrapie
PRR	Pattern recognition receptor
RBC	Red blood cell
RNI	Reactive nitrogen intermediates
ROI	Reactive oxygen intermediates
TLR	Toll-related receptor
Uncon CTL	Unconventional CTL

and gastrointestinal tracts. Extracellular bacteria often secrete proteins that penetrate or enzymatically cleave components of the glycocalyx covering cells of the mucosal epithelium, allowing the invaders access to interior tissue layers. A wide variety of extracellular bacteria gain access to the body via the M cells in the mucosal FAE, while others make use of complement receptors. The exploitation of various host components by pathogens is reviewed briefly in Box 22-2.

The major types of extracellular bacteria causing human disease are represented in the gram-positive cocci (e.g., *Staphylococcus*, *Streptococcus*), gram-positive bacilli (e.g., *Clostridium*, *Corynebacterium*), gram-negative cocci (e.g., meningococcus), and gram-negative bacilli (e.g., *Escherichia*) (Table 22-4). Many of the gram-positive cocci are *pyogenic*, meaning that they induce an acute, immunopathic inflammatory response accompanied by a high fever. In this situation, macrophages responding to the bacterium produce massive amounts of cytokines (particularly IL-1) that spill into the bloodstream and travel to the brain. The hypothalamus of the brain makes prostaglandins in response to these cytokines, and the prostaglandins act on the pituitary gland and influence it to raise the body's temperature.

i) Characteristics of Selected Species of Extracellular Bacteria

The most common extracellular bacteria causing acute human infections are species of *Streptococcus* and *Staphylococcus*. For example, *Streptococcus pneumoniae* (also known in short-hand as pneumococcus) infection can lead to pneumonia, while group

---

Box 22-2. Pathogen entry: exploitation of host components

Many pathogens have evolved devious means by which to circumvent the innate defenses of the host's body. Among the most popular devices are use of the M cells of the FAE and the subversion of complement receptors and RCA proteins.

M Cells

The reader will recall from Chapter 20 that the relatively rare M cells of the FAE are key mediators of the antigen sampling necessary for mucosal immunity. Unfortunately, M cells are also inviting portals of entry for a variety of pathogens. The structure of the M cell is such that it lacks the protective glycocalyx of other mucosal epithelial cells, and forms a pocket that extends deep into the underlying lamina propria. Furthermore, the well-developed endocytic apparatus of the M cell is designed to expedite the passage of foreign material through the cell and into the dome underlying the pocket. Invasive species of bacteria and viruses have exploited this transport mechanism as

an easy route to cross the otherwise resistant gut epithelium.

Binding to M cells has been demonstrated for (among several other bacterial species) *Salmonella typhimurium*, *Salmonella enterica*, *Yersinia enterocolitica*, and *Vibrio cholerae*, all of which cause enteric disease. (In contrast, the pathogenic strains of *E. coli* that cause similar symptoms attack gut enterocytes.) Only a few specific receptor–ligand pairs have been identified that exclusively facilitate bacterial entry through M cells. For example, the invasin proteins of *Yersinia* bind to β1 integrins on the M cell apical pole. Entry may be mediated more often by glycoproteins expressed on the apical surfaces of the M cells that can bind to particular bacterial adhesins. While some species of invading bacteria (e.g., *Salmonella*) are toxic to the M cells they have invaded, others are not. For example, *V. cholerae* does not cause disease until it reaches the small intestine, where it commences expression of proteins that allow it to adhere to intestinal epithelial cells. Only then does the bacterium

produce the cholera toxin that induces the exodus of chloride ions from the cell that leads to severe diarrhea.

Shigella flexneri also exploits M cells to cause enteric disease, but this intracellular bacterium accesses these cells by a different mechanism. In animal models, the M cells take up *S. flexneri* by what appears under the electron microscope to be a macropinocytotic event. In some cases, the bacteria cross the M cell without lysing the endocytic vacuole, travel to the deep end of the pocket, and infect intestinal epithelial cells by the basolateral (rather than the apical) pole (see Figure). In other cases, the *Shigella* are taken up by macrophages within the dome. The bacteria induce the apoptosis of these cells, and with this apoptotic death comes the release of more *Shigella* that can invade neighboring epithelial cells, as well as a flood of IL-1β into the local area that induces inflammation. This inflammation contributes to a destabilization of the epithelial cell layer that makes it easier for the *Shigella* to reach

Continued

Box 22-2. **Pathogen entry: exploitation of host components**–*cont'd*

fresh epithelial cells. *Yersinia* and *Salmonella* also make use of dome macrophages but in a different way, hiding in these cells without harming them while the cells migrate from the FAE throughout the body. The bacteria can then induce apoptosis of the macrophages at a later time and attack epithelial cells in new locations.

Several types of viruses also access the body via M cells. In mice, reovirus entering the gut via the digestive tract is proteolytically modified by host digestive such that the virus is activated. This activation induces the expression of an (unidentified) protein that allows the virus to adhere to M cells in the intestinal PPs and the respiratory tract. The virus uses the pocket of the M cell to access the interior of the follicle and infect FAE cells from the basolateral side. Viral progeny then go on to infect the preferred targets of this virus, the macrophages and neurons in the dome underlying the FAE. Interestingly, DCs are not productively infected by reovirus. Rather, DCs capture viral antigens from apoptotic or necrotic epithelial cells and present them to mucosal T cells to initiate an immune response (refer to Ch.20). Another virus that exploits M cells in a variety of sites in mice is the retrovirus called mouse mammary tumor virus (MMTV). Newborn mice may be infected with MMTV after consuming their infected mother's milk, and the virus then moves from the gut lumen to the M cells of the PPs. The virus spreads from there to the gut epithelial cells, the mesenteric lymph nodes and eventually to all lymphoid organs. In adult mice, MMTV enters the body via the M cells of the nasal or rectal epithelium. In humans, polio virus enters the body orally and passes through the M cells of the PPs on its way to its preferred target cells, the neuronal cells in the lamina propria. HIV is also thought to take advantage of M cells. HIV is frequently transmitted by anal intercourse but the virus cannot penetrate the glycocalyx of healthy rectal epithelial cells. Rather, the virus may enter the body via M cells present in FAE in the rectum, possibly by binding to galactosylceramine expressed on

the M cell surface. Studies in rabbits and mice have confirmed that HIV can bind to M cells of the PPs and be transcytosed *in vitro*. Transportation through the M cell into its pocket would give ready access to the T cells, DCs, and macrophages that are this virus's primary targets. More on HIV appears in Chapter 25.

Complement System Components

Once in the tissues, various pathogens have exploited numerous components of the complement system to either increase their adherence to host cell surfaces or to gain access to the interior of a host cell by receptor-mediated endocytosis. For example, pathogens as varied as HIV, *Legionella pneumophila*, *Mycobacterium tuberculosis*, and *Leishmania major* all express ligands that

allow them to bind to CR1, and several also make use of CR3. CR2 acts as a receptor for the Epstein-Barr virus that mediates infection of epithelial cells and B cells. Many picornaviruses such as the Coxsackie viruses express proteins that allow them to adhere to DAF. Host cell entry is not achieved, however, unless ICAM-1 is present to play an (unknown) coreceptor-like role. Various strains of *Escherichia coli* express different adhesin proteins (*Dr proteins*) that can also adhere to DAF, while *Neisseria gonorrhea*, *Neisseria meningitidis*, and *Helicobacter pylori* express pili that bind to MCP on host cells in target tissues. Some laboratory strains of measles virus and HHV-6 are thought to make similar use of MCP to effect host cell entry.

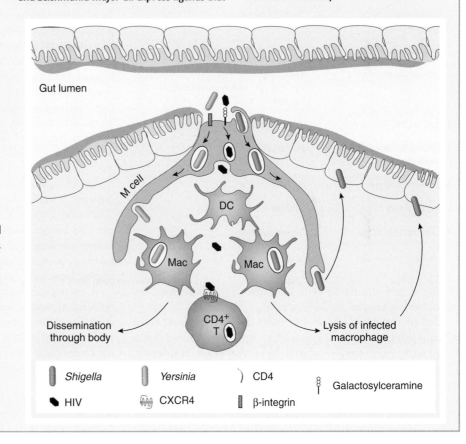

A streptococci are responsible for the severe throat infections commonly known as "strep throat." *Staphylococcus aureus* is responsible for many minor infections of the skin and mucosae, but also produces toxins causing food poisoning or toxic shock syndrome. A much more serious disease caused by either *Staphylococcus aureus* or group A streptococci alone, or by both in combination, is necrotizing fasciitis, more commonly known as "flesh-eating disease." Necrotizing fasciitis is characterized by the sudden onset of a rapidly spreading inflammatory infection that destroys the fascia, the layers of tissue covering the subcutaneous muscle layers. The skin and subcutaneous tissue become separated from the underlying muscles, and the patient experiences fever, severe pain, and a reddened swelling that spreads extremely quickly under the surface of the affected body part. Affected limbs may have to be amputated. Necrotizing fasciitis most often occurs in adults who have undergone surgery or experienced physical trauma of some sort, or who have compromised immune systems. Thankfully, this

Table 22-4 Examples of Extracellular Bacteria and the Diseases They Cause

Organism	Disease
Bacillus anthracis	Anthrax
Borrelia burgdorferi	Lyme disease
Clostridium botulinum	Botulism
Clostridium tetani	Tetanus
Corynebacterium diphtheria	Diphtheria
Escherichia coli O157:H7	Hemorrhagic colitis ("hamburger disease")
Helicobacter pylori	Ulcers
Leptospira (various)	Leptospirosis
Neisseria gonorrhea	Gonorrhea
Neisseria meningitidis (Meningococcus)	Bacterial meningitis
Salmonella enteritidis	Food poisoning (poultry)
Staphylococcus aureus	Food poisoning, toxic shock syndrome, eye and skin infections, necrotizing fasciitis
Streptococcus (Group A)	"Strep throat," necrotizing fasciitis
Streptococcus pneumoniae (Pneumococcus)	Pneumonia
Treponema pallidum	Syphilis
Vibrio cholerae	Cholera
Yersinia enterocolitica	Severe diarrhea

disease is rare (incidence of 0.3–0.7 per 100,000 persons per year), but it is fatal in 20–30% of cases.

One of the most feared (but rare) diseases resulting from a gram-positive bacillus infection is botulism, caused by a neurotoxin (see later) produced by *Clostridium botulinum*. *C. botulinum* is an anaerobic bacterium that commonly resides in the form of dormant spores in soil, fresh water, and house dust, and on the surfaces of many foods. Upon introduction into a suitable low oxygen habitat, the bacterium resumes replication and production of the neurotoxin. Unfortunately, such an environment is sometimes provided by home-canned or home-preserved foods that did not receive adequate heat treatment. Within 8–36 hours after consuming food contaminated by even a minuscule amount of *C. botulinum* neurotoxin, the patient experiences blurred or double vision, slurred speech, swallowing and breathing difficulties, and muscle weakness due to paralysis of the muscle caused by the neurotoxin. Recovery may take weeks or months, and 7–8% of patients die.

An example of a pathogenic gram-negative coccus is *Neisseria meningitidis*, a member of the meningococci. *N. meningitidis* causes bacterial meningitis, a severe inflamma-

tion of the membranes covering the brain and spinal cord (the meninges). About 10% of the general population carries these bacteria harmlessly in the nose and back of throat. However, in the rare event that these bacteria enter the blood, travel to the brain, and attack the meninges, toxins produced by these organisms induce inflammation that damages the brain. The patient complains of a severe headache, rash, lethargy, and a stiff neck, and avoids looking at bright lights. In some cases, the patient may enter a coma, which can be fatal. For unknown reasons, children under the age of 5 years and college and university students are at particularly high risk for contracting bacterial meningitis. Kissing and the sharing of items such as food and beverage utensils or even musical instrument mouthpieces can transmit the bacterium between individuals.

E. coli strain O157:H7 is a gram-negative bacillus that causes hemorrhagic colitis, more commonly known as "hamburger disease." *E. coli* O157:H7 lives harmlessly in the intestines of most food animals, including pigs, chickens, and cattle. During slaughter, the bacterium transfers to the exterior surfaces of the butchered meat. In the case of beef, the grinding of a cut of beef into hamburger distributes the bacterium throughout the meat. If this hamburger not cooked thoroughly and is eaten, the bacterium survives and commences replication in the human gut. The consumption of water contaminated by the manure of farm animals can also cause the disease. The patient experiences hemorrhagic colitis (bloody diarrhea), which may be accompanied by vomiting, abdominal cramps, and fever. A toxin (see later) produced by *E. coli* O157:H7 attacks the endothelial cells lining the body's small blood vessels, allowing the leakage of blood into the tissues. A potentially fatal complication of hemorrhagic colitis that occurs in 2–7% of infected patients (usually the young or very old) is hemolytic uremic syndrome. Because the toxin can kill the endothelial cells lining the tiny blood vessels of the glomerulus of the kidney, severe kidney damage can result such that the patient may permanently require dialysis after recovery. About 5% of hemolytic uremic syndrome cases are fatal.

A particularly nasty subgroup of extracellular gram-negative bacteria are the spirochetes, including members of the genera *Borrelia*, *Leptospira*, and *Treponema*. *B. burgdorferi* is the pathogen that causes Lyme disease. Lyme disease, named after the Connecticut town in which it was discovered, is characterized by a skin rash often accompanied by acute arthritis and/or carditis. The pathogen is transmitted to humans by way of tick bites. Various species of *Leptospira* cause leptospirosis, a disease characterized by acute fever that can lead to liver and kidney damage. Contact with urine from leptospiruric animals (including pet dogs) is the primary route of transmission to humans. Syphilis is a sexually transmitted disease caused by invasion of the genital mucosae by *Treponema pallidum*. The pathogen replicates rapidly at first but then can enter a latent phase. Recurrences erupting several years after the initial infection can lead to damage of the aorta, central nervous system, eyes, and ears.

An organism that has become a topic of much public concern lately is the organism that causes anthrax, *Bacillus anthracis*. This large, gram-positive extracellular bacterium is a common pathogen of hoofed livestock and range animals. It lives in the soil of farm and range lands and produces spores that can survive in soil for decades. While *B. anthracis* infection occurs commonly in livestock, it rarely infects humans. Indeed, prior to the bioterrorism attacks through the postal system in the United States in the fall of 2001, the last case of inhalation anthrax in the United States was recorded in 1976. When *B. anthracis* does infect a human, the resulting anthrax disease takes one of three forms: inhalation (the most lethal), cutaneous, and gastrointestinal. If left untreated, any type of anthrax can lead to septicemia and death. Cutaneous anthrax, the most common form, usually occurs after introduction into a skin abrasion of bacteria from contaminated meat, wool, hides, or leather of infected animals. Skin lesions develop within 12 days of exposure and become necrotic. Patients may also experience fever and lymph node swelling. Gastrointestinal anthrax occurs after consumption of raw or undercooked contaminated meat. Severe abdominal symptoms, including bloody diarrhea, vomiting, and fever, occur within a week of exposure. Inhalation anthrax is contracted by breathing aerosolized spores of *B. anthracis*. It was originally thought that thousands of spores had to be inhaled to cause disease, but that number has been revised downward in light of the deaths of mail workers exposed to more modest bacterial loads. Within 7 days of exposure to aerosolized *B. anthracis*, symptoms resembling a viral respiratory illness ensue that rapidly progress to shock, respiratory failure, and sometimes meningitis. Death occurs in 75% of cases, even with aggressive antibiotic therapy.

ii) Host Damage Caused by Extracellular Bacteria

Many of the disease symptoms caused by extracellular bacteria can be attributed to their production of toxins. *Exotoxins* are toxic proteins actively secreted by both gram-positive and gram-negative bacteria, while *endotoxins* are the lipid portion of lipopolysaccharides embedded in the walls of gram-negative bacteria. A given bacterial species may supply both exotoxins and endotoxins. For example, many strains of *E. coli* produce one or more exotoxins and contain the LPS endotoxin in their cell walls. *Pseudomonas aeruginosa* also contains LPS but secretes a different cytolytic exotoxin called exotoxin A.

iia) Direct toxicity caused by exotoxins. Different exotoxins cause disease by different means and in different locations, depending on the proclivities of the individual bacterial species. For example, infection with *V. cholerae* results in the local release of an exotoxin that binds to gut epithelial cells. A massive release of electrolytes and tissue fluids is induced that is manifested as the severe diarrhea that characterizes cholera. Although they are derived from extracellular bacteria, many bacterial exotoxins have the ability to translocate into mammalian cells and wreak havoc on intracellular processes. The diphtheria exotoxin secreted by *Corynebacterium diphtheriae* travels the body systemically and is absorbed by cells of the heart and peripheral nervous system. The toxin inhibits protein synthesis in these cells, leading to myocarditis and neuritis. The diphtheria exotoxin also promotes colonization of the throat by the bacterium, which provokes an acute inflammatory response resulting in severe respiratory obstruction. As mentioned previously, *C. botulinum* produces a neurotoxin that blocks the transmission of nerve impulses to the muscles, resulting in paralysis. This toxin is much feared as a potential biological weapon

because a dose of less than 1 μg is fatal to humans. Another *Clostridium* species, *Clostridium tetani*, synthesizes a neurotoxin that causes uncontrollable muscle contractions. Other exotoxins trigger specific host cell necrosis, such as the leukocidin produced by *S. aureus* that is toxic to granulocytes. Another example is the exotoxin produced by *E. coli* O157:H7, which causes severe hemorrhaging because it blocks protein synthesis within vascular endothelial cells and kills them. *B. anthracis* produces two exotoxins called *lethal toxin* and *edema toxin* that damage phagocytes in an unknown way. Lethal toxin is composed of a zinc protease called *lethal factor* and a protein called *protective antigen*, while edema toxin is composed of an adenylate cyclase called *edema factor* plus protective antigen.

iib) Immunopathic damage. Extracellular bacteria furnish many examples of situations in which the adaptive immune response to the infection results in immunopathic disease. In some cases, the destruction is caused by endotoxins, and in others an exotoxin is responsible. It is also becoming apparent that prolonged infection with many organisms can sometimes lead to the production of antibodies or activation of T cells that cross-react with self tissues, causing inflammation and possibly autoimmune disease. The concept of "molecular mimicry," in which a pathogen antigen resembles a self antigen enough to provoke an attack on self tissues by anti-pathogen antibodies and T cells, is explored fully in Chapter 29.

Damage to a host caused by an endotoxin is always immunopathic and occurs by the same mechanism in each case. As described in Box 22-1, the cell wall of a gram-negative bacterium contains lipopolysaccharides composed of the endotoxin Lipid A and one of several *O-linked polysaccharides* (O-PS). Not only are the O-PS moieties often specific to a particular bacterial species but they are also immunogenic, leading to their designation as "O-specific antigens." When the immune system responds to these antigens and destroys a gram-negative bacterium, its membrane disintegrates and Lipid A is released into the blood or gastrointestinal tract. The presence of Lipid A induces a massive release of cytokines (particularly TNF) that cause high fever and endotoxic shock (refer to Box 17-2). For example, in the case of *Helicobacter pylori* infection, Th1 cells responding to this organism produce large amounts of pro-inflammatory cytokines (principally TNF and IFNγ) that exacerbate the injury to the gastric lining and promote the formation of stomach ulcers. Thus, current treatments for ulcers include a course of antibiotics, to tackle the problem at its root.

An effect similar to endotoxic shock can result from the polyclonal activation of CD4$^+$ T cells by bacterial exotoxins that are superantigens (refer to Box 14-2). The reader will recall that T cell activation by a superantigen does not depend on either specific peptide or costimulation. The simultaneous activation of a wide variety of different T cell clones results in the production of huge quantities of pro-inflammatory cytokines. In addition, some of the activated T cell clones may be directed against self antigens. These clones would normally be held anergic in the presence of the self antigens in the host, but may go on to attack self tissues after non-specific activation by the bacterial exotoxin.

II. EFFECTOR MECHANISMS

The major mechanisms by which the immune system eliminates extracellular bacteria are summarized in Figure 22-2.

i) Humoral Defense

Extracellular bacteria by definition cannot routinely "hide" within host cells. Thus, the antibodies of the humoral response are generally highly effective against these species. Indeed, patients who are deficient in immunoglobulins (*agammaglobulinemic*) and infected with an extracellular bacterium can be successfully treated by a passive transfer of specific anti-bacterial antibodies. For a normal, healthy individual, the polysaccharides present in bacterial cell walls make perfect Ti antigens for B cell activation, while other bacterial components supply Td antigens that induce primarily a Th2 response and help for B cells recognizing bacterial Td antigens. Anti-bacterial antibodies of the polymeric IgM isotype dominate in the vascular system, while IgG antibodies, because of their smaller size, are able to access the tissues.

As we learned in the first section of this book, antibodies can neutralize bacteria by physically blocking them from attaching to host cell surfaces. Even though they do not need to enter host cells for replication, most extracellular bacteria adhere to host cells to avoid being swept off or out of the host by skin sloughing or movement of the intestinal contents. Small hair-like molecules called *pili* on the exterior surface of the bacterium allow it to stick to glycoproteins on host cells via a lectin-like mechanism. Different bacterial species bear different types of pili that interact with different ligands on host cells in different locations. Thus, the nature of the pili can determine where in the body the bacteria will attempt to establish colonization. Binding of antibody to a protein in the pili can inhibit bacterial adherence to a host cell, allowing the invader to be flushed away.

Antibodies can also serve as opsonins, coating the surface of the bacterium such that it becomes a more attractive target for phagocytosis mediated by neutrophils and macrophages that express Fc receptors. Extracellular bacteria are generally very vulnerable to the killing mechanisms that operate within a phagosome. Changes in pH within the phagosome, defensins, and the action of the reactive oxygen intermediates (ROIs) and reactive nitrogen intermediates (RNIs) all contribute to bacterial killing.

Antibodies are also made against the exotoxins used by bacteria to kill or disable host cells. *Antitoxins* are antibodies that bind to the toxin itself (not the bacterium producing it) and either cause its rapid removal or block its active site. Effective antitoxins are often of the IgG isotype because of the ability of this class to diffuse into the tissues. Antitoxins can also act as opsonins and enhance the clearance of the toxin by phagocytes. Sometimes secretory IgA can act as an antitoxin by preventing a toxin from making contact with and damaging a body tract such as the airway. If the toxin is the sole element causing disease in the host, the production of the antitoxin alone will be enough to restore health. For example, human resistance to tetanus relies on the presence of antibodies directed against the exotoxins produced by *Clostridium tetani*. The tetanus *toxoid*

vaccination that we receive as infants (followed by tetanus boosters every 10 years) is a preparation of inactivated tetanus exotoxin that serves to induce and maintain the production of neutralizing antitoxin antibodies (see Ch.23). Not surprisingly, antigenic variation can influence antitoxin effectiveness. For example, the composition of a bacterial exotoxin may vary slightly among strains such that an antitoxin generated in an exposure to one strain may not recognize the epitopes of exotoxin produced by a later infection with a second strain. The advantage of secondary response levels of antibody is lost, as the host is forced to mount a primary response to the new epitope.

ii) Complement

All three pathways of complement activation can be brought to bear on extracellular bacteria. The binding of antibody to a pathogen allows MAC-mediated destruction of some bacterial species by the classical pathway of complement activation. The alternative pathway can be activated in the absence of antibody by peptidoglycan in gram-positive cell walls or LPS in gram-negative cell walls. Because of their more delicate membrane structure, gram-negative bacteria usually succumb more rapidly to complement-mediated lysis than gram-positive bacteria. Lastly, the lectin pathway is activated by the binding of MBL to the distinctive sugars arranged on the surfaces of bacterial cells. As we saw in Chapter 19, complement-mediated lysis is particularly crucial for defense against the *Neisseria* group of gram-negative extracellular bacteria. Patients lacking the terminal complement components are still able to resist most bacterial infections, but not those initiated by these pathogens.

iii) Th Cells

Resolution of an infection by an extracellular bacterium cannot always be directly equated with elimination by the antibodies of a Th2 response. For example, infection with the extracellular bacterium *Yersinia enterocolitica* is associated primarily with a Th1 response, an interesting observation since this pathogen was originally thought to replicate intracellularly. Other particularly invasive or insidious extracellular pathogens, such as the spirochetes, provoke a combination of Th1 and Th2 responses. For example, immune responses to *B. burgdorferi* and *T. pallidum* have both humoral and inflammatory cell-mediated components. Even so, these organisms are difficult to eradicate, and development of a secondary response adequate to protect against re-infection develops only slowly, if at all. A clue to this intransigence may arise from the study of immune responses against *Borrelia recurrentis*, which causes recurring febrile episodes. Most of these bacteria are eliminated by the humoral response, but a few mutate just at the very end of the acute infection and escape immune system elimination.

iv) Mast Cells

Mast cells play a prominent role in defense against extracellular bacteria, in particular against gram-negative species expressing a bacterial adhesin protein called FimH. Mast cells lurking in the MALT both phagocytose the offending bacteria and release inflammatory mediators such as TNF, leukotrienes (especially LTβ4), IL-6, and histamine. One of

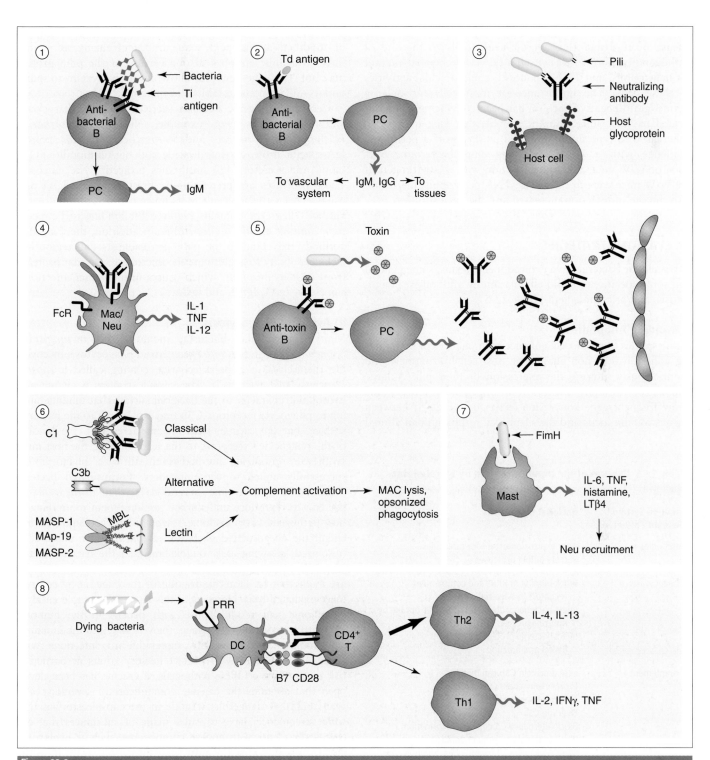

Figure 22-2

Major Mechanisms of Immune Defense against Extracellular Bacteria
(1) Polysaccharides on bacteria can act as Ti antigens and activate B cells that produce anti-bacterial IgM antibodies. (2) Other bacterial components can act as Td antigens and activate additional B cells that produce anti-bacterial IgM and IgG antibodies. (3) Neutralizing antibodies recognizing bacterial pili block access of these bacteria to host cell glycoproteins. (4) Neutralizing antibodies bound to bacteria or their components are recognized by FcR on macrophages and neutrophils. Once activated, these cells initiate receptor-mediated endocytosis of the pathogen and pro-inflammatory cytokine production. (5) A B cell recognizing a bacterial toxin produces neutralizing anti-toxin antibodies that prevent the toxin molecules from damaging the cell surfaces. (6) Bacteria that have been bound by antibody plus C1, C3b, or MBL activate complement via all three pathways. (7) Mast cells activated by an encounter with a FimH-expressing bacterium release histamine and pro-inflammatory cytokines that lead to the recruitment of neutrophils. (8) Bacterial components captured by DCs are processed and presented to CD4+ T cells. Th2 differentiation, which supports antibody production, is generally favored over Th1 differentiation.

the primary functions of these mediators is to promote the release of IL-8 and thus recruit neutrophils to the site of pathogen invasion. LTβ4 is itself a powerful chemoattractant for neutrophils, and TNF enhances the microbicidal activities of these cells. The importance of mast cells for defense against gram-negative bacteria has been convincingly demonstrated in W/W mice, mutant animals that lack mast cells. Whereas wild-type mice were able to shrug off a peritoneal challenge with a FimH-expressing enterobacterium, over 80% of W/W mice died after infection with this pathogen. If the W/W mice were injected with TNF, neutrophil emigration into the site of infection increased and the mice survived.

III. EVASION STRATEGIES

Extracellular bacteria have evolved mechanisms, some general and some specific, to avoid elimination by antibodies, phagocytosis, and/or complement (Table 22-5).

i) Avoiding Antibodies

Many species of gonococci enhance their chances of maintaining adhesion to a non-phagocytic host cell by routinely and spontaneously altering the expression of their pili. A large number of normally silent loci in the gonococcal genome allow frequent gene conversion (refer to Box 21-3) between the active pilin locus and the silent loci, resulting in an altered

nucleotide sequence that translates into an adhesion protein of modified amino acid sequence. Neutralizing antibodies directed against epitopes in one version of the pilin protein may not "see" the new pili, allowing the bacterium to maintain its adhesion and establish an infection.

Certain pathogenic bacteria secrete proteases that cleave antibody proteins. For example, *Neisseria gonorrhoeae*, *Haemophilus influenzae*, and *Streptococcus sanguis* express IgA-specific proteases that degrade both sIgA antibodies in the blood and secretory IgA antibodies protecting the mucosae. These proteases are preferentially directed against IgA1 and usually cleave this molecule in its hinge region. In this case, the antibody loses the domains required for binding to Fc receptors, limiting its effector functions. In addition, the truncated antibody may bind to the pathogen using its Fab region but lacks the ability to agglutinate the microbes. A similar protease from *Proteus mirabilis*, which causes urinary tract infections, can cleave IgA1, IgA2, and IgG.

ii) Avoiding Phagocytosis

Many extracellular bacteria, including *N. meningitidis*, *S. pneumoniae*, and *H. influenzae*, avoid phagocytosis by covering themselves in a polysaccharide coating called a *capsule* ("encapsulated bacteria"). The capsule confers a charge and hydrophilic character to the bacterial surface that inhibits binding to phagocyte receptors. C3b may still attach to the bacterial surface, but the capsule sterically interferes with the binding of the phagocyte's receptors to the C3b. Instead, the host must synthesize opsonizing antibodies to allow a phagocyte to successfully attach to these invaders. Encapsulated bacteria therefore have more time (several days) than unencapsulated bacteria to reproduce undisturbed, making them more dangerous pathogens. For example, alveolar macrophages cannot engulf the encapsulated bacterium *S. pneumoniae* unless it is opsonized, allowing early establishment of the bacteria in the lungs. Eventually, antibodies to components of the capsule are made that facilitate destruction of the bacteria by complement-mediated lysis. However, if this tactic is not rapid enough, pneumonia can result. Those with deficits in the humoral response, such as the very young, the very old, and the immunocompromised, are especially susceptible to infections with encapsulated bacteria. To protect healthy adults at particular risk of exposure to these pathogens, a vaccine has been developed that contains the capsule components of several species (see Ch.23). A dangerous wrinkle in the capsule story is that some streptococci have capsules made of substances that are chemically identical to human components, such as hyaluronic acid. Antibodies raised against such substances might have the unwelcome collateral effect of promoting autoimmunity. Such considerations have mandated caution in vaccination against these pathogens.

Some extracellular bacteria avoid capture by hostile phagocytes by entering non-phagocytes such as epithelial cells and fibroblasts. The pathogens inject bacterial proteins into the host cell that promote cytoskeletal rearrangements facilitating macropinocytosis. The bacteria are taken up by the non-phagocytes and thus gain unmolested access to the tissues. Receptor-mediated endocytosis can be exploited in a similar

Table 22-5 Evasion of the Immune System by Extracellular Bacteria

Immune System Element Thwarted	Bacterial Mechanism
Antibodies	Alter expression of surface molecules
	Secrete anti-Ig proteases
Phagocytosis	Block binding of PRRs and complement receptors to bacterial capsule
	Inject bacterial protein that promotes uptake by non-phagocytes
	Inject bacterial protein that disrupts phagocyte fusion and function
Complement	Lack a protein C3b can bind to
	Express long-chain O-antigen in lipopolysaccharide
	Degrade C3b
	Block alternative C3 convertase formation
	Block terminal component addition
	Shed activating surface moieties, immune complexes, or partially assembled MACs
	Capture host RCA proteins
	Alter expression of host RCA proteins and complement receptors
	Express molecules that mimic complement components or RCAs
	Inactivate anaphylatoxins
	Induce host production of poor complement-fixing Abs or Abs that block C3b binding
APCs	Block maturation or migration of DCs

way. For example, enteropathogenic *E. coli* injects the Tir (*translocation intimin receptor*) protein into the cytoplasm of epithelial cells. The act of injection induces a rearrangement of the host cell cytoskeleton that results in Tir being fixed in the plasma membrane, ready to act as a receptor for the surface protein intimin on the bacterium. Receptor-mediated endocytosis ensues, importing the bacterium into the non-phagocytic cell.

Some species of *Yersinia* have direct anti-phagocyte activity. *Y. enterocolitica* is an extracellularly replicating enteric pathogen that targets the FAE of the PPs and causes severe diarrhea. This bacterium forestalls phagocytosis by attaching to the exterior of a looming macrophage and injecting a set of so-called *Yop proteins* into the macrophage cytoplasm. Among the Yop proteins is YopH, a phosphatase that binds to certain tyrosine-phosphorylated proteins in the macrophage that are thought to be involved in intracellular signaling and the operation of its actin cytoskeleton. When YopH dephosphorylates these host proteins, disruptions in intracellular signaling block integrin-mediated phagocytosis of the bacterium. A different Yop protein appears to induce the apoptosis of macrophages, at least *in vitro*.

iii) Avoiding Complement

Evasion of the complement system is a time-honored strategy employed by a vast array of pathogens. Specific methods include avoiding recognition by complement proteins, removing any attached complement proteins, exploiting various host RCA (regulators of complement activation) proteins to either inactivate the complement cascade or gain access to a host cell, and expressing molecules that resemble host complement components or RCAs. Many anti-complement strategies employed by extracellular bacteria are shared by intracellular bacteria, viruses, and parasites. Box 22-3 explores in depth several of

---

Box 22-3. Avoidance of the complement system by pathogens

The complement system is a vital contributor to the defense of the host in both the innate and adaptive immune responses. Naturally, therefore, pathogens have devoted much time and energy to devising ways to confound the complement system, or to make use of it for their own ends. Pathogens can avert complement-mediated destruction by presenting a cell surface that does not permit deposition of C3b (avoiding recognition); by shedding incipient complexes as they form; by destroying or inhibiting various complement proteins; by acquiring host RCA proteins; and/or by expressing molecules that closely resemble host RCA proteins. At any one time, a given pathogen may be able to use just one or several of these mechanisms to protect itself.

A pathogen can avoid complement recognition if the surface moieties it expresses do not bind C3b or antibody well or at all. The surface glycoproteins of some parasites and capsules of some bacteria fit this description. For example, some species of gram-negative bacteria avoid detection by expressing LPS residues with unusual side chain sugars. These sugars mask the LPS determinants that would normally bind to C3b and activate complement via the alternative pathway. Another avoidance mechanism is the shedding of activatory surface moieties or immune complexes, as occurs for the parasites *Schistosoma mansoni* and *Trypanosoma cruzi*, respectively. Some organisms express molecules that disrupt one or more complement activation pathways. For example, Protein H expressed by *Streptococcus pyogenes* binds to C1 but does not allow the cascade to proceed, competitively inhibiting the classical pathway. Several members of the poxvirus and herpesvirus families secrete proteins that block

formation of the alternate C3 convertase. Even if either C3 or C1 has succeeded in initiating complement activation, many pathogens, such as the bacterium *Escherichia coli* and the parasite *Leishmania major*, respond by shedding the complex, particularly at the MAC formation stage. Proteolytic degradation and phosphorylation of complement components are also used by many organisms to derail MAC assembly before it can go too far. Some organisms (like certain *Borrelia* and *Salmonella* species) express proteins that inhibit the terminal complement proteins, so that C6 deposition is blocked or C9 polymerization is inhibited.

Another effective strategy for blocking complement-mediated destruction is to acquire the host RCA proteins that normally protect host cells. Budding viruses take sections of the host cell membrane with them as they emerge, and embedded in the membrane are host RCA proteins. DAF and MIRL are acquired by HIV and vaccinia in this way. Other organisms express molecules that adsorb soluble RCA proteins from the host plasma. The host RCA proteins retain their regulatory activity even when bound to a bacterial surface, so that the alternative pathway of complement activation is inhibited. For example, the fluid phase RCA proteins Factor H and C4bp can be bound to the M proteins of *Streptococcus pyogenes*. The C4bp–M protein binding is completely sufficient to halt complement-mediated opsonization and phagocytosis. C4bp has also been shown to bind to all known strains of *Bordetella pertussis*. In *Neisseria gonorrhea*, both surface sialic acids and an outer membrane protein called *porin* can bind Factor H.

Altering the expression of host RCA and CR proteins is another means of countering complement. Some intracellular pathogens induce new expression of RCA proteins to block complement-mediated destruction, thus preserving the pathogen's chosen home. For example, DAF and MCP expression are upregulated on cells infected with human CMV, and a modified form of DAF appears on cells infected with *Plasmodium berghei*. HIV downregulates the expression of CR1 and CR2 on the infected host cell, reducing opsonized phagocytosis. HIV also reduces the expression of C5aR by monocytes, decreasing the chemotaxis of these cells to sites of infection and inflammation. A related mode of pathogen defense is the expression of molecules that resemble host RCA proteins. Proteins with similar structures and functions to mammalian CR1, DAF, MCP, C4bp, and MIRL have been described in viruses such as vaccinia, KSHV, and HSV, as well as in parasites such as *Schistosoma*, *Entamoeba*, and *Trypanosoma*. Studies of *Herpesvirus saimiri* have shown that this virus has apparently acquired the cDNA for mammalian MIRL and incorporated it into the viral genome. The virus induces the expression of a GPI-anchored protein in the host cell membrane that closely resembles monkey MIRL and protects against MAC-mediated lysis.

Lastly, some bacteria and parasites proteolytically inactivate complement cascade products such as the anaphylatoxins C5a and C3a. A lack of these molecules dampens the inflammatory response and gives the pathogen more time to multiply or hide. For example, a membrane-bound enzyme that removes the C-terminus of C5a is expressed on the surface of streptococci.

these tactics from the point of view of a particular complement system component.

Perhaps the simplest approach to evading complement has been taken by the organism causing syphilis, *T. pallidum*. The outer membrane of this gram-negative bacterium is strikingly devoid of transmembrane proteins, offering almost no place suitable for the deposition of complement components. Other bacteria avoid complement-mediated lysis due to the nature of the O-PS of their cell wall lipopolysaccharides. Classical complement activation mediated by antibodies directed against the O-PS can eliminate a bacterium. However, if the polysaccharide chains of the O-PS are very long and project a great distance from the bacterial cell wall, the MAC cannot assemble on the bacterial surface and the invader is spared. Bacteria expressing O-PS with short polysaccharide chains are vulnerable to complement-mediated lysis.

Many extracellular bacteria secrete enzymes that inactivate various steps of the complement cascade. For example, group B streptococci, which cause neonatal meningitis, contain sialic acid in their cell walls, which can block alternative complement activation by degrading C3b that is attempting to attach to the pathogen surface. Several streptococci species also produce M proteins that can bind to the fluid phase RCA protein Factor H and fix it onto the bacterial surface. In its hijacked site, the recruited Factor H makes any C3b that has attached susceptible to degradation, and thus protects the microbial surface. Even if a host develops antibodies to a particular M protein, these antibodies tend not to be cross-reacting, and there are more than 80 different types of M proteins. This means that the host will likely still be vulnerable to a subsequent infection with a different streptococcal strain expressing a different M protein.

Gonococci and meningococci may have the most complex approach to avoiding death by complement. These organisms induce the host to produce "blocking antibodies" that recognize particular proteins on the surface of the bacteria but are of isotypes that are poor at fixing complement, such as IgA. The blocking antibodies compete with complement-fixing antibodies for binding to the bacterial surface, and prevent bacterial destruction by complement-mediated lysis. In addition, steric hindrance by the blocking antibodies interferes with the deposition of C3b and thus thwarts the alternative pathway of complement activation.

C. Immunity to Intracellular Bacteria

I. WHAT ARE INTRACELLULAR BACTERIA?

Intracellular bacteria have evolved the ultimate escape from phagocytes, complement, and antibodies: they move right inside the host cell and complete their reproduction out of reach of these host defenses. Some species infect non-immune system host cells such as hepatocytes and epithelial and endothelial cells, while others show a strong predilection for macrophages. While some intracellular bacterial species cannot survive outside of a host cell, others merely make intracellular

replication a preference. Like extracellular bacteria, most intracellular bacteria access the host via breaches in the mucosae and skin, but some are introduced directly into the bloodstream by the bites of vectors such as ticks, mosquitoes, and mites.

Intracellular bacteria generally enter the host cell by receptor-mediated endocytosis and are thus first confined to intracellular vacuoles. Some species remain in the vacuolar compartment, while others leave it to take up residence in the cytosol. Because of their desire to replicate within a host cell and keep it alive for this purpose, intracellular bacteria are generally not very toxic to the host and do not produce tissue-damaging bacterial toxins. Because of their low toxicity, infections with intracellular bacteria have extended incubation times. However, their intracellular lifestyle makes infections with these organisms difficult to resolve completely. These factors mean that infections with intracellular bacteria tend to result in chronic or recurrent rather than acute disease. Cell-mediated immunity is crucial for combating infections by intracellular bacteria. Thus, immunocompromised individuals who lack the ability to mount cell-mediated immune responses are particularly at risk for infection with these pathogens.

Important species of intracellular bacteria belong to the *Salmonella*, *Listeria*, *Brucella*, *Rickettsia*, and *Legionella* genera (Table 22-6). *Salmonella typhi* is an encapsulated, gram-negative enterobacterium that causes typhoid fever in humans. *Salmonella typhimurium* causes the equivalent of typhoid fever in mice. Humans can be infected by *S. typhimurium* if they eat improperly cooked meat from infected cattle. In this case, the infection results in food poisoning. Like all *Salmonella* species, *S. typhi* and *S. typhimurium* transcytose across M cells in the mucosae. However, unlike the extracellular *Salmonella enteritidis* bacteria, which cause poultry-related food poisoning, *S. typhi* and S. *typhimurium* invade macrophages to replicate and thus are intracellular pathogens. Another food-borne intracellular pathogen is *Listeria monocytogenes*, a gram-positive rod that infects the cytoplasm of macrophages and hepatocytes. *L. monocytogenes* causes meningeal or systemic

Table 22-6 Examples of Intracellular Bacteria and the Diseases They Cause

Organism	Disease
Brucella (various)	High fevers, brucellosis
Legionella pneumophila	Legionnaire's disease
Listeria monocytogenes	Listeriosis
Mycobacterium leprae	Leprosy
Mycobacterium tuberculosis	Tuberculosis
Rickettsia rickettsii	Rocky Mountain spotted fever
Salmonella typhi	Typhoid fever
Salmonella typhimurium	Typhoid fever-like disease in mice, food poisoning in humans
Shigella flexneri	Enteric disease

infections most often in sheep and cattle but sometimes in humans. Unlike most bacteria, *Listeria* can multiply quite successfully at low temperatures, meaning that the infection continues to grow in contaminated food even when it is refrigerated. *Listeria* infection has also been associated with miscarriages and stillbirths in humans and cattle. Mouse models of *Listeria* infection (listeriosis) have been very helpful in discovering how the immune system combats intracellular pathogens (see later). However, for all its use in experimental models, it is not yet clear how *L. monocytogenes* makes its way across the intestinal barrier into the liver and spleen and finally across the blood–brain barrier. Various *Brucella* species are transmitted to humans via contact with animals and result in high fevers. *Rickettsia rickettsii* is transmitted into the bloodstream via tick bites and proceeds to attack endothelial and smooth muscle cells, causing Rocky Mountain spotted fever. *Legionella pneumophila*, the relatively recently identified pathogen that causes the lung disorder known as "Legionnaires' disease," replicates only within macrophages.

More familiar examples of intracellular bacterial pathogens belong to the *Mycobacterium* genus. *M. tuberculosis* causes tuberculosis, and *M. leprae* infection leads to leprosy. The mycobacteria do not produce bacterial toxins and cause only mild inflammation on their own. However, these bacteria are extraordinarily hard to remove, and it is the prolonged host response to a mycobacterial infection that causes most of the tissue damage. Consider *M. tuberculosis*, a slow-replicating (12 hour) aerobic bacillus that gravitates to well-aerated tissues such as the lung. Tubercular disease associated with *M. tuberculosis* is caused not only by inflammatory damage in the lungs induced by the bacteria but also by the host's adaptive immune response. The persistence of the bacteria summons activated T cells that supply products and contacts that hyperactivate macrophages. The hyperactivated macrophages in turn secrete copious amounts of pro-inflammatory cytokines that result in lung damage. In addition, the macrophages may eventually form granulomas to wall off the bacteria. These granulomas are the "tubercles" (lumps or knobs) on the lungs that give the disease its name. Eventually, the inner circle of macrophages in the tubercles breaks down and the bacilli are released. If the body's defenses can arrest the disease effectively at this point, the lesion calcifies and becomes visible in X-rays as scarring on the lung. If the disease is not arrested, the bacilli persist and may become dormant for months or even years. At some point, when the immune system is weakened due to poor health, malnutrition, or other infections, the bacilli in the center of the tubercle may be able to proliferate again. However, in this situation, bacterial multiplication takes place <u>outside</u> macrophages for the first time. Should the tubercles rupture, millions of bacteria are released into the lungs. The patient becomes highly contagious at this point due to the release of bacilli during coughing episodes. From the lungs, the bacteria disseminate throughout the patient's body and induce the formation of seed-sized tubercles in infected tissues. Body defenses may be overwhelmed, leading to the weight loss, coughing (often with blood), and loss of vigor known collectively as "consumption."

Treatment of TB is a challenge, because the slow growth habit of the bacterium and its ability to hide in macrophages

or other locations for extended periods mean that the patient is often asymptomatic. Indeed, 90% of individuals infected with *M. tuberculosis* remain clinically healthy. An asymptomatic patient is sometimes reluctant to continue with the long-term course of antibiotics needed to successfully treat the disease. A classical "TB test," involving the cutaneous injection of tuberculin antigen (often called *purified protein derivative*; PPD) into the skin, indicates whether an individual is or has been infected with *M. tuberculosis* (and thus may still be harboring the bacterium) (Plate 22-1). A pronounced,

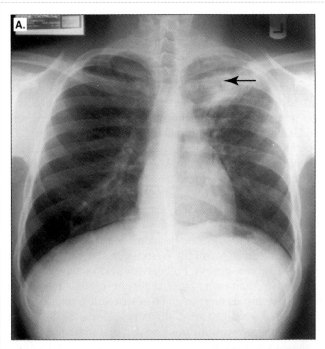

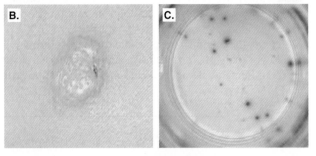

Plate 22-1

Tuberculosis
(A) Chest X-ray of an adolescent with tuberculosis showing disease of the upper lobe of the lung and early formation of a cavity (arrow). Courtesy of Dr. Ian Kitai, The Hospital for Sick Children, Toronto. (B) Tuberculin skin test based on a delayed type hypersensitivity in response to a crude mixture of *M. tuberculosis* proteins. (C) *Ex vivo* assay detecting secretion of IFNγ by individual peripheral blood T cells after stimulation with specific mycobacterial antigens. Courtesy of Dr. Ajit Lalvani, Nuffield Department of Medicine, University of Oxford, John Radcliffe Hospital, Oxford. [Lalvani A. *et al.* (2001) Rapid detection of *Mycobacterium tuberculosis* infection by enumeration of antigen-specific T cells. *American Journal of Respiratory and Critical Care Medicine* 163, 824–828.]

reddened reaction on the skin around the inoculation site after about 48 hours is a DTH response: memory T cells responding to the inoculated antigen activate macrophages that produce cytokines inciting local inflammation. In the very young, a positive TB test probably means that an active TB infection is occurring. In older people, a positive test can result from either an active infection, or from a previous infection that has been cleared, or from vaccination. In these cases, follow-up testing is done by X-ray examination of the lungs, which can distinguish between an active and a calcified lesion. There may also be an attempt to culture the bacterium from the patient. A new generation of TB tests under development includes variations of ELISA assays that measure the production of IgM antibodies or activation of T cells directed against epitopes specific to *M. tuberculosis*.

Healthy individuals living in generally sanitary environments (such as those in most of the developed world) are usually able to successfully fight off *M. tuberculosis*. However, the immune systems of those living in developing countries and socio-economically depressed areas of developed countries may be stressed just enough to put them at increased risk for TB. Attempts have been made to slow the spread of TB in such populations by antibiotic treatment and vaccination programs (see Ch.23). However, it takes at least 6 months of intense chemotherapy to cure a TB patient because of the limited effectiveness of available drugs against entrenched *M. tuberculosis*. Moreover, poor patient compliance with this stringent regime has led to the rise of multi-drug resistant strains of *M. tuberculosis*. In addition, administration of the bacillus Calmette-Guérin (BCG) strain of *Mycobacterium bovis* used as a vaccine against TB does not always induce complete cell-mediated immunity. These factors have led to a maddening persistence and recurrence of TB in susceptible populations. It is estimated that there are 2 billion people worldwide who are latently infected with *M. tuberculosis*, and that about 5–10% of these experience re-activation of the bacterium, leading to active disease. Tuberculosis kills about 1.6 million people per year worldwide.

M. leprae, the organism responsible for leprosy, grows even more slowly than *M. tuberculosis*, taking about 12 days to replicate in humans. Because the optimal growth temperature for this organism is 30 °C, *M. leprae* preferentially establishes infections just under the skin. The bacterium infects the peripheral nerves in this location and causes visible skin lesions. Two clinical types of leprosy have been identified based on disease phenotype and the strength of the cell-mediated immune response (Plate 22-2). "Tuberculoid leprosy," also called "neural leprosy," is associated with a vigorous cell-mediated immune response. The lesions, which are confined to the skin, contain moderate numbers of bacteria. However, the damage to peripheral nerves causes loss of nerve function that frequently leads to injury. Anti-mycobacterial antibody titers are low in these patients. However, if the infected patient mounts only a weak cell-mediated response to the organism, tuberculoid leprosy may become "lepromatous leprosy." This form of the disease is characterized by severe skin lesions containing large numbers of bacteria that can destroy underlying bone and cartilage, leading to loss of structures such as the fingers and the nose. Anti-mycobacterial antibody titers are high in these patients but are ineffective in controlling the disease.

II. EFFECTOR MECHANISMS

The major mechanisms by which the immune system eliminates intracellular bacteria are summarized in Figure 22-3.

i) Neutrophils
Early infections by intracellular bacteria are controlled by killing mechanisms associated with neutrophil phagocytosis. For example, infection of epithelial cells by *L. monocytogenes* generally induces the apoptosis of these cells, releasing chemoattractant molecules that draw neutrophils to the area. Bacteria released from the dying cells are phagocytosed by the neutrophils, whose defensins and powerful respiratory burst usually kill the invaders.

ii) Macrophages
If neutrophil killing does not suffice, the monocyte/macrophage arm of the innate response is activated. The ability of a macrophage to kill phagocytosed bacteria is influenced by

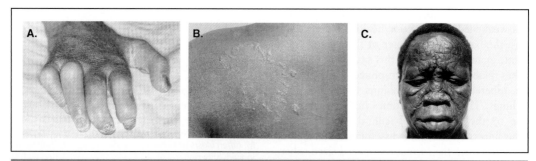

Plate 22-2

Leprosy
(A) Tuberculoid leprosy. (B) Leprosy characterized as the "middle of the spectrum." (C) Lepromatous leprosy.
Reproduced with permission from Emond R.T.D., Welsby P.D., and Rowland H.A.K. (2003) "Colour Atlas of Infectious Disease," 4th edn. Elsevier Science.

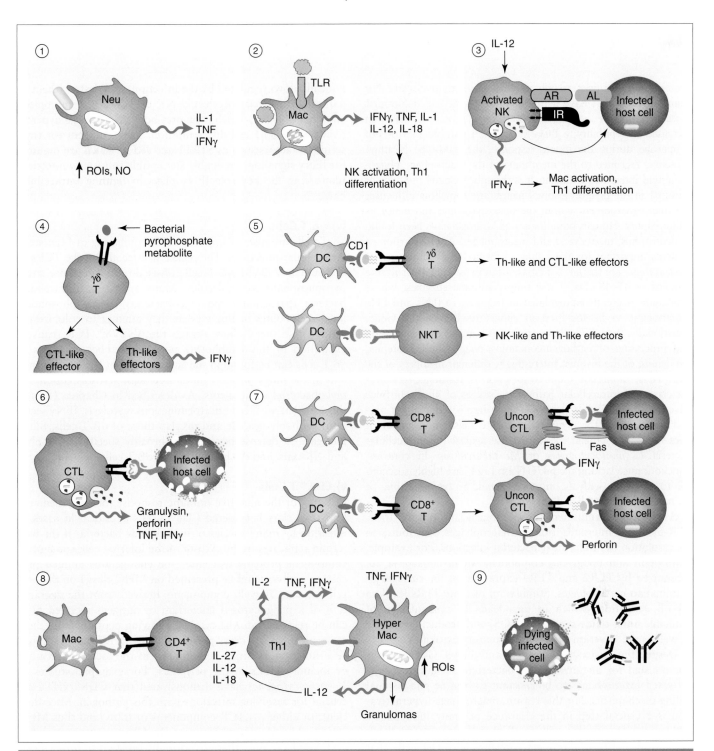

Figure 22-3

Major Mechanisms of Immune Defense against Intracellular Pathogens
(1) Phagocytosis of bacteria by neutrophils triggers phagosomal killing via respiratory burst and cytokine secretion. (2) Phagocytosis of bacteria by macrophages initiates phagosomal killing and secretion of cytokines that maintain inflammation, activate NK cells, and promote Th1 differentiation. TLR-mediated endocytosis also triggers cytokine secretion. (3) NK cells activated by IL-12 kill infected host cells by natural cytotoxicity and secrete IFNγ which activates macrophages and supports Th1 differentiation. (4) Phosphorylated metabolites released by a bacterium activate γδ T cells that generate T cell effectors. (5) Infected DCs present bacterial components on CD1 to γδ T cells and NKT cells. Once activated, these cells generate cytotoxic- and Th-like effectors. (6) CTLs recognizing bacterial peptides presented on MHC class I by an infected host cell kill the cell by perforin- and granzyme-mediated cytotoxicity. The CTL also secretes the anti-microbial molecule granulysin and pro-inflammatory cytokines. (7) Unconventional (Uncon) CTL subsets are activated by bacterial components presented on CD1 by infected DCs. One subset kills infected host cells by Fas killing, while another relies on perforin-mediated cytotoxicity. (8) Infected macrophages present bacterial peptides on MHC class II to CD4+ T cells. In the presence of IL-27, IL-12, and IL-18, Th1 effectors differentiate and supply cytokines that both support the CTL response and hyperactivate macrophages. Hyperactivated macrophages produce increased levels of pro-inflammatory cytokines and ROIs that increase killing. Granuloma formation may occur if these measures are insufficient. (9) Bacterial components released from a dying infected cell can activate B cells to produce neutralizing antibodies. These antibodies intercept any bacterium temporarily transiting the extracellular environment.

proteins that scientists are only now beginning to investigate. For example, mice that have a mutation of a gene called Nramp1 (*natural resistance associated macrophage protein* 1) have an increased susceptibility to BCG infections. Upon phagocytosis of a microbe during the innate response, the wild-type Nramp1 protein is recruited to the membrane of the macrophage phagosome and integrates within it. It is thought that the presence of Nramp1 in the phagolysosomal membrane somehow influences the microenvironment within the phagolysosome and hence its killing ability. Human homologues of Nramp1 have been found in neutrophils, monocytes, and macrophages. Another group of proteins important for the killing of intracellular pathogens (at least in mice) are the *p47GTPases*, a family of GTP-hydrolyzing proteins of 47–48 kDa. These transmembrane enzymes, whose expression is greatly upregulated in response to IFNs and LPS, are thought to be localized in either the ER or the Golgi. Functionally, the p47GTPases appear to regulate the maturation and processing of pathogen-containing phagosomes, hastening the demise of the invader. Interestingly, different members of the p47GTPase family are required for effective resistance to different classes of intracellular pathogens. Studies of knockout mice have shown that, while the p47GTPase Igtp is required for normal resistance to the protozoan parasite *Leishmania major*, it does not play a crucial role in defense against the intracellular bacteria *L. monocytogenes* or *M. tuberculosis*. In contrast, knockout mice lacking the p47GTPase Lrg47 are highly susceptible to infection with *L. monocytogenes*, *S. typhimurium*, or *M. tuberculosis* as well as *L. major*.

As well as engulfment of pathogens by phagocytosis, TLR-mediated internalization by macrophages contributes to the resolution of intracellular bacterial infections. For example, lipoprotein and lipoglycan components of mycobacteria are recognized by TLR2 and TLR4 expressed on the surfaces of mammalian macrophages. Stimulation of these TLRs leads to NF-κB activation, which in turn leads to upregulation of inducible nitric oxide synthase (iNOS) and production of NO as well as the secretion of pro-inflammatory cytokines.

Despite the existence of Nramp1 and the p47GTPases, it is not unusual for an intracellular bacterium phagocytosed by activated macrophages to be resistant to routine phagosomal killing mechanisms. For this reason, macrophage hyperactivation is a crucial step in the clearance of many intracellular pathogens. The reader will recall that a positive feedback mechanism exists between macrophages and Th1 cells that is required for macrophage hyperactivation. Macrophages activated by exposure to the pathogen secrete IL-12, which influences activated T cells to differentiate into Th1 effectors. The IFNγ secreted by the activated Th1 cells acts on the macrophages to hyperactivate them such that they gain the enhanced microbicidal powers detailed in Chapter 4. Foremost among these powers is the capacity to produce the large quantities of ROIs and RNIs that efficiently kill almost all intracellular pathogens.

iii) NK Cells

The IL-12 secreted by activated macrophages also activates NK cells. As described in Chapter 18, NK cells detect infected host cells by their lack of MHC class I expression (which is typically downregulated by the infection) and destroy them by natural cytotoxicity. Activated NK cells also secrete copious amounts of IFNγ, which promotes both macrophage hyperactivation and Th1 cell differentiation. Mice deficient for transcription factors such as STAT1 and IRF2 (which are required for IFNγ signaling) are highly susceptible to *Listeria* infection, reinforcing the key role IFNγ plays in fighting intracellular bacteria.

iv) γδ T Cells

γδ T cells are also of importance in combating at least some intracellular infections. The reader will recall that the TCRs of the human Vγ2Vδ2 γδ T cell subset directly recognize small pyrophosphate-like molecules. Many species of intracellular bacteria (particularly the mycobacteria) release phosphorylated metabolites of this type as they attempt to colonize the host. These metabolites engage the Vγ2Vδ2 TCR, causing expansion of this lymphocyte population. High numbers of γδ T cells can be found in the acute lesions of leprosy patients, and in the lungs of mice immunized with aerosols containing mycobacterial components. As described in Chapter 18, it is thought that γδ T cell effector functions (cytolysis, IFNγ secretion) probably precede and overlap those of αβ T cells, filling any gaps in defense between the broadly specific neutrophils and NK cells, and the highly specific αβ T cells.

v) CD8⁺ T Cells

In terms of the adaptive immune response, defense mediated by MHC class I-restricted CD8⁺ CTLs is critical in cases of infection by many species of intracellular bacteria. If the bacterium replicates in the cytosol of the infected cell, some of its component proteins will enter the endogenous antigen processing pathway and be presented on MHC class I on the host cell surface. As well, components derived from the degradation of a phagocytosed bacterium or antigens secreted by it can be released into the cytosol, allowing peptides from these molecules to be cross-presented on MHC class I. Such cells function as target cells for Tc in the local lymph node or memory CTLs in the periphery. For example, studies of *L. monocytogenes* have demonstrated that CD8⁺ CTLs are crucial for resolving infections with this pathogen. Mice deficient for either αβ TCR components or β2m (and thus MHC class I) show decreased resistance to *L. monocytogenes* infection. Similarly, mice that are normally resistant to tuberculosis can be rendered susceptible if CD8⁺ T cells are depleted by treatment with antibodies prior to infection. However, cytolysis via the perforin or Fas pathways does not appear to be a big factor in CD8⁺ T cell-mediated defense against intracellular bacteria (unlike antiviral defense; see later). Studies in knockout mice have shown that Fas-mediated cytolysis is not required for any stage of host defense against *L. monocytogenes* or *M. tuberculosis*, and perforin-mediated cytotoxicity is important only for containing disease in the chronic stages of *L. monocytogenes* and *M. tuberculosis* infections. Rather, CD8⁺ CTLs contribute to anti-bacterial defense primarily by releasing IFNγ and TNF, which are critical both for inflammatory cell influx and direct killing of infected cells, and by secreting *granulysin*, a molecule that has direct anti-microbial activity. Knockout mice deficient for the expression of either TNF or IFNγ or the receptors for these molecules show increased susceptibility to infections with *L. monocytogenes*. Furthermore, animals treated with anti-TNF or anti-IFNγ antibodies show increased susceptibility to listeriosis, while those treated with anti-IL-4 actually show increased resistance. Human patients lacking the IFNγ receptor are highly susceptible to certain types of mycobacterial infections.

vi) Unconventional CD8⁺ CTLs

Unconventional CTL subsets also contribute to defense against intracellular bacteria. Many types of intracellular bacteria secrete proteins that can be processed into peptides containing N-formyl-methionine. As we learned in Chapters 10 and 11, these peptides can be bound by non-classical MHC class Ib molecules (such as H-2M3) and presented to αβ CD8⁺ T cells. Several *Listeria* peptides can be presented by H-2M3 in mice, and a role for this type of presentation has been proposed for the early stages of the response to *Listeria* infection. However, this type of priming appears to induce only minimal memory responses. In the case of mycobacteria, infected DCs can present non-peptide glycolipid bacterial antigens on the non-polymorphic MHC-like molecule CD1b. Two subsets of CD1b-restricted CTLs have been defined in *M. tuberculosis*-infected mice. One subset uses Fas-mediated cytolysis to kill infected host cells (but not the pathogen), while the other subset uses perforin-mediated cytolysis to simultaneously eliminate both the infected host cell and the pathogen. Both subsets also secrete IFNγ. In humans, CTLs directed against glycolipid antigens have been isolated from tuberculosis patients, and these cells directly recognize and lyse infected macrophages *in vitro*. γδ T cells and NKT cells bearing CD1c- and CD1d-restricted TCRs may also play a role in anti-mycobacterial defense *in vivo* but the antigens recognized by these TCRs have yet to be identified.

vii) CD4⁺ T Cells

CD4⁺ T cells make a significant contribution to defense against intracellular bacteria not only because of the IL-2 they secrete to support Tc differentiation but also because these cells are required for the hyperactivation of macrophages. Bacterial antigens either secreted by the bacteria themselves or released by necrotic infected cells are taken up by professional APCs and presented to CD4⁺ T cells in association with MHC class II. The IFNγ produced by NK cells activated early in the infection favors the differentiation of Th1 effectors. Activated Th1 cells then supply the intercellular contacts (particularly CD40L) and TNF and IFNγ that drive the hyperactivation of macrophages and, if necessary, the formation of granulomas (see later).

While the IL-12 secreted by activated and hyperactivated macrophages and DCs is vital for the Th1 response to intracellular pathogens, it is not the only important cytokine. As discussed in Chapter 17, IL-18 acts synergistically with IL-12 to promote IFNγ production and Th1 differentiation. Resistance to infection by species of *Yersinia*, *Salmonella*, and *Cryptococcus* is enhanced by administration of IL-18 to wild-type mice, and IL-18⁻/⁻ mice show increased susceptibility to

infection with intracellular pathogens such as mycobacteria and *L. major*. IL-27 is required for the initiation (but not maintenance) of Th1 responses. Mice lacking IL-27R thus show decreased IFNγ production, which increases their susceptibility to *L. major* infection.

The importance of the Th1 response in defense against intracellular pathogens is nicely illustrated in human immune responses to *M. leprae* infection. When the T cells surrounding the lesions of the mildly affected tuberculoid leprosy patients described previously were isolated and tested for cytokine production, they were found to secrete IFNγ. It was hypothesized that these patients mounted a Th1 response that promoted cell-mediated immunity and ameliorated the disease. In contrast, the T cells surrounding the lesions in the more severely affected lepromatous patients were found to produce predominantly IL-4 and IL-10. These patients appeared to have mounted a Th2 response that favored humoral immunity (less effective against intracellular pathogens) over cell-mediated immunity. This difference is likely attributable to genetic differences between patients that affect the Th1/Th2 polarization of Th cells responding to this pathogen.

viii) Granuloma Formation

When an intracellular pathogen is able to resist killing even by CTLs and hyperactivated macrophages, the body takes the approach of "If you can't remove 'em, confine 'em." A group of hyperactivated macrophages fuses to form a granuloma that walls off the pathogen from the rest of the body (as described previously for chronic TB infection) (Fig. 22-4 and Plate 22-3). Scattered among the macrophages in the inner layer of the granuloma are numerous CD4⁺ T cells, while an outer ring of CD8⁺ T cells forms the exterior layer. Eventually the exterior of the granuloma becomes calcified and fibrotic, and cells in the center undergo necrosis. In some cases, all the pathogens trapped in the dying cells are killed and the infection and inflammation are resolved. In other cases, a few pathogens remain viable but dormant within the granuloma, causing it to persist. Granuloma persistence is an overt sign that the disease is becoming chronic. An event that results in the breakdown of the structure of the granuloma can release the trapped pathogens back into the body to commence replication anew. Should the host be immunosuppressed for some reason and unable to marshal the T cells and macrophages necessary to form new granulomas, the pathogen may be released into the bloodstream, from which it can infect other organs throughout the body and even cause death.

Cytokines play a critical role in granuloma formation. Sustained IFNγ production by Th1 cells and CTLs is required to maintain macrophage hyperactivation. TNF production by hyperactivated macrophages is crucial not only for early chemokine synthesis (to recruit fresh cells to the incipient granuloma) but also for aggregating these cells and establishing the "wall" around the invaders. Secretion of IL-4 and IL-10 by Th2 cells late in the adaptive response serves to control the formation of granulomas, damping them down as the bacterial threat is contained.

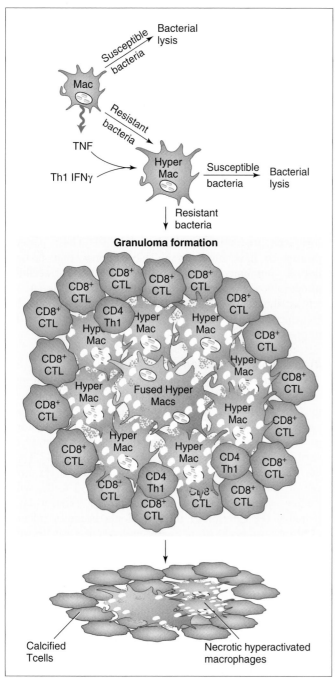

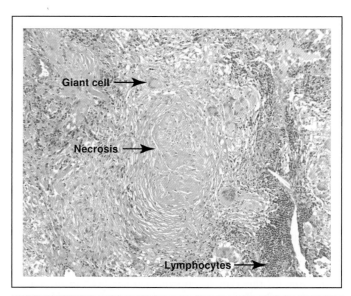

Granuloma

A central zone of necrosis is surrounded by activated macrophages and multinucleated giant cells (fused macrophages), with a rim of fibrosis and lymphocytes. Courtesy of Dr. David Hwang, Department of Pathology, University Health Network, Toronto General Hospital.

ix) Humoral Defense

Although it may seem counterintuitive at first, antibodies can contribute to defense against intracellular pathogens. Some antibodies have been shown to block the access of bacteria to host receptors used for cell entry. In addition, newly arrived bacterial cells or those transiting the extracellular space before invading the neighboring host cell can be opsonized or subjected to classical complement-mediated lysis. It is further speculated that antibodies taken into the phagosome with the bacterium can act in this location. For example, monoclonal antibodies to *listeriolysin O* (LLO), a bacterially produced molecule that facilitates the escape of *L. monocytogenes* from the phagosome into the host cytosol, have been shown to provide protection against *L. monocytogenes* infections in this way.

III. EVASION STRATEGIES

Strategies used by intracellular bacteria to evade the immune response are summarized in Table 22-7.

i) Avoiding Phagolysosomal Destruction

Evolution has confined intracellular pathogen killing to the phagosome of professional phagocytes. Consequently, internal damage to the phagocyte itself is prevented, and non-phagocytic cells (which do not create phagosomes) run no risk at all. In view of the compartmentalized nature of phagocytosis, intracellular bacteria can evolve in one of two ways to improve their chances of survival. They can either enter and replicate in non-phagocytic cells, or enter phagocytic cells and employ strategies to avoid phagolysosomal destruction. One example of an intracellular bacterium that targets a non-phagocytic cell is

Granuloma Formation

Most bacteria engulfed by activated macrophages succumb to phagosomal killing in these cells. However, bacteria resistant to this mode of destruction induce the macrophage to increase its secretion of TNF. In the presence of this TNF plus IFNγ produced by activated Th1 cells, the macrophage becomes hyperactivated and capable of killing most initially resistant bacteria due to its increased production of ROIs and RNIs. However, if any bacteria survive, the hyperactivated macrophages fuse to form the center of a granuloma. Surrounding the fused macrophages are additional hyperactivated macrophages and activated CD4+ and CD8+ T cells. Eventually, the center of the TNF-permeated granuloma becomes necrotic, killing most bacteria, and the surrounding T cells calcify to physically wall off any bacteria surviving this assault.

Table 22-7 Evasion of the Immune System by Intracellular Bacteria

Immune System Element Thwarted	Bacterial Mechanism
Phagolysosomal destruction	Reproduce in non-phagocytes Synthesize proteins blocking lysosomal fusion and/or phagosomal killing Recruit host proteins blocking lysosome function Reverse acidification of phagosome
Microbicidal action and hyperactivation of macrophages	Block intracellular signaling leading to expression of microbicidal genes and IFN-induced genes Produce phenolic glycolipid to neutralize ROI Produce enzymes breaking down ROI and H_2O_2
Antibodies	Spread to new host cell via pseudopod invasion
T cells	Anergize T cells via contact with LLO protein
APCs	Downregulate CD1 expression Block maturation or migration of DCs

M. leprae, which infects Schwann cells of the peripheral nervous system. The Schwann cells, which wrap around the axons of nerves and generate the myelin sheath, express a surface protein called α-*dystroglycan*. This protein normally has a role in early development and morphogenesis, but in this case acts as a receptor for *M. leprae*. The microbe enters the Schwann cells and is protected from phagocytosis. Similarly, *L. monocytogenes* expresses *internalin* proteins that bind to the host adhesion molecule E-cadherin, allowing the bacteria to enter non-phagocytes. *L. monocytogenes* also deliberately accesses phagocytes using the host cell Fc and complement receptors, but then employs survival tactics to prevent killing within the phagosome. These tactics include the synthesis of the listeriolysin O protein mentioned previously. LLO induces pore formation in the membrane of the phagolysosome, allowing the bacterium to escape into the relative safety of the cytoplasm.

M. tuberculosis has devised several different ways of avoiding phagolysosomal destruction. First, when this bacterium finds itself being engulfed in a macrophage phagosome, it recruits a host protein called TACO (tryptophan-aspartate-containing coat protein) to the cytoplasmic surface of the developing phagosome. The presence of TACO inhibits the fusion of the phagosome with lysosomes and other vesicles containing hydrolytic enzymes and specialized killing components. Secondly, mycobacteria also produce NH_4^+, which reverses the acidification of phagolysosomes and promotes fusion with harmless endosomes. Thirdly, *M. tuberculosis* infection of macrophages interferes with STAT1-mediated transcriptional responses to IFNγ signaling, blocking the expression of genes needed for microbicidal action and hyperactivation. As a result of these three strategies, mycobacteria can survive within host phagosomes for long periods. The bacteria may remain active, or enter a state of dormancy in which many of their metabolic pathways are downregulated.

Other intracellular bacteria focus on blocking the ROIs produced within the phagosome. The phenolic glycolipid found in *M. leprae* neutralizes ROIs, while other bacterial species produce superoxide dismutase and catalase that break down ROIs and hydrogen peroxide. Catalase may also inhibit the generation of RNIs.

S. typhi and *S. typhimurium* preferentially enter the body via the M cells of the PPs. Once in the dome below the FAE, these bacteria induce resident macrophages to capture them by macropinocytosis. Inside the macrophages, the bacteria secrete a protein called SpiC that efficiently blocks fusion of lysosomes with phagosomes, and other molecules that decrease the recruitment of NADPH oxidase to the phagolysosome. Still other bacterial products confer resistance to cationic peptides. These measures permit the *Salmonella* to survive within the macrophage for a relatively short period of time. However, because macrophages are mobile, the bacterial infection is quickly disseminated all over the body. In typhoid carriers, the disease becomes chronic, as the bacteria accumulate in the gall bladder and are shed for several months or even indefinitely into the feces. Transmission to fresh hosts is achieved under conditions of poor sanitation when the feces of an infected person enter the water supply consumed by non-infected individuals. An infected human does make antibodies to the capsule of *S. typhi*, and carriers have such antibodies in their plasma, but the antibody is not protective while the bacterium is hidden in the macrophages. Eventually, macrophages harboring *S. typhi* are induced to undergo apoptosis.

ii) Avoiding Antibodies

The ultimate escape for an intracellular pathogen is to make it into a new host cell without attracting the attention of the immune system at all. Ultrastructural studies have shown that, in host cells infected with *L. monocytogenes*, the bacterium induces the actin-based formation of a pseudopod that invaginates into a neighboring non-phagocytic cell. The neighboring cell engulfs the pseudopod (which contains the bacterium plus bits of the original host cell cytoplasm) and pinches it off to form a secondary vacuole surrounded by a double plasma membrane (Plate 22-4). The bacterium then uses LLO and phospholipases to forge its way through both sets of host membranes and enters the cytoplasm of the new host cell, free to start the cycle anew. The beauty of this tactic is that the bacterium is never extracellular, meaning that antibodies can never bind to it. Indeed, although low levels of natural IgM antibodies to *Listeria* have been detected in naive inbred mice, these antibodies do not play a significant role in resistance to natural *Listeria* infections.

iii) Avoiding T Cells

The LLO molecule of *L. monocytogenes* provides another means of defense in addition to its pore-forming capacity. It has been observed that contact with LLO irreversibly inactivates CD4+ T cells. In *in vitro* experiments, antigen-specific CD4+ T cells were exposed to APCs that were either infected with *L. monocytogenes* or treated with bacterial LLO peptides. T cells treated in this way became anergic and could not be activated by subsequent exposure to a known stimulatory

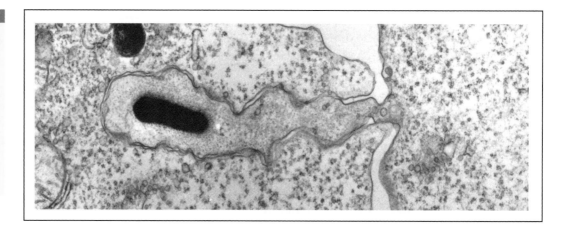

peptide, even after LLO was removed. The mechanism underlying this inactivation is unknown but may be related to the antagonism of TCRs by altered peptide ligands (see Ch.14).

M. tuberculosis escapes T cell attack by interfering with APC function. Infection of DCs by this pathogen promotes downregulation of the expression of MHC class I and II and CD1, decreasing the presentation of non-peptide antigens. The number of T cell clones that can be brought to bear on the pathogen is thus reduced.

D. Immunity to Viruses

I. WHAT ARE VIRUSES?

i) General Characteristics

Like many species of bacteria, viruses are obligate intracellular pathogens. However, unlike bacteria, viruses consist only of DNA or RNA genomes packaged in proteinaceous capsids. Viruses therefore do not have their own protein synthesis machinery and rely on subverting that of the host cell. The genomes of viruses are much smaller than those of other classes of pathogens, being limited to about 100 genes at most; the smallest viruses make do with four genes. Most of the genes in a virus genome encode structural proteins or are required for viral replication, but a surprisingly large number of genes in the more complex viruses encode proteins that undermine the host immune response or regulate host cell apoptosis. Many of these genes are dispensable for viral replication *in vitro*, emphasizing their function as countermeasures to the host immune response *in vivo*. Viruses with a small genome count on rapid replication and dissemination to new host cells to establish an infection before the immune system can respond. Viruses with larger genomes need more time to replicate and are transmitted more slowly; accordingly, these organisms have developed ways of interfering with various components of the host immune response that allow sufficient time for the establishment of an infection.

Viruses usually use a host receptor to bind to and enter the cell, followed by replication and virion assembly within the cell. Progeny virions released into the local area can infect neighboring host cells and initiate new replicative cycles that facilitate widespread dissemination of the virus. Progeny virions that reach the blood circulation are free to spread systemically. Different viruses have different relationships with their hosts. Viruses may vary with respect to the species they can infect (host range), the kinetics of their replication, and their propensity to cause either acute or chronic disease. Naturally, a successful virus is not so virulent that it wipes out all members of its target species, since that would prevent the virus's own future replication. Viruses that damage the host cell during infection are said to be *cytopathic*, while those that merely take over cell functions without damaging the host cell are *non-cytopathic*. In addition to virally induced cytopathy, it is not unusual for the host's own immune response to the virus to result in immunopathic damage to host tissues.

After an initial infection, some types of viruses remain in the body and establish a persistent infection. Unlike bacteria and parasites, which often give rise to chronic disease in cases of persistent infection, most viruses that persist in the body do so in a *latent* state. During latency, viral activity is held in check by cell-mediated immunity. The host experiences no symptoms and the virus remains in a non-infectious mode in which no new virus particles are assembled. However, as the host ages or faces immune system challenges such as other infectious agents, cancer, or immunosuppressive drug treatments, the cell-mediated response weakens. This weakening increases the likelihood that the latent virus will become reactivated, leading to recommencement of a productive infection with associated symptoms. A common example of this type of latent viral infection occurs with the virus that causes chicken pox, *varicella zoster virus* (VZV; also known as human herpesvirus 3, or herpes zoster). After an initial infection that results in the characteristic chicken pox rash, the virus moves up to the dorsal root ganglia of sensory nerves that reach into the skin and becomes latent, usually for years. Upon reactivation, the viral infection spreads down the affected nerves to the skin, causing the formation of vesicles called "shingles" that are very painful to the host and are teeming with infectious virus particles. *Herpes simplex virus* (HSV) acts in a similar way on nerves supplying the mouth area, giving rise to cold sores after a period of latency. *Cytomegalovirus* (CMV) persists in a latent infection that causes no symptoms and is not infectious as long as cell-mediated responses are intact and ongoing. If the virus becomes reactivated, patients may come down with pneumonitis or hepatitis and may shed infectious virus in their urine.

The oncogenic viruses are an important group of latent viruses. Both DNA and RNA viruses are represented in this heterogeneous group, their common characteristic being an association with malignancy in a small percentage of hosts. Infection with these viruses is usually inconsequential to everyday health, but, in some individuals, the replication of these pathogens disturbs the host cell cycle in such a way that the cell is transformed and becomes cancerous. The DNA oncogenic viruses include *human papillomavirus* (HPV; cervical cancer), *Epstein-Barr virus* (EBV; Burkitt's lymphoma and nasopharyngeal cancer), and *hepatitis B virus* (HepB; liver cancer). The RNA oncogenic viruses are all retroviruses, meaning that they possess a reverse transcriptase enzyme that allows them to make a cDNA copy of their RNA genome that can integrate into the host cell DNA. In humans, the retrovirus *human T cell leukemia virus* (HTLV) causes T cell leukemias and lymphomas (see Ch.30). The *human immunodeficiency virus* (HIV), which causes AIDS, is also a retrovirus (see Ch.25). HIV-infected individuals have a higher incidence of various lymphomas (see Ch.30). More on oncogenic viruses and immune responses to them appears in Chapter 26.

ii) Characteristics of Selected Viruses

Table 22-8 contains an overview description of important virus families. Table 22-9 contains a summary of illnesses caused by pathogenic viruses mentioned in this chapter. More detailed information on several well-studied and important human viruses appears in the following sub-sections. Additional viral human pathogens are discussed in Chapter 23 in the context of vaccination.

iia) Adenovirus. Adenovirus is a DNA virus that most often infects the upper respiratory tract of young children but that can also affect the eye and gastrointestinal tract. Disease is

---

Table 22-8 **Overview of Virus Families**

Virus Family	Envelope	Size (nm)	Important Members
DOUBLE-STRANDED DNA			
Adenoviridae	No	70–90	Adenoviruses
Hepadnaviridae	Yes	42	HepB
Herpesviridae	Yes	150–200	EBV, varicella, KSHV, HSV, CMV
Papovaviridae	No	40–57	Papillomavirus, polyoma, simian virus
Poxviridae	Yes	200–350	Vaccinia, variola, molluscum contagiosum
SINGLE-STRANDED RNA			
Arenaviridae	Yes	110–130	LCMV
Bunyaviridae	Yes	90–120	Hantavirus
Coronaviridae	Yes	80–160	Coronavirus, SARS virus
Filoviridae	Yes	80–1,400	Ebola virus, Marburg virus
Flaviviridae	Yes	40–50	HepC, yellow fever, dengue fever, West Nile virus
Paramyxoviridae	Yes	150–300	Parainfluenza virus, mumps, measles
Picornaviridae	No	28–30	HepA, enterovirus, rhinovirus, polio virus, Coxsackie virus
Rhabdoviridae	Yes	70–180	VSV, rabies virus
Togaviridae	Yes	60–70	Alphaviruses, rubella
SINGLE-STRANDED RNA IN SEGMENTS			
Orthomyxoviridae	Yes	80–200	Influenza viruses
DOUBLE-STRANDED RNA			
Reoviridae	No	60–80	Reovirus, rotavirus
Retroviridae	Yes	100–120	Lentiviruses (HIV), HTLV

Table 22-9 Examples of Viruses and the Diseases They Cause

Virus	Disease
Adenovirus	Acute respiratory infections
Epstein-Barr virus (EBV)	Infectious mononucleosis, Burkitt's lymphoma, nasopharyngeal cancer
Hepatitis A virus (HepA or HAV)	Hepatitis, cirrhosis, liver cancer
Hepatitis B virus (HepB or HBV)	Hepatitis, cirrhosis, liver cancer
Hepatitis C virus (HepC or HCV)	Hepatitis, cirrhosis, liver cancer
Herpes simplex (HSV)	Cold sores
Human cytomegalovirus (huCMV or HCMV)	Pneumonitis, hepatitis
Human immunodeficiency virus (HIV)	AIDS, lymphomas
Human papillomavirus (HPV)	Skin warts, asymptomatic genital infections, cervical cancer
Human T cell leukemia virus (HTLV)	T cell leukemias and lymphomas, chronic inflammation of the spinal cord
Influenza virus	Influenza (respiratory infection; often called the "flu")
Kaposi's sarcoma herpesvirus (KSHV)	Kaposi's sarcoma (associated with AIDS), B lymphoma
Lymphocytic choriomeningitis virus (LCMV)	Brain damage in mice, loss of motor control; viral meningitis in severe human infections
Measles virus (MV)	Measles
Molluscum contagiosum (MCV)	Benign skin tumors in humans
Mouse cytomegalovirus (muCMV or MCMV)	Pneumonitis, hepatitis in mice
Mouse mammary tumor virus (MMTV)	Mammary tumors in mice
Polio virus	Poliomyelitis, post-polio fatigue
Rabies virus	Rabies
Rhinovirus	Common cold
SARS virus	Severe acute respiratory syndrome
Varicella zoster virus (VZV)	Chicken pox, shingles (herpes zoster)
Variola virus	Smallpox
Vesicular stomatitis virus (VSV)	Mouth lesions and hoof loss in horses; flu-like symptoms in humans
West Nile virus (WNV)	Flu-like illness, encephalitis, fatigue; often fatal in horses and some species of birds

usually acute but mild, with symptoms ranging from those associated with the common cold to tonsillitis. If the virus attacks the lower respiratory tract, bronchitis and pneumonia can result. While 50% of adenovirus infections in otherwise healthy children are asymptomatic, adenovirus can kill immunocompromised individuals.

Adenovirus is one of the best-studied pathogens at the molecular level. The virus first expresses E1A genes that force the infected host epithelial cell to enter S phase. The viral DNA then replicates after about 7 hours. Various E1B, E2, E3, and E4 transcription units are subsequently activated that supply structural proteins for the virus and other proteins that control host cell functions and viral resistance to the immune response. Virions are assembled within 24 hours and are released from the lysed host cell by 48–72 hours.

iib) Cytomegalovirus. Cytomegaloviruses are β-herpesviruses, meaning that they are large, double-stranded DNA viruses. Human CMV is endemic in all human populations but causes only sub-clinical infections in immunocompetent hosts due to control by CD8$^+$ T cells. However, in immunocompromised hosts, such as fetuses, AIDS patients, and transplant recipients (who may acquire CMV from their transplants), the active virus can cause severe or even fatal disease. About 10–15% of CMV-infected fetuses suffer brain damage, while infected adults may come down with pneumonia, hepatitis, nephritis,

encephalitis, and/or an increased incidence of bacterial and fungal infections. CMV is also capable of adopting a latent state (see later) in which the inactive virus is found predominantly in the kidney, liver, and heart.

iic) Epstein-Barr virus. EBV is a DNA herpesvirus that preferentially infects B cells of young children by exploiting host CR2 molecules. In developing countries, EBV is endemic and almost all young children contract a mild infection. This early exposure confers a degree of resistance to EBV in later life, and adolescents and young adults are rarely ill with EBV infections. In developed countries, early childhood EBV infection is much less common, and an individual's first encounter with the virus may occur during adolescence. Such later infections produce a more serious glandular fever called *infectious mononucleosis* ("mono"), which is characterized by extreme fatigue. This type of EBV infection is particularly contagious, in part because the virus is very active in the salivary gland ("kissing disease"). In addition, as mentioned previously, EBV is an oncogenic virus. It persists in the body and may cause malignant transformation of cells in a small proportion of cases. Persistent EBV infection of tonsillar cells can result in nasopharyngeal cancer, while persistence of EBV in resting memory B cells leads to Burkitt's lymphoma. In the laboratory, EBV is often used to immortalize cultured B cells *in vitro* to establish B cell lines.

iid) Hepatitis B virus. The hepatitis B virus is an enveloped DNA virus that preferentially infects hepatocytes. Clinical symptoms of HepB infection include jaundice, nausea, vomiting, fatigue, and pain in the abdomen and joints. HepB infections may be either acute or chronic. Transmission is primarily by sexual contact or via contaminated needles used to inject intravenous drugs, although body fluids such as saliva and breast milk can be infectious as well. About 30% of acute infections are asymptomatic. When the virus adopts a latent state in which the viral genome integrates into the host genome, a chronic HepB infection is established and these persons become HepB carriers (who may or may not show symptoms). Chronic HepB infections may eventually lead to more serious conditions such as cirrhosis of the liver and hepatic cancer. The propensity for HepB infection to become chronic is greatest in the very young. That is, in 90% of individuals who are infected with the HepB virus when under the age of 1 year, the infection becomes chronic. In contrast, only 5% of those who are infected when over the age of 5 years continue to harbor the virus latently. The WHO estimates that, worldwide, 1.2 billion people are infected with HepB, and 350 million of these infections are chronic. About 15–25% of chronically infected persons die of HepB-related disease each year.

iie) Hepatitis C virus. The hepatitis C virus is a small, enveloped, single-stranded RNA virus with effects very similar to those of HepB, except that the infection becomes chronic at a much higher rate: fully 80% of all individuals infected are unable to clear the virus. The major mode of HepC transmission is via exposure to contaminated blood, either through blood transfusion (see Box 25-1 in Ch.25) or the sharing of contaminated needles. It remains controversial whether HepC is also transmissible through sexual contact. Worldwide, at least 3–4 million new HepC infections occur each year, and an estimated 3% of the world's population is currently infected. Of those chronically infected, 10–20% go on to develop cirrhosis of the liver or liver cancer, and an estimated 0.4% of these die each year. In the United States, chronic liver disease is the 10th leading cause of death among adults, and studies suggest that 40% of all such cases are due to HepC infection. In developed countries, chronic HepC infection is the most common reason for liver transplants.

The mechanism by which HepC establishes chronic infection is not yet understood. Most individuals attacked by HepC mount humoral and cell-mediated responses against the virus, but these responses are not usually sufficient to clear it. Moreover, HepC exhibits considerable genetic heterogeneity, with six major genotypes and over 100 strains. The envelope proteins of different HepC strains can diverge by as much as 50%, such that neutralizing antibodies that are effective against one strain may not be effective against another strain. More on the difficulties of dealing with HepC appears in Chapter 23.

iif) Human immunodeficiency virus (HIV). Because an extensive discussion of HIV and AIDS appears in Chapter 25, we include only a few brief comments here. HIV is a retrovirus that exhibits extreme *antigenic variation*. HIV infects both macrophages and CD4$^+$ T cells and destroys the latter.

The lack of CD4$^+$ T cells fatally cripples adaptive immune responses in these patients, and they usually die of either opportunistic infections or unusual tumors. The WHO estimates that, worldwide, over 42 million people are currently infected with HIV and that more than 3 million infected persons die each year of AIDS.

iig) Influenza virus. The influenza virus is an enveloped RNA virus with a segmented genome. As most of us know from personal experience, the "flu" is a nasty respiratory infection that initiates in ciliated epithelial cells but eventually produces symptoms that affect the entire body. Much of the malaise associated with the flu can be attributed to the large quantities of IFNα produced by immune system cells fighting the infection. Headaches, chills, high fever, muscle aches, weakness, and dry cough are common symptoms, most of which are resolved within 2 weeks. However, the virus abrogates the normal function of the ciliated epithelial cells lining the respiratory tract, decreasing the expulsion of pathogens in the mucus. Macrophage and neutrophil functions may also be suppressed in the local area of infection. As a result, opportunistic bacteria may become established in the respiratory tract and go on to cause potentially fatal pneumonia.

Part of the difficulty in combating influenza viruses is that a number of different strains exist in the human population. Because the principal proteins of these strains differ slightly (another example of antigenic variation), an immune response mounted against one strain does not guarantee protection against another strain. Thankfully, neither chronic nor latent infections occur with influenza virus. Influenza virus has been responsible in the past for several hemispheric and global epidemics in which the elderly and the immunocompromised were killed in large numbers. Even today, influenza virus infections cause significant morbidity and mortality, particularly among the aged and chronically ill. On a global scale, it is estimated that the annual bout of influenza infection causes 3–5 million cases of severe disease and 250,000–500,000 deaths.

iih) Measles virus. The measles virus is a cytopathic, enveloped RNA virus that initially attacks the upper respiratory tract but then spreads via the lymphatics and blood circulation to most other tissues. The characteristic red "measles spots" that appear all over the body are due to cell-mediated immune responses against infected host cells. Acute measles infections are highly contagious and the virus is easily spread to new hosts through respiratory secretions. Moreover, measles virus can survive for at least 60 minutes in aerosolized droplets, allowing it to spread through ventilation systems. Globally, an estimated 40 million new measles infections occur each year.

At the cellular level, the virus primarily targets monocytes, macrophages, lymphocytes, and DCs by binding to a membrane glycoprotein called SLAM (*signaling lymphocyte activation molecule*; CD150). However, the measles virus can also penetrate the walls of cerebral capillaries and enter the CNS, where it infects neurons. Due to the loss of function of multiple types of immune system cells, measles patients experience profound immunosuppression that makes them highly susceptible to secondary infections that may be severe or even fatal. In unvaccinated individuals, the disease can have significant mortality: 1–2 million deaths due to measles occur worldwide each year. While most survivors enjoy lifelong immunity to this virus, they are not entirely without risk. Some variant strains of the measles virus have defects in their surface proteins that block expression of viral antigens on the infected host cell surface, meaning that these cells escape the immune response. In a small percentage of individuals, these variants persistently infect the CNS and cause a chronic, progressive neurological disease called _subacute sclerosing panencephalitis_ (SSPE). More on the measles virus appears in Chapter 23.

iii) Polio virus. Polio virus is a small, non-enveloped RNA virus that replicates first in the intestinal tract. The virus is shed into the feces of an infected individual and spreads to new hosts via feces-contaminated food or water. In the later stages of an infection, the virus moves to the blood and spreads to the spinal cord and CNS. While most infected individuals recover completely, in a small number of patients the virus causes _poliomyelitis_, permanent muscle paralysis due to the destruction of the motor nerves of the CNS. In fact, in the pre-vaccine era, poliomyelitis was the leading cause of permanent disability in the United States. The disease was called "infantile paralysis" at that time because of the large number of children affected. When viral destruction of brain stem cells resulted in paralysis of the respiratory muscles, the patient had to be placed in an "iron lung" (artificial breathing apparatus). Even today, polio virus infection is fatal in 2–5% of poliomyelitic children and in 15–30% of adults. In aging survivors, the neurological damage suffered during an acute infection may be manifested as "post-polio fatigue," similar in symptoms to chronic fatigue syndrome. Polio is now almost unknown in the Americas, Europe, and the Antipodes due to a very successful vaccination campaign (see Ch.23), although it continues to plague some populations in Asia and Africa.

iij) Rabies virus. _Rabies_ is a disease characterized by hyperexcitability and spasms of the mouth and pharynx that occur when swallowing liquids. This latter feature can sometimes lead to "foaming at the mouth" and/or a fear of water ("hydrophobia"). The disease is caused by the rabies virus, an enveloped RNA virus. Upon entering the body, the virus replicates first in skeletal muscle and connective tissue. It then enters and spreads along the peripheral nerves to the spinal cord and CNS, where it eventually causes encephalitis. Initial symptoms are slow to develop, and may include irritability, fever, pain, and fatigue that can progress to hallucination and paralysis. The rabies virus readily infects cats, dogs, raccoons, foxes, bats, and skunks, all of which are creatures that interact quite frequently with humans. The bite of a rabid animal is usually fatal for humans unless a course of injections of protective anti-rabies antibodies is undertaken immediately (see Ch.23). The virus replicates in slow waves, meaning that the passive supply of relatively short-lived antibodies must be replenished several times to ensure elimination of the virus.

iik) Rhinovirus. Rhinovirus (rhino, "nose") is a non-enveloped, single-stranded RNA virus and a member of the picornavirus family. Over 100 rhinoviruses are known. About 50% of common colds are caused by some kind of rhinovirus. (About 10% of common colds are due to infection with a different type of virus, such as adenovirus, while the cause in 40% of cases is unknown.) Rhinoviruses thrive in the upper respiratory tract, particularly the nose and throat. The infection is thus characterized by the familiar symptoms of sneezing, excessive nasal secretion ("runny nose"), and congestion ("stuffy nose"). Complications of rhinovirus infection include spreading of the virus from the throat to the sinuses, lower respiratory tract, and middle ear, resulting in sinusitis, laryngitis, and otitis media, respectively.

iil) SARS virus. The virus that causes SARS (_severe acute respiratory syndrome_) is a novel member of the coronavirus family. Coronaviruses are enveloped, single-stranded RNA viruses with an asymmetric shape and surface spikes positioned so as to resemble a crown. Most human coronaviruses are associated with only mild respiratory disease. Thus, the world was caught off-guard in early 2003 by the severity of the atypical pneumonia caused by the newly emerged SARS coronavirus. Over 8000 cases of infection leading to more than 700 deaths were recorded in this initial outbreak. Scientists speculate that the SARS virus may have arisen in a sudden species jump from healthy civet cats to their owners in China in late 2002.

Symptoms of SARS are flu-like and include fever, chills, malaise, dry cough, shortness of breath, and headache. Liver damage and lymphopenia are also often present. Damage to the alveoli of the lungs is progressive and can be fatal. The virus spreads between humans by close contact, such as touching a contaminated surface or breathing in aerosol droplets created by the coughing or sneezing of an infected individual. At the cellular level, the SARS virus replicates chiefly in respiratory epithelial cells, although other tissues (such as eyes, heart, liver, kidney) and other cell types (such as macrophages) may also be targeted. Unlike most coronaviruses, the cytopathic SARS virus can be cultured _in vitro_, aiding the study of its infection mechanisms.

iim) Smallpox virus. The smallpox virus (variola) was introduced in Chapter 1. Although this virus has been eliminated globally in the wild, in this age of bioterrorism, it behooves us to briefly comment on it. Variola is a highly infectious member of the DNA poxvirus family. Smallpox disease is characterized by pain, fever, and a severe blistering of the skin that leaves disfiguring marks, particularly on the face of the infected individual (refer to Plate 1-1). Smallpox infection is fatal in 30% of cases. There is controversy over whether stocks of variola should be maintained for vaccine development. More on this issue appears in Chapter 23.

iin) Vaccinia virus. As recounted in Chapter 1, Jenner used the cowpox virus to vaccinate against smallpox back in the late 1700s. In modern times, the vaccinia virus is used. Vaccinia is an enveloped, double-stranded DNA virus and a member of the poxvirus family that includes cowpox and variola (smallpox). Curiously, it is not clear how the vaccinia virus arose. Some researchers have speculated that early attempts at developing a better smallpox vaccine resulted in a recombination of cowpox virus with another poxvirus that resulted in the vaccinia virus. While vaccinia does replicate in human cells, it usually causes relatively mild illness (rash, low fever, enlarged lymph nodes, papule at the site of vaccination) and is only very occasionally fatal. As described in more detail in Chapter 23, vaccinia is under intense investigation as a vector for DNA vaccines.

II. EFFECTOR MECHANISMS

It should be noted that not all of the effector mechanisms described in the following sections will apply to the resolution of all virus infections, depending on the type of virus and the particular host it is infecting at the time. However, contributions by CTLs and NK cells will likely play a role at some point in every infection, whether that contribution takes the form of cytolysis or cytokine secretion. Antibody defense, particularly neutralization, is crucial for protection against re-infection with many viruses. We also remind the reader that members of the TLR family of PRRs play a key role in both the innate and adaptive antiviral responses. TLR9 recognizes CpG-containing viral DNA, while TLR3 binds to double-stranded viral RNA, and TLR7 and TLR8 are triggered by single-stranded GU-rich viral RNA. TLR-stimulated macrophages and DCs produce large quantities of pro-inflammatory cytokines such as IFNα, IL-6, TNF, and IL-12. In the presence of these cytokines, TLR-stimulated DCs acquire viral components and are induced to mature, becoming competent APCs able to present viral antigens to naive T cells. The major antiviral effector mechanisms are summarized in Figure 22-5.

i) IFNs and the Antiviral State
Interferon secretion is one of the earliest means by which the body defends itself against virus infections. As we learned in Chapter 17, the type I interferons were first discovered on the basis of their antiviral activities. Host cells infected with a virus produce IFNα and IFNβ, which in turn initiate a series of metabolic events in neighboring cells that result in their adopting an *antiviral state*. The reader will recall from Chapter 17 that IFNα and IFNβ induce the synthesis of the host enzyme PKR, which is inactive until it encounters a dsRNA molecule such as a viral genome. Among other substrates, activated PKR phosphorylates first itself and then eIF2, a factor necessary for translation of both host and viral proteins. Since the virus relies on the host cell's translation machinery for viral protein synthesis, viral reproduction is halted. IFNs also induce expression of 2-5A synthetase, an enzyme activated by dsRNA that generates short 2′,5′ oligoadenonucleotides. These 2-5A molecules activate RNAse L, which degrades both viral and cellular RNAs, again blocking translation. Surprisingly, mice deficient for PKR are capable of mounting effective immune responses to almost all viruses (except *vesicular stomatitis virus*, VSV), suggesting that redundancy for PKR

function or a parallel antiviral pathway exists. These matters are under investigation.

As well as their involvement in the antiviral state, IFNs trigger the expression of multiple genes with profound effects on both the innate and adaptive immune responses (see Ch.17). These effects combine to control infection by a very broad spectrum of viruses. IFNs enhance phagocytosis, regulate the production of cytokines and NO, and promote the differentiation of monocytes into DCs. IFNγ is crucial for the mounting of a Th1 response necessary to support cell-mediated defense, consistent with the finding that mice lacking IFNγR are highly susceptible to vaccinia infection. IFNα has been used with success in the clinic against hepatitis infections, emphasizing the importance and utility of this family of molecules.

ii) NK Cells
Because viruses are intracellular pathogens, cell-mediated mechanisms are of key importance in their elimination. While CTLs are the prime mediators of antiviral immunity, there is often a 4–6 day delay before these cells can expand to sufficient numbers to successfully eliminate the virus-infected cells. In situations in which virus infection causes the downregulation of MHC class I on the host cell surface, direct cytolysis of infected cells by NK cells (via ADCC or natural cytotoxicity) and NK production of inflammatory cytokines become of paramount importance. For example, human patients lacking NK cells show increased susceptibility to virus infections, especially herpesviruses.

iii) CD8+ T Cells
The intracellular nature of viral replication means that viral antigens are processed by the host cell's endogenous antigen-processing pathway and peptides are displayed on MHC class I on the surface of an infected cell. The viral peptides that constitute the T cell epitopes may be derived from almost any internal, structural, or envelope protein of the virus, although nucleoproteins and matrix antigens are favored. CTLs kill most virus-infected cells using perforin-mediated cytotoxicity or by inducing Fas-mediated apoptosis, but also secrete the cytotoxic molecules granulysin, TNF, and IFNγ. The objective is to rapidly kill an infected host cell prior to virus assembly and thus block viral spread. Studies of knockout mice lacking either CD8 or β2m (and thus MHC class I) have shown that these animals are highly susceptible to virus infections.

An interesting twist to CTL-mediated elimination of viruses arises with respect to hepatocytes. It seems that, unlike the case for any other cell type, low levels of viruses infecting hepatocytes are controlled solely by cytokine action that does not lead to cytolysis. For example, LCMV and mouse CMV are cleared from murine liver cells primarily by the action of CTL-synthesized IFNγ, in sharp contrast to the cytolytic elimination of these viruses in other cell types. In a poorly defined mechanism, the binding of IFNγ to its receptor on an infected hepatocyte activates intracellular degradation of viral components, curing the cell without killing it. In the case of hepatocytes infected with HepB virus, both TNF and IFNγ must

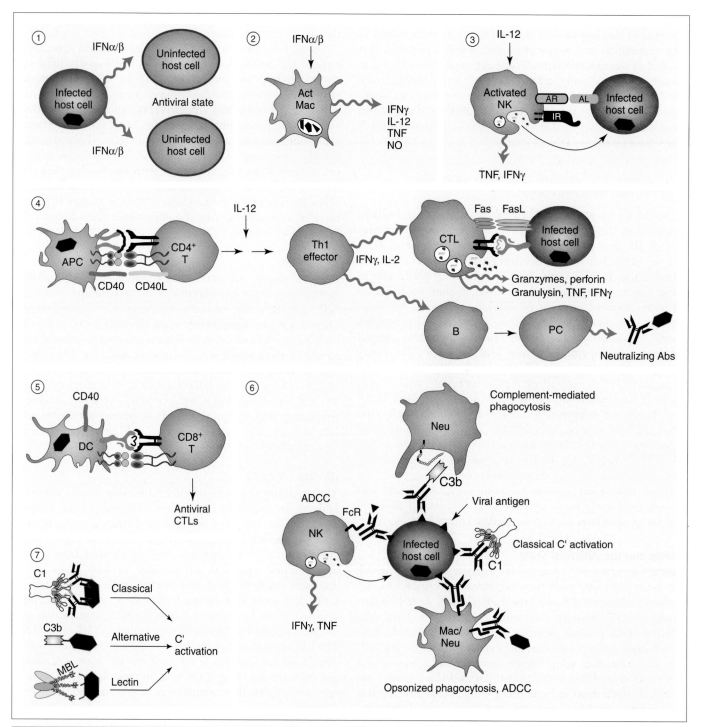

Figure 22-5

Major Mechanisms of Immune Defense against Viruses

(1) IFNα/β secreted by infected host cells induces an antiviral state in neighboring uninfected host cells. (2) IFNα/β also activates macrophage killing and production of pro-inflammatory cytokines and NO. (3) IL-12 produced by macrophages activates NK cells that secrete additional cytokines and use natural cytotoxicity to kill infected host cells that fail to express MHC class I. (4) Infected APCs present viral peptides to CD4⁺ T cells that supply CD40L contacts required for licensing APCs. The CD4⁺ T cells are activated by this encounter and generate Th1 effectors that supply cytokine help to Tc cells and B cells. Antiviral neutralizing antibodies block further spread of the virus, and antiviral CTLs secrete cytotoxic molecules and kill virus-infected host cells by Fas killing or perforin/granzyme-mediated cytotoxicity. (5) DCs infected by a "strong" virus upregulate their costimulatory molecules even in the absence of CD40 ligation, and activate Tc cells in absence of Th help. (6) Antibodies bound to a viral antigen on the surface of an infected host cell may engage the FcR on an NK cell, macrophage, or neutrophil and trigger ADCC or opsonized phagocytosis by these cells. Alternatively, the bound antibody may be bound by free C3b, facilitating opsonized phagocytosis of the infected cell. If the bound antibody binds C1, classical complement activation can lead to MAC-mediated destruction of the infected cell. (7) Complement may also be activated by the binding of C1, C3b, or MBL to the virus itself (as opposed to the infected host cell).

bind to their respective receptors on the hepatocyte surface to induce selective degradation of the viral genome as the virus replicates within the hepatocyte. Again, the threat is removed without killing the liver cell. Liver cells are killed when viral loads are high, because the heightened inflammatory response sparked by large numbers of virus particles recruits activated macrophages. In the cytokine-rich milieu of the inflamed liver, the macrophages commence expression of CD40L, which engages CD40 expressed on hepatocytes. This interaction induces FasL expression on hepatocytes, making them vulnerable to Fas-mediated apoptosis. Should FasL be engaged either by Fas on the surface of the same hepatocyte, or on an adjacent hepatocyte, cell death is induced, which can eventually lead to clinical liver damage. This type of liver pathology is prominent in severe cases of HepB and HepC infection in humans.

Immunopathology following CTL activation also arises with HTLV-1 infection. The vast majority of people infected with HTLV-1 go on to become carriers of the virus. In about 3% of carriers, the virus causes the adult T cell lymphoma/leukemia that gives this virus its name (see Ch.30). However, another 2.5% of carriers develop a chronic neurological disease called HAM/TSP (_HTLV-associated myelopathy/tropical spastic paraparesis_). The principal clinical feature of HAM/TSP is chronic inflammation of the spinal cord. It has been postulated that those individuals who mount a particularly strong CTL response restrict HTLV-1 spreading very efficiently, reducing the concentration of viral antigens that might otherwise induce inflammation. In an individual mounting a less effective CTL response, viral antigens are allowed to accumulate. A continuous cell-mediated response is stimulated that is accompanied by large amounts of pro-inflammatory cytokines that result in inflammation and damage within the CNS of the patient.

Some virus infections are combated by CTL responses that do not require CD4+ T cell help in either cytokine or intercellular contact form. DCs play prominent roles in such viral infections. For example, influenza virus on its own can activate DCs such that they strongly upregulate B7 costimulatory molecules even in the absence of CD40 stimulation. In other words, the DCs become fully mature APCs able to activate CD8+ T cells directly, without the need for help (CD40L engagement) from CD4+ T cells. A similar scenario holds for LCMV infection in mice. While antibody and CD4+ T cell responses do occur in response to LCMV infection, it is the CD8+ CTL cell response that is critical for defense. Studies have shown that strong primary CTL responses can be mounted to LCMV in the absence of CD4+ T cell help, reflecting the putative ability of this virus to access and directly activate APCs.

iv) Humoral Defense

Because a virus is an intracellular pathogen, it is often out of reach of antibodies during the primary adaptive response. Nevertheless, naive B cells may recognize viral components displayed on the surface of an infected cell or may encounter progeny virions as they are released from an infected host cell. In the presence of the appropriate T cell help, neutralizing antibodies are produced and released into the circulation and

memory B cells are generated. In a secondary response, the re-infection of the host can be rapidly blocked by the circulating neutralizing antibodies. These antibodies bar access of the virus to host cell receptors, and can initiate classical complement-mediated destruction of enveloped viruses. Virus-bound antibodies also act as opsonins and engage Fc receptors on phagocytes, facilitating internalization and phagosomal destruction of the invader. Similarly, antibodies that bind to viral components displayed on the surfaces of infected host cells can engage the Fc receptors of NK cells, neutrophils, and macrophages, inducing cellular and thus viral destruction by ADCC.

Antibodies directed against viral Ti antigens can be of great value in host defense. Because Ti responses involve only a B cell, they can be mounted more quickly than Td responses (which require a T cell and a B cell). Thus, B cells producing anti-Ti antibodies may act relatively early during an infection to minimize the spread of the virus until antibodies against viral Td antigens can be synthesized. Viruses such as VSV, LCMV, and Coxsackie virus have highly repetitive antigenic Ti structures on their surfaces that can activate virus-specific B cells without T cell help. Limited isotype switching may even occur in response to these antigens. For example, rotavirus infection in mice leads to mucosal production of virus-specific IgA even in the absence of CD4+ T cells. It is thought that cytokines such as IL-4, IFNγ, and TGFβ secreted by γδ T cells, mast cells, macrophages, and/or NK cells may be sufficient to support isotype switching in response to viral Ti antigens.

v) CD4+ T Cells

As was true for intracellular bacteria, whole virions or their components may be taken into a professional APC by receptor-mediated endocytosis or phagocytosis. The viral proteins are then subjected to the exogenous pathway of antigen processing, and viral peptides displayed on MHC class II can activate CD4+ T cells. Th cells are important for defense against most viruses because these cells both license APCs for naive CD8+ Tc cell activation and supply the T help required for humoral responses to viral Td antigens. For example, neutralizing antibodies play a key role in the elimination of many rhabdoviruses, and the production of these antibodies is completely dependent on CD4+ T cell help. Similarly, humoral and cell-mediated responses to adenovirus infections depend on CD4+ T cell help in the form of CD40L contacts. Mice deficient for CD40L are highly susceptible to adenovirus since the mutant animals fail to produce adequate titers of antibodies and mount only minimal CTL responses. Finally, as one might expect, CD4 knockout mice readily succumb to a broad range of virus infections.

vi) Complement

Surface components of virus particles can directly activate the lectin and alternative complement pathways, leading to the lysis of enveloped viruses. In addition, opsonization of viruses by complement components C3b and C3d promotes phagocytosis by neutrophils and macrophages that bear complement receptors. Simultaneous binding of a C3d-coated virus particle to the

CR2 B cell coreceptor (CD21) and the BCR of a B cell promotes B cell activation and production of virus-specific antibodies. As mentioned previously, such antibodies can neutralize virions and trigger classical complement-mediated destruction of either the virus or the infected host cell displaying viral antigens.

III. EVASION STRATEGIES

Viruses are the most devious of pathogens when it comes to evading the host immune response. Different viruses employ different tactics, and it is not unusual for a single virus to make use of several different mechanisms (Table 22-10). Viruses are not generally fazed by the barrier thrown up by intact skin or mucosae. Many viruses have evolved the capacity to bind to host cell receptors that facilitate viral internalization into epithelial cells. In addition, a broad spectrum of viruses comes equipped with enzymes that break down whole epithelial layers, allowing them access to deeper tissues. Viruses may also be introduced directly into the blood via insect or animal bites.

Once infection has occurred, many viruses hide from the immune system. Others confront the immune response head on by interfering with host cellular pathways. Indeed, it is hard to find a cellular pathway that is not subject to some kind of impairment by a virus. For example, each step of the MHC class I and class II antigen processing pathways can be disrupted, as can complement activation, host apoptotic pathways, antibody responses and cytokine/cytokine receptor signaling. Such strategies require substantial amounts of gene expression. For example, over 25% of the adenovirus genome is devoted to genes encoding proteins that block the host immune response. Some examples of viral strategies are discussed in detail in the following sections.

i) Latency

A virus that can adopt a latent state escapes the attention of the immune response, at least temporarily. Many viruses have the ability to persist and even multiply in cells in a defective form that renders them non-infectious for a period of time. The virus then waits until conditions in the host, such as drug-induced immunosuppression or immunodeficiency associated with disease or aging, provide better odds of survival and transmission to new hosts. In most cases, latency involves the inactivation of viral gene transcription needed for productive infection and the subsequent expression of new viral *latency-associated transcripts* (LATs). Reversal from latency back to productive infection requires some type of reactivation of the productive infection genes that can only occur when the immune response is weakened.

Different viruses achieve latency in different ways. The retroviruses, like HIV, become latent by integrating a cDNA copy of their RNA genome into the DNA of a host cell in such a way that there is limited transcription of viral genes (see Ch.25). The herpesviruses, which are large, double-stranded DNA viruses with genomes encoding numerous viral proteins, rely on other means of transcriptional control. For example, both VZV and HSV are herpesviruses that can establish latent infections in the neuronal ganglia. Neurons express little or no

Table 22-10 Evasion of the Immune System by Viruses

Immune System Element Thwarted	Viral Mechanism
Passive physical barriers	Breach epithelial layer via insects, animal bites, and viral hydrolases
Detection	Adopt latency
Antibodies	Undergo antigenic drift or shift Express viral FcR that blocks ADCC or neutralization Block B cell signaling
DCs	Block DC differentiation, maturation, or migration Inhibit DC expression of co-stimulatory molecules Upregulate surface FasL
CD8$^+$ T cells	Infect host cells with very low MHC class I expression Interfere with MHC class I presentation Express viral proteins that interfere with pMHC binding to CD8 and TCR Force internalization of pMHC
NK cells	Express viral homologue of MHC class I Increase synthesis of HLA-E Increase synthesis of MHC class I
CD4$^+$ T cells	Avoid infection of professional APCs Interfere with MHC class II presentation Express viral peptides that interfere with pMHC binding to CD4 and TCR
Complement	Express viral homologues of host RCA proteins Alter synthesis of host RCAs Bud through host membrane and acquire coat of RCAs Inhibit C9 polymerization
Antiviral state	Block secretion of IFNγ Interfere with PKR/2-5A pathway
Apoptosis	Block receipt of apoptotic signal Interfere with death receptor pathways Express enzymes that neutralize the effects of free radicals Manipulate the host cell cycle
Cytokines and chemokines	Express competitive inhibitors of cytokines and chemokines Block cytokine/chemokine transcription Downregulate host cell cytokine/chemokine receptor expression

MHC class I and so are relatively safe havens to which these viruses can retreat. The viral genome persists in only 0.01% of the neuron population but does not integrate into the host DNA. Rather, the viral genome forms a complex with host nucleosomal proteins that blocks transcription of productive infection genes. A new set of LATs is then transcribed to maintain latency. Another important human viral pathogen that adopts latency is CMV. Latent CMV takes up residence in leukocytes and endothelial cells concentrated around the kidneys, heart, and liver. It is not yet clear whether the CMV genome integrates into the host DNA during latency, or whether there is complete silencing of productive infection genes. There is some evidence for LAT-like transcripts in cells infected with latent CMV.

EBV and *Kaposi's sarcoma herpesvirus* (KSHV; formerly called *human herpesvirus-8*, HHV-8) are herpesviruses whose latency is associated with the development of host tumors. As mentioned previously, EBV infection is linked with B cell lymphomas and nasopharyngeal carcinomas, while KSHV appears to cause the Kaposi's sarcoma often found in AIDS patients. Unlike the oncogenic retroviruses, however, the genomes of these oncogenic DNA viruses do not integrate into the host DNA. Productive infection is halted via downregulation of productive infection genes and subsequent upregulation of LATs. For example, EBV latency correlates with the downregulation of the EBNA3 family of viral antigens. It is these proteins that normally supply immunodominant peptides for EBV-specific CTLs. The virus then upregulates expression of EBNA1 (*EBV nuclear antigen-1*), a protein essential for EBV latency. However, as is described in more detail later, the structure of the EBNA-1 protein blocks its presentation on MHC class I molecules, obscuring the virus from CTL recognition.

ii) Antigenic Variation

A common way for a virus to hide from the host immune system is to change its antigenic "stripe" over successive generations, expressing antigenically new forms of viral proteins that may not be recognized by existing memory lymphocytes or antibodies in populations previously exposed to the virus. Such antigenic variation as a means of viral survival is most likely to be important in hosts that are long-lived (such as humans). Host longevity means that multiple re-infections of an individual can occur, improving the chance that the virus will remain in circulation. This factor is particularly important if the virus lacks the ability to become latent.

Rapid modification of viral antigens through random mutations is known as *antigenic drift*. Antigenic drift usually involves proteins expressed on the surface of the virion that would normally be the target of neutralizing antibodies. Extensive antigenic drift is a hallmark of influenza virus, which lacks proofreading enzymes and thus experiences high rates of mutation of its RNA genome during replication. Minor mutations to the hemagglutinin and neuraminidase proteins present on the surface of the influenza virion arise at a rate of about 1 in 10^6 virus particles. These minor variants often replicate preferentially in the host, as they cannot be neutralized by existing antibodies raised against earlier strains. New strains of influenza arising due to continual antigenic drift spread fairly quickly through a population and are responsible for localized influenza outbreaks. A similar form of antigenic drift can be observed in human rhinoviruses as well as in foot-and-mouth disease virus. HIV undergoes very rapid antigenic drift (even within a given infected individual) due to the error-prone reverse transcriptase involved in the replication of its genome (see Ch.25).

Perhaps unique to the influenza virus is its ability to undergo *antigenic shift*, another type of variation that is less frequent but much more radical and dangerous. Antigenic shift can occur because the influenza virus genome exists as eight separate single-stranded RNA segments, each of which encodes a discrete protein involved in viral function. With such a genetic structure, two different influenza strains that simultaneously infect a single host cell can undergo a reassortment (sometimes inaccurately called "recombination") of their genomic segments (Fig. 22-6). Virus particles containing new combinations of parental RNA suddenly arise, dramatically changing the spectrum of protein epitopes presented to the immune system. Another factor favoring antigenic shift is the ability of many strains of influenza virus to infect an intermediate host (such as a pig or chicken) before being passed on to humans. These animal hosts are known as *animal reservoirs*. Reassortment of gene segments can also occur in an animal reservoir, again leading to new combinations of proteins expressed in the progeny viruses. An antigenically novel flu virus can then infect the human population, safe from antibodies and CTLs raised during previous exposure or vaccination.

Because influenza infections occur seasonally, the appearance of entirely new strains from one winter to the next seemed very sudden to early workers in the field, initiating the use of the term "shift." Antigenic shifts were responsible for the global or "pandemic" outbreaks of influenza that occurred in 1918, 1946, 1957 (Asian flu), and 1968 (Hong Kong flu). As an example of how severe the influenza virus can be, consider the pandemic of 1918. Even with the relatively slow nature of international travel in the early 20th century, this version of the influenza virus swept around the world and killed close to 20 million people within a few months.

iii) Interference with MHC Class I-Mediated Antigen Presentation

Antigen processing pathways offer a wealth of opportunities for a virus to sabotage the initiation of an adaptive immune response (Fig. 22-7). Without MHC class I expression, a peptide cannot be presented to either CD8$^+$ Tc cells or CTLs, so that many viruses have targeted various steps of MHC class I protein synthesis and peptide display. Some viruses (such as CMV) avoid MHC class I presentation of viral antigens by infecting mesenchymal or epithelial cells that have very low MHC class I expression. As a result, almost no viral antigens are displayed to alert passing T cells. Other viruses attempt to circumvent MHC presentation by producing viral proteins that resist proteolysis, meaning that peptides capable of fitting in the MHC binding groove are not generated. For example, as mentioned previously, an important protein expressed by latent EBV is EBNA-1, a molecule important for the maintenance of latent EBV in the host cell. The EBNA-1 protein contains about 200 Gly-Ala repeats that resist proteolysis, blocking the presentation of peptides from this antigen and thus the activation of T cells directed against it. Cytomegalovirus expresses a kinase called pp65 that phosphorylates a viral transcription factor such that it cannot be processed by the infected cell's antigen processing machinery.

Still other viruses prevent peptide loading by blocking the normal synthesis of MHC class I. Adenovirus makes a small protein called E19 that has a lysine-containing "ER retention motif." E19 binds to MHC class I molecules and the ER retention motif traps the complex within the ER, abrogating

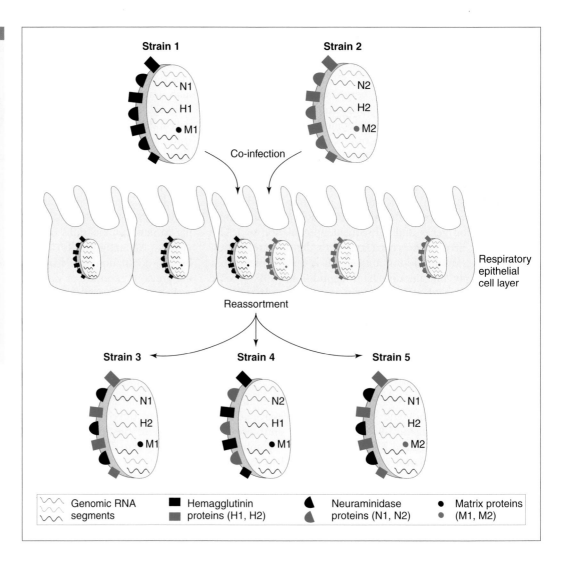

Figure 22-6

Antigenic Shift
The genome of the influenza virus is composed of eight segments that can reassort if two different viral strains infect the same cell. Progeny viruses acquiring various combinations of parental segments may express new constellations of proteins (not all possible combinations are shown here). Although internal genes like that encoding the matrix protein can also reassort, clinical immunologists define a particular antigenic shift by the identity of the hemagglutinin (H) and neuraminidase (N) molecules, since it is the presence or absence of B cell memory to these surface glycoproteins that influences the production of neutralizing antibodies.

peptide display on the cell surface so that CTL killing does not occur. In contrast, human CMV produces glycoproteins (such as the US3 protein) that induce the deglycosylation of some classes of newly synthesized MHC class I chains. The abnormal MHC class I proteins are then conveyed to the proteasomes for degradation. Similarly, the Vpu protein of HIV (see Ch.25) destabilizes newly synthesized MHC class I molecules, preventing them from ever reaching the surface. Some murine CMV proteins such as m06 bind tightly to MHC class I chains and act as lysosomal targeting sequences, causing the MHC class I chain to be destroyed in this organelle.

Another way to interfere with the MHC class I pathway is to alter the interaction of the peptide and the TAP transporter complex. As we learned in Chapter 11, peptides generated in the cytosol must be imported via the TAP machinery into the lumen of the ER for attachment to MHC class I and eventual presentation on the target cell surface. The process occurs in two steps: the ATP-independent binding of peptides to the TAP heterodimer and the ATP-dependent translocation of the peptides into the lumen of the ER. In EBV-infected B lymphoma cells, the transcription of both TAP-1 and TAP-2 is downregulated, reducing peptide loading and thus presentation of viral

antigens to CD8$^+$ T cells. Herpesviruses express small proteins that interfere with peptide binding to TAP on the cytosolic side of the ER. Other viruses express proteins that allow the peptide to bind to TAP but then trap the complex on the luminal side of the ER, blocking its release and subsequent interaction with MHC class I.

MHC class I presentation can also be inhibited by events closer to the cell surface. For example, a mouse CMV protein called m04 associates with mature peptide–MHC class I structures at the cell surface, interfering with TCR- and/or coreceptor-mediated interactions with CD8$^+$ T cells. Another example is the multifunctional Nef protein of HIV (see Ch.25). Certain amino acid motifs in Nef promote its association in the plasma membrane with the host clathrin proteins involved in receptor-mediated endocytosis. Other sequences in Nef allow it to bind to certain MHC class I molecules (among other proteins; see later). This Nef-mediated physical connection between MHC class I and clathrin forces the internalization of the MHC class I molecule. The lysosomal degradation of MHC class I follows, so that the surface level of MHC class I is drastically decreased.

One last note: in addition to sabotaging MHC class I antigen presentation, HIV attempts to avoid CTL attack by

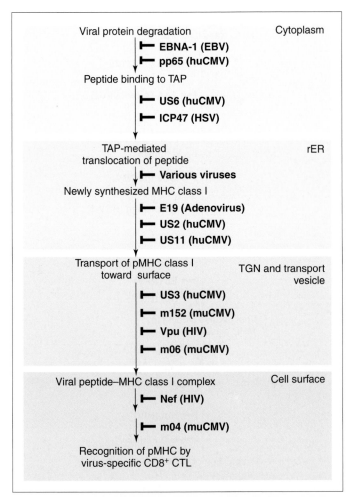

Viral protein degradation   Cytoplasm
⊢ **EBNA-1 (EBV)**
⊢ **pp65 (huCMV)**
Peptide binding to TAP

⊢ **US6 (huCMV)**

⊢ **ICP47 (HSV)**

TAP-mediated   rER
translocation of peptide
⊢ **Various viruses**
Newly synthesized MHC class I

⊢ **E19 (Adenovirus)**

⊢ **US2 (huCMV)**

⊢ **US11 (huCMV)**

Transport of pMHC class I
toward surface   TGN and transport
vesicle
⊢ **US3 (huCMV)**

⊢ **m152 (muCMV)**

⊢ **Vpu (HIV)**

⊢ **m06 (muCMV)**

Viral peptide–MHC class I complex   Cell surface

⊢ **Nef (HIV)**

⊢ **m04 (muCMV)**

Recognition of pMHC by
virus-specific CD8⁺ CTL

Figure 22-7

Viral Interference with the Endogenous Antigen Presentation Pathway
The boxes of increasingly darker shading represent the major
compartments in which viral peptides are processed and loaded onto
MHC class I for presentation on the infected cell's surface. Examples of
proteins that inhibit events occurring in each compartment are indicated in
bold, with the viruses from which they are derived shown in parentheses.
TGN, trans-Golgi network.

producing variant peptides that act as TCR antagonists for
HIV-specific CTLs (see Ch.25). Presentation of these peptides
blocks the activation of the CTL subsets that recognize them,
protecting the infected host cell from cytolysis.

iv) Fooling NK Cells

Because NK cells are activated in the absence of MHC class I,
they are crucial for the control of viruses that actively interfere
with the processing of antigen and attachment to MHC class I.
However, as one might expect, mechanisms exist that thwart this
means of host defense, too. Human CMV expresses an MHC
class I-like surface molecule called UL18 that engages the NK
inhibitory receptor and fools the cell into thinking it has detected
normal MHC class I (Fig. 22-8); the virus proceeds with its replication undisturbed. Poxviruses and herpesviruses also express
MHC class I homologues, many of which have deficits in peptide binding. We mentioned previously that the Nef protein of

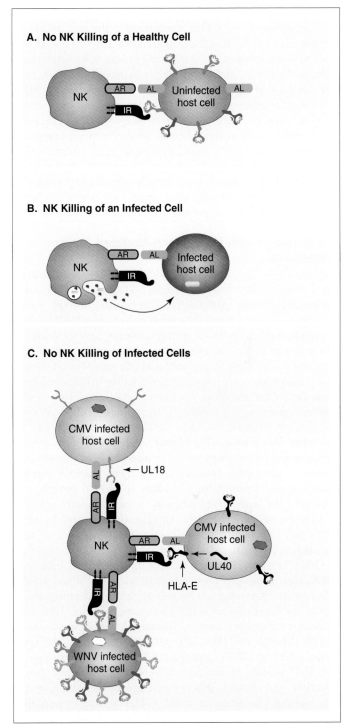

Figure 22-8

Fooling NK Cells
(A) An uninfected host cell expresses MHC class I molecules that
bind to the inhibitory receptors (IR) of the NK cell and block signals
mediated by the engagement of the activatory receptors (AR).
(B) Infected cells often lose expression of MHC class I, resulting in their
destruction by NK cells. (C) Viruses have developed ways of engaging
the IR and blocking NK killing. For example, human CMV induces both
expression of the MHC class I analogue UL18 that can bind the IR, and
the expression of UL40 that induces upregulation of HLA-E. In contrast,
West Nile virus (WNV) induces overexpression of conventional MHC
class I.

HIV depresses surface expression of MHC class I, blocking antigen presentation to CTLs. Interestingly, the expression of the non-classical MHC class I molecules HLA-C, HLA-E, and HLA-G is <u>not</u> affected by Nef, leaving at least some MHC class I-like molecules to bind to the inhibitory receptors of NK cells. The human CMV protein UL40 actually upregulates expression of HLA-E, increasing the numbers of non-classical MHC class I molecules able to bind NK inhibitory receptors. Curiously, some small viruses (such as the flavivirus West Nile virus) upregulate conventional host MHC class I expression in an infected cell in the hopes of staving off NK-mediated destruction. It is thought that this measure allows these rapidly reproducing viruses to complete their life cycle and disseminate long before the MHC class I-dependent CTL response can be brought to bear.

v) Interference with MHC Class II-Mediated Antigen Presentation

Because CD4+ Th cells support antiviral responses by delivering T help to antiviral B cells and CTLs, the MHC class II antigen presentation pathway has also become a target of viral evasion strategies. One of the easiest ways to block MHC class II presentation is to avoid getting antigens into professional APCs. For example, the rabies virus preferentially infects neurons and is very slow to replicate and lyse these cells, meaning that viral antigens are not easily collected by APCs until quite some time after the virus has entered the body. The natural immune response to rabies is thus considerably delayed. Interestingly, because the virus replicates relatively slowly, an individual bitten by a rabid animal can be successfully treated with a <u>post</u>-exposure vaccine, as an alternative to the administration of anti-rabies antibodies. The rabies vaccine contains an inactivated form of the virus that is injected directly into regions of the body where APCs are present, rapidly inducing an immune response (see Ch.23). Effectors activated by these APCs may then take steps to eliminate the rabies-infected neurons before the virus can spread further.

Other viruses take more indirect approaches to inhibiting MHC class II antigen presentation (Fig. 22-9). When a virus infects a cell, it produces IFNs that act on neighboring cells to induce the expression of many genes, among them MHC class II. As we learned in Chapters 10 and 17, IFNγ transduces its signals through the Jak-STAT pathway, activating the transcription factor CIITA required for MHC class II expression. The IE/E protein of human CMV is produced early during infection and blocks IFNγ signaling by destabilizing the Jak molecule, resulting in failed CIITA activation and thus reduced MHC class II expression. Proteins from adenovirus (e.g., E1A) can also affect MHC class II expression by influencing intracellular levels of activated Jak and STAT.

Although they differ in primary sequence, the secondary structures of newly synthesized MHC class I and II molecules are quite similar. This similarity allows the US2 molecule of human CMV to bind to MHC class II molecules and target them for proteosomal degradation before they can enter the endocytic compartment, as was true for MHC class I. In contrast, human herpes simplex virus-1 (HSV-1) expresses the gB viral protein that prevents MHC class II molecules from entering the endocytic system but does not

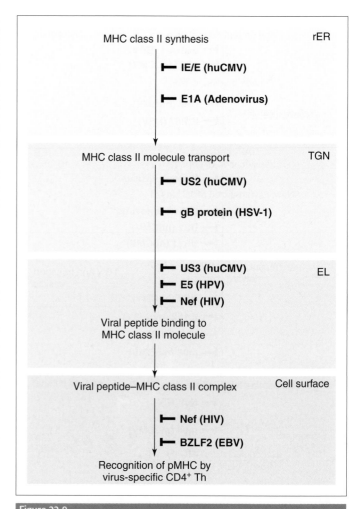

Figure 22-9

Viral Interference with the Exogenous Antigen Presentation Pathway
The boxes of increasingly darker shading represent the major compartments in which viral peptides are processed and loaded onto MHC class II for presentation on the infected cell's surface. Examples of viral proteins that inhibit events occurring in each compartment are indicated in bold, with the viruses from which they are derived shown in parentheses. TGN, trans-Golgi network; EL, endolysosome.

trigger their degradation. Other viruses interfere with MHC class II presentation after the MHC molecule has entered the endocytic system. The pathway to peptide loading of MHC class II molecules is so complex that there are many opportunities for viral interference. The US3 protein of human CMV competes with invariant chain for binding to MHC class II. Papillomaviruses express small proteins such as E5 that disrupt the acidification of the endocytic compartments, inhibiting proteolysis and the generation of antigenic peptides. Similarly, the talented Nef protein of HIV can interfere with the proton pump required to acidify endosomes.

MHC class II molecules that have reached the cell surface are still not safe from some viruses. In addition to its ability to associate with the proton pump, HIV Nef induces internalization of surface MHC class II molecules just as it does for MHC class I molecules, forcing them back into the endocytic system and to the lysosomes for destruction. EBV expresses a membrane glycoprotein called BZLF2 that interacts with MHC class II molecules intracellularly and on the cell surface. It is thought that this interaction inhibits antigen presentation to CD4$^+$ T cells, leading to a failure in T cell activation.

vi) Interference with DCs

Some viruses block the initiation of immune responses directed against them by inhibiting the survival or maturation of DCs. For example, HTLV-1 infects DC precursors and prevents their further differentiation into immature DCs capable of capturing antigen. The infection of immature DCs by HSV-1 and vaccinia precludes DC maturation in response to cytokines. Other viruses, such as vaccinia and canarypox virus, kill DCs by inducing apoptosis. Measles virus forces DCs to form large aggregations called *syncytia* in which the virus replicates freely but further DC maturation is stymied. When human CMV infects a DC, it blocks its upregulation of costimulatory molecules so that the DC may anergize, rather than activate, any naive T cell it encounters. Alternatively, virus infection may turn the DC into a T cell killer. Infection by human CMV or measles virus upregulates the expression of FasL on the DC surface, forcing it to kill Fas-bearing T cells with which it comes into contact.

vii) Avoiding Antibodies

Some viruses take the approach of interfering with B cell activation leading to antibody production. The nucleocapsid protein (NP) of the measles virus binds to the host FcγRIIB protein, which, as we learned in Chapter 9, exerts a negative regulatory effect on B cell activation. The binding of viral NP to FcγRIIB appears to enhance negative signaling such that B cell activation is dampened and antibody production is decreased. Other viruses (such as HSV-1) thwart antibody-mediated destruction by expressing viral Fcγ receptors on the infected host cell surface. These viral FcγRs are capable of binding to host antiviral IgG molecules that have bound viral antigen. Thus, when a host antibody binds to a viral antigen on an infected host cell surface in an effort to initiate complement-mediated lysis or ADCC, the viral FcγR binds to the Fc portion of the complexed IgG and makes it inaccessible. Neither ADCC nor classical complement activation can be triggered, and the infected host cell is not destroyed.

viii) Avoiding Complement

Viruses use many of the same mechanisms as other pathogens to avoid complement-mediated destruction, and the reader is therefore referred once again to Box 22-3. Briefly, many viruses express homologues of the RCA proteins that control complement activation pathways, and some viruses directly inhibit host RCA proteins or alter the expression of these molecules. Other viruses fool the immune system by budding through the host cell membrane, acquiring the surface phenotype of the host and its membrane-integrated RCA proteins.

ix) Counteracting the Antiviral State

Several viruses have developed mechanisms that disrupt the antiviral state. EBV blocks the initiation of the antiviral state by expressing the BARF1 gene. BARF1 encodes a soluble decoy

receptor for the monocyte/macrophage growth factor CSF-1. BARF1 thus decreases the concentration of CSF-1 available to stimulate monocytes and macrophages. Macrophage proliferation and monocyte secretion of IFNα are inhibited, blocking the delivery of signals that protect neighboring cells. When HSV infects a cell that has already established the antiviral state (i.e., eIF2 has been phosphorylated by PKR), the virus expresses phosphatases that dephosphorylate eIF2, allowing viral protein translation to resume. The vaccinia and HepC viruses synthesize proteins that compete with PKR as sites for phosphorylation, while EBV and adenovirus express RNAs that bind to PKR but inhibit its activation. HSV also produces inhibitory analogues of the 2-5A molecules that block RNAse L activation, permitting viral protein synthesis to proceed.

Other viruses block host transcription, leading to establishment of the antiviral state. KSHV produces proteins that are homologous to the host IRF transcription factors. These viral proteins bind to IRF-binding sites on the host DNA but do not permit transcriptional activation. As a result, expression of IFN-inducible genes does not occur and the antiviral response is blocked. Similarly, adenovirus E1A proteins interfere with the formation of transcription factors required for the transcription of IFN-inducible genes, inhibiting host cell responses to the IFNs.

x) Inactivation of CD4-Bearing Cells

Although HIV preferentially infects CD4$^+$ T cells, the very presence of CD4 appears to inhibit the release of newly assembled HIV virions from the T cell (see Ch.25). HIV thus has developed tactics for "going after" the CD4 molecule. The HIV Vpu protein promotes the degradation of newly synthesized CD4 molecules in the proteasome as it does for MHC class I, while the Nef protein promotes internalization and lysosomal destruction of surface CD4 as it does for surface MHC class I and II. Poxviruses and herpesviruses also produce proteins that trigger internalization and destruction of CD4. There is some debate as to whether the primary function of this mechanism is to promote the release of virions or to compromise T cell activation; both may be relevant.

xi) Manipulation of Host Cell Apoptosis

Host cell apoptosis prior to completion of replication is a distinct disadvantage for the more complex viruses with larger genomes, inspiring these viruses to develop means of blocking host cell death. Both genome size and tissue specificity of infection influence strategies to manipulate host apoptosis, since different apoptotic pathways may be triggered in different tissues by different stimuli. Host cell apoptosis is most commonly induced by CTL degranulation, NK cytolysis, Fas/FasL interaction, or the binding of TNF to TNFR. Many of the steps leading to apoptosis in each of these pathways are subject to viral interference (Fig. 22-10).

The extrinsic pathway of apoptosis induced by engagement of the death receptors Fas and TNFR is blocked by different viruses in different ways. Adenovirus expresses several E3 proteins that combine to form the *receptor internalization and degradation* (RID) complex. RID removes Fas and TNFR1 from the host cell surface by inducing the internalization of Fas and TNFR. These molecules are directed into the endosomes, followed by

Figure 22-10

Viral Evasion of Host Cell Apoptosis
A CTL often induces the death of an infected host cell via engagement of the death receptors Fas or TNFR1. Engagement of these receptors triggers the intrinsic and extrinsic pathways of caspase-mediated apoptosis. Examples of viral proteins that inhibit various steps of these pathways are shown in bold. See text for mechanisms and associated viruses.

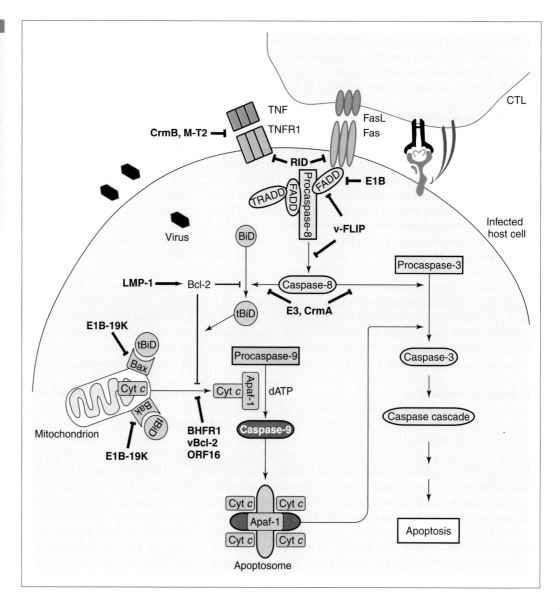

transport to the lysosomes, where they are degraded. Cell death induced by an encounter with a T cell or monocyte expressing TNF or FasL is thus avoided. Several poxviruses express soluble or membrane-bound homologues of TNFR (such as CrmB; *cytokine response modifier B*) that act as decoy receptors for TNF and related cytokines. As was discussed in Chapter 17, myxoma virus produces a secreted homologue to TNFR called M-T2 that blocks TNF-mediated apoptosis.

Other viruses have targeted signal transducers further down the TNFR and Fas signaling paths. An adenovirus E1B protein blocks the oligomerization of FADD necessary for the activation of caspase-8. Several herpesviruses and poxviruses express proteins called vFLIPs (*viral FLICE inhibitory proteins*) that contain DEDs (*death effector domains*) capable of binding to FADD and caspase-8. Association with vFLIP blocks the initiation of the caspase cascade. Caspase-8 activation is also inhibited by the direct binding of the CrmA protein of cowpox virus, the E3 protein of adenovirus, and similar proteins in

herpesviruses and vaccinia virus. These inhibitors are all *serpins*, members of the *serine protease inhibitor* superfamily.

We have mentioned in previous chapters the anti-apoptotic properties of the mammalian Bcl-2 protein. Bcl-2 blocks the intrinsic pathway of apoptosis by regulating the permeability of the mitochondrial membrane to ion traffic such that the release of cytochrome *c* from these organelles is inhibited. In the absence of cytochrome *c*, caspase-9 cannot be activated. Many viruses have sought to manipulate host cell apoptosis by either altering intracellular levels of host Bcl-2 or expressing Bcl-2 homologues. For example, herpesvirus samurai (HVS) expresses a homologue of Bcl-2 called ORF-16 that blocks mitochondria-mediated apoptosis. Two other examples can be found during EBV infection. In its productive infection phase, EBV expresses a Bcl-2 homologue called BHFR1 that blocks apoptosis induced by a variety of stimuli. However, during its latent phase, EBV expresses the LMP-1 protein instead. This molecule, which closely resembles a constitutively active CD40 molecule, is anti-apoptotic because it upregulates intracellular levels of host Bcl-2 and stimulates the activities of the host transcription factors NF-κB and AP-1; these transcription factors promote cell survival. This persistent stimulation by LMP-1 is thought to drive abnormal B cell survival and proliferation to the point of malignant transformation. A fourth example is the E1B-19K protein of adenovirus. E1B-19K binds to and inhibits those members of the Bcl-2 family that promote cell death, rescuing the host cell from apoptosis. Lastly, KSHV produces a protein called vBcl-2, which closely resembles human Bcl-2 and blocks apoptosis of infected host cells.

As well as by death receptor engagement, host cell apoptosis can be induced by external factors such as UV-irradiation (which generates free radicals) or the presence of chemically reactive molecules such as hydrogen peroxide. The genomes of several types of viruses contain genes whose products are homologous to cellular enzymes (such as glutathione peroxidase) and neutralize the effects of free radicals. Expression of such genes in transfected cells *in vitro* can protect them from cell death induced by UV-irradiation or hydrogen peroxide, but not apoptosis induced by TNF or Fas ligation. These anti-apoptotic mechanisms are likely to be relevant *in vivo* as well.

Sometimes a virus wants to induce the apoptosis of its infected host, especially a non-enveloped virus incapable of budding. Once the virus has completed replication and new virus particles have been assembled, the virus needs a way to get its progeny out of the cell. By inducing apoptosis of the infected cell, optimal dissemination of the progeny is achieved without the inflammation generated by necrotic death. An immune response is thus less likely to be mounted. Some examples of viruses that are thought to facilitate apoptosis in order to exit an infected cell include chicken anemia virus, Sindbis virus, HIV-1, and human parvovirus.

xii) Manipulation of the Host Cell Cycle

Many viruses express proteins that interfere with regulation of the host cell cycle and subvert cell functions to viral use. Products of the tumor suppressor genes p53 and Rb (see Ch.26) are favorite targets. (p53 induces cell cycle arrest or apoptosis in response to DNA damage, and Rb blocks the division of cells with genetic abnormalities.) Interestingly, different proteins in the same virus can have seemingly opposing effects. For example, the E1A proteins of adenovirus influence the regulatory functions of p53 and induce p53-dependent apoptosis, while other adenovirus proteins bind to p53 and trigger its destruction. Adenovirus E1A proteins also bind to members of the Rb group and the p300/CBP family of transcriptional co-activators, disrupting host cell cycling. Similarly, the large T antigen of SV40 binds to both p53 and Rb proteins and inactivates them. In HepB virus, the pX protein binds to host p53 and either promotes or inhibits p53-mediated apoptosis, depending on the immediate microenvironment. In contrast, the E2 protein of HPV regulates p53-mediated cell cycle arrest, rather than apoptosis, in epithelial cells. The HPV E6 protein binds to p53 protein and targets it for ubiquitination and proteasomal degradation. In this case, levels of p53 are kept low enough in the infected cell that the cell can continue to divide, allowing further propagation of the virus.

xiii) Interference with Host Cytokines

Early in viral infections, host cells are induced to produce copious quantities of cytokines with potent antiviral effects. Chemokines that govern the recruitment of infection-fighting leukocytes are also released. The inhibition of the production or action of cytokines and chemokines is thus a desirable strategy from the virus's point of view. For example, members of the poxvirus family evade immune surveillance by altering the local cytokine milieu and making it less favorable to the cellular cooperation that underpins an immune response. In particular, the IFNs, TNF, IL-1, IL-12, and various chemokines are targeted by different viruses. We have already mentioned that some viruses interfere with the pro-apoptotic signal transduction associated with TNF, and that both KSHV and adenovirus express proteins that inhibit IFN-inducible gene transcription. Poxviruses express a protein that blocks maturation of the precursor form of IL-1β, while a human CMV protein disrupts the transcription of chemokine genes. Other viruses have attempted to head off the host immune response by inducing the downregulation of cytokine receptor expression. For example, the surface expression of CXCR4 on CD4+ T cells is decreased following infection by several types of human herpesviruses. Chemotaxis is inhibited and the calcium flux that supports intracellular signaling is suppressed.

Inhibition of IL-12 production is a major goal of many viruses since this cytokine is crucial for Th1 differentiation. For example, EBV synthesizes a homologue of the IL-12p40 subunit that can bind to host IL-12p35. The imposter may competitively inhibit the activity of host IL-12, contributing to a dampening of Th1 responses. EBV also produces a protein called BCRF1 which is homologous to mammalian IL-10. Like IL-10, BCRF1 suppresses IL-12 production by macrophages and IFNγ production by lymphocytes, further downregulating Th1 responses. The measles virus takes a different approach to IL-12 inhibition, at least *in vitro*. When the laboratory strain of measles virus binds to the RCA MCP (CD46) on cultured DCs and macrophages, IL-12 production by these cells is inhibited. Studies done *in vitro* to ascertain the underlying mechanism of this inhibition have shown that IL-12 synthesis can be impaired even in uninfected cells by the cross-linking of FcγRs, CD46, or the complement receptor CR3 on the host cell surface. It is speculated that the aggregation of CD46 by the measles virus delivers signals that block the transcription of both IL-12 subunits.

As discussed in Box 17-3 of Chapter 17, certain viruses appear to have appropriated and subverted mammalian genes for cytokines and cytokine receptors, producing proteins that are homologous to host cytokines and cytokine receptors. In general, the viral cytokines compete with host cytokines for host receptors, while the viral cytokine receptors divert the cytokine away from the host receptor. These interactions generally result in abnormal intracellular signaling and altered chemotaxis. For example, poxviruses synthesize a CC chemokine homologue called MC148 that binds with high specificity to CCR8 and blocks the chemotaxis of lymphocytes, macrophages, and neutrophils *in vitro*. On the other hand, HHV-6 synthesizes a viral cytokine U83 that binds to host CC or CX3C chemokine receptors but increases the calcium flux in an infected host cell. Chemotaxis of mononuclear cells to the site of viral infection appears to be enhanced, providing the virus with a supply of new cells to infect. Sometimes a virus produces viral cytokines with seemingly opposite effects on chemotaxis. For example, KSHV produces a viral cytokine vMIP-II that antagonizes the CXCR5 and CCR5 receptors preferentially expressed on Th1 cells, suppressing their chemotaxis. However, KSHV also synthesizes another viral cytokine called vMIP-1 that binds to the CCR8 receptor expressed preferentially on Th2 cells, promoting the chemotaxis of these cells. Why KSHV should want to regulate Th cell behavior in this way is not known.

With respect to viral cytokine receptors, we have already mentioned the TNFR homologues produced by poxviruses and myxoma virus that block TNF from inducing host cell apoptosis. Similarly, vaccinia virus secretes an IFNα/βR homologue that intercepts IFNα, and an IFNγR homologue that binds IFNγ. Myxoma virus also produces huge amounts of an IFNγR-like viral cytokine receptor called M-T7 that binds host C, CC, and CXC chemokines, disrupting leukocyte extravasation into infected tissues. The p35 protein expressed by vaccinia and other poxviruses acts like a decoy chemokine receptor although it does not structurally resemble one. P35 binds to CC chemokines with higher affinity than does the host receptor, sequestering host chemokine molecules and reducing the efficiency of the immune response. The M-T1 protein of myxoma virus is another secreted molecule that binds and inhibits several chemokines *in vitro* and blocks the influx of monocytes into sites of acute infection *in vivo*. The opposite strategy has been adopted by KSHV. The suspected natural host cells of this virus are endothelial cells and B cells. KSHV induces the expression of a viral transmembrane receptor called ORF74 in the host membrane. ORF74 closely resembles CXCR2, an IL-8 receptor. However, unlike CXCR2, ORF74 is constitutively active, constantly stimulating signaling pathways associated with the synthesis of pro-inflammatory cytokines and *vascular endothelial growth factor* (VEGF). The increase in VEGF production is thought to trigger the formation of new sites of vascularization, facilitating the spread of the virus to fresh host cells. ORF74 may also regulate the migration of virus-laden B cells to lymphoid organs. Here, virions released from lysing host cells would find fresh, susceptible B cells to infect, enhancing propagation of the virus.

Other examples of virus-encoded homologues to mammalian immune system proteins may be found in the viral *semaphorins*. Mammalian semaphorins are signaling proteins that were first studied in neuronal cells but later identified in hematopoietic cells. The structurally diverse semaphorin superfamily includes secreted, GPI-linked, and transmembrane molecules, each of which contains a cysteine-rich "sema" domain of about 500 amino acids. The first recognized function of semaphorins was one of conveying signals guiding the growth of axons during embryonic neuronal development, but the fact that certain viruses have apparently acquired the genes for semaphorins implies the subversion of some kind of immunological function. The genome of vaccinia virus contains the gene for a small semaphorin called A39R that has a truncated sema domain. When A39R expression is engineered in human monocytes, this protein downregulates ICAM-1 expression, induces synthesis of IL-8, TNF, and IL-6, and inhibits migration in response to chemoattractants.

The relevance of these activities to *in vivo* viral infections is under investigation.

E. Immunity to Parasites

Parasites are some of the biggest killers in the pathogen pantheon and claim millions of lives every year, particularly in developing countries. An estimated 300–500 million people worldwide have contracted *malaria*, caused by the parasite *Plasmodium falciparum*. Over 1 million persons die of malaria each year, and many other infected individuals suffer greatly from severe lethargy that leaves them unable to work. Another 200 million people are infected with *Schistosomes* (blood flukes), and 800,000 die of these infections annually. *L. major*, which causes *leishmaniasis*, infects 2 million new hosts annually. The severity of the disease ranges from sub-clinical to mild to fatal even with treatment. Leishmaniasis was responsible for the deaths of 10% of the population of southern Sudan over 1995–2000. Even more worrying is the fact that this disease is on the rise, with a 500% increase in incidence reported in many endemic areas from 1993–2000. The development of clinical strategies to manage parasitic diseases occupies many medical researchers in the developing world.

I. WHAT ARE PARASITES?

As mentioned at the start of this chapter, the term "parasite" covers a vast range of organisms that differ considerably in size, complexity, form, replication mode, and pathogenicity. At one end of the scale are the protozoans. Parasitic protozoan species are single-celled organisms whose behavior resembles that of bacteria or viruses. They often replicate directly within host cells, including within leukocytes. Important protozoan parasites for humans are *L. major*, *P. falciparum*, and *Trypanosoma brucei*, which causes *African sleeping sickness* (Table 22-11). At the other end of the scale, we have the parasitic worms. Many helminth species, which include the trematodes (flukes and worms that attack the lungs and liver), cestodes (tapeworms in the gut lumen), and nematodes (roundworms and hookworms, primarily in the intestinal lumen), are also major human pathogens. These organisms are complex, multicellular beasts that reproduce inside a host's body but outside its cells, or outside the host entirely. However, in the latter case, millions of parasite eggs or larvae are often deposited in places (like water sources) where access to a host is made easy. The growth and maturation of the parasite then occur within the host.

Many parasites have multi-stage life cycles and operate through a *vector*, an intermediary organism that attacks the ultimate host. Indeed, each stage of a parasite may be able to infect a different host. This is a problem from a public health point of view, since a parasite that can make use of an invertebrate vector (such as a mosquito) or takes up residence in an animal reservoir (such as a raccoon) is much harder to deal with than a pathogen that infects humans only. As well as

Table 22-11 Examples of Parasites and the Diseases They Cause

Parasite	Disease
PROTOZOANS	
Entamoeba histolytica	Amebiasis (enteric disease)
Leishmania donovanii	Leishmaniasis in viscera
Leishmania major	Leishmaniasis in face, ears, and skin
Plasmodium falciparum	Malaria
Toxoplasma gondii	Toxoplasmosis
Trypanosoma brucei	African sleeping sickness
Trypanosoma cruzi	Chagas' disease
HELMINTH WORMS	
Ascaris (roundworm)	Ascariasis (lung damage)
Echinococcus (tapeworm)	Alveolar echinococcosis (liver and lung damage)
Onchocerca (filarial worm)	African river blindness (eye damage)
Schistosoma (blood fluke)	Schistosomiasis (liver damage)
Trichinella (roundworm)	Trichinosis (intestinal damage)
Wuchereria (filarial worm)	Elephantiasis (lower trunk swelling due to lymphatic blockage)

taking steps to fight pathogen infections of humans, measures to control the mosquito or raccoon population must be implemented. From an individual's point of view, that fact that some parasite stages may be intracellular while others are extracellular means that the infected individual must be able to mobilize both the cell-mediated and humoral arms of the immune response to eliminate the pathogen. The multiplicity of these factors makes it very difficult to design effective vaccines to combat parasites. We will now briefly describe some of the more important parasites and their modes of infection, before moving on to the immune responses required to control them.

The malarial parasite *P. falciparum* has a very complex life cycle (Fig. 22-11). When a female *Anopheles* mosquito bites an infected human to acquire a blood meal, it takes up RBCs bearing *P. falciparum* gametocytes. When the gametocytes reach the mid-gut of the mosquito, they commence maturation and 24 hours later combine to form worm-like zygotes that mature into *ookinetes*. Each ookinete penetrates into the mosquito's mid-gut wall, forms an oocyst, and starts to produce progeny called *sporozoites* within it. When an oocyst bursts, the sporozoites are released and migrate to the mosquito's salivary glands. When the infected mosquito bites a human, 5–20 sporozoites are injected into his or her subcutaneous tissues. Within minutes of injection, the sporozoites race through the blood to the liver and invade hepatocytes. In these cells, the sporozoites replicate asexually (and asymptomatically) for 2–10 days, producing thousands of *merozoite* progeny. The infected hepatocytes then lyse, each releasing 10,000–30,000

merozoites into the bloodstream where they invade erythrocytes and express parasite proteins on the RBC surface. Within the erythrocytes, the merozoites mature into *schizonts* that reproduce asexually and exponentially. The RBCs lyse 48 hours later, releasing new progeny merozoites into the blood, where they attack more erythrocytes. This synchronized rupture of the erythrocytes is felt as the cyclical constellation of chills, fever, and sweating that constitutes clinical malaria. These symptoms are due to the release of TNF and IL-1 by macrophages responding to the presence of lysed RBCs. However, not all RBCs are lysed by the merozoites. Within some infected RBCs, the merozoites develop into the sexual male and female gametocyte forms. These gametocytes can persist for years in the host. The cycle repeats when another mosquito bites the infected human and takes up RBCs infected with gametocytes.

Interestingly, individuals that suffer from sickle cell anemia are resistant to malaria. "Sickle cell" refers to the abnormal shape of the erythrocytes in these individuals. This shape is caused by an altered hemoglobin protein that arises from a particular mutation in the hemoglobin gene. In addition to inducing sickling of the erythrocytes, the mutated hemoglobin protein appears to hinder the intracellular replication of the parasite. The sickle cell mutation occurs most frequently in Africa, where the advantage it confers over malaria has resulted in its selection despite its associated anemia. In other words, the prevalence of sickle cell disease in Africa is mainly the result of evolutionary pressure exerted by malaria. Other mutations in individuals naturally resistant to malaria inhibit the penetration of the erythrocytes by the parasite.

Leishmania are obligate intracellular protozoan parasites that preferentially invade macrophages in both humans and animals. These parasites have a two-stage life cycle and operate through sand flies as a vector. These insects introduce the parasite into the host when taking a blood meal. The *promastigote* stage is the form transmitted by the invertebrate vector, while the *amastigote* stage is the intracellular form that establishes infection in the vertebrate host. Depending on the species of *Leishmania*, the skin or the viscera of the host is affected. *L. major* causes cutaneous leishmaniasis, characterized by skin lesions and erosion of cartilage in the face and ears. Death can result from asphyxiation or starvation if much of the nasopharyngeal cartilage is destroyed. The less common visceral leishmaniasis is a systemic disease caused by *Leishmania donovani*. Splenomegaly, release of TNF, and severely decreased levels of leukocytes are its signature features.

Toxoplasmosis is caused by the spore-forming protozoan *Toxoplasma gondii*. This organism primarily infects and reproduces in cats. The parasite causes no illness in the infected cat, but when the parasite enters its sexual stage in the feline intestine, oocysts containing sporozoites are formed that are then shed into the feces. These feces may then contaminate food, water, or soil consumed or handled by other animals or humans. If a mouse drinks water contaminated with *T. gondii* oocysts and an uninfected cat eats that mouse, the cycle repeats in the new host. If an oocyst is ingested by a human, the sporozoites are released from the oocyst and travel through the blood and lymph to various tissues, causing infections of the eye, muscle, and brain. In healthy adult humans, the symptoms of

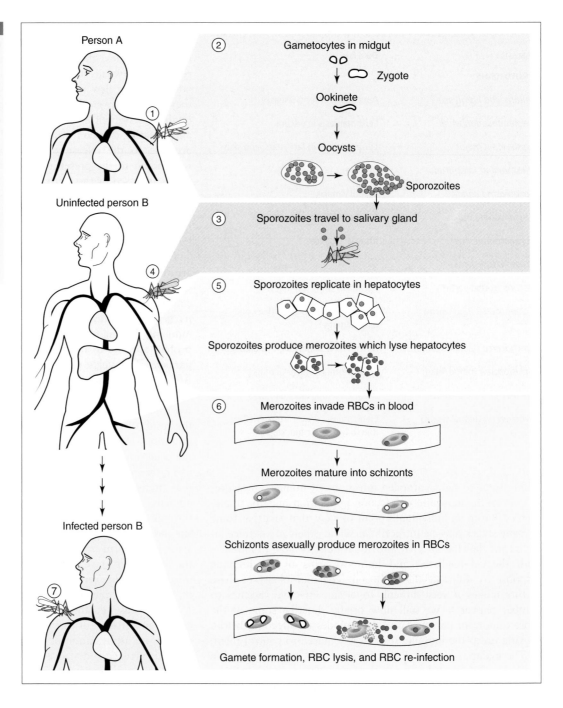

Figure 22-11

Malarial Infection
(1) Person A infected with *P. falciparum* is bitten by an uninfected mosquito that ingests RBC containing *P. falciparum* gametocytes. (2) Events in the mosquito gut. (3) Events in the mosquito salivary gland. (4) Infected mosquito bites uninfected Person B and injects sporozoites that travel to Person B's liver. (5) Events in person B's liver. (6) Events in person B's blood as he becomes infected. (7) Repeat of cycle when an uninfected mosquito bites infected person B.

Person A

Uninfected person B

Infected person B

② Gametocytes in midgut

Zygote

Ookinete

Oocysts

Sporozoites

③ Sporozoites travel to salivary gland

⑤ Sporozoites replicate in hepatocytes

Sporozoites produce merozoites which lyse hepatocytes

⑥ Merozoites invade RBCs in blood

Merozoites mature into schizonts

Schizonts asexually produce merozoites in RBCs

Gamete formation, RBC lysis, and RBC re-infection

T. gondii infection are usually mild and resemble mononucleosis at worst. However, in immunocompromised individuals (such as AIDS patients), sporozoites reaching the brain or eye can cause cerebral abscesses and neurological damage or retinitis leading to visual impairment. Pregnant women face the greatest danger from *T. gondii* infection (which is why mothers-to-be are warned not to change the litter boxes of their pet cats). Infection during pregnancy can cause severe brain damage or vision problems in the fetus or even stillbirth.

T. brucei is a flagellated protozoan parasite that uses the blood-sucking tsetse fly as a vector. Both humans and cattle can be bitten by the tsetse fly and thus be infected, causing devastation on two fronts in African villages. In the first stage of a human infection (which can extend for months or even years), the trypanosomes replicate freely in the blood and cause limited disease. Symptoms include fever, headaches, joint pain, and itchiness. However, in the second stage of the human disease, the parasite crosses the blood–brain barrier and attacks elements of the CNS. Victims show confusion, loss of appetite, lethargy, poor coordination, tissue wasting, and severe disruption of the sleep cycle (hence the name, African "sleeping sickness"). The disease is fatal without treatment, and irreversible neurological damage can occur if treatment is not started before the onset of the second stage. *T. brucei* can also cross the placenta and kill the developing fetus of an infected pregnant woman.

Another trypanosome, *Trypanosoma cruzi*, causes *Chagas' disease*. *T. cruzi*, which is carried by small blood-sucking insects called triatomines, is endemic in the southern United States, Central America, and South America. It is estimated that more than 20 million individuals are infected with *T. cruzi*, and that this parasite causes 50,000 deaths each year. The disease course has three stages: acute, indeterminate, and chronic. Only 1% of newly infected individuals show symptoms, including fatigue, fever, and a distinctive swelling at the site where the triatomine accessed the body. In most individuals, these difficulties are resolved without treatment. However, if *T. cruzi* is not completely cleared, the infection enters an indeterminate stage that can prevail for decades. The patient remains asymptomatic because the infection is kept under control by effective anti-parasite cell-mediated and humoral Th1 immune responses. About one-third of Chagas' disease patients go on to the chronic stage of the disease, in which much more serious symptoms appear 10–40 years after the initial infection. These patients exhibit a life-threatening inflammatory cardiomyopathy characterized by the infiltration of CD8+ T cells and macrophages and the presence of DTH lesions. The esophagus and colon also become enlarged. Interestingly, *T. cruzi* itself is not found in high numbers in the affected organs, leading some researchers to suspect that the symptoms of chronic stage Chagas' disease may be autoimmune in origin. That is, high levels of parasites present at the acute stage of the infection prime the immune system to cause damage to self tissues many years later (see Ch.29). Indeed, healthy mice that receive isolated CD4+ T cells from mice chronically infected with *T. cruzi* soon show myocardial pathology.

The *Schistosoma* worms (members of the trematode family) cause a disease called *schistosomiasis* in humans. The disease is associated with proximity to large, static bodies of water in developing countries because the intermediate host for these worms is a freshwater snail. Parasite eggs hatch in fresh water and penetrate into the snails. The larvae multiply in the snails, exit from the snails into the water, and attack and penetrate the skin of any nearby human. The immature schistosomes migrate first to the lungs and then to the liver, intestine, and blood. Eggs from reproducing schistosomes can get trapped in the walls of the intestine or the sinusoids of the liver, inducing a chronic inflammatory response. Products from the eggs can be directly toxic to liver cells, sometimes causing the host's death. To corral these egg products and block further development and dissemination of the parasite, granulomas form around the egg deposits. After several weeks, the granulomas are slowly replaced by collagen to form scars that can interfere with blood flow, particularly in the liver.

Echinococcus tapeworms, a genus of the cestodes, reproduce only in dogs but the eggs can be ingested by humans and cause alveolar *echinococcosis*. The eggs hatch in the human intestine and the larvae burrow through the intestinal wall into the lymphatic and blood circulatory systems. The larvae take up residence in various tissues and form large cysts, especially in the liver, kidney, lung, and brain. If a cyst ruptures, a sometimes-fatal IgE-mediated hypersensitivity reaction (see Ch.28) to the highly immunogenic contents of the cyst can occur.

A wide variety of helminth parasites is included in the nematode worm group, many of which cause well-known and severe diseases in humans. *Ascariasis* is an infection of the lung by members of the *Ascaris* (giant roundworm) genus. *Ascaris* eggs ingested by humans hatch into larvae that penetrate the intestinal wall, enter the blood, and travel to the lung. Severe inflammation in the lung can result from vigorous IgE-mediated attacks on *Ascaris* antigens in this location. *Trichinella spiralis* is another nematode whose larval form survives in cysts in uncooked meat. This parasite causes *trichinosis* when the meat is eaten and digested. The larvae develop into adults in the gut lumen, penetrate the intestinal mucosa, and produce more larvae in this location that enter the lymphatics and blood circulation. The larvae take up residence in muscles such as those of the tongue, diaphragm, and eye. *Onchocerca volvulus* is a nematode worm that causes *African river blindness*. Larvae are transmitted by African black fly bites to humans and the parasites migrate to the subcutaneous tissues. The larvae develop into adults that produce numerous microscopic thread-like progeny called *microfilariae*. The microfilariae can attack the eye, inducing inflammation that can rapidly lead to blindness. *Elephantiasis* is a severe swelling of the lower limbs and trunk that results from the blockage of the lower lymphatic vessels by the microfilariae of the worms *Wuchereria bancrofti* and *Brugia malayi* (Plate 22-5). These worms are transmitted by mosquitoes.

II. EFFECTOR MECHANISMS

Different parasites evoke different types of immune responses. Some are countered by both humoral and cell-mediated responses, while others are best contained by one or the other. Obviously, the cellularity of the beast, the stage of life cycle, and whether that stage is intra- or extracellular will have a bearing on the defense mechanisms used at any one time. In general, protozoan parasites tend to induce Th1 responses, while helminth worm infections are handled by Th2 responses.

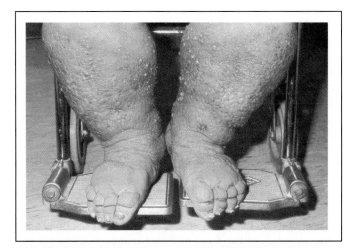

Plate 22-5

Elephantiasis
Reproduced with permission from Cooke R.A. and Stewart B. (2004) "Colour Atlas of Anatomical Pathology," 3rd edn. Elsevier Science, Amsterdam.

However, for unknown reasons, most anti-parasite immune responses are not 100% effective in ridding the body of parasites. The immune system then works to keep parasite numbers at a very low level such that the host does not experience disease. Should the immune system of the host later be compromised in any way, the residual parasites may be able to multiply freely, establishing a recurrence of symptoms.

i) Defense against Protozoans
ia) Humoral defense. Although most protozoan parasites adopt intracellular habitats for at least some of their life cycle, humoral immune responses to these organisms do occur. All the effector mechanisms ascribed to antibodies for defense against extracellular bacteria (refer to Fig. 22-2) apply to defense against small extracellular parasites. Both natural and antigen-specific antibodies may act as opsonins for the extracellular stage of protozoans which, unlike the helminths, are small enough to be phagocytosed. As well, ADCC, classical activation of the complement cascade, and neutralization have all been described as effective means of eliminating these organisms. For example, high titers of neutralizing antibodies that recognize epitopes of a large surface molecule called MSP1 expressed on the *P. falciparum* merozoite surface can block the infection of new erythrocytes by merozoites released from a lysing RBC. Some of these antibodies appear to result from CD1-mediated presentation of part of the MSP1 protein to NKT cells, which may supply help to B cells in the form of secreted IL-4. Complement-mediated lysis of merozoite-infected RBCs via the classical pathway also occurs. In addition to antigen-specific antibodies, there are abundant natural antibodies present in human serum that recognize widely distributed carbohydrate antigens. Parasites bearing these molecules in their cell walls can trigger both the classical and lectin pathways of complement activation.

ib) IFNγ, Th1 responses, and macrophage hyperactivation. Like many intracellular bacteria, some parasites (e.g., *L. major*) resist destruction within an ordinary phagosome and actually enjoy life within a macrophage. These protozoans are resistant to or fail to induce the ordinary respiratory burst in activated macrophages. Only in a hyperactivated macrophage are there sufficient levels of RNI to efficiently kill such parasites. TNF produced by hyperactivated macrophages also plays an important but ill-defined role in anti-*L. major* responses, since administration of anti-TNF antibodies to cultures of IFNγ-treated macrophages blocks their ability to kill *L. major*. If even hyperactivated macrophages cannot clear the infection, a granuloma is formed to contain the threat.

We saw for intracellular bacteria that the hyperactivation of macrophages requires a Th1 response. How does the anti-parasite Th1 response arise? In the case of *L. major*, the primary target of infection is the macrophage. However, *L. major* also infects Langerhans cells in the epidermis, which become the prime initiators of the T cell-mediated immune response to this organism. In mouse models, *L. major* infection of DCs immediately induces these cells to produce copious quantities of IL-12. This IL-12 drives activated anti-parasite Th0 cells along the Th1 path, an event absolutely crucial for defense against *L. major* because these cells secrete the IFNγ required to hyperactivate

macrophages. Mouse strains that are abnormally susceptible to *L. major* infection, such as the inbred strain BALB/c, are naturally incapable of inducing Th1 responses and mount ineffective Th2 responses instead. Th2 responses are unhelpful in this situation because Th2 cytokines such as TGFβ, IL-4, IL-10, and IL-13 inhibit IFNγ production and suppress the activation of iNOS. The same effect can be demonstrated in knockout mice lacking the genes for IFNγ itself, molecules regulating IFNγ production, or components of the IFNγ receptor.

In addition to macrophage hyperactivation, IFNγ has other powerful effects that make it the key molecule in anti-protozoan defense (Fig. 22-12). First of all, IFNγ is directly toxic to forms of many protozoan pathogens, including the sporozoite form of *P. falciparum*. Studies *in vitro* have shown that IFNγ eliminates sporozoites from infected hepatocytes in culture, and mice treated with IFNγ prior to infection with sporozoites are partially protected. Secondly, IFNγ establishes a positive feedback loop of Th1 differentiation because it stimulates IL-12 production by DCs and macrophages. This IL-12 in turn triggers additional IFNγ production by NK and NKT cells. In fact, mice and monkeys treated with IL-12 alone are able to resist sporozoite infections. Thirdly, IFNγ is a potent inducer of iNOS in infected macrophages. Induction of iNOS results in the production of intracellular NO, which eliminates either the parasite itself or the entire infected cell. As well, IFNγ upregulates the expression of p47GTPases localized in the ER or Golgi. The p47GTPase Lrg47 is essential for defense against *L. major*, *T. gondii*, and *T. cruzi*. Lastly, IFNγ upregulates the expression of Fas on the infected macrophage surface, rendering it susceptible to Fas-mediated apoptosis when it contacts a FasL-expressing T cell.

The CD40-CD40L intercellular contact is also crucial for anti-parasite Th1 responses, as evidenced by the increased susceptibility of CD40L$^{-/-}$ and CD40$^{-/-}$ mice to *L. major* infection. Engagement of CD40 on macrophages and DCs by CD40L on activated T cells promotes the production of IL-12 by these APCs. In the absence of such engagement and IL-12 production, the Th1 response and IFNγ production cannot be sustained, and the animals become vulnerable to the parasite. Injection of IL-12 into CD40L-deficient mice can partially protect them against *L. major* infection.

ic) CD8$^+$ T cells and γδ T cells. In infected cells in which protozoan parasites escape from the macrophage phagosome into the cytosol, parasite antigens may enter the endogenous antigen processing system and be presented on MHC class I. CD8$^+$ T cells are then activated and respond to the threat. Indeed, acute disease in experimental mice caused by *L. major* or *T. gondii* is exacerbated if the host is depleted of CD8$^+$ T cells by injection of anti-CD8 antibodies prior to infection. However, CTL-mediated cytolysis is not actually very effective against acute protozoan infections. Rather, it is the secretion of IFNγ by activated parasite-specific CTLs that is important. Similarly, IFNγ secretion by γδ T cells has been implicated at the sporozoite stage of malarial infections, since parasite loads are increased in animals lacking these cells. It should be noted that, while perforin-mediated cytotoxicity is not effective in the acute stage of *T. gondii* infection in mice, this mechanism is important for controlling the chronic stages of infection.

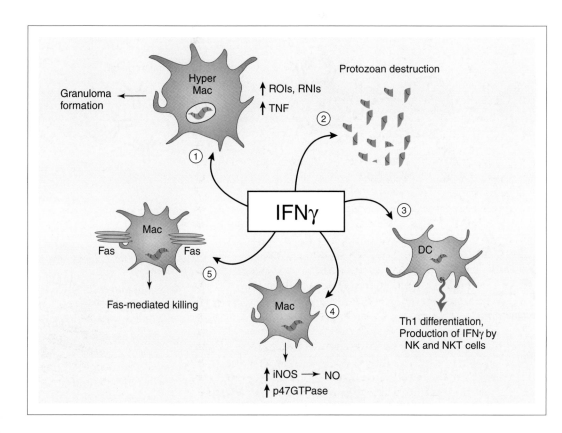

Figure 22-12

Role of IFNγ in Immune Defense against Protozoan Parasites
IFNγ produced by activated Th1 cells, CTLs, NK, and NKT cells is important for (1) hyperactivation of macrophages and granuloma formation that control resistant protozoans; (2) direct toxicity to many protozoan species; (3) stimulation of IL-12 production by infected DCs to sustain a feedback loop of Th1 differentiation; (4) induction of iNOS and increased NO production as well as increased p47GTPase activity in infected macrophages; and (5) induction of Fas expression on infected macrophages.

ii) Defense against Helminth Worms

iia) Unique Th2 responses. While Th1 responses are best for combating protozoan parasites, Th2 responses are vital for defense against large, multicellular helminth worms. In fact, *in vitro*, antigens from the surface of a worm egg can trigger an abrupt switch from an incipient Th1 response to a Th2 response. Humans naturally resistant to *Schistosoma mansonii* exhibit high levels of Th2 cytokines, whereas those susceptible to this pathogen have high levels of Th1 cytokines. Furthermore, susceptibility to *S. mansonii* infection in humans has been found to be under the control of a gene called SM1. Interestingly, this gene occurs in a chromosomal region (5q31-33) that encodes several molecules that regulate T cell differentiation. It is speculated that SM1 may be involved in influencing Th1/Th2 immune deviation.

The anti-helminth Th2 response requires the involvement of IgE, eosinophils, and mast cells, a combination that does not contribute significantly to defense against other types of pathogens (Fig. 22-13). Activated CD4$^+$ T cells are critical for anti-helminth defense because these cells differentiate into effectors supplying the Th2 cytokines and CD40L contacts required for isotype switching to IgE by B cells. Circulating IgE directed against worm surface molecules can bind to the pathogen, attracting the attention of circulating eosinophils expressing FcεRI molecules at a moderate density. Alternatively, prior to encountering specific antigen, the anti-parasite IgE may bind to mast cells expressing high levels of FcεRI, allowing these cells to "pre-arm" against the parasite. In either case, the interaction of parasite antigen, anti-parasite IgE, and FcεRI triggers intracellular signaling within eosinophils and mast cells that induces degranulation in close proximity to the worm. Eosinophil granules contain a variety of chemical substances that work directly and indirectly to kill the worm. Some molecules degrade the skin of the worm, creating an opening that allows other eosinophils to penetrate into its underlying tissues. These cells also degranulate and release additional toxic proteins and peptides that kill the worm. In addition, certain enzymes released from the eosinophil granule can further stimulate the release of histamine by activated mast cells. Histamine causes the contraction of host intestinal and bronchial smooth muscles that is associated with peristalsis and bronchospasm. The actions of these muscles can shake the parasite loose from its grip on the mucosae and expel it from the body. Histamine is also directly toxic to some helminth parasites.

iib) Roles of Th2 cytokines in anti-helminth defense. The Th2 cytokines IL-5, IL-4, and IL-13 are vitally important for defense against helminth worms. IL-5 strongly promotes the proliferation, differentiation, and activation of eosinophils, and supports the differentiation of plasma cells that have undergone isotype switching to IgA production. As we learned in Chapter 20, IgA is the precursor to secretory IgA that coats the mucosae and fends off parasite attachment, particularly that of schistosome worms. In humans, IL-5 expression is strongly associated with resistance to *S. mansonii* infection. IL-4 and IL-13 suppress macrophage production of IL-12, inhibiting IFNγ production and hence the development of a Th1 response. IL-13 is also required for the development

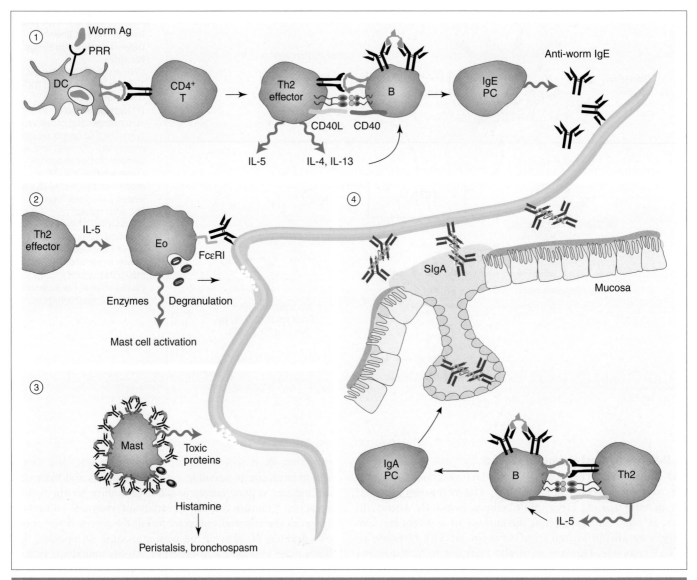

Figure 22-13

Major Mechanisms of Immune Defense against Helminth Worm Parasites
(1) DCs that have captured a worm antigen activate CD4⁺ T cells that are induced to undergo Th2 differentiation. Th2 effectors produce cytokines that induce isotype switching in activated B cells to IgE. (2) IL-5 produced by Th2 cells activates eosinophils that bind to worm-bound antibodies via their FcεRIs. Eosinophil cytotoxic granules contain molecules that directly damage the worm surface and stimulate mast cells to degranulate. (3) Mast cells pre-armed with anti-parasite IgE are activated by worm antigens and release histamine, which induces gut and airway spasms to eject the worm. Mast cells also release proteins that are directly toxic to the worm. (4) IL-5 produced by Th2 cells also induces isotype switching to IgA in mucosal anti-worm B cells. Secretory IgA (SIgA) blocks the worm from gaining a foothold on the mucosal surface.

of airway hyperresponsiveness and the alterations to gastrointestinal fluids and the functions of intestinal smooth muscle and epithelial cells that accompany GI pathogen expulsion. Both IL-4 and IL-13 are effective in inducing Th2 responses to nematode and schistosome infections, but the response to a specific parasite may rely more heavily on one cytokine than the other. For example, IL-13 is more important than IL-4 for the expulsion of *Nippostrongylus brasiliensis* from the gut of mice, but either will do for the elimination of *Trichurus muris*.

III. EVASION STRATEGIES

The fact that many parasites have multi-stage life cycles means that each pathogen has many different opportunities to thwart the immune response. Some important mechanisms underlying this evasion are summarized in Table 22-12. Because each stage of the parasite may employ multiple mechanisms and each parasite may have multiple life stages, natural immunity that successfully blocks parasite infection is difficult to establish in humans, and may take many years to develop even with ongoing exposure to the parasite.

Table 22-12 Evasion of the Immune System by Parasites

Immune System Element Thwarted	Parasite Mechanism
Specificity of T and B cells	Have a multi-stage, multi-host life cycle Lyse infected lymphocytes
Antibodies	Acquire host surface proteins that block binding of anti-parasite antibodies Shed parasite membranes bearing immune complexes Hide in macrophages Modify surface proteins to cause antigenic variation
Phagolysosomal destruction	Block fusion of phagosome to lysosome Escape from phagosome into cytoplasm Resist lysosomal enzymes Decrease respiratory burst Lyse resting granulocytes and macrophages
Complement	Modify cell membrane so alternative pathway cannot be activated Express RCA-like proteins Degrade attached complement components and release the terminal complex Force continuous complement activation so that components are exhausted
T cells	Induce accelerated apoptosis of memory T cells Secrete immunosuppressive peptides
APCs	Interfere with DC maturation and macrophage activation Promote IL-10 secretion by T cells to downregulate MHC class II Decrease IL-12 production to reduce Th1 differentiation

i) Avoiding Detection

The simplest evasion strategy is, of course, to avoid detection in the first place. *L. major* hides from the immune response by sequestering itself within host macrophages. Some schistosomes are masked from immune surveillance by the surface acquisition of host self antigens such as blood group glycolipids and forms of MHC class I and II molecules. The dense "forest" created by these host molecules blocks the host antibodies from binding to parasite surface antigens. Other schistosomes (and other protozoans) repel antibody attack by shedding their membranes, ejecting the immune complex of the parasite antigen and host antibody.

Like viruses, parasites use antigenic variation to avoid detection. For example, *T. brucei* spontaneously modifies its expression of surface molecules to forestall recognition by circulating antibody. There are multiple (perhaps as many as 1000) separate genes encoding versions of the dominant surface glycoprotein VSG (*variable surface glycoprotein*) covering the trypanosome. Each trypanosome expresses only one copy of the VSG gene at a time, but regularly shuts down that gene and switches to another that results in a glycoprotein coat of a slightly different amino acid sequence. Antibodies raised to the

first VSG protein may not recognize the second, allowing the parasite to escape recognition and thus destruction.

P. falciparum has evolved multiple means of antigenic variation to confound immune surveillance. First of all, different strains of *P. falciparum* express variant-specific forms of a large number of non-cross-reacting antigens on the surfaces of infected host erythrocytes. These strains tend to be endemic to a particular geographic region. An individual resistant to one strain of the parasite who travels to a new region will therefore likely be susceptible to the strain dominant in that region. Moreover, within the same host, successive populations of RBCs infected with the same strain of *P. falciparum* can express slightly different versions of certain parasite proteins. The expression of the novel antigens confounds antibodies previously raised against the antigens expressed in the first wave. In addition, many proteins necessary for parasite development are expressed only during the developmental stage in which they are required. This strategy minimizes the "window" of time during which a particular anti-parasite antibody can be effective. By the time antibodies to an early stage antigen are raised, the parasite may have either become intracellular or moved on in its life cycle such that the early antigen is no longer expressed.

ii) Avoiding Phagolysosomal Destruction

Most parasites have developed strategies to avoid destruction by the effector cells in which they have hidden. The macrophage phagosomes in which *T. gondii* is contained are unable to fuse with lysosomes, sparing this organism attack by hydrolytic enzymes. Like many intracellular bacterial species, the protozoan *T. cruzi* enzymatically lyses the phagosomal membrane prior to lysosomal fusion and escapes to the cytoplasm of the host cell. While a significant proportion of early stage *L. major* promastigotes succumb to lysosomal destruction of phagosomal contents, the later stage is somehow able to resist digestion. There is evidence that the receptor used by the later stage parasite to access macrophages may not trigger a complete respiratory burst, making the phagosome a less hostile environment for the late stage invader. Some parasites, such as the intestinal protozoan *Entamoeba histolytica*, turn the tables and lyse resting granulocytes, macrophages, and lymphocytes with which they come in contact.

iii) Avoiding Complement

Parasites have developed unique approaches to avoiding complement-mediated destruction. While the stages of many parasites that live in invertebrate vectors can be easily eliminated by MAC attack, those stages that infect vertebrate cells have modified their cell membranes so that these organisms no longer trigger the alternative complement activation pathway. Furthermore, there is evidence that the vertebrate stages of several parasites express a molecule that functionally resembles the mammalian RCA protein DAF. Still other parasites can proteolytically repel complement-activating molecules that have attached to their surfaces, or cleave the Fc portions of membrane-bound antibodies. For example, *L. major* can induce the release of the entire complement terminal complex from its surface. Some parasites can secrete molecules that force fluid phase complement activation, resulting in exhaustion of complement components.

iv) Manipulation of the Host Immune Response

Active manipulation of the host immune response is a tactic used by some sophisticated parasites, including *P. falciparum*. Certain epitope variants produced by this parasite promote secretion by T cells of the immunosuppressive cytokine IL-10 rather than IFNγ, resulting in downregulation of MHC class II expression, inhibition of NO production, and decreased expression of pro-inflammatory cytokines by T cells. In addition, *Plasmodium*-infected RBCs express a parasite-encoded adhesion protein that allows the RBCs to bind to integrins on the surfaces of DCs and macrophages. This binding blocks the activation of macrophages and interferes with the maturation of DCs, thus inhibiting APC function and T cell responses. The erythrocytic stage of malaria infection can also induce the accelerated apoptosis of antigen-specific memory T cells. *L. major* uses slightly different tactics to manipulate the host immune response. The reader will recall from our previous discussion of viral evasion strategies that the cross-linking of CD46, complement receptors, or FcγRs on macrophages by viral components inhibits IL-12 production at the transcriptional level. Similarly, molecules expressed by *L. major* promastigotes and amastigotes can bind to macrophage CR3 and FcγRs, respectively, and reduce IL-12 production. The Th1 response that would kill the parasite is thus inhibited. Lastly, some parasites secrete hormone-like peptides that appear to downregulate the host immune response.

F. Immunity to Fungi

I. WHAT ARE FUNGI?

Fungi tend to be either unicellular and yeast-like (spherical), such as *Candida* species, or multicellular and filamentous, like *Aspergillus*. Some fungi live commensally on the topologically external surfaces of the body, while others live most of their lives in the soil as a mass (*mycelium*) of thread-like processes (*hyphae*). *Dimorphic* fungi adopt a yeast-like form at one stage in their life cycle and a hyphal form at another stage. Fungal cells have a cell wall like bacteria but also a cell membrane like mammalian cells. However, the fungal cell wall lacks the peptidoglycans, teichoic acids, and lipopolysaccharide components of the bacterial wall, and the main component of the fungal cell membrane is ergosterol rather than the cholesterol found in mammalian cell membranes.

Most fungal species are not harmful to healthy humans. However, immunocompromised individuals can suffer from acute infections that sometimes go on to become persistent. Such is the case with many *Candida* and *Aspergillus* species, which are part of the normal host flora but which can start to become invasive if immune surveillance by phagocytes fails. Species of soil-living filamentous fungi produce progeny in the form of *conidia* (spores), which can become airborne and thus inhaled by a host. Within the host, the conidia develop into either hyphae or other specialized forms better suited for host tissue invasion. However a pathogenic fungus accesses the

body, its primary target tissue is usually the vascular system. Invasion of blood vessels by a growing fungus chokes off the blood supply to the host tissue, damaging or killing it. The exceptions are the *dermatophytes*, filamentous fungi that infect the skin, hair, and nails. These organisms, which include species of *Epidermophyton*, *Microsporum*, and *Trichophyton*, cannot penetrate the living, cellular tissue of a healthy host, and so are restricted to parts of the body that lack living cells, such as the keratinized outer layer of skin.

Diseases caused by fungal infections are called *mycoses*. Important fungal pathogens are *Blastomyces dermatitidis*, *Histoplasma capsulatum*, *Candida* species, *Aspergillus* species, *Cryptococcus neoformans*, and *Pneumocystis carinii* (Table 22-13). *Blastomycosis* occurs when conidia of the yeast-like fungus *B. dermatitidis* are inhaled. This organism replicates extracellularly to cause a pulmonary infection that spreads through the blood to the skin, bones, and male urogenital tract, but not the gut. In contrast, *H. capsulatum* is an intracellularly replicating yeast-like fungus that causes *histoplasmosis*. Inhaled microconidia develop into *Histoplasma* that take up residence preferentially in local respiratory macrophages. A progressive pulmonary disease resembling tuberculosis can spread to the secondary lymphoid organs, mucosae, gut, and adrenal glands. *Candida* species such as *C. albicans* and *C. tropicalis* lurk in the normal flora at the mucosae (but not in the skin) and cause disease only if these mucosal barriers are compromised. A deficiency of neutrophils in the host (*neutropenia*) leaves him or her especially vulnerable to *candidiasis*. Such *Candida* infections are usually fairly superficial in nature (such as vaginitis and cystitis) but can progress to infections of the eye, skin, and brain. Inhalation of spores of *Aspergillus* species causes a variety of diseases and can induce allergic responses. Conidia of three species, *A. fumigatus*, *A. flavus*, and *A. niger*, are particularly pathogenic for humans, causing invasive pulmonary infections that can be fatal if allowed to

Table 22-13 Examples of Fungi and the Diseases They Cause

Organism	Disease
Aspergillus species, including *A. fumigatus*, *A. flavus*, *A. niger*	Airway and pulmonary infections and allergic reactions
Blastomyces dermatitidis	Blastomycosis (pulmonary infection)
Candida sp., *C. albicans*, *C. tropicalis*	Yeast infections, vaginitis, cystitis
Cryptococcus neoformans	Cryptococcosis (pulmonary infections and meningitis)
Histoplasma capsulatum	Histoplasmosis (TB-like disease)
Pneumocystis carinii	Pneumonia and severe lung damage
Dermatophytes, including *Epidermophytyon floccosum*	Infections of skin, nails, and hair

entrench. Again, neutropenia is an aggravating factor. Clumps of filamentous *Aspergillus* readily colonize previously damaged airway tissues. *Aspergillus* species also produce *mycotoxins* that damage hepatocytes, macrophages, and CTLs. *C. neoformans* is a yeast-like fungus often present in pigeon droppings. When the unencapsulated spores of *Cryptococcus* are inhaled by a host, the parasite enters the lung and synthesizes a protective capsule that inhibits phagocytosis. If the infection becomes established, the result may be *cryptococcosis*, a syndrome of pulmonary infection accompanied by meningitis. *P. carinii* is a unicellular eukaryote that, although classified as a fungus, shares structural features with the protozoan parasites. *P. carinii* infection causes PCP (*P. carinii pneumonia*), a common feature in immunocompromised individuals (especially AIDS patients). Inhalation of *P. carinii* results in severe lung damage in hosts lacking cell-mediated immune responses. Spread of the infection to other tissues does not usually occur.

II. EFFECTOR MECHANISMS

Effector mechanisms used against various fungi are summarized in Figure 22-14.

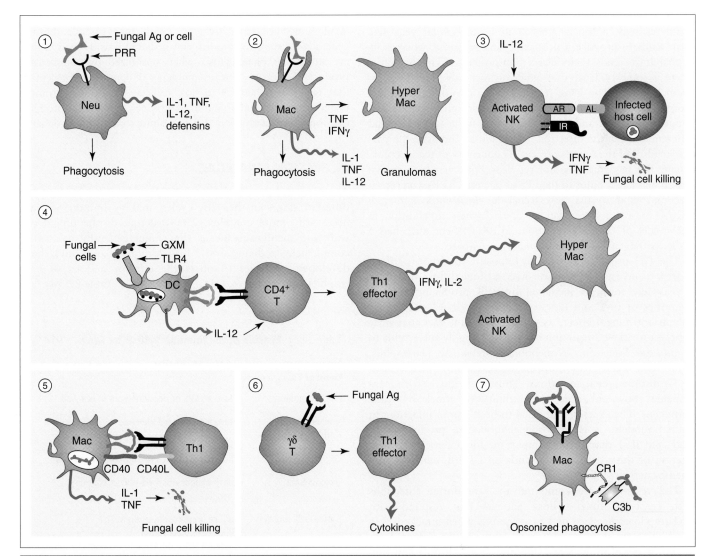

Figure 22-14

Major Mechanisms of Immune Defense against Fungi
(1) Neutrophils activated by PRR-mediated recognition of fungal PAMPs carry out phagocytosis and secrete cytokines. (2) Macrophages also carry out these functions plus can undergo hyperactivation and form granulomas to contain resistant fungi. (3) Activated NK cells kill fungi by secreting cytotoxic cytokines rather than by natural cytotoxicity. (4) Fungal TLR ligands such as GXM (glucuronooxylomannan) activate DCs that in turn initiate T cell activation and Th1 effector differentiation. These cells produce cytokines that stimulate macrophage hyperactivation and NK cell activation. (5) Infected macrophages that establish CD40-CD40L contacts with activated Th1 cells produce copious amounts of IL-1 and TNF that are directly toxic to fungal cells. (6) Mucosal γδ T cells activated by fungal products generate effectors that secrete additional cytokines supporting B cells. (7) Fungi coated in either anti-fungal antibody or C3b undergo opsonized phagocytosis by macrophages and neutrophils. Note that the structure of the fungal cell wall allows them to resist MAC lysis.

i) Cell-Mediated Defense

As mentioned previously, most fungi do not establish infections in humans with competent innate and adaptive immune responses. Cell-mediated innate immunity is the primary means by which fungus infections are controlled. Neutrophils and macrophages both carry out vigorous phagocytosis and produce powerful anti-fungal defensins. These defensins induce an osmotic imbalance in pathogens such as *Candida* and *Cryptococcus* that kills them. Neutrophils and macrophages also secrete copious quantities of IL-1, IL-12, and TNF. As we have seen, IL-12 stimulation activates NK cells that contribute to fungal cell killing via cytokine secretion (rather than natural cytotoxicity). IFNγ produced first by activated NK cells and later by activated Th1 cells also hyperactivates macrophages, which can initiate granuloma formation.

Interestingly, a fungus present in its unicellular yeast-like form tends to provoke a protective Th1 response, whereas its hyphal form tends to induce a non-protective Th2 response. There is some evidence that either distinct subsets of DCs, or distinct receptors on DCs, respond to the two different fungal morphologies. These DCs then proceed with phagocytosis and antigen processing and presentation, and influencing Th1/Th2 differentiation in the direction best suited to eliminate the particular form of the fungus present. The identity of the receptors driving this recognition is not yet clear, but it has been shown that, much like it binds to the LPS of bacteria, TLR4 can recognize and bind to the polysaccharide *glucuronoxylomannan* (GXM) present in the capsule of *C. neoformans*. In addition, in one model of *Candida* infection, TLR4-deficient mice had a higher pathogen burden than wild-type controls.

The Th1 response induced by exposure to airborne fungal spores, or invasion by skin or mucosal fungal flora that have a yeast-like form, is mediated by cells producing copious quantities of IL-2 and IFNγ. Accordingly, IL-12 production has an anti-fungal effect, as evidenced by the fact that mice deficient for the p40 subunit of IL-12 are highly susceptible to *C. albicans* infection. As mentioned previously, neutrophils, along with macrophages, appear to be the key producers of IL-12 during fungal infections. Thus, individuals with neutropenia show enhanced susceptibility to attack by these pathogens. The exception appears to be *P. carinii* infection. In animal models, it is macrophage/monocyte production of TNF and IL-1 that is most important for defense. The production of these cytokines depends on CD40L contacts supplied by activated T cells.

Th2 responses are comparatively rare during infections with yeast-like fungi. Those patients that respond with Th2 responses in place of Th1 responses when attacked by a yeast-like form show decreased resistance to the fungus. Such a bias appears in patients suffering from chronic mucosal *C. albicans* infections. While DTH responses in the skin counteract fungal invasion, the presence of Th2-associated antibodies such as IgE and IgG4 (in the absence of a DTH response) can actually predispose patients to recurrent fungal infections.

γδ T cells play a significant role in anti-fungal defense at the mucosae. Mice deficient for γδ T cells show enhanced susceptibility to *C. albicans* infections. In wild-type mice, polyclonal expansion of γδ T cells in the gastric mucosa occurs upon oral exposure to *C. albicans*. Protein antigens isolated from certain fungi can be used *in vitro* to stimulate human γδ T cells to produce factors promoting B cell differentiation.

ii) Humoral Defense

Other than the secretory IgA that defends the mucosae, antibodies are thought to contribute in only a limited way to defense against fungi. Antibody-mediated opsonization may promote phagocytosis, and thus contribute to the presentation of fungal antigens that activates Th1 cells. Mice deficient for B cells show a deficit in cell-mediated immunity that allows the establishment of infections with *Candida* species and *P. carinii*.

iii) Complement

While fungal cells can activate the complement cascade, they are generally resistant to complement-mediated lysis. However, they are subject to phagocytosis when opsonized by complement products. Fungi also express analogues of complement receptors that facilitate adherence to host cells and may also promote phagocytosis. Pro-inflammatory cytokines induced by products of complement activation also contribute to anti-fungal defense.

III. EVASION STRATEGIES

As mentioned previously, many fungi adopt different forms at different stages in their life cycles, making immune defense necessarily more complex. The structure of the fungal cell wall and membrane means that fungi generally can avoid complement-mediated lysis. In addition, many fungi have developed strategies to offset the effector actions of neutrophils, macrophages, CTLs, and NK cells (Table 22-14).

Table 22-14 Evasion of the Immune System by Fungi

Immune System Element Thwarted	Fungal Mechanism
PRR recognition	Have no LPS or peptidoglycan in cell wall
Specificity of T and B cells	Have a multi-stage life cycle
Complement	Block access to cell membrane via cell wall
Phagocytosis	Block phagocytosis via polysaccharide capsule
T and B cell function	Induce immune deviation to Th2 Block NF-κB activation Increase NO production to decrease lymphocyte proliferation Block phagocytosis Inhibit neutrophil migration Decrease IL-12 and B7 expression by monocytes Activate regulatory T cells via polysaccharide capsule component Produce melanin to decrease Th1 and Th2 responses Block TNF production

Many fungi produce toxins and other molecules that have immunosuppressive effects. Some appear to mediate immune deviation, biasing the host's immune reactions to ineffective Th2 responses at the expense of Th1 responses. Others have more direct inhibitory effects. For example, *gliotoxin*, a metabolite produced by *A. fumigatus*, inhibits the activation of NF-κB in stimulated T and B cells, blocking the transcription of genes needed for differentiation. A molecule called *protein 43* acts in an unknown way to suppress immune responses during murine *Candida* infections. *H. capsulatum* infection in mice results in NO production that downregulates lymphocyte proliferation. The GXM polysaccharide present in the capsule of *C. neoformans* spores not only blocks phagocytosis and inhibits neutrophil migration but also has downregulatory effects on adaptive immunity. In its free form, GMX blocks production of IL-12 by monocytes, downregulates B7 expression on macrophages, and activates regulatory T cells. In addition, melanin produced by *C. neoformans* inhibits both Th1 and Th2 responses in mice. The BAD1 molecule of *B. dermatitidis* binds to CR3 on macrophages and blocks the production of TNF by these cells.

G. The Mysterious Prions

For decades, scientists were puzzled by a number of related neurodegenerative diseases found in both humans and animals. These diseases came to be known as *spongiform encephalopathies* (SE) because they all caused CNS lesions that rendered the brain "sponge-like." Spongiform encephalopathies, which are invariably fatal, include variant Creutzfeldt-Jakob disease (vCJD) and kuru in humans, scrapie in sheep, and bovine spongiform encephalopathy (BSE or "mad cow disease") in cattle. Over the years, it was found that each of these diseases could be transmitted to experimental animals by injecting brain extracts from patients or animals that had died of SE. Indeed, D. Carleton Gajdusek was awarded a Nobel Prize in 1976 for his discovery in the early 1960s that kuru in humans represented a new form of infectious disease. However, while an infectious etiology had been clearly established, the actual agents involved remained mysterious. Analysis of the scrapie agent revealed it to be extremely resistant to heat, formalin, and both UV- and ionizing irradiation—all treatments that normally destroy infectious pathogens. Furthermore, scrapie, kuru, and vCJD did not provoke any of the normal immune responses expected during an infection. An infectious etiology was also difficult to reconcile with the earlier identification in the 1930s of a rare, familial form of vCJD that presumably had a genetic etiology. Finally, an SE would occasionally arise in a sporadic manner in a patient or animal with no history of familial or infectious transmission.

In the 1970s, Stanley Prusiner of the University of California at San Francisco attempted to purify the "scrapie agent" to determine its molecular nature. Surprisingly, protocols that destroyed nucleic acids had no effect on scrapie infectivity, while procedures known to destroy proteins reduced its infectious strength. By 1982, Prusiner concluded that the agent responsible for scrapie infectivity was a protein devoid of any

nucleic acid, an unprecedented discovery in biology. At this point, he coined the term "prion" to identify this novel type of pathogen that was both proteinaceous and infectious. The scrapie agent was therefore denoted as PrP^{sc} (*prion protein, scrapie*). Further work by Prusiner established that PrP^{sc} was actually a conformational isomer of a normal glycoprotein found on the surfaces of many mammalian cells. This normal glycoprotein was denoted PrP^c (*prion protein, cellular*). It turns out that, upon exposure to a PrP^{sc} molecule, a PrP^c molecule undergoes a dramatic change in conformation. An α-helical region of the native PrP^c protein spontaneously re-folds into the β-pleated sheet structure characteristic of PrP^{sc}. This refolding process appears to be facilitated by an as-yet-unidentified co-factor (perhaps a membrane lipid). It is also unclear exactly where the conversion occurs, although there is some evidence pointing to the minute infoldings of the plasma membrane generated during pinocytosis (called *caveolae*) as the site. In any case, when PrP^{sc} is introduced into a fresh host, it acts as a template for the refolding of existing host PrP^c molecules into additional copies of PrP^{sc}. In other words, the disease-causing prion can effectively "replicate" itself in a new host and a mass conversion of the host's PrP^c molecules to the PrP^{sc} conformation occurs (Fig. 22-15). Unlike PrP^c, the misfolded PrP^{sc} has profoundly altered physicochemical properties. When the PrP^{sc} protein invades cells of the CNS, it induces brain degeneration that is manifested as the clinical signs of SE. Intriguingly, no other part of the body appears to be affected by the presence of PrP^{sc}. For his discovery of prions and his investigations into the mechanisms by which they cause disease, Dr. Prusiner was awarded a Nobel Prize in Physiology or Medicine in 1997.

We emphasize that, in this infectious form of prion disease, there is no mutation of the PrP^c gene of the host, and no change in the amino acid sequence of the affected PrP^c proteins; the disorder is purely one of protein misfolding. Interestingly, it appears that misfolding of native proteins may be implicated in other neurodegenerative disorders such as Alzheimer's disease, Parkinson's disease, and frontotemporal dementia. Other forms of prion disease are inherited rather than infectious and do involve a mutation of the host's PrP^c gene. More than 20 such mutations have been identified, some of which may account for the sporadic cases of prion disease that have been reported over the years. Once a mutated PrP^c gene arises *de novo* in one individual, the pathogenic gene is passed down to his or her offspring.

The BSE epidemic in Great Britain in the 1990s was likely a case of infectious prion disease caused by the feeding of scrapie-infected sheep offal to cattle. A small number of humans who consumed beef processed from these cattle then acquired the PrP^{sc} protein and became ill with vCJD. The use of sheep offal as animal feed is now banned, but 1 million cattle were thought to have been infected with prions before this restriction came into place. Infectious prions are also responsible for the human disease kuru. Kuru was found almost exclusively among the Fore people of New Guinea, a tribe that used to routinely engage in ritualistic cannibalism. This practice included eating the brain tissue of dead relatives, some of whom had died of kuru. At the urging of missionaries,

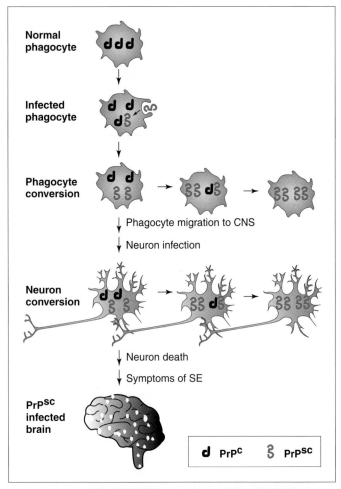

Figure 22-15

Infectious Prion Disease
A phagocyte in a healthy animal contains normally folded PrP^C molecules. Disease is initiated when the healthy animal ingests an abnormally folded prion PrP^{SC} molecule in contaminated food. A phagocyte takes up the PrP^{SC} molecule, which serves as a template for the misfolding of the cell's normal PrP^C molecules. Gradually, more and more normal PrP^C molecules are converted into PrP^{SC} molecules but the converted phagocyte does not die. Rather, it conveys the PrP^{SC} molecules to neurons in the CNS that are killed upon uptake of PrP^{SC} and conversion of neuron PrP^C molecules. Massive neuronal death causes the "spongy brain" and other symptoms of spongiform encephalopathy (SE).

Despite the preceding, work in experimental animals has shown that the PrP^C protein can be immunogenic. In so-called PrP$^{0/0}$ mice, in which the gene encoding PrP^C has been knocked out, the natural tolerance to the protein is absent and a normal, PrP^C-specific humoral response is seen after immunization with murine PrP^C in adjuvant. In another experiment, PrP$^{0/0}$ mice were immunized with plasmids expressing DNA (see Ch.23) encoding human prion proteins, inducing production of mouse anti-human PrP^C antibodies. Similarly, chickens immunized with either human or bovine PrP^C coupled to the carrier KLH generate antibodies directed against human or bovine PrP^C. Such anti-PrP^C antibody preparations are useful for both the study of prions and as potential tools for the diagnosis, prevention, and treatment of prion diseases.

In view of the natural tolerance to PrP^C discussed previously, one might assume that the immune system is neutral with respect to prion diseases. However, it appears that the immune system may actually promote the progress of prion disease both directly and indirectly. In a natural setting, prion infection occurs by a peripheral route, usually oral. Leukocytes are among the first cells to take up the prions, but the "replication" of the prions via PrP^C conversion causes no harm to these cells, in contrast to the severe damage done to neurons. However, leukocytes are apparently responsible for the subsequent introduction of prions into the CNS. In addition, it seems that the FDC in the lymphoid follicles is required for the replication of prions that precedes their spread into the nervous system. The mechanism remains obscure but is an absolute requirement for prion disease progression. There is also evidence that the activation of specific complement components plays a role in the initial trapping of prions in the lymphoid organs soon after infection. This too may relate to FDC function, since C3d/C4b-opsonized antigens interact with CR2/CR1 complement receptors on FDC and may therefore be important for the intracellular accumulation of prions in the FDC. Other studies have suggested that macrophages and DCs can be involved in prion disease, and that a microbial or viral pathogen may sometimes act as a triggering co-factor.

Indirect support for the involvement of the immune system in prion pathogenesis comes from studies of mouse models of scrapie infection. In these experiments, susceptibility to prion disease correlated with the functional status of the immune system. For example, while mitogenic stimulation of lymphoid cells enhanced prion disease susceptibility, lymphoid suppression achieved by either corticosteroid administration or splenectomy reduced it. In addition, studies of mice assessed at various time points after birth showed that increasing susceptibility to prion infection correlated with the progressive maturation of the immune system. Lastly, and perhaps most strikingly, *scid* mice (which cannot mount either T or B cell-mediated responses) were resistant to infection by the BSE or scrapie agents, while immunologically reconstituted *scid* mice were not. A more complete understanding of prion infection and disease processes will be critical to development of prevention strategies and therapies for these fascinating but deadly disorders.

This brings us to the end of our description of natural immune defense against pathogens. We move now to a discussion of "manufactured" immunity to pathogens, created by the techniques of vaccination.

the practice of cannibalism ended in the late 1950s, so that the incidence of kuru has declined steadily since that time.

Prion infection destroys the brain without inducing either a humoral or cell-mediated adaptive response. The host's T cells are usually tolerant to the PrP^{SC} protein, as it is merely a naturally occurring host protein with modified secondary structure. Thus, almost all peptides generated from newly formed PrP^{SC} will be seen as "self" peptides. By extension, in the absence of the activation of prion-specific T cells, no Td humoral response can be mounted. Furthermore, although the "foreign" conformation of PrP^{SC} might be recognized by the BCR of a B cell, the antigen itself cannot act as a Ti immunogen as it has neither the large size or multi-valency needed to stimulate B cells directly.

SUMMARY

Immune responses have evolved to combat the five major types of pathogens: extracellular bacteria, intracellular bacteria, viruses, parasites, and fungi. For all types of pathogens, the mechanisms of innate immunity offer an immediate response that either foils the establishment of infection or slows the infection down until adaptive immune mechanisms can target the pathogen more effectively. When an adaptive immune response is activated, the elements that will be most effective depend on whether the pathogen is extracellular or intracellular. Extracellular entities that are relatively small, such as extracellular bacteria, virus particles, protozoan parasites, and some fungi, can be targeted by antibody and then cleared effectively by antibody- and complement-mediated mechanisms that involve either direct lysis or phagocytic destruction. Protozoan worms are also subject to antibody assault because they are extracellular; however, these organisms are too large to be subsequently engulfed. Instead, specialized antibody isotypes such as IgA and IgE are produced to ensure that the worm does not become anchored in the host. In addition, mediators released by degranulating mast cells and eosinophils act on worm tissues to degrade them. Intracellular bacteria and replicating viruses, which are found inside host cells and cannot be targeted by antibody, must be eliminated by cytolytic mechanisms mediated by CTLs, NK cells, NKT cells, and $\gamma\delta$ T cells. A successful immune response against a given pathogen thus depends on cells and cytokines of the innate response inducing Th responses of the appropriate subtype, with Th1 responses being required for cell-mediated immunity against internal threats, and Th2 responses being needed for humoral responses against external threats. In the quest for continued survival, many pathogens have evolved complex evasion strategies intended to compromise the success of the immune response invoked.

Selected Reading List

Abel L. and Dessein A. J. (1997) The impact of host genetics on susceptibility to human infectious diseases. *Current Opinion in Immunology* 9, 509–516.

Bachmann M. and Kopf M. (2002) Balancing protective immunity and immunopathology. *Current Opinion in Immunology* 14, 413–419.

Bangham C. R. (2000) The immune response to HTLV-I. *Current Opinion in Immunology* 12, 397–402.

Barry M. and McFadden G. (1998) Apoptosis regulators from DNA viruses. *Current Opinion in Immunology* 10, 422–430.

Basta S. and Bennink J. R. (2003) A survival game of hide and seek: cytomegaloviruses and MHC class I antigen presentation pathways. *Viral Immunology* 16, 231–242.

Bertoletti A. and Maini M. K. (2000) Protection or damage: a dual role for the virus-specific cytotoxic T lymphocyte response in hepatitis B and C infection? *Current Opinion in Immunology* 12, 403–408.

CDC Public Health Emergency Preparedness and Response (http://www.bt.cdc.gov/index.asp).

Cohen J. I. (1999) The biology of Epstein-Barr virus: lessons learned from the virus and the host. *Current Opinion in Immunology* 11, 365–370.

Cossart P. and Bierne H. (2001) The use of host cell machinery in the pathogenesis of *Listeria monocytogenes. Current Opinion in Immunology* 13, 96–103.

Cuconati A. and White E. (2002) Viral homologs of BCL-2: role of apoptosis in the regulation of virus infection. *Genes and Development* 16, 2465–2478.

Dimopoulous G., Muller H.-M., Levashina E. A. and Kafatos F. C. (2001) Innate immune defense against malaria infection in the mosquito. *Current Opinion in Immunology* 13, 79–88.

Dussurget O., Pizarro-Cerda J. and Cossart P. (2004) Molecular determinants of *Listeria monocytogenes* virulence. *Annual Review of Microbiology* 58, 587–610.

Edelson B. T. and Unanue E. R. (2000) Immunity to *Listeria* infection. *Current Opinion in Immunology* 12, 425–431.

Finkelman F. D., Wynn T. A., Donaldson D. D. and Urban Jr. J. F. (1999) The role of IL-13 in helminth-induced inflammation and protective immunity against nematode infections. *Current Opinion in Immunology* 11, 420–426.

Flynn J. L. and Ernst J. D. (2000) Immune responses in tuberculosis. *Current Opinion in Immunology* 12, 432–436.

Flynn J. L. and Chan J. (2001) Immunology of tuberculosis. *Annual Review of Immunology* 19, 93–129.

Friedlander A. (2001) Microbiology: tackling anthrax. *Nature* 414, 160–161.

Good M. F. (1999) Immune effector mechanisms in malaria. *Current Opinion in Immunology* 11, 412–419.

Grandvaux N., tenOever B., Servant M. and Hiscott J. (2002) The interferon antiviral response: from viral invasion to evasion. *Current Opinion in Infectious Disease* 15, 259–267.

Grewal I. S., Borrow P., Pamer E. G., Oldstone M. B. and Flavell R. A. (1997) The CD40-CD154 system in anti-infective host defense. *Current Opinion in Immunology* 9, 491–497.

Hay S. and Kannourakis G. (2002) A time to kill: viral manipulation of the cell death program. *Journal of General Virology* 83, 1547–1564.

Hengel H. and Koszinowski U. H. (1997) Interference with antigen processing by viruses. *Current Opinion in Immunology* 9, 470–476.

Hilleman M.R. (2004) Strategies and mechanisms for host and pathogen survival in acute and persistent viral infections. *Proceedings of the National Academy of Sciences USA* 101(Suppl. 2), 14560–14566.

Hunter C. A. and Reiner S. L. (2000) Cytokines and T cells in host defence. *Current Opinion in Immunology* 12, 413–418.

Kemp A. and Bjorksten B. (2003) Immune deviation and the hygiene hypothesis: a review of the epidemiological evidence. *Pediatric Allergy and Immunology* 14, 74–80.

Kerksiek K. and Pamer E. G. (1999) T cell responses to bacterial infection. *Current Opinion in Immunology* 11, 400–405.

Klein M., Kaeser P., Schwarz P., Weyd H., Xenarios I. *et al.* (2001) Complement facilitates early prion pathogenesis. *Nature Medicine* 4, 488–492.

Lindahl G., Sjobring U. and Johnsson E. (2000) Human complement regulators: a major target for pathogenic microorganisms. *Current Opinion in Immunology* 12, 44–51.

Lingelbach K., Kirk K., Rogerson S., Langhorne J. *et al.* (2004) Molecular approaches to malaria. *Molecular Microbiology* 5, 575–587.

Louis J., Himmelrich H., Parra-Lopez C., Tacchini-Cottier F. and Launois P. (1998) Regulation of protective immunity against *Leishmania major* in mice. *Current Opinion in Immunology* 10, 459–464.

MacDonald A., Araujo M. and Pearce E. (2001) Immunology of parasitic helminth infections. *Infection and Immunity* 70, 427–433.

Maizels R. and Yazdanbakhsh M. (2003) Immune regulation by helminth parasites: cellular and molecular mechanisms. *Nature Reviews Immunology* 3, 733–744.

Malaviya R. and Abraham S. (2001) Mast cell modulation of immune responses to bacteria. *Immunological Reviews* 179, 16–24.

Mansour M. and Levitz S. (2002) Interactions of fungi with phagocytes. *Current Opinion in Microbiology* 5, 359–365.

McGuinness D., Dehal P., and Pleass R. (2003) Pattern recognition molecules and innate immunity to parasites. *Trends in Parasitology* 19, 312–319.

Mims C. (1982) "The Pathogenesis of Infectious Disease," 3rd edn. Academic Press, London.

Moll H. (2003) Dendritic cells and host resistance to infection. *Cell Microbiology* 5, 493–500.

Montagna L., Ivanov M. and Bliska J. (2001) Identification of residues in the N-terminal domain of the *Yersinia* tyrosine phosphatase that are critical for substrate recognition. *Journal of Biological Chemistry* 276, 5005–5011.

Moorthy V. and Hill A. (2002) Malaria vaccines. *British Medical Bulletin* 62, 59–72.

Morgan B. P. (1999) Regulation of the complement membrane attack pathway. *Critical Reviews in Immunology* 19, 173–198.

Mosser D. M. and Karp C. (1999) Receptor mediated subversion of macrophage cytokine production by intracellular pathogens. *Current Opinion in Immunology* 11, 406–411.

Neild A. L. and Roy C. R. (2004) Immunity to vacuolar pathogens: what can we learn from *Legionella? Cell Microbiology* 6, 1011–1018.

Neutra M. R., Pringault E. and Kraehenbuhl J.-P. (1996) Antigen sampling across epithelial barriers and induction of mucosal immune responses. *Annual Review of Immunology* 14, 275–300.

Palucka K. and Banchereau J. (2002) How dendritic cells and microbes interact to elicit or subvert protective immune responses. *Current Opinion in Immunology* 14, 420–431.

Pieters J. (2001) Evasion of host cell defense mechanisms by pathogenic bacteria. *Current Opinion in Immunology* 13, 37–44.

Plebanski M. and Hill A. V. (2000) The immunology of malaria infection. *Current Opinion in Immunology* 12, 437–441.

Ploegh H. (1998) Viral strategies of immune evasion. *Science* 280, 248–253.

Reddehase M. (2000) The immunogenicity of human and murine cytomegaloviruses. *Current Opinion in Immunology* 12, 390–396.

Reis e Sousa C., Sher A. and Kaye P. (1999) The role of dendritic cells in the induction and regulation of immunity to microbial infection. *Current Opinion in Immunology* 11, 392–399.

Rescigno M. (2002) Dendritic cells and the complexity of microbial infection. *Trends in Microbiology* 10, 425–431.

Roeder A., Kirschning C., Rupec R., Schaller M. and Korting H. (2004) Toll-like receptors and innate antifungal responses. *Trends in Microbiology* 12, 44–49.

Romani L. (1997) The T cell response against fungal infections. *Current Opinion in Immunology* 9, 484–490.

Romani L., Bistoni F. and Puccetti P. (2002) Fungi, dendritic cells and receptors: a host perspective of fungal virulence. *Trends in Microbiology* 10, 508–514.

Roulston A,. Marcellus R. and Branton P. (1999) Viruses and apoptosis. *Annual Review of Microbiology* 53, 577–628.

Sansonetti P. J. and Phalipon A. (1999) M cells as ports of entry for enteroinvasive pathogens: mechanisms of interaction, consequences for the disease process. *Seminars in Immunology* 11, 193–203.

Schluger N. and Rom W. (1997) Early responses to infection: chemokines as mediators of inflammation. *Currrent Opinion in Immunology* 9, 504–508.

Scholtissek C. (2002) Pandemic influenza: antigenic shift. In Potter C. W., ed. "Influenza." Elsevier Science B.V., Amsterdam, 87–100.

Segal S. and Hill A.V. (2003) Genetic susceptibility to infectious disease. *Trends in Microbiology* 11, 445–448.

Shen H., Tato C. M. and Fan X. (1998) *Listeria monocytogenes* as a probe to study

cell-mediated immunity. *Current Opinion in Immunology* 10, 450–458.

Spriggs M. K. (1999) Shared resources between the neural and immune systems: semaphorins join the ranks. *Current Opinion in Immunology* 11, 387–391.

Stenger S. and Modlin R. L. (1998) Cytotoxic T cell responses to intracellular pathogens. *Current Opinion in Immunology* 10, 471–477.

Szabo S., Sullivan B., Peng S. and Glimcher L. (2003) Molecular mechanisms regulating Th1 immune responses. *Annual Review of Immunology* 21, 713–758.

Szomolanyi-Tsuda E. and Welsh R. M. (1998) T-cell-independent antiviral antibody responses. *Current Opinion in Immunology* 10, 431–435.

Tatis N., Sinnathamby G. and Eisenlohr L. (2004) Vaccinia virus as a tool for immunologic studies. *Methods in Molecular Biology* 269, 267–288.

Taylor G., Feng C. and Sher A. (2004) p47 GTPases: regulators of immunity to intracellular pathogens. *Nature Reviews Immunology* 4, 100–109.

Tortora G., Funke B. and Case C. (2004) "Microbiology: An Introduction," 7th edn. Benjamin Cummings, San Francisco.

Tortorella D., Gewurz B. E., Furman M. H., Schust D. J. and Ploegh H. L. (2000) Viral subversion of the immune system. *Annual Review of Immunology* 18, 861–926.

Turner M.W. (2003) The role of mannose-binding lectin in health and disease. *Molecular Immunology* 40, 423–429.

Urban B. and Robers D. (2002) Malaria, monocytes, macrophages and myeloid dendritic cells: sticking of infected erythrocytes switches off host cells. *Current Opinion in Immunology* 14, 458–465.

Wold W. S., Doronin K., Toth K., Kuppuswamy M., Lichtenstien D. *et al.* (1999) Immune responses to adenoviruses: viral evasion mechanisms and their implications for the clinic. *Current Opinion in Immunology* 11, 380–386.

Wurzner R. (1999) Evasion of pathogen by avoiding recognition or eradication by complement, in part via molecular mimicry. *Molecular Immunology* 36, 249–260.

Yanagi Y. (2001) The cellular receptor for measles virus—elusive no more. *Reviews in Medical Virology* 11, 149–156.

Vaccines and Clinical Immunization

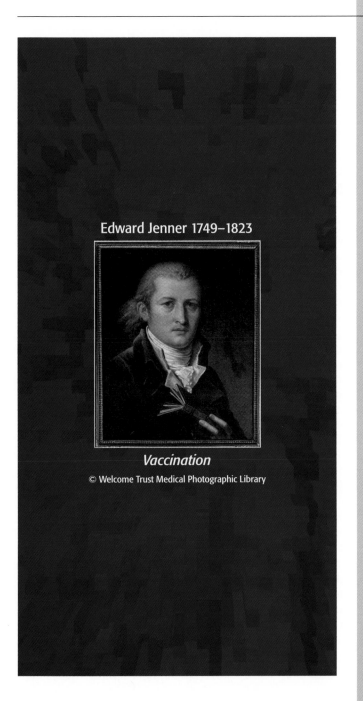

Edward Jenner 1749–1823

Vaccination

© Welcome Trust Medical Photographic Library

CHAPTER 23

A. VACCINATION: PUBLIC HEALTH SUCCESSES AND CHALLENGES

B. HISTORICAL NOTES

C. GENERAL PRINCIPLES OF VACCINE DESIGN

D. TYPES OF VACCINES

E. FACTORS AFFECTING VACCINATION

F. PROPHYLACTIC VACCINES

G. THE "DARK SIDE" OF VACCINES

H. PROPHYLACTIC VACCINES OF THE FUTURE

I. PASSIVE IMMUNIZATION

J. THERAPEUTIC VACCINES

"I hope that some day the practice of producing cowpox in human beings will spread over the world. When that day comes, there will be no more smallpox"—Edward Jenner

"Immunization" in the clinical sense is the process of manufacturing immune defense, that is, artificially helping the body to defend itself using the effectors and mechanisms of the immune system. Clinical immunization may be either *active* or *passive*. In *active* immunization, which we commonly call "vaccination," a modified form of the natural immunogen (the vaccine) is administered so that the host develops specific B and T memory cells that can act against the natural immunogen. These cells stand ready to be activated should the host later be exposed to the pathogen bearing the natural immunogen (Fig. 23-1). *Passive* immunization describes the situation in which exogenous antibodies directed against an antigen are administered to an individual exposed to the pathogen in question. Immediate protection is provided without the lag necessary for an immune response, but because the individual's own immune system has not been stimulated, no memory is generated.

The study of vaccines and vaccination, which involves aspects of immunology, microbiology, and cell biology, is called *vaccinology*. In most instances, we administer vaccines *prophylactically* in an effort to pre-empt disease. Because the vaccine has already induced a primary response, a subsequent natural exposure to the pathogen in question will protect the individual by inducing secondary response levels of antibodies and effector cells rather than primary response levels. *Therapeutic* vaccination is the administration of a vaccine to an individual to alter a pre-existing condition. For example, an individual suffering from a chronic infection might be vaccinated against the pathogen in an effort to give the immune system the extra boost it needs to finally clear the infection. We start this chapter with two sections placing vaccination into perspective, the first in modern times and the second, historically.

A. Vaccination: Public Health Successes and Challenges

Why vaccinate for prophylaxis? The commercialization of antibiotic use in the mid-20th century led to the mitigation of much bacterial disease in the developed world. However, drug-resistant strains of these pathogens are on the rise, and many infectious diseases are caused by classes of pathogens (such as viruses) that are not affected by antibiotics. In addition, in the developing world, access to sufficient amounts of necessary antibiotics remains problematic. Prophylactic vaccination has the potential to alleviate these concerns. For an individual, vaccination means diseases induced by pathogens may be prevented entirely or at least mitigated. At the population level, the pathogen causing a given disease may be eradicated, since few vulnerable hosts are available in which the pathogen can multiply. Today's best known vaccination success story is the campaign of the World Health Organization (WHO) to eradicate smallpox globally. Edward Jenner's discovery of the smallpox vaccine in 1798 led rapidly to smallpox immunization and control in most developed countries. However, 160 years later, smallpox was still a severe problem in many developing countries: 10 million cases and 2 million deaths per year worldwide were reported in the early 1960s. In 1967, the WHO launched the $300 million global smallpox eradication program which involved 200,000 health workers in 70 countries working for 10 years to vaccinate the world's population in its remotest corners. In 1976, a reward of $1000 was offered to anyone identifying a new case of smallpox. Over the next 3 years, only one case was recorded, leading to the declaration in 1980 that smallpox had been officially eradicated (Plate 23-1). A similar global

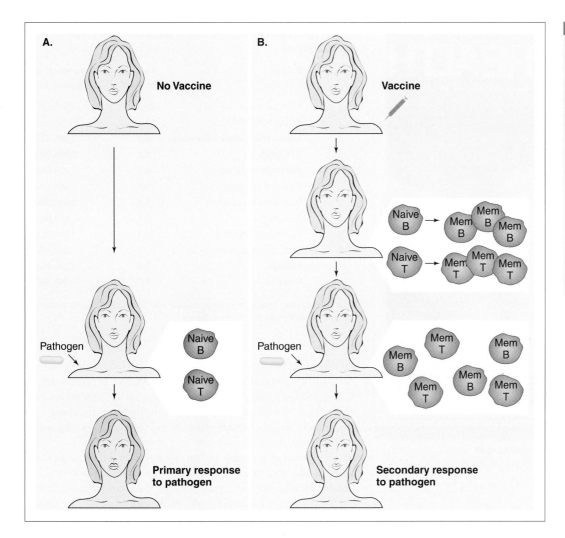

Figure 23-1

The Principle of Vaccination
In individuals who do not receive a vaccine (A), only naive B and T cells are present to combat an infecting pathogen. A primary immune response is mounted that is not usually sufficient to prevent disease in the infected individual. In individuals who receive a vaccine (B), the naive B and T cells are activated in advance of pathogen exposure, generating memory B cells and memory T. If pathogen exposure occurs, a secondary immune response is mounted that efficiently clears the pathogen before it can cause disease.

immunization program against polio virus is currently pushing this pathogen toward extinction. In addition, in the 200 years since Jenner's work with smallpox paved the way for vaccination to become a public health procedure, at least part of the world has seen the control of several other serious diseases, including diphtheria, tetanus, pertussis, measles, mumps, rubella, *Haemophilus influenzae* type b, and yellow fever. Vaccination is thus one of the world's few public health success stories, involving the application of research to the practical protection of many millions of the earth's inhabitants against devastating diseases.

Current vaccines are both clinically efficacious and cost-effective. For example, in the United States in 1920–1922 (the 3 years prior to the development of the diphtheria vaccine), there were on average 176,000 recorded cases of diphtheria per year. In 1998, there was one case. Similarly, the average annual number of cases of paralytic polio in the United States in 1950–1954 (prior to vaccine development) was about 16,000. In 1998, not a single case was recorded. While some may argue about the cost of mass vaccination programs, it should be noted that the cost of a population becoming infected is sure to be much higher. According to the WHO,

for every U.S. $1 million spent on global childhood vaccines, U.S. $4 million in long-term health care savings is realized. In the case of smallpox, not only the cost of infection but also the cost of universal vaccination itself has been eliminated by the eradication of the virus. The WHO has estimated that the investment in smallpox eradication, estimated to be about $300 million, is recouped every 26 days.

Costs and, more importantly, lives are saved with other vaccines as well. A single dose of oral polio vaccine costs about U.S. $0.10, and the total cost of vaccinating a child in a developing country, including transportation of clinic workers and the refrigeration of vaccines, is U.S. $1.00. Once polio virus has been globally eradicated, the annual savings gained by not having to vaccinate or care for and rehabilitate polio victims is projected to reach billions of dollars. Similarly, it has been estimated that every U.S. $1.00 spent on measles vaccine translates into health care savings of U.S. $13.50, and an average of U.S. $117 in direct medical costs is saved for each elderly person vaccinated against influenza. Vaccination can also have an impact on workforce productivity. One 1990 study of chicken pox infection in the United States estimated that, if the cost of the lost work time of parents caring for

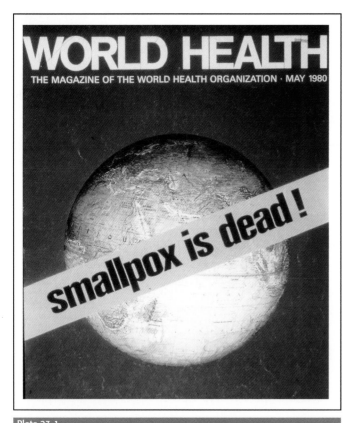

Plate 23-1

WHO Declaration of Smallpox Eradication in 1980
Courtesy of the Magazine of the World Health Organization.

Table 23-1 Global Annual Leading Causes of Death (WHO 2002 figures)

Rank	Cause	Percentage of Total	Number of Deaths
1	Cardiovascular disease	29.3	16,585,000
2	Malignant neoplasms	12.6	7,115,000
3	**Lower respiratory infections**	**6.8**	**3,871,000**
4	**HIV/AIDS**	**5.1**	**2,866,000**
5	Pulmonary disease	4.7	2,672,000
6	Perinatal conditions	4.4	2,504,000
7	**Diarrheal diseases**	**3.5**	**2,001,000**
8	**Tuberculosis**	**2.9**	**1,644,000**
9	**Childhood diseases***	**2.3**	**1,318,000**
10	Road traffic accidents	2.1	1,194,000
11	**Malaria**	**2.0**	**1,124,000**
12	Diabetes	1.6	895,000

* The WHO includes diphtheria, pertussis, tetanus, poliomyelitis, and measles in this category. Entries in bold are infectious diseases.

infected children was factored in, each U.S. $1.00 spent on universal vaccination of children against chicken pox would save U.S. $5 in total costs associated with the infection.

Despite these "good news stories," there are still major challenges in vaccinology, especially in the developing world. In 1974, only 5% of the world's children were being vaccinated against the six infectious diseases considered to be the biggest childhood killers: diphtheria, measles, pertussis (whooping cough), polio, tetanus, and tuberculosis. This number had increased to 60% by 1988 and to 80% by the late 1990s. However, of the 20% of children who are not vaccinated (mostly in the developing world), over 1 million die annually from these diseases. In some cases, logistical and social factors prevent the delivery of life-saving vaccines. For example, an excellent vaccine to combat measles exists, but because of difficulties in getting the vaccine to those who need it, hundreds of thousands of children still die of measles each year. Indeed, communicable diseases represent six of the top 12 causes of mortality worldwide, and account for over 20% of all the world's annual deaths (Table 23-1).

Progress on vaccine development and programs to combat diseases of the developing world such as malaria and African sleeping sickness has been frustratingly slow, partly because of economic factors and partly because of the slippery nature of the pathogens involved. The impact of these scourges should not be underestimated: malaria kills one child in Africa every 30 seconds and costs the African economy about U.S. $2 billion each year. In addition, HIV remains an elusive target worldwide, resisting the efforts of thousands of researchers to develop an effective vaccine, and leaving much societal and economic devastation in its wake. Close to 90% of HIV-positive persons go on to develop AIDS, indicating that, in the vast majority of cases, the immune response to natural HIV infection cannot prevent the virus from establishing a permanent foothold in the body (see Ch.25). The design of a vaccine that can induce a response more effective than the natural response, especially when the virus targets the all-important $CD4^+$ T cells, is a monumental task. Another relatively new pestilence is the hepatitis C virus (HepC), which causes liver failure and hepatic carcinoma. The HepC virus is extremely difficult to grow in culture, limiting the amount of pathogen material that can be generated for study. In addition, HepC does not replicate in mice so there is no small animal model for the disease. Thus, despite the intense efforts of a multitude of researchers worldwide, the development of a viable HepC vaccine remains elusive.

B. Historical Notes

Prior to the development of vaccines and antibiotics, infectious diseases were the leading killers of humans in every corner of the globe. The spread of "germs" was aided by a limited knowledge of the benefits of sanitation, so it was not uncommon for local or even global epidemics of diseases such as smallpox, diphtheria, and polio to take a horrendous toll. The first recognized instance of an effort to prevent disease by introducing a disease agent into a healthy individual was the variolation process described at the very beginning of this book. In use prior to 1796, variolation involved inoculating healthy individuals with material from smallpox lesions of infected patients, in an attempt to prevent the healthy individuals from contracting smallpox. Variolation was sometimes fatal to the inoculated person, but offered a 10-fold lower risk of dying than did a naturally contracted smallpox infection. In the 1790s, Jenner noticed that milkmaids, who were necessarily exposed to the cowpox virus when it attacked their charges, had a greatly reduced incidence of smallpox. Jenner then invented the first vaccination procedure in which smallpox was prevented by pre-emptively infecting a person with live cowpox virus. Of course, Jenner did not understand the mechanism underlying the success of his vaccine: the cross-reaction of anti-cowpox antibodies with epitopes on the smallpox virus. Nevertheless, the subsequent success of vaccination in preventing the devastation previously caused by smallpox epidemics convinced several European governments to make smallpox vaccination mandatory in the early 1800s.

In 1876, Robert Koch showed that inoculation of the bacterium *Bacillus anthracis* into an animal led to the disease anthrax, laying the groundwork for the germ theory of disease. In 1880, Louis Pasteur expanded this concept, and experimented with various pathogens in an attempt to replicate the vaccination protocol that had worked for smallpox. However, unlike the smallpox virus, not many other pathogens appeared to have cross-reactive counterparts (like the cowpox virus) that could infect the host and establish protection against the primary pathogen, but not cause disease themselves. Instead, Pasteur serendipitously stumbled on ways to *attenuate* a pathogen, i.e., treat the pathogen so that it remained viable (and immunogenic) but had lost its capacity to cause disease. For many pathogens, attenuation results when the pathogen is maintained in an alternative host species in which it is less virulent, or when the pathogen is cultured in the laboratory under sub-optimal growth conditions. With the emergence of these technologies, it became possible in 1881 to make an anthrax vaccine for animals, and in 1885 to produce a rabies vaccine for humans. By growing the rabies virus (which is highly virulent in humans) in rabbits, a less virulent form of the live virus was generated that could be used for vaccination. At about the same time, it was shown in birds that the vaccine organism did not have to be alive to achieve effective vaccination. Cholera bacteria that had been killed by exposure to very high temperatures still induced effective immunity in avian hosts subsequently challenged with the live pathogen. By the late 1880s, work with diphtheria infection had shown that even isolated pathogen products could induce immunity.

The definition of the nature of immune responses to microbes continued into the early 20th century. The bacillus Calmette-Guérin (BCG) strain of *Mycobacterium bovis* used as a vaccine for tuberculosis was discovered in the early 1900s. Also during this period, Emil von Behring and Shibasaburo Kitasato transferred serum from immune individuals to healthy recipients and found that protection against infection could sometimes be conferred for short periods. Unbeknownst to these researchers, their experiments were the first successful attempts at what later came to be known

as passive immunization, since the transferred immune serum contained specific antibodies. In the 1920s, pathogens inactivated by chemical treatment were found to induce immune responses, leading to the production in 1925 of a vaccine against diphtheria. Also in 1925, the first *toxoids* to protect against diphtheria and tetanus were devised. (Toxoids, which no longer cause disease but retain their immunogenic epitopes, are created by chemically inactivating the disease-causing toxins of bacteria with formalin.) In addition to determining what parts and products of a pathogen could be used as vaccines, researchers grappled with the problem of how to obtain vaccine materials from pathogens that could not be isolated in large amounts or cultured in the laboratory. In 1931, the technique of growing viruses in chick embryos (within the egg) was established, a method that remains in use today for many viruses. This breakthrough allowed the culture of sufficient virus materials to make a yellow fever vaccine in 1937 and an influenza vaccine in 1943. In addition to growth in embryonated hen's eggs, certain cell types such as chick embryo fibroblasts and rabbit kidney cells were found to support the growth of various viruses in tissue culture. For example, the vaccinia virus strains used for the smallpox eradication campaign in the 1960s and 1970s were both *in vivo* on live calf skin, as had been the primary method for many years, and expanded in bulk in hen's eggs and rabbit kidney cells.

By the late 1940s, clinical scientists had a wealth of appropriate vaccine materials to administer; however, there was still one hurdle to overcome. Researchers needed to find an effective means of delivering the vaccines into hosts so as to provoke protective immune responses. Injection of a modified pathogen alone into a host animal resulted in only low, transient levels of antibody production. It became apparent that the vaccine material was not being retained long enough in the body to induce a sustained immune response. To slow down dispersal of the vaccine, researchers mixed candidate vaccine immunogens with various particulate substances that acted as adjuvants. These mixtures were then injected into animals and the strength of the ensuing immune response and the incidence of adverse effects were monitored. The results were both good and bad. Many adjuvants promoted powerful immune responses to the test vaccines, but significant adverse reactions (particularly at the injection site) were also observed, making these adjuvants unsuitable for use in humans. Only a preparation of alum mixed with a vaccine gave an adequate immune response in animals with minimal adverse effects. Alum thus became the first adjuvant used in humans to enhance vaccination and remains the most widely used today, although other substances and methodologies have since been developed (see later in this chapter).

The 1950s saw the launch of large-scale culture methods for growing viruses *in vitro*, making it feasible to contemplate mass vaccination programs requiring large quantities of vaccine. Accordingly, in 1959, the World Health Assembly approved the plan described previously to eradicate the smallpox virus globally. A concentrated push in the 1970s saw the development of practical means for vaccinating individuals in isolated locations. Progress on the fight against other infectious diseases also took off during these three decades. Two types of vaccines to combat polio were developed in the mid-1950s: the *Salk vaccine*, which requires subcutaneous injection of killed polio virus, and the *Sabin vaccine*, which is a live, attenuated virus that can be delivered orally. By the late 1960s, vaccines against measles, mumps, and rubella were widely available in the developed world. In the late 1970s, researchers came up with a live, attenuated vaccine for typhoid fever that could be administered orally.

With the revolution of recombinant DNA technology in the 1980s, new approaches to developing vaccines became possible. Where pathogen protein isolation was a problem, the gene could be cloned and the protein produced using bacterial and yeast expression systems. For example, the hepatitis B virus cannot be grown in culture. However, recombinant DNA technology was used to produce a hepatitis B vaccine in 1986 and programs to mass-vaccinate infants and teenagers in the developed world are under way. Advances in molecular

chemistry solved other production problems. For example, the development of a vaccine against *Haemophilus influenzae* was stymied for a long time by the fact that the dominant epitope, called the *B antigen*, is a Ti-immunogen, meaning that no lasting memory is developed in response to B antigen administration. In 1988, a method of creating a covalently bonded conjugate between the carbohydrate epitopes of B antigen and a protein carrier was devised. The carrier half of this vaccine conjugate induces a Td response to B antigen and the generation of memory cells. In the last 15 years, numerous technologies have been developed to advance vaccine design and production. Site-directed mutagenesis of genes and improved transfection protocols have given rise to new methods of pathogen attenuation and vaccine delivery. The emerging sciences of genomics, transcriptomics, and proteomics (see Box 23-1) have made possible the predictive identification of epitopes that are most likely to induce protective responses. Edible vaccines, which need no injection,

---

Box 23-1. Genomics, transcriptomics, and proteomics as tools for vaccine development

Vaccine development has always followed the introduction of technological innovations. In this new millennium, we are poised for another revolution in vaccine design predicated on the prior identification of epitopes that induce protective immune responses against pathogens ("protective epitopes") using the new sciences of *genomics*, *transcriptomics*, and *proteomics*.

Genomics is the name given to the study of the entire genome of an organism, made feasible by recent monumental feats of gene sequencing in several organisms (including humans). Both coding and regulatory sequences can be studied, and how these sequences are organized and interact can be analyzed. Predictions can be made about the surface, cytoplasmic, or nuclear localization of a gene product, whether it is conserved among species, what sort of biological activity it might have based on putative domain structures, and where it is likely to be expressed *in vivo*. Transcriptomics is the study of all transcripts expressed by an organism. *DNA microarrays*, in which hundreds of cDNA copies of known transcripts are fixed onto a small glass chip, can identify and determine relative concentrations of transcripts expressed by an organism through hybridization experiments (see Plate). This method can be used to compare the transcripts expressed by virulent and benign strains of an organism, and to define specific genetic adaptations that have conferred virulence. Proteomics is the study of all the proteins

encoded in the genome of an organism, while "expression proteomics" is the study of all proteins actually expressed by an organism (usually determined by 2D gel electrophoresis and mass spectrometry). More than one protein may be derived from a single gene (by alternative mRNA splicing or post-translational modification) so that "knowing the genes" does not guarantee "knowing the proteins." Advances in technology allowing one to manipulate, extract, and identify proteins more easily are accelerating the identification of proteins in complex mixtures such as cell walls and membranes. The systematic analysis of the large bodies of information generated by these new sciences has been made possible by the development of sophisticated computer programs that can rapidly and automatically look for particular sequence motifs in a gene, or compare a novel protein sequence with others containing known functional domains.

Genomic, transcriptomic, and proteomic analyses of various pathogens have been undertaken in an effort to identify sequences that may represent novel protective epitopes. When found, these epitopes may form the basis for new vaccines against old foes such as malaria and TB. Such new vaccines will be needed to counteract the ever-increasing antigenic variation undertaken by pathogens, and the emergence of pathogen strains that are resistant to all known antibiotics. Furthermore, genomics has given us tools to search for antigens that can be used against

organisms that are hard to grow in the laboratory (such as *M. leprae*), or exist in strains that fall outside the purview of the existing vaccine (such as strains of *H. influenzae* B that lack polysaccharide capsules).

How does one find a candidate epitope in the sea of nucleotides that makes up a complete genome? How does one validate such an antigen once it is identified? Often the first step toward identifying a potential epitope is to search the genome for sequences that show homology to known immunogens in other strains or species, or are already components of an existing vaccine. Much success has been gained through these types of analyses of bacterial genomes. Bacterial genomes are relatively small (about 550–9200 kb), lack introns, and are easily mapped. Since antibodies recognize surface structures, examining the genome for sequences compatible with a localization on the cell surface may reveal a potential B cell epitope. The presence of a hydrophobic sequence, a transmembrane domain, or a glycosylation signal sequence can indicate a surface protein. Secretion signals may identify secreted antigens such as toxins that could be used to develop toxoid vaccines. T cell epitopes may be represented in potential antigens bearing peptide sequences known to bind to MHC molecules. In another approach, researchers compare gene expression patterns between an attenuated strain of a pathogen and its pathogenic parent. Genes whose expression

Continued

Box 23-1. Genomics, transcriptomics, and proteomics as tools for vaccine development—*cont'd*

is associated with pathogen virulence can be identified using these methods, giving the researcher more targets for attenuating mutations. Still another approach called *saturation mutagenesis* identifies virulence genes by systematically mutating each and every gene of a pathogen and testing the mutant for pathogenicity.

Proteomics has also provided several interesting insights into potential vaccine epitopes. Because of alternative splicing, alternative translation frame and post-translational modifications such as glycosylation, more than one protein may be derived from the same gene. As well, the same organism cultured under different conditions may produce different subsets of proteins. Recent advances in protein microsequencing have accelerated the identification of potential vaccine epitopes among all the potential candidates. In addition, the preferred peptide sequences bound by various MHC molecules have been characterized and mathematical models constructed to assess how well an unknown peptide might fit into the binding pocket of a particular MHC protein. Starting from a given protein sequence, all theoretical peptides can be identified using computer analysis, and the degree to which these peptide sequences match the "ideal" peptide can be calculated. These scores have been shown to correlate well with the binding affinity of known ligands for a particular MHC molecule. This technology can also be used to screen for proteins

(and thus peptides) that are differentially displayed by pathogenic and non-pathogenic strains of an organism, identifying potential T cell epitopes associated with pathogenicity.

How does one know that a viable vaccine candidate has been isolated? To test the epitope for its ability to induce protective immunity, a candidate gene encoding the protein of interest is expressed in *E. coli*, and the protein is purified and injected into experimental animals as a vaccination. At the appropriate time, the experimental animals are challenged with the pathogen and the development of infection and/or immune responses is monitored. The reader can see

that this *in vivo* testing for efficacy is still a labor-intensive, time-consuming process, but it is currently the only way to be sure that the epitope is indeed protective. Much current research is devoted to creating *in vitro* systems that will reliably identify protective epitopes.

With the new tools of genomics, transcriptomics, and proteomics, and the development of rapid ways to assess the biological relevance of an epitope, the process of vaccine antigen discovery should be accelerated. The rational design of new prophylactic and therapeutic vaccines will then perhaps become almost routine in the future.

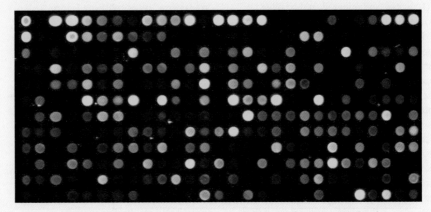

Gene expression in the liver of a rat treated with a toxic agent has been compared to that of an untreated control. Red signals indicate that the expression of the gene in that position on the gene chip is upregulated in the treated liver, while green signals indicate downregulated gene expression. Yellow signals indicate unchanged gene expression. Courtesy of Dr. Leming Shi, www.Gene-Chips.com.

cooking, or sophisticated storage, are in clinical trials (see Box 23-2), and new adjuvants and delivery vehicles that will make vaccination more efficient and less stressful are becoming available.

One aspect of vaccinology that is peculiar to our modern age is the economics of vaccine development and production. Some commentators have estimated that the cost of bringing a vaccine from the design stage through animal experiments and clinical trials and finally to the commercial market is U.S. \$200–400 million. Such an outlay means that most vaccines are made by only a very few companies in the Western world. Naturally these companies want a return on their investment, meaning they seek a large market for a fairly high-priced final product. Thus, diseases that affect only small numbers of people, or only people living in poorer parts of the world, are of considerably less interest to these manufacturers. In the interest of compassion, tiered pricing of certain

vaccines now exists, meaning that people in developing countries can pay substantially less for life-saving vaccines than people in the developed world. In 2000, the Global Alliance for Vaccines and Immunization (GAVI) was established to make new and future vaccines available to everyone, regardless of economic circumstances. In particular, GAVI has targeted the development of vaccines for HIV/AIDS, malaria, and TB as priorities. In a sign of more globally responsible times, Western private and corporate entities, as well as local and international politicians, have stepped forward to support GAVI's aims. In addition to GAVI, the Global Fund for Children's Vaccines (originally funded by the Bill and Melinda Gates Foundation) provides resources to build better immunization infrastructure where it is needed and to supply existing vaccines that have not been adequately distributed in the past. With these laudable initiatives, the complete eradication of many devastating diseases may finally be within our grasp.

Box 23-2. Edible vaccines

One of the greatest difficulties in vaccinating remote populations in the world is the preservation of the vaccine in an efficacious form. It is often difficult to maintain the "cold chain" (constant refrigeration of a labile vaccine) all the way to the intended vaccinees' villages. The other major stumbling block is the prohibitive cost of many vaccines for developing countries. To address both these difficulties, Charles Arntzen and colleagues at Texas A&M University in the United States pioneered the concept of *edible vaccines* in the early 1990s. Edible vaccines are produced within plants that are grown cheaply in the local area and can be processed into a form of food. Plants require only sunlight, nitrogen, CO_2, water, and minerals in the soil to grow, making them very cheap as a means of large-scale manufacturing compared to the fermentation facilities required to grow large quantities of recombinant yeast. Using genetic engineering techniques, transgenes for mammalian vaccine antigens can be introduced into the genomes of plants such as potatoes, bananas, lettuce, tomatoes, rice, wheat, soybeans, and corn. Plant regulatory sequences and/or expression vectors based on plant plasmids and plant viruses are used to ensure efficient expression of genes of the mammalian pathogen in the plant environment. If the transgenic plant synthesizes the vaccine protein in its edible parts, the protein is delivered to the vaccinee by the gut mucosal route when the raw plant is eaten. (The plant must be raw because cooking denatures the plant proteins, including the immunogen.)

Plants are particularly well-suited for the generation of edible vaccines for several reasons: (1) The eating of local plants obviates the need for sterile syringes and temperature-controlled transportation arrangements. (2) Transgenic plants can theoretically be grown year after year to generate more vaccine without the need for a manufacturing facility or the repeated purchase of seeds or plants. A cheap, local source of vaccine can thus be established wherever the plant can grow. (3) Being eukaryotes, plant cells faithfully express and appropriately glycosylate mammalian proteins. (4) Plant cells have thick walls that slow down the action of degradative enzymes in the mammalian gut. Vaccine immunogens contained in such cells are thus gradually released into the gut as the plant

cell wall is broken down, allowing a sustained delivery of antigen. (5) There is no chance of animal viruses contaminating the vaccine preparation, a constant worry with vaccines produced in human or animal cells or from blood.

The laboratories of Arntzen and others have fed experimental animals plant vaccines in which the genes encoding antigens from enteropathic *E. coli*, *V. cholerae*, or the Norwalk virus (all of which cause severe diarrhea) have been integrated into the genomes of tobacco, tomato, or potato plants. (Integration of exogenous genes into plant chromosomes is usually carried out using a naturally occurring soil bacterium called *Agrobacterium tumefaciens* that "transforms" plant cells; see Figure. The whole process takes anywhere from 3 to 9 months.) In several experiments, raw plant preparations fed to experimental animals have induced both systemic and mucosal immune responses to the original pathogens or their toxins. For example, mice fed extracts of tobacco leaves expressing the H protein of measles virus mounted serum and mucosal antibody responses that were able to neutralize measles virus *in vitro*. Buoyed by these successes in the mid-1990s, clinical trials were begun with human volunteers in the late 1990s, with encouraging results. For example, volunteers who ate raw potato transgenic for a component of the *E. coli* enterotoxin developed mucosal and systemic immune responses to the toxin. A hepatitis B antigen and a human CMV antigen have been expressed in tobacco, and a rabies antigen in tomatoes. Edible vaccines are now under construction to prevent several other animal and human disorders, including cholera and foot-and-mouth disease.

Besides their convenience, ease of administration, and indifference to cold chain considerations, a plant can be engineered to synthesize antigens from more than one transgene, meaning that several vaccine antigens might be delivered at one time by simply eating a hearty meal of the plant. For example, in one study, the efficacy of an edible combination vaccine in protecting mice from enteric disease was demonstrated. Transgenic potatoes expressing a stable oligomeric structure containing both the cholera toxin B subunit and a rotavirus epitope were fed to mice. When the mice were later challenged with cholera toxin and rotavirus, protective immune responses to

both entities were observed. A similar approach might be feasible in the future to feed multiple childhood vaccines to young children.

Despite their promise, several challenges remain to be solved with edible vaccines: (1) Very low concentrations of vaccine antigen are produced per transgenic plant, meaning that enormous quantities of a food might have to be consumed before an immunogenic dose was achieved, and frequent "boosters" would be required to maintain immunity. (2) Vaccine antigen concentration varies from plant to plant, making it difficult to calibrate a dose. (3) The production of large quantities of a foreign protein often stunts the growth of a plant. (4) Many ingested antigens promote oral tolerance to the antigen, not an immune response. All of these difficulties can theoretically be overcome using a variety of strategies: (1) Positive regulatory elements can be engineered into the plant genome. (2) Mammalian elements that interfere with gene expression in plants can be removed. (3) A gene encoding an adjuvant can be added. (4) A vaccine may be directed to the M cells of the host via co-transfection of the vaccine gene with a gene encoding a protein that enhances adhesion to M cells (such as the cholera toxin B subunit). (5) Regulatory elements that either inhibit synthesis of the vaccine antigen until the plant has reached a certain stage in its growth or respond to an external stimulus (like a growth factor) could be employed to encourage normal plant growth. (6) Plant viruses engineered to express the mammalian gene of interest can be used to produce up to 2 g/kg plant tissue of foreign protein, although the process involves the additional step of infecting the plant with the vaccine virus. Further purification to remove viral components is also required before serving.

In addition to the preceding considerations, each type of plant has its own advantages and disadvantages. For example, bananas can be eaten raw and are palatable to most people but the fruit rots relatively quickly. Banana plants also take some years to mature to the point of producing bananas. Tomatoes produce fruit much faster but they too spoil quite rapidly, as does lettuce. While the tobacco genome is well-known and the technology is in place to engineer these plants, tobacco is not generally eaten by humans and contains toxic substances such as nicotine. Exogenous

Continued

Box 23-2. Edible vaccines—*cont'd*

proteins expressed in rice grains are generally stable at room temperature, and rice is easy to store and transport. However, rice plants are slow to grow and can require greenhouse conditions in some locales. Potatoes are easily propagated and are resistant to spoilage but are not usually eaten raw. Corn is widely grown and has seeds that can be shipped over long distances at ambient temperature but the kernels require some cooking before consumption. Some of these difficulties may be circumvented by drying a fruit quickly after ripening, or partially cooking a transgenic potato so that it tastes more like dinner but maintains the immunogenicity of the vaccine antigens. Another approach may be to modify bacteria that occur naturally in human foods such that they express vaccine antigens. For example, *Lactobacillus* is a bacterial species that is used to make yogurt. When mice were intranasally vaccinated with a *Lactobacillus* strain engineered to express tetanus toxin antigens, mucosal responses and protection against a subsequent challenge with the toxin could be demonstrated. It remains to be seen whether oral consumption of yogurt vaccines will be effective.

One of the more intangible difficulties with edible vaccines is that these genetically engineered plants are negatively viewed by some as "frankenfoods," or *genetically modified organisms* (GMOs) that may be harmful. Of course, these types of plants should be grown under strictly controlled conditions that limit their unintended spread. One technology that may alleviate concerns about the latter possibility is *chloroplast transformation*. Like mitochondria in mammals, chloroplasts in most plant species contain their own genome and are inherited maternally. Thus, exogenous genes introduced into the chloroplast genome stay with the transgenic plant and are not packaged and distributed in its pollen. The risk of

transmission of the transgene beyond its prescribed borders is thus substantially reduced. Hopefully, sufficient clinical trial data can soon be accumulated that will demonstrate the efficacy and safety of

edible vaccines, allowing us to finally achieve the worthy goal of vaccinating all the world's children against a wide spectrum of devastating diseases both cheaply and painlessly.

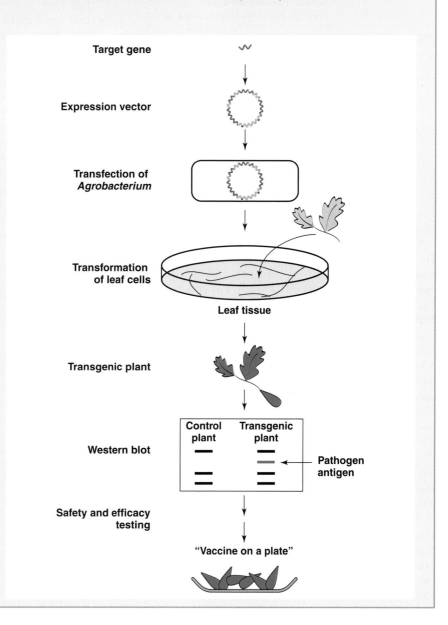

C. General Principles of Vaccine Design

In this section, we discuss what makes a good vaccine (Table 23-2). Because prophylactic vaccines are given to healthy people, usually the children of a population, it is extremely important that vaccines be both efficacious and safe. We start with a molecular description of what a vaccine is supposed to do.

I. BIOLOGICAL PURPOSE OF A VACCINE

The purpose of a vaccine is to stimulate a vigorous Td immune response that generates long-lasting B and T cell memory capable of counteracting the pathogen of interest. The vaccine should therefore contain a version of the pathogen or one of its epitopes that is immunogenic but not pathogenic, meaning that the vaccine induces an immune response in the host that is effective against the viable pathogen but does not itself

Table 23-2 Characteristics of a Successful Vaccine

Characteristic	Description
Pathogen	Undergoes little antigenic variation Does not have environmental or animal reservoir Causes acute rather than chronic disease Natural exposure leads to immunity Does not attack cells of the immune system
Vaccine efficacy	Stimulates vigorous Td responses leading to T and B memory Stimulates appropriate Td response; i.e., antibodies for extracellular pathogens, CTLs for intracellular pathogens Coverage of vaccinated population is 80–95%
Vaccine safety	No risk of causing disease Side effects do not outweigh severity of natural disease

cause disease in the host. At the very least, the health benefit of using the vaccine must outweigh the risk involved in its use. To be effective, the immune response induced by the vaccine must be appropriate for the elimination of the pathogen. For example, an extracellular pathogen is best countered by antibody, so that the vaccine must be capable of activating B cells. Furthermore, the antibodies produced must be of an isotype known to be effective against the particular pathogen: secretory IgA protects mucosal surfaces but IgG is needed to immobilize pathogens in the circulation. Similarly, if a pathogen is intracellular, it is vital that $CD8^+$ T cells be activated by the vaccine. As introduced in Box 23-1, epitopes that induce an effective immune response against a pathogen are known as *protective epitopes* and are derived from *protective antigens*. Although analysis of gene and protein sequences may give hints as to likely candidate epitopes, protection cannot be gauged from nucleotide or amino acid sequences alone. We reiterate that there is currently no reliable way to assess whether an epitope is protective or not other than by testing *in vivo*.

The pathogens most easily controlled by vaccination are those that infect only a single host species, meaning that they cannot escape into another species or an environmental "reservoir." A reservoir in this sense is a species or environmental niche outside of the human population in which a pathogen can survive. Pathogens that do not exhibit a high degree of antigenic variation are also good candidates for a successful vaccination program since memory B and T cells and their antibodies and effectors continue to recognize the pathogen in successive exposures. A vaccine is also more likely to be effective if the disease caused by the pathogen is acute rather than chronic, and if natural exposure to the pathogen induces robust immunity in an individual.

II. VACCINE EFFICACY AND SAFETY

A vaccine must be efficacious; that is, it must be effective in protecting the vaccinee from disease. Ideally, an animal model

that mimics the disease in humans should be available to judge the efficacy of the candidate vaccine prior to human testing. Efficacy in humans or animals is often judged by *seroconversion*, defined as a measurable increase in the serum level of specific anti-pathogen antibody that occurs after vaccination. However, seroconversion does not always correlate with protection from disease, especially for pathogens best combated by cell-mediated responses. It is also possible to judge whether a vaccine has worked by examining the immune response in the vaccinee for secondary response characteristics, such as the presence of IgG antibodies and/or the more rapid resolution of the infection. In population terms, the efficacy of a vaccine is often expressed as the percentage of individuals vaccinated who do not experience disease after exposure to the pathogen. No human vaccine has yet proved 100% effective due to the inherent genetic variation in human beings, but most childhood vaccines are effective in 80–95% of a given population; that is, the *coverage* of these vaccines is 80–95%. It is here that the concept of *herd immunity* comes into play. If a pathogen cannot gain a foothold in the majority of a population because these individuals respond to the vaccine, even those members of the population who are not vaccinated or who do not respond to the vaccine will be protected because the overall chances of any one individual being exposed to the pathogen are much reduced. More on herd immunity appears later in this chapter.

In addition to human variation, pathogen variation affects vaccine efficacy. Some viruses, such as influenza, evade the immune response by undergoing extensive antigenic variation, altering key epitopes so that the antibodies raised against one version of that epitope fail to recognize the altered form. HIV is particularly adept at antigenic variation, constantly changing the nucleotide sequences of its key antigens such that the viral population in a single infected individual may contain 15–20 different variants (see Ch.25). The search is ongoing for an HIV antigen that is immunogenic but does not undergo such variation, and thus might form the basis for a useful vaccine.

Safety of a vaccine refers to its associated incidence of detrimental side effects such as redness and tenderness at the injection site, high fever, seizures, pneumonia, encephalitis, and/or even death. Balancing the severity of vaccine side effects against the incidence and severity of the disease gives different results in different parts of the world. Where disease incidence is low, as in the developed world, a vaccine that results in relatively severe side effects is not used. In contrast, where disease incidence is much higher, as in the developing world, the risk of harm from the disease may outweigh the risk of harm due to vaccine side effects. These practical realities have the unfortunate corollary of establishing a double standard for vaccine use.

The safety and efficacy of a candidate vaccine are determined in a series of pre-clinical and clinical trials (Table 23-3). Pre-clinical trials involve rigorous testing of vaccine toxicity and teratogenicity first in non-primate animals and then most often in non-human primates. Testing for efficacy is also carried out where possible. One problem with testing

Table 23-3 **Pre-clinical and Clinical Trials**

Clinical Trial Stage	Protocol
PRE-CLINICAL	
Trial I	Test vaccine for toxicity and teratogenicity in non-primate animals
Trial II	Test vaccine for toxicity and teratogenicity in non-human primates
CLINICAL	
Phase I trial	Test vaccine for dose–response range and side effects in a small number of healthy human volunteers
Phase II trial	Test for efficacy against challenge with the pathogen in a small number of human volunteers; or vaccinate a small group of individuals living in an endemic area where natural exposure to the pathogen is likely
Phase III trial	Test for safety and efficacy against challenge with the pathogen in a large group of healthy human volunteers; or vaccinate a large group of individuals living in an endemic area
Phase IV trial	Monitor vaccinees in the "real world" for rare adverse events

safety and efficacy in animal trials is that some pathogens infect only humans, so that good animal models (in which immune responses and side effects are analogous to those in humans) may not be available. If, however, the vaccine candidate fares well in pre-clinical trials, phase I clinical trials are initiated to determine a dose range and assess the side effects of the candidate vaccine in a small number of healthy human volunteers. Antibody titers may also be measured to determine if the vaccine has induced an immune response. If no serious adverse effects are observed in this group, the candidate vaccine is formally tested for efficacy in phase II clinical trials. These trials generally involve the administration of either the candidate vaccine or a placebo to small groups of healthy volunteers that then undergo a challenge with the pathogen of interest. The outcome monitored is successful resistance to the disease. In situations in which it would be unethical to deliberately challenge a human volunteer with the infectious agent (such as malaria or HIV), a phase II clinical trial is set up in which the candidate vaccine is administered to individuals living in a geographic region where the disease of interest is endemic. The incidence and/or amelioration of disease in the vaccinated segment of the naturally exposed population can then be evaluated. In cases in which a live virus is being tested as a vaccine, the level of shedding of the virus into the vaccinee's feces or respiratory secretions may be monitored. In phase III clinical trials, large numbers of human volunteers are given the vaccine, and safety and efficacy following exposure to the natural pathogen are assessed in a randomized, controlled experimental design. Again, in the case of a particularly dangerous pathogen, large numbers of individuals living in an area where the pathogen is endemic can be vaccinated. Phase III trials are often of long duration and expensive to carry out. Should a vaccine pass its phase III trials, it can be licensed and distributed for use. Phase IV trials are then conducted, in which the real-life effectiveness of the vaccine in the general population is evaluated in case studies. Rare adverse consequences of vaccine use often turn up in phase IV trials.

D. Types of Vaccines

There are many different ways to construct a vaccine (Table 23-4). Traditional types of vaccines involve either whole organisms, either *live and attenuated* or *killed*, or their products (*toxoids*). More recently, innovative technologies have allowed the creation of *subunit vaccines* based on isolated pathogen components such as viral proteins or bacterial polysaccharides. The vaccine antigens in these vaccines may be purified from the natural organism or synthesized *in vitro* using chemical or recombinant DNA technology. These technologies can also be used *in vitro* to produce *peptide vaccines* in which the vaccine antigen is a small peptide. An *in vivo* approach is the use of recombinant DNA technology to make DNA vaccines in the form of either *recombinant vectors* or *naked DNA vaccines*. In recombinant vectors, the DNA encoding the vaccine antigen is incorporated into a vector that enters host cells and promotes translation of the vaccine antigen directly within it. In naked DNA vaccines, the vaccine antigen is encoded on a small bacterial plasmid (no vector) that is introduced directly into a vaccinee's body. Lastly, *anti-idiotypic antibodies* have been explored as vaccines in some limited circumstances. As is discussed in detail in the following sections, each vaccine type has its own drawbacks and benefits (Fig. 23-2), and types of vaccines that work best for one class of pathogen may not work at all for another.

I. LIVE, ATTENUATED VACCINES

i) What Is a Live, Attenuated Vaccine?
An attenuated vaccine consists of live, whole bacterial cells or viral particles that are treated in such a way that they have reduced virulence within the host but retain their ability to provoke an immune response. Attenuating mutations of a pathogen are often induced by culturing the pathogen for extended periods under less than optimal conditions (i.e., conditions very different from the physiological conditions

Table 23-4 **Types of Vaccines**

	Description
Live, attenuated	Whole pathogen treated to decrease pathogenicity but maintain immunogenicity; may still replicate
Killed or inactivated	Whole pathogen killed or inactivated to block replication but maintain immunogenicity
Toxoid	Formalin-inactivated toxin of pathogen
Subunit	Pathogen protein or polysaccharide purified from natural sources or synthesized using recombinant DNA methods
Peptide	Pathogen peptide purified from natural sources or synthesized using recombinant DNA methods
Recombinant DNA vector	Virus-based vector containing recombinant DNA of pathogen antigen Vaccinee is infected with the vector and the pathogen DNA is transcribed and translated within vaccinee's cells
Naked DNA	Small plasmid containing recombinant pathogen DNA Plasmid is injected into a vaccinee and the pathogen DNA is transcribed and translated within vaccinee's cells
Anti-idiotypic Ab	Antibodies directed against idiotypic differences of anti-pathogen Ab An animal is immunized with an anti-pathogen Ab and anti-idiotypic Abs are recovered from the animal and injected into a vaccinee

in its usual human host). This approach selects for organisms that change their growth habits such that they grow well in the sub-optimal setting but can no longer cause disease in the host (although they may still be able to replicate). A virus is often attenuated by introducing it into a species in which it does not replicate well (i.e., infection of an animal with a human virus), or forcing it to replicate repeatedly in tissue culture, a protocol called *passaging*. For example, influenza vaccine is prepared by passaging the influenza virus in embryonated chick eggs for an extended period, and the original Sabin polio vaccine was produced by passaging polio virus in monkey kidney epithelial cells. Attenuated strains of the viruses causing measles, rubella, mumps, and yellow fever have also been created by passage through non-human hosts or cells. Attenuation of live bacteria for vaccine development using the sub-optimal culture method has been much more limited, constrained by the unacceptably high rates of reversion of attenuating mutations in experimental strains.

Another way of attenuating a pathogen is to select mutant strains with growth temperature requirements that are different from the wild-type strain. Mutant strains that cannot grow at slightly elevated temperatures (at which the wild-type strain can still grow) are called *temperature-sensitive* strains, while those mutants that show improved growth at reduced temperatures are called *cold-adapted* strains. Temperature-sensitive and cold-adapted pathogens often show weaker growth *in vivo* in comparison to the corresponding parental strain, such that a level of attenuation suitable for vaccination is achieved. For example, in Russia, a cold-adapted influenza vaccine has been used extensively for the past several years, while a temperature-sensitive *respiratory syncytial virus* (RSV) vaccine has reached the point of clinical testing.

When a pathogen population is grown under sub-optimal conditions, selection occurs for an individual member that

has undergone a random mutation altering its growth habits. Often it is difficult to deduce precisely in which gene the mutation has occurred. Whether the virulence of the pathogen has actually been reduced by the attenuation procedure can be determined only by testing in experimental animals. The most modern methods of attenuation directly target genes and proteins known to contribute to pathogen virulence. Using recombinant DNA technology, virulence genes can be modified or deleted as appropriate. However, *in vivo* screening of the mutated pathogens is still required to identify those that are much less toxic to their hosts but continue to induce immune responses capable of recognizing the wild-type pathogen. In many cases, the safety of such vaccines is increased by deleting (rather than point-mutating) the gene in question because the chances of reversion are then virtually nil. For bacteria, reversion can theoretically occur by the random acquisition of DNA from plasmids transferred from commensal bacteria to vaccine bacteria in the site of vaccination. Engineering the blocking of more than one metabolic pathway in the vaccine bacteria prevents this type of rescue. Genetically engineered vaccines of this sort are under development for several pathogens, most notably *Vibrio cholerae*.

Sometimes nature creates a live, attenuated vaccine for us. As the reader has learned, the first form of live, attenuated vaccine was Jenner's cowpox virus. In this almost unique case, an intact, normal organism of moderate pathogenicity to humans was able to induce an immune response protecting against the much more virulent smallpox virus. This type of luck is not readily duplicated in nature, necessitating the development of less virulent, mutated vaccine organisms in the laboratory as described previously. For example, the BCG vaccine for TB results from a natural combination of the "related organism" and "mutation" approaches. *M. bovis* is not very pathogenic to humans but induces antibodies and

LIVE, ATTENUATED VACCINE

Pros

Whole organism supplies both T and B epitopes
Replication provides large quantities of immunogen
No need for adjuvant
Single dose often produces long-lasting immunity
Cost-effective and less need for boosters
Can be effective against intracellular organisms

Cons

Chance of reversion of attenuating mutation
Possible contamination with animal viruses
Can induce transient immunosuppression
Side-effects from non-pathogen parts of vaccine
Cold chain required for transport

TOXOID VACCINE

Pros

Avoids use of whole organism

Cons

Only effective if disease caused solely by bacterial exotoxin

PEPTIDE VACCINE

Pros

Avoids use of whole organism
Avoids purification from blood and tissue
May induce Tc response if peptide delivered to endogenous
   antigen processing system
Precise composition of single epitope known
Small enough to fit into most delivery vehicles
Very stable

Cons

Only small, linear, non-conformational epitopes
May possess epitopes recognized by only a small number of
   HLA molecules
May be perceived as hapten if not conjugated to carrier
Rapidly dissipated in tissues, thus requires highly effective
   adjuvant or delivery vehicle for immunogenicity
May be costly or difficult to identify and purify

NAKED DNA VACCINE

Pros

Vaccine plasmid contains bacterial sequences that can act as an
   adjuvant
Uptake by wide range of host cells and stable in tissues
Pathogen expressed in vaccinee's cells: Tc, Th, and B responses
   are mounted
Easier and less costly to construct than recombinant vector
   vaccines
Easy to modify DNA sequence encoding immunogen
Easy to add gene encoding helper molecule such as a cytokine

Cons

Need large amounts of plasmid for effective uptake, therefore
costly to produce
"Wrong" type of Th response may be induced depending on route
   of administration
CTL response may be low initially
Induced antibody response may show decreased titer, avidity,
   and longevity compared to natural infection or purified antigen
Plasmid integration into host genome could be tumorigenic

KILLED OR INACTIVATED VACCINE

Pros

Whole organism supplies both T and B epitopes
No possibility of reversion
Decreased chance of animal virus contamination
Quite stable, thus less need for cold chain

Cons

Cannot replicate: limited amount of immunogen
Multiple doses, adjuvants, and boosters required
May be ineffective against intracellular organisms

SUBUNIT VACCINE

Pros

Avoids use of whole organism
Avoids purification from human blood or tissue
Supplies multiple epitopes
Recombinant DNA technology allows manipulation of
   immunogen characteristics, including creation of conjugates
   to ensure T and B responses

Cons

Can be labor-intensive and costly to purify immunogen
Does not usually induce Tc responses
Possible lack of post-translational modification of pathogen
   protein may decrease immunogenicity
Possible alterations to pathogen protein conformation during
   purification may decrease immunogenicity
May require intact cold chain
Too large to fit into some delivery vehicles

RECOMBINANT VECTOR VACCINE

Pros

Avoids use of pathogen protein or whole pathogen
Large amounts of immunogen produced within vaccinee's cells
Can direct expression of immunogen to specific anatomical
   site by nature of vector
Proper post-translational modification and conformation
Can induce both Th and Tc responses by endogenous and
   cross-presentation of antigen

Cons

Vector may be associated with tumorigenicity
Replication of vector may induce side effects
Primary immune response mounted against vector proteins can
   generate anti-vector antibodies that block booster immunization

ANTI-IDIOTYPIC ANTIBODY VACCINE

Pros

Avoids risk of exposure to any and all pathogen components
No need to know anything about pathogen lifestyle or
   composition

Cons

Relatively complex and costly procedures required

Figure 23-2

The Pros and Cons of Various Types of Vaccines

CTLs that recognize the related, deadly human pathogen *Mycobacterium tuberculosis*. The BCG strain of *M. bovis* is a spontaneous mutant that is even less pathogenic to humans than the parent strain. Administration of BCG as a vaccine can be helpful in preventing TB in some (but not all) populations (see later in this chapter).

ii) Pros and Cons of Live, Attenuated Vaccines

Live, attenuated vaccines are usually very effective, and a single dose is often enough to induce long-lasting immunity. Because a viable, replicating pathogen is used, the innate immune system is triggered and cytokines are secreted that create an inflammatory milieu. Structures on the whole pathogen bind to PRRs on APCs and trigger the upregulation of costimulatory molecules. Most attenuated pathogens still supply both B and T epitopes, and consequently both humoral and cell-mediated responses are mounted. The fact that the attenuation process often preserves replication has another advantage in that large quantities of pathogen antigen are synthesized and accumulate in the host. Thus, only small amounts of the vaccine need to be administered to the vaccinee, maximizing cost-effectiveness. Even when boosting is required, a single dose is usually sufficient, an advantage in developing countries where returning to an immunization clinic may present insurmountable logistical barriers.

The principal danger with an attenuated vaccine is that the organism, because it is still alive, can sometimes recover its virulence and cause disease in the vaccinee. For example, in rare, unpredictable instances, one of the three attenuated viral strains comprising the Sabin oral polio vaccine reverts to virulence after passing through the human intestinal tract. The vaccinee may then develop *vaccine-associated paralytic polio* (VAPP), poliomyelitis caused by the vaccine. Another danger with live, attenuated vaccines is that, because they are grown in large quantities in culture vessels, contaminating organisms may invade the culture and become incorporated into the vaccine. For example, the monkey virus SV40 was detected in early preparations of the Sabin polio vaccine attenuated by passage through monkey kidney cells. Practices used to manufacture live vaccines can also introduce contaminants. A production process was to blame for two instances of contamination of a yellow fever vaccine with hepatitis B virus. The first occurred in Brazil from 1938 to 1940 and the second appeared among vaccinees in the U.S. military in 1942. In both these situations, pooled human serum had been used as a vaccine stabilizer, but the serum had been (unknowingly) contaminated with hepatitis B virus. Needless to say, this method of vaccine stabilization was discontinued soon after the 1942 outbreak.

A third difficulty with live, attenuated vaccines is that, even when attenuated, certain viruses apparently induce transient immunosuppression that leaves a vaccinee vulnerable to infections easily fended off by an unvaccinated individual. For example, vaccination with a certain attenuated strain of measles virus (no longer used) rendered vaccinees unusually susceptible to pneumonia, diarrhea, and parasitic infections. A last disadvantage to using a live vaccine is that it must be stored properly to be effective; that is, the cold chain of refrigeration from manufacturer to vaccinee must be maintained, a significant problem in many developing countries.

II. KILLED VACCINES

i) What Is a "Killed" Vaccine?

In a killed vaccine, a whole bacterial, parasitic, or viral entity is still involved but it is dead due to assault by γ-irradiation or a chemical agent such as formaldehyde. Used correctly, such procedures preserve the structure of the protective epitopes but remove the capacity of the organism to replicate or recover virulence. The treatment process also decreases the likelihood of contaminating organisms surviving in the vaccine preparation. The first killed vaccines were employed in the 1890s to prevent typhoid fever. Killed bacterial vaccines for *V. cholerae* (cholera) and *Yersinia pestis* (plague) may still be used in endemic areas today, and killed viral vaccines exist for polio (Salk vaccine), rabies, influenza, and Japanese B encephalitis.

ii) Pros and Cons of Killed Vaccines

The biggest advantage to using a killed vaccine is that the danger of reversion is removed. Killed vaccines are also generally more stable than live vaccines and less sensitive to disruptions in the cold chain. However, because they cannot replicate, larger amounts of these vaccines must be administered in the primary dose, raising costs. In addition, because they are dead and therefore trigger less intense "danger signals" than live vaccines, more frequent boosters are usually required to establish effective immunity. However, the greatest drawback to a killed vaccine is that it is often not effective against intracellular pathogens. Because the vaccine cannot actively penetrate host cells, the pathogen proteins do not make it to the endogenous antigen processing pathway for presentation on MHC class I. In addition, in the absence of significant "danger," cross-presentation by DCs is limited. These two factors mean that Tc activation seldom occurs and the cell-mediated response is limited. As a result, killed vaccines tend to induce predominantly systemic humoral responses consisting of the production of neutralizing antibodies.

iii) Considerations Related to Whole Organism Vaccines

One general difficulty with using whole organism vaccines (live or killed) is that the myriad components present may induce detrimental reactions in the host. For example, endotoxins in killed bacterial preparations can still induce toxic shock symptoms in vaccinees. Pertussis vaccines furnish us with another example. Although no definitive cause-and-effect relationship could be established, one in 300,000 children immunized with the original triple combination DTP vaccine (diphtheria toxoid, tetanus toxoid, killed whole *Bordetella pertussis* cells) experienced encephalopathy leading to permanent brain damage. Such problems have been alleviated by the use of an acellular pertussis vaccine (DTaP) that retains the protective pertussis epitopes but avoids the (unknown) components of the bacterial cell that may have induced the neurological symptoms.

Some vaccination strategies combine the best features of live and killed vaccines. For example, a North American adult who has previously received the Salk (killed) polio vaccine and intends to travel to an area in Africa where polio is endemic may be administered the Sabin (oral) vaccine. While the anti-polio antibodies and memory Th and B cells that were induced in the primary response to the Salk vaccine remain in this individual's circulation and tissues, the subsequent boosting with the live, attenuated virus of the Sabin vaccine induces strong, long-lasting CTL and B cell responses against the natural virus infection to which the individual may soon be exposed. Furthermore, if one of the attenuated viruses in the Sabin vaccine reverts, it should be neutralized by the anti-polio antibodies raised in the primary response to the Salk vaccine. The danger of VAPP is avoided.

III. TOXOIDS

As we learned in Chapter 22, the disease caused by a pathogen may be entirely due to its production of exotoxins. Thus, vaccinating a host against the exotoxin protects the host from disease (if not infection). As mentioned previously, *toxoids* are exotoxin molecules that have been chemically altered (usually by formalin treatment) such that they lose their toxicity but not their immunogenicity. Neutralizing antibodies generated in response to toxoid administration recognize and bind to the exotoxin and render it harmless. For example, vaccination or boosting with the tetanus toxoid protects against this disease for at least 10 years at a time. The DTaP vaccine cited previously contains both the diphtheria and tetanus toxoids. Toxoids are also sometimes used as carriers to increase the immunogenicity of Ti antigens such as bacterial capsule polysaccharides (see later).

IV. SUBUNIT VACCINES

Another way to avoid the problems of reversion and unwanted side effects associated with using whole bacterial cells or viral particles is to devise a vaccine consisting of purified components of pathogens. The idea is to use a component capable of inducing immune responses that protect against the whole pathogen. The immunological community tends to call vaccines based on whole macromolecules or large fragments of macromolecules *subunit vaccines*, although a biochemist might take issue with the inclusion of polysaccharide chains and polypeptide fragments in this definition. In any case, the pathogen component chosen to act as the vaccine must contain protective epitopes and be immunogenic, or must be modified to be so.

i) Viral Subunit Vaccines

The first attempts to develop sub-cellular vaccines involved the purification of whole macromolecules from pathogens. In particular, scientists used conventional protein chemistry techniques to purify the surface antigen prevalent on the surface of HepB virus particles (HBsAg). The rationale was that neutralizing antibodies raised during immunization with the purified protein should block infection by intact virus. Indeed, immunization with HBsAg purified from the plasma of infected patients provided vaccinees with protection against HepB infection. However, the labor-intensive effort required to purify sufficient amounts of HBsAg for vaccination made this method of production very costly.

Scientists then turned to recombinant DNA technology, a boon to those working with pathogens such as HepB or HepC that are difficult or impossible to grow *in vitro*, and/or are a challenge to purify in sufficient amounts from *in vivo* infections. Recombinant DNA technology allows the scientist to manipulate the DNA of pathogens, to clone genes of interest in vast quantities, mutate them to order, or fuse them to genes encoding components that are more easily purified and/or immunogenic. To construct a subunit vaccine using recombinant DNA technology, the DNA encoding the vaccine antigen is introduced into the genome of an organism such as yeast, *Bacillus subtilis*, or *Escherichia coli*. These microbes can easily be cultured in huge amounts in the laboratory, and the pathogen protein is synthesized in correspondingly high amounts along with the rest of the microbial proteins. A polypeptide fragment or peptide of interest can then be isolated from the newly synthesized protein. In addition, the regulation of expression of the pathogen gene in its microbial host can be altered such that the desired protein or peptide is overexpressed relative to other proteins, further simplifying purification. Lastly, the dangers of vaccine side effects due to co-purification of harmful pathogen components are avoided.

To produce the recombinant HBsAg vaccine for HepB virus, the gene encoding HBsAg was cloned into the yeast genome. However, rather than express HBsAg on its surface, the yeast produced copious amounts of the viral protein in its cytoplasm. High-pressure disruption of the yeast proved to be a relatively easy means of purifying the HBsAg protein. This subunit vaccine turned out to be so much more affordable than its conventionally prepared counterpart that the recombinant vaccine is now the standard. A similar strategy has been attempted with the hemagglutinin protein of influenza virus, but the extensive antigenic variation of this particular protein renders this subunit vaccine less effective than the killed, whole virus version. Recombinant veterinary vaccines are available for livestock protection against animal viruses such as that causing foot-and-mouth disease. Human recombinant vaccines utilizing various proteins from HIV, HepC, EBV, *Plasmodium falciparum*, and bacterial endotoxins are under development.

ii) Bacterial Subunit Vaccines

The HBsAg protein described previously contains both T and B cell protective epitopes, and so can function as subunit vaccine on its own. However, other macromolecules are less immunogenic and must be conjugated to carrier proteins in order to induce an effective immune response. The first attempts at bacterial subunit vaccines involved purifying the capsule polysaccharides from *H. influenzae* type b (Hib) and pneumococcus, because molecules in the capsules were known to be important for the pathogenicity of these organisms. Capsule polysaccharides are abundant on the bacterial surface and thus easy to purify, and do not have adverse side effects when injected into animals or humans. However, as we have learned, polysaccharides tend to be Ti antigens incapable of generating T cell responses or substantial memory. In addition, many polysaccharides do not induce immune responses of any type in children under 2 years of age. Vaccine efficacy in these cases has been greatly increased by conjugating the capsule polysaccharides to protein carriers capable of supplying T epitopes, creating a *conjugate vaccine*. The diphtheria and tetanus toxoids and modified versions of these proteins are frequently used as carriers in conjugate vaccines. Other common carriers include the *outer membrane proteins* (OMPs) of various bacterial species and *bovine serum albumin* (BSA). The *core antigen* of HepB virus (HBcAg) and the *nucleoprotein* of influenza virus (NP) have also been used as carriers because these molecules are localized in the interior of the virion and thus are likely to harbor T cell epitopes. HBcAg has the interesting property that, even with an attached molecule of interest, it attempts to assemble into a virion, making the composite structure even more provocative to the immune system.

The best example of a successful conjugate vaccine is the Hib vaccine given to infants. Conjugation of the Hib polysaccharide to diphtheria toxoid results in a vaccine that prevents the meningitis caused by *H. influenzae*. Bacterial colonization of the respiratory tract is also prevented, decreasing the spread of the bacteria to other children. Vaccines for infants based on the same principles have been developed for the prevention of pneumonia (which is often fatal in infants) and otitis media (middle ear infections). Because of their protein component, conjugate vaccines are more sensitive to cold chain disruption and are more costly to manufacture than simple polysaccharide vaccines. Recombinant DNA technology has assisted in reducing the costs of production of carriers.

iii) Limitations of Recombinant Subunit Vaccines

Despite the advantages cited previously, there are a few problems with the production of vaccine materials in recombinant organisms that must be kept in mind. First, production of a protein in this way may alter its conformation, such that stability may be affected or protective conformational epitopes do not form. Some sub-populations of neutralizing antibodies may not therefore be induced in the vaccinee. Secondly, it must also be determined whether post-translational modifications such as glycosylation are required for immunogenicity, since this process does not occur in a bacterial host. Production of a pathogen protein in need of glycosylation must be carried out in yeast, plant, or insect cells. Thirdly, because subunit vaccines involve purified macromolecules without the capacity to invade host cells, their presentation to the immune system occurs most often via APCs and the exogenous pathway. Thus, peptides from the protein are less likely to access MHC class I and induce a Tc response. This deficit is a major disadvantage if the pathogen of interest is intracellular.

V. PEPTIDE VACCINES

Reducing a pathogen to the fundamental elements required for immunogenicity eventually brings one to isolated peptides, some of which can function as protective T and B epitopes. There are several advantages to using a peptide as a vaccine: (1) The precise molecular composition of the vaccine antigen is known. (2) There is no possibility of reversion or contamination with (much larger) infectious agents or genomic material during purification. (3) Peptides are more stable than macromolecules or whole pathogens. (4) Peptides are small enough to fit easily into delivery vehicles designed to prolong antigen release within the body and direct the peptide toward antigen-processing pathways within the cell. Both natural and synthetic peptides have been explored as candidates for vaccines, and peptides have been produced both by standard purification methods and by recombinant DNA technology.

i) Identification of Peptide Vaccine Candidates

Merely deciding which peptides to pursue as vaccine candidates is a challenging task. Obviously, one wants to determine which segments of the amino acid sequence of the pathogen protein of interest induce the most effective immune response. As we learned in Chapter 6, most B cell epitopes are hydrophilic sequences, reflecting their positioning on the exterior of a molecule. Examining the binding kinetics of candidate peptides with antibodies from the serum of a patient with the disease of interest is often helpful in determining which of several peptides is the most immunogenic. For T cell epitopes, the problem is complicated by the fact that the MHC molecule and peptide are seen as a unit by the TCR. The peptide must therefore contain not only the epitope recognized by the TCR but also residues allowing it to bind to MHC. A complicating factor is that different MHC molecules may bind to the same peptide differently, causing variations in presentation to T cells and thus immune responses. In other words, a peptide vaccine representing a T cell epitope may work beautifully in individuals of a certain MHC haplotype but poorly in individuals of a different haplotype. Considerable work has been done on identifying peptides that can bind to a variety of MHC molecules (see Box 23-3).

Natural peptides have been tried as vaccines for influenza virus, pertussis, and cholera but only mixed results have been achieved. Where the natural version of a peptide is not immunogenic, modifications can be made in the laboratory that can cause it to become so. Such an approach has been taken with the *hemagglutinin* (HA) molecule of influenza virus. The natural HA molecule contains a peptide that allows the virus to bind to the sialic acid residues projecting from human cells. The natural peptide is not immunogenic, but modified synthetic versions of the peptide have been devised that induce neutralizing antibodies effective against several strains of influenza virus. Another way to increase the immunogenicity of a peptide may be to include a "supermotif" sequence that facilitates binding to the TAP antigen transporter. The presence of the supermotif should favor increased processing of the peptide, and thus promote immune responses against it.

One difficulty with the development of peptide vaccines is finding a peptide that will bind to (ideally) all the MHC molecules represented in a population; only by doing so can a large percentage of the population become successfully vaccinated. Well over 500 different MHC class I and II molecules have been identified in humans, many with distinct binding specificities. One approach has been to group these proteins into three large categories called *HLA supertypes*. (Note: "HLA supertypes" are not the same as the "HLA supratypes" (ancestral haplotypes) mentioned in Ch.10.) There are three HLA supertypes: A2, A3, and B7. All the proteins belonging to a supertype can recognize a shared or very similar motif such that the peptide-binding repertoires of these molecules overlap. For example, HLA supertype A2 includes at least nine HLA-A molecules that all bind to 9–10 amino acid peptides bearing hydrophobic residues at position 2 and at the C-terminus. The A3 supertype includes at least five HLA-A molecules that bind to 9–10 amino acid peptides bearing hydrophobic residues at position 2 but a Lys or Arg at the C-terminus.

The B7 supertype includes at least 11 HLA-B molecules that bind to peptides with Pro at position 2 and a hydrophobic residue at the C-terminus. Each HLA supertype is represented in about 30–55% of any human population, regardless of ethnicity. This frequency means that 83–89% of any population expresses an MHC protein that falls into one of the three supertypes. Thus, although a given HLA molecule might be rare in a particular population, the supertype to which it belongs will not be. Designing a vaccine that provides peptide epitopes corresponding to the three supertype binding specificities would therefore afford excellent vaccination coverage of any population in the world.

The promiscuous binding of peptides to different members of a supertype has been confirmed by X-ray crystallographic studies of peptide binding to pockets within HLA molecules. Examination of the features of the peptide-binding pockets among different HLA proteins has revealed that a consensus structure exists for the binding pockets of all MHC molecules within a supertype. At the cellular level, this supertype cross-reactivity

means that CTLs that recognize a particular peptide bound to, say, HLA-A03 can lyse target cells displaying the same peptide on, say, HLA-A11, because both proteins are members of the same supertype. Studies have confirmed that promiscuous binding of pathogen peptides occurs in patients. For example, a particular peptide of HIV reverse transcriptase that was displayed on one A2 supertype MHC molecule in one individual was bound to a different A2 supertype MHC protein in another HIV-infected individual. In another study, vaccination with an EBV peptide normally displayed on one supertype A2 protein triggered the generation of CTLs that could lyse infected target cells displaying the EBV peptide bound to a wide variety of different A2 supertype family members. Similar results have been obtained in studies of *P. falciparum* and HepB epitopes. Taken together, these results indicate that epitopes capable of binding to most members of a supertype will likely induce a response in a large proportion of a population, making them excellent candidates for vaccine development.

Sometimes a peptide is ineffective as a vaccine because it is subject to antigenic variation between pathogen strains. Researchers have thus attempted to identify sequences that are invariant in a large number of different pathogen isolates, and to generate synthetic versions of these peptides suitable for vaccines. In theory, immunization with an immunogenic invariant peptide should induce an immune response that is not confounded by antigenic variation. The synthetic invariant peptide vaccine approach is under evaluation for highly variable pathogens such as influenza, HIV, HepC, and *P. falciparum*.

ii) Limitations of Peptide Vaccines

Despite the attractive possibilities described previously, there are several difficulties with peptide vaccines. (1) Only small, linear, non-conformational epitopes can be represented by peptides. Those epitopes that are conformational in nature and depend on protein folding (like most B cell epitopes) simply will not be recognized by peptide-specific B cells. (2) Conjugation to a carrier protein may be required to circumvent the peptide being perceived as a hapten by the immune system (too small to be immunogenic). (3) Due to their small size, peptide vaccines dissipate in the tissues more quickly than the larger subunit or whole organism vaccines, sometimes too fast for an immune response to be mounted. It is then critical to use an adjuvant or delivery vehicle that offers maximal protection from degradation. (4) Unless

recombinant DNA technology can be used, there is a substantial cost attached to the purification of the large amounts of natural peptides needed for effective immunization, or chemical synthesis of a synthetic peptide.

VI. RECOMBINANT VECTOR DNA VACCINES

The vaccines discussed previously are all based on the administration of proteins (or parts thereof), be they prepared by conventional protein chemistry or via recombinant DNA technology. In either case, a large quantity of a purified, but inert, vaccine protein prepared *in vitro* must be injected into the vaccinee. In contrast, a recombinant DNA vector vaccine involves the administration of the DNA encoding the desired protein immunogen to a vaccinee. The rationale for DNA vaccines is that the vaccinee's own body makes the required immunogen, much as it would make a viral protein for a virus. The vaccine immunogen is thus presented to the immune system through the endogenous antigen processing pathway, rather than the exogenous pathway used to process injected proteins. Tc cells are activated in great numbers, a required step if the cell-mediated responses that can combat intracellular pathogens are to be invoked. Pathogens for which DNA vaccines are of particular interest are those that have resisted the development of vaccines by all other means: HIV, HepC, *Mycobacterium leprae*, and *P. falciparum*.

i) Production and Advantages of Recombinant DNA Vector Vaccines

Recombinant DNA vectors are typically attenuated viruses or bacteria that are unrelated to the pathogen of interest. These vectors can penetrate human cells and often replicate within them, but do not cause disease in the host. A selectable marker and the gene encoding the pathogen antigen of interest are incorporated into the vector using recombinant DNA technology. Since no other components of the pathogen of interest are present in the vaccinee after immunization, the danger of reversion is avoided. After the recombinant vector is injected into the vaccinee, it accesses the interior of host cells and the vaccine gene is transcribed like a viral gene. The pathogen antigen mRNA is then translated into picogram amounts of protein in the cytoplasm and glycosylated in the Golgi. Because the vaccine protein is synthesized and glycosylated by the host cell itself, it acquires the same glycosylation pattern as it would had it entered as part of the original infectious pathogen. The pathogen antigen therefore folds normally and assumes the conformation it would during a natural infection. Some molecules of the vaccine protein then enter the endogenous antigen presentation pathway as would any other host cell protein. Peptides of the pathogen antigen displayed on MHC class I activate Tc cells and initiate a cell-mediated response. Alternatively, the vaccine protein may be released from the synthesizing host cell and activate B cells directly, or may be taken up and processed by DCs to activate Th cells. The native conformation and glycosylation pattern of the vaccine protein ensure that the full range of antibodies and effector T cells is produced in a way that closely resembles a natural infection.

Another advantage to recombinant vector vaccines is that expression of the pathogen antigen can be directed to the location where it will induce the most effective immune response. For example, attenuated *Salmonella typhimurium* can be useful as a recombinant vector where immune defense at mucosal surfaces is key. As we have seen, *S. typhimurium* preferentially infects the M cells of the Peyer's patches. *V. cholerae* is a vicious pathogen that causes disease by attacking the gut mucosae. In one mouse model, an attenuated *S. typhimurium*-based recombinant vector containing sequences from *V. cholerae* antigens was able to induce the production of mucosal secretory IgA antibodies directed against *V. cholerae*. Because these antibodies were on guard exactly where *V. cholerae* accesses the body, they were capable of blocking a natural infection by this pathogen.

ii) Types of Recombinant Vectors

The organism most often used as a recombinant vector is the *vaccinia* virus. Vaccinia is a large complex pox virus, with many genes that are not essential for host cell invasion and replication. Consequently, these genes can be replaced in the vaccinia genome with foreign DNA encoding protective antigens from a pathogen of interest. High levels of the pathogen antigen are produced by the vaccinia virions, promoting presentation to both B and T cells. If the vaccine protein contains signals specifying localization in a membrane (such as a viral envelope protein), it will be directed to the host

cell membrane as would other vaccinia membrane-bound proteins. Vaccines based on highly attenuated vaccinia vectors are under study for the envelope glycoproteins of HIV and for proteins of the sporozoite stage of *P. falciparum*. HIV envelope glycoproteins have also been engineered into one of the attenuated polio virus strains used in the Sabin polio vaccine. The advantage of polio virus recombinant vectors is that they can be administered orally, resulting in the induction of both systemic and mucosal responses. However, the genome of polio virus can accept only a small amount of foreign DNA, and the genome of at least one polio virus strain is known to be genetically unstable.

Human *adenoviruses* have also been investigated as recombinant vectors. These viruses are attractive candidates because, although replication can result in the presence of high copy numbers of the viral genome in a host cell, the virus is naturally not very pathogenic to humans. In addition, adenovirus is easily cultured *in vitro*, and large amounts of foreign DNA can be inserted into the genome without affecting replication. Cell-mediated, humoral, and mucosal responses have been induced by recombinant adenoviruses expressing antigens from herpesvirus, measles virus, HepC virus, and rabies virus. Replication-deficient adenoviruses can also be used with no loss of immunostimulatory capacity. A disadvantage to using adenovirus as a recombinant vector is that this pathogen has been associated with tumorigenicity, although it may be possible to remove the genes conferring this property from the recombinant vector.

Alphaviruses such as *Semliki Forest virus* (SFV) are insect-borne viruses that infect a wide range of animal hosts. Because these viruses readily replicate in animal cells, they are being tested for use as recombinant vaccine vectors. Alphaviruses have a huge capacity for mRNA production, particularly of the viral structural genes. In animal studies, high levels of a vaccine protein have been detected within a vaccinee's body when the gene for a pathogen protein of interest has been used to replace a structural gene of an alphavirus. For example, mice infected with recombinant SFV expressing influenza NP showed significant and long-lived humoral and cellular responses directed against the NP protein. Avipox virus vectors (which infect birds) have been modified to encode the glycoproteins of various human pathogens, including HIV, rabies, and influenza. These vaccines induce both humoral and cell-mediated responses to pathogen antigens in experimental animals. Vectors employing the canarypox virus have induced particularly efficient immune responses in human clinical trials.

An insect virus called *baculovirus* has been used to generate HepC virion-like particles that may act as a potential vaccine. For unknown reasons, it is very difficult to induce expression of the envelope proteins of HepC in mammalian cells. However, HepC structural proteins are readily expressed in insect cells infected with a baculovirus bearing HepC cDNAs. In one study, HepC virion-like particles (which were non-infectious) were recovered from the insect cells and used to immunize mice. The animals mounted strong humoral and cell-mediated responses to HepC envelope proteins. Moreover, when the animals were later challenged with a vaccinia virus expressing HepC antigens, the infection was contained. Studies of this approach to HepC vaccination are under way in chimpanzees.

iii) Limitations of Recombinant Vector Vaccines

There are dangers associated with the use of replication-competent recombinant viruses for human vaccines. In particular, these vaccines pose special hazards for immuno-compromised individuals. The degree of attenuation of a live viral vector may not be sufficient to prevent rampant infection of an immunocompromised vaccinee, leading to serious side effects or even death. To address this concern, replication-deficient human vaccinia strains have been devised, and pox viruses of other species (which cannot replicate in human cells) have been investigated as recombinant vectors.

Another major problem associated with the use of recombinant vectors designed to elicit CTL responses is that an immune response is mounted not only to the pathogen protein of interest but also to the vector itself. Thus, such vaccines are generally used only for the primary immunization, since a booster dose of the recombinant vector meets a wall of neutralizing anti-vector antibodies. These antibodies bind to the vector before it can get into the host cell and initiate the expression of the vaccine antigen required to provoke the secondary CTL response. Instead, primary immunization with a recombinant vector vaccine is followed by boosting with a recombinant protein vaccine (minus the vector). For example, priming and boosting with a live, attenuated vaccinia vector encoding the HIV glycoprotein gp160 produced only marginal anti-HIV immune responses in humans. In contrast, both neutralizing antibodies and CTLs directed against gp160 were detected when subjects were primed with the gp160-vaccinia vector but boosted with purified gp160 alone. Alternatively, the primary immunization may involve one recombinant virus vector (say, vaccinia virus) encoding the antigen of interest, while the boost is carried out with a different recombinant vector (say, canarypox virus) encoding the antigen of interest. Antibodies raised against the first vector will not "see" the second vector, and the vaccine will avoid elimination in the tissues long enough to enter a cell, achieve expression, and provoke an effective secondary immune response.

VII. "NAKED DNA" VACCINES

The recombinant vector approach involves vaccination with an attenuated virus or bacterium expressing the genes for the pathogen protein as the vector infects a host cell. The most recent recombinant approach (still highly experimental) involves the direct injection into a vaccinee of a non-replicating plasmid containing DNA encoding a pathogen antigen. Surprising and imprecise as it may seem, this "naked DNA" technique (meaning that the vaccine DNA is not contained within a virion or bacterium) actually works quite well. The plasmid is taken up by host cells that synthesize the pathogen protein in such a way that protective immune responses are induced.

i) Preparation of Naked DNA Vaccines

How is a naked DNA vaccine prepared? The gene encoding the pathogen antigen of interest is cloned into a plasmid that can replicate to a high copy number in *E. coli* but not at all in human cells. The plasmid contains promoter and regulatory sequences (often derived from the CMV or SV40 viruses) that maximize the expression of the pathogen antigen gene in mammalian cells and stabilize its mRNA (Fig. 23-3). The plasmid also contains a drug selection marker, such as kanamycin resistance. Once the vaccine plasmid is constructed, it is used to transfect *E. coli* and large quantities of the bacteria are grown up under drug selection pressure. The vaccine plasmid is purified from the lysed bacteria and injected subcutaneously or intramuscularly into the vaccinee. Host cells of various types take up the injected DNA and commence production of the pathogen antigen. In rare cases, the plasmid is not degraded, meaning that production of the pathogen antigen can persist for months after injection. Studies of mice injected with a naked DNA vaccine against influenza have shown that most of the vaccine is taken up by non-lymphoid cell types such as muscle cells and keratinocytes. However, enough of the plasmid can be acquired by DCs to successfully initiate primary B and T cell responses. It is thought that these responses are then sustained by the presence of vaccine antigen produced and released by the transfected non-lymphoid cell types.

Interestingly, the bacterial DNA sequences making up the plasmid bearing the DNA of interest do not appear to interfere with naked DNA vaccination. Indeed, the unmethylated cytosine-phosphate-guanosine (CpG) motifs characteristically present in bacterial DNA can act as an effective adjuvant (see later).

ii) Immune Responses Induced by Naked DNA Vaccines

Naked DNA vaccines have been shown to induce either Th1 or Th2 responses in experimental animals, depending on the method of administration. When delivered by intramuscular injection that deposits the vaccine <u>outside</u> of cells, DNA vaccines tend to induce Th1 responses, as judged by the production of Th1-associated antibody isotypes. Scientists speculate that the CpG motifs in the bacterial plasmid induce a local inflammatory response, leading to the accumulation of copious amounts of pro-inflammatory cytokines (including IL-12) at the vaccination site. As we have learned, the presence of IL-12 generally biases the differentiation of any activated Th cells in the area toward Th1. Indeed, DNA vaccines have been used successfully in mice to prevent leishmaniasis, which is controlled by a Th1 response.

Naked DNA vaccine technology may be particularly useful for immunization against infection with respiratory syncytial virus (RSV), a pathogen for which it has been difficult to create an effective vaccine. Both natural RSV infection and immunization with formalin-inactivated RSV lead to a Th2 response that results in an undesired increase in the severity of lung disease. However, in one study, mice immunized intramuscularly with a recombinant plasmid expressing an RSV protein mounted a mixed Th1/Th2 response that resulted in the production of neutralizing antibodies

The Principle of Naked DNA Vaccines

The DNA encoding the pathogen immunogen chosen as the vaccine antigen is inserted into a plasmid bearing the indicated bacterial and viral sequences and an antibiotic resistance gene. The plasmid is transfected into *E. coli*, positive transfectants are selected by culture in the antibiotic, and large quantities of these transfectants are grown up. The plasmid is purified from lysed bacterial cells and used as the vaccine for injection into the host. The plasmid is taken up by host cells and mRNA of the vaccine antigen is transcribed and translated along with other host cell proteins. Presentation of endogenously processed peptides from the vaccine antigen on MHC class I activates Tc cells, while release of the vaccine protein (or its fragments) from host cells allows interaction with B cells. Receptor-mediated uptake of the vaccine protein by APCs followed by presentation via the exogenous processing pathway activates Th cells. Memory B, Th, and Tc cells are generated to protect the host from subsequent exposure to the pathogen from which the original vaccine antigen was derived.

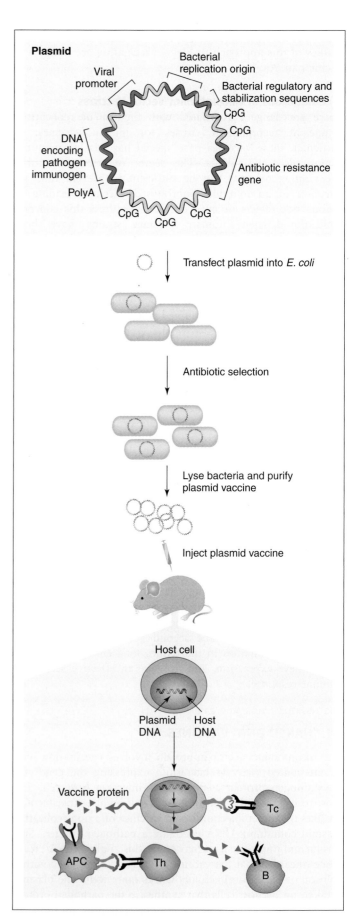

and CTLs without enhancing lung pathology. This mixed Th1/Th2 response, which was dominated by IFNγ, was achieved only by intramuscular delivery of the DNA vaccine; intradermal delivery gave the undesired Th2-type response.

What happens if the pathogen of interest is best managed by a Th2 response rather than a Th1 response? When a naked DNA vaccine is administered using a "gene gun" (see Box 23-4), the plasmid DNA is propelled directly into the target cells (typically skin, muscle, or liver cells), bypassing the induction of the local inflammatory response. In the ensuing absence of IL-12, it is possible that the DCs that acquire the vaccine DNA are influenced to prime the T cells they interact with for Th2 differentiation. In mice vaccinated via a gene gun, Ig isotypes associated with Th2 responses are preferentially synthesized, and IL-4 production increases while that of IFNγ decreases. Thus, the choice of method of DNA vaccine administration can have profound effects on the type of Th response induced, and therefore vaccine efficacy.

iii) Advantages of Naked DNA Vaccines

Naked DNA vaccines have the advantages of live, attenuated vaccines in that they induce both humoral and cell-mediated immune responses, but avoid the safety risks imposed by the use of a living, potentially dangerous organism. There are no worries about the effectiveness of attenuation of a viral/microbial vector, and the production and purification of large quantities of the desired DNA are technologically much easier and less costly than that of viral/microbial vectors, proteins, or peptides. Naked DNA is also more stable than proteins, peptides, or viral/microbial vectors, meaning that maintaining the cold chain is less important. It is also comparatively easy to modify the DNA sequence of an antigen, allowing one to optimize features to enhance immunogenicity, change intracellular localization, or remove immunosuppressive capacity. Combination DNA vaccines containing the genes for multiple antigens of one pathogen, or single antigens from more than one pathogen, are also relatively easy to prepare. A gene encoding a cytokine or

Box 23-4. Vaccination by bombardment: the "gene gun"

The "gene gun" is a unique means of introducing naked DNA vaccines directly into cells, bypassing the cell membrane and its receptors. The basic principle is the following: a sudden motive force propels minuscule plasmid-coated metal particles directly into the living tissue of interest, causing the particle to be deposited inside the cytoplasm without killing the cell. Expression of the transgene encoded on the plasmid ensues at a higher level than that achieved by other methods of inducing naked DNA uptake. The biggest advantage of the gene gun is its direct delivery: one does not have to isolate the target cells first, grow them *in vitro*, transfect them *in vitro*, and then implant them back into the recipient.

Several different types of bombardment devices have been developed and are routinely employed for transfections in various laboratories. Many systems use very small gold particles that are only 1–3 μm in diameter. About 10,000 copies of a 5 kb fragment of DNA can be precipitated onto a single particle. Tungsten particles (diameter of 3.5–4 μm) can also be used. The motive force can take one of several forms: a gunpowder charge, a high-voltage electrical charge, a helium gas charge, or a vaporized water droplet charge, among others. High-velocity delivery of the DNA-coated particles is essential, since lower speeds do not permit efficient penetration of the target cell and/or may damage it. The DNA loaded onto the particle remains in place and is not "rubbed off" as the particle penetrates into the tissue. The bombardment device can be hand-held, which facilitates delivery of the particles to tissues in intact animals (see Plate), or fixed like a microscope. The bombardment of the target tissue is sometimes carried out in a vacuum chamber.

The efficiency of DNA transfer by bombardment varies with the device used and the type of target cell, but frequencies in the range of one stably transfected cell in 1,700–60,000 have been reported. These frequencies are in line with other methods of cell transfection such as electroporation or calcium phosphate precipitation. The viability of bombarded cells (80–95%) is also equivalent to that of cells transfected by other methods. However, gene gun methods use far less material than an *in vivo* vaccination procedure: while conventional injection methods require 10–100 μg of a vaccine plasmid to induce an immune response in a rat, gene guns can accomplish the same task using only 0.1–1 μg plasmid. Experiments using luciferase reporter gene systems have shown that transient expression of the gene of interest can first be detected 1–2 days after tissue bombardment into the skin, or into surgically exposed liver or muscle tissues of living rats or mice. DNA transfer occurs about 10 times more readily to skin and liver than to muscle; about 10–20% of cells in bombarded skin and liver express the transgene. Transient gene expression can be detected for at least 4–24 days after transfer, depending on the cell type targeted. Attempts to transfect human mammary gland explants and human endothelial, fibroblast, lymphocytic, and epithelial cell lines have also been successful. However, in the limited human studies available, it appears that higher amounts (500–2500 μg) than those used in rodents are required for *in vivo* human DNA vaccination resulting in measurable antibody and CTL responses. One last note: gene gun technology was first developed for the transfection of plant cells, and is used to great advantage in this field. Gene guns have also been used to introduce DNA into mitochondria and chloroplasts.
[Plate courtesy of Dr. John O'Brien, www2.mrc-lmb.cam.ac.uk/personal/job.]

costimulatory molecule can be embedded in a DNA vaccine plasmid, supplying factors that can further enhance the immune response to the pathogen antigen.

As stated previously, DNA vaccines have the major advantage of generating endogenous antigens that induce CTL responses. How do these responses compare to those mounted against the natural pathogen? The evidence to date suggests that DNA vaccines can induce adequate levels of CTL activity to most intracellular pathogens in the primary response. The frequency of antigen-specific Tc cells in mice injected with DNA vaccines for the nucleoprotein of either influenza virus or Sendai virus was equal to that in mice naturally infected with either live virus. However, mice injected with a DNA vaccine against LCMV did not produce immediately detectable anti-LCMV CTL responses (unlike a natural virus infection), indicating a lower initial level of antigen-specific Tc cells. Nevertheless, these cells were able to expand and protect the animals against a later challenge with LCMV. The level of Tc cells initially induced by a DNA vaccine may be relevant because control of some particularly nasty intracellular pathogens (such as *P. falciparum*, HepC, and HIV) may depend on the rapid deployment of large numbers of effector CTLs. Initial examinations of the effectiveness of a DNA vaccine encoding an envelope protein of HepC showed that mice vaccinated with this construct mounted fairly strong Th1 responses but weak CTL responses. Stronger CTL responses in mice have been induced by vaccination with a plasmid encoding HepC non-structural proteins known to supply Th and CTL epitopes. Chimpanzees vaccinated using this approach that were later challenged with HepC initially developed mild hepatitis but were eventually able to clear the virus. Unvaccinated control animals went on to develop chronic HepC infection.

iv) Limitations of Naked DNA Vaccines

There are some caveats to naked DNA vaccines such that the approach remains unproven in humans. In animals, there is considerable variation in the adequacy of the protection induced by DNA vaccines, depending on the pathogen antigen and experimental system. In some studies, the titer, avidity, and longevity of antibodies induced by DNA vaccination were lower than those induced by either natural infection or a purified protein vaccine. In other situations, the DNA vaccine has persisted *in vivo* for as long as a couple of years; the possible side effects of such persistence are unknown. Some researchers have wondered whether the vaccine plasmid DNA might sometimes integrate into the host cell genome. No evidence for such an event has been obtained as yet, but the possibility remains. Depending on the site of integration, the vaccine DNA could disrupt a tumor suppressor gene or activate an oncogene, increasing the risk of tumor development in the vaccinee. Methodological concerns center around how to control which cell types take up the vaccine plasmid, and how to involve more professional APCs in DNA uptake. Lastly, the amounts of naked DNA that must be used to induce an immune response have so far proved to be comparatively large and very expensive to produce. Indeed, an experimental naked DNA vaccine generated to fight HIV was recently estimated to cost about U.S. $100 per single dose. Enhanced adjuvants and/or delivery vehicles are under development to reduce the amount of DNA needed to provoke an adequate immune response.

VIII. ANTI-IDIOTYPIC VACCINES

We learned in Chapter 5 about the existence of idiotopes, epitopes arising from subtle differences in the hypervariable region sequences of antibody antigen-binding sites. The reader will recall that the unique collection of idiotopes on a given antibody defines its *idiotype*. If an individual is immunized with a given purified antibody, at least part of the B cell response against this antibody will be directed against its idiotype, giving rise to *anti-idiotypic antibodies* (anti-Id Ab). Such anti-Id Ab may have binding sites that are complementary in shape to the binding site of the first antibody. Structurally, this makes such anti-Id binding sites equivalent to the antigen originally recognized by the first antibody. In other words, the anti-Id antibody is a molecular mimic of the original antigen. Thus, if the first antibody was specific for a nasty pathogen, this mimicry raises the possibility that an individual could be vaccinated with an anti-Id antibody that resembles an epitope on that pathogen, rather than vaccinated with the pathogen or pathogen product itself. The lymphocytes stimulated by the anti-Id antibody should cross-react

with the original pathogen antigen and protect the vaccinee without the risk of direct exposure to the pathogen or any of its components. In addition, this method of vaccination allows one to induce an effective immune response against a pathogen without having to isolate it or know very much about its lifestyle or composition.

Anti-idiotypic vaccines can be prepared in two ways: by isolating the anti-Id Ab itself, or by manipulating the DNA encoding it (Fig. 23-4). In the protein approach, mice are immunized with purified anti-pathogen antibody and hybridomas are established that produce anti-Id Ab. The desired anti-Id Ab-producing hybridoma is grown up in large quantities and the anti-Id Ab is purified. As with other protein-based vaccines, the purified anti-Id Ab may then be conjugated to an immunogenic carrier protein such as KLH and injected into the vaccinee. Alternatively, the purified anti-Id Ab can be injected into the patient along with GM-CSF. This growth factor acts as an adjuvant by increasing the numbers of DCs recruited to the local site that can take up and process the anti-Id Ab just as they would any other protein antigen. For the DNA approach to anti-Id Ab vaccination, the DNA sequences specifying the variable regions of both the light and heavy chains of the anti-Id Ab are inserted into a plasmid vector along with a linker peptide that ensures the assembly and folding of the truncated polypeptide chains into the correct Ig structure. The plasmid is grown up in a microbe, purified, and injected into the muscle cells of the patient. The muscle cells steadily synthesize the anti-Id Ab fragment along with the rest of their structural proteins, processing it through the endogenous pathway for display on MHC class I and releasing fragments of it to be taken up by APCs or recognized by B cells. The antibodies, effector T cells, and CTLs produced in these responses should recognize pathogen antigens in a subsequent natural infection and protect the vaccinee.

Do anti-Id Ab vaccines work? Anti-Id Ab vaccines have been investigated in rodents for many years, with protection noted against pathogens such as rabies virus, *Listeria monocytogenes*, and *Schistosoma mansonii*. In addition, early work with an anti-Id Ab raised against a first antibody specific for HBsAg demonstrated that chimpanzees vaccinated with the anti-Id Ab were protected from pathogenic doses of HepB. In contrast, anti-idiotypic vaccines for humans have yielded only modest results, with levels of protection no greater than those furnished by conventional or recombinant vaccines. Using the DNA strategy to make customized human anti-Id Abs greatly simplifies the process and decreases costs compared to the protein strategy, but the anti-Id Ab approach is still rarely used in general vaccination programs. Where anti-Id Abs may shine as clinical tools is in therapeutic vaccination against certain cancers. In some cancers, the antigens expressed by the tumor cells are unique tumor-specific antigens expressed only at low levels. For reasons detailed in Chapter 26, the immune system often fails to recognize these tumor cells as foreign. The theory is that the administration of large amounts of anti-Id Abs resembling these tumor-specific antigens may kick-start the immune system into mounting a more effective anti-tumor response *in vivo*.

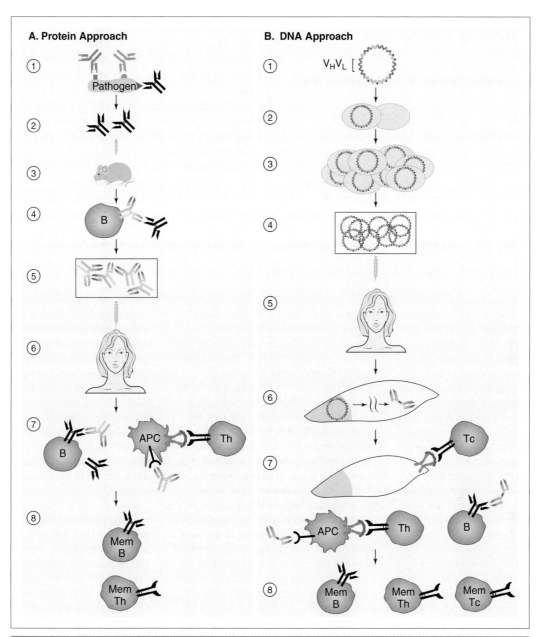

A. Protein Approach

B. DNA Approach

Figure 23-4

Two Ways to Prepare Anti-idiotypic Vaccines

(A) Protein approach. Infection by a pathogen leads to production of different antibodies responding to different epitopes (1). One of these anti-pathogen antibodies is purified as the primary antibody (2) and injected into an animal (3) to elicit production of secondary antibodies specific for the idiotype of the primary antibody (anti-Id Ab) (4). Hybridomas producing the anti-Id Ab are established and grown up in large quantities, and the anti-Id Ab protein is purified from these cultures (5). Vaccination of an individual (6) with the anti-Id Ab (in conjunction with a conjugate protein or adjuvant, as necessary) induces the activation of naïve B cells (7) that recognize the anti-Id Ab, a specificity that matches that necessary to recognize the original pathogen. Memory B cells of this specificity are generated (8). Uptake of the anti-Id Ab protein by APCs (7) and activation of naïve Th cells (7) ensures that memory Th cells will also be generated (8). (B) DNA approach. A plasmid is constructed bearing the gene segments encoding the variable region (VHVL) of an anti-Id Ab (1). The plasmid is transfected into yeast cells (2) that are grown up in large quantities under selection (3). The plasmid is purified from these cells (4) and injected into a vaccinee (5). Muscle cells take up the plasmid and transcribe and translate the anti-Id Ab fragment (6) as they would any other host cell protein. Because the anti-Id fragment is synthesized endogenously, it is presented on MHC class I to activate naïve Tc cells (7). When the fragment is released outside the host cell, it may activate naïve B cells (7) or be taken up by APCs and processed via the exogenous pathway for presentation to naïve Th cells (7). Memory B, Th, and Tc cells are generated (8) that can react rapidly to an invasion by the original pathogen.

E. Factors Affecting Vaccination

I. SCHEDULING OF VACCINE ADMINISTRATION AND BOOSTING

i) Age of Vaccinee

The age of a vaccinee is an issue because the immune system is immature during infancy and reacts differently to antigenic stimulation than does an adult immune system. Antibody responses are generally are weak in infants and Ti polysaccharide antigens do not induce immune responses at all in this age group. These factors are important because most vaccines are designed to be administered to young children (within 12–15 months after birth) to prevent the onset of disease during childhood. In addition, neonatal tolerance must be taken into account. The reader will recall that, while an adult will mount an effective immune response to a foreign antigen, a newborn will often develop tolerance to the same antigen. These factors are even more pertinent for naked DNA and recombinant vector vaccines, since the pathogen antigen is produced and processed endogenously in the host cell just as a self-antigen would be. Neonatal tolerance of a DNA vaccine encoding a malarial protein has been noted in vaccinated newborn mice. That is, these animals failed to mount B and T cell responses against *P. falciparum* when later challenged with the pathogen. In other words, the vaccination left the newborn animals more vulnerable to infection than if they had not been treated. The struggle to come up with an effective vaccine for malaria is outlined in Box 23-5.

Box 23-5. Toward a malaria vaccine

Malaria remains one of the greatest killers in the developing world, and the need for an effective vaccine is pressing. Malaria is caused by four species of the single-celled parasite *Plasmodium* (in order of their importance): *P. falciparum*, *P. vivax*, *P. ovale*, and *P. malariae*. However, despite millions of dollars spent and years of diligent research, *Plasmodium* has so far managed to outwit the vaccinology community. The complex, multi-host, multi-stage life cycle of this organism (refer to Fig. 22-14) is mostly to blame, since the immune system no sooner mounts a response to antigens of one developmental stage than the parasite moves on to the next completely different stage. Moreover, in the course of its life cycle, this parasite expresses 5000–6000 different proteins, some of which are highly polymorphic. Identifying immunogenic epitopes at each stage is time consuming and costly, and testing these epitopes for vaccine efficacy even more so. In addition, as described in Chapter 22, *Plasmodium* species have numerous means of evading the immune responses that are mounted against them. And if these problems were not enough, there is now evidence that *Plasmodium* strains are emerging that are resistant to the drugs used to treat acute malaria, and that the mosquitoes that carry *Plasmodium* are developing resistance to insecticides. There are difficulties to be overcome on the human end, too. Immune responses to *Plasmodium* antigens are highly dependent on HLA haplotype, and variations in RBC antigens and hemoglobin type are known to affect susceptibility to malaria. These factors also affect vaccine development.

Many adults who live in areas where malaria is endemic develop effective immunity to the disease after about 10 years' constant exposure. However, this protection requires reinforcement by continual re-exposure, and if the individual leaves the endemic area for a significant time, he or she is likely to become susceptible to the disease again. Researchers have attempted to improve on the natural immune response to *Plasmodium* by designing vaccines targeting one of three stages of its development: (1) the "pre-erythrocytic" stage, when the sporozoites migrate to the liver and develop into schizonts; (2) the "blood" or "erythrocytic stage," when the merozoites infect the RBCs; or (3) the "mosquito stage," when the gametes form ookinetes that are taken up by a mosquito. The reader should note that none of these strategies has yet given rise to a truly efficacious vaccine for malaria, although great strides are being made. The idea is that each of the proposed vaccine types would be appropriate for different human populations. For example, unexposed travelers from a non-endemic area would require a vaccine effective at the pre-erythrocytic stage, to prevent sporozoites from either reaching the liver or producing merozoites once they got there. No RBCs would be lysed, precluding the onset of clinical symptoms. Those already living in an endemic area could get away with a less protective vaccine that targets the erythrocytic stage and mimics the natural persistent infection experienced by most adults in the area. Merozoites would still be produced and invade some RBCs but these infected cells would be destroyed before the parasite could multiply or generate gametes.

Symptoms might still arise but they would be considerably milder than those of full-blown malaria. This type of vaccine would be of great benefit to infants and children in endemic areas, who can suffer greatly from malaria-induced anemia or coma. Mosquito stage vaccines would block the formation of the ookinete in the mosquito, thereby preventing transmission to the next host. This vaccine would not do the vaccinee much good but would reduce the spread of the disease to his or her family members or community.

• **Pre-erythrocytic stage vaccines:** These types of vaccines generally target sporozoite and schizont antigens. One candidate is *circumsporozoite protein* (CSP). CSP is a major surface protein first synthesized by sporozoites in the mosquito salivary gland. CSP is responsible for sporozoite binding to hepatocyte receptors. Experimental vaccination with this immunogen, especially when a prime-boost protocol is followed, induces both humoral and cell-mediated immune responses. Encouraging preliminary results have also been obtained in naive humans challenged with *P. falciparum* sporozoites. However, CSP sequences differ among *Plasmodium* strains so that protection against one strain does not guarantee protection against another. In addition, sporozoites entering the human body are present in the blood for only a few minutes before they hide in hepatocytes and become inaccessible to anti-CSP antibodies. Finally, CSP expression is low to absent on merozoites (depending on their stage of development) so that this symptom-causing stage of the parasite is ignored. A more

Continued

promising approach may be to target an antigen expressed by schizonts in the liver called *liver-specific antigen* (LSA). LSA does not vary significantly between different *Plasmodium* strains, and merozoites also express LSA. CTLs directed against LSA have been isolated from naturally infected patients, and experimental vaccination with LSA induces a detectable CTL response. Whether this response is sufficient to block subsequent infection is under investigation, as are various recombinant approaches to expressing LSA.

• **Erythrocytic stage vaccines:** These types of vaccines focus on the merozoites that attack RBCs. The reader will recall that, unlike most other body cells, RBCs do not express MHC class I, so that an infected RBC has no way of letting a Tc cell know it harbors a parasite. Immune defense against this stage of the infection is thus mediated primarily by two types of antibodies. First, antibodies recognizing surface antigens of merozoites bind to them and block their entry into RBCs. Secondly, antibodies recognizing parasite antigens expressed on infected RBC surfaces trigger ADCC and complement-mediated lysis of these cells before the merozoites can expand. One candidate erythrocytic vaccine consists of *merozoite surface protein 1* (MSP1), a relatively non-polymorphic glycoprotein that may facilitate merozoite invasion of RBCs. In one study, mice and monkeys vaccinated with MSP1 were protected against a later lethal challenge with *Plasmodium*. Another candidate vaccine is based on *apical membrane antigen 1* (AMA1) expressed by late stage schizonts. Mice vaccinated with AMA1 mounted both antibody and CD4+ T cell responses to this antigen. However, despite these encouraging observations, field

trials of both these vaccines in humans showed them to be of low efficacy and to provide only short-term protection against *P. falciparum*. A synthetic peptide vaccine derived from an erythrocytic stage antigen and administered with alum has been tested in endemic areas of South America and Tanzania but was found to be of minimal protective benefit.

• **Mosquito stage vaccines:** Mosquito stage vaccines are designed to forestall the development of the ookinete from the gametes and thus block transmission from an infected host to a new host. Immunization of animals with antigens derived either from *Plasmodium* gametes (pre-fertilization antigens) and ookinetes (post-fertilization antigens) induces antibody responses. These antibodies are then taken up by a mosquito feeding on the vaccinee. Once in the mid-gut of the mosquito, antibodies raised in response to a pre-fertilization antigen immediately destroy gametes and thus block fertilization. Antibodies raised in response to a post-fertilization antigen act within 12–24 hours to destroy ookinetes and thus block sporozoite formation. In both cases, the progress of the malarial cycle is halted. Unfortunately, these antibodies cannot reach gametes sequestered in the RBCs of the infected animal itself. Clinical trials in human volunteers of a vaccine based on a post-fertilization antigen (Pfs25) are under way.

It has become clear in recent years that only a vaccine incorporating elements of all three strategies described previously will succeed in generating effective protection against malaria. As well as the multiple stages of the *Plasmodium* life cycle, such a vaccine will also have to target multiple *Plasmodium* strains. In 1996, J. A. Tine and

colleagues inserted seven *Plasmodium* genes (including CSP, LSA, MSP1, AMA1, and Pfs25) into an attenuated vaccinia virus and vaccinated human volunteers with this vector (which they called NYVAC-Pf7). Although immune responses to the *Plasmodium* antigens could be detected, the NYVAC-Pf7 vaccine did not protect against *P. falciparum* infection. Indeed, thus far, the experience has been that DNA plasmids encoding *Plasmodium* antigens tend to generate antibody responses in animals but not in humans. Strategies in which priming with a plasmid bearing DNA for *Plasmodium* antigens is followed by boosting with a viral vector bearing DNA for the same *Plasmodium* antigens are currently being tested in several centers. In 2002, a collaborative international effort to develop a multi-stage DNA vaccine against malaria was initiated. This naked DNA vaccine, dubbed "MuStDO" (*multi-stage DNA-based vaccine operation*), consists of multiple plasmids bearing genes encoding sequences of 15 pre-erythrocytic and erythrocytic stage *Plasmodium* antigens. The vaccine is injected into volunteer vaccinees and the plasmids are taken up by muscle cells. A booster consisting of attenuated vaccinia virus bearing the same MuStDO gene sequences is subsequently administered. The idea is to induce both humoral and cell-mediated responses against infected hepatocytes and free sporozoites in the blood, and humoral responses blocking merozoite entry into RBCs, merozoite maturation within the RBCs, and the adhesion of infected RBCs to the vascular endothelium. Additional versions of MuSTDO that incorporate mosquito stage antigens are also under development. Hopefully, this strategy will finally allow clinicians to add malaria to the list of vaccine-preventable diseases.

An adaptation of the Recommended Childhood Immunization Schedule for the United States is shown in Figure 23-5. Ideally, primary doses and boosters of most vaccines are given in the first 15 months of life. In general, boosters are administered at least 4–8 weeks after the previous dose. Where more than three doses of a vaccine must be administered, the last dose is usually given 6–12 months after the penultimate dose. In some cases, later boosters around ages 5 and 15 years are also recommended. Although maternal antibodies that have been passed on to the infant either *in utero* or via breast feeding might be expected to interfere with the success of infant immunization by neutralizing the vaccines, they do not

appear to do so to any noticeable extent. It should be noted, however, that vaccination scheduling can be affected by the disease status or history of the mother. In the United States, neonates whose mothers are infected with HepB should receive recombinant HepB vaccine within 12 hours of birth (ignoring maternal antibodies), while those born to HBsAg-negative mothers can wait until 2 months of age. In addition, vaccination schedule times are not carved in stone, and one can often still "catch up" later in life on vaccinations missed in infancy. As we get elderly, however, standard vaccines delivered in standard ways usually offer less protection due to an age-associated decline in the immune system; in general, B cell

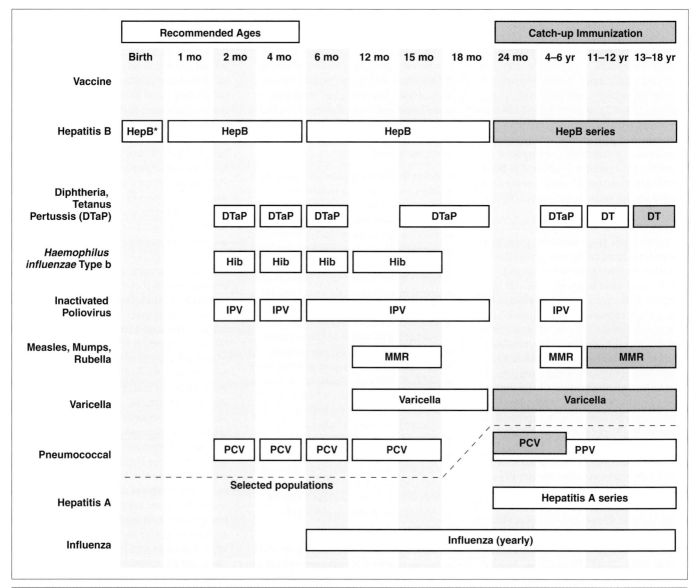

Adaptation of the Recommended Childhood and Adolescent Immunization Schedule
The age ranges at which the listed vaccines should be given are indicated by a white bar. The grey bars show where "catch-up" vaccinations should be given if the individual was not vaccinated during infancy. Vaccines below the dotted line are given only to selected populations (see text for conditions). DT, diphtheria/tetanus only; mo, months; y, years; *, only if the mother is infected with HepB; PCV, pneumococcal conjugate vaccine; PPV, pneumococcal polysaccharide vaccine. Adapted from United States National Immunization Program 2004, www.cdc.gov/nip/acip.

numbers are reduced and IL-2 secretion is decreased. Adjuvants and delivery vehicles can be of assistance in augmenting immune responses in aged individuals.

ii) Boosting

The second and subsequent doses of a vaccine have been dubbed *boosters* because they induce stronger, faster secondary immune responses. As noted previously, most vaccines are administered 2–3 times, with a gap of several weeks or months in between. When a booster is given can be influenced by the pathogen involved, its frequency in the environment of

the vaccinee, and the longevity of neutralizing antibodies. Some vaccines, particularly live, attenuated vaccines such as those for measles and mumps, induce very strong memory responses even after the primary dose, meaning that boosting may only be necessary at intervals of several years in order to sustain life-long immunity. In addition, if a pathogen is endemic in the local area, the vaccinee may encounter it naturally often enough to keep triggering memory responses without the need for more than a primary dose of vaccine. In cases of vaccines such as varicella that have been added relatively recently to the list of recommended childhood immunizations, the need for and timing of future boosters will only become clear with the aging of the first cohort of vaccinees.

Current research into vaccination techniques indicates that the booster dose does not have to involve the same vaccine formulation as the primary dose. As mentioned earlier, a live, attenuated vaccine might be administered in the primary immunization followed by a booster of killed or protein subunit vaccine, or the reverse strategy might be used in the case of a recombinant vector vaccine. With respect to naked DNA vaccines in mice, it is apparent that for some pathogens, particularly those requiring cell-mediated immune responses for elimination, boosting of the primary response induced by the DNA vaccine with a recombinant vector vaccine elicits better CTL-mediated protection than two rounds of naked DNA vaccination. Similarly, studies of DNA vaccines in monkeys have shown that boosting with a purified protein vaccine may be helpful for inducing full protection from some pathogens. The timing of DNA vaccine administration is also an issue, since multiple DNA vaccinations given relatively close together are less effective than fewer vaccinations given at longer intervals.

iii) Combination Vaccines

When vaccines against two different pathogens must be given to an individual, optimal immunization against both pathogens is achieved only if the vaccines are given either on the same day or at least 1 month apart, but not 2 days or 2 weeks apart. For example, if vaccinations for cholera and yellow fever are given about 3 weeks apart, the antibody responses to both pathogens are depressed. The immune system appears to require a certain "consolidation period" to regroup resources and/or expand important cell populations. One means of getting around the problems of multiple injections and/or repeated clinic visits is to combine several pathogen antigens into one vaccine. Although it has been speculated that competition might develop between antigens for APCs and/or cytokines that could affect the activation of the relevant lymphocytes for each antigen and thus the effectiveness of the immune responses, there is no evidence that combination vaccines overwhelm an infant's or child's immune system. Nevertheless, a combination vaccine has to be rigorously tested for safety and efficacy in clinical trials even if each of its component vaccines has already been tested. In addition, before licensing, it must be shown that the immune responses generated to each component of a combination vaccine occur at levels comparable to those seen when the corresponding vaccines are delivered separately. Examples of combination vaccines already in wide use in North America are those for the control of diphtheria, pertussis, and tetanus (the DTaP vaccine) and measles, mumps, and rubella (the MMR vaccine). More recently, a combination vaccine has been licensed in the United States that includes the DTaP plus HepB plus HibB conjugate vaccines. Other combinations that have been shown to have similar efficacy whether given alone or in combination are MMR and varicella (chicken pox); MMR, DTaP, and Sabin; HepB, DTaP, and Sabin; influenza and pneumococcus;

MMR, DTaP-Hib, and varicella; MMR and Hib; and DTaP and Hib.

Still more combination vaccines are under development, driven by our desire to protect our youngsters from as many pathogens as possible without further "crowding" of the childhood immunization schedule. Surveys have shown that today's children routinely receive as many as 20 shots by age 2 years to deliver 11 different vaccines, compared to 2-year-olds of a generation ago who received 8 shots to deliver 5 different vaccines. How many separate injections should a child be expected to endure, and how often can parents be persuaded to return to clinics? The more vaccines that can be combined into one injection, the greater the patient's comfort and compliance. From a purely logistical point of view, the merging of separate vaccines into one product reduces stocking and transportation costs, minimizes use of the cold chain, and decreases the need for sterilized injection supplies.

iv) Vaccinations for Special Situations

In the developed world, adults who have a specific disease or who run particular occupational hazards should be vaccinated against pathogens that might otherwise take advantage of their weakened immune systems or increased risk of exposure. For example, vaccination against influenza virus and *Streptococcus pneumoniae* is recommended for those who suffer from diabetes mellitus, alcoholism, or cardiovascular disease. Those with complement deficiencies should be vaccinated against meningococcal infection, and those who lack a spleen should seek out the Hib vaccine. Working with animal hides or hairs can increase one's likelihood of contracting anthrax, and so vaccination against this bacterial pathogen is urged. Military personnel who might be exposed to anthrax or smallpox during a biological warfare attack are vaccinated against these pathogens before entering the sphere of battle. Individuals working in nursing homes or hospitals are now routinely required to be vaccinated against influenza virus, and often rubella and varicella as well. Hepatitis A vaccine is recommended for people with chronic liver disease, household contacts of infected individuals, those such as hemophiliacs who regularly receive transfused blood products, travelers to hepatitis A-endemic areas, and children living in endemic areas of the United States. Besides being currently advised for all children under 18 years of age, hepatitis B vaccine is recommended for those living with chronically infected individuals, hemodialysis patients, injection drug users, health and safety workers, and those with multiple sex partners. Both HepA and HepB vaccines are recommended for researchers and clinicians handling human or non-human primate tissues or blood. Similarly, clinicians and researchers handling cultures or animals that might be infected with vaccinia or other pox viruses should be inoculated with the vaccinia-based smallpox vaccine. Sewage workers who were not vaccinated against diphtheria and tetanus as children are advised to obtain catch-up vaccinations. In addition to the preceding, North Americans may be advised to get vaccinations against cholera, typhoid, and yellow fever if they are traveling to regions where these pathogens are endemic.

II. ROUTES OF VACCINE ADMINISTRATION

Because of the adjuvant currently used with most human vaccines (see later), intramuscular injection is the most common route of administration. The thigh and deltoid muscles are the sites of choice, since injection into the buttocks has been known to cause damage to the sciatic nerve. The clinician must also take care that the injection is not lost in fat (especially in adults) instead of reaching muscle. Other routes of administration that are used routinely in animals, such as epidermal, intradermal, intravenous, intranasal, and oral, are being used more frequently in humans. The latter two routes are particularly attractive for two reasons. First, under the right circumstances, intranasal and oral regimens can induce both systemic and mucosal immune responses, enhancing protection at the interfaces where the vast majority of pathogens gain access to the body. Secondly, any method that obviates the need for sterile syringes and needles will be an immediate boon to the developing world. In the course of therapeutic injections as well as vaccinations, syringes and needles are commonly reused in poorer areas where sterilization is often inadequate. It has been estimated that, worldwide, 12 million new cases of HepB infection, 3.5 million new cases of HepC infection, and 120,000 new cases of HIV infection result annually from the reuse of syringes and needles in clinics.

In the case of oral vaccines, there remains the problem of how to avoid establishing oral tolerance to the vaccine antigen. The Sabin polio vaccine is one of the very few vaccines to be routinely administered by the oral route. In this case, the natural route of virus infection is through the digestive tract, so that attenuated polio virus vaccine delivered orally enters the body and stimulates the immune system in the same way as the natural pathogen. The polio vaccine attaches to polio virus receptors on intestinal cells and replicates in a way that induces both systemic and local immunity. In this case of targeted pathogen or vaccine entry, tolerance induction does not occur. Researchers have tried to avoid oral tolerance by delivering vaccine antigens to the MALT via M cells. As noted previously, several viruses, some species of *Salmonella*, *Yersinia*, and *Shigella*, and the exotoxins of enterotoxic *E. coli* and *V. cholerae* preferentially bind to M cells and are conveyed to the collection of immune system cells in the underlying dome. Mucosal and often systemic immune responses to these entities are then mounted. By attaching a vaccine antigen to the M cell-binding components of these exotoxins or organisms, the vaccine is delivered directly into the MALT where the natural infection would initiate. In mice, vigorous mucosal immune responses have been induced to various protein antigens introduced in this way. However, the experience with the mucosal delivery of DNA vaccines has been mixed, with mucosal administration sometimes, but not always, resulting in systemic antibody and/or cell-mediated responses.

III. ADJUVANTS AND DELIVERY VEHICLES

Live, attenuated vaccines and many recombinant vectors often retain the capacity of a pathogen to replicate in the host, penetrate the tissues, and induce inflammation. For these reasons, these types of vaccines are highly immunogenic. Killed vaccines, purified protein vaccines, and naked DNA vaccines generally lack these properties, meaning that they often need help to induce an effective immune response. Furthermore, proteins and DNA molecules run the risk of encountering extracellular proteases and nucleases that can degrade them before they can be taken up by APCs. Researchers have long experimented with protecting vaccine antigens from degradation (by the use of delivery vehicles) and enhancing their ability to induce an immune response (by the use of adjuvants). In the 1930s, J. Freund found that immunization of guinea pigs with killed mycobacteria antigens in paraffin oil resulted in very high anti-bacterial antibody titers after only a single injection. The mixture of paraffin oil and the aqueous suspension of mycobacterial antigen formed a viscous water-in-oil emulsion in which the antigens were found in the aqueous phase. The oil was included to slow the dispersion of the aqueous phase in the tissues. Freund and others went on to confirm that this emulsion caused a pronounced stimulation of the immune response to both the mycobacterial antigen and any other antigen of interest added to the aqueous phase. If the mycobacterial antigens were omitted from the immunizing mixture, a milder but still enhanced response to the protein antigens of interest could be obtained. By the 1940s and 1950s, the basic paraffin oil (usually mixed with a surfactant) used to emulsify the antigen of interest became known as *incomplete Freund's adjuvant* (IFA). If the mycobacterial antigens were included in the aqueous phase of the emulsion, it was called *complete Freund's adjuvant* (CFA). While both CFA and IFA quickly became indispensable in experimental immunization protocols, their use in clinical immunization of humans has always been impossible due to unacceptable side effects, most notably the formation of granulomas at the site of injection. The work of many vaccinologists is focused on developing adjuvants and delivery vehicles that are safe and effective in humans.

i) Adjuvants

The reader will recall from Chapter 6 that one of the key functions of an adjuvant is to induce local inflammation. Such inflammation recruits APCs and lymphocytes to the site of infection or inoculation, ensuring that antigens are presented to T cells in an activatory context. The precise mechanism underlying adjuvant activity is unknown and it is still not clear whether the adjuvant exerts its effects at the site of injection or in the local lymph node. What is clear is that the immunogen and the adjuvant must be injected together to achieve optimal immunogenicity: separate injections into the same animal are not enough. How then does an adjuvant work? One possibility is that the adjuvant promotes capture of the adjuvant-immunogen mixture by DCs in the tissues. These DCs become activated and transport the immunogen from the site of injection to the local lymph node and its resident naive T cells. Another hypothesis holds that an adjuvant represents "nonself" to the host innate immune system. In this context, the adjuvant works because it contains a molecule or sequence that binds to PRRs on innate immune system cells, initiating the inflammatory response. Alternatively, the adjuvant may cause sufficient damage to host cells at the injection site that stress molecules are released. In either case, the presence of inflammatory mediators (particularly IL-1) induces APCs to upregulate their costimulatory molecules, meaning that antigen presentation will be more likely to activate than anergize T cells. The adaptive immune response is thus enhanced and the vaccination is a success. However, it remains puzzling why killed virus particles and bacteria, which retain the motifs recognized by PRRs, still require adjuvants to induce an effective immune response. Moreover, host cell debris injected as an adjuvant has no immunostimulatory activity, while lipid-containing particles, which are thought to be innocuous, do have adjuvant activity.

Another explanation for adjuvant activity may be the "depot effect." Injection of adjuvant material often induces the formation at the injection site of a granuloma that allows only a slow release of the vaccine antigen. The life of the antigen in the tissues is therefore prolonged, and degradation is avoided until the antigen is taken up by APCs. This protective effect means that use of an adjuvant can reduce the amount of antigen needed to induce an immune response, an important factor when vaccines containing low amounts of antigen, such as purified protein vaccines or DNA vaccines, are used. We will now examine a selection of clinical and experimental adjuvants (Table 23-5).

- **Alum:** Alum, a gel composed of aluminum hydroxide or aluminum phosphate, was one of the first adjuvants ever tried and is still the only one licensed for routine use in humans. Despite its long history, it is still not understood exactly how alum enhances the immune response. There is some evidence that alum promotes Th2 responses at the expense of Th1 responses, resulting in a bias toward antibody production. This characteristic means that alum is not particularly good at supporting cell-mediated responses. An advantage of alum over protein adjuvants is that alum is more stable when the vaccine cold chain is broken. A disadvantage of alum is that it must be injected deep into muscle and not subcutaneously, increasing vaccinee discomfort. The presence of alum in subcutaneous tissues can cause necrosis.
- **Bacterial toxins:** Although not yet routinely used in humans due to safety concerns, some bacterial toxins or their components have proved to be useful adjuvants for immunization studies in experimental animals. Modified versions of cholera toxin have long been employed as adjuvants. The B subunit of natural cholera toxin directs the toxin to the M cells in the intestine. Natural cholera toxin can be altered by point mutation, or the A subunit can be deleted, to produce a molecule that is no longer toxic but still directs associated antigens to the mucosae. *Procholeragenoid* (PCG) is an immunogenic derivative of cholera toxin that is less expensive and easier to produce than natural cholera toxin B subunit. Another relatively recent toxin-derived adjuvant is *heat-labile enterotoxin* (HLT), a product of *E. coli*. Both PCG and HLT have substantial adjuvant activity when administered systemically or mucosally, and, when used at low concentrations, are not

toxic to mice. HLT is the more effective adjuvant for intranasal administration of a vaccine and induces the production of large quantities of pathogen-specific secretory IgA (as well as antibodies directed against HLT). One problem with extending the use of these adjuvants to humans is that there are few good animal models in which to investigate toxic side effects. For unknown reasons, animals tend to be much more resistant than humans to the neurological effects of these toxins. Nevertheless, clinical trials in 2000 demonstrated that small amounts of HLT could be used as an adjuvant for an intranasally administered influenza vaccine in humans without adverse consequences.
- **CpG motifs:** As mentioned previously, bacterial DNA often contains CpG motifs, short sequences of unmethylated Pu-Pu-CpG-Py-Py. These motifs are about 20-fold more common in bacterial DNA than in eukaryotic DNA, making them ideal targets for PRR-mediated recognition by the innate immune system. Indeed, NK cells encountering synthetic versions of such sequences are stimulated to secrete IFNγ, and activated B cells increase their proliferation and antibody secretion. Several lines of evidence suggest that the CpG motif can act as an effective adjuvant in mice. First, immune responses characterized by the secretion of IL-6, IL-12, TNF, and the IFNs are observed following immunization of mice with bacterial DNA containing CpG. Secondly, when an animal is immunized with CpG plus a purified protein antigen, antibody- and cell-mediated responses to the antigen are achieved that are equivalent to those observed when a DNA vaccine containing the gene for the same protein antigen is used. Thirdly, CpG is a powerful inducer of IFNα secretion by APCs, and IFNα is known to stimulate expansion of memory CD8$^+$ CTLs. CpG may thus indirectly enhance memory CTL responses. Unfortunately, synthetic CpG motifs that work well in mice are not very effective in humans. The design of a CpG sequence more stimulatory in humans, and the investigation of whether a CpG adjuvant effect applies in humans at all, are ongoing.
- **Cytokines:** We have learned in previous chapters that the presence of certain key cytokines can influence an immune response to take on either a Th1- or a Th2-like character. For some pathogens, it may be advantageous to encourage a Th2 response in place of a Th1 response, or vice versa. Co-administration of cytokines (particularly IL-12) with vaccine antigens can favorably skew the immune response in mice. While direct injection of cytokines into humans has toxic side effects, intranasal administration of cytokines such as IL-2, GM-CSF, and IL-12 in human volunteers has been shown to enhance the induction of mucosal immunity by an antigen. Work on defining safe conditions for the use of such adjuvants is under way.

ii) Delivery Vehicles

Significant increases in antibody titers and levels of relevant cytokines can be achieved when a vaccine is administered in a non-toxic delivery vehicle. The vehicle protects the vaccine molecules from protease or nuclease degradation (increasing its persistence in the tissues), and may also act as an adjuvant. Some vehicles facilitate the display of multiple vaccine antigen

Table 23-5 **Adjuvants and Delivery Vehicles**

Adjuvants	Characteristics	Pros	Cons
*Alum [aluminum hydroxide: Al(OH)₃; aluminum phosphate: AlPO₄]	Favors Th2 responses	Few side effects Stable	Does not induce Tc activation Requires painful injection Can cause necrosis
Bacterial toxins (derivatives of exotoxins)	Effective mucosal adjuvant	Not toxic at low concentrations	Possible neurological effects at high concentration
CpG motifs (component of bacterial DNA)	Favor Th1 responses and induce IFN-γ secretion by APCs, which promotes CTL expansion	Induces strong Ab and CTL responses in mice	Does not work as well in humans
Cytokines	Immune deviation	Promotes most effective immune response	Systemic toxicity

Delivery Vehicles	Mechanism	Pros	Cons
Liposomes (spherical phospholipid bilayer with aqueous center)	Phagocytosis by macrophages and uptake by other host cells induces Th, B and Tc responses	No immune response to liposome itself	Not very efficient, requires large amounts of Ag Possible toxic side effects at high liposome concentrations
Virosomes (IRIV) (liposome plus viral envelope glycoproteins)	Viral fusion to host cell receptors and cell entry like a virus provokes Tc response Phagocytosis by APCs provokes Th and B responses	Very effective and efficient with few side effects Can capture naked DNA, small peptides, and large proteins	
ISCOMs (soccer ball-like cage of cholesterol, phospholipid, and saponin surrounding Ag)	Presentation of multivalent antigen Phagocytosed by APCs to provoke strong Th1, B, and CTL responses Saponin stimulates APCs	Low amounts of Ag required Particularly good for lipid-bearing Ag Local and systemic responses Can be given by oral and intranasal routes Relatively stable	More difficult to make than IRIVs Possibly toxic if saponin impure
SMAAs (vaccine epitope bound to specific mAbs bound to microbeads)	Phagocytosed by APCs so that Tc, B, and Th responses induced	Can coat beads simultaneously with T and B epitopes	Labor-intensive to construct
Biodegradables (polymeric chitosan microspheres)	Slow degradation releases Ag over time; induces prolonged Ab response	Increased persistence of Ag	Continuous release of Ag may inhibit memory generation
DCs (immature DCs that have phagocytosed Ag)	Ag-loaded DCs activate naive Th and Tc cells	Efficient Provokes systemic and mucosal responses	Autologous DCs must be isolated for use in humans
Intracellular bacteria (attenuated intracellular bacteria transfected with naked DNA vaccines)	Macrophages phagocytose and lyse DNA-loaded bacteria, releasing Ag DNA Transcription and translation of Ag DNA occurs within host cell so a Tc response is induced	Relatively easy to construct	Possible integration of bacterial DNA into host genome Possible reversion of attenuating mutation
Gene gun (high-pressure blasting of DNA directly into cells)	Induces Tc response	Avoids degradation by nucleases	Complex apparatus required

*Alum is currently the only adjuvant licensed for standard patient treatment.

molecules on their surfaces, creating a multivalent form of the antigen that increases its immunogenicity. These properties have made delivery vehicles invaluable adjuncts for vaccination with subunit and DNA vaccines.

iia) Liposomes. Liposomes are one of the most versatile types of delivery vehicles. Liposomes are spherical structures composed of a bilayered membrane (often phospholipid) surrounding an aqueous center. A wide range of entities can be assembled that vary in lipid content, size, charge, and number of membranes. Typically the antigen is mixed with a suspension of phospholipids under conditions that favor the formation of a spherical structure that has its hydrophobic elements facing outward and hydrophilic elements facing inward. The vaccine antigen can be trapped within the liposome with or without an accompanying adjuvant, and the liposome can be administered by injection or mucosal absorption. Liposomes are enthusiastically phagocytosed by DCs and macrophages, ensuring fast uptake of the vaccine antigen by professional APCs. In addition, non-lymphoid host cells are able to take up liposomes, meaning that the antigen can be widely displayed on MHC class I for Tc cell activation. Fortunately, it appears that the human immune system does not usually mount significant immune responses to the types of phospholipids used in liposomes. One drawback to the use of liposomes is that delivery of the antigen is not specifically targeted and thus not overly efficient, meaning that relatively large amounts of antigen must be used. In addition, some researchers are concerned about the possible toxic effects of high concentrations of cationic lipids introduced into the body. These matters are under investigation.

iib) Virosomes. A virosome (or IRIV, *immunopotentiating reconstituted influenza virosome*) is a type of "artificial virus" that can be used to deliver vaccine antigens directly into a host cell. A virosome consists of a type of liposome that contains the hemagglutinin (HA) and neuraminidase (NA) envelope glycoproteins of influenza virus plus an immunogen from the pathogen of interest. A virosome is prepared by first solubilizing native influenza virus in detergent to remove the nucleocapsid. The viral envelope proteins are reconstituted into a liposome that retains the membrane fusion properties of the native virus as well as its conformational stability and ability to enter host cells. The resulting particle is spherical with a mean diameter of about 150 nm. Antigens of interest are either encapsulated within the lumen of the virosome or are chemically cross-linked to its surface.

The inclusion of the influenza envelope glycoproteins in the virosome promotes the fusion of the virosome to the membranes of non-phagocytic cells. The virosome is internalized and processed, and the vaccine antigen is directed into the MHC class I antigen processing pathway to provoke a specific anti-pathogen CTL response. However, if the virosome does not immediately fuse with the membrane of a non-phagocytic cell, the presence of HA in the virosome structure ensures that it maintains a compact conformation that is easily taken up by phagocytes. Furthermore, the HA contains residues that help target the virosome to sialic acid-containing receptors on the surfaces of APCs. The uptake of virosomes by APCs and their subsequent degradation in the endocytic system ensures rapid and efficient antigen presentation on MHC class II and the induction of a Th response. Thus, packaging a pathogen antigen in a virosome promotes both humoral and cell-mediated responses.

Virosomes have proved to be a very effective mode of vaccine administration with a low incidence of adverse events. For example, in the hopes of devising a better flu vaccine, scientists in the Netherlands confined a major epitope

of the influenza virus nucleoprotein (NP) to a virosome and used it to intramuscularly vaccinate mice that had been primed with infectious virus. (In this situation, the viral priming was done so that the response examined would be a secondary one and therefore of a magnitude that would be more easily measured; a primary response might not result in sufficient Ab production to be sure that the response had indeed occurred.) The virosomes induced powerful and specific memory CTL responses against influenza-infected cells. Subsequent intranasal administration of NP-containing virosomes to mice induced both mucosal and systemic responses. Virosome vaccines are also very efficient in that minute amounts of protein antigen can induce protective CTL responses. In one study of synthetic peptide vaccines, over 100-fold less peptide was required to induce CTL activity when the peptide was delivered in a virosome than when it was either mixed with adjuvant in an emulsion or complexed to inert latex beads and injected conventionally. Similar results were found in a mouse study that compared immune responses to diphtheria and tetanus toxoids delivered in virosomes to those induced by intramuscular injection of conventional toxoid preparations adsorbed to alum. Up to 3-fold higher anti-toxoid antibody titers were obtained after two immunizations with virosome vaccine compared to the corresponding toxoid-alum vaccine. Injection of toxoid-containing virosomes also caused fewer painful reactions at the injection site. Moreover, because delivery of the antigen was so efficient, the use of virosomes allowed for the injection of decreased amounts of toxoid, further reducing side effects in vaccinated mice. Research investigating the suitability of virosomes for the delivery of naked DNA vaccines is also under way.

The effectiveness of virosomes has been examined in a limited number of human clinical trials. In one study of influenza vaccines in geriatric human vaccinees, a virosome subunit influenza vaccine induced superior levels of anti-influenza antibodies compared to a conventionally injected influenza subunit vaccine. More importantly, compared to the latter, a greater proportion of the elderly individuals in the virosome group responded to the vaccine, meaning that the virosome vaccine provided better population coverage. Similarly, a virosome vaccine containing inactivated HepA virus was found to induce higher antibody titers and cause fewer side effects in healthy adult volunteers than a conventional HepA vaccine used with alum. Several virosome vaccines, including those directed against HepA and influenza, have been licensed for use in at least some jurisdictions.

iic) ISCOMs. An intact vaccine antigen can also be delivered to a broad range of host cells by packaging it in an *immunostimulating complex* (ISCOM). To form an ISCOM, the vaccine antigen is mixed with cholesterol, phospholipid, and the detergent saponin (from the bark of the tree *Quillaia saponaria* Molina). When saponin is mixed with cholesterol, stable rings are formed in aqueous solution that combine to form a soccer ball-like structure with a strong negative charge. Phospholipid is added to soften the rigidity of the cholesterol-saponin rings and to allow relatively bulky protein immunogens to insert into the cage-like structure. In this case, the vaccine antigen is usually an amphipathic protein: one with both hydrophobic and hydrophilic regions. Proteins containing transmembrane domains or significant amounts of lipid are most easily incorporated into ISCOMs. Immunogens that have internal hydrophobic regions can be denatured to expose them, or covalently bonded to strings of fatty acids; the latter approach works well for peptide antigens. Electrostatic interactions and chemical chelation methods can also be used to incorporate proteins into ISCOMs.

Overall, a complete ISCOM forms a spherical, hollow cage about 40 nm in diameter that closely resembles a virus particle in its size and orientation of surface proteins. To the immune system, the ISCOM resembles a multivalent antigen with a shape that invites phagocytosis by APCs. In addition, the saponin in the ISCOM is a powerful adjuvant and activator of APCs. When an APC engulfs an ISCOM, upregulation of MHC class II and secretion of IL-12 and IFNγ are induced. ISCOM immunization thus results in the generation of primarily Th1 effector cells and high titers of IgG2a antibodies that favor ADCC and complement fixation. ISCOMs can also induce a vigorous CTL response due to cross-presentation of the vaccine antigen by DCs stimulated by ISCOM components. CTL responses to ISCOM-delivered vaccine antigens can sometimes arise even in the absence of CD4[+] help, an important observation since it means that ISCOM vaccination may be able to protect immunocompromised individuals.

ISCOM immunization is highly efficient, as effective Th and CTL responses have been achieved using comparatively small amounts of immunogen. Another advantage of ISCOM vaccines is that they are stable for at least 18 months when refrigerated. However, compared to other types of novel delivery vehicles, ISCOMs are more difficult to make. In addition, the original experimental ISCOM vaccines carried the risk of toxicity because their saponin components were incompletely purified. (Saponin is not a pure substance; as many as 23 different closely related fractions have been identified.) The assembly of ISCOMs is now a much better controlled procedure and reproducible ratios of known saponins are used for construction.

In mice, ISCOMs have been used to induce both antibody and CTL responses to HIV envelope proteins, influenza virus glycoproteins, measles virus proteins, mycobacterial hsp60, SIV (*simian immunodeficiency virus*) proteins, and HPV proteins. Success with parenteral, intranasal, and oral modes of ISCOM administration has been reported, with both local and systemic antibody and CTL responses observed. Protective responses to many pathogens have also been induced in several other mammalian species, including CTL responses to SIV proteins and HepC core protein in monkeys. An ISCOM vaccine containing Newcastle disease virus antigens was recently successful in inducing high titers of protective antibodies in chickens. Through all these studies, ISCOM vaccines have appeared very safe. For example, over 1 million doses of an ISCOM vaccine for equine influenza have been sold in Sweden since 1989 and no adverse effects have been reported. ISCOM vaccines have reached the advanced clinical trial stage in humans. A test of a new ISCOM-delivered

influenza vaccine showed that it induced higher levels of anti-influenza antibody more rapidly than did a conventional influenza vaccine in human subjects. More importantly, specific anti-influenza CTL responses were induced in 50–60% of vaccinees, compared to only 5% of subjects vaccinated with the conventional vaccine.

iid) SMAAs. A very labor-intensive way to build a multivalent vaccine antigen is to construct a *solid matrix antibody-antigen* (SMAA) complex. Monoclonal antibodies to the vaccine epitope of choice are chemically bound to a solid support such as a microbead. The vaccine epitope is mixed with the bead and attaches to the mAb, fixing multiple vaccine epitopes on the surface. By coating the bead with mAbs to different epitopes, one can cover the microbead with both B and T cell epitopes. APCs readily take up the microbeads, and antibody and cell-mediated responses to antigens packaged in this way have been demonstrated in mice. An obvious disadvantage to this technique is that the antibodies to the desired epitopes may have to be generated *de novo*, potentially adding more steps to the vaccine production process.

iie) Biodegradables. There has been much interest in biodegradable materials that can be used to mediate the controlled release of vaccines. Particles called *microspheres* that are composed of polymeric molecules such as chitosan or poly[lactide-co-glycolide] can be slowly degraded to metabolizable products, releasing the vaccine antigen within. Extended antibody responses to toxoid vaccines delivered in this way have been reported in rabbits. Both mucosal and systemic responses to pathogen antigens have been observed in response to intraperitoneal or oral administration of chitosan-packaged DNA vaccines. One potential difficulty with this approach is that the antigen is released continuously rather than at defined intervals, which may affect memory cell triggering and generation.

iif) DCs as delivery systems. Vaccination of experimental animals using DCs loaded ("pulsed") with whole pathogens or their antigenic peptides has also been attempted, with some success. In one study, immature DCs were incubated in the presence of killed bacteria under conditions that stimulated these cells to take up bacterial proteins and commence production of IL-12. When the activated, antigen-pulsed DCs were administered intravenously to inbred mice, immune responses were mounted that efficiently protected the mice from a later challenge with the same bacteria. Protective responses against infection with *Chlamydia trachomatis*, *Borrelia burgdorferi*, *Leishmania major*, *M. tuberculosis*, LCMV, or *Toxoplasma* have also been obtained when DCs pulsed with soluble antigens from these pathogens were used for immunization. However, there is evidence that, to make the most of this approach, care must be taken to choose an antigen inducing immune deviation in the right direction. For example, while DCs loaded with killed, whole *Chlamydia* cells provoked protective Th1 responses in mice, DCs pulsed with only the outer membrane protein of *Chlamydia* elicited a non-protective Th2 response. Some researchers are loading DCs with pathogen RNA instead of protein. Experimental animals inoculated with DCs pulsed with *Candida* RNA were capable of fending off a later challenge with this fungus. *In vitro*, antigen-loaded DCs have also been used to elicit responses from human cells to EBV, influenza virus, and HIV antigens, as well as to antigens derived from melanoma, bladder cancer, colorectal cancer, and prostate cancer cells (see Ch.30). A major obstacle to using DC vaccination in humans is histoincompatibility: *autologous* (the vaccinee's own) DCs would have to be isolated in each case, a time-consuming and laborious process most likely suitable only for therapeutic vaccination.

iig) Other delivery systems. We would be remiss if we did not remind the reader of the gene gun delivery method for DNA vaccines. As described in Box 23-4, gene guns facilitate the introduction of a DNA vaccine directly into the vaccinee's cells, avoiding extracellular nucleases and initiating Tc responses. Intracellular bacteria have also been marshaled into DNA vaccine delivery. Attenuated strains of invasive bacteria such as *L. monocytogenes* and *Shigella flexneri* can be transiently transfected with naked DNA vaccines. The bacteria invade the host, are taken up by phagocytes, and are destroyed in the phagosome. The vaccine DNA is released and accesses the host cell cytoplasm, eventually initiating both antibody and cell-mediated responses. A significant caveat here is that, in one *in vitro* study, the DNA vaccine was found to have integrated into the host cell genome after delivery by attenuated *L. monocytogenes*. Another concern is that disease could result if the attenuating mutations of these invasive bacterial strains underwent reversion.

Other alternative delivery systems are mini-needles, needle-less injection, and vaccination patches. Mini-needles are an array of very tiny needles mounted onto an adhesive patch. The antigen is applied to the mini-needles in powdered form and delivered by patch pressure into the subcutaneous layer. For needle-less injection, a spring-powered device forces a liquid vaccine into the subcutaneous layer of the vaccinee without the need for piercing by a needle. The device can be adjusted to take the thickness of the skin of a vaccinee into account. Topical vaccine administration is also under trial. The antigen plus an adjuvant are contained in a patch that is pressed onto wet skin. The antigen is taken up by Langerhans cells in the skin, leading to measurable humoral and cell-mediated responses.

F. Prophylactic Vaccines

It may strike the reader as odd that we have left the discussion of vaccines available for human use to this point in the chapter. However, we felt it was important that the reader understand the nature of vaccine development and potential before delving into the current situation. We can expect great strides to be made in the early years of the new millennium that will expand the following list of diseases prevented or mitigated by vaccination.

At the time of writing, there are over 25 human infectious diseases for which safe and effective vaccines are available. Some are recommended for routine childhood vaccinations while others are recommended in certain situations for individuals at high risk. Note, however, that there are as yet no effective vaccines for parasite infections. In Table 23-6 and the sections that follow, we explore some of the most insidious diseases for which effective vaccines currently exist. We then discuss why vaccination programs are not 100% successful, and why an individual might not be vaccinated. Failure to vaccinate can be related to logistics, particularly in the developing world, or to a conscious decision by an individual not to vaccinate. Such a decision is sometimes based on the rare, but real, adverse side effects associated with some vaccines, but can be due to misconceptions about the nature of vaccination.

I. ANTHRAX

As discussed in Chapter 22, anthrax is caused by the extracellular, gram-negative bacterium *B. anthracis*, which occurs naturally in the soils of farms and woodlands. It commonly infects livestock and range animals but rarely humans. Even with the current concerns about bioterrorism, vaccination against anthrax is recommended only for members of the military who might encounter this organism in a battlefield context, researchers culturing this organism, those working with potentially infected animals in a region of high anthrax incidence, and those handling animals hides or wools imported from countries with lax standards for spore transfer prevention. Disease caused by this bacterium is due in part to the toxins it produces (which contain the *B. anthracis* proteins protective antigen [PA], lethal factor [LF], and edema factor [EF]; see Ch.22), and in part to other ill-defined bacterial functions. The current licensed vaccine for anthrax, referred to as the AVA vaccine, consists of a cell-free preparation of *B. anthracis* PA adsorbed to aluminum hydroxide. The AVA vaccine is given in a series of three subcutaneous injections 2 weeks apart, followed by three additional subcutaneous boosters at 6, 12, and 18 months. Annual boosters are recommended thereafter. This involved immunization schedule leads to problems with compliance among vaccinees. The most common clinical adverse effect of the vaccine is pain in the injected arm that restricts its capacity for lifting for a day or so. Other effects are generally mild and occur in about 30% of vaccinees.

New types of anthrax vaccines have been investigated in experimental animals. For example, in one gene gun study, mice were vaccinated with recombinant expression plasmids carrying fragments of the genes encoding either PA or LF or both. All mice vaccinated with any of these vectors survived a challenge with a fatal dose of purified *B. anthracis* lethal toxin. Interestingly, when the animals were vaccinated with both PA and LF, titers of anti-PA and anti-LF antibodies were increased 4- to 5-fold over titers achieved using either single vaccination.

Table 23-6 Examples of Current Vaccines

Disease Vaccine	Route of Administration	Minimum Vaccine Duration	Comments
Anthrax Acellular *B. anthracis* extract (AVA)	SC	1 yr	For military personnel and those who work with animal hides or wools
Cholera			For travelers to endemic areas
Killed bacteria	IM, ID, SC	6 mo	Significant side effects 50% of vaccinees may not be protected
Killed bacteria plus recombinant cholera toxin B	IM	3 yr	Few side effects
Live, attenuated bacteria	Oral	3 yr	Few side effects
Diphtheria Toxoid	IM	10–15 yr	Given in combination with tetanus and pertussis in the DTaP vaccine
***Haemophilus influenzae* B** (Hib) Bacterial capsule polysaccharide conjugated to toxoid or OMP	IM	<4 yr	Can be given in combination with DTaP
Hepatitis A Killed virus Live, attenuated virus	IM IM	10–15 yr 10–15 yr	For travelers in endemic areas or occupational exposure
Hepatitis B Recombinant HBsAg	IM	<5 yr	Given to infants soon after birth; catch-up recommended for adolescents and health care workers exposed to blood
Influenza virus			Given by November to ward off current season's "flu"
Killed virus	IM	1 yr	Significant side effects
Viral subunit	IM	<1 yr	Only low levels of Ab induced
Virosome	IM	1 yr	Most effective vaccine so far
Japanese B encephalitis			Possible neurological and allergic side effects
Inactivated	SC	3 yr	Derived from infected mouse brains
Live attenuated virus	IM	3–4 yr	For high-risk groups in Asia only Not available internationally
Measles Live, attenuated virus	SC	10–15 yr	Given in combination with mumps and rubella in the MMR vaccine; must be boosted in adolescence
Meningococcal meningitis Bacterial capsule polysaccharide	SC	3–5 yr	Not very effective in infants Group A and group C vaccines available
Polysaccharide conjugated to diphtheria toxoid	SC	10–15 yr	Effective in infants and adolescents Only group C vaccine available
Mumps Live, attenuated virus	SC	10–15 yr	Given as part of the MMR vaccine
Pertussis Several bacterial antigens (no whole cells)	IM	<5 yr	Given as part of DTaP in infancy; not given after age 10 yrs
Plague Killed bacteria	IM	6 mo	Given only for occupational exposure
Poliomyelitis Killed virus (Salk)	IM	5 yr	Requires regular boosters
Live, attenuated virus (Sabin)	Oral	10–15 yr	No booster required Oral vaccine has been associated with VAPP
Rabies Killed virus	IM	2 yr	Usually given as a therapeutic following an animal bite

Continued

Table 23-6 Examples of Current Vaccines—*cont'd*

Disease Vaccine	Route of Administration	Minimum Vaccine Duration	Comments
Rubella Live, attenuated virus	SC	10–15 yr	Given as part of the MMR vaccine
Streptococcus pneumoniae Bacterial polysaccharide (PPV)	IM	3–5 yr (children) 10 yr (adults) 5–10 yr (elderly)	Efficacious only after 24 months of age
Bacterial (PCV7) polysaccharides plus conjugate	SC or IM	<5 yr	Now part of routine infant immunization schedule
Tetanus Toxoid	IM	10 yr	Given as part of DTaP in infancy Regular boosters recommended for adults
Tuberculosis BCG vaccine (live, attenuated *M. bovis*)	ID or SC	10–15 yr	For persons in direct contact with patients having active TB infections
Typhoid Killed bacteria	IM	5 yr	For travelers to endemic areas
Live, attenuated (Ty21a)	Oral	2–3 yr	Significant side effects
Vi (purified capsule)	SC	2–3 yr	
Varicella (chicken pox) Live, attenuated virus	SC	Not known	Now recommended as part of childhood series Coverage >97% with one dose
Variola (smallpox) Live vaccinia virus	ID	10–20 yr	Significant side effects
Yellow fever Live, attenuated virus	SC	10 yr	For travelers to endemic areas

Ab, antibodies; Ag, antigen; ID, intradermal; IM, intramuscular; OMP, outer membrane protein (of bacteria); SC, subcutaneous; mo, months; yr, years.

II. CHOLERA

Cholera is a disease of debilitating diarrhea caused by the extracellular bacterium *V. cholerae*. Transmission is by water contaminated with fecal waste from infected humans, so that in regions where sanitation is good, disease incidence is minimal. Where sanitation is poor, as is the case in many developing countries, the disease is endemic. Sporadic cases of cholera can occur even in non-endemic areas when undercooked shellfish or raw fish that has come in contact with contaminated water (as in a harbor) is consumed. Ingested bacteria multiply rapidly in the gut and produce an exotoxin that induces relentless loss of water and ions from gut epithelial cells. Life-threatening dehydration can occur within hours of infection, particularly in very young children. Antibodies to the exotoxin mitigate but do not eliminate the disease. Rather, mucosal IgA that blocks the attachment of the bacteria to the gut epithelial cells is thought to be of key importance.

Cholera is an obvious candidate for a mucosally administered vaccine, but the only vaccine currently available in many countries is a killed whole cell preparation administered parenterally. Unfortunately, this vaccine is effective in only 50% of a given population and protection lasts only 6 months. Significant side effects, such as fever and debilitating malaise, may be experienced. This vaccine is really suitable only for foreign travelers to countries in which cholera is endemic. It is actually more cost-effective to improve the sanitation in an endemic area and implement rehydration therapy for cholera cases than to mass-vaccinate populations with this vaccine. New types of cholera vaccines that can be given orally are in use in some countries. A combination of killed whole *V. cholerae* cells and recombinant cholera toxin B subunit has had some success, as has an oral live, attenuated vaccine. Both these vaccines have fewer side effects than the original vaccine and confer longer-lasting protection (up to 3 years, at least for older children and adults).

III. DIPHTHERIA, TETANUS, AND PERTUSSIS

i) Diphtheria

Diphtheria is a devastating childhood disease caused by the exotoxin of the facultative anaerobic bacterium *Corynebacterium diphtheriae*. The exotoxin inhibits protein synthesis in cells of the heart and peripheral nervous system and also induces an inflammatory response in the throat that can obstruct breathing. Disease can be prevented by adequate levels of circulating antibodies to the exotoxin. The current vaccine is a formalin-inactivated exotoxin toxoid. Diphtheria toxoid is included with the tetanus toxoid and acellular pertussis in the combination DTaP vaccine (see later). From 1990 to 1999, the global incidence of diphtheria was reduced from about 22,400 cases to 3900 cases by increasingly widespread use of various diphtheria vaccines.

ii) Tetanus

Tetanus is caused by the exotoxin produced by the anaerobic extracellular bacterium *Clostridium tetani*. Because the disease is caused solely by the exotoxin and not by any other bacterial component, antitoxin antibodies raised in response to the exotoxin are protective. Tetanus is thus easily prevented by vaccination with tetanus toxoid, an inactivated form of the neurotoxin that induces the production of neutralizing antibodies. The global use of tetanus vaccination programs has reduced the incidence of tetanus from over 116,000 cases worldwide in 1980 to about 15,000 in 1999.

iii) Pertussis

Whooping cough is a highly contagious disease caused by the extracellular bacterium *B. pertussis*. *B. pertussis* gravitates to the mucosae of the bronchi and induces severe coughing that quickly debilitates young children. Almost half of *B. pertussis* infections occur in infants, and a significant proportion of these cases require hospitalization. Potentially fatal pneumonia occurs in about 15% of pediatric cases, while seizures and encephalopathy are observed in about 2% and 0.7% of cases, respectively. Permanent brain damage and even death can result from pertussis-induced encephalopathy in infants.

The natural immune response to *B. pertussis* is still not fully understood. The bacteria produce two exotoxins: tracheal cytotoxin and pertussin toxin. Tracheal cytotoxin destroys the cilia on cells in the throat, impeding the clearance of mucus, while pertussis toxin both promotes attachment of the bacteria to the respiratory epithelium and induces an influx of lymphocytes into the affected area. Neutralizing antibodies that prevent attachment of the bacteria to respiratory epithelial cells and antitoxin antibodies that bind to the exotoxins thus play key roles in combating this disease. In the late 1980s to the early 1990s, vaccination programs in the developing world began to make a serious dent in the global incidence of pertussis, reducing the number of reported cases from almost 2,000,000 in 1980 to about 443,000 in 1990 and 70,000 in 1999. For the reasons described earlier in this chapter, the current pertussis vaccination strategy is to include several isolated *B. pertussis* proteins (but no whole, killed *B. pertussis* cells) in the combination DTaP vaccine.

iv) Combination D, T, and P Vaccines

Childhood vaccination against diphtheria, pertussis, and tetanus employs the DTaP vaccine, consisting of *B. pertussis* proteins combined with diphtheria and tetanus toxoids. Interestingly, the inclusion of the pertussis components in the vaccine appears to enhance the antibody responses to the diphtheria and tetanus toxoids. Completion of a full series of DTaP doses while a child is young is important: 83% of vaccinees receiving three doses are completely protected from

these diseases, but only 36% are resistant to pertussis after one dose. In addition, the risk of adverse effects of the DTaP vaccine is greater when an individual receives his or her first dose as an adolescent or an adult. For both diphtheria and tetanus, immunity is known to wane with time and booster shots are necessary to maintain a high level of protection. Anti-tetanus protection drops within 10 years of the last booster. In the case of diphtheria, outbreaks among adults in certain parts of Europe and England in the 1980s were the first indication that adequate protection might not be in effect 20 years after immunization. As for the pertussis component of the DTaP vaccine, the acellular form of the vaccine has not been in use long enough to determine the duration of immunity induced. However, protection with the older whole cell vaccines was observed to decline within 10 years of immunization. Although pertussis boosters may be required for optimal long-term protection, adults and adolescents contracting *B. pertussis* infection experience far fewer ill effects than do children. Thus, boosters containing only diphtheria and tetanus toxoids are routinely recommended for adults every 10 years. Adults that increase their risk of tetanus exposure by (for example) cutting themselves on rusty metal are usually given a therapeutic tetanus booster as soon as possible.

IV. *HAEMOPHILUS INFLUENZAE* TYPE b

H. influenzae is a species of encapsulated extracellular bacteria. Different strains of this species bear different types of capsules. *H. influenzae* with the type "b" capsule (Hib) are particularly pathogenic to humans. These organisms preferentially colonize the mouth and throat, but the infection can disseminate throughout the body and cause fatal meningitis if the immune defenses are sufficiently depressed. Anti-bacterial antibodies provide very good defense against this pathogen, and the disease resistance of an individual correlates directly with his or her titer of circulating anti-bacterial antibodies. While maternal anti-Hib antibodies can protect infants for a few months, the young child is vulnerable until about age 3–4 years when sufficient natural antibodies recognizing Hib are produced. These cross-reactive antibodies are likely induced by the presence of benign bacteria bearing capsules containing the same or a very similar polysaccharide to that in the Hib capsule.

The first vaccines developed to combat Hib infection were produced in the early 1980s and were based on a Hib capsule polysaccharide called PRP (*polyribosylribitol phosphate*). However, PRP is a Ti antigen that is highly immunogenic in adults but only weakly so in infants. As a result, these first vaccines were not very effective in protecting the population most at risk. In the late 1980s, four different conjugate vaccines were developed in which PRP was combined with a carrier protein to form a Td antigen with improved immunogenicity in infants. In current vaccines, PRP is combined with diphtheria toxoid, tetanus toxoid, or the *outer membrane protein* (OMP) of *Neisseria*. The Hib conjugate vaccine is sometimes given in combination with the DTaP vaccine.

Hib vaccines have been very effective in the developed world. In 1986, the year prior to introduction of the vaccine, over 13,000 cases of invasive disease and over 8600 cases of meningitis caused by Hib were recorded in the United States; in 1998, only 228 cases of Hib infection were reported. However, the picture is much less rosy in the developing world, where the cost of the conjugate vaccines is prohibitive. The WHO has estimated that less than 2% of global Hib disease is prevented annually, which translates into about half a million infant deaths worldwide.

V. HEPATITIS A

The HepA virus is a non-enveloped single-stranded RNA virus that causes liver damage. Until the 1990s, passive immunization with immunoglobulin from plasma collected from thousands of donors was the only protection available against HepA-related disease. However, once researchers discovered how to grow the virus in cell lines, the *in vitro* production of vaccines began. Inactivated, whole virus vaccines are now licensed and available in many countries, while live, attenuated vaccines have been approved and used to a more limited extent. Studies of vaccine efficacy suggest vaccination is highly effective in protecting against HepA infection and that the immunity induced may last for decades. Vaccination is currently recommended for those 2 years and older who either live in or plan to travel to places where HepA is endemic. As mentioned previously, sewage workers as well as those who work in food service, day care centers, and health care facilities are also considered vaccination candidates. Prior to vaccine availability, the number of HepA cases reported in the United States reached a high of 35,000 per year. The annual infection rate had reached record lows by the late 1990s and is expected to decline even further.

VI. HEPATITIS B

As described in Chapter 22, the hepatitis B virus causes severe liver damage and hepatocarcinoma. HepB infection becomes chronic in about 90% of infected infants, and up to 25% of these will die of HepB-associated liver disease as adults. The virus is highly infectious, meaning that chronically infected individuals can easily pass on the virus to their contacts. As described earlier in this chapter, the current HepB vaccine consists of the HBsAg protein derived using recombinant DNA techniques. Protective levels of neutralizing antibodies are produced in 95% of infants receiving this vaccine, making it one of the most successful vaccinations ever devised. Furthermore, considering the costs to the health care system of later caring for those with HepB-induced liver cirrhosis or cancer, HepB vaccination is remarkably cost-effective.

In the early 1990s, some physicians raised concerns about a possible association between the HepB vaccine and the triggering of episodes of multiple sclerosis (MS; see Ch.29) in susceptible patients. However, infection with wild HepB is not a risk factor for MS. Moreover, in a well-controlled clinical study in 2001, no increased risk of MS was found in a large group of women who had received the HepB vaccine.

VII. INFLUENZA VIRUS

In humans, the influenza virus attacks and destroys respiratory epithelial cells and causes severe acute respiratory symptoms. Influenza virus infection is particularly problematic for very young and very old individuals whose immune systems are functioning at less than full strength. Persons suffering from underlying chronic respiratory disorders, such as cystic fibrosis, are also at increased risk for disease even though their immune systems may be normal. Three serotypes of influenza virus, A, B, and C, can be distinguished by the antigenic characteristics of their structural proteins. Influenza type A, which can infect both humans and birds, is the most pathogenic to humans. In a natural infection, neutralizing antibodies directed against the hemagglutinin (H) protein, which mediates attachment to host cells, and the neuraminidase (N) protein, an enzyme that assists in the release of progeny virions from the infected host cell, play key roles in immune defense. However, because of the propensity for influenza virus to undergo antigenic drift and shift (refer to Ch.22), it has been impossible so far to produce a vaccine that confers life-long protection against all strains of influenza. Consequently, annual vaccination programs are undertaken in developed countries just prior to "flu season." Previously, the elderly and workers in hospitals and geriatric facilities were the principal targets of these programs, but more recently all adults (and many children) have been encouraged to "get their flu shot" as winter approaches. How do the authorities know which variant of the virus to target for vaccine production? In February of any given year, the WHO studies which flu viruses are emerging around the world and chooses a combination of candidate strains for production of that year's vaccine. For example, for the 2001–2002 flu season, it was recommended that flu vaccines contain the New Caledonia/20/99 (H1N1)-like influenza A virus, the Moscow/10/99(H3N2)-like influenza A virus, and the Sichuan/379/99-like influenza B virus. However, if the strains chosen turn out not to be responsible for the majority of flu cases in that year, the vaccine will not be effective. It is estimated that this "mismatching" occurs once every 8–10 years.

Several forms of influenza vaccine exist. The original vaccine, based on formalin-inactivated whole virus, can result in unpleasant side effects. However, many subunit vaccines, which are safe to use in infants and the elderly, provide only limited protection against infection with a live virus because only low levels of antibodies are produced that have a short half-life. As discussed previously, a virosome vaccine bearing components of the influenza virus has superior immunogenicity to the subunit vaccine and causes fewer side effects than the whole virus vaccine. The virosome vaccine has proven effective in particularly vulnerable populations such as infants, the elderly, and CF patients. However, the problem of extreme antigenic variation associated with the influenza virus remains. Researchers are still struggling to identify an invariant influenza virus epitope capable of inducing a protective immune response.

VIII. JAPANESE B ENCEPHALITIS

The Japanese B encephalitis (JBE) virus is a mosquito-borne pathogen that is the leading cause of encephalitis and neurological infections in Asia. The virus is endemic in much of Asia and also infects pigs, allowing the virus to propagate readily in rural areas. At least 50,000 cases per year are formally reported, and 1 in 300 infections leads to symptomatic disease. Most infections occur in adolescents and children, and the severity of the infection depends on whether the virus reaches the neurons. The virus replicates very efficiently within these cells and destroys them, causing encephalitis. The early symptoms of JBE are high fever of sudden onset, headache, chills, vomiting, photophobia, areas of mild paralysis, and rigid neck. Respiratory symptoms, seizures, and abnormal reflexes often follow. No effective treatment exists, and 5–35% of symptomatic infections are fatal. Three vaccines exist: an inactivated JBE virus derived from mouse brain, an inactivated JBE virus derived from cell culture, and a live, attenuated JBE virus vaccine. The latter two vaccines are available only in Asia where the disease is endemic. Problems associated with these vaccines are high cost, limited production, and only short-term protection (3–4 years). In some jurisdictions (but not all), there have been reports of an increased frequency of neurological complications following administration of a JBE vaccine.

IX. MEASLES, MUMPS, AND RUBELLA

i) Measles

As noted in Chapter 22, the measles virus causes fever and a characteristic rash of red spots, and can induce temporary, general depression of both antibody and cell-mediated responses. Potentially fatal opportunistic infections may thus gain a foothold. Those individuals who survive a natural exposure to the measles virus usually acquire life-long protection against re-infection. Fortunately, unlike the influenza virus, the measles virus exhibits limited variation in its surface proteins. There are several different current vaccines for measles but all are based on live, attenuated viruses. A vaccine that contains live, attenuated measles virus alone is generally referred to as *measles-containing vaccine* (MCV). Live, attenuated measles virus also forms part of the MMR vaccine, discussed below.

ii) Mumps

"Mumps" was first described as a disease in the 5th century BC, and was named in a Scottish report published in 1790 by a scientist who used the name given to the affliction by the "common people." The mumps virus is an enveloped RNA paramyxovirus. The disease typically starts as a respiratory infection that often leads to viremia associated with fever, headache, muscle pains, and loss of appetite. Traveling in the blood, the virus readily infects many different glands but particularly the salivary glands. The signature symptom of mumps is a pronounced swelling of the parotid salivary gland (located in front of the ear). Mumps is generally a childhood disease, although those growing up in relatively isolated settings in the developed world may escape early exposure and thus be susceptible to infection as an adult. Outbreaks of mumps were recorded among American soldiers in France during World War I, when groups of young adults who had never experienced mumps were suddenly crowded together in military barracks. A demand for a vaccine against mumps arose from the dangers lurking in the severe complications that are associated with the disease. Sudden and permanent deafness in one or both ears can result, and meningoencephalitis occurs in 5% of all cases. In adults, inflammation of the testicles or breasts occurs in 30–40% of cases. The current vaccine is a live, attenuated form of the virus included in the MMR vaccine. Vaccination is very effective protection against mumps: the number of cases of mumps in the United States dropped from 152,000 in 1967 (just before vaccine use was initiated) to 2982 cases in 1985. Any subsequent upswing in mumps incidence in adolescents has been associated with a failure to vaccinate during childhood.

iii) Rubella

Rubella ("little red") is also known as "German measles," since the physicians who first distinguished this disease from other rash-related disorders were German. The rubella virus is a cuboidal, enveloped RNA virus and a member of the togavirus family (but is not transmitted by arthropods). In young children, rubella is a relatively mild disease characterized by a rash, low fever, some malaise, and mild conjunctivitis. Life-long immunity results from childhood infection. The danger in rubella lies in its teratogenic effects on the developing fetus of a woman who never had the disease as a child. Rubella viremia in naive pregnant women can lead to infection of the placenta, from which the virus enters the fetal circulation and infects fetal organs. If infection occurs within the first trimester, severe consequences for the fetus can result, including cataracts, heart disease, deafness, and sometimes hepatitis and mental retardation. Taken together, these symptoms have been termed *congenital rubella syndrome* (CRS).

Up until 1962, there was not much of a push for a rubella vaccine, since the disease was generally held to be fairly benign. However, in 1962–1964, a pandemic occurred in Europe and the United States in which 12,500,000 cases of rubella were recorded leading to 20,000 cases of CRS, 6000 spontaneous abortions, and 2000 stillbirths. Of the 20,000 surviving children affected by CRS, 11,600 were deaf, 3580 were blind, and 1800 were mentally retarded. It was estimated that the total economic burden of the pandemic was U.S. $1.5 billion. The impact of the pandemic was such that the rubella virus was quickly isolated and attenuated and a commercial vaccine developed by 1969. Outbreaks of the type described previously do not seem to occur in developing countries, perhaps because the disease is contracted by almost all children. As a result, almost no mothers are naive to this virus, and their fetuses are protected. The current vaccine for rubella is included as a live, attenuated virus in the MMR vaccine.

iv) MMR Vaccine

The combination MMR vaccine can confer life-long protection against measles, mumps, and rubella. This vaccine was introduced into the developed world in the late 1960s and immediately cut the incidence of measles, mumps, and rubella in these countries by about 98%. The current MMR vaccine is administered parenterally and is made up of live, attenuated versions of the measles, mumps, and rubella viruses. The vaccine is effective in inducing protective anti-measles antibody production in 98% of children vaccinated at age 15 months. Similar levels of protection are achieved for mumps and rubella. Both antibody and cell-mediated immune responses are mounted and long-lasting memory is induced.

It should be noted that maternal antibodies have a significant influence on the effectiveness of the MMR vaccine. Mothers who experienced a natural measles infection in their childhood pass on anti-measles antibodies that last for 11 months after birth in their infants. While these maternal antibodies protect the infant, they also bind to the vaccine before it can induce the infant's immune system to produce its own antibodies to the virus. The MMR vaccine was thus originally administered to infants at 15 months of age, to ensure that all maternal antibodies had disappeared. However, for unknown reasons, mothers who are immune to measles because of vaccination produce anti-measles antibodies that last only 9 months in their offspring. Under the original regimen, these infants were without antibody protection against measles for close to 6 months. Administration of the MMR vaccine is now recommended at 12–15 months, depending on maternal history. Despite these measures, local measles outbreaks occurred in the United States in the 1980s in children who had been vaccinated according to the recommended regimen. It was found that anti-measles antibody production is not initiated or is sub-optimal in 2–5% of 12-month-old infants receiving one dose of MMR vaccine. Current regimens recommend that a first dose of MMR vaccine be given at age 12–15 months followed by a second dose at age 4–5 years. Because measles is

now a relative rarity in the developed world, many North American parents have adopted a rather cavalier attitude toward it. It should be borne in mind that hundreds of thousands of children around the world still die from measles infection each year (Fig. 23-6). Even in North America, local outbreaks in the early 1990s caused the deaths of several unvaccinated high school students (see later).

Considering our previous discussion on the safety issues surrounding the use of live viruses as vaccines, why not use killed viruses in the MMR vaccine? In fact, the first versions of measles vaccines did contain inactivated virus. However, when exposed to the natural measles virus, some vaccinees developed "atypical measles" characterized by a different type of rash and lung inflammation. It was discovered that the effectiveness of the humoral anti-measles response rested on the induction of antibodies recognizing two key viral proteins. Vaccination with the inactivated virus failed to induce antibodies to one of these two viral proteins, leaving the vaccinee partly vulnerable to the natural virus. The success of the current MMR vaccine also works against the development of alternative measles vaccines, since there is little incentive for drug companies to put resources toward replacing a vaccine that is already statistically very safe. Indeed, in 1996, the WHO convened a consultative group that concluded that global measles eradication was feasible with the vaccines currently available. Nevertheless, research is proceeding aimed at developing a subunit or DNA measles vaccine that would overcome maternal antibody and allow vaccination at a younger age. Another goal is to produce a vaccine that allows simplified administration during mass immunization campaigns.

One last note on the MMR vaccine: in the late 1990s, concerns were raised about a possible association between this

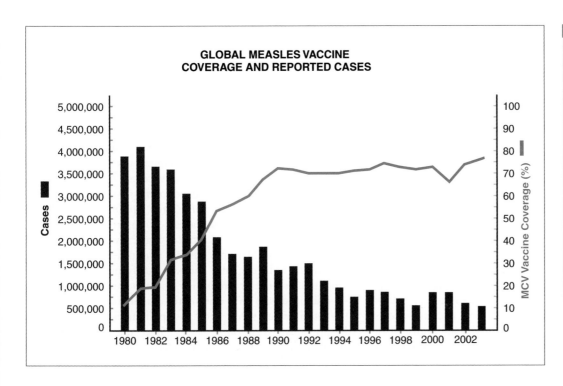

GLOBAL MEASLES VACCINE COVERAGE AND REPORTED CASES

Figure 23-6

An Example of the Effectiveness of Vaccine Coverage
The graph shows how disease incidence, in this case measles, can fall dramatically with increasing vaccine coverage. Note, however, that even at a global coverage rate of 70%, hundreds of thousands of individuals still come down with the disease. Note also that when MCV use declined briefly in 2000, there was a spike of increased disease incidence. Adapted from Vaccine, Immunization and Biologicals, WHO, 2003; www.who.int/vaccines-surveillance/graphics/htmls/measlescascov.htm.

vaccine and autism in developed countries. However, extensive studies and retrospective assessments have confirmed that MMR vaccination does not increase the risk of autism.

X. MENINGOCOCCUS

The encapsulated bacterium *Neisseria meningitidis* (commonly called meningococcus) causes bacterial meningitis characterized by severe inflammation of the membranes of the brain. The disease most often strikes children and adolescents, is readily transmitted through respiratory secretions, and is fatal in 5–15% of cases. Survivors may suffer from neurological defects leading to mental impairment, deafness, or seizures. Outbreaks can be severe, as evidenced by the 1996 epidemic in Western Africa that infected 250,000 people and killed 25,000 of them. Ninety percent of meningococcal disease is caused by three serogroups of meningococcus defined by their capsular polysaccharides: groups A, B, and C. Group A infections dominate in Africa, while group C infections are most prevalent in the Americas, Australasia, and some parts of Europe. Group B outbreaks have occurred most recently in Europe, Latin America, and New Zealand.

Both the capsular polysaccharide and the endotoxin produced by these bacteria contribute to their virulence, and humoral immunity is the key to successful host defense. Vaccines have been devised based on the capsular polysaccharides of group A and/or C bacteria alone, or on conjugates of the group C polysaccharide plus a protein carrier (such as diphtheria toxoid). Only the conjugate vaccines are effective in very young children, but all have a good safety record. Boosters are required every 3–5 years for the polysaccharide vaccines, but a single dose of a conjugate vaccine generates long-lasting protective memory. Vaccines based on group B polysaccharides are not efficacious to date, probably because this polysaccharide resembles an antigen prevalent in the human CNS. Unfortunately, vaccines based on group A or group C polysaccharides offer no protection against group B infection. In addition, the concept of herd immunity does not appear to apply to the group A/C vaccines, meaning that all (rather than most) members of a population must be immunized to prevent a wide outbreak. Mass pre-emptive vaccination against group C meningococcus in young children and adolescents has been instituted in several Western countries. Development of a cheap, effective group A conjugate vaccine for use in countries in the "African meningitis belt" is a priority of the WHO.

XI. PLAGUE

Plague is caused by the gram-negative bacillus *Y. pestis*. *Y. pestis* naturally infects rodents (especially rats) and their fleas, and is transferred to humans by flea bites. At the site of the bite, the skin becomes blistered and blackened (hence, "Black Death"). Within a week of the bite, the bacteria access the draining lymph node via the lymphatics and cause high fever and a large, painful swelling of the node known as a "bubo" (hence, *bubonic plague*). From the lymph node, the bacteria then spread to the blood and organs with disastrous speed. Once the bacteria access the blood, the plague is said to be *septicemic* in form. Septicemic plague can also result from the direct introduction of bacteria into the circulation. If the bacteria reach the lungs, pneumonia occurs and *pneumonic plague* is said to be present. Unfortunately, pneumonic plague is also readily spread by the inhalation of respiratory droplets expelled by the coughing or sneezing of an infected person. In the absence of antibiotic treatment, all three forms of plague have a high fatality rate: 50% for bubonic, 75% for septicemic, and 90% for pneumonic. Plague has been a major killer of human populations several times in the past. For example, this disease resulted in 100 million deaths over 50 years in the 6th century AD, and 25 million deaths in Europe over 5 years in the 14th century. Today, plague is endemic in many countries in Africa, Latin and South America, and Asia, but the risk to travelers to these areas is low as long as they avoid areas where rats congregate. The WHO records about 200 deaths from plague each year, the vast majority in Africa.

The current plague vaccine is based on formalin-killed whole bacterial cells. However, vaccination is recommended only for adult, high-risk occupational groups, such as health care workers in endemic areas, disaster relief workers, and laboratory personnel working directly with *Y. pestis*. Side effects of fever, headache, pain, and malaise are common and increase in severity with repeated doses. In addition, although various plague vaccines have been in use since the late 1800s, their efficacy has yet to be formally evaluated. Control of the local rodent population and the elimination of habitats that these animals favor is often a more cost-effective and efficient way of decreasing plague exposure. When vaccination is required, three doses are given over 6 months, with a booster every 6–12 months if potential exposure to *Y. pestis* continues.

XII. POLIO

As noted in Chapter 22, polio virus is the very infectious enterovirus that causes poliomyelitis. Polio virus occurs in three serotypes, called 1, 2, and 3, which are distinguished by their viral coat proteins; however, all three serotypes cause the same symptoms. In a natural infection, both antibodies and cell-mediated responses are needed to eliminate the virus. As mentioned earlier in this chapter, two polio vaccines are currently available: the Salk vaccine, more commonly known now as the <u>i</u>nactivated <u>p</u>olio virus <u>v</u>accine (IPV), and the Sabin vaccine, or <u>o</u>ral <u>p</u>olio virus <u>v</u>accine (OPV). IPV is administered by intramuscular injection, requires repeated boosters, and does not induce high levels of mucosal immunity in the GI tract. OPV is administered orally, does not require boosters, and induces gastrointestinal mucosal immunity. However, there is also a small chance (one case per 2–3 million doses) that one of the attenuated viruses in the OPV vaccine may revert and cause poliomyelitis in a vaccinee (the previously mentioned VAPP). Since the wild polio virus has been eliminated in North America, the risk of VAPP is greater than the risk of getting the disease, such that the OPV vaccine is no longer recommended for use in the United States. (The litigious nature of this country may also have something to do with it: a case of VAPP can lead to a $3 million lawsuit.) Rather, four doses of IPV spread over infancy to age 4–6 years is the preferred protocol in the United States. The OPV vaccine is still recommended for vaccination programs elsewhere in the world where the wild polio virus is endemic and sterile injections and repeated clinic visits are an issue. A plus to the OPV vaccine is that vaccinees in these areas who shed the attenuated polio virus in their feces can unwittingly immunize non-vaccinated individuals in their communities, extending vaccination coverage.

In 1988, when the WHO launched its global polio virus eradication effort, the wild virus was endemic on five continents and had caused the paralysis of 350,000 children. The number of reported cases of wild polio virus infection has since declined by 95% worldwide. The original objective of the campaign was to eliminate the virus completely by 2000. Although the virus has now been eradicated in more than 175 countries, including all of North and South America, Europe, and the Pacific Rim, it can still be found in parts of southern Asia and Africa. It is hoped that the virus will be wiped out completely in the wild by 2005. *National Immunization Days* (NIDs), country-wide blitzes in which outreach workers go house to house in an attempt to vaccinate each and every child in the country, are key tools in the fight to eradicate polio. For example, more than 470 million children under 5 years of age in 83 countries were vaccinated during a series of NIDs in 1998, including 134 million children immunized on a single day in India. Even in areas of armed conflict, cease-fires have been held to permit the vaccination of local children. In Somalia in 1998, no weapons were permitted in villages on NIDs and mines were removed from roads to facilitate access to outreach workers. These NIDs have other benefits. Outreach workers can monitor the children brought in for vaccination for other infectious diseases such as tetanus and measles, and administer vitamin supplements. Interaction with parents during the vaccination procedure can smooth the delivery of information, leading to better disease prevention in general.

XIII. RABIES

Rabies is caused by a slow-replicating virus that causes progressive encephalitis. The disease is inevitably fatal in humans if left untreated. Rabies kills 50,000 people worldwide (most in developing countries), and more than 10 million people each year exposed to the virus (usually in the form of a dog or bat bite) receive a combination of post-exposure vaccination plus a preparation of anti-rabies antibodies. In developed countries, rabies is kept under control by the vaccination of wildlife, domestic pets, and stock animals. In developing countries, however, vaccination programs are rare and bites by stray or feral dogs or wolves are common.

Prior to 1967, rabies vaccines were derived from crude preparations of nerve tissue. Administration of these vaccines required an extended and painful course of 23 injections, and neurological side effects were not infrequent. For these

reasons, these vaccines were only given post-exposure. Today's rabies vaccines are based on killed viruses grown up in human cell cultures or chick embryos, and are highly efficacious whether given pre- or post-exposure. For example, three doses of a modern rabies vaccine given prophylactically evoke an antibody response in 99% of vaccinees. In developed countries, only individuals who might expect to encounter rabies in their line of work, such as wildlife workers or veterinarians, receive the vaccine prophylactically. Travelers intending to go to countries where rabies is more common may also consider getting vaccinated. The cost of modern rabies vaccines inhibits widespread pre-emptive vaccination in developing countries. Patients in these regions receive a post-exposure administration of rabies vaccine that usually involves five doses given over 1 month. Serious side effects of rabies vaccines are extremely rare.

XIV. *STREPTOCOCCUS PNEUMONIAE*

One of the newer infant vaccinations to join the recommended list is the conjugate vaccine for *S. pneumoniae* (commonly known as pneumococcus). Pneumococci are encapsulated extracellular bacteria that initially colonize the upper respiratory tract. The capsule forestalls engulfment by pulmonary phagocytes such that the bacteria multiply in great numbers in the lung and then spread easily throughout the body. Thousands of cases of pneumonia, otitis media, bacteremia, and meningitis are attributed to this bacterium every year. The first vaccine developed against this pathogen was a preparation of the principal capsule polysaccharide. Modern versions of the polysaccharide vaccine (*pneumococcal polysaccharide vaccine*; PPV-23) contain the polysaccharides of the 23 strains of *S. pneumoniae* that represent 80–90% of disease-causing strains. However, as noted previously, polysaccharides are Ti antigens and do not induce memory responses in children under 24 months of age. To protect infants, a conjugate heptavalent vaccine called *pneumococcal conjugate vaccine-7* (PCV-7) was developed. PCV-7 contains the capsule polysaccharides of the seven most common strains of *S. pneumoniae* attacking infants plus modified diphtheria toxoid. When administered intramuscularly, this vaccine functions as a Td antigen and leads to effective primary and memory responses in about 90% of vaccinated infants. Unfortunately, the PCV-7 vaccine induces immunity to only 50% of the strains that cause disease in older children and adults. As a result, some children particularly susceptible to *S. pneumoniae* infections may need a booster of PCV-23 vaccine a little later in life.

XV. TUBERCULOSIS

We learned much about the intracellular bacterium *M. tuberculosis* in Chapter 22. Although the majority of adults and children that become infected with this organism develop no signs or symptoms at any time in their lives, TB is still one of the worst killers among the infectious diseases, responsible for more than a million deaths worldwide each year. The rise of multi-drug resistant strains of *M. tuberculosis* and the inadequacy of the existing vaccine prompted the WHO to declare TB to be a global health emergency in 1993.

It has been very difficult to develop an effective vaccine against TB because we still do not completely understand natural immunity to *M. tuberculosis* or why the Th1-dependent cell-mediated response that normally combats this pathogen is not always totally protective. To date, the only vaccine available to fight TB is the BCG vaccine. As noted earlier in this chapter, BCG is a live, attenuated form of *M. bovis* that shares many antigens with *M. tuberculosis*. The original BCG strain underwent a spontaneous deletion that removed genes conferring virulence without compromising immunogenicity. However, numerous mutations have occurred in the BCG genome over the many decades this organism has been maintained in the laboratory, meaning that some of the original immunogenic sequences may have been modified or lost. Eight major trials of BCG vaccine have been conducted over the years, with most involving children but some including vaccinees of any age. In these studies, cohorts of vaccinees were followed for periods ranging from 10 to 20 years. Protection levels in these trials ranged from zero to 80%. Many theories have been advanced to explain this amazing variation in efficacy, including problems with trial methodology, vaccine variations, and regional differences in BCG strains. One theory that has garnered much attention suggests that the presence of environmental mycobacteria (non-pathogenic species related to *M. tuberculosis*) may have a negative influence on vaccination results. This hypothesis is based on the observation that vaccination with BCG is highly effective (protects close to 80% of vaccinees) and confers life-long protection when the vaccinated population is from an area of the world where TB is uncommon (typically in prosperous areas of developed countries). However, in locations where infection with *M. tuberculosis* is endemic, such as in developing countries, BCG vaccination often fails to protect vaccinees against TB. The immune response induced against the environmental mycobacterial antigens may cross-react with antigens in the BCG vaccine, perhaps interfering with BCG replication and hence the efficacy of the vaccine. Exposure to environmental mycobacteria may also prime the individual to mount an unfavorable Th response. A Th1 response protects an individual from TB disease, whereas progressive TB is usually associated with a Th2 or a mixed Th1/Th2 response. Thus, if a local encounter with environmental mycobacteria has previously primed a Th2 response, the individual's subsequent response to BCG may be pre-disposed to be non-protective.

In developing countries with a high prevalence of infectious TB, the WHO recommends a single dose of BCG vaccine for newborns. Unfortunately, however, BCG does not induce life-long memory, and vaccinated children are protected against TB for only 10–15 years due to the progressive loss of TB-specific CD4[+] memory T cells. A booster strategy similar to that employed for tetanus and diphtheria is thus recommended. Routine vaccination for TB is not the norm in North America because the general population is at low risk of infection. In such regions, conscientious case detection and proper drug therapy are considered the most successful means of TB control. BCG vaccination should be considered for high-risk groups such as infants and children who have negative tuberculin tests but are continually exposed to people with untreated or ineffectively treated infectious pulmonary TB. In adults who have been neither infected nor immunized, the efficacy of current BCG vaccines is unknown. This is becoming an important question in developed countries where health care workers with no previous exposure are now dealing with TB cases among isolated groups at high risk (those in jails, hospitals, shelters for the homeless). Considering that more than 4 billion doses of BCG vaccine have been administered worldwide, it is surprising how many questions still remain about this vaccine. Much research is currently focused on developing a vaccine using *M. tuberculosis* itself. Some of the challenges facing scientists pursuing this goal are discussed in Box 23-6.

XVI. TYPHOID FEVER

Typhoid fever is caused by *Salmonella typhi*, a highly invasive intracellular bacterium whose only natural host is humans. (Typhoid fever was so-named because of the similarity of its symptoms to typhus; typhus is a distinct disease caused by infection with a rickettsial organism.) An acute infection is characterized by high fever, abdominal discomfort, malaise, and headache that can last for several weeks. Many (but not all) patients get a rash of salmon-colored spots (hence, *Salmonella typhi*). Life-threatening complications include intestinal perforation and hemorrhage. Prior to the discovery of antibiotics, typhoid fever was a severe disease, killing 10–20% of patients.

S. typhi is ingested via food and/or water contaminated with fecal matter from infected individuals. Ingested bacteria initially enter either enterocytes or M cells in the small intestine but eventually spread to macrophages throughout the body. In 2–5% of patients, bacteria reach the gallbladder and establish a chronic infection. These people become asymptomatic carriers who shed the bacteria in infectious form. The most famous carrier was "Typhoid Mary," a cook in New York who unknowingly infected 53 people and killed five in 1906. While typhoid fever is now a rarity in North America, where good water treatment prevails, the WHO estimates that more than 30 million cases of typhoid fever occur worldwide each year, resulting in about 500,000 deaths.

Inactivated, whole cell typhoid vaccines (developed originally for the military) have been administered parenterally since the late 1800s. This vaccine is reasonably efficacious (70%) and confers protection for 5 years, but often produces unpleasant local and systemic side effects. Two modern vaccines have been developed: Vi, a preparation of purified bacterial capsule given parenterally, and Ty21a, an attenuated strain of *S. typhi* given orally. Both these vaccines are less toxic than the original but are less efficacious and offer shorter protection (2–3 years). Development of additional vaccines against typhoid is fast becoming a key priority, because some of the over 100 known strains of *S. typhi* have become resistant to almost all known antibiotics. The complete genome of *S. typhi* has now been sequenced, facilitating the identification of new protective epitopes.

Box 23-6. **The challenges of tuberculosis vaccination**

New hope for TB vaccine development has surged with the complete sequencing of the *M. tuberculosis* genome (3924 genes). Functions could be ascribed to a majority of these genes and much research is now directed toward identifying genes that can be mutated to decrease the virulence or persistence of the organism without affecting its immunogenicity. While other vaccine strategies could be explored, the focus continues to be on development of live, attenuated *M. tuberculosis* vaccines. As we have learned, such vaccines supply many different antigens and replicate in the host (so that antigen persists), factors that favor a comprehensive and vigorous immune response characterized by long-lived memory. These characteristics are particularly useful in combating a pathogen such as *M. tuberculosis* that can lurk intracellularly for decades.

Researchers have developed several mutant *M. tuberculosis* strains that have lost the capacity to make their own essential amino acids or outer cell wall lipids required for bacterial survival in the lung. However, there appears to be a direct correlation between virulence and immunogenicity: the less virulent a bacterial strain, the less protective it is. One exception is the tryptophan auxotrophic mutant of *M. tuberculosis*, which requires exogenous supplies of tryptophan to survive. Mice vaccinated with this strain experienced fewer side effects and were better protected against a challenge with wild-type *M. tuberculosis* than mice that received the BCG vaccine.

While attempts to develop an improved live, attenuated vaccine for TB are ongoing, concerns about the general safety of such vaccines have spurred the search for *M. tuberculosis* antigens that can be used in subunit or DNA vaccines. However, these efforts have been hampered by the lack of a reliable *in vitro* system to test for the protective efficacy of an antigen. Given the long latency period of *M. tuberculosis* in humans (up to 20 years), clinical trials of immense length must be conducted to fully judge the efficacy of a vaccine. Even more frustrating is the fact that no one antigen appears to be able to protect against all stages of *M. tuberculosis* infection. In animals, some antigens that provoke almost no measurable response during the acute infection are very protective against re-infection, while antigens that induce a response during the acute phase often utterly fail to protect against re-infection. This phenomenon is no doubt why so many TB patients experience re-infection or reactivation

of disease long after the acute infection has resolved.

One parameter used to assess prospective vaccine epitopes *in vitro* is the determination of how vigorously the epitope stimulates cultured T cells to proliferate. *M. tuberculosis*-infected cells secrete several low molecular weight bacterial products into the culture medium that stimulate cultured T cells. Four of these molecules, called ESAT-1, ESAT-6, hsp60, and antigen 85A/B, are also strongly recognized during the course of a natural infection. When subunit vaccines were devised based on these molecules, protection against TB about equivalent to that provided by BCG vaccination was obtained in experimental animals. DNA vaccines employing these and other novel *M. tuberculosis* antigens have also had modest success in animal models. DNA vaccines are of particular interest for TB because they efficiently induce the CTL responses needed to eliminate intracellular bacteria. One area where subunit and DNA vaccines may prove especially useful is for boosting immunity in adults as the protective effect of the initial BCG vaccine given in childhood begins to fade. TB patients apparently retain both CD4$^+$ and CD8$^+$ T cell responses to ESAT-6 antigens for decades, and healthy individuals exposed to *M. tuberculosis* respond to ESAT-6 vaccine. A strategy in which the primary immunization consisted of a naked DNA vaccine encoding ESAT-6, followed by a booster with vaccinia virus vector expressing ESAT-6, induced substantial anti-mycobacterial CD4$^+$ and CD8$^+$ T cell responses in mice.

Another strategy to identify TB vaccine candidates has been to compare the proteomes of *M. tuberculosis* and *M. bovis* BCG, looking for antigens that are present only in the virulent *M. tuberculosis*. In theory, some of the 32 antigens found to be exclusive to *M. tuberculosis* could make good vaccine candidates, and several are being investigated for efficacy. One candidate of particular interest is the ESAT-6 antigen mentioned previously. Because ESAT-6 is not present in *M. bovis* BCG, it may be a virulence factor for *M. tuberculosis*. Over two dozen other putative T cell antigens (identified using mass spectrometry and an *in vitro* IFNγ assay) are also under investigation.

Some researchers have adopted another approach to TB vaccine development that is sometimes called *reverse vaccinology* or *epitope-driven vaccine design*. In this

approach, the researcher peruses the pathogen genome or proteome and tries to select sequences that are most likely to be epitopes that are "seen" by the infected patient's T cells. The researcher screens out those prospective T cell epitopes that (1) are homologous to already-known (and unsuccessful) epitopes; (2) are shared with non-pathogenic strains; or (3) bind only weakly to MHC proteins and so are unlikely to be recognized during infection *in vivo*. The remaining, smaller number of epitopes can then be screened for possible vaccine efficacy.

Unfortunately, a major stumbling block in developing any new TB vaccine has been the lack of a sufficiently accurate animal model to use for testing. Immune responses in general are well-characterized in mice, and immunological tools such as purified antibodies and cytokines are readily available. However, mice are not highly susceptible to *M. tuberculosis* and do not develop the severe granulomas and lesions seen in infected humans. In contrast, guinea pigs are very sensitive to *M. tuberculosis* and suffer lung lesions reminiscent of TB in infected humans. However, the relevant cytokines and antibodies needed to exploit this system have yet to be developed. Moreover, the acute course of disease in both these animal models does not parallel the chronic course of the human disease. Ideally, the disease should appear months or even years after infection in the animal. Because there are no adequate small mammal models for TB, researchers have had to turn to trials in primates in which the course of *M. tuberculosis* infection resembles that in humans. While primate trials are expensive and more difficult to carry out, it is hoped that improved *in vitro* methods for preliminary vaccine assessments will allow available resources to be concentrated on the candidates showing the greatest potential for success.

A last challenge to be faced in the development of TB vaccines is the current lack of optimal adjuvant options. The standard adjuvant alum promotes primarily Th2 responses, which actually enhance the detrimental effects of TB. Experiments with alternative delivery systems such as microspheres have been attempted, but none so far has offered superior protection to the venerable BCG vaccine. Millions around the world wait in hope that all these difficulties will soon be overcome and that prevention of this killer disease by an effective vaccine will become a reality.

XVII. VARICELLA ZOSTER (CHICKEN POX)

Chicken pox in children is characterized by fever and an outbreak of itchy red spots over most cutaneous and exterior mucosal surfaces. In its acute phase, the virus is highly contagious and transmission is via respiratory secretions and direct contact with skin lesions. As a result, almost all young children in endemic areas contract chicken pox at some point during their school years. In some cases, high fever, pneumonia, or encephalitis arises as a complication of varicella infection, necessitating hospitalization. In adults and adolescents, latent varicella from the original infection can become reactivated and cause *shingles*, a very painful resumption of viral activity in the skin. This reactivation is usually associated with extreme stress or immunosuppression and can occur in individuals vaccinated against chicken pox. This observation suggests that the virus is never totally cleared but only kept in check by the immune system. When immune responses are compromised, the virus resurges.

As noted earlier in this chapter, the development of the varicella vaccine was originally driven by North America's obsession with cost reduction. However, this vaccine is a great boon for the very small number of children with skin disorders in whom chicken pox would lead to unusually severe scarring. The current varicella vaccine, which contains live, attenuated virus, is highly immunogenic and one dose is sufficient to induce antibody production in 97% of school-aged vaccinees. However, the duration of immunity conferred by vaccination is currently unknown. Although studies indicate that vaccinees retain circulating anti-varicella antibody for several years after vaccination, it is unclear whether part of this persistence may be due to regular re-exposure to varicella zoster in the community. Such sporadic natural exposure, which may serve to keep protection levels high, may no longer exist once vaccination has become a widespread practice for a number of years. Thus, there are concerns that vaccinating young children against the disease may have the unwanted effect of delaying susceptibility to varicella infection to adolescence or adulthood, when the consequences of virus infection can be much more harmful. In addition, this scenario could lead to an increase in varicella-susceptible women becoming infected during pregnancy. In the first trimester, varicella can damage the CNS of the fetus, while in the third trimester, it can lead to severe maternal infection with the possibility of lethal spread to the newborn. Not surprisingly, the Food and Drug Administration (FDA) of the U.S. government has demanded the implementation of long-term studies to follow the duration of varicella immunity post-vaccination.

XVIII. VARIOLA (SMALLPOX)

Variola virus, which causes smallpox, is no longer endemic in any part of the world. However, the current fear of bioterrorism has spurred many nations to take a look at their preparedness in the event of a variola release into the public. Variola is a pathogen for which no effective drug exists. The length of protection afforded by the existing smallpox vaccine (based on live, attenuated vaccinia virus) is not known with certainty. However, minor smallpox outbreaks that occurred in the developed world from the mid-19th to the mid-20th century suggest that the risk of illness upon exposure to smallpox 10 to 20 years after vaccination is substantial, although mortality rates may be reduced. It is therefore sobering to realize that much of the world's population has not been immunized for more than 20 years, and that the younger generation has never been immunized. Moreover, the existing vaccine is unsafe for immunocompromised individuals who, due to HIV, are now present in great numbers around the world. Even among the general population immunized in the United States in the 1960s, hundreds of complications and several deaths were recorded, representing a safety record that would be unacceptable by today's vaccination standards. Many therefore believe that a new, safer smallpox vaccine should be developed.

Unfortunately, a major ethical hurdle that did not stand in Jenner's way is now a serious problem for modern vaccine researchers. In nature, smallpox is a strictly human disease, meaning that researchers lack an animal model to test the efficacy of any new candidate vaccines developed. Researchers are diligently trying to find a combination of test species and route of variola administration that will result in smallpox-like disease in an animal. In the meantime, most countries have opted to replenish their smallpox vaccine supplies with existing standard vaccine strains. The current thinking is that, rather than mass vaccinations, inoculation should be reserved for situations in which people are known to have been exposed to variola. Because the virus replicates slowly, exposed individuals can be vaccinated up to 4 days later and still develop the immune responses necessary to defeat the virus. The immediate contacts of infected individuals as well as health care workers involved in the case would also be vaccinated. This restriction of the size of the vaccinated population might make the inherent risks of smallpox vaccination more acceptable.

Where improved smallpox vaccines are being pursued, a practical concern is the availability of live variola for use in both animal trials and clinical trials. With the eradication of smallpox firmly established, the WHO had recommended that all remaining stocks of variola be destroyed by June 1999. However, it was decided to maintain two stockpiles of samples for research purposes at secure locations in the United States and Russia until 2002, at which time any remaining variola was to be destroyed. This decision was reversed in the autumn of 2001. In fact, among experts in the field, a fierce debate continues to rage concerning the fate of these stocks. Retention is favored by those who feel that the re-emergence of smallpox could occur due to accidental or deliberate release of undeclared stocks. They feel that having variola available for the testing of new, improved vaccines and possible anti-variola drugs warrants maintaining the virus under heavily controlled conditions. Those in favor of the destruction of the two official stocks argue that current vaccinia-based vaccines are adequately effective, and since the genomic DNA sequencing of representative variola strains has been completed, there is no scientific basis for the continued (potentially threatening) existence of the two stocks.

XIX. YELLOW FEVER

Yellow fever is caused by a small, round, enveloped RNA flavivirus that is mosquito-borne and can infect both monkeys and humans. This disease is prevalent in tropical climates where mosquitoes are endemic year-round. Different strains of the yellow fever virus are characterized by different envelope glyco-proteins and vary greatly in their pathogenicity to humans. Three levels of severity of human infections have been observed, determined by the pathogenicity of the infecting strain and the immune status of the host. Although some patients are asymptomatic, others experience moderate symptoms such as headache, fever, vomiting, and nosebleeds. In severe cases, patients have fever accompanied by hepatic, circulatory, and renal failure as well as severe jaundice (hence, "yellow" fever). Recovery from a moderate infection can be expected within a few days, but a high proportion of severe infections are fatal.

The yellow fever vaccine is a live, attenuated virus usually given in one subcutaneous dose. The vaccine virus was originally isolated from a patient in 1927 and was attenuated by passaging through chick embryos. Protection generally lasts for about 10 years. Vaccination programs have been quite successful, dropping the global incidence of yellow fever to about 100 cases in 1980. However, the disease resurged in Africa in the late 1980s to the early 1990s, peaking at over 10,600 cases and about 6000 deaths before receding again to about 300 cases in 1999. Such yellow fever outbreaks are usually due to upswings in the mosquito population, which is greatly influenced by climate (wet/dry seasons). The emergence of a highly pathogenic strain of yellow fever virus coupled with a comparatively wet year is likely to lead to an outbreak. Individuals living in temperate climates are not routinely vaccinated against yellow fever and receive the vaccine only when traveling to an area where the virus is endemic.

G. The "Dark Side" of Vaccines

Many laud vaccination as the "single greatest achievement of medicine," and indeed it has been responsible for the prevention of much misery and millions of premature deaths worldwide. Nevertheless, vaccines, like all powerful medicines, must always be assiduously tested for safety and administered with appropriate caution. In this section, we discuss some of the shortcomings, both real and perceived, of vaccination.

I. ADVERSE EFFECTS OF VACCINES

In general, because of extensive animal and cell culture testing, vaccination carries few risks of side effects. Mild reactions, such as redness or pain at an injection site, sneezing or nasal congestion after intranasal administration, fatigue, or headache, may be experienced by a small proportion of vaccinees. Oral vaccines have fewer side effects than parenterally administered vaccines. For example, only 1 in 50,000 persons vaccinated with the Ty21a typhoid vaccine experiences even mild and transient side effects.

Serious adverse effects of vaccines are rare. Several vaccines (for example, influenza and MMR) are produced in chick embryos, and contaminating chick egg proteins may be present in a vaccine preparation. For this reason, such vaccines should not be given to individuals with allergies to eggs. Allergies to other molecules used during vaccine preparation, such as preservatives or antibiotics, should also be taken into account. Contaminating viruses, bacteria, or chemical agents can also have detrimental effects on some vaccinees. For example, vaccines prepared in chick eggs must be carefully screened to rule out the presence of the avian leukosis virus, a bird pathogen that can be oncogenic in human cells. Ever-improving means of vaccine purification are steadily decreasing the frequency of events of this type. As a result, the risk of experiencing a severe reaction associated with a specific vaccine is usually much lower than the risk of severe consequences from the disease itself.

As mentioned previously, there are unique safety concerns associated with the use of live, attenuated vaccines. Such vaccines should not be given to pregnant women to avoid infection of the fetus. In addition, many live vaccines cause mild side effects such as fever or rash even in immunocompetent vaccinees. The live, attenuated virus used in the measles vaccine can induce moderate immunosuppression when first administered. This is not a problem for healthy adults, but infants and immunocompromised individuals may become vulnerable to opportunistic organisms causing diarrhea, pneumonia, or parasitic infections that an unvaccinated individual can resist. Mild arthritis is sometimes noted after rubella vaccination. The most serious example of damage caused by a live vaccine is VAPP caused by the reversion of the attenuated polio virus used in Sabin OPV.

In the 1960s, important lessons on vaccine design were learned from adverse events associated with the use of an inactivated vaccine for respiratory syncytial virus (RSV). RSV is the main cause of pneumonia and bronchial infections in pediatric populations, and respiratory tract illnesses in the elderly. In the United States, RSV is responsible for 4500 deaths annually. Natural infection induces immunity that rapidly decreases to levels that mitigate rather than prevent disease. A formalin-inactivated virus used as a vaccine in the 1960s induced the production of neutralizing antibodies in vaccinees but was unable to prevent disease. In fact, vaccinees contracted more severe cases of pulmonary disease than nonvaccinees, and some even died upon exposure to natural RSV. The reason for the exacerbation of disease by vaccination in this case remains obscure, but it is suspected that the symptoms occurred because the vaccine induced a Th2 type of immune response. As noted previously for TB, exacerbated lung pathology is observed when Th2 responses are mounted. In contrast, infection with wild-type, live RSV induces a Th1 response that is associated with a milder effect on the lungs. As mentioned previously, a DNA vaccine for RSV that induces Th1 responses is under development.

Another case of surprise adverse events arose with a rotavirus vaccine in the late 1990s. Rotavirus is the most important cause of severe diarrhea of infants and young children. A vaccine against rotavirus was developed at the National Institutes of Health (United States) in the early 1990s and was found to be safe and effective in phase III clinical trials. In 1998, this vaccine was recommended as one of the routine childhood vaccinations to be given at 2, 4, and 6 months of age. However, in July 1999, after about 1.5 million doses of this vaccine had been administered, the Centers for Disease Control in Atlanta suspended its use. A post-licensing surveillance of adverse events suggested that the vaccine was associated with an unacceptable frequency of intussusception (the inversion of one portion of the intestine within another). A safe and effective rotavirus vaccine has yet to be developed.

II. FAILURE TO VACCINATE AND OPPOSITION TO VACCINATION

In the developing world, the reasons why many people are not vaccinated are quite straightforward: the cost of vaccines, an unreliable cold chain, a lack of sterile syringes and needles, a shortage of immunization clinics and qualified personnel, and geographic distances from clinics. Wars and civil conflicts also block access to immunization clinics, or push vaccination programs to the bottom of a government's priority list. Localized infectious disease outbreaks are therefore frequent. For example, polio outbreaks occurred in Angola in 1999 and in Zaire in 1995 following military conflicts. Sometimes certain groups believe that vaccination is unnecessary or prohibited by their religion. With the fall of the Communist government in Bulgaria in 1991, many gypsy families mistakenly concluded that vaccination was no longer necessary. In Holland in 1992, a minority group refused to have their children vaccinated for religious reasons and suffered an outbreak of 71 cases of polio. India recorded 800 cases of polio in 1997 when religious concerns prevented many women from bringing their children to immunization clinics due to a dearth of male relatives willing to be escorts. A major outbreak of over 10,000 polio cases occurred in China in 1989–1990 when children who were not registered as part of the family planning strategy were not brought for vaccination.

When a booster is required for effective immunization, the preceding problems are compounded, especially if the individual erroneously believes that one dose should be enough. The latter applies even to many people in the developed world, often coupled with a belief that the risks of the disease are not that great. Some parents have also voiced the concern that the increasing number of vaccines available means that their children's immune systems are being exposed to too many antigens at a young age. In fact, the smallpox vaccine used a generation ago on its own contained 200 proteins, whereas all the vaccines in the current childhood schedule together now expose children to fewer than 130. In some ways, vaccination has been a victim of its own success, since vaccination has made personal experience with the ravages of diseases like diphtheria or pertussis comparatively rare. In addition, organized anti-vaccine groups continue to spread misinformation and myths via the Internet, and alternative medicine advocates argue against routine vaccination for childhood diseases. The very small chance of an adverse event is frequently blown out of proportion by the media, causing some parents to choose not to vaccinate their

children at all. Others fail to bring their infants in for the complete series of boosters. Such decisions led to the 1989–1991 measles epidemics in high schools in the United States. These outbreaks had severe effects on hundreds of students and killed 120 of them. To put the matter in perspective, encephalopathy following measles vaccination occurs at a rate of 1 in 10^6 vaccinees, while the risk of encephalomyelitis after natural measles virus infection is 1 in 10^3. In the same vein, epidemics of pertussis broke out in Japan, the United Kingdom, and Sweden in the early 1990s following parental decisions not to vaccinate. Diphtheria has also claimed the lives of several young unvaccinated children in North America in the past decade.

Those opposing vaccination often pose the question "This disease has been eradicated from my country, so why should I vaccinate my child against it?" Obviously, one answer is that the child may well travel to another country where the pathogen persists, or equally likely, that someone may travel from that country and unwittingly bring the pathogen with them. In an unvaccinated population, the pathogen could quickly take hold. A less obvious answer to the question is more altruistic, and lies in invoking herd immunity to protect those in the population who cannot be vaccinated because of allergies to the materials used to prepare the vaccine, or who have chronic diseases such as diabetes mellitus, chronic obstructive lung disease, or cardiovascular disease. In addition, a small percentage of any population simply will not respond to the vaccine. For these individuals, life within a "herd" of vaccinated individuals means that the pathogen may well die out before finding them. Herd immunity can protect the entire community (immune and non-immune individuals), but only if the overall vaccination rate is substantial.

One popular misconception used as an argument against vaccination is that the majority of people who get a disease have in fact been vaccinated. This is actually a trick of statistics, based on the fact that no vaccine is 100% effective. About 5–15% of any vaccinated population fails (for unknown reasons) to develop a protective immune response against the pathogen, meaning that they can become ill when exposed to the pathogen even though they are vaccinated. The sheer numbers of individuals vaccinated in a country mean that this pool of vaccinated but susceptible individuals will be sizeable. For example, suppose a population contains 100 individuals among whom there are 5 people who are not vaccinated to pathogen X, and 10 who are vaccinated but who do not respond to the vaccine. A total of 15 individuals therefore will get the disease caused by pathogen X, 10 (67%) of whom were vaccinated (hence the misconception). However, on an absolute basis, 100% of unvaccinated individuals get sick, whereas only 10/95 or about 10% of vaccinated individuals do. Without vaccination, the entire population of 100 souls might have fallen ill.

H. Prophylactic Vaccines of the Future

Much ongoing basic and clinical research is devoted to improving existing vaccines in the short term and designing new and more effective vaccines for the long term. An immediate goal is the production of cheaper, more stable vaccine materials that can be administered without injection but that will still induce effective immunity with fewer side effects and adverse reactions. Eliminating the need to reconstitute certain vaccines before use by providing vaccines as stable liquids, tablets, or powders is a key objective. Also required in many cases are simple but accurate tests for disease diagnosis under difficult conditions in the field.

New delivery methods, such as the edible plant vaccines described in Box 23-2, have the capacity to expand vaccination coverage globally, avoiding the difficulties associated with vaccine injection in developing countries. Mucosal administration via nasal sprays and skin creams may also become possible, once the underlying mechanisms of mucosal tolerance are elucidated. Further experimentation with timed-release biodegradable packaging may allow the administration of a single dose of vaccine that liberates its own booster several weeks later. In addition to these clinical goals, progress is being made on reducing the costs of vaccine production and distribution for developing countries. Admittedly, the basic research and clinical testing of vaccines are labor-intensive and hugely expensive procedures, but what does it gain the developed world if the developing world, which cannot afford the standard prices for vaccines, remains a reservoir for a deadly pathogen? Fortunately, we are starting to see special programs offered by the major multinational pharmaceutical companies, and manufacture of vaccines by generic companies in developing countries, that facilitate the production and distribution of vaccines in the developing world. The framework of guidelines for the ethical testing of vaccines for deadly pathogens (such as HIV) must also undergo refinement in this context.

More purified subunit, peptide, and DNA vaccines that induce cell-mediated immune responses with a lower risk of contaminants and unwanted side effects are on the horizon. Now that the complete nucleotide sequences of many pathogens are known, it should be easier to identify virulence genes and design vaccines to counteract their effects. Fast and easy methods to identify protective B and T epitopes and ways to test them quickly for immunogenicity are also under development. The design of strategies to circumvent the antigenic variation presented by pathogens such as HIV, influenza virus, and trypanosomes are additional priorities. Future vaccines may contain multiple protective epitopes from multiple antigens of a given pathogen, defeating antigenic drift. We may also see genes for specific cytokines and costimulatory molecules added to DNA vaccines to promote lymphocyte activation.

Elimination of disease via global programs using existing vaccines is a real, if somewhat distant, possibility for several infectious diseases. Eradication programs work best for those organisms that infect only humans and cannot take refuge in an animal reservoir. Diseases in this class are polio, measles, rubella, mumps, and Hib meningitis. More progress on vaccines for old diseases like TB and influenza is also needed, as well as new insights into respiratory infections caused by RSV and parainfluenza virus. With the new weaponry available, there is hope that high-profile diseases for which there are currently no vaccines, such as AIDS, HepC infection, and malaria, will yield to vaccination in the near future. Scores of other debilitating disorders, such as sexually transmitted diseases caused by bacteria or herpesviruses, childhood diarrhea due to enteroviruses, and the myriad afflictions caused by group A streptococci, should also soon become priorities for vaccine development.

I. Passive Immunization

Passive immunization is the administration of pre-formed antibodies directed against a particular pathogen to combat or prevent disease. Rather than waiting the required 7–10 days for one's own adaptive immune response to either a pathogen or vaccine immunogen, one is afforded immediate protection by the injected antibodies. A comparison of the features of active and passive immunization is given in Table 23-7. Passive immunization works because, depending on the isotype of the

Table 23-7 Passive versus Active Immunization

Feature	Passive	Active
Protection	Immediate	Requires 7–10 days
Duration	Days to months	Years
Host immune system	Not involved	Required
Elements of response	Anti-pathogen Ab only	Anti-pathogen Ab, CTLs, Th cells
Memory	No	Yes
Control of response	Can be tailored to be systemic or mucosal, or to recognize cryptic epitopes	Under influence of route of administration, immune response itself determines which epitopes are recognized and whether responses are mucosal or systemic
Uses	Immunocompromised individuals Unvaccinated individuals exposed to a dangerous pathogen Where no vaccine exists	Immunocompetent individuals Individuals not naturally exposed to pathogen Where a safe, efficacious vaccine exists

antibodies in the preparation, the pathogen may be blocked from entering the body or from attaching to host cells. Alternatively, the antibody may trigger activation of the complement cascade, or facilitate the activation of phagocytes and NK cells of the innate response that eliminate the invader by ADCC. The recipient's own immune system is generally not activated by passive immunization (unless a response is mounted against the injected antiserum). Protection is relatively short-lived (days to months) and is not characterized by memory.

Natural passive immunization occurs when maternal anti-pathogen antibodies pass through the umbilical circulation to the developing fetus, and later to the newborn in colostrum and breast milk. Transient neonatal protection against polio, tetanus, and diphtheria is provided in this way. Acquired passive immunization is very useful in situations in which no vaccine exists for a pathogen, the vaccine is not 100% efficacious, an unvaccinated individual has been or expects to be exposed to a certain pathogen, or an individual is immunodeficient or immunocompromised. For example, newborn babies of mothers infected with HepB are given both a HepB vaccine and an anti-HepB antibody preparation, to both induce protective immunity and immediately block transferred virus. Similarly, an anti-CMV antibody preparation is given to recipients of organs transplanted from CMV-positive donors. Individuals at high risk for complications of RSV or varicella infections, or those exposed to rabies, are treated with antibody preparations directed against these pathogens. Passive antibody treatment can also neutralize the effects of toxins such as those produced by black widow spiders, venomous snakes, and *Clostridium botulinum*.

Passive immunization is most often administered in the form of polyclonal antibodies, either as whole serum or as the gamma globulin fraction (predominantly IgG); the latter preparation is called an *immune globulin* (IG). Preparations of antibodies directed against toxins or venoms are IGs called *antitoxins* or *antivenins*, respectively. IGs can be raised in either an animal species or other humans; for obvious reasons, antitoxins and antivenins are almost exclusively of animal origin. Naturally, because antibodies are immunogens in their own right, the recipient's immune system may see a life-saving IG as foreign and produce antibodies and T cells recognizing species-specific or allotypic epitopes. If these antibodies happen to be of the IgE isotype, immune complexes of recipient antibody bound to the IG may induce widespread degranulation of FcεR-bearing mast cells. The resulting load of histamines may trigger *serum sickness*, characterized by fever, rash, and arthritis (see Ch.27), or anaphylaxis, a severe systemic allergic reaction that can be fatal (see Ch.28). To decrease the occurrence of these types of events, human IG preparations are used when at all possible. Such IGs are usually derived from the pooled serum of donors who have been selected for naturally high titers of the desired antibodies. Sometimes volunteers are repeatedly immunized with a pathogen antigen of interest to generate high levels of specific IgG antibodies that are recovered as the IG preparation. IGs are traditionally administered either intramuscularly (IM-IG) or intravenously (IV-IG). Preparations destined for intravenous

use are semi-purified to remove high molecular weight immune complexes that might activate complement in the circulation of the recipient. More recently, researchers have been experimenting with oral and mucosal passive immunization. Cows immunized with an antigen of interest produce antibodies that can be transferred to humans via consumption of milk, while mucosal antibodies can be applied in nasal sprays. Work in animal models and human volunteers suggests that these methods are efficacious and certainly less stressful to the patient. In mice, a slow-release delivery device that regularly administered a controlled dose of anti-HSV-2 antibody was implanted in the vagina. The animals were protected from vaginal HSV-2 infection.

In the past decade, the use of monoclonal antibodies for passive immunization has become feasible with successes achieved in generating "humanized" or fully human mAbs (see Ch.7). Why humanize? Humanizing an antibody decreases the danger of serum sickness and increases the longevity of the antibody in the host (it is not cleared as quickly from the circulation). Why monoclonals? Monoclonal antibodies are by definition specific for a particular epitope such that the exact target on the pathogen is known and the precise isotype of antibody best suited to eliminating the pathogen can be controlled. Secretory IgA antibodies can be generated for mucosal applications, while IgG would be the isotype of choice for systemic applications. Specific IgM antibodies would be desirable where multiple copies of an epitope are sterically accessible and multivalent interaction of the antibody with the target antigen can increase the duration of protection. For example, in one experiment in which rats were treated with either IgG or IgM mAbs against a pathogen and then challenged with the pathogen, the animals that had been treated with the IgM mAb were 1000× better protected than those treated with the IgG mAb. It turned out that the multivalent IgM antibody bound to the pathogen with 100× more avidity than the IgG mAb. In addition to these advantages, preparations of mAbs are generally extremely potent because only a highly effective antibody is chosen in the first place and every antibody molecule in a given dose is equally efficacious. Thus, protection can be achieved with a much smaller dose (up to 50× less) of a mAb than a conventional polyclonal IG preparation (which contains a mixture of antibodies of varying effectiveness). Indeed, treatment with mAbs is quickly rivaling treatment with pharmaceuticals for some pathogens. The inherent stability of antibody molecules (the half-life of IgG in the serum can be up to 20 days) means that a single dose per month of mAb preparation can be given to treat a persistent infection, compared to 1–3 doses per day of antibacterial or antiviral drugs. In addition, mAbs are manufactured in facilities free of pathogens, eliminating the danger inherent in the administration of IG preparations from human donors that an unrecognized blood pathogen might be passed on to the patient. The cost of producing mAbs on an industrial scale continues to plummet, making this mode of therapy economically viable as well. Moreover, mAbs are readily stored: freeze-dried antibodies are stable at room temperature for at least 2 years, and antibodies in solution retain substantial levels of protective activity even after several months at room temperature. At the end of 2000, at least seven mAbs were approved in the United States for therapeutic applications and more than 80 mAbs were in clinical trials.

An example of passive immunization that does not involve a pathogen is the prevention of Rh disease (Fig. 23-7). Rh is an antigen expressed on the surface of erythrocytes in many,

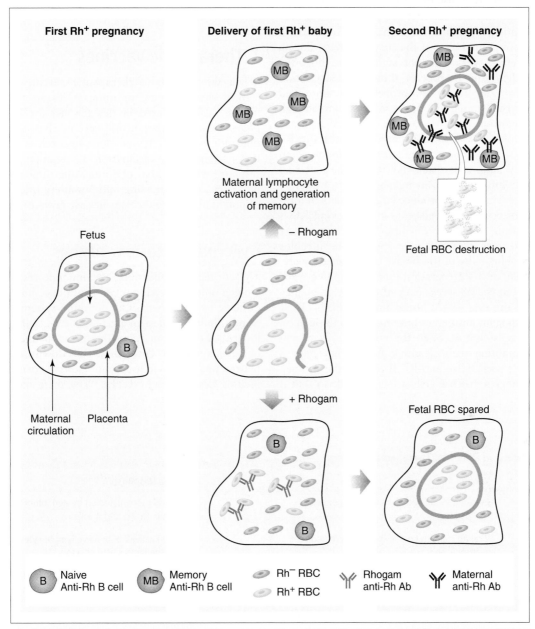

First Rh⁺ pregnancy

Delivery of first Rh⁺ baby

Second Rh⁺ pregnancy

Maternal lymphocyte activation and generation of memory

Fetal RBC destruction

− Rhogam

Fetus

Maternal circulation   Placenta

+ Rhogam

Fetal RBC spared

| B | Naive Anti-Rh B cell | MB | Memory Anti-Rh B cell | Rh⁻ RBC Rh⁺ RBC | Rhogam anti-Rh Ab | Maternal anti-Rh Ab |

Figure 23-7

Passive Immunization to Prevent Rh Disease

During the first pregnancy of an Rh⁻ mother carrying an Rh⁺ fetus, only the occasional fetal RBC carrying the Rh antigen accesses the maternal circulation. Few maternal anti-Rh B cells are present, and those that are present are naive. Little or no activation of these B cells occurs and virtually no memory anti-Rh B cells are generated. However, during delivery of the first Rh⁺ fetus, many fetal cells escape into the maternal circulation. In the absence of Rhogam (anti-Rh antibody), the maternal naive anti-Rh cells are activated and generate large numbers of memory B cells (top pathway). During a second Rh⁺ pregnancy, fetal RBCs escaping into the maternal circulation are much more likely to encounter and activate the memory anti-Rh B cells. Antibodies are generated that cross the placenta and destroy the RBCs of the second Rh⁺ fetus. However, if the Rh⁻ mother receives Rhogam during her first pregnancy (bottom pathway), these passively administered antibodies quickly bind to any fetal RBCs accessing her circulation during delivery and target them for disposal. Activation of naive anti-Rh B cells is thus avoided. Because memory anti-Rh B cells are not generated, little or no anti-Rh antibody production is induced by the RBCs of a second Rh⁺ fetus. Administration of Rhogam during the second and subsequent Rh⁺ pregnancies ensures that maternal anti-Rh B cells remain naive.

but not all, individuals. In "Rh disease," more properly known as *erythroblastosis fetalis*, the erythrocytes of the fetus are destroyed during the pregnancy of Rh⁻ mothers carrying their <u>second</u> (or later) Rh⁺ fetus. During a first pregnancy in which the Rh⁻ mother is carrying an Rh⁺ baby, very few fetal cells get into the mother's circulation prior to birth and only naive maternal anti-Rh lymphocytes are present. Any anti-Rh immune response mounted against the first Rh⁺ fetus during gestation is therefore weak and not sufficient to hurt the fetus. However, the traumatic process of birth and separation of the placenta from the uterine wall propels many more fetal cells into the maternal circulation. These fetal cells are enough to provoke a vigorous response by the maternal immune system that eliminates the fetal RBCs and generates anti-Rh memory cells. If a subsequent pregnancy also involves an Rh⁺ fetus, the few fetal cells that make it into the maternal circulation prior to birth will activate memory B cells. Maternal plasma cells are generated that produce large amounts of anti-Rh IgG capable of crossing the placenta and attacking RBCs in the fetus, causing hemolytic anemia. The infant suffers from a lack of oxygenation and may appear blue in color (hence, the term "blue baby"). Rh disease is prevented by giving an Rh⁻ woman an IG preparation (one well-known trade name is Rhogam) at the 28-week mark of the first pregnancy and again within the first 48–72 hours after the birth of her first child. The Rhogam antibodies bind to the Rh antigen on any fetal blood cells that accessed the maternal circulation during birth, promoting their clearance before they have a chance to interact with naive anti-Rh B cells in the mother. Virtually no memory anti-Rh B cells are generated, so that if

the fetus in the next pregnancy is also Rh⁺, he or she should be safe from harm. The mother should enjoy uneventful births if she continues to take Rhogam for all subsequent Rh⁺ pregnancies to maintain the naive state of her anti-Rh B cells.

J. Therapeutic Vaccines

When we talk about therapeutic vaccines, we mean substances that can cure or mitigate disease, rather than prevent it. Vaccines used in this manner are sometimes called *pharmaccines*. A therapeutic vaccine that is truly efficacious does not exist as yet, but much research is devoted to using the principles of vaccination to manipulate the immune response with the goal of eliminating tumors, curing chronic infections, or suppressing autoimmunity or allergy. Examples of potential therapeutic vaccines are given in Table 23-8.

I. VACCINES TO COMBAT TUMORS

The concept of unleashing the immune response against an antigen unique to a tumor cell is not new, but it has been very difficult to implement. Our immune system already has the responsibility for destroying tumor cells, self-cells that are no longer self due to mutations that promote uncontrolled growth or block apoptosis. Often these cells downregulate MHC class I, attracting lysis by NK cells, or express novel tumor antigens, inviting destruction by CTLs. Why these mechanisms fail to

Table 23-8 Examples of Potential Therapeutic Vaccines

Situation	Vaccine Agent	Expected Response
Tumor therapy	Cytokine-stimulated, inactivated tumor cells	Tumor Ags upregulated on treated tumor cells induce anti-tumor Th, Tc, and B cells
Chronic infection	Viral Ag contains cryptic epitope expressed in chronic phase	Enhanced activation of naive lymphocytes by cryptic epitopes generates new effectors
Indirect effects of infections	Pathogen components	Disease indirectly triggered by pathogens is mitigated. E.g., elimination of *H. pylori* infection reduces ulcers and MALT lymphomas
Autoimmunity	Self-Ag delivered orally	Oral tolerance to self-antigen decreases autoimmune response
	Self-Ag delivered with a cytokine favoring immune deviation	"Correct" Th response is induced to decrease autoimmune pathology
	Suppressive peptide of self-Ag	Suppressive properties of peptide block autoimmune response to whole self-antigen
Allergy	Allergen delivered with cytokines favoring Th1 response	Production of IgE anti-allergen Ab does not occur
Fertility	Reproductive protein conjugated to immunogenic carrier	Th cells recognizing carrier epitopes provide help to B cells producing Ab blocking function of reproductive antigen
Addiction	Cocaine	Anti-cocaine Ab blocks cocaine molecule from stimulating neurotransmitters and generating "high"
	Abzyme	Abzyme cleaves cocaine molecule so it cannot stimulate neurotransmitters

stop the growth of all tumors is the subject of much ongoing research (see Ch.26). Attempts have been made to therapeutically vaccinate individuals with their own modified tumor cells. The tumor cells are first stimulated *in vitro* with cytokines to promote MHC class I expression and antigen presentation, and are then inactivated. The treated cells are injected back into the patient, in the hope that the immunogenicity of antigens unique to the tumor cells will be enhanced without promoting tumor growth at the injection site.

Prophylactic vaccines against cancer development are also under exploration. Vaccination with a subunit or peptide vaccine directed against a pathogen associated with tumorigenesis might forestall cancer initiation. For example, cervical cancer is caused by papillomavirus. Subunit vaccines based on either two structural proteins or two oncogenes of the virus are now in clinical trials. Vaccination with antigens known to be associated with tumor development is also under investigation. For example, several biotechnology companies are making experimental vaccines directed against *mucins* (glycolipids) expressed specifically on tumor cells. Evaluations of the latest anti-cancer vaccines have indicated that immunization with these tumor-associated antigens is safe and induces humoral and cell-mediated responses that may be of significant clinical benefit. Immune responses to tumors are explored in depth in Chapter 26.

II. VACCINES TO CURE CHRONIC VIRAL DISEASES

The elimination of most viruses relies on a combination of effective humoral, Th, and CTL responses. In chronic viral diseases, including those associated with HepB, HepC, and HIV infections, these responses are simply not vigorous enough to completely clear the pathogen such that clinical symptoms are present on an ongoing or recurring basis. Thus, to be effective, a therapeutic vaccine for chronic viral infection must induce levels of all three types of responses that exceed those in a natural infection. A complicating factor is that the epitopes that are important during a chronic infection most likely are not those that are dominant during an acute infection. A therapeutic vaccine must therefore be aimed at enhancing the activation of naive T and B cells that recognize previously subdominant or cryptic epitopes. The vaccine must not induce anergy of responding T cells, meaning that the vaccine should contain sufficient viral antigen and an adjuvant to activate sufficient APCs (primarily DCs) to support CD4$^+$ and CD8$^+$ cell activation. However, as is true for prophylactic vaccines, the vaccine should not contain so much viral antigen that it induces a very rapid, vigorous response which leads to exhaustion of antiviral CD8$^+$ T cells due to AICD. The vaccine should also contain sufficient CD8$^+$ T cell epitopes to activate a polyclonal T cell response so that a single mutation in the virus does not allow it to escape all CTL effectors. Obviously, the correct balance of all these parameters is tricky to ascertain, making therapeutic vaccine design a challenge that may be only solved by empirical means. Unfortunately, although much research is under way, there are as yet no effective therapeutic vaccines for chronic viral diseases.

III. VACCINES TO MITIGATE INDIRECT EFFECTS OF INFECTIOUS DISEASE

As we learned in Chapter 22, pathogen infection underlies several diseases that were long thought not to involve a specific infectious agent at all. For example, it is now clear that many gastric ulcers, gastric carcinomas, and MALT lymphomas are promoted by *Helicobacter pylori*, a bacterium that invades the stomach mucosae and persists there for long periods. It is estimated that 80% of the adult population in developing countries is infected with this organism, and 30% in the developed world. Obviously, the natural immune response to *H. pylori* is not as efficient as it could be. Antibiotics have been very effective in resolving ulcers and even some MALT lymphomas driven by this organism, but concerns remain about the rise of drug-resistant *H. pylori* and the costs and side effects of prolonged antibiotic treatment. A vaccine containing components of *H. pylori* that would induce neutralizing antibodies capable of withstanding the gut environment and protecting the gut mucosae is therefore an attractive prospect. Recombinant vaccines based on the urease enzyme or heat shock proteins expressed by *H. pylori* have been tested in animal models with some success. Long-term (over 12 months) protection against challenge with *H. pylori* or closely related organisms has been achieved in mice given bacterial urease conjugated to modified cholera toxin. Indeed, there is some evidence that giving this vaccine to an infected animal therapeutically can mitigate its disease. A similar vaccine modified for human use might be used therapeutically to reverse ulcers in affected adults and perhaps forestall the development of gastric carcinomas and MALT lymphomas.

A parallel approach can be pictured for other diseases linked to prior pathogen infections. For example, many cases of juvenile diabetes are thought to be associated with infection by a particular enterovirus. Left untreated, this virus appears to induce the production of autoantibodies that destroy the insulin-producing β-islet cells of the pancreas. Theoretically, once the enterovirus is identified in a patient at particular risk, vaccination against the pathogen could be initiated to clear it and perhaps prevent the development of diabetes.

IV. VACCINES TO MITIGATE AUTOIMMUNITY

The ingestion of many immunogens induces oral tolerance to the immunogen rather than an immune response. The underlying mechanisms of oral tolerance remain to be completely defined, but some researchers have already started exploiting the concept of oral vaccines to suppress autoimmune disease. That is, if an animal is fed self-antigen under the right conditions, it should develop tolerance to that antigen. If that antigen is driving an autoimmune response, the induction of tolerance to the antigen should mitigate the disease symptoms. For example, William Langridge and colleagues at the Loma Linda University School of Medicine, United States, have used this approach to combat diabetes in mice. The two proteins that appear to be the prime targets of autoimmune responses in the diabetes-susceptible NOD (*non-obese diabetic*; see Ch.29) strain of mice are insulin and *glutamic acid decarboxylase* (GAD), both expressed by pancreatic β-islet cells. The researchers fed NOD mice with potatoes containing GAD linked to the B subunit of cholera toxin (to direct the vaccine to the M cells). The vaccinated mice developed oral tolerance to GAD and downregulated the autoimmune response attacking the pancreatic β-islet cells; the onset of clinical diabetes was inhibited. If the concentration of self-antigen made by such plants can be scaled up, this approach might provide a viable therapy for some cases of autoimmune diabetes in humans.

Vaccination to induce immune deviation has also been considered as a strategy to treat autoimmune diabetes in experimental animals. It is thought that diabetes usually results from the mounting of Th1 responses in which autoaggressive CTLs attack the pancreas (see Ch.29). The induction of a Th2 response instead would favor the activity of regulatory T cells that might suppress the CTLs and thus prevent the development of diabetes. Research is proceeding into the design of DNA vaccines that induce immune deviation and promote Th2 responses over Th1 responses. It may also be possible to use conventional vaccination approaches to decrease autoimmunity. Certain regions of some proteins suppress immune responses rather than induce them. For example, immunization with the N-terminal region of lysozyme actually suppresses an animal's immune response to the whole lysozyme molecule. Once such a suppressive peptide is identified, it could form the basis for a vaccine that might prevent autoimmune attacks on the tissue expressing the molecule containing the peptide. More on the treatment of autoimmunity appears in Chapter 29.

V. VACCINES TO MITIGATE ALLERGY

The reader will note that we have not yet discussed the underlying mechanisms of allergy in this book. Suffice it to say here that allergic symptoms are caused by degranulation of mast cells that are activated by the binding of specific antigen (the allergen) to IgE already bound to FcεR on the mast cells. Not all individuals produce antibodies of the IgE isotype to such antigens, which is why only some among us have hay fever or cannot eat shellfish. Because generation of the IgE isotype is promoted by cytokines synthesized by Th2 cells, one could imagine that a vaccine designed to preferentially induce a Th1 response to the allergen might reduce allergic symptoms. Indeed, the oral administration of naked DNA vaccines has been used to alleviate allergy in mice. For example, the *arah2* gene encodes the principal allergen in peanuts. The feeding of a DNA vaccine containing the *arah2* gene to susceptible mice decreased their levels of circulating IgE and protected them from anaphylaxis in a subsequent challenge with peanuts. It is not yet clear whether the DNA vaccine exerted its effects by influencing fully differentiated Th2 cells, or by directing newly produced uncommitted Th cells down the Th1 path, or by a mechanism yet to be elucidated. More on allergy and immunotherapy to treat it appears in Chapter 28.

VI. VACCINES TO SUPPRESS FERTILITY

An application that is not strictly speaking therapeutic but of benefit to humans is the use of vaccines against reproductive hormones or gamete antigens to control fertility. While pharmaceutical contraceptives have been successful in aiding women in developed countries to avoid unwanted pregnancies, this option is often not practical or desirable in developing countries. The large Western pharmaceutical companies have not pursued fertility vaccines in any measure but much is being done in the developing world to create vaccines that can be used to reversibly inhibit fertility. Targets for immunization have included antigens specific to sperm, the *zona pellucida* (ZP) of the egg, and reproductive hormones.

One problem with fertility vaccines is that the target antigens are usually self proteins of the reproductive system and thus are not perceived as foreign by the vaccinee's immune system. That is, tolerance to these components must be overcome in order to provoke an immune response that would block fertility. Generally speaking, tolerance in this situation is broken by abnormal delivery of the reproductive antigen: a small reproductive protein (such as a hormone or hormone subunit) or peptide antigen is chemically linked to a protein carrier (such as tetanus toxoid). Use of such a hapten-carrier system can recruit Th cells (recognizing carrier epitopes) that provide help to antibody-producing B cells (recognizing peptide or hormone epitopes). A corollary to breaking this tolerance, however, is that the potential for antigen-specific autoimmunity exists, because one is attempting to induce an immune response to a self protein. Another concern is that some reproductive hormones have functions in other body systems, making their total elimination unwise. Moreover, since large amounts of these hormones are often constitutively produced, the mounting of an immune response against them might cause the accumulation of detrimental amounts of immune complexes in the tissues. Reversibility of the anti-fertility immune response is also an issue. Many women want to delay, not eliminate, childbearing, so that a vaccine that resulted in an anti-fertility response throughout the childbearing years would not be accepted.

Rather than reproductive hormones, the gametes, the trophoblast of the conceptus, and proteins secreted by these structures have become favored sources of fertility vaccine epitopes. For example, immunization of female baboons with a peptide derived from the sperm-specific form of lactate dehydrogenase conjugated to diphtheria toxoid decreased fertility by 75%. The resistance to pregnancy lasted for 12 months, making this vaccine effectively reversible. Other vaccines have used proteins expressed specifically on the sperm surface or the ZP of the egg to successfully induce humoral and cell-mediated immune responses leading to infertility in some mammalian species (but not others). Unfortunately, vaccination with anti-ZP or anti-sperm protein vaccines has sometimes led to irreversible ovarian or testicular damage in experimental animals.

The most promising human anti-fertility vaccine is that targeting the pregnancy hormone hCG (*human chorionic gonadotropin*). hCG is a heterodimeric glycoprotein secreted by cells of the trophoblast prior to implantation of the blastocyst in the uterus. hCG is vital for gestation because it stimulates the production of progesterone required to maintain the earliest stages of pregnancy. Various vaccines directed against a unique peptide of hCG conjugated to tetanus toxoid have been tested in human clinical trials. The results of early trials were very promising, in that neutralizing anti-hCG antibodies were produced and pregnancy blocked in 80% of the women vaccinees. Furthermore, the effect either reversed spontaneously after several months or could easily be induced to do so by treatment with progesterone. It was speculated that the 20% of women who did not generate protective levels of anti-hCG antibodies may have expressed MHC haplotypes that could not bind to the vaccine peptide. In an effort to overcome this problem, researchers in India immunized mice with a cocktail of hCG peptides conjugated to three different pathogen peptides known to bind promiscuously to a broad range of MHC proteins. The anti-hCG responses in the animals immunized with the cocktail (plus alum adjuvant) exceeded those in mice immunized with each individual peptide conjugate. Furthermore, four different inbred mouse strains were rendered infertile. If a similar cocktail approach to human hCG vaccination can be devised, these results hold out the hope that almost any woman could be protected against pregnancy, if she so desired. Other studies have experimented with alternative adjuvants and delivery vehicles and hCG produced by recombinant DNA techniques. Oral vaccination with recombinant vaccinia virus engineered to express the β subunit of hCG induced both protective antibody and T cell responses in vaccinees. Studies to evaluate the long-term safety and efficacy of such vaccines are in progress.

Veterinary applications of anti-fertility immunization can be used to control the reproduction of both domestic livestock and "pest" animals in the wild. Damage to sensitive environments caused by the uncontrolled breeding of wild horses and elephants has been mitigated by vaccinating the females of these species with fertility antigens. The vaccines are administered by the shooting of darts, much as a tranquilizer dart is delivered in the wild. In these situations, irreversibility of the anti-fertility response is a therapeutic goal.

VII. VACCINES TO SUPPRESS ADDICTION

Drug addicts, such as those abusing cocaine, often find it difficult to comply with regimes designed to "break their habits." Moreover, the drugs used to replace the need for cocaine are sometimes almost as addictive as cocaine itself. Cocaine works by inhibiting neurotransmitters in the brain that downregulate the "reward" sensation pathway. As a result, the reward pathway fires continuously, generating a "high." Vaccination of experimental rodents with cocaine metabolites results in a decrease in the level of cocaine in the CNS and in cocaine-associated behavior. High titers of long-lasting anti-cocaine antibodies appear in the serum. However, this strategy is unlikely to work in humans because the addicts can merely overload on cocaine in an attempt to swamp the antibodies. Another approach is to immunize the addict with an antigen that induces the production of an abzyme (refer to Ch.7) capable of cleaving the cocaine molecule. In this situation, the catalytic antibody continuously inactivates large numbers of cocaine molecules such that a "high" is not attained. Addictive behavior is no longer "rewarded" in this situation, allowing the addict to finally break his or her habit.

We have come to the end of our discussion of vaccines. Vaccinology is a young science—just over 200 years old. We can expect many advances in the future that will greatly expand the scope and reach of vaccinology around the world, and hopefully reduce the incidence and severity of many devastating infectious diseases to a point at which doctors regard their presentation as a rarity. We move now to a discussion of primary immunodeficiencies, defects in immunity that arise from congenital genetic abnormalities.

SUMMARY

A prophylactic vaccine is given to an individual prior to pathogen exposure to establish expanded clones of long-lived, pathogen-specific memory lymphocytes. In a subsequent exposure to a pathogen, the individual mounts a secondary rather than primary response and is thus far less likely to become seriously ill or die. The major types of vaccines include attenuated, inactivated, toxoid, subunit, recombinant vector, naked DNA, and anti-idiotypic vaccines. Each type of vaccine offers particular advantages and disadvantages, but all must meet a common set of requirements if they are to be deemed successful. First, a vaccine must be capable of inducing a protective response against the pathogen of interest. In most cases, such a response is not induced in all members of the vaccinated population, but if a large enough proportion of vaccinees respond, the spread of the pathogen will be successfully halted. Secondly, the vaccine must be shown to be safe, such that the risk of harm from any associated side effects is far outweighed by the risk of harm due to the disease itself. Vaccine efficacy and safety must be formally tested in a rigorous series of pre-clinical and clinical trials before government approval is granted for medical use. Many factors affect the outcome of immunization with a particular vaccine, including the age of the vaccinee, the number of times the vaccine is administered, and the route of immunization. Also critical is the choice of adjuvant used to enhance the vaccine's immunogenicity, and the delivery vehicle used to protect the vaccine from degradation *in vivo*. Passive immunization with preformed antibodies recognizing a pathogen can be given to forestall disease. Unlike prophylactic vaccines, which are given to prevent illness prior to exposure to a pathogen, therapeutic vaccines are administered in an attempt to contain the impact of an acute infection with a slow-replicating pathogen. Therapeutic vaccines are also under investigation for the improvement of chronic conditions such as chronic infections, cancer, autoimmunity, allergy, infertility, and addiction.

Selected Reading List

Allwin R. and Doerr H. (2002) The "influenza vaccine"—benefit, risk, costs. *Medical Microbiology and Immunology* **191**, 183–185.

Andersen P. (2001) TB vaccines: progress and problems. *Trends in Immunology* **22**, 160–168.

Andre F. (2001) The future of vaccines, immunisation concepts and practice. *Vaccine* **19**, 2206–2209.

Arkema A., Huckriede A., Schoen P., Wilschut J. and Daemen T. (2000) Induction of cytotoxic T lymphocyte activity by fusion-active peptide-containing virosomes. *Vaccine* **18**, 1327–1333.

Barry M., Howell D., Andersson H., Chen J. and Singh R. (2004) Expression library immunization to discover and improve vaccine antigens. *Immunological Reviews* **199**, 68–83.

Bendandi M. (2000) Anti-idiotype vaccines for human follicular lymphoma. *Leukemia* **14**, 1333–1339.

Berthet F.-X., Coche T. and Vinals C. (2001) Applied genome research in the field of human vaccines. *Journal of Biotechnology* **85**, 213–226.

Betts J. (2002) Transcriptomics and proteomics: tools for the identification of novel drug targets and vaccine candidates for tuberculosis. *IUBMB Life* **53**, 239–242.

CDC. Web sites on vaccines: http://www. cdc.gov http://www.cdc.gov/node.do/ id/0900f3ec8000e2f3

Combination vaccines for childhood immunization: recommendations of the Advisory Committee on Immunization Practices (ACIP), the American Academy of Pediatrics (AAP), and the American Academy of Family Physicians (AAFP). (1999) *Pediatrics* **103**, 1064–1076.

Cohen I., Quintana F. and Mimran A. (2004) T_{regs} in T cell vaccination: exploring the regulation of regulation. *Journal of Clinical Investigation* **114**, 1227–1232.

Conne P., Gauthey L., Vernet P., Althaus B., Que J. *et al.* (1997) Immunogenicity of trivalent subunit versus virosome-formulated influenza vaccines in geriatric patients. *Vaccine* **15**, 1675–1679.

Cryz S. J. (1999) BERNA: a century of immunobiological innovation. *Vaccine* **17**, S1–S5.

Cusi M. and Gluck R. (2000) Potential of DNA vaccines delivered by influenza virosomes (letter to the editor). *Vaccine* **18**, 1435.

Cusi M., Lomagistro M., Valassina M., Valensin P. and Gluck R. (2000) Immunopotentiating of mucosal and systemic antibody responses in mice by intranasal immunization with HLT-combined influenza virosomal vaccine. *Vaccine* **18**, 2838–2842.

Doherty T. and Andersen P. (2002) Tuberculosis vaccine development. *Current Opinion in Pulmonary Medicine* **8**, 183–187.

Durbin J. and Durbin R. (2004) Respiratory syncytial virus-induced immunoprotection and immunopathology. *Viral Immunology* **17**, 370–380.

Feng H., Sandlow J., Sparks A. and Sandra A. (1999) Development of an immunocontraceptive vaccine: current status. *Journal of Reproductive Medicine* **44**, 759–765.

Forns X., Bukh J. and Purcell R. (2002) The challenge of developing a vaccine against hepatitis C virus. *Hepatology* **37**, 684–695.

Gluck R., Mischler R., Durrer P., Furer E., Lang A. *et al.* (2000) Safety and immunogenicity of intranasally administered inactivated trivalent virosome-formulated influenza vaccine containing *Escherichia coli* heat-labile toxin as a mucosal adjuvant. *Journal of Infectious Diseases* **181**, 1129–1132.

Green B. and Baker S. (2002) Recent advances and novel strategies in vaccine development. *Current Opinion in Microbiology* **5**, 483.

Gupta A., Pal R., Ahlawat S., Bhatia P. and Singh O. (2001) Enhanced immunogenicity of a contraceptive vaccine using diverse synthetic carriers with permissible adjuvant. *Vaccine* **19**, 3384–3389.

Gurunathan S., Klinman D. M. and Seder R. A. (2000) DNA Vaccines: immunology, application, and optimization. *Annual Review Immunology* **18**, 927–974.

Hayes S. and World M. (2000) Adverse reactions to anthrax immunization in a military field hospital. *Journal of the Royal Army Medical Corps* **146**, 191–195.

Heath W., Belz G., Behrens G., Smith C., Forehan S. *et al.* (2004) Cross-presentation, dendritic cell subsets, and the generation of immunity to cellular antigens. *Immunological Reviews* **199**, 9–26.

Heathcote E. (2004) Prevention of hepatitis C virus-related hepatocellular carcinoma. *Gastroenterology* **127**(Suppl. 1), S294–302.

Hilleman M. R. (2000) Overview of vaccinology with special reference to papillomavirus vaccines. *Journal of Clinical Virology* **129**, 79–90.

Howie D., Sayos J., Terhorst C. and Morra M. (2000) The gene defective in X-linked lymphoproliferative disease controls T cell dependent immune surveillance against Epstein-Barr virus. *Current Opinion in Immunology* **12**, 474–478.

Klein K. and Diehl E. (2004) Relationship between MMR vaccine and autism. *Annals of Pharmacotherapy* **38**(7–8), 1297–1300.

Knaust A. and Frosch M. (2004) Genome-based vaccines. *International Journal of Medical Microbiology* **294**, 295–301.

Kutzler M. and Weiner D. (2004) Developing DNA vaccines that call to dendritic cells. *Journal of Clinical Investigation* **114**, 1241–1244.

Lajeunesse M., Zhang Q. and Finn A. (2004) Mucosal immunity to infections and its importance in future vaccinology. *Advances in Experimental Medicine and Biology* **549**, 13–22.

Lambkin R., Novelli P., Oxford J. and Gelder C. (2004) Human genetics and responses to influenza vaccination: clinical implications. *American Journal of Pharmacogenomics* **4**, 293–298.

Langridge W. (2000) Edible vaccines. *Scientific American* **283**(3), 66–71.

Levine M., Campbell J. and Kotloff K. (2002) Overview of vaccines and immunisation. *British Medical Bulletin* **62**, 1–13.

Li X., Sambhara S., Xin Li C., Ewasyshyn M., Parrington M. *et al.* (1998) Protection against respiratory syncytial virus infection by DNA immunization. *Journal of Experimental Medicine* **188**, 681–688.

Lindblad E. (2004) Aluminium adjuvants—in retrospect and prospect. *Vaccine* **22**(27–28), 3658–3668.

Lutwick L. (1999) Unconventional vaccine targets; immunization for pregnancy, peptic ulcer, gastric cancer, cocaine abuse and atherosclerosis. *Infectious Disease Clinics of North America* **13**, 245–264.

Makela P. H. (2000) Vaccines, coming of age after 200 years. *FEMS Microbiology Reviews* **24**, 9–20.

McDevitt H. (2004) Specific antigen vaccination to treat autoimmune disease. *Proceedings of the National Academy of Sciences USA* **101**(Suppl. 2), 14627–14630.

Mims C. (1982) "The Pathogenesis of Infectious Disease," 3rd edn. Academic Press, London.

Moingeon P., de Taisne C. and Almond J. (2002) Delivery technologies for human vaccines. *British Medical Bulletin* **62**, 29–44.

Moll H. (2004) Antigen delivery by dendritic cells. *International Journal of Medical Microbiology* **294**, 337–344.

Moore S., Surgey E. and Cadwgan A. (2002) Malaria vaccines: where are we and where are we going? *Lancet Infectious Diseases* **2**, 737–743.

O'Hagan D., Singh M. and Ulmer J. (2004) Microparticles for the delivery of DNA vaccines. *Immunological Reviews* **199**, 191–200.

Palucka K. and Banchereau J. (2002) How dendritic cells and microbes interact to elicit or subvert protective immune responses. *Current Opinion in Immunology* **14**, 420–431.

Peltola H. (2000) Worldwide *Haemophilus influenzae* type b disease at the beginning of the 21st century: Global analysis of the disease burden 25 years after the use of the polysaccharide vaccine and a decade after the advent of conjugates. *Clinical Microbiology Reviews* **13**, 302–317.

Piyasirisilp S. and Hemachudha T. (2002) Neurological adverse events associated with vaccination. *Current Opinion in Neurology* **15**, 333–338.

Price B., Liner A., Park S., Leppla S., Mateczun A. *et al.* (2001) Protection against anthrax lethal toxin challenge by genetic immunization with a plasmid encoding the lethal factor. *Infection and Immunity* **69**, 4509–4515.

Recommended Childhood Immunization Schedule (2004), Advisory Committee on Immunization Practices (www.cdc.gov/nip/acip), The American Academy of Pediatrics (www.aap.org) and The American Academy of Family Physicians (www.aafp.org).

Richie T. and Saul A. (2002) Progress and challenges for malaria vaccines. *Nature* **415**, 694–701.

Rocha C., Caetano B., Machado A. and Bruna-Romero O. (2004) Recombinant viruses as tools to induce protective cellular immunity against infectious diseases. *International Microbiology* **7**, 83–94.

Rolph M. and Ramshaw I. A. (1997) Recombinant viruses as vaccines and immunological tools. *Current Opinion in Immunology* **9**, 517–524.

Schaad U., Buhlmann U., Burger R., Ruedeberg A., Wilder-Smith A. *et al.* (2000) Comparison of immunogenicity and safety of a virosome influenza vaccine with those of a subunit influenza vaccine in pediatric patients with cystic fibrosis. *Antimicrobial Agents and Chemotherapy* **44**, 1163–1167.

Schijns V. E. (2000) Immunological concepts of vaccine adjuvant activity. *Current Opinion in Immunology* **12**, 456–463.

Sette A. and Sidney J. (1998) HLA supertypes and supermotifs: a functional perspective on HLA polymorphism. *Current Opinion in Immunology* **10**, 478–482.

Sjolander A., Cox J. and Barr I. (1998) ISCOMs: an adjuvant with multiple functions. *Journal of Leukocyte Biology* **64**, 713–723.

Sjolander A., Drane D., Maraskovsky E., Scheerlinck J.-P., Suhrbier A. *et al.* (2001) Immune responses to ISCOM formulations in animal and primate models. *Vaccine* **19**, 2661–2665.

Snowden S. and Langridge W. (2003) Plant-based mucosal immunization. *Biotechnology & Genetic Engineering Reviews* **20**, 165–182.

Stephenson I., Nicholson K., Wood J., Zambon M. and Katz J. (2004) Confronting the avian influenza threat: vaccine development for a potential pandemic. *Lancet Infectious Diseases* **4**, 499–509.

Sturniolo T., Bono E., Ding J., Raddrizzani L., Tuereci O. *et al.* (1999) Generation of tissue-specific and promiscuous HLA ligand databases using DNA microarrays and virtual HLA class II matrices. *Nature Biotechnology* **17**, 555–561.

Subbarao K. and Katz J. (2004) Influenza vaccines generated by reverse genetics. *Current Topics in Microbiology and Immunology* **283**, 313–342.

Sukumaran B. and Madhubala R. (2004) Leishmaniasis: current status of vaccine development. *Current Molecular Medicine* **4**, 667–679.

Tjalma W., Arbyn M., Paavonen J., van Waes T. and Bogers J. (2004) Prophylactic human papillomavirus vaccines: the beginning of the end of cervical cancer. *International Journal of Gynecological Cancer* **14**, 751–761.

Tortora G., Funke B. and Case C. (2004) "Microbiology: An Introduction," 7th edn. Benjamin Cummings, San Francisco, CA.

Tregoning J., Maliga P., Dougan G. and Nixon P. (2004) New advances in the production of edible plant vaccines: chloroplast expression of a tetanus vaccine antigen, TetC. *Phytochemistry* **65**, 989–994.

Vazquez M. (2004) Varicella infections and varicella vaccine in the 21st century. *Pediatric Infectious Disease Journal* **23**, 871–872.

Walmsley A. M. and Arntzen C. J. (2000) Plants for delivery of edible vaccines. *Current Opinion in Biotechnology* **11**, 126–129.

Webster D., Thomas M., Strugnell R., Dry I. and Wesselingh S. (2002) Appetising solutions: an edible vaccine for measles. *Medical Journal of Australia* **176**, 434–437.

WHO. WHO Vaccine Preventable Diseases Monitoring System; 2000 global summary. Department of Vaccines and Biologicals, WHO, Geneva.

WHO. Home Web site: http: //www.who.int

WHO. (1999) Global Polio Eradication Initiative, Advocacy: A practical guide with polio eradication as a case study. Geneva, Switzerland.

Williams R. S., Johnston S. A., Riedy M., DeVit M., McElligott S. G. *et al.* (1991) Introduction of foreign genes into tissues of living mice by DNA-coated microprojectiles. *Proceedings of the National Academy of Sciences USA* **88**, 2726–2730.

Yang N.-S., Burkholder J., Roberts B., Martinelli B. and McCabe D. (1990) In vivo and in vitro gene transfer to mammalian somatic cells by particle bombardment. *Proceedings of the National Academy of Sciences USA* **87**, 9568–9572.

Young D. (2000) Current tuberculosis vaccine development. *Clinical Infectious Diseases* **30**, S254–S256.

Yu A., Cheung R. and Keeffe E. (2004) Hepatitis B vaccines. *Clinical Liver Diseases* **8**, 283–300.

Zajac A. J., Murali-Krishna K., Blattman J. N. and Ahmed R. (1998) Therapeutic vaccination against chronic viral infection: the importance of cooperation between CD4$^+$ and CD8$^+$ T cells. *Current Opinion in Immunology* **10**, 444–449.

Zeitlin L., Cone R., Moench T. and Whaley K. (2000) Preventing infectious disease with passive immunization. *Microbes and Infection* **2**, 701–708.

Zimmerman R. and Ahwesh E. (2000) Adult vaccination, part 2: vaccines for persons at high risk. *Journal of Family Practice* **49**, S51.

Zimmerman R. and Burns I. T. (2000) Child vaccination, part 1: routine vaccines. *Journal of Family Practice* **49**, S22–S32.

Ziv E., Daley C. and Blower S. (2004) Potential public health impact of new tuberculosis vaccines. *Emerging Infectious Diseases* **10**, 1529–1535.

Zurbriggen R. and Gluck R. (1999) Immunogenicity of IRIV- versus alum-adjuvanted diphtheria and tetanus toxoid vaccines in influenza-primed mice. *Vaccine* **17**, 1301–1305.

Primary
Immunodeficiencies

Robert Good 1922–2003

Clinical and experimental immunodeficiencies

© AAAAI

24

CHAPTER 24

A. GENERAL CONCEPTS

B. PRIMARY IMMUNODEFICIENCIES DUE TO DEFECTS IN ADAPTIVE IMMUNE RESPONSES

C. PRIMARY IMMUNODEFICIENCIES DUE TO DEFECTS IN THE INNATE IMMUNE RESPONSE

"That was David. He was born into a bubble and he died at the age of 12 in 1984 undergoing an experimental bone marrow transplant, he's the symbol of our patients"–Marcia L. Boyle

Much of this book has been devoted to demonstrating the essential nature of our immune system. Without its vigilance, our bodies fall prey to pathogen infections and tumors. In the next two chapters, we explore the consequences of a failure in some aspect of the immune system, a situation known as *immunodeficiency*. Immunodeficiencies (IDs) can be either *acquired* or *primary*. In *acquired* immunodeficiencies, the patient is born with normal immune responses but later experiences an event that damages the immune system in some way. The most notorious example is, of course, acquired immunodeficiency syndrome (AIDS) caused by human immunodeficiency virus (HIV) infection. However, acquired immunodeficiency can be a secondary consequence of other events, such as severe infections and treatments to combat them, cancers and cancer therapies, and malnutrition. Acquired immunodeficiencies are discussed in Chapter 25.

This chapter is concerned with *primary immunodeficiencies* (PIs) in which the patient is born with a genetic mutation that results in a defect in either the innate or adaptive immune response. Mutations have been identified that affect various aspects of the differentiation and/or effector functions of B, T, or NK cells or phagocytes, the function of the complement system, and anti-bacterial defenses. Some mutations alter the activity of an enzyme or factor that is relevant to more than one cell type. Other mutations cause defects in cellular interaction or regulation that have effects on immunity. For example, the name "hyper-IgM syndrome" looks like a misnomer at first, since the phenotype is the exact opposite of "deficiency." However, in most cases of this disorder, the defect lies in the reduced expression of CD40L by helper T cells. The intercellular contacts required for isotype switching are not supplied to B cells, and the outcome is elevated (rather than diminished) IgM production.

The vast majority of the more than 100 distinct PIs that have been defined are rare (<100 patients worldwide), and the total global incidence of PIs is about 1 in 5000 live births. The most common PI is *selective IgA deficiency* (see later),

which has been reported in the United States at 1 in 700 live births. We start our discussion of PIs with an examination of some general concepts related to diagnosis and treatment of PIs. We move then to detailed descriptions of PIs due to defects in the adaptive immune response, since these are the disorders with which the reader is most likely to be familiar. The "combined immunodeficiencies" (both T and B cell functions are affected), T cell-specific IDs, B cell-specific IDs, IDs due to defects in DNA repair, lymphoproliferative IDs, and other adaptive immunodeficiencies are included in this group. We conclude this chapter with a description of PIs arising from defects in the innate immune response, including phagocyte response disorders and complement deficiencies. Table 24-1 summarizes the defining characteristics of these categories of PIs.

A. General Concepts

I. DIAGNOSIS OF PIs

While most PIs do not manifest symptoms until well into a patient's adult years, others are diagnosed in infancy. Protection furnished by maternal antibodies lasts for 6–9 months after birth, but when these antibodies are finally depleted, the young patient shows signs of PI. The principal clinical phenotype of most PIs is increased susceptibility to infections with organisms that are of low pathogenicity in a normal human. Different types of PIs are associated with different types of infections. For example, enterovirus infections are commonly seen in patients with antibody deficiencies but not those with phagocyte defects or complement deficiency. Similarly, infection with *Salmonella typhi* or various fungi is a sign of a defect in cell-mediated immunity. As well as multiple and recurrent infections, PI patients often present with weight loss, failure to thrive,

Table 24-1 General Categories of PIs

PI Category	Defining Characteristics
SCID	Low numbers or absence of T cells and sometimes B cells; lack of B cell function may be due to lack of T cell function
T-PI	Normal numbers of T, B, and NK cells but T cells are non-functional; B cell function may be compromised
B-PI	B cells are absent or non-functional; T and NK cells are normal
DNA repair defects	Failure in DNA repair in immune system cells (as well as other body cells); lead to low lymphocyte numbers
Lymphoproliferative disorders	Mutations in death receptors or ligands, or in molecules of the caspase cascade; lead to uncontrolled proliferation of T and B cells
Phagocyte response deficiencies	Defects in extravasation, activation, or function of phagocytes
IL-12/IFNγ axis	Failure in macrophage hyperactivation
Autoinflammatory syndromes	Severe local inflammation and prolonged periodic fevers with no obvious cause; lymphocytes are normal
Neutropenias	Reduced levels of neutrophils in the circulation
Complement deficiencies	Lack of complement activation; T and B cells and phagocytes are normal

dermatitis, and enlargement of lymph nodes, tonsils, and other organs. Autoimmunity, malignancies, and allergic diseases can also appear as PI symptoms depending on the defect in the immune system. The severity of PIs ranges from relatively benign to life-threatening. In the worst cases, the failure in the immune response allows chronic infections to take hold that can cause irreversible and sometimes fatal damage to vital organs. Because this deterioration can occur rapidly, a diagnosis of a severe PI constitutes a pediatric emergency that should be treated as quickly as possible.

Precise diagnosis of a PI can be a challenge, since similar clinical pictures can result from very different defects, and very similar defects can produce dramatically different clinical signs. Indeed, affected members of the same family, who all carry the same mutation, can show striking variability in the severity and range of clinical symptoms. Studies of several PIs of this sort have led to the conclusion that the clinical phenotype of a PI caused by a specific mutation is often influenced by environmental factors. Clinicians usually commence a diagnosis by ruling out factors that might result in acquired ID, such as recent surgery, immunosuppressive treatment for a transplant, protein insufficiency disorders, or tumorigenesis. HIV screening may also be carried out. A complete medical and family history is obtained, noting the incidence of infection in other family members, the possibility of consanguinity (because many PIs are autosomal recessive disorders), and whether the patient has experienced any unusual events following childhood vaccinations that might suggest a compromised immune system. Consultation with experts in microbiology, virology, and immunology is often necessary to nail down the diagnosis. Immunological screening tests for antibody production, DTH reactions, complement function, lymphocyte surface markers, lymphocyte proliferation *in vitro*, and neutrophil function can be used to help identify the defect.

Obviously, patients confirmed to have an immunodeficiency (either acquired or primary) should be vaccinated with live, attenuated vaccines only after careful consideration, for fear that the degree of attenuation may not be sufficient to make the pathogen controllable by a weakened immune system. For the same reason, ID patients should be prevented from coming into contact with vaccinees immunized with attenuated pathogens (e.g., polio) that remain highly contagious and are easily shed in body secretions.

II. TREATMENT OF PIs

i) Intravenous Immunoglobulin Replacement Therapy (IV-IG)

The deficit in Ig production associated with many PIs can be treated on an ongoing basis with *intravenous immune globulin* (IV-IG) *replacement therapy*, a technique established in the 1980s to treat hypogammaglobulinemia. The IV-IG preparation most often used contains 10% IgG (of all specificities) in a form that resists aggregation. The protein is stabilized, partially digested, or acidified so that the chances of complement activation following intravenous injection are reduced. A dose of about 400–600 mg IV-IG/kg/month is often sufficient to prevent a PI patient from experiencing recurrent infections. The reader will note that IV-IG therapy is a more general form of the passive immunization technique described in Chapter 23 in which patients are administered preparations of specific antibody directed against a single pathogen.

ii) Enzyme Replacement

Some PIs are due to an enzyme deficiency and can be treated by enzyme replacement therapy in which the patient is injected with a stabilized form of the missing enzyme. This strategy works only if enzyme function in the blood is sufficient to mitigate the disease, and if the enzyme does not actually have to enter the hematopoietic cells affected. This therapy is discussed later in this chapter in the context of the diseases to which it is applied.

iii) Bone Marrow/Hematopoietic Cell Transplantation

Increasingly, many types of PI patients are being cured with a *bone marrow transplant* (BMT; see Ch.27) or a *hematopoietic cell transplant* (HCT; see Ch.27) from a histocompatible donor. When successful, this procedure allows a patient to completely replace his or her immune system and subsequently lead a normal life. Indeed, prenatal diagnosis of many PIs has made it possible to carry out a BMT or HCT *in utero*, and the earlier the transplant is undertaken, the greater the chance of success. Because identical HLA matches among humans are rare, allogeneic BMT/HCT from siblings or other close relatives is most often attempted. Up to 60–80% of PI patients treated with BMT/HCT survive and go on to develop new T cells after 3–4 months. However, this delay in re-establishing the immune system can lead to the death of a transplant recipient due to infections or *graft-versus-host disease* (GvHD). (In GvHD, lymphocytes in the donated graft attack normal tissues in the recipient and cause severe clinical complications. Depletion of T cells from the donor marrow prior to transplantation can reduce the chance of GvHD; see Ch.27). In a proportion of the BMT/HCT cases that survive, T cell function is only partially restored and B cell production and/or function is still lacking. This deficit in Ig production can be treated on a continuing basis with IV-IG therapy.

iv) Gene Therapy

If the underlying genetic defect of a PI is known, it may become possible to attempt "gene therapy," in which a normal copy of the faulty gene is introduced into the somatic cells of the patient. To date, gene therapy protocols for PIs have not been as practical or successful as once hoped. Attempts at gene therapy are covered in this chapter in the context of the diseases to which they have been applied.

III. FOCUS OF PI RESEARCH

Much research has been devoted to identifying the mutated genes responsible for various PIs. Knowing the genetic defect allows a precise diagnosis, and thus a prediction of what the prognosis may be for the patient and the chances of disease for other family members. Genetic counseling of identified carriers may be appropriate. Prenatal diagnosis of a PI also becomes possible, allowing parents to make informed reproductive choices. In the future, when gene therapy moves from an experimental protocol to a practical treatment, knowing which gene to replace is obviously of paramount importance. From the basic research point of view, the workings of a biological system are most often revealed when a mutation causes a disruption to that system. Indeed, several experts in the field have noted that PIs are a prime example of William Harvey's 1657 definition of "Experiments of Nature," as quoted by Lord Garrod in 1924: "Nature is nowhere accustomed more openly to display her

secret mysteries than in cases where she shows traces of her working apart from the beaten path; nor is there any better way to advance the proper practice of medicine than to give our minds to discovery of the usual law of nature by careful investigation of cases of rarer forms of disease." (Garrod, AE. *The debt of science to medicine.* The Harveian Oration of 1924, Royal College of Physicians. Oxford: Clarendon Press, 1924.) Analysis of the mutations in PIs has revealed much about heretofore hidden aspects of the human immune system, much as knockout mice have served to define the roles of countless proteins involved in immunity in that species. However, comparison of human PI patients with their counterpart mouse mutants reminds us that there are distinct differences between human and murine immune responses, and what holds true in mice may be completely false in humans. Basic research also yields knowledge that may have therapeutic application, as in the deciphering of the mechanisms of an inflammatory PI called *TNF-receptor-associated periodic syndrome* (TRAPS; see later). It was established in the laboratory that TRAPS was caused by a surfeit of TNF signaling; treatment in the clinic with TNF blocking agents has proved to be an effective therapy for TRAPS in preliminary trials.

B. Primary Immunodeficiencies Due to Defects in Adaptive Immune Responses

Defects in T cell and B cell biology can lead to immunodeficiency. A mutation may affect immune responses by just T or just B cells, or may affect responses by both (a combined immunodeficiency). Because of its shared developmental path with T cells, NK cell function is also be affected by many of these mutations.

I. COMBINED IMMUNODEFICIENCIES

Some of the best known immunodeficiencies are the SCIDs, standing for *severe combined immunodeficiency diseases*. SCID cases represent about 20% of PIs, and the global frequency of all forms of SCID is about 1 in 80,000–100,000. The "Boy in the Bubble" story from in the 1970s awoke the general public to the plight of children with SCID, and introduced many to the science of immunology for the first time (see Box 24-1). In all forms of SCID, T cell development and/or function is compromised, such that cell-mediated as well as Ig immune responses are impaired (hence the term "combined"). In some SCID cases, the B cells may have an intrinsic defect, whereas in other cases, the B cell defects are secondary to a lack of T cell help caused by non-functional or absent T cells. NK development and function are also impaired in some forms of SCID. No matter what their genetic defect, patients with SCID generally present with chronic diarrhea, failure to thrive, and severe opportunistic infections. Untreated by BMT/HCT, SCID is inevitably fatal, usually in early childhood but sometimes in adolescence. There are over 400 SCID patients worldwide who have been successfully treated with BMT/HCT.

Box 24-1. Lessons in immunodeficiency: "The Boy in the Bubble"

On September 21, 1971, the birth of David Vetter marked the beginning of an unprecedented experiment that gave doctors, scientists, and laypeople dramatic insight into the consequences of immunodeficiency. Even before David was born, his parents had made special plans for his entry into the world at Texas Children's Hospital in Houston. While the Vetters had a healthy 3-year-old daughter, a son had died after being born with SCID. Because their genetic defect leaves them without B or T cell functions, SCID patients rapidly succumb to infections that would not be threatening to those with functioning immune systems. For this reason, as a precautionary measure, David was delivered by Caesarean section under "germ-free" conditions and was immediately placed in a sterile "isolator." This procedure had been followed in several cases in which a newborn was known to have a family history of SCID. In all of these earlier cases, the babies were quickly found to have fully functioning immune systems, and they were able to leave their sterile environments almost immediately. However, when David was born, his parents became the first ever to face the more difficult outcome—their baby had the disease. The doctors offered an alternative to removing the growing baby from the sterile incubator and waiting for infection to claim his life: David could live in a larger, specially designed sterile environment or "bubble" indefinitely. With time, a cure might be found for SCID or David's immune system might show some latent development, allowing him to "outgrow" the disease. In the meantime, researchers would seek to learn more about SCID by studying David on an ongoing basis. David's family agreed to this plan, and David became known worldwide as "the Boy in the Bubble."

Although David lived most of the time at home in his bubble, he was occasionally studied by doctors and researchers at the Clinical Research Center of Texas Children's Hospital. One of the first lessons gained from study of David was that, despite previous assumptions that SCID patients have no lymphocytes, B cells were present in David's blood. These B cells expressed but did not secrete Ig proteins. David showed no antibody response to challenges with KLH or isolated typhoid antigens, and his lymphocytes could not be stimulated in culture with mitogens. Another lesson was

learned when, between 2.5 and 4 years of age, David suddenly developed T cells. Enough T cells accumulated to bring David's B:T cell ratio into the normal range, but the absolute numbers of both cell types remained below normal. Furthermore, while some activity of B cells was eventually observed, the T cells remained non-functional.

Other aspects of David's life showed better progress. Doctors found that David's growth was normal if not advanced, despite the fact that babies born with SCID who are forced to live in a normal, pathogen-laden environment usually have stunted development. This observation suggested that there is no intrinsic growth defect in SCID; rather, the lack of growth is likely secondary to the repeated infections sustained by these children. David's early years also represented a triumph for those who argued that a person could develop in a normal way both intellectually and emotionally despite living in isolation. By 3 years of age, tests measuring intellectual development showed him to be 1 to 2 years beyond his chronological age. By age 5, he showed above-average speech and language abilities and attended school very successfully by telephone. People pointed out that, rather than wanting to escape from his bubble, he seemed to regard it as something to protect. In 1977, NASA presented him with a special suit that allowed him to experience what it was like

to move around outside his bubble. However, as David neared his adolescent years, he expressed a wish to live a normal life in the outside world and experience such simple pleasures as feeling the grass on his bare feet.

The only hope for a cure during the first years of David's life was a bone marrow transplant (BMT) from a genetically matched donor. In this procedure, the hematopoietic cells from a healthy donor establish themselves in their new host and give rise to the lymphocytes necessary for productive immune responses. In the late 1970s, such transplants were only possible if exact genetic matches could be made between the recipient and donor. Almost invariably, this meant the donor had to be a sibling, and unfortunately, David's older sister was not an exact match. However, by the early 1980s, advances in BMT technology had allowed some success with inexact donor/recipient matches. By this point, David and his family were prepared to take a gamble if it meant he might have his freedom. At the age of 12, David received a BMT from his 15-year-old sister. He remained in his sterile environment while the doctors waited to see if his body would accept the transplanted cells. While there were no signs of rejection, he began to show signs of illness, including fever, diarrhea, and vomiting. These symptoms suggested that David was suffering from a case of "graft

Continued

The reader may recall that *scid* mice have been mentioned in earlier chapters. These mutant animals suffer from severe combined immunodeficiency but differ from their human counterparts in the underlying defect. *Scid* mice have a mutation in the DNA-dependent protein kinase required for V(D)J recombination in T and B cells, a defect yet to be reported in human PI. Thus, *scid* mice are not particularly relevant for studying human SCID per se. However, they are frequently used in reconstitution experiments in which their immune systems are restored using human stem cells. Human hematopoiesis and diseases of various cell lineages can then be studied in the chimeric animals.

We will now discuss each of the various types of human SCID in detail. These disorders are summarized in Figure 24-1 and Table 24-2.

i) X-linked SCID

X-linked SCID (XSCID), the type of SCID suffered by David Vetter (the Boy in the Bubble; refer to Box 24-1), accounts for about half of all SCID cases. XSCID is caused by mutation of the gene encoding the common γ chain (γc). The γc gene is located on the X chromosome. As we learned in earlier chapters, the γc protein is a subunit in the IL-2, IL-4, IL-7, IL-9, IL-15, and IL-21 receptors and transduces intracellular signaling initiated by the binding of the appropriate cytokine to its receptor (Fig. 24-2A). Because both IL-7 and IL-15 signaling are disrupted in the absence of γc, the development of T and NK cells (but not B cells) is disrupted (Fig. 24-2B). Boys affected with this disorder have few if any T and NK cells but normal or even increased numbers of circulating B cells. This form of SCID is therefore sometimes referred to as T⁻B⁺NK⁻ SCID, or the X-linked form of "SCID hyper-B." XSCID B cells may not be entirely normal, however, and exhibit defects in isotype switching that are only partly due to an absence of T cell help. Abnormalities in IL-2 and IL-4 signaling are thought to contribute to the additional defects in B cell function. A point mutation of γc that affects how this receptor chain interacts with its signal transducer Jak3 results in an extremely rare, milder form of XSCID called XCID (i.e., an X-linked combined immunodeficiency that is not "severe"). XCID males survive into adulthood because the mutation allows partial T cell development. About 25–50% of the normal number of circulating T cells can be

found in these patients; however, these T cells tend to be non-functional.

In contrast to human SCID, knockout mice lacking γc have some mature T cells but conventional B cells are almost undetectable. These discrepancies may be attributable to the differential importance of IL-7 signaling in the human and murine immune systems (see later).

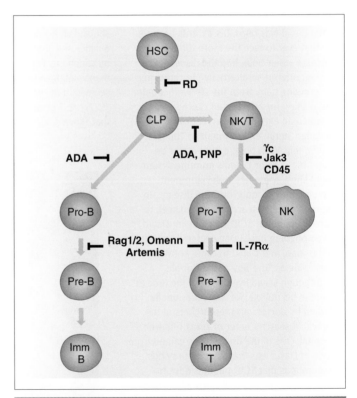

Figure 24-1

Defects Leading to Human SCID
SCID diseases are characterized by the loss of function of both B and T cells. Defects at the indicated points during B and T cell development result in the indicated types of SCID. In some cases of SCID, loss of B cell function is due to a lack of T cell help. NK cells may also be affected by SCID mutations. RD, reticular dysgenesis; ADA, adenosine deaminase; PNP, purine nucleoside phosphorylase; Imm, immature.

Table 24-2 **Types of SCID**

Name	Defective Gene	T Cells	B Cells*	NK Cells
X-linked SCID (XSCID)	γc chain	−	+	−
Jak3 SCID	Jak3	−	+	−
IL-7Rα SCID	IL-7Rα	−	+	+
CD45 SCID	CD45	−	+	−
ADA SCID	ADA	−	−	−
PNP SCID	PNP	−	+	−
RAG SCID	RAG-1 or RAG-2 (<1% function)	−	−	+
Omenn syndrome	RAG-1 or RAG-2 (1–25% function)	±	−	+
Artemis SCID	Artemis (VDJ recombination)	−	−	+
Reticular dysgenesis	?	−	−	−

*Even if B cells are present, they do not secrete Igs.

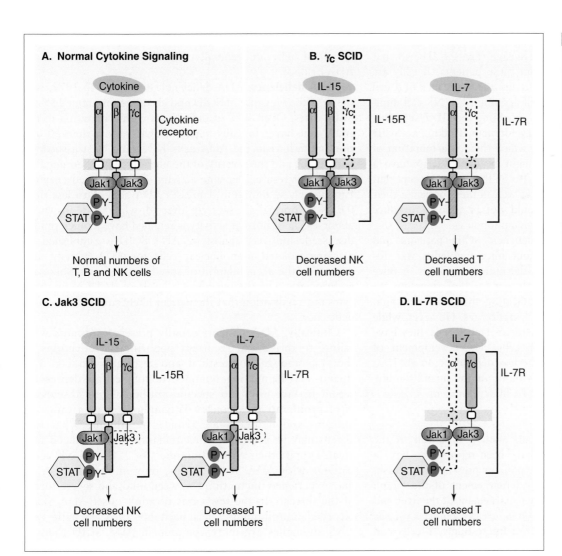

Figure 24-2

Defects in Cytokine Receptor Signaling that Lead to SCID
(A) Many cytokine receptors are heterodimers or heterotrimers whose cytoplasmic regions associate with Jak kinases that in turn activate STAT transcription factors. Normal function of all these components results in normal numbers and function of T, B, and NK cells. (In parts B, C, and D, mutated molecules are indicated with a dashed outline.) (B) Mutations of the γc chain, which is a component of the IL-15 and IL-7 receptors, prevent the development of mature NK and T cells, respectively. B cells are non-functional due to the lack of T cell help. Note that mutations of the γc chain also prevent signaling by IL-2, IL-4, IL-9, and IL-21. Loss of these intracellular messages may independently affect the normal functions of many hematopoietic cell types. (C) Mutations in Jak3 kinase have the same effects as γc chain mutations. (D) Mutations in the IL-7Rα chain affect only IL-7 signaling and impair T cell development.

ii) Jak3 SCID

A disease clinically indistinguishable from XSCID (but much more rare) can result from mutations in Jak3, the lymphoid-specific kinase responsible for downstream transduction of signals from γc (Fig. 24-2C). Because the gene encoding Jak3 is not on the X chromosome, this disease has an autosomal recessive inheritance pattern and is sometimes called "autosomal recessive SCID hyper-B." Again, T and NK cells are missing, but B cell numbers are normal or increased. However, the B cells are non-functional due to both a lack of T cell help and a lack of cytokine signaling.

Jak3$^{-/-}$ knockout mice have a phenotype that is indistinguishable from that of γc$^{-/-}$ animals. Unlike their human counterparts, Jak3$^{-/-}$ animals have some T cells. However, these T cells are highly proliferative *in vivo*, have activated or memory surface markers, show a decreased resistance to apoptotic stimuli, and are non-functional in *in vitro* assays. Experimental gene replacement therapy in Jak3$^{-/-}$ mice has led to mitigation of the immunodeficiency symptoms, suggesting that gene therapy for the human disorder may be possible.

iii) IL-7Rα SCID

This type of SCID can be thought of as T$^-$B$^+$NK$^+$ SCID. A mutation in the gene encoding the IL-7Rα chain affects the development of only the T lineage (Fig. 24-2D), so that mature T cells are missing from these patients. B cells are present, but are unable to produce Igs due to the lack of T cell help (so that the deficiency is still "combined"). NK cell numbers and function are normal, confirming that IL-7 signaling is not required for human NK development. To date, no SCID patients have been identified in which there is a mutation of the IL-7 gene. Such patients might be expected to have a phenotype similar to that of IL-7Rα SCID patients except that a BMT/HCT might not "take" as well in these patients; their bone marrow stromal cells would not be able to produce the IL-7 needed to support early lymphopoiesis.

A comparison of IL-7Rα-deficient SCID patients and IL-7$^{-/-}$ and IL-7Rα$^{-/-}$ knockout mice highlights the differential importance of IL-7 to the immune systems of mice and humans. Knockout mice deficient for IL-7 signaling are missing both T and B cells, indicating that IL-7 is a vital growth factor for both types of lymphocytes. However, while human XSCID and Jak3 SCID patients lack T cells, they have at least normal numbers of B cells. The development of human B cells must therefore be less dependent on IL-7, so that a mutation in either the cytokine or the ligand-binding chain of its receptor affects only T cell development.

iv) CD45 SCID

A very small number of patients with a mutation in the regulatory phosphatase CD45 have been reported to have a form of T$^-$B$^+$NK$^-$ SCID. These patients present with a rash, pneumonitis, *lymphoadenopathy* (enlargement of the lymph nodes), and *hepatosplenomegaly* (enlargement of the liver and spleen). CD45 surface expression is greatly reduced on the surfaces of all types of leukocytes. Interestingly, T cells and NK cells in these patients are non-functional and profoundly decreased in number but B cell numbers are normal or increased. However, despite the apparent normal B cell development, production of both IgM and IgA is greatly decreased in these patients and B cell proliferative responses are impaired *in vitro*. It remains unclear how the mutation in CD45 blocks surface expression, and how B cells, but not NK or T cells, can compensate for the loss of an enzyme important in all three cell types. Mice lacking CD45 show immunodeficiencies very similar to those observed in CD45-deficient humans.

v) Adenosine Deaminase and Purine Nucleoside Phosphorylase SCIDs

A form of T$^-$B$^-$NK$^-$ SCID arises from autosomal recessive mutations in the gene encoding *adenosine deaminase* (ADA). Mutations in the gene for *purine nucleoside phosphorylase* (PNP) result in a similar SCID that affects T and NK cells but not B cells. The ADA and PNP genes encode enzymes required for the salvage pathway of purine metabolism (see later). ADA and PNP are normally present in all cells but are most highly expressed in hematopoietic cells and the kidney. Because lymphocytes are a particularly dynamic and actively proliferating cell population, the accumulation of toxic metabolites that occurs in the absence of ADA or PNP is devastating to the survival and/or differentiation of lymphoid cells. Non-lymphoid cells are generally not affected by mutations in ADA or PNP.

va) ADA deficiency. ADA deficiency, discovered in 1972, was the first known cause of SCID and accounts for about 15% of SCID cases. Over 80% of individuals with a mutation in the ADA gene have clinically apparent SCID. More than 50 distinct mutations in the ADA gene have now been associated with SCID, and the severity of the phenotype is determined by the level of residual enzyme activity of the mutated enzyme. Patients whose mutation depresses ADA activity to less than 0.05% of normal are severely affected, while those in whom about 1–5% of normal activity is retained have a much milder disease. Definitive diagnosis of ADA SCID is established by assaying isolated mononuclear cells for ADA activity and measuring the accumulation of specific purine metabolites in erythrocytes. The finding of high levels of dATP in erythrocytes is an indication that the patient likely suffers from ADA deficiency.

Clinically, ADA patients usually present as infants with failure to thrive and recurrent opportunistic infections of the skin and gastrointestinal and respiratory tracts. The thymus may be missing, lymphocyte numbers are depressed, serum Igs are low, and specific antibodies are absent. *In vitro* proliferative responses of lymphocytes to a range of antigens and mitogens are reduced or absent. What few T cells may be present in the circulation show reduced survival. Despite their reduced numbers of T, B, and NK cells, patients with ADA SCID may be only mildly affected, or may experience fluctuations between mild and severe forms of the disease. In the few cases of adult onset ADA SCID recorded, autoimmunity has been observed. Hepatic and CNS anomalies are more common in ADA SCID than in other forms of SCID.

ADA is a monomeric enzyme present in all cells that converts adenosine and 2′-deoxyadenosine molecules into inosine and 2′-deoxyinosine, respectively (Fig. 24-3). If not metabolized, adenosine and 2′-deoxyadenosine accumulate to toxic levels both within cells and in the circulation. Adenosine and 2′-deoxyadenosine are generated via the breakdown of DNA and RNA during cell turnover. These molecules can pass through the cell membrane into the extracellular fluid and are

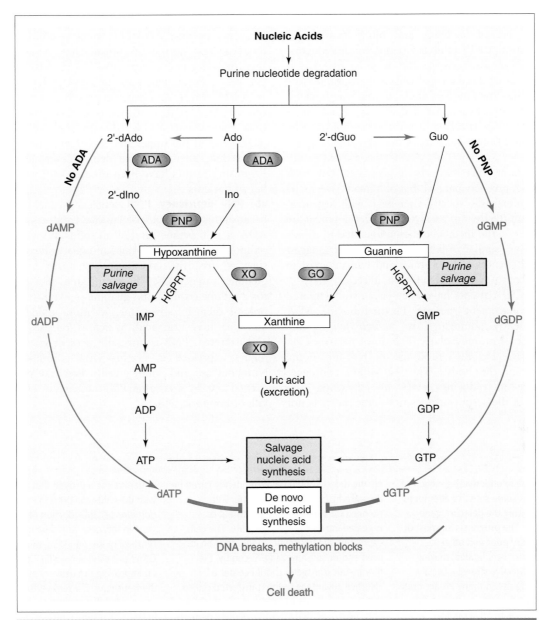

Figure 24-3

SCID Associated with Defects in Purine Salvage
Normal nucleic acid degradation leads to an accumulation of purine nucleotides that are broken down into adenosine (Ado) and deoxyadenosine (2′-dAdo), and guanosine (Guo) and deoxyguanosine (2′-dGuo). ADA converts Ado and 2′-dAdo to inosine (Ino) and 2-deoxyadenosine (2′-dIno), respectively. PNP converts Ino and 2′-dIno to hypoxanthine, and Guo and 2′-dGuo to guanine. These molecules can enter the purine salvage pathway via HGPRT and are eventually converted back to ATP and GTP that can be recycled into new nucleic acids. Alternatively, hypoxanthine and guanine are converted by xanthine oxidase (XO) and guanine oxidase (GO), respectively, to xanthine. Xanthine is then converted by XO to uric acid, which is excreted. In the absence of ADA, Ado and 2′-dAdo accumulate and are metabolized into dATP. Similarly, loss of PNP function results in the accumulation of dATP and dGTP. These toxic molecules induce breaks in DNA, block normal DNA methylation, and interfere with the *de novo* pathway of DNA synthesis such that cell death is induced. Rapidly proliferating cells such as lymphocytes are most affected, and ADA and PNP mutations result in a variable loss of T and B cells. ADA, adenosine deaminase; PNP, purine nucleoside phosphorylase; HGPRT, hypoxanthine-guanine phosphoribosyl transferase; IMP, inosine monophosphate.

taken up by cells using their salvage pathways to synthesize new purines in preparation for cell division. Thus, ADA is crucial for the recycling of elements of old purines into new purines, particularly in tissues in which cell division occurs at a rapid pace. As we have seen, such tissues include the bone marrow, thymus, and lymph nodes. Without ADA function, high levels of dATP accumulate in lymphoid cells, a situation that can induce breaks in DNA and block methylation reactions. In addition, dATP inhibits ribonucleotide reductase activity so that DNA synthesis via the *de novo* pathway is impaired. Apoptosis of the damaged cell then ensues, leading to lymphocytopenia (low numbers of circulating lymphocytes) and SCID. In addition, accumulated dATP can block other enzymatic reactions (not involved in purine metabolism) that lead to the buildup of other substances that promote lymphocyte apoptosis.

ADA$^{-/-}$ knockout mice have been generated but their phenotype does not precisely parallel that of human ADA SCID patients. These animals die shortly after birth with degeneration of the liver, but the thymus and lymphoid development are normal. Nevertheless, the metabolic disruptions due to a lack of ADA function are similar in both species. The data suggest that important differences between mice and humans exist in how purine metabolism is handled in lymphoid cells.

Many ADA-SCID patients have been successfully treated using enzyme replacement therapy. Purified bovine ADA protein is covalently conjugated to *polyethylene glycol* (PEG) and injected intramuscularly 1–2 times per week on a continuing basis. (The PEG prolongs the half-life of the injected enzyme in the body.) The ADA-PEG conjugate is very effective in reducing levels of adenosine and deoxyadenosine in the serum (it does not enter cells), is not toxic, and is well tolerated by patients. Within 3–4 months of treatment, thymus development often resumes and B and T cell numbers slowly increase. Antibody responses (including that against ADA-PEG) become detectable, although T cell function may still be impaired. Adverse effects are few and mild, and most patients gain considerable health benefits if they persist with treatment for more than 6 months. However, because ADA-PEG must be continuously supplied to avoid resumption of clinical signs, enzyme replacement therapy is extremely expensive (>U.S. $100,000 per patient per year). Permanent cure of ADA-SCID can be achieved with BMT/HCT, and gene therapy has been attempted with several ADA-SCID patients (see Box 24-2). In the latter case, the patient is injected with a plasmid containing the ADA gene in the hopes that expression of the gene in cells will reduce adenosine and deoxyadenosine concentrations. Promising results have been obtained for a very small number of patients.

vb) PNP deficiency. PNP deficiency can also lead to SCID because this homotrimeric enzyme catalyzes the step after the ADA step in purine catabolism (refer to Fig. 24-3). PNP acts on inosine and deoxyinosine, and guanosine and deoxyguanosine, to free the core base structures hypoxanthine and guanine, respectively. These molecules are in turn substrates for xanthine oxidase and guanine oxidase, respectively, which process them into xanthine. Xanthine oxidase then converts xanthine into uric acid, which is released into the blood and excreted. PNP is normally present in all cells but most highly expressed in erythrocytes, granulocytes, peripheral lymphocytes, and kidney cells. Prenatal diagnosis can be carried out by analyzing PNP activity in cultured chorionic villi or amniotic cells.

Box 24-2. Somatic gene therapy for primary immunodeficiencies

Because PIs are due to underlying genetic mutations, it makes some sense to attempt to cure them by replacing the defective gene with a normal one. The prognosis for most of these diseases is very poor and other treatment options are very limited, making gene therapy an attractive strategy. Luckily, the mutations giving rise to many PIs are quite well-defined so that there is no issue about which gene to replace. In addition, the mutations resulting in PIs generally affect a relatively discrete number of cells, meaning that somatic gene therapy is practical. In view of our limited understanding of the effects of altering genomic DNA, somatic modification (which cannot be passed on to the next generation) rather than germline gene therapy is preferred.

Pre-clinical investigations of gene therapy for PIs have helped to define what makes a good therapeutic transgene and when it is appropriate to attempt gene therapy. First of all, the defect should be permanently cured by the introduction of the transgene, with no need for supplemental treatment or repeated introduction of the gene. The cells expressing the transgene should thus be long-lived and/or self-renewing. Secondly, the therapeutic transgene should confer a selective advantage to the transfected cell so that it can outgrow its defective counterparts and "take over" the relevant cellular compartment. Thirdly, the therapeutic gene in question must have the potential to be expressed at the correct time in all tissues in which loss of the corresponding natural gene causes a significant physiological defect. However, lineage-specific and stage-specific controls must be used when expression or overexpression in particular cell types or at particular stages of maturation would be detrimental. Finally, the disease for which the gene therapy route is being attempted should not be caused by a mutation capable of generating a protein that can interfere with the wild-type protein (*negative transdominance*). Introduction of the wild-type gene in this case would be useless without removal of the offending mutant protein. Other pre-clinical investigations have focused on improving the delivery and expression of the transgene. Efficient vector delivery systems are needed that can support gene expression in HSCs or resting lymphocytes. Ways to remove negative regulatory sequences in murine retroviral vectors or to add enhancers or demethylating fragments have been devised. It may be possible to include cytokine genes in the vector to stimulate proliferation, or to replace native promoters with those favoring expression in HSC and/or resting lymphocytes.

In 1990, ADA deficiency became the first PI to be treated experimentally with somatic gene therapy. In earlier mouse experiments, murine stem cells had been transfected

Continued

Box 24-2. Somatic gene therapy for primary immunodeficiencies—*cont'd*

in vitro (a process called *transduction* by clinicians in this field) with a retrovirus vector carrying the human ADA gene. These cells gave rise to circulating lymphoid cells that expressed the transfected gene. Based on this success, scientists began to explore similar vectors that could be used in human HSCs. In theory, the transplantation of HSCs transfected *in vitro* with modified retrovirus expressing the normal ADA gene should support normal thymopoiesis in the ADA patient and lead to normal thymic selection and the development of a normal mature T cell repertoire. In practice, the first trials of this approach showed insufficient rates of transfection (1–10%) of the HSCs, preventing the rebuilding of the immune system after transfer to the ADA patient. Scientists also tried transfecting T cells with the ADA gene and infusing them into ADA patients (who discontinued their ADA-PEG treatment). Although these T cells initially proliferated well in the patient, expressed normal ADA, and appeared to have normal functions, the impact of these relatively short-lived cells was not enough to ameliorate the metabolic defect. In 2002, researchers in Italy reported improved techniques for HSC transfection and transplantation that resulted in marked clinical improvement in the two ADA patients in which the approach was tried. These patients were not receiving ADA-PEG treatment (due to a lack of availability in their home countries) so that the effect of the gene therapy alone could be monitored. A combination of a more efficiently transfected HSC population plus a less severe bone marrow ablation technique allowed the transfected HSCs to successfully engraft in the patient bone marrow in high numbers. Multiple lineages of hematopoietic cells, including myeloid, erythroid, T, B, and NK cells, were produced that stably expressed normal ADA. Lymphocyte numbers, circulating antibody levels, and immune responses were increased in these patients, while the levels of toxic metabolites were decreased. One year after the transplant, the respiratory infections and chronic diarrhea that had plagued these patients had disappeared in one of them. To date, these patients are free of malignancies.

Like ADA SCID, PNP SCID is a reasonable candidate for gene therapy because many of its effects are exerted on hematopoietic cells. Model systems have been devised in which PNP-deficient human fibroblasts and

T lymphoma cell lines have been transfected with retroviral vectors expressing PNP. The transfected cells do indeed express PNP and are able to withstand experimental exposure to deoxyguanosine. PNP-deficient peripheral T cells transfected with such vectors regain the ability to proliferate. Whether this approach to therapy, or a version of the transfected HSC approach described previously, will succeed in PNP SCID patients remains to be determined.

Gene therapy is also an attractive approach for the treatment of XSCID (γc-deficiency) and Jak3 SCID because the underlying defects cannot be cured simply by introducing a normal protein into the blood. Moreover, it was shown in one XSCID child that a natural reversion of the γc mutation in a single T cell was sufficient to rebuild the T cell repertoire to about 1% of normal. This observation gave researchers hope that, even if the transfection efficiency of a therapeutic transgene was very low, it might be sufficient to correct the disease *in vivo*. Several XSCID patients have been treated by transplantation of γc-expressing human HSC. Early results were very promising in that, although B and NK cell numbers remained low, transfected T cells were present in significant numbers at about 2 months post-transplant and T cell numbers were generally restored to normal by 6 months. By 12 months post-transplant, T, B, and NK responses had recovered enough to decrease the number and severity of infections in these patients and greatly improve their rate of growth and overall health. Researchers began to muse that the *in vivo* correction might be permanent, an improvement over the conventional treatment of SCID by BMT/HCT in which new T cells are produced for only about 14 years. Unfortunately, the optimism of the gene therapists turned out to be premature: 3.5 years after treatment, two out of the 10 children treated with γc-transfected T cells in this trial developed T cell leukemia. It seems that the retroviral vector used to deliver the γc gene inserted itself into the genome of one founder T progenitor right at the site of a known oncogene. The oncogene was activated and induced the malignant transformation of the progenitor. This T cell gained a selective advantage and became an aggressive leukemic clone that "took over" the T cell compartment in the patients. With low numbers of NK cells, the patients may

not have been able to eliminate the transformed cells promptly enough to avoid establishment of the leukemia. Although oncogenesis has generally been a rare event in clinical trials of gene therapy, this experience with XSCID patients has raised concerns about the safety of transfection mediated by retroviral vectors, particularly in immunocompromised patients. Ways of reducing the transforming potential of retroviral vectors, such as inserting DNA sequences to block the transcription of neighboring genes, or ensuring that the therapeutic gene is expressed only from an internal promoter, are under investigation.

Lineage-specific gene therapy, controlled by promoters that are active only in certain cell types, will likely be required to cure forms of immunodeficiency caused by mutations of cell-type specific genes. For example, the RAG genes are expressed only in lymphocytes, so that expression of RAG1/2 in non-lymphoid cells induced during global RAG gene therapy might be toxic. Similarly, T-PIs caused by a lack of CD3 components or ZAP-70 might be addressed by transfection of genes encoding these proteins, but without lineage-specific expression controls, their overexpression in B and myeloid cells might be problematic. Temporal control of transcription may also be needed in some cases. For example, the expression of the transcription factor CIITA is tightly controlled throughout T cell development. Patients with CIITA deficiency would therefore likely require treatment with a genetically engineered construct that restricted expression of CIITA to only the proper stage of T cell development.

In addition to lineage-specific and temporal controls on gene expression, the amount of protein produced may also have to be regulated to achieve safe gene therapy. Two examples are Wiscott-Aldrich syndrome, a PI in which a protein called WASP (*Wiscott-Aldrich syndrome protein*) is missing or mutated, and a form of X-linked hyper IgM syndrome in which the CD40L gene is mutated (see main text). In mice, the overexpression of WASP or CD40L has been associated with undesirable secondary effects such as cytolysis or lymphoma development. The design of vectors for gene therapy in humans in these cases may therefore require much prior experimentation to determine safe levels of protein expression.

Because as little as 1.5% of normal PNP activity (as measured in erythrocytes) is sufficient for survival, PNP SCID occurs much less often than ADA SCID and constitutes only 4% of all SCID cases. Profound deficiency for PNP is fatal due to overwhelming infections. In patients homozygous for a PNP mutation, T cell function may be normal at birth, accounting for the virtually normal humoral responses seen in very young PNP patients. However, as the patient grows, T cell function declines or may fluctuate between normal and impaired, adversely affecting B cell responses. PNP patients often first come to clinical attention because of varicella infections that are widely disseminated due to the lack of functioning T cells. Recurrent infections of the sinuses, urinary tract, and middle ear, and diarrhea and pneumonia are common. Hypouricemia (a reduction in the level of urate in the blood) may also be present due to the block in the production of substrates for xanthine oxidase. Defects in the CNS leading to skeletal anomalies affecting motor function are also frequent. Autoimmunity occurs more frequently in PNP SCID than in ADA SCID, affecting 30% of PNP SCID patients.

Immunodeficiency arises in PNP deficiency because, like dATP, dGTP can inhibit ribonucleotide reductase and DNA replication. In the absence of purine nucleotide processing by PNP, dGTP accumulates in cells and poisons them, resulting in lymphocytopenia. However, damage by dGTP occurs only in T and NK cells and not in B cells. Due to intrinsic differences in the activities of nucleotide kinases in T/NK and B lymphocyte progenitors, dGTP accumulates to high levels in T/NK lineage cells but not in B lineage cells. As a result, B cell numbers are normal in these patients.

Mice with chemically induced point mutations of PNP show a phenotype that is similar to that of PNP-deficient humans and correlates well with the amount of residual PNP activity retained in the mutant animal. Accumulation of dGTP was found to occur preferentially in the mitochondria of T cells, leading to defects in mitochondrial DNA repair. It is speculated that this lack of repair renders the murine PNP-deficient T cells hypersensitive to spontaneous mitochondrial DNA damage. This damage induces the T cells to undergo apoptosis in droves, leading to lymphocytopenia.

Enzyme replacement using a PEG-conjugated form of purified PNP can be used to treat human PNP deficiency. The circulating PNP-PEG processes any inosine, deoxyinosine, guanosine, and deoxyguanosine released from cells into the blood, reducing the chance of uptake by lymphocytes. As is true for ADA-PEG, this method of therapy is safe and efficacious but very expensive. Interest in maximizing the stability of injectable PNP has led to investigation of the hexameric PNP of *Escherichia coli* as a substrate for conjugation with PEG. The bacterial PNP enzyme is more stable at 37° than the human, murine, or bovine PNP enzymes, and its hexameric structure allows the conjugation of more molecules of PEG. In treatment trials in mice, bacterial PNP-PEG showed decreased immunogenicity but increased half-life in the blood and the same enzymatic activity as the murine PNP-PEG. Gene therapy has also been attempted for PNP deficiency but only at an experimental level (refer to Box 24-2).

vi) RAG SCID

A form of SCID sometimes called *alymphocytosis* results from null mutations of either the RAG-1 or RAG-2 genes. These mutations cause defects in V(D)J recombination of the gene segments of the TCR and Ig genes, impairing antigen receptor expression. Mature T and B lymphocytes are absent, but NK cells and other hematopoietic lineages are present, leading to the occasional designation of this disease as $T^-B^-NK^+$ SCID.

RAG-1$^{-/-}$ and RAG-2$^{-/-}$ mice show identical, severe blocks in early lymphocyte development. Development of T cells halts at the pro-T cell stage and that of B cells at the pro-B cell stage. Thus, like human RAG SCID patients, no mature circulating T or B cells are found in these animals. Animals carrying mutations of other genes involved in V(D)J recombination show similar phenotypes to RAG-1$^{-/-}$ and RAG-2$^{-/-}$ mice. Natural or engineered mutations in mice or horses have been identified for Ku70, Ku80, DNA-PK, XRCC4, and DNA ligase IV. Since these molecules are involved in DNA double-strand break repair (see later), the mutant animals show phenotypic features such as increased sensitivity to ionizing radiation in addition to defects in V(D)J recombination. To date, no human SCID patients with mutations in Ku70, Ku80, DNA-PK, XRCC4, or DNA ligase IV have been identified.

vii) Omenn Syndrome

Omenn syndrome is caused by amino acid substitution mutations in the RAG genes that reduce their activity to 1–25% of normal. Children with Omenn syndrome fail to thrive and suffer greatly from opportunistic infections, but also display scaly rashes, alopecia (hair loss), and diarrhea. These patients have extremely low levels of circulating B cells but a variable number of T cells that infiltrate multiple organs, leading to lymphadenopathy and hepatosplenomegaly. The skin and GI tract also experience T cell infiltration. Germinal centers (GCs) are absent from the spleen and lymph nodes. High numbers of eosinophils are present and (for unknown reasons) serum IgE is elevated. Omenn syndrome is often fatal unless corrected by BMT/HCT.

The presence of T cells, but dearth of B cells, in Omenn syndrome patients indicates that V(D)J recombination in B cells requires a higher level of RAG activity than in T cells. Omenn syndrome T cells bear the surface markers of activated Th2 cells and express a restricted TCR repertoire associated with autoreactivity. It is unclear why these T cells are autoreactive: autoimmunity is not observed in *scid* mice with leaky mutations of DNA-PK that allow small numbers of T cells to be produced.

viii) Artemis SCID

The most recently discovered kind of SCID is caused by mutations of the Artemis gene on human chromosome 10p. Artemis encodes a factor that belongs to the metallo-β-lactamase superfamily and is important for both V(D)J recombination and DNA repair. In the absence of Artemis, T and B cells cannot repair the double-stranded DNA cuts made by RAG-1 and RAG-2 during V(D)J recombination. Thus, like RAG SCID, Artemis SCID (sometimes called

Athabascan SCID for the native North American population in which it was first identified) is characterized by a lymphocyte profile of T⁻B⁻NK⁺ SCID. However, unlike RAG SCID cells, Artemis-deficient skin fibroblasts and bone marrow cells show increased sensitivity to radiation. Symptoms and treatment of Artemis SCID are the same as for RAG SCID.

ix) Reticular Dysgenesis

Reticular dysgenesis (RD) is perhaps the most severe type of SCID but the least understood. This autosomal recessive disorder is extremely rare, comprising only 1–3% of SCID cases. The exact genetic defect remains a mystery, and indeed, even the precise cell types affected are obscure. From examination of RD patients' blood cell populations, it appears that the defect disrupts stem cell maturation early in hematopoiesis, resulting in variable deficits in T, B, NK, and other hematopoietic lineages. Some observers have therefore called this disease "inherited bone marrow failure," although some HSCs appear to be present. Most striking in RD are the very low levels of leukocytes in patients' blood smears. Erythrocytes are normal, and platelets are normal or slightly decreased, but monocytes and neutrophils are severely depleted. Interestingly, despite the deficit in monocyte numbers, macrophages are present in normal numbers in the dermis and (abnormal) lymphoid tissues. This observation implies that there must be a residual population of monocytes (or some other hematopoietic cell type) that can differentiate into at least some subsets of macrophages. In addition, the residual monocytes must be able to differentiate successfully into osteoclasts since RD patients do not suffer from the increased bone density. In contrast, Langerhans cells (LCs), which are also derived from blood monocytes, are completely absent in the epidermis of RD patients. A clue to this puzzle may lie in the fact that monocytes isolated from RD patients are able to differentiate into immature DCs *in vitro*. Thus, it is thought that the defect in RD affects a yet-to-be-identified bone marrow-derived factor required to support the differentiation of some, but not all, myeloid and/or lymphoid

progenitors. Short term treatment with GM-CSF can ameliorate RD by stimulating stem cell differentiation, but unless a BMT/HCT can be arranged quickly, death occurs within 3 months due to overwhelming infections.

II. T CELL-SPECIFIC IMMUNODEFICIENCIES

T cell-specific PIs (T-PIs) are those in which the gene affected has its predominant impact on T cells as opposed to B cells or NK cells. Patients with these disorders generally have normal numbers of T cells but these lymphocytes show defects in activation or function (Table 24-3). Clinically, T-PI patients often show defects in the CTL response manifested as increased susceptibility to infections with viruses, intracellular bacteria, *Pneumocystis carinii*, and other fungi. Infections may be mild or severe. Levels of Igs in these patients are usually low but may be normal. Autoimmunity is common. These patients may also experience a higher incidence of tumors, some of which may be induced by infection with oncogenic viruses.

i) MHC Deficiencies

MHC deficiencies were introduced in Chapter 10. Cells in MHC-deficient patients fail to express the MHC molecules necessary for antigen presentation to T cells, causing a failure in the cell-mediated response and impairment of the humoral response. Mutations in the TAP transporter result in impaired expression of MHC class I on all tissue cells. As a result, Tc cells cannot be activated. Autosomal recessive mutations in the transcription factor CIITA, or in any of three RFX proteins required for the regulation of MHC class I and II transcription, result in a lack of MHC class I and II expressed on APCs (including B cells; hence the historical name "bare lymphocyte syndrome"). Due to the lack of MHC class II, Th cells cannot be activated. Patients with MHC deficiencies present with multiple severe infections that are usually fatal at a young age. BMT/HCT is the treatment of choice for MHC deficiencies.

Table 24-3 Types of T-PI

Name	Defective Protein	Functional Defect
MHC deficiencies "Bare lymphocyte syndrome"	TAP	Reduced MHC class I expression
	CIITA	Reduced MHC class I and II expression
	RFX	Reduced MHC class I and II expression
HIGM1	CD40L	Lack of T cell co-stimulation blocks isotype switching in B cells; lack of DC licensing prevents Tc activation
Hemophagocytic syndrome	Perforin Exocytosis protein Munc 13-4 CHS1/LYST	Defective granule exocytosis; uncontrolled infiltration of CTLs into tissues; phagocytosis of RBCs by activated macrophages
TCR signaling defects	CD3γ	No mature CD4⁺ or CD8⁺ T cells
	CD3ε	No mature CD4⁺ or CD8⁺ T cells
	Lck	No mature CD4⁺ or CD8⁺ T cells
	Ca²⁺ flux	No mature CD4⁺ or CD8⁺ T cells
	ZAP-70	No mature CD8⁺ T cells

ii) Hyper IgM Syndromes

At first glance, "hyper IgM" suggests a disease that should be categorized as a B cell defect. Patients with *hyper IgM syndrome* (HIGM) have normal or very high levels of circulating IgM but very low or absent levels of all other isotypes. There are six types of HIGM, labeled types 1–5, plus HIGM-NEMO (Table 24-4). Two of these, HIGM1 and HIGM-NEMO, are X-linked, while HIGM2, HIGM3, HIGM4, and HIGM5 have autosomal recessive inheritance patterns. HIGM2, HIGM3, HIGM4, HIGM5, and HIGM-NEMO are indeed due to intrinsic B cell defects and will be discussed later in the section on B cell-specific PIs. However, HIGM1 is the most common form of the syndrome, and in this case, the defect is one of isotype switching due to a disruption of help supplied by T cells. For all types of HIGM, the lack of isotypes such as IgG and IgA and the presence of severe neutropenia means that the patients suffer from frequent infections with opportunistic organisms such as *Pneumocystis carinii* and *Cryptosporidium parvum*.

In HIGM1, the mutation lies in the CD40L gene, which is located on the X chromosome. As we have learned, interaction between CD40L on a Th cell and CD40 on a B cell is vital for B cell costimulation, isotype switching, and GC formation. The B cells of HIGM1 patients, although intrinsically normal, cannot proliferate and GCs do not form in the lymphoid organs. Memory B cell generation is also impaired. Curiously, despite the lack of CD40-CD40L signaling, a limited amount of somatic hypermutation is able to occur in the few memory B cells present in HIGM1 patients, unlike the case for B cells of HIGM3 patients (who lack CD40 expression; see later). The mechanism underlying this limited somatic hypermutation in HIGM1 B cells is unknown. As well as disrupting isotype switching and GC formation, the CD40L defect in HIGM1 impairs the licensing of DCs, leading to deficits in the cell-mediated response. Thus, in addition to the opportunistic fungal and parasitic infections that plague other types of HIGM patients, HIGM1 patients suffer from recurrent infections with intracellular bacteria. For unknown reasons, HIGM1 patients also run an increased risk of liver disease and liver cancer such that only one-quarter of patients survive beyond 25 years of age. Autoimmunity has been reported in some HIGM1 patients.

IV-IG replacement therapy can help prevent infections normally addressed by the humoral response, but has no effect on those pathogens usually combated by cell-mediated immunity. In particular, CD40L normally expressed on the surface of biliary epithelial cells is thought to be vital for CD40L–CD40 interaction required to control protozoan parasites such as *C. parvum*. BMT/HCT is the only cure for HIGM1.

iii) Hemophagocytic Syndrome (Hemophagocytic Lymphohistiocytosis)

The name *hemophagocytic syndrome* (HPS) sounds like this inherited disease ought to be caused by a defect in innate immunity. However, HPS turns out to be due mostly to a defect in granule exocytosis by CTLs. For unknown reasons, this defect leads to uncontrolled infiltration of activated polyclonal CTLs into various organs (including the brain), and the phagocytosis of erythrocytes by activated macrophages. Both T cells and macrophages secrete copious amounts of cytokines that cause immense tissue damage leading to organ failure and death in the absence of immunosuppressive therapy. However, B cell numbers and activation appear to be normal in HPS patients and there are no clinical signs of autoimmunity.

Mutations in three separate genes have been associated with HPS: the gene encoding the perforin protein, an unknown gene required for the process of exocytosis itself, and the Munc13-4 gene. The Munc13-4 protein is thought to be involved in the docking of cytotoxic granules to the CTL plasma membrane after conjugate formation between the CTL and the target cell. Mutations of the *CHS1* gene that causes *Chédiak-Higashi syndrome* (see later) can also lead to HPS because this mutation impairs the packaging of granule contents. Curiously, unlike humans, mice deficient for perforin do not spontaneously develop HPS even when maintained in a non-pathogen-free facility. However, when these mutant animals are infected with LCMV or herpes simplex virus, HPS-like symptoms develop that kill the mice within a few weeks. A similar course has been noted in other mice bearing mutations in other parts of the granule exocytosis pathway. Why impaired CTL granule exocytosis should lead to uncontrolled T cell infiltration is not clear.

Table 24-4 **Types of HIGM Syndrome**

Type	Defective Gene	Chromosome	Defect in Isotype Switching	Defect in Somatic Hypermutation	Defect in Cell-Mediated Immunity
HIGM1	CD40L	X	Yes	Partial	Yes
HIGM2	AID	12	Yes	Yes	No
HIGM3	CD40	20	Yes	Yes	Yes
HIGM4	Unknown	Not X	Yes	No	Yes
HIGM5	UNG	12	Yes	Partial	No
HIGM-NEMO	NEMO	X	Yes	No	Yes

iv) Defects in TCR Signaling

PI patients have been described in whom various steps of the TCR signaling pathway have been impaired. Defects have been identified in the CD3ε and CD3γ chains of the TCR complex, in Lck kinase that transduces downstream signaling, and in the Ca²⁺ flux that follows TCR triggering. Neither CD4⁺ nor CD8⁺ mature T cells are present in these patients. An interesting group of PI patients lacks ZAP-70 function. Although SP CD8⁺ T cells fail to mature in these patients, mature CD4⁺ T cells are present. It is unclear why human CD4⁺ T cells can compensate for ZAP-70 deficiency but CD8⁺ T cells cannot. It is possible that human CD4⁺ T cells express higher levels of the ZAP-70-related kinase Syk. Mice deficient for ZAP-70 lack both CD4⁺ and CD8⁺ SP T cells.

III. B CELL-SPECIFIC IMMUNODEFICIENCIES

In this section, we discuss PIs due to mutations that affect B lymphocytes only (Fig. 24-4 and Table 24-5), as opposed to those affecting both B and T cells, or those in which the Ig deficit is caused by lack of T cell help. Those born with defects specifically impairing the humoral response account for 70% of all PIs and are relatively easy to diagnose and treat. Patients are usually free of infection until age 7–9 months due to the presence of maternal antibodies. As these antibodies decrease in concentration, the patient manifests low levels of one or more serum Igs and an increased susceptibility to infections with enteroviruses, parasites, and encapsulated bacteria (such as *Streptococcus pneumoniae* and *Haemophilus influenzae* b) that cause pyogenic infections. The stunted growth often evident in patients with T-PIs is not usually seen in patients with B-PIs. Two broad classes of antibody deficiencies exist: patients who do not have mature B cells, and those who do. XLA is the prototypical example for the first category, while diseases such as CVID, IgAD, and HIGM2 (all defined later) fall into the second category. Most humoral immunodeficiencies are treated with IV-IG replacement therapy that must be continued for the life of the patient.

i) Congenital Agammaglobulinemia

ia) Bruton's XLA. <u>X-linked agammaglobulinemia</u> (XLA) is a rare disorder (about 1 in 200,000) in which affected boys lack mature B cells and all antibody isotypes, and thus are unusually susceptible to pyogenic bacterial infections and enteroviral disease. XLA is caused by a defect in B cell maturation that is most often due to a mutation in the cytoplasmic tyrosine kinase *Bruton's tyrosine kinase* (Btk). Over 300 different mutations (mostly amino acid substitutions) of this enzyme have been reported in various XLA patients. Btk is expressed in B cells and myeloid cells but not in T cells. Btk signaling appears to be critical for the continued maturation of pre-B cells, and has been implicated in both intracellular signal transduction affecting calcium mobilization and transcription factor activation. Although pre-B cells are normal in the bone marrow of XLA patients, immature B cells are greatly reduced in the lymphoid organs and periphery. The secondary lymphoid tissues are reduced in size.

Myeloid cells are apparently not affected by the loss of Btk. Interestingly, the same Btk mutation may result in different clinical phenotypes in different members of the same family. This observation implies that other modifying genes or environmental factors may contribute to the clinical manifestation of XLA.

It is still not clear why Btk deficiency has such a devastating effect on B cell development in humans. Gene-targeted Btk⁻/⁻ mice, and mice bearing the natural mutation *xid* (an amino acid substitution in Btk), show much less severe defects in B cell development. Both strains of mutant mice retain up to 50% of normal B cell numbers, although the cells

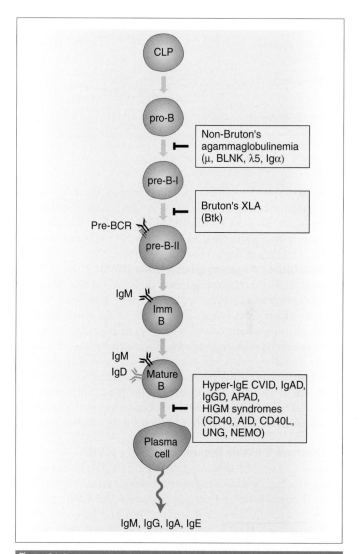

Figure 24-4

Defects Leading to B-PIs

Defects at the indicated points during B cell development give rise to immunodeficiencies characterized by loss of B cell function only. Genes found to be mutated in various B-PI patients are shown in parentheses. XLA, X-linked agammaglobulinemia; CVID, common variable immunodeficiency; IgAD, selective IgA deficiency; IgGD, selective IgG deficiency; APAD, anti-polysaccharide antibody deficiency; HIGM, hyper IgM; NEMO, NF-κB essential modulator; AID, activation-induced deaminase; UNG, uracil DNA glycosylase.

Table 24-5 Types of B-PI

Name	Defective Gene	Functional Defect
Bruton's XLA	Btk	Pre-B cell maturation blocked
Non-Bruton's agammaglobulinemia	Igh-Cμ BLNK λ5 Igα	Pro-B cell maturation blocked
CVID	ICOS ?	Plasma cells produce only low levels of all Ab isotypes
IgAD	?	Plasma cells do not produce IgA (other isotypes are normal)
IgGD	Igh-Cγ ?	Plasma cells do not produce IgG (other isotypes are normal)
Anti-polysaccharide Ab deficiency	?	Plasma cells do not produce Ab to polysaccharides in capsules of bacteria
HIGM syndromes	AID, CD40, ?, UNG, NEMO	Plasma cells produce normal IgM but decreased amounts of all other isotypes due to defects in isotype switching and/or somatic hypermutation
Hyper-IgE syndrome	?	Plasma cells produce very high levels of IgE, possibly due to increased Th2 cytokine secretion or a block in IgE catabolism

have an immature phenotype and reduced survival. Serum concentrations of IgM and IgG3 are decreased but other Ig isotypes are normal. Curiously, although they fail to make antibodies to Ti antigens, *xid* and Btk$^{-/-}$ mice can respond to Td antigens, unlike human XLA patients.

ib) Non-Bruton's agammaglobulinemia (NBA). About 10% of patients with the clinical features of XLA do not have Btk mutations. Instead, these patients have mutations in the μ heavy chain gene, the λ5 component of the surrogate light chain, the Igα chain, or BLNK, a signaling adaptor acting downstream of BCR engagement. These defects lead to a block in the pro-B to pre-B cell transition, a step earlier than the defect in Bruton's XLA patients. Because the genes mutated in NBA are autosomal, this disorder can affect both boys and girls. Again, defects in B cell development are much milder in knockout mice lacking λ5 than in their human counterparts.

ii) Common Variable Immunodeficiency (CVID)

Common variable immunodeficiency (CVID) is the name given to a family of heterogeneous diseases characterized by a general impairment of humoral responses. All CVID patients show profoundly decreased levels of IgA and IgG, and about 50% also lack IgM. Circulating mature B cells are present but plasma cell differentiation and antibody production are impaired. Memory B cells may also be absent. The frequency of CVID is about 1 in 30,000, making it among the most prevalent of all PIs. Onset may occur in childhood, adolescence, or adulthood. The types of recurrent bacterial infections observed in CVID are similar to those in XLA but the secondary lymphoid tissues are frequently enlarged, particularly the Peyer's patches. Virus-associated liver disease occurs in 10% of CVID patients. About 24% of patients die of chronic pulmonary disease or B cell lymphoma, and

autoimmunity occurs in 22% of patients. While most CVID patients are treated using IV-IG, the condition of some has been improved by low dose IL-2 administration.

Some cases of CVID are believed to be the result of (as yet undefined) intrinsic B cell defects. Partial defects in isotype switching and somatic hypermutation have been found in some CVID B cells, and these cells respond only weakly to antigenic stimulation *in vitro*. Other CVID B cells have shown defective PKC activation, impaired DNA repair, abnormal protein tyrosine phosphorylation, or upregulated Fas expression. However, other CVID cases appear to be secondary to T cell defects. Mutations in a poorly defined part of the MHC class III region close to the TNF and LT genes appear to increase susceptibility to CVID. Moreover, about 30% of CVID patients have abnormally low numbers of CD4$^+$ T cells, and 40% of CVID patients show decreased CD40L expression on activated T cells. These findings have led to speculation that CVID can arise from defects in required interactions between B and Th cells, or between Th cells and APCs. CVID has also been associated with homozygous deletions of the ICOS gene. The reader will recall from Chapter 14 that ICOS is an inducible costimulator that contributes to T cell activation, germinal center formation, and isotype switching. While these ICOS-deficient CVID patients had apparently normal T cell subset numbers and functions, they exhibited profound deficits in numbers of naive and memory B cells as well as B cells that had undergone isotype switching.

iii) Selective Ig Deficiencies

In patients with selective Ig deficiencies, antibodies of a particular isotype, or antibodies directed against a particular pathogen structure, are missing. The total level of circulating B cells is usually normal. A few of these cases are caused by straightforward deletions or mutations in the Igh locus that affect the constant region exons for the IgA or IgG subclasses. Other mutations that affect the expression of the membrane-bound form of an Ig protein result in impaired synthesis of the soluble antibody form. Defects in the regulatory and switch regions of the Ig heavy chain exons can also result in selective Ig deficiency.

iiia) Selective IgA deficiency (IgAD). As mentioned previously, selective IgA deficiency is the most common PI at a frequency of 1 in 700 people. Antibodies of both the IgA1 and IgA2 subclasses are missing. All other antibody isotypes are usually synthesized in normal amounts, although some patients may gradually go on to show signs of CVID. T cell function is normal. Gastrointestinal and respiratory infections are the most common clinical signs of IgAD, although the disease is asymptomatic in many people. Autoimmunity is not uncommon and allergy is frequent. Sporadic, autosomal dominant, and autosomal recessive forms of IgAD have been described but the underlying genetic defects are unknown. A particular MHC haplotype (HLA-DR3, -B8, -A1) found in people of Northern European heritage has been linked to IgAD. Susceptibility to IgAD may be associated with the same locus responsible for susceptibility to CVID. Antibiotic use to combat infections is usually adequate treatment for IgAD.

iiib) Selective IgG deficiency (IgGD). Clinical immunodeficiency can also result from the selective loss of the ability to produce one of the IgG subclasses. IgGD is characterized by recurrent upper and lower respiratory tract infections. Inheritance of these disorders is autosomal recessive. While the underlying molecular defects are unknown in many cases, some may be due to mutations in the Igh locus. For example, in one patient, a mutation upstream of the $\gamma3$ exon affected switching to this isotype and resulted in selective IgG3 deficiency. In another family, two cases of selective IgG2 deficiency were found to be due to an insertion mutation in a $\gamma2$ sub-exon. Infections due to IgGD can be prevented with IV-IG replacement therapy. For unknown reasons, these PIs are sometimes associated with allergy, autoimmunity, and malignancy.

iiic) Anti-polysaccharide antibody deficiency (APAD). Some young children have normal levels of serum Igs but suffer recurrent infections because they cannot make antibodies to the polysaccharides of encapsulated bacteria such as pneumococci or meningococci. This condition, which is sporadic in occurrence, often comes to light after vaccination with a polysaccharide vaccine. The underlying molecular defect is unknown. Treatment is by IV-IG replacement therapy.

iv) B Cell-Intrinsic HIGM Syndromes

The most common form of hyper-IgM syndrome (HIGM1) was discussed previously in the section on T cell disorders because it is due to defects in the CD40L signaling pathway in T cells. There are five other HIGM syndromes caused by various defects in B cell signaling pathways involved in isotype switching.

iva) HIGM2. HIGM2 is caused by autosomally inherited mutations of the AID gene (refer to Table 24-4). The reader will recall from Chapter 9 that AID (*activation-induced deaminase*) is a putative RNA- or DNA-editing enzyme that is selectively expressed in GC B cells and critical for both isotype switching and somatic hypermutation. AID expression and activation in B cells is induced by CD40 signaling. Thus, like HIGM1 patients, HIGM2 patients exhibit normal or elevated serum IgM but sharply reduced serum IgG, IgA, and IgE. However, unlike HIGM1 patients, antibody diversity due to somatic hypermutation is completely absent in HIGM2 patients. Abnormal giant GCs filled with proliferating B cells form in the secondary lymphoid organs, causing them to enlarge. The B cells are mature, indicating that loss of AID does not affect B cell development. However, many of these cells express IgD as well as IgM, an unusual surface phenotype in GC B cells. CD40L expression is normal in HIGM2 patients as is T cell function, meaning that these patients do not suffer the same range of opportunistic infections as HIGM1 patients. Phenotypic features identical to HIGM2 can be found in knockout mice deficient for AID.

ivb) HIGM3. HIGM3 is caused by autosomal recessive mutations of the CD40 gene. A lack of this molecule blocks CD40-CD40L signaling on the B cell end, resulting in an autosomal disorder that is very similar to X-linked HIGM1. As expected, isotype switching and memory B cell generation are impaired. However, the defect in somatic hypermutation is more profound in HIGM3 patients than in HIGM1 patients. Like HIGM1 patients, HIGM3 patients show a deficit in cell-mediated immunity that leaves them vulnerable to a broad range of recurrent bacterial and opportunistic infections. Symptoms of HIGM3 patients are recapitulated in the phenotype of CD40$^{-/-}$ mice.

ivc) HIGM4. HIGM4 patients show a phenotype similar to, but slightly milder than, that of HIGM2 patients. IgM is not as highly elevated and more IgG can be detected in the blood. However, antibody responses are impaired (especially those directed against polysaccharide antigens) and HIGM4 patients therefore suffer greatly from opportunistic, bacterial, and mycobacterial infections. While isotype switching is impaired in HIGM4 B cells, somatic hypermutation is not. The CD40, CD40L, AID, UNG (see below), and NEMO (see below) genes are not mutated in HIGM4 patients, and the formation of double-stranded DNA breaks necessary for isotype switching is normal in HIGM4 B cells. The specific defect causing this disease thus is thought to lie well downstream of CD40 engagement and AID signaling. An element of the DNA repair machinery that is required to stitch together isotype-switched exons may be affected. Alternatively, the delivery of survival signals to B cells that have completed isotype switching may be faulty.

ivd) HIGM5. Patients with HIGM5 show symptoms that most closely resemble those of HIGM2 patients: highly elevated IgM, a near-absence of IgG, enlarged lymphoid tissues, and greatly increased susceptibility to bacterial (but not opportunistic) infections. However, the AID gene is normal in HIGM5 patients. Rather, the autosomal recessive defect lies in the gene encoding *uracil DNA glycosylase* (UNG). It is thought that UNG may act downstream of AID, removing uracil residues deaminated by AID from the affected DNA. This removal ultimately leads to the double-stranded breaks in DNA necessary for isotype switching. HIGM5 B cells also show a partial defect in somatic hypermutation. Knockout mice lacking UNG show an impairment of isotype switching that is less severe than that in human HIGM5 B cells and a skewed spectrum of somatic hypermutations.

ive) HIGM associated with X-linked hypohydrotic ectodermal dysplasia (HIGM-NEMO). A mild form of HIGM occurs in association with an X-linked disorder of ectodermal tissue development. These patients have sparse hair and abnormal or missing teeth, and lack sweat glands. Humoral responses to antigens are impaired, particularly to polysaccharide antigens. Cell-mediated immunity is also apparently affected. As a result, these patients are subject to frequent opportunistic, bacterial, and mycobacterial infections. In these HIGM cases, the mutation lies in the X-linked gene encoding NEMO (*NF-κB essential modulator*). NEMO is part of the enzymatic IKK (*IκB kinase*) complex that phosphorylates IκB, the inhibitor that binds to the transcription factor NF-κB and holds it inactive in the cytoplasm. Phosphorylated IκB is ubiquitinated and degraded, freeing NF-κB to translocate to the nucleus and initiate transcription. Mice lacking subunits of NF-κB itself have a phenotype similar to that of HIGM-NEMO patients. These observations imply that NF-κB is involved in CD40-mediated isotype switching, but how CD40 stimulation triggers activation of the IKK complex is unclear.

v) Hyper-IgE Syndrome

The hallmark of patients with hyper-IgE syndrome is, naturally enough, very high levels of serum IgE. IgD may also be moderately increased. Affected children have characteristic coarse facial features and suffer very early in life from dermatitis and pyogenic infections, particularly recurrent abscesses caused by staphylococci that attack the lungs, joints, and skin. The skin abscesses usually localize on the face, neck, and scalp, especially in infants and young children. The abscesses in these patients can be very severe, giving this disorder its alternative name "Job's syndrome" (after the Biblical sufferer). Interestingly, despite the presence of elevated IgE, the dermatitis in these patients is quite distinct from that of patients suffering from the atopic dermatitis associated with severe allergy (see Ch.28). Indeed, respiratory allergies are rare in hyper-IgE patients. In addition, although some symptoms of hyper-IgE patients resemble those of patients with *chronic granulomatous disease* (CGD; see later), the infections are by different organisms and occur in anatomically different sites. Hyper-IgE patients are also often distinguished by their double rows of teeth. The primary teeth are not lost in these youngsters due to a failure in tooth root resorption. In general, bone density is reduced in hyper-IgE patients and relatively minor traumas frequently result in fractures. There have been rare reports of autoimmunity and malignancy.

With respect to hematopoietic cells, numbers of neutrophils and T and B lymphocytes are normal in hyper-IgE patients but eosinophils are greatly elevated in both blood and sputum. It is thought that these eosinophils or their products may contribute to the tissue destruction that accompanies lung abscess formation. Secondary humoral responses are impaired to a variable degree, and in some cases, secondary cell-mediated responses are also affected. For a small proportion of patients,

neutrophil phagocytosis and phagosome killing are normal but neutrophil chemotaxis *in vitro* is defective.

Hyper-IgE syndrome sometimes runs in families, showing a variable pattern of autosomal dominant inheritance, but the molecular defect remains unclear. In at least some hyper-IgE patients, the accumulation of IgE in the blood appears to be due to a defect in IgE catabolism that blocks normal turnover of the IgE protein. Researchers speculate that other patients might be suffering from an imbalance in Th1/Th2 cytokine secretion, such that excessive IL-4/IL-13 signaling might promote abnormal levels of isotype switching to IgE in B cells. Indeed, a viral infection that induces vigorous production of Th1 cytokines appears to temporarily ameliorate the symptoms of hyper-IgE syndrome. In addition, studies *in vitro* of lymphocytes isolated from one group of hyper-IgE patients revealed a decrease in the ability of these cells to respond to IL-12 and secrete IFNγ. However, while some hyper-IgE patients have decreased levels of IFNγ and TNF in their sera, levels of IL-4 are not generally elevated. Limited information exists with respect to IL-13, but the proportion of CD4$^+$ T cells expressing IL-13 was increased in one study of hyper-IgE patients. It is possible that a mutation of the IL-4Rα chain (which is part of both IL-4R and IL-13R) that results in upregulated or constitutive IL-4 or IL-13 signaling could account for at least some cases of hyper-IgE syndrome. One problem with the Th1/Th2 imbalance hypothesis is that serum cytokine levels and IgE levels are not positively correlated. Indeed, in some older patients, serum IgE levels decline suddenly without explanation and without any improvement in symptoms. In addition, mice that overexpress IL-4 (or lack IFNγ expression) do not show any of the non-immunological abnormalities of human hyper-IgE syndrome.

Antibiotics to combat staphylococcal infections are an effective treatment for hyper-IgE syndrome, although surgery may be required to remove entrenched abscesses. Treatment with cyclosporin A, thought to shift immune responses away from the Th2 type, has been used to improve the condition of some young patients. IFNγ treatment improved the *in vitro* chemotaxis of the neutrophils of several hyper-IgE patients but did not bring clinical relief. To date, BMT/HCT has not been successful in correcting this disorder.

IV. ADAPTIVE IMMUNODEFICIENCIES DUE TO DEFECTS IN DNA REPAIR

Immunodeficiency is also a prominent clinical feature in a group of diseases that are caused by defects in genes encoding proteins responsible for the repair of damage to cellular DNA (Table 24-6). These genes are expressed in other cell types in addition to hematopoietic cells and so have effects outside the immune system. Mutations in three pathways of DNA repair have been associated with ID: *non-homologous end joining* (NHEJ), *homologous recombination repair* (HRR), and *nucleotide excision repair* (NER) (Fig. 24-5). The NHEJ pathway is responsible for repairing small *double-stranded breaks* (DSB) in DNA, most often caused by insults such as ionizing radiation. NHEJ simply ligates the ends of the broken DNA strands back together to reform the original strands. When a double-stranded DNA gap is too large for repair by NHEJ, it is often fixed by HRR. The HRR pathway uses a different collection of enzymes and a homologous DNA strand to restore DNA integrity in a templated, "error-free" fashion. UV-irradiation alters nucleotides such that they become "bulky" and disrupt DNA strand matching. In this case, the DNA is repaired by the NER pathway, which

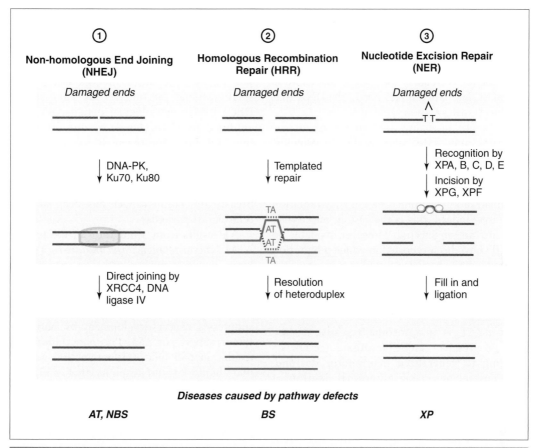

Figure 24-5

PIs Associated with Defects in DNA Repair Pathways
Damage to cellular DNA can be repaired by the three pathways shown involving the indicated molecules. NHEJ is used when a double-stranded break has occurred with no significant loss of nucleotides. HRR can repair larger double-stranded gaps using a homologous strand as a template. NER repairs UV-induced damage to nucleotides and removes cross-strand linkages. Ataxia telangiectasia (AT) results from mutation of the ATM protein that is able to sense DNA damage that can be repaired by NHEJ. Mutations of the nibrin protein that functions downstream of ATM during NHEJ lead to Nijmegen breakage syndrome (NBS). Bloom's syndrome (BS) is due to mutations of the BLM helicase involved in HRR. Mutations in multiple genes involved in the NER pathway lead to xeroderma pigmentosum (XP). It remains a mystery why defects in DNA repair lead to immunodeficiency. Adapted from Gennery A. R. *et al.* (2000) Immunodeficiency associated with DNA repair defects. *Clinical and Experimental Immunology* **121**, 1–7.

Table 24-6 PIs Associated with DNA Repair Defects

Name	Defective Gene	Functional Defect
AT	ATM	Failure of ATM to phosphorylate p53 and stabilize it so it can arrest the cell cycle and allow repair of DSB by NHEJ; unknown how ATM impairs immunity
NBS	NBS1	Failure of nibrin to act in same pathway as ATM to stabilize p53 and facilitate DNA repair by NHEJ; nibrin may be involved in complex supporting V(D)J recombination
Bloom syndrome	BLM	Putative helicase enzyme may be involved in DNA repair by HRR; unknown how lymphocytes affected
XP	XPA-XPG	Failure to repair DNA by NER; unknown how lymphocytes affected

snips out the offending nucleotides and carries out fill-in and ligation reactions to restore the double-stranded DNA.

In PIs associated with defects in DNA repair, the patient's cells are hypersensitive to DNA-damaging agents and show chromosomal anomalies, instabilities, and breakages that disrupt cell cycle progression. Lymphocytes appear to be particularly sensitive to these mutations, and both humoral and cell-mediated immune responses are impaired. For reasons that are unclear, autoimmunity is common in one of these disorders (ataxia-telangiectasia) but not in other disorders of this group. Malignancy is frequent in all these diseases due to the accumulation of unrepaired mutations in the DNA that can lead to cancerous transformation.

The reader may notice in Figure 24-5 that some of the enzymes used to carry out NHEJ, such as the Ku proteins, DNA-PK, XRCC4, and DNA ligase IV, were also cited as being important for V(D)J recombination in the antigen receptor genes. Thus, one might expect that mutations affecting these enzymes would impair both resistance to ionizing radiation and immune responses. Indeed, this is the case in mice when mutations of this sort are engineered. However, to date, no cases of human SCID have been reported in which a mutation of any of the preceding genes has been found.

i) Ataxia-Telangiectasia

Ataxia-telangiectasia (AT) is an autosomal recessive disease characterized by progressive cerebellar ataxia (cerebellar degeneration leading to generalized neuromotor dysfunction); oculocutaneous telangiectasia (abnormal dilation of blood vessels in the conjunctiva of the eyes; Plate 24-1); and a variable CID. Both the cell-mediated and humoral immune responses are affected, but responses to bacterial infections can range widely among patients. Recurrent severe lung and sinus infections are common and almost half of AT patients die of pulmonary failure due to pneumonia. Hypogammaglobulinemia, particularly of IgA, IgE, IgG2, and IgG4, is observed due to defective B cell maturation, and development of the thymus is impaired. T cell responses to viral antigens and mitogens are decreased and peripheral blood memory T cells are reduced in number. Patients are hypersensitive to ionizing radiation (such as that associated with sun exposure or cancer treatment) and run a 100-fold increased risk of developing lymphomas and acute lymphocytic leukemias by the age of 15 years. Autoimmunity is also common in these patients. Some female AT patients fail to ovulate and experience delayed puberty and premature menopause, while males may have small testicles. Secondary sex characteristics are normal. Due to their neurological deficits, AT patients are usually confined to a wheelchair by age 20 and die of cancer or lung infections before age 30. No effective treatment for AT is available because the radiation required to prepare the patient's bone marrow for BMT/HCT is too toxic for these ultra-sensitive patients. Increased incidence of lymphoma, breast, and colon cancer has been noted in heterozygous relatives of patients.

AT results from a mutation of the tumor suppressor gene ATM (*ataxia telangiectasia mutated*). The ATM gene is very large, being composed of 66 exons spread over 150 kb. It encodes a 350 kDa member of the PI3K family of kinases, enzymes that are characteristically involved in cell cycle control and DNA damage repair. The normal ATM protein suppresses cancer development by phosphorylating the key tumor suppressor p53. Phosphorylation of a specific serine residue in p53 blocks its interaction with Mdm2, a protein that binds to unphosphorylated p53 and targets it for ubiquitin-mediated degradation. Thus, a major function of ATM is to stabilize p53 in response to DNA damage such that p53 can arrest the cell cycle and promote repair of DSB in DNA via NHEJ. In other words, ATM can function as an early sensor of DNA damage. Classical AT patients typically have frameshift mutations that prevent production of the ATM protein. In its absence, spontaneous chromosomal anomalies such as characteristic translocations can occur (Plate 24-2). More rarely, one may observe dicentric, multicentric, and ring chromosomes in the cells of AT patients. Much more mildly affected AT variant patients have missense mutations that preserve a small percentage of normal ATM activity. Despite the identification of the genetic defect underlying AT, it is not known why ATM mutations impair immunity. Cells of AT patients can successfully carry out NHEJ *in vitro* and no deficits in V(D)J recombination have been reported. However, lymphocytes of AT patients often have translocations involving the TCR and Ig loci, events that could substantially interfere with lymphocyte maturation. There is little evidence that these translocations are linked to tumorigenesis.

ATM$^{-/-}$ mice have a phenotype similar in some respects to that of human AT patients. The mutant animals show immunological deficits, radiation sensitivity, and increased incidence of tumors, especially thymic lymphomas. Female mice do not ovulate and males lack sperm. However, ATM$^{-/-}$ mice do not show ataxia or other defects in the cerebellum.

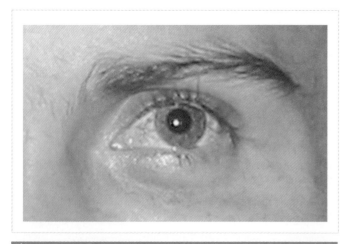

Plate 24-1

Eye Phenotype of Ataxia-Telangiectasia
Courtesy of the Kindregan family.

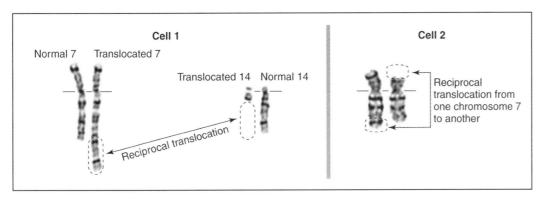

Plate 24-2

Chromosomal Abnormalities in Ataxia-Telangiectasia
Examples of the chromosomal rearrangements involving T cell receptor loci seen in a fraction of stimulated lymphocytes from AT patients. Courtesy of the Cytogenetics Laboratory, The Hospital for Sick Children, Toronto.

ii) Nijmegen Breakage Syndrome

Nijmegen breakage syndrome (NBS) is another rare PI caused by a defect in NHEJ. This autosomal recessive disease is named for the location in the Netherlands where an affected family was first studied. NBS patients can be distinguished from AT patients at birth by their bird-like facial features and microcephaly (Plate 24-3). Later on, NBS patients develop mild to moderate mental retardation that is not present in AT patients. In addition, NBS patients do not show the ataxia or telangiectasia characteristic of AT. However, like AT patients, NBS patients are of noticeably short stature, have a truncated life span, and experience recurrent, severe sinopulmonary infections due to a variable CID. Hypogammaglobulinemia is common and antibody responses to antigen are defective. *In vitro* proliferation of T cells in response to mitogens is impaired. Cells subjected to ionizing radiation display the same spectrum of chromosomal rearrangements seen in irradiated cells from AT patients. NBS patients also have an increased risk of cancer development, particularly B cell lymphomas. Defects in primary and secondary sex characteristics are more severe in female NBS patients than in female AT patients, but NBS males are sexually normal. In mice (unlike humans), homozygous null mutations of NBS1 are embryonic lethal at an extremely early stage, consistent with a key role for this protein in maintaining genomic stability in this species.

NBS arises from mutations of the NBS1 gene, which encodes nibrin, a 754 amino acid protein homologous to proteins known to be involved in cell cycle control. It is thought that nibrin participates in a complex that acts on the tumor suppressor p53 in the same DNA damage repair pathway as ATM, which may explain some of the overlapping clinical features of these disorders. Indeed, *in vitro*, ATM has been observed to phosphorylate nibrin in response to DNA-damaging agents. In normal individuals, the complex containing nibrin moves rapidly to the sites of DSB, presumably to support DNA repair. In cells from NBS patients subjected to DNA-damaging agents, cell cycle progression is halted, usually at the G1/S checkpoint. NBS1 has also been implicated in the maintenance of telomeres and DNA replication. With respect to immune responses, the nibrin-containing complex may help to open the DNA hairpins formed during V(D)J recombination in T and B cells. A defect in this process could theoretically result in the CID noted in these patients but has yet to be identified *in vivo*.

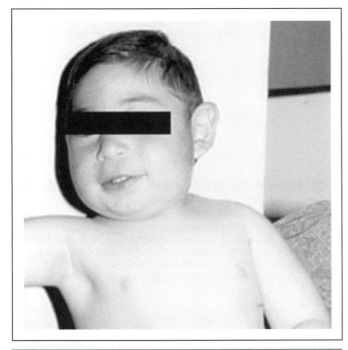

Plate 24-3

Nijmegen Breakage Syndrome
Reproduced with permission from Marcelain K. *et al.* (2004) Clinical, cytogenetic and molecular characterization of a new case of Nijmegen breakage syndrome in Chile. *Revista Medica de Chile* **132**, 211–218.

iii) Bloom Syndrome

Bloom syndrome (BS) is a rare autosomal recessive disorder characterized by genomic instability and CID. Spontaneous chromosomal anomalies appear in lymphoblasts and other cell types. Hypogammaglobulinemia, including low levels of IgM, is common. As a result, BS patients experience recurrent respiratory infections that can lead to chronic lung disease. BS patients are also short in stature and are particularly sensitive to the sun, breaking out in facial erythema upon exposure (Plate 24-4). Due to their genomic instability, BS patients have a higher risk of developing lymphomas or lymphoid leukemias in their 20s, and other solid tumors in later decades. Unfortunately, because of the nature of their mutations, these patients are particularly vulnerable to the potential adverse effects of radiation or chemotherapeutic cancer treatments.

BS results from a mutation in the BLM gene acting in the HRR pathway. BLM encodes a helicase enzyme that is thought to resolve DNA replication forks that have stalled due to DNA damage. The spectrum of chromosomal anomalies in BS is therefore different from that in AT or NBS, being dominated by abnormal replication intermediates such as sister chromatid exchanges (Plate 24-5) and, more rarely, quadriradial chromosomes. How a defect in the BLM helicase results in CID is under investigation.

iv) Xeroderma Pigmentosum

Xeroderma pigmentosum (XP) is a rare disorder (1 in 250,000 live births) characterized by extreme sensitivity to the sun and a marked predisposition to the development of skin cancers. Variable immunodeficiency occurs in some, but not all, XP patients. The immune system defect is manifested in cell-mediated but not humoral responses. Defects in NK-mediated natural cytotoxicity have been reported, as has decreased

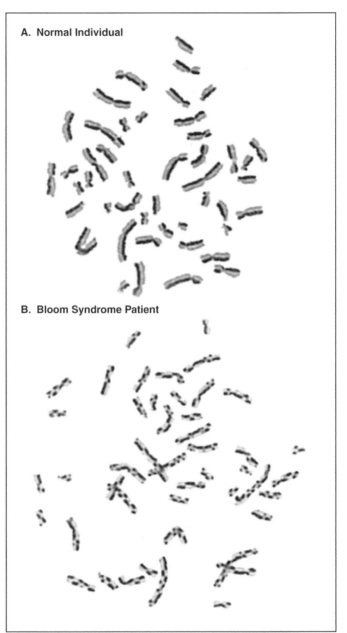

A. Normal Individual

B. Bloom Syndrome Patient

Plate 24-5

Bloom Syndrome Chromosomal Abnormalities
Chromosomes are stained such that genetic material from sister chromatids can be distinguished. Sister chromatid exchange is rare in a lymphocyte from a normal individual (A) but occurs at high frequency in a lymphocyte from a patient with Bloom syndrome (B). Courtesy of the Cytogenetics Laboratory, The Hospital for Sick Children, Toronto.

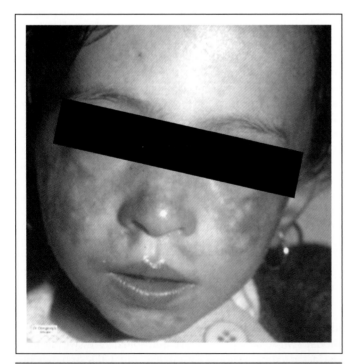

Plate 24-4

Bloom Syndrome
Courtesy of Dr. Dengvezli (www.atlas-dermato.org).

IFNγ production. Cells from XP patients are highly sensitive to UV *in vitro*.

XP is caused by mutations in one of seven genes (called XPA–XPG, respectively) involved in the NER pathway. Mice with a mutated XPA gene have impaired immune responses following UV-irradiation, but how mutations in these molecules lead to immunodeficiency is currently unknown.

V. LYMPHOPROLIFERATIVE IMMUNODEFICIENCY SYNDROMES

Although it sounds counterintuitive, lymphoproliferative syndromes exist that are associated with immunodeficiency. Both X-linked and autosomal inheritance patterns have been observed. We will discuss two of these disorders that have very different manifestations and that are due to distinct types of underlying mutations (Table 24-7).

i) X-Linked Lymphoproliferation

X-linked lymphoproliferation (XLP) is a very curious disease of young boys in which the primary clinical manifestation is an inappropriate response to EBV infection. When XLP males are infected with EBV (which is fairly endemic everywhere), they develop infectious mononucleosis that can be fatal due to fulminant liver necrosis. Excessive numbers of anti-EBV lymphocytes responding to the infection apparently contribute to the pathophysiology. Uncontrolled proliferation of B cells, lymphoadenopathy, hepatosplenomegaly, hypogammaglobulinemia, aplastic anemia, and fever are often observed, and CD8$^+$ T cells infiltrate extensively into the liver. Most strikingly, XLP patients have high serum levels of IFNγ and their immune responses to acute EBV infection are dramatically skewed to the Th1 type. The high levels of Th1 cytokines observed in these patients may contribute to the widespread tissue necrosis that develops. If a patient does not succumb to the EBV infection, he usually develops either a selective Ig deficiency or Burkitt's lymphoma. (Autoimmunity is not associated with this syndrome.) Death often occurs at a young age (and always by age 50 years) due to hepatic necrosis and/or bone marrow failure.

XLP has now been described in close to 90 families in the developed world. For unknown reasons, virtually no pathogens other than EBV appear to trigger XLP disease in susceptible patients. These boys start out with apparently intact immune systems, mounting normal responses to viruses such as varicella prior to infection with EBV. It was originally thought that the disease could not occur in the absence of EBV infection, but more recent work has shown that individuals negative for anti-EBV serum antibodies and

EBV DNA can develop XLP. Indeed, antibody deficiencies and lymphomas have been found in XLP patients prior to EBV infection. Now that the mutation underlying XLP has been identified (see later), gene therapy is under investigation to replace the current preferred treatment of BMT/HCT. IV-IG replacement therapy can be helpful when the patient's clinical picture includes hypogammaglobulinemia.

What causes XLP? It is thought that EBV infection itself is not to blame but rather is the trigger that causes the defect to show itself. At the genetic level, the defect is a deletion or point mutation of the SAP gene (*SLAM-associated protein*; also called SH2D1A or DSHP) located on the X chromosome. SAP is a small, single SH2 domain-containing molecule that associates with SLAM, the protein we encountered in Chapter 22 as the physiological receptor for measles virus. SLAM is a transmembrane protein that acts as a minor costimulator on activated B and T cells and is also expressed on DCs. The precise biological role of SAP is unclear but its dominant function appears to be negative regulation of IFNγ secretion by T cells. In the absence of SAP, phosphorylation of SLAM on T cells leads to CD28-independent costimulation, activation, and uncontrolled polarization toward Th1 differentiation. The production of excessive amounts of IFNγ follows. SAP also interacts with the phosphatases SHP-2 and SHIP2 and the adaptor proteins Dok1 and Dok2, but it is unclear what roles these molecules play. Finally, SAP is expressed in NK cells, and studies *in vitro* have shown that NK cells from XLP patients cannot kill EBV-infected B cells. Taken together, these observations may account for the inability of XLP patients to successfully eliminate EBV. However, the reasons why excessive IFNγ production leads to the XLP phenotype, and why a Th1/Th2 imbalance in an XLP patient should only manifest itself clinically during EBV infection, remain obscure.

SAP-deficient knockout mice show the same skewing to the Th1 phenotype and increased levels of IFNγ as seen in XLP patients, and antibody responses to infection with *Leishmania major* or various viruses are depressed. Study of these animals is expected to clarify the chain of events leading to XLP.

Table 24-7 **Lymphoproliferative Syndromes**

Name	Defective Gene	Functional Defect
XLP	SAP	Failure to negatively regulate IFNγ production by T cells; excessive Th1 differentiation results
ALPS Ia	Fas	Failure of activated lymphocytes to undergo apoptosis
ALPS Ib	FasL	Failure of activated lymphocytes to undergo apoptosis
ALPS II	Caspase-10	Failure of caspase cascade after Fas triggering
ALPS III	?	Failure of caspase cascade after Fas triggering

ii) Autoimmune Lymphoproliferative Syndromes

Autoimmune lymphoproliferative syndromes (ALPS; refer to Table 24-7) are generally characterized by hyperproliferation of B cells (mostly the CD5+ subpopulation), CD8+ T cells, and DN thymocytes. For unknown reasons, the proliferation of CD4+ T cells and CD5− B cells is not affected. The expanding lymphocytes accumulate in the lymphoid organs in a non-malignant way, causing lymphoadenopathy and hepatosplenomegaly. These diseases are sometimes diagnosed within the first year of life and almost always by age 5 years. Due to the B cell hyperproliferation, levels of serum IgG and IgA are increased in ALPS patients. Autoantibody production is found in almost all affected individuals, with anti-erythrocyte antibodies leading to anemia, and anti-platelet antibodies promoting thrombocytopenia (platelet deficiency). ALPS patients also have an increased risk of developing tumors of various types, including breast, colon, and lung cancers and lymphomas. BMT/HCT appears to be the only practical treatment for ALPS.

ALPS are caused by defects in lymphocyte apoptosis, particularly in the Fas-mediated pathway. Accordingly, lymphocytes from ALPS patients show defective cell death *in vitro* in response to various apoptotic stimuli. ALPS patients have been classified by their underlying genetic mutations. ALPS Ia patients (the majority) have a mutation in the gene encoding the Fas death receptor (CD95), while ALPS Ib patients have a mutation in its ligand FasL. ALPS II patients have a mutation in caspase-10, a caspase vital in the apoptotic cascade downstream of Fas ligation. The gene mutated in ALPS III patients remains to be identified.

The autosomal inheritance of ALPS Ia is dominant in most cases, an interesting phenomenon that may be related to the fact that the functional Fas receptor occurs in trimeric form on the cell surface. FasL cannot bind to a receptor in which one of the component monomers is mutated. Thus, if one Fas allele is mutated, only one out of eight Fas trimers on the cell surface will be functional, apparently not sufficient for a normal phenotype. This type of autosomal dominance associated with a trimeric receptor also holds true for TNFR1 in the PI called TRAPS (see later).

VI. OTHER ADAPTIVE IMMUNODEFICIENCIES

Two other adaptive PIs that do not fit into the previous categories are detailed in the following sections and summarized in Table 24-8.

i) DiGeorge Syndrome

DiGeorge syndrome (called by some the "DiGeorge anomaly") is a complex disorder in which the thymus often does not develop fully or at all. These patients have characteristically abnormal facial features, including a long narrow face, small mouth, prominent nose, hooded or full upper eyelids, and low-set, cupped ears (see Plate 24-6). The hands and feet of these individuals tend to be small, and they have an abundance of hair on their heads. Cardiovascular and renal abnormalities are frequently present, and the patients suffer from parahypothyroidism and recurrent infections. Cleft palate is common and can lead to difficulties in early feeding and speech. Neural tube defects have also been reported. However, the disease is clinically very heterogeneous, and not all patients show all symptoms. Modern diagnostic techniques have put the incidence of DiGeorge syndrome (and closely related syndromes) at 1 in 4000 live births. Autosomal dominant, autosomal recessive, and X-linked modes of inheritance of DiGeorge syndrome have been reported.

The precise molecular defect underlying DiGeorge syndrome has yet to be identified, but it appears to affect a subset of neural crest cells that is destined to populate the third and fourth pharyngeal pouches during embryogenesis. In normal embryos, these structures eventually develop into the thymus, parathyroid glands, the aortic arch arteries and the outflow tract of the heart, and parts of the face. In most (but not all) DiGeorge syndrome patients, translocations or large (up to 2 megabases) deletions involving chromosome 22q11 are observed. Disruption of 22q11 is associated with impaired embryonic cell migration patterns such that the thymus (among other tissues) fails to develop or is smaller than normal. Phenotypes typical of DiGeorge syndrome have been associated with external events during embryogenesis, such as maternal diabetes or exposure of the fetus to high levels of alcohol or other chemicals.

With respect to immunodeficiency in DiGeorge syndrome, the extent of the defect is governed by the extent of the thymic hypoplasia. It has been found that surprisingly little thymic tissue is sufficient for normal T cell immunity. When a thymus is present in a DiGeorge patient, it is often normal in architecture (if small in size) and the differentiation of the cortex and medulla and Hassall's corpuscles is normal. However, the thymus may be abnormally positioned, shifting from its normal site under the breastbone to a site in the neck. Numbers of T cells vary in DiGeorge thymi but can range from almost nil to almost normal. However, T cells from at least some DiGeorge patients show increased Fas and FasL expression

---

Table 24-8 Other Adaptive PIs

Name	Defective Gene	Functional Defect
DiGeorge Syndrome	Unknown; may be associated with deletion or translocation of chromosome 22q11 in some cases	Failure to develop thymus plus T cell abnormalities
WAS	WASP	Defects in actin cytoskeleton reorganization; unknown defect impairs T cell development, survival, and/or activation

Plate 24-6

DiGeorge Syndrome
Courtesy of Andrea Shugar, Division of Clinical and Metabolic Genetics, The Hospital for Sick Children.

A murine equivalent to at least some aspects of DiGeorge syndrome is the *nude* mouse, which is hairless and lacks a thymus. As mentioned in Chapter 13, these animals have an autosomal recessive mutation on chromosome 11 that results in thymic hypoplasia and a failure to develop hair follicles. As is true in DiGeorge patients, the degree of thymic hypoplasia correlates with a variable CID. However, mice heterozygous for the *nude* mutation are normal, unlike heterozygous DiGeorge patients. Another mouse model that addresses some aspects of DiGeorge syndrome is the $Crkol^{-/-}$ knockout mouse. When homozygous, loss of this gene results in embryonic lethality with defects in the thymus, craniofacial structures, aortic arch arteries, and cardiac outflow tract. Again, however, heterozygous $Crkol^{+/-}$ mice are normal. $TBX1^{-/-}$ knockout mice also show most features of DiGeorge syndrome, including thymic hypoplasia, cleft palate, and facial and cardiac abnormalities. Heterozygous $TBX1^{+/-}$ animals have defects in the fourth pharyngeal arch.

Treatment of DiGeorge syndrome is necessarily complex, given the extent of the abnormalities and the several different organ systems involved. The hypoparathyroidism can be treated by administering vitamin D and calcium. Heart defects must be surgically repaired. The CID can be mitigated by administration of thymic hormones or by thymic transplant. In 2004, successful thymic tissue transplantation (coupled with immunosuppression) was carried out to ameliorate the immunodeficiency phenotype in several DiGeorge babies who completely lacked a thymus.

ii) Wiscott-Aldrich Syndrome

Wiscott-Aldrich syndrome (WAS) is an X-linked recessive, clinically heterogeneous disorder caused by mutations in the WAS protein (WASP). WASP is expressed constitutively by all hematopoietic cells, and may be involved in the growth and survival of HSC. In 27% of WAS patients ("classical" WAS), the mutated WASP is aberrantly expressed in lymphocytes and megakaryocytes and a characteristic constellation of clinical features results: CID, eczema, and thrombocytopenia. Patients are highly susceptible to viral, pyogenic, and opportunistic infections, and lymphomas, autoimmunity, and allergy are common in this group. Other WAS patients have only selected disease features: 20% have platelet defects without ID (also called X-linked thrombocytopenia; XLT), and 5% have ID without platelet defects. It has been proposed that classical cases of WAS are due to an abrogation of major functions of WASP, and that non-classical cases of WAS may be due to mutations that affect only a particular domain of the multi-functional WASP protein.

WASP appears to be crucial for platelet survival and quite important for lymphocyte survival. In classical WAS patients, platelet formation is severely disrupted and only low numbers of small, non-functional platelets are found. T cell numbers steadily decline during early childhood, and those few cells that are left function abnormally in that *in vitro* proliferative and DTH responses are impaired. This progressive lymphopenia leads to deficits in both the cell-mediated and humoral responses. B cell numbers are essentially normal but antibody responses to both protein and polysaccharide antigens

and an enhanced tendency to undergo spontaneous apoptosis. Such a tendency would reduce numbers of circulating mature T cells. The T cells that are present often fail to proliferate in response to mitogens in *in vitro* assay systems. NK numbers and distribution are normal, and B cells are normal and may even be elevated in number. However, the lack of T cells decreases the amount of T cell help available to B cells, often compromising the humoral response. In DiGeorge patients with a profound T cell defect, the first sign is usually a persistent oral thrush infection. Recurring severe infections typical of other CIDs soon follow. Rheumatoid arthritis and Graves' disease (hyperthyroidism; see Ch.29) have been reported in some DiGeorge patients, suggesting that deletions of 22q11 may predispose humans to autoimmunity in some cases.

Extensive microdeletion experiments have failed to pinpoint a gene or group of genes whose loss directly results in DiGeorge syndrome, although the critical region has been narrowed to about 250 kb. Disease severity does not correlate with size of deletion, suggesting that the genes responsible are not contiguous. Recent evidence has suggested that the DiGeorge deletions may remove the gene encoding the transcription factor TBX1 required for normal aortic arch formation, and perhaps the gene encoding the adaptor protein *Crkol*, which functions in growth factor signaling pathways. It should be noted, however, that only some humans with point mutations of TBX1 have DiGeorge syndrome. Furthermore, the TBX1 locus is not included in the deletion of 22q11 present in many DiGeorge patients, and 10% of DiGeorge patients do not have any deletion of 22q11. Besides *Crkol* and TBX1, other candidates include the GSCL gene ("goosecoid-like"), which is a regulator of embryonic neurogenesis, and HIRA, a corepressor involved in histone transcription during the embryonic development of the heart, cranium, and pharyngeal arches.

are impaired. While serum IgG is normal, levels of IgM are reduced and those of IgA and IgE are increased. Numbers of monocytes and neutrophils in the blood of classical WAS patients are normal, but the chemotaxis of these cells *in vitro* is impaired. BMT/HCT is the treatment of choice for WAS, although antibiotic treatment, IV-IG replacement therapy, splenectomy (to decrease thrombocytopenia), and platelet transfusions are often helpful.

WASP knockout mice have a phenotype that shares similarities with the human situation but also exhibits striking differences. Mild thrombocytopenia and lymphopenia are present in WASP$^{-/-}$ animals and T cell activation is defective. However, the eczema found in many human patients is missing, as are hematopoietic malignancies. More dramatically, WASP$^{-/-}$ mice suffer from chronic colitis while WASP-deficient humans do not. As in humans, B cell activation is normal in WASP$^{-/-}$ mice, but unlike humans, antibody responses to Td and Ti antigens are normal in the mutant animals, at least under pathogen-free conditions.

The precise mechanism by which mutation of WASP leads to WAS remains unclear. WASP is a multi-functional protein containing six domains. WASP is thought to play a role in both signal transduction mediated by SH3 domain-containing proteins and the polymerization of the actin cytoskeleton in response to cellular activation. With respect to signal transduction, WASP has domains capable of interacting with SH3-containing signaling molecules such as Src family kinases, the Grb2 adaptor protein, PLCγ, and Btk. Depending on the site of the mutation, one or more of these domains may be affected, and a mutated WASP protein may lead to classical WAS, XLT, or ID alone. With respect to the actin cytoskeleton, WASP physically clusters with filamentous actin and can act as a direct effector of the GTPase Cdc42, a crucial regulator of cytoskeletal organization. As we learned in Chapter 14, changes to the actin cytoskeleton of a T cell are required when the cell undergoes activation. A lack of WASP could therefore halt responses induced by engagement of the TCR, leading to blocks in T cell development and activation. Indeed, T cells deficient for WASP cannot respond to the proliferative signals delivered by the engagement of the TCR by anti-CD3 antibody. Normal differentiation of megakaryocytes also depends on the interaction of WASP with actin filaments, as does the general motility of immune system cells. The chemotaxis of macrophages and DCs of WAS patients is markedly decreased due to the failure to develop the polarized extensions of the cell membrane and cytoplasm required for cell movement.

C. Primary Immunodeficiencies Due to Defects in the Innate Immune Response

Compromised immunity to pathogens can also be caused by defects in the innate response. Most of these disorders arise from defects in phagocytes, although some stem from complement deficiencies as described in Chapter 19. Autoimmunity and malignancy are infrequently associated with innate response PIs.

I. IMMUNODEFICIENCIES AFFECTING PHAGOCYTE RESPONSES

About 9% of PIs are due to defects in phagocyte extravasation, activation, or function, particularly that of neutrophils (Table 24-9). These PIs manifest at an early age with recurrent fevers and infections by normally non-pathogenic organisms. Severe gingivitis and periodontitis are common, as are hepatomegaly and lymphoadenopathy. Lung, liver, and bone infections may occur. Prompt antibiotic treatment is essential to avoid fatality.

i) Leukocyte Adhesion Deficiencies
ia) Leukocyte adhesion deficiency type I. *Leukocyte adhesion deficiency type I* (LAD-I) is characterized by recurrent non-pyogenic infections of the mucosal surfaces by *Staphylococcus aureus* and *Aspergillus* and *Candida* species. LAD-I results from autosomal recessive mutations in CD18, the common subunit of many β2 integrins (such as LFA-1). Without functional integrins, leukocytes such as neutrophils cannot leave the circulation and extravasate into sites of infection or injury. Numbers of circulating neutrophils in LAD-I patients are thus almost twice those in normal individuals, and few neutrophils are found in the inflammatory sites where they are needed. *In vitro*, LAD-I neutrophils are unable to aggregate normally or bind to endothelial cells. Depending on the mutation and on how badly it depresses CD18 expression, the mucosal infections can be mild to severe. Serious periodontitis and tooth decay are also common in affected individuals. Children with the most devastating mutations (essentially no expression of CD18) die by age 10 years, while milder cases (low expression of CD18) may survive into their 40s. Variant LAD-I patients retain significant amounts of CD18 expression, but the protein has a structural or signaling defect that blocks intracellular pathways leading to neutrophil mobilization.

ib) Leukocyte adhesion deficiency type II. *Leukocyte adhesion deficiency type II* (LAD-II) is extremely rare and is characterized by defective neutrophil chemotaxis. Patients with LAD-II carry autosomal recessive mutations of a fucose transporter protein responsible for translocating fucose from the cytoplasm to the lumen of the Golgi. As a result, the glycoprotein sialyl Lewis X, which is expressed on the neutrophil surface, is not properly fucosylated. Sialyl Lewis X is a ligand for the selectin molecules expressed by endothelial cells that mediate extravasation of neutrophils into sites of injury or infection. The failure in extravasation means that high levels of neutrophils remain in the circulation rather than enter the tissues under attack, leading to recurrent bacterial infections. LAD-II patients also have dysmorphic facial features and suffer growth deficits and mental retardation due to the effect of the fucosylation defect on

Table 24-9 PIs Associated with Defects in Phagocytes

Name	Defective Protein	Functional Defect
LAD-I	CD18	Non-functional integrins inhibit neutrophil extravasation
LAD-II	Fucose transporter	Lack of fucosylation of neutrophil ligand blocks selectin-mediated extravasation
Rac2 deficiency	Rac2	Defective cytoskeleton function leads to defective extravasation; failure of neutrophils to degranulate
CGD	NADPH oxidase subunits Associated GTPase	Failure in phagosomal killing
CHS	CHS1/LYST	Defect of protein involved in intracellular vacuolar trafficking, lysosomal fusion, phagolysosomal killing, and granule function
IL-12/IFNγ axis	IFNGR1, IFNGR2 IL-12p40 IL-12Rβ1 STAT1	Defect in macrophage hyperactivation due to decreased IL-12 and/or IFNγ signaling
FMF	Marenostrin	Monocytes and granulocytes fail to regulate inflammatory response; excessive chemotaxis of neutrophils
HIDS	Mevalonate kinase	Dysregulation of inflammation and IgD production by unknown mechanism
TRAPS	TNFR1	Blockage in TNFR1 cleavage leads to uncontrolled TNFR1 signaling
SCN	Neutrophil elastase	Constant severe decrease in neutrophil numbers caused by increased apoptosis of neutrophil precursors in the bone marrow
X-SCN	WASP	Constitutive activation of WASP destabilizes neutrophil progenitors and induces apoptosis
CyN	Neutrophil elastase	Cyclical decrease in neutrophil numbers due to block in neutrophil precursor differentiation

glycoproteins essential for metabolism outside the immune system. Infections and fevers in LAD-II patients can be reduced by treatment with oral fucose.

ic) Rac2 deficiency. Rac2 deficiency was first described in 2000 as a new type of leukocyte adhesion deficiency. The affected patient was diagnosed at the age of 5 weeks with severe infections and elevated numbers of neutrophils in the circulation. Rac2 is an important GTPase that is crucial for cytoskeletal function in neutrophils. A mutation of this protein rendered the patient's neutrophils unable to respond to chemotactic stimuli or to degranulate properly.

ii) Chronic Granulomatous Disease

Chronic granulomatous disease (CGD) is a clinically heterogeneous disorder caused by failures in phagosomal killing. Patients generally suffer from recurrent and sometimes life-threatening infections with catalase-positive fungi and bacteria such as *S. aureus* and *Aspergillus* species. (Catalase is an enzyme that degrades hydrogen peroxide, H_2O_2.) In contrast, infections with catalase-negative microbes (such as *S. pneumoniae*) are infrequent in CGD patients. The lungs, lymph nodes, and skin are particularly affected in CGD and show prominent granulomatous lesions (sites of excessive formation of granulomas in the tissues). Obstruction of the ureters or bowel can result from chronic granuloma formation. Patients often exhibit severe facial acne and gingivitis.

CGD is caused by mutation of any one of the four subunits of the NADPH oxidase enzyme positioned in the phagosome membrane. Without a functional NADPH oxidase, there is a sharp decrease in H_2O_2 production during the respiratory burst, and killing within the phagosome is diminished. X-linked CGD is the most common and most severe form, affecting about 70% of patients. The NADPH oxidase subunit mutated in this case is encoded by the gp91[phox] gene. A milder (and less common; about 30% of cases) form of CGD is due to autosomal recessive mutations in the p47[phox] gene, which encodes a cytosolic component of NADPH oxidase. In this case, onset of disease is later than in X-linked CGD, the incidence of infections and granulomas is reduced, and fatalities are less frequent. Rarely, mutations have been found in the genes encoding the p22[phox] and p67[phox] subunits, and in a gene encoding a GTPase that associates with the NADPH oxidase complex.

While a BMT/HCT is needed to effect a permanent cure for CGD, prophylactic treatment with IFNγ or antibiotics can reduce the frequency of life-threatening infections. It has been shown in female carriers of X-linked CGD that 10% of normal NADPH oxidase activity is sufficient to protect against the recurrent infections characteristic of this disease. As 10% is a readily attainable level of enzyme activity, gene therapy has been pursued in the hope that a reconstituted neutrophil compartment would be able to protect the patient. Trials of the transfer of HSC transfected with NADPH oxidase subunit genes into mutant mice were carried out with reasonable success, but early clinical trials using a similar approach in humans yielded only transient detection of myeloid progenitors bearing the transfected gene. Some investigators fear that it may be difficult to get gene therapy to work in this situation because NADPH oxidase is expressed only in differentiated granulocytes and thus cannot provide a survival advantage for the transfected HSC over the deficient progenitors already present. In addition, granulocytes have a half-life of only 2 days, meaning that a constant large volume of transfected cells would have to be produced to do the patient any good. Work is under way to develop gene transfer vectors and protocols that can achieve clinical success for CGD.

iii) Chédiak-Higashi Syndrome

Chédiak-Higashi syndrome (CHS) is a disease of intracellular vacuolar and granule fusion caused by autosomal recessive mutations to the *CHS1* gene (the human equivalent of the mouse *LYST* gene). CHS has both hematopoietic (first described by Chédiak) and neurological (first described by Higashi) manifestations. Clinical symptoms include recurrent infections with organisms normally eliminated by phagocytosis, peripheral neuropathy, partial oculocutaneous albinism, slight mental retardation, platelet dysfunction, severe periodontitis, and a sometimes fatal infiltration of lymphocytes and macrophages into various organs. Neither autoimmunity nor malignancy is increased in CHS. CHS is usually fatal at a young age, although some patients can survive until age 20 or 30 years. These survivors are usually confined to a wheelchair by the symptoms of peripheral neuropathy. BMT/HCT is the only effective treatment for CHS.

What causes CHS? The *CHS1/LYST* gene encodes a cytoplasmic protein that is expressed in a wide variety of cell types and is frequently found in association with microtubules. Although its precise function is unknown, the CHS1/LYST protein is thought to be involved in the fusion/fission events that affect the size and intracellular trafficking of lysosomes and related organelles (such as lytic granules within CTLs, neutrophils, and NK cells and MHC class II endosomal compartments in APCs). Thus, the fusion of intracellular structures, such as primary with secondary lysosomes, and endocytic vesicles with phagosomes, depends on *CHS1/LYST* function. The intracellular localization of molecules such as perforin and granzymes in CTL granules, the negative regulator CTLA-4 in T cells, and MHC class II molecules in B cells are thus affected by alterations to CHS1/LYST. Under the light microscope, the lysosome-related organelles in neutrophils, platelets, melanocytes, and some neurons of CHS patients are all substantially increased in size due to abnormal fusions with primary lysosomes. Histologists say that these cells are filled with "giant granules." In phagocytes of CHS patients, the giant granules cannot fuse properly with phagosomes so that phagosomal bacterial killing is impaired. In addition, CHS neutrophils lack the granule proteins cathepsin G and elastase, and show defects in chemotaxis *in vitro*. NK cells in CHS patients also show functional defects and CTL killing is reduced, most likely due to the inability of the giant granules to release their lytic contents. The cause of the neuropathy in CHS remains obscure, but unusually large lysosomes have been observed in the Schwann cells of CHS patients. It is speculated that these lysosomes either cause some kind of direct injury to the nerve cells, or induce lymphocytic infiltration that results in nerve cell damage. In mice, the natural *beige* mutant strain has an alteration to the *LYST* gene. These animals have abnormal pigmentation, giant lysosomes in leukocytes, and deficiencies in NK function that make them a reasonably accurate model of human CHS.

iv) Defects of the IFNγ/IL-12 Axis

As described in Chapter 22, the control of many intracellular pathogens depends on granuloma formation by hyperactivated macrophages. Macrophages become hyperactivated in response to IFNγ produced by activated Th1 cells, and, in a positive feedback loop, Th1 differentiation depends largely on IL-12 produced by activated macrophages and DCs (Fig. 24-6). Patients with autosomal recessive null mutations of the IFNGR1 or IFNGR2 chains, the p40 subunit of IL-12, the IL-12Rβ1 chain, or the signal transducer STAT1, begin to show ID symptoms in early infancy. These patients show an unusual constellation of infections with strains of *nontuberculous mycobacteria* (NTM) and *Salmonella* that are normally not very pathogenic to humans. Other opportunistic infections are not observed. Due to the lack of granuloma formation, NTM infections are sometimes fatal in these patients, particularly in those deficient in IFNγ signaling. Unfortunately, death can also result when individuals with undiagnosed IFNγR deficiency are (unknowingly) vaccinated against TB with the BCG strain of *Mycobacterium bovis*. While exogenous IFNγ has been effective in preventing NTM infections in IL-12-deficient children, even high doses of IFNγ (obviously) cannot help patients with null mutations of the IFNγR chains.

A milder disease phenotype has been observed in patients with a partial deficiency of IFNγR expression, and some of these individuals can respond to very high doses of exogenous IFNγ. In addition, some people born with defects in IL-12 signaling can still make some IFNγ, possibly due to the action of IL-18. When NTM infections strike these individuals, a few mature granulomas are formed that afford some degree of infection control. Curiously, while humans with defects in the IFNγ/IL-12 axis do not generally succumb to viral infections, IFNγ-deficient mice are highly susceptible to a broad spectrum of viruses. The mechanisms underlying this difference are under investigation.

v) Autoinflammatory Syndromes

Patients with autoinflammatory syndromes suffer bouts of severe local inflammation and prolonged, periodic fevers in the absence of any obvious pathogenic cause. No significant increases in autoantibodies or autoreactive T cells are present. Both autosomal recessive and autosomal dominant syndromes have been reported, but the underlying mechanisms are generally obscure. We shall describe three of the better-studied autoinflammatory syndromes: *familial Mediterranean fever* (FMF), hyper-IgD with periodic fever syndrome (HIDS), and

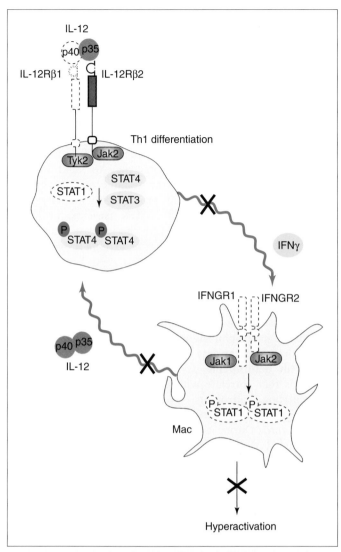

Figure 24-6

PIs Associated with Defects in the IFNγ/IL-12 Axis
Macrophage hyperactivation may be defective due to any one of a number of deficiencies in the IL-12 and IFNγ signaling axis. Dashed outlines indicate molecules in which mutations are known to cause primary immunodeficiency. Mutations of the p40 subunit of IL-12 or the IL-12Rβ1 chain inhibit Th1 differentiation, reducing the production of IFNγ. Loss of STAT1 function needed for IL-12 and IFNγR signaling also compromises the Th1 response. Mutation of either chain of the IFNγR prevents a macrophage from receiving the signals necessary to undergo hyperactivation or produce IL-12. Mutations of other molecules shown in this figure, such as IL-12p35 and IL-12Rβ2, have yet to be associated with PI. The reader is referred to Fig. 17-5 and Fig. 17-30 for depictions of the normal IFNγ and IL-12 signaling pathways, respectively.

TNF-receptor-associated periodic syndrome (TRAPS) (refer to Table 24-9).

va) Familial Mediterranean fever (FMF). FMF is an autosomal recessive disease that occurs preferentially in people of Middle Eastern ethnic backgrounds. FMF is caused by mutations in an inflammatory regulator called *marenostrin* (or *pyrin*), which is expressed almost exclusively by activated

monocytes and granulocytes. Patients suffer from periodic fevers that last 1–3 days accompanied by abdominal pain, pleurisy, limited arthritis, rash, and myalgia. The clinical picture can be influenced by the expression of specific modifier alleles of genes that are associated with the occurrence of *amyloidosis*. (Amyloidosis, which is a feature of numerous diseases in addition to FMF, is the extracellular deposition of a waxy, starch-like glycoprotein called *amyloid* in the tissues. Amyloid is derived from the acute phase protein *serum amyloid A*, which is produced by hepatocytes in response to the inflammatory cytokines TNF and IL-6. Deposition of amyloid in a tissue can destroy the structure of the affected organ.) Treatment with the alkaloid drug colchicine is often helpful for FMF patients because colchicine decreases amyloidosis. Colchicine also inhibits microtubule formation and may reduce neutrophil chemotaxis, which would in turn decrease inflammation.

vb) Hyper-IgD with periodic fever syndrome (HIDS). HIDS was named for the original observation that serum levels of IgD were elevated in affected patients. This manifestation has now been found to be secondary to the primary clinical signs of fever and inflammation. Indeed, some FMF and TRAPs patients also show elevated IgD. While HIDS patients have most of the same clinical signs as FMF patients, HIDS differs from FMF in that it occurs primarily in people of Northern European ethnic backgrounds, the periodic fever lasts for 3–7 days, and the associated arthritis affects multiple sites. No amyloidosis has yet been noted in HIDS patients but lymphoadenopathy is common. HIDS is caused by mutations in *mevalonate kinase*, an enzyme involved in cholesterol biosynthesis, but how one gets from impaired cholesterol synthesis to overproduction of IgD is a mystery. In addition, it is unclear why some mutations in this gene result in HIDS, and others result in a severe metabolic disorder that affects brain function. Treatment with anti-inflammatory drugs, either steroidal (prednisone) or non-steroidal (aspirin), can be beneficial in reducing the severity, if not the frequency, of the fever attacks.

vc) TNF-receptor-associated periodic syndrome (TRAPS). TRAPS is an autoinflammatory disease of autosomal dominant inheritance that occurs in many different ethnic groups. TRAPS is characterized by severe localized inflammation and a prolonged fever that lasts anywhere from 7–30 days. Peritoneal and pulmonary inflammation, abdominal pain, arthritis, a spreading rash, conjunctivitis, and edema in the eye are the major clinical signs. Amyloidosis is seen in about 25% of TRAPS patients.

TRAPS is caused by mutations of TNFR1 that result in uncontrolled stimulation of the TNFR1 signaling pathway. As we learned in Chapter 17, most of the pro-inflammatory effects of TNF are mediated via TNFR1. In normal cells, TNF signaling is negatively regulated at least in part by cleavage of the TNFR1 receptor protein. When TNF binds to trimeric TNFR1, part of the extracellular domain of the receptor is clipped off by a metalloprotease. This freed TNFR1 domain can then compete with intact receptors for incoming TNF, blunting the promotion of the inflammatory response. Without cleavage of the extracellular domain, TNF signaling rages uncontrollably. Patients with TRAPS have mutations in the first extracellular domain of the TNFR1 protein that block its cleavage, leading to increased levels of intact TNFR1 on their leukocytes. It is not clear exactly how the mutations block cleavage. None of the documented mutations in patients are actually at the site where the metalloprotease clips the TNFR1 protein. It is therefore speculated that TRAPS mutations may generate an unfavorable overall conformation of the receptor that prevents the metalloprotease from binding to it. TRAPS is thought to be autosomal dominant for the same reason as ALPS due to Fas mutation: the homotrimeric nature of TNFR1. Receptor complexes containing just one mutated TNFR1 chain appear to be completely resistant to cleavage, amplifying the effect of the mutated allele.

Conventional treatment of TRAPS involves corticosteroids with anti-inflammatory effects. Experimental treatment of TRAPS patients with TNF-blocking agents such as *etanercept*, a fusion protein containing TNFR type II sequences, or a humanized anti-TNF mAb, have been successful in reducing fever and inflammatory attacks. More on these types of treatment appears in Chapter 29.

vi) Congenital Neutropenia

Neutropenia is the term used to describe extremely low numbers of neutrophils present in the circulation. Because neutrophils are vital for innate defense against a wide range of organisms, a drop in their numbers is an invitation for pathogen establishment. Neutropenia can be constant (*severe congenital neutropenia*; SCN) or intermittent (*cyclic neutropenia*; CyN), but it turns out that mutations in the neutrophil elastase gene involved in extracellular matrix dissolution can underlie both phenotypes. Severe congenital neutropenia can also be caused by gain-of-function mutations in the WASP gene. These diseases are included in the summary in Table 24-9.

via) Severe congenital neutropenia (SCN and X-SCN).
SCN is characterized by dramatically reduced levels of neutrophils in the circulation and recurrent bacterial and fungal infections. Peritonitis, stomatitis, and meningitis caused by *S. aureus* are common clinical signs. The extremely low level of neutrophils does not fluctuate, unlike the case with cyclic neutropenia described later. The vast majority of cases of SCN are caused by mutations in the gene encoding *neutrophil elastase* (NE), an enzyme used by neutrophils to blast a path through the extracellular matrix in response to inflammatory stimuli. Both autosomal dominant and autosomal recessive inheritance patterns of SCN have been reported. Curiously, patients with elastase mutations leading to SCN also have an increased risk of leukemia, unlike patients with neutrophil elastase mutations that lead to CyN (see later).

Neutrophil elastase is a serine protease derived from a gene expressed only at the promyelocyte stage of granulopoiesis. (The gene is silenced in mature neutrophils.) The NE protein made during granulopoiesis is stored in the cytoplasmic granules, to be employed later by the mature neutrophil. It remains unclear how the mutated elastase protein is responsible for decreasing neutrophil numbers in SCN, but it seems that mutations in the enzyme somehow induce inappropriate apoptosis of neutrophil precursors in the bone marrow. Some NE mutations alter the glycolytic processing and subsequent intracellular storage of the NE protein, while other mutations cause the NE protein to alter its substrate specificity or acquire resistance to natural serine protease inhibitors. It is possible that leakage of the abnormal NE protein into the cytosol kills the cell. A few cases of SCN have been described that have an X-linked pattern of inheritance. These X-SCN cases have been attributed to mutations of the WASP gene that cause its protein product to become constitutively active. It is speculated that, for unknown reasons, a hyperactive WASP enzyme destabilizes the cytoskeleton specifically in neutrophil progenitors, inducing their apoptosis. In both SCN and X-SCN patients, prolonged treatment with recombinant G-CSF can increase the survival of neutrophil progenitors such that neutrophil levels rise and infections decrease.

Again, mouse modeling has highlighted differences between the human and murine immune systems. Knockout mice completely deficient for murine NE, although at increased risk for infections, do not develop a SCN phenotype. Their neutrophil numbers remain at normal levels but show a decreased capacity to kill pathogens.

vib) Cyclic neutropenia.
CyN is characterized by recurrent infections that correlate with cyclic fluctuations in numbers of circulating neutrophils. Over a 21-day cycle, neutrophil counts in CyN patients can range from near-normal to virtually zero. During the 3–6 days when neutrophil counts are at their lowest, the patient slips from apparently normal health to a state of severe infection that leads to death in 10% of cases. Like SCN, CyN results from mutations of the NE gene. The block in neutrophil precursor differentiation occurs at the same stage in CyN as in SCN, and CyN can also be treated successfully with G-CSF administration. However, unlike SCN, inheritance of CyN is always autosomal dominant. It should be noted that the exact same mutation may give rise to either CyN or SCN, indicating that other modifying factors are at work in these disorders.

II. COMPLEMENT DEFICIENCIES

Deficiencies for elements of the various complement activation pathways were covered in detail in Chapters 4 and 19 and thus will be reviewed only briefly here (Table 24-10). Mannose-binding lectin (MBL) is important for the opsonization of pathogens as well as for triggering the lectin pathway of complement activation. Mutations in the MBL gene are usually autosomal dominant and relatively common, affecting 16–29% of a population depending on ethnic background. Patients with very low levels of serum MBL suffer from recurrent infections and (for unknown reasons) are at increased risk for the development of rheumatoid arthritis. This PI has its greatest impact in early infancy, before the patient's adaptive immune system is mature enough to respond and after maternal antibodies have dissipated.

Children lacking the earliest components (C1 and C4) of the classical complement activation pathway suffer from

Table 24-10 Complement Deficiencies

Missing Complement Component	Functional Defect
MBL	Failure in opsonization and lectin complement activation pathway leads to recurrent infections with a wide range of organisms; increased risk of RA
C1, C4	Failure in classical complement activation pathway leads to pyogenic infections; increased risk of SLE
C2	Failure in classical complement activation pathway leads to non-pyogenic infections; increased risk of SLE
C3, Factor H, Factor I, properdin	Failure in alternative complement activation pathway leads to pyogenic infections
C5–C8	Failure in MAC formation leads to *Neisseria* infections

RA, rheumatoid arthritis; SLE, systemic lupus erythematosus (see Ch.29).

recurrent pyogenic infections with encapsulated bacteria. Infections in patients lacking C2 tend not to be pyogenic. For unknown reasons, patients lacking C1, C2, or C4 also have an increased incidence of systemic lupus erythematosus (SLE; see Ch.29). Patients lacking C3, or the Factor H, Factor I, or properdin components of the alternative pathway, also suffer from severe recurrent infections with pyogenic encapsulated bacteria but do not show signs of SLE. Increased infection with *Neisseria* (only) occurs in adult patients lacking any of the terminal complement components C5, C6, C7, and C8. Patients deficient for C9 are generally asymptomatic. These complement deficiencies are all rare autosomal recessive disorders and account for about 1% of all PIs. Administration of prophylactic antibiotics and specific immunization against encapsulated organisms are effective treatments.

This concludes our discussion of primary immunodeficiencies, defects of the immune system that are congenital. We move now to a discussion of acquired immunodeficiencies, defects of the immune system that are secondary to an external event in the patient's life. Naturally, this chapter is dominated by a description of AIDS caused by HIV, a pathogen that was unknown until the 1980s.

SUMMARY

Primary immunodeficiencies (PIs) are rare congenital diseases caused by mutations in genes that affect the generation or function of either lymphocytes, NK cells, phagocytes, or the complement system. The defects in cellular and molecular function vary widely among PIs but include developmental arrest, blockage of metabolic pathways, cytokine abnormalities, reduced MHC expression, disrupted signaling pathways, impaired DNA repair mechanisms, and failure of normal apoptotic pathways. In the case of lymphocyte malfunctions, some PIs are either T cell- or B cell-specific, while others result in severe combined immunodeficiency (SCID) in which the functions of both T and B cells are compromised and the patient is left with virtually no adaptive immunity. Many PIs are diagnosed in infancy when the protection of maternal antibodies begins to wane and the affected individual becomes susceptible to infection. In some cases, due to deregulation of lymphocyte growth and differentiation, autoimmunity, allergy, and lymphoproliferative disorders may also occur. The severity of PIs may range from mild to life-threatening. Treatment strategies include intravenous immunoglobulin replacement therapy, enzyme replacement therapy, or bone marrow/hematopoietic cell transplantation. The use of gene therapy in appropriate PI cases holds great promise.

Selected Reading List

Aiuti A., Slavin S., Aker M., Ficara F., Deola S. *et al.* (2002) Correction of ADA-SCID by stem cell gene therapy combined with nonmyeloablative conditioning. *Science* **296**, 2410–2413.

Altare F., Jouanguy E., Lamhamedi S., Doffinger R., Fischer A. *et al.* (1998) Mendelian susceptibility to mycobacterial infection in man. *Current Opinion in Immunology* **10**, 413–417.

Aprikyan A. and Dale D. (2001) Mutations in the neutrophil elastase gene in cyclic and congenital neutropenia. *Current Opinion in Immunology* **13**, 535–538.

Berger M. (2004) Subcutaneous immunoglobulin replacement in primary immunodeficiencies. *Clinical Immunology* **112**, 1–7.

Bleesing J., Straus S. and Fleisher T. (2000) Autoimmune lymphoproliferative syndrome;

A human disorder of abnormal lymphocyte survival. *Pediatric Clinics of North America* **47**, 1291–1309.

Buckley R. (2001) The hyper-IgE syndrome. *Clinical Reviews in Allergy and Immunology* **20**, 139–154.

Buckley R. (2002) Primary cellular immunodeficiencies. *Journal of Allergy and Clinical Immunology.* **109**, 747–757.

Bzowska A., Kulikowska E. and Shugar D. (2000) Purine nucleoside phosphorylases: properties, functions, and clinical aspects. *Pharmacology and Therapeutics* 88, 349–425.

Carney J. (1999) Chromosomal breakage syndromes. *Current Opinion in Immunology* 11, 443–447.

Cavazzana-Calvo M., Hacein-Bey-Abina S. and Fischer A. (2002) Gene therapy of X-linked severe combined immunodeficiency. *Current Opinion in Allergy and Clinical Immunology* 2, 507–509.

Chinen J. and Puck J.M. (2004) Successes and risks of gene therapy in primary immuno-deficiencies. *Journal of Allergy and Clinical Immunology* 113, 595–603.

Conley M. (1999) Genetic effects on immunity; New genes—how do they fit? *Current Opinion in Immunology* 11, 427–430.

Conley M. and Cooper M. (1998) Genetic basis of abnormal B cell development. *Current Opinion in Immunology* 10, 399–406.

Delpech M. and Grateau G. (2001) Genetically determined recurrent fevers. *Current Opinion in Immunology* 13, 539–542.

de Saint Basile G. and Fischer A. (2001) The role of cytotoxicity in lymphocyte homeostasis. *Current Opinion in Immunology* 13, 549–554.

Desiderio S. (1997) Role of Btk in B cell development and signaling. *Current Opinion in Immunology* 9, 534–540.

Di Renzo M., Pasqui A. and Auteri A. (2004) Common variable immunodeficiency: a review. *Clinical and Experimental Medicine* 3, 211–217.

Durandy A. (2001) Terminal defects of B lymphocyte differentiation. *Current Opinion in Allergy and Clinical Immunology* 1, 519–524.

Durandy A. and Honjo T. (2001) Human genetic defects in class-switch recombination (hyper-IgM syndromes). *Current Opinion in Immunology* 13, 543–548.

Emile J.-F., Geissmann F., de la Calle Martin O., Radford-Weiss I., Lepelletier Y. *et al.* (2000) Langerhans cell deficiency in reticular dysgenesis. *Blood* 96, 58–62.

Erlewyn-Lajeunesse M. (2000) Hyperimmunoglobulin-E syndrome with recurrent infection: a review of current opinion and treatment. *Pediatric Allergy and Immunology* 11, 133–141.

Ferrari S. and Plebani A. (2002) Cross-talk between CD40 and CD40L: lesson from primary immunodeficiencies. *Current Opinion in Allergy and Clinical Immunology* 2, 489–494.

Fischer A. (1998) Genetic effects on immunity; editorial overview. *Current Opinion in Immunology* 10, 397–398.

Fischer A. (2001) Immunogenetics; editorial overview. *Current Opinion in Immunology* 13, 533–534.

Fischer A. (2004) Human primary immunodeficiency diseases: a perspective. *Nature Immunology* 5, 23–30.

Fischer A., Hacein-Bey S., Le Deist F., Soudais C., Di Santo J. *et al.* (2000) Gene therapy of severe combined immunodeficiencies. *Immunological Reviews* 178, 13–20.

Galon J., Aksentijevich I., McDermott M., O'Shea J. and Kastner D. (2000) TNFRSF1A mutations and autoinflammatory syndromes. *Current Opinion in Immunology* 12, 479–486.

Gatti R., Becker-Catania S., Chun H., Sun X., Mitui M. *et al.* (2001) The pathogenesis of ataxia-telangiectasia: learning from a Rosetta stone. *Clinical Reviews in Allergy and Immunology* 20, 87–108.

Gennery A., Cant A. and Jeggo P. (2000) Immunodeficiency associated with DNA repair defects. *Clinical and Experimental Immunology* 121, 1–7.

Grimbacher B., Belohradsky B. and Holland S. (2002) Immunoglobulin E in primary immunodeficiency diseases. *Allergy* 57, 995–1007.

Hershfield M. (1998) Adenosine deaminase deficiency: clinical expression, molecular basis, and therapy. *Seminars in Hematology* 35, 291–298.

Hershfield M. (2003) Genotype is an important determinant in adenosine deaminase deficiency. *Current Opinion in Immunology* 15, 571–577.

Heyworth P., Cross A. and Curnutte J. (2003) Chronic granulomatous disease. *Current Opinion in Immunology* 15, 578–584.

Hong R. (2001) The DiGeorge anomaly. *Clinical Reviews in Allergy and Immunology* 20, 43–60.

Howe S. and Thrasher A. (2003) Gene therapy for inherited immunodeficiencies. *Current Hematology Reports* 2, 328–334.

Jones A. and Gaspar H. (2000) Immunogenetics: changing the face of immunodeficiency. *Journal of Clinical Pathology* 53, 60–65.

Leonard W. (1997) Protein tyrosine kinases: disease loci for inherited immunodeficiencies. *Current Opinion in Immunology* 9, 525–527.

Leonard W. (2000) Genetic effects on immunity: editorial overview. *Current Opinion in Immunology* 12, 465–467.

Lieber M. (2000) Antibody diversity: a link between switching and hypermutation. *Current Biology* 10, R798–R800.

Lim M. and Elenitoba-Johnson K. (2004) The molecular pathology of primary immunodeficiencies. *Journal of Molecular Diagnostics* 6, 59–83.

Mackay I. and Rosen F. (2000) Immunodeficiency diseases caused by defects in phagocytes. *Advances in Immunology* 343, 1703–1714.

Masternak K., Barras E., Zufferey M., Conrad B., Corthals G. *et al.* (1998) A gene encoding a novel RFX-associated transactivator is mutated in the majority of MHC class II deficiency patients. *Nature Genetics* 20, 273–277.

Matheux F. and Villard J. (2004) Cellular and gene therapy for major histocompatibility complex class II deficiency. *News in Physiological Sciences* 19, 154–158.

Matsuda S., Suzuki-Fujimoto T., Minowa A., Ueno H., Katamura K. *et al.* (1999) Temperature-sensitive ZAP70 mutants degrading through a proteasome-independent pathway. *Journal of Biological Chemistry* 274, 34515–34518.

Nonoyama S. and Ochs H. (1998) Characterization of the Wiskott-Aldrich syndrome protein and its role in the disease. *Current Opinion in Immunology* 10, 407–412.

Notarangelo L., Villa A. and Schwarz K. (1999) RAG and RAG defects. *Current Opinion in Immunology* 11, 435–442.

Parkman R., Weinberg K., Crooks G., Nolta J., Kapoor N. *et al.* (2000) Gene therapy for adenosine deaminase deficiency. *Annual Review of Medicine* 51, 33–47.

Puel A. and Leonard W. J. (2000) Mutations in the gene for the IL-7 receptor result in T-B+NK+ severe combined immunodeficiency disease. *Current Opinion in Immunology* 12, 468–473.

Riminton D. and Limaye S. (2004) Primary immunodeficiency diseases in adulthood. *Internal Medicine Journal* 34, 348–354.

Rosenzweig S. and Holland S. (2004) Phagocyte immunodeficiencies and their infections. *Journal of Allergy and Clinical Immunology* 113, 620–626.

Schinke M. and Izumo S. (2001) Deconstructing DiGeorge syndrome. *Nature Genetics* 27, 238–240.

Schwartz S. (2000) Intravenous immunoglobulin treatment of immunodeficiency disorders. *Pediatric Clinics of North America* 47, 1355–1369.

Simonte S. and Cunningham-Rundles C. (2003) Update on primary immunodeficiency: defects of lymphocytes. *Clinical Immunology* **109**, 109–118.

Snapper S., Rosen F., Mizoguchi E., Cohen P., Khan W. *et al.* (1998) Wiskott-Aldrich syndrome protein-deficient mice reveal a role for WASP in T but not B cell activation. *Immunity* **9**, 81–91.

Sullivan J. (1999) The abnormal gene in X-linked lymphoproliferative syndrome. *Current Opinion in Immunology* **11**, 431–434.

Tchilian E., Wallace D., Spencer Wells R., Flower D., Morgan G. *et al.* (2001) A deletion in the gene encoding the CD45 antigen in a patient with SCID. *Journal of Immunology* **166**, 1308–1313.

Ten R. (1998) Primary immunodeficiencies. *Mayo Clinic Proceedings* **73**, 865–872.

Thomis D. (1997) The role of Jak3 in lymphoid development, activation and signaling. *Current Opinion in Immunology* **9**, 541–547.

Wahn V. (2003) Primary immunodeficiencies caused by defects of cytokines and cytokine receptors. *Methods in Molecular Biology* **215**, 3–12.

Wright E. (1999) Inherited and inducible chromosomal instability: a fragile bridge between genome integrity mechanisms and tumourigenesis. *Journal of Pathology* **187**, 19–27.

HIV and Acquired Immunodeficiency Syndrome

25

Peyton Rous 1879–1970

Retroviruses
Nobel Prize 1966

©Nobelstiftelsen

CHAPTER 25

A. HISTORICAL NOTES

B. WHAT IS HIV?

C. STRUCTURE AND FUNCTION OF HIV PROTEINS

D. HIV INFECTION AND AIDS

E. THE IMMUNE RESPONSE DURING HIV INFECTION

F. HOST FACTORS INFLUENCING THE COURSE OF HIV INFECTION

G. EPIDEMIOLOGY AND SOCIOLOGY OF HIV INFECTION

H. ANIMAL MODELS OF AIDS

I. HIV VACCINES

J. TREATMENT OF HIV INFECTION WITH ANTI-RETROVIRAL DRUGS

"A warless world will come as men develop warless hearts"
—Charles Wesley

In the last chapter, we discussed primary or congenital immunodeficiencies (IDs), defects in the immune system that are inborn. Acquired or secondary IDs are those that are not inborn but are due to external interference in the immune system. Severe metabolic disturbances caused by nutritional deficiencies, cancer, and trauma due to burns, injuries, or surgery can all cause secondary IDs, as can immunosuppressive drugs taken to combat cancer (Ch.26), circumvent transplant rejection (Ch.27), or treat hypersensitivities (Ch.28) or autoimmune diseases (Ch.29). Some types of infectious diseases, such as measles or malaria, can also lead to immunodeficiency although the mechanisms involved are poorly understood. Patients with such secondary IDs are left vulnerable to opportunistic bacterial and viral infections that can be lethal. However, despite the potential seriousness of these conditions, their overall health impact pales in comparison to that of a relatively new secondary ID that is the subject of the remainder of this chapter, namely *acquired immunodeficiency syndrome* (AIDS) due to infection with *human immunodeficiency virus* (HIV).

AIDS is one of the top five leading causes of death of young adults around the world. Indeed, AIDS is now the leading cause of all adult deaths due to infectious disease, surpassing malaria in 1999. The situation is worst in sub-Saharan Africa, where AIDS accounts for 20% of all young adult deaths. Much has been learned about the biology of HIV over the past 20 years, but, for a variety of reasons detailed later, this virus still frustrates the best efforts of researchers to produce an effective vaccine. We will discuss the biology of HIV infection, the structure of the virus, epidemiological aspects of HIV infection, and the societal and economic devastation wrought by this virus. Animal and tissue models of HIV infection, experimental animal and human vaccines, and anti-retroviral drug treatments are also described. We commence with a brief history of the discovery of HIV and AIDS.

A. Historical Notes

In 1981, Michael Gottlieb and colleagues in Los Angeles, California reported an unusual cluster of cases of pneumonia due to *Pneumocystis carinii* in young homosexual men. This observation was puzzling because *P. carinii* is an opportunistic organism that becomes a problem only when the cell-mediated immune response is compromised. Similar cases were then reported in gay communities in other cities, and it was noted that many of these patients also had a rare tumor called *Kaposi's sarcoma* (Plate 25-1). The occurrence of Kaposi's sarcoma is also strong

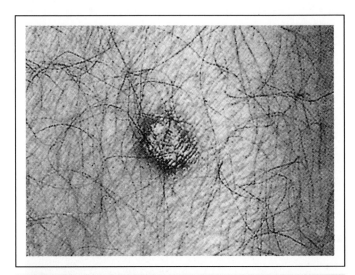

Plate 25-1

Kaposi's Sarcoma
Reproduced with permission from Emond R. T. D., Welsby P. D. and Rowland H. A. K (2003) "Colour Atlas of Infectious Diseases," 4th edn. Elsevier Science, London.

evidence of a failure in cell-mediated immunity. Up to this point, these types of afflictions had been associated almost exclusively with IDs secondary to the use of immunosuppressive or chemotherapeutic drugs in transplant or cancer patients. The sudden, unexplained loss of immune system function in ostensibly healthy young men was a new finding. With the demonstration that these "gay plague" patients had suffered a loss of $CD4^+$ T cells, the scientific community was galvanized into tracking down the agent responsible for the disease entity that came to be known as "acquired immunodeficiency syndrome" or "AIDS." Two factors prompted researchers to wonder whether this agent could be a new human retrovirus. First, a feline retrovirus (*feline leukemia retrovirus*; FeLV) was known to profoundly impair immune responses in cats. Secondly, the first human retrovirus isolated, HTLV-1, had been shown to preferentially infect human T cells. In 1983–1984, prodigious efforts in the laboratories of Luc Montagnier of the Pasteur Institute in Paris, Robert Gallo of the National Cancer Institute in Bethesda, and Jay Levy at the University of California in San Francisco resulted in the independent isolation of LAV (*lymphadenopathy-associated virus*), HTLV-III (*human T cell lymphotrophic virus III*), and ARV (*AIDS-related virus*), respectively, from the cells of AIDS patients. After extensive study showed that the nucleotide sequences of all three virus isolates differed by only 1–2%, an international committee recommended in 1986 that the new human retrovirus be named the "human immunodeficiency virus".

As the AIDS epidemic spread around the world in the late 1980s, it became apparent that, in addition to male-to-male sexual relations, heterosexual intercourse and intravenous drug use were major modes of HIV transmission. In addition, blood transfusion recipients were getting the disease, confirming that transmission occurred primarily via transfer of body fluids. It was also becoming devastatingly clear that the disease was almost invariably fatal after a lengthy latency of 7–12 years. A graphic demonstration of the cause and effect relationship between HIV-1 and AIDS occurred in 1988 when a laboratory worker working with HIV was accidentally infected. The worker became ill with AIDS, and the strain of HIV isolated from this victim was essentially identical to the strain the worker had been handling at the time of the accident.

Where did HIV originally come from? There are two known types of HIV: HIV-1 and HIV-2. HIV-1, the virus that is prevalent in the Western hemisphere, can be found anywhere in the world. HIV-2 is of much more restricted geographic distribution, being found in few places outside Western Africa. There is good genetic evidence that HIV-2 infection of humans may have originated with a cross-species jump by SIV_{smm}, a *simian immunodeficiency virus* that infects sooty mangabey monkeys. The origin of HIV-1 is less well-defined. Chimpanzees can be infected with a virus called SIV_{cpz}, which is closely genetically related to HIV-1. However, chimpanzees infected with either HIV-1 or SIV_{cpz} do not get any illness approximating AIDS, despite the fact that chimps are 98% genetically identical to humans. To resolve this apparent paradox, researchers examined the mitochondrial DNA of three geographically isolated chimp populations, each of which was naturally infected with a distinct substrain of SIV_{cpz}. By determining how much the sequences of the mitochondrial DNA in these chimp

populations had diverged, and by comparing the sequences of the three SIV_{cpz} isolates, it was determined that SIV_{cpz} most likely first infected chimpanzees more than 10,000 years ago. Such a lengthy time span has allowed for considerable co-evolution of SIV_{cpz} and chimpanzees. The present-day SIV_{cpz} virus has thus been selected for decreased pathogenicity to chimps, preserving the host for ongoing rounds of virus replication.

SIV_{cpz} may have made the jump to humans to become HIV-1 fairly recently, perhaps when humans disrupted the African jungle environment that is home to large populations of wild chimps. A retrospective study of serum samples from young women who had been living in Zaire in Africa in the early 1970s showed that HIV-1 was present in 0.25% of them. Going back even farther, a frozen blood sample originally drawn in Zaire in 1959 was found in the 1980s to contain anti-HIV antibodies. Similarly, tissue samples preserved from an English seaman whose death in 1959 was attributed to "immunological collapse" contained HIV-1 DNA. Tellingly, the seaman's wife and a child conceived since the seaman's return to England also died with similar symptoms. These data suggest that the latest time HIV may have evolved from SIV_{cpz} is in the early 1950s. Unlike SIV in chimps, HIV has not yet had over 10,000 years of selection to adapt to our species and become less pathogenic. As a result, most HIV-infected humans die.

One of the most tragic aspects of the current AIDS epidemic is that transmission of the virus can be halted by relatively easy means. The use of condoms and single-use syringes stops sexual and drug-related transmission, and the feeding of babies with formula instead of breast milk prevents maternal–child transmission. Careful screening of blood products dramatically reduces transmission via the blood supply. However, human behavioral or cultural patterns are sometimes very difficult to change, even in the face of overwhelming evidence. Indeed, after a downswing in the incidence rate of new cases in the late 1990s in North America, the greater availability and effectiveness of antiretroviral drugs has had the counterproductive effect of decreasing the motivation of some gay men in the Western world to use safer sex practices. As a result, the incidence of HIV infection in developed countries climbed again in the early 2000s.

Despite their success in slowing the onset of AIDS, anti-retroviral drugs are not the final answer. Computer modeling studies have estimated that at least 60 years of anti-retroviral therapy that completely suppresses viremia would be required to eradicate HIV from an infected person. Furthermore, these drugs are expensive and not readily available in the developing world, where the vast majority of HIV-infected individuals live. More frightening is the fact that drug-resistant isolates of HIV are becoming common. All of these factors have spurred immense amounts of research into HIV vaccines. However, here we come to the key conundrum: high levels of anti-HIV CTLs and neutralizing antibodies contain the virus for a few months or years, but then apparently lose their effectiveness as the virus adapts to and escapes these responses. This is a very worrying observation because it leads to the question of whether HIV vaccination will do any good: if a vigorous natural immune response cannot do the job, how can we develop a vaccination response (typically weaker than that induced by natural infection) that can stop the virus? The answer to this question remains unknown, and despite the identification of the causative agent, the purification of its proteins, and the cloning of its genes, many obstacles to the development of an efficacious HIV vaccine still exist.

Let's take a closer look at this insidious agent of immune system destruction, the human immunodeficiency virus.

B. What is HIV?

As previously mentioned , there are two known human immunodeficiency viruses: HIV-1 and HIV-2. Both HIV-1 and HIV-2 are very complex, cytopathic retroviruses belonging to the lentivirus class. A typical lentivirus persists in its host despite antibody and cell-mediated immune responses directed against it, and clinical disease is detected only after a very long latency period. Unlike many retroviruses, however, HIV-1 and HIV-2 are incapable of cellular transformation leading to cancer. Instead, both HIV-1 and HIV-2 cause AIDS, the constellation of clinical symptoms that occurs after wholesale destruction of the CD4$^+$ T cell population is nearly complete. However, AIDS caused by HIV-1 is more aggressive in onset and spreads more rapidly than AIDS caused by HIV-2. Indeed, the genetic similarity between HIV-1 and HIV-2 is in the range of only 40–50%, and the envelope protein surrounding the HIV-2 capsid is quite

different from that protecting the HIV-1 capsid. We shall concentrate on describing the better-studied HIV-1 virus, noting differences in HIV-2 where appropriate.

A short overview of the life cycle of HIV-1 is provided to place the molecules discussed later in their proper context. We then describe in detail the functions of the HIV RNA genome and important structural and regulatory proteins.

I. OVERVIEW OF THE HIV-1 LIFE CYCLE

An HIV virion most often accesses the body by breaching the mucosae during sexual contact or via intravenous drug injection. It can also be transmitted from mother to child before or during birth or by breast-feeding. Once past the mucosal barrier, the virus binds to certain proteins (including CD4, CXCR4, and CCR5) on the surfaces of host macrophages and T cells using its *glycosylated protein* 120 (gp120) molecule (Fig. 25-1). The virion is then internalized and its outer protective layers stripped away by cellular enzymes. Once the viral genome and its associated proteins are exposed in the cytosol, a viral enzyme called *reverse transcriptase* (RT) commences reverse transcription of the viral RNA genome into a DNA copy. This viral DNA becomes associated with a viral *integrase* enzyme, and both are transported from the cytosol into the host cell nucleus where the integrase mediates insertion of the

---

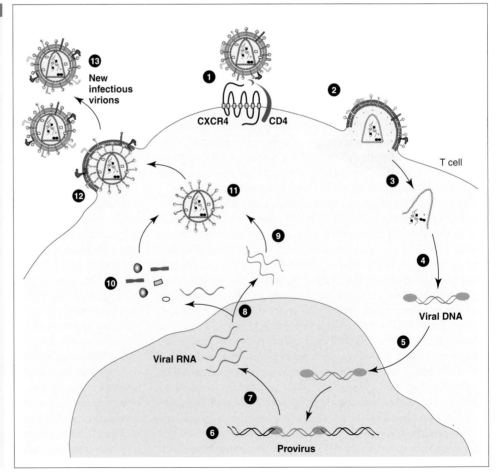

Figure 25-1

HIV Life Cycle
(1) The HIV virion (shown here in cross-section) uses its gp120 protein to bind to CXCR4 and CD4 on the surface of a host T cell. (2) The viral envelope fuses to the host cell membrane and the capsid is everted into the cytoplasm, where it is uncoated by host cell proteases (3). The viral genomic RNA undergoes reverse transcription to produce the viral DNA, which associates with viral integrase proteins (4). The viral DNA is imported into the nucleus (5), where the integrase mediates its integration into the host cell genome to form the provirus (6). When the cell is activated, the provirus is transcribed to generate viral RNA (7). The viral RNA is exported out of the nucleus (8) and meets two fates. Some becomes the genomic RNA of new progeny viruses (9), while some is translated in the host cell cytoplasm to generate the structural, regulatory, and accessory proteins required for progeny virus production (10). The newly transcribed genomic RNA is packaged with the newly translated viral proteins into the capsid to form the progeny virions (11). The progeny virions bud through the host cell membrane to acquire an envelope containing not only viral proteins but also many host proteins (12). The progeny virions (13) then seek out fresh host T cells to infect. The original host T cell undergoes lysis after many rounds of virus replication. See Figure 25-2 for a more detailed view of HIV structure.

viral DNA into the host cell genome. This integrated form of the viral DNA is called *proviral DNA* or the *provirus*. The provirus may remain untranscribed in the host cell genome (a state called *preactivation*) for a considerable time until the infected cell is activated by a stimulus such as TCR engagement or cytokine binding. Virus production is then also activated in this cell because the proviral promoter contains the binding motifs for the NF-κB and NF-AT transcription factors that transcribe many host genes during T cell activation. Transcription of the integrated proviral DNA results in the production of new copies of the RNA genome, some of which are translated to generate the structural proteins necessary to assemble progeny virions. The progeny virions bud through the host cell membrane to acquire their envelopes and proceed to spread through the body in search of fresh host CD4$^+$ T cells to infect (Plate 25-2). The CD4$^+$ T cell that has supported the generation of these virions is lysed in the process. Importantly, other <u>uninfected</u> CD4$^+$ T cells are also killed by HIV via mechanisms that remain unclear (see later). The result is a devastating systemic loss of CD4$^+$ T cells that leaves the victim vulnerable to tumorigenesis and fatal opportunistic infections.

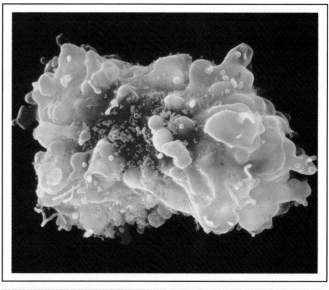

Plate 25-2

HIV Budding from an Infected T Cell
Budding HIV virions appear light blue. Courtesy of Boehringer Ingelheim Pharma KG, photo Lennart Nilsson, Albert Bonniers Förlag AB.

II. OVERVIEW OF HIV STRUCTURE

HIV-1 has a multi-layered structure consisting of (from the inside out) the genome, the capsid, the matrix, and the envelope (Fig. 25-2). The genomic RNA with its associated proteins, the capsid, and the matrix constitute the *core* of the HIV virion. The entire virion particle has a diameter of 100–120 nm.

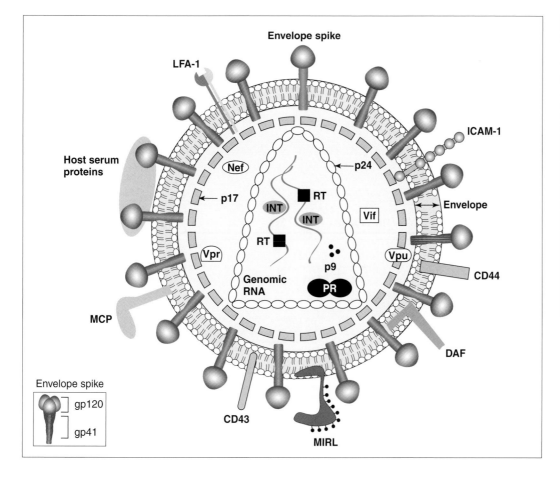

Figure 25-2

Structure of the HIV Virion
The HIV genome consists of two strands of +RNA which are associated with reverse transcriptase (RT) and integrase (INT) molecules. These elements plus the nucleocapsid protein p9 and the viral protease (PR) are surrounded by a capsid made up of p24 protein. The capsid is nestled in a matrix of p17 protein. The matrix also contains the accessory proteins Nef, Vpr, Vif, and Vpu. The matrix is surrounded by a lipid bilayer envelope containing the viral envelope spike and various host surface proteins. The inset shows the composition of the HIV envelope spike.

i) The Genome

The HIV genome is a "plus" (coding) ssRNA molecule present in two copies. The ssRNA molecule is 9.4 kb long and gives rise to double-stranded viral DNA that encodes nine partially overlapping genes (Fig. 25-3). As is true for all retroviruses, the HIV genome contains the three principal genes *gag*, *pol*, and *env*, which encode polyproteins that are subsequently cleaved into multiple shorter functional proteins. Like all lentiviruses, the HIV genome contains genes encoding the regulatory factors Tat and Rev. The *tat* and *rev* genes (two exons each) substantially overlap the *env* gene but are translated in different reading frames. Finally, the HIV genome contains four small genes encoding the accessory proteins Nef, Vpr, Vpu, and Vif. The *vpu* gene partially overlaps the *env* gene, and the *vif* gene overlaps the *pol* gene. After integration, the proviral DNA contains long terminal repeat (LTR) sequences that differ slightly between the 5' and 3' ends. The 5' LTR contains an active promoter that binds host RNA polymerase II to initiate transcription, as well as regulatory sites (including the κB motif) that control transcription in response to signals received from either the host cell or viral proteins. The 5' LTR also contains a binding site for Tat, the importance of which is made clear later. The 3' LTR contains sequences directing host RNA processing enzymes to cleave the primary transcript and add the polyA tail to form the translatable mRNA.

ii) The Capsid

The two RNA molecules making up the viral genome are contained in a cone-shaped *capsid*. The capsid is composed of about 1200 molecules of the internal structural protein p24 (refer to Fig. 25-2). Within the capsid, each RNA molecule is associated with several molecules of the viral RT and integrase. Viral protease and nucleocapsid protein p9 are also present.

iii) The Matrix

Surrounding the capsid is a spherical layer called the *matrix*, composed primarily of the viral structural protein p17. p17 is anchored to the internal surface of the viral envelope and appears to supply the scaffolding around which the envelope is wrapped (refer to Fig. 25-2). Within the matrix are the important accessory proteins Nef, Vif, Vpu, and Vpr.

iv) The Envelope

The HIV core is surrounded by an envelope made of a phospholipid bilayer acquired when the progeny virion budded out through membrane of the host cell infected by its parent. Embedded in the envelope are various viral and host cell proteins (refer to Fig. 25-2). Most prominent in the HIV-1 envelope is the viral envelope spike, a structure made up of three copies of a heterodimeric transmembrane glycoprotein. The heterodimer results from a non-covalent association between two viral proteins: gp41, which anchors the heterodimer in the membrane, and gp120, which projects from the virion surface. Both of these proteins are derived from the gp160 polyprotein precursor encoded by the *env* gene. It is estimated that each virion is covered with 7–14 envelope spikes, so that 21–42 gp120/gp41 molecules are present on each virion's surface. It is the carbohydrate-rich gp120 protein that binds to host cell receptors and contains the immunodominant epitopes of HIV-1. It should be noted that HIV-2 also has an envelope protein but that this molecule is unrelated in sequence to that of HIV-1. In both cases, the HIV envelope contains low amounts of host cell surface proteins such as the adhesion molecules ICAM-1, LFA-1, CD43, and CD44, and the complement regulatory proteins MCP, MIRL, and DAF.

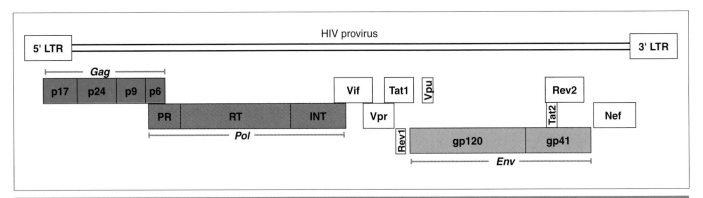

Figure 25-3

Proteins Encoded by HIV Genes
The HIV proviral DNA contains genes encoding the Gag, Pol, and Env polyproteins, plus genes encoding the Vif, Vpr, Tat, Rev, Vpu, and Nef proteins. Tat is encoded by the Tat1 and Tat2 exons while Rev is encoded by the Rev1 and Rev2 exons. Note the overlap among some of the gene sequences. Adapted from Girard M. and Excler J. L. (1999) Human immunodeficiency virus. In Plotkin S. A. and Orenstein W. A., eds. "Vaccines," 3rd edn. W.B. Saunders Co., Philadelphia, PA.

C. Structure and Function of HIV Proteins

We will now describe in detail the structure and function of the important HIV proteins, grouping them by the gene from which they are derived (Table 25-1).

It is reiterated here that, as illustrated in Figure 25-3, the *gag* and *pol* genes physically overlap in the same piece of viral DNA. The mRNAs transcribed from this stretch of DNA are translated either in one frame to give rise to the Gag polyprotein, or in another frame to give rise to the Pol polyprotein. Each of these polyproteins is then cleaved into smaller viral components.

Table 25-1 **HIV Proteins and Their Functions**

Protein	Name	Function
GAG POLYPROTEIN		
p17	Matrix protein	Acts as scaffolding for envelope bilayer
p24	Capsid protein	Protects RNA genome
p9	Nucleocapsid protein	Stabilizes DNA and RNA during replication
p6	Budding protein	?
POL POLYPROTEIN		
RT (p66)	Reverse transcriptase catalytic subunit	Transcribes viral RNA into viral DNA
RT (p51)	Reverse transcriptase support subunit	Assists in transcription of viral RNA
INT (p32)	Integrase	Inserts viral DNA into host DNA to form provirus
PR (p10)	Protease	Cleaves Gag and Pol polyproteins into mature components Promotes host cell apoptosis and inhibits cell survival
ENV POLYPROTEIN		
gp160		
gp41	Transmembrane envelope glycoprotein	Anchors envelope "spike" in lipid bilayer of envelope; essential for viral entry via fusion of envelope and host cell membrane
gp120	Non-transmembrane envelope glycoprotein	Binds to host receptors to initiate viral entry Promotes host cell apoptosis
REGULATORY PROTEINS		
Tat (p14)	Transactivator	Recruits cyclin-T to prevent premature transcription termination Promotes host cell apoptosis and inhibits cell survival
Rev (p19)	Regulator of viral protein expression	Binds to nascent viral mRNAs and promotes transport from nucleus into cytoplasm for translation
ACCESSORY PROTEINS		
Nef (p27)	"Negative factor"	Connects host surface proteins to clathrin to force internalization Interferes with host endosome acidification Promotes cellular activation that supports viral DNA synthesis and transport to the nucleus Promotes host cell apoptosis
Vpr (p15)	Viral protein "r"	Facilitates transport of viral DNA into nucleus Induces host cell cycle arrest in G2 Promotes host cell apoptosis
Vpu (p16)	Viral protein "u"	Binds to newly synthesized CD4 and targets it for ubiquitination May form channels in host cell membrane to promote budding Blocks MHC class I synthesis Enhances susceptibility of host cell to apoptosis
Vif (p23)	Virion infectivity factor	Stabilizes viral DNA and promotes nuclear transport Promotes viral DNA integration by counteracting host CEM-15

I. VIRAL PROTEINS DERIVED FROM THE *gag* GENE

The *gag* gene encodes a precursor protein of about 53 kDa that is cleaved by the viral protease (see later) into the matrix protein p17, the capsid protein p24, the nucleocapsid protein p9, and a small protein called p6. The p9 protein is a multifunctional nucleic acid chaperone that aids in stabilizing RNA and DNA structures during replication and viral packaging. The precise function of p6 is not known but HIV mutants lacking this protein cannot bud properly.

II. VIRAL PROTEINS DERIVED FROM THE *pol* GENE

i) Reverse Transcriptase

Reverse Transcriptase (RT) is essential for HIV replication because the viral RNA genome on its own is highly susceptible to degradation by intracellular RNases. RT rapidly makes a much more nuclease-resistant double-stranded DNA copy of the RNA template that later integrates to form the proviral DNA. HIV RT is a heterodimer composed of a 66 kDa subunit (p66) and a 51 kDa subunit (p51) created by cleavage of a separate molecule of p66. All the catalytic activity of HIV RT is attributable to p66, while p51 supports the functions of p66. As well as its DNA polymerase function, RT has an RNase H function that degrades the RNA template used to make the viral DNA. The complex method by which the RT synthesizes the viral DNA from the genomic ssRNA is described in Figure 25-4.

HIV RT is responsible for much of the antigenic variation of HIV that confounds both the natural immune response and vaccine development. HIV RT lacks the proof-reading capabilities inherent in cellular polymerases, meaning that its duplication of the HIV genome is highly error-prone. Mutations due to uncorrected RT activity appear in the HIV genome at a rate of about 1 in every 1500–4000 nucleotides

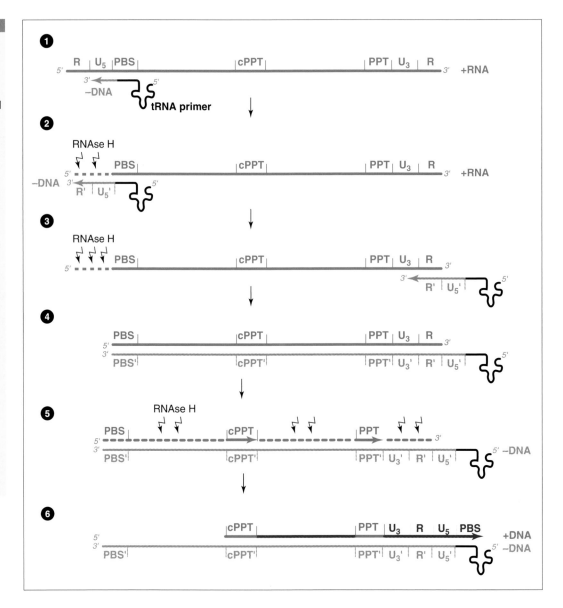

Figure 25-4 Part A

Synthesis of HIV DNA from Genomic RNA

(1) A cellular tRNA^{lys3} molecule hybridizes with the primer binding site (PBS) on the genomic +RNA strand and HIV RT commences synthesis of the negative strand DNA (−DNA) in the 3′ direction. (2) As the −DNA elongates, the 5′ section of genomic +RNA used as the template is degraded by the RNase H activity of RT. (3) The newly synthesized fragment of −DNA "jumps" to the 3′ end of the genomic RNA where the R′ sequence in the −DNA hybridizes to the R sequence in the +RNA. (4) Synthesis of the −DNA strand then continues in the 5′ to 3′ direction until it is completed. (5) The genomic +RNA is degraded by RT RNase H activity with the exception of two poly-purine tracts (cPPT and PPT) that remain hybridized to the −DNA. Using the −DNA as the template and both PPTs as primers, the positive strand DNA (+DNA) is elongated toward its 3′ end until a PBS site is created (6). Please see part B to continue.

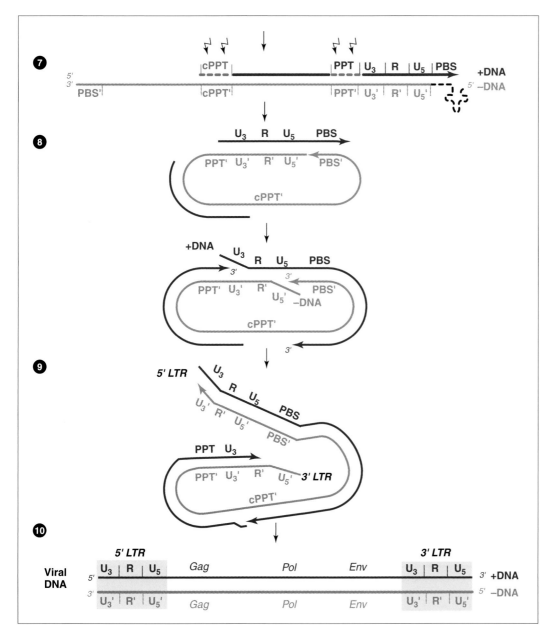

Figure 25-4 Part B

Synthesis of HIV DNA from Genomic RNA
(7) The tRNA^lys3 at the 5′ end of the −DNA strand and the PPTs are degraded. (8) Hybridization of the PBS′ with the PBS site results in circularization of the − and + DNA strands and elongation of the +DNA. (9) The U3′-R-U5′ sequence of the −DNA is used as a template to complete the 3′ end of the +DNA strand and form the 3′ LTR of the proviral DNA. Elongation of the −DNA strand through to its 3′ end copies the 5′ +DNA to produce the 5′ LTR of the provirus. Elongation of the +DNA strand ends at a termination site 3′ of the cPPT′. The resulting overlap in the +DNA strand is not shown in the viral DNA. (10) Note that R-U5 and U3-R from the genomic RNA have now both become U3-R-U5 in the viral DNA, giving it the required LTR at each end. The R site in the 3′ LTR contains a cleavage site and a polyadenylation site used for the transcription of viral mRNA. Adapted from Gotte M. et al. (1999) *Archives of Biochemistry and Biophysics* **365**(2), 199–210.

per replication cycle. As a point of comparison, the average rate of mutation of the human cellular genome is 1 in 10^7–10^8 base pairs.

ii) Integrase

HIV integrase (INT; 32 kDa) is responsible for inserting the viral DNA randomly into the host cell genome. This multimeric enzyme carries out integration in a multi-step process that is described in detail in Figure 25-5.

iii) Protease

HIV protease (PR; 10 kDa) is a homodimeric aspartyl protease. This enzyme is vital for cleaving other polyprotein precursors that give rise to mature viral components. For example, PR is required to cleave the Gag polyprotein into various viral capsid, matrix, and other structural proteins. It

should be noted that PR is not responsible for cleaving newly synthesized gp160 into viral envelope proteins; that job is performed by a host protease. In an infected individual, PR can contribute directly to the demise of infected T cells because it both cleaves caspase-8, triggering apoptosis, and cleaves Bcl-2, blocking survival signaling.

III. VIRAL PROTEINS DERIVED FROM THE ENV GENE

The *env* gene encodes a glycosylated polyprotein of about 160 kDa (gp160) that is subsequently cleaved by a host cell protease to generate the viral envelope proteins gp120 and gp41. These two chains associate non-covalently in trimers to form the viral envelope spikes extending from the virion envelope. As mentioned previously, gp41 is a transmembrane

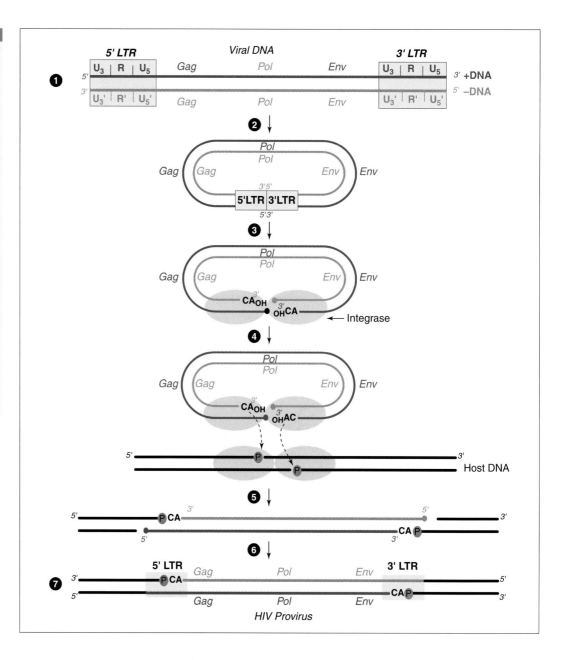

Figure 25-5

Integration of the HIV DNA into the Host Genome to Form the Provirus
Newly synthesized viral DNA (1) circularizes such that the LTRs come together (2). HIV integrase first recognizes sequences in the LTRs and nicks the viral DNA adjacent to conserved CA dinucleotides in the 3′ ends of both strands (3). Two nucleotides are removed per strand and a 3′ OH residue is exposed. The integrase then makes a staggered cut in the cellular DNA, facilitating the attack by the viral 3′ OH ends on cellular DNA phosphate bonds (4). The integrase covalently joins the 3′ ends of the viral DNA to the cellular DNA, leaving short gaps between the 5′ ends of the viral DNA and the cut host cell DNA (5). The gaps are repaired by the host cell DNA repair machinery to complete the integration of the viral DNA to form the provirus (6).

protein that anchors the spikes in the lipid layer of the envelope. gp41 both holds the spike together and is crucial for the fusion between viral and host cell membranes that permits viral entry. gp120 is a non-transmembrane protein containing high mannose and sialic acid-bearing carbohydrates. The gp120 molecule is made up of multiple domains bounded by many internal disulfide bridges. Five hypervariable regions called V1–5 are formed, of which V3 constitutes a major epitope identified by the TCRs of certain laboratory T cell lines directed against HIV. The V3 hypervariable region is critical for HIV binding to host cell receptors. The extensive glycosylation of gp120 helps to sterically obscure its binding sites, protecting them from assault by the immune system. As well as its structural function in the virion, gp120 promotes the intrinsic and extrinsic apoptosis of infected host cells in several ways (see later).

IV. REGULATORY PROTEINS

Tat and Rev are regulatory proteins expressed by all lentiviruses but by no other retroviruses.

i) Tat

The Tat (*transactivator*) molecule is a regulatory protein of 14 kDa that acts in the nucleus of the host cell as a potent activator of viral gene transcription. Tat is highly conserved and also appears in HIV-2 as well as in the related SIVs. Tat is crucial for HIV replication because the HIV promoter is more sensitive than host gene promoters to inhibition of transcription initiation by various cellular regulators. To overcome this sensitivity, viral Tat binds to a host cell protein called *cyclin-T*. Cyclin-T is the regulatory subunit of *cyclin-dependent kinase 9* (Cdk9), an enzyme required in host cells for the continued elongation of transcripts initiated by RNA polymerase II complexes. Once bound to cyclin-T, Tat is able to bind with greater affinity to a stem-loop structure called the TAR (*trans-acting responsive*) element at the 5′ end of all nascent viral RNA molecules. By linking cyclin-T to the TAR, Tat recruits Cdk9 activity to the nascent viral RNA and prevents premature termination of its transcription. This action, plus the actions of other host cell factors that associate with Tat, increases by several hundred-fold the number of viral transcripts that are produced for translation. In the absence of Tat/cyclin-T interaction, HIV transcription driven by the LTRs is crippled and viral replication is blocked. Indeed, mouse cells that are transgenically engineered to express human CD4 and other proteins serving as viral receptors can be infected by HIV but do not support replication because mouse cyclin-T cannot interact with HIV Tat. As well as its regulatory function during viral replication, Tat has direct cytotoxic effects on infected host cells because it promotes extrinsic and intrinsic apoptosis and interferes with Bcl-2-mediated survival signaling (see later).

ii) Rev

Rev (*regulator of expression of virion proteins*) is a regulatory protein of 19 kDa that promotes the exit of viral transcripts from the host cell nucleus into the cytoplasm for translation. Rev is vital for the expression of the structural, enzymatic, and accessory proteins necessary to complete the virus's life cycle. How does Rev act? Newly transcribed mRNAs for HIV structural, enzymatic, and accessory proteins (but not the regulatory proteins) contain *nuclear retention sequences* that force them to remain in the host cell nucleus. Rev counteracts these sequences by binding to "Rev-responsive sites" on these nascent mRNAs. Rev then interacts with nuclear export factors and enzymes that facilitate expulsion of the viral mRNAs into the cytoplasm. Until sufficient Rev is present to bind to the Rev-responsive sites, structural viral mRNAs are trapped in the nucleus and are not translated, inhibiting progeny virion production. In the presence of adequate Rev, translation of mRNAs for structural and other viral components proceeds, virion assembly is completed, and the progeny particles bud through the host cell membrane. The expression of Rev from the original proviral DNA is then downregulated, which in turns shuts off the production of structural and other proteins.

V. ACCESSORY PROTEINS

i) Nef

Nef stands for *negative factor*, because it was erroneously thought at first that Nef inhibited HIV transcription. We have encountered the Nef protein before, in our discussions of viral evasion mechanisms in Chapter 22. In that chapter, we learned that the 27 kDa Nef protein physically connects certain MHC class I, class II, and CD4 molecules on the host cell surface with the clathrin-coated pits involved in receptor-mediated endocytosis. This connection forces the internalization of surface molecules followed by lysosomal degradation. The expression of CD4 as well as HLA-A and HLA-B (but not HLA-C or HLA-E) is thus drastically decreased on the surface of an infected cell. The loss of these molecules has multiple negative effects on the host immune response and positive effects on virus production. The reduction in host cell surface CD4 somehow promotes the release of progeny HIV virions (see later). The decrease in surface MHC class I blocks CTL recognition and killing, while the decrease in MHC class II inhibits antigen presentation to Th cells. At the same time, the fact that HLA-C and HLA-E on the infected cell surface are spared prevents killing by NK cells.

Nef has multiple additional functions that promote viral reproduction or, alternatively, enhance host cell destruction: (1) Nef interferes with the host cell proton pump required to acidify endosomes, blocking endosomal destruction of HIV virions. (2) Nef can stimulate various signaling pathways within an infected cell, activating it such that it supports viral DNA synthesis. (3) Nef can upregulate the expression of FasL and activate caspase-3, promoting host cell apoptosis. (4) Nef can downregulate expression of Bcl-2 and Bcl-xL, blocking cell survival signaling.

All these factors make Nef a key culprit behind the generation of high viral loads in AIDS patients and thus a principal virulence factor *in vivo*. Certain AIDS patients that have mild, slowly progressive disease carry an HIV virus in which the *nef* gene has undergone deletion, confirming the key role *nef* plays in viral replication and CD4+ T cell loss. Monkey and mouse models (see later) in which the animals are infected with viruses lacking *nef* also experience less serious disease than their counterparts infected with intact virus.

ii) Vpr

Vpr (*viral protein "r"*) is required for productive HIV infections. The main function of the 15 kDa Vpr protein is to get the viral DNA to the host cell nucleus. Vpr collaborates with the matrix protein and interacts with the nuclear importation proteins *importin-α* and *nucleoporin* to direct the viral DNA to the nucleus. Vpr also alters phosphorylation patterns of enzymes crucial for progress through the G2 phase of the cell cycle, causing dividing cells to arrest at this stage. Vpr can also promote the intrinsic apoptosis of infected host cells by promoting mitochondrial cytochrome *c* release. In HIV-2 and SIV, the *vpr* gene is replaced with the very similar *vpx* (viral protein "x") gene.

iii) Vpu

Vpu (*viral protein "u"*) is a protein of 16 kDa that is also required for productive HIV infections. Vpu binds to newly synthesized CD4 in the host cell cytoplasm and targets it for ubiquitin-mediated proteasomal degradation, thereby decreasing surface CD4 expression. There is also some evidence that Vpu can promote the release of virions from an infected cell by forming ion channels in the host cell membrane. In addition, Vpu interferes with an early step in MHC class I biosynthesis, blocking its expression on the cell surface. Finally, Vpu appears to enhance the susceptibility of infected host cells to Fas-mediated apoptosis.

iv) Vif

Vif (*virion infectivity factor*) is required for the preservation of viral DNA. In HIV strains lacking Vif, either reverse transcription is not completed or the newly made DNA does not survive long enough to generate progeny viruses. Vif stabilizes newly synthesized viral DNA and associates with host cytoskeleton filaments in such a way that transport of the viral DNA to the nucleus is facilitated. Vif is expressed at high levels in the cytoplasm of infected cells and increases the efficiency of HIV infection *in vitro*. Work in 2003 showed that the mechanism by which Vif protects viral DNA is by foiling a host molecule called CEM-15 (also called APOBEC3G) that attempts to block HIV replication (see later).

D. HIV Infection and AIDS

In this section, we offer first a clinical view of HIV infection followed by a description of some of its underlying molecular events.

I. CLINICAL VIEW OF HIV INFECTION

i) Clinical Course of HIV Infection

As stated previously, HIV is most often introduced into the body via intravenous injection or by sexual contact. The rectal and vaginal mucosae appear to be particularly vulnerable to infections with lentiviruses such as HIV. In the earliest, asymptomatic stages of a primary HIV infection, HIV infects DCs, macrophages, and CD4$^+$ T cells in the lamina propria just under the rectal or vaginal epithelium. Infected DCs and T cells carry the virus to additional resting naive CD4$^+$ T cells located in the local lymph node, and viral replication proceeds at an exponential rate such that the population doubles every 6–10 hours. Within 2–6 weeks of exposure, the individual may experience acute fever and an illness similar to infectious mononucleosis characterized by rash, lymphadenopathy, and fatigue (Fig. 25-6). High viral loads (10^5–10^7 virions/ml blood) are present in the circulation and significant levels of the HIV structural protein p24 can be detected in the blood. The trafficking patterns of infected and uninfected cells through the lymphoid recirculation system may be altered. Infected lymphocytes increasingly convey virions to additional lymphoid tissues where macrophages and T cells can be found in abundance. It is at this point that the individual, who may not suspect HIV infection at all, is most contagious. Unfortunately, the levels of anti-HIV antibodies in the infected individual at this stage cannot be detected by standard serological screening procedures. At about 4–8 weeks post-infection, the ratio of CD4$^+$ T cells to CD8$^+$ T cells is reduced drastically from its normal value of approximately 2:1 due to the killing of CD4$^+$ T cells in large numbers and the peaking of the anti-HIV CTL response. The infection is partially contained (but not totally cleared from the body), the clinical symptoms abate, and viremia subsides by about 2 months. However, this defense has come at considerable cost to the immune system, as large numbers of HIV-specific CTL clones are lost to clonal exhaustion.

Circulating anti-HIV neutralizing antibodies finally appear at detectable levels at about 2 months post-infection. At this point, the patient is seropositive for anti-HIV antibodies recognizing the viral Gag and Env proteins as well as some other non-structural proteins. *Seroconversion* is said to have occurred. However, because the virus undergoes tremendous antigenic variation, many of the antibodies that would previously have been neutralizing now fail to recognize the viral epitopes. Worse, the antibodies that can bind to virions may in turn be bound by macrophage Fc receptors that internalize the virions directly into a cell population that serves as a reservoir for the virus. Large numbers of live viruses integrate as proviruses in both macrophages and resting T cells, and the individual enters a period of chronic infection but clinical latency. As soon as these CD4$^+$ T cells and macrophages are activated, the virus re-commences its error-prone replication. However, HIV-specific CTLs and antibodies hold the viremia to a manageable level and other clinical signs are not apparent. As the virus spreads to more and more new T cells in the lymphoid organs, it replicates and releases progeny to the tune of 10^9–10^{11} new virions/day. Viral loads in the blood of clinically latent patients can rise to 10^7 HIV RNA molecules/ml plasma (equivalent to 5×10^6 virions/ml blood) but the patient does not become overtly ill. Complement components and antibodies that have bound to virions are in turn bound to complement receptors and Fc receptors expressed on FDC surfaces. Some virions thus end up "trapped" on the exterior surfaces of the FDC in the germinal centers of the lymph nodes. Hyperplasia of the FDC network ensues, and germinal center macrophages and antigen-specific T cells that make contact with the FDC find themselves infected, thus becoming unwilling contributors to further viral replication. Billions of T cells, as many as 2×10^9/day, are destroyed. However, only an estimated 1% of the killed CD4$^+$ T cells are actually infected with the virus. The bulk of the collapse in the CD4$^+$ T cell population during the chronic infection phase is due to the loss of uninfected cells.

During the latency phase, the drastic assault on CD4$^+$ T cells can be partially compensated for by the tremendous power of the bone marrow to produce new hematopoietic cells. As long as the "T4 count," the number of CD4$^+$ T cells in a patient's blood (normally 800–1100 CD4$^+$ T cells per mm^3), is maintained above 200 CD4$^+$ T cells/mm^3 blood, the infected individual usually shows no clinical signs. Clinical latency may last only a few months or extend to

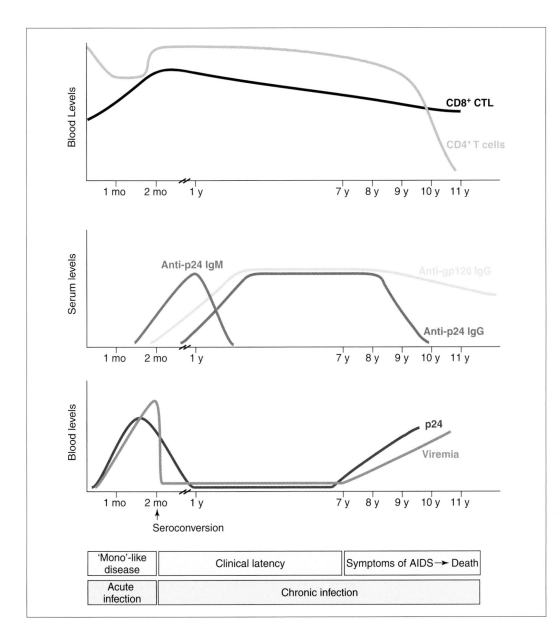

Figure 25-6

Clinical and Immunological Course of HIV Infection
The three graphs show the progression of several clinical and immunological parameters over a generic time course of 1 month to 11 years. Top graph, relative levels of CD8+ and CD4+ T cells. Middle graph, relative levels of antiviral IgM and IgG Ab directed against p24 and gp120. Bottom graph, relative levels of p24 viral protein and virus particles in the blood (viremia). The progression and clinical features of the acute and chronic phases of HIV infection are indicated below the bottom graph.

over 16 years. Interestingly, how rapidly the disease will progress in a given patient can be predicted from the plasma viral load at 6 months post-infection. Those with fewer than 10^5 HIV RNA molecules/ml plasma at this point are more likely to survive for at least 5 years before developing the clinical symptoms of AIDS, while those with more than 10^5 HIV RNA molecules/ml plasma are 10 times as likely to fall ill within 5 years.

Clinical latency ends when the host's immune system has been so disrupted that he or she can no longer fend off opportunistic pathogens. Anti-HIV antibody titers drop, numbers of anti-HIV CTLs decline, viral p24 again appears in the blood, and the plasma viral load increases sharply as up to 10^{10} virions/day are produced. As the disease approaches its late stages, the FDC network within the lymph nodes breaks down and is replaced with fibrotic and/or fatty cellular infiltrations. Lymph node architecture degenerates, and the

virus that was trapped in immune complexes on the surfaces of the FDC earlier in the infection is freed. Increased viremia is observed, and the T4 cell count falls below 200 CD4+ T cells/mm³ blood. With the loss of the FDC, the lymph node microenvironment, and CD4+ T cells, cytokine profiles are altered and the capacity to mount adaptive immune responses is drastically compromised. When a patient has a T4 count of <50 CD4+ T cells/mm³ (i.e., close to 90% of the normal T cell population has been destroyed), death is imminent in the absence of anti-retroviral drug therapy.

ii) Classification of HIV-Infected Persons

Originally, clinical AIDS was subdivided into three broad categories based on number and severity of symptoms, the mildest form of clinical disease being lymphoadenopathy, followed by "AIDS-related complex" (ARC), and finally full-blown AIDS.

These days, both the Centers for Disease Control in the United States and the WHO categorize HIV-infected persons using classification schemes that have both an immunological status component and a clinical presentation component. As shown in Table 25-2, there are three categories of immunological status based on CD4$^+$ T cell count. (In some countries, total T lymphocyte counts are used instead. In addition, it should be noted that a low CD4$^+$ T cell count is not enough to establish a diagnosis of AIDS; the person must also be shown to be HIV$^+$.) There are also three levels of clinical AIDS status based on the opportunistic diseases and cancers diagnosed. Clinical category A patients are HIV-infected but asymptomatic, or may show generalized persistent lymphoadenopathy. Clinical category B patients have opportunistic infections that are slightly less devastating than those in category C. Treatment strategies are then custom-designed for each patient depending on his or her classification. For example, an A1 patient (HIV-infected but asymptomatic) can afford to wait a little longer to start anti-retroviral therapy than a B2 patient. Patients who fall into A3, B3, C1, C2, or C3 should start anti-retroviral therapy as soon as possible along with prophylactic treatment against *P. carinii* infection.

A signature clinical feature of C3 HIV patients is Kaposi's sarcoma, an unusual tumor of blood vessel tissue that was originally associated with aging in some parts of the world. Kaposi's sarcoma is caused not by HIV but by *Kaposi sarcoma herpesvirus* (KSHV; also known as *human herpesvirus-8* [HHV8]). It is thought that the immune deficits caused by HIV allow KSHV to take hold in AIDS patients. Another feature of many AIDS cases is HIV encephalitis. HIV-infected macrophages are somehow able to penetrate the blood–brain barrier and infiltrate into the cerebrum. Characteristic histological features of AIDS encephalitis are gliosis and the presence of multinucleated giant cells. Clinically, patients can present with "AIDS dementia" and/or paralysis. The precise mechanism by which HIV affects the CNS has yet to be established.

II. MOLECULAR EVENTS UNDERLYING HIV INFECTION

i) Viral Tropism

While the predominant target of HIV is the CD4$^+$ T cell population, the virus can infect a wide range of other cell types that also express CD4, including macrophages, Langerhans cells, dermal DCs, and microglia of the CNS. However, *in vitro*, HIV can also infect cells such as mammary cells, NK cells, and brain endothelial cells that do not express CD4. The latter observations were among the first clues that viral entry could involve receptors other than CD4. It turns out that the entry of HIV into cells is most often mediated by the co-binding of the virus to a chemokine receptor as well as to CD4. The major chemokine receptors that help CD4 to facilitate HIV entry are CCR5 and CXCR4. CCR5 is expressed on both CD4$^+$ T cells and macrophages, while CXCR4 is expressed on CD4$^+$ T cells but not on macrophages.

In vivo, an HIV clone that has penetrated the mucosal barrier rapidly generates divergent strains within a single individual due to the error-prone DNA polymerase activity of its RT enzyme. Early in the infection, most of these HIV substrains are able to bind to both CCR5 and CXCR4 and so infect macrophages as well as CD4$^+$ T cells (Fig. 25-7). (Historically, these substrains were known as "macrophage- or *M-tropic* strains" to emphasize this ability. Today, these substrains are known as *R5 viruses* because they can bind to CCR5.) Unlike T cells, macrophages are quite resistant to the cytopathic effects of HIV and are not killed in great numbers during the course of the disease. For comparison, macrophages infected with HIV are thought to have a half-life of up to 4 weeks, while HIV-infected naive T cells have a half-life of only 1–6 days. Macrophages infected with an R5 substrain thus represent a significant reservoir where HIV can reproduce out of reach of the immune system. With time and the accumulation of mutations in the viral genome (particularly those affecting the V3 hypervariable region of gp120), the R5 substrain often loses its ability to bind to CCR5 but retains the capacity to bind to CXCR4. As a result, the new substrain, which is known as an *X4 virus*, can no longer infect macrophages and preferentially attacks T cells. (Historically, these X4 substrains were known as "T cell- or *T-tropic* viruses.") X4 viruses usually replicate more rapidly than R5 viruses and also gain the ability to infect immortalized T cell lines in culture, a capacity generally lacking in R5 HIV. The transition from R5 to X4 HIV in a patient is typically associated with an accelerated loss of CD4$^+$ T cells that occurs about 18 months before the development of full-blown AIDS. (Note, however, that this transition does not occur in all patients and AIDS can develop in patients infected

Table 25-2 Classification of HIV-Infected Persons

Immunological Status (T4 Count)	Clinical Status		
	Category A: Asymptomatic or Mild Disease	Category B: Moderate Disease	Category C: Severe Disease
Category 1: ≥500 CD4$^+$ T cells/mm^3 blood	A1	B1	C1
Category 2: 200–499 CD4$^+$ T cells/mm^3 blood	A2	B2	C2
Category 3: <200 CD4$^+$ T cells/mm^3 blood	A3	B3	C3

Examples of category A diseases: persistent generalized lymphoadenopathy, acute HIV infection with mononucleosis-like disease, fever, fatigue.
Examples of category B diseases: thrush, vulvovaginal candidiasis, cervical dysplasia, shingles, listeriosis, pelvic inflammatory disease, peripheral neuropathy.
Examples of category C diseases: pulmonary TB, invasive cervical cancer, recurrent pneumonia, candidiasis of bronchi, trachea, or lungs, cytomegalovirus infection, Kaposi's sarcoma, wasting syndrome, *Pneumocystis carinii* pneumonia, Burkitt's lymphoma, chronic herpes simplex ulcers, recurrent *Salmonella* septicemia, brain toxoplasmosis, brain lymphoma.

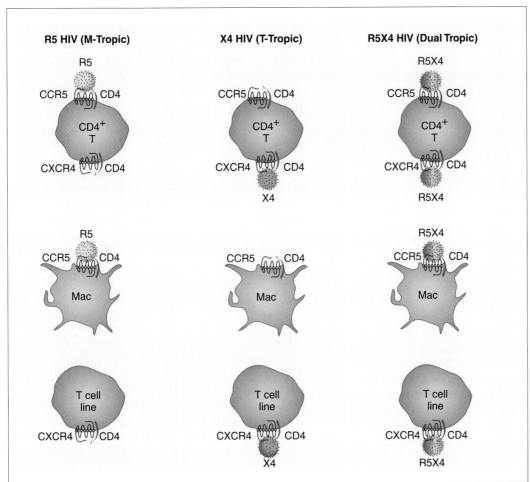

Figure 25-7

Progression of HIV Tropism
Early in the infection, R5 (M-tropic) substrains of HIV dominate. These viruses can bind to both CCR5 and CXCR4 and can infect primary T cells and macrophages. Despite their expression of CXCR4, T cell lines in culture are not infected by R5 HIV isolates. As the infection progresses *in vivo*, the R5 substrains lose their ability to bind to CCR5 but retain the ability to bind to CXCR4. These viruses (T-tropic) thus infect primary T cells but not macrophages. X4 viruses also gain the capacity to infect T cell lines in culture. *In vitro* (and in some patients), HIV may pass through a "dual tropic" stage, in which it can bind to both CCRs and infect all three cell types.

with R5 HIV.) Both *in vitro* and in a minority of patients, the evolution of R5 to X4 HIV passes through a middle stage called *dual tropism*. Dual-tropic strains reproduce quickly and can infect macrophages, primary T cells, and immortalized T cell lines in culture. Characteristics of R5, X4, and dual tropic viruses are summarized in Table 25-3.

We mention here a phenomenon called *syncytia formation* that is associated primarily with *in vitro* studies of HIV. Some HIV substrains (which can be either R5 or X4) can induce the formation of giant, multinucleated structures in which an HIV-infected cell "glues together" a large number of uninfected cells (up to 500 *in vitro*). These syncytia produce huge amounts of virus for a short time (2 days) before dying off. *In vivo*, documented instances of syncytia formation have been few except in cases of overwhelming SIV infection of monkeys. In human AIDS

Table 25-3 Characteristics of HIV Tropism

	R5 HIV (M-Tropic)	X4 HIV (T-Tropic)	R5X4 (Dual Tropic)
Can infect	Macrophages Primary T cells	Primary T cells Immortalized T cell lines in culture	Macrophages Primary T cells Immortalized T cell lines in culture
Cannot infect	Immortalized T cell lines in culture	Macrophages	
Binds to CD4	Yes	Yes	Yes
Preferred chemokine receptor	CCR5 CXCR4	CXCR4	CCR5 CXCR4
Replication rate	Slow	Fast	Fast
Induces T cell syncytia *in vitro*	Few substrains	Many sub-strains	Many sub-strains

patients, only very rare occurrences of syncytia in lymph nodes and brain have been reported.

ii) Viral Entry

When HIV gp120 binds to CD4, the conformation of gp120 is altered such that the V3 hypervariable region capable of binding to chemokine receptors is exposed. The interaction of the CD4–gp120 complex with a chemokine receptor further adjusts the conformation of the envelope spike such that the fusogenic domain in the N-terminus of the gp41 molecule is brought into contact with the host cell membrane. The fusogenic domain of gp41 is glycine-rich and hydrophobic, properties that promote its insertion into the host cell membrane and ultimately the fusion of the two membranes. The virus is then efficiently transferred into the cell's interior. *In vitro* experiments with R5 HIV have confirmed that gp120, CD4, and CCR5 can participate in a trimolecular complex, while gp120, CD4, and CXCR4 can do the same for X4 HIV. The formation of these trimolecular complexes *in vitro* can be blocked by adding the chemokines RANTES or MIP-1α (natural ligands for CCR5) in the first case, or SDF (natural ligand for CXCR4) in the second. Residues in all four extracellular domains of CCR5 are required for interaction with gp120 and CD4, while residues in the first and second extracellular loops of CXCR4 perform the same function for this receptor. While the signaling functions of CCR5 and CXCR4 do not appear to be required for entry of HIV into host cells, signaling triggered by engagement of these receptors may contribute to the destruction of host cells, particularly infected neurons. *In vitro* studies of neuronal cell lines have shown that engagement of CXCR4 by the Env polyprotein induces apoptosis of these neurons. The relevance of this observation to other cell types is under investigation.

At least *in vitro*, HIV strains that bind to CD4 are able to make use of a wider range of chemokine receptors, including CCR2B and CCR3, than strains that cannot bind to CD4. How these other chemokine receptors contribute to HIV infection *in vivo* remains unclear. *In vitro* studies of HIV infection have also shown that, in the absence of CD4, the virus can exploit mannose receptors on astrocytes, proteoglcycans on endothelial cells, and galactosyl ceramide on neurons. However, *in vivo*, there is considerable selection against HIV substrains that use these methods because the regions of the envelope spike binding to these non-chemokine receptors are comparatively exposed. Neutralizing antibodies directed against these epitopes can thus offer effective protection against infection. In contrast, the V3 region of gp120 is hidden from antibodies until this viral protein binds to CD4 plus CCR5 or CXCR4, by which point it is too late for the antibodies to block viral entry. HIV substrains using the CD4-chemokine receptor route of host cell entry thus soon dominate the infection.

As well as by host receptor-mediated infection, HIV can be transferred into T cells by another mechanism called *carriage*. During carriage, infected APCs that encounter CD4+ T cells in the lymph node can form a bicellular conjugate called an *infectious synapse* that allows rapid transfer (by an unknown mechanism) of the virus from the infected APCs to the T cells. The infected T cells recirculate to other secondary lymphoid tissues, allowing the virus to establish multiple reservoirs from which to disseminate to other body systems and vital organs.

iii) Activation of Viral Replication

After fusion of the HIV envelope with the host membrane, the viral core is everted into the host cell cytoplasm and the capsid protecting the viral genome is removed by host cell proteases. The replication cycle commences within a *preintegration complex* consisting of the remaining elements of the viral core. Within the preintegration complex, HIV RT associated with the viral RNA rapidly synthesizes the viral DNA. The matrix proteins present in the preintegration complex possess nuclear localization signals that facilitate the transport of the preintegration complex containing the newly synthesized viral DNA, integrase, and other viral proteins through a nuclear pore into the nucleus. As noted previously, the HIV Vpr protein is especially important for this process. Once in the nucleus, the integrase inserts a single copy of the viral DNA into the host cell genome to form the provirus. In the mucosal DCs that are often the first targets of HIV attack, the production of progeny virions cannot proceed because DCs lack expression of the vital transcription factor Sp1. Similarly, in infected resting T cells, there are only maintenance levels of certain host transcription factors important for viral replication, particularly the Rel protein subunits of NF-κB. Infected T cells in this "preactivation" stage do not transcribe the proviral genome to any meaningful extent and very few progeny virus particles are formed. However, if the T cell becomes activated in response to either TCR engagement or stimulation by TNF, IL-6, or GM-CSF or a mitogen, host cell transcription involving NF-κB activation is induced (Fig. 25-8). Levels of cyclin-T, the host regulatory protein that binds to Tat and sustains viral gene transcription, are upregulated. Because the proviral DNA contains NF-κB binding sites in its LTRs, the viral regulatory genes start to be expressed, followed shortly thereafter by the structural, enzymatic, and accessory genes. In this state of HIV activation, new progeny virions are assembled and released to commence new infections, and infected T cells are lysed.

iv) Antigenic Variation

HIV infections are characterized by extreme antigenic variation. Although a single HIV clone is passed on to initiate a new infection, distinct HIV isolates can soon be found within an individual patient. With time, a range in RNA sequence of as much as 2–3% can be found among HIV viral clones isolated from a single individual. Thus, anatomical structures that are situated physically quite close to each other may harbor HIV clones that are surprisingly different in sequence. The rate of mutation of the virus can vary substantially among infected persons, although HIV isolates obtained from people with a link to the same viral strain (e.g., transmission from mother to child, between sex partners, among addicts sharing needles) are more likely to be

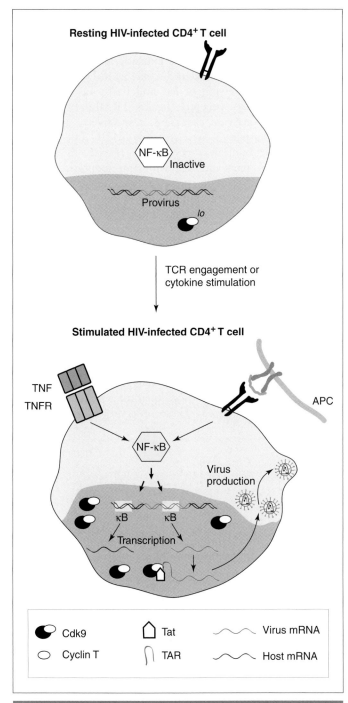

Figure 25-8

Activation of HIV Replication
In the pre-activation stage, CD4$^+$ T cells infected with HIV are in a resting state and no virus is produced. Only low levels of NF-κB and other transcription factors are present and NF-κB is held inactive in the cytoplasm by binding to IκB. However, if the T cell is stimulated by either cytokine binding or interaction with an APC, the cell enters the proviral activation stage. In response to cytokine or TCR signaling, NF-κB is activated and enters the nucleus, where it binds to κB motifs in both cellular and proviral DNA. Host and viral genes are transcribed and cyclin T, the regulatory subunit of Cdk9, is upregulated. Transcription of viral RNA is sustained by the binding of cyclin T to Tat associated with the TAR sequences of viral RNAs. Progeny virus production can then occur in the cytoplasm.

similar than those obtained from completely independent sources.

Originally, three broad groups of antigenically distinct types of HIV-1 were defined based on nucleotide differences in the *env* and *gag* genes: the "majority" group (group M), the "outlier" group (group O), and the "non-M, non-O" group (group N). Within group M, 10 subtypes called *clades* A–J were defined based on sequence diversity. Members of the same clade differed in sequence by less than 10%, whereas different clades differed in sequence by 15% or more. Group O HIV-1 was established when the Env protein sequence of this particular strain of HIV-1 was found to differ from that of group M by almost 47%. Until recently, different group M clades tended to have a particular geographical occurrence. For example, the most common HIV-1 clade in North America and Western Europe was the B clade, while clades A and D were found primarily in infected individuals in central Africa, and clade E predominated in Southeast Asia. However, by 2001, recombination between the clades (achieved by dual infection of an individual) had started to become more prevalent, blurring the distinctions between the clades and their geographic distributions. Viral isolates with characteristics of clades A and J have been identified in Cameroon, and a clade C/D hybrid has turned up in Tanzania. Group O HIV-1 was originally found almost exclusively in Cameroon and Gabon but it too is now spreading. At the molecular level, the clades and recombinant subtypes differ in the nucleotide sequences of several HIV-1 proteins, but the most variable is gp120, particularly its V3 hypervariable region. At the clinical level, some minor differences in transmissibility and disease progression have been noted among HIV-1 clades, but all cause premature death in the absence of anti-retroviral drug treatment. With respect to HIV-2, antigenic variation among substrains is also substantial, but the differences are spread more widely across the proteome of the virus and are not concentrated in the Env protein.

Three factors contribute to the generation of HIV antigenic variability. First, as mentioned previously, the error-prone RT enzyme introduces mutations when it copies the RNA genome. Secondly, the very high production rate of the virus provides plenty of replication cycles and thus chances for mutations to be introduced. Thirdly, the ability of viruses of different clades to recombine if they co-infect the same individual suddenly generates a new viral genome with sequence characteristics of both parental viruses. These new HIV strains that evade the immune response by altering their antigenicity are called *genetic escape mutants*. The modified antigenic epitopes in these mutated substrains allow them to avoid neutralizing antibodies and specific CTLs. Anti-retroviral drugs that act on particular residues in the active sites of HIV-1 enzymes may not be able to recognize their targets in the variant viruses, although the enzymes themselves may remain functional. In addition, some mutated HIV virions have been observed to interfere (by an unknown mechanism) with the recognition of the original HIV virions by specific CTLs. In other words, both the mutant and the original HIV virions escape CTL-mediated destruction and persist in the infected individual.

E. The Immune Response during HIV Infection

HIV has been monumentally successful as a pathogen for two reasons. First, HIV acts slowly on its hosts, a characteristic that has allowed the virus time to spread widely throughout the world's population. Secondly, HIV employs sophisticated mechanisms that permit it to both evade the immune response and destroy immune system cells. Several of HIV's evasion mechanisms were described in Chapter 22. Here we expand on responses by elements of the immune system to HIV infection and the effects of HIV on these elements. How HIV devastates the immune system is summarized in Table 25-4.

I. Th RESPONSES

At least in the initial stages of an HIV infection, when infected macrophages, DCs, and T cells are still targets visible to the immune system, strong responses by HIV-specific Th cells are correlated with reduced viral loads. Cytokine help supplied by HIV-specific Th cells is crucial for supporting the differentiation of the HIV-specific CTLs that reduce viremia in the first few weeks after infection. However, in most HIV-infected individuals, the virus soon evades the CTL response

and proceeds to infect the Th population in droves. The virus further adapts to replicate efficiently in activated Th cells, and to decimate this population. As mentioned previously, several HIV proteins promote both the extrinsic and intrinsic pathways of apoptosis in infected T cells: gp120, Nef, and Tat upregulate the expression of Fas and FasL; Tat upregulates expression of death receptor ligand TRAIL; Tat and PR upregulate and activate caspase-8, respectively; gp120 and Nef increase caspase-3 activity; and gp120, Tat, and Vpr promote mitochondrial cytochrome *c* release. In addition, various HIV proteins interfere with anti-apoptotic proteins: gp120, Tat, and Nef inhibit the expression of Bcl-2; Nef also blocks Bcl-xL expression; and PR cleaves Bcl-2. Those few Th cells that survive these direct cytotoxic assaults are often unable to proliferate *in vitro* in response to specific antigen, implying that some kind of anergization has occurred. *In vivo*, only weak anti-HIV Th responses can eventually be mounted (if any), compromising the CTL response. Persistent viremia is then observed. In most HIV-infected persons, this viremia can only be reduced to tolerable levels by the use of anti-retroviral drugs. However, certain HIV-infected individuals called *long-term non-progressors* (LTNPs; see later) appear to be able to control their viremia without any help from anti-retroviral drugs. Many of these LTNPs retain large populations of HIV-specific Th cells capable of vigorous Th1 responses in which copious quantities of IFNγ and CC chemokines are produced. These observations suggest that,

Table 25-4 Effects of HIV on Host Immune Responses

Immune Response Element	Effect of HIV infection
Primary Th response	Destruction of infected and uninfected CD4+ T cells Interference with cell survival pathways Anergization of CD4+ T cells
Memory Th response	Anergization of memory T cells Destruction of lymph node follicles Impaired antigen presentation by infected FDC
CTL response	Compromised naive CD8+ T cell response due to lack of CD4+ T cell help Compromised recognition of target cells because Nef and Tat induce downregulation of MHC class I Destruction of CD8+ T cell clones due to induction of CD4 expression Induction of Fas expression on CD8+ T cells and FasL on CD4+ T cells leading to death of CD8+ T cells Development of memory CTL response blocked by genetic escape mutants
Antibody response	Compromised naive B cell response due to lack of CD4+ T cell help gp120 acts as a superantigen causing polyclonal activation of B cells and clonal exhaustion Antibodies produced may not neutralize sub-strains of HIV arising by antigenic variation
Cytokines	Abnormal cytokine production, especially increased TNF Cytokines favoring immune deviation to Th2 accelerate progression to AIDS; ↑ IL-4 and IL-10, ↓ IFNγ in blood Compromised cytokine and chemokine production due to CD4+ T cell destruction
CEM-15 (APOBEC3G)	Vif blocks packaging of CEM-15, thereby preventing lethal hypermutation of viral DNA
NK cells	Free Tat interferes with natural cytotoxicity killing
Complement	Capture of host RCA proteins in viral envelope while budding Recruitment of Factor H to virion surface Downregulation of expression of CR1, CR2

under the right circumstances, Th responses can play a direct role in containing HIV infection. The mechanism underlying the resistance of LTNPs is under investigation.

Memory Th responses are greatly impaired in AIDS patients, judging by defects in skin tests measuring DTH reactions. It is unclear whether this impairment is due to anergy of memory CD4$^+$ T cells, destruction of lymph node follicles by HIV, or lack of antigen presentation by FDCs. *In vitro* studies have indicated that infection of APCs by HIV compromises the ability of these cells to present antigen to Th cells, and that these Th cells fail to proliferate when re-challenged with an antigen in a memory response situation. *In vivo*, an antiviral Th memory response may still be detected in the form of weak cytokine production, although the memory Th cells cannot proliferate.

What about the billions of uninfected CD4$^+$ T cells that are lost during HIV infection? Several poorly understood mechanisms have been proposed to account for this killing. First, under the influence of gp120, Nef, Vpu, and/or Tat, HIV-infected T cells upregulate their expression of FasL, allowing them to kill neighboring uninfected CD4$^+$ T cells that express Fas. Nef expression appears to protect the infected cell itself from undergoing Fas-mediated death. Secondly, uninfected bystander CD4$^+$ T cells sometimes bind soluble gp120 shed into the extracellular fluid by HIV virions. T cells that display gp120 in this way may become targets for anti-gp120 antibodies or gp120-specific CTLs, or may be triggered to undergo apoptosis after TCR ligation. Thirdly, soluble gp120 may interfere with T cell maturation by binding to CD4 on developing thymocytes and inhibiting positive selection. Lastly, an HIV gene product may be acting as a superantigen, inducing the deletion of all CD4$^+$ T cells bearing TCRs containing the Vβ region that binds to the HIV protein.

II. CTL RESPONSES

Several lines of evidence indicate that HIV-1 replication and viremia are held at bay in the initial stages of an HIV infection by a vigorous CTL response. In animal models, chronically SIV-infected macaques experimentally depleted of CD8$^+$ T cells experience an increase in viral load. In humans, some individuals who remain uninfected with HIV despite repeated exposure have been found to retain strong HIV-specific CTL responses. *In vivo* studies have shown that the CTLs generated in the initial stages of an HIV infection recognize epitopes derived from both structural and accessory HIV proteins. For example, HIV gp160 contains several epitopes recognized by CTLs in the context of various HLA molecules. One reasonably well-studied epitope recognized in association with HLA-A2 occurs in the V3 loop of gp120. Other CTL epitopes are derived from the highly conserved HIV Pol, Nef, and Gag proteins. The p24 capsid protein also appears to be highly immunogenic. CTL clones recognizing HIV epitopes have been found in the lymph nodes, skin, semen, and peripheral blood of infected individuals.

How do CTLs attempt to defend the body against HIV? *In vitro* (and presumably *in vivo*), HIV-specific CTLs can destroy HIV-infected cells using perforin/granzyme-mediated cytolysis. However, the cytotoxic cytokines usually secreted by CTLs do not appear to play a significant role in the killing of HIV-infected cells. Indeed, *in vitro*, TNF can actually stimulate the replication of the virus in an infected cell. A novel mechanism of CTL-mediated host cell defense may lie in the actions of CC chemokines secreted by these cells. CC chemokines can interfere with viral binding to the chemokine receptors used by HIV to facilitate host cell entry, and at least *in vitro*, CC chemokines inhibit HIV replication in cultured human cells. Similar observations have been made in animal cell studies, in that molecules such as MIP-1α, MIP-1β, and RANTES inhibit SIV replication *in vitro*. However, whether chemokine-mediated interference contributes to CTL-mediated antiviral defense in infected humans is not known.

The CTL response eventually fails in most HIV-infected individuals primarily because of the catastrophic loss of Th cells. A general decrease in CTL production is observed, and cytolytic activity is reduced. In addition, the continuously mutating virus generates escape mutants that present new epitopes, requiring the activation of more and more naive CD8$^+$ T cells. In the absence of sufficient Th help, these cells or their newly generated effectors may not be adequately activated. Indeed, in a study of AIDS patients with rapidly progressing disease, CTL clones that could readily be expanded *in vitro* could not do so *in vivo*, a finding correlated with a lack of HIV-specific Th cells in these patients.

Other mechanisms contribute to inadequate CTL responses during HIV infection. First, HIV infection of an individual can induce the expression of CD4 on the surfaces of naive CD8$^+$ T cells, marking them for infection and destruction by HIV. Secondly, the presence of the virus upregulates Fas expression on CTLs, making them vulnerable to apoptosis induced by FasL-expressing CD4$^+$ T cells. Viral accessory and regulatory proteins also conspire to dampen the anti-HIV CTL response. Nef downregulates the expression of surface MHC class I and thus inhibits antigen recognition by CD8$^+$ T cells. The Tat protein may also contribute to a lack of cell surface MHC class I by suppressing transcription from MHC class I promoters.

III. ANTIBODY RESPONSES

As we saw in Figure 25-6, humoral responses against HIV do occur. However, they are very weak compared to antibody responses to other viral infections and are not very effective. Several weeks after viral replication is activated and the viral p24 antigen is produced, the IgM antibody response finally kicks in (seroconversion), followed a few weeks later by the IgG response. Judging by serum antibody titer, the most immunogenic molecules of HIV are the proteins making up its envelope. (These proteins can also appear in the membranes of infected cells.) In the case of X4 HIV-1, the gp120 protein supplies both pan-specific and strain-specific epitopes. Epitopes shared by different strains of HIV-1 tend to be derived from a conserved region of gp120 containing the CD4 binding site. The strain-specific epitopes are supplied primarily by the

V2 and V3 loops of gp120. The highly variable V3 loop is the immunodominant epitope of X4 HIV-1, meaning that antibodies to this epitope are made by most individuals either infected with X4 HIV-1 or immunized with gp120/gp160. *In vitro* at least, anti-V3 neutralizing antibodies can block fusion of either the virus to the host cell membrane, or an uninfected cell to an infected cell. The situation is slightly different for R5 HIV-1 in that V3 is not a key epitope for neutralizing antibodies and the known dominant epitopes tend to be more non-linear and conformational. A similar situation holds for HIV-2 epitopes.

Although antibodies to various internal and external HIV epitopes can be detected in infected individuals, and many show neutralizing activity against a patient's own substrain of HIV *in vitro*, these Igs do not seem to be able to eliminate the virus *in vivo*. It is suspected that key antibody epitopes are sterically obscured in the envelopes of primary isolates of HIV. (In this context, a "primary" HIV strain is one that has been isolated directly from a patient, as opposed to a strain cultured for a period in the laboratory.) In addition, even when the epitopes are accessible, the extreme antigenic variability of the virus is likely to confound recognition by the first sets of antibodies produced. Indeed, as HIV infection progresses, the viral substrains that emerge show evidence of having been selected for complete resistance to neutralization.

How else is humoral immunity compromised during HIV infection? Although HIV does not replicate in B cells, the virus has a severe impact on their function. The gp120 protein of HIV binds to a site in the V region of the mIg heavy chain and acts as a superantigen, non-specifically activating multiple clones of B cells. Indeed, only about 20% of the antibodies produced in response to HIV infection are actually directed against HIV. Clinically, HIV-infected persons exhibit B cell hyperplasia and polyclonal activation, circulating immune complexes, defects in immune responses to both Td and Ti antigens, and autoimmunity. Eventually, B cell numbers decline drastically, presumably due to clonal exhaustion. In addition, the decimation of the CD4+ Th cell population impairs the activation of any remaining clones of naive anti-HIV B cells. In patients in the latest stages of AIDS, it may not be possible to detect any anti-HIV antibodies in peripheral blood.

While detectable levels of antibodies to most other pathogen antigens appear within the first 2 weeks after infection, the lag between HIV infection and the appearance in the blood of measurable levels of anti-HIV IgM antibodies is much longer (about 8 weeks). Even after seroconversion, titers of anti-HIV antibodies remain very low in comparison to antibodies raised during immune responses to other pathogens. A similar lag in the appearance of anti-SIV antibodies has been observed in macaques infected with virulent forms of SIV. The delay in the human antibody response means that the standard ELISA screening test for specific antibody may give a false negative result in the early stages of the disease. For this reason, a person should not consider himself or herself free of HIV until he or she has tested negative in two serological assays 3 months apart. When it is necessary to determine immediately whether an individual has contracted the virus, PCR determinations for the detection of HIV proviral DNA or ELISA assays for the presence of HIV proteins can be performed. The current generation of rapid screening tests uses ELISA determinations of both viral antigens and antiviral antibodies to achieve early detection.

IV. CYTOKINES

HIV infection results in an abnormal profile of cytokine secretion that contributes to many AIDS symptoms. In the early stages of the disease, HIV-infected persons have elevated blood levels of TNF, IL-1, IL-2, IL-6, and IFNα. Some researchers have gone so far as to call AIDS a "TNF disease," because this cytokine is such a powerful inducer of NF-κB and thus an activator of integrated HIV genomes. In addition, TNF is known to induce the *cachexia* (wasting) associated with AIDS (Plate 25-3) and to promote deterioration of the CNS. IL-1 has been associated with AIDS-associated persistent fever and dementia. *In vitro*, IL-1, TNF, and IL-6 all induce the proliferation of Kaposi's sarcoma cells and promote HIV replication in peripheral blood cells.

The balance of Th1/Th2 cytokines may also be relevant to the progress of HIV infection. *In vitro*, the R5 HIV isolates prevalent early during infection can replicate in either Th1 or Th2 cells, but the X4 HIV isolates that dominate later in the infection prefer Th2 or Th0 cells. Thus, over time, a shift in a patient's cytokine profile to favor the differentiation of Th2 cells spurs progression to clinical AIDS. In patients with advanced AIDS, increased IL-4 and/or IL-10

Plate 25-3

Wasting Effects of AIDs on a Patient
Reproduced with permission from Emond R. T. D., Welsby P. D. and Rowland H. A. K. (2003) "Colour Atlas of Infectious Diseases," 4th edn. Elsevier Science, London.

in the blood may replace declining levels of IFNγ and IL-2. What few Th0 cells might be left are influenced to differentiate into the Th2 phenotype, expanding the target population for X4 infection and promoting the less effective antibody response at the expense of the cell-mediated response. The reduction in IFNγ also impedes the DTH response, leaving the patient ever more vulnerable to intracellular pathogens such as *P. carinii*.

V. CEM-15 (APOBEC3G)

Host cells under attack from HIV produce an RNA/DNA editing enzyme called CEM-15 or APOBEC3G (*apolipoprotein B mRNA editing enzyme, catalytic polypeptide-like 3G*). CEM-15 is structurally related to AID, the RNA/DNA editing enzyme associated with somatic hypermutation in B cells. CEM-15 functions by inducing lethal mutations into HIV DNA, an action counter-attacked by the HIV Vif protein. To illustrate, let us suppose that a host cell has been infected with a strain of HIV that cannot express the Vif protein. As this virus replicates and starts to make the proteins required to assemble progeny HIV virions, CEM-15 binds to newly synthesized HIV Gag-Pol polyprotein molecules and is packaged along with them into the progeny virions. When one of these Vif-deficient progeny virions infects a new host cell, the virus attempts to synthesize viral DNA by reverse transcription of its RNA genome. However, the CEM-15 protein that was packaged into the virion now deaminates cytidine residues in the negative strand of the newly synthesized single-stranded DNA intermediate, converting them to uracils. As a result, the plus strand of the viral DNA is filled with G to A mutations. (CEM-15 does not appear to hypermutate viral genomic RNA.) It is thought that the hypermutated viral DNA is unstable and fails to integrate into the host genome properly. Even if the viral DNA is successfully integrated, the viral proteins produced from it are riddled with deleterious mutations. As a result, little or no infectious virus is produced. Wild-type HIV confounds this ingenious method of innate defense by synthesizing Vif. Vif prevents the packaging of CEM-15 into progeny virions by blocking the association of CEM-15 with Gag-Pol. Vif also recruits a ubiquitination enzyme complex that ubiquitinates CEM-15 and thus promotes its degradation by the ubiquitin-proteasome pathway.

Other host proteins with activities similar to those of CEM-15 have been identified, including APOBEC3B and APOBEC3F. All these enzymes appear to be coordinately expressed and may synergize in inducing hypermutation that inhibits HIV replication. Interestingly, APOBEC3F is partially resistant to the effects of HIV Vif, perhaps evidence of evolution in action.

VI. NK CELLS

HIV also has detrimental effects on NK cells. Extracellular Tat protein released by HIV-infected cells can interfere with natural cytotoxicity exerted by an NK cell once it has bound to a target cell. The extracellular Tat binds to the calcium channels of NK cells and inhibits the intracellular signaling necessary to trigger degranulation. Thus, even though HIV downregulates the expression of many MHC class I molecules on infected T cells (which should act as a call to arms for NK cells), the warriors are disabled and the infected T cells escape death.

VII. COMPLEMENT

HIV blocks complement-mediated defense in several ways: (1) As new progeny viruses bud through the host cell membrane, they incorporate host RCA proteins into their envelopes and can thus forestall the deposition of the MAC. (2) The HIV envelope has multiple binding sites that recruit Factor H to the virion surface. As the reader will recall from Chapter 19, Factor H is a soluble host regulatory protein that inhibits alternative complement activation by promoting the cleavage of C3b to iC3b. (3) HIV infection downregulates expression of both CR1 and CR2 on host cells by unknown mechanisms, decreasing the likelihood of complement component binding to infected cells. (4) *In vitro*, C5aR expression is decreased upon exposure to gp120, inhibiting the receipt of pro-inflammatory signals.

F. Host Factors Influencing the Course of HIV Infection

The previous section described the general course of immune responses to HIV in most individuals. In this section, we discuss host factors that can influence whether an exposed individual will actually be infected by HIV or not, and whether an infected person progresses slowly or rapidly to AIDS. These factors are summarized in Table 25-5.

Table 25-5 **Host Factors Affecting Speed of Progression to AIDS**	
Phenotype	**Associated Host Factor**
HIV resistance	Deletion of 32 bp in CCR5 gene Expression of MHC class I HLA-A2/6802 supertype Expression of MHC class II alleles DRB1*0101 or *0102
Long-term non-progressor	Expression of HLA-A2, B27, B51, or B57 Complete heterozygosity for all MHC class I loci R5 HIV prevalence Younger age when infected
Rapid progressor	Expression of HLA-B35, Cw4, A*2301 Infection with other pathogens X4 HIV prevalence Increased TNF in the blood

I. RESISTANCE TO HIV INFECTION

Two groups of people appear to be able to resist HIV infection upon direct exposure to the virus. The first group includes individuals carrying mutations in host genes required for entry of the virus into host cells. For example, a mutated allele of CCR5 called CCR5Δ32 has a 32 bp deletion in the receptor nucleotide sequence that renders it non-functional. The vast majority of individuals homozygous for CCR5Δ32 are not susceptible to HIV-1 infection and indeed seem to suffer no health deficits at all in the absence of the receptor. Furthermore, many CCR5/ CCR5Δ32 heterozygotes experience at least a 2 year delay in the onset of AIDS after HIV infection. The CD4$^+$ T cells of these individuals show reduced expression of CCR5 on the cell surface and viral replication *in vivo* is decreased. The CCR5Δ32 allele is present at its highest frequency in northern European populations and at its lowest frequency in Africa, Asia, and Oceania. Other mutations in the promoter of CCR5 have been linked to slowed, but not abolished, disease progression.

The association of other chemokine receptor mutations with resistance to HIV infection is less clear. The evolutionary conservation of the CXCR4 gene is considerably greater than that of the CCR5 gene. Very few polymorphisms of the CXCR4 gene have been identified and none has been unequivocally associated with resistance to HIV-1 infection. A point mutation of the CCR2B gene has been linked to slowed progression of AIDS in some studies but not in others. CCR2B expression is not reduced on cells bearing the mutated CC2RB allele and levels of CCR5 and CXCR4 are also normal, so how the CC2RB mutation might block HIV-1 infection remains unknown. AIDS progression has also been associated with a particular mutation in the SDF-1 gene. However, while some studies show a positive association of this particular SDF-1 mutation and delayed AIDS progression, other studies have shown the opposite. Such results underscore the complex nature of HIV infection and viral interaction with host receptors.

Individuals in the other HIV-resistant group have normal CCR5 (and other chemokine receptor) genes but express certain HLA proteins. Among MHC class I molecules, proteins belonging to a subset of the HLA-A2 supertype called HLA-A2/6802 have been linked to HIV resistance. (The reader will recall from Box 23-3 that an HLA supertype is a collection of HLA proteins that are structurally and functionally related (i.e., present the same or very similar peptides).) The frequency of particular MHC proteins constituting a supertype can differ in different populations, although the overall frequency of the supertype tends to be the same. For example, the overall frequency of the A2/6802 supertype is about 40% in both Africans and Asians, but the HLA-A*0206 allele is expressed in 10.8% of Asians and only 0.6% of Africans, while the HLA-A*0202 allele is expressed in 0.2% of Asians but 8.7% of Africans. (For a short explanation of how HLA proteins and alleles are named, see Box 25-1. For a description of how they are typed in the laboratory, see Ch.27.) In several different studies, individu-

als expressing the HLA-A2/6802 supertype have remained seronegative and uninfected by HIV after documented exposure (Fig. 25-9A). These individuals have therefore been dubbed *persistently seronegative* (PSN) persons. HLA-A2/6802 is often observed in uninfected newborns of HIV$^+$ mothers, uninfected sexual partners of infected people, and in a well-studied group of repeatedly exposed but uninfected prostitutes in Nairobi, Kenya. It is not yet clear why the A2/6802 supertype is protective. One theory holds that HLA-A2/6802 proteins are able to present peptides that are derived from parts of the HIV virion that are shared between substrains. Indeed, sustained cell-mediated responses and even some helper T cell responses directed against viral Pol, Nef, and Env proteins have been observed in some of these individuals. In these cases, the immune system may be able to rapidly mount a protective CTL response that permanently blocks establishment of the virus. Another theory suggests that member molecules of this supertype are more flexible than usual about the peptides they present. Indeed, cross-clade recognition has been documented in some A2/6802 individuals.

Among MHC class II molecules, there is some quite convincing evidence that expression of DRB1*0101 and DRB1*0102 may have a protective effect against HIV infection (Fig. 25-9B). MHC class II molecules containing these β chains are very similar in structure and present the same peptides. Vaccinologists are investigating the possibility that highly conserved HIV epitopes presented by the A2/6802 supertype or MHC molecules containing DRB1*01 chains may be strongly immunogenic and induce vigorous Th and CTL responses. Such responses could serve as the basis for a widely effective anti-HIV vaccine.

II. CLINICAL COURSE VARIABILITY

Once an individual has become infected with HIV, additional host-related factors may have a positive effect on symptom-free survival and clinical latency. As introduced above, HIV-infected individuals who resist clinical disease for more than 10 years are called long-term non-progressors (LTNP) while those that succumb within 2 years of infection are called rapid progressors (RP). What determines who will be a LTNP and who will be an RP is not precisely known, but several elements have been identified as having some influence. First, LTNPs exhibit relatively stable CD4$^+$ T cell counts and ongoing polyclonal anti-HIV CTL responses directed against multiple HIV antigens. In contrast, RPs show more fluctuation in their CD4$^+$ T cell counts and the expansion of only one or two Vβ families of CTL clones.

Some people who are HIV-infected but AIDS-free for extended periods may have their MHC haplotype to thank. HLA molecules belonging to the A2, B27, B51, and B57 families are associated with a more favorable clinical outcome. In addition, individuals fully heterozygous at all HLA class I loci enjoy longer clinical latency. These people have an advantage

Box 25-1. **Nomenclature of HLA alleles**

In Chapter 10, we learned that the MHC locus in humans consists of the HLA-A, B, and C regions containing genes encoding the class I α chains, and the DP, DR, and DQ regions containing genes encoding the class II α and β chains. Within each of the HLA-A, B, and C regions, one gene specifies the corresponding α chain that pairs with β2-microglobulin to form the complete heterodimeric HLA-A, B, or C molecules. There are no HLA-A, B, or C pseudogenes. The situation is more complicated for the class II genes. Within the DP region, the DPA1 gene specifies the DPα chain that combines with the DPβ chain specified by the DPB1 gene to form the HLA-DP molecule. The "1" is included in the designations DPA1 and DPB1 to distinguish them from the unexpressed pseudogenes (DPA2, DPA3, DPB2) in the DP region of the MHC. Similarly, in the DQ region, only one each of the collection of DQA and DQB genes present is expressed (DQA1 and DQB1); the others are pseudogenes (DQA2, DQB2, DQB3). In the DR region, there is a single functional DRA gene but no closely related pseudogenes (so that there is no need for a "1" in this gene's label). However, there are four functional DRB genes (DRB1, 3, 4, and 5) and five non-functional pseudogenes (DRB2, 6, 7, 8, 9).

Another level of complexity in naming arises because each HLA gene has many different alleles that can be expressed in a population. These alleles, which have been identified by their gene sequences, are indicated by an asterisk and a specific multi-digit number, for example, HLA-DRB1*0305. This designation contains the following information about this MHC class II chain: HLA-DRB1 indicates that it is a β chain derived from the DRB1 gene; the "03" signifies that, like other members of the 03 group, its presence in a class II molecule defines a group of serological epitopes collectively known as DR3; and the "05" indicates that this particular allele is the fifth to be discovered of the DR3 group.

Specialists in HLA nomenclature recognize several enhancements of this system that provide even more detail. For example, the designation HLA-DRB1*0305N indicates that the allele is null; that is, the mutation in its nucleotide sequence renders the resulting protein chain non-functional.

The alleles HLA-DRB1*030501 and HLA-DRB1*030502 differ from HLA-DRB1*0305 by two different synonymous mutations; that is, the nucleotide sequence has changed in each case but the amino acid sequence has not. These proteins are functional. Should there in future be a mutation to the HLA-DRB1*030502 allele that is outside the coding region for the β chain, it would be designated as HLA-DRB1*03050201. If this mutation has occurred outside the coding region but has still caused the β chain to become non-functional, it would be indicated as HLA-DRB1*03050201N. As the reader can see, many happy hours can be spent deciphering the meaning of HLA allele names.

Nomenclature	Interpretation
HLA	The allele is encoded by the human MHC
HLA-DRB1	The allele is encoded by the HLA locus DRB1
HLA-DRB1*03	The allele is one of a group that encodes the serologically determined DR3 antigen
HLA-DRB1*0305	The allele is the fifth member of the above group
HLA-DRB1*0305N	A mutation in the 0305 allele has rendered it null
HLA-DRB1*030501	The allele differs from the 0305 allele by a synonymous mutation
HLA-DRB1*03050101	The allele differs from the 0305 allele by a mutation outside the coding region
HLA-DRB1*03050101N	The mutation outside the coding region of allele 030501 has rendered it null

Modified from the original prepared by The HLA Informatics Group, http://www.anthonynolan.org.uk/HIG/, with the kind permission of Dr. Steven Marsh, Anthony Nolan Research Institute.

in antigen presentation over their more homozygous counterparts, and thus may be able to mount broader CTL responses against HIV. Interestingly, HLA-B35 and HLA-Cw4, which have been linked to decreased NK activity against HIV, are associated with increased susceptibility to rapid AIDS onset. Increased susceptibility to HIV infection in a given population has also been associated with expression of HLA-A*2301. Researchers have speculated that MHC molecules containing this chain may either present an antagonist HIV peptide, or be relatively intolerant of mutations to a favored epitope. It is also possible that HLA-A*2301 is genetically linked to another polymorphic gene that increases susceptibility to HIV infection.

HIV-infected persons who can avoid infections with pathogens that tend to upregulate CC chemokine receptor expression (such as *Mycobacterium tuberculosis* and mycoplasmas), or viruses that encode viral receptors serving as conduits for HIV entry (such as human CMV), have a longer latency period. In addition, as mentioned previously, patients in whom the predominant HIV isolates remain R5 in phenotype appear to stay healthier longer than those in which the switch to X4 is made rapidly (although it is not clear whether this switch is the cause or consequence of disease progression). Some researchers believe that since the R5-to-X4 switch is influenced by cytokine production (particularly TNF), individuals who naturally have less TNF in their blood may slow their progression to AIDS. Lastly, the younger a patient is when infected with HIV, the longer he or she may expect to survive, possibly because the immune system is inherently more robust at a younger age.

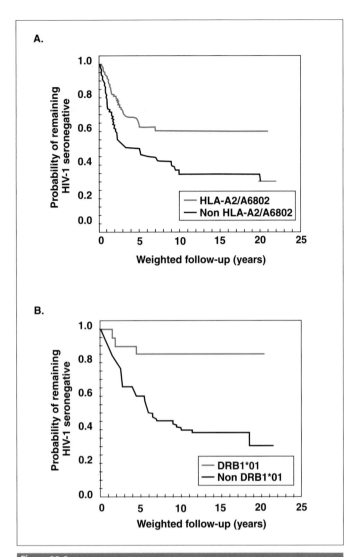

Figure 25-9

Influence of HLA on HIV Seroconversion
Upon exposure to HIV, individuals expressing either (A) MHC class I molecules belonging to the HLA-A2/A6802 supertype or (B) MHC class II molecules containing a member of the DRB1*01 family of β chains are more likely to resist HIV infection and remain seronegative. Reproduced with permission from MacDonald K. S. *et al.* (2000) Influence of HLA supertypes on susceptibility and resistance to human immunodeficiency virus type 1 infection. *Journal of Infectious Diseases* **181**, 1581–1589.

G. Epidemiology and Sociology of HIV Infection

I. TRANSMISSION OF HIV

HIV is not easy to get. Infection requires transfer of the virus into a body fluid by sexual contact, breast-feeding, the mixing of blood during transfusions or trauma (including during birth), or the sharing of contaminated needles. Merely touching, kissing, or being sneezed on by an HIV-infected person is not enough to transmit the virus. The probability of transmission for each unprotected encounter with HIV-1 is 1 in 1000, depending on the dose of the virus in the transferred body fluid. Transmission is rare if the number of copies of viral RNA/ml of blood is below 1500. Minimum RNA levels for other body fluids are unknown, but viral replication has been detected in the rectal mucosae of individuals with undetectable plasma levels of the virus.

It is clear that HIV can be halted in its tracks in most cases if elementary precautions are taken, such as the use of condoms, gloves, screened or cloned blood products, and clean needles and syringes. Since the tragedy resulting from HIV contamination of donated blood supplies (see Box 25-2), almost all countries have adopted rigorous programs for screening blood donors that have greatly reduced the chances of contracting HIV by this means (risk of 1 in 15 million in the United States). Unfortunately, Africa still lags behind in this regard, screening only 75% of its donated blood for HIV. The current AIDS epidemic is therefore not due so much to a highly contagious virus (which HIV is not) as it is to faulty human behavior and cultural intransigence. Particularly risky behaviors are having multiple sexual partners, engaging in anal intercourse, having unprotected sex, having sex with prostitutes, abusing intravenous drugs, and getting a tattoo via a needle. If exposed to HIV, uncircumcised males are at higher risk of infection, as are individuals who already have another sexually transmitted disease. A health worker in the developed world has a risk of about 0.3% of acquiring HIV from a needle stick injury. About 300,000 new cases of HIV-1 infection occur each year globally due to maternal–newborn transmission. Transfer of the virus can occur during the second half of pregnancy, at delivery (where most cases occur), or via breast-feeding. Treatment of the mother with anti-retroviral drugs during pregnancy, delivery, and after the birth can substantially reduce viremia in the mother and thus the risk of transmission to the child. The infant also receives anti-retroviral drugs for the first 6 weeks of life and can be given baby formula to avoid the risk of breast-feeding.

HIV-2 is even harder to transmit than HIV-1. It is estimated that it is three times more difficult to acquire HIV-2 by sexual contact than HIV-1, and 10 times more difficult for a mother to pass HIV-2 on to her infant. Infection with HIV-2 does not appear to prevent a subsequent infection with HIV-1.

II. EPIDEMIOLOGY

The numbers associated with the AIDS epidemic are truly staggering (Fig. 25-10). It has been estimated that over 20 million people around the world have died of the disease, and another 40 million are living with the infection. Almost 5 million new HIV infections were reported globally in 2004, equivalent to over 13,000 new HIV-1 infections each day. Most of these occur in developing countries, more than half affect persons younger than 25 years of age, and more than half occur in women. More than 3 million people a year die of the disease, including over 1 million women and 600,000 children.

Box 25-2. **The tainted blood scandal**

One of the greatest tragedies in the HIV/AIDS story is the infection of thousands of individuals due to HIV contamination of blood products used for transfusions and the treatment of hemophilia. Tragedy turned to scandal when it was revealed that health officials and drug companies could have screened and treated blood products to guard against the possibility of transmitting HIV but failed to do so in a timely manner. It has been estimated that 90% of the 8000–10,000 hemophiliacs living in the United States in the early 1980s were infected with HIV, and that most of them have now died. Similar sobering statistics hold for most other developed countries. The only good coming out of this debacle is that blood supplies in Western countries are now screened rigorously, giving hope that future contamination of the blood supply by a new scourge will be quickly detected and rectified.

The roots of the tainted blood scandal can be traced back to the suffering of hemophiliacs. Hemophilia arises from a defect in the clotting of blood that causes the victim to bleed for extended periods of time. The most common form of the disease, hemophilia A, is due to an X-linked genetic defect in the production of clotting factor VIII. Hemophilia B is due to an X-linked defect in the gene for clotting factor IX and is 20× more rare than hemophilia A. Approximately 17,000 persons in the United States suffer from some form of hemophilia. The physical consequences of this disease are severe. Everyday activities are fraught with the peril of sustaining a minor cut, and contact sports are a high-risk activity. Bruising or internal bleeding from a ruptured blood vessel leads to the pooling of blood in the joints, tissues, internal organs, and muscles. Intense pain results from the pressure of the pooled blood on local nerves, and the affected area swells alarmingly.

Prior to the 1960s, most hemophiliacs died before they became adults, and the few that survived to adulthood were often confined to a wheelchair. The repeated incidents of internal bleeding could only be stopped by transfusions of large volumes of human plasma, which had to be stored frozen until use. In 1968, a great boon was granted to hemophiliacs in the form of a freeze-dried, concentrated version of plasma and its clotting factors called *clotting factor concentrate* (CFC). The ability to merely "add water and stir" to produce clotting factors freed hemophiliacs to travel and live life in a way that before had been impossible. However,

the seeds of tragedy lay in this same life-giving product because it was produced from plasma drawn from tens of thousands of people. Many hemophiliacs required injections of CFC every 1–2 weeks, meaning that their intake of blood products was enormous. The population of volunteer donors was simply not large enough to provide the huge quantities of plasma required, so that American drug companies turned to paying blood donors. However, these donors were not screened, so that blood was accepted from many persons with less than healthy lifestyles.

The drug companies were not entirely ignorant of the need to purify the blood donations they collected. For decades, human albumin supplies had been treated to inactivate hepatitis viruses. In the late 1970s, a German company developed a heat pasteurization process that could inactivate various viruses in CFC, but the concentrate lost much of its potency in the process. As a result, the cost per unit of the treated product was much higher than that of the untreated, a factor that caused some health officials to balk at recommending implementation of heat treatment. At the time, it was thought that even if a virus (most likely a hepatitis virus) did infect the blood supply, it would at worst cause a chronic disease that could be managed. Governments and health officials weighed the known pain and difficulties to hemophiliacs if they stopped using CFC against the perceived likelihood that the patients would contract a manageable virus from the blood supply, and decided in favor of untreated CFC. Hemophiliacs would later charge that the officials should have gone ahead with heat treatment if only to inactivate hepatitis viruses.

In 1983, HIV was discovered to be the cause of AIDS, and the deadly nature of the virus was confirmed as hundreds of gay men and IV drug users began to die of the disease. However, there was as yet no reliable way to test a blood sample for the presence of HIV. Some health professionals in the United States and Europe urged that blood donors be screened for AIDS-associated lifestyle factors such as male-to-male sex and IV drug use. However, most officials demurred at asking blood donors the necessary hard questions about their habits that could have barred them from donating. Non-profit and commercial operators of blood collection centers alike were loath to narrow the sources feeding their already precarious blood supplies,

fearing the expense of testing and the increased cost per unit of CFC. Officials also did not want high risk individuals masquerading as blood donors to obtain free AIDS tests. At the time, even some doctors believed that hemophiliacs might be infected with HIV but would not come down with AIDS. Thus, it was a minority of blood collection centers that believed the HIV threat was real.

Governments, health organizations, and doctors continued to tell hemophiliacs well into 1984 that using CFC was still their best option. By this time, American blood products companies had come up with alternative heat treatments that preserved a greater percentage of clotting factor activity. However, this treatment did not in fact kill the virus, and inadequately treated CFC was sold throughout North America for another 2 years. In 1985, a test for HIV antibody finally became commercially available, and it was shown that the previously cited "one in a million" risk of getting HIV from donated blood was actually closer to 1 in 700 in large urban areas. It has been estimated that about 10,000 transfusion-related HIV transmissions had occurred in the United States by 1985. By the late 1980s, it was clear that an epidemic of HIV infection in hemophiliacs and blood transfusion recipients was under way.

The tainted blood scandal was an international phenomenon. In Canada, more than 2000 hemophiliacs contracted HIV and 20,000 more were infected with HepC. The Canadian Red Cross (now defunct) delayed HIV testing until 8 months after a reliable test was commercially available and did not test for hepatitis C until 4 years after an indirect test for the virus had been developed. In 1997, the Krever Commission concluded that the Canadian Red Cross, all levels of the Canadian government, and the American blood products company that was a major CFC supplier at the time were all to blame. In late 2002, criminal charges were laid against two federal health officials, a senior Canadian Red Cross official, and the vice-president of the American blood products company. In Germany, it was known in 1983 that there were problems with imported American CFC, and it was recommended that local, heat-treated products be used instead. However, German federal officials continued to allow the import of American CFC, and the German Red Cross refused to either dispose of its inventory of these products or screen them

Continued

for HepB. Public health insurers declined to pay for safer, heat-treated CFC, leading to the infection of half of German hemophiliacs. In 1987, substantial compensation was paid to most of them. In France, an infected hemophiliac laid criminal charges against government officials in that country. As a result of these proceedings, the government was ousted from power and the blood supply infrastructure underwent a major restructuring. In Norway, compensation was paid to the 20% of hemophiliacs that were infected with HIV and their spouses. Infection rates were much lower in this country because the CFC used was made from the blood of much smaller

numbers of local donors. In Japan, CFC was not subjected to heat treatment until long after American products were. More than 1800 Japanese hemophiliacs contracted HIV, and the Japanese officials responsible were convicted of criminal negligence. Substantial financial compensation was awarded to each victim.

The residue of the tainted blood scandal is still with us. Individuals who contracted HepC prior to 1985 have a good chance of developing chronic liver disease and/or hepatocarcinoma later in life. Fortunately, HepC transmission via sexual activity is rare, and the virus can be slowed by treatment with a combination of IFNα and a nucleoside

analogue called *ribavirin*. Nevertheless, the care of these individuals is expected to consume prodigious amounts of health care dollars in the future. Sadly, hemophiliacs that were infected with HIV between 1980 and 1985 have grimmer prospects and many are already dead. Compensation lawsuits are still dragging through the courts. On the positive side, many valuable lessons were learned from this tragedy that spurred concrete measures to protect the blood supply (see Ch.27). Recombinant forms of clotting factors VIII and IX were also rapidly developed in the wake of the scandal, providing a much safer means of treatment for today's hemophiliacs.

Figure 25-10

Total Number of Persons Living with HIV/AIDS in 2004
Regions of the world where HIV infection occurs have been ranked in order of severity of impact. With information from the Coordinating Committee of the Global HIV/AIDS Vaccine Enterprise (2005). The Global HIV/AIDS Vaccine Enterprise: Scientific Strategic Plan. *Public Library of Science Medicine* **2**(2), e25.

	Region	Epidemic started	Total number of persons living with HIV/AIDS in 2004	New HIV infections during 2004	Deaths due to AIDS in 2004
1	Sub-Saharan Africa	Late '70s - early '80s	25.4 million	3.1 million	2.3 million
2	South & South East Asia	Late '80s	7.1 million	890,000	490,000
3	Latin America	Late '70s - early '80s	1.7 million	240,000	95,000
4	Eastern Europe & Central Asia	Early '90s	1.4 million	210,000	60,000
5	East Asia & Pacific Rim	Late '80s	1.1 million	290,000	51,000
6	North America	Late '70s - early '80s	1.0 million	44,000	16,000
7	Western and Central Europe	Late '70s - early '80	610,000	21,000	6,500
8	North Africa & the Middle East	Late '80s	540,000	92,000	28,000
9	Caribbean	Late '70s - early '80s	440,000	53,000	36,000
10	Oceania	Late '70s - early '80s	35,000	5,000	700
	Total		40 million	4.9 million	3.1 million

i) Southern Africa

Southern Africa, especially the sub-Saharan region, is the area of the world hit hardest by AIDS. It has been estimated that over 75% of deaths attributable to the AIDS epidemic have occurred in this region. Over 25 million people are infected and heterosexual intercourse is the dominant mode of viral transmission. About 3 million new infections per year occur in these countries, and the risk of a newborn being infected by his or her mother is about 35–50%. Over 1.1 million children under 15 years of age were living with HIV in sub-Saharan Africa in 2000. In some areas, 40% of pregnant women and 80% of female prostitutes are infected with HIV, and rates in the general adult population can reach 20%. In Botswana, 39% of the adult population is estimated to be infected with HIV, the highest rate in the world (in 2002). Swaziland, Lesotho, and Zimbabwe, all at about 31–33%, are not far behind.

Several factors contribute to the high rate of HIV infection in Southern Africa. Large numbers of workers in this part of the globe are migratory and their movement enhances the spread of the virus. Civil and international conflicts, social upheaval, and political instability also contribute both to population migration and to major difficulties in establishing the infrastructure required to fight the AIDS epidemic. However, there have been some public policy successes, particularly in encouraging young mothers to come to antenatal clinics. As a result, the infection rate is declining in certain regions of countries such as Zambia and Tanzania. Botswana has implemented a national program to reduce maternal transmission by 50–80% by treating all HIV-positive pregnant women with anti-retroviral drugs. One particularly bright spot is Senegal, which has maintained effective prevention programs from the start. The rate of HIV infection in the Senegalese general population has been held to less than 2%.

ii) South and South East Asia

AIDS in Asia has the potential to be as devastating as in Africa. Over 7 million people were estimated to be living with HIV in this region in 2004, and there were an estimated 890,000 new HIV infections in 2004 alone. The first AIDS cases identified in Thailand in the late 1980s were associated with IV drug use, but female prostitutes soon became major sources of the virus. Programs promoting condom use have been successful in Thailand in that the rate of infection of young adult men has declined since 1995. In India, however, HIV is spreading relatively unchecked from the urban centers to the rural areas. Increases in IV drug use and unprotected heterosexual and homosexual activity are driving the growing epidemic in this country. An outbreak of HIV-2 occurred in prostitutes in Bombay, India in the late 1990s, indicating that this virus has jumped continents.

iii) Central and South America and the Caribbean

Overall, there are an estimated 2.3 million people living with HIV in Central and South America and the Caribbean, and the overall new infection rate is about 300,000 people per year. AIDS mortality overall in this region has declined due to the advent of anti-retroviral drugs. The prevalence of the virus in the adult populations of individual countries varies widely. For example, 5% of the adult population of Haiti was infected in 1999, whereas Brazil has benefited from aggressive prevention and drug therapy programs and has a relatively low incidence. Heterosexual transmission is the dominant mode of viral transfer in the Caribbean, while IV drug use and homosexual activity are the main contributors to HIV incidence in Central and South America.

iv) Eastern Europe and Central Asia

The AIDS epidemic was slow to start in Eastern Europe and Central Asia but gained alarming speed in 2000. It was estimated in 2004 that about 1.4 million people were living with HIV in this region. A 7-fold increase in HIV infections was noted in the Russian Federation in 2000 compared to 1999, and the virus has now reached almost all sections of the country. The predominant victims are IV drug users, but the infection is expected to spread rapidly to the sexual partners of these people and prostitutes.

v) North America, Australia, New Zealand, and Central and Western Europe

HIV-1 infection was originally most prevalent in North America, Australia, New Zealand, and Central and Western Europe, where established communities of gay and bisexual men and intravenous drug users existed. Blood transfusion recipients and hemophiliacs then also became victims, as did heterosexuals with multiple partners. Indeed, heterosexual transmission is now the most common mode of viral transfer in several European countries. Prevention campaigns, needle exchange programs, and the availability of anti-retroviral drugs slowed the overall rate of HIV infection in these regions of the world in the mid-1990s. However, as mentioned previously, the success of new drug regimens had the unfortunate side effect of inducing a false sense of security in some young gay men and a belief that HIV infection was no longer a death sentence. Unprotected sex was again on the rise in 2001 in this population, as was the HIV infection rate.

In these areas of the world, an estimated 70,000 people a year are newly infected, and the combined total number living with HIV has now reached about 1.7 million. Due to the implementation of prevention programs, the risk of a newborn being infected by his or her HIV$^+$ mother is about 12–15%. Proportionally, HIV infection rates are increased in ethnic minority and socioeconomically disadvantaged populations, probably because prevention and drug therapy programs are not accessed as readily by these groups. Cultural resistance to the introduction of anti-HIV programs can dramatically affect the HIV infection rate. In the United States, where the majority of the public opposes needle exchange programs for IV drug users, the HIV infection rate among urban IV drug users is 18–30%. In Australia, which implemented needle exchange programs in the early 1990s, the infection rate in IV drug users is 0.5–2%. From 2000 to 2003, the estimated number of HIV-infected persons in Australia and New Zealand stayed stable at about 18,000. An international study comparing 29 cities that had needle exchange programs with 52 cities that did not have such a program found that HIV infection decreased by 6% per year in the former and increased by 6% per year in the latter. Neither drug use nor crime rates increased in cities that implemented needle exchange programs.

vi) East Asia and the Pacific Rim

HIV infection and deaths due to AIDS remained relatively low up until the early 2000s in East Asia and the Pacific Rim. Despite the huge populations of China and Japan, a total of 1.1 million people (just under 0.1% of the adult population) was estimated to have HIV in 2003. However, the rising incidence of IV drug use and unprotected sex is expected to rapidly increase these rates of infection. The sale of contaminated blood is also a major contributor.

vii) North Africa and the Middle East

North Africa and the Middle East have been more successful than Southern Africa in containing the spread of HIV. In 2004, it was estimated that about 540,000 people were living with HIV in this region. However, new infections are increasing at a significant pace. Data on infection rates are not easy to acquire in these countries due to cultural sensitivities. A report from one antenatal clinic in Algeria stated that about 1% of pregnant women who came to the clinic were infected with HIV.

III. SOCIETAL AND ECONOMIC IMPACT OF AIDS

The cost to societies of the AIDS epidemic, in both human and economic terms, is already immense and predicted to increase astronomically. The long lag time between infection and disease, and the rapid spread of the virus over the past two decades, means that the worst is yet to come, even if new drugs do prove to be financially accessible and clinically successful. Some commentators have likened the effect of the AIDS epidemic to an "erosion of social capital": the disease deals a crushing blow to the young adults of a society, those on whom the very young and elderly depend for financial and emotional support. Both the wealthy and the poor of a society are affected, both the well-educated and the ignorant. Life expectancies in the most affected areas of developing countries have sunk from 63 years to 36 years, a level not seen since the early 1950s. In addition, as women of child-bearing age and their offspring are increasingly infected, a crippling decline in population growth looms. The disaster will be averted only if governments and individuals take action now to interrupt HIV transmission on a population-wide scale.

AIDS also hammers the economy of a country because its most productive work force is decimated by the disease. One study of 15 countries with high rates of HIV infection concluded that deaths due to AIDS would reduce the overall work force in these countries by 24 million people by 2020. Economists have noted a direct relationship between the rate of AIDS infection in a country and a decline in that country's gross domestic product (GDP): a 5% increase in HIV prevalence corresponds to a 0.2% decline in GDP growth rate per capita. It has been estimated that, by 2010, families in South Africa with an HIV-infected member will be 13% poorer than if there had been no AIDS to deal with.

When a family member has AIDS, the whole family suffers. The loss of work of the infected family member takes a toll, and other family members may have to stop their own work to look after him or her. Medical and legal costs take another bite from the family income. Expenditures on food, housing, school fees, and health care for other family members may be curtailed. Children may be required to quit school and work in the family fields or care for dying parents. In countries where livestock are a mainstay of family economics, children orphaned by loss of their parents to AIDS may lack the skills to look after the livestock, or the livestock may have to be sold to cover funeral and other expenses. The long term prospects for the family fall dramatically with the loss of their main income generator. In Africa, where AIDS is increasingly becoming a "women's disease," women are responsible for 80% of the food production. Thus, a whole community may face famine as well as devastating illness when its women are infected. The vicious circle is continued when infected, starving individuals migrate from the countryside to the cities to barter sexual favors for jobs and food.

Children orphaned by AIDS now number about 13.2 million worldwide, a figure expected to reach about 21 million by 2010. These children suffer from a lack of family stability, interrupted education, and gloomy future prospects. Teachers are more often lost to AIDS than to retirement in some areas of sub-Saharan Africa, crippling the education system. The fate of this lost generation of children may in turn seal the fate of their countries. The dearth in skilled (and unskilled, for that matter) workers will become a major problem for employers. Already employers bear significant costs associated with the loss of employees to AIDS: medical and health insurance payments, funeral contributions, and increased hiring and re-training costs. Some mining companies in South Africa have found it cost-effective to sponsor and implement their own AIDS prevention and drug treatment programs for their workforces. One study of one company's efforts to decrease risky behaviors in its workforce found that for every $1 spent on intervention, $25 in AIDS-related expenses down the road were avoided.

Despite the very discouraging picture just described, many countries are slowly taking a pro-active stance and implementing programs to combat the AIDS epidemic. Over the past two decades, certain strategies have emerged as being more effective than others. A national will to combat AIDS must be present, implemented in a uniform way across the country. In 2002, the President of Uganda called on all Ugandan citizens to fight AIDS as a patriotic duty, and this call to action has helped to substantially reduce the rate of HIV infection in young Ugandans. However, societies must also become accepting of HIV-infected persons and those most at risk of infection, providing them with the support they need to seek help when required. AIDS is often the result of risky behaviors indulged in by an individual, but for the sake of the country's future as a whole, communities must strive to build a society in which these behaviors are no longer acceptable or, at least, are drastically decreased. Easy access to condoms and clean needles is helpful, and ready access to non-judgmental medical care is essential. Voluntary testing and counseling have proved cost-effective measures for reducing the spread of the virus. Antenatal clinic care of pregnant women can prevent transmission to the next generation. Sexual and HIV prevention education for children and adolescents is a must. Cultural resistance to the implementation of these measures must be overcome, and myths regarding who can and cannot be infected with the virus must be exploded. The pragmatic must win over the idealistic, if any real progress is to be made on a global scale.

H. Animal Models of AIDS

While *in vitro* studies have revealed important information about the HIV virion, studies done solely in culture dishes cannot reproduce many important aspects of *in vivo* infection and disease progression, and vaccine candidates cannot be evaluated. A major challenge in studying AIDS has been the development of an adequate animal model in which to examine immune responses, test anti-retroviral drugs, and assess vaccination strategies.

Clinical researchers have compiled a "wish list" of features that a system ideally should have to be a useful animal model of AIDS (Table 25-6). Most importantly, the disease caused in the animal following virus infection should closely resemble human AIDS in terms of clinical features. Routes of transmission of the virus and the range of organs, tissues, and cells infected in the animal should approximate those in humans. However, the disease course should be significantly shorter, since experiments with a 10–15 year lag phase before clinical signs appear are costly to maintain. With respect to the model animal itself, it should be readily available in adequate numbers, not endangered or the subject of heated ethical or legal debate, and reasonably affordable. Several different types of animal models have been established to study HIV infection, including some in primates and others in mice. Each model has its advantages and drawbacks, but no single one includes all the ideal features listed previously.

Table 25-6 Features of the Ideal Animal Model of AIDS

Animal should be naturally susceptible to HIV infection

Animal should develop AIDS symptoms, including attack on lymphoid tissues and cells

HIV transmission should be achieved either naturally via copulation or artificially by IV injection

Disease course should be accelerated to weeks/months instead of years

Disease course should include HIV antigenic variation

Animal should not be endangered species

Animal should not be costly or difficult to acquire, or hard to maintain

I. PRIMATE MODELS OF AIDS

Models for the study of AIDS have been developed in several primate species. As mentioned previously, chimpanzees are the only animal that can be naturally infected with HIV-1. However, these animals fail to develop AIDS, are costly and difficult to obtain, and are an endangered species. Models based on monkeys are attractive because, unlike the apes, the animals are readily available, are not endangered, are comparatively affordable, and breed well in captivity. However, monkeys are not susceptible to HIV-1 infection. Several species of monkeys are susceptible to infection with SIV_{mac} (SIV that infects macaque monkeys) and SIV_{smm}, monkey lentiviruses related to HIV-2 in nucleotide sequence, but only macaques infected with SIV_{mac} get sick. The more distantly related SIV_{agm} chronically infects another simian species, the African green monkey, but does not cause AIDS-like symptoms. Another strain of SIV has been isolated from mandrills but has not been investigated extensively. Monkey infection models involving SIV_{smm} and SIV_{agm} have been used much less frequently than those involving SIV_{mac} infection.

i) HIV-1 Infection of Chimpanzees

As in humans, HIV-1 uses CD4, CCR5, and CXCR4 as its principal entry receptors for infection of chimpanzee cells. This observation is not surprising since these molecules are almost identical in chimps and humans. In the infected chimp, neutralizing antibodies are generated that fail to clear the virus, but, unlike the human situation, disease does not develop. Infected animals do not lose their CD4$^+$ T cells and never develop immunodeficiency. It had been hoped that the means by which chimpanzees are able to prevent HIV-1 from replicating and causing disease could be defined by studying these animals, but the mechanisms have remained elusive thus far. However, other aspects of disease pathogenesis have been successfully studied in this model; for example, HIV-1 transmission via the mucosal route was first identified in chimp studies. Because of the genetic similarity of humans and chimps, the chimpanzee/HIV-1 model has also been among the most useful for determining the effects of various drug treatments and for the development of vaccination and passive immunization protocols intended for human treatment.

ii) SIV$_{mac}$ Infection of Macaques

The model that most closely approximates a human HIV-2 infection is that of SIV_{mac} infection of macaque monkeys. Macaques infected with SIV_{mac}, or HIV-2 itself, develop disease in which CD4$^+$ T cells are depleted and immunodeficiency eventually ensues. These clinical signs may take weeks, months, or years to develop, a lag that depends partially on the virulence of the infecting SIV strain. Paralleling the anti-HIV response in humans, macaques infected with SIV_{mac} produce ineffective antibodies but efficient cell-mediated responses that can curtail virus replication. In addition, being a lentivirus, SIV_{mac} has an RT that is inherently error-prone, meaning that the virus undergoes mutations both within and between experiments, just as in a natural HIV infection in humans. Because of these similarities, the study of the macaque/SIV$_{mac}$

model has furnished useful information since 1985 on HIV transmission, development of organ-specific disease, histological changes to lymphoid tissues, and the roles of chemokines in HIV infection. The fact that host determinants influence the course of disease has also been elegantly demonstrated in rhesus monkeys infected with SIV_{mac}. In this case, the virus was molecularly cloned so as to generate only one isoform of the virus and eliminate intrastrain variation. Monkeys infected with the same cloned virus showed two different phenotypes: they either contracted AIDS within 6 months of infection, or staved it off for 2–3 years. It is unclear precisely what differences between the host monkeys were responsible for this variation.

iii) Chimeric Virus Infection of Macaques

Chimeric viruses are derived from laboratory-engineered recombination of two different strains of the same virus or even of two different viruses. By including or excluding defined nucleotide sequences in a given chimera, researchers have been able to identify important epitopes inducing humoral or cell-mediated immune responses and sequences that contribute to the infectivity of the virus. A useful chimeric virus in the context of AIDS research is SHIV (*simian-human immunodeficiency virus*), a chimera of SIV_{mac} and HIV-1 that infects macaques. The prototypic SHIV virion (dubbed SHIV89.6P) has the *env, tat, rev,* and *vpu* genes of HIV-1, but all its other genes (including *gag/pol*) are derived from SIV_{mac}. The first strains of SHIV created did not cause immunodeficiency in monkeys. To obtain highly virulent strains of SHIV, the virus was repeatedly passaged through monkeys so as to select for the virus strains with the fastest replication kinetics in this animal's tissues. Speed is known to be important for virulence, because establishment of SHIV (or HIV, for that matter) infection can be aborted if an anti-retroviral drug is employed at the earliest stages of viral attack.

When macaques are infected with a virulent strain of SHIV, the animals lose their $CD4^+$ T cells and develop immunodeficiency over a period of weeks to months, offering a practical, accelerated model for human AIDS. As is true for the SIV_{mac} model, organ-specific diseases and histological changes to lymphoid tissues associated with SHIV infection parallel those seen in human AIDS. Moreover, because SHIV carries the *env* gene of HIV-1, the envelope of SHIV is the same as that of HIV. This model can thus be used to assess the efficacy of vaccinations or passive immunizations based on gp120, making it directly relevant to human HIV-1 infection. It should be noted, however, that anti-gp 120 vaccines administered to macaques have produced mixed results with respect to protection against subsequent SHIV infection (see later).

Other chimeric viruses containing only one HIV-1 gene can be used to pursue very narrow areas of study. For example, a virus in which only the reverse transcriptase is derived from HIV-1 (SHIV-RT) has been used to study inhibitors of reverse transcriptase as anti-retroviral drugs. Chimeras have also been made from simian viruses other than SIV_{mac}. SIV_{agm} has been combined with HIV-1 but not investigated to the same extent as SHIV.

II. MOUSE MODELS OF AIDS

Much valuable information on various aspects of AIDS has been gained from the study of mouse models. The main advantage of mouse models over primate models is the increased speed of the experimental outcome. More rapid first stage screening of AIDS vaccine candidates for toxicity or the establishment of dose–response curves can thus be carried out. Mice are readily available and are not an endangered species, and mouse trials are significantly cheaper than similar evaluations in monkeys or full-scale clinical trials in humans.

Mouse models have been used primarily to study HIV infection of human lymphoid tissue transplanted into mutant mice. This system provides a further advantage over primate models because, despite the many physiological similarities between monkeys and humans, lentivirus infection of monkey lymphoid tissues differs significantly from HIV infection of human lymphoid tissues. To more closely approximate the earliest stages of HIV infection of humans, human PBL or lymphoid tissues are transplanted into *scid* mice to create chimeric living models of human lymphoid tissue. As we have seen, *scid* mice lack their own functional T and B cells due to a defect in DNA-PK activity but can be reconstituted by the introduction of exogenous hematopoietic tissue or cells.

i) Hu-PBL-scid Mouse

The Hu-PBL-*scid* mouse is created by injecting human PBL (including lymphocytes) donated by healthy adults into the peritoneal cavity of a *scid* mouse. Human APCs, $CD4^+$, and $CD8^+$ T cells readily spread throughout the mouse's body and take up residence in its spleen, liver, lymph nodes, and peripheral blood. Although these cells may attack the "foreign" mouse tissue and induce immunopathological damage (which decreases the useful life span of the model), the donated human lymphocytes remain capable of responses to specific antigen. When these Hu-PBL-*scid* mice are infected with HIV, loss of $CD4^+$ T cells occurs rapidly, just as is observed in HIV-infected humans. In one study designed to test the effectiveness of HIV envelope proteins as vaccines, Hu-PBL-*scid* mice were created using PBL donated by adult humans immunized with either gp120 or gp160 from a particular strain of HIV. When the chimeric animals were injected with the same strain of HIV, immune responses (including secondary humoral responses) were mounted that were able to subdue the viral attack and sustain the animals.

A model of pediatric AIDS has been established by implanting neonatal *scid* mice with human cord blood PBLs (Hu-PBL-*nscid* mice) and then infecting the animals with HIV. Features of this model parallel aspects of neonatal infection with HIV-1 quite nicely, allowing the evaluation of therapy and vaccination strategies for human newborns. Other researchers have recreated HIV encephalitis by injecting HIV and human PBLs into the cerebral regions of *scid* mice. It is hoped that study of the resulting pathology in these animals, which resembles that seen in the brains of AIDS patients, will yield clues as to the mechanism underlying HIV encephalitis.

ii) Hu-Thy/Liv-scid Mouse

The Hu-Thy/Liv-*scid* mouse (also known as the *scid*-hu mouse) is created by implanting human thymus and fetal liver fragments under the kidney capsule of a *scid* mouse. The implant acts as a single new organ in which the liver supplies hematopoietic stem cells and the thymus provides the correct stromal microenvironment for T cell differentiation. Normal human T cells derived from the implant are exposed to a combination of human and mouse self proteins during development such that they become "mouse-tolerant." Low numbers of human T cells leave the artificial organ and soon populate the periphery of the chimeric mouse. Human T cells and macrophages colonize the murine lymph nodes, and human IgG-secreting B cells appear in the circulation. An advantage of the *scid*-hu model is that the human tissue is fairly stable in the mouse, allowing prolonged examination (up to a month) of the response of human lymphoid tissue to injected HIV. HIV invades the implant much as it would natural thymic tissue, and human CD4$^+$ SP and DP thymocytes in the implant (which, like natural thymocytes, express high levels of CXCR4) are severely depleted within weeks of infection. T cells and monocyte/macrophages in lymph nodes also become infected and are killed by CTL-mediated cytotoxicity. Work in the *scid*-hu model has shown that HIV infection also has effects on other hematopoietic lineages because human CD34$^+$ stem cells (which express CXCR4 and CCR5) can become infected. However, at least some CD34$^+$ stem cells appear to escape viral attack and retain their pluripotency. Thus, if *scid*-hu mice are infected with HIV but are also treated with anti-retroviral drugs to contain viral replication, the surviving human CD34$^+$ stem cells can regenerate human CD4$^+$ T cells. The *scid*-hu model has been used to determine which viral genes are required for HIV virulence *in vivo* (as opposed to *in vitro*) by infecting *scid*-hu mice with viruses lacking *nef*, *vif*, or *vpu*. In addition, the *scid*-hu model has been used to evaluate drugs as candidates for immunomodulation, and to assess the promise of various exogenous cytokine therapies.

iii) Natural Mutant Mouse Strains

One issue that niggles in mouse models of the *scid* background is that murine innate immunity is intact in these animals. This factor reduces the percentage of human tissue that successfully "takes" during implantation, and may affect the immune responses observed in response to HIV infection in experiments. Some researchers have turned to the natural *beige* mouse mutant in an effort to resolve this difficulty. The reader will recall from earlier chapters that the *beige* mouse carries a mutation of the LYST gene that affects cytoplasmic granule function such that NK activity is abolished. The crossing of *beige* to *scid* mice generates an animal in which the efficiency of human tissue implantation is dramatically increased. Further efficiencies have been achieved by crossing *scid* mice to NOD mice (*non-obese diabetic* mice; see Ch.29). As well as lacking T and B cells, NOD/*scid* mice are also deficient for murine complement, NK, and macrophage functions. NOD/*scid* mice can then be reconstituted with human PBL as described previously for Hu-PBL-*scid* mice. These mouse models are expected to yield much information about HIV pathogenesis and the merits of various anti-retroviral drug candidates.

iv) Transgenic Mouse Models

Several different strains of transgenic mice expressing various HIV or SIV genes have been generated to study the function of these genes in a whole animal. The viral gene is integrated into the mouse genome and can be expressed freely in a wide variety of tissues. The reader should note, however, that the effects seen are (by definition) due to a gene unnaturally expressed outside the context of a whole virus. Nevertheless, much valuable information on the functions of *tat*, *env*, and *nef* has been gained by studying mice transgenic for HIV proviruses containing altered versions of these viral genes or the genes alone. For example, mice transgenic for an HIV provirus containing an intact *nef* gene (under the control of human promoters) both downregulate CD4 expression on cells and lose CD4$^+$ T cells, while those carrying a provirus in which *nef* is mutated do not. Mice transgenic for SIV *nef* show a similar phenotype. Some (but not all) strains of mice transgenic for *tat* develop skin lesions resembling those associated with Kaposi's sarcoma. Mice in which expression of an HIV provirus is restricted to neurons develop neurodegenerative changes, and mice transgenic for *env* alone show evidence of toxicity in several cell types in the CNS.

I. HIV Vaccines

Since HIV has such a devastating impact and is seemingly impossible to cure, the best way to head off further misery is to vaccinate the world's population against the virus. In the late 1980s and early 1990s, there was much optimism that an effective vaccine could be readily produced using the techniques of modern biotechnology because the virus had been identified so rapidly. Unfortunately, HIV's combination of antigenic diversity and multiple strategies for evading and destroying the human immune system have greatly hampered vaccine development. It remains unclear which epitopes of the virus should be included in a vaccine, what type of response (humoral or cell-mediated or both) will be most effective, where (mucosally or systemically) the response will have to be induced, and how strong the response will have to be.

I. BARRIERS TO HIV VACCINE DEVELOPMENT

Obviously, the destruction of the very cells responsible for responding to a vaccine antigen constitutes a huge hurdle to AIDS vaccine development, but there are other barriers (Table 25-7). The extreme variability of HIV presents enormous difficulties. An HIV immunogen has yet to be identified that induces antibodies or CTLs capable of recognizing a broad range of primary viral isolates from a multitude of patients. Extensive *in vitro* studies have shown that neutralizing antibodies recognizing one clade usually do not recognize

Table 25-7 Barriers to HIV Vaccine Development

Targeting by HIV of immune system cells necessary for response to vaccine

Extreme antigenic variation of HIV within individuals and between populations

Non-cross-reaction of antibodies recognizing different clades

Promotion of genetic escape mutants by vaccine focus on one epitope

Low immunogenicity of the most conserved HIV epitopes

High susceptibility to antigenic drift of the most immunogenic HIV epitopes

Enhanced infectivity promoted by some anti-gp160 antibodies

Differences in behavior of laboratory-cultured and primary HIV strains

Lack of adequate *in vitro* system based on human cells for testing

Lack of animal AIDS model that is truly parallel to the human situation

Lack of effective mucosal vaccine administration protocol to induce Ab and CTLs at site of HIV entry

another, and that anti-HIV-1 antibodies rarely cross-neutralize HIV-2. Cross-reactivity between HIV-specific CTL clones tends to be broader than that among antibody clones, but focusing on one epitope has tended to foster the emergence of genetic escape mutants. Moreover, it seems that the most conserved epitopes in HIV tend to be the least immunogenic, and those that are immunogenic are subject to extensive antigenic drift. These factors do not bode well for producing a vaccine that will induce protective immunity in most individuals within a population, let alone between populations. Having to make multiple vaccines for multiple populations drives up the cost of vaccination, a significant barrier to vaccination programs in the developing world.

With respect to the antibody response, researchers must ensure that a prospective HIV vaccine does not induce the production of antibodies that will do more harm than good. In a study of gp160-immunized human volunteers, some serum samples contained antibodies that actually <u>enhanced</u> the infectivity of the virus *in vitro*. Virions complexed to these antibodies bound more readily to Fc or complement receptors on macrophages, facilitating their internalization by macrophages and thus the entry of the virus into this reservoir population. Antibodies with this effect were identified among those recognizing gp41 and the V3 loop of gp120. The effects of such enhancing antibodies will have to be carefully considered in vaccine design.

Another technical difficulty in AIDS vaccine development is related to the animal models described in the last section. While monkey, mouse–human tissue, and chimeric virus models have been very helpful in investigating isolated aspects of HIV biology, infections in these models differ significantly in key areas from natural HIV infections of humans. Moreover, laboratory-maintained strains of HIV differ significantly from primary HIV isolates in several characteristics. These factors combine to

make extensive clinical trials of vaccine candidates in humans unavoidable, despite their inherent risks and considerable expense. The development of a laboratory system using isolated human cells that could identify the most promising vaccine candidates before the clinical trial stage would be of tremendous value to scientists, patients, and governments; unfortunately, such a system has remained elusive.

The route of administration of a prospective AIDS vaccine is an issue. Most HIV vaccines are administered by intramuscular injection, a route that primarily induces circulating antibodies. What is desperately needed is a protocol that can trigger both neutralizing antibody and vigorous CTL-mediated responses against HIV at the mucosae, so as to block the virus at entry and prevent its establishment in sub-mucosal tissues. Vaccines deliverable by various mucosal routes (nasal, oral, vaginal, rectal) are under investigation. Some of the efforts to induce effective mucosal anti-HIV responses in animals are outlined in the next section.

There is one last factor that may have to be addressed in HIV vaccine design: the matter of distinguishing those that are infected with HIV from those who have been vaccinated with HIV proteins. Individuals may, for legal or insurance reasons, need to be able to prove that they have been vaccinated against HIV rather than infected with the virus. Individuals are commonly tested for HIV infection by screening their blood for the presence of anti-HIV antibodies. Such antibodies are identified using prepared Western blots containing regions of various HIV proteins. Deliberately deleting one of these regions from the HIV proteins or genes used to constitute the vaccine will result in a lack of antibodies recognizing this region in the vaccinee's serum. A negative result for this region will appear on the standard Western screening blot, indicating to the health professional that the individual has been vaccinated against HIV and is not infected with the virus.

II. EXPERIMENTAL AIDS VACCINES IN ANIMALS

Much of our current knowledge of HIV vaccines has been garnered from studies of the animal models described previously, particularly the HIV-1/chimpanzee and SHIV/macaque models. Studies in chimpanzees have confirmed that more than one vaginal exposure to HIV is required to achieve uptake of the virus and initiate a productive infection. Thus, a single vaccination via the vaginal route is unlikely to introduce sufficient viral antigens to induce the mucosal immunity required to later ward off a bona fide HIV-1 attack. In the SHIV system, vaccines based on live, attenuated virus, killed whole virus, viral DNA, recombinant subunits, and live vaccinia vectors expressing SHIV proteins have been evaluated. Those based on live, attenuated virus are generally the most effective in protecting against subsequent SHIV infection, although results have been mixed.

We now present some general conclusions about manufactured immunity to HIV that have emerged from studies of SIV and SHIV in vaccination trials.

1. The identity of a deleted gene may dictate the degree of attenuation of the virus. In studies designed to test the pathogenicity of mutated SIV strains, it was found that

strains in which only *vpr* or *vpx* was deleted remained pathogenic, while those in which *vif* was deleted were not infectious. Multiple deletions had varying effects on resulting CD4$^+$ T cell counts and disease progression in animals infected with the attenuated strain.

2. The more attenuated an SIV strain, the less effective it is as a vaccine. This fact has profound implications for the balance of vaccine safety versus efficacy, as the reader will recall from Chapter 23. Using the SIV$_{mac}$ model, it has been shown that a SIV$_{mac}$ virus attenuated by deletion of its *nef* gene (SIV$_{mac}$Δ*nef*) is able to protect macaques from infection with wild-type SIV$_{mac}$. In addition, at least in a proportion of experimental animals, SIV$_{mac}$Δ*nef* vaccination protects macaques against subsequent infection with virulent chimeric SHIV, a virus even more closely related to HIV-1. However, a problem with applying this approach to humans is that attenuation by this method may not offer a sufficient guarantee of vaccine safety. In some studies, the onset of immunodeficiency in macaques was only delayed by vaccination with attenuated SIV, not eliminated, implying the persistence of the virus after the conclusion of immune responses intended to clear it. Indeed, there are some HIV-infected patients in whom natural mutations of certain HIV genes, principally *nef* or the regulatory region of the LTR, have occurred. Unfortunately, these patients do not produce antibodies capable of halting AIDS progression: while deterioration is delayed, it is still inevitable. Even more worrying is the fact that some attenuated SIV strains have shown a remarkable talent for reversion, repairing engineered deletions in genes by use of their error-prone reverse transcriptases. It has become clear that deletions used to attenuate SIV (and thus potentially HIV) strains must be large and totally out-of-frame if the attenuated virus is to be used safely as a vaccine. Attenuation by simultaneous deletion of several genes, such as *vpr*, *vpu*, and the LTRs in addition to *nef*, may remedy this problem. Elimination of proviral integration by anti-retroviral drug treatment in combination with vaccination using an attenuated virus may also be a viable strategy.

3. The longer the interval between vaccination with attenuated SIV and challenge with virulent virus, the greater the protection against infection. In one study, full protection against virulent SIV was achieved only at 20 months post-vaccination, with very limited protection apparent at 2 months post-vaccination. Other studies using other attenuated viruses have shown varying intervals before full protection is achieved, but none has been shorter than 10 weeks.

4. How quickly full protection is achieved is influenced by the route of vaccine administration and the route of viral challenge, and how closely related the vaccine virus is to the challenge virus. The more similarity there is in the *gag-pol* genes (rather than the *env* genes) of the two viruses, the more likely vaccination with one will protect against the other.

5. In the early days of HIV vaccine development, researchers focused on Env as an immunogen because it seemed logical that this surface protein should be able to induce anti-retroviral neutralizing antibodies. Unfortunately, these antibodies proved not to be protective. Indeed, in most animal models, the production of neutralizing antibodies is not essential for protection against subsequent SIV infection and does not even contribute to protection in many cases. It has thus far proved impossible to develop HIV immunogens and vaccination protocols that induce broadly effective neutralizing antibodies. Those antibodies that are produced tend to be too strain-specific for use in a population in which each individual may harbor a slightly different strain of HIV. There are immunogens that elicit antibodies protective against infection by laboratory isolates of HIV, but these antibodies are generally ineffective against viral isolates from patients.

6. The preceding observations imply that effective immune responses against SIV, SHIV, and HIV will depend primarily on the recognition of internal viral epitopes rather than the surface envelope epitopes. CTLs directed against internal viral proteins other than those derived from *gag-pol* have been explored as vaccine candidates in SHIV models. Rev is found primarily in the nucleus of the infected cell and is not highly immunogenic, precluding its use as a vaccine. Tat in native and toxoid forms has been tried as a vaccine candidate, but results have been mixed. When Tat protein was the basis of a vaccine administered to macaques, there was some success in reducing SHIV replication. However, when the *tat* gene was used as a DNA vaccine, no protective effect was observed.

7. Protection against disease in SHIV- or SIV-infected macaques can be achieved even in the presence of replicating virus. In one study, the vaccination did not completely eliminate the virus from the body but mitigated the viral load such that fewer CD4$^+$ T cells were lost. The vaccinated animals did not develop overt signs of immunodeficiency. The bottom line seems to be that current SIV/SHIV vaccination protocols can protect macaques from the development of AIDS-like disease but cannot prevent SIV/SHIV infection.

8. There are many reports of vaccination approaches that induce production of SIV-specific CTLs but still fail to protect the animal from infection. Animals in which the vaccination protocol induces an early and vigorous SIV-specific CTL response are more likely to be protected against subsequent viral challenge than are animals whose immune systems are slower to respond. Similar results have been obtained for vaccination studies involving SHIV.

9. Recombinant viral vectors have been used with promising results in SIV/SHIV infection experiments. For example, adenovirus vectors are highly immunogenic in monkeys and target their APCs. Macaques and chimpanzees infected intranasally with replication-competent adenovirus carrying HIV or SIV genes have been shown to mount detectable systemic and mucosal immune responses against the viruses. Similarly, replication-defective alphaviruses, which also target APCs, have been used to vaccinate macaques against SIV challenge with modest results.

Recombinant live, attenuated polio virus carrying SIV genes has induced both systemic and mucosal anti-SIV immune responses in macaques, while vaccination with a recombinant, replication-defective, attenuated HSV carrying the SIV *env* gene resulted in a persistent anti-SIV antibody response.

In one particularly successful study, researchers first primed naive macaques with a recombinant vaccinia virus-based vector expressing SIV and HIV-1 structural proteins. Doses of this vaccine were given at the start of the experiment and then 8 weeks later. The animals were boosted at 20 and 28 weeks by injection of chemically inactivated, intact SIV and HIV virion particles. On week 46 after the start of the experiment, these macaques were challenged with SHIV. Control, unvaccinated monkeys immediately underwent complete loss of their CD4$^+$ T cells and developed AIDS-like disease within 5 months. In contrast, although the vaccinated monkeys were infected, viremia was low, CD4$^+$ T cells were largely spared, and the animals stayed healthy for at least 15 months after the infection. Both neutralizing antibody and CTL antiviral responses could be demonstrated in these animals.

10. Vaccination of primates with naked DNA alone (by gene gun or other means) fails to result in the vigorous immune responses often noted in rodent systems (see Ch.23), and primates are thus usually not protected against subsequent viral challenge. Similarly, in human volunteers, intramuscular injection of a plasmid encoding Env and Rev from HIV-1 induced minimal and inconsistent T cell responses. As we saw in Chapter 23, DNA vaccination can be augmented by later administration of the antigen in protein form. Briefly, the vaccinee is given a DNA vaccine followed by boosting with the recombinant protein or vector bearing the gene encoding the protein of interest. Protective and potent T cell responses (but not antibody responses) have been obtained when macaques were immunized with a DNA plasmid vaccine carrying both SIV genes and HIV-1 *env* followed by boosting with a recombinant vaccinia vector bearing SIV *gag-pol* and HIV-1 *env*. Similar results have been obtained using a modified approach in rhesus monkeys. In this case, the monkeys were co-injected four times over 40 weeks with both a DNA vaccine plasmid encoding SIV *gag* and either a plasmid encoding IL-2 fused to the Fc regions of human IgG or the IL-2/Ig fusion protein itself. (IL-2 enhances the activation of the responding APCs and Th1 cells, and fusion of IL-2 to the IgG Fc region increases the half-life of the IL-2 in the blood.) Effective SHIV-specific CTL responses were achieved by this strategy. Furthermore, in contrast to animals receiving only the SIV *gag* DNA, animals receiving both SIV *gag* DNA plus IL-2/Ig protein (or DNA) showed good control of viremia and had not developed clinical signs of immunodeficiency 5 months after challenge with virulent SHIV. By 6 months, however, one of the monkeys in this trial showed a resurgence of the virus that led to the animal's death by 12 months. It was found that the SHIV strain in this animal had undergone mutation of a key CTL epitope, permitting it to genetically escape the CTL response. Nevertheless, at the time of writing, the remaining seven monkeys in this trial remain free of both detectable virus and immunodeficiency disease.

III. EXPERIMENTAL AIDS VACCINES IN HUMANS

The small number of people who have been continually exposed to HIV but never infected, and LTNPs who have the virus but experience a long delay in developing AIDS, represent hope for the eventual development of an effective AIDS vaccine. These populations have been extensively investigated in an effort to determine why they have been able to resist the virus, and a variety of mechanisms have been elucidated. As discussed previously, some individuals express a mutated version of the CCR5 protein used as a coreceptor for viral entry. Others have high levels of anti-HIV neutralizing antibodies that are unusually effective (including mucosal IgA responses), while others display vigorous anti-HIV CTL responses. In other individuals, the mechanism remains to be defined. The emerging fields of genomics, transcriptomics, and proteomics may assist us in identifying genes and proteins that confer HIV resistance, and in defining polymorphisms associated with increased susceptibility to infection.

As of early 2002, about 20 candidate HIV vaccines were under evaluation in close to 75 phase I, five phase II, and two phase III human clinical trials, so far without significant adverse events. Most of these trials have involved individuals that have already been infected with HIV, but uninfected volunteers have participated in some. In 2003, the results of the first phase III efficacy trial of a gp120 vaccine tested in 5000 at-risk volunteers were released: sadly, no protective effect was found. The various vaccines under investigation have generally involved envelope proteins, particularly gp120 and gp160, produced in mammalian cells, insect cells, or yeast by recombinant DNA technology. Fusion proteins and peptides of different HIV proteins delivered in various immunogen delivery systems have also been tested, as have naked DNA and viral and bacterial vector-based vaccines. The effectiveness of co-administration of plasmids expressing cytokines or immunostimulatory DNA sequences has been assessed. Unconventional routes of vaccine administration such as via mouth and rectum have also been explored. In general, only minimal neutralizing antibody responses have been elicited by the approaches tried to date. Furthermore, most of these candidate vaccines induced the production of antibodies that recognized only the laboratory-cultivated strain of HIV used to prepare the vaccine, not primary HIV strains isolated from patients. Vaccine-induced HIV-specific CTL responses, while more promising in that they recognized more than one strain of HIV, have been inconsistent.

A more promising AIDS vaccine approach involves priming with a live recombinant canarypox virus vector vaccine containing Env, Gag, Pol, and Nef domains from HIV followed by boosting with a recombinant subunit vaccine containing purified HIV Env protein. In at least some human volunteers, there was induction of HIV-1 neutralizing antibody responses coupled with CTL responses capable of recognizing different HIV-1 clades. Priming with naked DNA vaccines followed by boosting with recombinant viral vectors expressing HIV proteins has also yielded encouraging preliminary results. In addition, researchers have identified three human monoclonal antibodies, called 2F5, 2G12, and IgG1b12, which in combination can neutralize multiple HIV strains *in vitro*. The 2F5 antibody recognizes an epitope located in the extracellular region of gp41, close to the transmembrane domain. The 2G12 antibody recognizes an epitope near the base of the V3 loop of gp120, while IgG1b12 (also known as F105) recognizes a discontinuous epitope near the CD4 binding site of gp120. Investigations are under way to devise a vaccine that can elicit the production of antibodies with these specificities *in vivo*.

In 2003, The Global HIV/AIDS Vaccine Enterprise was established "as an alliance of independent organizations committed to accelerating the development of a preventive vaccine for HIV/AIDS through implementation of a shared scientific strategic plan, mobilization of additional resources, and a greater collaboration among HIV vaccine researchers world-wide" [Klausner R.D. *et al.* (2004) *Science* 303, 1293]. The Enterprise involves scientists from private companies and government-funded institutes, public health officials, members of advocacy groups, and representatives of agencies and organizations that secure funding for HIV/AIDS research. The scientific strategy blueprint was drafted with the participation of more than 140 scientists and other interested individuals from 17 countries and international agencies. The plan seeks to standardize methodologies and results assessments for each phase of vaccine discovery, as well as promote an unprecedented sharing of information and resources. The Enterprise is also actively pursuing increased funding to support HIV/AIDS vaccine research, and finding ways to overcome regulatory obstacles and increase the complement of trained personnel in developing countries.

IV. PASSIVE IMMUNIZATION

Passive immunization experiments in animal models have been used by researchers in an attempt to establish what sorts of antibodies at what sorts of concentrations might confer protection against SIV/SHIV. In early trials, passive transfer of serum from vaccinated animals usually could not protect recipients against viral challenge. While parenteral passive immunization of macaques with anti-SHIV antibodies protected recipients from subsequent challenge with virulent SHIV, the levels of antibody introduced to achieve this protection were much higher than those induced by standard vaccination protocols, and indeed higher than those present in immune animals. Furthermore, the anti-SHIV antibodies were unable to protect the macaques if they were administered even as soon as 2 hours after viral infection; only administration before infection was effective.

Refined approaches to passive immunization in the early 2000s have been more successful. Passive immunization of infected macaques with physiological levels of exogenous anti-HIV antibodies has resulted in a modest reduction in viremia and resistance to further loss of CD4$^+$ T cells. Other animals that received neutralizing IgG monoclonal antibodies directed against HIV-1 epitopes have been able to repel both oral and vaginal challenges with SHIV, and the relevant mAb was found to be localized to mucosal surfaces. If one could achieve the same effect in humans (assuming the anti-HIV mAb used recognized the HIV strain dominant in a particular individual), HIV infection through the mucosae, or transmission deeper into the body, might be prevented. In an effort to find ways of blocking maternal transmission of HIV to the neonate during the birth process, researchers have treated pregnant monkeys with a combination of the three human IgG1 neutralizing mAbs mentioned previously (which have also been shown to protect cultured cells from SHIV infection). When the baby monkeys were born, they were treated with the mAb triple combination and then orally challenged with SHIV. The majority of the neonates were protected for at least several months. Even neonates that received the triple mAb combination only after birth were able to resist SHIV infection or at least maintain normal CD4$^+$ T cell counts. Significantly, newborn monkeys that were first challenged with oral SHIV and then given the triple mAb combination were also protected.

J. Treatment of HIV Infection with Anti-retroviral Drugs

Until an effective HIV/AIDS vaccine is developed, the best weapon the world has against HIV is anti-retroviral drug therapy. In 1996, there were dramatic reductions in AIDS progression and AIDS-related deaths in the developed world following the introduction of effective anti-retroviral drugs. Current regimens of *anti-retroviral therapy* (ART) or *highly active anti-retroviral therapy* (HAART) are even more successful. HAART features combinations ("cocktails") of at least two different types of drugs that go after the virus in different ways. For example, an inhibitor of HIV RT may be combined with an inhibitor of HIV protease or viral entry to limit viral replication for extended periods. Suppression of viral replication has two effects. First, by reducing the number of replicative cycles, the chance of generating a mutated virus whose protease or RT is resistant to the HAART drugs is decreased. Secondly, depending on when it is administered, HAART can give the patient's own immune response some breathing room such that at least some restoration of HIV-specific T cell responses (and responses against other pathogens such as CMV) can occur. For example, when HAART is started in the acute stages of an HIV infection, strong, persistent HIV-specific Th responses are mounted that help to control viremia. However, if HAART is not started until after the disease has become chronic, the prospects are not quite as good. The production of new, naive CD4$^+$ T cells is increased following therapy, and Th responses specific for CMV may still be recovered, but HIV-specific Th responses are rarely restored. In addition, for reasons that remain unclear, late-starting HAART can be associated with loss of CTL numbers and inhibition of CTL function. Current guidelines specify that HAART should be started in asymptomatic HIV$^+$ persons when their T cell counts drop to 200–350 cells/mm^3. However, HAART may be commenced at a higher CD4 T cell count if the HIV RNA level is high (i.e., >100,000 copies/ml blood).

Table 25-8 **Anti-retroviral Drugs**

Inhibitor Type	Mechanism	Examples
Protease inhibitors	Small molecules competitively bind to active site of viral protease	Indinavir, nelfinavir, saquinavir
Nucleoside RT inhibitors	2'3'-dideoxynucleoside analogues competitively inhibit 2'-deoxynucleosides and block viral DNA synthesis by RT	AZT, abacavir, didanosine, 3TC
Non-nucleoside RT inhibitors	Molecules of diverse structure induce conformational changes to RT that inactivate it	Nevirapine, efavirenz
Fusion inhibitors	Peptides derived from gp41 block virus envelope fusion to host cell membrane	Enfuvirtide

There are at present four classes of licensed anti-retroviral drugs that act on proteins essential for viral spread: protease inhibitors, which act on HIV protease, nucleoside and non-nucleoside RT inhibitors, which act on HIV reverse transcriptase, and fusion inhibitors, which interfere with viral envelope fusion to the host cell membrane and thus block viral entry (Table 25-8). However, these drugs are effective only on actively replicating virus and do not eliminate latent virus in resting cells. In addition, the highly mutable nature of HIV means that strains are constantly evolving that are resistant to existing drugs. New classes of drugs are therefore under investigation, such as those that can degrade viral RNA, prevent viral DNA integration, or block chemokine receptor access. It is hoped that some of these new drugs will be more effective in eliminating the virus wherever it hides while generating fewer detrimental side effects. Cytokines are also being explored as therapeutics. Early results of treatment regimens in which patients were repeatedly given IL-2 showed that plasma viral loads did not decrease but numbers of CD4$^+$ T cells rose. In addition, IL-2 may stimulate the activation of latent proviruses, forcing them to replicate and, with any luck, become susceptible to one of the anti-retroviral drugs described later. Trials are under way to assess the effect of including IL-2 as an anti-HIV drug in combination therapies. We will now describe some examples of existing anti-retroviral drugs in more detail.

I. PROTEASE INHIBITORS

Protease inhibitors such as indinavir, nelfinavir, saquinavir, and ritonavir are small molecules that work by competitively binding to the active site of the viral protease. Without protease function, the production of progeny viruses is stymied because the post-translational processing and cleavage of the large structural core proteins of the virus do not occur. Protease inhibitors are highly effective medicines and were responsible for the first major drops in morbidity and mortality due to HIV infection in the developed world in 1996. However, this type of therapy entails taking numerous pills that have unpleasant side effects, decreasing patient compliance. Diarrhea, hyperlipidemia, rash, dry skin, nausea, and vomiting are all commonly experienced side effects.

II. NUCLEOSIDE RT INHIBITORS

The nucleoside RT inhibitors work on the principles of competitive inhibition and premature DNA chain termination. The inhibitors are 2'3'-dideoxynucleoside analogues of the usual 2'-deoxynucleosides that the RT joins together to synthesize DNA. If the 3' hydroxy group of a nucleoside is replaced with an azido group or a hydrogen molecule, the elongation of the growing viral DNA chain is interrupted because there is no 3' OH onto which the next nucleoside can be joined. Moreover, the substrate binding site of the RT protein is firmly occupied by the inhibitor. Lastly, there may be so many molecules of the inhibitor present that the RT enzyme may simply have a hard time finding its natural substrate. Commonly used nucleoside inhibitors are zidovudine (commonly known as AZT, *azidothymidine*), stavudine, lamivudine (3TC), abacavir, and didanosine. Unfortunately, patients taking RT inhibitors often experience unpleasant side effects such as lipodystrophy, lactic acidosis, bone marrow suppression, peripheral neuropathy, myopathy, mouth ulcers, nausea, and vomiting.

III. NON-NUCLEOSIDE RT INHIBITORS

Unlike nucleoside inhibitors, the non-nucleoside inhibitors act on the RT molecule at locations distant from the active site. These inhibitors are diverse in structure but share the characteristic of inducing major conformational changes to the RT molecule that disrupt its enzymatic activity. Common non-nucleoside inhibitors are nevirapine and efavirenz, drugs that are better tolerated by patients than either protease inhibitors or nucleoside RT inhibitors. Side effects include hyperlipidemia, lipodystrophy, and diabetes mellitus. One disadvantage to non-nucleoside RT inhibitors is that they can be too specific: in one study, the RT of HIV-1, but not that of HIV-2, was inactivated *in vitro* by a particular inhibitor of this class.

IV. FUSION INHIBITORS

Enfuvirtide is the only licensed fusion inhibitor to date. This relatively new class of anti-retroviral drug works by blocking viral entry into host cells. Enfuvirtide is a 36 amino acid synthetic peptide based on the extracellular region of the transmembrane protein gp41. This drug blocks the conformational change of the HIV envelope spike required to expose the fusogenic domain of gp41. The virus cannot fuse with the host cell membrane and thus cannot enter the cell. Disadvantages of enfuvirtide are that it is currently very expensive and difficult to manufacture, and must be delivered by subcutaneous injection. Enfuvirtide is recommended for HAART in patients whose viral isolates show resistance to other classes of inhibitors. Other fusion inhibitors are undergoing clinical testing.

V. IMMUNE RESPONSES AND ANTI-RETROVIRAL THERAPY

Because HIV mutates so readily, point mutations to the protease or RT or *gag* genes can easily thwart all four classes of inhibitors described previously. Effective treatment thus requires that combinations of anti-HIV drugs be given rather than just one type. The original anti-HIV drug cocktail regimens called for the use of three different nucleoside RT inhibitors, or two nucleoside RT inhibitors plus one or more protease inhibitors, or two nucleoside RT inhibitors plus one non-nucleoside RT inhibitor. As of 2003, fusion inhibitors are being added to these cocktails.

Anti-HIV cocktails given at the appropriate time for each patient (based on viral load) can block viral replication, maintain low viral loads, and allow restoration of CD4$^+$ T cell numbers, including measurable numbers of HIV-specific CD4$^+$ T cells. Therapeutic vaccination may then become feasible. For example, in one animal model, T cell responses were examined in macaques that were infected with SIV and treated 2 weeks later with HAART alone, attenuated vaccinia virus expressing SIV genes alone (as an immunogen), or both HAART and attenuated vaccinia/SIV. Only those animals in which viral loads were reduced by HAART were able to respond to the immunogen by producing virus-specific T cells; immunization itself was not enough in the face of massive viremia. However, anti-retroviral drugs do not entirely eliminate an HIV or SIV infection because the drugs do not affect viruses lying dormant in resting cells. When the drug regimen is stopped, any virus that is activated in these cells can replicate freely. In the macaque experiment described previously, it was found that prior HAART, but not immunization, had a profound effect on the ability of the animal to cope with the infection. Viremia eventually rebounded sharply in animals that had only been immunized, but was less intense in macaques, immunized or not, that had originally been treated with HAART.

These observations may lead to concrete benefits for HIV patients on HAART. Because the virus reappears with a vengeance if drug therapy is discontinued, a patient must continually take the drug cocktail to suppress the viral load, sometimes an issue because of the sheer numbers of pills involved. Indeed, as many as 40–50% of AIDS patients do not fully comply with their drug regimens due to the arduous nature of the therapy. Consequently, these individuals will not show a significant decrease in viral load even after 1 year of treatment. Work is under way to develop combination pills to reduce the number of pills a patient has to take. Researchers have also used primate AIDS models to experiment with different drug treatment

schedules in an effort to find a protocol that minimizes the time of drug treatment without compromising suppression of viral replication. These protocols have been dubbed *structured treatment interruptions* (STI) or *cycling HAART*. In one study, macaques were infected with virulent SIV and treated 6 weeks later with HAART. One group received HAART continuously, while another group underwent a "3 weeks on, 3 weeks off" schedule of HAART. It was found that viremia was better controlled in the animals that had undergone the interrupted HAART schedule than in animals that had experienced continuous HAART. It is speculated that the cessation of HAART after suppression of the viral load allowed the restoration of immune responses that were able to act against the virus. In addition, continuous HAART can lead to the selection of drug-resistant HIV mutants, while on-again/off-again drug regimens tend to maintain the sensitivity of the virus to the drugs. Such cycling HAART regimens are now being used for treatment of human HIV patients, with clear gains in compliance and marked decreases in both expense and toxic side effects.

VI. OTHER DRUG THERAPY ISSUES

The expense of Western anti-retroviral drugs has become a major issue for developing countries, which contain the populations that need the drugs the most but can least afford to buy them. Different countries have responded to this difficulty in different ways. Some countries have patent laws that allow companies to manufacture any patented drug as long as the process used is different from that specified in the original patent. The Brazilian public health system produces its own generic anti-retroviral drugs that allow it to treat AIDS sufferers at a fraction of the cost of treatment with patented medicines. Indeed, in Brazil, access to anti-retroviral therapy is enshrined in the national constitution. Citing similar patent laws, a company in India called CIPLA has offered to supply the international charity Medicins sans Frontières (Doctors without Borders) with a generic version of a triple anti-retroviral drug cocktail at 1/20th the cost of the brand name version in developed countries. In an international initiative, a preferential pricing agreement has been reached between the WHO and five major brand name drug manufacturers to supply anti-HIV drugs at substantially reduced prices to developing countries. Under this agreement, the annual cost of treating an HIV patient in Senegal would be about $1000, as compared to the approximately $15,000 charged in North America and Europe. (It has been pointed out, however, that $1000 is still beyond the reach of most infected people in Africa.) Other agreements have mobilized funding for drugs specifically to block maternal–child transmission in developing countries. For example, a regimen that has been cost-effective and clinically beneficial in Uganda has been a short course of perinatal therapy with non-nucleoside RT inhibitors. HIV-infected pregnant women are treated orally with a single dose of a non-nucleoside RT inhibitor at the onset of labor, and their new babies are treated orally with a single dose of the same drug within 72 hours of birth. This practice can easily and cheaply decrease the risk of HIV transmission from mother to child by 50%.

Some observers have noted that merely supplying low-cost anti-retroviral drugs to combat HIV, while a great start, will not be enough to contain this pandemic. Funding (and political will) must also be directed toward health care delivery infrastructure and education programs. However, this point of view has been challenged by those who say that it is the high cost of drugs that discourages any expenditure on infrastructure: if the anti-retroviral drugs were readily available at almost no cost, the building of a health care system to distribute those drugs would suddenly become worthwhile. Citing a similar strategy that has been successful in treating patients with drug-resistant TB in developing countries, various health care officials have challenged the CEOs of brand name drug companies to make their anti-retroviral medications available to developing countries essentially for free. Until 2003, however, a major obstacle to this and other cost-relief proposals was the Trade Related Aspects of Intellectual Property Rights (TRIPS) agreement established in 1997 by the World Trade Organization (WTO). This agreement specified that developing countries should enforce patent protection of brand-name pharmaceuticals. However, a meeting of the WTO in 2003 smoothed a legal path for developing countries to import cheaper generic copies of patented drugs for catastrophic diseases affecting the developing world (including AIDS, TB, and malaria) if these countries could not manufacture the medicines themselves.

Another issue associated with anti-HIV drug treatment is that we do not yet know the side effects of long term use of these drug cocktails. Patients may end up having to choose between physical detriments due to the virus itself and those due to HAART. Some relief might be gained if drugs that specifically block access to CCR5 or CXCR4 are successfully developed. The apparent good health of individuals who naturally lack expression of CCR5 and their resistance to HIV infection make this protein a highly promising target. Indeed, the chemokine ligands of CCR5 and CXCR4 have anti-HIV effects *in vitro*, most likely because binding of a chemokine to its receptor induces downregulation of receptor expression.

Specific chemokine receptor antagonists have been generated by truncating or making N-terminal modifications to RANTES, the ligand for CCR5. *In vitro*, such antagonists are powerful inhibitors of R5 HIV infection of lymphocytes and macrophages. The modified RANTES molecule interferes with the internalization and recycling of CCR5, rapidly reducing surface expression of the receptor on lymphocytes, monocytes, and macrophages. These cells are then protected from HIV infection. Similarly, peptide inhibitors of CXCR4 have been described that can block cell fusion and infection mediated by X4 HIV *in vitro*. Unfortunately for AIDS patients, none of these inhibitors has turned out to be sufficiently metabolically stable for therapeutic use. More promising results (in the form of reduced viral loads) have been obtained in early clinical trials of a small molecule inhibitor of CXCR4. Evaluations of the efficacy of this class of drugs and their potential side effects are ongoing.

Despite the successes achieved with HAART, clinicians have come to the realization that it will be extremely difficult to completely eliminate HIV from a patient's body. HIV can maintain latency inside resting T cells for years, and also hides in other body reservoirs safe from inactivation by current HAART agents. Such reservoirs include the CNS and semen, in which any virus present evolves independently of virus isolates in blood cells such as macrophages. The patient thus harbors multiple strains of the virus, increasing the likelihood that at least one will mutate to a drug-resistant form. Moreover, HAART apparently does not stop <u>all</u> viral replication, and a low cryptic level continues that cannot be detected by current clinical assays. New approaches are needed that can completely suppress viral replication and attack viruses in reservoirs without harming normal cells.

This concludes our description of HIV and AIDS. We move now to Chapter 26 and a discussion of how the healthy immune system strives to fight cancer development, and how tumors evade both immune surveillance and elimination by immune system effector cells.

SUMMARY

Human immunodeficiency virus (HIV) is a cytopathic RNA retrovirus causing acquired immunodeficiency syndrome (AIDS) in human hosts. Globally, AIDS has become the leading cause of adult death due to infectious disease, and approximately 40 million people worldwide are currently infected. HIV is most commonly transmitted during sexual contact or intravenous drug use involving contaminated needles. Within the host, the interaction of the viral envelope protein gp120 with CD4 and host chemokine receptors allows the virus to enter both macrophages and CD4$^+$ T cells. Viral reverse transcriptase transcribes the viral RNA into a viral DNA that is integrated into the host genome to become the provirus. When the infected cell is stimulated, viral transcription and translation begin and copious amounts of newly synthesized virions bud from the host cell prior to its eventual lysis. In a newly infected individual, the virus can infect DCs, macrophages, and T cells. However, the virus sustains small mutations to its gp120 protein that usually result in a restricted tropism for T cells as the infection progresses. HIV also kills uninfected CD4$^+$ T cells via an unknown mechanism. In terms of immune responses to HIV, B cells and CD8$^+$ T cells respond to the viral attack but the HIV-specific antibodies and CTLs produced are effective only for a brief period, if at all. Clinically, the infected individual first experiences a mononucleosis-like illness followed by a period of viral and clinical latency that may last for several years. Upon reactivation of the virus, massive loss of CD4$^+$ T cells results in the profound immunodeficiency characteristic of AIDS, and infected individuals succumb to a variety of opportunistic infections or certain malignancies. The only effective form of therapy is life-extending treatment with anti-retroviral drugs. Many barriers continue to thwart development of a successful HIV vaccine, including the extreme antigenic variation of the virus, the incapacitation of immune system cells following infection, and the lack of adequate *in vitro* and *in vivo* models of HIV infection.

Selected Reading List

Altmeyer R. (2004) Virus attachment and entry offer numerous targets for antiviral therapy. *Current Pharmaceutical Design* **10**, 3701–3712.

Andrake M. and Skalka M. (1996) Retroviral integrase, putting the pieces together. *Journal of Biological Chemistry* **271**, 19633–19636.

Benito J., Lopez M. and Soriano V. (2004) The role of CD8$^+$ T-cell responses in HIV infection. *AIDS Reviews* **6**, 79–88.

Brander C. and Walker B. (1999) T lymphocyte responses in HIV-1 infection: implications for vaccine development. *Current Opinion in Immunology* **11**, 451–459.

Busch M., Chamberland M., Epstein J., Kleinman S., Khabbaz R. et al. (1999) Oversight and monitoring of blood safety in the United States. *Vox Sanguinis* **77**, 67–76.

Centers for Disease Control. 1993 revised classification system for HIV infection and expanded surveillance case definition for AIDS among adolescents and adults. CDC web site, http: //www.cdc.gov/epo/mmwr/preview/ mmwrhtml/00018871.htm.

Chinen J. and Shearer W. (2002) Molecular virology and immunology of HIV infection. *Journal of Allergy and Clinical Immunology* **110**, 189–198.

Coordinating Committee of the Global HIV/AIDS Vaccine Enterprise (2005) The Global HIV/AIDS Vaccine Enterprise: Scientific Strategic Plan. *Public Library of Science* **2**(2), e25.

Delamothe T., Swan N., Bono A., Westin S., Ruiz M. et al. (1990) Blood, HIV and compensation. *British Medical Journal* **300**, 67–68.

Douaisi M., Dussart S., Courcoul M., Bessou G., Vigne R. et al. (2004) HIV-1 and MLV Gag proteins are sufficient to recruit APOBEC3G into virus-like particles. *Biochemical and Biophysical Research Communications* **321**, 566–573.

Fauci A. (2003) HIV and AIDS: 20 years of science. *Nature Medicine* **9**, 839–843.

Galel S., Lifson J. and Engleman E. (1995) Prevention of AIDS transmission through screening of the blood supply. *Annual Review of Immunology* **13**, 201–227.

Garber M. and Jones K. (1999) HIV-1 Tat: coping with negative elongation factors. *Current Opinion in Immunology* **11**, 460–465.

Girard M. and Excler J.-L. (1999) Human immunodeficiency virus. In Plotkin S. A. and Orenstein W. A., eds. "Vaccines," 3rd edn. W.B. Saunders Co., Philadelphia, PA.

Gotte M., Li X. and Wainberg M. (1999) HIV-1 reverse transcription: a brief overview focused on stucture-function relationships among molecules involved in initiation of the reaction. *Archives of Biochemistry and Biophysics* **365**, 199–210.

Gougeon M-L. (2003) Apoptosis as an HIV strategy for escape immune attack. *Nature Reviews Immunology* **3**, 392–404.

Greene W. (2004) The brightening future of HIV therapeutics. *Nature Immunology* **5**, 867–871.

Harris R. and Liddament M. (2004) Retroviral restriction by APOBEC proteins. *Nature Reviews Immunology* **4**, 868–877.

Horuk R. (1999) Chemokine receptors and HIV-1: the fusion of two major research fields. *Immunology Today* **20**, 89–94.

Joag S. (2000) Primate models of AIDS. *Microbes and Infection* **2**, 223–229.

Johnson R. P. and Desrosiers R. C. (1998) Protective immunity induced by live attenuated simian immunodeficiency virus. *Current Opinion in Immunology* **10**, 436–443.

Kitchen S. and Zack J. (1998) HIV type I infection in lymphoid tissue: natural history and model systems. *AIDS Research and Human Retroviruses* **14**, S235–S239.

Klein H. (2001) Will blood transfusion ever be safe enough? *Transfusion Medicine* **11**, 122–123.

Lee C. (1995) Hepatitis C and haemophilia. *British Medical Journal* **310**, 1619–1620.

Lekkerkerker A., Van Kooyk Y. and Geijtenbeek T. (2004) Mucosal-targeted AIDS vaccines: the next generation? *Trends in Microbiology* **12**, 447–450.

Lemckert A., Goudsmit J. and Barouch D. (2004) Challenges in the search for an HIV vaccine. *European Journal of Epidemiology* **19**, 513–516.

Marsh S., Albert E., Bodmer W., Bontrop R., Dupont B. et al. (2002) Nomenclature for factors of the HLA system, 2002. *Tissue Antigens* **60**, 407–464.

Mascola J. and Nabel G. (2001) Vaccines for the prevention of HIV-1 disease. *Current Opinion in Immunology* **13**, 489–495.

McMichael A. and Hanke T. (2003) HIV vaccines 1983–2003. *Nature Medicine* **9**, 874–880.

Michael N. (1999) Host genetic influences on HIV-1 pathogenesis. *Current Opinion in Immunology* **11**, 466–474.

Nabel G. (2001) Challenges and opportunities for development of an AIDS vaccine. *Nature* **410**, 1002–1007.

Pierson T., McArthur J. and Siliciano R. (2000) Reservoirs for HIV-1: mechanisms for viral persistence in the presence of antiviral immune responses and antiretroviral therapy. *Annual Review of Immunology* **18**, 665–708.

Piguet V. and Sattenau Q. (2004) Dangerous liaisons at the virological synapse. *Journal of Clinical Investigation* **114**, 605–610.

Pillay D. (2004) Current patterns in the epidemiology of primary HIV drug resistance in North America and Europe. *Antiviral Therapy* **9**, 685–702.

Piot P., Bartos M., Ghys P., Walker N. and Schwartlander B. (2001) The global impact of HIV/AIDS. *Nature* **410**, 968–973.

Sacks S., Griffiths P., Corey L., Cohen C., Cunningham A. et al. (2004) Lessons from HIV and hepatitis viruses. *Antiviral Research* **63**(Suppl. 1), S11–S18.

Sheehy A., Gaddis N., Choi J. and Malim M. (2002) Isolation of a human gene that inhibits HIV-1 infection and is suppressed by the viral Vif protein. *Nature* **418**, 646–649.

Smith, K.A. (2003) The HIV vaccine saga. *Medical Immunology* **2**, 1–7.

Steinhart C. (2004) Recent advances in the treatment of HIV/AIDS. *Expert Review of Anti-Infective Therapy* **2**, 197–211.

Tatt I., Barlow K., Nicoll A. and Clewley J. (2001) The public health significance of HIV-1 subtypes. *AIDS* **15**, S59–S71.

Tortorella D., Gewurz B. E., Furman M. H., Schust D. J. and Ploegh H. L. (2000) Viral subversion of the immune system. *Annual Review of Immunology* **18**, 861–926.

Trimble J., Salkowitz J. and Kestler H. (2000) Animal models for AIDS pathogenesis. *Advances in Pharmacology* **49**, 479–514.

Van den Brink M, Alpdogan O. and Boyd R. (2004) Strategies to enhance T-cell reconstitution in immunocompromised patients. *Nature Reviews Immunology* **4**, 856–867.

Willemot P. and Klein M. (2004) Prevention of HIV-associated opportunistic infections and diseases in the age of highly active antiretroviral therapy. *Expert Review of Anti-Infective Therapy* **2**, 521–532.

Zhang H., Yang B., Pomerantz R., Zhang C., Arunachalam S., et al. (2003) The cytidine deaminase CEM-15 induces hypermutation in newly synthesized HIV-1 DNA. *Nature* **424**, 94–98.

Zhang Z-Q., Schuler T., Zupancic M., Wietgrefe S., Staskus K. et al. (1999) Sexual transmission and propagation of SIV and HIV in resting and activated CD4$^+$ T cells. *Science* **286**, 1353–1357.

Tumor Immunology

William Coley 1862–1936

Bacterial extracts for cancer therapy

© Archives of the Cancer Research Institute (NY)

26

CHAPTER 26

A.   HISTORICAL NOTES

B.   TUMOR BIOLOGY

C.   IMMUNE RESPONSES TO CANCER

D.   CANCER THERAPY

"We must accept finite disappointment, but we must never lose infinite hope"
—Dr. Martin Luther King Jr.

Immunologists have been fascinated for decades with the idea that our immune systems may protect us against cancer by acting as a tumor surveillance mechanism. The great hope is that it may be possible to one day manipulate the immune system to kill tumor cells. Many immunologists have pursued this concept with the goal of finding a means of supplementing or replacing the standard cancer treatments of radiation therapy and chemotherapy. The importance of this quest to human health cannot be overstated, as more than 8 million new cases of cancer occur annually throughout the world. In North America, only heart disease claims more lives than cancer (refer to Table 23-1). In the United States, cancer is responsible for one of every four deaths. In 2004, over 1,386,000 new cases of cancer were diagnosed in that country, and over 500,000 people died of some form of this disease (about 1500 per day) (Table 26-1). As well as the cost in lives, cancer imposes a huge economic burden. The direct annual costs associated with the treatment of cancer patients in the United States (including drugs, hospital care, and home care) exceed U.S. $50 billion, even before the lost years of productivity of stricken workers are taken into account.

In this chapter, we first review the biology of tumor development at the molecular and cellular levels, and discuss the antigens that are associated with cancer cells. We then outline the evidence suggesting that immune responses can keep tumorigenesis in check to a point, and describe the roles of immune system cells in this task. Included in this discussion is a delineation of the mechanisms that a tumor can use to thwart immune responses. The last section of this chapter describes the various strategies of immunotherapy that have been devised to combat cancers.

A. Historical Notes

The medical literature is sprinkled with cases of "spontaneous regression"; that is, situations in which a tumor disappears apparently on its own. Even early on, this phenomenon was attributed to a successful immune response against an unknown antigen on the tumor cell. However, these cases were difficult to fully document and the hypothesis impossible to prove. About a century ago, scientists began formally investigating the possibility that the immune system could reject transplanted tumors. Their interest was awakened by the discovery that a spontaneous tumor from one experimental mouse could be transplanted into another mouse, but that the tumor was soon rejected by a host response mechanism. This success was embraced enthusiastically by researchers who were not yet aware of the allogeneic immune response that is provoked between different outbred mice. These workers mistakenly believed that the rejection of the transplanted tumor was triggered by the presence of specific tumor antigens. These results prompted a flood of hasty attempts in the late 1940s–1950s to create cancer vaccines based on what turned out to be an overly optimistic interpretation of the rodent experiment results: approaches that had apparently worked in mice turned out to be ineffective in the clinic. More in-depth laboratory studies then made it clear that the observed rejection of transplanted tumors was actually due to the recognition of the inherent differences in cell surface antigens expressed on <u>all</u> tissues of different mouse strains. Indeed, further exploration into the nature of these antigens led to the discovery of the MHC genes by Peter Gorer and George Snell and their colleagues (refer to Ch.10). Scientists were then prompted to develop inbred and congenic strains of mice with defined MHC backgrounds. Once the normal histocompatibility barriers were removed by the use

Table 26-1 **Estimated New Cancer Cases and Deaths for 2004 in the United States**

Cancer Type	New Cases			Deaths		
	Total	Male	Female	Total	Male	Female
All types	1,386,030	699,560	668,470	563,700	290,890	272,810
Lung/bronchial	173,770	93,110	80,660	160,440	91,930	68,510
Colorectal	146,940	73,620	73,320	56,730	28,320	28,410
Breast	217,440	1,450	215,990	40,580	470	40,110
Pancreatic	31,860	15,740	16,120	31,270	15,440	15,830
Prostate	230,110	230,110	N/A	29,900	29,900	N/A
Leukemia	33,440	19,020	14,420	23,300	12,990	10,310
Non-Hodgkin's lymphoma	54,370	28,850	25,520	19,410	10,390	9,020
Ovarian	25,580	N/A	25,580	16,090	N/A	16,090
Liver	18,920	12,580	6,340	14,270	9,450	4,820
Bladder	60,240	44,640	15,600	12,710	8,780	3,930
Myeloma	15,270	8,090	7,180	11,070	5,430	5,640
Melanoma	55,100	29,900	25,200	7,910	5,050	2,860
Mouth	10,080	5,410	4,670	1,890	1,070	820
Thyroid	23,600	5,960	17,640	1,460	620	840
Hodgkin's lymphoma	7,880	4,330	3,550	1,320	700	620
Bone and joint	2,440	1,230	1,210	1,300	720	580
Testicular	8,980	8,980	N/A	360	360	N/A
Eye	2,090	1,130	960	180	110	70

Not all types of cancers are listed. With information from Table 1 of "Estimated New Cancer Cases and Deaths for 2004 in the USA," Jemal *et al.* American Cancer Society: Cancer Statistics, 2004. *Cancer J Clin* **54**, 8–29. Cancers are listed in order of decreasing number of annual deaths. N/A, not applicable.

of inbred mice, a tumor transplanted from one mouse could often grow in another mouse of the same strain. Moreover, the malignant cells survived for prolonged periods in the syngeneic hosts without rejection. While these developments heralded a new era of understanding of the genetic control of the immune response, they also temporarily scuttled the theory that tumors expressed antigens that the host immune system could be encouraged to attack.

The discovery of true tumor-related antigens came in the mid-20th century from the work of Ludwig Gross and others. A popular model used at that time involved the transplantation among inbred mice of tumors induced by a chemical carcinogen called *methylcholanthrene* (MCA). For example, a mouse treated with MCA would develop tumors from which tumor cells could be recovered (Fig. 26-1). If these tumor cells were injected into a naive mouse of the same inbred strain, the tumor cells would grow, as expected. On the other hand, if the tumor in this second mouse was excised and the mouse was allowed to recover, it would later reject a second chal-

lenge with the tumor cells, demonstrating that an immune response had been mounted against the tumor. However, it was not possible to conclude definitively from this result that the response was directed at tumor-related antigens, because there was a chance that residual heterozygosity existed within the mice of this inbred strain. To rule out this possibility, it was shown that a mouse immunized with normal cells from another mouse did not mount a response that would protect against a later tumor challenge. Conversely, inoculation of a mouse with tumor cells did not result in a response that would subsequently mediate rejection of normal skin grafts. These results suggested that the inbred mice used were indeed homozygous. The final, direct proof that an immune response could be mounted specifically to a tumor-related antigen came from an experiment in which an MCA-induced tumor was completely removed from a mouse and, after a recovery period, this same mouse (the "autochthonous" host) was challenged again with the same tumor. (The tumor was kept growing in another mouse during the recovery period.) It was

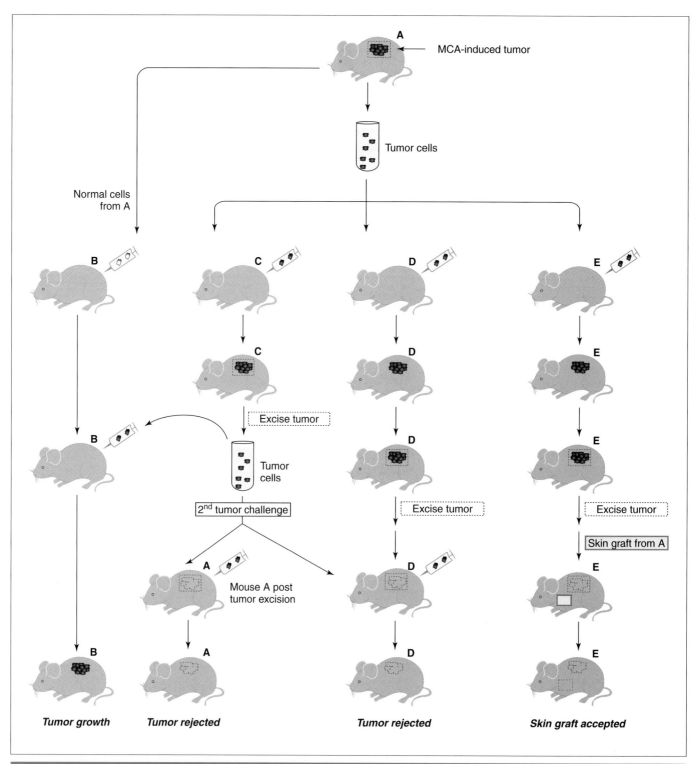

Figure 26-1

Demonstration of the Existence of Tumor-Related Antigens
Illustration of an experiment carried out using a strain of inbred mice. Mouse A is treated with methylcholanthrene (MCA) to induce tumor formation. When tumor cells are excised from A and injected into another mouse (C, D, or E), tumor growth is observed. However, when mouse D has its tumor excised and is allowed to recover, it can then reject a secondary challenge with the tumor cells that have been harvested from mouse C, demonstrating that an immune response to the tumor cells has occurred. Priming of mouse B with normal cells from mouse A does not illicit anti-tumor protection, and the response of mouse E to the tumor does not prevent it from later accepting a skin graft from mouse A, suggesting that residual allelic differences between the inbred mice are not responsible for the observed response against the tumor cells. As a final confirmation that the response is truly directed against tumor-specific antigens not found on normal cells, the original tumor host (mouse A) is re-challenged with the same tumor and the tumor is rejected. Adapted from Schreiber H. (2003) Chapter 48. In Paul W., ed. "Fundamental Immunology," 5th edn. Lippincott Williams and Wilkins. Philadelphia.

found that this animal was able to reject cells from its own tumor. Taken together, these experimental results confirmed that the immunity induced by the tumor cells was directed against antigens found on the tumor but not on normal cells.

Similar experimental results were obtained for tumors induced by other chemical agents or UV-irradiation. Researchers then concluded that tumors had to contain specific antigens that were not present in normal tissues and could be recognized by the immune system. These observations encouraged investigators in their belief that one of the normal functions of the immune system was to monitor the body for the rise of tumor cells and to eliminate them before they could cause disease. Subsequent investigations implicated T cells as the agents of tumor rejection. It was therefore puzzling when Osias Stutman reported that *nude* mice (which lack a thymus and thus most mature T cell subsets) did not have an elevated incidence of tumors. However, it was subsequently demonstrated that *nude* mice do have some T cells derived by extrathymic means, and also possess NK cells capable of producing IFNγ and carrying out the perforin-mediated cytolysis of targets such as tumor cells. STAT1$^{-/-}$ mice, which have defects in both innate and adaptive immunity, show a significantly increased frequency of tumor development. These observations have spurred some researchers to continue their attempts to identify antigens capable of provoking anti-tumor responses in normal animals.

Another milestone in tumor immunology came with the recognition of the role of pathogens in some types of cancer. It had long been noted that patients with inherited immuno-deficiencies had an elevated risk of developing certain rare cancers such as Burkitt's lymphoma and Kaposi's sarcoma. With the rise of AIDS in the 1980s, it became clear that acquired immunodeficiency also left a patient vulnerable to these same kinds of tumors. Transplant recipients, who were deliberately immunosuppressed to allow their transplants to "take," were also sometimes victims of these rare cancers. Some scientists took these observations as evidence that, in healthy individuals, the immune system routinely seeks out and destroys tumor cells, a concept referred to as "immunosurveillance." However, detractors of the immunosurveillance theory pointed to the fact that the more common malignancies such as colon, breast, lung, and prostate tumors did not occur at increased frequency in immunodeficient or immunosuppressed individuals. Debate see-sawed back and forth until the late 1990s, when the sum of the accumulated evidence indicated that the increased frequency of only some types of tumors in immunosuppressed individuals was most likely due to associations between particular pathogen infections and the development of certain tumors. For example, it was found that EBV infection was associated with Burkitt's lymphoma, HepB and HepC with liver cancer, KSHV with Kaposi's sarcoma, HPV with cervical cancer, and *Helicobacter pylori* with stomach cancer. In each of these cases, the compromised immune response in an immunodeficient individual gave the pathogen free rein, increasing the potential for it to pursue its transformative course.

Although the preceding revelations have potentially clarified the origins of some tumors, the role of the immune system in dealing with malignant transformation remains obscure. To this day, tumor immunology is a field in which much is still a "black box." Researchers continue to struggle to understand how malignancies arise, why the immune response does not eliminate all tumors promptly, and why some cancers appear to induce tolerance to themselves rather than an immune response. We move now to a discussion of basic tumor biology and the underlying molecular changes that drive malignant cell transformation and proliferation.

B. Tumor Biology

I. WHAT IS A TUMOR AND WHAT IS A CANCER?

Throughout our lives, our cells divide, differentiate, and die in a way that is carefully controlled. Organ-specific stem cells in various adult tissues (like the HSC of the hematopoietic system) give rise to progressively more differentiated forms, each of which strikes a balance between proliferation, which allows the tissue to grow, and ever more restricted differentiation, which allows the tissue to function. In other words, the more differentiated a cell, the slower its rate of proliferation. The requirements for growth and differentiation of different tissues vary over time from prenatal life and birth through childhood, puberty, reproductive maturity, and lastly, the senescence of old age. A complicated and fascinating genetically controlled program quietly governs this complex process such that we are not aware of it on a day-to-day basis. Sometimes, the cells of a tissue will undergo more specialized cell division or differentiation. Such normal (but not routine) tissue changes include *hyperplasia* (more cells are produced than lost), *metaplasia* (a different cell type is produced), and *dysplasia* (undifferentiated or partially differentiated cells appear with fully differentiated cells). Examples are the hyperplasia seen in the breast tissue of a pregnant woman and the metaplasia seen in the bronchus of a smoker, where, due to damage being incurred, cells begin to differentiate into squamous rather than typical respiratory epithelial cells. These types of changes are normal responses to specific stimuli and demonstrate reversibility when the stimulus is removed. On rare occasions, however, the cells of a tissue undergo division that is unusual and serves no useful function for the host. This proliferation creates an abnormal tissue mass called either a *neoplasm* ("new growth") or *tumor* ("swelling"). It is important to note that, although the term "tumor" makes one think of a mass of rapidly dividing cells, the majority of neoplastic cells are not dividing faster on an individual basis than normal cells of the same type and stage of differentiation. Rather, it is the higher than normal proportion of cells dividing, coupled with a decreased rate of programmed cell death, that causes the abnormal accumulation of cells that constitutes a tumor.

i) Classification of Tumors

All tumors can be classified as either *benign* or *malignant*. A benign tumor is relatively slow-growing because it contains cells that are well differentiated and well organized so that the tumor is very much like the normal tissue from which it originated (Plate 26-1). Like a healthy tissue, the tumor is surrounded by a stromal cell framework and is nourished by blood vessels. Factors secreted by the stromal cells then perpetuate the proliferation of the neoplastic cells. However, a benign tumor is securely encapsulated and the altered cells cannot break away from the cell mass and circulate in the body. In clinical nomenclature, many benign tumors are denoted by the suffix *oma*, which is added to the root term identifying the tissue or organ involved. For example, a *lipoma*, a *neuroma*, and an *adenoma* are benign tumors of fat tissue, neural tissue, and glandular tissue, respectively. These tumors do not normally cause death, but if they do, they do so by indirect means. For example, a benign tumor of the adrenal medulla may produce epinephrine the way normal cells of this tissue would. The resulting oversupply of epinephrine may cause an increase in blood pressure that is life-threatening. Benign tumors may also cause illness or death if the pressure they exert due to their abnormal size or location compresses or damages adjacent tissues and organs, particularly the brain.

In contrast to benign tumors, malignant tumors are directly lethal to the host unless they are completely removed or killed. The level of differentiation of cells in a malignant tumor is defined by where the original transformed cell was on the continuum from stem cell (completely undifferentiated) to mature cell (completely differentiated). Tumors in which the cells are poorly differentiated (*anaplastic*) tend to be more aggressive and associated with poor prognoses. Clinicians sometimes refer to these tumors as being *high grade* malignancies. Conversely, a *low grade* tumor is one that is not growing aggressively and whose component cells are fairly well differentiated. In general, the cell mass of a malignant tumor is disorganized and rarely encapsulated. Thus, these tumors are able to become *invasive*, moving into and destroying the healthy architecture of nearby organs. It is this characteristic that gave cancer its name. In

ancient Greece, postmortem examinations sometimes revealed that normal tissues had been invaded by lines of hard gray tissue that bore a resemblance to the legs of a crab. This appearance inspired physicians of the day to call this condition *cancer*, from the Latin word meaning "crab." In addition to being anaplastic and invasive, malignant tumors are frequently *metastatic*, meaning that cells from the original tumor mass (the *primary tumor*) can break away and spread via the blood to secondary sites at nearby or distant locations in the body. These secondary tumors are called *metastases* and the process of their establishment is called *metastasis* (from the Greek for "change in position"). Metastasis is not necessarily a late-stage feature of tumorigenesis: for example, certain small breast cancers become metastatic very early in their development and rapidly disseminate to distant tissues.

Malignant tumors are named using terms that describe the tissue of origin and the cell type in that tissue from which the tumor arose. When the suffix *carcinoma* appears in the tumor name, the malignancy has arisen from epithelial cells. For example, an "adenocarcinoma" is a malignant tumor of glandular epithelial origin, while a "carcinoma of the lung" is derived from the epithelium of lung. The suffix *sarcoma* indicates that the malignancy arose in one of the numerous types of cells supporting the body, such as cells of connective tissue, muscles, bone, or cartilage. For example, an "osteosarcoma" is a malignant tumor of bone, and a "rhabdomyosarcoma" is a malignant tumor of the skeletal muscles or tendons. The suffix *blastoma* is used for highly malignant childhood tumors that are so undifferentiated that the cells resemble those in a blastocyst. Three tumors in this category are neuroblastoma, retinoblastoma, and nephroblastoma, which are derived from neuroblasts, retinal cells, and embryonic kidney cells, respectively. It should be noted that there are many exceptions to these nomenclature guidelines, some of the better known being melanoma (malignancy of skin melanocytes responsible for producing skin color pigment), hepatoma (malignancy of hepatocytes), leukemia (liquid malignancy of leukocytes in the

---

Plate 26-1

Melanoma
(A) Benign melanoma showing lack of mitosis and atypical cells. (B) Malignant melanoma showing mitosis and atypical cells. Courtesy of Dr. Danny Ghazarian, Princess Margaret Hospital, University Health Network, Toronto.

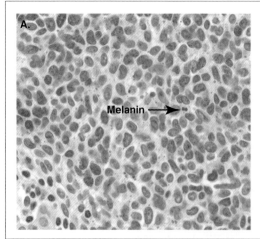

A.

Melanin →

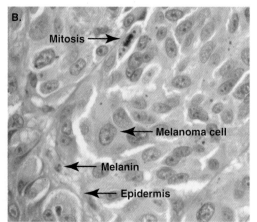

B.

Mitosis →

← Melanoma cell

← Melanin

← Epidermis

blood), and lymphoma (solid malignancy of lymphoid cells in lymphoid tissues). Examples of various benign and malignant tumors are given in Table 26-2.

ii) Morbidity and Mortality

How does a cancer cause morbidity and mortality? Although one might assume that a direct effect of malignant growth would be solid organ failure, this is rarely a direct cause of death. The vital organs are very resilient, and one can often survive with only a fraction of an organ in functional

condition. Indeed, a person with a solid organ tumor may be unaware of the impending disaster for quite a long time. Eventually, the stretching of the organ capsule may lead to pain that will bring attention to the condition. Abnormal bleeding and pain due to penetration of the tumor deeper into the wall of a body tract may also signal the presence of cancer. In hollow areas of the body, such as the GI tract and various ducts, a cancer will often cause an obstruction that leads to symptoms of illness and sometimes death. For example, total obstruction of the descending colon causes death quickly unless surgical removal of the offending mass is carried out immediately upon diagnosis.

Despite the seriousness of the preceding effects, the major causes of death in cancer patients are cachexia and infection. The cachexia seen in cancer cases is often described as "terminal wasting" and is marked by weight loss and muscle wasting resulting in a form of starvation. While one might think that this wasting is due to a redirecting of nutrients to the growing tumor, it is actually due to a general upset in metabolism caused by biologically active substances released from the tumor, the best-recognized being TNF. The weakness associated with cancer-induced cachexia also renders the patient immunodeficient, so that, in most terminal cancer patients, the immediate cause of death is pneumonia or septicemia caused by opportunistic microbes.

iii) Stages of Cancer

Oncologists use a nomenclature system to describe the seriousness of a given cancer. Depending on this assessment, the appropriate combination of treatment options can be selected. The parameters considered are the primary tumor size, how many lymph nodes have been infiltrated, and whether metastasis to distant tissues has occurred. The size of the primary tumor (T) can range from T1 (small) to T4 (large). Similarly, lymph node involvement (N) ranges from N0 (no involvement), to N1 and N2 (progressively more lymph nodes infiltrated), to N3 (extensive nodal involvement). (This travel of cancerous cells to the draining lymph node occurs via the afferent lymphatic and is considered by solid tumor oncologists to be "nodal metastasis.") Metastasis to distant tissue sites via the blood is symbolized by "M," with M0 meaning no metastasis, and M1 meaning that secondary tumors in distant organs or tissues (other than lymph nodes) are present. Thus, a cancer described as T2N1M0 is a medium-sized primary tumor that has spread to one set of lymph nodes, but has not metastasized to distant tissues. A cancer described as T4N3M1 consists of a large primary tumor with extensive lymph node involvement and at least one metastasis in a distant tissue.

Once the TNM characteristics of a cancer are defined, it is classified as being at a particular stage of cancer progression for that particular tumor type. Based on the stage of the cancer, the oncologist then decides on the combination of surgical excision (if possible), chemotherapy, and irradiation (localized or whole body) that best suits the control of the tumor for that particular patient. In general, the malignant tumor in a stage I cancer patient is small and quite confined, and often easily removed by surgery. A stage II cancer is also considered operable but the risk of the disease spreading further is significant. A stage III

Table 26-2 Examples of Benign and Malignant Tumors

Tumor Name	Tissue of Origin
BENIGN TUMORS	
Adenoma	Glandular cells
Angioma	Cells of blood vessels, lymphatic vessels
Chondroma	Cartilage cells
Fibroma	Connective tissue fibroblasts
Lipoma	Fat cells
Neuroma	Nervous system cells
MALIGNANT TUMORS	
Adenocarcinoma	Epithelial cells of the adrenal gland
Astrocytoma	Low grade tumor of astrocytes (star-shaped brain cells)
Basiloma	Carcinoma of basal cells in skin
Carcinosarcoma	Mixture of epithelial and connective tissue cells
Chondrosarcoma	Cartilage cells
Fibrosarcoma	Connective tissue fibroblasts
Glioblastoma	High grade tumor of astrocytes
Hepatoma	Hepatocytes
Leiomyosarcoma	Smooth muscles of digestive tract, uterus, bladder, etc.
Leukemia	Leukocytes
Lymphoma	Lymphocytes
Melanoma	Melanocytes
Nephroblastoma	Embryonic kidney cells
Neuroblastoma	Neuroblasts
Osteosarcoma	Connective tissue of bone
Retinoblastoma	Retinal cells
Rhabdomyosarcoma	Skeletal muscles and tendons
Seminoma	Testis

tumor has invaded deeply into surrounding structures such that it cannot be operated on easily. A stage IV cancer shows metastatic spread of tumor cells from the primary tumor to remote sites in the body. Some stage IV cancers can be treated successfully, such as many cases of testicular and colorectal cancers and Hodgkin's lymphomas. However, the care of most other stage IV cancer patients is usually palliative; that is, designed to relieve pain and provide comfort to the patient rather than totally eliminate his or her widely spread cancer. While the surgical removal of isolated metastases coupled with chemotherapy and radiation treatment may improve the survival of stage IV patients (at least for a short time), they may also decrease the "quality of life" of the patient's remaining days. The intensity of any therapy must therefore be carefully managed.

iv) Paraneoplastic Syndromes

Sometimes a cancer affects organs or tissues that are quite distant from the site of the actual tumor. When these effects result in clinical symptoms, the patient is said to have a *paraneoplastic syndrome* because the condition is caused by the neoplasm in an indirect manner. Some examples of paraneoplastic syndromes include hypercalcemia (abnormally high calcium in the blood), the inappropriate secretion of anti-diuretic hormone (which normally reduces the amount of water eliminated by the body), and a range of skin disorders. An intriguing subgroup of these diseases causes dysfunction within the central nervous system that is often of equal or greater concern than the primary cancer because the disruption can result in sudden, severe disability. For example, some cases of cerebellar degeneration, dementia, amyotrophic lateral sclerosis, and myasthenia gravis (see Ch.29) can be attributed to the presence of a malignant tumor. For a given type of cancer, the range of CNS-related paraneoplastic disorders triggered may be quite limited or quite broad. For example, while breast and ovarian cancer are most likely to trigger cerebellar degeneration, a thymoma has the potential to initiate more than 10 different nervous system disorders, including encephalitis (inflammation of the brain), polyneuritis (inflammation of the peripheral nerves), and chorea (involuntary irregular movements of the head and extremities). The onset of the CNS symptoms is often what prompts a person to seek medical attention in the first place, because the cancer itself is asymptomatic. Further investigation then leads to the diagnosis of the causative tumor.

So how does a tumor actually cause dysfunction at a remote site in the body? Researchers currently studying CNS paraneoplastic syndromes believe that they are the result of cancer-initiated autoimmunity. A newly transformed tumor cell may inappropriately express proteins that would normally be found exclusively on CNS tissue at non-immunogenic levels. Such abnormal expression of these proteins on the tumor may then trigger both humoral and cell-mediated responses against the antigens (see later). While this immune response is thought to slow tumor growth, it has the unfortunate side effect of targeting and damaging vital CNS tissues, leading to CNS malfunction. Evidence that CNS paraneoplastic syndromes are indeed autoimmune-mediated includes the presence in patients of high titer antibodies specific for both the tumor and CNS tissue. In addition, antigen-specific T cells can be found in the blood, brain, and cerebrospinal fluid. Once a paraneoplastic syndrome is diagnosed, timely identification and treatment of the causative tumor constitute the most effective therapeutic approach. While a reversal of the CNS symptoms may not be possible, the progression of the syndrome may at least be halted. In some cases, immunosuppression may also lead to some clinical improvement.

II. CARCINOGENESIS

The malignant transformation of a cell is due to a multi-step process called *carcinogenesis* that culminates in the deregulation of cellular growth. Since cellular growth patterns are controlled genetically, it follows that the deregulation observed in cancer cells is a direct result of the inappropriate expression of various genes. The human genome is normally very stable, maintained by multiple mechanisms entrusted with the rapid and accurate repair of mutations, or the induction of cell cycle arrest or death of cells with gross genetic abnormalities that cannnot be repaired. Cancer results only when these genome stabilizing or elimination mechanisms fail, or environmental factors increase the rate of mutation such that the repair mechanisms cannot keep up. Deleterious mutations can then accumulate, leading to inappropriate gene expression that drives abnormal cellular growth.

Some scientists believe that at least some cancers establish and expand due to the transformation of stem cells, be they hematopoietic stem cells (HSCs) or organ-specific stem cells. In this scenario, the transformed stem cells driving tumorigenesis constitute only a very small proportion of the cells making up the malignancy. Indeed, if one isolates the leukemic blood cells from patients with a leukemia known as *acute myelogenous leukemia* (AML; see Ch.30) and tests various fractions of these cells for their ability to initiate tumors in cancer-prone mice, one finds that only a small fraction of undifferentiated cells is able to do so. Moreover, the leukemia that develops in these mice shows a range of differentiated cell types that parallels the situation in the human patients. This result suggests that AML is driven by a rare subset of malignant stem cells with the power to both self-renew and potentially give rise to more differentiated progenitors. Similar results were found in one study of human breast cancer, in that only a small minority of cells within the breast tumor had the capacity to successfully establish tumors in cancer-prone mice. While examples of this sort are still limited in number, the transformed stem cell hypothesis is gaining adherents because it could neatly account for the continuous proliferation and undifferentiated nature of many tumor cells. In addition, stem cells have comparatively long life spans and self-renew constantly, providing ample opportunity for tumorigenic mutations to accumulate. Nevertheless, despite the intuitive appeal of this hypothesis, it remains to be shown that it applies in general to human cancers. In addition, it cannot yet be ruled out that the transformation event actually occurs in partly or fully differentiated cells that then undergo some kind of de-differentiation process to acquire their proliferative powers. De-differentiation of committed progenitors has been demonstrated in *Drosophila melanogaster*. If indeed the stem cell theory of tumorigenesis is demonstrated more widely, our concepts of cancer treatment may have to undergo a major shift. Treatments designed to cure cancer would necessarily become more focused on targeting the malignant stem cell component.

We note here that the nature of the mutation in a human cancer classifies it as either *sporadic* or *familial*. Most human cancers are sporadic in that the tumorigenic mutations occur in a somatic cell of a tissue and do not affect his or her germ cells. In rare cases, the genetic alterations that increase the chances of malignant transformation occur in the germ cells of an individual. In this situation, the genetic predisposition toward malignancy becomes a heritable trait that may lead to the increased incidence of particular cancers among descendants. Such familial cancers are thus also called "inherited" or "hereditary" malignancies. In laboratory animals, a *spontaneous* cancer is one that develops in the absence of any experimental manipulation.

i) Four Steps to Carcinogenesis

The steps of carcinogenesis necessary to establish a primary tumor are defined as *initiation, promotion, progression*, and *malignant conversion* (Fig. 26-2). In the initiation step, the DNA in the nucleus of a cell experiences either a strand break or, more commonly, a nucleotide alteration. In most cases, the altered nucleotide (known as an *adduct*) either has no effect on the cell or is repaired such that there are no alterations to the structure or function of any cellular proteins. Similarly, the majority of strand breaks are repaired enzymatically with no negative consequences for the cell. In some cases, however, the alteration or break introduces an error that is not repaired and alters the sequence of a protein involved in growth regulation. In a subset of these cases, the cell in which the genetic event occurred may eventually gain a growth advantage. Such a cell is said to be the *target* cell of a tumor, and the target cell is said to have undergone *initiation*. Because the genetic change that has occurred within the genome of the target cell is permanent, initiation is irreversible.

Promotion, the next step in carcinogenesis, involves the exposure of the initiated target cell to a stimulus that allows the selective proliferation of this cell. An expanding *preneoplastic clone* of genetically altered cells then develops. However, unlike initiation, promotion is completely reversible, so that if the promoting stimulus is removed, the clone will undergo *regression* (resolution of the tumor) to a cell number in keeping with that present in neighboring healthy tissue.

In the penultimate step of carcinogenesis, known as *progression*, additional genetic events occur in one of the cells of the preneoplastic clone. Genetic instabilities introduced into regulatory genes during the initiation and promotion stages may predispose the genome of the cell to such further alterations. These secondary mutations may arise spontaneously, or may be caused by environmental influences impinging on the cell. At this point, the preneoplastic clone becomes a *neoplastic clone* with a significant growth advantage over normal cells. In some cases, this growth advantage depends on the presence of a hormone or pathogen, in that the tumor regresses when the hormone or pathogen is removed. In other cases, the growth

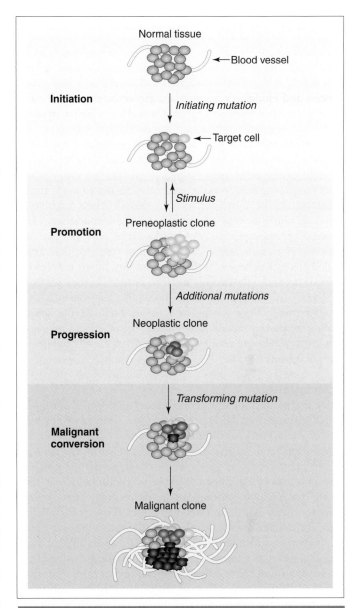

Figure 26-2

Four Steps of Carcinogenesis
Initiation: A normal cell becomes a target cell when it undergoes a mutation that damages its DNA such that it cannot be repaired properly and confers a growth advantage. Promotion: In the presence of a stimulus, the target cell proliferates to become a preneoplastic clone. At this point, the enhanced growth is reversible if the stimulus is removed. Progression: A cell of the preneoplastic clone accumulates additional mutations further driving growth such that it proliferates to become a neoplastic clone. Malignant conversion: A cell of the neoplastic clone acquires a transforming mutation that causes it to become fully malignant and allows it to proliferate uncontrollably. Fed by newly formed blood vessels, the tumor grows aggressively and becomes invasive.

advantage is due solely to the accumulated mutations and is irreversible. These clones are on the threshold of malignancy.

Carcinogenesis is complete when the *malignant conversion* stage is reached. One of the neoplastic cells in a state of progression acquires additional mutations, including the transforming mutation that finally pushes the cell to adopt the invasive and metastatic growth pattern characteristic of malignancy. There may be structural changes to cell morphology and architecture, and biochemical changes to metabolic pathways, that enhance the tumor cell's resistance to normal apoptosis-inducing signals. The highly mutated and malignant cell then grows in a totally deregulated manner, generating a clone of continually dividing abnormal cells within the original tissue. Even within this clone, there is a constant competition for nutritional resources. The most aggressive cells that acquire new mutations conferring enhanced survival are selected in a Darwinian manner, leading to cell overgrowth and formation of the primary tumor. However, to grow beyond a minimal size, the primary tumor must induce the formation of new blood vessels capable of bringing nutrients into the interior of the tumor mass. This process of *angiogenesis* depends in large part on *vascular endothelial growth factor* (VEGF), a molecule that acts as an endothelial cell mitogen. Usually the primary tumor induces the normal stromal cells surrounding it to upregulate their secretion of VEGF and other angiogenic molecules, but sometimes the proliferating cells of the primary tumor itself are the source. With a supporting vasculature in place, the malignant neoplasm can grow without restraint and eventually disrupts neighboring normal tissues.

ii) Metastasis

After the establishment of the primary tumor, the malignant cells continue to acquire new mutations, some of which may confer metastatic competence. Metastasis requires that the transformed cells gain the ability to detach from the primary tumor mass, move on their own through the tissues, and survive in a new environment. Metastatic tumor cells first bind to the basement membrane of the tissue layer in which the primary tumor has established and punch through it, invading the extracellular matrix. The cells pass through the extracellular matrix and invade the endothelial cell layer lining a local blood vessel. Once inside the blood vessel, the metastatic cells circulate around the body until they adhere to the surface of a blood vessel in another part of the body. The metastatic cells then reverse the process, extravasating out of the blood and into the extracellular matrix. A secondary tumor can then start to grow in this location.

What characterizes a metastatic cell? Such a cell must acquire altered properties of adhesion, migration, and survival to be able to detach from the primary tumor, evade the immune system during migration, and establish in a secondary site. Scientists are still deciphering the very complex mechanisms underlying metastasis, but a molecule that appears to be of key importance is CD44. We have encountered this protein before in the context of T cell activation and homing, but it is also expressed on most other hematopoietic cells and on many non-hematopoietic cell types. In the context of metastasis, variant isoforms of CD44 are overexpressed on many types of metastatic tumor cells. These isoforms appear to be derived from alternative splicing of the 20-exon CD44 gene. Only 10 exons are used in normal CD44, leaving 10 others that are differentially included in the variant CD44 isoforms associated with tumorigenesis. One theory is that, in contrast to the adhesion function of normal CD44, these variant CD44 isoforms may be anti-

adhesive in nature, facilitating the escape of the affected cell from the primary tumor. In CD44 knockout mice lacking CD44 exon 6 (one of the exons found frequently in tumor-associated variant CD44 molecules), primary tumors grew unimpeded following induction but metastasis was inhibited.

Another molecule that may be crucial for metastasis is RhoC, a member of the Rho subfamily of the Ras superfamily of signaling molecules. RhoC is overexpressed on highly metastatic melanoma cells and breast and pancreatic cancer cells. In addition, the engineered overexpression of RhoC by a poorly migrating melanoma cell can increase its metastatic capacity. Conversely, inhibition of RhoC function can decrease melanoma cell metastasis. Preliminary studies have shown that conditional knockout mice lacking RhoC expression in the mammary gland show decreased metastasis of mammary tumors compared to controls. The mechanism by which RhoC might influence metastasis is under investigation.

The migration of metastatic tumor cells is not random. Different types of cancer cells migrate to and establish secondary tumors in different tissues. The factors governing this migration are only now being uncovered, and it should not surprise the reader to learn that the expression of chemokines and their receptors are of key importance. For example, cells in some primary breast cancers have been found to express increased levels of the chemokine receptor CXCR4 compared to normal epithelial cells. Researchers speculate that, should a cancerous cell break free of the tumor mass, the presence of this receptor on the cell surface might allow it to home to tissues in which there are large quantities of the SDF-1 (CXCL12) ligand for this receptor (Fig. 26-3). SDF-1 is produced selectively by resting cells in the lungs, liver, bone marrow, and lymph nodes, and it is these tissues that happen to be the most common sites of breast cancer metastasis. Breast cancer cells also express high levels of CCR7, a receptor whose chemokine ligand SLC (CCL21) is expressed almost exclusively in the lymph nodes. Once in the secondary sites, adhesion molecules come into play that promote the extravasation of the tumor cells from the bloodstream into the target tissue. Melanoma cells also express CCR7, meaning that cells of this type of tumor are similarly drawn into the lymph nodes. In addition, melanoma cells express CCR10. Secretion of the chemokine CCL28 (a ligand for CCR10) by skin cells promotes the migration of melanoma cells to various locations in the skin, establishing metastases at distant sites in this tissue (Plate 26-2).

III. TUMORIGENIC GENETIC ALTERATIONS

While many genes in a malignant cell may undergo mutation, it is the modification of those involved in the control of cell growth that are at the root of cancer development. The major classes of genes involved in controlling cell growth that are modified in cancer are the *DNA repair genes*, *oncogenes*, and *tumor suppressor genes* (TSG). In some cancers, it is a DNA repair gene that is mutated first during the initiation phase of carcinogenesis. The cell's genetic stability is thus weakened and it becomes predisposed to acquiring additional mutations that may activate oncogenes or inactivate TSGs. Mutation

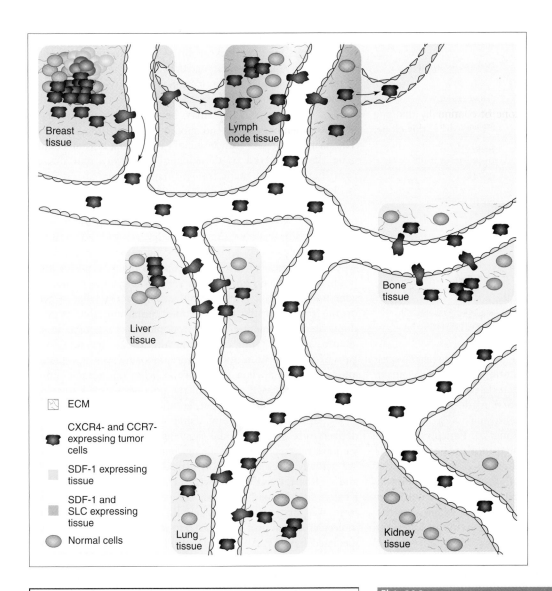

Figure 26-3

The Role of Chemokines in Metastasis
In this example, malignant transformation of a breast cell has given rise to a primary tumor (upper left) in which the component cells express CXCR4 and CCR7. These cells traverse the extracellular matrix (ECM) and either access the lymphatic vessels draining the breast tissue or force their way through the endothelium lining a local blood vessel to enter the circulation. The metastatic primary tumor cells are drawn to leave the blood or lymphatic circulation wherever the cells of a tissue express high levels of the chemokines SDF and/or SLC which are ligands for CXCR4 and CCR7, respectively. The lymph nodes, liver, bone, and lung are such tissues and thus are the primary sites of breast cancer metastasis. The kidney does not express significant levels of either SDF or SLC and thus escapes invasion by this type of tumor cell.

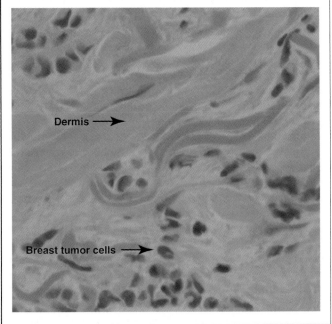

Plate 26-2

Metastatic Breast Tumor Infiltrating the Dermis of the Skin
Courtesy of Dr. Danny Ghazarian, Princess Margaret Hospital, University Health Network, Toronto.

of an oncogene or TSG greatly increases the likelihood that the cell will go on to malignant transformation. In other cancers, the sequence of tumorigenic mutations is completely unknown.

i) DNA Repair Genes

The protein products of DNA repair genes detect and fix damage that other genes sustain due to spontaneous mutation, nucleotide mismatching, or the oxidation and alkylation that occur sporadically in mammalian cells. Thus, although proteins involved in DNA repair are not directly involved in cell cycle regulation, if they do not function properly, the chances are increased that genes directly involved in proliferation

Table 26-3 DNA Repair Genes Associated with Cancer

DNA Repair Pathway	Gene Name	Disease Association
HRR	BRCA1 BRCA2 XRCC3	Breast cancer Breast cancer Melanoma
MMR	MSH1 MSH2 MSH6	HNPCC HNPCC HNPCC
NER	XRCC1 XPC	Breast cancer Melanoma in xeroderma pigmentosum
NHEJ	ATM NBS1	Breast cancer, lymphoma, rhabdomyosarcoma Leukemia, breast cancer, multiple myeloma

HNPCC, hereditary non-polyposis colorectal cancer; HRR, homologous recombination repair; MMR, mismatch repair; NHEJ, non-homologous end-joining pathway; NER, nucleotide excision repair.

control will acquire deregulating mutations that lead to cancer (Table 26-3). The connection between defects in DNA repair and increased carcinogenesis was first made over 20 years ago in studies of xeroderma pigmentosum (XP), a primary immunodeficiency described in Chapter 24. Patients with XP had long been observed to have a high incidence of skin cancers, but it was unknown why. Later investigations at the molecular level revealed that XP patients have deficiencies in the nucleotide excision pathway that repairs DNA damaged by exposure to UV radiation. A more recent example is the discovery that individuals in some families with an inherited predisposition to colorectal cancer carry mutations in the DNA repair gene MSH2. MSH2 participates in a complex that is responsible for the repair of nucleotide mismatches. The loss of function of the MSH2 protein results in what is called a "mutator phenotype," meaning that the subsequent string of mutations required to achieve colorectal carcinogenesis (see Box 26-1) can occur more easily in a person bearing an MSH2 mutation than in a normal individual.

Two important DNA repair genes involved in carcinogenesis are BRCA1 (*breast cancer associated-1*) and BRCA2. The products of these genes are crucial for the proper functioning of the homologous recombination DNA repair pathway (refer to Ch.24). BRCA1 is involved in both regulating transcription and responding to signals from upstream molecules that DNA has been damaged, while BRCA2 recruits a key DNA repair enzyme to the site of DNA damage. About 5–10% of all breast cancers are hereditary, and a large percentage of these cases are due to a mutation in either the BRCA1 or BRCA2 gene. Cells in which these proteins no longer function experience genomic instability, resulting in gross chromosomal aberrations such as truncations and ring chromosomes. Knockout mice totally deficient for either Brca1 or Brca2 (the murine equivalents of BRCA1 and BRCA2) die very early during embryonic development. In conditional knockout mice lacking expression of Brca1/2 only in their mammary glands, the animals successfully reach adulthood but develop mammary tumors as they age. Interestingly, mice heterozygous for the knockout Brca1 or Brca2 mutations show no increased risk of developing cancer, indicating that one copy of these genes is sufficient (at least in mice) to avoid an increased incidence of unrepaired genetic aberrations.

ii) Oncogenes

The term oncogene ("cancer gene") refers to a normal cellular gene that is altered or deregulated in a way that directly contributes to malignant transformation. Once an oncogene is identified, its normal cellular counterpart is referred to as a *proto-oncogene*. Most proto-oncogenes are positive regulators of cell growth, such as growth factors, growth factor receptors, intracellular signal transducers, and transcription factors (Table 26-4). A proto-oncogene gains oncogene status when it becomes constitutively activated due to translocation, point mutation, proximity to a retroviral gene (see later), or gene amplification. (Gene amplification occurs when a gene, or the region of the chromosome containing that gene, is duplicated to generate multiple copies of that gene in the genome.) These types of genetic alterations are said to be *gain-of-function* mutations because oncogenes are more active and/or functional than their normal proto-oncogene counterparts. The increased activity of the oncogene drives sustained intracellular signaling and perpetual cell division.

A particularly important family of human oncogenes is derived from the intracellular signaling transducer Ras. Although we have referred to Ras as a single entity up to this point, there are actually three distinct Ras genes, known as H-Ras, K-Ras, and N-Ras (H for its discoverer Harvey, K for its discoverer Kirsten, and N for "neuronal," the cell type in which this isoform was first identified). Each Ras gene encodes a 21 kDa molecular switch protein involved in relaying intracellular signals mediated by GTP. About 30% of all human tumors, including colorectal, pancreatic, breast, and skin cancers, exhibit oncogenic mutations in one of the three Ras proteins. Interestingly, regardless of whether the targeted protein is H-Ras, K-Ras, or N-Ras, the oncogenic mutation is found exclusively in codons 12, 13, or 61, with the most frequent changes occurring in codon 12.

Most other oncogenes are rogue forms of nuclear transcription factors. The best-studied example is c-myc, first isolated as the mammalian homologue of the transforming sequence of an avian tumor virus. As we have learned, one of the normal functions of c-myc is to promote proliferation. When c-myc is overexpressed due to a gain-of-function mutation or gene amplification, carcinogenesis is favored due to the resulting abnormally rapid cell division. c-myc amplifications have been found in human breast, colon, and lung carcinomas. More on c-myc and malignant transformation in leukemias and lymphomas can be found in Chapter 30.

Another cause of oncogene activation is retroviral integration (at least in mice and chickens). In fact, oncogenes were first discovered during studies of retroviral replication. Retroviruses are generally categorized as either *exogenous* or *endogenous* viruses. An exogenous retrovirus can survive for lengthy periods outside of host cells in the intercellular spaces of the tissues. An endogenous retrovirus is maintained as a provirus in the host cell genome for a long time and passes most of its existence as

Intense interest in the pathogenesis of colorectal cancer is hardly surprising from a clinical point of view: it is the second leading cause of cancer mortality in North America, exceeded only by lung cancer. However, colorectal cancer has also drawn the attention of researchers hoping to better understand the molecular nature of tumorigenesis. This malignancy lends itself to such analysis because, unlike other common human tumors, samples of colorectal cancers from very early through late stages of development are easily obtained from large numbers of patients. As a result, colorectal cancer has emerged over the last 10–15 years as

an excellent model of how an accumulation of genetic mutations can lead to tumor formation.

The clinical progression of colorectal cancer occurs in a well-defined series of steps. First, a normal colonic epithelial cell gives rise to early adenoma cells, which in turn generate intermediate and then late adenomas. The final transition is seen when one or more cells of the late adenoma undergoes malignant transformation to colorectal carcinoma. Researchers have found that, while tumors at each of these stages may show a range of genetic alterations, there are a few mutations that are present at very high frequency at a particular stage. These mutations are

responsible for the growth advantage and expansion of cells that facilitate continued development of the tumor, as shown in Figure A. Interest in this model of colorectal tumorigenesis continues to grow, as it is now thought that many other carcinomas, such as those in the lung and pancreas, may arise via a similar developmental pathway.

What are the mutations that drive the development of colorectal cancer? The earliest event is often the inactivation of a tumor suppressor gene (TSG) called APC (*adenomatous polyposis coli*). APC normally acts as a "gatekeeper" controlling cell division in the colon-rectum. More specifically, the

A. Progression of Colorectal Cancer

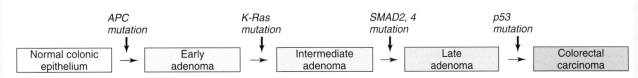

protein product of the APC gene is required to regulate the normal process by which epithelial cells in the intestinal crypts switch from proliferation to differentiation. In cells that have lost the function of both APC alleles, the balance is tipped abnormally toward proliferation, and the switch is made from a normal epithelial cell to an early adenoma. The transition to an intermediate adenoma is often associated with mutations in the K-Ras proto-oncogene that lead to its inappropriate activation. Deregulated K-Ras is thought to constitutively activate the MAPK and PI3K-PKB/Akt pathways responsible for further proliferation of the adenoma as well as the inhibition of apoptosis. Progression to the late adenoma stage shows a high correlation with mutations that affect the TSGs SMAD2 and SMAD4. Without the SMAD2 and SMAD4 proteins, the transcription of TGFβ is disrupted and the growth inhibition that this cytokine normally exerts in the colon-rectum is lost. The final step from late adenoma to carcinoma is often associated with mutations that inactivate the TSG p53. The general genomic instability caused by p53 inactivation (see later in this chapter) opens the door to the subsequent

mutations necessary for the cell to progress to the metastatic state.

While colonic epithelial cells in any individual may accumulate the multiple alterations necessary to proceed through to the cancerous state, the chances are significantly higher in an individual who inherits a mutation that predisposes his or her genome to instability. Two forms of such hereditary colorectal cancer exist: *familial adenomatous polyposis* (FAP) and *hereditary nonpolyposis colorectal cancer* (HNPCC) (see Figure B). In FAP, a germline APC mutation is inherited from one parent. This individual is thus born with the first of the "two hits" necessary to completely inactivate the APC locus. When the second APC allele is mutated, the individual develops thousands of benign, early adenomas at an accelerated rate compared to the general population or in individuals on their way to developing the sporadic form of colorectal cancer. However, each of these adenomas still requires the accumulation of the mutations described previously to progress to malignancy. While the chances of these mutations occurring on a cell-by-cell basis are the same in a FAP patient

as in an individual without a germline mutation, the elevated number of adenomas in the former increases the overall likelihood that a malignancy will ultimately arise. That is, the rate of tumor initiation is increased. The situation is slightly different for HNPCC, in which an individual inherits a germline mutation of one of several genes involved in DNA mismatch repair (MMR). The loss of one allele of an MMR gene (such as MSH2) greatly increases the chances that a sporadic mutation will result in total inactivation of that MMR gene. For unknown reasons, however, HNPCC patients show no more early adenomas than the general population (or those with developing sporadic cancers). Instead, the genomic instability generated by mutation of the MMR gene leads to accelerated tumor progression from the intermediate adenoma stage on. As a result, despite their differing mechanisms, FAP and HNPCC patients both develop colorectal cancer at the same median age of 42 years, which is 25 years earlier than patients with sporadic colorectal cancer.
Courtesy of Dr. Danny Ghazarian, Princess Margaret Hospital, University Health Network, Toronto.

B. Inherited Colorectal Cancer

FAP

APC germline mutation	Tumor initiation	Tumor progression

HPCC

MMR germline mutation	Tumor initiation	Tumor progression

Colorectal carcinoma

42 years

Table 26-4 Oncogenes

Oncogene*	Name or Function of Normal Gene	Tumor Association
Bcl-2	Anti-apoptosis	Lymphomas
Bcr-Abl fusion gene	Non-receptor tyrosine kinase	Chronic myelogenous leukemia
ErbB (EGFR)	Epidermal growth factor receptor tyrosine kinase	Glioblastoma
H-, N-, and K-Ras	GTP-binding proteins	Colorectal, pancreatic, mammary gland, and skin cancers
Her-2/neu	Epidermal growth factor family receptor tyrosine kinase	Schwann cell and breast cancers
Hst	Fibroblast growth factor	Stomach cancer
Int-2	Fibroblast growth factor	Breast cancer
Jun	Transcription factor	Osteosarcoma, skin cancer
Met	Hepatocyte growth factor receptor	Multiple sarcomas, renal carcinoma
Mdm-2	Negative regulator of p53	Multiple sarcomas, colorectal cancer
Myb	Transcription factor	Myeloblastosis
Myc	Transcription factor	Neuroblastoma
PKC	Serine-threonine kinase	Skin cancer
Raf	Serine-threonine kinase	Liver, lung cancers
Sis	Platelet-derived growth factor	Multiple sarcomas
Src	Non-receptor tyrosine kinase	Multiple sarcomas
VEGF	Vascular endothelial growth factor	Colorectal, gall bladder cancers

*C-prefixes, i.e., as in c-myc, have been omitted for simplicity.

a silent agent of infection. If the provirus has inserted into the genome in close proximity to a proto-oncogene and viral replication is triggered, the LTR of the provirus (which contains a transcriptional promoter) may induce the inappropriate activation of the proto-oncogene and cause it to act as an oncogene. Significantly, endogenous retroviruses can be induced to replicate by exposure of infected cells to carcinogens such as chemicals or radiation.

On rare occasions, an integrated retrovirus can "capture" and incorporate a copy of a proto-oncogene permanently into the viral genome. When this incorporation occurs, the gene in question may become converted into a retroviral oncogene that is able to transform new host cells upon infection. Retroviral oncogenes are often named after the virus in which they occur. For example, in 1911, when Peyton Rous discov-

ered a sarcoma-inducing retrovirus, the virus was named *Rous sarcoma virus* (RSV) and the oncogene responsible for transformation was named "Src," for "sarcoma." To distinguish between the viral oncogene and the cellular proto-oncogene, the former was called v-Src (viral Src) and the latter c-Src (cellular Src). Because of their ability to transform cells through oncogene activation, retroviruses can be considered carcinogens (see later). While retroviral infection remains a relatively minor cause of natural malignancies in most mammals, study of the behavior of the murine c-Src oncogene has helped scientists to gain much general knowledge about human oncogenes.

iii) Tumor Suppressor Genes

Unlike oncogenes, TSGs are usually negative regulators of cell growth or survival (Table 26-5). That is, the normal function of

Table 26-5 Tumor Suppressor Genes

Gene	Name or Function of Normal Gene	Tumor Association
APC	Regulates intestinal epithelial cell proliferation and differentiation	Familial adenomatous polyposis (colorectal cancer)
NF1	Inhibits Ras	Neurofibromatosis, glioma
p53	Transcription factor regulating cell growth arrest or apoptosis in response to DNA damage	Li-Fraumeni syndrome,* multiple advanced sporadic cancers
PTEN	Phosphatase negatively regulating PKB-mediated survival signaling	Multiple advanced cancers
Rb	Binds to transcription factor E2F to maintain cells in G_0 of cell cycle	Retinoblastoma, osteosarcoma
VHL	Decreases transcription and promotes degradation of cell cycle molecules	Renal cell carcinoma, von Hippel-Lindau syndrome˅
WT1	Transcription factor required for kidney development	Wilm's tumor#

* Li-Fraumeni syndrome: an inherited increased risk of cancer during childhood and young adulthood.
˅von Hippel-Lindau syndrome: an inherited disorder in which abnormal blood vessel growths appear in the retina, brain, spinal cord, and/or adrenal glands.
#Wilm's tumor: nephroblastoma, usually in children.

the product of a TSG discourages tumor development. TSGs become implicated in malignant transformation when they undergo *loss-of-function* mutations such that the restraining influence of the TSG product is lost. Point mutations, abnormal methylation, or partial or complete gene deletions can lead to lack of TSG function. Because diploid cells have two copies of any TSG, mutation of one of them is not usually sufficient to cause malignant transformation. The normal copy continues to produce the regulatory protein encoded by the TSG, and cell growth and survival remain under control. However, if this copy of the TSG is also inactivated (referred to as *loss of heterozygosity*; LOH), the individual becomes predisposed to tumor formation.

Perhaps the best known example of a TSG is *p53*. Mutations of p53 have been found in close to 50% of all human cancers, making it the gene most often involved in human neoplastic transformation. Germline mutations of p53 are associated with *Li-Fraumeni syndrome*, an inherited susceptibility to the development of various cancers. The p53 gene encodes a transcription factor that functions as a master regulator of cell cycle progression and cell death. Indeed, p53 has been called the "guardian of the genome" by some researchers. The functions of p53 are activated by phosphorylation via the DNA repair kinases ATM and Chk2. (ATM is the gene mutated in families with ataxia-telangiectasia [refer to Ch.24] and Chk2 is another signaling kinase in the same pathway.) By a mechanism that remains a mystery, ATM "senses" when the DNA of a cell has been damaged and activates the cell's DNA repair programs via p53 phosphorylation. p53 either causes the arrest of the cell cycle so that DNA repair enzymes have a chance to repair the error, or forces the affected cell into apoptosis if the damage cannot be repaired. The vigilance of p53 ensures that few mutated cells survive and proliferate to become tumors. Conversely, in the absence of p53 function, the survival of the deregulated cells becomes possible and their unrestrained growth promotes the accumulation of further mutations that cannot be repaired. As a result, the eventual malignant conversion of these cells becomes highly probable.

Examination at the nucleotide level has shown that more than 1700 different types of p53 mutations occur in human cancers. In many tumor cells, one p53 allele has experienced a missense mutation while the other allele has been lost by deletion. Substitutions of amino acids in the DNA-binding domain of the remaining allele are the most frequent alterations associated with transformation because these defects break the contact of p53 with the genomic DNA and block p53-mediated induction of repair processes. Curiously, in some tumor cells, there appears to be a mutation in only one allele of p53. For other TSGs, the presence of one functioning copy of the gene would be enough to prevent increased genetic deregulation within the cell. However, heterozygosity for certain p53 mutations is devastating to the function of the p53 protein because of the homotetrameric structure of this molecule. For example, suppose an individual has one normal p53 allele and one allele in which a mutation drastically changes the conformation of the p53 chain. Should a homotetramer form in which three chains are normal but the fourth has the altered conformation, this one chain may be enough to prevent the function of the entire tetramer. Moreover, such an altered tetramer effectively sequesters normal p53 chains and decreases the formation of normal tetramers. The conformational change that mediates such disruption can result from as little as a single point mutation in the p53 gene.

PTEN ("phosphatase and tensin homologue located on chromosome 10") is a TSG whose normal function is to control signal transduction by the PI3K pathway. The reader will recall that one of the downstream targets of PI3K signaling is PKB/Akt, a kinase critical for cell growth and survival. PTEN negatively regulates the activation of PKB/Akt, thereby preventing this enzyme from prolonging the survival of cells that are scheduled to die. PTEN is the second most commonly mutated gene in human cancers, and germline mutations of PTEN have been found in families with *Cowden syndrome*, another disease in which there is an inherited susceptibility to the development of various cancers. A high frequency of mutations leading to impaired PTEN function has also been associated with sporadic brain, breast, prostate, and skin tumors.

The TSG known as Rb was named for retinoblastoma (RB), the cancer in which this gene was first identified. Retinoblastoma is a malignancy that affects cells in the retinas of very young children (incidence 1 in 250,000 in the United States). These young patients rapidly become blind and, if left untreated, soon die. About 40% of RB cases are familial, leaving about 60% of cases to arise sporadically. It was largely studies of RB inheritance that led Alfred Knudson of the Fox Chase Cancer Center in Pennsylvania, United States to the conclusion that the process of tumorigenesis required at least two "hits" to the genome. In families with the inherited form of RB, a germline mutation of the Rb TSG already exists (the "first hit"). However, the disease does not show up until a second, somatic mutation of the other Rb allele occurs (the "second hit"). Because the first allele is already mutated, the frequency of RB in these families is much higher than the frequency of sporadic RB in the general population (which requires two somatic mutations). The normal function of the Rb protein is to act as a cell cycle control checkpoint, blocking the division of cells with genetic abnormalities.

IV. CARCINOGENS

A carcinogen is any substance or agent that significantly increases the incidence of malignant tumors. Carcinogens promote tumorigenesis by inducing the mutation or deregulation of oncogenes, TSGs, and DNA repair genes. In the majority of cases, a given carcinogen does not affect every stage of malignant transformation, and different carcinogens contribute to different stages. However, a "complete carcinogen" is one that can drive initiation, promotion, and progression right through to malignant conversion. A complete carcinogen may play all these roles during a single exposure, or may influence the cell over the course of multiple exposures.

The role of carcinogens in malignant transformation may account at least partially for the well-established observation that human cancer incidence increases dramatically in the sixth and seventh decades of life. Since tumorigenesis is a multi-step process, the required mutations need time to accumulate before complete transformation is achieved. However, it can also be argued that the naturally weaker immune system present in an older individual cannot eliminate tumors as easily as the more robust immune system in a younger person, once the malignancy has developed. Both factors are likely important.

The three major categories of carcinogens are chemicals, radiation, and pathogens. We now offer a brief description of the functions of these agents in carcinogenesis.

i) Chemical Carcinogens

External agents were first recognized as possibly playing a role in tumorigenesis in 1775 in England. A correlation was made between the intense exposure of chimney sweeps to soot and the prevalence of scrotal cancer among these men. In the 20th century, as more and more synthetic chemicals came into use, the prevailing dogma was that the majority of these agents were likely to be carcinogenic. Those man-made substances whose chemical structures appeared to predispose them to deregulating DNA replication came under particular suspicion, since such deregulation would result in a loss of cell growth control. However, when tested for their ability to transform cells, only 25–30% of these "high-risk" synthetic chemicals were found to be carcinogenic. We now know that many chemicals found in nature can be equally, if not more, carcinogenic than synthetic ones. For example, cigarette smoke created by burning the leaves of tobacco plants contains scores of natural chemicals sadly proven to be carcinogenic in the bronchi of lung cancer patients, as well as in the esophagus and bladder. Examples of chemicals demonstrated to be carcinogenic in humans are given in Table 26-6.

ii) Radiation

Ultraviolet rays from the sun are a common source of potentially carcinogenic radiation. The correlation of overexposure to UV radiation with skin cancer has led to public health campaigns to reduce harmful exposure to the sun. Ionizing radiation from X-rays is also carcinogenic, as demonstrated by the prevalence of skin cancer in early radiologists. These pioneering clinicians constantly exposed their hands to X-rays without the shielding that is standard practice today. Nuclear detonations have also confirmed that radiation exposure can be carcinogenic. People who survived the atomic explosions in Japan in 1945 and the Chernobyl nuclear reactor meltdown in 1986 have a much higher incidence of cancer (particularly leukemias; see Ch.30) than the general population. Examples of radioisotopes known to be carcinogenic in humans are given in Table 26-7.

iii) Pathogens

About 15% of human cancers appear to be linked to pathogen infection (Table 26-8). For example, *H. pylori* infection is associated with certain gastric carcinomas, and HepB and HepC infections are seen in a relatively high number of liver cancer patients. Similarly, EBV infection correlates closely with the appearance of Burkitt's lymphoma in African children (see Ch.30). However, there is no solid evidence that these pathogens are directly responsible for the genetic deregulation seen in malignant transformation. For one thing, not every individual infected with *H. pylori*, HepB, HepC, or EBV develops a malignancy. Some researchers interpret this observation to mean that prior genetic and/or environmental factors

Table 26-6 Chemical Carcinogens

Chemical	Tumor Association
Aflatoxin	Liver
Alcohol	Liver, esophagus, pharynx, larynx
Aluminum	Lung
Aminobiphenyl compounds	Bladder
Arsenic compounds	Lung, skin
Asbestos	GI tract, peritoneum, lung
Benzene	Leukemia
Benzidine	Bladder
Beryllium compounds	Lung
Cadmium compounds	Lung
Diethylstilbestrol	Cervix, vagina, breast, testis
Naphthylamine	Bladder
Nickel compounds	Lung, nasal sinus
Silica crystals	Lung
Soot	Lung, skin
Tobacco smoke	Lung, esophagus, pharynx, larynx, pancreas, liver
Vinyl chloride	Brain, lymphoma, liver, lung

Table 26-7 Radioactive Carcinogens

Radioisotope	Tumor Association
Plutonium-239 and decay products	Bone, liver, lung
Radium-224, -226, -228 and decay product	Bone
Radon-222 and decay products	Lung
Thorium-232 and decay products	Leukemia, lung
Iodine-131	Breast, thyroid, leukemia

Table 26-8 Pathogens Associated with Carcinogenesis

Pathogen	Putative Tumor Association
EBV	Burkitt's lymphoma, nasopharynx, Hodgkin's lymphoma
Helicobacter pylori	Stomach
HepB	Liver
HepC	Liver
HPV	Cervix, penis
HTLV-1	Adult T cell leukemia/lymphoma
KSHV	Kaposi's sarcoma
*Opisthorchis viverri**	Liver

*Liver fluke (trematode).

must have been involved that constituted a predisposition to pathogen-associated malignancy. Other researchers believe that it is chronic inflammation triggered by these persistent pathogens that pushes cells at the initiation stage of carcinogenesis to the promotion stage (see Box 26-2). However, still other scientists view the presence of such pathogens in a cancer patient as coincidental, a consequence of their relatively widespread prevalence and the possibly depressed immunocompetence of the affected patients.

Because of the propensity of retroviruses for insertional mutagenesis, a more direct link might be expected between tumorigenesis and infection with these viruses. Studies *in vitro* have shown that the random insertion of a retroviral provirus into or near a cellular gene can disrupt either its structure or the control of its expression, potentially deregulating cell growth and differentiation. Indeed, historically, many of the murine tumors that were studied to determine the nature of tumorigenesis were directly or indirectly induced by retroviruses, including *Moloney murine leukemia virus* (MoMuLV)

and *murine mammary tumor virus* (MMTV). Furthermore, studies of the infection of experimental animals with oncogenic viruses have shown that immune responses that eliminate normal cells harboring these viruses can prevent later tumorigenesis. Many researchers thus believe that T cell-mediated immunity can control the formation of at least certain tumors by targeting components of the oncogenic virus linked to its occurrence. However, as mentioned previously, retroviruses are associated with only a minority of human cancers. An example of a human retrovirus-associated cancer is *adult T cell leukemia* (ATL; see Ch.30), which is linked to infection with the exogenous retrovirus HTLV-1. Unlike the situation in mice and chickens, in which the development of malignancy is often associated with insertion of the retrovirus into the same site in different animals, human ATL is not associated with a recurring insertion site. There have also been reports of endogenous retroviruses in human tumor samples; however, it is not known whether the proviral insertion event actually caused the cancer in these cases.

Box 26-2. Links between chronic inflammatory disease and cancer?

Many patients with chronic inflammatory diseases go on to develop cancers. Scientists have speculated that perhaps the inflammatory milieu, with all its cytokine mediators, induces cells in the area to undergo mitosis at an increased rate. This enhanced proliferation may increase the chance that the genome of one of these cells might undergo an initiating mutation, making it a target cell. The "cytokine storm" that ensues during chronic inflammation may also induce changes to cellular differentiation or apoptosis programs in the target cell. Most importantly, macrophages activated by pro-inflammatory cytokines produce ROI and RNI that can be taken up by the target cell. Once within this cell, these intermediates could cause additional damage to its genomic DNA. The combination of these factors may be sufficient to push the target cell down the carcinogenesis pathway.

Several lines of evidence link inflammatory cytokines and cancer. The progression of breast carcinomas is enhanced by M-CSF, and sustained exposure to IL-15 promotes the development of T or NK leukemias. A pro-inflammatory cytokine called *macrophage inhibitory factor* (MIF) has been shown to directly interfere with p53 function. The induction of skin cancers or lymphomas by certain chemical agents absolutely requires the activities of the pro-inflammatory cytokines TNF and IL-6. As we have seen, TNF signaling leads to NF-κB activation, which in turn promotes cell survival and proliferation.

Indeed, mice lacking the IKKβ subunit of the kinase required for NF-κB activation show decreased tumorigenesis in a colitis-associated cancer model. Genetically engineered mice in which both inflammatory bowel disease and colon carcinomas develop under normal, non-sterile animal colony conditions had both disorders resolve when the animals were maintained under special germ-free conditions. The theory here is that removal of the pathogen that was sustaining the inflammation resulted in abatement of the inflammatory response and its generation of mediators promoting carcinogenesis. Similarly, if mice that are prone to colon cancer are treated with a non-steroidal anti-inflammatory drug, their rate of tumorigenesis is decreased. Intriguingly, the same observation has been made in human patients with familial colon cancer.

Another line of evidence suggesting that inflammatory mediators may be linked to cancer comes from studies of *Cox-2 inhibitors*. Cox-2 stands for "cyclo-oxygenase-2," a pro-inflammatory enzyme involved in the production of prostaglandins (refer to Ch.4). In a normal, healthy tissue, Cox-2 is undetectable. However, this enzyme is induced in response to inflammatory cytokines, growth factors, carcinogens, and oncogene activation. Experimentally, upregulated Cox-2 expression has been linked to constitutive NF-κB activation. In several model systems, the inhibition of Cox-2 has led to a decrease in cancer-promoting factors

or in tumorigenesis itself. For example, the Her-2/neu strain of transgenic mice spontaneously develops mammary tumors at about 8–9 months of age. When Her-2/neu mice were fed the Cox-2 inhibitor Celecoxib starting at age 1 month, significantly fewer cancers developed as the mice aged. These results have inspired clinical researchers to investigate the use of Cox-2 inhibitors in women at high risk for breast cancer. Other studies have implicated various arachidonic acid and prostaglandin metabolites in the progression of prostate cancer, and Cox-2 activity in gastric cancer cells has been associated with their proliferation. Perhaps the strongest association of Cox-2 activity and tumorigenesis lies in the development of lung cancer. In one lung cancer cell line, elevated Cox-2 activity was associated with enhanced production of angiogenic chemokines and factors promoting tumor cell proliferation. In another study, a Cox-2 inhibitor called deguelin was able to induce the apoptosis of cells of a lung cancer cell line *in vitro*. With respect to colorectal cancer, daily consumption of the humble aspirin (which is also a Cox-2 inhibitor) has been associated with a decreased incidence of this disease. The targeting of Cox-2 may thus provide substantial benefits to a wide range of cancer patients. It should be noted, however, that high doses of the newer Cox-2 inhibitors have recently been associated with an increased risk of heart attack or stroke, necessitating an analysis of the balance of risk for each patient.

Another group of viruses known as the *DNA tumor viruses* has a more clearly defined role in promoting cancer in humans and animals. Although they did not understand the mechanism by which transformation occurred, early investigators made opportune use of DNA tumor viruses such as adenovirus, SV40, and murine polyoma to induce tumors in mice. Over the years, it became clear that certain proteins of these viruses, such as the E1A protein of adenovirus, are able to bind to TSG proteins such as p53 and inactivate their regulatory functions. Mutations can then accumulate in the infected cells that lead to malignant transformation. Similarly, several human DNA tumor viruses have been directly implicated in human cancers. For example, certain strains of *human papilloma virus* (HPV) cause cervical cancers, and the herpesvirus KSHV is a known contributor to the development of Kaposi's sarcoma in AIDS patients (refer to Ch.25).

C. Immune Responses to Cancer

Mutations associated with carcinogenesis, be they translocations, inversions, point mutations, or deletions, can lead to the production of new molecular structures. These structures can take the form of a near-normal protein of altered conformation, a protein with a subtle or not-so-subtle amino acid substitution, a protein that has suffered a deletion or internal rearrangement, or, in the case of certain chromosomal translocations, the fusion of part of one protein with part of another. For immunologists attempting to manipulate the immune system to fight cancer, the most important question is "What determines whether these structural changes are recognized by the immune system?" Indeed, very few of the many mutations identified in cancer cells lead to the expression of antigens that can provoke tumor rejection. Conversely, many of the antigens that do induce tumor rejection in mice, or constitute high affinity targets recognized by human anti-tumor T cells, result from somatic mutations in proteins that were not predicted to be of relevance to tumorigenesis. This section of the chapter deals with the expression of tumor antigens, and the body's attempts to mobilize cells of the immune system to attack cells that were originally "self" but that are now the enemy.

I. THE CONCEPT OF IMMUNOSURVEILLANCE

As mentioned in the Historical Notes section, the concept that the immune system can prevent cancer originated over 100 years ago and is still held dear by some (but not all) immunologists. Modern studies of knockout mice deficient for various components of the immune system have supported the concept of immunosurveillance. Increased frequencies of spontaneous and carcinogen-induced cancers in various tissues have been noted in some studies of mice deficient for STAT1, IFNγ, RAG2, perforin, Fas, FasL, or IL-12 (although some of these results remain controversial). Animals with defects in the TCR genes such that they lack NKT cells or γδ T cells also

show a greater susceptibility to carcinogens. However, the fact that cancers develop frequently in immunocompetent animals clearly establishes that immunosurveillance fails on a regular basis. One theory for this failure holds that the tumors caught by immunosurveillance are those expressing a particularly immunogenic antigen, such as might be produced in large quantities by an oncogenic virus. In contrast, tumors that carry a weakly immunogenic tumor antigen, or fail to express any tumor antigen at all, may develop quietly under the radar of the normal immune response.

Robert Schreiber and his colleagues at Washington University in St. Louis have gone further and suggested that, while immunosurveillance destroys some cancer cells, it may play an active role in promoting the growth of others. According to this view, immunosurveillance by both innate and adaptive immune mechanisms places a selection pressure on a growing tumor that favors survival of neoplastic cells of reduced immunogenicity. Such "immune selection" or "immunoediting" of the repertoire could actually increase the chance of tumor survival. In a perfect world, an initial immune response mounted against tumor cells would eliminate them completely and the immunoediting process would proceed no further. However, the immune response and the tumor cells may settle instead into a state of equilibrium in which tumor growth is simply held in check. During this phase, the immune system exerts various anti-tumor mechanisms that apply constant selection pressure to the genetically unstable tumor cells. Eventually this selection pressure may promote the emergence of tumor mutants that escape the immune response via any one of a number of different mechanisms (see later), including loss of expression of the provoking antigen or the MHC on which its peptide is presented. Thus, in this scenario, a failure of immunosurveillance is not due to insufficient strength of the immune response itself, but rather to phenotypic adaptations on the part of the tumor that allow it to evade immunity in a manner reminiscent of some pathogens. Other researchers feel that, rather than increasing escape from immune effector mechanisms, this type of selection pressure simply favors clones with increased malignant growth potential; that is, immune selection promotes the predominance of those cells that can grow so quickly that they "outrun" the immune response.

II. TUMOR ANTIGENS

For our immune systems to detect a tumor cell, it must be recognized as "foreign" in some way. The expression by a tumor cell of a macromolecule that is abnormal in appearance, concentration, or location, or is expressed at an unusual time during development, should draw the attention of innate and adaptive immune system cells. In general, there are two types of such tumor-related antigens: *tumor-associated antigens* (TAAs) and *tumor-specific antigens* (TSAs). Protein TAAs (some are carbohydrates) are encoded by normal cellular genes that have become dysregulated in some way such that their expression is altered inappropriately. Genes encoding protein TAAs can be found in both normal and cancerous cells. In contrast, protein TSAs are encoded by mutated cellular genes, or by viral oncogenes. The products of these genes are foreign to a normal host right from the start. Genes encoding TSAs are not expressed by non-transformed cells.

In the early days of tumor antigen research, it was hoped that there might be a master TAA or TSA shared by all (or at least most) malignancies that could serve as the basis of a vaccine to defend against or provide therapy for all types of cancers. However, as discussed previously, advances in the understanding of tumor biology have made it clear that a broad range of genes influences malignant transformation, and that a wide variety of genetic aberrations may cause the deregulation of any one of these genes. Thus, any antigen inducing an anti-tumor response would likely be restricted to expression in an individual tumor, rather than being shared among all tumors or even among tumors of a given type. Indeed, tumor rejection studies have confirmed this hypothesis. For example, a mouse vaccinated with cells of a given tumor may later mount a rejection response to a challenge with cells derived from the same tumor. However, if challenged with cells from the same type of tumor from a different source, or with a different type of tumor, the mouse does not respond. Even when the same carcinogen is used to induce tumors of the same histological type in the same organ system of animals of the same strain, a different range of tumor antigens is often induced. This diversity is not totally unlimited, however, as cross-reacting antigens have been identified that are shared by different mouse melanomas (see later).

It should also be noted that the initial assault by immune system cells on a tumor mass may physically expose new antigens as the cells and surrounding tissues are degraded. Some of these antigens, which may have previously been hidden from the immune system, may be taken up, processed, and presented as new pMHC epitopes on the surfaces of APCs. The presence of these novel epitopes draws new clones of naive T cells to the fight. This broadening of the immune response as a result of initial immune recognition is called *epitope spreading*, a topic that is discussed in detail in Chapter 29. Epitope spreading is thought to be important for sustaining an immune response against a tumor long enough for regression to be achieved.

i) Tumor-Associated Antigens (TAAs)

For a molecule to be viewed as a TAA, it must be a normal protein or carbohydrate expressed in the tumor in a way that is abnormal relative to its status in the healthy, fully differentiated cells in the surrounding tissue of origin. In other words, a TAA is almost always a case of the "right molecule expressed at the wrong concentration, place, or time." Sometimes the expression of a TAA is in keeping with the lineage of the target cell, but under other circumstances, the malignant cells express normal genes that are characteristic of a completely unrelated tissue. This is especially true when the malignant cells become metastatic. When a TAA corresponds to a normal cellular antigen that is expressed only in a rare cell type, the extensive proliferation of the malignant cells exposes the immune system to an unusually high abundance of this constituent macromolecule. The increased concentration of the macromolecule may then make it immunogenic.

TAAs have been broadly categorized into four classes: *normal cellular antigens, differentiation antigens, tissue-specific antigens, embryonic antigens,* and *idiotypic antigens* (Fig. 26-4).

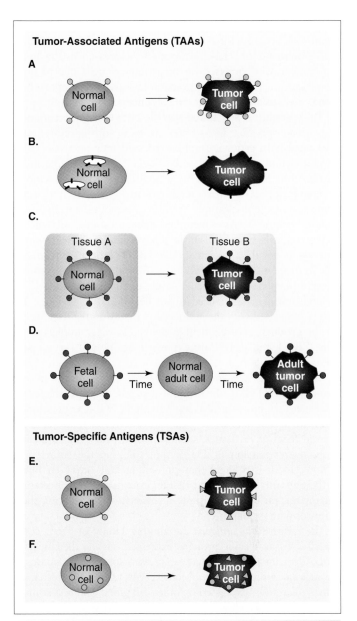

Figure 26-4

Types of Tumor-Related Antigens
TAAs are encoded by normal genes but are expressed at the wrong concentration, place, or time, while TSAs are encoded by mutated genes. Examples of TAAs: (A) a protein expressed at a low level on a normal cell is highly overexpressed on a tumor cell; (B) a protein normally expressed within an organelle is expressed outside that organelle in an inappropriate location in the tumor cell; (C) a protein whose expression is normally restricted to cells of Tissue A is expressed inappropriately by cells of Tissue B; (D) a protein whose expression is normally restricted to fetal cells becomes reactivated in adult cells that have undergone transformation. Examples of TSAs: An abnormal protein is expressed either (E) on the surface or (F) in the interior of a tumor cell.

ia) Normal cellular antigens. Genetic modifications that deregulate the expression of a normal gene can result in a tumor cell that expresses a normal protein at much higher levels (up to 100-fold) than a healthy cell. An example of a normal protein that is carcinogenic when overexpressed is the Her-2 protein. Her-2 (also known as Neu and c-erbB-2) is a transmembrane tyrosine kinase receptor that binds human epidermal growth factor. Her-2 is normally expressed in the heart but can be overexpressed on breast cancer cells due to gene amplification. Her-2 overexpression has also been reported in some prostate, ovary, lung, and GI cancers.

ib) Differentiation and tissue-specific antigens. Differentiation and tissue-specific TAAs are proteins or carbohydrates associated with genes normally expressed only at a particular developmental stage or in a particular tissue or subcellular localization. The expression by a tumor of such a molecule in an inappropriate context tends to make the tumor more immunogenic. For example, certain melanocyte-specific proteins constitute TAAs when they appear on the surfaces of melanoma cells, instead of in the internal structures with which they are normally associated. The TRP-1 antigen (*tyrosine-related protein-1*; also known as gp75) and gp100 are proteins normally expressed in the membrane of the *melanosome*. The melanosome is a specialized organelle present in melanocytes that is associated with the endocytic pathway and is responsible for the production of pigment proteins. However, in melanoma cells, TRP-1 and gp100 are expressed on the cell surface and act as TAAs.

Another example of a normal protein expressed in the wrong place is EGFR, the receptor for *epidermal growth factor* (EGF) and TGFα. In a healthy individual, EGFR expression is confined predominantly to the skin. However, many cancers derived from other tissues, such as breast, ovary, and prostate, show overexpression of EGFR molecules on the surfaces of tumor cells. It is thought that circulating EGF or TGFα binds to the abnormally expressed EGFR proteins present on premalignant cells in breast, prostate, or ovarian tissue, driving their inappropriate proliferation.

The *cancer-testis antigens* are another family of TAAs composed of tissue-specific proteins expressed outside their normal range. Proteins that are usually expressed solely in spermatogonia and spermatocytes can become TAAs when transformation causes them to be expressed on other cell types. The best-studied examples of this class of TAAs are the MAGE (melanoma antigen gene) proteins expressed by melanomas. The MAGE proteins were initially thought to be TSAs because CTL responses were mounted *in vitro* to MAGE peptides presented on syngeneic MHC class I, suggesting a response against a non-self entity. However, it was later discovered that the MAGE proteins are normal proteins that are transcriptionally repressed by methylation in most normal cell types but expressed at low levels in testicular cells such as spermatogonia and spermatocytes. Some clinicians thus refer to the MAGE proteins as *oncospermatogonal* antigens. During melanoma formation, the DNA of the MAGE genes (among others) is somehow demethylated, leading to abnormal overexpression of MAGE proteins on melanoma cells. Interestingly, mRNAs encoding most of the cancer-testis antigens are expressed by mTECs in the thymus, suggesting that high affinity T cell clones recognizing these antigens may have been deleted during the establishment of central tolerance (see Ch.16). Such deletion would account for the fact that anti-cancer vaccination (see later) with many of these antigens usually provokes only a limited anti-tumor response *in vivo*.

ic) Embryonic antigens. The embryonic class of TAAs comprises proteins or carbohydrates expressed on normal embryonic tissues; these genes are silent in the normal cells of the adult. Malignant transformation can cause resumption of expression of these TAAs in adult cells in a type of de-differentiation called *retrodifferentiation*. Two prominent examples of embryonic TAAs are *carcino-embryonic antigen* (CEA) and *alpha feto-protein* (AFP). CEA, which is normally expressed only in the liver, intestines, and pancreas of the human fetus, is highly associated with colon cancer in the adult, as well as with other adult tumor types to a lesser degree. Similarly, AFP, which is normally expressed only in fetal liver and yolk sac cells, is strongly associated with hepatocarcinoma and testicular cancer in the adult, among other tumor types. The monitoring of CEA and AFP levels has been successfully used to detect the recurrence of cancers associated with these markers.

id) Idiotypic antigens. The last class of TAAs, the idiotypic antigens, is specific to T and B cell leukemias and lymphomas and thus is discussed in more detail in Chapter 30. The reader will recall from Chapters 5 and 12 that the very subtle differences between the variable regions of the Ig and TCR molecules expressed by different B and T cell clones are called "idiotypic differences." Because the lymphoid malignancies are also clonal in nature, the unique idiotypic structure of the Ig or TCR expressed by a particular lymphocytic tumor cell can be used to detect the recurrence of the leukemia or lymphoma. Accordingly, mAbs directed against specific Ig idiotypes have been used successfully for the treatment of a small number of patients with B cell malignancies (refer to Chs. 23 and 30). Similarly, antibodies against TCRαβ V region sequences have been used to treat a few patients with T cell leukemia or lymphoma.

ii) Tumor-Specific Antigens (TSAs)

TSAs are new macromolecules that are unique to the tumor and are not produced by any type of normal cell. Many TSAs are the protein products of normal genes altered by point mutations, deletions, or chromosomal translocations during the transformation process. One of the best-known human TSAs is a fusion protein produced when the Bcr (*breakpoint cluster region*) gene on chromosome 9 is translocated into the Abl gene (a tyrosine kinase) on chromosome 22, creating the *Philadelphia chromosome* (see Box 30-2 of Ch.30). This translocation is found frequently in cases of a myeloid cell malignancy called *chronic myelogenous leukemia* (CML) (see Ch.30). Unlike most TSAs, the fusion protein derived from the Philadelphia chromosome is almost identical in the tumor cells of different CML patients. This happy fact means that, if a way of inducing an immune response against the fusion protein or its novel component peptides could be devised, the strategy could be applied to the treatment of most CML patients. We note here that, while a genetic deletion can theoretically also result in a new fusion protein, the excision points would likely be different in every patient. As a result, any new structures generated would be specific to the individual tumor and thus not as useful as targets for developing a general therapeutic strategy.

Some TSAs are known to be derived from point mutations of key genes controlling cell growth or apoptosis. The most important of these is p53, many alterations of which result in a non-functional protein that cannot induce cell cycle arrest or apoptosis following DNA damage. Another example occurs in certain patients with head and neck cancer whose tumor cells show a mutation in a stop codon of the caspase-8 gene. This mutation results in the addition of 88 amino acids to the caspase-8 protein, an alteration that presumably interferes with the pro-apoptotic function of this enzyme. The tumor cells in some melanoma patients have a mutation in the Cdk4 gene that prevents the Cdk inhibitor p16 from binding to it, resulting in uncontrolled cell cycling. Mutated forms of EGFR are expressed by several different types of cancer cells, and abnormal signaling driven by the binding of EGF or TGFα to this receptor drives tumor growth. Other TSAs are abnormal carbohydrates, produced when an enzyme involved in carbohydrate biosynthesis has undergone a transforming mutation. Certain mucins (such as MUC-1) and gangliosides, which are altered in some types of melanomas and brain, breast, ovary, and lung cancers, are considered to be TSAs. Still other TSAs may be derived from DNA sequences that are not mutated but that are not transcribed in the normal way. For example, new proteins can be derived from the use of a cryptic transcription or translation start site, an alternative reading frame, transcription of a gene in the opposite direction, or transcription/translation of a pseudogene sequence. Finally, the viral proteins expressed on tumor cells that originated from a target cell infected with an oncogenic virus can also be considered TSAs.

Because of their nonself nature, TSAs should constitute bona fide immunogens that provoke an immune response. However, for reasons that are elaborated later, very few of them do. The cancer therefore continues to grow unchallenged as if it were "self." The reader is cautioned here that, just because a given molecule is apparently detected only on tumor cells, it does not necessarily mean that the antigen is truly unique to that tumor. Some putative TSAs (for example, certain MUC-1 derivatives) may in fact be normal proteins that are merely expressed at undetectably low levels on normal cells (i.e., are TAAs). Such an expression pattern could account for the lack of an immune response against what initially appears to be a nonself antigen.

The lack of natural immune responses against most TAAs and TSAs has meant that the development of effective treatments for cancers using immunological approaches has not progressed as quickly as once hoped. However, breakthroughs were achieved in the late 1990s in studies of melanoma regression. When patients were treated with purified TSAs from a mixture of melanoma cells, specific immune responses were detected in some of them that led to tumor regression. Similarly, in the early 2000s, melanoma patients treated with highly purified autologous T cells recognizing a TAA overexpressed on melanoma cells experienced regression of their disease. These results suggest that cautious optimism may be warranted, and the efforts of many laboratories are focused on identifying TAAs and TSAs and using them to devise protocols that can be used to induce effective anti-tumor immunity.

III. IMMUNE RESPONSES TO TUMOR CELLS

Despite the difficulties described previously in identifying TAAs and TSAs, various components of both the innate and adaptive arms of the immune system have been shown to respond to tumor cells (Fig. 26-5). These responses have been implicated in both immunosurveillance guarding against the growth of incipient tumors and in the rejection of established tumors.

i) NK Cell Responses

Perhaps one of the best-studied effectors thought to mediate anti-cancer cell immunity is the NK cell. As described in Chapter 18, NK cells are activated when their activatory receptors are engaged but their inhibitory receptors are not. Tumor cells often lose expression of MHC class I, meaning that they cannot engage most of the inhibitory receptors on NK cells and thereby prevent a cytotoxic "hit." Moreover, many human tumor cells express MICA, the stress molecule described in Chapter 18 as a ligand for the activatory receptor NKG2D expressed by NK cells. In mice, many cancer cells express two stress molecules called Rae-1 and HP60 that resemble MHC class I in structure but function as activatory ligands because they bind to NKG2D. In normal mice, Rae-1 is expressed in embryonic but not adult tissues. HP60 may appear in normal adult spleen but is most often found in embryonic tissues. Thus, tumor cells induce the expression of NK activatory ligands that do not normally appear on healthy adult cells, and in so doing, become obvious targets for NK-mediated cytolysis. With respect to *in vivo* work, there have been conflicting reports as to links between decreased NK cell cytotoxicity and human cancer incidence. Scientists are still struggling to generate a mouse that is specifically deficient in NK cells and so could be used to test the contribution of NK cells to the fight against tumorigenesis *in vivo*.

As described in Chapters 3 and 18, NK cells that have been treated *in vitro* with IL-2 become large, more powerfully cytolytic LAK cells. At least *in vitro*, LAK cells have the ability to lyse cells belonging to a wide range of tumor cell lines. However, it has not been possible to demonstrate any effectiveness of LAK cells against cancer cells *in vivo*. Several clinical trials in the 1980s in which cancer patients were treated with autologous LAK cells had disappointing results.

ii) Inflammatory Cell Responses to Tumors

In Box 26-2, we explored the concept that chronic inflammation can promote tumorigenesis. In contrast, acute inflammation has anti-tumorigenic effects, and activated neutrophils, eosinophils, and macrophages appear to play key roles in tumor rejection. The reader will recall that neutrophils and macrophages are activated by "danger signals" in the immediate microenvironment. In a cancer situation, the "danger signals" may arise from molecules associated with pathogens that have penetrated the local area, or from molecules released from the debris of necrotic tumor cells. In addition, it has been well documented that the mere presence of premalignant or malignant cells can induce inflammation. Some researchers believe that, in some cases, there may be new molecules expressed on these cells that are capable of binding to the

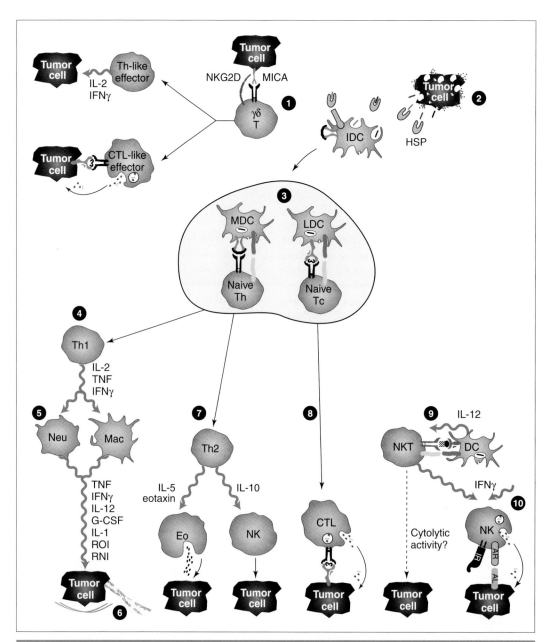

Figure 26-5

Immune Responses Against Tumor Cells

Tumor cells often express stress ligands, such as MICA in humans (or Rae-1 in mice) that are recognized by the TCRs and NKG2D receptors of γδ T cells (1). Once activated by these or other ligands, γδ T cells generate anti-tumor Th-like and CTL-like effector cells. Tumor cells also shed TAA/TSAs that are acquired by heat shock proteins (HSPs) (2) and conveyed to immature DCs (IDC), which mature as they migrate to the local lymph node. In this location (3), mature DCs (MDC) activate naive anti-tumor Th cells while licensed DCs (LDC) activate naive anti-tumor Tc cells via cross-presentation. Th1 effectors (4) migrate to the tumor site and release cytokines activating neutrophils and macrophages (5), which in turn produce a spectrum of anti-tumor molecules that carry out direct cytolysis, induce collagen capsule formation, or damage tumor vasculature (6). Meanwhile, Th2 effectors (7) that have migrated from the lymph node to the tumor site secrete cytokines that either activate eosinophil-mediated cytolysis or increase the vulnerability of tumor cells to NK cell killing. In addition CTL effectors (8) migrating from the lymph node carry out perforin/granzyme-mediated cytolysis of tumor cells. IDCs that capture tumor antigens and subsequently present them on CD1d (9) activate NKT cells that produce copious amounts of IFNγ and may kill tumor cells by cytolysis. NK cells (10) activated by engagement of activatory receptors (including NKG2D) and stimulated by IFNγ destroy tumor cells by natural cytotoxicity.

PRRs of immune system cells. Once activated, these cells release inflammatory cytokines (such as IFNγ, IL-12, and TNF) and reactive molecules (such as ROI and RNI) that can lyse cancer cells and damage tumor vasculature.

iii) Anti-tumor Cytokine Responses

The most important anti-tumor cytokines identified to date are IFNγ and IL-12, two cytokines that promote Th1 responses and CTL generation. Knock-out mice deficient for IFNγR or IL-12 show a greatly increased incidence of spontaneous tumors. IFNγ and IL-12 also have powerful anti-angiogenic effects that stifle the growth of blood vessels feeding a tumor mass. In addition, IFNγ induces the formation of a collagen capsule around a tumor mass, a natural barrier against metastasis. There is also some evidence that IL-10 may sometimes suppress tumor expansion (rather than supporting it, as described later). For example, in *in vitro* experiments, the addition of IL-10 enhances the vulnerability of tumor cells to NK-mediated cytolysis. Similarly, the transfection of tumor cell lines with the IL-10 gene reduces their ability to expand and increases their immunogenicity. Other cytokines produced locally at the tumor site may also play roles in controlling the growth of cancer cells. In human cancer patients, IL-2, GM-CSF, and IL-15 are found at higher levels in melanomas undergoing regression than in those expanding aggressively. Indeed, direct injection of exogenous IL-2 into tumors can induce regression. *In vitro*, CTLs whose function has been blocked by IL-10 can be rescued by IL-2 administration. G-CSF, IL-1, and IL-7 have also been implicated in containing tumor expansion, but it is not clear whether these cytokines block the growth of a malignancy directly by interfering with tumor cell proliferation, or indirectly by stimulating DCs and/or lymphocytes with specificity for various TSAs and TAAs, or both.

As alluded to in Box 26-2, the cytokines, chemokines, and inflammatory mediators secreted by immune system cells in response to the presence of premalignant or malignant cells can sometimes stimulate the growth of the tumor. For example, activated macrophages drawn to the tumor site secrete growth factors such as PDGF and TGFβ that stimulate angiogenesis supporting the expansion of the tumor. Thus, an immune response to a premalignant or malignant cell can simultaneously promote both regression and progression of tumors. This conundrum serves as a warning to those pursuing various cancer vaccination strategies (see later), because the induction of an anti-tumor immune response could potentially stimulate cancer growth rather than inhibit it.

iv) Conventional αβ T Cell Responses

The role of the adaptive response in combating tumor cells in experimental models is reasonably well established. There have been conclusive demonstrations in animals that CD4+ Th cells are involved in immunosurveillance activities and contribute to the elimination of tumors in experimental settings. In particular, mice deficient for CD4 or depleted of CD4+ T cells show reduced tumor rejection capacity. Since most cancer cells are derived from cell types that normally do not express MHC class II, how are anti-tumor CD4+ T cells

activated? It is speculated that macrophages and DCs that become activated in the tumor site take up antigens shed by tumor cells, process them, and present peptides on MHC class II to either memory CD4+ T cells resident in the affected tissue or naive CD4+ T cells in the local lymph node. Once activated, the primary role of CD4+ T cells is to supply help, via cytokines and/or CD40-CD40L licensing of APCs, to CD8+ T cells. B cells do not appear to play a significant anti-cancer role (see later), so that no CD4+ help is needed here. However, the Th1 effectors generated from activated CD4+ T cells also supply cytokines needed for the hyperactivation of macrophages. In addition, *in vitro* studies suggest Th1 effectors also play a role in the production of NO and ROI by macrophages needed to kill tumor cells. Cytokines secreted by Th2 effectors are required for the recruitment and activation of eosinophils that take localized anti-tumor cell action. As well, as mentioned previously, IL-10 produced by Th2 cells may promote tumor cell killing by NK cells.

The role of CD8+ T cells in anti-tumor responses has been more clearly defined. Numerous studies have shown that tumor-specific CD8+ CTLs can lyse cancer cells directly *in vitro* and reject transplanted tumor cells in experimental animals. Moreover, when mice are depleted of CD8+ T cells, anti-tumor responses are greatly reduced. Of note is the observation that many CD8+ T cells express NKG2D, meaning that they can be stimulated by the binding of the stress ligand MICA present on many tumor cells. In human cancer patients, CTLs can often be found within the tumor tissue itself, leading to the designation of these cells as *tumor-infiltrating lymphocytes* (TILs) (Plate 26-3). However, it is still not clear whether TILs play a role in controlling human cancers. Research is under way in many cancer settings with the goal of designing

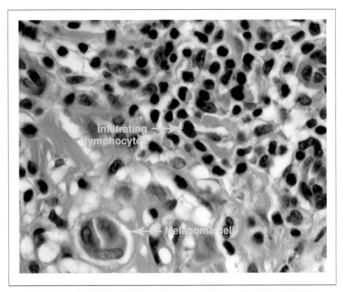

Plate 26-3

Tumor Infiltrating Lymphocytes in Melanoma
Courtesy of Dr. Danny Ghazarian, Princess Margaret Hospital, University Health Network, Toronto.

procedures to stimulate the TILs present within a tumor. In particular, patients with melanomas and kidney cancers are under study for TIL exploitation since regression of these tumors can be monitored relatively easily.

What epitopes do anti-tumor T cells see? CD4$^+$ T cells that respond to antigens associated with lymphomas, melanomas, or cervical, colon, or breast tumors have been isolated from various cancer patients. However, it has been difficult to identify the precise antigens supplying the stimulatory epitopes recognized by these T cells. It is known that a TAA called NY-ESO-1 associated with esophageal cancer and the MAGE-3 TAA of melanoma produce epitopes that are recognized by CD4$^+$ T cells. In the case of virus-associated tumors, viral antigens expressed in the course of infection and replication may appear in their native forms on the tumor cell surface, or may be presented as peptides on pMHC. MHC class I-restricted CTLs directed against peptides from different virally encoded proteins have been detected in tumors associated with EBV, HepB, HepC, KSHV, or HPV infection. In general, these epitopes are derived from products of the early region genes and viral capsid proteins. CTLs directed against these epitopes can successfully induce rejection of the cancerous cells. Consequently, several of these viral proteins are candidates for vaccination against these cancers (see later). In studies of individual cancer patients with unusually favorable clinical courses, researchers have been able to characterize the TSAs recognized by tumor-specific CTLs that appeared to be conferring protection. For example, in one melanoma patient, the TSA was derived from a protein encoded by a mutated gene homologous to RNA helicases. Similarly, in one patient with human large cell lung carcinoma, the TSA was derived from a mutated version of the actin cross-linking protein α-actinin-4. In another lung carcinoma patient, a mutation in the gene encoding the mitochondrial protein malic enzyme was the source of the TSA. In all three cases, CTLs specific for the relevant TSA were present at high frequency in the blood of the patient (for extended periods) and were able to specifically lyse cells expressing that TSA *in vitro*.

v) NKT Cell Responses

It is early days yet in the exploration of the contribution of NKT cells to immune responses against cancer, but the available evidence from animal models suggests that these cells do have anti-tumor effects. As described in Chapter 18, NKT cells express a semi-invariant αβ TCR that binds to CD1d molecules presenting glycolipid or non-peptidic antigens. Once activated, NKT cells mediate both T helper-like and NK-like effector functions. Activated NKT cells rapidly express a collection of Th0 cytokines, including IFNγ, that stimulates the anti-tumor activities of NK cells. As well, activated NKT cells can carry out perforin-dependent cytolysis of tumor cells (at least *in vitro*), although it is not yet clear how important this capacity of NKT cells is *in vivo*. Experiments in animal models have shown that tumor metastasis is blocked if mice are injected with an α-galactosylceramide ligand known to bind to CD1d and to activate NKT cells. This anti-metastatic effect is mediated by an early burst of IFNγ production by activated NKT cells, followed by prolonged IFNγ production by NK cells. Consistent with this observation, mice lacking NKT cells show increased susceptibility to chemical carcinogens. The observed anti-tumor effect of IL-12 is also thought to be due at least in part to the stimulation of NKT cells by this cytokine. Direct contact between DCs and NKT cells via CD40-CD40L interactions stimulates IL-12 production by the DCs, and this IL-12 promotes NKT activation. Although likely, it remains to be proven *in vivo* that endogenous glycolipids expressed by tumor cells can be presented by CD1d on APCs to activate NKT cells.

vi) γδ T Cell Responses

The reader will recall from Chapters 18 and 20 that the IELs (intraepithelial lymphocytes) present in the epithelia of the body carry out important sentinel functions in eliminating pathogens before they can gain a firm foothold in deeper tissue layers. Since most common adult cancers (such as tumors of the lung, breast, colon, and prostate) arise from transformation of an epithelial target cell, it is fair to wonder whether IELs also have an immunosurveillance function to guard against malignancies. As we have seen, the IEL population in humans is composed of 50% γδ T cells and 50% αβ T cells, while mouse IELs are almost all γδ T cells. It is significant, then, that knockout mice lacking γδ T cells are one of the very few types of knockout mice that spontaneously and rapidly develop tumors. Mutant mice lacking these cells showed a higher level of skin cancer formation when a chemical carcinogen was topically applied to the skin, or when MCA was injected to induce fibrosarcomas.

How would γδ T cells be activated by tumor cells? γδ T cells express both γδ TCRs that bind non-peptidic ligands and the NK activatory receptor NKG2D that binds to MICA and MICB in humans (Rae-1 and H60 in mice). In *in vitro* experiments, mouse γδ T cells whose NKG2D receptor was engaged by Rae-1 on a tumor cell (and whose TCR was engaged by a tumor-related antigen) could directly kill the tumor cell via cytotoxic mechanisms. In this context, the NKG2D molecule was thought to be acting as either a coreceptor or a costimulatory molecule for γδ TCR-mediated activation. In addition, γδ T cells preferentially express the CD8αα form of the CD8 coreceptor. CD8αα does not function well as a standard coreceptor since it interacts poorly with classical MHC class I molecules. Instead, CD8αα binds strongly to an MHC class Ib protein called *thymus leukemia antigen* (TL). TL was originally identified as an antigen expressed by the leukemic cells of patients with the lymphoid cell malignancy *acute lymphocytic leukemia* (ALL). In healthy individuals, TL is expressed almost exclusively by epithelial cells of the small intestine. TL does not bind peptide ligands and does not interact with CD8αβ heterodimers, but does bind to CD8αα. Engagement of CD8αα by TL can increase cytokine production by activated (but not resting) CD8αα-expressing γδ T cells, indicating that this interaction is regulatory in nature. *In vivo*, it may be that the engagement of the γδ TCR plus NKG2D and/or CD8αα signals to the γδ T cell that a premalignant cell has been encountered that should be eliminated. It should be noted that NKG2D is also expressed by αβ IELs in humans, but whether it fulfils the same costimulatory function for these cells is unknown. Finally, some scientists believe that γδ T cell responses to tumors can also be activated by heat shock proteins (HSPs). The reader will recall from previous chapters that the expression of HSP occurs in response to stress. *In vitro*, the engagement of γδ TCRs by HSP has been shown to lead to γδ T cell activation. More on the role of HSP in anti-tumor responses appears later.

vii) B Cell Responses

It remains unclear whether B lymphocytes actively participate in immunosurveillance and tumor rejection. Anti-tumor antibodies are not often made in cancer patients, and when they are, they are usually not effective. However, if tumor cells are isolated from a mouse and injected back into the same host (i.e., the mouse receives autologous tumor cells), high titers of anti-tumor IgG antibodies can be detected. The exploration of the use of induced or exogenous anti-cancer antibodies in cancer therapy is ongoing (see later).

viii) The Role of Heat Shock Proteins in Anti-tumor Responses

Originally, the involvement of HSPs in tumor immunogenicity was puzzling. Scientists observed that HSPs isolated from tumor cells could induce anti-tumor immune responses, but the same HSPs isolated from normal cells could not. No differences in nucleotide or amino acid sequence were found between the tumor cell and normal cell versions of HSPs. It was then surmised that the difference likely stemmed from the capacity of HSPs to bind to denatured proteins and peptides from necrotic cells; that is, the peptides associated with HSPs isolated from tumor cells were different from those associated with HSPs from healthy cells. *In vivo*, a positive correlation has been observed in rat carcinomas between the expression of HSPs by a tumor clone and immune system-mediated rejection of that clone. Conversely, loss of HSP expression has caused the tumor clone to revert to a non-immunogenic phenotype. In humans, immune responses against breast cancer cells have been associated with the formation of intracellular complexes between HSP70 and activated mutated forms of p53. Other studies have shown that the expression of HSP60 or HSP70 on the surfaces of tumor cells stimulates the proliferation of human γδ T cells directed against those cells. However, despite the preceding observations, the direct involvement of HSPs in anti-tumor responses awaits more definitive confirmation.

As well as directly activating γδ T cells, HSPs may be involved in supplying peptides for the activation of αβ T cells via both the exogenous and endogenous antigen processing pathways. For example, HSPs released from a necrotic cell and bound to a tumor antigen peptide may convey that peptide to a professional APC, which in turn may present it to a tumor-specific T cell on MHC class II. Cross-presentation (described in Ch.11) may also come into play, with HSP–peptide complexes being taken up by APCs but presented on MHC class I. Indeed, when purified complexes of peptide-bound HSPs are administered to experimental animals, strong CD8+ (rather than CD4+) T cell responses

directed against the relevant pMHC on the tumor cell are mounted. In these cases, the role of the HSPs may be to act as a chaperone and enhance cross-presentation of tumor antigens.

IV. HURDLES TO EFFECTIVE ANTI-TUMOR IMMUNITY

Why does the immune system not prevent all tumorigenesis? For those tumors overexpressing TAAs but not TSAs, central and peripheral T cell tolerance will have already been established and the tumor cell will be seen as self. This tolerance will have to be broken in order for these TAAs to become visible to the immune system, with the added complication that self-tissues expressing that TAA may also then come under attack. Because cancer cells are genetically unstable, it might be logical to think that numerous mutations could occur that might generate new antigenic structures constituting targets for adaptive immunity. Nevertheless, it is also possible that the genetic alterations acquired by a tumor cell might allow it to escape immunosurveillance, suppress the initiation of immune responses, and/or thwart the effector actions of immune system cells. In other words, the tumor may essentially induce peripheral tolerance to itself. In fact, some scientists feel the immune response to a tumor may favor the emergence of such variants, resulting in the "immunoediting" process described previously that may reduce tumor immunogenicity.

What is the evidence that a tumor can actively evade or block immune responses? First, most cancer patients are highly prone to opportunistic infections, some of which can be life-threatening. However, within weeks of the removal of the tumor (by surgery, chemotherapy, or radiation treatment), the immune system of the patient is often restored to full function. Secondly, it has been shown in mouse models that animals bearing tumors respond poorly to antigens that normally induce vigorous immune responses in healthy animals. We will now review several mechanisms by which tumor cells evade the immune system (Table 26-9).

i) Secretion of Immunosuppressive Cytokines

Tumor cells often release immunosuppressive cytokines such as IL-10 and TGFβ, as well as prostaglandins, chemokines, and other unknown factors that can downregulate immune responses. As a result, any TAAs or TSAs acquired by an immature DC for cross-presentation are taken up in a microenvironment that does not support DC maturation. T cells that encounter these DCs are thus likely to be anergized rather than activated. Such an effect may explain why patients with colorectal cancer or lymphoma often have a worse prognosis if they also have elevated plasma IL-10 levels. *In vitro*, IL-10 treatment decreases MHC class I and adhesion molecule expression on melanoma cell lines, inhibits TAP function, suppresses DC activity, and blocks CTL-mediated lysis of tumor cells. Similarly, elevated TGFβ has been associated with various human cancers, including lung, colorectal, pancreatic, stomach, and breast cancers. TGFβ, which was first named for its ability to promote the

Table 26-9 Mechanisms by Which Tumors Thwart the Immune System

Secrete immunosuppressive cytokines such as IL-10 and TGFβ that block effector cell functions and promote T_{reg} differentiation

Induce downregulation of molecules in the TCR signaling path

Develop disorganized vasculature lacking appropriate integrin expression for lymphocyte infiltration

Lose or mutate expression of TAAs/TSAs

Downregulate expression of MHC class I, co-stimulatory molecules, and PAMPs

Express HLA-E or HLA-G to inhibit NK cells and CTLs

Express TAAs/TSAs not presented efficiently by immunoproteasomes

Induce tolerance by anergizing anti-tumor T cells

Constitutively activate STAT3 to block pro-inflammatory cytokine expression

Shed copious amounts of TAAs/TSAs to the point of T cell clonal exhaustion

malignant transformation of fibroblasts, also promotes angiogenesis and invasion of the extracellular matrix in cultured tissue explants. *In vitro*, TGFβ blocks the functions of CTLs and NK cells and suppresses Ig synthesis.

Tumors may also contribute indirectly to the establishment of an immunosuppressive cytokine environment by failing to produce pro-inflammatory cytokines and chemokines. Many tumor cells appear to be able to resist the extracellular signals that would normally induce them to express these molecules because the malignant cells show constitutive activation of the transcription factor STAT3. Such resistance can be overcome *in vitro* via a STAT3 blockade.

ii) Promotion of Regulatory αβ T Cell Responses

Evidence is accumulating that the weak response of T cells to many TAAs may be due in some cases to suppression by $CD4^+CD25^+$ T_{reg} cells. As described in Chapter 16, T_{reg} cells are able to suppress the effector functions of activated conventional T cells. When T_{reg} cells are depleted from mice transplanted with tumors, the rejection response to the tumors is enhanced. Studies *in vitro* have suggested that T_{reg} cells act in a non-antigen-specific way to inhibit the anti-tumor responses of NK cells as well as those of helper T cells and CTLs. It is not yet clear whether this inhibition is cytokine- or cell contact-mediated. How are T_{reg} cells activated in a tumor situation? It has been speculated that the large quantities of immunosuppressive cytokines (such as TGFβ) secreted by tumor cells may block the maturation of local DCs that have taken up tumor antigen. These immature DCs may then promote the activation of T_{reg} cells while being unable to activate conventional T cells. Both measures would blunt anti-tumor immune responses.

iii) Inhibition of T Cell Signaling

Another means of derailing anti-tumor immune responses involves a mechanism by which cancer cells are able to downregulate the expression of CD3ζ, ZAP-70, and Lck in peripheral T cells that have infiltrated the cancer. With this blockage of the TCR-initiated signaling pathway, an anti-tumor T cell cannot be activated to attack its target. It is unknown how the tumor accomplishes this feat, but the establishment of a negatively regulating intercellular contact is suspected.

iv) Avoidance of Recognition

In addition to inactivating the immune system cells they encounter, tumors are often able to evade immune scrutiny in the first place. At the anatomical level, the capillaries that wind in and out of the tumor mass and support malignant cell growth are disorganized and often lack the expression of the appropriate integrins to facilitate T cell extravasation into the tumor site. An anti-tumor T cell thus may not be physically able to reach its tumor cell target. At the molecular level, there may simply be no nonself tumor epitopes to be presented. A tumor may never express a TAA or a TSA in the first place, or may spontaneously lose expression of its TAA or TSA. The tumor then is no longer a target for lymphocytes recognizing that antigen or peptides derived from that antigen. These variants are called *tumor escape mutants*, and these cells pose a significant problem for clinical researchers who design vaccines based on a single tumor epitope (see later).

A major mechanism by which cancer cells avoid immune detection is by downregulating or altering MHC class I expression on their surfaces. Mutation of the MHC class I genes themselves, mutation of β2m, alteration of glycosylation patterns, or mutation of genes required for antigen processing can all lead to reductions in surface pMHC and thus a lack of recognition by $CD8^+$ T cells and CTLs. The increased c-myc transcription present in some tumor cells can induce methylation of the promoters of certain HLA genes, blocking their transcription. Some tumor cell clones lose their responsiveness to IFNs and fail to upregulate their existing low levels of MHC class I in response to these cytokines, while other tumor cells actually downregulate their MHC class I expression. Although low levels of classical MHC class I should invite recognition and elimination by NK cells, some tumor cell clones replace them with the non-classical HLA-E, HLA-F, or HLA-G molecules. HLA-E is able to send a strong inhibitory signal to NK cells by binding to the NK inhibitory receptor CD94/NKG2A. Similarly, HLA-G expressed on tumor cells can bind to the inhibitory receptors ILT2 and KIR2DL4 on NK cells and deliver an inhibitory signal able to override the activatory signal mediated by MICA (at least *in vitro*). NK-mediated destruction is thus blocked. Since some CTLs also express ILT2, the expression of HLA-G on a malignant cell helps to protect it from CTL-mediated cytolysis as well. Significantly, the expression of HLA-G on tumor cells is upregulated by IL-10. Little is known about HLA-F but it also appears to bind to ILT2.

v) Immunoproteasome Bias

An intriguing barrier to CTL recognition of tumor cells is the inefficient processing of some TAAs by the immunoproteasome. The reader will recall from Chapter 11 that there are two types of proteasomes. The standard proteasome, the form present in most resting cells, is responsible for the routine protein degradation that generates the peptides presented on MHC class I molecules. The immunoproteasome, the form induced in cells exposed to IFNγ and constitutively present in APCs, is active in supplying peptides for MHC class I antigen presentation in tissues mounting immune responses. Different subsets of peptides can be produced from the same protein, depending on whether the proteasome involved in protein degradation is of the standard or immuno-type. Thus, the efficiency of presentation of a given peptide may be highly dependent on the type of proteasome involved in its production. For example, immunoproteasomes are better at processing and presenting the immunogenic peptide that is derived from the MAGE-3 TAA. In this case, the suppressive cytokine milieu of the tumor may favor maintenance of the standard proteasome, decreasing the efficiency of MAGE-3 peptide presentation to CD8$^+$ T cells. On the other hand, in one study, melanoma cells expressing a TAA called Melan-AMART1 were recognized by CTLs specific for certain peptides generated efficiently from this TAA by the standard proteasome. When the melanoma cells were treated for 8 days with IFNγ, the same CTLs could no longer recognize the melanoma cells because the IFNγ had induced the predominance of immunoproteasomes over standard proteasomes. Translating this observation to an *in vivo* situation, the exposure of tumor cells to a cytokine milieu that favors one type of proteosome over another might bias the spectrum of peptides presented and help the tumor to avoid detection by the immune system.

D. Cancer Therapy

The first step in cancer therapy is almost always the surgical removal of the complete primary tumor mass, if possible. This approach is clearly not an option for very diffuse cancers, such as leukemias, in which the tumor cells circulate throughout the body, nor for metastatic cancers, in which many small and hidden sites may be affected. Neither is surgery advisable when the tumor is located in such a way that the efforts to remove it would irreparably damage a vital body structure, as in many types of brain cancers. Until relatively recently, the only next step an oncologist could take after surgery was either performed or ruled out was the implementation of a course of conventional radiation therapy and/or chemotherapy. Both of these techniques exploit the biology of the cancer cell to kill it, and in both cases, the therapy predominantly affects rapidly dividing cells while sparing the resting cells that comprise the majority of normal cells in the body. However, these therapies also have detrimental side effects that often have a significant impact on the quality of life of the cancer patient. Today, thanks to 50 years of intensive

research, we stand on the threshold of a new era of cancer therapy that exploits our growing understanding of both cancer cell biology and the roles of the innate and adaptive immune responses in immunosurveillance and tumor rejection. The challenge now is to translate the knowledge and experimental success gained in animal models into practical and effective therapies for human cancers.

In this section, we briefly describe the long-standing therapies of radiation and chemotherapy, including mention of some of the novel chemotherapeutic drugs developed as a result of our increased understanding of tumor biology. We then delve more deeply into the possibilities of *immunotherapy*, the manipulation of the immune system to fight cancer. It is the fervent hope of many researchers that immunotherapy will complement or even replace surgical, chemotherapeutic, or radiation treatments, or at least be useful for eliminating those last few cancer cells that remain after more conventional therapies have been attempted.

I. RADIATION THERAPY

The use of radiation therapy to treat cancer has a very long history, beginning in the late 1800s. However, "cures" of only a limited list of tumors could be achieved up until the 1950s, when the introduction of cobalt-based irradiation machines expanded the utility of this form of cancer treatment. Further gains in precision and dose control were made after the discovery of linear particle accelerators in the 1960s. Today, radiation oncologists are able to deliver "hits" of damaging energy to tumors that are hidden deep within the body with a minimum of damage to more superficial tissues. Nevertheless, in most cases, radiation therapy is carried out in concert with chemotherapy and surgery to ensure that as few cancer cells as possible escape elimination.

Ionizing radiation, as opposed to UV-irradiation or microwaves, is used for radiation therapy because of its high energy. While ionizing radiation includes γ-irradiation and X-rays, only γ-irradiation is used for cancer treatment. The sources of radiation used are the forms or *isotopes* of elements that undergo radioactive decay. Depending on the isotope used (Table 26-10), the energy is delivered either as photons from γ-irradiation, or as sub-nuclear particles emitted during radioactive decay. These sub-nuclear particles are protons, α-particles (helium nuclei), or β-particles (electrons). Isotopes differ with respect to their half-lives, and the half-life of an isotope determines how long it is effective. Isotopes also vary with respect to the amount of energy carried by their emitted particles and the distance these particles can travel within a tissue. These factors must be carefully weighed when choosing an isotope for treatment. Different isotopes are used for treating a superficial versus a deep-seated tumor, or a tumor positioned very closely to a vital organ that cannot tolerate much radiation.

The principle of radiation therapy is that photons and sub-nuclear particles cause damage to the DNA of cells with which they interact. The molecules of an irradiated cell begin to eject electrons from their atomic nuclei, releasing large quantities of

Table 26-10 **Common Radioisotopes Used for Radiation Therapy**

Radioisotope	Half-life	Cancer Treated
213-Bismuth	46 min	RIT: leukemias, colon cancer
137-Cesium	30 years	Implant: uterine, vaginal cancers
60-Cobalt	5.3 years	Whole body or restricted field: various cancers Implant: ocular cancer, melanoma
198-Gold	2.7 days	Implant: prostate cancers
125-Iodine	60 days	Implant: prostate, colon, rectal, brain, and ocular cancers; melanoma
131-Iodine	8.2 days	RIT: lymphomas and leukemias; prostate
192-Iridium	74 days	Implant: brain, breast, prostate cancers
32-Phosphorus	14 days	IP: early stage ovarian cancer; reduce pain from bone metastases
256-Radium*	1622 years	Implant: breast, head and neck, uterine, vaginal cancers
153-Samarium	46 hours	IV: bone metastases in breast, lung, prostate cancer
89-Strontium	50.5 days	IV: bone metastases in breast, lung, prostate cancer
90-Strontium	28.8 years	Implant: ocular cancer
90-Yttrium	64 hours	RIT: lymphomas and leukemias; ovarian, breast, pancreatic, thyroid

*Used less often now because of fears of radon gas escaping the implant.
RIT, radioimmunotherapy; see later in this chapter. IP, intraperitoneal injection; IV, intravenous injection.

energy that damage DNA directly or indirectly. The direct effect occurs when the energy of an ejected electron is absorbed by the DNA itself, breaking the strand. The indirect effect occurs when the energy of the ejected electron is absorbed by a water molecule within the cell, and free radicals are generated that bind to DNA and break its bonds. Both single-strand and double-strand breaks may be induced. In both cases, if the damage is severe enough that it cannot be fixed by the DNA damage repair mechanisms normally operating within the cell, the cell dies when it attempts to initiate DNA replication. Thus, it is rapidly dividing cells like cancer cells that succumb to radiation therapy. For unknown reasons, hematopoietic cells (whether resting or dividing) are also inherently highly sensitive to irradiation. The bone marrow is therefore particularly sensitive to collateral damage during radiation therapy.

Radiation therapy can be delivered either externally as a focused beam targeting the tumor in a particular part of the patient's body, or internally as a metal "seed" containing the radioisotope. The latter approach is often called *brachytherapy* or *implant radiation*. The advantage of brachytherapy is that the radioisotope can be placed directly in the tumor, or at least in close proximity to it. Very high doses of radiation can thus be delivered very precisely for shorter periods of time than by external therapy, decreasing detrimental side effects on patients. In general, the radioisotopes used for brachytherapy have short half-lives and exert high energy over a short distance, so that they kill only the tumor cells while sparing surrounding normal cells, and their effects are not prolonged. These considerations also ensure that implanted patients can be discharged relatively quickly without irradiating everyone else they meet. Brachytherapy has been successfully used for treatment of cancers in accessible body cavities such as the uterus, bronchus, vagina, and esophagus. As well, implantation of radioisotope seeds directly into tumors of the prostate, thyroid, brain, breast, and bladder has yielded clinical benefits.

Despite the fact that radiation therapy is not usually delivered systemically, it can have wide-reaching and serious side effects on patients. In general, radiation side effects are classified as *early* or *late reactions*. Early reactions are due to the irradiation of fast-growing normal tissues, such as the mucosae of the GI and the skin. Accordingly, nausea, vomiting, hair loss, and bone marrow suppression are all commonly observed. Inflammation of mucosae and skin may appear just after the completion of treatment but is usually resolved without intervention by 2–4 weeks after the last irradiation session. Late reactions, which occur 6–12 months after cessation of radiation treatment, are more serious and are associated with damage to tissues directly within the irradiated field. For example, there may be ongoing inflammation of the spinal cord, or necrosis of brain cells. Cataracts or ulcerations may develop in the eyes of patients being treated for ocular tumors. Irradiation of a lung tumor can lead to pneumonitis and fibrosis of surrounding healthy tissue, while irradiation of a stomach cancer can result in ulcers or perforation of the normal stomach wall. Surgical removal of the deteriorating tissue, if possible, is often attempted. The reader should also keep in mind that the *threshold tolerance* of an organ to irradiation (the dose before significant damaging side effects occur) is decreased when the patient is also being treated with chemotherapy. Chemotherapeutic drugs such as bleomycin and doxorubicin (see later) increase the sensitivity of all cells to irradiation, most likely by inhibiting the repair of sub-lethal DNA damage. This factor must be taken into account when choosing the radiation dose.

II. CHEMOTHERAPY

Chemotherapy is the use of drugs to kill tumor cells in a cancer patient. It is a mode of treatment that originated over 50 years ago as a "blunt force" instrument for the treatment of advanced or metastatic cancers only. These days, chemotherapy is tailored to the cancer patient and his or her disease, and is used at much earlier stages. In addition, because combinations of two or more chemotherapeutic drugs often work synergistically, patients are usually treated with "cocktails" containing anywhere from two to six agents, depending on the type and stage of the tumor. Considerably more benefit is accrued to the patient than if any one of the cocktail drugs is used alone. Furthermore, the use of more than one drug greatly reduces the chance of the tumor becoming resistant to drug treatment.

Chemotherapy is usually applied systemically, particularly when the cancer has spread beyond the reach of the surgeon's scalpel. Because cells throughout the body are affected, the *therapeutic index* of each drug used in the cocktail must be taken into account when planning treatment. The therapeutic index of a drug is the dose range of that drug that benefits the patient, i.e., the dose is high enough to kill the cancer cells but not so high that it kills (or severely damages) the patient. This dose range is usually very narrow and is affected by factors such as the route of administration, how the drug is metabolized, the overall health of the patient, the stage of the cancer, and the sensitivity of the cancer cells to that particular drug. The effective concentration of the drug that actually reaches the tumor site must also be taken into account as it may be comparatively low due to systemic dilution. Moreover, the administration of a single dose of any drug or combinations of drugs (at their therapeutic indexes) does not kill every single cancer cell in the patient but only a proportion of them, meaning that cyclical rounds of chemotherapy are needed to systematically reduce the overall number of cancer cells in the patient's body. Whether the chemotherapy will eventually be successful in reducing this number to zero depends in part on whether any tumor cells (because of their genetic instability) become resistant to the drugs, or whether any tumor cells escape to areas of the body not penetrable by the drugs.

The underlying principle of chemotherapy is that a chemotherapeutic agent primarily affects only those cells that are growing faster than most normal cells, or those that have a metabolic imbalance. For the most part, chemotherapeutic drugs work by directly or indirectly damaging the replicating DNA of the dividing cell, inhibiting DNA synthesis, preventing cell division, or blocking access of the tumor cells to a necessary growth factor. Most chemotherapy drugs are alkylating agents, antibiotics, anti-metabolites, glucocorticoids, or plant alkaloids. Some of these drugs attack cells at a particular stage of the cell cycle (such as during mitosis or S phase), while others can damage cells at any stage. Examples of conventional chemotherapeutic drugs and the mechanisms by which they operate are given in Table 26-11.

Unfortunately, by its very nature, chemotherapy is not very specific, and fast-growing normal cells such as those in bone marrow are also damaged. Immunity to pathogens therefore plunges. Cells of the GI tract are also highly proliferative, so that chemotherapy almost inevitably causes nausea and vomiting. In addition, because the liver and kidneys are sites that accumulate these powerful drugs and their metabolites, hepatic and nephrotic toxicity are common side effects. As a result, many researchers are dedicated to developing new, more specific targeted chemotherapies that will preferentially affect tumor cells and thus cause fewer detrimental side effects in patients (Table 26-12). For example, several of these new drugs, such as Letrozole and Iressa, focus on blocking the synthesis or function of estrogen, removing the essential growth factor driving many breast cancers. Others, such as Gleevec and Bortezomib, exploit the molecular defect leading to the rise of the cancer itself. The mechanisms by which these drugs exert their effects are explored in Chapter 30.

A relatively recently devised strategy for combating cancer cells involves the use of *ribozymes*. Ribozymes are catalytic RNA molecules that bind specifically to mRNAs of complementary sequence and cleave them, thus acting as "molecular scissors." The idea is to prevent the protein expression of key molecules driving cancer cell proliferation, drug resistance, or metastasis. For example, Angiozyme, a ribozyme that targets the mRNA encoding VEGFR (the receptor for VEGF), is under investigation in human trials as an anti-angiogenic therapy for breast and colon cancer. Similarly, Herzyme is a ribozyme that cleaves Her-2 mRNA and thus may be of value for Her-2-dependent breast and ovarian cancers. Hepatocarcinomas driven by HepC may be resolved in the future by administration of Heptzyme, a ribozyme that targets HepC viral mRNA.

III. TUMOR HYPOXIA

Tumors can become resistant to both chemotherapy and radiation therapy if they grow so fast or become so large that the growth of blood vessels required to supply the malignant cells with oxygen and nutrients can no longer keep pace. The tumor is then said to have "outgrown" its blood supply and parts of the tumor experience low oxygen pressure (*hypoxia*). Hypoxia can also arise when the abnormal structure of the tumor vasculature has led to "blind ends" or tortuous blood vessel shapes that restrict blood flow. Where the loss of oxygen is severe, the tumor shows areas of necrosis. However, the loss of oxygen has a detrimental effect from the cancer therapy point of view. Without oxygen for oxidation, the free radicals generated by ionizing radiation cannot have their full effect and the DNA damage to the tumor cells may not be permanent. For this reason, hypoxic cells are $2-3\times$ less sensitive to radiation damage than normally oxygenated cells. Hypoxic tumor cells are also resistant to most chemotherapeutic agents, partly because hypoxic cells proliferate less and do not respond to signals to undergo p53-mediated apoptosis, and partly because hypoxia upregulates normal cellular genes conferring drug resistance. These factors mean that the core of a solid tumor that experiences hypoxia cannot be eliminated by conventional therapies alone. Moreover, despite suffering decreased oxygen, hypoxic tumor cells do not die and continue to accumulate mutations that allow them to become metastatic. Indeed, hypoxia appears to select for tumor cells with an accelerated rate of mutation, and thus ultimately for a more aggressive cancer. Hypoxia also induces the expression of the transcription factor HIF-1 (*hypoxia-inducible factor-1*). HIF-1 drives the expression of genes containing an HRE (*hypoxia-response element*) in their promoters. One such gene is VEGF, which induces new angiogenesis in an attempt to increase oxygenation. These new blood vessels support tumor growth and thus contribute to disease progression. The

Table 26-11 **Common Chemotherapeutic Drugs**

Drug	Mechanism	Cancer Treated	Side Effects
ANTIBIOTICS			
Adriamycin (Doxorubicin)	Intercalates into DNA and breaks strands; free radical formation	Breast, lung, gastric, ovarian, thyroid, liver, testicular cancers; soft tissue sarcomas, lymphomas, myelomas, leukemias	Bone marrow suppression, cardiac anomalies, hair loss
Bleomycin	Intercalates into DNA and breaks strands; free radical formation	Testicular cancer, lymphomas	Fever, hypertension, liver anomalies
ANTI-METABOLITES			
Azacytidine	Ribonucleotide analogue, inhibits RNA synthesis	Adult non-lymphocytic leukemias	Nausea, vomiting, bone marrow suppression
Cladribine (2-CdA)	Purine analogue, inhibits DNA synthesis	CLL, HCL	Kidney and nerve damage, bone marrow suppression, nausea, vomiting, diarrhea
Cytarabine (Arabinosyl-cytosine; Ara-C)	Pyrimidine analogue, inhibits DNA synthesis	Brain, myeloid leukemias, lymphomas	Nausea, vomiting, bone marrow suppression, liver damage
Deoxycoformycin	Purine analogue, inhibits DNA synthesis	HCL	Bone marrow suppression, thrombocytopenia, nausea, vomiting
Fludarabine	Purine analogue, inhibits DNA synthesis	CLL	Fever, chills, nausea, vomiting, chest pain, swollen glands
Fluoruoracil (5-FU)	Pyrimidine analogue, inhibits DNA synthesis	Breast, GI, head and neck, liver cancers	Mild bone marrow suppression, oral and GI sores (mucositis)
Gemcitabine	Pyrimidine analogue, inhibits DNA synthesis	Pancreas, lung cancers	Nausea, vomiting, rash, fluid retention, tingling and numbness of extremities
Hydroxyurea	Inhibitor of nucleotide reductase, inhibits DNA synthesis	CML	Nausea, vomiting, bone marrow suppression, convulsions
Mercaptopurine	Inhibits *de novo* purine synthesis, inhibits DNA synthesis	Adult non-lymphocytic leukemias	Bone marrow suppression, liver damage
Methotrexate	Blocks dihydrofolate reductase, inhibits DNA synthesis	Brain, breast, bladder, ovarian, head and neck cancers; lymphomas; bone sarcomas	Nausea, vomiting, bone marrow suppression, dermatitis, hair loss
Thioguanine	Inhibits *de novo* purine synthesis, inhibits DNA synthesis	Adult non-lymphocytic leukemias	Nausea, vomiting, bone marrow suppression
ALKYLATING AGENTS			
Busulfan	Alkylates DNA	CML	Diarrhea, weight loss, fatigue, hair loss
Carmustine	Cross-links and breaks DNA strands	Brain, GI cancers; melanoma, lymphomas, myelomas	Acute leukemia, kidney, CNS and peripheral nerve damage, hair loss
Chlorambucil	Alkylates DNA	CLL, lymphomas	Bone marrow suppression
Cisplatin	Binds DNA and blocks replication	Testicular, ovarian, bladder, lung, head and neck cancers etc.	Nausea, vomiting, platelet and kidney damage, nerve damage
Cyclophosphamide	Binds DNA and blocks replication	Breast, lung cancers, soft tissue sarcomas, myelomas, lymphomas	Nausea, vomiting, immunosuppression, sterility

Continued

Table 26-11 Common Chemotherapeutic Drugs—*cont'd*

Drug	Mechanism	Cancer Treated	Side Effects
ALKYLATING AGENTS—*cont'd*			
Dacarbazine	Alkylates DNA	Melanoma, soft tissue sarcomas, lymphomas	Nausea, vomiting, bone marrow suppression
Mechlorethamine	Alkylates DNA	CML, CLL, HL, NHL, polycythemia vera, mycosis fungoides	Nausea, vomiting, bone marrow suppression, weakness, diarrhea, fever
Melphalan	Alkylates DNA	Myeloma, ovarian cancer	Nausea, vomiting, bone marrow suppression, fever, chills, joint pain
Procarbazine	Alkylates DNA	Lymphomas	Nausea, vomiting, bone marrow suppression, flu-like symptoms
GLUCOCORTICOIDS			
Dexamethasone	Binds to glucocorticoid receptors inactivates transcription factors, activates caspase-9	Myeloma	Nausea, vomiting, paranoia, muscle wasting, diabetes, infection, hypertension, osteoporosis, heart disease
Prednisolone	Binds to glucocorticoid receptors inactivates transcription factors	ALL, CLL, myeloma, lymphomas	Facial weight gain, stomach irritation, skin changes, mood swings
PLANT ALKALOIDS			
Etoposide	Causes DNA strand breaks during metaphase	Lung, testicular cancers	Nausea, vomiting, leukopenia
Irinotecan (CPT-11)	Topoisomerase inhibitor, inhibits DNA synthesis	Metastatic colon and rectal cancers	Diarrhea, neutropenia
Taxol (Paclitaxel)	Stabilizes tubulin polymerization, blocks mitosis	Breast, ovarian cancers	Leukopenia, anaphylaxis, cardiac anomalies, hair loss
Vinblastine	Binds tubulin in microtubules, arrests mitosis	Testicular, lung cancers; lymphomas, Kaposi's sarcoma	Nausea, vomiting, leukopenia, skin necrosis, loss of reflexes, muscle wasting, hair loss
Vincristine	Binds tubulin in microtubulin, arrests mitosis	Leukemias, myelomas, lymphomas	Nausea, vomiting, leukopenia, skin necrosis, loss of reflexes, muscle wasting, hair loss
SYNTHETIC DRUGS			
Bisphosphonate (zoledronate, pamidronate)	Pyrophosphate analogue, block secretion of IL-6, reduces reduces osteoporosis	Bone disease in myeloma	Fever, kidney dysfunction, esophageal irritation (oral form)
Thalidomide	Organic chemical, decreases angiogenesis, activates caspase-8, suppresses TNF production	Myeloma	Sedation, neuropathy, deep vein thrombosis (teratogenic)

Hematopoietic cancers are discussed in Chapter 30. These include the following: ALL, acute lymphoblastic leukemia; CLL, chronic lymphocytic leukemia; CML, chronic myelogenous leukemia; HCL, hairy cell leukemia; HL, Hodgkin's lympoma; NHL, Non-Hodgkin's lymphoma; With information from G. R. Weiss, Table 14-1 of "Clinical Oncology," Prentice-Hall International Inc. Norwalk, CT & www.chemocare.com.

stability of the HIF-1 protein is greatly increased under hypoxic conditions, and the expression of the HIF-1 gene is upregulated in cells bearing a Ras, Src, or Her-2 oncogene. As a result, HIF-1 is active in most tumor cells but not in most normal cells. Experimental tumor therapy approaches that directly target HIF-1 have been investigated that prevent this transcription factor from activating its target genes, destabilize the HIF-1 protein, or interfere with HIF-1 transcription or translation. Ways of maximizing the efficiency of delivery of such inhibitors are under development.

Table 26-12 Examples of Targeted Chemotherapeutic Drugs

Drug Name (Other Names)	Mode of Action	Cancers Treated
ATRA [All-trans retinoic acid] (Vesanoid, Tretinoin)	Retinoid binds retinoic acid nuclear receptors, restores the differentiation of the leukemic cells	Acute promyelocytic leukemia
Bortezomib (Velcade)	Boronic acid dipeptide inhibits proteasomes, blocks IL-6 production, activates caspases-8 and -9	Myeloma
Exemestane (Aromasin)	Aromatase inhibitor blocks synthesis of estrogen, inhibits estrogen-driven tumors	Breast cancer
Gleevec (Imatinib mesylate)	Protein tyrosine kinase inhibitor inhibits activity of Bcr-Abl kinase, blocks signal transduction driving survival	CML, certain gastrointestinal stromal cell tumors
Iressa (Gefitinib)	Anilino-quinzoline inhibits tyrosine protein kinases, including EGFR	Metastatic lung cancer
Letrozole (Femara)	Aromatase inhibitor blocks synthesis of estrogen, inhibits estrogen-driven tumors	Breast cancer
Tamoxifen (Novaldex)	Estrogen receptor antagonist blocks estrogen receptor, inhibits estrogen-driven tumors	Metastatic breast cancer, ovarian cancer
Toremifene (Fareston)	Estrogen receptor antagonist blocks estrogen receptor, inhibits estrogen-driven tumors	Breast cancer
Angiozyme	Ribozyme specifically cleaves VEGFR mRNA, inhibits tumor angiogenesis	Breast cancer, colon cancer

CML, chronic myelogenous leukemia (see Ch.30).

One ongoing goal of cancer researchers has been to devise methods of reducing hypoxia in a tumor's core, thereby increasing its sensitivity to conventional cancer therapies. However, these attempts have generally not been very successful. Researchers have now turned to novel approaches that exploit the fact that hypoxia is rare in normal tissues but common in tumors. The idea is to use this specificity to establish a complementary therapy that eliminates those parts of the tumor that are resistant to radiation and chemotherapy. For example, a cancer patient can be given an inactive drug precursor (a *pro-drug*) that remains inactive as long as normal oxygen levels are present. However, in a hypoxic environment, the drug is chemically activated to a cytotoxic form that kills cells. One such drug, called *tirapazamine* (TPZ), not only kills hypoxic cells but also interacts directly with cisplatin (refer to Table 26-11) to potentiate its anti-cancer effects. Significant success in the treatment of small cell lung cancer has been observed in clinical trials of this combination approach. A related hypoxia-based strategy (still experimental) involves gene therapy, in which a vector is constructed such that a gene encoding a cytotoxic drug is placed under the control of the HRE motif. As mentioned previously, expression of a gene containing HRE in its promoter is activated only under conditions of hypoxia. Thus, in theory, a hypoxic tumor cell that takes up this vector is killed as soon as HRE activation drives transcription tumor translation of the cytotoxic gene. However, as was true for the gene therapy approaches for primary immunodeficiencies discussed in Chapter 24, a major hurdle is the development of vectors that can efficiently transfer the cytotoxic gene into the tumor cells of interest.

IV. IMMUNOTHERAPY

While chemotherapy and radiation therapy have improved the survival of many cancer patients, there are several problems with these approaches. The high levels of drugs and radiation that are applied in an effort to achieve sufficient killing power at the tumor site may encourage the rise of drug-resistant and radiation-resistant populations of tumor cells. In addition, systemic or local toxicity can be high, because there is often no way to confine the drug or radiation treatment precisely to the tumor site. This problem is compounded several-fold if the tumor has metastasized. Clinical researchers have thus turned to the immune system to devise therapies that kill cancer cells as specifically as possible. The idea is to either create an immune response against the tumor, or augment an existing effector mechanism attempting to eliminate the tumor. In some cases, the agents used for immunotherapy belong to a class of therapeutic molecules often referred to as *biologicals*. "Biologicals" are natural or engineered proteins whose purpose is to disrupt the proliferation, metabolism, or function of pathogenic host cells.

The field of cancer immunotherapy has had its share of highs and lows. For example, by the late 1970s, it seemed that the recognition of the existence of TAAs and TSAs on tumors combined with the advent of monoclonal antibody technology would result in a new era of cancer treatment. Indeed, researchers of the day confidently suggested that a unique TSA would exist for each tumor, and that all one would have to do was identify the TSA, raise a specific mAb against it, and administer this "magic bullet" to the patient. In theory, all cancer cells would be killed by this customized treatment while all healthy cells would be spared. In practice, very few hybridomas generating mAbs recognizing TSAs were ever successfully generated. Several factors have contributed to the loss of popularity of this approach. First, it has proved extremely challenging to identify non-viral TSAs, with the only true examples being those described previously for melanomas. Secondly, mAbs generally recognize conformational epitopes, but many TSAs likely exist only as small mutated peptides. Thirdly, while TAAs might be used as alternative targets, there remains the potential for a TAA-specific mAb to cross-react with healthy tissues since TAAs are, by definition, "self."

With the great strides made in the 1970s and 1980s in understanding T cell recognition of antigens, immunotherapy mediated by T cells became a focus of cancer therapy research. Again, mixed results have been obtained. It has been demonstrated that T cells directed against tumor cells exist, and that anti-tumor T cells introduced into a patient home to the tumor mass. Moreover, vaccinating a patient with a purified TAA or TSA induces the expansion of T cells recognizing that antigen (see later). However, there seems to be no correlation *in vivo* between the presence of proliferating anti-tumor T cells and clinical outcome. To date, a clinical response to tumor antigen vaccination has been observed in just 20–40% of patients, and tumor regression in only a small minority. Even these very modest "successes" must be interpreted with caution, since several tumor types have relatively high rates of spontaneous regression. For example, 10–20% of melanomas and 5% of renal cell carcinomas will regress in the absence of any treatment, making it difficult to determine whether a vaccine is in fact doing anything. The current limited response rate to vaccines may indicate that there is more heterogeneity among tumor cells than originally thought, and/or that the protocols used to induce immune responses against a specific tumor antigen have merely promoted the selection of tumor cells that fail to express that antigen.

Despite the many difficulties encountered, constant progress in the understanding of both tumor biology and immune responses to tumors has allowed improved therapeutic manipulation of the immune system to fight cancer (Table 26-13). Ironically, the humoral arm of adaptive immunity, which does not appear to combat tumors when they first arise, has been exploited successfully to treat certain kinds of established malignancies. As well as antibody-based therapies, researchers are testing targeted cytokine secretion and immunotherapeutic agents designed to block the supply or action of a required growth factor, or disrupt the angiogenesis supporting tumor growth. Refined anti-tumor vaccination protocols are also under development. Among other approaches, the delivery of

Table 26-13 General Approaches to Cancer Immunotherapy

Administer a pure cytokine with anti-tumor effects

Block metastasis with a chemokine receptor antagonist

Administer autologous tumor cells modified to express a cytokine with anti-tumor effects

Administer an anti-TSA/TAA mAb that
- induces tumor cell killing via ADCC, complement activation, apoptosis
- activates NK cells
- blocks receipt of a tumor growth factor

Administer mAb that damages tumor vasculature

Administer an anti-TSA/TAA mAb conjugated to
- a toxin (immunotoxin)
- a radioisotope (RIT)
- an enzyme (ADEPT)
- a cytokine (ICK)

Administer conjugates of toxins and tumor growth factors

Administer a vaccine composed of
- whole tumor cells expressing TSA/TAA
- purified TSA/TAA, plus adjuvant, cytokine, or HSP
- GM-CSF-stimulated DCs pulsed with TSA/TAA peptide, protein, or RNA
- DCs pulsed with α-galactosylceramide to stimulate NKT cells

Administer anti-CTLA-4 Ab to inhibit negative regulation of activated anti-tumor T cells

Administer agonist costimulatory Ab to stimulate anti-tumor T cells

genetically engineered tumor cells, the administration of DCs loaded with purified tumor peptides, and the administration of purified peptides or macromolecules derived from TAAs or TSAs have all been used in a bid to kick-start the immune system into controlling the malignancy.

i) Cytokine-Based Therapies

It has been difficult to correlate either systemic or intratumoral levels of particular cytokines or classes of cytokines with clinical outcome. For example, high plasma levels of Th2 cytokines have been observed in patients with advanced tumors and poor prognosis, but also in patients with earlier stages of cancer and a more favorable prognosis. Similarly, although IFNγ is most often thought of as an anti-tumor cytokine, it may have a pro-tumorigenic effect or no effect depending on the situation. In the next section, we discuss the effects of several cytokines in relation to tumorigenesis and the clinical outcomes when these cytokines are injected systemically into patients. A discussion of new approaches that promise more directed introduction of cytokines into tumor masses follows.

ia) Pure cytokine administration. From what we have learned throughout this book on the nature and function of cytokines, it would be logical to assume that at least some of them might have anti-tumor effects that could be manipulated, at least under some circumstances. Unfortunately, it has been difficult to define and exploit those circumstances. The reader might wonder in particular about the therapeutic use of IFNγ, since this cytokine promotes the CTL response, activates macrophages, and has anti-angiogenic effects. Moreover, as stated earlier in this chapter, IFNγR knockout mice spontaneously develop tumors. However, while high IFNγ levels have been found in spontaneously regressing melanomas, they have also been detected in developing renal carcinomas and in tumor cells isolated from ovarian and breast cancers. In the clinic, administration of IFNγ seems to enhance the effectiveness of various tumor cell vaccination procedures (see later). On the other hand, treatment of metastatic melanoma patients with exogenous IFNγ alone has offered little clinical benefit. Similar difficulties have come to light with the use of IL-10, which can either promote or suppress tumor cell growth *in vitro* depending on the experimental system used. Administration of exogenous IL-10 has induced regression of melanoma and breast cancer metastases in some animal models but has had no effect in others.

IL-2 promotes the expansion of tumor-specific T cells *in vitro* and counteracts the effects of immunosuppressive cytokines. However, intravenous injections of IL-2 have induced clinical tumor regression in only 20% of treated renal carcinoma and melanoma patients. Moreover, the dose of IL-2 that can be safely administered to humans is quite limited due to toxicity. Even doses that are sub-toxic can have unexpected effects. In one clinical trial, melanoma patients received a vaccine containing an epitope of the melanoma TAA gp100 plus a substantial (but non-toxic) dose of IL-2. In theory, the IL-2 was supposed to promote the activation and survival of gp100-specific T cells. However, those patients showing a clinical response actually had <u>fewer</u> gp100-specific CTLs in the tumor site. It was speculated that, rather than homing to the tumor site, the gp100-specific CTLs migrated to other areas of the body because the relatively high dose of IL-2 used altered the characteristics of the blood vessels in the tumor site. It remains unclear why these patients enjoyed a clinical response despite the faulty migration of the tumor-specific CTLs.

IL-12 has considerable anti-tumor activity due to its ability to promote NO production by macrophages, stimulate NK and NKT cells, and drive IFNγ secretion. Indeed, a major part of IL-12's anti-tumor activity is due to its induction of IFNγ. IL-12 also induces the local secretion of chemokines such as IP-10 and MIG that contribute to the destruction of tumor vasculature. Interestingly, the ability of IL-12 to drive the generation of Th1 effectors does not appear to be crucial for its anti-tumor activity. Studies have shown that IL-12 can induce tumor rejection in *nude* mice despite the fact that these animals lack mature T cells and thus cannot generate Th1 effectors. On the clinical side, systemic injections of IL-12 into human cancer patients have yielded mixed results, with the induction of tumor cell necrosis in kidney cancer patients but no appreciable benefit for other types of tumors. Large systemic doses of IL-12 are toxic, causing hematological abnormalities and the loss of liver function.

The verdict is also mixed on the potential efficacy of IL-4 as a therapy. On the one hand, exogenous IL-4 can block the expansion of various tumor cell lines *in vitro*. As well, there is a correlation between the lack of IL-4 expression by a given human tumor sample and the failure of anti-tumor immune responses raised against it. However, other studies have found no differences in IL-4 expression between normal PBLs and tumor-associated TILs. At the clinical level, treatment with exogenous IL-4 has not improved patient outcomes.

Equivocal results have also been obtained for GM-CSF therapy. GM-CSF is a powerful stimulator of DC maturation and function, and enhances ADCC-mediated tumor cell killing by granulocytes *in vitro*. However, systemically injected GM-CSF is cleared very rapidly from the circulation, limiting its ability to access tumor sites *in vivo*. As we shall see in the next section, strategies that introduce exogenous GM-CSF directly into a tumor site have had more promising results.

Cytokine therapy has also been employed to block metastasis. Because metastasis appears to depend on chemokine gradients, some researchers are experimenting with chemokine receptor blockades to stop the spreading of cancers. For example, small molecule antagonists of CXCR4 (developed to block HIV infection of T cells; refer to Ch.25) are in clinical trials to assess their potential to block breast cancer metastasis.

The reader may wonder why, with its name "tumor necrosis factor," TNF is not the most obvious choice for anti-tumor cytokine therapy. Unfortunately, when clinical trials of the systemic administration of recombinant TNF were carried out in patients with advanced cancer, the patients were unable to tolerate a large enough dose of TNF to have any effect on their tumors. Severe side effects, including liver toxicity, organ failure, and hypotension, were observed. Better results have been obtained using localized application of TNF, which has been used in conjunction with IFNγ and chemotherapy to treat certain sarcomas and inoperable melanoma metastases.

We include one last note on IFNα therapy. While the administration of this cytokine (Intron-A, Roferon-A) has been very useful for the treatment of certain leukemias (see Ch.30), it has not proved as beneficial for therapy of non-hematopoietic tumors. Indeed, because IFNα upregulates expression of HLA-G, IFNα may actually promote the escape of certain tumor cells from NK-mediated cytolysis. In one clinical study, the condition of patients with an HLA-G-expressing melanoma deteriorated following treatment with IFNα.

ib) Engineered cytokine secretion. Some researchers believe that pure cytokines have frequently not proven effective in controlling tumor growth because the dose of cytokine has not been delivered directly to the tumor in a sustained manner. Several laboratories have attempted to remedy this difficulty using a transfection approach. Tumor cells are isolated from an experimental animal and genetically engineered *in vitro* to express a cytokine gene. When these tumor cells are returned to the same animal, the cells home back to the primary tumor and commence secreting the cytokine directly into the tumor microenvironment. For example, mice injected with tumor cells engineered to express GM-CSF develop strong anti-tumor immune responses characterized by enhanced DC activity. Both Th1 and Th2 responses are induced in mice by the introduction of GM-CSF-secreting tumor cells. The production of NO and anti-angiogenic factors by activated macrophages is also enhanced.

In human melanoma patients, trials have been carried out in which autologous tumor cells were isolated, engineered *in vitro* to express GM-CSF, and returned to each patient. The homing of these modified cells back to the primary tumor created a local milieu of GM-CSF that induced the infiltration of the tumor by activated lymphocytes, neutrophils, and eosinophils that worked together to destroy the tumor vasculature. In addition, vigorous antibody responses against melanoma surface antigens and CTL responses against intracellular epitopes were induced. Encouraging results have also been obtained in cases of lung and prostate cancer where patients have been treated with autologous GM-CSF-secreting lung or prostate carcinoma cells, respectively.

Other cytokines have been used to modify tumor cells. When tumor cells are isolated, engineered to express IL-12, and returned to the original tumor-bearing animal, regression of the tumor can be induced. Moreover, healthy animals treated with IL-12-expressing tumor cells are able to resist a later challenge with cells from the original, non-transfected cancer. The expression of IL-2 by engineered tumor cells has also been used to reverse the effects of immunosuppressive cytokines. One *in vivo* study examined the effect of IL-2 on a murine sarcoma cell line that was able to resist rejection due to its substantial production of TGFβ. When these cells were transfected with IL-2 and introduced into mice, the animals were able to reject the sarcoma cells. Similarly, mouse renal tumor cells transfected with the IL-4 gene do not grow once transplanted *in vivo*, and transfection of mouse colon carcinoma cells with the IL-10 gene reduces their malignancy. Treatment of mice with mouse neuroblastoma cells engineered to express IFNγ leads to rejection of non-transfected neuroblastoma cells in a later challenge. These results strongly suggest that engineered cytokine secretion can be a useful mode of cancer immunotherapy.

In the hopes of streamlining the time-consuming preparation of modified tumor cells, some investigators have explored the possibility of infecting autologous tumor cells with retroviral or adenoviral vectors expressing the cytokine gene. While this type of measure would speed up the gene transfer step, fears remain that these viruses might not truly be inactivated (refer to Ch.23). These concerns have limited the enthusiasm of researchers to translate this approach to humans.

ii) mAb-Based Therapies

Several novel cancer therapy techniques have taken advantage of the specificity and ease of manipulation of monoclonal antibodies. When employed on its own, an exogenous mAb binds to a TAA/TSA on the surface of the tumor cell and is in turn bound by the FcR of macrophages, neutrophils, or eosinophils. The tumor cell is then destroyed by ADCC. mAb binding may also trigger classical complement activation, leading to the deposition of C3b on the tumor cell surface and elimination by effector cells expressing CRs. Certain mAbs have also been observed to induce the apoptosis of tumor cells on their own. Finally, antibodies may be used to sequester growth factors

needed by the tumor to continue its expansion, or to interfere with angiogenic factors and thus reduce the blood supply to the tumor mass.

Some mAb therapies are based on *immunoconjugates* in which a tumor-specific mAb is covalently linked to a cytotoxic molecule that can kill a cell it contacts. In theory, use of a TSA-directed mAb conjugated to such a molecule would ensure that normal cells are spared while the drug, toxin, or cytokine is targeted to the cancer cells in either the primary tumor site or areas of metastasis. Such targeting should decrease systemic toxicity while increasing the effective concentration of the cytotoxic agent at the tumor site, thus improving efficacy.

We now discuss the therapeutic potential of conventional mAbs and immunoconjugates based on conventional mAbs. As well as pursuing these two avenues, several clinical researchers have investigated ways of manipulating the structure of the Ig molecule itself, generating derivatives that may have advantages over conventional mAb structures (see Box 26-3).

iia) Unconjugated mAb administration. With respect to employing unconjugated antibodies to combat cancer *in vivo*, in those rare cases when a TSA or TAA can be identified on the tumor cell, a mAb can be raised against it in mice, "humanized" (as described in Ch.7), and used in the clinic (Table 26-14). One of the best-known anti-tumor humanized mAbs, called rituximab, is directed against CD20. CD20 is a surface antigen found on pre-B cells and resting and mature B cells, but not plasma cells, pro-B cells, or HSC. (The ligand and function of CD20 are unknown, and CD20$^{-/-}$ mice appear to have perfectly normal immune responses.) Anti-CD20 mAb treatment is now widely used in conjunction with chemotherapy to control certain B cell lymphomas in which the cancerous lymphocytes express significant levels of CD20 (see Ch.30). Because CD20 is not expressed on plasma cells, these cells are spared and the patient is able to maintain adequate levels of circulating Igs to fight infection. In addition, the chimeric antibody protein does not provoke the production of large quantities of antibodies directed against its mouse sequences.

A similar approach has been successful for the 25% of breast cancer patients in whom the Her-2 gene is amplified. Anti-Her-2 mAbs (drug name Herceptin), which bind to the Her-2 receptor, have been used very successfully in combination with anti-cancer drugs to promote tumor regression. It is thought that Herceptin works in three ways: (1) the Herceptin mAb binds to Her-2 and interferes with Her-2-mediated signaling, (2) the Her-2/Herceptin complex binds to FcR on NK cells and triggers tumor cell destruction by ADCC, and (3) Herceptin blocks (by an unknown mechanism) the repair of DNA damage induced by chemotherapeutic agents. It should be borne in mind, however, that Her-2 is also expressed in normal heart tissue, so that cardiotoxicity can be a serious complication of Herceptin treatment. Another TAA that has been targeted with mAbs is EGFR, overexpressed in many breast, prostate, ovary, and colon cancers. The most effective anti-EGFR mAbs appear to work by blocking the binding of the ligands EGF and TGFα to the receptor, inhibiting EGFR phosphorylation and downstream signal transduction. Receptor internalization is also induced, decreasing the number

Box 26-3. Use of structural derivatives of mAbs for immunotoxins

The reader will recall from Chapter 5 that a conventional immunoglobulin molecule (150 kDa) consists of two identical heavy chains and two identical light chains bound together by disulfide bonds. The N-terminal V domains of each pair of heavy and light chains come together to form two identical antigen-binding sites (Fab regions), while the three (or four) C_H domains of the heavy chains mediate antibody effector functions (Fc region). Researchers have been able to use the techniques of genetic engineering to manipulate the various Ig domains and generate antibody fragments useful for various applications. For example, a *minibody* (80 kDa) is a truncated IgG molecule in which the V domains are joined to a shortened Fc region composed solely of the C_H3 domain (see Figure). This smaller structure is more rapidly cleared from the circulation than the complete antibody, an advantage when the mAb is conjugated to a highly toxic molecule or radioisotope that would cause unacceptable damage to normal cells if allowed to stay in the tissues for a prolonged period. In addition, the minibody's reduced size means that it is able to penetrate better into solid tumors, slipping between the tight junctions holding the malignant cells together. The time for the toxin or radioisotope to reach its target tumor cells is thus much reduced.

The smallest mAb derivative to date that can still bind the original antigen is called the "single chain Fv" molecule (scFv). This structure (25–30 kDa) is composed of a single V_H domain linked to a single V_L domain via a peptide. scFv molecules are readily linked to other molecules or radioisotopes to form immunoconjugates, and their very small size means that any scFv that does not home rapidly to the tumor site is quickly taken up and eliminated by the kidneys, minimizing damage to normal tissues. When it is desirable to extend the half-life of the scFv in the tissues and circulation (as when the scFv is intended to deliver a toxin or radioisotope to a large tumor mass), the scFv can be dimerized into a *diabody* (50 kDa). A diabody contains two Fab sites linked to each other by disulfide bonds and a peptide. Because of their linked V domains, diabodies show superior avidity of binding to the relevant tumor antigens compared to scFv molecules. This avidity can be increased still further if the diabody is joined to an additional scFv to form a *triabody* (90 kDa), or if two diabodies each containing two antigen-binding sites are joined together by a peptide linker and disulfide bonds to generate a *tetrabody* (120 kDa; also known as a *tandem diabody*). These modified structures show the ideal combination of high binding avidity, moderate resistance to kidney clearing, and ease of tumor mass penetration.

Sometimes it is desirable to use an antibody fragment to link two different epitopes together that are located on either the same or different cells. A *bi-specific Fab* or even a *tri-specific Fab* can be created using chemical joining of two or more scFv molecules of different specificity. For example, one bi-specific anti-cancer antibody derivative was designed using an scFv molecule with specificity for CD3 (so that it bound to T cells), and another scFv molecule with specificity for the TAA CEA (so that the T cell would be brought into immediate proximity with a CEA-expressing tumor cell). A toxin molecule can also be brought directly to a tumor cell surface using a bi-specific Fab in which one antigen-binding site recognizes an epitope on the target cell surface, and the other site recognizes an epitope on the toxin molecule. The ability to manipulate Ig structures in this way holds out great hope for the increasingly precise use of such "magic bullets."

With information from von Mehren M. *et al.* (2003) Monoclonal antibody therapy for cancer. *Annual Review of Medicine* **54**, 343–369.

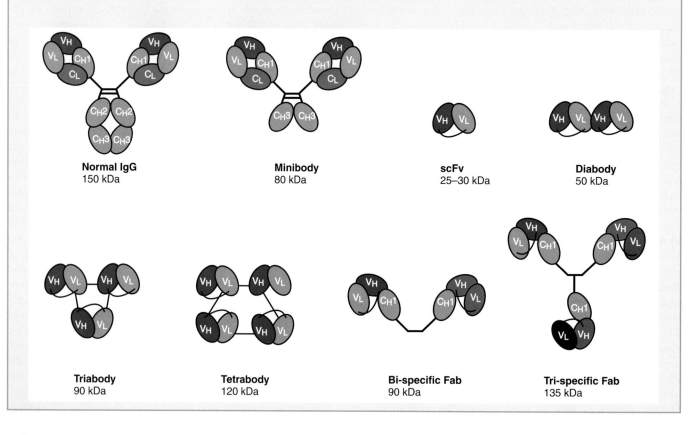

Normal IgG
150 kDa

Minibody
80 kDa

scFv
25–30 kDa

Diabody
50 kDa

Triabody
90 kDa

Tetrabody
120 kDa

Bi-specific Fab
90 kDa

Tri-specific Fab
135 kDa

Table 26-14 Examples of Monoclonal Antibody-Based Immunotherapeutics

Drug Name (Other Names)	Mode of Action	Cancers Treated
Avastin (Bevacizumab)	mAb binding to VEGF receptor stops tumor angiogenesis promoted by VEGF	Metastatic colorectal cancer
Erbitux (Cetuximab)	mAb binding to EGFR facilitates killing of EGFR-bearing cancer cells	Metastatic colorectal cancer, advanced head and neck cancer
Herceptin (Trastuzumab)	mAb binding to Her-2 receptor blocks signaling driving proliferation, promotes NK activity	Breast cancer
Rituximab (Rituxan)	mAb binding to CD20 facilitates killing of CD20-bearing B cells	Certain NHL

NHL, non-Hodgkin's lymphoma (see Ch.30).

of EGFR proteins available for ligand binding. Encouraging results have been obtained using this approach in combination with radiation treatment in patients with metastatic colorectal cancer and advanced head and neck cancers.

Some researchers have explored the possibility that anti-tumor responses might be improved by depleting patients of CD25-bearing cells, the idea being to remove the T_{reg} population that downregulates the responses of conventional effector cells. However, since T_{reg} cells are also responsible for controlling autoimmunity, there is a danger of inducing autoimmune responses along with anti-tumor immunity. In addition, activated conventional T cell effectors express CD25, and if they were also depleted, the patient could be left vulnerable to pathogens. Nevertheless, in mice treated with anti-CD25 mAbs, there has so far been little evidence of autoimmunity or undue pathogen infection. The translation of this strategy to the human situation is under investigation.

A problem with administering purified mAbs on their own as therapeutic agents is that it is often difficult for the mAb molecule to penetrate into the tumor mass. The encapsulated structure of most solid tumors, the tight junctions between component malignant cells, and the disordered and decreased vasculature in the core of the tumor reduce the number of mAb molecules that are actually able to bind to the cancerous cells. (Such problems do not arise when mAbs are used to treat leukemias [see Ch.30], since these tumor cells are dispersed in the circulation.) Some investigators have tried a different approach in which the mAbs are directed against the endothelial cells lining the blood vessels feeding the tumor, rather than against the tumor mass itself. The rationale is that interference with endothelial cell function should choke off the tumor's blood supply and kill the cancer. The tumor vasculature is much more accessible to exogenous proteins than is the encapsulated malignant mass, meaning that more of the mAb should be able to reach the targeted tumor cells. In addition, the genomes of endothelial cells are much more stable than those of tumor cells, meaning that fewer mutations conferring resistance to mAb binding can occur. One promising molecule targeted using this type of approach is VEGFR-2 (*vascular endothelial growth factor receptor 2*), which is expressed at higher levels on endothelial cells of tumor blood vessels relative to normal blood vessels. The ligand for VEGFR-2 is VEGF, which, as we have seen, is highly expressed by many cancers to stimulate tumor angiogenesis. Anti-VEGFR-2 mAbs have been used to block the binding of VEGF to VEGFR-2, with some clinical success noted in some patients with breast cancers or renal cell carcinomas. In one clinical trial, the administration of an anti-VEGFR-2 mAb called Avastin used in combination with the pyrimidine analogue 5-FU extended the lives of advanced colon cancer patients by 4 months.

iib) Immunoconjugates. As noted previously, *immunoconjugates* are chimeric proteins in which mAbs are linked to other molecules either chemically or at the DNA level (Fig. 26-6). When the molecule linked to the mAb is a toxin, the immunoconjugate is known as an *immunotoxin*. When a radioisotope is linked to the mAb, the result is an *immunoradioisotope*. Enzymes needed to metabolize and activate a precursor form of an anti-cancer drug, and cytokines with anti-tumor properties, have also been conjugated to mAbs. In all of these instances, the idea is to count on the specificity of mAb binding to bring the toxin, radioisotope, enzyme, or cytokine to the tumor site alone, sparing normal tissues. In some cases, the molecule conjugated to the mAb is internalized and exerts its effects on intracellular pathways crucial for the cell's survival. In other cases, the effector molecule is delivered within the tumor cell mass or its vasculature but remains external to the individual tumor cell.

In the case of immunotoxins and immunoradioisotopes, the immunoconjugate kills tumor cells because the toxin or radioisotope component is internalized following mAb binding. Once inside the tumor cell, the toxin or radioisotope inhibits protein or nucleic acid synthesis or damages DNA. Many immunotoxins are extremely potent, requiring only one molecule per cell to kill. This efficacy means that very few immunoconjugate molecules have to reach the tumor site to have a therapeutic effect. Unfortunately, as should be clear from our discussions in this chapter, there are very few known TSAs to raise a mAb against, making it difficult to construct a truly tumor-specific vector. As well, it has been difficult to control the immunogenicity and half-life of immunotoxins, and to achieve adequate penetration of solid tumor sites. Despite these impediments, researchers have developed mAbs conjugated to plant toxins such as ricin, and to bacterial toxins such as *Pseudomonas* exotoxin (PE) and diphtheria toxin (DT). Monoclonal antibodies have also been conjugated to anti-cancer drugs such as doxorubicin and highly potent antibiotics such as members of the enediyne family. Unfortunately, however, only limited clinical success

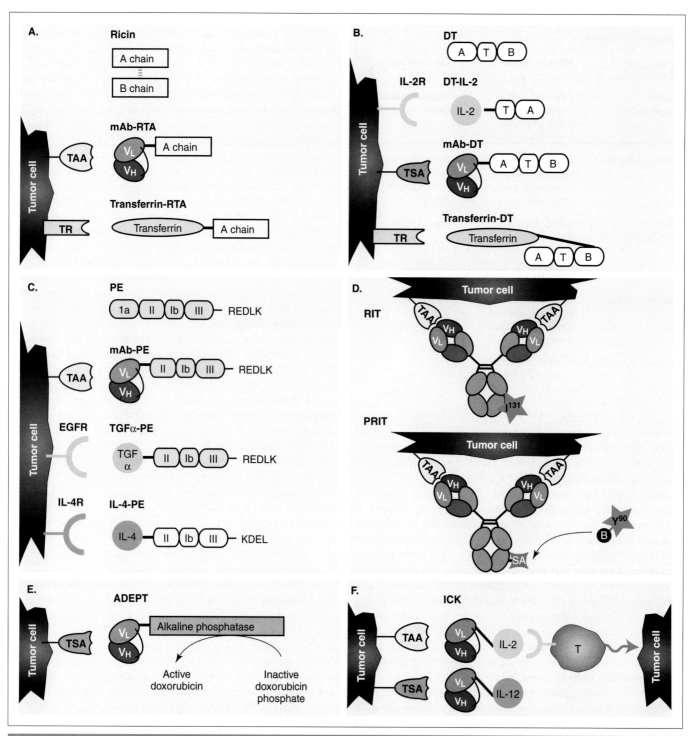

Figure 26-6

Immunoconjugates

An effector molecule such as a toxin or cytokine is conjugated to a molecule conferring binding specificity such as a mAb (or its scFv derivative) or a receptor ligand. (A) The plant toxin ricin is composed of the A and B chains. The cytotoxic A chain (RTA) can be conjugated to a mAb recognizing a specific TAA (or TSA), or to transferrin, which will bind to transferrin receptors (TR) inappropriately expressed on tumor cells. (B) The bacterial diphtheria toxin (DT) is composed of the A, T, and B subunits. Various combinations of these subunits can be conjugated to a cytokine such as IL-2, to a TAA/TSA-specific mAb, or to transferrin. (C) Components and variants of the bacterial *Pseudomonas* exotoxin (PE) can also be conjugated to mAbs or cytokines. (D) Radioimmunotherapy (RIT) is performed by conjugating a toxic radioisotope such as I131 to a TAA/TSA-specific mAb. In pretargeted RIT (PRIT), the TAA/TSA-specific mAb conjugated to streptavidin (SA) localizes in the tumor first, followed by binding of biotin (Ⓑ) conjugated to a toxic radioisotope (Y90 in this case). (E) Antibody-directed enzyme/prodrug therapy (ADEPT) is performed using TAA/TSA-specific mAb conjugated to an enzyme that then activates a chemotherapeutic drug locally. (F) Immunocytokines (ICK) are TAA/TSA-specific mAbs conjugated to cytokines that either stimulate immune system cells or kill tumor cells directly. REDLK, KDEL, single letter amino acid code. With information from Kreitman R. J. (1999) Immunotoxins in cancer therapy. *Current Opinion in Immunology* **11**, 570–578.

using the immunotoxin approach has been achieved in cancer patients. Very few instances of tumor regression have been recorded after immunotoxin treatment, and systemic symptoms such as nausea and gastritis are often significant.

In a variation on the immunotoxin approach, a protein other than a mAb is sometimes used to direct toxin binding to tumor cells. For example, bladder carcinoma cells often express high levels of EGFR. As mentioned previously, one ligand for EGFR is TGFα. In one animal study, when PE was conjugated to TGFα and introduced directly into a bladder tumor, some regression was observed. The direct administration used in this trial appeared to avoid systemic effects on liver cells, which normally express EGFR. Brain cancer can also be treated in this way, at least in experimental animals. The transferrin receptor is expressed by normal hepatic cells and cancerous brain cells, but not by normal brain cells. Substantial regression of brain tumors results in mice when a conjugate composed of transferrin and ricin is introduced directly into brain tumors. Cytokines may also be conjugated to toxins to direct the toxins to cells expressing particular cytokine receptors. Again, however, despite the encouraging results obtained in pre-clinical studies, the use of these agents in cancer patients has rarely yielded the hoped-for results.

There has been better luck in the use of immunoradioisotopes to treat cancer patients, a strategy often called *radioimmunotherapy* or RIT. Two β-particle-emitting isotopes are commonly used for RIT: the traditional iodine-131 and the newer and less volatile yttrium-90. When mAbs specific for the epithelial mucin TAA MUC-1 were conjugated to yttrium-90 and infused into patients with advanced breast cancer, clinical responses were observed in half of the cases. Remission of ovarian cancer has also been reported following a combination therapy based on intraperitoneal administration of an immunoradioisotope recognizing MUC-1. Other immunoradioisotopes include a yttrium-90-labeled mAb binding to the CEA protein often expressed by colon, pancreatic, breast, ovarian, and thyroid cancers, and an iodine-131-labeled mAb recognizing a transmembrane phosphoprotein expressed by renal carcinoma cells. The targeting of these immunoconjugates to tumor cells has been encouragingly precise, and clinical responses have been observed in significant numbers of patients. One difficulty with RIT is that the body (and particularly the bone marrow) is exposed to the radioisotope while the mAb is searching out its target. In an effort to reduce systemic effects of RIT, a variation called *pretargeted RIT* (PRIT) has been devised. The cancer patient is first given the required mAb conjugated to streptavidin (SA) to form a mAb–SA complex. After a period in which the mAb–SA is allowed to localize in the tumor site, unbound mAb–SA is allowed to clear from the body, and the radioisotope conjugated to biotin is administered. Because of the enormous avidity of biotin-SA binding, the radioisotope is quickly concentrated at the tumor site and does minimal damage to normal tissues.

Another type of immunoconjugate is used for ADEPT: *antibody-directed enzyme/pro-drug therapy*. The idea here is to conjugate the specific anti-tumor mAb to an enzyme capable of converting an inert pro-drug into an active cyto-

toxic drug. For example, alkaline phosphatase conjugated to a mAb is capable of converting inactive doxorubicin phosphate to the active anti-cancer drug doxorubicin, or etoposide phosphate to the pro-apoptotic drug etoposide. The immunoconjugate is administered to the patient in advance of the pro-drug, giving the mAb–enzyme complex time to specifically localize in the tumor site and to be cleared from other areas of the body. The pro-drug is then administered such that its activation into the toxic metabolite is (theoretically) confined to the tumor mass where the mAb-enzyme is bound. In practice, it is sometimes difficult to clear the mAb–enzyme complex from all non-cancerous tissues. Should any of the mAb–enzyme complex be left behind, or should there be a natural source of endogenous enzyme in normal tissues or the blood, unwanted activation of the pro-drug in non-cancerous sites may be initiated. There may even be a systemic effect. For this reason, it may be necessary to administer a secondary antibody (directed against the first antibody) that is conjugated to galactose. The galactose moiety prevents the secondary antibody from penetrating into the tumor mass while expediting the clearance of circulating mAb–enzyme complexes from non-cancerous tissues. Another problem with mAb–enzyme immunoconjugates is that the complexes can be immunogenic, inducing an immune response that inactivates the enzyme before it can do its job. Moreover, immune complexes may form that can clog the tissues and induce hypersensitivity reactions (see Ch.28).

Some immunoconjugates are *immunocytokines* (ICK), chimeric proteins in which a whole anti-TAA/TSA mAb (or a structural derivative of that mAb) is linked to a cytokine that either has anti-tumor properties or can stimulate immune system cells in the tumor site. Here, the plan is to prolong the half-life of the cytokine and to bring it directly into the malignant mass by virtue of the specificity of the mAb. Within the tumor, the modified cytokine binds to the appropriate cytokine receptors and triggers the usual signaling pathways. (ICKs are not internalized like immunotoxins.) Obviously, the mAb must be tumor-specific and not directed against an antigen expressed on large numbers of normal cells, and the local concentration of the cytokine the mAb brings along as a passenger must not reach a level sufficient to induce endotoxic shock. Several ICKs involving IL-2 have been constructed and are under investigation in animal models and clinical trials. For example, an ICK composed of IL-2 fused to a mAb directed against the GD2 glycolipid antigen found on the surfaces of melanoma and neuroblastoma cells has been evaluated in phase I clinical trials. The chimeric protein retained IL-2 biological activity, was able to bind to the tumor cells, and accumulated in the tumor site. When mice were transplanted with human melanoma cells and then treated with the IL-2/anti-GD2 immunoconjugate, metastasis of the melanoma was curtailed. In a similar experiment, an ICK was created in which IL-2 was coupled to a mAb recognizing an epithelial adhesion molecule called EpCAM. EpCAM is upregulated on many carcinomas, and the chimeric IL-2/anti-EpCAM protein was able to control colonic metastases in both humans and mice. It is thought that the IL-2 portion of the immunoconjugate may have stimulated T, B, and NK cells and/or macrophages in the tumor site that were already attempting to eliminate the tumor. ICKs containing IL-2 have worked well in some patients with early stage adult T cell leukemia.

ICKs have been constructed with other cytokines. A chimeric protein composed of IL-12 and anti-EpCAM can block the metastasis of human prostate cancer cells and murine colon cancer cells. A double whammy of IL-12 and IL-2 hooked up to anti-EpCAM is even more effective. An ICK of GM-CSF plus anti-GD2 mAb enhances ADCC-mediated killing of neuroblastoma cells by granulocytes. Clinical researchers have also attempted to deliver ICKs based on TNF, LTα, IL-1, IL-7, and G-CSF directly into sites of solid tumor growth but the results have been unremarkable to date.

An emerging problem in clinical research on all types of immunoconjugates is that the preparatory work in animal models generates results that frequently overestimate the efficacy and underestimate the toxicity of a drug or toxin in humans. In the purebred animals used for pre-clinical studies, the antigen recognized by the mAb partner of the immunoconjugate may be a TSA. However, in humans, the antigen is often more broadly expressed and constitutes a TAA. Monoclonal antibody binding in humans is thus dispersed more widely than desired and unwanted killing of normal cells occurs. This relative lack of specificity in immunoconjugate targeting must be kept in mind when weighing the benefits and detriments of immunoconjugate treatment of patients.

iii) Cancer Vaccines

The reader will recall from Chapter 23 that there are two types of vaccines: prophylactic vaccines, given to healthy individuals to prevent pathogen infection, and therapeutic vaccines, given to infected individuals to boost the immune response against established disease. In the context of cancer treatment, there are several therapeutic vaccine approaches being studied. With respect to cancer prevention, the only prophylactic vaccines currently available are those that prevent infection by pathogens associated with the development of specific cancers. By blocking the infection, the subsequent development of cancers associated with these pathogens is thwarted. Work in mouse models on prophylactic cancer vaccines able to block tumor challenges has raised the hope that humans predisposed to cancer development might one day be vaccinated with a tumor antigen that can prime an anti-cancer immune response capable of blocking the expansion of any tumor cell that progresses to full malignancy.

iiia) Pathogen cancer vaccines. The most common pathogen cancer vaccines are those directed against the oncogenic viruses. While there is still much debate about whether these viruses are themselves agents of cancerous transformation, it is generally agreed that infected cells expressing oncoviral antigens often later become malignant. Thus, these antigens are logical targets for vaccine development. HepB proteins are already important reagents used for conventional vaccination against HepB infection. In geographic areas where HepB virus is endemic, the implementation of HepB vaccination programs correlates with a reduced incidence of hepatocarci-

noma in the populations of these areas. Similarly, the incidence of cancers associated with HPV infection has been reduced by inducing protective immunity against this virus. Researchers are also developing a vaccine to block *H. pylori* infection, with the goal of reducing the incidence of gastric lymphomas and carcinomas as well as ulcers.

iiib) Non-pathogen cancer vaccines. Two possible non-viral TSAs that have been investigated as components of a non-pathogen cancer vaccine are the proteins encoded by the oncogene Ras and the TSG p53. In theory, distinctive peptides from the mutated versions of these proteins presented on the surface of cancer cells should activate $CD8^+$ T cells. Indeed, *in vitro* assays have demonstrated the presence in mice of CTLs capable of responding to purified mutated Ras peptides. Unfortunately, however, these anti-Ras CTLs did not respond to intact tumor cells. In the same vein, while it has been possible to demonstrate immune responses against mutated p53 peptides in experimental animal models, the same has not been true in human cancer patients.

What about TAAs? The reader might wonder whether immunotherapy against TAAs would work at all, since any therapy designed to induce an immune response against tumor cells bearing self antigens might also do unacceptable damage to normal tissues. In reality, many TAAs are expressed in non-vital organs, so that limited destruction of healthy cells in these locations does not threaten the overall survival of the host. For example, melanoma patients immunized with melanoma TAAs often benefit from some regression of their tumors, but they also experience a skin condition called *vitiligo*. Vitiligo is a progressive loss of pigment from the skin and hair, in this case resulting from an autoimmune attack on normal melanocytes by T cells directed against the melanoma cell TAAs. Thankfully, while vitiligo may be unsightly, it is not life threatening.

Several mouse models have yielded encouraging results with respect to the potential of non-pathogen TAA cancer vaccines. For example, transgenic mice overexpressing rat Neu (the oncogenic form of Her-2) under the control of the MMTV promoter are highly susceptible to mammary carcinogenesis. When these animals were immunized with DNA or plasmids (refer to Ch.23) encoding whole or partial sequences of the Neu gene, mammary gland tumor formation was blocked or slowed in these animals because the vaccine induced anti-Her-2 antibodies that blocked Her-2-mediated proliferative signaling. Anti-Her-2 CTLs did not appear to be generated. A caveat with this system, however, is that it is not very physiological, being based on expression of a rat oncogene in mice under the control of a viral promoter. In another approach, transgenic mice engineered to express human CEA were immunized with either DNA encoding CEA or a vaccinia virus vector bearing CEA sequences. The idea was to test whether the tolerance developed to human CEA in the transgenic mouse would preclude immune responses against a later challenge with a CEA-bearing tumor. It was found that vaccination with CEA could indeed overcome tolerance and provoke the mounting of a CTL-mediated immune response that rejected the tumor challenge. Significantly, there was no clinical evidence of an attack by these CTLs on "normal" CEA-expressing tissues of the transgenic mouse. Similar results have been obtained in mice transgenic for the human TAA MUC1, a mucin antigen associated with epithelial carcinomas.

Although much useful information has been gleaned from the study of potential TAA/TSA vaccines in mice, the reader is cautioned that there are limitations to their applicability to human cancer. Many of these studies are carried out using transgenic mouse strains engineered to generate cells expressing the TAA or TSA at artificially high levels. In other cases, tumor growth in the subject animal is induced by inoculation of transformed cell lines that cause tumor formation within days or weeks of the procedure. These scenarios are at considerable odds with the much longer natural history of tumorigenesis in humans and the much lower natural expression of (mostly unknown) TAAs and TSAs on human cancer cells.

Most clinical research to date on vaccination of humans with TAAs has been done with melanoma antigens and cancer-testis antigens (Table 26-15). For example, melanoma patients have received peptide vaccines (refer to Ch.23) derived from the MAGE, tyrosinase, gp100, or Melan-AMART1 TAAs. Immune responses against the appropriate peptides were detected, and tumor regression occurred in some cases. Similarly, antigen-specific CTLs were generated in esophageal cancer patients who received a peptide vaccine derived from the NY-ESO-1 cancer-testis antigen, and some patients experienced regression of their tumors. Peptide-specific immune responses have also been noted in some patients with breast or ovarian cancer who received the Her-2 vaccine plus GM-CSF, and in prostate cancer patients vaccinated with *prostate-specific antigen* (PSA). Finally, positive early results have been obtained in a trial of vaccination of colon cancer patients with CEA. These findings bode well for the future development of at least some TAA cancer vaccines. Moreover, for both melanoma and prostate cancer vaccinations, the relevant TAAs are expressed by the tumor cells of most patients with the disease, making the design of a generic therapy feasible.

While the preceding is very encouraging, there are still significant hurdles to be overcome before therapeutic cancer vaccines become a reality. The proportion of patients that currently gains clinical benefit from any given cancer antigen vaccination approach is usually small (5–20%). This high rate of failure is perhaps unsurprising since most patients who have been treated with experimental cancer vaccines to date suffer from advanced tumors which are refractory to other modes of therapy, and thus represent the greatest possible clinical challenge. More puzzling perhaps is the lack of correlation between the T cell response generated and the clinical outcome: some patients with very low numbers of antigen-specific CTLs experience tumor regression, while others in whom high numbers of antigen-specific CTLs are induced by vaccination show no clinical improvement. However, there is growing evidence that TAA vaccination may be very useful for the prevention of relapses following the surgical removal of primary tumors. The theory is that even a weak immune response mounted against the tumor peptide antigens in the vaccine might be sufficient to eliminate residual cancer cells. Indeed, in one study of melanoma patients whose tumors were surgically removed, those who also received a MART-1 peptide vaccine enjoyed a longer respite before relapse. Even more promising results have

been obtained with a melanoma cancer vaccine originally developed in the 1980s by Donald Morton and colleagues at the John Wayne Cancer Institute in California. This vaccine was prepared by irradiating whole melanoma cells from three allogeneic melanoma cell lines. Together, these cell lines expressed more than 20 highly immunogenic melanoma-specific and general TAAs. The vaccine was given at several points over a 5 year period to stage II melanoma patients whose primary tumors and any metastases had been previously excised. In comparison to a control group receiving an alternative treatment, the vaccinees had a better survival rate. Moreover, this improved survival correlated well with tumor-specific cellular and humoral immune responses. It is believed that the immune response induced against antigens expressed by the allogeneic

Table 26-15 Potential Cancer Vaccine Antigens

Name of Antigen	Tumor Association
PATHOGEN ANTIGENS DERIVED FROM	
EBV	Burkitt's lymphoma, nasopharynx
Helicobacter pylori	Stomach
HepB	Liver
HepC	Liver
HPV	Cervix, vagina, penis
TSAs	
Bcr-Abl fusion protein	CML, ALL
Mutated Cdk4	Melanoma
Mutated Cdc27	Prostate
Mutated EGFR	Breast
Mutated p53	Colorectum, bladder, head and neck, others
Mutated Ras	Pancreas, colon, lung
Ig idiotype	Follicular B cell lymphoma
TCR "idiotype"	T cell lymphoma
TAAs	
AFP	Liver
CD20	Lymphomas
CEA	Colorectum, lung, breast
GM-2, GD2, GD3 gangliosides	Melanoma
Gp100	Melanoma
Her-2	Breast
MAGE-1, MAGE-3	Melanoma, colorectum, lung, stomach
Melan-A^{MART1}	Melanoma
MUC-1	Colorectum, pancreas, lung, ovary
NY-ESO-1	Esophagus
PSA	Prostate
PAP	Prostate
Prostate-specific membrane antigen	Prostate
Thyroglobulin	Thyroid
Tyrosinase	Melanoma

ALL, acute lymphoblastic leukemia; CML, chronic myelogenous leukemia.
(See Ch.30.)

melanoma cells generated lymphocytes capable of cross-reacting with antigens expressed on any residual tumor cells remaining in the patient after excision. This vaccine, whose commercial name is Canvaxin, is one of the few to make it to a full-fledged phase III clinical trial (begun in July 2003).

A key difficulty with most cancer vaccines is that many TAAs and TSAs simply are not very immunogenic. Researchers are therefore exploring the hypothesis that one must also induce inflammation or increase immunogenicity by some other means to obtain an effective anti-tumor response. In fact, the use of a cancer vaccine of low immunogenicity may have deleterious consequences, as suggested by a small number of studies in which administration of a TAA vaccine actually enhanced tumor cell growth in some patients. The growing speculation is that, if the vaccine does not succeed in driving APC maturation such that the TAA is presented in an activatory context, anergization (rather than activation) of responding TAA-specific lymphocytes may be induced. Such a scenario would promote progression of the malignancy rather than rejection.

Two other factors that should not be overlooked when considering TAAs for therapy are proteasome preference and HLA preference. Since some epitopes are generated from TAAs by standard proteasome digestion, and others by immunoproteasome digestion, the administration strategy should take into account factors such as whether the whole protein or just the peptide of interest should be given, whether DCs should be involved, and whether IFNγ should be co-administered. In some cases, it may be more efficient to carry out the immunization with the specific peptide only (rather than the complete TAA protein), as the correct peptide may not be generated from the whole protein if the appropriate proteasome is not present. With respect to HLA selection, the TAA peptide chosen should be able to bind to at least one HLA molecule expressed in the potential vaccinee. For example, HLA-A2 is expressed by 25–50% of the populations of most parts of the world. Choosing a peptide that binds to this MHC molecule would ensure that the vaccine would be of potential benefit to a large number of cancer patients.

Rather than isolated TAAs or TSAs, whole autologous tumor cells can be administered as a vaccine. A wide variety of relevant epitopes from both surface and intracellular antigens are acquired by APCs and presented to lymphocytes. Although the procedure is technically demanding, laborious, and expensive, it has also proved thus far to be the most effective way to get specific destruction of a patient's cancer. Modification of the autologous cells to express an appropriate cytokine(s) as described previously can ensure that these epitopes are presented to T and B cells in an activatory context. Several laboratories have investigated the option of using modified allogeneic cells for introduction into patients, but so far, anti-tumor immunity induced by this approach has fallen short of that induced by the autologous approach.

iiic) Heat shock proteins as cancer vaccine adjuvants. The reader will recall from Chapter 16 that one way to overcome low immunogenicity of an antigen is to boost the response to it with an adjuvant. Several researchers have explored the potential of HSPs as adjuvants for tumor-specific immunity (Fig. 26-7). Tumor cells subjected to stress (such as radiation

treatment) increase their intracellular levels of HSPs, which in turn spur the presentation of tumor antigens on MHC class I. As well as conveying proteins and peptides to the antigen-processing machinery, HSPs are associated with necrotic (but not apoptotic) cell death and thus serve as "danger signals" that promote the activation and maturation of DCs. In addition, the release of HSP–tumor antigen complexes from necrotic or stressed tumor cells induces neutrophils to secrete pro-inflammatory cytokines such as IL-6 and TNF, which further stimulate DC maturation. This DC activation is thought to break peripheral tolerance to TAAs.

From the clinical point of view, a major advantage to the HSP approach is that HSPs are not particularly specific for the peptides and proteins they chaperone, meaning that they can convey a wide variety of potential epitopes to APCs. This non-specificity means that one can lyse whole tumor cells, combine the lysate with HSPs, and use the mixture as a multi-epitope vaccine. Alternatively, one can isolate whole tumor cells, engineer them to overexpress HSPs (if they do not already do so), and re-introduce the engineered cells back into a tumor-bearing host. In either case, one does not have to know the precise identity of which TAA or TSA peptides associate with the HSPs to form the complexes provoking the immune responses: the mere presence of the HSPs helps to ensure the uptake of any TAAs/TSAs by APCs and the mounting of adaptive responses directed against the tumor cells. In addition, should the tumor lose expression of one antigen, it will remain under attack due to the presence of other tumor peptide–HSP complexes in the vaccine. Work in mouse models has validated the HSP vaccine concept. For example, mice vaccinated with tumor cells engineered to express HSP72 were later able to reject tumor cells that did not express HSP72, confirming that the HSP72 was not itself an antigen but that it promoted an immune response against (unknown) tumor antigens. Similarly, if HSP65 was introduced into melanoma cells and these cells were used to vaccinate a mouse, tumor-specific T cell-mediated responses were mounted that did not arise when non-engineered melanoma cells were used for vaccination.

In humans, therapeutic vaccination with peptides complexed to HSP70 or the ER-chaperone molecule gp96 (refer to Ch.11) has been attempted in cases of pancreatic, breast, colorectal, and gastric cancer, among others. Tumor cells from a given patient are lysed and the cell lysates are mixed with HSPs. Each patient is then vaccinated with his or her autologous lysate–HSP mixture. In one study of this type, about half of the vaccinated patients developed expanded CTL populations specific for the autologous tumors, and some also showed increases in peripheral blood NK cells. There were no significant detrimental side effects associated with the procedure and autoimmunity did not develop. A similar approach was used to treat cases of advanced melanoma. Patients were vaccinated with lysates of autologous melanoma cells mixed with HSP70. Increased numbers of melanoma-specific T cells were observed, and several vaccinated patients enjoyed clinical improvement. Clinical trials of gp96 as an adjuvant for the vaccination of renal cell carcinoma patients are under way.

iiid) Modified DCs as cancer vaccines. The DCs naturally present at a tumor site do not seem to be able to prime naive

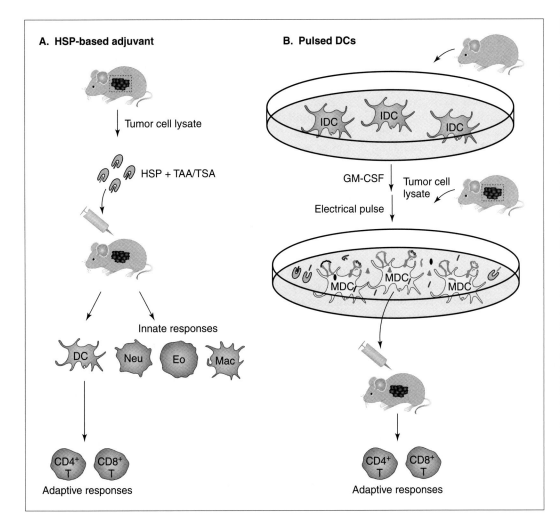

A. HSP-based adjuvant

Tumor cell lysate

HSP + TAA/TSA

Innate responses

DC   Neu   Eo   Mac

CD4⁺ T   CD8⁺ T

Adaptive responses

B. Pulsed DCs

IDC   IDC   IDC

GM-CSF   Tumor cell lysate

Electrical pulse

MDC   MDC   MDC

CD4⁺ T   CD8⁺ T

Adaptive responses

Figure 26-7

Two Novel Approaches to Anti-cancer Therapeutic Vaccination
(A) HSP-based adjuvant. The tumor is excised from the host, its component cells lysed, and the lysate mixed with purified HSPs. HSPs bound to TAA/TSAs are injected into a syngeneic host bearing the same tumor. The presence of the HSPs enhances the uptake of TAA/TSAs by host DCs and thus promotes adaptive and innate responses. (B) Pulsed DCs. Immature DCs (IDC) are recovered from a mouse and cultured *in vitro*. DC maturation is induced by the addition of GM-CSF and these cells acquire TAA/TSAs derived from the lysate of a tumor from a syngeneic mouse. Mature DCs loaded with TAA/TSA peptides presented on MHC class I and II are injected into a syngeneic tumor-bearing mouse to induce adaptive anti-tumor responses.

T cells to mount effective anti-tumor responses, most likely because of the lack of "danger signals" and the presence of immunosuppressive cytokines secreted by the tumor cells. How then to induce DC maturation and/or activation so that these cells can initiate anti-tumor immune responses? One approach in animal models has been to recover DCs from one inbred mouse, treat the DCs *in vitro* with GM-CSF to trigger maturation, "load" the stimulated DCs with some form of tumor antigen, and introduce the DCs into a tumor-bearing, syngeneic mouse as a therapeutic vaccine (refer to Fig. 26-7). (This loading of DCs is carried out by electroporation and the procedure is sometimes called *pulsing*.) The source of tumor antigens used to load the DCs may be whole tumor cell lysates, purified native TAA/TSA proteins (if known), or peptides released by acid treatment from MHC class I molecules on the surfaces of tumor cells. The electroporation process introduces proteins and peptides directly into the cell cytoplasm, allowing peptides derived from the pulsing to provoke a CD8⁺ CTL response due to presentation on MHC class I. Some *in vitro* uptake of antigen via endocytosis may also allow presentation of relevant peptides on MHC class II, facilitating CD4⁺ T cell stimulation as well.

Does the loaded (pulsed) DC approach work? When activated DCs pulsed with a tumor peptide mixture were injected into syngeneic animals, multiple T cell clones directed against various tumor peptides in the mixture were detected. It did not seem to matter whether the loaded DCs were injected intradermally, intravenously, or intranodally (into the lymph nodes), offering hope that this strategy might offer general protection. In some cases, loading has been accomplished by allowing antibody–tumor epitope complexes to be taken up by DCs *in vitro* via FcR-mediated endocytosis. Tumor peptides appeared on both MHC class I (by cross-presentation) and MHC class II on the DC surface, and both CTLs and Th cells directed against the original tumor were generated *in vitro*. Sometimes the loaded DCs are modified to supply their own activatory cytokines. In one mouse model of melanoma, retroviruses were used to engineer overexpression of GM-CSF, TNF, or CD40L in DCs. When these modified DCs were pulsed with a MAGE-1 peptide and introduced into melanoma-bearing mice, effective *in vivo* immune responses against the tumors were observed.

Several studies have extended the pulsed DC approach to treatment of human cancer. Unfortunately, while the side effects of this type of treatment have been few, the actual clinical benefits of this approach have also been minimal. Human melanoma patients were given intranodal injections of autologous DCs that had been induced to mature *in vitro* and pulsed with melanoma peptides. In almost all cases, there was a strong induction of CTL clones directed against several melanoma peptides. Indeed, in one study, these clones could still be detected 2 years later. However, at the clinical level, melanoma regression occurred only to a minor degree in a small proportion of patients. Limited clinical results were also obtained for patients with metastatic renal cell carcinoma who were given stimulated DCs loaded with renal carcinoma peptides. Better clinical responses were obtained when stimulated DCs pulsed with CEA were used to treat a group of patients with colon or small cell lung cancers. Specific anti-tumor CTL responses were also induced in half of a group of patients with advanced breast or ovarian cancer that were treated with stimulated DCs pulsed with Her-2 peptides. Interestingly, in one study, researchers loaded stimulated DCs with peptides from mouse *p*rostatic *a*cid *p*hosphatase (PAP) and injected them into human prostate cancer patients. All patients developed immune responses against mouse PAP, as expected, but about half also mounted responses against the human PAP homologue. Clinical benefit was shown in several individuals of the latter group. Much effort is currently devoted to improving the isolation, *in vitro* maturation, and peptide/protein loading of DCs. Researchers are also seeking new methods of administering the pulsed cells, and developing means of accurately monitoring and correlating immunological and clinical responses.

Some investigators have loaded stimulated DCs with tumor RNAs instead of tumor peptides or proteins. The idea is that a DC takes up an exogenous RNA encoding a tumor antigen, translates that RNA as it would any of its own mRNAs, and processes the resulting protein for presentation on MHC class I. The use of RNA has several advantages over the use of either protein or DNA. First, RNA is much cheaper to engineer and produce in large quantities than peptides or protein. Secondly, RNA cannot integrate into mammalian genomes like DNA can, meaning that the vaccine cannot inadvertently be passed on to the next generation of cells. Thirdly, the RNA is only transiently present in the DC, just long enough for protein production and processing: there is no danger of uncontrolled protein production some time in the future. The disadvantage of the RNA approach is that the sequence of the TAA or TSA under investigation must be known, and that only a CTL response is induced. Nonetheless, clinical trials using RNA encoding PSA have had encouraging results. A group of men with metastatic prostate cancer were injected with DCs that had been pulsed with PSA RNA and then induced to mature *in vitro*. All patients showed increased IFNγ production and several exhibited anti-PSA CTL responses. A decrease in the number of circulating tumor cells was also observed in several patients.

It may also be possible to use DCs to kick-start endogenous NK- and NKT-mediated anti-tumor responses. We noted previously that the administration of α-galactosylceramide to mice induces an anti-metastatic effect dependent on IFNγ produced by both NKT cells and NK cells. If the α-galactosylceramide is packaged into DCs prior to administration, the anti-metastatic effect is significantly heightened. For example, if α-galactosylceramide-loaded DCs were given to mice in which a melanoma had already been established for at least 1 week, metastasis of the tumor was greatly inhibited. The

potential use of α-galactosylceramide-pulsed DCs to combat tumors in humans is under investigation in clinical trials. Early results are promising, with sustained decreases in serum tumor markers observed in the handful of adenocarcinoma or renal cell carcinoma patients examined.

iv) Manipulation of T Cell Costimulatory and Regulatory Molecules

Tumor immunologists have been interested in the relevance of T cell costimulation to anti-tumor immunity since the early 1990s. The hypothesis is that tumor immunosurveillance sometimes fails because a malignant clone does not express the appropriate costimulatory molecules required for T cell activation. Indeed, if tumor cells are manipulated to express high levels of B7, potent Tc responses are induced in animals transplanted with these cells and tumor regression is sometimes observed. Unfortunately, this approach has not translated well to the clinic, prompting researchers to search for other means of boosting T cell costimulation and overcoming tolerance to tumor cells.

The reader will recall from previous chapters that CTLA-4 is a negative regulator of T cell activation that competes with CD28 for binding to B7-1/B7-2. In the course of a normal immune response, CTLA-4 curtails the activation of Th cells whose CD28 molecules interact with B7-1/B7-2 on an activated APC. Similarly, the costimulation of naive Tc cells whose CD28 molecules interact with B7-1/B7-2 on a licensed DC is controlled by CTLA-4. Several investigators have explored the possibility that inhibition of these negative effects of CTLA-4 on T cell activation might allow a stronger or sustained T cell response against tumor cells. If the balance can be tipped away from Tc tolerance toward Tc activation through increased DC licensing and Th activation, then increased numbers of tumor-specific CTLs will be generated and go on to kill their targets in their usual B7-independent manner. Indeed, studies have shown that inhibition of CTLA-4 can lead to enhanced anti-tumor responses in experimental animals. For example, injection of a soluble anti-CTLA-4 antibody that inhibits the action of this molecule establishes a *CTLA-4 blockade* that appears to prevent the negative regulation of anti-tumor Th and Tc cells. Furthermore, tumor cells lacking B7-1/7-2 expression are rejected by a mechanism that requires both CD4$^+$ and CD8$^+$ T cells. In the small number of human patients examined, CTLA-4 blockade can lead to the regression of tumors but autoimmunity is also induced (see Ch.29). It remains unclear whether the CTLA-4 blockade in these cases is directly affecting effector T cell functions, or if the effect is due to an impact of the blockade on T$_{regs}$.

CTLA-4 blockade is used most often to boost the effectiveness of other treatment modalities. In one study, mice treated with a combination of CTLA-4 blockade plus either low dose chemotherapy or surgical excision of the primary malignancy experienced significant tumor regression and enhanced survival. In another study, mice were vaccinated with tumor cells that had been engineered to express GM-CSF. The efficacy of this vaccine was synergistically enhanced when anti-CTLA-4 antibody was also administered. Similarly, CTLA-4 blockade has been combined with malignant cell vaccination to reduce tumor incidence in a transgenic mouse model of prostatic cancer. A triple combination approach involving CTLA-4 blockade has shown even more promise. A liposomal vaccine containing a peptide from a lymphoma TSA was used to immunize mice. Along with the vaccine, the mice received both anti-CTLA-4 antibody plus agonistic anti-CD40 antibody (binds to CD40 on DCs and activates them). Dramatic increases in tumor-specific CTL activity were obtained that allowed the mice to survive a subsequent challenge with the original lymphoma cells.

What about manipulating other costimulatory or regulatory molecules? As described in Chapter 14, OX40 is a TNFR-related molecule with modest T cell costimulatory activity. Upregulated OX40 expression has been found on TILs, leading researchers to attempt to stimulate TILs by administering an OX40L-Ig fusion protein or agonist anti-OX40 antibody. Indeed, these strategies can promote the regression of established tumors in mice. Another TNFR-related molecule that may be relevant in anti-tumor immunity is a lymphotoxin-like protein called LIGHT. LIGHT was first identified when its expression on tumor cells was found to induce their apoptosis *in vitro*. This apoptosis depended on the co-expression by the tumor cells of two cell surface receptors: a widely expressed molecule called HVEM (*herpesvirus entry mediator*) and the LTβR described in Chapter 17. However, LIGHT has some other anti-tumor effects. First, LIGHT, which structurally resembles LTαβ, induces the production of chemokines such as SLC (CCL21) that recruit inflammatory cells. Secondly, when LIGHT engages HVEM expressed on T cells (which do not normally express LTβR), T cell activation and Th1 differentiation are costimulated. In one study, when tumor-bearing mice were vaccinated with autologous tumor cells that had been manipulated to express LIGHT, anti-tumor CD4$^+$ and CD8$^+$ responses were enhanced *in vivo* and tumor regression was observed. (The reader may wonder why we did not define LIGHT previously: the reason is the sheer bulk of its name. LIGHT stands for "homologous to lymphotoxins, exhibits inducible expression, and competes with HSV glycoprotein D for HVEM, a receptor expressed by T lymphocytes".)

v) Adoptive Transfer of Anti-tumor Leukocytes

Adoptive transfer of exogenously activated or manipulated immune system cells is also an option for tumor therapy. Studies in experimental animals have shown that CTL clones directed against a known TSA can be adoptively transferred into mice bearing tumors expressing that TSA. The transferred cells multiply and survive for at least a week, a period that is doubled if the recipient is also given a low dose of IL-2 at the time of cell transfer. The anti-tumor cells home to the tumor site and infiltrate among its cells, killing those expressing the TSA. In human cancer patients, autologous CTLs activated *in vitro* have been adoptively transferred back into the patients from whom they were isolated. Useful anti-tumor effects have been observed in certain cases, particularly in patients suffering from melanoma or renal cell carcinoma. Similarly, some patients with Burkitt's lymphoma and nasopharyngeal carcinoma (also linked to EBV infection) have improved following infusion of preparations of T cells enriched for EBV-specific CTLs.

Despite the preceding, the adoptive transfer approach has not yielded an effective enough killing response to make a clinical difference in most cases. It is difficult to expand populations of tumor-specific cells *in vitro* to the number that appears to be required to have a detectable effect on the cancer *in vivo*. In addition, it seems that accurate homing of adoptively transferred anti-tumor cells to the tumor site is not guaranteed. Even if the transferred cells succeed in homing to the tumor site, they may be stymied in this location by the suppressive cytokine milieu and the evasion mechanisms discussed previously.

We have come to the end of our discussion of whether and how the immune system fights cancer. There is still much to be explored in this area, and hopefully many clinical benefits to be reaped in the future. We move now from a field in which researchers try to induce immune responses to sustain life to one in which they try to suppress them to sustain life: human tissue transplantation.

SUMMARY

Tumors are masses of clonally derived cells that form in tissues due to abnormal cell division. A benign tumor contains well-differentiated cells, grows slowly, and is not usually life threatening. A malignant tumor or "cancer" is composed of poorly differentiated cells, grows in an uncontrolled and invasive way, and is lethal if not treated. The spreading of a cancer to distant locations in the body is called metastasis. A cancer arises due to a multi-step process in which a target cell accumulates genetic mutations. Inappropriate gene expression results that ultimately allows irreversible deregulation of cell growth. Carcinogens, such as certain chemicals, radiation sources, or pathogens, promote this mutation process. The mutations most directly connected to malignancy occur in DNA repair genes, oncogenes, or tumor suppressor genes. Mutations that cause the deregulation of cell growth do so through altered protein expression that in turn can trigger recognition of the tumor by the immune system. Such "immunosurveillance" may involve the detection of either tumor-associated antigens (TAAs), which are normal cellular proteins expressed at abnormal levels, tissue locations, or developmental stages, or tumor-specific antigens (TSAs), which are macromolecules unique to the tumor and not expressed on any normal cells. Immune defense against tumors includes the actions of NK, NKT and T cells, IELs, inflammatory cells, and cytokines such as IFNγ and IL-12. Unfortunately, the genetic instability of tumor cells gives them an inherent ability to evade or suppress immune responses. While radiation therapy and chemotherapy still form the cornerstones of cancer therapy, the goal of much current research is the development of immunotherapy techniques that can specifically kill tumor cells or induce an augmented anti-tumor immune response. These techniques include the manipulation of TAAs/TSAs, mAbs, heat shock proteins, DCs, and other elements of the immune system to devise agents that can bind to and disable tumor cells directly, block angiogenesis or hormone secretion supporting a tumor, increase costimulation or decrease negative regulation of anti-tumor T cells, or act as a cancer vaccine.

Selected Reading List

Beachy P., Karhadkar S. and Berman D. (2004) *Nature* **432**, 324–331.

Bellacosa A. (2003) Genetic hits and mutation rate in colorectal tumorigenesis: versatility of Knudson's theory and implications for cancer prevention. *Genes, Chromosomes and Cancer* **38**, 382–388.

Berthet F.-X., Coche T. and Vinals C. (2001) Applied genome research in the field of human vaccines. *Journal of Biotechnology* **85**, 213–226.

Berzofsky J., Terabe M., Oh S., Belyakov I., Ahlers J. *et al.* (2004) Progress on new vaccine strategies for the immunotherapy and prevention of cancer. *Journal of Clinical Investigation* **113**, 1515–1525.

Blattman J. and Greenberg P. (2004) Cancer immunotherapy: a treatment for the masses. *Science* **305**, 200–205.

Brown J. and Wilson W. (2004) Exploiting tumour hypoxia in cancer treatment. *Nature Reviews. Cancer* **4**, 437–447.

Brown M., Lipscomb J. and Snyder C. (2001) The burden of illness of cancer: economic cost and quality of life. *Annual Review of Public Health* **22**, 91–113.

Brutkiewicz R. and Sriram V. (2002) Natural killer T (NKT) cell and their role in antitumor immunity. *Critical Reviews in Oncology and Hematology* **41**, 287–298.

Carding S. and Egan P. (2004) Gammadelta T cells: functional plasticity and heterogeneity. *Nature Reviews Immunology* **2**, 336–345.

Coulie P., Karanikas V., Lurquin C., Colau D., Connerotte T. *et al.* (2002) Cytolytic T-cell responses of cancer patients vaccinated with a MAGE antigen. *Immunological Reviews* **188**, 33–42.

Davis C. and Gillies S. (2003) Immunocytokines: amplification of anti-cancer immunity. *Cancer Immunology and Immunotherapy* **52**, 297–308.

Dela Cruz J., Huang T., Penichet M. and Morrison S. (2004) Antibody-cytokine fusion proteins: innovative weapons in the war against cancer. *Clinical & Experimental Medicine* **4**, 57–64.

Fearon E. and Vogelstein B. (1990) A genetic model for colorectal tumorigenesis. *Cell* **61**, 759–767.

Feldmeier J. (1993) Radiation oncology. In Weiss G. (ed.) "Clinical Oncology." Prentice-Hall International Inc., Norwalk CT.

Fong L. and Engleman E. (2000) Dendritic cells in cancer immunotherapy. *Annual Review of Immunology* **18**, 245–273.

Gallimore A. and Sakaguchi S. (2002) Regulation of tumor immunity by CD25$^+$ T cells. *Immunology* **107**, 5–9.

Garcia-Lora A., Algarra I. and Garrido F. (2003) MHC class I antigens, immune surveillance, and tumor escape. *Journal of Cellular Physiology* **195**, 346–355.

Godfrey D. and Kronenberg M. (2004) Going both ways: immune regulation via CD1d-dependent NKT cells. *Journal of Clinical Investigation* **114**, 1379–1388.

Harris M. (2004) Monoclonal antibodies as therapeutic agents for cancer. *Lancet Oncology* **5**, 292–302.

Houghton A., Gold J. and Blachere N. (2001) Immunity against cancer: lessons learned from melanoma. *Current Opinion in Immunology* **13**, 134–140.

Jager E., Jager D. and Knuth A. (2002) Clinical cancer vaccine trials. *Current Opinion in Immunology* **14**, 178–182.

Kalos M. (2003) Tumor antigen-specific T cells and cancer immunotherapy: current issues and future prospects. *Vaccine* **21**, 781–786.

Kinzler K. and Vogelstein B. (1996) Lessons from hereditary colorectal cancer. *Cell* **87**, 159–170.

Klein G., Sjogren H., Klein E. and Hellstrom K. (1960) Demonstration of resistance against methylcholanthrene-induced sarcomas in the primary autochthonous host. *Cancer Research* **20**, 1561–1572.

Lejeune F., Ruegg C. and Lienard D. (1998) Clinical applications of TNF-alpha in cancer. *Current Opinion in Immunology* **10**, 573–580.

Liu B., DeFilippo A. and Zihai L. (2002) Overcoming immune tolerance to cancer by heat shock protein vaccines. *Molecular Cancer Therapeutics* **1**, 1147–1151.

Mahnke Y., Speiser D., Luescher I., Cerottini J. and Romero P. (2005) Recent advances in tumour antigen-specific therapy: in vivo veritas. *International Journal of Cancer* **113**, 173–178.

McKinnell R. (1998) "The Biological Basis of Cancer." Cambridge University Press, Cambridge, UK.

Mocellin S., Wang E. and Marincola F. (2001) Cytokines and immune response in the tumor microenvironment. *Journal of Immunotherapy* **24**, 392–407.

Moingeon P. (2001) Cancer vaccines. *Vaccine* **19**, 1305–1326.

Murphy P. (2001) Chemokines and the molecular basis of cancer metastasis. *New England Journal of Medicine* **345**, 833–835.

Offringa R., van der Burg S., Ossendorp F., Toes R. and Melief C. (2000) Design and evaluation of antigen-specific vaccination strategies against cancer. *Current Opinion in Immunology* **12**, 576–582.

Pardoll D. (2003) Does the immune system see tumors as foreign or self? *Annual Review in Immunology* **21**, 807–839.

Posner J. (2003) Immunology of paraneoplastic syndromes. *Annals of the New York Academy of Sciences* **998**, 178–186.

Rammensee H., Weinschenk T., Gouttefangeas C. and Stevanovic S. (2002) Towards patient-specific tumor antigen selection for vaccination. *Immunological Reviews* **188**, 164–176.

Schreiber H., Wu T., Nachman J. and Kast W. (2002) Immunodominance and tumor escape. *Seminars in Cancer Biology* **12**, 25–31.

Schuler G., Schuler-Thurner B. and Steinman R. (2003) The use of dendritic cells in cancer immunotherapy. *Current Opinion in Immunology* **15**, 138–147.

Seliger B., Abken H. and Ferrone S. (2003) HLA-G and MIC expression in tumors and their role in anti-tumor immunity. *Trends in Immunology* **24**, 82–87.

Smyth M., Crowe N., Hayakawa Y., Takeda K., Yagita H. *et al.* (2002) NKT cells—conductors of tumor immunity? *Current Opinion in Immunology* **14**, 165–171.

Soloski M. (2001) Recognition of tumor cells by the innate immune system. *Current Opinion in Immunology* **13**, 154–162.

Spence R. and Johnston P. (2001) The molecular biology of cancer. In "Oncology." Oxford University Press, Oxford.

Steinman R. and Mellman I. (2004) Immunotherapy: bewitched, bothered and bewildered. *Science* **305**, 197–200.

Van den Eynde B. and Morel S. (2001) Differential processing of class-I-restricted epitopes by the standard proteasome and the immunoproteasome. *Current Opinion in Immunology* **13**, 147–153.

Van Der Bruggen P., Zhang Y., Chaux P., Stroobant V., Panichelli C. *et al.* (2002) Tumor-specific shared antigenic peptides recognized by human T cells. *Immunological Reviews* **188**, 51–64.

Velders M., Schreiber H. and Kast M. (1998) Active immunization against cancer cells: impediments and advances. *Seminars in Oncology* **26**, 697–706.

von Mehren M., Adams G. and Weiner L. (2003) Monoclonal antibody therapy for cancer. *Annual Review of Medicine* **54**, 343–369.

Weiss G. (1993) Chemotherapy. In Weiss G. (ed.) "Clinical Oncology". Prentice-Hall International., *Norwalk*, CT.

Wells A. and Malkovsky M. (2000) Heat shock proteins, tumor immunogenicity and antigen presentation: an integrated view. *Immunology Today* **21**, 129–132.

Yee C., Riddell S. and Greenberg P. (2001) In vivo tracking of tumor-specific T cells. *Current Opinion in Immunology* **13**, 141–146.

Transplantation

Joseph Murray 1919–

Human organ transplantation
Nobel Prize 1990

© Nobelstiftelsen

CHAPTER 27

A. HISTORICAL NOTES

B. THE MOLECULAR BASIS OF ALLORECOGNITION

C. MINOR HISTOCOMPATIBILITY ANTIGENS

D. TYPES OF CLINICAL REJECTION AND THEIR MECHANISMS

E. HLA TYPING

F. IMMUNOSUPPRESSION

G. INDUCTION OF GRAFT TOLERANCE

H. XENOTRANSPLANTATION

I. BLOOD TRANSFUSIONS

J. HEMATOPOIETIC CELL TRANSPLANTATION (HCT)

K. GENE THERAPY IN TRANSPLANTATION

"Beware of Greeks bearing gifts"
—the story of the Trojan horse

When the casual reader thinks of transplantation, life-saving surgeries come to mind: kidney transplants, heart and lung transplants, and skin grafts for burn victims. These procedures are called "solid organ transplants." However, transfers of whole bone marrow or isolated hematopoietic cells (BMT/HCT; refer to Ch.24) are also tissue transplants, as are blood transfusions. Research and clinical experimentation on these different types of tissue transfers have historically been very separate, and it has only been in recent times that these fields have started to converge and take advantage of their pooled knowledge to enhance benefits to patients.

The reader should keep in mind that, with the exception of the fetus (refer to Ch.16), nature never intended a part of one human's body to be introduced into another human's body. Unless the donor and recipient (sometimes called the *host*) of a transplant are genetically identical, such transfers of donor tissue are regarded by the immune system of the recipient as the appearance of foreign material that should be repelled. That is, an immune response against the grafted tissue cannot be avoided unless the donor and recipient are fully *histocompatible*. In fact, for reasons that are elaborated later, an immune response to a transplanted tissue is much more vigorous than a response to a pathogen. Without immunosuppression, the donor tissue is destroyed by recipient effector cells and mechanisms in a process called *graft rejection* (Plate 27-1). The opposite situation can also arise. When donor lymphocytes in the transplant mount an immunological attack on recipient tissues, the resulting damage is called *graft versus host disease* (GvHD). The mechanisms behind these immunological reactions are discussed in detail in this chapter, as are the various clinical interventions designed to forestall them.

A few more definitions may be helpful at this point. The terms *syngeneic* (identical at all genetic loci) and *allogeneic* (having allelic variation at some genetic loci) were discussed in Chapters 10 and 16 with reference to histocompatibility and peripheral tolerance, respectively. In the simplest of transplant scenarios, the donor and recipient are the same patient,

as in the case of a burn victim whose own healthy skin is used to replace a damaged section of skin elsewhere on his or her body. This type of syngeneic, self–self tissue transfer is called an *autologous* graft. Syngeneic grafts that involve the transfer of tissue between two genetically identical individuals (such as between human identical twins or two inbred mice) are called *isografts*. Conversely, tissue transfers between genetically different members of the same species (such as two humans who are not identical twins) are called *allografts*. X*enografts* are tissue transfers between members of two different species, such as a pig organ transplanted into a human.

A. Historical Notes

Damaged noses seem to be the first recorded targets of attempted transplants. Hindu surgeons are credited with performing the first autologous grafts in 700 BC when they used the forehead skin of patients to repair their damaged or missing noses. The Italians reported similar clinical treatments starting in the 1400s. One suspects that many attempts were made to use another person's skin for some of these repairs, but graft rejection would have made them unsuccessful and perhaps not deemed worthy of reporting. It is known that skin grafting was quite common by the late 1800s, although its many practitioners did not distinguish between the use of autologous and allogeneic skin. There is also a dearth of information on the long term prospects of these grafts. John Hunter in England pioneered the transplantation of teeth and also demonstrated (unfortunately for his patients) that diseases of the tooth donor could be passed on to the recipient. Hunter's work also led to the first ethical questions about transplantation, as subordinates were forced to give up teeth for transplantation into their superiors. In 1900, Karl Landsteiner discovered the existence of human blood group antigens and antibodies, work that garnered him the

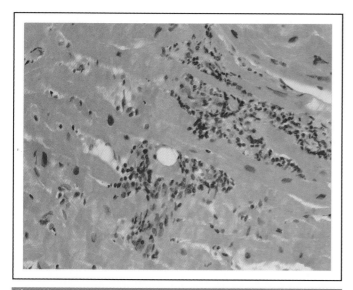

Plate 27-1

Graft Rejection
Mononuclear cells have infiltrated the muscle tissue of a transplanted heart.
Courtesy of Dr. Jagdish Butany, University Health Network/Toronto Medical
Laboratories, Toronto General Hospital.

Nobel Prize in 1930. The early 1900s also saw improvements in techniques for surgically joining blood vessels so that the transplantation of solid organs within the body became theoretically possible. In the mid-1930s, Russian surgeon Yu Yu Voronin was the first to take the kidney of a recently deceased human (a *cadaver*) and transplant it into a patient with kidney failure. Despite the initial success of the operation, the grafted kidney did not survive. Many subsequent attempts were made to transplant solid organs between humans, between different members of various animal species, and between animals and humans, but the transplants failed to survive for reasons that were mysterious to researchers of the day.

No real improvements in transplant success or in understanding mechanisms of organ rejection were made until the mid-1940s. At this point, as we learned in Chapter 16, Ray Owen reported that non-identical twin calves were natural chimeras that accepted each other's blood types. The unique exchange of blood cells that occurs in the placenta during fetal development allowed the immune system of each calf to become tolerant to the other's blood cell antigens. At around the same time, Peter Medawar and colleagues were carrying out clinical work with severely burned World War II pilots. These researchers noticed that a badly burned patient given skin grafts from the same donor on two different occasions rejected the second graft more rapidly than the first graft, suggesting that immunological memory might be involved. Medawar then turned his attention to transplantation studies in laboratory mice and rabbits, and clearly demonstrated that the increased speed of rejection of a second graft occurred only if the second graft was from the same donor as the first graft. In other words, the graft rejection responses he observed were exhibiting both specificity and memory, well-established hallmarks of adaptive immune responses. Medawar's exper-

iments in graft rejection thus forged the first link between the fields of immunology and clinical transplantation. Rupert Billingham capitalized on Owen's and Medawar's work when he showed that non-identical twin calves that shared a single placental circulation were tolerant to each other's skin grafts, indicating that natural tolerance of allogeneic tissue as well as blood could be established, at least *in utero*. Further extensions of Medawar and Billingham's work demonstrated that rejection could be adoptively transferred in animals if the animals received the sensitized blood cells, but not the serum, of the original graft recipient. These studies both laid the foundation for our knowledge of immune tolerance and gave hope to patients and researchers that the immune system might be manipulated some day to permit tolerance of foreign tissue in the complete absence of drug-induced immunosuppression. Indeed, the latter concept has come to be called the "Holy Grail of Transplantation."

In the light of this new understanding of the immunological basis of transplantation, surgeons under the leadership of Dr. Joseph Murray at the Brigham and Women's Hospital in Boston carried out the first successful kidney transplant between living, identical twins in 1954. This success was rapidly replicated with other sets of identical twins, confirming that a tissue transplant could survive for lengthy periods and function normally if immune responses to it were absent. Concurrently, the work of many researchers contributed to the demonstration that, as in other species, it was differences in MHC types (*HLA mismatches*) between humans that led to the most powerful form of graft rejection. Because there was a significant demand for kidney transplants in patients who did not have an identical twin, researchers pursued the use of human cadavers as kidney donors. The brave recipients of these HLA-mismatched organs underwent total body irradiation (TBI), thymectomy, thoracic duct drainage, and steroid treatment in an effort to suppress immune responses against the transplanted tissue, but the end result was usually death. Research into immunosuppressive drugs proceeded in earnest in the 1950s. At about the same time, it was discovered that dialysis could extend the lives of kidney failure patients, fostering hope that these patients might survive until a reliable means of immunosuppression was discovered. In 1959, the immunosuppressive properties of 6-mercaptopurine and its derivative azathioprine (AZA) were reported. These anti-metabolic drugs were able to blunt certain recipients' immune responses to a graft while generating an acceptable level of side effects. When these drugs were used in combination with the steroid prednisone, the transplantation of kidneys from unrelated cadavers became a real possibility. Improvements in surgical techniques and in tissue typing for closer MHC matches then contributed to a growing success rate in renal transplants. Transplantation of other solid organs such as liver (1963), lung (1963), pancreas (1966), and heart (1967) was also undertaken, with limited success in most cases. Meanwhile, refinements in the technique of BMT were also being developed under the leadership of E. Donnell Thomas. It was found that the bone marrow (BM) of a patient with a blood disorder or blood cancer could be destroyed by a combination of chemotherapeutic drugs and irradiation, and his or her immune system rebuilt by what we now know to be the small percentage of HSCs present in BM obtained from a suitable donor. The first BMT between HLA-identical siblings was performed in 1968. Subsequent BMT in situations of HLA mismatch unfortunately gave ample demonstration of the dangers of GvHD.

By the late 1960s, it was known that lymphocytes were the linchpin of graft rejection. Clinicians thus boldly attempted to increase transplant success rates by pre-treating their recipients with anti-lymphocyte globulin, a preparation of antibodies that specifically attacked lymphocytes. During this period, physicians also agreed upon a definition of "brain death" based on neurological activity assessments that expanded the population of potential organ donors (see Box 27-1). The ability to artificially maintain the respiration of an individual whose brain function had totally ceased meant that organs could be kept healthy for a longer time, giving transplant teams a chance to notify families of donors, line up recipients, and recover healthy organs.

The 1970s and 1980s saw great strides made in the elucidation of lymphocyte development and function, particularly regarding the central importance of T cells in immune responses. After the advent of monoclonal antibody technology, a mAb called OKT3 was identified that specifically recognized T cells. Clinicians began to pre-treat their transplant recipients with OKT3 to deplete their T cells prior to organ grafting. Success rates for kidney transplants increased appreciably, but those for other organs still lagged behind. A huge breakthrough for transplantation occurred in 1972 with the discovery by Jean Borel in Switzerland of *cyclosporine A* (CsA), a fungal molecule that inhibits T cell activation (see later). Administration of CsA as an immunosuppressive drug to recipients of all types of transplants increased success rates

---

Box 27-1. Precious gifts: types of organ donation

Originally, organs could be harvested only from persons who had died a natural death due to irreversible cessation of heart and lung function. Such dead humans are termed "cadavers," and those from whom organs are harvested are called "cadaveric" or "non-heart-beating" donors. Unfortunately, the organs of a cadaver remain healthy for only a very short period (measured in minutes to hours) before the lack of oxygen kills their component cells. Cadaveric organs therefore are often rendered unusable before a suitable recipient can be notified and reach the hospital. In a development that increased the number of transplantable organs, physicians of the 1960s devised a series of tests of neurological function that would establish if "brain death" had occurred prior to the cessation of heart and lung function. Typically, brain death results from trauma to the brain, loss of oxygen flow to this organ, or the

rupture of a blood vessel in the brain. A patient who has undergone brain death loses all brain function required to sustain life. However, his or her heart and lungs are unaffected and their function can be maintained artificially using a respirator. Brain-dead donors are thus known as "heart-beating donors." The organs of brain-dead patients on life support continue to receive oxygen and remain healthy throughout the processes of obtaining family consent to organ donation, HLA matching, and transfer of the organ to a suitable recipient.

Because only a small percentage of deaths involve a brain death scenario, clinicians are keen to improve the efficiency of cadaveric organ donation and thus increase the number of available transplantable organs. However, the question which then arises is "When is one legally dead?" In other words, how

rapidly can organs be taken out of a newly deceased individual? Within a minute of the official pronouncement? Within two minutes? Some commentators fear that the rapid deterioration of organs after heart/lung death may spur some transplant teams into harvesting organs before the individual has completely succumbed. Legal experts are currently working with clinicians to try to establish a clear framework for the harvesting of cadaveric organs.

Because the number of patients on transplant waiting lists greatly exceeds the supply of available organs, clinicians have increasingly reached out to "living related" donors, since family members would be expected to share at least some MHC alleles with the prospective recipient. The Figure illustrates the concept of "shared haplotypes" and how their determination can be used to find a potentially suitable donor. For example,

Types of Organ Donors

Donor Type	Brain Function	Heart/Lung Function	Timing of Transplant	Caveats
Cadaveric (non-heart-beating)	No	No	Within minutes	Determination of legal time of death and rapidity of organ removal
Brain-dead (heart-beating)	No	Yes, with respirator	Within hours	Trauma for family of disconnecting respirator
Living related	Yes	Yes	Within hours of prearranged start	Surgical risk to donor; altruism must be motive
Living non-related	Yes	Yes	Within hours of prearranged start	Risk to donor; demand for financial or other compensation
Living heart donor	Yes	Heart: yes; Lungs: only with medical support	"Domino transplant" of healthy heart within hours	Donor trades perfectly functional heart for a donated one

Continued

Box 27-1. **Precious gifts: types of organ donation**–*cont'd*

individual A has two copies of chromosome 6 bearing the MHC and its component loci HLA-DP, -DQ, -DR, -A, -B, and -C: one inherited from her father and one from her mother. Individual A thus has the two MHC haplotypes indicated in the Figure as haplotypes 1 and 2. Let us suppose that individual A marries individual B who has inherited haplotypes 3 and 4 from his parents. The couple has five children named P, Q, R, S, and T. Each child will receive a copy of chromosome 6 from individual A and a second copy from individual B. The MHC tends to be inherited as a block, with no crossing over within the locus. Thus, when the spectrum of haplotypes is examined among the children, it can be seen that four possible haplotype combinations exist: (1,3), (1,4), (2,3), and (2,4). Suppose now that child P is in need of a kidney transplant. The most

suitable donor for child P would be child Q, because these two individuals share two haplotypes (1 and 3). Child R and child S are less suitable matches because they share only 1 haplotype, while child T shares no haplotypes with child P and would thus be the least suitable donor.

Because healthy humans can survive with only one kidney, the risk of the surgery to the donor in this type of transplant is considered acceptable. As well, the amazing powers of regeneration of the liver have meant that individuals can donate a portion of their livers to a relative in need. In general, living related donors participate out of concern and love for their stricken family member and do not expect to be compensated financially. Indeed, many jurisdictions expressly forbid financial compensation for organ donation so that the poor, ill and desperate will not

be drawn into selling their organs. Clinicians and ethicists differ on whether a living non-related donor should be compensated by means other than financial, and whether a person in a position of subordination (for example, a trusted employee of the prospective recipient) should be permitted to donate organs to his or her superior. Because of these concerns, many transplant centers do not accept organ donations from living, non-related donors.

The reader might be startled to learn that there are also "living heart donors." These individuals, who have diseased lungs but healthy hearts, are actually recipients of a combined heart-lung transplant. The healthy hearts of these recipients are then transferred to another recipient (with healthy lungs) who is in need of a heart transplant, a procedure termed a "domino transplant."

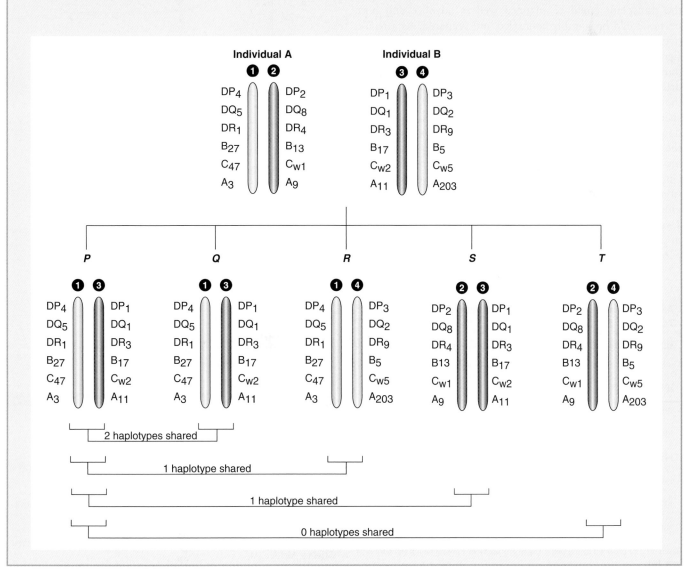

dramatically in the mid-1980s. Heart and liver transplants in particular became realistic prospects rather than just wishful thinking. Organizations were established in most developed countries to help match organ donors with those in need of transplants, and clinical standards for organ harvesting and HLA matching were drafted.

With the preceding advances, transplantation became the preferred treatment for kidney failure. As a result, the demand for replacement organs soon outstripped the supply available from cadavers and brain-dead individuals. In the late 1980s, clinicians desperate for new sources of organs began to consider the possibility of taking one kidney from a living, related family member (a *living related donor*; refer to Box 27-1) and transplanting it into the stricken patient. Because a normal healthy individual can get by quite adequately on 15% of the function of a single kidney, this procedure was not considered to be particularly risky for the donor. There were two huge advantages of using kidneys from living related donors: (1) the MHC match was almost always guaranteed to be better with a family member than with someone unrelated, and (2) because both the donor and recipient were known, the transplant could be planned well in advance and both parties could be suitably counseled and prepared. As a result, over the 1990s, the number of available organs increased significantly and the relative dependence on cadaveric donors diminished (Table 27-1). Several new immunosuppressive drugs offering various advantages were also developed such that even greater numbers of mismatched organs could be successfully transplanted. This combination of improved donor selection and better immunosuppression led to a rise in the rate of successful kidney transplants in the 1990s, increasing the mean life span of a kidney transplant patient in the United States from 12.7 years in 1988 to 21.6 years in 1996 (if a living related donor was used). Overall, the percentage of kidney grafts in the United States that survived for at least 1 year increased from 45% in 1975 to 80% in 1994. Indeed, in 2001, the one-year patient survival rate for kidney transplants was over 95% in the United States. Significant gains in patient survival have also been achieved for other types of organ transplants such that the majority of patients can expect to survive at least 3 years post-transplantation (Fig. 27-1 and Table 27-2). In many non-Western countries, transplantation progress has proceeded steadily but more slowly. A brief outline of the state of transplantation medicine in several of these countries is included in Box 27-2.

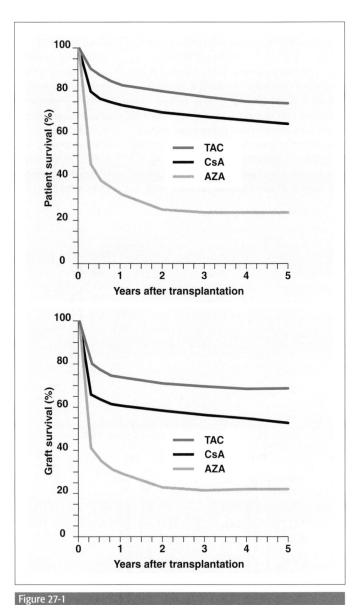

Figure 27-1

Comparison of Liver Transplantation Outcomes Using Various Immunosuppressive Drugs
Recipients of liver transplants were treated with tacrolimus (TAC), cyclosporine A (CsA), or azathioprine (AZA). (See text for details on modes of action of these drugs.) TAC is the immunosuppressant that most effectively shuts down graft rejection while inducing the least toxic side effects. Accordingly, this drug results in the highest survival percentages for both patients (top panel) and grafts (bottom panel). Note that a control curve (no drug) does not appear here because it would be unethical to withhold drug treatment. Adapted from Ginns L. C., Cosimi A. B. and Morris P. J., eds. (1999) "Immunosuppression in Transplantation." Blackwell Science, Malden, MA.

Table 27-1 **Decreased Reliance on Cadaveric Donors for Kidney Transplants in the United States**

Year	Total Donors	Cadaveric Donors	
		Number	Percentage Total Donors
1988	8828	7062	80
1990	9387	7322	78
1992	9731	7201	74
1994	10608	7638	72
1996	11356	7722	68
1998	12334	8017	65
2000	13239	8076	61

Based on data provided by the UNOS Organ Procurement Transplantation Network as of June 2001.

Table 27-2 Organ/Recipient Survival for Non-kidney Transplants in the United States

Transplant Type	Organ Survival (%)			Patient Survival		
	1 month	1 year	3 years	1 month	1 year	3 years
Pancreas	87	72	51	99	94	88
Lung	90	75	55	90	76	56
Liver	90	80	69	94	86	76
Heart and heart/lung	92	84	77	92	85	77

With information from Shannon J. B., ed. (2002) "Transplantation Sourcebook," 1st edn. Omnigraphics, Inc., Detroit.

Modern Perspective

As we start the 21st century, our enhanced ability to suppress immune responses has made transplantation the treatment of choice for many different diseases. In addition, researchers are developing tolerance induction strategies in both experimental animals and patients suffering from autoimmune diseases that should be applicable in the clinical transplantation setting. Cell isolation techniques have been refined so that whole BM no longer has to be used as a source of HSCs for the rebuilding of a recipient's immune system. Patients with blood cancers or other hematopoietic disorders can be treated with a *hematopoietic cell transplant* (HCT) in which peripheral blood or (less often) umbilical cord blood is used as the source of stem cells (see Section H). However, with the ever-increasing application of transplantation as a cure for disease, the problem of the shortage of suitable human tissue donors is exacerbated. Many more people are on transplant waiting lists than there are tissues to transplant. While dialysis can extend the time that patients are able to wait for a kidney transplant, those in need of new hearts, lungs, and livers do not have an equivalent form of therapy that can support them until an appropriate organ becomes available. This situation has led to research work on artificial organs (a topic beyond the scope of this book) and *xenotransplantation* (addressed later in this chapter). The possibility of obtaining organs from living, *non-related* donors has also been raised (refer to Box 27-1), although there are considerable ethical concerns over whether such individuals should be compensated in some way, financial or otherwise, for their organs. In poorer societies, an individual may be willing to sell his or her healthy organ to a "transplantation tourist" from a developed country for thousands of U.S. dollars, a practice many condemn as exploitative.

Box 27-2. Transplantation: a global endeavor

Transplantation is slowly becoming an accepted means of saving lives in most parts of the world. A transplant registry was started in Asia in 1989 to keep track of the numbers of centers performing transplants, numbers of patients on waiting lists, numbers of patients going abroad for transplants, the specific organ types undergoing transplant, and ethical and social issues associated with transplantation. Member countries of this registry now reply to regular surveys to contribute data. Transplantation rates vary widely by country: for example, only 85 kidney transplants were carried out in Bangladesh from 1995 to 1999, whereas China performed 14,717 in the same period. While all member countries carried out kidney transplants and several countries did liver transplants, only China and Thailand performed (a very few) pancreas transplants in 1995–1999. Issues surrounding definitions of brain death are being debated in many member countries. For example, a brain death law was implemented in Japan only in 1997. In several countries, the donor must conform to very strict criteria, the donor's family must approve, and the donor must have signed a written living will indicating an intention to donate organs. Partly because of a lack of cadaveric donors, a relatively modest total of 540 heart transplants and 71 lung transplants were carried out in 11 member countries between 1995 and 1999. Interestingly, the annual number of heart transplants in Asia actually dropped between 1998 and 1999 due to a decrease in donors following the introduction of mandatory motorcycle helmet use in Taiwan. Acceptable sources of donated organs vary in member Asian countries, with living, related kidney donors being much more common in India than in China. Numbers of living liver donors were also increasing in the late 1990s. Waiting lists are an issue in Asia as they are in Western countries. Prospective recipients wait almost 5 years for a kidney transplant in the most advanced of member countries.

A Latin American Transplant Registry with 18 member countries has also been established relatively recently. As is true in Asia, rates and successes of transplants vary widely by country. Use of immunosuppressive drugs is becoming more widespread. Brazil and Mexico performed the most kidney transplants in 1999, the majority from living donors. Brazil also leads the way in heart and liver transplants. Kidney transplantation frequency was stable or on the rise for 13 member countries but decreased (compared to 1998 figures) in five countries. A significant factor in the increase in transplantation in several countries has been the return of native surgeons trained elsewhere who have brought their enhanced skills home.

Table 27-3 The Organ Matching Process in the USA

Step	What Happens
1. Organ becomes available	Local *Organ Procurement Organization* (OPO) sends information on the organ size, condition, blood type, and tissue type to UNOS (United Network for Organ Sharing)
2. Local potential recipients are ranked by UNOS	Ranking factors include blood type, tissue type, clinical condition, time spent on waiting list
3. Organ is offered	UNOS offers the organ to the appropriate transplant center through their OPO
4. Transplant team considers offer (**1 hour allowed to reach decision**)	Factors include organ compatibility, organ and candidate condition, staff and patient availability
5. Organ accepted or declined	If accepted, surgery proceeds; if rejected, the offer procedure shifts to next transplant candidate on the UNOS list

At a societal level, many developed countries are emphasizing public education programs to make people aware of the precious gifts represented by their organs and those of family members, and encouraging them to give consent for such organs to be used when death occurs. In addition, there has been an increase in the instruction of physicians and nurses in the proper way to approach families of the recently deceased on the topic of organ use, to ensure that no chance to harvest an organ is missed through lack of information or tact. In many jurisdictions in the United States, organ retrieval exchange programs facilitate the sharing of information about organs as they become available. In addition, a coordinated, standardized organ matching process establishes equitable treatment for those on waiting lists and increases the chances of finding a suitable recipient for an organ before it deteriorates (Tables 27-3 and 27-4).

The ability to give a wide range of patients a second chance at life through organ transplantation is a medical feat made possible by a century of progress in understanding immunology, the perfection of surgical techniques, the discovery of efficacious pharmaceuticals, and a deep consideration of ethical concerns. In 1990, the Nobel Prize in Physiology or Medicine was jointly awarded to Joseph Murray and E. Donnell Thomas for their very different contributions to transplantation medicine, reflecting the multi-faceted nature of this field. The work of both these pioneers also led to an appreciation of the importance of HLA typing as well as an evolving use of various immunosuppressive techniques.

We move now to an in-depth discussion of topics relevant to transplantation immunology, starting with an examination of the molecular basis of why a transplant is viewed as foreign. We then describe the various types of graft rejection and their underlying mechanisms, followed by HLA typing and immunosuppressive drugs. Methods for inducing tolerance to grafts are discussed, as are the pros and cons of xenotransplantation. We conclude the chapter with a discussion of issues relevant to HCT.

Table 27-4 Time Constraints on Organ Viability

Organ	Post-harvest Viability (hr)
Kidney	48–72
Liver	12–24
Pancreas	12–24
Heart	4–6
Heart–lung	4–6

With information from Shannon J. B., ed. (2002) "Transplantation Sourcebook," 1st edn. Omnigraphics, Inc., Detroit.

B. The Molecular Basis of Allorecognition

Transplant rejection happens when the recipient's immune system responds to antigenic differences between self and the incoming donor tissue. Antigenic differences encoded by genes in minor histocompatibility loci (outside the MHC) result in slow, mild responses that contribute to graft rejection but are not usually its primary cause (see later). It is the antigenic differences encoded by the MHC genes themselves that result in the strong, rapid responses that are principally responsible for transplant rejection. Thus, while responses to differences in other polymorphic proteins are technically included in the definition, in practice, the term *allorecognition* as commonly used refers to immune responses to MHC-encoded differences (allo-MHC). How are these differences manifested? For B cells, the rationale for the anti-donor response is straightforward. Donor allo-MHC molecules often have a slightly different conformation than recipient MHC molecules so that the recipient's B cells recognize them as foreign, as they would any incoming nonself protein. In contrast, the physiological role of a T cell is to recognize foreign antigen in the context of *self*-MHC. It may therefore be initially more difficult to see why T cells should be stimulated so vigorously by the presence of *allo*-MHC molecules in an incoming graft. There are two mechanisms that account for these T cell responses: *direct allorecognition* and *indirect allorecognition*.

I. DIRECT ALLORECOGNITION

In direct allorecognition, the recipient's Tc and Th cells recognize peptide/allo-MHC class I and peptide/allo-MHC class II epitopes, respectively, presented on the surfaces of donor cells in the graft. What is the basis for this recognition? As we have learned, T cells are selected in the thymus through the mechanisms of central tolerance such that cells released to the periphery recognize (self-MHC + X) complexes, where X is a nonself-peptide. Any T cells recognizing (self-MHC + Y), where Y is a self-peptide, are either eliminated in the thymus or silenced in the periphery by the mechanisms of peripheral tolerance. In a transplant situation, the cells of an allogeneic graft express hundreds of thousands of allo-MHC molecules loaded with a wide variety of peptides. Most of these peptides are considered "self" with respect to both the donor and recipient because they are derived from monomorphic proteins that are identical throughout a given species. Thus, from the recipient's point of view, the peptide/MHC complexes in the transplanted tissue can be described as (allo-MHC + Y). However, T cells do not distinguish the individual components of such complexes; they recognize the overall shape. If the overall conformation of an (allo-MHC + Y) epitope looks like some combination of (self-MHC + X), recipient T cells specific for (self-MHC + X) will be activated by the donor cells presenting (allo-MHC + Y) (Fig. 27-2A). In other words, the response of the recipient's T cells to the graft is really due to cross-reactivity in which peptide/allo-MHC complexes look like nonself-peptides presented on self-MHC. These findings were confirmed in 2000 by crystal structure analysis of an alloreactive TCR bound to an allogeneic MHC class I molecule. In some instances of direct allorecognition, the identity of the peptide Y may be irrelevant because the allo-MHC molecule itself bears resemblance to (self-MHC + X). Indeed, some allo-CTLs have been shown to recognize empty allo-MHC molecules *in vitro*. A series of H-2 mutant mice was instrumental in elucidating the mechanism underlying direct allorecognition, as described in Box 27-3.

The complete understanding of direct allorecognition was not possible until the peptide-presenting function of MHC molecules was elucidated in the 1980s. For many years before that, immunologists had struggled to explain why T cell responses to allogeneic cells were so much more intense than responses to pathogens. However, in direct allorecognition, a vast number of self-peptides are presented simultaneously and continuously by the allo-MHC molecules of the graft, leading to the simultaneous and continuous stimulation of a wide range of T cell clones. In contrast, a pathogen attack results in the presentation of a relatively small number of different peptides, stimulating only a small number of T cell clones. In fact, it has been estimated that while 1 in 10,000 T cells is activated in a conventional immune response to a foreign antigen, 1 in 50 T cells is activated in response to allogeneic cells. Because of these increased numbers of activated T cells, direct allorecognition leads to rejection of the graft that is much more violent and swift than a conventional immune response.

II. INDIRECT ALLORECOGNITION

In indirect allorecognition, recipient CD4+ T cells are stimulated by recipient APCs that have acquired peptides derived from proteins (predominantly MHC molecules) of the donor. When cells of an allogeneic graft die, some of their component proteins are shed into recipient tissues and are taken up by recipient APCs, or recipient APCs may enter the graft itself. Again, most of these donor proteins are encoded by genes that are monomorphic within a species, so that peptides of these donor proteins are seen as "self." If acquired by recipient APCs, these self-peptides are presented on self-MHC and do not provoke an immune response because T cells recognizing this combination were deleted during the establishment of central tolerance. However, because the MHC molecules of the recipient and donor are usually different, nonself-peptides derived from donor MHC molecules will be generated. If these peptides are acquired by recipient APCs, they are presented on self-MHC as nonself (self-MHC + X), and an immune response against them is mounted (Fig. 27-2B). (There may be other non-MHC proteins that differ between donor and recipient, but these are usually far less abundant than the MHC proteins, resulting in far fewer nonself-peptide/self-MHC complexes.) Because donor proteins are taken up exogenously, the processing of allo-MHC molecules leads to presentation by recipient MHC class II molecules and activation of CD4+ T cells. The activated CD4+ T cells secrete cytokines that damage the graft and support the activation of B cells producing alloantibodies directed against the donor protein in the graft. In addition, the presence of cytokines produced by the recipient CD4+ T cells may non-specifically support or stimulate recipient CD8+ T cells that recognize graft epitopes by direct allorecognition. Indirect allorecognition by CD8+ T cells may also occur due to cross-presentation of exogenous donor MHC peptides on MHC class I of recipient APC (refer to Ch.11). CD8+ T cells activated in this way can mediate cytotoxicity and express damaging cytokines that contribute to graft rejection.

From a timing point of view, direct and indirect allorecognition appear to make major contributions to graft rejection at different stages of the response. For example, soon after a transplant is introduced, donor cells are present in abundance and are able to present peptides on allo-MHC to recipient Tc and Th cells. Direct allorecognition is thus likely dominant in the early stages. As donor cells die and shed donor antigens, these proteins are taken up more often by recipient APCs and presented to recipient Th cells. Indirect allorecognition may therefore play an increasingly important role in graft rejection later in the process.

C. Minor Histocompatibility Antigens

We learned in Chapter 10 about the MHC, the major histocompatibility complex, and how these genes were identified by their very rapid induction of graft rejection. In this section, we examine the minor histocompatibility genes, which are

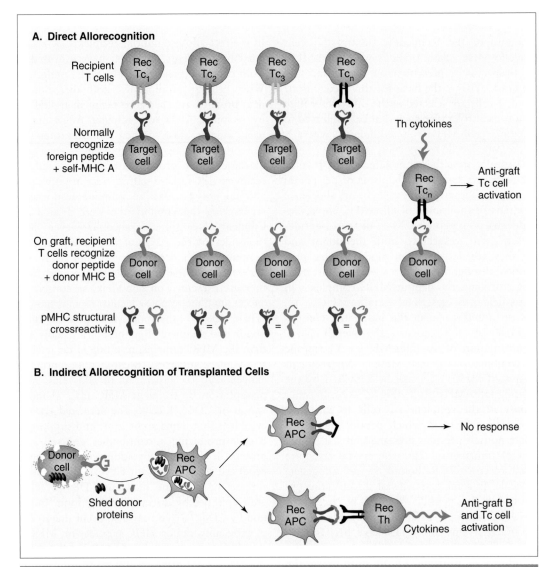

A. Direct Allorecognition

Recipient T cells

Rec Tc₁ Rec Tc₂ Rec Tc₃ Rec Tcₙ

Normally recognize foreign peptide + self-MHC A

Target cell Target cell Target cell Target cell

Th cytokines

Rec Tcₙ → Anti-graft Tc cell activation

On graft, recipient T cells recognize donor peptide + donor MHC B

Donor cell Donor cell Donor cell Donor cell Donor cell

pMHC structural crossreactivity

B. Indirect Allorecognition of Transplanted Cells

Donor cell → Rec APC → Rec APC → No response

Shed donor proteins

Rec APC — Rec Th → Anti-graft B and Tc cell activation

Cytokines

Figure 27-2

Direct and Indirect Allorecognition

(A) Direct allorecognition. Prior to any transplant, the T cell population in a recipient has been "educated" by central tolerance mechanisms to recognize (self-MHC A + X), where X is any peptide derived from nonself sources such as pathogens. However, when a graft is introduced into this recipient, a given recipient T cell will cross-react with any donor cell bearing a combination of (donor MHC B + donor peptide Y) that structurally resembles (self-MHC A + X). Direct allorecognition of pMHC class I by recipient Tc cells is shown here, but the same holds true for direct allorecognition by recipient Th cells of pMHC class II presented by donor APCs. For complete activation, recipient Tc cells must receive costimulation from licensed DCs (not shown). (B) Indirect allorecognition. Spent or damaged donor cells shed proteins that may be monomorphic (dark grey) or polymorphic (blue; such as allogeneic MHC) into the graft and surrounding tissue. These proteins are taken up by recipient APCs and presented on MHC class II. If the peptide is derived from a monomorphic protein, it is viewed as "self" by recipient Th cells and there is no response. However, if the peptide is derived from a molecule that is not expressed in the recipient, the recipient pMHC combination is seen as "nonself" by recipient Th cells. These cells are activated and supply help to alloreactive B cells as well as Tc cells that recognize targets by direct allorecognition. Recipient APCs cross-presenting nonself-peptides on MHC class I may also activate recipient CD8⁺ T cells (not shown).

generally associated with a slower rejection of the graft when mismatched. Subtle differences in the nucleotide sequence of a minor histocompatibility gene give rise to a *minor histocompatibility antigen* (MiHA). When a transplant between HLA-identical siblings is rejected, it is due to differences in MiHAs.

MiHAs were first identified in experiments carried out by field pioneer George Snell of the Jackson Laboratory in Bar Harbor, United States. Snell exchanged skin grafts or tumors between different strains of inbred mice that were matched at the MHC but otherwise had different genetic backgrounds. Snell tracked the crosses in which rejection of the transplant

Box 27-3. Lessons from MHC class I mouse mutants

As immunologists struggled to understand the role of MHC molecules in T cell recognition during the 1970s and 1980s, some valuable and interesting insights came out of the study of certain H-2 mutant mouse strains. These strains had actually been developed years before by D. W. Bailey and H. I. Kohn (1965) and R. W. Egorov (1967) as a means of investigating transplantation genetics. In these experiments, H-2 mutant mice were identified using a selection system based on skin graft rejection. (Skin graft rejection is considered to be the most sensitive indicator of MHC incompatibility. This sensitivity arises because skin is less tolerogenic than most other tissues due to its lack of vascularization and the presence of large numbers of highly costimulatory Langerhans cells ready to act as APCs.) Two well-defined parental strains of inbred mice were crossed to generate F1 hybrid offspring. When the skin of either parent was later grafted onto an F1 mouse, no rejection was expected and usually none was observed. However, in rare cases, an F1 mouse would reject the parental graft vigorously, indicating that a mutation had arisen in the MHC of the F1 mouse. This mutant F1 mouse was then back-crossed to one of the parental strains, and progeny with the mutation were inbred by brother–sister matings until a congenic line bearing the mutation on the parental background was established. The scope of this pioneering work was staggering: approximately 100,000 mice were tested in order to identify fewer than 30 mutants. Appreciation of the mutants outside the field of transplantation genetics grew in the 1970s when the direct involvement of MHC molecules in immune responses was first established, and immunologists began to study the mutants for clues as to the immunological role of the MHC.

The majority of the H-2 mutations identified in the F1 mutants had occurred in mice derived from the C57BL/6 mouse strain (H-2^b). More specifically, researchers found that, although the H-2D^b and H-2L^b loci were sometimes involved, it was the H-2K^b locus that was most often affected. These "K^b mutants" were considered remarkable because they occurred spontaneously at an unusually high rate. When tested for reactivity against mouse tissues bearing the parental K^b molecule, it was found that, in addition to provoking vigorous skin graft responses, the mutant H-2 molecules were also responsible for cell-mediated cytotoxicity and reactivity in MLR assays. During the 1970s, amino acid sequencing of the K^b mutants revealed the surprising fact that, in most cases, these highly allogeneic MHC class I molecules differed from the wild-type K^b molecule by only one or two amino acids. For the most part, single base changes at the DNA level were responsible for the amino acid substitutions observed. Indeed, the mutant and parental K^b molecules could seldom be distinguished serologically. How could the mutant and wild-type molecules be so similar structurally and yet provoke such intense immune responses? What could they tell immunologists of the day about the role of the MHC proteins in immune responses?

Perhaps the first mutant-derived clue about the function of MHC class I molecules came when it was found that all the amino acid changes in the K^b mutants were localized to the N-terminal half of the molecule. Further answers took shape as structural and functional studies of the wild-type molecules revealed the secondary structure of the MHC class I molecules. By 1981, Stanley Nathenson and his colleagues had established that the extracellular region of the MHC class I molecule was divided into three globular domains of about 90 amino

acids each. They called these domains α1, α2, and α3. A comparison of the domain sequences and immune function data between wild-type mice and K^b mutant mice led to the conclusion that the α1 and α2 domains of the MHC class I molecule were those involved in immune responses. By the mid-1980s, other experiments (previously described in Ch.10) had established that the MHC class I protein was actually a peptide presentation molecule. Structural data from the mutants led to the prediction that amino acids 70–90 (α1) and 150–180 (α2) formed the TCR recognition site. These proposals were largely verified in 1987 with the publication of the crystal structure of a wild-type MHC class I molecule. It became clear that α1 and α2 comprised a superdomain with two alpha helical regions (aa 58–84 and 138–180) that formed the sides of an apparent peptide-binding groove. Further evidence that the TCR actually recognized peptides in the groove and not just the MHC itself came from the observation that several K^b mutants evoked strong alloreactivity despite the fact that their mutations were in the bottom of the MHC groove and caused no alteration to the α-helices.

With the structure and function of the MHC class I molecule fully elucidated, it was possible to rationalize how subtle sequence changes in the K^b mutants could lead to their discrimination by alloreactive T cells. We now know that as little as a single amino acid change in the groove of the MHC molecule can affect the manner in which it binds the host's array of self-peptides. Because these peptides sit in the groove of the mutant or allogeneic MHC in an altered way, the structures mimic various "self-MHC + foreign peptide" structures, provoking the intense alloreactive response originally observed by Bailey.

---

occurred, and noted that some genetic loci were associated with transplant rejection like the major histocompatibility genes but were far more numerous and segregated away from the MHC in genetic analyses. Because of the milder rejection associated with these non-MHC loci, they were called the "minor" histocompatibility genes. The nature of the proteins encoded by these genes was puzzling for a long time because, unlike other antigens, antibodies could not be raised against them. Later, when antigen processing and presentation were understood at the molecular level, it became clear that MiHAs are not recognized as whole proteins. Instead, they give rise to peptides called "minor H peptides" that recipient

T cells recognize as foreign in the context of MHC molecules. Both syngeneic and allogeneic MHC molecules will combine with a minor H peptide to form pMHC structures that can be recognized by the recipient's T cells, so a response by recipient T cells is provoked against an MiHA difference regardless of whether the recipient and donor are MHC-matched.

What exactly are MiHAs? Some MiHAs result from allelic variation in "housekeeping" genes within a population. Housekeeping genes are expressed in virtually all cell types of a species and are either non-polymorphic genes or genes of very low polymorphism. For example, many housekeeping genes encode proteins that maintain basal metabolic functions.

When the protein encoded by the housekeeping gene of the tissue donor differs slightly from the form of that protein in the recipient, the difference constitutes an MiHA, a variant minor H peptide is presented, and a (mild) T cell-mediated response is mounted (Fig. 27-3). Other MiHAs result from differences in the expression of tissue-specific genes (such as proteins that differ between males and females of a species). B cells do not respond to MiHA, either because the protein is only expressed intracellularly or because the polymorphism involved does not result in a conformational change that can be detected by BCRs.

Why is the response to MiHA weak compared to that invoked by an MHC difference? As discussed previously, when an allogeneic MHC molecule is expressed by a donor cell, the vast majority of peptides that combine with that MHC are likely to form pMHC structures that look foreign to the recipient, stimulating a relatively large proportion of the recipient's T cell population. On the other hand, any given MiHA difference expressed by a donor cell will give rise to a very limited number of peptides with the potential to form pMHC structures that recipient T cells will then recognize. In fact, the proportion of recipient T cells that respond to the MiHA difference would be no different than that expected to respond to a given pathogen-derived protein.

Over 50 murine MiHA genes have been identified to date using skin grafting and mapping of congenic strains of mice that differ only in particular segments of the genome (Table 27-5). In theory, any genetic polymorphism that affects a peptide displayed on self-MHC could constitute an MiHA. Indeed, minor histocompatibility genes are scattered throughout the genome (except in the MHC region). Sometimes much biochemical information on the nature of a minor H peptide will be available but its gene will be a mystery. In other cases, transplant rejection attributable to minor histocompatibility differences will be observed, but the minor H peptide and its source will resist identification. Researchers are investing considerable time and energy in the study of the TCRs that recognize MiHA in the hopes of gaining new tools to reduce graft rejection.

The first MiHA genes to be definitively identified were two murine mitochondrial genes in which single amino acid changes led to the generation of minor H peptides and thus histoincompatibility. Another prominent murine MiHa is β2-microglobulin, which has several allelic variants. However,

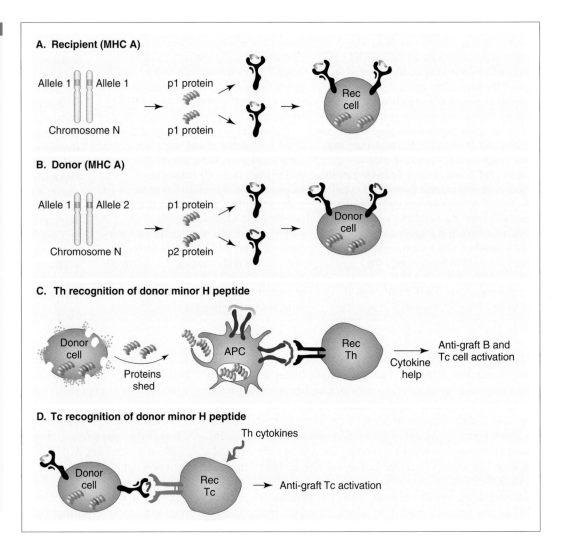

Figure 27-3

Immune Response Resulting from Differences in Minor Histocompatibility Antigens
In this example, chromosome N contains a housekeeping gene displaying co-dominant expression. (A) A transplant recipient expresses MHC class I molecule A and is homozygous for the housekeeping gene. Allele 1 gives rise to p1 proteins that are processed into p1 peptides (grey) and presented on recipient MHC A. (B) The transplant donor also expresses MHC A but expresses both allele 1 and allele 2 of the housekeeping gene of interest on chromosome N. In the heterozygote, protein is expressed from both alleles, and p1 (grey) and p2 (blue) peptides are presented on MHC A donor cells. (C) Provision of cytokine help for a graft rejection response arises when recipient Th cells recognize a p2-derived minor H peptide presented by MHC class II on either a donor or a recipient APC. (D) In the presence of cytokine help, recipient Tc cells that recognize a different minor H peptide derived from p2 presented on MHC class I are activated and generate CTLs that attack the graft.

A. Recipient (MHC A)

Allele 1   Allele 1   p1 protein

Chromosome N   p1 protein   Rec cell

B. Donor (MHC A)

Allele 1   Allele 2   p1 protein

Chromosome N   p2 protein   Donor cell

C. Th recognition of donor minor H peptide

Donor cell   Proteins shed   APC   Rec Th   Cytokine help   Anti-graft B and Tc cell activation

D. Tc recognition of donor minor H peptide

Th cytokines

Donor cell   Rec Tc   Anti-graft Tc activation

Table 27-5 **Examples of Murine and Human Minor Histocompatibility Antigens**

MiHA	Protein/Gene	Localization
MOUSE		
COI	Cytochrome oxidase	Mitochondria
ND-1	NADH dehydrogenase	Mitochondria
H-YK^k	SCMY	Y chromosome
H-YD^k	SCMY	Y chromosome
H-YD^b	UTY	Y chromosome
β2m	β2-microglobulin	Chromosome 2
B6^{dom}	Unknown	Unknown
Mx1	Unknown	Unknown
H-13	Unknown	Chromosome 2
HUMAN		
HY-B7	SMCY	Y chromosome
HY-A1	DFFRY	Y chromosome
HY-B8	UTY	Y chromosome
HY-DQ5	DBY	Y chromosome
HY-DRB3	RP54Y	Y chromosome
HA-1	KIAA0223	Chromosome 19
HA-2	Class I myosin	Chromosome 6
HA-8	KIAA0020	Chromosome 19
HB-1	HB1	Chromosome 5
UGT2B17	UGT2B17	Chromosome 4
BCL2A1	BCL2A1	Chromosome 15

the best-studied MiHAs in both mice and humans are the male-specific H-Y antigens encoded by genes on the Y chromosome. Because the Y chromosome does not appear in females, if an inbred female mouse receives a graft from a male of the same inbred strain, the only T cell responses by the female will be those provoked by epitopes composed of H-Y minor H peptides bound to female self-MHC.

It has been estimated that up to 720 MiHAs may exist in mice: an equivalent number in humans would be a transplant clinician's worst nightmare. In reality, only a small number of MiHAs seem to be immunodominant; that is, the majority of immune responses are directed against only a small number of minor H peptides. The other MiHAs are said to be either "dominated" or "cryptic." Indeed, if one suppresses the responses to the immunodominant specificities, responses to the dominated MiHA will emerge. Immunodominance is thought to arise when the interaction between the responding CTL and the donor cell is of particularly high avidity, or when the minor H peptide is presented at particularly high frequency. *In vitro*, CTL responses against cryptic MiHAs are not observed unless there are no T cells present that can recognize dominant MiHAs, or the donor cell presents only the cryptic peptide on its MHC molecules. In humans, only a small number of immunodominant MiHAs appear to be clinically relevant, and only a small number of T cell clones are activated (i.e., an oligoclonal rather than a polyclonal response).

D. Types of Clinical Rejection and Their Mechanisms

Transplantation is a traumatic event from the body's point of view. Surgery is required, tissues are resected, and blood vessels are damaged. All these events initiate responses from innate immune system cells even in a syngeneic transplant situation. Adhesion molecules are upregulated in the injured endothelium of the graft, promoting the infiltration of activated recipient macrophages and other inflammatory cells that produce IL-1 and IL-6. Because these events are nonspecific and do not usually result in loss of the grafted tissue, they are not considered graft rejection. (They may, however, play an important role in setting the stage for graft rejection; see later.) Graft rejection proper is caused by adaptive immune responses against antigens in the incoming organ (Fig. 27-4), and may be *hyperacute, acute,* or *chronic.* Here we will discuss the clinical features of various forms of graft rejection and the mechanisms proposed to account for them (Table 27-6). In addition, we describe modern treatments designed to promote acceptance of the graft by the recipient.

I. HYPERACUTE GRAFT REJECTION

Hyperacute rejection (HAR) is the destruction of a graft almost immediately after transplantation. Solid organs are vascularized structures and HAR is usually caused by the presence in the recipient of pre-formed antibodies that recognize antigens on the endothelial cells of blood vessels within the transplanted organ. When these preformed antibodies bind to these antigens, they rapidly initiate complement activation via the classical pathway such that the vascular network of the graft is destroyed within minutes by MAC deposition. Clinically, the graft succumbs to microvascular thrombosis (blockage of small blood vessels) and interstitial hemorrhage (bleeding in the extracellular matrix between cells). In general, no cellular infiltration is seen.

In practice, the chance of HAR is greatest in the presence of anti-ABO blood group antibodies in a recipient (see Section H, 'Blood Transfusion,' later in this chapter). Although ABO antigens are defined as antigenic structures present on an individual's red blood cells, they are also expressed on vascular endothelial cells, making them a major target within donated tissue. HAR can also be triggered by pre-formed antibodies recognizing nonself-MHC. Such antibodies are common in the circulation of prospective recipients who have been previously exposed to foreign MHC molecules, as in the case of a previous graft, blood transfusion, or pregnancy. The presence of anti-HLA antibodies has been documented to cause HAR of kidney, heart, lung, and pancreas transplants. The natural antibodies described in Chapter 6 are another type of preformed antibody causing HAR. These Igs, which are normally directed against common pathogen determinants, are thought to cross-react with donor antigens on an incoming graft. In the transplantation world, individuals who harbor these antibodies likely to cause HAR are said to be "highly sensitized" or "hyperimmunized."

Figure 27-4

Types of Clinical Graft Rejection

Four types of graft rejection may destroy a transplanted organ such as the donated kidney shown here in blue (KC = kidney cells). Hyperacute rejection (HAR), which occurs within minutes of transplantation, is due to attack by preformed antibodies recognizing structures (often allo-MHC molecules) on the vasculature of the donated organ. Acute graft rejection occurs within days or weeks of transplantation. Acute cellular rejection (ACR) is due to direct allorecognition of KC pMHC by recipient Tc cells. Acute humoral rejection (AHR) is thought to involve indirect allorecognition of shed donor allo-MHC by recipient Th cells. Th cytokine help supports the activation of B cells producing alloantibodies. Chronic graft rejection (CGR) is seen several months after transplantation and is usually due to local ischemia within the graft. Recipient alloreactive Th cells induce macrophages and endothelial cells to produce cytokines and growth factors that alter the structure and function of graft blood vessels. IGF, insulin-like growth factor.

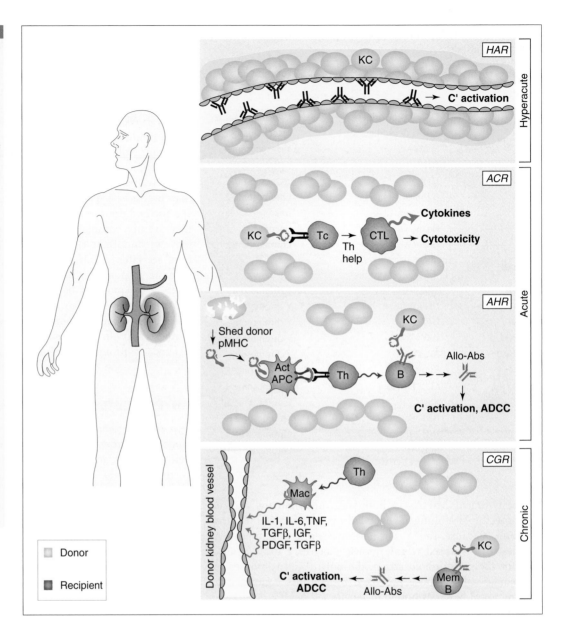

The chance of HAR of a graft can be greatly decreased by careful screening of the recipient and potential donor for blood group mismatches. In addition, a patient's blood is tested for the presence of antibodies specific for the HLA expressed on donor cells, a mandatory pre-transplant test called *cross-matching*. A patient whose blood contains antibodies to a particular HLA molecule expressed on the donor organ is said to exhibit a *positive cross-match* for that HLA determinant, and another donor must be found. If a transplant must be made between mismatched donors, the patient may undergo pre-transplant plasmapheresis or immunoadsorption to remove the alloantibodies from his or her blood. Even with immunosuppressive drugs, the survival of a graft in the face of HAR is guaranteed to be extremely short. An exception to this rule is the liver, which can be transplanted with little regard for cross-matching. The lack of organ rejection in these cases is not fully understood, but some

researchers suspect that the tremendous regenerative powers of liver tissue are responsible. The relatively low level of MHC class I expression on hepatocytes may also spare the transplanted organ from other types of immune system attack (see later).

II. ACUTE GRAFT REJECTION

There are two types of acute graft rejection, *acute cellular* and *acute humoral*, that generally occur within a few days or weeks of transplantation. These types of rejection are distinguished by their underlying mechanisms.

i) Acute Cellular Graft Rejection

Acute cellular rejection (ACR) is almost always the result of direct allorecognition of mismatched donor MHC by recipient

Table 27-6 **Types of Allograft Rejection**

Property	Hyperacute (HAR)	Acute Cellular (ACR)	Acute Humoral (AHR)	Chronic (CGR)
Time for graft rejection	Within minutes	Within days or weeks	Within days or weeks	Within months
Targets on donor cell	ABO blood group Ag Allogeneic donor pMHC Donor Ag that resembles pathogen Ag	Allogeneic donor pMHC	Allogeneic donor pMHC	Allogeneic donor pMHC
Mechanism	Preformed Ab that induce lysis by MAC formation	Direct allorecognition by recipient CTLs that exert cytotoxicity and secrete cytokines	Indirect allorecognition by recipient CD4$^+$ T cells induces alloAb production	Indirect allorecognition; cytokine-induced ischemia
Cellular infiltration	None	Lymphocytes	Neutrophils	Lymphocytes and neutrophils
C4d staining	Yes	No	Yes	Yes
Fibrosis	No	No	Yes	Yes
Treatment	None effective	Immunosuppressive drugs; steroids	Immunosuppressive drugs; plasmapheresis	None effective; experimental treatments with CTLA-4 and TGFβ inhibitors

alloreactive CTLs. The greater the number of MHC differences, the more effector clones are mobilized and the faster the rejection. Clinically, ACR is characterized by the presence of an infiltrate of mononuclear cells into the transplanted organ and inflammation of the endothelial lining of its blood vessels that can lead to necrosis. Infiltrating CTLs destroy the graft both by secreting destructive cytokines such as TNF and by initiating perforin- and granzyme-mediated killing. CTLs also secrete high levels of IFNγ, which may promote DTH reactions that compromise the graft.

The normal roles of CD4$^+$ and CD8$^+$ T cells sometimes become blurred in a transplant situation. CD8$^+$ cells secreting cytokines have been encountered in studies of graft rejection, as have CD4$^+$ cytotoxic T cells. Indeed, when RAG-deficient mice were reconstituted with various purified lymphocyte populations, acute rejection of an allogeneic heart could be mediated by purified alloreactive CD4$^+$ T cells, 90% of which were exercising direct allorecognition. In other experiments involving mouse skin grafts, treatment of the recipients with antibodies to remove CD8$^+$ T cells did not significantly enhance the survival of grafts. However, treatment with anti-CD4 antibodies extended graft survival moderately, and treatment with both anti-CD8 and anti-CD4 antibodies prolonged graft survival significantly.

The reader may wonder if NK cells play a role in graft rejection. NK cells are activated by the absence of MHC class I and thus are not generally a factor in solid organ rejection. However, NK cells are capable of rejecting donor BM cells which express mismatched, little, or no MHC class I (see later in this chapter).

Immunosuppression using the drugs discussed later to deactivate or eliminate lymphocytes is the principal means of preventing ACR, and a patient who stops using these drugs is left open to ACR soon after. Steroid treatment and

T cell depletion can be helpful in reversing an acute episode of ACR.

ii) Acute Humoral Graft Rejection

Acute humoral rejection (AHR) can occur rapidly but is usually seen several weeks or months after transplantation and so is sometimes called "delayed graft rejection" or "delayed vascular rejection." In these cases, rejection is due to indirect recognition of alloantigen by recipient CD4$^+$ T cells. These cells produce effectors that support the *de novo* production of antibodies directed against allo-MHC expressed on the graft. The clinical features of graft rejection caused by *de novo* alloantibodies are distinct from those of HAR and ACR and include neutrophil infiltration into capillaries in the graft, inflammation of the vascular system, fibrosis, and necrosis of blood vessel walls. In contrast to ACR, infiltration by mononuclear cells is not usually observed. AHR can also be diagnosed by staining for C4d, a degradation product of the complement component C4b. C4d tends to accumulate on capillaries in a graft undergoing AHR, whereas ACR is generally negative for C4d staining. In addition to MAC attack, AHR can sometimes be mediated by ADCC. Donor cells that have bound alloantibodies are destroyed after these alloantibodies are in turn bound by the FcR of cytolytic effectors.

Steroid treatment and anti-lymphocyte therapies are generally not very effective in reversing AHR. Typically, over 50% of grafts that undergo this type of rejection do not survive. However, if patients are treated post-transplant with plasmapheresis (to remove newly synthesized alloantibodies) and with two of the newer immunosuppressive drugs (TAC and MMF; see later), alloantibody concentrations can be greatly reduced and the incipient graft loss due to AHR can be reversed. In transplantation lingo, such recipients are said to be "desensitized" by the procedure.

III. CHRONIC GRAFT REJECTION

Chronic graft rejection (CGR) of solid organs (also called *chronic allograft dysfunction*, CAD) is defined as the loss of allograft <u>function</u> several months after transplantation. The transplanted organ may still be in place, but persistent immune system attacks on its component cells have gradually caused the organ to cease functioning. While immunosuppressive drugs have greatly improved the 1 year graft survival rate for many organ types, they are less effective in extending the life of a graft after 12 months. To illustrate, although 90% of transplanted kidneys are alive at 1 year post-transplant, more than 50% of them are lost by 10 years post-transplant. As a result, over 20% of kidney transplants in the United States are "re-transplants," the grafting of new organs in patients whose first transplants have failed due to CGR. Chronic graft rejection thus remains a major hurdle in the quest for transplant permanence. Finding ways to decrease CGR will minimize the need for re-transplants and conserve valuable donated organs.

Precisely what triggers CGR is unknown. The most important clinical predictor of CGR is an episode of acute rejection (that has been reversed). If this episode of acute rejection occurs after the first 3 months, the long-term prospects of the transplant are particularly poor. The occurrence of several such episodes is associated with dramatically decreased graft survival. Most researchers believe that, like acute rejection, chronic rejection is related to the presence of alloantigen. Indeed, comprehensive surveys of transplant recipients have shown that for every HLA mismatch, a 5% decline in long term graft survival results.

Both alloantibodies and cell-mediated responses are involved in CGR, and the indirect pathway of allorecognition appears to be particularly important. As is true for AHR, recipient T cells differentiate in response to allogeneic donor peptide presented on recipient APCs. These effectors produce cytokines that either directly damage the grafted tissue or constitute T cell help delivered to recipient B cells producing alloantibodies that recognize donor MHC class I and II expressed on cells of the graft. Over half of chronically rejected organs exhibit C4d deposition in the capillaries of the transplant.

In a critical part of the CGR response that is characteristic but poorly understood, activated recipient T cells also induce monocytes/macrophages that have infiltrated into the graft, and endothelial cells in the walls of its blood vessels, to produce growth factors and cytokines. Molecules such as *platelet-derived growth factor* (PDGF), *insulin-like growth factor* (IGF), *epidermal growth factor* (EGF), IL-1, IL-6, TNF, and TGFβ are secreted and bind to receptors present on smooth muscle cells lining the arteries and arterioles of the graft. Smooth muscle cell proliferation and the synthesis of extracellular matrix proteins are increased. The proliferating smooth muscle cells shrink the lumens of the graft blood vessels, inhibiting blood flow and causing *ischemia* (local loss of oxygen to a tissue). The resulting pathology is called *chronic graft vasculopathy* or *chronic graft vascular disease*. The increase in extracellular matrix proteins causes fibrotic changes that also promote deterioration of the graft and compromise its function. Thus, CGR is clinically characterized by fibrosis, collagen deposition, and loss of blood flow within the graft.

The phenomenon of epitope spreading introduced in Chapter 26 is sometimes associated with CGR. At the start of a CGR response, only a small number of different T cell clones with distinct TCRs may recognize a small number of allo-epitopes. However, as the graft deteriorates, more allo-epitopes become physically exposed and accessible for recognition. New naive recipient T cell clones can then be activated and the response "spreads" from the original epitope to new ones. The attack on the graft is thus perpetuated by the continuous activation of new T cell clones. Epitope spreading is discussed in detail in Chapter 29.

A growing number of researchers believe that CGR may be related to events during or following the transplantation procedure that have little to do with alloantigen *per se*. In this regard, CMV infection, delayed or poor function of the donor organ, and pathological lipid metabolism have been identified as risk factors for CGR. These factors have more and more influence as the taking of organs from "marginal donors"—elderly, diabetic, or hypertensive individuals—is increased due to the pronounced shortage of transplantable tissues.

Chronic graft rejection is best prevented by treating the patient with combinations of immunosuppressive drugs that block acute rejection episodes. However, while these therapies tend to downregulate inflammatory Th1 responses, they can upregulate Th2 responses that promote fibrosis associated with chronic rejection. While there is currently no truly effective treatment for CGR, promising results have been obtained from experimental strategies based on using the negative regulator CTLA-4 to block costimulation and induce T cell anergy (see later). CGR has been either prevented or interrupted by the application of soluble forms of CTLA-4 in several animal allograft rejection models. Another therapeutic thrust takes aim at blocking the fibrosis associated with chronic rejection. The high levels of TGFβ found in transplant recipients have been implicated in promoting fibrosis, and TGFβ inhibitors are being explored as adjunct therapies following transplantation.

IV. THE ROLE OF CYTOKINES AND CHEMOKINES IN GRAFT REJECTION

As mentioned previously, cytokines have direct and indirect roles in transplant rejection. Pro-inflammatory cytokines spark inflammation and infiltration of recipient leukocytes into the graft. Cells drawn to the rejection site produce IFNα, β, and γ, which promote upregulated MHC expression, and IL-2, which supports T cell proliferation and the generation of effector CTLs. TNF produced by activated macrophages and CTLs damages grafted tissues directly (particularly pancreatic islets), and IL-1, TGFβ, and PDGF produced by activated macrophages promote smooth muscle cell proliferation and upregulated synthesis of extracellular matrix proteins, contributors to CGR. Pro-inflammatory cytokines (particularly TNF) also induce the synthesis of chemokines that may play a more direct role in graft damage.

Chemokines are, by definition, involved in leukocyte chemotaxis, migration, and adhesion, and thus play important (if ill-defined) roles in allograft rejection. In particular, IL-8 appears to be of great relevance to human transplantation because the presence of this chemokine has been correlated with graft failure. For example, during CGR of lung transplants, increased levels of IL-8 as well as neutrophil infiltration are observed in the grafted lungs. IL-8 also induces the firm adhesion of monocytes to vascular endothelial cells, facilitating damage to the graft's circulatory network. Researchers have begun systematic study of the expression of various other chemokines during graft rejection in rodents. Studies of heart allografts in mice and rats have shown that the first chemokines expressed by infiltrating mononuclear cells in the grafted tissue are MIP-1α (CCL3) and MIP-1β (CCL4) (which have functions similar to human IL-8), followed about 8 days later by IP-10 (CXCL10; induced by IFNγ), Mig (CXCL9), and RANTES (CCL5). In contrast, while MCP-1 and MIP-1β are prominent in kidney transplants undergoing rejection, RANTES and MIP-1α are not. Macrophages infiltrating lung transplants in humans express high levels of IP-10 and other CXCR3 ligands. Liver allografts undergoing rejection are permeated with high levels of TNF. TNF induces the production of "cytokine-induced neutrophil chemoattractant" (CINC), a member of the IL-8 superfamily. It is thought that TNF produced by monocytes and macrophages involved in the liver graft rejection response induces autocrine production of CINC, which in turn recruits neutrophils into the graft.

Of the chemokine receptors, two of the most important molecules associated with allograft rejection are CXCR3 and CCR5. These receptors are highly expressed by effector T cells, macrophages, and NK cells, all cell types prominent in graft infiltration. For example, CXCR3-expressing T cells have been found in human lung, cardiac, and renal transplants undergoing acute or chronic rejection. Significantly, mice lacking CXCR3 (whose ligands include IP-10 and Mig) are able to tolerate a cardiac allograft for over 50 days without immunosuppression (rather than the usual 7 days exhibited in wild-type mice). Similarly, blockade of CXCR3 by neutralizing mAb can reduce the rejection of cardiac allografts in mice. In contrast, CCR5 is the most important receptor for the rejection of renal allografts in rodents. CCR5 (whose ligands include RANTES, MIP-1α, and MIP-1β) is not expressed on healthy kidneys but CCR5 mRNA can be found in mononuclear cells infiltrating rejected kidneys. Chronic renal allograft rejection can be inhibited by either gene targeting of CCR5 or the use of an analogue of RANTES that functions as a CCR5 antagonist. This strategy may be of future benefit to human kidney transplant patients, since the expression of both CCR5 and its ligands is elevated in human renal allografts undergoing acute rejection. Retrospective studies of patient groups have been done to correlate polymorphism in the CCR5 gene with the incidence of graft rejection. Kidney transplant survival was examined in a large group of patients,

1.7% of whom were found to express the non-functional CCR5Δ32 allele discussed in Chapter 25. There was no change in the incidence of acute graft rejection frequency among these patients, but 5- and 10-year survival rates were significantly better in this group than in patients bearing the normal CCR5 allele. In mice, CCR5 also mediates the rejection of pancreatic β-islet cell allografts, which may be of relevance for human diabetic patients. A lesser role in allograft rejection in rodents is played by CCR1 (whose ligands include RANTES, MIP-1α, and several other chemokines). Cardiac allografts transplanted into CCR1$^{-/-}$ mice show increased survival times. Furthermore, given a very low dose of an immunosuppressive drug, CCR1$^{-/-}$ animals can permanently accept their allografts, whereas the wild-type mice cannot. Another chemokine receptor with minor involvement is CCR2 (whose ligands include MCP-1 and MCP-3). Expression of CCR2 has been implicated in human renal allograft rejection and pancreatic β-islet cell rejection in mice.

Much remains to be determined about the functions of cytokines, chemokines, and their receptors in allograft rejection, particularly for chronic rejection. Progress in understanding the roles of chemokines and their receptors in clinical transplantation is hampered by the fact that these molecules are quite different among species, necessitating cautious interpretation of studies in animal models. Some human chemokines (like IL-8) have no structural homologue in rodents, and other chemokines bind to additional or different receptors in different species. In addition, the relevance of a given chemokine receptor or its ligand to rejection varies by the organ being transplanted. Another difficulty is that the number of truly useful mAbs directed against chemokines or their receptors is currently limited. Finally, while studies in knockout mice lacking a particular chemokine receptor have been extremely useful for examining acute rejection, the complete absence of the receptor in question could obscure a role for that receptor in attracting regulatory cells later on that could mitigate the rejection response. These considerations have slowed the development of clinical therapies to reduce graft rejection based on the manipulation of chemokines or their receptors.

V. A ROLE FOR TLRs IN GRAFT REJECTION?

What about a role for innate immunity in transplant failure? Prior to the early 2000s, there was little solid evidence that innate immunity was involved in allograft rejection. However, studies of knockout mice lacking the MyD88 adaptor protein crucial for TLR signaling have indicated that TLR engagement may play an indirect role in at least some cases of allograft rejection. The reader will recall from previous chapters that, in addition to being bound by pathogen PAMPs, TLRs can be engaged by host-derived heat shock and other stress ligands. Immature DCs expressing TLRs engaged by these ligands are induced to mature, produce pro-inflammatory cytokines, and express the costimulatory molecules necessary to activate naive T cells. The trauma of transplantation may cause the local release of stress ligands that subsequently induce TLR-mediated DC maturation. Using the HY MiHA system, it was shown that female MyD88-deficient mice could not reject skin grafts from littermate male mice. The mutant females showed reduced numbers of mature DCs in their draining lymph nodes and decreased numbers of anti-HY CD8$^+$ T cells in their spleens. The production of Th1 cytokine IFNγ, but not the Th2 cytokine IL-4, was impaired in the draining lymph nodes of these mice. In humans, support for a role of TLRs in allograft rejection comes from studies of lung transplants in patients heterozygous for polymorphisms of TLR4. These polymorphisms affected the function of the TLR4 molecules such that the patients exhibited a reduced responsiveness to endotoxin (the pathogen-derived LPS ligand of TLR4). In the lung transplant situation, recipients who expressed one of two such TLR4 functional polymorphisms were much less likely to experience acute rejection of the transplanted organ. Another aspect of TLR biology that may be relevant to allograft rejection is the expression of these receptors on non-immune system cells. For example, tubular cells in kidney epithelium express TLR1, -2, -3, -4, and -6, and TLR4 has been found on airway epithelial cells. It is possible that stress ligands released during transplantation may activate these cells to secrete pro-inflammatory cytokines that contribute allograft rejection.

VI. GRAFT-VERSUS-HOST DISEASE (GvHD) IN SOLID ORGAN TRANSPLANTS

Transplantation, although a life-saving procedure, can be a double-edged sword. When immune system cells associated with a transplanted donor tissue cause damage to recipient tissue, "graft-versus-host disease" is said to have occurred. The epithelial cells of the skin, liver, and intestine of the transplant recipient are the prime targets of GvHD. If the latter two tissues come under attack, the recipient may start to lose weight and experience toxicity that decreases the function of the liver, lungs, or gut. Onset can be acute (and sometimes fatal) or chronic. Not surprisingly, the incidence of GvHD depends on the tissue transplanted and how many donor lymphocytes have traveled with the graft. While GvHD has been reported in cases of liver and intestine transplants, it rarely occurs in transplants of other solid organs. In contrast, GvHD is very prominent in HCT since, by definition, the transplant is of donated immunocompetent cells. GvHD is therefore discussed in detail later in Section I on HCT.

E. HLA Typing

The reader is by now well aware of the polymorphism of the MHC loci. Only 1 in 400,000 unrelated persons are fully matched at the major HLA loci. Such rarity dictates that almost all transplants (except those involving twins or siblings) will be done under conditions of at least some HLA mismatching. Depending on the type of transplant being performed (solid organ versus BMT/HCT; what type of solid

organ is required) and the alternatives open to the patient (i.e., dialysis in the case of kidney failure), the doctors involved decide what degree of mismatch, if any, is acceptable before proceeding with the transplant. For example, heart, lung, liver, and, to a great extent, kidney transplants can now be done under conditions of significant HLA mismatching due to the control of graft rejection by immunosuppressive drugs and the generally rare occurrence of GvHD in solid organ transplantation. HAR is more of a threat to these types of transplants, and so ABO blood group matching is considered more important than HLA matching in these cases.

To determine the degree of HLA mismatching that exists between a prospective donor and recipient, these individuals may be subjected to extensive *HLA* or *tissue typing*. In general, several pieces of HLA-related information are required for a complete evaluation of tissue compatibility. Initially, the recipient's HLA haplotype is determined using their leukocytes as assay material. In addition, the recipient's serum is tested for the presence of pre-formed anti-HLA antibodies that would automatically exclude the use of some donor organs. These first two tests can be carried out as soon as a patient's need for transplantation has been established. Once an organ becomes available, the HLA haplotype of the donor is determined so that mismatches with the recipient's HLA haplotype can be evaluated. In addition, organs that will be an obvious positive cross-match with recipient anti-HLA antibodies can be ruled out for use at this point. Finally, for donors still under consideration, HLA cross-matching of greater specificity is performed by testing the recipient's serum for reactivity against cells that are actually taken from the prospective donor. Methods used to obtain information about a recipient's and donor's HLA compatibility status are outlined in the next section.

The speed of the typing procedure can be an issue in establishing the HLA profile of a donor, given that some organs made available by sudden death are viable for transplantation for only a short time (refer to Table 27-4). Modern methods of tissue typing have made it possible to determine a complete HLA profile in less than 3 hours. The reader should note, however, that the same time pressures do not exist for BMT or HCT; in these situations the donor is always living and the exact timing of the transplant is known in advance. There is usually ample time to search bone marrow donor databases, secure consent for the procedure, and carry out repeated testing of the prospective donor's blood for histocompatibility at the HLA and other loci. A sample of the donor BM itself may also be examined in advance to ensure that there is no incipient leukemia lurking that has not yet manifested itself in the blood.

Although we discuss tissue typing in the context of transplantation in this chapter, the reader should be aware that determining HLA haplotypes can also be crucial for the gathering of forensic evidence and for studies of anthropology, disease incidence, and drug reactions. Research into mechanisms underlying cancer and autoimmune diseases also benefits from knowledge of the HLA haplotypes of the affected individuals. It will likely be possible in the not-too-distant future to predict an individual's susceptibility to disease or his or her chances of keeping a transplanted tissue based on an exquisitely detailed HLA profile.

I. SEROLOGICAL TECHNIQUES

i) Complement-Dependent Cytotoxicity (CDC)

Before the 1980s, when serology was the primary means by which the major alleles of the principal HLA classes were distinguished, an assay method called complement-dependent cytotoxicity (CDC) was developed. In this assay, leukocytes from blood samples of the individual being typed are screened for reactivity using panels of antibodies obtained from individuals exposed to allogeneic cells via a previous transplant, transfusion, or pregnancy. The peripheral blood leukocytes (PBLs) being typed are mixed with separate antibody samples in a parallel series of microwells (Fig. 27-5). Also added to the wells are complement and a visible dye (such as trypan blue) that is excluded from viable cells. In any given well, if the cells express HLA recognized by a given anti-HLA antibody, they are lysed by the complement and take up the dye. Mismatching of the donor and recipient at the various HLA loci is thus revealed by different patterns of reactivity across the same antibody panel. Although labor-intensive, CDC assays have the advantage of rapidity in that results can be obtained within 3 hours.

Originally, polyclonal alloantisera were used in CDC assays, with each antiserum often cross-reacting with more

than one MHC epitope. Thus, a disadvantage of the original CDC technique was that it did not detect subtle polymorphisms between some HLA alleles. Today, the use of mAbs allows more precise serological analyses to be carried out while maintaining assay speed (see later). However, there are difficulties even with mAb CDCs in distinguishing between certain human MHC class II alleles. Most recently, reliance on serological typing by CDC has waned as newer, DNA-based typing techniques have provided more sensitive results (see later).

ii) Panel Reactive Antibody (PRA)

Just matching the tissue types of a donor and a recipient is sometimes not enough to ensure the acceptance of a grafted tissue. The chances of a transplant surviving are greatly increased when the recipient's serum has also been screened for the presence of anti-MHC alloantibodies that could potentially mediate HAR. Serological screening is particularly important for kidney transplants and platelet transfusions, and essential in cases in which the prospective recipient has been pregnant or has had a blood transfusion. Because many MHC alleles share epitopes that trigger alloantibody production, it is more efficient to screen recipient serum against panels of pooled cells expressing a broad spectrum of MHC molecules (hence the name "panel reactive antibody"). One currently used PRA assay method involves mixing the prospective recipient's serum with pools of standard cells

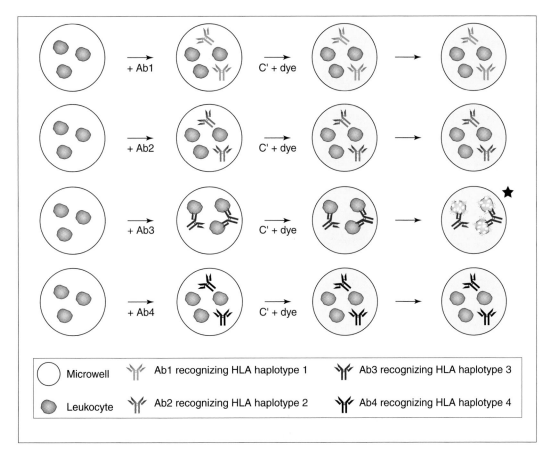

Figure 27-5

Complement-Dependent Cytotoxicity (CDC) Tissue Typing
In this example, leukocytes from an individual being typed are mixed (separately) with purified antibodies recognizing different HLA haplotypes. Rabbit complement and trypan blue dye are added to each microwell. Where the Ab fails to bind to the leukocytes, complement is not activated; the cells remain intact and exclude the blue dye. Where the Ab binds to the leukocytes, the cells are permeabilized by complement activation and take up the blue dye. Here we see that only the addition of Ab3 results in blue cells, meaning that this individual expresses HLA haplotype 3 but not haplotypes 1, 2, or 4.

◯ Microwell   Ab1 recognizing HLA haplotype 1   Ab3 recognizing HLA haplotype 3

⬡ Leukocyte   Ab2 recognizing HLA haplotype 2   Ab4 recognizing HLA haplotype 4

(either cell lines or PBLs), and then adding exogenous human complement as well as a labeled monoclonal antibody that detects the complement product C4d (Fig. 27-6). Where recipient alloantibody binds to the test cells, complement is activated and C4d is deposited on the test cell surface, allowing the labeled anti-C4d mAb to bind. Flow cytometry is then used to detect the level of C4d on each test cell sample. A positive result indicates that alloantibodies to standard cells are present in the recipient's serum, and that precise HLA cross-matching of the prospective recipient and donor is critical. If the recipient's serum contains antibodies that react with the donor's cells, the transplant should not proceed.

An advantage of flow cytometry screening for the preceding tests is that hundreds of samples can be examined in a day; however, its resolution is considered to be low. Many HLA typing labs use a combination of methods: large numbers of samples are pre-screened by flow cytometry, but only those that are positive are subjected to ELISA to precisely determine the MHC specificity of the alloantibody in the sample.

II. TYPING BY CELLULAR RESPONSE: MLR

Another means of investigating histocompatibility is to conduct a one-way mixed lymphocyte reaction (MLR). The reader will recall from Chapter 10 that the MLR takes place between a set of stimulator T cells and a set of responder T cells. In the transplantation context, T cells from a potential organ donor (the stimulators) are irradiated (to block their responses) and mixed with T cells from the recipient (the responders) (Fig. 27-7). The mixed cells are cultured for 4–7 days to allow a response to alloantigens to develop. ^{3}H-thymidine is added to the culture at the end of this period to measure the proliferation of responding T cells. If the donor and recipient are histocompatible, very little proliferation is observed. Vigorous proliferation is a sign of histoincompatibility. This assay is sensitive but not very specific as well as technically demanding and time-consuming. For these reasons, it is considered historical and is rarely used today for tissue typing.

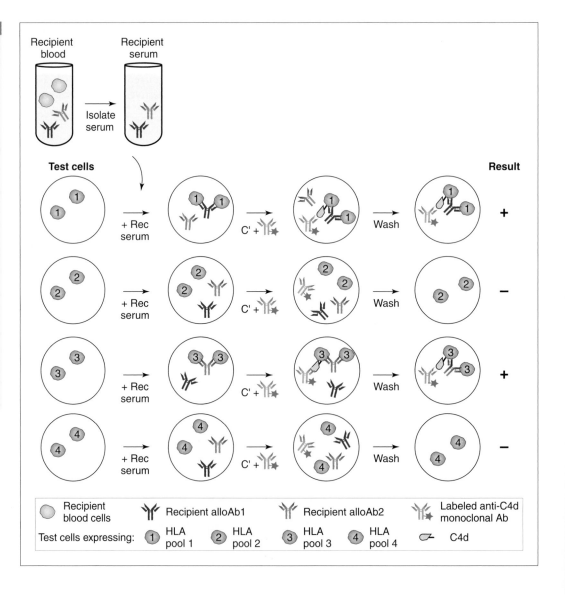

Figure 27-6

Panel Reactive Antibody (PRA) Typing

To determine if a prospective transplant recipient has pre-existing alloreactive antibodies in his or her circulation, samples of recipient serum are mixed with pooled test cells expressing different HLA haplotypes. Rabbit complement is added along with a tagged mAb recognizing the complement activation by-product C4d. After incubation to allow binding, a washing step removes unbound mAb and cell-bound tagged anti-C4d is measured in each well. Here we see that the recipient's serum contains alloantibodies recognizing test cells from HLA pool 1 or pool 3 but not HLA pool 2 or pool 4. The test indicates that the prospective transplant recipient should not receive tissue from a donor expressing alleles from HLA pools 1 or 3.

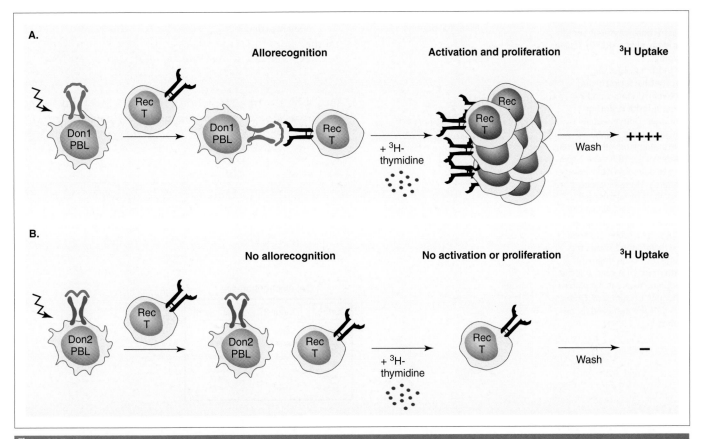

Figure 27-7

Mixed Lymphocyte Reaction (MLR) Tissue Typing
In this example, donor 1 and donor 2 are screened for suitability as tissue sources for a recipient. PBLs (the "stimulators") isolated from each donor are irradiated (jagged arrow) and mixed with recipient PBLs (the "responders"). (A) T cells from the recipient are activated by direct allorecognition of donor cells and incorporate ³H-thymidine into their DNA as they proliferate. High levels of radioactivity are detected in the cell culture indicating MHC mismatching. (B) The recipient T cells are not activated by donor 2 cells so negligible incorporation of ³H-thymidine occurs, indicating MHC matching. Even if there had been minor histocompatibility differences between donor 2 and the recipient, little ³H-thymidine uptake would be observed as the response to MiHA is relatively weak.

III. TYPING AT THE DNA LEVEL

i) Fragment-Based Methods

In the 1980s, after the discovery of restriction enzymes that cut DNA in sequence-specific ways, new tissue typing techniques were developed for use in both research labs and transplantation clinics. The reader will recall that, when a given gene is digested with a particular restriction enzyme, it is cleaved in a sequence-dependent manner into "restriction fragments" of distinctive length. The fragments can then be separated by size on a DNA gel and visualized by hybridization to a tagged probe specific for the gene of interest (Fig. 27-8). In the case of polymorphic genes, the number and location of restriction enzyme cleavage sites can vary between alleles so that the lengths of fragments generated by one allele are often slightly different from those generated by another allele. Such a difference is referred to as a *restriction fragment length polymorphism* (RFLP). In order to establish the HLA haplotype of an individual, the RFLP pattern of their HLA genes can be compared to those of individuals known to express particular MHC alleles. Also

in the 1980s came improvements in protein analysis by 2D-electrophoresis, which allowed accurate comparison of component peptides of MHC proteins. Differing patterns between individuals in this situation could also be used to assess MHC matching. However, although sensitive, RFLP and 2D protein analyses are labor-intensive and relatively slow methodologies, so they are now rarely used in the context of tissue matching for transplantation.

ii) PCR-Based Methods

The late 1980s saw the development of PCR techniques that allowed researchers to greatly amplify specific DNA sequences (refer to Box 8-3). With the large quantities of the DNA of interest that could be generated from the smallest of samples, it was possible to accurately determine the nucleotide sequences of various HLA alleles and identify the specific variations that made them unique. The sequences of hundreds of human MHC class I and II alleles have been resolved, giving rise to truly allele-specific tissue typing. Synthetic, labeled oligonucleotide probes complementary to

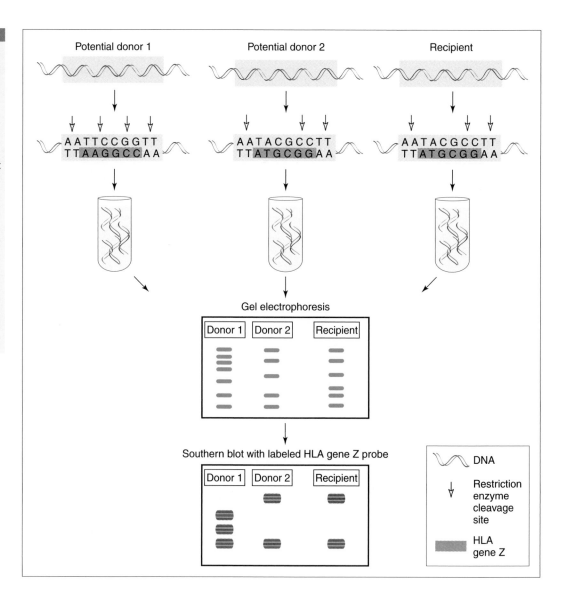

Figure 27-8

Restriction Fragment Length Polymorphism (RFLP) Tissue Typing
DNA from cells of the prospective recipient and potential donors 1 and 2 is digested with restriction enzymes. Differences in the nucleotide sequences surrounding or within different alleles of the HLA gene Z give rise to different DNA cleavage patterns. These differences are visualized by agarose gel electrophoresis followed by Southern blotting using an HLA gene Z probe. Here we see that donor 2 shows the same restriction fragment pattern for HLA gene Z as the recipient, whereas the pattern of donor 1 is different. Donor 2 is thus the preferred tissue donor.

these identifying sequences are hybridized with the DNA of individuals of unknown HLA types to define their HLA profiles. The increased sensitivity of these more modern techniques is considerable. For example, in one study in which an antibody detected a single DR4 allele, the MLR test detected six alleles and oligonucleotide hybridization detected 11 alleles. In another study of a population of donors and recipients that had been typed serologically for identity at HLA-A, -B, and -DR, fully 45% showed mismatches of one or more of these alleles when typed at the DNA level.

Several different variations on PCR techniques were developed in the late 1990s that were designed to give high-, medium-, or low-resolution analysis of MHC alleles. Naturally, the higher the resolution, the slower the speed of results and the greater the cost. While medium-resolution analysis is generally sufficient for the typing of solid organs for transplantation, high-resolution analysis is required for an unrelated HCT. There are currently two main categories of modern PCR techniques: the *probe hybridization* methods and the *direct amplicon* methods.

For probe hybridizations, total DNA is extracted from the cells of the individual to be typed and PCR is carried out using primers that allow generic amplification of all HLA alleles (Fig. 27-9A). The amplified HLA sequences are then dot-blotted onto a membrane and the membrane hybridized to specific single-stranded probes representing diagnostic sequences from specific MHC alleles. Both radioactive and non-radioactive probes have been used successfully. This technique, originally dubbed "PCR followed by sequence-specific oligonucleotide probing" (PCR-SSOP), can give results within 4–5 hours.

In contrast to probe hybridization, direct amplicon analyses rely on initial amplification of a selected region of the DNA of the individual to be typed. The region amplified is determined by the PCR primers used. For a technique called PCR-RFLP, PCR primers are chosen so as to amplify a region of the HLA gene known to be polymorphic and therefore differentially sensitive to particular restriction enzymes in an allele-dependent way. After the amplified patient DNA is cut with the relevant restriction enzymes, the resulting

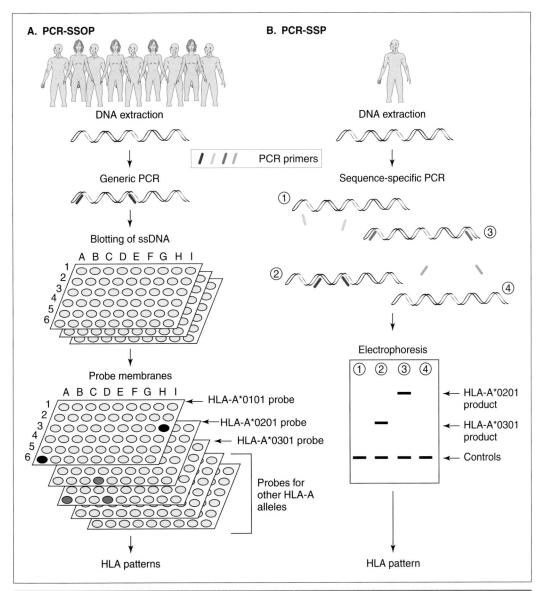

A. PCR-SSOP

DNA extraction

↓

Generic PCR

↓

Blotting of ssDNA

A B C D E F G H I
1
2
3
4
5
6

↓

Probe membranes

A B C D E F G H I
1
2
3 ← HLA-A*0101 probe
4 ← HLA-A*0201 probe
5 ← HLA-A*0301 probe
6

Probes for other HLA-A alleles

↓

HLA patterns

B. PCR-SSP

DNA extraction

↓

Sequence-specific PCR

① ② ③ ④ PCR primers

↓

Electrophoresis

① ② ③ ④

← HLA-A*0201 product
← HLA-A*0301 product
← Controls

↓

HLA pattern

Figure 27-9

PCR-SSOP and PCR-SSP Tissue Typing

(A) PCR-SSOP is generally used for the screening of large numbers of samples. In the example shown, DNA from each individual is extracted and hybridized separately to generic primers that allow amplification of all HLA alleles. The ssDNA PCR products are replicatively dot-blotted onto membranes and each membrane is hybridized to a tagged probe specific for a different HLA allele. The hybridization pattern at a given coordinate on multiple membranes yields the HLA haplotype for a given individual (54 individuals in example shown). For example, we see here that individual A6 expresses HLA-A*0101 and HLA-A*0301 but not HLA-A*0201. In contrast, individual D6 expresses HLA-A*0201 and HLA-A*0301 but not HLA-A*0101. (B) In PCR-SSP, samples of the DNA of an individual are hybridized to PCR primers specific for individual HLA alleles (in this case, HLA-A*0101, *0201, *0301, and *0401). The PCR products are mixed with ethidium bromide and fractionated on an agarose gel to reveal a pattern of bands of expected sizes for particular alleles. Here we see that the individual being typed shows bands indicating expression of HLA-A*0201 and HLA-A*0301. Adapted from Welsh K. and Bunce M. (2001) New methods in tissue typing. In Thiru S. and Waldmann H., eds. "Pathology and Immunology of Transplantation and Rejection." Blackwell Scientific Ltd., Oxford.

---

fragment pattern is matched with those for known alleles. While PCR-RFLP is highly sensitive and accurate, it is technologically complex to perform and relatively costly. PCR-RFLP has therefore largely been supplanted by PCR-SSP (PCR using sequence specific primers). In this approach, allele-specific primers are used to amplify patient DNA and the amplification products are characterized by gel electrophoresis and ethidium bromide staining (Fig. 27-9B). PCR-SSP is highly sensitive such that tests for multiple alleles can be conducted in the same sample. Moreover, accurate tissue typing results can be obtained in about 2.5 hours, making this technique suitable for those situations in which a cadaveric solid organ donor must be typed. Variations on the PCR-SSP technique have allowed clinical researchers to distinguish between closely related, highly polymorphic alleles. In addition, because the two initiating PCR primer sequences must be located on the same chromosome in order to drive any potential amplification, this method can be used to show that multiple sites of polymorphism are linked on the same chromosome, allowing determination of whole MHC haplotypes.

A disadvantage of the PCR-SSP technique is that it is too unwieldy to use if the complete screening of an entire MHC haplotype must be accomplished in a short time. In addition, PCR-SSP depends on known sequence information to produce PCR primers, meaning that new alleles may be mistyped or missed entirely. To fill this gap, some tissue typers have turned to sequencing the individual nucleotides of the amplified DNA generated by PCR-SSP. Sequence-based typing results in very precise allele determinations but requires expensive equipment and a comparatively long time to type just one individual. In addition, sequencing errors (particularly GC inversions) are not uncommon.

F. Immunosuppression

As we have learned earlier in this chapter, graft rejection can be combated by immunosuppressive drugs designed to derail the alloreactive response to graft antigens. These modern miracles of biochemistry have made transplants between unrelated individuals possible, saving the lives of countless patients. In addition, the broad panel of drugs and techniques now available for immunosuppression means that therapy can be customized to take into account the individual patient's clinical course during transplantation. In general, primary immune responses are more sensitive to immunosuppressive agents than are secondary responses, so immunosuppressive treatment should be initiated as soon as possible before the clinical manifestation of immmunological rejection symptoms. However, the reader should bear in mind that immunosuppression is not a panacea. The same immune system machinery blocked to avoid graft rejection is also responsible for dealing with pathogen infections and for tumor surveillance. Thus, an immunosuppressed patient is liable to come down with opportunistic infections or develop malignancies. In addition to these risks, the various drugs used for immunosuppression may have serious complications, including diabetes, nephrotoxicity, neurotoxicity, hepatotoxicity, thrombocytopenia, leukopenia, hyperglycemia, diarrhea, hyperuricemia, and hyperlipidemia leading to cardiovascular disease. Physicians must attempt to maintain a balance between transplant survival and the patient's tolerance of detrimental side effects. As survival rates and life spans of transplant patients increase, these quality-of-life issues become more important.

We now briefly describe the most commonly used immunosuppressive drugs, how they work, and their associated side effects.

I. AZATHIOPRINE

The anti-metabolic drug 6-mercaptopurine was one of the first compounds found to have immunosuppressive characteristics. However, because this drug could only be administered parenterally, it could not be used by transplant patients once they returned home from the hospital. A derivative of 6-mercaptopurine called azathioprine (AZA) was then developed that could be taken orally and thus outside of a hospital setting. Because it is a purine analogue, AZA disrupts both the *de novo* and salvage pathways of nucleotide synthesis (refer to Fig. 24-3). The S phase of the cell cycle is thus shut down in all dividing cells, including those lymphocytes undergoing antigen-driven proliferation. AZA's ability to block cellular proliferation makes it a powerful bone marrow suppressant. However, should this suppression be excessive, bone marrow aplasia can result. Hepatotoxicity is also a significant side effect in some cases. In addition, for unknown reasons, AZA is not effective in all individuals. As a result, when AZA was first introduced, up to 50% of recipients treated with this drug lost their grafts. Clinicians then used AZA in combination with various steroids (particularly prednisone) with great success, and this regimen became the standard treatment for immunosuppression during transplantation in the l960s. Indeed, despite intensive research, no better drug was discovered until cyclosporine A in 1972.

II. CYCLOSPORINE A

Cyclosporine A (CsA) is a compound produced by the fungus *Tolypocladium inflatum gams*. The clinical drug developed from this compound was used for transplants of all types of organs in the late 1970s and early 1980s. Indeed, successful liver and heart transplants and HCTs became possible only after the advent of CsA. CsA works by inhibiting the phosphatase calcineurin required for the dephosphorylation of the cytoplasmic component of the transcription factor NFAT. Following dephosphorylation, this component can be translocated into the nucleus, where it participates in the initiation of transcription of many interleukin genes, especially IL-2. As we have learned, IL-2 plays an important (but not exclusive) role in the activation and expansion of T cell clones. Thus, without calcineurin activity, allorecognition may occur but the intensity of the response is severely curtailed. Because T cell help is reduced, CsA also helps to prevent or reduce the production of alloantibodies that mediate AHR. The clinical side effects associated with inhibition of calcineurin can be severe and include atherosclerotic cardiovascular disease, nephrotoxicity, neurotoxicity, diabetes, excessive hair growth or loss, and swelling of the gums.

Despite the short term benefits of IL-2 inhibition in warding off graft rejection, the use of CsA (or any other agent blocking IL-2) may actually work against the long term establishment of graft tolerance for two reasons. First, as described in Chapter 15, IL-2 triggers the expression of Fas and FasL on T cell surfaces and downregulates expression of the anti-apoptotic molecule FLIP. AICD signaling is thus initiated that induces activated T cells to undergo apoptosis when their contribution is no longer needed. This means that the use of CsA ultimately protects graft-specific T cells from deletion. If use of the drug is subsequently discontinued, T cell clones capable of attacking the graft will still be present and ready for activation. Secondly, IL-2 appears to be essential for the generation and maintenance of T_{reg} cells that can shut down alloreactive T cells (see later). Without T_{reg} cells present, the remaining alloreactive T cells that escape AICD can plunge wholeheartedly into graft rejection when immunosuppression is lifted. Thus, the high dose of CsA given to prevent short term graft rejection may also block the establishment of long term graft tolerance due to impaired development of T_{reg} cells.

The importance of IL-2 for allograft tolerance has been demonstrated in studies of IL-2$^{-/-}$ mice. Immunosuppression strategies (non-cyclosporine-based) that allowed long term survival of pancreatic islet and cardiac grafts in wild-type mice were totally unsuccessful in IL-2$^{-/-}$ animals. Although T cells could still be activated in these mice in the absence of IL-2, AICD was not initiated and clones reacting to the graft were not shut down. In addition, although they were not examined in this particular study, it has been shown in other work that T_{reg} cells are absent in IL-2$^{-/-}$ mice. These findings support the hypothesis that IL-2 is required to establish natural limits for T cell proliferation and effector and memory cell generation and function during immune responses, including those directed against allografts.

III. TACROLIMUS

A more recently developed calcineurin inhibitor is _tacrolimus_ (TAC; also known as FK506), isolated originally from the bacterium _Streptomyces tsukubaensis_. Tacrolimus is considered by many to be a more effective immunosuppressive agent than CsA because it extends the life of kidney grafts to 14 years over the 8–9 years routinely experienced with CsA. It has also been used extensively in liver transplants to reduce acute rejection. Side effects of TAC are similar to those of CsA, although TAC appears to be associated more often with hair loss and less often with increased cholesterol.

IV. MYCOPHENOLATE MOFETIL

Mycophenolate mofetil (MMF) is a derivative of mycophenolic acid that can be given orally. MMF blocks T and B lymphocyte proliferation by inhibiting the enzyme inosine monophosphate dehydrogenase required for _de novo_ purine synthesis (refer to Fig. 24-3). Because lymphocytes use this pathway almost exclusively during proliferation (i.e., do not use the salvage pathway), DNA replication is blocked, lymphocytes cannot proliferate, and the immune response is shut down. Other cell types can get by with the salvage pathway, accounting for the milder side effects attributable to MMF in comparison to AZA (or CsA). In addition to its direct effects on DNA, studies of blood samples from MMF-treated animals and healthy human volunteers have shown that MMF increases lymphocyte apoptosis and blocks expression of adhesion molecules, costimulatory molecules, and cytokine receptors. MMF causes B cells to decrease antibody production _in vitro_ and inhibits humoral responses _in vivo_, decreasing the chance of AHR. When MMF treatment is combined with administration of a particular vitamin D3 derivative (which inhibits DC maturation), transplant tolerance is enhanced in conjunction with the appearance of increased numbers of T_{reg} cells. MMF has also proven to be of use in combating CGR. MMF prevents the structural changes associated with CGR of transplanted kidneys because MMF blocks the expression and glycosylation of the adhesion molecules that allow lymphocytes and monocyte/macrophages to stick inappropriately to endothelial cells and the extracellular matrix. MMF also suppresses the expression of several chemokines that promote the late infiltration of these cells into an allograft. Finally, MMF may increase graft survival because it appears to boost the effectiveness of drugs administered to combat CMV infection (a prime contributor to late graft loss).

In humans, MMF is very effective in reducing the risk (by more than 50%) of acute rejection episodes within the first year of transplantation, and also significantly decreases CGR. MMF used in combination with TAC has been highly successful in very high risk kidney and liver transplant cases. MMF is not diabetogenic, neurotoxic, or nephrotoxic, but patients may experience some nausea, vomiting, diarrhea, leucopenia, and thrombocytopenia.

V. SIROLIMUS

Sirolimus (also known as rapamycin) is an intracellular signaling inhibitor with antibiotic properties. This drug was originally derived from _Streptomyces hygroscopius_. Sirolimus binds to a cytoplasmic protein called _FK-binding protein_ (FKBP) and forms a complex that inhibits a cytosolic enzyme called "mammalian target of rapamycin signaling" (mTOR). During G1 of the cell cycle, mTOR promotes the growth and proliferation of lymphocytes induced by engagement of IL-2R. Inhibition of mTOR blocks intracellular signaling leading to the production of proteins required for cell cycle progression. Thus, unlike CsA, sirolimus blocks only proliferation induced by IL-2 signaling and not AICD. Sirolimus therefore continues to promote tolerance of the graft by allowing AICD of peripheral clones recognizing the grafts. The effects of sirolimus on T_{reg} development and function are unclear.

As well as influencing T cells, sirolimus inhibits B cell activation and antigen uptake by DCs. Sirolimus can prevent CGR because it blocks expression of IGF, PDGF, and pro-inflammatory cytokines and adhesion molecules that promote the proliferation and "stickiness" of fibroblasts and endothelial cells. Use of sirolimus in transplantation has been associated with anemia, thrombocytopenia, leukopenia, and hyperlipidemia, but not nephrotoxicity or cardiac atherosclerosis.

VI. MALONONITRILAMIDES (MNAs)

Malononitrilamides, such as *leflunomide* (LFM), are a relatively new class of immunosuppressive drug. These low molecular weight molecules inhibit the dihydroorotate dehydrogenase enzyme required for pyrimidine biosynthesis and thus block the proliferation of both T and B cells following activation. In addition, MNAs block signaling through Btk in B cells and therefore shut down alloantibody production. MNAs also affect cytokine-triggered signal transduction pathways such that the expression of E- and P-selectins by activated endothelial cells is suppressed. Furthermore, MNA-mediated inhibition of STAT4 signaling results in defects in NK maturation and activation. The MNAs have exceptional oral bioavailability, meaning that they can be swallowed without losing significant potency. In addition, LFM has an extraordinarily long half-life in the blood, being detectable for at least 2 weeks after administration. Derivatives of LFM are being developed that have shorter half-lives more compatible with variable dose adjustments. MNAs are well-tolerated by animals and humans and do not have the toxic side effects characteristic of many other immunosuppressants.

MNAs have been shown to suppress GvHD, allograft rejection, and even xenograft rejection in experimental animals. Significantly, MNAs are one of the few immunosuppressive drugs to block a phenomenon called *primary non-function* (PNF). PNF is a poorly understood type of graft failure (which is <u>not</u> graft rejection) exemplified by the way in which pancreatic β-islet cells are often destroyed immediately upon transplantation. The trauma of transplantation may lead to the activation of local macrophages, and pancreatic β-islet cells are particularly sensitive to the nitric oxide, free radicals, and cytotoxic cytokines secreted by infiltrating activated macrophages. A combination of MNAs and CsA has been shown to reduce the expression of TNF and IL-1 within grafted tissue such that PNF does not occur. Clinical trials are ongoing to assess the possible interactions of LFM with other immunosuppressive drugs such as TAC.

VII. FTY720

This still investigational drug has shown early promise as an immunosuppressant to prevent acute and chronic rejection of kidney transplants. FTY720 is a synthetic analogue of a fungal toxin called *myriocin*, produced by *Isaria sinclairii*. Myriocin, which structurally resembles sphingosine, is an inhibitor of the serine palmitoyltransferase enzyme required for the *de novo* synthesis of sphingolipids from sphingosine. When administered to rodents, FTY720 becomes phosphorylated and acts as an agonist binding to the sphingosine 1-phosphate receptor (S1P-R) expressed on T cells. Signaling initiated by engagement of S1P-R alters the homing of peripheral T cells, bringing them from the spleen and peripheral blood preferentially to the lymph nodes and PPs. The recirculation of T cells (including those bent on graft destruction) to the peripheral blood and tissues is thus reduced, and fewer alloreactive T cells infiltrate the graft. No other T cell functions appear to be impaired, such that proliferation and cytokine production are normal in T cells treated *in vitro* with this drug. In addition, unlike many other immunosuppressive drugs, FTY720 has no effect on cells of the innate immune system, meaning that patients treated with FTY720 suffer from fewer opportunistic infections than those treated with other immunosuppressive drugs. Furthermore, to date there have been no reports of side effects of FTY720 used in the liver, bone marrow, kidney, or pancreas. When FTY720 is used in combination with certain other immunosuppressants, a synergistic reduction in the graft rejection rate is observed, and side effects are not increased.

VIII. FLUDARABINE

Fludarabine is a nucleoside drug precursor (a "pro-drug") that is converted to a purine analogue (F-ara-A) upon injection into a human or animal. F-ara-A is freely able to enter cells where it accumulates in phosphorylated form (F-ara-ATP). F-ara-ATP inhibits cell proliferation primarily by inhibiting ribonucleotide reductase and DNA polymerase. F-ara-ATP incorporation into DNA blocks DNA ligase, while incorporation into RNA inhibits transcription. Interestingly, because of its similarity to the dATP required for caspase-9 activation, F-ara-ATP may also be able to induce the apoptosis of a cell it has gained access to. Fludarabine was originally developed as an anti-cancer drug for the treatment of low-grade leukemias and lymphomas (see Ch.30). With respect to transplantation, fludarabine can enhance graft acceptance when used in combination with low doses of other immunosuppressive drugs. Although quite effective when administered intravenously, fludarabine is less potent when given orally. Toxicity appears to be low but is still under investigation.

IX. ANTI-LYMPHOCYTE ANTIBODIES

Once it had become clear in the late 1960s that lymphocytes were responsible for graft rejection, researchers pounced on the idea of using specific antibodies to delete a patient's lymphocytes and forestall the recognition of foreign tissue that provoked the allograft response. In 1967, researchers in both the United States and France produced anti-lymphocyte globulin preparations that attacked all lymphocytes. A more refined approach later became possible with a preparation called thymoglobulin that was made up of polyclonal antibodies directed against T cells. However, as mentioned at the start of this chapter, the biggest breakthroughs came out of the development of mAb technology in the 1980s and the isolation of the OKT3 mAb that recognizes CD3 (originally called T3) and thus all T cells. Much experimentation ensued in which anti-CD3 antibodies were conjugated to mutated diphtheria toxin molecules to form anti-CD3 immunotoxins (refer to Ch.26). When administered to rhesus monkeys repeatedly over the week prior to transplantation, these T cell-toxic antibodies induced profound T cell depletion in the recipient's blood and lymph nodes for about a month. Apparent tolerance to the alloantigens of a mismatched kidney graft resulted and was maintained in the absence of additional immunosuppression.

Another more modern humanized mAb that has been highly useful in depleting T cells for immunosuppression during both solid organ transplantation and HCT is CAMPATH-1H. CAMPATH-1H is an anti-CD52 mAb originally developed to kill lymphoid cancer cells (see Ch.30). CAMPATH-1H binds to the GPI-anchored receptor CD52 expressed on the surfaces of most lymphocytes (alloreactive or not), setting them up for ADCC- or complement-mediated destruction. CAMPATH-1H treatment is well-tolerated and allows the use of lower, less toxic maintenance doses of other immunosuppressive drugs. Antibodies directed against CD4, CD8, CD18, or CD45RB (with or without toxin conjugation) have also been used successfully in conjunction with low levels of an immunosuppressive drug to permit organ or skin transplantation across MHC barriers, at least in the short-term.

Another approach is to use mAbs such as daclizumab and basiliximab that recognize the IL-2R expressed by activated T cells and block the clonal expansion induced by IL-2. Due to the use of techniques described in Chapters 7 and 26 that "humanize" mouse mAbs, these antibodies retain only the minimal murine sequences necessary for specific recognition of the target antigen. The bulk of the antibody is made up of human IgG sequences and thus provokes negligible immune responses to itself. These antibodies generally have long half-lives in the blood and can be administered intravenously in almost unlimited quantities every 2 weeks. Thus, these antibodies can significantly reduce graft rejection with minimal side effects. Indeed, in one study, daclizumab contributed to a reduction in the incidence of acute graft rejection from 40 to 20%.

X. CYTOKINES

We noted previously that, because of their redundancy of function, no one cytokine is essential for graft rejection responses. However, can any cytokine protect against such a response? IL-10 and TGFβ were known early on to down-regulate immune responses *in vitro*, an observation that led transplantation researchers to investigate whether cytokine administration could prevent graft rejection. Indeed, isolated organs that have been transfected with vectors expressing TGFβ or IL-10 seem to have improved survival rates. However, the administration of mammalian IL-10 to mice *in vivo* accelerated graft loss rather than inhibited it. In this particular study, it appeared that the IL-10 was promoting the generation of anti-graft CTLs. Other transplantation researchers have examined polymorphisms in the TGFβ gene that lead to high or low expression of this cytokine. One study concerned patients expressing a TGFβhi allele who underwent lung transplantation. While the elevated TGFβ present in the blood of these patients may have suppressed graft rejection responses, the fibrogenic properties of this cytokine led to an increased incidence of fibrosis in the grafts.

Although administration of a single cytokine may not be able to block graft rejection, creation of an appropriate cytokine milieu may promote graft survival. Animal studies of allografts across MiHA barriers have shown that graft rejection can be reduced and tolerance induced if the cytokine milieu is biased toward Th2 conditions. Consistent with these observations, patients bearing a polymorphism of the TNF gene that resulted in low blood levels of TNF (which is considered a Th1 cytokine) were less likely to acutely reject an allogeneic heart, kidney, or liver transplant than patients with a TNFhi gene. Investigators have concluded from these studies that it is crucial to gain early control of the production of pro-inflammatory cytokines induced by the insult of transplantation surgery. Regimens that enhance Th2 differentiation at the expense of Th1 cells are under exploration as a means of promoting the development of tolerance (see later) and prolonging graft survival.

Because certain cytokines appear to provoke graft rejection, the reader might wonder whether agents that block cytokine action might be useful immunosuppressants. Much work has been done to study the effects of anti-cytokine antibodies, anti-cytokine receptor antibodies, and soluble cytokine receptor antagonists on graft acceptance. However, contrary to expectations, many of these molecules enhance the effects of cytokines rather than suppress them. For example, in the case of anti-interleukin antibodies, the binding of the antibodies to IL-3, IL-4, IL-6, or IL-7 stabilized the cytokine proteins and prolonged their detrimental activities.

G. Induction of Graft Tolerance

We stated at the start of this chapter that the ultimate goal of transplantation researchers is to induce permanent tolerance of an allogeneic graft in a recipient in the absence of immunosuppression. Immunosuppressive drugs are helpful in the short term, but they have toxic side effects and leave the patient vulnerable to pathogen attack and tumor development. Similarly, because reagents like anti-CD3 mAbs are non-specific, all T cells in the recipient (including regulatory T cells) are affected regardless of antigenic specificity. A more refined approach is to induce antigen-specific tolerance, anergizing or deleting only those T cells that would attack the graft. To achieve such tolerance, one must find ways to delete the cells causing rejection mediated by direct allorecognition, and to regulate the cells causing rejection by indirect allorecognition.

Researchers have been studying the few examples of natural graft tolerance for clues on how to induce it clinically. Spontaneous tolerance to transplanted organs has been observed in certain patients who were weaned off immunosuppressive drugs and whose liver or kidney grafts have persisted intact for 5–30 years. It is speculated that cells from the graft may have migrated throughout the periphery of a recipient's body and mixed with the recipient's own cells, and, reciprocally, that cells from the recipient may have entered the graft. Precisely how this limited "chimerism" might induce tolerance of the graft in the recipient is not fully understood, but researchers have attempted to replicate it in transplantation situations in various ways (see later). In addition, several experimental tolerance induction protocols have originated from studies of maternal–fetal tolerance. As described in Chapter 16, the growing fetal tissue resembles an allograft because half its genes are paternal and therefore potentially foreign to the mother. However, fetal tissues express almost no MHC class I, meaning that even if maternal alloantibodies or CTLs manage to access the amniotic sac, an attack on fetal cells will not be mounted. As we have learned, several other mechanisms contribute to the maintenance of localized tolerance to fetal tissue in the uterine environment. Researchers are studying these processes in the hopes of finding a means of translating them to transplantation.

Many models of transplantation tolerance have been established in animals, including in rodents, pigs, dogs, and non-human primates, and much of the information presented in the following sections is derived from studies of these models. Unfortunately, results gleaned from one species are often not applicable to another. In particular, rodent and primate immune systems differ in important (mostly unknown) ways such that tolerance is much more easily induced in rodents. The reader is thus cautioned, as he or she contemplates the experimental results presented here, that primates require considerably more onerous protocols for tolerance induction, and human transplant recipients are likely to have even more complex requirements.

We will now address some experimental approaches by which clinical researchers have sought to encourage true tolerance of a graft by an HLA-mismatched recipient. Protocols aimed at developing allograft-specific tolerance are tested in experimental animals and judged successful if a second graft from an original donor is accepted for the long term by the treated recipient, but a graft from a third party is rejected. We describe two strategies based on manipulation of primary lymphoid organs, followed by discussion of a raft of molecular strategies designed to exploit mechanisms underlying peripheral tolerance.

I. BONE MARROW MANIPULATION

If thymocytes with the potential to recognize tissue from a given donor could be negatively selected in the recipient's thymus, central tolerance to that particular donor's tissue could be established. In experimental animals, the thymic environment required to establish such donor tolerance can be contrived by seeding the recipient's hematopoietic system with donor BM progenitors that will give rise to thymic DCs expressing donor MHC. The recipient must first undergo a regimen to partially empty the hematopoietic compartment in preparation for a BMT/HCT from the organ donor. Such treatment might include administration of a chemotherapeutic agent, or immunosuppression via anti-T cell antibody, CsA, or prednisone and is called *non-myeloablative conditioning*. While not usually considered necessary, aggressive

chemotherapy and TBI may be used to achieve *myeloablative conditioning*, so that immune system cells are virtually eliminated from the body.

After the conditioning step, and within a day or so of the solid organ transplant, the recipient is intravenously injected with a *donor cell infusion* consisting of donor BM (which naturally lacks mature effector cells) or purified donor HSCs. The idea is that the donor stem cells will travel to the recipient's BM and engraft in this location. The BM of the recipient thus becomes *chimeric*; that is, it contains cells of both recipient and donor origin. The donor stem cells make a significant contribution to hematopoiesis in the recipient, including the generation of thymic progenitors. These progenitors seed the emptied recipient thymus, giving rise to thymic DCs that express donor MHC alongside thymic DCs expressing recipient MHC. Donor MHC molecules on the thymic DCs are loaded with recipient self-peptides, so that developing thymocytes undergo negative selection following encounters with both donor and recipient pMHCs. In other words, the thymocytes become "educated" to recognize both donor and recipient MHC molecules as self. Thus, when lymphocytes in the reconstituted recipient encounter the solid organ from the same donor, they will tolerate the MHC differences of the donor cells. The chance of graft rejection is thus decreased. In theory, tolerance of the grafted tissue should be maintained as long as there are sufficient donor APCs being generated by the chimeric BM. In practice, however, this chimerism is usually only transient and the donor cells are lost. Other means of establishing tolerance must then take over if the graft is to achieve long term survival. This approach has progressed from testing in solid organ transplantation in large animal models to limited clinical trials in humans.

Several terms are used to describe the degree of chimerism of a recipient's hematopoietic compartment after transplantation. *Full chimerism* occurs in cases of BMT/HCT in which myeloablative conditioning has been used to destroy all recipient hematopoietic cells. After the transplant, 100% of hematopoietic cells in the recipient are necessarily of donor origin while all non-hematopoietic cells are of recipient origin. *Mixed chimerism* results after non-myeloablative conditioning followed by donor cell infusion since the surviving recipient and donor hematopoietic cells co-exist. In this situation, the donor cells can constitute anywhere from 1 to 100% of the total hematopoietic cell population. In monkey transplantation, the establishment of mixed chimerism, even if only temporary, is required for the induction of allograft tolerance. Donor lymphocytes may represent as little as 1.5% of the lymphocyte population in a recipient's peripheral blood, but this number can be sufficient to protect the allograft at least in the short term. *Microchimerism* refers to a situation in which donor hematopoietic cells that happen to have been carried along with a solid organ transplant spontaneously establish in the recipient. In this case, the donor hematopoietic cells represent <1% of all hematopoietic cells in the recipient. It remains controversial whether microchimerism is of much help in promoting graft tolerance.

II. THYMIC MANIPULATION

As well as manipulating the bone marrow, some transplantation researchers have tried to induce central T cell tolerance to alloantigens by introducing donor cells directly into the thymus of an animal whose mature T cells have been depleted in the periphery by irradiation or anti-lymphocyte antibody treatment. The presence of donor pMHC structures on cells in the thymus should result in deletion of thymocytes with the potential to mediate allorecognition of a graft from this same donor. In addition, some researchers believe that the presence of alloantigen in the thymus induces the generation of T_{reg} cells recognizing alloantigenic peptides presented by recipient APCs. There is some evidence that these allo-specific T_{reg} cells can subsequently migrate to the periphery and may be able to suppress rejection of grafts expressing the alloantigen, at least temporarily. (More on the role of T_{reg} cells in allograft tolerance appears later.) In rats, the thymic manipulation approach was successful in inducing tolerance to pancreatic β-islet cells transplanted from a particular donor. Moreover, after the transplant, β-islet cells introduced peripherally from a different donor were rejected, demonstrating that donor-specific tolerance had been induced. Unfortunately for humans, unlike rodents, our thymus involutes as we approach adulthood and the stage in our lives when most transplants are required. The chances of influencing human central tolerance in this way are therefore limited. The possibility of administering hormones that can reverse thymic involution (refer to Box 13-2) is under investigation.

III. COSTIMULATORY BLOCKADE

We learned in Chapter 16 that peripheral tolerance can be induced by anergizing T cells, and anergy is achieved when T cells are activated in the absence of costimulation. Artificial interference with the delivery of costimulatory signals is known as establishing a *costimulatory blockade*. The advantage of a costimulatory blockade as a therapy is that, in the long term, only T cells activated by encounter with specific antigens of the transplant are anergized. (Resting T cells are not affected, and a patient with an obvious infection would not be treated with a costimulatory blockade so that there would be no worries about anti-pathogen T cells also being anergized.) Accumulated experience in both the laboratory and the clinic suggests that the costimulatory blockade is a promising approach for the induction of permanent, or at least sustained, tolerance to grafts. There are two ways to block costimulatory pathways, both of which have been successful in increasing graft survival in various experimental transplantation settings. The first employs antibodies that specifically bind to costimulatory binding partners, and the second employs soluble forms of the negative T cell regulators CTLA-4 and PD-1.

i) Anti-costimulatory Antibodies

The binding of inhibitory antibodies directed against B7-1, CD40L (CD154), or ICOS can prevent the establishment of vital contacts between T cells and APCs needed for complete T cell activation and thus graft rejection. Both direct allorecognition (important for acute rejection) and indirect allorecognition (implicated in chronic rejection) can be blocked. Inhibition of B7 function on donor APCs blocks activation of recipient Th cells capable of direct allorecognition (Fig. 27-10). In addition, work on heart allografts in mice has shown that B7 expression on recipient APCs is crucial for the graft rejection response. When wild-type allogeneic hearts are transplanted into mice deficient for either B7-1 or B7-2, the grafted organs are not rejected and show long term survival. These types of results have encouraged researchers to develop anti-CD80 mAbs to target B7-1 during transplantation. (Particularly in large mammals, B7-1 is more important than B7-2 for costimulation in sites distant from draining lymph nodes, such as a graft would tend to be.) Similarly, anti-CD40L (anti-CD154) antibodies have been used to block the CD40-CD40L costimulatory pathway. A failure to engage CD40 on recipient DCs prevents upregulation of B7 and thus the "licensing" of the DC. This DC cannot support the activation of recipient alloreactive Tc cells (leading to a dearth of CTL effectors) and DC production of IL-12 and IFNγ is decreased. Furthermore, Th cells interacting with such DCs may be induced to differentiate into some type of regulatory T cell capable of suppressing the activities of alloreactive T cells and thus promoting tolerance (see later). A lack of CD40 signaling also appears to depress the pro-inflammatory activities of macrophages, and endothelial cells fail to upregulate the expression of adhesion molecules that draw leukocytes into sites of inflammation. The use of anti-ICOS mAbs reduces cytokine and chemokine production within an allograft, decreases the expression of chemokine receptors such as CCR5 and CXCR3 (see later), and inhibits the infiltration of T cells and macrophages into an allograft.

Does the costimulatory blockade strategy work *in vivo*? Anti-CD40L treatment of primates delays the rejection of fully mismatched kidney transplants, while anti-CD80 mAb has been used in conjunction with low dose CsA in HCT situations to reduce rejection in primates. Unfortunately, neither of these approaches on its own is sufficient to induce permanent graft tolerance. However, a combination of these therapies is synergistic in that decreases in both acute and chronic graft rejection have been observed. More promising is the anti-ICOS strategy, since allograft survival is prolonged and chronic rejection does not set in. Anti-ICOS plus administration of CsA or anti-CD40L can lead to permanent graft tolerance in rodents. Additional improvements are achieved if the anti-costimulatory antibody strategy is used in conjunction with injection of donor cells prior to transplantation of MHC-mismatched skin, heart, or pancreatic islets. It is speculated that the injection of the donor cells stimulates the recipient alloreactive T cells, which, in the absence of costimulation due to the blockade, become tolerized to the donor antigens. Alternatively, the presence of the donor cells may activate the recipient T cells to the point of AICD in a way that does not occur with the donated tissue itself.

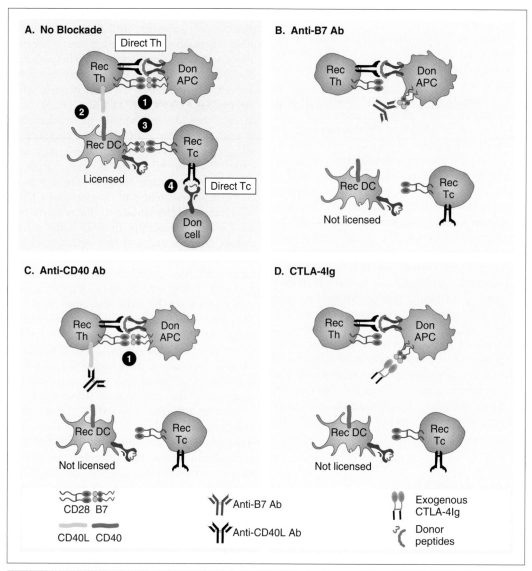

CD28 B7

CD40L CD40

Anti-B7 Ab

Anti-CD40L Ab

Exogenous
CTLA-4Ig

Donor
peptides

Figure 27-10

Example of Costimulatory Blockade to Reduce Graft Rejection
(A) During direct Th allorecognition, a recipient Th cell encounters a donor APC presenting an allo-pMHC combination
that cross-reacts with the recipient TCR. If the donor APC also expresses B7, costimulation occurs (1) that leads to
activation of the recipient Th cell. CD40L is upregulated on the activated recipient Th cell surface and binds to CD40 on a
recipient DC (2). The DC becomes "licensed" and upregulates its expression of B7, which interacts with CD28 on an
alloreactive recipient Tc cell (3), inducing costimulation. When the TCR of the Tc cell binds to allo-pMHC on a donor
target cell (4), activation is triggered by direct Tc allorecognition and graft rejection is initiated. (B) Anti-B7 Ab binds to B7
on the donor APC and blocks costimulation of the recipient Th cell. No CD40L upregulation occurs, so that the recipient
DC is not licensed and the alloreactive recipient Tc is not activated due to a lack of costimulation. (C) Anti-CD40L Ab
blocks the interaction of CD40L on an activated recipient Th cell with recipient DCs. Again, the recipient DC is not
licensed and the alloreactive recipient Tc is not activated. (D) CTLA-4Ig has the same blockade effect as anti-B7 Ab. The
unlicensed DCs in panels B, C, and D may also induce the differentiation of recipient Th cells to regulatory T cells that
suppress the alloreactive response (not shown).

ii) CTLA-4 and PD-1

Reductions in graft rejection have also been achieved experimentally by simultaneously blocking costimulation and
downregulating T cell activation with the negative regulators
CTLA-4 and PD-1. As we have learned, 3–4 days after T cell
activation, endogenous CTLA-4 displaces CD28 on the T cell

from its B7 binding partners on the APC and blocks costimulation. By introducing exogenous CTLA-4 into a transplantation situation before recipient T cell activation can occur,
the chance of transplant rejection can be decreased. In theory
at least, the TCR of an alloreactive recipient T cell may be
engaged by alloantigen but signal two is never received and

the cell is anergized (refer to Fig. 27-10). The production of IFNγ needed to stimulate expression of chemokines within the allograft that attract infiltrating immune system cells is thus inhibited. A soluble fusion protein consisting of the extracellular domain of CTLA-4 conjugated to the constant region of IgG (CTLA-4Ig) has been tested for its ability to prevent graft rejection. Transient, systemic administration of CTLA-4Ig has proved effective in some rodent transplantation models, decreasing the incidence of acute rejection and extending graft survival. Furthermore, second grafts from the same donor are accepted by animals that were previously treated with CTLA-4Ig, whereas third party grafts are rejected. More precisely targeted administration of CTLA-4Ig has also been attempted in rodents. Researchers infected liver grafts with an adenovirus vector expressing the CTLA-4Ig gene. The infected liver cells began to secrete soluble CTLA-4Ig, which inactivated any recipient antigen-specific T cells that entered the graft. The grafts showed prolonged survival in allogeneic recipients. Islet cell, small bowel, and heart transplants in rodents have also been improved by targeted CTLA-4Ig treatment, and extended survival of xenogeneic organ transplants expressing CTLA-4Ig has been documented. Unfortunately, in this type of approach, the expression of CTLA-4Ig decreases with time, and the graft is eventually rejected. Ways of manipulating CTLA-4Ig expression with the goal of inducing true and permanent graft tolerance are under investigation.

There is an interesting twist to the CTLA-4Ig story. Some researchers believe that CTLA-4Ig has an additional tolerogenic activity mediated by its effects on APCs. The engagement of B7 on a donor DC induces it to secrete IFNγ, which in turn can act in autocrine fashion on the DC to induce upregulation of an enzyme called *indeolamine 2,3 dioxygenase* (IDO). IDO is essential for the catabolism of tryptophan, an amino acid that T cells need for cell cycling. In the absence of sufficient tryptophan, T cells undergo apoptosis. Thus, treatment with CTLA-4Ig may induce a local breakdown of tryptophan in the vicinity of the donor APC, killing whatever recipient T cells have come to inspect the antigens presented by that APC. Interestingly, IDO expression is increased during pregnancy and correlates with the establishment of maternal–fetal tolerance. It remains to be determined whether natural, cell-bound CTLA-4 has the same effect on APC IDO expression as CTLA-4Ig.

In the early 2000s, researchers began investigating the potential of a second negative regulator of T cell activation, PD-1, to reduce allograft rejection in rodents. The reader will recall from Chapter 14 that PD-1 binds to two ligands, PDL1 and PDL2. An agonistic PDL1-Ig fusion protein was generated that was able to stimulate PD-1 and block the proliferation of anti-CD3-stimulated CD4$^+$ and CD8$^+$ T cells *in vitro*. When PDL1-Ig (alone) was given to wild-type mice that had received a cardiac allograft, no improvement of graft survival was seen. However, PDL1-Ig treatment greatly increased cardiac allograft survival time when given to CD28$^{-/-}$ allograft recipients, or when combined with administration of either CsA or anti-CD40L mAb. In fact, permanent engraftment was achieved in some of these cases. In another study, a combination of PDL1-Ig and anti-CD40L

mAb induced long term survival of pancreatic islet allografts in mice. These results suggest that the manipulation of PD-1 signaling can help to promote allograft tolerance *in vivo*, at least in rodents.

IV. REGULATORY T CELLS

As described in Chapter 16, CD4$^+$ and CD8$^+$ regulatory T cell subsets are thought to play an active role in suppressing immune responses, particularly those directed against autoantigens. However, evidence is also accumulating that these cells can suppress alloreactive responses in a transplantation situation. While the mechanisms underlying the differentiation and functions of regulatory T cells remain to be fully elucidated, transplantation researchers are working on experimental strategies designed to manipulate these cells in the hopes of increasing graft acceptance.

i) Regulatory CD4$^+$ T Cells (T$_{reg}$, Tr1, and Th3)

It is assumed (but not yet proven) that the same CD4$^+$CD25$^+$ T$_{reg}$ cells that act to prevent autoimmunity can also induce tolerance to allografts that, unlike pathogens but like self tissues, constitute a persistent antigen that is not cleared. In the case of a transplant, it appears that T$_{reg}$ cells activated by alloantigens act to inhibit the activation of other alloreactive T cell clones that would otherwise damage the graft. For example, in mice, it was shown using a TCR transgenic model that T$_{reg}$ cells specific for a given skin graft antigen were able to reduce the proliferation of naive alloreactive T cells and block their rejection of an established skin graft. Alloreactive T$_{reg}$ cells have also been identified in cases in which spontaneous tolerance of a liver transplant has been observed, and in mice tolerant to transplanted rat islets (a case of xenograft tolerance). T$_{reg}$ cells are present in tolerant hosts in very low numbers and are concentrated in the graft itself, although some of these cells may migrate to the periphery (principally the spleen) after a couple of months. T$_{reg}$ cells appear to exercise the indirect pathway of allorecognition, recognizing peptides of allogeneic donor cell proteins presented on self-APCs. Both CD4$^+$ and CD8$^+$ effector T cells that encounter antigen-specific T$_{reg}$ cells are induced to decrease cytokine production, downregulate expression of costimulatory and adhesion molecules, stop proliferating, and sometimes become anergic. Some of these cells may even be converted to additional T$_{reg}$ cells via infectious tolerance (see later). Both naive and memory alloreactive T cells can be controlled by T$_{reg}$ cells.

The mechanism by which T$_{reg}$ cells suppress alloreactive T cells is not yet clear. T$_{reg}$ cells constitutively express CTLA-4, and blockade of this molecule prevents suppression of allograft rejection by T$_{reg}$ cells. However, how CTLA-4 exerts its effects has yet to be defined. The TNFR family member CD30 may also play a role in the regulatory mechanism, particularly for the T$_{reg}$-mediated induction of the apoptosis of alloreactive memory CD8$^+$ T cells. Mice lacking CD30 generated T$_{reg}$ cells that were not able to block the rejection of cardiac allografts. Some researchers have postulated that, rather than acting directly on alloreactive T cells,

T_{reg} cells may act indirectly on APCs, inducing them to down-regulate costimulatory molecules so that alloreactive T cells encountering these modulated APCs are anergized rather than activated.

The reader will recall from Chapter 16 that (at least) two other types of regulatory CD4+ T cell exist: Th3 cells, which secrete large amounts of TGFβ, and Tr1 cells, which preferentially secrete IL-10 plus low amounts of TGFβ. *In vitro*, both of these subsets mediate specific inhibition of immune responses in a manner that does not depend on intercellular contacts. *In vivo*, it can be challenging to distinguish among the various regulatory T cell subsets. Several studies have shown that the administration of anti-IL-10 antibody can block the suppression of allograft rejection mediated by "regulatory CD4+ T cells." Similarly, *in vivo* administration of TGFβ can block the induction of unresponsiveness to alloantigens. Nevertheless, it has been difficult to clarify the roles of cytokines in transplantation using *in vitro* experimentation: cytokines that are important *in vivo* for regulatory effects seem to lose their relevance *in vitro*.

Regulatory CD4+ T cells (which may be T_{reg}, Tr1, or Th3 cells) are also thought to be responsible for two aspects of tolerance we addressed in Chapter 16: infectious tolerance and linked suppression. The reader will recall that "infectious tolerance" refers to the situation in which T cells tolerized to accept an allograft can seemingly "pass on" this tolerance to normal alloreactive T cells introduced into the graft site. Both naive and effector T cells can apparently be tolerized by encounters with these tolerant cells so that they also fail to initiate graft rejection. "Linked suppression" refers to the observation that tolerance can extend to antigens in close proximity on a protein, or to antigens that are present as components of the same graft. Both direct contact by T_{reg} cells and cytokines (IL-10 and/or TGFβ) secreted by Tr1 and/or Th3 cells have been implicated in linked suppression. The ability of CD4+ regulatory T cells to promote these phenomena may prove invaluable for the prevention of graft rejection in future transplant recipients. In theory, the patient could be exposed to just one of the donor MHC antigens prior to the transplant. The presence of this alloantigen would spur the development of regulatory T cells that would shut down not only recipient cells directed against this antigen but also those directed against any other closely associated MHC proteins in the graft.

ii) Regulatory CD8+ T Cells

In Chapter 16, we described a subset of regulatory CD8+ T cells called T suppressor (Ts) cells that have a surface phenotype of CD8+D57+CD28−. When a Ts cell binds to specific pMHC presented by an APC, this interaction suppresses the expression of B7 molecules by the APC and upregulates expression of the ITIM-bearing receptors ILT3 and ILT4. Antigen-specific CD4+ Th0 cells interacting with such tolerogenic APCs are anergized because costimulation is inhibited, promoting allograft tolerance. Furthermore, at least some of these anergized CD4+ Th0 cells are induced to become antigen-specific CD4+ regulatory T cells. Importantly, the suppression of allograft rejection induced by ILT3/ILT4 upregulation is alloantigen-specific. *In vivo*, heart transplant patients that do not reject their new organs show greater numbers of Ts cells in their blood and higher levels of ILT3/ILT4 expression in the fraction of blood cells containing APCs.

It has also been noted that certain donor CD8+ T cells with the marker profile CD2+CD16+CD38− have a form of "veto power" *in vivo* that allows them to block graft rejection by neutralizing recipient T cells attacking allogeneic donor tissue. This mysterious veto power does not appear to be a form of anergy as we currently understand it, and research into the underlying mechanism continues. Some researchers have postulated that it may be possible to manipulate CD8+ veto cells to protect transplant recipients against graft rejection. In theory, CD34+ HSCs isolated from a living donor (using a procedure described in Section H of this chapter) could be cultured *in vitro* with Flt3-L, SCF, and thrombopoietin, a procedure that generates substantial numbers of veto cells. The infusion of the expanded donor veto cells along with the transplanted organ into the recipient might help protect the graft from attack by recipient T cells. This hypothesis has been tested in mice and rhesus monkeys, and those animals that received an infusion of donor CD8+ veto cells following transplantation of an allograft did indeed show reduced graft rejection. For example, in one study, monkeys underwent renal transplantation followed 12 days later by an infusion of a donor T cell preparation enriched for the CD8+ veto cell population. Graft rejection was significantly reduced in these animals compared to those who received a T cell preparation that did not include the veto cell fraction. Long term (greater than 1 year) graft survival occurred in more than 50% of the monkeys that received veto cells. While anti-alloantibody production could not be stopped in these survivors, anti-donor CTL-mediated cytotoxicity was shown to be suppressed.

V. REGULATION BY NKT CELLS

The reader will recall from Chapter 18 that NKT cells are a lymphoid lineage cell that expresses a semi-invariant TCR bearing the Vα14/Jα18 chain. The NKT TCR recognizes glycolipid or lipid antigens presented on CD1d. When activated, NKT cells respond immediately (without the need for effector cell differentiation) with the production of a broad range of Th1 and Th2 cytokines, including IL-2, IL-4, IL-13, and IL-10 and IFNγ. This cytokine production by NKT cells has been implicated in the state of immune privilege (refer to Ch.16) that prevails in the anterior chamber of the mammalian eye. Unlike wild-type mice, mice lacking NKT cells are unable to suppress the inflammatory response that is initiated by the introduction of a foreign antigen into the eye. It is speculated that IL-10 secreted by activated NKT cells induces the generation of antigen-specific CD8+ regulatory T cells that suppress responses to the foreign antigen. Accordingly, corneal allografts that are readily accepted by wild-type mice are rejected by NKT-deficient mice. NKT-mediated regulation of tolerance may extend beyond immune-privileged sites, as various conditioning regimens (such as costimulatory blockade) that reduce the rejection of cardiac allografts in wild-type mice do not do so in NKT-deficient mice. In addition, NKT cells have been shown to be crucial for the acceptance of xenografts, since a regimen in which rat pancreatic β-islet cells were successfully transplanted into wild-type mouse livers failed when the recipients lacked NKT cells. Finally, NKT cells have been implicated in the inhibition of GvHD in mice. Whether these tolerogenic effects of NKT cells apply in humans as well as in rodents is under investigation.

VI. TOLEROGENIC DCs

As mentioned previously, immature DCs that interact with CD8$^+$ Ts cells such that the DCs are induced to express ILT3/ILT4 become tolerogenic. In addition, as the reader will recall from Chapter 16, DCs become tolerogenic if exposed to high levels of IL-10 and/or TGFβ. Such modulated DCs no longer produce IL-12 and anergize, rather than activate, naive T cells with which they interact. Modulated DCs can also induce the apoptosis of some of these cells, or trigger their differentiation into T$_{reg}$ cells. Scientists have attempted to exploit these effects to reduce allograft rejection. DC precursors that are cultured *in vitro* under conditions of low GM-CSF, high IL-10, or TGFβ remain immature when exposed to foreign antigens. Similarly, DC maturation can be prevented by prior exposure in culture to pharmacological agents such as aspirin, corticosteroids, or CsA that interfere with NF-κB signaling. Co-culture of such DCs with naive alloreactive T cells results in the generation of alloreactive T$_{reg}$ cells. Moreover, in the presence of moderate immunosuppression, the introduction of donor DCs tolerized *in vitro* in this way into a mouse that also receives a fully MHC-mismatched β-islet cell graft results in acceptance of the graft. Genetic engineering of DCs to enhance their tolerizing powers is discussed later in Section J.

Some researchers have pursued the tack that particular subsets of DCs are intrinsically tolerogenic. For example, murine CD8$^+$ conventional DCs express FasL and induce apoptosis of T cells, suggesting that they may contribute to the maintenance of peripheral tolerance to self antigens (refer to Ch.16). This DC subset also expresses IDO, the tryptophan-metabolizing enzyme required for T cell cycling as mentioned previously. *In vitro*, DCs that transgenically express IDO can suppress alloreactive T cell proliferation. In humans, plasmacytoid DCs expressing ILT3/ILT4 have been shown to anergize CD4$^+$ T cells *in vitro*. Confirmation of the tolerogenicity of plasmacytoid DCs *in vivo* is awaited.

Two strategies have been designed to deliver donor antigens to DCs *in vivo* without triggering the maturation of these cells. The idea is that the processing of these antigens and their presentation to recipient T cells under non-inflammatory conditions may anergize the T cells and prevent an alloreactive response. First, mAbs that recognize DC-specific markers (such as DEC-205) can be biochemically linked to a donor antigen of interest. Intravenous or subcutaneous administration of this modified mAb conveys the donor antigen to DCs

in lymph nodes and spleen and allows antigen internalization without triggering maturation. This strategy has been shown to induce the deletion of donor antigen-specific CD8$^+$ T cells resulting in tolerance in mouse model systems. Alternatively, since DCs routinely process apoptotic cells to acquire self antigens without triggering maturation, isolated apoptotic donor cells can be injected intravenously into mice to achieve the acquisition of donor antigens. Apoptotic cells carry specific molecules on their cell surfaces that minimize the production of pro-inflammatory cytokines by DCs taking up the dead cells. Administration of apoptotic donor cells to mice has been demonstrated to induce T cell non-responsiveness to the donor antigen and to improve bone marrow cell engraftment. The generation of T$_{reg}$ cells has been implicated in this effect.

VII. CAVEATS

There are several caveats to be borne in mind when considering the use of tolerance induction as a means of decreasing graft rejection. First, a protocol that completely shuts down T cell activation by anergy will also shut down AICD and may block the differentiation of regulatory T cells mediating tolerogenic effects. As mentioned previously in the discussion of CsA in transplantation, without AICD, true peripheral tolerance to the graft may never be established because the alloreactive clones will still be present. For example, T cells of transgenic mice that overexpress Bcl-xL (an anti-apoptotic member of the Bcl-2 family) do not undergo AICD, and allograft tolerance cannot be established in these animals even in the presence of a block in costimulation. Such results further support the idea that complete suppression of graft-specific T cell function is detrimental in the long run. Conversely, a level of costimulatory and regulatory blockade that reduces proliferation of alloreactive cells but does not block their AICD (or the differentiation or function of regulatory T cells) can lead to permanent allograft acceptance (at least in mice). Secondly, the relative importance of various costimulatory pathways likely varies with tissue type, the nature of the provocation to the T cells (pathogen versus autoimmunity versus graft), the disease setting, and the state of the reactive T cells. Custom-tailored costimulatory blockades may have to be designed to achieve optimum results.

A third caveat pertains to the fact that it is much harder to induce tolerance in primates than in rodents. Regimens that induce graft acceptance in rodents will thus likely have to be modified to succeed in primates and humans. For example, while the survival of an allograft for more than 100 days is considered to be evidence for tolerance induction in mice, this time span falls far short of what would be considered adequate for the survival of a primate or human transplant. More trials of tolerance induction strategies carried out in primates would help to establish optimal conditions for these species, and give insight into approaches to be tested in human recipients in clinical trials. Nevertheless, for practical and ethical reasons, strategies and results obtained with primates may not be strictly transferable to human transplantation situations.

A fourth issue is the development of standard measures of tolerance induction and loss. That is, clinical researchers need to be able to assess the strength of the recipient's immune response against a graft in such a way that the effects of various regimens on the success of human transplantation can be monitored and compared. Such data will be critical for determining when immunosuppression can be safely withdrawn from a recipient who may have developed tolerance to his or her transplant.

Finally, the safety aspects of tolerance induction regimens have yet to be fully explored. Quite unexpectedly, the use of anti-CD40L antibody and CTLA-4Ig treatments in human clinical trials has led to several episodes of unexplained thromboembolism. Moreover, the long term effects of tolerance induction regimens are unknown. Do these treatments also block chronic rejection (which may be due in part to antigen non-specific mechanisms)? Is the risk of infection or malignancy increased, and does the presence of infection or malignancy affect allograft rejection? Would a strategy that induces tolerance to donor MHC also induce tolerance to a latent or undetected pathogen carried by the donor or recipient? These questions must be resolved before tolerance induction in human transplantation becomes routine.

H. Xenotransplantation

The success of transplantation strategies over the past 15 years has led to them being considered the cure of choice for an ever-growing list of disorders. As a result, patients with failing organs that used to be condemned to certain death are now placed on transplantation waiting lists. Despite efforts to educate the public and physicians about organ donation, the number of organs available for transplantation remains stuck

at roughly one-quarter to one-third of the number of waiting patients (Table 27-7). While some countries have "opt-out" organ donation policies in place, according to which a deceased person's organs are considered available for donation unless that person has explicitly declined, many other jurisdictions have "opt-in" policies stating that a deceased person must have explicitly given prior consent for his or her organs to be harvested. Even then, if a family member objects, the would-be donor can be overruled in some cases. The organ shortage can be eased somewhat by family members who altruistically serve as living related donors, but not all patients have access to such a donor. The ethical difficulties involved with using elderly or unrelated living donors were outlined at the start of this chapter. In addition, although immunosuppression techniques are improving all the time, it is still not yet possible to take just any organ and put it in just any patient. As a result, the waiting time for a kidney transplant now stands at over 800 days in the United States, and many patients die before they receive a transplant. Similarly, 550,000 new heart failure cases are diagnosed annually in the United States. Almost 38% of these patients die within 2 years, and it is estimated that as few as 5% of those who could benefit from cardiac transplantation actually receive a donated heart each year. While progress has been made in the design of artificial hearts, these devices are not yet as functional as a human heart, and no artificial lung equivalent exists for those in need of a lung transplant. The anatomical complexities of real hearts and lungs have also made the transplantation of cardiac myoblasts or stem cells (grafted in the hopes of regenerating a healthy organ) a goal for the future. Xenotransplantation is thus a practical attempt to increase the survival of patients on the waiting list by temporarily installing an animal organ until a suitable donated human organ becomes available. Transplantation specialists also dream of the day when they can fearlessly implant

Table 27-7 Trends in Transplantation Frequency in the United States (1988–2001)*

Organ Type	Transplants Performed			Number on Waiting List	Percentage of Patients Receiving Transplants
	1988	1998	2001	2001	2001
Kidney	8,873	12,166	14,184	49,935	22
Liver	1,713	4,487	5,158	18,353	22
Heart	1,677	2,345	2,202	4,207	34
Kidney–pancreas	170	973	886	2,502	23
Lung	34	862	1053	3,773	22
Pancreas	79	248	466	1,132	29
Intestine	0	69	111	174	39
Heart–lung	73	47	27	215	11
Total number	**12,619**	**21,197**	**24,087**	**80,291**	**23**

With information from Shannon J. B. (2002) "Transplantation Sourcebook," 1st edn. Omnigraphics, Inc., Detroit.
*Total donors, both living and cadaveric.

pig-derived β-islets into diabetics, neuronal cells into Parkinson's disease patients, and liver and kidney cells into individuals suffering failures of these organs. Indeed, because the transplantation of cells raises fewer obstacles (both ethical and immunological) than the transplantation of whole animal organs, much of current xenotransplantation is focused on these strategies.

Transplantation between species has historically been the stuff of myths. For example, for the ancient Greeks, the centaur sported the upper body of a man and all but the head and neck of a horse. The Minotaur was half bull and half man, while the Harpies were savage birds with women's heads. In the mid-20th century, researchers started to experiment with organ transplants between animal species but severe vascular thrombosis (which blocks the small blood vessels feeding the transplanted organ) caused failure in the vast majority of cases. In the early 1960s, prior to the development of dialysis for the maintenance of kidney failure patients, there were also several xenotransplantation attempts in humans. However, when chimpanzee kidneys were transplanted into renal failure patients, the majority of them succumbed either to graft rejection or to pathogens allowed to run rampant by high levels of immunosuppression. As immunosuppression techniques became more effective, xenotransplantation was placed on the back burner in favor of allogeneic transplantation. However, the growing organ donor shortage in the 1980s spurred a resurgence of interest in this field. In 1984, the attention of the media and the general public was captivated by an infant named Baby Fae who received a baboon heart (see Box 27-4). The animal organ sustained Fae for only 2 weeks, but this was considered great progress because other baboon organ transplants had rapidly failed due to the very vigorous rejection reaction mounted by the human recipients.

I. CHOICE OF SPECIES FOR XENOTRANSPLANTATION

The large number of antigenic differences between members of two different species results in a much more vigorous graft rejection response than occurs between two unrelated members of the same species. Indeed, the more unrelated two species are, the faster the graft rejection. In the case of two widely divergent species (such as pigs and humans), the species combination is said to be *discordant* and immediate rejection of the xenograft occurs due to an HAR response. When the species are more closely related (such as chimpanzees and humans), the species combination is said to be *concordant*. The graft may survive for a short period but is eventually lost to acute or chronic rejection.

Assuming that one intends to fight rejection with large quantities of immunosuppressive drugs, which species should one consider for transplantation to humans? One might logically turn first to chimpanzees because of their close genetic match to humans. However, chimpanzees are an endangered species, are not readily available, and are expensive to maintain. Other primate species, such as baboons, have been ruled out in recent years because of fears that a primate pathogen

might be transferred to humans. In contrast, so-called "mini" pigs (a strain of small-sized pigs) are not endangered, are available in large numbers, can be raised in a specific-pathogen-free environment, and are easy to feed and maintain. These animals mature and breed quickly and produce large litters of piglets. Most importantly, despite their significant genetic differences, the physiology of mini-pigs and humans is surprisingly similar. Mini-pig organs are comparable in size to human organs, meaning that the transplanted organs fit well into the recipient's body cavities with minimal disruption to surrounding structures. Moreover, many pig hormones (e.g., insulin) are also active in humans. Theoretically, if the problem of HAR can be overcome (see later), mini-pigs maintained in a specific-pathogen-free environment could furnish an unlimited supply of organs for human transplantation. To further reduce the risk of rejection, isolated animal cells (rather than whole organs) may be transplanted after "packaging" in a collagen capsule that inhibits contact with the patient's immune system. Limited successes in the transplantation of isolated cells of normal pigs into humans have already been reported in the following situations: pig hepatocytes into acute liver failure patients, fetal pig brain cells into Huntington's and Parkinson's disease patients, and pig pancreas β-islet cells into diabetics.

II. XENOGRAFT REJECTION

Since pigs are physiologically the best candidates for organ transplantation to humans, most studies have been carried out on these animals. Our discussion of xenotransplantation is thus necessarily focused on the difficulties and corrective strategies involved in pig–human transplantation.

i) Types of Rejection

Hyperacute (or *immediate*) *xenograft rejection* is an HAR response mediated by human natural IgM antibodies that cross-react with galactose-α (1-3) galactose. This sugar molecule appears at the termini of glycolipids and glycoproteins expressed on the surfaces of pig cells but not human cells. Humans probably have natural antibodies recognizing this sugar because galactose-α (1-3) galactose is structurally very similar to a sugar expressed by certain human gut bacteria. In the case of a pig–human transplant, the human natural anti-Gal antibodies immediately recognize the galactose-α (1-3) galactose expressed on the pig endothelial cells and activate the recipient's complement system, destroying the graft via HAR.

*Acute vascular rejection (*also known as *delayed xenograft rejection)* results in the loss of the xenograft in a few days or weeks after transplantation. Rejection in this case can be attributed to hyperactivation of the vascular endothelium of the graft. Analysis of rejected tissue has shown that, in this type of rejection, the anti-Gal antibody binds to the endothelium but complement deposition does not necessarily occur to a meaningful extent. Instead, in a mechanism that remains unclear, the antibody induces constant activation of the vascular endothelium, which leads to elevated expression of

Box 27-4. The story of Baby Fae

Because xenotransplantation involves the mixing of animal and human tissues, its clinical use raises serious ethical issues in many people's minds. A case that graphically demonstrated both the benefits and pitfalls of xenotransplantation was that of Baby Fae, an infant who lived with a transplanted baboon heart for a short period in 1984. Baby Fae was born 3 weeks prematurely with a fatal congenital condition called "hypoplastic left heart." In this disorder, which affects 1 in 12,000 infants in the United States, the left ventricle and the aorta fail to form properly during fetal development. As a result, the main pumping machinery of the heart is absent and the infant falls critically ill and dies within 2 weeks of birth unless a heart transplant can be performed. However, pediatric heart donors are notoriously scarce, such that a new human organ is rarely found before the infant succumbs. In Baby Fae's case, matters were made worse by the fact that she was only 5 pounds in weight at the age of 2 weeks and thus required an extraordinarily small heart.

Baby Fae's cause was taken up by pediatric surgeon Leonard Bailey at Loma Linda Medical Center in California. Dr. Bailey had experimented extensively with cardiac xenotransplantation among various animal species. He transplanted the walnut-sized heart of a 7-month-old female baboon named Goobers into Baby Fae in order to give the little girl some chance of survival. Copious amounts of CsA were used in an attempt to block graft rejection. At first, the procedure appeared to be working. The baboon heart was beating normally and Baby Fae was able to breathe without supplementary oxygen and to feed via a bottle rather than by intravenous drip at about 10 days post-transplant. Unfortunately, Baby Fae's body started to reject the baboon heart at about 14 days. At 21 days post-transplant, she died of heart failure preceded by kidney failure. It was not clear why Baby Fae's kidneys failed; one of the side effects of CsA is nephrotoxicity, and the dose given to the infant to ward off graft rejection may have been dangerously high. However, the surgery itself may have been too traumatic, or graft rejection may have set in even in the presence of CsA. Despite the ultimate failure of the procedure, Baby Fae's survival for more than 2 weeks longer than any other animal heart recipient was considered a milestone in transplantation research.

On the other side of the coin, how Baby Fae's baboon heart transplant was carried out raised several serious ethical questions. Although Baby Fae's parents were required to give informed, written consent twice over a period of 24 hours, some observers questioned whether the parents really understood all that they were consenting to. Was the potential success of the baboon transplant procedure exaggerated? Were the parents told that the transplantation of a human infant heart was more likely to succeed? In response, Baby Fae's mother replied that even if a human heart and a baboon heart had both been available, she would have picked the organ that was most compatible with her daughter's body regardless of its species of origin. The availability of a human heart later became an issue in this story, because it seems that the hospital did not try to find one. To be fair, past experience had shown that a suitable infant heart was almost never around when it was needed, one of the key reasons Dr. Bailey had embarked on his experiments with animal hearts. Baby Fae was critically ill and needed intervention immediately, precluding even a short wait for an infant organ. Unbelievably, the heart of a 2-month-old infant in fact became available later in the day of Baby Fae's transplant surgery. The organ may have still been too large for Baby Fae's tiny body, and the histocompatibility testing might have taken too long, but the hospital later admitted that it was totally focused on the baboon transplant option. Another alternative that was perhaps underplayed was the Norwood procedure, a tricky method of surgically correcting the defect in Baby Fae's own heart. In the Norwood procedure, the blood vessels feeding the heart are reorganized and reattached such that the right ventricle takes over the pumping function normally done by the left ventricle. This surgery had been successful in about 40% of attempts up to that time but very few surgeons had been trained to perform it. Critics claimed that the success rate and risks of the Norwood procedure had been misrepresented to bias the parents toward proceeding with the baboon heart transplant.

Numerous experts stated in the media that the human–baboon experiment was doomed from the start because it is not possible to accurately assess histocompatibility in baboons using human reagents (although an attempt was made in this case). Baboons are more distantly related to humans than are the great apes, and close homologies are few and far between. Others pointed out that, even if MHC antigens could be matched in some way, blood group antigens certainly could not be. Indeed, preformed antibodies that recognized baboon blood antigens were present in samples of Baby Fae's serum examined after her death. Some claim that severe rejection was therefore bound to lead to the loss of the organ despite the use of immunosuppressive drugs, and that Baby Fae's death was only being prolonged by transplantation of the baboon heart. However, other clinicians argued that baboon hearts could be acceptable as stop-gap measures, keeping infant patients alive until a human heart became available. Other commentators focused on Goobers and whether it was ethical to "play God" and kill a healthy, living creature just to save a human. Many decried the "ghoulish tinkering" that resulted in a human with a simian heart. One very encouraging offshoot of the Baby Fae story was that public awareness of the shortage of organs for transplantation was radically increased, leading to a sharp expansion in the supply of donated infant tissues.

Perhaps we should end this box by thinking about what Baby Fae's mother believed in the face of this controversy. To paraphrase her statements to the media at the time: "The heart is just a pump; the seat of the soul is in the brain, and how can it be wrong to try to save a little girl's life?" Many are still wrestling with the implications raised by the Baby Fae story.

---

procoagulant proteins, cell adhesion molecules, and cytokines. Clinically, microvascular thrombosis results in *focal ischemia* (local loss of blood supply); no cellular infiltrate is observed. Nevertheless, analyses at the cellular level have implicated both direct and indirect allorecognition as mechanisms of acute xenograft rejection.

Cellular (or *chronic*) *xenograft rejection* of a pig–human transplant is due to the vigorous attack of human CTL and NK cells on the pig endothelium such that the graft is lost several weeks after transplantation. The response is one of unusual strength, perhaps because pig endothelial cells, unlike human endothelial cells, express a form of the costimulatory molecule B7-2 which can bind to CD28 on human Tc cells. In addition, human NK cells have (unknown) activatory receptors that recognize galactose-α-1,3-galactose, precipitating the death of pig cells by natural cytotoxicity.

ii) Strategies to Prevent Hyperacute Xenograft Rejection

Several approaches have been tried to prevent HAR in pig–human transplants. Even in the presence of high levels of immunosuppressive drugs, substantial amounts of natural antibodies continue to be made. The simplest therapeutic approach is to remove these antibodies by either plasma-pheresis or immunoabsorption with sugars cross-reacting with the anti-galactose-α-1,3-galactose antibodies. In other cases, the recipient receives a preparation of peptides with binding specificities similar to those of the xenoreactive IgM. The innocuous peptides then compete with the antibodies for binding to the offending sugars and prevent the activation of complement. Administration of complement activation inhibitors such as cobra venom factor and a soluble form of CR1 can also prevent HAR and extend xenograft survival for up to 6 weeks, but these strategies are rarely used because this practice leaves the transplant recipient very vulnerable to pathogens.

Another approach is to engineer the overexpression of another sugar (not recognized by human natural antibodies) that competes with galactose-α-1,3-galactose for the transferase required for glycoprotein synthesis. In theory, this sugar would replace galactose-α-1,3-galactose on the pig vascular endothelium and avoid the triggering of complement-mediated destruction. More practical, perhaps, is engineering the expression on pig cells of human complement regulatory proteins such as DAF, CD59, and MCP. As we have learned, these RCA proteins can disrupt the complement cascade and block deposition of the MAC. When transplanted into non-human primates, pig organs expressing human RCAs show improved survival (up to 2 months). A danger exists in this scheme, however. Because DAF is a receptor for echovirus and MCP is a receptor for measles virus, transgenic pig cells expressing these proteins could theoretically become infected with these viruses, perhaps ultimately transmitting them to the human transplant recipient.

In 2002, transplantation researchers were finally able to develop a genetically engineered mini-pig that does not express galactose-α-1,3-galactose on its vascular epithelium. Deletion of the gene (GGTA1) encoding the α-1,3-galactosyl-transferase enzyme responsible for the attachment of galactose-α-1,3-galactose onto proteins had been achieved much earlier in knockout mice, but several technical hurdles in the manipulation of mini-pig ES cells had to be cleared before the first knockout pig could be generated. As of 2004, it has been possible to demonstrate the survival of GGTA1$^{-/-}$ pig hearts in baboons for more than 2 months. Investigation of the use of the organs of GGTA1$^{-/-}$ mini-pigs

for the prevention of HAR in human xenotransplantation is now under way. Unfortunately, even if the problems of hyperacute and acute vascular xenograft rejection are solved, the problems posed by cellular xenograft rejection remain. These difficulties must also be overcome before the xeno-transplantation of whole mini-pig organs becomes a standard option.

III. TRANSMISSION OF ZOONOTIC DISEASES

Another concern related to the use of animal organs in humans is the possibility that animal pathogens might be transmitted to humans, an event called a *zoonosis*. In the case of transplantation of a primate organ, the risk of trans-mission of a virus or intracellular parasite that can attack human tissues is significant. The inadvertent introduction into a recipient of a known or unknown virus would be a tragedy, not just for the individual but perhaps for the human species as well. Transplant patients are particularly at risk for zoonoses because the skin barrier is deliberately breached during transplantation surgery, and any pathogen riding along unnoticed in the graft is introduced directly into the recipient's tissues. In addition, transplant patients are usually on immunosuppressive drugs, decreasing their ability to fight pathogens. Thus, in the case of pig–human transplantation, it is critically important that the pigs intended for xenotrans-plantation be born and raised in as clean and contained an environment as possible. Screening piglets for known pig pathogens such as porcine CMV or porcine lymphotropic herpesvirus can alleviate some concerns, but pig pathogens may exist for which there are no simple means of diagnosis. In addition, organisms or viruses that are harmless in a pig could turn out to be deadly in a human.

Particularly worrying among pig pathogens are the *pig endogenous retroviruses* (PERVs) present in all pig organs. The DNA of the average pig contains 50 integrated PERVs of varying degrees of replication competence. There are three major classes of PERVs, termed PERV-A, PERV-B, and PERV-C. Although all PERVs integrate into a pig cell's DNA and remain quiescent for long periods, the various classes differ with respect to their envelope protein sequences (which define their host cell range) and their tendency to undergo recombination. A virus resulting from the recombination of PERV-A and PERV-C has been found to be able to infect humans cells *in vitro* because it acquired the gene encoding the *env* receptor domain of PERV-A as well as sequences pro-moting infectivity from PERV-C. Fortunately, PERV infection of human transplant recipients has not been demonstrated *in vivo* although more than 200 patients have received a pig organ. It is also somewhat reassuring that baboons and macaques that have received pig xenografts carrying PERV do not become infected with the virus. However, it remains theoretically possible that a replication-deficient PERV travel-ing with the transplanted pig organ might recombine with another PERV to generate a replication-competent virus that could infect a human cell. This recombinant virus might then be able to recombine with one of its human counterparts, the normally quiescent HERVs (*human endogenous retroviruses*) integrated in the human genome, to create a hybrid virus that might be deadly for humans. The specter of the cross-species jump that gave rise to HIV starkly demonstrates that much caution is warranted. The PERV transmission issue is also of concern with respect to engineered pigs transgenic for human RCAs. PERVs budding from the cells of such pigs would carry human RCAs on their envelopes, and recipients of such organs might have to take anti-retroviral drugs as a precaution. In this regard, it is worrying to note that, of all the HIV RT and protease inhibitors tested, only AZT inhibited PERV replication at a drug concentration that could be achieved *in vivo*.

IV. REGULATORY AND LEGAL OBSTACLES

Aside from the physiological difficulties with xenotransplantation, significant regulatory and legal obstacles exist, at least in developed countries. The usual means of testing a drug or procedure for human safety is to try it out in non-human primates first. However, there is considerable evidence to suggest that primate immune systems may react differently than human immune systems to the presence of a pig organ, meaning that primate trial results (obtained with a considerable expenditure of time and money) could not be extrapolated to the human situation. Moreover, in the case of pig organs genetically designed for human use (say, organs expressing human RCAs), no meaningful results would be obtained because RCAs are species-specific and the human proteins would not function in a primate's circulation. Finally, diagnostic technologies related to the transplantation procedure that have been optimized for use in humans may not function properly in primates.

Other non-immunological barriers slowing research into xenotransplantation arise due to issues of concern to commercial interests involved. For example, although the widespread use of xenotransplants would generate increased revenues in the form of money spent on surgery, medical treatments, and immunosuppressive drugs, worries about liability for graft failure or inadvertent pathogen transmission, and the need to do much preparatory and follow-up work on patients, has made companies think twice about investing in such ventures. The need for collaboration may also impede commercial progress. In fact, several biotechnology companies in the developed world have each generated their own genetically engineered pigs designed to get around HAR. However, one approach alone is unlikely to be sufficient to make xenotransplantion a routine procedure, and these companies have yet to agree on a way to combine information and share possible future profits. In many Western countries, government- and foundation-sponsored research into organ xenotransplantation has also slowed, further reducing the funding available for the expensive and lengthy clinical trials that are needed. Interest in the xenotransplantation of isolated cells for treatment of Parkinson's disease and Huntington's chorea continues on a reduced scale.

V. ETHICAL AND MORAL CONSIDERATIONS

While some people have no problem with using an animal organ to sustain the life of a human, others feel it is objectionable to use animals in this way. Still others feel that sacrificing pigs for such purposes can be justified, but that the apes and monkeys are too much like humans to be so used. Many people have religious prohibitions about eating or otherwise incorporating the parts of certain animals into their bodies. Other ethical concerns arise regarding how clinical trials of xenotransplantation are conducted, with some companies attempting to recruit subjects in developing countries where few or no guidelines on xenotransplantation exist. In May 2004, the World Health Organization published a series of recommendations that may lead to the establishment of global standards for xenotransplantation.

The combination of all the preceding factors means that xenotransplantation research, while proceeding steadily, is doing so circumspectly and at different paces in different jurisdictions.

I. Blood Transfusions

One of the most common tissue transplants is blood transfusion, where components of blood donated by healthy volunteers are injected into the bloodstream of injured or diseased individuals or those undergoing surgery. Blood donations are typically purified into three "blood products": erythrocytes, platelets, and plasma. Which product a recipient receives depends on his or her particular need. For example, those lacking erythrocytes receive new ones, and those lacking clotting factors receive concentrated plasma components. In the United States in 1 year, close to 4 million people are transfused with over 26 million units of blood products. The human error that results in transfusion of incompatible blood is minimized by standard procedures but still occurs in up to 1 in 12,000 transfusions. Such errors are usually due to a clerical mistake in labeling or a rushed assessment of an unconscious emergency room patient. In addition, although we will discuss the risk of disease transmission associated with the blood supply, it should be kept in mind that blood transfusions are generally lifesavers and that patients are far more likely to succumb to complications arising from surgery or infections acquired in the hospital than to pathogens passed on in transfused blood.

I. ALLOREACTIVITY IN BLOOD TRANSFUSIONS

If antibodies specific for donor blood cells are present in a recipient's circulation, they may immediately attack transfused RBCs and induce *acute intravascular hemolysis*, the complement-mediated destruction of blood cells within the vessels. Such transfusion reactions may have systemic consequences ranging from mild to severe. In mild cases, high fevers called "febrile transfusion reactions," or chills, nausea, headache, wheezing, and skin rashes may result. In more severe cases, activation of clotting factors and fibrinolytic enzymes may proceed uncontrolled throughout small blood vessels, resulting in the tissue necrosis and bleeding that are characteristic of *disseminated intravascular coagulation* (DIC). In the most severe cases, falling blood pressure triggers vasoconstriction leading to renal failure and shock. These immunopathological consequences are fatal in 10% of cases (about 1 patient death per 500,000 units transfused in the United States). Transfusion reactions are best treated by prompt recognition of the initial mild symptoms and termination of the transfusion. Hydration of the patient is then required to maintain blood pressure and renal function. Products to encourage blood clotting may also be given to control symptomatic bleeding. Within the first few hours of receiving a mismatched transfusion, a patient must also be watched for respiratory failure. Complement-mediated destruction of endothelial cells in the lung capillaries may lead to lung injuries and pulmonary edema.

The majority of severe transfusion reactions result when blood of the incorrect ABO blood type is transfused into a recipient. The ABO blood type antigens are glycoproteins expressed predominantly on the surface of red blood cells. Every individual produces a precursor form of this glycoprotein to which terminal sugars are added by a series of enzymes (Fig. 27-11). Every individual also expresses a fucose transferase enzyme encoded by the H gene. The form of the glycoprotein that results from the addition of the fucose by the H enzyme is called the *H antigen*. Individuals of blood type A express an additional transferase enzyme from the A gene that attaches an N-acetylgalactosamine residue to the H antigen. The resulting glycoprotein is called the A antigen. In contrast, individuals of blood type B express a different transferase that attaches a galactose residue to the H antigen, creating the B antigen. Individuals of blood type AB possess both A and B transferases and produce both A and B antigens. Individuals of type O do not express either the A or B transferase and only antigen H is found on their erythrocytes (Table 27-8).

Because of the differences in terminal sugars on the A and B antigens, these proteins constitute Ti antigens in a person of the wrong blood type. (In the world of blood typing, the blood group antigens are known as *agglutinins* whereas the antibodies recognizing them are called *isohemagglutinins*.)

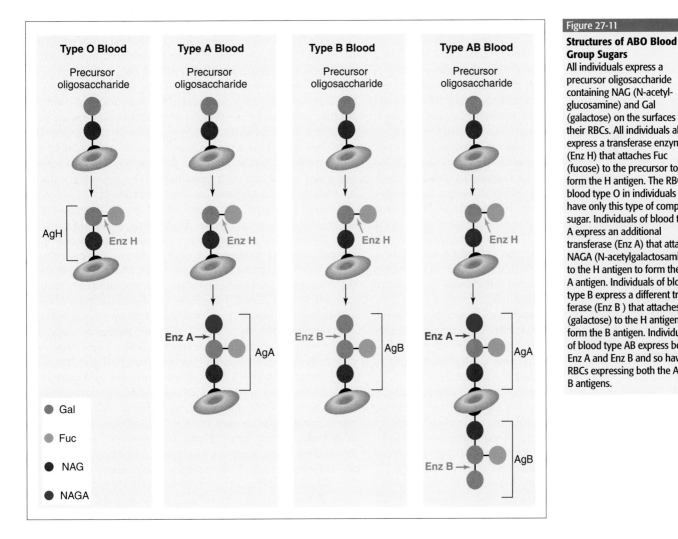

Figure 27-11

Structures of ABO Blood Group Sugars
All individuals express a precursor oligosaccharide containing NAG (N-acetyl-glucosamine) and Gal (galactose) on the surfaces of their RBCs. All individuals also express a transferase enzyme (Enz H) that attaches Fuc (fucose) to the precursor to form the H antigen. The RBCs of blood type O in individuals have only this type of complex sugar. Individuals of blood type A express an additional transferase (Enz A) that attaches NAGA (N-acetylgalactosamine) to the H antigen to form the A antigen. Individuals of blood type B express a different transferase (Enz B) that attaches Gal (galactose) to the H antigen to form the B antigen. Individuals of blood type AB express both Enz A and Enz B and so have RBCs expressing both the A and B antigens.

Table 27-8 Blood Group Antigens and Antibodies

Blood Type of Recipient	Genotype	Sugars Added by Transferases	Antigens on RBC	Anti-ABO Antibodies in Circulation	Blood Type of Compatible Donor(s)
A	AA or AO	Fucose; NAGA	A	Anti-B	A, O
B	BB or BO	Fucose; GAL	B	Anti-A	B, O
AB	AB	Fucose; NAGA; GAL	A and B	None	A, B, AB, O
O	OO	Fucose	H	Anti-A and anti-B	O

NAGA, N-acetylgalactosamine; GAL, galactose.

During the development of central tolerance, an individual of blood type A loses B cells recognizing blood antigen A (due to negative selection) but retains B cells recognizing blood antigen B. Unfortunately, certain common intestinal bacteria express glycoproteins with epitopes closely resembling the A and B antigens. Thus, long before any transfusion, the B cells in the type A individual that recognize B antigen are activated by cross-reaction with the bacterial epitopes, and anti-B antibodies are synthesized and released into the circulation. These cross-reactive antibodies are then maintained at significant levels by the ongoing presence of the bacteria. Thus, when a transfusion of blood of type B is attempted in the individual of blood type A, the preformed cross-reactive antibodies in the recipient's circulation immediately attack the transfused RBCs. Individuals of blood type O express neither A nor B antigens on their RBCs and were dubbed "universal donors" in the past because their antigen H-expressing RBCs would not be targeted by anti-A or anti-B antibodies in any recipient. However, modern protein chemistry has revealed the presence of other RBC antigens that may induce a response in an insufficiently matched recipient; the term "universal donor" has therefore been dropped. Type O individuals produce both anti-A and anti-B antibodies, meaning that they can receive blood only from other O donors. Individuals of blood type AB have RBCs that bear both A and B antigens. They used to be called "universal recipients" because neither anti-A nor anti-B antibodies are present, so the individual may receive blood from A, B, or O individuals (barring mismatch at other RBC antigens). On the other hand, AB donors may donate their blood only to other AB individuals.

We have focused on the reactions caused by recipient antibodies against incoming blood cells. What about the reaction of donor antibody against recipient blood cells? For example, if type O individuals possess antibodies to both A and B antigens, and if type O blood is transfused into an A, B, or AB recipient, one might expect these antibodies to attack recipient RBCs expressing one or the other of these antigens. If whole type O blood is transfused, this outcome is definitely a possibility. However, a modern blood transfusion usually consists of the cellular fraction of blood without the serum fraction, meaning that the antibodies have been removed.

In addition to screening for A and B blood antigens, prospective transfusion recipients are also evaluated for Rh positivity. The reader will recall from Chapter 23 that women who do not express Rh antigen may be primed against Rh antigen during pregnancy with an Rh$^+$ fetus. If this woman later carries a second Rh$^+$ fetus, a memory response of maternal anti-Rh antibodies may destroy the erythrocytes of the growing fetus, causing erythroblastosis fetalis. A severe anti-Rh Ab response may also occur in a blood transfusion situation if the recipient is Rh$^-$ but the donated blood is Rh$^+$.

There are other *minor blood group antigens* that may be mismatched between donor and recipient but are not typically assessed during routine blood screening. Antibodies to these antigens do not pre-exist in the recipient. Instead, exposure of the recipient to such alloantigens in a first transfusion results in a state called *alloimmunization* or *sensitization*, in which the recipient produces antibodies against the donated antigens. While alloimmunization may be not be clinically obvious at the time of the first transfusion, the activation of memory B cells results in the production of alloantibodies upon a subsequent blood transfusion. The recipient may then experience a delayed hemolytic reaction manifested by a drop in hemoglobin about 2 weeks after the transfusion. These reactions are much more common than the severe reactions induced by ABO or Rh mismatches but they are rarely fatal.

Several "systems" of minor blood group antigens exist for erythrocytes, including the Duffy, Kell, and Kidd systems. The *Duffy blood group system* contains five antigens that are polymorphic forms of the Duffy protein, an acidic transmembrane glycoprotein expressed by erythroid cells and the endothelial cells of capillaries and postcapillary venules (as well as by cerebellar neurons and epithelial cells of lung and kidney). In contrast, the *Kell blood group system* consists of over 20 different antigens that are primarily (but not exclusively) expressed on erythroid cells. There are two physically linked transmembrane proteins, Kell and XK, that carry between them all the Kell antigens. The Kell protein is a membrane glycoprotein with zinc endopeptidase activity. The function of XK has yet to be defined. The *Kidd blood group system* involves two antigens derived from co-dominant alleles of a single gene. These antigens vary in frequency depending on ethnic group. Functioning as a urea transporter, the transmembrane Kidd protein likely helps to maintain the osmotic stability of erythrocytes and kidney endothelial cells. Polymorphic forms of antigens expressed by platelets or granulocytes may also give rise to alloimmunization.

It should be noted that transfusion-related GvHD is also known to occur, in which donor leukocytes transferred with the blood product attack recipient tissues. Although transfusion-related GvHD causes death in 5% of cases, its occurrence is now very rare in developed countries thanks to the implementation of irradiation protocols that deplete the leukocytes from the donated blood.

II. ENSURING THE SAFETY OF THE BLOOD SUPPLY

It is not just the matching of antigens that is crucial for the safety and efficacy of blood transfusions; one must also be as sure as possible that no pathogens are transferred to the patient in the donated blood. Because of the tainted blood scandal associated with AIDS in the 1980s (refer to Box 25-1), governments in developed countries have set up elaborate measures to protect their blood supplies from specific threats such as HIV and HepC, as well as the more general danger of unknown pathogens. These measures include detailed donor screening and education programs, exclusion of high risk donors, a return to voluntary donations rather than paid donors, quarantining of donated blood until HIV antibody testing is completed, screening for increased numbers of viral markers, trace-back labeling of donations, and recipient notification when a problem has been identified. Since the full implementation of these programs in the United States, fewer than 45 people have been infected with HIV as a result of a blood transfusion, and the overall risk of HIV infection by this means is measured in cases per million units transfused. Not a single new case of HepC infection via blood transfusion has been recorded by the U.S. Centers for Disease Control (CDC) surveillance program for this pathogen since 1994. The only reason that HIV can still be transmitted via the blood supply in Western countries is that some individuals may unknowingly donate blood during their initial "window period" of HIV infection when no illness is apparent and molecular markers of viral infection such as anti-HIV antibodies and HIV antigens are at levels below current assay detection limits. Ever more sensitive assay systems are under development to address this problem, as are methods to inactivate viruses and sterilize blood products without diminishing their activity. Although artificial blood substitutes are currently under development by Western drug companies, these products are unlikely to replace more than a fraction of the volume of donated blood needed around the world.

Hemophiliacs in the United States enjoy an extra layer of surveillance in addition to the measures described previously. In 1998, the CDC set up the Universal Data Collection Project (UDC) for persons with bleeding disorders. Participating hemophiliacs undergo a free annual clinical exam and screening of their blood for HepA, HepB, HepC, and HIV. Portions of these blood samples are retained in case they are needed for future blood safety investigations. Health officials also counsel patients on ways to prevent transmission of HIV or HepC to family members. As of February 2002, there had been no new cases of HepC or HIV infection found in UDC participants.

Despite new screening precautions, it is inevitable that previously unknown pathogens will arise in the human blood supply. Like HIV, these agents may cause devastation before they are identified or before screening techniques are developed to guard against them. A case in point is West Nile virus. It was discovered in 2002 that this mosquito-borne pathogen could be transmitted via blood transfusion. Unfortunately, the current test for this virus takes 2 weeks to carry out, whereas the donated blood must be used within 5 days of collection. Faster tests are under feverish development. Malaria, Chagas disease, hepatitis G, and KSHV are other pathogens of great concern since conventional methodologies are inadequate to ensure that these pathogens are not present in donated blood. One promising possibility for the detection of difficult pathogens is *nucleic acid amplification testing* (NAT), which uses PCR to detect pathogen-specific nucleic acids. NAT tests sensitive enough to detect HIV, HepB, and HepC during the initial window period of infection are approaching commercial availability. Unfortunately, NAT testing will not aid in the early detection of protein-based pathogens such as the prion that causes variant Creutzfelt-Jakob disease (vCJD) in humans (refer to Ch.22). Some scientists suspect that vCJD may be transmitted by the consumption of beef contaminated with the prion that causes bovine spongiform encephalopathy (BSE), an outbreak of which killed over 100 people and devastated the British beef industry in the 1980s. Due to a lack of tests that can detect prions in living humans, blood supply managers have resorted to rejecting blood from persons who have spent a total of 6 months in the UK since 1986.

Even with progress toward improved screening tests, such technologies are likely to benefit only populations in developed countries. Sophisticated blood screening practices are generally out of reach in cost and feasibility for many countries in the developing world. It has been estimated that, each year, over 13 million blood donations in developing countries are not tested for hepatitis viruses or HIV. It can only be hoped that a concerted bid by the developed countries to fight AIDS globally will include resources directed at the screening of all blood supplies.

J. Hematopoietic Cell Transplantation (HCT)

Hematopoietic cell transplants are carried out when a patient has a disease involving hematopoietic cells, such as a primary immunodeficiency or a blood cell cancer, that calls for the replacement of the patient's hematopoietic system. At one time, all hematopoietic cell transplants involved the transplantation of whole bone marrow, by necessity. This procedure requires that the bone marrow of the donor be physically penetrated, entailing general surgery and the risks of anesthesia, not to mention the considerable pain involved. The advent of HCT based on the collection of HSCs from peripheral blood (or umbilical cord blood) and the ability to manipulate HSC populations *in vitro* have conferred great benefits on HCT recipients and have made life infinitely easier for HCT donors. When the stem cell donor is the patient himself or herself, the patient is said to receive an *autologous* HCT. When the stem cell donation has come from a non-syngeneic donor, the HCT is *allogeneic* because, even though the donor and recipient will have been matched at the HLA loci as much as possible, allogeneic differences will still exist for the minor histocompatibility loci.

How are donor HSCs, which are not normally present in peripheral blood, harvested for an HCT? Prior to the scheduled transplant, the donor repeatedly injects himself or herself subcutaneously with G-CSF (or sometimes GM-CSF) to stimulate the proliferation of his or her BM HSCs. These cells then "spill out" from the BM into the peripheral blood and are said to be "mobilized." The HSCs in the peripheral blood are then harvested by a procedure called *leukapheresis* (from "leukocyte" and "aphairesis," a Greek word meaning "to take away"), in which the peripheral blood passes out of the donor's body through a machine that sorts and collects the leukocytes (including the all-important HSCs). The RBCs and other blood components are returned to the donor's circulation. The leukocytes harvested in this way are then carefully frozen down until the day of the transplant. Prior to freezing or before administration, the harvested leukocytes may be manipulated so as to deplete donor T cells or enrich for the CD34$^+$ fraction that contains a high proportion of HSCs. This preparation is then administered to the patient as a donor cell infusion. Much useful information on the behavior and requirements of HSCs for HCTs has come from model systems based on the transplantation of human HSCs into mice (see Box 27-5).

I. GRAFT REJECTION IN HCT

While HCT has become the preferred treatment for hematopoietic disorders such as leukemias and lymphomas (see Ch.30), in almost 40% of cases, a suitable donor cannot be found despite the existence of BM registries to facilitate the search. HCT ideally requires HLA <u>identity</u> since, even with standard immunosuppression, transplants expressing less than identical HLA markers are often rejected. As little as a single amino acid difference in a single HLA molecule can cause HCT failure, and almost 30% of HCT with two or more differences in HLA-A, -B, and/or -C alleles fail. For this reason, HCT donors are most often found among the patient's HLA-matched siblings (refer to Box 27-1).

The reader should note that, whereas solid organ transplants "take" much better if there is ABO blood group compatibility between donor and recipient as well as HLA compatibility, the same is not true for BMT or HCT. Indeed, about one-third of allogeneic BMT and HCT are performed under conditions of ABO incompatibility. To decrease the

---

Box 27-5. Mouse models of human HSC behavior

With HCT becoming the treatment of choice for so many hematological disorders, and with the potential of mixed chimerism for improving solid organ transplantation, it is essential that we know as much as possible about how human HSCs behave during and after transplantation procedures. To generate *in vivo* models of the behavior of grafted human HSCs, scientists have turned to transplanting human HSCs into immunodeficient mice. The most popular recipients in such experiments are either *scid/*NOD mice or *bnx* mice (*bnx* mice result from the interbreeding of *beige, nude*, and *xid* mice). Human hematopoietic cells are injected into either the tail vein or peritoneum of the mouse, the animal is treated in various ways, and the fate of the human HSC within its body is monitored. This type of work has identified different subsets of HSC with reconstitutive powers, and has distinguished between cell populations that engraft easily (those CD34$^+$ cells expressing a marker called AC133) and those that do not (CD34$^+$ AC133$^-$).

Requirements for the homing of HSCs to the murine bone marrow have also been elucidated. The HSCs must express CXCR4 in order to find BM stromal cells secreting the chemokine SDF-1 and to be retained in the BM. If the HCT preparation is treated with antibodies directed against CXCR4 prior to transplantation, the graft does not "take" in the murine BM. CXCR4 expression on HSCs is upregulated if the HCT is treated with IL-6 or SCF prior to transplantation. Once the HCT is engrafted in the murine BM, treatment of the animal with G-CSF results in the fanning out of progenitor human hematopoietic cells into the peripheral blood. Work on defining the importance of adhesion molecules to human HSC engraftment has shown that BM HSCs (but, for unknown reasons, not cord blood HSCs) must associate with a stromal layer to be able to sustain hematopoiesis in the long term. HSCs cultured *ex vivo* on stromal layers engraft better than those cultured in suspension with cytokines because expression of the vital adhesion molecule VLA-4 on HSCs (upon which successful engraftment depends) is markedly decreased in HSCs cultured in suspension. Engraftment is further improved if the stromal layer produces Flt3 ligand.

Some surprising observations on cytokine involvement in HSC biology have also come out of these studies. IL-3 turns out to be a species-specific factor required to sustain hematopoiesis from human HSCs transplanted into a mouse. The generation of hematopoietic progenitors from the transplanted human HSCs could only be sustained if the stem cells were co-transplanted with human BM mesenchymal cells that had been transfected with the human IL-3 gene. Another lesson learned in studies of several other cytokines is that the *in vitro* activities established for some cytokines are often not representative of the *in vivo* situation.

Other investigations have focused on ways to increase HSC numbers prior to transplantation while preventing premature differentiation. If isolated HSCs are cultured *in vitro* with a combination of TGFβ (which blocks the cell cycle) and two bone morphogenetic proteins, resting HSC survival is improved without affecting engraftment capacity. When the HSCs are then transferred to serum-free medium lacking TGFβ, relatively rapid expansion leading to HSC differentiation ensues. These insights into how HSC behavior can be manipulated are starting to be translated into concrete clinical benefits for patients.

chance of graft rejection, the donated BM or stem cell preparation is often purged of donor RBCs prior to the transplant, removing the target antigens of recipient anti-ABO antibodies. In addition, recipient B cells producing these antibodies appear to be tolerized over time, reducing the production of fresh anti-ABO antibodies. As a result, a BMT or HCT patient can walk into the hospital with blood type "O" but leave with blood type "B." Happily for these recipients, the rate of patient survival associated with ABO-incompatible HCTs is equivalent to that of ABO-compatible HCTs.

II. GRAFT-VERSUS-HOST DISEASE (GvHD) IN HCT

As mentioned earlier in this chapter, graft-versus-host disease occurs when donor cells in the transplanted tissue attack recipient cells and tissues. Because the transplanted tissue in an HCT consists of competent immune system cells, GvHD is a significant issue for HCT recipients. HLA mismatching in HCT thus leads not only to graft rejection but also to acute and severe GvHD that can be fatal.

i) Occurrence of GvHD

Acute GvHD occurs in 75% of HCT recipients with a mismatch at a single MHC locus, and in 80% of recipients with three mismatches. Interestingly, those with disparities in MHC class II alleles are at higher risk for developing GvHD than are those with mismatched MHC class I alleles. Even among recipients completely matched at all MHC loci, about 30–50% will develop clinically significant GvHD due to MiHA mismatches.

ii) Mechanisms of GvHD

Once activated by recognition of an MHC or MiHA mismatch, alloreactive donor T cells undergo clonal expansion, and pro-inflammatory cytokines and chemokines are released that recruit donor and recipient macrophages, NK cells, and other cell types that do not discriminate between donor and recipient targets. TNF secreted by these cells is thought to be responsible for the wasting associated with some cases of GvHD. Accordingly, treatment with anti-TNF antibodies can be helpful in reducing GvHD-mediated tissue damage. However, the bulk of tissue destruction in GvHD is caused by Fas- or perforin-mediated cytotoxicity exerted by donor CTLs. We know these mechanisms are crucial for GvHD because grafts from Fas-deficient or perforin-deficient mice cannot initiate GvHD. In transplant models, treatment with neutralizing anti-FasL antibodies (in conjunction with anti-TNF antibodies) can be quite effective in reducing tissue damage due to GvHD. Helper Th1 responses directed against MiHA presented on MHC class II may also be involved in GvHD. Several studies have shown that Th1 cells secreting IFNγ and IL-2 tend to promote GvHD, while Th2 cells secreting IL-4 and IL-10 usually prevent it (although there are contradictory data). In patients, the higher the level of IL-10 in a recipient's blood prior to the HCT, the lower this recipient's chances of experiencing GvHD. Similarly, in experimental animals, grafts treated *ex vivo* with IL-10 and TGFβ prior to transplant experience lower levels of GvHD.

Unfortunately, GvHD is promoted by myeloablative conditioning, which completely wipes out the recipient's immune system and is traditionally used to prepare recipients for HCT. The tissue damage caused by the high doses of chemotherapy or irradiation used to achieve myeloablation promotes an inflammatory response and the production of cytokines (especially TNF) that both damage tissues and encourage donor T cells to attack recipient tissues. In theory, non-myeloablative conditioning, which is not as damaging to recipient tissues, should decrease these activatory signals. However, in HCT cases in which non-myeloablative conditioning has been applied (mostly hematologic malignancies), although the patients appear to recover more quickly from the transplantation procedure and show fewer side effects, there has been little reduction in GvHD.

iii) Strategies to Reduce GvHD

The blunt instrument method of preventing GvHD in HCT is to deplete the donor cell population of all mature T cells. Lymphocyte-depleting antibodies directed against certain T cell surface antigens have been used in leukemia patients to decrease GvHD. The systemic administration of a chemical inhibitor of protein kinase C has also shown promise. However, the same mature T cells that carry out GvHD are also responsible for the beneficial *graft-versus-leukemia* (GvL) effect (see later). The desire to dampen but not obliterate T cell function in the graft has led to a variety of different approaches that have some influence on GvHD. For example, lengthening the delay between non-myeloablative conditioning and donor cell infusion can reduce the incidence and severity of GvHD. The degree of mixed chimerism achieved in a graft recipient is improved and stronger responses to the alloantigens necessary for the GvL effect are mounted. Other strategies are designed to block donor T cell recognition of recipient antigens or to anergize the donor T cells. For example, infusion into a recipient mouse of a synthetic peptide known to bind to a wide range of recipient MHC class I molecules blocked the presentation of recipient MiHA to donor T cells, reducing the incidence of GvHD. Similarly, less GvHD occurred in rodents treated prior to transplant with synthetic soluble analogues of CD8 and CD4, which blocked TCR-MHC class I and TCR-MHC class II interactions, respectively.

The observation that CD28$^{-/-}$ mice experience greatly reduced GvHD has encouraged researchers to pursue anti-costimulatory approaches to combat human transplant GvHD. The graft rejection reduction agent CTLA-4Ig was tested in clinical trials of human HCT to see if it could reduce GvHD. Peripheral blood leukocytes were collected from prospective transplant recipients, irradiated, and cultured *in vitro* with donor bone marrow and CTLA-4Ig prior to transplantation of the mixture back into the recipient. The donor T cells in the transplant were anergized as a result and the incidence of GvHD was indeed reduced. Encouragingly, the responses *in vitro* of isolated recipient T cells to "third party" cells were retained.

Certain regulatory T cell subsets may also be useful for reducing GvHD. In one study, isolated mouse CD4$^+$CD25$^+$

cells were cultured in the presence of IL-2 plus an allogeneic antigen to generate an expanded and stimulated T_{reg} cell population. These T_{reg} cells were then infused back into the animal, which later received an allogeneic transplant. GvHD was decreased, suggesting that the infused T_{reg} cells were able to suppress donor effector T cells in the transplanted tissue. With respect to the human situation, peripheral blood lymphocytes have been isolated and cultured with TGFβ to expand the T_{reg} population *in vitro*. Increased numbers of T_{reg} cells result that are able to suppress immune responses *in vitro*. However, it is still unknown whether human regulatory T cells generated *in vitro* can survive in a patient after infusion, and whether these cells retain sufficient regulatory activity against allogeneic donor cells to be useful.

GvHD has been reduced in humans using a "megadose" HSC transplant strategy. A group of heavily conditioned leukemia patients received both a BMT and a peripheral blood HCT from HLA-mismatched donors that had been treated with G-CSF prior to the transplant procedure. T cells were depleted as much as possible from both the BM and the peripheral blood stem cell preparations. These mega-grafts established successfully in over 80% of the recipients despite the significant MHC barriers, and only 20% of recipients experienced GvHD. Additional studies using large doses of purified CD34+ HSC have replicated this effect. As a result, clinicians have been able to reduce immunosuppressive pre-treatments for leukemia patients, mitigating the impact of toxic side effects.

Since much of the damage of GvHD is due to effector cytokines and other molecules released by both donor and recipient immune system cells, scientists have attempted to block their production or action with various antibodies or antagonists. Modest reductions in GvHD in animals have been observed when TNF, Fas, or perforin is neutralized. Some clinical researchers have taken advantage of the observation that Th1 cytokines increase GvHD while Th2 cytokines decrease it. For example, if mobilized stem cells recovered from the peripheral blood of a mouse are exposed to a cocktail of growth factors and cytokines, multiple types of hematopoietic progenitors are generated. If these progenitors are then treated with G-CSF, IL-12 production is blocked and DCs developing in this population are biased toward secreting cytokines favoring Th2 development. Immune deviation thus occurs that results in a greater number of Th2 cells at the expense of Th1 cells. In addition, G-CSF induces the production of soluble TNF receptors and IL-1 receptor antagonist. These molecules block TNF- and IL-1-mediated inflammatory responses and reduce the induction of adhesion and costimulatory molecules on APCs. As a result, GvHD in mice receiving infusions of these G-CSF-treated progenitor cells is diminished.

A novel approach to combating GvHD is the removal of donor T cells once GvHD has started. If donor T cells are manipulated *ex vivo* so as to introduce an inducible suicide vector, administration of the trigger molecule to the recipient causes all the donor T cells in his or her body to self-destruct. Thus, the donor T cells would be allowed to carry out their anti-leukemia function (see later) but could be eliminated if they started to initiate GvHD. This strategy has been tried in leukemia patients given donor material containing T cells carrying the gene for <u>h</u>erpes <u>s</u>implex <u>v</u>irus <u>t</u>hymidine <u>k</u>inase (HSV TK). When the patients started to show signs of GvHD, they were given the antiviral drug gancyclovir. Gancyclovir is a guanosine analogue that is processed into a toxic metabolite by HSV TK. Thus, donor cells bearing HSV TK were killed by the application of gancyclovir, halting the onslaught of GvHD.

III. GRAFT-VERSUS-LEUKEMIA (GvL) EFFECT

It has been repeatedly observed that leukemia patients receiving an allogeneic HCT suffer a relapse of their disease if all the donor T cells are eliminated. It seems that, if a fraction of the donor T cells is retained in the HCT, these cells are able to attack any leukemic cells arising in the recipient after the HCT. This phenomenon is called the "graft versus leukemia" (GvL) effect. The physician must then weigh the balance of preventing GvHD against the value of the GvL effect. It is thought that GvL is usually due to responses against MiHA differences between hematopoietic cells of the graft donor and the recipient, rather than responses against specific tumor antigens. GvL does not occur in identical twins or in transplants between totally syngeneic donor–recipient pairs. Clinicians have attempted to establish selective responses to MiHA differences in a procedure called *adoptive cancer immunotherapy*. The idea is to preserve donor T cells responding to MiHA differences between donor and recipient <u>hematopoietic</u> cells (maintaining GvL) while eliminating or anergizing donor T cells responding to MiHA differences between donor and recipient <u>epithelial</u> cells (preventing GvHD). In practice, this approach has yielded only equivocal results thus far.

IV. BENEFICIAL EFFECTS OF ALLOGENEIC NK CELLS

Another way of trying to reduce GvHD while maintaining GvL takes advantage of the properties of NK cells. Most transplants are necessarily carried out with some degree of HLA mismatching between the recipient and donor. As we learned in Chapter 18, NK cells are activated by an absence of a self-MHC class I ligand that binds to their inhibitory receptors. A donor NK cell finding itself in a recipient of mismatched MHC class I might therefore be activated, since some of its inhibitory receptors might not be bound by the correct MHC class I ligand. Although one might expect activated donor NK cells to contribute to GvHD, they do not appear to mount attacks on the recipient epithelial cells whose destruction is the hallmark of GvHD. Instead, NK activity is focused on recipient hematopoietic cells. Indeed, scientists postulate that donor NK cells may actually reduce GvHD by attacking and destroying recipient APCs (Fig. 27-12). Why would the destruction of recipient APCs decrease GvHD? In a GvHD situation, donor lymphocytes exit the graft and invade the recipient tissue. Through direct allorecognition, donor Th cells become activated when they encounter recipient APCs expressing allogeneic MHC class II. If donor NK cells kill these recipient

the chance that recipient Th cells will be activated by indirect allorecognition.

The postulated beneficial effects of allogeneic NK cells on transplantation have been confirmed in mice. When donor NK cells were adoptively transferred into an allogeneic recipient mouse prior to an HCT, the risks of both GvHD and graft rejection were reduced due to the NK-mediated killing of recipient APCs and disabling of donor T cells. Indeed, prior infusion of HLA-mismatched donor NK cells has permitted the use of milder myeloablative pre-transplant conditioning regimens and has increased the survival of recipient animals in experimental transplant protocols. Taken together, these results suggest that the transfer of allogeneic NK cells may be of great benefit to leukemia patients because these cells increase GvL while simultaneously decreasing graft rejection and GvHD.

Can the NK cell-mediated GvL effect be applied to the containment or elimination of solid tumors? There is some evidence in animal models that HCT can help to clear metastatic breast cancers or prevent spontaneous sarcomas, what we might call a "graft versus tumor" (GvT) effect. For example, in one study, mice that received stimulated, allogeneic NK cells survived for longer and had fewer lung metastases of a colon adenocarcinoma than control animals. However, preliminary studies in breast cancer patients subjected to nonmyeloablative conditioning and allogeneic HCT (containing NK cells) have shown only modest signs of a GvT effect. Researchers are following up these studies with more extensive clinical trials.

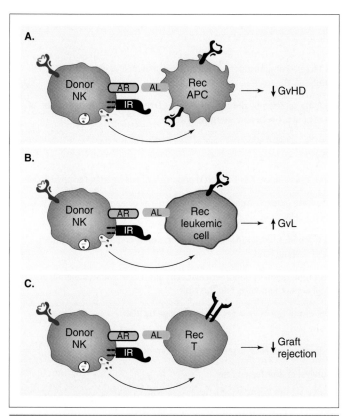

Figure 27-12

Beneficial Effects of Allogeneic Donor NK Cells
(A) GvHD is reduced because donor NK cells activated by an HLA mismatch kill recipient APCs. Donor alloreactive Th cells are not activated and cannot provide cytokine help for donor alloreactive Tc cells that would otherwise attack recipient epithelial cells due to direct allorecognition. (B) GvL is increased because donor NK cells destroy recipient leukemic cells expressing little or mismatched HLA. (C) Graft rejection is decreased because donor NK cells kill recipient T cells. In addition, the destruction of recipient APCs in (A) decreases allorecognition by recipient T cells, further reducing graft rejection.

APCs before they can activate donor Th cells, donor Th cell activation is decreased. Consequently, cytokines necessary to support the activation of donor Tc cells are not produced, and GvHD due to the action of donor CTLs against recipient epithelial cells is blocked. Fortunately, the donor NK cells also mediate a considerable GvL effect. In a study of human leukemia patients who underwent an allogeneic HCT, two sub-groups were established based on whether or not their HLA molecules were recognized by the donor's NK inhibitory receptors. Clinicians found that there was a lower risk of leukemia relapse in the sub-group whose HLA molecules did not inhibit donor NK activity. Presumably the destruction of residual leukemic cells in this sub-group was more complete due to the action of the donor NK cells. Finally, the targeting of recipient hematopoietic cells by allogeneic donor NK cells may reduce graft rejection in two ways. First, donor NK cells can directly kill recipient Th and Tc cells that express mismatched MHC. Secondly, the destruction of recipient APCs by donor NK cells reduces

V. INFECTION CONTROL

A major problem in HCT is that the immunosuppression given to reduce rejection and prevent GvHD leaves the recipient wide open to infections. Reconstitution of immune responses in HCT patients is relatively slow, such that they run a significant risk of contracting a fatal infection within the first few weeks of transplantation. It takes about 12 months for the immune system to be fully reconstituted, and if the patient can survive to this point, he or she is unlikely to succumb to infection. Researchers are looking for ways to speed up reconstitution, including the prompt cessation of G-CSF treatment so that APCs can again produce IL-12 and mount immune responses directed against pathogens. Another tactic may be to adoptively transfer donor T cells directed against specific pathogens that commonly attack transplant recipients such as *Aspergillus* and CMV.

K. Gene Therapy in Transplantation

We first encountered techniques of gene therapy in our examination of primary immunodeficiencies in Chapter 24. Transplantation researchers have taken a similar tack in trying

Table 27-9 **Potential Gene Therapy Approaches in Transplantation**

Gene Introduced	Tissue Manipulated	Rationale
TGFβ, vIL-10	Donor heart, liver, pancreatic islets	Donor cells secrete immunosuppressive cytokines that shut down recipient T cell attacking graft
CTLA-4Ig	Donor heart, islets	Soluble CTLA-4Ig secreted by donor cells blocks CD28-mediated co-stimulation of alloreactive recipient T cells and anergizes them
CD40L-Ig	Donor heart, islets	Soluble CD40-Ig secreted by donor cells blocks contact with CD40L on T cells and thus activation
FasL	Donor kidney, islets	FasL on donor cells induces Fas-mediated apoptosis of Fas-bearing recipient T cells
IL-10, TGFβ, CTLA-4Ig, FasL, TRAIL, PDL1, IDO	Donor DCs	Modified DCs have enhanced powers to induce anergy or apoptosis of alloreactive T cells, or generate T_{reg} cells
Anti-sense ICAM	Donor heart	Anti-sense nucleotides block ICAM expression on donor endothelium in graft so recipient T cells cannot bind
iNOS	Donor heart	NO prevents proliferation of donor vascular cells and T cells within the graft; also decreases leukocyte chemotaxis into the graft
Copper-zinc superoxide dismutase	Donor liver	Expression of dismutase in liver decreases damage by ROI
Catalase	Donor liver	Expression of catalase in liver degrades H_2O_2 and reduces damage by ROI
Bcl-2	Donor endothelial cells, islets	Increases survival of transfected cells
Donor MHC	Recipient fibroblasts	Recipient fibroblasts expressing donor MHC infused back into recipient prior to heart transplant anergize alloreactive recipient T cells because fibroblasts do not express co-stimulatory molecules
Donor MHC	Recipient BM cells	Developing progenitor T cells in the recipient are "educated" to see donor MHC as self
Donor MHC	Recipient DCs	Modified DCs expressing donor MHC anergize alloreactive recipient T cells

to modify gene expression in recipients and/or donated grafts in an effort to decrease graft rejection and promote true tolerance to the grafted tissue (Table 27-9). Transplantation is actually an ideal venue for gene therapy because it already calls for manipulation of tissue outside the body, and the introduction of a gene or gene regulator into isolated donor tissue can be readily accomplished.

Gene therapy has been applied in several ways in attempts to suppress recipient immune responses leading to graft rejection. For example, plasmids expressing TGFβ or vIL-10 (a viral cytokine closely resembling mammalian IL-10) have been introduced into mouse heart allografts. Similarly, isolated human hearts, livers, and pancreatic islets have been infected with an adenovirus vector expressing vIL-10. In both humans and mice, the presence of an immunosuppressive cytokine delayed the rejection of the transplanted tissue in the recipients, but only slightly.

Genetic engineering has also been used to enhance the tolerizing properties of donor DCs. The hope is that, in an allograft recipient, modified donor DCs will traffic to the lymph node and express their tolerizing products in the site where naive alloreactive T cells are being activated, reducing the need for systemic immunosuppression. For example, the survival of mouse heart and pancreatic islet allografts has been improved to a significant degree by infecting donor DCs with an adenovirus vector expressing soluble CTLA-4Ig. As described previously, CTLA-4Ig blocks the engagement of CD28 on recipient T cells by B7 and thus inhibits their activation, decreasing graft rejection. A similar adenovirus approach has been tested in rats using CD40-Ig protein to block CD40L engagement on recipient T cells by APCs within the graft. In this case, acute allograft rejection was prevented but chronic alloreactivity developed. Transfection of murine islet or kidney grafts with the FasL gene can also lead to better transplant survival due to the induced apoptosis of Fas-bearing alloreactive T cells. Tolerogenic DCs that overexpress TGFβ, PDL1, TRAIL, and IDO have also been generated.

Despite these successes, early results suggest that more than one immunosuppressive gene may have to be introduced into a graft to have a meaningful clinical effect on rejection. In addition, FasL transfection has to be carefully managed because the introduction of FasL into rodent liver grafts can lead to inflammation and neutrophil infiltration or even fatal hepatocyte apoptosis. It must also be kept in mind that the manipulation of donor DCs will not modify the indirect allorecognition response, since this depends on recipient DCs. Instead, recipient DCs can be engineered to express donor MHC class I. In one study of heart allograft rejection, treatment with a combination of anti-CD4 mAb plus donor MHC-expressing DCs resulted in improved allograft survival in mice. Experiments exploring the potential of *in vitro* modified DCs in larger mammals are under way.

Another application of gene therapy may be to introduce genes that protect donor endothelial cells (the main target of rejection responses) from damage. At the most fundamental level, it may be theoretically possible to prevent the attacking recipient T cells from making contact with graft endothelial cells by introducing "anti-sense" oligonucleotides that inhibit the expression of the important adhesion molecule ICAM-1 in the donor cells. (Anti-sense oligonucleotides use their complementarity to bind to a DNA sequence in a particular gene and block its transcription.) Indeed, chronic graft vasculopathy is decreased in transplanted rat hearts transfected with anti-sense ICAM-1. Alternatively, the arteriosclerosis that is the hallmark of chronic heart transplant rejection can be suppressed by introducing the iNOS gene into the cardiac tissue. *In vitro*, nitric oxide prevents the proliferation of vascular smooth muscle cells and T cells, and inhibits leukocyte chemotaxis. The loss of islet or hepatocyte transplants is often due to localized inflammation accompanying the transplant surgery and the accumulation of toxic oxygen and nitrogen radicals. In one study, rat liver cells were transfected with an adenovirus vector bearing a gene encoding copper-zinc superoxide dismutase (which neutralizes oxygen radicals). Expression of this gene in transplanted livers dramatically reduced tissue necrosis within the graft and prolonged its survival. Similar experiments have been carried out with the transfected catalase gene (degrades H_2O_2). Another approach to increasing graft survival has been transfection of the anti-apoptotic gene Bcl-2. Expression of Bcl-2 can protect isolated human endothelial cells from CTL attack, and has enhanced the insulin production of transplanted, transfected pancreatic islets in mice.

Bone marrow chimerism has also been duplicated using gene transfer. In this case, instead of donor and recipient MHC molecules being expressed by two different sources of BM cells, they were co-expressed on the surfaces of recipient BM cells. The experiment was conducted by introducing DNA encoding donor MHC class I antigens directly into recipient BM cells prior to transplantation. Although expression of the allo-MHC was quite low, acceptance of skin grafts expressing the same allo-MHC molecule was increased. Progeny of the transfected stem cells presumably seeded the thymus such that developing thymocytes were tolerized to the allo-MHC. Another possibility is that the expression of the allogeneic MHC class I transgene was sufficient to provoke regulatory T cell development.

Finally, a strategy in which recipient mouse fibroblasts were transfected with genes encoding donor MHC molecules has been shown to prolong heart allograft survival in mice. The engineered fibroblasts were infused back into the recipient mouse several days prior to heart transplantation. The survival of the heart allograft was improved over non-transfected controls, and non-specific immunosuppression did not occur. Presumably an encounter by the recipient's T cells with the fibroblasts bearing donor MHCs (cells that do not generally express key costimulatory molecules) anergized the attacking CTLs and thus protected the incoming heart tissue to a degree.

Challenges remain in the application of gene therapy to eliminating graft rejection. The ideal vector will be safe, have a high transfection rate, will integrate stably, will not be restricted to a particular tissue, will not be immunogenic, will not damage the infected cell, and will produce high levels of the desired protein. Overall, levels of protein currently expressed from transfected genes tend to be quite low *in vivo*. In particular, in the case where a gene is introduced into a graft *ex vivo*, it may take a day or more for significant amounts of the desired protein to be produced. Keeping the donor tissue at its peak for this interval is an issue. Indeed, just getting the gene into the right cell at the right concentration continues to be difficult. Retroviruses can be very useful vectors, but these viruses only infect replicating cells. Thus, the human HSC that one would often want to manipulate cannot be transfected easily by most members of this virus class. HSCs resist entering the cell cycle *in vitro* and appear to be impervious to the cytokines effective with other cell types. Some researchers have experimented with reducing levels of intracellular cell cycle inhibitors to stimulate HSCs to enter the cell cycle. The cycling HSCs are then transfected with a vector based on the Moloney murine leukemia virus (MoMuLV). This vector has been shown to integrate stably into the HSC genome and does not disrupt either HSC homing to the BM or HSC differentiation into both myeloid and lymphoid cells. Most importantly, the MoMuLV virus has proved safe over many years of research. An alternative to inducing the cycling of HSCs is to infect them with a member of the lentivirus subclass (retroviruses that can infect resting cells). However, the lentivirus subclass includes HIV, and few patients would serenely contemplate being transplanted with even the smallest of vectors based on sequences that could be related to HIV. Despite these concerns, the potential of gene therapy to assist transplantation efforts is huge and hopefully will be realized in the not-too-distant future.

We have come to the end of our discussion of transplantation. We move now to Chapter 28 and an exploration of allergy and hypersensitivity. The reader will find that many of the concepts learned in Chapters 26 and 27 overlap those relevant to the control of allergy and hypersensitivity and even autoimmunity (Ch.29). Indeed, it is the hope of many clinical researchers that lessons learned in one discipline will have direct application to the others.

SUMMARY

Transplantation is often the only clinical option for organ failure. However, with rare exceptions, a transplant recipient and an organ donor will not be syngeneic at the MHC. The resulting immune response mounted by the recipient against allogeneic MHC and minor histocompatibility molecules in the donor tissue invariably leads to graft rejection in the absence of some type of control measure. Histoincompatibility between a recipient and donor triggers recipient T cell activation by both direct allorecognition, in which donor pMHC complexes are recognized by cross-reacting TCRs of recipient Th and Tc cells, and indirect allorecognition, in which shed donor pMHC complexes are taken up, processed, and presented by recipient APCs to recipient Th cells. Activated recipient anti-graft Th cells provide T help for both the anti-graft Tc cells that mediate acute cellular rejection, and the anti-graft B cells whose alloantibodies mediate acute humoral rejection. Both cell-mediated and alloantibody-induced damage can result in the local production of growth factors and chemokines leading to chronic graft rejection and loss of graft function. The recipient may also experience graft versus host disease, in which donor lymphocytes in the graft attack recipient tissues and cause potentially life-threatening illness. To optimize graft and patient survival, the number of MHC mismatches may be minimized by using various tissue typing techniques to define the MHC genotypes of prospective recipients and donors in advance. ABO blood group matching of donor and recipient as well as cross-matching of recipient serum and donor leukocytes is also done to avoid hyperacute graft rejection. This type of rejection occurs when the recipient's circulation contains preformed antibodies that recognize allogeneic MHC or ABO blood group antigens. Vast improvements in long term allograft acceptance have been achieved through the routine use of immunosuppressive drugs. However, such drugs are not completely effective and have deleterious side effects. A key goal of transplantation researchers is thus to find ways of inducing specific graft tolerance in the recipient.

Selected Reading List

Akalin E. and Murphy B. (2001) Gene polymorphisms and transplantation. *Current Opinion in Immunology* 13, 572–576.

Bach J.-F. and Chatenoud L. (2001) Cellular and molecular basis of immunosuppression. In Thiru S. and Waldmann H., eds. "Pathobiology and Immunology of Transplantation and Rejection." Blackwell Scientific, Oxford.

Bagley J. and Iacomini J. (2003) Gene therapy progress and prospects: gene therapy in organ transplantation. *Gene Therapy* 10, 605–611.

Baid S., Saidman S., Tlkoff-Rubin N., Williams W., Delmonico F. *et al.* (2001) Managing the highly sensitized transplant recipient and B cell tolerance. *Current Opinion in Immunology* 13, 577–581.

Bell E. (2002) Mismatch advantages. *Nature Reviews Immunology* 2, 302–303.

Chiffoleau E., Walsh P. and Turka L. (2003) Apoptosis and transplantation tolerance. *Immunological Reviews* 193, 124–145.

Dallman M. (2001) Immunobiology of graft rejection. In Thiru S. and Waldmann H., eds. "Pathobiology and Immunology of Transplantation and Rejection." Blackwell Scientific, Oxford.

Dao M. and Nolta J. (1999) Immunodeficient mice as models of human hematopoietic stem cell engraftment. *Current Opinion in Immunology* 11, 532–537.

Delpin E. and Garcia V. (2001) The 11th report of the Latin America Transplant Registry: 62,000 transplants. *Transplantation Proceedings* 33, 1986–1988.

Editorial. (1985) The case of Baby Fae. *Journal of the American Medical Association* 254, 3358–3360.

Fiebig E. (1998) Safety of the blood supply. *Clinical Orthopaedics and Related Research* 357, 6–18.

Finger E. and Bluestone J. (2002) When ligand becomes receptor—tolerance via B7 signaling on DCs. *Nature Immunology* 3, 1056–1057.

Furukawa H. and Todo S. (2004) Evolution of immunosuppression in liver transplantation: contribution of cyclosporine. *Transplantation Proceedings* 36, 274S–284S.

Gandhi V. and Plunkett W. (2002) Cellular and clinical pharmacology of fludarabine. *Clinical Pharmacokinetics* 41, 93–103.

Ginns L., Cosimi A. and Morris P. (1999) "Immunosuppression in Transplantation." Blackwell Science, Malden, MA.

Gojo S., Cooper D., Iacomini J. and LeGuern C. (2000) Gene therapy and transplantation. *Transplantation* 69, 1995–1999.

Gonwa T. (2000) Transplantation. *American Journal of Kidney Diseases* 35, S153-S159.

Halloran P. (2004) Immunosuppressive drugs for kidney transplantation. *New England Journal of Medicine* 351, 2715–2729.

Hancock W., Wang L., Ye Q., Han R. and Lee I. (2003) Chemokines and their receptors as markers of allograft rejection and targets for immunosuppression. *Current Opinion in Immunology* 15, 479–486.

Housset D. and Malissen B. (2003) What do TCR-pMHC crystal structures teach us about MHC restriction and alloreactivity? *Trends in Immunology* 24, 429–437.

Karre K. (2002) A perfect match. *Science* 295, 2029–2031.

Kawai T., Sachs D. and Cosimi A. (1999) Tolerance to vascularized organ allografts in large-animal models. *Current Opinion in Immunology* 11, 516–520.

Knechtle S. (2000) Knowledge about transplantation tolerance gained in primates. *Current Opinion in Immunology* 12, 552–556.

Locatelli F., Rondelli D. and Burgio G. (2000) Tolerance and hematopoietic stem cell transplantation 50 years after Burnet's theory. *Experimental Hematology* 28, 479–489.

Logan J. (2000) Prospects for xenotranplantation. *Current Opinion in Immunology* 12, 563–568.

Menikoff J. (2002) The importance of being dead: non-heart-beating organ donation. *Journal of Issues in Law and Medicine* 18, 3–20.

Mollness T. E and Fiane A. E. (1999) Xenotransplantation: how to overcome the complement obstacle? *Molecular Immunology* 36, 269–276.

Morelli A. and Thomson A. (2003) Dendritic cells: regulators of alloimmunity and opportunities for tolerance induction. *Immunological Reviews* 196, 125–146.

Munson R. (2002) "Raising the Dead: Organ Transplants, Ethics, and Society." Oxford University Press, Oxford.

O'Neill J., Taylor D. and Starling R. (2004) Immunosuppression for cardiac transplantation—the past, present and future. *Transplantation Proceedings* 36, 309S-313S.

Ota K. (2001) Asian transplant registry. *Transplantation Proceedings* **33**, 1989–1992.

Perez-Simon J., Caballero D., Mateos M. and San Miguel J. (2004) Graft vs. host disease and graft vs. myeloma effect after non-myeloablative allogeneic transplantation. *Leukemia & Lymphoma* **45**, 1725–1729.

Perreault C., Roy D. C. and Claudette F. (1998) Immunodominant minor histocompatibility antigens; the major ones. *Immunology Today* **19**, 69–74.

Petersdorf E., Anasetti C., Martin P., Woolfrey A., Smith A. *et al.* (2001) Genomics of unrelated-donor hematopoietic cell transplantation. *Current Opinion in Immunology* **13**, 582–589.

Pierson III, R. (2004) Xenotransplantation at the cross-roads. *Xenotransplantation* **11**, 391–393.

Platt J., DiSesa V., Gail D. and Massicot-Fisher J. (2002) Recommendations of the National Heart, Lung, and Blood Institute Heart and Lung Xenotransplantation Working Group. *Circulation* **106**, 1043–1047.

Reiser J., Gregoire C., Darnault C., Mosser T., Guimezanes A. *et al.* (2002) A T cell receptor CDR3beta loop undergoes conformational changes of unprecedented magnitude upon binding to a peptide/MHC class I complex. *Immunity* **16**, 345–354.

Reiser J., Darnault C., Guimezanes A., Gregoire C., Mosser T. *et al.* (2000) Crystal structure of a T cell receptor bound to an allogeneic MHC molecule. *Nature Immunology* **1**, 291–297.

Reisner Y. and Martelli M. (2000) Tolerance induction by 'megadose' transplants of CD34$^+$ stem cells: a new option for leukemia patients without an HLA-matched donor. *Current Opinion in Immunology* **12**, 536–541.

Salama A., Remuzzi G., Harmon W. and Sayegh M. (2001) Challenges to achieving clinical transplantation tolerance. *Journal of Clinical Investigation* **108**, 943–948.

Salomon B. and Bluestone J. (2001) Complexities of CD28/B7: CTLA-4 costimulatory pathways in autoimmunity and transplantation. *Annual Reviews of Immunology* **19**, 225–252.

Sandrin M. and McKenzie I. (1999) Recent advances in xenotransplantation. *Current Opinion in Immunology* **11**, 527–531.

Schorlemmer H., Bartlett R. and Kurrle R. (1998) Malononitrilamides: a new strategy of immunosuppression for allo- and xenotransplantation. *Transplantation Proceedings* **30**, 884–890.

Shannon J. (2002) "Transplantation Sourcebook." Omnigraphics, Inc., Detroit, MI.

Simpson E., Roopenian D. and Goulmy E. (1998) Much ado about minor histocompatibility antigens. *Immunology Today* **19**, 108–112.

Slavin S. (2000) New strategies for bone marrow transplantation. *Current Opinion in Immunology* **12**, 542–551.

Soderhall K., Iwanaga S. and Vasta G., eds. (1996) "New Directions in Invertebrate Immunology." SOS Publications, Fair Haven, NJ.

Takeuchi Y. and Weiss R. (2000) Xenotransplantation: reappraising the risk of retroviral zoonosis. *Current Opinion in Immunology* **12**, 504–507.

Waaga A., Gasser M., Laskowski I. and Tilney N. (2000) Mechanisms of chronic rejection. *Current Opinion in Immunology* **12**, 517–521.

Waer M. (2001) Induction of xenotransplantation tolerance: a privileged role for the malononitrilamides. *Transplantation Proceedings* **33**, 2429–2430.

Waldmann H. (2001) Therapeutic approaches for transplantation. *Current Opinion in Immunology* **13**, 606–610.

Watschinger B. (1999) Indirect recognition of alloMHC peptides – potential role in human transplantation. *Nephrology Dialysis Transplantation* **14**, 8–11.

Weber M., Deng S., Olthoff K., Naji A., Barker C. *et al.* (1998) Organ transplantation in the twenty-first century. *Urologic Clinics of North America* **25**, 51–61.

Welsh K. and Bunce M. (2001) New methods in tissue typing. In Thiru S. and Waldmann H., eds. "Pathobiology and Immunology of Transplantation and Rejection." Blackwell Scientific, Oxford.

Wood K. and Sakaguchi S. (2003) Regulatory T cells in transplantation tolerance. *Nature Reviews Immunology* **3**, 199–210.

Allergy and Hypersensitivity

Charles Richet 1850–1935

Anaphylaxis
Nobel Prize 1913

© Nobelstiftelsen

CHAPTER 28

A.   HISTORICAL NOTES

B.   TYPE I HYPERSENSITIVITY: IMMEDIATE OR IgE-MEDIATED

C.   TYPE II HYPERSENSITIVITY: DIRECT Ab-MEDIATED CYTOTOXICITY

D.   TYPE III HYPERSENSITIVITY: IMMUNE COMPLEX-MEDIATED INJURY

E.   TYPE IV HYPERSENSITIVITY: DELAYED-TYPE OR CELL-MEDIATED HYPERSENSITIVITY

"In war, there are no unwounded soldiers"
—José Narosky

In general, we view the immune response as a helpful entity that functions to protect us from pathogen attack. The secondary response to a pathogen is generally so effective that we do not get sick at all, the signature manifestation of protective immunity. In this chapter, we examine what happens when the primary response is followed by a secondary response that <u>hurts</u> the individual rather than helping him or her. In this context, the primary exposure to the antigen in question has <u>sensitized</u> the individual such that a subsequent exposure results in disease rather than immunity. These immunopathological responses, whose root causes remain unknown for the most part, cause *allergies* and other *hypersensitivities*. We offer now a brief historical introduction to the subject.

A. Historical Notes

In the late 1800s, when immunology was still a relatively new science, the immune response was thought of as a flawless mechanism protecting self from nonself. Immunologists were reluctant to consider the possibility that the immune system—so finely evolved to maintain the health of the individual—could actually harm as well as help a host. Thus, in 1891, when Robert Koch found that tuberculous animals experienced tissue damage after receiving fresh injections of tubercle bacilli, he assumed the effects were due to a local excess of toxin from the pathogen. Soon after, Emil von Behring found that animals immunized with diphtheria toxin often became ill or even died after a second dose of the toxin, even if the doses were reduced to amounts that had no effect on unimmunized animals. He called this increased susceptibility "hypersensitivity" and reasoned that it was due to a direct effect of the toxin on the animal. In 1898, Charles Richet and Jules Hericourt reported a similar hypersensitivity in experimental animals upon their second exposure to eel toxin. These workers also concluded that the

toxin itself was somehow acting in a more potent fashion in the primed animals.

It was not until the first decade of the 20th century that the immune response itself began to be suspected as the cause of hypersensitivity reactions. In quick succession, three separate discoveries were made that confirmed this counterintuitive hypothesis. The first evidence emerged from experiments in physiology conducted in 1902 by Charles Richet and Paul Portier. These researchers were attempting to determine the strength of a toxin released from the tentacles of a particular marine invertebrate. After establishing a lethal dose per unit weight by injecting varying amounts of the toxin into dogs, they decided to use some of the surviving animals in further experiments. Several weeks had elapsed since the initial exposure to the toxin and these dogs now appeared normal and healthy. However, even the smallest dose of toxin caused an immediate clinical shock syndrome in these animals that led to their deaths from severe gastrointestinal and respiratory symptoms. Rather than concluding that the toxin caused this effect directly, Richet surmised that the dogs were in a hypersensitive state somehow caused by being primed immunologically to the toxin. He argued that the immune response to the priming dose of antigen had laid the groundwork for the catastrophic secondary response. Most importantly, Richet showed that this hypersensitivity could be transferred to a naive animal via the antibodies present in the serum of a primed animal.

Richet's interpretation of his findings was the first recognition that antibodies could harm as well as protect. Richet therefore called the hypersensitivity response he observed "anaphylaxis," from the Greek "ana," meaning "opposite," and "phylaxis," meaning "protection." In 1903, on the heels of the discovery of anaphylaxis, physiologist Maurice Arthus found that repeated injections of horse serum into the dermis of a rabbit resulted in neutrophil infiltration, hemorrhaging, and tissue death at the injection site—a clinical constellation later called the *Arthus reaction*. In light of the emerging view of hypersensitivity as a form of immunopathology, pediatricians Clemens von Pirquet and Bela Schick re-examined earlier reports that some patients receiving anti-diphtheria or anti-tetanus antibodies as part of a passive immunization protocol subsequently became ill with unusual local and systemic symptoms. In a 1906 publication, von Pirquet and Schick named this phenomenon "serum sickness" and were the first to demonstrate that it was due to immune responses in the immunized patients. Hypersensitivity reactions were then generally accepted as being due to immune responses and the door opened wide to their study. In 1913, in honor of his seminal work in this area, Charles Richet was awarded the Nobel Prize in Physiology or Medicine.

Around this time, von Pirquet gave the scientific community a new medical term that has since become a household word. As he analyzed the mechanisms underlying serum sickness, von Pirquet confirmed that immunity and hypersensitivity were basically due to the same responses of the immune system. In striving for linguistic clarity, von Pirquet set out to introduce a generalized term that would refer to any immunological response, no matter what its clinical outcome. To this end, he coined the term "allergy" from allos ("other") and ergon ("work"), hoping to capture the general concept of "changed reactivity" that occurs whenever the immune system is subjected to antigenic challenge. However, over the first half of the 1900s, the term "allergy" came to be exclusively associated with immune reactivity causing inflammation or tissue damage, that is, only with hypersensitivity. Simultaneously, in developed countries, there began to be a greater scientific and societal focus on the hypersensitivities that were increasingly observed in clinical practice, namely asthma, hay fever, eczema, and hives, as well as adverse reactions to insect stings and foods. As a result, by the 1950s, these clinical conditions had become virtually synonymous with "allergy" in the common parlance of patients, clinicians, and many scientists. By the same token, antigens that induced allergic responses came to be called "allergens."

On the biochemical front, progress was being made in elucidating the underlying mechanisms of some types of hypersensitivity. For example, in 1967, two independent research groups in Sweden and the United States showed that the molecule in serum that was capable of transferring sensitivity to an allergen to an unsensitized recipient was IgE. Over the next several years, Kimishige Ishizaka and Teruko Ishizaka elegantly demonstrated that, in the presence of allergen, IgE's role was to bind to mast cells and trigger their degranulation. The contents of the mast cell granules then caused the allergic symptoms.

In the late 1960s, Philip Gell and Robin Coombs published their classic book 'Clinical Aspects of Immunology,' in which they established a logical framework for classifying various hypersensitivities based on their underlying mechanisms. According to their scheme (which is still in use today), there are four types of hypersensitivity: *type I, IgE-mediated hypersensitivity; type II, direct antibody-mediated cytolytic hypersensitivity; type III, immune complex-mediated hypersensitivity; and type IV, delayed type cell-mediated hypersensitivity.* All these hypersensitivities develop in two stages: the *sensitization* stage, and the *effector* or *elicitation* stage. The sensitization stage is basically a primary immune response, while the effector stage is a secondary immune response. In this context, hypersensitivity is defined as any excessive or abnormal secondary immune response to a sensitizing agent.

Table 28-1 Outstanding Queries in Hypersensitivity

Why is one person primed and another sensitized by an encounter with a given antigen?

Why do some allergens induce localized reactions while others have systemic effects?

Why are type 1 hypersensitivity reactions mounted against inert antigens and not against pathogens?

Why do only some people respond with IgE production to an antigen?

Why do different people develop different atopic responses to the same allergen?

Why is a person allergic to only a subset of highly related antigens?

How does IgE (which does not bind to pIgR) get into body secretions?

What genes and/or alleles of genes are responsible for hypersensitivity?

Since its first delineation, the Gell–Coombs classification scheme has been reinforced by the detailed characterization of lymphocyte subsets and their roles in immunity. Critical interactions among lymphocytes and accessory cells have been defined, and the functions of an ever-increasing number of cytokines have been elucidated. However, the precise reasons why one individual may be sensitized to an antigen (and thus will experience a hypersensitivity reaction upon a secondary exposure) and another person only primed by the antigen (will mount a normal secondary immune response) remain obscure. Other outstanding queries concerning hypersensitivity are included in Table 28-1 for the reader to keep in mind as he or she digests this chapter.

We will now examine each of the four types of hypersensitivity (HS) in turn, describing the immune system mechanisms involved, the resulting clinical pathology, and the therapies available to those who experience these reactions. These features are summarized in Table 28-2. It should be noted that this classification scheme is helpful in broadly categorizing HS reactions but does not take into account the subtle shadings that may place them in more than one or none of these classes. In addition, many hypersensitivies are components of autoimmune diseases, because the HS reactions in these cases are mediated by antibodies and effector T cells directed against self antigens. Additional discussion of some of these disorders can therefore be found in Chapter 29 on 'Autoimmunity'.

B. Type I Hypersensitivity: Immediate or IgE-Mediated

I. WHAT IS TYPE I HS?

Type I hypersensitivity (HS) is what most people think of as "allergy." Hay fever, eczema, hives, and asthma are all considered type I hypersensitivities. As introduced in Chapter 5,

Table 28-2 Types of Hypersensitivity and Their Key Characteristics

Characteristic	Type I HS	Type II HS	Type III HS	Type IV HS
Common name(s)	Immediate IgE-mediated Atopy Allergy	Direct antibody-mediated	Immune complex-mediated	Delayed type Cell-mediated
Antigen	Soluble	Cell-bound	Initially soluble	Soluble
Immune system mediator	Anti-IgE antibody	IgG or IgM antibody	IgG or IgM antibody	Effector Th1 cell (T_{DTH})
Time to symptoms	<1 min–30 min	5–8 hr	4–6 hr	24–72 hr
Mechanism	Anti-IgE antibodies cross-link IgE bound on mast cells and basophils and induce degranulation	IgG/IgM antibodies cross-link cell-bound antigen; cell is destroyed by phagocytosis, complement activation, or ADCC	Immune complexes trigger complement activation and FcR aggregation of phagocytes producing inflammatory mediators	T_{DTH} cells produce IFNγ and other cytokines promoting hyperactivation of macrophages that release pro-inflammatory mediators
Examples	Asthma Hay fever Eczema Urticaria Food allergies Anaphylaxis	Drug-induced hemolytic anemia Autoimmune hemolytic anemia Goodpasture's syndrome	Arthus reaction Serum sickness Aspects of rheumatoid arthritis and SLE	Granulomatous disease Contact hypersensitivity Hypersensitivity pneumonitis

allergies occur in individuals who express IgE antibodies directed against certain common antigens in the environment. Type I HS is known as "immediate" HS because the secondary response to the allergen is generally very rapid, occurring within 30 minutes of the encounter. The antigens are typically soluble proteins that are components of larger particles such as pet dander or flower pollen (Table 28-3). Most people encountering antigens derived from bee venom or ragweed pollen produce IgM, IgG, or IgA antibodies to clear them; no ill effects on health are experienced. In those making IgE antibodies to these antigens, however, reactions are triggered that lead to side effects that can range from itching and swelling to breathing difficulties and even shock or death. It remains unknown why only some people produce IgE antibodies to allergens. Another major puzzle in type I HS is that these reactions occur in response to inert antigens and not to pathogens; that is, one may be allergic to penicillin but not to *Escherichia coli*. It is tempting to speculate that there is something about the presence of a whole pathogen that triggers the proper controls on the immune response such that the excesses of HS reactions are prevented. In the case of an inert antigen, these controls may never be activated or may be easier to derail.

Allergic diseases have been recognized as clinically distinct entities since ancient times and were described in Chinese and Greek writings of those days. Accordingly, another term for allergy is *atopy*, a word meaning "strange disease" derived from the Greek for "out of place." Atopy was the term originally used to describe the combination of asthma, eczema, and hay fever that commonly runs in families. Now, any HS characterized by elevated IgE in the serum and a positive skin test for anti-allergen IgE is considered a case of atopy. The term is often used as an adjective, as in "atopic asthma."

Table 28-3 Examples of Common Allergens

Allergen Name	Scientific Name of Source	Common Name of Source
Act c 1	*Actinidia chinensis*	Kiwi fruit
Amb a 2	*Ambrosia artemisifolia*	Ragweed
Api m 1	*Apis mellifera*	Bee venom
Ara h 2	*Arachis hypogea*	Peanuts
Asp f 1	*Aspergillus fumigatus*	Mold spores
Bet v 1	*Betula verrucosa*	Birch tree pollen
Can f 1	*Canis familiaris*	Dog dander
Der p 1	*Dermatophagoides pteronyssinus*	House dust mite
Fel d 1	*Felis domesticus*	Cat dander
Gly m 2	*Glycine max*	Soybean
Hev b 7	*Hevea brasiliensis*	Rubber tree sap (latex)
Lol p 2	*Lolium perenne*	Rye grass
Pen a 1	*Penaeus aztecus*	Shrimp
Per a 1	*Persea americana*	Avocado
Phl p 5	*Phleum pratense*	Timothy grass
Pru a 1	*Prunum avium*	Sweet cherry

Allergens are named according to the first three letters of the genus of the organism from which the antigen is derived combined with the first letter of the species and a number indicating order of discovery. For example, "Amb a 2" is the second allergen derived from *Ambrosia artemisifolia* (ragweed).

There are two types of atopic reactions, systemic and local. A systemic atopic response is called *anaphylaxis* and affects the entire body (see later). Overwhelming symptoms result that may lead to a catastrophic drop in blood pressure and sometimes death. In a local atopic reaction, the allergic symptoms depend on the anatomical location of the affected cell type and are generally confined to that organ system. For example, *asthma* is the term used to describe inflammation of the airway and lungs, and *atopic asthma* is airway inflammation resulting from IgE-mediated responses to inhaled allergens. Similarly, *atopic dermatitis* is the formal term for allergic eczema and describes an inflammatory reaction in the skin due to an IgE-mediated response. Hay fever is the better known common name for *seasonal atopic rhinitis*, meaning an IgE-mediated inflammatory reaction in the nose that occurs in response to an allergen that appears at a particular time of year. It should be noted here that a HS that has a seemingly uniform presentation may be heterogeneous in its etiology. For example, asthma is not always caused by an IgE-mediated mechanism. The 10–30% of asthmatics who have normal serum IgE are said to have *intrinsic asthma*. The classic wheezing symptoms in these cases are triggered by the effects of cold air, vigorous exercise, or respiratory viral infections on a sensitive airway.

Atopy is a disease of developed cultures. About 20% of the population of North America is considered atopic. In Britain and Australia, it has been estimated that 20% of children under 14 years of age have atopic dermatitis while 25% are asthmatics. In Swedish children, the prevalence of asthma, allergic rhinitis, and eczema doubled over 12 years in the late 1980s–1990s. Most of the morbidity and mortality associated with allergy in developed societies can be attributed to allergic rhinitis and asthma. Because there is currently no cure for atopy, billions of dollars are spent in treating symptoms in both children and adults, and the loss of workforce productivity related to atopy is considerable. In Japan, the total cost of diagnosis and treatment of allergic rhinitis has been estimated to exceed U.S. $1.5 billion. In the United States, over 5000 deaths and 500,000 hospitalizations are attributed annually to asthma attacks. The annual cost of caring for asthmatic patients in the United States alone is estimated to be over U.S. $6 billion, and the global market for asthma medication is over U.S. $5 billion per year.

II. MECHANISMS UNDERLYING TYPE I HS

i) Sensitization Stage

The sensitization stage of a type I HS reaction is initiated when an allergen penetrates a skin or mucosal barrier, is collected by an immature DC or other APC in the immediate vicinity, and is conveyed to the local lymph node (Fig. 28-1). Within the node, the now mature DC presents processed allergen to a naive Th cell, activating it. Naive allergen-specific B lymphocytes in the node are also activated when intact or fragmented allergen binds to their BCRs. At this point, the reader may well ask "But if allergens are generally innocuous, what supplies the 'danger signal' that induces DCs to mature and activate T cells?" Researchers speculate that allergens in fact are associated with

some kind of cellular stress or damage that provokes a response in everyone. However, in non-atopic individuals, this response is mediated asymptomatically by IgG antibodies. Another unsolved mystery relates to the association of atopy with immune deviation to the Th2 phenotype. Indeed, some commentators have defined allergy as a "Th2-associated reaction to a common environmental protein." It seems that local high concentrations of IL-4 and IL-6 in an atopic person's lymph node induce naive T cells to differentiate into Th2 effectors upon activation. These cells then secrete the Th2 cytokines that profoundly influence the effector actions of other immune system cells responding to the allergen.

Allergen-activated B cells and Th2 cells in the lymph node commence expression of tissue-specific homing receptors that lead them back to the target tissue into which the allergen first entered. In this site, the allergen-specific B and T cells cooperate in a primary response designed to eliminate the allergen. The activated Th2 cells produce copious amounts of IL-4, IL-5, and IL-13 that influence the isotype switching that occurs in the B cells. In atopic individuals, IgE (rather than IgG or IgA) anti-allergen antibodies are produced late in the primary response. Because the allergen is usually pretty much cleared at this point by effectors of the innate response, the IgE antibodies are free to float about and become bound to high-affinity FcεRI expressed on the surfaces of mast cells or basophils in the tissues or blood, respectively. These mast cells and basophils are then armed or sensitized (coated with allergen-specific IgE) and become bombs waiting to explode. Moreover, the binding of IgE to high-affinity FcεRI appears to deliver a survival signal to the armed mast cell or basophil. In the absence of the allergen, the armed mast cells and basophils can remain quiescent but on guard for up to 3 months.

ii) Effector Stage

If an allergen enters the body a second time during the period that the mast cells and basophils are armed, the effector phase is triggered. The allergen binds to the IgE molecules fixed on the mast cell and basophil cell surfaces and, in so doing, aggregates FcεRI molecules. Intracellular signaling is triggered that causes immediate degranulation of the cell with the release of symptom-causing histamines, proteases, and proinflammatory molecules. The rapid onset of these symptoms constitutes the *early phase reaction* of the effector stage. Leukocytes drawn to the site of allergen penetration by early phase mediators then release cytokines and other factors that stimulate inflammation in the target tissue; this later release of inflammatory mediators is called the *late phase reaction*. We will now discuss the early and late phase reactions in more detail.

iia) Early phase reaction. The early (or acute) phase reaction of type I HS is mediated primarily by the degranulation of mast cells in the target tissue (Fig. 28-2). Mast cells are abundant in the skin, in the loose connective tissue surrounding blood vessels, nerves, and glandular ducts, and in the mucosae. In the lungs, mast cells are found in the bronchial connective tissues and alveolar spaces. Mast cell trafficking is still poorly understood, but IL-8, SCF, SDF, TGFβ, RANTES, and eotaxin have all been implicated. As mentioned in Chapters 5 and 22, mast

Sensitization Stage of Type I Hypersensitivity
Allergen breaches the mucosal barrier (1) and is taken up by an immature DC (IDC) (2), which conveys it to the local lymph node (3). Naive T cells recognizing pMHC (derived from the allergen) presented by the now mature DC are activated and, in the appropriate cytokine milieu, differentiate into Th2 effectors that supply help to allergen-specific B cells. Activated B and Th2 cells travel via the thoracic duct (4) and blood and home back to the site where allergen entered the body (5). The Th2 cells supply cytokines (6) that influence differentiating plasma cells to switch to IgE production (7). IgE enters the lymphatics and recirculates through the blood (8), encountering basophils in the blood (9) and mast cells in various tissues (10). The IgE binds to FcεR on these cells, sensitizing them in readiness for the effector stage of type I hypersensitivity (see Fig. 28-2).

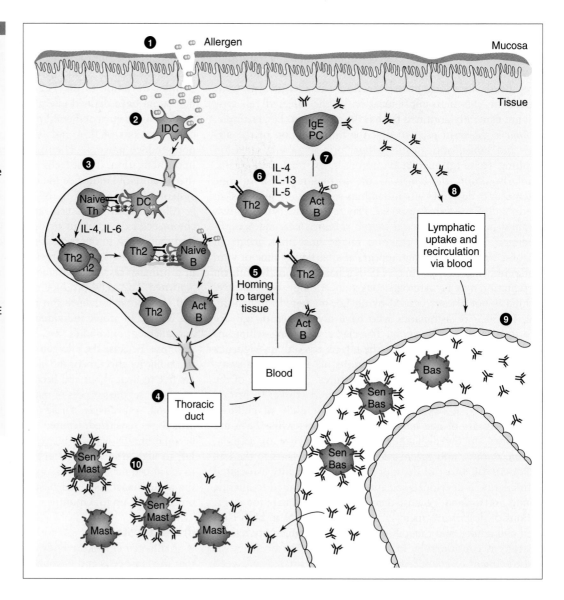

cell degranulation (whether IgE-mediated or not) most likely evolved to combat parasites, and the symptoms induced by the granule contents (coughing, sneezing, tearing of the eyes, scratching of the skin, cramping of the gut, and diarrhea) are designed to expel these types of pathogens. The primary mediator of allergic symptoms in the early phase of the effector stage is histamine, although serotonin and certain proteases are also important (Table 28-4). The fact that these fast-acting mediators are <u>preformed</u> and stored in mast cell granules accounts for the "immediate" nature of type I HS responses. After degranulation, the mast cell starts to break down. Phospholipase A in the plasma membrane of the deteriorating mast cell generates PAF and arachadonic acid from lipid molecules in the membrane. As mentioned in Chapter 4, arachadonic acid is digested by lipoxygenase to produce leukotrienes, and by cyclooxygenase to produce prostaglandins. These secondary mediators sustain the allergic response for several hours. Histamine and PAF bind to their specific receptors on the smooth muscle cells supporting the blood vessels and induce them to relax, expand-

ing the diameter of the blood vessel lumen (vasodilation) and increasing blood flow to the local area. Simultaneously, histamine and leukotrienes induce the contraction of the endothelial cells lining the blood vessels, creating opportunities for plasma proteins and cells in the blood to leak out of the circulation into the tissues (increased tissue permeability). As a result, fibrinogen leaking from the blood into the tissues induces thrombin activation, which causes local deposition of fibrin and tissue swelling. The action of histamine on sensory nerves causes the itching of eczema and hives and the sneezing of allergic rhinitis. Histamine also induces the increased mucus secretion in the bronchioles that is characteristic of asthma. At the molecular level, histamine has been implicated in the increased expression of adhesion molecules such as ICAM-1, LFA-1, and E-selectin in allergic individuals and the production of IL-8 and IL-6 by endothelial cells.

The events described previously are the common underpinnings for all type I HS reactions. The symptoms seen vary widely because the mediators leaking out of the circulation

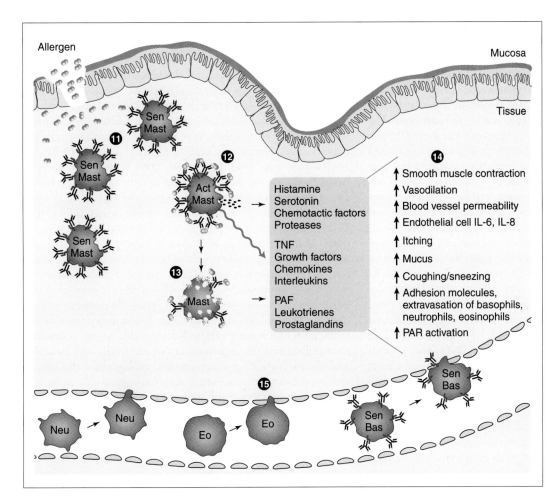

Early Phase Reaction of Effector Stage of Type I Hypersensitivity
Continuing on from Figure 28-1, sensitized mast cells in a tissue that encounter allergen (11) are induced to degranulate by the cross-linking of their FcεR (12), releasing histamine and other mediators. In addition, the activated mast cells secrete multiple cytokines, including TNF. A spent mast cell then breaks down (13), and enzymatic digestion of its membrane components generates mediators such as PAF. Together, all these molecules combine to produce the indicated physiological signs of an allergic response (14). Chemotactic factors released during mast cell activation encourage the extravasation of additional immune system cells from the circulation (15), including eosinophils and sensitized basophils. PAR, protease-activated receptor.

Table 28-4 **Major Mediators Contributing to Type I HS**

Mediator	Cellular Source	Promotes
Histamine	Mast cell and basophil granules	Vasodilation; increased vessel permeability; bronchial smooth muscle contraction; mucus production; sensory nerve stimulation (itching, sneezing)
Chemotactic factors	Mast cell and basophil granules	Chemotaxis of eosinophils and neutrophils
Proteases	Mast cell and basophil granules	Mucus production; basement membrane digestion; activation of receptors that facilitate cytokine secretion; increased blood pressure
Serotonin	Mast cell and basophil granules	Vasodilation; bronchial smooth muscle contraction
Pro-inflammatory cytokines and growth factors	Mast cells, basophils	Mobilization and activation of immune system cells; initiation of inflammatory response
Major basic protein	Eosinophil granules	Mast cell degranulation; smooth muscle contraction; death of respiratory epithelial cells
Platelet activating factor (PAF)	Mast cell membranes	Platelet aggregation and degranulation; pulmonary smooth muscle contraction
Prostaglandins	Mast cell membranes	Platelet aggregation; pulmonary smooth muscle contraction
Leukotrienes	Eosinophil granules, mast cell membranes	Vasodilation; increased vessel permeability; bronchial smooth muscle contraction; mucus production
Eosinophil cationic protein	Eosinophil granules	Death of respiratory epithelial cells
Eosinophil-derived neurotoxin	Eosinophil granules	Ribonuclease activity

affect different cell types in different locations. In addition, different tissues contain different resident mast cell subsets that can be distinguished by the granule proteases they release. Some mast cell granules contain only tryptase (MC_T cells), while others synthesize both tryptase and chymase (MC_{TC} cells). MC_T cells are the dominant subset in the airway and small intestine mucosa, while MC_{TC} cells predominate in the skin and the sub-epithelial layers of the small intestine. Tryptase potentiates histamine-induced smooth muscle contraction, and both tryptase and chymase activate proteins involved in tissue destruction or reorganization (*tissue remodeling*). Mast cell proteases also cleave and consequently activate certain *protease-activated receptors* (PARs) on the surfaces of eosinophils and epithelial and endothelial cells. Activation of these PARs leads in an unknown way to enhanced expression of cytokines and adhesion molecules that promote the selective recruitment of basophils and eosinophils. In contrast, cleavage of PARs expressed by neurons stimulates the release of neuropeptides related to vasodilation. In addition to mediators contained in their granules, mast cells secrete a very large number of cytokines (IL-1, -2, -3, -4, -5, -6, -13, -16, TNF, and LTα), chemokines (IL-8, MIP-1α, exotaxin, MCP-1), and growth factors (GM-CSF, PDGF, and FGF) that have profound effects on immune system cells. In particular, unlike most other cell types, mast cells maintain stored, intracellular pools of TNF that can be released almost instantaneously, contributing to the rapidity of type I HS reactions. The same storage strategy may also apply for many other mast cell cytokines.

Basophils are not as well studied as mast cells. Unlike resting mast cells, which reside for long periods just under mucosal epithelial surfaces, near peripheral nerves and around the blood vessels, resting basophils are routinely found in the blood circulation and must be recruited into the tissues by inflammatory mediators. The same is true for basophils armed with IgE in the sensitization response to an allergen. The release of chemokines by triggered mast cells draws the sensitized basophils into the tissue sites where allergen is present. When the allergen aggregates the FcεRI of an armed basophil, the cell degranulates and releases many of the same mediators as a triggered mast cell. Activated basophils also secrete large quantities of IL-4 and IL-13. Interestingly, a study comparing the contents of basophil granules of atopic and non-atopic individuals found that tryptase, chymase, and carboxypeptidase A were present only in the basophil granules of the atopic subjects.

iib) Late phase reaction. About 4–6 hours after the initiation of a type I HS reaction, the accumulation of chemotactic factors released at the site of allergen exposure during the early phase reaction induces the expression of new adhesion molecules on the activated endothelium of local blood vessels. A mob of immune system cells expressing the appropriate counter-receptors exits the circulation and infiltrates the allergen-polluted tissue (Fig. 28-3). Neutrophils and eosinophils make up most of the infiltrate, but Th2 lymphocytes, mast cells, basophils, and macrophages are also present. Cytokine and chemokine secretion by these cells creates a milieu in which mast cells and basophils are induced to degranulate via both IgE-dependent and IgE-independent mechanisms. Local

concentrations of IL-8 stimulate neutrophils, while IL-5, IL-3, and GM-CSF promote the activation and differentiation of eosinophils.

During the late phase of the effector stage, stimulated immune system cells collectively release more leukotrienes, enzymes, and cytokines that directly or indirectly cause clinical damage to tissues. This damage has been labeled the *late phase reaction*, and eosinophils appear to be the most important players in its generation. Eosinophils bear FcεRs, FcαRs, and FcγRs that can bind IgE-, IgA-, and IgG-coated allergen, respectively. Upon the stimulation of any of these receptors, the eosinophils degranulate and flood the surrounding tissue with inflammatory mediators such as leukotrienes, PAF, IL-4, IL-10, major basic protein, eosinophil-derived neurotoxin, and eosinophilic cationic protein. *Major basic protein* has potent activity against parasites such as helminth worms, but is also toxic to respiratory epithelial cells and can induce asthma-like symptoms in monkeys. *Eosinophil-derived neurotoxin* and *eosinophilic cationic protein* both have ribonuclease activity that is effective against viruses but also damages the airway epithelium. Molecules released during this tissue destruction then induce further degranulation of mast cells and basophils and the activation of macrophages. Eosinophils also release and/or activate peroxidases, collagenases, and proteases such as lysozyme that cause still more tissue damage.

NKT cells appear to play an important, if murky, role in the late phase reaction. As described in Chapter 18, NKT cells express a semi-invariant TCR that recognizes glycolipids presented on CD1d. During an allergic inflammatory response, self-glycolipids may become exposed or expressed, activating local NKT cells that produce large amounts of IL-4 and IL-13. Indeed, mice lacking either NKT cells or CD1d show reduced allergen-induced airway inflammation. In one experiment, the re-introduction of wild-type NKT cells into NKT-deficient mice restored the airway damage, while the transfer of NKT cells from IL-4$^{-/-}$ IL-13$^{-/-}$ double mutant mice did not. The precise mechanism by which NKT cells mediate airway damage is unclear. Some researchers speculate that NKT cells may help to supply the IL-4 required to support the Th2 differentiation of allergen-specific memory T cells. These Th2 effectors then produce Th2 cytokines that activate the eosinophils. It is also possible that the IL-4 and IL-13 secreted by activated NKT cells stimulate mast cells and basophils.

III. EXAMPLES OF TYPE I HS

i) Localized Atopy

As explained previously, most people experience allergies as a localized type I HS reaction affecting a specific target tissue such as the skin or bronchial passages. In these situations, the sensitized mast cells that are triggered lurk among the epithelial cells lining the target organ. Following mast cell activation, the target tissues exhibit the cellular and biochemical changes characteristic of atopic inflammatory infiltration: increased numbers of eosinophils and activated macrophages, increased Th2 cells but decreased Th1 cells, degranulated mast cells, elevated histamine, prostaglandins, and other

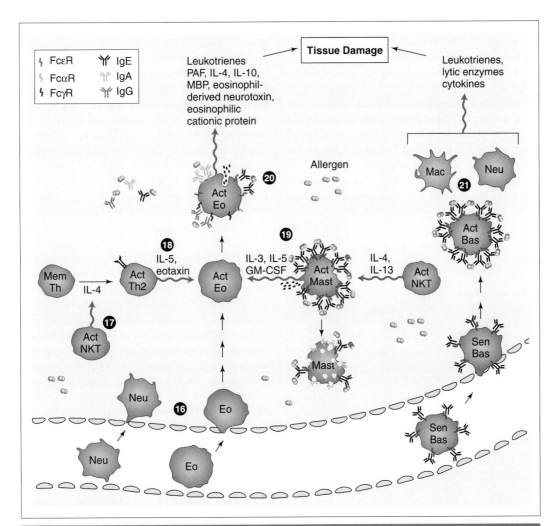

Figure 28-3

Late Phase Reaction of Effector Stage of Type I Hypersensitivity
Continuing on from Figure 28-2, chemotactic factors released during the early phase of the hypersensitivity reaction draw immune system cells into the allergen-polluted tissue (16). Eosinophils are particularly important players at this stage. Activated NKT cells (17) may support the differentiation of Th2 effectors that secrete cytokines inducing eosinophil differentiation (18). NKT-produced cytokines may also stimulate newly activated mast cells, which in turn secrete cytokines and growth factors contributing to eosinophil differentiation (19). Several types of FcRs are upregulated on the eosinophil cell surface, and these cells are activated upon FcR cross-linking by allergen–antibody complexes (20). Eosinophil cytokine secretion and degranulation release the indicated mediators, which damage host cells. Other immune system cells (21), such as sensitized basophils activated by encounter with allergen and macrophages and neutrophils stimulated by pro-inflammatory cytokines, also produce mediators that damage host cells.

mediators, increased IgE, and enhanced IL-4, IL-5, IL-13, GM-CSF, and various chemokines. Where these symptoms occur defines the nature of the allergy.

ia) Hay fever. Allergic rhinitis is the prototypical example of a type I HS disease and one of the first to be described. Scientists of the early 1800s reported a disorder associated with haying season, and soon identified pollen grains as the agent causing the allergic symptoms. Allergic rhinitis results when airborne allergens such as ragweed pollen or mold spores are inhaled. Antigens derived from these entities bind to allergen-specific IgE fixed on sensitized mast cells resident in the upper respiratory tract, the conjunctiva of the eyes, and the nasal mucosae. The release of pro-inflammatory mediators in these locations causes the characteristic symptoms of hay fever: coughing, tearing and itching of the eyes, sneezing, and the blockage of nasal passages. About 20% of the populations of developed countries suffer from allergic rhinitis. Onset usually occurs at age 7 years.

ib) Asthma. Atopic asthma is a type I HS reaction that occurs in the lower respiratory tract of 10–20% of children and adults of developed countries. Inhalation of antigen triggers degranulation of sensitized mast cells resident in the nasal or bronchiolar mucosae. The release of pro-inflammatory mediators in these locations results in the production of copious amounts of mucus, which constricts the bronchioles (sometimes severely). The patient soon complains of tightness in the chest, and begins to wheeze or gasp for air. Asthma can be fatal if an acute attack results in total blockage of the airways. The airways of asthmatics are often hyper-responsive to non-specific stimuli, a phenomenon known as _airway hyper-responsiveness_ (AiHR). AiHR can also lead to narrowing of the airways and increased breathing difficulties. Airways of asthmatics are also subject to a "tissue remodeling" process in which the top layers of the bronchial epithelium are stripped away and the submucosae are thickened due to the deposition of collagen beneath the basement membrane. The smooth muscles of the bronchioles tend to be enlarged, as are the goblet cells in the area. At the cellular level, the airways of asthmatic patients are chronically inflamed with an infiltration of eosinophils, mast cells, lymphocytes, and neutrophils (Plate

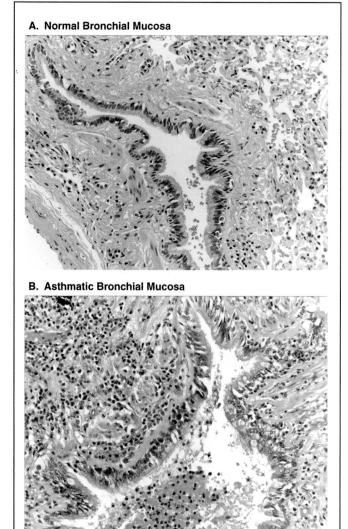

A. Normal Bronchial Mucosa

B. Asthmatic Bronchial Mucosa

Plate 28-1

Normal versus Asthmatic Bronchial Mucosae
Compared with normal bronchial mucosa (A), asthmatic bronchial mucosa (B) shows increased numbers of goblet cells and inflammatory cell infiltration, including numerous eosinophils. Courtesy of Dr. David Hwang, Toronto General Hospital.

28-1). Eosinophils are particularly prominent in the smooth muscle bundles of asthmatic bronchi, although increased numbers of mast cells may also be present. Over 50 separate inflammatory mediators have been associated with asthma symptoms, and high levels of inflammatory cytokines are found in the lung secretions of asthmatic patients. The prevalence and severity of asthma and its resistance to easy treatment have spurred researchers to develop animal asthma models for both drug testing and investigation of underlying mechanisms, but, to date, none has completely replicated the human situation (see Box 28-1).

ic) Urticaria (hives). Urticaria is a type I HS reaction in which sensitized skin mast cells degranulate and release mediators that cause swollen, reddened patches on the skin known as _wheal and flare_ eruptions (Plate 28-2). The whitish wheal in the center of the hive is composed of leukocytes that have escaped the blood vessels due to the increased "leakiness" of these channels. The flare is the ring of redness seen surrounding the wheal due to increased blood flow into this area. There is intense itching in the site of allergen exposure and pain caused by stimulation of skin nerve endings by histamine. Urticaria is frequently accompanied by prominent swelling beneath the mucosal and cutaneous layers in the area of exposure. Allergies to latex and some drugs are often associated with acute urticaria.

id) Atopic dermatitis (eczema). Atopic dermatitis (AD) is the most common type I HS reaction in the skin of those living in developed countries. Clinically, the skin tends to be excessively dry and the itchy rash is more scaly than in urticaria. Eczemic lesions affect different parts of the body at different ages, being most prominent on the face and trunk of infants but more prevalent on the limbs of older children and adults. Eczemic itching is worst at night, so that the sufferer's sleep patterns are often disrupted. AD tends to be more chronic in nature than urticaria and is often associated with respiratory allergies later in life. Individuals with AD also tend to be more susceptible to skin infections because the barrier function of the skin is compromised by the eczemic lesions.

ie) Food allergies. Food allergies result from IgE-mediated reactions to allergens in consumed foods. Although any food can cause an allergic reaction, 90% of food allergies have been linked to peanuts, soy, milk, eggs, wheat, or fish. Food allergies are manifested in two different ways. Oral food allergy is the burning or itching sensation that a person allergic to a particular fresh fruit or vegetable experiences when this morsel touches his or her lips. This reaction, which is a type of contact urticaria usually confined to the tongue, lips, and throat, can cause severe mucosal edema of the mouth and pharynx. Other food allergens (such as milk, eggs, cereal grains, and fish) slide safely down the esophagus but induce the release of inflammatory mediators from mast cells resident in the gut. These mediators then act on the smooth muscles in this location, causing them to contract and produce vomiting, nausea, abdominal pain and cramping, and/or diarrhea. Large numbers of eosinophils may be induced to infiltrate into the gastric and intestinal walls. In extreme cases, the mediators may increase the permeability of the gut mucosal layer such that a food

Box 28-1. Mouse models of asthma

Mouse models of asthma have been under investigation for some time in order to elucidate the underlying mechanisms of the disease and identify possible treatment strategies. For example, the administration of ovalbumin (OVA) to C57BL6 mice causes both early and late phase type I HS symptoms, including airway eosinophilia and AiHR. Studies of normal and IL-4-deficient mice treated with OVA showed that IL-4 was essential for the onset of OVA-specific IgE production and AiHR. STAT6 was also required for the development of OVA-induced atopy. Exposure of mice lacking mast cells or MHC class II to OVA under atopy-inducing conditions showed that the eosinophilic influx into the airways could occur in the absence of mast cell degranulation but not in the absence of APCs. These observations were instrumental in confirming the importance of Th2 cells to atopic inflammation. This model has also demonstrated the importance of genetic background to the manifestation of atopy. C57BL6 mice lacking IL-5 subjected to the OVA protocol do not develop AiHR, but IL-5$^{-/-}$ Balb/c mice do.

Wild-type mice treated with IL-13 also develop symptoms strikingly reminiscent of human asthma, including mucus overproduction and AiHR. However, there is a mystery here because mouse T cells do not express IL-13R. It is speculated that other cells bearing the receptor (perhaps mast cells?) may be activated in response to IL-13 binding and secrete other cytokines that drive mouse Th2 differentiation. As well, mice that transgenically overexpress IL-9 specifically in the lung exhibit AiHR and changes to airway

morphology similar to those in asthma patients. IL-9 is a powerful inducer of mast cell differentiation.

Knockout mice lacking expression of the transcription factor T-bet also constitute a model of human asthma. The reader will recall from Chapter 15 that T-bet is expressed exclusively in Th1 cells and induces transcription of the IFNγ gene. Forced expression of T-bet can also convert Th2 cells producing IL-4 and IL-5 into fully functional Th1 cells that suppress production of IL-4 and IL-5 and commence synthesis of IFNγ instead. T-bet$^{-/-}$ mice spontaneously (i.e., in the absence of any known allergen or trigger event) develop classic symptoms of asthma. The airway is hyper-responsive to stimuli and the bronchioles show infiltration by eosinophils and lymphocytes. Airway remodeling in the form of collagen deposition below the epithelium is present. Heterozygous and homozygous T-bet mutant mice are affected equally, showing that half of normal T-bet expression is not sufficient to preserve normal airway physiology. Significantly, T-bet expression is dramatically decreased in the lungs of many asthmatic humans. Two puzzles remain, however. First, if there is a global lack of T-bet function in T-bet mice, why do the atopic symptoms only occur in the airway? Secondly, unlike T-bet$^{-/-}$ mice, IFN$\gamma^{-/-}$ mice do not show any signs of spontaneous asthma. These observations imply that T-bet is responsible for the expression of genes in addition to IFNγ that are required to avoid the onset of asthma-like symptoms.

Despite their usefulness in dissecting components of human asthma, mouse

models of asthma differ from the human situation in several important ways that make the extrapolation of results from these models risky. Moreover, in contrast to the human situation, no generally accepted standard for measuring lung function prevails in mice, and there is wide variation in sensitization and allergen challenge protocols between laboratories. Differences that have been observed between mouse models and human asthma include: (1) The doses per body weight of substances used to induce allergy in mice are significantly greater than the levels of allergen a human is exposed to in his or her environment. (2) The timing and tissues affected in the inflammatory response associated with asthma symptoms are not strictly parallel in mice and humans. (3) Mouse eosinophils have many fewer granules than human eosinophils and lack expression of FcϵR and FcαR. Agents that provoke massive degranulation of eosinophils in human asthma challenge tests incite very little eosinophil degranulation in mice. (4) The conventional knockout mice used in many asthma studies by definition have permanently lost the gene of interest right from birth, whereas a given gene may be switched on or off inappropriately in asthmatic humans. To address this last issue, researchers are engineering conditional knockout mutant mice in which the loss of expression of a specific atopic gene can be triggered at will during the animal's life. Hopefully these sorts of modifications will bring us closer to a more accurate mouse model of human asthma.

---

allergen enters the circulation. Depending on where it ends up, the allergen may then induce an asthmatic response or urticaria in a site distant from the original site of exposure. The only proven therapy for a food allergy is elimination of that food from the diet; however, children and adults often grow out of mild food allergies with time. Note that a food allergy is not a food intolerance: the latter results from a non-immunological response to a food component. One of the best known food intolerances is lactose intolerance, in which a deficiency of the lactase enzyme leads to an inability to digest lactose-containing milk products.

ii) Systemic Atopy: Anaphylaxis

As introduced previously, anaphylaxis is a type I HS response with systemic consequences. Clinically, anaphylaxis is a form of extreme shock (hence, "anaphylactic shock") that can kill within minutes of exposure to the antigen. Some individuals

are so sensitive that even the smell of the offending allergen can trigger the reaction. Anaphylactic shock is most frequently observed in individuals sensitized to insect stings, peanuts, seafood, or penicillin, although hypersensitivity to latex (as in latex gloves) is on the rise in health workers. In the United States each year, anaphylactic reactions to food allergens account for about 29,000 emergency room visits, 2000 hospitalizations, and 150 deaths. Interestingly, anaphylaxis (like atopy in general) is not common in developing countries.

During anaphylaxis, large quantities of inflammatory mediators and vasodilators are released into the circulation by activated mast cells and basophils, causing rapid dilation of blood vessels throughout the body. Respiration immediately becomes difficult and is followed by a dramatic drop in the victim's blood pressure and extensive edema in the tissues. Patients have been known to report a "feeling of

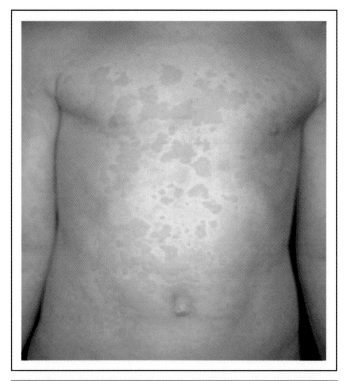

Plate 28-2

Urticaria
Typical wheal and flare reaction of urticaria (hives). Courtesy of Dr. Thomas Habif, and reproduced with permission from Merck manual of Diagnosis and Therapy, © Merck & Co., Inc.

doom" at this point. The lungs may fill with fluid, the heart may beat irregularly, and control of the smooth muscles of the gut and bladder is often lost. Constriction of the bronchioles may cause lethal suffocation of the victim unless treatment with epinephrine (adrenaline) is started immediately. Antihistamines, often used to treat atopy (see later), are useless here because they cannot halt the extremely rapid onset of symptoms. Epinephrine counters the actions of histamines because it immediately increases heart action, stimulates the circulatory system and the muscles supporting the capillaries, and relaxes the constricted bronchioles. These measures help to restore normal blood pressure and combat shock. Severely allergic individuals should never be without their "Epi-pens," a syringe-like device that even the most panicked of bystanders can use to inject epinephrine into the stricken victim.

The clinical course of anaphylaxis is biphasic in one-third of patients, a factor that has led to some fatalities due to premature release from the emergency ward. Typically, these patients experience severe initial symptoms that then appear to resolve for 1–3 hours. Symptoms then return with a vengeance and can kill if treatment is not sought again immediately. Anaphylactic patients should therefore always be observed for at least 4 hours after the initial resolution of symptoms.

IV. ROLES OF FCεR AND IgE IN TYPE I HS

i) FcεRI (High-Affinity IgE Receptor)

Type I HS reactions depend on the triggering of the high-affinity FcεRI. In rodents, FcεRI occurs as a heterotetramer of structure $\alpha\beta\gamma_2$. Expression of this receptor is confined to mast cells and basophils. In humans, there are two isoforms of FcεRI: a heterotetramer of structure $\alpha\beta\gamma_2$ and a lower affinity heterotrimer of $\alpha\gamma_2$ (Fig. 28-4). Both isoforms appear on mast cells and basophils, and low levels of the $\alpha\gamma_2$ isoform can be found on Langerhans cells (LCs) and DCs of normal individuals, and on eosinophils and monocytes of atopic individuals. However, the $\alpha\beta\gamma_2$ receptor is the isoform most closely associated with human atopy.

The α chain of FcεRI is responsible for binding to the Fc region of IgE, while the γ chains contain two ITAMs each that mediate intracellular signaling. The β chain also contains two ITAMs and serves as a signal amplifier. The α chain is normally heavily glycosylated, a status that prevents spontaneous aggregation of FcεRI receptors. However, when a multivalent allergen cross-links two FcεRI-bound anti-allergen IgE molecules, the resistance to interaction due to α chain glycosylation is overcome and a conformational change occurs that is somehow sensed by the β and γ chains. Multiple phosphorylation and dephosphorylation events then occur that propagate intracellular signaling leading to mast cell activation.

In a sensitized human mast cell, Lyn kinase is preferentially associated with the β chain of FcεRI. Upon allergen-mediated cross-linking of IgE bound to FcεRI, Lyn is activated (possibly by CD45-mediated dephosphorylation) and phosphorylates the ITAMs in both the β and γ chains (Fig. 28-5). In a signaling pathway reminiscent of that in activated B cells, FcεRI ITAM phosphorylation recruits Syk kinase to the γ chain via SH2 domain interactions and Lyn proceeds to phosphorylate and activate Syk. Syk activation then leads to LAT phosphorylation, PI3K and PLCγ activation, the generation of the second messengers DAG and PIP3, and the release of calcium from intracellular stores. As we have seen, calcium is absolutely required for the nuclear translocation of the transcription factor NF-AT and the induction and maintenance of cytokine synthesis and secretion. As well, DAG facilitates the recruitment of PKC, which leads to NF-κB activation. The adaptor protein Grb2 is also recruited to the FcεRI complex. Much as is true in antigen receptor signaling, Grb2 recruits the adaptors SLP-76 and Sos which in turn recruit Vav1. Vav1 is responsible for cytoskeletal rearrangements and acts as an exchange factor for GTPase signaling via Ras and Rac. The GTPase pathways triggered by these proteins activate ERK, MAPK, and SAPK/JNK signaling pathways that ultimately lead to the formation of the transcription factor AP-1. In addition, the GTPase pathways promote the release of preformed mediators from the mast cell granules and morphological changes in the mast cell membrane associated with activation.

The FcεRIβ chain has multiple roles in FcεRI signaling. The FcεRIβ chain stabilizes the receptor complex, significantly increasing the number of FcεRI structures on the mast cell surface. In the absence of the FcεRIβ chain, there is a 5- to 7-fold drop in Lyn recruitment to the receptor complex, Syk activation,

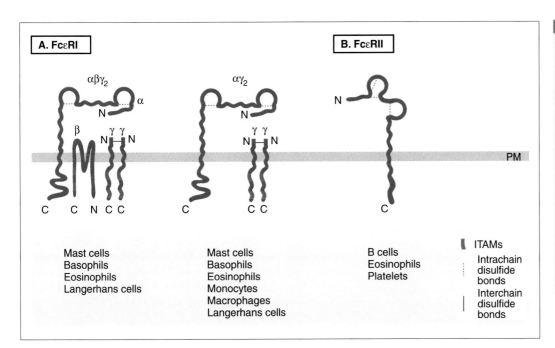

Figure 28-4

Human FcεR Complexes
Structural isoforms of FcεRI and FcεRII and their expression by immune system cells are shown. The αβγ₂ isoform of FcεRI is the high-affinity IgE receptor involved in atopy. Signaling via FcεRII on B cells has a negative effect on atopy because IgE synthesis is downregulated. PM, plasma membrane. Adapted from Oliver J. M. *et al.* (2000) Immunologically mediated signaling in basophils and mast cells: finding therapeutic targets for allergic diseases in the human FcεR1 signaling pathway. *Immunopharmacology* **48**, 269–281.

calcium release, and production of IL-6. However, the FcεRIβ chain may also be responsible for negative control of FcεRI signaling. The phosphatases SHP1 and SHP2, which dephosphorylate crucial signaling mediators, have been shown to bind to sites in the FcεRIβ chain. SHP1 and SHP2 may also be recruited to the FcεRI complex in the wake of co-ligation of FcεRI with the inhibitory receptor FcγRIIβ. We have seen this type of mechanism before, in the negative regulation of BCR signaling via the co-ligation of mIg with FcγRIIβ by IgG-bound antigen (see Ch.5). There is some *in vitro* evidence that IgG bound to allergen can

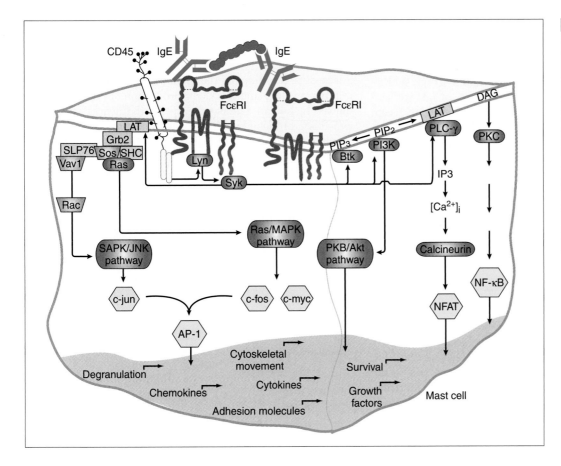

Figure 28-5

Signaling via the High-Affinity IgE Receptor
Allergen-induced cross-linking of FcεRI-bound IgE on a mast cell results in the activation of CD45. CD45 is thought to activate Lyn associated with the β chain of FcεRI. Lyn in turn phosphorylates Syk, which leads to activation of LAT, PI3K, PLCγ, and PKC. Downstream signaling via the indicated pathways induces cytoskeletal reorganization, degranulation, and the synthesis of various molecules required for the early phase reaction.

simultaneously bind to FcγRIIβ and IgE-bound FcεRI, triggering the phosphorylation of ITIMs and the recruitment of phosphatases. It is possible that, in an individual able to make IgG antibodies to an allergen, these enzymes act to depress the intracellular signaling that would otherwise lead to mast cell or basophil degranulation and atopic symptoms.

Another interesting facet of FcεRI biology is that receptor expression on mast cells and basophils is regulated by the serum IgE concentration in the allergic individual. For example, IgE-deficient mice possess mast cells and basophils that show an 80% reduction in FcεRI expression compared to the wild-type. The control of FcεRI expression by IgE is thought to be exerted in two ways. First, binding of IgE blocks FcεRI internalization and degradation, increasing the number of receptors maintained on the mast or basophil cell surface. Secondly, the more these initial FcεRI are occupied by IgE, the more new FcεRI molecules are synthesized by the cell. Unfortunately for the patient, the greater the surface expression of FcεRI, the more severe the clinical consequences when the receptor–IgE complexes are triggered by allergen binding.

ii) CD23 (FcεRII)

As mentioned previously, CD23 (FcεRII) is the low-affinity IgE receptor (refer to Fig. 28-4). While the binding affinity of FcεRI for IgE is close to $K_a = 10^9$ M^{-1}, that of CD23 for IgE is only about 10^6–10^7 M^{-1}. In addition, whereas FcεRI is expressed at high levels on mast cells and basophils, CD23 is expressed preferentially on B cells, platelets, and eosinophils. For these two reasons, CD23 is not associated with incidence of atopy. Indeed, studies of knockout mice lacking CD23 have indicated that this receptor serves as a negative regulator of IgE synthesis because these mutant animals have increased serum levels of IgE. The mechanism by which CD23 exerts this effect is unclear. CD23-mediated inhibition of IgE synthesis appears to kick in only when serum levels of IgE are very high; that is, sufficient to bind to this lower affinity IgE receptor. Cross-linking of CD23 by IgE may then deliver an inhibitory signal that shuts down IgE synthesis. One can imagine that an allergen that somehow interfered with this cross-linking such that the inhibitory signal was not delivered would result in increased IgE production and exacerbated atopy. For example, consider Der p I, an allergen derived from the house dust mite *Dermatophagoides pteronyssinus*. Der p I is a protease that has the ability to selectively cleave CD23 on the B cell surface. Without the negative regulatory signal delivered via CD23, the activated allergen-specific B cell is spurred to increase its output of IgE and atopy is enhanced.

iii) IgE

We have discussed the biology of the receptor to which IgE binds, but how and why is this IgE produced? The reader will recall from Chapter 5 that IgE has a very short half-life (1–5 days) and is present at vanishingly small amounts in the serum of normal individuals. Even in severely atopic individuals in which the serum concentration of IgE is 1000× normal, IgE is still present at a concentration below the normal baseline levels of other antibody isotypes. As we have learned, isotype switching to IgE production in an activated B cell results

from signals delivered via ICOS, CD40, and IL-4 or IL-13. Th2 cells are thought to be the primary sources of cytokines responsible for isotype switching to IgE, although both IL-4 and IL-13 can also be produced by NKT cells, mast cells, basophils, and eosinophils, and IL-13 by NK cells.

Studies of the Th1/Th2 balance in mice and humans have confirmed that Th2 responses are more likely to be associated with type I HS reactions than Th1 responses (Table 28-5). Indeed, the presence of the Th2 cytokines IL-4, IL-5, and IL-13 can account for all the major features of asthma: isotype switching to IgE, mucus production, and eosinophil activation and differentiation. For example, in adoptive transfer experiments in mouse asthma models, animals that received allergen-specific Th2 cells soon exhibited infiltration of eosinophils into the airways, AiHR, and excessive mucus secretion, whereas mice that received allergen-specific Th1 cells did not. Studies of intracellular signaling associated with Th2 responses have identified STAT6 as a transcription factor important in several ways for IgE production and atopy. As we learned in Chapter 17, the engagement of a naive Th0 cell's IL-4 receptors triggers Jak1 and Jak3 activation. Activated Jak1 in turn phosphorylates STAT6, which activates expression of GATA-3 and Th2 effector differentiation. STAT6 then cooperates with GATA-3, c-maf, NF-AT, and AP-1 to induce the expression of Th2 cytokines such as IL-4, IL-5, and IL-13. The expression of Th1-specific genes is suppressed while that of chemokine receptors (CCR4, CCR8) important for Th2 chemotaxis is increased. In other leukocytes, STAT6 activation promotes the synthesis of chemokines (TARC, RANTES) that are relevant for Th2 cell trafficking. In B cells activated by allergen, activated STAT6 promotes the production of the germline transcripts of the Cε Ig exon that are required for isotype switching to IgE synthesis. Studies of STAT6$^{-/-}$ mice in asthma models have shown that a lack of STAT6 impairs the inflammatory infiltration of Th2 cells into the lung and blocks the production of excessive airway mucus. In human asthmatic patients, the density of STAT6-expressing bronchial epithelial cells is increased in the airway. In addition, high levels of GATA-3 mRNA have been found in the bronchial tissues of asthmatics.

Atopy may also result in part from a failure in negative controls exerted by Th1-related mechanisms. In mouse models in which immune deviation to Th1 is induced, IFNγ and IL-2 produced by Th1 cells act directly on B cells to suppress IgE production. IL-12 and IL-18 produced by NK cells and macrophages also suppress IgE production, most likely because these cytokines promote Th1 differentiation and the secretion of IFNγ. Mechanistically, IFNγ inhibits IgE production by

Table 28-5 Th1/Th2 Balance in Atopy

Promote Atopy	Inhibit Atopy
Th2 response	Th1 response
IL-4, IL-5, IL-13	IFNγ, IL-2, IL-12, IL-18
STAT6, GATA-3	Socs-1, Bcl-6
Isotype switching to IgE	Isotype switching to IgG
CCR3, CCR4, CCR8	CCR1, CCR5, CXCR3

inducing expression of the suppressive factor Socs-1 discussed in Box 15-1. Socs-1 binds to Jak proteins and inhibits STAT activation. Another molecule important for limiting IgE production is Bcl-6, the transcriptional repressor we first encountered in Chapter 9 in our discussion of plasma cell differentiation. Knockout mice lacking Bcl-6 show type I HS-like disease in that eosinophils and other immune system cells infiltrate multiple organs in a massive inflammatory response. Moreover, the number of B cells producing IgE is markedly increased in these mutants. Bcl-6 appears to compete with STAT6 for binding to STAT6 motifs in the promoters of Th2-related genes.

One last detail of IgE synthesis in atopy remains a mystery. Particularly in asthmatic patients, IgE is often present in the airway mucus. We have learned in previous chapters that the entry of Igs (primarily IgA and IgM) into body secretions is facilitated by pIgR. However, IgE is not polymeric and is not transported across epithelial cells by pIgR; how it enters the mucus is not known. It may be that yet-to-be-characterized binding factors or transporter receptors exist in the mucosae for this purpose.

V. ALLERGEN BIOLOGY

i) What Makes an Allergen?

Despite intensive study, no one has been able to identify a single characteristic common to all allergens. They are diverse in their structure and biochemical properties, enter into the body in different ways, act at different concentrations, and interact with different molecules or cell types once within the body. Many allergens are small airborne glycoproteins derived from organisms such as house dust mites and plant pollens, but certain antibiotics, perfumes, and metalloproteins can also do the job. Some allergens have proteolytic activity that enables them to damage the mucosae and penetrate into the body more easily. Others have characteristics that aid them in becoming airborne. However, it is not only the nature of the allergen itself that is relevant. Some individuals are genetically predisposed to allergies (see later), and the circumstances under which an individual is first exposed to an allergen (i.e., dose, route of entry) appear to be important. Furthermore, allergies are not necessarily a scourge for life. Children often "outgrow" food allergies, possibly because the mucosal barrier of the gut gets tougher with age and resists penetration. Still, it remains unknown why one individual might be allergic to cat dander but not dog dander, nor why ragweed pollen is more allergenic than the pollen of other abundant plants, nor why allergy to birch pollen is common but that to pine pollen is extremely rare. Nor is it understood why some antigens cause localized reactions while others have systemic effects, nor why some people develop urticaria in response to a given allergen while others develop asthma.

In some types of chronic atopy, the nature of the allergen can change due to a mechanism called *epitope spreading* (see Ch.29). For example, in acute AD, the type I HS response is clearly directed against an exogenous allergen. However, in patients in which the inflammatory response and thus the dermatitis become chronic, one starts to see the emergence of circulating IgE antibodies that recognize epitopes on human

self protein. It is thought that the tissue damage associated with the dermatitis may expose new proteins and thus provide immunogenic epitopes that otherwise would have been hidden from the immune system. Thus, while the inflammatory response is initially induced by a bona fide allergen, it is later maintained by an autoreactive mechanism. Another autoimmune mechanism seen in chronic urticaria, asthma, and AD involves the cross-linking of the FcεRI on mast cells by an IgG autoantibody directed against the FcεRI, or against the IgE bound to the FcεRI. In rare cases, a second IgG autoantibody may be directed against the first IgG antibody binding to the IgE (Fig. 28-6).

ii) Determinants Associated with Type I HS
iia) Genetic determinants. Although we still do not know exactly why type I HS reactions occur in only some people, it has been clear for a long time that both genetic and environmental determinants play a role (Fig. 28-7). Indeed, the familial nature of asthma incidence was noticed as early as 1860. Families in which both parents are atopic have a 50% chance of having an allergic child, while children in a family with one non-atopic parent have a 30% chance of being allergic. Only 19% of children in a family with no history of atopy develop allergies. It seems that, while an excessive IgE response underlies almost all cases of atopy, other genes must be involved in determining the specific clinical manifestations of that response.

Family studies and cloning techniques have identified 13 chromosomal regions and more than 20 genes that may contribute to the development of allergic diseases (Table 28-6). The human chromosomal region 5q31–33 is of special interest because it contains a cluster of genes that encode Th2 cytokines, including IL-4. Patients with certain polymorphisms in this region have more Th2 cells in the relevant tissues than non-allergic individuals, and these Th2 cells produce greater than normal amounts of IL-4. In addition, particular polymorphisms in the IL-4Rα gene have been correlated with elevated total serum IgE levels as well as the occurrence of severe AD and asthma. Certain genetic alterations in the IL-13 gene have also been associated with human asthma, and IL-9 (whose gene is also in this cluster) is markedly reduced in mice showing AiHR. Other genes located in this chromosomal region that may have links to atopy include those encoding GM-CSF, TGFβ, CD14, a glucocorticoid receptor, and a β-adrenergic receptor.

Specific alleles of the HLA-D region genes on chromosome 6p21 also figure prominently in atopy. In particular, inheritance of HLA-DR4 and HLA-DR7 appears to predispose one to allergic disease, and HLA-DR2 has been linked quite strongly to ragweed allergy. In contrast, some HLA-DQ molecules have been found to confer resistance to the development of asthma. Other obvious candidate genes are those encoding the FcεRI subunits. There is some evidence suggesting that a gene on chromosome 11q, possibly the β subunit of FcεRI, may be linked to atopy. In particular, polymorphisms in intron 5 and exons 6 and 7 of the FcεRIβ gene have been associated with AiHR in some populations. Other studies have suggested that two other genes linked to the FcεRIβ gene may influence AiHR and serum IgE concentration.

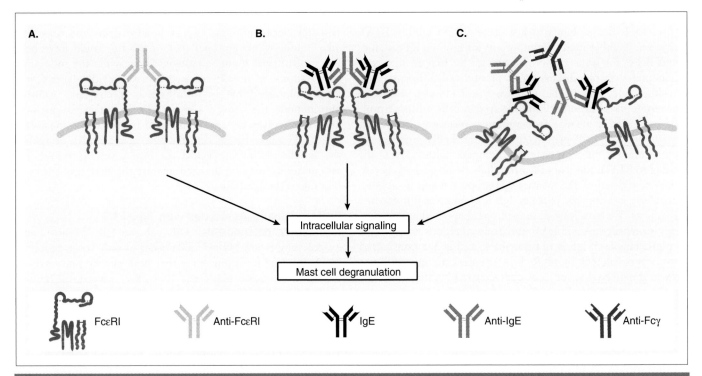

Figure 28-6

Examples of Autoimmunity in Atopy

Autoantibodies can promote atopic responses when they facilitate the cross-linking of FcεRI on mast cells. For example, in (A), an autoantibody recognizes a site on the FcεRI molecule itself and induces aggregation leading to intracellular signaling and degranulation even in the absence of allergen. In (B), an autoantibody recognizes a site on the Fc portion of IgE molecules on a sensitized mast cell. The autoantibody, rather than the allergen, cross-links the IgE molecules to initiate mast cell activation and triggers the allergic response. In (C), a second autoantibody recognizes the Fc region of a first IgG autoantibody that has bound to the Fc region of IgE. In cases in which the first autoantibody is not able to sufficiently cross-link the IgE molecules, the second autoantibody completes the link and triggers the allergic response. With information from Marone *et al.* (1999) The anti-IgE/anti-FcεRIα autoantibody network in allergic and autoimmune diseases. *Clinical and Experimental Allergy* **29**, 17–27.

Figure 28-7

Genetic and Environmental Determinants Favoring Atopy

The indicated genes and environmental factors may work alone or in concert to predispose an individual to atopy. Exposure to one or more of the indicated triggering events can result in a clinically relevant atopic response to an allergen that breaches the mucosal defenses.

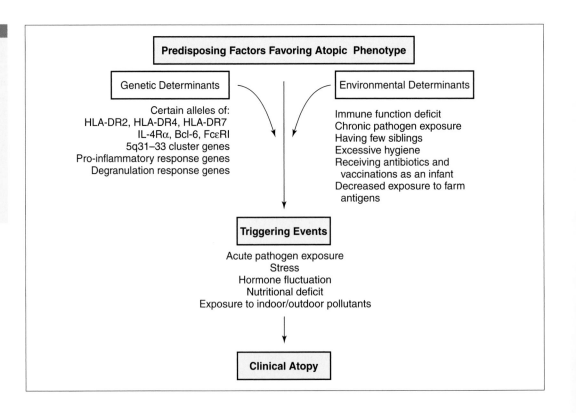

Table 28-6 Human Genes Potentially Contributing to Atopy

Chromosomal Region	Candidate Gene(s)	Proposed Role
3q27	Bcl-6	Negative regulation of Th2 responses
5q31–33	IL-3, IL-4, IL-5, IL-9, IL-13, TGFβ, CD14, GM-CSF, gene for IL-12 responsiveness	Immune deviation to Th2; isotype switching to IgE; activation of eosinophils, basophils, and mast cells
6p21–22	MHC genes, TAP1 TNF	Antigen processing and presentation Inflammation
11q13	FcεRI	Mast cell degranulation
12q14–24	IFN-γ, SCF, NF-YB iNOS, leukotriene A4 hydrolase, mast cell growth factor	Regulation of IL-4 transcription Pro-inflammatory response; stimulation of mast cells
14q11–13	TCRA, TCRD NF-κB	TCR genes Transcription of pro-inflammatory genes
16p	IL-4Rα	Component of IL-4 receptor

Certain regions of human chromosomes 1, 2, 4, 7, 8, 13, 17, 19, and 21 have also been implicated in atopy, but candidate genes in these locations have yet to be identified.

Another interesting candidate gene is the Bcl-6 gene located in chromosomal region 3q27. As mentioned previously, Bcl-6 regulates Th2 responses such that knockout mice lacking Bcl-6 have a phenotype strikingly reminiscent of atopy. For similar reasons, mutations of genes that affect IL-12 activity may also contribute to atopy. On human chromosome 5q31, a gene for IL-12 responsiveness has been found that appears to regulate responses to infectious organisms and to allergens. In addition to these reasonably well-defined candidates, one might suspect that genes encoding mast cell mediators, histamine receptors, chemokines and their receptors, antigen processing molecules, and even the TCR loci might be relevant to atopic disease.

The matter of identifying genes causing atopy is made more complicated by the fact that the expression of some of these genes is subject to maternal inheritance patterns or maternal "imprinting" based on specific modifications of chromatin components. In such cases, an allele inherited from the mother is more likely to be expressed, while the paternal allele remains silent. An individual may therefore possess an allergenic allele of a candidate atopy gene, but the likelihood of it being expressed is much higher if it was inherited from his or her mother. As a result, the presence of the allergenic allele in an individual's genotype will not necessarily correlate with the atopic phenotype, making it more difficult to definitively conclude that the genetic locus in question plays a role in the allergic response. Another difficulty is that some alleles appear to be relevant in only some ethnic groups. For example, an allele of the FcεRIβ gene that was linked to asthma in a white South African population was not associated with disease in a parallel black South African population. Indeed, black asthmatics carried an entirely different allele of the FcεRIβ gene. As the reader can see, much remains to be elucidated in the field of allergy genetics.

iib) Environmental determinants. Despite the clear involvement of genetic determinants in atopy, they are not the whole story. Environment and/or early allergen exposure must also play a role because identical twins do not always share allergies.

In fact, both twins are allergic in only 60% of cases. Clinicians have noted that chronic illnesses, acute viral, fungal, or bacterial infections, emotional stress, fluctuating hormone levels, nutritional deficits, and the presence of pollutants all seem to be associated with the onset of type I HS reactions. The common factor here may be a weakening of the immune system that leaves the body more accessible to allergen entry and sensitization. Alternatively, a pathogen molecule itself may be acting as an allergen in that the infected individual mounts an IgE response to the offending protein. For example, many flare-ups of AD are associated with the presence of certain skin fungi, and IgE antibodies directed against proteins from these organisms are found in the sufferer's circulation. Another organism almost always found in AD skin lesions is *Staphylococcus aureus*. It is thought that some of the toxins secreted by this bacterium may be acting as superantigens inducing polyclonal activation of T cells. A sizeable number of chronic AD patients have circulating IgE antibodies that recognize *S. aureus* superantigens, and the severity of the dermatitis correlates with the level of superantigen-specific IgE. Interestingly, AD patients also often have reduced numbers of CD8[+] T cells and NK cells, and are generally more susceptible to viral infections than non-allergic people. It may be that the original immune function deficits leave these sufferers more vulnerable than normal individuals to infections that can trigger allergy.

iii) Why Is Atopy Increasing?

Over the past two decades, the frequency of atopy has increased exponentially in North America, Europe, Australia, and New Zealand. As a result, over 130 million people now suffer from asthma in the developed world. In contrast, the prevalence of atopic disease remains very low in developing countries and the incidence of other types of hypersensitivity (types II, III, and IV) has been stable. Moreover, despite the allocation of an ever-increasing chunk of the health care budget to treatment of allergic diseases, morbidity and mortality associated with these disorders is mounting steadily. The reasons why atopy should increase in a country as its living standard rises are not clear. One factor sometimes noted is the relatively recent emphasis in Western cultures on energy conservation in the home. The upgrading of windows, insulation, and carpeting may have decreased draftiness but has also tremendously reduced the exchange of inside and outside air. As a result, occupants of such homes may suffer increased exposure to house dust mites and other allergens.

In almost any country, the frequency of atopy is higher in urban versus rural areas, an observation perhaps related to the higher numbers of motor vehicles in urban settings. It is thought that pollutants in automotive exhaust that increase lung responsiveness may promote allergen-specific IgE production and increase sensitization rates. Cigarette smoking and exposure to indoor pollutants or agents in the "blue collar" workplace may also be relevant here. Urban pollutants can also increase the bioavailability of some allergens. For example, certain pollen grains that would remain intact in a clean environment are activated by pollutants such that they release their allergens more easily. Even within a given rural area, environmental differences have an impact on atopy. Children raised on farms in a rural area are less likely to suffer from allergies than are children living in non-farm households in the same rural area. One theory holds that the former are exposed from a young age to animal antigens and higher concentrations of bacterial endotoxin than are their non-farm peers, and that this exposure to a broad range of pathogens is protective. Certain viral infections do seem to be directly associated with atopy. Respiratory viruses that target the <u>lower</u> respiratory tract can cause symptoms reminiscent of asthma in children, if the infection is severe. Moreover, there is some evidence that repeated or severe childhood infections with these viruses can predispose that child to asthma later in life. However, other viral infections may be protective. For example, infection in the first year of life with a virus that targets the <u>upper</u> respiratory tract appears to decrease the risk of subsequently developing hay fever or asthma.

Based on the preceding observations, several investigators have proposed that allergies result from a failure to encounter sufficient pathogens in infancy, a situation most likely to be achieved in highly protective families in wealthy, developed countries. The theory is that a lack of early exposure to pathogens may lead to a lack of "training" of the immature immune system in the infant, and consequently an imbalance in Th1/Th2 responses. The imbalance then leads to inappropriate responses to normally innocuous antigens. This concept is called the *hygiene hypothesis* and is discussed in detail in Box 28-2.

VI. DIAGNOSIS AND THERAPY OF TYPE I HS

i) Diagnosis

"Allergy testing" in common parlance is the process by which the allergen irritating a particular patient is identified. The procedure is often a progressive series of tests, starting with the taking of a blood sample for a RIST (*radioimmunosorbent test*). The serum of the individual patient is analyzed for the presence of total circulating IgE antibodies. A positive result means the patient may be atopic but does not identify the antigen. (It should be noted that high levels of serum IgE are also present in

other disease conditions, such as helminth worm infections and Hodgkin's lymphoma, so that the mere presence of elevated IgE does not necessarily confirm the presence of atopic disease.) Several of the most common allergens have been purified for use in more refined tests, and recombinant DNA technology has allowed the synthesis of allergens that have been hard to purify by conventional means. The RAST (*radioallergosorbent test*) uses these purified allergens in a panel to screen serum samples from the patient and identify the allergen by binding to circulating IgE. These methods of diagnosis are the safest for the patient because he or she is never directly exposed to the allergen, thereby avoiding any possibility of anaphylaxis. However, some patients do not exhibit circulating IgE (it may be present only attached to mast cells or present in mucosal secretions). In these cases, a *skin prick* test must be carried out. A sterile needle is dipped into a preparation of diluted allergen and the dermis of the forearm or back is pricked such that the allergen is introduced into the dermis. A control prick of solution without allergen is also applied nearby. If a hive appears within 20–30 minutes on the skin at the site where the allergen was applied, it can be concluded that the individual is allergic to the applied allergen (Plate 28-3). The *skin patch* test is similar. Surface skin cells are gently rubbed away from a small area and a patch soaked in diluted allergen is applied. The patch is covered and the reaction allowed to proceed for about 48 hours. A pronounced eruption of the skin at this site is evidence of allergy to the applied allergen. A disadvantage of these more invasive methods is that, by introducing a battery of new allergens into the body of a potentially atopic person, that person may indeed develop an allergy to a new allergen. The results of such tests should also be interpreted with caution for another reason: many people that show a positive skin test response to an allergen do not show a clinically significant allergic reaction to it. Needless anxiety is caused in patients that focus on the skin test result rather than the clinical reality.

For similar reasons, the diagnosis of a food allergy by skin testing should be confirmed with a DBPCFC test (*double-blind placebo-controlled food challenge*). DBPCFCs give clearer results than a one-time skin prick test because they eliminate observer bias and control for the variability of chronic effects such as urticaria or exacerbations due to an interruption in medication. In a DBPCFC, the suspected food is eliminated from the patient's diet 7–14 days before the test and antihistamines and other medications are minimized. The fasting patient is fed a low dose (25–50 mg) of a freeze-dried version of the suspect food in capsule or liquid form. Such doses are unlikely to provoke symptoms. The dose is then slowly increased every 15–60 minutes up to a dose of 10 g of freeze-dried food. The patient is observed for up to 2 hours to detect clinical signs of reactivity to the food. To validate this type of test, the patient must also be given an equal number of challenges with a placebo food. Such trials should be conducted in a random order such that neither the administering doctor nor the patient knows which dose is which. If the patient fails to react to 10 g of the suspect food, an *open feeding test* is done in which the doctor observes the patient unrestrictedly eating the food in question. A failure to react in this situation means that the patient's symptoms were not related to this food.

Box 28-2. **The hygiene hypothesis**

Urban areas of developed countries show the highest frequency of type I hypersensitivity. In contrast, citizens of less-developed countries with more primitive medical systems enjoy comparative freedom from atopic disorders. One theory advanced to account for this phenomenon is the *hygiene hypothesis*. This theory posits that, in more advanced countries, childhood infections are prevented or rapidly resolved with vaccination or antibiotics, and personal and food hygiene are fastidious. The immune system therefore gets fewer chances to mount the Th1 responses necessary to combat intracellular pathogens. As a result, T cell differentiation in this "sheltered" individual may be biased toward Th2 development, possibly predisposing him or her to atopic disease. Consistent with this idea, in both developed and less-developed countries, a history of substantial childhood infections (rather than the presence of specific pollutants or allergens) correlates strongly with resistance to atopic disease. In particular, infection with HepA virus has been implicated in protecting against the development of skin allergies and asthma. HepA infection used to be endemic in North American and European populations before improved hygiene reduced its incidence to less than 30%. An inverse correlation now exists between HepA prevalence and the development of atopy. Similarly, early infection with measles virus appears to reduce the chance of atopy, perhaps because this virus strongly induces a Th1 response.

Situations that expose a growing child to the possibility of infection reduce the chance of atopy. A child who is one of several siblings, or who attends a day care center at an early age, has a decreased likelihood of developing skin allergies, hay fever, or asthma. On the other hand, a child who receives copious quantities of antibiotics and/or is vaccinated during early childhood is at increased risk for allergies. In one study of Swedish children, two populations were compared for frequency of atopy. The population that had the highest frequency of atopy (25%) also had the highest rate of childhood vaccination and antibiotic use. In the population that rarely used antibiotics and was vaccinated only against tetanus and polio, the frequency of atopy was only 13% and asthma incidence was reduced almost 4-fold. Another example often cited to demonstrate a link between higher standards of living and atopy is the surge in atopy frequency in the former East Germany after the fall of the Berlin Wall in 1989. People of this region rapidly increased their standard of living but also increased their risk for several atopic diseases.

What is the underlying basis for the hygiene hypothesis? As we learned in Chapter 16, the placenta is a unique microenvironment that secretes Th2 cytokines and other factors designed to maintain tolerance of the growing fetus. As a result, T cells in the cord blood of newborns are largely biased to a Th2 phenotype. Serum IFNγ levels are reduced and Th1 cells are decreased in frequency. With normal exposure to pathogens, the Th1 population is built up at the expense of the Th2 population and the finely tuned balance between these cell subsets necessary for comprehensive immune responses is restored. In the absence of significant infection by pathogens (due to antibiotics or vaccinations), the balance remains tipped in favor of Th2 responses and thus atopy.

As attractive as the hygiene hypothesis is, it is likely too simplistic. There must be more to the story than merely steering immune responses toward Th1 because those individuals infected with helminth worms, which strongly induce Th2 responses characterized by IgE production and eosinophilia, do not suffer from an increase in atopic disease. A striking demonstration of this principle was observed when serum IgE levels, parasite burdens, and allergies were compared in children born and raised in West Germany versus those growing up in East Germany before 1989. Increased parasite infections, high levels of serum IgE (not associated with specific allergens), and decreased atopy were found in the East German children compared to the West German children. Similarly, in one study in Africa, even when an individual did show an atopic response to an allergen when tested with a skin prick, clinical symptoms of allergy did not appear if the individual also suffered from a significant helminth infection. In other words, even though these individuals had IgE-producing B cells sensitized to the allergen, their mast cells did not degranulate and inflammatory infiltration of target organs did not occur. (It is important to note that light helminth infections, of the sort that occur in developed countries where prompt medical attention is available, were not protective against allergy.)

Another line of evidence that argues against the hygiene hypothesis is that a simple shift in the Th1/Th2 balance should be associated not only with an increase in Th2-mediated atopy but also with a decrease in Th1-mediated autoimmune diseases. Such a decrease has not been observed; indeed, as described in detail in Chapter 29, the incidence of Th1-associated autoimmune diseases such as type I diabetes and inflammatory bowel disease is on the rise. Like atopy, autoimmune diseases are more prevalent in developed countries than in developing ones, and in urban as opposed to rural environments. In countries where a cereal diet predominates, low rates of both atopy and type I diabetes are observed, and one study on breastfeeding noted protective effects against the incidence of both asthma and type I diabetes.

The preceding observations have prompted some scientists to propose that it is not the development of Th1 responses that is important for suppressing atopy so much as the generation of adequate capacity to produce anti-inflammatory cytokines. Indeed, helminth-infected individuals have elevated levels of IL-10, while allergic individuals show decreased IL-10. IL-10 and TGFβ are known to suppress both cytokine secretion and proliferative responses of antigen-specific T cells, and IL-10 also inhibits mast cell degranulation. It may be that numerous childhood infections help to establish a powerful anti-inflammatory response capable of dampening allergic responses. If this anti-inflammatory capacity is not fully developed during childhood, a predisposition to allergy might arise. It is tempting to speculate that pathogens such as HepA might induce the development of Th3 cells that express large amounts of IL-10 and mediate anti-inflammation. Significantly, Th3 cells have been shown to decrease the incidence of type I diabetes in mouse models of the disease. It is hypothesized that a greater exposure to microbes in developing countries might spur the development or activation of Th3 cells, whose production of anti-inflammatory cytokines would depress the activation of both Th1 and Th2 cells and consequently reduce the incidence of both atopy and autoimmune disease.

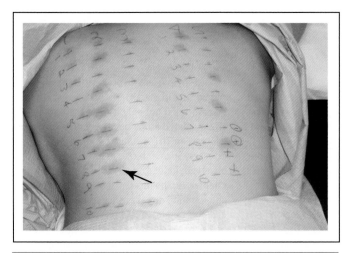

Skin Prick Test for Allergy
An example of a positive reaction to a test allergen is indicated by the arrow. Courtesy of Dr. William Howland III (www.allergy-asthma-austin.com).

ii) Allergy Therapy

Allergies often have different manifestations in different individuals, and even different manifestations at different phases of the response in the same individual. As a result, the appropriate therapy will depend on the individual's circumstances and exposure to the allergen. We discuss both the more conventional treatments in common use (Table 28-7) as well as some experimental approaches (Table 28-8).

iia) Basic approaches. Prevention of an allergic response by minimizing contact with the allergen is usually a beneficial approach but is not always possible if the allergen is ubiquitous. Antibiotics can reduce the presence of pathogens that trigger or exacerbate flare-ups of AD, while emollients can help to rehydrate dry skin and reduce the chance of allergen entry. Asthmatics should do their best to avoid breathing cold air or exercising too vigorously.

iib) Antihistamines. Because the principal mediator of symptoms released during the early phase of an allergic response is histamine, molecules that mitigate its effects are often used to treat mild allergies such as hives and hay fever. Antihistamines

Table 28-7 Conventional Allergy Therapies

Type	Rationale	Drawbacks
Minimize contact	Immune response of sensitized individual is not provoked	Hard to do if allergen is ubiquitous
Antihistamine (H1 receptor antagonist)	Blocks H1 receptors on cells of target organs and prevents histamine binding	May cause drowsiness, nausea, dry mouth, constipation Does not block late phase reaction
Antihistamine (H2 receptor antagonist)	Blocks H2 receptors on skin cells and prevents some of the effects of histamine binding	Limited to relief of urticaria
Epinephrine	Reverses mechanisms of anaphylactic shock	Major effects on circulatory system that are detrimental if sustained
Lipoxygenase antagonist	Blocks formation of leukotrienes and thus eosinophil influx	Churg-Strauss syndrome (rare)
Bronchodilator (β-adrenergic)	Blocks metabolism of cAMP in mast cells to prevent degranulation; opens K^+ channels and induces smooth muscle relaxation so airway expands	Does not treat underlying allergic inflammation
Corticosteroid	Binds to glucocorticoid receptors that inhibit transcription factors; reduces cytokine production and adhesion molecules	Long term use or high doses are associated with localized damage to skin or eyes; systemic effects on water retention, blood pressure; growth stunting in children; immunosuppression
Cromone	May block Ca^{2+} influx or alter Cl^- channels; may have effects on cytoskeletal proteins required for mast cell degranulation (mast cell stabilizer); inhibits late phase reaction in asthma	Minimal
Immunosuppressive drugs	Block T cell activation and function	Overly broad suppressive effect
Hyposensitization	Anergizes allergen-specific Th2 cells by repeated low doses of allergen; increases Th3 population; promotes isotype switching to IgG4	Requires repeated injections; anaphylaxis possible
Anti-IgE Ab	Inactivates circulating IgE and downregulates FcϵRI	Expensive; requires frequent injections; limited efficacy

Table 28-8 **Experimental Allergy Therapies**

Type	Rationale	Drawbacks
Lyn or Syk inhibitor	Blocks FcεRI signaling	Potentially non-specific effects
Co-ligation of FcεRI with FcγRIIB	Natural negative regulatory mechanism	Unknown as yet
IL-12	Promotes differentiation of allergen-specific Th1 rather than Th2 cells	Not in use; severe metabolic disruption
IFNγ	Th1 cytokine; blocks Th2 differentiation	Not in use; IFNγ promotes eosinophil survival and activation
IL-10	Suppresses anti-allergen IR	Not in use; no clinical benefit seen
Soluble IL-4R, anti-IL-4R Ab	Compete for IL-4 and block B cell isotype switching to IgE	Unknown as yet
Anti-IL-5 Ab	Blocks IL-5-mediated stimulation of eosinophil activation/survival	Unsuccessful in clinical trials to date
Anti-βc Ab	Blocks IL-3R, IL-5R, and GM-CSFR signaling and thus eosinophil differentiation and activation	Unknown as yet
Anti-TNF Ab or inhibitors of enzymes required for TNF synthesis	Prevents TNF from activating transcription factors contributing to HS response	Unknown as yet
IL-1Ra	Blocks inflammatory effects of IL-1	Unknown as yet
Socs-1, small molecule STAT6 inhibitor	Block STAT6 activation during IL-4 and IL-13 signaling and thus Th2 differentiation	Unknown as yet
Small molecule NF-κB, AP-1 inhibitors	Block transcription of IL-4- and IL-13-dependent genes	Potentially non-specific effects
Ribozymes, anti-sense oligonucleotides	Specifically prevent translation of proteins associated with atopy	Unknown as yet
Anti-CCR3 Ab; modified RANTES	Blocks receipt of chemokine signals by eosinophils, Th2 cells, and basophils	Unknown as yet
CTLA4-Ig, anti-B7-2	Costimulatory blockade that anergizes allergen-specific Th2 cells	Also affects allergen-specific Th1 cells
Anti-allergy vaccine	Induces pre-emptive Th1 response to allergen with bacterial product co-administration	Potential risk of endotoxic shock

work by binding to histamine receptors on target organs, preventing histamine released by degranulating mast cells from triggering symptoms. If an individual knows that he or she is about to be exposed to an allergen, the prior use of antihistamines can prevent symptom development.

Histamine was discovered in 1911, and the first generation of useful antihistamine drugs (including chlorpheniramine and promethazine) was developed in the mid-1940s. These molecules block the effects of many of the effects of mast cell degranulation by antagonistically binding to H1 histamine receptors on target cells. Unfortunately, these antagonists are also able to cross the blood–brain barrier and thus cause significant drowsiness as well as unwanted side effects such as dry mouth, constipation, cough, nausea, and vomiting. A second generation of H1 antihistamines was then developed in which modifications to the basic chemical structure of the H1 drugs prevented them from crossing the blood–brain barrier. These drugs, which include cetirizine,

fexofenadine, ebastine, and loratadine, thus relieve allergy symptoms without sedating the patient. While these second generation H1 antagonists are useful for treating rhinitis and the itching of AD, they (like first generation H1 antagonists) are ineffective in alleviating the late phase symptoms characteristic of asthma. There are three other types of histamine receptor, H2, H3, and H4, that bind to histamine in different locations in the body. H2 receptors primarily mediate the induction of gastric acid secretion by histamine but are also expressed in the skin. Drugs designed to antagonize H2 receptors, such as cimetidine and ranitidine, can relieve many of the symptoms of urticaria. The H3 receptors are involved in neurotransmitter release in the CNS, while the H4 receptors appear to function intracellularly to promote wound healing. Drugs antagonizing these receptors have not been shown to be useful for atopy to date.

As mentioned previously, antihistamines are no match for the rapidity of the early phase reaction in anaphylaxis.

Epinephrine is the only effective treatment but it must be given immediately. Anaphylactic individuals are well-advised to be vigilant and avoid all contact with the offending allergen. At the moment, there is no preventive treatment for anaphylaxis.

iic) Lipoxygenase antagonists. Leukotrienes generated by the action of lipoxygenase mediate or promote many symptoms of atopy, particularly bronchoconstriction and eosinophil infiltration. As a result, the inhalation of lipoxygenase-blocking agents such as zileuton, zafirlukast, pranlukast, and montelukast has been used to obtain relief from asthma symptoms and improve pulmonary function. These agents can be given orally and usually have minimal side effects, making them very useful for small children who have difficulty with inhalers (see later). Because lipoxygenase antagonists combat inflammation in a different way from corticosteroids (see later), current asthma treatment plans often include a combination of these agents; the presence of the lipoxygenase antagonist allows the dose of the corticosteroid to be reduced. Lipoxygenase antagonists are also effective in treating allergic rhinitis. One concern with some lipoxygenase antagonists is *Churg-Strauss syndrome*, a rare disease characterized by flu-like symptoms and inflammation of pulmonary blood vessels. Although this syndrome can be life-threatening, it is usually rapidly resolved with corticosteroid treatment. Some investigators feel that the lipoxygenase antagonist does not cause Churg-Strauss per se but rather causes the pre-existing condition to be unmasked, because the antagonist allows the corticosteroid dose keeping the Churg-Strauss under control to be decreased.

iid) Bronchodilators, corticosteroids, and cromones. Many symptoms of a type I HS response are due to molecules other than histamine (principally cytokines), so that most antihistamines are not of particular use for cytokine-dominated disorders such as asthma and AD. Indeed, some antihistamines can even be harmful to an asthmatic because these agents cause the mucus in the airway to thicken and further shrink the airway passages. Current treatment of an acute asthma attack often entails the administration of rapidly acting bronchodilators ("puffers"). These agents are usually aerosolized β-adrenergic drugs (such as theophylline, salbutamol, salmeterol, or formoterol) that block the phosphodiesterase enzyme responsible for metabolizing cAMP in mast cells. The accumulation of cAMP then blocks mast cell degranulation. β-adrenergic drugs also open potassium channels, inducing smooth muscle relaxation that relaxes bronchoconstriction. However, bronchodilators do not address the underlying inflammatory response responsible for atopic symptoms.

Corticosteroids have powerful anti-inflammatory effects because they bind to glucocorticoid receptors and inactivate transcription factors such as NF-κB, AP-1, NF-AT, and the STATs. Both cytokine production and the expression of adhesion molecules required for the entry of inflammatory cells into target tissues are thus inhibited. Low dose corticosteroids are often used in a cream form to treat AD and as nose drops for rhinitis. Inhaled aerosolized corticosteroids are often highly effective for the treatment of asthma, although an estimated 10% of children and 25% of adolescents with asthma do not respond to inhaled corticosteroids. Corticosteroids are associated with very few side effects when used at a low dose. However, high dose, long term, or oral use of corticosteroids has been linked to nasty local side effects such as dermal atrophy, glaucoma, and optic nerve damage and systemic side effects such as elevated blood pressure, water retention, and increased excretion of potassium and calcium. Growth stunting in children and immunosuppression may also be a concern in some situations. Researchers are in the process of designing modified corticosteroids that minimize systemic side effects, focusing on corticosteroid-based molecules that are inactivated when they enter the blood.

The cromones (including nedocromil sodium and sodium cromoglycate) are effective anti-inflammatory drugs that have fewer systemic effects than corticosteroids. When intranasally or ophthalmologically applied, cromones are effective in reducing symptoms of allergic rhinitis. Intranasal cromones can also alleviate asthma symptoms without significant side effects. Oral cromones are useful for treating food allergies in some (but not all) patients. The mechanism of action of these compounds is unclear but it is speculated that they may either block Ca^{2+} influx into mast cells or influence the chloride channels that figure prominently in mast cell degranulation. Other evidence has indicated that sodium cromoglycate alters a key cytoskeletal protein required for mast cell degranulation. Because they prevent mast cells from degranulating, these agents are often known as "mast cell stabilizers." Cromones have also been found to inhibit the macrophages and eosinophils driving the late phase reaction of asthma.

iie) Immunosuppressive drugs. The immunosuppressive drugs described in Chapter 26 and 27 have been used to decrease the functions of activated allergen-specific Th2 cells. However, these agents also affect allergen-specific Th1 cells and thus do not restore a normal Th1/Th2 balance in atopic patients. The development of regulatory T cells thought to mitigate the allergic inflammatory response is also inhibited by these drugs, making them counterproductive in this situation.

iif) Hyposensitization. The side effects of many drugs used to treat symptoms of allergy have bolstered the search for clinical approaches that would block the initiation of the atopic response itself. *Hyposensitization* (or *desensitization*) is the name given to a procedure in which the allergy sufferer receives subcutaneous injections of ever-increasing amounts (5–20 μg) of the purified allergen every week or month for a period of up to 3–5 years. In many cases, the atopic individual eventually ceases to respond to the allergen. Hyposensitization is not new: the first efforts at this type of treatment were undertaken by Noon and Freeman in 1911. Hyposensitization works well for those who are allergic to insect venom or whose response to the allergen takes the form of rhinitis or urticaria but is not as effective for those suffering from asthma or eczema. Hyposensitization is almost always started only after a child has passed his or her 5th birthday.

The mechanism underlying hyposensitization is still not completely understood. It is thought that the continuous dosing with increasing allergen may have three major effects (Fig. 28-8). First, the differentiation of allergen-specific Th1 cells may be promoted at the expense of Th2 cells. Cytokines (such as IFNγ) produced by Th1 cells may inhibit isotype switching to IgE. Secondly, allergen-specific Th2 cells may be anergized by the repeated stimulation with sub-immunogenic doses of allergen under conditions in which costimulation is minimal. Thirdly, but perhaps most importantly, CD4⁺ regulatory Th3 cells (which express high levels of IL-10 and TGFβ) may be increased in number or activated. *In vivo* studies have shown that the blood of allergic individuals contains lower levels of Th3 cells than the blood of non-allergic controls. Work in animal asthma models has supported the idea that Th3 cells may be helpful in modulating the allergic inflammatory response. For example, when T cells engineered to overexpress TGFβ or IL-10 ("artificial" Th3 cells) were adoptively transferred into mice with allergic airway symptoms, the treated animals showed greatly diminished AiHR and airway inflammation. In allergic humans in whom hyposensitization is successful, serum levels of IL-10 (likely produced by activated Th3 cells) are sharply elevated. High levels of IL-10 have been shown to suppress T cell activation, decrease mast cell numbers, and depress eosinophil function, ameliorating the atopic symptoms.

Another end result of hyposensitization is that both IL-10 (and TGFβ) induce a patient's B cells to favor the production of IgG₄ antibody instead of IgE. IgG₄ does not bind to any FcRs on mast cells, so the presence of the allergen will not trigger mast cell degranulation. In addition, direct competition by the anti-allergen IgG₄ will prevent the allergen from binding to any IgE still present on mast cells. There is also some evidence that IgG₄ anti-allergen antibodies may displace the IgE from circulating IgE–allergen complexes, so that there is less binding of IgE–allergen complexes to FcγRII expressed on B cells. Fewer IgE–allergen complexes are internalized by B cells, reducing their ability to present allergen epitopes to memory T cells. Interestingly, IgG₄–allergen complexes do not appear to bind to the low-affinity FcγR expressed on B cell surfaces.

A significant danger with hyposensitization is that the process may induce an anaphylactic response to the allergen. To avoid this problem, researchers have been working on various ways to modify the native allergen and minimize systemic reactions. The earliest efforts involved pre-treating allergen extracts with denaturing agents such as formaldehyde, creating *allergoids*. In more recent years, peptide fragments of the principal allergen have been given to experimental animals to induce hyposensitization. In the case of mice repeatedly injected with a peptide derived from Fel d 1, production of IgE and IL-2 by lymphocytes isolated from these animals was decreased. Human CD4⁺ T cells repeatedly exposed to modified allergenic T cell peptide epitopes

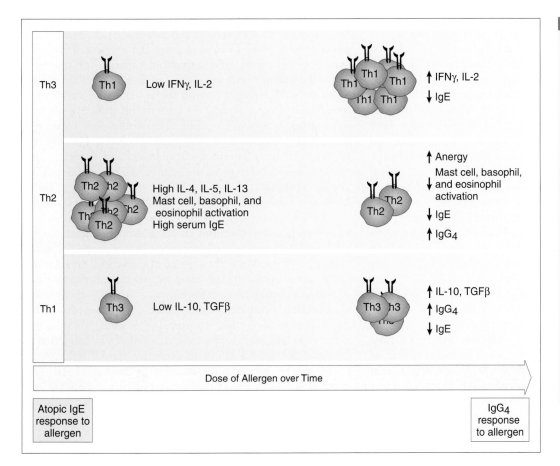

Figure 28-8

Hyposensitization
The end result of hyposensitization is the conversion of a harmful IgE response to an allergen to a harmless IgG4 response. The individual is given ever-increasing doses of the allergen over a lengthy period of time. The number of allergen-specific Th1 cells is increased at the expense of the Th2 population, promoting the production of cytokines that inhibit isotype switching to IgE. Concomitantly, allergen-specific Th2 cells that secrete cytokines favoring mast cell activation and isotype switching to IgE are anergized by the constant exposure to sub-immunogenic doses of allergen. Most importantly, hyposensitization gradually increases the number and/or activation of Th3 cells. These cells secrete cytokines suppressing mast cell activation and promoting isotype switching to IgG4.

in vitro also showed evidence of non-responsiveness. It is speculated that these peptides were of sufficient size to induce the anergy of allergen-specific T cell clones but were too small to cross-link IgE and induce the release of histamine from mast cells and basophils. Results have not been as favorable in the clinic. In one study, allergic patients injected four times with peptide from either cat or ragweed allergens showed some improvement in symptoms but also unacceptable side effects. Other work has shown very little effect of peptide injection on symptoms and significant numbers of adverse events.

Hyposensitization with plasmid DNA expressing a gene encoding an allergen has been modestly successful in experimental animals. In theory, the allergen protein would be continuously produced after its introduction, avoiding the need for repeated "allergy shots." In the case of rats injected with DNA encoding the Der p 5 allergen from the house dust mite, subsequent inhalation of the allergen did not trigger an anti-allergen IgE-mediated response. A similar approach has been tried in mice sensitized to the principal peanut allergen Ara h 2. DNA encoding this allergen was complexed to a chitosan delivery vehicle and fed to the mice. The allergen was expressed in the intestine, and anaphylactic responses to the oral allergen were blocked. However, as outlined in Chapter 23, the adaptation of DNA vaccination technology to humans requires further optimization.

iig) Interference with IgE/FcεRI signaling. An obvious way to stop early phase reactions is to interfere with FcεRI signaling, either by blocking the binding of IgE to FcεRI or by disrupting signaling events downstream of this binding. An allergy patient can be passively immunized with a recombinant anti-IgE humanized mAb (such as omalizumab; also called Xolair) that selectively binds to circulating IgE and prevents it from binding to mast cell and basophil FcεRI. Not only does this strategy directly interfere with cellular activation but the reduction in circulating IgE induces the downregulation of FcεRI expression on these cells. In early clinical trials, administration of anti-IgE ameliorated both the early and late phase responses in asthma patients and decreased the accumulation of eosinophils in their airways. No adverse side effects were reported. However, subsequent larger scale clinical trials of blocking anti-IgE mAbs have failed to demonstrate broad clinical efficacy in atopic patients. In addition, this type of therapy is expensive and requires frequent injections. At the moment, this approach is offered only to those over 12 years of age whose moderate or persistent asthma is not adequately controlled by corticosteroid inhalation.

Other mAbs have been isolated that bind to FcεRI and activate Lyn but do not induce Syk activation, leading to failed signal transduction. A small molecule inhibitor has also been identified that specifically inhibits Syk activation. However, the problem with these types of therapies is that, as we have learned, Syk kinase is vital for BCR signaling, so that some way will have to be found to selectively block Syk function in mast cells and basophils. Another means of modifying FcεRI signaling may be to promote the natural negative regulation exerted by FcεRI/FcγRIIB co-ligation described earlier in this chapter. Indeed, the success of hyposensitization may

be due in part to such a mechanism since this therapy deliberately increases the production of allergen-specific IgG molecules. *In vitro*, co-ligation of FcεRI and FcγRIIB on the surfaces of basophils and mast cells reduces their production of allergy-related mediators.

Studies of the crystal structures of FcεRI–IgE complexes have suggested other ways in which the triggering of the atopic signaling cascade can be prevented. Peptide blockers have been designed that interfere with the binding of IgE to FcεRI, or upset the interactions between the component chains of the FcεRI, but these are still very much experimental in nature.

iih) Deviation of T cell responses. With the observation that many allergy sufferers show a bias toward Th2 responses, clinical researchers have investigated ways of shifting their immune responses back toward the Th1 side. An obvious approach is administration of IL-12, the primary cytokine required for Th1 differentiation. In one study, mice vaccinated with a fusion protein of IL-12 plus allergen successfully mounted a Th1 response against the allergen rather than a Th2 response with its associated IgE production. However, in another study of rodent models of asthma, the transfer of allergen-specific Th1 cells to susceptible animals exacerbated rather than ameliorated airway inflammation. In humans, the administration of IL-12 to the airway results in unacceptably severe side effects. One might then turn to administration of IFNγ, since this key Th1 cytokine inhibits Th2 development. However, in one important study of asthmatics, IFNγ was <u>elevated</u> (not decreased) in the serum and bronchial fluids of patients undergoing an acute asthma attack. In addition, IFNγ has been found to prolong the activation and survival of eosinophils *in vitro*. In view of these factors, and the possible correlation of atopy with a reduced capacity for mounting anti-inflammatory responses (refer to Box 28-2), the clinician might be well-advised to avoid the induction of an allergen-specific Th1 response and concentrate on inducing an anti-inflammatory response instead. The reader will recall that IL-10 has powerful anti-inflammatory effects, blocking the synthesis of many enzymes, chemokines, and cytokines that promote tissue infiltration. Indeed, in sensitized animals, an allergic response can be blunted by treatment with IL-10. Moreover, some studies have found a defect in IL-10 transcription in the macrophages of asthmatic patients. However, the administration of exogenous IL-10 to atopic patients has yet to show any clinical effect. It may be that endogenous induction of IL-10 directly in the site and at the time of allergen sensitization is required for blocking the development of a Th2-mediated anti-allergen response.

iii) Blocking cytokines. Agents that can prevent IL-4 and IL-13 from binding to IL-4R and IL-13R on B cells and thus block switching to IgE production are under investigation as therapies for type I HS. For example, aerosolized, soluble forms of the extracellular domain of human IL-4R (one example is Nuvance) have been given to asthma patients with the goal of sopping up IL-4 before it can reach B cells and induce them to produce IgE. An advantage to this agent is that, because its amino acid sequence is identical to that of the extracellular domain of native IL-4R, the solubilized decoy should be non-immunogenic (unlike a humanized mAb). So far, modest efficacy in relieving asthma symptoms has been achieved with no serious side effects. This success has spurred a search for similar soluble antagonists of IL-13. In addition, antibodies directed against IL-4, IL-9, and IL-13 are under study to determine their effects on asthma symptoms. In a related approach, antibodies that block IL-4R have been shown to reduce AiHR and eosinophil infiltration in a mouse model of asthma.

Particularly in cases of asthma, infiltration by eosinophils is a major contributor to pathogenesis. Cytokines such as IL-5, GM-CSF, and IL-3 promote eosinophil survival, activation, and differentiation, and elevated levels of these cytokines are present in the lungs of asthmatics undergoing an attack. Accordingly, some clinical researchers have explored blocking these cytokines as a treatment for asthma. In experimental mouse models of asthma, the administration of neutralizing antibodies against IL-5 prevented the influx of eosinophils into the airways and decreased AiHR. While similar results were obtained in guinea pig models of asthma, protection in experimental monkeys was not as complete. Unfortunately, clinical trials of an anti-IL-5 mAb as a treatment for asthmatic humans have also failed to show significant efficacy. Eosinophils were reduced in the blood and sputum but there was no improvement in either AiHR or the late phase reaction to inhaled allergen. Other researchers have targeted the relevant cytokine receptors. Knockout mice lacking the βc subunit shared by the IL-5, GM-CSF, and IL-3 receptors show significantly decreased numbers of eosinophils and fail to mount an eosinophilic influx in response to an allergen. To imitate this situation, a monoclonal antibody called BION-1 has been generated that binds to βc and stops signal transduction through it, blocking the action of all three cytokines. Experimental mice treated with BION-1 showed reduced eosinophil numbers in the bone marrow. A humanized version of this mAb is under development.

Another major contributor to allergy is TNF, which amplifies atopic responses by activating NF-κB, AP-1, and other transcription factors. Treatment with a blocking anti-TNF antibody called infliximab has improved the clinical symptoms of patients with other TNF-mediated diseases, suggesting that this agent might be given to patients whose atopic disease is not significantly ameliorated by corticosteroids. Inhibitors of enzymes required for TNF synthesis are also under investigation. Because IL-1 is another major inflammatory mediator, clinicians have tested the administration of the soluble IL-1R antagonist IL-1Ra for treatment of asthma symptoms in experimental animals. There is some evidence that IL-1Ra can reduce AiHR in these models.

The effects of cytokines can also be abrogated by blocking their downstream signaling. A prime target is STAT6 function, since this transducer is required for both IL-4 and IL-13 signaling. Small molecule inhibitors of STAT6 have been shown to reduce IL-4-dependent gene expression *in vitro*. However, because it remains a challenge to deliver an exogenous inhibitor intracellularly, scientists are investigating the manipulation of the endogenous STAT6 inhibitor Socs-1. Further

downstream, small molecule inhibitors of NF-κB and AP-1 have been devised that reduce IL-13 and eotaxin levels as well as AiHR in mouse asthma models. A possible drawback to these agents may be the inhibition of expression of many other "desirable" genes.

As well as targeting molecules contributing to allergy at the protein level, researchers have invented ways to intercept them at the RNA level. An advantage of an RNA approach over the protein- and small molecule-based strategies described previously is that the offending protein is never produced. One strategy is to use a *ribozyme* (introduced in Ch.26) to cleave the mRNA encoding a protein driving atopy. This cleavage blocks the translation of the mRNA, leading to a decrease in the protein (but not gene) expression of the molecule of interest. Ribozymes targeting sequences in the mRNAs for IL-5, ICAM-1, and NF-κB are under investigation in mouse asthma models. Another way of preventing the translation of specific mRNAs is to use an *anti-sense oligonucleotide*. These oligonucleotides are synthesized to order *in vitro* and are designed to bind to a complementary sequence (the "sense strand") in the primary RNA transcript or mRNA encoding a protein of interest. The duplex mRNA that forms may block the splicing of the primary transcript, arrest translation of the mRNA, disrupt ribosomal assembly, or induce mRNA cleavage by RNase H activation. Several agents of this type have been formulated for use in animal asthma models, and the molecules targeted have included IL-4, IL-5, IL-5Rα, βc chain, Syk, Lyn, GATA-3, and STAT6. Promising reductions in outcomes associated with asthma (AiHR, lung inflammation, switching to IgE) have been noted in several cases. When an anti-sense oligonucleotide is designed to be inhaled directly into the lungs, it is called a RASON (*respirable anti-sense oligonucleotide*). In this case, the anti-sense oligonucleotide is associated with a lipid surfactant that increases the uptake of the oligonucleotide and its distribution in the lung tissue. For example, Durason is a 21-nucleotide RASON that inhibits the expression of the adenosine A1 receptor in bronchial smooth muscle. In an asthmatic person or animal, engagement of this G-protein-coupled receptor promotes lung inflammation and bronchorestriction. In experimental rabbits, Durason inhibited expression of the A1 receptor in bronchial smooth muscle and decreased airway obstruction and AiHR. In early clinical trials in patients with moderate to severe asthma, inhalation of Durason significantly reduced airway obstruction. The effects were specific and long-lasting, and accompanied by minimal systemic side effects. Clinical investigation of these and other RNA-based agents is ongoing.

iij) Blocking chemokines. The infiltration of leukocytes in the late phase reaction of type I HS is dependent on certain chemokines. Accordingly, these chemokines and their receptors have been examined as potential therapeutic targets. Eosinophils, Th2 cells, and basophils express high levels of the chemokine receptor CCR3. This receptor binds multiple ligands necessary for chemotaxis and tissue infiltration, including eotaxin, RANTES, MCP-3, and MCP-4. An anti-CCR3 antibody and a modified RANTES molecule that block the binding of all of these ligands to the receptor have been shown to inhibit eosinophil chemotaxis *in vitro*. However, for unknown reasons, CCR3$^{-/-}$ mice show <u>increased</u> infiltration of mast cells into bronchial epithelium and enhanced AiHR. In addition, studies of eotaxin-deficient mice have demonstrated that eotaxin is not essential for the development of allergic airway responses. These findings cast doubt on the potential therapeutic value of strategies blocking either of these molecules *in vivo*.

Greater success in treating asthma may come from blocking CXCR2 function. CXCR2 (whose major ligand is IL-8) is expressed on neutrophils, activated monocytes and macrophages, and activated eosinophils. CXCR2 function and IL-8 overexpression have both been associated with the increased production of mucus that is a hallmark of asthma. This mucus overproduction results from hyperplasia of the airway goblet cells. In CXCR2$^{-/-}$ mice, goblet cell hyperplasia is reduced and mucus production is significantly decreased. Conversely, the introduction of chemokine ligands for CXCR2 into the airways of mice induces goblet cell hyperplasia. Researchers believe that the engagement by chemokine ligands of CXCR2 (and possibly other chemokine receptors such as CCR2) on macrophages, eosinophils, and neutrophils triggers the release of mediators such as TGFα and *epidermal growth factor* (EGF) that stimulate airway goblet cells to proliferate and produce copious amounts of mucus. CXCR2 and CCR2 may thus represent additional therapeutic targets for human asthma treatment.

iik) Costimulatory blockade. The costimulatory blockade strategies described in Chapters 26 and 27 have shown some promise for atopy treatment in experimental settings. These approaches aim to disrupt T cell activation by blocking the interaction of the CD28 and B7 molecules expressed on T cells and APCs, respectively. Both CTLA4-Ig fusion protein and anti-B7-2 antibodies have been shown to reduce the development of allergen-specific IgE, pulmonary eosinophilia, and AiHR in mouse models of asthma.

iil) Anti-allergy vaccines. It may be possible in the future to reduce atopy in babies or young children at particularly high risk by administering vaccines that induce immune deviation favoring Th1 responses. This idea is related to the hygiene hypothesis discussed in Box 28-2. The theory is that, if an individual is treated so as to set up overall Th1-inducing conditions in the body (prior to exposure to the allergen), when his or her allergen-specific naive T cells encounter the allergen, they will be induced to develop into relatively harmless Th1 effectors rather than symptom-generating Th2 effectors. Evidence for the feasibility of this approach has been gained from animal studies. For example, if a mouse is injected with killed bacteria or bacterial products 2 weeks prior to administration of an allergen, anti-allergen Th1 responses are favored, the production of allergen-specific IgE is reduced, and symptoms of atopy (such as eosinophilia and AiHR) are reduced. A parallel phenomenon has been observed in some human populations, in that subsets of individuals receiving the BCG vaccine (live attenuated bacteria; refer to Ch.23) occasionally show a decreased tendency to be atopic. Many researchers now envision an approach in which vaccination with a bacterial product induces the production of IL-12 and IFNγ at the probable sites of allergen entry into the body and forestalls a Th2 response.

One promising vaccination strategy involves the use of bacterial unmethylated CpG motifs. We have encountered bacterial CpG before as the ligand for TLR9 (refer to Ch.4) and as an adjuvant for DNA vaccines (refer to Ch.23). Engagement of TLR9 on a macrophage by CpG activates it and upregulates the expression of MHC class II and costimulatory molecules. Most relevant to our discussion of allergy, Th1 cytokine production is also induced. When mice were given intranasal or intraperitoneal injections of CpG along with an allergen known to induce asthma-like symptoms, both eosinophil infiltration into the airway and AiHR were reduced compared to controls that did not receive CpG. Subsequent studies confirmed that the effects were long-lasting (a number of weeks), and the mice were capable of mounting Th1 memory responses in a subsequent exposure to allergen. IFNγ was identified as important (but possibly not essential) for the replacement of Th2 responses to the allergen with Th1 responses.

There are considerable hurdles to be overcome before the preceding strategy becomes truly workable in humans. First of all, the approach rests on the ability of a killed bacterium or bacterial product to induce a vigorous Th1 response capable of suppressing Th2 responses, and such vigor might require a considerable quantity of the bacterium or product. The presence of large amounts of such materials in a tissue might itself induce severe inflammation and immunopathology. Some progress has been made in experimental animals in reducing the amount of bacterial product necessary to achieve the desired response. Smaller amounts of CpG can be used if they are directly conjugated to the allergen rather than merely mixed with it. Another potential problem with this approach is that all possible effects of a bacterial product must be considered. In some animal studies, small amounts of LPS induce IL-12 production and reduce airway constriction arising from allergen challenge. However, LPS in sufficient quantity can induce the production of TNF and other cytokines causing endotoxic shock. CpG has also been reported to cause endotoxic shock in mice. Given that most human allergies are not life-threatening, the potential risk of endotoxic shock must be reduced to a negligible level before these types of treatments could be considered sensible for human allergy therapy.

iim) Gene therapy? The reader may be wondering about gene therapy for treatment of atopy. As we have seen, atopy is not a single gene disorder, in which the addition of a functional copy of a mutated gene has the potential to save the day (if it is expressed in the right place). Thus, the modification of the expression of one or a few atopy-related genes is unlikely to be effective. Engineered overexpression of an anti-inflammatory gene (such as IL-10) might be helpful, but the trick will be getting the expression to occur in the right place at the right time and in a sustained manner.

C. Type II Hypersensitivity: Direct Antibody-Mediated Cytotoxicity

I. WHAT IS TYPE II HS?

A type II hypersensitivity is said to occur when damage to the host tissues is caused by cellular lysis induced by the direct binding of antibody to cell surface antigens. While the antibodies involved in type I HS are of the IgE isotype, those involved in type II HS reactions are mainly of the IgM or IgG isotype. In some cases of type II HS, the pathological antibodies attack leukocytes or red blood cells, so-called "mobile cells." In other cases, the antibodies bind to cells that are "fixed" as part of a solid tissue. With respect to the antigenic specificity of the pathological antibody, some recognize a foreign entity that has "stuck" on the surface of a host cell in some way. Others may be autoantibodies that, due to a failure in tolerance mechanisms, are free in the periphery to bind to self-epitopes on host cells. Examples of type II HS include some forms of anemia, blood transfusion reactions, certain platelet disorders, and some types of tissue transplant rejection. Type II HS reactions are also frequently components of autoimmune diseases, and thus more discussion of the mechanisms and treatment of these disorders appears in Chapter 29. As for all hypersensitivities, some type II HS reactions are induced by drugs. The mechanisms responsible for drug hypersensitivities are discussed in Box 28-3.

II. MECHANISMS UNDERLYING TYPE II HS

The mechanisms of cell lysis and tissue damage involved in type II HS are the same as those triggered when IgG or IgM antibodies bind to cellular pathogens. ADCC, complement activation, and opsonized phagocytosis all take their toll (Fig. 28-9). To recap briefly: ADCC of a target cell is triggered by the binding of the Fab portion of the pathological antibody to the target cells, followed by the binding of the Fc region of the pathological antibody to FcR-bearing cytotoxic cells such as NK cells, neutrophils, eosinophils, and macrophages. Classical complement activation is initiated by the binding of the pathological antibody to the antigen on the target cells. C3b is deposited, which facilitates either opsonized phagocytosis by CR1-bearing cells such as macrophages or neutrophils, or MAC assembly and lysis of the target cell due to pore formation. (Presumably the RCA proteins attempting to protect the cell surface from MAC assembly are overwhelmed in hypersensitive individuals.) Interestingly, the damage mediated by opsonization depends on whether the target cell is mobile or fixed in a tissue. If the target is mobile, the cell can sometimes be successfully phagocytosed "on the spot" by the phagocyte. Alternatively, the antibody-bound target cell makes its way to the liver or spleen for phagocytosis in that location. However, if the target cell is a fixed component of a larger tissue or organ, a phagocyte may be unable to engulf the cell. The "frustrated"

phagocyte then releases the contents of its lysosomes externally, leading to localized tissue damage. NK cells contribute to the destruction by initiating ADCC of an antibody-bound target cell. This damage is exacerbated by the local presence of the complement activation by-products C3a and C5a. These anaphylatoxins set up a chemotactic gradient that draws additional neutrophils into the area and encourages basophils and mast cells to degranulate, further damaging the affected tissues.

III. EXAMPLES OF TYPE II HS: CYTOTOXICITY AGAINST MOBILE CELLS

Specific examples of type II HS reactions are summarized in Table 28-9.

i) Hemolytic Anemias

Anemia is a general term referring to any condition in which the level of RBCs in an individual is below normal. Many cases of anemia are the secondary consequences of non-immunological disease. However, when a pathological antibody has mediated the lytic destruction of erythrocytes in a type II HS reaction, the resulting disease is called *hemolytic anemia* or *immune hemolytic anemia*. The actual hemolysis often takes place extravascularly, with the location determined by the predominant antibody isotype involved. In the case of RBCs coated with IgG antibodies, the cells are destroyed by FcγR-bearing phagocytes in the spleen and very little intravascular complement activation is triggered. When IgM antibodies are involved, however, complement is readily activated by the classical pathway such that some RBCs are lysed "on the spot," resulting in a small amount of intravascular hemolysis. More commonly, IgM-bound RBCs bearing surface C3b and iC3b acquired during complement activation are conveyed to the liver, where they are destroyed by CR-bearing Kupffer cells.

ia) Alloimmune hemolytic anemias. Alloimmune hemolytic anemias occur when the individual is exposed to allogeneic erythrocytes such that antibodies directed against the foreign RBC antigens are generated. Should the interaction of these antibodies with the foreign antigen induce too vigorous an inflammatory response, the resulting tissue damage constitutes a type II HS reaction. Two of the best known examples of type II HS fall into this category: the *transfusion reaction* that occurs when a patient is transfused with ABO-incompatible donor blood, and *Rh disease* (also known as hemolytic disease of the newborn or erythroblastosis fetalis), which occurs when Rh-positive RBCs of a fetus are destroyed by maternal anti-Rh IgG antibodies. These disorders were discussed in detail in Chapters 27 and 23, respectively.

ib) Autoimmune hemolytic anemias. Autoimmune hemolytic anemias occur when the individual makes antibodies directed against epitopes on his or her own RBC. The autoantibodies involved are classified as being either "warm" or "cold" depending on the temperature at which they show optimal affinity and reactivity *in vitro*. Warm autoantibodies are most potent around 37 °C and have reduced effects at lower temperatures,

Box 28-3. Drug hypersensitivities

Although drugs are tested rigorously for patient safety during both pre-clinical and clinical trials, adverse drug reactions remain a serious medical problem. It is estimated that such reactions account for 2–5% of hospital admissions and thousands of deaths per year in the United States. Thus, while the vast majority of adverse drug reactions are mild and easily reversed by discontinuing use of the drug, the more serious reactions are the fourth most common cause of death in the United States. In addition to threatening the patient's health directly, drug reactions also harm patients indirectly by preventing them from receiving potentially useful therapeutic treatments.

Approximately 80% of adverse drug reactions occur because the pharmacological effect of the drug is too powerful for a given individual. These are known as "type A" or "augmented" drug reactions and do not involve the immune response in any way. A typical example would be bleeding due to administration of anticoagulants. Dose reduction is often enough to render the drug safe, and serious life-threatening outcomes are rare in these cases. The remaining 20% of adverse drug reactions cannot be predicted from the known pharmacology of the drug and are called "type B," "bizarre," or "idiosyncratic" reactions. Type B reactions tend to be more severe, accounting for a disproportionately high number of drug reaction deaths. In the majority of cases, these reactions are immune system-mediated and are therefore called *drug hypersensitivities*. (Those type B reactions that are non-immunological in origin are sometimes called "metabolic idiosyncrasies.")

The ultimate pathology caused by a drug HS may be due to the involvement of type I, II, III, or IV HS reactions or a combination of more than one of these immunological mechanisms. However, the initiation of the immune responses to the various offending drugs is thought to always occur via the same mechanism. This mechanism is referred to as the *hapten hypothesis of immune recognition of drugs*. Most drugs are chemically inert in their original form but become *bioactivated* during the normal process of drug metabolism. This metabolism involves catalysis of the drug by the cytochrome P450 enzyme present in hepatocytes as well as in many other organs in the body. In most individuals, the potential toxicity of the reactive metabolites created by bioactivation is counteracted by the presence of intracellular antioxidants and detoxification enzymes. However, due to infection, interference of other drug metabolites, genetic predisposition, or other unknown factors, some individuals have an imbalance in this process that leaves them with a surplus of reactive metabolites. In such cases, the reactive metabolites are free to bind covalently to serum or cellular proteins. Because the reactive metabolites are often of low molecular mass, their binding to the host proteins creates an array of immunogenic hapten–carrier complexes that can be processed and presented to Th cells. Following activation of the Th cells, both antibody and cell-mediated immune responses are mounted, potentially triggering both Ig-mediated (types I, II, and III) and cell-mediated (type IV) HS reactions in host tissues containing the drug-host protein conjugates.

Like all HS reactions, drug hypersensitivities require a sensitization phase and an effector phase. The primary sensitization to a drug takes place over an induction period of 2–6 weeks as the bioactive metabolites haptenate host proteins and activate naïve Th cells. Continued exposure to the drug or re-introduction of the drug at a later time then leads to the secondary response (effector phase) and elicitation of the adverse reaction. Clinically, patients with drug hypersensitivities present in many different ways. The most commonly affected organ is the skin, with most reactions being attributable to sulfonamides, penicillins, anticonvulsants, and non-steroidal anti-inflammatory drugs. While the majority of skin reactions are mild and pass quickly upon discontinuation of the causative drug, 1 in 1000 patients with cutaneous drug hypersensitivities experiences a life-threatening condition known as *toxic epidermal necrolysis* (TEN). Severe, painful blistering of the skin occurs followed by loss of the epidermis. Depending on the extent of skin sloughing, fluid loss and electrolyte imbalance can be significant. Mortality is approximately 30% for severe cases of TEN, while a slightly milder form of the reaction called *Stevens-Johnson syndrome* (SJS) has a mortality rate of 5%. For both TEN and SJS, the main cause of death is sepsis due to loss of the protective skin barrier. The underlying HS in these conditions is thought to be a type IV response mediated by CD4+ and CD8+ T cells.

The liver is also a common target of drug HS. An example is *halothane hepatitis*, in which inhalation of the analgesic halothane leads to production of drug-specific antibodies and mononuclear cell infiltrates in the liver. These features are suggestive of both type II and type IV HS responses in action. Other drugs cause a liver HS known as *immuno-allergic hepatitis*. Reactions to amodiaqine (anti-malarial), carbamazepine (anti-convulsant), diclofenac (non-steroidal anti-inflammatory), and dihydralazine (anti-hypertensive) are the most common causes of this disorder.

Some drugs induce immune system-mediated destruction of one or more types of blood cells, leading to dangerous deficits. Such reactions, while quite rare, can have severe consequences for the patient. For example, destruction of pluripotent stem cells results in *aplastic anemia*; a loss of neutrophils leads to *agranulocytosis*; and decimation of erythrocytes causes *hemolytic anemia*. The drugs aminophenazone (anti-pyretic and analgesic), amodiaquine (anti-malarial), mianserin (antihistamine), and clozapin (anti-psychotic) have all been implicated in cases of agranulocytosis. Widespread destruction of neutrophils induced by these drugs releases pyrogens that cause acute fever. Sore throat and a variety of infections then follow. In the case of clozapine, the incidence of agranulocytosis has been high enough (0.8%) to warrant the restriction of its use and vigilant monitoring of the neutrophil counts of patients who are prescribed this drug. Indeed, the serum from some individuals experiencing this type of drug reaction can lyse blood cells in culture. The isotypes responsible appear to be IgG and IgM antibodies, suggesting that type II HS is involved.

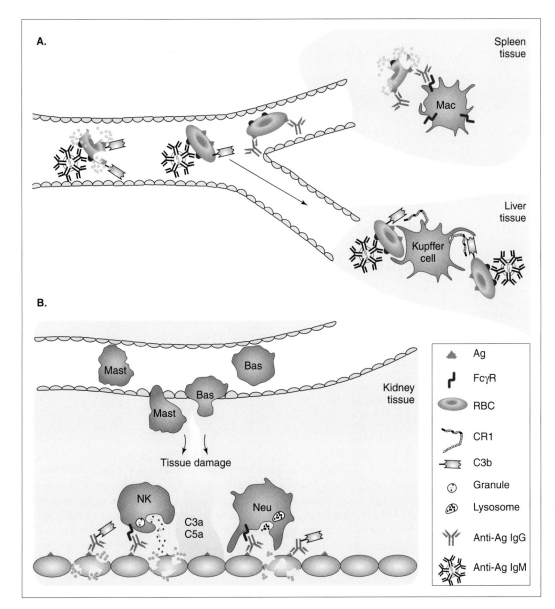

A.

Spleen
tissue

Mac

Liver
tissue

Kupffer
cell

B.

Mast   Bas

Bas

Mast

Kidney
tissue

Tissue damage

NK   Neu

C3a
C5a

	Ag
	FcγR
	RBC
	CR1
	C3b
	Granule
	Lysosome
	Anti-Ag IgG
	Anti-Ag IgM

Figure 28-9

Type II Hypersensitivity
Type II HS is mediated by non-IgE antibodies directed against antigens (which may be self antigens) on either mobile (A) or fixed (B) host cells. In (A), the target cells are host RBCs, such that the HS reaction leads to hemolytic anemia. In some cases, an anti-IgM antibody directed against an antigen on the RBCs binds to these cells, promoting complement-mediated destruction by MAC assembly that can occur directly within the blood vessel itself. Alternatively, RBCs complexed to either IgM or IgG travel to the liver or spleen, where destruction is mediated by resident phagocytes via C3b-opsonized phagocytosis or MAC assembly, or FcR-mediated ADCC. In (B), kidney cells expressing an antigen recognized by a pathological IgG antibody are fixed in the tissue and cannot be engulfed. Instead, "frustrated" phagocytes and NK cells release lysosomal enzymes and granule contents that damage the kidney cells. In addition, the local activation of complement establishes a chemotactic gradient that encourages the influx of mast cells and basophils that are triggered to degranulate by IgE-independent mechanisms, causing additional tissue damage.

while cold autoantibodies bind effectively only at temperatures below 37 °C. This difference in molecular behavior is also manifested *in vivo* in the patient.

Between 50 and 70% of autoimmune hemolytic anemias are due to warm autoantibodies. The presence of these antibodies is often secondary to an autoimmune or lymphoproliferative disorder, with the incidence of associated hemolytic anemia rising in those over 40 years of age. The onset of the anemia may be gradual or acute. In gradual onset cases, symptoms such as fatigue, dizziness, weakness, and breathing difficulty upon exertion first appear. Fever, coughing, abdominal pain, and weight loss may also be experienced. In these cases, the intensity of the symptoms often varies over time. In acute onset cases, the anemia strikes in a sudden, potentially life-threatening way. The patient experiences an intense wave of hemolysis that leads to all the preceding symptoms plus jaundice, pallor, edema, hemoglobinuria (hemoglobin in the urine), and swelling of the spleen, liver, and lymph nodes. If large amounts of hemoglobin

are released into the blood, kidney toxicity may result. Treatment of warm autoimmune hemolytic anemias usually involves glucocorticoids to inhibit the inflammatory response. As well, these disorders tend to involve antibodies of the IgG isotype; cases not controlled with drug therapy may be improved by splenectomy. (Splenectomy removes the main organ in which the extravascular hemolysis occurs.)

Two major types of cold autoimmune hemolytic anemia are relevant clinically: *cold agglutinin syndrome* (CAS), also known as *cold hemagglutinin disease* (CHD), and *paroxysmal cold hemoglobinuria* (PCH). CAS accounts for about 20–30% of all autoimmune hemolytic anemias, while PCH is the underlying cause in less than 10%. CAS is generally a disease of older adults, with the majority of patients around 70 years of age. The intensity of symptoms varies with the ambient temperature because of the nature of the cold autoantibodies involved. For example, patients may simply be pale and experience mild fatigue in temperate surroundings or seasons,

Table 28-9 Type II HS Reactions

Examples	Pathological Ab Directed Against	Tissue Destroyed
MOBILE CELLS		
Alloimmune hemolytic anemia in the form of blood transfusion reactions	Allo-Ag on donated RBCs	Donated RBC
Alloimmune hemolytic anemia in the form of erythroblastosis fetalis	Rh Ag on fetal RBCs when the mother is Rh$^-$	Fetal RBC
Warm and cold autoimmune hemolytic anemia	Self-Ag on host's RBC	Host's own RBC
Alloimmune thrombocytopenia in the form of neonatal thrombocytopenia	PLA1 on fetal platelets when mother is PLA$^-$	Fetal platelets
Alloimmune thrombocytopenia in the form of post-transfusion purpura	PLA1 on donated platelets when the recipient is PLA$^-$	Donated platelets
Acute and chronic autoimmune thrombocytopenia	Self-Ag on host platelets	Host's own platelets
FIXED TISSUES		
Hyperacute graft rejection, acute humoral graft rejection, chronic graft rejection	Foreign Ag on cells of a solid organ graft	Component cells of a grafted tissue
Goodpasture's syndrome	Collagen in basement membranes of host's kidneys and alveoli	Host's own lung and kidney epithelia and endothelia
Pemphigus	Desmogleins mediating adhesion of host's skin and mucosal epithelial cells	Skin and mucosae

while the transition to more northern locations or the onset of winter may trigger episodes of acute hemolysis. Intensity of symptoms also varies anatomically. Because the blood in the extremities is particularly subject to temperature change, antibody-mediated damage of RBCs may be more prevalent in the fingers, toes, wrists, and ankles. These anatomical structures may feel extremely cold and take on a bluish tint due to the reduced numbers of RBCs available to transport oxygen.

The primary therapy for people with CAS is avoidance of exposure to cold. Immunosuppressants are used only in the most severe cases. Unlike IgG-mediated warm autoimmune hemolytic anemias, splenectomy is not generally helpful in CAS because the condition is caused by IgM autoantibodies that induce hemolysis in the liver. Treatment with rituximab (anti-CD20 antibody; refer to Ch.26) can be effective in cases of CAS. This therapeutic mAb binds to B cell surfaces (including those producing the pathological antibodies) and promotes their destruction by complement-mediated lysis and ADCC.

Unlike CAS, PCH mainly affects children and is one of the leading causes of hemolytic anemia in this age group. Interestingly, when PCH was first described in 1904, 90% of cases were secondary to syphilis infection, making it a chronic, relapsing disease affecting the adult population. After the advent of antibiotics brought effective treatment for syphilis, the epidemiology of PCH changed dramatically and the disease became an acute, non-recurring condition seen almost exclusively in the pediatric population. Children typically experience PCH in an acute form that comes and goes after an infection with the measles, mumps, EBV, varicella, or influenza viruses.

Exposure to cold temperatures may also trigger the onset of PCH. Sudden high fever, chills, back and leg pain, abdominal cramping, and hemoglobinuria are common symptoms, with headache, nausea, vomiting, and diarrhea experienced somewhat less frequently. Generally, a PCH attack subsides within hours, although in some cases the extent of hemolysis can be life-threatening such that the patient requires several weeks for recovery.

PCH is caused by an IgG autoantibody called the *Donath-Landsteiner antibody*. *In vitro*, this antibody binds to a polysaccharide *P antigen* on RBCs at low temperatures and then activates complement when the cells are returned to a 37°C environment. A similar phenomenon occurs in patients exposed to cold and then warm temperatures. Treatment for PCH therefore revolves around keeping the patient warm (to prevent autoantibody binding) and managing the symptoms. In severe cases, plasmapheresis can be used to remove the autoantibodies and reduce the intensity of the hemolysis.

ii) Thrombocytopenia

A patient with thrombocytopenia has an abnormally low number of platelets in the blood. Thrombocytopenia is the most common cause of abnormal bleeding and is caused by decreased platelet production, increased platelet destruction, or abnormal distribution of platelets within the body. Type II HS reactions can contribute to increased platelet destruction and thus cause immune system-mediated thrombocytopenia. As with hemolytic anemia, type II HS thrombocytopenia may be due to either alloimmune or autoimmune antibodies.

iia) Alloimmune thrombocytopenia. *Neonatal thrombocytopenia* and *post-transfusion purpura* (PTP) are two types of thrombocytopenia that are considered to be type II HS. Both conditions are caused by alloantibodies specific for the platelet surface antigen PLA1. In the case of neonatal thrombocytopenia, a pregnant, PLA1-negative woman becomes sensitized by the PLA1-positive fetus she is carrying. The mother produces PLA1-specific antibodies of the IgG isotype that can cross the placenta and destroy fetal platelets, causing thrombocytopenia at the fetal and neonatal stages. Fortunately, since 98% of humans are PLA1$^+$, neonatal thrombocytopenia is rare and occurs only once in 5000 deliveries. Affected neonates develop areas of purplish or brownish red discoloration on the skin (*purpura*) caused by leakage of blood (*hemorrhage*) into the skin layers. Very small red spots called *petechia* may also be seen. While these symptoms are mild and of no lasting consequence, a very serious complication is the potential for intracranial bleeding affecting the brain. Standard therapy for neonatal thrombocytopenia usually involves steroid treatment to reduce inflammation and inhibit the function of phagocytic cells. Platelet transfusion may also be used to bring relief from symptoms and reduce the chance of intracranial hemorrhage. In addition, if neonatal thrombocytopenia is diagnosed during pregnancy, early delivery by caesarean section may be recommended to decrease exposure of the baby to the damaging alloantibodies.

PTP occurs when PLA1$^-$ individuals receive PLA1$^+$ platelets through the therapeutic transfusion of blood or blood products. While some PLA1$^-$ individuals harbor natural antibodies recognizing PLA1, PTP occurs more often in patients who have experienced multiple pregnancies or multiple transfusions. The anti-PLA1 antibodies initiate platelet destruction in the patient, resulting in the sudden onset of thrombocytopenia. Purpura and petechia are seen in the skin and mucosae and, in some cases, nasal, gingival, gastrointestinal, or urogenital tract bleeding may occur. Luckily, PTP is usually self-limited but, as with neonatal thrombocytopenia, intracranial hemorrhage is a concern. When warranted, therapy for PTP involves steroids or plasmapheresis.

iib) Autoimmune thrombocytopenia. Autoimmune thrombocytopenia is caused by an autoantibody attack on one's own platelets. The disorder may be either acute or chronic. In both cases, the binding of autoantibodies directed against platelet antigens triggers platelet destruction via phagocytosis. Both IgG and IgM bind to platelets in patients with autoimmune thrombocytopenia, although IgG appears to predominate. Analogous to the hemolytic anemias, IgG autoantibodies cause the lysis of platelets in the spleen, while IgM autoantibodies target the platelets for destruction in the liver.

Acute autoimmune thrombocytopenia (AAT) is seen most often in children 2–6 years old and in young adults. In 80% of patients, there is a history of a recent viral infection involving EBV, rubella, varicella, or rubeola, or one of several respiratory viruses. Indeed, the incidence of AAT follows that of viral infections in general, reaching its peak in the winter and spring months. Symptoms are similar to those described previously for PTP, with the hallmark again being abnormal bleeding in the skin and at various mucosal surfaces. In about 90% of acute cases, symptoms resolve on their own within 6 months and require no

intervention. Although intracranial hemorrhage can be a serious complication, it occurs in 1% or less of all acute cases.

Chronic autoimmune thrombocytopenia (CAT) affects mainly adults but is also seen in children over 8 years old. While symptoms are similar to those of acute AAT, the condition lasts longer than 6 months. Patients experience bleeding sporadically, with each episode lasting days or weeks. In contrast to AAT, CAT is not associated with prior viral infections. Steroids form the main basis of therapy when required, but splenectomy is also an option if the patient does not respond to steroids over the course of several weeks.

IV. EXAMPLES OF TYPE II HS: CYTOTOXICITY AGAINST FIXED TISSUES

i) Antibody-Mediated Rejection of Solid Tissue Transplants

A well-known example of type II HS against a fixed cellular target is *hyperacute graft rejection*. As discussed in Chapter 27, HAR occurs within minutes or hours of organ transplantation when the recipient has in his or her circulation pre-existing alloantibodies directed against donor MHC. Such pre-existing alloantibodies are most likely to be found in patients who have been exposed to allogeneic cells during previous pregnancies, organ or bone marrow transplantation procedures, or blood transfusions. As we have seen, alloantibody-mediated destruction of the grafted tissue proceeds quickly and irreversibly during HAR. Antibody-mediated destruction of fixed cells also occurs to varying degrees in acute humoral graft rejection and chronic graft rejection, although there is usually concurrent T cell-mediated damage. These somewhat slower forms of graft rejection are thus examples of the simultaneous occurrence of type II and IV HS reactions.

ii) Goodpasture's Syndrome

As is described in more detail in Chapter 29, Goodpasture's syndrome (GS) is caused by autoantibodies that recognize the C-terminal domain of collagen IV protein found in the basement membranes of lung and kidney cells (Plate 28-4). A type II HS reaction is initiated by these autoantibodies that damages the epithelia and endothelia of the lungs and kidneys, causing lung hemorrhage and *glomerulonephritis* (inflammation of the renal glomeruli). The nephritis is often severe and progressive and may result in renal failure within weeks if left untreated. A patient who is simultaneously suffering from GS and a renal infection may experience a collapse of kidney function within days or even hours. Therapy for GS usually involves some combination of plasmapheresis, cyclophosphamide administration, and corticosteroid treatment. A therapy period of about 3 months provides the immunosuppression necessary to overcome GS since the causative autoantibodies are produced for a relatively short time.

iii) Pemphigus

Pemphigus is an autoimmune disease characterized by potentially fatal blistering of skin and mucosal surfaces. This disorder is caused by autoantibodies (usually IgG or IgA) that attack

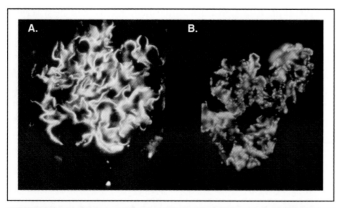

Plate 28-4

Immunofluorescence of Type II and Type III Hypersensitivity in the Kidney

(A) Typical "linear" immune deposits of IgG along the glomerular basement membrane characteristic of Goodpasture's syndrome, a type II hypersensitivity in which autoantibodies are directed against collagen IV. (B) Typical "granular" immune deposits within the glomerular capillary wall characteristic of systemic lupus erythematosus, a type III hypersensitivity in which immune complexes formed elsewhere in the body are trapped between the endothelium or epithelium and the basement membrane. Courtesy of Dr. Michael Madaio, Renal Electrolyte and Hypertension Division, University of Pennsylvania and Dr. William Couser, Vanderbilt Medical Center.

the various cadherin-like proteins involved in adhesion between epithelial cells. In particular, proteins called *desmogleins* seem to be targeted. Autoantibody binding to the desmogleins causes the epithelial cells to lose contact with each other and with the lamina propria, resulting in the lifting away of large areas of cells (*acantholysis*). Although pemphigus mainly affects the skin, the blisters are also found on mucosal tissues in the mouth, nose, conjunctiva, genitals, and throat. Patients are usually middle-aged or elderly. Systemic immunosuppression is generally required to manage this disease, with corticosteroids being the standard treatment. Indeed, before corticosteroids were available, the most common form of the disease (*pemphigus vulgaris*) was almost invariably fatal due to the dehydration and infections that resulted from the blistering. With treatment, the mortality rate for this disease is now lower than 10%. In the most severe cases, other immunosuppressants such as azathioprine or cyclophosphamide may be of benefit. More on pemphigus can be found in Chapter 29.

D. Type III Hypersensitivity: Immune Complex-Mediated Injury

I. WHAT IS TYPE III HS?

Type III hypersensitivity is also known as *immune complex injury*. The damaging inflammatory reaction is triggered by a soluble antigen capable of forming large insoluble immune complexes (IC) with IgM or IgG antibodies in the circulation.

These complexes are too large to phagocytose, so, rather than being cleared, the ICs are deposited either at a single site or in various locations in the body (often in the walls of vessels). The presence of the ICs in the tissues initiates immune responses that damage surrounding cells, leading to localized pain, edema, and inflammation. Histologically, type III HS is distinguished by an accumulation of neutrophils at the site of tissue injury at about 4–6 hours after exposure to the antigen.

As we learned in Chapter 7, antigen and antibody naturally come together to form immune complexes during an adaptive immune response. Normally, these complexes do not become large and insoluble because the Fc regions of the antibodies facilitate the binding of complement C1q. In addition to triggering the classical complement cascade that leads to clearance of the antibody-bound antigen, the binding of C1q interferes with the growing antigen–antibody lattice structure. The ICs are thus maintained at a size that remains soluble in the circulation. So, when do ICs become a problem that triggers hypersensitivity reactions? Some type III HS reactions result when the individual has a complement deficiency that leads to inefficient removal of ICs (see Ch.19). In other cases, when the individual has an otherwise competent immune system, the sheer quantities of antibody and antigen present may generate damaging ICs. Many bacteria, parasites, and viruses supply antigens that remain after the main pathogen threat has been resolved, and it is these "persistent antigens" that cause type III HS reactions. The ICs formed by the persistent antigen and anti-pathogen antibodies circulate in the blood and are deposited in various tissues, causing symptoms that are distinct from those due to the pathogen itself. In these cases, the type III HS reaction is a clinical complication of the pathogen infection. In similar fashion, some cancer patients make antibodies to tumor antigens shed into the blood by cancer cells. Unless the tumor resolves naturally or is forced to do so by medical intervention, exposure to such antigens is continuous and relatively long term. Thus, ICs may form between the antibodies and the tumor antigens and cause type III HS symptoms. Some drug "allergies" (refer to Box 28-3) may also be due to type III HS reactions. In this case, the symptoms persist as long as the drug the individual is taking induces the problem antibodies. Type III HS is also often found in patients expressing autoantibodies. The autoantibodies combine with specific autoantigens, which might be proteins, glycoproteins, or even DNA, that are naturally present at all times. Large ICs are formed that then lead to inflammatory autoimmune diseases (see Ch.29).

II. MECHANISM UNDERLYING TYPE III HS

How does an IC cause inflammatory disease? Once ICs become extensive enough to become insoluble, they accumulate in narrow blood vessels or body channels in the tissues, where they physically "get stuck" (Fig. 28-10). In the case of a blood vessel, the presence of the accumulated ICs induces inflammation such that the endothelial cell layer becomes leaky, allowing the ICs to penetrate into the tissues beneath. The presence of the ICs in the tissues triggers the activation of complement via

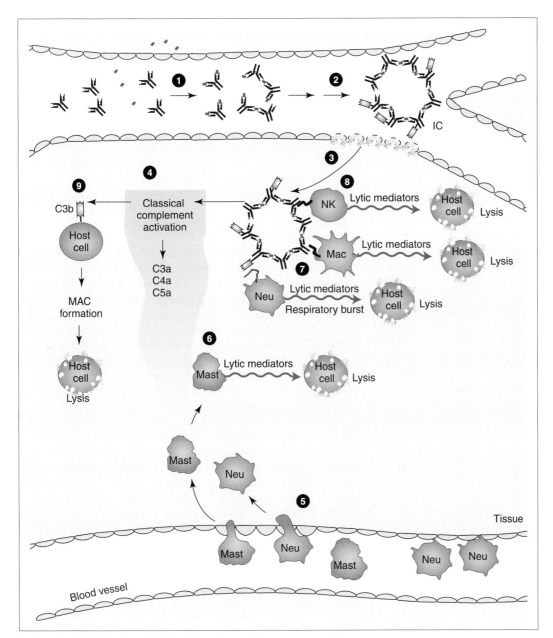

Type III Hypersensitivity
Type III HS is mediated by inflammatory responses to immune complexes that accumulate in narrow channels in the body. In this example, antigen–antibody pairs in the blood (1) cross-link to form an insoluble immune complex (IC) that "gets stuck" in a narrow capillary (2). The deposited IC activates complement, initiating C3b deposition on the IC and inflammation that damages the endothelial cells such that the IC can enter the underlying tissue (3). The presence of the IC induces further complement activation (4), generating anaphylatoxins that summon mast cells and neutrophils (5) to the site of IC deposition. Host cells are then destroyed by mediators released when the cytokine milieu induces mast cell degranulation (6). In addition, tissue-damaging molecules are released from "frustrated" neutrophils and macrophages (7) that bind to the IC via either C3b-CR1 or Ig-FcR interactions but are unable to phagocytose the IC due to its large size. NK cells triggered by Ig/FcR interactions also release lytic mediators (8). Finally, complement activation initiated by the IC results in the coating of nearby host cells with C3b (9) and their destruction by MAC formation.

the classical pathway. The anaphylatoxins released in the course of the complement cascade draw neutrophils and mast cells to the site of IC deposition. Cellular damage then ensues via the same mechanisms mediating type II HS reactions against fixed tissues. That is, the mast cells and neutrophils degranulate and release enzymes and mediators that both lyse local tissue cells and have chemotactic and pro-inflammatory effects. Surrounding blood vessels are further induced to dilate, increasing the access of the ICs into the tissues and the influx of leukocytes into the site. Complement activation also results in the deposition of C3b on the ICs and surrounding tissue cells. Since neutrophils carry C3b receptors, they attempt to phagocytose cells fixed in the solid tissues at the IC deposition site. When engulfment fails, the "frustrated" neutrophils release lytic products that cause the damage seen as hypersensitivity. In addition, the presence of C3b on host cells

facilitates MAC formation, which results in the lysis of those cells. As one would expect for mechanisms dependent on classical complement activation, specific antibody isotypes, namely IgM, IgG3, and IgG1 in humans and IgM, Ig2a/b, and IgG3 in mice, are the principal culprits in type III HS.

Also involved in type III HS reactions are cells activated by the binding of their FcγRs to the numerous immobilized Fcs in deposited ICs. Aggregation of FcγRI and FcγRIII on macrophages, DCs, LCs, NK cells, neutrophils, and mast cells at the local site is readily achieved, triggering intracellular signals that result in secretion of pro-inflammatory cytokines, respiratory burst, and degranulation (depending on the cell type bearing the receptor). Studies of knockout mice lacking Fc receptors have shown that FcR aggregation plays an important role in the induction of type III HS reactions. For example, when FcγR$^{-/-}$ mice were subjected to experimental protocols that normally invoke type III HS reactions, the mice showed significant reductions in edema, hemorrhaging, and neutrophil infiltration.

III. EXAMPLES OF TYPE III HS

Type III HS reactions are classified by their clinical manifestations, which affect either a single site (localized type III HS) or multiple sites (systemic type III HS) (Table 28-10).

i) Localized Type III HS

In a localized type III HS reaction, the ICs involved are deposited only where antibody and antigen first encounter one another; the ICs do not get a chance to circulate systemically. Only local tissue damage characterized by pain, redness, and swelling is observed in what clinicians call an *Arthus reaction*. Much of what is known about the cellular mechanisms underlying type III HS has been defined by noting what happens at a local site when an Arthus reaction is induced experimentally. An animal is primed systemically with a known antigen and later challenged subcutaneously with the same antigen. On the skin at the site of the secondary injection, an Arthus reaction develops that is relatively controlled and easy to observe. Scientists can then manipulate conditions under which the reaction develops. Researchers can also induce an experimental form of type III HS called the *reverse passive Arthus reaction*. In this situation, mice are injected with specific antibody at the precise site (usually lung, skin, or peritoneum) where the researcher wants to initiate the inflammation. (In other words, the mice do not actively make their own antibodies to the antigen.) The antigen of interest is then injected intravenously such that it circulates around the body until it meets up with the antibody in the tissue site. Immune complexes are then formed in this location. Under these controlled conditions, the effect of the presence or absence of a particular component or agent on the occurrence of type III HS can be determined.

ii) Systemic Type III HS

Most type III HS cases seen in the clinic involve systemic reactions affecting those parts of the body that tend to inadvertently trap large ICs circulating in the blood. Thus, the site at which symptoms appear may be far removed from the original site of antigen–antibody contact, and multiple sites may be affected simultaneously. Systemic type III HS often occurs in sites where the local blood vessels have a narrow diameter, or where there is circulatory turbulence due to changes in direction of blood flow at the site. More specifically, joints, arteries, and the renal glomeruli are particularly susceptible to IC deposition, such that type III HS disease often involves some combination of arthritis, vasculitis, and nephritis.

Serum sickness (introduced at the start of this chapter) is the prototypical systemic type III HS. The patients in whom this disease was originally observed had been given large quantities of anti-tetanus or anti-diphtheria antitoxins that had been raised in animals. The recipients mounted antibody responses against non-human serum proteins in the antitoxin preparations, and these antibodies formed ICs that accessed the circulation. These ICs then traveled around the body and became trapped in the tissues, where they triggered an inflammatory response. Today, serum sickness is more often associated with repeated exposure to an environmental or drug antigen. Serum sickness in response to a pathogen is usually fleeting unless the infection becomes prolonged. The clinical signs of acute serum sickness include fever, weakness, rash with swelling, arthritic pain in the joints, and (frequently) glomerulonephritis. Repeated or persistent exposure to the offending antigen can cause the illness to become chronic, resulting in recurring damage to the kidneys, arteries, joints, and lungs. Treatment of chronic serum sickness caused by an environmental or drug antigen requires the identification of the offending antigen and limitation of the patient's exposure to that substance.

As well as being provoked by foreign antigens, systemic type III HS can result from the accumulation of ICs involving autoantibodies recognizing soluble self antigens. In these cases, the type III HS reaction is just one manifestation of the autoimmune disease. For example, patients with systemic lupus erythematosus (SLE) produce many different types of autoantibodies, including those directed against DNA, nucleoproteins, cytoplasmic antigens, leukocyte antigens, and clotting factors. Similarly, in rheumatoid arthritis (RA), autoantibodies to the patient's own IgG molecules can be found in the circulation. In both SLE and RA, the deposition of ICs occurs systemically so that, along with their other symptoms, patients experience joint inflammation, vasculitis and, in the case of SLE, kidney disease. It should be noted that both Goodpasture's syndrome, discussed previously as a prototypic type II HS disease, and SLE have glomerulonephritis as disease hallmarks. However, while the glomerulonephritis in GS is characterized by the binding of antibodies directly to cellular basement membranes, the kidney damage observed in SLE is caused by the deposition of ICs in the glomeruli. The two conditions can be clearly distinguished by the

Table 28-10 **Type III HS Reactions**

Example	Pathological Ab Directed Against	Deposition of Immune Complexes
Arthus reaction	Subcutaneously injected Ag	Localized
Serum sickness	Toxins, environmental Ag, drug Ag, Ag of persistent pathogen	Systemic
SLE	Host DNA, nucleoproteins, clotting factors, other self-proteins	Systemic
RA	Host IgG	Systemic

immunofluorescent staining pattern of the antibodies involved (refer to Plate 28-4). The underlying mechanisms and treatment of these disorders are discussed in Chapter 29.

E. Type IV Hypersensitivity: Delayed-Type or Cell-Mediated Hypersensitivity

I. WHAT IS TYPE IV HS?

Type IV HS is immunopathological damage that occurs at about 24–72 hours after exposure of a sensitized individual to an antigen (hence, "delayed-type" hypersensitivity or DTH). Histologically, type IV HS reactions are distinguished from other HS reactions by the infiltration of basophils and mononuclear cells (particularly Th1 lymphocytes and macrophages) that occurs at the site of exposure. Indeed, unlike type I, II, and III HS (which are all antibody-mediated), type IV HS results primarily from the actions of effector T cells and macrophages. Landsteiner and colleagues were among the first to demonstrate that type IV HS reactions were cell-mediated rather than antibody-mediated. These researchers transferred peritoneal exudate cells from a host sensitized with antigen A to a naive recipient. If the recipient was then challenged with antigen A, the animal showed signs of a type IV HS reaction. However, no anti-A DTH response occurred if serum from the sensitized animal was transferred instead. In this section, we discuss three well-known manifestations of type IV HS: a chronic form of the *DTH reaction* we have encountered in Chapters 15 and 22, *contact hypersensitivity*, and *hypersensitivity pneumonitis*.

It should be noted that some clinicians consider certain forms of chronic graft rejection to be type IV HS reactions because a normal cellular immune response precipitates immunopathological damage to the recipient. For the same reason, the acute damage done by infiltrating CTLs in response to some pathogens is sometimes deemed to be type IV HS. An experimental example is the destruction of brain tissue resulting from the CTL response to LCMV infection in mice. Finally, cellular responses against certain autoantigens play a role in various autoimmune diseases, contributing a type IV HS component to the observed pathology (see Ch.29). Type IV HS reactions are summarized in Table 28-11.

II. CHRONIC DTH REACTIONS

DTH reactions are usually initiated by antigens derived from intracellular pathogens that resist elimination by the usual immune response measures (Fig. 28-11). For example, the reader will recall from Chapter 22 that persistent *Mycobacterium tuberculosis* is eventually walled off from the rest of the body by the formation of a granuloma. As we have seen, granuloma formation depends on activated Th1 cells and hyperactivated macrophages. The hypersensitivity part comes in when these cells do not "back off" after the pathogen has been sequestered. Sustained release of the pro-inflammatory cytokines and factors driving granuloma formation starts to damage the surrounding healthy tissue. The DTH response then becomes chronic and constitutes a type IV HS reaction. The delay in this type of HS (compared to antibody-mediated types of HS) is due to the time required for T cell activation and differentiation, cytokine and chemokine secretion, and for the accumulation of macrophages at the site of exposure.

The clinical signs of a DTH reaction are influenced by the site of antigen entrenchment. The skin over the site of exposure may appear abnormally red about 2–3 days after exposure. Calcification, caseation necrosis (a destruction of the tissue that leaves it looking like cheese curds), and cavity formation are common clinical findings. Infiltrating mononuclear leukocytes are a characteristic feature of DTH responses associated with infectious diseases (tuberculosis, leprosy, leishmaniasis), reactions to non-infectious agents (silicosis, berylliosis), and reactions to unknown agents (Crohn's disease, sarcoidosis). When a granuloma finally becomes encapsulated in a layer of fibroblasts and connective tissue (refer to Fig. 22-4), it can cause a lesion that may interfere with the function of an organ such as the liver or lung. Chronic liver disease or respiratory failure can result. With respect to treatment, corticosteroids can be used to block cytokine production by effector T cells in type IV HS reactions. If the reaction is confined to a local area, topical creams may be used. More extensive tissue involvement may require systemic corticosteroid administration.

As mentioned in Chapter 22, DTH reactions have been exploited as a means of determining whether an individual has been previously exposed to a pathogen. The skin prick test for tuberculosis is an example of such a test, in which redness and swelling at the site of an injection of a small amount of *M. tuberculosis* antigen indicates that the individual has previously been infected with the bacterium. Similar tests can be

Table 28-11 **Type IV HS Reactions**		
Examples	**Pathological T Cells Directed Against Peptides of**	**Tissue Damage Observed**
Chronic DTH	Conventional foreign antigen	Granuloma formation
Contact HS	Neo-Ag of chemically reactive hapten bonded to skin self-proteins	Itching and rash
HS pneumonitis	Neo-Ag of inhaled inorganic hapten bonded to lung self-protein	Granuloma and fibrosis in lungs
Chronic graft rejection	Allo-Ag in donor tissue	Destruction of donor cells, fibrosis, collagen deposition, restricted blood flow to transplanted organ
Autoimmunity	Auto-Ag in target tissue	Destruction of self-cells, fibrosis

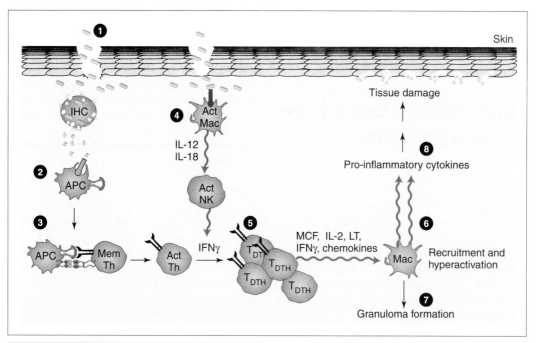

Figure 28-11

Type IV Hypersensitivity: Chronic DTH Reaction
In this example, a pathogen has penetrated the skin (1) and infected a host cell. Pathogen antigens released by the infected host cell (IHC) are taken up (2), processed, and presented by host APCs to memory Th cells (3). At the same time, macrophages (4) activated by the pathogen produce cytokines that activate NK cells to secrete IFNγ. This IFNγ promotes the differentiation of T$_{DTH}$ cells (5) producing chemokines and cytokines that recruit macrophages to the site and drive their hyperactivation (6). Granulomas may be formed to contain pathogens that cannot be eliminated (7). Hypersensitivity arises when the macrophages continue to secrete pro-inflammatory mediators that damage host cells (8).

used to determine prior infection with organisms causing diphtheria or brucellosis. The DTH reaction can also be used to test the functionality of an individual's T cells. The yeast *Candida albicans* is so prevalent in our environment that virtually everyone has experienced at least one infection with this organism in childhood. Thus, a DTH test using *Candida* antigen should provoke redness and swelling at the site of testing in virtually all individuals. A patient that fails to mount a DTH response during this test likely has a deficit in T cell function and may be suffering from an acquired immunodeficiency.

III. CONTACT HYPERSENSITIVITY (CHS)

Contact hypersensitivity, sometimes called *allergic contact dermatitis*, is an immune response to a chemically reactive hapten that has bound covalently to self proteins in the uppermost layers of the skin. Examples of contact HS include the patchy rash and intense itching that follow a plunge into a patch of poison oak or poison ivy, and the local skin irritations experienced by individuals sensitive to chemicals such as paint thinner or cosmetics. The alteration of self proteins by the binding of the CHS antigen generates a "nonself" entity that can be thought of as a *neo-antigen* ("new" antigen). Most neo-antigens are thus basically hapten–carrier complexes of the type discussed in Chapter 5. The small molecules playing the part of the hapten are usually derived from *xenobiotics*, a class of non-living entities with bio-

logical effects. Examples of xenobiotics can be found among drugs, metals, and industrial and natural chemicals. For example, poison ivy contains the xenobiotic urushiol in the surface oils of its leaves. Upon contact with human skin, urushiol molecules penetrate the protective epithelial layers and bind covalently to reactive groups present on local self proteins. Other xenobiotics alter self proteins by oxidation. When the xenobiotic is a metal, it may form stable metalloprotein complexes that are particularly provocative to macrophages. Sometimes a xenobiotic will have to be metabolized in the liver before the reactive component is available for neo-antigen formation. Certain cell types, such as LCs, dermal DCs, neutrophils, and macrophages, can also metabolize xenobiotics into reactive entities that participate in neo-antigen formation. This metabolic process is a form of bioactivation. Genetic polymorphisms in the enzymes responsible for hepatic metabolism or bioactivation may affect an individual's chances of developing type IV HS to a xenobiotic.

Many cell types play a role in CHS reactions (Fig. 28-12). As well as forming neo-antigens, xenobiotics are able to stimulate keratinocytes to upregulate ICAM-1 expression and produce pro-inflammatory cytokines (IL-1, TNF, and IL-6) and chemokines (IP-10, MIG, RANTES, TARC, CCL18) that draw APCs, T cells, neutrophils, monocytes, and other immune system cells into the affected region of the skin. Expression of these molecules at the site of xenobiotic exposure has been detected at about 12 hours after contact. In such a milieu, local LCs are stimulated to capture the neo-antigen and cross-present

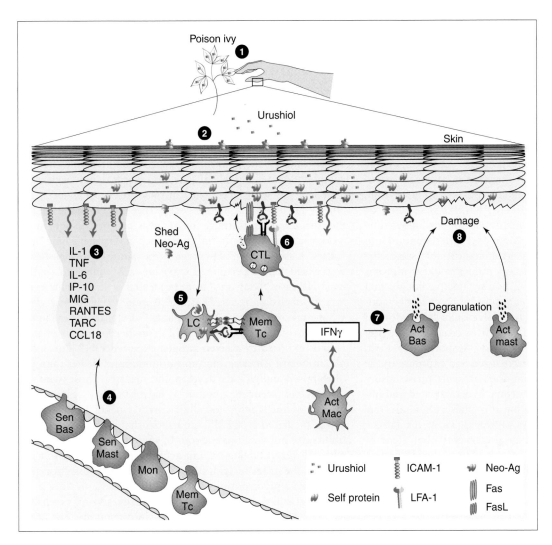

Figure 28-12

Type IV Hypersensitivity: CHS
In this example, a misguided individual has touched a poison ivy plant (1), allowing urushiol to bind to self proteins in the skin (2) and form a neo-antigen. The neo-antigen induces skin cells to upregulate the expression of adhesion molecules and release numerous chemokines and cytokines (3) that draw immune system cells from the circulation (4). Neo-antigen shed by skin cells is taken up by Langerhans cells (LC) and peptides derived from it are cross-presented to memory Tc cells (5). CTL effectors are generated that damage host skin cells directly via degranulation and/or Fas-mediated apoptosis (6), and also produce IFNγ. The high level of IFNγ in the milieu stimulates the degranulation of mast cells and basophils (7) and the release of lytic mediators that contribute to host cell damage (8).

it to local CD8$^+$ memory T cells generated during the sensitization stage. CTL effectors are generated that both secrete IFNγ and turn their cytotoxic activity onto keratinocytes displaying the appropriate pMHC. That this cytotoxicity is integral to CHS reactions was demonstrated by the fact that mice lacking Fas, FasL, or perforin were unable to mount CHS responses, although they did possess hapten-specific CD8$^+$ T cells. IFNγ produced by responding CTLs (and other cell types) upregulates the expression of Fas on keratinocytes, making them more susceptible to the cell death that is a key manifestation of CHS. IFNγ also stimulates the production of chemokines, histamine, serotonin, and vasodilators by basophils and mast cells. Mast cells play a major role in CHS, since natural mouse mutants lacking mast cells show impaired CHS reactions.

Peptides from a neo-antigen are also presented on MHC class II by LCs to local memory CD4$^+$ T cells, but these cells appear to play only a minor role, if any, in CHS. Studies of CHS in MHC class II-deficient mice have shown that LCs in these animals are just as efficient as LCs from MHC class II-competent mice at inducing CHS, whereas LCs from MHC class I-deficient mice are unable to facilitate CHS. Indeed, presentation of neo-antigen peptides on MHC class II may

normally induce the generation of CD4$^+$ Th3 cells that are able to damp down a CHS response due to their ability to secrete large quantities of IL-10. Significantly, when the blood of individuals reactive to nickel is compared with that of non-reactive individuals, the former have lower levels of Th3 cells.

Neo-antigen formation may also give rise to autoimmunity that is CHS-mediated. The physical alteration of a self protein by the addition of a xenobiotic may expose cryptic epitopes in the self protein (for which lymphocytes were not deleted during the establishment of central tolerance). New peptides (which are not haptenated) may be presented that activate a whole new spectrum of T cells. For example, if mice are treated with mercury, they begin (for unknown reasons) to present novel peptides of the nucleolar protein fibrillarin. Autoantibodies soon appear that react to the fibrillarin epitopes. Presumably the structure of the fibrillarin molecule was altered in some way by the presence of the mercury atom such that previously cryptic peptides became exposed for presentation.

As might be expected, the primary mode of treatment of CHS is avoidance of exposure to the inciting antigen. Corticosteroid creams may provide some welcome relief from symptoms.

IV. HYPERSENSITIVITY PNEUMONITIS (HP)

Hypersensitivity pneumonitis, also called *extrinsic allergic alveolitis*, is a type IV HS reaction in the lung caused by prolonged exposure to an inhaled antigen. A wide range of antigens can trigger HP, but the resulting response in the lung usually takes the same form and occurs in three distinct clinical phases: the acute phase, the subacute phase, and the chronic phase. When a sensitized individual inhales the offending antigen, macrophages are activated in the lungs either because these cells ingest the antigenic particles directly or because of complement activation. Within 48 hours, chemokines (IL-8, MIP-1α, RANTES) secreted by the macrophages promote an influx of neutrophils followed by T lymphocytes. CD4$^+$ Th0 cells commence differentiation into Th1 effectors secreting copious cytokines. At this stage, the affected individual likely experiences influenza-like symptoms such as fever, chills, muscle pain, non-productive cough, and breathing difficulties. If further exposure to the causative antigen is avoided, symptoms usually resolve quickly. If a formal diagnosis is made, a short round of corticosteroids may be prescribed to help dampen the inflammatory response. However, if the acute stage of HP goes unrecognized and exposure to the antigen continues or occurs again, the sub-acute phase ensues during which the macrophages become hyperactivated and initiate granuloma formation. During this phase, there are few overt symptoms for several days or weeks. However, fatigue and cough then reappear and progress subtly but steadily. Again, corticosteroids can be of great benefit to the HP patient at this point. However, if there is no intervention and exposure continues, the chronic phase of HP sets in and leads to fibrosis of lung tissue via the same mechanisms seen in the chronic DTH responses described previously. The activated alveolar macrophages secrete large amounts of TGFβ, further encouraging the fibrosis. HP sufferers may also produce specific antibodies that contribute type II and III HS reactions to the observed pathology. For example, antibody may either bind to tissues in which the inhaled antigen has lodged, or form ICs whose deposition promotes lung inflammation. As chronic HP progresses to its end stages, patients show increasingly labored breathing and inexorable weight loss. Unfortunately, this phase of the disease is easy to confuse with other lung diseases involving fibrosis, and the damage to the lungs may be irreversible and eventually fatal despite the use of corticosteroids.

HP is associated with a wide range of conditions that tend to be named for a defining antigen. These antigens are generally proteins derived from microbes, fungi, plants, or animals. In some instances, the inducing substance is an inorganic chemical compound that is capable of activating specific lymphocytes only after it reacts with lung tissue. In these cases, the induction of type IV HS in HP resembles an occurrence of CHS in the lung. In the past, exposure to triggering antigens occurred mainly in the workplace, giving rise to afflictions such as "farmer's lung" and "cheese washer's lung." However, as the nature of these disorders was elucidated, occupational exposure was controlled so that most HP cases today arise from residential or leisure-time exposure to inhaled antigens. Thus, one is more likely to hear about "sauna-taker's disease," "hot-tub lung," or "bird-fancier's lung."

Studies have shown that only a small proportion of individuals exposed to a given antigen experience HP. For example, in North American farming communities, fewer than 6% of exposed individuals develop the disease. Those who are disease-free seem to mount a harmless IgG response to inhaled antigens with no cellular involvement. As in the case of IgE-mediated type I HS, the factors that control HP susceptibility are not well understood. In some instances, individuals with HP recall having had an acute respiratory infection just prior to the onset of symptoms, suggesting that infection may have been a predisposing factor.

This concludes our discussion of allergy and hypersensitivity. We move now to a discussion of autoimmunity, one of the great mysteries of immunology. Why does one's immune system attack one's own body, and what can be done to prevent it from doing so? The knowledge gained by the reader from the previous 28 chapters of this book may help him or her to assist in the ongoing search for the answers to these questions.

SUMMARY

A hypersensitivity response is a secondary immune response that is deleterious to the host rather than beneficial. Four categories of hypersensitivity are defined based on their mechanisms and immunopathology. Type I hypersensitivity, known as allergy or atopy, is mediated by IgE antibody specific for antigens that are normally non-pathogenic (allergens), such as plant pollens. In an atopic individual exposed to allergen, B cells produce allergen-specific IgE, which binds to FcεRs of mast cells, "arming" or "sensitizing" these cells. Re-exposure to the allergen immediately triggers the sensitized mast cells to degranulate and release soluble inflammatory mediators that cause local symptoms such as dermatitis, urticaria, asthma, and rhinitis. Anaphylaxis is a systemic atopic response that can be life-threatening and requires immediate epinephrine administration. Allergy symptoms may be treated with antihistamines, mast cell stabilizers, anti-inflammatory or immunosuppressive drugs, or a hyposensitization regimen that encourages the production of IgG instead of IgE. Many experimental therapies that target the initiation of the atopic response itself are under development. Type II hypersensitivity responses are due to the direct Ab-mediated cytotoxicity that occurs when IgG or IgM antibodies bind to antigenic epitopes on cells, triggering ADCC, complement activation, and phagocytosis. Targets include mobile cells such as red blood cells or neutrophils, and cells fixed in tissues such as keratinocytes or epithelial cells. In type III hypersensitivity, IgG or IgM antibodies in the circulation bind to antigen and form large, insoluble immune complexes. These complexes become lodged in narrow channels in the body, triggering inflammatory responses that damage underlying tissues. Type IV hypersensitivity, as exemplified by chronic DTH reactions, contact hypersensitivity, and hypersensitivity pneumonitis, is delayed in character and is primarily due to the reactivity of Th1 cells and macrophages that infiltrate a site of antigen exposure and induce tissue damage.

Selected Reading List

Akdis D., Akdis M., Trautmann A. and Blaser K. (2000) Immune regulation in atopic dermatitis. *Current Opinion in Immunology* **12**, 641–646.

Barnes P. (1999) Therapeutic strategies for allergic diseases. *Nature* **402**, B31–B38.

Barnes P. (2004) New drugs for asthma. *Nature Reviews Drug Discovery* **3**, 831–844.

Bataille V. (2004) Genetic factors in nickel allergy. *Journal of Investigative Dermatology* **123**, xxiv–xxv.

Behrendt H. and Becker W.-M. (2001) Localization, release and bioavailability of pollen allergens: the influence of environmental factors. *Current Opinion in Immunology* **13**, 709–715.

Black P. (2001) Why is the prevalence of allergy and autoimmunity increasing? *Trends in Immunology* **22**, 354–355.

Buhl R. (2004) Anti-IgE antibodies for the treatment of asthma. *Current Opinion in Pulmonary Medicine* **11**, 27–34.

Burks W. (2002) Current understanding of food allergy. *Annals of the New York Academy of Science* **964**, 1–12.

Busse W. and Lemanske R. (2001) Asthma. *The New England Journal of Medicine* **344**, 350–362.

Caramori G., Ito K. and Adcock I. (2004) Targeting Th2 cells in asthmatic airways. *Current Drug Targets Inflammation and Allergy* **3**, 243–255.

Celedon J. and Weiss S. (2004) Use of antibacterials in infancy: clinical implications for childhood asthma and allergies. *Treatments in Respiratory Medicine* **3**, 291–294.

Corry D. and Kheradmand F. (1999) Induction and regulation of the IgE response. *Nature* **402**, B18–B23.

Crispo J., James J. and Rodriquez J. (2004) Diagnosis and therapy of food allergy. *Molecular Nutrition and Food Research* **48**, 347–355.

Dombrowicz D. and Capron M. (2001) Eosinophils, allergy and parasites. *Current Opinion in Immunology* **13**, 716–720.

Finotto S., Neurath M., Glickman J., Qin S., Lehr H. A. *et al.* (2002) Development of spontaneous airway changes consistent with human asthma in mice lacking T-bet. *Science* **295**, 336–338.

Girolomoni G., Sebastiani S., Albanesi C. and Cavani A. (2001) T-cell subpopulations in the development of atopic and contact allergy. *Current Opinion in Immunology* **13**, 733–737.

Godfrey D. and Kronenberg M. (2004) Going both ways: immune regulaton via CD1-dependent NKT cells. *Journal of Clinical Investigation* **114**, 1379–1388.

Griem P., Wulferink M., Sachs B., Gonzalez J. and Gleichmann E. (1998) Allergic and autoimmune reactions to xenobiotics: how do they arise? *Immunology Today* **19**, 133–145.

Gutermuth J., Ollert M., Ring J., Behrendt H. and Jakob T. (2004) Mouse models of atopic eczema critically evaluated. *International Archives of Allergy and Immunology* **135**, 262–276.

Kawaguchi M., Adachi M., Oda N., Kokubu F. and Huang S. K. (2004) IL-17 cytokine family. *Journal of Allergy and Clinical Immunology* **114**, 1265–1273.

Kay A. (2000) Allergy and allergic diseases: with a view to the future. *British Medical Bulletin* **56**, 843–864.

Kemp J. (2003) Recent advances in the management of asthma using leukotriene modifiers. *American Journal of Respiratory Medicine* **2**, 139–156.

Kimber I. and Dearman R. (2002) Allergic contact dermatitis; the cellular effectors. *Contact Dermatitis* **46**, 1–5.

Kips J., Tournoy K. and Pauwels R. (2000) Gene knockout models of asthma. *American Journal of Respiratory Care Medicine* **162**, S66–S70.

Kobayashi K., Kaneda K. and Kasama T. (2001) Immunopathogenesis of delayed-type hypersensitivity. *Microscopy Research and Technique* **53**, 241–245.

Kohl J. and Gessner J. (1999) On the role of complement and Fc gamma receptors in the Arthus reaction. *Molecular Immunology* **36**, 893–903.

Leung D. and Soter N. (2001) Tacrolimus ointment: advancing the treatment of atopic dermatitis. *Journal of the American Academy of Dermatology* **44**, S1–S12.

Maddox L. and Schwartz D. (2002) The pathophysiology of asthma. *Annual Reviews of Medicine* **53**, 477–498.

Marone G., Spadaro G., Palumbo C. and Condorelli G. (1999) The anti-IgE/anti-Fc epsilon RI alpha autoantibody network in allergic and autoimmune diseases. *Clinical and Experimental Allergy* **29**, 17–27.

Meenu S. (2004) Newer drugs for asthma. *Indian Journal of Pediatrics* **71**, 721–727.

Miller A. and Lukacs N. (2004) Chemokine receptors: understanding their role in asthmatic disease. *Immunology and Allergy Clinics of North America* **24**, 667–683.

Novak N., Kraft S. and Bieber T. (2001) IgE receptors. *Current Opinion in Immunology* **13**, 721–726.

Oliver J., Kepley C., Ortega E. and Wilson B. (2000) Immunologically mediated signaling in basophils and mast cells: finding therapeutic targets for allergic diseases in the human Fc epsilon R1 signaling pathway. *Immunopharmacology* **48**, 269–281.

Ono S. (2000) Molecular genetics of allergic diseases. *Annual Reviews of Immunology* **18**, 347–366.

Pernis A. and Rothman P. (2002) JAK-STAT signaling in asthma. *Journal of Clinical Investigation* **109**, 1279–1283.

Platts-Mills T. (2001) The role of immunoglobulin E in allergy and asthma. *American Journal of Respiratory Care Medicine* **164**, S1–S5.

Popescu F. (2003) New asthma drugs acting on gene expression. *Journal of Cellular and Molecular Medicine* **7**, 475–486.

Ramshaw H., Woodcock J., Bagley C., McClure B., Hercus T. R. *et al.* (2001) New approaches in the treatment of asthma. *Immunology and Cell Biology* **79**, 154–159.

Ring J., Kramer U., Schafer T. and Behrendt H. (2001) Why are allergies increasing? *Current Opinion in Immunology* **13**, 701–708.

Robinson D., Larche M. and Durham S. (2004) Tregs and allergic disease. *Journal of Clinical Investigation* **114**, 1389–1397.

Sampson H. (2000) Food anaphylaxis. *British Medical Bulletin* **56**, 925–935.

Schmid-Grendelmeier P., Simon D., Simon H.-U., Akdis C. and Wuthrich B. (2001) Epidemiology, clinical features, and immunology of the "intrinsic" (non-IgE-mediated) type of atopic dermatitis (constitutional dermatitis). *Allergy* **56**, 841–849.

Schwartz R. (2002) A new element in the mechanism of asthma. *New England Journal of Medicine* **346**, 857–858.

Sheikh A. (2002) Itch, sneeze and wheeze: the genetics of atopic allergy. *Journal of the Royal Society of Medicine* **95**, 14–17.

Simons F. (2004) Advances in H1-antihistamines. *New England Journal of Medicine* **351**, 2203–2217.

Turner H. and Kinet J.-P. (1999) Signalling through the high-affinity IgE receptor Fc epsilon RI. *Nature* **402**, B24–B30.

Wedemeyer J. and Galli S. (2000) Mast cells and basophils in acquired immunity. *British Medical Bulletin* **56**, 936–955.

Weiss K. and Sullivan S. (2001) The health economics of asthma and rhinitis. I. Assessing the economic impact. *Journal of Allergy and Clinical Immunology* **107**, 3–8.

Wills-Karp M. (2001) IL-12/IL-13 axis in allergic asthma. *Journal of Clinical Immunology* **107**, 9–18.

Wohlleben G. and Erb K. (2001) Atopic disorders: a vaccine around the corner? *Trends in Immunology* **22**, 618–626.

Yazdanbakhsh M., Kermsner P. and van Ree R. (2002) Allergy, parasites, and the hygiene hypothesis. *Science* **296**, 490–494.

Autoimmune Disease

Henry Kunkel 1916–1983

Rheumatoid factor a autoantibody

© Rockefeller Archive Center

CHAPTER 29

A.   OVERVIEW

B.   EXAMPLES OF HUMAN AUTOIMMUNE
     DISEASES

C.   ANIMAL MODELS OF AUTOIMMUNE DISEASE

D.   DETERMINANTS OF AUTOIMMUNE DISEASE

E.   MECHANISMS UNDERLYING AID

F.   THERAPY OF AID

G.   RELATIONSHIP BETWEEN AID AND CANCER

"Men are at war with each other because each man is at war with himself"
—Francis Meehan

Autoimmunity arises when immune responses mounted in the host are directed against self-components. *Autoimmune diseases* (AID) are pathophysiological states in which the host's own tissues are damaged as a result of these immune system attacks. Compared to infectious diseases, AID are much less common but they impose a significant burden of suffering for the 3–5% of populations in developed countries that suffer from them. Indeed, AID cause 33% of premature deaths in the developed world, and are a leading cause of death among women under 65 years of age in the United States. In contrast, AID are rarely diagnosed in infants, perhaps because their immune systems are still immature and subject to the robust mechanisms of neonatal tolerance. Curiously, like allergy and hypersensitivity, AID are diseases of higher socio-economic strata, and are rare in developing countries. In this chapter, we will examine several aspects of AID, including their discovery, clinical examples, related animal models, predisposing and triggering factors, cellular mechanisms, and conventional and immunotherapeutic treatments. We conclude this chapter with a brief discussion of the possible relationship between AID and cancer.

A. Overview

I. HISTORICAL NOTES

For a long time, immunologists doubted that self-reactive immunity existed. Paul Ehrlich and others held at the beginning of the 1890s that the immune system would naturally be incapable of mounting anti-self immune responses. This concept was labeled "*horror autotoxicus*," meaning that the immune system should have a "horror of" (and therefore avoid) being "autotoxic." However, even as far back as the early 1900s, there were isolated reports suggesting that autoreactivity, although unusual, could definitely occur. For example, anti-self agglutinins in humans were reported by J. Donath and K. Landsteiner and their colleagues at the turn of the 20th century. A short while later, it was demonstrated that rabbits could mount immune responses against their own ocular lens proteins. In 1906, it was found that human antibodies directed against the organism that causes syphilis (*Treponema pallidum*) could cross-react with an antigen (later identified as *cardiolipin*) in normal tissues, damaging the latter. Indeed, the action of this antibody on normal cells induced the release of more of the self antigen, feeding the disease cycle. An unexpected discovery in the 1930s provided a marker for what became one of the best-studied AID. Quite serendipitously, patients with *rheumatoid arthritis* (RA; see later) were found to have a factor in their serum that was capable of agglutinating indicator (Ig-coated) sheep RBCs in complement fixation tests. This "rheumatoid factor" was shown much later to be an IgM antibody directed against the Fc region of IgG molecules.

Clear clinical and experimental evidence for autoimmunity was finally obtained in the early 1950s when William J. Harrington and co-workers transferred serum from patients suffering from *idiopathic thrombocytopenic purpura* (ITP; see later) to healthy volunteers (themselves). Recipients of the ITP patient serum soon experienced a dramatic decrease in platelet counts, suggesting that the transferred serum contained antibodies directed against the platelets. Since platelets are a normal body tissue, these antibodies that destroyed a self tissue were called *autoantibodies*, the first formal use of the term. Soon after, researchers such as Peter Miescher and Henry Kunkel showed that the disease known as *systemic lupus erythematosus* (SLE) was associated with production of antibodies directed against components of the nucleus such as DNA, RNA, and various nuclear proteins. In addition, in 1956, Noel Rose and his colleagues showed that immunization of a healthy rabbit with an extract of its own thyroid induced a form of autoreactivity against this organ that led to thyroiditis (inflammation of the thyroid gland).

One of the most important animal models for research into autoimmunity was discovered in the 1960s when New Zealand researchers Marianne and Feliz Bielschowsky intercrossed two black-colored littermates that turned up in a colony of normally white mice being maintained for inbred strain development. Successive backcrossing produced an inbred strain of black mice called New Zealand Black (NZB) that died prematurely of an AID that resembled autoimmune hemolytic anemia in humans. When NZB mice were crossed to other mouse strains, different AID symptoms developed in the progeny, such as autoimmune damage to the kidney or thymus. Several different NZB hybrid substrains were subsequently developed, and, as is described in more detail later in this chapter, the study of these mice has underpinned many of our modern concepts of AID. Nevertheless, consistent with the mysteries that still surround AID today, the precise genetic defect leading to the predisposition to autoimmunity in NZB mice has yet to be elucidated.

Two of the discoveries most germane to understanding autoimmunity were the elucidation of the roles in adaptive immune responses of T and B cells in the 1960s, and the MHC in the 1970s. On the technological side, in the 1960s–1970s the use of fluorescence microscopy become routine for the detection of autoantibodies on cell surfaces and in serum samples. For example, in the mid-1970s, G. F. Bottazzo and colleagues, as well as W. J. Irvine and associates, independently demonstrated the presence of anti-pancreatic β-islet cell antibodies in insulin-dependent diabetes. In the 1980s–1990s, the characterization of Th cell subsets and DC functions plus investigations of peripheral tolerance shed much light on possible mechanisms underlying AID. The dissection of apoptotic mechanisms in the 1990s and the recognition of the importance of these pathways for controlling autoimmunity have also been important contributions to this field. However, despite all this progress, we still do not know in most cases precisely what sparks an autoimmune attack.

II. WHAT CHARACTERIZES AN AUTOIMMUNE DISEASE?

The first book on AID was published in 1963 by Ian Mackay and F. Macfarlane Burnet. This volume included discussions of several diseases that were thought to be clear examples of autoimmunity, such as thyroiditis, ITP, SLE, RA, and multiple sclerosis, as well as some suspect disorders that only later were confirmed to be AID, such as rheumatic fever, ulcerative colitis, and myasthenia gravis. Many of these diseases had been previously classified as "idiopathic" (occurring without known cause) or "of unknown etiology." The current list of recognized AID contains over 80 distinct disorders that vary widely in incidence and phenotype. A summary of some of the better-studied AID is presented in Figure 29-1, and the reader is encouraged to refer back to this chart frequently.

AID		Incidence/100,000/yr	Dominant sex	Disease pattern	Tissue(s) affected
APECED	Autoimmune polyendocrinopathy candidiasis ectodermal dystrophy	1	F	C	Multiple endocrine glands (thyroid, adrenals, etc.)
APS	Anti-phospholipid syndrome	5	F	C	Blood clots at multiple sites
AS	Ankylosing spondylitis	10	M	C	Tendons, ligaments, bone
CD	Crohn's disease	2	M = F	R/R C	Walls of colon and small intestine
GBS	Guillain-Barré syndrome	1–2	M = F	A	Peripheral nerves
GD	Graves' disease (hyperthyroidism)	50	F	R/R C	Thyroid gland
GS	Goodpasture's syndrome	1	M	A	Kidney and lung

Figure 29-1

Examples of Human Autoimmune Diseases
Several features of AID covered in this chapter are summarized. "Incidence" refers to the number of new cases that occur per 100,000 population in 1 year. "Dominant sex" indicates whether males (M) or females (F) are affected most often (*, indicated sex is only slightly predominant). "Disease pattern" describes whether the course of the disease is acute (A), is chronic (C), or follows a relapsing/remitting pattern (R/R) (some AID are manifested in more than one pattern). "Tissues affected" specifies those tissues that characteristically come under autoimmune attack.

Continued

	AID	Incidence/100,000/yr	Dominant sex	Disease pattern	Tissue(s) affected
HT	Hashimoto's thyroiditis (hypothyroidism)	500	F	C	Thyroid gland
IPEX	Immunodysregulation, polyendocrinopathy, enteropathy, X-linked syn.	Unknown, very rare	M	C	Variety of tissues and organs
ITP	Immune (idiopathic) thrombocytopenia purpura	7	M = F (children) F (adults)	A (children) C (adults)	Platelets
KD	Kawasaki disease	10–70 (USA) 134 (Japan)	M*	A	Mucosae, lymph nodes, heart, vasculature
MG	Myasthenia gravis	2	F (30–50 yrs) M (70–80 yrs)	R/R	Neuromuscular junctions
MS	Multiple sclerosis	7	F	R/R C	Brain, spinal cord
PG	Pemphigus	0.5–3.2	M = F	C	Skin
PM	Polymyositis	1	F	R/R C	Voluntary muscle fibers
PS	Psoriasis	1000–3000	M = F	R/R C	Skin
RA	Rheumatoid arthritis	1000	F	C	Joints, muscle, connective tissue
RF	Rheumatic fever	200	F*	R/R	Heart muscle and valves, kidney, CNS
SD	Scleroderma	0.2–0.4	F	C	Skin, bones, connective tissue
SLE	Systemic lupus erythematosus	40	F	R/R	Skin, joints, kidney, lung, heart, GI tract
SS	Sjögren syndrome	20	F	C	Exocrine glands (lacrimal, salivary)
T1DM	Type 1 diabetes mellitus	35 (Finland) 0.7(China)	M = F	C	β-islet cells of pancreas
TTP	Thrombic thrombocytopenia purpura	0.1–0.3	F	A R/R	Platelets
UC	Ulcerative colitis	7	M = F	R/R C	Inner wall of colon

Figure 29-1 *cont'd*

How does an AID arise? We learned in Chapter 13 about T cell development and the processes of negative selection in the thymus that are designed to remove autoreactive clones and establish central tolerance. We then learned in Chapter 16 that, despite these measures, anti-self clones do escape to the periphery where they are normally controlled by the mechanisms of peripheral tolerance. Most AID develop when these mechanisms fail or are not triggered in the first place. (The rare occurrence when an AID results from a defect in central tolerance is discussed in Box 29-1.) Failures in peripheral control mechanisms may be linked to allelic variation in genes encoding proteins important for immune system function. That is, an individual may be genetically predisposed to developing AID. As well, environmental triggers such as exposure to particular pathogens, chemicals, or toxic agents are often associated with the induction of episodes of AID. Altered expression of certain self proteins may also contribute to autoimmune diseases under some circumstances. An autoreactive response may thus be initiated in a genetically predisposed individual when an autoreactive clone encounters a suitable antigen in the right cytokine milieu (Fig. 29-2). In a healthy individual, even if an activated autoreactive cell proliferates and differentiates into effector cells, intact mechanisms of peripheral tolerance ensure that the response cannot proceed much further, and the chance of damage to the tissues is minimal. The clinical damage of AID occurs when the activation of autoreactive clones is not properly held in check.

AID have traditionally been categorized into two broad classes: (1) *organ-specific autoimmunity*, in which an anatomically-specific site is targeted for immune destruction, and (2) *systemic autoimmunity*, in which the immune response is not restricted to a specific organ or tissue. However, it should be noted that the categorization of an AID as organ-specific or systemic is based primarily on clinical observations rather than the expression pattern of the self antigen that appears to be targeted in the attack. In some instances, an antigen may be ubiquitously expressed but an autoimmune response to it may occur only in one organ such that the manifestation of the AID is organ-specific. For example, in a disease called *primary biliary cirrhosis* (PBC), the small bile ducts in the liver are destroyed by an inflammatory response. Autoantibodies recognizing mitochondrial and nuclear components are present in these patients, causing researchers to believe PBC is an AID. However, why the disease manifestation is restricted to the liver is unknown. Some researchers consider Sjögren syndrome (see later) to be a similar case, because the AID-mediated damage appears primarily in the exocrine glands, but the nuclear proteins against which the autoantibodies are directed occur in almost all body cells. Conversely, in some systemic AID, tissue damage may occur at sites where the autoantigen is not expressed

---

Box 29-1. AID resulting from a defect in central tolerance

In very rare cases, human autoimmunity arises from a defect in central tolerance. *Autoimmune polyendocrinopathy candidiasis ectodermal dystrophy* (APECED) is an AID in which multiple endocrine glands sustain autoimmune attack, including the pancreas and the thyroid and adrenal glands. Patients thus present with a complex array of clinical signs generally associated with Hashimoto's thyroiditis (HT), insulin-dependent diabetes mellitus (T1DM) (see text for details), and a very rare AID specifically affecting the adrenal gland, *Addison's disease*. (In Addison's disease, autoimmune destruction of the adrenal gland leads to deficient production of cortisol and aldosterone, two hormones with critical roles in stress responses and metabolic regulation.)

Unlike most AID, which involve contributions from many genes, APECED results from a single gene defect: an autosomal recessive mutation in the AIRE gene (discussed in Ch.13). The reader will recall that AIRE is a transcriptional co-activator in thymic epithelial cells that may play a key role in establishing central tolerance. In most cases, the mutation causing APECED prevents the normal localization of the AIRE protein in the nucleus. It is theorized that sub-optimal expression of peripheral self antigens (such as insulin and thyroglobulin) in thymic epithelial cells then results. In the absence of normal presentation of these antigens, developing thymocytes recognizing them escape negative selection and are released to the periphery in numbers that overwhelm normal peripheral tolerance mechanisms. Indeed, in knockout mice lacking AIRE, negative selection cannot occur in the thymus and large numbers of autoreactive T cells can be detected in the periphery. Like AIRE-deficient humans, the mutant mice develop multi-organ autoimmunity.

Clinically, the young children that make up the bulk of APECED patients often first present with either a chronic mucosal yeast infection or a neurological seizure due to low calcium. This latter deficit is due to a failure in the parathyroid gland and the seizures can be fatal if the cause is not recognized and treated promptly. Hypothyroidism, anemia, and defects in the nails and tooth enamel are also common features. Patients may fail to mature sexually. Treatment of APECED primarily involves the relief of symptoms. Vitamin D and calcium are given to correct for the parathyroid hormone deficiency, along with hydrocortisone supplements to compensate for the failure of the adrenal gland. Anti-fungal drugs are used to control candidiasis.

No other examples of central tolerance failure leading to AID have been found to date in humans, but a murine model demonstrating the principle can be found in the development of T1DM in *non-obese diabetic* (NOD) mice. If these animals are transgenically engineered to express high levels of insulin in the thymus as well as in the periphery, autoreactive clones directed against insulin are eliminated during thymic negative selection and the incidence of diabetes is reduced. Thus, it is hypothesized that NOD mice may develop diabetes in part because insulin may be sub-optimally expressed in the thymus, leading to a failure in negative selection of autoreactive clones directed against this self antigen. The consequent release of these autoreactive cells to the periphery, coupled with a failure of peripheral tolerance mechanisms to control them, leads to attacks on the β-islet cells and symptoms of diabetes.

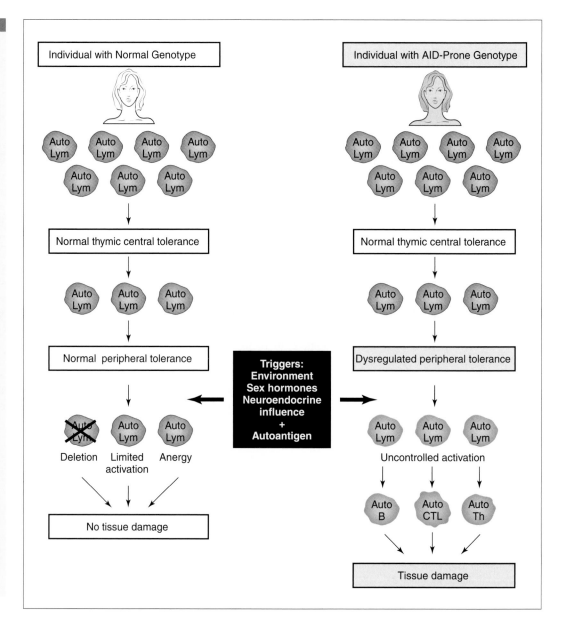

How Dysregulation of Peripheral Tolerance Leads to Autoimmune Disease
In an individual with a normal genotype (left), most autoreactive lymphocytes are eliminated by negative selection during the establishment of central tolerance. Those autoreactive cells that escape into the periphery are kept under control by the mechanisms of normal peripheral tolerance. An encounter with autoantigen, even in the presence of one of the indicated triggers, almost always results in clonal deletion or anergization. Even if an autoreactive lymphocyte achieves limited activation, it is held in check by regulatory mechanisms that ensure that the cell does not continue to respond to the autoantigen and damage self tissues. In an individual with an AID-prone genotype (right), normal central tolerance is usually in place but the mechanisms of peripheral tolerance are compromised. Autoreactive lymphocytes that escape to the periphery and encounter autoantigen in the presence of a trigger are activated uncontrollably. Effector functions of the autoreactive cells result in tissue damage and clinical symptoms that constitute "autoimmune disease."

because antibody–antigen complexes formed elsewhere travel in the circulation and lodge in these locations. A summary of self antigens that come under autoimmune attack in various AID is presented in Table 29-1.

How do activated autoreactive cells cause the tissue damage associated with AID? In some cases, upon activation by binding to self antigen, autoreactive Th cells release cytokines such as TNF and IFNγ that can damage host tissues or supply T help for autoantibody production. Autoantibodies that have bound to their self antigens can induce classical complement activation, facilitate opsonized phagocytosis, and form immune complexes with self antigens that are either soluble or present on the surfaces of healthy host cells. In the case of localized damage, autoanti-

bodies that bind to cellular targets may directly and locally impede the normal function of a tissue and disrupt its contribution to the physiology of the host. Alternatively, as noted previously, systemic damage can be caused by immune complexes formed between autoantibodies and soluble self antigens that block the body's narrow channels. Such antibody-mediated AID mechanisms represent physiological examples of the type II and III hypersensitivity reactions described in Chapter 28. In addition to the destruction caused by autoantibodies, autoreactive Tc cells receiving help from autoreactive Th cells generate CTLs that kill healthy host cells expressing self antigens via perforin- and granzyme-mediated cytolysis. These CTLs also release damaging pro-inflammatory cytokines.

Table 29-1 **Autoreactivity to Self Antigens in Selected Human AID**

Acronym	Disease	Self Antigen
APS	Anti-phospholipid syndrome	Glycoproteins of the prothrombin activator complex
AS	Ankylosing spondylitis	Fibrocartilage-derived Ag HLA-B27
GBS	Guillain-Barré syndrome	Gangliosides, heparan sulfate
GD	Graves' disease	TSH receptor, thyroid peroxidase, thyroglobulin
GS	Goodpasture's syndrome	Collagen in basement membrane
HT	Hashimoto's thyroiditis	TSH receptor, thyroid peroxidase, thyroglobulin
IPEX	Immunodysregulation, polyendocrinopathy, enteropathy X-linked syndrome	Antigens of pancreatic islets, thyroid and adrenal glands, smooth muscle, intestine, kidney
ITP	Immune thrombocytopenic purpura	vW receptor, fibrinogen receptor, other platelet antigens
LA	Lyme arthritis	LFA-1
MG	Myasthenia gravis	Nicotinic acetylcholine receptor
MS	Multiple sclerosis	Myelin basic protein, myelin oligodendrocyte protein, proteolipid protein
PG	Pemphigus	Desmoglein-3, -1
PM	Polymyositis	Aminoacyl tRNA synthetases
RA	Rheumatoid arthritis	IgG, citrullinated polypeptides
RF	Rheumatic fever	Cardiac myosin
SD	Scleroderma	Topoisomerase I, RNA polymerases, centromeric proteins, collagen, heat shock proteins, phospholipids, nuclear components, IgG
SS	Sjögren syndrome	Muscarinic acetylcholine receptor, α-fodrin, IgG, nuclear components, ICA69, plasma membrane protein, proteasome protein
SLE	Systemic lupus erythematosus	Nuclear components
T1DM	Insulin-dependent diabetes mellitus	Glutamic acid decarboxylase, insulin, ICA512
TTP	Thrombotic thrombocytopenia purpura	vW-cleaving protease

ICA, islet cell antigen; TSH, thyroid-stimulating hormone; vW, von Willebrand.

III. OUR APPROACH TO DISCUSSING AUTOIMMUNITY

Autoimmunity is a fascinating and complex subject, and there are as many ways to approach its varied topics as there are immunologists to study it. We have chosen to start this chapter by giving short clinical descriptions of 20 reasonably well-researched AID (Section B), with the rationale that the reader will better appreciate how a contributing trigger or mechanism might work if the basic features of the disease in question are understood beforehand. Similarly, the relevance of a mouse model of an AID (Section C) might be easier to grasp if the reader is first familiar with the features of the human disease. We then move to a discussion of the genetic, environmental, and hormonal determinants of various human AID, pointing out how these factors contribute to various AID (Section D). The mechanisms underlying AID remain to be fully defined but it is believed that many AID share at least

some of them, so that collective discussions of these subjects follow in Section E. As well, the therapeutic approaches to treating AID patients (Section F), including those based on immunological strategies, can be more readily appreciated when the features of the disease to be treated are understood.

B. Examples of Human Autoimmune Diseases

Because SLE is one of the best-studied AID and is often considered a prototype of human autoimmunity, we start with a discussion of this disorder, and proceed through a list of several other important human AID in an order that generally reflects our state of knowledge about them.

I. SYSTEMIC LUPUS ERYTHEMATOSUS (SLE)

We first encountered SLE in Chapter 16, where we discussed peripheral tolerance. SLE is a systemic AID that initially affects the skin, joints, kidney, lung, heart, and gastrointestinal tract. A pattern of relapse and remission is common, with an unpredictable frequency of "flare-ups." With time, SLE may become chronic in nature. Patients may initially present with alopecia (hair loss), fatigue, musculoskeletal symptoms, vascular abnormalities, photosensitivity, and a characteristic rash. It is this rash that gave the disease its name: the rash is red in color (erythematous) and is concentrated in the malar region of the face (on the cheeks), giving many patients a "wolfish" appearance ("lupus" is Latin for "wolf") (Plate 29-1). More severe SLE cases may involve arthritis, inflammatory serositis (inflammation of the lining of organs such as the lungs and heart), glomerulonephritis (inflammation of the glomeruli of the kidney) or renal failure, *autoimmune hemolytic anemia* (AHA), thrombocytopenia, lymphopenia, and neurological problems. Multiple elements of the immune system may be disrupted in SLE patients, including the complement system (see Ch.19). SLE patients are thus usually highly vulnerable to opportunistic infections. The incidence of SLE is estimated to be 40 cases per 100,000 population per year. People of non-Caucasian descent are disproportionately affected, as are women, especially those of childbearing age.

A signature feature of SLE (especially during the active phase of the disease) is the production of high levels of autoantibodies referred to as *anti-nuclear antibodies* (ANAs). The ANAs are directed against the host's double-stranded DNA (dsDNA) and small nuclear ribonucleoproteins (snRNPs) exposed by cellular breakdown. The *Smith* (Sm) *antigen* associated with SLE is one such snRNP. While levels of anti-DNA antibodies vary with disease flare-ups, the titers of autoantibodies directed against snRNPs tend to remain constant. Variable levels of antibodies against other nuclear self antigens such as histone proteins, RNA binding proteins, and DNA polymerase components may also be present, as may autoantibodies directed against soluble cytoplasmic and membrane components such as phospholipids, IgG, and certain coagulation factors and complement components. These autoantibodies form immune complexes that accumulate first in the blood and eventually in the target organs, triggering inflammation that exacerbates the disease. Some circulating autoantibodies may also bind directly to autoantigens in the kidney and form immune complexes in this organ. The extent of abnormal elevation of each type of autoantibody varies greatly from individual to individual, but the profile tends to remain relatively constant within a given patient. This finding suggests that the genetic make-up of an individual plays an important role in determining the pattern of autoreactivity against these host proteins and nucleic acids. Why a high percentage of autoantibodies is directed against nuclear-associated components is not understood.

Examination of B cells isolated from SLE patients has shown that the activation of these cells is abnormal. Increased numbers of activated B cells at all stages of differentiation can be found in the circulation of SLE patients, and intracellular calcium flux within these cells is elevated. SLE B cells also tend to be more sensitive than normal B cells to the effects of cytokines. SLE patients have higher levels of serum IL-10 than do normal individuals, which is relevant because IL-10 can stimulate B cell proliferation and differentiation. Peaks in serum IL-10 levels have been reported to correlate with upswings in ANA titers in SLE patients. IL-12, which favors Th1 responses and thus may have inhibitory effects on antibody production, is produced at abnormally low levels by monocyte/macrophages in SLE patients. The fact that the autoantibodies in SLE are of high affinity suggests that affinity maturation has occurred in the self-reactive B cells, which in turn implies that T cells capable of delivering the intercellular signals to B cells that promote affinity maturation may play a role in the development of the disease. Indeed, there is some evidence that the autoreactive T cells in SLE are activated, as judged by their enhanced expression of surface activation markers. However, the specific self-epitopes recognized by such T cells have yet to be identified. In addition, the relative importance of Td autoantibodies (as opposed to Ti autoantibodies) in the development of SLE is still not clear: it is likely that both types of autoantibodies make a contribution.

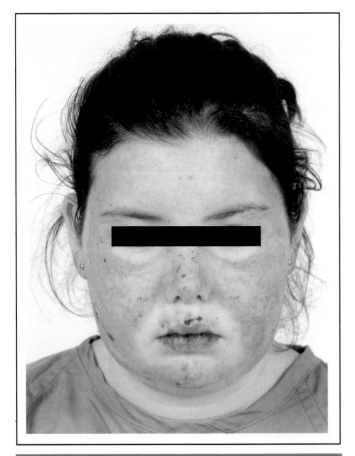

Malar Rash of SLE
Courtesy of Dr. Rae Yeung, The Hospital for Sick Children, Toronto.

II. RHEUMATOID ARTHRITIS (RA)

Rheumatoid arthritis (RA) is an organ-specific AID that typically attacks the synovial tissue (called the *synovium*) and cartilage of the joints (Fig. 29-3). The terms "rheumatic" and "rheumatoid" thus refer to diseases in which the joints, muscles, or connective tissues of the body come under inflammatory attack. There are two types of RA that show both similarities and differences: *juvenile RA* (JRA) and adult RA (usually referred to as just "RA"). It remains unclear whether these AID have separate causes, but their HLA associations (see later) differ, as do their phenotypic presentations. JRA strikes children under 10 years of age, while RA affects adults up to 70 years of age. The incidence of RA is a shocking 1 case per 100 population per year, while JRA is about 10 times less frequent. RA is less clinically and genetically heterogeneous compared to JRA, but even in RA the disease presentation and prognosis vary dramatically among patients. In the early phase, the dominant features are morning stiffness in the affected joints, which commonly include the knees, hips, wrists, fingers, and ankles. Severe or untreated disease may lead to cartilage destruction and bone erosion (Plate 29-2), or inflammation of the cervical spine that results in a chronic crippling disease. In the most severe cases, patients can be disabled and unable to perform daily functions. On average, RA patients have a life span that is decreased from the norm by 3–7 years, although those with extreme cases may succumb 10–15 years early. JRA and RA are similar in that both occur about three times more often in females than in males. However, JRA tends to have a better long term outcome compared to RA.

At the cellular level, examination of the inflamed joints of RA (and JRA) patients reveals signs of inflammatory infiltration. The blood vessels in the synovial layer assume the characteristics of HEVs, facilitating the extravasation of activated macrophages and DCs into the joint. It is these cells that appear to produce the bulk of the pro-inflammatory cytokines, including TNF, IL-1, and GM-CSF, that are present in RA tissue. Collagen and fibronectin in the ligaments, tendons, and bone may also suffer some degradation due to the action of matrix metalloproteinases secreted by the activated macrophages. Additional cytokines secreted by macrophages and T cells induce angiogenesis and the enhanced expression of adhesion molecules, further promoting inflammation. At some point during disease progression, low numbers of CD4$^+$ and CD8$^+$ T cells infiltrate RA joints. These cells produce cytokines such as IL-2 and IFNγ that help to activate B cells and macrophages, perpetuating the inflammation. Indeed, T cells isolated from affected sites may show specificity against antigenic components (such as collagen) in the diseased tissues. In addition to their production of cytokines, activated T cells present in an inflamed joint express costimulatory and regulatory molecules that may help to sustain the autoimmune response.

With respect to the humoral arm, RA synovial tissues are characterized by the presence of ectopic germinal centers

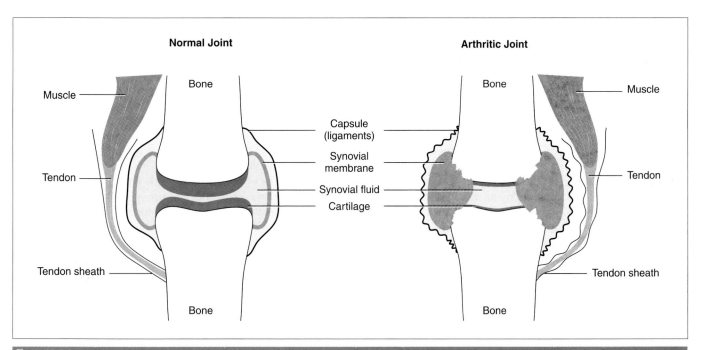

Figure 29-3

How Rheumatoid Arthritis Affects Joints
A normal joint consists of two bones held together by ligaments that form a capsule around the bone ends. A "cushion" of cartilage between the bone ends prevents the ends from grinding against each other. A synovial membrane surrounding a small volume of synovial fluid lubricates the joint. In RA, autoantibodies attack self antigens in the synovial membrane and the cartilage. The inflammation that accompanies this attack then destroys the cartilage and sometimes the bone itself. The lubrication and cushioning in the joint is lost, leading to stiffness or immobility of the joints.

vessel lining can cause narrowing and blockage of the vessels that results in *ischemia* (restricted blood supply). In turn, ischemia can induce damage to other organs.

A distinctive serological sign associated with RA (but not JRA) is the presence of *rheumatoid factor* (RhF or RF). RhF is not a single entity but rather a collection of autoantibodies that are directed against the patient's own IgG molecules. However, RhF is also found in patients with AID other than RA, in the serum of normal individuals suffering from chronic infections with viruses and bacteria, and even in about 15% of healthy seniors. The normal function, if any, of RhF is not known, but these autoantibodies have been postulated to target immune complexes and clear them from the body via the mononuclear phagocyte system. The CD5$^+$ B1 subset of B cells is thought to be a major source of RhF. The reader will recall from Chapter 9 that CD5$^+$ B cells arise early during fetal development and are concentrated in the peritoneal and pleural cavities. These cells produce the low-affinity IgM natural antibodies that constitute part of innate immunity. However, at least some CD5$^+$ clones appear to be positively selected by self antigens during development, meaning that an abnormality in the regulation of these cells could lead to their persistent activation and the production of autoantibodies. Elevated numbers of CD5$^+$ B cells have been found in patients with a variety of organ-specific and systemic AID.

A relatively newly recognized and highly specific serological marker for RA (but not JRA) is the presence of autoantibodies directed against *citrullinated polypeptides* (anti-CPP antibodies). *Citrullination* is a post-translational modification process in which arginine residues in a protein are deiminated (i.e., =NH is removed) by the enzyme *peptidylarginine deiminase* (PAD) to form citrulline. While citrullination of cellular proteins can occur during apoptosis, few proteins in normal cells contain citrulline. An exception is the *filaggrin* protein expressed in terminally differentiated epidermal epithelial cells. Anti-CPP antibodies in the serum of RA patients were identified early on by their reaction with citrullinated forms of this protein. However, it is unlikely that this skin protein is an autoantigen in a joint disease such as RA. Significantly, the fibrin protein present in joints is citrullinated in the inflamed synovial tissue of at least some RA patients. Such a finding is consistent with observations that the synovium of RA patients contains plasma cells that locally produce anti-CPP antibodies. Anti-CPP antibodies have turned out to be a useful diagnostic marker for early RA, since they may appear in a patient's serum before any clinical manifestations. In addition, the presence of these antibodies has been correlated with higher levels of joint damage as the disease progresses, making their evaluation of prognostic value as well.

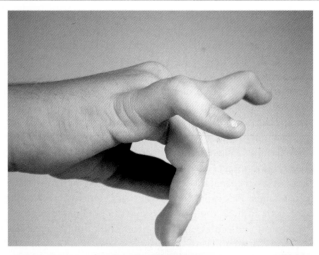

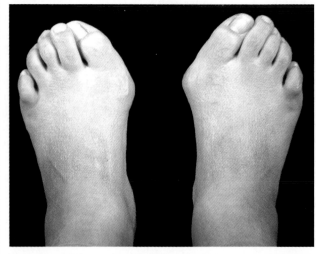

Plate 29-2

RA Joints
Courtesy of Dr. Rae Yeung, The Hospital for Sick Children, Toronto.

populated by plasma cells. It is speculated that something in these affected tissues, perhaps inflammatory cytokines or unknown intercellular signals delivered by synovial cells, induces the local differentiation of B cells that produce autoantibodies directed against antigens in the synovial membrane and cartilage. As well as being sources of autoantibody production, these abnormally situated germinal centers can seriously damage the normal architecture of the tissue in which they develop.

In addition to direct attack on the joints, RA has effects on other organs. For example, RA may be associated with inflammation of the lungs that results in shortness of breath, inflammation of various parts of the eye, and skin ulceration. Another complication that occurs in RA (but rarely in JRA) is *vasculitis*, an intense inflammation of the endothelial lining of the blood vessels. Initially the inflammatory damage may be caused by IC deposition in the vessel walls and an influx of neutrophils, but later stages may be characterized by lymphocyte infiltration. The inflammatory damage to the blood

III. RHEUMATIC FEVER (RF)

Despite its similarity in name, rheumatic fever (also known as *acute rheumatic fever*; ARF) is not directly related to rheumatoid arthritis and patients are not positive for rheumatoid factor. This AID primarily attacks cells in the heart muscle, heart valves (Plate 29-3), kidney, and CNS, but also affects

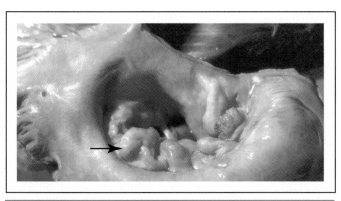

Plate 29-3

RF Heart Damage
Mitral valve in the heart of a person who had acute rheumatic fever as a child. The opening of the valve is greatly reduced in diameter due to the thickening of the mitral valve cusps (arrow). Reproduced with permission from Cooke R. A. and Stewart B. S. (2004) "Colour Atlas of Anatomical Pathology," 3rd edn. Churchill Livingstone, Edinburgh.

the joints. Unlike most AID, RF has its highest incidence (200 cases per 100,000 population per year) in developing countries. Children 5–15 years of age are the most frequent victims but adults can get RF as well. The disease affects girls slightly more often than boys, and Blacks more frequently than Caucasians. RF follows a wax-and-wane pattern, with acute episodes of RF lasting 6–14 weeks, interspersed with longer periods of disease remission. The major clinical manifestations of RF are fever, a distinctive rash, carditis, arthritis, and neurological effects (such as uncontrolled movements). Some patients present with acute cardiac failure or tachycardia (very rapid heartbeat). Indeed, RF has a high mortality rate due to its effects on the heart, and for persons under 40 years of age in developing countries, carditis associated with RF is the leading cause of heart-associated deaths.

In a large number of RF patients, AID symptoms appear 2–6 weeks after infection with certain strains of Group A streptococcal bacteria, such as those causing "strep throat." These bacteria express the M antigen, a bacterial cell wall protein that confers pathogenicity. Some researchers believe that the M protein sparks the autoreactive response that establishes the AID through *molecular mimicry* (see later, Section D). That is, lymphocytes that originally were activated by epitopes of the streptococcal M protein cross-react with epitopes on host tissues in the heart. Indeed, antibodies recognizing the M protein are present at high levels in patients with acute RF, and relapses of symptoms may occur following a subsequent infection with *Streptococcus*. This connection with bacterial infection may explain why RF is seen most often today in developing countries: the antibiotics necessary to treat streptococcal infections are not always readily available, and poor hygiene and housing conditions may encourage the growth of *Streptococcus* in the local environment. Interestingly, there has been a recent upswing in the number of RF cases seen in the developed world that some scientists suspect may be related to widespread prescribing and/or inappropriate use of antibiotics.

This practice may have favored the selection of antibiotic-resistant M strains of Group A *Streptococcus* in these countries. There is also almost certainly a genetic component to this AID, as family members of RF patients have an increased risk of getting the disease regardless of environmental factors.

At the cellular level, organs affected in acute RF show elevated numbers of CD4$^+$ T cells, decreased CD8$^+$ T cells, and increased levels of IL-6, IL-8, and TNF. These cytokines decline to normal levels as the disease subsides and rise again during relapses. Subsets of T cells isolated from patients with severe RF have been shown to recognize certain peptides from the M protein. In addition, streptococcal M protein is toxic for macrophages such that, although these cells can still secrete IL-1 and TNF, they can no longer undertake their normal function of clearing immune complexes. These complexes then accumulate in the circulation and lodge in the heart and joints, provoking intense inflammation.

IV. TYPE 1 DIABETES MELLITUS (T1DM)

The sugar glucose is the main energy source for mammalian cells and as such drives all basic metabolic processes. The hormone insulin is produced in response to food intake and is the master regulator of glucose metabolism. Insulin is synthesized by the β cells of the islets of Langerhans in the pancreas (often called "β-islet cells"). Insulin facilitates the entry of glucose from the blood into cells and thereby regulates blood sugar levels. *Diabetes* results when the body either does not produce insulin or its cells cannot respond to it normally. In the absence of insulin, glucose cannot get into cells and blood sugar levels rise catastrophically. The patient can fall into a life-threatening "diabetic coma," a condition that results from *ketoacidosis*, a buildup in the blood of acid and ketone products of abnormal cellular metabolism. A diabetic coma is fatal if blood sugar levels are not reduced immediately by insulin treatment. Less dramatic symptoms of diabetes include excessive hunger and thirst, increased urination, weight loss, fatigue, and blurred vision. Slow wound healing and increased susceptibility to infections may also be present in some patients. If diabetes is controlled inadequately for an extended period, the elevated concentration of sugar in the blood (hyperglycemia) damages the vascular endothelium in multiple tissues. This damage can result in blindness, heart attacks and strokes, nerve damage, kidney failure, or tissue necrosis necessitating limb amputation. Diabetics often die of the heart attacks and strokes that are vascular complications of this disease. As a result, diabetes is the sixth leading cause of death in the United States.

There are two major types of diabetes mellitus: (1) *type 1 diabetes mellitus* (T1DM), formerly known as insulin-dependent diabetes mellitus (IDDM), juvenile onset diabetes, or autoimmune diabetes, and (2) *type 2 diabetes mellitus* (T2DM), formerly known as non-insulin-dependent diabetes mellitus (NIDDM). About 80% of diabetes patients have T2DM, which is not an AID. T2DM generally affects overweight adults in middle age but can also strike obese children. In these "insulin-resistant" cases, insulin is produced normally but cells in the patient stop responding properly to the hormone and cannot make full use of the glucose in the blood. The pancreas then increases its production of insulin to compensate, but much of it is removed by the abundant fat cells, leading to effective insulin deficiency and a slow rise in blood sugar levels. Eventually the pancreas may lose its ability to produce insulin. T2DM occurs more frequently in certain ethnic groups, including African Americans, the Inuit, and native North Americans.

T1DM is an organ-specific AID that preferentially occurs in young children of Caucasian ethnicity. For example, the incidence of T1DM is 54 cases per 100,000 population per year in Finland, but only 0.7 cases per 100,000 population per year in China. Overall, T1DM is the most common chronic disease in children and adolescents, and equal numbers of boys and girls are affected. Insulin deficiency itself may have an acute, abrupt onset at birth, but progression to clinical diabetes may take several years, depending on the rate of β-islet cell destruction. Usually by 10 years of age, the secretion of insulin by the pancreas ceases completely and the patient requires multiple daily injections of exogenous insulin to survive. The patient's autoimmune response specifically destroys his/her β-islet cells: no other endocrine cells in the pancreas are affected. Clinically, T1DM is a heterogeneous disease, but most patients fall into one of two distinct clinical categories. The first is a condition called *insulitis*, in which histological examination of the pancreas shows that the islets have been invaded by inflammatory cells (including macrophages, B cells, and CD4$^+$ and CD8$^+$ T cells) (Plate 29-4). These patients may have autoreactive T and B cells that can be shown to recognize β-islet antigens, but the actual destruction of the β-islet cells (and thus overt diabetes) does not occur. The second presentation is overt diabetes, in which the invasion of the pancreas results in T cell-mediated destruction of the β-islet cells and an absolute insulin deficiency.

Both humoral and cell-mediated responses directed against several β-islet cell antigens can be detected in T1DM patients. One such self antigen is a cellular enzyme called *glutamic acid decarboxylase* (GAD). GAD is expressed primarily in neuroendocrine tissues but also in pancreatic β-islet cells. In neurons, GAD catalyzes the synthesis of the important inhibitory neurotransmitter GABA (gamma-aminobutyric acid), but its role in the islets is unclear. Autoantibodies reactive to different epitopes, one central and one C-terminal, of the GAD protein have been found in T1DM patients. T cell epitopes have also been identified in these regions of GAD. Another candidate self antigen may be insulin or its precursor proinsulin. Autoantibodies against insulin are often present in T1DM patients, and a particular polymorphism in the insulin promoter has been found more often in T1DM patients than in normal individuals. ICA512 (islet cell antigen 512; also known as IA-2), a tyrosine phosphatase localized in the β-islet cell secretory granules, is also the target of specific autoantibodies. However, it is not yet clear whether these antibodies precede or arise in response to islet destruction. Large international trials are under way to resolve this issue.

Some researchers theorize from observational studies that T1DM may somehow be linked to the early exposure of an immature immune system to intact foreign proteins. Very young babies who are genetically susceptible to T1DM (have a family history of the disease) and who are prematurely switched from breast-feeding to formula containing intact cow's milk proteins often go on to develop T1DM. Decreased rates of diabetes development have been observed in young, genetically susceptible rats if the animals are fed hydrolyzed formula (in which the foreign proteins are broken down to amino acids) rather than formula with intact foreign proteins. A large international study known as TRIGR (Trial to Reduce IDDM in the Genetically at Risk) is currently comparing the rates of T1DM development in cohorts of 6-month-old genetically susceptible babies who are switched from breast-feeding to either standard cow's milk formula or formula containing hydrolyzed cow proteins.

V. MULTIPLE SCLEROSIS (MS)

Multiple sclerosis (MS) is an organ-specific AID that primarily affects the brain and spinal cord. The *myelin sheath* surrounding the nerve axons (Plate 29-5), and brain cells called *oligodendrocytes* that make the myelin, come under autoimmune attack. Axon function may be lost due to impaired myelination, ultimately leading to the inhibition of nerve impulse transmission. Damage to the axonal components underlying the myelin may also occur. Eventually, myelinated axons in the peripheral nervous system may also be affected. MS patients thus suffer from widespread motor weakness,

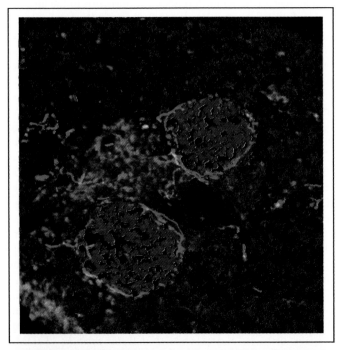

Plate 29-4

T Cell Infiltration of the Pancreas
Two pancreatic islets of Langerhans in early stages of insulitis. Insulin is stained in blue, cells enveloping the pancreatic islets are stained in green, and the infiltrating T cells are stained in red. Courtesy of Dr. Hubert Tsui and Dr. H.-Michael Dosch, University of Toronto.

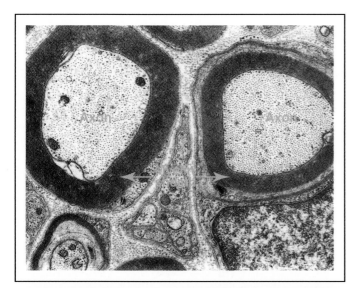

Plate 29-5

Myelinated Axon
Electron micrograph of normal myelinated peripheral nerve axon.
Reproduced with permission from Gartner L. P and Hiatt J. L. (2001) "Color
Textbook of Histology," 2nd edn. W. B. Saunders Company, Philadelphia.

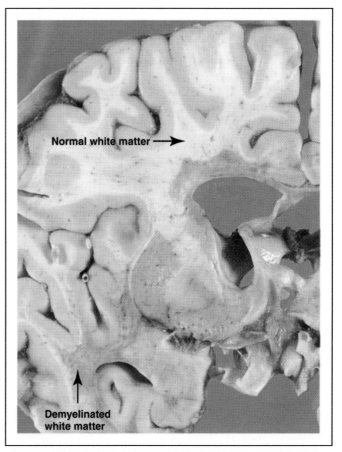

Plate 29-6

Multiple Sclerosis
Post-mortem brain slice from a person with MS showing characteristic
demyelination of the white matter. Reproduced with permission from
Cooke R. A. and Stewart B. S. (2004) "Colour Atlas of Anatomical Pathology,"
3rd edn. Churchill Livingstone, Edinburgh.

spasticity, and visual, auditory, and cognitive impairments. Analysis of the nerve fibers in the brains and spinal cords of MS patients reveals areas of hardened scar tissue called plaques (*sclerosis*), which are largely responsible for the localized loss of neurological function (Plate 29-6). The incidence of MS is 7 cases per 100,000 population per year, and an estimated total of 2.5 million people around the world suffer from this AID. MS preferentially strikes people between the ages of 15 and 40 and is the most common cause of neurological dysfunction in young adults. MS occurs twice as often in women as in men and shows a bias for women of Northern European heritage.

The disease course of MS varies widely among patients, such that some show very few symptoms and are able to continue all normal daily activities, while others become severely disabled within months. In 70% of MS patients, the disease follows a relapsing/remitting pattern characterized by acute attacks that may last for days or weeks followed by remission periods of months to years. Complete recovery is usually experienced after the first few flare-ups but progressively increasing functional deficits remain after later relapses. In 30% of MS patients, the disease starts with a relapsing/remitting pattern but then progresses to a second stage of chronic progressive disability that has no remission periods for relief. These patients eventually lose their ability to walk and are confined to a wheelchair.

The pathogenesis of MS is thought to stem from the activation (by unknown means) of autoreactive T cells which can recognize pMHC involving peptides derived from *myelin basic protein* (MBP), the principal protein composing the myelin sheath (Fig. 29-4). The activation of T cells recognizing other myelin sheath components, including *myelin oligodendrocyte glycoprotein* (MOG) and *proteolipid protein*

(PLP) has also been observed. These cells (particularly CD8$^+$ T cells) start to express the VLA-4 integrin needed to bind to VCAM-1 upregulated on CNS endothelial cells and cross the blood–brain barrier. MBP-specific T cells also produce *matrix metalloproteinases* (MMPs) that degrade both the basement membrane of the endothelium and the extracellular matrix components in the brain, allowing the T cells direct access to CNS neurons and their axons. Autoreactive CTLs responding to various peptides presented by oligodendrocytes produce still more MMPs and cytokines (such as LTα and TNF) that damage the myelin sheath. In addition, CTLs have been shown to kill neurons *in vitro*, suggesting that CTLs may directly eliminate neurons by perforin/granzyme-mediated mechanisms.

The destruction of the blood–brain barrier and brain matrix induced by autoreactive T cells facilitates an influx of inflammatory cells into the brain. *Microglia* (macrophage-like cells resident in the brain) and infiltrating neutrophils are induced by T cell-derived cytokines to secrete pro-inflammatory mediators and cytokines that further contribute to myelin destruction. Among these mediators is *osteopontin*, a cytokine that primarily

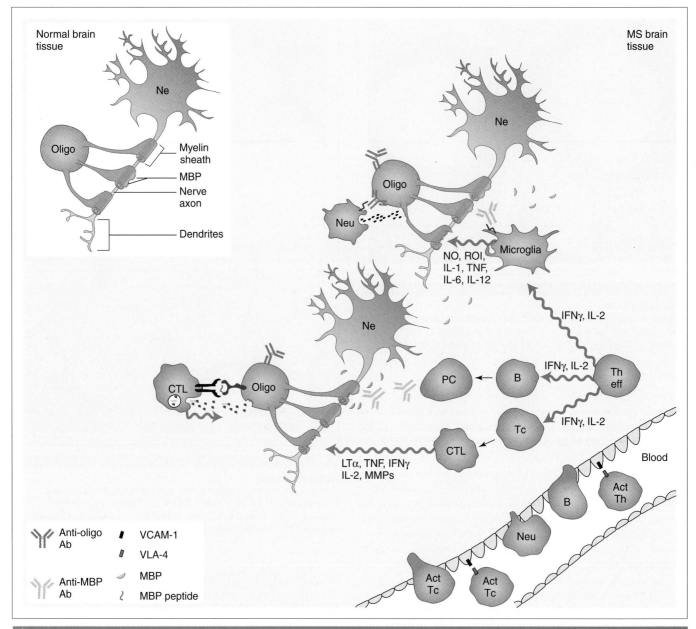

Figure 29-4

Putative Pathogenesis of Multiple Sclerosis

In the normal brain, the nerve axons are surrounded by a myelin sheath that facilitates normal nerve impulse transmission. Myelin basic protein (MBP) is a key component of the myelin sheath and is made by oligodendrocytes (Oligo). In MS brain tissue, activated autoreactive T cells (lower right) upregulate VLA-4 and bind to VCAM-1 on the endothelium of the blood–brain barrier and enter the CNS, along with other inflammatory cells. Within the CNS, cytokines and matrix metalloproteinases (MMPs) produced by these cells lead to damage of the myelin sheath and exposure of MBP. Autoreactive Th effectors (Th eff) supply help for autoreactive B cell and Tc activation. Plasma cells (PC) produce autoantibodies recognizing MBP or other autoantigens expressed by the oligodendrocytes, and CTLs secrete cytokines and exert cytotoxicity that damages oligodendrocytes. Progressive damage to the myelin sheath and oligodendrocytes increasingly impedes nerve impulse transmission.

stimulates osteoclasts to break down bone but also promotes the secretion by various cell types of pro-inflammatory cytokines such as IFNγ and IL-12. In addition, the NO produced by phagocytes blocks nerve conduction pathways and contributes to the irreversible structural damage to the nerves. Sites of demyelination also contain elevated levels of IgG autoantibodies and complement products, indicating that

autoreactive B cells accompany autoreactive T cells into the CNS such that the humoral response contributes to MS pathogenesis. Among the targets of these autoantibodies are antigens on oligodendrocytes.

Remission of MS disease is thought to occur when anti-inflammatory cytokines and growth factors produced by cells in the inflammatory infiltrate offset the autoimmune attack and allow the oligodendrocytes to remyelinate the damaged nerves. However, over time, the buildup of the scar tissue around the nerves, coupled with the accumulating assaults on the oligodendrocytes, may prevent remyelination, so that the patient is not able to fully recover from an MS episode. If the sclerosis becomes severe, the patient enters the chronic progressive stage of MS.

VI. ANKYLOSING SPONDYLITIS (AS)

Ankylosing spondylitis (AS) is a chronic inflammatory disease of bone and joints, particularly of the spine. The terms "anky-losing" and "spondylitis" are derived from Greek words meaning "fusing together" and "inflammation of the vertebrae," respectively. AS affects about 10 individuals per 100,000 population per year, predominantly males between the ages of 20 and 40. The course of AS disease can vary greatly but is generally more severe in men than in women. Affected individuals first show symptoms of chronic lower back and hip pain that can persist for years. The normally elastic tissue in the tendons and ligaments of the joint is eroded and then replaced first by fibrocartilage and finally by bone. When such infiltration and bone replacement occurs in the spine, the lower vertebrae fuse together, causing irreversible and serious damage to the spinal column. In the most severe cases, vertebral fusion progresses all the way up the spinal column, leaving the patient with a "bamboo spine." The patient suffers from significant back and neck pain and joint stiffness, and his/her mobility is markedly compromised. The chest expansion may be reduced and the patient's posture is often characteristically altered (Plate 29-7). The patient has to be constantly aware of the increased fragility of his/her rigid spine: if trauma occurs in the neck region, the patient may become a quadriplegic. Because AS progression is insidious and a definitive diagnosis is difficult to ascertain in the early stages of the disease, patients have often suffered irreversible damage by the time they are properly diagnosed. The growing use of *magnetic resonance imaging* (MRI) analysis to detect early AS-linked changes in the sacroiliac region should increase the number of patients who receive timely treatment.

The target tissue that undergoes autoreactive attack in AS appears to be the fibrocartilage supporting the sites where tendons and ligaments attach to the bones of the joints. Affected AS joints contain elevated numbers of many cell types, including plasma cells, macrophages, lymphocytes, and mast cells. A particularly large increase in CD8$^+$ T cells has been noted. HLA-B27 expression is strongly associated with AS, but the mechanism by which HLA-B27 confers susceptibility to AS has not been defined (see later). Some researchers believe that conformational epitopes in HLA-B27 molecules expressed by fibrocartilage cells may come under autoantibody attack. On the biochemical side, AS patients often have elevated serum IgA, IL-10, and acute phase proteins, but autoantibodies are notable for their absence.

A disease related to AS is *reactive arthritis*, described as the development of arthritis in association with an infection that occurred in the 1–4 weeks prior to the onset of the arthritis. Bacteria belonging to the *Salmonella* and *Yersinia* genera have

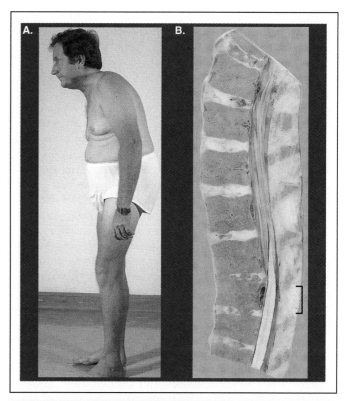

Plate 29-7

Ankylosing Spondylitis
(A) AS patient demonstrating posture imposed by a rigid spinal column. (B) Cross-section of an AS spine showing calcification of the interspinous ligaments. An example is indicated by the bracket. Reproduced with permission from Cooke R. A. and Stewart B. S. (2004) "Colour Atlas of Anatomical Pathology, 3rd edn. Churchill Livingstone, Edinburgh.

been particularly implicated, but the pathogen itself is not present by the time the diagnosis is reached. The development of arthritis is thus thought to be an indirect, rather than direct, result of the infection. There is also an element of genetic predisposition, as 50% of reactive arthritis cases are associated with expression of HLA-B27. Clinically, the arthritis develops primarily in the knees, ankles, and feet although the sacroiliac region may also be affected. A mucosal or skin rash may occur. Antibodies against *Salmonella* and *Yersinia* can sometimes be detected in the sera of reactive arthritis patients. A serendipitous demonstration of the link between reactive arthritis and bacterial infection occurred in 1984 in Toronto, Canada following a papal visit. Hundreds of police officers on security duty suffered food poisoning due to *Salmonella typhimurium*, and about 5% of them showed signs of acute arthritis within 3 months of the incident. The disease resolved satisfactorily in one-third of these officers, but the remaining two-thirds (18 individuals) developed reactive arthritis.

VII. SJÖGREN SYNDROME (SS)

Sjögren syndrome is a systemic AID with an incidence of 20 cases per 100,000 population per year. SS primarily strikes women in their 40s to 50s. SS can be present either on its own (primary SS), or as SS secondary to another AID such as SLE, RA, or scleroderma (see later). Autoantibodies produced in SS patients primarily attack exocrine glands such as the lacrimal glands in the eyes and the salivary glands in the mouth, causing these structures to stop production of their usual secretions. As a result, the signature clinical signs of SS are a constellation of complaints known as *sicca symptoms* (*sicca*, meaning "dry"): dry eyes (*keratoconjunctivitis sicca*; KCS), leading to blurry vision; dry mouth (*xerostomia*) and dry throat, leading to swallowing difficulties; and dry nose and skin. SS is considered a rheumatic disease in part because the arthritis associated with this disease resembles RA, and SS often overlaps with other connective tissue disorders. Numbers of CD5$^+$ B cells are often increased and RhF is usually present. SS patients also exhibit hypergammaglobulinemia and the presence of two types of anti-nuclear antibodies called SS-A and SS-B. These particular ANAs are characteristic of SS but occasionally turn up in other AID such as SLE and RA. Anti-thyroid antibodies are also common in SS and thyroiditis is frequently observed.

Although SS is usually a relatively benign disorder, patients can present with a lymphoproliferative disease that progresses to a B cell lymphoma 5% of the time. These lymphomas tend to affect mucosal tissues and contain large numbers of CD27$^+$ B lymphocytes that, contrary to the normal situation, may be either naive or memory B cells. The fact that SS patients show lymphoproliferation and sometimes develop lymphomas has raised suspicions that the underlying defect lies in the regulation of lymphocyte proliferation or apoptosis. SS patients exhibit extensive infiltration of activated CD4$^+$ T cells into the lacrimal and salivary glands, and these cells are important drivers of the disease. B cells also infiltrate the lacrimal and salivary glands, and the presence of IgA$^+$ plasma cells and ectopic germinal centers in these tissues is diagnostic for SS. These B cell anomalies may be linked to the ectopic expression of the B cell survival/proliferation factor BAFF and the B cell chemokine BCA-1 (CXCL-13) in the salivary glands of SS patients.

Several potential autoantigens associated with SS have been studied in some depth. T cells recognizing an epitope derived from the muscarinic M3 acetylcholine receptor have been identified in SS patients. The M3 receptor is upregulated on the surfaces of acinar cells in the salivary glands of some SS patients. IgA antibodies directed against the M3 receptor have been detected in the saliva of SS patients and may be responsible for the xerostomia associated with this disease. Indeed, clinical trials of a drug that binds to the M3 receptor showed that this agent provided some relief from xerostomia in SS patients. Another autoantigen that may be associated with SS is an antigen called ICA69. ICA69 (islet cell antigen 69) was originally identified in the pancreatic islets but is also expressed in the salivary and lacrimal glands. Both T and B cell responses to ICA69 have been detected in SS patients. Autoantibodies against a ubiquitously expressed membrane protein called α-fodrin have also been found in large numbers of SS patients. Evidence in animal studies has suggested that if this protein is not properly processed and presented on MHC

class II, abnormalities in salivary and lacrimal function can arise. It may be that these abnormalities expose autoantigens that then precipitate the autoimmune attack.

VIII. AUTOIMMUNE THYROIDITIS

There are two major types of autoimmune thyroiditis: *Hashimoto's thyroiditis* and *Graves' disease*. To understand how these AID develop, a brief review of the biology of the thyroid gland may be helpful. The thyroid gland, which surrounds the trachea in the throat, produces a hormone called *thyroxine*. Thyroxine governs the metabolic rate of all cells in the body, helps to regulate body temperature and energy use, and contributes to the normal functions of the CNS and cardiovascular system. When the level of circulating thyroxine falls below an optimal level, the hypothalamus in the brain releases *thyrotropin-releasing hormone* (TRH; also known as *thyroid-releasing hormone*), which acts on the pituitary gland at the base of the brain. The anterior pituitary then produces *thyroid-stimulating hormone* (TSH; also known as *thyrotropin*), which binds to TSH receptors (TSHR) on thyroid follicular cells (Fig. 29-5). These cells routinely make *thyroglobulin* (TG), the inactive protein precursor of thyroxine, and store it in the central *colloid* regions of the thyroid follicles. On the luminal surface of a follicle, molecules of iodine are added to the thyroglobulin by the enzyme *thyroid peroxidase (TPO)*. In response to TSH signaling, the iodinated thyroglobulin is imported back into the follicular cell via the *sodium iodide symporter* and cleaved by lysosomal enzymes to produce thyroxine. The thyroxine molecules then diffuse out of the follicular cell and enter the circulation.

i) Hashimoto's Thyroiditis (HT)

Hashimoto's thyroiditis is an AID in which the autoreactive response is directed against thyroid antigens. Thyroid follicular cells are destroyed such that insufficient levels of thyroxine are produced, leading to hypothyroidism. HT has an incidence of 500 cases per 100,000 population per year and typically affects women over the age of 50, but younger patients have also been reported. Although many individuals who technically have HT show no symptoms, others present with clinical signs typical of metabolic deceleration: depression, forgetfulness, muscle weakness, brittle hair, sensitivity to cold and stress, fatigue, weight gain, and dry, rough skin. The menstrual cycle and ovulation may also be affected, increasing the chance of miscarriage. In more severe cases, the thyroid gland is greatly enlarged in size (*goiter*) as it attempts to make more thyroglobulin to compensate for the thyroxine deficiency. The hyperproliferation of the glandular cells makes HT patients vulnerable to tumorigenesis in the thyroid.

The thyroid gland of HT patients shows infiltration by plasma cells and T cells (Plate 29-8), and ectopic germinal centers may be present. Autoantibodies directed against TSHR, thyroglobulin, and thyroid peroxidase are found in many HT patients, and some also have antibodies recognizing the sodium iodide symporter. Because these anti-thyroid antibodies tend to be of isotypes that can fix complement, it

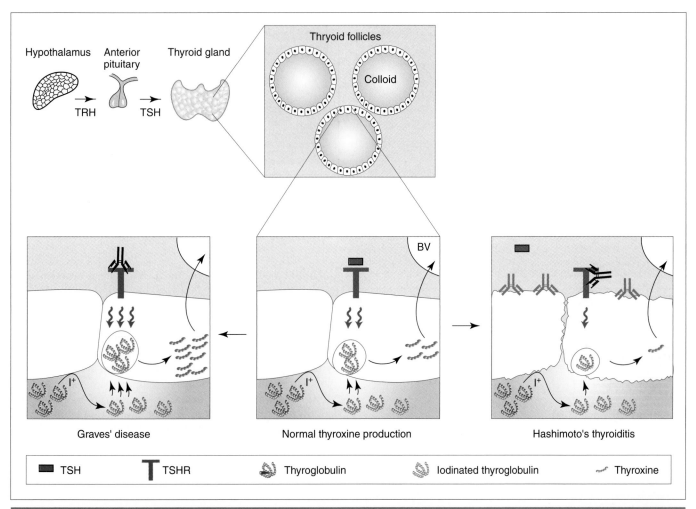

Figure 29-5

Putative Pathogenesis of Autoimmune Thyroiditis
In a normal individual, the hypothalamus in the brain produces thyrotropin-releasing hormone (TRH), which acts on the anterior pituitary gland and induces it to produce thyroid-stimulating hormone (TSH). TSH binds to thyroid-stimulating hormone receptors (TSHR) expressed by the follicular cells of the thyroid gland. Thyroid follicular cells routinely make the inactive precursor thyroglobulin and store it in the colloid of the thyroid follicles (lower center panel). Iodine is then added to produce iodinated thyroglobulin. In response to TSHR signaling, iodinated thyroglobulin is imported back into the follicular cell, followed by cleavage to produce thyroxine. Thyroxine is released into the circulation (blood vessel, BV) to control the body's metabolic rate. In GD patients (left panel), autoantibodies overstimulate TSHR signaling, resulting in overproduction of thyroxine and clinical hyperthyroidism. In HT patients (right panel), autoantibodies, CTLs, and cytokines all play a role in damaging thyroid follicular cells. This destruction results in insufficient thyroxine production and clinical hypothyroidism. In some cases, autoantibodies may also inhibit TSHR signaling.

would be logical to think that they contribute to the ensuing destruction of thyroid follicular cells. Indeed, the early stages of MAC assembly can be seen on the membranes of these cells. However, thyroid follicular cells also express high levels of RCAs which are further upregulated in response to cytokines secreted by the infiltrating macrophages and T cells. With such a high level of RCA expression, MAC assembly is rarely completed. A more likely antibody-mediated mechanism of destruction in HT may be ADCC, perhaps carried out by the significant numbers of NK cells that infiltrate the thyroid. CTL cytotoxicity may also take a toll, as the HT thyroid contains large concentrations of activated T cells. Indeed, experimental thyroiditis can be induced in naive animals by

transferring lymphocytes from an affected animal. *In vitro*, T cell clones isolated from the thyroid glands of HT patients proliferate in response to thyroglobulin or thyroid peroxidase, indicating that epitopes derived from these proteins can activate autoreactive T cells. The precise epitopes appear to vary between patients and no one dominant peptide has been identified. Once activated, these T cells produce IFNγ, IL-2, IL-6, and TNF. Fas-mediated apoptosis may also be an important mechanism of thyroid follicular cell destruction. Thyroid follicular cells constitutively express FasL, and Fas expression is upregulated on the surfaces of these cells in response to IL-1α, a cytokine that has been found at high levels in the thyroid glands of HT patients.

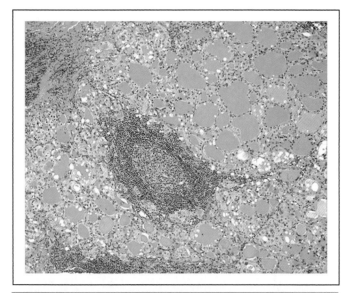

Plate 29-8

Hashimoto's Thyroiditis
A section of the thyroid gland of an HT patient showing infiltration by T cells and plasma cells, and an ectopic germinal center. Courtesy of Dr. Sylvia Asa, University Health Network, Toronto.

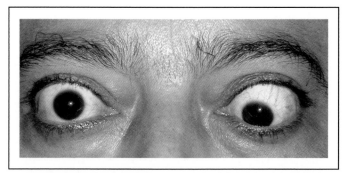

Plate 29-9

Graves' Disease Ophthalmopathy
Courtesy of Dario Surace and S.I.F.I. S.p.A.

ii) Graves' Disease (GD)

In Graves' disease, the autoreactive response against the thyroid gland has the opposite effect: the gland becomes hyperactive and overproduces thyroxine. GD has an incidence of 50 cases per 100,000 population per year and primarily affects women over 20 years of age. Onset may be acute or gradual, and the course may adopt a relapsing/remitting pattern. The symptoms of GD include hand tremors, brittle hair, insomnia, irritability, paranoia, weight loss, sensitivity to heat, increased perspiration, changes to hair structure, muscle weakness, anxiety, and rapid heart beat. Goiter can be present due to expansion of the hyperactive thyroid gland. GD is sometimes accompanied by a distinctive bulging of the eyes known as *Graves' ophthalmopathy*. The eyes and the tissues around and behind them become inflamed and swollen due to the autoimmune response, forcing the eyeball out of its orbit (Plate 29-9). Double vision, restricted eye movement, and blurred vision may occur. Some GD patients also develop a painless but lumpy, reddish thickening of the skin on the shins and feet called *Graves' dermopathy*. This thickening is due to the abnormal synthesis of collagen by underlying local fibroblasts.

The autoantibodies prominent in GD are often directed against the TSHR but bind to a different region than those in HT. Rather than blocking thyroxine production, engagement of these receptors by GD anti-TSHR antibodies stimulates the thyroid gland to produce more thyroxine. The multiple B cell epitopes involved appear to be conformational. Autoantibodies against thyroglobulin and thyroid peroxidase may also be present but do not seem to reduce the excess thyroxine being produced. Nor do the immune complexes formed between the thyroid autoantigens and their respective autoantibodies appear to be much of a clinical problem since they activate complement only weakly and are easily cleared. The thyroid tissue in a GD patient shows T and B lymphocyte infiltration and the formation of ectopic germinal centers. The infiltrating activated T cells, which respond to peptide epitopes derived from the same thyroid autoantigens recognized by the autoantibodies, produce Th1 and Th2 cytokines. These cytokines induce thyroid cells to upregulate expression of MHC class I and II, various adhesion molecules, and CD40. The thyroid cells also increase their production of inflammatory cytokines such as IL-6, IL-8, and IL-12. Activated T cells that infiltrate the muscles of the eye and the dermis of the skin mediate autoimmune responses that lead to Graves' ophthalmopathy and dermopathy, respectively.

IX. AUTOIMMUNE THROMBOCYTOPENIC PURPURA

A patient with *thrombocytopenia* has abnormally low platelet numbers in the blood. Platelets play a critical role in the blood clotting system such that abnormal bleeding occurs in their absence. Patients with thrombocytopenia often develop *purpura*, a purplish-brown skin rash resulting from inadequate clotting of blood that leaks from the capillaries into the skin. There are two types of thrombocytopenic purpura that are associated with AID: *immune* (or *idiopathic*) *thrombocytopenic purpura* (ITP) and *thrombotic thrombocytopenic purpura* (TTP).

i) Immune (or Idiopathic) Thrombocytopenic Purpura (ITP)

ITP is caused by an autoantibody attack on platelets. Two groups of people appear to get ITP: young children and young adults. In children, boys and girls are equally affected. In adults, about three times more women than men are affected. The overall incidence of ITP is estimated to be about 7 cases per 100,000 population per year. ITP occurs in both acute and chronic forms. In acute cases (most often children), the patient experiences sudden bleeding into the skin, nose, mouth, or digestive or urinary tracts due to leakage of blood from capillaries feeding these sites. In more severe cases, the bleeding may occur in the brain, lungs, or other vital organs.

When the platelet count falls from the normal 100×10^9/liter to below 5×10^9/liter, intracranial bleeding becomes a real threat. However, in 80% of children with ITP, the disease is self-limited and requires no treatment. Onset in chronic ITP cases (most often adults) is gradual and may be asymptomatic, discovered only in the course of blood testing for other reasons. Alternatively, the occurrence of purpura, heavy menstrual periods, and/or unusually easy bruising may cause a patient to seek medical attention.

The IgG autoantibodies in ITP are directed against several platelet membrane glycoproteins, including the fibrinogen receptor and the receptor for *von Willebrand factor* (vWF). (vWF is a carrier protein produced by endothelial cells that stabilizes clotting Factor VIII in the blood and helps platelets stick to each other and blood vessel walls.) The binding of the autoantibodies marks the platelets for Fc-mediated or possibly CR-mediated destruction by phagocytes in the spleen or liver, causing platelet levels to fall precipitously. It remains unclear whether the underlying defect lies in the autoreactive B cells, or whether events associated with T cell activation or regulation have gone awry.

ii) Thrombotic Thrombocytopenic Purpura (TTP)

In contrast to ITP, where platelet numbers are low due to lysis, circulating platelets in TTP patients are low because they aggregate abnormally and form blood clots (*thromboses*) in capillaries throughout the body. The kidney and brain are most severely affected, followed by the liver and lungs. There is no inflammation, but the partial blockage of the blood vessels damages erythrocytes trying to circulate through and results in their lysis. Initial symptoms of TTP may include malaise, fever, and diarrhea, followed by bruising or bleeding as the platelets are effectively removed. When the platelets block capillaries, patients may show signs of confusion, headache, temporary paralysis, speaking difficulties, and numbness. Anemia may result if damaged and lysed erythrocytes are removed from the circulation faster than they can be replaced.

Autoimmune TTP is a disease of adults, whereas a form of the disease that occurs in children is usually due to an inherited deficiency of an enzyme called *vWF-cleaving protease*, which is required for normal clotting. Adult TTP occurs at an incidence of 0.1–0.3 cases per 100,000 population per year and is most common in women of 20–40 years of age. More than 80% of adult TTP cases appear spontaneously, but about 20% have been associated with pregnancy, infection, use of anti-clotting drugs or oral contraceptives, BMT or HCT, or cancer. The disease pattern of adult TTP is usually acute, although one-third of patients experience at least one relapse.

Platelets aggregate in adult TTP because of the presence in the blood of abnormally large multimers of the clotting factor vWF. In normal individuals, megakaryocytes and endothelial cells synthesize vWF as an extremely large molecule composed of many identical subunits. This extremely large multimer is broken down into a collection of merely large multimers by vWF-cleaving protease. The large multimers are "sticky" and bind to platelets and damaged endothelial cell walls at sites of injury to promote the clotting process. In adult TTP, autoantibodies directed against the vWF-cleaving protease prevent the cleavage of the vWF multimer, leaving a very sticky, mammoth molecule to circulate in the blood and clump platelets. These clumps then promote the formation of dangerous blood clots, particularly in the capillaries of the brain and kidneys.

X. SCLERODERMA (SD)

Scleroderma (also known as *systemic sclerosis*, SSc) is an extremely rare chronic AID with an incidence of 0.2–0.4 cases per 100,000 population per year. The term "scleroderma" is derived from the Greek words *sklerosis* (meaning "hardening") and *derma* ("of the skin"). This phenotype results from the overproduction of collagen by fibroblasts. SD patients are usually women between the ages of 30 and 50 years. Like RA and SLE, SD can be considered a rheumatic disease since it involves inflammation of the skin, bones, or connective tissues.

SD occurs in two major sub-types: *localized SD* and *systemic SD*. Symptoms of localized SD are restricted to the skin and underlying tissues, whereas systemic SD affects blood vessels and multiple organs, including the heart, lungs, and kidneys. Localized SD is more common in Caucasians than Blacks, but the reverse is true for systemic SD. In localized SD, the skin of the chest, stomach, back, face, arms, and legs develops one or more oval-shaped reddish patches with whitish centers. The skin in these patches is thickened and tight (Plate 29-10). In a sub-type of localized SD called *linear SD*, the disease produces a disfiguring, single straight line of hardened skin on the arm, leg, or forehead. Patients with localized SD symptoms may find it hard to carry out routine activities if the skin on their hands or feet becomes tight and hard. Patients with systemic SD show a much broader range

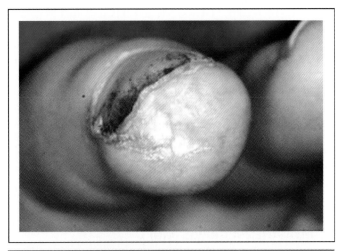

Plate 29-10
Fingertip of Scleroderma Patient
Reproduced with permission from 2002 *Dermatology Online Journal* **8**(1), 3, Department of Dermatology, Leipzig, Germany.

of symptoms. In addition to the symptoms of local SD, these individuals exhibit calcium deposits in connective tissues; *Raynaud's phenomenon*, in which small blood vessels in the limbs contract abnormally in response to cold temperatures or stress; impaired movement of the smooth muscles in the esophagus, leading to swallowing difficulties; thick and tight skin on the fingers; and skin telangiectasias (red spots due to capillary swelling). Internally, the thickening of connective tissue around the heart, lungs, and kidneys may compress these organs and impair their function. Fatigue, loss of appetite, weight loss, and joint swelling are often present. SD can be severely debilitating or even lethal, but also may resolve after 3–5 years and leave the patient with either near-normal skin, or thin, delicate skin lacking hair and sweat glands.

At the laboratory level, the serum of most SD patients contains many autoantibodies found in the sera of patients with other AID. Such antibodies include anti-IgG (RhF), anti-ssRNA, anti-snRNP, anti-histones, anti-phospholipids, anti-collagen, anti-HSP, SS-A, and SS-B, among others. Autoantibodies unique to SD (so far) include anti-topoisomerase I (an enzyme needed for DNA replication), anti-RNA polymerase I, and anti-RNA polymerase III. Anti-centromere protein antibodies have been found in the blood of 90% of patients with one sub-type of systemic SD. The chromosomes in lymphocytes and fibroblasts of SD patients show an abnormally high frequency of breaks, deletions, and acentric fragments, implying that SD lymphocytes may be unusually sensitive to DNA damage.

It is not clear how SD fibroblasts are induced to overproduce collagen. However, endothelial cells in the capillaries of SD patients show enhanced expression of adhesion molecules such as MAdCAM-1 and ICAM-1 that draw lymphocytes and inflammatory cells into the tissues. SD endothelial cells also undergo abnormal apoptosis, which may contribute to the release of self antigens. Having been drawn to skin capillaries, CD4$^+$ cells infiltrate the tissues and surround the blood vessels. These cells (plus smaller numbers of CD8$^+$ T cells, monocytes, and NK cells) produce cytokines, including IL-1, -2, -4, 6, -8, -10, and -13, TGFβ, PDGF, TNF, and IFNγ, as well as soluble forms of CD4 and IL-2R, that are present at high concentrations in the circulation of SD patients. Whether fibroblasts near the affected blood vessels are induced to overproduce collagen by these cytokines, or by intercellular contacts established with the invading inflammatory cells, remains under investigation. It is noteworthy that TGFβ is a powerful inducer of collagen gene promoter activation. In addition, SD fibroblasts show decreased activity of the collagenase enzyme that normally balances collagen synthesis. Finally, SD fibroblasts express lower levels of the α-integrins that allow these cells to use a feedback mechanism to assess how much collagen they have made.

XI. MYASTHENIA GRAVIS (MG)

Myasthenia gravis is a rare AID (incidence 2 cases per 100,000 population per year) that results in severe muscle weakness. The name of the disease is derived from Greek and Latin words: "my" refers to "muscle," "asthenia" means "weak-

ness," and "gravis" means "severe." MG affects more women than men in the third to fifth decades of life, but more men than women in the seventh and eighth decades. Early and late onset forms of MG exist and the clinical presentation is heterogeneous in both. A relapsing/remitting pattern of disease is common, and early onset MG is often accompanied by symptoms of other AID. MG patients usually complain of specific muscle weakness rather than generalized fatigue. In Asian populations at least, the early onset form is often restricted to the eye. Weakness of the facial muscles may cause a patient to have difficulty smiling, chewing, talking, and swallowing. Neck muscle weakness can lead to drooping of the head, and breathing difficulties (which can be life-threatening) can stem from weakness of the respiratory muscles.

In 85% of MG cases (both early and late onset), the symptoms are caused by autoantibodies that attack the nicotinic acetylcholine receptors (AChRs) of the muscle at the neuromuscular junctions (Fig. 29-6). Acetylcholine (ACh) is the chemical that carries the electrical signal across the synapse from a neuron to a muscle cell, stimulating the muscle to move. This signaling is controlled by cholinesterase enzymes that break down the acetylcholine and prevent constant firing of the signal. MG autoantibodies bind to the extracellular regions of AChRs on the muscle cells and block ACh access to AChRs. Ig deposits, complement components and MAC assembly in the affected junctions may also interfere with nerve impulse transmission. As a result, the muscles are immobilized and the patient feels weak and fatigued. In the remaining 15% of MG cases, the autoantibodies are directed against muscle-specific kinase, another neuromuscular junction protein.

One striking feature of early onset MG is that it is associated with the presence of thymoma (thymic lymphoma) in 10% of patients, and with hyperplastic changes in this organ in another 70%. Ectopic germinal centers containing T and B cells can be seen in the thymus of many MG patients. Some researchers speculate that the thymus is the site where MG autoreactivity first develops, since the thymus contains APCs, T cells, and muscle-like cells that express AChR. Indeed, when thymic tissue (containing B cells) from MG patients is transplanted into immunodeficient mice, it triggers production of anti-AChR antibodies. Researchers have also been able to isolate T cell clones from MG thymi that are specific for AChR epitopes. Curiously, despite similarities in other clinical signs, thymic abnormalities are not usually present in late onset MG.

XII. KAWASAKI DISEASE (KD)

Kawasaki disease is an AID that affects the mucosal surfaces, lymph nodes, vasculature, and heart. It was first described in Japan and its incidence (134 cases per 100,000 population per year) remains highest in that country. In contrast, the incidence of KD in the United States is 10–70 cases per 100,000 population per year. KD is a disease of young children, with the vast majority of diagnoses being made in patients under 5 years of age. Boys are affected slightly more

A. Normal neuromuscular junction

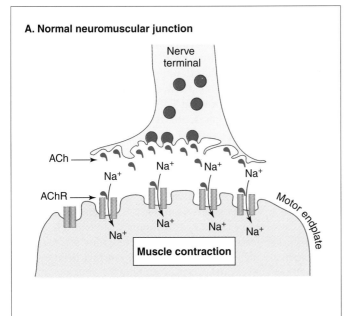

B. Myasthenia gravis

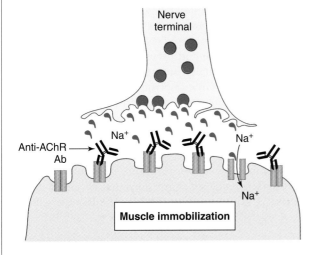

Figure 29-6

Putative Pathogenesis of Myasthenia Gravis
(A) In normal neuromuscular junctions, the chemical neurotransmitter acetylcholine (ACh) is released from a nerve terminal and binds to acetylcholine receptors (AChR) on the motor endplate of a muscle. This binding opens channels for the influx of Na^+ ions, which induces contraction of the muscle. (B) In an MG patient, autoantibodies attack AChR molecules, blocking the receipt of ACh signals. Affected Na^+ channels remain closed and the muscle fails to contract in response to nerve impulses.

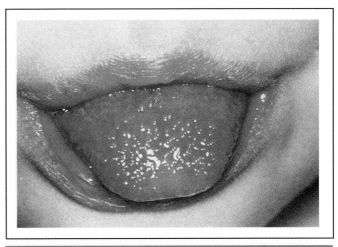

Plate 29-11

Strawberry Tongue of Kawasaki Disease Patient
Reproduced with permission from Emond R. T. D., Welsby P. D. and Rowland H. A. K. (2003) "Colour Atlas of Infectious Diseases," 4th edn. Mosby, Edinburgh.

body may be covered in a measles-like rash, and the reddened skin on the swollen hands, feet, and genitals may slough off in sheets. The patient may also complain of swelling and pain in the joints. Less obvious externally but more dangerous are the cardiac complications of KD that occur in about 30% of patients. Indeed, KD is now considered a leading cause of acquired heart disease in children in developed countries. Vasculitis of the coronary arteries can cause *aneurysms* (areas of local blood vessel weakening and dilation) that sometimes lead to heart attack. These events are fatal in about 2% of KD patients. Even when a child appears to have recovered completely from KD, an undetected aneurysm can sometimes cause sudden death in adulthood.

Although the high fever associated with KD suggests the involvement of a pathogen as a triggering agent, none has been consistently linked to KD so far and it is unknown what initiates the disease process. Whatever the causative event, it apparently triggers an autoimmune response in which high levels of cytokines are produced. Researchers suspect that these cytokines damage mucosal cells and the vascular endothelium, exposing previously sequestered antigens that may activate ignorant autoreactive B cells. Autoantibodies are then produced and attack the mucosal surfaces and arteries. With respect to laboratory findings, acute phase proteins, serum IgE, and platelets are often elevated in KD patients.

XIII. POLYMYOSITIS (PM)

Polymyositis is a rare organ-specific AID (incidence 1 case per 100,000 population per year) in which the fibers of the voluntary muscles come under inflammatory attack. Women are affected twice as often as men, and Blacks considerably more often than whites. Onset of PM occurs gradually in the second decade of life and initially affects the proximal

often than girls. Symptoms include high fever persisting for up to 2 weeks that does not abate when treated with antibiotics. Irritability is often increased out of proportion to the degree of fever. The lymph nodes swell (particularly in the neck) and the eyes and mouth appear red and inflamed. The lips are intensely red in color and chapped. The underside of the tongue also appears bright red and may display red bumps, a sign sometimes called "strawberry tongue" (Plate 29-11). The

muscles (closest to the trunk of the body). The patient complains mainly of muscle weakness. Weight loss, fatigue, and a low-grade fever may be observed. Over weeks or months, the disease spreads slowly to distal muscles so that eventually patients have difficulty kneeling, standing up, lifting objects, reaching overhead, and managing stairs. The joints may also be painful. In the worst cases, the muscles needed for swallowing are affected and the patient may suffer from malnutrition and complications from choking on food. Pneumonia, lung disease, and myocarditis are serious complications. Respiratory failure can occur due to progressive weakness of the chest wall muscles. A condition closely related to PM is *dermatomyositis* (DM), in which patients show PM symptoms accompanied by a characteristic purple-red rash over the face, chest, and hands.

At the biochemical level, the muscle enzyme creatine (or creatinine) kinase is elevated in the serum of PM patients and myoglobin may be present in the urine. PM is sometimes diagnosed in conjunction with SLE, SD, or RA, and the RhF and ANAs characteristic of these disorders are present in about half of PM patients. While 50% of all PM patients recover fully, residual weakness remains in 30% of treated PM patients, and as many as 20% have disease that resists all attempts at therapy. The overall 5-year mortality rate for PM and DM is 20%. Some surviving DM patients go on to develop cancers such as lymphomas and breast, lung, ovarian, or colon carcinomas.

Autoantibodies directed against aminoacyl tRNA synthetases are frequently found in a PM patient's serum. Aminoacyl tRNA synthetases are enzymes critical for protein synthesis because they attach the correct amino acid to the specific tRNA responsible for conveying that amino acid to a growing protein chain on a ribosome. A different aminoacyl tRNA synthetase exists for each amino acid. Autoantibodies directed against asparaginyl, lysyl, glycyl, histidyl, alanyl, threonyl, and isoleucyl tRNA synthetases have been found in various PM patients. Anti-Jo-1 antibodies (directed against histidyl tRNA synthetase) are the most common, being present in 20–30% of PM patients. Anti-PL-7 and anti-PL-12 antibodies (directed against threonyl and alanyl tRNA synthetases, respectively) are each present in 3% of PM patients, while anti-EJ and anti-OJ antibodies (directed against glycyl and isoleucyl tRNA synthetases, respectively) are each present in 2% of PM patients. A major puzzle in PM disease is why, since tRNA synthetases are ubiquitously expressed, the inflammatory damage is seen only in muscle tissue.

XIV. GUILLAIN-BARRÉ SYNDROME (GBS)

GBS is a rare AID (incidence 1–2 cases per 100,000 population per year) resulting from acute autoantibody attacks on the peripheral nerves. Onset of GBS occurs equally in men and women of age 30–50 years and is sometimes associated with another illness such as SLE, AIDS, or Hodgkin's lymphoma. GBS results when IgM and IgG autoantibodies directed against gangliosides, heparan sulfate, and other glycolipids attack neurons in the peripheral nerves, inducing acute inflammation that demyelinates the nerve fibers and reduces electrical impulse

transmission. The patient may first notice tingling in the feet or hands that rapidly (within hours) spreads up or down the body. Some patients experience symptoms on only one side of their bodies. Blurred vision, clumsiness, fainting, and swallowing difficulties may occur. Within days or weeks, muscle weakness or cramping can progress to extensive paralysis. In some cases, the axons are destroyed, leading to cessation of nerve impulses entirely. In the worst cases, the respiratory muscles are paralyzed and the patient has to be put on a respirator. Histological examination shows that the spinal roots and peripheral nerves of GBS patients are infiltrated with lymphocytes and macrophages. Serum TNF is usually elevated, with higher values present in patients showing the greatest demyelination. Despite these serious clinical findings, a vast majority of GBS patients recover on their own without treatment. The typical disease course features worsening symptoms for 3 weeks after onset, then a widely variable period of no further deterioration, followed by recovery over days or months.

In 66% of GBS cases, onset occurs following the recovery of the patient from a GI or respiratory infection with *Campylobacter jejuni*, *Mycoplasma pneumoniae*, HSV, EBV, or CMV. Some GBS patients have apparently experienced the onset of their disease following surgery, or after vaccination for polio, influenza, swine flu, rabies, or tetanus. It is unknown precisely how these events are linked to the generation of anti-ganglioside and other autoantibodies. Moreover, only 0.1% of individuals contracting *C. jejuni* infection are later diagnosed with GBS, implying that there is a large environmental and/or genetic component involved (see Section D).

XV. PSORIASIS (PS)

Psoriasis (from a Greek word meaning "the itch") is an AID characterized by reddened skin lesions that develop a covering of silver scaly skin cells (Plate 29-12). The lesions (which are

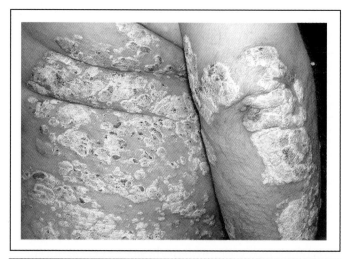

Plate 29-12

Psoriasis
Skin lesions of severely affected patient. Courtesy of Phototherapy Education and Research Centre, Women's College Hospital, Toronto.

called *plaques*) either occur in localized patches or cover large sections of the body. The patient may suffer from persistent itching and flaking of the abnormal skin. While not a great threat to physical health, PS can cause the sufferer much discomfort and sometimes emotional or self-esteem problems. The incidence of PS is 1000–3000 cases per 100,000 population per year and has two peaks: 16–22 years of age and 57–60 years. Men and women are affected equally. The skin symptoms can occur in a relapsing/remitting pattern but do not usually increase in intensity with each flare-up or spread more widely with age. Flare-ups may have no apparent cause. Patients with severe PS may be of any age, and 10% of these individuals go on to develop *psoriatic arthritis* (PsA).

The scaling of psoriasis is due to the presence in the affected region of abnormal keratinocytes with an accelerated growth rate. The cause of these skin cell defects is unknown. Some researchers suspect that a T cell-mediated autoreactive response is to blame, in that these cells might release growth-promoting cytokines. Abnormal regulation of TNF and IL-1 has been cited as being associated with the development of PS, particularly in patients with PsA. Autoantibodies directed against keratinocyte proteins or IL-1 have been found in a few PS patients but the significance of these antibodies remains unclear.

XVI. ANTI-PHOSPHOLIPID SYNDROME (APS)

APS (also abbreviated as APLS and known as Hughes syndrome) is an AID in which autoantibodies attack components of the *prothrombin activator complex* (PAC) that is required for the blood coagulation cascade. When components of the PAC are bound to autoantibody, unwanted coagulation occurs. Thromboses can block blood vessels in various regions of the body, giving rise to variable clinical symptoms including headache, migraine, deep vein thrombosis, stroke, heart attack, and pulmonary embolus (blood clot in the lung). Memory loss, seizures, and speech difficulties may ensue, and a net-like red rash may appear on the wrists and knees. Recurring miscarriage is a common occurrence because blood clots form in the placenta and block the blood supply to the fetus. APS occurs with incidence of about 5 cases per 100,000 population per year and affects more women than men. Patients are usually between 20 and 60 years of age, and some may also have SLE. Interestingly, some patients with APS have been misdiagnosed as having MS because several APS symptoms overlap those of MS. Indeed, in one study in the UK, one-third of patients diagnosed with MS actually had APS, a finding with serious implications because therapy options for these two disorders are very different in scope and cost. Heart attack or stroke in a person younger than 50 years of age may be an indicator of APS.

APS was so named because it was originally thought that the autoantibodies present in these patients bound to membrane phospholipids, such as the mitochondrial and cell surface phospholipid cardiolipin. (Hence, there are many references to "anti-cardiolipin antibodies" in the older literature.) Further analysis revealed that the autoantibodies in fact were directed against PAC glycoproteins (such as β₂-glycoprotein I or prothrombin) that can associate with anionic phospholipids once they are exposed on the cell surface. Such exposure might occur following damage to the endothelial layer. It is not known exactly how the resulting surface-bound immune complexes promote coagulation. One hypothesis is that they interfere with phospholipid-dependent anti-coagulation pathways. Another theory is that the immune complexes induce the activation of platelets such that thrombin production is enhanced. In any case, there is some evidence that the autoantibodies directed against β₂-glycoprotein I or prothrombin are natural antibodies normally present at low levels in the blood of healthy adults. It is speculated that AID may arise when the production of these natural antibodies is abnormally stimulated, perhaps due to an alteration in the costimulatory context in which β₂-glycoprotein I and prothrombin are encountered.

XVII. INFLAMMATORY BOWEL DISEASE (IBD): CROHN'S DISEASE (CD) AND ULCERATIVE COLITIS (UC)

Inflammatory bowel disease is actually a family of autoimmune disorders affecting the large and small intestines. IBD affects both sexes equally and preferentially occurs in persons 15–35 years of age, although children may also be affected. The two major types of IBD are *Crohn's disease* and *ulcerative colitis*. CD affects all layers of the wall of the colon and the small intestine, while UC affects only the innermost layer of the wall of the colon. In CD, there may be healthy patches of bowel interspersed with diseased patches, while in UC, the entire colonic lining is often affected. The incidence rates of CD and UC are 2 and 7 cases per 100,000 population per year, respectively. Both CD and UC are characterized by chronic or relapsing/remitting inflammation of the intestinal lining that causes loss of appetite, weight loss, diarrhea, pain, and fever. The scarring and swelling associated with IBD can cause obstructions that lead to abdominal cramps and vomiting. Ulcers arising from IBD can penetrate the gut wall and cause fistulas (abnormal connections) between the gut and the bladder, vagina, or rectum. Some of the fistulas may cause bleeding or form abscesses. Anemia, malnutrition, and growth retardation are other debilitating side effects of IBD. IBD usually has a relapsing/remitting pattern and can be severe and debilitating but is rarely fatal.

One hypothesis to account for IBD is that a mutation in a gene governing mucosal inflammation drives a damaging and inappropriate inflammatory response to unknown antigens present in the normal gut flora. The action of the pro-inflammatory cytokines on the sensitive gut mucosae then causes the symptoms in the patient. Interestingly, CD4⁺ T cells isolated from the intestinal lamina propria of CD patients tend to secrete high levels of Th1 cytokines such as IL-12, IFNγ, and TNF, while those from UC patients produce elevated levels of Th2 cytokines such as IL-5 (but not necessarily IL-4). Although autoantibodies are sometimes detected in IBD patients, B cells do not appear to play a large part in this family of diseases.

XVIII. PEMPHIGUS (PG)

Pemphigus is an AID characterized by blistering of the mucosae and skin that results from an autoantibody attack on desmoglein proteins. This disease was discussed in Chapter 28 as an example of type II (antibody-mediated) hypersensitivity. In the skin and mucosae, respectively, the desmogleins "glue" keratinocytes or mucosal epithelial cells together to form the intact upper epidermal or mucosal layers. The action of autoantibodies on the desmogleins not only causes *acantholysis* (separation of the epidermal cells), but also allows the release of a protease that causes skin blisters. These blisters are exceedingly painful and just touching the skin can be enough to cause it to peel off. If the barrier of intact skin is lost, innate immunity is breached and the patient becomes vulnerable to infections. Indeed, prior to the advent of modern antibiotics, pemphigus was fatal 99% of the time. Currently, mortality has been reduced to about 10%. Even so, treated patients do not ever fully recover and the disease becomes chronic in nature.

The overall incidence of PG is about 0.5–3.2 cases per 100,000 population per year. The disease strikes men and women of any age equally, but is rarely seen in children. There are three clinically distinct types of PG: *pemphigus vulgaris* (PV), the most common form characterized by severe blistering of the skin and mouth (Plate 29-13); *pemphigus foliaceus* (PF), in which itchy (rather than painful) blisters occur on the face, scalp, back, and chest but not in the mouth; and *paraneoplastic pemphigus*, a very rare and painful form of the disease that affects the lips, mouth, and throat of individuals already suffering from cancer.

XIX. GOODPASTURE'S SYNDROME (GS)

GS was also discussed in Chapter 28 as an example of type II hypersensitivity. GS is caused by autoantibody attack on

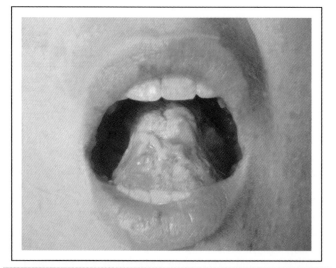

Plate 29-13

Mouth Blisters of Pemphigus Vulgaris
Courtesy of Dr. Vijay Chaddah, Grey Bruce Health Services, Owen Sound, Ontario.

collagen proteins that are components of the basement membranes of both the glomeruli in the kidney and the alveoli in the lungs. As a result, immune complexes and complement are deposited in the basement membranes, and patients present with progressive kidney dysfunction, bleeding in the lungs, cough, and blood in the sputum and urine. The incidence of GS is 1 case per 100,000 population per year, and the disease affects primarily young men (median age 20 years; range 16–61 years). The disease has a variable duration, lasting anywhere from a few weeks to about 2 years. While permanent lung damage is rare, kidney damage can be severe and long-lasting. Rapid kidney failure is seen in some cases, accompanied by iron deficiency anemia. The worst cases may be fatal due to pulmonary hemorrhage and respiratory failure.

XX. IMMUNODYSREGULATION, POLYENDOCRINOPATHY, ENTEROPATHY X-LINKED (IPEX) SYNDROME

IPEX is an extremely rare X-linked disease characterized by autoimmune inflammation. IPEX is first observed in boys when they are very young and is often fatal. Patients present with eczema, failure to thrive, chronic wasting, severe diarrhea, early onset diabetes, AHA, thrombocytopenia, lymphoadenopathy, enteropathy, and hypothyroidism. Arthritis and increased susceptibility to infection are observed in a minority of cases. At the biochemical and cellular levels, increased serum IgE is frequently present as is eosinophilia. Numbers of CD4$^+$ T cells may be slightly increased. Many (but not all) patients have autoantibodies directed against self antigens in a variety of tissues, including the pancreatic islets, thyroid and adrenal glands, smooth muscle, intestine, and kidney. Deposits of immune complexes have been found in the kidneys of IPEX patients, as well as the infiltration of immune system cells into the skin, pancreas, and intestinal mucosae.

Unlike most AID, IPEX results from the disruption of a single gene. More than 90% of IPEX patients examined to date exhibit a mutation of some sort in the X-linked FOXP3 gene. FOXP3 is a member of the FOX family (<u>F</u>orkhead <u>Box</u>) of transcription factor genes. The human FOXP3 gene encodes the "scurfin" protein, named for the natural *scurfy* mouse mutant in which the original mutation was identified (see Section C). The normal function of scurfin is not yet known, but it appears to be a transcriptional repressor important for the function of the regulatory T cells that contribute to the maintenance of peripheral tolerance.

C. Animal Models of Autoimmune Disease

Because the complex series of biological and physiological changes involved in the development of any AID cannot be recreated *in vitro*, the establishment of animal models is essential for the *in vivo* dissection of cellular and molecular events leading to the initiation and maintenance of these ailments. Historically, investigators took advantage of several mouse strains that spontaneously developed AID. Much of our current knowledge of AID was then gained from studies of susceptible small animal strains in which AID could be induced by various treatments. Most recently, researchers have generated transgenic and knockout mouse models of several AID. However, although these latter mutants are very useful, it should be borne in mind that they probably reflect only an approximation of the genetic and cellular events that occur in humans who develop AID. As a result, numerous therapy strategies that have succeeded in preventing the development of AID in animal models have often failed when translated to human AID. These disappointing results emphasize the complexity of human AID etiology.

Scientists have developed two ways to study the genetic basis of AID in animal models: mutate a gene and see if it causes AID (the "forward genetics approach"), and examine the genes in an animal with AID and determine which genes have sustained mutations (the "reverse genetics approach"). The latter method has been used in an attempt to understand the mechanisms underlying the AID that spontaneously develops in various natural mouse mutants. Scientists have painstakingly sifted through the genomes of mouse strains prone to developing diabetes or SLE, in their search for distinctive genotypes associated with AID development. However, in these and most other cases, AID development has proved to be polygenic. Using the forward genetics approach, over 40 genes have been identified in mice whose manipulation or deletion appears to induce autoimmunity. These genes include those encoding complement components, costimulatory molecules, cytokine receptors, Fc receptors, intracellular signaling proteins, and pro- and anti-apoptotic molecules. It is also thought that alterations in genes that determine the inherent resistance of a tissue to lymphocyte attack, or control the access of lymphocytes to a tissue, may contribute to AID.

Curiously, most mouse AID models in which a spontaneous or engineered mutation exists result in a limited range of AID phenotypes rather than a spectrum of AID. For example, there are many knockout and spontaneous mouse mutants that show an SLE-like or IBD-like phenotype, but very few that show diabetes or an RA-like disease. Generally speaking, the genes whose altered expression can lead to SLE-like symptoms fall into three groups: genes affecting the general clearance of apoptotic cells (e.g., C1q, IgM), those affecting lymphocyte apoptosis (e.g., Fas, Bcl-xL), and those involved in lymphocyte intracellular signaling (e.g., Src kinases, CD45). Similarly, alterations to the expression of certain cytokines (e.g., IL-7, IL-10), cytokine receptors (e.g., IL-2R), or downstream signaling intermediaries (e.g., STAT-4) tend to lead to IBD-like symptoms. Why the mutation of so many different genes should result predominantly in an SLE-like or IBD-like phenotype remains unknown. Some researchers have speculated that a skewing of the B and/or T cell repertoires toward the recognition of SLE- or IBD-associated antigens has taken place in

these mice. Genetic background may come into play here, since most of these studies have been done in inbred mice of only one strain. Studies in mice of other inbred backgrounds are needed to clarify whether an SLE-like or IBD-like phenotype always results when these genes are affected. It is also unclear how or whether mutations of these various genes in mice relate to the (mostly unknown) genetic alterations underlying human AID.

We will now describe several spontaneous, induced, and genetically engineered mouse models that have been used to study autoimmunity (Tables 29-2 and 29-3). Other knockout and transgenic mice that have AID symptoms are included in Section D. Two AID models in other animal species are briefly described in Box 29-2.

Table 29-2 Selected Animal Models of AID

Animal Model	Description	Human AID Mimicked
	NATURAL MUTATIONS	
NZB/W F1 mouse	Sle1, 2, 3, 6 genes plus predisposing H-2 alleles	SLE
MRL/lpr mouse	Fas	SLE, ALPS, SS
gld mouse	FasL	SLE, ALPS, SS
NOD mouse	*iddm* genes (over 20)	IDDM, SS
Scurfy mouse	Foxp3	IPEX
motheaten mouse	Shp1	SLE
BSXB mouse	Unknown gene(s)	SLE
BB-DP rat	Ian5	IDDM
OS chicken	Several loci	HT
	INDUCED	
EAE mouse	MOG peptide + CFA	MS
CIA mouse	Collagen II + CFA	RA
EAU mouse	IRBP (whole protein or peptides) + CFA	Autoimmune uveitis
EAT mouse	TG or TPO (whole protein or peptides) + CFA	HT
EAMG mouse	AchR (whole protein or peptides) + CFA	MG
EAM mouse	Cardiac myosin (whole protein or peptides) + CFA; infection with Coxsackie B3 virus	Autoimmune myocarditis
Thymectomized animals	Thymectomy 2–5 days after birth	Autoimmune gastritis, oophoritis

AChR, acetylcholine receptor; IRBP, interphotoreceptor retinoid-binding protein; MOG, myelin oligodendrocyte glycoprotein; TG, thyroglobulin; TPO, thyroid peroxidase.

Table 29-3 Selected Engineered Animal Models of AID

Mouse Model	Human AID Mimicked
KNOCKOUT	
AIRE$^{-/-}$	APECED
Cbl-b$^{-/-}$	Multi-organ inflammation
IL-10$^{-/-}$	IBD
TGFβ$^{-/-}$	Multi-organ inflammation
IL-2Rβ$^{-/-}$	IBD
IκBα$^{-/-}$	IBD
C1q$^{-/-}$	SLE
Lyn$^{-/-}$	SLE
CD45$^{-/-}$	SLE
IL-2$^{-/-}$	ALPS (see Chapter 24)
CTLA-4$^{-/-}$	ALPS
ICOS$^{-/-}$ (+ MOG peptide)	MS (EAE)
TRANSGENIC	
Bcl-2 Tg	SLE
IL-7 Tg	IBD
STAT4 Tg	IBD
Bcl-xL Tg	SLE
BAFF Tg	SLE
TNF Tg	RA (CIA)
HLA-DR3 Tg	Autoimmune uveitis (EAU)

MOG, myelin oligodendrocyte glycoprotein.

I. NZB/W F1 MICE

As introduced previously in the 'Historical Notes' of this chapter, the New Zealand Black (NZB) strain of mice carries a constellation of susceptibility alleles (mostly unknown) that spontaneously leads to polyclonal B cell activation, the production of anti-erythrocyte autoantibodies, and lethal hemolytic anemia. Quite independently, a strain of New Zealand White (NZW) mice was developed by Walter H. Hall in New Zealand in 1952; this strain carries a different subset of susceptibility alleles. NZW mice appear healthy for most of their lives and have phenotypically normal B cells but develop ANA and nephritis (inflammation of the kidney) late in life. The interesting part comes when NZB mice are crossed to NZW mice, as demonstrated in the 1960s by F. M. Burnet and his colleagues at the Walter and Eliza Hall Institute in Melbourne, Australia. Within their first year, over 90% of the F1 hybrid progeny mice [designated NZB/W F1 or (NZB × NZW)F1 mice] develop AHA and *lupus nephritis*, a fatal SLE-like disease accompanied by kidney damage. Moreover, just as in human SLE patients, it is predominantly female mice that develop the AID. NZB/W F1 mice thus constitute one of the oldest and best-studied animal models for systemic AID in general and SLE in particular.

NZB/W F1 mice have hyperactive B cells that produce high titers of autoantibodies directed against nuclear components. However, autoreactive T cells are believed to play a major role in the disease because NZB/W F1 mice depleted of αβ T cells develop only attenuated AID. Autoantigens identified to date as being recognized by the autoreactive T cells in NZB/W F1 mice include peptides derived from nuclear and cytoplasmic antigens. NZB/W F1 mice show a complex cytokine profile in the blood that does not reflect a simple Th1 or Th2 bias. Of note, TNF is elevated in the blood of these animals.

Multiple genes are known to be involved in the predisposition of NZB/W F1 mice to lupus nephritis. A combination of five susceptibility loci has been identified as being particularly important for the loss of self-tolerance: *Sle1* on mouse chromosome 1, *Sle2* on chromosome 4, *Sle3* on chromosome 7, *Sle6* on chromosome 5, and certain alleles of the H-2 locus (MHC) on chromosome 17. The *Sle1* sequence is a single nucleotide polymorphism of the shared gene encoding CR1 and CR2. This alteration introduces a novel glycosylation site into these complement receptors that compromises their function. Defects in the clearance of immune complexes and apoptotic cells may result that lead to the abnormal release of nuclear antigens. These antigens could then spark the production of ANAs. However, a mouse bearing the *Sle1* allele experiences only minimal tissue damage in the absence of the other susceptibility alleles. Similarly, the presence of only *Sle2* (which affects B cell activation) or only *Sle3* (which affects T cell activation) results in immunological abnormalities but not autoimmune lupus nephritis. The presence of the *Sle1*, *Sle2*, and *Sle3* alleles, plus mutations in Fas, Lyn, or SHP-1, result in generalized immune system dysfunction and expansion of the ANA response. However, full-blown lupus nephritis, arthritis, and vasculitis do not occur unless the *Sle6* allele (of unknown function) is also present. Polymorphisms of the FcγRIII gene have also been implicated in lupus nephritis induction in NZB/W F1 mice.

Studies of crosses among different substrains of NZB and NZW mice have shown that, in addition to genes promoting AID, there are loci in the mouse genome that appear to suppress AID. In particular, several modifying genes seem to mitigate the severity of the lupus nephritis in NZB/W F1 mice and prevent it from causing death. These genes have been designated *Sles 1–4*, for *SLE suppressor*. For example, expression of *Sles1* is sufficient to block the autoreactivity to chromatin that is associated with *Sle1* expression.

II. *MRL/lpr* AND *gld* MICE

The *MRL/lpr* and *gld* mouse strains represent two other well-studied animal models for systemic AID. (*MRL/lpr* mice are known as *lpr* mice outside the AID field.) Both of these strains of mice spontaneously develop an SLE-like disease characterized by severe lymphadenopathy and glomerulonephritis accompanied by hypergammaglobulinemia and autoantibody production. The genetic defects of these animals are related, since *MRL/lpr* mice have a mutation of the Fas gene, and *gld* mice have a mutation of the FasL gene. These genetic deficiencies are thought to prevent the apoptosis of autoreactive T and B cells implemented through the Fas/FasL

Some of the earliest animal models of AID to be studied were spontaneous mutants in species other than mice. Among these are the BB-DP rat and the OS chicken.

• BB-DP Rat

The acronym "BB-DP" stands for "Bio-Breeding Diabetes-Prone" rat. This strain, which was discovered in the 1970s in a Canadian commercial rat breeding facility, spontaneously develops a T cell-dependent AID strikingly similar to T1DM in humans. Both male and female BB-DP rats develop diabetes at age 8–12 weeks. Prior to this time, insulitis occurs that gradually affects each islet in the pancreas. Histologically, the insulitis in BB-DP rats does not involve as many lymphoid cells as does the insulitis in the diabetes-prone NOD mouse (see main text). Unlike human diabetics, BB-DP rats are lymphopenic from birth. It is not clear how this lymphopenia relates to diabetes development, but adoptive transfer of histocompatible, normal T cells can prevent the development of diabetes in BB-DP rats. A related strain of rats called BB-DR (Bio-Breeding Diabetes-Resistant) does not spontaneously develop diabetes when maintained in a pathogen-free environment (unlike BB-DP rats) and is not lymphopenic. However, these animals can develop diabetes in a conventional animal colony if they are forcibly depleted of their lymphocytes (including regulatory T cells; see later) by the use of mAbs or sublethal irradiation. Diabetes also develops in BB-DR rats if they are infected with *Kilham's rat virus* (KRV) or injected with polyinosinic–polycytidylic acid (poly-IC; mimics viral double-stranded RNA and

induces IFNα expression). This induction can occur in either conventional or pathogen-free environments.

Two loci conferring susceptibility to diabetes development have been identified in BB-DP rats: *Iddm1* on chromosome 4 and *Iddm2*, which maps to the rat MHC. The locus affected by the Iddm1 mutation was originally called Lyp (for "lymphopenia"). Mutations of the Lyp gene decrease the life span of naive T cells by 5- to 10-fold, resulting in a severe reduction in peripheral T cell numbers. Further investigation of the Iddm1/Lyp locus has shown that the lymphopenia in BB-DP rats is due to a frameshift deletion in a gene called IAN5 (*immune-associated nucleotide gene*-5). This mutation, which results in a severe truncation of the protein product, does not appear in BB-DR rats. IAN5 is a GTP-binding protein that may be involved in T cell development and maturation. In particular, a rat regulatory T cell population known as RT6+ T cells may be affected. BB-DP rats have very low numbers of these cells, and diabetes development can be prevented by the transfer of RT6+ T cells from BB-DR mice. If BB-DR mice are depleted of RT6+ T cells, they soon develop diabetes.

• OS Chicken

The OS (obese strain) chicken was recognized in the 1960s as spontaneously developing autoimmune thyroiditis closely resembling HT in humans. The disease is primarily manifested in females, and the affected chickens show physical signs typical of hypothyroidism, including altered body size, altered feather texture, fat deposits in the abdomen and under the skin, high levels of

lipid in the serum, and sensitivity to temperature changes. The ability of OS chickens to lay eggs is compromised. At the cellular level, from day 7 after hatching, OS chicks show an infiltration of lymphoid cells, including T cells and plasma cells, into the thyroid gland. The infiltrating T cells exhibit the marker expression pattern of activated CD4+ T cells. The thyroid becomes filled with structures resembing germinal centers and high levels of anti-thyroglobulin antibodies are produced. The normal architecture of the thyroid is almost obliterated by 1–2 months of age. As well as effects on the thyroid, OS chickens show general hyper-reactivity to immunogens. Th cells exit prematurely from the thymus to the periphery during development and regulatory T cell populations are depressed in number. Before hatching, OS chick embryos show elevated levels of IFNγ, IL-1, -2, -6, -8, -15, and -18 in their thyroids compared to control strain CB chickens. IL-15 is also upregulated in the spleens of both OS chick embryos and 5-day-old OS chicks. It has been speculated that IL-15 overproduction may be making a major contribution to the AID in these birds by driving lymphocyte infiltration into the thyroid.

Genetic studies in the OS chicken are limited, but several loci associated with the development of AID in this bird have been tentatively identified. Environment can have a significant influence on OS AID development, since the severity of the disease is increased if the iodine level in the chickens' food is increased. Conversely, studies have shown that anti-thyroid drugs that reduce thyroidal iodine levels or iodine metabolism can suppress or delay the onset of OS AID.

---

pathway. In addition, these animals accumulate large numbers of CD3+CD4−CD8−B220+ T cells. Members of this unusual T cell subset, which is thought to be derived from CD8+ T cells, are anergic. Their role in the pathology of these mutant mice is unknown, as it is the CD4+ T cell population that is believed linked to the observed autoantibody production and lymphadenopathy.

The reader should note that, although *MRL/lpr* and *gld* mice are convenient models of AID, they do not represent ideal models for human SLE disease. Fas and FasL expression are normal in human SLE patients, lymphocyte apoptosis is normal, and no overt lymphoproliferation is seen. A better human parallel for *MRL/lpr* and *gld* mice is subtype I of the *autoimmune lymphoproliferative syndromes* (ALPS) described in Chapter 24. These patients carry autosomal dominant mutations of Fas or FasL and show excessive lymphocyte

proliferation and accumulations of lymphocytes in the lymphoid organs. Autoantibodies directed against erythrocytes and platelets can lead to anemia or thrombocytopenia, respectively. ALPS patients also have an increased risk of developing various tumors, as discussed later in this chapter (see Section F). Finally, *MRL/lpr* and *gld* mice have been used to explore certain aspects of human SS, since lymphocytes from patients with this AID are resistant to apoptosis.

III. NOD MICE

The NOD (*non-obese diabetic*) inbred mouse strain is an important model for human T1DM. This strain arose spontaneously in 1980 in a Japanese mouse colony that was being screened for animals showing high blood glucose after fasting.

NOD mice have proved invaluable as a tool for exploring organ-specific AID. NOD mice express a single MHC class II molecule, I-A^{g7}, which is homologous in structure and function to the human MHC class II molecule HLA-DQ8 highly associated with T1DM risk. The physical symptoms of diabetic disease in NOD mice parallel those in humans, except that females are affected more often than males, and the animals are deaf. In general, by the age of 6 weeks, most NOD mice spontaneously show a progressive accumulation of DCs, macrophages, and NK, T, and B cells around the pancreatic islets, a condition called *peri-insulitis*. No destruction of the islet tissue is observed at this point. However, by about age 10–12 weeks, pancreatic insulitis is observed in which the infiltrating immune system cells start to destroy the β-islet cells. By the age of 30 weeks, 80–90% of females and up to 50% of males develop T1DM. It should be noted that the incidence of diabetes in different colonies of NOD mice can vary considerably, and that some NOD mice also develop autoimmune thyroiditis. Interestingly, unlike most humans, NOD mice can survive without insulin administration for several weeks.

Both autoreactive T and B cells contribute to T1DM development in NOD mice. With respect to T cells, both CD4$^+$ and CD8$^+$ clones recognizing peptides from various islet proteins (including insulin) have been found in these animals. With respect to B cells, insulitis and diabetes do not develop in either newborn NOD mice treated with anti-IgM antibodies or in NOD mice that lack B cells because they carry a knockout mutation of the Ig μ gene. Autoantibodies recognizing various ICA, insulin, and GAD homologues as well as other autoantibodies can be detected in the affected tissue, although not to the same levels as in human T1DM patients. However, these antibodies may be a consequence rather than a cause of T1DM in NOD mice. The injection of GAD45 or insulin does not lead to insulitis or diabetes in NOD mice, as might be expected if a causal relationship existed. Indeed, young NOD mice that transgenically express TNF (TNF-NOD mice) show accelerated development of diabetes, and this acceleration is observed whether or not B cells are present.

Close to 20 T1DM susceptibility loci (denoted *Idd* loci) have been defined in NOD mice, but few genes in these loci have been conclusively identified. As noted previously, the greatest risk of diabetes development maps to the I-A^{g7} gene. I-A^{g7} has a critical amino acid substitution that directly corresponds to a substitution in the human HLA-DQ8 molecule. A candidate for an underlying driver of the disease is the IL-2 gene. The onset of NOD insulitis can be adoptively transferred to irradiated normal recipients if both CD4$^+$ and CD8$^+$ T cells from a diabetic NOD mouse are transplanted, and the most potent initiators appear to be Th subsets that secrete high levels of IL-2. Researchers have attempted to identify other components that drive T1DM development by treating NOD mice with various proteins that might either block an immune response, or serve as antagonists of molecules known to promote immune responses. Knockout and transgenic mouse models have also been developed to explore these issues. However, the results have been perplexing since the altered expression of over 150 molecules has been documented to affect the development of T1DM in NOD mice in one way or another. Many of these genes are of uncertain relevance to the human situation, but one example of a molecule that appears to play a role in T1DM in both NOD mice and humans is IL-12 p40. NOD mice treated with antagonists of IL-12 do not develop diabetes, and polymorphisms of the IL-12B gene encoding the IL-12 p40 subunit have been found in some diabetic humans (see Section D).

Certain aspects of SS have also been explored in NOD mice. As mentioned previously, human SS patients show T and B cell responses to ICA69. NOD mice with an intact ICA69 gene spontaneously lose function of their salivary glands, but NOD mice in which the ICA69 locus is experimentally disrupted show a decrease in salivary and lacrimal gland disease. Researchers have taken the NOD genomic region known to be responsible for salivary gland function loss and introduced it into wild-type C57BL/6 mice (which are not predisposed to developing AID). The resulting mutant mice constitute a useful model of SS in that the animals develop salivary gland abnormalities but no signs of insulitis or diabetes.

IV. *Scurfy* MICE

Scurfy mice bear a natural mutation of the Foxp3 gene whose human homologue FOXP3 is mutated in the vast majority of IPEX patients. As in humans, the murine Foxp3 gene is X-linked so that the *scurfy* phenotype appears almost exclusively in males. *Scurfy* mice suffer from runting, diarrhea, wasting, rough skin, anemia and thrombocytopenia, lymphoadenopathy, and a heightened susceptibility to infections. High levels of IL-2, -4, -5, -6, -10, IFNγ, and TNF are present in the circulation and skin of affected animals. Death invariably occurs by 3–4 weeks of age. As expected, a very similar phenotype occurs in gene-targeted mice lacking Foxp3.

The autoimmune inflammation in *scurfy* mice has been linked to the presence of unusual autoreactive CD4$^+$ T cells that do not require normal levels of CD28-mediated costimulation and can be activated via the TCR more easily than normal CD4$^+$ T cells. *Scurfy* T cells are also resistant to control by CsA. Evidence is accumulating that the scurfin protein encoded by the Foxp3 gene is involved in the generation and/or function of regulatory T cells. *Scurfy* mice that receive an injection of normal total T cells are "cured," implying that the abnormal T cells in these animals can still respond to regulation but that this regulation is lacking due to the mutation. In normal mice, expression of Foxp3 is highest in T$_{reg}$ cells and low or absent in non-regulatory CD4$^+$ and CD8$^+$ T cells. In one study, non-regulatory CD4$^+$ T cells that were engineered to express a Foxp3 transgene took on a T$_{reg}$ phenotype, and gained the ability to establish suppressive intercellular contacts. *In vivo*, the Foxp3-transfected T cells were able to suppress inflammation and autoimmunity in both *scurfy* and Foxp3$^{-/-}$ mice. Consistent with its putative negative regulatory function, overexpression of scurfin decreases immune responses and reduces the cellularity of the lymphoid organs in mice.

V. THYMECTOMY OF NEONATAL MICE

Certain mouse strains, including BALB/c, develop spontaneous autoimmunity when thymectomized during a narrow window extending between 2 and 5 days after birth. Organ-specific ailments such as gastritis and oophoritis (inflammation of the ovaries) tend to be the dominant manifestations. T cells infiltrate the affected organ and autoantibodies directed against that organ are present in the animal's blood. The phenotype can be strikingly similar to that of human autoimmune gastritis. The genes that predispose these strains to AID have yet to be identified, but work on the progeny of crosses of susceptible mice to resistant strains (such as C57BL/6J) has suggested that the MHC genes are not responsible. Significantly, the narrow window resulting in AID development corresponds to the time in neonatal mouse development when the first T_{reg} cells migrate from the thymus to the periphery. The theory is that, in the absence of these cells, the other unknown defects in tolerance mechanisms present in susceptible strains are sufficient to break peripheral tolerance and induce AID. Indeed, injection of purified normal T_{reg} cells into a thymectomized BALB/c mouse by day 5 can prevent the onset of AID symptoms.

VI. IBD MODELS

There are many animals models for IBD, some in which the mucosal inflammation is spontaneous, others in which it is induced by chemical agents, and still others in which it results from gene deletion or transgenesis. As is true for human IBD, both an underlying genetic defect and an environmental stimulus (such as a particular type of bowel bacterium) must be present for IBD-like disease to be observed. IBD-prone mice kept in a sterile environment do not develop IBD, and IBD-prone mice kept in a conventional environment but treated with antibiotics experience less severe disease. These observations have convinced some researchers that the development of IBD is a case of molecular mimicry in which T cells originally directed against antigens of enteric bacteria attack cross-reactive self antigens (see Section E). Interestingly, different species of bacteria induce IBD in different mouse strains, demonstrating the importance of the genetic component of this disease.

STAT4 transgenic mice, mice with defects in NF-κB activation (such as the IκBα knockout mouse), and mice transgenic for IL-7 develop clinical features of IBD. Studies of these animals have revealed that the infiltration of autoreactive CD4⁺ cells into the mucosal lamina propria is a key driver of intestinal inflammation. Where the autoreactive cells are of the Th1 subtype, TNF and IFNγ are produced and directly damage the delicate mucosal surface. Macrophages recruited and activated by these inflammatory cytokines then secrete additional TNF, IL-1, and IL-6, enhancing tissue destruction. These cells also produce IL-12, which drives additional Th1 cell differentiation and induces symptoms that resemble human CD. Significantly, treatment of these mutant animals with anti-CD4 or anti-IL-12 mAbs blocks development of the disease. Autoreactive Th2 cells secreting IL-4 can also cause

IBD, although the inflammation is not as intense as in the case of Th1 infiltration and the disease symptoms more closely resemble human UC than CD. Although CD8⁺ T cells are also present in the lamina propria, these cells do not appear to contribute to IBD, at least in animals.

Defects in the production of immunosuppressive cytokines such as IL-10 and TGFβ may also contribute to the manifestation of IBD symptoms in mouse models. The theory here is that these cytokines are produced alongside the pro-inflammatory cytokines in a form of self-limiting regulation. If the immunosuppressive cytokines are not produced, the autoreactive Th cells are not restrained and more and more pro-inflammatory cytokines are secreted. For example, knockout mice lacking TGFβ expression die by age 3–4 weeks with generalized autoimmune inflammation. Similarly, IL-10⁻/⁻ mice or mice with defects in the IL-10R spontaneously develop severe Th1-mediated inflammation of the mucosae in both the large and small intestines. It is also likely that IL-10 is important for the development of regulatory T cell subsets (particularly Th3 cells) that may play a role in subduing the autoreactive Th1 cells causing the IBD.

VII. EXPERIMENTAL AUTOIMMUNE ENCEPHALITIS (EAE)

EAE is a model of human MS that can be induced in certain susceptible strains of rodents. In mice, the most commonly used strains are C57BL/6 or SJL. The animal is injected with a mixture of complete Freund's adjuvant plus MOG 35-55, a peptide from the myelin ogligodendrocyte glycoprotein made by oligodendrocytes. The immunized mouse develops MS-like symptoms with a relapsing/remitting pattern of disease that ultimately results in death in most cases. IL-23, the cytokine related to IL-12 that shares its IL-12 p40 subunit, has been shown to be the main driver of EAE in susceptible mice. Susceptible mice treated with anti-IL-12 p40 mAb (which blocks both IL-12 and IL-23) showed much less AID development than mice treated with anti-IL-12 p35 mAb (which blocks only IL-12). Abnormal predominance of inflammatory Th1 responses is thus highly implicated in EAE. The pro-inflammatory cytokine osteopontin (OPN) was noted previously as possibly being involved in MS development. To examine the role of this protein in EAE, the *opn* gene was knocked out in a susceptible strain of mice. These OPN⁻/⁻ mice still developed EAE upon immunization with MOG 35-55 plus CFA, but symptoms were less severe and remissions more frequent. The mutant animals also did not die as frequently as the controls. OPN⁻/⁻ T cells recognizing MOG 35-55 produced more IL-10 and less IFNγ and IL-12 compared to controls, suggesting immune deviation to a less damaging Th2 response.

Mast cells may play a role in the development of EAE because W/Wᵛ mice, which are natural mouse mutants lacking the c-kit receptor required for normal mast cell development, experience much less severe EAE when treated with the MOG peptide protocol. When normal mast cells were introduced into W/Wᵛ mice, it became possible to induce severe EAE in these mutants. It is speculated that some of the mediators contained in the mast cell granules, particularly histamine, may affect the blood–brain barrier such that T cells gain easier access to the CNS. Indeed, drugs that inhibit mast cell degranulation can delay the onset of EAE in mice. With respect to humans, mast cells have been found in sites of demyelination in MS patients, and anti-histamines can ameliorate MS symptoms. Interestingly, mast cells have also been found in the salivary glands of SS patients and in the synovial fluid of the joints of RA and SLE patients, suggesting that these cells may be involved in other AID.

Several other factors may contribute to EAE development and by extension, perhaps to human MS. For unknown reasons, knockout mice deficient for the ICOS costimulatory molecule are highly susceptible to EAE development. The minor costimulatory molecule OX40 may also be involved, as this member of the TNFR superfamily is upregulated on autoreactive T cells prominent in EAE. In addition, APCs expressing high levels of OX40L (the ligand for OX40) infiltrate the CNS during acute EAE episodes. Polymorphisms of various MCP chemokines that attract monocytes and macrophages have also been linked to EAE.

VIII. COLLAGEN-INDUCED ARTHRITIS (CIA)

CIA is used by investigators as a mouse model of human RA. Inoculation of certain substrains of DBA/1J mice with type II collagen (CII) plus CFA leads to an autoimmune response that results in severe arthritis in multiple joints within 4–6 weeks of injection. Both cartilage and bone are eroded and the synovial tissues are inflamed. Damage is primarily due to the effects of TNF and complement-mediated destruction induced by the binding of autoantibodies to CII present in both bone and cartilage. The administration of GM-CSF exacerbates this damage. Interestingly, however, it seems that neither T nor B cells are absolutely required for CIA development. When CII is injected into RAG-deficient mice (which lack T and B cells), the animals still exhibit severe inflammation in the joints. Thus, CIA may be a disease primarily of innate autoimmunity. Despite these findings, an accelerated induction of CIA that does not depend on genetic background can be achieved by treating mice with a cocktail of four purified mAbs directed against arthritogenic B cell epitopes identified in the CII protein. Amino acid sequences in the CII molecule that give rise to pMHC epitopes recognized by autoreactive T cells have also been identified. Administration of peptide antagonists of these sequences can prevent the development of CIA in susceptible mice.

Like human RA, CIA is promoted by the expression of certain MHC class II alleles, namely the H-2^q and H-2^r haplotypes. There is also evidence that polymorphisms of a gene on chromosome 2 (possibly C5 of the complement cascade) and another gene on chromosome 1 (possibly FcγRIIB) can contribute to arthritis development in mice. With respect to environmental triggers, the involvement of a pathogen in CIA progression is suspected because co-administration of purified LPS with CII exacerbates the arthritic symptoms.

Mice that transgenically overexpress TNF have a phenotype very similar to that of CIA mice. Arthritis develops rapidly and spontaneously in the joints of these animals in a process that does not require T or B cells. Curiously, despite the presence of the TNF transgene in a variety of cell types in the mutant animal, TNF protein is expressed at high levels only in the cells of the synovial membranes in the joints. It remains a mystery why other tissues do not express the TNF transgene in the same way and thus do not show AID symptoms.

IX. EXPERIMENTAL AUTOIMMUNE MYASTHENIA GRAVIS (EAMG)

EAMG is a very useful small animal model of human MG and perhaps the best-studied example of an antibody-mediated AID. To induce disease, purified AChR (or its peptides) from the electric organ of the Pacific electric ray *Torpedo californica* are combined with CFA and used to immunize genetically susceptible strains of mice, rats, or rabbits. Anti-AChR CD4$^+$ T cells and IgG1 and IgG2a autoantibodies are generated that cross-react with epitopes of AChR molecules in the mammalian skeletal muscle end plate. Complement activation is induced, mediating tissue destruction that in turn interferes with muscle contractile function. Damage to the neuromuscular end plate and its junctional folds can be detected about 2 weeks after AChR immunization. Macrophages are observed in the neuromuscular junction. The animals experience rapid onset muscle weakness and fatigue, much like human MG patients, followed by a chronic phase of muscle dysfunction. The disease can be passively transferred to naive susceptible animals by administration of anti-AChR antibodies or lymphocytes isolated from EAMG animals.

Mice are much less prone to EAMG induction than rats or rabbits. Nevertheless, certain strains (such as C57BL/6), which tend to mount Th1 rather than Th2 responses to antigens, are susceptible. C57BL/6 mice develop severe muscle weakness and sometimes die in response to AChR administration. In contrast, mouse strains such as BALB/c, which preferentially mount Th2 responses, are usually resistant to EAMG. Consistent with this observation, administration of IFNγ or IL-12 exacerbates EAMG symptoms. Researchers have studied the contributions of various cytokines and signaling molecules to EAMG development using knockout mice. For example, studies of IL-4$^{-/-}$ and IFNγ$^{-/-}$ C57BL/6 mice have confirmed earlier investigations indicating that Th1 cells and IFNγ play a crucial role in anti-AChR antibody production. While IL-4$^{-/-}$ C57BL/6 mice develop EAMG at the same rate as wild-type C57BL/6 mice, IFNγ$^{-/-}$ C57BL/6 mice are resistant to EAMG and show decreased levels of circulating anti-AChR antibodies. In rats, it has been shown that a key function of IFNγ in EAMG is the induction of MCP-1 production by skeletal muscle cells. MCP-1 promotes the trafficking of inflammatory cells, especially macrophages, into the neuromuscular junction. Another cytokine critical for EAMG is IL-6, since far fewer IL-6$^{-/-}$ C57BL/6 mice develop the disease compared to the wild-type. IL-6$^{-/-}$ C57BL/6 mice show decreased levels of IgG (but not IgM) anti-AChR antibodies, suggesting a defect in T cell help needed for isotype switching. Circulating levels of complement component C3 are reduced *in vivo* in the absence of IL-6, and lymphocyte proliferation and IFNγ production in response to AChR are decreased *in vitro*. These findings are consistent with the observation that human AChR-specific lymphocytes express elevated levels of IL-6 mRNA. In a mouse study examining signaling molecules, EAMG development in BALB/c mice lacking either STAT6 (promotes Th2 responses) or STAT4 (promotes Th1 responses) was compared. The results showed that an EAMG-resistant strain would develop EAMG if Th2 development was hampered in the animal (due to STAT6 deficiency) such that a relative abundance of Th1 cells prevailed. In contrast, STAT4$^{-/-}$ BALB/c mice, in which Th1 cell development is inhibited, remained resistant to EAMG.

Studies in various animals have shown that the peptides represented by AChR amino acids 146–162, 181–200, and 360–378 are highly immunogenic and immunodominant in establishing EAMG. Significantly, nasal administration of synthetic versions of these AChR epitopes prior to immunization with whole AChR plus CFA can forestall the development of EAMG in C57BL/6 mice. The *in vitro* responsiveness of anti-AChR CD4$^+$ T cells was reduced by this tolerization regimen and the *in vivo* production of anti-AChR antibodies was inhibited. In a similar study in a rat EAMG model, a shift from a Th1 to a Th2 cytokine profile was noted upon AChR peptide administration, as well as a downregulation in the expression of lymphocyte costimulatory molecules. The potential application of a mucosal AChR peptide administration strategy to induce tolerance in human MG patients is in under investigation.

X. EXPERIMENTAL AUTOIMMUNE THYROIDITIS (EAT)

EAT is a cell-mediated AID in mice that serves as a model of HT in humans. If genetically susceptible strains of mice are immunized with CFA plus either whole thyroglobulin or TG peptides, the thyroid gland becomes infiltrated with mononuclear cells that gradually destroy the gland. Granulomatous lesions in the thyroid may be present in severe cases. EAT can also be induced by the transfer of lymphocytes isolated from TG-immunized donors and subsequently activated *in vitro* with TG and IL-12. About 3 weeks after the transfer of such cells to susceptible mice, the disease becomes chronic with prominent fibrosis in a subset of the recipients. CD4$^+$ T cells are particularly important for EAT development in this system, since *nude* mice (which lack a thymus) are resistant. Both Th1 and Th2 effectors have been implicated. Immunization with thyroid peroxidase (TPO) in CFA can also induce symptoms of EAT and the production of anti-TPO antibodies. At about 2–3 months post-immunization with TPO, mice show inflammation and CD4$^+$ T and B lymphocyte infiltration into the thyroid and destruction of thyroid follicles.

Strains of mice (such as CBA/J) expressing the H-2^k haplotype are particularly susceptible to EAT induction mediated by TG immunization. Multiple pathogenic peptides of TG have been identified, most of which are clustered at the C-terminal end of this very large protein. The majority of these peptides

have been shown to bind to H-2A^k *in vitro*. Indeed, the administration of anti-H-2A^k antibodies can prevent EAT development in these animals. Current research is focused on determining the effects of various cytokines on EAT development, and searching for an immunodominant peptide of TG or TPO that can be used for tolerance studies. Treatment of susceptible mice with TG prior to immunization with TG plus adjuvant induces the generation of T$_{reg}$ cells that can suppress EAT.

XI. EXPERIMENTAL AUTOIMMUNE UVEORETINITIS (EAU)

EAU is a model of human *autoimmune uveitis*, an organ-specific, inflammatory AID of the neural retina of the eye. In human uveitis, T cells (particularly Th1 cells) and monocytes breach the blood–retina barrier and damage the photoreceptor cell layer of the retina. Impaired vision or blindness can result. Patients with uveitis also have antibodies to retinal antigens such as *interphotoreceptor retinoid-binding protein* (IRBP) and *arrestin* (also known as *retinal soluble antigen*, or S-Ag), although it is not clear whether these antibodies are a cause or effect of the disease. There is a strong genetic component to human uveitis, and disease development is associated with expression of HLA-DR3, -DR4, -DQ6, or -DQ8.

EAU can be induced in rats, guinea pigs, monkeys, and mice either by immunization with purified retinal antigen (or its fragments) plus adjuvant, or by transfer of lymphocytes (but not serum) from donor animals previously immunized with retinal antigens. Different antigens work best in different species. For example, both S-Ag and IRBP, as well as three other retinal antigens, induce EAU in susceptible rats, while only IRBP is effective in susceptible mice. Cells with a Th1 phenotype have been most strongly implicated in EAU development, but NK cells play a role in increasing the severity of the disease. In animals of any genotype, the early response to retinal antigens tends to be characterized by the production of both Th1 and Th2 cytokines and antibody isotypes. However, as the response persists, the Th2 phenotype gradually predominates if the animal is genetically resistant to EAU, while a Th1 phenotype is observed in animals with a susceptible genotype. In susceptible mice (such as the B10.A strain), treatment with IL-10 plus IL-4 suppresses Th1 effector generation and function and decreases EAU symptoms. Conversely, mAb-mediated inhibition of endogenous IL-10 increases EAU severity. BALB/c mice, which naturally tend to mount Th2 responses to antigens, are resistant to EAU.

The EAU model has been "humanized" by transgenically introducing human HLA alleles associated with human uveitis susceptibility into mice that would normally be resistant to EAU. Transgenic mice lacking endogenous murine MHC class II molecules but expressing human HLA-DR3, -DR4, -DQ6, or -DQ8 developed EAU in response to IRBP immunization. Anti-human MHC class II antibodies were able to block EAU induction in this model, confirming that the disease was due to presentation by the human MHC molecules. As in standard EAU models, the disease could be transferred to naive recipients by administration of isolated Th1 lymphocytes (but not serum) from the transgenic mice. Significantly, transgenic mice expressing HLA-DR3 could also develop EAU in response to S-Ag, a protein that does not normally induce EAU in mice. This humanized model may thus provide a more accurate model of human uveitis relative to the standard mouse model.

In addition to their use in mechanistic studies, EAU-susceptible rodents have been valuable in investigations of potential therapies for human uveitis. For example, it is known that the neuropeptide *vasoactive intestinal peptide* (VIP) is present in lymphoid tissues and has anti-inflammatory effects. Intraperitoneal administration of VIP to EAU-susceptible mice prior to injection of IRBP decreased the frequency and severity of EAU. It is thought that VIP may have induced the generation of regulatory T cells able to inhibit EAU development.

XII. EXPERIMENTAL AUTOIMMUNE MYOCARDITIS (EAM)

EAM is a model of human *autoimmune myocarditis*, a condition that can lead to dilated cardiomyopathy. In patients with dilated cardiomyopathy, the ventricular diameter of the heart becomes enlarged and the heart muscle is weakened such that it no longer pumps blood efficiently. Fatal organ damage can result. Dilated cardiomyopathy is a major cause of heart failure in adolescents and young adults, and often necessitates a heart transplant. Acute myocarditis leading to chronic myocarditis and/or dilated cardiomyopathy arises in some patients who have been infected with Coxsackie virus. Similarly, infection of genetically susceptible strains of mice (such as A/J and BALB/c) with Coxsackie B3 virus induces acute myocarditis (EAM) strikingly similar to the human disease. In EAM mice, the autoimmune response is mediated primarily by CD4$^+$ T cells and directed against the heavy chain of cardiac myosin. Significantly, immunization of susceptible mice with cardiac myosin itself (or its immunodominant peptides) plus CFA also induces acute myocarditis.

Cytokines are important for the induction of EAM, and both Th1 and Th2 responses play a role. The heart lesions of EAM mice are filled with eosinophils, and autoantibodies of the IgG1 and IgE isotypes predominate in the serum. These features imply that the disease may be linked to a Th2 autoimmune response. Indeed, anti-IL-4 mAb treatment of susceptible mice immunized with cardiac myosin leads to reduced EAM and a shift in antibody isotype to IgG2a. The production of IL-4, IL-5, and IL-13 is decreased in these animals while that of IFNγ is increased. Similar results have been obtained using IL-4$^{-/-}$ susceptible mice. On the other hand, IL-6$^{-/-}$ susceptible mice also show disease of decreased frequency and severity, indicating that this pro-inflammatory cytokine is important for EAM development. The expression of complement component C3 is compromised in IL-6$^{-/-}$ susceptible mice compared to their wild-type susceptible littermates, and T cells isolated from the IL-6$^{-/-}$ animals are no longer able to cause EAM when adoptively transferred into IL-6$^{+/+}$ naive, susceptible hosts. The general inflammatory environment promoted by IL-6 may be critical to disease outcome, as humans with heart dis-

eases and high plasma levels of IL-6 have a poorer prognosis than patients with lower IL-6 levels. Interestingly, while both IL-12 and TNF promote heart damage in EAM, IFNγ appears to protect against, rather than exacerbate, the disease. In one study, IFNγ$^{-/-}$ susceptible mice sustained more heart damage than IFNγ$^{+/+}$ susceptible mice. The induction of iNOS was impaired in the hearts of IFNγ$^{-/-}$ animals and anti-myosin T cell responses were enhanced. It is not clear whether IFNγ itself can protect against EAM (perhaps by eliminating the Coxsackie virus promptly), or whether the presence of IFNγ in wild-type mice allows another protective molecule to show its effects. It may be that the prime role of IL-4 in EAM development is to suppress IFNγ and abrogate its protective effects.

XIII. OTHER POTENTIAL AID MODELS

Several other mouse strains spontaneously develop systemic AID, and thus may serve as additional models for human systemic AID. For example, the *motheaten* mouse strain (which has patchy fur) suffers early mortality due to severe, generalized autoimmune inflammatory lesions accompanied by hypergammaglobulinemia and autoantibody production. *Motheaten* mice carry a mutation of phosphatase gene SHP-1 (also called PTPIC). The reader will recall from Chapter 9 that SHP-1 functions as a negative regulator of BCR signaling (see Section E). The BXSB mouse strain also has a curtailed life span due to the rapid development (in males) of SLE-like disease characterized by B cell hyperactivity, T cell dysfunction, and monocyte proliferation. The genetic defect in BXSB mice is not known.

In addition to the genetically engineered transgenic and gene-targeted mutant mice described previously, several other lines of manipulated mice may furnish useful models of AID. For example, mice whose lymphocytes are transgenic for the anti-apoptotic genes Bcl–2 and Bcl-xL develop an SLE-like disease. Massive lymphoproliferation and lymphocytic infiltration into multiple organs reminiscent of ALPS occurs in mice deficient for CTLA-4. The reader will recall that CTLA-4 is a negative regulator of T cell activation that also plays a role in the generation and/or function of T$_{reg}$ cells. A knockout mutation of the Cbl-b gene encoding an E3 ubiquitin ligase also results in spontaneous autoantibody production and the infiltration of hyperactivated T cells into multiple organs in mice. Interestingly, a strain of rats prone to the spontaneous development of diabetes has been shown to have a nonsense mutation in the Cbl-b gene. As discussed in Chapter 14, Cbl-b negatively controls signaling pathways downstream of both the TCR and BCR. The evidence suggests that Cbl-b may regulate calcium flux in T cells subjected to anergizing signals.

We note here that, just because an autoantigen has been identified for an AID, it is not always possible to generate a useful animal model for the disease. For example, despite the cloning of autoantigens associated with Graves' disease, an accurate animal model of this disease still does not exist. Some potential GD models lack lymphocyte infiltration into the thyroid, while others show no hyperthyroidism. The best approximations have come from the immunization of mice with fibroblasts transfected with the TSHR. Studies of these animals have suggested that regulatory T cells may play a role in controlling hyperthyroidism in mice.

D. Determinants of Autoimmune Disease

The development of an AID is a very complex process involving both environmental and genetic factors. For many AID, the break in peripheral self-tolerance leading to an anti-self immune response is linked to an encounter with a particular pathogen, chemical, drug, toxin, or hormone. However, the single most important factor contributing to AID is the genetic make-up of the host. A complex constellation of AID susceptibility alleles and haplotypes exists that determines the ongoing deregulation of self-tolerance mechanisms.

I. GENETIC PREDISPOSITION

The most convincing evidence that vulnerability to AID is heavily dependent on the inheritance of disease susceptibility genes comes from studies of identical (monozygotic; MZ) twins. When one MZ twin develops AID, there is a reasonable chance that the other twin will develop the same ailment. The *concordance rate*, a measure of the frequency of the same disease developing in two twins, has been reported to range from 12 to 60% for MZ twins, depending on the specific AID (Table 29-4). In contrast, the concordance rate of an AID developing in two fraternal (dizygotic) twins is about 5%. It should be noted that concordance in MZ twins will never be 100% due to two factors: the random V(D)J recombination that occurs independently in the lymphocyte repertoires of each twin as he/she grows, and the differing "immune system histories" of each twin that are shaped by the different pathogens and environmental agents each encounters. The fact that MZ concordance in situations of AID is so much lower than 100% confirms that non-genetic factors, such as environmental triggers, are also important.

Inherited predisposition to AID can also be seen in families without twins. Strikingly, in these cases, the actual disease entities may be different in different members of the same family. This observation is reminiscent of atopy (refer to Ch.28), in which family members share a general predisposition to allergy but may display very different forms of it, such as hives, eczema, or asthma. It is also very common, especially with systemic autoimmunity, for patients suffering from one type of AID to develop another AID. For example, patients with "mixed connective tissue disease" have the combined symptoms of SD, PM, and SLE, and patients with T1DM are significantly more prone to developing MS than would be expected from a random association. These observations suggest that what is often inherited is a general susceptibility to AID rather than a single defect that is restricted to a particular manifestation.

While studies of NOD mice and other animal models have taught us much about the underpinnings of animal AID, the number and identity of the genes involved in human predisposition to AID are not known. People are not inbred animals, so that there is wide individual variation in the sets of specific predisposing alleles that are assembled from the perhaps hundreds of genes that can affect immune responses. Because of this influence of the total genetic background, a given human ends up with a genotype that confers either increased risk or resistance to AID. Thus, not every human who has a particular predisposing allele gets AID, and the course of a given AID varies greatly from person to person. For example, human T1DM has a particularly complex inheritance pattern: there is an increased frequency of the disease in close relatives of a diabetic individual, suggesting a genetic basis, but the disease can skip whole generations. In addition, some AID appear to concentrate in particular ethnic groups. For example, Eastern European Jews and people of Mediterranean extraction have a higher incidence of PG, and the incidence of CD is increased in Caucasians of Jewish descent. How scientists attempt to establish the genetic basis of an AID is outlined in Box 29-3.

Due to their direct involvement in T cell responses, the most important genes that predispose both humans and animals to AID are the MHC genes (Table 29-5). Perhaps the best illustration of AID–HLA association in humans can be found in AS. Over 90% of Caucasians with AS express an allele belonging to the HLA-B27 family. AS is thus one of the few AID to show a predominant linkage to an MHC class I molecule rather than to an MHC class II molecule. The only difference between the HLA-B27 alleles associated with AS and B27 alleles that are not associated with the disease is the identity of two amino acids in the bottom of the peptide-binding groove. The importance of HLA-B27 to AS development has been confirmed in transgenic rats expressing a complex of human HLA-B27 plus human $\beta 2m$. These animals spontaneously develop an inflammatory autoimmune disease with features of AS. Other AID also show strong associations with specific HLA allele families. For example, expression of HLA-DR2 and HLA-DR3 predisposes an individual to developing SLE, while T1DM has particularly strong links to HLA-DR3, -DR4, -DQ2, and -DQ8. Individuals expressing certain alleles of HLA-DR4 are especially prone to RA or JRA, while primary SS and PM are associated with HLA-DR3 in some populations. HT has been linked to HLA-DR3 and

Table 29-4 **Concordance of Selected AID in Monozygotic Twins**	
AID	**Concordance Rate (%)**
CD	40
GD	20
T1DM	20–30
SLE	24–49
MS	18
RA	15

Like many complex disorders, AID cluster in families, indicating a genetic component in the etiology of these diseases. The extent to which genes are involved in a given AID is often estimated by comparing the prevalence of the disease among siblings with the prevalence in the general population. The ratio of these prevalence values is called the *relative sibling risk* and is denoted λ_s, where $\lambda_s = 1.0$ indicates no familial clustering and $\lambda_s > 1.0$ indicates a genetic influence is present. For many AID, λ_s ranges between 10 and 20, which means that siblings of an affected individual have a 10–20 times greater risk of having the disease compared to members of the general population. When the siblings involved are monozygotic twins, the relative risk factor is denoted λ_{MZ}.

Once it is established that a given AID is influenced by genetics, the next challenge is to identify the genetic loci involved and the allelic variants of each gene whose expression predisposes an individual to developing the disease. In addition to aiding in disease detection and screening for disease susceptibility, such information may point to the cause of the disease and thus possible therapies. The process of identifying such AID susceptibility genes is not easy and progress in this field has been slow. More than one genetic locus is almost always involved, and the influence attributable to a single gene of a collection may be weak and difficult to detect. Two approaches are generally used to identify susceptibility alleles: the more traditional *association studies*, and *linkage analysis*, which has developed in step with technology facilitating large-scale DNA analyses.

To carry out association studies for a given AID, one must first choose a candidate gene (which must be polymorphic) that might logically be expected to contribute to disease pathogenesis. Such candidate genes traditionally are those that encode proteins either important for the immune response, such as the MHC class II genes, or for the biological function of the target tissue, such as the insulin gene in T1DM. Alleles of the candidate gene must then be defined to allow genotyping of the relevant locus in a large group of unrelated individuals recruited for study. To determine if a given allele A predisposes an individual to the AID, genotypes are obtained from both affected *(case)* and unaffected *(control)* individuals (which is why such analyses are

often called *case–control studies*). If allele A is significantly over-represented among the affected individuals relative to controls, then allele A is deemed a "risk allele" or "susceptibility allele." With the discovery in the 1970s of the role of MHC class II molecules in antigen presentation, various MHC alleles became obvious candidates for risk association. Indeed, over the last 30 years, case–control studies have been very successful in identifying HLA alleles that predispose individuals to particular AID. In contrast, although some non-HLA risk alleles have been identified this way, results have generally been far less conclusive, perhaps because these genes have a much weaker effect on an individual basis. In addition, differences in allele frequencies between patient and control populations due to ethnic backgrounds have resulted in false positive results in some cases.

Unlike association studies, which begin with an assumption concerning the possible pathogenic role of a candidate gene, linkage analysis is an unbiased method that identifies candidate alleles regardless of their biological connection to the disease. The basis for linkage analysis is a general genotype comparison between affected relatives (usually siblings). The goal is to spot any polymorphism (called in this context an SNP, for *single nucleotide polymorphism*) that is shared above chance expectation. On average, when comparing two human genomes, SNPs can be found every 1000 to 2000 base pairs, which translates into about 3–10 million SNPs across the genome. In reality, far fewer SNP markers need to be analyzed due to the genetic phenomenon known as *linkage disequilibrium*, in which genetic variants co-segregate in the population when they are fairly closely spaced within the genome (within approximately 60,000 bp). Thus, a genome-wide scan actually analyzes about 300 evenly spaced polymorphic loci, allowing the identification of chromosome "blocks" or haplotypes inherited by affected individuals with a higher than random frequency. Within an identified haplotype, a number of possible risk alleles may be present so researchers must then choose candidate alleles from among them. The significance of these potential risk alleles is then evaluated by carrying out association studies. A potential aid to linkage studies began development in 2002 with initiation of a collaborative project known as the International HapMap Project. Geneticists in Canada, Japan, China, the United

States, and Nigeria are working to develop a haplotype map of the human genome in which SNPs strongly associated with particular human haplotypes will act as markers of haplotype location. Such a map will allow researchers doing linkage studies of autoimmunity or other diseases to bypass looking at the 99.9% of the human genome that is invariant, so that they can focus their analyses on the 0.1% of the genome that might show variation associated with disease.

Since the introduction of genome-wide linkage analysis in the early 1990s and its first application to human T1DM, the technique has been used to analyze many other human AID, including MS, RA, SLE, and CD. While initial enthusiasm for the approach was high, relatively few significant linkages outside of the HLA loci have been found. The ability of this technique to correctly identify a risk allele increases with both the degree of influence of the allele itself and with the number of *affected sibling pairs* (ASPs) studied. However, with the exception of HLA involvement, a given AID usually has risk alleles at multiple loci, each having a weak effect. Thus, on a statistical basis, researchers estimate that a minimum of several hundred, if not thousands, of ASPs are required to detect weak but meaningful allele associations via linkage analysis. Since most preliminary studies were relatively small in scope, few risk alleles have exerted significant enough effects to be detected. The lesson learned is that with most AID being relatively rare, identifying and recruiting the required number of families with two or more affected members is not feasible unless researchers studying a given AID pool their data, an approach that is now being actively pursued. For example, the North American Rheumatoid Arthritis Consortium (NARAC), which represents a collaborative effort among a large group of U.S. researchers, aims to carry out genomic screening of 1000 or more ASPs with RA. Similarly, several groups studying T1DM are collaborating to assemble genotype information from 4000 ASPs. One caveat to this data pooling approach is that among genetically diverse populations, different combinations of weak genetic effects may lead to a given AID phenotype. As a result, when data are combined, a locus that has a very weak effect in a number of populations will appear significant, while a locus that has a strong effect in a single population may not be detected.

Table 29-5 HLA Alleles Associated with Selected Human AID

AID	HLA Molecule
AS	HLA-B27 (Caucasians)
Autoimmune uveitis	HLA-DR3, DR4, DQ6, DQ8
CD	HLA-DRB1*07
GD	HLA-DR3
GS	HLA-DR2
HT	HLA-DR3, DR5
T1DM	HLA-DR3, DR4, DQ2, DQ8
Lyme arthritis	HLA-DR4, DR1
MG	HLA-B8, HLA-DR3 (early onset)
	HLA-B7 and HLA-DR2 (late onset)
	HLA-B46 (Asian)
MS	HLA-DR2
PM	HLA-DR3
PS	HLA-B27, Cw6
RA	HLA-DR4
	HLA-DRB1*04 (Caucasians)
SD	HLA-DR5, DR3, DR11, DR52
SLE	HLA-DR2, DR3
SS (primary)	HLA-DR3

Table 29-6 Non-HLA Genes Associated with Selected Human AID

AID	Non-HLA Genes
AS	Unknown genes in chromosomal regions 1p, 2q, 6p, 9q, 10q, 16q, and 19q
CD	CARD15, cation transporters
GBS	FcγRII, TNF
GD	CTLA-4
HT	CTLA-4
T1DM	Insulin, CTLA-4, IL-12 p40
PS	Gene on chromosome 17 containing a Runx-binding site
RA	IL-1β, IL-1R, MBL, TNFR2, ICAM-1, IFNγ, various FcRs IL-6 (juvenile) Gene on chromosome 17 containing a Runx-binding site
SD	C4, C2, Factor B, HSP70, TNF, fibrillin (one population)
SLE	CR1, HSP70, FcγII, FcγIII, IL-6, IL-10, TNF, TNFR2, PD-1 Gene on chromosome 2 containing a Runx-binding site C1, C4, or C2 (5% of cases)
SS (primary)	IL-10 promoter (not all groups)

HLA-DR5 in some groups, while PS is often associated with HLA-B27 and HLA-Cw6. Predisposition to GS appears to be inherited in association with HLA-DR2, and many CD patients have been found to express the HLA-DRB1*07 allele.

It should be noted, however, that an association between an AID and a particular HLA molecule may not hold true for all populations. For example, the HLA-DRB1*04 allele is tightly linked to RA development in Caucasians but not in Hispanic or African populations. Similarly, different HLA alleles are linked to SD in different ethnic groups, and ocular MG is associated with expression of HLA-B46 but only in Asian populations. Sometimes the HLA association may differ between clinical subsets of a given AID. For example, while the early onset form of MG is associated with expression of HLA-B8 and HLA-DR3, the late onset form is associated with HLA-B7 and HLA-DR2. Finally, to complicate matters further, there is now evidence that some HLA proteins confer protection against the development of AID even when another HLA molecule linked to increased AID susceptibility is present. For example, HLA-DQ6 is strongly protective against the development of T1DM. How this protection works is not known.

What about alleles of genes outside the MHC? In humans, certain alleles of several polymorphic genes involved in inflammation, complement activation or function, clearance of antigen, and T cell regulation have been linked to AID (Table 29-6). Indeed, in 95% of SLE patients, one observes the expression of the AID-associated alleles of a minimum of four "susceptibility genes," only some of which are HLA genes. Other polymorphisms linked to SLE development

occur in the CR1, HSP70, FcRγII, FcRγIII, IL-6, IL-10, TNF, and TNFR2 genes and in the gene encoding the negative T cell regulator PD-1. In human RA patients, polymorphisms of the IL-1β, IL-1R, MBL, TNFR2, ICAM-1, IFNγ, and various FcR genes have been identified. A polymorphism in the IL-6 gene has been associated with an increased risk of developing JRA. Polymorphisms of the C4, C2, Factor B, HSP70, and TNF genes have been identified in various SD patients. In addition, in one population of Native American Indians, SD was associated with mutations in the fibrillin gene encoding a glycoprotein component of the extracellular microfibrils. In mice, mutations in the fibrillin gene cause an SD-like disease. Polymorphisms of the T cell regulatory molecule CTLA-4 have been implicated in some patients with GD, HT, or T1DM. Human T1DM has also been associated with certain polymorphisms of the insulin promoter and the IL-12 p40 subunit gene (in some ethnic groups). A particular polymorphism of the promoter of the IL-10 gene has been noted in some groups of SS patients but not others. In GBS, the expression of certain polymorphisms of FcγRII and TNF have been associated with susceptibility to this AID. (We caution the reader here that, despite the wealth of detail cited, it has yet to be definitively proved that any of these polymorphisms actively contribute to the development of any AID.)

Some of the most detailed information on non-HLA gene association with AID has come from studies of CD patients. A key non-HLA gene associated with CD development is CARD15 (*caspase recruitment domain-containing protein* 15; also known as NOD2; refer to Ch.4). About 10–20% of CD patients in North America exhibit frameshift or missense mutations of this gene, and those homozygous for loss of CARD15 function invariably develop CD. (Curiously, however, CD patients in Japan do not show an association with CARD15 although the clinical phenotype is identical.) The CARD15 gene is located on human chromosome 16, and its protein product belongs to the Apaf-1 family of pro-apoptotic molecules. The CARD15 protein is preferentially expressed as an intracellular protein in monocytes. As noted in Chapter 4, CARD15 contains an LRR sequence similar to that found in TLRs and the plant R proteins discussed in Chapter 21. This LRR serves as an intracellular sensor of bacterial peptidoglycan. *In vitro*, CARD15 facilitates the activation of NF-κB induced by peptidoglycan or bacterial LPS. Mutations of CARD15 associated with CD result in a protein that can no longer recognize peptidoglycan but which can still respond to LPS. Different mutations of CARD15 have different effects on LPS-induced NF-κB activation: some enhance it while others inhibit it. It can be imagined that a mutated CARD15 protein that was constitutively active might incite the constant production of inflammatory cytokines that could damage the gut. Alternatively, the mutation may affect the domain in the CARD15 protein that allows it to interact with caspases. An inactive CARD15 protein might inhibit apoptosis and thus permit the inappropriate continued survival of T cells producing inflammatory cytokines. In addition to CARD15, a region on human chromosome 5 contains two genes associated with predisposition to CD. These genes encode transmembrane proteins responsible for organic cation transport across the gut wall. Polymorphisms of these transporter genes that result in decreased cation transport *in vitro* are found at increased frequency in CD patients.

A non-HLA gene of particular relevance to several other AID is Runx-1, which encodes the CBF2A subunit of a heterodimeric transcription factor called CBF. A DNA motif that specifically binds to the CBF2A half of CBF occurs throughout the human genome, and mutations in this sequence (termed by workers in the field as "Runx-1 binding sites") have been identified as AID susceptibility loci. For example, mutations in a Runx-1 binding site on chromosome 2 are associated with SLE, and mutations of a different Runx-1 binding site on chromosome 17 have been linked to RA and PS. Cell biologists know the CBF2A/Runx-1 gene as a gene required for the development and proliferation of myeloid cells. Intriguingly, mutations in the Runx-1 gene itself are found in patients with acute myeloid leukemia (see Ch.30), explaining why researchers in this field call the Runx-1 gene AML1.

Sometimes a non-HLA genetic susceptibility region of a chromosome may have been identified for an AID but the genes remain a mystery. For example, in addition to the association of HLA-B27 with AS, specific genetic susceptibility regions have been mapped to chromosomes 1p, 2q, 6p, 9q, 10q, 16q, and 19q. Similarly, in addition to the genes described previously, more than 10 other chromosomal loci associated with CD development have been identified in families with multiple affected members. Research is under way to identify the relevant genes in these regions.

Occasionally, despite intense research, information on the genetic determinants associated with a particular AID remains surprisingly limited. For example, MS occurrence is strongly associated with family groups and ethnic background, suggesting

an important role for genetic determinants. However, aside from an association with HLA-DR2, no other major susceptibility genes have been definitively associated with this disease. Similarly, while non-HLA loci are suspected to be involved in the development of HT, none has been conclusively linked to it. The same holds true for the other autoimmune thyroiditis: GD is weakly associated with expression of HLA-DR3, and this MHC molecule binds with high affinity *in vitro* to the peptides isolated as T epitopes from GD patients. However, no significant associations with polymorphisms in cytokine genes or the TSHR gene have been found. With respect to RF, no definitive genetic marker has yet been identified, and there have been only conflicting results with respect to HLA association.

II. ENVIRONMENTAL TRIGGERS

Environmental stimuli, including chemical agents and pathogens, show tantalizing links to AID onset or flare-ups in both humans and animal models. For example, in studies of animal AID based on adoptive transfer, the environmental stimulus has been shown to be critically important because autoreactive responses do not arise if the potentially autoreactive cells transferred into a genetically susceptible recipient are in the resting state. It is only if these cells are activated *in vitro* prior to adoptive transfer that AID develops. Despite this intriguing evidence, however, definitive proof that an encounter with an environmental stimulus actually triggers the initial onset of human AID is lacking.

i) Chemical Agents

Certain chemical and pharmaceutical agents have been linked to the onset of particular systemic AID symptoms. For example, toxins such as the heavy metal mercuric-chloride or polyvinyl-chloride can precipitate immune complex nephritis, systemic sclerosis, or the development of autoantibodies. Smoking, use of hair dyes (which contain aromatic amines), glue-sniffing, or exposure to silica dust (as occurs in many types of manufacturing and mining jobs) or other toxins can bring on an episode of RA, SLE, HT, GD, or SD. Heavy exposure to solvents such as paint thinners and removers has been linked to SD and some cases of MS, while a weak association has been noted between pesticide exposure and RA. Workers in industries such as furniture re-finishing, spray-painting, or perfume or cosmetic manufacturing also have a slightly increased risk of developing AID. Excess iodine intake may be a precipitating factor for HT or GD in some populations, and an episode of allergic rhinitis may bring on GD in certain patients. Skin injuries such as cuts and burns seem to initiate some cases of PS, and scratching and sun exposure can exacerbate PS symptoms. Some PS patients report that drinking alcohol aggravates their disease. Exposure to UV radiation, particularly UV-B rays, has been linked to a physical (rather than strictly chemical) insult that results in flare-ups of SLE. *In vitro* studies suggest that exposure of DNA and snRNPs to UV-B results in changes to the conformation and location of these molecules that increase their chances of activating an autoreactive lymphocyte. Accumulations of nuclear and cytoplasmic components appear on the membranes of apoptotic keratinocytes that have been subjected to UV radiation.

Some AID may initiate in response to drug treatment. For example, thiol-containing drugs and sulfonamide derivatives, as well as certain antibiotics and non-steroidal anti-inflammatory drugs, appear to trigger the onset of PG. Several pharmaceuticals have been associated with the initiation of an SLE-like syndrome in a small proportion of patients. Drugs such as hydralazine and procainamide or similar aromatic amine drugs prescribed to combat high blood pressure and irregular heartbeat, respectively, can induce SLE-like symptoms such as arthritis, pleuropericarditis, and myocarditis. Fortunately, the disease is often mild and of short duration, and usually resolves when the drug is stopped. The use of TNF inhibitors and IFNγ to treat AID (see Section F) has been associated with the induction of anti-nuclear antibodies in a high percentage of patients but actual drug-induced SLE appears to be rare.

How do chemical agents trigger AID? Silica exposure has been the best-studied agent in model systems. If silica particles of a size that can be inhaled are phagocytized by alveolar macrophages *in vitro*, the cells become activated and secrete high levels of TNF, IL-1, and chemokines. Lysosomal enzymes and ROIs are released, all of which can contribute both to the destruction of self cells (exposing self antigens) and the recruitment of additional APCs and lymphocytes. The pro-inflammatory milieu serves to activate any APCs in the immediate area, leading to enhanced presentation of the exposed self antigens to T cells. *In vivo*, treatment of animals with aerosol particles of silica has led to the accumulation of silica-laden macrophages in lymph nodes and elevated autoantibody production. With respect to T cells, if one incubates a population of purified T cells with silica particles *in vitro*, polyclonal activation is observed. This result indicates that self-reactive T cells in the population that would normally be anergized or ignorant may be non-specifically activated by silica. Moreover, exposure to silica biases the differentiation of activated T cells toward Th1, increasing IFNγ production and prolonging the inflammatory response. Macrophages hyperactivated by the IFNγ produced by the activated self-reactive Th1 cells continue to accumulate silica to the point where the release of internal proteases causes the death of the macrophages by necrosis. The silica particles strewn in the wreckage of the dying macrophages are phagocytosed by fresh macrophages, and the cycle is perpetuated.

ii) Infections

Infections with certain viruses, bacteria, and mycoplasmas appear to provoke the initiation of systemic AID in genetically predisposed individuals. Moreover, a severe bacterial or viral infection may trigger an increase in autoreactive antibodies or CTLs that leads to a flare-up of quiescent AID or an exacerbation of existing symptoms. With respect to viruses, the onset of various AID has been variably associated with infection by HSV-1, Coxsackie virus, EBV, HIV, HPV, or influenza virus. In particular, viral infections have been closely associated with flare-ups of SLE. Similarly, the development of GBS may follow infection with HSV, EBV, or CMV, and the onset of acute ITP may be preceded by varicella infection.

Infections with respiratory viruses may trigger GD or acute ITP. Viral infections may also be a factor in SD, since autoantibodies in some SD patients cross-react with sequences in the HIV Tat protein and the SV40 large T antigen. Based on the vacuolar appearance of PM muscle cells, some researchers speculate that an unknown virus can trigger this AID.

For many AID, the causal involvement of virus infections is still under considerable debate. For example, while pathogens such as EBV, HepC, and HTLV-1 have been proposed by some as triggers of SS, these viruses can cause sicca symptoms in the absence of AID, leaving the issue of causality unresolved. Some investigators believe that T1DM is triggered by a viral infection (particularly by the rubella virus or Coxsackie B4 virus) and that insulitis may be the body's response to these pathogens, but this theory is controversial. Whether MS is triggered by a viral infection or vaccination is also under debate. About 30% of new cases of MS appear to emerge following infection with some kind of virus, and relapses in some MS patients seem to be triggered by adenovirus or GI infections. However, no one viral infection has been consistently associated with this subset of MS cases. An association between MS onset and the HepB vaccine has been found in some studies but not others. Among the former, some scientists believe a component in the vaccine may actually cause the disease, while others speculate that vaccination merely accelerates onset in patients that would have eventually developed MS anyway.

Infections with various bacterial species have also been associated with AID. The most striking example is the development of RF following recovery from infection with a virulent member of the Group A streptococci. Another close link is that between the onset of GBS and *C. jejuni* infection. Antibodies directed against the LPS of *C. jejuni* that cross-react with human nerve gangliosides have been isolated from GBS patients. Episodes of AS also often follow bouts of *C. jejuni* infection. GD has been linked to infection with *Yersinia enterocolitica*, and mycoplasma infections are thought to sometimes precipitate the onset of RA, CD, or SD. A connection between *Helicobacter pylori* and ITP has been proposed because some ITP patients infected with this organism show improvement in their ITP symptoms following antibiotic treatment. In children and adolescents with PS, infection of the throat or upper respiratory tract with streptococci can cause a flare-up. Certain GI infections (particularly with *Salmonella* or *Yersinia*) can trigger episodes of reactive arthritis in susceptible individuals, and infection with *Klebsiella pneumoniae* or *Escherichia coli* is believed by some to be a risk factor for AS.

Finally, for a few AID, there appears to be no consistent link to pathogen exposure. For example, researchers have been unable to establish an association between a specific pathogen infection and the later development of RA, HT, or PG. Similarly, although the clinical course of KD suggests a pathogen infection, and various groups have implicated various organisms in small groups of cases, no one pathogen has been consistently associated with onset of this disease. In SS patients, alterations of the normal microbial flora of the mouth are often seen, but whether these changes cause SS or are consequences of the reduced salivation associated with this disease has yet to be determined.

III. HORMONAL INFLUENCES

Many AID show a gender bias in incidence. For example, women make up 88% of SLE patients and 90% of SS patients, and RA affects three times as many women as men. Among other AID, patients with MS, RF, SS, HT, GD, ITP, PM, and APS are most often females, while those with AS, KD, or GS are usually males. These findings suggest that sex hormones can play a major role in inducing AID onset in genetically predisposed individuals. In general, females are more susceptible to systemic AID such as SLE, whereas males have a greater chance of developing an organ-specific AID such as AS. Researchers hypothesize that the expression of hormones or factors associated with the development of sex-specific organs can activate previously tolerant or ignorant lymphocytes. Indeed, in a mouse model of SLE, the administration of estrogen upregulated Bcl-2 in B cells and blocked B cell tolerization. Disease symptoms were exacerbated in the estrogen-treated animals. In another study, chronic exposure of very young mice to estrogen altered thymic development and blocked the establishment of tolerance. Estrogen metabolism is often abnormal in SLE patients, and flare-ups of SLE may on occasion be associated with changes in hormonal status, such as during pregnancy or the initiation of hormone replacement therapy.

The two types of autoimmune thyroiditis, HT and GD, also occur predominantly in women. Significant numbers of HT and GD patients first develop their disease in the postpartum period, suggesting that major hormonal changes can precipitate onset. In animal models of hypothyroidism, estrogen exacerbates HT symptoms while testosterone reverses it. As well as a surfeit of female hormones, a deficiency of male hormones may be linked to AID. In several studies of SS, female patients had lower than normal levels of the male androgen hormones that regulate lipid production in mammalian glands.

Another hormonal influence that may be relevant to AID etiology is the hypothalamic–pituitary–adrenal (HPA) axis, introduced in Box 20-2. The reader will recall that the HPA axis is the neural circuit responsible for the "fight or flight" response to acute stress. Animals with defects in their HPA axis show increased susceptibility to AID, implying that stress-induced increases in glucocorticoids such as corticosterone and cortisol are required to restrain autoreactive lymphocytes. *MRL/lpr* mice respond to IL-1 stimulation (stress) by producing lower serum levels of corticosterone compared to wild-type animals. Similarly, in untreated SLE patients, the induction of hypoglycemia (stress) results in decreased production of serum cortisol compared to normal individuals. Psychological stress can also bring on symptoms of AID. For example, marital difficulties, job stress, and economic hardship have all been tentatively linked to the onset of RA in men and women. Stress can also bring on PG. The speculation is that stress causes the release of hormones or neurotransmitters that can influence the immune system. Indeed, it has been shown that stress alters the body's release of insulin, a factor that may be relevant to the onset of some cases of T1DM. This issue is currently under long term study.

IV. REGIONAL/ETHNIC DIFFERENCES

As mentioned previously, many AID appear to vary in incidence by region or by ethnic group, although such data have been relatively hard to come by and are not consistent for all AID or countries. In some cases, this variation may be due to the uneven prevalence of an HLA allele linked to a particular AID (due to ethnic differences) or of a triggering pathogen or chemical agent (due to geographic or environmental differences). In other cases, the reasons for variation in AID incidence among countries or ethnic groups are not obvious. It has been observed that Southern European and Asian countries have a lower incidence of T1DM and MS than do northern European countries, and that SLE incidence is higher in natives of African and Caribbean countries than in Europeans and North Americans. Within a single country such as the United States, African-Americans and Hispanics have a higher risk of developing SLE and SD than do Caucasians, but the risk of T1DM and MS in the African-American population is lower. In contrast, the rate of RA incidence in the United States does not vary by ethnic group. In Britain, immigrants from Africa, Asia, and the Caribbean show a decreased incidence of MS but an increased incidence of SLE compared to the general population.

Tied to the regional/ethnic issue is the observation that the incidence in developed countries of certain prototypical AID, including CD, T1DM, and MS, has rapidly increased over the past 30 years (Fig. 29-7). Some researchers believe that

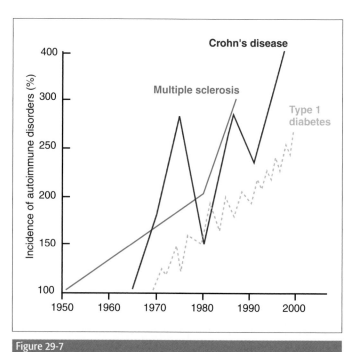

Figure 29-7

Increasing Autoimmune Disease Incidence in the Developed World
The incidence in developed countries of several common AID, including MS, CD, and T1DM has increased dramatically over the past five decades. For each curve, the starting incidence has a value of 100%. Incidence in later years is expressed as a percentage of the starting incidence. Adapted from Bach J. F. (2002) The effect of infections on susceptibility to autoimmune and allergic diseases. *New England Journal of Medicine* **347**, 911–920.

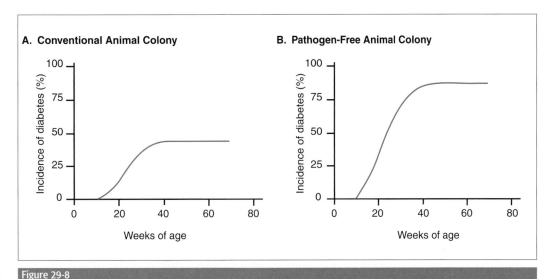

Figure 29-8

Animal Model Support for the Hygiene Hypothesis of Increased AID Incidence
(A) Diabetes-prone NOD mice born and maintained in a conventional animal colony such that they encounter normal levels of pathogens develop diabetes at a predictable frequency (about 40%). (B) NOD mice born and maintained in a sterile environment such that they do not encounter pathogens develop diabetes at a higher frequency (about 75%) and at a younger age. Adapted from Bach J. F. (2002) The effect of infections on susceptibility to autoimmune and allergic diseases. *New England Journal of Medicine* **347**, 911–920.

this increase may be related to the "hygiene hypothesis" (discussed in Box 28-2, in reference to the increased incidence of type I hypersensitivities in developed countries). To review, the hygiene hypothesis posits that excessive zeal in maintaining sanitary living conditions during infancy and childhood may result in an immune system that is not properly primed and/or does not develop adequate anti-inflammatory capacity. In this light, the observation that SLE incidence is much higher in Black Americans than in West Africans (who share ethnicity but not environment) might make sense. Another study has demonstrated that, just as is true for asthma, the incidence of T1DM in young children is lower in those who have multiple siblings or attend day care centers, compared to only children or children reared at home. However, contradictory evidence also exists, in that the incidence of T1DM and MS is unexpectedly high in areas of lower socioeconomic development in Sardinia and Puerto Rico.

Some animal studies have supported the hygiene hypothesis with respect to AID, while others have weighed against it. For most of the animal models described in Section C, individuals of a susceptible strain who are born and reared in a pathogen-free environment show earlier and more severe signs of AID than do animals of the same strain raised in a conventional animal colony (Fig. 29-8). Indeed, the development of diabetes in NOD mice can be prevented by prior infection with various pathogens, including LCMV, mycobacteria, or filarial parasites. Similarly, the development of SLE-like disease in NZB/W F1 mice can be forestalled by prior infection with *Plasmodium berghei*. However, the reactive arthritis that develops in transgenic rats expressing

human HLA-B27 is markedly reduced if the animals are reared in a pathogen-free environment.

E. Mechanisms Underlying AID

The vast majority of AID stem from abnormalities in the mechanisms of peripheral tolerance that fine-tune the repertoires of mature T and B peripheral lymphocytes. (APECED, the one defect known to be due to a failure in central tolerance, was described in Box 29-1.) We reiterate that the mere presence of autoreactive lymphocytes in an individual's repertoire is not enough to trigger AID: it only predisposes that individual to developing AID. Indeed, many perfectly healthy individuals have demonstrable (low) levels of autoantibodies and autoreactive T cells in their tissues. For AID to develop, a stimulus that activates the autoreactive cells must be present, and mechanisms designed to regulate autoreactive lymphocyte responses must fail.

How is the peripheral tolerance that controls autoreactive cells broken? In most cases, it is an unusual encounter between a self antigen and host lymphocytes that induces the autoreactive response. In most of these cases, regulatory cells and cytokines in the surrounding microenvironment ensure that the autoreactive response is mild and self-limiting and does not lead to clinical AID. However, in some situations, a cyclical accumulation of anti-self responses may occur that overwhelms peripheral tolerance mechanisms. For example, imagine that a particular self antigen is present at a higher concentration than usual, or is abnormally exposed, or is present in an unaccustomed inflammatory context. Molecules of this antigen may be recognized by an autoreactive B cell. If this B cell not only binds to the autoantigen but also internalizes it, processes it, and presents epitopes to T cells, autoreactive T cells recognizing epitopes from the antigen may also be activated if the surrounding microenvironment permits. Such autoreactive T cells then furnish more T help to the B cell and promote the expansion of the autoreactive clone. Cumulative cycles of this sort in the appropriate cytokine milieu can ultimately disrupt the maintenance of peripheral self-tolerance, and the tissues start to sustain immunopathic damage. Importantly, because the antigens recognized by the autoreactive cells are often routinely expressed by large numbers of self cells, the anti-self response does not resolve as a normal immune response would when the inciting pathogen is finally eliminated. This persistence of autoreactive clone activation may eventually be manifested as chronic autoimmunity.

We will now discuss several mechanisms, some of which remain controversial, that are believed to contribute to the development of AID in susceptible individuals.

I. PATHOGEN-RELATED MECHANISMS

As stated earlier in this chapter, the onset or flare-ups of many AID appear to be triggered by particular pathogens. Several hypotheses advanced to account for these observations are discussed in the following sections. However, the reader should keep in mind that, even if one or more of these hypotheses are proved correct, there must be other factors involved in AID development because, as once noted by Robert Inman and Lori Albert at the University of Toronto, "Infection is common, autoimmunity is not." Millions of people experience pathogen infections, many of them very serious, but only a small fraction of infected individuals develop AID. What determines which individuals will recover from an infection only to find they have developed an autoimmune disease? Can a preventive course of antiviral or antibacterial treatment avert AID in susceptible individuals? These matters remain under investigation.

i) Molecular Mimicry

The first pathogen-related hypothesis, called *molecular mimicry* (or *antigenic mimicry*), holds that autoreactive lymphocytes in the periphery are sometimes activated by cross-reacting pathogen antigens (Table 29-7). That is, an epitope derived from a pathogen may resemble an epitope derived from a self antigen closely enough to activate the autoreactive lymphocyte, if the appropriate cytokine microenvironment is present. This lymphocyte then mounts an attack not only on the pathogen but also on self tissues expressing the self antigen (Fig. 29-9). Because there is usually plenty of self antigen continuously present (because the antigen is part of a self tissue), the immune response cannot mop it up the way it does a pathogen, and the unwanted attack by autoreactive T and B cells is perpetuated. A few examples of rodent and human AID in which the

Table 29-7 Examples of AID Potentially Linked to Molecular Mimicry

Pathogen Antigen	Cross-reacting Mammalian Self Antigen	AID
	HUMAN	
Streptococcus cell wall M protein	Myosin, other heart valve proteins	RF
Peptides of EBV, influenza virus, HPV, measles virus, HHV-6	Myelin basic protein	MS
Borrelia burgdorferi, OspA protein	LFA-1	Lyme arthritis
Proteins of *Klebsiella pneumoniae* or *Shigella flexneri*	HLA-B27	AS
Proteins of *Salmonella typhimurium* or *Yersinia enterocolitica*	HLA-B27	Reactive arthritis
P2-C protein of Coxsackie virus	Glutamic acid decarboxylase	T1DM
LPS of *Campylobacter jejuni*	Peripheral nerve gangliosides	GBS
Protein of *Yersinia enterocolitica*	Thyrotropin receptor	GD
Lipoprotein of *Treponema pallidum*	Fibronectin, collagen	Vasculitis associated with syphilis
B13 protein of *Trypanosoma cruzi*	Cardiac myosin	Chagas heart disease
Protein of Coxsackie B3 virus	Cardiac myosin	Autoimmune myocarditis
	MOUSE	
Protein of *Chlamydia pneumoniae*	Myelin basic protein	EAE
Mycobacterial HSP60	Mammalian HSP60	IBD
UL6 protein of HSV-1	Corneal antigen	Autoimmune keratoconjunctivitis
Proteins of Coxsackie B3 virus	Cardiac myosin	Autoimmune myocarditis
P2-C protein of Coxsackie B4 virus	Glutamic acid decarboxylase	NOD diabetes
Protein of *Trypanosoma cruzi*	Cardiac myosin	CHD-like disease

A. T cell molecular mimicry

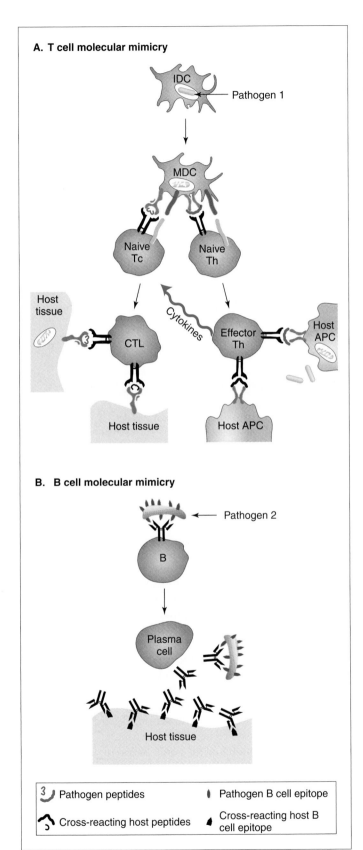

Figure 29-9

Theory of Molecular Mimicry
Both T cell- and B cell-mediated autoreactive responses may be triggered by molecular mimicry. (A) T cell molecular mimicry. An immature DC is induced to mature by its acquisition of pathogen antigens. The mature, licensed DC presents pathogen peptides on both MHC class I and class II. Naïve Tc and Th cells with TCRs that recognize these pMHC epitopes are activated and generate effectors that eliminate the pathogen. However, self-peptides presented by MHC class I and II on host tissues may mimic the pathogen pMHC epitopes. Damage to host cells ensues as the effector T cells attack healthy host tissues presenting the cross-reacting pMHC. (B) B cell molecular mimicry. A pathogen may supply B cell epitopes that activate naïve B cells. The plasma cell progeny produce antibodies that eliminate the pathogen but may also recognize similar-looking B cell epitopes on host tissues. Host cells are damaged as the antibodies attack healthy host tissues presenting the cross-reacting epitope.

In humans, molecular mimicry has been proposed to account for several prominent AID:

- Perhaps the best studied of these examples is RF, in which antibodies generated against the cell wall M protein present in Group A *Streptococcus* species cross-react with the structural protein myosin in the human heart. Immune complexes that are formed in the heart induce an inflammatory response and cause tissue damage. Anti-M protein T cell clones isolated from RF patients have been shown to cross-react with both human myosin and other heart valve proteins *in vitro*.

- While a century of study has failed to definitively identify a particular virus as the cause of MS, viral antigens from EBV, influenza virus, measles virus, and HPV (among others) can provoke the production of antibodies and T cells that cross-react with MBP *in vitro*. Most MS patients go through relapsing/remitting cycles of their disease. In some cases, relapses appear to follow episodes of viral infection, suggesting that the disease mechanisms are reactivated when memory anti-self responses are triggered. The prevailing theory is that the HLA-DR2 molecule closely associated with MS incidence presents a viral peptide in such a way that this peptide–MHC complex closely resembles a self-epitope involving the MBP peptide. Antiviral T cells are then continually activated, inducing inflammation and the destruction of myelin.

- *Lyme disease* caused by *Borrelia burgdorferi* infection is frequently associated with severe arthritis in the joints (*Lyme arthritis*). *B. burgdorferi* expresses a surface protein called OspA, which contains an epitope that is remarkably similar to a peptide in the human LFA-1 molecule. In many Lyme disease patients with severe joint disease, the subpopulation of Th1 cells that infiltrates the joint and dominates the inflammatory response is specific for an OspA peptide. It is thought that these Th1 cells later initiate an autoimmune attack on self cells in the joint presenting peptides derived from LFA-1. The high concentrations of IFNγ produced by the OspA-reactive Th1 cells may upregulate ICAM-1 expression on local synoviocytes and synovial fibroblasts and draw large numbers of LFA-1-expressing

B. B cell molecular mimicry

Pathogen peptides	Pathogen B cell epitope
Cross-reacting host peptides	Cross-reacting host B cell epitope

sequences of potentially corresponding self and cross-reactive pathogen T cell peptide epitopes have been compared are summarized in Figure 29-10.

AID	Cross-reacting sequences	Origin of peptide
EAE	YGSLPQKSQRTQDEN	Rat MBP
	YGCLLPRNPRTEDQN	*Chlamydia pneumoniae* protein
IBD	NIISDA	Mouse HSP60
	NAASIA	Mycobacterium HSP60
Lyme arthritis	VVLEGTLTA	Human LFA-1
	YVIEGTSKQ	*Borrelia burgdorferi* OspA protein
MS	VVHFFKNIV	Human MBP
	VYHFVKKHV	EBV protein
RF	QKMRRDLEE	Human myosin
	KGLRRDLDA	*Streptococus* cell wall M protein

Figure 29-10

Examples of Proposed Cross-reaction between Pathogens and Mammalian Peptides
Several AID in rodents and humans may arise from molecular mimicry between pathogen peptides and host peptides of similar sequence. Cross-reacting sequences are given in single-letter amino acid code. With information from Rohm A. P. *et al.* (2003) Mimicking the way to autoimmunity: an evolving theory of sequence and structural homology. *Trends in Microbiology* **11**, 101–105; Gross D. M. *et al.* (1998) Identification of LFA-1 as a candidate autoantigen in treatment-resistant lyme arthritis. *Science* **281**, 703–706; Fraser C. M. *et al.* (1997) Genomic sequence of a Lyme disease spirochaete, *Borrelial burgdorferi. Nature* **350**, 580–586; Quinn A. *et al.* (1998) Immunological relationship between the class I epitopes of streptococcal M protein and myosin. *Infection and Immunity* **66**, 4418.

cells into the joint. Increased presentation of LFA-1 peptides could then occur, providing continuous stimulation for the OspA-primed Th1 cells. In mouse models of Lyme disease, animals that mount a Th1 response characterized by CTL generation and the production of high concentrations of IFNγ have more severe arthritis than mice that mount a Th2 response. Depletion of CD8+ T cells also ameliorates arthritis in these animals.

- The onset of reactive arthritis in HLA-B27-expressing individuals is often preceded by infection with an enteric bacterium such as *Salmonella typhimurium* or *Yersinia enterocolitica*. These bacteria may express B cell conformational epitopes that resemble those on the HLA-B27 molecule whose expression is associated with this AID. Antibodies synthesized in a response to the bacteria during the acute infection may persist in the body and later attack subsets of healthy host tissue cells expressing HLA-B27.

- About 30% of patients with chronic Chagas disease (resulting from infection with the protozoan *Trypanosoma cruzi*; see Ch.22) develop *Chagas heart disease* (CHD). CHD takes the form of acute or chronic myocarditis that can progress to potentially fatal dilated cardiomyopathy. Autoimmune responses to cardiac myosin (among other self antigens) can be demonstrated in CHD patients, and T cell clones that recognize both human myosin and *T. cruzi* B13 protein have been found in the hearts of CHD patients. It should be noted, however, that autoimmune responses to cardiac myosin can be induced in the absence of any pathogen, such as by the surgical trauma associated with heart transplantation. Whether the autoimmune response to myosin is the true cause of CHD remains under debate.

- Autoimmune myocarditis leading to chronic heart disease also occurs in some patients that have recently recovered from a Coxsackie B3 viral infection. Again, the principal autoantigen appears to be cardiac myosin.

- T1DM has been linked in some cases to infection with Coxsackie B4 virus. The replicative enzyme P2-C of this virus shows striking sequence similarities to the human GAD protein.

Evidence for molecular mimicry has also come from mouse models.

- HSV-1 produces a viral protein called UL6 that closely resembles part of an antigen present in the mouse cornea. Anti-UL6 antibodies are able to cross-react with the murine corneal antigen *in vitro*, and HSV-1 infected mice develop autoimmune keratoconjunctivitis *in vivo* that closely resembles a T cell-mediated AID of the human eye called *herpes stromal keratitis*. Herpes stromal keratitis is triggered by HSV-1 infection of human eyes.

- NOD mice infected with Coxsackie B4 virus generate T cells directed against the P2-C protein of the virus. These T cells also appear to recognize GAD peptides presented by insulin-producing pancreatic β-islet cells.
- The infection of genetically susceptible wild-type mice (such as the A/J or BALB/c strains) with Coxsackie B3 virus induces autoimmune myocarditis in which the principal autoantigen is the heavy chain of cardiac myosin.
- As in humans, the infection of susceptible mice with *T. cruzi* causes CHD-like disease. The mice develop autoantibodies recognizing cardiac myosin.
- The cross-reaction of CD8$^+$ T cells directed against mycobacterial HSP60 with murine HSP60 can lead to the onset of IBD in genetically predisposed mice.

The reader is again cautioned, while anti-pathogen antibodies and CTLs can be shown to respond to purified human self proteins and cells expressing these proteins *in vitro*, it remains to be conclusively demonstrated that this type of attack is responsible for clinical disease in AID patients.

ii) Induction of Inflammation and DC Maturation

Infection by a pathogen induces inflammation, supplying "danger signals" and a cytokine milieu that favors DC maturation and activation. Many researchers believe that it is this inflammation-induced maturation of DCs that may be the key link between pathogen infection and autoimmunity, the so-called "adjuvant effect." Indeed, in animal models of T1DM and EAE, if the accumulation of DCs in the target tissue is blocked, clinical signs of AID do not develop. The hypothesis is that bacterial DNA, bacterial components, and mammalian DNA released upon pathogen-induced cell death are particularly potent adjuvants because they engage the TLRs of immature DCs. As we have learned, in the absence of TLR engagement by a PAMP, a DC remains immature and does not upregulate costimulatory molecules. Autoreactive T cells that interact with such DCs are then either anergized or induced to undergo a short burst of proliferation followed by cell death. In addition, immature DCs may induce the differentiation of regulatory T cells with immunosuppressive properties. In contrast, following TLR engagement, DCs are induced to mature and upregulate their expression of costimulatory molecules. When such mature DCs encounter autoreactive T cells in the lymph node, activation leading to an autoimmune response may result if the pMHC derived from a pathogen or self antigen is recognized by the T cell. In addition, activated mature DCs secrete proinflammatory cytokines, particularly IL-6, which tends to make activated T cells resistant to the suppressive functions of T$_{reg}$ cells. Thus, autoreactive T cells that might have been held quiescent due to a lack of costimulation and/or the effector actions of T$_{reg}$ cells regain their capacity for activation. T cell effectors are then generated that migrate back to tissue expressing the self antigen and destroy its cells.

The hypothesis of AID induction via inflammation fits nicely with studies of β-islet cell destruction in rodent models of T1DM. The development of diabetes is accelerated in TNF-NOD mice, which overexpress the inflammatory cytokine TNF. More evidence has come from studies of EAM in mice. Only infection of genetically susceptible mice with live Coxsackie virus (which induces significant inflammation) is able to induce autoimmune myocarditis: injection of killed virus cannot. Moreover, resistant mouse strains can be rendered susceptible to EAM induction if LPS is injected along with the virus. It is believed that the binding of the LPS to TLR4 on innate immune system cells triggers production of IL-1, TNF, and other pro-inflammatory cytokines that promote DC maturation. In humans, increased numbers of DCs can be found in the cellular infiltrates affecting the target tissues in several AID, including GD, HT, RA, T1DM, SLE, and SS. These DCs appear to be mature in phenotype, although it is not clear whether they arrive in the lesions as mature cells or are induced to mature once they arrive. Indeed, the synovial fluid from RA-affected joints and the serum from SLE patients contain unknown factors that can induce DC differentiation from HSC and the maturation of these DCs *in vitro*.

Pathogen-induced inflammation and the release of PAMPs from infected host cells may not be the only way to drive DC maturation leading to AID. Cells that have become necrotic due to mechanical injury, transformation, or other forms of stress may release host stress molecules with effects on DCs. For example, *in vitro* studies have shown that stress molecules such as HSP70, HSP60, and gp96 can mimic the effects of cytokines on DCs, inducing these cells to produce proinflammatory cytokines and chemokines and to upregulate their expression of MHC class II and B7. There is also now *in vivo* evidence obtained using a transgenic model of mouse diabetes that HSP70 can act on DCs to increase their T cell stimulatory function, overcoming established tolerance to a self antigen and invoking CTL-mediated autoimmunity. While the precise mechanism by which DC function is enhanced by HSP70 remains to be clarified (there was no upregulation of the usual collection of DC costimulatory molecules), these results suggest a means by which AID can be induced by endogenous host stress molecules in the absence of pathogen infection.

iii) Microbial Superantigens

Another theory to account for at least some episodes of pathogen-linked AID involves microbial superantigens. The reader will recall that these molecules can non-specifically activate a large number of different T cell clones by binding directly to particular TCR Vβ sequences. We emphasize that superantigens have been implicated in relapses of AID or the exacerbation of existing AID, but they do not appear to be able to <u>initiate</u> AID. Convincing evidence for autoimmune damage mediated by pathogen superantigens has emerged from mouse models. For example, in one study of EAE induction in mice, the affected animals did not show any relapse of AID symptoms following recovery from the acute episode. However, if these animals were injected with the superantigen

SEB after recovery, relapse of the disease was observed. This study also showed that administration of SEB alone could not induce EAE. In fact, injection of SEB into naive mice prior to the induction of EAE prevented the development of AID features, most likely because the binding of SEB to the potentially autoreactive T cells induced their clonal deletion. Similar results have been obtained in a mouse CIA model. If mice that experience an acute episode of autoimmune arthritis are allowed to recover and are then treated with a superantigen called MAM (derived from *Mycoplasma arthritidis*), a relapse of severe and persistent arthritis occurs. Treatment with SEB following recovery from an acute CIA episode can also induce a relapse. However, if the mice are treated with SEB prior to the administration of CII, the degree of arthritis developing in the initial acute episode is markedly diminished.

In humans, there is evidence that a bacterial superantigen from an unknown species (possibly one of the commensal bacteria in the intestine) may be a factor in CD. Molecular analyses of the lesions in CD patients have revealed the presence of DNA sequences derived from the gene encoding a bacterial transcription factor called I2. Normal tissues of CD patients did not contain this DNA. *In vitro*, I2 is capable of binding non-specifically to the TCRs of $V\beta5^+$ T cells, activating them. Whether this mechanism operates *in vivo* is under investigation. Researchers have also noted that certain TCR $V\beta$ T cell subsets are elevated in cases of KD and PS. Indeed, T cells whose TCR $V\beta$ regions are recognized by Group A streptococcal superantigens have been isolated from PS skin lesions.

II. INHERENT DEFECTS IN IMMUNE SYSTEM CELLS

i) Abnormalities in B Cells

The reader will recall from Chapter 9 that there are several checkpoints during B cell maturation that could affect the survival of autoreactive B cells generated by random V(D)J recombination in the bone marrow. The majority of autoreactive B cells undergo negative selection via mIg- or Fas-mediated apoptosis to establish central B cell tolerance. Autoreactive B cells that escape negative selection in the bone marrow and make it to the periphery undergo receptor editing in a second effort to generate a non-autoreactive BCR. Those B cells that fail this step usually undergo apoptosis, but some may either remain ignorant (due to sequestration of the relevant autoantigen), or are anergized. In the case of human SLE, peripheral B cells are hyper-reactive *in vitro* and show evidence of abnormal receptor editing and selection in the periphery. In mice, a broad spectrum of mutations affecting B cell survival, development, and activation gives rise to an AID phenotype (Table 29-8). For example, SLE-like symptoms develop in mice that bear mutations decreasing the expression of Fas, or in transgenic mice that overexpress either Bcl-2 or BAFF.

We learned in Chapter 9 that when the TNF-related molecule BAFF binds to its primary receptor BAFF-R on the B cell surface, intracellular signaling is triggered that is critical for continued maturation of transitional B cells. BAFF has a

Table 29-8 Inherent B Cell Defects Leading to AID in Mice

Molecule	Normal Function	AID Results from
BAFF	Essential for mature B cell survival	↑ Expression
Bcl-2	Promotes hematopoietic cell survival	↑ Expression
Cbl-b	Negative modulator of B cell signaling	Loss of function
CD19	BCR co-receptor	↑ Expression
CD22	Negative modulator of BCR signaling	↓ Expression
CD40/40L	Costimulatory molecules, isotype switching	↑ Interactions
CD45	Activates Src kinases	↑ Expression
Fas/FasL	Required for B cell AICD	↓ Expression
FcγRIIB	Negative modulator of BCR signaling	Loss of function
Lyn	Required for positive and negative BCR signal transduction	↓ Expression or function
SHP-1	Inactivates BCR signal transduction	Loss of function
TACI	Negative modulator of BAFF signaling	↓ Expression

direct connection to AID because overexpression of BAFF can result in the survival of autoreactive B cell clones that normally would have been removed by negative selection. Consistent with this observation, SLE-prone NZB/W F1 mice have elevated serum levels of BAFF. In addition, transgenic mice overexpressing BAFF develop AID resembling SLE and SS. Interestingly, mice deficient for TACI, an alternative receptor that binds to BAFF, have highly elevated B cell numbers and die rapidly of fatal autoimmune glomerulonephritis. It has thus been proposed that BAFF engagement of TACI (rather than BAFF-R) sends signals that negatively regulate B cell homeostasis and balance the effects of BAFF/BAFF-R binding. Presumably, TACI present on the B cell surface competes with BAFF-R for whatever BAFF is available. Accordingly, in a mouse model of SLE, the administration of TACI–Fc (a soluble fusion protein composed of TACI and the Fc region of Ig) to block BAFF/BAFF-R interaction reduced the production of anti-dsDNA autoantibodies and prevented Ig deposition in the kidney. A similar protective effect in mouse models of SLE, CIA, and EAE has been found for chimeric receptors containing BCMA, a third receptor that can bind BAFF and draw it away from BAFF-R. In humans, elevated levels of BAFF have been found in the sera of SLE, RA, and SS patients. Polymorphisms of the BAFF gene have been associated with AID in some populations but not others. In addition to BAFF, several other members of the TNF/TNFR superfamily are thought to make contributions to AID. Some examples are briefly discussed in Box 29-4.

Mutations that directly affect B cell signaling also lead to SLE-like disease. For example, AID in mice is associated with

Box 29-4. **TNF/TNFR-related proteins and AID**

We have learned previously in this book that various members of the TNF/TNFR superfamilies of receptors and ligands collectively influence the generation, activation, survival, and function of lymphocytes. Accordingly, altered expression of these molecules can sometimes result in defects predisposing an individual to autoimmunity. The postulated role of the TNF/TNFR family member BAFF (a B cell survival factor) in AID appears elsewhere in this chapter. Here we briefly outline several other examples of links between TNF/TNFR-related molecules and various AID in animal models and humans.

• LTαβ

The reader will recall from Chapter 17 that the normal architecture of the lymphoid tissues depends on the engagement of LTβR on supporting non-lymphoid cells by the TNFR-related heterodimer LTαβ expressed on lymphoid cells. Both the LTα and LTβ subunits are required for normal primary follicle and germinal center formation. Unfortunately for AID patients, the normal functions of LTαβ expressed by infiltrating autoreactive T cells may also contribute to the formation of the ectopic germinal centers common in AID, and thus the production of autoantibodies. Indeed, in a mouse model of MG, susceptible mice deficient for LTα did not develop AID symptoms. Similarly, inhibition of LTβR function or treatment with a soluble LTβR decoy receptor can protect NOD mice from developing diabetes.

• LIGHT

We learned in Chapter 26 that LIGHT is a molecule of unknown function that is structurally related to LTαβ. LIGHT can bind to LTβR and is expressed by both DCs and activated T cells, although its presence on the latter cells is normally transient. Overexpression of LIGHT in transgenic mice leads to AID characterized by lymphadenopathy, splenomegaly, and severe inflammation in the intestinal mucosae mediated by increased numbers of activated CD4$^+$ and CD8$^+$ T cells. In a mouse model of UC, administration of a decoy receptor that blocked LIGHT binding to T cells was able to prevent development of the disease. Similarly, in NOD mice, interception of LIGHT by use of a decoy receptor reduced the incidence of diabetes.

Some scientists have speculated that LIGHT is involved in negative selection in the thymus, since inhibition of LIGHT signaling can block the deletion of thymocytes recognizing thymic self antigens, but this role remains to be definitively established.

• TRAIL

The TNF superfamily member TRAIL (*TNF-related apoptosis-inducing ligand*) has been investigated as a potential molecular treatment for AID. As its name indicates, TRAIL is a widely-expressed molecule related to TNF that can induce apoptosis under certain circumstances *in vitro*. However, the physiological function of TRAIL is unknown. TRAIL can bind to several receptors, including the two death receptors DR4 and DR5. DR5 is highly expressed on activated but not resting lymphocytes, and activated DR5-expressing normal human T and B cells can be induced to undergo apoptosis *in vitro* by treatment with TRAIL. B cells from SLE patients show elevated expression of DR5, and *in vitro* treatment of these cells with TRAIL induces their death by apoptosis in excess of that seen in normal lymphocyte cultures. These observations have led some researchers to theorize that autoreactive T and B cells in patients might be induced to undergo AICD by treatment with TRAIL. Unfortunately, the administration of high doses of TRAIL to humans may be toxic due to the induction of liver cell necrosis. In mice, TRAIL does indeed protect against AID, since blockade of TRAIL in mouse EAE models exacerbates disease. However, the underlying mechanism appears to be different from that originally proposed. Rather than engaging DR5 and inducing apoptosis, elevated levels of TRAIL appear to promote the expression of the Cdk inhibitor p27 in murine lymphocytes, inducing cell cycle arrest and halting the proliferation of autoreactive cells.

• GITR

The reader will recall from Chapter 16 that GITR (glucocortocoid-induced TNFR) is a TNFR-related surface molecule constitutively expressed by T$_{reg}$ cells. Its ligand GITR-L is expressed by the CD5$^+$ B cell subset associated with autoimmunity. Upon engagement of GITR by GITR-L, the suppressive function of T$_{reg}$ cells is inhibited. One can imagine that autoreactive T cell clones that had been suppressed by T$_{reg}$

cells might then be susceptible to activation, resulting in autoimmunity. However, there may be a more direct effect involved. GITR expression can be induced on activated conventional T cells in the spleen, lymph nodes, and thymus. *In vitro*, engagement of GITR by GITR-L sends inhibitory signals that block AICD. If this phenomenon occurs *in vivo* (where the required GITR-L would presumably be supplied by CD5$^+$ B cells), autoreactive T cells might be able to resist AICD. Of interest, the T cells of GITR-deficient mice exhibit hyperactivation, elevated IL-2 and IL-2R expression, increased proliferation, and heightened sensitivity to AICD.

• 4-1BB

4-1BB (CD137) is a TNFR-related receptor with minor costimulatory function. 4-1BB is expressed primarily on activated CD4$^+$ and CD8$^+$ T cells and on NK cells, while its TNF-related ligand, 4-1BBL, is expressed by activated APCs. Engagement of 4-1BB on T cells stimulates proliferation independently of CD28 and promotes IL-2 and IFNγ secretion by Th effector cells and CTLs, respectively. Like Fas but independently of it, 4-1BB may contribute to the AICD of activated T cells. However, there is also evidence that 4-1BB may downregulate the functions of Th cells, either by anergizing them directly or by promoting the generation of T$_{reg}$ cells. These observations have led researchers to test anti-4-1BB mAbs as therapy for AID. In a mouse EAE model, treatment of the animals with an agonistic anti-4-1BB mAb stimulated the proliferation of autoreactive T cells and drove them to AICD more quickly, ameliorating symptoms of the disease. Treatment with this mAb also forestalled the development of lymphoproliferative disease in *MRL/lpr* mice due to the deletion of autoreactive B and T cells. However, when NOD mice were engineered to transgenically express anti-4-1BB mAb, the development of autoimmune diabetes was accelerated rather than inhibited. With respect to AID in humans, it has been shown that activated T cells can release a soluble form of 4-1BB, and that levels of soluble 4-1BB are increased in the serum of RA patients. How these observations are related to the findings in mice is unclear, and much remains to be investigated about 4-1BB function before experimental therapies in humans can be attempted.

Continued

• CD30

CD30 is a TNFR-related surface protein expressed by activated (but not resting) T and B cells, while its ligand CD30L (CD153) is a TNF-related transmembrane protein primarily expressed on activated T and B cells, eosinophils, and macrophages. CD30 was originally identified as a surface marker of the tumor cells of Hodgkin's lymphoma, an unusual B cell-like subset called "Reed-Sternberg" cells (see Ch.30). The expression of CD30 and CD30L has subsequently been found on several other types of tumors. With respect to AID, a soluble form of CD30 has been found at high levels in the serum of patients with RA, SS, and SLE. In animal models of AID, engagement of CD30 blocks the proliferation of autoreactive CD8$^+$ T cells and prevents the development of AID via adoptive transfer of these cells. There is also some evidence that CD30 signaling can dampen Th1-driven inflammation.

• CD27

As outlined in Chapter 14, CD27 is a TNFR-related membrane protein expressed on T cells that functions as a minor costimulatory molecule when engaged by its TNF-related ligand CD70. Both molecules are expressed on resting T cells and are upregulated in response to T cell activation. However, the signaling contributed by CD27/CD70 appears to be more important for the functions of mature effector T cells and the establishment of memory than for the initial stages of naive T cell activation. There is also some evidence that CD27/70 signaling enhances the effects of CD40 signaling on plasma cell differentiation. In mouse models of EAE, inhibition of CD27/CD70 signaling via anti-CD70 mAb treatment prevented the development of disease. In humans, primary SS patients show increased serum levels of soluble CD27 that correlate positively with

their increased serum levels of IgG. Similarly, T cells from SLE patients overexpress CD70 and readily support the overproduction of IgG by autoreactive B cells.

• OX40

OX40 is a TNFR-related molecule whose expression is upregulated on naive and effector T cells upon activation. The importance of OX40 signaling to AID has been demonstrated in animal studies. For example, when OX40 signaling was artificially sustained by treating mice with agonistic anti-OX40 mAb, T cell tolerance that had been established to a self antigen was reversed. In another study, researchers genetically manipulated mice to contain an autoreactive T cell population that did not induce clinically detectable disease. When these mice were treated with an agonist anti-OX40 mAb, they died of a severe inflammatory AID within 12 days.

---

mutations that increase BCR signaling, such as upregulated expression of CD19 (the B cell coreceptor) or downregulated expression of CD22 (negative modulator of BCR signaling). Loss of function of FcγRIIB (negative regulator of BCR signaling), SHP-1 (dephosphorylates and inactivates downstream signal transducers), or Cbl-b (ubiquitin ligase controlling downstream signal transducers) also leads to sustained B cell activation and SLE-like symptoms. CD45 is normally required for the activation of the Src kinases controlling signal transduction downstream of antigen receptors and its overexpression is also associated with autoreactivity in transgenic mouse models. Some of the Src family kinases controlled by CD45, particularly the Lyn kinase associated with the BCR, have multiple functions such that they not only promote signal transduction favoring cellular activation (as detailed in Ch.9) but also signal transduction that has negative regulatory effects on the same cell. For this reason, mutations that decrease the function or expression of Lyn have the net effect of hyperactivating B cells. All of these types of mutations might allow previously ignorant or anergized anti-self B cells to break peripheral tolerance.

CD40–CD40L costimulation, the process critical for effective T–B cell cooperation, is another suspect. Indiscriminant signaling via CD40 could lead to increased autoantibody production. For example, active flare-ups in SLE patients have been linked to spontaneous CD40–CD40L interactions that drive the increased activation, proliferation, and differentiation of autoreactive B cells. Similarly, the induction of CIA in mice is exacerbated if the animals also receive an agonistic anti-CD40 mAb (which acts like CD40L). Conversely, NOD, *MLR/lpr*, and NZB/W F1 mice treated with an antagonistic anti-CD40L mAb (which blocks CD40–CD40L interaction)

all develop AID at a reduced frequency. Anti-CD40L mAb treatment also delays the onset of EAE or CIA in mouse models. Dysregulation of T–B cooperation due to abnormalities in CD40–CD40L interaction could also lead to abnormal isotype switching. A change in the isotype of an autoantibody usually present at clinically insignificant levels might alter its trafficking or stability such that it becomes capable of mediating clinically detectable autoimmunity.

An intriguing means by which autoreactive B cells might be stimulated to proliferate and produce autoantibodies involves TLR9. The reader will recall from earlier chapters that this innate immune system receptor initiates signaling leading to phagocyte activation when it is engaged by the unmethylated CpG motif commonly found in bacterial DNA. TLR9 is also expressed by B cells, although its precise function here is unknown. While some TLR9 appears on the B cell surface, other TLR9 molecules are localized in the membrane of the endocytic compartment. This intracellular TLR9 appears to be able to bind to mammalian chromatin despite the fact that most CpG sequences in mammals are methylated. The following scenario has thus been proposed to account for some cases of autoantibody production. Let us suppose that the B cell repertoire of the mammalian host contains an autoreactive B cell directed against self antigen A, but that this cell is anergized in the periphery. Let us now suppose that a cell in our host dies and starts to release intracellular contents before it can be properly cleared away by scavenging macrophages. Chromatin released in the cellular debris attaches to self antigens, including self antigen A. The autoreactive B cell that recognizes self antigen A binds the chromatin/self antigen A complex to its BCR and internalizes it. In the endocytic compartment, the chromatin moiety of the complex binds to the transmembrane TLR9, triggering intracellular signaling in the cytoplasm that overcomes the ignorant or anergized state of the autoreactive B cell. Thus, the chromatin/TLR-induced activation of the autoreactive B cell results in the production of anti-self antigen A autoantibodies. Evidence to support this hypothesis has come from the study of murine autoreactive B cells producing anti-IgG2a antibody (a model for RF-producing B cells in human RA). These cells were activated much more efficiently *in vitro* when the IgG2a recognized by the autoreactive B cells was complexed to chromatin, and this enhanced activation was dependent on TLR9.

The "dual engagement" of the BCR and TLR9 by chromatin may explain why anti-DNA antibodies are such a common feature of AID. Autoreactive B cells specific for chromatin would receive two sets of activatory signals: those engendered by chromatin binding to the BCR on the cell surface, and those triggered by internalized chromatin binding to intracellular TLR9. The activation of these autoreactive B cells would thus be more efficient compared to that of autoreactive B cells without a means of stimulating TLR9. The result would be a predominance of anti-DNA antibody production.

ii) Abnormalities in Conventional T Cells

Abnormalities in T cell signaling and activation can be associated with AID. For example, the vital signal transducer Lck is reduced in SLE T cells due to increased ubiquitination that is independent of T cell activation. Mutations in other molecules that alter Lck localization and/or function in the membrane rafts important for TCR signaling may also be associated with SLE, including CD45, Csk (kinase needed to balance CD45 action), Cbl-b, and c-Cbl (another ubiquitin ligase). In activated SLE T cells, the proportion of the plasma membrane devoted to lipid rafts is increased, CD45-mediated activation of signaling molecules within the lipid rafts occurs more rapidly than usual, and CD45 is excluded from the rafts more slowly. In addition, a mutation that increases the expression of the raft-associated ganglioside GM1 is thought to alter Lck interaction with CD45. Precisely how these defects lead to clinical AID remains under investigation, but it may be that SLE T cells are in a pre-primed state (despite their reduced Lck expression) and can be activated more easily than normal T cells.

Defects in regulators of T cell activation may also lead to AID. For example, should an autoreactive T cell be activated in an individual, the negative regulators CTLA-4, Cbl-b, c-Cbl, and PD-1 should kick in to terminate the response before too long. Loss of the function of these regulators could lead to AID. Indeed, gene-targeted mice lacking PD-1 or Cbl-b show SLE-like symptoms.

iii) Abnormalities in Regulatory T Cells

Abnormalities in regulatory T cells have also been implicated in the development of certain AID. As noted in Section B, the first evidence that regulatory T cells could suppress AID came from studies of the neonatal thymectomy of certain AID-prone mouse strains. These animals soon developed organ-specific AID such as autoimmune gastritis or oophoritis. Similarly, *nude* mice (which are naturally athymic) were found to produce autoantibodies that formed elevated levels of immune complexes causing renal disease. Subsequent investigations showed that, in many of these cases, the AID induced by a lack of thymic function could be suppressed by adoptive transfer of CD4$^+$CD25$^+$ T cells (containing the T$_{reg}$ subpopulation) from normal animals. Since IL-2 (and perhaps CTLA-4) are essential for the generation and/or function of T$_{reg}$ cells, the lymphoproliferation and autoimmunity that occur in mutant mice with defects in IL-2, IL-2Rα/β, and CTLA-4 can also be reduced by adoptive transfer of normal CD4$^+$CD25$^+$ T cells. As was noted in our discussion of IPEX, there is now good evidence at the molecular level that the FOXP3 transcription factor is implicated in the generation and/or function of T$_{reg}$ cells that control AID.

Work in several other experimental systems has demonstrated the importance of T$_{reg}$ cells to AID suppression. In mouse models of T1DM, EAE, CIA, and IBD, the elimination of T$_{reg}$ cells speeds up disease progression. In neonatal NOD mice, TNF injection may exacerbate disease not only because it promotes inflammation but also because it decreases the T$_{reg}$ cell population. In NOD mice deficient for either CD28 or B7-1/B7-2, an exacerbation of diabetes correlates with a deficiency in T$_{reg}$ cells (which depend on CD28 signaling for their homeostasis). In *MLR-lpr* mice, the differentiation of anti-dsDNA B cells depends on help provided by CD4$^+$ T cells, and this help is not delivered if the animals are given exogenous preparations of normal T$_{reg}$ cells.

iv) Abnormalities in APCs

As well as defects in effector lymphocytes and regulatory T cells, aberrations in the generation and/or function of APCs have been associated with AID. Abnormalities in the development of macrophages and DCs from bone marrow precursor cells have been found in the BB-DP and NOD rodent models of T1DM. In these animals, macrophages are increased in number while DCs are decreased, and DC and macrophage functions are altered. For example, macrophages in NOD mice show enhanced powers of migration and pro-inflammatory secretion. It is speculated that these alterations may increase the non-specific damage done to self tissues by responding macrophages, and that this damage causes the release of self antigens. DCs in NOD mice show reduced expression of MHC class II and costimulatory molecules, a factor that prevents them from activating naive T cells efficiently. While this feature might first be seen as helpful (since the DC would become tolerogenic), it also means that this DC can no longer activate the regulatory T cells thought to be crucial for blocking the initiation of diabetes. Indeed, the onset of diabetes in NOD mice and BB-DP rats is hastened, rather than suppressed, if the interaction between CD28 and B7 is blocked. Conversely, NOD mice or BB-DP rats treated with agonistic CD28 mAbs do not develop diabetes. *In vitro*, isolated regulatory T cells from BB-DP rats cannot expand normally when cultured with defective BB-DP DCs. In humans, corresponding defects in the development and maturation of DCs and macrophages have been noted in patients with T1DM and HT.

v) Abnormalities in CD1-Restricted Lymphocytes

CD1-restricted lymphocyte subsets such as γδ T cells and NKT cells may also play a part in controlling autoreactive αβ T cell responses, most likely by cytokine secretion. For example, if γδ T cells are depleted from *MRL/lpr* mice, the SLE-like symptoms of these animals are exacerbated, and the high levels of anti-snRNP autoantibodies normally present in these mutant mice are further increased. In humans, studies of AID onset in sets of MZ twins have yielded results that are consistent with these findings. Where one twin has diabetes and the other does not, the diabetic twin has much lower levels of CD1-restricted T cells. Why CD1-restricted T cells are reduced in these patients is unknown, but the deletion of the homologous CD1 locus in mice has been shown to hasten the development of diabetes in NOD mice. However, contradictory results have been obtained in studies of other AID models, with some results supporting a role for γδ T cells in prevention of disease, others finding little or no effect, and still others implicating γδ T cells in the exacerbation of disease. Indeed, accumulations of γδ T cells have been found in the inflamed synovia of RA patients. Time course studies of γδ T cell depletion have suggested that a particular subset of these cells initially exerts a pro-inflammatory effect that helps to initiate and support αβ T cell responses (including autoreactive ones), but that later on, either the same set or a different subset of γδ T cells switches gears to anti-inflammatory functions that can suppress AID.

Controversy remains as to the precise role of NKT cells in AID in both humans and mice. The reader will recall from Chapter 18 that most NKT cells bear semi-invariant TCRs that recognize ligands resembling α-galactosyl ceramide (αGalCer) presented on CD1d. Upon activation, NKT cells rapidly secrete IFNγ, IL-4, GM-CSF, and IL-13, among other cytokines. Because of this mixed Th1/Th2 cytokine expression profile, NKT cells can have both stimulatory and suppressive effects on AID, depending on the specific disease and the circumstances involved. For example, NOD mice have reduced numbers of NKT cells. When wild-type NKT cells are adoptively transferred into these animals, these cells can apparently inhibit the development of diabetes if they secrete predominantly Th2 (rather than Th1) cytokines in response to stimulation by β-islet cell antigens. A similar effect is seen if NOD mice are injected with αGalCer (to activate residual NKT cells). An even greater benefit is realized if DCs pulsed with αGalCer are used for treatment, most likely because delivery of the stimulatory ligand is streamlined. Other studies have reported that, in addition to polarizing the cytokine response of autoreactive T cells to the less harmful Th2 type, NKT cells can induce the anergy of IFNγ-secreting Th1 cells directed against islet antigens. In contrast, conflicting results have been obtained with respect to a role for NKT cells in protection against EAE. In situations where NKT-mediated protection did occur, a shift in the production of cytokines by autoreactive T cells from a Th1 to a Th2 profile was observed.

What about the function of NKT cells in human AID? In some (but not all) studies, lower numbers of NKT cells were found in the blood of T1DM patients compared to healthy individuals, and the ability of patient NKT cells to secrete IL-4 was impaired. Decreases in NKT cell numbers have also been found in the peripheral blood of SS, RA, and MS patients. However, NKT cell numbers vary widely even in healthy humans, and it is not clear that values obtained for peripheral blood populations are representative of numbers in the affected tissues. In any case, deficits in NKT function do not appear to be a universal mechanism underlying AID because patients with GD or MG have normal numbers and activity of NKT cells.

vi) Abnormalities in Phagocytes

Abnormalities in mechanisms controlling the prompt removal and disposal of self antigens may also underlie some AID. For example, in some SLE patients, phagocytes fail to properly take up immune complexes due to reduced numbers and function of CR1 receptors on their cell surfaces. This observation may be the basis for the relationship between complement component deficiencies (see Ch.19) and SLE. Other SLE patients express alleles of the FcγRs that do not bind to antibody-bound antigen as efficiently as they should, reducing phagocytosis of IgG-containing immune complexes.

III. ALTERATIONS TO CYTOKINE EXPRESSION

Studies in mice and in human tissues have indicated that over-expression or deficiency of some cytokines as well as defects in cytokine signaling pathways can lead to AID (Table 29-9). IL-12 is a particularly important cytokine for AID development, since it drives Th1 responses leading to damaging inflammation and promotes the development of CTL effectors. Abnormally high levels of IL-12 production by APCs can predispose a susceptible mouse to developing an

Table 29-9 Defects in Cytokine Production or Signaling Associated with Murine or Human AID Symptoms

Disease/Model	Cytokine or Cytokine Receptor Status	
	Overexpression	**Deficiency**
CIA	IL-17, IL-6	
EAE	IL-23, osteopontin*	IFNγ (late stages)
IBD	IL-2, IL-2R, IL-7, IL-10, TNF, STAT4	IL-2R, IL-10R, STAT3, TGFβ
T1DM	IL-12, IL-27, TNF (early onset)	IL-6, TNF (late onset)
MS	IL-17, IL-6	
RA	TNF, IL-15, osteopontin	IL-1RA
SLE	IL-12, IL-27, IFNγ (in skin)	TNF, TGFβ, TGFβR (in T cells)
Vasculitis	TNF	

*Normal expression required.

organ-specific AID such as T1DM, while moderate IL-12 upregulation has been associated with a systemic AID resembling SLE. As noted previously, the IL-12-related cytokine IL-23 is linked to the development of EAE and therefore possibly MS. Another IL-12-related cytokine, IL-27, may influence the phenotype of glomerulonephritis developed in SLE patients. Some RA patients show elevated levels of TNF and IL-15 in inflamed synovial tissues as well as increased levels of the pro-inflammatory chemokine osteopontin. In the EAE mouse model, animals lacking osteopontin develop less severe disease. In the CIA mouse model, IL-17 and IL-6 are elevated in the sclerotic lesions of affected animals, and inhibition of endogenous IL-17 or deficiency of IL-6 blocks CIA development. IL-17 and IL-6 are also increased in lesions of human MS patients.

The influence of a cytokine can be different in different AID or can change as the AID progresses. For example, rather than exacerbating diabetes in NOD mice (as might be expected for a pro-inflammatory cytokine), overexpression of IL-6 slows diabetes development in these animals. In the case of transgenic TGFβ expression, autoimmunity is dampened in rodent AID models at the initiation of the disease (as expected for an anti-inflammatory cytokine), but fibrosis of the organs is promoted later on. Overexpression of TNF in neonatal NOD mice exacerbates disease, but if TNF expression is delayed until adulthood, this cytokine can have a protective effect. It seems that the same proinflammatory cytokines that inflict damage on tissues also trigger homeostatic mechanisms that slowly gain control over the inflammatory response. For example, while IFNγ initially promotes a Th1 response with its inherent tissue damage, this cytokine may eventually promote the apoptosis of the offending autoreactive T cells, mitigating disease. Indeed, in IFNγ-deficient mutants of susceptible mouse strains, the development of EAE (a Th1 disease) is ultimately enhanced rather than suppressed. A similar scenario may explain the protective effect of TNF on diabetes development in adult NOD mice, in that TNF may induce the apoptosis of CD8$^+$ T cells directed against β-islet cells in these animals. Alternatively, it may be that both IFNγ and TNF eventually trigger anti-inflammatory mechanisms that suppress or counter the actions of autoreactive effector lymphocytes. These observations have important implications for the treatment of AID patients. A therapy that initially damps down an autoreactive response may later promote tissue damage under certain circumstances. For example, a proportion of RA patients treated with anti-TNF antibodies to block the effects of this pro-inflammatory cytokine in their joints later developed anti-nuclear antibodies and even SLE on rare occasions. Such observations mandate caution in the prolonged treatment of AID (or any) patients with cytokines.

IV. DEFECTS IN THE COMPLEMENT SYSTEM

As we learned in Chapter 19, the complement system is crucial for the efficient disposal of immune complexes, including those containing self antigens. As a result, patients with mutations of various complement components exhibit SLE-like symptoms. The rationale is that, without C1, C4, or C2 function, levels of C3b and C3d are sub-optimal and clearance of immune complexes containing self antigens is reduced. Self antigens thus persist and participate in the formation of large networks of immune complexes that can accumulate at numerous sites in the body (particularly its narrow channels) and trigger inflammation. AID symptoms are then induced or perpetuated. Other proposed mechanisms by which complement deficiency might lead to AID, including the "waste disposal hypothesis" of apoptotic cell clearance, were detailed in Box 19-3.

V. EPITOPE SPREADING

"Epitope spreading" was briefly discussed in Chapters 26 and 27. This term is used to describe the phenomenon in which the immune system appears to expand its response (no matter how triggered) beyond the immunodominant epitopes first recognized by T and B cells to include new, cryptic, non-cross-reactive epitopes that are recognized only much later. The appearance of responses to these later epitopes underlies the progression of AID and characterizes its chronic phase. The theory is that the actions of anti-self effector cells and cytokines on cells or self antigens bearing the original immunodominant epitopes may have effects beyond the first, acute anti-self response. This first response is usually self-limiting and does not lead to clinical AID. However, the effector actions unleashed during this response to the original epitope may disrupt cell structure or protein conformation to expose either entirely new self antigens (*intermolecular epitope spreading*) or new epitopes on existing self antigens (*intramolecular epitope spreading*). These epitopes would previously have been either totally sequestered from the immune system, or processed in amounts insufficient to activate T cells, so that lymphocytes recognizing them would not have been deleted during the establishment of central tolerance or anergized in the periphery. Upon exposure, molecules containing the cryptic epitopes become available for the first time, or in increased amounts, for uptake by DCs. In addition, tissue damage from the original autoreactive response may result in the release of stress signals that can induce DC maturation. Within this inflammatory milieu, autoreactive T and B lymphocytes directed against epitopes of the newly exposed proteins become activated and expand the autoreactive attack to additional self tissues where these proteins are expressed. There might then be virtually no limit to the duration of the autoimmune response, because new cryptic epitopes would be continually revealed as tissue destruction proceeded. In addition, the autoreactive lymphocytes might generate memory cells to all epitopes, initial and cryptic. These memory cells would lie in wait and might be responsible, at least in part, for subsequent flare-ups of the disease.

Evidence to support the importance of epitope spreading in AID has been obtained from several mouse models. In NOD mice, the initial damage to the pancreatic islets arises from an autoimmune attack on GAD. However, as the disease progresses, lymphocytes directed against β-islet cell proteins such as insulin and carboxypeptidase H are activated. If, instead of an immune response, tolerance is induced to GAD in NOD mice, the animals do not develop diabetes and no responses against cryptic self antigens are subsequently mounted. Similarly, in mice susceptible to EAE induction, anti-self immune responses mounted against a series of epitopes occur in a reproducible sequence that can be attributed to epitope spreading. These responses cannot occur until after the initial attack on MBP has been sustained; that is, until after the cryptic epitopes are exposed by the initial tissue damage. Indeed, the late phase of EAE disease in these animals is demonstrably due to epitope spreading.

Proven examples of epitope spreading in humans are few. In PG, blistering of the mouth almost always precedes blistering of the skin, and mouth blisters are associated with the presence of autoantibodies directed against the desmoglein-3 protein. It is not until the attacks on desmoglein-3 expose epitopes on the related protein desmoglein-1 that autoantibodies directed against this latter protein are produced and skin blistering commences. This chain of events constitutes an example of intermolecular epitope spreading. PF furnishes an example of intramolecular epitope spreading. PF patients (who have skin blisters but not mouth blisters) exhibit only anti-desmoglein-1 antibodies. Before the onset of clinical signs of the disease, patients predisposed to PF produce anti-desmoglein autoantibodies that are directed against an epitope in the C-terminus of the desmoglein-1 protein. With the onset of clinical disease, autoantibodies directed against epitopes in the N-terminus of the desmoglein-1 protein can be found in the serum. It is speculated that the sub-clinical response against the C-terminal epitopes exposes the N-terminal epitopes and invokes the production of skin-blistering antibodies. Epitope spreading has also been invoked to account for the polyclonal lymphocyte activation evident in SLE. For example, multiple T and B cell clones reactive to different epitopes of snRNP can be found at different stages of AID progression in SLE patients. A similar phenomenon occurs during the development of SLE-like disease in *MRL/lpr* mice.

Some scientists speculate that, as well as exacerbating AID, epitope spreading can sometimes control it. If an epitope exposed as a result of the autoimmune attack preferentially induces a Th2 (rather than Th1) response, inflammation may be downregulated. With decreased inflammation should come reduced tissue damage, decreased activation of APCs, and reduced exposure of new epitopes. Such a process might contribute to the resolution of the original autoimmune response without provoking any further pathology. This scenario could account for the fact that some AID that appear in acute fashion (such GBS) disappear on their own in a subset of patients.

F. Therapy of AID

In this section, we review both conventional and immunotherapeutic strategies applied to the treatment of AID. Some of these therapies are aimed at alleviating symptoms only, while others are intended to eliminate, either specifically or non-specifically,

the autoreactive lymphocytes or autoantibodies causing the damage. Many of the same immunosuppressive drugs used to inhibit lymphocyte proliferation or function described in earlier chapters are often used, raising the same concerns with respect to long term side effects. Immunotherapeutic approaches to the treatment of AID have thus been developed that resemble the new strategies intended to prevent transplant rejection and allergy as discussed in Chapters 27 and 28, respectively. All have the objective of blocking unwanted immune responses, be they directed toward allogeneic antigens, allergens, or self antigens. As applied to therapy for AID, these strategies can be classified in three groups: (1) those that block the innate immune response, with a view to reducing the inflammation that stimulates APCs and damages host tissues; (2) those that alter the adaptive immune response—that is, block the production of autoantibodies or alter cytokine or chemokine secretion patterns by inducing immune deviation to a less harmful T cell response; and (3) those that establish tolerance to the autoantigen. We also briefly discuss the possibilities of HCT and β-islet cell transplantation as treatments for particular AID.

One major problem with research into experimental therapies for AID is that results obtained in the mouse models used to explore the initial hypotheses are often not translatable to human AID patients. Lack of efficacy, or, even more menacingly, toxicity and exacerbation of AID, have often been reported. For example, treatment with anti-CD40 mAb is highly effective in preventing lupus nephritis in mice, but can cause thromboembolism in human SLE patients. Unfortunately, there just is no substitute for human clinical trials despite their length and expense. Even the results of human clinical trials must be cautiously interpreted, because many human AID have a relapsing/remitting disease pattern. The observed resolution or amelioration of an AID symptom may or may not be the direct result of the candidate treatment. Such hurdles make the development of new therapeutics for AID a slow process.

I. CONVENTIONAL TREATMENTS

i) Anti-inflammatory Agents

Although there is no known cure for the autoimmune arthritis associated with rheumatic diseases such as RA, RF, or SD, or the spinal fusion of AS, various treatments aimed at relieving the inflammation driving the bone and joint symptoms can be applied. Traditionally, such patients are treated with anti-inflammatory agents such as aspirin and ibuprofen. Aspirin is also given to KD, TTP, and APS patients to prevent excessive blood clotting.

The administration of corticosteroids (prednisone or prednisolone), which block the transcription of pro-inflammatory cytokines (TNF, IL-1) and destructive enzymes (collagenase, elastase), can decrease the number of MS relapses, improve skin and serosal inflammation in SLE, ameliorate the muscle weakness of MG or PM, relieve the eczema of IPEX, control bleeding in the lungs caused by GS, and soothe the blistering of PG. Corticosteroids can also be injected directly into an affected joint in cases of RA or AS, or applied as a topical cream for patients with PS. Steroidal anti-inflammatories may also be of help in cases of SS, ITP, and TTP. However, prolonged use of large doses of these drugs can have multiple detrimental side effects. Indeed, corticosteroids are not given to KD patients since they appear to increase heart damage. In recent years, the use of the newer Cox-2 inhibitors described in Chapter 26 has increased because these drugs are effective but gentler on the GI tract. In 2004, however, concerns arose that at least some of these drugs may increase the risk of heart attack. The reader will recall that Cox-2 inhibitors block cyclooxygenase-2, an enzyme involved in the production of the prostaglandins, which are key components of inflammation.

ii) Strategies to Reduce Pathogen Infections

Because of the striking link between RF and streptococcal infections, some RF patients are prescribed a continuous course of prophylactic antibiotics to prevent re-infection and thus RF relapses. In a parallel approach, researchers are working on a vaccine representing a collection of M protein-bearing strains of *Streptococcus* in an effort to prevent the triggering infection. Because these bacteria concentrate in the throat, an oral vaccine that stimulates mucosal production of anti-bacterial IgA antibodies is favored. Antibiotics have also improved the symptoms of some CD patients, at least temporarily. However, despite the evidence for pathogen involvement in AID such as reactive arthritis and AS, antibiotics have been of no value in treating these disorders to date.

iii) Disease-Specific Symptom Relief

Several AID have symptoms that are unique. For example, to reduce the painful muscle spasticity of MS, anti-spasmodic drugs such as diazepam, dantilene, or baclofen are often prescribed. Clinical management of the sicca symptoms of SS relies primarily on drugs that induce moisture in the eyes and mouth. In adults with ITP, a splenectomy may help maintain adequate platelet levels by removing the site where spent and excess platelets are normally destroyed. Compounds currently under development may ameliorate ITP by directly boosting the production of thrombocytes. MG can be effectively treated using cholinesterase inhibitors that cause an accumulation of ACh in the junction and sustain any ACh signaling that is getting through to the muscle cells. Because of their severe blistering, PG sufferers are treated like burn patients and vigorous anti-bacterial and anti-fungal measures are instituted. Intravenous feeding and fluids may be necessary if mouth blistering is severe. Dialysis may be required for GS patients with severe kidney damage. The excessive blood clotting of APS may be abated through administration of anti-coagulants such as warfarin or heparin. Exercise programs are helpful to maintain the mobility, breathing, and posture of AS patients. Physical therapy is encouraged to avoid muscle atrophy in PM, RA, and MS patients, and to restore muscle function in patients recovering from GBS. In some AS and RA cases, surgery to repair joint damage or replace a joint is warranted. Similarly, some CD patients benefit from surgical removal of the affected part of the bowel. *Ileostomy*, a surgical procedure by which a passage is created from the small intestine through the abdominal wall to the exterior of the body, may

be used to divert fecal contents and allow the inflamed bowel to rest.

A novel approach directed at relieving the bone-related symptoms of RA is based on OPGL administration. In Chapter 17 we learned that the TNF-related molecule OPGL (also known as TRANCE or RANKL) is the master regulator of osteoclastogenesis. The receptor for OPGL is RANK, a molecule that is highly expressed on osteoclast progenitors. T cells of RA patients show elevated expression of OPGL, and the bone destruction mediated by abnormal numbers of osteoclasts generated in the wake of this signaling is thought to play a major role in RA. Thus, the decoy receptor OPG that blocks OPGL binding to RANK may be of value in reducing bone loss in RA patients as well as in patients suffering from other rheumatic diseases. Indeed, in a rat model of arthritis characterized by severe bone destruction and joint inflammation, administration of OPG at the onset of AID symptoms did not reduce inflammation but was able to block bone degradation.

While some cases of HT require no therapy, patients with significant symptoms may find they are alleviated by the oral administration of synthetic thyroxine. In contrast, treatment of GD relies on the use of drugs such as beta-blockers that counteract the symptoms of too much thyroxine, and anti-thyroid medications that decrease the ability of the thyroid to produce thyroxine. For both HT and GD, patients with severe goiter or signs of lymphoproliferative disease may undergo either surgery to remove the abnormal thyroid gland, or irradiation with radioactive iodine (which naturally concentrates in the thyroid gland) to kill thyroid cells. The patient then proceeds with life-long use of replacement thyroxine therapy. Treatment to lengthen the eye muscles or reduce the swelling behind the eyes can alleviate the bulging eye symptom of GD.

Phototherapy in the form of controlled exposure to UV-A (long wave) or UV-B (short wave) rays can help clear up more widely spread PS lesions. Systemic medications are used to treat severe PS cases. Psoralen is an orally administered drug extracted from plants that is effective in clearing PS when used in combination with UV-A phototherapy, a protocol often referred to as *PUVA therapy*. Derivatives of Vitamin A, such as Tegison, can slow the growth of abnormal PS keratinocytes. When used in combination with phototherapy, Tegison is helpful for some (but not all) PS patients.

The Raynaud's phenomenon associated with SD can be treated with calcium channel blockers that prevent the blood vessels from contracting too much. However, there is currently no known way to stop the fibroblasts in SD patients from overproducing collagen. Clinical trials are under way to test the efficacy of halofugione, a drug that inhibits the synthesis of type I collagen.

Similarly, there is no simple means of arresting the progression of T1DM once it has advanced beyond a certain point. Traditional treatment focuses on day-to-day management of glucose levels by the patient. Through carefully calibrated injections of insulin to reduce hyperglycemia, and prompt sugar intake to combat hypoglycemia (low blood sugar), the patient strives to achieve normal blood glucose levels at all times. Many patients now wear a small insulin pump with implanted catheters that allows the individual to administer the hormone in small increments without syringes and needles. A potential cure for T1DM is the transplantation of normal (usually allogeneic) β-islets into the patient's liver. This procedure, called the "Edmonton Protocol," was pioneered by James Shapiro in 2000 at the University of Alberta in Edmonton, Canada. To date, 50% of patients treated with this procedure have not required injection of exogenous insulin for over a year (although they have had to take immunosuppressive drugs). While promising, several unsolved problems curtail the usefulness of this approach. Because donated islets are generally obtained from cadavers, the supply is extremely limited. As well, the immunosuppression necessary to control graft rejection and autoimmunity has had serious side effects. Whether the grafted islets will survive in the long term is also unknown. As an alternative, researchers are working on ways to stimulate β-islet cell regeneration in the patient, or to encourage a patient's own tissue stem cells to differentiate into β-islet cells *in vitro* so that they might later be introduced back into the patient without the need for immunosuppression.

We note here a case of disease-specific exacerbation by potential treatment. While platelet infusions are useful for most types of thrombocytopenia because they boost flagging platelet levels, they can increase the detrimental effects of TTP. Because of the clotting abnormalities in these patients, any additional platelets introduced into the patient in the donated blood product may cause additional blood clots in the body's capillaries.

iv) Immunosuppressive Drugs

Many AID, including severe cases of RA, SD, MG, PM, PS, SS, GS, and PG, are treated with anti-proliferative drugs such as methotrexate, azathioprine, or cyclophosphamide (refer to Chs.26 and 27) to non-specifically inhibit the expansion of autoreactive lymphocytes. However, the reader will recall that these drugs are non-specific in their effect, and all rapidly dividing cells, including non-autoreactive lymphocytes, may be affected. Thus, prudent use of these immunosuppressive therapies is required to avoid increasing the risk of life-threatening infections in AID patients. Tacrolimus has proved especially helpful in cases of IPEX, and very promising results have been obtained in RA patients treated with leflunomide. Leflunomide improves patient symptoms without increasing the opportunistic infections, bone marrow toxicity, and mucosal inflammation associated with the use of other nucleotide synthesis inhibitors. Curiously, the same immunosuppressive agents that have been effective in RA treatment do not seem to help AS patients. Similarly, immunosuppressive drugs are only modestly effective in ITP patients, although cyclosporin can be helpful in increasing the platelet count. Some children in whom T1DM has been diagnosed early have been treated with immunosuppressive drugs to reduce the attack on the islets and preserve any remaining function. However, the deleterious side effects of these treatments, particularly on the kidney and liver, are considerable.

v) Strategies to Non-specifically Control Autoantibodies

The reader will recall from Chapter 27 that plasmapheresis is a procedure in which a machine is used to remove antibody proteins, including both anti-pathogen antibodies and alloantibodies, from the blood. The cleansed blood is then returned to the patient. Patients with MS, GBS, MG, ITP, GS, and PG have shown modest clinical improvement after this procedure has been used to remove autoantibodies. Where plasmapheresis has been of the greatest benefit is in cases of TTP: 80% of treated patients now survive, whereas this disease was almost always fatal prior to the advent of plasmapheresis. A type of plasmapheresis that removes granulocytes (key sources of TNF and other cytokines that might be driving keratinocyte proliferation) has been helpful in reducing joint pain in some PsA patients.

Infusion of intravenous immunoglobulin (IV-IG) is useful for the treatment of many AID, including MS, MG, GBS, and subsets of ITP and PM. Indeed, the current standard treatment for children with KD is infusion of high doses of IV-IG to prevent permanent damage to the coronary arteries and the heart. The reader will recall from Chapter 23 that IV-IG is a preparation of antibodies of all specificities that have been pooled from a group of healthy donors. Exactly how IV-IG reduces autoimmune symptoms is unclear, but it is speculated that the heterologous antibody preparation may interfere with autoantibody access to targets, block complement activation, inhibit pro-inflammatory cytokine function and/or production, and/or modulate FcR expression and function. IV-IG also contains antibodies that bind to receptors (including Fas) on T and B cells, and may thus have moderating effects on the activation and differentiation of autoreactive lymphocytes. For many AID patients, relief from symptoms gained via IV-IG therapy may last as long as 2 months.

II. IMMUNOTHERAPEUTICS

The therapies described in this section tend to be of more recent development than those described in the preceding section; indeed, some are still primarily experimental. Most either attempt to block only the aspect of the immune system causing the problem, or use the specificity of the immune system to target the autoreactive lymphocytes (Table 29-10).

i) Strategies to Block the Innate Immune Response

In the past decade, new types of agents have been developed that block the harmful inflammation driving many AID while causing fewer side effects than traditional anti-inflammatory or immunosuppressive drugs. RA patients have been among the first to enjoy the fruits of research into these new biological therapeutics. Blockade of TNF by anti-TNF mAb (infliximab), when used alone or in combination with immunosuppressive drugs, has been of real clinical benefit for the 70% of RA patients and 80% of AS patients that respond to this strategy. Patients experience decreased inflammation in their affected joints and their quality of life improves. Similar results have been achieved for RA using etanercept, a soluble recombinant fusion protein of TNFRII coupled to the Fc region of an Ig.

Etanercept has also been used successfully for AS treatment and to reduce PS flare-ups. There is some evidence that the mechanisms of action of infliximab and etanercept are similar but not identical. Infliximab is effective in CD patients because, as well as neutralizing TNF, this drug is able to bind to lamina propria T cells and increase (by an unknown mechanism) their susceptibility to apoptosis. Infliximab also inhibits GM-CSF production. Since GM-CSF prolongs the survival of neutrophils that have infiltrated into the inflamed mucosae of CD patients, infliximab treatment likely reduces this inflammatory component as well.

Table 29-10 **Immunotherapeutic Strategies**

Strategy	Treatment	Examples of AID Where Applied†
I. BLOCK INNATE IMMUNE RESPONSES		
TNF blockade	Anti-TNF mAb, TNFRII–Ig	RA, AS, PS, CD, MS
IL-1 blockade	IL-1RA (IL-1R antagonist)	RA†
GM-CSF blockade	Anti-GM-CSF mAb, anti-TNF mAb	CIA, RA, CD, EAE (MS?)
ROI blockade	iNOS inhibitor	RA
Chemokine blockade	CCR2 signaling inhibitor	RA
Modulate complement activation	Factor B, C3, C5, C5aR inhibitors	CIA
II. ALTER ADAPTIVE IMMUNE RESPONSES		
Bias response to Th2 Downregulate MHC class II, inhibit MMPs Block DC maturation	IL-4 IFNβ IL-10, TGFβ	EAE, CIA, PS MS PS
Bias response to Th2	Autoreactive T cells transfected with IL-4, IL-10, TGFβ DCs transfected with IL-4	EAE, CIA, T1DM CIA
IL-2 blockade	Anti-CD25 mAb	PV
Interfere with T cell function	Anti-CD3 or anti-CD4 mAbs	RA, T1DM
Induce T cell apoptosis	LFA-1/IgG fusion protein	PS
Kill lymphocytes	Anti-CD52 mAb (CAMPATH-H1)	RA, ITP, MS,† AHA
Alter T cell trafficking	Anti-LFA-1 mAb, anti-VLA-4 mAb	PS, EAE, MS, UC, CD
OX40 co-stimulatory blockade	Anti-OX40 mAb, anti-OX40L mAb, OX40-Ig	EAE, IBD, CIA
Block binding of self-peptide to relevant MHC molecules	Co-polymer of L-alanine, L-lysine, L-tyrosine, L-glutamic acid (Copaxone)	MS
Enhance negative T cell regulation	CTLA-4Ig	RA, PS, SLE (mice)
Kill B cells	Anti-CD20 mAb, anti-CD22 mAb	RA, MG, ITP, AHA
CD40 co-stimulatory blockade	Anti-CD40L mAb Anti-CD40 mAb	SLE, ITP EAE
Reduce B cell survival	Anti-BAFF mAb	SLE, RA†
Block formation of ectopic GCs	Soluble LTβR decoy receptor	T1DM (mice)
LIGHT blockade‡	Soluble LIGHT decoy receptor	UC (mice), IDDM (mice)
Induce cell cycle arrest via p27 upregulation‡	TRAIL	EAE
Accelerate AICD‡	Agonistic anti-4-1BB mAb	EAE, AID in *MRL/lpr* mice
CD70 blockade‡	Anti-CD70 mAb	EAE

Continued

Table 29-10 Immunotherapeutic Strategies—cont'd

Strategy	Treatment	Examples of AID Where Applied[†]
III. ESTABLISH TOLERANCE		
Induce oral tolerance	Feed self antigen	MS, RA[†]
Induce non-oral tolerance	Inject continuous low doses of self antigen	T1DM[†]
DNA vaccination	Inject DNA encoding self antigen	EAE
Anergize or exhaust autoreactive T cells	Tolerogenic peptides (APLs)	M, T1DM
IV. HEMATOPOIETIC CELL TRANPLANTATION		
	Ablate patient's immune system and reconstitute with autologous or allogeneic stem cells	SLE, CD, MS, LA, SD, IPEX, ITP

AHA, autoimmune hemolytic anemia; MMPs, matrix metalloproteinases; ROI, reactive oxygen intermediates; Tg, transgenic.
[†] Little or no success to date.
[‡] Discussed in Box 29-4.

The 30% of RA patients that fail to respond to TNF blockade have only low levels of TNF in their affected joints. It is speculated that their disease is driven by cytokines other than TNF, most likely IL-1 and/or GM-CSF. Kineret (anakinra) is a recombinant version of the physiological IL-1RA protein that antagonizes IL-1R. In general, treatment with kineret is considerably less effective than treatment with therapeutics based on anti-TNF strategies, possibly due to the short half-life (<6 hours) of kineret in the plasma. Blocking GM-CSF with anti-GM-CSF mAbs may be more worthy of consideration as a therapy. The use of a blocking anti-GM-CSF mAb in the CIA mouse model of RA decreases the severity of AID developing in susceptible animals. In cultures of human synovial tissues from RA patients, the co-administration of anti-GM-CSF mAb plus anti-TNF mAb decreases inflammation. Anti-GM-CSF mAbs can also blunt symptoms of AID in the EAE model, suggesting that MS patients might one day benefit from an agent targeting GM-CSF.

One potential problem with long-term TNF blockade therapy is that this treatment may increase the risk of eventually developing cancer or another AID. Chronic exposure to anti-TNF in mice of a susceptible genotype increases T cell responses to antigens in general, including autoantigens. Indeed, anti-inflammatory strategies have been shown to backfire on occasion. For example, in one study of potential MS therapies, agents that antagonized TNF and should have reduced the inflammation driving demyelination actually increased MS relapses and plaque formation. Researchers speculate that the relatively bulky anti-TNF agents used could not cross the blood–brain barrier, and thus could not reach sites in the brain where autoreactivity was ongoing. It should also be borne in mind that, as well as reducing autoreactive responses, anti-TNF strategies compromise host immune responses to pathogens. Increased susceptibility to infection becomes a serious consideration in these patients.

What about interfering with other aspects of the innate response to mitigate autoimmunity? Various groups have tried inhibiting iNOS activity to reduce the presence of damaging ROIs, blocking chemokines that attract lymphocytes, and modulating complement activation. For example, CCR2 expressed by monocytes, macrophages, and T cells allows these cells to follow a gradient of MCP-1 emanating from an inflamed joint in an RA patient. A novel drug that specifically blocks the MCP-1/CCR2 signaling pathway is currently in clinical trials. Complement-mediated damage is another key contributor to RA pathology. Inhibitors that act on Factor B, C3, C5, or the C5a receptor have been tested successfully in the CIA mouse model. Translation of these inhibitors to human treatment is under way.

ii) Strategies to Alter the Adaptive Immune Response
iia) Approaches involving cytokines. As we have learned, subsets of T helper and cytotoxic T cells that secrete distinct sets of cytokines can influence the type of immune response that is mounted (e.g., cell-mediated vs. humoral). Such biases can alter the course of an AID. With respect to an AID predominantly characterized by inflammation, a deviation to a Th2 response can decrease the level of Th1 cytokines, thereby reducing the severity of the disease. For example, NOD mice that overexpress IL-4 do not get diabetes, and CD8$^+$ T cells that are directed against β-islet cell antigens that are adoptively transferred into NOD mice lacking IFNγ lose their ability to invade pancreatic islets. The ability of IL-4 to bias responses away from the Th1 type has also been used to block the development of disease in EAE and CIA mouse models. Some of these effects have been translatable to humans. For example, patients with PS (a Th1-driven disease) who were treated with recombinant IL-4 showed an amelioration of disease symptoms owing to the presence of increased numbers of Th2 cells. On the other hand, for disorders such as SLE, in which autoantibodies are the driving force, researchers have tried to augment the Th1 response in the hopes of minimizing the Th2 response. In theory, autoantibody production might be reduced and the patient's prognosis might improve with time. In practice, this strategy has achieved only mixed results.

Significant success has been achieved for large numbers of MS patients treated with IFNβ, and this treatment is fast becoming the standard of care in many jurisdictions. The original rationale for this approach was that IFNβ would control the unknown virus that was thought to be triggering MS onset. However, it appears now that IFNβ reduces relapse rates by countering the effects of pro-inflammatory cytokines and downregulating expression of MHC class II that promotes presentation of autoantigens. In addition, IFNβ is a powerful inhibitor of the matrix metalloproteinases used by the activated T cells to invade brain tissue. Some researchers have investigated the use of IFNα as a therapy for AID. As was mentioned in Chapter 26 and is described in more detail in Chapter 30, IFNα has been used very successfully for the treatment of some types of leukemia. The mechanism of IFNα action in cancer treatment is not clear, but the antiproliferative properties of this cytokine and/or its ability to inhibit the expression of other cytokines have been implicated. Unfortunately, despite the fact that both leukemia and AID involve proliferating lymphocytes, the use of recombinant IFNα for treatment of AID has not been effective. Researchers speculate that IFNα may promote the differentiation of autoreactive B cells into autoantibody-secreting plasma cells, meaning that it would exacerbate any AID (such as RA and SLE) in which the underlying mechanism was primarily humoral.

Another approach to reduce AID with purified cytokines has been the administration of the immunosuppressive cytokines IL-10 and TGFβ. These cytokines block the maturation and migration of DCs and inhibit the expression of pro-inflammatory cytokines by DCs as well as by other APCs. For example, PS patients who received IL-10 showed suppression of inflammation in the skin and the restoration of normal keratinocyte growth and cytokine/chemokine production. The interval between PS relapses was lengthened and the number of relapses was reduced.

Rather than systemic administration of a purified cytokine, some researchers have experimented with the use of engineered murine T cells to deliver a cytokine capable of either immune deviation or immunosuppression directly to the site of the autoimmune attack. For example, primary CD4[+] T cells specific for the relevant autoantigen are isolated from a mouse with an experimental AID such as EAE. The cells are stably transfected with a construct containing the gene encoding the cytokine of interest (usually IL-4, IL-10, or TGFβ) plus a gene encoding a luminescent marker protein (to follow cell trafficking *in vivo*). When the modified autoreactive T cells are returned to the original animal, they home rapidly and preferentially to the site of autoimmune attack and commence secretion of the transfected cytokine. In situations where the damage has been caused by a Th1 response, the presence in the milieu of copious amounts of this cytokine may tip the balance back to a less damaging Th2 response. For example, the administration of autoantigen-specific CD4[+] T cells modified to express IL-4 can prevent the development of diabetes in NOD mice, and block the development of arthritis in a CIA mouse model. Antigen specificity is the key to the success of this approach, since the modified

CD4[+] T cells able to prevent EAE were ineffective in ameliorating symptoms in CIA mice.

A similar approach has been tried with modified DCs. Immature DCs transfected with a construct driving expression of IL-4 were adoptively transferred into CIA mice. The idea here was to bias the differentiation of autoreactive T cells away from Th1 and towards Th2. Indeed, mice with established CIA who received IL-4-expressing DCs showed suppression of disease symptoms for several months. Moreover, if IL-4-expressing DCs were administered to susceptible mice prior to injection of collagen, the onset of CIA could be prevented.

iib) Approaches targeting autoreactive T cells. Other immunotherapeutic approaches have taken direct aim at autoreactive T cells (keeping in mind that activated, non-autoreactive T cells are often affected as well). Treatment with humanized anti-CD25 mAbs such as daclizumab and basiliximab, which block the effects of IL-2, have proved useful for the treatment of PV. Modified anti-CD3 and anti-CD4 mAbs that block T cell functions but do not deplete the T cell population have shown some clinical efficacy for treatment of RA and T1DM. In the latter case, newly diagnosed patients were treated for 12 days with a modified anti-CD3 mAb called (for the moment) "hOKT3γ1 (ala-ala)." These patients enjoyed a marked clinical benefit in terms of increased insulin production and better control of glucose levels. This mAb is now under investigation as an adjunct treatment for diabetics who have received pancreatic β-islet cell transplants (see later). Other mAbs tried as therapies for AID include a soluble LFA-1/IgG fusion protein (alefacept). This molecule binds to both CD2 receptors on T cells and FcγRs on phagocytes and NK cells, allowing the induction of ADCC of the T cells. In one study of PS patients, administration of alefacept led to clinical improvement in skin lesions that correlated with a reduction in the memory T cell population. Improvement in PS symptoms has also been observed following treatment with an anti-LFA-1 mAb, most likely due to alterations to the trafficking and costimulation of autoreactive T cells.

CAMPATH-1H, the humanized anti-CD52 mAb originally developed for killing lymphoid cancer cells (see Ch.30), has been successful in inducing clinical responses in some AID patients. The reader will recall from Chapter 27 that CAMPATH-1H binds to the GPI-anchored receptor CD52 expressed on the surfaces of most lymphocytes (autoreactive or not), setting them up for ADCC- or complement-mediated destruction. The symptoms of AID as varied as AHA, RA, and ITP have been ameliorated following treatment with CAMPATH-1H. However, it seems that treatment with this mAb alone is not the whole answer. In one study of MS, a group of patients with progressive disease was treated with CAMPATH-1H. Inflammation within the cerebrum was reduced and new plaques stopped forming in the CNS. However, the axons in these patients continued to deteriorate and their disabilities worsened. The results of this trial graphically illustrate that our knowledge both of autoimmune mechanisms and how best to derail them remains incomplete. In addition to these difficulties, CAMPATH-1H treatments have been associated with a prolonged and detrimental depression in circulating T cells.

A novel mAb-based approach towards the elimination of autoreactive T cells focuses on OX40 (CD134), a minor costimulatory molecule transiently upregulated on activated naive and effector T cells only in inflammatory sites. For example, OX40 is upregulated on autoreactive T cells prominent in EAE mice, and is also elevated in T cells isolated from inflammatory sites in human patients with RA, MS, PS, UC, or CD. The advantage of using anti-OX40 mAbs is that, unlike mAbs targeting broadly expressed T cell markers such as CD3 or CD52, an anti-OX40 mAb spares the vast majority of peripheral T cells (except those responding to pathogens, which would also upregulate OX40). Anti-OX40 mAb-based immunotoxins have been investigated in animal models to determine whether the elimination of OX40-expressing cells can ameliorate AID. In a rat EAE model, an anti-OX40 mAb immunotoxin was successful in preventing the onset of EAE in naive animals if it was given on the same day as the adoptive transfer of EAE-causing T cells. If administration of the immunotoxin was delayed until the first signs of paralysis, the disease did not progress further. The number of MBP-specific T cells in the spinal cords of treated rats was also profoundly decreased. In another approach, researchers targeted the interaction of OX40 with its ligand OX40L (CD134L), which is expressed on B cells, activated macrophages, activated endothelial cells, and some activated DCs. Administration of a blocking anti-OX40L mAb or an OX40-Ig fusion protein in animal models of IBD, CIA, and EAE had positive effects if the inhibitory agent was applied fairly early in the AID induction process. Levels of pro-inflammatory cytokines were reduced, and the adhesion of autoreactive T cells to endothelial cells in inflammatory sites (which is partly mediated by OX40/OX40L interaction) was decreased. Unfortunately, however, interference with anti-OX40/OX40L interaction did not induce tolerance, and AID symptoms returned in treated animals shortly after the cessation of anti-OX40L therapy.

A strategy for preventing AID based on blocking T cell access to target tissues has been explored in an EAE mouse model. Activated autoreactive T cells express VLA-4, which allows them to bind to VCAM expressed on activated endothelial cells, including those forming part of the blood–brain barrier. EAE-susceptible mice that were treated with an agent that blocked VLA-4 function showed a decrease in clinical paralysis. Presumably the activated autoreactive T cells could no longer penetrate the blood–brain barrier, preventing further relapses of the disease. Clinical trials of humanized anti-VLA-4 antibodies (natalizumab) have shown promise for the treatment of not only MS but also UC and CD.

A reduction in MS relapse rates has been observed if the drug Copaxone is given after the first MS episode. Copaxone, a random co-polymer of L-alanine, L-lysine, L-tyrosine, and L-glutamic acid, was originally called "co-polymer I" (generic name, glatiramer acetate). Copaxone inhibits autoreactive T cell activation by binding to the HLA-DR proteins associated with the incidence of MS. The binding of the pMHC containing the MBP to the TCRs of autoreactive T cells is thus blocked. There is some evidence that Copaxone may also induce immune deviation to a less damaging Th2 response.

The reader will recall from Chapter 27 that CTLA-4Ig is a soluble agent that can bind to B7 and deliver a negative regulatory signal that damps down T cell activation. Thus, for any disease in which autoreactive T cells are thought to be playing a primary role, the induction of the anergy of autoreactive T cells via CTLA-4Ig may be helpful. Indeed, sustained administration of CTLA-4Ig can prevent the onset of AID symptoms in both NZB/W F1 and *MRL/lpr* mice. In humans, CTLA-4Ig has been found to be of promising clinical benefit to RA and PS sufferers.

A caveat associated with the approaches described here is that, as well as affecting autoreactive T cells, the activation or function of regulatory T cells may also be impeded. Since regulatory T cells normally inhibit autoreactive CD4$^+$ and CD8$^+$ T cells, the AID may be exacerbated rather than mitigated.

iic) Approaches targeting autoreactive B cells. Depletion of autoreactive B cells that would otherwise differentiate into plasma cells producing pathological antibodies has been tried for several AID. Humanized mAbs directed against B cell markers induce the destruction of B cells (both normal and autoreactive) by either complement-mediated cytolysis, ADCC, or induction of apoptosis. Accordingly, the administration of the anti-CD20 mAb rituximab (refer to Ch.26) or an anti-CD22 mAb called epratuzumab has been of clinical benefit for patients with AHA, or refractory RA, MG, or ITP. For example, in one study of rituximab efficacy for refractory RA, all treated patients experienced a reduction in synovial inflammation 6 months after infusion of the mAb. Most patients enjoyed significant clinical benefit for at least 1 year, and side effects (including infections) were mild. While circulating B cells were almost entirely eliminated and RhF levels were profoundly decreased, total serum Ig levels underwent only a small reduction. In another study of RA that examined the effectiveness of rituximab in combination with conventional drugs, methotrexate plus rituximab gave five-fold better results compared to methotrexate treatment alone. Clinical benefits were retained for at least 6 months. Rituximab has shown similar efficacy for ITP treatment. A small group of ITP patients, each of whom had suffered a relapse after a first round of a conventional treatment, received a month-long course of rituximab. Platelet counts were restored to near normal in almost half of the treated patients and stayed that way for at least a month. Most of the remaining patients showed a partial response. Again, only mild side effects were observed, including dizziness, fever, and chills. Increased infection due to possible general impairment of the humoral response did not appear to be a problem.

Another experimental therapy that is being tested for its effects on autoreactive B cells is anti-CD40L mAb treatment. CD40/CD40L engagement is critical for the formation of the ectopic germinal centers that facilitate plasma cell differentiation and thus drive autoantibody production in various AID.

Indeed, SLE patients treated with anti-CD40L mAb show reduced titers of circulating autoantibodies and decreased glomerulonephritis. Anti-CD40L mAbs have also been used to ameliorate symptoms of ITP. Interestingly, and perhaps due to differences in treatment regimens, there was no sign of the thrombosis that accompanied the use of anti-CD40L mAb to induce tolerance during transplantation (described in Ch.27). The opposite tack, treatment with anti-CD40 mAb, has been tried in animals. Administration of this mAb can prevent AID development in a monkey model of EAE.

An approach that is still highly experimental is based on an anti-BAFF mAb that blocks the binding of BAFF to all three of its receptors. In preclinical testing, this mAb was able to block B cell proliferation *in vitro* and to reduce numbers of CD20+ B cells in mice and monkeys. The hope is that anti-BAFF mAb treatment may be of particular benefit to human SLE and RA patients. Approaches involving the BMCA decoy receptor or the agonistic anti-TACI mAb described previously may also be worth pursuing for human therapy.

iii) Strategies to Induce Tolerance

There were originally high hopes for therapeutic approaches based on vaccination with autoantigens. The theory was that exposure to high concentrations of the offending autoantigen (delivered in a benign context or via an unusual route of administration) should anergize or perhaps exhaust the autoreactive lymphocyte clones. However, it seems that approaches that are successful in inducing tolerance to self antigens in rodents (such as the feeding of an antigen to induce oral tolerance) are often ineffective in humans. For example, no efficacy was demonstrated in a group of MS patients given oral MBP, and disappointing results were obtained when oral collagen was administered to a group of RA patients. The tolerizing approach has also been tried as a means of preventing the development of an AID. The Diabetes Prevention Trial in the United States was a large-scale examination of whether small, regular doses of insulin (delivered either orally or by injection) could prevent the development of diabetes proper in healthy individuals ("pre-diabetics") who were relatives of T1DM patients and who already possessed autoantibodies directed against β-islet cell antigens. The hope was that the continuous administration of insulin might induce tolerance to it, and thus reduce β-islet cell destruction. Unfortunately, although this strategy had had encouraging results in rodents, it was reported in 2003 that neither parenteral nor oral administration of insulin could prevent or delay the onset of diabetes in humans at high risk of developing the disease.

DNA vaccination has been also been explored as a possible AID therapy. In one EAE mouse model, paralysis in affected animals was reversed if they were given DNA encoding an MBP peptide plus IL-4. The translation of these animal results to human clinical trials is now under investigation. Other researchers have tried vaccination with purified "altered peptide ligands" (APLs) that are tolerogenic versions of peptides associated with induction of autoimmunity. For example, a group of MS patients was injected with various doses of an MBP-derived APL. Low doses of MBP–APL were associated with modest clinical improvement and a shift away from the damaging Th1 response to a Th2 response. However, for unknown reasons, high doses of MBP–APL caused allergic reactions in some patients. In others, the disease was exacerbated due to the cross-reaction of T cells directed against MBP–APL with native MBP, indicating that the anticipated anergization or exhaustion of the autoreactive T cells had failed. Vaccination with an APL derived from the insulin protein is under investigation for T1DM. When tested in NOD mice, inoculation of this peptide (in IFA) significantly reduced T1DM incidence. Trials are ongoing in humans.

One reason for the discrepancies between the mouse and human results is that, in many mouse model systems, the AID is specifically induced *de novo* at a particular time for a limited duration. One self antigen, or at most a very few self antigens, triggers the autoimmune response. In contrast, in human AID patients, by the time the AID symptoms are clinically important, a very active autoimmune response is fully under way. A cytokine environment is created in which immunomodulation techniques that work in naive animals may have only a limited effect. In addition, epitope spreading has already likely begun such that numerous unknown epitopes are being sequentially exposed. Without knowing what these epitopes are, it is not possible to inject the self antigens from which they are derived (or the DNA encoding them) to induce specific tolerance.

We offer one last note on tolerance induction to treat AID: strategies that promote the differentiation of regulatory T cells capable of responding to autoantigens could theoretically be helpful in inducing tolerance, since these cells produce cytokines and deliver intracellular contacts that shut down autoreactive T cells. However, to date, the reliable induction of these cells in numbers sufficient to be useful remains problematic.

iv) Hematopoietic Cell Transplantation

HCT has been attempted as a last resort for some patients with severe AID. The theory is that if the autoreactive immune system is eliminated (or at least severely repressed) using irradiation or immunosuppressive drugs, a new immune system with normal self-tolerance can be established from the transplanted hematopoietic cells. The initial defect that caused the AID may either not occur or may be bypassed, even in the case of an autologous transplant. Transplantation of autologous stem cells has been tried for several AID, with encouraging results. In one phase I trial of autologous HCT for 18 severely affected SLE patients, 16 of the transplant recipients enjoyed a prolonged respite from AID symptoms (over 18 months) and were able to stop treatment with corticosteroids during this time. Similar success has been achieved for small numbers of patients with refractory CD, MS, RA, SD, or ITP. Allogeneic HCT has been tried in animal models to see if the donor lymphocytes respond to HLA differences on residual autoreactive lymphocytes in the host. The hope is that a "graft vs. autoimmunity" effect may exist akin to the "graft vs. leukemia" (GvL) effect seen during allogeneic HCT treatment of leukemia (refer to Ch.27). Preliminary results suggest that this approach may be useful for treatment of human AID such as IPEX. However, as was true for the transplant patients discussed in Chapter 27, the immunosuppression administered to patients entering HCT trials for AID will have to be monitored very closely to avoid the onset of life-threatening opportunistic infections. In addition, the use of allogeneic HCT raises the specter of graft vs. host disease (GvHD). The long term safety of HCT as a treatment for AID remains to be established.

G. Relationship between AID and Cancer

Patients with AID have an increased risk of developing cancer compared to the general population. While this association between autoimmunity and malignancy has long been recognized, it is still not understood. One of the first connections came out of the 1964 observation by Norman Talal and Joseph Bunim that SS patients had a dramatically increased risk (44-fold) of developing non-Hodgkin's lymphoma. Other AID are also associated with a higher risk of tumor incidence in the affected individual. For example, vasculitis is associated with an increased risk of renal or lung cancer, hairy cell leukemia (see Ch.30), or non-Hodgkin's lymphoma; RA is associated with Hodgkin's lymphoma; SLE, SS, and chronic thyroiditis are associated with non-Hodgkin's lymphoma; SD is associated with breast and lung cancers and non-Hodgkin's lymphoma; and PM is associated with breast, lung, and GI cancers. Some of these cancers may be first detected as long as 20 years after the onset of the AID. Similar findings have been made in animal models of AID; that is, certain strains of mice that are predisposed to developing autoimmunity also have a higher risk of developing cancers. For example, NZB mice frequently succumb to B cell neoplasms resembling human chronic lymphocytic leukemia (see Ch.30).

Patients with various subtypes of the primary immunodeficiency ALPS described in Chapter 24 and mentioned earlier in this chapter furnish one of the best examples of the link between AID and tumorigenesis. These patients not only develop an autoimmune lymphoproliferative disorder and display elevated autoantibody production but also have an increased incidence of tumors, including a 50-fold and 10-fold higher risk of developing Hodgkin's and non-Hodgkin's lymphomas, respectively. The autosomal dominant mutations of Fas, FasL, or caspase-10 in these patients are believed to cause a failure in apoptosis, including the AICD of activated lymphocytes. Thus, in addition to the lack of death of tumorigenic cells, autoreactive T and B cells remain active, resulting in the continuous production of autoantibodies. Curiously, the equivalent single gene defects in inbred mice lead to AID but not malignancy. For example, *MRL/lpr* and *gld* mice, which have defects in Fas and FasL, respectively, develop AID symptoms and a lymphoproliferative disorder but not lymphomas. A similar scenario holds true for *scurfy* mice, which have AID due to

a Foxp3 mutation but do not develop malignancies. (The incidence of tumorigenesis in human IPEX patients has yet to be reported.) The reasons for these interspecies discrepancies remain unclear, but the outbred nature of the human population suggests that genetic background is also important in determining whether malignancy will accompany an AID.

The relationship between autoimmunity and malignancy is bidirectional, because cancer patients often show features of AID months or years after the diagnosis of the tumor. Patients with hematopoietic cell cancers have been variously diagnosed as also having SS, SLE, HT, GD, ITP, or AHA. Moreover, the anti-nuclear antibodies that are a dominant feature of SLE, SS, and PM are found in the serum of many cancer patients. RhF, anti-histone, and anti-phospholipid antibodies may also be present. In rare situations, the production of these autoantibodies may be due to the malignant transformation of an autoreactive B cell. For example, if a B cell with specificity for self-IgG is activated, the cancer patient may show clinical signs of RA.

How could AID lead to cancer? As discussed in Box 26-2, the destructive chronic inflammation associated with many AID may increase the level of carcinogenic ROIs in the microenvironment. These reactive molecules could act on a nearby cell in the early stage of carcinogenesis and push it down the path to malignant conversion. Alternatively, the rapid proliferation of an autoreactive lymphocyte might provide more opportunities for oncogenic mutations that could lead to its transformation and the development of a lymphoma.

How might cancer lead to AID? Antibodies directed against tumor-related antigens are frequently found in the serum of patients with lymphomas or breast, ovarian, lung, or colon tumors. These antibodies include those directed against oncogene products such as c-myc, tumor suppressors such as p53, cell cycle control proteins such as cyclin D1, and the MAGE cancer-testis antigens. As we saw in Chapter 26, while some of these antigens may be true tumor-specific antigens (TSAs), in that they are mutated proteins that are not expressed by normal cells, others are tumor-associated antigens (TAAs) that are not mutated and are immunogenic because they are inappropriately expressed. In these latter cases, the antibodies produced may attack normal tissues as well as the cancer. In addition, the physical disruption associated with the growing tumor may expose new autoantigens in a context of inflammation. The fact that vasculitis often develops in small blood vessels adjacent to a tumor site suggests that the presence of malignant cells can facilitate an autoreactive response. The paraneoplastic syndromes discussed in Chapter 26 are also evidence of a link between cancer and AID. The reader will recall that, in some of these cases, the presence of the tumor appears to spark the development of an AID such as MG or PV. In all of these examples, the cancer itself may be initially quite asymptomatic, but complaints stemming from the AID bring the patient to his/her physician, whereupon a diagnosis of cancer is made.

The therapies used to rid cancer patients of their tumors can sometimes induce the onset of AID after the cessation of the therapy. For example, at a point 2 to 12 months after the completion of chemotherapy, breast cancer patients treated with cocktails containing cyclophosphamide, fluorouracil, and/or doxorubicin occasionally develop the arthritic joints characteristic of RA. RA symptoms have also been reported in patients with ovarian cancer or non-Hodgkin's lymphoma treated with high-dose chemotherapy. Similarly, many patients with testicular cancer who receive bleomycin therapy later show signs of the Raynaud's phenomenon characteristic of SD. In addition to induction, if the cancer patient has pre-existing rheumatic symptoms, a course of chemotherapy may exacerbate them. As noted earlier in this chapter, there also appears to be a connection between the IFN regimens used to treat cancer patients and the later appearance of symptoms of RA, SLE, and HT. The mechanisms underlying the induction of AID in these cases are not clear.

An intriguing interaction at the molecular level has led some researchers to postulate the existence of a direct connection between cancer and AID via the BAFF system. The reader will recall that BAFF is a cytokine promoting the survival and maturation of B cells that have avoided negative selection. BAFF induces this survival via binding to its receptor BAFF-R expressed on immature B cells. BAFF also binds to the TACI and BMCA receptors but signaling induced by these interactions may have a negative modulatory effect. In any case, when BAFF is overexpressed in transgenic mice, autoreactive B cells receive abnormally strong or sustained survival signals and are able to avoid negative selection. The hyperproliferation of autoreactive B cells then appears to cause SLE-like disease. Interestingly, high levels of BAFF can be found in the serum of patients with non-Hodgkin's lymphomas, and the cancerous cells of some leukemias and myelomas show abnormal expression of BAFF and/or BCMA. It may be that this involvement of the BAFF signaling system protects a premalignant cell from undergoing apoptosis, thus contributing to tumorigenesis. Another connection to cancer may lie in the fact that, as mentioned in Chapter 17, a BAFF-related molecule called APRIL can serve as a ligand for BCMA and TACI. APRIL was originally identified as a soluble factor secreted at high levels by many types of tumors. Indeed, administration of APRIL can promote the survival and growth of tumor cells in culture and in immunodeficient mice. However, thus far, APRIL has not been associated with AID. How BAFF, APRIL, AID, and cancer might be interrelated is under investigation.

Some scientists have wondered whether the relationship between autoimmunity and cancer might be exploited to improve disease outcomes. An intriguing hint lies in the fact that the spontaneous regression of cancers such as melanomas and renal cell carcinomas is often associated with the autoimmune manifestation vitiligo (described in Ch.26). Understanding this phenomenon may assist researchers in their intense efforts to find a means of inducing immune responses able to trigger cancer regression.

We have come to the end of our discussion of autoimmunity, hopefully having enlightened the reader as to the nature of these diseases, as well as to the scope of the mysteries surrounding them. We move now to another topic in which there are still many unanswered questions, cancers of hematopoietic cells. In Chapter 26, we outlined tumor biology and described the mechanisms thought to underlie tumorigenesis. In that chapter, the immune system and its component hematopoietic cells were discussed from the point of view of their roles in fighting cancer. In Chapter 30, we examine situations in which hematopoietic cells themselves become the targets of malignant transformation, leading to the development of leukemias, myelomas, and lymphomas.

SUMMARY

While most autoreactive lymphocytes are removed from the immune response repertoire by central tolerance mechanisms, a certain number inevitably escape into the periphery. If the mechanisms of peripheral tolerance cannot adequately control the activity of such clones, autoimmune disease (AID) will result. A large number of AID have been identified that vary widely in their effects on patients, since symptoms depend on the identity and location of self antigen under attack in each case. Underlying the diversity of these diseases are common adaptive and innate immune system defects that occur alone or in concert to deregulate peripheral tolerance. Since many immune response mechanisms subject to dysregulation involve proteins encoded by polymorphic genes, an individual's genotype, particularly the identity of their HLA alleles, may predispose them to autoimmunity. Ultimately, in an individual genetically prone to autoimmunity, the regulatory balance may be tipped toward AID by various triggers such as pathogen infection, toxin exposure, sex hormones, and neuroendocrine influences. Some AID therapies are simply aimed at alleviating disease symptoms, while the objective of more ambitious therapeutic approaches is the control or elimination of either the disease-causing autoreactive lymphocytes or the APCs that lead to their inappropriate activation. Many conventional therapeutics target lymphocytes regardless of specificity, an approach that can lead to generalized immunosuppression and increased vulnerability to infection. In contrast, newer therapeutics are designed to target autoreactive clones more specifically. Both the study of AID mechanisms and the development of effective AID therapies have relied heavily on the use of animal models because the complexities of these diseases cannot be duplicated *in vitro*. Such model systems are either naturally occurring, induced, or genetically engineered.

Selected Reading List

Abu-Shakra M., Buskila D., Ehrenfeld M., Conrad K. and Shoenfeld Y. (2001) Cancer and autoimmunity: autoimmune and rheumatic features in patients with malignancies. *Annals of the Rheumatic Diseases* **60**, 433–440.

Adorini L., Gregori S. and Harrison L. C. (2002) Understanding autoimmune diabetes: insights from mouse models. *Trends in Molecular Medicine* **8**, 31–38.

Albert L. J. and Inman R. D. (1999) Molecular mimicry and autoimmunity. *New England Journal of Medicine* **341**, 2068–2074.

Anderson M. (2002) Autoimmune endocrine disease. *Current Opinion in Immunology* **14**, 760–764.

Andreakos E., Foxwell B., Brennan F., Maini R. and Feldmann M. (2002) Cytokines and anti-cytokine biologicals in autoimmunity: present and future. *Cytokine and Growth Factor Reviews* **13**, 299–313.

Arnout J. and Vermylen J. (2002) Current status and implications of autoimmune antiphospholipid antibodies in relation to thrombotic disease. *Journal of Thrombosis and Haemostasis* **1**, 931–942.

Bach J.-F. (2002) The effect of infections on susceptibility to autoimmune and infectious diseases. *New England Journal of Medicine* **347**, 911–920.

Burt R., Traynor A. E., Craig R. and Marmont A. M. (2003) The promise of hematopoietic stem cell transplantation for autoimmune diseases. *Bone Marrow Transplantation* **31**, 521–524.

Compston A. and Coles A. (2002) Multiple sclerosis. *Lancet* **359**, 1221–1231.

Cooper G. and Stroehla B. (2003) The epidemiology of autoimmune diseases. *Autoimmunity Reviews* **2**, 119–125.

Curotto de Lafaille M. and Lafaille J. (2002) CD4+ regulatory T cells in autoimmunity and allergy. *Current Opinion in Immunology* **14**, 771–778.

Davidson A. and Diamond B. (2001) Autoimmune diseases. *New England Journal of Medicine* **345**, 340–350.

Dedhia H. and DiBartolomeo A. (2002) Rheumatoid arthritis. *Critical Care Clinics* **18**, 841–854, ix.

Doolye M. and Hogan S. (2003) Environmental epidemiology and risk factors for autoimmune disease. *Current Opinion in Rheumatology* **15**, 99–103.

Gavin M. and Rudensky A. (2003) Control of immune homeostasis by naturally arising regulatory CD4+ T cells. *Current Opinion in Immunology* **15**, 690–696.

Granger S. and Ware C. (2001) Turning on LIGHT. *Journal of Clinical Investigation* **108**, 1741–1742.

Green E. and Flavell R. (1999) The initiation of autoimmune diabetes. *Current Opinion in Immunology* **11**, 663–669.

Gregersen P. (2003) Teasing apart the complex genetics of human autoimmunity: lessons from rheumatoid arthritis. *Clinical Immunology* **107**, 1–9.

Hamilton J. (2002) GM-CSF in inflammation and autoimmunity. *Trends in Immunology* **23**, 403–408.

Hansen A., Lipsky P. E. and Dorner T. (2003) New concepts in the pathogenesis of Sjögren syndrome: many questions, fewer answers. *Current Opinion in Rheumatology* **15**, 563–570.

Haustein U.-F. and Anderegg U. (1999) Pathophysiology of scleroderma: an update. *European Academy of Dermatology and Venereology* **11**, 1–8.

Kalman B., Albert R. and Leist T. (2002) Genetics of multiple sclerosis: determinants of autoimmunity and neurodegeneration. *Autoimmunity* **35**, 225–234.

Kazatchkine M. and Kaveri S. (2001) Immunomodulation of autoimmune and inflammatory diseases with intravenous immune globulin. *New England Journal of Medicine* **345**, 747–755.

Klareskog L. and McDevitt H. (1999) Rheumatoid arthritis and its animal models: the role of TNF-alpha and the possible absence of specific immune reactions. *Current Opinion in Immunology* **11**, 657–662.

Kohm A., Fuller K. G. and Miller S. D. (2003) Mimicking the way to autoimmunity: an evolving theory of sequence and structural homology. *Trends in Microbiology* **11**, 101–105.

Lam-Tse W., Lernmark A. and Drexhage H. A. (2002) Animal models of endocrine/organ-specific autoimmune diseases: do they really help us to understand human autoimmunity? *Springer Seminars in Immunopathology* **24**, 297–321.

Liblau R., Wong F. S., Mars L. T. and Santamaria P. (2002) Autoreactive CD8 T cells in organ-specific autoimmunity: emerging targets for therapeutic intervention. *Immunity* **17**, 1–6.

Mackay F. and Ambrose C. (2003) The TNF family members BAFF and APRIL: the growing complexity. *Cytokine and Growth Factor Reviews* **14**, 311–324.

Mackay I. and Burnet F. (1963) "Autoimmune Diseases: Pathogenesis, Chemistry and Therapy." Charles C. Thomas, Springfield, IL.

Millar D., Garza K., Odermatt B., Elford A., Ono N. *et al.* (2003) HSP70 promotes antigen-presenting cell function and converts T-cell tolerance to autoimmunity in vivo. *Nature Medicine* **9**, 1469–1476.

Mok C. and Lau C. (2003) Pathogenesis of systemic lupus erythematosus. *Journal of Clinical Pathology* **56**, 481–490.

Morahan G. and Morel L. (2002) Genetics of autoimmune diseases in humans and in animal models. *Current Opinion in Immunology* **14**, 803–811.

Myers L., Rosloniec E. F., Cremer M. A. and Kang A. H. (1997) Collagen-induced arthritis: an animal model of autoimmunity. *Life Sciences* **61**, 1861–1878.

Nepom G. (2002) Therapy of autoimmune diseases: clinical trials and new biologics. *Current Opinion in Immunology* **14**, 812–815.

Newman B. and Siminovitch K. (2003) Inflammatory bowel disease: Crohn's disease and the success of NODern genetics. *Clinical & Investigative Medicine—Médecine Clinique et Expérimentale* **26**, 303–334.

Onengut-Gumuscu S. and Concannon P. (2002) Mapping genes for autoimmunity in humans: type 1 diabetes as a model. *Immunological Reviews* **190**, 182–194.

Pasare C. and Medzhitov R. (2003) Toll-like receptors: balancing host resistance with immune tolerance. *Current Opinion in Immunology* **15**, 677–682.

Patel D. (2002) B cell-ablative therapy for the treatment of autoimmune diseases. *Arthritis and Rheumatism* **46**, 1984–1985.

Powell A. and Black M. (2001) Epitope spreading: protection from pathogens, but propagation of autoimmunity? *Clinical and Experimental Dermatology* **26**, 427–433.

Raman K. and Mohan C. (2003) Genetic underpinnings of autoimmunity—lessons from studies in arthritis, diabetes, lupus and multiple sclerosis. *Current Opinion in Immunology* **15**, 651–659.

Ramsdell F. and Ziegler S. (2003) Transcription factors and autoimmunity. *Current Opinion in Immunology* **15**, 718–724.

Rose N. and Mackay I., eds. (1998) "The Autoimmune Diseases," 3rd edn. Academic Press, San Diego, CA.

Sanders S. and Harisdangkul V. (2002) Lefluonomide for the treatment of rheumatoid arthritis and autoimmunity. *American Journal of Medical Science* **323**, 190–193.

Sieper J., Braun J,. Rudwaleit M., Boonen A. and Zink A. (2002) Ankylosing spondylitis: an overview. *Annals of Rheumatic Disease* **61** (Suppl. III), iii8–iii18.

Silverman G. and Weisman S. (2003)
Rituximab therapy and autoimmune
disorders: prospects for anti-B cell
therapy. *Arthritis and Rheumatism* **48**,
1484–1492.

Steinman L. (2004) Immune therapy for
autoimmune diseases. *Science* **305**,
212–216.

Stollerman G. (2001) Rheumatic fever. *Clinical
Infectious Diseases* **33**, 806–814.

Strober W., Fuss I. J. and Blumberg R. S. (2002)
The immunology of mucosal models of
inflammation. *Annual Review of Immunology*
20, 495–549.

Tarner I., Slavin A., McBride J., Levicnik A.,
Smith R. *et al.* (2003) Treatment of
autoimmune disease by adoptive cellular
gene therapy. *Annals of the New York
Academy of Sciences* **998**, 512–519.

Torres B. and Johnson H. M. (1998)
Modulation of disease by superantigens.

Current Opinion in Immunology **10**,
465–470.

Turley S. (2002) Dendritic cells: inciting and
inhibiting autoimmunity. *Current Opinion in
Immunology* **14**, 765–770.

Tuscano J., Harris G. and Tedder T. F. (2003)
B lymphocytes contribute to autoimmune
disease pathogenesis: current trends and
clinical implications. *Autoimmunity Reviews*
2, 101–108.

Vincent A. (2002) Unravelling the pathogenesis
of myasthenia gravis. *Nature Reviews.
Immunology* **2**, 797–804.

Vyse T. and Todd J. (1996) Genetic analysis of
autoimmune disease. *Cell* **85**, 311–318.

Wakeland E., Wandstrat A., Liu K. and Morel L.
(1999) Genetic dissection of systemic lupus
erythematosus. *Current Opinion in
Immunology* **11**, 701–707.

Weinberg A. (2002) OX40: targeted
immunotherapy—implications for

tempering autoimmunity and enhancing
vaccines. *Trends in Immunology* **23**,
102–109.

Wildin R., Smyk-Pearson S. and Filipovich A. H.
(2002) Clinical and molecular features of
the immunodysregulation, polyendocrinopathy,
enteropathy, X-linked (IPEX) syndrome.
Journal of Medical Genetics **39**, 537–545.

Wong F. and Janeway C., Jr. (1999) Insulin-
dependent diabetes mellitus and its animal
models. *Current Opinion in Immunology* **11**,
643–647.

Wucherpfennig K. (2001) Mechanisms for the
induction of autoimmunity by infectious
agents. *Journal of Clinical Investigation* **108**,
1097–1103.

Yamada A. (2002) The role of novel
T cell costimulatory pathways in
autoimmunity and transplantation. *Journal
of the American Society of Nephrology* **13**,
559–575.

Hematopoietic Cancers

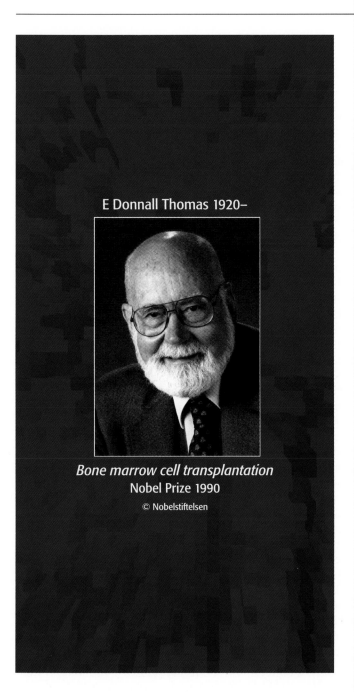

E Donnall Thomas 1920–

Bone marrow cell transplantation
Nobel Prize 1990

© Nobelstiftelsen

CHAPTER 30

A. HISTORICAL NOTES

B. OVERVIEW OF HEMATOPOIETIC CANCER BIOLOGY

C. TERMS USED IN CLINICAL ASSESSMENT AND TREATMENT OF HC

D. LEUKEMIAS

E. PLASMA CELL DYSCRASIAS

F. LYMPHOMAS

"Now this is not the end. It is not even the beginning of the end. But it is, perhaps, the end of the beginning"—Sir Winston Churchill

I n Chapter 29, we saw how an unhealthy immune system can cause autoimmune disease. Immune system cells that have undergone malignant transformation can also fail the body. We have chosen to call these cancers of immune system cells *hematopoietic cancers* (HCs), to distinguish them from the *non-hematopoietic cancers* (NHCs) we discussed in Chapter 26. HCs account for about 8–10% of all cancer diagnoses in the developed world and a similar percentage of cancer deaths. The tumor biology we described in Chapter 26 remains relevant here despite the fact that cancers of the immune system concern hematopoietic cells that are inherently mobile and not fixed like those of body organs. However, there are unique aspects to the biology of HCs that make them a fascinating area of study for clinical oncologists. In this chapter, we discuss the main categories of HCs, how they arise, and their clinical features. Conventional treatments and immunotherapies that have shown promise for the treatment of specific HCs are also described.

A. Historical Notes

Malignancies of the hematopoietic system were first recognized clinically in the mid-1800s. The German physician Rudolf Virchow observed greatly increased numbers of white blood cells in the blood smear of one of his patients, and coined the term *leukemia* (meaning "white blood") to describe this phenomenon. Clinicians of the day subsequently reported that although leukemia was rare, its disease course was often very rapid and severe and caused death within weeks or months of diagnosis. These reports led to the first suspicions that leukemia might be a blood cell cancer. In 1853, Thomas Hodgkin described a solid mass made up of white blood cells that behaved as a tumor in one of his patients. With the identification of similar masses in other patients, *Hodgkin's disease* (HD) became recognized as a clinical entity.

For almost a century, HD was thought to be associated with or caused by a pathogen infection that was drawing inflammatory leukocytes into the affected tissue. This belief persisted until the 1950s, when lymphocytes were discovered. It was then determined that the cells making up the solid mass in HD were indeed lymphocytes, and that HD was a hematopoietic malignancy rather than an infectious disease. A solid mass of malignant lymphocytes became known as a *lymphoma*, and HD is now referred to as *Hodgkin's lymphoma* (HL).

In the mid-20th century, chemical staining techniques and microscopic methods that highlighted morphological differences among cells were used to demonstrate the existence of many subtypes of leukemias and lymphomas. In the 1950s–1960s, the new science of cytogenetics revealed the presence of prominent chromosomal abnormalities (particularly translocations) in leukemic cells. Perhaps the most dramatic story arose out of the observation in 1960 by Peter Nowell and David Hungerford of the Wistar Institute in Philadelphia that tumor cells from most patients with *chronic myelogenous leukemia* (CML) possessed abnormally short versions of chromosome 22. (CML is characterized by the presence of increased numbers of myeloid-like cells in the blood; see later in this chapter.) This truncated chromosome 22, which came to be called the *Philadelphia chromosome*, turned out to be a defining characteristic of this type of leukemia. The hunt was then on to determine where the missing chromosome 22 DNA had gone and how this defect might be related to CML. In the early 1970s, Janet Rowley in Chicago used new staining techniques that highlighted chromosomal band patterns to determine that chromosome 9 in CML patients was increased in length. She proposed that a translocation between chromosome 22 and chromosome 9 was responsible for the appearance of the abnormally long chromosome 9 and the Philadelphia chromosome in CML patients. Technical advances in the 1980s confirmed this hypothesis and identified the specific genes affected by the translocation. Rowley was also among the first to note that most of the cancerous cells in a group of patients with a different type of leukemia involving lymphocyte-like cells had common recurring chromosomal breakage patterns, but that these translocations were different from that seen in CML. It then became possible to correlate specific abnormal chromosomes with particular leukemia and lymphoma subtypes. These findings not only confirmed the clonal nature of HCs but also ultimately helped to reveal the molecular pathogenesis of many of them. As is detailed later in this chapter, this precision of knowledge has spurred research into the development of novel drugs for specific HCs.

B. Overview of Hematopoietic Cancer Biology

I. WHAT ARE HEMATOPOIETIC CANCERS?

Hematopoietic cancers arise from the malignant transformation of hematopoietic cells. For the purposes of this book, our discussion of HCs will focus on the *leukemias, myelomas,* and *lymphomas* (Fig. 30-1). As can be seen from Tables 30-1, -2, -3, -4, and -5, almost any hematopoietic cell type at any stage of development or differentiation can undergo malignant transformation, giving rise to a wide range of cancers. (The definitions of the terms used in these tables are revealed as this chapter unfolds, and the reader is encouraged to refer back to these tables frequently.)

Leukemias are "liquid" tumors that arise from the transformation of a hematopoietic cell in the bone marrow (BM). The cancerous progeny of this cell then make their way into the blood. Leukemias are therefore usually manifested as greatly increased numbers of myeloid, lymphoid, or (more rarely) erythroid lineage cells in the blood and BM. Leukemias represent about 29% of HCs. Myelomas are tumors of fully differentiated plasma cells that are present either as solid masses or as dispersed clones in the BM. Unlike normal plasma cells, which do

not divide after differentiation, myeloma cells continue to proliferate in an uncontrolled way and synthesize large amounts of Ig chains. Myelomas (and related plasma cell tumors) constitute 14% of hematopoietic malignancies. Lymphomas, which represent 57% of HCs, are solid tumors composed of distinct masses of lymphocytes. For reasons that are described later in this chapter, lymphomas are generally classified as either Hodgkin's lymphomas (HLs) or *non-Hodgkin's lymphomas* (NHLs). HLs represent about 12.5% of all lymphomas and are distinguished by the presence of a unique cell type. The vast majority of NHLs are derived from the transformation of B lineage cells but some are derived from T lineage cells. Lymphomas generally arise in a lymphoid tissue such as a lymph node, the thymus, or the spleen but do not usually involve the blood. Certain malignancies, such as Burkitt's lymphoma, exist in both a liquid and solid state and thus behave as both a leukemia and a lymphoma.

As was true for the NHCs described in Chapter 26, the specific cell that undergoes malignant transformation is known as the *target* cell. For HCs, the target cell may be either a developing hematopoietic precursor or a mature cell type. For example, in a *myelogenous* (myeloid) leukemia, the target cell may be a myeloblast precursor, a slightly more differentiated monocyte/macrophage precursor, or a fully differentiated monocyte or macrophage. For the lymphoid leukemias, clinicians use the term *lymphoblastic* to describe disease arising from the transformation of a T or B precursor cell, and *lymphocytic* if the target cell appears to have been a mature peripheral B or T cell. In rare cases, the origin or lineage of the cancer cannot be clearly defined and the specific transformation event remains a mystery. It is then assumed that the target cell is an HSC or a very early descendant that has not yet differentiated into a recognizable lineage-specific precursor. As described in Chapter 26, some investigators believe that all tumors (including NHCs) are derived from cells with HSC-like properties called *cancer stem cells*. The hypothesis is that these rogue stem cells undergo transformation but retain their powers of self-renewal and unlimited proliferation such that tumorigenesis is promoted. In addition, these cells can still differentiate so that the cancerous subpopulations that

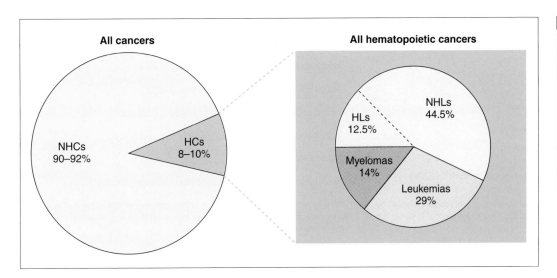

Figure 30-1

Relative Frequencies of Hematopoietic Cancers in North America Hematopoietic cancers (HCs) comprise 8–10% of all cancers. The three major classes of HCs are leukemias, myelomas, and lymphomas. Lymphomas are sub-classified as either Hodgkin's lymphomas (HLs) or non-Hodgkin's lymphomas (NHLs). NHCs, non-hematopoietic cancers.

Table 30-1 Classification of Selected Hematopoietic Cancers: Myeloid Lineage

ACUTE MYELOID LEUKEMIAS (AML)

Acute myeloid leukemia with the t(8;21) (q22;q22) translocation

Acute myeloid leukemia with the t(6;16) (p13;q22) translocation or inv(16)(p13q22)

Acute myeloid leukemia with the t(5;17) (q22;q12) translocation

Acute myeloid leukemia with 11q23 abnormalities

Acute myeloid leukemia with multilineage dysplasia

Acute myeloid leukemia, minimally differentiated

Acute myeloid leukemia, without maturation

Acute myeloid leukemia, with maturation

Acute myelomonocytic leukemia

Acute erythroid leukemia

Acute megakaryoblastic leukemia

Acute basophilic leukemia

CHRONIC MYELOPROLIFERATIVE DISEASES (MPD)

Chronic myelogenous leukemia (CML) with the t(9;22)(q34;q11) translocation

Atypical chronic myelogenous leukemia with the t(9;12)(q34;p13) translocation

Polycythemia vera

Essential thrombocythemia

Chronic idiopathic myelofibrosis

Chronic neutrophilic leukemia

Chronic eosinophilic leukemia

MYELODYSPLASTIC/MYELOPROLIFERATIVE DISEASES

Chronic myelomonocytic leukemia

Juvenile myelomonocytic leukemia

MYELODYSPLASTIC SYNDROMES (MDS)

Refractory anemia

Refractory anemia with ringed sideroblasts*

Refractory cytopenia with multi-lineage dysplasia

Refractory anemia with excess blasts

*Ringed sideroblasts are abnormal erythroid lineage cells in which the nucleus is surrounded by a ring of iron particles.

With information from Harris N. L. *et al.* (2000) Summary of WHO classification of tumors of hematopoietic and lymphoid tissues. *The Hematology Journal* **1**, 53–66. Not all neoplasms are included.

Table 30-2 Classification of Selected Hematopoietic Cancers: B Cell Lineage

PRECURSOR B CELL LYMPHOBLASTIC LEUKEMIAS/LYMPHOMAS

Precursor B lymphoblastic leukemia/lymphoma (B-LBL)

Childhood acute B cell lymphoblastic leukemia (B-ALL)

Adult acute B cell lymphoblastic leukemia

MATURE B CELL LEUKEMIAS

Chronic B lymphocytic leukemia (B-CLL)

Hairy cell leukemia (HCL)

B cell prolymphocytic leukemia (B-PLL)

PLASMA CELL DYSCRASIAS

Monoclonal gammopathy of undetermined significance (MGUS)

Waldenström's macroglobulinemia (WM)

Myeloma

Solitary plasmacytoma of bone

Extramedullary plasmacytoma

NON-HODGKIN'S LYMPHOMAS (NHL)

Mantle cell lymphoma (MCL)

B cell chronic lymphocytic lymphoma (BCLL)

Follicular lymphoma (FL)

Mucosa-associated lymphoid tissue lymphoma (MALT-L)

Diffuse large cell lymphoma (DLCL)
Mediastinal large B cell lymphoma
Intravascular large B cell lymphoma

Burkitt's lymphoma (BL)

With information from Harris N. L. *et al.* (2000) Summary of WHO classification of tumors of hematopoietic and lymphoid tissues. *The Hematology Journal* **1**, 53–66. Not all neoplasms are included.

Table 30-3 Classification of Hodgkin's Lymphomas

CLASSICAL HL

Nodular sclerosis Hodgkin's lymphoma (NSHL)

Mixed cellularity Hodgkin's lymphoma (MCHL)

Lymphocyte-depleted Hodgkin's lymphoma (LDHL)

Lymphocyte-rich classical Hodgkin's lymphoma (LRCHL)

NON-CLASSICAL HL

Nodular lymphocyte predominant Hodgkin's lymphoma (NLPHL)

With information from Harris N. L. *et al.* (2000) Summary of WHO classification of tumors of hematopoietic and lymphoid tissues. *The Hematology Journal* **1**, 53–66. Not all neoplasms are included.

Table 30-4 **Classification of Selected Hematopoietic Cancers: T Cell Lineage**

PRECURSOR T CELL LYMPHOBLASTIC LEUKEMIA/LYMPHOMAS

Precursor T cell acute lymphoblastic leukemia

Precursor T cell lymphoblastic leukemia/lymphoma (T-LBL)

Childhood acute T cell lymphoblastic leukemia (T-ALL)

Adult acute T cell lymphoblastic leukemia

MATURE T CELL LEUKEMIAS

T cell prolymphocytic leukemia (T-PLL)

Large granular lymphocytic leukemia of T cell origin (T-LGLL)

MATURE T CELL LYMPHOMAS

Peripheral T cell lymphoma, unspecified

Adult T cell leukemia/lymphoma (ATL)

Anaplastic large cell lymphomas (ALCL)

Angioimmunoblastic T cell lymphoma (AITL)

Mycosis fungoides

Sezary syndrome

Hepatosplenic T cell lymphoma

With information from Harris N. L. *et al.* (2000) Summary of WHO classification of tumors of hematopoietic and lymphoid tissues. *The Hematology Journal* **1**, 53–66. Not all neoplasms are included.

Table 30-5 **Classification of Selected Hematopoietic Cancers: Other Lineages**

Large granular lymphocytic leukemia of NK origin (NK-LGLL)

Aggressive NK cell leukemia

Macrophage/histiocytic sarcoma

Langerhans cell sarcoma

FDC sarcoma

DC sarcoma

Mast cell leukemia

Mast cell sarcoma

Cutaneous mastocytosis

Systemic mastocytosis

With information from Harris N. L. *et al.* (2000) Summary of WHO classification of tumors of hematopoietic and lymphoid tissues. *The Hematology Journal* **1**, 53–66. Not all neoplasms are included.

Thus, in a cancerous B lineage cell, a lack of somatic mutations in its Ig genes may indicate that the transformation event likely occurred at or before the centroblast stage. Conversely, the presence of somatic mutations in the Ig genes suggests that the cell reached the more advanced centrocyte stage before transformation took place.

II. HEMATOPOIETIC CANCER CARCINOGENESIS

The development of a leukemia is called *leukemogenesis*, while that of a myeloma is called *myelomagenesis*, and that of a lymphoma is called *lymphomagenesis*. Like NHCs, HCs arise when a target cell accumulates genetic changes in oncogenes and tumor suppressor genes. While about half of HCs show relatively subtle genetic aberrations such as small chromosomal deletions or point mutations, an estimated 50% of leukemias and lymphomas are associated with major chromosomal disruptions. These disruptions frequently take the form of *recurring chromosomal translocations* (i.e., the same translocation appears in many different patients with the same type of leukemia or lymphoma).

Recurring chromosomal translocations involve the abnormal exchange of genetic material between two different chromosomes. Not surprisingly, recurring translocations associated with HC development affect chromosomes in regions where genes regulating cellular growth, differentiation, and apoptosis are located. Mechanisms designed to preserve genetic stability in the cell may be compromised, opening the door to mutations that accumulate and promote malignant transformation as described in Chapter 26. Sometimes a reciprocal translocation results in a remote effect that alters the function and/or regulation of a transcription factor or enzyme required for growth control. In other cases, the exchanged chromosomal fragments come together in such a way that a new genetic entity (which may be a functional gene) is created (Fig. 30-2). Occasionally, the product of such an entity is a fusion protein with novel functions that disrupt the control of cellular homeostasis.

We remind the reader here that geneticists use a special form of notation to describe translocations. For example,

are generated may represent a range of more mature phenotypes. Some of the first evidence supporting the cancer stem cell hypothesis was garnered from studies of *acute myeloid leukemia* (AML). It was shown in these experiments that only a small minority of cells in the leukemic population was capable of establishing new tumors when introduced into in cancer-prone mice, and that these cells could give rise to the range of more differentiated cell types found in the original leukemia. Scientists continue to debate whether mature cell types are transformed themselves, or whether a cancer stem cell has differentiated into mature cancerous progeny.

Within an HC of a particular lineage, a malignancy may be distinguished by features that are characteristic of a specific stage of differentiation. Consider malignancies of mature B cells. The reader will recall from Chapter 9 that the primary follicles of the secondary lymphoid tissues are filled with naive B cells that are able to come into contact with both antigen and T helper cells in this location. These B cells are often called "pre-GC cells." When an activated B cell commences proliferation, it migrates to the germinal center (GC) of the follicle and begins to generate centroblasts. These centroblasts then mature to become centrocytes, which eventually exit the GC to become either memory B cells or plasma cells producing specific antibody. These B cells are sometimes called "post-GC cells." It is not until the centrocyte stage that somatic hypermutation and affinity maturation take place.

Figure 30-2

Reciprocal Chromosomal Translocation
In this example, normal chromosomes 1 and 2 (A; one chromosome of each pair shown at metaphase) sustain breaks in their DNA (B) that allow the reciprocal transfer of genetic material from chromosome 2 to chromosome 1 and vice versa (C). The resulting translocated chromosomes (D) may contain existing genes in new positions or new genes composed of a fusion of sequences from both chromosomes. In (E), the translocation positions a hypothetical transcription factor originally on chromosome 1 downstream of a hypothetical enhancer normally situated on chromosome 2. The translocation may also result in fusion of these sequences. Deregulated expression of the transcription factor induced by the enhancer may help to drive tumorigenesis.

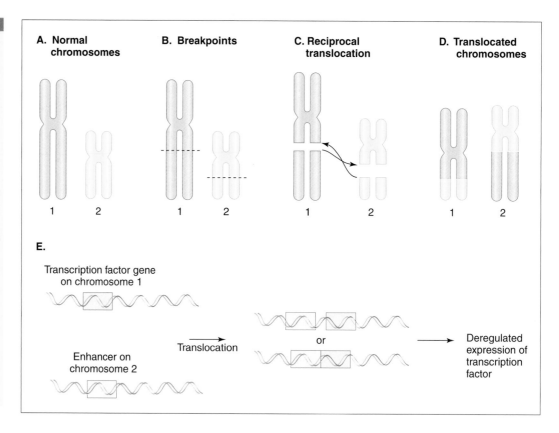

A. Normal chromosomes   B. Breakpoints   C. Reciprocal translocation   D. Translocated chromosomes

E.

Transcription factor gene on chromosome 1

Enhancer on chromosome 2

Translocation

or

Deregulated expression of transcription factor

---

consider the designation t(12;22)(p13;q11). The "t" indicates "translocation," while the "12" and "22" in the first set of parentheses refer to the two chromosomes involved in the exchange. By convention, the chromosome with the lower number is listed first in the first set of parentheses. In the second set of parentheses, p13 indicates that the break in chromosome 12 took place in the 1st block, 3rd band (originally defined by staining methods) of the short ("p," for the French "petit," meaning small) arm of the chromosome. Similarly, q11 indicates that the break in chromosome 22 took place in the 1st block, 1st band of the long ("q") arm. These breakpoints are sometimes expressed in shorthand as 12p13 and 22q11. The gene whose disruption does <u>not</u> appear to be primarily responsible for the development of the HC is often called the *partner gene*. A summary of recurring translocations leading to HCs discussed in this chapter appears in Table 30-6.

In addition to translocations, exposure to environmental carcinogens, radiation, or chemicals can promote HC development (just as was true for NHCs). For example, both cigarette smoking and exposure to industrial solvents such as benzene have been documented to increase the risk of leukemogenesis in later life. A much higher incidence of leukemia has also been recorded in populations living close to the sites of the atomic bomb explosions in Japan in 1945 and the nuclear reactor meltdown in Chernobyl in 1986. Sadly, leukemias can develop as secondary tumors following intensive radiation or chemotherapy applied as a treatment for a primary cancer (which could be an NHC or another HC), or in preparation for a bone marrow transplant (BMT) or a hematopoietic cell transplant (HCT).

C. Terms Used in Clinical Assessment and Treatment of HC

At this point, the definition of a few terms related to HC clinical assessment and treatment may be helpful for the reader. Just like the NHCs described in Chapter 26, HCs are often categorized by their grade and whether they cause *acute* or *chronic disease*. *High grade* HCs are generally composed of transformed <u>precursor</u> cells of a particular hematopoietic lineage. These malignant cells grow aggressively and cause acute disease that kills rapidly. The median survival of untreated patients with high grade HC is less than 2 years, with many patients succumbing within months. *Medium* and *low grade* HCs feature cells that are more differentiated or even mature, allowing median survival times (in the absence of treatment) of 2–5 years and greater than 5 years, respectively. These grades of HC cause chronic disease that is characterized by a slower course and symptoms that may appear to come and go.

Standard chemotherapy and radiation protocols (refer to Ch.26) are the first modes of treatment used for newly diagnosed HC patients. The response to such treatment is assessed by examining smears of cells taken from the patient's blood and/or BM. Malignant hematopoietic cells often resemble the blast-like (immature) stages of hematopoietic cells. For example, in an untreated patient suffering from an acute leukemia, over 20% of the cells observed in a blood or BM smear may be blast-like cells (Plate 30-1). In a healthy individual, such blast-like cells account for 5% of cells developing

Table 30-6 **Examples of Translocations Leading to HC**

Translocation	Genes Involved		Translocation	Genes Involved	
AML			**CLL**		
t(8;21)(q22;q22)	ETO	CBF2A	t(14;19)(q32;q13)	Igh	Bcl-3
t(16;16)(p13;q22)	MYH11	CBFB	t(14;18)(q32;q21) (rare)	Igh	Bcl-2
t(3;21)(q26;q22)	EVI-1	CBF2A	**B-PLL**		
t(12;22)(p13;q11)	TEL	MN1	t(11;14)(q13;32)	Bcl-1	Igh
t(4;11)(q21;q23)	AF4	MLL	**T-PLL**		
t(11;19)(q23;p13.3)	MLL	ENL	t(X;14)(q28;q11)	MAGE	TCRαδ
t(11;19)(q23;p13.1)	MLL	ELL	**MYELOMA**		
t(1;11)(q21;q23)	AF1q	MLL	t(11;14)(q13;q32)	Cyclin D1	Igh
t(11;22)(q23;q13)	MLL	P300	t(6;14)(p21;q32)	Cyclin D3	Igh
t(9;11)(q22;q23)	AF9	MLL	t(4;14)(p16;q32)	FGFR3	Igh
APL			t(14;16)(q32;q23)	Igh	c-maf
t(15;17)(q22;q21)	PML	RARα	t(6;14)(p25;q32)	IRF-4	Igh
t(11;17)(q23;q21)	PLZF	RARα	**NHL-MCL**		
CML			t(11;14)(q13;q32)	Bcl-1	Igh
t(9;22)(q34;q11)	Abl	Bcr	**NHL-FL**		
t(9;12)(q34;p13)	Abl	TEL	t(14;18)(q32;q21)	Igh	Bcl-2
t(3;21)(q26;q22)	EVI-1	CFB2A	**NHL-MALT-L**		
B-ALL			t(11;18)(q21;q21)	IAP-2	MALT1
t(2;8)(p12;q24)	c-myc	Igk	t(1;14)(p22;q32)	Bcl-10	Igh
t(8;14)(q24;q32)	c-myc	Igh	**NHL-DLCL**		
t(8;22)(q24;q11)	c-myc	Igl	t(3;14)(q26;q32)	Bcl-6	Igh
t(9;11)(q22;q23)	AF9	MLL	t(3;22)(q26;q11)	Bcl-6	MN1
t(9;22)(q34;q11)	Abl	Bcr	**NHL-BL**		
t(11;19)(q23;p13)	MLL	ENL	t(8;14)(q24;q32)	c-myc	Igh
t(12;21)(p13;q22)	TEL	CBF2A	t(2;8)(p11–12;q24)	Igk	c-myc
t(4;11)(q21;q23)	AF4	MLL	t(8;22)(q24;q11)	c-myc	Igl
t(1;19)(q23;p13)	PBX1	E2A	**NHL-ALCL**		
t(17;19)(q22;p13)	HLF	E2A	t(2;5)(p23;q35)	ALK	nucleophosmin
T-ALL					
t(1;7)(p32;q35)	TAL1	TCRβ			
t(11;14)(p13;q11)	LMO2	TCRαδ			
t(7;10)(q35;q24)	TCRβ	Hox11			
t(8;14)(q24;q11)	c-myc	TCRαδ			

in BM but they are virtually absent from the blood, so that the presence of cells of this morphology in the blood is a sign of cancer. After a HC patient has been treated, if more than 20% of cells in a blood or BM smear are still blast-like, the treatment has failed and there has been *no clinical response*. A *partial response* to treatment means that between 5 and 20% of cells in the blood or BM smear are blast-like. A *complete response* (CR) means that the number of blast-like cells in the patient's post-treatment blood smear has been reduced to almost zero, or that ≤5% blast-like cells remain in his or her BM smear. If these levels hold over a period of at least 4 weeks following treatment, the patient's disease is said to be

in remission. Remission can be induced either by a natural waning of the HC or by a treatment-induced CR. A *relapse* is the reappearance of disease in a patient who was previously in remission. A patient who has not suffered a relapse of the same tumor for a period of 5 years is said to be *cured, clinically disease-free*, or a *long term survivor*. Nevertheless, the "cured" patient may be harboring *residual disease* in the form of a few remaining malignant cells. These cells are usually identified at the genetic level either by karyotyping or by PCR detection of abnormal transcripts. Presumably the immunosurveillance mechanisms discussed in Chapter 26 are operating sufficiently well in these patients to keep these few

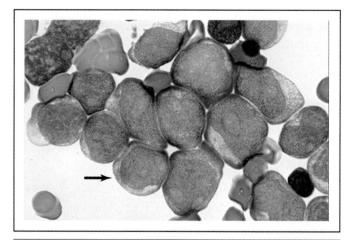

Bone Marrow Aspirate Showing the Diagnosis of Hematopoietic Cancer
In a bone marrow sample of a patient with acute leukemia, far more than 5% of cells are blast-like (arrow). Reproduced with permission from Tkachuk D. C., Hirschmann J. V. and McArthur J. R. (2002) "Atlas of Clinical Hematology." W.B. Saunders Company, Philadelphia.

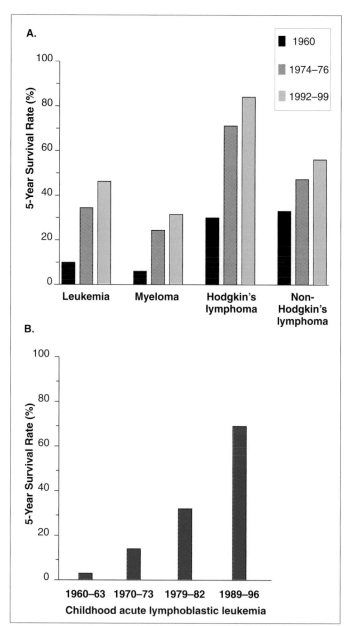

Five-Year Survival Rates for Hematopoietic Cancer Patients from 1960–1999
(A) The percentage of patients with leukemia, myeloma, HL, or NHL that survive for at least 5 years has steadily increased over the past 40 years due to improved therapies. (B) Particularly spectacular progress has been achieved for childhood acute lymphoblastic leukemia. Adapted from Surveillance, Epidemiology and End Results (SEER) Program 1975–2000 and 1973–1996, National Cancer Institute: Myeloma Biology and Management, 2nd edn. Oxford University Press, 1998.

malignant cells under control. Patients whose cancer resists treatment with conventional chemotherapy or irradiation are said to be *refractory* to treatment. *Salvage therapy* is then attempted on these individuals, often employing experimental treatment regimens in a last-ditch effort to defeat their disease by novel means.

The survival of HC patients has improved greatly over the past four decades (Fig. 30-3). This happy result can be attributed to a combination of aggressive chemotherapy and radiation treatments and highly accurate *immunophenotyping* methods. Immunophenotyping involves making microscope slide smears or flow cytometric preparations of cells from BM and peripheral blood and staining them with labeled specific antibodies to define marker expression. This type of molecular analysis allows an oncologist to more precisely identify the cell type that originally underwent transformation. Another tool of the modern HC diagnostic trade is *fluorescence in situ hybridization* (FISH), used for the examination of tumorigenic chromosomal translocations (see Box 30-1). One or more fluorescently tagged probes are applied to whole cells or a chromosomal spread on a microscope slide to visualize one or more whole chromosomes or a specific region of a single chromosome. Major aberrations become clearly visible, pointing the observer to a probable diagnosis. In the 2000s, RNA profiling of gene expression using high-density DNA "chip" microarrays has increasingly been used to definitively classify malignant subtypes. All these techniques help the physician to arrive at a more accurate diagnosis of the HC, allowing a tailoring of treatment that is appropriate to an individual case. More precise therapy may increase the chance of remission and prolong survival. At the basic science level, these types of in-depth studies of HC tumor cells have generated insights that have furthered our understanding of the etiology of cancers in general.

Hematopoietic cell transplants (refer to Ch.27) offer hope to HC patients because they provide a cure in many cases. For an autologous HCT, the patient undergoes an initial round of non-myeloablative chemotherapy that often (but not always) leads to the first remission. The patient is also treated with G-CSF to mobilize HSCs into the peripheral blood. The peripheral blood is collected from the treated patient, and leukocytes (including HSC) are purified by leukapheresis. To further reduce the number of tumor cells in the harvest, the

Box 30-1. **FISHing for answers: Detection of chromosomal aberrations**

Many HCs are associated with particular chromosomal translocations, insertions, or deletions. The identification of these anomalies in the cells from a patient's blood or BM smear allows the oncologist to give a more accurate diagnosis and prognosis to the patient, and thus facilitates the optimization of a corresponding treatment regimen. For these reasons, cytogenetic analysis is an integral part of the modern management of hematopoietic malignancies. A technique that was first introduced in the 1980s and went on to revolutionize this critical area of clinical work is *fluorescence in situ hybridization* (FISH). In this approach, an easily detected fluorochrome is incorporated into a DNA sequence of interest to make a specific probe. This probe is incubated with denatured patient DNA and given a chance to hybridize. After a washing step to remove unhybridized probe, the pattern of fluorescence is inspected to identify large or small chromosomal aberrations.

Prior to the advent of FISH, chromosome visualization relied on chemical staining techniques. However, these agents bind only to metaphase chromosomes fixed on a microscope slide, meaning that only dividing cells can be analyzed. In addition, the DNA has to be isolated from high quality cell samples; that is, samples in which the cells are capable of dividing in culture for 24 hours so that numerous metaphase cells are generated. Such constraints made the chemical analysis of chromosomal aberrations difficult in HC cases because cell samples obtained from these patients are often of poor quality (do not proliferate well in culture). The transforming mutation itself may be to blame, or the chemotherapeutic drugs used to treat the patient (which by definition kill dividing cells). With FISH, sample quality is not an issue because the DNA can be obtained not only from dividing cells but also from resting, interphase cells found in BM, peripheral blood, or lymph nodes. No *in vitro* growth or lysis of cells is necessary. Information is thus gleaned from all cells in a sample (increasing sensitivity) rather than just from the small fraction that happens to be in metaphase at any given time.

While chemical staining methods are adequate for detecting gross chromosomal abnormalities, they do not allow specific localization and identification of aberrant genetic sequences. In contrast, FISH reveals the number, size, and location of particular DNA segments at the single cell level. By combining FISH with immunophenotyping, correlations between chromosomal abnormalities and cell type or lineage can be established. FISH allows the rapid screening of large numbers of samples, aiding the busy clinic, and helps the clinician to quantitate the number of malignant cells in a given patient. Such information is critical in tracking disease progression and identifying residual malignant cells after treatment.

The fluorochromes most commonly used to label DNA probes for FISH include fluorescein (green), tetramethylrhodamine (red), and aminomethylcoumarine (blue). In some cases, multi-color FISH using two or more probes labeled with different fluorochromes can simultaneously detect different genetic sequences within one cell or one chromosome. Compared to the radioactive labels that were originally used to visualize chromosomes in hybridization assays, fluorochromes offer the same advantages exploited by scientists performing immunoassays with labeled antibodies (refer to Ch.7). Fluorochromes are much safer and more stable than radioisotopes, and much less time is required to generate results because no film exposure is necessary. The binding of a fluorochrome-labeled probe to the sample DNA can be detected almost immediately by fluorescence microscopy. Not surprisingly, FISH continues to improve as advances are made in microscope optics and computerized image processing.

Like immunoassay techniques, FISH may involve either direct or indirect detection methods. In direct methods, the probe is tagged with the fluorochrome and binding to DNA is detected immediately after the hybridization step. In an indirect assay, an untagged probe is used for the initial hybridization step followed by detection of this binding by a fluorochrome-tagged agent that binds to the untagged probe. In one popular indirect FISH method, the patient DNA is hybridized to a biotinylated probe that is then visualized using fluorochrome-labeled avidin or streptavidin. Another common approach is to employ dioxigenin-bound probes followed by detection with fluorochrome-conjugated anti-dioxigenin antibodies.

Three different types of DNA probes are commonly used in FISH (see figure). *Repeated sequence* or *centromeric probes* target highly repeated sequences characteristic of the centromere of a given chromosome. While these probes provide a simple, rapid method for detecting abnormal numbers of a given chromosome, they cannot detect genetic aberrations in non-centromeric DNA. Thus, in tumor cell analysis, centromeric probes are most often used to assess *aneuploidy* (incorrect number of chromosomes) in interphase nuclei. The second class of FISH probes is comprised of *whole chromosome probes* (WCP) or chromosome "paints." Each WCP is a mixture of sequences derived from the entire length of the chromosome of interest. A WCP cannot detect deletions, duplications, or inversions of short genetic sequences but is useful for identifying chromosomes affected by large additions or deletions in metaphase nuclei. Members of the third class of FISH probes are known as either *locus-specific probes, unique sequence probes,* or *single copy probes* (SCP). An SCP is complementary to a genetic sequence that is present in the genome in only one copy. SCPs allow the detection of translocations, deletions, and inversions regardless of size. These probes can be used in both interphase and metaphase nuclei.

FISH is now used for the clinical diagnosis and management of many HCs. In one of its first clinical applications, FISH was used to detect the Philadelphia chromosome in patients with chronic myelogenous leukemia (CML) (see plate). These days, FISH routinely and reliably diagnoses the presence of the t(9;22)(q34;q11) translocation responsible for CML in interphase cells from patient BM smears. In fact, in one study, when FISH was applied to samples from CML patients previously diagnosed as Ph-negative by traditional cytogenetic staining techniques, 5% were found to be Ph-positive by FISH analysis. FISH analysis has also distinguished between CML patients with and without a detectable deletion at the breakpoint site in chromosome 9. In those with the deletion, median survival time is 36 months and response to IFNα therapy is poor, while those with the deletion benefit from IFNα therapy and have a median survival time of more than 90 months.

FISH can also help basic researchers understand mechanisms underlying transformation. For example, the third most

Continued

frequent site of rearrangement associated with certain types of lymphomas is 3q26, the locus containing the Bcl-6 gene. Most translocations resulting in alterations in 3q26 are too minor to be detected by standard techniques but are easily visualized with FISH. At least 33 different partner sites participate in translocations affecting Bcl-6, leading scientists to conclude that it is alterations in Bcl-6 (rather than the partner genes) that are important for development of these lymphomas.

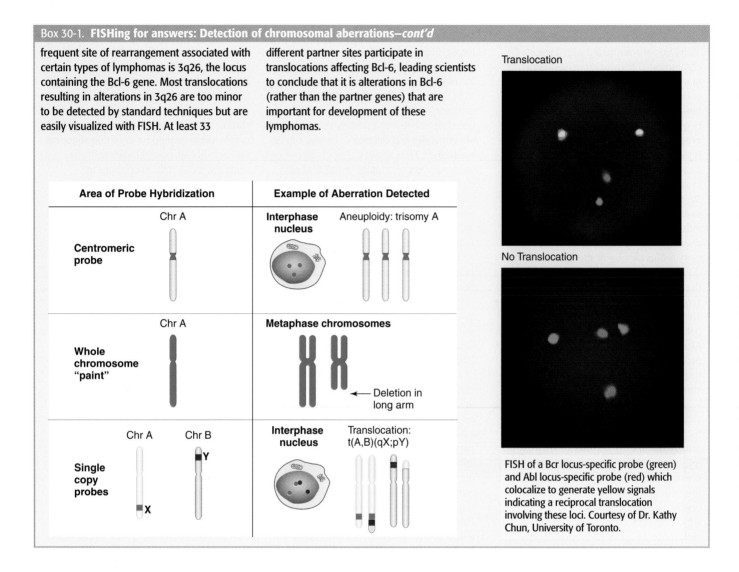

FISH of a Bcr locus-specific probe (green) and Abl locus-specific probe (red) which colocalize to generate yellow signals indicating a reciprocal translocation involving these loci. Courtesy of Dr. Kathy Chun, University of Toronto.

CD34$^+$ HSC population may be isolated *in vitro* by passage over anti-CD34 affinity columns (most tumor cells are CD34$^-$). Following a round of myeloablative chemotherapy to destroy the patient's bone marrow, the autologous stem cell preparation is infused back into the patient to reconstitute a hopefully cancer-free immune system.

An HCT or BMT using the peripheral blood or BM of an HLA-matched donor can also serve as a source of stem cells. These types of transplants are considered "allogeneic," since matching is carried out for HLA markers but allogeneic differences still exist for the minor histocompatibility loci. Because of these differences, the stem cell preparation infused into the BM-ablated patient contains immunocompetent donor lymphocytes that can help destroy any residual cancer cells (the *graft versus leukemia* or *graft versus lymphoma* effect). As a result, relapse rates are lower in patients who receive an allogeneic HCT compared to those who receive an autologous HCT. However, because the donor lymphocytes also attack non-cancerous cells in the recipient, the trade-off is a higher chance (up to 40%) of GvHD, which itself can be fatal. Allogeneic HCT is thus not considered a routine treat-

ment for HCs outside of specialized cancer care centers, and the balance of risk factors has to be carefully evaluated for each patient. A current major goal for onco-immunologists is to design an allogeneic HCT strategy that minimizes the relapse rate without increasing GvHD.

Progress in treating hematopoietic malignancies has also come from the ever-increasing use of *biologicals*. As defined in Chapter 26, biologicals are natural or engineered proteins that a clinician applies to disrupt the proliferation, metabolism, or function of tumor cells. For example, IFNα (which has potent anti-proliferative activity) is now routinely added to standard treatment regimens for several leukemia and lymphoma subtypes. In addition, monoclonal antibodies (mAbs) directed against molecules expressed predominantly on the surfaces of tumor cells are now used on a regular basis for the treatment of certain HCs. For example, the tumor cells of many B cell lymphomas and leukemias show elevated expression of CD19, CD20, and/or CD22, so that mAbs directed against these molecules (particularly the anti-CD20 mAb rituximab) can be used to induce death by either complement activation or ADCC-mediated destruction. Conjugation of these antibodies to toxins to form immunotoxins (refer to Ch.26) can also be used to kill cells bearing the appropriate markers. Similarly, diphtheria toxin has been conjugated to IL-2 for the destruction of IL-2R-bearing malignant T cells in cases of T cell lymphoma. The radiolabeling of mAb proteins with iodine-131 or yttrium-90 (refer to Ch.26) has also been explored for HCs, and has proved helpful in treating lymphomas that have resisted elimination by chemotherapy or standard irradiation protocols. An advantage of these lymphocyte-specific therapies is that relatively modest side effects are observed *in vivo*, at least in comparison to chemotherapy or radiation treatment.

With our increased knowledge of the molecular pathogenesis of hematopoietic diseases, other novel treatments are being contemplated. For example, it was discovered in the late 1990s that most cases of HL show deregulation of IL-13 signaling (see later). It is hoped that a means of blocking this cytokine can be developed that will ameliorate the disease in refractory patients. The reader should note, however, that the long term effects of the use of biologicals are still unknown, and diligent follow-up of patients treated with these agents is essential.

We will now discuss the clinical features, underlying molecular defects, and treatment of several important leukemias, myelomas, and lymphomas.

D. Leukemias

Leukemias are broadly classified by whether they are acute or chronic in onset (Fig. 30-4). Acute leukemias can strike both children and adults, while the chronic leukemias tend to arise in individuals 50–60 years of age. The overall incidence of leukemia is relatively low compared to NHCs and accounts for only 3% of all malignancies. Nevertheless, acute leukemia is the leading fatal cancer in persons under 20 years of age. The transformed cells in acute leukemias usually resemble undifferentiated blasts, while those in chronic leukemias are of a more mature stage. The acute and chronic leukemias can be further classified into numerous subtypes defined by cell lineage and stage of differentiation (refer to Tables 30-1, -2, -3, and -4). The reader is cautioned here that, while an elevated number of leukocytes in the blood is suggestive of a leukemia, the definitive diagnosis requires examination of the BM.

I. ACUTE MYELOID LEUKEMIA (AML)

AML is an acute cancer of the myeloid lineage in which the target cells are early stage myeloblasts (Plate 30-2). The overall incidence of AML is 2.5–4.0 per 100,000. While AML mainly affects patients over the age of 50 years, it also accounts for 20% of childhood leukemias. When AML arises directly from certain chromosomal aberrations (detailed later), it is referred to as *primary AML*. The disease onset in these cases is rapid and life-threatening. *Secondary AML*, which has a slower course, sometimes arises as a consequence of cyto-toxic chemotherapy in patients undergoing treatment for other

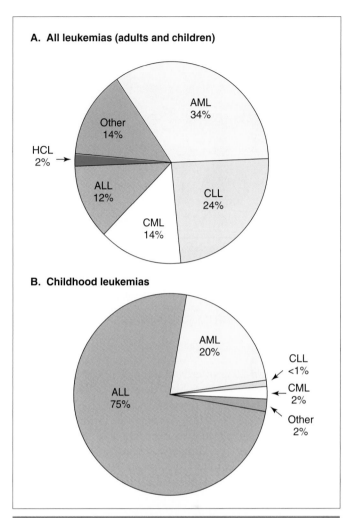

A. All leukemias (adults and children)

AML 34%
Other 14%
HCL 2%
ALL 12%
CML 14%
CLL 24%

B. Childhood leukemias

AML 20%
ALL 75%
CLL <1%
CML 2%
Other 2%

Figure 30-4

Relative Frequencies of Leukemias in North America
(A) The relative frequencies of the major types of leukemia are shown for both adults and children. (B) The relative frequencies of these disorders in the pediatric population are shown. AML, acute myelogenous leukemia; CLL, chronic lymphocytic leukemia; CML, chronic myelogenous leukemia; ALL, acute lymphocytic leukemia; HCL, hairy cell leukemia.

types of malignancies. However, it may also occur due to a progression from a less severe hematologic disorder such as bone marrow hyperplasia, abnormal myelopoiesis, or one of the *myelodysplastic syndromes* (MDS) (refer to Table 30-1). The MDS, which are considered less severe than the myelogenous leukemias, are a group of hematopoietic disorders characterized by increased infection, anemia, and hemorrhage. These symptoms arise because hematopoiesis in an MDS patient is "ineffective" and produces abnormal myeloid, erythroid, and megakaryocytic precursors. The numbers of these abnormal precursors in the blood and BM of MDS patients are not as high as in leukemia patients.

i) Clinical and Biochemical Features
Like most leukemia sufferers, AML patients present with symptoms of malaise, fatigue, fever, increased susceptibility to infection, bone pain, and weight loss. AML is diagnosed

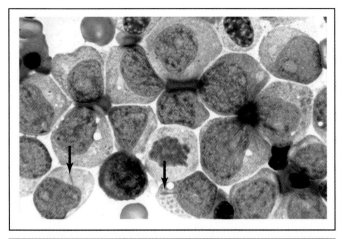

Plate 30-2

Acute Myeloblastic Leukemia
Bone marrow aspirate of an AML patient. Arrows point to two myeloblasts containing needle-like Auer rods. Reproduced with permission from Tkachuk D. C., Hirschmann J. V. and McArthur J. R. (2002) "Atlas of Clinical Hematology." W.B. Saunders Company, Philadelphia.

when more than 20% of cells in the patient's BM sample are leukemic myeloblasts. The high numbers of leukemic cells disrupt hematopoiesis in the BM such that the BM is said to have undergone "failure." Anemia and thrombocytopenia result that are manifested as bleeding and easy bruising. Immune responses are compromised such that upper and lower respiratory infections are common. Hepatosplenomegaly arises when leukemic cells accumulate in the liver and spleen.

The tumor cells may also migrate to *extramedullary tissues*, a term used by hematologists to refer to tissues and organs in the body that are neither lymphoid tissues nor BM. For example, the infiltration of leukemic cells into the gums occurs in some AML patients. When the infiltrating AML cells form a lump in an extramedullary tissue, the structure is called a *granulocytic sarcoma*. Because AML is generally very aggressive, patients have a poor prognosis and untreated AML patients have a median survival time of only 2 months.

While the WHO classification of AML subtypes (refer to Table 30-1) has superseded older schemes to some extent, the reader may find it helpful to be familiar with the traditional FAB (French, American, British) scheme. According to the FAB classification (Table 30-7), there are eight subtypes of AML, designated M0–7, which are distinguished primarily by tumor cell morphology, stage of differentiation, surface markers, and the activity of an enzyme called *myeloperoxidase* (MPO) that is a signature feature of granulocytic precursors. *Auer rods* may also be present in the leukemic cells. (Auer rods are finely granular flat bodies that appear in the cytoplasm of myeloid lineage leukemia cells. Auer rods are usually positive for acid phosphatase activity.)

ii) Genetic Aberrations

About half of AML cases show gross chromosomal abnormalities that are easily detected by cytogenetics. In some of these patients, a whole chromosome has been gained or lost, as in the case of AML associated with trisomy 8. (Trisomy 8 is also associated with other types of lymphoid and myeloid HCs.) Why these aberrations result in AML and which genes are involved are not known. Better success in deducing a disease

Table 30-7 FAB Classification of AML

FAB	Subtype	MPO	Morphology	CD13	CD14	CD15	CD33	CD34	Others
M0	AML no maturation	−	Myeloblasts (no granules)	+	−	−	+	+	
M1	AML minimum maturation	+/−	Myeloblasts (some granules)	+	−	−	+	−/+	
M2	AML with maturation	++	Myeloblasts (Auer rods)	+	−	+	+	+/−	
M3	Acute promyelocytic (APL)	+++	Promyelocytes (Auer rods)	+	−	+	+	−	
M4*	Acute myelomonocytic	+	Myeloblasts Promyeloblasts Monoblasts Monocytes	+	+	+	+	−	CD11b
M5	Acute monoblastic	+/−	Promyelocytes Monoblasts Monocytes	−/+	−/+	+	+	−	CD11b
M6	Erythroleukemia	+/−	Myeloblasts Megaloblasts	−	−	−	+	−	CD36
M7	Megakaryocytic	−	Megakaryoblasts	−	−	−	−	−	CD41 CD61

MPO, myeloperoxidase; *, includes "M4 eosino," which is acute myelomonocytic leukemia with eosinophilia.

mechanism has come from studying the many AML patients who show one of several recurring chromosomal translocations. The most common of these translocations is t(8;21)(q22;q22), which causes leukemia because it disrupts the function of a transcription factor called *core binding factor* (CBF). CBF regulates many genes involved in myeloid cell development, proliferation, and function, including M-CSF, GM-CSF, G-CSF, IL-3, and neutrophil elastase. How does this disruption occur? The heterodimeric CBF molecule is composed of two subunits called CBF2A and CBFB. The gene encoding CBF2A (also known as the AML1 gene) is located at 21q22. When the translocation occurs, part of the CBF2A gene is fused to part of a partner gene called ETO, located at 8q22 (Fig. 30-5). ETO binds to histone deacetylases in the nucleus and acts as a co-repressor of gene expression. The

translocation results in the expression of a chimeric CBF2A-ETO fusion protein that retains the DNA-binding domain of CBF2A but not its C-terminus. It is believed that this fusion protein acts as a dominant negative version of the CBF transcription factor and thus represses the expression of CBF-regulated genes (such as those encoding GM-CSF and M-CSF) that are necessary for myeloid cell development. Alternatively, or perhaps in addition, the disruption of the normal repressor activities of ETO may contribute to the defect. In any case, as a result of ETO-CBF2A expression, developing myeloblasts are "stuck" at an early stage of differentiation and never receive the appropriate signals to stop proliferating.

The gene encoding the CBFB subunit of CBF can also be altered by chromosomal rearrangement in AML leukemic cells. In this case, the CBFB gene located at 16q22 is translocated to

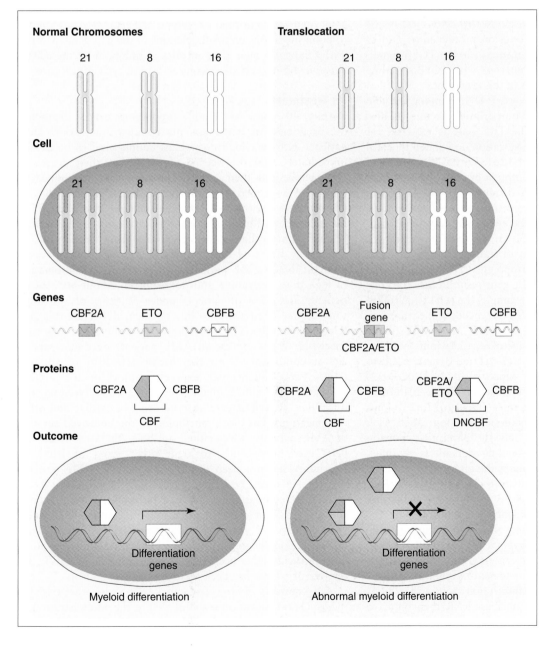

Normal Chromosomes

21   8   16

Cell

Translocation

21   8   16

Genes

CBF2A   ETO   CBFB

CBF2A   Fusion gene   ETO   CBFB

CBF2A/ETO

Proteins

CBF2A   CBFB

CBF

CBF2A   CBFB

CBF

CBF2A/ ETO   CBFB

DNCBF

Outcome

Differentiation genes

Myeloid differentiation

Differentiation genes

Abnormal myeloid differentiation

Figure 30-5

How the t(8;21)(q22;q22) Translocation Leads to Acute Myelogenous Leukemia
Normal chromosomes 21 and 16 contain the genes encoding the CBF2A and CBFB subunits of the CBF transcription factor. Chromosome 8 contains the ETO gene encoding a transcriptional repressor. In a normal cell, CBF drives the transcription of genes required for the normal differentiation of myeloid cells. The t(8;21)(q22;q22) reciprocal translocation results in the exchange of genetic material such that a fusion gene is created containing sequences from the CBF2A and ETO genes. Expression of this gene results in the translation of a mutated CFB2A subunit that combines with normal CBFB to produce a mutated transcription factor (DNCBF) that has a dominant negative effect on CBF-mediated transcription. The expression of genes required for normal myeloid differentiation is therefore blocked, leading to the development of AML.

the MYH11 locus (16p13) encoding the smooth muscle form of the myosin heavy chain. This genetic event is mediated either by a conventional chromosomal translocation designated t(16;16)(p13;q22), or by an inversion within one copy of chromosome 16, designated inv (16)(p13;q22). The chimeric CBFB-MYH11 fusion protein may bind to CBFA2 and prevent it from joining to normal CBFB, blocking the formation of functional CFB. Again, the transcription of myeloid differentiation genes regulated by CBF is disrupted.

Still another AML-related translocation involving CBF is designated t(3;21)(q26;q22). In this case, part of the CFBA2 gene at 21q22 is translocated into the EVI-1 locus at 3q26. The EVI-1 gene encodes a zinc finger DNA-binding protein that can interact with various co-activators and co-repressors governing gene expression. EVI-1 function is necessary for normal hematopoietic cell generation, and inappropriate EVI-1 expression blocks terminal differentiation of granulocytes and erythrocytes but promotes megakaryocytic differentiation. EVI-1 may also block cellular responses to cytokines. In the case of AML, the CFBA2/EVI-1 fusion protein may bind to Smad3, a downstream signaling mediator in the TGFβ signaling pathway. This binding may interfere with TGFβ signaling and thus the inhibitory functions of this regulator.

The TEL gene, which encodes a DNA-binding protein belonging to the Ets family of transcription factors, is also disrupted in cases of AML. The TEL gene is required for hematopoiesis in the BM. A translocation designated t(12;22)(p13;q11) fuses the TEL gene at 12p13 to the MN1 gene at 22q11. MN1 encodes a nuclear protein of unknown function.

Several additional chromosomal translocations resulting in AML involve the MLL (*mixed lineage leukemia*) gene located at 11q23. The MLL gene encodes a very large transcription factor of 431 kDa that is known to be important for the regulation of the Hox family of transcription factors and other genes required for hematopoiesis. The chromosomal translocations disrupting the MLL gene result in fusion to one of many different partners. For example, the t(4;11)(q21;q23) translocation fuses MLL with the gene encoding the transactivator AF4. Similarly, t(11;19)(q23;p13.3) joins MLL to the transcription factor ENL; t(11;19)(q23;p13.1) joins MLL to the transcription factor ELL; t(1;11)(q21;q23) fuses MLL to AF1q, a protein of unknown function; t(11;22)(q23;q13) fuses MLL to p300, a protein involved in the regulation of differentiation and cell cycling; and t(9;11)(q22;q23) joins MLL to a nuclear protein of unknown function called AF9. In all these chimeric proteins, only the N-terminal portion of MLL containing the DNA-binding domain is retained. The C-terminus of MLL containing the zinc finger domains involved in binding to and regulating other proteins is lost. These fusion events may therefore result in proteins that activate gene expression in an uncontrolled manner.

A subtype of AML associated with two unique genetic aberrations is *acute promyelocytic leukemia* (APL or M3 type AML). APL leukemic cells carry various translocations involving the gene encoding *retinoic acid receptor-α* (RARα). (RARα is one of a family of transcription factors that binds *retinoic acid* [RA] entering the nucleus. RAR then binds to response elements in RA-responsive genes and alters gene expression patterns.) For example, the t(15;17)(q22;q21) translocation fuses the RARα gene to the gene encoding the zinc finger protein PML at 15q22. A different translocation t(11;17)(q23;q21) joins RARα to another zinc finger protein called PLZF at 11q23. The resulting chimeric proteins appear to have dominant negative effects that block the normal function of RARα. As a result, genes normally involved in the progression of myeloid differentiation are suppressed.

What about the 50% of AML cases in which the tumor cells display normal karyotypes? One subgroup of these AML cases is associated with hereditary defects in genes responsible for DNA repair. For example, AML is frequently found in patients with Bloom syndrome or ataxia-telangiectasia (refer to Ch.24). These primary immunodeficiency patients have germline mutations of the BLM gene (a DNA helicase) or the ATM gene (a DNA damage sensor kinase), respectively. Patients with *Li-Fraumeni syndrome*, who have germline mutations of the tumor suppressor p53, also have AML associated with a normal karyotype. Other AML cases are due to point mutations or small deletions of the p15 gene. p15 is a TGFβ-inducible gene that encodes an inhibitor of the *cyclin-dependent kinases* (Cdks) required for cell cycle progression.

iii) Treatment

Despite 40 years of steadily improving cancer therapies, the prognosis for AML patients remains quite poor. AML patients with very high myeloblast counts need to be treated immediately, and the lineage of the blasts, their degree of maturation, and their numbers must all be determined to select the appropriate treatment. In the 1960s, AML was usually treated with the pyrimidine analogue cytarabine. (The reader is referred back to Table 26-11 for capsule descriptions of many chemotherapeutic drugs and their mechanisms of action.) A CR was induced in 20–30% of AML patients treated. Today, AML patients usually receive a combination of high dose cytarabine and doxorubicin. About 50% of all patients and 75% of patients under 60 years of age show some degree of response to this regimen. Reflecting this age-related difference in clinical response is the number of long term survivors. While almost all AML patients over 60 years of age succumb quickly to the disease, 30–40% of younger (<60 years) patients can expect to enjoy life for another 10 years. Better clinical management of anemia, bleeding, and infections, as well as careful monitoring of cardiac and other complications, has also contributed to the improved survival of AML patients. Monoclonal antibody therapy, which has been used successfully to treat other types of leukemia (see later), has not been of clinical benefit to AML patients to date. AML can be treated successfully by allogeneic HCT if an HLA-compatible donor can be found. Indeed, in one study, up to 50% of AML patients who received an allogeneic HCT achieved 5 years disease-free survival post-transplant. However, complications due to GvHD were significant. Some AML patients have been treated with autologous HCT, but the rate of relapse has been relatively high.

APL patients whose leukemic cells carry the t(15;17)(q22;q21) translocation respond well to the administration of trans-retinoic acid (ATRA; *all trans-retinoic acid*). ATRA is an oral agent that counters the effect of the chimeric RARα-PML protein and rescues the maturation of hematopoietic cells. Indeed, a combination of ATRA plus the anthracycline idarubicin allows 95% of t(15;17)(q22;q21) APL patients to achieve a CR and 80% to survive for at least 5 years disease-free. Interestingly, the clinical condition of APL patients with the t(11;17)(q23;q21) translocation does not improve with ATRA. Studies at the molecular level have indicated that this resistance to ATRA results from the recruitment of histone deacetylase by the fusion protein. These patients do respond to treatment with histone deacetylase inhibitors.

II. CHRONIC MYELOGENOUS LEUKEMIA (CML)

Chronic HCs involving cells of the myeloid lineage are called *myeloproliferative diseases* (MPDs; refer to Table 30-1). These disorders, which tend to affect older adults, arise from the malignant transformation of various myeloid cells in the BM. However, each disorder is due to a different collection of tumorigenic mutations and has a different predominant manifestation. For example, in *polycythemia vera*, the majority of the proliferating leukemic cells are erythroid in lineage, while in *essential thrombocythemia*, there is an excess of platelets. *Myelofibrosis* is manifested as replacement of the normal bone marrow stroma with a fibrotic mix of procollagen, collagen, and fibronectin. Anemia, neutropenia, and thrombocytopenia are all evident, but the nature of the original transformed cell is not known. The most common MPD is *chronic myelogenous leukemia* (CML), which is characterized by increased numbers of mature and immature granulocytes, with a predominance of the latter. Although CML accounts for a mere 0.3% of all cancers, it represents about 14% of adult leukemias (overall incidence of 1.3 per 100,000). This malignancy usually strikes adults with a median age of 50 years, and men are affected almost twice as often as women.

i) Clinical and Biochemical Features

The onset of CML is relatively benign and the disease can remain "silent" in many patients for 3–5 years. This stage of the disease is known as the *chronic phase* (Plate 30-3). However, the disease eventually progresses to an *accelerated phase* and finally to a *blast crisis*. In the accelerated phase, there is a sharp increase in the numbers of immature leukemic cells in the blood or BM. In the blast crisis, CML blast cells make up more than 30% of cells in the blood or BM. Indeed, phenotypically, blast crisis CML bears a striking resemblance to AML. Once the blast crisis has started, the leukemic cells start to infiltrate into extramedullary tissues and an untreated CML patient survives only a few months.

Most CML patients are diagnosed towards the end of the chronic phase. However, in about 10% of CML patients, the accelerated phase or blast crisis has already begun by the time of diagnosis. Clinical symptoms of blast crisis CML include weight loss, abdominal discomfort, bleeding, weakness, lethargy, and night sweats. Splenomegaly is present in 80% of CML patients. In some cases, the swollen spleen is large enough to induce splenic infarction (severance of the

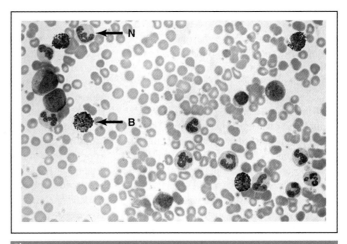

Plate 30-3

Chronic Myelogenous Leukemia: Chronic Phase
Peripheral blood smear showing increased numbers of precursor and mature granulocytes, including neutrophils (N) and basophils (B). Reproduced with permission from Tkachuk D. C., Hirschmann J. V. and McArthur J. R. (2002) "Atlas of Clinical Hematology." W.B. Saunders Company, Philadelphia.

local blood supply). Alkaline phosphatase activity in total leukocytes is low, a feature that helps to distinguish CML from several other similar-looking diseases and infections. Histological examination of the blood and BM reveals anemia with increased numbers of mostly granulocyte precursors but also B lymphoid (rarely T lymphoid), erythroid, and megakaryocytic lineage cells, plus connective tissue-forming cells. This spectrum of leukemic cell types indicates that the original transformation event must have taken place in a very early progenitor in the BM that retained its ability to divide and differentiate into multiple hematopoietic lineages. As a result, despite the name of this disorder, cells of either the myeloid or lymphoid lineage may predominate in the blast crisis phase.

ii) Genetic Aberrations

The tumor cells in over 90% of CML patients possess the Philadelphia (Ph) chromosome. As mentioned previously and as detailed in Box 30-2, the Ph chromosome is the shortened version of chromosome 22 that results from the t(9;22)(q34;q11) reciprocal translocation fusing the Bcr gene on chromosome 22 with the Abl gene on chromosome 9. The chimeric protein encoded by the Bcr–Abl fusion gene has constitutive Abl tyrosine kinase activity due to abnormal regulation and oligomerization mediated by the Bcr moiety. In particular, while the normal Abl enzyme plies its tyrosine kinase activity mainly in the nucleus, the Bcr–Abl fusion protein mediates tyrosine kinase activity in the cytoplasm. The activated Abl kinase moiety of Bcr–Abl directly and indirectly activates downstream cytoplasmic and nuclear targets, including several mediators affecting cell–cell interaction. The effects of Bcr–Abl on a plethora of mediators, such as Ras, PI3K, and NF-κB, result in altered cell division, cell survival, and myeloid differentiation. A less frequent

Box 30-2. **The Philadelphia chromosome**

The Philadelphia (Ph) chromosome is a cytogenetic anomaly that is manifested as a shortened version of human chromosome 22. Ph chromosomes are present in over 90% of chronic myelogenous leukemia (CML) patients and also in 25% of adult acute lymphocytic leukemia (ALL) patients. The Ph chromosome is derived from a reciprocal translocation between the long (q) arms of chromosomes 9 and 22. In general, the break in chromosome 9 occurs at q34 such that a 3′ portion of the gene encoding the Abl tyrosine kinase is transferred to chromosome 22. (The Abl gene was originally identified as a cellular gene that was "captured" by the A̲belson l̲eukemia virus; i.e., a copy of the human gene sequence was incorporated into the viral genome.) The break in chromosome 22 occurs at q11 at one of several points in a gene called Bcr, for b̲reakpoint c̲luster r̲egion (not "B cell receptor," BCR). In each case, only the 5′ portion of the Bcr gene is retained. The remaining 3′ portion of the Bcr gene goes to chromosome 9 to complete the reciprocity of the translocation (9q$^+$).

During the translocation process, the 3′ fragment of the Abl gene is joined to the 5′ fragment of the Bcr gene to form a fusion Bcr–Abl gene. This gene is transcribed and translated to produce a chimeric Bcr–Abl protein that is localized in the cytoplasm and has constitutive tyrosine kinase activity. In different patients, slightly different breakpoints in the Abl and Bcr genes occur, creating fusion mRNAs of slightly different lengths. As a result, fusion proteins of slightly different sizes and possibly functions are created. Most breaks in the Abl gene occur just 5′ of Abl exon 2, while those in the Bcr gene occur between Bcr exons 12 and 16. This region of the Bcr gene is known as the *major breakpoint cluster region* (M-bcr). Bcr–Abl proteins of 210 kDa are derived from fusion in this region. Similarly, fusion proteins of 190 kDa and 230 kDa are generated by breakpoints in the *minor breakpoint cluster region* (m-bcr) between Bcr exons 1 and 2, and the *micro breakpoint cluster region* (μ-bcr) between Bcr exons 19 and 20, respectively. (The relevant exons of the Bcr and Abl genes are shown in the figure as black and blue bars, respectively.)

In vitro studies have shown that the expression of Bcr–Abl in a myeloid cell promotes its uncontrolled proliferation and cancerous transformation. *In vivo*, the transforming activity of Bcr–Abl fusion proteins has been confirmed in several animal models. In one study, murine HSCs were transfected with the Bcr–Abl gene. These altered HSCs were then infused back into the original mice (which had been irradiated to destroy their immune systems). The majority of the treated animals went on to develop CML-like disease.

How do the fusion Bcr–Abl proteins actually cause transformation? The tyrosine kinase function of the normal Abl protein is regulated by the presence in its N-terminal end of an SH2 domain and an SH3 domain.

In particular, the SH3 domain has a negative regulatory effect, and when the function of this domain is blocked, Abl tyrosine kinase activity is abnormally elevated and promotes transformation. At its C-terminal end, Abl contains so-called DB domains that facilitate binding to nuclear proteins, DNA, and actin. Turning to the normal Bcr protein, we find that its N-terminus contains an oligomerization domain (OD), a serine-threonine kinase domain, and numerous adaptor recruitment domains, including Grb2 and SHC. The C-terminus of the normal Bcr protein acts as a GTP exchange factor and is important for controlling Ras activation. In the Bcr–Abl fusion protein, the N-terminal domains of the Bcr protein induce oligomerization of the Abl tyrosine kinase domains. Most importantly, the N-terminal Bcr region interferes with the negative regulatory function of Abl's SH3 domain, resulting in constitutive activation of Abl's tyrosine kinase activity. Abl binding to F-actin is also enhanced. The adaptor recruitment domains in the Bcr moiety of the fusion protein allow the recruitment of Grb2, which in turn allows the binding of the Ras activator Sos. As we have learned, Ras controls one of the most important signaling pathways driving cellular proliferation and differentiation. Sos-mediated activation of Ras is critical for the transforming activity of Bcr–Abl. Other adaptor proteins such as CRKL and SHC that are recruited to Bcr–Abl promote the phosphorylation and activation of ERK, SAPK/JNK, PI3K, PKB/Akt, STAT5, c-

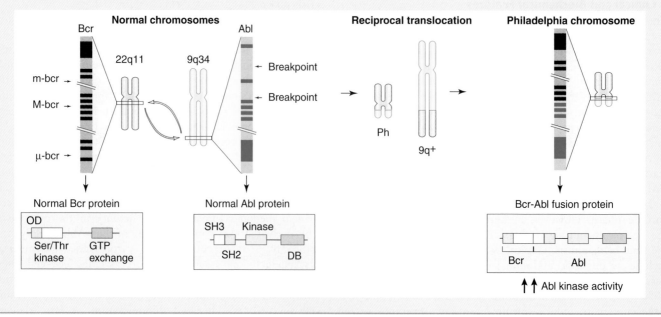

Continued

Box 30-2. **The Philadelphia chromosome**–*cont'd*

myc, and NF-κB, all molecules that promote signaling associated with cell survival and/or proliferation. In addition, Bcr–Abl expression appears to induce the proteasomal degradation of cell cycle inhibitors such as the Cdk inhibitor p27. Without such inhibitors, cell cycling is unrestrained. CML leukemic cells bearing the Ph chromosome are also able to ignore local stromal cell-derived cytokine signals that are

meant to curb cell division. Conversely, high quantities of IL-1β appear in the serum of advanced stage CML patients, which help to drive the continued proliferation of the leukemic cells. As well as affecting proliferation and survival, Bcr–Abl fusion proteins suppress apoptosis via upregulation of Bcl-2 and Bcl-xL expression. *In vitro*, CML tumor cells are less sensitive to death signaling induced by irradiation or cytotoxic

drugs. Finally, Bcr–Abl signaling disrupts integrin-mediated adhesion between developing leukemic myeloid progenitors and the surrounding stromal cells. The leukemic cells are thus released prematurely into the blood and gain enhanced powers of tissue infiltration. Figure adapted from Faderl S. *et al.* (1999) The biology of chronic myeloid leukemia. *New England Journal of Medicine* **341**, 164–172.

---

recurring chromosomal translocation found in *atypical CML* is the t(9;12)(q34;p13) translocation. This fusion event joins the Abl kinase gene with the TEL gene described previously in our discussion of AML.

The preceding genetic alterations are found in leukemic cells of the chronic and accelerated phases of CML. In the blast crisis phase, new genetic aberrations can appear such as trisomy 8, trisomy 19, or an inversion of chromosome 17. Double Ph chromosomes and the t(3;21)(q26;q22) translocation involving CFB2A and EVI-1 (as described previously for AML) have also been observed. More subtle mutations found in blast crisis CML cells affect the genes encoding Ras, p53, p15, and p16 (another Cdk inhibitor).

iii) Treatment

The prognosis of CML patients varies according to the phase of the disease at diagnosis. Conventional treatment for CML patients involves the administration of busulfan or hydroxyurea. Hydroxyurea is associated with longer patient survival (median 4 years) than busulfan (median 3 years). While both of these agents have toxic side effects, they control leukemic cell numbers such that patients may experience up to an 80% hematological remission. Cytogenetic remission, defined as the elimination of Ph chromosome-bearing cells, may also be achieved. IFNα treatment has been effective in over 50% of CML patients, with complete cytogenetic remission often achieved within 2 years of starting therapy. How IFNα works in this situation is not known, but mechanisms involving antiproliferation, anti-angiogenesis, or some other immunomodulatory function (such as inhibiting the expression of other cytokines) are suspected. *In vitro* studies of CML cells have confirmed that regulatory T cells can specifically suppress the proliferation of CML cells, and it may be that IFNα supports these T cell subsets in some way. In one study, it was shown that IFNα restored normal intercellular adhesion properties to leukemic cells, inhibiting their tendency to overgrow normal cells.

In the early 2000s, the treatment of CML patients whose tumor cells have the Ph chromosome was advanced with the introduction of the drug Gleevec (imatinib mesylate). This compound inhibits the activity of the Abl kinase driving the transformation of the leukemic cells. Gleevec has several other advantages in addition to its specificity: it can be administered orally; it has few systemic side effects (save

occasional thrombocytopenia); it does not have a significant impact on normal tissues; and most CML patients respond to it to some degree.

As was the case for AML, CML can be treated with an allogeneic HCT. For the 30% of chronic phase CML patients who are lucky enough to have HLA-matched donors, the prognosis of the disease dramatically improves, particularly if the patient is under 30 years of age. More than half of chronic phase CML patients who receive such transplants achieve over 5 years disease-free survival. If the disease has progressed to the accelerated phase or blast crisis stage, however, the chance of HCT success is reduced to about 15%. The mortality rate due to GvHD is also considerable, running at about 10–12% of all treated CML patients. When HLA matching has been particularly poor, the death rate due to GvHD can reach as high as 50%. Depletion of donor T cells from the stem cell preparation can reduce the chance of GvHD, but these patients also experience an increased rate of CML relapse due to a decreased GvL effect.

III. ACUTE LYMPHOBLASTIC LEUKEMIA (ALL)

ALL is characterized by increased numbers of usually B (but sometimes T) lineage cells in the blood and BM. The overall incidence of ALL is 1.3 per 100,000, accounting for 1.5% of all cancers. ALL can affect adults or children, but is most common in children of age 3–5 years. Indeed, the disease represents close to 75% of childhood leukemias, making ALL the most frequent form of cancer in children. A second peak of ALL incidence occurs in individuals over the age of 65 years, but only 20% of all adult leukemias are ALL. Males get ALL more frequently than females, and the disease incidence is higher in whites than blacks. The genetic factors that result in these biases are not known. Environmental exposure to ionizing radiation, electromagnetic fields, industrial pollutants, car exhaust, and tobacco smoke or the use of amphetamines, certain diet pills, or hallucinogenic drugs by women before or during pregnancy have also been linked to ALL.

i) Clinical and Biochemical Features

In general, ALL patients first present with fever, fatigue, weight loss, infections, hemorrhages, dizziness, easy bruising, and joint pain. High numbers of lymphoblast-like leukemic cells are present in the blood, BM, and extramedullary tissues. Lymphadenopathy and hepatomegaly may be present. Patients with T cell ALL (T-ALL) may have a mass in the *mediastinum* (i.e., among the organs, including the heart and thymus, that are positioned between the sternum and the spine). A mediastinal mass can lead to wheezing and cardiac complications if it is sufficiently large. B-ALL patients often have masses in the abdomen that can compromise kidney function. Central nervous system (CNS) involvement, characterized by leukemic cell invasion of the spinal cord and brain, is rare at ALL presentation but often occurs during relapses and in advanced stage disease (particularly in cases of B-ALL). Symptoms of CNS involvement include headaches, nausea, lethargy, and irritability.

A diagnosis of ALL is reached when analysis of the patient's blood smear reveals that 20% of cells present are leukemic lymphoblasts. Under the microscope, cases of B-ALL and T-ALL are morphologically very similar. However, ALL leukemic lymphoblasts are quite different from AML myeloblasts in appearance, in that the former are smaller, have more regular nuclei with agranular cytoplasm, and have clumpier chromatin (Plate 30-4). In contrast to AML, Auer rods are absent. Historically, morphological, biochemical, and cytochemical criteria were used by the FAB working group to distinguish three subtypes of ALL, called L1, L2, and L3. However, this scheme did not take hematopoietic lineage and stage of differentiation into account. The WHO has since come up with new criteria that emphasize immunophenotyping, the analysis of genetic aberrations, and the determination of certain enzyme activities. Immunophenotyping for the expression of surface Ig and multiple other markers is used to determine whether the ALL is, to name just a few subtypes, pro-B, pre-B, mature B, pro-T, pre-T, cortical T, or mature T (Table 30-8). Most cases of ALL involve B lineage cells that express CD19 and cytoplasmic CD22. Depending on the state of maturity of the leukemic blasts, CD10 (a peptidase called "common ALL antigen"), CD20, CD79a (Igα), cytoplasmic

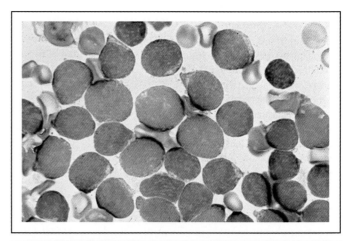

Plate 30-4

Acute Lymphoblastic Leukemia
Bone marrow aspirate of ALL patient showing lymphoblasts with large nuclei and minimal cytoplasm. Reproduced with permission from Tkachuk D. C., Hirschmann J. V. and McArthur J. R. (2002) "Atlas of Clinical Hematology." W.B. Saunders Company, Philadelphia.

Ig μ chain, and/or surface Ig may also be expressed. The less common forms of ALL involve cells of the T lineage that express CD2, CD5, and CD7. In some cases, CD1, CD3, CD4, CD8, and even the B cell markers CD10 and CD21 can be found on T-ALL cells. Mature B cell ALL is Tdt-negative, whereas precursor subtypes of B cell ALL and all T-ALLs are TdT-positive. In contrast to the medium/high levels of MPO activity seen in AML, ALL cells tend to show only very low levels of MPO activity.

ii) Genetic Aberrations

Some of the genetic changes leading to ALL development are similar to those found in AML and CML patients. The tumor

Table 30-8 Immunophenotyping of ALL

Subtype	CD2	CD3	CD5	CD7	CD10	CD19	CD22	CD79a	Cyt Ig*	mIg	MHC class II	TdT
Pro-B ALL	−	−	−	−	−	+	+	+	−	−	+	+
Common ALL	−	−	−	−	+	+	+	−	−	−	+	+
Pre-B ALL	−	−	−	−	+	+	+	−	−	−/+	+	+
Pre-B ALL with cyt mu	−	−	−	−	+/−	+	+	−	+	−	+	+
Mature B ALL	−	−	−	−	−	+	−/+	−	−	+	+	−
Pro-T ALL	−	−	−	+	−	−	−	−	−	−	−/+	+
Pre-T ALL	+	−	+	+	−	−	−	−	−	−	−/+	+
Cortical T ALL	+	−	+	+	−	−	−	−	−	−	−	+
Mature T ALL	+	+	+	+	−	−	−	−	−	−	−	+

*Cyt Ig, cytoplasmic Ig μ chain.

cells in about 25% of adult and 5% of childhood ALL cases exhibit the Ph chromosome first described for CML. However, the precise breakpoint of the translocation is usually slightly different in ALL, leading to the production of an oncoprotein of different size. The Abl tyrosine kinase is still deregulated, however, and its abnormal activation drives the development of aggressive pre-B ALL, in which the tumor cells express high levels of CD10. The prognosis for this subtype of ALL is one of the least favorable, although the success in the early 2000s of treatment with a combination of Gleevec and chemotherapy has brought new hope to these patients.

About 30% of childhood ALL cases show translocations in which the gene encoding the CBF transcription factor subunit CBF2A at 21q22 is translocated to the TEL transcription factor locus at 12p13 [t(12;21)(p13;q22)] (refer to Table 30-6). This translocation is rare in adult ALL. The prognosis for children with pre-B ALL driven by this translocation is quite good. Another pair of chromosomal translocations found in ALL involves the MLL gene and the AF9 nuclear protein [t(9;11) (q22;q23)], or the MLL gene and the ENL transcription factor [t(11;19) (q23;p13)]. The t(4;11)(q21;q23) translocation leading to activation of the AF4 transactivator is found in 80% of infant ALL cases but only 5% of adult ALL patients. Rearrangements involving 11q23 generally result in pro-B ALL characterized by low levels of CD10. This subtype of ALL generally has a poor prognosis.

Several recurring chromosomal translocations leading to ALL involve the introduction of gene fragments into either the Ig loci or the TCR loci. These loci are a common target not only in lymphoid leukemias but also in myelomas and lymphomas (refer to Table 30-6). This increased vulnerability to translocation may reflect the inherent genetic instability of the antigen receptor genes. It is normal for the Ig and TCR genes to undergo DNA breakage and rejoining events during somatic recombination, and the Ig genes experience further modifications due to somatic hypermutation and isotype switching. Moreover, a gene introduced into these loci is likely to come into proximity with an active Ig or TCR enhancer that may activate or upregulate transcription of the translocated gene. A gene frequently translocated and activated in this way in mature B cell ALL is that encoding the transcription factor c-myc. The t(8;14)(q24;q32) translocation introduces the c-myc gene at 8q24 into the Igh locus at 14q32 (80% of chromosome 8 translocations), while the t(8;22)(q24;q11) translocation fuses c-myc to the Igl locus at 22q11 (15%), and the t(2;8)(p12;q24) translocation joins c-myc to the Igk locus at 2p12 (5%). C-myc inserted into an Ig locus becomes constitutively activated, causing deregulation of the expression of genes usually controlled by another transcription factor called MAX. MAX normally homodimerizes or forms heterodimers with another protein called MAD. However, when c-myc is overexpressed due to its translocation into an Ig locus, MAX heterodimerizes with c-myc instead, reducing the number of MAX-MAX or MAD-MAX dimers. The altered transcriptional complex changes the transcription of genes involved in cell proliferation, apoptosis, and differentiation. How these events lead to malignant transformation is under investigation.

Several different genes introduced into the TCR loci result in cases of T-ALL (refer to Table 30-6). The TCRαδ locus is located at chromosome 14q11, while the TCRβ locus is at 7q35. The promoters of partner genes introduced into these TCR loci become activated, leading to the expression of proteins that are not normally present during T cell development (at least not at abundant levels). For example, t(1;7)(p32;q35) inserts the gene encoding the TAL1 transcription factor into the TCRβ locus, while t(11;14)(p13;q11) joins the gene encoding the zinc finger protein LMO2 with the TCRαδ locus. Both TAL1 and LMO2 are normally expressed at high levels during erythropoiesis rather than T cell development and influence the activity of the E2A transcription factor. Similarly, t(7;10)(q35;q24) joins the HOX11 gene to the TCRβ locus, leading to HOX11 overexpression during T cell development. The cycling of T lineage cells is disrupted and programmed cell death is blocked. Parallel to B-ALL, the t(8;14)(q24;q11) translocation associated with T-ALL inserts the c-myc gene into the TCRαδ locus, deregulating c-myc activation and promoting T cell proliferation.

A translocation so far unique to ALL is the t(1;19)(q23;p13) translocation common in childhood pre-B cell ALL. This genetic aberration introduces the transcription factor E2A gene located at 19p13 into the PBX1 gene located at 1p23. The reader will recall from Chapter 9 that E2A is important for the regulation of B cell development. When the transactivating domain of E2A is fused to PBX1, a chimeric protein is generated that can strongly activate genes not usually expressed in pre-B cells. The activation of these genes may drive the malignant transformation of the pre-B cells. A second chromosomal translocation involving E2A is found in a small subset of pre-B cell ALL cases. The t(17;19)(q22;1p13) translocation encodes a fusion protein that induces anti-apoptosis in pre-B cells. The partner at 17q22 is a transcription factor called HLF.

Deletions of 9p21 are found in about 20% of B-ALL cases and in 80% of T-ALL patients. The 9p21 chromosomal region contains the genes encoding the Cdk inhibitors p15 and p16. Malfunctions of these genes promote uncontrolled cell proliferation. Childhood B-ALL associated with the 9p21 abnormality is aggressive, but T-ALL is not (unless the deletion is homozygous). Point mutations of the p15 and p16 genes, as well as inactivation of these genes due to methylation, have also been found in ALL leukemic cells. Interestingly, some T-ALL leukemic cells show inactivation of the ATM kinase mutated in families with ataxia-telangiectasia. Lastly, patients with trisomy 21 have a 20-fold greater risk of developing ALL. It is not known what gene(s) on chromosome 21 are responsible for the transformation in these cases.

iii) Treatment

While the therapy regimens for the treatment of ALL are similar in children and adults, the outcomes are strikingly different. Over 90% of childhood ALL cases achieve a CR and 75–80% of these patients are cured, making the treatment of childhood ALL one of the modern success stories in cancer therapy. However, the prognosis for adult ALL is gloomy by comparison. While 65–80% of adult ALL patients achieve a CR, only 20–30% are cured. This difference in clinical outcome correlates with different genetic alterations associated with childhood and adult ALL. The poorer outcome of treatment in adult ALL may also be due in part to the greater number of mutations that older leukemic patients may have accumulated, or the inability of older patients to tolerate the high doses of chemotherapy used successfully in young children.

The standard treatment for ALL has four stages: (1) CR induction, (2) CNS prophylaxis, (3) consolidation, and (4) maintenance. In the induction phase, a CR is achieved with a combination of the anti-cancer drugs vincristine, prednisolone, doxorubicin, methotrexate, asparaginase, and cyclophosphamide. Slightly different combinations of chemotherapeutic drugs may be used depending on whether the disease is B-ALL or T-ALL. Growth factors such as G-CSF may also be administered to stimulate neutrophil production and stave off infections. The second stage of ALL treatment is CNS prophylaxis, necessary to prevent invasion of the CNS and brain. *Intrathecal* chemotherapy (in which the drug is introduced just under the sheath covering the spinal cord) or localized irradiation of the CNS and brain is implemented. Methotrexate and cytarabine are the drugs of choice for this stage of treatment because many other drugs are toxic to the CNS. In the consolidation phase, additional drugs are administered that help to prevent relapses. These drugs may include cytarabine given in combination with anthracycline, methotrexate, or thioguanine. Without the consolidation phase, the relapse rate of most subtypes of ALL patients is very high. In the maintenance phase, patients are given methotrexate and mercaptopurine for an additional 2–3 years. (It should be noted that maintenance therapy is not usually necessary for patients with mature B-ALL.) If a patient suffers a relapse of his or her ALL disease despite the preceding treatments, an allogeneic HCT is a viable option for those with an HLA-matched donor. Autologous HCT has been tried but is no more effective than chemotherapy in prolonging life.

What about biologicals? As was the case for AML, clinical benefit has yet to be demonstrated using mAb therapy against ALL leukemic cells. Alternative immunotherapy strategies to fight ALL are being actively pursued. In one approach, peptide sequences have been isolated from mutated oncogenes and chromosomal translocations present in ALL cells. Engineered presentation of these peptides to Tc cells has been used experimentally in an attempt to stimulate CTL cytotoxicity against the leukemic cells. The idea is to expand the anti-leukemic CTLs *in vitro* and return them to the patient using adoptive transfer (refer to Ch.26). Successful translation of these approaches to the clinic may provide new options for ALL patients.

IV. CHRONIC LYMPHOCYTIC LEUKEMIA (CLL)

CLL arises from the transformation of a peripheral blood lymphocyte. As its name suggests, CLL is a chronic form of lymphocytic leukemia characterized by a prolonged disease course. CLL mostly affects individuals over 60 years old and occurs about twice as often in men as in women. The overall incidence of CLL is the highest among all the leukemias, at 5.2 per 100,000 in people older than 50 years. Curiously, while CLL represents 30% of adult leukemias in the United States, only 2.5% of Asian adult leukemias are CLL.

i) Clinical and Biochemical Features

CLL progression has been divided into five stages according to the *Rai classification* system, which is based on the severity of the disease and other medical complications. Stage 0 CLL patients often have no clinical symptoms at diagnosis, although the BM is invariably involved. These patients are considered at "low risk" and generally survive at least another 10 years. Stage I and II CLL patients show lymphadenopathy, splenomegaly, and sometimes hepatomegaly, and are considered to be at "intermediate risk." Stage I and II patients live an average of 7 years following diagnosis. Stage III and IV patients have more advanced disease and are considered "high risk." As well as lymphadenopathy, splenomegaly, and hepatomegaly, about 20% of high risk CLL patients develop clinical hemolysis, and a very small number show reduced numbers of erythrocytes or platelets secondary to bone marrow failure. Infections with *Staphylococcus aureus*, *Streptococcus pneumoniae*, and *Haemophilus influenzae* are frequent because of compromised immune responses. In addition, the risk of a CLL patient developing an NHC is twice that of the general population. As a result of all these difficulties, CLL patients with advanced disease can expect to live only another 2–4 years after diagnosis. This course may be further shortened if the leukemic lymphocytes infiltrate the BM in a broadly diffuse manner, as opposed to forming nodules or remaining contained in the interstitial spaces. In 10% of CLL patients, the leukemia becomes an aggressive B cell lymphoma of the DLCL ("diffuse large cell lymphoma") subtype (see later). The patient experiences a sudden fever, enlargement of the lymph nodes, and increased hepatosplenomegaly. Other symptoms characteristic of lymphoma are also often present. In these cases, death usually results within 5 months, with or without treatment.

CLL is mainly a disorder of B lineage differentiation and proliferation. The leukemic cells often express CD23, CD25, and CD71 as well as the pan-B markers CD19, CD20, and MHC class II. Most B-CLL leukemic cells also express CD5. Normal B cell production in the BM and function in the periphery are disrupted with deleterious consequences. Alterations to the normal levels of many cytokines, including IL-2, -4, -7, -8, -10, -11, -12, and -13 as well as TNF, IFNγ, GM-CSF, and TGFβ, have been noted in B-CLL patients. Perhaps as a result of these cytokine abnormalities, helper T cell responses to antigens are also impaired. All of these factors culminate in reduced antibody responses in CLL patients.

To add insult to injury, about 20% of CLL patients develop some type of autoimmune disease such as autoimmune hemolytic anemia or ITP. The fact that CD5 is a marker present on B-CLL leukemic cells may not be a coincidence. As we learned in Chapters 16 and 29, CD5+ cells are often elevated in patients with autoimmune diseases, and activated CD5+ cells secrete broadly reactive autoantibodies. One emerging theory to account for the marker expression pattern of B-CLL cells is that the target cell may have been a self-reactive CD5$^+$ B cell. In a normal individual, this cell would have been anergized. However, the leukemic transformation event may have overcome the anergy and triggered inappropriate activation, allowing uncontrolled clonal proliferation and autoimmunity.

About 5% of CLL cases involve transformed mature T lymphocytes. The leukemic cells in these cases express CD2, CD3, CD4, and CD7. Patients with this HC respond only weakly to chemotherapy and thus have a very poor prognosis.

ii) Genetic Aberrations

Leukemic cells in about 80% of CLL cases show overt cytogenetic alterations. The most common mutation, found in over 50% of CLL patients, is a deletion at 13q14. Within this region are two *micro-RNA genes* called *miR15* and *miR16*. Micro-RNA genes constitute a large family of highly conserved non-coding genes that are transcribed and processed into short (21–25 nucleotide) RNA molecules. Micro-RNAs are thought to be involved in regulating the expression of tissue-specific and stage-specific genes during mammalian development. Interestingly, the V gene segments of the Ig loci of this sub-group of CLL patients tend to have elevated levels of somatic mutations. The 13q14 mutation is associated with the least aggressive course of CLL, as expected for leukemic cells with a more mature post-GC phenotype. Another chromosomal deletion found in 20% of CLL cases occurs at 11q23. The Ig genes in these leukemic cells are not mutated, suggesting that the leukemic cells are derived from more primitive B cell precursors. The disease course in these cases is more aggressive. About 15% of CLL patients have trisomy 12. Again, the Ig genes in the leukemic cells of this sub-group lack somatic mutations and the disease course is quite rapid. In half of trisomy 12 CLL cells, a translocation at t(14;19) (q32;q13) joins the Igh locus to the Bcl-3 gene. Bcl-3, which is a member of the IκB family of proteins, acts as a transcriptional co-activator of NF-κB. Presumably, overexpression of the Bcl-3 gene driven by an Igh enhancer results in excessive expression of NF-κB-activated survival genes. Overexpression of the anti-apoptosis gene Bcl-2 occurs in the leukemic cells of 70% of older CLL patients, but only 5–10% of these cases can be accounted for by translocations of Igh locus into the Bcl-2 gene [t(14;18)(q32;q21)].

Atypical CLL occurs in about 15% of patients. In this situation, the CLL is converted into a very aggressive disease called *Richter syndrome*. The CLL tumor cells acquire new mutations that drive the conversion of the disease into a large cell lymphoma. Patient prognosis is poor. This group of patients exhibits abnormalities of chromosome 17, some of which affect the p53 tumor suppressor gene located at 17p13.

iii) Treatment

Untreated CLL patients survive for an average of 6 years following diagnosis. Thus, as CLL symptoms are usually relatively mild, and the patients it affects are generally over 60 years of age, it is important to weigh the benefits of treatment against the significant side effects of chemotherapy.

Infections due to chemotherapy-induced immunosuppression and myelosuppression pose a real threat to the older patient. Consequently, one of the most important management issues of CLL is to decide when to simply watch disease progression and when to medically intervene.

When CLL requires treatment, chemotherapy with an alkylating agent (chlorambucil or cyclophosphamide) or a purine analogue (fludarabine or cladribine) is commonly used as a first approach. Unfortunately, the response of CLL leukemic cells to this panel of drugs is often underwhelming, and there is no guarantee that a CR will be induced. Partial remission occurs in 70% of CLL patients treated with purine analogues, while 30% achieve a CR. When a standard cocktail of alkylating agents called CHOP (*cyclophosphamide, [H]-doxorubicin, [O]-vincristine,* and *prednisolone*) is used, 40–50% of CLL patients experience partial remission, with 30–40% achieving a CR. If a CLL patient has proved refractory to treatment with alkylating agents or suffers a relapse, administration of a purine analogue can have some clinical benefit about 40% of the time. Research is ongoing into the potential therapeutic effectiveness of cisplatin, cytarabine, arsenic, cyclin inhibitors, kinase inhibitors, and anti-angiogenesis compounds as treatments for CLL.

At this time, biological therapeutics are not commonly used as treatments for CLL. IFNα, which works well for some leukemias (e.g., CML, hairy cell leukemia; see later) appears to be of minimal benefit to CLL patients. Some trials have suggested that a combination of IFNα and alkylating agents may be more effective. The most useful biological agent currently used for CLL cases is the CAMPATH-H mAb described in Chapter 29. The reader will recall that CAMPATH-H recognizes CD52, a GPI-anchored receptor expressed on the surfaces of most lymphocytes. About 30% of CLL patients who fail to respond to chemotherapy will respond to CAMPATH-H therapy. Another subset will respond to high doses of rituximab (anti-CD20) despite the fact that CLL tumor cells express only a low level of this surface antigen. HCT is rarely attempted as a treatment for CLL because the occurrence of severe GvHD is frequent. In addition, most CLL patients are at an age when they are less likely to survive the HCT procedure. However, some younger CLL patients have enjoyed prolonged remission following successful HCT.

V. OTHER LEUKEMIAS

i) Hairy Cell Leukemia (HCL)

HCL is a rare type of B cell leukemia that takes its name from the "hairy-looking" cytoplasmic projections extending off the transformed lymphocytes (Plate 30-5). HCL represents 2% of all leukemias in the United States. The vast majority of patients are older males. The cause of HCL is unknown, and no association with viruses, exposure to chemicals or radiation, or recurring genetic abnormalities has been observed. Individuals with HCL present with splenomegaly, hepatomegaly, cytopenia, anemia, neutropenia, and infections. Lymphadenopathy is rare. The malignant cells express the monocytic marker CD11c as well as the B cell markers

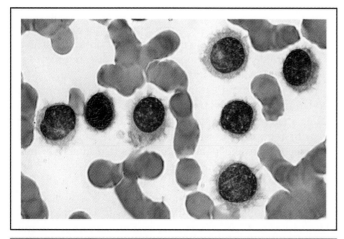

Plate 30-5

Hairy Cell Leukemia
Peripheral blood smear of leukemic cells featuring hair-like projections of cytoplasm. Reproduced with permission from Tkachuk D. C., Hirschmann J. V. and McArthur J. R. (2002) "Atlas of Clinical Hematology." W.B. Saunders Company, Philadelphia.

CD19, CD20, and CD103. Staining for CD5 is negative. HCL leukemic cells can also be distinguished from other types of leukemic blasts by their characteristic positive staining for TRAP (*tartrate-resistant acid phosphatase*). Happily for these patients, HCL is one of the most treatable leukemias. Up to 80% of HCL patients given purine analogues such as deoxycoformycin or cladribine respond clinically, and many of these can achieve a CR. IFNα treatment also results in a high rate of remission.

ii) Leukemia Associated with Lymphoma
Some types of B cell non-Hodgkin's lymphomas can progress to a leukemic phase. These lymphomas include *mantle cell lymphoma* (MCL), *MALT lymphomas*, and *follicular lymphomas* (FL), all described in detail later in this chapter. Attempts to treat these types of leukemias have not been very successful and the median survival rate of these patients is only about 2–3 years.

iii) B Cell Prolymphocytic Leukemia (B-PLL)
Despite its name, B cell prolymphocytic leukemia is not a cancer of a precursor B cell. The "lymphocytic" in the name of this disorder signifies that B-PLL cells appear to be more mature than the cancerous cells of "lymphoblastic" leukemias. B-PLL is considered to be a mature B cell leukemia because this disease sometimes arises from CLL, even though the leukemic cells of CLL appear to be more differentiated than those of B-PLL. Unlike CLL cells, B-PLL cells tend to be negative for CD5 expression. Some B-PLL cells contain somatic mutations in the Igh V gene segments.

B-PLL occurs at about one-tenth the frequency of CLL and, like CLL, usually strikes elderly men. B-PLL patients present with splenomegaly but minimal lymphadenopathy. A chromosomal abnormality associated with B-PLL is the t(11;14)(q13;q32) translocation, which brings the Cμ enhancer of the Igh locus at 14q32 into contact with the Bcl-1 locus (encoding the cyclin D1 protein) at 11q13. As a result, the Bcl-1 gene is transcriptionally activated and its cyclin D1 product pushes the affected cell into the cell cycle. Mutations of p53 appear in the cells of 50% of PLL patients. The prognosis of B-PLL patients is quite poor (less than 3 years) even with treatment. Chemotherapy is not successful in most cases, with only 25% of patients achieving partial remission.

iv) T Cell Prolymphocytic Leukemia (T-PLL)
T-PLL is a rare but serious disease characterized by very high numbers of leukemic T cells in the blood and BM. For the same reason that B-PLL is classed among the mature B cell leukemias, T-PLL is considered a mature T cell leukemia. T-PLL leukemic cells express T cell surface proteins such as CD3, CD4, and CD8 as well as CD5 and CD7. Lymphadenopathy and splenomegaly are the main clinical features of T-PLL patients, with tumor cell infiltration into the skin in a subset of patients. Chromosomal aberrations associated with this disorder involve the TCRαδ locus at 14q11: for example, an inversion described as inv(14)(q11q32) and a translocation with the X chromosome designated as t(X;14)(q28;q11). The Xq28 region contains several melanoma-related MAGE genes. Median survival in the absence of treatment is only 7 months after diagnosis. While the disease is resistant to most chemotherapeutic agents, clinical trials of newer regimens involving CAMPATH-H have had more promising results, with 60–70% of patients showing a clinical response.

v) Large Granular Lymphocytic Leukemias (LGLLs)
Two rare subtypes of leukemia feature large granular lymphocytic cells with the properties of either the T or NK lineages. Accordingly, these leukemias are called T-LGLL and NK-LGLL. Both these disorders have an *indolent* course, meaning that the transformed cells grow relatively slowly and disease progression is gradual. In general, LGLLs tend to strike middle-aged rather than older people, and women are affected more often than men. In T-LGLL, most of the large granular lymphocyte-like cells present in the blood express CD3 and CD8, while the rest are CD3⁺CD4⁻CD8⁻. T-LGLL leukemic cells also express V gene segments of the TCR genes. The mechanism of transformation is not known. In NK-LGLL cases, the large granular lymphocyte-like leukemic cells do not express CD3. Patients with either T-LGLL or NK-LGLL eventually develop splenomegaly, neutropenia, and infections. Interestingly, for unknown reasons, about a third of these patients also have rheumatoid arthritis. Some success in the treatment of LGLL has been obtained with purine analogues (deoxycoformycin or cladribine), corticosteroids, alkylating agents, cyclosporin A, or IFNα.

E. Plasma Cell Dyscrasias

Plasma cell dyscrasias are HCs of plasma cells, the terminally differentiated B cells that secrete antibody in response to antigen. Normal plasma cells cannot divide and so die after doing their job of secreting antibody for a few days. Transformed plasma cells continue to divide and may overexpress whole antibodies or single Ig chains whether antigen is present or not. The abnormal Ig protein produced by the transformed plasma cell is often called the *M protein* (for "monoclonal") or the *paraprotein* (as in "IgM paraprotein") to distinguish it from the conventional antibodies produced by normal plasma cells. Although large amounts of these Ig paraproteins may be produced, they appear to be non-functional. Moreover, paraprotein synthesis can create an imbalance of normal isotypes of light and heavy Ig chains that disrupts the humoral response and leads to clinical and biochemical abnormalities. It should be noted that the plasma cell dyscrasias are not considered leukemias because, although their paraprotein products may enter the blood, the transformed plasma cells themselves are (at least initially) confined to the BM and do not enter the circulation until the late stages of the disease.

There is a broad spectrum of plasma cell dyscrasias whose severity ranges from the almost asymptomatic to life-threatening. These disorders are sometimes named for the location of the tumor. For example, a patient showing a single plasma cell tumor within the BM is diagnosed with *solitary plasmacytoma of bone*. A patient whose cancerous plasma cells have moved outside the BM and infiltrated other tissues is said to have *extramedullary* or *extraosseous plasmacytoma*. We offer brief descriptions of two fairly mild plasma cell dyscrasias before concentrating on the most serious members of this HC family, the myelomas.

I. MONOCLONAL GAMMOPATHY OF UNDETERMINED SIGNIFICANCE (MGUS)

Individuals with this colorfully named syndrome are living demonstrations that the presence of transformed plasma cells is often of little clinical consequence. MGUS is the most common plasma cell dyscrasia, with an incidence of 1% in persons over 40 years of age. About 70% of MGUS patients show no clinical symptoms at all, and the finding of elevated levels of an Ig paraprotein in the blood usually occurs by chance during blood testing initiated for other reasons. If symptoms do occur, they take the form of mild anemia and usually disappear with time. There may be abnormally low levels of normal plasma cells in the BM, but this tissue has not suffered lesions or sustained failure as occurs in more serious plasma cell dyscrasias. MGUS patients do not show impaired immune responses. As a result, most MGUS patients are monitored but not treated. About 25–30% of MGUS patients go on to develop myelomas or B cell lymphomas.

II. WALDENSTRÖM'S MACROGLOBULINEMIA (WM)

More serious than MGUS is Waldenström's macroglobulinemia (WM). In this disorder, the cancerous plasma cells in the BM secrete large quantities of IgM paraprotein that makes its way into the blood. A patient's serum may be more viscous than normal due to the presence of the bulky IgM paraprotein molecules. This hyperviscosity can lead to altered vision and neurological problems. Accumulation of the tumor cells in the BM and infiltration of the lymph nodes and spleen are common, causing enlargements of these organs in 40% of patients. Because the malignant cells replace normal ones in the BM, hematopoiesis is compromised and anemia results. Hemorrhaging may occur due to the binding of para-IgM to clotting factors and fibrinogen. Cutaneous lesions are rare in WM but, when present, are a specific indicator of the disease. Examination of these lesions shows them to be filled with either the malignant cells themselves or deposits of para-IgM. The cancerous plasma cells causing WM do not generally secrete factors inducing osteoclast activation (unlike myeloma cells; see later), and bone lesions associated with bone pain are therefore seen in only 10% of WM patients. WM patients generally respond well to treatment with purine analogues such as cladribine.

III. MYELOMA

Myeloma is a very serious form of plasma cell cancer. Myelomas often arise from pre-existing MGUS disease when a minimally transformed plasma cell acquires additional mutations that cause it to divide aggressively and uncontrollably and synthesize high levels of a specific paraprotein. In its most advanced stages, large numbers of para-Ig-secreting tumor cells can be detected in multiple body sites in these patients, such that this disease is often referred to as *multiple myeloma*.

The reader will recall that we have encountered myeloma cells before. The structure of antibody proteins was elucidated by studying myeloma paraproteins (refer to Ch.5), and myeloma cells were used to develop hybridoma and monoclonal antibody technology (refer to Ch.7). It was the property of immortalization of the myeloma cell that allowed Köhler and Milstein to fuse it to a normal B cell and produce a long-lived hybrid cell secreting large amounts of specific antibody.

How do myeloma cells cause disease? The increased numbers of myeloma cells in the BM affect hematopoiesis such that the production of erythrocytes is decreased, resulting in anemia. Normal B cell production may be impaired by the presence of the malignant cells in the BM, leading to a state of immuno-suppression, which leaves the patients very vulnerable to infections. The physical presence of high concentrations of para-Ig in the blood can cause hypercoagulation and circulatory difficulties, and may decrease pulmonary function. Myeloma disease can also affect the CNS and lead to neurological symptoms. In many cases of myeloma, the transformed cells produce an abundance of free Ig light chains. When secreted into the urine, these free Ig light chains are called *Bence Jones proteins*, after their co-discoverers. The Bence Jones proteins bind to a particular glycoprotein in the urine called the *Tamm-Horsfall protein* to form multimeric aggregates. These aggregates can induce hypercalcemia and hypercalciuria (elevated calcium in the blood and urine, respectively). The deposition of the Bence Jones aggregates plus calcium in the kidney leads to a form of interstitial nephritis called "myeloma kidney" that can culminate in kidney failure.

i) Clinical and Biochemical Features

Myeloma occurs at an annual incidence of about 4 persons per 100,000 population. About 1% of all malignancies in the United States are myelomas, accounting for about 14,000 new diagnoses and 11,000 deaths per year. Myeloma is one of the few cancers whose incidence in developed countries has been steadily rising over the last few decades. One reason for this observation is that over 80% of myeloma cases occur in persons over 60 years of age, and this demographic stratum has been proportionally increasing in the populations of these countries. Myeloma incidence is slightly higher in men than women, and twice as many blacks develop the disease as whites. Social and economic factors have been ruled out as reasons for these differences, suggesting that, although no studies have conclusively linked hereditary factors to myeloma predisposition, genetic determinants likely play a contributing role. Indeed, some studies have indicated that expression of HLA-Cw2 favors myelomagenesis. Another major factor is exposure to ionizing radiation, which is an important driving force behind the mutational events leading to myelomagenesis. For example, epidemiological studies of individuals exposed to atomic radiation in Hiroshima and Nagasaki demonstrated a 5-fold increase in myelomagenesis beginning about two decades after exposure. Exposure to environmental carcinogens or certain chemicals also increases the rate of myeloma development.

Myeloma disease is classified according to a staging system (Table 30-9) based on the histology of the malignant cells (Plate 30-6), their pattern of distribution in the BM, and blood levels of hemoglobin, calcium, and paraproteins. The normal B cell markers CD19 and CD20 are not usually present on myeloma cells, although CD10 may appear on cells in advanced stage disease. An evaluation of these factors can give a good indication of whether the patient is likely to suffer a mild or aggressive disease course. However, regardless of histology, patients over 65 years of age have slightly lower survival rates.

In stage I myeloma disease, the myeloma clone remains partially reliant on bone marrow stromal cell factors for its survival and so is confined to the BM. Serum hemoglobin and calcium levels are normal, and only a few malignant

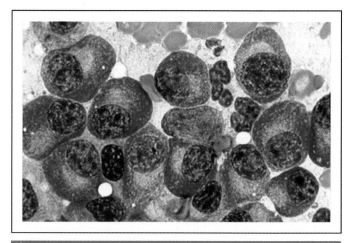

Plate 30-6

Multiple Myeloma
Bone marrow aspirate of a patient with multiple myeloma showing a predominance of plasma cells. Reproduced with permission from Tkachuk D. C., Hirschmann J. V. and McArthur J. R. (2002) "Atlas of Clinical Hematology." W. B. Saunders Company, Philadelphia.

plasma cells are observed in the BM. Any solid masses made up of myeloma cells are small. Bence Jones chains are present in the urine, but the overall production of Ig paraproteins is relatively low. Clinical symptoms are often mild and may go unnoticed. These patients are considered at low risk and are likely to survive at least 5 years. Stage II myeloma patients show subnormal hemoglobin, elevated serum calcium, and moderate levels of Ig paraprotein and Bence Jones proteins in the blood. In 70% of cases, the paraprotein is a form of IgG, with para-IgA making up another 20% of cases. Para-IgD and para-IgE are present in 2% and 0.1% of myeloma patients, respectively. Any masses of myeloma cells present in the BM are medium-sized. Stage II patients are considered to be at medium risk and can expect to live another 3 years after diagnosis. Patients with stage III myeloma disease are in the highest risk group and have a life expectancy of only 1 year. At this point, the myeloma cells have accumulated additional oncogenic mutations such that the clone loses its dependence on the bone marrow stroma. The transformed cells migrate out of the BM, infiltrating a variety of extramedullary tissues and disrupting their functions. Serum calcium is highly elevated and hemoglobin is severely decreased. Anemia and renal complications are sometimes present. Levels of Ig paraproteins and Bence Jones proteins in the serum and/or urine may be three times those in stage I patients. Radiography of the bones reveals multiple lesions so that most patients present with bone pain and exhibit the simultaneous presence of osteopenia (abnormally low numbers of bone cells) and osteoporosis. Myeloma cell masses present in the BM can be large. The degree of dispersion of myeloma cells in a stage III patient is a prognostic indicator: the more widely distributed the malignant cells, the worse the prognosis.

One of the biochemical consequences of myelomagenesis is the deregulation of several cytokines and chemokines, the

Table 30-9 Stages of Myeloma

	Stage I	Stage II	Stage III
Location	Bone marrow	Bone marrow	Bone marrow and extramedullary tissues
Serum Hb and Ca²⁺	Normal	↓ Hb; ↑ Ca²⁺	↓ Hb; ↑ Ca²⁺
Bence Jones protein	Low	Moderate	High
Ig paraprotein	Low	Moderate	High
Size of myeloma cell mass (if present)	Small	Medium	Large
Bone pain	No	No	Yes
Survival (untreated)	5 yr	3 yr	1 yr

most prominent of which is IL-6. IL-6 overexpressed either by the myeloma cells themselves or by stromal cells in the BM appears to drive the development and maintenance of the cancer. Indeed, the binding of myeloma cells to receptors on bone marrow stromal cells stimulates the synthesis of IL-6 by the latter. As well as its pro-inflammatory functions, IL-6 can act as a proliferation and differentiation factor for B lineage cells. IL-6 is known to trigger cell survival signaling through STAT3, which leads to the upregulation of the anti-apoptotic gene Bcl-xL and another gene called MCL1 (*myeloid cell factor 1*). The precise function of MCL1 is unknown but its expression appears to be essential for the survival of myeloma cells. IL-6 also induces the expression of *vascular endothelial growth factor* (VEGF) by myeloma cells that supports the construction of new blood vessels to support tumor growth. VEGF also promotes the migration of myeloma cells and inhibits antigen presentation by DCs.

Myeloma cells produce other cytokines and chemokines in addition to IL-6, with damaging consequences. A combination of IL-6, MIP-1α, IL-1β, and TNF secreted by the myeloma cells triggers bone marrow stromal cells to release OPGL, the master regulator of osteoclastogenesis (refer to Ch.17). Osteoblast production is decreased, osteoclasts are activated, and bone lesions are created, which produce severe bone pain. Additional cytokines implicated in myeloma disease progression include IL-2, IL-7, IL-11, LTα, and GM-CSF. The upregulation of these cytokines and their receptors tends to suppress Ig production as well as helper T and NK cell functions, thereby further increasing the chances of infection in these patients. TGFβ secreted by myeloma cells further stimulates IL-6 production. Lastly, the growth factor IGF-1 (*insulin-like growth factor 1*) secreted by bone marrow stromal cells may play a role in myelomagenesis, since myeloma patients with lower levels of this molecule in their serum have a better prognosis than those with higher levels. *In vitro*, IGF-1 treatment of isolated myeloma cells induces activation of the PI3K-PKB/Akt survival pathway and inhibition of apoptosis.

ii) Genetic Aberrations

The genetic aberrations leading to myelomagenesis are complex and have been much harder to identify than those associated with leukemias. There are some recurring chromosomal translocations associated with this disease (see later) but no one genetic change appears to be dominant or essential. Trisomy 3, 5, 6, 7, 9, 11, 15, or 19 has been reported in various myeloma clones. Interestingly, aberrations of chromosomes 6 and 9 are associated with the most favorable disease outcomes. In contrast, deletion of the chromosome 13q region is associated with a very poor prognosis. It is unknown which gene(s) in this region function to ward off myelomagenesis. At the molecular level, about 70% of myeloma patients have mutations in the Cdk enzymes themselves or in the p15 and p16 Cdk inhibitors required to control cell cycling. Hypermethylation of these genes (which leads to their inactivation) has also been found in some myeloma cells. Activation of Ras oncogenes and alterations to c-myc are other mutations reported in myeloma patients, especially those with stage II and III disease. Similarly, mutations of p53

are rare in early stage myeloma cells but are common in cells of relapse patients.

Translocations that affect the Igh locus appear to be an early event in the carcinogenic process that leads to stage III myeloma. One partner gene participating in an Igh translocation is the fibroblast growth factor receptor III (FGFR3) gene located at 4p16 (refer to Table 30-6). The fusion product derived from this union is thought to block apoptosis mediated by caspase-3. The IRF4 gene at 6p25, the cyclin D3 gene at 6p21, and the c-maf transcription factor gene at 16q23 have also been identified as partners in Igh translocations in various myeloma cases. Activation of the Bcl-1 gene (11q13) via translocation has been reported in myeloma cells, although the clinical significance of this activation is not clear. Translocations of Bcl-1 can also be found in some MGUS patients; it is rare that the MGUS cells in these cases convert to myelomas.

Despite their varied genetic defects, one interesting feature shared by the majority of myeloma cell clones is upregulation of the anti-apoptotic genes Bcl-2 and Bcl-xL. The overexpression of these cell survival proteins is also found in certain lymphomas (see later) due to a t(14;18) translocation between the Igh locus and the Bcl-2 gene. However, these types of translocations are not usually found in myeloma cells. While overexpression of IL-6 as described previously may account for the elevated Bcl-xL, why Bcl-2 should be upregulated in myeloma cells is unclear.

The most aggressive myeloma cases are associated with p53 mutations. As many as 50% of myeloma patients exhibit p53 mutations by the late stages of their disease, and this finding predicts a poor outcome. Chromosomal deletions at 13q14, the region containing the Rb tumor suppressor gene, are also associated with an unfavorable prognosis.

iii) Treatment

Myelomas are among the hardest HCs to treat. The 5-year survival rate for treated myeloma patients is only about 30%, despite recent advances in therapy methods. Radiographic examination of the myeloma patient is used to determine which modality of treatment to use. Patients in the early stages of an indolent myeloma generally do not need treatment for at least 2 years, unless the tumor starts to compress adjacent organs or the spinal cord. At later stages or in a more aggressive case, magnetic resonance imaging (MRI) may be helpful in evaluating the tumor load (numbers and locations of myeloma cells present). For patients with isolated myeloma masses, localized irradiation can be used to reduce the tumor load. As was true for CLL treatment, the timing and choice of myeloma therapy are especially important for older patients, for whom the consequences of medical intervention have to be balanced against quality of life.

Chemotherapy is the first treatment employed in myeloma patients with progressing disease. In 1962, Daniel E. Bergsagel introduced the use of the alkylating agent melphalan for the treatment of myeloma and it remains the standard chemotherapeutic drug. A combination of melphalan and prednisolone can induce a partial response in over 50% of myeloma patients, but very few achieve a CR. Ironically, a rapid response to this regimen is an unfavorable prognostic indicator, as it shows that the tumor cells are multiplying very rapidly. Several other chemotherapeutic approaches have been tried, including various combinations of cyclophosphamide, prednisolone, melphalan, carmustine, and/or doxorubicin. However, none has proved more effective clinically than the melphalan–prednisolone strategy. For patients refractory to melphalan–prednisolone or for relapse cases, an alternative regimen that includes dexamethasone, vincristine, and adriamycin has been implemented. Over half of refractory myeloma patients treated with this combination experience a partial response. Dexamethasone is thought to decrease IL-6 levels and also to inhibit NF-κB, thereby blocking the expression of anti-apoptotic and survival genes.

The drug *thalidomide*, which was originally prescribed for relief of nausea in pregnant women, became infamous for its devastating effects on fetuses. In the 21st century, this drug is finding new uses as a chemotherapeutic agent, particularly for treatment of myeloma. Thalidomide activates caspase-8 and promotes the apoptosis of both myeloma cells and cells making up the tumor vasculature. In addition, this drug blocks secretion of the IL-6 and angiogenic factors that support myeloma growth. However, the side effects of thalidomide use are significant. Researchers are now investigating several less toxic thalidomide analogues for efficacy. A more promising approach may involve a proteasome inhibitor called *bortezomib* (Velcade). Bortezomib prevents the degradation of ubiquitinated proteins, including the IκB inhibitor bound to NF-κB. If IκB is not degraded, NF-κB cannot be activated, and the abnormal upregulation of IL-6 expression that drives myeloma cell proliferation is blocked. Bortezomib also directly induces the apoptosis of myeloma cells and decreases VEGF-associated angiogenesis. Bortezomib treatment has been shown to induce clinical responses in a subset of refractory myeloma patients, prolonging their lives. Turning to the biologicals, IFNα has been shown to prolong remission in myeloma patients but does not significantly increase their life spans.

The bone pain experienced by myeloma patients is caused by excessive osteoclast activation, so that treatment with a bisphosphonate is usually of great benefit. This class of drugs, which includes zoledronate and pamidronate, blocks osteoclast activity and so prevents osteoporosis. Either oral or parenteral administration is effective. An experimental treatment for myeloma-associated osteoporosis involves the decoy receptor OPG discussed in Chapter 17. OPG downregulates the signaling initiated by OPGL binding that drives osteoclastogenesis.

Autologous HCT is now a realistic therapy option for myeloma patients, especially for those younger than 70 years of age. The overall survival of a myeloma patient after autologous HCT is about 5 years on average, and patients remain disease-free for about 3–4 years. Some patients have received "back-to-back transplants" (a second autologous HCT soon after the first) in an attempt to reduce residual tumor cells as much as possible and further increase survival. Allogeneic HCTs have also been tried, to take advantage of the postulated *graft versus myeloma* effect. However, the morbidity and mortality in myeloma patients that undergo HCT are much worse than those in leukemia patients that receive similar transplants. The more advanced age of myeloma patients at diagnosis and

their inherent increased susceptibility to infection may account for this difference. There is some evidence that a non-myeloblative approach coupled with allogeneic HCT may be of clinical benefit to some myeloma patients.

F. Lymphomas

As stated at the start of this chapter, lymphomas are solid cancers that generally initiate from the malignant transformation of a single lymphocyte resident in an organized lymphoid tissue such as the spleen, thymus, or a lymph node. When the malignancy arises in a diffuse lymphoid tissue, such as that associated with the GI tract, the lymphoma is referred to as *extranodal*. The transformed lymphocyte proliferates to establish a clone that remains dependent on the surrounding stromal cells for growth and survival factors and vital intercellular contacts. As a result, lymphoma cells do not usually migrate into the blood. However, if a lymphoma cell undergoes a mutation that allows it to proliferate in the absence of stroma, or to survive in the presence of growth/survival factors in the blood, the lymphoma cell may travel in the circulation and become a leukemic cell.

We noted previously that lymphomas are broadly classified into *Hodgkin's lymphomas* (HLs) and *non-Hodgkin's lymphomas* (NHLs). In HL, the tumor mass is made up of a so-called *reactive infiltrate* composed of non-transformed lymphocytes, histiocytes, fibroblasts, and other cell types. Only a tiny percentage of cells in the tumor mass are in fact cancerous, and these cells are a peculiar B lineage-like cell type called *Reed-Sternberg cells* (see later). In NHL, the solid mass consists almost entirely of malignant lymphocytes. The cancerous clone is usually derived from a peripheral B cell but sometimes from a peripheral T cell. In some situations, the cellular origin of the lymphoma cannot be clearly defined.

The progression of any lymphoma can be described in four stages (Table 30-10). In stage I lymphoma, one or more lymph nodes enlarge in a single, localized region. In stage II lymphoma, two or more lymph node regions that are adjacent to each other are involved, but on only one side of the diaphragm. One additional organ close to the original affected node may also be infiltrated. In stage III lymphoma, two or more lymph nodes on both sides of the diaphragm are affected. Tumor cells may be found in the spleen or other organs near an affected lymph node. In stage IV lymphoma, the tumor mass disseminates widely into multiple lymph nodes and the BM as well as into several organs and tissues.

We will now discuss the clinical, histological, and biochemical features of HL, its subtypes, and treatment, before doing the same for NHL.

I. HODGKIN'S LYMPHOMA (HL)

The Reed-Sternberg (RS) cells that define an HL tumor were first identified through microscopic examinations of abnormal cells in the enlarged nodes of Hodgkin's disease patients by Carl

Table 30-10 Stages of Lymphoma

Stage	Diagnostic Features
I	One or more enlarged lymph node(s) in a single lymph node region, lymphoid structure, or extralymphoid site
II	Two or more enlarged lymph node regions on the same side of diaphragm; tumor cells may also be present in a single organ near an affected node
III	Two or more enlarged lymph nodes on both sides of the diaphragm; tumor cells may also be present in the spleen and/or another organ near an affected node
IV	Wide dissemination of tumor cells into multiple lymph nodes, BM, liver, and multiple organs

Sternberg in 1898 and Dorothy Reed in 1902. These researchers found that each affected lymph node contained a few large, multi-nucleated cells within an infiltrate of normal-looking lymphocytes, fibroblasts, and other cell types. These cells became known as "Reed-Sternberg cells" in honor of these investigators. RS cells were then demonstrated to be clonal in their growth, suggesting that they were transformed. Indeed, HL was subsequently shown to be a true hematopoietic cancer in which the RS cells are the tumor cells of the HL mass. Tissues showing the infiltration of an HL mass are said to be "HL-involved."

i) Clinical and Histological Features

HL is a rare disorder, representing 0.7% of all malignancies in the United States with an annual incidence of 3 per 100,000 population. Unlike most solid cancers, HL usually affects relatively young patients between the ages of 15 and 35 years. Males are affected slightly more often than females. Most patients initially present with a lump in the neck region due to one or more enlarged lymph nodes, although mediastinal masses may appear upon chest radiography. Patients also experience a constellation of systemic complaints known as *B symptoms* that are defined as unexplained rapid weight loss, fatigue, cyclic bouts of fever, and night sweats frequently accompanied by chest pain. (Patients that have tumor cells

but none of these systemic symptoms are said to be "A" patients.) *Pruritis* (intense itching) is present in about 20% of patients. HL patients are highly vulnerable to fungal and viral infections, particularly with varicella.

Biopsies of the enlarged lymph nodes of HL patients reveal that the multi-nucleated RS cells are sparsely dispersed among lymphocytes and other cells in the tumor mass at a frequency of 0.1–1% (Plate 30-7). As well as their multiple nuclei, RS cells are identified by their distinctive shape and expression of CD30. Expression of CD15, CD25, CD40, CD95, ICAM-1, and HLA-DR on the surfaces of RS cells is also common. What is the significance of CD30 expression on RS cells? The RS cells found in most HL tumors are thought to be derived from B cell precursors at an early stage of differentiation. Accordingly, the Ig genes are often rearranged in RS cells, but Ig protein is not usually expressed. As we learned in Chapter 9, in normal developing B cells, a lack of surface Ig expression leads to apoptosis. CD30 is a transmembrane glycoprotein found mainly on activated T cells whose engagement leads to NF-κB activation and the expression of survival genes such as FLIP. When CD30 on an RS cell is engaged, the resulting expression of FLIP blocks FADD-mediated signal transduction and allows RS cells to resist Fas-mediated apoptosis. Indeed, isolated RS cells in which the expression of FLIP protein is inhibited succumb to both Fas- and TNF-induced apoptosis, despite the expression of other survival genes. Thus, even though they do not express surface Ig, RS cells employ means such as CD30-mediated overexpression of FLIP to escape death.

One curious feature of HL is that the RS cells appear to travel via the lymphatics in a sequential fashion from one lymph node to the next lymph node in the anatomic chain. It is sometimes said that a "chain" or "cluster" of HL-involved tissues results from the "spreading of the disease to contiguous lymph nodes." It is not understood how this unique pattern of HL spreading occurs. It may be that a particular microenvironment is needed to sustain the growth of RS cells, and that this microenvironment must be in place before a migrating RS cell can proliferate in the new location. Since cytokines are such a prominent feature of HL (see later), some scientists have speculated that cytokine molecules accumulating in an affected lymph node spill down the efferent lymphatic leaving

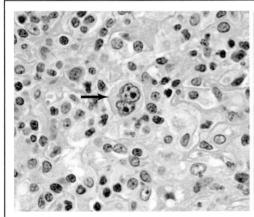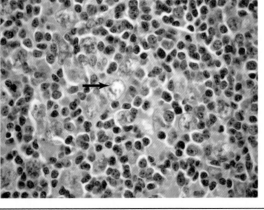

Plate 30-7

Reed-Sternberg and Popcorn Cells in Hodgkin's Lymphoma
Left: Arrow shows a binucleate Reed-Sternberg cell in the lymph node of a patient with classical HL. The Reed-Sternberg cell has multiple red-staining nucleoli. Right: Arrow shows a popcorn cell in a lymph node of a patient with non-classical HL. Courtesy of Dr. Doug Tkachuk, Princess Margaret Hospital, UHN, Toronto.

that node and enter the next node in the chain. These cytokines may then alter conditions among the cells in the second node such that a reactive infiltrate is recruited or established. When an RS cell does break away from the original node and reaches the second node, conditions may then be ripe for its continued survival and proliferation. If this hypothesis is true, it would account for the fact that RS cells in the early stages of HL do not appear to be able to establish just anywhere in the body. Histological studies have confirmed that lymph nodes near an HL-involved lymph node frequently contain reactive infiltrate but no RS cells. In the later stages of the disease, however, the RS cells acquire additional mutations that allow them to become independent of the support of cytokines and the reactive infiltrate. The disease then spreads to additional lymph nodes in a non-contiguous fashion, and invades non-lymphoid organs and tissues.

ii) Viral Involvement?

The exact etiology of HL is unknown, but some scientists theorize that infection by EBV, a virus capable of transforming lymphocytes, is a causative factor in at least some cases. The LMP1 (*latent membrane protein 1*) and EBNA2 (*Epstein Barr nuclear antigen 2*) molecules of EBV can be found on the surfaces of RS cells in 40% of HL tumors, suggesting that these viral proteins might have a role in the development of this lymphoma. LMP1 is known to act through the CD40 signaling pathway and to trigger NF-κB-mediated induction of Bcl-2 expression. Significantly, LMP1 and activated NF-κB have been found in other tumors in which EBV is believed to have a role in transformation, such as nasopharyngeal carcinoma and Burkitt's lymphoma. Moreover, clinical studies have found that HL is 2- to 5-fold more frequent in patients that have had infectious mononucleosis (which is caused by EBV infection). However, fully 60% of HL cases appear to be independent of EBV. Many clinical researchers therefore believe that EBV does not play any role in the transformation process leading to HL, and that the presence of EBV proteins on RS cells is simply a reflection of the coincidental presence of this virus in many individuals. It is estimated that over 80% of humans have been exposed to EBV at some point in their lives. However, only a very small number of them develop HL. In addition, while epidemiological studies have shown that HL is more common in developed than developing countries, EBV is found more frequently among HL patients in the lower socio-economic strata of Western cultures than in wealthier groups. Moreover, women in the highest socio-economic stratum, who are the group most frequently infected by EBV, have the lowest incidence of HL. If HL does develop in one of these women, it manifests as the subtype associated least frequently with the presence of EBV.

iii) Genetic Aberrations

The precise genetic aberrations leading to HL transformation are not known. Unlike the leukemias and NHL subtypes (see later), recurring chromosomal translocations are not common in HL. It has been difficult to ascertain what types of mutations are associated with HL because of the challenges in obtaining adequate amounts of relatively pure populations of fresh RS cells for biochemical and molecular analyses. Studies to date have shown that the tumor cells of between 25 and 75% of HL patients have cytogenetic anomalies, including trisomy 1, 2, 5, 12, or 21. A t(14;18) translocation that activates Bcl-2 was identified in one study, but its clinical significance remains questionable. P53 mutations have been found in only a small percentage of HL cases, and no abnormalities associated with c-jun, Raf, Ras, or c-myc have been found in HL tumors. Most researchers believe that the unknown genetic changes underlying HL somehow lead to the constitutive activation of NF-κB in RS cells. Research is doggedly ongoing to determine the exact nature of these mutations.

iv) Role of Cytokines in HL

Whatever the defect in the genome of an RS cell, the abnormality has a profound effect on the regulation of cytokines and their receptors. This dysregulation appears to be a crucial factor in HL disease. HL patients succumb readily to fungal and viral infections, indicating that Th1 responses are impaired. However, antibody responses to pathogens are functional. This reduction in cell-mediated immunity coupled with normal humoral immunity suggests that immune deviation to a Th2 response may be occurring during HL development. The pruritis observed in HL patients might then be due to the triggering of IgE production by a deregulated Th2 cytokine.

A multitude of cytokines has been implicated in HL in various studies, including TNF, TGFβ, IL-1, IL-5, IL-6, IL-9, and IL-10. However, work at the Ontario Cancer Institute in Toronto by Ursula Kapp and Brian Skinnider indicated that it is IL-13 that plays the most important role in HL pathogenesis. Studies of both primary RS cells and cultured RS cell lines showed that only IL-13 is consistently and abundantly expressed by RS cells in different HL patients. The IL-13 receptor is also upregulated on RS cells, establishing an autocrine loop of IL-13R engagement. It is known that IL-13 and CD40 signaling promote the survival of normal B cells by inducing expression of Bcl-xL. In RS cells, persistent IL-13 signaling results in the constitutive activation of STAT6, which in turn activates transcription of genes that drive proliferation and survival. Indeed, the anti-apoptotic gene Bcl-xL is highly expressed by primary RS cells in many cases of HL. Moreover, when antibody capable of blocking IL-13 signaling is added to cultured RS cell lines, the proliferation of the transformed cells is blocked. Some RS cells deprived of IL-13 even undergo spontaneous apoptosis. These results suggest that it is autocrine IL-13 signaling that permits the transformed B cell to become a full-fledged RS cell. IL-13 and other cytokines frequently elevated in HL may also contribute to the recruitment of the cells composing the reactive infiltrate, and this infiltrate may secrete still more cytokines and/or supply intercellular contacts that promote RS survival and proliferation. For example, T cells in the reactive infiltrate express CD40L, which engages CD40 on the RS cells, while eosinophils express CD30L, which binds to CD30. A complicated series of positive feedback cytokine loops may be generated that perpetuate the disease.

v) Subtypes of HL

Five subtypes of HL are defined by the morphology and marker expression of the transformed cells in the HL tumor, the relationship of these cells to the surrounding cells of the reactive infiltrate, and the lineage and morphology of the reactive infiltrate cells (Fig. 30-6). Four of these subtypes comprise 95% of HL tumors and are considered *classical HL* in that they all contain RS cells that generally have a surface immunophenotype of CD30$^+$, CD15$^+$, CD20$^-$, CD45$^-$. These subtypes are *nodular sclerosis HL, mixed cellularity HL, lymphocyte-rich classical HL,* and *lymphocyte-depleted HL.* The fifth subtype, *nodular lymphocyte-predominant HL,* is considered to be a *non-classical HL* and comprises the remaining 5% of HL tumors. The malignant cells in this subtype are not RS cells but a different cell type called a *popcorn cell* or *L&H cell* (for *lymphocytic and histiocytic*) (refer to Plate 30-7). Popcorn cells generally have a surface immunophenotype of CD30$^-$, CD15$^-$, CD19$^+$, CD20$^+$. Unlike the Ig genes in RS cells, the Ig genes of popcorn cells show evidence of considerable somatic hypermutation.

va) Nodular sclerosis Hodgkin's lymphoma (NSHL).
NSHL is the most common subtype of HL, accounting for about 73% of all cases. The majority of patients are young adults. Patients usually present with enlarged lymph nodes in the mediastinal and supra-diaphragmatic areas. The B symptoms of weight loss, fever, and night sweats are common, as are pruritus and a transient feeling of pain in the lymph nodes after alcohol consumption. Within the NSHL tumors, the RS cells are surrounded by normal lymphocytes, fibroblasts, histiocytes, eosinophils, plasma cells, and neutrophils. NSHL acquired its name because the tumor is typically divided into nodules by collagenous bands, thought to be synthesized in response to TGFβ secreted by RS cells and surrounding eosinophils. Significantly, IL-13 is a potent stimulator of TGFβ production and fibrosis.

vb) Mixed cellularity Hodgkin's lymphoma (MCHL).
MCHL accounts for about 20% of all HL cases. Unlike NSHL, this disease can strike patients of any age. MCHL patients often present with enlarged lymph nodes in the spleen and abdomen, although involvement in the mediastinal area can also occur. The tumor lacks the nodules and fibrous bands of NSHL. The RS cells in these tumors are nested among normal lymphocytes, epithelial cells, histiocytes, eosinophils, and plasma cells.

vc) Lymphocyte-depleted Hodgkin's lymphoma (LDHL).
LDHL is a relatively rare form of classical HL with an incidence of about 1% of all HL cases. The disease usually strikes elderly individuals, and patients are often diagnosed when the cancer is quite advanced. The spleen, abdomen, and BM are often involved, and fibrosis and necrosis may be present. LDHL is distinguished from other classical HL subtypes by the dearth of cells surrounding the RS cells in LDHL tumors. Indeed, LDHL can sometimes be mistaken for non-Hodgkin's lymphoma (which lacks RS cells) if careful histology and immunophenotyping are not performed. Interestingly, LDHL is the most common type of HL in patients with immunodeficiency, such as AIDS sufferers.

vd) Lymphocyte-rich classical Hodgkin's lymphoma (LRCHL).
This subtype of classical HL occurs in a very small fraction of HL patients and is the least aggressive form of the disease. The RS cells in these patients are surrounded by higher numbers of normal lymphocytes in the reactive infiltrate than in the other classical subtypes, and the clinical behavior of the tumor is more like the non-classical nodular lymphocyte-predominant HL type than a classical HL tumor.

ve) Nodular lymphocyte predominant Hodgkin's lymphoma (NLPHL).
NLPHL, the sole non-classical HL subtype, has an incidence of about 5% of all HL cases. The median age of NLPHL patients is 30 years, and men are affected three times as often as women. Most patients present during the early stages of the disease with enlarged lymph nodes but few other symptoms. As noted previously, rather than CD30$^+$ RS cells, NLPHL tumors contain the CD19$^+$CD20$^+$ popcorn cells. Interestingly, EBV proteins are also found on NLPHL

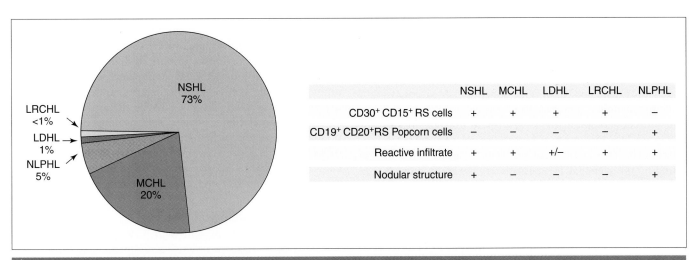

	NSHL	MCHL	LDHL	LRCHL	NLPHL
CD30$^+$ CD15$^+$ RS cells	+	+	+	+	–
CD19$^+$ CD20$^+$RS Popcorn cells	–	–	–	–	+
Reactive infiltrate	+	+	+/–	+	+
Nodular structure	+	–	–	–	+

Figure 30-6

Relative Frequencies and Characteristics of Hodgkin's Lymphoma Subtypes in North American Adults
Five subtypes of HL are defined by the shape and marker expression of their tumor cells, the presence and composition of the reactive infiltrate, and the overall morphology of the tumor mass. NSHL, nodular sclerosis HL; MCHL, mixed cellularity HL; LRCHL, lymphocyte-rich classical HL; LDHL, lymphocyte-depleted HL; NLPHL, nodular lymphocyte predominant HL; RS, Reed-Sternberg.

popcorn cells. As suggested by the name of this disease, the popcorn cells are surrounded by large numbers of lymphocytes as well as a few epithelioid histiocytes. In about 10% of NLPHL patients, the disease progresses to NHL (see later).

vi) Treatment of HL

The effective management of HL began in the 1940s when R. Gilbert and M.V. Peters started to irradiate lymphatic tissues adjacent to affected lymph nodes. Henry Kaplan of Stanford University refined this approach through clinical trials designed to define minimum effective doses and appropriate fields of radiation. The collective efforts of these workers led to improved regimens that allowed 80% of HL patients whose disease was caught in the early stages to survive for at least 10 years. By the 1960s, the addition of chemotherapy to the armamentarium began to make a clinical difference in patients with late stage HL disease. As many as 50% of late stage patients were "cured" by treatment with a combination of drugs called MOPP (*mechlorethamine, [O]-vincristine, procarbazine,* and *prednisolone*). Subsequently, a version of this cocktail with fewer detrimental side effects was developed called ABVD: *adriamycin (doxorubicin), bleomycin, vinblastine,* and *dacarbazine*.

The prognosis of all HL patients was improved considerably in the early 1980s by the discovery that HL disease spread through the lymphatics to adjacent lymph nodes. Prophylactic irradiation of lymphatic channels in the immediate area of the affected node, a procedure called *extended field radiation*, became routine and increased the survival of many HL patients. Today, while some early stage HL patients may require only radiation treatment, chemotherapy is advised for patients with more aggressive subtypes of the disease. Patients with advanced HL undergo multiple rounds of ABVD chemotherapy combined with aggressive local irradiation. These regimens, coupled with good clinical management to control infections, have enabled about 80% of both early and later stage HL patients to enjoy long term disease-free survival. The exact treatment regimen may vary by HL subtype. For example, because NLPHL tumors are usually highly localized, they can sometimes be treated by surgical excision alone, or with surgery plus *involved field radiation* (only the tumor region is irradiated). Anti-CD20 mAb therapy has also shown promising preliminary results in NLPHL patients because, unlike RS cells, popcorn cells express CD20.

Unfortunately, the increased doses of chemotherapy and radiation required to achieve these excellent results for advanced HL are associated with several serious secondary effects. The impact of these problems is made more devastating by the fact that HL patients are generally quite young (20–40 years of age). Sterility and hypothyroidism are not uncommon after extensive MOPP or ABVD treatment, and pulmonary and cardiac complications such as heart disease and coronary and carotid artery occlusion are frequent. However, the most sobering unintended consequence of aggressive HL treatment is a high rate of secondary tumorigenesis. AML, NHL, and even some NHCs such as lung and breast cancers have appeared in HL patients within 5–15 years after initial treatment. Indeed, about 15% of conventionally treated HL patients can expect to develop a secondary cancer

within 20 years, meaning that they may survive their HL at age 20 only to be felled by AML at age 40.

It is the desire to eliminate these secondary cancers and severe complications that drives the development of alternative therapies for HL. Patients whose disease resists aggressive chemotherapy and irradiation, or who want to avoid the side effects of conventional treatment, now have several other options, some of which are still experimental in nature. First, autologous HCT has been used to cure some HL patients, and allogeneic HCT is under investigation. Secondly, preclinical trials are ongoing to evaluate the effectiveness of immunotoxins directed against the CD30 marker found on the RS cells of the majority of HL cases. Promising results have been obtained in studies of cultured RS cell lines and mice inoculated with HL tumors. Finally, the administration of neutralizing antibodies against IL-13, the growth factor postulated to drive RS proliferation and survival, is also being investigated as a novel therapy for HL.

II. NON-HODGKIN'S LYMPHOMAS

The NHLs are a family of heterogeneous cancers. Unlike HL, the cancerous mass is composed almost entirely of malignant lymphoid lineage cells. In addition, NHLs differ from HL in their lack of unique tumor cell types such as RS cells or popcorn cells. NHLs also pose a more significant health problem than HLs, at least in developed societies. NHLs account for about 3–5% of all human cancers, occurring twice as often in men as in women. The median age of NHL patients is 65 years, considerably older than that for HL patients; however, significant numbers of young people between the ages of 30 and 40 also succumb to NHL disease. Interestingly, the incidence of NHL has doubled over the last three decades. Although a small percentage of this "epidemic" is due to a 10-fold increase in various lymphomas in AIDS patients, an across-the-board elevation of NHL incidence has been found for all subtypes. This higher incidence has been attributed to a combination of factors, including (i) an expanding population of older individuals, (ii) new imaging technologies that can detect previously unnoticed lymphomas, and (iii) the reclassification of some HL malignancies under the NHL umbrella. While specific genetic determinants promoting the development of NHL have yet to be definitively identified, there is accumulating evidence that prolonged exposure to pollution, pesticides or herbicides, paints, wood preservatives, oil or rubber products, chemical hair dyes, or the highly toxic chemical defoliant "Agent Orange" (used in the 1970s during the Vietnam War) may promote lymphomagenesis.

i) Clinical and Histological Features

NHL patients usually present with B symptoms (unexplained weight loss, fever, night sweats) accompanied by fatigue and pain in the bones, chest, and abdomen. Physical examination may reveal lymphadenopathy, hepatomegaly, and/or splenomegaly. The lymphadenopathy may "wax and wane" over time, meaning that the lymph nodes swell and resolve repeatedly. Sometimes the NHL presentation is extranodal, in that the lymph nodes are not obviously affected and the suspicious lumps are present in the skin, GI tract, salivary gland, eye, cervix, or tonsils. A late diagnosis of NHL can usually be confirmed by BM biopsy, as most NHLs go on to involve the BM. Radiography may reveal the spread of the disease to the stomach, kidneys, ovaries, spleen, or liver.

Unlike HL, NHL does not usually spread sequentially from lymph node to lymph node in a contiguous fashion. Rather, NHL cells may migrate from the initial affected lymph node and travel via the blood and lymphatics to scattered nodes in different regions of the body. While this mechanism is just like that underlying NHC metastasis, some clinical researchers refer to this process as "spreading to non-contiguous lymph nodes" because the primary and secondary sites are both lymph nodes. The invasion of a different tissue, such as breast or lung, by an NHL tumor cell would be considered NHL metastasis by these clinicians. It is postulated that NHL does not exhibit the contiguous spreading pattern of HL because NHL tumor cells that have left the original lymph node are no longer reliant on a particular microenvironment, making local and distant lymph nodes equally suitable for invasion.

ii) Viral Involvement
While the cause of most cases of NHL is unknown, infection by a transforming virus has been linked to several types of NHLs. Different NHLs are thus prevalent in different geographic areas, depending on the global distribution of the relevant transforming virus. For example, the greatest numbers of T cell

lymphoma cases occur in Japan, the Mediterranean, and the Caribbean islands, where there is a high rate of infection by HTLV-1. As mentioned in Chapter 26, HTLV-1 is an oncogenic retrovirus capable of inducing T cell malignancies (both leukemias and lymphomas) several decades after infection. Similarly, Burkitt's lymphoma (linked to EBV infection) is most often found in equatorial Africa, where this virus is highly prevalent. Interestingly, Italy has the highest incidence of gastric lymphoma, which may be linked to the increased prevalence of *H. pylori* bacteria in that country.

Consistent with the involvement of a pathogen in NHL pathogenesis, immunodeficiency (either inherited or acquired) appears to contribute to some cases of NHL development. For example, patients with AIDS, SCID, CVID, hypogammaglobulinemia, Wiskott-Aldrich syndrome, or ataxia-telangiectasia all have a dramatically enhanced risk of developing NHL. In most of these situations, EBV has been strongly implicated as the transforming agent.

iii) Subtypes of NHL
The NHLs cover a wide range of lymphoproliferative malignancies with very different histological and clinical features and incidence (Fig. 30-7). The World Health Organization (WHO), in conjunction with European and American lymphoma working groups, has worked out the REAL (*revised European-American classification of lymphoid neoplasms*) classification scheme of NHLs based on a complex portfolio of morphological and immunophenotypic criteria, clinical

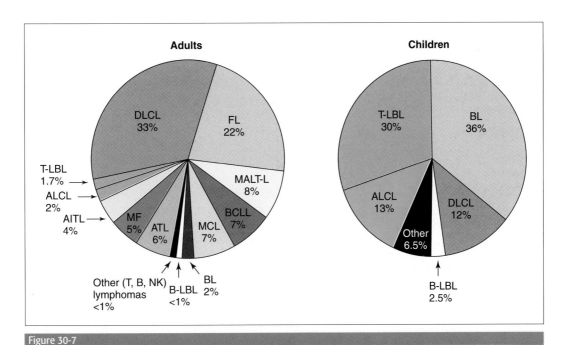

Figure 30-7

Relative Frequencies of Non-Hodgkin's Lymphoma Subtypes in North American Adults and Children
B cell NHLs are shown in shades of grey, while the major T cell NHLs are shown in shades of blue. DLCL, diffuse large cell lymphoma; FL, follicular lymphoma; MALT-L, mucosa-associated lymphoid tissue lymphoma; BCLL, B cell chronic lymphocytic lymphoma; MCL, mantle cell lymphoma; BL, Burkitt's lymphoma; B-LBL, precursor B cell lymphoblastic leukemia/ lymphoma; ATL, adult T cell lymphoma; MF, mycosis fungoides (includes Sezary syndrome cases); AITL, angioimmunoblastic T cell lymphoma; ALCL, anaplastic large cell lymphoma; T-LBL, precursor T cell lymphoblastic leukemia/lymphoma.

features, and genetic aberrations. However, even within a given subtype, there are variations of disease with features that may substantially overlap another subtype. As noted previously, in the majority of NHLs, the target cell is a peripheral B cell, but some are derived from a transformed peripheral T cell. Some oncologists roughly sub-classify B cell NHLs into "pre-GC NHLs" and "post-GC NHLs" (Fig. 30-8), depending on whether the Ig genes of the tumor cells show somatic

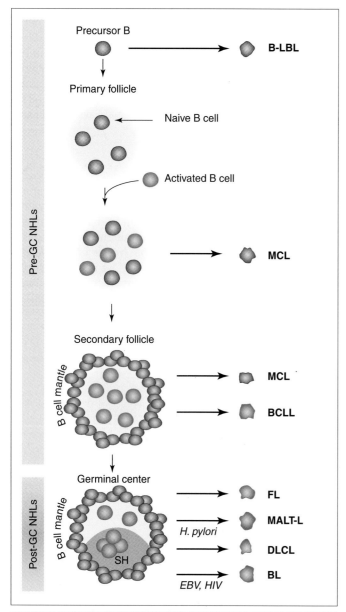

Figure 30-8

Development of B Cell Non-Hodgkin's Lymphomas
B cell NHLs are sub-classified as either pre-GC (germinal center) or post-GC diseases depending on whether the target cell acquires somatic hypermutations (SH) prior to transformation. Pathogens that have been associated with the development of some (but not all) cases of a particular lymphoma subtype are indicated. B-LBL, precursor B cell lymphoblastic leukemia/lymphoma; MCL, mantle cell lymphoma; BCLL, B cell chronic lymphocytic lymphoma; FL, follicular lymphoma; MALT-L, mucosa-associated lymphoid tissue lymphoma; DLCL, diffuse large cell lymphoma; BL, Burkitt's lymphoma.

hypermutations. The T cell NHLs are classified according to tumor cell morphology and surface marker expression pattern (Fig. 30-9). We continue with a brief description of some of the more important NHLs.

iiia) Precursor B lymphoblastic leukemia/lymphoma (B-LBL). B-LBL is a very aggressive lymphoma with biological and clinical similarities to ALL. Children are stricken more frequently than adults, with B-LBL representing about 2.5% of childhood NHLs but less than 1% of adult NHLs. Males are affected twice as often as females. Onset is rapid, such that more than 70% of patients present with stage IV disease. Lymphadenopathy is obvious and may compress nearby anatomical structures. Mediastinal masses are not usually present (unlike cases of precursor T lymphoblastic leukemia/lymphoma, T-LBL; see later). Patients complain of B symptoms and immune responses are impaired so that opportunistic infections are frequent. CNS involvement is common (33% of cases) and patients sometimes have neurological deficits. Histologically, the lymph nodes, BM, spleen, and/or CNS show infiltration by immature B cells. The invasion of the BM by the malignant cells leads to ineffective hematopoiesis in many cases. The rapidly dividing tumor cells are large with little cytoplasm and odd-looking nuclei (Plate 30-8). The nucleoli are indistinct and the chromatin looks "dusty." B-LBL tumor cells are usually positive for expression of TdT, CD79a, CD5, CD19, and CD43, but low or negative for CD10, CD20, and CD23 (Table 30-11). The Igh genes may be rearranged. The underlying genetic aberrations have yet to be defined. Mortality is high unless immediate chemotherapy (including intrathecal therapy) is instituted. Although relapses are frequent, chemotherapy is generally quite successful. Five year survival rates in treated children are much better (85%) than in treated adults (50%). Autologous or allogeneic HCTs are equally beneficial for adult patients younger than 60 years who lack CNS involvement and are in their first remission.

iiib) Mantle cell lymphoma (MCL). MCLs comprise about 7% of NHL and occur 4–6 times more often in men than in women. The disease usually strikes adults over 50 years of age. The disease course is often aggressive, and by the time most MCL patients seek out a physician, their disease has already disseminated and can be classified as stage IV. At presentation, common findings are generalized lymphoadenopathy and involvement of the BM, spleen, and liver. Strikingly, the colon is an involved extranodal site in a large proportion of MCL patients. About half of MCL patients present with B symptoms. When anemia and thrombocytopenia are present, they are usually mild.

The component cells of an MCL are derived from naive B cells making up either a primary follicle, or from cells in the follicular mantle (sometimes called the *mantle zone*) of a secondary follicle. MCL tumor cells are primarily small- to medium-sized, irregularly shaped B cells at a pre-GC stage and show little evidence of Ig somatic hypermutation. In general, these tumor cells multiply to fill an expanded mantle area surrounding the (benign) germinal center (Plate 30-9). The pattern of growth may be nodular, diffuse, or a combination of the two. In the late stages of the disease, the tumor

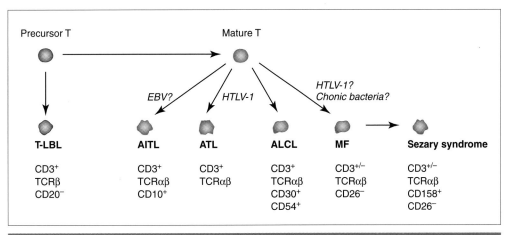

Figure 30-9

Distinguishing Features of T Cell Non-Hodgkin's Lymphomas
T-LBL arises from precursor T cells in either the bone marrow or the thymus while mature T cells give rise to the other forms of T cell NHL. In each case, the tumor cells express or lack expression of the indicated markers. Only the TCRβ gene is rearranged in T-LBL, while both the TCRα and TCRβ genes are rearranged in mature T cell NHLs. Pathogens that have been associated with the development of some (but not all) cases of a particular lymphoma subtype are indicated. T-LBL, precursor T cell lymphoblastic leukemia/lymphoma; AITL, angioimmunoblastic T cell lymphoma; ATL, acute T cell leukemia/lymphoma; ALCL, anaplastic large cell lymphoma; MF, mycosis fungoides.

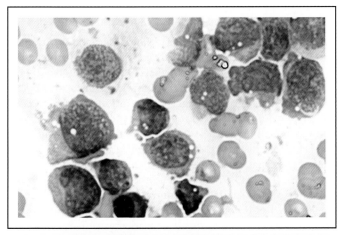

Plate 30-8

Precursor B Lymphoblastic Leukemia/Lymphoma
Bone marrow aspirate from a B-LBL patient showing a predominance of B lymphoblasts. Courtesy of Dr. Doug Tkachuk, Princess Margaret Hospital, UNH, Toronto.

cells obliterate the germinal centers and interfollicular areas. MCL cells express the B cell markers CD19, CD79a, CD20, and CD22, and are positive for CD5 and CD43 but negative for CD10, CD23, and TdT. Surface IgM and IgD are present and usually associated with the λ light chain rather than the κ light chain. Progression of MCL to DLCL (see later) is rare.

The characteristic chromosomal anomaly associated with MCL is the t(11;14)(q13;q32) translocation resulting in Bcl-1 activation. In cases of MCL lacking this chromosomal translocation, other genetic modifications appear to force the overexpression of cyclin D1. For example, mutations of the cell cycle regulators p15 and p16 have been associated with MCL, as has trisomy 12. Some aggressive MCL tumor cells have alterations to the p53 gene along with decreased expression of the Cdk inhibitor p27.

MCL is generally considered to have a poor prognosis, with or without treatment. About 40–50% of MCL patients treated with CHOP chemotherapy achieve a CR. However, the rapid and persistent proliferation of the tumor cells means

Table 30-11 Immunophenotyping of B Cell NHLs

Lymphoma	CD5	CD10	CD19	CD20	CD22	CD23	CD43	CD79a	TdT
B-LBL	+	Lo or −	+	Lo or −	+	−	+	+	+
MCL	+	−	+	+	+	−	+	+	−
BCLL	+	−	+	+	−	+	+	+	−
FL	−	+/−	+	+	+	+/−	−	+	−
MALT-L	−	−	+	+	+	−	+/−	+	−
DLCL	+/−	+/−	+	+	+	−	−	+	−
BL	−	+	+	+	+	−	−	+	−

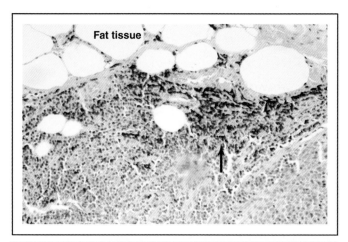

Fat tissue

Plate 30-9

Mantle Cell Lymphoma
Lymph node of MCL patient showing disruption of the lymph node capsule and invasion of surrounding fat tissue by lymphoma cells (arrow). Courtesy of Dr. Doug Tkachuk, Princess Margaret Hospital, UNH, Toronto.

that few patients will still be in remission 12 months after treatment. As a result, the median survival rate of this disease is 3 years. Anti-CD20 mAb and IFNα treatments can induce partial remission in relapsed patients. For younger MCL patients, an autologous or allogeneic HCT may be an attractive option (even considering the risk of GvHD) because of the short life expectancy associated with this type of lymphoma.

iiic) B cell chronic lymphocytic lymphoma (BCLL). BCLL (also known as *small lymphocytic lymphoma*) can exist as a leukemia, a lymphoma, or both, depending on the circumstances. In its leukemia form, BCLL is very similar to CLL. In its lymphoma form, BCLL comprises about 7% of all NHL. The disease primarily strikes adults over 60 years of age, affecting twice as many men as women. Patients present with generalized lymphoadenopathy and infiltration of the spleen, BM, liver, and sometimes peripheral blood. The small round B cells making up the tumor are positive for the B cell markers CD19, CD20, and CD79a and CD23 as well as CD5 and CD43, but negative for CD10, CD22, and TdT. Most cases of BCLL are pre-GC diseases with no somatic mutations present in the Ig genes. A minority of BCLL cases are post-GC in phenotype, indicating that B cells at various stages of differentiation may undergo transformation to produce this disease.

The genetic aberrations associated with some cases of BCLL are similar to those found in MCL. Chromosomal rearrangements and other more subtle genetic modifications may lead to activation of the Bcl-1 gene and hence overexpression of cyclin D1 protein. Abnormalities on chromosome 13q are also found in about 25% of BCLL cases.

With respect to treatment, about 40% of patients with BCLL in its leukemia form show a partial response to fludarabine, with a CR being achieved in about 30% of cases. Over 50% of those achieving a CR survive for several years after treatment. When the BCLL is present as a lymphoma, local

irradiation is added to the treatment plan. Anti-CD20 mAb and CAMPATH-H have also been employed with some success. HCT as a therapy for this disorder is currently being attempted.

iiid) Follicular lymphoma (FL). FL is the second most common NHL after DLCL, accounting for about 22% of cases. FL is a slow-growing, mostly post-GC B cell lymphoma that affects mainly older adults. Patients often present at late disease stages with extensive infiltration of transformed B cells into the BM and spleen as well as lymph nodes. The involvement of the peripheral blood or extranodular tissues is rare. As the name implies, FL tumor cells are derived from B cells originally residing in the centers of lymphoid follicles. The tumors are composed of a mixture of centroblasts and centrocytes, with the latter cell type predominating. The aggressiveness of the tumor correlates with an increased proportion of centroblasts. All FL tumor cells express the B cell markers CD19, CD20, CD22, and CD79a, over half of them express CD10, and some express CD23. However, FL tumor cells are negative for CD5, CD43, TdT, and CD11c. In addition, most FL tumor cells are positive for surface Ig, with about half expressing IgM and the other half displaying IgG. IgA expression is rare on FL cells but can occur. The Ig molecules expressed by FL cells have a high frequency of somatic hypermutation.

With respect to genetic aberrations, the hallmark translocation leading to FL is t(14;18)(q32;q21) in which the Igh locus is translocated into the Bcl-2 locus. The Bcl-2 gene is activated by this event and promotes cell survival. In some FL tumor cells, the zinc finger transcriptional repressor Bcl-6 may also be upregulated in the nucleus.

FL is one of the hardest NHLs to treat. A few patients with very localized disease respond to targeted irradiation, but only half of these patients can expect to enjoy disease-free survival of up to 10 years. Whether chemotherapy supplies additional benefits to these patients is controversial. In most FL cases, the disease has become disseminated by the time of diagnosis, reducing the median survival of patients to 7–8 years even when heavy doses of chemotherapy are applied. Overall, about 50% of FL patients achieve some degree of remission following chemotherapy with either a single agent or a combination of drugs. However, the median survival rate does not improve significantly. More aggressive chemotherapy regimens are under examination in clinical trials. Treatment of FL with anti-CD20 can induce a partial response or even a CR in about 50% of patients. Similarly, a response rate of 50% can be attained with IFNα therapy, an approach that is not very effective for other NHLs. Autologous and allogeneic HCTs have been successful in prolonging life in some FL cases. However, while the graft versus lymphoma effect of an allogeneic HCT can block progression of the tumor and promote a lower relapse rate, the risk of mortality due to GvHD is significant.

A novel treatment for FL may be idiotypic therapeutic vaccination (refer to Chs.23 and 26). This approach can be contemplated for this type of lymphoma because the idiotypic determinants of the Ig expressed on the surface of the tumor cells are known to be bona fide tumor-specific antigens. In one clinical study, when the Ig from the FL cells of a given patient was purified and injected back into the same patient, a humoral response was induced that was specific for the idiotypic determinant, i.e., directed against the tumor cells. Several FL patients first given high dose chemotherapy and then vaccinated in this way have experienced extended remission of their disease. (The high rate of somatic hypermutation of the Ig molecules on FL cells did not appear to hamper the treatment.) Clinical trials are ongoing to determine how broadly applicable this labor-intensive therapy may be to cases of FL.

iiie) Mucosa-associated lymphoid tissue lymphoma (MALT-L). MALT lymphomas are extranodal NHLs located in the diffuse lymphoid tissues underlying the mucosae (particularly in the GI tract). MALT lymphomas constitute about 8% of NHLs. A MALT lymphoma is usually a post-GC disease, with tumors composed of a mixture of small mature B lymphocytes, centrocytes, *monocytoid* B cells (phenotypic B cells expressing monocyte markers), and plasma cells. The appearance of centroblasts in MALT tumors is rare, but when they are present, the prognosis is poor. In a minority of cases, MALT-L progresses to DLCL (see later). MALT-L cells carry the B cell markers CD19, CD20, CD22, and CD79a, have variable expression of CD43, and are negative for CD5, CD10, CD23, and TdT. Surface IgM is generally expressed, but IgG or IgAl is displayed in a few cases. Like the Ig genes in FL tumor cells, the Ig genes in MALT-L cells show a high degree of somatic hypermutation.

Clinicians recognize two types of MALT lymphomas, those that appear to depend on the presence of a bacterial antigen or an autoantigen (Ag-dependent), and those that are clearly independent of any apparent antigen (Ag-independent). The Ag-dependent MALT lymphomas are almost always found at sites of infections or autoantigen expression or exposure. Indeed, most MALT lymphomas occur in areas of the stomach where *H. pylori* bacteria are also present. MALT lymphomas develop at lower frequencies in the diffuse lymphoid tissues of the intestine, lung, thyroid, and salivary glands. In addition, patients with autoimmune diseases such as Hashimoto's thyroiditis or Sjögren syndrome (refer to Ch.29) can develop Ag-dependent MALT lymphomas. The constant presence of a pathogen antigen or autoantigen appears to be necessary to drive the proliferation of an Ag-dependent MALT-L cell, suggesting that these neoplastic cells are still one step away from complete malignant transformation and total growth independence. However, the unregulated proliferation of these cells provides ample opportunity for additional carcinogenic mutations to arise. Furthermore, the neutrophils that flock to the gut (drawn by the bacteria) may release ROIs that promote further genetic alteration. If a mutation occurs that allows the B cell to divide regardless of the presence of the bacterial antigen or autoantigen, the lymphoma cell becomes Ag-independent and completely transformed.

Antigen independence of a MALT lymphoma is associated with two rare genetic aberrations that do not occur in other types of NHL and do not lead to activation of Bcl-1, Bcl-2, Bcl-6, or c-myc. The t(11;18)(q21;q21) translocation, found in 50% of Ag-independent MALT lymphomas, brings

an anti-apoptotic gene called IAP-2 into the MALT1 locus. The t(1;14)(p22;q32) translocation, found in 10% of Ag-independent MALT lymphomas, brings the gene encoding the signal transducer Bcl-10 into the Igh locus. MALT1 encodes a protein that has a death domain, two Ig domains, and a domain that resembles a caspase (so that MALT1 is considered to be a "paracaspase"). The Bcl-10 protein contains a <u>c</u>aspase <u>r</u>ecruitment <u>d</u>omain (CARD) that allows the binding of caspases and paracaspases. Work at the Ontario Cancer Institute in Toronto led by Jurgen Ruland showed that both MALT1 and Bcl-10 are involved in the signaling pathway linking BCR and TCR engagement by antigen to the activation of NF-κB. It is thought that the preceding translocation events inappropriately activate MALT1 or Bcl-10 in a MALT-L B cell, leading to constitutive NF-κB signaling. The B cell proceeds to proliferate independently of either pathogen or autoantigen, setting the stage for lymphomagenesis.

Treatment of MALT lymphomas depends on whether the lymphoma is Ag-dependent or Ag-independent, and the nature of the initiating antigen. About 70% of MALT-L patients who are positive for *H. pylori* will respond to antibiotics. As the bacteria are killed, the pathogen antigen driving lymphomagenesis disappears and the tumor slowly resolves. Experience has shown that the tumor cells of MALT-L patients who do not respond to antibiotic treatments tend to have one of the two translocations activating Bcl-10 or MALT1. For these "bacteria-negative" patients, radiation therapy is the preferred treatment, but good results have also been obtained with chemotherapy. In some situations, surgery can be effective. If the disease progresses despite these treatments, patients undergo therapy similar to that applied to FL or DLCL.

iiif) Diffuse large cell lymphoma (DLCL). DLCL is the most common NHL subtype in adults, comprising 33% of these cancers (12% of childhood NHLs). Patients with DLCL usually present with rapidly expanding lymph nodes as well as extra-nodular growths in the GI tract, bones, and/or CNS. DLCL acquired its name both from its diffuse growth pattern and the large size of its tumor cells. Morphologically, DLCL tumor cells most closely resemble centrocytes, although centroblast-like cells may also be present. Somatic hypermutations are generally found in the Ig genes so that DLCL is considered to be a post-GC disease. DLCL tumor cells are usually positive for the B cell markers CD19, CD20, CD22, and CD79a but negative for CD23, CD43, and TdT. Cells in some tumors express CD5 or CD10.

There are several morphological variants of DLCL, including *centroblastic* DLCL (in which the tumor cells most closely resemble centroblasts); *immunoblastic* DLCL (in which the tumor cells are large, rapidly proliferating lymphocytes with large nuclei, clumped chromatin, distinct nucleoli, and basophilic cytoplasm); and *anaplastic* DLCL (in which the tumor cells behave like undifferentiated stem cells and pile up on each other). Other DLCL variants appear to have T cells, histiocytes, or granulocytes mixed in with the cancerous B cells. *Mediastinal* DLCL arises from the transformation of rare B cells residing in the thymic medulla. As its name indicates, *intravascular* DLCL is characterized by the adhesion of transformed B cells to the lumenal walls of the capillaries and post-capillary venules. These B cells can proliferate within these vessels and accumulate such that the vessels are distended. Two extranodal variants of DLCL are *CNS lymphoma* and *testicular lymphoma*. The tumors in CNS lymphoma are very aggressive and rapidly infiltrate the brain, meninges, and eyes. This NHL is difficult to treat because the combination of irradiation and chemotherapy effective against the tumor cells is also neurotoxic. Testicular lymphoma is rare and affects men over 60 years of age. An unwelcome complication of this otherwise localized tumor is its propensity for relapsing into the CNS following standard chemotherapy/radiation regimens. Thus, as is true for ALL and B-LBL treatment, preventive intrathecal chemotherapy is usually applied.

Genetic aberrations associated with DLCL include chromosomal rearrangements or mutations that activate the Bcl-6 gene at 3q26. The Bcl-6 gene is known to be a "promiscuous translocator," and translocations of Bcl-6 into a dozen different chromosomal sites have been found in DLCL tumor cells. The most common partners are the Igh locus at 14q32 and the MN1 gene at 22q11. In other DLCL cases, promotor methylations or gene truncations that lead to deregulation of Bcl-6 have been observed. Translocations leading to the activation of the Bcl-2 or c-myc genes can also give rise to DLCL. In these cases, the prognosis is worse than for disease derived from Bcl-6 activation. Mutations in p53 are found in some DLCL tumor cells and are associated with a less favorable disease outcome.

Patients with stage I and II DLCL do relatively well with standard CHOP chemotherapy, with or without localized irradiation of the involved tissue. However, those with the stage III and IV disease require more aggressive chemotherapy. One commonly employed drug regimen features a combination of cyclophosphamide, etoposide, anthracyclins, methotrexate, and vincristine. A clinical response to this regimen (and other similar drug combinations) occurs in a majority of patients. Long term follow-up of these patients is required, because relapses can occur many years after treatment has ended. When chemotherapy has failed, several options remain. A few DLCL patients show long-lasting responses to IFNα treatment, and about 40% of patients (usually those with small tumors) benefit to some degree from treatment with anti-CD20. Finally, autologous HCT can be efficacious in some DLCL cases.

iiig) Burkitt's lymphoma (BL). About 2% of all NHLs are cases of BL, named for Dennis Burkitt, who discovered them while examining jaw tumors in African children in 1958. Most cases of BL occur in children (36% of childhood NHLs), and the kidney (rather than the jaw) is the primary site of involvement. In adults, BL is found most often in the lymph nodes or small intestine. In its most advanced stages, BL may become a leukemia. BL occurrence is strongly associated with EBV infection in Africa (most likely due to the immunosuppression induced by malarial infection) but not in North America. Similarly, most AIDS-associated BLs are EBV-positive in Africa but EBV-negative in North America. It is not known why EBV should be so prevalent in African AIDS sufferers but not in other immunosuppressed individuals.

Many cases of BL resemble immunoblastic DLCL. The tumors are composed of medium-sized B cells with round nuclei, multiple well-defined nucleoli, and basophilic cytoplasm. BL cells undergo both tremendous proliferation and monumental cell death, and large numbers of (normal) macrophages are present in the tumor mass that may function to remove the apoptotic BL cells. The presence of these vacuolated cells gives the affected tissue a "starry sky" appearance (Plate 30-10). BL tumor cells are generally negative for TdT, CD5, CD23, and CD43 expression but positive for CD10, CD19, CD20, CD22, CD79a, and CD45R. BL is a post-GC disease, and studies of the Ig genes in the tumor cells suggest that the original target cells may be memory B cells.

Recurring chromosomal translocations into the Igh, Igl, and Igk loci that activate c-myc are found in the vast majority of BL cases. About 80% of patients have tumor cells bearing the t(8;14)(q24;q32) translocation described previously for ALL, while about 15% show t(2;8)(p11–12;q24) and about 5% have t(8;22)(q24;q11). However, it is suspected that mutations of other genes controlling the cell cycle may also contribute to some BL cases. Two such candidates are the related tumor suppressors Rb2 and p107. In the case of AIDS-associated BL, it is postulated that the Tat protein of HIV binds to Rb2 and inactivates it. The expression of numerous cell cycle genes is then inappropriately induced. Mutations of p53 have been found in 35% of BL samples.

Although BL is considered a high grade lymphoma, chemotherapy can be remarkably effective if implemented soon after diagnosis. In cases of BL in African children, one dose of cyclophosphamide can be sufficient to induce resolution of the cancerous lump, although the full course of treatment is needed to prevent relapse. BL in AIDS patients is more difficult to treat, most likely due to underlying effects of HIV infection. Radiation therapy as well as high doses of combination chemotherapy (including cyclophosphamide, doxorubicin, vincristine, methotrexate, cytarabine, and

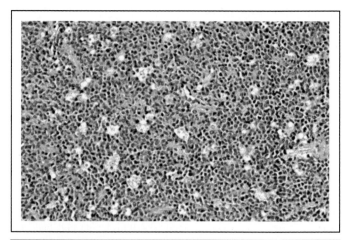

Plate 30-10

Burkitt's Lymphoma
Lymph node biopsy of a BL patient showing the 'starry sky' appearance caused by the presence of vacuolated macrophages that have phagocytosed tumor cell debris. Courtesy of Dr. Doug Tkachuk, Princess Margaret Hospital, UNH, Toronto.

eptoposide) are applied. Intrathecal chemotherapy is used to prevent spread of the disease to the brain and spinal cord.

iiih) Precursor T lymphoblastic leukemia/lymphoma (T-LBL). T-LBL represents 1.7% of all NHLs in adults but close to 30% of NHLs in children. The clinical features of T-LBL are essentially identical to those of B-LBL described previously except that a mediastinal mass is more common in T-LBL patients. Teenagers are the most frequent victims, with boys being affected twice as often as girls. T-LBL tumor cells are morphologically indistinguishable from B-LBL tumor cells and are positive for Tdt expression but also carry the markers CD1, CD2, CD3, CD5, and CD7. CD20 expression is absent. The TCRβ chain (only) shows rearrangement in 95% of T-LBL cases. The underlying genetic aberrations are unknown. Treatment and survival expectations are as for B-LBL.

iiii) Adult T cell leukemia/lymphoma (ATL). ATL is the general name given to a heterogeneous group of lymphomas that collectively account for about 6% of all NHLs. There is often a leukemic component to ATL. ATL usually strikes older individuals and is most common in Japan, the Mediterranean, and the Caribbean, where the instigating HTLV-1 virus is prevalent. ATL starts off as a relatively slow-growing malignancy but becomes aggressive towards the end of its course. Patients with ATL suffer from anemia, hepatosplenomegaly, and infiltration of the leukemic cells into the bone, skin, and CNS. Infections are common due to impaired immune responses. The disease also has an unfortunate knack of recurring long after treatment has been deemed successful. It is believed that the target cells of ATL are mature T cells circulating in the peripheral blood. Morphologically, ATL tumors consist of a mixture of large and small atypical lymphocytes that may be CD4$^+$ or CD8$^+$ or CD4$^-$CD8$^-$. The tumor cells may also express CD2, CD3, CD5, or CD7 on the cell surface, and both the TCRα and TCRβ genes (and sometimes the TCRγ gene) are usually rearranged. Complex karyotypic abnormalities have been identified in ATL cells, but no one event leading to specific oncogene activation has been consistently identified.

Traditionally, treatment of ATL has been similar to that used to combat DLCL. However, survival following conventional CHOP chemotherapy is poor, with only 50% achieving a CR. Current regimens include therapy with IFNα and the anti-retroviral drug zidovudine (AZT; refer to Ch.25) to combat HTLV-1. This strategy greatly improves survival rates. HCT is recommended for younger ATL patients and some success has been reported.

iiij) Mycosis fungoides and Sezary syndrome. The name "mycosis fungoides" looks like it should be attached to a fungal infection, and indeed, that is what the tissues of patients suffering from this rare cutaneous T cell lymphoma look like. A peripheral CD4$^+$ T cell located in or near the skin undergoes cancerous transformation and establishes a lymphoma in or on the skin. *Mycosis fungoides* (MF) represents 5% of all NHLs, affects older individuals (>50 years), and occurs more often in blacks than whites. Patients usually present with a rash or scaly patches on the skin that are often misdiagnosed as eczema or psoriasis (Plate 30-11). These symptoms may persist for years as the disease pursues its

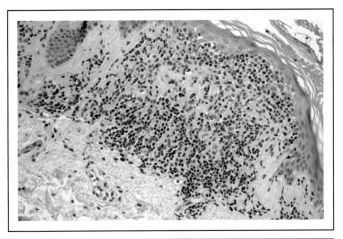

Mycosis Fungoides
Skin biopsy showing aggressive infiltration by malignant T cells. Courtesy of Dr. Doug Tkachuk, Princess Margaret Hospital, UHN, Toronto.

indolent course. In about 5% of MF patients, the disease presents in an advanced form and is called *Sezary syndrome* (SS). These patients have more severe, widespread reddening of the skin that has led to a nickname for this disease of "Red Man Syndrome." Intense itching and exfoliation of the skin are common. SS patients show lymphoma cells in the blood and metastases to lymph nodes and other organs.

The cause of MF is unknown, but it has been speculated that individuals with chronic bacterial infections in the skin are at greater risk. HTLV-1 has also been associated with this disease in some populations. Exposure to the defoliating chemical "Agent Orange" has been anecdotally linked to an increased risk of developing MF in war veterans. Modern methods of diagnosis of these disorders now include examination of the TCR genes, as there are often abnormal TCR rearrangements in the transformed T cells. A lack of CD26 expression and diminished CD3 expression also distinguish tumor cells from MF and SS patients. (CD26 is a widely expressed surface protein with peptidase activity.) Expression of the NK inhibitory receptor KIR3DL2 (CD158) and abnormalities in STAT5 function have been reported for SS tumor cells in several cases.

Patients with early stage MF can be treated with a combination of UVA light exposure and topical chemotherapy with carmustine, mechlorethamine, or steroids. Disease that has spread to the dermis may require treatment by exposure to a penetrating electron beam. Intensive therapy of this sort can lead to superficial burning of the skin (like a bad sunburn). Another curious (and sometimes appreciated) side effect is that the hair of an older, grey-haired individual that grows back in the region under treatment recovers its original pigment. For disease that has spread to the lymph nodes and internal organs, systemic chemotherapy is required. As well as conventional therapies, biologicals have been explored as treatments for MF and SS. IFNα administered subcutaneously is helpful for MF, as is administration of CAMPATH-1. Since about half of the malignant cells in MF and SS patients express IL-2R (CD25), an immunotoxin called DAB-IL2 derived from the fusion of the receptor binding domain of IL-2 to diphtheria toxin has been investigated as a potential therapy, with some success in clinical trials.

iiik) Anaplastic large cell lymphoma (ALCL). ALCL comprises about 2% of NHLs overall but 13% of childhood lymphomas. ALCL tumor cells have a distinct morphology and are large lymphoblast-like cells with multiple nuclei. In some (but not all) cases, the tumor cells express T cell markers such as CD3 as well as CD25, CD45, and CD53. (CD53 is a poorly characterized transmembrane protein involved in leukocyte signaling pathways.) The TCR genes are rearranged in a majority of ALCL cases, indicating that the target cells that undergo transformation are mature T cells. Interestingly, many ALCL cells also express CD30, the marker found on the RS cells of HL. Some cases of ALCL also show architectural features of HL tumors, leading to their occasional designation as "ALCL Hodgkin's related disorders."

The tumor cells of 30–50% of ALCL patients show the t(2;5)(p23;q35) translocation that fuses the *anaplastic lymphoma kinase* (ALK) gene on chromosome 2 to the nucleophosmin gene (involved in nucleolar assembly and ribosome assembly and transport) on chromosome 5. ALK is a tyrosine kinase thought to activate PLCγ, among other molecules. Occasionally, this translocation turns up in the tumor cells of other NHLs when very sensitive PCR detection methods are used.

The prognosis of ALCL is very good when the strategies used to treat DLCL are used. Pediatric cases are especially responsive, and for children with the ALK rearrangement, over 90% show a CR and enjoy long term survival. Adults with this genetic modification have a cure rate of about 70%. However, for ALCL patients without the ALK chromosomal translocation, prospects are not as good: only one-third attain long-term survival, even with treatment.

iiil) Angioimmunoblastic T cell lymphoma (AITL). AITL accounts for 4% of all NHLs and 20% of T cell lymphomas. This lymphoma is an aggressive disease affecting adults and has a median survival rate of only about 18 months even with treatment. The tumor mass is composed of large numbers of small- to medium-sized CD4$^+$ T cells and clustered FDCs. Scattered CD8$^+$ T cells, early stage B cells, and plasma cells may be present as well. The TCR genes are rearranged in most of the T lineage tumor cells. Some cases of AITL involve trisomy 3, trisomy 5, or both, and EBV infection has been found in a significant number of patients.

iiim) Peripheral T cell lymphoma, unspecified. About 12% of NHLs are lumped into a group called "peripheral T cell lymphomas, unspecified." These lymphomas, which typically affect adults rather than children, are generally characterized as diffuse with no one distinguishing feature. The transformed cells, which are immunophenotypically mature T cells, may be present in a spectrum of abnormal morphologies and sizes. Genetic aberrations are variable, as is expression of CD7 and CD30. Because this is a heterogeneous group, some of its member disorders are aggressive in nature while others are indolent.

We have come to the end, not only of this chapter, but also of this book. Hopefully we have achieved our goal of giving the reader a sound foundation in basic and clinical immunology, and have perhaps inspired him or her to pursue further study in this field. Many intriguing problems in immunology remain to be resolved. More importantly, the growing links between the immune system and the nervous system, and between immune responses and cancer, mean that a solid understanding of immunology is more useful than ever before. Multidisciplinary approaches in both the laboratory and the clinic are the way of the future, and immunologists can expect to be valuable contributors to these endeavors.

SUMMARY

Hematopoietic cancers (HCs) are malignancies of immune system cells that account for approximately 10% of all cancers occurring in the developed world. Like non-hematopoietic cancers (NHCs), HCs develop due to an accumulation of genetic changes in DNA repair genes, oncogenes, and tumor suppressor genes. However, unlike NHCs, HCs are much more commonly associated with gross chromosomal abnormalities such as recurring translocations. The three major types of HC are the leukemias, the myelomas, and the lymphomas. Leukemias are "liquid" tumors in the blood derived from the transformation of either a hematopoietic precursor in the bone marrow or a mature hematopoietic cell circulating in the blood. Leukemias are classified by whether their onset is acute or chronic, and subtypes are defined based on the lineage and state of differentiation of the transformed cell. In myelomas, the transformed cell is a fully differentiated plasma cell. The proliferating malignant clone may be present either in dispersed form or as a solid mass in the bone marrow. In lymphomas, the transformed cell is a lympho-cyte resident in a lymphoid tissue outside the bone marrow. The proliferating cells create a solid mass in the affected tissue. A lymphoma is classified as either a Hodgkin's lymphoma (HL) or a non-Hodgkin's lymphoma (NHL). In an HL, a reactive infiltrate of non-transformed cells is drawn to form a mass around a malignant clone of Reed-Sternberg cells. In an NHL, the entire cancerous mass develops from a transformed B or T lineage cell. Subtypes of HL and NHL are defined based on cell morphology, state of differentiation, surface marker expression, and genetic aberrations. For all HCs, immunophenotyping methods often allow precise identification of the transformed cell type, while fluorescence *in situ* hybridization (FISH) techniques can define the genetic aberrations that have occurred in the transformed cell. Such information guides clinicians in optimizing the radiation and chemotherapy protocols used as standard treatment for HC patients. Improvements in diagnosis and therapy (including immunotherapeutic approaches) have led to increased patient survival over the last four decades.

Selected Reading List

Beachy P., Karhadkar S. and Berman D. (2004) Tissue repair and stem cell renewal in carcinogenesis. *Nature* **432**, 324–331.

Bergsagel P. and Kuehl W. (2001) Chromosome translocations in multiple myeloma. *Oncogene* **20**, 5611–5622.

Brentjens R. and Sadelain M. (2004) Somatic cell engineering and the immunotherapy of leukemias and lymphomas. *Advances in Pharmacology* **51**, 347–370.

Callander N. and Roodman G. (2001) Myeloma bone disease. *Seminars in Hematology* **38**, 276–285.

Cameron E. and Neil J. (2004) The Runx genes: lineage-specific oncogenes and tumor suppressors. *Oncogene* **23**, 4308–4314.

Chaganti R., Nanjangud G., Schmidt H. and Teruya-Feldstein J. (2000) Recurring chromosomal abnormalities in non-Hodgkin's lymphoma: biologic and clinical significance. *Seminars in Hematology* **37**, 396–411.

Chiorazzi N. and Ferrarini M. (2003) B cell chronic lymphocytic leukemia: lessons learned from studies of the B cell antigen receptor. *Annual Review of Immunology* **21**, 841–894.

Crawford D. and Talbot S. (2004) Viruses and tumours—an update. *European Journal of Cancer* **40**, 1998–2005.

Dono M., Cerruti G. and Zupo S. (2004) The CD5$^+$ B cell. *International Journal of Biochemistry and Cellular Biology* **36**, 2105–2111.

Du M.-Q. and Isaccson P. (2002) Gastric MALT lymphoma: from aetiology to treatment. *Lancet Oncology* **3**, 97–104.

Evans L. and Hancock B. (2003) Non-Hodgkin lymphoma. *Lancet* **362**, 139–146.

Faderl S., Jeha S. and Kantarjian H. (2003) The biology and therapy of adult acute lymphoblastic leukemia. *Cancer* **98**, 1337–1354.

Farag S. and Caligiuri M. (2004) Cytokine modulaton of the innate immune system in the treatment of leukemia and lymphoma. *Advances in Pharmacology* **51**, 295–318.

Ferrando A. and Look A. (2000) Clinical implications of recurring chromosomal and associated molecular abnormalities in acute lymphoblastic leukemia. *Seminars in Hematology* **37**, 381–395.

Goldman J. and Melo J. (2003) Chronic myeloid leukemia—advances in biology and new approaches to treatment. *New England Journal of Medicine* **349**, 1451–1464.

Greiner T. (2004) mRNA microarray analysis in lymphoma and leukemia. *Cancer Treatment Research* **121**, 1–12.

Harris N., Jaffe E., Diebold J., Flandrin G., Muller-Hermelink H. *et al.* (2000) The World Health Organization classification of neoplasms of the hematopoietic and lymphoid tissues: report of the Clinical Advisory Committee meeting—Airlie House, Virginia, November, 1997. *The Hematology Journal* **1**, 53–66.

Hayashi Y. (2000) The molecular genetics of recurring chromosome abnormalities in acute myeloid leukemia. *Seminars in Hematology* **37**, 368–380.

Houghton A., Gold J. and Blachere N. (2001) Immunity against cancer: lessons learned from melanoma. *Current Opinion in Immunology* **13**, 134–140.

Janz S., Potter M. and Rabkin C. (2003) Lymphoma- and leukemia-associated chromosomal translocations in healthy individuals. *Genes, Chromosomes and Cancer* **36**, 211–223.

Lamy T. and Loughran T. J. (2003) Clinical features of large granular lymphocyte leukemia. *Seminars in Hematology* **40**, 185–195.

Mathas S., Lietz A., Anagnostopoulos I., Hummel F., Wiesner B. *et al.* (2004)

c-FLIP mediates resistance of Hodgkin/Reed-Sternberg cells to death receptor–induced apoptosis. *Journal of Experimental Medicine* **199**, 1041–1052.

Minden M., Imrie K. and Keating A. (1996) Acute leukemia in adults. *Current Opinion in Hematology* **3**, 259–265.

Najfeld V. (2003) Diagnostic application of FISH to hematological malignancies. *Cancer Investigation* **21**, 807–814.

Reichardt V. and Brossart P. (2004) DC-based immunotherapy of B-cell malignancies. *Cytotherapy* **6**, 62–67.

Richardson P., Hideshima T., Mitsiades C. and Anderson K. (2004) Proteasome inhibition in hematologic malignancies. *Annals of Medicine* **36**, 304–314.

Saglio G. and Cilloni D. (2004) Abl: the prototype of oncogenic fusion proteins. *Cellular and Molecular Life Science* **61**, 2897–2911.

Seidl S., Kaufmann H. and Drach J. (2003) New insights into the pathophysiology of multiple myeloma. *Lancet Oncology* **4**, 557–564.

Sirohi B. and Powles R. (2004) Multiple myeloma. *Lancet* **363**, 875–887.

Skinnider B. and Mak T. (2002) The role of cytokines in classical Hodgkin lymphoma. *Blood* **99**, 4283–4297.

Stone R. (2002) Treatment of acute myeloid leukemia: state-of-the-art and future directions. *Seminars in Hematology* **39**, 4–10.

Waters J., Barlow A. and Gould C. (1998) Demystified . . . FISH. *Journal of Clinical Pathology* **51**, 62–70.

Werner M., Wilkens L., Aubele M., Nolte M., Zitzelsberger H. *et al.* (1997) Interphase cytogenetics in pathology: principles, methods, and applications of fluorescence in situ hybridization (FISH). *Histochemistry and Cell Biology* **108**, 381–390.

Yung L. and Linch D. (2003) Hodgkin's lymphoma. *Lancet* **361**, 943–951.

Appendix: CD Molecules

With information from www.hlda8.org and Coligan J.E. *et al*. (1996) "Current Protocols in Immunology," John Wiley & Sons, Inc., New York.

CD Number*	Common Names	Gene Name	Chromosomal Location	Cellular/Tissue Expression	Function**
CD1a	T6/leu-6, R4, HTA1	CD1A	1q22-q23	Cortical thymocytes but not mature T cells; B cells, DCs, LCs, other APCs	Integral membrane protein with structural similarity to MHC class I Associates with β2-microglobulin Presents nonpeptide antigens such as lipids and glycolipids to T cells
CD1b	R1	CD1B	1q22-q23	Cortical thymocytes but not mature T cells; B cells, DCs, LCs, other APCs	Integral membrane protein with structural similarity to MHC class I Associates with β2-microglobulin Presents nonpeptide antigens such as lipids and glycolipids to T cells
CD1c	M241, R7	CD1C	1q22-q23	Cortical thymocytes but not mature T cells; B cells, DCs, LCs, other APCs	Integral membrane protein with structural similarity to MHC class I Associates with β2-microglobulin Presents nonpeptide antigens such as lipids and glycolipids to T cells, especially γδ T cells
CD1d	R3	CD1D	1q22-q23	Cortical thymocytes but not mature T cells; B cells, DCs, intestinal epithelial cells	Integral membrane protein with structural similarity to MHC class I Presents glycolipid antigens to NKT cells
CD1e	R2	CD1E	1q22-q23	Cortical thymocytes but not mature T cells; B cells, DCs, LCs, other APCs	Integral membrane protein with structural similarity to MHC class I Presents nonpeptide antigens such as lipids and glycolipids to T cells
CD2	**LFA-2** Lymphocyte function associated antigen 1, T11, Tp50, sheep red blood cell (SRBC) receptor	CD2	1p13	Most T cells, thymocytes, NK cells, some early developing B cells	Integral membrane glycoprotein and member of Ig superfamily Adhesion molecule binding to LFA-3 (CD58) Has minor T cell costimulatory function
CD3d	**CD3δ** CD3 delta, T3, Leu4	CD3D	11q23	T lineage cells	Glycosylated Ig superfamily member forming part of CD3 complex associating non-covalently with TCR chains Forms ITAM-containing heterodimer with CDε Required for cell surface expression and signal transduction of TCR
CD3e	**CD3ε** CD3 epsilon	CD3E	11q23	T lineage cells	Non-glycosylated Ig superfamily member forming part of CD3 complex associating non-covalently with TCR chains Forms ITAM-containing heterodimer with CDδ or CDγ Required for cell surface expression and signal transduction of TCR

continued overleaf

CD Number*	Common Names	Gene Name	Chromosomal Location	Cellular/Tissue Expression	Function**
CD3g	**CD3γ** CD3 gamma	CD3G	11q23	T lineage cells	Glycosylated Ig superfamily member forming part of CD3 complex associating non-covalently with TCR chains. Forms ITAM-containing heterodimer with CDε. Required for cell surface expression and signal transduction of TCR
CD4	OKT4, Leu3a, L3T4, T4	CD4	12pter-p12	Subsets of thymocytes and T cells (especially Th cells), some monocytes and macrophages	Type I transmembrane protein and member of Ig superfamily. Coreceptor for MHC class II antigen-restricted T cell recognition and activation. Required for thymocyte development and effector T cell differentiation. Binds to HIV gp120 protein and facilitates viral entry into CD4$^+$ T cells, macrophages
CD5	Tp67;T1, Ly1, Leu-1	CD5	11q13	T lineage cells, B-1 B cell subset, neonatal B cells, B cell CLL	Type I transmembrane monomeric glycoprotein and member of scavenger receptor family. Binds to CD72 on B cells. Forms complexes with TCR and BCR and is phosphorylated upon antigen stimulation. Promotes proliferation of activated T cells but may oppose B cell activation. CD5$^+$ B cells are implicated in autoimmunity
CD6	T12	CD6	11q13	Thymocytes, T cells, B cell CLL, brain	Type I integral monomeric glycoprotein and member of scavenger receptor family. Adhesion function mediated by binding to CD166 expressed by activated leukocytes. Also mediates adhesion between thymocytes and thymic epithelial cells. Cross-linking on T cells by anti-CD6 antibody promotes T cell proliferation
CD7	Leu 9, 3A1, gp40, Tp41, p41, T cell leukemia antigen	CD7	17q25.2-q25.3	Thymocytes, peripheral T cell subset, HSCs; upregulated on activated T cells	Homodimeric glycoprotein and member of Ig superfamily. Associates with TCR complex. May modulate cell adhesion or signaling. No ligand demonstrated as yet
CD8a	OKT8, Leu2, Lyt2, T8	CD8A	2p12	Subsets of thymocytes and T cells (especially Tc cells and CTLs), some γδ T cells and NK cells, intestinal iIELs	Type I transmembrane protein and member of Ig superfamily. Part of heterodimeric coreceptor required for MHC class I-restricted T cell recognition and activation. Required for thymocyte development and effector T cell differentiation. CD8αα homodimers promote memory Tc cell differentiation
CD8b	Lyt3, Ly3	CD8B1	2p12	Subsets of thymocytes and T cells (especially Tc cells and CTLs)	Type I transmembrane protein and member of Ig superfamily. Part of heterodimeric coreceptor for MHC class I-restricted T cell recognition and activation. Required for thymocyte development and effector T cell differentiation

CD Number*	Common Names	Gene Name	Chromosomal Location	Cellular/Tissue Expression	Function**
CD9	Drap-27, MRP-1, p24, leukocyte antigen MIC3	CD9	12p13.3	Endothelium, platelets, megakaryocytes, stromal cells, early B cells, activated T cells, eosinophils, basophils	Type III transmembrane protein and member of tetraspanin family Modulates cell adhesion and migration Involved in platelet aggregation and tumor metastasis Required for gamete fusion
CD10	**CALLA** Common acute lymphoblastic leukemia antigen, neutral endopeptidase, membrane metallo-endopeptidase	MME	3q25.1-q25.2	Pre-B cells, germinal center B cells, fetal thymocytes, neutrophils, granulocytes, bone marrow stromal cells, some epithelial cells	Type II membrane zinc metalloproteinase with neutral peptidase activity May carry out cleavage to regulate the effects of biologically active peptides May regulate B and T development and neutrophil activity
CD11a	**LFA-1 α chain** Lymphocyte function associated antigen 1, gp180/95, LeuCAM	ITGAL	16p11.2	Neutrophils, monocytes, macrophages, lymphocytes	Type I transmembrane glycoprotein and member of CAM family Complexes with CD18 to form the β2-integrin LFA-1 Binds to ICAM1, ICAM2, ICAM3, ICAM4 on resting and activated endothelial cells Has adhesion and signal transduction functions important for inflammation, ADCC, and CTL killing
CD11b	**CR3 α chain** **Mac-1 α chain** gp155/95, LeuCAMb, alpha chain of C3bi receptor	ITGAM	16p11.2	Neutrophils, monocytes, macrophages, NK cells	Type I transmembrane glycoprotein and member of CAM family Complexes with CD18 to form the β2-integrin Mac-1 (complement receptor 3; CR3) Binds to iC3b, ICAM1, CD23, and other ligands important for inflammation Has adhesion and signal transduction functions
CD11c	**CR4 α chain** gp150/95	ITGAX	16p11.2	Neutrophils, monocytes, macrophages, NK cells, DCs, activated T and B cells	Type I transmembrane glycoprotein and member of CAM family Complexes with CD18 to form the β2-integrin p150, 95 (complement receptor 4; CR4) Binds to iC3b, LPS, ICAM-1, fibrinogen, and other ligands important for inflammation Has adhesion and signal transduction functions
CDw12	p90–120			Monocytes, granulocytes, NK cells, platelets	90–120 kDa protein of unknown structure and function
CD13	**ANPEP** Aminopeptidase N, APN, gp150, EC 3.4.11.2	ANPEP	15q25-q26	Granulocytes, monocytes, bone marrow stromal cells, fibroblasts, osteoclasts, intestinal epithelium	Type II transmembrane glycoprotein with zinc metalloproteinase activity Trims peptides bound to MHC class II and may regulate activities of biologically active peptides Receptor for coronaviruses
CD14	LPS receptor	CD14	5q31.1	Monocytes, macrophages, granulocytes, microglia, uterus, adipose tissue	GPI-anchored membrane protein with leucine-rich repeats (also soluble forms) Cooperates with MD-2 and TLR4 to form complex transducing LPS signaling; i.e., binds to serum LBP complexed to LPS

continued overleaf

CD Number*	Common Names	Gene Name	Chromosomal Location	Cellular/Tissue Expression	Function**
CD14 (cont'd)					Promotes inflammation, cytokine secretion, and upregulation of adhesion molecules in innate responses Mediates endotoxic shock
CD15	**Lewis X** Le-X, X-Hapten, CD15u (sulfated Lewis X), CD15s (sialyl Lewis X)		Carbohydrate antigen	Neutrophils, eosinophils, monocytes, cells of HEVs and lymph nodes, Reed-Sternberg cells in Hodgkin's lymphomas	Pentasaccharide linked to various lipid or protein structures Various forms bind to CD62P, CD62E, or CD62L to mediate leukocyte adhesion, extravasation, and activation
CD16a	**FcγRIIIa** Low-affinity IgG Fc receptor	FCGR3A	1q23	NK cells, macrophages activated monocytes, mast cells	Type I transmembrane glycoprotein and member of Ig superfamily Associates with CD3ζ chain or FcεRIγ chain to form type III Fc receptor for IgG Mediates phagocytosis and ADCC
CD16b	**FcγRIIIb** Low-affinity IgG Fc receptor	FCGR3B	1q23	Neutrophils, activated eosinophils	Ig superfamily member that is GPI-anchored to membrane (also a soluble form) Does not mediate phagocytosis or ADCC but rather serves as a trap for immune complexes
CDw17	**LacCer** Lactosylceramide		Carbohydrate antigen	Neutrophils, monocytes, granulocytes, basophils, platelets, subset of B cells, some leukemic cells	Phosphatidyl inositol-anchored membrane antigen Cross-linking by anti-CD17 antibody leads to calcium flux, upregulation of adhesion molecules Binds to bacteria and may function in phagocytosis
CD18	**Mac-1** Macrophage antigen 1, integrin β2 β chain, LAD, LCAMB, MF17	ITGB2	21q22.3	Broadly expressed on leukocytes	Type I transmembrane glycoprotein Associates non-covalently with CD11a, b, or c to form the adhesion molecules LFA-1, Mac-1, and p150,95, respectively Facilitates binding to ICAMs, complement components, fibrinogen, Factor X Important for the cell adhesion to matrix proteins necessary for innate and adaptive immune responses
CD19	Bgp95, B4	CD19	16p11.2	All B lineage cells (except plasma cells), FDC, malignant B cells	Type I transmembrane protein and Ig superfamily member Associates with CD21 and CD81 in BCR complex and promotes signal transduction Functions as coreceptor for B cell activation
CD20	B1, membrane-spanning 4-domains, subfamily A, member 1	MS4A1	11q12	All B lineage cells, including plasma cells	Non-glycosylated member of the tetraspanin family Membrane-spanning domains form ion channel May be involved in regulating calcium flux Required for mitogen-induced B cell proliferation and B cell differentiation Therapeutic target for non-Hodgkin's lymphomas
CD21	**CR2** Complement receptor 2, C3d receptor, EBV receptor, gp140	CR2	1q32	B cells, FDCs, DCs, epithelial cells, T cells, thymocytes	Type I glycoprotein and member of RCA family Associates with CD21 and CD81 in BCR complex and promotes signal transduction during B cell activation Complement receptor for C3d Serves as receptor for Epstein-Barr virus

CD Number*	Common Names	Gene Name	Chromosomal Location	Cellular/Tissue Expression	Function**
CD22	BL-CAM, Bgp135, Siglec2, sialic acid binding lectin-2	CD22	19q13.1	Surface expression by mature B cells (but not plasma cells), cytoplasmic expression in early B cells	Type I transmembrane glycoprotein and member of Ig superfamily Contains both ITAMs and ITIMs Adhesion and signaling molecule Binds to sialic acid residues on numerous glycoproteins, including CD45 on T cells May regulate B cell activation May be involved in B cell localization in lymphoid tissues
CD23	FcεRII Low-affinity IgE receptor, B6, Leu-20, BLAST-2	FCER2	19p13.3	B and T cells, monocytes, FDC, thymocytes, eosinophils, LCs, thymic epithelial cells	Type II transmembrane glycoprotein expressed as two splice variants Low-affinity receptor for Fc of IgE Negative feedback regulation of IgE synthesis Triggers NO production and cytokine release by monocytes
CD24	HSA heat stable antigen, BA-1	CD24	6q21	Mature granulocytes, most B lineage cells (except plasma cells), immature and activated T cells, tumor cells	Small GPI-anchored membrane glycoprotein Primary ligand is unknown but can bind to P-selectin (CD62P) May contribute to thymocyte selection, T cell costimulation, and B cell development May modulate B cell activation
CD25	IL-2Rα Interleukin-2 receptor α chain, TAC	IL2RA	10p15-p14	Activated T and B cells, monocytes, some thymocytes, T$_{reg}$ cells	Type I transmembrane glycoprotein containing sushi repeats Induced by cellular activation Associates with IL-2Rβ (CD122) and IL-2Rγ (CD132) to form the high-affinity IL-2 receptor Has a short cytoplasmic tail and is involved in ligand binding rather than signaling
CD26	Dipeptidyl peptidase (DPPIV), adenosine deaminase (ADA)-complexing protein, Ta1	DPP4	2q24.3	Activated T and B cells, some monocytes, some epithelial cells, some tumor cells	Type II transmembrane glycoprotein (also soluble form) Associates with ADA in T cells Has peptidase activity that promotes cell migration across endothelia Associates with CD45 and binds collagen May contribute to T cell costimulation May regulate glucose homeostasis
CD27	T14, S152	TNFRSF7	12p13	Most T and B cells, NK cells, upregulated on activated T cells	Type I glycosylated transmembrane protein and member of TNFR superfamily (soluble form in serum) Homodimer is ligand for CD70 Important for thymocyte development and cell survival during primary T cell responses Critical for memory CTL responses Also regulates NK cytotoxicity and B cell activation
CD28	Tp44, T44	CD28	2q33	T cell subset, activated B cells	Type I membrane protein that forms disulfide-linked homodimers Structurally related to CTLA4 (CD152) but functionally distinct Ligand for B7-1 (CD80) and B7-2 (CD86) expressed by APCs

continued overleaf

CD Number*	Common Names	Gene Name	Chromosomal Location	Cellular/Tissue Expression	Function**
CD28 (cont'd)					Mediates T cell costimulation and stimulates T cell survival, proliferation, production of IL-2 and other cytokines TCR engagement in the absence of CD28 signaling can lead to anergy
CD29	**VLA β chain** Very late antigen beta chain, integrin β1 chain, platelet GPIIa	ITGB1	10p11.2	Most leukocytes but weak on neutrophils; adipocytes, hepatocytes, smooth muscle cells; not RBCs	Type I transmembrane protein that associates with CD49$_{a-f}$ chains to form VLA molecules Mediates cellular adhesion to VCAM-1, MAdCAM-1, and matrix proteins such as collagen, laminin, fibronectin Critical for leukocyte transendothelial migration, embryogenesis, and HSC differentiation
CD30	Ki-1, Ber H2, D1S166E, TNFRSF8	TNFRSF8	1p36	Activated B, T, and NK cells, some monocytes, Reed-Sternberg cells in Hodgkin's lymphomas	Type I transmembrane glycoprotein and member of TNFR superfamily (also soluble form) Affects gene expression by activating NF-κB May regulate development of thymocytes and proliferation of activated lymphocytes May promote apoptosis and limit autoimmunity Upregulated on T cells from allergic and HIV-infected persons
CD31	**PECAM-1** Platelet/endothelial cell adhesion molecule-1, endocam, platelet GPIIa	PECAM1	17q23	Vascular endothelium, granulocytes, monocytes, platelets, B and T cell subsets	Type I transmembrane glycoprotein and member of CAM family (also soluble forms in serum) Contains six ITIMs Concentrated at endothelial cell junctions to facilitate leukocyte transmigration Promotes adhesion through homophilic and heterophilic binding but ligands (other than CD31 and CD38) are not clear
CD32	**FcγRII** gp40	FCGR2A	1q23	Monocytes, neutrophils, B and T cell subsets, eosinophils, NK cells, platelets	Type I membrane protein and member of Ig superfamily Low-affinity receptor for Fc of IgG Preferentially binds IgG2 Various isoforms promote inhibition or activation of B cell, NK cell, and mast cell responses
CD33	My9, gp67, p67	CD33	19q13.3	Myeloid progenitors, monocytes, macrophages, mast cells, LCs, neutrophils (weak); not HSCs	Type I membrane protein and member of sialoadhesin family Binds to α-2, 3 linked sialic acids and contains two ITIMs Interacts with SHP-1 and SHP-2 upon tyrosine phosphorylation and so may inhibit signal transduction
CD34	My10, gp105-120	CD34	1q32	Hematopoietic precursor cells, bone marrow stromal cells, capillary endothelium	Type I transmembrane glycoprotein of sialomucin family Serves as a marker for pluripotent HSCs May mediate adhesion between hematopoietic precursors and bone marrow extracellular matrix or stromal cells Physiological ligand in bone marrow unknown but can bind to L-selectin (CD62L)

CD Number*	Common Names	Gene Name	Chromosomal Location	Cellular/Tissue Expression	Function**
CD35	CR1 Complement receptor 1, C3b/C4b receptor, immune adherence receptor	CR1	1q32	Erythrocytes, neutrophils, FDC, monocytes, eosinophils, T and B cell subsets; not platelets	Type I transmembrane glycoprotein and monomeric member of RCA family Receptor for C3b and C4b bound to immune complexes Mediates opsonized phagocytosis, immune complex clearance
CD36	Scavenger receptor class B member 3, platelet GPIV, PASIV, FAT, OKM-5 SCARB3, platelet collagen receptor	CD36	7q11.2	Platelets, mature monocytes, macrophages, some endothelial cells	Transmembrane glycoprotein and scavenger receptor Binds to broad range of ligands, including LDL, collagen, long-chain fatty acids, platelet-agglutinating protein, malaria-infected RBCs, thrombospondin, anionic phospholipids Mediates phagocytosis of apoptotic cells as well as platelet adhesion and aggregation May function in leukocyte adhesion May regulate fatty acid transport
CD37	gp40-52	CD37	19p13-q13.4	Strong expression on most B cells (not plasma cells), weak on other leukocytes	Integral membrane glycoprotein and member of tetraspanin family Function is unclear, but may be involved in T-B interaction under suboptimal costimulatory conditions
CD38	T10, gp45, ADP-ribosyl cyclase	CD38	4p15	Developing B and T cells, activated T cells, germinal center B cells, plasma cells, platelets, some hematopoietic progenitors	Type II transmembrane glycoprotein (soluble form in serum) Adhesion molecule binding to PECAM-1 (CD31) Also an ectoenzyme with NAD^+ glycohydrolase, ADP-ribosyl cyclase, and ADP-ribose hydrolase activities Anti-CD38 ligation induces lymphocyte activation and proliferation or death
CD39	gp80, ATPDase, ectonucleoside triphosphate diphosphohydro-lase 1	ENTPD1	10q24	Mantle zone B cells, activated T cells, NK cells, macrophages, LCs, DCs	Potential transmembrane homodimeric glycoprotein May mediate homotypic adhesion Also has ecto-apyrase activity (hydrolysis of ATP to give AMP)
CD40	TNF Receptor 5, Bp50	TNFRSF5	20q12-q13.2	Mature B cells, some pre-B cells, monocytes, DCs, T cell subsets	Type I integral membrane glycoprotein and member of TNFR superfamily Binds to CD40L (CD154) on Th cells to costimulate B cell activation, GC formation, and isotype switching Also important for DC licensing
CD41	Platelet glycoprotein IIb, αII β integrin, alpha subunit of platelet membrane adhesive protein receptor complex	ITGA2B	17q21.32	Platelets, megakaryocytes	Type I transmembrane protein of integrin family Present as heterodimer of two subunits generated by post-translational cleavage of protein precursor Associates with CD61 to mediate platelet activation and aggregation The CD41/CD61 complex binds to several ligands, including fibrinogen, fibronectin, plasminogen, prothrombin

continued overleaf

CD Number*	Common Names	Gene Name	Chromosomal Location	Cellular/Tissue Expression	Function**
CD42a	Platelet glycoprotein IX, GPIX	GP9	3q21	Platelets, megakaryocytes	Type I membrane glycoprotein with leucine-rich repeat Participates in CD42 complex (von Willebrand factor receptor) essential for platelet adhesion to sites of injury and clotting
CD42b	CD42b, GPIb-α	GP1BA	17pter-p12	Platelets, megakaryocytes	Type I membrane glycoprotein with leucine-rich repeats Participates in CD42 complex (von Willebrand factor receptor) essential for platelet adhesion to sites of injury and clotting Disulfide-linked to CD42c Contains binding site for von Willebrand factor and thrombin
CD42c	CD42c, GPIb-β	GP1BB	22q11.21	Platelets, megakaryocytes	Type I membrane glycoprotein with leucine-rich repeat Participates in CD42 complex (von Willebrand factor receptor) essential for platelet adhesion to sites of injury and clotting Disulfide-linked to CD42b
CD42d	CD42d, GPV	GP5	3q29	Platelets, megakaryocytes	Type I membrane glycoprotein with leucine-rich repeats Participates in CD42 complex (von Willebrand factor receptor) essential for platelet adhesion to sites of injury and clotting
CD43	Leukosialin, LSN, gp115 sialophorin	SPN	16p11.2	Many leukocytes, highly expressed on T cells and thymocytes; a pro-B cell marker	Type I membrane glycoprotein Possible ligands are ICAM-1 (CD54), hyaluronic acid, selectins, and Siglec-1 (CD169) CD43 is removed from the T cell surface at the APC–T cell interface during T cell activation, and so may have a negative regulatory effect
CD44	pgp1, gp80-95, Hermes antigen, ECMR-III, MC56, MDU2, MIC4 HUTCH-I	CD44	11p13	Leukocytes, erythrocytes; variant forms are widely expressed on endothelia and epithelia	Type I transmembrane glycoprotein that is highly heterogeneous due to extensive alternative splicing and post-translational modifications Binds hyaluronic acid, collagen, laminin, fibronectin with its N-terminus; and ANK, the ERM proteins and NF2 with its C-terminus Mediates cell–cell and cell–matrix adhesion and signaling Important for lymphocyte activation, homing, recirculation Certain CD44 isoforms may be involved in tumor metastasis
CD45	**B220** Leukocyte common antigen, LCA, T200, protein tyrosine phosphatase receptor type C	PTPRC	1q31-q32	All leukocytes and hematopoietic progenitors	Type I transmembrane glycoprotein that undergoes alternative splicing to generate several isoforms (including CD45RA and CD45RO) differentially expressed according to developmental/activation status Tyrosine phosphatase (EC 3.1.34) required for signal transduction

CD Number*	Common Names	Gene Name	Chromosomal Location	Cellular/Tissue Expression	Function**
CD45 (cont'd)					Required for antigen receptor-mediated lymphocyte activation and T cell development
CD45RA	Restricted T200, gp220, isoform of leukocyte common antigen	PTPRC	1q31-q32	Naive T cells, B cells, medullary thymocytes	Isoform of CD45
CD45RO	Restricted T200, gp180, isoform of leukocyte common antigen	PTPRC	1q31-q32	Memory T cells, activated T cells, cortical thymocytes, monocytes	Isoform of CD45 CD45RA$^+$ cells lose RA and acquire RO upon activation
CD46	**MCP** Membrane cofactor protein	MCP	1q32	Most leukocytes, endothelial and epithelial cells, tumor cells, placental trophoblasts, sperm	Type I transmembrane glycoprotein and member of RCA family Controls complement activation Cofactor for Factor I-mediated proteolytic cleavage of C3b and C4b May prevent complement activation at the maternal–fetal interface May be involved in fusion of sperm with oocyte during fertilization
CD47	**IAP** Integrin-associated protein, ovarian carcinoma antigen OA3	CD47	3q13.1-q13.2	Most hematopoietic cells; epithelial and endothelial cells, fibroblasts, sperm	Integral membrane protein Binds to inhibitory receptor PTPNS1 (SIRPα1; CD172a) May mediate calcium flux in response to binding of adhesion molecules to extracellular matrix Important for neutrophil activation and migration in innate responses Expression prevents premature clearance of self RBCs
CD48	BLAST-1, CD2 ligand, Hulym3, OX45, BCM1	CD48	1q21.3-q22	Most leukocytes (not neutrophils or platelets), some endothelial and epithelial cells	GPI-anchored membrane protein and member of Ig superfamily Ligand for CD2 and so a minor costimulator of T cell responses
CD49a	**VLA 1 α chain** Very late antigen α1, integrin α1	ITGA1	5q11.2	Activated T cells, monocytes, cultured neuronal cells, melanoma cells	Type I transmembrane glycoprotein and member of integrin family Associates with CD29 (integrin β1 chain) to form the VLA-1 adhesion receptor for collagen and laminin-1 Upregulated during inflammation
CD49b	**VLA-2 α chain** Integrin α2 chain, platelet GPIa	ITGA2	5q23-q31	Platelets, megakaryocytes, activated T cells, B cells, monocytes, epithelial and endothelial cells, fibroblasts	Type I transmembrane glycoprotein and member of integrin family Associates with CD29 (integrin β1 chain) to form the VLA-2 receptor Mediates cell and platelet adhesion to collagen, laminin, fibronectin, E-cadherin Promotes wound healing
CD49c	**VLA-3 α chain** Integrin α3, GAPB3, VCA-2, MSK18	ITGA3	17q21.33	B cells, adherent cell lines	Type I transmembrane glycoprotein and member of integrin family Cleaved into heavy and light chains that then are disulfide-linked Associates with CD29 (integrin β1 chain) to form the VLA-3 receptor

continued overleaf

CD Number*	Common Names	Gene Name	Chromosomal Location	Cellular/Tissue Expression	Function**
CD49c (cont'd)					Mediates adhesion to laminin-5, fibronectin, invasin, laminin-1, collagen, entactin, thrombospondin
CD49d	VLA-4 α chain Integrin α4	ITGA4	2q31-q32	Most leukocytes (not platelets or neutrophils)	Type I transmembrane glycoprotein and member of integrin family Cleaved into heavy and light chains that then are disulfide-linked Associates with either CD29 (integrin β1 chain) or the integrin β7 chain to form the VLA-4 receptor Mediates adhesion to VCAM-1, MAdCAM-1, fibronectin, thrombospondin Important for inflammatory responses and differentiation and migration of HSCs May play a role in tumor progression and metastasis
CD49e	VLA-5 α chain Integrin α5, FNRA	ITGA5	12q11-q13	Variety of adherent and non-adherent cells, monocytes, NK cells, DCs, osteoblasts	Type I transmembrane glycoprotein and member of integrin family Cleaved into heavy and light chains that then are disulfide-linked Associates with CD29 (integrin β1 chain) to form the VLA-5 receptor Receptor for fibronectin and fibrinogen
CD49f	VLA-6 α chain Integrin α6, platelet GPIc	ITGA6	2q31.1	Platelets, megakaryocytes, monocytes, T cells, thymocytes, many cultured adherent cell lines	Type I transmembrane glycoprotein and member of integrin family Cleaved into heavy and light chains that then are disulfide-linked Associates with CD29 (integrin β1 chain) or the integrin β4 chain to form the VLA-6 receptor Receptor for laminins, invasin, merosin Mediates interaction between epithelial cells and basement membrane during wound healing Involved in tumor metastasis
CD50	ICAM-3 Intercellular adhesion molecule 3	ICAM3	19p13.3-p13.2	Most leukocytes	Type I membrane glycoprotein and member of Ig superfamily Binds to LFA-1 and contributes to APC adhesion and costimulation during T cell activation Regulates LFA/ICAM1 and integrin β1-dependent adhesion
CD51	Integrin α chain, vitronectin receptor α chain, VNRA, MSK8	ITGAV	2q31-q32	Endothelial cells, monocytes, macrophages, platelets, activated T cells, some B cells, osteoclasts	Type I transmembrane protein which is cleaved into heavy and light chains that then are disulfide-linked Associates with CD61 to form receptor binding to extracellular matrix proteins, vitronectin, von Willebrand factor Mediates cellular adhesion to extracellular matrix
CD52	CAMPATH-1 HE5	CDW52	1p36	Lymphocytes, monocytes, eosinophils, male reproductive tract epithelium, mature sperm	Small GPI-anchored membrane protein Physiological function unknown Anti-CD52 cross-linking induces calcium flux and activates T cells, but decreases monocyte activity Anti-CD52 mAbs are used therapeutically to treat leukemia and reduce GvHD

CD Number*	Common Names	Gene Name	Chromosomal Location	Cellular/Tissue Expression	Function**
CD53	MRC-OX44, MOX44, TSPAN25	CD53	1p13	Thymocytes, T cells, B cells, monocytes, granulocytes, osteoblasts, osteoclasts; not RBCs	Integral membrane protein and member of tetraspanin family Tyrosine phosphatase activity Anti-CD53 cross-linking induces calcium mobilization in lymphocytes, NO release in macrophages
CD54	**ICAM-1** Intercellular adhesion molecule 1	ICAM1	19p13.3-p13.2	Activated (but not resting) T cells, B cells, monocytes, endothelial cells	Type I glycoprotein and member of Ig superfamily Adhesion molecule binding to LFA-1, Mac-1, fibrinogen, hyaluronan, CD43 Major signaling and adhesion molecule in inflammatory and adaptive immune responses Receptor for rhinovirus
CD55	**DAF** Decay accelerating factor	DAF	1q32	Most hematopoietic cells	GPI-anchored membrane protein and member of the RCA family Protects against inappropriate complement activation Binds C3b and C4b to inhibit C3 convertase formation Binds C3bBb and C4b2a to accelerate decay of C3 convertases Receptor for echovirus
CD56	**NCAM** Neural cell adhesion molecule, NKHI, Leu-19	NCAM1	11q23.1	Various isoforms present on NK cells, T cell subsets, neural cells, muscle cells, embryonic tissue, tumors	GPI-anchored type I membrane glycoprotein and member of Ig superfamily; numerous splice variants Adhesion molecule in neuronal tissue and possibly for NK cells Function for T cells not known
CD57	HNK1, Leu-7	CD57	11q12-qter	Subsets of NK, NKT, and T cells (especially CD8+ Ts regulatory cells), neural tissue	Trisaccharide epitope of a myelin-associated glycoprotein May contribute to neural cell adhesion Function in hematopoietic cells is unknown but may have a role in anti-tumor immunity
CD58	**LFA-3** Lymphocyte function associated antigen-3	CD58	1p13	Leukocytes, erythrocytes, endothelial and epithelial cells, fibroblasts	GPI-anchored or transmembrane protein and member of Ig superfamily Ligand for CD2 and so contributes to costimulation during T cell activation Mediates adhesion between Th cells and APCs, Tc cells and target cells, thymocytes and thymic epithelial cells
CD59	**MIRL** MACIF, protectin, P-18	CD59	11p13	Widely expressed on hematopoietic cells	GPI-anchored membrane glycoprotein and member of RCA family Protects cells against complement-mediated lysis in a species-specific manner by binding to C8 and/or C9 and inhibiting the final step of MAC formation Essential for RBC survival Can bind to CD2 and stimulate T cell activation Incorporated by budding viruses into their envelopes as an evasion tactic

continued overleaf

CD Number*	Common Names	Gene Name	Chromosomal Location	Cellular/Tissue Expression	Function**
CD60	GD3 (CD60a), 9-0-acetyl GD3 (CD60b), 7-0-acetyl GD3 (CD60c)		Carbohydrate antigens	Subset of activated Th cells, B cells, monocytes, some thymocytes, platelets, thymic epithelium, smooth muscle cells, activated keratinocytes	GD3 ganglioside antigens Anti-CD60 cross-linking of CD60a and CD60 is costimulatory for T cells CD60c may be involved in apoptosis induction CD60 is upregulated by LFA-3 engagement High numbers of CD60+ T cells are present in synovial fluid
CD61	Glycoprotein IIIa, β3, GP3A; integrin, vitronectin receptor β chain	ITGB3	17q21.32	Platelets, megakaryocytes, endothelium, smooth muscle, some B cells, monocytes, macrophages	Type I membrane protein and member of integrin family Common β subunit of CD41/61 and CD51/61 CD41/61 mediates attachment to matrix proteins such as fibronectin, laminin, matrix metalloproteinase-2, osteopontin, prothrombin, vitronectin Mediates platelet activation and aggregation
CD62E	**E-selectin** LECAM-2, ELAM-1	SELE	1q22-q25	Activated endothelial cells, activated platelets, megakaryocytes	Type I transmembrane glycoprotein with N-terminal C-type lectin domain and Sushi repeats Adhesion molecule required for leukocyte extravasation and activation during inflammatory responses Binds glycoproteins and glycolipids containing sialyl Lewis-X (CD15s) Ligands include PSGL-1 (CD162)
CD62L	**L-selectin** LAM-1, Mel-14, lymph node homing receptor	SELL	1q22-q25	Neutrophils, monocytes, NK cells, memory T cells	Type I transmembrane glycoprotein with N-terminal C-type lectin domain and Sushi repeats Adhesion molecule required for leukocyte extravasation and activation during inflammatory responses Mediates lymphocyte adherence to HEVs in lymph nodes Binds glycoproteins and glycolipids containing sialyl Lewis-X (CD15s) or CD155 epitopes Also binds certain polyanionic molecules, including heparin sulfate
CD62P	**P-selectin** GMP-140, granule membrane protein-140,	SELP	1q22-q25	Megakaryocytes, activated platelets, activated endothelium	Type I transmembrane glycoprotein with N-terminal C-type lectin domain and Sushi repeats Adhesion molecule required for leukocyte extravasation and activation during inflammatory responses Binds to glycoproteins and glycolipids with CD15s (sialyl Lewis-X) epitopes, including CD162 (PSGL-1) Binds to certain polyanions—endothelial CD62P is involved in the rolling reaction that precedes leukocyte extravasation
CD63	LIMP, gp55, LAMP-3, neuroglandular antigen, melanoma-associated antigen ME491, PTLGP40, granulophysin	CD63	12q12-q13	T cells, some B cells, activated platelets, monocytes and macrophages, degranulated neutrophils	Lysosomal integral membrane glycoprotein and a member of tetraspanin family Strongly expressed on early primary melanoma cells Found with P-selectin and von Willebrand factor in secretory granules of platelets May play a role in regulating hematopoietic cell growth

CD Number*	Common Names	Gene Name	Chromosomal Location	Cellular/Tissue Expression	Function**
CD64	**FcγR1** High-affinity Fcγ receptor, FCR1	FCGR1A	1q21.2-q21.3	Monocytes, macrophages, neutrophils, DC subsets	Type I membrane protein and a member of Ig superfamily Binds the Fc region of IgG molecules with high affinity Carries out receptor-mediated phagocytosis of immune complexes, ADCC Transferred across placenta
CD65	Ceramide dodeca-saccharide 4c, VIM2		Carbohydrate antigen	Granulocytes, some monocyte and DC subsets	18-residue sugar moiety; also has sialylated form (CD65s) Physiological function unknown but anti-CD65 cross-linking induces calcium flux and respiratory burst Potential ligand for CD62P and CD62E May be involved in phagocyte activation
CD66a	Carcinoembryonic antigen-related cell adhesion molecule 1; BGP, biliary glycoprotein, NCA-160	CEACAM1	19q13.2	Granulocytes, epithelial cells	Type I membrane protein of CEA family and member of Ig superfamily (also secreted isoforms) Mediates homophilic and heterophilic adhesion to E-cadherin Capable of activating neutrophils and endothelial cells Receptor for *N. meningitis, N. gonorrhoeae*
CD66b	Previously CD67; CGM6, NCA-95	CEACAM8	19q13.2	Granulocytes	GPI-anchored glycoprotein of CEA family and member of Ig superfamily Mediates heterophilic adhesion to CD66c Capable of activating neutrophils Released during granulocyte activation
CD66c	NCA, nonspecific cross-reaction antigen, NCA-50/90	CEACAM6	19q13.2	Granulocytes, epithelial cells	GPI-anchored glycoprotein of CEA family and member of Ig superfamily May mediate homophilic adhesion Capable of activating neutrophils
CD66d	CGM1	CEACAM3	19q13.2	Granulocytes	Type I membrane protein of CEA family and member of Ig superfamily Adhesion and signaling leading to phagocytosis of microbes binding to CEACAM Important for innate immunity
CD66e	CEA, carcinoem-bryonic antigen	CEACAM5	19q13.1-q13.2	Epithelial cells of normal fetal colon, adult adenocarcinomas	GPI-anchored glycoprotein of CEA family and member of Ig superfamily Mediates homophilic and heterophilic adhesion (to endothelial E-selectin) Marker of colon cancer that may be involved in tumor metastasis
CD66f	**PSBG** Pregnancy-specific β-glycoprotein, Sp-1	PSG1	19q13.2	Secreted protein produced by placental syncytiotrophoblasts, fetal liver, myeloid cell lines	Secreted glycoprotein of CEA family and member of Ig superfamily Precise function unknown but appears to protect fetus from maternal immune system
CD67	CD67 has been renamed CD66b				

continued overleaf

CD Number*	Common Names	Gene Name	Chromosomal Location	Cellular/Tissue Expression	Function**
CD68	gp110, macrosialin, SCARD1, scavenger receptor class D member 1	CD68	17p13	Monocytes, macrophages, neutrophils, basophils, eosinophils, mast cells, osteoclasts, DCs, subset of CD34+ bone marrow progenitors	Type I transmembrane glycoprotein and member of scavenger receptor family Localized primarily in endosomes and lysosomes but also on cell surfaces Binds to selectins and lectins, including those expressed on pathogens May play roles in macrophage phagocytosis and lysosomal metabolism
CD69	**AIM** Activation inducer molecule, MLR3, EA1, VEA, Leu-23	CD69	12p13-p12	Activated T and B cells, thymocytes, NK cells, neutrophils, eosinophils, platelets	Type II transmembrane glycoprotein with C-type lectin domain Expressed as disulfide-linked homodimer Earliest inducible cell surface marker seen during lymphocyte activation Involved in signal transduction, calcium flux leading to lymphocyte proliferation Anti-CD69 cross-linking on platelets results in aggregation and activation
CD70	CD27 ligand, CD27L, KI-24 antigen	TNFSF7	19p13	Activated T and B cells	Type II transmembrane protein and member of TNF superfamily Likely expressed as homotrimer Binds to CD27 to contribute to costimulation of T and B cells Enhances CTL generation and memory responses
CD71	**Transferrin receptor** T9, p90	TFRC	3q29	All proliferating cells	Type II transmembrane protein occurring as disulfide-linked homodimer Binds transferrin and mediates iron uptake Required for RBC and nervous system development Also binds to HLA-DR1 Receptor for HepB virus, canine/feline parvoviruses
CD72	Lyb-2, Ly-19.2, Ly32.2	CD72	9p13.3	All B cells (except plasma cells), macrophages (weak); not monocytes	Type II membrane protein occurring as a disulfide-linked homodimer with a C-lectin domain Contains ITIMs and acts as a negative regulator of B cell responsiveness Binding of CD72 to CD100 turns off negative signaling by CD72
CD73	Ecto-5'-nucleotidase	NT5E	6q14-q21	Subsets of T and B cells, FDCs, epithelial and endothelial cells; not NK cells	GPI-anchored disulfide-linked homodimer Dephosphorylates extracellular ribo- and deoxyribonucleoside monophosphates to generate membrane-permeable nucleosides
CD74	**Invariant chain** Ii; MHC class II–associated invariant chain	CD74	5q32	MHC class II-positive cells, including B cells, other APCs; activated T cells	Type II membrane glycoprotein with several isoforms Occurs as nonamer of three α, β, γ heterotrimers Required for stabilization of peptide-free MHC class II molecules in the ER and their intracellular transport to endolysosomal compartments for peptide loading Critical for MHC class II antigen presentation Surface function not known

CD Number*	Common Names	Gene Name	Chromosomal Location	Cellular/Tissue Expression	Function**
CD75	Lactosamine		Carbohydrate antigen	Strong on mature B cells (including GC cells), weak on extrafollicular and mantle zone B cells, erythrocytes, T cell subset, various B cell malignancies	No known function; may be ligand for sialic acid-binding lectins
CD75s	α2,6-sialylated lactosamines (formerly CDw75 and CDw76)		Carbohydrate antigen	Mature B cell subsets (not GC B cells or plasma cells), T cell subsets, some B cell malignancies	Carbohydrate antigen produced by α2,6-sialyltransferase Present on wide variety of glycoproteins and glycolipids No known function: may be a ligand for sialic acid-binding lectins
CDw76	CDw76 has been renamed CD75s				
CD77	Gb3, Pk blood group antigen, BCA, Burkitt's lymphoma associated antigen		Carbohydrate antigen	GC B cells (especially centroblasts), endothelia, epithelia, strongly expressed on Burkitt's lymphoma cells	Neutral glycolipid hapten Binds to bacterial pili, Shiga toxin, and verotoxin Also binds to CD19 and may promote GC formation May mark centroblasts for apoptosis unless rescued by binding to antigens on FDC
CDw78	Deleted from CD classification				
CD79a	Igα Immunoglobulin receptor α chain, MB1, IGA	CD79A	19q13.2	All mature B cells and plasma cells	Type I membrane glycoprotein containing ITAM and Ig domains Associates with Igβ (CD79b) in a heterodimer that participates in the BCR complex to carry out signal transduction Essential for BCR function, B cell development, and B cell effector functions
CD79b	Igβ Immunoglobulin receptor β chain, B29	CD79B	17q23	All mature B cells but not plasma cells	Type I membrane glycoprotein containing ITAM and Ig domains Associates with Igα (CD79a) in a heterodimer that participates in the BCR complex to carry out signal transduction Essential for BCR function, B cell development, and B cell effector functions
CD80	**B7-1** B7.1, B7, BB1	CD80	3q13.3	B cells, monocytes, macrophages, DCs, activated T cells	Type I transmembrane glycoprotein and member of Ig superfamily B7-1 binding to CD28 is major mechanism of naive T cell costimulation B7-1 binding to CTLA-4 (CD152) has negative effect on T cell activation Highly induced on stimulated APCs
CD81	**TAPA-1** Target of anti-proliferative antibody, M38	CD81	11p15.5	Most leukocytes (especially B cells), endothelia	Integral membrane protein and member of tetraspanin family Participates in signal transduction complex with CD19 and CD21 that may amplify BCR signaling Anti-TAPA cross-linking induces cellular aggregation and growth inhibition Also binds to HepC glycoprotein and facilitates malarial infection of hepatocytes

continued overleaf

CD Number*	Common Names	Gene Name	Chromosomal Location	Cellular/Tissue Expression	Function**
CD82	R2, IA4, 4F9, C33, KAI1, kangai 1	KAI1	11p11.2	Many types of activated hematopoietic cells (not RBCs), epithelial and endothelial cells, fibroblasts	Integral membrane glycoprotein and member of tetraspanin family Function unknown but appears to suppress metastasis of some types of tumor cells
CD83	HB15, BL-11	CD83	6p23	DCs, LCs, mitogen-activated and GC lymphocytes	Monomeric type I membrane glycoprotein and member of Ig superfamily Function unknown but may be involved in antigen presentation or events following lymphocyte activation
CD84	p75, GR6	CD84	1q24	Mature and memory B cells, NK cells, monocytes, T cell subsets; low levels on neutrophils, platelets	Type I integral membrane glycoprotein and member of Ig superfamily Function unknown but structural similarities to CD2 and CD48 suggest a role in intercellular interaction and signaling
CD85a	**ILT5** Ig-like transcript 5, leukocyte Ig-like receptor subfamily B, LIR3, HL9	LILRB3	19q13.4	NK cells, B cells, DCs, monocytes, plasma cells	Type I membrane protein containing ITIMs and member of Ig superfamily May bind MHC class I and regulate NK cell recognition of self cells
CD85d	**ILT4** Ig-like transcript 4, LIR2; MIR10	LILRB2	19q13.4	Monocytes, macrophages, DCs, B cells, NK cells	Type I membrane protein containing ITIMs and member of Ig superfamily Binds to broad spectrum of HLA-A, -B, -C, and -G alleles and downregulates adaptive responses Upregulated on tolerogenic APCs following interaction with regulatory Ts cells Also binds to FcγRIa and inhibits protein phosphorylation and calcium flux
CD85j	**ILT2** Ig-like transcript 2, LIR-1, MIR7	LILRB1	19q13.4	Monocytes, NK cells, T cells, B cells, other peripheral blood leukocytes, placenta, lung, liver	Type I membrane protein containing ITIMs and member of Ig superfamily Binds to broad spectrum of HLA-A, -B, -C, and -G alleles and downregulates adaptive responses Receptor for CMV protein
CD85k	**ILT3** Ig-like transcript 3, LIR5, HM18	LILRB4	19q13.4	Monocytes, DCs, macrophages	Type I membrane protein containing ITIMs and member of Ig superfamily Binds to broad spectrum of HLA-A, -B, -C, and -G alleles and downregulates adaptive responses Upregulated on tolerogenic APCs following interaction with regulatory Ts cells Interferes with CD40 signaling and NK-κB activation Negative regulatory effect on calcium mobilization
CD86	**B7-2** B7.2, B70	CD86	3q21	Professional APCs, (especially DCs); constitutively expressed on resting B cells; induced on eosinophils by IL-3	Type I membrane protein and member of Ig superfamily B7-2 binding to CD28 is major mechanism of naive T cell costimulation B7-2 binding to CTLA-4 (CD152) has negative effect on T cell activation May preferentially support Th2 differentiation

CD Number*	Common Names	Gene Name	Chromosomal Location	Cellular/Tissue Expression	Function**
CD87	**PLAUR** Plasminogen activator, urokinase receptor; UPAR, Mo3	PLAUR	19q13	Macrophages, monocytes, neutrophils, DCs, activated T cells, fibroblasts, endothelial cells, smooth muscle	GPI-anchored monomeric glycoprotein (also secreted isoform) Binding of CD87 to urokinase plasminogen activator localizes it on the cell surface and induces tyrosinase phosphorylation Localizes and promotes plasmin formation during inflammation Also binds to vitronectin and $\beta 1$ and $\beta 2$ integrins Promotes neutrophil recruitment into lungs in response to bacteria
CD88	**C5aR** C5a receptor	C5R1	19q13.3-q13.4	Most leukocytes, epithelial and endothelial cells, hepatocytes, astrocytes	Integral seven-transmembrane protein and member of G protein-coupled receptor family Binds anaphylatoxin C5a and mediates its pro-inflammatory effects Stimulates granule release, chemotaxis, reactive anion production Critical for lung mucosal defense
CD89	**FcαR** IgA receptor	FCAR	19q13.2-q13.4	Neutrophils, monocytes, macrophages, activated eosinophils	Type I transmembrane glycoprotein and member of Ig superfamily (several splice variants) Binds Fc region of IgA Induces phagocytosis, respiratory burst, degranulation, ADCC, release of inflammatory mediators and cytokines
CD90	**Thy1**	THY1	11q22.3-q23	Thymocytes, T cells, some hematopoietic progenitors, neurons, fibroblasts	GPI-anchored membrane protein and member of Ig superfamily May contribute to regulation of proliferation and differentiation of hematopoietic cells Anti-CD90 cross-linking stimulates T cells May be involved in cell–cell adhesion and interaction in brain Putative tumor suppressor gene for ovarian cancer
CD91	α2-macroglobulin receptor, LDL receptor-associated protein	LRP1	12q13-q14	Monocytes, macrophages, erythroblasts, reticulocytes, liver, brain, lung	Type I transmembrane protein and member of LDLR family Originates as large precursor cleaved into α and β subunits that remain non-covalently associated Mediates plasma clearance of chylomicron remnants and activated α2-macroglobulin Regulates fibroblast proliferation and internalization of pro-urokinase and thrombospondin Binds to heat shock proteins and so may be an inflammatory sensor for necrotic cell death Low CD91 expression correlates with early Alzheimer's disease onset May be involved in tumor metastasis
CDw92	p70, choline-transporter-like protein-1	CDW92	9q31.2	B cells, most T cells, some NK cells, monocytes, granulocytes, fibroblasts, endothelial cells	Membrane protein similar to choline transporter; has a potential ITIM Function unknown but is downregulated upon DC/monocyte activation

continued overleaf

CD Number*	Common Names	Gene Name	Chromosomal Location	Cellular/Tissue Expression	Function**
CDw93	GR11			Monocytes, granulocytes AML blasts, myeloid lines, endothelial cells	Sialoglycoprotein of unknown function
CD94	Killer cell lectin-like receptor subfamily D member 1, kP43	KLRD1	12p13	NK cells, some $\gamma\delta$ T cells, some $\alpha\beta$ CD8$^+$ T cells	Type II transmembrane glycoprotein containing C-lectin domain and ITIM Associates with NKG2A (CD159a) and NKG2C (CD159c) to form inhibitory receptors that bind to HLA-E; in the absence of this binding, NK cytotoxicity is activated Also associates with NKG2D to form an activatory receptor promoting expression of inflammatory cytokines
CD95	**Fas** APO-1, TNFR superfamily member 6	TNFRSF6	10q24	Activated T and B cells	Type I transmembrane glycoprotein and member of TNF superfamily Upon binding of FasL (CD178), Fas delivers signals initiating apoptosis Important for maintenance of peripheral tolerance and homeostasis
CD96	Tactile, T cell activation increased late expression	CD96	3q13.13-q13.2	Peripheral T and NK cells (weak), some AML cells; strongly upregulated on T, NK, and B cells in late stages of activation	Type I transmembrane protein forming disulfide-linked homodimers May be involved in adhesion during late stages of immune responses
CD97	BL-KDD/F12	CD97	19p13	Activated leukocytes, smooth muscle cells; upregulated in inflamed synovium; various carcinomas	Large precursor protein that is proteolytically cleaved into an alpha subunit with an extracellular region containing EGF-like domains that interact with DAF (CD55), and a beta subunit consisting of a seven-span transmembrane G protein-coupled receptor May function in adhesion, signaling, and/or migration following cellular activation
CD98	4F2, FRP-1, solute carrier family 3 member 2, fusion regulatory protein 1	SLC3A2	11q13	Activated and transformed cells, including non-hematopoietic cells	Type II transmembrane glycoprotein acting as the heavy chain of a heterodimeric receptor Disulfide-bonded to nonglycosylated light chain Potential amino acid transporter strongly associated with actin Involved in normal and cancer cell survival or growth
CD99	MIC2, E2	CD99	Xp22.32; Yp11.3	Highest on T cells, thymocytes; weakly on other leukocytes	Type I transmembrane glycoprotein with proline repeats Mediates homotypic and heterotypic adhesion Antigen identified by mAbs 12E7, F21, O13
CD100	**SEMA4D** Semaphorin 4D, BB18, collapsin 4	SEMA4D	9q22-q31	Wide variety of hematopoietic and non-hematopoietic cells; upregulated on activated T and B cells	Type I membrane protein and member of semaphorin family Homodimer binds to plexin B1 to promote changes to actin cytoskeleton, invasive growth, cell migration, axon guidance
CD101	V7, P126, immunoglobulin superfamily member 2	IGSF2	1p13	Monocytes, granulocytes, DCs, LCs, activated T cells	Type I transmembrane glycoprotein and member of Ig superfamily Accessory molecule for T cell costimulation

CD Number*	Common Names	Gene Name	Chromosomal Location	Cellular/Tissue Expression	Function**
CD102	**ICAM-2**	ICAM2	17q23-q25	Leukocytes (not neutrophils); strongest on vascular endothelium	Type I transmembrane glycoprotein and member of Ig superfamily Binds to LFA-1 (CD11a/CD18) to mediate intercellular adhesion and signal transduction important for T cell interactions, lymphocyte recirculation, NK cell migration
CD103	Integrin αE subunit, human mucosal lymphocyte antigen 1, HML-1	ITGAE	17p13	Intraepithelial lymphocytes (especially γδ T cells), some circulating lymphocytes, activated T and mast cells, macrophages	Type I transmembrane glycoprotein that is cleaved into two subunits that remain disulfide-linked Forms a complex with integrin β7 that binds to E-cadherin on epithelial cells May promote homing of iIELs to GALT
CD104	Integrin β4 subunit, TSP-1180, gp150	ITGB4	17q25	Epithelial cells, in hemidesmosomes	Type I transmembrane glycoprotein with fibronectin domains; member of β integrin family Associates with integrin α6 subunit (CD49f) to form laminin receptor Mediates adhesion to epithelial layers Promotes invasiveness of carcinomas
CD105	Endoglin, TGFβ receptor	ENG	9q33-q34.1	Endothelial cells, activated macrophages, bone marrow stromal cells, some leukemic cells	Type II transmembrane glycoprotein and disulfide-linked homodimer Component of multi-chain TGFβ receptor May play role in vascular morphogenesis
CD106	**VCAM-1** Vascular cell adhesion molecule-1, INCAM-110	VCAM1	1p32-p31	Endothelial cells, DCs, FDCs; upregulated on surface and secreted by activated vascular endothelium	Type I transmembrane sialoglycoprotein and member of Ig superfamily (also soluble form) Binds to integrins VLA-4 and α4/β7 to mediate leukocyte extravasation for immune responses and into sites of inflammation Required for marginal B cell localization and normal heart development
CD107a	**LAMP-1** Lysosome associated membrane protein 1	LAMP1	13q34	Lysosomal protein; low expression on cell membrane; granulocytes, activated platelets, tonsillar epithelium, melanoma cells, T cells, macrophages, endothelial cells	Type I transmembrane sialoglycoprotein Shuttles between lysosomes, endosomes, plasma membrane Constitutively expressed Ligand for galectin and furnishes carbohydrate ligands for selectins Increased expression on tumors and may be associated with metastasis
CD107b	**LAMP-2** Lysosome associated membrane protein 2, indoleamine pyrole 2,3 dioxygenase	LAMP2	Xq24	Lysosomal protein; LAMP-2A is highest in placenta, lung, liver; LAMP-2B is highest in skeletal muscle, fibroblasts	Type I transmembrane sialoglycoprotein with two variants Shuttles between lysosomes, endosomes, and possibly plasma membrane Inducibly expressed Ligand for galectin and furnishes carbohydrate ligands for selectins May import proteins into lysosomes
CD108	John-Milton-Hagen (JMH) human blood group antigen, GPI-gp80	SEMA7A	15q22.3-q23	Erythrocytes	GPI-anchored glycoprotein and member of semaphorin family May function in nervous and immune system modulation

continued overleaf

CD Number*	Common Names	Gene Name	Chromosomal Location	Cellular/Tissue Expression	Function**
CD109	Platelet activation factor, 8A3, E123, Gov a/b alloantigen	CD109	6q14	Platelets, monocytes, granulocytes, activated T cells, some bone marrow progenitors, umbilical vein endothelial cells	GPI-anchored glycoprotein and member of α2-macroglobulin family May mediate cell–substrate or cell–cell interaction
CD110	**Thrombopoietin receptor** c-mpl, myeloproliferative leukemia virus oncogene	MPL	1p34	Platelets, HSCs, megakaryocytes	Type I membrane protein of type I cytokine (WSXWS) receptor family Receptor for thrombopoietin Regulates megakaryocyte and platelet formation
CD111	**PRR1** Polio virus receptor related 1 protein, Nectin 1, Hve C1	PVRL1	11q23	Broadly expressed by both hematopoietic and non-hematopoietic cells	Type I transmembrane protein and member of Ig and polio virus receptor families; also secreted form Probably an adhesion molecule Receptor for cell entry of HSV-1, HSV-2
CD112	**PRR2** Polio virus receptor related 2 protein, Nectin 2, Hve B	PVRL2	19q13.2-q13.4	Broadly expressed by both hematopoietic and non-hematopoietic cells	Type I transmembrane glycoprotein and member of Ig and polio virus receptor families; two isoforms Probably an adhesion molecule Receptor for cell entry of HSV-1, HSV-2
CDw113	**PRR3** Polio virus receptor-related 3, Nectin 3	PVRL3	3q13	Epithelial cells, neurons, placenta, testis	Transmembrane protein and member of Ig and polio virus receptor families Binds to afadin, an actin filament-binding protein that connects nectins to the actin cytoskeleton Contributes to intercellular junctions Required for normal numbers and sizes of synapses between hippocampal neurons
CD114	**G-CSFR** Granulocyte colony stimulating factor receptor, HG-CSFR, CSFR3	CSF3R	1p35-p34.3	Granulocytes, monocytes and their precursors; mature platelets, endothelial cells, placenta, trophoblasts	Type I membrane protein of type I cytokine (WSXWS) receptor family Receptor for G-CSF Regulation of granulocyte differentiation and proliferation Stimulates mobilization of HSCs from bone marrow into blood Used to boost granulocyte production following chemotherapy
CD115	**M-CSFR** Macrophage colony stimulating factor receptor, CSF-1, c-Fms oncogene	CSF1R	5q33-q35	Myeloid cells and progenitors	Type I membrane protein of type III tyrosine kinase growth factor receptor family Receptor for M-CSF Regulation of monocyte/macrophage differentiation and proliferation (including osteoclasts) Promotes adhesion to bone marrow stroma
CD116	**GM-CSFRα** Granulocyte/macrophage colony stimulating factor receptor alpha subunit, GMR α	CSF2RA	Xp22.32; Yp11.3	Macrophages, neutrophils, eosinophils, DCs, myeloid progenitors, erythroid progenitors	Type I membrane protein of type I cytokine (WSXWS) receptor family On its own, binds to GM-CSF with low affinity but associates with βc (CD131) to bind to GM-CSF with high affinity Promotes differentiation, proliferation, and activation of myeloid and erythroid cells

CD Number*	Common Names	Gene Name	Chromosomal Location	Cellular/Tissue Expression	Function**
CD117	c-kit SCFR, stem cell factor receptor, steel factor receptor	KIT	4q11-q12	HSCs and downstream progenitors, mast cells, melanocytes, neurons	Type I membrane protein of type III tyrosine kinase growth factor receptor family; also Ig superfamily Receptor for SCF Binding to SCF leads to CD117 dimerization, autophosphorylation, and interaction with signaling molecules
CD118	LIFR Leukemia inhibitory factor receptor, SJS2, STWS	LIFR	5p13-p12	Epithelial cells, placenta, bone, brain, spinal cord	Type I transmembrane protein of type I cytokine (WSXWS) receptor family; also a soluble form Associates with gp130 (CD130) to form heterodimeric receptor for LIF Binding to LIF stimulates differentiation, proliferation, and survival of many hematopoietic cell types (including embryonic stem cells) as well as some epithelial cells
CD119	IFNγRα IFN gamma receptor α subunit	IFNGR1	6q23-q24	Ubiquitous except RBCs	Type I membrane protein of type II cytokine receptor family; also Ig superfamily Associates with IFN-γRβ subunit to form complex transducing IFN-γ signaling Triggers numerous antiviral, antiproliferative, and immunomodulatory effects during innate and adaptive responses
CD120a	TNFR1 Tumor necrosis factor receptor 1, TNFRp55	TNFRSF1A	12p13.2	T and B cells; upregulated on most other cell types; highest expression in thymus, spleen, PBL, placenta	Type I membrane protein containing death domain; prototypical member of TNFR superfamily Soluble form produced by proteolytic cleavage of membrane form; used as a therapeutic for inflammatory diseases Binding of TNF leads to homotrimerization of TNFR1 and signaling responsible for most TNF activities TNF promotes inflammation, fever, shock, tumor necrosis, cell proliferation, differentiation, and apoptosis Also binds to LTα with similar but milder effects
CD120b	TNFR2 Tumor necrosis factor receptor 2, TNFRp75	TNFRSF1B	1p36.3-p36.2	Most hematopoietic cells, especially myeloid cells; highly expressed on activated T and B cells	Type I membrane protein containing death domain and member of TNFR superfamily; Soluble form produced by proteolytic cleavage of membrane form Binding of TNF leads to homotrimerization of CD120b Signaling by CD120b is not systemic (unlike CD120a signaling) and is limited to the immune system Primarily promotes survival of hematopoietic cells (occasionally necrosis) Also binds to LTα with similar but milder effects
CD121a	IL-1RI Type I IL-1 receptor, IL1RA	IL1R1	2q12	Almost ubiquitous	Type I transmembrane glycoprotein with TIR domain and member of Ig superfamily

continued overleaf

CD Number*	Common Names	Gene Name	Chromosomal Location	Cellular/Tissue Expression	Function**
CD121a (cont'd)					Associates with IL-1RAcP to form complex binding to IL-1α, IL-1β, and IL-1Ra (IL-1 receptor antagonist) Mediates pro-inflammatory activities of IL-1
CD121b	**IL-1RII** Type II IL-1 receptor, IL1RB	IL1R2	2q12-q22	B cells, monocytes, neutrophils, some epithelial cells	Type I transmembrane glycoprotein and member of Ig superfamily Decoy receptor that binds well to IL-1β but poorly to IL-1α, IL-1Ra Inhibits pro-inflammatory effects of IL-1
CD122	**IL-2Rβ** IL-2 receptor β chain, p75	IL2RB	22q13	Constitutive low levels on T, B, and NK cells; upregulated by activation	Type I transmembrane glycoprotein of type I cytokine (WSXWS) receptor family Associates with CD132 (γc; IL-2Rγ) to form low-affinity IL-2 receptor Associates with CD25 (IL-2Rα) and CD132 to form high-affinity IL-2 receptor Binding to IL-2 promotes lymphocyte proliferation, differentiation, and regulation contributing to peripheral tolerance
CD123	**IL-3Rα** IL-3 receptor α chain	IL3RA	Xp22.3; Yp11.3	HSCs and downstream progenitors, NK cells, some myeloid and B lineage subsets	Type I transmembrane glycoprotein of type I cytokine (WSXWS) receptor family Associates with βc (CD131) to form IL-3 receptor Mediates signaling influencing growth and differentiation of erythroid, platelet, neutrophil, eosinophil, basophil, and monocyte lineages
CD124	**IL-4R** IL-4 receptor chain	IL4R	16p11.2–12.1	T and B cells, hematopoietic precursors, fibroblasts, endothelial cells	Type I transmembrane glycoprotein of type I cytokine (WSXWS) receptor family Associates with γc (CD132) to form high-affinity receptor for IL-4 IL-4 binding initiates signals promoting growth of B and T cell Required for Th2 differentiation Regulates IgE production and allergic inflammation Also associates with IL-13Rα (CD213) to form IL-13 receptor
CD125	**IL-5Rα** IL-5 receptor α chain	IL5RA	3p26-p24	Eosinophils, basophils	Type I transmembrane glycoprotein of type I cytokine (WSXWS) receptor family Associates with βc (CD131) to form IL-5 receptor Promotes eosinophil generation and activation
CD126	**IL-6Rα** IL-6 receptor α chain	IL6R	1q21	Mature T cells, some B cell subsets, activated B cells, most blood leukocytes, epithelial cells, liver	Type I transmembrane glycoprotein of type I cytokine (WSXWS) receptor family; also soluble form Associates with gp130 (CD130) to form IL-6 receptor Stimulates acute phase response in liver, regulates hematopoiesis Mediates growth signals for myelomas
CD127	**IL-7Rα** IL-7 receptor α chain	IL7R	5p13	Pro- and pre-B cells, thymocytes	Type I transmembrane glycoprotein of type I cytokine (WSXWS) receptor family Associates with γc (CD132) to form IL-7 receptor

CD Number*	Common Names	Gene Name	Chromosomal Location	Cellular/Tissue Expression	Function**
CD127 (cont'd)					Promotes T and B lymphopoiesis Essential for γδ T cell development and survival
CDw128a	**CXCR1** IL-R8 type I, IL-8 receptor α, IL-8RA	IL8RA	2q35	Neutrophils, basophils, T cell subsets, monocytes, fibroblasts	G protein-anchored transmembrane protein and member of chemokine receptor family Binds to IL-8 (CXCL8), a chemotactic factor preferentially drawing neutrophils to inflammatory sites IL-8 binding activates neutrophils
CDw128b	**CXCR2** IL-R8 type II, IL-8 receptor β, IL-8RB	IL8RB	2q35	Neutrophils, basophils, T cell subsets, monocytes, fibroblasts	G protein-anchored transmembrane protein and member of chemokine receptor family Binds to IL-8 (CXCL8), a chemotactic factor preferentially drawing neutrophils to inflammatory sites IL-8 binding activates neutrophils Mediates angiogenic activity of CXC chemokines in lung carcinoma models
CD129	**IL-9Rα** IL-9 receptor α chain	IL9R	Xq28; Yq12	Activated T cells, B cells, mast cells, macrophages, erythroid and myeloid precursors, neutrophils of asthmatics	Type I transmembrane glycoprotein of type I cytokine (WSXWS) receptor family Associates with γc (CD132) to form IL-9 receptor Promotes growth of mast cells and erythroid and myeloid progenitor cells
CD130	**gp130** IL-6 receptor β chain, IL-6 signal transducer, IL-6ST	IL6ST	5q11	Almost ubiquitous	Type I transmembrane glycoprotein of type I cytokine (WSXWS) receptor family Common signaling chain for receptors binding IL-6, oncostatin M, LIF, IL-11, IL-27, or cardiotrophin-1 Does not bind to cytokines itself
CD131	**βc** Common β chain	CSF2RB	22q13.1	HSCs and downstream progenitors, myelomonocytic lineages	Type I transmembrane glycoprotein of type I cytokine (WSXWS) receptor family Common signaling chain for receptors binding IL-3, IL-5, or GM-CSF Does not bind to cytokines itself Upregulated in response to inflammatory cytokines
CD132	**γc** Common γ chain, interleukin 2 receptor gamma chain	IL2RG	Xq13.1	T, B, and NK cells, monocyte lineage cells, glioma cells	Type I transmembrane glycoprotein of type I cytokine (WSXWS) receptor family Common signaling chain for receptors binding IL-2, 4, 7, 9, 15, and 21 Does not bind to cytokines itself
CD133	PROML1, prominin, AC133	PROM1	4p15.32	HSCs in adult bone marrow and blood, fetal liver, and cord blood; epithelial cells, endothelial precursors, pancreas, placenta, brain, retina, some tumor cells	Five-transmembrane glycoprotein and member of prominin family Function unknown Marker for myogenic progenitors May be a marker for brain cancer stem cells
CD134	**OX40** TNFRSF4 ACT35, TXGP1L	TNFRSF4	1p36	Activated T cells, B cells, fibroblasts, hematopoietic precursors	Type I transmembrane glycoprotein and member of TNFR superfamily Binds to OX40L (CD252) to promote

continued overleaf

CD Number*	Common Names	Gene Name	Chromosomal Location	Cellular/Tissue Expression	Function**
CD134 (cont'd)					T cell interaction with APCs, including B cells
					Triggers PKB signaling required for survival of activated T cells
					Also mediates adhesion of activated T cells to endothelium
CD135	**FLT3** Fms-related tyrosine kinase3, STK-1, flk-2	FLT3	13q12	Hematopoietic progenitors	Type I transmembrane glycoprotein of type III receptor tyrosine kinase family; also member of Ig superfamily
					Receptor for FLT3 ligand
					Important for hematopoiesis and differentiation, especially for NK cells
CDw136	**MSP-R** Macrophage stimulating protein receptor, Ron	MST1R	3p21.3	Peritoneal macrophages, osteoclasts, keratinocytes, megakaryocytes, epithelial cells	Type I membrane glycoprotein precursor that is post-translationally cleaved into two heterodimeric, disulfide-linked subunits; member of the tyrosine protein kinase family
					Receptor for MSP
					Has tyrosine protein kinase activity
					Ligation induces cellular activation, migration, proliferation, invasion
					May play roles in bone resorption and wound healing
CDw137	**4-1BB** ILA, Ly63	TNFRSF9	1p36	Activated T cells, B cells, IELs, monocytes	Type I transmembrane glycoprotein and member of TNFR superfamily
					Present as disulfide-linked homodimer; splice variant is soluble
					Binds to 4-1BB ligand (TNFSF9) present on B cells, macrophages
					Costimulates T cell activation and promotes T cell proliferation
CD138	**SDC1** Syndecan-1, heparan sulfate proteoglycan, B-B4	SDC1	2p24.1	Pre-B cells, plasma cells, mammary epithelial cells, skin, brain, mammary carcinoma cells	Type I membrane protein and member of syndecan proteoglycan family
					Cell surface proteoglycan that links the cytoskeleton to extracellular matrix
					Can modulate neutrophil influx into sites of injury
					Can modulate Wnt signaling during mammary tumorigenesis in mice
CD139	B-031	Not assigned	Not known	B cells, monocytes, granulocytes, FDCs; weak on RBCs and smooth muscle cells of some blood vessels	>200-kDa protein of uncharacterized structure
					Function unknown
CD140a	**PDGFα** Platelet-derived growth factor receptor type α; α subunit, PDGFR2	PDGFRA	4q11-q13	Fibroblasts, smooth muscle cells, liver endothelial cells, mesothelial cells	Type I transmembrane glycoprotein of type III protein kinase receptor family and member of Ig superfamily
					Forms homodimers and heterodimer with PDGFRβ
					Receptor for PDGFA and PDGFB
					Regulates PDGF-induced chemotaxis, cell growth
					Important for wound healing and fetal development
					Receptor for adeno-associated virus

CD Number*	Common Names	Gene Name	Chromosomal Location	Cellular/Tissue Expression	Function**
CD140b	**PDGFβ** Platelet-derived growth factor receptor type β; β subunit, PDGFR1	PDGFRB	5q31-q32	Endothelial cells, stromal cells, fibroblasts, neurons, meningeal cells, smooth muscle cells	Type I transmembrane glycoprotein of type III protein kinase receptor family and member of Ig superfamily Forms homodimers and heterodimer with PDGFRα Receptor for PDGFB only Regulates PDGF-induced chemotaxis, cell growth Important for wound healing and fetal development Receptor for adeno-associated virus
CD141	**Thrombomodulin** TM, fetomodulin	THBD	20p12-cen	Endothelial cells, plasma, urine	Type I transmembrane glycoprotein with C-type lectin and EGF-like domains Receptor for thrombin Thrombin complexed to thrombomodulin has increased activity over thrombin alone Complex rapidly converts protein C to activated protein C, reducing thrombin generation and promoting anti-coagulation Inhibits thrombin-mediated clotting of fibrinogen and platelet activation
CD142	**Coagulation factor III** Tissue factor, thromboplastin	F3	1p22-p21	Activated monocytes, endothelial and epithelial cells, fibroblasts, vascular smooth muscle cells	Type I transmembrane glycoprotein of type II cytokine receptor family and member of Ig superfamily Associates with Factor VII to form an enzyme complex which initiates the extrinsic coagulation cascade
CD143	**ACE** Angiotensin 1 converting enzyme, peptidyl dipeptidase A-1	ACE	17q23	Vascular endothelium of arterioles, capillaries; epithelial cells in lung, kidney, testis	Type I transmembrane protein with Zn metallopeptidase activity; also soluble form in plasma, body fluids Two isoforms: one active in somatic cells and the other active in sperm only Converts angiotensin I to angiotensin II, increasing vasoconstriction Also inactivates bradykinin Important for regulation of blood pressure and electrolyte balance, and for fertility
CD144	**VE-cadherin** Vascular endothelial cadherin, cadherin-5	CDH5	16q22.1	Cell–cell and cell–matrix junctions involving endothelial cells	Type I membrane protein with five cadherin domains Ca^{2+}-dependent adhesion protein Predominantly homotypic binding Organizes adhesion junctions for endothelial cells Important for contact inhibition of cell growth
CDw145	None	Not assigned	Unknown	Endothelial cells, including those in umbilical vein, kidney, lung, spleen, liver	Identified by immunoprecipitation bands at 251, 110, and 90 kDa Function unknown
CD146	**MCAM** Melanoma cell adhesion molecule, Muc 18, Mel-CAM, S-endo	MCAM	11q23.3	Endothelial cells, some activated T cells, smooth muscle, melanoma	Type I transmembrane glycoprotein resembling NCAM (CD56) and member of Ig superfamily Probable adhesion function but ligand unknown

continued overleaf

CD Number*	Common Names	Gene Name	Chromosomal Location	Cellular/Tissue Expression	Function**
CD146 (cont'd)					May play a role in tumor growth and metastasis by promoting melanoma cell interaction with vascular cells
CD147	**BSG** Basigin, M6, extracellular metalloproteinase inducer, EMMPRIN, OK blood group	BSG	19p13.3	Most leukocytes, endothelial cells, platelets, RBCs	Type I membrane protein and member of Ig superfamily Forms homo-oligomers on the plasma membrane Stimulates fibroblasts to produce matrix metalloproteinases (MMPs) May recruit monocarboxylate transporters to retinal cell plasma membranes Important for spermatogenesis, embryo implantation, neural network formation Forms a complex with matrix metalloproteinase-1 (MMP1) on tumor cell surfaces that promotes progression
CD148	DEP-1, HPTP-η; protein tyrosine phosphatase receptor type J	PTPRJ	11p11.2	Most leukocytes, platelets, fibroblasts, some neurons	Type I transmembrane glycoprotein and member of receptor protein tyrosine kinase phosphatase type III family Anti-CD148 ligation leads to cytokine secretion but natural ligand is unknown Deleted in many tumors May contribute to tumor suppression by arresting cell growth at confluence
(CD149)	Not in use				
CD150	**SLAM** Signaling lymphocyte activation molecule, IPO-3	SLAMF1	1q22-q23	Activated T and B cells, thymocytes, DCs, memory T cells, endothelial cells	Type I transmembrane glycoprotein and member of Ig superfamily; also secreted and cytoplasmic forms Exhibits weak homophilic binding Adaptor protein that interacts with SAP in cytoplasm Minor costimulator of T and B cell activation, leading to enhanced proliferation, inflammatory cytokine secretion, Ig production Physiological receptor for measles virus
CD151	**PETA-3** Platelet-endothelial tetraspan antigen 3, SFA-1, Mer2	CD151	11p15.5	Most tissues (not brain), highly expressed in megakaryocytes, RBCs	Transmembrane glycoprotein and member of tetraspanin family Binds to several β-integrins in hemidesmosomes and may modify their function or signaling Also binds to CD9 and CD81 Essential for signaling leading to normal platelet function and activation
CD152	**CTLA-4** Cytotoxic T lymphocyte associated protein 4	CTLA4	2q33	Activated T and B cells, regulatory T cells	Type I transmembrane protein and member of Ig superfamily Shows sequence similarity to CD28 High avidity receptor for CD80/CD86 Downregulates T cell activation and regulates Th2 differentiation Contributes to maintenance of peripheral tolerance and homeostasis
CD153	**CD30L** CD30 ligand	TNFSF8	9q33	Activated T cells and macrophages, neutrophils, eosinophils, normal and malignant B cells	Type II transmembrane glycoprotein and member of TNF superfamily Binds to CD30

CD Number*	Common Names	Gene Name	Chromosomal Location	Cellular/Tissue Expression	Function**
CD153 (cont'd)					CD153 signaling has been associated with stimulation of T–B cell interaction, cellular proliferation, and apoptosis Precise function is unclear
CD154	**CD40L** CD40 ligand, TRAP-1, TNF-related activation protein-1 T-BAM, gp39	TNFSF5	Xq26	Activated T cells (especially CD4$^+$), mast cells, NK cells, granulocytes, monocytes, activated platelets	Type II transmembrane glycoprotein and member of TNF superfamily Binds to CD40 Provides major costimulatory signal for B cell activation and a survival signal to GC B cells Also stimulates monocytes, T cells Required for isotype switching, DC licensing, CD8$^+$ memory T cell generation, and regulation of autoreactive T cells
CD155	**PVR** Polio virus receptor	PVR	19q13.2	Monocytes	Type I transmembrane glycoprotein and member of polio virus receptor family and Ig superfamily; also secreted isoforms Normal function unknown but is expressed in embryonic structures giving rise to neurons Receptor for polio virus
CD156a	**ADAM 8** A disintegrin and metalloproteinase domain 8, MS 2	ADAM8	10q26.3	Neutrophils, monocytes, macrophages, some B cell lines	Type I transmembrane protein containing metalloproteinase and disintegrin domains Binds Zn Possible role in leukocyte extravasation
CD156b	**ADAM 17** TACE, tumor necrosis factor α-converting enzyme, snake venom-like protease, CSVP	ADAM17	2p25	Monocytes, macrophages, neutrophils, T cells, endothelial cells, myocytes	Type I transmembrane protein containing metalloproteinase and disintegrin domains Binds Zn Cleaves other membrane-anchored proteins (including TNFRI, TNFRII, IL-1RII, L-selectin) such that they "shed" their extracellular domains Converts membrane-bound precursors of TNF and CX3CL1 to their mature soluble forms Activates the Notch1 pathway
CDw156c	**ADAM10** A disintegrin and metalloproteinase 10, kuz, MADM	ADAM10	15q22	Broadly expressed, especially in Golgi-derived vesicles	Type I transmembrane protein containing metalloproteinase and disintegrin domains Binds Zn Contributes to neuronal α-secretase activity Converts membrane-bound precursor of TNF to its mature soluble form Cleaves ephrin-A2 to regulate axon signaling Contributes to normal cleavage of cellular prion protein May be involved in Notch signaling during embryogenesis
CD157	**BST-1** Bone marrow stromal antigen-1, BP-3/IF7, Mo5	BST1	4p15	Monocytes, granulocytes, lymphoid progenitors, bone marrow stromal lines, vascular endothelial cells, FDCs, pancreatic islets	GPI-anchored homodimeric protein with ADP-ribosyl cyclase activity Generates cyclic ADP-ribose that stimulates Ca^{2+} release from intracellular pools Stromal cell expression of BST-1 supports pre-B cell growth

continued overleaf

CD Number*	Common Names	Gene Name	Chromosomal Location	Cellular/Tissue Expression	Function**
CD158e	KIR3DL1 Killer cell Ig-like receptor three domains long cytoplasmic tail 1	KIR3DL1	19q13.4	NK cells, subset of CD8$^+$ T cells	Type I transmembrane glycoprotein and ITIM-containing member of Ig superfamily NK inhibitory receptor Binds to HLA-B on potential target cells and blocks NK cytotoxicity
CD158i	KIR2DS4 Killer cell Ig-like receptor two domains short cytoplasmic tail 4	KIR2DS4	19q13.4	NK cells	Type I transmembrane glycoprotein and member of Ig superfamily (no ITIM) Potential NK activatory receptor Binds to HLA-C on potential target cells but does not block NK cytotoxicity
CD158k	KIR2DL2 Killer cell Ig-like receptor two domains long cytoplasmic tail 2	KIR2DL2	19q13.4	NK cells	Type I transmembrane glycoprotein and ITIM-containing member of Ig superfamily NK inhibitory receptor Binds to HLA-C and inhibits NK cytotoxicity
CD159a	NKG2A Natural killer gene 2A, killer cell lectin-like receptor subfamily C member 1	KLRC1	12p13	NK cells, subset of CD8$^+$ T cells	Type II transmembrane glycoprotein with C-type lectin domains and ITIMs Associates with CD94 to form NK inhibitory receptor binding to HLA-E Blocks NK cytotoxicity
CD159c	NKG2C Natural killer gene 2C	KLRC2	12p13	NK, NKT, and γδ T cells	Type II transmembrane glycoprotein with C-type lectin domains Associates with CD94 to form NK activatory receptor binding to HLA-E May activate NK, NKT, and γδ T cells
CD160	NK1, NK28, BY55	CD160	1q21.1	NK cells, cytotoxic T cells, iIELs	GPI-anchored membrane protein and member of KIR superfamily Occurs as disulfide-linked homomultimer Binds a range of classical and non-classical MHC class I molecules May be involved in costimulation
CD161	NKR-P1A, killer cell lectin-like receptor subfamily B member 1	KLRB1	12p13	Most NK cells, some T cell subsets, NKT cells	Type II transmembrane glycoprotein with C-type lectin domains Occurs as disulfide-linked homodimer Binds to oligosaccharides May be NK activatory receptor since anti-CD161 cross-linking on NK cells induces IFNγ secretion and cytotoxicity Also induces NKT IFNγ production
CD162	PSGL-1 P-selectin glycoprotein ligand 1	SELPLG	12q24	Neutrophils, monocytes, granulocytes, some B cells, some CD34$^+$ bone marrow cells	Type I transmembrane sialomucin protein occurring as disulfide-linked homodimer Binds to P-, E-, and L-selectins (CD62) Mediates leukocyte tethering and rolling on activated endothelium or activated platelets
CD162R	PEN5		Carbohydrate antigen	NK subset	Unknown function
CD163	GHI/61, D11, RM3/1, M130, hemoglobin scavenger receptor	CD163	12p13.3	Tissue macrophages	Type I transmembrane protein and member of scavenger receptor superfamily; alternative splice variants Binds to haptoglobin–hemoglobin complexes May play role in anti-inflammatory responses of monocytes

CD Number*	Common Names	Gene Name	Chromosomal Location	Cellular/Tissue Expression	Function**
CD164	**MGC 24** Multi-glycosylated core protein 24, MUC-24, sialomucin	CD164	6q21	Bone marrow stromal cells, myelomonocytic progenitors, monocytes, epithelial cells	Type I integral membrane glycoprotein and member of the mucin family Mediates adhesion of CD34+ progenitors to bone marrow stroma and negatively regulates their growth Associated with various carcinomas
CD165	AD2, A108, gp37	Not assigned	Unknown	Thymocytes, thymic epithelial cells, platelets, monocytes, CNS neurons, some lymphocytes	Membrane glycoprotein that mediates adhesion between thymocytes and thymic epithelial cells
CD166	**ALCAM** Activated leukocyte cell adhesion molecule, KG-CAM	ALCAM	3q13.1	Neurons, activated T cells, activated monocytes, epithelial cells, fibroblasts	Type I membrane protein and member of Ig superfamily Mediates T and B cell adhesion to other leukocytes by binding to CD6 and to itself Also involved in neurite extensions by neurons
CD167	**DDR1** Discoidin receptor	DDR1	6p21.3	B cells, immature DCs, epithelial cells, epithelial tumors	Type I transmembrane glycoprotein with discoidin domain; member of protein tyrosine kinase family Binds to collagen and has kinase activity May play role in epithelial tumor invasion and metastasis
CD168	**RHAMM** Receptor for hyaluronan involved in migration and motility	HMMR	5q33.2-qter	Thymocytes, lymphocytes, B cell malignancies	Surface protein lacking a transmembrane domain or signal sequence Binds to hyaluronan Part of a larger complex involved in cell motility May also be involved in oncogenic transformation and metastasis Intracellular protein in human breast cancer cells
CD169	**Sialoadhesin** Siglec-1, Sialic acid binding Ig-like lectin 1	SN	20p13	Macrophages, monocytes (weak), DCs, lymphocytes	Type I transmembrane glycoprotein and member of Ig superfamily; soluble variant Adhesion molecule mediating intercellular interaction by preferentially binding to glycolipids and glycoproteins with terminal α-2 sialyl residues May play a role in hematopoiesis
CD170	**Siglec-5** Sialic acid binding Ig-like lectin 5	SIGLEC5	19q13.3	Lymphocytes, monocytes, DCs, myeloid leukemic cells	Type I transmembrane protein and ITIM-containing member of Ig superfamily Occurs as homodimer May mediated adhesion by binding to α-2 sialyl residues
CD171	Neuronal adhesion molecule, LI	L1CAM	Xq28	Lymphocytes, monocytes, APCs, some endothelial and neuronal cells	Type I transmembrane glycoprotein with fibronectin domains; member of Ig superfamily Mediates homotypic adhesion Also binds to laminin, integrins, axonin, and proteoglycans containing chondroitin sulfate Important for neuron migration and differentiation

continued overleaf

CD Number*	Common Names	Gene Name	Chromosomal Location	Cellular/Tissue Expression	Function**
CD172a	SIRPα Signal inhibitory regulatory protein a or alpha	PTPNS1	20p13	Monocytes, hematopoietic precursors, neuronal cells	Type I transmembrane glycoprotein and ITIM-containing member of Ig and SIRP superfamilies Receptor-type adhesion molecule that induces translocation of binding partners from cytosol to plasma membrane Can deliver an inhibitory signal influencing phagocytosis, activation of mast cells and DCs Involved in growth factor receptor signaling and cerebellar neuron adhesion Binding to CD47 (IAP) blocks DC maturation and cytokine secretion
CD172b	SIRPβ Signal regulatory protein beta 1	SIRPB1	20p13	DCs, monocytes	Type I transmembrane glycoprotein and ITIM-containing member of Ig and SIRP superfamilies Receptor-type adhesion molecule SIRPβ lacks the cytoplasmic domain of SIRPα allowing recruitment of SHP-2 and instead associates with ITAM-containing DAP12 to stimulate cellular activation Recruits Syk kinase
CD172g	SIRPγ Signal regulatory protein beta 2 or gamma	SIRPB2	20p13	T cells, activated NK cells, placenta	Type I transmembrane glycoprotein and ITIM-containing member of Ig and SIRP superfamilies Structure similar to SIRPβ; function unknown
CD173	Blood group H2		Carbohydrate antigen	CD34+ hematopoietic precursors, carcinomas, endothelial cells, erythrocytes	Precursor to ABO blood group carbohydrate antigens and Lewis antigens Product of fucosyl transferase-1
CD174	Lewis Y blood group, LeY, fucosyltransferase 3	FUT3	19p13.3	CD34+ hematopoietic precursors, erythrocytes	Type II membrane protein; also in Golgi membrane Transfers fucose to polysaccharide precursors to create Lewis blood antigens May also act as cofactor for procoagulant activity of cancer cells Possible role in early commitment to apoptosis
CD175	Tn antigen, T-antigen novelle		Carbohydrate antigen	CD34+ hematopoietic precursors, RBCs, various T, B, and myeloid leukemias, epithelial tumors	O-linked monosaccharide Precursor for ABO blood group antigens and TF antigen (CD176)
CD175s	Sialyl-Tn		Carbohydrate antigen	Same as CD175	Sialylated form of CD175
CD176	Thomsen-Friedenreich antigen, TF		Carbohydrate antigen	CD34+ hematopoietic precursors, endothelial cells, erythrocytes; a sialylated form is ubiquitous	Function unknown Appears on most carcinomas
CD177	NB1, PRV1, polycythemia rubra vera 1, HNA2A	PRV1	19q13.2	Neutrophil subset, myelomonocytic precursors	GPI-anchored glycoprotein Upregulated during actin redistribution
CD178	FasL Fas ligand, CD95L, CD95 ligand	TNFSF6	1q23	T and NK cells; may be expressed more widely but shed through proteolysis	Type II transmembrane glycoprotein and member of TNF superfamily; soluble form generated by proteolysis

CD Number*	Common Names	Gene Name	Chromosomal Location	Cellular/Tissue Expression	Function**
CD178 (cont'd)					Occurs as trimer
					Binds to Fas (CD95) to induce apoptosis
					Important for maintenance of peripheral tolerance
CD179a	V pre-β	VPREB1	22q11.22	Pro-B, pre-B cells	Atypical member of Ig superfamily with homology to Ig V region
					Associates with lambda V chain (CD179b) to form surrogate light chain required for pre-BCR expression and early B cell development
CD179b	Lambda 5	IGLL1	22q11.23	Pro-B, pre-B cells	Member of Ig superfamily with homology to Ig lambda J region and constant region
					Associates with V pre-β chain (CD179a) to form surrogate light chain required for pre-BCR expression and early B cell development
CD180	RP105, Bgp95	LY64	5q12	Mantle zone B cells, marginal zone B cells, GC B cells (weak)	Type I transmembrane protein and member of TLR family
					Associates with MD-1 to form receptor complex that works with TLR4 to regulate B cell recognition and responses to LPS
					Anti-CD180 ligation leads to upregulation of CD80/CD86 and increased cell size
CD181	CXCR1 IL-8RA, IL-8R1	IL8RA	2q35	Neutrophils, some T cells, Fallopian tube epithelial cells	Integral membrane protein and member of G protein-coupled receptor and rhodopsin superfamilies
					Chemokine receptor binding to IL-8 only
					Mediates IL-8 signaling promoting neutrophil activation and chemotaxis
CD182	CXCR2 IL-8RB, IL-8R2	IL8RB	2q35	Neutrophils, granulocytes, monocytes, some T cells	Integral seven-span membrane protein and member of G protein-coupled receptor and rhodopsin superfamilies
					Chemokine receptor binding to IL-8 plus two related growth factors
					Mediates IL-8 signaling promoting neutrophil activation and chemotaxis
					Transduces signals for chemokine-mediated angiogenesis, and possibly metastasis
CD183	CXCR3 GPR9, IP-10 receptor, Mig receptor	CXCR3	Xq13	Activated T cells, NK cells, Th1 effectors, leukemic B cells (not normal B cells)	Integral seven-span membrane protein and member of G protein-coupled receptor and rhodopsin superfamilies
					Chemokine receptor binding primarily to chemokines IP-10, Mig, I-TAC
					Facilitates T cell homing to the lung
CD184	CXCR4 Fusin	CXCR4	2q21	Widely expressed in hematopoietic cells vascular endothelial cells, neural tissue, breast cancer cells	Integral seven-span membrane protein and member of G protein-coupled receptor and rhodopsin superfamilies
					Chemokine receptor binding primarily to chemokine SDF-1
					Directs T cells to T cell-rich zones of secondary lymphoid tissues
					Coreceptor for HIV infection
CD185	CXCR5 Burkitt's lymphoma receptor 1	BLR1	11q23.3	B cells	Integral seven-span membrane protein and member of G protein-coupled receptor and rhodopsin superfamilies

continued overleaf

CD Number*	Common Names	Gene Name	Chromosomal Location	Cellular/Tissue Expression	Function**
CD185 (cont'd)					Chemokine receptor binding primarily to chemokine BCA-1
					Directs B cells to B cell-rich zones of secondary lymphoid tissues
					Deficiency results in a lack of GC formation in mice
CDw186	**CXCR6** BONZO	CXCR6	3p21	Some T and B cells	Integral seven-span membrane protein and member of G protein-coupled receptor and rhodopsin superfamilies
					Chemokine receptor binding primarily to chemokines CXCL15 and CXCL16
					Preferentially expressed by Th1 cells
					Facilitates homing to sites of inflammation
(CD187–190)	In reserve				
CD191	**CCR1** RANTES receptor	CCR1	3p21	Activated T cells, macrophages, neutrophils, basophils, myeloid precursors, B cell subsets, stem cell subset	Integral seven-span membrane protein and member of G protein-coupled receptor and rhodopsin superfamilies
					Chemokine receptor binding primarily to chemokines RANTES, MIP-1α, MCP-3, MCP-4, Lkn-1
					Facilitates early events in inflammation
CD192	**CCR2** MCP-1 receptor	CCR2	3p21	Macrophages, monocytes, activated T cells, basophils, endothelial cells	Integral seven-span membrane protein and member of G protein-coupled receptor and rhodopsin superfamilies
					Chemokine receptor binding primarily to chemokines MCP-2, -3 , -4
					Required for normal trafficking of APCs capable of inducing IFNγ production by T cells
					Functions as a minor HIV coreceptor
CD193	**CCR3** Eotaxin receptor	CCR3	3p21.3	Eosinophils, some DCs, activated T cells, Th2 effectors, basophils, airway epithelial cells	Integral seven-span membrane protein and member of G protein-coupled receptor and rhodopsin superfamilies
					Chemokine receptor binding primarily to chemokines eotaxin, MCP-2, -3, -4
					Required for eosinophil trafficking
					Facilitates allergic reactions
					Functions as a minor HIV coreceptor
(CD194)	Reserved for **CCR4**	CCR4	3p23	Activated T cells, Th2 effectors, basophils	Integral seven-span membrane protein and member of G protein-coupled receptor and rhodopsin superfamilies
					Chemokine receptor binding primarily to chemokines TARC, MDC, RANTES, MIP-1α, MIP-1β
CD195	**CCR5**	CCR5	3p21	Lymphocytes, monocytes, macrophages, DCs	Integral seven-span membrane protein and member of G protein-coupled receptor and rhodopsin superfamilies
					Chemokine receptor binding primarily to chemokines MIP-1, MIP-1β, RANTES
CD196	**CCR6** GPR29	CCR6	6q27	Activated T cells, B cells, immature DCs, CLA$^+$ memory T cells	Integral seven-span membrane protein and member of G protein-coupled receptor and rhodopsin superfamilies
					Chemokine receptor binding primarily to chemokine MIP-3α and β-defensins
					Induces PLC-mediated calcium flux

CD Number*	Common Names	Gene Name	Chromosomal Location	Cellular/Tissue Expression	Function**
CD196 (cont'd)					Facilitates recruitment of immature DCs and T cells to sites of microbial invasion
CD197	**CCR7** EBV-induced gene 1, EBI1	CCR7	17q12-q21.2	Naive T and B cells, mature DCs, Th1 effectors, central memory T cell subset	Integral seven-span membrane protein and member of G protein-coupled receptor and rhodopsin superfamilies Chemokine receptor binding primarily to chemokines SLC, ELC, MIP-3β Directs homing to splenic PALS and lymph nodes Directs activated B cells to move to T cell zones of secondary lymphoid tissues
CDw198	**CCR8**	CCR8	3p22	Activated T cells, NK cells, Th2 effectors, monocytes	Integral seven-span membrane protein and member of G protein-coupled receptor and rhodopsin superfamilies Chemokine receptor binding primarily to chemokine I-309 Required for Th2 cytokine responses
CDw199	**CCR9**	CCR9	3p21.3	DCs, immature and mature thymocytes, CLA⁻ mucosal memory T cells	Integral seven-span membrane protein and member of G protein-coupled receptor and rhodopsin superfamilies Chemokine receptor binding primarily to chemokines MEC, TECK Facilitates homing to the mucosa of small intestine
CD200	MRC OX 2, MOX1, MOX2	CD200	3q12-q13	B cells, neurons, some DCs	Type I transmembrane glycoprotein and member of Ig superfamily May regulate cells of the macrophage lineage in various tissues
CD201	EPCR Endothelial protein C receptor	PROCR	20q11.2	Endothelial cells in skin, lung, heart	Type I transmembrane glycoprotein and member of CD1/MHC family Receptor for and activator of activated protein C (involved in coagulation pathway)
CD202b	**TEK** TIE2	TEK	9p21	Endothelial cells, subset of CD34⁺ HSCs	Type I transmembrane protein with fibronectin domains; member of Ig superfamily Receptor tyrosine kinase that binds to angiopoietin 1 Important for regulation of endothelial cell proliferation, differentiation, and vascular patterning during angiogenesis
CD203c	E-NPP3, PDNP3, PD-1β	ENPP3	6q22	Basophils, mast cells and their precursors, uterus	Type II transmembrane glycoprotein Also an ectoenzyme with ATPase and ATP pyrophosphatase activity
CD204	**SR-A** MSR, macrophage scavenger receptor class A	MSR1	8p22	Macrophages	Type II transmembrane protein and member of scavenger receptor family; splice variants Occurs as homotrimer Scavenger receptor mediating uptake of negatively charged macromolecules such as oxidized LDL Important for innate defense
CD205	**DEC-205**	LY75	2q24	DCs, LCs, thymic epithelium, T and B cells (weak)	Type I transmembrane glycoprotein with C-lectin and fibronectin domains Endocytic receptor that directs captured antigen to endocytic compartments specialized for antigen presentation

continued overleaf

CD Number*	Common Names	Gene Name	Chromosomal Location	Cellular/Tissue Expression	Function**
CD206	**MR** Mannose receptor type 1, macrophage mannose receptor, MMR	MRC1	10p12.33	Macrophages	Type I transmembrane glycoprotein with C-lectin and fibronectin domains Receptor that binds to oligomannose-containing molecules on pathogens and induces their phagocytosis
CD207	Langerin	CD207	2p13	LCs, some activated DCs	Endocytic receptor with C-type lectin activity May be involved in internalization, processing, and presentation of mannose-bearing antigens Expressed in Birkbeck granules
CD208	DC-LAMP	LAMP3	3q26.3-q27	Activated DCs	Type I transmembrane glycoprotein with C-type lectin domain Found in lysosomes, endosomes, MHC class II compartments
CD209	**DC-SIGN**	CD209	19p13	DCs	Type II transmembrane protein with C-type lectin domain Binds to mannose oligosaccharides on wide variety of pathogens to initiate internalization for antigen presentation Binds to ICAM-3 (CD50) to mediate DC–T cell interaction, or DC transendothelial migration Also binds to HIV gp120 *in vitro*
CDw210a	**IL-10R1** IL-10RI IL-10 receptor α	IL10RA	11q23	T, B, and NK cells, monocytes, macrophages	Type I transmembrane protein and type II cytokine receptor with fibronectin domains Extracellular domain resembles that in IFN receptors Binds to IL-10R2 (CDw210b) to form high-affinity IL-10 receptor Mediates suppression of inflammatory cytokine synthesis Promotes survival of myeloid progenitors
CDw210b	**IL-10R2** IL-10RII IL-10 receptor	IL10RB	21q22.11	T, B, and NK cells, monocytes, macrophages	Type I transmembrane protein and type II cytokine receptor with fibronectin domains Binds to IL-10R1 (CDw210a) to form high-affinity IL-10 receptor Also participates in heteromeric receptors for IL-22, -26, -28A, -28B, and -29
(CD211)	Not assigned				
CD212	**IL-12Rβ** IL-12 receptor β chain	IL12RB1	19p13.1	Activated T and NK cells, some B cells, some DCs	Type I transmembrane protein of type I cytokine (WSXWS) receptor family; has fibronectin domain Binds via disulfide link to IL-12Rα chain to form heterodimeric IL-12 receptor IL-12 promotes IFNγ production that directs immune responses preferentially toward Th1 type
CD213a1	**IL-13Rα** IL-13 receptor α1	IL13RA1	Xq24	Vascular endothelial cells, monocytes, macrophages, B cells, broadly in non-hematopoietic tissues, Reed-Sternberg cells of Hodgkin's lymphomas	Type I transmembrane protein of type I cytokine (WSXWS) receptor family; has fibronectin domain On its own, binds to IL-13 with low affinity Binds to IL-4R (CD124) to form high-affinity IL-13 receptor

CD Number*	Common Names	Gene Name	Chromosomal Location	Cellular/Tissue Expression	Function**
CD213a1 (cont'd)					Functions of IL-13 overlap with those of IL-4 (inhibition of Th1 cytokine production) but exclude a role for IL-13 in Th2 differentiation Promotes isotype switching to IgE
CD213a2	IL-13 receptor α2	IL13RA2	Xq13.1-q28	B cells, monocytes, fibroblasts, endothelial cells, placenta	Type I transmembrane protein of type I cytokine (WSXWS) receptor family; has fibronectin domain Decoy receptor that binds IL-13 (but not IL-4) with high affinity but does not transduce signaling
(CD214-CD216)	Not assigned				
CDw217	**IL-17R** IL-17 receptor	IL17R	22q11.1	Broad tissue distribution, cord blood lymphocytes, peripheral blood lymphocytes, thymocytes	Type I transmembrane glycoprotein with no homology to other cytokine receptors Binds IL-17 with low affinity IL-17 causes stromal cells to secrete pro-inflammatory cytokines and chemokines
CDw218a	**IL-18R** IL-18Rα IL-18 receptor alpha chain, IL-1RRP	IL18R1	2q12	Broad tissue distribution, leukocytes, lung, spleen, liver, thymus, intestine	Type I transmembrane protein and member of IL-1 receptor family Associates with IL-18RAcP chain (CD218b) to form high-affinity receptor for IL-18 IL-18 binding subunit IL-18 synergizes with IL-12 in inflammatory responses; sustains Th1 response Signaling activates NF-κB leading to IL-1 and TNF production
CDw218b	**IL18RAcP** IL-18 receptor accessory protein, IL-18Rβ, ACPL	IL18RAP	2p24.3-p24.1	Broad tissue distribution, T and B cells, neutrophils, some DCs, lung, spleen, colon	Type I transmembrane protein and member of IL-1 receptor family Associates with IL-18R chain (CD218a) to form high-affinity receptor for IL-18 Signal transduction subunit
(CD219)	Not assigned				
CD220	**Insulin receptor**	INSR	19p13.3-p13.2	Ubiquitous	Type I membrane precursor protein that is proteolytically cleaved into α and β chains that are disulfide-linked Splice variants exist in different tissues Tetramer of 2 α and 2 β chains forms the insulin receptor α subunit binds insulin β subunit is a transmembrane tyrosine kinase Insulin controls cell utilization of glucose
CD221	**IGF1R** Insulin-like growth factor receptor-1, type I IGF receptor	IGF1R	15q26.3	Ubiquitous	Type I membrane precursor protein that is proteolytically cleaved into α and β chains that are disulfide-linked Tetramer of 2 α and 2 β chains forms receptor binding to IGF1 with high affinity, IGF2 with low affinity α subunit binds IGFs β subunit is a transmembrane tyrosine kinase Anti-apoptotic effects Critical role in transformation

continued overleaf

CD Number*	Common Names	Gene Name	Chromosomal Location	Cellular/Tissue Expression	Function**
CD222	IGF2R Insulin-like growth factor II receptor, mannose-6-phosphate receptor	IGF2R	6q26	Ubiquitous	Type I transmembrane protein with collagen-binding domains; predominantly located intracellularly Binds to IGF2 and mannose-6-phosphate-containing proteins such as lysosomal hydrolases Involved in transport of phosphorylated lysosomal enzymes from surface and Golgi to lysosomes Regulates TGFβ activity by binding to LAP May also bind to plasminogen, CD87, retinoic acid
CD223	LAG-3 Lymphocyte activation gene 3	LAG3	12p13.32	Activated T cells, NK cells	Type I transmembrane protein with homology to CD4; member of Ig superfamily Binds to MHC class II Induces DC maturation and activation
CD224	GGT γ-glutamyl transferase, γ-glutamyl transpeptidase	GGT1	22q11.23	Broadly expressed in tissues involved in absorption and secretion, some leukocytes, some hematopoietic precursors	Type II transmembrane glycoprotein Heterodimer of heavy and light chains (latter is catalytic) Also an ectoenzyme that transfers the glutamyl moiety of glutathione to amino acids and dipeptide acceptors Involved in extracellular glutathione breakdown, maintenance of intracellular glutathione, and cell antioxidant defense
CD225	Leu-13, interferon-induced transmembrane protein 1	IFITM1	11p15.5	Peripheral blood leukocytes, endothelial cells	Type I transmembrane glycoprotein Associates with the CD21/CD19/CD81 complex involved in cellular activation (especially B cells) Implicated in cell growth control
CD226	DNAM-1 DNAX accessory molecule-1, DTA-1	CD226	18q22.3	NK cells, platelets, monocytes, T and B cell subsets	Type I transmembrane glycoprotein and member of Ig superfamily Mediates adhesion of CTLs and NK cells to unknown ligand on target cells Anti-CD226 ligation induces intracellular signaling and activation
CD227	MUC1 Mucin 1, epithelial membrane antigen, EMA	MUC1	1q21	Widely expressed on leukocytes and epithelia of glands and ducts, aberrantly expressed in breast cancer	Type I transmembrane phosphoglycoprotein with numerous tandem 20 residue repeats and variable glycosylation; also secreted forms C-terminal peptide is cleaved and bound to remainder in heterodimer Protects and lubricates epithelial surfaces May be involved in intercellular adhesion, links to cytoskeleton Can bind to ligands on pathogens
CD228	Melano-transferrin, p97	MFI2	3q28-q29	Melanoma cells	GPI-anchored membrane glycoprotein belonging to transferrin superfamily Presumed iron uptake function May also bind Zn
CD229	Ly9 Lymphocyte antigen 9, SLAMF3	LY9	1q21.3-q22	T and B cells, thymocytes	Type I transmembrane glycoprotein and member of Ig superfamily Function unknown, but similarity to CD2 and CD48 suggests an adhesion function

CD Number*	Common Names	Gene Name	Chromosomal Location	Cellular/Tissue Expression	Function**
CD229 (cont'd)					May be involved in homophilic interactions between T cells and APCs
CD230	**PrP^c** Prion protein cellular, Creutzfeldt-Jakob disease	PRNP	20pter-p12	Brain, T and B cells, monocytes, DCs	GPI-anchored membrane glycoprotein with alpha-helical structure Function of normal PrP^c protein is unknown; may bind copper PrP^sc is abnormal form with β-pleated sheets that tend to aggregate into rod-like structures Aggregates bind to cholesterol-rich phospholipid membranes in brain and are cytotoxic PrP^sc is infective agent in spongiform encephalopathies
CD231	**TSPAN7** Tetraspanin 7, TALLA-1, TM4SF2	TSPAN7	Xp11.4	Neuronal tissue, neuroblastoma, T-ALL	Transmembrane glycoprotein and member of tetraspanin family May control neurite outgrowth Forms complexes with integrins
CD232	**Plexin C1** VESPR, virus encoded semaphorin protein receptor	PLXNC1	12q23.3	Monocytes, some B cells, brain, lung, spleen, heart, placenta	Transmembrane glycoprotein containing sema domain; member of plexin superfamily Receptor for virally encoded semaphorin
CD233	Band 3, AE1, anion exchanger 1, Diego blood group antigen, solute carrier family 4 member 1	SLC4A1	17q21-q22	Erythrocytes	Integral membrane glycoprotein and member of anion exchanger family N-terminal cytoplasmic domain binds ankyrin and links membrane to RBC cytoskeleton C-terminal membrane domain mediates exchange of chloride and bicarbonate anions across plasma membrane Implicated in RBC aging and attachment of malarial parasites to RBCs
CD234	**Duffy blood group antigen** DARC, Fy glycoprotein	FY	1q21-q22	Erythrocytes, endothelial cells	Integral seven-span membrane glycoprotein and member of G protein-coupled chemokine receptor superfamily Receptor for chemokines including IL-8, GRO, RANTES, MCP-1, TARC Also binds to malarial parasites
CD235a	Glycophorin A	GYPA	4q28.2-q31.1	Erythrocytes	Type I transmembrane glycoprotein Major RBC sialoglycoprotein determining MN blood group Also binds influenza virus Mediates resistance to malaria
CD235b	Glycophorin B	GYPB	4q28-q31	Erythrocytes	Type I transmembrane glycoprotein Major RBC sialoglycoprotein determining Ss blood group
CD236c	Glycophorin C, encoded by GYPC exons 1–4	GYPC	2q14-q21	Erythrocytes	Type III sialoglycoprotein linked to membrane via Band 4.1 Little homology to glycophorins A and B Minor presence on RBCs Regulates RBC mechanical stability Receptor for malarial merozoites

continued overleaf

CD Number*	Common Names	Gene Name	Chromosomal Location	Cellular/Tissue Expression	Function**
CD236d	Glycophorin D, encoded by part of GYPC exon2 and GYPC exons 3 and 4; Webb and Duch blood group antigens	GYPC	2q14-q21	Erythrocytes	See CD236c
CD236r	Mutations of GYPC giving rise to Gerbich and Yus blood group antigens	GYPC	2q14-q21	Erythrocytes	See CD236c
(CD237)	Not assigned				
CD238	Kell blood group antigen	KEL	7q33	Erythroid cells, myeloid progenitors, testis; low levels in brain, heart, skeletal muscle	Type II transmembrane glycoprotein disulfide-linked to XK protein of cytoskeleton Binds to Zn and has endopeptidase activity Cleaves endothelin-3 to give bioactive form
CD239	**B-CAM** B cell adhesion molecule, Lutheran blood group glycoprotein	LU	19q13.2	Erythrocytes, broad range of non-hematopoietic cells	Type I transmembrane glycoprotein and member of Ig superfamily Receptor for laminin Mediates cell–cell and cell–matrix adhesion and possibly signaling May be involved in erythropoiesis May play role in epithelial cancer
CD240CE	Rh blood group system, Rh30CE	RHCE	1p36.11	Erythroid cells	Transmembrane glycoprotein with extracellular palmitolylation; member of Rh antigen family Maintains erythrocyte membrane integrity May participate in complex acting as ammonium transporter Rh-positive persons express both RHCE (Cc and Ee polypeptides) and RHD (CD240D; D polypeptide)
CD240D	Rh blood group system, Rh30D	RHD	1p36.11	Erythroid cells	Transmembrane glycoprotein with extracellular palmitolylation; member of Rh antigen family RhD is highly immunogenic Expression of RhD is absent in Rh-negative persons Rh incompatibility leads to erythroblastosis fetalis
CD241	Rh blood group associated glycoprotein, RhAg, Rh50	RHAG	6p21.1-p11	Erythroid cells	Transmembrane glycoprotein belonging to Rh antigen family Associates with Rh30 blood group antigen polypeptides Can function as an ammonium transporter
CD242	**ICAM-4** LW blood group, Landsteiner-Wiener blood group antigens	ICAM4	19p13.2-cen	Erythrocytes, expression associated with Rh antigens (CD240)	Type I transmembrane glycoprotein and member of Ig and ICAM families; also secreted form Binds to LFA-1 (CD11a/CD18) Mediates adhesion of RBCs to leukocytes

CD Number*	Common Names	Gene Name	Chromosomal Location	Cellular/Tissue Expression	Function**
CD243	MDR-1, P-glycoprotein, pgp 170, multidrug resistance protein I	ABCB1	7q21.1	Kidney, liver, gut, HSCs, some malignant cells, normal hematopoietic cells (weak)	12-span transmembrane glycoprotein and member of ABC family of ATP-binding transporters Carries out ATP-dependent pumping of small molecules (including drugs) out of cells Activity leads to anti-cancer drug resistance Also transporter across blood–brain barrier
CD244	**2B4** NK receptor 2B4	CD244	1q23.3	NK cells, monocytes, some T cells	Type I transmembrane glycoprotein and member of Ig superfamily Related to CD2 and binds to CD48 Able to recruit SAP to prevent its binding to SHP-2 Modulates intracellular signaling to increase activation Ligation of CD244 triggers NK cytotoxicity but not CTL cytotoxicity
CD245	p220/240, DY12, DY35	Not assigned	Unknown	T, B, and NK cells, monocytes, granulocytes, platelets; not RBCs or thymocytes	Transmembrane protein with weak phosphatase activity May be involved in late stages of activatory signal transduction in T and NK cells Ligation by anti-CD245 enhances proliferation induced by other activators
CD246	**ALK** Anaplastic lymphoma kinase	ALK	2p23	Neuronal tissue, intestine, testis; not normal lymphoid cells	Type I transmembrane glycoprotein and receptor protein tyrosine kinase Important role in brain development Translocation-mediated fusion of 3′ half of ALK with 5′ half of nucleophosmin creates oncogene frequently found in anaplastic large cell lymphomas
CD247	**CD3ζ** CD3 zeta, T cell receptor ζ chain	CD3Z	1q22-q23	All T lineage cells	Type I transmembrane protein and member of Ig superfamily Non-glycosylated ITAM-containing partner in CD3 complex associating non-covalently with TCR chains Participates in CD3ζζ homodimer, CD3ζη heterodimer, or CD3ζ–FCεRIγ heterodimer Required for cell surface expression and signal transduction of TCR Also interacts with HIV Nef protein
CD248	**Endosialin** TEM1, CD164 ligand, sialomucin-like 1	CD164L1	11q13	Stromal fibroblasts, tumor (but not normal) adult endothelium, vasculature of developing embryo	Type I membrane glycoprotein with C-type lectin and Sushi domains Resembles sialomucin and thrombomodulin May be involved in regulating blood coagulation May function as complement receptor May promote angiogenesis and tumor progression

continued overleaf

CD Number*	Common Names	Gene Name	Chromosomal Location	Cellular/Tissue Expression	Function**
CD249	**Glutamyl aminopeptidase A** BP-1, gp160	ENPEP	4q25	Capillary endothelial cells; epithelial cells of placenta, intestine, and kidney tubules; early B cells, stromal cells of thymus and bone marrow	Type II transmembrane glycoprotein Occurs as disulfide-linked homodimer Binds to Zn Ectopeptidase activity that releases N-terminal glutamate or aspartate from peptides May regulate early B cell differentiation and growth Mediates resistance of renal carcinoma cancer cells to anti-proliferative effects of IFNα
(CD250-251)	Reserved				
CD252	**OX40L** OX40 ligand, CD134L, gp34	TNFSF4	1q25	Activated B cells, macrophages, DCs, endothelial cells, spleen, thymus, heart, HTLV-infected cells	Type II membrane protein belonging to TNF superfamily; also cleaved to give secreted cytokine Occurs as homotrimer binding to OX40 Mediates adhesion of T cells to APCs or endothelial cells Mediates CD28-independent T cell costimulation leading to production of pro-inflammatory cytokines
CD253	**TRAIL** TNF-related apoptosis-inducing ligand	TNFSF10	3q26	Broadly expressed on leukocytes, including activated T and B cells, activated monocytes	Type II membrane protein belonging to TNF superfamily; also cleaved to give secreted cytokine Occurs as homotrimer binding to TRAILR family of receptors Engagement induces apoptosis of transformed cell lines and normal brain cells, thymocytes, and hepatocytes
CD254	**TRANCE** OPGL, TNF-related activation induced cytokine, RANKL, ODF	TNFSF11	13q14	Lymph node and bone marrow stromal cells, activated T cells, osteoblasts, chondrocytes, primitive mesenchymal cells	Type II membrane protein belonging to TNF superfamily; also cleaved to give secreted cytokine Occurs as homotrimer binding to RANK (CD265; osteoprotegerin; OPG) Osteoclast activation and differentiation factor Promotes bone disease in myeloma cases Increases T cell survival and ability of DCs to activate naive T cells
(CD255)	Reserved for **TWEAK** TNF-related weak inducer of apoptosis, APO3L, TL1A	TNFSF12	17p13.3	Myeloid cells, including macrophages; broad tissue expression	Type II membrane protein belonging to TNF superfamily; also cleaved to give secreted cytokine Binds to TWEAK-R (CD266) to promote cell–matrix interaction and endothelial cell growth and migration Weakly induces apoptosis and IL-8 production by several cell types
CD256	**APRIL** A proliferation-inducing ligand, TALL2	TNFSF13	17p13.1	Myeloid cells, including macrophages, DCs; bone marrow, spleen, tumor cell lines (strong)	Precursor type II membrane protein belonging to TNF superfamily; cleaved by furin to yield secreted APRIL Binds to TACI (CD267) and BMCA (CD269) In normal cells, APRIL downregulates Td antibody responses and promotes isotype switching to IgA Promotes B cell development, and T and B cell proliferation

CD Number*	Common Names	Gene Name	Chromosomal Location	Cellular/Tissue Expression	Function**
CD257	**BAFF** B cell activating factor, BlyS, TALL1, THANK	TNFSF13B	13q32-34	Myeloid cells, including monocytes macrophages; spleen, bone marrow	Type II membrane protein belonging to TNF superfamily; also cleaved to give secreted cytokine Major biological effects mediated by binding to BAFFR (CD268) but also binds to TACI (CD267) and BMCA (CD269) Upregulated on monocytes by IFNγ Essential for T1 to T2 transition during B cell development Stimulates B and T cell function and humoral immunity
CD258	**LIGHT** Homologous to lymphotoxins exhibits inducible expression and competes with HSV glycoprotein D for HVEM a receptor expressed by T lymphocytes	TNFSF14	19p13.3	Activated T cells, immature DCs, stromal cells, spleen, brain	Type II membrane protein belonging to TNF superfamily; also cleaved to give secreted cytokine Binds to HVEM as well as LTβR Activates NF-κB Stimulates T cell proliferation Induces SLC production to recruit inflammatory cells Induces apoptosis of tumor cells expressing it
(CD259–260)	Reserved				
CD261	**TRAIL-R1** DR4, APO2	TNFRSF10A	8p21	Normal leukocytes (weak); tumor cells (strong)	Type I transmembrane receptor with intracellular death domain; member of TNFR superfamily Receptor for TRAIL (CD253) that can interact with TRADD and RIP Induces FADD-independent, caspase-mediated apoptosis upon engagement
CD262	**TRAIL-R2** DR5, TRICK2, TRICKB, KILLER	TNFRSF10B	8p22-p21	Ubiquitous but highest in PBL, spleen, ovary	Type I transmembrane receptor with intracellular death domain; member of TNFR superfamily Receptor for TRAIL (CD253) that can interact with TRADD and RIP Induces FADD-dependent, caspase-mediated apoptosis upon engagement Frequent site of mutations associated with head and neck tumors
CD263	**TRAIL-R3** TRID, LIT, DCR1	TNFRSF10C	8p22-p21	Broadly but weakly expressed; highest in PBL, spleen, lung, placenta	GPI-anchored glycoprotein and member of TNFR superfamily Lacks death domain due to absence of cytoplasmic region Decoy receptor binding to TRAIL Competitive inhibitor blocking TRAIL-R-mediated apoptosis
CD264	**TRAIL-R4** DCR2, TRUNDD	TNFRSF10D	8p21	Ubiquitous	Type I transmembrane receptor with truncated death domain; member of TNFR superfamily Decoy receptor binding to TRAIL Blocks TRAIL-induced apoptosis

continued overleaf

CD Number*	Common Names	Gene Name	Chromosomal Location	Cellular/Tissue Expression	Function**
CD265	**TRANCE-R** **RANK** **OPG** ODFR	TNFRSF11A	18q22.1	Broadly expressed; highest in osteoclasts, DCs	Type I transmembrane receptor and member of TNFR superfamily Binds to TRANCE (CD254; RANKL) Regulates interactions between DCs and naive T cells Promotes DC survival Regulates osteoclastogenesis and lymph node development
CD266	**TWEAK-R** DR3, APO3, FN14	TNFRSF12A	16p13.3	Broadly expressed; highest in heart, placenta, kidney, smooth muscle cells; elevated in liver cancers	Type 1 transmembrane receptor and member of TNFR superfamily Binds TWEAK to promote cell–matrix interactions Regulation of hepatocyte growth and liver neoplasia Regulation of endothelial cell growth and migration
CD267	**TACI** Transmembrane activator and CAML (calcium modulator and cyclophilin interactor	TNFRSF13B	17p11.2	Activated T cells, B cells (weak), spleen, thymus, small intestine	Type III transmembrane receptor and member of TNFR superfamily Binds to BAFF (CD257) but not essential for T1 to T2 transition during B cell development Also binds APRIL (CD256) Negative regulator of B cell homeostasis May protect against autoimmunity mediated by pathogenic autoantibodies
CD268	**BAFFR** B cell activating factor receptor, BLyS receptor	TNFRSF13C	22q13.1-q13.31	B cells, spleen, lymph nodes	Type III transmembrane receptor and member of TNFR superfamily Binds to BAFF (CD257) and is essential for T1 to T2 transition during B cell development Does not bind to APRIL (CD256)
CD269	**BCMA** B cell maturation antigen	TNFRSF17	16p13.1	B cells	Type III transmembrane receptor and member of TNFR superfamily Binds to BAFF (CD257) but not essential for T1 to T2 transition during B cell development Also binds APRIL (CD256) May be involved in B cell terminal differentiation
(CD270)	Reserved for **LIGHTR** TNFR superfamily member 14, HVEM, herpes virus entry mediator	TNFRSF14	1p36.3-p36.2	T cells, lung, spleen, thymus	Transmembrane receptor and member of TNFR superfamily Binds to LIGHT Binds to herpesvirus envelope glycoprotein D (gD) to facilitate viral entry into activated T cells
CD271	**NGFR** Nerve growth factor receptor, p75	NGFR	17q21-q22	Neurons, Schwann cells, stromal cells, FDCs, testis, heart	Type I transmembrane glycoprotein and member of TNFR superfamily Occurs as disulfide-linked homodimer Binds neurotrophic factors including NGF, BDNF, NT-3, NT-4 Participates in complex that can induce apoptosis through the sphingomyelin pathway

CD Number*	Common Names	Gene Name	Chromosomal Location	Cellular/Tissue Expression	Function**
CD271 (cont'd)					Can also promote cell survival Promotes myelination of developing and regenerating peripheral nerves Receptor for rabies virus
CD272	BTLA B and T lymphocyte attenuator	BTLA	3q13.2	Activated T cells, Th1 (but not Th2) effectors, lymph nodes, spleen	ITIM-containing transmembrane glycoprotein; member of Ig superfamily Binds to B7H4 and associates with SHP-1 and SHP-2 Inhibits antigen-induced IL-2 production and proliferation by T cells
CD273	PDL2 Programmed cell death 1 ligand 2, B7DC	PDCD1LG2	9p24.2	Activated T cells, IFN-γ-stimulated monocytes and DCs; highest in placenta, heart, lung, pancreas, liver	Type I transmembrane protein and member of Ig and B7 families Binds to T cell-negative regulator PD-1 (CD279) Inhibits T cell proliferation and cytokine production Regulates asthmatic response
CD274	PDL1 Programmed cell death 1 ligand 1, B7H1	PDCD1LG1	9p24	IFN-γ-stimulated monocytes and DCs, activated B cells, resting T and B cells (weak); highest in heart, skeletal muscle, placenta, lung	Type I transmembrane protein and member of Ig and B7 families Different isoforms for plasma and intracellular membranes Binds to T cell-negative regulator PD-1 (CD279) Inhibits T cell proliferation and cytokine production Regulates asthmatic response
CD275	ICOSL Inducible T cell costimulator ligand, B7H2, B7RP1	ICOSL	21q22.3	Monocytes, macrophages, some stimulated DCs, T and B cells (weak); highest expression in brain, heart, kidney, liver, PBL	Type I transmembrane protein and member of Ig and B7 families Specific ligand for inducible costimulator ICOS (CD278); does not bind to CTLA-4 (CD152) Upregulated by inflammatory cytokines Promotes proliferation and cytokine production of antigen-stimulated T cells Plays roles in T–B interaction, GC formation, IgG1 and IgE production May be most relevant for secondary responses
CD276	B7H3 B7 homologue 3	CD276	15q23-q24	Highest expression in heart, liver, placenta; not detected in resting PBL; can be induced on DCs, monocytes, and T, B, and NK cells	Type I transmembrane protein and member of Ig and B7 families Ligand currently unknown Binds to activated T cells Negative regulator of proliferation and cytokine production by Th1 effectors (in mice)
CD277	BT3.1 butyrophilin 3	BTN3A1	6p22.1	Highest expression in heart, pancreas, lung; lower in placenta, liver, muscle; T, B, and NK cells, monocytes, DCs, stem cell subset	Type I transmembrane protein and member of Ig and butyrophilin families May be involved in lipid metabolism May regulate T cell activation and function
CD278	ICOS Inducible T cell costimulator	ICOS	2q33	Activated T cells, Th2 cells, thymocyte subsets	Type I transmembrane glycoprotein and member of CD28 and Ig superfamilies

continued overleaf

CD Number*	Common Names	Gene Name	Chromosomal Location	Cellular/Tissue Expression	Function**
CD278 (cont'd)					Occurs as a disulfide-linked homodimer binding to ICOSL (CD275) on APCs
					Induced on activated T cells and contributes to T cell costimulation for proliferation, cytokine secretion
					Necessary for Td antibody responses
					Required for T-B cooperation, GC formation, isotype switching
					Promotes Th2 responses, including upregulation of IL-4, IL-10, and IL-17 production but not IL-2
					Contributes to IgE production and inflammatory cell recruitment in allergic reactions
CD279	**PD1** Programmed cell death 1	PDCD1	2q37.3	Activated T and B cells, pro-B cells, $\gamma\delta$ thymocyte subset, DN $\alpha\beta$ thymocytes	Type I transmembrane glycoprotein and member of Ig superfamily
					Induced in response to certain apoptotic signals
					Negative regulator of T cell activation
					Required to avoid development of autoimmunity
					May play a role in pro-B cell differentiation
CD280	**ENDO180** TEM22, mannose receptor C type 2, uPARAP	MRC2	17q23.3	Stromal fibroblasts, macrophages, some endothelial cell subsets, sites of cartilage deposition	Transmembrane glycoprotein with C-type lectin domain
					Non-phagocytic mannose receptor that also binds to collagen
					May be involved in extracellular matrix degradation and remodeling
					May link urokinase-type plasminogen activator and its receptor for directional sensing during chemotaxis
CD281	**TLR1** Toll-like receptor 1	TLR1	4p14	PBL, monocytes, neutrophils, macrophages, DCs, keratinocytes	Type I transmembrane glycoprotein of TLR family; contains TIR domain
					Occurs in plasma membrane and phagosomes
					Pattern recognition receptor triggering antibacterial innate response
					Cooperates with TLR2 to bind to lipotechoic acid and peptidoglycan in gram-positive bacterial cell walls
CD282	**TLR2** Toll-like receptor 2; TIL4	TLR2	4q32	Myelomonocytic cells, particularly monocytes, neutrophils, granulocytes, macrophages, DCs; heart, brain, muscle	Type I transmembrane glycoprotein of TLR family; contains TIR domain
					Pattern recognition receptor triggering antibacterial innate response
					Cooperates with TLR1 or TLR6 to bind to lipotechoic acid and peptidoglycan in gram-positive bacterial cell walls
					Stimulates respiratory burst and production of NO and IL-12 by macrophages
					Induces apoptosis in response to bacterial lipoprotein
CD283	**TLR3** Toll-like receptor 3	TLR3	4q35	DCs, fibroblasts, placenta, pancreas	Type I transmembrane glycoprotein of TLR family; contains TIR domain
					Pattern recognition receptor triggering IRF3-dependent innate antiviral response (IFNα/β production)
					Binds to double-stranded viral RNA

CD Number*	Common Names	Gene Name	Chromosomal Location	Cellular/Tissue Expression	Function**
CD284	**TLR4** Toll-like receptor 4	TLR4	9q32-q33	Myelomonocytic cells, endothelial cells, B cells, cardiomyocytes, placenta	Type I transmembrane glycoprotein of TLR family; contains TIR domain Pattern recognition receptor triggering antibacterial innate response via PKR Binds to LPS of gram-negative bacterial cell walls (in conjunction with CD14 and LBP) and induces phagocytosis and production of inflammatory cytokines Used by pathogenic bacteria to induce apoptosis of infected cells May also contribute to antiviral innate response
(CD285)	Reserved for **TLR5** Toll-like receptor 5, TIL3	TLR5	1q32.3-q42	Myelomonocytic cells, prostate, ovary, liver, lung	Type I transmembrane glycoprotein of TLR family; contains TIR domain Pattern recognition receptor triggering innate immune responses Binds to bacterial flagellin Mutations in TLR5 are associated with susceptibility to Legionnaire's disease
(CD286)	Reserved for **TLR6** Toll-like receptor 6	TLR6	4p16.1	Myelomonocytic cells, lung, ovary	Type I transmembrane glycoprotein of TLR family; contains TIR domain Pattern recognition receptor triggering innate immune responses Cooperates with TLR2 to bind to lipotechoic acid and peptidoglycan in gram-positive bacterial cell walls
(CD287)	Reserved for **TLR7** Toll-like receptor 7	TLR7	Xp22.3	Plasmacytoid DCs, lung, placenta, spleen, brain, small intestine	Type I transmembrane glycoprotein of TLR family; contains TIR domain Pattern recognition receptor triggering innate immune responses Binds to single-stranded GU-rich viral RNA Active in endosomal compartment
(CD288)	Reserved for **TLR8** Toll-like receptor 8	TLR8	Xp22	Lung, PBL, brain, liver, heart	Type I transmembrane glycoprotein of TLR family; contains TIR domain Pattern recognition receptor triggering innate immune responses Binds to single-stranded GU-rich viral RNA Active in endosomal compartment
CD289	**TLR9** Toll-like receptor 9	TLR9	3p21.3	Monocytes, macrophages and DCs, spleen, lymph node, bone marrow, PBL	Type I transmembrane glycoprotein of TLR family; contains TIR domain Pattern recognition receptor triggering innate immune responses Binds to the unmethylated CpG nucleotides common in bacterial DNA Localized in endoplasmic reticulum and moves to CpG-containing intracellular vesicles
(CD290)	Reserved for **TLR10** Toll-like receptor 10	TLR10	4p14	Spleen, thymus, lymph node, tonsil, lung	Type I transmembrane glycoprotein of TLR family; contains TIR domain Pattern recognition receptor triggering innate immune responses May recognize unmethylated CpG

continued overleaf

CD Number*	Common Names	Gene Name	Chromosomal Location	Cellular/Tissue Expression	Function**
CD291	Reserved for **TLR11** Toll-like receptor 11	TLR11	Not yet identified	Macrophages of liver, kidney, bladder	Type I transmembrane glycoprotein of TLR family; contains TIR domain Pattern recognition receptor triggering innate immune responses Ligand unknown Prevents infection by uropathogenic bacteria in mice
CD292	**BMPR1A** Bone morphogenetic protein (BMP) receptor type IA, ALK3	BMPR1A	10q22.3	Bone progenitors, skeletal muscle	Type I transmembrane protein and member of activin receptor serine/threonine kinase superfamily Forms heterodimer with type II BMP receptor that binds to BMP-2 and BMP-4 Plays role in cartilage and bone formation Required for regression of embryonic female reproductive organs in developing males Required for control of intestinal stem cell renewal
CDw293	**BMPR1B** Bone morphogenetic protein receptor type IB, ALK6	BMPR1B	4q22-q24	Bone progenitors, prostate, brain	Type I transmembrane protein and member of activin receptor serine/threonine kinase superfamily Forms heterodimer with type II BMP receptor that binds to BMPs Plays role in cartilage and bone formation
CD294	**CRTH2** Chemoattractant receptor-homologous molecule expressed on Th2; PGRD2, G protein-coupled receptor 44	GPR44	11q12-q13.3	Th2 effectors, eosinophils, basophils (not B or NK cells), adult brain, thymus, fetal liver	G protein-coupled seven-span transmembrane receptor Binds to prostaglandin D2 to induce intracellular calcium mobilization in Th2 cells, and chemotaxis of Th2 cells, eosinophils, and basophils May play a role in allergic inflammation
CD295	**LEPR** Leptin receptor, OBR	LEPR	1p31	Broadly expressed, high levels in hypothalamus of brain; some endothelial cells and monocytes	Type 1 transmembrane protein related to gp130 chain Homodimerizes upon binding by leptin, an adipocyte-specific hormone that controls body weight by regulating adipose tissue mass via effects on the hypothalamus Associates with SH2-containing signal transducers Can stimulate neoangiogenesis
CD296	**ART1** ADP-ribosyltransferase 1	ART1	11p15	Epithelial cells, neutrophils, T and NK cell subsets; highly expressed in skeletal, cardiac muscle	GPI-anchored membrane ectoenzyme Catalyzes ADP-ribosylation of arginine residues in proteins, thus post-translationally modifying their functions Associates with PDGFβ (CD140b), integrins, and defensin
CD297	**ART4** ADP-ribosyltransferase 4, Dombrock blood group glycoprotein	DO	12q13.2-q13.3	T cells, RBCs, activated monocytes, spleen, fetal liver	GPI-anchored membrane protein that may have ectoenzyme activity similar to ART1 (i.e., Arg-specific ADP-ribosylation) Dombrock blood group antigen

CD Number*	Common Names	Gene Name	Chromosomal Location	Cellular/Tissue Expression	Function**
CD298	ATP1B3 Na/K ATPase beta3 subunit	ATP1B3	3q23	Broadly expressed; highest in testis	Type II membrane glycoprotein Non-catalytic subunit of integral Na/K ATPase responsible for transporting Na and K ions across cell membranes
CD299	L-SIGN CLEC4M, C-type lectin domain family 4 member M, DCSIGN-related, DCSIGNR, CD209 antigen-like, CD209L	CLEC4M	19p13	Some endothelial cells, some DCs, lung, placenta, endometrium, liver, lymph node	Type II transmembrane protein containing C-lectin domains and related to DC-SIGN (CD209) Binds carbohydrates of ICAM-3 (CD50) and HIV gp120; enhances HIV infection of T cells Alternate receptor for SARS, Ebola, and Sindbis viruses Promotes hepatocyte infection by HepC
CD300a	CD300A antigen CMRF-35H, IRC1, IRP60	CD300A	17q25.1	DCs, monocytes, macrophages, some T and B cells	Type I transmembrane glycoprotein and member of Ig superfamily One of a family of type I integral membrane antigens recognized by anti-CMRF35 antibody CMRF35 is related to the poly-Ig receptor CD300A is distinct from CMRF35 CD300A contains four ITIMs and so may be inhibitory
CD300c	CD300C antigen CMRF-35A, CMRF35, LIR	CD300C	17q25.1	Monocytes, neutrophils, granulocytes, some DCs; T, B, and NK cells	Type I transmembrane glycoprotein and member of Ig superfamily One of a family of type I integral membrane antigens recognized by anti-CMRF35 antibody CD300C contains a charged residue that may allow it to associate with ITAM-containing proteins; may be activatory
CD300e	CD300E antigen CMRF35L1	Not assigned	Likely 17q25.1	Monocytes, macrophages, DC subset	Type I transmembrane glycoprotein and member of Ig superfamily One of a family of type I integral membrane antigens recognized by anti-CMRF35 antibody
CD301	MGL1 Macrophage galactose lectin, C-type lectin superfamily member 10A, CLECSF14, HMGL, HML2	CLECSF10A	17p13.1	Immature DCs, inducible on activated monocytes and macrophages	Type II transmembrane C-type lectin receptor Binds to galactose- or N-acetylgalactosamine-bearing antigens and facilitates their endocytosis Involved in cell adhesion, migration, recognition Also binds to Tn Ag (CD175; human carcinoma-associated epitope)
CD302	DCL1 CLEC13A, BIMLEC	CD302	2q24.2	DCs, monocytes, macrophages, granulocytes, some B cell lines	Type I transmembrane C-type lectin receptor Binds to glycosylated antigens and facilitates their endocytosis Part of a novel fusion protein (with DEC-205; CD205) expressed by Reed-Sternberg cells in some Hodgkin's lymphomas
CD303	BDCA2 Blood dendritic cell antigen 2, C-type lectin	CLEC4C	12p13.2-p12.3	Plasmacytoid DCs in peripheral blood; tonsils, lymph nodes	Type II transmembrane protein with C-type lectin domain Mobilizes intracellular calcium May facilitate antigen internalization

continued overleaf

CD Number*	Common Names	Gene Name	Chromosomal Location	Cellular/Tissue Expression	Function**
CD303 (cont'd)	domain family 4 member C, CLECSF11, CLECSF7				Anti-CD303 ligation triggers protein phosphorylation Anti-CD303 ligation inhibits IFNα/β induction and thus may promote Th2 responses
CD304	BDCA4 Neuropilin-1, NRP1, vascular endothelial growth factor 165 receptor, VEGF-165R	NRP1	10p12	Plasmacytoid and monocyte-derived DCs in peripheral blood, endothelial cells, some T cells, neurons; high in heart and placenta, low in brain	Type I membrane-bound coreceptor for a plexin tyrosine kinase receptor binding to both VEGF and semaphorin; also soluble forms Originally identified as an antigen of developing neuronal axons Also facilitates angiogenesis, embryonic blood vessel development, tissue invasion, cell survival and migration May also promote DC–T cell interaction leading to T cell activation Soluble isoforms may act as regulatory antagonists
CD305	**LAIR1** Leukocyte-associated Ig-like receptor 1	LAIR1	19q13.4	T, B, and NK cells, macrophages, monocytes, intraepithelial T cells, some thymocytes	Type I transmembrane ITIM-containing glycoprotein and member of Ig superfamily Distinct from KIR, ILT, and NKG2 families Physiological ligand unknown; does not bind to MHC class I Associates with SHP-1 and positions it near the plasma membrane Inhibits NK and CTL cytotoxicity, B cell activation, and proliferation of T, NK, and myelomonocytic cells
CD306	**LAIR2** Leukocyte-associated Ig-like receptor 2	LAIR2	19q13.4	Monocytes, T cells	Potential secreted protein homologous to LAIR1 Lacks transmembrane and cytoplasmic domains Physiological ligand unknown May regulate mucosal tolerance May inhibit cellular activation and inflammation
CD307	**FCRL5** Fc receptor-like 5, IRTA2, Immunoglobulin superfamily receptor translocation associated 2, BXMAS1	FCRL5	1q21	B cell subsets, especially GC centrocytes and marginal zone B cells, lymph node, spleen, bone marrow, small intestine, tonsil	Glycoprotein with Ig domains Multiple isoforms exist: some secreted, some GPI-anchored, and some containing transmembrane domains with ITIMs Homologous to Fc receptors and inhibitory receptors Implicated in B cell development and activation Chromosomal aberrations involving the IRTA2 gene are implicated in the development of some Burkitt's lymphomas
(CD308)	Reserved for **VEGFR1** FLT1, fms-related tyrosine kinase 1	FLT1	13q12.2	Endothelial cells and monocytes; highest in lung, lower in kidney, placenta, liver, heart, brain	Type I membrane tyrosine kinase protein receptor Binds to VEGF, VEGFB, and PGF Not expressed in tumor cell lines
CD309	**VEGFR2 KDR** Kinase insert domain receptor	KDR	4q11-q12	Endothelial cell precursors, some stem cells, some tumor cells	Type III receptor tyrosine kinase Binds strongly to VEGF Promotes angiogenesis, including tumor angiogenesis

CD Number*	Common Names	Gene Name	Chromosomal Location	Cellular/Tissue Expression	Function**
CD309 (cont'd)					Required for development of retinal vascular system
(CD310)	Reserved for **VEGFR3** Fms-related tyrosine kinase 4, PCL	FLT4	5q34-q35	Endothelial cells	Structurally related to VEGFR1 and VEGFR2 Binds to VEGFC but not VEGF Marker for high endothelial venules Stimulates growth of lymphatic vessels during embryogenesis
(CD311)	Reserved for **EMR1** EGF-like module containing mucin-like hormone receptor-like 1	EMR1	19p13.3	Broad expression, high in peripheral blood mononuclear cells	G protein-coupled seven-transmembrane glycoprotein receptor containing EGF-like repeats and a mucin-like stalk May be involved in cell–cell interactions
CD312	**EMR2** EGF-like module containing mucin-like hormone receptor-like 2	EMR2	19p13.1	Macrophages, monocytes, neutrophils, some DCs, activated lymphocytes, spleen, lymph nodes	G protein-coupled seven-transmembrane glycoprotein receptor containing EGF-like repeats and a mucin-like stalk Cleaved into non-covalently linked heterodimer Related to CD97 but does not bind DAF Binds to chondroitin sulfate found on some B cells May promote intercellular interactions between T cells and DCs, or macrophages and B cells May play role in adhesion and migration during phagocytosis
(CD313)	Reserved for **EMR3** EGF-like module containing mucin-like hormone receptor-like 3	EMR3	19p13.1	Leukocytes	G protein-coupled seven-transmembrane glycoprotein receptor containing EGF-like repeats and a mucin-like stalk Recognizes unknown ligand on macrophages and activated neutrophils Does not bind DAF Differs from EMR2 in its extracellular domain
CD314	**NKG2D** KLRK1, killer cell lectin-like receptor subfamily K member 1	KLRK1	12p13.2-p12.3	NK cells, $\gamma\delta$ T cells, some CD8$^+$ $\alpha\beta$ T cells	Type II transmembrane glycoprotein with C-type lectin domains and ITIMs Homodimer associates with DAP10 to form NK activatory receptor Binds to stress antigens MICA, MICB Activates NK cytotoxicity and costimulates T cells
CD315	**PTGFRN** Prostaglandin F2 receptor negative regulator, FPRP, CD9P1, CD9 partner	PTGFRN	1p13.1	Activated monocytes, some B cells, epithelial and endothelial cells, keratinocytes (weak); high expression in ovary, lung	Type I transmembrane protein and member of Ig superfamily Present primarily in endoplasmic reticulum and trans-Golgi network Associates with prostaglandin F2α receptor and inhibits binding of prostaglandin F2α Also associates with CD9 and CD81 Upregulated in cancer cell lines
CD316	**IGSF8** Immunoglobulin superfamily, member 8, CD81	IGSF8	1q23.1	T and B cells, NK cells (weak), hepatocytes	Type I transmembrane protein and member of Ig superfamily Contains a Glu-Trp-Ile (EWI) motif not seen in other Ig proteins

continued overleaf

CD Number*	Common Names	Gene Name	Chromosomal Location	Cellular/Tissue Expression	Function**
CD316 (cont'd)	partner 3, EWI2, prostaglandin regulatory-like, PGRL				Associates specifically with CD9 and CD81 Precise function unclear; may be required for normal CD81 and/or CD9 functions May be involved in cell migration
CD317	**BST2** Bone marrow stromal cell antigen 2	BST2	19p13.2	Bone marrow stromal cells, fibroblasts, some B, T, and NK cells, some DCs	Type II transmembrane protein May promote growth of pre-B cells
CD318	**CDC1P** CUB domain-containing protein 1, SIMA135	CDC1P	3p21.31	Some stem cells, epithelial cells, skeletal muscle, colon	Type I transmembrane glycoprotein with CUB domains May be involved in cell–cell or cell–matrix adhesion Engagement can stimulate the generation of erythroid colonies Overexpressed in colorectal, breast, and lung cancer; also myeloid leukemic blasts
CD319	**CRACC** CD2-like receptor activating cytotoxic cells, SLAM family member 7, 19A24, CS1	SLAMF7	1q23.1-q24.1	NK cells, CD8$^+$ CTLs, activated B cells, mature DCs, some CD4$^+$ T cells, spleen, lymph node	Transmembrane glycoprotein related to CD2; member of Ig superfamily Functions as NK activatory receptor May regulate cytotoxic T cell functions and lymphocyte adhesion
CD320	**CD320 antigen** 8D6	CD320	19p13.3-p13.2	Germinal center FDC	Transmembrane glycoprotein similar to LDLR family Facilitates FDC–B cell interaction that stimulates GC formation Promotes cell cycling of GC B cells Enhances B cell proliferation and Ig secretion May be involved in B lymphomagenesis or metastasis
CD321	**JAM1** Junctional adhesion molecule 1, JAM-A, F11 receptor, F11R, PAM1	F11R	1q21.2-q21.3	Monocytes, neutrophils, platelets, DCs, some T and B cells, epithelial cells, liver, kidney, lung, intestine, placenta, pancreas; high in leukemia cell lines	Type I transmembrane glycoprotein and member of Ig superfamily Binds to PAR3 expressed by epithelial cells in area of tight junctions Binds to LFA-1 to promote transendothelial migration of T cells and neutrophils Also binds to antibody F11 that activates platelets; may be involved in platelet aggregation Acts as reovirus receptor
CD322	**JAM2** Junctional adhesion molecule 2, JAM-B, VEJAM	JAM2	21q21.2	HEV endothelial cells, some T and B cells, heart, placenta, lymph node	Type I transmembrane glycoprotein and member of Ig superfamily Homophilic interactions Also binds to JAM3 on T and NK cells and DCs and promotes transendothelial migration of circulating leukocytes May mediate cell–cell adhesion, particularly between Sertoli cells and spermatids
(CD323)	Reserved for **JAM3** Junctional adhesion molecule 3,	JAM3	11q25	Endothelial cells, activated T and NK cells, DCs, platelets; placenta, kidney, brain	Type I transmembrane member of Ig superfamily Binds to JAM2 and promotes transendothelial migration of circulating leukocytes

CD Number*	Common Names	Gene Name	Chromosomal Location	Cellular/Tissue Expression	Function**
CD323 (cont'd)	JAM-C				Essential for round polarization of spermatids
CD324	**E-Cadherin** Epithelial cadherin, cadherin 1, CDHE, ECAD, UVO, LCAM	CDH1	16q22.1	Epithelial cells, erythroblasts, some stem cells, keratinocytes, trophoblast cells, platelets	Type I transmembrane calcium-dependent adhesion molecule containing cadherin domains Occurs as disulfide-linked homodimer Binds to itself and to integrin αE/β7 (CD103) Required for cell–cell adhesion Required for epithelial-mesenchymal transitions during embryogenesis Low expression correlates with tumor cell invasiveness and T cell malignancies Binds to *Listeria monocytogenes* internalin protein to facilitate pathogen entry
CDw325	**N-Cadherin** Neuronal cadherin, cadherin 2, NCAD	CDH2	18q11.2	Neurons, endothelial cells, some stem cells, brain, skeletal and cardiac muscles	Type I transmembrane calcium-dependent adhesion molecule containing cadherin domains Binds to itself and other ligands Required for gastrulation and left-right symmetry Required for pre-synaptic/post-synaptic adhesion in some cases Required for retinal cell patterning Low expression correlates with tumor cell invasiveness
CD326	**Ep-CAM** Epithelial cellular adhesion molecule, tumor-associated calcium signal transducer 1, MIC18, TROP1, M4S1, MK-1	TACSTD1	2p21	Epithelial cells, normal colon, many carcinoma cells	Type I transmembrane glycoprotein related to thyroglobulin-1 Binds to itself in a calcium-independent manner to promote cell–cell adhesion Also binds to LAIR-1 (CD305) and LAIR-2 (CD306) to inhibit cellular activation and inflammation May act as a growth factor receptor Upregulated on cells of colorectal, ovarian, renal cell, breast, and pancreatic carcinomas, lung adenocarcinoma, retinoblastoma Higher expression correlates with increased tumor invasiveness
CDw327	**Siglec6** Sialic acid binding Ig-like lectin 6, CD33L, CD33-like antigen, OBBP1	SIGLEC6	19q13.3	High in placental trophoblasts, circulating B cells; low expression in spleen, small intestine	Type I transmembrane ITIM-containing protein of sialoadhesin and Ig superfamilies Also secreted isoform Closely related to Siglec3 (CD33) Promotes adhesion by binding to Neu5Ac (alpha)2-6GalNAc(alpha) (sialyl-Tn) Also binds to leptin
CDw328	**Siglec7** Sialic acid binding Ig-like lectin 7, AIRM1, adhesion inhibitory	SIGLEC7	19q13.3	Strongly on NK cells (strong) weakly on some T cells, monocytes, granulocytes, placenta, liver, spleen	Type I transmembrane ITIM-containing protein of sialoadhesin and Ig superfamilies Closely related to Siglec3 (CD33) Associates with CD3 Inhibitory in function due to ITIMs that recruit SHP-1 upon phosphorylation

continued overleaf

CD Number*	Common Names	Gene Name	Chromosomal Location	Cellular/Tissue Expression	Function**
CDw328 (cont'd)	receptor				Binds cell surface sialic acids but not MHC class I Inhibits the proliferation and differentiation of myelomonocytic cells
CDw329	**Siglec9** Sialic acid binding Ig-like lectin 9, OBBP-like	SIGLEC9	19q13.3	Neutrophils, NK cells, monocytes, bone marrow cells, hepatocytes, placenta, spleen, fetal liver	Type I transmembrane ITIM-containing protein of sialoadhesin and Ig superfamilies Closely related to Siglec7 (CDw328) Binds to both α-2-3- and α-2-6-linked sialic acids Inhibitory function that modulates cellular actions during immune responses
(CD330)	Reserved for **Siglec10** Sialic acid binding Ig-like lectin 10	SIGLEC10	19q13.3	Eosinophils, monocytes, some NK cells, spleen, lymph nodes, PBL, bone marrow, small intestine	Type I transmembrane ITIM-containing protein of sialoadhesin and Ig superfamilies Binds to sialic acids and may have inhibitory function
CD331	**FGFR1** Fibroblast growth factor receptor 1, FLT2, FLG	FGFR1	8p11.2-p11.1	Epithelial cells, fibroblasts; broad expression in adult and fetal tissues; overexpressed in some breast cancers	Type I transmembrane protein with unusual acidic region and intracellular tyrosine kinase domain; member of Ig superfamily Binds to FGFs and heparin with high affinity Required for recruitment of epithelial precursors during organogenesis Also important for wound healing Implicated in tumor expansion and angiogenesis
CD332	**FGFR2** Fibroblast growth factor receptor 2, keratinocyte growth factor receptor, KGFR, BEK, TK14	FGFR2	10q26	Epithelial cells, fibroblasts, skin, bone, brain, liver, prostate, kidney, lung, spinal cord	Type I transmembrane protein with unusual acidic region and intracellular tyrosine kinase domain; member of Ig superfamily; also secreted isoforms Two isoforms generated by alternative splicing: KGFR (skin) and BEK (bone) Bind to FGFs with high affinity KGFR prominent in skin development BEK prominent in osteogenesis Mutations in FGFR2 are linked to several craniosynostosis syndromes (premature suturing of skull bones)
CD333	**FGFR3** Fibroblast growth factor receptor 3, CEK2, ACH	FGFR3	4p16.3	Epithelial cells, fibroblasts, brain, kidney, adult testis, fetal small intestine	Type I transmembrane protein with unusual acidic region and intracellular tyrosine kinase domain; member of Ig superfamily; also secreted isoforms and splice variants Preferentially binds to FGF1 Mutations in FGFR3 are linked to skeletal abnormalities as well as bladder and cervical carcinomas
CD334	**FGFR4** Fibroblast growth factor receptor 4, TKF	FGFR4	5q35.1-qter	Epithelial cells, fibroblasts, lymphocytes, macrophages, lung	Type I transmembrane protein with unusual acidic region and intracellular tyrosine kinase domain; member of Ig superfamily Tissue expression pattern differing from other FGFRs

CD Number*	Common Names	Gene Name	Chromosomal Location	Cellular/Tissue Expression	Function**
CD334 (cont'd)					Preferentially binds to FGF19 and acidic (but not basic) FGF Required for endoderm and skeletal muscle development Mutations in FDFR4 are linked to accelerated progression of breast and colon carcinomas, and may be associated with leukemia/lymphoma
CD335	**NKp46** NCR1, natural cytotoxicity triggering receptor 1, Ly94	NCR1	19q13.42	Resting and activated NK cells, spleen	Type I transmembrane glycoprotein and member of Ig superfamily Lacks ITAMs and associates with CD3ζ or FcεRI-γ chains for signaling Binds to a non-MHC ligand Activates and/or enhances NK cytotoxicity, cytokine production Especially effective against intracellular bacteria
CD336	**NKp44** NCR2, natural cytotoxicity triggering receptor 2, Ly95	NCR2	6p21.1	Activated NK cells, some activated $\gamma\delta$ T cells	Type I transmembrane glycoprotein and member of Ig superfamily Lacks ITAMs and associates with DAP12 chain for signaling Binds to a non-MHC ligand Increases efficiency of NK cytotoxicity, cytokine production
CD337	**NKp30** NCR3, natural cytotoxicity triggering receptor 3, 1C7	NCR3	6p21.3	Resting and activated NK cells, spleen	Type I transmembrane glycoprotein and member of Ig superfamily Lacks ITAMs and associates with CD3ζ chain for signaling Binds to a non-MHC ligand Increases efficiency of tumor cell killing by activated NK cells
CDw338	**ABCG2** MRX, MXR, ABCP, BCRP	ABCG2	4q22	Endothelial cells, brain, placenta, lactating mammary gland	Transmembrane glycoprotein and member of White subgroup of ABC family of ATP-binding transporters Concentrates xenobiotics and toxins in mammalian milk Implicated in multi-drug resistance of breast cancer cells to chemotherapeutic agents
CD339	**Jagged-1** JAG1, JAGL1 HJ1, AGS	JAG1	20p12.1-p11.23	Broadly expressed, particularly in ovary, prostate, placenta, heart, kidney; embryonic progenitor cells, epithelial cells, stromal cells, carcinoma cells	Type I transmembrane protein Ligand for Notch1, Notch2, and Notch3 cell fate determination molecules Involved in cellular differentiation and proliferatiion, especially in embryonic heart

* A "w" in a CD number indicates "workshop" and that the designation for this molecule is not yet fully confirmed.
** Selected abbreviations: ADCC, antibody-dependent cell-mediated cytotoxicity; ALL, acute lymphoblastic leukemia; APC, antigen-presenting cell; B-CLL, chronic lymphocytic leukemia, B cell type; BCR, B cell receptor; BMP, bone morphogenetic protein; CLL, chronic lymphocytic leukemia; BMP, bone morphogenetic protein; CNS, central nervous system; DC, dendritic cells; DN, double negative (CD4$^-$CD8$^-$); EC, extracellular; FDC, follicular dendritic cells; GM-CSF, granulocyte-macrophage colony stimulating factor; GPI, glycosyl phosphatidylinositol; GSL, glycosphingolipid; GvHD, graft-versus-host disease; HB-EGF, heparin-binding epidermal growth factor-like growth factor; HEV, high endothelial venule; HUVEC, human umbilical vein endothelial cells; HVEM, herpesvirus entry mediator; IL, interleukin; LAP, latency associated peptide; LDL, low-density lipoprotein; LIF, leukemia inhibitory factor; LPS, lipopolysaccharide; MCF, macrophage colony stimulating factor; MHC, major histocompatibility complex; NGF, nerve growth factor; NHL, non-Hodgkin's lymphoma; NK, natural killer; PBL, peripheral blood lymphocytes; PBMC, peripheral blood mononuclear cells; PMN, polymorphonuclear lymphocytes; PTK, protein tyrosine kinase; RBC, red blood cell; SAP, SLAM associated protein; SCF, stem cell factor; SLC, secondary lymphoid tissue chemokine; SRBC, sheep red blood cell; TCR, T cell receptor; TNF, tumor necrosis factor; TNFR, tumor necrosis factor receptor; VEGF, vascular endothelial growth factor.

Glossary

Term	Definition
αβTCR *checkpoint*	Second major checkpoint in T cell development. *Positive selection* of *DP thymocytes* expressing a fully functional αβ TCR promotes the survival of thymocyte clones recognizing self-MHC alleles with moderate *avidity*, while *negative selection* deletes *autoreactive* thymocyte clones.
ABO antigens	Family of glycoprotein antigens expressed on the surfaces of RBCs that define the ABO blood group system. Differences in ABO blood types can cause *hyperacute graft rejection* and severe blood *transfusion reactions*.
abortive infection	An infection in which the *pathogen* enters the body but is unable to successfully propagate.
ABVD chemotherapy	Cocktail of Adriamycin (doxorubicin), Bleomycin, Vinblastine, and Dacarbazine.
abzyme	Antibody that cross-reacts with and stabilizes a transition state species in an enzymatic reaction, thereby acting as a catalyst.
accessory cells	Cells with effects on lymphocyte proliferation and differentiation, or on antigen processing and presentation; includes APCs.
accessory molecules	Molecules promoting or supporting adaptive immune responses. Includes the CD4 and CD8 coreceptors, CD3 components of the TCR, adhesion molecules, and costimulatory molecules.
acoelomates	Animals without a body cavity. See *lower invertebrates*.
acquired immunodeficiency	Failure of some component of the immune system due to an external factor such as a nutritional deficiency, cancer, drug, or pathogen.
activation-induced cell death (AICD)	Apoptotic death of lymphocytes subjected to prolonged antigenic stimulation. A means of eliminating effector cells after they are no longer needed. Often mediated by the *Fas pathway* or *TNFR1 pathway*.
activation-induced deaminase (AID)	A putative RNA- or DNA-editing enzyme critical for both *isotype switching* and *somatic hypermutation*. Selectively expressed in germinal center B cells.
active immunization	See *vaccination*.
acute cellular rejection (ACR)	A form of *acute graft rejection* in which allogeneic graft cells are destroyed by effector functions of recipient *CTLs* and *DTH reactions*.
acute graft rejection	Graft rejection occurring within days or weeks of a transplant. Can be cell-mediated (*acute cellular rejection*) or antibody-mediated (*acute humoral rejection*).
acute humoral rejection	A form of *acute graft rejection* due to *de novo* production of antibodies directed against allo-MHC in the graft. Characterized by vascular inflammation and neutrophil infiltration. Also called "delayed graft rejection" or "delayed vascular rejection."
acute infection	An infection causing disease symptoms that appear rapidly but remain for only a short time.
acute lymphoblastic leukemia (ALL)	*Leukemia* characterized by an acute increase in numbers of usually B (or sometimes T) lineage cells in the blood and bone marrow.
acute myelogenous leukemia (AML)	An acute cancer of the *myeloid* lineage in which the target cells are early stage *myeloblasts*.
acute phase proteins	Early inflammatory proteins made by hepatocytes in response to TNF and IL-6. Acute phase proteins (particularly *C-reactive protein*) bind to cell wall components of microbes and activate complement. See also *inflammation*.
adaptive immunity (also *specific* or *acquired immunity*)	Highly specific recognition of a non-self entity by lymphocytes whose activation leads to elimination of the entity and the production of specific memory lymphocytes. Because these memory lymphocytes forestall disease in subsequent attacks by the same pathogen, the host immune system has "adapted" to cope with the entity.
adaptor proteins	Non-enzymatic proteins that serve as a scaffold for other proteins, or recruit or link together other proteins. Often contain specific amino acid motifs that promote physical interaction of signaling proteins.

Term	Definition
ADCC	= Antibody-dependent cell-mediated cytotoxicity. The lysis of a cell that occurs when the Fc region of surface-bound antibody interacts with FcR expressed on lytic cells, such as eosinophils and NK cells, triggering their degranulation. Macrophages, neutrophils, and monocytes may also mediate ADCC.
adenoids	See *tonsils*.
adenosine deaminase (ADA) deficiency	Subtype of *SCID* caused by mutation of the gene encoding adenosine deaminase required for the *salvage pathway of nucleic acid synthesis*.
ADEPT	= Antibody-directed enzyme/pro-drug therapy. Therapy employing an immunoconjugate in which the mAb is linked to an enzyme capable of converting an inert pro-drug into an active cytotoxic drug. Used for anti-cancer therapy.
adjuvant	A substance that, when mixed with an isolated antigen, increases its immunogenicity. Adjuvants provoke local inflammation, drawing immune system cells to the site and triggering maturation of DCs. See *Freund's adjuvant*.
adoptive transfer	A form of transplantation in which lymphocytes from a donor are transferred to and function in a recipient whose own lymphocytes are non-functional, greatly reduced in number, or absent. Used for reconstitution of the immune system.
afferent lymphatic vessel	A vessel that conveys *lymph* into a *lymph node*.
affinity	A measure of the strength of the association established at single point of binding between a receptor and its ligand. For *antibody*, the strength of the non-covalent association between a single antigen-binding site on the antibody and a single *epitope* on an *antigen*.
affinity chromatography	Isolation of a protein from a mixture in solution based on its ability to bind to a particular ligand. A solution containing the protein of interest is passed over a column to which the ligand has been attached, and bound protein is then released from the column by adding a high concentration of soluble ligand or changing the pH. In the immunological context, an antigen from a protein mixture can be isolated by passage over a column to which antibody of appropriate specificity has been attached.
affinity maturation	Positive selection of developing B cells with *BCRs* that have undergone *somatic hypermutation* resulting in increased *affinity* for specific antigen. Memory cells retaining this increased affinity for antigen are generated.
agammaglobulinemia	Lack of any antibodies in an individual. Congenital defects in B cell signaling result in several *primary immunodeficiencies* characterized by agammaglobulinemia.
agglutination	Aggregation or "sticking together" of cells to form a visible particle. Used in some assays to detect antigen–antibody binding.
agonist ligand	A ligand that binds to an antigen receptor and has a positive effect on downstream signaling and thus cellular function.
AICD	See *activation-induced cell death*.
AIDS	= Acquired immunodeficiency syndrome. Failure of adaptive immunity due to T cell destruction caused by HIV infection.
airway hyper-responsiveness	A narrowing of the airways in an individual with *asthma* that occurs in response to non-specific stimuli.
airway remodeling	In an individual with *asthma*, the stripping away of the top layers of the bronchial epithelium and the thickening of the submucosae due to the deposition of collagen beneath the basement membrane.
ALL	See *acute lymphoblastic leukemia*.
alleles	Two slightly different sequences of the same gene. Proteins produced from alleles have the same function.
allelic exclusion (Ig locus)	The production of a functional μ heavy chain from the Igh locus on one chromosome inhibits further V(D)J recombination and production of μ chains from the other Igh allele.
allelic exclusion (TCR locus)	The production of a functional TCRβ chain from the TCRβ locus on one chromosome inhibits further V(D)J recombination and production of TCRβ chains from the other TCRβ allele, as well as V(D)J recombination in the TCRγ locus on both chromosomes.

Term	Definition
allergen	An antigen that is innocuous in most individuals but provokes *type I hypersensitivity* in some.
allergy	Clinical manifestation of *type I hypersensitivity*. Mediated by mast cells armed with allergen-specific IgE.
alloantibodies	Antibodies distinguishing between the different forms of a protein encoded by different alleles of a given gene. Often refers to antibodies specific for *MHC* molecules.
alloantiserum	An *antiserum* containing *alloantibodies*.
allogeneic	Having different alleles at one or more loci in the genome compared with another individual of the same species.
allograft	Tissue transplanted between genetically different (*allogeneic*) members of the same species.
alloimmune hemolytic anemia	*Hemolytic anemia* that occurs when an individual is exposed to allogeneic RBCs such that antibodies directed against the foreign RBC antigens are generated. Examples are *transfusion reaction*s to unmatched ABO blood group antigens and *Rh disease*.
alloimmune thrombocytopenia	*Type II hypersensitivity* caused by *alloantibodies* specific for *platelet* surface antigens.
alloimmunization	Exposure of an individual to cells expressing an allogeneic protein resulting in establishment of memory B lymphocytes specific for this alloantigen. May occur during transfusion, transplantation, or pregnancy. Upon subsequent exposure to the alloantigen, rapid production of *alloantibodies* occurs that may have systemic consequences. May also be called "sensitization."
alloreactivity	Immune responses resulting from lymphocyte recognition of genetic differences between individuals. Most often refers to responses against MHC-encoded differences.
allorecognition	Recognition of allelic differences expressed by cells of one individual by the lymphocytes of another individual. Most often refers to recognition of MHC-encoded differences.
allotopes	Differences in the constant region structure of an Ig that makes that Ig immunogenic in some members of the same species.
allotype	Collection of *allotopes* found in a particular Ig constant region.
alpha-beta T cells	T lymphocytes bearing an αβ TCR, including populations of Th, Tc, and regulatory T cells. Principal mediators of highly specific, but very diverse, adaptive immune responses.
ALPS	See *autoimmune lymphoproliferative syndromes*.
alternative complement activation	Activation of the *complement* cascade initiated by direct binding of complement component C3b* to a stabilizing ligand on a microbe. Involves cleavage and activation of factors B and D and properdin.
alternative splicing	Mechanism by which the RNA splicing machinery may process primary transcripts of a given gene at splice acceptor sites 3′ of different exons giving rise to mRNA populations containing different subsets of exons. Exons whose splice acceptor sites are spliced out of the mRNA so their corresponding amino acid sequences are absent from the resulting protein. A single gene can therefore give rise to two or more versions of the same protein.
AML	See *acute myelogenous leukemia*.
amoebocytes	Mobile phagocytic cells in the tissues of primitive organisms lacking a circulatory system. Responsible for host defense via *phagocytosis* as well as digestion and excretion.
amphipathic	An *amphipathic* molecule has both hydrophobic and hydrophilic regions.
anaphylatoxins	Complement component cleavage products that have pro-inflammatory and chemoattractant effects; includes C3a, C4a, and C5a. See *complement*.
anaphylaxis	Systemic *type I hypersensitivity* response that may be fatal due to a catastrophic drop in blood pressure induced by the release of large quantities of inflammatory mediators. Associated most often with insect stings, peanuts, seafood, or penicillin.
anaplastic	Anaplastic cells are poorly differentiated and, in a tumor, are associated with aggressive growth.
ancestral haplotype	An MHC *haplotype* shared among a large number of families that likely have common ancestors. Also known as an "HLA supratype" or a "common extended haplotype."
anergize	To induce a state of *anergy* in another cell.

Term	Definition
anergized	To have induced or to be in a state of *anergy*.
anergy	State of lymphocyte non-responsiveness to specific antigen. Induced by an encounter of the lymphocyte with cognate antigen under less than optimal conditions, i.e., in the absence of *costimulation*. See *tolerization* and *peripheral tolerance*.
aneuploidy	Having a chromosome number that is not a multiple of the haploid number for the species.
angiogenesis	Process by which new blood vessels are formed.
animal reservoir	Intermediate animal host for a pathogen whose primary host is another species. See *reservoir*.
ankylosing spondylitis (AS)	*Autoimmune* chronic inflammatory *disease* of bone and joints, particularly of the spine. Linked to HLA-B27 expression.
antagonist ligand	A ligand whose binding to a receptor blocks the normal function of another ligand for the same receptor.
anti-allotypic antibodies	Antibodies directed against the *allotopes* of an Ig of a member of the same species. For example, antibodies raised in inbred mouse strain A after immunization with Igs from inbred mouse strain B.
antibody	Secreted *immunoglobulin* that is produced by B lineage *plasma cells* and binds to specific antigens. Able to recognize antigens that are either soluble or fixed in a tissue or on a cell surface.
antigen	Entity (such as an element of an infectious pathogen, cancer, or inert injurious material, or of a self-tissue) that can bind to the antigen receptor of a T or B cell. This binding does not necessarily lead to lymphocyte activation.
antigen clearance	The removal of an antigen by phagocytosis, cytolysis, or complement-mediated destruction.
antigenic determinant	See *epitope*.
antigenic drift	Subtle modification of pathogen antigens through point mutations. Usually involves surface proteins that would normally be the target of neutralizing antibodies.
antigenic shift	Dramatic modification of viral antigens due to reassortment of genomic segments of two different strains of a virus (that simultaneously infect the same individual) to generate progeny virions with new combinations of genome segments and thus new proteins.
antigen presentation	Cell surface display of peptides in association with either MHC class I or II molecules allows T cell recognition of antigenic peptides on the surface of target cells or APCs, respectively.
antigen-presenting cell (APC)	Cell expressing MHC class II and thus capable of presenting peptides to $CD4^+$ Th cells. Professional APCs such as DCs, B cells, and monocyte/macrophages express high levels of MHC class II. Non-professional APCs such as keratinocytes and some epithelial, endothelial, and mesenchymal cells can act as APCs once activated by inflammation.
antigen processing	Degradation of a protein into peptides suitable for binding in the peptide binding grooves of either MHC class I or II molecules.
anti-histamine	Molecules that bind to histamine receptors on target organs, preventing *histamine* released by degranulating mast cells from triggering symptoms. Used to treat mild *allergies*.
anti-idiotypic antibodies	Antibodies directed against the *idiotopes* of an Ig of an individual. For example, antibodies raised in individual X of inbred mouse strain A after immunization with Igs from individual Y of inbred mouse strain A.
anti-idiotypic vaccine	*Vaccine* based on an *anti-idiotypic antibody* whose binding site resembles an epitope on the original pathogen or tumor cell.
anti-isotypic antibodies	Antibodies raised in one species that are directed against determinants found exclusively on one immunoglobulin isotype of another species. For example, rabbit anti-mouse IgM antibodies.
anti-microbial peptides	Small peptides that damage bacteria or fungi by permeabilizing their cell walls. In mammals, these peptides are produced by keratinocytes or intestinal epithelial cells under pathogen attack, and some can be found in neutrophil granules. Includes the defensins and the cathelicidins. In lower and primitive animals, anti-microbial peptides include the magainins, attacins, agglutinins, cecropins, and drosomycin. In plants, anti-microbial peptides include defensins, thionins, and phytoalexins.
anti-nuclear antibodies	Antibodies directed against self double-stranded DNA and small nuclear ribonucleoproteins (snRNPs) exposed by cellular breakdown.

Term	Definition
anti-phospholipid syndrome (APS)	*Autoimmune disease* in which *autoantibodies* attack components of the prothrombin activator complex required for the blood coagulation cascade. Abnormal blood clots block blood vessels and may induce migraine, stroke, heart attack, pulmonary embolus, or recurrent miscarriage.
anti-retroviral drugs (HIV)	Drugs used to inhibit replication and spread of HIV. Include protease inhibitors, nucleoside and non-nucleoside reverse transcriptase inhibitors, and fusion inhibitors that interfere with viral envelope fusion to the host cell membrane.
anti-sense oligonucleotide	Oligonucleotide designed to disrupt synthesis of a given protein by specifically binding to a complementary sequence (the "sense strand") in a primary RNA transcript or mRNA encoding the protein of interest. The duplex mRNA that forms may block the splicing of the primary transcript, arrest translation of the mRNA, disrupt ribosomal assembly, or induce mRNA cleavage by RNase H activation.
antiserum	The clear liquid (serum) fraction of clotted blood containing antibodies produced when an individual or animal is exposed to a foreign substance or infectious agent.
antitoxins	Antibodies made against bacterial *exotoxins* and *endotoxins*. Historically, the term *antitoxin* was used for antibodies that could clump and lyse bacteria, and precipitate and neutralize toxins and viruses.
antivenin	Antibodies made against snake venom. May be commercially prepared and used for *passive immunization* of snake bite victims.
antiviral state	A metabolic state of viral resistance induced in a cell by exposure to type I IFN released by neighboring virus-infected cells. Depends on IFN-induced degradation of viral RNA and inhibition of viral transcription and translation.
APC licensing	Concept that full activation of an antigen-stimulated Tc cell is made possible through costimulation by a DC that has upregueated its costimulatory molecules in response to prior CD40–CD40L-mediated contact with an antigen-activated Th cell.
APECED	See *autoimmune polyendocrinopathy candidiasis ectodermal dystrophy.*
apical	The *apical* surface of an epithelial cell faces the lumen rather than the tissue.
apoptosis	The controlled death of a cell mediated by certain intracellular proteases (including *caspases*) that cause the orderly breakdown of the cell nucleus and its DNA. Death occurs without the release of internal contents and without triggering inflammation. The *intrinsic apoptotic pathway* involves the release of mitochondrial cytochrome *c*, while the *extrinsic apoptotic* pathway is triggered by the engagement of *death receptors* such as *Fas* and *TNFR*. Apoptosis is also called "programmed cell death."
Artemis SCID	Subtype of *SCID* caused by mutations of the Artemis gene that encodes a factor important for both *V(D)J recombination* and *DNA repair*.
arthritis	*Inflammation* of a joint. May cause localized pain and immobility.
Arthus reaction	Localized *type III hypersensitivity* characterized by redness and swelling in the site of *immune complex* deposition.
armed CTL	A mature CTL that has yet to encounter antigen but that has synthesized the chemical mediators that will be used to carry out target cell destruction.
armed mast cell	A mast cell that has bound *allergen*-specific IgE to its FcεR but has yet to encounter the allergen.
ascites fluid	Fluid in the peritoneal cavity.
asthma	*Inflammation* of the bronchi in the lungs. Characterized by *airway hyper-responsiveness*. Often *atopic* (due to IgE-mediated responses to inhaled *allergens*) but also intrinsic (independent of IgE).
ataxia-telangiectasia (AT)	Autosomal recessive disease characterized by progressive cerebellar ataxia (cerebellar degeneration leading to generalized neuromotor dysfunction); oculocutaneous telangiectasia (abnormal dilation of blood vessels in the conjunctiva of the eyes); increased risk of leukemia; and variable hypogammaglobulinemia. Due to mutation of the *tumor suppressor gene* ATM involved in *non-homologous end-joining (NHEJ) pathway* of *DNA repair*.
atopic	Involving symptoms or reactions caused by *type I hypersensitivity*.

Term	Definition
atopic dermatitis	Inflammation of the skin due to *type I hypersensitivity*. Characterized by a scaly itchy rash and dry skin. Also called "eczema."
atopic rhinitis	Inflammation of the nasal mucous membranes due to *type I hypersensitivity* to an inhaled *allergen*. Characterized by runny nose and itchy, watery eyes. Also called "hay fever."
atopy	Any *type I hypersensitivity* response. Commonly called *allergy*.
ATRA	= All *trans*-retinoic acid. An oral agent that counters the effect of the chimeric protein in acute promyelocytic leukemia and rescues the maturation of hematopoietic cells.
attenuation	Weakening of a pathogen by sub-optimal culture or chemical or genetic modification so that it remains viable and immunogenic but loses its capacity to cause disease.
autoantigen	Self antigen inducing an *autoreactive* response.
Auer rods	Finely granular flat bodies in the cytoplasm of *myeloid* lineage *leukemia* cells.
autocrine	A molecule that acts in an autocrine fashion affects only the gland or cell that produced it.
autoimmune disease	Pathophysiological state in which the host's own tissues are damaged as a result of *autoimmunity*.
autoimmune hemolytic anemia	*Hemolytic anemia* that occurs when an individual makes antibodies directed against epitopes on his or her own RBC.
autoimmune lymphoproliferative syndromes (ALPS)	Family of *PIs* caused by defects in lymphocyte apoptosis, particularly in the *Fas*-mediated pathway.
autoimmune myocarditis	*Inflammation* of the heart due to an autoimmune response to myocardial tissues. Cardiac myosin may be the principal autoantigen. Can lead to chronic heart disease or *dilated cardiomyopathy*.
autoimmune polyendocrinopathy candidiasis ectodermal dystrophy) (APECED)	*Autoimmune disease* characterized by chronic mucosal yeast infection and attacks on multiple endocrine glands (thyroid, pancreas, adrenal). Due to a mutation in AIRE, a transcriptional co-activator required for expression of peripheral self-antigens in thymic epithelial cells. AIRE deficiency results in a defect in *central tolerance*.
autoimmune thrombocytopenia	Acute or chronic *type II hypersensitivity* caused by platelet-specific *autoantibodies* that trigger platelet destruction via phagocytosis. Includes immune thrombocytopenic purpura (ITP) and thrombotic thrombocytopenic purpura (TTP).
autoimmune uveitis	Organ-specific, inflammatory *autoimmune disease* of the neural retina of the eye characterized by infiltration of Th1 cells and monocytes across the blood–retina barrier. *Autoantibodies* to retinal antigens are present.
autoimmunity	The response of an individual's immune system against self tissues. May cause *autoimmune disease* if unrestrained by mechanisms of *peripheral tolerance*.
autoinflammatory syndromes	PI characterized by bouts of severe local inflammation and prolonged, periodic fevers in the absence of any obvious pathogenic cause.
autologous graft	Transplantation of tissue from one part of an individual's body to another part of his or her body.
autoreactive	Lymphocytes that recognize and are activated by self-antigens are *autoreactive*.
avidity	A measure of the total strength of all the associations established between a multivalent receptor and its multivalent ligand.
avr-R system	Immune response mechanism of plants involving the disease resistance (R) genes of plants and avirulence (*avr*) genes of microbes. Only when the *R* product of the plant and the *avr* product of the attacking pathogen "match" can the plant respond by walling off the pathogen and degrading it. See *hypersensitive response* and *local acquired resistance*.
β_2-*microglobulin* ($\beta 2m$)	Invariant chain of *MHC class I proteins*; encoded by a gene outside the MHC.
β-*selection*	First major checkpoint in T cell development. Process in which the TCRβ chain expressed by a DN3 thymocyte participates in a *pre-TCR* that allows the thymocyte to receive a survival/ proliferation signal (*pre-TCR activation*) and become committed to the $\alpha\beta$ T lineage.
bacillus Calmette-Guérin	Strain of *Mycobacterium bovis* used as a vaccine against tuberculosis.
bacteria	Microscopic, prokaryotic single-celled organisms in which the nucleus lacks a nuclear membrane and the genetic material is usually contained in a single linear chromosome.

Term	Definition
BALT	See *bronchi-associated lymphoid tissue* and *MALT*.
bare lymphocyte syndrome	Lack of expression of MHC molecules on cells (including lymphocytes) due to deficiency of the RFX or CIITA components of the *SXY-CIITA regulatory system*.
basement membrane	A layer composed of collagen, laminin, heparan sulfate, and glycosaminoglycan that supports an overlying layer of epithelial or endothelial cells.
basolateral	The *basolateral* surface of an epithelial cell faces the tissue rather than the lumen.
basophil	Circulating granulocyte with an irregularly shaped nucleus. Contains granules that stain a dark blue color with basic dyes. Basophil granules contain heparin and vasoactive amines, as well as many enzymes capable of promoting inflammation.
B cell receptor (BCR) complex	Antigen receptor complex of B lineage cells. Composed of a *membrane-bound Ig* (mIg) monomer plus the accessory *Igα/Igβ* complex required for *intracellular signaling pathway*.
BCG vaccine	See *bacillus Calmette-Guérin*.
beige mouse	Natural mouse mutant with defect in granule function. Sometimes used as an NK cell-deficient mouse model. Also a model of *Chédiak-Higashi syndrome*.
Bence Jones proteins	Soluble Ig light chains found in the urine of individuals with *myeloma*.
benign tumor	A mass formed by abnormally dividing cells that are well differentiated and well organized. Resembles the normal tissue from which it originated and is securely encapsulated and relatively slow growing. Causes death only by indirect means.
binder-ligand assay	An assay in which one of two binding partners is labeled with a measurable *tag* so it can be used to detect the presence of its untagged binding partner. Examples are *ELISA* and *RIA*.
biodegradables	In *vaccine* design, *delivery vehicles* made of microspherical particles composed of polymers such as chitosan that can be degraded to metabolizable products. The vaccine antigen in the particle is released continuously.
biologicals	Natural or engineered proteins whose purpose is to activate or disrupt the proliferation, metabolism, or function of normal cells or pathogenic host cells such as cancer cells.
biotin-avidin system	A sensitive method of amplifying the detection of binding pairs in a *binder-ligand assay*. Biotin is a small co-factor molecule that is easily coupled to proteins such as antibodies via amide linkages. Avidin and streptavidin are tetrameric basic glycoproteins, each of which can form an extremely high affinity non-covalent bond with four molecules of biotin. After allowing a biotinylated antibody or antigen to bind to its corresponding binding partner in a binder-ligand assay, the binding pair can be detected by adding tagged avidin/streptavidin, or untagged avidin/streptavidin followed by tagged biotin.
blast crisis (in CML)	The late phase of *CML* in which blast-like leukemia cells make up more than 30% of cells in the patient's blood or bone marrow. May infiltrate into *extramedullary tissues*.
blastoma	In general, highly *malignant* childhood *tumors* that are so undifferentiated that the cells resemble those in a blastocyst.
blood–thymus barrier	The inability of blood cells to pass readily through the endothelial cells lining the vessels in the thymic cortex.
Bloom syndrome (BS)	Rare autosomal recessive *PI* characterized by genomic instability and *CID*. Caused by mutation of the BLM gene involved in the *homologous recombination pathway* of DNA repair.
BMT	See *bone marrow transplant*.
bone marrow	The *primary lymphoid tissue* (in the adult) in which all lymphocytes and other hematopoietic cells arise and B cells mature. *Red marrow* is the hematopoietically active tissue in which HSCs either self-renew or differentiate into red and white blood cells of all lineages. Fibroblasts, fat cells, endothelial cells, and fibers form a stromal framework supporting the developing hematopoietic cells. *Yellow marrow* is hematopoietically inactive but contains significant numbers of adipocytes that act as an energy reserve.

Term	Definition
bone marrow transplant (BMT)	Replacement of a patient's dysfunctional immune system by *transplantation* of healthy whole *bone marrow* from a donor. A high degree of *histocompatibility* between recipient and donor is necessary. In some cases, bone marrow cells may be isolated from a patient during disease remission and held in storage for a future *autologous* BMT in which the patient receives his or her own cells.
booster	A second or subsequent immunization (*vaccination*) to stimulate *memory lymphocyte* production.
brachytherapy	Implantation of a metal "seed" containing a radioisotope to mediate *radiation therapy* internally. Also called "implant radiation."
bradykinin	See *kinins*.
brain death	Permanent loss of brain function in the absence of trauma to other body systems. The respiration of an individual who has suffered brain death can be artificially maintained to preserve healthy organs for transplantation.
Brambell receptor	See *FcRn*.
BrdU labeling	Incorporation of bromo-deoxyuridine into the DNA of dividing cells. Used to measure cellular proliferation.
bronchi-associated lymphoid tissue (BALT)	Lymphoid patches and cells in the mucosae of the trachea and lungs. See *MALT*.
bronchodilators	Rapidly acting, aerosolized β-adrenergic drugs used to block *mast cell* degranulation and relax bronchoconstriction during episodes of acute *asthma*.
brush border	A layer of dense microvilli that cover the *apical* surfaces of gut epithelial cells, greatly increasing the surface area.
B symptoms	Constellation of systemic complaints experienced by *lymphoma* patients. Includes unexplained rapid weight loss, fatigue, cyclic bouts of fever, and night sweats frequently accompanied by chest pain.
bursa of Fabricius	Unique organ in birds responsible for B cell production; site where B cells were first identified.
bystander activation	Activation of a B cell with T cell help supplied by a Th cell responding to an unrelated antigen.
C1inh	= C1 inhibitor protein. An inhibitor of *classical complement activation* that acts as a circulating decoy substrate for the activated C1r and C1s serine proteases within the C1 complex.
C3 convertase	Enzyme composed of complement components C4b2a (classical) or C3bBbP (alternative). Cleaves C3 into C3b and C3a.
C5 convertase	Enzyme composed of complement components C4b2a3b (classical) or C3bBbPC3b (alternative). Cleaves C5 into C5b and C5a.
cachexia	Wasting of the body due to uncontrolled cellular catabolism. Cachexia in cancer patients is induced by their high levels of TNF.
cadaveric organ	Transplantable organ recovered from a recently deceased human (a cadaver).
CAMPATH-1H	Anti-CD52 mAb originally developed to promote the killing of lymphoid cancer cells.
cancer	See *malignant tumor*.
cancer-testis antigens	Proteins that are normally expressed solely in spermatogonia and spermatocytes but become *TAAs* when transformation causes them to be expressed on other cell types. See *differentiation* and *tissue-specific antigens*.
canonical structure	A commonly observed molecular sequence or molecular arrangement.
capsid	Protein coat of a *virus*.
capsule	See *encapsulated bacteria*.
carcinogen	Any substance or agent that significantly increases the incidence of *malignant tumors* by mutating or deregulating *oncogenes*, *tumor suppressor genes*, or *DNA repair* genes.
carcinogenesis	Multi-step process by which malignant transformation of a cell culminates in cell growth deregulation. Steps are *initiation*, *promotion*, *progression*, and *malignant conversion*.
carcinoma	A *malignant tumor* derived from epithelial cells.
carriage (HIV)	Rapid transfer of HIV from infected APCs to lymph node T cells by way of a bicellular conjugate called an "infectious synapse." The infected T cells recirculate to other *secondary lymphoid tissues*, establishing multiple reservoirs from which the virus disseminates through the body.

Term	Definition
carrier (person)	A person with a *latent infection* that can be transmitted unknowingly to others.
carrier (protein)	Large macromolecule to which a *hapten* is conjugated to stimulate an anti-hapten antibody response. Provider of T cell epitopes required for B-T cooperation.
carrier effect	Phenomenon in which a secondary antibody response to a hapten is seen only if the hapten is conjugated to the <u>same</u> carrier for both the primary and secondary immunizations.
caspase	Intracellular protease involved in *apoptosis*.
caspase cascade	Series of proenzyme cleavage events mediated by cysteine-aspartic acid proteases called "caspases." The key "executioner" caspases are caspase-3, caspase-6, and caspase-7. These enzymes break down numerous intracellular substrates, leading to apoptosis. Caspase-3 initiates the cascade and is activated by either caspase-8 in the *extrinsic apoptotic pathway* or caspase-9 in the "apoptosome" of the *intrinsic apoptotic pathway*.
cathepsins	Proteases in the endolysosomes that progressively degrade proteins. Cathepsins are important for degrading *Ii* chains into small fragments, generating *CLIP*. Integral part of *exogenous antigen processing and presentation*.
CD1 molecules	Unconventional *antigen presentation* molecules that present non-peptide antigens (lipid or glycolipid antigens) to subsets of αβ and γδ T cells and *NKT* cells. Include CD1c and CD1d. See *MHC-like proteins*.
CD3 chains	*ITAM*-containing accessory chains necessary for TCR signaling: CD3γ, CD3δ, CD3ε, CD3ζ, and CD3η. Also facilitate insertion of the TCR complex in the membrane. See *TCR complex*.
cell-mediated immunity	Originally, adaptive immune responses mediated by effector actions of cytotoxic T lymphocytes. Now includes effector actions of NK and NKT cells.
cellular adhesion molecules (CAMs)	Molecules facilitating cell–cell and cell–matrix adhesion as well as extravasation. Includes members of the ICAM, *selectin*, and *integrin* protein families. See *homing receptors* and *vascular addressins*.
CEM-15	Host RNA/DNA editing enzyme that binds to *HIV* polyproteins and is packaged into virions. CEM-15 hypermutates replicating HIV DNA, blocking integration into the host cell genome.
central tolerance	Mechanism that eliminates most *autoreactive* T and B cells during lymphocyte development. Established by *negative selection* processes in the thymus and bone marrow. Autoreactive cells that escape central tolerance are prevented from attacking self cells by the mechanisms of *peripheral tolerance*.
centroblasts	Rapidly proliferating, antigen-activated B cells that fill the *dark zone* of a GC and undergo *somatic hypermutation*.
centrocytes	Smaller, non-dividing cells arising from *centroblasts* that migrate to the *light zone* of the GC and undergo *affinity maturation* and *isotype switching*. Give rise to *plasma cells* and memory cells.
CFA	See *Freund's adjuvant*.
Chédiak-Higashi syndrome (CHS)	Complex disease with a *PI* component caused by defects in intracellular vacuolar and *granule* fusion. CHS has both hematopoietic and neurological manifestations. Caused by mutations to the *CHS1* gene involved in intracellular fusion/fission events.
chemokine receptors	Receptors expressed on leukocytes that allow them to follow appropriate chemokine gradients for homing. For example, *Th1* cells express CCR1, CCR5, and CXCR3, which direct cells to tissue inflammatory sites. *Th2* cells express CCR4, CCR3, and CCR8, which direct cells to the *mucosae*. Naive lymphocytes express CCR7, which directs cells to *lymph nodes*.
chemokines	Chemotactic cytokines. Contain intramolecular disulfide bridges involving conserved N-terminal cysteine residues. Structural families include the C, CC, CXC, and CX3C subgroups. See also *chemotaxis*.
chemokinesis	The stimulation of random leukocyte movement by chemical signals.
chemotactic factors	Factors derived from either host or bacterial metabolism that can induce *chemotaxis*. Includes *chemokines*, kallikrein, *PAF*, *thrombin*, and the *anaphylatoxin* C5a.
chemotaxis	The directed movement of cells along the concentration gradient of a *chemotactic factor*. Serves to draw neutrophils, lymphocytes, and other leukocytes from the circulation into an injured or infected tissue.

Term	Definition
chemotherapy	Use of chemical drugs to kill tumor cells. Chemotherapeutic agents generally affect only those cells that are growing faster than most normal cells, or those that have a metabolic imbalance.
chimeric protein	A chimeric protein consists of parts of two or more different proteins.
chimerism	The co-existence of tissues of two or more genotypes in a single organism. Chimerism in transplantation refers to the presence of donor HSCs in a recipient. See also *full*, *mixed*, and *microchimerism*.
CHOP therapy	*Chemotherapy* cocktail of the anti-cancer agents: cyclophosphamide, [H]-doxorubicin, [O]-vincristine, and prednisolone.
chromatin remodeling	Alteration of chromatin structure of a gene to enhance or suppress its accessibility for transcription.
chromatin remodeling complexes (CRCs)	Large, variable multi-component enzymatic complexes capable of reorganizing and bending DNA such that its transcription is either enhanced or repressed. See *chromatin remodeling*.
chromosomal translocation	The abnormal exchange of genetic material between two different chromosomes or two different regions on the same chromosome.
chronic disease	A disease in which symptoms are experienced on an ongoing or recurring basis. See *persistent infection*.
chronic graft rejection (CGR)	Immune response against an *allograft* that occurs several months after transplantation and leads to loss of allograft function. Characterized by fibrosis, collagen deposition, and *chronic graft vasculopathy*.
chronic graft vasculopathy	Recipient T cells activated by *allorecognition* induce monocytes/macrophages present in a graft, and endothelial cells in the walls of its blood vessels, to produce growth factors and cytokines that induce smooth muscle cell proliferation in arteries and arterioles of graft. The diameter of the graft blood vessels decreases, causing *ischemia*. See *chronic graft rejection*.
chronic granulomatous disease (CGD)	Clinically heterogeneous *PI* of the innate response caused by a failure in *phagosomal* killing. Due to mutations in NADPH oxidase involved in the *respiratory burst*.
chronic lymphocytic leukemia (CLL)	Chronic form of *leukemia* arising from the transformation of a peripheral blood lymphocyte (usually a B cell). Characterized by a prolonged disease course.
chronic myelogenous leukemia (CML)	Most common *myeloproliferative disease. Leukemia* characterized by increased numbers of mature and immature *granulocytes*. Includes chronic phase, accelerated phase, and *blast crisis*.
chronic xenograft rejection	Loss of a xenograft several weeks after transplantation due to the attack of CTLs and NK cells on the donor endothelium. Also known as "cellular xenograft rejection."
CIA	See *collagen-induced arthritis*.
CID	See *combined immunodeficiency*.
CIIV	Class II vesicle. See *MIICs*.
clades	Original subtypes of *HIV* based on sequence diversity. Members of the same clade differed in sequence by less than 10%, whereas different clades differed in sequence by 15% or more.
classical complement activation	Initiated by the Fc region of an antigen-bound antibody binding to C1q, followed by recruitment and cleavage of C4, C2, C3, and assembly of the terminal complement components. See *complement*.
classical Hodgkin's lymphoma	Comprises 95% of *Hodgkin's lymphoma* tumors. Characterized by CD30$^+$, CD15$^+$, CD20$^-$, CD45$^-$ *Reed-Sternberg cells*. Includes subtypes "nodular sclerosis HL," "mixed cellularity HL," "lymphocyte-rich classical HL," and "lymphocyte-depleted HL."
clathrin	A protein component of the microtubule network involved in *receptor-mediated endocytosis*.
clinical trials	A series of controlled tests of a drug, vaccine, or treatment in human volunteers that is used to determine if its safety and efficacy warrant licensing and use in the general population. Generally involves four phases from I to IV.
CLIP	= Class II invariant chain peptide. A small fragment of *invariant chain* that sits in the MHC class II groove and prevents the premature binding of rER peptides by steric hindrance.

Term	Definition
clonal deletion	The induction of apoptosis in B or T lymphocytes that have bound their cognate antigen. In the establishment of *central tolerance*, clonal deletion of T or B cells with antigen receptors that recognize self antigens with high affinity underlies *negative selection*. In the establishment of *peripheral tolerance*, clonal deletion of mature naive T cells that meet their cognate self peptide/MHC presented by immature DCs results in elimination of T cell clones that might otherwise cause autoimmunity.
clonal exhaustion	An activated lymphocyte divides so quickly in the face of persistent antigen that its progeny reach the point at which they can no longer divide (*replicative senescence*) before memory cells have been generated.
clonal expansion	The proliferation of only that lymphocyte clone that was originally activated by recognition of specific antigen.
clonal selection	The activation of only those clones of lymphocytes bearing receptors specific for a given antigen.
cluster of differentiation (CD)	A CD number is the unique designation given to a cell surface protein identified serologically by the binding to that protein of a cluster of different antibodies. That is, the CD number points to the specific collection of serological epitopes furnished by the protein.
CML	See *chronic myelogenous leukemia*.
CNS prophylaxis	*Intrathecal chemotherapy* or localized irradiation of the CNS and brain are implemented to prevent invasion by cancer cells.
coagulation	In invertebrate immunity, refers to a mechanism in *higher invertebrtates* that combines wound healing and host defense. *LPS* binding to a *PRR* activates a zymogen cascade of three proteases resulting in cross-linking of coagulogen protein in *hemolymph*. A clot is formed that prevents further leakage of body fluids and walls off any pathogen that has entered the wound.
coding joint	Junction of coding sequences of *gene segments* in the Ig and TCR loci following *V(D)J recombination* and *NHEJ repair*.
co-dominance	Expression of a given gene from both the maternal and paternal chromosomes. If the genetic locus is heterozygous, both alleles of the gene are expressed.
coelomates	Animals with a body cavity. See *higher invertebrates* and *vertebrates*.
cognate antigen	An antigen known to be recognized by a given lymphocyte antigen receptor because it was used for the original activation of that lymphocyte.
cold chain (vaccination)	Constant refrigeration of a labile *vaccine* from the point of manufacture and shipping to the point of use in the field.
collagen-induced arthritis (CIA)	Mouse model of human *rheumatoid arthritis*. Injection of substrains of DBA/1J mice with type II collagen plus CFA leads to *arthritis* in multiple joints due to local induction of *autoimmunity*.
collectins	Soluble *pattern recognition molecules* that clear pathogens by *opsonization*, *agglutination*, or the *lectin pathway of complement activation*. Includes MBL and lung surfactant proteins A and D.
colony forming units (CFU)	A measure of the number of hematopoietic progenitors present in a sample that have the capacity to generate a colony composed of hematopoietic cells of a specified type or types. For example, CFU-GEMM is a measure of the precursors in a sample that can generate granulocytes, erythrocytes, monocytes, or macrophages.
colony-stimulating factors (CSF)	Proteins that promote the growth and differentiation of hematopoietic cells in the bone marrow, and the formation of hematopoietic cell colonies *in vitro*. For example, granulo-monocyte-CSF (GM-CSF) induces a *common myeloid progenitor (CMP)* to produce granulocyte/monocyte precursors.
combined immunodeficiency (CID)	A relatively mild disorder due to a deficit in the presence and/or function of both B and T lymphocytes.
commensal organisms	Beneficial microbes that normally inhabit the surfaces of skin and body tracts and provide innate immune protection against pathogens. Also known as *normal flora*.
common gamma chain (γc chain)	A signal transduction protein containing *ITAMs* that functions as a subunit in the receptors for IL-2, IL-4, IL-7, IL-9, IL-15, and IL-21. See *cytokine receptors*.

Term	Definition
common lymphoid progenitor (CLP)	Early descendant of HSCs that gives rise to *lymphoid cells*.
common mucosal immune system (CMIS)	A system of extended mucosal defense established when lymphocytes activated in one mucosal *inductive site* migrate to a large number of *effector sites* in various mucosal tissues. See *MALT, BALT, NALT, GALT*.
common myeloid progenitor (CMP)	Early descendant of HSCs that gives rise to *myeloid* lineage *cells*.
common variable immunodeficiency (CVID)	A family of heterogeneous *PIs* characterized by decreased IgA and IgG, and sometimes IgM. Circulating mature B cells are present but *plasma cell* differentiation and *antibody* production are impaired. *Memory B cells* may also be absent. Due to intrinsic B cell defects or defects in delivery of *T cell help*.
competitive binder-ligand assay	An assay in which *tagged* and untagged forms of the component of interest compete for binding to the specific binding partner.
complement	System of over 30 soluble and membrane-bound proteins that act through a tightly regulated cascade of pro-protein cleavage and activation to mediate cell lysis through assembly of the *membrane attack complex* (*MAC*) in a target cell membrane. Intermediates in the complement cascade can opsonize cells, and peptide by-products of complement component cleavage (*anaphylatoxins*) promote inflammation and serve as chemoattractants. Complement activation also solubilizes *immune complexes* and promotes their clearance and enhances antigen presentation and B cell activation and memory.
complementarity determining region (CDR)	See *hypervariable region*.
complementation	Process by which a hybrid cell acquires the abilities of two different parental cells to survive under different sets of selective conditions.
complement control proteins (CCP)	See *regulators of complement activation (RCA)*.
complement-dependent cytotoxicity test	Cells are tested for the expression of particular surface molecules by incubation *in vitro* with antibodies of the appropriate specificity, followed by the addition of complement. Cells positive for antigen expression are assessed by the uptake of dye that occurs upon cell lysis.
complement fixation	The consumption of complement components following complement activation. Used in assays to detect specific antibody.
complement fixing antibodies	Antibodies that efficiently trigger the classical complement activation pathways. In humans: IgM, IgG1, IgG2, and IgG3.
complete carcinogen	Carcinogen which can drive all four stages of *carcinogenesis*.
complete clinical response (hematopoietic cancer)	The number of blast-like cells in the patient's post-treatment blood or bone marrow smear has been reduced to almost zero or ≤5%, respectively.
concordance rate	A measure of the frequency of the same disease developing in two monozygotic or dizygotic twins.
conditional knockout mouse	Mutant mouse in which the gene of interest is rendered non-functional ("knocked out") only at a particular developmental stage or in a particular cell type or tissue upon induction by a specific stimulus. See *knockout mouse*.
conformational determinant	B cell *epitope* in which the contributing amino acids are located far apart in the linear sequence but which become juxtaposed when the protein is folded in its native shape.
congenic	Two individual animals that carry identical *alleles* at all but one locus. Created by backcrossing and selecting for the desired trait for several generations.
congenital neutropenia	*PIs* of the *innate immunity* that lead to *neutropenia* that is constant (severe congenital neutropenia; SCN) or intermittent (cyclic neutropenia; CyN). Caused by mutations in the *neutrophil* elastase gene involved in extracellular matrix dissolution.
conjugate	A structure resulting from the physical joining of two other structures. Two cells may form a conjugate, such as Th and B cells during B–T cooperation, or *CTLs* and target cells during cytolysis. Two molecules may form a conjugate, such as the covalent joining of a *hapten* to a *carrier* macromolecule to induce an immune response.

Term	Definition
conjugate vaccine	*Vaccine* based on the covalent linkage of a pathogen carbohydrate epitope to a protein carrier such that it induces a B cell response to *Td antigens* rather than *Ti antigens*.
constant domain	A domain of an Ig or TCR chain that is encoded by the corresponding *constant exon*. The constant domains have very little amino acid variability and they define the C-terminal *constant regions* of the Ig and TCR chains.
constant exon	Exon encoding the C-terminal portion of either an Ig or a TCR protein. A C exon is spliced at the mRNA level to a rearranged *variable (V) exon* to produce a transcript of a complete Ig or TCR gene. The C exons of the Ig heavy chain gene locus are composed of *sub-exons*.
constant region	The relatively invariant C-terminal portion of an Ig or TCR molecule comprised of *constant domains*.
contact hypersensitivity	*Type IV hypersensitivity* caused by an immune response to a chemically reactive *hapten* that has bound covalently to self proteins in the uppermost layers of the skin to form a *neo-antigen*.
core (HIV)	For *HIV*, the genomic RNA with its associated proteins, the *capsid* and the *matrix*.
coreceptor	Protein that facilitates or enhances the binding of a primary receptor to a ligand. The T cell coreceptors CD4 and CD8 bind to non-polymorphic sites on MHC class II and I, respectively, that are outside the peptide-binding groove. This binding stabilizes the contact between the pMHC and the TCR, and also recruits PTKs that initiate intracellular signaling. The CD19–CD21–CD81 complex functions as a *BCR* coreceptor. *NKG2D* may function as a $\gamma\delta$ *TCR* coreceptor.
cortex	The outer layer of an organ such as the *thymus*. The thymus possesses an outer and inner cortex.
cortical thymic epithelial cells (cTECs)	See *thymic epithelial cells*.
corticosteroids	Molecules that mediate powerful anti-*inflammatory* and *immunosuppressive* effects when they bind to glucocorticoid receptors and inactivate *transcription factors* such as NF-κB, AP-1, NF-AT, and the STATs.
costimulation	The second signal required for completion of lymphocyte activation and prevention of *anergy*. Supplied by engagement of CD28 by B7-1/7-2 (T cells), and of CD40 by CD40L (B cells). Costimulatory signaling lowers the activation threshold of a lymphocyte, supports *chromatin remodeling*, and promotes and sustains TCR or BCR signaling.
costimulatory blockade	Artificial interference with the delivery of *costimulation* signals such that antigen-specific T lymphocytes are anergized.
costimulatory molecules	Molecules involved in mediating the *costimulation* of a cell.
coverage (vaccine)	See *efficacy*.
Cowden syndrome	Inherited susceptibility to the development of various cancers caused by germline mutations of the *tumor suppressor gene PTEN*.
C-reactive protein	An *acute phase protein*.
Crohn's disease	Chronic *inflammatory bowel disease* affecting all layers of the wall of the colon and small intestine. Healthy patches of bowel may be interspersed with diseased patches. Thought to be caused by an autoimmune response.
cross-linking	Cross-linking occurs when multiple identical molecules are linked together by the binding of multivalent ligands such as antibodies. Both soluble and cell-surface structures may be cross-linked.
cross-presentation	Presentation of peptides from exogenous antigens on the MHC class I molecules of DCs and macrophages. May involve the fusion of a phagosome containing exogenous antigen with an ER-derived vesicle containing proteins of the endogenous antigen processing machinery. May also occur via *peptide regurgitation* or *peptide interception*.
cross-priming	Activation of a CD8$^+$ Tc response through *cross-presentation* of exogenous antigen by a DC or macrophage.
cross-reactivity	Recognition by a lymphocyte or antibody of an antigen other than the *cognate antigen*. Cross-reactivity results either when the same *epitope* is found on two different antigens, or when two epitopes on separate antigens are similar.

Term	Definition
cSMAC	= Central SMAC. Innermost ring of the *SMAC*. Contains triggered TCRs, pMHC, *coreceptors* and associated kinases, and *lipid rafts*.
C-type lectin	A *lectin* that requires calcium to bind to carbohydrate ligands.
cutaneous immunity	Immune response mediated by diffuse collections of APCs and lymphocytes in the *SALT* that can respond to pathogens attacking the skin without necessarily involving the draining lymph node.
CVID	See *common variable immunodeficiency*.
cytokine receptors	Families of cell surface receptors that bind to specific *cytokines* and initiate *intracellular signaling pathway*. Cytokine receptors are generally multimeric. "Public" subunits are chains shared among a group of cytokine receptors and usually carry out signal transduction. "Private" subunits are unique chains that usually dictate binding specificity.
cytokines	Low molecular weight, soluble proteins that bind specific cell surface receptors whose engagement leads to activation, proliferation, differentiation, effector action, or death of the cell. Synthesized by leukocytes and non-hematopoietic cells under tight regulatory controls. Exert their effects in a *paracrine* or *autocrine* fashion.
cytolytic T lymphocytes (CTLs)	*Effector* T cell progeny of an activated Tc cell. CTLs recognize and destroy target cells displaying foreign peptide complexed to MHC class I. Target cell killing occurs via *cytotoxic cytokine* secretion and *perforin/granzyme-mediated cytotoxicity*.
cytopathic virus	A *virus* that damages the host cell during infection.
cytotoxic cytokines	Cytokines that directly induce the death of cells. For example, TNF and LTα.
danger signals	Molecules released by stressed or dying cells of host or pathogen origin that bind to *PRMs* of innate response cells and induce inflammation. Immature DCs exposed to danger signals undergo maturation and upregulate *costimulatory molecules*, allowing them to activate, rather than *tolerization*, T cells. Signals include bacterial CpG DNA, ds and ss viral RNA, *inflammatory cytokines* and *chemokines*, internal *ROI*, products of *complement* activation, and *heat shock proteins*.
dark zone	Region of a *germinal center* where *somatic hypermutation* occurs.
death by neglect	Process by which DP thymocytes with non-functional TCRs undergo apoptosis. Also called "death by non-selection."
death receptor	Cell surface receptor whose engagement triggers *apoptosis*. Examples include *Fas, TNFR1, DR5*.
delayed type hypersensitivity (DTH)	Immunopathological damage occurring 24–72 hours after exposure of a sensitized individual to an antigen. Characterized by the infiltration of Th1 cells, macrophages, and basophils, and *granuloma* formation. See also *type IV hypersensitivity*.
delayed xenograft rejection	Loss of a *xenograft* a few days or weeks after transplantation. Caused by *ischemia* resulting from hyperactivation of the graft vascular endothelium induced by the binding of anti-galactose-α (1-3) galactose antibodies. Also known as "acute vascular rejection."
deletional joining	Occurs during *V(D)J recombination* when the two *gene segments* to be brought into apposition are in the same transcriptional orientation.
delivery vehicle	Inert, non-toxic structure designed to protect *vaccine* antigens from nuclease or protease-mediated degradation. May also act as an *adjuvant* or increase antigen display. Includes *liposomes, virosomes, ISCOMs, SMAAs*, and *biodegradables*.
dendritic cells (DCs)	Irregularly shaped phagocytic leukocytes with finger-like processes resembling the neuronal dendrites (except plasmacytoid subset). DC subsets arise from both the myeloid and lymphoid lineages and include conventional and plasmacytoid DCs. Immature DCs in the tissues capture antigen. DCs activated by pro-inflammatory cytokines migrate to the draining lymph node and mature, upregulating expression of costimulatory molecules. Mature DCs are the only APC capable of activating naive T cells. *Modulated DCs* mediate *peripheral tolerance*.
de novo pathway of nucleic acid synthesis	Biosynthetic process by which new nucleotides are built from amino acids; inhibited by aminopterin.
depot effect	The ability of an adjuvant to induce granuloma formation that allows a slow release of a vaccine antigen.

Term	Definition
dermis	Lower layer of skin beneath the *epidermis* and *basement membrane* containing nerve fibers, blood vessels, lymphatic vessels, and scattered fibroblasts, macrophages, DCs, mast cells, and αβ T cells. The papillary dermis lies just below the basement membrane and the reticular dermis lies below the papillary dermis. Both layers are formed from networks of collagen and elastin fibers embedded in a hyaluronic acid matrix.
desmogleins	Protein components of the *desmosomes*.
desmosomes	Specialized junctions between *keratinocytes* that physically connect these cells and ensure that each layer of keratinocytes divides and migrates upward as a unit.
DETC	= Dentritic epidermal T cells. See *intraepithelial lymphocytes*.
determinant selection model	Model that holds that variations in immune responsiveness to a given antigen in different individuals correspond to the ability of each individual's MHC alleles to effectively bind and present determinants from that antigen.
deuterostomes	*Coelomate*–derived animals that became the *vertebrates*.
diabetes mellitus	Metabolic disorder resulting when the body does not produce insulin or its cells cannot respond to insulin normally. In the absence of insulin, glucose cannot get into cells and blood sugar levels rise. Symptoms such as excessive thirst, excessive hunger, emaciation, and weakness can result. The patient can eventually fall into a "diabetic coma" brought on by *ketoacidosis*. See *type 1* and *type 2 diabetes*.
diapedesis	Second step of *extravasation* in which flattened leukocytes squeeze between the endothelial cells of a post-capillary venule and through their basement membranes, emigrating into the tissues.
differential RNA processing	Mechanism at the polyadenylation level by which two structural versions of a protein can be derived from a single gene. Multiple polyadenylation sites are positioned after two or more exons. Depending on which site is chosen for cleavage and polyadenylation, primary RNA transcripts of different lengths (including different subsets of exons and introns) will be generated, leading to proteins with alternative C-terminal amino acid sequences.
differentiation	Functional specialization of a developing precursor cell or an activated cell.
differentiation antigen	A protein or carbohydrate antigen associated with only a particular developmental stage. Inappropriate appearance in a tumor cell can make it a *tumor-associated antigen*.
DiGeorge syndrome	Clinically heterogeneous disorder including a *PI* component. The thymus does not develop fully or at all. Some DiGeorge cases may involve mutations in the chromosome 22q11 region.
dilated cardiomyopathy	Cardiac disease in which the ventricular diameter of the heart becomes enlarged and the heart muscle is weakened such that it no longer pumps blood efficiently. Fatal organ damage or heart failure can result.
direct allorecognition	A transplant recipient's T cells recognize peptide/allo-MHC epitopes presented on the surfaces of *allogeneic* donor cells in the graft. See *allorecognition*.
direct amplicon analysis	In *tissue typing*, the characterization of an unknown HLA allele by determining the specific *PCR* primers that are able to amplify a region of the gene known to be *polymorphic*. Detection of amplification products is by gel electrophoresis and ethidium bromide staining.
direct tag assays	Single step procedures in which a *tagged* antigen (or antibody) is used to detect the presence of its untagged antibody (or antigen) binding partner.
DN thymocytes	= Double negative *thymocytes*. Thymocytes that express neither CD4 nor CD8 molecules, giving them a CD4⁻CD8⁻ surface phenotype. Encompasses four subpopulations: DN1–4.
DNA binding motif	Binding site for a nuclear *transcription factor* found at various places in genomic DNA. Examples include κB (which binds NF-κB) and ISRE (interferon-stimulated response element; binds IRFs).
DNA repair	Removal of mutations in DNA to maintain genomic stability. DNA repair is mediated in large part by the *NHEJ*, *homologous recombination*, and *nucleotide excision pathways*.
dome	Region of mucosal lymphoid follicle overlying a germinal center. Contains DCs, macrophages, CD4⁺ T cells, regulatory T cells, and mature B cells.

Term	Definition
dominant negative transgene	A transgene expressing a non-functional form of a protein that interferes with the function or expression of the endogenous protein. See *transgenic mouse*.
donor cell infusion	Donor BM or a purified donor HSC preparation is administered to a *solid organ transplant* recipient to promote *chimerism* and acceptance of donor cells.
DP pause	Stage of T cell development when the TCRα genes in the earliest *double positive* (DP) *thymocytes* are still undergoing TCRα gene rearrangement on both chromosomes. Over a period of 3–4 days, a DP cell produces candidate TCRα chains and carries out *productivity testing* with the functional TCRβ chain it already expresses.
DP thymocytes	= Double positive *thymocytes*. Thymocytes that express both CD4 and CD8 molecules, giving them a CD4$^+$CD8$^+$ surface phenotype.
DRiP pathway	= Defective ribosomal products pathway. *Antigen processing* pathway in which polypeptides that are misfolded are rapidly ubiquitinated and digested by standard *proteasomes* such that the resulting DRiP peptides become associated with MHC class I molecules.
drug hypersensitivity	*Hypersensitivity* induced by immune responses to *conjugates* of host proteins with drug metabolites.
DTH reaction	Delayed type *hypersensitivity*. See *type IV hypersensitivity*.
dual tropic HIV	*HIV* strains that can infect macrophages, primary T cells, and immortalized T cell lines in culture.
EAE	See *experimental autoimmune encephalitis*.
EAM	See *experimental autoimmune myocarditis*.
EAMG	See *experimental autoimmune myasthenia gravis*.
early phase reaction	In the *elicitation* phase of *type I hypersensitivity*, the rapid onset of clinical symptoms induced by preformed inflammatory mediators immediately released via *mast cell* degranulation.
early signalosome	The early complex of TCR signal transduction molecules that can participate in either the *immunosome* or the *tolerosome*.
EAT	See *experimental autoimmune thyroiditis*.
EAU	See *experimental autoimmune uveoretinitis*.
ectopic	A protein or structure is *ectopic* if it is expressed or found in an abnormal location in the body.
eczema	See *atopic dermatitis*.
edible vaccine	Component of a plant that has been genetically manipulated to express a pathogen antigen. Can be ingested when the plant is processed into a form of food.
effector B cell	See *effector lymphocytes (effector cells)*.
effector T cell	See *effector lymphocytes (effector cells)*.
effector functions	The actions *effector cells* take to eliminate foreign entities. For example, the effector functions of Th1/Th2 cells, CTLs, and plasma cells are cytokine secretion, cytotoxicity, and antibody production, respectively.
effector lymphocytes (effector cells)	The differentiated progeny of an activated leukocyte that act to eliminate a nonself entity. *Plasma cells, Th cells,* and *CTLs* are effector cells that differentiate from proliferating B, Th, and Tc cells, respectively. These effector cells generally have lower *costimulation* requirements than corresponding resting naive cells and express different *homing* and *chemokine receptors* that allow access to inflammatory sites. *NKT* and *NK* cells can act as effectors without having to proliferate.
effector site	Remote mucosal location where lymphocytes activated in a mucosal *inductive site* differentiate into effector cells and exert effector actions (such as antibody production).
effector stage (hypersensitivity)	See *elicitation*.
efferent lymphatic vessel	A vessel that conveys *lymph* away from a *lymph node*.
efficacy (vaccine)	The ability of a vaccine to effectively protect individuals from disease. Expressed as the percentage of individuals vaccinated that develop immune responses to the pathogen. Often judged by post-vaccination *seroconversion*. Also called "coverage."
electrostatic bond	A bond formed by the attraction between charged residues with opposite polarities.

Term	Definition
elicitation	The excessive, abnormal secondary response to a *sensitizing agent* that results in inflammatory tissue damage. Also called "effector stage." See *hypersensitivity* and *type I–IV hypersensitivities*.
ELISA	= Enzyme-linked immunosorbent assay. *Binder-ligand* assay in which the antibody or antigen used is linked to an enzyme. The presence of binder-ligand pairs can be determined by adding a chromogenic (color-producing) substrate whose enzymatic conversion causes a detectable color change.
embryonic antigens	A protein or carbohydrate antigen normally expressed solely in the embryo. Inappropriate expression in a tumor cell can make it a *tumor-associated antigen*.
embryonic stem (ES) cells	Early embryonic cells with the potential to give rise to any cell type.
emerging infectious disease	A disease with the potential to have great impact on the global population because it is caused by a newly developed pathogen or one that is changing in character.
encapsulated bacteria	Bacteria with a polysaccharide coating (capsule) that inhibits phagocytosis.
encapsulation	A mechanism mediated by *PRRs* in *higher invertebrates* by which pathogens too large to be phagocytosed are surrounded by many phagocytic cells, sealing them off from the hemolymph. Encapsulated pathogens are then killed by phagocyte ROI and lysosomal enzymes. Enhanced by molecules produced by *granular cells*.
endocrine	A molecule that acts in an *endocrine* fashion affects cells systemically or over substantial distances.
endocytic pathway	Intracellular system of membrane-bound vesicles, including endosomes and endolysosomes, that contain hydrolytic enzymes and other molecules and substances necessary for the digestion of internalized materials. Responsible for *exogenous antigen processing and presentation*.
endogenous antigen	An antigen that originates within a cell in the host, as in a protein synthesized in a cell infected by a virus or intracellular bacterium.
endogenous antigen processing and presentation	A mechanism by which endogenous antigens in the cytosol are degraded into peptides via *proteasomes* and complexed to MHC class I in the rER. The peptide–MHC class I complex is then displayed on the cell surface. This pathway operates in almost all nucleated cell types. Also known as the "cytosolic antigen processing pathway."
endotoxic shock	A sometimes fatal collapse of circulatory and metabolic systems induced by overwhelming amounts of cytokines (particularly IL-1 and TNF) released into the circulation by macrophages in response to infection with *gram-negative bacteria* or their components (especially *LPS*). Also known as "septic shock."
endotoxin	Toxins released from the cell walls of damaged *gram-negative bacteria*. Endotoxins induce the production of cytokines that cause fever, increase capillary permeability, and may lead to *endotoxic shock*. The lipid A portion of *lipopolysaccharide* (*LPS*) is the most significant endotoxin.
enhanceosome	A multi-protein transactivator complex that binds to an active promoter to initiate or support gene transcription. See *SXY-CIITA regulatory system*.
enhancer	Short region of DNA containing a collection of *DNA-binding motifs*. Nuclear transcription factors that bind to an enhancer may influence the transcription of a target gene on the same chromosome positively or negatively. Enhancers can be positioned external to or within a gene, and may be active only within certain cell types or at certain times.
enterocytes	Gut epithelial cells that actively absorb nutrients. Express *Toll-like receptors* and produce *cytokines* that support *iIEL* activation.
envelope (in HIV)	Phospholipid bilayer surrounding a viral *core*. In the case of *HIV*, the envelope contains spikes made up of the HIV proteins *gp41* and *gp120*.
enzyme replacement therapy	Treatment in which a genetically deficient individual, such as a *PI* patient, is injected with a stabilized form of a missing enzyme whose activity in the blood is sufficient to mitigate disease.
eosinophils	Connective tissue granulocytes with bilobed nuclei and granules that stain reddish with acidic dyes. The granules contain highly basic proteins and enzymes effective in the killing of larger parasites. Eosinophils also play a role in allergy.
epidermis	Top layers of skin that contain mainly *keratinocytes* plus elements of *SALT*, but no blood vessels.

Term	Definition
epitope	The small region of a macromolecule that specifically binds to the antigen receptor of a B or T lymphocyte. B cell epitopes can be comprised of almost any structure, including native and denatured proteins, chemical groups, lipids, carbohydrates, nucleic acids, etc. T cell epitopes are a complex of antigenic peptide bound to either MHC class I or II.
epitope spreading	An immune response against one epitope causes tissue destruction that exposes previously hidden epitopes, activating additional lymphocyte clones. Spreading may be inter- or intramolecular. Plays a role in *tumor regression*, *chronic graft rejection*, and autoimmune responses.
equilibrium dialysis	An experimental method that can be used to measure the *affinity* of an antibody for an antigen. The antibody is placed at a fixed concentration in a series of dialysis bags that are then suspended in solutions containing known concentrations of tagged antigen. The amount of antigen that has diffused into each bag at equilibrium can be used to calculate the antibody affinity.
erythrocytes	Red blood cells.
erythroblastosis fetalis	See *Rh disease*.
erythropoietin (EPO)	Growth factor promoting the differentiation of CFU-E precursors and eventually mature erythrocytes.
evasion strategies	Pathogen mechanisms designed to avoid or compromise the host's immune response.
exocytosis	Process by which the membrane of an exocytic vesicle fuses with the plasma membrane, everting the contents into the extracellular fluid.
exogenous antigen	An antigenic protein that originates outside the cells of the host, as in a bacterial toxin.
exogenous antigen processing and presentation	*Exogenous antigens* are internalized into an APC, degraded within the *endocytic pathway*, and complexed to MHC class II in an endolysosomal vesicle. The peptide–MHC class II complex is then displayed on the cell surface. This pathway operates almost exclusively in APCs.
exotoxin	Toxic protein actively secreted by a gram-positive or gram-negative *bacterium*.
experimental autoimmune encephalitis (EAE)	Mouse model of human *multiple sclerosis*. Can be induced by injection of C57BL/6 or SJL mice with CFA plus MOG (a peptide from the myelin oligodendrocyte glycoprotein).
experimental autoimmune myasthenia gravis (EAMG)	Animal model of human *myasthenia gravis*. Induced by immunizing various genetically susceptible strains of mice, rats, or rabbits with purified acetyl choline receptor protein (or its peptides) from the electric organ of the pacific electric ray *Torpedo californica* plus *CFA*.
experimental autoimmune myocarditis (EAM)	Mouse model of human *autoimmune myocarditis* (inflammation of the heart). Induced by infection of A/J and BALB/c mice with Coxsackie B3 virus. Immunization of susceptible mice with cardiac myosin also results in EAM.
experimental autoimmune thyroiditis (EAT)	Mouse model of *Hashimoto's thyroiditis* in humans. Induced by immunizing substrains of CBA/J mice with CFA plus either whole thyroglobulin protein or its peptides.
experimental autoimmune uveoretinitis (EAU)	Animal model of human *autoimmune uveitis*. Induced by immunizing various genetically susceptible strains of rats, guinea pigs, monkeys, or mice with purified retinal antigen (or its fragments) plus adjuvant.
experimental tolerance	Lack of an immune response to a foreign antigen induced by treatment of a mature animal with either a non-immunogenic form of the antigen via the usual route of immunization, or an immunogenic form of the antigen via a non-immunogenic route of administration.
extracellular antigen	See *exogenous antigen*.
extracellular pathogen	A pathogen that does not enter host cells, but reproduces in the interstitial fluid, blood, or lumens of the respiratory, urogenital, and gastrointestinal tracts.
extramedullary plasmacytoma	A *plasma cell dyscrasia* in which the cancerous cells have moved outside the bone marrow and infiltrated other tissues.
extramedullary tissues	Tissues and organs that are neither *lymphoid tissues* nor *bone marrow*.
extranodal lymphoma	*Lymphoma* arising in a diffuse *lymphoid tissue*, such as that associated with the GI tract.
extrathymic T cell development	Development and maturation of T cells in tissues outside the thymus.
extravasation	Exit of leukocytes from the blood circulation into the tissues in response to inflammatory signals. The two steps of this process are *margination* and *diapedesis*.

Term	Definition
extrinsic apoptotic pathway	See *apoptosis*.
F1 hybrid resistance	F1 hybrid resistance is said to occur when two inbred strains of mice are crossed and a parental bone marrow graft from either of the parents is subsequently rejected by the F1 offspring. Due to variation in *NK inhibitory receptor* expression by F1 *natural killer (NK) cells*.
Fab fragment	= Fragment, antigen binding. N-terminal portion of Ig molecule left after digestion with papain. Contains the two antigen-binding sites of the antibody.
FAE	See *follicle-associated epithelium*.
familial cancer	Cancer that develops with increased frequency in related individuals. Established when a genetic alteration that increases the chances of *malignant transformation* occurs in a germ cell. This genetic predisposition toward malignancy then becomes a heritable trait that leads to increased incidence of particular cancers among descendants. Also called "inherited" or "hereditary" cancer.
familial Mediterranean fever (FMF)	Autoinflammatory *PI* due to inherited mutations of the inflammatory regulator marenostrin.
Fas pathway	An apoptotic pathway induced by ligation of the death receptor Fas by FasL, resulting in death of the Fas-expressing cell. Signal transduction is mediated by FADD, caspase-8, and the *caspase-3 cascade*.
Fc fragment	= Fragment, crystallizable. C-terminal portion of Ig molecule left after digestion with papain; crystallizes at low temperature. Contains the *constant region* of the antibody.
Fc receptor (FcR)	Family of structurally diverse receptors that bind to the Fc regions of a specific antibody isotypes. Includes isoforms of FcαR, FcγR, and FcεR. FcRs are expressed on effector cells and trigger signaling that mediates *receptor-mediated endocytosis, phagocytosis*, and *ADCC*, as well as *neutrophil* and *mast cell* degranulation and the release of inflammatory mediators and cytokines. Many FcRs are members of the *immunoglobulin(Ig) superfamily* and contain *ITAMs*.
FcRn (neonatal)	FcR expressed on cells of the neonatal gut that binds the IgG in breast milk, mediating its passage into the neonatal gut lumen. Also facilitates IgG transport in adult intestine and the placenta. Originally known as the Brambell receptor.
FDCs	See *follicular dendritic cells*.
fetal-maternal tolerance	Tolerance of a mammalian mother for her fetuses despite their expression of allogeneic paternal MHC. A fetus implanted in the uterus resembles a tissue graft that is not rejected. Maintained by *immune privilege* and other mechanisms.
fetal thymic organ culture (FTOC)	*In vitro* culture of the excised thymus of a fetal mouse. The cultured thymus remains capable of supporting T cell development.
FISH	See *fluorescent* in situ *hybridization*.
flow cytometry	Use of a specialized electronic instrument (flow cytometer) to count or otherwise evaluate individual cells based on various physical and chemical characteristics such as size, granularity, and binding of fluorochrome-tagged antibodies of defined specificity.
fluorescence-activated cell sorting (FACS)	The fluorescence detectors of a flow cytometer are connected to computer-controlled electromagnetic deflector plates that are programmed to direct a cell with a fluorescent signal (due to binding by fluorochrome-tagged antibody) of a given wavelength (color) and intensity to a particular collection tube.
fluorescent in situ *hybridization (FISH)*	Assay in which a fluorescently tagged DNA probe is applied to whole cells or a chromosomal spread on a microscope slide to visualize a whole chromosome or a specific region of a chromosome. Often used for the examination of tumorigenic *chromosomal translocations* since major aberrations become clearly visible. Multiple probes labeled with distinct fluorochromes may be used simultaneously.
fluorochromes	Molecules that fluoresce, i.e., they absorb light at a particular wavelength and subsequently emit that light at a longer wavelength (i.e., at a lower energy) of distinctive color.
follicle-associated epithelium (FAE)	Epithelium lying directly over single or aggregated lymphoid follicles that is specialized for *transcytosis* due to the presence of *M cells*.
follicular B cells	Mature B cells found in lymphoid follicles of *secondary lymphoid tissues*. Respond to Td antigens and generate *memory* B cells, and both short- and long-lived *plasma cells*.

Term	Definition
follicular dendritic cells (FDC)	Distinct lineage of DCs found in B cell-rich areas of lymphoid organs. FDC do not internalize antigen and do not function as APCs but rather trap antigen–antibody complexes on their cell surfaces and display them for extended periods.
food allergies	*Type I hypersensitivity* to *allergens* in consumed foods. Usually peanuts, soy, milk, eggs, wheat, or fish.
forward genetics approach	Process of establishing the role of a gene in a given disease by mutating it in an experimental animal and observing if the mutation causes the disease of interest.
framework regions	Relatively invariant parts of the *V domain* of an Ig or TCR chain that are outside the *hypervariable regions*.
Freund's adjuvant	Complete Freund's adjuvant (CFA): antigen is emulsified in paraffin oil plus an aqueous suspension of mycobacterial antigen. Incomplete Freund's adjuvant (IFA): antigen is emulsified in paraffin oil plus surfactant (no mycobacterial antigen). Both CFA and IFA are used in experimental animals only.
frustrated phagocytosis	A process in which a *phagocyte* that cannot engulf an entity because it is too large or is fixed in a tissue releases the contents of its lysosomes externally, leading to localized tissue damage.
full chimerism	Situation in which 100% of hematopoietic cells in an individual are of donor origin. May be established by BMT/HCT in which *myeloablative conditioning* has been used prior to transplantation to destroy all recipient hematopoietic cells.
fungi	Eukaryotes that can exist comfortably outside a host but that will invade and colonize a host if given the opportunity. May be single-celled and spherical (yeast) or multicellular and filamentous (bread molds).
GALT	See *gut-associated lymphoid tissue* and *MALT*.
gamma-delta (γδ) *T cells*	T lymphocytes bearing γδ *TCRs* that recognize broadly expressed pathogen antigens in their natural, unprocessed forms. γδ T cells respond more rapidly than αβ T cells to antigen, proliferate less, and require little or no conventional costimulation. Considered sentinels of innate immunity and a bridge to the adaptive immune system.
gamma-delta (γδ) *TCRs*	Heterodimer of TCRγ and δ chains plus the *CD3 complex*. Antigens include *MICA/MICB*, Hsps, T10/22 proteins, modified phospholipids, lipoproteins, phosphorylated oligonucleotides, pyrophosphate-like epitopes, alkyl amines, pathogen peptides, or whole proteins. Some γδ TCRs recognize *CD1c*. See *TCR complex*.
gamma globulins	The third of three fractions of plasma proteins (α, β, γ) that are defined by their electrophoretic mobility. Most immunoglobulins are found in the gamma globulin fraction.
gene amplification	A gene or its chromosomal region is duplicated one or more times to generate multiple copies of the gene in the genome.
gene gun	Bombardment device that applies a sudden motive force to propel miniscule plasmid coated gold particles directly into the cytoplasm of living cells. The antigen encoded by the plasmid is then synthesized by the cells. Often used in experimental *vaccination* protocols.
gene segment	A short, germline sequence of DNA from either the variable (V), diversity (D), or joining (J) families that randomly joins via *V(D)J recombination* with one or two other gene segments to complete a *V exon* in either the Ig or TCR loci. Gene segments lack RNA splice donor and acceptor sites.
gene therapy	Treatment of a genetic defect by introduction of a normal copy of the faulty gene into somatic cells.
genetic escape mutant	A pathogen that has evaded the immune response or drug therapy by mutating a gene encoding a protein that normally confers either antigenicity or drug susceptibility, respectively.
genomics	Study of the entire genome of an organism.
germinal centers (GC)	Aggregations of rapidly proliferating B cells and differentiating memory B and *plasma cells* that develop in the *secondary lymphoid follicles*. Site of *isotype switching, somatic hypermutation,* and *affinity maturation*.
germ theory of disease	Concept that pathogenic microbes invisible to the naked eye (bacteria, viruses, fungi, and parasites) are responsible for specific illnesses.

Term	Definition
gld mouse	Natural mouse mutant with *FasL* mutation. Develops *autoimmune disease* with aspects of human *SLE* and *ALPS*.
Gleevec	A drug (imatinib mesylate) that inhibits the activity of several kinases, including the chimeric Bcr–Abl kinase in *CML*.
glycocalyx	A thick *glue*-like layer of mucin-related molecules anchored in the apical membrane of the *brush border of gut epithelial cells*. Contains hydrolytic enzymes and has a negative charge that repels pathogens.
goblet cells	Mucus-producing cells in the gut epithelium that contribute to *mucosal immunity*.
goiter	Enlarged *thyroid gland*.
Goodpasture's syndrome	An *autoimmune* disease involving *type II hypersensitivity* mediated by *autoantibodies* directed against collagen epitopes found in the basement membranes of lung and kidney cells.
gp41	An HIV glycoprotein that anchors *gp120* in the *HIV envelope*.
gp120	An HIV glycoprotein that is the principal envelope protein of *HIV*. Binds to host cell receptors and contains *immunodominant epitopes*.
graft rejection	Attack by recipient immune system on transplanted donor tissue (a graft). See *hyperacute, acute*, and *chronic graft rejection*.
graft versus host disease (GvHD)	Attack by immunocompetent cells in transplanted donated tissue (a graft) on recipient tissues due to *MHC* or *MiHA mismatching*. Most common in *BMT/HCT*, but rare in *solid organ transplants*.
graft versus leukemia (GvL) effect	In BMT/HCT for cancer treatment, destruction of residual recipient *leukemia* cells by T cells from an *allogeneic* donor. A highly desirable consequence of *MiHA* and *MHC* mismatches between recipient and donor.
gram-negative bacteria	Bacteria that have thin cell walls containing peptidoglycan and *LPS* such that they stain red with gram staining procedures.
gram-positive bacteria	Bacteria that have thick cell walls containing peptidoglycan and lipotechoic and techoic acids such that they stain purple with gram staining procedures.
granular cells	*PRR*-expressing *hemocytes* in *higher invertebrates* that mediate wound healing and *coagulation* to trap pathogens. Also produce bactericidal proteins and *opsonins* that assist phagocytic cells in *encapsulation*. Also called "hemostatic cells."
granule exocytosis pathway	A mechanism used by armed CTLs to kill target cells. After conjugate formation and TCR triggering, the CTL reorients its Golgi so that its pre-formed cytotoxic granules fuse with the CTL membrane. The contents of the granules, chiefly perforin and the granzymes, are directionally exocytosed and released toward the target cell membrane. The target cell undergoes apoptosis. See *perforin/granzyme mediated cytotoxicity*.
granulocytes	*Myeloid* leukocytes that harbor large intracellular granules containing microbe-destroying hydrolytic enzymes. Include *neutrophils, basophils*, and *eosinophils*.
granuloma	Structure formed by a group of *hyperactivated macrophages* that fuse together to wall off a persistent *pathogen* from the rest of the body. Also contains CD4$^+$ and CD8$^+$ T cells. Formation depends on TNF produced by activated *Th1* effectors.
granzymes	Serine esterases released from the granules of cytotoxic lymphocytes that induce apoptosis upon entry into the cytosol of a target cell. Also called "fragmentins." See *perforin/granzyme-mediated cytotoxicity*.
Graves' disease	*Autoimmune disease* in which the *thyroid gland* becomes hyperactive and overproduces *thyroxine*, causing hyperthyroidism. Autoantibodies against thyroid antigens, including thyroid-stimulating hormone receptor, are present.
growth factors	Messenger molecules produced inducibly or constitutively by leukocytes and non-leukocytes rather than by glands. Often present at significant levels in the circulation. Influence a broad range of cell types to initiate and/or sustain proliferation.
Guillain-Barré syndrome (GBS)	*Autoimmune disease* resulting from acute autoantibody attacks on antigens of peripheral nerves. Characterized by demyelination and reduced electrical impulse transmission leading to muscle weakness and paralysis. Usually self-limiting. Linked to recovery from certain pathogen infections.

Term	Definition
gut-associated lymphoid tissue (GALT)	The Peyer's patches, the appendix, and diffuse collections of immune system cells in the linings of the small and large intestine. See *MALT*.
H-2 complex	Murine *major histocompatibility complex*. Contains the K, P, A, E, S, D, Q, T, and M loci, each containing multiple genes.
HAART	= Highly active anti-retroviral therapy. A drug treatment regimen for HIV patients that combines at least two different types of *anti-retroviral drugs* in a cocktail.
haemolin	*Higher invertebrate* protein occurring both as a soluble molecule in the *hemolymph* and as a membrane-bound protein on the surface of phagocytic hemocytes. Soluble haemolin opsonizes whole pathogens or LPS and promotes *phagocytosis*.
haplotype	An array of linked genes on one chromosome. An MHC haplotype is the set of MHC alleles contained on a single chromosome of an individual.
hapten	A small molecule that can bind an antigen receptor but is not immunogenic on its own. Hapten-specific B and T cell responses can be induced if the hapten is conjugated to a *carrier* macromolecule.
HAR	See *hyperacute graft rejection*.
Hashimoto's thyroiditis (HT)	*Autoimmune disease* in which a cell-mediated response destroys thyroid follicular cells such that insufficient levels of *thyroxine* are produced, leading to hypothyroidism. Autoantibodies against thyroid antigens are present. See *thyroid gland*.
HAT selection	Culture of cells in medium containing hypoxanthine, aminopterin, and thymidine to select cells expressing both hypoxanthine-guanine phosphoribosyl transferase (HGPRT) and thymidine kinase (TK). Used in *somatic hybrid selection* for *hybridoma* production.
HCT	See *hematopoietic cell transplant*.
heat shock proteins (HSPs)	Family of proteins whose expression is upregulated in cells subjected to environmental stresses such as heat, inflammation, or cancerous transformation. Within cells, HSPs chaperone peptides from the *proteasome* to *TAP* during *endogenous antigen processing*. HSPs can also be released externally from stressed cells as "danger signals" binding to TLRs. May promote *exogenous antigen processing* and *cross-presentation*.
hemagglutination	An *agglutination* assay in which the target cells to be clumped are RBCs.
hematopoiesis	The generation of *hematopoietic cells* from HSCs in the bone marrow or fetal liver. Involves a continuum of identifiable intermediates called progenitor and precursor cells. These intermediates progressively lose their capacity to self-renew and to differentiate into multiple cell types.
hematopoietic cancers	Immune system cells that have undergone *malignant transformation*. See *carcinogenesis*.
hematopoietic cells	Red and white blood cells.
hematopoietic cell transplant (HCT)	Replacement of a patient's damaged immune system by transplantation of isolated healthy *HSCs* from the peripheral blood or (less often) umbilical cord blood of a *histocompatible* donor. Ideally requires HLA identity, but ABO compatibility is not an issue.
hematopoietic stem cell (HSC)	The common pluripotent hematopoietic precursor that can either self-renew, or commit to differentiating into the *common myeloid progenitor* (CMP) that gives rise to cells of the myeloid lineage (*myelopoiesis*), or the *common lymphoid progenitor* (CLP) that gives rise to cells of the lymphoid lineage (*lymphopoiesis*).
hemocytes	Leukocyte-like cells present in the *hemolymph* of invertebrates.
hemolymph	Fluid in the circulatory system in *invertebrates* that ferries nutrition and waste products around the body.
hemolytic anemia	*Type II hypersensitivity* in which a pathological antibody mediates the lytic destruction of RBCs.
hemophagocytic syndrome (HPS)	PI due mostly to a defect in the *granule exocytosis pathway* of CTLs. Also called "hemophagocytic lymphohistiocytosis."
hepatosplenomegaly	Enlargement of the liver and spleen due to increased numbers of lymphocytes.

Term	Definition
herd immunity	Protection of non-immune individuals in a population from a given pathogen due to effective vaccination of the majority of the population. The non-immune individuals may be unvaccinated or may not respond to the vaccine. The overall chances of such vulnerable individuals being exposed to the pathogen are much reduced because the pathogen cannot gain a foothold in the vaccinated population.
heterokaryon	Single large cell resulting from fusion of two or more cells that are not genetically identical. Contains one cytoplasm but multiple nuclei.
HEV	See *high endothelial venules*.
high endothelial venules (HEV)	Specialized post-capillary venules in most secondary lymphoid tissues that allow the extravasation of lymphocytes from the blood into these sites. HEV endothelial cells have a plump, cuboidal shape and express adhesion molecules that facilitate the adherence and transendothelial migration of naive lymphocytes.
higher invertebrates	Annelids (segmented worms), molluscs (such as clams, snails, mussels, squid, octopus), arthropods (such as insects, crustaceans).
higher vertebrates	Agnatha (jawless fish such as lampreys); Chondrichthyes (cartilaginous jawed fish such as skates, sharks, and rays); Osteichthyes (bony jawed fish including lobe-finned, ray-finned, and advanced bony fish such as salmon); amphibians (such as frogs, toads, salamanders); reptiles (such as crocodiles, snakes); birds (such as ducks, robins); mammals (such as mice, humans).
high grade tumor	Aggressively growing tumor containing poorly differentiated cells. Usually derived from the *malignant transformation* of a precursor cell. Generally associated with a poor prognosis.
high zone tolerance	*Experimental tolerance* induced by a very large dose of an immunogen.
hinge region	Site in the Ig monomer where the *Fab* region joins the *Fc* region.
histamine	*Vasoactive amine* in mast cell granules. When released by degranulation of the cell, it causes vasodilation and contraction of intestinal and bronchial smooth muscles. Directly toxic to some *parasites*. See *inflammation*.
histiocytes	Macrophages in connective tissues.
histocompatibility	Ability of a recipient to accept a tissue graft from another individual.
HIV	See *human immunodeficiency virus*.
HL	See *Hodgkin's lymphoma*.
HLA complex	Human *major histocompatibility complex* (MHC).
HLA supertype	Defines a group of human MHC alleles whose peptide binding repertoires overlap. Three HLA class I supertypes have been identified and are denoted A2, A3 and B7.
Hodgkin's lymphoma	*Lymphoma* in which the tumor mass is made up of a reactive infiltrate of non-transformed lymphocytes, histiocytes, fibroblasts, and other cell types plus scattered, malignant *Reed-Sternberg cells* that are characteristic of the disease.
"hole in the repertoire" model	Model that holds that a non-responder to a given antigen is missing the required T cell specificity due to tolerance mechanisms. It is assumed that the relevant foreign peptide/self MHC combination very closely resembles a self peptide/self MHC combination in this particular individual, such that any T cell clones that would respond to the antigen are deleted as autoreactive during *negative selection* in the thymus or are tolerized in the periphery. The individual thus has a "hole" its T cell repertoire relative to the repertoire of a responder.
homeostasis	The tendency of a biological system to maintain a steady state. For example, in a healthy individual, a balance of cell proliferation and cell death maintains a constant physiological level of each cell type.
homing receptors	Receptors expressed by lymphocytes that bind to specific *vascular addressins* and direct lymphocyte trafficking. For example, naive lymphocytes expressing L-selectin bind to GlyCAM-1 on *high endothelial venules*.
homokaryon	Multinucleate cell resulting from the fusion of two or more genetically identical cells.

Term	Definition
homologous recombination	The reciprocal exchange of genetic material between two DNA segments of highly similar sequence. For example, integration of engineered plasmid DNA into a gene of interest due to inclusion in the plasmid of sequences homologous to the gene of interest. Used to introduce a mutated version of a gene into mouse *embryonic stem (ES) cells* to generate a *knockout mouse* or a mouse expressing a *knock-in transgene*.
homologous recombination pathway	*DNA repair* pathway that uses a homologous DNA strand to repair large gaps in double-stranded DNA in an "error-free" fashion. Involves the BLM gene (see *Bloom syndrome*).
hormones	Messenger molecules induced by specific stimuli and synthesized in specialized glands. Tend to operate in an *endocrine* fashion and influence only a very limited spectrum of cells.
horror autotoxicus	The concept put forward by Ehrlich that the immune system would "have a horror" of being toxic to its host (i.e., would avoid *autoreactivity*).
HR	See *hypersensitive response*.
HSC	See *hematopoietic stem cell*.
HSC mobilization	Prior to a scheduled HCT, a living donor receives G-CSF or GM-CSF to stimulate *HSC* proliferation in the bone marrow. These cells then "spill out" from the bone marrow into the peripheral blood and can be collected by *leukapheresis*.
human immunodeficiency virus (HIV)	Human retrovirus that destroys T cells and causes acquired immunodeficiency syndrome (AIDS). See also *anti-retroviral drugs, carriage, core, envelope, matrix, Nef, protease, Rev, Tat, Vif, Vpr,* and *Vpu*.
humanized antibodies	Engineered antibodies in which the *variable domains* of non-human mAbs that recognize a human antigen of interest are combined genetically with the *constant domains* of human antibodies, decreasing the *anti-isotypic* response and allowing optimal Fc-mediated effector functions. Alternatively, the *hypervariable* regions of non-human mAbs are engineered into a human antibody framework.
humoral immunity	Adaptive immune responses mediated by B cells that differentiate into plasma cells producing antibodies.
hybridoma	An immortalized cell secreting large amounts of pure *monoclonal antibody* of a single known specificity; created by fusion of a *myeloma* cell with an activated B cell producing antibody of known specificity.
hydrogen bonding	A bond established when the positive charge surrounding a hydrogen atom belonging to a residue in one molecule is shared by an electronegative chemical group of a residue in another molecule.
hydrophobic bond	A bond formed in aqueous solution when polar water molecules force hydrophobic, non-polar chemical groups together in an effort to generate the minimum non-polar surface area possible (and maximize the entropy of the water molecules).
hygiene hypothesis	The theory that excessive zeal in preventing exposure to pathogens in infancy leads to a lack of activation of the immature immune system. The resulting bias toward Th2 responses may predispose an individual to *hypersensitivity* and/or *autoimmunity*.
hyperactivated macrophage	See *macrophage*.
hyperacute graft rejection (HAR)	Extremely rapid destruction of a graft almost immediately after transplantation. Mediated by classical *complement* activation initiated by preformed antibodies recognizing *allogeneic* epitopes on the graft vasculature. A form of *type II hypersensitivity*.
hyperacute xenograft rejection	*HAR* response mediated by human *natural antibodies* that cross-react with galactose-α (1-3) galactose on glycolipids and glycoproteins expressed by pig endothelial cells in a xenograft.
hyper IgD with periodic fever syndrome	Autoinflammatory *PI* characterized by elevated serum IgD. Caused by mutations in mevalonate kinase, an enzyme involved in cholesterol biosynthesis.
hyper IgE syndrome	*PI* of unknown cause characterized by very high levels of circulating IgE.
hyper IgM syndrome (HIGM)	Family of *PIs* characterized by normal or very high levels of circulating IgM but very low or absent levels of all other *isotypes*.

Term	Definition
hyperimmunized individual	Individual who has developed *alloantibodies* to an antigen due to prior exposure, such as a previous graft, blood transfusion, or pregnancy. Occasionally hyperimmunization can be due to the presence of *natural antibodies* that are normally directed against common pathogen determinants but which cross-react with *allogeneic* donor antigens on an incoming graft.
hypersensitive response (HR)	In plants, an anti-pathogen response induced by activation of the *avr–R system* resulting in changes to Ca^{2+} flux, MAPK activation, and NO and ROI production. Plant cells in immediate contact with the pathogen undergo rapid *necrosis*, preventing pathogen spread and releasing hydrolytic enzymes that damage pathogen structures.
hypersensitivity	Excessive immune reactivity to a generally innocuous antigen that results in inflammation and/or tissue damage. Includes *type I–IV hypersensitivities*. Occurs in two stages: *sensitization* and *elicitation*.
hypersensitivity pneumonitis	*Type IV hypersensitivity* in the form of chronic *DTH* in the lungs caused by inhaled antigens.
hypervariable region	Region of extreme amino acid variability in the *V domain* of an Ig or TCR chain. The hypervariable regions largely form the antigen-binding site. Also known as *complementarity determining regions* (CDRs).
hypogammaglobulinemia	Very low levels of antibodies in an individual, resulting in a reduced *gamma globulin* fraction.
hyposensitization	Allergy therapy in which the patient receives subcutaneous injections of ever-increasing amounts of purified *allergen* regularly for 3–5 years. Production of allergen-specific IgE is converted to production of allergen-specific IgG4, which does not trigger *mast cell* degranulation.
hypothalamic-pituitary-adrenal (HPA) axis	The neural circuit responsible for the "fight or flight" response to acute stress. High stress stimulates HPA activation which in turn results in excess glucocorticoid production, suppressed *inflammatory responses*, and increased susceptibility to infections.
H-Y system	The H-Y antigens are encoded on the Y chromosome so they are "self" if expressed in males but "non-self" if expressed in females that are otherwise genetically identical. Used in *negative selection* experiments.
iccosome	A small immune complex-covered particle formed by the "budding off" of a piece of FDC membrane that has trapped multiple immune complexes. Iccosomes released by the FDC can bind to BCRs by virtue of the presence of specific antigen. Iccosomes can be internalized by a B cell and the component antigen presented on MHC class II to a Th cell.
idiotope	*Epitope* formed by an Ig or TCR antigen binding site that is unique to an individual and can induce the production of antibodies when introduced into another member of the same species. Within an individual, an idiotope may represent a *tumor-associated antigen* if it is expressed by the dominant clone of a *leukemia* or *lymphoma*. This type of TAA is called an "idiotypic antigen."
idiotype	The collection of *idiotopes* on a given Ig or TCR.
I exons	Leader-like sequences in the C_H region of the Igh locus where the transcription of germline C_H transcripts necessary for *switch recombination* initiate.
IFA	= Incomplete Freund's adjuvant. See *Freund's adjuvant*.
Igα/Igβcomplex	Accessory heterodimer within the *BCR* complex. Required to transduce intracellular signaling initiated by *mIg* engagement by antigen. Cytoplasmic tails of Igα and Igβ contains *ITAMs*.
Ii	See *invariant chain*.
iIELs	= Intestinal IELs. See *intraepithelial lymphocytes*.
IL-1Ra	= IL-1 receptor antagonist. A soluble protein that binds to IL-1R but blocks signal transduction so inflammation is inhibited. Combats *endotoxic shock*.
immediate hypersensitivity	See *type I hypersensitivity* and *hypersensitivity*.
immune complex	Lattice-like structure composed of interlinked antigen–antibody complexes. Large immune complexes are insoluble and can become trapped in vessel walls or narrow body channels, provoking inflammation.
immune complex glomerulonephritis	Inflammation caused by large circulating antigen–antibody complexes that become trapped in the glomeruli of the kidneys during the renal filtration process.

Term	Definition
immune deviation	Conversion of an adaptive immune response that is harmful (such as *autoimmunity* or *hypersensitivity*) to a less harmful response. Most often seen in the context of a switch from a *Th1* to *Th2* response, or vice versa. Can appear to engender *peripheral tolerance* to an antigen.
immune exclusion	A form of *neutralization* in which a *secretory antibody* binds to a pathogen and "excludes" it from making contact with the *mucosae*.
immune privileged sites	Anatomical sites in which immune responses are actively or passively suppressed by physical barriers, hormone secretion, low DC numbers, *immune deviation*, *immunosuppressive cytokines*, or *Fas* killing.
immune response	A coordinated action by numerous cellular and soluble components in a network of tissues and circulating systems that combats pathogens, injury by inert materials, and cancers. Immune responses are characterized by the recognition stage (identification of the entity as nonself or dangerous to self) and the effector stage (elimination of the entity).
immune response (Ir) genes	Term formerly used to describe genes of the *major histocompatibility complex* because studies using inbred animals showed that the alleles expressed at these loci affected the strength of an individual's immune response to a given antigen.
immune system	A coordinated system of cells, tissues, and soluble products that constitute the body's defense against invasion by nonself entities, including infectious and inert agents and tumor cells.
immunity	The ability to rid the body successfully of a foreign entity.
immunization	Process of manufacturing immune defense; that is, artificially helping the body to defend itself using the effectors and mechanisms of the immune system. May be *active* or *passive*.
immunoconjugate	*Chimeric protein* in which a whole *mAb* (or a structural derivative of that mAb) is linked to another molecule either chemically or at the DNA level. See *immunocytokine*, *immunoradioisotope*, and *immunotoxin*.
immunodiffusion	*Precipitin*-based assay in which antigen and antibody diffuse toward each other within a semi-solid medium such as an agar gel. Analysis of location of resulting precipitin lines allows qualitative or quantitative identification of an unknown.
immunodominant epitope	An *epitope* against which the majority of antibodies is raised, or to which the majority of T cells responds, even though there may be numerous epitopes within the immunogen.
immunodysregulation, polyendocrinopathy, enteropathy X-linked (IPEX) syndrome	X-linked *autoimmune* disease characterized by *inflammation* of multiple tissues and the presence of autoantibodies against antigens of pancreatic islets, thyroid and adrenal glands, smooth muscle, intestine, and kidney. Caused by a mutation of the FOXP3 *transcription factor* required for IL-2 expression and T_{reg} *cell* generation and function.
immunocytokine (ICK)	*Immunoconjugate* in which the mAb (or mAb derivative) is linked to a cytokine. Used in cancer therapy to deliver cytokines with anti-tumor properties to the tumor site.
immunoelectrophoresis	Assay that combines the separation of proteins by non-denaturing electrophoresis with the determination of protein identity by *immunodiffusion*.
immunofluorescence	*Immunohistochemical* assay in which specific antibody *tagged* with a *fluorochrome* binds to an antigen on a cell surface or within an immobilized cell or tissue section and is detected by use of a fluorescence microscope.
immunogen	An antigen that can induce lymphocyte activation; that is, elicits an adaptive immune response.
immunoglobulin (Ig)	Y-shaped antigen binding protein expressed by B lineage cells. An Ig monomer is composed of two identical light and two identical heavy chains (H_2L_2). In its plasma membrane-bound form, an Ig is the antigen-binding component of the BCR complex. In its secreted or *secretory* form, an Ig is an antibody. Igs were originally named for both their involvement in immunity and their presence in the *gamma globulin* fraction of plasma proteins.
immunoglobulin (Ig) domain	A protein domain found in all immunoglobulin molecules as well as many other molecules of the immune and nervous systems. Defined by a sequence of about 70–100 amino acids that forms a characteristic structure known as the "Ig fold" or "Ig barrel." Mediates inter- and intramolecular interactions.
immunoglobulin (Ig) fold	See *Immunoglobulin domain*.

Term	Definition
immunoglobulin (Ig) monomer	The basic Ig unit, composed of two heavy chains and two light chains (H_2L_2).
immunoglobulin (Ig) subclasses	Subgroups of the major Ig *isotypes:* e.g., in humans: IgA1, IgA2, IgG1, IgG2, IgG3, and IgG4.
immunoglobulin (Ig) superfamily	Any protein with 15% amino acid homology to Ig proteins and containing one or more *immunoglobulin domains* is a member of this large family.
immunohistochemistry (immunostaining)	Visualization of cellular structures or molecules of interest within tissues using *tagged* antibodies that are easily identified using the appropriate type of microscopy and detection system.
immunological ignorance	Apparent peripheral tolerance to selfcomponents present in such low amounts that immature DCs rarely acquire and present them to cognate T cells.
immunological memory	Hallmark of adaptive immunity critical in preventing disease. During the *primary response* to a pathogen, each antigen-specific lymphocyte clone activated generates many memory lymphocytes of identical specificity and greater affinity. When the same pathogen attacks the body a second time, it is eliminated more rapidly and efficiently by the *secondary response* mediated by the activation of these memory lymphocytes.
immunological synapse	See SMAC (*supramolecular activation complex*).
immunology	The study of the immune system. That is, the study of the activation, development, and defense functions of adaptive and innate immune responses and the investigation of the cells, genes, and proteins underlying these interactions.
immunopathic damage	Collateral damage to host tissues caused by cytokines and effector cells deployed during an immune response.
immunophenotyping	Analysis of microscope slide smears or flow cytometric preparations of cells by staining them with tagged specific antibodies to define cell surface protein ("marker") expression. This term is used most often in the context of hematopoietic cell cancers where cells of peripheral blood and bone marrow are examined.
immunoprecipitation	Use of specific antibody to isolate or detect antigens present in complex mixtures of proteins in solution. In most cases, either before or after antigen binding, the antibody is complexed to an insoluble particle that allows isolation of antibody–antigen complexes by centrifugation.
immunoproteasome	See *proteasome.*
immunoradioisotope	An *immunoconjugate* in which the mAb (or mAb derivative) is linked to a radioisotope that can kill a cell on contact. Used for *radioimmunotherapy.*
immunosome	The entire membrane complex of antigen-stimulated TCR, CD3 chains, costimulatory and adhesion molecules, coreceptors, and other signaling molecules associated with the *lipid rafts* in an activated T cell.
immunosuppression	The reduction or elimination of immune responses. Can be mediated by *immunosuppressive cytokines, immunosuppressive drugs,* anti-lymphocyte antibodies, or anti-cytokine antibodies. Leaves an individual vulnerable to infection.
immunosuppressive cytokines	Cytokines (e.g., IL-10, TGFβ) that damp down immune system cell activation by interfering in intracellular signaling pathways.
immunosuppressive drugs	Drugs that decrease immune responses by interfering with either lymphocyte *homing receptors, costimulation,* proliferation, or *AICD,* or leukocyte function or trafficking. May also suppress *cytokine* or *chemokine* expression or antigen uptake by DCs. These drugs have significant detrimental side effects.
immunosurveillance	Concept that the immune system monitors the body for pathogens and tumor cells and destroys them before they can establish a significant presence.
immunotherapy	The manipulation of the immune system to fight diseases.
immunotoxin	An *immunoconjugate* in which the mAb (or mAb derivative) is linked to a *toxin* that can kill a cell it contacts.
inbred mice	Strains of mice in which all members are genetically identical at virtually all loci and each mouse is homozygous at all loci. Created by sibling mating for multiple generations.

1145

Term	Definition
incubation time	Interval between initial infection and onset of disease.
indirect allorecognition	Occurs in a transplant situation when peptides derived from allogeneic proteins of the graft (often MHC molecules) are presented to recipient T cells (mostly CD4$^+$ T cells) by recipient APCs.
indirect tag assays	Two-step procedure in which antigen–antibody binding is detected by tagging a third component that binds to the unlabeled antigen–antibody pair. Such components are typically *secondary antibodies*, *protein A* or *G*, or the *biotin-avidin* system.
indolent	An indolent cancer is one in which the transformed cells grow relatively slowly and disease progression is gradual.
induced fit	Hypothesis that an antigen can influence the conformation of the antigen binding site of an antibody or TCR such that it better accommodates the antigen. Also applies to peptide binding in the grooves of MHC class I and II proteins.
inducible nitric oxide synthetase (iNOS)	Enzyme induced mainly in phagocytes by the presence of microbial products or pro-inflammatory cytokines. Converts arginine to citrulline and *nitric oxide*, which is toxic to endocytosed pathogens.
inductive site	A local area of the *mucosae* where antigen is encountered and a primary adaptive response initiates.
infection	The attachment and entry of a *pathogen* into a host's body such that the organism successfully avoids innate defense mechanisms and is able to reproduce; not synonymous with disease.
infectious tolerance	Phenomenon in which induced *peripheral tolerance* to a given antigen is "passed along" from one population of *anergic* T cells to a normal T cell population, rendering the normal cells also tolerant to the same antigen. May be mediated by *regulatory T cells*.
inflammation	A local response at a site of infection or injury initiated by an influx of innate response leukocytes that fight infections using broadly specific recognition mechanisms. Inflammatory cells phagocytose antigens, produce chemical signals that promote wound healing, and secrete cytokines and chemokines that attract and regulate lymphocytes. Clinically, inflammation is characterized by heat and pain as well as swelling and redness due to vasodilation and increased vascular permeability. These effects are due to inflammatory mediators such as *kinins, acute phase proteins, leukotrienes, prostaglandins*, and *vasoactive amines* including *histamine*. Inflammation is an integral part of both innate and adaptive immunity.
inflammatory bowel disease (IBD)	Family of inflammatory *autoimmune diseases* affecting the large and small intestines. See *Crohn's disease* and *ulcerative colitis*.
inflammatory response	See *inflammation*.
initiation	First step of *carcinogenesis*. A mutation in a cell is not repaired and gives the cell an irreversible growth advantage. This cell is said to be the "target cell" for eventual *malignant transformation*.
innate immunity	Non-specific and broadly specific mechanisms that deter entry or result in elimination of nonself entities. Innate immunity is mediated by (1) physical, chemical, and molecular barriers that exclude antigens in a totally non-specific way, and (2) receptors (*PRMs* and *PRRs*) that recognize a limited number of molecular patterns that are common to a wide variety of pathogens. There is no memory development, so that responses during the first and subsequent exposures to a given antigen have the same kinetics and magnitude.
insulitis	Infiltration of the pancreas by inflammatory cells and other leukocytes (including macrophages, B cells, and CD4$^+$ and CD8$^+$ T cells). Insulin-producing β-islet cells are not destroyed. See *type I diabetes*.
integrase	Enzyme of retroviruses (e.g., *HIV*) that mediates insertion of the viral DNA into the host cell genome.
integrin	See *cellular adhesion molecule*.
interferons (IFN)	Family of *cytokines* produced by cells upon infection by virus. Type I IFNs (IFN α, β, and others) are important in inhibiting proliferation and establishing an *antiviral state* in bystander cells. Type II IFN (IFNγ) has multiple immunoregulatory and pro-inflammatory effects in addition to its anti-proliferative and anti-viral activities, and is essential for a *hyperactivated macrophage*.
interfollicular region	Tissue between lymphoid follicles positioned in a group. Characterized by the presence of high concentrations of mature T cells surrounding HEVs.

Term	Definition
interleukins (IL)	Family of over 25 secreted proteins that convey information between leukocytes and trigger signaling that regulates the growth, differentiation, and function of hematopoietic cells.
intestinal cryptopatches	Small regions of lymphoid tissue in the *lamina propria* of the crypts of the small intestine. Site of *extrathymic* lymphopoiesis.
intestinal follicles	Lymphoid follicles located in the intestinal *lamina propria* either grouped in *Peyer's patches* or the appendix, or scattered singly. Composed of a *germinal center* containing B cells and *FDCs* topped by a *dome* containing DCs, macrophages, CD4$^+$ T cells, regulatory T cells, and mature B cells. Surrounded by mature CD4$^+$ Th and CD8$^+$ Tc cells.
intracellular antigen	See *endogenous antigen*.
intracellular pathogen	Pathogen that spends a significant portion of its life cycle within a host cell; reproduction is usually within the host cell.
intracellular signaling pathway	The means by which the binding of a ligand to its receptor on a cell surface eventually triggers new patterns of gene expression in the cell nucleus. Ligand-receptor binding initiates a series of interactions between various proteins, including enzymes and adaptors, which culminate in the activation of transcription factors that enter the nucleus and alter the transcription patterns of genes controlling cellular proliferation, differentiation, and effector functions.
intraepithelial lymphocytes (IEL)	$\alpha\beta$ and $\gamma\delta$ T cells dispersed among the epithelial cells lining a body tract. High concentrations occur in the intestinal epithelium (intestinal IELS; iIELs) and top skin layers (dendritic epidermal T cells; DETCs). May also include intraepithelial *natural killer (NK) cells*.
intraepithelial pocket	Pocket within the *FAE* created by invagination of the *basolateral* surface of an *M cell*. Location of subpopulations of CD4$^+$ memory T cells, naive B cells, DCs, and macrophages. Provides access to antigens that have been transported through the *M cell* by *transcytosis*.
intrathecal chemotherapy	Introduction of a chemotherapeutic drug just under the sheath covering the spinal cord.
intrinsic apoptotic pathway	A mechanism that triggers apoptosis in mammalian cells by inducing changes in the mitochondrial membrane that cause it to release cytochrome *c*. Cytochrome *c* combines with caspase-9 and Apaf-1 in the apoptosome, which in turns triggers the *caspase-3 cascade*.
invariant chain (Ii)	Transmembrane protein that binds to peptide-binding groove of newly synthesized *MHC class II molecules* and chaperones them out of the rER into the endocytic system. Upon cleavage, Ii gives rise to *CLIP*.
invasive pathogen	A pathogen that is very adept at entering the body, even when surface defenses are intact.
invasive tumor	A tumor that is able to move into and destroy the healthy architecture of nearby organs.
inversional joining	Occurs during *V(D)J recombination* when the two *gene segments* to be joined are in the opposite transcriptional orientation.
invertebrates	Animals without a true backbone. See *higher* and *lower invertebrates*.
in vitro	*In vitro* ("in glass") experiments are conducted in tissue culture plates or test tubes.
in vivo	*In vivo* ("in life") experiments are conducted using live animals or human subjects.
ionic bonds	See *electrostatic bonds*.
IPEX	See *immunodysregulation, polyendocrinopathy, enteropathy X-linked (IPEX) syndrome*.
ischemia	Local loss of oxygen to a tissue, due to constriction or blockage of blood vessels.
ISCOM	= Immunostimulating complex. A vaccine in which a pathogen antigen is mixed with cholesterol, phospholipid, and saponin to form a soccer ball-shaped structure with a strong negative charge. The vaccine antigen is usually an *amphipathic* protein. See *delivery vehicle*.
isograft	Tissue transplanted between two genetically identical individuals.
isotypes	Classes of immunoglobulins defined on the basis of the amino acid sequences of their constant regions. There are five different Ig heavy chain constant regions in mice and humans, Cμ, Cδ, Cα, Cγ, and Cε, which define the five antibody isotypes IgM, IgD, IgA, IgG, and IgE. Each antibody isotype differs in physical and functional characteristics, in anatomical distribution, and in the ability to polymerize. There are also light chain isotypes defined by either a kappa light chain constant region (Cκ) or one of several lambda light chain constant regions (Cλ).

Term	Definition
isotype switching	Mechanism by which a B cell producing immunoglobulin of one isotype can then switch to producing immunoglobulin of the same variable region but different isotype. Involves rearrangement of immunoglobulin heavy chain constant region genes in the B cell. See *switch recombination*.
ITAM	= Immunoreceptor tyrosine-based activation motif. A sequence present in the cytoplasmic tails of many activatory transmembrane receptors that allows recruitment of *protein tyrosine kinases (PTKs)*. Phosphorylation of ITAMs by PTKs triggers or upregulates intracellular signaling pathways leading to new gene transcription and cellular activation.
ITIM	= Immunoreceptor tyrosine-based inhibition motif. A sequence present in the cytoplasmic tails of many inhibitory transmembrane receptors that allows recruitment of phosphatases that counter the actions of PTKs. ITIM signaling also interferes with intracellular Ca^{2+} mobilization. Intracellular signaling pathways are downregulated, suppressing or altering gene transcription and inhibiting cellular activation.
IV-IG	Intravenously administered preparation of pooled antibodies of all specificities collected from volunteers. Used to prevent recurrent infections in patients with *hypogammaglobulinemia* or *primary immunodeficiency*. When antigen-specific antibodies are pooled, the preparation is used for *passive immunization*.
Jak kinases	= Janus kinase or "just another kinase." Family of non-receptor tyrosine kinases that associate with cytokine receptor tails and phosphorylate the *STAT* transcription factors.
J chain	= Joining chain. A small acidic polypeptide that can bond to the tail pieces of α and μ Ig heavy chains, stabilizing polymeric IgA or IgM formation, respectively.
junctional diversity	Variation in the amino acid sequences of Igs and TCRs that arises during V(D)J *recombination* in the Ig or TCR loci, respectively. Due to imprecise joining of *gene segments*, deletion of nucleotides in the joint region, and/or the addition of P and/or N *nucleotides*.
kallidin	See *kinins*.
Kaposi's sarcoma	An unusual tumor of blood vessel tissue that is common in immunodeficient individuals such as those with AIDS. Associated with infection by Kaposi's sarcoma herpesvirus (KSHV).
kappa-lambda exclusion	Process by which the production of a functional κ light chain from the Igk locus on one chromosome shuts down further V(D)J rearrangement at the other Igk allele, and also prevents V(D)J rearrangement of either Igl allele.
Kawasaki disease (KD)	An *autoimmune disease* affecting mucosal surfaces, lymph nodes, vasculature, and heart. Characterized by persistent high fever and a signature rash. May induce vascular aneurysms.
keratin layer	Tough outer layer of skin that resists penetration by inert stimuli as well as microbes. Composed of filaments of keratin protein made by *keratinocytes*.
keratinocytes	Specialized epidermal squamous epithelial cells that are generated in waves, are connected by *desmosomes*, and produce keratin. Continual migration of cells upward ensures constant regeneration of the skin. The coordinated death of a wave of these cells forms the *keratin layer*. Keratinocytes in lower levels of the epidermis can secrete pro-inflammatory *cytokines* and *complement* components. Can also act as anergizing *APCs*.
ketoacidosis	A build-up in the blood of acid and ketone products due to abnormal cellular metabolism. See *diabetes mellitus*.
killed vaccine	*Vaccine* based on a pathogen inactivated by irradiation or chemical treatment. Incapable of replication but retains immunogenicity.
kinins	Family of small peptide inflammatory mediators (including *bradykinin* and *kallidin*) that circulate in inactive form in the blood until cleaved by blood clotting enzymes or enzymes released from damaged cells. These peptides cause increased vascular permeability, smooth muscle contraction, and pain. See also *inflammation*.
knock-in transgene	A transgene bearing a mutation of interest is flanked by sequences that are homologous to the endogenous locus and thus ensure that the transgene integrates in its natural position. See *transgenic (Tg) mouse*.
knockout mouse	Mouse strain in which a single gene is heterozygously or homozygously deleted or rendered defective by genetic engineering techniques in embryos. See *homologous recombination*.

Term	Definition
Kupffer cells	Macrophage-like cells in liver sinusoids.
lactoferrin	Iron-chelating protein that inhibits bacterial growth; found in neutrophil granules.
lamina propria	The layer of loose connective tissue between the *basolateral* surface of the *mucosal* epithelium and the underlying muscle layer. Major site of *plasma cell* concentration.
Langerhans cells	Epidermal DCs that function as APCs and secrete cytokines drawing lymphocytes to the skin.
LAR	See *local acquired resistance.*
latent infection	A *persistent infection* in which a pathogen is present for an extended time but is non-infectious and does not cause clinical symptoms. Achieved by integration of the viral genome into the host DNA, or by alterations to viral gene expression.
late phase reaction	In the *late phase reaction* of *type I hypersensitivity*, leukocytes are drawn to the site of *allergen* penetration by mediators released in the *early phase reaction*. These leukocytes release additional cytokines, enzymes, and inflammatory mediators that do further damage. Dominated by *eosinophils.*
lectin	A protein that binds to particular carbohydrate moieties on membrane glycoproteins or glycolipids on cell surfaces.
lectin-mediated complement activation	Complement activation initiated primarily by the binding of *mannose-binding lectin* (MBL) to monosaccharides on a pathogen surface. Involves MASP1/2, MAp19, and the cleavage and activation of C4, C2, and C3. See *complement.*
leukapheresis	The selective removal of leukocytes from blood drawn from a donor, followed by return of the remaining blood products to the donor. When carried out in preparation for an HCT, the collected donor leukocytes may be enriched for HSCs by prior treatment of the donor to mobilize HSCs from the bone marrow into the peripheral blood.
leukemia	Liquid malignancy of *leukocytes*. Manifested as greatly increased numbers of *myeloid*, *lymphoid*, or erythroid lineage *cells* in the blood and bone marrow. Arises from the *malignant transformation* of a hematopoietic cell in the bone marrow.
leukocyte adhesion deficiencies	PIs of innate responses caused by mutations in *CAMs* that alter the *extravasation* of leukocytes (especially *neutrophils*).
leukocytes	White blood cells, including lymphocytes, granulocytes, monocytes, macrophages, NK, and NKT cells.
leukotrienes	Lipid inflammatory mediators whose formation is initiated when phospholipids in the membranes of macrophages, monocytes, neutrophils, and mast cells are degraded and converted to arachidonic acid. Metabolism of arachidonic acid via the lipoxygenase pathway then produces leukotrienes. See also *inflammation.*
Li-Fraumeni syndrome	An inherited susceptibility to the development of various cancers. Most cases of this syndrome are due to germline mutations of *p53*.
light zone	Region of the *germinal center* where *isotype switching* occurs, and where B cells whose immunoglobulin genes have undergone *somatic hypermutation* are either negatively selected to establish peripheral B cell tolerance, or selected for increased affinity for antigen (*affinity maturation*).
limiting dilution analysis	Cells of interest are diluted to a known concentration so low that when the cell suspension is dispensed into multi-well culture plates, individual wells are unlikely to contain more than a single cell each.
linear determinant	B cell *epitope* defined by a series of consecutive amino acids.
linkage disequilibrium	The tendency of genetic loci to co-segregate in a population because they are fairly close together within the genome.
linked recognition	The requirement that a hapten and a carrier be physically linked, not just mixed together, to induce a secondary response to the hapten.
linked suppression	Phenomenon in which *peripheral tolerance* to one antigen spreads to other antigens in close physical proximity within the same host. May be mediated by *regulatory T cells.*

Term	Definition
lipid rafts	Small areas of the fluid bilayer of the plasma membrane rich in glycosphingolipids and cholesterol. In a triggered T cell, proteins required for TCR signaling associate with the rafts and form a large, stable structure (the *immunosome*) that recruits and activates various additional signaling molecules. The rafts plus signaling molecules then associate with the cytoplasmic portion of the triggered TCR and transduce TCR signaling to the nucleus.
lipopolysaccharide	See *LPS*.
liposome	Spherical structure composed of a phospholipid bilayer surrounding an aqueous center. May be used as a vaccine *delivery vehicle* if a *vaccine* antigen is incorporated in the aqueous solution.
live attenuated vaccine	*Vaccine* based on a weakened version of the pathogen that has reduced virulence within the host but retains its ability to provoke an immune response. Can replicate to some extent but does not cause overt disease. See *attenuation*.
local acquired resistance (LAR)	Metabolic changes occurring in plant cells surrounding an *HR* site. Glycoproteins and other molecules are deposited and cross-linked to fortify the walls of the plant cells, trapping the pathogen. *Anti-microbial peptides* are induced, as are plant *regulatory peptides* that in turn induce the *pathogenesis-related (PR) proteins*.
locus accessibility	With respect to Ig and TCR genes, the degree to which the chromatin structure of a given antigen receptor gene allows it to serve as a substrate for the recombinase complex. Governed by *enhancers*.
long-term non-progressors (LTNPs)	*HIV*-infected individuals whose viremia remains under control without the use of *anti-retroviral drugs*.
long terminal repeat (LTR)	Repeated sequence in retroviruses at the 5′ and 3′ ends of the *provirus*. For *HIV*, the 5′ LTR contains an active promoter that initiates transcription, and regulatory sites that control transcription in response to signals received from either the host cell or viral proteins. The 3′ LTR contains sequences controlling cleavage of the primary transcript and poly-A tail addition.
lower invertebrates	Coelenterates (such as corals and jellyfish), platyhelminths (flatworms), nematodes (roundworms), nemertines (ribbonworms).
low grade tumor	Tumor that does not grow aggressively and contains fairly well-differentiated cells. Usually derived from the *malignant transformation* of a mature cell. Generally associated with a favorable prognosis.
low zone tolerance	*Experimental tolerance* induced by repeated administration of a very small dose of an immunogen over a long period of time.
LPS	= *Lipopolysaccharide*. Component of *gram-negative* bacterial cell walls that generates *endotoxin* and induces *endotoxic shock*. Binds to *TLR4*.
lupus nephritis	A fatal SLE-like disease in mice characterized by kidney damage.
lymph	The nutrient-rich interstitial fluid that bathes all cells in the body. Consists of blood plasma that, under the pressure of the circulation, leaks from the capillaries into spaces between cells. Ninety percent of this fluid returns to the circulation via the venules, but 10% filters slowly through the tissues, collecting antigen, and eventually enters the *lymphatic system* via tiny lymphatic vessels.
lymphatic system	A network of lymphatic capillaries, larger lymphatic vessels, and two lymphatic trunks through which lymph and lymphocytes are taken up and recirculate. The lymphatic trunks connect with the subclavian veins of the blood circulation.
lymph nodes	Bean-shaped, encapsulated *secondary lymphoid tissues* containing the concentrations of T and B lymphocytes, APCs, and other accessory cells required for the activation of naive lymphocytes and the mounting of primary adaptive immune responses. Lymph nodes occur along the entire length of the lymphatic system but are clustered in key regions.
lymphoadenopathy	Enlargement of the lymph nodes due to increased lymphocyte proliferation.
lymphoblastic leukemia	*Leukemia* arising from the transformation of a T or B precursor cell.

Term	Definition
lymphoblasts	Resting lymphocytes activated by engagement of their TCRs or BCRs become *lymphoblasts* within 18–24 hours. Lymphoblasts are larger and display more cytoplasmic complexity than resting cells. Lymphoblasts undergo rapid cell division and differentiate into memory and effector lymphocytes.
lymphocyte	A class of leukocyte generally concentrated in the *secondary lymphoid tissues* while in the resting state. Morphologically, lymphocytes are small round cells with a large nucleus surrounded by a narrow rim of cytoplasm containing few intracellular organelles. T cells and B cells are lymphocytes and can be distinguished phenotypically by the cell surface expression of *TCRs* and *BCRs*, respectively.
lymphocyte activation	Occurs when multiple copies of a specific antigenic *epitope bind* to multiple copies of the TCR or BCR of a resting lymphocyte stimulating it to become a *lymphoblast* and undergo cell division (proliferation). For naive cells, *costimulation* and cytokines are also required. The transcription of numerous genes is triggered, causing the progeny cells to undergo morphological and functional changes leading to the generation of *memory* and *effector lymphocytes*.
lymphocyte activation threshold	The number of antigen receptors that have to be aggregated and triggered (in conjunction with costimulatory signaling) to deliver a proliferative signal.
lymphocyte precursors	Populations of immature cells in the bone marrow or thymus that arise from HSCs and are committed to developing into lymphocytes. Includes pro-B, pre-B, pro-T, pre-T, and immature B and T cells.
lymphocyte recirculation	Continual migration of lymphocytes (but not other leukocytes) from the tissues back into the blood via the *lymphatic system*, followed by return to the tissues via *extravasation* from the circulation.
lymphocyte specificity	The highly restricted and unique range of *epitopes* potentially bound by a lymphocyte due to its expression of a single type of randomly generated antigen receptor gene.
lymphocyte tolerization	Process by which a lymphocyte is rendered *anergic*. For B cells, tolerization often results from the engagement of the BCR in the absence of T cell help, or from *receptor blockade*. For T cells, tolerization results from the interaction of the lymphocyte with an immature DC, a *modulated DC*, or a *regulatory T cell*. Inadequate *costimulation* and/or the induction of suppressive *intracellular signaling* may underlie tolerization.
lymphocytic leukemia	*Leukemia* characterized by the abnormal proliferation of transformed T or B lineage cells that appear to have a mature peripheral cell phenotype.
lymphoid cells	Cells that develop from the CLP, including *T and B lymphocytes, NK cells*, and *NKT cells*.
lymphoid follicles	Organized clusters of lymphocytes.
lymphoid organs	Groups of lymphoid follicles encapsulated by specialized supporting tissues and membranes.
lymphoid patches	Groups of lymphoid follicles that are not encapsulated.
lymphoid tissue	A tissue in which lymphocytes are found, including diffuse arrangements of individual cells, follicles, patches, and encapsulated organs.
lymphokine-activated killer (LAK) cells	*NK cells* cultured in high levels of IL-2 are induced to differentiate into LAKs, which have even broader powers of target cell recognition and cytolysis. It is unclear if LAKs exist *in vivo*.
lymphoma	Solid malignancy of lymphoid cells arising in a structured or diffuse *secondary lymphoid tissue* rather than in the blood or bone marrow. Malignant cells are not usually found in the blood. Classified as either a *Hodgkin's lymphoma* (HL) or a *non-Hodgkin's lymphoma* (NHL).
lymphopenia	Greatly reduced numbers of lymphocytes in the circulation.
lymphopoiesis	The process of HSC differentiation into CLPs and descendant cells of the lymphoid lineage.
lysozyme	Protease that digests a common component of bacterial cell walls. Found in neutrophil granules and in tears and other body secretions.
mAb	See *monoclonal antibodies*.

Term	Definition
macrophage	Powerful *phagocyte* that also secretes a large array of proteases, cytokines, and growth factors and can act as an *APC*. Resting or immature macrophages receiving an inflammatory signal become responsive to chemotactic signals. Cytokines, in particular IFNγ, cause the responsive macrophage to become "activated" or "primed" and capable of antigen presentation and costimulation. In the presence of high levels of IFNγ plus bacterial endotoxin, the activated cell becomes a "hyperactivated" or "triggered cytolytic" *macrophage*, exhibiting enhanced anti-pathogen activities and the capacity to kill tumor cells. These cells are incapable of proliferation and have a reduced ability to present antigen.
macropinocytosis	A mechanism of antigen uptake in which a cell internalizes extracellular fluid containing soluble macromolecules by engulfing droplets and forming *macropinosomes*. The macropinosomes then enter the *endocytic pathway*. Also called "cell drinking."
macropinosome	Membrane-bound vesicle formed by plasma cell invagination around a droplet of extracellular fluid. See *macropinocytosis*.
major histocompatibility complex (MHC)	Region of the genome containing genes encoding the chains composing the *MHC class I, class II*, and *class III proteins*. Originally defined as a cluster of genes encoding proteins controlling tissue compatibility between individuals. The function of the proteins encoded in the MHC class I and class II regions is to combine with antigenic peptides, both self and nonself, and display them on the surface of host cells for perusal by T cells. MHC class I molecules are expressed on most nucleated cells and present peptides to CD8-expressing Tc cells. MHC class II molecules are expressed only on APCs that present peptides to CD4-expressing Th cells. The class I and class II MHC genes are highly polymorphic in many species, including mice and humans. The class III MHC genes encode various proteins important in complement activation, inflammation, and stress responses.
malignant conversion	Fourth stage of *carcinogenesis*. One of the cells in a *neoplasm* accumulates additional deleterious mutations such that it becomes a *malignant tumor* with completely deregulated growth. It may also become an *invasive tumor* and may undergo *metastasis*.
malignant transformation	See *carcinogenesis*.
malignant tumor	A mass formed by abnormally dividing cells that appears anywhere from poorly to completely differentiated. The tissue mass is disorganized, rarely encapsulated and subject to *metastasis*. Directly lethal to the host unless removed or killed. Also called "cancer."
MALT	See *mucosa-associated lymphoid tissue*.
mannose-binding lectin (MBL)	A serum *collectin that* specifically binds to distinctive mannose structures on microbial pathogens in the blood. Engagement of MBL triggers the *lectin-mediated complement activation*.
MAPK	= Mitogen-activated protein kinase. See *Ras/MAPK signaling pathway*.
marginal zone	See *spleen*.
marginal zone B cells	A subset of B cells found in the marginal zone surrounding the lymphoid follicles in the *spleen*. Important in the earliest stages of the adaptive response to blood-borne antigens, especially Ti antigens. Give rise to short-lived *plasma cells* in the spleen but not memory cells.
margination	First step of *extravasation*. Leukocytes adhere to endothelial cells in the post-capillary venules in three phases: selectin-mediated tethering, integrin-mediated activation, and activation-dependent arrest and flattening.
mast cells	Granule-containing leukocytes with non-lobed nucleus derived from a lineage separate from that giving rise to other *granulocytes*. The cytoplasmic granules of mast cells stain like those of basophils and also contain heparin and *histamines*, but are more numerous and smaller than those of basophils. Mast cells are found in both mucosal and non-mucosal tissues and their degranulation is important for *inflammation* and *allergy*.
maternal imprinting	Specific modification of the chromatin of the maternal *allele* of a gene in an individual such that it is more likely to be expressed than the paternal allele. Affects the correlation between possession of a disease-related allele and disease incidence.
matrix (HIV)	Supplies the scaffolding around which the viral *envelope* is wrapped.

Term	Definition
M cells	= Membranous cells. Large epithelial cells with an *intraepithelial pocket* that *transcytose* antigens from a body tract lumen across the epithelial layer to its basolateral surface. M cells lack a *glycocalyx* and *brush border*, facilitating antigen capture. Many M cells overlie *intestinal follicles* and have long processes that extend into the dome covering the follicle.
mediastinal mass	A mass situated among the organs, including the heart and thymus, that are positioned between the sternum and the spine.
medulla (*thymus*)	Inner region of the *thymus*.
medullary thymic epithelial cells (*mTECs*)	See *thymic epithelial cells*.
megakaryocytes	Multinucleate *myeloid cells* lineage leukocytes from which platelets are derived.
melanization reaction	See *ProPO system*.
melanocyte	Skin cell responsible for producing skin pigment melanin.
melanoma	Malignancy of skin *melanocytes*.
melanosome	Specialized organelle present in *melanocytes* that is associated with the endocytic pathway and is responsible for the production of pigment proteins.
membrane attack complex (*MAC*)	Pore-shaped structure assembled in the membrane of a pathogen or target cell as a consequence of *complement* activation. Facilitates osmotic imbalance and lysis of the pathogen or cell.
membrane-bound Ig (*mIg*)	Form of cell-surface Ig molecule containing a transmembrane region, extended C-terminal region, and no tail piece. Antigen-binding moiety of the *BCR*.
membrane nibbling	Acquisition by a DC of a portion of the membrane of a live, whole infected cell.
memory	See *immunological memory*.
memory B cells	See *memory cells (memory lymphocytes)*.
memory T cells	See *memory cells (memory lymphocytes)*.
memory cells (*memory lymphocytes*)	Lymphocytes generated during a *primary response*. Remain in a quiescent state until fully activated by a subsequent exposure to specific antigen (a *secondary response*). Memory cells are activated and generate effector cells more quickly and with less *costimulation* than naive lymphocytes. Memory cells express *homing* and *chemokine receptors* that allow them to return to the site of first antigen exposure.
memory response	See *immunological memory*.
mesangial phagocytes	Macrophages in the kidney.
metastases	Secondary tumors established by *metastasis* of a primary tumor to sites in the same or a different organ or tissue.
metastasis	Process by which malignant cells break away from a *primary tumor* and spread via the blood to secondary sites at nearby or distant locations.
MGUS	See *monoclonal gammopathy of undetermined significance* (*MGUS*).
MHC class I proteins	Cell surface heterodimeric proteins composed of the polymorphic transmembrane MHC class I α chain in non-covalent association with the invariant β_2-*microglobulin* ($\beta2m$) chain. The MHC class I α chain is encoded by genes in the K and D loci in mice, and by genes in the HLA-A, -B, and -C loci in humans, while $\beta2m$ is encoded outside the MHC. Bind 8–10 amino acid peptides of endogenous origin and present them for recognition by CD8$^+$ T cells. Expressed on almost all nucleated cell types.
MHC class Ib and IIb proteins	Non-classical MHC proteins with restricted polymorphism. Most are not involved in antigen *presentation*. Class Ib proteins are encoded by genes in the Q, T, and M loci in mice, and by genes in the HLA-X, -E, -J, -G, and -F and HFE loci in humans. Class IIb proteins are encoded by genes in the P locus in mice, and by genes in the DM and DO loci in humans.
MHC class II proteins	Cell surface heterodimeric proteins composed of polymorphic MHC class II α and MHC class II β chains. The α and β chains are encoded by genes in the A and E loci in mice, and by genes in the DP, DQ, and DR loci in humans. Bind 13–18 amino acid peptides of exogenous origin and present them to CD4$^+$ T cells. Expressed on APCs.

Term	Definition
MHC class III proteins	Proteins encoded by the S region genes in mice, and the MHC class III region genes in humans. Include many proteins involved in immune response functions such as complement proteins, heat shock proteins, TNF, and LT.
MHC-like proteins	Non-polymorphic proteins encoded outside the MHC that have structural and functional similarities to MHC proteins. Include CD1 molecules.
MHC restriction	The principle that a T cell that recognizes a given antigenic peptide when presented by a particular MHC molecule will not recognize the same peptide if presented by a different MHC molecule. Most commonly observed when antigen-specific T cells derived from one individual will not respond to the antigen presented by APCs or target cells from an individual that is *allogeneic* at the MHC.
MICA, MICB	= MHC class I chain related A and B. Human stress ligands upregulated in response to heat or other environmental stresses, virus infection, or tumorigenesis. Bind to NK activatory receptor NKG2D and to Vγ1Vδ1 TCRs. Murine equivalents, Rae-1 and H60, bind to NKG2D. See *NK activatory receptors* and *gamma-delta TCRs.*
microchimerism	Occurs when donor hematopoietic cells that happen to have been carried along with a *solid organ transplant* become established in the recipient.
microglia	Cells in the brain that are derived from the same precursor as macrophages and have phagocytic function.
mIg	See *membrane-bound Ig.*
MiHA	See *minor histocompatibility antigens.*
MIICs	= MHC class II compartments; also called CIIVs, "class II vesicles." Specialized late endosomal compartments that are part of the *exogenous (endocytic) antigen processing* pathway. MHC–CLIP complexes from endolysosomes enter MIICs and undergo *CLIP* exchange and peptide loading.
mini pigs	A strain of small-sized pigs bred specifically to provide organs and tissues to use as *xenografts* in humans. Mini pig organs are comparable in size to human organs.
minor blood group antigens	Non-*ABO* antigens on RBCs, including the Duffy, Kell, and Kidd systems. *Alloimmunization* to these antigens causes delayed, mild *transfusion reactions.*
minor histocompatibility antigens (MiHA)	Proteins that exist in a small number of different allelic forms in the population (dimorphic, trimorphic, etc.) so that in a transplant situation, peptides derived from allogeneic MiHA of the donor are recognized as nonself by the recipient's T cells. Generally invoke a slower, weaker *graft rejection* response than MHC incompatibilities.
minor H peptides	Peptides of donor *minor histocompatibility antigens* presented by APCs to recipient T cells.
"missing self" hypothesis	Concept that *NK cell* function is triggered by interaction with cells lacking surface expression of self MHC class I, i.e., tumor cells, virus-infected cells, and *allogeneic* cells.
mitogen	Molecule that non-specifically stimulates DNA synthesis and cell division (mitosis). In the case of lymphocytes, the mitogen-binding site is distinct from the antigen-binding site of the BCR or TCR.
mixed chimerism	An HCT recipient receives *non-myeloablative conditioning* followed by *donor cell infusion.* Surviving recipient HSCs and donor HSCs co-exist and give rise to cells of the myeloid and lymphoid lineages.
mixed leukocyte reaction (MLR)	Assay used to detect MHC differences between individuals. Based on the proliferation of *allogeneic* T cells in response to these differences. Two-way MLR: lymphoid cells of two individuals are incubated together and total proliferation is measured. One-way MLR: "stimulator" lymphoid cells from one individual are inactivated to prevent proliferation. "Responder" lymphoid cells from a second individual proliferate if MHC differences exist.
modulated DC	A DC that *anergizes* rather than activates naive T cells, and/or induces generation of T_{reg} cells. *In vitro*, modulation may be induced by encounter of the DC with *Ts cells*, or culture in conditions of low GM-CSF, high IL-10, or TGFβ, or in the presence of pharmacological agents interfering with NF-κB signaling (aspirin, corticosteroids, cyclosporine A). Also called a "tolerized" or "tolerogenic DC."

Term	Definition
molecular mimicry	Concept that a pathogen epitope may resemble a self-epitope closely enough to activate an *autoreactive* lymphocyte, if the appropriate *cytokine* microenvironment is present. This lymphocyte mounts an attack not only on the pathogen but also on self tissues expressing the self epitope.
monoclonal antibodies (mAb)	Antibodies produced by a single B cell clone such that the antibody preparation contains a single type of antibody protein with a single, known specificity. See *hybridoma*.
monoclonal gammopathy of undetermined significance (MGUS)	Mild *plasma cell dyscrasia* in which elevated levels of an Ig *paraprotein* appear in the blood. Associated with few or no clinical consequences.
monocytes	*Myeloid cells* in the blood. Monocytes circulate in the blood for approximately 1 day before entering the tissues and serous cavities and maturing further into *macrophage*s.
monomorphic	A given gene is monomorphic if only one nucleotide sequence (one allele) exists for this gene in the population.
MOPP chemotherapy	Cocktail of mechlorethamine, [O]-vincristine, procarbazine, and prednisolone. Used for cancer therapy.
moth-eaten mice	Natural mouse mutant with patchy fur. These animals suffer early mortality due to inflammatory *autoimmune disease*. Carry mutation of gene encoding SHP-1, a phosphatase that normally functions as a negative regulator of BCR signaling.
MRL/lpr mice	Natural mouse mutant with a *Fas* mutation. Also known as "lpr mice." These animals develop an *autoimmune disease* with aspects of human *SLE* and *ALPS*.
MS	See *multiple sclerosis*.
mucosae	Mucosal epithelial layers that cover the topologically exterior surfaces of the gastrointestinal, respiratory, and urogenital tracts. Also called "mucous membranes."
mucosa-associated lymphoid tissue (MALT)	Diffuse collections of APCs and lymphocytes scattered in the *mucosae*. Includes the *gut-associated lymphoid tissue (GALT)*, the *bronchi-associated lymphoid tissue (BALT)*, and the *nasopharynx-associated lymphoid tissue (NALT)*.
mucosal immunity	Immunity mediated by physical barriers, substances (such as *mucus*), and cells in the *MALT* that can respond to antigen attacking the mucosae without necessarily involving the draining lymph node.
mucosal tolerance	Concept that an individual continually exposed to a modest amount of an antigen through a mucosal route becomes tolerant to that antigen systemically, in that responses to the same antigen later introduced by a non-mucosal route are abrogated.
mucous membranes	See *mucosae*.
mucus	Sticky, viscous fluid that coats the luminal surface of cells of the *mucosae*. A product of glandular secretion. Contains *secretory antibodies* and anti-microbial molecules. Element of *mucosal immunity*.
multiple myeloma	Advanced *myeloma* in which the cancerous cells have spread to multiple body sites.
multiple sclerosis (MS)	*Autoimmune disease* affecting the brain and spinal cord. Believed to be due at least in part to autoimmune attacks on the myelin sheath surrounding the nerve axons and on oligodendrocytes (the brain cells that make myelin). Characterized by inhibition of nerve impulse transmission.
multiplicity	With respect to the genome, refers to the existence of multiple independent genes encoding proteins of identical function.
myasthenia gravis	*Autoimmune disease* leading to severe muscle weakness that may interfere with breathing if respiratory muscles are affected. Caused by autoantibody attack on the nicotinic acetylcholine receptors (AChR) of the muscle at the neuromuscular junctions.
mycoses	Diseases caused by infection with *fungi*.
mycosis fungoides	Cutaneous T cell *non-Hodgkin's lymphoma*.
myeloablative conditioning	Elimination of a patient's hematopoietic cells in the *bone marrow* using aggressive *chemotherapy* and *TBI*, leading to eventual depletion of immune system cells from the peripheral blood and all *secondary lymphoid tissues*. Used to prepare the patient for immune system reconstitution via HCT or bone marrow transplantation.

Term	Definition
myeloblast	An immature hematopoietic cell of the *myeloid* lineage that is produced by and found in the bone marrow but not normally in peripheral blood. Gives rise to *granulocyte* precursors. Has non-granular basophilic cytoplasm.
myelodysplastic syndromes (MDS)	A group of hematopoietic disorders characterized by increased infection, anemia, and hemorrhage. Arise from "ineffective" hematopoiesis, which produces abnormal myeloid, erythroid, and megakaryocytic precursors. May progress to *acute myeloid leukemia (AML)*.
myelogenous leukemia	*Leukemia* of a *myeloid* lineage *cell*.
myeloid cells	Cells that develop from the *common myeloid progenitors (CMP)*, including erythrocytes, *neutrophils*, *monocyte/macrophages*, *eosinophils*, *basophils*, and megakaryocytes.
myeloma	*Plasma cell dyscrasia* in which highly malignant plasma cells occur either as solid masses or as dispersed clones in the bone marrow. Large quantities of an Ig *paraprotein* are secreted as well as *Bence Jones proteins* (free Ig light chains).
myelopoiesis	The process of HSC differentiation into *common myeloid progenitors (CMPs)* and descendant cells of the *myeloid* lineage.
myeloproliferative diseases (MPD)	*Chronic* hematopoietic cancers of *myeloid* lineage *cells*. Includes *chronic myelogenous leukemia (CML)*, as well as polycythemia vera, essential thrombocythemia, and myelofibrosis.
naive lymphocytes	Resting B and T cells that have never interacted with specific antigen in the periphery. Also called "virgin" or "unprimed."
naked DNA vaccine	*Vaccine* based on an isolated DNA plasmid (no vector) encoding the vaccine antigen. The plasmid is introduced directly into the vaccinee's body, where it is taken up by host cells that then synthesize the pathogen protein.
NALT	See *nasopharynx-associated lymphoid tissue* and *MALT*.
nasopharynx-associated lymphoid tissue (NALT)	Mucosal lymphoid tissues including the *tonsils* and adenoids in the nasopharynx and diffuse collections of lymphocytes resident in the upper respiratory epithelium. See *MALT*.
natural antibodies	*Polyreactive antibodies* (mostly IgM) that arise in the body independently of external antigenic stimulation. Includes antibodies binding to various components of commensal microbes as well as to self-elements such as nucleic acids, erythrocytes, insulin, and other cellular components. Natural antibodies are produced by CD5$^+$ B cells and considered part of *innate immunity*.
natural cytotoxicity	*Perforin/granzyme-mediated cytotoxicity* of target cells carried out by NK cells.
natural killer (NK) cells	Lymphoid lineage cells that recognize nonself entities with broad specificity. NK cells are activated when a target cell expresses ligands that bind to *NK activatory receptors* but lacks MHC class I to engage *NK inhibitory receptors*. NK cells bear cytoplasmic granules that allow them to kill targets such as virus-infected and tumor cells by *natural cytotoxicity*. NK cells also secrete inflammatory cytokines and carry Fc receptors mediating *ADCC*. Considered sentinels of innate immunity.
natural killer gene complex (NKC)	Cluster of *NK inhibitory/activatory receptor* genes on mouse chromosome 6 and human chromosome 12. Includes genes for Ly-49, CD94, CD69, NKG2, and NKRP1 (NK1.1) molecules.
natural killer T (NKT) cells	Lymphoid lineage cells with combined features of T cells and NK cells. NKT cells carry a semi-invariant *TCR* recognizing glycolipid or lipid antigens presented on *CD1d molecules*. Once activated, NKT cells quickly secrete Th0 cytokines that support the activation and differentiation of B and T cells. NKT cells may also carry out target cell cytolysis like NK cells.
necrosis	Sudden, uncontrolled cell death due to infection or trauma. The cell loses the integrity of its nucleus and spills its contents into the surrounding milieu, releasing "danger signal" mediators that trigger inflammation.
Nef (HIV)	In an HIV-infected cell, this multi-functional *HIV* protein (1) forces clathrin-mediated internalization of certain MHC class I, class II, and CD4 molecules; (2) interferes with the proton pump required to acidify endosomes; (3) stimulates intracellular signaling pathways such that the infected cell supports viral DNA synthesis; (4) upregulates FasL and activates caspase-3; and (5) downregulates survival signaling.

Term	Definition
negative selection	A central process that removes *autoreactive* cells from the lymphocyte population destined for the periphery. In the bone marrow, developing B cells whose *BCRs* have high *affinity/avidity* for self antigens present on bone marrow stromal cells are induced to undergo *apoptosis*. In the thymus, developing thymocytes whose *TCRs* have high affinity/avidity for self-peptide complexed to self MHC presented by thymic epithelial cells are induced to undergo *apoptosis*. *NKT cells* also undergo negative selection in the thymus.
neo-antigen	A foreign antigen generated by the binding of a *xenobiotic* to a self protein. See *contact hypersensitivity*.
neonatal immunity	Immunity in the newborn due to maternal circulating antibodies that were passed on to the fetus via the placenta, or maternal secretory antibodies consumed by the newborn in breast milk. A form of *passive immunization*.
neonatal tolerance	Phenomenon that tolerance to an antigen is established more easily in neonatal than mature animals. Due to functional immaturity and low numbers of neonatal T and B cells, DCs, macrophages, and FDCs, and altered *lymphocyte recirculation* patterns.
neoplasm	An abnormal tissue mass, i.e., a tumor. May be *benign* or *malignant*.
neoplastic clone	See *neoplasm*.
neutralization	Ability of an antibody to bind to an antigen and physically prevent it from binding to and infecting a host cell.
neutropenia	Greatly reduced numbers of *neutrophils* in the circulation.
neutrophilia	Greatly increased numbers of *neutrophils* in the circulation.
neutrophils	Most common leukocytes in the body. Respond immediately in great numbers to tissue injury or pathogen attack. Neutrophils are both *granulocytes* and *phagocytes*, and are distinguished by their irregularly shaped multi-lobed nuclei and cytoplasmic granules that stain neutrally. Also called "polymorphonuclear cells" (PMNs).
NF-κB	A family of heterodimeric *transcription factors*. NF-κB activation is induced by engagement of the BCR, TCR, and many cytokine/growth factor receptors. In resting cells, NF-κB is held inactive in the cytoplasm by binding to the IκB inhibitor. Receptor engagement stimulates intracellular signaling that activates the IKK kinase, which phosphorylates IκB, triggering its degradation. Free NF-κB enters the nucleus and binds to the κB *DNA binding motif*, initiating new gene transcription.
NHEJ pathway	See *non-homologous end joining pathway*.
Nijmegen breakage syndrome (NBS)	*PI* that includes variable hypogammaglobulinemia and *CID*. Patient cells subjected to ionizing radiation show abnormal levels of chromosomal defects. Caused by mutations in the gene encoding nibrin, a protein involved in *NHEJ pathway* of *DNA repair*.
nitric oxide (NO)	Reactive nitrogen species that interferes with the citric acid cycle and microbial enzymes containing iron or sulfur atoms. Also inhibits virus replication.
NK	See *natural killer cells*.
NK activatory receptors	Receptors bearing *ITAMs* whose engagement in the absence of inhibitory receptor engagement induces *natural cytotoxicity* and *cytotoxic cytokine* secretion by *NK cells*, resulting in the death of *target cells* lacking self MHC class I molecules. Include NCRs, KARs, NKG2D, -C, and -E, and Ly-49D, -H, -P, and -W.
NK inhibitory receptors	Receptors bearing *ITIMs* whose engagement by self MHC class I molecules on a potential target counteracts the effects of *NK activatory receptor* engagement, preventing target cell destruction. Include KIRs; Ly-49A, -B, -C, -E, -F, -G, Ly-49I; NKG2A and -B; and ILT2.
NKT	See *natural killer T cells*.
NK tolerance	Established during NK development by the selection of only those NK clones that express *NK inhibitory receptors* able to recognize self-MHC and send a strong enough signal to override host cell-induced *NK activatory receptor* signaling.
NK/T precursor	*CLP*-derived precursor that can develop into T or NK cells but not B cells.
N nucleotides	= Non-templated nucleotides. Nucleotides that are added randomly by TdT onto the ends of nicked hairpins between two antigen receptor *gene segments* undergoing *V(D)J recombination*. Occurs before final repair and ligation.

Term	Definition
NOD mice	= Non-obese diabetic mice. An inbred mouse strain that develops *autoimmune* diabetes. An important model for human *T1DM*.
NOD proteins	= Nucleotide-binding oligomerization domain proteins. Family of PRMs that operate in the cytoplasm to detect products of intracellular pathogens. Structurally related to the TLRs. Engagement of NOD proteins induces expression of pro-inflammatory cytokines.
non-classical Hodgkin's lymphoma	Also known as "nodular lymphocyte-predominant HL." Comprises 5% of HL tumors. Tumor cells are CD30$^-$ CD15$^-$ CD19$^+$ CD20$^+$ *popcorn cells*.
non-cytopathic virus	*Virus* that takes over cellular functions and reproduces without damaging the host cell.
non-Hodgkin's lymphoma	*Lymphoma* in which the solid tumor mass consists almost entirely of malignant lymphocytes. Usually derived from a transformed peripheral B cell but sometimes from a transformed peripheral T cell. This heterogeneous family of cancers includes precursor B lymphoblastic leukemia/lymphoma, mantle cell lymphoma, B cell chronic lymphocytic lymphoma, follicular lymphoma, mucosa-associated lymphoid tissue lymphoma, diffuse large cell lymphoma, Burkitt's lymphoma, precursor T lymphoblastic leukemia/lymphoma, adult T cell leukemia/lymphoma, mycosis fungoides, Sézary syndrome, anaplastic large cell lymphomas, angioimmunoblastic T cell lymphoma, and peripheral T cell lymphoma (unspecified).
non-homologous end-joining (NHEJ) pathway	*DNA repair* pathway present in all cells. Acts to repair double-stranded DNA breaks without the need for DNA sequence homology between the ends to be joined; i.e., the DNA ends are simply ligated back together. Involves some of the same proteins involved in *V(D)J recombination*.
non-myeloablative conditioning	Less aggressive regimen of *chemotherapy* designed to only partially deplete the *bone marrow* in preparation for *transplantation*.
non-responder	An individual that fails to mount an immune response to a foreign protein that provokes a strong immune response in other individuals of the same species. Due to lack of expression of an MHC allele with the ability to present an immunogenic peptide from the foreign protein.
normal cellular antigen (class of TAA)	Normal macromolecule that is overexpressed in a tumor. Often caused by *gene amplification*. See *tumor-associated antigen*.
normal flora	See *commensal organisms*.
Northern blotting	Laboratory technique in which RNA transcripts are separated by agarose gel electrophoresis, transferred to a nitrocellulose blot, and detected by hybridization to a labeled cDNA probe of complementary sequence.
Notch1	Cell fate determination molecule that regulates transcription by associating directly with nuclear transcription factors. Promotes (1) T cell development at the expense of B cell development, (2) the *DN* to *DP thymocyte* transition, and (3) Th1 differentiation at the expense of Th2 differentiation.
nucleotide excision pathway (NER)	*DNA repair* pathway involving products of the XPA-XPG genes that excise nucleotides damaged by UV irradiation.
nude mice	Natural mouse mutant with an autosomal recessive mutation that both blocks hair follicle development and prevents or greatly reduces thymus development. *Nude mice* are thus hairless and lack T cells. Partial model for *DiGeorge syndrome* in humans.
nurse cells	See *thymus*.
NZB/W F1 mice	F1 progeny of a cross of NZB and NZW mouse strains. Progeny develop *autoimmune hemolytic anemia* and *lupus nephritis* and are used as an animal model for human *SLE*.
Omenn syndrome	*SCID* caused by amino acid substitution mutations in the RAG genes that reduce the activity of the corresponding proteins to 1–25% of normal.
oncogene	A gene whose deregulation is associated with *carcinogenesis*. Oncogenes often encode positive regulators of cell growth. Activation may result from mutation or retroviral integration.
opportunistic pathogen	A pathogen that does not cause disease unless offered an unexpected opportunity by a failure in host defense.
opsonin	A host protein that coats a pathogen or macromolecule such that the entity binds more easily to phagocyte receptors, enhancing *phagocytosis*. Includes immunoglobulins of certain isotypes, the complement intermediates C3b, iC3b, and C4b, and the wound healing protein fibronectin.

Term	Definition
opsonization	Enhanced phagocytosis of a pathogen or macromolecule due to the binding of molecules that interact with cell surface receptors on phagocytes.
oral tolerance	*Experimental tolerization* due to feeding of an immunogen.
organ-specific autoimmunity	An *autoimmune* response in which a specific anatomical site is targeted.
original antigenic sin	Phenomenon in which B cell clones triggered by an original antigen are reactivated in response to a new cross-reactive antigen. This antigen contains novel B cell epitopes as well as T cell epitopes found on the original antigen.
osteoclast	Large, macrophage-like cell in the bone responsible for bone resorption.
p53	A *tumor suppressor gene* that functions as a gatekeeper of genomic stability. Induces cell cycle arrest to allow *DNA repair*, or induces *apoptosis* of cells in which the damage cannot be repaired. Gene most often mutated in human cancers.
PAMPs	See *pathogen-associated molecular patterns*.
panel reactive antibody (PRA) test	A laboratory technique used for *tissue typing*. A prospective transplant recipient's serum is screened against panels of pooled cells expressing a broad spectrum of MHC molecules. Used to identify preformed alloantibodies in the recipient that could mediate *hyperacute graft rejection* of tissue from certain *allogeneic* donors.
Paneth cells	Located at the bottom of *intestinal crypts*. Produce anti-microbial proteins.
paracrine	A molecule acting in paracrine fashion affects only cells located a short distance away.
paralyzed TCRs	TCRs that have bound but cannot release an *antagonist ligand*. Also known as "spoiled" TCRs.
paraneoplastic syndrome	Clinical symptoms arising from the indirect effect of a *malignant tumor* on remote organs or tissues.
paraprotein	Immunoglobulin secreted in high amounts by a clone of transformed *plasma cells*. Detected as an abnormal level of structurally homogeneous Ig in the blood. See *plasma cell dyscrasias*.
parasite	A pathogen that depends on a host organism for both habitat and nutrition at some point in its life cycle, usually doing damage to the host but not killing it. Parasites are eukaryotic organisms with several chromosomes contained in a membrane-bounded nucleus. Includes single-celled *protozoa* and multicellular helminth worms. Ranges in size from only a few micrometers to several meters.
parenteral	Administration of a substance by a non-oral route, i.e., by injection.
passive immunization	Transfer of *antibodies* from one or more immune individuals to a non-immune recipient to provide immediate protection against a particular pathogen. No memory is generated.
pathogen	An organism that causes disease in its host as it attempts to reproduce. The five major types of pathogens are *extracellular antigens, intracellular antigens, viruses, parasites,* and *fungi*.
pathogen-associated molecular patterns (PAMPs)	Structural patterns present in components or products that are common to a wide variety of microbes but are not usually present in host cells. Ligands for *pattern recognition molecules (PRMs)*.
pathogenesis-related (PR) proteins	Chitinases, β1,3 glucanases, peroxidases, RNases, proteases, and enzyme inhibitors released by plant cells that have developed *LAR*. The PR proteins degrade pathogen components, resulting in the release of small degradation products that induce additional plant responses.
pattern recognition molecules (PRMs)	Proteins that recognize *PAMPs*. PRMs can be soluble or membrane-bound. Soluble PRMs include the collectins (including *MBL*), *acute phase proteins, natural antibodies,* and *NOD proteins*. Membrane-bound PRMs are called *pattern recognition receptors (PRRs)*.
pattern recognition receptors (PRRs)	Widely distributed membrane-bound *PRMs* fixed in either the plasma membrane of a cell or in the membranes of its endocytic vesicles. Include CD14, scavenger receptors, and the *Toll-like receptors*. NK activatory receptors, the γδ TCR, and the NKT semi-invariant TCR can also be considered PRRs. Engagement of PRRs induces the expression of pro-inflammatory cytokines.
pemphigus	*Autoimmune type II hypersensitivity* characterized by blistering of skin and mucosal surfaces. Caused by *autoantibodies* (usually IgG or IgA) that attack the *desmogleins* required for adhesion between epithelial cells.

Term	Definition
peptide interception	Potential mechanism of *cross-presentation*. An *MIIC* containing exogenous peptides fuses with a vesicle containing MHC class I recycling from the cell surface. Low pH in the MIICs forces the endogenous peptide out of the groove of the recycling MHC class I molecule and allows the TAP-independent loading of an exogenous MIIC peptide onto MHC class I.
peptide regurgitation	Potential mechanism of *cross-presentation*. Extracellular proteins are internalized and processed to peptides as usual within the DC endosomes, but the peptides are then released ("regurgitated") back into the extracellular environment. The extracellular exogenous peptide then mediates a peptide exchange at the cell surface independently of intracellular processing. Peptides already associated with MHC and displayed on the surface are displaced by the exogenous peptide of interest.
peptide vaccine	A *vaccine* that uses a small antigenic peptide for immunization.
perforin	See *perforin/granzyme-mediated cytotoxicity*.
perforin/granzyme-mediated cytotoxicity	Mechanism of *target cell* destruction triggered when CTLs, NK or NKT cells degranulate to release both granzymes, which are proteases, and perforin, which is a pore-forming protein. These molecules are taken up by the target cell and perforin then facilitates the egress of granzyme from target cell endosomes. Perforin and granzyme are synthesized in advance and are stored in cytoplasmic granules in both cell types.
periarteriolar lymphoid sheath (PALS)	A cylindrical lymphoid tissue surrounding each splenic arteriole like a sleeve. The PALS are populated primarily by mature T cells but also contain some B cells, plasma cells, macrophages, and interdigitating DCs.
peripheral tolerance	Functional silencing or deletion of self-reactive lymphocytes that escaped elimination during the establishment of *central tolerance*. Established by a collection of mechanisms that act outside the thymus and bone marrow, including peripheral *clonal deletion*, *anergization*, *clonal exhaustion*, *immunological ignorance*, *immune privilege*, *immunosuppressive cytokines*, *immune deviation*, and *regulatory T cells*.
periphery	Tissues and organs outside the primary lymphoid organs (bone marrow and thymus).
persistent infection	A pathogen remains in the host's body for prolonged periods or throughout life. Persistent infections may be *latent* or cause *chronic disease*.
Peyer's patches	*Secondary lymphoid tissue* in the intestine. See *intestinal follicle*.
phage display	A phage display library of *Fab* fragments is constructed by joining each of the *variable region* gene sequences of a collection of B cells to a gene encoding a bacteriophage minor coat protein, pIII. Each phage then expresses a fusion protein on its coat surface consisting of an Ab variable domain joined to the N-terminus of the pIII coat protein. The desired Ab fragment is selected by "panning" the recombinant phage library over the immobilized antigen of interest.
phagocyte	Cell capable of carrying out *phagocytosis*.
phagocytosis	Process by which a cell captures ("eats") particulate pathogens or organic or inorganic debris by membrane-mediated engulfment. Ligands on the particle bind to multiple receptors on the phagocyte surface in a "zippering" manner and induce actin polymerization, plasma membrane invagination, and the sequestration of the particle into an intracellular vesicle called a *phagosome*. If the particle is a microbe, it will be killed by *reactive oxygen intermediates* and *reactive nitrogen intermediates* species within the phagosome. The phagosome then enters the endocytic processing pathway and undergoes step-wise maturation to form a *phagolysosome*.
phagolysosome	See *phagocytosis*.
phagosome	See *phagocytosis*.
Philadelphia (Ph) chromosome	Shortened chromosome 22 resulting from the t(9;22)(q34;q11) reciprocal translocation fusing the Bcr gene on chromosome 22 with the Abl gene on chromosome 9. The chimeric protein encoded by the Bcr–Abl fusion gene has constitutive Abl tyrosine kinase activity due to abnormal regulation and oligomerization capability mediated by the Bcr moiety. Prominent in many CML and some ALL patients.
PI	See *primary immunodeficiency*.
plantibodies	Antibodies produced by transgenic plants.

Term	Definition
plasmablasts	Proliferating progeny of an activated B cell. Destined to become *plasma cells* and so have increased rER, Golgi, and ribosomes.
plasma cell dyscrasias	Hematopoietic cancers of *plasma cells*.
plasma cells	Terminally differentiated B cells that secrete antibody. Plasma cells do not express MHC class II or mIg and can no longer receive T cell help. "Short-lived" plasma cells differentiate rapidly without undergoing isotype switching or somatic hypermutation and thus secrete low-affinity IgM antibodies. "Long-lived" plasma cells differentiate after *isotype switching* and *somatic hypermutation* and thus secrete high-affinity antibodies of diverse effector functions.
plasmapheresis	An individual's blood is withdrawn and passed through a machine designed to remove antibody proteins. The machine then returns the treated blood to the patient.
platelet-activating factor (PAF)	A lipid inflammatory mediator that activates *platelets*.
platelets	Small, colorless, irregularly shaped non-nucleated cells in blood derived from megakaryocytes. Activated platelets aggregate to promote blood clotting and secrete cytokines (from preformed mRNAs) that influence the functions and migration of other leukocytes during inflammation.
pleiotropic	A pleiotropic molecule (e.g., cytokine) is one that acts on many different cell types, or has multiple effects on the same cell type.
pluripotency	The capacity to differentiate (specialize) into any one of a variety of other cell types.
P nucleotides	If the hairpin loops joining *gene segments* undergoing *V(D)J recombination* are nicked in the intervening DNA rather than at the precise ends of the coding sequences, a recessed strand end and an overhang are generated. The nucleotides added to fill the gaps on both strands are considered P nucleotides.
polyclonal activator	An entity that can activate lymphocytes non-specifically. For example, an antibody that binds to mIgM can non-specifically activate multiple B cell clones regardless of their antigen receptor specificity. Similarly, for T cells, an antibody that binds to *CD3 chains* non-specifically activates many different T cell clones.
polyclonal antiserum	An antiserum that contains antibodies produced by many different B cell clones in response to an antigen. These antibodies are raised in response to different epitopes on the antigen.
poly-Ig receptor	= Polymeric immunoglobulin receptor. Ig domain-containing transmembrane receptor positioned on the basolateral surface of mucosal epithelial cells. Binds to the *J chain* in secreted Ig and facilitates *transcytosis* of the Ig and its delivery into the external secretions. Cleavage of the poly-Ig receptor occurs during this process, leaving a portion called *secretory component* attached to the Ig molecule.
polymerase chain reaction (PCR)	Amplification of a DNA sequence of interest. Known oligonucleotide primers flanking the gene fragment of interest are incubated with the gene under appropriate conditions with a heat-stable DNA polymerase. Multiple cycles in which the DNA is denatured, annealed with the primers, and copied are required.
polymeric Ig	Ig macromolecule composed of multiple identical H_2L_2 monomers joined by a J chain. Examples include pentameric IgM [$(H_2L_2)5$], which has ten antigen-binding sites, and dimeric IgA [$(H_2L_2)2$], which has four.
polymorphic	A given gene is polymorphic if different *alleles* exist for this gene in the population.
polymorphonuclear (PMN) cells	See *neutrophils*.
polymyositis (PM)	A form of *organ-specific autoimmunity* in which the fibers of the voluntary muscles come under inflammatory attack. Slowly progresses from distal to proximal body muscles. Characterized by autoantibodies against aminoacyl tRNA synthetases.
polyreactive antibody	An individual Ig recognizes and binds to several different antigens of substantially different structure with varying affinity.
popcorn cells	Malignant cells of *non-classical HL*. Surface phenotype of CD30⁻ CD15⁻ CD19⁺ CD20⁺. Ig genes of such cells show *somatic hypermutation*.

Term	Definition
positive cross-match	The laboratory finding demonstrating that an individual's blood contains preformed antibodies to one or more HLA molecules expressed on an *allogeneic* donor organ.
positive selection	Process in the *thymus* that promotes the survival and maturation of developing αβ thymocytes that bind to self-peptide complexed to self MHC with only low *affinity/avidity*, i.e., those clones that are most likely to recognize nonself-peptide complexed to self MHC with high affinity. *NKT cells* also undergo positive selection in the thymus. It is controversial whether developing B cells with low affinity for self-antigen are positively selected, i.e., receive survival signals.
post-GC	Term used to describe a *leukemia/lymphoma* in which the transformed cells have the phenotype of an activated B cell or memory B cell. The genome shows a high frequency of somatic hypermutations.
preactivation	The state of a provirus that remains untranscribed in the host cell genome due to the absence of host cell stimulation.
pre-BCR	Complex composed of *SLC*, a candidate μ chain, and the Igα/Igβ heterodimer. The pre-BCR is inserted transiently in the membrane of a developing B cell for *productivity testing* of a particular heavy chain VDJ combination. A pre-BCR cannot respond to antigen but may bind to stromal ligands such that an intracellular signal is delivered indicating that the μ chain is functional.
precipitin reaction curve	Large antigen–antibody complexes precipitate out of solution. A graph showing the amount of antibody precipitated out of solution by varying amounts of antigen yields a precipitin reaction curve.
pre-clinical trials	Testing of an entity for toxicity and teratogenicity (the tendency to induce fetal abnormalities) in non-primate animals and non-human primates. Testing for *efficacy* is also carried out when possible.
pre-GC	Term used to describe a *leukemia/lymphoma* in which the transformed cells have the phenotype of a naive B cell. The genome shows few somatic hypermutations.
preintegration complex	A complex in the host cell cytoplasm containing all the elements of the *HIV core* except the capsid. Within this complex, HIV RT rapidly synthesizes the viral DNA. The matrix proteins of the complex then direct it into the nucleus.
premalignant clone	A clone of genetically altered cells at a stage prior to *malignant conversion*. Arises from a *target cell* during the *progression* step of *carcinogenesis*.
preneoplastic clone	A clone of genetically altered cells at a stage prior to becoming a *neoplasm*. Arises from a *target cell* during the *promotion* step of *carcinogenesis*.
presentosome	Collection of molecules on both sides of the ER membrane that facilitate *endogenous antigen processing*. Peptides derived by proteasomal degradation are complexed with HSP70, HSP90, or HSP110 in the cytosol for transport to TAP. On the other side of TAP in the ER, the peptides are complexed to gp96 or PDI.
pre-T alpha chain (pTα)	Invariant transmembrane TCRα-like chain expressed only in *DN thymocytes*. Used for *productivity testing* of candidate TCRβ chains. Not required for γδ T cell development. See *pre-TCR*.
pre-TCR	Transient complex composed of a candidate TCRβ chain plus the *pTα* chain plus the *CD3 chains*. Enables *productivity testing* of a particular V(D)J rearrangement in the TCRβ gene. A pre-TCR cannot respond to antigen but may bind to stromal ligands such that an intracellular signal is delivered indicating that the TCRβ chain is functional.
pre-TCR activation	The rapid proliferation during β-selection of only those thymocyte clones with functional TCRβ chains. Induced by *pTα* signaling triggered by an unknown ligand.
pre-vertebrates	Animals evolving from a *deuterostome* ancestor but lacking a true backbone. Include echinoderms (starfish, sea urchins) and protochordates (ascidians and tunicates, such as sea squirts).
primary follicles	Spherical aggregates of (mainly) resting mature B lymphocytes, macrophages, and FDC within the B cell-rich regions of *secondary lymphoid tissues* such as spleen, lymph nodes, and Peyer's patches.
primary immunodeficiency (PI)	Failure of a component of the immune system due to an inborn genetic mutation.

Term	Definition
primary lymphoid tissue	Lymphoid tissues (*bone marrow* and *thymus*) where lymphocytes are generated and mature.
primary response	The immune response mounted upon a first exposure to a nonself entity. Activated T and B lymphocytes that recognize epitopes on the entity proliferate and differentiate into antigen-specific *effector lymphocytes* and *memory cells*. The primary response is slower and weaker than secondary (or subsequent) responses.
primary tumor	The original tumor mass established by the first transformed cell.
prime-boost strategy	Vaccination approach in which primary *vaccination* with a *recombinant vector* or *naked DNA vaccine* is followed by a *booster* with a recombinant protein *subunit vaccine*.
priming	First encounter of a naive lymphocyte with specific antigen. Leads to a *primary response*.
prions	Proteins that are infectious despite being non-living entities lacking any type of associated nucleic acid. Possess an abnormal conformation that induces *spongiform encephalopathies*. Prions spread disease by altering the conformation of their normal protein counterparts in the brain of the infected host.
probe hybridization (tissue typing)	A laboratory technique used for *tissue typing*. Total DNA is extracted from an individual's cells and *PCR* is carried out using primers that allow generic amplification of all HLA alleles. Sequences are detected by hybridization to allele-specific probes.
productive infection	An infection in which the pathogen multiplies despite immune responses against it.
productivity testing	The testing by a developing B or T cell of the functionality of a particular VDJ combination (in the H or β chain, respectively) by transcribing the newly assembled gene and translating its mRNA. In combination with accessory proteins of the *pre-BCR* or *pre-TCR*, the candidate chain is expressed on the cell surface and delivers a signal back down to the cell that the recombination has been successful. See also *surrogate light chain (SLC)* and *pTα*.
proenzyme	Inactive precursor of an enzyme; also called a "zymogen." Often activated by enzymatic cleavage in a cascade, e.g., complement activation.
professional APC	See *antigen presenting cell*.
programmed cell death	See *apoptosis*.
progression	Third stage of *carcinogenesis*. Additional genetic events occur in one cell of a *preneoplastic clone* that give it a significant growth advantage. This cell generates a *neoplasm (neoplastic clone)* that is a *premalignant clone*.
promiscuous binding	A binding site that accommodates several different ligands with similar affinity exhibits promiscuous binding.
promotion	Second step of *carcinogenesis*. A *target cell* formed by *initiation* is exposed to a stimulus that allows its selective proliferation to form a *preneoplastic clone*. Reversible with removal of the stimulus.
prophylactic vaccination	A *vaccine* given to pre-empt disease. See *vaccination*.
ProPO system	= Prophenoloxidase activating system. Operates only in *higher invertebrates*. Microbial ligands activate the *coagulation* cascade, in turn activating serine proteases in the *hemolymph*. These serine proteases cleave inactive, circulating prophenoloxidase (ProPO) and generate active phenoloxidase (PO). PO is an oxidoreductase that oxidizes phenols to produce quinones. The quinones polymerize nonenzymatically to form melanin that coats and paralyzes the microbe in a process called the *melanization reaction*.
prostaglandins	Lipid inflammatory mediators. Phospholipids in the membranes of macrophages, monocytes, neutrophils, and mast cells are degraded and converted to arachidonic acid. Metabolism of arachidonic acid via the cyclooxygenase pathway produces prostaglandins. See also *inflammation*.
protease (HIV)	*HIV* enzyme that cleaves polyprotein precursors to give rise to mature viral components.

Term	Definition
proteasome	Huge, cylindrical, non-lysosomal cytoplasmic organelle composed of multiple proteases that digest proteins into peptides via multiple catalytic activities. Integral component of the *endogenous antigen processing and presentation* pathway. The standard proteasome dominates in resting cells in which mostly unwanted self proteins are digested. The immunoproteasome dominates in DCs and in other cells during inflammation, meaning that the digestion of foreign proteins is a major function.
protective epitopes	In the context of *vaccination*, epitopes from a pathogen that induce an immune response that will prevent subsequent infection by that pathogen. Derived from "protective antigens."
proteins A and G	Proteins found naturally on the surface of *Staphylococcus aureus* bacteria that are used in laboratory assays because of their ability to bind the Fc region of antibody proteins.
protein tyrosine kinase (PTK)	Kinases (phosphorylating enzymes) involved in *intracellular signaling pathways* leading to transcriptional activation. Activated PTKs phosphorylate the tyrosine residues of substrate proteins, initiating a GTP-dependent phosphorylation/dephosphorylation cascade that involves other kinases and additional substrates. Examples include the Src and Syk PTK families.
proteomics	The study of all the proteins encoded in the genome of an organism. "Expression proteomics" is the study of all proteins actually expressed by an organism.
proto-oncogene	The normal cellular counterpart of an *oncogene*.
protostomes	*Coelomate*-derived animals that then became *higher invertebrates*.
protozoans	Single-celled *parasites* that often replicate directly within host cells.
proviral DNA	See *provirus*.
provirus	Viral DNA that has been integrated in the host cell genome.
pruritis	Intense itching.
pseudogene	DNA sequence that resembles a functional gene but contains a defect (such as a premature stop codon) such that it cannot be expressed.
pSMAC	= Peripheral *SMAC*. Inner pSMAC surrounds the *cSMAC* and contains CD2/LFA-3 pairs. Outer pSMAC surrounds the inner pSMAC, contains LFA-1/ICAM-1 pairs, and is linked to the the actin cytoskeleton.
psoriasis (PS)	*Autoimmune disease* characterized by reddened skin lesions that develop a covering of silver scaly skin cells. Due to abnormal function of keratinocytes.
PTEN	Tumor suppressor gene that negatively regulates the activation of the cell survival kinase PKB/Akt induced by PI3K signaling. Second most commonly mutated tumor suppressor gene in human cancer.
pulsing	Loading of APCs with antigen by electroporation.
purine nucleoside phosphorylase (PNP) deficiency	Subtype of *SCID* caused by mutation of purine nucleoside phosphorylase required for the *salvage pathway of nucleic acid synthesis*.
purpura	Areas of purplish or brownish red discoloration on the skin caused by leakage of blood (hemorrhage) into the skin layers.
pus	Cream-colored substance at a site of injury or infection. Consists of an accumulation of leukocytes that have died fighting infection.
pyogenic infections	Infections that induce an acute, immunopathic inflammatory response accompanied by high fever.
R5 viruses	Strains of *HIV* that bind to CCR5 and infect macrophages as well as CD4$^+$ T cells. Originally called "M-tropic viruses."
radiation therapy	Use of ionizing radiation to kill tumor cells. Can be delivered either externally as a focused beam or via *brachytherapy*.
radioimmunoassay (RIA)	*Binder-ligand assay* in which the antibody or antigen molecule is *tagged* with a radioisotope.
radioimmunotherapy (RIT)	Use of *immunoradioisotopes* for cancer therapy.

Term	Definition
RAG blastocyst complementation	A technique often used to examine the function of a gene when its null mutation in a whole animal is embryonic lethal. RAG mice lack T and B cells. When *embryonic stem (ES) cells* from a mutant of interest (with intact RAG genes) are introduced into the blastocyst (8-cell stage) of a developing RAG$^{-/-}$ embryo, the embryo absorbs the new cells and continues normal development after implantation. Different tissues are derived from different parts of the chimeric embryo, meaning that some reconstituted mice will be viable (due to a lack of expression of the mutation in vital organs) but have a lymphoid system expressing the mutation of interest. The effect of this mutation on lymphocyte development and function can then be examined.
RAG recombinases	= Recombination activation gene recombinases. RAG-1 and RAG-2 are the key enzymes mediating *V(D)J recombination* of *gene segments* in the Ig and TCR loci. The RAG genes are expressed only in developing B and T lymphocytes.
Ras/MAPK signaling pathway	Intracellular signaling pathway involving the membrane-associated signal transducer Ras, which in turn activates the *MAPK* enzyme ERK1 (extracellular signal regulated kinase 1). ERK1 activates c-fos, a component of the AP-1 *transcription factor*. Ras/MAPK signaling is activated downstream of TCR engagement.
Raynaud's phenomenon	Small blood vessels in the limbs contract abnormally in response to cold temperatures or stress. May be observed in autoimmune diseases such as *scleroderma (SD)* and *SLE*.
reactivation	When a *latent infection* undergoes *reactivation*, it re-commences replication, leading to a *productive infection* with associated disease symptoms.
reactive arthritis	*Autoimmune* development of arthritis in the knees, feet, and sacroiliac region 1–4 weeks after an acute bacterial infection.
reactive nitrogen intermediates	Nitrogen-derived free radicals (such as nitric oxide) that kill microbes within *phagosomes*.
reactive oxygen intermediates (ROI)	Oxygen-derived free radicals that kill microbes within phagosomes. Includes singlet oxygen, H_2O_2, hydroxyl radicals. Also called "reactive oxygen species" (ROS).
reaginic antibodies	Prior to the molecular characterization of IgE, pathogenic antibodies directed against *allergens* were called *reaginic antibodies*. Binding of these antibodies by FcεR on a mast cell stimulates release of inflammatory mediators.
receptor blockade	Anergization of a B cell due to very large amounts of antigen that persistently occupy the BCRs without cross-linking them. Receptor-associated tyrosine kinase activation is decreased, inhibiting signal one.
receptor editing	Secondary round of gene rearrangement in an autoreactive immature mIg$^+$ B cell. If self-antigen binds to the BCR during development in the bone marrow, the cell is given a brief opportunity to rearrange its light chain again in an attempt to stave off apoptosis by altering its antigenic specificity.
receptor-mediated endocytosis	Process in which binding of a soluble macromolecule to its complementary cell surface receptor induces internalization via *clathrin* polymerization. Invagination of the clathrin-coated pit captures the receptor and bound macromolecule in a clathrin-coated vesicle. The vesicle then enters the *endocytic* processing *pathway*.
recombinant vector vaccine	*Vaccine* in which the DNA encoding the vaccine antigen is incorporated into a vector that enters host cells and promotes translation of the vaccine antigen directly within them.
recurring chromosomal translocation	A *chromosomal translocation* that appears in many different patients. Most often observed in hematopoietic cancers, and usually associated with malignancies of the same type of cell.
red marrow	See *bone marrow*.
red pulp	See *spleen*.
Reed-Sternberg cells	The tiny percentage of cancerous cells present in a classical *Hodgkin's lymphoma* tumor mass. Large, CD30$^+$, multi-nucleate B lineage-like cells with a distinctive shape and variable somatic hypermutation in the Ig genes.
refractory cancer	A malignancy that resists treatment with conventional *chemotherapy* or *radiation* treatment.
regulators of complement activation (RCA)	Family of structurally homologous soluble and membrane proteins that bind to C4 and C3 products and interfere with *complement* activation. Also known as *complement control proteins (CCP)*. Include DAF, C4bp, CR1, factors H and I, MCP, vitronectin, clusterin, MIRL, and HRF.

Term	Definition
regulatory peptides (plants)	Induced during the establishment of *LAR* in a plant. Include salicylic acid (SA), ethylene (ET), and jasmonates (*JA*). ET and JA are plant hormones analogous to mammalian inflammatory mediators.
regulatory T cells	T cells that inhibit the responses of other immune system cells. Include CD4$^+$ subsets such as T_{reg}, *Tr1*, and *Th3 cells*, and CD8$^+$ subsets such as *Ts* cells. *In vitro*, inhibition is mediated by intercellular contacts and/or *immunosuppressive cytokine* secretion. Regulatory T cell populations may be induced or pre-existing.
relapse	The reappearance of clinical disease in a patient who was previously in *remission*.
relative sibling risk	A ratio comparing the prevalence of a disease among siblings and its prevalence in the general population.
remission	A patient's disease is in remission if it cannot be detected clinically. In the case of hematopoietic cancer, a remission is defined as a *complete clinical response* that holds for at least 4 weeks following treatment.
repertoire	The pool of antigenic specificities represented in the total population of T or B lymphocytes, as in "the T cell repertoire" or the "B cell repertoire."
replicative senescence	The death of a cell once it has reached the end of its pre-determined number of cell divisions.
reservoir	A non-human species or environmental niche in which a *pathogen* that normally infects humans can survive.
respiratory burst	After *neutrophil* degranulation, NADPH-dependent oxidases attached to the granule membranes stimulate both the generation of inorganic radicals and myeloperoxidase action resulting in the formation of hypochlorous acid, which oxidizes nucleic acids, amino acids, and thiols of the microbe. The significant increase in oxygen utilization by the NADPH oxidases necessary to sustain this antimicrobial mechanism is the *respiratory burst*.
restriction fragment length polymorphism (RFLP)	Differences in fragment lengths generated when DNA sequences surrounding different *alleles* of a polymorphic gene are digested by a certain restriction enzyme. Differences are due to variation in the number and location of restriction sites among the alleles. May be resolved by gel electrophoresis.
reticular dysgenesis (RD)	A rare, severe subtype of *SCID* that may result from a defect in *HSC* maturation. Characterized by variable deficits in T, B, NK, and other hematopoietic lineages.
retroviral oncogene	When an integrated *retrovirus* incorporates a copy of a *proto-oncogene* permanently into its genome, adjacent regulatory sequences may convert the proto-oncogene into a retroviral oncogene able to transform new host cells upon infection. See also *oncogene*.
retrovirus	*Virus* characterized by expression of *reverse transcriptase* and *integrase* enzymes that allow the synthesis of a DNA copy of the viral RNA genome and its integration into the host cell DNA.
reverse genetics approach	Process of establishing the role of a gene in a given disease by examining the genome of an animal with the disease and determining which of its genes have sustained mutations.
reverse transcriptase	Enzyme of retroviruses (e.g., *HIV*) that makes a nuclease-resistant double-stranded DNA copy of the viral RNA genome. A relatively high frequency of mutations can be introduced into the DNA due a lack of proofreading capability.
reverse vaccinology	Design of a *vaccine* antigen by study and selection of sequences from a pathogen's genome or proteome that are most likely to be *epitopes* recognized by T cells. Also called "epitope-driven vaccine design."
Rev (HIV)	Retroviral regulatory protein that promotes the exit of viral transcripts from the host cell nucleus into the cytoplasm for translation.
Rh disease	Also called "erythroblastosis fetalis." The erythrocytes of the fetus are destroyed during the pregnancy of an Rh$^-$ mother carrying her second (or subsequent) Rh$^+$ fetus. Due to the response of maternal anti-Rh memory B cells originally primed upon exposure to Rh$^+$ cells during delivery of the first Rh$^+$ baby. Prevented by *passive immunization* of the mother in advance with anti-Rh *IV-IG* to prevent memory cell generation during each pregnancy with an Rh$^+$ fetus.
rheumatic	Term indicating a disease in which the joints, muscles, or connective tissues of the body come under inflammatory attack.

Term	Definition
rheumatic fever	*Autoimmune disease* attacking the joints, heart muscle, heart valves, kidney, and central nervous system. Characterized by fever and prior streptococcal infection. Antibodies against streptococcal M protein are often present.
rheumatoid	See *rheumatic*.
rheumatoid arthritis (RA)	*Organ-specific autoimmunity* in which the synovial membranes in joints are attacked. Characterized by the presence of *rheumatoid factor*.
rheumatoid factor	An IgM *autoantibody* directed against the Fc region of host IgG molecules. Present in several *autoimmune diseases* but particularly in *rheumatoid arthritis*.
Rh factor	Antigen expressed on RBCs that defines the Rh blood group system. Differences in Rh positivity between a pregnant woman and the fetus she is carrying can lead to *Rh disease*.
RIA	See *radioimmunoassay*.
ribozyme	A catalytic RNA molecule that binds specifically to mRNAs of complementary sequence and cleaves them, thus acting as "molecular scissors." Used therapeutically to prevent expression of specific proteins.
RSS	= Recombination signal sequence. Composed of a conserved heptamer and a conserved nonamer separated by a non-conserved spacer of either 12 bp (12-RSS) or 23 bp (23-RSS). RSSs flank germline V, D, and J *gene segments*. The complementary nature of the 12-RSS and 23-RSS is responsible for the *12/23 rule* of RAG-mediated gene segment joining during *V(D)J recombination*.
SALT	See *skin-associated lymphoid tissue*.
salvage pathway of nucleic acid synthesis	Biosynthetic process by which new nucleotides are built from the routine degradation of spent nucleic acids. Involves adenosine deaminase and *purine nucleoside phosphorylase (PNP)*.
salvage therapy	In cancer therapy, an experimental or novel treatment regimen tried as a last resort for a patient with *refractory cancer*.
SAPK/JNK signaling pathway	= Stress-activated protein kinase/jun N-terminal kinase signaling pathway. Activated directly in response to extracellular stresses such as osmotic shock and chemical agents. Also activated indirectly downstream of TCR engagement. Culminates in activation of c-jun, a component of the AP-1 *transcription factor*.
sarcoma	A *malignant tumor* derived from cells of connective tissue, muscles, bone, or cartilage.
SCID	= Severe combined immunodeficiency disease. Family of *PIs* characterized by loss of both T and B cell functions due to various genetic defects affecting the generation and/or function of both T and B cells. NK cell function may also be affected.
Scid mice	Natural mouse mutant that lacks T and B cells due to a defect in DNA-PK function required for *V(D)J recombination*.
scleroderma (SD)	*Autoimmune disease* characterized by hardening of the skin and thickening of connective tissue caused by overproduction of collagen by fibroblasts.
Scurfy mouse	Natural mouse mutant showing rough skin, runting, *thrombocytopenia, lymphoadenopathy*, and increased susceptibility to infections. T_{reg} *cells* do not function due to mutation of the FoxP3 gene encoding the scurfin *transcription factor*. Model of human *IPEX*.
secondary antibody	An antibody with specificity for the Fc region of an untagged primary antibody that has been allowed to bind to an antigen of interest. By tagging the secondary antibody and adding it to the assay system after unbound primary antibody is washed away, formation of primary antibody–antigen pairs can be detected.
secondary follicles	Once the *primary follicles* are infiltrated by activated T and B cells, they become *secondary follicles* that form *germinal centers* and foster terminal differentiation of activated B cells into *memory* B and *plasma cells*.
secondary lymphoid tissues	Peripheral lymphoid tissues inhabited by mature lymphocytes. Secondary lymphoid tissues trap antigen in the circulation (*spleen*), in the lymphatic system (*lymph nodes*), at the mucosal membranes (*MALT*), or in the skin (*SALT*). *Adaptive immune responses* are initiated in these sites.

Term	Definition
secondary response	A *secondary response* is mounted by antigen-specific *memory* lymphocytes activated by a subsequent exposure to a given nonself entity. Memory lymphocytes require less *costimulation* than naive lymphocytes. In addition, due to clonal expansion during the *primary response*, the frequency of antigen-specific lymphocytes able to respond in the secondary response is high. The secondary response is therefore faster and stronger than the primary response, allowing it to forestall disease.
secreted antibody	Soluble form of Ig serving as circulating antibody in the blood. Contains a short *tail piece* and no transmembrane region.
secretory antibody	*Secreted antibodies* that undergo post-translational modifications that allow them to enter body secretions such as tears and mucus. Generally contain a *tail piece*, *J chain*, and *secretory component*.
secretory component (SC)	Polypeptide associated with secreted *polymeric Igs*. SC is a protein fragment of the *poly-Ig receptor* that remains associated with the Ig after cleavage of the receptor during *transcytosis* of the Ig through a mucosal epithelial cell.
selectin	See *cellular adhesion molecules (CAMs)*.
self-tolerance	Lack of an immune response to self antigens in an individual's tissues. See *peripheral tolerance*.
sensitization (hypersensitivity)	An abnormal *primary response* to a *sensitizing agent* such that, in a subsequent exposure, the individual mounts an excessive or abnormal secondary response that causes disease rather than immunity. Associated with *immune deviation* to *Th2*. See *elicitation* and *hypersensitivity*.
sensitization (transplantation)	See *alloimmunization*.
sensitizing agent	A generally innocuous *antigen* inducing *hypersensitivity*. Includes *allergens*.
septic shock	See *endotoxic shock*.
serial TCR triggering model	A model based on the hypothesis that a relatively low number of pMHC complexes can successively engage sufficient TCRs to generate the sustained signaling necessary for complete activation of a naive T cell.
seroconversion	A measurable increase in the serum level of specific anti-pathogen antibodies.
serological epitope	An antigenic *epitope* recognized by an antibody or group of antibodies.
serology	The study of the *antibodies* present within a given *antiserum*.
serum antitoxins	A historical term used to describe *antibodies* that could clump and lyse bacteria, and precipitate and neutralize toxins and viruses.
serum sickness	*Systemic type III hypersensitivity* response against a normally innocuous antigen introduced into the body. Most often associated with exposure to foreign serum components, or an environmental or drug antigen. Antibodies form *immune complexes* with the antigen that circulate and then become trapped in multiple body sites, triggering *inflammation*.
Sézary syndrome	Advanced form of *mycosis fungoides*, a T cell *non-Hodgkin's lymphoma*.
shared haplotypes	MHC *haplotypes* shared by two siblings due to genetic inheritance from the same parents. The haplotypes on the maternal and paternal chromosomes of each parent can reassort among offspring in four different patterns, such that a given pair of siblings shares zero, one, or two haplotypes. Siblings thus have a far greater chance of sharing one or two haplotypes than unrelated individuals.
SHIV	= Simian-human immunodeficiency virus. A chimera of SIV_{mac} and HIV-1 that infects macaques. Has the *Env, Tat, Rev*, and *Vpu* genes of HIV-1, but all its other genes are from SIV_{mac}.
Shwarztman reaction	Local or systemic response to endotoxin induced experimentally in animals by two sequential exposures to LPS or IL-1. Localized (subcutaneous) endotoxin exposure followed by systemic (i.v.) administration results in a localized vasculitis and tissue necrosis called the "Shwartzman phenomenon." If both exposures to the inducing endotoxin are systemic, a systemic response called the "generalized Shwartzman reaction" occurs, resulting in *anaphylaxis*.
sicca symptoms	A constellation of symptoms marked by dry eyes leading to blurry vision, dry mouth (xerostomia), and dry throat leading to swallowing difficulties, and dry nose and skin.
signal joint	The DNA sequence formed by the joining together of blunt *RSS* ends following *V(D)J recombination* of *gene segments*.

Term	Definition
silencer	An *enhancer* that suppresses transcription of a target gene.
single nucleotide polymorphism (SNP)	Single nucleotide variation in a gene that generates an amino acid difference in an *allele* of the corresponding protein.
SIV$_{mac}$	Simian immunodeficiency virus that infects macaque monkeys.
Sjögren syndrome (SS)	*Autoimmune disease* caused by autoantibodies that attack exocrine glands and block production of body secretions. Characterized by *sicca symptoms* as well as *rheumatic* symptoms.
skin-associated lymphoid tissue (SALT)	The diffuse collections of epidermal αβ and γδ T cells (*DETCs*) and *DCs* (*Langerhans cells*), and dermal αβ T cells, fibroblasts, DCs, macrophages, and lymphatic vessels.
SLE	See *systemic lupus erythematosus*.
SMAA	= Solid matrix antibody–antigen complex. A vaccine in which mAbs to a vaccine epitope are chemically bound to a microbead and the vaccine epitope is then attached to the mAb. By coating the bead with mAbs to different epitopes, the microbead can display both B and T cell epitopes. See *delivery vehicle*.
SMAC	= Supramolecular activation complex. The interface or contact zone between a T cell and an APC where stabilized *lipid rafts* and triggered TCRs accumulate. Also called the "immunological synapse." The SMAC has a concentric triple ring structure made up of the *cSMAC*, inner *pSMAC*, and outer *pSMAC*, where "c" denotes central and "p" denotes peripheral.
solid organ transplant	Transfer of a solid tissue such as a kidney, heart, lung, liver, or skin from a donor to a recipient.
solitary plasmacytoma of bone	*Plasma cell dyscrasia* in which a single plasma cell tumor occurs within the bone marrow.
somatic cell hybrid	Cell in which the multiple nuclei of a *heterokaryon* have fused, combining the two (or more) sets of chromosomes in one large nucleus.
somatic gene conversion	Homologous recombination in which an acceptor gene is altered by the copying of multiple nucleotides from a homologous donor gene (templated alteration), but the donor gene sequence is not altered (non-reciprocal). A mechanism by which germline diversity of Ig genes is created in some species.
somatic hybrid selection	Use of *complementation* to select *somatic cell hybrids* with desired characteristics.
somatic hypermutation	A process by which single or double nucleotide substitutions are introduced at an unusually high frequency into the V exons of Ig genes. Results in very high numbers of random point mutations in the V regions of Ig proteins. Does not occur in the TCR genes.
somatic recombination (somatic gene rearrangement)	See *V(D)J recombination*.
Southern blotting	A laboratory method in which DNA fragments are separated by agarose gel electrophoresis and transferred to a nitrocellulose blot. DNA fragments of interest are then detected by hybridization to a labeled cDNA probe of complementary sequence.
spleen	An organ in the abdomen containing *secondary lymphoid tissue*. Traps blood-borne *antigens* introduced by an insect bite or due to spillover of a tissue-based infection into the blood. Splenic *white pulp* contains the *PALS* and follicles containing T and B cells. The *marginal zone* surrounding the follicles contains B cell subsets. Splenic *red pulp* consists of *splenic cords* and venous sinuses.
splenic cords	A collagen-containing lattice surrounding collections of erythrocytes, reticular cells, fibroblasts, macrophages, and lymphocytes in the *spleen*.
split tolerance	(1) An experimental situation in which different measures of immune reactivity give different results; e.g., an antigen induces a humoral but not a cell-mediated response. May be due to *immune deviation*. (2) Tolerance is induced to epitope A but not epitope B of the same molecule, or to some antigens present on a cell but not others. (3) T cell tolerance to a tolerogen is invoked but B cell reactivity is maintained.
spongiform encephalopathies	Fatal diseases caused by *prions*. Characterized by CNS lesions that render the brain "sponge-like." Include variant Creutzfeldt-Jakob disease (vCJD) in humans, scrapie in sheep, and bovine spongiform encephalopathy (BSE or "mad cow disease") in cattle.

Term	Definition
spontaneous cancer	A *malignant tumor* that develops in a laboratory animal in the absence of any experimental manipulation.
sporadic cancer	A *malignant tumor* in humans that is caused by transformation of a somatic cell of a tissue, rather than a germ cell.
SP thymocytes	= Single positive *thymocytes*. Thymocytes that express either CD4 or CD8 but not both, giving them a CD4$^+$CD8$^-$ or CD4$^-$CD8$^+$ surface phenotype.
stages of cancer	Stage I: *tumor* is small, confined, and often easily removed by surgery. Stage II: tumor is operable, but risk of disease spread is significant. Stage III: tumor has invaded deeply into surrounding structures and cannot be operated on easily. Stage IV: tumor has metastasized.
stages of lymphoma	Stage I: one or more lymph nodes enlarged in a single, localized region. Stage II: two or more lymph node regions that are adjacent to each other are involved, but on only one side of the diaphragm. One additional organ close to the original affected node may also be infiltrated. Stage III: two or more lymph nodes on both sides of the diaphragm are affected. Tumor cells may be found in the spleen or other organs near an affected lymph node. Stage IV: the tumor mass disseminates widely into multiple lymph nodes and the BM as well as into several organs and tissues.
STAT transcription factors	= Signal transducer and activator of transcription. Heterodimeric *transcription factors* that move into the nucleus once activated by *Jak* kinases previously activated by *cytokine receptor* engagement.
stochastic models	Models in which events happen by chance without external direction or regulation.
sub-exon	In the Igh locus, a small exon containing splice donor and acceptor sites that is spliced at the RNA level with other small sub-exons to form a complete *constant exon*.
subunit vaccine	*Vaccine* based on an isolated pathogen component such as a viral protein or bacterial polysaccharide. Whole macromolecules or large fragments of macromolecules containing *protective epitopes* are used.
superantigen	Protein that binds simultaneously to an invariant site on the MHC protein outside the peptide binding groove and to the CDR2 of certain TCRβ chains. Non-specifically activates a large number of T cell clones bearing TCRs with the appropriate Vβ segments.
surface marker	Cell surface protein whose presence identifies a cell type or a stage of *differentiation*.
surrogate light chain (SLC)	Two polypeptides called V$_{preB}$ and λ5 assemble non-covalently to form the SLC. V$_{preB}$ supplies a sequence homologous to the V regions of λ and κ chains, while λ5 supplies a constant region-like sequence. The SLC participates in the *pre-BCR* for μ chain *productivity testing*.
switch recombination	Mechanism underlying immunoglobulin *isotype switching*. Depends on the pairing of switch regions found 5′ of each C$_H$ exon (except Cδ). The switch region of the C$_H$ exon that was originally part of the VDJ-C gene pairs with the switch region of a downstream C$_H$ exon such that the intervening C$_H$ exons are excised. The VDJ exon is then joined to the new C$_H$ exon. Specific cytokines (TGFβ, IFNγ, or IL-4) and germline Ig transcripts influence which C$_H$ exon is selected. See *I exons*.
SXY-CIITA regulatory system	Governs expression of *MHC class I and II molecules*. SXY is a regulatory motif in the promoters of genes required for MHC class I and II expression. SXY allows the assembly of the RFX/X2BP/NF-Y *enhanceosome* on the promoter. CIITA (class II transactivator) is a transcription factor that binds to the RFX/X2BP/NF-Y complex.
syncytia formation	The formation of giant, multinucleated structures in which a virus-infected cell "glues together" a large number of uninfected cells.
syngeneic	Individuals are *syngeneic* if they have the same alleles at all genetic loci. In some cases, "syngeneic" is used to refer to genetic identity at the MHC loci only.
systemic acquired resistance (SAR)	A state in which an entire plant is immune to attack by a broad range of pathogens. SAR is mediated by elevated concentrations of NO, salicylic acid, and *pathogenesis-related (PR) proteins*.
systemic autoimmunity	An autoimmune response in which the destruction is not restricted to a specific organ or tissue.
systemic lupus erythematosus (SLE)	Systemic AID in which the skin, joints, kidney, lung, heart, and gastrointestinal tract come under inflammatory attack. Characterized by the presence of *anti-nuclear antibodies*.

Term	Definition
systemic type III hypersensitivity	*Type III hypersensitivity* caused by the deposition in the tissues of circulating immune complexes that then trigger inflammatory mechanisms. Symptoms may appear at a site far removed from the original site of antigen–antibody contact, and multiple sites may be affected simultaneously.
T1DM, T2DM	See *type 1* and *type 2 diabetes*.
T4 count	Clinical term indicating the number of CD4$^+$ T cells in an individual's blood (normally 800–1100 CD4$^+$ T cells/mm^3).
TAA	See *tumor-associated antigen*.
tag	A labeling molecule used to make unitary antigen–antibody pairs detectable. Tags are typically radioisotopes, enzymes, *fluorochromes*, or electron-dense materials that can be covalently bound to either the antigen or the antibody.
tail piece	Short C-terminal region of *secreted antibody*.
Tamm-Horsfall protein	Urine glycoprotein to which *Bence Jones proteins* bind to form multimeric aggregates in *myeloma* patients. These aggregates can lead to excess calcium in the blood (hypercalcemia) or urine (hypercalciuria).
TAP	= Transporter of antigen processing. Heterodimeric protein positioned in the membrane of the rER that transfers HSP-chaperoned peptides from the cytosol into the rER lumen for loading onto MHC class I molecules. Integral component of *endogenous antigen processing and presentation* pathway.
tapasin	rER protein that uses two separate binding sites to link *TAP* and the *MHC class I* α chain. May stabilize empty MHC class I heterodimers in a conformation suitable for peptide loading.
target cell (cancer)	The first cell that undergoes a tumorigenic mutation during the *initiation* step of *carcinogenesis*.
target cell (cytolysis)	An altered self cell, such as an infected cell, a cancer cell, or a graft cell, that is destroyed by cytolysis or cytokine secretion mediated by *CTLs* or *NK cells*.
Tat (HIV)	= Transactivator protein. Activates transcription of *proviral DNA* by preventing premature transcription termination. Also promotes apoptosis of infected host cells and interferes with survival signaling.
TBI	See *total body irradiation*.
Tc cells	Cytotoxic T cells that recognize nonself-peptide presented on MHC class I molecules expressed by a target cell. Upon activation, Tc cells differentiate into *CTL effector cells* that kill target cells by perforin/granzyme-mediated cytotoxicity or by secretion of cytotoxic cytokines. Tc cells generally express the CD8 *coreceptor*.
T cell help	Provision of cytokines or costimulatory contacts by *Th cells* to B and *Tc cells* to support their activation. The different cytokines secreted by Th1 and Th2 cells result in the promotion of switching to different antibody isotypes in B cells.
T cell receptor (TCR)	Heterodimeric antigen receptor expressed by T cells. May be composed of an alpha and beta chain, or a gamma and delta chain.
TCRαβ transgenic mouse (TCR Tg)	A *transgenic (Tg) mouse* created by introducing into its germline DNA the rearranged genes encoding a particular TCRα and β chain that form a TCR recognizing a specific pMHC. Instead of a varied repertoire of TCR specificities, a large percentage of the T cells in each TCR Tg mouse expresses the same TCR protein and so exhibits the same antigenic specificity.
TCR complex	Complete antigen receptor of T lineage cells. For αβ T cells, the complex contains the αβ TCR plus CD3γε, CD3δε, and CD3ζζ, CD3ζη, or CD3ζ-FcεRIγ. For γδ T cells, the complex contains a γδ TCR, two CD3γε heterodimers, and CD3ζζ, CD3ζη, or CD3ζ-FcεRIγ.
TCR tickling	Continual but subtle stimulation of the TCRs of resting naive T cells by self-pMHC complexes presented in the lymph nodes. Blocks apoptotic death normally induced in naive mature T cells in the absence of specific antigen. May also maintain memory T cell population.
Td antigens	See *T-dependent (Td) antigens*.
T-dependent (Td) antigens	Td antigens can bind to the BCRs of B cells to initiate activation but cannot induce B cell differentiation or Ig production in the absence of direct intercellular contact with a primed helper T cell. Td antigens are proteins that have both B and T cell epitopes.

Term	Definition
terminal complement components	Complement components C5–9. Required for *MAC* formation.
tertiary lymphoid tissue	Inflamed peripheral tissues where post-capillary venules have been altered to permit an influx of leukocytes, especially effector and memory lymphocytes.
Th cells	Helper T cells that recognize nonself-peptide presented on MHC class II molecules expressed by an APC. Effector Th cells are generated that may be either Th1 or Th2 in phenotype. Th1 cells secrete IFNγ and IL-2, generally combat intracellular bacteria and viruses, and induce isotype switching in humans to IgG1 and IgG3. Th2 cells secrete IL-4, IL-5, and IL-10, act against extracellular bacteria and parasites, and induce isotype switching to IgA, IgE, and IgG4. Th cells generally express the CD4 *coreceptor*.
Th1/Th2 differentiation	Process by which an activated Th cell proliferates to generate IFNγ- and IL-4-secreting Th0 cells, which then differentiate into either Th1 or Th2 effectors depending on the presence of other transcription factors and cytokines. Under the influence of T-bet and IL-12, Th0 cells differentiate into Th1 *effector cells*. Under the influence of GATA-3 and IL-4, Th0 effectors differentiate into Th2 effectors. See *Th cells* and *transcription factors*.
Th3 cells	CD4$^+$ *regulatory T cells* that secrete large amounts of TGFβ.
12/23 rule	An observed pattern in which *V(D)J recombination* occurs only between *gene segments* whose apposition brings together a *12-RSS* and a *23-RSS*. Only pairing of opposing types of RSSs allows gene segments to be recognized by *RAG recombinases*.
therapeutic vaccination	A *vaccine* given to alter or ameliorate a pre-existing disease or condition such as cancer, allergy, or autoimmunity. See *vaccination*.
three signal model of lymphocyte activation	A model based on the observation that most lymphocytes require three coordinated signals for complete activation, namely (1) antigen recognition, (2) costimulation, and (3) cytokine signaling. In general, for T cells, signal one is TCR engagement by pMHC; signal two is costimulation via engagement of CD28 by B7-1/B7-2; and signal three is cytokine signaling (usually by IL-2). For B cells, signal one is BCR engagement by antigen; signal two is costimulation via engagement of CD40 by CD40L on an activated, antigen-specific Th cell; and signal three is cytokine signaling delivered by cytokines secreted by the Th cell.
thrombin	An enzyme necessary for blood clotting and tissue repair.
thrombocytopenia	An abnormally low number of *platelets* in the blood.
thymic epithelial cells (TECs)	Stromal epithelial cells in the *cortex* (cTECs) or *medulla* (mTECs) of the thymus. *TECs* present self ligands thereby mediating *positive selection* and *negative selection* in the thymus.
thymic involution	After puberty, a block in *DN thymocyte* proliferation causes the thymus to progressively lose both its tissue mass and its functionality. Thymic tissue is largely replaced by fat deposits.
thymocytes	T cell precursors developing in the thymus. Progenitors proceed through a double negative (DN; CD4$^-$CD8$^-$) stage, to a double positive (DP; CD4$^+$CD8$^+$) stage, and finally to a single positive (SP; CD4$^+$CD8$^-$or CD4$^-$CD8$^+$) stage. SP thymocytes enter the periphery to become mature CD4$^+$ or CD8$^+$ T cells.
thymoma	*Neoplasm* of thymic epithelial cells.
thymus	A small bilobed organ located above the heart, consisting of the *medulla*, the inner and outer *cortex*, and the subcapsule. The thymus is the primary lymphoid tissue in which thymocytes develop into mature T cells. Thymocyte development is supported by *thymic epithelial cells (TECs)* and specialized epithelial cells in the outer cortex called *nurse cells*.
thyroid gland	An endocrine gland surrounding the trachea that produces *thyroxine*, a hormone involved in regulating metabolic rate.
thyroxine	Hormone regulating cell metabolism, body temperature, energy use, and normal CNS and cardiovascular functions. Produced by the *thyroid gland*.
tight junction	Joint between epithelial cells at their *apical* poles. Inhibits the passive diffusion of pathogens, macromolecules, and peptides. Can open in response to cytokine signaling.
Ti antigens	See *T-independent (Td) antigens*.
TILs	See *tumor-infiltrating lymphocytes*.

Term	Definition
T-independent (Ti) antigens	Ti antigens stimulate B cells in the absence of T cell help. However, due to this lack of Th help, these responses involve only very limited isotype switching and are not characterized by somatic hypermutation or memory B cell generation. The result is a primary IgM response of relatively low affinity. Ti-1 antigens, which are generally large polymeric proteins, bind to BCRs to fully activate B cells in the absence of any type of help. Ti-2 antigens, which are generally polysaccharides containing repetitive elements, bind to BCRs but cannot fully activate B cells without cytokine help from bystander cells.
TIR domain	= Toll/IL-1R domain. Found in the cytoplasmic tails of TLRs, IL-1R, and certain plant *PRMs*. The TIR does not have intrinsic tyrosine kinase activity but recruits the MyD88 *adaptor protein*.
tissue-specific antigen	A protein or carbohydrate antigen associated with only a particular tissue or subcellular location. Inappropriate expression of the antigen in a tumor cell can make it a *tumor-associated antigen*.
tissue typing	Identification of *HLA* alleles expressed on an individual's cells, usually with the goal of establishing the level of HLA mismatching between a donor and recipient in a transplant situation. Techniques used include *complement-dependent cytotoxicity test, mixed eukocyte reaction, restriction fragment length polymorphisms*, and the *PCR*-based *probe hybridization* and *direct amplicon analysis* methods. *Panel reactive antibody* may also be considered a tissue typing technique.
titer	Concentration of antigen-specific antibodies within a sample of *antiserum*. Usually determined by serially diluting samples of the antiserum until binding to a standard amount of specific antigen can no longer be detected.
TNF	See *tumor necrosis factor*. TNFα is often known simply as TNF. TNFβ is the TNF-related cytokine known as lymphotoxin α (LTα).
TNF receptor-associated periodic syndrome (TRAPS)	Autoinflammatory *PI* characterized by repeated bouts of severe localized inflammation and prolonged fever. Caused by mutations of TNFR1 that result in uncontrolled stimulation of the *TNFR1 pathway*.
TNFR1 pathway	= Tumor necrosis factor receptor-1 pathway. An *intracellular signaling pathway* in which a ligand binds to TNFR1 and induces either cell death or cell survival depending on the signaling components involved. Death signaling is mediated by TRADD, FADD, caspase-8, and the *caspase-3 cascade*. Survival signaling is mediated by TRADD, TRAF2/RIP, and the cIAPs leading to activation of the *NF-κB* and *SAPK/JNK signaling pathway*s.
tolerance	The absence of an immune response to a given antigen. See also *self-tolerance, experimental tolerance, central tolerance*, and *peripheral tolerance*.
tolerization	See *lymphocyte tolerization* and *experimental tolerance*.
tolerogen	An experimental foreign antigen that is recognized by a T or B lymphocyte but induces lymphocytes to become refractory to activation.
tolerogenic DC	See *modulated DC*.
tolerosome	The altered membrane complex of antigen-stimulated TCRs, CD3 chains, costimulatory and adhesion molecules, coreceptors, and other signaling molecules associated with the *lipid rafts* in an *anergized* T cell. The component signaling molecules differ from those in the *immunosome* of an activated T cell.
Toll-like receptors (TLR)	An evolutionarily conserved family of membrane-bound *pattern recognition receptors*. Ligands for specific TLRs include host stress ligands, LPS, peptidoglycan, bacterial flagellin, viral dsRNA, viral ssRNA, and bacterial CpG nucleotides. Some TLRs (e.g., TLR2, TLR4) are fixed on the cell surface, while others (e.g., TLR7, TLR9) reside in the membranes of endosomal vesicles. TLR engagement triggers DC maturation and the production of defensins by epithelial cells.
tonsils	A network of reticular cells that supports *lymphoid follicles* and *interfollicular regions* that are part of the *NALT*. Five tonsils form a structure in the nasopharynx called Waldeyer's ring: the nasopharyngeal tonsil (commonly referred to as "the adenoids") in the roof of the nasopharynx, two lingual tonsils at the back of the tongue, and two palatine tonsils at the sides of the back of the mouth. Tonsils serve as both *inductive* and *effector sites* of adaptive immunity.
total body irradiation (TBI)	A *hematopoietic cell transplant* recipient receives sufficient ionizing radiation over his or her whole body to kill developing hematopoietic cells in the *bone marrow*.

Term	Definition
toxic shock syndrome	Potentially lethal systemic cytokine induction mediated by the polyclonal activation of T cells by bacterial superantigen TSST1. Associated with improper use of feminine hygiene products.
toxin	Molecule produced by a *pathogen* that kills host cells, inhibits host cell metabolism, or alters host immune responses against the pathogen. See *exotoxin* and *endotoxin*.
toxoid	Chemically inactivated pathogen *toxin*.
Tr1 cells	CD4$^+$ *regulatory T cells* that preferentially secrete IL-10 plus low amounts of TGFβ.
transcription factors	Proteins that bind to motifs in promoters and initiate, support, or inhibit gene transcription. Examples include NF-κB, NF-AT, CIITA, IRF, AP-1 (c-fos/c-jun), c-myc, CREB, CBF, Ikaros, PU.1, Tcf-1, Lef-1, GATA-3, and T-bet.
transcriptomics	Study of all RNA transcripts produced by an organism.
transcytosis	Transport of a molecule from one surface of a cell to its opposite surface, via an intracellular transport vesicle.
transformation	*Malignant conversion.* See *carcinogenesis*.
transfusion reaction	Destruction of transfused blood cells by pre-formed recipient antibodies specific for donor blood group antigens.
transgenic (Tg) mouse	A mutant mouse created by the introduction of an isolated gene (the "transgene") into the genome of a whole mouse embryo. The transgene is expressed along with the recipient's own genes. Expression of a transgene may be controlled either by its natural transcriptional promoters and enhancers, or by exogenous regulatory elements engineered into it.
transplantation	The transfer of living tissue or cells between individuals, or from one site to another in a given individual.
transplant rejection	See *graft rejection*.
T_{reg} *cells*	CD4$^+$ CD25$^+$ *regulatory T cells* that express CTLA-4 constitutively but do not secrete significant amounts of cytokines. Can *anergize* T cells non-specifically via intercellular contacts. May also cause APCs to become *tolerogenic* by inducing downregulation of their *costimulatory molecules*.
Ts cells	CD8$^+$D57$^+$CD28$^-$ *regulatory T cells*. Thought to have a downregulating effect on immune responses. *In vitro*, the binding of a Ts cell to pMHC on an APC tolerizes the APC by suppressing B7 expression and upregulating ILT3 and ILT4. Antigen-specific CD4$^+$ Th0 cells interacting with these APCs may be anergized and/or become T_{reg} *cells*. Activated Ts cells secrete a Th0-like cytokine profile.
TSA	See *tumor-specific antigen*.
tumor	See *neoplasm*.
tumor-associated antigen (TAA)	Normal protein or carbohydrate expressed in a tumor at a concentration, location, or time that is abnormal relative to its status in the healthy, fully differentiated cells in the tissue of origin. Includes *normal cellular proteins, differentiation antigens, tissue-specific antigens, embryonic antigens*, and idiotypic antigens (see *idiotope*). TAAs are encoded by normal cellular genes that have become dysregulated.
tumor hypoxia	Low oxygen pressure (hypoxia) that occurs in a part of a tumor that has outgrown its blood supply. Hypoxic tumor cells are less sensitive to *radiation* and *chemotherapy*.
tumor-infiltrating lymphocytes (TILs)	CTLs found within a tumor.
tumor necrosis factor (TNF)	First identified as a soluble molecule capable of inducing the hemorrhagic necrosis of tumors in mice. The major inflammatory mediator induced by *gram-negative* bacteria and their components and thus a principal contributor to *endotoxic shock*. Depending on its concentration and the signal transduction induced downstream of its binding to *TNFR*, TNF also has potent immunoregulatory, cytotoxic, antiviral, and pro-coagulatory activities. It may also promote cell survival and has effects on *hematopoiesis*.
tumor regression	Disappearance of a *malignant tumor*. May be induced by anti-cancer treatment, or occur spontaneously (without treatment).

Term	Definition
tumor-specific antigen (TSA)	A macromolecule that is unique to a tumor and not produced by any type of normal cell. TSAs are encoded by mutated cellular genes, or by viral oncogenes.
tumor suppressor gene	Gene encoding a protein whose absence promotes *carcinogenesis*. Often negative regulators of cell growth or survival. Includes *p53* and *PTEN*.
two-signal paradigm	A model based on the concept that a lymphocyte that does not receive signals from both antigen receptor engagement and costimulation is rendered unresponsive.
type 1 diabetes mellitus (T1DM)	A form of *diabetes mellitus* in which a loss of insulin production is associated with *autoimmunity* that damages pancreatic β-islet cells. May be characterized by *insulitis* and antibodies directed against pancreatic β-islet cell antigens. Also known as "insulin-dependent diabetes mellitus" (IDDM).
type 2 diabetes mellitus (T2DM)	A form of *diabetes mellitus* in which insulin is produced normally but host cells stop responding to it. Not an autoimmune disease. Also known as "non-insulin-dependent diabetes mellitus" (NIDDM).
type I hypersensitivity	*Hypersensitivity* arising from the synthesis of IgE antibodies directed against an antigen (sensitizing agent). These antibodies arm *mast cells* via FcεR binding. Upon re-exposure to the same antigen, rapid mast cell degranulation occurs, triggering allergic symptoms. Also called "IgE-mediated" or "immediate" hypersensitivity. See *allergy* and *anaphylaxis*.
type II hypersensitivity	Hypersensitivity arising from direct antibody-mediated cytotoxicity. Damage to host tissues is caused by cellular lysis induced by the direct binding of IgM or IgG antibody to cell surface antigens. Target cells may be mobile (blood cells) or fixed as part of a solid tissue. If *autoantibodies* are involved, the reaction may be a component of an *autoimmune disease*.
type III hypersensitivity	*Immune complex*-mediated *hypersensitivity*; also known as "immune complex injury." A soluble antigen forms large insoluble immune complexes with IgM or IgG antibodies in the circulation. Deposition of these complexes at single or multiple body sites (often in the walls of vessels) triggers a damaging *inflammatory* response. May be *localized* or *systemic*. If *autoantibodies* are involved, the reaction may be a component of an *autoimmune disease*.
type IV hypersensitivity	Cell-mediated hypersensitivity in which Th cells activate Tc cells and macrophages that then damage or destroy host cells. Includes chronic *delayed type hypersensitivity (DTH)* reactions.
ubiquitination	Process in which ubiquitinating enzymes covalently bind polymers of ubiquitin to a protein, making it a target for degradation by the standard *proteasome*.
ulcerative colitis	Chronic *inflammatory bowel disease (IBD)* affecting only the innermost layer of the wall of the colon. The entire colonic lining is often affected.
urticaria	*Type I hypersensitivity* in which sensitized skin mast cells degranulate and release mediators that cause swollen, reddened, itchy patches on the skin known as *wheal and flare* eruptions. Also known as "hives."
vaccination	Administration of a non-infectious form of a pathogen or its components (the *vaccine*) to prime an *adaptive* response and induce the generation of antigen-specific *memory T cells* and *memory B cells* directed against the pathogen. A natural exposure to the pathogen then triggers a *secondary*, rather than *primary*, adaptive *response* and protects against overt disease.
vaccine	A modified, non-pathogenic form of a natural immunogen. May be a killed, inactivated, or attenuated form of the pathogen, or composed of pathogen proteins, DNA, or other molecules. See *vaccination*.
vaccinia virus	Member of the poxvirus family that may have arisen from recombination of cowpox virus with another poxvirus. Used for vaccination against *variola* (smallpox).
Van der Waals forces	A bond established when polarities oscillate in the outer electron clouds of two neighboring atoms, creating attractive/repulsive forces between them.
variable (V) domain	N-terminal domain of an Ig or TCR chain that is encoded by the corresponding *variable (V) exon*. The variable domains of an Ig heavy and light chain or a TCR α and β chain define the variable regions and antigen-binding sites of the Ig and TCR molecules.

Term	Definition
variable (V) exon	Exon encoding the variable domain of an Ig or TCR protein. The V exons in the Igk, Igl, TCRα, and TCRδ loci are randomly assembled from V and J *gene segments*, while the V exons in the Igh, TCRβ, and TCRγ loci are composed of V, D, and J gene segments.
variable lymphocyte receptors (VLRs)	GPI-anchored cell surface receptors that are expressed by primitive lymphocytes in Agnatha (see *higher vertebrates*) and are capable of somatic variation.
variable (V) region	The N-terminal portion of Ig or TCR proteins that is characterized by a high level of amino acid variability and is responsible for antigen recognition.
varicella	Chicken pox virus.
variola	Smallpox virus.
variolation	Historic method of inducing immunity to smallpox by inhalation of material from smallpox pustules or insertion of the material into superficial wounds. Although it appeared to be effective in some cases, the outcome was unreliable and potentially dangerous.
vascular addressins	*Cellular adhesion molecules* that are expressed on *HEVs* and help mediate lymphocyte *extravasation* at particular sites in the body. Different lymphoid organs and tissues display different addressins on their HEVs, attracting only those lymphocytes with the appropriate complementary *homing receptors*.
vasoactive amines	Compounds contained in *mast cell* granules that, upon degranulation-mediated release, induce vasodilation and increase vascular permeability, facilitating the influx of inflammatory leukocytes into the tissues. Include *histamine* and 5-hydroxytryptamine. See also *inflammation*.
V(D)J recombination	Site-specific recombination at the DNA level of pre-existing V, D, and J *gene segments* in the Ig and TCR loci to generate unique *variable (V) exons*. Mediated by recognition, cutting and rejoining of the heptamer-nonamer *recombination signal sequences (RSSs)* flanking those gene segments. When three segments are used, the D and J gene segments are joined first, and V is then joined to the DJ unit.
vertebrates	Animals with a true backbone. See *higher* and *lower vertebrates*.
Vif (HIV)	*HIV* protein that stabilizes newly synthesized HIV DNA and facilitates its transport to the nucleus. Blocks packaging of *CEM-15* into virions.
virosome	A vaccine composed of a spherical "artificial virus" used to introduce *vaccine* antigens directly into a host cell. Made of a *liposome* that contains the vaccine antigen and the hemagglutinin and neuraminidase proteins of influenza. Retains the native virus's membrane fusion properties, conformational stability, and ability to enter host cells. The antigen is either encapsulated within the lumen of the virosome or chemically cross-linked to its surface.
virulence	Ability of a *pathogen* to invade host tissues and cause disease. A virulent pathogen is highly pathogenic.
virus	Very small acellular particles consisting of a capsid (protein coat) encasing a genome of either DNA or RNA. To propagate, a virus must enter a host cell and exploit its protein synthesis machinery.
vitiligo	Progressive loss of pigment from the skin or hair. May occur in melanoma patients as a result of *autoimmunity* that damages melanocytes. Mediated by T cells directed against the melanoma cell TAAs.
Vpr (HIV)	HIV protein that helps to direct the *HIV* DNA to the nucleus. Promotes G2 arrest and activation of the *intrinsic apoptotic pathway*.
Vpu (HIV)	HIV protein that binds to newly synthesized CD4 in the host cell cytoplasm and targets it for ubiquitin-mediated degradation by *proteasomes*. Forms ion channels in the host cell membrane for virion release. Interferes with *MHC class I* synthesis. Enhances susceptibility to *Fas*-mediated apoptotic *pathway*.
Waldenström's macroglobulinemia (WM)	*Plasma cell dyscrasia* in which cancerous cells in the bone marrow secrete large quantities of IgM *paraprotein* that enter the blood, causing hyperviscosity. Malignant cells also infiltrate secondary lymphoid tissues. Hematopoiesis is compromised.

Term	Definition
waste disposal hypothesis	A model proposing that some cases of *autoimmunity* may be due to a failure to clear apoptotic cells. These cells could then accumulate in the tissues, releasing autoantigens (such as double-stranded DNA) and triggering autoimmune reactions.
Western blotting	A laboratory method in which proteins are separated by SDS-PAGE and transferred onto a nitrocellulose membrane. Proteins of interest are then detected by incubation of the blot with specific *tagged* antibody.
wheal and flare	See *urticaria*.
white pulp	See *spleen*.
Wiscott-Aldrich syndrome (WAS)	X-linked recessive, clinically heterogeneous disorder with a *PI* component (CID). Caused by mutations in the WAS protein (WASP) that may be involved in survival and/or function of *HSCs*, *platelets*, and *lymphocytes*.
X4 viruses	Strains of *HIV* that can no longer infect macrophages and preferentially attack CD4$^+$ T cells. Originally called "T-tropic viruses."
xenobiotic	Class of non-living entities usually derived from drugs, metals, or industrial or natural chemicals. Bind to self proteins to form *neo-antigens*.
xenogeneic	Genetic relationship between two individuals of different species.
xenograft	Tissue or cells transplanted between members of two different species.
xeroderma pigmentosum (XP)	Rare disorder characterized by extreme sensitivity to sun and variable *CID*. Caused by mutations to XPA-XPG genes involved in *nucleotide excision pathway* of *DNA repair*.
X-linked lymphoproliferation (XLP)	*PI* characterized by CTL-mediated *immunopathic damage* in response to EBV infection. Caused by mutation of the SAP gene responsible for negative regulation of T cell IFNγ secretion.
X-ray crystallography	A technique used to analyze protein structure at the atomic level. A crystallized protein is bombarded with monochromatic X-rays from either a rotating anode X-ray generator or a synchrotron source. As the X-rays strike the crystal, most go right through, but some strike the atoms making up the protein and are deflected into a detectable and reproducible pattern. This diffraction pattern can be analyzed to glean information about the position and orientation of the atoms in the crystal.
yellow marrow	See *bone marrow*.
zone of antibody excess	Area of the *precipitin reaction curve* where the number of antibody-combining sites exceeds the number of antigenic epitopes. Few immune complexes can form so there is little or no precipitate.
zone of antigen excess	Area of the *precipitin reaction curve* where the number of antigenic epitopes exceeds the number of antibody-combining sites. Few immune complexes can form so there is little or no precipitate.
zone of equivalence	Area of the *precipitin reaction curve* where the number of antigenic epitopes and antibody-combining sites is approximately equal. The formation of extended immune complexes is optimal so that maximum precipitation is observed.
zoonosis	An animal disease that can be transmitted to humans.
zymogen	See *proenzyme*.

Index

Page numbers in italics refer to figures and tables; page numbers in bold refer to focus boxes.

A

ABF-1, B cell activation modulation, **224**
Abzyme, overview, **149**
Acquired immunodeficiency syndrome (AIDS),
   see also Human immunodeficiency virus
  animal models
    chimpanzee, 813
    macaque, 813–814
    mouse, 814–815
    overview, 813, *813*
  blood transfusion recipients, **809–810**
  cachexia, 804, *804*
  history of study, 786–788
  societal and economic impact, 812–813
Activation-induced cell death (AICD)
  definition, 424
  modulation to induce autoimmunity, 1007
  pathways, *425*
    extrinsic pathway, 424–426
    intrinsic pathway, 426
  pro-/anti-apoptotic balance, 426–427, *427*
  susceptibility of T cells, 426
Activation-induced cytidine deaminase (AID)
  immunodeficiency, 764,767
  isotype switching role, 239
  somatic hypermutation role, 233–234
Acute lymphoblastic leukemia (ALL)
  clinical features, 1041–1042
  epidemiology, 1041
  genetic aberrations, 1042–1043
  immunophenotyping, *1042*
  treatment, 1043–1044
Acute myeloid leukemia (AML)
  classification, 1036, *1036*
  clinical features, 1035–1036
  epidemiology, 1033
  genetic aberrations, 1036–1038, *1037*
  treatment, 1038–1039
Adaptive immunity
  diversity, 20–21
  division of labor, 21, *22*
  host defense phases, 11–12, *12*
  immunogens, 24
  comparative study, *see* Comparative immunology
  innate immunity relationship, 12–13, *18*
  memory, 19–20, *20*
  overview, 7, 10, *10*, 19
  primary versus secondary immune responses, 31–32, *32*
  specificity, 19, *19*
  tolerance, 21
ADCC, *see* Antibody-dependent cell-mediated cytotoxicity
Adenosine deaminase, deficiency, 758–760, *759*
Adenovirus
  description, 665–666
  vaccine vector, 712
Adhesion molecules
  B cell–T helper cell interactions, 225–227, *226*
  immunodeficiency, 776–777
  inflammatory response, 83–84, *83*
  lymphocyte homing, 64–65, *65*
  naive and effector T cells, 422–423
  T cell–antigen-presenting cell interactions, 375–378, *376*
Adjuvant
  alum, 723
  bacterial toxins, 723
  complete Freund's adjuvant, 722
  CpG motifs, 723
  cytokines, 723
  incomplete Freund's adjuvant, 722

  mechanisms of action, 130, 722–723
  types, *724*
Adult T cell leukemia/lymphoma, 1061
Affinity chromatography, principles, 173–175, *174*
Affinity maturation
  nature of, 234
  memory B cells, 243
Agglutination
  assays, 159
  types, *160*
AICD, *see* Activation-induced cell death
AID, *see* Activation-induced cytidine deaminase
AIDS, *see* Acquired immunodeficiency syndrome
Aiolos
  B cell development role, **224**
  T cell development role, **353**
ALL, *see* Acute lymphoblastic leukemia
Allelic exclusion
  in B cells, 192–193, *193*
  in T cells, 326
Allergy, *see* Hypersensitivity, type I
Alloreactivity
  immune responsiveness
    haplotype correlation, 273, *273*
    models of non-responsiveness, 274, *274*
  major histocompatibility complex role, 273
  transplantation role, 880
ALPS, *see* Autoimmune lymphoproliferative syndrome
Alum, vaccine adjuvant, 723
AML, *see* Acute myeloid leukemia
Anaphylaxis, 933–934
Anaphylatoxins, **559**
Anaplastic large cell lymphoma, 1062
Anergy
  anti-HIV response, 802
  B cells, 447
  Dysregulation leading to autoimmunity, 968
  induction to prevent graft rejection, 901, 905
  induction to prevent GvHD, 915
  induction to treat autoimmunity, 1019
  T cell peripheral tolerance
    demonstrations, 441–442, *442*
    induction, 443–444
    signaling, 443–444, *443*, *444*
Angioimmunoblastic T cell lymphoma, 1062
Ankylosing spondylitis
  clinical features, 977, *977*
  pathophysiology, 977
Anthrax, vaccination, 727
Anti-allotypic antibodies, **102**
Antibody, *see* Immunoglobulin
Antibody-dependent cell-mediated cytotoxicity (ADCC)
  anti-tumor response, 857–859, 864
  as a result of CAMPATH-1H treatment, 899
  graft rejection role, 886–887
  induction to treat autoimmunity, 1017
  induction to treat hematopoietic cancers, 1034
  mechanism, 114–115, *115*
  natural killer cells, 522–523, *522*
  type II hypersensitivity role, 949–952; also *see* Hypersensitivity
Antigen
  antibody interactions, *see* Antigen–antibody interaction
  B cell recognition, 26, *26*
  definition, 7, 24, 122
  epitope, 25, *25*, 137
  T-dependent antigens(Td antigen)
    adjuvants, 130
    anti-bacterial defense, 652

    anti-viral defense, 671
    B cell activation, 124, *125*
    immunogenicity factors, 125, *125*
    properties
      nature of, 122
      dosage, 128–129, *128*
      foreigness, 125–126
      molecular complexity, 126–128, *126*, **127**
      route of entry or administration, 129–310, *129*
    vaccine design, 703–704
  T-independent antigen (Ti antigen)
    anti-viral defense, 671
    classes, 123–124
    limitations in vaccines, 710
    nature of, 122
  types, 7, 122, *123*
Antigen–antibody interaction
  affinity
    definition, 140
    determination, **141–142**
    distribution curve for polyclonal antibody response, 143, *143*
    kinetics, 140, *140*
  avidity, 143, *143*
  B epitope structural requirements, 137
  complementarity-determining regions, 137–139
  cross-reactivity, 143–145
  hapten–antibody reaction specificity, *138*
  non-covalent forces, 139–140, *140*
Antigen-presenting cell (APC)
  antigen processing, *see* Antigen processing and presentation
  autoimmune disease abnormalities, 1009
  B cells, 288–289
  dendritic cells, 285–286
  functional overview, 27–28
  macrophages, 288
  monocytes, 288
  T cell activation, *see* T cell activation
  types, 282, *282*
Antigen processing and presentation
  CD1 molecules and non-peptide antigen presentation, 306–308, *307*
  cross-presentation
    mechanism, 302–304, *303*
    minor pathways
      peptide interception, 305, *305*
      peptide regurgitation, 304–305, *305*
    peripheral tolerance role, 439
    rationale, 304
  endogenous processing pathway
    MHC class I molecule α-chain and chaperones, 296, *296*, **301**
    peptide loading onto MHC class I molecules, 302
    proteasome generation of peptides
      core proteasome, 296–297
      immunoproteasome, 298–299
      sampling of organelle proteins, 299
      standard proteasome, 297–298, *297*
    TAP transporters, 299–300, 302
    target cells, 295
  exogenous processing pathway
    antigen-presenting cells
      B cells, 288–289
      dendritic cells, 285–286
      macrophages, 288
      monocytes, 288
      types, 282, *282*
    factors affecting, 294–295

Antigen processing (*Continued*)
history of study
endogenous processing, 295
exogenous processing, 281–282
general, 249, 251–252, 254
non-classical MHC proteins and antigen presentation, 305–306
overview, 27–28, *28*
processing mechanism
CLIP exchange and peptide loading, *292*
HLA-DM-H-2M, 293, *294*
HLA-DO-H-2O, 293–294
Ii digestion, 291–293
minor pathways of MHC class II peptide loading, 294
exogenous peptide generation, 290–291, *293*
invariant chain, 289–290
viral interference with MHC antigen presentation
MHC class I, 673–675, *675*
MHC class II, 676–677, *676*
Antigen recognition, *see also* Epitope
by B cells, 122
by αβ T cells, 337–338
by γδ T cells, 539
Antihistamines, allergy treatment, 942–944
Anti-idiotypic antibodies, **102**, 716
Anti-idiotypic vaccine, *see* Vaccination
Anti-isotypic antibodies, **102**
Anti-phospholipid syndrome
clinical features, 985
pathophysiology, 985
Anti-polysacharide antibody deficiency (APAD), 767
Antiserum
advantages and disadvantages in immunoassays, 149–150
definition, 149
APAD, *see* Anti-polysacharide antibody deficiency
APC, mutation in cancer, **87**
APC, *see* Antigen-presenting cell
APOBEC3G, *see* CEM-15
Apoptosis, *see also* Activation-induced cell death
CD4/CD8 commitment, 365
central B cell tolerance, 215
cytotoxic T lymphocyte induction via Fas pathway, 422
homeostasis role, 50–51, **51**
mature naive B cells, 218
mature naive T cells, 368
phases, 424
thymic selection, 357–359
virus manipulation, 677–679, *678*
Artemis
function, 198–199
mutation in mice, 200
severe combined immunodeficiency, 762–763
Asthma, 931–932, *932*, **933**
AT, *see* Ataxia-telangiectasia
Ataxia-telangiectasia (AT), 770, *770*, *771*
Atopy, *see* Hypersensitivity
Autoimmune disease, *see also specific diseases*
animal models
BB-DP rat, **989**
collagen-induced arthritis, 992
experimental autoimmune encephalitis, 991–992
experimental autoimmune myasthenia gravis, 992–993
experimental autoimmune myocarditis, 994
experimental autoimmune thyroiditis, 993
experimental autoimmune uveoretinitis, 993–994
gld mouse, 988–989
inflammatory bowel disease, 991
MTL/lpr mouse, 988–989
NOD mouse, 989–990
NZB/W F1 mouse, 988
OS chicken, **989**
overview, 986–987, *987*, *988*
prospects, 994–995
scurfy mouse, 990
thymectomy, 991

antigen-presenting cell abnormalities, 1009
B cell abnormalities, 1006, 1008–1009
cancer association, 832, 1020–1021
chemical agents as triggers, 998–999
complement abnormalities, 1011
cytokine abnormalities, 1010–1011
danger signal model, 438
epitope spreading, 1011–1012
examples, 965–*966*, 969–986
γδ T cell abnormalities, 1010
genetic predisposition
introduction, 274–275
specific diseases, 995, *995*, **996**, 997–998, *997*
geographic and ethnic differences, 1000–1001, *1000*
history of study, 964–965
hormonal influences, 1000
infection as trigger
epidemiology, 999
mechanisms
dendritic cell maturation induction, 1005
inflammation induction, 1005
molecular mimicry, 1002–1005, *1002*, *1003*
superantigens, 1005–1006
organ-specific versus systemic autoimmunity, 967
phagocyte abnormalities, 1010
regulatory T cell abnormalities, 1009
self antigens, *969*
T cell abnormalities, 1009
tissue damage mechanisms, 968
tolerance defects, 436, 967, **967**, *968*
treatment
anti-inflammatory agents, 1012–1013
disease-specific symptom relief, 1013–1014
immunosuppressants, 1014
immunotherapy
autoreactive B cell targeting, 1018–1019
autoreactive T cell targeting, 1017–1018
cytokines, 1016–1017
hematopoietic cell transplantation, 1019–1020
innate immune response blocking, 1014–1016
overview, 1015–*1016*
tolerance induction, 1019
infection prevention, 1013
intravenous immunoglobulin, 1014
Autoimmune lymphoproliferative syndrome (ALPS), 774
Azathioprine, immunosuppression in transplantation, 896

B

B7
antibody targeting and graft tolerance induction, 901
blockade in allergy treatment, 948
costimulation role, 393–396, *393*, *395*, 397
isoforms, 393–394
knockout mice, 394
ligands, 394
T helper cell activation, 414
T helper cell differentiation role, 408
Bacteria, *see* Extracellular bacteria; Intracellular bacteria
BAFF
B cell development role, 217
expression, 501
functions, 501
receptor, 501
BALT, *see* Bronchi-associated lymphoid tissue
Basophil
compartmentalization, *39*
functional overview, 39
histology, *38*
type I hypersensitivity role, 929–930
BB-DP rat, **989**
B cell, *see also* Memory B cell; Plasma cell
activation, *see* B cell receptor
affinity maturation, 234–235
antigen presentation, 288–289
autoimmune disease
abnormalities, 1006, 1008–1009
autoreactive B cell targeting, 1018–1019

compartmentalization, *39*
differentiation, *see* Memory B cell; Plasma cell
discovery, 6
division of labor, 21, *22*
effector functions, 29–30
functional overview, 13, *13*
germinal center formation, 230–231, **230**, *232*
histology, *37*
identification, 43
lymphoblast production, 41
maturation
mig+ B cells
central B cell tolerance, 215
coexpression of IgM and IgD, 216–217, *216*
peripheral B cell maturation, 217
receptor editing, 215, *216*
pre-B cells
pre-B-I cells, 213–214
pre-B-II type 1 cells, 214, *214*
pre-B-II type 2 cells, 214
pre-B-II type 3 cells, 214–215
surrogate light chain, 213
pro-B cells
bone marrow stromal cells in development, 211–213, *213*
markers, 211, *212*
ontogeny overview, 210–211, *211*
priming, 31
repertoire and memory, 44, 46
resting cells, 41
T cell cooperation in humoral immune response
hapten–carrier complexes *in vivo*, 134–136
hapten–carrier experiments
carrier effect, 132, *132*
hapten–carrier system, 131, *133*
linkage of B and T cell determinants, 132–133, *133*
rationale for linked recognition, 136–137, *136*
reconstitution experiments, 130–131
T cell ontogeny comparison, 344, 347
tolerance, *see* Peripheral tolerance
tumor immune response, 849
B cell chronic lymphocytic lymphoma, 1058
B cell Prolymphocytic leukemia (B-PLL), features, 1046
B cell receptor (BCR), *see also* Immunoglobulin; Immunoglobulin genes
activation
cellular interactions
follicular dendritic cells (FDC), 229
T cells, 229
T helper subtype influence, 230
cytokine role, 227–229, *228*
overview, 25, *25*, 218, *219*
signal transduction, 218–224
T helper costimulation, 224–226
antigen complex, 108, *108*
antigen recognition, 26, *26*
density, 44
gene rearrangement, *see* V(D)J recombination in B cells
Bcl-6
plasma cell differentiation, 239–240
T helper cell differentiation role, **410**
BCR, *see* B cell receptor
Bence Jones proteins, 1047
Beta-selection of αβ T cells
developmental checkpoint, 348, 355
Overview, 325
Biotin-avidin assay system, *see* Tag assays
Blood transfusion
ABO blood types, 911–912, *911*, *912*
acquired immunodeficiency syndrome in recipients, **809–810**
alloreactivity, 910–913
frequency, 910
minor blood group antigens, 912
safety of blood supply, 913
Bloom syndrome, 771–772, *772*

Bone marrow
 anatomy, 53, *54*
 B cell development role, 211–213
 cellular composition, 53
 T cell development role, 347
Bone marrow transplantation
 autoimmune disease and hematopoietic cell
  transplantation, 1019–1020
 graft rejection, 914–915
 graft tolerance induction, 900
 graft-versus-host disease, *see* Transplantation
 graft-versus-leukemia effect, 916–917, 1034
 hematopoietic stem cell transplantation overview,
  913–914
 infection control, 917
 mouse models, **914**
 primary immunodeficiency management, 754
B-PLL, *see* B cell Prolymphocytic leukemia
Bronchi-associated lymphoid tissue (BALT)
 antigen sampling, 593
 composition, 52, 591–593, *591*, *592*
Bronchodilators, allergy treatment, 944
Bruton's agammaglobulinemia, 765–766
Burkitt's lymphoma
 Association with Epstein-Barr virus (EBV), 840
 description, 1060–1061, *1060*

C

CAMPATH-1H
 immunosuppression in autoimmunity, 1017
 immunosuppression in transplantation, 899
Cancer, *see* Hematopoietic cancers; Tumors
Caspases, *see* AICD, Apoptosis
Catalytic antibody, *see* Abzyme
CD markers
 definition, 44, **45**
 list of CD molecules, see Appendix I
CD1
 antigen presentation to γδ T cells, 539
 antigen presentation to NKT cells, 548
 non-peptide antigen presentation, 306–308, *307*
CD2
 natural killer cell costimulation, 529
 T cell–antigen-presenting cell interactions, 378
CD3
 as part of NCR NK activatory receptor, 528
 DN2 thymocytes, 352
 CD4/CD8 commitment, 366
 embryological expression, 369
 functions, 331
 genes, 322
 knockout mouse studies, 331–332
 NK cell expression, 520
 pre-TCR formation and signaling, 355–356
 structure, 330–331
 T cell receptor complex, *318*
 thymic selection role, 360–361
 TCR triggering, 379
CD4
 discovery, 332
 functional overview, 334–336
 peptide–MHC complex interactions, *335*
 structure, 334, *334*
 T cell coreceptor, 29, *29*, 30–31
 T cell receptor signaling role, 391
 thymic selection role, 362
 transgenic mice, 332
CD5
 association with autoimmunity, 972
 B cell development, 218
 marker of B-CLL, 1044
CD8
 discovery, 332
 functional overview, 334–336
 peptide–MHC complex interactions, *335*
 structure, 334, *334*
 T cell coreceptor, 29, *29*, 30–31
 T cell receptor signaling role, 391

 thymic selection role, 362
 transgenic mice, 332
CD14
 pattern recognition, 85
 TLR signaling, 476–476
CD19
 B cell activation modulation, 221, 224
 expression in leukemias and lymphomas, 1034
CD20, monoclonal antibody specific for
 autoimmunity treatment, 1018
 cancer immunotherapy, 859, 1034
 type II hypersensitivity treatment, 952
CD27
 autoimmune disease role, **1008**
 costimulation role, 398
 memory T cell generation, 428
CD28
 absence of signaling in T cell anergy, 441–442
 costimulatory effects of signaling, *393*
 cytokine induction role, 393
 cytotoxic T lymphocyte-associated molecule-4 control
  of signaling, 396, *396*
 signal transduction, 394–396, *395*
 T helper cell activation, 414
 T$_{reg}$ cell generation, 452
CD29, role in intercellular adhesion, 408
CD30
 autoimmune disease role, **1008**
 expression by Reed-Sternberg cells, 1051
 T helper cell expression, 413
CD40
 absence in B cell tolerance, 448
 B cell effector differentiation, 239–240
 B cell–T helper cell interactions, 226–227
 DC licensing model, **375**
 germinal center formation, 231
 isotype switching, 237
 somatic hypermutation, 231
 Th1 differentiation, 408
 thymic selection role, 362
CD44
 intercellular adhesion role, 211
 memory T cell marker, 428
 T cell activation, 377
 thymocyte development, 347
CD45
 activation of Src kinases
  in B cells, 220
  in T cells, 384, 388
 severe combined immunodeficiency, 758
 thymic selection, 361
CD55, Decay accelerating factor (DAF), complement
  fixation regulation, 567
CD59, Membrane inhibitor of reactive lysis
  (MIRL),complement fixation regulation, 571–572
CD69, natural killer cell costimulation, 529
CD70
 costimulation role, 398
 memory T cell response, 428
CD80, *see* B7 (B7–1)
CD86, *see* B7 (B7–2)
CD94, component of NK receptors, 526–527, 530
CD137 (4–1BB)
 autoimmune disease role, **1007**
 T cell costimulation, 392, 398
CD154 (CD40L), antibody targeting and graft tolerance
  induction, 901
CDRs, *see* Complementarity-determining regions
C/EBPβ, T helper cell differentiation role, **410**
CEM-15, human immunodeficiency virus immune
  response, 805
Centroblast
 affinity maturation, 234–235
 proliferation, 231–232
CGD, *see* Chronic granulomatous disease
Chédiak-Higashi syndrome, 778
Chemokines, *see also specific chemokines*
 allergy treatment and blocking, 947–948

 allograft rejection role, 888–889
 chemotaxis in inflammation, **81–82**
 expression, 507
 functions, 505–507, *506*
 human immunodeficiency virus entry via receptors,
  800, 806
 metastasis role, 834, *835*
 nomenclature, *506*
 receptors and signaling, 507–508, *508*, *509*
 T helper cell expression of receptors, 412–413
Chicken pox, vaccination, 738
Cholera
 involvement of exotoxin, 651
 vaccination, 729
Chronic granulomatous disease (CGD), 777–778
Chronic lymphocytic leukemia (CLL)
 clinical features, 1044–1045
 epidemiology, 1044
 genetic aberrations, 1045
 treatment, 1045
Chronic myelogenous leukemia (CML)
 clinical features, 1039
 epidemiology, 1039
 genetic aberrations, 1039, **1040–1041**, 1041
 treatment, 1041
CLL, *see* Chronic lymphocytic leukemia
Cluster of differentiation, *see* CD markers
Clusterin, complement fixation regulation, 571
CML, *see* Chronic myelogenous leukemia
c-myb
 B cell development modulation, **222**
 T cell development role, **353**
Coagulation, comparative immunology, 621
Collectins
 pathogen defense mechanisms, 90
 structure, 89, *89*
 types, 89
Colony-stimulating factors (CSF); see also G-CSF,
 GM-CSF, M-CSF
 anti-tumor treatment, 858–859, 867
 autoimmunity treatment, 1015–1016
 DC tolerization, 905
 eosinophil differentiation and activation, 947
 functions, 502–505
 hematopoiesis, 48, 50
 HSC mobilization, 914, 1032
 link to type I hypersensitivity, 939
 primary immunodeficiency treatment, 780
 secreted by Th cells, 416
Combinatorial diversity
 B cell receptor, 201–202
 T cell receptor, 328
Common variable immunodeficiency (CVID), 766
Comparative immunology
 cell types, 626
 evolution
  animal kingdom, 612, *613*, 614
  pressures in immune recognition, 614–615, 617
 gut-associated lymphoid tissue (GALT)
  amphibians, 627–628
  B cell receptor, 628
  birds, 628
  bony fish, 627
  cartilaginous fish, 627
  jawless fish, 626–627
  reptiles, 628
 immunoglobulin
  heavy chains, *629*
   amphibians, 630
   birds, 630, 632
   bony fish, 630
   cartilaginous fish, 629–630
   reptiles, 630
  light chains, 631
 innate immunity
  anatomical and physiological barriers, 21
  anti-microbial proteins and peptides, 623, 625
  blood cells, 518, *618*

Comparative immunology (*Continued*)
    coagulation, 621
    complement, 622–623
    components across animal types, *616–617*
    cytokines, 620–621
    pattern recognition molecules and Toll-like receptors, 615, 619, *619–620*, 620
    phagocytes, 621
    prophenoloxidase system, 621–622, *622*
  lymphoid tissues
    amphibians, 627–628
    birds, 628
    bony fish, 627
    cartilaginous fish, 627
    jawless fish, 626–627
    primitive animals, 626
    reptiles, 628
  major histocompatibility complex
    amphibians, 633–634
    birds, 634
    bony fish, 633
    cartilaginous fish, 633
    evolutionary rationale, 632–633
    jawless fish, 633
    primitive animals, 632
    reptiles, 634
  mutational mechanisms, *631*
  plants, 624–*625*
  T cell receptor
    amphibians, 635
    birds, 635
    bony fish, 635
    cartilaginous fish, 634–635
    reptiles, 635
Complement
  activation pathways
    alternative pathway
      alternative C3 convertase, 562
      alternative C5 convertase, 562
      C3, 562
      Factor B, 562
      Factor D, 562
      principles, 561–562, *561*
      properdin, 562
      regulation, 570, *570*
    classical pathway
      C1, 557–558, *557*
      C2, 559
      C3, 559–561, *560*
      C4, 558–559
      principles, 557, *558*
      regulation, 567, 568, *569*
    lectin pathway
      C-reactive protein, 563–564
      early steps, *564*
      mannose-binding lectin, 562–563, *563*
      regulation, 571
    overview of pathways, 555–556, *556*, *567*
    terminal components
      C5, 564
      C6, 564–565
      C7, 564–565
      C8, 565
      C9, 565, *566*
      regulation, 571–572, *571*
  amplification and sensitivity, 556
  anaphylatoxins, **559**
  autoimmune disease abnormalities, 1011
  comparative immunology, 622–623
  deficiencies
    alternative pathway components, 578
    C1, 576–577
    C2, 576–577
    C3, 577–578
    C4, 576–577
    complement receptors, 579
    lectin pathway components, 578
    phenotypes, 575–576, *576*, 780–781, *781*

      regulatory proteins, 578–579
      terminal components, 578
    developmental functions, 579
    evasion by extracellular bacteria, 655–656, **655**
    extracellular bacteria elimination, 652
    fixation assays, 159–162, *161*
    fungus elimination, 690
    history of study, 554–555
    human immunodeficiency virus immune response, 805
    infection barrier, 73, **74**
    membrane attack complex
      formation, *565*
      targets, 566
    nomenclature, 556–557
    parasite evasion, 687
    pathogen clearance mechanisms
      antigen presentation enhancement, 116
      B cell activation enhancement, 116, 574
      immune complex clearance from circulation, 116, 573–574
      inflammation induction, 116
      lysis, 115
      opsonization, 115–116, 573
      overview, 115, *115*
    pathology, **568**
    peripheral tolerance role, **577**
    receptors and pathogen entry mechanisms, 648, **649**
    receptors
      C1q receptors, 575
      C3a receptors, 575
      C4a receptors, 575
      C5a receptors, 575
      classification, 572, *573*
      CR1
        antigen processing enhancement, 574
        immune complex clearance, 573–574
        opsonized phagocytosis, 573
      CR2
        B cell activation, 574
        B cell memory development, 574
      CR3, 574–575
      CR4, 575
    sources, 555
    virus avoidance, 677
    virus elimination, 671–672
Complementarity-determining regions (CDRs), antigen–antibody interaction, 137–139
Complement-dependent cytotoxicity, tissue typing, 891, *891*
Complete Freund's adjuvant, 722
Congenic mouse, generation, **257–258**
Contraception, vaccines, 746–747
Corticosteroids
    allergy treatment, 944
    autoimmunity treatment, 1012
    TRAPS treatment, 780
CpG motif
    ligand binding to TLR, 86, 646
    possible role in autoimmunity, 1008
    vaccine adjuvant, 723, 948
C-reactive protein (CRP)
    complement activation, 563–564
    inflammation initiation role, 80
    pattern recognition, 90
Crohn's disease
    clinical features, 985
    pathophysiology, 985
Cross-presentation, *see* Antigen processing and presentation
CRP, *see* C-reactive protein
CTL, *see* Cytotoxic T lymphocyte
CTLA-4, *see* Cytotoxic T lymphocyte-associated molecule-4
CVID, *see* Common variable immunodeficiency
Cyclosporine A, *see* Immunosuppressive drugs
Cytokines, *see also Interferons, Interleukins, Lymphotoxins, Tumor Necrosis Factor and specific cytokines*
    allergy treatment and blocking, 946–947
    allograft rejection role, 888–889

      autoimmune disease
        abnormalities, 1010–1011
        management, 1016–1017
      B cell activation role, 227–229, *228*
      bone remodeling role, *513*
      cancer immunotherapy
        engineered cytokines, 858–859
        purified cytokines, 857–858
      comparative immunology, 620–621
      cytolytic cytokines, 422
      functional classification, *511*
      general properties, *465*
        complexity rationale, 466–467
        functional categories, 466
        half-lives, 465–466
        structure, 465, *466*
      hematopoiesis regulation, 50, *50*, *512*
      history of study, 464–465
      Hodgkin's lymphoma abnormalities, 1052
      human immunodeficiency virus immune response, 804–805
      immunosuppression in transplantation, 899
      inflammatory cytokines, *511*
      parasite evasion, 685–686
      pathogen response
        extracellular pathogens, *512*
        intracellular pathogens, *513*
      pleiotropic effects, 22
      receptors
        complexes, *470*
        shared subunits, 467–469, *469*
        structural families, 467
      T cell activation role, 398–400
      T cell immunosuppression, 446–447
      T helper cell differentiation, 409, 412
      tumors
        immune response, 847
        immunosuppressive cytokine secretion, 849–850
      types and sources, *23*
      vaccine adjuvant, 723
      viral homologs, **509–510**
      virus interference, 679–680
Cytomegalovirus, 666
Cytotoxic T cell (Tc), *see* T cell, CD8+ cytotoxic T cell
Cytotoxic T lymphocyte (CTL)
    adoptive transfer in cancer treatment, 869
    assays of function, **250–251**
    human immunodeficiency virus immune response, 803
    intracellular bacteria elimination, 660–661
  mucosa-associated lymphoid tissue, 597
    target cell destruction mechanisms
      cytolytic cytokines, 422
      Fas pathway, 422
      granule exocytosis
        granzymes, 421
        perforin role, 420–421
      overview, 419, *420*
      self-destruction prevention, 422
    virus elimination, 669, 671
Cytotoxic T lymphocyte-associated molecule-4 (CTLA-4)
    anergy induction role, 443
    blockade
      allergy treatment, 948
      autoimmunity treatment, 1018
      cancer treatment, 868–869
    control of CD28 and T cell receptor signaling, 396, *396*
    graft tolerance induction, 902–903

D

DAF, *see* CD55, Decay accelerating factor
Danger signal model, 437, **438**, 439
DC, *see* Dendritic cell
Decay accelerating factor (DAF), complement activation regulation, 567, 569
Delayed type hypersensitivity, *see* Hypersensitivity
Dendritic cell (DC)
    activation, 46–47
    antigen processing and presentation, 285–286, *286*

cancer vaccines, 866–868
compartmentalization, *39*, 46
functional overview, 13, *13*, **283**
graft tolerance induction, 905
gut-associated lymphoid tissue, 590
histology, *37*
immature versus mature cells, *288*
licensing model, **375**
markers, 46, *284*, *285*
maturation, 286–288
migration, 46, 286–288
pathogens and autoimmune disease, 1005
regulatory T cell generation role, 452–453, *453*
subtypes and development
  conventional dendritic cell, 283–285
  plasmacytoid dendritic cell, 285
  thymic dendritic cell, 283
T cell peripheral tolerance role, 437, **438**, 439
vaccine delivery, 726–727
virus interference, 677
De novo pathway of nucleotide synthesis, 151
Determinant selection model, MHC and immune
  responsiveness, 274, *274*
Diabetes mellitus
  type I
    clinical features, 974
    pathophysiology, 974, *974*
  types, 973
Diffuse large cell lymphoma, 1059–1060
DiGeorge syndrome, 774–775, *775*
Diphtheria
  involvement of exotoxin, 651
  vaccination, 729–730
Disease
  carriers, 645
  chronic disease, 645
  definition, 644
  emerging infectious diseases, *645*
  immunopathic disease symptoms, 644, *645*
DNA repair pathways
  defects leading to immunodeficiencies, 768–774, *769*
  defects associated with AML, 1038
  description, 768–769, *769*
  involvement in cancer, 834–839
DNA vaccine, *see* Vaccination
Duffy blood group system, 912

E

E2A
  B cell development modulation, **222–223**
  thymocyte development role, 353
Early B cell factor (EBF), B cell development
  modulation, **223**
EBF, *see* Early B cell factor
EBV, *see* Epstein-Barr virus
Eczema, 932
Electron dense material, *see* Tag assays
Elephantiasis, 683, *683*
ELISA, *see* Enzyme-linked immunosorbent assay
Endocytic processing pathway, *see also* Antigen
  processing and presentation
  model, 76–78, *77*
Endotoxic shock, pathophysiology, 475–**476**
Enzyme-linked immunosorbent assay (ELISA),
  see also Tag assays
  principles, 168, *169*
Eosinophil
  compartmentalization, *39*
  functional overview, 39
  histology, *38*
  importance in hypersensitivity, 929–930
Epitope
  B cell epitope
    agglutination, 159
    cross-reactivity, 144
    immunodiffusion, 156
    introduction, 25
    nature of, 122
    recognition by polyclonal antisera, 97

structural requirements, 127, 137
  definition, 25, *25*, 122
  protective epitope, 704
  T cell epitope
    introduction, 26
    nature of, 249–251, 313
    structure of TCR/pMHC complex, 337–338
Epitope spreading
  autoimmune disease, 1011–1012
  tumor immune response, 843
Epstein-Barr virus (EBV)
  Burkitt's lymphoma association, 829, 1052
  features, 666, 773
  Hodgkin's lymphoma role, 1052
  XLP association, 773
ERK, *see* extracellular signal regulated kinase
ERM, T helper cell differentiation role, **410**
Erythrocyte
  compartmentalization, *39*
  functional overview, 39
  histology, *38*
  target of malarial infection, 681
Experimental autoimmune models, see Autoimmune
  disease, animal models
Extracellular bacteria
  bacteria classification, *643–644*
  definition, 647–648
  effector mechanisms in elimination
    complement, 652
    humoral defense, 652
    mast cells, 652, 654
    overview, *653*
    T helper cells, 652
  entry mechanisms, 648, **648–649**
  evasion strategies
    antibody avoidance, 654
    complement avoidance, 655–656, **655**
    overview, *654*
    phagocytosis avoidance, 654–655
  host damage mechanisms
    exotoxins, 651
    immunopathic damage, 651
  species and diseases, 648–651, *650*
Extracellular signal regulated kinase (ERK)
  CD4/CD8 commitment, 366
  thymic selection, 361
  TCR signaling, 390–391
Exogenous processing, see Antigen processing and
  presentation

F

Familial Mediterranean fever, 779
Fas
  activation-induced cell death role,
    424–426
  defect leading to ALPS, 774
Fc receptors (FcRs)
  effector functions, 110–111, *111*
    FcαR, 112
    FcγR, 111–112
    FcεR, 112
    FcμR, 112
    transporter receptors, 112–113
  functional overview, 108
  nomenclature, 108–109
  signaling pathways, 110, *110*, 113
  structure, 109–110, *109*
  type I hypersensitivity roles
    FcεRI, 934–936, *935*
    FcεRII, 936
FcRs, *see* Fc receptors
FDC, *see* Follicular dendritic cell
Fetal–maternal tolerance, 457–459, *458*
Fetal thymic organ culture (FOTC)
  nature of, 358
  thymic selection studies, 363
FISH, *see* Fluorescence in situ hybridization
FLIP, description, 427
Flow cytometry, principles, 171–172, *172*

Flt3 ligand, 532
Fludarabine, *see* Immunosuppressive drugs
Fluorescence in situ hybridization (FISH), chromosomal
  translocations, **1033–1034**
Fluorochromes, *see* Tag assays
Follicular dendritic cell (FDC)
  B cell interactions, 229
  functional overview, 47
Follicular lymphoma, 1058–1059
Food allergy, 932–933
FOTC, *see* Fetal thymic organ culture
FoxP3
  IPEX syndrome, 986
  scurfy mice, 990
  T$_{reg}$ cell generation, 451
FTY720, see Immunosuppressive drugs
Fungi
  definition, 688
  effector mechanisms in elimination
    cell-mediated defense, 690
    complement, 690
    humoral defense, 690
    overview, *689*
  evasion strategies, 690–691, *690*
  species and diseases, 688–689, *688*

G

GALT, *see* Gut-associated lymphoid tissue
GATA-3
  CD4/CD8 T cell lineage commitment role, 367
  knockout mouse studies, **344**
  T cell development role, **353**
  T helper cell differentiation role, **410**
Gamma-delta (γδ)T cell
  activation, 541–542, *542*
  anatomical distribution, 539
  antigens
    binding, *541*
    CD1c, 540–541
    host stress antigens, 540
    MICA, 540
    MICB, 540
    pathogen-derived antigens, 540
    recognition, 539
  anti-tumor response, 848
  autoimmune disease abnormalities, 1010
  comparison with αβ T cells, *538*
  definition, 537–538
  development, *546*
    embryology, 544
    selection and tolerance, 547
    T cell receptor gene rearrangement, 544–547
  effector functions, 542–544, *543*
  functional overview, 46
  intracellular bacteria elimination, 660
  markers, *521*, 538–539
  memory, 544
  pathology, **545**
  pattern recognition by receptors, 88
  tumor immune response, 848–849
G-CSF, *see* Granulocyte colony-stimulating factor
Gene gun, vaccination, **715**, 727
Gene therapy
  allergy treatment prospects, 948
  primary immunodeficiency, **760–761**
  primary immunodeficiency management, 754
  transplantation application prospects,
    917–919, *918*
Genetic predisposition, *see* Autoimmune disease
Germinal center
  defects in immunodeficiencies, 762, 766
  formation, 230–231, **230**, *232*
gld mouse, 988–989
GM-CSF, *see* Granulocyte-macrophage colony-stimulating
  factor
Goodpasture's syndrome
  Categorization as type II hypersensitivity, 953
  clinical features, 953, 986
  pathophysiology, 986

Graft-versus-host disease (GvHD), see Transplantation
Graft-versus-leukemia effect, 916–917, 1034
Granulocyte colony-stimulating factor (G-CSF)
    expression, 504–505
    functions, 504, *505*
    receptor, 505
Granulocyte-macrophage colony-stimulating factor
    (GM-CSF)
    cancer immunotherapy, 858–859
    expression, 504
    functions, 504, *505*
    receptor, 504
Granuloma
    formation, 661, *662*
    in type IV hypersensitivity, 957
Granzymes, granule exocytosis, 421
Graves' disease
    clinical features, 980, *980*
    pathophysiology, *978*, 980
Grb2, T cell receptor signaling, **384**
Guillain-Barré syndrome
    clinical features, 984
    pathophysiology, 984
Gut-associated lymphoid tissue (GALT)
    antigen sampling
        dendritic cells, 590
        enterocytes, 590
        follicle-associated epithelium, 589–590, *589*
        M cells, 589–590
        macrophages, 590–591
    comparative immunology, 626–628
    composition
        gut epithelium, 586–587
        lamina propria, 588
        overview, 52, *587*, *588*
GvHD, see Transplantation, graft-versus-host disease

H

H-2 complex in mice, see Major histocompatibility complex
Haemophilus influenzae, vaccination, 730–731
Hairy cell leukemia (HCL), features, 1045–1046, *1046*
Hapten–carrier complex
    experiments
        carrier effect, 132, *132*
        hapten–carrier system, 131, *133*
        linkage of B and T cell determinants, 132–133, *133*
    in vivo complexes, 134–136
Hashimoto's thyroiditis
    clinical features, 978
    pathophysiology, 978–979, *978*, *979*
Hay fever, 931
HCL, see Hairy cell leukemia
Heat shock proteins (HSPs)
    antigen processing chaperones, **301**
    association with autoimmunity, 997, 1002, 1005
    cancer vaccine adjuvants, 866, *867*
    tumor immune response, 849
Helicobacter pylori, vaccination, 745
Helios, T cell development role, **353**
Hemagglutination assay, principles, 159
Hematopoiesis
    apoptosis, 50–51
    cytokine regulation, 50, *50*
    cytokine roles, *512*
    models, 49, *50*
    sites, 47, *47*
Hematopoietic cancers, *see also specific diseases*
    carcinogenesis, 1029–1030, *1030*
    chromosomal translocations, 1030, *1030*, *1031*,
        1033–1034
    classification, 1027, *1028*, 1029, *1029*
    clinical assessment and treatment terminology,
        1030–1032, 1034–1035
    frequencies, 1027, *1027*
    history of study, 1026–1027
    survival rates, *1032*
Hematopoietic stem cell (HSC)
    abundance, 49

B cell development role, 210–211
differentiation, 49, *48*
discovery, **48**
NK cell development role, 532
pluripotency, 47
self-renewal, 49, *49*
T cell development role, 347
transplantation, *see* Bone marrow transplantation
Hemolytic anemia
    alloimmune anemias, 949
    autoimmune anemias, 949, 951–952
Hemophagocytic syndrome, 764
Hepatitis A, vaccination, 731
Hepatitis B
    vaccination, 731
    virus, 667
Hepatitis C virus, 667
HEV, *see* High endothelial venule
HIDS, *see* Hyperimmunoglobulin D with periodic fever
    syndrome
High endothelial venule (HEV)
    lymphocyte migration, 63–64
    role in mucosal immunity, 588–589
High zone tolerance, 455
HIGM, *see* Hyperimmunoglobulin M syndrome
HIV, *see* Human immunodeficiency virus
HLA, *see* Human leukocyte antigen; Major
    histocompatilibity complex
Hodgkin's lymphoma
    clinical and histological features, 1051–1052, *1051*
    cytokine abnormalities, 1052
    genetic aberrations, 1052
    history of study, 1050–1051
    subtypes
        frequencies, *1053*
        lymphocyte-depleted Hodgkin's lymphoma, 1053
        lymphocyte-rich classical Hodgkin's lymphoma, 1053
        mixed cellularity Hodgkin's lymphoma, 1053
        nodular lymphocyte predominant Hodgkin's
            lymphoma, 1053–1054
        nodular sclerosis Hodgkin's lymphoma, 1053
        overview, 1052–1053
    treatment, 1054
    viruses in etiology, 1052
Homologous restriction factor (HRF), complement fixation
    regulation, 571–572
HRF, *see* Homologous restriction factor
HSC, *see* Hematopoietic stem cell
HSPs, *see* Heat shock proteins
Human immunodeficiency virus (HIV), *see also* Acquired
    immunodeficiency syndrome
    anti-retroviral drugs
        costs, 821–822
        fusion inhibitors, 820–821
        immune responses, 821
        overview, 819–820, *820*
        protease inhibitors, 820
        reverse transcriptase inhibitors, 820
        safety, 822
    epidemiology
        Africa, 811–812
        Americas, 811
        Asia, 811
        Asia, 811–812
        Europe, 811–812
        overview, 667, 808, *810*
    history of study, 786–788
    host factor effects on course
        human leukocyte antigen haplotypes, 806–807,
            807, *808*
        overview, 805, *805*
        resistance, 806
    immune response
        CEM-15, 805
        complement, 805
        cytokines, 804–805
        cytotoxic T lymphocytes, 803
        humoral immunity, 803–804

natural killer cells, 805
overview, 802, *802*
T helper cells, 802–803
infection
    antigenic variation, 800–801
    course, 796–797, *797*
    replication activation, 800, *801*
    staging, 797–798, *798*
    viral entry, 800
    viral tropism, 798–800, *799*, *799*
latency, 822
life cycle, 788–789, *788*, *789*
proteins and functions
    env gene-derived proteins, 793–794
    gag gene-derived proteins, 792
    pol gene-derived proteins
        integrase, 793, *794*
        protease, 793
        reverse transcriptase (RT) and DNA synthesis,
            792–783, *782*, *783*
    Nef, 795
    Rev, 795
    Tat, 794–795
    Vif, 796
    Vpr, 795
    Vpu, 795–796
serotypes, 788
structure
    capsid, 789
    envelope, 789
    genome, 790, *789*
    matrix, 789
    overview, 789, *789*
transmission, 808
vaccine
    animal models, 816–818
    barriers, 815–816, *816*
    clinical trials, 818–819
    development, 698, 700, 704
    passive immunization, 819
Human leukocyte antigen (HLA) complex, *see also* Major
    histocompatibility complex
    autoimmune disease susceptibility alleles, 995,
        997, *997*
    genes, *see* Major histocompatibility complex, HLA
        complex
    haplotype matching, *see* Transplantation, HLA
        matching
    human immunodeficiency virus infection and
        haplotypes, 806–807, **807**, *808*
    supertypes and vaccination, 710, **711**
Humoral immunity, *see also* B cells, effector function;
    Immunoglobulin
    introduction, 5, 7, 25
Hybridoma
    generation for monoclonal antibody production, 151,
        152, 153
    somatic cell hybrid generation, 150–151, *150*, *151*
Hygiene hypothesis, **941**
Hyperimmunoglobulin D with periodic fever syndrome
    (HIDS), 779
Hyperimmunoglobulin E syndrome, 768
Hyperimmunoglobulin M syndrome (HIGM)
    B cell-intrinsic syndromes, 767–768
    types, 764, *764*
Hypersensitivity
    allergy, see type I hypersensitivity
    atopy, see type I hypersensitivity
    history of study, 924–925
    overview of types, 925, *926*
    research prospects, 925, *925*
    type I hypersensitivity (allergy; atopy)
        allergens
            characteristics, 937, *938*
            common allergens, 926, *926*
            environmental determinants, *938*, 939
            genetic determinants, 937, *938*, 939
        anaphylaxis, 933–934

asthma, 931–932, *932*, **933**
candidate genes, *939*
definition, 925–926
diagnosis, 940, *942*
eczema, 932
effector stage
early phase reaction, 927–928, *929*, 930
mediators, 928, *929*
late phase reaction, 930, *931*
Fc receptor roles
FcεRI, 934–936, *935*
FcεRII, 936
food allergy, 932–933
hay fever, 931
immunoglobulin E role, 936–937
localized atopy, 927, 930–931
prevalence trends and hygiene hypothesis, 939–940, **941**
sensitization stage, 927, *928*
systemic atopy, 933–934
T helper balance, *936*
treatment of allergy
antihistamines, 942–944
bronchodilators, 944
chemokine blocking, 947–948
corticosteroids, 944
costimulatory blockade, 948
cromones, 944
cytokine blocking, 946–947
gene therapy, 948
hyposensitization, 945–947, *946*
immunoglobulin E signaling interference, 946
immunosuppressants, 944
lipoxygenase antagonists, 944
overview, 942, *942*
T cell response deviation, 946
vaccines, 948
urticaria, 932, *934*
type II hypersensitivity
antibody-mediated rejection of solid tissue transplants, 953
definition, 949
drug hypersensitivities, **950**
Goodpasture's syndrome, 953
hemolytic anemias
alloimmune anemias, 949
autoimmune anemias, 949, 951–952
mechanisms, 949, *951*
pemphigus, 953–954
thrombocytopenia
alloimmune thrombocytopenia, 953
autoimmune thrombocytopenia, 953
type III hypersensitivity
definition, 954
localized reactions, 956, *956*
mechanisms, 954–956, *955*
systemic reactions, 956–957, *956*
type IV delayed type hypersensitivity
chronic reactions, 957–958, *958*
contact hypersensitivity, 958–959, *959*
definition, 957
hypersensitivity pneumonitis, 960
T helper cell role, 417
H-Y system
as a minor histocompatibility antigen, 885
nature of, 358
T cell tolerance studies, 460

I

ICOS
antibody targeting and graft tolerance induction, 901
costimulation role, 396–397, *397*
deletion in CVID, 766
involvement in EAE, 992
T helper cell activation, 414
Id proteins, T cell development role, **353**
IFN-γ, *see* Interferon-γ
Ig, *see* Immunoglobulin

Ikaros
B cell development modulation, **222**
T cell development role, **353**
IL-1 through IL-28, *see* Interleukins
Ileostomy, 1013
Immunization, *see* Vaccination
Immunoassays, *see specific assays*
Immunoconjugates, cancer immunotherapy, 861, *862*, 863–864
Immunodeficiency, *see also specific diseases*
B cell-specific immunodeficiencies, 765–768, *765*, *766*
complement deficiencies, see Complement, deficiencies
diagnosis of primary immunodeficiency, 752–753
DiGeorge syndrome, 774–775, *775*
DNA repair defects and adaptive immunodeficiencies, 768–774, *769*
lymphoproliferative syndromes, 773–774, *773*
phagocytes, 776–780, *777*
primary versus acquired immunodeficiency, 752
severe combined immunodeficiency, see Severe combined immunodeficiency
T cell-specific immunodeficiencies, 763–765, *763*
treatment of primary immunodeficiency
bone marrow transplantation, 754
enzyme replacement, 753–754
gene therapy, 754
intravenous immunoglobulin, 753
Wiscott-Aldrich syndrome, 775–776
Immunodiffusion assays, principles, 156–157, *157*
Immunodysregulation, polyendorinopathy, enteropathy X-linked (IPEX) syndrome, features, 986
Immunoelectron microscopy, *see also* Tag assays
principles, 169–170, *170*
Immunoelectrophoresis
principles, 157–158, *158*
two-dimensional, *158*, 159
Immunofluorescence assays, *see also* Tag assays
principles, 170–171, *170*
Immunogen, *see also* Antigen
adjuvants, 130
definition, 24, 122
tolerogen comparison, *455*
types, 122
Immunoglobulin (Ig)
anti-immunoglobulin antibodies, **102**
antigen interactions, *see* Antigen–antibody interaction
antigen recognition, *see also* Epitope, B cell
introduction 26–27, *26*
biosynthesis, 242–243, *242*
carbohydrate content, 103
comparative study, see Comparative immunology, immunoglobulin
constant regions, 98–99, 101
cross-reactivity, 143–145
discovery, 5–6
diversity
chain pairing, heavy and light, 204
constant region, see Isotype switching
variable region, see V(D)J recombination in B cells;
Somatic hypermutation
effector functions
antibody-dependent cell-mediated cytotoxicity, 114–115, *115*
complement-mediated clearance of antigen, 115–116, *115*
neutralization, 113, *114*
opsonization, 113–114, *114*
extracellular bacteria
avoidance, 654
elimination, 652
Fc receptors, *see* Fc receptors
fragments, 94, *96*
fungus elimination, 690
genes, see Immunoglobulin genes
heavy chains, 97–98, *98*
history of study, 94, **95**, 96–97, 180–181

human immunodeficiency virus immune response, 803–804
immunoglobulin E and type I hypersensitivity
role, 936–937
therapeutic targeting, 946
intracellular bacteria
elimination, 662
evasion, 663
isotypes
classification, 29–30
functional properties
IgA, 118–119
IgD, 118
IgE, 119–120
IgG, 118
IgM, 116–117
overview, 116, *117*
heavy chain isotypes, 103, *104*
light chain isotypes, 101, 103, *103*
switching, *see* Isotype switching
light chains, 97–98, *98*
monoclonal antibody, see Monoclonal antibody
mucosa-associated lymphoid tissue secretory antibodies, 595–597, *595*, *596*
myeloma, see Myeloma
natural antibodies and pattern recognition, 90–91
parasite elimination, 684
plasma cell secretion, 26, 29
polyclonal nature of natural antisera, 97
structural isoforms, *105*
membrane-bound immunoglobulins, 105–106, 240–242
secreted immunoglobulins, 106–108, *107*, 240–242
structure, 29, *30*, 97–99, 101, 103
superfamily molecules, **99–100**
variable regions, 98–99, 101, *101*
virus
elimination, 671
evasion, 677
Immunoglobulin genes
allelic exclusion, 192–193, *193*
fine structure
heavy chain gene, 186–187, *187*
light chain gene, 185–186
general organization, 184–185, *185*, *186*
kappa/lambda exclusion, 193–194
loci, 183–184, *184*
productivity testing, 191–192
rearrangement, *see* V(D)J recombination in B cells
structure and splicing, **181–182**
transcriptional control sequences
enhancers, 204–206, *206*
general characteristics, 204
identification, **205**
transcription factor binding sites
nuclear factor-κB, 207, *207*
OCTA-binding transcription factors, 207
Immunohistochemistry, principles, 168
Immunology
balance between health and disease, 15–16, *15*
definition, 4, 7
historical perspective, 4–7, **8–9**
Nobel Prize winners, **9**
Immunoprecipitation, principles, 173, *173*
Immunoreceptor tyrosine-based activation motifs (ITAMs)
B cell activation, 220, **221**
NK cell activation, 523, 526–527
T cell activation, 318, 331
Immunoreceptor tyrosine-based inhibitory motifs (ITIMs)
Inhibitory NK receptors, 526–527, 531
regulation of B cell activation, 220, **221**
regulation of T cell activation, 396
Immunostimulating complex (ISCOM), vaccine delivery, 725–726
Immunosuppressive drugs
azathioprine, 896
CAMPATH-1H, 899
cyclosporine A, 896–897

Immunosuppressive drugs (*Continued*)
cytokines, 899
fludarabine, 898
FTY720, 898
leflunomide, 898
mycophenolate mofetil, 897
OKT3, 898
sirolimus, 897
tacrolimus, 897
treatment of autoimmune disease, 1014
treatment of hypersensitivity, 944
Inbred mouse, generation, **257–258**
Incomplete Freund's adjuvant, 722
Infection, *see also* all of Chapter 22
anatomical and physiological barriers, 70–73, *71*, *72*
definition, 10–11
host defense phases, 11–12, *12*
pathogen types, 11, *11*
persistent, 645
Infectious tolerance, 453–454
Inflammatory bowel disease
animal models, 991
clinical features, 985
pathophysiology, 985
Inflammatory response
adhesion molecules, 83–84, *83*
auxiliary cells, 84
cellular response and chemotaxis, *80*, **81–82**
clinical signs, 78, 80
complement induction, 116
definition, 14
initiation, 80, 83
mediators, *82*
tissue repair, 84
Influenza
vaccination, 731
virus, 667
Innate immunity
adaptive immunity relationship, 12–13, *18*
anatomical and physiological barriers to infection, 70–73, *71*, *72*
cellular internalization mechanisms, 73, 76–78
comparative study, *see* Comparative immunology, innate immunity
host defense phases, 11–12, *12*
inflammation, *see* Inflammatory response
overview, 7, 10, *10*, 18–19
pattern recognition, 84–91
Interferon, *see also specific interferons*
classification, 470
common functions, 470–471, *472*
expression, 472
history of study, 469–470
receptors
interferon-α/β receptor, 472–473
interferon-γ receptor, 474
signaling, 473–474, *473*
virus elimination, 669
Interferon α/β
as a danger signal, 438–439
induction of Ts cells, 452
NK cytotoxicity effect, 522
Interferon-γ (IFN-γ)
functions, 471–472
isotype switching role, 237, *238*
MHC class I molecule induction, 271
NK cell priming, 522
parasite elimination, 684, *685*
production by γδ T cells, 543
production by NK cells, 524
production by NKT cells, 548
receptor, 474
signaling defects and primary immunodeficiency, 778, *779*
stimulation of NK anti-tumor response, 549
T helper cell
differentiation role, 409–410
expression of receptor, 413
Th1/Th2 cross-regulation, 417

Interferon regulatory factor-4 (IRF-4), B cell activation modulation, **224**
Interleukin-1 (IL-1)
as a danger signal, 438
effect on Th2 cells, 415
expression, 474, 476
functions, 474, *474*
inflammatory role overview, 80
receptor and signaling, 476–477, *477*
receptor antagonist, 477
T helper cell activation, 415
Interleukin-2 (IL-2)
anergy role, 441, 443–444
cancer immunotherapy, 858
CTL generation, 418–419
expression, 478
FLIP downregulation, 427
functions, 477–478, *478*
memory T cell generation, 428
NK cell development, 532–533
NK cell priming, 522
production by NKT cells, 549
receptor, 399
receptor chains
α chain, 478
β chain, 478
γ chain, 478
T cell activation role, 398–400
Th0 cells, 406
Th1 cytokine secretion, 416–417
transcriptional control, 399–400, *399*
Treg cells, 452
Interleukin-3 (IL-3)
expression, 479
functions, 479, *479*
receptor, 479
Interleukin-4 (IL-4)
B cell activation role, 219–220
cancer immunotherapy, 858
expression, 480
functions, 479–480, *480*
gamma-delta T cell development, 543
infectious tolerance, 453
isotype switching role, 237, *238*
production by NKT cells, 549
receptor and signaling, 480–481, *480*
T helper cell differentiation role, 409, 412
Th0 cells, 406–407
Th1/Th2 cross-regulation, 417
Th2 cytokine secretion, 416
Th3, Tr1 differentiation 452
Interleukin-5 (IL-5)
effect on eosinophils, 417
expression, 481
functions, 481, *481*
isotype switching role, 237, *238*
production by Th2 cells, 410, 416
receptor, 481
type I hypersensitivity role, 927, 930, 933
Interleukin-6 (IL-6)
as a danger signal, 438
CTL differentiation, 418–419
expression, 481
functions, 481, *481*
inflammatory role overview, 80
receptor and signaling, 481–482, *482*
T helper cell differentiation role, 410,412
Th2 cytokine secretion, 414, 416
Interleukin-7 (IL-7)
CD4/CD8 lineage commitment, 365
expression, 483
functions, 482–483, *482*
gamma-delta T cell development, 544–545
memory T cell generation, 428
memory T cell life span, 430
receptor, 483
receptor and severe combined immunodeficiency, 758
T cell development, 347, 351–352
thymic involution, 345–346

Interleukin-8 (IL-8)
expression, 483
functions, 483, *483*
inflammatory role overview, 81
receptor, 483
Interleukin-9 (IL-9)
expression, 484
functions, 483–484, *484*
receptor, 484
Interleukin-10 (IL-10)
B cell activation role, 228
B cell effector differentiation, 239
expression, 485
functions, 484–485, *484*
immunosuppression, 446
infectious tolerance, 453
inhibition of Th1 development, 409
isotype switching role, 238
production by NKT cells, 549
receptor, 485
T helper cell differentiation role, 409–410
Th1/Th2 cross-regulation, 417
Th2 cytokine secretion, 414–416
Th3, Tr1 differentiation, 452
Interleukin-11 (IL-11)
expression, 485
functions, 485, *485*
receptor, 485
Interleukin-12 (IL-12)
cancer immunotherapy, 858
CTL differentiation role, 418–419
expression, 486
functions, 486, *486*
macrophage hyperactivation role, 417
NK cytotoxicity role, 522
NKT cell activation, 548
receptor and signaling, 486, *493*
signaling defects and primary immunodeficiency, 778, *779*
Th0 cells, 406–407
Th1 cytokine secretion, 416–417
Th1 differentiation, 409–410
Interleukin-13 (IL-13)
expression, 487
functions, 486–487, *487*
production by NKT cells, 549
receptor, 487
Th1/Th2 cross-regulation, 417
Th2 cytokine secretion, 414, 416
Interleukin-14 (IL-14)
expression, 487
functions, 487
receptor, 487
Interleukin-15 (IL-15)
expression, 488
functions, 487–488, *488*
memory T cell life span and proliferation, 430
NK cell development, 532–534
NKT cell development, 550
receptor and signaling, 488, *489*
Interleukin-16 (IL-16)
expression, 488–489
functions, 488, *489*
receptor, 489
Interleukin-17 (IL-17)
expression, 490
functions, 489–490, *490*
receptor, 490
Interleukin-18 (IL-18)
as a danger signal, 438–439
effect on NK cells, 522
effect on Th1 cells, 415
expression, 491
functions, 490–491, *490*
NKT cell activation, 549
receptor, 491
T helper cell activation, 415
Interleukin-19 (IL-19)
expression, 491

functions, 491
receptor, 491
Interleukin-21 (IL-21)
expression, 492
functions, 491, *491*
receptor, 492
Interleukin-22 (IL-22)
expression, 492
functions, 492, *492*
receptor, 492
Interleukin-23 (IL-23)
expression, 492
functions, 492, *492*
memory effector T cell differentiation, 430
receptor and signaling, 492, *493*
Interleukin-24 (IL-24)
expression, 493
functions, 493
receptor, 493
Interleukin-25 (IL-25)
expression, 493
functions, 493, *493*
receptor, 493
Interleukin-26 (IL-26)
expression, 494
functions, 493
receptor, 494
Interleukin-27 (IL-27)
expression, 495
functions, 494–495, *494*, *495*
initiation of Th1 differentiation, 407, 419–410
receptor, 495
T helper cell differentiation role, 409
Interleukin-28 (IL-28)
expression, 495
functions, 495
receptor, 496
Intracellular bacteria
bacteria classification, 643–644
definition, 656
effector mechanisms in elimination
CD4$^+$ T cells, 661
CD8$^+$ T cells, 660–661
$\gamma\delta$ T cells, 660
granuloma formation, 661, *662*
humoral defense, 662
macrophages, 658, 660
natural killer cells, 660
neutrophils, 658
overview, *659*
evasion strategies
antibody avoidance, 663
overview, *663*
phagocytosis avoidance, 663–664
T cell avoidance, 663–664
species and diseases, 656–658, *656*,
657, *658*
Intravenous immunoglobulin (IVIG)
autoimmune disease management, 1014
primary immunodeficiency management, 753
IPEX syndrome, *see* Immunodysregulation,
polyendorinopathy, enteropathy X-linked
syndrome
IRF-4, *see* Interferon regulatory factor-4
ISCOM, *see* Immunostimulating complex
Isotype switching
activation-induced cytidine deaminase role, 239
cytokine roles, 237–238, *238*
defect in immunodeficiency, 764, 767
differential requirements, 238–239
example, *236*
germline CH transcription, 237–238, *238*
overview, 30, 103–104, *105*, 235, *235*
switch regions, 235–237
ITAMs, *see* Immunoreceptor tyrosine-based activation
motifs
ITIMs, *see* Immunoreceptor tyrosine-based inhibitory
motifs
IVIG, *see* Intravenous immunoglobulin

J
Jak
Jak3 severe combined immunodeficiency, 758
signal transduction, *468*
Japanese B encephalitis, vaccination, 732
Jenner, Edward, smallpox immunization, 4–5, *5*, **6**
Junctional diversity
B cell receptor, 202, *203*
T cell receptor, 328–329

K
Kaposi's sarcoma, 786–787, *786*
Kawasaki disease
clinical features, 982–983, *983*
pathophysiology, 983
Kell blood group system, 912
Kidd blood group system, 912
Killed vaccine, *see* Vaccination
Knockout mouse
B7 studies, 394
CD3 studies, 331–332
GATA-3 studies, 344
generation, **194–195**
RAG studies, 344
somatic recombination studies, 194, 196
T cell receptor genes, 327
Koch, Robert, germ theory of disease, 5

L
LAD, *see* Leukocyte adhesion deficiency
LAG-3, *see* Lymphocyte activation gene-3
LAK cells, *see* lymphokine-activated killer cells
Large granular lymphocytic leukemia (LGLL), features, 1046
LEF-1
B cell development role, **223**
T cell development role, 353–354
Leflunomide, *see* Immunosuppressive drugs
Leishmaniasis, 680–681
Leprosy, 658, *658*
Leukemia, *see also specific leukemias*
Definition, 1027
Leukocyte adhesion deficiency (LAD)
Rac2 deficiency, 777
type I, 776
type II, 776–777
Leukocyte extravasation, 61–62, *62*
Leukotriene
as a chemokine, 81
definition and function, 82–83
LFA-1, T cell–antigen-presenting cell interactions, 376–377
LGLL, *see* Large granular lymphocytic leukemia
LIGHT
autoimmune disease role, **1007**
manipulation in cancer treatment, 869
Linked suppression
decrease in graft rejection role, 904
definition, 453–454
Liposome, vaccine delivery, 724–725
Lipoxygenase
antagonists in allergy treatment, 944
Involvement in leukotriene production, 83
Live, attenuated vaccine, *see* Vaccination
Low zone tolerance, 456
LTs, *see* Lymphotoxins
Lymph, collection in peripheral tissues, 56, *56*
Lymph nodes
afferent lymphatic vessels, 57–58
anatomy, 57, *57*, *58*
cells, 14
follicles, 58–59
site of B and T cell interaction, 229
site of T cell activation, 375–376
Lymphocyte, see also B cells; T cells
discovery, 5–6
homing, 64–66, *64*, **66**
recirculation, 62–64, *63*
Lymphocyte activation gene-3 (LAG-3)
natural killer cell costimulation, 528–529
T helper cell expression, 413

Lymphocyte-depleted Hodgkin's lymphoma, 1053
Lymphocyte-rich classical Hodgkin's lymphoma, 1053
Lymphokine-activated killer (LAK) cells, 522
Lymphoid tissue
bone marrow, 53
bronchi-associated, see Bronchi-associated lympoid
tissue
comparative immunology, 616
gut-associated, see Gut-associated lymphoid tissue
leukocyte extravasation, 61–62
lymphatic system, 56–59
lymphocyte homing, 64–66
lymphocyte recirculation, 62–64
mucosa-associated, see Mucosa-associated lymphoid
tissue
nasopharynx-associated, see Nasopharynx-associated
lymphoid tissue
organization, 51–52, *52*
primary versus secondary, 14, *14*, 52, *52*, *53*
skin-associated, see Skin-associated lymphoid tissue
spleen, 60–61
thymus, 53–55
Lymphoma, *see also specific lymphomas*
classification, 1050
definition, 1027
leukemia association, 1046
progression, 1050, *1051*
Lymphotoxins (LTs)
LTα
expression, 499
function, 498–499
LT$\alpha\beta$
expression, 499
function, 499
receptors and signaling, 499, *499*

M
mAb, *see* Monoclonal antibody
MAC, *see* Membrane attack complex
Macrophage
activation, 40, **42–43**
antigen presentation, 288
compartmentalization, *39*, 40, *41*
functional overview, 13, *13*, 36, 40
gut-associated lymphoid tissue, 590–591
histology, *37*
hyperactivation by T helper cells, **42–43**, 417
intracellular bacteria elimination, 658, 660
parasite elimination, 684
Macrophage colony-stimulating factor (M-CSF)
expression, 505
functions, 505, *505*
receptor, 505
Macropinocytosis, mechanism, 73, *75*
Major histocompatibility complex (MHC), *see also* Human
leukocyte antigen
alleles and disease, 274–275, *275*
alloreactivity, 273
antigen binding, 26–27, *27*
antigen processing and presentation, *see* Antigen
processing and presentation
classes of molecules encoded
MHC class I molecules
peptide-binding site, 261–262, *262*
polypeptides, 260–261
structure, *260*, *261*
MHC class II molecules
peptide-binding site, 263–264, *264*
polypeptides, 263
signaling, **287**
structure, *260*, *263*
MHC class III molecules, introduction, 255
MHC-like molecules, description, **266**
Non-classical MHC molecules, description, **266**
overview, 26, 254–255
comparative study, *see* Comparative immunology, MHC
deficiencies and primary immunodeficiency, 763
definition, 248
epidemiology, 680

Major histocompatibility complex (*Continued*)
  expression of MHC proteins
    class I genes, 271
    class Ib genes, 272
    class II genes, 271
    regulatory elements, *270*
    SXY-CIITA regulatory system, 270–271
  H-2 complex organization in mice, *255*, *265*
    class I genes, 264
    class Ib genes, 264
    class II genes, 264–265
    class IIb genes, 265
    class III genes, 265, 267
    β2-microglobulin, 264
    non-MHC genes, 265, 266–267
  haplotypes, 259–260, *259*, 268, *268*, 272–273
  history of study
    antigen presentation, 249, 251–252, 254
    discovery, 248–249
    restriction discovery, **253–254**
    T cell recognition, 249
  HLA complex organization in humans, *255*, *265*
    class I genes, 267–268
    class Ib genes, 268
    class II genes, 268
    class IIb genes, 268–269
    class III genes, 269
    β2-microglobulin, 268
    non-MHC genes, 269
  loci in humans and mice, *255*
    comparison of products, *269*
    multiplicity, 255–256, *256*
    organization, 255, *255*
    polymorphisms, 256–257
Malaria
  parasite life cycle, 681, *682*
  vaccine development, 718, **718–719**
MALT, *see* Mucosa-associated lymphoid tissue
Mancini assay, principles, 156, *157*
Mannose-binding lectin (MBL)
  complement activation, 562–563, *563*
  immunodeficiency, 780
Mantle cell lymphoma, 1056–1058, *1058*
MAPK, *see* Mitogen-activated protein kinase
Mast cell
  compartmentalization, *39*
  differentiation, 50
  extracellular bacteria elimination, 652, 654
  functional overview, 39
  histology, *38*
  inflammation role, 84
  type I hypersensitivity role, 927–928
  type II hypersensitivity role, 951
  type III hypersensitivity role, 955
MBL, *see* Mannose-binding lectin
M cell, pathogen entry mechanisms, 648, **648–649**
M-CSF, *see* Macrophage colony-stimulating factor
Measles
  vaccination, 732–734, *733*
  virus, 667–668
Megakaryocyte
  functional overview, 39
  histology, *38*
Membrane attack complex (MAC)
  formation, *565*
  targets, 566
Memory B cell
  fate decisions, 239, *240*
  general characteristics, 243
  surface markers, 243, *244*
Memory T cell
  activation, 429–430
  differentiation and effector function, 430
  distribution, 429
  generation, 428
  life span, 430–431
  markers, 428
Meningococcus, vaccination, 734

Metastasis, *see* Tumors
MG, *see* Myasthenia gravis
MHC, *see* Major histocompatibility complex
MiHA, *see* Minor histocompatibility antigen
Minor histocompatibility antigen (MiHA), 881–885,
  884, 885
MIRL, *see* CD59
Mitogen-activated protein kinase (MAPK)
  B cell receptor signaling, 220
  CD4/CD8 T cell lineage commitment signaling, 366
  disruption in signaling leading to anergy, 444
  T cell receptor signaling, 390–391, *390*
  thymic selection role, 361
Mitogens
  T cell receptor signaling, **386–387**
  Ti antigens, 123
Mixed cellularity Hodgkin's lymphoma, 1053
Mixed lymphocyte reaction (MLR)
  Description, 250
  tissue typing, 892, *893*
MLR, *see* Mixed lymphocyte reaction
Monoclonal antibody (mAb)
  advantages and disadvantages in immunoassays,
    154–155
  humanization, 153, **153**
  hybridoma generation, 151, *152*, 153
  phage display of Fab fragments, **153–154**
  plantibodies, **154**
  use as treatment; *see also* CTLA-4 blockade
    autoimmunity, 991, 1015
    cancer
      immunoconjugates, 861, *862*, 863–864, 1034
      unconjugated antibodies, 859, **860**, 861, *861*
    HIV infection, 819
    to increase graft tolerance, 899
    TRAPS, 780
    type I hypersensitivity, 946–947
    type II hypersensitivity, 952
Monoclonal gammopathy of undetermined significance
  (MGUS), features, 1047
Monocyte
  antigen presentation, 288
  compartmentalization, *39*
  functional overview, 36
  histology, *37*
MS, *see* Multiple sclerosis
MRL/lpr mouse, 988–989
Mucosa-associated lymphoid tissue (MALT), *see also*
  Bronchi-associated lymphoid tissue; Gut-associated
  lymphoid tissue; Nasopharynx-associated lymphoid
  tissue
  common mucosal immune system evidence, 597,
    599, *599*
  composition, 52, 59–60, 584
  cytotoxic T lymphocytes, 597
  distribution, *59*
  inductive versus effector sites, *585*
  minor tissue immune responses
    ear, 600
    eye, 600
    urogenital tract, 599–600
    neurological components, **603–604**
  poly-immunoglobulin receptor and pathogen
    neutralization, **598**
  secretory antibodies, 595–597, *595*, *596*
  Th2 responses, 593–595
Mucosa-associated lymphoid tissue lymphoma,
  1059
Mucosal epithelium, *see also* BALT, GALT, MALT,
  NALT
  anatomy, 70, *71*
  barrier to infection, 71–72
Multiple myeloma, *see* Myeloma
Multiple sclerosis (MS)
  clinical features, 974–975
  experimental autoimmune encephalitis, 991–992
  pathophysiology, 975–977, *975*, *976*
Mumps, vaccination, 732–734

Myasthenia gravis (MG)
  clinical features, 982
  experimental autoimmune myasthenia gravis, 992–993
  pathophysiology, 982, *983*
Mycophenolate mofetil, *see* Immunosuppressive drugs
Mycosis fungoides, 1061–1062, *1061*
Myeloma
  clinical and biochemical features, 1048–1049, *1048*
  definition, 1027
  genetic aberrations, 1049
  overview, 1047
  stages, *1048*
  treatment, 1049–1050

N

Naked DNA vaccines, *see* Vaccination
NALT, *see* Nasopharynx-associated lymphoid tissue
Nasopharynx-associated lymphoid tissue (NALT)
  antigen sampling, 593
  composition, 52, 591–593, *591*, *592*
Natural antibodies
  defense against parasites, 684
  pattern recognition, 90–91
  production triggered by complement, 574
  production by CD5⁺ B cells, 218
Natural killer (NK) cell
  allogeneic cell benefits in hematopoietic stem cell
    transplantation, 916–917, *917*
  compartmentalization, *39*
  costimulation, 528–529
  development, 532–534, *533*
  F1 hybrid resistance, 534–535, *536*
  functions
    cytokine secretion and immunoregulation, 524, *524*
    cytolysis
      antibody-dependent cell-mediated cytotoxicity,
        523–524, *523*, *524*
      Fas-mediated killing, 524
      natural cytotoxicity, 521–522
    overview, 46, 520–521
  histology, *37*, 520
  history of study, 518–520
  human immunodeficiency virus immune response, 805
  intracellular bacteria elimination, 660
  markers, 520, *521*
  pattern recognition by receptors, 88
  priming and activation, *522*
  receptors
    activatory receptors
      costimulators, 528–529
      natural cytotoxicity receptors, 528
      NKG2, 528
      overview, 524, *524*, **526–527**
    inhibitory receptors
      CD94, **530**, 531
      expression, 534–535, *535*
      ILT2, **530**, 531–532
      KIR molecules, **530**, 531
      Ly-49 family, **530**, 531
      overview, *526*, 529–530, **530**
  tolerance
    anergy model, 536–537
    at least one model, 535–536
    receptor calibration model, 536, *537*
  tumor immune response, 845
  virus
    elimination, 669
    evasion, 676
Natural killer T (NKT) cell
  activation, 548–549, *548*
  antigen recognition, 548
  compartmentalization, *39*
  development, 549–550, *550*
  effector functions, 549
  functional overview, 46, 547–548, *548*
  graft tolerance induction, 904–905
  histology, *37*
  markers, *521*

pattern recognition by receptors, 88–89
type I hypersensitivity role, 930
tumor immune response, 848
Nck, T cell receptor signaling, **385**
Neonatal tolerance, 459–461, *459*
Neutropenia
cyclic, 780
severe congenital, 780
Neutrophil
bystander damage to host tissue, 40
chemotaxis, 39
compartmentalization, *39*
degranulation, 39
functional overview, 13, *13*, 36, 39
histology, *38*
intracellular bacteria elimination, 658
respiratory burst, 40
NF-AT, T helper cell differentiation role, **411**
NF-κB, *see* Nuclear factor-κB
NHL, *see* Non-Hodgkin's lymphoma
Nijmegen breakage syndrome, 771, *771*
NK1.1, natural killer cell costimulation, 529
NK cell, *see* Natural killer cell
NKT cell, *see* Natural killer T cell
Non-obese diabetic (NOD) mouse, 989–990
NOD proteins, pattern recognition, 91
Nodular lymphocyte predominant Hodgkin's lymphoma,
1053–1054
Nodular sclerosis Hodgkin's lymphoma, 1053
Non-Bruton's agammaglobulinemia, 766
Non-Hodgkin's lymphoma (NHL)
clinical and histological features, 1054–1055
development, 1056, *1056*
pathogen induction, 1055
subtypes
adult T cell leukemia/lymphoma, 1061
anaplastic large cell lymphoma, 1062
angioimmunoblastic T cell lymphoma, 1062
B cell chronic lymphocytic lymphoma, 1058
Burkitt's lymphoma, 1060–1061, *1060*
diffuse large cell lymphoma, 1059–1060
follicular lymphoma, 1058–1059
frequencies, *1055*
immunophenotyping, *1057*
mantle cell lymphoma, 1056–1058, *1058*
mucosa-associated lymphoid tissue lymphoma, 1059
mycosis fungoides, 1061–1062, *1061*
overview, 1055–1056
peripheral T cell lymphoma, 1062
precursor B lymphoblastic leukemia/lymphoma,
1056, *1057*
precursor T lymphoblastic leukemia/lymphoma, 1061
Sezary syndrome, 1061–1062
Notch1
B cell development, 223
CD4/CD8 commitment, 366–367
NK cell development, 532
signaling in cell fate determination, **350**
T lineage commitment, 349–350
Th2 differentiation, 412
Nuclear factor-κB (NF-κB), gene transcription regulation
in somatic recombination, 207, *207*
NZB/W F1 mouse, 988

O

OKT3, immunosuppression in transplantation, 898
Omenn syndrome, 762
Oncogenes, 836, 838, *838*
OPGL, *see* RANK ligand
Opsonization
as a function of CR2, 574
complement effector function, 115–116, 554
increase phagocytosis role, 560
inhibition by C4bp, 655
mechanism, 76, *76*, 113–114, *114*
via MBL or CRP, 563–564
Organ transplantation, *see* Transplantation
OS chicken, **989**

Ouchterlony assay, principles, 156, *157*
OX40
autoimmune disease role, **1008**
manipulation in cancer treatment, 869

P

PAMP, *see* Pathogen-associated molecular pattern
Panel reactive antibody (PRA), tissue typing,
891–892, *892*
Paraneoplastic syndrome, *see* Tumors
Parasites
effector mechanisms in elimination
helminth worms, *686*
cytokines, 685–686
T helper cells, 685
protozoans
humoral defense, 684
interferon-γ, 684, *685*
macrophages, 684
T cells, 684
T helper cells, 684
evasion strategies
complement avoidance, 687
detection avoidance, 687
host immune response manipulation, 688
overview, 686, *687*
phagolysosomal destruction avoidance, 687
malaria, *see* Malaria
species and diseases, 680–683, *681*, *683*
Pasteur, Louis, vaccine development, 5
Pathogens
autoimmune disease role
infection epidemiology, 999
mechanisms
molecular mimicry, 1002–1005, *1002*, *1003*, *1003*
inflammation induction, 1005
dendritic cell maturation induction, 1005
superantigens, 1005–1006
bacteria, *see* Extracellular bacteria; Intracellular
bacteria
carcinogenesis, 840–842, *840*
fungi, *see* Fungi
innate defenses, 645–647, *647*
parasites, *see* Parasites
types, 642, 644
viruses, *see* Viruses
Pathogen-associated molecular pattern (PAMP), 84–85, 647
Pattern recognition receptors (PRRs)
CD14, 85
overview, 84–85, 85, *86*
scavenger receptors, 87–88
soluble molecules, 89–91
T cell receptors, 88–89
Toll-like receptors, 85–87
Pax5, B cell development modulation, **223**
PCR, *see* Polymerase chain reaction
PD-1
defect leading to autoimmunity, 1009
graft tolerance induction, 902–903
negative regulation of T cell activation, 397
Pemphigus
categorization as a type II hypersensitivity, 953
clinical features, 953–854, 986, *986*
pathophysiology, 986
Peptide vaccine, *see* Vaccination
Perforin, granule exocytosis, 420–421
Peripheral T cell lymphoma, 1062
Peripheral tolerance
B cells
anergy, 447
mechanisms, 448–450, *449*
evidence, 436–437
history of study, 434–436, *435*
T cells
anergy
demonstrations, 441–442, *442*
induction, 443–444
signaling, 443–444, *443*, *444*

cytokine immunosuppression, 446–447
danger signal model, 437, **438**, 439
exhaustion, 447
immune deviation, 447
immune privilege, 446
immunological ignorance, 445–446
peripheral clonal deletion, 439–440
two-signal paradigm, 436, **437**
Pertussis, vaccination, 730
Phage display, Fab fragments, **153–154**
Phagocyte
autoimmune disease abnormalities, 1010
comparative immunology, 621
discovery, 5, **7**
primary immunodeficiencies, 776–780, *777*
respiratory burst, 77, **79**
Phagocytosis
evasion by extracellular bacteria, 654–655
intracellular bacteria evasion, 663–664
mechanism, 73, *75*, 76
Phagolysosome, maturation, 77, *78*
Philadelphia chromosome, 1039, **1040–1041**, 1041
Phosphatidylinosotol 3-kinase (PI3K), T cell receptor
signaling, 389–390
Phototherapy, 1013
PI3K, *see* Phosphatidylinosotol 3-kinase
pIgR, *see* Poly-immunoglobulin receptor
PKC, *see* Protein kinase C
Plague, vaccination, 734
Plants, comparative immunology, *624–625*
Plasma cell
antibody secretion, 26, 29
dyscrasias, 1046–1050
fate decisions, 239, *240*
general characteristics, 240
immunoglobulin synthesis, 242–243, *242*
membrane versus secreted immunoglobulin secretion,
240–242, *241*
Platelet, histology, *38*
Polio
vaccination, 734–735
virus, 668
Poly-immunoglobulin receptor (pIgR)
generation of secretory antibodies, 107, 113, 596
pathogen neutralization, **598**
Polymerase chain reaction (PCR)
detection of V(D)J recombination, *188–189*
tissue typing techniques, 893–896, *895*
Polymorphonuclear leukocyte, *see* Neutrophil
Polymyositis
clinical features, 983–984
pathophysiology, 984
Popcorn cells, in Hodgkin's lymphomas, 1053
PRA, *see* Panel reactive antibody
Pre-B cell
pre-B-I cells, 213–214
pre-B-II type 1 cells, 214, *214*
pre-B-II type 2 cells, 214
pre-B-II type 3 cells, 214–215
surrogate light chain, 213
Precipitin curve, generation, 155–156, *155*
Precipitin ring assay, principles, 156, *157*
Precursor B lymphoblastic leukemia/lymphoma,
1056, *1057*
Precursor T lymphoblastic leukemia/lymphoma, 1061
Pre-TCR
developmental checkpoint, 348
overview, 325–326
Primary immunodeficiency, *see* Immunodeficiency
Prion
diseases, 691
immune response, 692
infection mechanism, 691–692, *692*
Pro-B cell
bone marrow stromal cells in development, 211–213,
213
markers, 211, *212*
Prophenoloxidase system, 621–622, *622*

Prostaglandin, definition and function, 82–83
Proteasome
    core proteasome, 296–297
    immunoproteasome, 298–299
    sampling of organelle proteins, 299
    standard proteasome, 297–298, *297*
Protein A, antibody binding in immunoassay, 162, *164*
Protein G, antibody binding in immunoassay, 162, *164*
Protein kinase C (PKC), B cell peripheral tolerance
    signaling, 450
Protein tyrosine kinase (PTK)
    B cell activation, 219–220, *220*
    CD4/CD8 T cell lineage commitment signaling, 366
    signal transduction, 23, *24*
    T cell receptor signaling, 384, 387–388
PRRs, *see* Pattern recognition receptors
Psoriasis
    clinical features, 984–985, *984*
    pathophysiology, 985
PTK, *see* Protein tyrosine kinase
PU.1
    B cell development modulation, **222**
    T cell development role, **353**
Purine nucleoside phosphorylase, deficiency, 760, 762

R

Rabies
    vaccination, 735
    virus, 668
Rac2, deficiency, 777
Radioimmunoassay, *see also* Tag assays
    principles, 167–168
RAG proteins, *see* Recombination activation gene proteins
RANK ligand
    expression, 500
    functions, 499–500, *500*
    knockout mice, 499
    receptor, 500–501, *500*
Ras
    B cell receptor signaling, 220
    T cell receptor signaling, 390–391, *390*
    thymic selection role, 361
Receptor-mediated endocytosis, mechanism, 73, *75*
Recombinant vector DNA vaccine, *see* Vaccination
Recombination activation gene (RAG) proteins
    knockout mouse studies, **344**
    severe combined immunodeficiency, 762
    V(D)J recombination, 197
Recombination signal sequence (RSS)
    function in B cell V(D)J recombination 196–197,
        197, 198
    function in T cell V(D)J recombination, 322–323
Reed-Sternberg cells, in Hodgkin's lymphomas, 1051
Regulatory T cell
    autoimmune disease abnormalities, 1009
    CD4$^+$CD25$^+$ T cells, 450–452
    CD8$^+$ suppressor T cells, 452–453
    dendritic cells in generation, 452–453, *453*
    graft tolerance induction
        CD4$^+$ cells, 903–904
        CD8$^+$ cells, 904
    history of study, 450
    infectious tolerance role, 453–454
    linked suppression role, 453–454
    Th3 cells, 452
    Tr1 cells, 452
    tumors and response promotion, 850
    types, *451*
Rejection, *see* Transplantation
Restriction fragment length polymorphism (RFLP), tissue
    typing, 893, *894*
Reticular dysgenesis, 763
Rev, human immunodeficiency virus, 795
Reverse transcriptase, human immunodeficiency virus,
    792–783, *782, 783*, 820
RF, *see* Rheumatic fever
RFLP, *see* Restriction fragment length polymorphism
Rh disease, 743–744, *743*

Rheumatic fever (RF)
    clinical features, 972–973, *973*
    pathophysiology, 973
Rheumatoid arthritis (RA)
    clinical features, 970, *971*
    collagen-induced arthritis, 992
    pathophysiology, 971–972, *971*
Rhinovirus, 668
Rocket electrophoresis, principles, 158–159, *158*
RSS, *see* Recombination signal sequence
Rubella, vaccination, 732–734
Runx-3, CD4/CD8 T cell lineage commitment role, 367

S

SALT, *see* Skin-associated lymphoid tissue
Salvage pathway of nucleotide synthesis
    immunodeficiency, 758–759
    nature of, 151
SAPK/JNK, *see* stress-activated protein kinase/jun kinase
SARS virus, *see* Severe acute respiratory syndrome virus
Scavenger receptors
    classes and structures, 87–88, *87*
    signaling, 88
SCF, *see* Stem cell factor
Schistosomiasis, 683
SCID, *see* Severe combined immunodeficiency
Scleroderma
    clinical features, 981–982, *981*
    pathophysiology, 982
Scurfy mouse, 990
Selectins
    leukocyte extravasation role, 61–61
    T cell activation role, 376
Selective immunoglobulin A deficiency, 767
Selective immunoglobulin G deficiency, 767
Semliki forest virus, vaccine vector, 712
Severe acute respiratory syndrome (SARS) virus, 668
Severe combined immunodeficiency (SCID)
    case example, **755–756**
    epidemiology, 754
    gene therapy, **760–761**
    types
        adenosine deaminase deficiency, 758–760, *759*
        Artemis disease, 762–763
        CD45 disease, 758
        interleukin-7 receptor disease, 758
        Jak3 disease, 758
        Omenn syndrome, 762
        overview, *756, 757*
        purine nucleoside phosphorylase deficiency, 760,
            762
        RAG disease, 762
        reticular dysgenesis, 763
        X-linked disease, 756, *757*
Sezary syndrome, 1061–1062
Signal transducer anad activator of transcription (STAT)
    signal transduction, *468*
    T helper cell differentiation role, **410**
Sirolimus, see Immunosuppressive drugs
Sjögren syndrome
    clinical features, 977–978
    pathophysiology, 978
Skin, *see also* Skin-associated lymphoid tissue (SALT)
    anatomy, 70, *71*
    barrier to infection, 71
Skin-associated lymphoid tissue (SALT)
    composition
        basement membrane, 605
        dermis, 605
        keratin layer, 601
        lower epidermis, 601–603
        neurological components, **603–604**
        overview, 52, 60, 584, 600–601, *601*
    distribution, 59
    immune responses, 605–608, *606*
    structure, *602*
SLE, *see* Systemic lupus erythematosus
SLP76, T cell receptor signaling, **385**

SMAA, *see* Solid matrix antibody-antigen complex
SMAC, *see* Supramolecular activation complex
Smallpox
    eradication, 696–697, *698*
    immunization, 4–5, *5*, **6**
    vaccination, 738
    virus, 668
Solid matrix antibody-antigen complex (SMAA), vaccine
    delivery, 726
Somatic gene rearrangement, *see* V(D)J recombination in
    B cells; V(D)J recombination in T cells
Somatic hypermutation
    activation-induced cytidine deaminase role, 233–234
    association with affinity maturation, 234–235
    definition, 232–233, *233*
Somatic recombination, *see* V(D)J recombination in B
    cells; V(D)J recombination in T cells
Southern blot, somatic recombination detection, 188–189
Sox-4, B cell development modulation, **223**
Spi-B, B cell activation modulation, **224**
Spleen
    anatomy, 60–61, *60*
    cellular composition, 61
    circulation, 60–61
Split tolerance, 457
Src kinases
    B cell receptor signaling, 219–221
    T cell receptor signaling, 331, 384
    regulation by CD45, 220, 388
STAT, *see* Signal transducer and activator of transcription
Stem cell factor (SCF)
    expression, 503
    functions, 503, *503*
    receptor and signaling, 503–504, *504*
Streptococcus pneumoniae, vaccination, 735
Stress-activated protein kinase/jun kinase (SAPK/JNK)
    signaling pathway, 383, 390
    T cell costimulation role, 393–394
    Th1 differentiation, 415
Subunit vaccine, *see* Vaccination
Superantigens
    pathogens and autoimmune disease, 1005–1006
    T cell receptor signaling, **386–387**
Suppressor T cells, see Regulatory T cells
Supramolecular activation complex (SMAC), formation,
    380–381, *382*
Systemic lupus erythematosus (SLE)
    clinical features, 970, *970*
    pathophysiology, 970

T

Tacrolimus, see Immunosuppressive drugs
Tag assays, *see also Enzyme-linked immunosorbent assay
    (ElISA); Immunoelectron microscopy;
    Immunofluoresence; Radioimmunoassay (RIA)*
    direct tag assays, 162
    indirect tag assays
        blocking, 166
        standard curves, 167, *167*
    tags
        biotin–avidin system, 162–164
        electron-dense materials, 166
        enzymes, 164–165
        fluorochromes, 165–166, *166*
        molecular tags, *165*
        protein A or G, 162, *164*
        radioisotopes, 164
        secondary antibodies, 162
        third components, *163*
    washing, 166–167
TAP, *see* Transporter associated with antigen processing
Tat, human immunodeficiency virus, 794–795
T cell, *see also* Cytotoxic T lymphocyte (CTL); γδ T cell;
    Memory T cell; Natural killer T cell; Regulatory T cell
    accessory molecules, 29, *29*
    activation, see T cell activation
    activation-induced cell death, *see* Activation-induced
        cell death

autoimmune disease
  abnormalities, 1009
  autoreactive T cell targeting, 1017–1018
B cell ontogeny comparison, 344, 347
CD4⁺ T helper cell (Th)
  activation
    antigen-presenting cells, 413–414
    assays, **250–251**
    costimulation, 414–415
    T cell receptor signaling, 415
  atopy
    balance, *936*
    therapeutic targeting
    B cell costimulation, 224–226
  cooperation in humoral immune response
    hapten–carrier complexes *in vivo*, 134–136
    hapten–carrier experiments
      carrier effect, 132, *132*
      hapten–carrier system, 131, *133*
      linkage of B and T cell determinants, 132–133, *133*
      rationale for linked recognition, 136–137, *136*
      reconstitution experiments, 130–131
  differentiation
    history of study, 404–405
    Th0, 406
    Th1/Th2
      antigen effects, 408
      antigen-presenting cell type effects, 408–409, 412
      costimulation effects, 408
      cytokine effects, 409, 412
      overview, 406–408, *407, 407*
      surface markers, 412–413, *413*
      transcription factors, 410–**411**
  effector functions
    cytokine secretion, 416–417
    macrophage hyperactivation and delayed type hypersensitivity, role in, **42–43**, 417
    overview, 30–31, *30*, 405–406, *414*, 415–416, *416*
    Th1/Th2 cross-regulation, 417, *417*
  extracellular bacteria elimination, 652
  fungus elimination, 690
  human immunodeficiency virus immune response, 802–803
  intracellular bacteria elimination, 661
  mucosa-associated lymphoid tissue responses, 593–595
  parasite
    elimination, 684
    evasion, 685
  subsets, 31, *405*
  Th3 cells as regulatory T cells, 452
  virus elimination, 671
CD8⁺ cytotoxic T cell (Tc), *see also* Cytotoxic T lymphocyte (CTL)
  activation, 418–419, *419*
  differentiation, 418
  effector functions, 30–31, *30*, 417–418
  suppressor T cells, 452–453
compartmentalization, *39*
coreceptors, *see* CD4; CD8
costimulatory molecules, 29, *29*
cytotoxicity, *see* Cytotoxic T lymphocyte (CTL)
development
  bone marrow, 347
  differences over life stages, *347*
  double-negative phase, *352*
    DN1, 351
    DN2, 351–353
    DN3, 352, 354–356
    DN4, 356
  DP phase, 356–364
  embryology, 368–370, *369*
  maturation in thymus, 53–55, 347–349, *348*, 351
  Notch1 signaling in cell fate determination, **350**
  overview, 346–347
  peripheral cell fates, 368

SP phase and CD4/CD8 lineage commitment, 364–368
TCRα locus rearrangements, 356
TCRαβ pool expansion, 356
thymic selection, *see* Thymic selection
discovery, 6
division of labor, 21, *22*
effector functions, general,13, *13*, 30–31, *30*
exhaustion, 427
helper function, see T cell, CD4⁺ T helper cell (Th)
histology, *37*, 41
history of study, 342–344
identification, 43
intracellular bacteria evasion, 663–664
lineage commitment, CD4/CD8
  models, 365–366, *365*
  signaling, 366–367
  T cell receptor V segment usage, 367–368
  transcription factors, 367
lymphoblast production, 41
memory, *see* Memory T cell
parasite elimination, 684
repertoire and memory, 44, 46
receptor, *see* T cell receptor
resting cells, 41
signaling, *see* T cell receptor
tolerance, *see* Peripheral tolerance; Tolerance
trafficking and adhesion of naive and effector cells, 422–423
tumor immune response, 847–848, *847*
virus inactivation, 677
T cell activation, *see also* T cell receptor
  adhesion molecules in T cell-APC interactions, 375–378, *376*
  antigen-presenting cell licensing model, see Dendritic cell, licensing model *375*
  assays, **250–251**
  costimulation
    CD27/CD70, 398
    CD28-B7, 393–396, *393, 395*
    ICOS-ICOSL, 396–397, *397*
    minor interactions, 398
    overview, 392–393, *392*
    PD-1/PDL, 397
    time course of gene expression, *398*
    tumor necrosis factor-related molecules, 397–398
  cytokines
    interleukin-2, 399–400
    overview, 398–399
  cytotoxic T lymphocytes, 418–419, *419*
  gamma-delta (γδ? T cells, 541–542, *542*
  memory T cells, 429–430
  naive and effector T cells, 423–424
  overview, 374–375
  signaling
    adaptor proteins, 384–387
    complexity rationale, 391–392
    coreceptor roles, 391
    overview, 383–384, *383*
    phosphatidylinosotol 3-kinase, 389–390
    protein tyrosine kinases, 384, 387–388
    Ras, 390–391, *390*
    Vav1 and cytoskeletal reorganization, 388–389, *389*
    ZAP-70, 388
  triggering
    oligomerization model, 378–379
    overview, 378
    serial triggering model, 379–380, *380*
    supramolecular activation complex formation, 380–381, *382*
T cell prolymphocytic leukemia (T-PLL), features, 1046
T cell receptor (TCR)
  binding with peptide–MHC complexes, 336
  CD3 complex, *318*
  CD4/CD8 T cell lineage commitment signaling, 366
  comparative study, see Comparative immunology, T cell receptor
  density, 44
  discovery, 312–315

diversity
  chain pairing, 329
  combinatorial diversity, 328
  junctional diversity, 328–329
  multiplicity of gene segments, 327
  overview, *328*
downregulation, 382–383
genes, *see* T cell receptor genes
pattern recognition by TCRγδ, 88–89
protein engineering, 336
signaling, *see* T cell activation
signaling defects and primary immunodeficiency, 765
staining, **316**
structure, 317–319, *317, 318*
tickling, 368
triggering, *see* T cell activation
tumor inhibition, 850
X-ray crystal structures
  contacts with ligands, 337–338, *337*
  ligand recognition stringency, 338
T cell receptor genes
  enhancers and gene expression regulation, 329–330, *330*
  gamma-delta (γδ) T cell receptor genes, *see also* γδ-T cell
    discovery, 315, 317
    structure, 317–319, *317, 318*
  gene cloning, 313–315, *314*
  gene rearrangement, *see* V(D)J recombination in T cells
  genomic organization
    TCRα gene
      human, 319–320, *319*
      mouse, *319*, 320
    TCRβ gene
      human, 320, *320*
      mouse, *320*, 321
    TCRγ gene
      human, 321, *321*
      mouse, 321, *321*
    TCRδ gene
      human, 322, *322*
      mouse, 322, *322*
  V(D)J recombination, see V(D)J recombination in T cells
TCR, *see* T cell receptor
T-dependent antigen, *see* Antigen
Tetanus, vaccination, 730
TGF-β, *see* Transforming growth factor-β
T helper cell (Th), see T cell, CD4⁺ T helper cell (Th)
Thrombocytopenia
  alloimmune thrombocytopenia, 953
  autoimmune thrombocytopenia, 953
Thrombocytopenic purpura
  idiopathic thrombocytopenic purpura, 980–981
  thrombotic thrombocytopenic purpura, 981
Thymic selection
  affinity/avidity model, 357, *360*
  antigen processing molecules, 362–363
  coreceptors, 362
  definition, 356
  experimental systems, **358–359**
  peptide–MHC role, 363
  processes
    positive and negative selection, 357, 359
    thymocyte death by neglect, 357
  rationale, 356–357
  signaling
    CD3 chains, 360–361
    costimulatory signaling, 362
    protein tyrosine kinases, 361
    Ras, 361
    T cell receptor chains, 360, *360*
  stromal cell interaction with thymocytes
    negative selection, 364
    positive selection, 363
Thymocyte
  lineage restriction, *354*
  markers, 347–348
  populations in adult thymus, *370*
  T cell development, *see* T cell; Thymic selection

Thymus
  anatomy, 53, *55*
  involution, 55, **345–346**
  microenvironment in thymocyte development,
    348–349, *349*, 351
  T cell maturation, 53–55, 347–349, 351
T-independent antigen, *see* Antigen
Tissue typing, see Transplantation, HLA matching
TLRs, *see* Toll-like receptors
TNF, *see* Tumor necrosis factor
TNFR, *see* Tumor necrosis factor receptor
Tolerance, *see also* Peripheral tolerance
  autoimmune disease defects, 967, **967**, *968*
  central versus peripheral, 21
  degree and persistence, 457
  experimental tolerance
    characteristics, 454
    genetic susceptibility, 457
    oral tolerance, 456–457, *456*
  fetal–maternal tolerance, 457–459, *458*
  γδ T cells, 547
  induction therapy in autoimmune disease, 1019
  natural killer cells
    anergy model, 536–537
    at least one model, 535–536
    receptor calibration model, 536, *537*
  neonatal tolerance, 459–461, *459*
  split tolerance, 457
  tolerogen characteristics
    administration routes, 456–457
    dose, 455–466
    immunogen comparison, *455*
    molecular nature, 454–455
    size, 455
Toll-like receptors (TLRs)
  allograft rejection role, 889–890
  comparative immunology, 615, 619, *619–620*, 620
  discovery, 85–86
  effector mechanisms against bacteria, 647,
    659–660
  effector mechanisms against fungi, 689
  effector mechanisms against viruses, 669
  endotoxic shock, 475–476
  ligand specificity, 86–87
  signaling, 87, 475
Toxicoplasmosis, 681–682
Toxoid, *see* Vaccination
T-PLL, *see* T cell prolymphocytic leukemia
TRAIL
  autoimmune disease role, **1007**
  nature of, 227
  upregulation by HIV Tat, 802
Transforming growth factor-β (TGF-β)
  B cell development role, 228
  cancer promotion, 850
  DC modulation for transplantation, 905
  decrease in GvHD role, 915
  defect leading to autoimmunity in mice, 991
  expression, 502
  functions, 501–502, *502*
  graft rejection role, 888
  immunosuppression, 446, 451
  induction of NK inhibitory receptors, 530
  infectious tolerance, 453
  isotype switching role, 237, *238*
  myeloma survival role, 1049
  peripheral T cell tolerance role, 447
  possible role in hypersensitivity, 945
  possible treatment for autoimmunity, 1015, 1017
  receptors and signaling, 502, *503*
  Th3, Tr1 differentiation, 452
Transgenic mouse
  construction, **332–333**
  coreceptor studies, 332
  TCR transgene
    nature of, 358
    Thymic selection studies, 362

Transplantation
  allorecognition
    direct, 881, *882*
    indirect, 881, *882*
    overview, 880
  blood transfusion, *see* Blood transfusion
  bone marrow transplantation, *see* Bone marrow
    transplantation
  demand in United States, 906, *906*
  gene therapy prospects, 917–919, *918*
  global perspective, **878**
  graft types, 784
  graft-versus-host disease (GvHD)
    bone marrow transplantation
      incidence, 915
      mechanisms, 915
      prevention strategies, 915–916
    organ transplants, 890
  graft-versus-leukemia effect, 916–917, 1034
  historical perspective, 874–876, 878–880, **883**
  HLA matching
    complement-dependent cytotoxicity, 891, *891*
    mixed lymphocyte reaction, 892, *893*
    overview, **876–877**, 890–891
    panel reactive antibody, 891–892, *892*
    polymerase chain reaction techniques,
      893–896, *895*
    restriction fragment length polymorphism,
      893, *894*
    process in United States, *880*
  immunosuppression, *see* Immunosuppressants
  minor histocompatibility antigens, 881–885, *884*, *885*
  organ donors, **876–877**
  rejection
    acute graft rejection
      acute cellular rejection, 886–887
      acute humoral graft rejection, 887
      antibody-mediated rejection of solid tissue
        transplants, 953
    chemokine roles, 888–889
    chronic graft rejection, 888
    cytokine roles, 888–889
    hyperacute graft rejection, 885–886
    overview of types, 885, *886*, 887
    Toll-like receptor roles, 889–890
  survival of solid organ recipients, *878*, *879*
  tolerance induction
    bone marrow manipulation, 900
    caveats, 905–906
    costimulatory blockade
      antibody targeting, 901
      CTLA-4, 902–903
      PD-1, 902–903
    dendritic cells, 905
    natural killer T cells, 904–905
    regulatory T cells
      CD4$^+$ cells, 903–904
      CD8$^+$ cells, 904
    thymic manipulation, 901
  viability of organs, *880*
  xenotransplantation
    case example, **908**
    ethical and moral considerations, 910
    rationale, 906–907
    regulatory and legal obstacles, 910
    rejection
      prevention, 909
      types, 907–909
    species selection, 907
    zoonotic disease transmission risk, 909–910
Transporter associated with antigen processing (TAP)
  proteins, antigen processing, 299–300, 302
TRAPS, *see* Tumor necrosis factor-associated periodic
  syndrome
Tuberculosis
  clinical features, 657–658, *657*
  vaccination, 735–736, **737**

Tumor necrosis factor (TNF)
  activation of HIV, 800
  autoimmunity association, 968
  blockade to treat rheumatoid arthritis, 1014–1016
  deficiency effects, *496*
  excess effects, *497*
  expression, 498
  functions
    antiviral effects, 498
    coagulation, 498
    cytotoxic effects, 498
    hematopoiesis, 498
    immunoregulation, 497
    inflammation, 496–497
  graft rejection promotion, 888–889
  GvHD promotion, 915
  increased blood levels and cachexia in AIDS, 804
  introduction, 80
  MHC class I upregulation, 271
  receptors, 498
  related costimulatory molecules, 397–398
  see also TRAPS
  type I hypersensitivity promotion, 930, 947
Tumor necrosis factor-associated periodic syndrome
  (TRAPS), 779–780
Tumor necrosis factor receptor (TFNR)
  activation-induced cell death role, 426
  knockout mice, 498
  removal by virus, 677
  see also TRAPS
  superfamily
    introduction **226**, **227**
    association with autoimmunity, 1007
Tumors
  autoimmune disease induction, 1020–1021
  cancer
    epidemiology, 26, *27*
    staging, 831–832
  carcinogenesis steps, 832–834, *833*, **837**
  carcinogens
    chemicals, 840, *840*
    pathogens, 840–842, *840*
    radiation, 840, *840*
  classification, 829–831, *830*, *831*
  definitions, 829
  evasion of immune response
    antigen recognition avoidance, 850
    immunoproteasome bias, 851
    immunosuppressive cytokine secretion, 849–850
    overview, 849, *850*
    regulatory T cell response promotion, 850
    T cell signaling inhibition, 850
  gene mutations
    DNA repair genes, 835–836, *836*
    oncogenes, 836, 838, *838*
    tumor suppressor genes, 838–839, *838*
  history of tumor immunology, 826–827, 829
  hypoxia and therapeutic targeting, 853, 855–856
  immune response
    B cells, 849
    cytokines, 847
    γδ T cells, 848–849
    heat shock proteins, 849
    inflammatory cells, 845, 847
    natural killer cells, 845
    natural killer T cells, 848
    overview, 845, *846*
    T cells, 847–848, *847*
  immunosurveillance, 842
  metastasis, 834, *835*
  morbidity and mortality, 831
  paraneoplastic syndromes, 832
  treatment
    adoptive transfer of cytotoxic T lymphocytes, 869
    cancer vaccines
      dendritic cells, 866–868
      heat shock proteins as adjuvants, 866, *867*

non-pathogen vaccines, 864–866, *865*
pathogen vaccines, 864
principles, 744–745
treatment of follicular lymphoma, 1058
chemotherapy, 852–853, 854–*855*, *856*
immunotherapy
approaches, 857
cytokines, 857–859
historical perspective, 856–857
monoclonal antibodies, 859, **860**, 861, *861*, *862*, 863–864
radiation therapy, 851–852, *852*
T cell costimulatory and regulatory molecules, 868–869
tumor-related antigens
classification, 842, *843*
discovery, 827, *828*, 842–843
tumor-associated antigens
differentiation and tissue-specific antigens, 844
embryonic antigens, 844
idiotypic antigens, 844
normal cellular antigens, 843–844
tumor-specific antigens, 844–845
Tumor suppressor genes, 838–839, *838*
Typhoid fever, vaccination, 736

U
Ulcerative colitis
clinical features, 985
pathophysiology, 985
Urticaria, 932, *934*

V
Vaccination
addiction suppression, 747
adjuvants
types, *724*
mechanisms of action, 722–723
incomplete Freund's adjuvant, 722
complete Freund's adjuvant, 722
alum, 723
bacterial toxins, 723
CpG motifs, 723
cytokines, 723
administration routes, 722
allergy mitigation, 746, 948
autoimmunity mitigation, 745–746
boosting, 720–721
chronic viral diseases, 745
clinical trials, 704–705, *705*
combination vaccines, 721
communicable disease mortality, 698, *698*
cost-effectiveness, 697–698
delivery vehicles
biodegradables, 726
dendritic cells, 726–727
immunostimulating complexes, 725–726
liposomes, 724–725
overview, 723–724, *724*
solid matrix antibody-antigen complex, 726
virosomes, 725
design principles
biological purpose, 703–704
successful vaccine criteria, *704*
edible vaccines, **702–703**
efficacy, 697, 704–705
examples
anthrax, 727
chicken pox, 738
cholera, 729
diphtheria, 729–730
Haemophilus influenzae, 730–731
hepatitis A, 731
hepatitis B, 731
influenza, 731
Japanese B encephalitis, 732

measles, 732–734, *733*
meningococcus, 734
mumps, 732–734
overview, 727, 728–729
pertussis, 730
plague, 734
polio, 734–735
rabies, 735
rubella, 732–734
smallpox, 738
Streptococcus pneumoniae, 735
tetanus, 730
tuberculosis, 735–736, **737**
typhoid fever, 736
yellow fever, 739
fertility suppression, 746–747
genomics, transcriptomics, and proteomics in development, **700–701**
historical perspective, 5, **6**, 696–700
mitigation of indirect effects of infectious disease, 745
opposition and failure to vaccinate, 740
passive immunization, 741–744, *741*, *743*
principles, 696, *697*
prospects, 740–741
safety and adverse effects, 704–705, 739–740
scheduling, 718–720, *720*
smallpox eradication, 696–697, *698*
special patient populations, 721
therapeutic versus prophylactic vaccination, 696
tumor vaccines, *see* Tumors
types
anti-idiotypic vaccines, 716, *717*
comparison of types, *707*
killed vaccines, 708–709
live, attenuated virus, 705–706, 708
naked DNA vaccines
advantages, 714–716
gene gun, **715**, 727
immune response, 713–714, *714*
limitations, 716
preparation, 713
overview, 705, *706*
peptide vaccines
candidate identification, 710–711, **711**
limitations, 711
recombinant vector DNA vaccines
advantages, 712
limitations, 713
pathogens, 711
production, 712
recombinant vector types, 712–713
subunit vaccines
bacterial, 709–710
limitations, 710
viral, 709
toxoids, 709
Vaccinia virus
nature of, 668–669
vaccine vector, 712–713
Varicella zoster, *see* Chicken pox virus
Variola, *see* Smallpox virus
Vav1
T cell development, 361
T cell receptor signaling, 388–389
V(D)J recombination in B cells
allelic exclusion, 192–193, *193*
control by enhancers, 204
diversity generation mechanisms
combinatorial diversity, 201–202
junctional diversity, *203*
deletion, 202
N nucleotide addition, 202
P nucleotide addition, 202
multiplicity of germline gene segments, 201
overview, 201, *201*
detection, **188–189**

gene rearrangement
at *Igh* locus, *190*
at *Igk* locus, 190–191, *191*
at *Igl* locus, 190–191, *192*
gene segment numbers, *186*
gene splicing, **181–182**
history of study, 180–181, 183, **183**
kappa/lambda exclusion, 193–194
knockout mouse studies, 194–196
molecular mechanisms
coding joint, 198–199
deletional joining, 199, *200*
D-J joining, example, *199*
inversional joining, 200, *200*
mutant studies, 200–201
recombination activation gene proteins, 197
recombination signal sequences, 196–197, *197*, *198*
signal joint, 198
synapsis, 198
mutations, 200
overview, 183–184
productivity testing, 191–192
the 12/23 rule, 197
V(D)J recombination in T cells
developmental aspects
TCRα rearrangement, 326–327
TCRβ rearrangement, 325–326, *326*
TCRγ rearrangement, 327
TCRδ rearrangement, 327
knockout mouse studies, 327
diversity generation in T cell receptors
chain pairing, 329
combinatorial diversity, 328
junctional diversity, 328–329
multiplicity of gene segments, 327
overview, 327, *328*
gene transcription and protein assembly, 324–325, *325*
mechanism, 322–324, *323*, *324*
Virosome, vaccine delivery, 725
Virulence, definition, 644
Viruses
characteristics
adenovirus, 665–666
chicken pox virus (varicella zoster), 738
cytomegalovirus, 666
Epstein-Barr virus, 666
general characteristics, 664–665
hepatitis A virus, 731
hepatitis B virus, 667, 731
hepatitis C virus, 667
human immunodeficiency virus, 667
influenza virus, 667, 731
measles virus, 667–668, 732
mumps virus, 732
polio virus, 668, 734
rabies virus, 668, 735
rhinovirus, 668
rubella virus, 732
severe acute respiratory syndrome virus, 668
smallpox virus (variola), 668
vaccinia virus, 668–669
yellow fever virus, 739
disease examples, *666*
effector mechanisms in elimination
CD4$^+$ T cells, 671
CD8$^+$ T cells, 669, 671
complement, 671–672
humoral defense, 671
interferons, 669
natural killer cells, 669
overview, *670*
evasion strategies
antibody avoidance, 677
antigen presentation interference
effect on MHC class I function, 673–675, *675*
effect on MHC class II function, 676–677, *676*

Viruses (*Continued*)
   antigenic variation, 673
   antiviral state counteraction, 677
   apoptosis manipulation, 677–679, *678*
   complement avoidance, 677
   cytokine interference, 679–680
   dendritic cell interference, 677
   host cell cycle manipulation, 679
   latency, 672–673
   natural killer cell evasion, 676
   overview, 672, *672*
   T cell inactivation, 677
   families, *665*
Vitronectin, complement fixation regulation, 571

W
Waldenström's macroglobulinemia,
   features, 1047
Wiscott-Aldrich syndrome, 775–776
Western blot, principles, 175–176, *175*

X
Xenotransplantation, *see* Transplantation
Xeroderma pigmentosum, 772, *772*
X-linked lymphoproliferation, 773
X-ray crystallography
   Fab–lysozyme complex, *139*
   principles, **127–128**

MHC class I and II molecules, 260–261
MHC-like molecules, 266–267
T cell receptor
   contacts with ligands, 337–338, *337*
   ligand recognition stringency, 338

Y
Yellow fever, vaccination, 739

Z
ZAP-70
   T cell development, 361
   T cell receptor signaling, 388